Evolution and Selection of Quantitative Traits

Evolution and Selection of Quantitative Traits

Bruce Walsh
University of Arizona

Michael Lynch
Arizona State University

Sinauer Associates

NEW YORK OXFORD
OXFORD UNIVERSITY PRESS

Cover Photo Credits (top to bottom)

- *Lithophane leeae* Walsh 2009 (Noctuid moth). Photo by Bruce Walsh.
- Chuck (Angora rabbit). Photo by Lee Fulmer.
- Maize diversity. Photo by Nick Kaczmar.
- Bony armor variation in *Gasterosteus aculeatus* (the threespine stickleback). Photo by Rowan Barrett.

OXFORD
UNIVERSITY PRESS

Great Clarendon Street, Oxford, OX2 6DP,
United Kingdom

Oxford University Press is a department of the University of Oxford. It furthers the University's objective of excellence in research, scholarship, and education by publishing worldwide. Oxford is a registered trade mark of Oxford University Press in the UK and in certain other countries

First Edition published in 2018
Reprinted with corrections 2018

Impression: 2

Published in the United States of America by Oxford University Press
198 Madison Avenue, New York, NY 10016, United States of America

British Library Cataloguing in Publication Data
Data available

Library of Congress Control Number: 2017054086

ISBN 978-0-19-883087-0

DOI: 10.1093/oso/9780198830870.001.0001

Printed in Great Britain by
Bell & Bain Ltd., Glasgow

Seek simplicity and distrust it. Whitehead (1920)

No efforts of mine could avail to make the book easy reading.
Fisher (1930. *The genetical theory of natural selection*)

ADVANCE PRAISE FOR THIS VOLUME

"Their 1998 book has been an essential and popular teaching and reference text for quantitative geneticists for 20 years. At last the long promised and awaited second volume is available, structured to provide both an introduction for the graduate student and a reference volume for the seasoned professional. The authors display both an impressive breadth and depth of knowledge and ability to explain it."

William G. Hill, University of Edinburgh, UK

"The book is a fabulous resource; it will be the go-to place to learn about a broad variety of topics in population and quantitative genetics, from measuring differences among populations to estimating selection from genomic sequence data. A real strength of the book is the expert overview of the topics alongside pointers to the key papers for more details."

Sarah P. Otto, University of British Columbia, Canada

"This long-awaited volume gives an exceptionally comprehensive and thoughtful overview of the field, and will be invaluable to all those who are trying to understand the genetics of complex traits. As well as covering quantitative genetics, it gives a thorough treatment of the most recent methods for making inferences from DNA sequence data."

Nicholas H. Barton, Institute of Science and Technology, Austria

"This book is a virtuoso synthesis of the theory and application of selection and evolution of complex traits. Much more than an advanced textbook, it is the most comprehensive treatment and synthesis of the theory and analysis of selection and evolution of quantitative traits to date."

Peter M. Visscher, University of Queensland, Australia

"A masterful synthesis of the growing points in selection analysis."

Stevan J. Arnold, Oregon State University, USA

"Decades in the making, the somewhat audacious conceit of collecting everything that we know about evolutionary genetics into a single tome pays dividends in the form a book that will serve as a central resource in the library of every evolutionary biologist."

Patrick C. Phillips, University of Oregon, USA

"This book has the same outstanding merits as the previous volume. It is authoritative, comprehensive (covering an extremely wide range of topics) and beautifully written. It will remain as the definitive account of the material that it considers for many years to come."

Warren J. Ewens, University of Pennsylvania, USA

Contents

PREFACE

One hundred years ago, Sir Ronald Aylmer Fisher (1918), in a paper that was rejected by the Royal Society of London, laid the foundations for the field that has come to be known as quantitative genetics. This paper, and subsequent work by Fisher and others (often in the search of biologically motivated problems), also gave birth to much of modern statistics. The past century has seen quantitative genetics flourish, and indeed, the field has never been stronger or more vibrant as it moves seamlessly into the post-genomics era with an ever-broadening reach. If Fisher were alive today, we believe that he would be fascinated, and somewhat pleased, both at how much the field has grown, and also at how much it remains very familiar to him, even after all these years.

This text is the second part (out of three) of our treatment of the modern field of quantitative genetics from its humble, but path-breaking, beginnings 100 years ago. Some 20 years ago, we remarked in the introduction to the first part of our treatment (Lynch and Walsh 1998) that "today's quantitative genetics is not the science that it was 25 (or even 10) years ago," a statement that is even truer today. Our further comment that "the current machinery of quantitative genetics stands waiting (and its practitioners willing) to incorporate the fine genetic details of complex traits being elucidated by molecular and developmental biologists" has certainly been proven true over the past two decades. Modern quantitative genetics is the glue that connects many disciplines, especially given its ability to model uncertainty while fully accounting for known genetic and genomic features.

The first part of our treatment (Volume 1) dealt with genetics, which is the underpinning of complex traits. While we also remark on some recent developments in this area in this volume, our main focus is on the evolution of such traits, with applied evolution (plant and animal breeding) as an important application. To address these concerns, we have attempted to merge modern population-genetics theory with quantitative genetics and genomics. As a result, some of our treatment is more technical than was presented in Volume 1, as it provides a much fuller development of mathematical models of evolution. We have strived to present a holistic treatment, showing the interplay between theory and data, with extensive discussions on statistical issues relating to the estimation of the biologically relevant parameters for these models. We view quantitative genetics as the bridge between complex mathematical models of trait evolution and real-world data, and have tried to frame our treatment as such.

As with our first volume, a key goal in producing this volume is to increase communication among the various disciplines that make up the very diverse field of quantitative genetics. Twenty years ago, this included plant and animal breeders, evolutionary biologists, human geneticists, and statisticians. This list has only grown with the advent of genomics, which attracts an ever-widening array of those who use quantitative genetics (many of whom may not even be aware of it).

Given a project of this size and scope, and despite our own best efforts, excellent copy editors, and wonderful colleagues, it is likely that errors have escaped scrutiny. Hopefully, most will be obvious and trivial, but the posterior for this process often has a long tail, and, therefore, the potential for some interesting outliers. Please report any errors you find to BW (**jbwalsh@email.arizona.edu**). Likewise, we have a facebook group, **Quantitative Genetics Book**, whose file section will contain updated pdf files for current errata, as well as a forum for discussing issues related to the book. Finally, the final part of our treatment, which deals largely with multivariate issues, will be forthcoming in a more timely manner than this volume!

Acknowledgments

As numerous seasoned readers will know, various components of this book have been in the works for over three decades. For a project of this magnitude, there is a host of people to thank, and we have undoubtedly forgotten to credit individuals who provided us with critical feedback in the earliest years. We profusely apologize for any such slights and blame our slowly fading memories.

As we moved into the field of quantitative genetics in earnest in the 1980s, critical foundational support in the development of our ideas was provided by our colleagues and teachers Stevan Arnold, Reinhard Burger, James Crow, Joe Felsenstein, Wilfried Gabriel, Daniel Gianola, Thomas Hansen, William Hill, Elizabeth Housworth, Alex Kondrashov, Russell Lande, Bill Muir, Tom Nagylaki, Tim Prout, Monty Slaktin, Michael Turelli, and Michael Wade.

We have also greatly benefited from all past and current members of our labs for their comments over the years. These especially include Ph.D. students: Matthew Ackerman, Desiree Allen, Stephan Baehr, David Butcher, Chi Chun Chen, Hong-Wen Deng, Suzanne Estes, Allan Force, Kyle Hagner, Parul Johri, Vaishali Katju, Weiyi Li, Timothy Licknack, Samuel Miller, Kendall Morgan, Angela Omilian, Michael Pfrender, Taylor Raborn, Sarah Schaack, Ryan Stikeleather, Ken Spitze, and Larry Weider; and post-doctoral associates: Charles Baer, Megan Behringer, Melania Cristescu, Dee Denver, Thomas Doak, Jeffrey Dudycha, Suzanne Edmands, Jean Francois Gout, Kevin Higgins, David Houle, Sibel Kucukyildirim, Niles Lehman, Hongan Long, Takahiro Maruki, Martin O'Hely, Susan Ratner, Barrie Robison, Stewart Schultz, Douglas Scofield, and John Willis.

BW also wishes to thank the thousands of students who have attended his short courses on various aspects of quantitative genetics over the past two decades, during which he has taught over 5000 students from roughly 60 countries at various locations around the world (25 countries at last count). This international interaction and feedback have greatly sharpened our ability to present the material in this book to a truly universal audience.

We have been exceptionally blessed with outstanding feedback throughout this project from a long list of colleagues, who directly engaged wih the manuscript. Sally Otto and Michael Morrissey read whole parts of the manuscript, while critical comments on specific chapters were provided by Steve Arnold, Nick Barton, Piter Bijma, Mark Kirkpatrick, Bill Muir, Shinichi Nakagawa, Guilherme Rosa, Ruth Shaw, John Storey, and Michael Turelli.

We especially want to acknowledge Bill Hill, who has been a major force throughout this project, and in many ways could be considered its third author. Bill read all of both Volume 1 (Lynch and Walsh) and the current volume, offering very detailed and critical comments, and saving us from much embarrassment. Our goal throughout the project was to present the field with the breadth and depth that Bill carries around in his head. He also kept us going by constantly ending his emails with the reminder to "keep writing!" Hence, we are very pleased to codedicate this volume to Bill.

With respect to the production end, we have greatly enjoyed working with the highly professional staff at Sinauer Associates through the bulk of the editing process, and then with their new colleagues at Oxford during the final phase. Copy editor Nicole Balant and production editor Martha Lorantos faced the thankless task of trying to catch typos in a 1400-page technical manuscript where the lead author is rather dyslexic.

Elizabeth Morales and Ann Chiara did a wonderful job with the illustrations and patiently dealt with us, as we frequently changed our minds on how to best illustrate various concepts, while Michele Beckta dealt with copyright issues regarding the figures. Last, but certainty not least, production manager Chris Small thoughtfully guided us through the typesetting of both this and the previous volume (both of which were typeset by BW using Textures). Any less-than-professional appearance of aspects of this volume are there despite the best efforts of Chris!

We save very special thanks for our publisher and friend, Andy Sinauer. Andy initially started this project in the mid-1980s when we (BW and ML) discovered (via Andy) that we were both working on separate books on the same subject and he logically queried whether it might be better if we worked together. Indeed it was! Andy has shown incredible patience over the past three decades on the pace of this project, and we hope that he will be pleased with the final outcome, despite its long gestation. We are thus extremely pleased to codedicate this volume to Andy for his long and outstanding service as one of the preeminent publishers of biological science textbooks. Upon his retirement, he has left a gap that will be hard to fill, and his impact on a number of fields, particularly evolutionary biology, is hard to overstate.

In closing, ML is especially grateful to the administration and colleagues at the University of Illinois, his place of first employment, for providing him with a safe environment for making an early (and risky) career transition from limnology to evolutionary genetics. And we both thank NSF and NIH for continuing support for research related to the subject matter of the book.

Finally, and especially, we wish to thank our incredible wives, Emília Martins (ML) and Lee Fulmer (BW), for their enduring patience during this long project.

Bruce Walsh, Tucson

Michael Lynch, Tempe

March 2018

I

Introduction

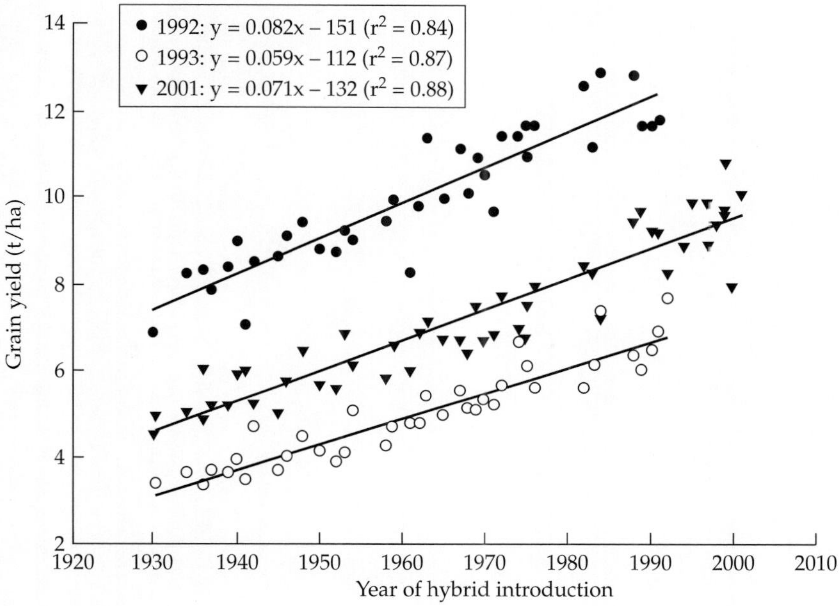

1

Changes in Quantitative Traits Over Time

The tendency of modern scientific teaching is to neglect the great books, to lay far too much stress upon relatively unimportant modern work, and to present masses of detail of doubtful truth and questionable weight in such a way as to obscure principles. Fisher and Stock (1915)

Quantitative traits—be they morphological or physiological characters, aspects of behavior, or genome-level features such as the amount of RNA or protein expression for a specific gene—usually show considerable variation within and among populations. A central (perhaps *the* central) question in evolution is the nature of the underlying forces generating this variation, be it the standing variation within a population or the divergence between species. Fully interwoven into this question is the corollary of how best to achieve targeted changes in organisms exploited for human welfare, namely breed improvement (Figures 1.1 and 1.2). The importance of this applied aspect cannot be overstated. The most important technology developed by humans is agriculture, as all of our other impressive advances rest upon the foundation of a stable and sustainable food system. At present, despite significant improvements in yield (e.g., Evenson and Gollin 2003; Thornton 2010), there is considerable uncertainty about our ability to keep pace with projected global needs (Ray et al. 2012, 2013).

It is against this background that we attempt to present a modern, unified theory of quantitative genetics. Such a theory must draw heavily from population genetics, statistical theory, mathematical modeling, genetics, and genomics. It must also be built upon empirical observations from a very wide range of fields. As such, it must consider changes in complex traits in populations of organisms ranging from long-term laboratory studies on viral and bacterial evolution, much shorter-term selection experiments on metazoans, the exhaustive experience of breeders of a wide range of domesticated plants and animals, recent information from human biology, and a growing number of model (and also non-model) species in wild populations. Quantitative-genetics theory also needs to consider the full range of mating systems: inbred and outbred populations of sexual species and the clonal populations of asexual species, and it must also be applicable to haploids, diploids, and polyploids. It should be flexible and extensible, being able to incorporate new sources of information as they become available (e.g., functional-genomics features). Finally, such a theory should be empirically testable, making predictions (such as selection response) based on quantities that we have some hope of estimating.

Given the vast diversity of stakeholders in quantitative genetics, it is not surprising that very useful ideas, results, and machinery developed in one subfield may only slowly (if at all) migrate out into the broader community of users. A critical goal of this book is to facilitate this dissemination. For example, it is not fully appreciated that modern quantitative genetics is now fully merged with population genetics and is a growing part of genomics. Indeed, despite concerns of being outdated and in need of major revision (e.g., Nelson et al. 2013), the field is extremely robust and quite healthy. Modern quantitative genetics is the glue that connects many disciplines, especially given its ability to model uncertainty while fully accounting for known genetic and genomic features.

This introductory chapter starts with a historical overview of considerations of the evolution of quantitative traits. We then briefly comment on the theoretical foundations of evolutionary biology as they relate to qualitative traits (addressing some common misconceptions). We conclude with an overview and road map of this volume.

Evolution and Selection of Quantitative Traits. Bruce Walsh & Michael Lynch.
Published 2018 by Oxford University Press.
DOI 10.1093/oso/9780198830870.001.0001

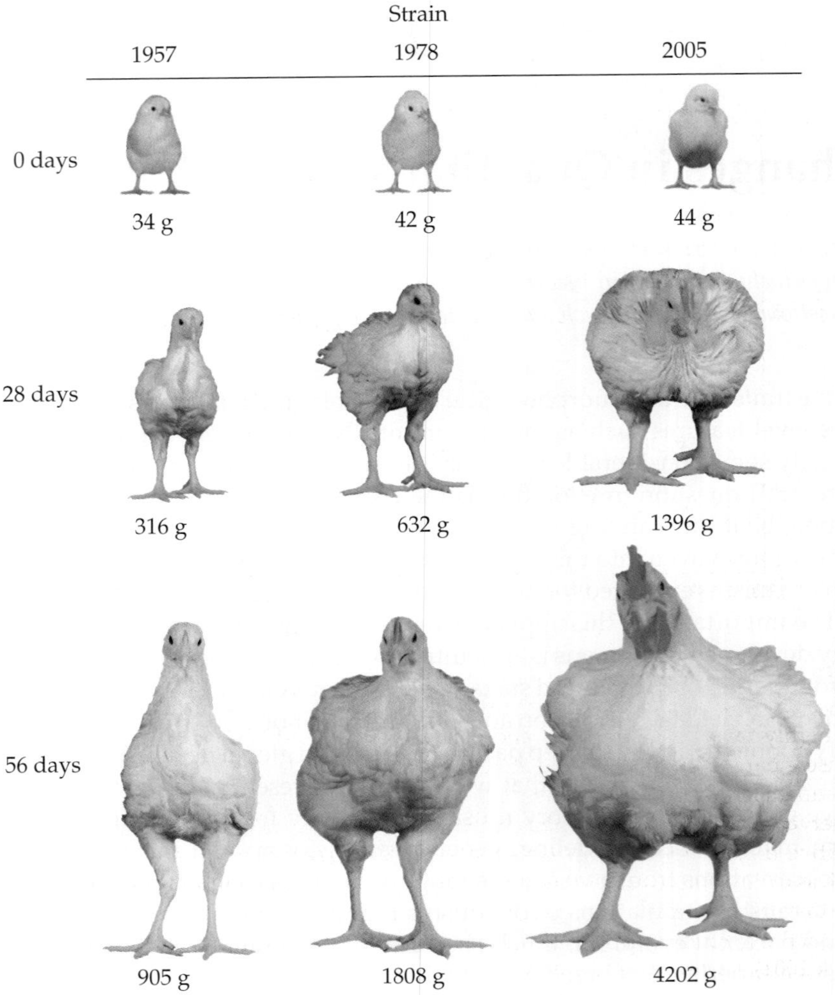

Figure 1.1 A striking example of the power of artificial selection is seen in lines of broilers, chickens selected for meat production. The figure compares the performance of a representative base line from 1957, an improved line from 1978, and a recent line from 2005, all raised in the same environment. Besides these extremely impressive differences in growth, there are equally important improvements in feed-to-protein conversion efficiency. (After Zuidhof et al. 2014.)

A BRIEF HISTORY OF THE STUDY OF THE EVOLUTION OF QUANTITATIVE TRAITS

The Fusion of Population and Quantitative Genetics

Although the histories of population and quantitative genetics are highly intertwined, they have experienced long periods of apparent separation. Their initially separate trajectories trace back to the bitter debate between the **biometricians** and the **Mendelians** following the rediscovery of Mendel in the early 1900s (Provine 1971, 2000; Tabery 2004). Motivated by Mendel's work on the inheritance of major genetic factors, the Mendelians believed that evolution was saltational: moving only by major leaps via the appearance of new mutations. The biometricians (supporters of Darwin and the founders of modern statistics) introduced concepts such as regressions and correlations to quantify continuous variation, and they thought that evolution occurred in very small steps by exploiting this variation. This initial schism between inheritance (Mendel) and evolution (Darwin) was largely driven by antagonistic personalities in the different camps (Provine 1971; although see Hogben 1974 for a

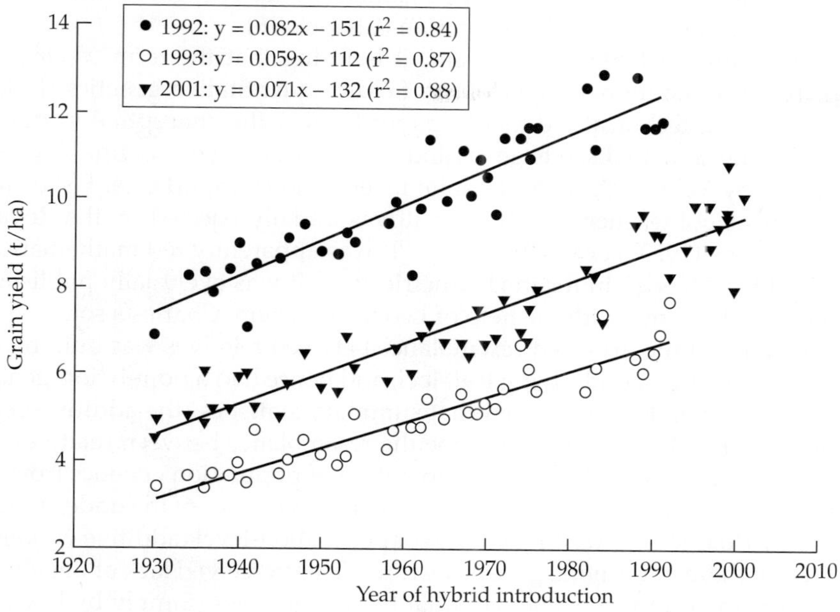

Figure 1.2 Yield (measured in tons per hectare) in maize hybrid lines as a function of year of release. Using remnant seed, all lines were grown in the same set of years, with 1992 being highly favorable, 1993 cool and extremely wet, and 2001 hot and dry. Note that the response is parallel over the three different environments (years), suggesting little genotype × environment interaction. Such "common-garden" experiments are the cleanest way to separate an observed gain into genetic versus environmental components (Chapters 19 and 20 present alternative mixed-model approaches that can also accomplish this goal). This separation is critical, as a yield improvement over time could simply reflect improved agronomic practices, rather than genetic gain. In maize, hybrid improvement accounts for between 50 and 70% of the total improvement in yield, with the remainder due to improved farming practices (Duvick 2001). (After Duvick 2005.)

different perspective), and significantly delayed the modern synthesis—the fusion of Darwin's evolution by natural selection with Mendelian inheritance. Vestiges of this difference between the gene-based focus of the Mendelians and the continuous-trait focus of the biometricians still persist today.

The first population-genetics paper, which predated the rediscovery of Mendel (and hence was published well before any considerations of the actual dynamics of genes), was concerned with a quantitative-genetics question. Fleeming Jenkin, Regius Professor of Engineering at the University of Edinburgh, was asked to write a review of Darwin's *Origin of Species*. Jenkin was a polyglot, and, among other things, invented the cable car and was an actor and an artist. Jenkin (1867) pointed out that, under blending inheritance, half of any variation is removed in each generation. Thus, any change achieved via natural selection would quickly be diluted away by the assumed nature of inheritance (also see Bulmer 2004). Phrased in terms of modern statistics, and recalling that the variance of a product of a constant (a) times a random variable (x) is $\sigma^2(ax) = a^2\sigma^2(x)$, his argument was that

$$\sigma^2(z_o) = \sigma^2\left(\frac{z_f + z_m}{2}\right) = \frac{\sigma^2(z_f)}{2^2} + \frac{\sigma^2(z_m)}{2^2} = \frac{\sigma^2(z)}{2}$$

where z_f and z_m are the (assumed to be uncorrelated) paternal and maternal trait values, z_o is the offspring value, and we assume that the trait variance is the same in both sexes, so that $\sigma^2(z_f) = \sigma^2(z_m) = \sigma^2(z)$. Mendelian genetics provides the solution to this quandary of reduced variation following reproduction, as the segregation variance (variation generated by the transmission of alternative alleles in the gametes of heterozygous parents) completely restores the genetic variance in each generation (Chapter 16).

Given the nature of Jenkin's argument, it is fitting that the term *variance* was first introduced in the foundational paper of the field of quantitative genetics (Fisher 1918). Fisher's brilliant (yet technically difficult) paper formed the conceptual bridge between Mendelian inheritance with discrete genes and evolution acting on continuous traits (ideas hinted at earlier by Yule 1902, 1906). Despite its enormous importance, Fisher wrote this paper as a high school teacher in 1916, and it was initially rejected by the Royal Society of London. As Crow (1972) succinctly stated, "It was apparently too mathematical for the Mendelists and too Mendelian for the biometricians." It was eventually published by the Royal Society of Edinburgh with the help of Leonard Darwin, Charles's son.

Fisher's theory of the expected resemblance between relatives was built by considering the sharing of alleles at underlying trait loci, and hence had a population-genetics core. Operationally, however, Fisher showed that summary statistics (the additive-genetic and other variance components) largely describe the resemblance between relatives. This was the undercurrent for much of the passive separation of population genetics from quantitative genetics. Although the latter was built around the dynamics of the underlying genes, it focuses on individual-level (breeding values) or population-level (additive variances) summary statistics of these underlying genetic features. This perceived lack of detailed genetics is both the strength and the weakness of quantitative genetics. Simply by having phenotypic measures of a trait on a set of known relatives, a number of short-term predictions (such as the response to selection and the expected resemblance between different sets of relatives) can be made in the absence of any other genetic information.

Immediately following Fisher's paper, major contributions started to appear from the two other founders of population genetics, Sewall Wright and J. B. S. Haldane, who also helped solidify the foundation of quantitative genetics initially laid out by Fisher. Despite this trinity all making deep contributions to the founding of both fields, one source of the passive separation of quantitative from population genetics was the multilocus nature of the former and the initial focus of the latter on single-locus models (Chapters 2 and 5).

No explicit considerations of multiple-locus models of selection were made until the late 1950s (Kimura 1956; Lewontin and Kojima 1960; Chapter 5). The complexity of the dynamics under even two-locus selection models seemed to prevent an easy generalization to the types of genetic architectures thought to be typical of quantitative traits, namely, a large number of loci. However, as detailed in Chapter 5, most of the complications in multilocus systems arise because of selection-generated gametic-phase (linkage) disequilibrium (abbreviated LD). If LD is assumed to be fairly weak (in other words, recombination is strong relative to selection on a given locus), single-locus results (for additive loci) can be used in models predicting selection responses (Chapters 5, 24–28). An important extension of this simplifying assumption was advanced by Bulmer (1971b, 1974a), who showed that the cumulative effects of even small amounts of selection-generated LD have a significant impact on the genetic variance when a large number of loci underlie a trait under selection. Fortunately, the expected LD-generated change in the additive variance is largely predictable (at least over a few generations) given the value of the **genic variance**, σ_a^2 (the additive variance in the absence of LD; Chapters 16 and 24).

Given the assumption of weak selection (on any given locus, but not necessarily on the trait as a whole) and loose linkage, much of the theory of short-term selection response in quantitative traits is built around the **infinitesimal model**, a simplification wherein a large number of loci, each of small effect and usually additive, are assumed to underlie a trait (Chapter 24). In this setting, selection-induced allele-frequency changes at the underlying loci are assumed to be small for several generations, allowing response (both in terms of the mean and any LD-induced changes in the variance) to be largely predictable from information on the resemblances between relatives in the base (i.e., unselected) population. This setting, which is based on parent-offspring regressions, is the foundation for much of modern breeding theory (Chapters 5, 6, 13–23).

Starting in the late 1980s, more realistic, yet partly tractable, models of multilocus selection beyond the infinitesimal model started to appear (Barton and Turelli 1987; Keightley and Hill 1987; Turelli 1988; Turelli and Barton 1990; Bürger 1991a; Chevalet 1994; Chapter 24). To date, the issues with any multiple-locus model are twofold. First, the dynamics can be very complex, making generalizations difficult (but see Chapters 6 and 24). Second, and more critically, predicting the long-term behavior of such multiple-locus genetic systems requires, not just manageable theory, but also a detailed knowledge of the underlying genetic architecture of a trait (number of genes, allelic frequencies, and effect sizes) that is largely unavailable at the present, and likely to remain so into the rather far future (but see Visscher 2016 for a different perspective).

Finally, the foundations for the field of molecular population genetics were laid by Kimura (1955a, 1955b, 1957), and started with Kimura and Crow's (1964) infinite-alleles model (the first attempt to fully model mutation at the DNA-sequence level; Chapter 2). Although this field initially appeared to be very tangential to quantitative genetics, even here there was a strong initial connection with quantitative-genetics theory, as Kimura (1965a) used a modification of the infinite-alleles model to examine the expected distribution of allelic effects at a locus underlying a trait under stabilizing selection (Example 28.4).

Realizing that stochastic forces are extremely important in molecular evolution, Kimura marshaled the power of diffusion approximations (Appendix 1) to examine the interactions of evolutionary forces in finite populations, culminating with his **neutral theory of molecular evolution** (Kimura 1968b, 1983). Again, as with other fields that at first were perceived as only tangential, there were early applications of this molecularly driven theory to quantitative genetics. Robertson (1960a) realized that Kimura's (1957) results on the fixation probability of an allele under drift and selection were directly applicable to breeding. This led to his remarkably simple expression for the expected limit to the selection response: the expected total long-term response (ignoring new mutations) is bounded above by $2N_eR(1)$, where N_e is the effective population size (Chapter 3) and $R(1)$ is the expected response to selection in the first generation (Chapters 13 and 26).

A watershed event in molecular population genetics was Kingman's (1982a, 1982b) introduction of the powerful concept of the **coalescent** (treating a sample of alleles from a population as a genealogy, which eventually coalesces to a common ancestral sequence at some point in the past; Chapter 2). This approach allowed for the development of statistical tests on the nature of forces shaping sequence evolution (Chapters 4 and 8–10). The very slow trickle of molecular data on which to apply this theory in those early days has now become a veritable tsunami of whole-genome sequences at the population level. It is this very flood of population-genomics data that is now driving the fusion of molecular and quantitative genetics. Much of the theoretical foundations for this fusion derive from results originally developed for, and envisioned as being restricted to, molecular population genetics.

The Ongoing Fusion of Molecular and Quantitative Genetics

In large part, the interaction between so-called classical genetics and quantitative genetics that played out during the rise of molecular genetics parallels much of the debate between the early Mendelians and biometricians, but with less rancor. During the rise of molecular genetics in the 1950s (nicely reviewed by Judson 1979 and Mukherjee 2016), the notion of a gene grew from a fairly nebulous concept into a rather concrete, well understood, object. This resulted in any statistical description of inheritance from quantitative genetics (variances of traits instead of following genes) becoming far less appealing. Indeed, to many geneticists, quantitative genetics appeared as an anachronism, a crutch that biologists no longer needed. As massive advances were made in the dissection of specific, individual genes with discrete, and highly reproducible, phenotypes, the result was a diminished interest in continuous traits. Indeed, Fisher's notion of an infinitesimal model—a large number of loci, each with very small effects—was anathema to this molecular way of thinking.

By the early 1980s, one of the byproducts of this revolution was access to an ever-growing number of molecular-based markers. Their availability opened up the potential for quantitative trait locus (QTL) mapping, which involves the isolation of small chromosomal regions that influence a trait of interest by applying linkage analysis to line-cross derivatives (LW Chapters 14 and 15). Such studies routinely found relatively small (~10s of megabases) chromosomal segments that appeared to account for nontrivial fractions of the trait variation (5% or more; e.g., LW Figure 15.13). At odds with infinitesimal-like model thinking, this finding reinforced the view of many molecular geneticists on the importance of single genes.

The apparent presence of such major genes suggested that population-genetic models working with these genes (once they have been isolated) could largely supplant standard quantitative-genetics approaches. However, the actual isolation of such candidate genes proved extremely elusive. Eventually, this led to the realization that the large estimated effects of such genes were very often the result of significantly upward estimation bias and the presence of tightly linked clusters of QTLs that fractionate upon finer mapping (Chapter 24).

By the early 2000s, as molecular-marker technology continued to advance, the ability to quickly and cheaply score tens of thousands of markers allowed for association mapping (LW Chapter 16). The foundation of this approach rests on the use of linkage disequilibrium in random samples of unrelated individuals from a large population to fine-map genes (an idea deeply rooted in coalescent theory). As detailed in Chapter 24, one generalization from these studies was that only a very tiny fraction of the variance (typically far less than 1%) of a typical trait can be ascribed to any particular genomic region (with resolution now on the kilobase scale). The use of such dense marker coverage showed that the response to artificial selection is usually the result of changes over a large number of sites (Chapter 25), and their availability has facilitated tests of selection in any given region within a genome (Chapters 9 and 10).

The culmination of these association studies, involving up to millions of scored single-nucleotide polymorphisms (SNPs) on tens of thousands of individuals, was the crisis of apparently "missing heritability" (Manolio et al. 2009). The concern among many human geneticists was that, for all of their hard work on highly heritable traits such as height, detected SNP markers could only account for ~10% of the heritability found by standard biometrical analysis (i.e., resemblance between relatives). The result was a near-hysteria on the part of some, as they felt that because the molecular results *must* be correct, something fundamental was being overlooked in quantitative genetics. In reality this disconnect was simply a manifestation of the narrow focus that arose from being trained for decades in a single-gene culture. As pithily remarked by Sheldon (2014),

> There is something simultaneously remarkable and encouraging about the fact that a centuries-old method requiring no more than a ruler, a pencil and (I suppose) a slide rule outperformed, by an order of magnitude, the fruits of the genomic revolution.

As detailed in Example 24.1, such a pattern of "missing heritability" is exactly what is expected when each underlying locus only explains a small fraction of trait variation (namely, a large number of underlying loci in partial linkage disequilibrium with the scored markers).

Thus, we have come full circle, with the genomic (i.e., association) data providing strong support for an infinitesimal-like model for the genetic architecture of many traits. However, the subtle distinction from the standard infinitesimal model is that while any particular site accounts for only a very small fraction of the *variance* of a trait, rare segregating alleles having large individual *effects* can be found (Figure 28.8). A quantitative-genetic framework is thus required when using association data for genomic prediction (the framework for genomic selection and genomic-based individualized medicine). A quick glance at any of the front-line journals in genomics highlights this growing merger with quantitative genetics, as many functional-genomic features—such as expression levels of RNAs, proteins, and metabolites—are now routinely examined in a quantitative-genetics framework.

A final example highlighting the connectedness between genomics and quantitative genetics concerns the **Hill-Robertson effect**. In contemplating the consequences of linkage for applied breeding, Hill and Robertson (1966) noticed that the presence of linkage reduced the chances of the joint fixation of linked beneficial alleles (Chapters 3, 7 and 8). While their work was regarded as entirely theoretical at the time (they never had any expectations of having the marker density required to test it), their idea that selection reduces the effective population size at linked sites is now a key organizing principle in the evolutionary analysis of genomic sequence data (Chapters 3 and 8).

The Common Thread Between Breeding and Evolution in Natural Populations

The connection between trait change through artificial selection in domesticated organisms and trait evolution in natural populations was one of the pillars of evidence offered by Darwin in support of his theory of evolution by natural selection. It is thus not surprising that the subfields of breed improvement (which is simply applied evolution) and evolutionary biology have historically had considerable interactions (Hill and Kirkpatrick 2010; Hill 2014; Gianola and Rosa 2015). Both fields are concerned with predicting an expected change in a trait (or vector of traits) given a specific pattern of selection. Evolutionary biologists face the additional, and daunting, challenge of trying to estimate the nature and target (or targets) of selection (i.e., assigning fitnesses to specific trait combinations; Chapters 20, 29, and 30). We consider this last topic separately, after first examining the common connections between the work of breeders and evolutionary biologists.

The introduction of the full power of quantitative genetics into breeding is in large part due to Jay Lush, and his extremely influential book, *Animal Breeding Plans* (Lush 1937). In particular, he proposed the classical breeder's equation, $R = h^2 S$, for predicting the response to a single generation of selection. Lush was heavily influenced by Sewall Wright, as in 1931 he commuted weekly from Iowa State to the University of Chicago to attend Wright's statistical genetics course. He was also influenced by R. A. Fisher, who lectured at Iowa State during the summers of 1931 through 1936. The extension from the univariate breeder's equation to the response to selection on multiple traits, in the form of selection on a weighted index of trait values, was influenced by both Lush and Fisher. Smith (1936), who acknowledges Fisher's help, and Lush's colleague Hazel (1945) both examined the expected selection response and optimal weights of such an index of trait values. Lush (1947) applied the Smith-Hazel results to obtain the optimal weighting of within- and among-family information for artificial selection programs. Smith and Hazel's work also introduced the very powerful concept of genetic correlations.

The breeder's equation is the cornerstone for much of the predicted response to selection in natural populations, and its multivariate extension, which is due to Lande (1976), builds heavily on the ideas of Smith and Hazel, as well as Pearson (1903). Hence, ideas that were initially developed for evolution (Wright's interests) were translated into animal breeding by Lush, and then subsequently found their way back into the evolutionary literature for modeling the expected reponse in natural populations (Chapter 20; Slaktin 1970; Roughgarden 1972; Lande 1976).

An alternative to the breeder's equation for predicting selection response was proposed by Alan Robertson (1966a) in a very opaque paper on the culling process in cattle. He proposed that the selection response in a trait could be expressed as the covariance between the breeding value of that trait and relative fitness, $R = \sigma(A_z, w)$. As this expression was based on an extension of **Fisher's fundamental theorem of natural selection** (which states that the rate of response in fitness is simply the additive variance in fitness; Chapter 6), it is often called the **secondary theorem** (Orr 2009 suggested that Robertson's theorem is actually more fundamental, and that Fisher's theorem follows as a special case). Price (1970) independently suggested a completely general covariance-based expression for the response to *any* system of selection (Equation 6.8). While both the Robertson and Price results are widely used in evolutionary biology (Chapters 20 and 22; Frank 1995), they have only recently made their way back into breeding, most notably in the analysis of associative effects (Chapter 23; Bijma et al. 2007a).

Until very recently, most selective breeding in outbred populations was based on **best linear unbiased predictor (BLUP)** selection (Chapters 18 and 20), an approach introduced by Charles Henderson (1975). In keeping with the focus on breeding values seen in Robertson's secondary theorem, Henderson noted that one can significantly improve the estimate of individual breeding value by considering *all* of the known relatives in a sample. This is accomplished using linear mixed models (Chapters 19 and 20; LW Chapters 26 and 27) to return BLUPs of individual breeding values. The individuals with the highest BLUPs are then chosen as the parents for the next generation, with the expected selection response simply being the average breeding values of the selected parents. As this same approach yields a **restricted maximum likelihood** estimator (or **REML** estimate) of the additive-genetic variance, evolutionary biologists realized that mixed models could be used to estimate trait heritabilities in natural populations using a set of known relatives (Shaw 1987; Cheverud and Dittus 1992; Kruuk 2004). As detailed in Chapter 20, there was some initial overreach in the early 2000s in an attempt by evolutionary biologists to use BLUPs to assess direct targets of selection, but the BLUP framework still remains a powerful tool for evolutionary analysis.

The most recent chapter in the development of breeding methods is the use of dense marker information. Initially motivated by the apparent ubiquity of QTLs of major effects, breeders flirted with **marker-assisted selection** (**MAS**) schemes, wherein a few well-chosen markers were used to augment phenotypic information when selecting individuals (Lande and Thompson 1990). However, in response to a renewed appreciation for the infinitesimal model, molecular-based breeding has moved to **genomic selection** (Meuwissen et al. 2001), wherein mixed models are used to assign weights to tens of thousands of SNP-tagged segments that span the genome. The simplest variant of this approach recovers BLUP selection (Chapters 13 and 19), but now with the **A** matrix of all pairwise relationships between individuals in the sample being replaced by a marker-based estimate of the fraction of shared markers (**G-BLUP**, for genomic-BLUP). With the increasing availability of dense SNP markers (and now whole-genome sequences) for even nonmodel organisms, marker-based estimations of relationships are now widely used for REML estimates of additive variances, and also (to a lesser extent) for association mapping, in natural populations (Chapter 20).

Detecting Selection in Natural Populations

Given the biometrician focus on evolution acting on continuous variation, early attempts to measure natural selection in the wild predated Fisher's (1918) treatment of complex trait genetics. Although this focus on detecting selection on metric traits was entirely concerned with the phenotypic level, it marked the beginning of evolutionary quantitative genetics. Two early such attempts were particularly noteworthy. The first was due to Weldon (1895, 1899), who examined morphological features in adult and juvenile crabs (*Carcinus maenas*) in Plymouth Sound and in the Bay of Naples, generating data that enticed Karl Pearson (1903) to start a serious consideration of the statistics of selection. The second was the classic study by Herman Bumpus (1899). During a severe ice storm in February 1898, Bumpus collected 136 immobilized house sparrows (*Passer domesticus*) from several locations around Providence (in the US state of Rhode Island). Of these, only 72 survived. For all 136 individuals, Bumpus recorded sex and male maturity state (adult vs. juvenile) and presented meticulous measurements of their body weight and eight skeletal features. As reviewed in Chapter 30, Bumpus's results continue to be examined today, illustrating the power of carefully collected, and widely disseminated, datasets.

While numerous studies during the early to mid-1900s focused on the fitness consequences of discrete traits, such as various color morphs (reviewed by Endler 1986; Reznick and Travis 1998), the explosion of studies on fitness surface estimation (obtaining an expression for $W(z)$, the expected fitness of an individual with trait value z) began with the method proposed by Lande and Arnold (1983), which built on Pearson (1903). As detailed in Chapters 29 and 30, Lande and Arnold proposed fitting a quadratic regression of fitness as a function of the phenotypes of candidate traits. Although purely phenotypic in nature, this approach has deep connections, not only with Pearson's work, but also with the index-selection results of Smith (1936) and Hazel (1943). While most estimates of the nature of selection on a particular trait (or combinations of traits) focus on morphological characters, nothing about this approach prevents it from being applied to *any* quantitative trait. Hence, measures of physiological features (such as the levels of target hormones) or genomic-level characters (such as levels of RNAs, proteins, and metabolites) all hold as appropriate traits. The only challenge (which is far from trivial) is their measurement in a meaningful way in natural populations where individual fitnesses have been assessed.

Both Lande-Arnold fitness estimation, and the subsequent methods it spawned, require data on individual fitness, which is often difficult to collect (Chapters 29 and 30). A related question of deep evolutionary interest is whether an observed pattern of trait divergence (either between populations or over time within a single population) is due to selection. Such questions are typically framed using the null hypothesis of whether this observed pattern is consistent with drift alone. The roots of many of the current approaches for testing patterns of trait divergence trace back to Lande (1976), in which he explored whether an observed amount of phenotypic divergence in a specific trait in the fossil record of a species exceeded that expected by drift alone. Unlike methods that estimate whether a particular trait is *currently* under selection, Lande's (1976) approach does not require estimates of individual fitness, but does require models for the expected change in the trait over time under drift. Chapter 12 examines a number of such approaches for constructing null models, be it for a fossil sequence, the trait divergence between extant populations, or genomic data. Given the reckless abandon with which some biologists suggest adaptive scenarios, null models are unquestionably required to judge the significance of a particular pattern of trait divergence. This is especially true for functional-genomic features, such as the levels of mRNA expression among sibling species (Chapter 12).

A final approach for detecting selection, which follows as one of the fruits of the genomic revolution, involves tests for selection on specific genomic locations (Chapters 8–10). These tests have the advantage of being trait-independent (no specific candidate traits are assumed). Further, contemporaneous fitness data are not needed, and indeed such genomic signatures of selection can be a consequence of a past history (perhaps long ago) of selection that may not be presently ongoing. The downside of these purely marker-based approaches is that it may be rather difficult to associate a genomic region with a change in a specific trait, leaving the target of selection unclear.

THE THEORETICAL FOUNDATIONS OF EVOLUTIONARY BIOLOGY

Lewin's maxim that "there is nothing more practical than a good theory" (Lewin 1943; McCain 2015) is highly apropos for evolution. The basic idea of evolution is so inherently simple—organisms change over time from preexisting ones—that it appears to be the ultimate play-at-home game. Despite the care with which most biological studies are presented, often the evolutionary comments are wildly speculative, with little apparent grounding in modern theory. Even worse are the sporadic, and usually rather forceful, comments that routinely appear in the (nonevolutionary) literature that suggest something is woefully wrong, and/or missing, in the current theory. We conclude our introductory remarks by offering some commentary on these issues.

The Completeness of Evolutionary Theory

Without a theoretical framework, science is simply a fact-establishing enterprise. The emergence of facts from consistent observations is progress, but theory provides a mechanistic explanation of the material basis of facts. A theoretical framework also provides the basis for making predictions in areas where observations have not previously been made, not simply based on statistical extrapolation but rather on arguments from first principles. Mathematical theory is particularly desirable, as it allows the development of logical arguments from well-defined assumptions, whereas verbal theorizing can easily go awry in the analysis of complex systems. Fortunately, evolutionary biology has a well-established framework of principles from which to draw.

As noted above, starting with Fisher's seminal 1918 paper, which both founded the field of quantitative genetics and established a number of principles upon which modern statistics relies, the field of population genetics has experienced a rich history. Indeed, the field of theoretical population genetics, which is at least as well-grounded as any other area of quantitative biology (including biophysics), forms the formal foundation for all of evolutionary theory.

The earliest period of theoretical development (1918–1950) substantially preceded any knowledge of the molecular basis of genes. Starting in the 1950s, dramatic new findings in the field of molecular genetics emerged, including the discovery of DNA as the ultimate genetic material, the mechanisms of recombination, the basic structure of genes and their component parts, the central roles of transcription and translation (and downstream modifications), the existence of mobile genetic elements, and various nuances related to epigenetic inheritance. Yet none of these discoveries led to any alteration in the basic structure of evolutionary theory. Such robustness in the face of revolutionary changes in our understanding of genetics at the molecular level speaks volumes about the generality of our current theoretical framework. Important specific applications may remain to be developed, but the theoretical foundations of evolutionary biology are remarkably accepting of new observations.

Nonetheless, periodic claims are made that the field of evolutionary biology is in a phase of upheaval, with the harbingers of such messages seldom offering the much-needed solution to the (in their view) previously unappreciated problem. Without exception, these episodes have gone badly. Perhaps the most notable of these excursions was Goldschmidt's (1940) argument that large changes in evolution are products of macromutations with coordinated developmental effects (a flashback to the Mendelians), and Lysenko's rejection of natural selection in favor of the inheritance of acquired characteristics, which led to a decades-long demise of genetics in the Soviet Union.

One of the more recent alarmist exercises involves a tiny but vocal community clamoring for an "extended evolutionary synthesis" (EES), and claiming that "The number of biologists calling for change in how evolution is conceptualized is growing rapidly," and that there is currently a "struggle for the very soul of the discipline" (Pigliucci and Müller 2010; Shapiro 2011; Laland et al. 2014, 2015). It is interesting that the list of individuals engaged in this struggle appears not to extend beyond the authors themselves, and the level of discourse is reminiscent of the distant "bean-bag genetics" diatribe of Mayr (1959, 1963), which was promptly disemboweled by Haldane (1964). No glaring errors in contemporary evolutionary theory have been pointed out by the EESers, no evidence of familiarity with current theory has been provided, and no novel predictions have been offered (Welch 2017). There is only a warning that once qualified theoreticians come on board, the revolution will begin.

One of the more dramatic claims is that the discovery of various epigenetic effects amounts to a game-changer in evolutionary biology, imposing the need to revamp our general understanding of inheritance and its evolutionary implications (Jablonka and Lamb 2005; Caporale 2006; Shapiro 2011). Highlighted phenomena include base modifications on DNA, histone modifications on nucleosomes, and mechanisms of gene regulation by small RNAs, all of which can, in principle, have transgenerational effects without direct changes at the level of genomic DNA. Advocates of epigenetic inheritance have generally argued that such phenomena respond in beneficial ways to environmental induction, which then allows for an acceleration in the rate of adaptive phenotypic evolution. This echo of Lysenkoism, in effect, resurrects the concept of the inheritance of acquired characteristics.

The logic underlying the entire subject of epigenetic contributions to evolution has been masterfully dismantled by Charlesworth et al. (2017), and here just two points are made. First, to appreciate the implausibility of a long-term contribution of nongenetic effects to phenotypic evolution, one need only recall the repeated failure of inbred (totally homozygous) lines to respond to persistent selection, with many such experiments dating back to the beginning of the twentieth century (Lynch and Walsh 1998). The absence of genetic variation does not imply an absence of environmental or epigenetic sources of phenotypic variation, yet there is no permanent response to selection within populations unless there is variation at the DNA level.

Second, with respect to the claim that evolutionary theory is incapable of addressing the matter of epigenetic inheritance, one need only turn to models for the inheritance of environmental maternal effects developed well before the discovery of the molecular basis of any epigenetic effects. Existing theory readily demonstrates that variance in maternal effects can indeed contribute to the response to selection, but unless such effects reside at the DNA level, the response is bounded, owing to the fact that transgenerational effects are progressively diluted out (Chapters 15 and 22). The response to selection on environmental (epigenetic) maternal effects is also transient, and it decays away if the selective pressure is eliminated. Finally, if epigenetic effects are sufficiently stochastic, they will reduce, rather than enhance, the response to selection, owing to the reduction in the correspondence between genotype and phenotype.

A persistent claim to novelty by the EESers is the idea that the environmental induction of a trait in a novel situation can enhance the exposure of the trait to selection, thereby magnifying the response to selection. This, however, is by no means a new insight. Rather, such effects are central to the concept of genotype × environment interaction, the theory of which dates back decades. Indeed, breeders have long exploited this concept to determine the optimum environmental setting in which to select for particular phenotypes. Thus, the idea that evolutionary theory needs to be remodeled to account for phenotypic plasticity is also without merit, and simply reflects a basic lack of familiarity of the existing evolutionary quantitative-genetics literature. Hopefully, the material presented in the following pages will provide some aid in this regard.

Another argument raised against the adequacy of evolutionary theory involves the idea that the field is incapable of dealing with the possibility that evolution "is guided along specific routes opened up by the processes of development" (Laland et al. 2014). This claim again belies a lack of awareness of long-existing evolutionary theory. Although models concerned with the role of mutation in phenotypic evolution commonly assume a symmetrical distribution of mutational effects, it is relatively easy to modify any quantitative-genetic model to allow for biased mutational effects. Indeed, the classical house-of-cards model has a structure in which mutation bias is intrinsic—here there is a fixed set of possible allelic effects, so there is always a directional bias to mutation unless the mean genotypic value is positioned at exactly the intermediate location along the axis of possibilities. As for the role of developmental bias in evolution, one need only consider the substantial theory devoted to multivariate evolution that is explicitly focused on pleiotropy and the ways in which this influences the evolutionary trajectories of complex traits (e.g., Jones et al. 2003, 2007, 2012).

Most remarkable is the EESer claim that the key flaw of contemporary evolutionary theory is the assumption that change in allele frequencies is a necessary component of the response to selection, their counterview being that "the direction of evolution does not depend on selection alone, and need not start with mutation" (Laland et al. 2014). Whereas it has long been appreciated that evolution can, and sometimes does, occur in the absence of selection (for example, by random genetic drift of neutral traits), we await an explanation as to how any sustained form of evolution (aside from cultural evolution) can occur in the absence of genetic variation. Technically speaking, evolution can occur in the absence of allele-frequency change, but only via changes in the form of allelic associations across loci (e.g., via LD, which necessarily implies *genotype*-frequency change), and this does not appear to be what the dissenters have in mind.

Finally, we note that unlike the laws of physics and the features of chemical elements, biology is subject to historical contingencies, and for virtually every set of general observations, one can find some kind of exception. Aficionados of such exceptions are sometimes tempted to claim that their observations are sufficient to dismantle the previous theoretical framework for broadscale patterns. However, more often than not, a deeper look almost always reveals explanations for exceptions that are fully compatible with the rules of life. The discovery of mitochondrial DNA and nuclear epigenetic effects did not alter our basic understanding of maternal effects, and the discovery of transposable elements did not alter our appreciation of the mutational process. Such observations simply provided a deeper molecular explanation of modes of production of phenotypic variation. Because evolution is a stochastic process, no theoretical framework can ever be expected to predict the exact trajectories of evolution at the molecular, cellular, or developmental levels. As Haldane (1964) pointed out, if population genetics could make such specific predictions, it would not be a branch of biology—it would be the entirety of biology.

Far from providing a weak and/or incomplete caricature of evolving genetic systems, population- and quantitative-genetic theory has provided powerful, general, and sometimes unexpected mechanistic explanations for trait variation and phenotypic evolution. We note several of these in Table 1.1, and emphasize that few of these issues would have ever been revealed and/or resolved with simple verbal arguments. Indeed, as noted above, it was Fisher's (1918) paper that rescued the previously verbalized theory of evolution from the high seas of obfuscation. Having resisted quantitative thinking derived from first principles in genetics, developmental biology, from which many of the pleas for novel theory emanate, remains in many respects in a pre-population-genetics mode of confusion.

Although our specific railing on the EES ideology may be offensive to some or seem like pandering to trivia to others, the implication that a century's worth of theoreticians have been woefully misled is a gross misrepresentation of the facts. Repeated peddling of this view by individuals with no intention of confronting the facts belies reality. As outlined in Table 1.1 and expanded upon in the following chapters, the structure of evolutionary theory that has been developed over the past century has, to this point, found no boundaries in terms of applications, has not been challenged by novel findings in the molecular era, and continues to make predictions that are bolstered by empirical observations. While conflict and cooperation are the engines that keep science running, conflict engineered under false pretenses and incessantly repeated with no evidence is generally motivated by nonscientific goals.

Nonadaptive Hypotheses and our Understanding of Evolution

Darwin's and Wallace's grand views of selection as a natural force for the emergence of adaptive change marked a key moment in the history of biology. So convincing were they, and their popular-science writing disciples such as Dawkins, that most people (including most biologists) view all aspects of biology to be necessary products of natural selection. As it channels the entire field away from a landscape of unbiased study, this dogmatic adherence to the unbounded power of natural selection constitutes one of the most significant problems in evolutionary science.

Whereas natural selection is one of the most pervasive forces in the biological world, it is not all-powerful. As is evident throughout the following pages, the genetic paths open to exploitation by selection are strongly influenced by another pervasive force—the noise in the evolutionary process imposed by random genetic drift which is caused by both the fact that there are finite numbers of individuals within populations and the physical linkage of different nucleotide sites on chromosomes. If the power of selection is weak relative to that of drift, as is often the case at the molecular level, evolution will proceed in an effectively neutral manner. Biased mutation pressure can also play a role here if it is sufficiently strong relative to selection, steering evolution in the direction of mutation bias.

Even if selection were to be pervasive, in order to understand the degree to which natural selection molds the features of populations, it is essential to know what to expect in the absence of selection. It is for this reason that neutral models have been repeatedly exploited in population- and quantitative-genetic analyses. Under such models, random genetic drift, mutation, and recombination are the sole evolutionary forces, with the resultant formulations then providing null models for testing for natural selection. As noted above, such

Table 1.1 A few key areas in evolutionary biology where the mathematical/statistical theory of population and quantitative genetics has enhanced our understanding of the mechanistic basis of trait variation and provided novel predictions. This list is by no means complete, and many of the references are simply exemplars, rather than a full literature citation.

Topic	References
Quantitative-trait variation	
Phenotypic resemblance between relatives, and how this scales with the degree of relationship.	Fisher 1918; Kempthorne 1954 LW Chapter 7
Inbreeding depression, and how this scales with parental relatedness.	Crow 1948 LW Chapter 10
Quasi-inheritance of familial (including maternal) effects, and transient selection response.	Willham 1963; Falconer 1965 LW Chapter 23
Expression of all-or-none traits as a function of underlying determinants.	Wright 1934a, 1934b LW Chapter 25
Pleiotropy and the genetic correlation between traits.	Mode and Robinson 1959 LW Chapter 21
Long-term patterns of evolution	
Sudden (saltational) transitions from one discrete character state to another.	Lande 1978
Rates and patterns of evolution in the fossil record.	Charlesworth et al. 1982; Charlesworth 1984b
Rapid evolution across adaptive valleys by stochastic tunneling.	Lynch 2010b; Weissman et al. 2010
Mutation bias and the inability of a mean phenotype to attain an optimal state.	Lynch 2013; Lynch and Hagner 2014
Spatial variation in genotypic values in the absence of underlying ecological variation	Higgins and Lynch 2001
Genome evolution	
The fate of duplicate genes.	Walsh 1987, 1995; Force et al. 1998; Lynch and Force 2000
Conditions for the spread of mobile elements.	Charlesworth and Charlesworth 1983 Charlesworth and Langley 1986
Evolution of codon bias.	Bulmer 1991
Evolution of transcription-factor binding sites.	Lynch and Hagner 2014
The illusion of evolutionary robustness.	Frank 2007; Lynch 2012b
Evolution of the genetic machinery	
Evolution of the mutation rate.	Lynch 2011; Lynch et al. 2016
Evolutionary consequences of sexual reproduction.	Kondrashov 1988; Charlesworth 1990; Otto and Barton 2001
Evolutionary deterioration of sex chromosomes.	Charlesworth and Charlesworth 2000
Evolutionary features of alleles	
Low probability of fixation of mutant alleles.	Kimura 1962
Ages of alleles.	Kimura and Ohta 1973
Conditional time to fixation of deleterious mutations equaling that of beneficial mutations.	Maruyama and Kimura 1964
Enhanced evolutionary divergence under uniform selection.	Cohan 1984b; Lynch 1986
Doubling the effective population size by equalizing family sizes.	Crow and Kimura 1970

models have been particularly useful in attempts to understand long-term phenotypic changes recorded in the fossil record, where seemingly dramatic changes that have elicited selection arguments are found to be less than impressive when evaluated in the proper context of drift and mutation (Lande 1976; Lynch 1990; Chapter 12). Neutral models are particularly easy to develop for DNA-level features, as mutation can be explicitly defined in terms of the possible nucleotide substitutions. Although neutral models can become more challenging in the case of more complex traits, where the baseline features of mutations can be more difficult to define, this is not a justification for ignoring the matter.

Some investigators (Pigliucci and Kaplan 2000; Hahn 2008) have suggested that so much evidence for selection has emerged that we should abandon the use of neutral theory, with Hahn going so far as to argue that "the implications of our continued use of neutral models are dire," and "can positively mislead researchers and skew our understanding of nature." No one argues that selection is unimportant, but the proposition of a selection theory as a null model for hypothesis testing has no logical basis. When properly constructed, neutral models make very explicit predictions, and without such null expectations, arguments for the role of natural selection in evolution are reduced to qualitative hand-waving. Formal rejection of a neutral model, along with measurement of the deviations between observations and expectations, yields a deeper and more defensible understanding of evolutionary processes.

The second problem with invoking the need for a "selection theory" as the null model is that one can concoct a selection argument for essentially any observed pattern, leaving no room for falsifying the null hypothesis of the universal power of selection. If one form of selection does not adequately fit the data, then one can try another, and failing that, still another, but without ever abandoning the pan-selection view. Fluctuating selection (wherein the direction of selection wanders around a mean of zero) can even yield a population-genetic history that closely mimics the expectation under random genetic drift alone, with no discrimination between models being possible without a knowledge of effective population sizes.

A third problem with criticisms of neutral theory is the reliance on incorrect assumptions to reach questionable conclusions. For example, Lewontin (1974) long ago suggested that the fact that standing variation in natural populations is only weakly associated with effective population size (N_e) suggests a violation of the neutral theory, as standing levels of variation at silent sites in diploid populations should scale with $4N_e u$, where u is the mutation rate per nucleotide site (Chapter 4). Although this argument continues to be made (Hahn 2008), it ignores the fact that mutation rates evolve to be inversely correlated with N_e, which naturally leads to weak dependence of standing variation on N_e (Lynch et al. 2016). In this particular case, the discrepancy between data and the neutral theory is greatly diminished once the appropriate null model is constructed.

The call for a selection theory of evolution, like the call for an extended evolutionary synthesis, has not resulted in any theoretical upheaval. No offering of what a novel theory of selection might be has been presented, and none is likely to emerge for the very simple reason that we already have such a theory. From the very beginning, population- and quantitative-genetic theory has fully embraced selection as a central force in evolution. Indeed, the numerous accomplishments of selection theory are summarized, explained, and celebrated in the remainder of this book.

OVERVIEW AND PATHWAYS THROUGH THIS VOLUME

Given the scope (and length) of this volume, we conclude our introduction by providing a brief roadmap and overview of the material presented in the following chapters. We envision the users of this book to be a very diverse community: animal, crop, and tree breeders; evolutionary biologists, human geneticists, and population and statistical geneticists. Given this eclectic audience, our challenge has been to present both the necessary material a user

Table 1.2 Suggested topic-specifc pathways through this book.

Topic	Suggested Chapters
Basic population genetics	2–5, 7–10, 27, Appendix 1
Applied breeding	3, 6, 13–19, 21–23, 25–26, Appendix 5
Evolutionary quantitative genetics	6, 13–20, 22, 27–30, Appendicies 5, 6
Population genetics of selection response	5–10, 24–28, Appendix 1
Detection of selection with genomic data	2–5, 7–12, 27, Appendices 1–4

might be searching for, while also highlighting connections with other areas that may not initially have been considered relevant.

We assume that the reader has some basic knowledge of the foundations of the genetics of complex traits, which is covered in our first volume (Lynch and Walsh 1998). Given that we extensively refer back to this volume, for brevity it is denoted throughout the current volume by LW. While LW focused on the genetics of complex traits, the current volume examines the forces behind the evolution of such traits: drift, mutation, recombination, and selection. Our treatment of the field will conclude in our final (third) volume, which largely deals with multivariate traits, plus a few specialized, but important, topics (such as genotype $\times$ environment interactions, inbred-line development, and breeding for heterosis). Table 1.2 offers some suggested pathways through this volume, and our overview below is organized by the major sections of this book.

Evolution at One and Two Loci

The next nine chapters develop the population-genetic theory that underpins most analyses of the evolution of quantitative traits, serving as the foundation for the remainder of this volume. Chapter 2 begins this excursion by examining drift and mutation-drift interaction, including important concepts such as the infinite-sites and infinite-alleles models of sequence evolution and the extremely powerful tool of the coalescent for treating drift.

Chapter 3 introduces the concept the genetic effective size of a population. One of the major extensions of this powerful idea over the last few decades was the realization that different regions of the genome likely experience *different* effective population sizes. This is the Hill-Robertson effect: selection acting on a particular locus reduces, perhaps quite dramatically, the effective population size experienced by linked sites. An unresolved debate is whether the majority of this interfering selection simply involves the removal of new deleterious mutations (background selection) or the fixation (or at least increase in frequency) of newly-favorable alleles (hitchhiking and selective sweeps).

Chapter 4 examines the relative strengths of the nonadaptive forces of evolution: mutation, drift, and recombination. All three of these forces can vary substantially among phylogenetic lineages and therefore can substantially influence modes of evolution. Genomics has painted a much fuller picture of these processes than was present even a decade ago, and we review results from numerous whole-genome sequencing studies.

Chapter 5 covers the theory of one and two-locus selection, concluding with a derivation from population-genetic principals of Lush's classic breeder's equation, $R = h^2S$ (which is the foundation for much of current selection-response theory). We also bridge population and quantitative genetics by showing how selection on a trait translates into the nature and amount of selection on underlying loci.

Chapter 6 is essentially the gene-free counterpart of Chapter 5, as it examines general theories of selection response, such as Price's equation, Fisher's fundamental theorem, and both versions of Robertson's secondary theorem. Price's equation is a general expression for any type of selection response, expressed in terms of variances and covariances, which we use to examine the robustness of the breeder's equation.

Chapter 7 focuses on the interactions of drift and mutation with selection, highlighting some of the less intuitive, and sometimes counterintuitive, features that can arise when these forces interact. Together with Chapter 5, the machinery developed in this chapter is used extensively in Chapters 24–28 to build the population-genetic theory of the selection response of a quantitative trait.

Chapters 8–10 apply the concepts developed in these early chapters in the search for signals of both recent and past selection in genomic data. Chapter 8 develops the theory of selective sweeps, namely the nature of genomic signatures generated by relatively recent selection events. Chapter 9 relates these results to the myriad of tests of very recent (or ongoing) selection using a sample from a population (polymorphism-based tests). Chapter 10 examines approaches using divergence data (differences between two, or more, distantly related populations or species) to search for repeated episodes of selection over a genomic region. Genomic-based searches for selection are an extremely fashionable area of evolutionary and human genetics, and, when handled appropriately, offer important tools for breeders searching for genetic footprints from domestication events. Chapters 9 and 10 are rather detailed, as numerous tests have appeared in the literature, and our goal is to provide the reader with an overview of all relevant approaches. Both of these chapters also stress the important point that all such methods have critical, indeed often fatal, flaws raising caveats to consider when applying these approaches to genomic scans for selection.

Drift and Quantitative Traits

Paralleling our development of population-genetics theory (where we first considered the roles of drift and mutation before examining the impact of selection), before we turn to selection acting on traits, Chapters 11 and 12 examine trait evolution under drift and mutation alone. Chapter 11 examines the impact of drift and mutation on the additive-genetic variance within a population. In order to fully address this issue, it also develops the rather intricate covariances between relatives that arise under general schemes of inbreeding.

Chapter 12 examines the among-population divergence in trait means under drift and mutation. While this topic may seem rather esoteric to some, especially to those with functional-genomics data, this theory is actually very relevant. Modern omics data (such as expression levels of RNAs, proteins, or metabolites) are very much quantitative traits, influenced by (potentially) numerous genes and the environment. Chapter 12 develops a number of tests for whether an observed pattern of change (be it between a set of natural populations, a temporal sequence of morphological phenotypes in the fossil record, the pattern of allele-frequency change over a set of QTL or GWAS-linked markers, or differences in gene-expression levels between species) is consistent with drift. Such drift-based models provide critical null models for tests of selection-driven divergence.

Short-term Response on a Single Character

The remainder of this volume examines the role of selection in changing the distribution of trait values for a quantitative character. We start by considering the short-term changes in a single trait. By short-term, we mean the first few generations, where selection-induced changes in allele frequencies at underlying loci are sufficiently small to be ignored.

Chapter 13 further explores Lush's classical breeder's equation, $R = h^2 S$, for the expected single-generation change in the mean of a quantitative trait. We examine extensions of this result to a number of settings, including selection schemes based on other estimates of an individual's breeding value beyond just their own phenotypic value (e.g., also using the phenotypic values of relatives). We also introduce the multivariate breeder's equation (which is extensively covered in Volume 3).

Chapter 14 considers truncation selection and the response in threshold traits, wherein a binary (presence/absence) trait results from the transformation of some underlying continuous variable (the liability). We also briefly introduce logistic regression and log-linear models, two approaches for the statistical modeling of discrete quantitative-genetic traits.

Chapter 15 extends of the breeder's equation to scenarios where the selection response in the mean has a transient component that changes over time. This can occur when additive

$\times$ additive epistasis underlies a trait, for dominance in a tetraploid species, or when there are environmental effects that are partly shared between parent and offspring. The last scenario serves as a brief introduction to maternal effects, which are more extensively covered in Chapter 22 (and especially in Volume 3). The key concept of Chapter 15 is that such transient components of response decay away under random mating when selection is stopped, leaving a reduced permanent response in the mean given by an appropriate interpretation of the breeder's equation.

Chapters 13–15 are entirely concerned with predictable changes in the mean, while Chapters 16 and 17 examine the impact of short-term selection on the variance. Chapter 16 examines Bulmer's theory (and its extensions) on the role of gametic-phase disequilibrium (or linkage disequilibrium [LD] for brevity, even though results also apply to unlinked loci) in changing the additive-genetic variance. The key concept is that the additive variance can be decomposed as $\sigma_A^2 = \sigma_a^2 + d$. The genic variance, σ_a^2, is entirely a function of the allele frequencies and represents the additive variance in the absence of LD. The disequilibrium contribution, d, is generated by LD, and eventually decays to zero (via recombination) under random mating when selection is relaxed. We use Bulmer's result to examine both the change in σ_A^2 under directional selection and the change in σ_A^2 when selection is largely focused on the population variance (e.g., stabilizing or disruptive selection).

Chapter 17 examines the short-term change in another variance component, the environmental variance, σ_E^2. A relatively recent appreciation is that σ_E^2 can vary over genotypes, and, as such, can potentially respond to selection. We examine different parameterizations for modeling σ_E^2 in terms of heritable components (i.e., breeding values for environmental variances), and the expected selection response under directional, as well as stabilizing and disruptive selection.

Chapters 18 and 19 cover the analysis of selection experiments, including the very important case of assessing the selection response (e.g., progress) in a breeding program. Chapter 18 reviews least-squares models of analysis, while Chapter 19 examines more recent approaches based on mixed-models (i.e., BLUP) and Bayesian methods (Appendices 2 and 3).

Chapter 20 examines the expected selection response in natural populations. Here, a significant confounding factor is the inability to precisely know the target of selection. We examine both BLUP-based approaches and those based on Robertson's secondary theorem (introduced in Chapter 6), the latter focusing on the nature of selection acting directly on the breeding value of a trait. A key finding is that numerous traits measured in natural populations show at least moderate heritabilities and apparently experience strong selection (so that $h^2 S$ is fairly large), yet show little to no response. We examine a variety of reasons for the apparent failure of the breeder's equation in these settings.

Selection in Structured Populations

Chapters 21–23 conclude our treatment of short-term response on a single trait, considering situations other than mass selection (wherein individuals are selected based on their phenotypic value alone) under random mating. Chapter 21 examines family-based selection schemes, wherein information from families—such as family means over a set of environments or deviations within a given family—are used in selection decisions.

Chapter 22 introduces the important, and much underappreciated, topic of associative effects. Here, individuals interact, and the phenotype of a focal individual is a function of both its own genetic and environmental values, as well as the values of the interacting individuals. This generates a selection response on both a direct and a social (or associative) component, and can lead to situations wherein selection to improve a trait actually results in a reduction in the mean. For example, when one selects the best individuals for egg production in a cage, often these individuals are the most aggressive (and hence able to garner the most resources). While they may have superior genes that directly influence egg production, they also have genes that reduce production in other cage members. A cage full of such birds can easily show a significant *reduction* in egg production. Associative effects also allow us to model maternal effects (as a special case), and place the concepts of kin and group selection into a single, unified, framework that involves components that can be estimated in standard breeding designs.

Chapter 23 concludes our discussion of structured populations by considering the impact of inbreeding on selection. This theory forms the basis for much of modern plant breeding. We also review some of the theory on the evolution of selfing rates in monoecious species.

Population-genetic Models of Trait Response

Chapters 13–23 consider short-term response, wherein either allele-frequency change is sufficiently small to be ignored, or is modified in very predictable ways, largely independent of the selection scheme (such as the decline in σ_A^2 under drift). Building on the results from Chapters 5 and 7, Chapters 24–28 examine the theory of long-term selection response in the face of significant allele-frequency change. Long-term response is far less predictable than short-term response. Indeed, it is entirely *unpredictable* from knowledge of just the base population genetic variances alone.

Chapter 24 examines the classical infinitesimal model, wherein a large number of loci, all of small effect, underlie a trait. We then consider departures from this basic model, including the continuum-of-alleles model, and the impact of LD and allele-frequency change in driving the distribution of breeding values away from a normal (Gaussian) distribution. Parts of this chapter are fairly technical.

Chapter 25 examines the deterministic theory of long-term response, as well as models with a major gene and background polygenes. In addition to theory (largely based on results from Chapter 5), we also examine the empirical data on long-term selection, and review the common features seen in many experiments.

Chapter 26 examines the impact of adding drift and mutation to the deterministic results from Chapter 25. In particular, we develop Robertson's classical result that the expected long-term selection response (in the absence of mutation) is usually bounded above by the product of $2N_e$ (twice the effective population size) and the response in the first generation. This key result implies that a higher short-term response (which results in a reduced N_e) comes at the expense of a reduced long-term response.

Chapter 27 examines a much longer-term view of selection response by considering adaptive walks, namely the sequence of successive fixations during adaptation over evolutionary time scales. One exciting aspect of this work is that some of the theory (which is rather technical and detailed in places) can be tested in microbial systems, and we review some of these empirical results, comparing them with their theoretical expectations.

Chapter 28, perhaps the most technical chapter in this volume, examines the extensive population-genetics modeling that has been performed on possible explanations for the maintenance of quantitative-genetic variation. Here, we are in the rather embarrassing position of having a wealth of highly detailed models, none of which can account for the data given our current understanding of strengths of selection in natural populations and details of the underlying genetics (number of loci and average mutation rates).

Measuring Selection on Traits

We conclude with the estimation of individual fitnesses in natural populations and the search for specific traits, or trait-combinations, currently under selection. In many ways, these last two chapters are the complement of Chapters 9 and 10. The tests from these previous chapters are trait-independent and require only sequence information, while the methods in Chapters 29 and 30 require estimates of the fitnesses of individuals (which are far from trivial to obtain), as well as the phenotypic values of candidate traits. The search for traits under selection in natural populations is a major focus of much of evolutionary and ecological genetics.

Chapter 29 focuses on the measurement of fitness, and how to partition these estimates across specific episodes of selection. A key concept is the opportunity for selection, I, the variance in relative fitness. This sets an upper limit on the maximum amount of change in any particular trait over a single generation of selection. We also examine measures of sexual selection, differences in the ability to attract mates. We then turn to estimation of, and various descriptive statistics for, the fitness surface, $W(z)$. This surface describes the expected fitness of an individual whose phenotypic value is z. Most of the methods for estimating fitness-trait associations assume some simple distribution, such as a normal, binomial, or Poisson for fitness (total number of offspring left by an individual). In reality, fitness distributions are expected to be very complex, in part because we expect them to have a large point mass at zero, representing the fraction of individuals leaving no offspring. We conclude by introducing the important Aster model approach, which allows for far more realistic distributions of fitnesses to be constructed.

Chapter 30 extends the univariate analysis of fitness-trait associations to multivariate settings and examines the resulting geometry of multivariate fitness surfaces. We also review the empirical data on the strength of selection in natural populations, introduce path-analytic models for the analysis of selection, and discuss the estimation of selection that occurs on levels above the individual (such as on groups or populations).

Appendices

Our six appendices introduce and review important mathematical and statistical tools used throughout this volume. Appendix 1 reviews diffusions, a class of approximations from probability theory widely used in population genetics to handle complex models with random features, notably the interaction of drift and selection.

Appendices 2 and 3 introduce, respectively, Bayesian methods and the Markov Chain Monte Carlo (MCMC) approaches used to obtain their resulting posterior distributions. Bayesian methods are widely applied in genetics, in large part because of their ability to fully account for all the potential sources of uncertainty when estimating model parameters. The resulting models are analytically rather complex, but with the development of computer resampling methods (MCMC approaches), they can be analyzed in rather straightforward ways. Bayesian methods are ideal for handling the high-dimensional datasets generated by genomics, in which massive numbers of features are measured on a much more modest number of actual samples (e.g., 30,000 measures of gene expression on, say, 100 individuals).

Appendix 4 examines several aspects of multiple comparisons, moving beyond simple Bonferroni corrections to the concept and application of the **false-discovery rate**. The latter is a powerful alternative to the more classical stringent control of experiment-wide type I error, as in genomic studies, we *expect* numerous tests to reject the null, and our concern is more about the quantification of those we accept (**discoveries**) rather than trying to control against *any* false positives (the Bonferroni setting). It also examines another powerful multiple comparison setting, **meta-analysis**, the comparison of results over a large number of studies, showing how to place this into a mixed-model framework.

Appendix 5 presents useful results from matrix algebra, and in particular, the geometry of a matrix, such as eigenstructure and matrix decompositions. While these methods are used throughout the book, they are especially important when describing the curvature of a multivariate fitness surface (Chapter 30).

Appendix 6 concludes with a few brief results on the calculus of matrix and vector derivatives.

Volume 3

The advanced reader might have noticed that a number of important topics are not covered in this volume. As mentioned at the start of this chapter, our treatment of selection response on quantitative traits concludes in Volume 3, which covers important topics in applied breeding such as the development and selection of pure lines and selection programs for hybrid breeding. The bulk of Volume 3, however, is multivariate, examining multivariate selection response in detail. We consider both the change in the mean vector and in the $\mathbf{G}$

matrix of the additive-genetic variances and covariances for the vector of traits of interest and examine methods for comparing the stability of **G** over populations. We conclude with a variety of applications of multivariate methods, such as the theory of index selection, BLUP and genomic selection, dealing with genotype $\times$ environment interactions, and selection on longitudinal traits (those whose phenotypes change over time, such as a growth curve or milk yield). The latter are often called function-valued traits, and can they be modeled by random-regressions which is an extension of BLUP methodology.

Notation

Finally, we need to briefly comment on the bane of almost every theoretical treatment: notation. One of our greatest challenges in trying to forge a cross-disciplinary synthesis spanning the full breadth of quantitative genetics was dealing with the considerable amount of overlapping notational symbols found in the literature. For example, θ is widely used for very different quantities in different subfields of quantitative and population genetics. While we could have created new notation for the various quantities with overlapping symbols, this would have introduced an additional level of complexity. Hence, we have tried (as much as possible) to follow the common usage in the relevant literature for a given field at the expense of some notational overlap. Hopefully, any slight ambiguity introduced by the recycling of some of the notational symbols across different chapters can be distinguished from the context of the presentation.

II

Evolution at One and Two Loci

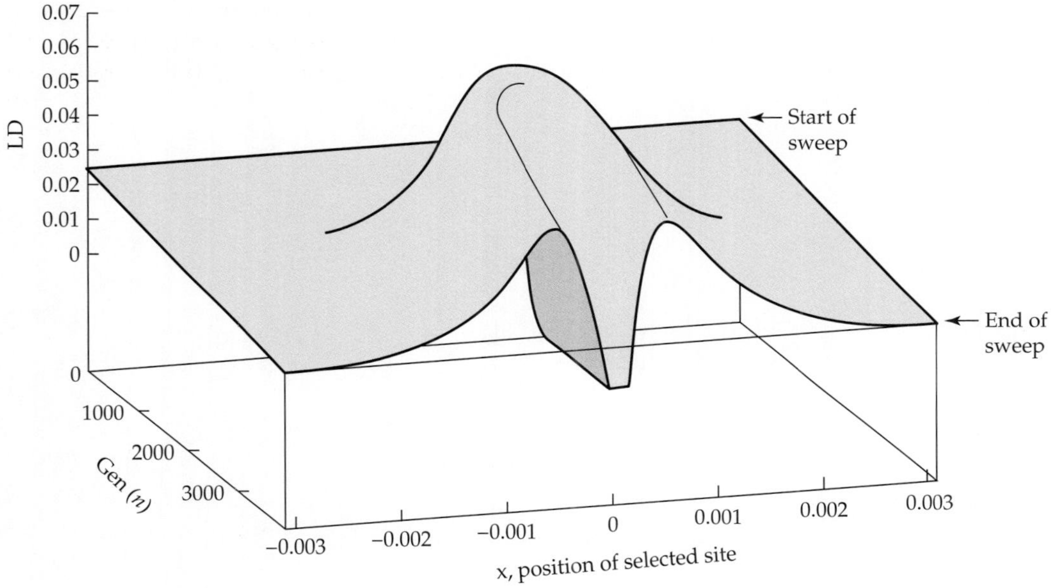

2

Neutral Evolution in One- and Two-Locus Systems

Variations neither useful nor injurious would not be affected by natural selection, and would be left either a fluctuating element, as perhaps we see in certain polymorphic species, or would ultimately become fixed, owing to the nature of the organism and the nature of the conditions.
Charles Darwin (Chapter 4, 1859)

Although most of the research in evolutionary biology is focused on issues related to natural selection, the rigor of all such analyses depends critically on our understanding of expected patterns of evolution in the absence of selection. The simple reasoning here is that if we are to have much confidence in any adaptive argument, it ought to be possible to firmly reject a simpler, neutral hypothesis. Thus, prior to exploring various population-genetic models for adaptive evolution, we first embark on a broad overview of neutral models of evolution. The theory underlying such models brings us into immediate contact with the issue of **genetic drift**, a commonly misunderstood factor in evolutionary biology, but which is nothing more than the random fluctuations in allele frequencies that necessarily result from sampling finite numbers of gametes in each generation. The magnitude of such allele-frequency fluctuations increases with decreasing population size, and combined with the input of new alleles by mutation, the incessant stochasticity of the process ensures that populations will evolve even in the absence of selection (Kimura 1983).

Consider, for example, a single heterozygous *Bb* parent that produces two progeny. There is a 50% probability that one offspring will inherit the *B* allele and the other the *b* allele, in which case no net change in allele frequency has been transmitted from parent to offspring. However, there is also a 50% probability that both offspring will inherit the same allele. In extremely large populations, these random changes resulting from gamete sampling tend to average out, leaving the allele frequency in the offspring population very close to that in the parental generation. However, over sufficiently long time scales, the cumulative effects of even small single-generation changes in allele frequencies can become quite pronounced. As we will show, if the time scale of interest (t, in generations) is much less than the average number of reproductive adults in the population (N), random fluctuations in allele frequencies can usually be ignored. This justifies the assumption of an effectively infinite population size as a good first approximation in many applications of population and quantitative genetics. However, for situations in which t is on the order of N or larger, evolution can no longer be viewed as a strictly deterministic process. Rather, any observed evolutionary change must be viewed as one realization of many possible outcomes.

In the following pages, it will be shown that even though finite population size induces stochastic evolutionary change, genetic drift has several *predictable effects*. First, even if mating is completely random, there will still be some long-term trend toward matings among relatives. Because all members of a population must ultimately descend from a narrow ancestral base, the smaller the population size, the greater this tendency will be. Thus, in a tiny **dioecious** (separate-sex) population with a stable adult number of two, all matings must be between full sibs, even though the reproductive pair itself may be a random draw from a larger progeny pool in the preceding generation. It follows that the genetic consequences of finite population size must be similar to those of inbreeding—the average homozygosity at a locus is expected to increase with smaller N, although as noted below, this can be offset in part by replenishment from mutationally derived variation. Second, gamete sampling causes allele frequencies to gradually drift toward zero or one, with the probability of ultimate fixation of any particular allele in the absence of selection being equal to its initial frequency. Third, subdivision of a population into isolated demes

Evolution and Selection of Quantitative Traits. Bruce Walsh & Michael Lynch.
Published 2018 by Oxford University Press.
DOI 10.1093/oso/9780198830870.001.0001

results in allele-frequency divergence among demes. The greater the degree of isolation of the subgroups, the more pronounced this differentiation will be.

This and the following two chapters provide a formal basis for these ideas. We first consider matters of one- and two-locus evolution in the context of a population with an idealized mating system and no influence from selective forces. As we will show in subsequent chapters, such models are often of great utility even when selection is operating, provided that the forces of selection are weaker than those associated with random genetic drift. In addition, such theory provides the underlying logic for the development of molecular-marker methods for estimating the power of mutation, recombination, and random genetic drift. These methods, the subject of Chapter 4, are the primary ways we have of describing the past population-genetic environment. Chapter 3 provides a critical link between Chapters 2 and 4 by demonstrating how results derived under the assumption of an ideal random mating population can be extended to a variety of alternative reproductive systems and population structures. In subsequent chapters, the one- and two-locus results introduced here will be used to develop neutral models for formal tests of adaptive evolution within a genomic region of interest (Chapters 8–10) and for the evolution of quantitative traits (Chapter 12).

THE WRIGHT-FISHER MODEL

Because the number of possible types of population structure is literally infinite (involving, for example, various degrees of local inbreeding, geographic subdivision, and age-specific mortality and fecundity), and temporal variation in population size is also common, it is impossible for us to consider the dynamics of neutral alleles in a fully general sense. Instead, we will focus initially on single finite populations of constant size within which mating is random. Even this simple structure admits to many possible variants, depending, for example, on whether there are separate sexes, whether there is variation in family size, and whether generations overlap. These additional layers of complexity will be taken up in Chapter 3, where it will become clear that, with an appropriate redefinition of the concept of population size, most of the results in the current chapter often still hold.

Perhaps the most frequent description of drift is the **Wright-Fisher model**, whose roots trace to Fisher (1922) and Wright (1931). We assume here a diploid population with a fixed number (N) of **monoecious** (hermaphroditic) adults, random mating (including the possibility of self-fertilization), and discrete generations, and we follow the number ($0 \leq i \leq 2N$) of copies of a given allele. The gamete pool produced by the adults is assumed to be effectively infinite, such that the $2N$ gametes that actually contribute to the next generation can be viewed as being sampled with replacement.

Consider a locus with two alleles, *B* and *b*, with neither having a selective advantage with respect to the other. If there are i copies of allele *B* in generation t, the probability P_{ij} that the number in generation $t+1$ is equal to j follows the binomial distribution. Assuming the Wright-Fisher model, each of the $2N$ sampled gametes has probability $i/(2N)$ of being *B* and probability $[1-(i/2N)]$ of being *b*, yielding

$$P_{ij} = \binom{2N}{j} (i/2N)^j [1-(i/2N)]^{2N-j} \tag{2.1}$$

where the first term in large parentheses is the binomial coefficient. This expression holds for all possible values of $i, j = 0, 1, \ldots, 2N$. Note that throughout we will be referring to a diploid population of size N, which requires $2N$ gametes for replacement; this same expression applies to a haploid population of size N if the 2 is deleted.

Letting $\mathbf{P}$ be the $(2N+1) \times (2N+1)$ matrix of all the P_{ij}, the probability distribution of the number of copies of allele *B* in a population can then be expressed succinctly as

$$\mathbf{x}(t+1) = \mathbf{x}(t)\,\mathbf{P} \tag{2.2a}$$

where the elements of the row vector $\mathbf{x}(t)$ are the probabilities that the allele is present in $i = 0, 1, \ldots, 2N$ copies in generation t. Note that P_{ij}, which refers to the element in row i and column j in matrix **P**, is the probability that a population makes a transition from i copies of B to j copies, conditional on starting at i. If the transition matrix **P** remains constant from generation to generation, as it does under the assumptions given previously, Equation 2.2a generalizes to

$$\mathbf{x}(t) = \mathbf{x}(0)\,\mathbf{P}^t \tag{2.2b}$$

This is an example of a **Markov chain** (considered in more detail in Appendix 3).

When considering a single population starting with an allele frequency of $i/2N$, all of the entries in the initial vector $\mathbf{x}(0)$ are equal to zero, except $x_i(0) = 1.0$ (corresponding to i copies being present). Equation 2.2b then yields the evolution of the probability distribution of allelic copy number over time. That is, the elements of $\mathbf{x}(t)$ denote the frequencies of hypothetical replicate populations at time t that are expected to have exactly i copies of the focal allele. The first ($i = 0$) and final ($i = 2N$) elements of $\mathbf{x}(t)$ are of special interest, as they are **absorbing states**—once an allele becomes **lost** ($i = 0$) or **fixed** ($i = 2N$) in a population, it remains at that state indefinitely (barring reintroduction via mutation or migration). As t increases in Equation 2.2b, all of the interior elements of $\mathbf{x}(t)$ eventually converge on zero, and the sum $x_0(t) + x_{2N}(t)$ converges to one. The ultimate probability of fixation of allele B is given by $x_{2N}(\infty)$, whereas the ultimate probability of loss of allele B (or equivalently, of fixation of allele b) is $x_0(\infty)$.

From the elements of $\mathbf{x}(t)$, it is straightforward to compute the expected allelic copy number, the variance in copy number among replicate populations, the probability of fixation by generation t, etc. This **transition-matrix** approach is exact, but many useful approximations have been developed for it (e.g., Gale 1990; Ewens 2004). Some of these approaches will be discussed later, with a powerful alternative method, the **diffusion approximation**, being covered extensively in Appendix 1. In a diffusion, the focus shifts from copy number to allele frequency.

It should be noted that the Wright-Fisher model, in which all individuals synchronously turn over, is just one of many possible conceptual frameworks for approximating a randomly mating population. For example, Moran (1962) developed a treatment whereby a single random individual is chosen to reproduce at each point in time, with a single random individual then being chosen to die. Because allele frequencies can change by only single steps during each time interval under this scenario, the **Moran model** turns out to be more analytically tractable than the Wright-Fisher model, although it is restricted to haploid populations.

Example 2.1. Consider an initially heterozygous individual *Bb* in a self-fertilizing line maintained by single-progeny descent. With $N = 1$, the only three possible allele-frequency states in the population are zero, one, or two B alleles. Denoting the initial state of the population by $\mathbf{x}(0) = [0, 1, 0]$, the probability that the population is in states 0, 1, or 2 at some future generation t is given by Equation 2.2b with

$$\mathbf{P} = \begin{pmatrix} 1 & 0 & 0 \\ 0.25 & 0.50 & 0.25 \\ 0 & 0 & 1 \end{pmatrix}$$

Although the numerical values for the elements of **P** can be obtained directly from Equation 2.1, for this simple example they can also be arrived at intuitively. For example, the elements in the first row of **P** denote the probabilities that the population will be in states $j = 0, 1, 2$ in generation $t + 1$ given that it is in state 0 in generation t. The only nonzero element in this row is $P_{00} = 1$. It is nonzero because the $i = 0$ state is absorbing, i.e., once the population enters this state, it remains there indefinitely.

The probability of being in any particular allele-frequency category in generation t, which follows from Equation 2.2b, is a function of $\mathbf{P}^t$, so for example,

$$\mathbf{P}^2 = \begin{pmatrix} 1 & 0 & 0 \\ 0.375 & 0.250 & 0.375 \\ 0 & 0 & 1 \end{pmatrix}, \quad \mathbf{P}^5 = \begin{pmatrix} 1 & 0 & 0 \\ 0.48438 & 0.03125 & 0.48438 \\ 0 & 0 & 1 \end{pmatrix}$$

and so on. With the initial vector $\mathbf{x}(0) = (0, 1, 0)$, only the middle row of $\mathbf{P}^t$ is relevant, giving the following table for the progression of the elements of $\mathbf{x}(t)$ over time:

t	BB $x_0(t)$	Bb $x_1(t)$	bb $x_2(t)$
0	0.00000	1.00000	0.00000
1	0.25000	0.50000	0.25000
2	0.37500	0.25000	0.37500
3	0.43750	0.12500	0.43750
4	0.46875	0.06250	0.46875
5	0.48438	0.03125	0.48438
...	...	...	...
10	0.49951	0.00098	0.49951
...	...	...	...
15	0.49998	0.00003	0.49998
...	...	...	...
∞	0.50000	0.00000	0.50000

The first and last elements of $\mathbf{x}(t)$, respectively, denote the probabilities that the line will have become fixed for the *B* or the b allele by time t. Thus, for this particular case, the line eventually becomes completely monomorphic for either the *B* or the b allele with equal probability. This meets our intuitive expectations for a neutral locus—in the absence of any directional forces, the two probabilities of fixation are equal to the initial frequencies of the respective alleles.

LOSS OF HETEROZYGOSITY BY RANDOM GENETIC DRIFT

The **sampling variance of an allele frequency** provides one way to succinctly define the stochastic effects of random genetic drift. Consider a large pool of gametes, a fraction p of which carry the *B* allele, and let $2N$ gametes be randomly drawn to produce a new generation of N individuals. Defining the expected frequencies of genotypes *BB*, *Bb*, and *bb* in the progeny generation by the Hardy-Weinberg proportions p^2, $2p(1-p)$, and $(1-p)^2$, the expected number of *B* alleles contained in a random offspring is simply $(2 \cdot p^2) + [1 \cdot 2p(1-p)] + [0 \cdot (1-p)^2] = 2p$. The expected square of the number of *B* alleles carried per individual is $(2^2 \cdot p^2) + [1^2 \cdot 2p(1-p)] + [0^2 \cdot (1-p)^2] = 2p(1+p)$. Thus, the variance (the mean squared value minus the square of the mean) of the number of *B* alleles carried by an individual is $2p(1+p) - (2p)^2 = 2p(1-p)$, whereas the variance of the total number of *B* alleles carried in the offspring generation is N times this, $2Np(1-p)$. Because the frequency of allele *B* is the number of copies divided by $2N$, the sampling variance of the frequency (a second-order moment) is $2Np(1-p)/(2N)^2 = p(1-p)/(2N)$, which is directly proportional to the heterozygosity (the fraction of *Bb* individuals in the population) and inversely proportional to the population size. The expression $p(1-p)/(2N)$ defines the dispersion in allele frequency resulting from a single generation of gamete sampling, conditional on allele frequency p in the parental population.

In the absence of any counteracting evolutionary forces, the dispersive effects of genetic drift will continue in each generation, leading to a progressive erosion of population-level heterozygosity until all loci have eventually become fixed for just a single allele. To evaluate the long-term impact of finite population size on the expected heterozygosity of a locus, we make use of the properties of the **inbreeding coefficient**, f, which denotes the probability that two alleles at a locus in an individual are **identical by descent (IBD)** (LW Chapter 7).

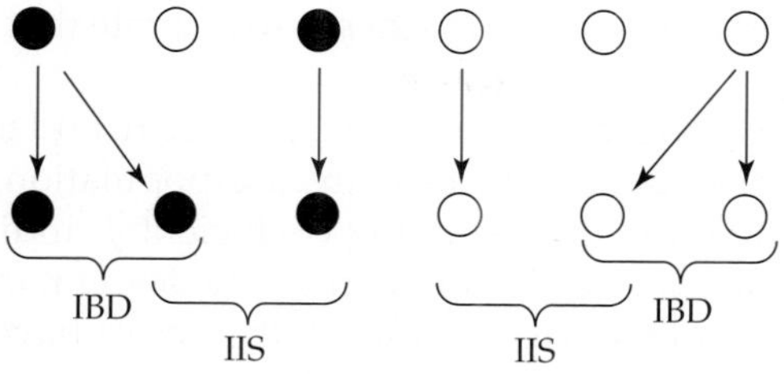

Figure 2.1 Distinction between alleles that are identical in state (IIS) and identical by descent (IBD). The simple example is a case in which the population consists of just three individuals, so that the total number of gametes transmitted randomly per generation is $2N = 6$. A polymorphism exists in the time-zero population (black and white alleles), so not all alleles are initially IIS. After one generation of sampling, two pairs of alleles are IBD because the two copies are direct copies of a parental allele.

Because all gene copies at a locus must ultimately trace back to a single, remote common ancestor, it is essential to start with an appropriate reference generation, which we here take to be the current generation, and then simply query forward in time as to the probability that any random pair of genes traces back to a specific copy of one of the $2N$ genes present at the locus in the reference generation. Under this framework, a pair of genes that are **identical in state** (**IIS**; also know as **alike in state, AIS**) need not be IBD, but barring mutation, genes that are IBD must also be IIS (Figure 2.1).

There is always a small chance that uniting gametes will derive from related individuals, even in a randomly mating population. For example, in a monoecious population containing only two individuals, there are only four genes residing at each locus, so the probability that one gamete will randomly unite with another containing a direct descendant of the same parental gene is 1/4. With four individuals, there are eight gene copies, and this probability becomes 1/8. Thus, under the idealized Wright-Fisher model with population size N, the probability that two direct copies of any parental gene will randomly unite in an offspring is $1/(2N)$. Barring a rare mutation, all such offspring are homozygotes.

Although the quantity $1/(2N)$ may be thought of as the new inbreeding that is incurred in each generation, this does not fully describe the buildup of homozygosity in a population. For even if uniting gametes do not carry genes that are direct copies of a parental gene, they may still be identical by descent through inbreeding in a previous generation. Under random mating, the probability of the latter event is simply the inbreeding coefficient of the parental generation. Thus, because the probability of drawing genes that are not direct copies of the same parental gene is $[1 - (1/2N)]$, the expected inbreeding coefficient in generation t is

$$f_t = \frac{1}{2N} + \left(1 - \frac{1}{2N}\right) f_{t-1} \tag{2.3}$$

Subtracting both sides from one yields the recursion formula

$$(1 - f_t) = \left(1 - \frac{1}{2N}\right)(1 - f_{t-1}) \tag{2.4a}$$

which generalizes to

$$(1 - f_t) = \left(1 - \frac{1}{2N}\right)^t (1 - f_0) \tag{2.4b}$$

and finally to

$$(1 - f_t) = \left(1 - \frac{1}{2N}\right)^t \tag{2.4c}$$

if we assume a noninbred base population ($f_0 = 0$). Again, we see the central role that population size plays in the dynamics of genetic variation. As $t \to \infty$, the fraction of the population that is not inbred, $1 - f_t$, approaches zero at a rate that is inversely proportional to N.

To see the connection between the inbreeding coefficient and the expected heterozygosity in a population, consider a diallelic locus with base-population heterozygosity $2p(1-p)$. In the descendant population with inbreeding coefficient f, individuals can only be heterozygotes if they carry alleles that are not identical by descent, the probability of which is $(1 - f)$. If two alleles are not identical by descent, they must have been acquired independently, so the probability that a genotype containing a pair of such alleles is a heterozygote is $2p(1-p)$. Thus, the expected heterozygosity of a population with inbreeding coefficient f and initial allele frequency p is $2p(1-p)(1-f)$. This shows that the fractional reduction in heterozygosity relative to the base population is equal to f. Because this argument applies regardless of the initial heterozygosity (and regardless of the number of segregating alleles), Equation 2.4c may be rewritten to describe the expected population heterozygosity at time t,

$$H_t = H_0 \left(1 - \frac{1}{2N}\right)^t \tag{2.5}$$

This **rate of decay of heterozygosity** of $1/(2N)$ was first obtained by Wright (1931). It may be a source of encouragement to the nonmathematically inclined that the brilliant Fisher (1922), using a rather different approach, obtained the wrong answer.

The time course for the loss of heterozygosity can be clarified by using an exponential approximation to Equation 2.5. Because $(1-x)^t \simeq e^{-xt}$ for $|x| \ll 1$, for N greater than 10 or so,

$$H_t \simeq H_0 e^{-t/(2N)} \tag{2.6a}$$

Rearrangement then leads to the expected time to reach a certain reduction in heterozygosity,

$$t = -2N \ln(H_t/H_0) \tag{2.6b}$$

which shows that the heterozygosity is reduced to half of its initial value in $\sim 1.4N$ generations and to 5% of H_0 in $\sim 6N$ generations. Thus, a population twice the size of another requires twice the number of generations to reach the same expected state.

With a temporally varying population size, Equation 2.6a becomes

$$H_t = H_0 \prod_{i=1}^{t} \left(1 - \frac{1}{2N_i}\right) \simeq H_0 \exp\left[-\sum_{i=1}^{t} 1/(2N_i)\right] \tag{2.7}$$

where the $\prod$ sign denotes a product of terms, and N_i is the population size in generation i. This expression illustrates an important point. Because each of the generation-specific terms, $[1 - (1/2N_i)]$, is necessarily less than one, Equation 2.7 shows that an expansion of population size can reduce the rate of erosion of heterozygosity, but does not eliminate it.

One significant limitation of the preceding expressions is that they only provide information on the behavior of the *expected* heterozygosity in a population. In reality, fluctuations in allelic copy number resulting from random genetic drift ensure that variation in heterozygosity will arise among loci that start in the same state. In a finite population of size N, the heterozygosity of a diallelic locus can take on $N+1$ discrete values: $0, 2(1/2N)[1-(1/2N)], \cdots, 2(N/2N)[1-(N/2N)]$. Using the transition-matrix approach (Equations 2.2a and 2.2b), one can obtain the exact probability distribution of heterozygosity for a locus starting with allele frequency $i/2N$, using the fact that $x_j(t) + x_{2N-j}(t)$ is the probability that the population has heterozygosity $2(j/2N)[1-(j/2N)]$.

An alternative approach was developed by Kimura (1955b), who used diffusion theory (Appendix 1) to obtain an analytical expression for the probability density of allele frequency at time t, given the starting value p_0,

$$\varphi(p_t|p_0) = p_0(1-p_0)\sum_{i=1}^{\infty} i(2i+1)(i+1)\cdot$$
$$F(1-i, i+2, 2, p_0)\cdot F(1-i, i+2, 2, p_t)\cdot e^{-i(i+1)t/(4N)} \tag{2.8}$$

where $F(1-i, i+2, 2, p_0)$ and $F(1-i, i+2, 2, p_t)$ are specific variants of the hypergeometric function (Equation 15.1.1 in Abramowitz and Stegun 1972). When we use this expression, $[\varphi(p_t|p_0)+\varphi(1-p_t|p_0)]$ is the probability of heterozygosity $2p_t(1-p_t)$ at time t. We will make more use of Equation 2.8 in the next sections, illustrating in particular its implications for the dispersion of allele frequencies among isolated populations. The utility of the diffusion approximation is that it yields to numerous closed-form mathematical approximations, some of which we note below.

PROBABILITIES AND TIMES TO FIXATION OR LOSS

Because Equation 2.8 denotes the probability density of allele frequency p given that the population is still polymorphic,

$$\Omega(p_0, t) = \int_{1/(2N)}^{1-1/(2N)} \varphi(p_t|p_0)dp_t \tag{2.9a}$$

is the probability that both alleles are still present in generation t. The probability that an allele with initial frequency p_0 has been fixed by generation t is

$$p_f(p_0, t) = p_0 + p_0(1-p_0)\sum_{i=1}^{\infty}(2i+1)(-1)^i$$
$$\cdot\, F(1-i, i+2, 2, p_0)\cdot e^{-i(i+1)t/4N} \tag{2.9b}$$

whereas the probability of loss of the allele, $p_l(p_0, t)$, is given by Equation 2.9b with $(1-p_0)$ exchanged for p_0 (Kimura 1955b). Summing up,

$$\Omega(p_0, t) + p_f(p_0, t) + p_l(p_0, t) = 1 \tag{2.10}$$

As can be seen from the negative exponential terms in the previous expressions, as $t \to \infty$, $\Omega(p_0, t) \to 0$, $p_f(p_0, t) \to p_0$, and $p_l(p_0, t) \to (1-p_0)$. Thus, under neutrality, the probability that a particular allele will become fixed is simply equal to its initial frequency, p_0. It follows that, in averaging over a very large number of replicate populations, the expected allele frequency will remain constant at p_0, with a fraction p_0 of all replicates ultimately reaching allele frequency 1, and the remaining fraction $1-p_0$ having allele frequency 0.

An issue of special interest is the mean time until an allele is absorbed into either state $p = 0$ or $p = 1$. Using diffusion approximations (Appendix 1), Kimura and Ohta (1969a) obtained expressions for both quantities, and Kimura (1970) presented a description of the entire probability distributions for absorption times. The following example uses a somewhat simpler approach to arrive at results identical to those of Kimura and Ohta (1969a), and provides yet another illustration of how the effects of random genetic drift scale with population size.

Example 2.2. Ewens (2004) used the following line of reasoning to derive the expected time to absorption of a neutral allele under the Wright-Fisher model. Letting δp denote the change in allele frequency in one unit of time, the mean time to absorption for an allele with frequency p may be rewritten as

$$\bar{t}_a(p) = E[\,\bar{t}_a(p+\delta p)] + 1$$

where E denotes an expected value. In words, this expression states that the mean absorption time starting at frequency p is equal to the mean absorption time one time unit later when the allele frequency is $p + \delta p$, plus one. Approximating $\overline{t}_a(p + \delta p)$ by the first three terms in its Taylor series (see LW Equation A1.2) and then taking expectations (only the δp are random terms; the rest are fixed constants), gives

$$E[\,\overline{t}_a(p + \delta p)] \simeq E\left[\,\overline{t}_a(p) + \delta p\,\frac{\partial\,\overline{t}_a(p)}{\partial p} + \frac{(\delta p)^2}{2}\frac{\partial^2\,\overline{t}_a(p)}{\partial p^2}\right]$$

$$= \overline{t}_a(p) + E[\delta p]\,\frac{\partial\,\overline{t}_a(p)}{\partial p} + \frac{E[(\delta p)^2]}{2}\frac{\partial^2\,\overline{t}_a(p)}{\partial p^2}$$

Hence, we have

$$\overline{t}_a(p) \simeq \overline{t}_a(p) + E[\delta p]\frac{\partial\,\overline{t}_a(p)}{\partial p} + \frac{E[(\delta p)^2]}{2}\frac{\partial^2\,\overline{t}_a(p)}{\partial p^2} + 1$$

Under neutrality, the expected change in allele frequency is $E(\delta p) = 0$, and as derived previously, the expected variance in allele-frequency change is $E[(\delta p)^2] = p(1-p)/(2N)$. Substituting into our approximation and rearranging gives

$$\overline{t}_a(p) - \overline{t}_a(p) - 1 \simeq \frac{p(1-p)}{2\cdot 2N}\frac{\partial^2\,\overline{t}_a(p)}{\partial p^2}$$

which implies the differential equation

$$\frac{\partial^2\,\overline{t}_a(p)}{\partial p^2} \simeq -\frac{4N}{p(1-p)}$$

Performing the double integration with respect to p leads to the solution

$$\overline{t}_a(p_0) \simeq -4N[p_0 \ln(p_0) + (1-p_0)\ln(1-p_0)] \tag{2.11a}$$

which is the mean time until an allele with initial frequency p_0 is either lost or fixed in a population.

A similar approach can be used to estimate the mean time to fixation for the subset of alleles that specifically become fixed, $\overline{t}_f(p_0)$. The essential modification here is that in estimating $\overline{t}_f(p_0)$, $E(\delta p)$ is no longer equal to zero, because in order for an allele to become fixed, at least one copy must be produced in each generation. That is, in the case of conditional fixation, of the $2N$ genes drawn in each generation, one is definitely a *B* allele, whereas the remaining $2N-1$ genes can be viewed as random, leading to $E(\delta p) = \{(1/2N) + [1-(1/2N)]p\} - p = (1-p)/(2N)$. Similarly, because the states of only $2N-1$ genes are random, $E[(\delta p)^2] = \{(1\cdot 0) + [(2N-1)p(1-p)]\}/(2N)^2$. Unless N is very small, the approximation $E[(\delta p)^2] = p(1-p)/(2N)$ still holds quite well, and following the procedures utilized previously, we then have

$$\left(\frac{2}{p}\right)\frac{\partial \overline{t}_f(p)}{\partial p} + \frac{\partial^2 \overline{t}_f(p)}{\partial p^2} \simeq -\frac{4N}{p(1-p)}$$

The solution of this second-order differential equation requires several steps, which we omit, the final result being

$$\overline{t}_f(p_0) \simeq -\frac{4N(1-p_0)\ln(1-p_0)}{p_0} \tag{2.11b}$$

The mean time to loss of an allele conditional upon loss is identical to the previous expression, but with $(1-p_0)$ interchanged with p_0,

$$\overline{t}_l(p_0) \simeq -\frac{4Np_0\ln(p_0)}{1-p_0} \tag{2.11c}$$

These expressions show that the conditional time to fixation asymptotically approaches $4N$ generations for rare alleles, with the more general fate of such alleles being loss from the population in $\ll 4N$ generations (Figure 2.2).

Finally, because the probability of ultimate fixation of a neutral allele is equal to its initial frequency (p_0) and the probability of ultimate loss is $(1 - p_0)$, it follows that

$$\bar{t}_a(p_0) = p_0\bar{t}_f(p_0) + (1 - p_0)\bar{t}_l(p_0) \tag{2.11d}$$

Example A1.8 (in Appendix 1) uses diffusion theory to obtain results identical to those just presented.

THE AGE OF A NEUTRAL ALLELE

A classic result from standard theory relates the age of a neutral allele to its frequency (Kimura and Ohta 1973; Maruyama 1974; Watterson 1976; Slatkin and Rannala 1997, 2000). In particular, the expected age t (in generations) of an allele with current frequency p is

$$E(t) = -\frac{4Np\,\ln(p)}{1-p} \tag{2.12}$$

assuming a constant population size during the allele's sojourn through the population (Kimura and Ohta 1973). As shown in Figure 2.3, alleles at higher frequency are expected to be older. Most notably, as $p \to 1$, $E(t) \to 4N$ generations, which is consistent with the result in the previous example showing that the mean time to fixation of a rare (e.g., new mutant) neutral allele is $4N$ generations. This result is more than just an esoteric finding, as one class of tests for selection evaluates whether an allele, given its frequency, is too young to be compatible with neutrality (Chapter 9).

Equation 2.12 is not, however, the whole story, as a very old allele can sometimes be at low frequency, having transiently drifted up to a high frequency before drifting back toward zero frequency by chance. Thus, we expect the confidence interval for allele-age estimates to be highly asymmetric about the mean. Slatkin and Rannala (2000) provide an approximation for the cumulative probability for the age of an allele, given a frequency of p in a random sample of n alleles,

$$\Pr(t \leq \tau) = (1-p)^{-1+[n/(1+nN\tau)]} \tag{2.13}$$

One final caveat is in order with respect to Equation 2.12. Unless the sample size is very large, the estimated frequency of a rare allele can be quite misleading. Consider, for example, a singleton with a sample frequency of $1/n$. Application of Equation 2.12 would imply an estimated age of $4N\ln(n)/[n(1-1/n)]$ generations. If ten diploid individuals were sampled, the minimum allele frequency would be 0.05, so the estimated age of a singleton would be $\sim 0.84N$ generations. As very rare alleles, with true frequencies $\ll 1/n$ will either be recorded as singletons or not at all in a sample of size n, it is clear that Equation 2.12 yields upwardly biased estimates of the ages of rare alleles unless the sample size is large enough that the estimate of frequency p is highly accurate.

Example 2.3. The mutation *CCR5-δ32* destroys the human *CCR5* receptor, which is used by the HIV virus to enter the cell, leading to significant resistance against HIV infection. This deletion occurs at frequencies up to 14% in Eurasians, but is absent in Africans, Native Americans, and East Asians. Assuming a frequency of $p = 0.10$ and an effective population size $N = 5000$ for Caucasians, Stephens et al. (1998) used Equation 2.12 to estimate the age of this allele (under the assumption of neutrality) to be

$$\widehat{t} = -\frac{4 \cdot 5000 \cdot 0.1\,\log(0.1)}{0.9} = 5116 \text{ generations}$$

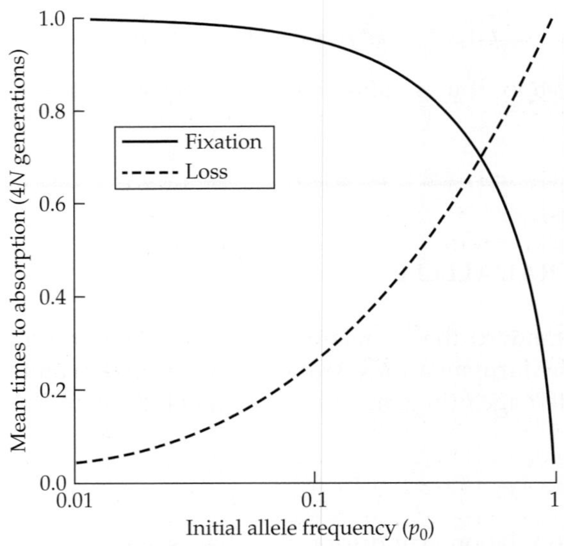

Figure 2.2 Mean times to fixation and loss of neutral alleles with starting frequency p_0 (from Equations 2.11b and 2.11c). The times are scaled in units of $4N$ generations, and thus need to be multiplied by $4N$ to obtain absolute numbers of generations.

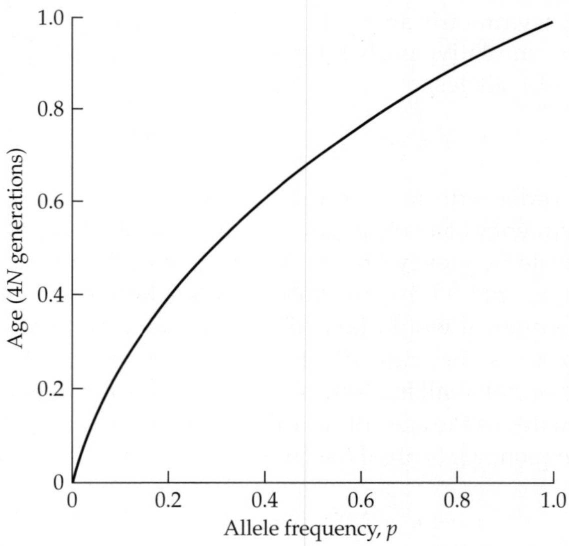

Figure 2.3 Expected age of a neutral allele, given its frequency p (Equation 2.12). Time is scaled in units of $4N$ generations.

However, an independent (and more direct) estimate of the allele's age can be obtained by considering the variation in haplotypes among all sequences carrying this mutation. The $\delta 32$ mutation is in strong linkage disequilibrium with allele *215* at the *AFMB* marker (a highly variable tandem-repeat locus), to the extent that 84.8% (39 of 46) of sampled $\delta 32$ mutations have the *δ32-215* haplotype. Clearly, the initial $\delta 32$ mutation at *CCR5* must have arisen on a chromosome carrying the *215* allele. The recombination fraction between *CCR5* and *AFMB* was estimated by Stephens et al. (1998) to be $c = 0.006$. The probability of the *δ32-215* haplotype remaining intact after t generations (i.e., experiencing no recombinational breakdown) is just $\pi = (1-c)^t$, which rearranges to

$$t \simeq \ln(\pi)/\ln(1-c) = 27.4 \text{ generations}$$

Stephens et al. (1998) took these great disparities between age estimates as an indicator that strong selection has promoted the $\delta 32$ mutation much more rapidly than would be likely under a pure drift model. Assuming $\delta 32$ originated as a single mutation, they estimated the selection coefficient to be between 0.2 and 0.4, depending on assumptions about dominance. We will revisit the age of this mutation in Example 9.14.

Example 2.4. Consider a situation in which a random sample of 500 gametes from a population yields 350 copies of a particular allele A, and hence an estimated p of 0.7. Under the joint assumptions of neutrality and constant population size, what is the 95% confidence interval for the estimated age of this allele? From Equation 2.12, the expected age is

$$\widehat{t} = -\frac{(4N)\,0.7\,\ln(0.7)}{1-0.7} = 1.45N$$

An approximate confidence interval using the approximation given by Equation 2.13 is obtained as follows. To account for N, we first rescale time so that $\tau = 2N$ generations, and after solving for the critical values of τ on this scale, we convert it back to generations. Define τ_α as satisfying $\Pr(t \leq \tau_\alpha) = \alpha$. The 95% confidence interval for allelic age t is given by $(\tau_{0.025}, \tau_{0.975})$. From Equation 2.13, $\tau_{0.025}$ must satisfy

$$(0.3)^{-1+[500/(1+250\tau_{0.025})]} = 0.025$$

the solution of which is $\tau_{0.025} = 0.49$, or $0.49 * (2N) = 0.98N$ generations (Equation 9.38d gives an approximation for τ when n is large). The same procedure can be used to find that $\tau_{0.975} = 3.9N$, showing that the confidence interval about the mean value is very asymmetric.

ALLELE-FREQUENCY DIVERGENCE AMONG POPULATIONS

A natural consequence of allele-frequency drift within populations is the divergence of isolated replicate populations. Suppose a monoecious base population with allele frequency p_0 is suddenly split into several completely isolated subpopulations, each of size N, with random mating within each subpopulation and no selection, migration, or mutation. The variance in allele frequency among subpopulations in generation t is

$$\sigma_p^2(t) = E(p_t^2) - E^2(p_t)$$

Adding and subtracting $E(p_t)$,

$$\begin{aligned}\sigma_p^2(t) &= [E(p_t) - E^2(p_t)] + [E(p_t^2) - E(p_t)] \\ &= E(p_t)[1 - E(p_t)] - E[p_t(1-p_t)]\end{aligned}$$

Because there are no systematic forces causing the allele frequency to increase or decrease, $E(p_t) = p_0$, so the first quantity on the right is $p_0(1-p_0)$. The quantity $E[p_t(1-p_t)]$ is half the expected heterozygosity in a population in generation t, which was already defined in Equation 2.5. Thus

$$\sigma_p^2(t) = p_0(1-p_0)\left[1-\left(1-\frac{1}{2N}\right)^t\right] \tag{2.14a}$$

which is well approximated by

$$\sigma_p^2(t) \simeq p_0(1-p_0)(1-e^{-t/2N}) \tag{2.14b}$$

for $N > 10$. This shows that the among-population variance asymptotically approaches $p_0(1-p_0)$, which is half the heterozygosity in the base population—over time, the allelic variance within populations (half of the quantity in Equation 2.6a) is transformed into among-population variance, with the total of the two remaining constant. An alternative way to envision this asymptotic result is to note that at fixation the allele frequency has a value of 1.0 with probability p_0, and otherwise is zero, giving $E(p_0) = 1 \cdot p_0$ and $E(p_0^2) = 1^2 \cdot p_0 = p_0$. Hence, the among-population variance when all alleles are fixed is just $E(p_0^2) - [E(p_0)]^2 = p_0(1-p_0)$.

Although Equations 2.14a and 2.14b deal with the expected allele-frequency variance, they do not describe the actual form of the distribution of population allele frequencies. However, all of this information is contained in the formulations presented previously on the probability distribution of allele frequencies within populations. For example, the transition-matrix approach (Equations 2.2a and 2.2b) and the diffusion approximation (Equation 2.8) yield the expected temporal dynamics of the distribution of allele frequencies in different replicate populations, all starting from an identical frequency, p_0 (Figure 2.4).

Summing up the calculations to this point, five significant conclusions can be gleaned with respect to neutral alleles. First, with increasing time, the total probability mass for the allele-frequency distribution at a locus declines because only *segregating* alleles are considered, i.e., the proportions of populations that have experienced gene fixation or loss are ignored. Second, regardless of the starting condition, the distribution becomes flatter with increasing time, such that the frequency of a sufficiently old segregating allele is equally likely anywhere over the (0,1) interval. Third, high-frequency alleles are generally expected to be old. Fourth, the distributions in Figure 2.4 can be interpreted in two different ways: as the probability distribution of allele frequencies over a very large number of replicate populations over time, all starting at an identical state, or as the expected distribution of allele frequencies for the subset of loci with identical starting frequencies within a single population. Finally, the expected allele-frequency distribution is a function of t/N generations, as can be seen from the exponential terms in Equation 2.8. As should be clear by now, this scaling of the temporal dynamics of random genetic drift to the reciprocal of population size is a natural consequence of the fact that the variance of allele-frequency change is inversely proportional to N.

BURI'S EXPERIMENT

Because all populations are finite in size, the theory of random genetic drift is of central significance to all areas of population genetics. It may therefore come as a surprise that highly replicated experiments examining the chance dynamics of allele-frequency change are extremely rare. However, the results of one massive experiment nicely affirm the theoretical expectations outlined above, while making one additional important point. Starting with two homozygous lines of *Drosophila melanogaster*, one of which was fixed for allele bw^{75} and the other for allele bw at the brown locus, Buri (1956) established 212 F_1 hybrid populations, with initial frequency 0.5 for both alleles. For the following 19 generations, he randomly mated eight males and eight females within each population and monitored the changes in allele frequencies in each subline. This could be done in the premolecular era because the genotype at the brown locus determines eye color: $bw^{75}bw^{75} =$

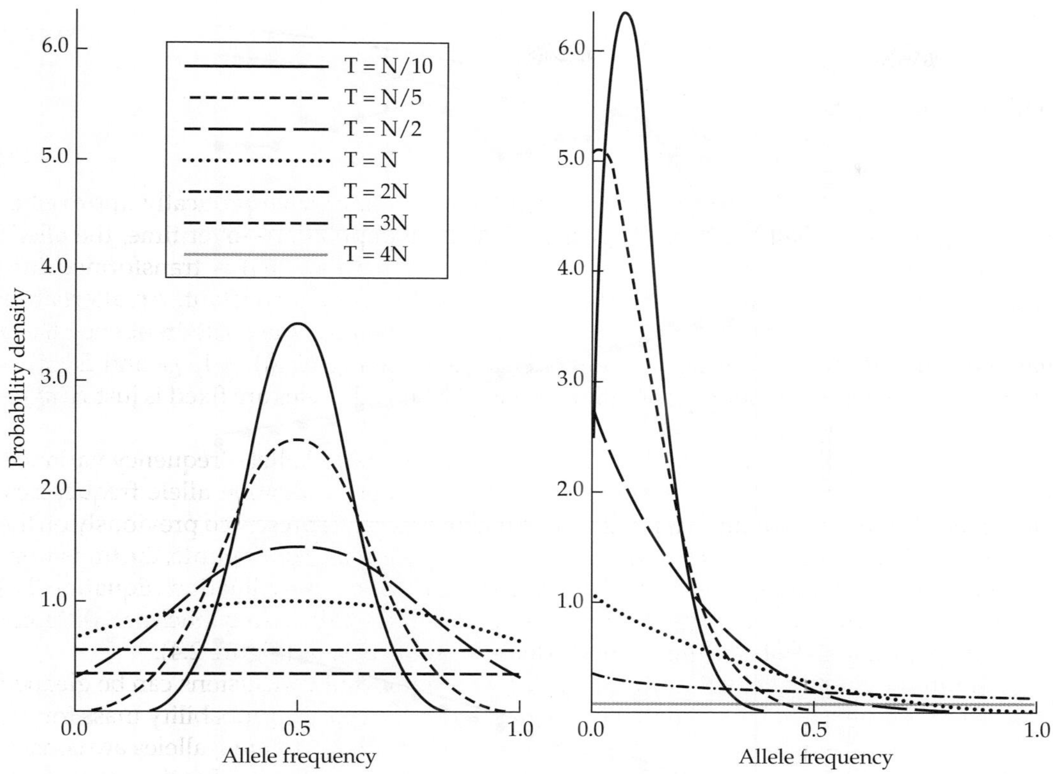

Figure 2.4 Expected probability distributions for the frequencies of *segregating* neutral alleles in replicate, randomly mating populations of size N after t generations of divergence (fixed alleles are ignored). The initial allele frequency in the base population is 0.5 on the left and 0.1 on the right. The abscissa is the population allele frequency, whereas the ordinate is proportional to the probability of occurrence of that frequency. Note that the time scale is in units of N generations, where N is the population size, so that $t = N$ generations implies 100 generations for a population of size 100 and 10,000 generations for a population of size 10,000. (From Kimura 1955b.)

bright red-orange, $bw^{75}bw$ = deep red-brown, and $bwbw$ = white. (Two separate experiments were performed, one with 107 and the other with 105 populations, but the results are so similar that they have been pooled in the following analysis.)

To evaluate the results in the light of the preceding theory, it is first necessary to demonstrate that the bw^{75} and bw alleles are indeed neutral with respect to each other. This can be done as follows (Figure 2.5, top). In the absence of selection, the expected frequency of the bw^{75} allele averaged over all populations should equal its initial frequency, 0.50, in all generations. Nevertheless, just as the frequency within any population is expected to deviate from 0.5 because of drift, the mean allele frequency in the total aggregate of populations will also vary slightly because the number of populations is finite. The sampling variance of the overall mean frequency is equal to the sum of the expected within- and among-population allele-frequency variances divided by the number of populations, 212. The latter quantity was already defined in Equation 2.14, whereas the former is the expected binomial sampling variance divided by the sample size ($2N$), or $p_0(1-p_0)[1-(1/2N)]^t/(2N)$. Figure 2.5 shows that although the frequency of the bw^{75} allele averaged over all populations increased to

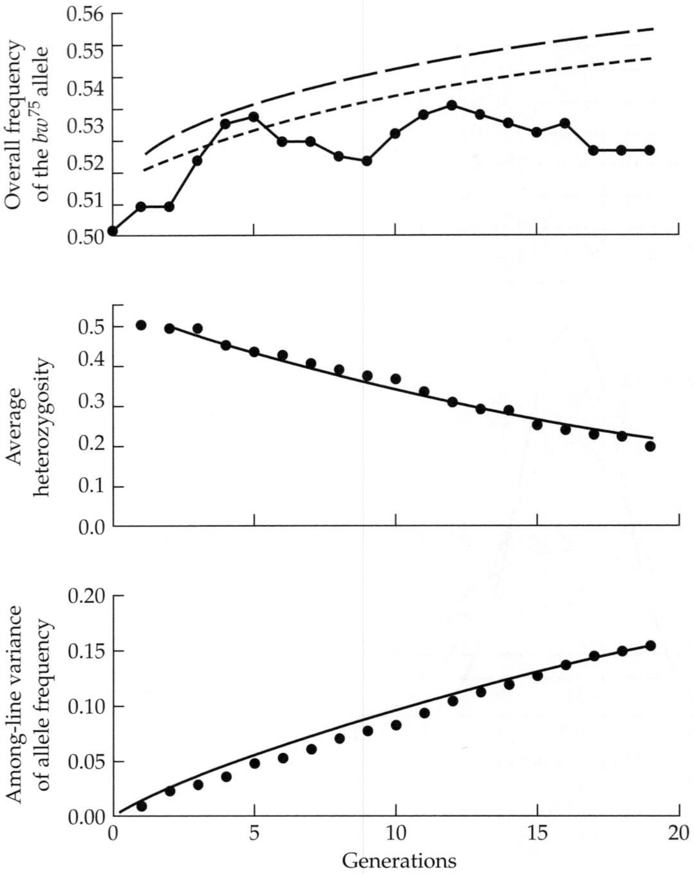

Figure 2.5 Patterns of change in the frequencies of the bw^{75} allele in 212 isolated populations of *Drosophila melanogaster,* each consisting of eight breeding males and eight females. (**Top**) The average allele frequency over the entire pool of populations. The dotted and solid lines, respectively, denote upward deviations of two standard errors from the expected value of $p_0 = 1/2$ under the assumption of effective population sizes of 16 and 10.2 individuals. (**Middle**) Mean observed heterozygosity compared to the expectations assuming an effective population size of 10.2. The expected heterozygosity is 0.5 in generations 1 and 2 because the base population (generation 0) consisted entirely of heterozygotes, and with separate sexes, an additional generation is required for the unification of alleles that are identical by descent. (**Bottom**) Among-line variance of allele frequencies compared with their expectations assuming an effective population size of 10.2. (After Buri 1956.)

0.525, it generally remained within two standard errors of the expectation under pure drift. The overall pattern of change in mean allele frequency is therefore compatible with the expectations for a neutral locus subject to random genetic drift.

The dynamics of the among-population divergence (Figure 2.6) are qualitatively very similar to the expected pattern illustrated in the left panel of Figure 2.4 (corresponding to $p_0 = 0.5$). As the population allele frequencies diverge, the initial bell-shaped distribution does indeed become flatter, eventually acquiring a U-shape as populations that are fixed for the bw^{75} or bw alleles accumulate. Had the experiment been extended further in time, the distribution would have eventually consisted of only two classes, populations fixed for bw^{75} and those fixed for bw, with nearly equal frequencies.

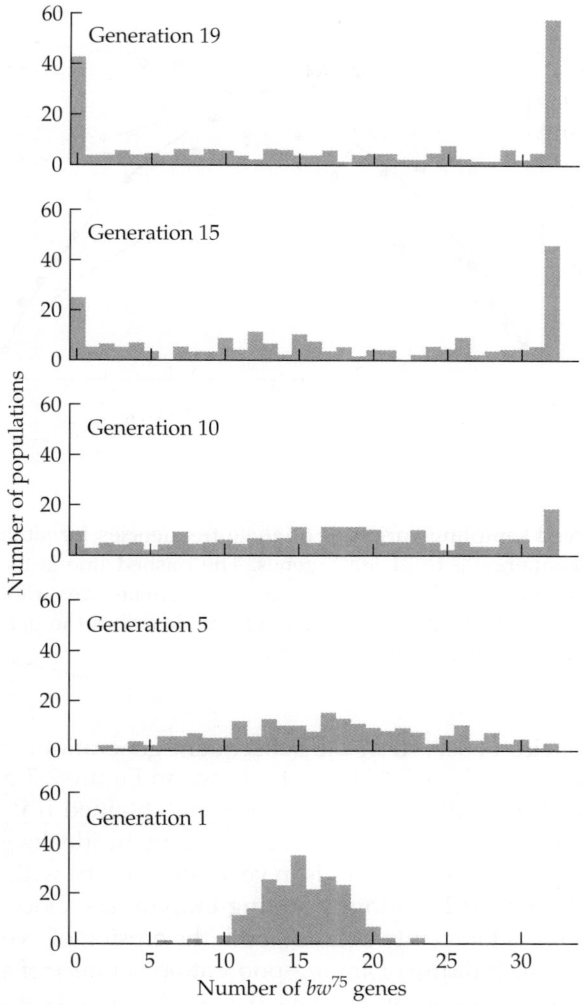

Figure 2.6 Distribution of the number of bw^{75} alleles in 212 populations of *D. melanogaster* each initiated with a frequency of 0.5. Two features are represented in this temporal series: the distribution of frequencies for segregating alleles (1 to 31 copies), the expected form of which is given in Figure 2.4, and the accumulation of fixed alleles (0 or 32 copies). (From Buri 1956).

Despite the qualitative agreement with theoretical expectations, the rate of divergence illustrated in Figure 2.6 is somewhat greater than that expected for randomly mating populations of 16 individuals. However, this does not necessarily invalidate the theory outlined previously, as it is possible that not all 16 potential parents reproduced each generation, and/or that the distribution of family sizes deviated from randomness. Either condition would cause the populations to behave genetically as though they were smaller than its actual size (Chapter 3). With the massive amount of data in Buri's experiment, it is possible to obtain an empirical estimate of this **effective population size** in the following way.

Not including fixed classes, there are 31 possible allele frequencies in Buri's populations (1/32 to 31/32), each of which was observed at various times in one or more of the 212 populations. Focusing on any one allele-frequency class, the single-generation sampling variance conditional on the initial allele frequency for this class (p) can then be calculated

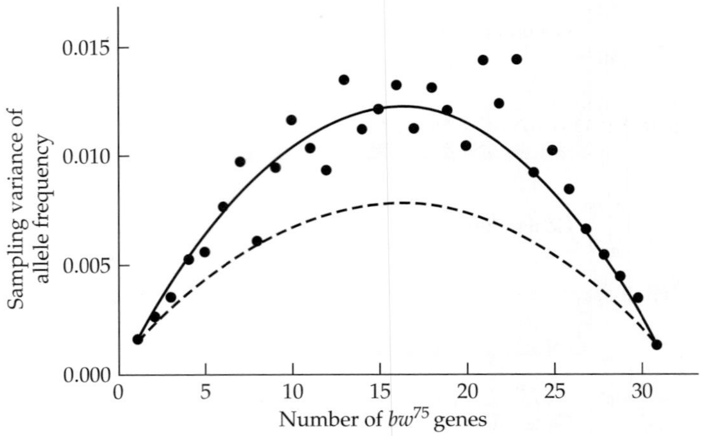

Figure 2.7 Observed sampling variances of allele frequencies for situations in which the donor population contained 1 to 31 bw^{75} genes. The dashed line is the expected pattern, $p(1-p)/2N$, if the actual populations of 8 males and 8 females were randomly mating with equal chances of contributing offspring. The solid line describes the pattern for an average effective population size of 10.2. (From Buri 1956.)

from the allele frequencies observed in the subsequent generation, and compared to the expected value of $p(1-p)/(2N)$. The 31 points shown in Figure 2.7 provide an empirical description of this function, with an excellent fit being obtained if it is assumed that the average effective population size was $N \simeq 10.2$ rather than the idealized 16. In other words, the sampling variance of allele frequencies is in very close accord with that expected for an average ideal population of 10.2 randomly mating individuals. Once this change in scale from $N = 16$ to $N = 10.2$ is taken into account, both the erosion of average heterozygosity within populations and the buildup of among-population variance of allele frequencies are quite consistent with the theory outlined above (Figure 2.5, middle and bottom).

HIGHER-ORDER ALLELE-FREQUENCY MOMENTS

In the previous sections, we evaluated the expected values of various population features under neutrality. However, as just noted, in applying such expressions to empirical studies, it is important to keep in mind that the random sampling of allele frequencies across generations will cause the realized behavior of any particular population or group of populations to deviate from the expected pattern. Thus, there is a practical need for expressions for the variance of various population parameters that result from genetic sampling. This in turn requires an understanding of the behavior of higher-order allele-frequency moments. For example, although the *expected* heterozygosity is a function of $2p(1-p) = 2(p-p^2)$, as will be shown later, its *variance* depends on p^3 and p^4.

For a population obeying the features of the idealized Wright-Fisher model, useful expressions can be obtained by noting that the expected value of an allele-frequency moment in generation $t+1$ can be written as a conditional function of the allele frequency p_t in the previous generation. For example, letting δp denote the change in allele frequency in the previous generation resulting from gamete sampling, the behavior of the first moment (the mean) can be written as

$$E[p_{t+1}|p_t] = E[(p_t + \delta p)|p_t]$$

In the case of neutrality, there are no directional forces, so $E(\delta p) = 0$, and the expected frequency of an allele remains perpetually at its initial value (p_0),

$$E(p_t) = p_0 \tag{2.15}$$

The second moment is obtained by noting that

$$E[p_{t+1}^2|p_t] = E[(p_t^2 + 2p_t\delta p + \delta p^2)|p_t]$$

Because $E(\delta p) = 0$ and $E(\delta p^2|p_t) = p_t(1-p_t)/(2N)$ under binomial sampling,

$$E[p_{t+1}^2|p_t] = E\left(p_t^2 + \frac{p_t - p_t^2}{2N}\right)$$

Letting $\lambda_1 = 1 - (1/2N)$, and noting that $E(p_t) = p_0$, this expression can be rearranged to give the recursion equation

$$E[p_{t+1}^2|p_t] - p_0 = [E(p_t^2) - p_0]\lambda_1$$

the general solution of which is

$$E(p_t^2) = p_0 - [p_0(1-p_0)]\lambda_1^t \tag{2.16a}$$

The general starting expression, which can be extended to all higher-order moments, is

$$\begin{aligned} E[p_{t+1}^k|p_t] &= E[(p_t + \delta p)^k|\, p_t] \\ &= \sum_{i=0}^{k} \binom{k}{i} p_t^i\, E\left[(\delta p)^{k-i}|\, p_t\right] \end{aligned} \tag{2.16b}$$

where the summation gives the terms in the polynomial expansion. For binomial sampling theory, expressions are available for all expected values of powers of δp (e.g., Johnson et al. 2005), so Equation 2.17 can be solved recursively starting with the lower-order moments.

Using expectations for higher-order δp^k terms, expressions for all higher-order moments can be acquired, two of which prove to be particularly useful (Crow and Kimura 1970):

$$E(p_t^3) = p_0 - \frac{3}{2}p_0(1-p_0)\lambda_1^t - \frac{1}{2}p_0(1-p_0)(2p_0-1)(\lambda_1\lambda_2)^t \tag{2.16c}$$

$$\begin{aligned} E(p_t^4) = p_0 &- \frac{18N-11}{10N-6}p_0(1-p_0)\lambda_1^t - p_0(1-p_0)(2p_0-1)(\lambda_1\lambda_2)^t \\ &+ p_0(1-p_0)\left(p_0(1-p_0) - \frac{2N-1}{10-6}\right)(\lambda_1\lambda_2\lambda_3)^t \end{aligned} \tag{2.16d}$$

where $\lambda_i = 1 - (i/2N)$. Modifications for these expressions for populations with separate sexes and 1:1 sex ratios are given by Lynch and Hill (1986).

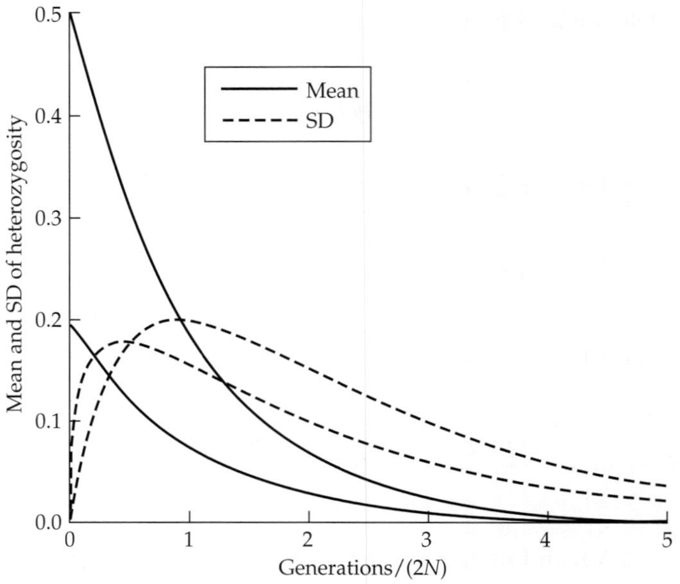

Figure 2.8 Mean heterozygosity and its standard deviation (SD) among replicate populations as a function of time (scaled in units of $2N$ generations). The upper curves assume an initial heterozygosity of $H_0 = 0.5$ and the lower curves of $H_0 = 0.2$. A diallelic locus is assumed, and new variation generated by mutation is ignored. Experimental error resulting from sampling of a finite number of individuals is ignored as well, i.e., we consider only the variance of true population-level heterozygosities resulting from gamete sampling. The expected heterozygosities (solid lines) are obtained with Equation 2.5, whereas the standard deviations (dotted lines) follow from Equation 2.17.

Example 2.5. The preceding expressions can be used to derive the evolutionary (or drift) variance of heterozygosity at a locus under the assumption of Hardy-Weinberg equilibrium, provided there are only two alleles segregating at the locus. Letting $H_t = 2p_t(1 - p_t)$ denote the heterozygosity at generation t, the expected variance of heterozygosity is

$$\begin{aligned}\sigma^2(H_t) &= E\{[2p_t(1-p_t)]^2\} - \{E[2p_t(1-p_t)]\}^2 \\ &= E(4p_t^2) - E(8p_t^3) + E(4p_t^4) - [E(2p_t - p_t^2)]^2\end{aligned}$$

A solution is obtained by substituting Equations 2.16a–2.16d for the expectations of allele-frequency moments, with further simplification made possible by using the approximations $\lambda_1 \simeq e^{-t/2N}$, $\lambda_2 \simeq e^{-2t/2N}$, and $\lambda_3 \simeq e^{-3t/2N}$, yielding

$$\sigma^2(H_t) \simeq H_0 \left[\frac{2}{5} e^{-t/2N} + \left(H_0 - \frac{2}{5} \right) e^{-3t/N} - H_0 e^{-t/N} \right] \tag{2.17}$$

This quantity can be viewed as either the variance in heterozygosity that develops at a particular neutral locus among replicate populations starting from the same initial allele frequencies or as the variance in heterozygosity among a pool of loci within the same population with identical initial allele frequencies. As with the expected heterozygosity, the temporal dynamics of the evolutionary variance of heterozygosity scale inversely with the size of the population. Moreover, the variation in heterozygosity resulting from genetic drift can be quite high, with the standard deviation always exceeding the expected heterozygosity once $t > 2N$ generations (Figure 2.8). With more than two alleles per locus, the preceding expressions would need to be modified to account for the negative drift sampling covariance between different alleles at the locus.

LINKAGE DISEQUILIBRIUM

In the study of multilocus traits, we are naturally interested in combinations of alleles among loci. If the alleles at two loci are independently distributed, the expected frequency of each gamete type can be predicted from the products of the allele frequencies at the two loci. For example, with two alternative alleles (*A* and *a*) at one locus having frequencies p and $1-p$, and those (*B* and *b*) at another locus having frequencies q and $1-q$, the expected frequencies of gametic types *AB*, *Ab*, *aB*, and *ab* are pq, $p(1-q)$, $(1-p)q$, and $(1-p)(1-q)$, respectively, under the assumption of independence. A natural measure of the deviation of the frequency of a gametic type from such expectations is the **coefficient of linkage disequilibrium**

$$D_{\text{AB}} = p_{AB} - pq \tag{2.18}$$

where p_{AB} denotes the observed frequency of the *AB*th gamete.

Here a word on notation is in order. Throughout, we will use italicized letters to designate alleles associated with a particular locus, which will in turn be denoted with an uppercase nonitalicized letter. Thus, Equation 2.18 defines the linkage disequilibrium at loci A and B, using alleles *A* and *B* to define the deviation from random expectations.

This definition of D_{AB} has the useful feature of being equivalent to the covariance of the distribution of alleles *A* and *B* in the same gametes. To see this, let the random variable x take on a value of one when the allele at the first locus is *A* and zero otherwise, and likewise let y equal one when the allele at the second locus is *B* and zero otherwise. Then, $E(xy) = p_{AB} \cdot 1$, $E(x) = \text{freq}(A) \cdot 1 = p$, and $E(y) = \text{freq}(B) \cdot 1 = q$, giving the covariance between allele presence at the two loci as $E(xy) - E(x)E(y) = p_{AB} - pq$.

In the absence of selection, there will be no tendency for the alleles at different loci to be associated positively versus negatively. However, forces such as migration or nonrandom mating may cause some such correlations. Letting D_0 denote an initial level of disequilibrium, c denote the frequency of recombination between loci, and $\lambda_1 = 1 - (1/2N)$, the expected disequilibrium (under random mating) resulting from the joint forces of recombination and gametic sampling is

$$\begin{aligned} E(D_t) &= [(1-c)\lambda_1]^t D_0 \\ &\simeq D_0 e^{-(2Nc+1)t/(2N)} \end{aligned} \tag{2.19}$$

(Hill and Robertson 1966), showing that disequilibrium declines toward zero in the absence of any replenishing forces.

In contrast, the *variance* of D can be quite substantial even when its expected value is zero. The problem can be evaluated by use of the following set of recursion equations for fourth-order moments of allele frequencies,

$$\begin{pmatrix} E[p(1-p)q(1-q)] \\ E[D(1-2p)(1-2q)] \\ E(D^2) \end{pmatrix}_{t+1} = \lambda_1 \cdot \begin{pmatrix} \lambda_1 & \lambda_1(1-c)/(2N) & 2(1-c)^2/(4N^2) \\ 0 & \lambda_2^2(1-c) & 4\lambda_2(1-c)^2/(2N) \\ 1/(2N) & \lambda_1(1-c)/(2N) & [\lambda_2^2 + (1/4N^2)](1-c)^2 \end{pmatrix} \cdot \begin{pmatrix} E[p(1-p)q(1-q)] \\ E[D(1-2p)(1-2q)] \\ E(D^2) \end{pmatrix}_t \tag{2.20}$$

where, as previously, $\lambda_i = 1 - [i/(2N)]$ (Hill and Robertson 1968). The evolutionary variance of D associated with drift among replicated populations or among loci starting from the same allele frequencies is

$$\sigma^2(D) = E(D^2) - E^2(D) \tag{2.21}$$

If $D_0 = 0$, then $E(D_t) = 0$, and $\sigma^2(D_t) = E(D_t^2)$. Ohta and Kimura (1969a; their Equations 20–25) obtained a closed-form solution to this expression.

Hill and Robertson (1968) introduced a standardized measure of linkage disequilibrium, often referred to as the squared within-gamete correlation of allele frequencies at two loci

$$r^2 = \frac{D^2}{p(1-p)q(1-q)} \tag{2.22}$$

This is a natural definition in the sense of a correlation coefficient being generally defined as the ratio of a covariance (in this case, D) and the square root of the product of the variances (in this case, $p[1-p]$ and $q[1-q]$ for the allelic variances at each locus). There are, however, conceptual limitations to this measure, as it can only achieve a maximum value of 1.0 if the heterozygosities at both loci are equal. To make matters worse, in many applications of Equation 2.22, investigators have computed the ratio of average values of the numerator and denominator for large numbers of pairs of loci, which is not equivalent to the squared gametic correlation averaged over pairs of loci. This difference in definition can lead to up to 100-fold differences between estimated values of r^2 and the true correlation if allele frequencies are extreme, as is often the case with neutral alleles (Song and Song 2007).

Example 2.6. One common setting generating linkage disequilibrium involves a new allele arising as a single copy on a particular background. Let A denote the derived allele and assume it arose on a B background at a second locus. Initially, all copies of A are associated with B (there are no Ab gametes). Because the sum of the AB and aB gamete frequencies is just the frequency q of B, the resulting 2×2 gametic contingency table is

	A	a
B	p	$q-p$
b	0	$1-q$

and the resulting initial values for D and r^2 become

$$D_{\text{AB}} = p_{AB}p_{ab} - p_{aB}p_{Ab} = p(1-q) - 0 \cdot (q-p) = p(1-q)$$

$$r^2 = \frac{D^2}{p(1-p)q(1-q)} = \frac{[p(1-q)]^2}{p(1-p)q(1-q)} = \frac{p(1-q)}{(1-p)q}$$

For this example, $r^2 = 1$ only when $p = q$ (Sved 1971), or when there are no aB gametes (A and B always co-occur), in which case the 2×2 gamete contingency table becomes

	A	a
B	p	0
b	0	$1-p$

, or equivalently

	A	a
B	q	0
b	0	$1-q$

For a newly arising mutation, $p = 1/(2N)$, so initially $D_{\text{AB}} = (1-q)/(2N)$ and $r^2 \simeq (1-q)/(2Nq)$.

MUTATION-DRIFT EQUILIBRIUM

In the preceding pages, we were largely concerned with the dynamics of gene-frequency change owing to the effects of random genetic drift alone. Under this model, finite population size eventually results in the complete loss of genetic variation (and covariation) within populations, at which point all loci are fixed for ancestral alleles with probabilities equal to their initial frequencies. In reality, however, mutation will always introduce variation at a low rate, which not only offsets some of the loss resulting from drift, but also ensures that neutral loci will continue to diverge among isolated populations. If the time scale of the problem under consideration is short ($t \ll 2N$) and the initial level of within-population variation is high (relative to the mutational rate of production of new heterozygosity per generation), the contribution from mutation will be negligible, and the preceding expressions will be quite adequate. However, for longer-term evolutionary issues, such as the maintenance of variation in natural populations and interspecific divergence, mutation cannot be ignored.

The incorporation of mutation into a neutral model of evolution is relatively straightforward. For example, suppose there are k possible alleles at a locus, each with a mutation rate u per generation (the **k-exchangeable alleles model**). The dynamics of heterozygosity can then be obtained by recalling from above that the expected frequency of heterozygotes in generation $t+1$ in the absence of mutation is $[1-(1/2N)]H_t = \lambda_1 H_t$, whereas the expected frequency of homozygotes is $1-\lambda_1 H_t$. Following mutation, the heterozygous state will be retained if: (1) neither allele mutated, the probability of which is $(1-2u)$, ignoring the very small probability of double mutations to the same state; or (2) one of the alleles mutated to a different state than the other, the probability of which is $[2u(k-2)/(k-1)]$, assuming that all allelic types are equally mutationally exchangeable. On the other hand, homozygotes will be mutationally converted to heterozygotes at a rate of $2u$. Thus, the expected dynamics of heterozygosity can be expressed as

$$H_{t+1} = H_t\lambda_1\left((1-2u)+\frac{2u(k-2)}{k-1}\right)+2u(1-\lambda_1 H_t) \tag{2.23}$$

Setting $H_{t+1} = H_t$, the expected value of heterozygosity under drift-mutation balance is found to be

$$E(H) = \frac{\theta}{1+[\theta k/(k-1)]} \tag{2.24a}$$

where $\theta = 4Nu$, a result first given by Kimura (1968a). Note that θ has the pleasing interpretation of being the ratio of the rates of mutational production of heterozygotes from homozygotes ($2u$) and the rate of loss of heterozygosity by drift ($1/2N$). If a large number of alternative alleles ($k \gg 1$) is assumed, as is reasonable when the unit of analysis is an entire gene, Equation 2.24a reduces to

$$E(H) \simeq \frac{\theta}{1+\theta} \tag{2.24b}$$

which is equivalent to the **infinite-alleles model** of Kimura and Crow (1964). On the other hand, if the unit of analysis is a nucleotide site, then $k=4$, and

$$E(H) = \frac{\theta}{1+(4/3)\theta} \tag{2.24c}$$

where u is now the mutation rate per nucleotide site. Equation 2.24a needs to be modified if alleles mutate at different rates or are not equally mutationally accessible (Kimura 1983; Nei and Kumar 2000), but provided $2Nu \ll 1$, as seems to be generally the case (Chapter 4), then $E[H] \simeq \theta$ regardless of the model assumed. As $2Nu \to \infty$, the infinite-alleles model implies $E(H) \to 1.0$, whereas the $k=4$ model implies $E(H) \to 0.75$. The latter result is a simple consequence of four segregating alleles with equal frequencies (0.25) when the power of drift is overwhelmed by mutation. A number of other mutational models are possible, and a very general treatment is given by Cockerham (1984c), who also considered the transient approach to equilibrium.

As these drift-mutation models play a central role in the neutral theory of molecular evolution (Kimura 1983), substantial attention has been given to additional details that are obscured by the summary statistic of average heterozygosity. For example, θ is the expected average heterozygosity over a large number of loci; this is not likely to be exactly realized at any particular site. Because of the stochastic nature of both mutation and drift, the allele frequencies at any neutral locus are expected to wander stochastically over time, with some loci being transiently fixed for one particular allele, and others being distributed over the remaining spectrum of allele frequencies. Kimura (1968a) obtained an expression analogous to Equation 2.8 for the complete probability distribution of the frequency of an allele under the symmetric mutation model described above, starting from an arbitrary allele frequency. Although this expression is quite complicated, a highly useful result is that regardless of the starting point, a steady-state distribution of allele frequencies p is eventually attained

$$\phi(p) = \frac{\Gamma(\theta+\beta)}{\Gamma(\theta)\Gamma(\beta)}(1-p)^{\theta-1}p^{\beta-1} \tag{2.25a}$$

where $\theta = 4Nu$, $\beta = 4Nu/(k-1)$, with

$$\Gamma(\alpha) = \int_0^\infty x^{\alpha-1}e^{-x}\,dx \tag{2.25b}$$

being the **gamma function** (Appendix 2). Equation 2.25a may be viewed as either the expected distribution of allele frequencies over all neutral loci within a single population in mutation-drift equilibrium or as the distribution of allele frequencies at a particular locus among replicate populations (or species) with identical N and u. Nei and Li (1976) presented the theory necessary for predicting the approach to the equilibrium state.

The expected value of any function of population allele frequencies (e.g., homozygosity) can be obtained by simply integrating the function over the density distribution $\phi(p)$. It is useful that Equation 2.25a defines a **beta distribution** (Appendix 2), as many of its properties are already well known. For example, the mean of the distribution, which is the expected allele frequency, is

$$E(p) = \frac{\beta}{\theta+\beta} = \frac{1}{k} \tag{2.26a}$$

and the allele-frequency variance among replicates is

$$\sigma^2(p) = \frac{\theta\beta}{(\theta+\beta)^2(\theta+\beta+1)} = \frac{k-1}{k^2[2Nuk(k-1)+1]} \tag{2.26b}$$

Expressions for the variance of heterozygosity for a population en route to equilibrium were derived by Li and Nei (1975) and Lessard (1981), and at equilibrium

$$\sigma^2(H) = \frac{2\theta[1+(\theta/\ell)]}{[1+\theta+(\theta/\ell)]^2[2+\theta+(\theta/\ell)][3+\theta+(\theta/\ell)]} \tag{2.27a}$$

where $\ell = k-1$ (Stewart 1976). Note that as $k \to \infty$ (the infinite-alleles model),

$$\sigma^2(H) = \frac{2\theta}{(1+\theta)^2(2+\theta)(3+\theta)} \tag{2.27b}$$

which reduces to about $\theta/3$ for $\theta \ll 1$.

Although the preceding results lead to predicted equilibrium allele frequencies, heterozygosities, etc., when viewed in isolation, such results obscure the long-term dynamics of neutral mutations. Given a population of size N, on average $2Nu$ new mutations arise per nucleotide site per generation, each with initial frequency $1/(2N)$. As noted earlier, the probability of fixation of a neutral allele is simply equal to its initial frequency, so that in the long run, at a neutral site, there is an average turnover rate of $2Nu \cdot 1/(2N) = u$ mutations per generation. This simple but powerful result tells us that the long-term rate of nucleotide substitution at a neutral site is equal to the mutation rate, *regardless of the size of the population,* a hallmark of the **neutral theory of molecular evolution** (Kimura 1983). This long-term flux occurs in the face of the maintenance of quasi-steady-state within-population heterozygosity, which arises as a consequence of the per-generation balance between the loss of variation by fixation and replenishment by recurrent mutation.

Finally, Ohta and Kimura (1971) and Hill (1975) obtained expressions for the expected values of the two-locus moments described in Equation 2.20, under the infinite-alleles model at stochastic drift-mutation equilibrium. Letting $\rho = 4Nc$ and $\theta = 4Nu$, where c is the recombination rate between sites, and u is the mutation rate per site, Hill's expressions reduce to

$$\begin{aligned} E[p(1-p)q(1-q)] &= M(22 + 13\rho + 32\theta + \rho^2 + 6\rho\theta + 8\theta^2) \\ E[D(1-2p)(1-2q)] &= 8M \\ E[D^2] &= M(10 + \rho + 4\theta) \end{aligned} \tag{2.28a}$$

where

$$M = \theta^2/[(\theta+1)(18 + 13\rho + 54\theta^2 + \rho^2 + 19\rho\theta + 40\theta^2 + 6\rho\theta^2 + 8\theta)], \tag{2.28b}$$

and the standardized linkage disequilibrium is given by

$$E(r^2) = \frac{10 + \rho + 4\theta}{22 + 13\rho + 32\theta + \rho^2 + 6\rho\theta + 8\theta^2} \tag{2.29a}$$

As will be seen in Chapter 4, for individual nucleotide sites θ is generally substantially smaller than one, whereas for sites separated by hundreds of base pairs or more, ρ is generally greater than one. Under such conditions,

$$E(r^2) \simeq \frac{10 + \rho}{22 + 13\rho + \rho^2} \tag{2.29b}$$

which asymptotically approaches $1/\rho$ for large ρ. Additional theoretical points of interest were developed in Strobeck and Morgan (1978) and Golding and Strobeck (1980). A more daunting problem is obtaining expressions for the evolutionary variances of these statistics. As the components in Equation 2.28a already involve fourth-order moments of allele frequencies within and between loci, their variances are functions of moments up to the eighth order. Hill and Weir (1988) tackled this problem.

Example 2.7. A widely cited expression for $E(r^2)$ originates with Sved (1971; Feldman and Sved 1973), which we will refer to as

$$E(r^2_{\text{IBD}}) = \frac{1}{1 + 4Nc} = \frac{1}{1 + \rho}$$

Note that this expression is quite different from Equation 2.29a (as no terms for mutation appear), and also different from the approximation given by Equation 2.29b. To understand this discrepancy, it is useful to reflect on Sved's simple derivation, which focuses on a different measure of correlation than that outlined earlier. Conditional on two random gametes being identical by descent (IBD) from common ancestry at locus A, Sved wished to know the probability Q_{AB} that the alleles at locus B on both gametes are also IBD owing to the absence of recombination between the sites during the entire two pathways back to the common ancestor. With this definition in hand, Sved showed that $Q_{\text{AB}} = r^2_{\text{IBD}}$.

Let Q_t denote the value of joint IBD in generation t, and consider how this relates to Q_{t-1}, one generation in the past. For a population with size N, at any particular site a random pair of gametes will be direct copies of a gamete in the preceding generation with probability $1/(2N)$, and IBD will exist at both sites provided no recombination has occurred between them, the probability of which is $(1-c)^2$, where c refers to the recombination rate. Alternatively, the two sampled gametes will be drawn from different gametes leading to the preceding generation, the probability of which is $1 - 1/(2N)$. In the latter case, joint IBD will still exist if it happened to have been present for the two chromosomal segments in the preceding generation with no recombination again occurring in either. Summing up, the recursion equation for joint IBD is

$$Q_t = \frac{1}{2N}(1-c)^2 + \left(1 - \frac{1}{2N}\right)(1-c)^2 Q_{t-1}$$

Setting $Q_t = Q_{t-1}$ gives the equilibrium solution as

$$\widetilde{Q} = \frac{1/(2N)}{1 - [1 - 1/(2N)](1-c)^2} \simeq \frac{1}{1 + 4Nc}$$

Although this is an equilibrium solution like Equations 2.29a and 2.29b, Sved's Q is not equivalent to the measure r^2 in these previous formulae, which are concerned with the long-term average disequilibrium associated with identity in state (IIS). Whereas IIS is a directly observable quantity, IBD is not, and hence there are interpretative problems when applying Sved's expression to empirical data. This is because parallel mutation can cause IIS for alleles that are not IBD, and secondary mutations can eliminate IIS for pairs of alleles that are otherwise IBD. Thus, although Sved's expression is widely used, apparently because of its simplicity, Equations 2.29a and 2.29b appear to be more appropriate for practical applications to observed molecular variation.

THE DETAILED STRUCTURE OF NEUTRAL VARIATION

Although heterozygosity provides a robust measure of genetic variation, as a summary statistic it obscures the details of the underlying structure of this variation, e.g., the number of alleles and their frequencies. Fortunately, at mutation-drift equilibrium, most of these features are just relatively simple functions of θ (Ewens 2004). There are, however, two alternative ways of thinking about molecular variation, each appropriate in a different context, and care must be taken in the interpretation of the mutation rate in each model. In one case, the **infinite-alleles** (or **infinitely many alleles**) **model**, the unit of observation is the locus (a stretch of DNA), whereas in the **infinite-sites** (or **infinitely many sites**) **model**, it is the nucleotide site. Results developed here are extensively used in Chapters 8 and 9 in methods for detecting departures from the **equilibrium neutral model**.

The Infinite-alleles Model and the Associated Allele-frequency Spectrum

The infinite-alleles model (briefly introduced earlier) was developed prior to the DNA-sequencing era, but was motivated by emerging knowledge on the structure of DNA sequences. In today's world, different alleles under this model are typically viewed as different sequences (**haplotypes**) over a region of L nucleotide sites, and the general assumption is that L is large enough that each mutation generates a new haplotype (not preexisting in the population), but small enough that recombination can be ignored (so that mutation is the sole generator of novel sequences). Considering the five short sequences in Figure 2.9, under the infinite-alleles framework, there are three different alleles, although there are only two segregating sites. With allele frequencies 0.4 (AAGACC), 0.4 (AAGGCC), and 0.2 (AAGGCA), the allelic heterozygosity of the sample is 0.64. This is, of course, a rather crude perspective, as it ignores the ways in which alleles differ from each other (in this example, two pairs of alleles differ at a single site, whereas one pair differs at two sites).

A A G A C C
A A G G C C
A A G A C C
A A G G C C
A A G G C A

Figure 2.9 An example of the difference between the infinite-alleles and infinite-sites models. Five sequences (horizontal rows) scored at six nucleotide sites (vertical columns) are sampled from a population. Three of these five sequences are different and are scored as three alleles (or haplotypes) under an infinite-alleles framework (from top to bottom, sequences 1 and 3; 2 and 4; and 5). Conversely, only two of the six sites are segregating (from left to right, columns 4 and 6), giving two polymorphic sites under an infinite-sites framework.

A key parameter for the infinite-alleles model is the per-*locus* population mutation rate, $\theta_L = 4NuL$, which we distinguish from the more commonly used per-*site* measure, $\theta = 4Nu$, where u is the mutation rate per site. As noted previously, under neutrality, provided that the population has been at constant size long enough to be in mutation-drift equilibrium, the expected heterozygosity is given by Equation 2.24b, with $E(H) \approx \theta_L$ for $\theta_L \ll 1$. If the expected heterozygosity is small, a sample will often be **monomorphic**, consisting of only a single allele, or **dimorphic**, consisting of just two alleles. However, at higher levels of heterozygosity, multiple alleles can be expected, and a natural measure of variation is the number of different alleles in a sample. Insight into this quantity under drift-mutation equilibrium is given the probability of having k different alleles in a sample of n genes, which is

$$\Pr(k \,|\, \theta_L, n) = \frac{S_n^k \, \theta_L^k}{S_n(\theta_L)} \tag{2.30a}$$

where

$$S_n(\theta_L) = \theta_L(\theta_L + 1)(\theta_L + 2) \cdots (\theta_L + n - 1) \tag{2.30b}$$

and S_n^k is the coefficient on the θ_L^k term obtained by expanding the polynomial in Equation 2.30b (Ewens 1972). This formula opened up the field of formal statistical tests for whether a pattern of allelic variation is consistent with the equilibrium neutral model. For example, using Equation 2.30a, one can ask if an observed estimate of θ_L (obtained in this case as the allelic heterozygosity) is consistent with the observed number k of different alleles (Chapter 9).

Several useful results follow from Equation 2.30a. First, the probability of a monomorphic sample is

$$\Pr(k = 1) = \frac{(n-1)!}{(\theta_L + 1)(\theta_L + 2) \cdots (\theta_L + n - 1)} \tag{2.31a}$$

Second, a bit of algebra gives the mean and variance for the number of alleles in a sample as

$$E(k) = 1 + \theta_L \cdot \sum_{j=2}^{n} \frac{1}{\theta_L + j - 1}, \qquad \sigma^2(k) = \theta_L \cdot \sum_{j=1}^{n-1} \frac{j}{(\theta_L + j)^2} \tag{2.31b}$$

An even more complete description of the segregating allelic variation is given by the

allele-frequency spectrum, which describes the joint probability distribution of the *number of alleles* in the sample and their *frequencies*. Given that the numbering of alleles is arbitrary, the convention is to consider the vector $(n_1, \cdots, n_n)$, where n_i denotes the number of alleles that have exactly i copies in the sample. If the sample is monomorphic, then $n_n = 1$, whereas if all n alleles are unique (**singletons**), $n_1 = n$. For the example data set in Figure 2.9, one allele appears as a singleton, whereas the other two alleles both appear as two copies, giving $n_1 = 1, n_2 = 2$, and $n_3, n_4, n_5 = 0$. The constraint on the n_i is that

$$\sum_{i=1}^{n} i \cdot n_i = n \tag{2.32}$$

A very powerful result, due to Ewens (1972) and Karlin and McGregor (1972), is that the joint probability distribution of $n_1, \cdots, n_n$ and k is given by the **Ewens sampling formula** (Ewens 1972)

$$\Pr(n_1, n_2, \cdots, n_n, k \,|\, n) = \frac{n!\theta_L^k}{S_n(\theta_L)\,(1^{n_1} 2^{n_2} \cdots n^{n_n})n_1!n_2! \cdots n_n!} \tag{2.33a}$$

Only nonzero values of n_i are included. The probability that the sample is monomorphic, Equation 2.31a, directly follows by setting $k = 1, n_1 = n$. Equations 2.30a and 2.33a show that the conditional distribution of $n_1, \cdots, n_n$ given k is

$$\Pr(n_1, n_2, \cdots, n_n \,|\, n,\, k) = \frac{n!}{S_n^k\,(1^{n_1} 2^{n_2} \cdots n^{n_n})n_1!n_2! \cdots n_n!} \tag{2.33b}$$

Note that the righthand side in Equation 2.33a is *independent* of θ_L. This property arises because k is a sufficient statistic for θ_L under the equilibrium neutral model, so that conditioning on k removes any dependence on θ_L. It is this independence that forms the basis of the Ewens-Watterson test of neutrality considered in Chapter 9.

The Infinite-Sites Model and the Associated Site-frequency Spectrum

An alternate framework for summarizing molecular variation is embodied in the **infinite-sites** model, which treats a region as a series of L sites. Each new mutation is again viewed as unique, but now occurring at a novel (not currently segregating) *site*, and as with the infinite-alleles model, recombination within the region is assumed to be insignificant. Because this model allows only a single mutation per site, a particular variant at a polymorphic site is either **ancestral** (original) or **derived** (mutated), and each segregating site is treated as biallelic. With sequence data from one or more outgroups, one can often **polarize** the nucleotides at any particular site, determining which is derived by assuming the outgroup harbors the ancestral state. In the absence of such information, the **minor-allele frequency**, the frequency of the rarest nucleotide, is reported for each site. In the following expressions, we will make use of both measures of the population mutation rate noted above, with $\theta_L = 4NuL = \theta L$.

The infinite-sites model offers a much richer set of information than can be achieved with the infinite-alleles model. One measure is the **number of segregating sites** S, i.e., the number of polymorphic sites in a sample (the loose analog to number of alleles). A second is the **nucleotide diversity** π, the average per-site heterozygosity within the region of interest. Finally, one can consider the **site-frequency spectrum**, the counterpart of the *allele*-frequency spectrum. In this case, instead of counting the number of times each allele appears in the sample, one considers the number of *sites* s_j in the sample with j copies of a nucleotide. The **unfolded frequency spectrum** refers to polarized nucleotides, with s_j being the number of sites with $j \leq n$ copies of the *derived* nucleotide. For a **folded frequency spectrum**, s_j is the number of sites with j copies of the minor nucleotide, with $j \leq n/2$.

For the example data set in Figure 2.9, four of the six sites are monomorphic, whereas site 4 has 3 Gs and 2 As, and site 6 has 4 Cs and an A. For a folded frequency spectrum, this gives four sites in class 0, one site in class 1, and one site in class 2 (i.e., $s_0 = 4, s_1 = 1, s_2 = 1$). If the nucleotides were polarized, so that (for example) the ancestral states at the six sites were (respectively) A-A-G-G-G-A, the unfolded frequency spectrum would be three in class 0 (sites 1, 2, 3), one in class 2 (site 4), one in class 4 (site 6), and one in class 5 (site 5) (i.e., $s_0 = 3, s_2 = 1, s_4 = 1, s_5 = 1$). Many of the analyses using site-frequency spectrum data condition on only sites that are polymorphic in the sample.

For a very long region ($L \gg 1$) (again assuming neutrality, and mutation-drift equilibrium), the fraction of sites in the entire *population* that have a derived nucleotide at frequency x (for $0 < x < 1$) is given by the (unfolded) **Watterson** (1975) **distribution**,

$$\phi(x) = \frac{\theta}{x} \quad \text{for} \quad \frac{1}{2N} \leq x \leq 1 - \frac{1}{2N} \tag{2.34a}$$

This tells us that under the neutral model most sites are expected to have a low frequency of derived nucleotides in the population. Reconfiguring Equation 2.34a for unpolarized alleles, so that $0 < x \leq 0.5$ where x is the minor-allele frequency, gives the **folded Watterson distribution**,

$$\phi(x) = \frac{\theta}{x} + \frac{\theta}{1-x} = \frac{\theta}{x(1-x)} \quad \text{for} \quad \frac{1}{2N} \leq x \leq 1/2 \tag{2.34b}$$

The folded and unfolded frequency spectra are very similar over the range where both are defined ($0 < x \leq 0.5$), as high-frequency derived nucleotides are rare under the equilibrium neutral model, due to the fact that most new mutations are lost by drift.

The site-frequency spectrum for a *sample* is not the same as that for the entire population, but it follows from the Watterson distribution. In the case of an unfolded frequency spectrum, for a sample of size n, the number s_i of the L total sites with i derived nucleotides has the expected value

$$E(s_i) = \frac{\theta_L}{i}, \quad \text{for} \quad 1 \leq i \leq n-1 \tag{2.35a}$$

(Fu 1995; Ewens 2004). Because $\theta_L = i\,E(s_i)$, Equation 2.35a motivates several infinite-site estimators of θ developed in Chapters 4 and 9. By using different regions of the site-frequency spectrum (i.e., different ranges for i) to estimate θ, various assumptions of the standard neutral model (e.g., constant population size, and absence of selection) can be tested (Chapter 9). Similarly, for a folded frequency spectrum, where s_i now denotes the number of sites with minor-nucleotide frequency i/n,

$$E(s_i) = \frac{\theta_L}{i} + \frac{\theta_L}{n-i} = \frac{\theta_L n}{i(n-i)}, \quad \text{for} \quad 1 \leq i \leq [n/2] \tag{2.35b}$$

where $[n/2]$ denotes the largest integer $\leq n/2$.

Although the expectations given by Equations 2.35a and 2.35b can be used for method-of-moments estimators of θ, likelihood estimators (LW Appendix 4) require the full distribution within a sample, not just the expected value. The probability of seeing exactly k derived nucleotides at a site with allele-frequency x follows from the binomial,

$$\Pr(k \,|\, x, n) = \binom{n}{k} x^k (1-x)^{n-k} \tag{2.36a}$$

In addition, the probability that a sample is polymorphic at a site with allele-frequency x is just one minus the probability of a sample monomorphic for either the derived (x^n) or the ancestral $(1-x)^n$ nucleotide,

$$\Pr(1 \leq k \leq n-1) = 1 - x^n - (1-x)^n \tag{2.36b}$$

The probability of seeing k derived nucleotides at a random site in a sample is the average of Equation 2.36a over the possible x values,

$$\Pr(k \,|\, n) = \binom{n}{k} \int_{1/(2N)}^{1-1/(2N)} x^k (1-x)^{n-k} \phi(x) dx \tag{2.36c}$$

where $\phi(x)$ is defined by Equation 2.34a. This formula, which forms the null model for several of the likelihood-based tests for a selective sweep (Chapters 8 and 9), is the infinite-sites analog to Ewens' sampling formula for the infinite-alleles model (both of which are functions of n and θ). Ewens' formula (Equation 2.30a) gives the probability that k alleles are seen in a sample of size n, whereas Equation 2.36c gives the probability that a randomly-chosen site is segregating k copies of a derived nucleotide. The latter formula also defines the probability of k copies of a minor nucleotide under the folded spectrum if Equation 2.34b is used for $\phi(x)$.

THE GENEALOGICAL STRUCTURE OF A POPULATION

The preceding analyses showed that a number of summary statistics, such as average levels of heterozygosity and linkage disequilibrium, are defined by the processes of drift, mutation, and recombination in predictable ways, at least for neutral sites. Provided we retain our focus on neutral regions of the genome, it is possible to go quite a bit further, even to the extent of predicting the expected *genealogical* relationships among different sequences sampled within populations. The basic theory, first laid out by Kingman (1982a 1982b) and now called **coalescent theory**, provides an elegant and powerful approach for solving problems in population genetics and molecular evolution. Kingman (2000) reviewed the historical origins of this approach, and detailed overviews can be found in Hudson (1990), Donnelly and Tavaré (1995), Fu and Li (1999), Nordborg (2001), Stephens (2001), Rosenberg and Nordborg (2002), Hein et al. (2005), and Wakeley (2006).

Because all of the genes within a population are direct products of past gametic sampling, they are all ultimately related in a genealogical sense. Thus, if one were to sample two alleles in a current population and then follow them back in time, both copies would eventually be traced to a single copy in an ancestral individual, at which point the two alleles are said to have **coalesced.** A key principle is that the form of the expected gene genealogy for neutral genes, in particular the expected coalescence time, is ***completely independent of the mutational process***.

Consider a random sample of n alleles drawn from a current population, assumed to obey all the properties of the idealized Wright-Fisher model, and with no recombination within alleles. Focusing initially on just two of the sampled alleles, we first evaluate the probability that both members of the pair are direct copies of a single allele in the preceding generation. Assuming that each individual produces a large number of gametes, because there are $2N$ gene copies in the population in each generation, this probability is simply $1/(2N)$, whereas $\lambda_1 = 1-(1/2N)$ is the probability that coalescence occurred at some earlier generation. Conditional on coalescence not having occurred in generation one, the probability of coalescence one further generation in the past is again equal to $1/(2N)$, yielding $\lambda_1(1/2N)$ as the unconditional probability of coalescence two generations back. This simple rule can be generalized to give the probability of coalescence exactly t generations in the past,

$$P_c(t) = \lambda_1^{t-1}(1/2N) \tag{2.37}$$

which defines a **geometric distribution**, with the sum of $P_c(t)$ over the interval $t = 1$ to ∞ being equal to one. One simple related point is that the probability that the **most recent common ancestor** (**MRCA**) between two sampled alleles occurred within the last t generations is $1-\lambda_1^t \simeq 1-e^{-t/2N}$, namely one minus the probability of no common ancestor over the first t generations into the past.

The mean coalescence time for two randomly sampled genes is simply

$$\bar{t}_c(2) = \sum_{t=1}^{\infty} t \cdot P_c(t) = 2N \tag{2.38}$$

Thus, the expected number of generations required for any two random alleles to trace back

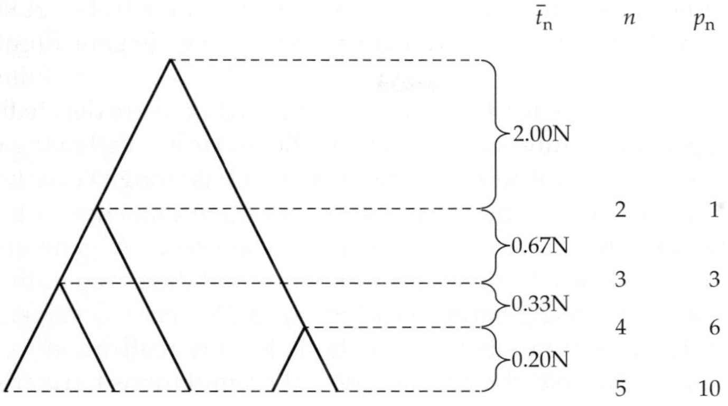

Figure 2.10 Expected coalescence times, $\bar{t}_n$, for a sample of $n = 5$ neutral genes taken from an idealized Wright-Fisher population of size N. The number of gene pairs in each consecutive step of the coalescent process is denoted by p_n, and the expected times to coalescence at each step are equal to $2N/p_n$ generations. The particular lineages that join during each step are arbitrary. Note that over half of the coalescent time for the total lineage of five samples involves the coalescent event between the final two branches.

to an ancestral copy is simply equal to twice the population size (more precisely, twice the effective population size, as will be defined in Chapter 3).

The logic used to derive this result is easily extended to the entire sample of n gene copies. There are $p_n = n(n-1)/2$ possible pairs of n copies, each of which will or will not coalesce in the preceding generation with respective probabilities $1/(2N)$ and $[1-(1/2N)]$. If the sample size is much smaller than the population size, the probability of coalescence for any pair in the sample in the preceding generation is simply the product $p_n/(2N)$. Thus, the probability distribution for the coalescence time of one pair within a set of n sequences is

$$P_c(p_n, t) = [1 - (p_n/2N)]^{t-1}[p_n/(2N)] \tag{2.39}$$

Namely, a geometric random variable with success parameter $(p_n/2N)$. The mean time to coalescence of the first pair is then $2N/p_n$ generations (as opposed to $2N$ generations with a single pair). Because at this point two copies have coalesced into one, the sample size has been reduced by one, and the mean time to coalescence of the next pair is found by resetting p_n to $p_{n-1} = (n-1)(n-2)/2$. This procedure can be followed recursively down to the final pair ($p_n = 1$), which again has an expected coalescence time of $2N$ generations (Figure 2.10). The implication of these results is that the expected time for merging n random lineages into $n-1$ lineages,

$$\bar{t}_n = 2N/p_n = \frac{4N}{n(n-1)} \tag{2.40}$$

increases with decreasing sample size.

The total expected genealogical depth of a sample, obtained by summing the expectations of each coalescence event, is

$$\bar{t}_c(n) = \sum_{i=2}^{n} \frac{4N}{i(i-1)} = 4N\left(1 - \frac{1}{n}\right) \tag{2.41}$$

Thus, under neutrality, the expected time to the most recent common ancestor of all alleles residing at a locus is $\simeq 4N$ generations. This is equivalent to the mean time to fixation of a neutral mutation, as can be verified by substituting $p_0 = 1/(2N)$ into Equation 2.11b.

Notably, the expected distance between the final two nodes in a neutral coalescent tree, $2N$, is *at least half the coalescence time for the entire sample*. Unfortunately, this fundamental issue is commonly ignored by those who invoke deep splits in a gene genealogy as evidence of an adaptive event.

It is important to note that all of the results in this section were derived without regard to any underlying genetic features of the sampled alleles. However, having determined the expected genealogical features of neutral gene sequences, it is straightforward to incorporate genetic issues, as mutations will arise randomly along the branches of the genealogy in numbers proportional to time. For example, given the average $2N$ generations separating two randomly sampled alleles, the average number of mutations separating such genes is $2 \cdot 2N \cdot u = 4Nu$, the two arising because each copy is $2N$ generations removed from the common ancestor. The actual number of mutations for any realization follows a Poisson distribution, so for this example, the probability that no mutations have arise is e^{-4Nu}.

One of the primary uses of the coalescent derives from its ability to efficiently generate sample distributions of quantities of interest (e.g., the expected pattern of molecular variation among neutral alleles), which provides a formal basis for statistical tests of various evolutionary models (Hudson 2002), including those involving selection (Chapter 9). Although all of the preceding results simply refer to the *expected* coalescent times, each individual coalescence time has considerable evolutionary variance, being equal to the square of its expected value (a feature of geometric distributions). As will be discussed in Chapter 9, a number of useful results have been obtained on the sampling distributions of coalescents, but it is also straightforward to obtain such information by using computer simulations to construct random genealogies. For each simulated sample, one starts with n distinct lineages, picks two at random, and generates a value for t_n by randomly drawing from an exponential distribution with mean value given by Equation 2.40. After these two samples are joined in the coalescent tree, the remaining sample of $n - 1$ distinct lineages is treated in the same way to generate a value of t_{n-1}, and so on, until the last two remaining lineages coalesce to yield t_2. For each branch of the resultant tree, the number of mutations is then drawn using a Poisson distribution with an expectation equal to the product of the mutation rate and the length of the branch (in generations), e.g., the two branches emanating from the first node have independent numbers of mutations with expectation ut_n. By repeating such **coalescent simulations** several thousands of times, a distribution of interallelic variation under mutation and drift can then be acquired for any mutational model of interest, with the resultant data providing a null model for various tests of selection based on the pattern of variation in an actual sample (Chapter 9).

MUTATION-MIGRATION-DRIFT EQUILIBRIUM

Populations are often structured in space, with distinct **demes** connected by migration to form a **metapopulation**. The joint forces of mutation, migration, and drift structure the genetic variation both within and among demes, and if kept constant eventually lead to equilibrium values. The equilibrium neutral results serve as the framework for tests of abnormally high or low amounts of among-population variation (Chapter 9), corresponding respectively to diversifying and stabilizing selection.

Quantifying Population Structure: F_{ST}

The classic measure of population structure, F_{ST}, was defined by Wright (1943, 1951) as the correlation (identical by descent status) between alleles in different individuals from the same subpopulation. F_{ST} is a measure of the amount of inbreeding introduced by the population structure, relative to what would be expected in one large panmictic population. Effectively, F_{ST} measures the fraction of total genetic variance due to differences between subpopulations (indeed, S stands for subpopulation and T for total population). This directly follows from the standard ANOVA (analysis of variance) identity that the among-group component of variance equals the within-group covariance (LW Chapter 18), with the latter equal to the correlation among group members times the total variance. In particular, consider the distribution of the allele frequencies for a biallelic locus (with alleles *B* and *b*) over a set of populations. If p_0 denotes the average frequency of *B* over this set, the allelic variance for the total metapopulation is just $p_0(1-p_0)$. F_{ST} is then defined as the fraction of the total variance attributable to the variance in the frequency of allele *B* among demes, $\sigma^2(p)$,

$$F_{ST} = \frac{\sigma^2(p)}{p_0(1-p_0)} \tag{2.42}$$

Wright was somewhat ambiguous in his use of F_{ST}, and some confusion has surrounded various interpretations of its true meaning. These issues were nicely cleaned up by Balding (2003).

Recalling Equation 2.14, which gives the expected variance in the allele frequency between two populations separated from a common ancestor t generations in the past (ignoring mutation and assuming no migration among groups), we have that

$$F_{ST} = \left[1 - \left(1 - \frac{1}{2N_e}\right)^t\right] \simeq \frac{t}{2N_e} \quad \text{for} \quad t \ll N_e \tag{2.43}$$

Under this model, F_{ST} eventually increases to one, as in the absence of mutation, drift eventually removes all variation within groups, with the different demes becoming randomly fixed for alternative alleles.

With recurrent mutation and gene flow among demes, however, neither the within- nor the among-population variation is ever expected to reach absolute zero, and F_{ST} will be in the (0,1) interval, with its magnitude depending on the relative impact of the three contributing forces (mutation, migration, and drift). A variety of methods, including extensions to multiple alleles, have been developed for estimating F_{ST} from samples of alleles from multiple subpopulations under the assumption of the infinite-alleles model (Nei and Chesser 1983; Weir and Cockerham 1984; Weir and Hill 2002; Balding 2003), and others allow for highly mutable alleles with a significant chance of back mutation (Slatkin 1995a; Goodman 1997).

A result of great conceptual utility is that F_{ST} values are closely related to the average coalescence times of alleles within subpopulations, $\bar{t}_0$, and of alleles drawn randomly from the entire metapopulation, $\bar{t}$,

$$F_{ST} = \frac{\bar{t} - \bar{t}_0}{\bar{t}} = 1 - \frac{\bar{t}_0}{\bar{t}} \tag{2.44}$$

(Slaktin 1991). If $\bar{t}_0$ is very close to $\bar{t}$, there is little among-group differentiation, as pairs of alleles within groups have essentially the same amount of time to diverge mutationally as those among groups. Because the coalescent structure of a population simply represents a genealogical sampling process and is independent of the mutations that incidentally arise, Equation 2.44 shows that the equilibrium level of population subdivision is also independent of the mutation process.

Mutation-migration-drift Equilibrium Values of F_{ST}

When migration rates are sufficiently high that essentially all demes exchange at least one individual per generation, the physical structure of the metapopulation has no consequences for population subdivision at the genetic level, i.e., $F_{ST} \simeq 0$. When the forces of mutation, migration, and drift operate simultaneously at moderate levels, some intermediate equilibrium level of among-population differentiation is reached such that the loss of new variants within demes by drift is balanced by the spread of novel variants across demes by migration. As one might expect, the resultant equilibrium F_{ST} value is a function of the population size within each deme and the pattern of migration over demes, so there are essentially endless numbers of possible patterns of spatial structure and migration. To make some general points, we will consider two commonly envisioned situations (Figure 2.11).

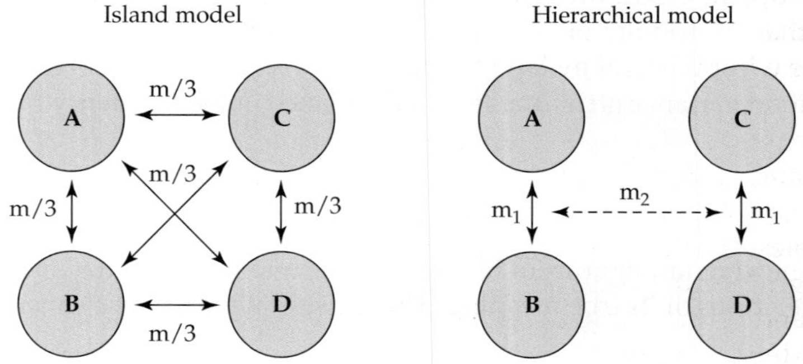

Figure 2.11 (**Left**) The island model. Most mating occurs within each subpopulation/deme, but some small amount of equally partitioned migration m occurs between them. Hence, a migrant from deme A is equally likely to end up in demes B through D. (**Right**) A hierarchically-structured population. Here, intragroup migration (between A and B, and between C and D, m_1) occurs at a much higher level than migration between these two groups (m_2).

The simplest structure is Wright's (1951) **island model**, wherein the population consists of d demes, each containing N breeding individuals. Each generation, each deme contributes a fraction m of its genes to a migrant pool, yielding an expected migration rate from any deme to any other of $m/(d-1)$. A remarkable feature of this model is that the equilibrium amount of genetic variation expected within demes is independent of the level of population subdivision (assuming $m > 0$), a feature known as the **geographic invariance principle** (Maruyama 1971; Nagylaki 1982). Provided there is some potential migratory route between all demes, regardless of the level of migration, the mean coalescence time between random pairs of genes within demes is

$$\bar{t}_0 = 2Nd \tag{2.45a}$$

generations, i.e., twice the sum of the demic population sizes (Li 1976; Strobeck 1987; Hey 1991; Nagylaki 2000). Under this model, the independence of the within-deme variation of the migration rate arises because a lower rate of migration encourages a higher rate of allelic divergence, resulting in a balance between these two opposing factors (frequency and impact of immigrant alleles).

On the other hand, the mean coalescence time for two genes randomly drawn from the *entire* metapopulation is

$$\bar{t} = 2Nd + \frac{(d-1)^2}{2dm} \tag{2.45b}$$

(Li 1976; Slatkin 1991; Nei and Takahata 1993), which is the sum of the previous term and the additional time required for alleles to coalesce into the same deme. Unlike the amount of variation *within* demes, the differentiation *among* demes clearly depends on m. Substituting Equations 2.45a and 2.45b into Equation 2.44 yields

$$F_{ST} = \frac{1}{1 + 4N\dfrac{md^2}{(1-d)^2}} \simeq \frac{1}{1+4Nm} \quad \text{for} \quad d \gg 1 \tag{2.46}$$

This again shows that the equilibrium level of population subdivision is completely independent of the mutation rate and largely independent of the number of demes (d), provided the latter is at least moderately large.

This type of framework for interpreting F_{ST} is readily extended to populations with a **hierarchical** structure (Figure 2.11). In the simplest hierarchical model, the metapopulation is arranged into a series of g groups, each of which is further subdivided into d demes. Within each group, the demes exchange migrants according to the island model with total rate of m_1 (so that $m_1/[d-1]$ is the rate at which each deme sends migrants to any other particular deme within its group). By definition, there is less frequent exchange of migrants between the different groups, i.e., $m_2 < m_1$. With this type of structure, it is necessary to consider the degree to which the total genetic variation partitions into three components: within demes, among demes within a group, and among groups.

Each of these components can again be expressed in terms of coalescence times. Let $\bar{t}_0$ be the mean coalescence time for two individuals from the same deme, $\bar{t}_1$ for two individuals from the same group but different demes, and $\bar{t}_2$ for two individuals from different groups. As with the island model, the geographic invariance principle gives the mean coalescence time for two individuals from the same deme as $\bar{t}_0 = 2Ngd$, i.e., twice the total population size. In order for two individuals in the same group but different demes to coalesce, they must first trace back to the same deme, which for $m_1 \ll m_2$ requires approximately $t_1' \simeq g(d-1)/(2m_1)$ generations (from the second term in Equation 2.45b), and then take an additional $\bar{t}_0$ generations to coalesce within that deme, giving $\bar{t}_1 = t_1' + \bar{t}_0$ (Slatkin and Voelm 1991). Finally, for individuals from different groups to coalesce, they first must trace back to the same group, which requires an average of $t_2' \simeq (g-1)/(2m_2)$ generations, then trace back to the same deme within a group (t_1'), and finally coalesce within that deme, giving $\bar{t}_2 = t_2' + t_1' + \bar{t}_0$.

Using these results, three hierarchical F statistics define the partitioning of the total population variation: F_{DG} among demes within groups, F_{GT} among groups within the total population, and F_{ST} among demes within the total population:

$$F_{DG} = \frac{\bar{t}_1 - \bar{t}_0}{\bar{t}_1}$$

$$F_{GT} = \frac{\bar{t}_2 - \bar{t}_1}{\bar{t}_2}$$

$$F_{ST} = \frac{\bar{t}_2 - \bar{t}_0}{\bar{t}_2} \tag{2.47a}$$

(Slatkin and Voelm 1991; Excoffier et al. 2009b). Substituting in the values for the various mean coalescence times noted previously, and assuming d and g moderately large and $m_2 \ll m_1$, results in further simplification,

$$F_{DG} \simeq \frac{1}{1 + 4Nm_1}$$

$$F_{GT} \simeq F_{ST} \simeq \frac{1}{1 + 4Ndm_2} \tag{2.47b}$$

Note that these expressions are equivalent in form to that for the simple island model, in this case with the demic population size (N) determining the variation of demes within groups and group size (Nd) determining that for groups within the total population. Such coalescence approaches are readily extended to much more complex situations (e.g., Nordborg 1997; Nagylaki 1998; Pannell and Charlesworth 2000; Pannell 2003).

3

The Genetic Effective Size of a Population

However, any theory which one finds uncomfortable but for which one cannot say exactly why, deserves serious consideration. Such an uncomfortable feeling signals a challenge to one or more of one's unexamined, and perhaps unjustified, assumptions. Van Valen (1976)

Throughout the preceding chapter, we assumed a population with an idealized set of Wright-Fisher features, including random mating within a homogeneous group of monoecious, self-compatible individuals with equal expected family sizes, discrete generations, and an absence of density fluctuations. Because almost all populations deviate from this ideal structure in one or more ways, and often substantially so, the relevance of the resultant theory might then seem in doubt. In fact, much of the theory of inbreeding and random genetic drift can be generalized to other types of population structures to a good approximation in a relatively simple manner. To accomplish this task, instead of relying on the total number of adult individuals (N) as a measure of population size, we construct a surrogate index that takes into account the deviations from the ideal model from a genetic perspective. Following the influential work of Wright (1931, 1938a, 1939), such an index has become widely known as N_e, the **effective population size**. With such a reparameterization, essentially all of the results in the preceding chapter hold when N_e is substituted for N. Buri's experiment (Chapter 2) is a case in point—a set of populations with actual size $N = 16$ exhibited allele-frequency dynamics closely approximated by the expectations for an idealized Wright-Fisher population with an effective size of just ten individuals.

Because of its central role in defining levels of variation within populations, the rate of divergence among populations, and the efficiency of natural selection, N_e is one of the most important parameters in population genetics. Although N_e is not as easily measured as the total (census) population size, as will be seen in the following sections, it is, at least in part, defined by observable demographic and mating-system properties of populations (Latter 1959; Lande and Barrowclough 1987; Crow and Denniston 1988). A central goal of this chapter is to illustrate that nearly every violation of the assumptions underlying the Wright-Fisher model leads to a reduction in N_e relative to N, thereby implicating a stronger role for genetic drift in evolution than might be surmised from estimates of total population sizes. We will progressively consider aspects of the mating system (including the possibility of self-fertilization, variation in the sex ratio, and variance in family size), age structure, temporal variation, and spatial structure.

In addition, we will show that the effective size of a population is often strongly influenced by the structural aspects of genomes, independent of population demographic features. The physical linkage of genes on chromosomes ties their mutual fates together, creating stochastic fluctuations of allele frequencies out of chance associations with other loci under selection. Thus, chromosomally regional variation in recombination rates and selection has the effect of creating variation in N_e among different loci in the same population. Tight linkage to a deleterious mutation can result in the loss of an otherwise neutral or beneficial allele from the population, whereas tight linkage to a beneficial mutation can result in hitchhiking to fixation.

This chapter focuses entirely on the concept of N_e from a theoretical perspective, with a goal of providing the reader with a qualitative understanding of the mechanistic determinants of this key population-genetic parameter. Chapter 4 will provide an overview of methods for the estimation of N_e using molecular markers in natural populations, which require no knowledge of a population's demographic features. A more thorough review of a number of the topics that we touch upon can be found in Caballero (1994), and although

Evolution and Selection of Quantitative Traits. Bruce Walsh & Michael Lynch.
Published 2018 by Oxford University Press.
DOI 10.1093/oso/9780198830870.001.0001

we focus on autosomal loci, the general principles are readily extended to sex-linked loci (Caballero 1995; Nagylaki 1995; Wang 1997; Charlesworth 2001).

GENERAL CONSIDERATIONS

An appreciation for the concept of effective population size can be gained by recalling that all members of an ideal monoecious population (consisting of hermaphrodites capable of random self-fertilization) contribute equally to the total gamete pool, with each successful gamete uniting randomly with another gamete derived from the total population of N individuals. Under these conditions, the probability that two uniting gametes are derived from the same parent is simply $P = 1/N$. However, many factors, including self-incompatibility, limited dispersal, differential productivity of gametes, and selection, can cause P to deviate from $1/N$. To account for the joint influence of all of these factors and many others, we define P to be the reciprocal of the effective population size.

Using this definition of $N_e = 1/P$, many of the results in the previous chapter can be generalized, at least to a first-order approximation. Consider, for example, the expected dynamics of the inbreeding coefficient for a diploid, monoecious population. The probability that two uniting gametes are derived from the same parent is now $1/N_e$, in which case there is a 50% chance that they each carry copies of the same gene (i.e., they are identical by descent, one generation removed) and a 50% chance that they carry copies of genes from different parental chromosomes. In the latter case, the uniting genes may still be identical by descent from previous inbreeding with probability f_{t-1}. Finally, there is a $1 - (1/N_e)$ probability that the uniting gametes are derived from different parents, in which case there is again a probability f_{t-1} that they are identical by descent from previous inbreeding. (Here, we are assuming a population without spatial structure, so that the mean degree of inbreeding does not depend on the parental source of alleles.) Summing up the three ways by which identity-by-descent can arise between uniting gametes,

$$
\begin{aligned}
f_t &= \left(\frac{1}{N_e}\right)\left(\frac{1}{2}\right) + \left(\frac{1}{2}\right)\left(\frac{1}{N_e}\right) f_{t-1} + \left(1 - \frac{1}{N_e}\right) f_{t-1} \\
&= \frac{1}{2N_e} + \left(1 - \frac{1}{2N_e}\right) f_{t-1} \qquad (3.1)
\end{aligned}
$$

This expression is identical in form to Equation 2.3, but with N_e replacing N.

Under the above interpretation, N_e is the size of an ideal population that would exhibit the same amount of inbreeding as the population under consideration. Defined in this way, N_e is the **inbreeding effective size**. There are, however, numerous additional ways to define the effective size of a population. One can, for example, define the **variance effective size** as the N that, when applied to Equation 2.14a, yields the temporal variance in allele-frequency change exhibited by a nonideal population. Crow (1954) emphasized that the inbreeding effective size is most closely related to the number of parents (or the number of grandparents if selfing does not occur) because it is based upon the probability of uniting gametes coming from the same ancestor. In contrast, the variance effective size, which is associated with allele-frequency drift resulting from gamete sampling, is primarily a function of the number of offspring produced. Thus, in an expanding or declining population, the rates of inbreeding and allele-frequency drift can differ in any particular generation (Templeton 2006). However, for populations with stable size, inbreeding and variance effective sizes are generally equivalent, and in the long run this is even true for fluctuating populations because both measures depend on the same sequence of adult population sizes (Caballero 1994; Whitlock and Barton 1997). More discourse on these issues may be found in Crow and Morton (1955), Kimura and Crow (1963b), Crow and Kimura (1970), Crow and Denniston (1988), and Caballero (1994). Unless stated otherwise, the following discussion will assume populations of constant size.

Before proceeding, it bears emphasizing that random genetic drift is a result of two stochastic processes. First, in sexual species, segregation during meiosis leads to the random transmission of alleles from heterozygous parents, as there is a 50% chance for each alternative allele to be inherited by any given offspring. Second, variation in family size encourages lineages of alleles that happen to be contained within large families to expand at the expense of others. As we will see in the following sections, such variation arises by numerous mechanisms, including simple sampling of gametes, spatial population structure, variation among the sexes, and trapping in various genetic backgrounds by linkage.

MONOECY

To illustrate the mathematical approach to deriving expressions for N_e, we first generalize the concept of a monoecious, self-compatible population to allow for arbitrary gamete production by different individuals. As a reminder, monoecious (a botanical term) individuals are equivalent to hermaphrodites (a zoological term), with both terms referring to the situation in which individuals produce male and female gametes.

Let k_i be the number of gametes that the ith parent contributes to offspring that survive to maturity, μ_k and σ_k^2 be the mean and variance of successful gamete production per individual, and N_{t-1} be the number of reproducing parents. Assuming that mating is random and **isogamous** (so that there is no distinction between male and female gametes), there are $k_i(k_i - 1)$ ways in which the gametes of parent i can unite with each other, and summing over all parents, $\sum_{i=1}^{N_{t-1}} k_i(k_i - 1)$ total ways by which gametes can unite by self-fertilization. Because a total of $N_{t-1}\mu_k$ successful gametes are produced, the expected fraction of zygotes derived from the same parent is

$$P_t = \frac{1}{N_e} = \frac{\sum_{i=1}^{N_{t-1}} k_i(k_i - 1)}{N_{t-1}\mu_k(N_{t-1}\mu_k - 1)} \tag{3.2}$$

where the denominator is the total number of pairs of uniting gametes necessary to produce the next generation. This expression can be simplified greatly by noting that $\sum_{i=1}^{N_{t-1}} k_i(k_i - 1)/N_{t-1} = E(k^2) - \mu_k = \sigma_k^2 + \mu_k(\mu_k - 1)$, and that because all zygotes are derived from two gametes, $N_{t-1}\mu_k = 2N_t$. Substituting into Equation 3.2 and inverting,

$$N_e = \frac{2N_t - 1}{(\sigma_k^2/\mu_k) + \mu_k - 1} \tag{3.3}$$

This shows that for a randomly mating monoecious population with discrete generations, the effective population size is a function of three measurable quantities: the actual population size (N_t) and the mean (μ_k) and variance (σ_k^2) of successful gamete production. All other things being equal, variance in gamete production causes a reduction in N_e, as this inflates the representation of a fraction of the population in the descendant gene pool. Because such variation is expected to be the rule in natural populations occupying environments that are heterogeneous with respect to resource availability, on this basis alone, we can generally expect N_e to be less than the number of reproducing adults.

Equation 3.2 simplifies greatly under a number of conditions. For example, populations that are stable in size have, on average, two successful gametes per parent ($\mu_k = 2$), leading to

$$N_e = \frac{4N - 2}{\sigma_k^2 + 2} \tag{3.4}$$

If we further assume that each parent produces the same number of *potential* gametes (returning us to the ideal random-mating population), an explicit statement can also be made about σ_k^2. In this case, for any particular draw from the gamete pool, the variance in the number of gametes derived from a particular parent (0 or 1) is $(1/N)[1 - (1/N)]$

(from the properties of a binomial distribution), and because a total of $2N$ gametes are drawn, $\sigma_k^2 = 2[1 - (1/N)]$. Note that this result is very close to the Poisson expectation of $\sigma_k^2 = 2$ with completely random gamete production (because the variance of a Poisson equals the mean), with the slight deviation resulting because we assume a fixed population size. Substitution of the exact expression into Equation 3.4 yields $N_e = N$, showing that the inbreeding effective size of an ideal random-mating population is indeed equal to the number of reproductive adults in the previous generation.

In contrast, in the opposite, and extreme, situation in which all parents produce *exactly* two progeny (such that $\sigma_k^2 = 0$ and $\mu_k = 2$), $N_e = 2N - 1 \simeq 2N$. This shows that the elimination of variance in family size results in the effective population size being twice the actual number of breeding adults, a feature that is often exploited in breeding schemes to minimize the amount of inbreeding.

There are a number of ways in which the breeding systems of a monoecious species can deviate from the assumptions made in the preceding derivations. Consider, for example, species with self-incompatibility, in which case identity-by-descent for pairs of uniting gametes comes through grandparents rather than parents. If we now let k_i be the number of successful gametes for individual i in generation $t - 2$, and again assume random mating, there are $2k_i(k_i - 1)$ ways in which pairs of genes from i can unite through matings in the parental generation $t - 1$ (the 2 arising because we assume that each individual can serve as a mother or father). Because there are $N_{t-2}\mu_k/2$ parents in generation $t - 1$, there are $2(N_{t-2}\mu_k/2)[(N_{t-2}\mu_k/2) - 1]$ ways of drawing different parents, and $4 \cdot 2(N_{t-2}\mu_k/2)[(N_{t-2}\mu_k/2) - 1]$ ways of drawing gene pairs (the 4 because each parent carries two genes). Therefore, the probability that a generation-*t* individual carries a pair of genes derived from the same grandparent is

$$P_t = \frac{1}{N_e} = \frac{\sum_{i=1}^{N_{t-2}} k_i(k_i - 1)}{N_{t-2}\mu_k(N_{t-2}\mu_k - 2)} \tag{3.5}$$

Employing the same kinds of substitutions used for Equation 3.3,

$$N_e = \frac{2(N_{t-1} - 1)}{(\sigma_k^2/\mu_k) + \mu_k - 1} \tag{3.6}$$

and for constant N (which implies $\mu_k = 2$),

$$N_e = \frac{4(N - 1)}{\sigma_k^2 + 2} \tag{3.7}$$

For populations that are moderately large and stable in size, Equations 3.4 and 3.7 give essentially the same answer, $N_e \simeq 4N/(\sigma_k^2 + 2)$, demonstrating that the prohibition of selfing has a negligible influence on N_e unless the total population size is tiny. The reason for this is that under random mating the increment in inbreeding resulting from self-fertilization is a transient event that can be completely undone in the following generation.

A second potential complication is that in most hermaphroditic species, there is a distinction between male and female gametes (**anisogamy**), so that even with selfing, only a fraction of potential gamete pairs are capable of spawning a successful zygote. When mating is random but selfing is prohibited, the effective population size is the same under isogamy and anisogamy, and Equation 3.6 still applies (Crow and Denniston 1988). However, with selfing permitted,

$$N_e = \frac{N_{t-1}}{(4\sigma_{o,p}/\mu_k^2) + 1} \tag{3.8}$$

where $\sigma_{o,p}$ is the covariance of the numbers of successful male (p, pollen) and female (o, ovule) gametes per parent (Crow and Denniston 1988). If $\sigma_{o,p}$ is positive, as might be expected in a spatially heterogeneous environment where some individuals acquire more resources than others, the effective population size will be less than the observed size. However, if $\sigma_{o,p}$ is negative, as might be expected when there is a trade-off between male and female function, N_e can exceed N_{t-1}. This results because a negative covariance in male and female gamete production reduces the variance in family size.

Example 3.1. Hedgecock (1994) suggested that marine organisms with high fecundities and broadcast spawning may have effective population sizes that are orders of magnitude smaller than the absolute number of potential breeders. This situation can arise if vagaries in oceanographic conditions are such that only a small fraction of adults produce gametes at points in time and space that allow recruitment to the next generation. Suppose the total adult population size is N, whereas only N_p individuals contribute equally to the breeding pool. Such a situation is sustainable if reproductive adults can individually produce an average of $2N/N_p$ gametes (and many marine species are capable of producing many tens of thousands of gametes). Given such a situation, N_p individuals have expected family sizes of $2N/N_p$, whereas $(N - N_p)$ have zero expected reproductive success, which results in an expected family-size variance among all N individuals of $4[(N/N_p) - 1]$ (Hedrick 2005). Using the logic outlined in the paragraph below Equation 3.4, the additional variance in reproductive success among spawning individuals resulting from random gamete sampling is equal to

$$\left\{\left(\frac{N_p}{N}\right)\cdot 2N\cdot\left(\frac{1}{N_p}\right)\left[1-\left(\frac{1}{N_p}\right)\right]\right\}+\left\{\left[1-\left(\frac{N_p}{N}\right)\right]\cdot 2N\cdot 0\right\}=2\left[1-\left(\frac{1}{N_p}\right)\right]$$

Summing these two sources of family-size variance gives $\sigma_k^2 = [(4N-2)/N_p] - 2$, and substituting into Equation 3.4, we obtain $N_e = N_p$. Thus, provided a species has a high enough gamete production to generate N surviving progeny from a small number of adults, the effective population size can be only a tiny fraction of N.

Example 3.2. Heywood (1986) estimated that σ_k^2/μ_k^2 for *seed* production is on the order of 1 to 4 in a number of annual plants (including self-compatible species). Unfortunately, the value of σ_k^2 for *total* gamete production requires additional information on successful pollen production, which is extremely difficult to acquire due to problems in ascertaining paternity. For heuristic purposes, however, let us assume a stable monoecious population. This necessarily implies mean seed and pollen production are both equal to one, and $\mu_k = \mu_o + \mu_p = 2$, as each parent must produce two successful gametes (on average, one male and one female). We will also assume a three-fold higher standard deviation for successful pollen relative to seed production, so that $\sigma_p = 3\sigma_o$, and a perfect correlation between ovule and pollen production. Because the correlation between the number of female and male gametes produced per individual is defined to be $\sigma_{o,p}/(\sigma_o\sigma_p)$, the latter assumption implies $1 = \sigma_{o,p}/[\sigma_o \cdot 3\sigma_o]$. Assuming random mating, and substituting $\sigma_{o,p} = 3\sigma_o^2$ into Equation 3.8, we obtain $N_e = N/[(12\sigma_o^2/\mu_k^2) + 1]$. Thus, for σ_o^2/μ_k^2 in the range of 1 to 4, N_e is between 2% and 8% of the census number (N).

DIOECY

As in the case of monoecy with self-incompatibility, when the sexes are separate, inbreeding always needs to be defined with reference to the grandparent generation, which is the earliest point back to which the two genes of an individual can coalesce. Separate sexes also introduce the possibility of different levels of inbreeding through males and females, as might be expected, for example, in polygynous species in which most females mate with a relatively small segment of the male population.

If O is the offspring of interest, with M and F being its mother and father, there are two ways by which O may derive two genes from the same grandparent: (1) M and F may share the same mother (with probability $1/N_{ef}$, where $1/N_{ef}$ is the effective number of females); or (2) M and F may share the same father (with probability $1/N_{em}$, where $1/N_{em}$ is the effective number of males). In either case, because each parent transmits to O a gene from the shared ancestor with probability 0.5, the probability that O inherits both genes from the shared grandparent is 1/4. Thus, the total probability that O inherits two genes from the same grandparent is

$$P = \frac{1}{N_e} = \frac{1}{4N_{em}} + \frac{1}{4N_{ef}} \tag{3.9}$$

What do we mean by the effective numbers of males and females? Assuming random mating (including no prohibition of mating between sibs), the effective number of each sex can be derived by the same method used to obtain Equation 3.3. Skipping the intermediate steps, we simply note that

$$N_{es} = \frac{\mu_{sk} N_{s,t-2} - 1}{(\sigma^2_{sk}/\mu_{sk}) + \mu_{sk} - 1} \tag{3.10}$$

where s denotes the sex (m or f), and μ_{sk} and σ^2_{sk} are the mean and variance of gamete production by sex s (Crow and Denniston 1988). Latter (1959) provided a more elaborate expression for N_{es} that explicitly accounts for the variance and covariance of male and female progeny production. Letting ϕ be the sex ratio (proportion of females),

$$N_{em} = \frac{4N_{m,t-2}}{2 + \sigma^2_{mm} + \frac{2(1-\phi)}{\phi}\sigma_{mm,mf} + \left(\frac{1-\phi}{\phi}\right)^2 \sigma^2_{mf}} \tag{3.11a}$$

$$N_{ef} = \frac{4N_{f,t-2}}{2 + \sigma^2_{ff} + \frac{2\phi}{1-\phi}\sigma_{fm,ff} + \left(\frac{\phi}{1-\phi}\right)^2 \sigma^2_{fm}} \tag{3.11b}$$

where for male parents, σ^2_{mm} is the variance of male progeny number, σ^2_{mf} is the variance of female progeny number, and $\sigma_{mm,mf}$ is the covariance of male and female progeny number, with similar definitions for female parents. There are a variety of situations in which these types of specifications may be useful. For example, if parents produce a fixed number of offspring, the covariances $\sigma_{mm,mf}$ and $\sigma_{fm,ff}$ between numbers of sons and daughters must be negative, whereas these terms can be positive if parents differ in the resources available for overall progeny production.

Further simplification of Equation 3.10 is possible when certain assumptions are met. Consider, for example, the case in which members of the same sex produce equal numbers of potential gametes, such that $\sigma^2_{sk} = \mu_{sk}$, and the variation in family size is a simple consequence of the random union of gametes. It then follows from the development of the monoecy model that $N_{em} = N_{m,t-1}$ and $N_{ef} = N_{f,t-1}$. Rearranging Equation 3.9, assuming constant population features and dropping the designation of time, and noting that $N_m = (1-\phi)N$ and $N_f = \phi N$,

$$N_e = \frac{4N_m N_f}{N_m + N_f} = 4\phi(1-\phi)N \tag{3.12}$$

In this case, N_e attains a maximum of N when the sex ratio is balanced ($\phi = 0.5$), but with skewed sex ratios, N_e is influenced much more strongly by the density of the rarer sex. For example, in a highly polygynous species, as $\phi \to 1$, $N_e \to 4(1-\phi)N \simeq 4N_m$, namely four times the number of males.

In natural populations, where individuals inhabit different environments that influence the availability of resources and mates, it is likely that the variance in progeny production will exceed the mean (i.e., $\sigma_k^2 > \mu_k$), in which case N_e will be less than that predicted by Equation 3.12. For example, in a summary of data on lifetime reproductive success in female birds, Grant (1990) found that σ_{fk}^2/μ_{fk} ranged from 1.2 to 4.2. Assuming a stable population size ($\mu_{fk} = 2$) and substituting into Equation 3.10, the female effective population sizes for these species are found to be 40% to 90% of the actual number of females. Nonrandom variation in family sizes appears to be the rule even in laboratory populations. For example, caged populations of *Drosophila* typically exhibit effective sizes on the order of 10% of the census size of the adult population (Briscoe et al. 1992). Observations from natural populations of other animals with separate sexes suggest an average N_e/N ratio for single generations on the order of 0.7 (Crow and Morton 1955; Nunney and Elam 1994).

Finally, the results for dioecy can be linked to those for the monoecy model in the following informative way. The mean gamete production for the whole population is $\mu_k = (1-\phi)\mu_{mk} + \phi\mu_{fk}$, or equivalently, because all individuals have a father and a mother, $\mu_k = 2(1-\phi)\mu_{mk} = 2\phi\mu_{fk}$. The variance of gamete production across the entire population is $\sigma_k^2 = (1-\phi)\sigma_{mk}^2 + \phi\sigma_{fk}^2 + \phi(1-\phi)(\mu_{mk} - \mu_{fk})^2$. Using these expressions, Equation 3.9 is essentially equivalent to Equation 3.7 (Kimura and Crow 1963b), showing that the effective size of an ideal population with separate sexes is the same as that for a monoecious, self-incompatible population with the same population properties μ_k and σ_k^2.

AGE STRUCTURE

Because the previous formulae were obtained under the assumption of discrete generations, they provide estimates of N_e for explicit generational intervals. Such expressions are reasonable for organisms such as annual plants (ignoring the problem of seed banks; Nunney 2002) or univoltine insects, but for species that reproduce at different ages, as is the case for most vertebrates and perennial plants, the overlapping of generations raises additional complications. Nevertheless, as first pointed out by Hill (1972e, 1979), there is a simple correspondence between the effective sizes of populations with and without age structure.

In the previous formulations, N was the number of potential reproductive individuals entering the population in each generation. For age-structured populations, we must consider instead N_b, the total number of newborns entering the population during each unit of time, as well as the number of time units per generation. The latter quantity, known as the **generation time** (T), is the average age of parents giving birth, which in turn is a function of the age-specific schedules of survival and reproduction. For an ideal monoecious population,

$$T = \frac{\sum_{i=1}^{\tau} i\ell_i b_i}{\sum_{i=1}^{\tau} \ell_i b_i} \tag{3.13}$$

where ℓ_i is the probability of surviving to age i, b_i is the expected number of offspring produced by parents of age i, and τ is the maximum reproductive age. The quantity $\ell_i b_i$ denotes the expected number of births by an individual of age i, discounting for prior mortality. For a dioecious population, T is further complicated by the need to average over males (m) and females (f),

$$T = \frac{T_{mm} + T_{mf} + T_{fm} + T_{ff}}{4} \tag{3.14}$$

where T_{mf}, for example, is the average age of male parents of daughters. The average generation length is the natural time scale for the evolutionary analysis of age-structured populations. Letting $N_e = N_{eb}T$, with $N_{eb} = 4\phi_b(1-\phi_b)N_b$ being the effective size of the newborn age class and ϕ_b being the sex ratio of newborns, all of the preceding formulae for discrete generations apply provided the structure and size of the population are stable.

However, we are still left with the rather substantial problem of estimating σ_k^2, which now depends on variation in longevity as well as variation in fertility.

Felsenstein (1971), Johnson (1977b), and Emigh and Pollak (1979) showed how the variance in offspring production can be expressed in terms of the age-specific parameters ℓ_i and b_i. Again making the assumption that the population is stable in terms of size, sex ratio, and age composition, the effective size of an age-structured population with separate sexes is

$$N_e = \frac{N_{eb}T}{1 + (1-\phi_b)\sum_{i=1}^{\tau_f}\left(\frac{1}{\ell_{i+1}^f} - \frac{1}{\ell_i^f}\right)\left(\sum_f\right)^2 + \phi_b \sum_{i=1}^{\tau_m}\left(\frac{1}{\ell_{i+1}^m} - \frac{1}{\ell_i^m}\right)\left(\sum_m\right)^2} \tag{3.15}$$

where $(\sum_s)^2 = \left(\sum_{j\geq i+1}^{\tau_s} \ell_j^s b_j^s\right)^2$, τ_s is the maximum age of reproduction for sex s, and $s = m$ or f (Emigh and Pollak 1979). An analogous expression is available for monoecious populations (Felsenstein 1971). While the derivations underlying these expressions rely on the assumption that gametes are drawn randomly from the members within age classes, no assumptions are made with regard to the preference of matings between age classes.

Despite their complicated structure, demographic formulae such as Equation 3.15 are useful for analyzing the sensitivity of a population's effective size to modifications in the life-history schedule. Nevertheless, the Emigh-Pollak equation has some practical difficulties. First, it rests on the assumption of a stable population structure. Such situations are rare in nature because of temporal changes in the environment. Johnson (1977b) and Choy and Weir (1978) derived dynamical equations to resolve these difficulties, and the entire subject was reviewed by Charlesworth (1994a). Second, Equation 3.15 was derived under the assumption that the age-specific mortality and birthrates of individuals are uncorrelated, i.e., that individuals with an elevated likelihood of survivorship do not have elevated or reduced birthrates. This will not be true for populations in which energetic trade-offs exist between different life-history characters. The problem needs further investigation.

Substantial simplification of Equation 3.15 can be achieved under some conditions. For example, if year-to-year survival is age independent, and if the mating system can be described in simple terms, N_e can be defined as a function of a small number of parameters, thus eliminating the need for refined age-specific schedules of survivorship and fecundity. Using this approach, Nunney (1993) concluded that N_e in animals with overlapping generations is typically on the order of $N/2$ to N, although his analysis ignores the important influence of variation in N across generations (as will be seen in a following section).

Example 3.3. While complete age-specific survivorship and reproductive schedules are available for the females of many natural populations, male promiscuity often imposes enormous practical difficulties for ascertaining paternity. Thus, the variance in male reproductive success is generally unknown. However, a long-term study on the behavior and demography of the red deer (*Cervus elaphus*) by Clutton-Brock et al. (1982) allows at least a crude estimate of N_e by use of the Emigh-Pollak equation, as shown in the table below. The study population was roughly constant in density for two decades, and observations on known individuals provide information on the age-specific rates of mortality and reproduction for both sexes. The sex ratio at birth (ϕ_b) averaged 0.43 over several years, so $N_{eb} = 0.98N_b$.

The age-specific survival rates, ℓ_i, in the following table were extracted directly from Clutton-Brock et al. (1982), while the age-specific reproductive schedules, b_i^f and b_i^m, were estimated from behavioral and demographic observations of the authors and adjusted downward to maintain a stable population size. The columns marked (1) and (2) are $[(1/\ell_{i+1}^s) - (1/\ell_i^s)]$ and $(\sum_{j>i+1}^{\tau_s} \ell_j^s b_j^s)^2$, and column (3) is the product of (1) and (2); all of these are deployed in Equation 3.15.

Age	Females					Males				
i	ℓ_i^f	b_i^f	(1)	(2)	(3)	ℓ_i^m	b_i^m	(1)	(2)	(3)
1	1.00	0.00	0.33	0.25	0.08	1.00	0.00	0.45	5.97	2.69
2	0.75	0.00	0.12	0.25	0.03	0.69	0.00	0.22	5.97	1.31
3	0.69	0.00	0.02	0.25	0.01	0.60	0.00	0.03	5.97	0.18
4	0.68	0.18	0.04	0.24	0.01	0.59	0.00	0.03	5.97	0.18
5	0.66	0.26	0.07	0.22	0.02	0.58	0.00	0.03	5.97	0.18
6	0.63	0.33	0.05	0.21	0.01	0.57	0.34	0.03	5.92	0.18
7	0.61	0.34	0.06	0.19	0.01	0.56	0.26	0.03	5.88	0.18
8	0.59	0.40	0.06	0.17	0.01	0.55	0.60	0.10	5.59	0.56
9	0.57	0.42	0.03	0.16	0.01	0.52	0.53	0.12	5.30	0.64
10	0.56	0.34	0.07	0.14	0.01	0.49	0.79	0.40	3.94	1.58
11	0.54	0.46	0.03	0.13	0.00	0.41	0.53	0.42	3.11	1.31
12	0.53	0.42	0.20	0.08	0.02	0.35	0.45	1.69	1.00	1.69
13	0.48	0.45	0.19	0.04	0.01	0.22	0.08	2.60	0.63	1.64
14	0.44	0.40	0.23	0.01	0.00	0.14	0.20	3.97	—	—
15	0.40	0.25	0.20	—	—	0.09	—	—	—	—
16	0.37	0.00	—	—	—	0.05	—	—	—	—
					0.23					12.32

The summations in the denominator of Equation 3.15 reflect the variation in lifetime reproductive success of females and males. As outlined in the table, these terms are equal to 0.23 and 12.32, respectively, indicating a great inequity between the reproductive properties of the sexes. This results because male red deer appropriate harems, and older males are much more successful at doing so than young ones. The few males that live to an old age may father up to two dozen offspring in their lifetimes, whereas males that die before the age of five (~40% of newborn males) have no reproductive success at all. On the other hand, almost all females reproduce to some degree once they have attained reproductive maturity.

Substituting the sums from the table into Equation 3.15, the effective population size is found to be $0.98N_bT/[1+(1-0.43)(0.23)+0.43(12.32)] = 0.15N_bT$. Thus, the effective size of this population is ~15% of the number of offspring produced by the population per generation. The mean generation time through females and males is 9.47 and 9.18 years, so $T \simeq 9.32$, and the annual number of offspring produced by the population is $N_b \simeq 270$. Thus, $N_e \simeq 0.15 \times 270 \times 9.32 = 377$.

VARIABLE POPULATION SIZE

Most populations vary in density from generation to generation, often dramatically so, and this raises practical problems in the implementation of the previous theory. As noted in Equation 2.7, with variable population size, the expected loss of heterozygosity over t generations is no longer $[1-(1/2N_e)]^t$ but rather is now a product of t terms, each incorporating the effective population size of a particular generation such that

$$H_t = H_0 \prod_{i=0}^{t-1} \left(1 - \frac{1}{2N_{e,i}}\right) \tag{3.16}$$

Thus, it is informative to evaluate the size of an ideal population of constant size with the same expected heterozygosity after t generations as a population with variable size over the same period. An approximate answer can be obtained by noting that with a moderately large $N_{e,i}$, Equation 3.16 simplifies to

$$H_t \simeq H_0 \exp\left(-\sum_{i=0}^{t-1} \frac{1}{2N_{e,i}}\right) \tag{3.17}$$

which may be compared to

$$H_t \simeq H_0 e^{-t/2N_e}$$

for the ideal case of constant effective size. Equating the exponents of these two expressions,

$$N_e^* \simeq \frac{t}{1/N_{e,0} + 1/N_{e,1} + \cdots + 1/N_{e,t-1}} \tag{3.18}$$

Thus, the long-term effective size N_e^* is approximately equal to the harmonic mean of the generation-specific effective sizes. An asterisk is placed on N_e to remind the reader that the inbreeding projected by N_e^* strictly pertains to generation t. Other generations may exhibit more or less loss of variation than anticipated depending upon the actual pattern of temporal changes in $N_{e,i}$.

Example 3.4. To see how population bottlenecks have especially pronounced effects on N_e^*, consider a population whose effective size regularly fluctuates between 10 and 100. From Equation 3.18, $N_e^* = 2/(0.1 + 0.01) = 18.2$. Thus, the total loss of heterozygosity from this population every two generations is equivalent to that expected for an ideal random-mating population with a constant effective size of 18, which is much closer to the expectation for a constant population size of 10 than 100. Frankham (1995) and Vucetich et al. (1997) showed that N_e^* is frequently in the range 10% to 20% of N_e for a diversity of natural populations of animals. In the extreme case where the effective population size is effectively infinite for $t-1$ generations and N_e for one generation, $N_e^* = tN_e$.

PARTIAL INBREEDING

Although most of the previous formulations assumed a random union of gametes, the frequency of mating between relatives often exceeds that expected under random mating. Many plants, for example, produce a significant proportion of offspring by self-fertilization. If the total population size were infinite, a fixed proportion of matings between relatives would simply lead to an equilibrium condition wherein the production of new inbreeding each generation is balanced by the breakdown of old inbreeding through outcrossing (Wright 1951, 1969; Hedrick 1986b; Hedrick and Cockerham 1986). However, such an equilibrium does not exist for finite populations, where allele frequencies are subject to random genetic drift. Here we consider the consequences of partial selfing in monoecious populations and of partial full-sib mating in populations with separate sexes, in both cases assuming an otherwise randomly mating population. Both subjects were evaluated in considerable detail by Caballero and Hill (1992a) and Wang (1996).

Assuming a constant number of adults, for a population in which a proportion β of progeny from each family is a product of self-fertilization,

$$N_e = \frac{2(2-\beta)N}{\sigma_k^2 + 2(1-\beta)} \tag{3.19}$$

(Crow and Denniston 1988; Caballero and Hill 1992a; Wang 1996). Further assuming that the numbers of outcrossed and selfed progeny per parent are independent variables, the variance in gamete transmission (σ_k^2) can be simplified by noting the logic below Equation 3.4: for the outcrossed progeny, there are $2N(1-\beta)$ random gametes drawn, each with variance $\simeq 1/N$ per focal parent, yielding a variance of $\simeq 2(1-\beta)$; for the selfed progeny, there are $N(1-\beta)$ parents to draw from, but because each successful draw leads to two transmitted gametes, the variance per draw is $\simeq 4/N$, yielding a total variance contribution $\simeq 4\beta$; the total variance is then $\sigma_k^2 \simeq 2(1+\beta)$, and Equation 3.19 reduces to

$$N_e = \frac{(2-\beta)N}{2} \tag{3.20}$$

If mating is random, so that $\beta = 1/N$, then $N_e = N - 0.5$, a result that can also be obtained directly from Equation 3.3 by letting $\sigma_k^2 \simeq \mu_k \simeq 2$. Under obligate self-fertilization, a mode of reproduction in some plants and hermaphroditic animals, $\beta = 1$ and $N_e = N/2$. Self-fertilization results in a reduction in the effective population size because the nonindependence imposed by inbreeding reduces the effective number of alleles per locus within individuals.

For the case of species with separate sexes, with β being the fraction of offspring derived by full-sib mating, and assuming equal numbers of males and females and Poisson-distributed family sizes,

$$N_e \simeq \frac{(4-3\beta)N}{4-2\beta} + 1 \tag{3.21}$$

(Wang 1995). Three special cases are of interest here. For a population derived entirely by full-sib mating, $\beta = 1$ and $N_e = N/2$, as in the case of complete selfing. With complete avoidance of sib mating, $\beta = 0$ and $N_e = N + 1$. Finally, with the type of population structure assumed, there are $N/2$ families, so under random mating, the probability of full-sib mating is simply $\beta = 2/N$, which implies $N_e \simeq N + 0.5$.

SPECIAL SYSTEMS OF MATING

It is important to bear in mind that, when applied to matters of genetic variation, the equations for N_e given above are appropriate for predicting the *expected* loss of heterozygosity resulting from inbreeding at the population level. When variation in pedigree structure exists among individuals, as will almost always be the case in nature, the actual degree of inbreeding will generally *vary* among loci within individuals as well as among individuals within the population. Given a cumulative level of inbreeding (loss of heterozygosity) at a locus equal to f, identity by descent will be binomially distributed among individuals with mean f and variance $\sigma_f^2 = f(1-f)$. With completely linked loci, this is also the total variance in f, as there will be no variation in f among loci. However, for unlinked loci, the realized inbreeding at each locus need not be the same. Weir et al. (1980) found that the coefficient of variation of $(1-f)$ among individuals is approximately $(3N_e)^{-1/2}$ for randomly mating monoecious populations, $(6N_e)^{-1/2}$ for randomly mating but monogamous, dioecious populations, and $(12N_e)^{-1/2}$ for monoecy with selfing excluded and for dioecy with random mating. These asymptotic values are reached in only a few generations. Thus, provided the population size and number of constituent loci are moderately large, the variation in inbreeding is negligible for most practical purposes (see also Franklin 1977; Cockerham and Weir 1983).

In contrast to the usual situation, one can also envision (and implement) systems of mating involving fixed relationships such that all members of the population have exactly the same average inbreeding coefficient over all loci (Wright 1921b). Consider first the most extreme form of inbreeding—obligate self-fertilization, a mode of reproduction in some plants and hermaphroditic animals. Because under this scheme of mating all individuals are reproductively isolated, a collection of such lines is equivalent to a series of populations, each consisting of a single individual, and after t generations of selfing, the expected fraction of heterozygotes at any locus is reduced to

$$H_t = H_0(1/2)^t \tag{3.22}$$

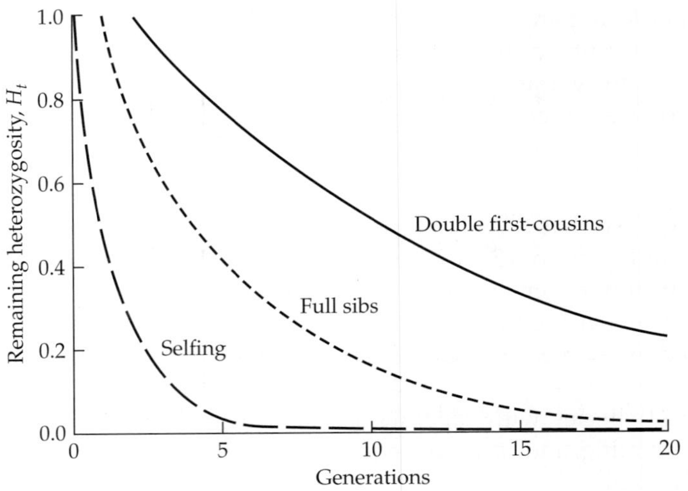

Figure 3.1 Erosion of expected heterozygosity under continuous breeding schemes involving self-fertilization, full-sib mating, and double first-cousin mating.

where H_0 denotes the initial level of variation. After $t = 3$ generations, only 12.5% of the initial heterozygosity remains (Figure 3.1).

The next most intense system of inbreeding involves continuous brother-sister mating. Starting with unrelated parents, it takes a generation of full-sib mating before alleles identical by descent can appear in the same individual. Written in terms of the inbreeding coefficient, such that $H_t = (1 - f_t)H_0$, the exact recursion equation under full-sib mating is

$$f_t = \frac{1}{4}(1 + 2f_{t-1} + f_{t-2}) \tag{3.23a}$$

and in one of the first applications of matrices in population genetics, Haldane (1937a) provided the approximation

$$H_t \simeq H_0(0.81)^t \tag{3.23b}$$

Thus, starting from a non-inbred base population, 12 generations of full-sib mating will result in a loss of 90% of the initial heterozygosity.

Moving on, with a constant population size of four breeding adults, the minimum relationship between individuals is that of double first-cousins (Figure 3.2, left). Starting with four unrelated individuals, it then takes three generations for alleles that are identical by descent (IBD) to appear in the same individual, and thereafter

$$H_t \simeq H_0(0.92)^t \tag{3.24}$$

(Wright 1921b). The number of generations required for the loss of 90% heterozygosity is now 30 (Figure 3.1).

These types of results are of special interest to managers of small, captive populations of endangered species and/or breeding stocks viewed as genetic resources for the future. Here we consider just one of the many practical questions that arise in these areas. Given a limited number of founders and an upper ceiling on the number of individuals that can be maintained, what is the optimal breeding scheme for minimizing the erosion of genetic variation? Wright (1921b) suggested that the best way to minimize the loss of heterozygosity from a small population would be to restrict matings to pairs of individuals with the

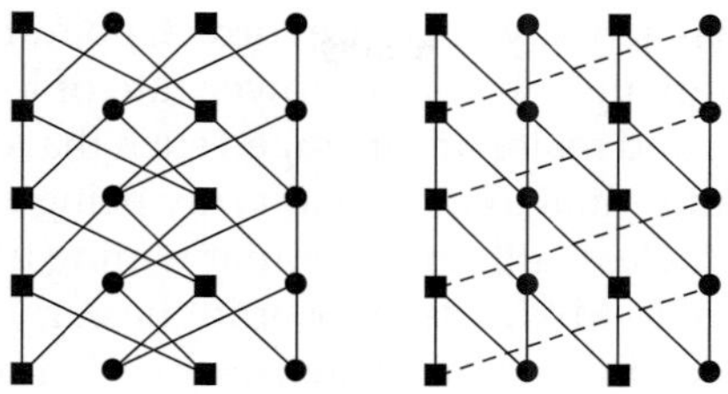

Figure 3.2 Mating schemes under continuous double first-cousin mating (left) and under circular mating with four individuals (right). Genes that are identical by descent do not appear in the same individual for three and four generations, respectively, under these two schemes. Males and females are denoted by separate symbols (squares and circles), and lines connect parents (above) and offspring (below).

least degree of relatedness. Such a breeding scheme, known as **maximum avoidance of inbreeding (MAI)**, is exemplified by all three of the special mating systems just noted—in each case, matings occur between the most distantly related individuals within each line. An added advantage of MAI is that for a population size of $N = 2^m$, m generations will pass before any inbreeding occurs at all. For example, with $N = 64$, $m+1 = 7$ generations would pass before two copies of a founding gene could appear in the same individual under a maximum avoidance scheme. Once the inbreeding begins, the proportion of heterozygosity lost each generation is very nearly constant, approaching an asymptotic value of $1/(4N - m - 1)$ (Robertson 1964), which, with $N = 4$ and $m = 2$ under double first-cousin mating, equals 0.08, giving the fraction retained as 1 −0.08 = 0.92, recovering Equation 3.24.

Note that when N is large, $m \ll N$ and the asymptotic rate of loss of heterozygosity is $\simeq 1/(4N)$ per generation under MAI. Comparing this expression with Equation 2.4c, it can be seen that this mating scheme has the same effect as doubling the size of a random-mating population. This result is not strictly a consequence of the avoidance of inbreeding but, again, is the outcome of all families producing equal numbers of offspring. In fact, even under a random-mating scheme, if family sizes are equilibrated, provided $N \geq 4$, the erosion of heterozygosity is

$$H_t \simeq H_0 \left(1 - \frac{1}{4N}\right)^t \tag{3.25}$$

where t is the number of generations after the onset of inbreeding (Wright 1951). This can be seen by returning to many of the formulae in the earlier sections of this chapter and setting $\sigma_k^2 = 0$. Under the idealized scheme of random mating discussed earlier, variances in allele frequencies arise from variance in the number of progeny left by each individual and from segregational variance resulting from the sampling of alleles within individuals. For randomly mating populations of even moderate size, about half of the total sampling variance of allele frequency arises from each source, so that equilibration of family size reduces the total sampling variance by 50%.

The central point of the preceding discussion is that, for a fixed census size, three factors can potentially be manipulated to reduce the rate of inbreeding: avoidance of matings among relatives; equilibration of family sizes; and a sex ratio as close to one as possible. In some domesticated animals, the latter is a problem because females have only one or two offspring per year. In such cases, Gowe et al. (1959) suggested that when the sex ratio of contributing parents is r females to each male, the loss of genetic variance will be minimized if every male contributes exactly one son and r daughters, and every female leaves one daughter and also contributes a son with probability $1/r$. Wang (1997) improved on this scheme with the constraint that a female contributing a son does not contribute a daughter and another female from the same male family instead contributes two daughters.

Kimura and Crow (1963a) noted that Wright's intuition that MAI minimizes the long-term loss of genetic variation is actually not quite correct, pointing out that a **circular mating** (CM) scheme (Figure 3.2) ultimately leads to a lower rate of loss of heterozygosity. Under this breeding design, females and males are arranged such that each of them is mated to two "neighbors," with the last individual in the linear array being mated with the first, thereby completing the circle. Nevertheless, although circular mating ultimately reduces the rate of loss of heterozygosity relative to MAI, it is inferior in the early generations of mating, and even with small N, it may take more than 100 generations before its superiority is realized. Thus, because most of the initial genetic variation in a population will generally have been lost by this time, the practical advantages of circular mating are actually quite negligible.

The major limitation of both the MAI and CM schemes is that they only impede the loss of genetic variation—ignoring new mutations, any randomly mating finite population will ultimately become homozygous at every locus. However, Robertson (1964) obtained the more general (and counterintuitive) result that the rate of loss of overall genetic variation from a population actually *declines* as the relatedness between mates increases. In the extreme, genetic diversity can be preserved indefinitely by subdividing a population into several isolated lines. Although the individual lines are all expected to become homozygous eventually, different lines will become fixed for different sets of genes, with the overall level of preservation of genetic diversity being defined by the number of inbred lines. In effect, complete inbreeding gives rise to a condition equivalent to each family having preserved the equivalent of one gamete from the base population. For example, for a locus with initial allele frequency p_0, assuming a large number of families, the allele frequency in the total collection of lines would remain close to p_0, so that subsequent random mating of the lines would render the heterozygosity close to its original state, $2p_0(1-p_0)$. It must be emphasized, however, that these arguments assume that intense inbreeding in small lines has no consequences that might endanger the line's survival. In reality, however, very small lines are likely to die out occasionally just by accident, and extreme inbreeding also often has serious deleterious effects on fitness (LW Chapter 10). The gradual replacement of extinct lines by members of surviving lines will lead to further loss of variation.

POPULATION SUBDIVISION

Although few species exhibit internal isolation as extreme as that just noted, many (probably most) species occupy spatially structured environments. Such structure causes local inbreeding, as mates are more related than random pairs from the entire population. One of the simplest models of population structure is Wright's (1951) island model, introduced in Chapter 2 (Figure 2.11). Here, the metapopulation consists of d demes, each containing a fixed number N of randomly mating individuals with idealized Wright-Fisher properties, and each deme contributes an equal fraction m of its genes to the migrant pool, which is then equally distributed among the remaining demes. Under this model, an equilibrium level of population divergence is eventually reached (Equation 2.46), at which point the increase in divergence resulting from within-deme genetic drift is balanced by the exchange of alleles by migration.

Recalling from Chapter 2 that the mean coalescence time for an ideal, random-mating population is $2N$ generations, Equation 2.45b (giving the mean coalescence time under this population structure) implies an effective size for the overall metapopulation of

$$N_e = Nd + \frac{(d-1)^2}{4dm} \tag{3.26}$$

In this simplest case, we see that the effective size of a metapopulation exceeds the sum of the demic effective sizes (Nd) by an amount approaching $d/(4m)$ when the number of demes is large. With low migration rates ($m \ll 1$), this inflation can be substantial. For example, if $m < 1/(4N)$, N_e exceeds twice the total number of breeding adults ($2Nd$). This confirms Robertson's (1964) argument that population subdivision can reduce the overall rate of loss of variation by drift as unique alleles are sequestered within individual demes. However, numerous authors have pointed out that this inflation of N_e in the ideal island model is a special consequence of the absence of variation in deme productivity (analogous to the consequences of constant family sizes within a single population, noted above).

The next simplest type of island model allows for extinction and recolonization of the individual demes. In each generation, a fraction e of the demes goes extinct, but immediate recolonization ensures the maintenance of a fixed number (d) of demes. Tracing back to Slatkin (1977) and Maruyama and Kimura (1980), most attempts to model this process have assumed that the newly colonized deme is immediately restored to size N in a single generation, with recolonization involving k immigrants either derived from a single random deme (the **propagule-pool model**) or from a random pool of migrants from the entire metapopulation (the **migrant-pool model**). This simple modification results in a reduction in N_e for the metapopulation by inducing variation in productivity among the demes surviving in each generation—demes that contribute to a colonization event experience a burst of productivity relative to demes that do not.

Using the logic noted above, approximate expressions for mean coalescence times for metapopulations experiencing extinction and recolonization (Pannell and Charlesworth 1999) yield formulae for the metapopulation N_e. Such expressions are functions of the relative rates of extinction and migration and the size and type of colonizing pool, and we only give two examples for the propagule-pool model. If the extinction rate is smaller than the migration rate ($e \leq m$) and much smaller than the relative size of the colonizing pool ($e \ll k/N$),

$$N_e \simeq Nd\,\frac{[4m + (1/N)]}{4(e+m)} \tag{3.27a}$$

a result also obtained by Maruyama and Kimura (1980) and Whitlock and Barton (1997). Under these conditions, local extinctions are sufficiently rare that within-deme variation is able to recover substantially by migration between bottleneck events, and although $N_e < Nd$, it approaches the latter value as $e \to 0$ unless N is tiny. On the other hand, if the extinction rate is relatively high, such that $e \gg m$ and $e \gg k/N$,

$$N_e \simeq \frac{d}{4e} \tag{3.27b}$$

In this case, extinctions are so frequent that local demes (reestablished from a small number of colonists and experiencing little immigration) are almost completely inbred, and the total effective size is independent of the number of individuals per deme and simply defined by deme number and average deme longevity (e^{-1} generations).

A more general expression for the effective size of a metapopulation under the island model was derived by Whitlock and Barton (1997). Assuming large d, so that $d \simeq d-1$ as a first-order approximation,

$$N_e \simeq \frac{d(1+4Nm)}{4m + 2\sigma_K^2(1+2m)} \tag{3.28}$$

where σ_K^2 denotes the among-deme variance in the number of gametes contributing to the next generation (the analogue for the variance σ_k^2 in individual contributions). Comparing this expression to Equation 3.26, it can be seen that $\sigma_K^2 > 0$ will always cause a reduction in N_e. In addition, for N_e to be less than the total metapopulation size $2Nd$, σ_K^2 need only be larger than $1/(2N)$. This amount of among-deme variance in gamete production is trivial, as even under ideal conditions in which individual family sizes are Poisson distributed, the variance in total deme productivity will be on the order of $2/N$. This follows from the fact, noted above, that within an ideal random-mating deme, $\sigma_k^2 = 2$, so that at the deme level, $\sigma_K^2 = 2/N$. Thus, it appears that population subdivision will almost always result in a reduction in N_e, provided the individual demes are not completely isolated, a point first made by Wright (1940; see also Nunney 1999).

SELECTION, RECOMBINATION, AND HITCHHIKING EFFECTS

Up to now, we have assumed that alleles are immune to selective processes, which is, of course, unrealistic in many cases. In fact, selection generally causes a still further reduction in N_e by inflating the among-family variance in offspring production. Evaluating the magnitude of such effects is complicated by the fact that unlike family-size variation induced by environmental heterogeneity, which can be erased in a single generation, genetic variation in fitness is sustained across generations. Such heritable transmission will elevate the genetic representation of some individuals in future generations beyond the expectations under drift alone. This phenomenon was initially mentioned by Morley (1954), who noted in sheep flocks exposed to selection that "the genetically superior individuals will tend to be most inbred." The processes that we will examine are analogous to those that occur in spatially structured populations with random extinction, except that now specific alleles can become trapped in genetic backgrounds that are destined for elimination or fixation.

Before proceeding, it must be emphasized that because N_e is defined in the context of hypothetically neutral loci that serve as benchmarks for the pure drift process, our concern here is not so much with the specific loci under selection, but with the effects of such selection on the dynamics of neutral-allele frequencies elsewhere in the genome. Because the long-term effects of selection depend on the frequency of recombination between selected loci and their associated neutral markers as well as on the mode of selection, the issues are quite technical. Our goal is simply to provide a heuristic overview of why the effects of selection almost always lead to a substantial reduction in N_e. Chapter 8 examines many of these issues in greater detail.

Effects From Selection at Unlinked Loci

Robertson (1961) first considered the influence of a constant selection regime on the long-term dynamics of a neutral locus assumed to be entirely unlinked to any selected loci. In addition to any baseline variance in gamete production among individuals that might exist for environmental reasons (our previous σ_k^2), in the first generation of selection there will also be an among-family genetic variance in relative fitness, σ_w^2, associated with the differential contributions of individual families. Here the relative fitness (w_i) of the ith family is simply the expected contribution to the next generation relative to the average in the population, such that the mean relative fitness $\sum_{i=1}^{N/2} w_i/(N/2) = 1$, where we assume a balanced sex ratio and $N/2$ families. For populations with features in accordance with the standard Wright-Fisher model, a single generation of selection will then reduce the effective population size to

$$N_e \simeq \frac{4N}{\sigma_k^2 + 2 + 4\sigma_w^2} \tag{3.29a}$$

which is identical in form to Equation 3.4 except for the additional variance associated with selection in the denominator. Note that because σ_w^2 is the variance in relative fitness among families and the average family size is two, $4\sigma_w^2$ is the genetic variance in actual family size. This expression demonstrates that the random association of neutral alleles with families with different genetic endowments has the same qualitative effect as environmental differences in family sizes (σ_k^2).

Of course, it would be a rare situation in which selection operated only for a single generation, and Robertson (1961) had the additional insight that with subsequent generations of selection, new stochastic associations between selected and unselected loci will arise each generation, while old associations are lost at rate 0.5 with free recombination. This yields a long-term cumulative contribution to the among-family genetic variance in relative fitness proportional to $[1+(1/2)+(1/4)+(1/8)+\cdots]^2\sigma_w^2 = 4\sigma_w^2$, which (as above) is then further multiplied by 2^2 to translate relative fitness into the genetic variance in family size. Although this result ignores the fact that the stochastic effects of selection will dissipate over time as favorable alleles at the loci under selection go to fixation, this additional layer of complexity is readily incorporated. Letting L denote the per-generation fractional loss of additive genetic variance at selected loci, the preceding series simply has terms in powers of $(1 - L)/2$ instead of $1/2$, and again after converting the variance in relative fitness to the absolute scale, the long-term effective population size becomes

$$N_e \simeq \frac{4N}{\sigma_k^2 + 2 + 4[2/(1+L)]^2\sigma_w^2} \tag{3.29b}$$

(Santiago and Caballero 1995). If we further assume that baseline variation in family sizes unassociated with selection simply reflects random gamete sampling, then from above $\sigma_k^2 \simeq 2$, and under Robertson's assumption of no reduction in variance by selection ($L = 0$), the long-term effective size becomes

$$N_e = \frac{N}{1 + 4\sigma_w^2} \tag{3.29c}$$

Equations 3.29b and 3.29bc are quite general in the sense that they apply to any scheme of selection. However, they are also a bit opaque in that the mechanistic determinants of σ_w^2 and L are not defined and the substantial effects of linkage are omitted. The remainder of this chapter is focused on the removal of these limitations.

Example 3.5. The genetic variance for relative mean-family fitness is a function of the intensity of selection and the heritability of the selected traits. For example, in the case of truncation selection on a single trait (Chapter 14), $\sigma_w^2 = \bar{\imath}^2 t_{FS}$, where $\bar{\imath}$ is the standardized selection differential (the change in mean phenotype imposed by selection in units of phenotypic standard deviations, Equation 14.3a), and t_{FS} is the phenotypic correlation among full sibs (Milkman 1978), which is equivalent to half the heritability for an ideal trait with an additive genetic basis (LW Chapter 18). For situations in which the most extreme 1% to 10% of the phenotypic distribution is selected, $\bar{\imath}$ is in the range of 2.7 to 1.8 (LW Chapter 2), and t_{FS} takes on a maximum value of 0.5 when the heritability of the trait is equal to 1.0. Thus, recalling Equations 3.29a through 3.29c, with very strong truncation selection on a highly heritable trait, σ_w^2 may take on high enough values to reduce N_e several-fold relative to the expectation in the absence of selection, even when the selected loci are unlinked. Chapter 26 examines the reduction in N_e due to truncation selection on a trait in more detail.

Selective Sweeps and Genetic Draft

The effects of linked loci on N_e are substantially greater than those from unlinked loci for the simple reason that chromosomally juxtaposed sites are necessarily mutually influenced by each other's fitness attributes for extended periods. For example, a neutral allele linked to a site under positive selection can **hitchhike** to high frequencies (and even fixation) if the force of selection is strong relative to the recombination rate between the sites (Maynard Smith and Haigh 1974). One direct consequence of this reduction in N_e is a depressed amount of molecular variation at neutral sites in regions of low recombination, an expectation that is in agreement with many empirical observations (Chapter 8). However, as will be emphasized in the next section, the periodic fixation of favorable alleles is just one potential explanation for this kind of observation, an alternative hypothesis being **background selection**, the constant purging of new deleterious mutations (Charlesworth 2012).

A simple way of evaluating the effects of **selective sweeps** of beneficial mutations on variation at completely linked neutral loci was presented by Gillespie (2000). Recall from Chapter 2 that the variance of neutral allele-frequency change from generation to generation in a Wright-Fisher diploid population is equal to $p(1-p)/(2N)$, where p is the current allele frequency. Now imagine that this locus is completely linked to other genomic sites incurring beneficial mutations that collectively cause rapid fixations at an average rate of δ per generation. Because such mutations arise independently of the allele at the linked neutral locus, such selective sweeps will result in the fixation of neutral alleles with probabilities proportional to their current frequencies, in this case p and $(1-p)$ at the neutral focal locus. If we assume that selective sweeps cleanse a population of linked variation essentially instantaneously (or at least rapidly relative to the usual rate of genetic drift), then conditional on a sweep occurring, the variance in allele-frequency change will be $p(1-p)$. Thus, for a neutral locus in an ideal randomly mating population subject to periodic selective sweeps, the total variance in allele-frequency change is the weighted sum of both contributions $\sim p(1-p)\{[(1-\delta)/(2N)]+\delta\}$. Because this expression applies to all initial allele frequencies, equating the term in braces to $1/(2N_e)$ yields

$$N_e \simeq \frac{N}{1+2N\delta} \tag{3.30a}$$

a result also obtained by Maruyama and Birky (1991) by a different method. Here and below, it is appropriate to view N as the effective size of a population based solely on the demographic considerations noted earlier in this chapter.

When selective sweeps are rare relative to the strength of random genetic drift, such that $\delta \ll 1/(2N)$, $N_e \simeq N$, but as $N \to \infty$, $N_e \to 1/\delta$, showing that even populations with enormous numbers of reproductive adults may approach an asymptotic upper limit to N_e defined, not by genetic drift, but by **genetic draft** (Gillespie 2000)—the stochastic result of hitchhiking effects that inevitably arise in linked genomes. That is, when N is very large, the effective size of a population can be more strongly influenced by the physical (i.e., linkage) features of the genome than by demographic factors. In principle, the frequency of selective sweeps may increase with N, as larger populations provide more opportunities for rare beneficial mutations, so strong linkage may even lead to the potential situation in which N_e eventually scales negatively with absolute population size (Lynch 2007).

The preceding result applies to the extreme case of complete linkage. If, instead, a significant amount of recombination occurs between a neutral marker and the selected locus while the latter is proceeding toward fixation, then a selective sweep is not expected to completely remove the variation at the marker locus. As detailed in Chapter 8, the extent to which a neutral locus can free itself of stochastic associations with newly arising beneficial mutations depends on the rate of the sweep (which in turn is a function of the relative power of selection and drift, $s/[1/(2N)] = 2Ns$) as well as on the relative power of recombination and selection (c/s, where c is the rate of recombination between the two loci).

Incorporation of these technical issues by Wiehe and Stephan (1993) led to an expression identical in form to Equation 3.30a, with $2N\delta$ being replaced by a term that is smaller in absolute value. Gillespie (2000) expressed this influence of recombination as

$$N_e \simeq \frac{N}{1+2Nf_s\delta} \tag{3.30b}$$

where f_s is the probability that no recombination occurs between the selected site and the marker locus under consideration during the sweep. Following the completion of the sweep, f_s is equivalent to the probability that a random individual will contain two IBD copies of the original neutral allele on the gamete in which the new favorable mutation arose. With free recombination $f_s \simeq 0$, but with complete linkage, $f_s = 1$, returning us to Equation 3.30a. The general form for f_s is

$$f_s \simeq (4Ns)^{-c/s} \tag{3.31}$$

which will be further discussed in Chapter 8.

Background Selection

We now turn to the influence of selection against recurrently appearing deleterious mutations, which cause a still further reduction in N_e as a consequence of induced variation in family size. Contrary to the situation with selective sweeps, which are sporadic and chromosomally restricted in scope, the effects of recurrent deleterious mutations are expected to be persistent across the entire genome for the simple reason that the vast majority of mutations are deleterious (LW Chapter 10). Here we attempt to provide a heuristic understanding of the effects of such background selection by considering separately the effects of unlinked and linked deleterious mutations, relying on a simple model in which interfering loci harbor two alternative allelic types (beneficial and deleterious).

If the beneficial allele at a locus mutates to a defective type at rate u per generation, with the latter causing a fractional reduction in heterozygote fitness equal to s, the equilibrium frequency of the deleterious allele is equal to u/s (provided that s is substantially stronger than the power of drift and mutation) (Chapter 7). The genetic variance in relative fitness associated with this locus then has an expected value close to $2us$. This result can be obtained by noting that the additive genetic variance for a single locus is equal to $2a^2p(1-p)$ (LW Chapter 4), where a is the difference in phenotype between adjacent genotypic classes and p is the allele frequency. In this case, $a = s$, and because u/s is small, $p(1-p) \simeq p = u/s$. Summing over all n loci capable of mutating to deleterious alleles, the total genetic variance in fitness among individuals is $2nus = Us$, where $U = 2nu$ is the diploid deleterious mutation rate.

This result can be used to evaluate the overall effect of unlinked background selection by noting that unless the number of chromosomes is very tiny, the vast majority of pairs of genes within genomes will be unlinked (with x chromosomes of equal size, the fraction of linked pairs will be $< 1/x^2$ because genes located on opposite ends of chromosomes are effectively unlinked). Thus, we can make use of Robertson's equation (3.29c), noting that the variance of mean family fitness is $Us/2$ after discounting by averaging over both parents. This shows that in the absence of any linkage, deleterious mutations are expected to cause a relatively small reduction in N_e,

$$N_e \simeq \frac{N}{1+2Us} \simeq Ne^{-2Us} \tag{3.32}$$

with the exponential approximation applying under the assumption that $Us \ll 1$. This assumption is justified by numerous observations suggesting that U is on the order of 0.1 to 1.0 and s is on the order of 0.01 (LW Chapter 12).

Some impression of the impact of linkage follows from the logic used to obtain Equations 3.29b and 3.29c. As noted above, for unlinked loci, the initial stochastic associations of neutral alleles and selected loci last for an average of $\sum_{i=0}^{\infty}(1/2)^i = 2$ generations, giving a total contribution to the variance proportional to 2^2. Letting the recombination rate between loci be c, this expression generalizes to $\sum_{i=0}^{\infty}(1-c)^i = 1/c$ generations, and hence a contribution of $1/c^2$ to the variance. Thus, a single selected locus is expected to reduce the variation at a linked neutral locus, which in effect causes a local depression in the effective population size to

$$N_e \simeq Ne^{-us/c^2} \tag{3.33}$$

(Barton 1995a). With $c \ll 0.5$, the absolute value of the exponential term can be considerably larger than that for unlinked loci, $4us$. The challenge is to determine the joint effects of the full spectrum of all linked and unlinked loci surrounding the neutral reference locus.

Insight into the overall power of selection on linked deleterious mutations can be gleaned by considering the extreme case of a completely nonrecombining, but otherwise sexual, genome, i.e., allowing only for segregation during gamete production. Assuming a total of n selected loci, for which the mutant alleles have identical and multiplicative effects on fitness, the average number of deleterious alleles in a gamete is $nu/s = U/(2s)$, where U is again the deleterious mutation rate per diploid genome. Assuming large enough N_e that deleterious mutations do not go to fixation, with this type of genomic architecture (no recombination), only those gametes that are free of deleterious mutations can contribute to the future genetic constitution of the population. Because the number of deleterious mutations per gamete is Poisson distributed in sufficiently large populations with multiplicative selection (Kimura and Maruyama 1966), the frequency of such mutation-free gametes is simply $e^{-U/(2s)}$, leading to the conclusion that with complete linkage, selection against segregating deleterious mutations leads to

$$N_e = Ne^{-U/(2s)} \tag{3.34}$$

(Charlesworth et al. 1993a). This shows that background selection has the potential to cause a dramatic reduction in N_e in a nonrecombining (but segregating) population. For example, with $U = 0.1$ and $s = 0.01$, $e^{-U/(2s)} = 0.0067$. This expression also applies to a nonrecombining chromosomal region if U is redefined to be the deleterious mutation rate for the region under consideration.

Hudson and Kaplan (1994) extended this result to allow for recombination, assuming that the latter operates at uniform rates per physical distance over chromosomal regions. Their results show that

$$N_e \simeq Ne^{-U/(2s+C)} \tag{3.35a}$$

where C denotes the rate of recombination between the ends of the region. Because $1/(2s + C) \gg s$, Equation 3.35a predicts a much smaller N_e than that obtained for freely recombining loci (Equation 3.32), showing that the total contribution from interference from unlinked loci (which is embedded in Equation 3.35a) is relatively minor relative to that from loci in the immediate vicinity of the neutral locus. Moreover, because s is expected to be $\ll 1$, and for an entire chromosome, C is of order 1.0, Equation 3.35a can be roughly approximated as

$$N_e \simeq Ne^{-U/C} \tag{3.35b}$$

where U/C is equivalent to the diploid deleterious mutation rate per unit of recombination (Hudson and Kaplan 1994, 1995). This result, which has been obtained by several different methods (Barton 1995a; Nordborg et al. 1996; Santiago and Caballero 1998), shows that the impact of segregating deleterious mutations on N_e is largely independent of the mutational effect s. As will be seen in Chapter 4, the ratio of mutation and recombination rates can be estimated from molecular polymorphism data, so if the fraction of mutations that are deleterious is known, U/C is also estimable.

Finally, it is worth noting that some conditions exist under which selection may actually promote an increase in N_e, the most obvious involving increases in the coalescence times for linked alleles in a chromosomal region under balancing selection. Pálsson and Pamilo (1999) also found that with very strong linkage and a low efficiency of selection ($2Ns < 1$), repulsion disequilibrium can build up between simultaneously segregating deleterious mutations, leading to a form of associative overdominance (LW Chapter 10) and an elevation of N_e. Although this condition arises in the absence of direct balancing selection on any specific site, it remains unclear whether the special requirements necessary for such an outcome are very common. Santiago and Caballero (2005) also found that in a subdivided population, selective sweeps within demes can sometimes lead to an increase in N_e for the total metapopulation, as the migration of a sweeping chromosomal region drags new variation into a recipient deme, leading to an overall effect akin to balancing selection.

With these exceptions aside, there are two general lessons to be learned from all of the preceding discussion. First, although the individual demographic and genetic effects that influence N_e may appear to be only moderate in nature, their cumulative effects can easily depress N_e below the actual number of reproductive individuals by several orders of magnitude, and second, although population geneticists often develop analytical descriptions of various processes under the assumption of an effectively infinite population size, the physical linkage of the genome ensures that even populations with extraordinarily large N need not be immune to drift-like processes imposed by hitchhiking effects. These points will be made clearer in Chapter 4 as we explore the direct manifestation of such effects on standing variation in natural populations.

Example 3.6. Because natural populations are subject to both positive and negative selective forces, the total influence of selection on N_e must reflect *both* background selection and selective sweeps. This necessarily raises even more technical issues than were outlined above. Significant progress was made by Kim and Stephan (2000), and here we simply outline the basic result. If background selection operates as an essentially continuous process resulting from the recurrent introduction of deleterious alleles, the depressive effects of both forms of selection, as well as baseline demographic effects, may be treated as largely independent. The reduction in N_e resulting from background selection can then be obtained by use of one of the above expressions, e.g., Equation 3.35b as a first-order approximation for a sexual population (with N already taking into consideration demographic effects).

Consider a large monoecious population of constant breeding size and variance in family size $\sigma_k^2 = 4$. Based on demographic considerations alone, from Equation 3.4, $N_e \simeq 2N/3$. Letting $U = 1$ and $C = 1$, Equation 3.35b implies that background deleterious mutations further reduce N_e to $(2N/3)e^{-1} \simeq 0.25N$. The effective population size dictated by these demographic and deleterious-mutation processes further defines the background N_e' within which occasional beneficial mutations arise and sweep to fixation, so that the effective population size resulting from the joint operation of all three effects can be approximated by substituting N_e' for N in Equation 3.30b. Supposing a complete sweep occurs every 10,000 generations (so that $f_s = 1$ and $\delta = 0.0001$), then $N_e = 0.25N/[1 + (0.50N \cdot 0.0001)]$. With $N = 10^4$, 10^6, and 10^8, this implies $N_e/N \simeq 0.17, 0.0049$, and 0.00005, respectively.

In general, the joint operation of background selection and selective sweeps will reduce N_e more than either does alone, although it is, at least in principle, possible for background selection to reduce the influence of selective sweeps in regions of very low recombination by depressing N_e, which in turn will reduce the fixation probability of favorable alleles (Chapters 7 and 8). The simultaneous operation of positive selection on multiple loci (which was ignored in the derivation of Equations 3.30a and 3.30b) can also slightly alleviate the overall effects of selection on N_e as simultaneously segregating mutations interfere with each other's fixation, thereby reducing the incidence of complete selective sweeps (Kim and Stephan 2003). These issues are examined in greater detail in Chapter 8.

4

The Nonadaptive Forces of Evolution

I may be wrong but I doubt it. Charles Barkley

Although natural selection plays a major role in the evolution of many traits, three additional factors determine the patterns of genetic variation within and among populations. We refer to these factors—-mutation, recombination, and random genetic drift—as the **nonadaptive forces of evolution** because their operation is generally independent of the specific selective factors operating on the extrinsic phenotypes of individuals. Migration (briefly touched upon in Chapters 2 and 3) might be added to this list, although we regard this added complexity as being independent of the internal genetic machinery of a population. As will become clear in the following chapters, the three nonadaptive forces together comprise the **population-genetic environment**, which defines the paths of evolutionary change that are open vs. closed to natural selection.

Knowledge of the magnitude of the nonadaptive forces of evolution should be sufficient to arrive at a full description of the dynamics of allele- and gamete-frequency change within populations in the absence of external forces of selection. Moreover, this logic works in reverse—under certain assumptions, observed patterns of variation in neutral genomic regions can be used to infer the magnitude of the evolutionary forces responsible for such patterning.

The goals of this chapter are, therefore, three-fold. First, we will consider how observations on putatively neutral molecular markers can be used to estimate rates of mutation, recombination, and random genetic drift. Second, we will summarize the existing data resulting from such analyses, providing information that will play a central role in subsequent chapters. As a consequence of the recent emergence of new technologies for high-throughput genomic sequencing, this is a rapidly developing area that will undoubtedly experience additional refinements in the near future. Finally, after showing that the intensities of the nonadaptive forces of evolution vary by orders of magnitude among species in fairly predictable manners, we will summarize existing theory that helps explain such patterns.

Although our ultimate desire is to obtain accurate estimates of the individual forces of mutation, recombination, and drift, as will be seen below, it is often much easier to obtain ratios of these features than to measure them individually. Fortunately, this is not always an undesirable situation, for as we saw in Chapter 2, in the absence of selection, the ratio of the power of mutation (u) and the power of drift ($1/2N_e$) defines the level of heterozygosity in a population, and the ratio of the recombination rate per nucleotide site (c_0) and the power of drift defines the magnitude of linkage disequilibrium. Therefore, before summarizing the approaches for estimating N_e, u, and c_0 separately, we will first consider methods for estimating the composite population parameters $\theta = 4N_e u$ and $\rho = 4N_e c_0$. As will be seen below, accurate estimates of N_e are particularly difficult to achieve directly, especially for large populations. However, by using combined estimates of θ, ρ, u, and/or c_0, obtaining approximate measures of long-term N_e is sometimes possible.

Throughout this chapter, we will assume that we are dealing with molecular markers known in advance to behave in an effectively neutral fashion. Numerous methods to test this hypothesis will be discussed in Chapters 9 and 10. We will largely focus on measures at the level of individual nucleotide sites, as it is now routine to obtain large quantities of DNA-sequence data, and per-nucleotide site measures are readily extrapolated to larger units of analysis such as genetic loci. Thus, u and c_0 will, respectively, denote the mutation rate per nucleotide site and the recombination rate between adjacent sites.

Evolution and Selection of Quantitative Traits. Bruce Walsh & Michael Lynch.
Published 2018 by Oxford University Press.
DOI 10.1093/oso/9780198830870.001.0001

RELATIVE POWER OF MUTATION AND GENETIC DRIFT

In Chapter 2, it was demonstrated that if the forces of drift and mutation remain constant for a sufficiently long time, the level of heterozygosity at a neutral nucleotide site (with four possible allelic states) will stochastically wander around an expected equilibrium value of $\sim 12N_e u/(3+16N_e u)$, where u is the mutation rate per gamete per nucleotide site (assuming all nucleotides mutate at the same rate). As will be seen below, the average heterozygosity per neutral nucleotide site is far below 1.0 in all phylogenetic groups, so the preceding expression is generally closely approximated by $4N_e u$. This particular measure has great practical utility. Because $2u$ is the mutation rate per site per diploid genome, $4N_e u$ is equivalent to the ratio of the power of mutation to the power of random genetic drift, $1/(2N_e)$. For haploid species, the expected nucleotide diversity at a neutral site is $2N_e u$.

Nucleotide Diversity

Suppose a population sample of n random sequences has been obtained for a particular genomic region. In principle, such a stretch of DNA might consist of intronic or intergenic sequence or of the subset of **silent** (**synonymous**) sites in one or more coding regions. Letting k_{ij} be the number of site-specific differences between observed sequences i and j, and L be the number of sites per sequence, the average fraction of pairwise differences between the sampled sites,

$$\widehat{\theta}_\pi = \frac{2}{n(n-1)} \sum_{i=1}^{n} \sum_{j>i}^{n} k_{ij}/L \tag{4.1}$$

yields a heterozygosity-based estimate of $\theta = 4N_e u$ (Tajima 1983). This formulation, frequently called the **Tajima estimator**, is often denoted by π in the literature.

If we are to confidently use $\widehat{\theta}_\pi$ as an estimator of θ, aside from knowing whether the assumptions of neutrality and equilibrium are valid, it is critical to know the sampling variance of $\widehat{\theta}_\pi$. Such variance results from two sources of uncertainty. First, heterozygosity is subject to **evolutionary variance**, which results from the natural fluctuations of nucleotide frequencies generated over time by the stochastic forces of mutation and drift (Chapter 2). Although this source of variation is not easily observed directly, assuming a population in drift-mutation equilibrium, the expected evolutionary variance of the true population value of θ based on L independent (effectively unlinked) sites is $\simeq (\theta/3)[(2\theta/3) + (1/L)]$ (Tajima 1983). This source of variance is intrinsic to the features of the population, independent of the sample taken. Second, **sampling variance** results from the use of a finite number of sampled sequences to estimate θ_π. For an equilibrium population, this variance is $\simeq \{2\theta/[3(n-1)]\}\{[(2n+3)\theta/3n] + (1/L)\}$, where n is the number of sequences sampled per site (Tajima 1983).

Summing over these two sources of variance, for sites in stochastic drift-mutation equilibrium, the expected total variance of heterozygosity-based estimates of θ is

$$\sigma^2(\widehat{\theta}_\pi) \simeq \frac{\theta}{3(n-1)} \left(\frac{2(n^2+n+3)\theta}{3n} + \frac{n+1}{L} \right) \tag{4.2}$$

(Pluzhnikov and Donnelly 1996). With increasing numbers of sampled alleles per site, i.e., as $n \to \infty$, the total variance of estimates of θ based on nucleotide diversity approaches a minimum equal to the evolutionary variance. Even in this case, and with an enormous amount of sequence per individual (large L), the sampling coefficient of variation of $\widehat{\theta}_\pi$ is $\simeq \sqrt{2/9} \simeq 0.47$. Adding more individuals or sites to a survey will not alter this minimum.

It is worth reemphasizing that Equation 4.2 is an appropriate estimator of the variance of a nucleotide-diversity estimate only if the latter is based on neutral sites in drift-mutation equilibrium (the **standard neutral model**). Even for assuredly neutral sites, this expression will not apply for *nonequilibrium* situations, e.g., populations that have experienced relatively recent expansions or contractions. Under the latter conditions, the variance of heterozygosity must be evaluated more directly from the spectrum of allele frequencies across all sites and from their higher-order moments. A number of related technical issues are covered and general expressions derived in Nei and Roychoudhury (1974), Nei (1978), Nei and Tajima (1981a, 1983), Nei and Jin (1989), and Lynch and Crease (1990).

Finally, now that high-throughput sequencing has become routine for entire diploid genomes, it is possible to estimate the average nucleotide diversity over millions of putatively neutral sites, yielding per-individual measures with a high degree of precision. Because most pairs of sites are on different chromosomes, a full survey of even a single individual from a random-mating population should provide a very accurate description of the average per-site diversity across the entire population. Moreover, when a survey of two random individuals is possible, the covariance of heterozygosity within sites provides a direct estimate of the evolutionary variance of heterozygosity among sites, which, as noted above, should closely approximate $(\theta/3)[(2\theta/3) + (1/L)]$ under the assumption of drift-mutation equilibrium (Lynch 2008a). These observations are now quite salient, as Pluzhnikov and Donnelly (1996) showed that for a fixed amount of resources for sequencing Ln total bases, the optimal strategy for obtaining minimal-variance estimates of θ is generally to sample no more than two or three individuals, putting the effort instead into sampling more sites, i.e., maximizing L at the expense of n.

Number of Segregating Sites

Although nucleotide diversity is the most transparent means of estimating θ, it is by no means the only, or even the most efficient, approach. Watterson (1975) pointed out an alternative statistical measure of allelic diversity—the total number of segregating sites (S) in the region analyzed over the full set of n sequences. Because a segregating site is any nucleotide position that harbors two or more variants, S clearly increases with the length L of the sequence and the number of individuals assayed. Watterson (1975) showed that under the assumptions of neutrality and drift-mutation equilibrium, an unbiased estimator of the per-site parameter $\theta = 4N_e u$ is

$$\widehat{\theta}_S = S/(L\, a_n) \tag{4.3a}$$

where

$$a_n = \sum_{j=1}^{n-1} 1/j \tag{4.3b}$$

By rearranging, it can be seen that Equation 4.3a relates directly to the expected site-frequency spectrum for a sample under drift-mutation equilibrium with a known value of θ (Equation 2.35a). A central point here is that when the nucleotide sites surveyed are neutral and in drift-mutation equilibrium, like the Tajima estimator (Equation 4.1), the **Watterson estimator** provides a separate estimate of θ. In Chapter 9, we will see that when the assumptions of neutrality and/or equilibrium are violated, the values of $\widehat{\theta}_\pi$ and $\widehat{\theta}_S$ deviate from each other in ways that yield insight into past population-genetic processes.

The sampling variance for the Watterson estimator, analogous to Equation 4.2 and again under the assumptions of neutrality and equilibrium, is

$$\sigma^2(\widehat{\theta}_S) \simeq \frac{\theta}{a_n}\left(\frac{\theta\, b_n}{a_n} + \frac{1}{L}\right) \tag{4.4a}$$

where

$$b_n = \sum_{j=1}^{n-1} 1/j^2 \tag{4.4b}$$

For sample sizes smaller than ten, the Tajima and Watterson estimators have similar expected sample standard deviations, but with larger n, the latter can be up to two-fold smaller

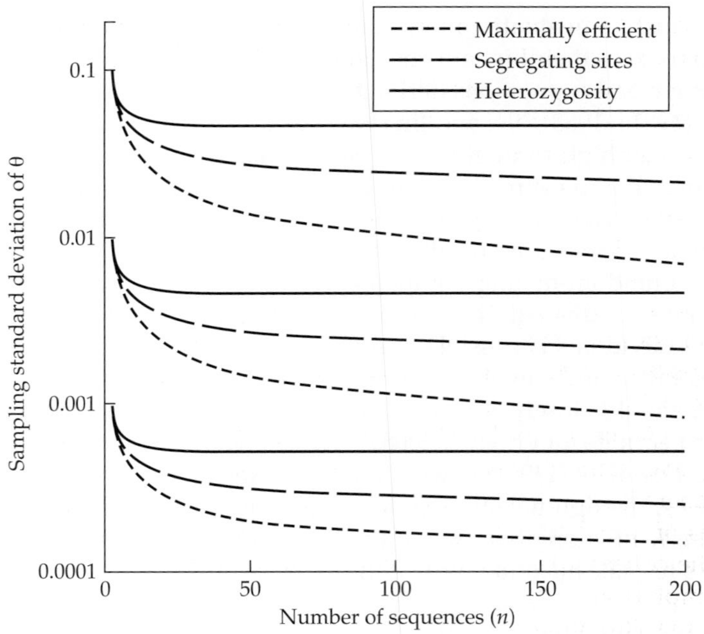

Figure 4.1 Expected sampling standard deviations for estimates of θ from sequences assumed to be neutral, in drift-mutation equilibrium, and experiencing no intragenic recombination. Results are derived from Equations 4.2 (Tajima estimator based on heterozygosity, solid line), 4.4 (Watterson estimator based on segregating sites, long-dashed line), and 4.5 (maximally efficient, short-dashed line), for $\theta = 0.1$, 0.01, and 0.001 in descending order. The assumed number of sites is $L = 10,000$ in all cases.

than the former, although there is little to be gained with either approach once n exceeds 50 or so (Figure 4.1). It should, however, be emphasized that both Equation 4.2 and Equation 4.4a were derived under the assumptions of sequences experiencing *negligible recombination*. The necessary modifications to allow for intragenic recombination, derived in Pluzhnikov and Donnelly (1996; their Equations 6 and 7), play a role in some methods for estimating the population recombination rate, as described in the following section.

One significant issue that arises with the use of S to estimate θ in the modern era of high-throughput sequencing involves the introduction of upward bias from sequencing errors. With large numbers of sites and individuals, errors will inevitably appear as singletons but nonetheless enter the estimate of S. Such effects can be quite deceptive in population-genetic analyses because rare alleles are expected to be common under the neutral hypothesis. Johnson and Slatkin (2008), Kang and Marjoram (2011), and Keightley and Halligan (2011) suggested methods for eliminating the bias from S when an accurate estimate of the sequencing-error rate is available. An alternative approach relaxes this constraint by estimating the error rate from the data themselves (Lynch 2009).

Alternative Approaches

Felsenstein (1992) pointed out that neither of these approaches is likely to provide the most efficient estimates of θ (i.e., to yield estimates with minimum sampling variance), as they do not utilize all of the information in the sample of sequences. In particular, both approaches ignore the genealogical relationships of sequences (i.e., the coalescent structure of the sample), although under neutrality, the expected contribution to variation from each genealogical branch can be expressed in terms of θ (Equation 2.33b).

To evaluate how much improvement might be achieved by exploiting such information, Fu and Li (1993a) derived a maximum-likelihood estimator of θ for the extreme situation in which one knows with certainty the genealogical relationships of the sequences and the numbers of mutations and generations on each branch of the genealogy. The expected sampling variance of this estimator is

$$\sigma^2(\widehat{\theta}_{ML}) = \frac{\theta}{a_n}\left(\frac{\theta a_n}{n-1} + \frac{1}{L}\right) \tag{4.5}$$

A comparison of this expression and Equations 4.2 and 4.4 illustrates that there is substantial room for improvement in the estimation of θ over the traditional heterozygosity and segregating-sites methods, provided the number of sequences exceeds five or so, and assuming a reasonably accurate gene genealogy can be obtained (Figure 4.1).

Gene genealogies cannot be constructed without error. However, by using information on the expected coalescence times of samples of neutral sequences, Fu (1994a, 1994b) developed several generalized least-squares estimators that account for the sampling variances and covariances of mutations on different branch segments. Several of these estimators, which utilize the concepts of the site-frequency spectrum (Fu 1995; Li and Fu 1999; see Chapter 2), asymptotically perform in a near-optimal manner as the sample size increases, again provided the sites are neutral and the population is in drift-mutation equilibrium.

As one or both of the latter two assumptions (neutrality and equilibrium) are likely to be violated to unknown degrees in many natural settings, having an estimator with minimum sensitivity to both problems would be highly useful. In fact, just such an approach can be extrapolated from Watterson's estimator (Equation 4.3). The basis for this strategy follows from the property for neutral alleles that in an equilibrium population, the number S_j of derived single-nucleotide variants found j times in a sample of size n has the expected value $L\theta/j$ (Watterson 1975; Fu 1995). Because the total number of segregating sites, S, has an expected value of $L\theta a_n$, it follows that Watterson's estimator is equivalent to an average of estimates of θ, each weighted by the inverse of the number of observations.

The simplest estimate of θ, based only on singletons ($j = 1$), is then

$$\widehat{\theta}_1 = S_1/L \tag{4.6a}$$

which is also equivalent to the number of mutations (per site) on the external branches of a gene genealogy (Fu and Li 1993b). Such an estimator is attractive for two reasons. First, the singletons in a sample are a function of the very recent past, especially when the overall sample size is large, and hence are not expected to be influenced by distant periods of population-size change. Second, because the dynamics of rare alleles are primarily governed by the drift process, singleton frequencies are expected to most closely reflect the pattern expected under neutrality even when such mutations are nonneutral (Messer 2009). The sampling variance of the singleton-based estimator is

$$\sigma^2(\widehat{\theta}_1) = \frac{\theta}{n}\left(\frac{n-1}{L} + \frac{\theta[2a_n(n-1)-1]}{n}\right) \tag{4.6b}$$

Considering just the sampling variance of the estimators of θ to this point, as $n \to \infty$, those for $\widehat{\theta}_W$ and $\widehat{\theta}_{ML}$ are $\theta/(15.4L)$, whereas that for $\widehat{\theta}_\pi$ is $\theta/(3L)$, and that for $\widehat{\theta}_1$ is θ/L. Thus, although the singleton-based estimator is likely to have the smallest amount of bias associated with selection, a focus on only a fraction of the segregating sites results in higher sampling variance.

Empirical Observations

Estimates of θ, mostly derived as silent-site heterozygosity from protein-coding genes using Equation 4.1, have been summarized for a wide range of species across the Tree of Life by Lynch (2007) and Leffler et al. (2012), and more specifically on metazoans and land plants by Romiguier et al. (2014) and Corbett-Detig et al. (2015). Across a diverse assemblage of more than 100 eukaryotic and prokaryotic species, there is an inverse relationship between organism size and θ_π, with estimates for prokaryotes falling in the broad range of 0.007 to 0.388, with an average value of 0.104 (and a large standard deviation of 0.111). The average values for unicellular eukaryotes (mean = 0.057, SD = 0.078) and invertebrates (mean = 0.026, SD = 0.015) are 50% to 75% lower, and estimates for land plants (mean = 0.015, SD = 0.013) and vertebrates (mean = 0.004, SD = 0.003) are still smaller. Because the numbers of independent studies contributing to these estimates are in the range of 15 to 50, the cited means approach a level of reliability (with some caveats given below), but because of sampling error at the gene, individual, and population levels, the standard deviations likely overestimate the true evolutionary variance.

For both of the unicellular groups, silent-site heterozygosity measures are likely to be downwardly biased estimators of $4N_eu$ ($2N_eu$ for haploids), for at least two reasons. First, most recorded studies of microbial species are derived from surveys of pathogens, whose N_e may be abnormally low because of the restricted distributions of their multicellular host species, and second, silent-site variation will underestimate the neutral expectation if such sites experience some form of purifying selection. Such conditions can arise for a variety of reasons: (1) translation-associated selection when certain tRNAs have higher affinities for certain alternative codons (often referred to as **codon bias**); (2) selection on sites involved in splice-junction identification for species with introns; (3) secondary selection against codons that are one mutational step from termination codons; and (4) inhibition of double-strand break repair between highly divergent alleles. The molecular biological underpinnings of some of these factors, as well as their potential population-genetic consequences, are reviewed in Lynch (2007). Because all of these forms of selection are expected to be quite weak, they will be most effective in populations with very large N_e. Thus, although θ_π may underestimate $4N_eu$ ($2N_eu$ for haploids) in some microbial species by as much as ten-fold, the bias may be minor in multicellular eukaryotes. Many uncertainties remain, however, and we return to the topic in Chapter 8.

With these caveats in mind, the existing data make a compelling statement with respect to the relative power of mutation and random genetic drift—in essentially no eukaryotic species is there evidence that the former exceeds the latter (as this would cause $4N_eu > 1$), and in large multicellular land plants and vertebrates, the ratio is almost always on the order of 0.03 or much smaller. Thus, drift appears to be a more powerful force than mutation at the nucleotide level in all species, except perhaps the smallest microbes. As the absolute population sizes of many species (certainly microbes) can exceed $1/u$ by orders of magnitude (see below), these observations clearly support the idea introduced in Chapter 3 (and detailed in Chapter 8) that N_e is usually substantially smaller than the actual number of reproductive individuals in a population, and that this is largely a consequence of selection on linked sites, especially in large populations. Analyses of data on metazoans and land plants by Corbett-Detig et al. (2015) suggest that selection on linked sites typically reduces silent-site heterozygosity below the expected value $4N_eu$ by $\sim$10%, although there are a few cases involving small, wide-ranging species where the downward bias is as great as 70%.

RELATIVE POWER OF RECOMBINATION AND GENETIC DRIFT

As will be seen in subsequent chapters, recombination plays an important role in evolution because the physical scrambling of linked genes increases the ability of natural selection to promote or eliminate mutations on the basis of their individual effects. On the other hand, high rates of recombination can often inhibit the establishment of pairs of mutations with favorable epistatic effects.

Two general approaches provide insight into the level of recombination per physical distance along chromosomes. Genetic maps, generally derived from controlled crosses, are based on observations on the frequency of meiotic crossovers between informative markers

(LW Chapter 14), whereas studies of linkage disequilibrium (LD) in natural populations use the theoretical concepts introduced in Chapter 2 to indirectly infer the relative magnitudes of the historical forces of random genetic drift and recombination. High-density genetic maps have the power to yield accurate estimates of average recombination rates over fairly long physical distances (usually with markers being separated by millions of nucleotide sites, which typically corresponds to >1% recombination per generation). However, because patterns of LD are generally outcomes of many thousands of generations, they have the potential to reveal much more refined (kilobase scale) views of the recombinational landscape. For a mapping cross involving n gametes with a recombination frequency c between marker sites, the expected number of recombinants is nc, so for sufficiently close sites, the typical outcome will be a complete absence of recombinants. On the other hand, if n random chromosomes are sampled from a natural population with a mean coalescence time between random alleles of $\bar{t} = 2N_e$ generations (Chapter 2), the expected number of recombination events is $2\bar{t}nc = 4N_enc$. Thus, this chapter will focus on the use of LD, rather than directly observed meiotic crossover events, to derive inferences about recombination.

Before proceeding, we again remind the reader that we use c_L to denote the recombination rate between sites separated by a particular distance L, with c_0 denoting the recombination rate between adjacent sites. When referring to population-scaled recombination rates, we will use separate notations of $\rho_L = 4N_ec_L$ and $\rho = 4N_ec_0$. Although it is tempting to assume that $c_L = c_0L$ and $\rho_L = \rho L$, as will be outlined below, such a linear transformation is not generally valid.

Recall from Chapter 2 that $\rho_L = 4N_ec_L$ is the effective number of recombination events between sites (separated by L base pairs [bp]) per generation at the entire population level, which is also equivalent to the ratio of the power of recombination to the power of drift. Just as the amount of segregating variation at neutral sites provides insight into the population mutation rate $\theta = 4N_eu$, the amount of standing LD is a function of the population recombination rate, ρ_L. Although a wide variety of methods for estimating the latter parameter have been proposed, the challenges to obtaining accurate measures are substantial. The markers employed must not only have at least moderate frequencies (to ensure accurate estimates of gamete frequencies and reasonable likelihoods of observing recombination events), but must also behave neutrally (to ensure the validity of the application of drift-recombination theory). Moreover, most of the proposed estimators rely on the assumption of drift-mutation-recombination equilibrium, while also suffering from very high sampling variance, which demands substantial replication over independent pairs of sites.

Number of Recombinational Events in a Sample of Alleles

The power of using population-level data in the detection of historical recombination rates is limited by the fact that recombination events only leave a trace if they involve pairs of doubly heterozygous chromosomes. Moreover, there is no way to directly determine whether multiple recombinants in a sample are a result of parallel recombination events or intact descendants of the same events. Thus, to obtain unbiased estimates of ρ_L, we require a method for converting the *observed* number of recombinant events in a sample to the *actual* number that likely occurred (R). This is not unlike the challenge in genetic-map production of converting observed into actual numbers of recombination events between markers (LW Chapter 14). We start with a description of methods involving shorts spans of DNA, e.g., single genes with **phased haplotypes** (such as complete sequences for each of the two alleles within diploid individuals). Chromosomal regions of such small size will often have recombination rates between their boundaries $\ll 0.01$, and hence would have no chance of revealing recombinants in simple mapping crosses.

For a set of sequences with the most extreme distance between polymorphic sites being L nucleotides, assuming a population in drift-mutation-recombination equilibrium, the expected number of recombination events, R, in a sample of n sequences is equal to $\rho_L a_n$, where a_n is given by Equation 4.3b (Hudson and Kaplan 1985). Thus, a potential estimator for ρ_L is

$$\widehat{\rho}_L = \widehat{R}/a_n \tag{4.7}$$

where $\widehat{R}$ is the estimated number of recombinational events that have occurred between the maximum span of polymorphic sites in the history of the sequences within the sample. Note the similarity of the form of this expression to that relating the number of segregating mutations to θ (Equation 4.3a).

The primary impediment to applying this expression is the estimation of R. One approach, proposed by Hudson and Kaplan (1985), starts with the **four-gamete test**, which asserts that any pair of heterozygous sites exhibiting four gametic haplotypes must reflect the prior action of at least one recombination event, assuming an absence of parallel mutations. Under this view, starting with a fixed gamete of the form *AB*, a single mutation will create either an *aB* or *Ab* gamete, resulting in two gametic types in the population. This is a noninformative situation because recombination between the ancestral (*AB*) and derived (*aB*) haplotypes cannot generate a novel haplotype. If, however, prior to fixation of the first mutation, a mutation arises at the remaining homozygous site, there will be three haplotypes (e.g., *AB*, *Ab*, and *aB*), with the fourth type (*ab*) arising only by subsequent recombination between *aB* and *Ab* haplotypes. Judiciously applying this criterion to all pairs of segregating sites in a sample of sequences and ensuring that the same event is not counted more than once, it is possible to estimate R_{min}, the minimum number of crossover events in the history of the sample (Hudson and Kaplan 1985). More complex approaches attempt to derive information from the complete haplotype structure in a sample (Myers and Griffiths 2003; Liu and Fu 2008).

In principle, with knowledge of the expected fraction of detectable recombination events, d_r, one could extrapolate the observed R_{min} to an estimate of the actual value R. Assuming conditions of drift-mutation equilibrium, Stephens (1986) found approximate lower and upper bounds to d_r giving rise to observable, nonparental haplotypes,

$$d_{r,\text{min}} = 1 - [2\ln(1+\Theta)]/\Theta + [1/(1+\Theta)] \tag{4.8a}$$

$$d_{r,\text{max}} = 1 - [2(1-e^{-\Theta})]/\Theta + e^{-\Theta} \tag{4.8b}$$

where $\Theta = 4N_e uL$ is the population mutation rate for the stretch of DNA being surveyed. These two limits are respectively approached as $c \to 0.0$ (complete linkage) and $c \to 0.5$ (free recombination). As $\theta = \Theta/L$ is generally on the order of 0.001 to 0.01 for neutral sites, unless the segments being analyzed have lengths in excess of 1000 nucleotides, the majority of recombination events will simply reproduce parental gamete types, and hence not be scored as recombinants (Figure 4.2).

Given an estimate of Θ, Equations 4.8a and 4.8b can be used to approximate the total number of recombination events in the sample as $R_{\text{min}}/\,\overline{d}_r$, where $\overline{d}_r$ is the average of $d_{r,\text{min}}$ and $d_{r,\text{max}}$. However, even this approach is not fully adequate because only a subset of the recombinant gametes that are nonparental with respect to markers are also novel with respect to the entire population, i.e., the fraction of uniquely detectable recombination events in the population is even lower than suggested by Equations 4.8a and 4.8b.

An empirical approach to this problem was suggested by Zietkiewicz et al. (2003; see also Lefebvre and Labuda 2008). Letting p_i denote the frequency of the *i*th haplotype in a sample, an estimator for the fraction of detectable (but not necessarily unique) recombinant alleles is

$$\widehat{d}_r = \sum_{i=1}^{L}\sum_{j>i}^{L} 2p_i p_j L_{\text{max},ij}/L \tag{4.9}$$

where $L_{\text{max},ij}$ is the distance between the maximally separated heterozygous sites in the *ij*th comparison. Through simulations, one can establish the fraction of potentially informative recombination events that would indeed produce novel haplotypes in the sample, thereby converting $\widehat{d}_r$ to $\widehat{d}'_r$, the fraction of recombination events that lead to uniquely observable

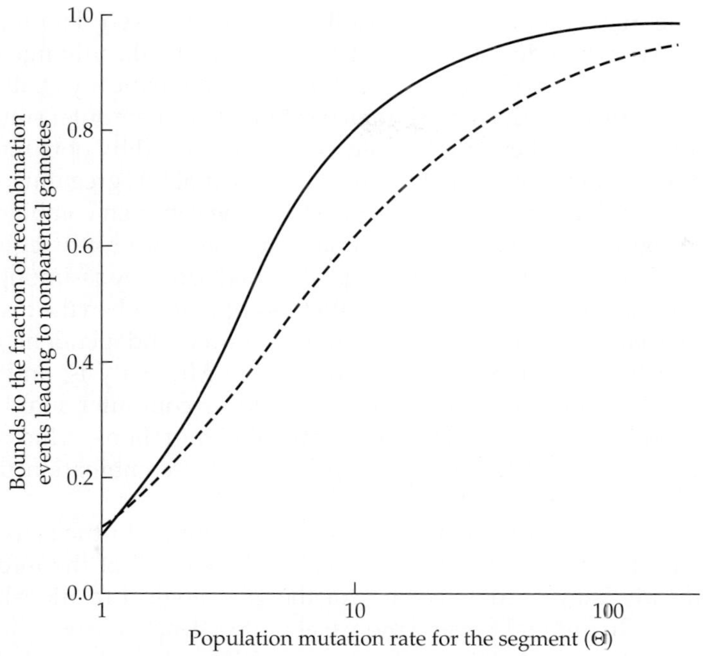

Figure 4.2 Approximate upper and lower bounds on the fraction of recombination events that produce nonparental gametes among two or more neutrally evolving sites (from Equations 4.8a and 4.8b). $\Theta = L\theta$ is the product of the population mutation rate per site (θ) and the length of the segment (L in base pairs).

recombinants. Recalling Equation 4.7, a method-of-moments estimator for the population recombination rate is then

$$\widehat{\rho}_L = \widehat{R}_{\min}/(\widehat{d'_r} a_n) \tag{4.10}$$

Other Approaches for Narrow Genomic Intervals

An alternative method-of-moments approach to estimating ρ_L was suggested by Hudson (1987), who noted that the variance of pairwise measures of neutral sequence divergence is expected to decline with increasing levels of recombination. (With strong linkage disequilibrium, some random pairs of haplotype blocks will be identical over all polymorphic sites, while others will differ at all such sites.) This approach requires an estimate of the average number of nucleotide differences between random sequences of length L, $\Theta_\pi = \theta_\pi L$, as well as the observed variance of pairwise divergence,

$$\widehat{\sigma}_k^2 = \frac{2}{n(n-1)} \sum_{i=1}^{n} \sum_{j>i}^{n} (k_{ij} - \Theta_\pi)^2 \tag{4.11}$$

where k_{ij} is the number of sites at which sequences i and j differ, and n is the number of chromosomes scored in the sample. Wakeley's (1997) Equation 15 allows one to estimate ρ_L as a function of Θ_π, $\widehat{\sigma}_k^2$, and n.

Fuller use of the information in sample data can be achieved by considering the probabilities of various sample counts of the four gametic types at two loci or nucleotide sites (i.e., *AB*, *Ab*, *aB*, and *ab*) assumed to be biallelic, neutral, and in drift-mutation-recombination equilibrium (Hudson 2001). For any hypothetical combination of the parameters θ, ρ_L, and sample size n, one may compute the probability of the observed data for each pairwise combination of markers (Golding 1984; Ethier and Griffiths 1990), although obtaining exact probabilities of two-locus sampling configurations is mathematically challenging, and for large sample sizes, approximations must often be obtained by computer simulation (but see Jenkins and Song 2009). Further simplification can be achieved by obtaining probabilities of sampling configurations conditional on two alleles actually segregating at both sites, as this eliminates the dependence on θ (Hudson 2001). One can then combine the likelihood estimates with respect to ρ_L over all nonoverlapping pairs of linked segregating sites to obtain a global estimate of ρ (Hudson 2001). (Usually, this is done by assuming that $\rho = \rho_L/L$, although as noted below there are problems with this approach when the distances between sites are highly variable.) Because the data are not entirely independent, this **composite likelihood approach** is just an approximation to a full ML analysis, and the confidence limits for the resultant estimates can only be achieved by computer simulations. McVean et al. (2002) extended this approach to allow for parallel mutations, which in species with high mutation rates, can lead to the false appearance of recombination under the usual assumptions of the four-gamete test.

The efficiency of all of these methods can be questioned in the sense that they use summary statistics that do not necessarily make full use of all of the information in the sample. Most notably, they do not account for the genealogical relationships among the sampled haplotypes. To this end, several more elaborate ML approaches and their Bayesian extensions go well beyond the method of Hudson (2001) (e.g., Kuhner et al. 2000; Nielsen 2000; Fearnhead and Donnelly 2001). As the number of genealogies consistent with any given set of mutational and recombinational parameters is enormous, exact solutions are not possible with these computationally intensive approximations. Moreover, although one would expect estimates derived in an explicit likelihood framework to perform better than the ad hoc procedures outlined above, it remains unclear whether that is the case for the sample sizes (n and L) that have been typically applied to date, as all existing estimators appear to be biased, have very large sampling variances, and rely on the assumption of an equilibrium population (Wall 2000).

Large-scale Analysis

The methods outlined in the preceding paragraphs were developed largely for analyzing sequences at the level of gene-sized fragments. However, with the sequencing of entire genomes of multiple individuals now becoming routine, genome-wide profiles of LD can be obtained. One limitation of genome-sequencing technologies is that sequence read lengths remain small (often on the order of 100–200 bp), so that unlike the situation when individual alleles are cloned and sequenced, the phases of haplotypes are not certain for double heterozygotes at distant pairs of sites. However, unambiguous haplotypes can still be inferred from information contained within singly heterozygous individuals, with the resultant frequency estimates enabling one to compute the full slate of LD statistics. Moreover, mean read lengths are rapidly expanding, so this will soon be a minor consideration.

One approach to estimating ρ from whole-genome sequencing relies on data from just a single individual (Lynch 2008a). This maximum-likelihood method estimates the correlation Δ of "zygosity" (heterozygosity and homozygosity) of pairs of sites separated by specific distances (L) across the genome to obtain disequilibrium measures that are nearly unbiased with minimal sampling variance. Spatial patterns of heterozygosity arise because recombination causes variation in coalesence times among chromosomal regions. In effect, this leads to clustering of heterozygous sites in long stretches of DNA that by chance have experienced little recombination and have long coalescence times. For any distance L (in nucleotides) between sites, Δ_L is defined as the deviation of the frequency of pairs of nucleotide sites with mixed zygosities from the random expectation

$$\Delta_L = 1 - \frac{H_{1d}}{2\pi(1-\pi)} \tag{4.12}$$

with H_{1d} denoting the fraction of pairs of sites at distance L containing one heterozygote and one homozygote, and $2\pi(1-\pi)$ being the expected fraction of such mixed pairs under a random distribution given an average level of heterozygosity π.

For the situation in which the genome-wide patterns of variation are largely driven by mutation, recombination, and genetic drift, and the population is in equilibrium, by using expressions from Ohta and Kimura (1969b) for the two-allele model, it can be shown that

$$E(\Delta_L) \simeq \frac{\theta(1+2\theta)(18+\rho_L)}{2(1+\theta)A} \tag{4.13a}$$

where

$$A = 9 + 6.5\rho_L + 0.5\rho_L^2 + 19\theta\rho_L + 12\theta^2\rho_L + \theta\rho_L^2 + 54\theta + 80\theta^2 + 32\theta^3 \tag{4.13b}$$

(Lynch et al. 2014), where $\rho_L = 4N_e c_L$ is the scaled population recombination rate for sites separated by distance L (and having recombination rate c_L). Note that as $\rho_L \to 0$, $E(\Delta_L) \to \theta(1+2\theta)/(1+7\theta)$, which is closely approximated by θ when $\theta \ll 1$ (which, as noted above, is generally the case). As $\rho_L \to \infty$, $E(\Delta_L) \to \theta(1+\theta)/\rho_L \simeq \theta/\rho_L$. Thus, given an estimate of θ, with estimates of average Δ_L for neutral sites separated by $L = 1, 2, 3, \ldots$ sites, each based on thousands to millions of pairs of sites, the decline in Δ_L with L can be used to infer the distance dependence of ρ_L.

Another potentially powerful method for estimating ρ with population-genomic data takes advantage of the standardized linkage disequilibrium (r^2) introduced in Chapter 2. For neutral sites in drift-mutation equilibrium, Equation 2.29a gives a full expression for r_L^2 in terms of θ and ρ_L. However, provided $\theta \ll 1$ (which is always the case) and $\rho_L \gg \theta$ (which, as shown below, is generally the case for physically distant sites), Equation 2.29a simplifies to

$$r_L^2 \simeq \frac{10+\rho_L}{(11+\rho_L)(2+\rho_L)} \simeq \frac{1}{2+\rho_L} \tag{4.14}$$

The simplification to the right of this equation (Hill 1975; McVean 2002), which causes no more than 10% bias in estimating ρ_L, is often relied on in the literature (Hayes et al. 2003; Tenesa et al. 2007). We noted in Example 2.7 that another commonly used approximation, $r_L^2 \simeq 1/(1+\rho_L)$, has a more restricted meaning, which limits its use with molecular data.

As the sampling variance for r_L^2 for single pairs of polymorphic sites is generally very high, the usual strategy is to procure a large number of estimates for different pairs of informative markers separated by a certain window of physical distance, and then to pool these into a single estimate for that distance. Subtracting an expected contribution $1/n$ to r_L^2 resulting from finite sample size (Weir and Hill 1980), and rearranging Equation 4.14, leads to the estimator for sites separated by distance L,

$$\widehat{\rho}_L = \frac{1}{\widehat{r_L^2} - (1/n)} - 2 \tag{4.15}$$

A significant problem, often unappreciated, is that estimates of r_L^2 can be substantially biased if sample sizes are small or allele frequencies are extreme (Song and Song 2007).

Before proceeding, it is useful to review the specific mechanics of recombination between nucleotide sites, as we have not yet clarified how the recombination rate scales with distance L. Although it is often assumed that the recombination rate is simply equal to the crossover rate between sites, this is generally not true for closely spaced sites. Recombination events nearly always involve heteroduplex formations between homologous chromosomes, i.e., the temporary physical annealing of homologous regions of complementary strands (usually no more than a few hundred base pairs). When such heteroduplexes contain heterozygous sites, the nonmatching sites have to be resolved by **gene conversion**. Inclusion of these processes in the interpretation of the recombination rate is essential because although recombination events result in the potential for gene conversion, not all gene conversion events are accompanied by crossovers. Because gene-conversion tracts are relatively short, when sites are far apart, most recombination events result from crossing over, but when sites are close together, recombination mostly results from the conversion of single sites.

To understand this in a more quantitative way, let c_0 be the total rate of initiation of recombination events per nucleotide site (with or without crossing over), L be the number of sites separating the two focal positions (with $L = 1$ for adjacent sites), and x be the fraction of recombination events accompanied by crossing over. Using Haldane's (1919) mapping function (LW Equation 14.3), which assumes random and independent recombination at all sites, the crossover rate can be represented as $0.5(1-e^{-2c_0xL})$, which is $\simeq c_0xL$ for $c_0xL \ll 1$, and asymptotically approaches 0.5 for large c_0xL. In the following, we assume distances between sites that are small enough that the crossover rate $\simeq c_0xL$.

As noted by Andolfatto and Nordborg (1998), from the perspective of two sites, a gene conversion event has consequences equivalent to a crossover if the conversion tract encompasses just one of the sites. Under the assumption of an exponential distribution of tract lengths with mean length T (in bp), the total conversion rate per site is $(1-x)c_0T(1-e^{-L/T})$ (Langley et al. 2000; Frisse et al. 2001; Lynch et al. 2014). The total recombination rate between sites separated by distance L is then

$$c_L \simeq c_0[xL + (1-x)T(1-e^{-L/T})] \tag{4.16a}$$

For $L \ll T$,

$$c_L \simeq c_0 L \tag{4.16b}$$

whereas for $L \gg T$,

$$c_L \simeq c_0 Lx \tag{4.16c}$$

These results show that the simple division of an estimate of ρ_L by L to obtain an estimate of the per-site parameter $\rho = 4N_ec_0$, a common practice, may yield rather different answers depending on the distance between sites; at large distances ρ specifically measures the population crossover rate between sites.

Empirical Observations

Applying the preceding methods to population samples, many attempts have been made to measure the population recombination rate $\rho = 4N_ec_0$, usually first estimating ρ_L at various distances between sites, and then dividing by L under the assumption that $c_L = c_0L$, i.e., assuming a linear relationship between the recombination rate and physical distance between sites. As noted above, this is a reasonable approximation provided the distance between sites is less than the average length of a conversion tract but will lead to an underestimate of ρ (by a factor of $1/x$) when greater distances are relied upon. Using this procedure, all estimates of the per-site parameter $4N_ec_0$ are smaller than 0.1, with many falling below 0.01 (Table 4.1). Because the fraction of recombination events resulting in crossing over (x) is typically in the range of 0.05 to 0.25 (as reviewed below), these general observations provide strong support for the idea that random genetic drift is generally a much more powerful force than recombination at the level of individual nucleotide sites.

By dividing estimates of $4N_ec_0$ by parallel estimates of $\theta = 4N_eu$, the effective population size cancels out, yielding an estimate of the ratio of recombination and mutation rates at the nucleotide level (c_0/u). All such estimates are smaller than 5.0, and nearly half are smaller than 1.0, implying that the power of recombination between adjacent sites is generally of the same order of magnitude or smaller than the power of mutation (Table 4.1). The average estimate of c_0/u for *Drosophila* is ~2.7, whereas that for humans is ~0.8. Average c_0/u for 14 land plants is 1.1 (SD = 1.2), although this may somewhat underestimate the average for purely outcrossing species because several of the taxa included in the survey (e.g., *Arabidopsis* and *Oryza*) are predominantly self-fertilizing, which reduces the effective amount of recombination (Hagenblad and Nordborg 2002).

It is notable that even though prokaryotes do not engage in meiosis, estimates of c/u for such species are generally of the same order of magnitude as those for eukaryotes (Lynch

Table 4.1 Estimates of the per-site population recombination rate ($\rho = 4N_e c_0$) and the ratio of the per-site recombination and mutation rates (c_0/u, obtained by dividing estimates of ρ by estimates of $\theta = 4N_e u$). All estimates are derived from population surveys of nucleotide variation at silent sites in protein-coding genes.

Species	ρ	c_0/u	References
Animals:			
Drosophila melanogaster	0.05846	3.545	Hey and Wakeley 1997 Andolfatto and Przeworski 2000
Drosophila pseudoobscura	0.08655	1.360	Hey and Wakeley 1997
Drosophila simulans	0.09720	3.306	Andolfatto and Przeworski 2000
Homo sapiens	0.00060	0.770	Frisse et al. 2001; Ptak et al. 2004 Lefebvre and Labuda 2008
Land plants:			
Arabidopsis thaliana	0.00160	0.193	Kim et al. 2007
Brassica nigra	0.00602	0.330	Lagercrantz et al. 2002
Cryptomeria japonica	0.00046	0.118	Fujimoto et al. 2008
Helianthus annuus	0.05280	4.100	Liu and Burke 2006
Hordeum vulgare	0.00080	1.417	Morrell et al. 2006
Oryza rufipogon	0.00003	0.006	Mather et al. 2007
Oryza sativa	0.00004	0.021	Mather et al. 2007
Persea americana	0.00338	0.582	Chen et al. 2008
Pinus sylvestris	0.01452	2.855	Pyhäjärvi et al. 2007
Pinus taeda	0.00175	0.266	Brown et al. 2004
Solanum chilense	0.02380	1.122	Arunyawat et al. 2007
Solanum peruvianum	0.03480	1.392	Arunyawat et al. 2007
Sorghum bicolor	0.00041	0.130	Hamblin et al. 2005
Zea mays	0.02840	2.176	Tenaillon et al. 2004

2007). This suggests that, relative to the background rate of mutation, recombination at the nucleotide level is not exceptionally low in prokaryotes, although the downward bias in estimates of θ for this group (noted above) may lead to inflated estimates of c_0/u.

Applying the single-individual estimator (i.e., the correlation of zygosity) and fitting the generalized recombination function (Equation 4.16a) clarifies a number of features of recombination in mammalian species (Lynch et al. 2014). First, estimates of $4N_e c_0$ in mammals are generally in the range of 0.001 to 0.005, again implying a substantially higher power of genetic drift than of recombination at the single-site level.

Second, the fraction of recombination events resulting in crossovers is generally in the range of $x = 0.05$ to 0.25. This is consistent with empirical work suggesting $x \simeq 0.30$ in the budding yeast *S. cerevisiae* (Malkova et al. 2004; Mancera et al. 2008), and $x = 0.15$ in the fly *D. melanogaster* (Hilliker et al. 1994). Indirect LD-based analyses have also led to estimates of $x \simeq 0.14$ in humans (Frisse et al. 2001; Padhukasahasram and Rannala 2013), $x = 0.08$ in *D. melanogaster* (Langley et al. 2000; Yin et al. 2009), $x = 0.05$ in the plant *A. thaliana* (Yang et al. 2012), and $x = 0.06$ to 0.16 in wild barley (Morrell et al. 2006). Thus, observations in a variety of organisms consistently point to the fact that the vast majority of recombination events are simple local gene-conversion events unaccompanied by crossovers, raising questions about the frequently used assumption that $c_L = c_0 L$.

Third, based on single-individual LD analysis, the inferred average lengths of conversion tracts in mammals are typically in the range of $T = 10^3$ to 10^4 bp, which is consistent with more direct observations made in other species: $T \simeq 400$ bp in bacteria (Santoyo and Romero 2005); $T = 500$ to 4000 bp in *S. cerevisiae* (Ahn and Livingston 1986; Judd and Petes 1988; McGill et al. 1990); $T = 400$ to 1400 bp in *D. melanogaster* (Hilliker et al. 1994; Preston and Engels 1996; Miller et al. 2012); and $T = 200$ to 3000 bp in mammals (Chen et al. 2007; Paigen et al. 2008; Rukść et al. 2008). Referring to Equations 4.16b and 4.16c, this suggests that the assumption of $c_L = cL$ is generally approximately valid provided $L < 500$ bp, whereas for distances > 5000 bp, the recombination rate primarily reflects the crossover rate, i.e., $c_L \simeq c_0 Lx$.

Finally, the results in Lynch et al. (2014) suggest that the level of LD at closely spaced sites (< 200 bp) in vertebrates is generally much higher than can be accounted for by the standard neutral model outlined above, a conclusion that was reached in a number of other studies: *Drosophila* (Andolfatto and Przeworski 2000); humans (Przeworski and Wall 2001); sorghum (Hamblin et al. 2005); and *Arabidopsis* (Kim et al. 2007). There are at least three reasons why unusually high levels of LD may exist at closely spaced sites, all associated with the nonindependence of mutational and/or recombinational events. First, new mutations arise in a significantly clustered manner on spatial scales of $\sim$100 bp, possibly as a consequence of an occasional defective polymerase engaging at origins of replication or of the localized deployment of error-prone polymerases in DNA repair (Schrider et al. 2011; Harris and Nielsen 2014). Second, recombination and double-strand-break repair are mutagenic, violating the usual assumption of the independence of these two processes (Hicks et al. 2010; Malkova and Haber 2012; Arbeithuber et al. 2015). Third, nonhomologous gene conversion can introduce excess LD at individual sites (Walsh 1988; Mansai and Innan 2010).

EFFECTIVE POPULATION SIZE

Although the theory outlined in Chapter 3 suggests numerous ways in which the effective size of a population might be estimated from demographic data, such information is often difficult to come by, except in carefully controlled breeding populations. Moreover, estimates of N_e based on demography alone generally do not incorporate the long-term effects of selection on linked chromosomal regions, and certainly not selective sweeps or background selection.

Nevertheless, there are several ways in which inferences about N_e can be made without direct demographic observation. From the standpoint of natural populations, two approaches harbor the most promise—monitoring temporal changes in putatively neutral allele frequencies, and ascertaining genome-wide patterns of LD, in both cases back-calculating the value of N_e that best explains the data (reviewed by Wang 2005).

Temporal Change in Allele Frequencies

Consider a **single nucleotide polymorphism** (usually abbreviated as a **SNP**) sampled on two occasions separated by t generations, with initial frequency p_0, and recall from Chapter 2 that the expected variance in allele-frequency change after t generations is $p_0(1-p_0)(1-e^{-t/(2N_e)}) \simeq p_0(1-p_0)t/(2N_e)$ for small $t/(2N_e)$. This represents only the true population variance (the evolutionary variance in the preceding parlance), to which the sampling variance associated with errors in *observed* allele-frequency estimates (owing to finite sample size) must be added. Summing these two sources of stochasticity yields an overall estimate of the expected variance of allele-frequency change of $p_0(1-p_0)[t/(2N_e) + 1/(2n_0) + 1/(2n_1)]$ between two time points, where n_0 and n_1 denote the number of individuals (assumed to be diploid) genotyped in the two generations. Letting $\widehat{p}_0$ and $\widehat{p}_1$ be the estimated allele frequencies in the two generations, an estimate of the observed variance in allele-frequency change across generations can be written as $(\widehat{p}_1 - \widehat{p}_0)^2$ because $E(\widehat{p}_1 - \widehat{p}_0) = 0$ under neutrality.

Krimbas and Tsakas (1971) suggested that by equating the observed and expected variance of allele-frequency change and rearranging, the effective population size can be estimated from observations over two consecutive generations ($t = 1$)

$$\widehat{N}_e = \frac{1}{2\widehat{F}_1 - (1/n_0) - (1/n_1)} \tag{4.17a}$$

where

$$\widehat{F}_1 = \frac{(\widehat{p}_0 - \widehat{p}_1)^2}{\widehat{p}_0(1 - \widehat{p}_0)} \tag{4.17b}$$

is a measure of the standardized variance of allele-frequency change. Provided $t/(2N_e) \ll 1$, the same expression applies when samples are made t generations apart, if t is substituted for one in the numerator of Equation 4.17a. (Note that the definition of F_1 is identical in form to the population-subdivision statistic F_{ST}, presented as Equation 2.42, except that the latter is concerned with spatial rather than temporal variation.)

Despite their intuitive nature, Equations 4.17a and 4.17b yield biased estimates because the contributions of the sampling variance (and in some cases, covariance) of allele frequencies to F_1 are not fully accounted for (Pamilo and Varvio-Aho 1980; Nei and Tajima 1981b; Pollak 1983; Tajima and Nei 1984; Waples 1989a). Additional limitations are that $\widehat{F}_1$ is undefined if $\widehat{p}_0 = 0$, and that Equations 4.17a and 4.17b do not immediately allow for the incorporation of multiple alleles. An alternative estimator that deals with these problems is

$$\widehat{N}_e = \frac{t - 2}{2\widehat{F} - (1/n_0) - (1/n_1)} \tag{4.18a}$$

where $\widehat{F}$ is calculated by either

$$\widehat{F}_2 = \frac{1}{k} \sum_{i=1}^{k} \frac{(\widehat{p}_{0i} - \widehat{p}_{1i})^2}{[(\widehat{p}_{0i} + \widehat{p}_{1i})/2] - \widehat{p}_{0i}\widehat{p}_{1i}} \tag{4.18b}$$

(Nei and Tajima 1981b), or

$$\widehat{F}_3 = \frac{1}{k} \sum_{i=1}^{k} \frac{(\widehat{p}_{0i} - \widehat{p}_{1i})^2}{(\widehat{p}_{0i} + \widehat{p}_{1i})/2} \tag{4.18c}$$

(Pollak 1983), where k is the number of alleles. The details leading up to these alternative expressions can be found in the primary references, but it is notable that because $(\widehat{p}_{0i} + \widehat{p}_{1i})/2$ is generally much larger than $\widehat{p}_{0i}\widehat{p}_{1i}$, both estimators usually lead to very similar results (Waples 1989a). One drawback of Equation 4.18a is that it requires an interval of at least three generations.

More refined measures of F can be obtained by averaging estimates of F_1, F_2, or F_3 over multiple loci, and Pollak (1983) derived a generalized estimator that allows for sampling across more than a single time interval. All of these approaches assume that the sampling of individuals at the beginning of an interval has no effect on the allele-frequency variance, which is reasonable when samples constitute a minor fraction of the population or are taken in a nondestructive manner or following reproduction. An additional concern is that the sampling scheme for allele frequencies, which is straightforward in a synchronized population with discrete generations but potentially problematical in species with overlapping generations. In the latter case, the contributions of sampled individuals to the overall allele-frequency estimates need to be weighted by the reproductive values of various age classes (Waples and Yokota 2007), a difficult enterprise with species with poorly understood life histories. Attention to these issues is provided in Nei and Tajima (1981b) and Waples (1989a).

Estimates of N_e derived by these method-of-moment estimators generally have substantial sampling variances, and negative estimates of N_e are even possible. Clearly, if $t/(2N_e) \ll 1/(2n_0) + 1/(2n_1)$, observed fluctuations in allele frequencies will be largely a consequence of sampling error, so the utility of the overall approach becomes diminishingly small in populations with large effective sizes. Assuming equal sample sizes for each locus, the sampling variance of $\widehat{N}_e$ is

$$\text{Var}(\widehat{N}_e) \simeq \left(\frac{8N_e^4}{t^2 M}\right) \left(\frac{1}{4N_e^2} + \frac{1}{N_e t\tilde{n}} + \frac{1}{t^2\tilde{n}^2}\right) \tag{4.19}$$

where M denotes the number of independent allelic comparisons and $\tilde{n}$ is the harmonic mean of the per-locus sample sizes in the two generations (Pollak 1983). In general, M, t, and $\tilde{n}$ will be under the control of the investigator, so Equation 4.19 provides a useful basis for designing an optimal sampling strategy. For example, a doubling of M will reduce the sampling variance by one half, whereas a doubling of the sampling interval (t), which may often be less costly, has a much greater effect.

The sampling distribution of the composite function $M[\widehat{F}/E(F)]$, where $E(F)$ denotes the expected value, is expected to be approximately χ^2 in form, with M degrees of freedom (Lewontin and Krakauer 1973; Nei and Tajima 1981b), and this fact can be used to construct confidence intervals for N_e by substituting the critical χ^2 values for $\widehat{F}$ into Equation 4.18a, e.g., using the values of F at the 2.5% and 97.5% cumulative probability levels to yield 95% confidence limits. However, using computer simulations, Goldringer and Bataillon (2004) found that the χ^2 assumption can be significantly violated when there is a minor allele with frequency < 0.1, there is large number of alleles (as with microsatellites) or the number of generations between sampling times is large. In Chapter 9, the issue of temporal change in allele frequencies will be revisited from a different perspective—testing the hypothesis that an observed magnitude of change is inconsistent with random genetic drift for an assumed value of N_e, or equivalently estimating the largest value of N_e that is consistent with the observed change being entirely due to drift.

Because of their simple heuristic interpretations, method-of-moments estimators, like those just noted, are highly popular approaches for estimating population parameters. However, on a single summary statistic, such methods do not fully utilize the information in a set of samples. A more powerful approach to estimating N_e from sequential samples involves the use of ML procedures (and their Bayesian extensions) to yield estimates that best explain the entire distribution of observed allele frequencies conditional on sample sizes (Williamson and Slatkin 1999; Anderson et al. 2000; Berthier et al. 2002). These methods are highly demanding computationally, a demand that increases with N_e, although Wang (2001), Beaumont (2003), Tallmon et al. (2004), Anderson (2005), and Bollback et al. (2008) have presented computationally efficient approximations.

Single-sample Estimators

Because of the practical difficulties in obtaining temporal sequences of samples, especially in species such as vertebrates and land plants with long generation times, a number of methods have been developed for estimating N_e from the information contained in just a single sample. Only a brief overview of such methods will be provided here. One of the most commonly applied single-sample estimators is the LD method, already outlined above. Under the assumption of drift-mutation-recombination equilibrium, estimates of $\rho_L = 4N_ec_L$ can be obtained, so if the recombination rate between the loci under consideration is known, then $\widehat{\rho}_L/(4c_L)$ will provide an estimate of N_e (Hill 1981). Likewise, if the mutation rate per nucleotide site (per generation) is known, any estimate of $\theta = 4N_eu$ can converted to an estimate of long-term N_e using $\widehat{\theta}/(4u)$. Again, being based on simple summary statistics, these estimators do not utilize all of the information inherent in a sample of alleles, and hence are not likely to provide the most efficient estimates of N_e. Alternative, highly computationally intense coalescent sampling methods have been developed to estimate various population-genetic parameters, including N_e, mutation and recombination rates, and other demographic parameters (e.g., population growth rates and degree of subdivision), using the genealogical information inherent in samples of population sequences (Kuhner 2008).

A second approach, applicable only to randomly mating species, relies on observed amounts of excess heterozygosity relative to Hardy-Weinberg expectations. The basis of this procedure is the random deviation in allele frequency that develops among the two sexes as a consequence of stochastic sampling of gametes in the preceding generation. In effect, the two sexes are being viewed as two random samples of gametes here—and the smaller the value of N_e, the larger the expected deviation between the sexes (Robertson 1965;

Pudovkin et al. 1996; Luikart and Cornuet 1999). Such variation among the sexes causes excess heterozygosity in the progeny generation by elevating the likelihood that each sex will contribute an alternate allele to the offspring.

A third single-sample method attempts to estimate the fraction of pairs of randomly sampled offspring from a population that are either full or half-sibs (Wang 2009). This method relies on statistical procedures for deriving estimates of relatedness with molecular markers. As in the case of the heterozygosity-excess approach, information on a large number of informative markers is required, and a random sampling scheme is essential. These last two approaches are restricted to very small population sizes (on the order of 100 reproductive adults or fewer), which are required to generate detectable deviations from Hardy-Weinberg expectations and detectable numbers of sib pairs.

Empirical Observations

In Chapter 3, we found that the numerous demographic factors influencing the effective size of a population almost always do so in a downward direction. Applications of the methods outlined above provide some indication as to the magnitude of this reduction relative to the actual size of a population (N). Because the temporal-fluctuation method requires a small enough N_e to yield meaningful results on a reasonable time scale, not surprisingly, almost all estimates using this technique derive from large-bodied (relatively small N) vertebrate species. In a survey of studies on mostly low-fecundity species, Frankham (1995) found an average N_e/N of $\sim$0.11, whereas a subsequent study with a much larger sample obtained an average of 0.14 (Palstra and Ruzzante 2008).

It is likely that the $\sim$90% reduction in N_e suggested by these studies is a considerable *underestimate* of the situation for many nonvertebrate species and even many vertebrates. For example, as noted in Chapter 3, high-fecundity fish in spatially variable environments appear to have $N_e/N < 0.001$ (Hedrick 2005). In addition, many unicellular species have conspicuous phases of asexual reproduction that can encourage the rapid proliferation of a small number of clones, generating N_e/N ratios much lower than 0.001 via genetic hitchhiking. Strongly inbreeding species (e.g., self-fertilizing plants) may also approach such extremes. Finally, one of the major shortcomings of the temporal-fluctuation approach to estimating N_e may be its tendency to overlook rare, but quantitatively significant, phases in which genomic regions are exposed to strong selective sweeps at linked loci (Chapter 8).

Example 4.1. Hill (1981) noted that estimates of N_e based on the amount of standing LD between tightly linked markers are more a function of the long-term population history while LD measures between more loosely linked markers are reflective of recent history. Hayes et al. (2003) seized upon this observation to suggest that by using estimates of $\rho_L = 4N_e c_L$ for different values of c_L (i.e., known genetic-map distances between sites), one could, in effect, estimate the effective population sizes at different times in the past. In particular, for a model of linear population-size change (growth or decline), they suggested that LD between markers with recombination frequency c_L yields estimates of N_e at roughly $1/(2c_L)$ generations in the past.

To apply this approach, Tenesa et al. (2007) scored $\sim 10^6$ SNPs to examine LD at various intrachromosomal distances for four different human populations, and Figure 4.3a shows the result for a Utah population of European ancestry. For given slices of time, the various points indicate the separate estimates based on each of the 22 autosomes. Note both the consistency of estimates over autosomes and the very recent expansion of population size. Similar studies in humans were performed by Sved et al. (2008) and McEvoy et al. (2011). In contrast, when Hayes et al. (2003) and Flury et al. (2010) applied this approach to modern dairy cattle, they concluded that N_e had dramatically *declined* from historical values, presumably reflecting the bottlenecking effects of selection for improved milk production (Figure 4.3b).

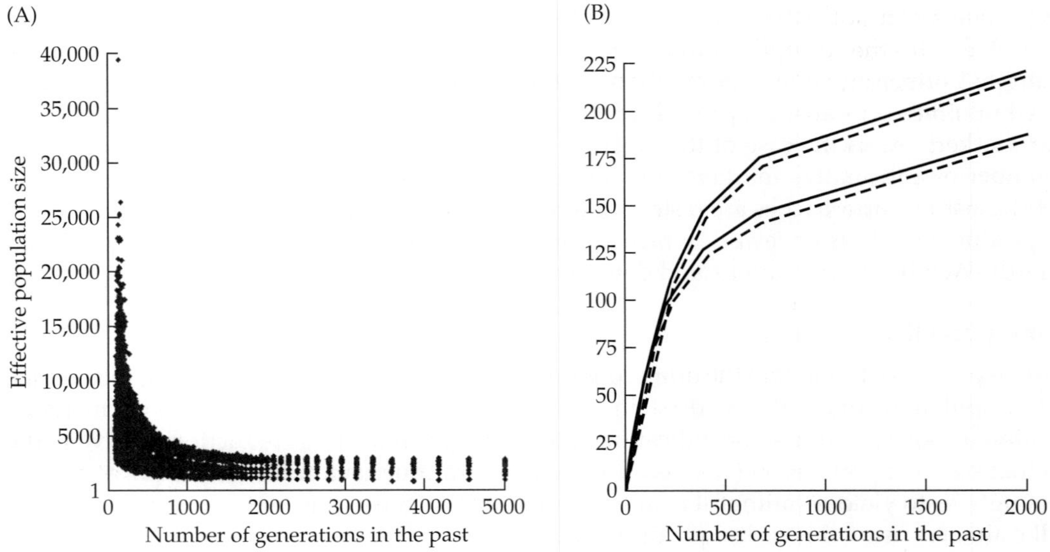

Figure 4.3 Estimates of historical values of N_e using linkage-disequilibrium between large numbers of pairs of markers with different genetic-map distances. The estimates were pooled into categories with different values of c (between markers), with the bin-specific values of $1/(2c)$ serving as estimates of the time in the past (in generations) for which the categories provide estimates of N_e. The latter is calculated by using the simplified version of Equation 4.14 given the average estimate of r^2 and c for the bin. **A:** Estimates of historical changes in N_e for a Utah population of European extraction. For a given generation time slice, the points represent an estimate based on the markers from each of the 22 human autosomes. Note the rapid increase in N_e in the recent past. (After Tenesa et al. 2007.) **B:** Estimates for the Swiss Eringer breed of cattle. Here, the different curves represent different assumptions used to correct estimates of ρ for sampling effects and different estimates of the fine-scale recombination rates. Regardless of the assumptions involved, it is clear that in contrast to the results for the human population, N_e has dramatically declined over the past 500 years. (After Flury et al. 2010.)

MUTATION RATE

The long-term evolution of complex traits ultimately depends on the input of new variation via mutation, which is a function of the rate at which new mutations arise at the DNA level and their influence at the phenotypic level, the combined effects defining the overall rate of polygenic mutation (LW Chapter 12). Here, we continue to focus specifically on the DNA-sequence level, with u being defined as the rate of mutation per nucleotide site per generation. Because mutations arise at an extremely low rate at most nucleotide sites, the direct estimation of u is formidably challenging, with most approaches relying on procedures that enrich the pool of experimentally derived mutations in an effectively neutral fashion (so that selection does not bias the outcome). Here, we review the two most commonly used methods of enrichment: (1) long-term genome-wide accumulation of mutations in isolated lineages with tiny effective population sizes; and (2) short-term isolation of conspicuous mutants at single marker loci from large populations raised on selective media.

Divergence Analysis

The most conceptually simple approach to estimating u, which is frequently applied to multicellular organisms with fairly long generation times, is to perform a mutation-accumulation experiment (LW Chapter 12), whereby a set of initially genetically identical (and usually homozygous, if not clonal) lines are passed through repeated population bottlenecks. For example, with the self-fertilizing nematode *Caenorhabditis elegans* and the plant *Arabidopsis thaliana*, an ancestral line can be repeatedly selfed to ensure homozygosity, with the progeny of one parent being used to synchronously initiate a set of parallel lines, each subsequently maintained by single-progeny descent. With each line having an effective population size of just one individual under this design, essentially all mutations that do not cause lethality or complete sterility (the vast majority of mutations) will accumulate independently at a rate u per site, in accordance with the neutral theory (Chapter 2). Under self-fertilization, newly arisen mutations are fixed or lost in just two generations on average, so after several dozens to hundreds of generations of mutation accumulation, nearly all mutations can be detected as fixed homozygotes by sequencing a subset of lines. Typically, nearly all lines will be identical at individual nucleotide sites (reflecting the ancestral state), with mutations in specific lines appearing as single-line outliers.

Letting n denote the number of sites surveyed, L the number of lines, T the average number of generations per line, and m the number of observed mutations summed over lines and sites, the mutation rate per site is estimated as

$$\widehat{u} = m/(nLT) \tag{4.20a}$$

with sampling variance of

$$\sigma^2(\widehat{u}) \simeq \widehat{u}/(nLT) \tag{4.20b}$$

The latter expression implies a coefficient of sampling variation for $\widehat{u}$ of $(unLT)^{-1/2}$, which is the inverse of the square root of the expected number of observed mutations in the assay.

Example 4.2. A commonly used variant of the laboratory mutation-accumulation experiment for estimating mutation rates exploits the information inherent in natural populations, relying on presumptively neutral sequences from isolated but closely related species. Recall from Chapter 2 that the long-term rate of nucleotide substitution at neutral sites is equal to the mutation rate regardless of N_e, and from above that the average nucleotide heterozygosity of random sites within a species has expected value $4N_eu$. Thus, for two sister taxa that became isolated t generations in the past, the expected divergence of orthologous neutral sequences (number of substitutions per site) is $d = 2tu + 4N_eu$, assuming equal N_e in both taxa. At $t = 0$, $d = 4N_eu$ (the average divergence of randomly sampled alleles in the ancestral population), whereas as $t \to \infty$, $d \simeq 2tu$ (a widely used approximation in applications of molecular clocks for dating evolutionary events). Rearranging, and using $\overline{\theta}_H$, the average within-species nucleotide diversity at silent sites as the estimate of $4N_eu$, we obtain an estimator for the mutation rate, $\widehat{u} = (\widehat{d} - \overline{\theta}_H)/(2t)$.

Nachman and Crowell (2000) used this approach to obtain an estimate of the mutation rate for humans from sequences of 12 unexpressed pseudogenes in human and chimpanzee. Because they are nontranscribed, such stretches of DNA are expected to fulfill the assumptions of neutrality. The average number of substitutions per site separating the two species was $\widehat{d} = 0.0133$. A broad geographic survey of within-species variation in 49 noncoding (and presumably largely neutral) regions yielded estimates of 0.00087 for human and 0.00134 for chimpanzee (Yu et al. 2003), implying $\overline{\theta}_H = 0.00110$. Nachman and Crowell assumed a divergence time of 5 million years, and an average generation time of 20 years, implying $t \simeq 250,000$ generations. Substitution into the preceding expression then gives an estimated mutation rate of 2.44×10^{-8} per site per generation for base-substitution mutations. Strictly speaking, this is an average over the chimpanzee and human lineages.

Short-term Enrichment

The preceding approach employs a strategy of augmenting the pool of observable mutations by passing lines through a large number of generations. The advantage of such a protocol is that mutations are equally enriched throughout the genome, minimizing the chances that the mutational profile will be biased by making observations at any particular target locus. However, with genome-wide mutation-accumulation analysis, an enormous number of sites (typically many tens of millions) need to be searched to obtain just a few dozen mutations.

An alternative approach, which is widely applied to microbial cultures, focuses on reporter constructs (specific marker loci at which at least a subset of mutations causes obvious phenotypic changes). Here, the emphasis is on the efficient screening of a very large pool of cells in a relatively short period of time for a small subset of mutations, e.g., exponentially growing an initially nonmutant stock to a population size in excess of the reciprocal of the mutation rate (so there will be more than one mutational event in the culture), and then isolating the subset of cells that have acquired a mutation at a locus that is nonessential in the background environment but permits subsequent growth on a selective medium (Luria and Delbrück 1943). From estimates of the total number of mutant and nonmutant cells in the culture, it is then possible to determine the mutation rate per cell division.

For the marker approach to yield reliable estimates of u, a good deal of knowledge must exist on the molecular features of the target locus. Because mutant cells reproduce during culture expansion, the relationship between the number of mutant cells observed in a population and the actual number of mutational events that produced them is generally not one-to-one. Thus, the first challenge is to convert the observed number of mutant cells to the number of mutations leading to them (m). In addition, because not all mutations produce an observed phenotype, the second challenge is to determine the fraction of mutations that are detectable at the target locus (d). The true number of mutations is estimated by m/d. Finally, in order to determine the mutation rate per nucleotide site, one must know the mutational target size (n, in base pairs).

Several methods exist for estimating the number of unique mutational events from the observed numbers of mutant and nonmutant cells in short-term experiments, with broad overviews provided by Rosche and Foster (2000) and Angerer (2001a, 2001b). Suppose a large series of replicate cultures is developed, and one then simply scores the fraction of cultures at the end point that are completely free of mutations (p_0). Assuming that the number of mutational events per culture is Poisson distributed with expectation m, the expected frequency of mutation-free cultures is then simply

$$E(p_0) = e^{-m} \tag{4.21}$$

Rearrangement leads to the estimator $\widehat{m} = -\ln(p_0)$, which ignores the sampling bias resulting from the error in estimating p_0. This approach works well when m is on the order of 0.5 to 2.5, but with more extreme values, p_0 will be close enough to 0.0 or 1.0 that meaningful estimates are not possible unless the number of cultures is enormous. A second disadvantage of this approach is its failure to use most of the information in the set of cultures, as the distribution of mutant numbers among replicate cultures is completely ignored. Full use of such information can be incorporated into a maximum-likelihood framework (e.g., Lea and Coulson 1949; Sarkar et al. 1992).

Example 4.3. The mutation rate in an exponentially growing culture can be estimated by considering the expected temporal dynamics of the frequency of mutant cells in the population. Letting f_0 be the initial frequency of mutations, r be the rate of exponential growth of the numbers of cells in the culture (assumed to be identical for cells that are mutant and nonmutant at the marker locus), and u_o be the rate of mutation to an observable phenotype per cell division, the expected frequency after t time units is

$$f_t = f_0 + (1 - f_0)(1 - e^{-u_o rt}) \tag{4.22a}$$

This follows from the fact that $e^{-u_o rt}$ is the probability that a descendant of a nonmutant cell has not acquired a detectable mutation after rt cell divisions. Note that if one starts with a mutation-free culture ($f_0 = 0$) and the cumulative probability of mutation ($\simeq u_o rt$) is $\ll 1$, the expected fraction of mutant cells will increase in an essentially linear fashion at a rate of $u_o r$.

Because of the stochastic nature of mutations, results from single cultures are not reliable with this approach. Thus, motivated by the original design of Luria and Delbrück (1943), most studies of microbial mutation grow a moderate number of initially (putatively) mutation-free cultures up to an arbitrarily large population size and then survey the frequency of mutants at the end point of each culture. Rearrangement of the preceding expression yields the relevant point estimator of the mutation rate to observable phenotypes,

$$\widehat{u}_o = -\frac{\ln[(1 - f_0)/(1 - f_t)]}{rt} \tag{4.22b}$$

Because $N_t = N_0 e^{rt}$ under exponential growth, where N_0 and N_t are the total numbers of cells in the culture at times 0 and t, so long as the observed mutant frequencies are < 0.1, so that $\ln(1 - f) \simeq -f$, Equation 4.22b further simplifies to

$$\widehat{u}_o \simeq \frac{f_t - f_0}{\ln(N_t/N_0)} \tag{4.22c}$$

which is simply the rate of accumulation of observable mutations per cell division.

Drake (1991) argued that this essentially deterministic view of the rate of increase of mutants is unlikely to hold very well until a culture has reached a large enough size to harbor at least some mutations, which is expected to take several generations. Taking the view that a reasonable benchmark is the point at which a culture is expected to contain a single mutant, which implies $u_o N = 1$, $f_0 = u_o$ and $N_0 = 1/u_o$ can be used as an arbitrary starting point, which, after substitution into Equation 4.22c, leads to

$$\widehat{u}_o \simeq \frac{f_t - \widehat{u}_o}{\ln(\widehat{u}_o N_t)} \tag{4.22d}$$

Given just the total number of cells, N_t, and the frequency of mutants at the end point, f_t, this expression can be solved recursively to obtain the estimate $\widehat{u}_o$. When data are available from multiple cultures, f_t is generally taken to be the *median* frequency of mutants, as the mean can be strongly biased if the sample includes any "jackpot" cultures that happened to have acquired a mutation during an early cell division.

Conversion of the rate of origin of *observable* mutations, u_o, to an estimate of the mutation rate at the nucleotide level requires that the fraction of detectable mutations at the marker locus (d) be known. Many mutations have no phenotypic effects, e.g., because they arise at silent sites or at amino-acid replacement sites that have no substantive effect on the causal locus. To determine the fraction of undetectable mutations, a large number of independent mutant cells can be sequenced to ascertain the molecular basis of the changes at the target locus and the degree to which these are concentrated at particular sites. Generally, because the mutation rate per nucleotide site is quite low, no more than a single change is found within any particular sequenced gene, so there is little ambiguity as to the identity of causal mutations.

For base-substitutional mutations, Drake (1991) made the following argument for obtaining an estimate of d. Assuming that all mutations causing premature translation termination (so-called nonsense mutations) cause functional changes that are detectable, and then letting n_n denote the number of such mutations observed in the sequenced sample, the expected total number of base-substitutional mutations per sequence in the sample (whether recorded as mutants or not) is $64n_n/3$. This follows from the fact that of the 64 possible triplet codons, three encode for chain termination (in most species), and it assumes random mutation to all 64 codons. Thus, letting n_o denote the total number of observed base-substitutional mutations in the set of sampled sequences (missense and nonsense mutations), $\widehat{d} = n_o/(64n_n/3)$ provides an estimate of the fraction of base-substitutional mutations that are detectable (if all detected base-substitutional mutations were to termination codons, implying no effects of missense mutations, $n_o/n_n = 1$, and $\widehat{d} = 3/64$). If n is the length of the target sequence (in base pairs) over which mutations are detectable (generally assumed to be the length of the coding region, which could be an overestimate), an estimator for the base-substitution mutation rate per nucleotide site is then

$$\widehat{u} = \frac{\widehat{u}_o}{\widehat{d}n} \tag{4.22e}$$

Example 4.4. To indirectly estimate the human mutation rate, Kondrashov (2003) took advantage of records on genetic pathologies attributable to dominant mutations at known causal loci. The population frequency of genetic disorders (I, incidence) caused by dominant autosomal mutations provides a simple basis for estimating the mutation rate to defective alleles. This is because the expected frequency of a dominant deleterious allele under selection-mutation balance is $p \simeq u/s$, where u is the mutation rate to defective alleles (per gene copy), and s is the selective disadvantage of affected (heterozygous) individuals (Equation 7.6b). For a severe disorder, the frequency of the deleterious allele will be so small that essentially all affected individuals are heterozygotes, implying an incidence of the disorder very close to $2p(1-p) \simeq 2p = 2u/s$. Thus, the mutation rate to dominant defective alleles can be estimated as $sI/2$. (For a dominant mutation that leads to complete loss of reproductive fitness, $s = 1$, and the incidence is simply equal to $2u$, as each functional parental allele has a probability u of mutating to a defective product.)

The remaining challenge is to convert the total rate of observed mutations at a locus to the underlying rate at the level of individual nucleotide sites. This can be accomplished by employing a strategy similar in spirit to that advocated by Drake (1991). For each disorder in the survey of Kondrashov (2003), a large sample of affected individuals (whose parents were known to be nonmutant) had both of their alleles sequenced to identify the nature of the newly arisen, causal mutations. Assuming all insertion and deletion mutations had detectable effects, the total detectability of mutations could then be calculated from the incidence of chain-terminating base-substitutional mutations, as outlined in Example 4.3. Although Kondrashov's (2003) survey involved 20 different genetic disorders (each determined by a unique locus), we will simply present the calculations for one such analysis, and conclude with a summary of all results.

Familial adenomatous polyposis is a genetic disorder known to be caused by dominant mutations in the adenomatous polyposis coli (APC) tumor-suppressor gene, arising at an estimated rate of $u_o = 7 \times 10^{-6}$ per gene copy per generation. Of the 799 mutations validated by sequencing and deemed to be causal, 202 involved nonsense base substitutions, with the remaining 597 being associated with major lesions, insertions, and deletions of various sorts. Assuming that the total number of base substitutional mutations (when extrapolated to unaffected mutants) is $202 \times (64/3)$, and that all insertions and deletions are detectable, the overall detectability is estimated as $799/[597+(202\times 64/3)] = 0.163$. From the pool of affected individuals subjected to sequencing, a fraction 0.325 exhibited no causal mutation (presumably because the mutation resided outside of the sequenced target exons, which summed to 4803 sites). The estimated total mutation rate at the locus is therefore $(7 \times 10^{-6}) \times 0.675/(4803 \times 0.163) = 6.0 \times 10^{-9}$ per site per generation, a fraction of which, $1 - \{597/[597 + (202 \cdot 64/3)]\} = 0.878$, involves base-substitutional mutations.

When these approaches are extended to the remaining 31 loci, the estimated average total mutation rate to base-substitutional changes is 1.70×10^{-8} per site per generation, averaged over both sexes. A subsequent estimate involving a larger number of loci underlying human genetic disorders and somewhat different assumptions yielded an estimate of 1.29×10^{-8} (Lynch 2009b).

More recently, direct estimates of the human mutation rate have been generated by whole-genome sequencing in known lines of descent. For example, from information on portions of Y chromosomes separated by 13 generations of paternal-line descent, Xue et al. (2009) obtained a base-substitutional mutation rate estimate of 1.73×10^{-8} after scaling across the sexes to account for the lower rate of mutation in females. Three additional studies, involving autosomal sequences of parent-offspring trios, all yield sex-averaged estimates close to 1.2×10^{-8} (Conrad et al. 2011; Campbell et al. 2012; Kong et al. 2012).

Taken together, these estimates point to a sex-averaged base-substitutional mutation rate of $\sim 1.4 \times 10^{-8}$ for humans, which is significantly lower than the phylogenetic estimate reported in Example 4.2 (2.44×10^{-8}). A number of factors might account for the elevated rate based on interspecies divergence: an incorrect estimate of the time of divergence between the human and chimpanzee lineages; an incorrect estimate of the amount of heterozygosity at initial divergence; inaccurate estimates of average generation times since the time of divergence; an elevated rate of mutation in the chimpanzee lineage; a recent decline in the human mutation rate; the operation of some selection on the sites analyzed in the comparative study; etc. The main point is that estimates of the mutation rate derived from phylogenetic data are subject to numerous sources of potential error, the magnitude of which is generally unknown (and in some cases unknowable). Short-term studies are less vulnerable to these uncertainties.

Evolution of the Mutation Rate

Whole-genome sequence analyses of mutation-accumulation lines have made clear that substantial variation in the mutation rate exists among species (Lynch et al. 2016). In all organisms, the bulk of small-scale mutations involve single base-substitutions, with the ratio of insertion and deletion mutations to the former typically being on the order of 0.1 (Lynch et al. 2016). Base-substitutional mutation rates in bacterial species are generally in the range of 10^{-10} to 10^{-9} per nucleotide site per generation, with a mean of 4×10^{-10}. Rates in unicellular eukaryotes are lower, in the range of 8×10^{-12} to 5×10^{-10} per nucleotide site per generation, with a mean of 2×10^{-10} (Table 4.2). On a per-generation basis, base-substitutional mutation rates are substantially higher in multicellular species, averaging 3.6×10^{-9} per nucleotide site in invertebrates, 1.3×10^{-8} per nucleotide site in the great apes, and 4.5×10^{-9} per nucleotide site in land plants. Thus, it appears that mutation rates are higher in multicellular than in unicellular species, and that among unicellular lineages, eukaryotes have higher levels of replication fidelity than do bacteria.

What are the likely mechanisms driving these sorts of differences? One obvious distinction among the above-mentioned groups is that multicellular species experience multiple germline cell divisions per generation, e.g., $\sim$10 for *C. elegans,* 36 for *D. melanogaster,* 40 for *A. thaliana,* and 200 for *H. sapiens* (Drost and Lee 1995; Kimble and Ward 1998; Crow 2000; Lynch 2010a), whereas there is one cell division per generation in unicellular species. If most mutations arise as replication errors, one would then expect the per-generation mutation rate to scale across yeast : *C. elegans* : *D. melanogaster /A. thaliana* : human in an $\sim$1 : 10 : 38 : 200 ratio. However, the per-generation mutation-rate scaling implied by the results given above is less extreme, approximately 1 : 6 : 25 : 57. Mutation rates of microsatellite loci, which mutate via changes in nucleotide-motif repeat numbers, are also magnified with the level of multicellularity, but the ratio of per-generation mutation rates for such loci, $\sim$1 : 50 : 13,400 for unicellular eukaryotes, invertebrates, and mammals (Seyfert et al. 2008), is much more extreme than the scaling of germline-cell division number. Thus, it appears that additional factors, including those independent of replication, must be responsible for the pattern exhibited in Table 4.2.

As with all phenotypic traits, the rate of mutation is subject to the forces of natural selection (Baer et al. 2007). However, selection on the mutation rate is unusual in that the fitness effects associated with a mutator or antimutator allele are generally manifested only indirectly through the mutational changes induced at other fitness-related loci. This raises the question as to whether mutation rates are typically held at optimum intermediate levels by stabilizing selection so as to somehow maximize the long-term rate of adaptive evolution, or simply pushed to their physiologically defined lower limits so as to minimize deleterious-mutation accumulation. If replication-error rates are maintained at higher levels than can be explained by constraints on cellular processes, the next obvious question is why dramatically higher mutation rates would be selectively promoted in multicellular relative to unicellular species, despite the fact that most mutations are deleterious (LW Chapter 12).

The central difficulty with arguments that invoke long-term benefits of elevated mutation rates is that high mutation rates are much more likely to evolve in predominantly asexual populations (the situation in many unicellular species, but not multicellular taxa), as an absence of recombination is essential if novel mutator alleles are to be pulled to fixation via linkage to induced beneficial mutations (Johnson 1999a; Sniegowski et al. 2000; Wilke et al. 2001; André and Godelle 2006; Denamur and Matic 2006). Yet, as noted above, it is among sexually reproducing multicellular eukaryotes that the highest mutation rates are consistently observed.

Despite substantial theoretical research, it has proven quite difficult to avoid the conclusion that mutation rates are predominantly driven downwardly by the transient linkage of mutator alleles to their recurrent deleterious side effects (Sturtevant 1937; Leigh 1970, 1973; Johnson 1999b; Lynch et al. 2016). Occasionally, a mutator allele may be brought to high frequency by hitchhiking with a tightly linked beneficial mutation (Clune et al. 2008; Desai and Fisher 2011), but such events are expected to be transient, as they are quickly followed by loss of the mutator phenotype by either recombinational decoupling or reversion of the mutation rate.

To see why recurrent deleterious mutation imposes selection against mutator alleles, note that any allele that magnifies the mutation rate (hereafter, designated as a mutator allele) will necessarily generate statistical associations with deleterious germline mutations induced at linked and unlinked loci. The duration of such disequilibria will depend on the rate of recombination between the mutator and affected loci, but because new associations will arise recurrently in each generation by mutation, an equilibrium background mutation load will eventually be reached, with alleles that impose higher mutation rates developing a higher associated deleterious load.

Consider a locus relevant to fitness that recombines at rate c with respect to the mutator locus. If, in the heterozygous state, the mutator induces deleterious mutations at the fitness locus at an elevated rate Δu per gene with a reduction in fitness equal to hs per induced mutation, the selective disadvantage of the mutator allele induced by linkage disequilibrium with this particular fitness locus is

$$s_d \simeq \frac{hs \cdot \Delta u}{1 - (1 - hs)(1 - c)} \tag{4.23a}$$

assuming $\Delta u \ll hs$ (Kimura 1967; Dawson 1999). An intuitive feeling for the structure of this equation can be obtained by noting that hs is the rate of removal of a deleterious mutation by selection, and c is the rate of recombination between the mutator allele and the fitness locus. The denominator is then equivalent to the per-generation rate of loss of associated fitness load for the mutator allele by either selection or recombination. The numerator is the rate of input of new fitness load per generation, so the selective disadvantage at equilibrium is simply the ratio of the rates of gain and loss of associated fitness load per generation. For unlinked loci ($c = 0.5$), this expression reduces to

$$s_d \simeq \frac{2hs \cdot \Delta u}{1 + hs} \tag{4.23b}$$

whereas in the absence of recombination ($c = 0.0$),

$$s_d = \Delta u \tag{4.23c}$$

These results indicate that the strength of selection opposing the downward drive of mutation rate is much weaker in sexual than in asexual species.

The total disadvantage of a mutator allele is obtained by summing the correlated load across all fitness-related loci. Thus, from Equation 4.23c, provided the equilibrium load associated with selection-mutation balance is reached, the total magnitude of selection against a mutator allele in an asexual population is simply equal to the elevation in the genome-wide deleterious mutation rate (ΔU, summed over all fitness-relevant loci), independent of the effects of the mutations. However, for a sexual species, the total disadvantage of a mutator must take into consideration mutations arising both on the chromosome carrying the mutator and on all other unlinked loci, as only tightly linked loci remain in association with the mutator for more than a few generations. Assuming L chromosomes, each measuring one Morgan in length (below), and a haploid genome-wide increase in the deleterious mutation rate of ΔU, after accounting for the spatial distribution of random mutations, the total induced selection coefficient against the mutator allele is found to be

$$s_{d,T} \simeq \frac{2hs \cdot \Delta U(L - 1 + \phi)}{L(1 + hs)} \tag{4.24a}$$

where

$$\phi = 1 + \ln\left(\frac{1 + hs - (1 - hs)e^{-1}}{2hs}\right) \tag{4.24b}$$

is the approximate elevation in the average induced fitness effect of mutations on the mutator-bearing chromosome relative to that on the other $L - 1$ unlinked chromosomes (Lynch 2008b). For $0.001 < hs < 0.1$, which fully covers the range of average mutational effects found in empirical studies (LW Chapter 12), ϕ is in the range of 2 to 7. Thus, the selective disadvantage of a mutator allele in a sexual species is close to twice the product of the heterozygous fitness effect of new mutations (hs) and the haploid genome-wide increase in the deleterious mutation rate (ΔU) unless the chromosome number is very small, and even then not likely to be much more than a few-fold higher. The factor by which $s_{d,T}$ exceeds $hs \cdot \Delta U/(1+hs)$ is equivalent to the average number of generations that an induced deleterious mutation remains associated with the mutator responsible for its origin (and as can be seen from Equation 4.23b, this factor is two for unlinked loci).

Because single amino-acid substitutions in DNA-processing proteins may have arbitrarily small effects on the mutation rate, and because existing mutation rates are already so low that there is little further room for improvement (the maximum possible reduction being the mutation rate itself), these results imply that the long-term selective disadvantage of many mutator alleles may be sufficiently small (relative to the power of genetic drift) to render them immune to the eyes of natural selection (Chapter 7). Thus, because there is a substantial decline in N_e from microbes to small invertebrates to vertebrates and large land plants (Lynch 2007), it is plausible that the elevation of mutation rates in multicellular lineages is not simply an inevitable outcome of an inherent physiological limitation in such species, but rather a consequence of the diminished ability of natural selection to enhance the level of replication fidelity in small-N_e species (Lynch 2011; Jain and Nagar 2013).

Several observations are consistent with this **drift-barrier hypothesis**. First, for the set of species with adequate data, there is an inverse relationship between the mutation rate per nucleotide site per generation (u) and N_e (Sung et al. 2012). Second, empirical observations on the molecular machinery involved in DNA replication and repair indicate that these processes are indeed more error-prone in taxa with higher overall mutation rates (Lynch 2008a, 2008b, 2011). Third, u is also inversely proportional to the number of a functional genes in a genome (Drake et al. 1998; Massey 2008; Ness et al. 2012; Sung et al. 2012; Lynch et al. 2016). The latter relationship is expected because, as noted above, selection operates on the total rate of deleterious-mutation production across the genome, which increases with the number of functionally relevant nucleotides in the genome. Finally, long-term laboratory-evolution experiments starting with mutator strains of microbes often reveal a gradual reduction in the mutation rate resulting from the spontaneous accumulation of changes at diverse genomic locations (Tröbner and Piechocki 1984; Notley-McRobb et al. 2002; Herr et al. 2011; Weigloss et al. 2013; Williams et al. 2013). Such observations clearly demonstrate that, even in microbes where the efficiency of selection is expected to be strong, the loci underlying replication fidelity have not been driven to a point where further improvement is no longer possible. The central point here is that one of the primary determinants of

Table 4.2 Base-substitution mutation rates (u, in units of 10^{-9} per nucleotide site per generation) for a diversity of eukaryotic species derived from whole-genome sequencing of mutation-accumulation lines or parent-offspring trios.

Species	u	References
Unicellular eukaryotes:		
Chlamydomonas reinhardtii	0.515	Sung et al. (2012); Morgan et al. (2014)
Paramecium tetraurelia	0.019	Sung et al. (2012)
Rhodosporidium toruloides	0.242	Long et al. (2016)
Saccharomyces cerevisiae	0.263	Lujan et al. (2014); Serero et al. (2014); Zhu et al. (2014)
Schizosaccharomyces pombe	0.200	Farlow et al. (2015)
Tetrahymena thermophila	0.008	Long et al. (2016)
Land plants:		
Arabidopsis thaliana	6.850	Ossowski et al. (2010); Yang et al. (2015)
Oryza sativa	2.150	Yang et al. (2015)
Invertebrates:		
Apis mellifera	6.800	Yang et al. (2015)
Caenorhabditis elegans	1.450	Denver et al. (2012)
Daphnia pulex	5.690	Keith et al. (2015)
Drosophila melanogaster	5.165	Schrider et al. (2013)
Heliconius melpomene	2.900	Keightley et al. (2014)
Pristionchus pacificus	2.000	Weller et al. (2014)
Mammals:		
Homo sapiens	13.513	Conrad et al. (2011); O'Roak et al. (2011, 2012); Campbell et al. (2012); Kong et al. (2012)
Pan troglodytes	12.000	Venn et al. (2014)

the evolutionary features of a population, the mutation rate itself, is subject to substantial evolutionary modification, with the effective population size and the functionally effective genome size dictating the degree to which selection can reduce the replication-error rate and/or levels of repair efficiency.

RECOMBINATION RATE

Although it is extraordinarily difficult to estimate recombination rates at specific nucleotide sites, some compelling general statements can be made about *average* levels of recombination over entire genomes. Such information derives from high-density genetic maps constructed from observed rates of meiotic crossing-over between molecular markers, now available for hundreds of eukaryotes thanks to the widespread availability of highly variable markers such as microsatellites. Genetic maps are based on mapping functions that attempt to convert observed recombination frequencies into the expected numbers of crossover events between pairs of markers (LW Chapter 14). Strictly speaking, such maps measure the frequency of crossover events, and generally do not include the added contributions of gene conversion, which can cause the recombination rate between very closely spaced sites to exceed by several-fold the expectation based on distant markers that are predominantly rearranged by crossovers (Equations 4.16a through 4.16c). Chromosome lengths are generally reported in units of **Morgans** (one Morgan equaling one crossover), with the sum of these lengths over all chromosomes giving the total map length.

Although eukaryotic genome sizes (total numbers of nucleotides) vary by four orders of magnitude, the range of variation in genetic-map lengths among species is only about ten-fold, with the averages for various phylogenetic groups deviating by only five-fold (Table 4.3). A simple physical constraint explains such behavior. During meiosis, there are

Table 4.3 Basic features of the physical and genetic maps of various eukaryotic groups, derived from a large survey of mapping studies involving high-density molecular markers. The grouping "Other unicellular species" includes algae, apicomplexans, ciliates, kinetoplastids, and oomycetes. Numbers in parentheses denote standard errors, and n denotes the number of species surveyed. Map lengths and mean chromosome (Chr.) sizes are in units of Morgans (M).

Group	Total Map Length	Genome Size (Mb)	Haploid Chr. No.	Mean Chr. Size (M)	n
Fungi	18.3 (2.2)	36.4 (3.2)	11.9 (1.2)	1.86 (0.36)	19
Other unicellular sps.	10.9 (1.2)	80.9 (23.3)	12.9 (1.2)	0.96 (0.18)	11
Arthropods	18.1 (3.7)	679.6 (172.4)	16.1 (3.4)	1.20 (0.18)	15
Mollusks	9.2 (1.1)	1270.7 (177.2)	13.3 (1.6)	0.71 (0.09)	6
Nematodes	4.5 (1.2)	97.6 (2.5)	7.3 (1.3)	0.59 (0.05)	3
Fish	16.0 (2.3)	1185.4 (190.5)	25.1 (0.6)	0.63 (0.08)	15
Birds	23.1 (5.4)	1334.0 (48.6)	39.6 (0.4)	0.58 (0.14)	5
Mammals	23.9 (2.5)	3222.0 (108.1)	22.1 (2.2)	1.10 (0.07)	19
Angiosperms	15.9 (1.6)	2020.3 (434.2)	13.2 (0.9)	1.19 (0.07)	44

typically no more than two crossover events per chromosome (one per arm), so that average chromosome lengths are generally on the order of one Morgan, regardless of chromosome size. Thus, because phylogenetic increases in genome size are generally associated with increases in chromosome size rather than chromosome number (Table 4.3), there is little variation in the total amount of meiotic crossing over per genome across a vast swath of life.

These observations lead to a simple structural model for the average recombination rate per physical distance across a genome ($\overline{c}$). Letting G be the total number of bases per haploid genome, and N be the haploid number of chromosomes per genome, G/N is the mean physical length of chromosomes. Letting C be the average number of crossovers (Morgans) per chromosome per meiosis, then $\overline{c} \simeq CN/G$, assuming that C is independent of chromosome size. If this model is correct, a regression of $\overline{c}$ on G on a log scale should have a slope not significantly different from -1.0, with the vertical distribution (residual deviations) around the regression line being defined largely by variation in CN (the total number of crossovers per genome). The data closely adhere to this predicted pattern, with the smallest genomes of microbial eukaryotes having recombination rates per physical distance that are $\sim$1000 times greater than those for the largest multicellular land plants (which have $\sim$1000 times larger genomes but approximately the same numbers of chromosomes) (Figure 4.4). Over this entire gradient, a smooth, overlapping decline in recombination intensity across unicellular species, invertebrates, vertebrates, and land plants reflects the general increase in genome sizes across these eukaryotic domains (Lynch 2007).

These observations suggest that the vast majority of the variance in the average recombination rate among eukaryotic species is simply due to variation in genome size and chromosome number. It should be noted, however, that even in the highest-density genetic maps, adjacent markers are generally separated by tens of thousands to millions of base pairs, so that measures of *average* levels of recombination for particular chromosomes can obscure fine-scale features. Indeed, up to 100-fold differences in recombination rates can exist among regions within chromosomes, with highly localized **recombinational hotspots** existing in well-studied species (Petes 2001; de Massy 2003; Jeffreys et al. 2004; Myers et al. 2005; Arnheim et al. 2007; Coop et al. 2008; Mancera et al. 2008).

Evolution of the Recombination Rate

As in the case of the mutation rate, considerable effort has been devoted to understanding how selection might favor recombination modifiers in various contexts (e.g., Feldman et al. 1996; Barton and Otto 2005; Keightley and Otto 2006; Barton 2010; Hartfield et al. 2010). As

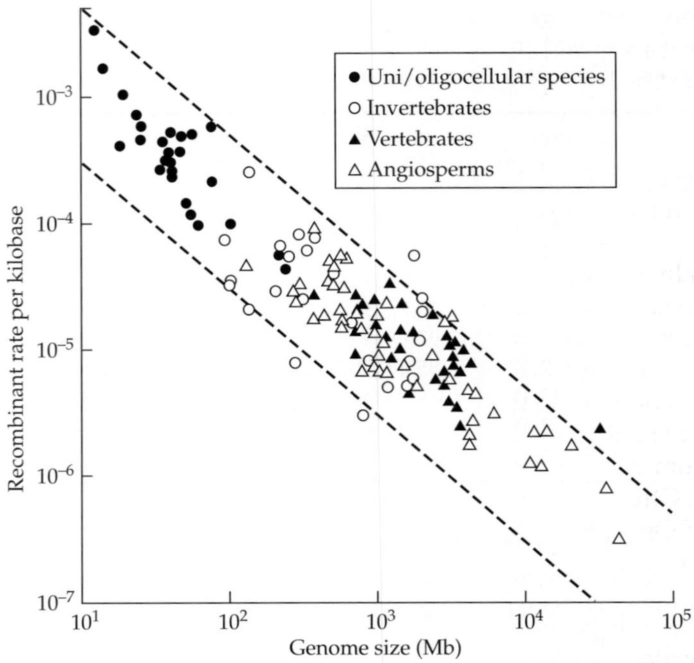

Figure 4.4 Average rates of recombination per physical distance for four major groupings of eukaryotes, determined from information on total physical and genetic map sizes. The two dashed lines have slopes of -1.0 in accordance with the theory discussed in the text. Letting C be the average number of crossovers, and N be the number of chromosomes, the top line assumes $CN = 50$, i.e., 50 chromosomes with an average length of 1.0 Morgans, 25 with average lengths of 2.0 Morgans, 100 with average lengths of 0.5 Morgans, etc. The lower line assumes $CN = 3$. For the plotted species, C is in the range of 0.3 to 3.1 (with one exception) and N is in the range of 3 to 44. (From Lynch et al. 2011.)

just noted, however, the fact remains that the vast majority of eukaryotic variation in the genome-wide amount of recombination per physical distance can be explained by a simple and largely invariant physical model of meiosis, leaving very little residual variation to be potentially assigned to mechanisms of adaptive fine-tuning. Thus, with a near-universal rule of approximately one crossover per chromosome arm, one could argue that if selection is involved at all in recombination-rate evolution, it generally operates in a way to minimize the amount of meiotic recombination across the genome. Dumont and Payseur (2007) find that variation in recombination rates across mammalian species evolves in a manner that cannot even be discriminated from the expectations of a neutral model.

Because it minimally involves three-locus dynamics in finite populations (one for the recombination modifier and two for the relevant loci under selection), most population-genetic theory on the evolution of recombination-rate modifiers has no simple analytical solutions available and so relies heavily on computer simulations. The basic motivation underlying all such work is the general principle that natural selection often encourages the build-up of repulsion disequilibria between alleles affecting fitness, i.e., the joint accumulation of gametes with different constitutions but essentially equivalent total fitness (Chapters 5 and 16). In principle, recombinational release of such hidden genetic variance can lead to more efficient selection for joint combinations with high fitness (Chapter 7). Two features of genetic systems might encourage such behavior.

First, **synergistic epistasis** (with fitness declining at an increasing rate with increasing numbers of deleterious alleles) tends to promote a selective advantage for recombination, as this will enhance the rate of production of double mutants and their more efficient (two in single events) promotion/elimination by selection (Eshel and Feldman 1970; Kondrashov 1988; Charlesworth 1990; Barton 1995b). In contrast, **diminishing-returns epistasis** (with fitness declining at a diminishing rate with increasing numbers of deleterious alleles) has the opposite effect, encouraging reduced recombination rates. As the evidence on the general incidence of these two forms of epistasis is mixed at best (Chapter 7) and the effectiveness of synergistic epistasis is greatly diminished when the single-locus effects of mutations are unequal (an issue ignored in most theory, but certainly the case in reality; Butcher 1995), the role of epistasis in the evolution of recombination rates remains unclear from an empirical perspective.

Second, as already noted in Chapter 3 and further elaborated on in Chapter 7, even in the absence of nonadditive gene action, linkage reduces the efficiency of selection on multi-locus systems, although the effect is expected to be more pronounced in larger populations harboring larger numbers of cosegregating loci. Plausible arguments have been made that the power of selection of modifiers increasing the recombination rate by this mechanism (i.e., improving the efficiency of selection among linked loci) may substantially outweigh that resulting from epistasis, even when synergistic effects are common (Felsenstein and Yokoyama 1976; Otto and Barton 2001; Pálsson 2002; Otto and Lenormand 2002; Iles et al. 2003; Barton and Otto 2005; Keightley and Otto 2006; Roze and Barton 2006).

What remains unclear is the extent to which modifiers of the recombination rate ever arise with substantial enough effects to be promoted by these kinds of associative effects. Most attempts to study the matter theoretically have focused on rather extreme situations in which either selection coefficients or the magnitude of the modifier's effect on the recombination rate are very large. Some approximations suggest that even under these conditions, the selective advantage of the modifier can be quite small (Barton and Otto 2005), perhaps too small to overcome the likelihood of being lost by drift in most cases. Nevertheless, empirical observations suggest that strong directional selection in artificial selection programs sometimes leads to the evolution of higher recombination rates (Barton and Otto 2005), and recombination-modifier models may be relevant to the more general issue of the adaptive significance of sexual versus asexual reproduction, where the former entails segregation of unlinked loci as well as recombination among linked loci.

GENERAL IMPLICATIONS

The results summarized above allow for several general statements about the intensities of mutation, recombination, and random genetic drift experienced by natural populations. As these three features define the population-genetic environment within which selection processes occur, such knowledge provides a powerful resource for understanding the limits of molecular, genomic, and phenotypic evolution and how these vary across phylogenetic lineages.

First, although the direct estimation of N_e in large populations is essentially impossible with current techniques, from information on within-population variation at putatively neutral sites, there are a number of ways to estimate the composite parameter $\theta = 4N_e u$ (or $2N_e u$ for haploids), which is equivalent to the ratio of the magnitudes of the power of mutation and drift. With direct estimates of the mutation rate (u) now available for a number of taxa, it is then also possible to estimate the long-term effective population size of a species by factoring the latter out from estimates of θ. For example, noting that the average estimate of θ for unicellular eukaryotes is 0.057 and that the average estimate of u for base-substitutional mutations in such species is $\sim 2 \times 10^{-10}$ (Table 4.2), the average N_e for such species appears to be on the order of 6×10^8 individuals if haploidy is assumed (and half that if diploidy is assumed). These estimates are likely to be somewhat downwardly biased as selection can reduce variation at silent sites in large microbial populations. When using an average θ of 0.026 and u of 3.6×10^{-9} for invertebrates, average N_e for this grouping is

$\sim 2 \times 10^6$. Likewise, when using $\theta = 0.0011$ (Example 4.1) and $u = 1.4 \times 10^{-8}$ (Example 4.4), long-term N_e for the human population is ~20,000.

Similar indirect inferences can be made from estimates of $\rho = 4N_e c_0$. For example, from Table 4.1, the average estimate of ρ for *Drosophila* species is 0.0807, whereas that for humans is ~0.0006, and for annual plants and long-lived trees is 0.0134 and 0.0050, respectively. From the genetic map data contributing to Figure 4.4, average c_0 ($\times 10^{-8}$ per site per generation, based on crossovers alone) is 2.14 for *Drosophila,* 1.28 for humans, 1.59 for annual plants, and 2.93 for trees. These results imply average values of N_e of $\sim 10^6$ for *Drosophila,* 12,000 for humans, 210,000 for annual plants, and 43,000 for trees. The consistency of the results when both approaches are applied to *Drosophila* and humans is compelling.

These estimates of N_e should be considered simply as broad indicators, as θ and ρ (and therefore N_e) can vary by at least an order of magnitude among species within major phylogenetic groups and probably within species as well, owing to long-term temporal fluctuations (Lynch 2006). Moreover, because the mean coalescence time for a random pair of alleles is $2N_e$ generations in a diploid species (Chapter 2), polymorphism-based estimates of N_e are expected to be reasonable approximations of the average conditions experienced over only the past $\sim 2N_e$ generations. Nevertheless, several general conclusions can be made: (1) the magnitude of the power of random genetic drift increases by a factor of 10^4 from unicellular eukaryotes to large multicellular species; (2) long-term effective population sizes are generally orders of magnitudes smaller than the actual numbers of breeding adults within species, probably largely as a consequence of the effects of selection on mutations physically linked on chromosomes (Chapters 3 and 8); and (3) it is possible that no eukaryotic species, even the most enormous microbial populations, has ever had a long-term N_e much beyond 10^{10}, owing to the stochastic effects of selective sweeps and background selection.

Second, a long-standing puzzle in evolutionary genetics has been that the within-species level of variation at putatively neutral sites is nearly independent of actual population sizes (Lewontin 1974). Given that such variation is expected to scale with N_e and that the numbers of individuals in bacterial species are many orders of magnitude greater than those for species of vertebrates and land plants, Lewontin dubbed this observation the **paradox of variation**. We now know that a strict linear increase in θ with absolute population size is unexpected owing to the effects of selection acting on linked loci (Chapters 3 and 8). Nevertheless, given the estimates of N_e just presented, one might still expect an increase of θ on the order of 10^4 over this gradient of organisms. However, the observed range is only two orders of magnitude (Nei 1983; Lynch 2007; Leffler et al. 2012).

The reason for this discrepancy is made clear by the preceding summary. The mutation rate u is not independent of N_e, but instead strongly declines with increasing N_e, thereby partly compensating for the direct influence of N_e on θ. As a consequence, it appears that in no species does the power of mutation exceed that of random genetic drift (i.e., θ is always much smaller than 1.0). Moreover, because estimates of ρ are also always well below 1.0, the same conclusion can be drawn with respect to the relative magnitudes of the power of recombination per nucleotide site and that of random genetic drift.

Third, it can be concluded that the ratio of the power of mutation to that of recombination increases substantially with genome size (which exhibits a strong increasing gradient with organismal size, from unicellular eukaryotes to vertebrates and land plants; Lynch 2007). Using the regression relationship in Figure 4.4 and additional information on mutation rates and genome sizes (Lynch 2010a), for eukaryotic genomes of size 10^1, 10^2, 10^3, and 10^4 Mb, average u/c_0 is ~0.00076, 0.028, 1.01, and 36.5, respectively. These extrapolations are consistent with the indirect (polymorphism-based) estimates of u/c_0 implied in Table 4.1, which are subject to substantial sampling error but nonetheless fall in the range of 1 to 100 for animals and land plants (with genome sizes in the range of 100 to 10^4 Mb).

These ideas need to be tempered by the fact that for closely spaced sites, gene conversion causes the recombination rate to be elevated relative to that expected on the basis of crossing over alone. From Equations 4.16b and 4.16c, the degree of inflation is $\simeq 1/x$, where x is the fraction of recombination events resulting in a crossover. As x is typically in the neighborhood of 0.1 (Lynch et al. 2016), this implies that the effective value of u/c_0 may be as much as 10 times lower than the values suggested above. On the other hand, with the emerging data suggesting that most recombination events are concentrated at a small number of hotspots, the recombination rate at most nucleotide sites will be much lower than the average, implying that except near hotspots, u/c_0 will be higher than implied with the use of average c_0 values.

5

The Population Genetics of Selection

Theoretical population genetics is surely a most unusual subject. At times it appears to have little connection with the parent subject on which it must depend, namely observation and experimental genetics, living an almost inbred life of its own. Warren Ewens (1994, p. 186)

Selection is the focus of much of this book, and here we lay the foundations for the response to selection on quantitative traits by first considering scenarios involving one or two loci. There are two fundamental reasons for starting here. First, in some settings, the trait of interest is indeed largely controlled by a single major gene, in which case the models introduced here are directly applicable. Second, these relatively simple population-genetic models form the foundation for models of the selection response when trait variation is controlled by multiple loci. Short-term prediction of the response of a quantitative trait to selection is done under the assumption of either constant genetic variances (Chapter 13) or predictable changes in the genetic variance from linkage disequilibrium (Chapter 16). Over longer time scales, allele-frequency change alters the genetic variances, and one- and low-locus population-genetic models are central to evaluating how these changes are influenced by the underlying genetic architecture of a trait (Chapters 24–28).

One key assumption of this chapter is that population size is effectively infinite, meaning that there is no effect of drift. A second assumption is that precise fitness values can be assigned to individual genotypes—one knows W_g, the average fitnesses for all genotypes g at the locus (or loci) of interest. Conversely, in the typical quantitative-genetic setting, fitness is defined for phenotypes, not genotypes, with $W(z)$ denoting the average fitness of individuals with phenotypic value z, with little regard for their underlying genotypes. We connect these different views of fitness at the end of the chapter, showing how selection on phenotypes maps into selection on an underlying locus, and forging a fundamental connection between the population-genetic and quantitative-genetic views of selection.

We start with a review of single-locus selection theory, highlighting how the dynamical equations for allele-frequency change can also be expressed in terms of quantitative-genetic parameters (average excesses) for fitness. While single-locus theory is essentially complete, when two or more loci are involved, gametic-phase disequilibrium is usually generated. If this occurs, single-locus equations for allele-frequency change no longer hold and no completely general statement can be made about the behavior of a system under selection.

This lack of a general theory can be addressed in one of two ways. First, exact results are available for particular two-locus models, and we present these to offer some guidance for the behavior of models with more loci. Second, rather than following the underlying changes in gametic frequencies (as required for multilocus models), we instead can approximate their impact on the expected change of the mean of a trait. We touch upon this trait-based approach at the end of this chapter by showing how single-locus approximations lead to the classic breeder's equation. General approximations for the behavior of traits under selection are further developed in Chapter 6.

SINGLE-LOCUS SELECTION: TWO ALLELES

Consider the simplest selection model: one locus with two alleles (A, a) and constant genotypic fitnesses of W_{AA}, W_{Aa}, and W_{aa}. The analysis of selection on such systems dates back to a series of papers by Haldane from 1924 to 1932 (summarized in Haldane 1932b; see Clark 1984 and Crow 1992 for overviews of Haldane's fascinating life and legacy). We

Evolution and Selection of Quantitative Traits. Bruce Walsh & Michael Lynch.
Published 2018 by Oxford University Press.
DOI 10.1093/oso/9780198830870.001.0001

Table 5.1 Genotype frequencies after viability selection. Here, p is the frequency of allele A, genotypes are in Hardy-Weinberg frequencies before selection, and $\overline{W}$ is the mean population fitness (Equation 5.1a).

Genotype	AA	Aa	aa
Frequency before selection	p^2	$2p(1-p)$	$(1-p)^2$
Fitness	W_{AA}	W_{Aa}	W_{aa}
Frequency after selection	$p^2\,\dfrac{W_{AA}}{\overline{W}}$	$2p(1-p)\,\dfrac{W_{Aa}}{\overline{W}}$	$(1-p)^2\,\dfrac{W_{aa}}{\overline{W}}$

deal first with **viability selection,** in which case W is the probability of survival from birth to reproductive age. Under this model, once adults reach reproductive age, there is no difference in mating ability and/or fertility between genotypes. Differential survival changes p, the initial frequency of allele A, to a new frequency, p', in prereproductive (but postselection) adults. Under the assumption of an effectively infinite population size, random mating then ensures that the offsprings' genotypic frequencies are in Hardy-Weinberg proportions. However, as we will show below, if parental genotypes differ in fertility, offspring genotypes will generally not be in Hardy-Weinberg proportions.

Viability Selection

Consider the change in the frequency p of allele A over one generation, $\Delta p = p' - p$. As shown in Table 5.1, the number of AA individuals following selection is proportional to $p^2 W_{AA}$, the frequency of AA genotypes before selection multiplied by their genotypic fitness. To ensure that post-selection frequencies sum to one, we divide this proportion by a normalization constant, the **mean population fitness** (the average fitness of a randomly chosen individual),

$$\overline{W} = p^2\, W_{AA} + 2\,p\,(1-p)\,W_{Aa} + (1-p)^2 W_{aa} \tag{5.1a}$$

Proceeding similarly for the other genotypes allows us to fill out the entries in Table 5.1.

From these new genotypic frequencies, the frequency of A after selection is

$$p' = \text{freq}(AA\text{ after selection}) + \frac{1}{2}\,\text{freq}(Aa\text{ after selection})$$

and applying the results in Table 5.1 gives the expected change in the frequency of A as

$$\Delta p = p' - p = p\left(p\,\frac{W_{AA}}{\overline{W}} + (1-p)\,\frac{W_{Aa}}{\overline{W}} - 1\right) \tag{5.1b}$$

This equation can also be expressed using the **relative fitnesses** $W_{ij}/\overline{W}$, abbreviated as w_{ij}, with mean fitness then scaling to $\overline{w} = 1$. Throughout, we will adhere to a notation whereby upper-case W corresponds to some absolute measure of fitness, and lower-case w corresponds to relative fitness.

If we assign the genotypes $aa : Aa : AA$ fitnesses of $1 : 1+s(1+h) : 1+2s$, Equation 5.1b becomes

$$\Delta p = \frac{sp(1-p)[\,1 + h(1-2p)\,]}{\overline{W}} \tag{5.1c}$$

As shown in Figure 5.1, a graph of Δp as a function of p provides a useful description of the allele-frequency dynamics under selection. In particular, allele frequencies that satisfy

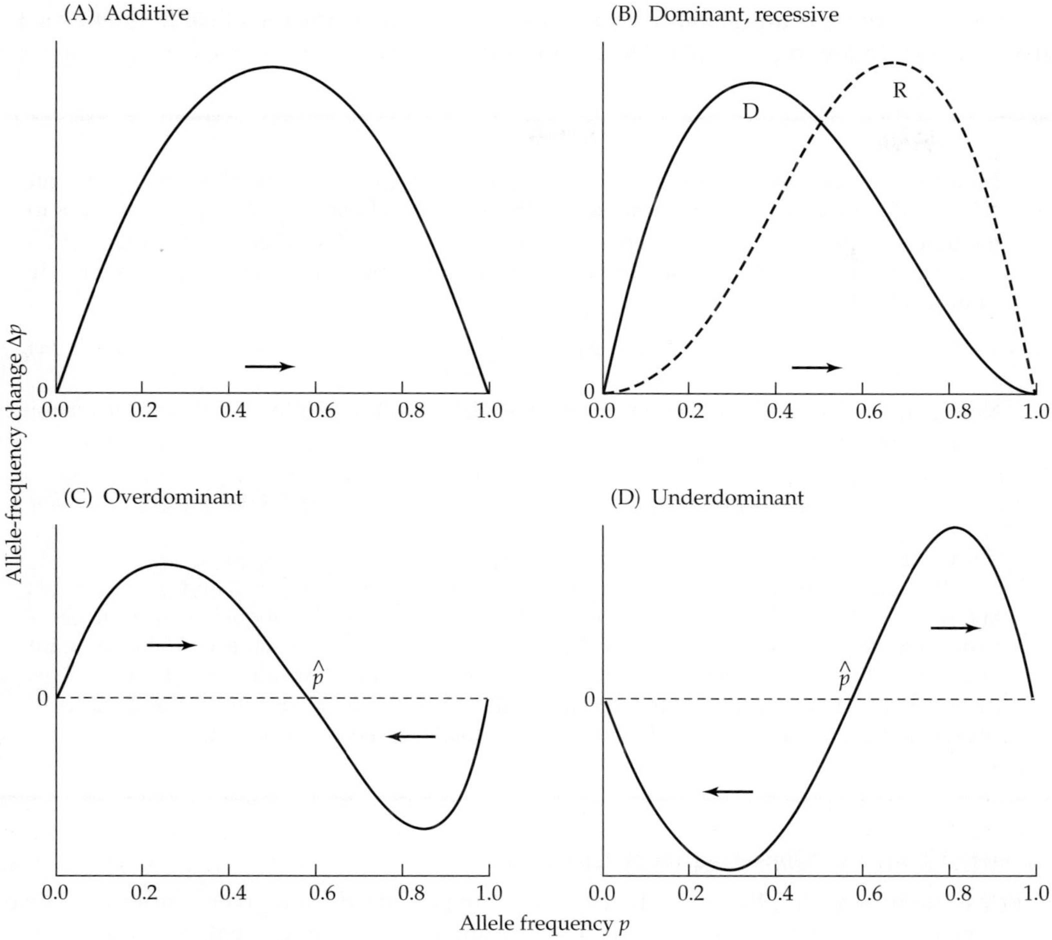

Figure 5.1 A plot of allele-frequency change Δp as a function of p is a useful device for examining how frequencies change under selection. If $\Delta p > 0$, the frequency of A increases (moves to the right), as indicated by the right-pointing arrow. If $\Delta p < 0$, the frequency of A decreases (left-pointing arrow). If $\Delta p = 0$, the allele frequencies are at equilibrium. **(A) Directional selection** with additive fitnesses favoring allele A. For $p \neq 0, 1$; $\Delta p > 0$, and p increases to one, with the rate of change becoming symmetric about $p = 1/2$. **(B)** Directional selection with dominance and allele A favored. Curve D corresponds to allele A dominant, and curve R to A recessive. Here the response is *asymmetric* about $p = 1/2$. In both cases, $\Delta p > 0$ (provided $p \neq 0, 1$), and the frequency of A increases to 1. **(C) Overdominant selection**, where the heterozygote is more fit than either homozygote (Example 5.4), has an internal equilibrium frequency of $\widehat{p}$. For frequencies above the equilibrium, $\Delta p < 0$ and the frequency decreases to $\widehat{p}$; whereas if p is less than $\widehat{p}$, $\Delta p > 0$ and the allele frequency increases to $\widehat{p}$. Thus, $\widehat{p}$ is a *stable equilibrium*. **(D)** With **underdominant selection**, the heterozygote is less fit than either homozygote. Again, there is an internal equilibrium allele frequency, but in this case it is *unstable*. If $p < \widehat{p}$, p decreases toward zero, while if $p > \widehat{p}$, p increases toward one. The result is fixation of either A or a, depending on the starting allele frequency.

$\Delta p = 0$ (i.e., no allele-frequency change after selection) are called **equilibrium frequencies**, which we denote by $\widehat{p}$. Regardless of the values of s or h, trivial **boundary equilibria** exist when only one allele is present ($\widehat{p} = 0$ or 1). Equation 5.1c shows that an **internal equilibrium**, in which both alleles are segregating, requires $1 + h(1 - 2\widehat{p}) = 0$, which yields $\widehat{p} = (1 + h)/(2h)$. Thus either $h > 1$ (**overdominance**) or $h < -1$ (**underdominance**) is required to ensure $0 < \widehat{p} < 1$. However, the equilibrium behavior is very different in these two cases. The situation $h > 1$ represents a **stable equilibrium**, in which following a small perturbation from $\widehat{p}$, selection returns the allele frequency to $\widehat{p}$ (Figure 5.1C). In contrast, with $h < -1$, there is an **unstable equilibrium**, selection sends the allele frequency *away* from $\widehat{p}$ which following a small perturbation (Figure 5.1D).

Example 5.1. Letting p = freq(A), what is Δp when $W_{AA} = 1 + 2s$, $W_{Aa} = 1 + s$, and $W_{aa} = 1$? These are **additive fitnesses**, with each copy of allele A adding an amount s to the fitness. In this case, mean fitness simplifies to $\overline{W} = 1 + 2sp$ (there is an average of $2p$ A alleles per individual, each of which increments fitness by s), and applying Equation 5.1c (setting $h = 0$) yields

$$\Delta p = \frac{sp(1-p)}{1+2sp} \tag{5.2a}$$

Noting that a first-order Taylor expansion (in s) yields $1/(1+2sp) = 1 - 2sp + O(s^2)$, for $|s|$ small, we have

$$\Delta p = sp(1-p)\left[1 - 2sp + O(s^2)\right] = sp(1-p) + O(s^2) \tag{5.2b}$$

The notation $O(s^2)$ denotes terms of order s^2, which are negligible for $|s| \ll 1$.

The only equilibrium allele frequencies under this model are $\widehat{p} = 0$ and $\widehat{p} = 1$. If A is favored by selection ($s > 0$), then $\Delta p > 0$ for $0 < p < 1$ and the frequency of A increases to one, regardless of the starting point (Figure 5.1A). Here $\widehat{p} = 1$ is a stable equilibrium point while $\widehat{p} = 0$ is unstable, because if even a few copies of A are introduced, selection drives them to fixation. In contrast, if allele a is favored ($s < 0$), the frequency of allele A declines to zero regardless of the starting point and $\widehat{p} = 0$ is stable, while $\widehat{p} = 1$ is unstable.

Expected Time for Allele-frequency Change

A key issue in selection theory is the expected time required for a given amount of allele-frequency change. Assuming that s and sh are small (i.e., there is weak selection, so that $\overline{W} \simeq 1$), we can ignore $\overline{W}$ in Equation 5.1.c as a first-order approximation (see Example 5.1). Equation 5.1c then shows that the change in allele frequency p under weak selection is approximated by the differential equation

$$\frac{dp}{dt} = sp(1-p)[\,1 + h(1-2p)\,] \tag{5.3a}$$

For additive selection ($h = 0$), this has a simple solution of

$$p_t = \frac{p_0}{p_0 + (1-p_0)e^{-st}} \tag{5.3b}$$

where p_t is the frequency of allele A at time t. Often of greater interest is t_{p_t,p_0}, the expected time required to move from initial frequency of p_0 to target value of p_t. From Equation 5.3a, this is given by the integral

$$t_{p_t,p_0} = \frac{1}{s}\int_{p_0}^{p_t} \frac{dx}{x(1-x)[1+h(1-2x)]} \tag{5.3c}$$

Crow and Kimura (1970) presented explict results for several important cases. If fitnesses are additive ($h = 0$)

$$t_{p_t,p_0} \simeq \frac{1}{s}\ln\left(\frac{p_t\,(1-p_0)}{p_0\,(1-p_t)}\right) \tag{5.3d}$$

whereas if A is recessive ($h = -1$)

$$t_{p_t,p_0} \simeq \frac{1}{2s}\left[\ln\left(\frac{p_t\,(1-p_0)}{p_0\,(1-p_t)}\right) - \frac{1}{p_t} + \frac{1}{p_0}\right] \tag{5.3e}$$

and finally, if A is dominant ($h = 1$)

$$t_{p_t,p_0} \simeq \frac{1}{2s}\left[\ln\left(\frac{p_t\,(1-p_0)}{p_0\,(1-p_t)}\right) + \frac{1}{1-p_t} - \frac{1}{1-p_0}\right] \tag{5.3f}$$

Example 5.2. Consider the time for a favored allele to move from a frequency of 0.1 to a frequency of 0.5. For an additive allele, Equation 5.3d yields

$$t \simeq s^{-1}\ln\left(\frac{0.5\,(1-0.1)}{0.1\,(1-0.5)}\right) = \frac{2.2}{s}\ \text{generations}$$

On the other hand, from Equations 5.3e and 5.3f, $t \simeq 1.5/s$ generations when A is dominant, and $t \simeq 5.1/s$ generations when A is recessive. The faster rate of response for a rare dominant allele occurs because A is fully exposed in heterozygotes ($W_{Aa} > W_{aa}$), while its fitness effects are completely hidden in heterozygotes when recessive ($W_{AA} > W_{Aa} = W_{aa}$). Conversely, this same feature slows down the rate of response of a dominant when A is common, as only rare aa homozgyotes will be selected against.

Differential Viability Selection on the Sexes

Up to now we have assumed there is equal selection operating on both sexes, but this need not be the case. To accommodate this complication, again assume there is random mating, an autosomal locus, and viability selection, but let x be the current frequency of allele A in males and y be the current value in females. The genotype frequencies following random mating and their fitnesses can be represented as

Genotype	AA	Aa	aa	Mean
Frequency	yx	$x(1-y)+y(1-x)$	$(1-x)(1-y)$	
Male fitness	W_{AA}	W_{Aa}	W_{aa}	$\overline{W}$
Female fitness	V_{AA}	V_{Aa}	V_{aa}	$\overline{V}$

As in Table 5.1, the frequencies of surviving genotypes in males and females are equal to the product of their starting values and relative fitnesses. For example, $yx\,W_{AA}/\overline{W} = yx\,w_{AA}$ and $yx\,V_{AA}/\overline{V} = yx\,v_{AA}$ are the frequencies of AA among surviving males and females, respectively. The frequency of A in males after selection is the sum of the postselection frequency of AA plus half that of Aa, giving the recursion equation for the allele frequency in males as

$$x' = \frac{xy\,W_{AA} + (1/2)[x(1-y)+y(1-x)]\,W_{Aa}}{\overline{W}} \tag{5.4a}$$

$$= xy\,w_{AA} + (1/2)[x(1-y)+y(1-x)]\,w_{Aa} \tag{5.4b}$$

where

$$\overline{W} = xy\,W_{AA} + [x(1-y)+y(1-x)]\,W_{Aa} + (1-x)(1-y)\,W_{aa} \tag{5.4c}$$

with an analogous expression for y' in females obtained by replacing W by V. These new values (x', y') are then used for the next iteration. Kidwell et al. (1977) explored the conditions under which differential selection in the sexes can maintain variation (i.e., support a stable polymorphism). For additive selection, they found that **antagonistic selection** (meaning that the sign of the selection coefficients differs between sexes, with A favored in one sex and a in the other), can stably maintain variation *only* if the absolute values of selective differences are fairly close to each other. Strong disproportional selection in one sex will remove variation. See Kidwell et al. (1977) for an analysis of more complex cases.

Frequency-dependent Selection

Although we have been assuming that the genotypic fitnesses W_{ij} are constants, this need not be the case. The fitness of a genotype may be a function of the other genotypes in the population, giving rise to **frequency-dependent selection**. For example, when a rare genotype has a selective advantage due to avoidance of a search image by a predator, its fitness declines as its frequency increases. Alleles at self-incompatibility loci in plants also have a fitness advantage when rare because successful gametes must fuse with those carrying different alleles.

If genotype fitness varies with allele frequencies, Equation 5.1 still holds, provided we replace the constant values of W_{ij} by the functions $W_{ij}(p)$. One interesting feature of frequency-dependent selection is that mean population fitness need not increase over time. Indeed, Wright (1948a) provided a simple two-allele example where mean fitness strictly decreases over time.

Frequency-dependent selection can maintain a polymorphism when rare alleles have the highest fitness. Such a situation is often called **balancing selection**, but some caution is in order when using this term as it is also used for constant-fitness overdominance. Wright and Dobzhansky (1946) noted just how subtle this distinction can be, showing that both fitness models (frequency-dependence and overdominance) can generate identical allele-frequency dynamics. Thus, the two models cannot be distinguished from allele-frequency data alone. Indeed, Denniston and Crow (1990) and Lachmann-Tarkhanov and Sarkar (1994) showed that for *any* set of constant fitnesses, there is always an alternative frequency-dependent fitness set that generates the same exact allele-frequency dynamics.

Making a case for balancing selection via rare-genotype advantage requires making direct estimates of genotype fitnesses at different allele frequencies. Genotype fitnesses are expected to be constant under overdominance, but they change under frequency-dependence. An example of this approach was provided by Fitzpatrick et al. (2007), who examined the *foraging* gene of *Drosophila melanogaster* and found that the alternative *sitter* and *rover* alleles have their highest fitnesses when rare.

Finally, a (somewhat) related topic is the fate of alleles whose selection coefficients randomly fluctuate over generations. The important feature in this case is that the allele with the highest (*arithmetic*) mean absolute fitness, $\mu_s = E[s]$, is not necessarily the winner. Rather, the allele with the highest *geometric* mean fitness wins (Dempster 1955a; Haldane and Jayakar 1963; Gillespie 1973, 1977; Orr 2007). The geometric mean can be approximated as $\mu_s - \sigma_s^2/[2\mu_s]$, so that the variance matters as well. Hence an allele that is less fit on average (i.e., has a smaller μ_s), can still win if it has a lower variance.

Fertility/Fecundity Selection

We have also been assuming no differential fertility/fecundity (we treat these two terms as synonymous), meaning that all combinations of genotypic pairs produce, on average, the same number of offspring. Obviously, this is often not true. To treat this problem formally, the average number of offspring produced by the (ordered) cross of an A_iA_j male with an A_kA_l female is denoted by the **fertility fitness**, f_{ijkl}. In this fully general case, it is no longer sufficient to simply follow allele frequencies. Rather, we must follow *genotypic* frequencies, and the resulting dynamics can quickly become very complex. For example, mean viability can easily decrease if the genotypes with low viability have sufficiently high fertility.

Bodmer (1965) and Kempthorne and Pollak (1970) further explored some of the consequences of fertility selection. A key result was that if the fertility fitnesses are multiplicative,

$$f_{ijkl} = f_{ij} \cdot f_{kl}$$

meaning that the average fertility of the cross is simply the product the fertility fitnesses for each genotype (as opposed to a specific value for each cross), then if w_{ij} is the viability fitness, the evolutionary dynamics proceed as with viability selection with fitness $w_{ij}f_{ij}$.

Sexual Selection

A final complication is **sexual selection**, non-random mating based on traits involved in mate choice (Chapter 29). In many species, mate choice is at least partly based on trait values, either through male-male competition for access to females or through female choice of specific males. Sexual selection for particular traits can result in very interesting evolutionary dynamics, especially when sexually preferred trait values conflict with natural selection (viability and/or fertility selection).

Example 5.3. An interesting example of the consequences of sexual selection was presented by Muir and Howard (1999). As exotic genes are introduced into domesticated species to create transgenic organisms, there is a biosafety concern in the potential genetic risk of the introduced transgene. If the gene "escapes" into a wild population, will it increase in frequency, be neutral, or quickly be lost by negative selection? Muir and Howard (1999, 2001) and Howard et al. (2004) developed population-genetic models to assess such risk and used them to understand the fate of a transgenic strain of the Japanese medaka fish (*Oryzias latipes*). After insertion of a human growth-hormone gene under a fish-specific promoter, the resulting transgenic fish grows faster and to a much larger size than a normal medaka. While such a genetic transformation may be a boon for aquaculture, what would happen if the growth-hormone gene found its way into natural medaka populations? Muir and Howard found that transgenic fish have only 70% of the survival rate of normal fish. Based on this strong viability selection, one might think that any transgenes that escape would quickly be lost. However, Muir and Howard found that larger fish have a roughly four-fold mating advantage relative to smaller fish. Based on these parameter values, any escaped transgene will spread, as the mating advantage more than offsets the survival disadvantage. However, simulation studies find a potentially more ominous fate under these parameter values. The transgene not only spreads, but it eventually may drive the population to extinction as a consequence of the reduction in viability. Muir and Howard coined the term **Trojan gene** for such settings. Such genes may also arise naturally. A potential example was given by Dawson (1969), who found that a newly arisen eye color mutation in *Tribolium castaneum* rapidly increased in frequency in tandem with the rate at which the line went extinct in a competition experiment.

WRIGHT'S FORMULA

A more compact and insightful way to express allele-frequency change was presented by Sewall Wright, one of the founding fathers (with Fisher and Haldane) of modern selection theory. Because

$$\begin{aligned}\frac{d\overline{W}}{dp} &= \frac{d\left(p^2 W_{AA} + 2p(1-p)W_{Aa} + (1-p)^2 W_{aa}\right)}{dp} \\ &= 2pW_{AA} + 2(1-2p)W_{Aa} - 2(1-p)W_{aa} \end{aligned} \tag{5.5a}$$

a little algebra shows that Equation 5.1b can be written as

$$\Delta p = \frac{p(1-p)}{2\overline{W}} \frac{d\overline{W}}{dp} = \frac{p(1-p)}{2} \frac{d\ln \overline{W}}{dp} \tag{5.5b}$$

The last step follows from the chain rule for differentiation,

$$\frac{d\ln f(x)}{dx} = \frac{1}{f(x)} \frac{df(x)}{dx}$$

Equation 5.5b is **Wright's formula** (1937), which holds provided the genotypic fitnesses are constant and **frequency independent** (not themselves functions of allele frequencies, which can be formally stated as $\partial W_{ij}/\partial p_k = 0$ for all i, j, and k). Normal derivatives are used in Equation 5.5b as there is just a single variable, the allele frequency p.

Example 5.4. Consider a locus with two alleles and genotypic fitnesses

$$W_{AA} = 1 - t, \quad W_{Aa} = 1, \quad \text{and} \quad W_{aa} = 1 - s$$

Letting $p = \text{freq}(A)$, Wright's formula can be used to find Δp and the equilibrium allele frequencies. Here mean fitness is given by

$$\begin{aligned}\overline{W} &= p^2(1-t) + 2p(1-p)(1) + (1-p)^2(1-s)\\ &= 1 - tp^2 - s(1-p)^2\end{aligned}$$

Taking derivatives with respect to p results in

$$\frac{d\overline{W}}{dp} = 2[s - p(s+t)]$$

which, upon substituting into Wright's formula, results in

$$\Delta p = \frac{p(1-p)[s - p(s+t)]}{1 - tp^2 - s(1-p)^2}$$

Alternatively, substituting these fitnesses into Equation 5.1b recovers the same result.

Setting $\Delta p = 0$ yields three equilibrium solutions: $\widehat{p} = 0, \widehat{p} = 1$, and, most interestingly,

$$\widehat{p} = s/(s+t)$$

which corresponds to $d\overline{W}/dp = 0$, a necessary condition for a local extremum (maximum or minimum) in $\overline{W}$. Recall from calculus that this extremum is a maximum when $d^2\overline{W}/dp^2 = -2(s+t) < 0$, and a local minimum when this second derivative is positive. With selective overdominance ($s, t > 0$), the heterozygote has the highest fitness, implying $\Delta p > 0$ when $p < \widehat{p}$, and $\Delta p < 0$ when $p > \widehat{p}$ (Figure 5.1C). Thus, *selection retains both alleles in the population*, as first shown by Fisher (1922). With overdominance, $\widehat{p}$ is also the allele frequency that maximizes $\overline{W}$.

With selective underdominance ($s,\ t < 0$), the heterozygote has lower fitness than either homozygote. Although there is still an equilibrium, $\widehat{p} = s/(s+t)$ corresponds to a local *minimum* of $\overline{W}$ (as $d^2\,\overline{W}/dp^2 > 0$) and is therefore unstable—if p is the slightest bit below $\widehat{p}$, it decreases to zero, while if p is the slightest bit above $\widehat{p}$, it increases to one (Figure 5.1D). In contrast to selective overdominance, *selective underdominance removes, rather than maintains, genetic variation*, with the initial starting frequencies determining which alelle is fixed.

Example 5.5. A classic example of selective overdominance is sickle-cell anemia, a disease due to a recessive allele at the beta hemoglobin locus. *SS* homozygotes suffer periodic life-threatening health crises due to their red blood cells being sickle-shaped. *SS* individuals often have near-zero fitness (due to their low survival to reproductive age), and ordinarily this would be expected to result in a very low frequency of the *S* allele. However, in malaria-infested regions, *SN* heterozygotes (*N* denoting the "normal" allele) have increased resistance to malaria relative to *NN* homozygotes.

A sample of 12,387 West Africans yielded 9365 *NN*, 2993 *NS*, and 29 *SS* individuals (Nussbaum et al. 2004), giving a frequency of *S* as

$$\frac{(1/2)\cdot 2993 + 29}{12,387} = 0.123$$

Assuming the frequency of S is at its selective equilibrium, what is the strength of selection against *NN* individuals due to malaria? Writing the fitnesses of the *SS*, *SN*, and *NN* genotypes as $1-t$, 1, and $1-s$ respectively, the result from Example 5.4 is an equilibrium frequency of $s/(s+t)$ for allele S. Setting this equal to 0.123 implies that

$$t = \frac{(1-0.123)}{0.123}\, s = 7.120\, s \quad \text{or} \quad s = 0.140\, t$$

If *SS* individuals are either lethal ($t = 1$) or have only 10% fitness ($t = 0.9$), then s = 0.140 and 0.126, respectively. In other words, relative to *NN* individuals, heteroyzgotes have a 13% to 14% fitness advantage due to increased malaria resistance.

Example 5.6. A common model in evolutionary genetics considers a trait (determined by a number of loci) that is experiencing stabilizing selection, with some intermediate phenotypic value θ favored (Chapters 28–30). Given that intermediate phenotypes are favored, naively one might think that heterozygotes (whose phenotypes are between those of the two homozygotes) are favored, resulting in selective overdominance. However, the application of Wright's formula to the allelic dynamics at one of the underlying loci shows that this is *not* the case.

To demonstrate this, we need to express the mean fitness $\overline{W}$ in terms of the allele frequency p_i at some focal locus. To do so, we start with a standard model for stabilizing selection, a **Gaussian fitness function**, with the expected fitness of an individual with trait value z given by

$$W(z) = e^{-(z-\theta)^2/\omega^2}, \quad \text{with} \quad \omega^2 > 0$$

Akin to the variance in a Gaussian (i.e., normal) distribution, ω^2 measures the strength of selection against the trait, with larger values indicating weaker selection (fitness declining more slowly as we move away from θ). Letting $s = 1/\omega^2$, Barton (1986) showed that if phenotypes are normally distributed with a mean of μ and a variance of σ^2, then (assuming weak selection, $\sigma^2 \ll \omega^2 = 1/s$) the mean fitness is

$$\overline{W} \simeq e^{-[\sigma^2+(\mu-\theta)^2]/(2\omega^2)}, \qquad \text{implying} \qquad \ln \overline{W} \simeq -s[\,\sigma^2 + (\mu-\theta)^2\,]/2 \tag{5.6a}$$

Because the mean and variance are functions of the allele frequencies over all loci, our concern is now the partial derivative with respect to the allele frequency p_i at our focal locus. Applying the chain rule,

$$\begin{aligned} \frac{\partial \ln \overline{W}}{\partial p_i} &= -(s/2)\frac{\partial[\sigma^2 + (\mu-\theta)^2]}{\partial p_i} \\ &= -(s/2)\left[\frac{\partial \sigma^2}{\partial p_i} + 2(\mu-\theta)\,\frac{\partial \mu}{\partial p_i}\right] \end{aligned} \tag{5.6b}$$

Now suppose that n diallelic fully additive loci underlie this character (no dominance or epistasis), with the genotypes b_ib_i, B_ib_i, and B_iB_i at each locus i having effects 0, a_i, and $2a_i$. Letting p_i be the frequency of allele B_i, the trait mean is some baseline value m plus the genetic contributions, while the trait variance is the additive-genetic plus environmental variances,

$$\mu = m + 2\sum_{i=1}^{n} a_i\, p_i \qquad \text{and} \qquad \sigma^2 = 2\sum_{i=1}^{n} a_i^2\, p_i\,(1-p_i) + \sigma_E^2 \tag{5.6c}$$

where the additive-genetic variance expression (LW Equation 4.23b) assumes no linkage disequilibrium. From Equation 5.6c,

$$\frac{\partial \mu}{\partial p_i} = 2\, a_i \qquad \text{and} \qquad \frac{\partial \sigma^2}{\partial p_i} = 2\, a_i^2\,(1-2p_i) \tag{5.6d}$$

Wright's formula (Equation 5.5b) returns the expected change in the frequency of allele B_i as

$$\Delta p_i = \frac{p_i(1-p_i)}{2} \left(\frac{\partial \ln \overline{W}}{\partial p_i} \right)$$

which, upon substituting Equation 5.6d into Equation 5.6b, yields

$$\Delta p_i = s\, a_i \left(\frac{p_i(1-p_i)}{2} \right) [\, a_i\,(2p_i - 1) + 2(\theta - \mu)\,] \tag{5.6e}$$

Thus, even if the population mean μ coincides with the phenotypic optimum θ, there remains the potential for selection on the underlying loci, as in this special case Equation 5.6e reduces to

$$\Delta p_i = a_i^2\, s\, p_i(1-p_i)(p_i - 1/2) \tag{5.6f}$$

This is a form of selective *underdominance*, with $\widehat{p}_i = 1/2$ being unstable, as $\Delta p_i < 0$ for $p_i < 1/2$, while $\Delta p_i > 0$ for $p_i > 1/2$. Hence, selection for an optimum phenotype tends to drive allele frequencies toward fixation, *removing*, rather than retaining, variation at underlying loci (Robertson 1956).

Adaptive Topographies and Wright's Formula

The surface $\overline{W}(p)$ of mean population fitness as a function of allele frequency forms an **adaptive topography**, showing which p value or values maximize mean fitness. Examples are shown in Figure 5.2, which plots $\overline{W}(p)$ for the same settings as in Figure 5.1. Note that stable equilibria correspond to local maxima (Figure 5.2C) and unstable equilibria correspond to local minima in mean fitness (Figure 5.2D). Because $p(1-p) \geq 0$, Wright's formula implies that the sign of Δp is the same as the sign of $d \ln \overline{W}/dp$, implying that *allele frequencies change to locally maximize mean fitness*. Wright's formula thus suggests a powerful geometric interpretation of the mean-fitness surface $\overline{W}(p)$—the local curvature of the fitness surface largely describes the behavior of the allele frequencies. In a random-mating population with constant W_{ij}, allele-frequency changes move the population toward the nearest local maximum on the fitness surface.

However, evolution toward maximum fitness is only guaranteed when the assumptions underlying Wright's formula hold (e.g., single-locus viability selection with constant genotypic fitnesses). Further, in a strict mathematical sense, mean fitness need not increase to a local maximum. If the initial allele frequencies are such that mean population fitness is exactly at a local minimum, allele frequencies do not change, as $d \ln \overline{W}/dp = 0$ (Example 5.4). However, this case is biologically trivial, as the resulting equilibrium is unstable—any amount of genetic drift moves allele frequencies away from this minimum, with mean fitness subsequently increasing to a local maximum.

SINGLE-LOCUS SELECTION: MULTIPLE ALLELES

Extending single-locus models from two to multiple alleles is a straightforward process and also reveals connections between quantitative-genetic concepts and the behavior of population-genetic models. In particular, multiple-allele models can be framed in terms of the average excess (LW Chapter 4) of each allele on fitness, which then leads into a discussion of how the additive genetic variance in fitness influences the selection response (allele-frequency change).

Marginal Fitnesses and Average Excesses

For a locus with n alleles under viability selection and random mating, the frequencies of the A_iA_j heterozygotes and A_iA_i homozygotes after selection are $2\,p_ip_jW_{ij}/\overline{W}$ and $p_i^2\,W_{ii}/\overline{W}$.

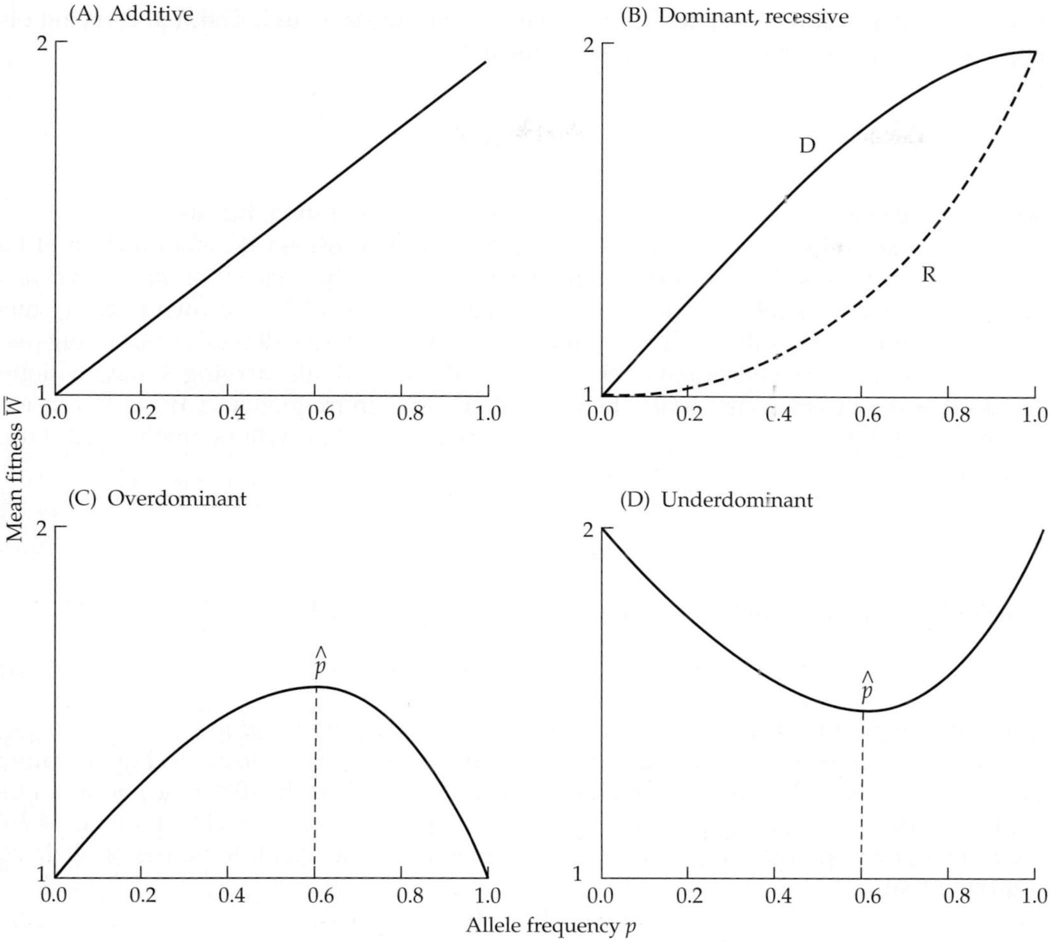

Figure 5.2 Plots of mean population fitness, $\overline{W}(p)$, as a function of allele frequency p (as in Figure 5.1). In all cases, the best genotype has a fitness of 2.0 and the worst has a fitness of 1.0. **(A)** Directional selection with additive fitness, with allele A favored. **(B)** Directional selection with dominance, with allele A favored. The upper curve (D) is for A dominant and the lower (R) is for A recessive. In both **(A)** and **(B)**, mean fitness is maximized at the stable equilibrium point ($p = 1$). **(C)** With overdominant selection, fitness is maximized at the stable equilibrium point $\widehat{p}$. **(D)** With underdominant selection, fitness is *minimized* at the unstable internal equilibrium point $\widehat{p}$.

As in Table 5.1, for a biallelic locus, this follows from weighting the (random-mating) frequency of a genotype before selection by its fitness, with

$$\overline{W} = \sum_{i=1}^{n}\sum_{j=1}^{n} p_i\, p_j W_{ij}$$

The frequency of allele A_i in survivors is the postselection frequency of the A_iA_i homozygote plus half the postselection frequencies of all A_iA_j heterozygotes, a sum that simplifies to

$$p_i' = \frac{p_i}{\overline{W}} \sum_{j=1}^{n} p_j W_{ij} = p_i \frac{W_i}{\overline{W}} \tag{5.7a}$$

where

$$W_i = \sum_{j=1}^{n} p_j W_{ij} \tag{5.7b}$$

is the **marginal fitness** of allele A_i, i.e., the expected fitness of an individual carrying one copy of A_i and a second randomly chosen allele. Further,

$$\overline{W} = \sum_{i=1}^{n} p_i W_i$$

so one can also express mean fitness as the average of the marginal fitnesses.

Note that, unlike *genotypic* fitness (W_{ij}), the marginal fitness W_i is a function of the allele frequencies and hence *is expected to change over time.* The concept of marginal fitness can be understood as follows: under random mating, if one allele is known to be A_i, then with probablity p_j, the other will be A_j and the resulting fitness will be W_{ij}. Summing over all possible alleles recovers Equation 5.7b. If $W_i > \overline{W}$ (individuals carrying A_i have a higher fitness than a random individual), then A_i will increase in frequency. If $W_i < \overline{W}$, A_i will decrease in frequency. Finally, if $W_i = \overline{W}$, the frequency of A_i will be unchanged. From Equation 5.7a, the expected allele-frequency change is

$$\Delta p_i = p_i' - p_i = p_i \frac{W_i - \overline{W}}{\overline{W}} \tag{5.7c}$$

which implies that at a polymorphic equilibrium (e.g., $\widehat{p_i} \neq 0, 1$),

$$W_i = \overline{W} \text{ for all } i \text{ with } 0 < \widehat{p_i} < 1 \tag{5.7d}$$

Thus, at an equilibrium, *all segregating alleles have the same marginal fitness.*

Marginal fitnesses provide a direct connection between single-locus and quantitative-genetic theory. Recalling that the **average excess** of allele A_i is the difference between the mean of individuals carrying a copy of A_i and the population mean (LW Equation 4.16), we immediately see that $(W_i - \overline{W}\,)$ is the average excess in absolute fitness of allele A_i, implying that

$$s_i = (W_i - \overline{W}\,) / \overline{W} = w_i - 1 \tag{5.8a}$$

is the average excess in *relative* fitness. Like W_i, s_i is a function of allele frequencies, and thus changes as these change. Equation 5.7c can therefore be expressed as

$$\Delta p_i = p_i\, s_i \tag{5.8b}$$

Thus, at equilibrium, *the average excess in fitness of each allele equals zero.* As there is then no variation in the average excesses, it immediately follows that the *additive genetic variance in fitness is also zero* at the equilibrium allele frequencies, as a nonzero additive genetic variance requires variation in the average excesses (see LW Equation 4.23a).

Equilibrium Frequencies With Multiple Alleles

As with a single locus with two alleles, mean fitness never decreases with n alleles at a single locus (again assuming constant fitnesses, viability selection, and random mating). The classical short proof for this was given by Kingman (1961a). A more interesting question involves the number of polymorphic equilibria that exist with two (or more) segregating alleles. In particular, how many equilibria with all n alleles polymorphic are possible? Kingman (1961b) showed that such a system either has no such equilibria, exactly one, or an infinite number (a line or hyperplane of solutions). We can see this from Equation 5.7d, as the equilibrium marginal fitnesses for all segregating alleles are identical, e.g., $\widehat{W_i} = \widehat{W_1} = \overline{W}$ for $i = 2, \cdots, n$. Because each marginal fitness is a linear function of the n equilibrium allele frequencies (Equation 5.7b), there are $n - 1$ linear equations in terms of the equilibrium frequencies for the n alleles (as the allele frequencies are constrained to sum to one), with

$$\widehat{W_i} = \sum_{j=1}^{n} W_{ij}\, \widehat{p_j} = \widehat{W_1} = \sum_{j=1}^{n} W_{1j}\, \widehat{p_j} \text{ for } i = 2, \cdots, n$$

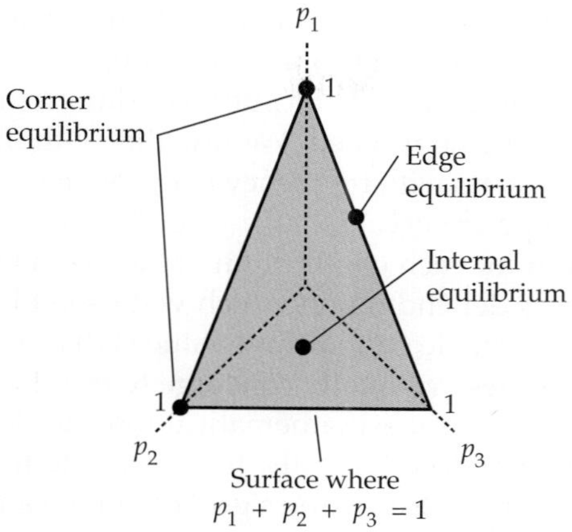

Figure 5.3 The simplex for three alleles, namely the space of all possible allele frequencies, subject to the constraint that they must sum to 1. Note that the plane of possible values intersects each allele-frequency axis at a value of 1 for that allele, and 0 for all others. Within the simplex, three types of equilibria are possible. Corner equilibria occur when one allele has frequency 1; these are monomorphic equilibria, with no genetic variation. Polymorphic equilibria can either be edge equilibria, when at least two (but not all) allele frequencies are nonzero (here alleles 1 and 3 are segregating while allele 2 is absent), or internal equilibria, wherein all alleles are segregating.

With a linear system of $n-1$ equations and $n-1$ unknowns, there will be either zero, one, or infinitely many solutions. An example of the latter is that when all the $W_{ij} = 1$, any set of allele frequencies is stable, as this is just the neutral condition.

A more profound result obtained by Kingman is that the existence of a single internal (and stable) equilibrium for all n alleles requires the fitness matrix **W** (whose ijth element is W_{ij}) to have exactly one positive and at least one negative eigenvalue (Appendix 5). More generally, if **W** has m positive eigenvalues, then, at most, $n - m + 1$ alleles can be jointly polymorphic.

Internal, Corner, and Edge Equilibria and Basins of Attraction

With more than two alleles, a number of different types of equilibria are possible, and some notation is helpful for characterizing these types. With n possible alleles, the space of potential allele frequencies is given by the **simplex** defined by the constraint $\sum_i^n p_i = 1$. Figure 5.3 shows this for the three-allele case, which is a section of a two-dimensional plane. With n alleles, the resulting simplex is a section of an $n-1$ dimensional hyperplane. We can distinguish between three types of equilibria based on their location on the simplex. **Corner equilibria** are those for which the frequency of one allele is 1, and hence all others are 0, corresponding to a corner of the simplex (Figure 5.3). With n alleles, there are n corner equilibria. With **edge equilibria**, the values of one (or more) of the allele frequencies are 0, while the rest are nonzero, i.e., two (or more) alleles are segregating in the population. Finally, we can have an **internal equilibrium**, in which all alleles are segregating (all $\widehat{p}_i > 0$). Thus, polymorphic equilibria correspond to either edge equilibria (not all alleles are segregating) or internal equilibria (all are segregating). Kingman's (constant-fitness) result states there is either no internal equilibrium, a single unique one, or a surface (such as a line or plane) embedded within the simplex.

Far more important than the existence of equilibria is their stability. As noted above, when allele frequencies at an unstable equilibrium are perturbed, they depart the neighborhood of this equilibrium value. Conversely, departures from the nearby vicinity of a stable equilibrium are followed by returns to the equilibrium. If we have a surface of equilibria (as might occur if two or more alleles have identical fitnesses), then we can also have a surface of **neutrally stable equilibria**. In such cases, provided we perturb the allele frequencies along the equilibrium surface, the subsequent allele frequencies do not change over time (the neutral Hardy-Weinberg condition is one such example).

When multiple stable equilibria exist, the initial conditions (history) of the process have a great influence on the final value reached. We saw this with underdominance (Example 5.4) where, if the population starts with frequency in the open interval $(0, \widehat{p})$, $p \to 0$, while if the population starts in the open interval $(\widehat{p}, 1)$, $p \to 1$. Thus, with multiple stable equilibria, there is a **basin of attraction** for each equilibrium value. Akin to rainfall over a wide area ending up in different rivers depending on which watershed basin it originally fell into, the domain of attraction for a stable equilibrium value is that region in the simplex within which starting allele frequencies eventually converge to the stable equilibrium of interest. In very special situations, one can use mathematical tools to determine such basins (e.g., Hofbauer and Sigmund 1988). More typically, however, one must systematically sample starting values within the simplex (e.g., using a grid of points) and then numerically iterate the equations of response to determine where the frequencies eventually converge.

Wright's Formula With Multiple Alleles

With only two alleles, Equation 5.5b (under the assumption of frequency-independent fitnesses) completely describes the evolutionary dynamics in terms of a single variable (the frequency of either allele). To express Equation 5.7c in a form analogous to Equation 5.5b, we again assume that $\partial W_{ij}/\partial p_k = 0$ for all i, j, and k (i.e., frequency-independent fitnesses). Taking the partial derivative of mean fitness with respect to p_i, the frequency of allele A_i,

$$\frac{\partial \overline{W}}{\partial p_i} = \frac{\partial}{\partial p_i}\left(\sum_j^n \sum_k^n p_j\, p_k\, W_{jk}\right) = 2\sum_k^n p_k W_{ki} = 2W_i \tag{5.9a}$$

Hence,

$$W_i = \frac{1}{2}\frac{\partial\,\overline{W}}{\partial\,p_i} \tag{5.9b}$$

Further, note that

$$\overline{W} = \sum_{j=1}^n p_j W_j = \frac{1}{2}\sum_{j=1}^n p_j \frac{\partial\,\overline{W}}{\partial\,p_j} \tag{5.9c}$$

Hence,

$$W_i - \overline{W} = \frac{1}{2}\left(\frac{\partial\,\overline{W}}{\partial\,p_i} - \sum_{j=1}^n p_j \frac{\partial\,\overline{W}}{\partial\,p_j}\right) \tag{5.9d}$$

Substituting Equation 5.9d into Equation 5.7c gives the allele-frequency change as

$$\Delta p_i = \frac{p_i}{2\overline{W}}\left(\frac{\partial\,\overline{W}}{\partial\,p_i} - \sum_{j=1}^n p_j \frac{\partial\,\overline{W}}{\partial\,p_j}\right) \tag{5.10}$$

This is the multiple-allele version of Equation 5.5b.

It is important to stress that Wright (1937) himself presented a different (and incorrect) version of his formula for multiple alleles, namely

$$\Delta p_i = \frac{p_i(1-p_i)}{2\overline{W}}\frac{\partial\,\overline{W}}{\partial\,p_i}$$

which appears widely in the literature. Comparing this with the two-allele version (Equation 5.5b), it is easy to see how Wright became seduced by this extension of his (correct) two-allele formula. In various subsequent descriptions of the multiple-allele version, Wright attempted to justify his 1937 expression by suggesting that it was not a normal partial derivative, but rather a measure of the gradient in mean fitness along a direction in which the relative proportions of the other alleles do not change (Wright 1942, 1955). However, Edwards (2000) showed that even this interpretation is not correct and presented the correct expression for Wright's (1942, 1955) later interpretation (Edwards' expression still differs from Equation 5.10 and lacks a transparent interpretation).

For a compact way to write Equation 5.10, recall that $(\partial\,\overline{W}/\partial\,p_i)(1/\overline{W}) = \partial\,\ln(\overline{W})/\partial\,p_i$, which yields

$$\Delta p_i = \sum_{j=1}^{n} G_{ij} \cdot \frac{\partial\,\ln\overline{W}}{\partial\,p_j} \tag{5.11a}$$

where

$$G_{ij} = \begin{cases} p_i(1-p_i)/2 & i = j \\ -p_i p_j/2 & i \neq j \end{cases} \tag{5.11b}$$

Equation 5.11a can be written in matrix form as

$$\Delta\mathbf{p} = \frac{1}{\overline{W}}\,\mathbf{G}\,\boldsymbol{\nabla}\overline{W} = \mathbf{G}\,\boldsymbol{\nabla}\ln(\overline{W}) \tag{5.12a}$$

where $\Delta\mathbf{p}$ is the vector of all allele-frequency changes,

$$\boldsymbol{\nabla}\overline{W} = \begin{pmatrix} \partial\,\overline{W}/\partial\,p_1 \\ \vdots \\ \partial\,\overline{W}/\partial\,p_n \end{pmatrix} \tag{5.12b}$$

is the gradient vector of all first partial derivatives, and the elements of the $n \times n$ **genetic variance-covariance matrix G** are given by Equation 5.11b.

Recall from vector calculus (Appendix 6) that the greatest local change in the value of $f(\mathbf{x})$ occurs by moving in the direction given by $\boldsymbol{\nabla} f$. Thus, $\boldsymbol{\nabla}\overline{W}$ (and hence $\boldsymbol{\nabla}\ln[\overline{W}]$) is the direction of allele-frequency change that maximizes the local change in mean fitness. However, the *actual* vector of change in allele frequencies is rotated away from $\boldsymbol{\nabla}\ln(\overline{W})$ by the matrix **G**. The genetic matrix **G** thus constrains the rate of selection response, a prelude to the theme of genetic constraints that arises in multivariate trait selection (Chapter 13).

It can be shown that Equation 5.12a implies $d\overline{W}/dt \geq 0$ (see Example A6.7), but unlike the diallelic case, the sign of Δp_i need not equal the sign of $\partial\,\ln\overline{W}/\partial\,p_i$. Alleles with the largest values of $p_i(1-p_i)\,|\,\partial\,\ln\overline{W}/\partial\,p_i\,|$ dominate the change in mean population fitness and hence dominate the allele-frequency dynamics. As these initially dominating alleles approach their equilibrium frequencies under selection (values where $|\,\partial\,\ln\overline{W}/\partial\,p_i\,| \simeq 0$), other alleles begin to dominate the dynamics of $\overline{W}$, with their frequencies changing in a way that continues to increase mean population fitness.

Changes in Genotypic Fitnesses, W_{ij}, When Additional Loci are Under Selection

All of the preceding results for allele-frequency change at a locus under selection assume that the genotypic fitnesses W_{ij} remain constant over generations. Changes in the environment can obviously compromise this assumption, as can frequency-dependent effects (such as rare-genotype advantage). A more subtle issue arises when additional loci influence fitness. Because the fitness of $A_i\,A_j$ is the average fitness over all multilocus genotypes containing these alleles, when linkage disequilibrium is present, correlations between alleles within gametes can create a dependency between the average fitness value of $A_i\,A_j$ and the frequency of at least one allele at this locus (Example 5.7). In this case, the assumption that $\partial\,W_{ij}/\partial\,p_k = 0$ can fail and Wright's formula no longer holds. A complete description of the allele-frequency dynamics then requires following *all* loci under selection. As we will show, this is not trivial, as it requires modeling more than just the allele-frequency changes at each locus. Selection generates nonrandom associations (linkage, or gametic-phase, disequilibrium) among loci, requiring us to model the dynamics of multilocus *gamete* frequencies to account for both the frequencies of alleles over all loci and the disequilibria between them. If gamete-frequency changes due to recombination occur on a much shorter time scale than changes due to selection, linkage disequilibrium is expected to be negligible, and Wright's formula can be applied as a good approximation to certain quantitative-genetic problems (e.g., Barton 1986; Barton and Turelli 1987; Hastings and Hom 1989).

Example 5.7. Consider two diallelic loci with alleles *A*, *a* and *B*, *b*, and let p = freq (*A*) and q = freq (*B*). The frequency of the gametes *AB* and *Ab* are $pq + D$ and $p(1- q) - D$, respectively, where *D* is the linkage disequilibrium between these two loci (Equation 2.18). The marginal (or **induced**) fitness W_{AA} of *AA* individuals (the fitness of *AA* averaged over all genetic backgrounds) is

$$W_{AA} = W_{AABB} \cdot \Pr(AABB\,|AA) + W_{AABb} \cdot \Pr(AABb\,|AA) + W_{AAbb} \cdot \Pr(AAbb\,|AA)$$

These conditional weights, such as $\Pr(AABB\,|AA)$, are the conditional probabilities of the two-locus genotype (here $AABB$) given a single-locus genotype (here AA), and are obtained from the standard formula for conditional probability, $\Pr(x\,|\,y) = \Pr(x,y)/\Pr(y)$, as follows. Under random mating $\Pr(AA) = p^2$, while $\Pr(AABB)$ is the probability of getting AB gametes from both parents; or, from above, $\text{Freq}(AB)^2 = (pq + D)^2$, giving $\Pr(AABB\,|AA) = (pq + D)^2/p^2$. Similar results for the remaining two *B* locus genotypes results in

$$W_{AABB}\,\frac{(pq + D)^2}{p^2} + W_{AABb}\,\frac{2(pq + D)\,[\,p\,(1 - q) - D\,]}{p^2} + W_{AAbb}\,\frac{[\,p(1 - q) - D\,]^2}{p^2}$$

In the absence of linkage disequilibrium (D = 0), the marginal fitness reduces to

$$W_{AA} = W_{AABB} \cdot q^2 + W_{AABb} \cdot 2\,q\,(1 - q) + W_{AAbb} \cdot (1 - q)^2$$

which is independent of p, the frequency of *A*. In this special case, even though the marginal fitness of *AA* changes as the frequency q of allele *B* changes, Wright's formula (Equation 5.5b) for the change in the frequency p of allele *A* still holds, as the fitness of *AA* does not depend on p. However, when $D \neq 0$, W_{AA} is a complex function of p, q, and D, so that $\partial\, W_{AA}/\partial\, p \neq 0$ and Wright's formula does not hold. It is worth nothing that even if $D = 0$ initially, disequilibrium is typically built up by selection (Chapters 16 and 24), although it may be sufficiently small for Wright's formula to serve as a good approximation.

SELECTION ON TWO LOCI

When fitness is influenced by n biallelic loci, we cannot generally predict how genotype frequencies will evolve by simply considering the n sets of single-locus allele-frequency change equations (Equation 5.7a). The major complication is *gametic-phase disequilibrium*, which (if present) thwarts the prediction of gamete frequencies from simple allele frequencies alone (Chapter 2; LW Chapter 5). In addition, the marginal fitnesses W_{ij} associated with any one of the loci can themselves be functions of the frequencies of alleles at other loci (see Example 5.7). These complications necessitate following *gamete* rather than *allele* frequencies, requiring many more equations. Further, when disequilibrium and/or epistasis in fitness occur, complicated multiple equilibria can result. Although most forms of selection generate some disequilibrium even between unlinked loci (Chapter 16), if selection is weak relative to recombination, disequilibrium is often very small. However, as we will see in Chapter 16, even small disequilibrium values can be cumulatively rather significant when summed over a large number of loci underlying quantitative-trait variation.

We focus here on the simplest case of two diallelic loci (alleles A, a and B, b) with random mating and frequency-independent viability selection. Even in this case, the general behavior with constant fitnesses has not been solved outside of a few special cases, and the development of theory beyond two loci is still in a rather embryonic stage (but see Kirkpatrick et al. 2002). Our purpose is simply to introduce some of the complications that arise due to gametic-phase disequilibrium, rather than to examine the theory in detail. For comprehensive reviews, see Karlin (1975), Nagylaki (1977a, 1992a), Hastings (1990b, 1990c), Bürger (2000), Christiansen (2000), and Ewens (2004).

Dynamics of Gamete-frequency Change

Denote the frequencies of the four different gametes by x_i, where

$$\begin{array}{ll} \text{freq}(g_1) = \text{freq}(AB) = x_1 & \text{freq}(g_2) = \text{freq}(Ab) = x_2 \\ \text{freq}(g_3) = \text{freq}(aB) = x_3 & \text{freq}(g_4) = \text{freq}(ab) = x_4 \end{array}$$

Under random mating (the random union of gametes), the frequency of the different (unordered) genotypes is given by

$$\text{freq}(g_i\, g_j) = \begin{cases} 2x_ix_j & \text{for } i \neq j \\ x_i^2 & \text{for } i = j \end{cases}$$

Let the fitness of an individual formed from gametes g_i and g_j be $W_{g_ig_j} = W_{g_jg_i}$ (we use the $g_i\, g_j$ subscript notation to stress that these fitnesses are for specific *gametic*, as opposed to *allelic*, combinations). $W_{g_1g_4}$ and $W_{g_2g_3}$ are of special note, being the fitness of ***cis*** (AB/ab) and ***trans*** (Ab/aB) double heterozygotes, respectively. One would normally expect these two genotypes to have equal fitness, but certain genetic interactions (such as position effects) can sometimes complicate matters. In addition, if the two loci being considered are themselves in gametic-phase disequilibrium with other loci affecting fitness, *cis* and *trans* fitnesses will appear to differ due to fitness associated with loci not considered (Turelli 1982). Denoting the gamete frequencies after selection by x_i', with constant viability selection ($W_{g_ig_j}$ constant), no *cis-trans* effect ($W_{g_1g_4} = W_{g_2g_3}$), and discrete nonoverlapping generations, the gametic recursion equations become

$$x_1' = [\, x_1\, W_{g_1} - c\, D\, W_{g_1g_4}\,]/\overline{W} \tag{5.13a}$$
$$x_2' = [\, x_2\, W_{g_2} + c\, D\, W_{g_1g_4}\,]/\overline{W} \tag{5.13b}$$
$$x_3' = [\, x_3\, W_{g_3} + c\, D\, W_{g_1g_4}\,]/\overline{W} \tag{5.13c}$$
$$x_4' = [\, x_4\, W_{g_4} - c\, D\, W_{g_1g_4}\,]/\overline{W} \tag{5.13d}$$

where c is the recombination fraction between loci, $D = x_1\, x_4 - x_2\, x_3$ is a measure of gametic-phase disequilibrium, and W_{g_i} is the average fitness of a g_i-bearing individual (the marginal fitness of gamete g_i), with

$$W_{g_i} = \sum_{j=1}^{4} x_j\, W_{g_ig_j} \quad \text{and} \quad \overline{W} = \sum_{i=1}^{4} x_i\, W_{g_i} \tag{5.13e}$$

These equations are due to Kimura (1956) and Lewontin and Kojima (1960). Observe that selection can change gamete frequencies by changing allele frequencies and/or by changing the amount of gametic-phase disequilibrium. Because the four gamete frequencies sum to one, Equations 5.13a–5.13d can be expressed in terms of three equations. Alternatively, one can also parameterize this set of equations using another set of three variables, namely, the allele frequencies p and q for the two loci and their disequilibrium, D.

Equations 5.13a–5.13d are similar in form to the multiple-allele equation (and identical to Equation 5.7a when c or D equals zero). Unlike allele frequencies (which do not change in the absence of selection under our assumption of infinite population size), gamete frequencies *can* change from generation to generation due to changes in D from recombination, even in the absence of selection (Chapter 2; LW Chapter 5). If D is zero and remains zero after selection (as occurs when fitnesses are **multiplicative** across loci, so that $W_{ijkl} = W_{ij}W_{kl}$), then the new gamete frequency is simply given by the product of the new allele frequencies, e.g., $x'_1 = p'_A\, p'_B$, and the dynamics can be followed by considering each locus separately (i.e., following Δp_A and Δp_B). However, except in this and a few other special cases, these two-locus equations turn out to be extremely complex. Indeed, as we detail below, there is no general analytic solution for even the simple model of constant fitnesses and viability selection, except when certain symmetry patterns (such as additivity) hold among the fitnesses (Table 5.2).

Example 5.8. If loci have effects on both fitness and on a character not under selection, an incorrect picture as to which characters are under selection can result. The following example, modified from Robertson (1967), illustrates some of the problems that can arise. Let loci **A** and **B** affect fitness (perhaps through some unmeasured character) in addition to influencing character z, which is not itself under selection, with the following fitnesses and character values:

Fitness	*AA*	*Aa*	*aa*
BB	1.0	1.1	1.0
Bb	1.1	1.2	1.1
bb	1.0	1.1	1.0

Character z	*AA*	*Aa*	*aa*
BB	2	2	1
Bb	2	2	1
bb	1	1	0

Alleles A and B are dominant for the character z, while fitness increases with the number of loci that are heterozygous.

Assume that there is gametic-phase equilibrium and that the frequencies of alleles A and B are both $2/3$, in which case $\overline{W} = 1.089$, $\mu_z = 1.78$, and the expected fitnesses for each phenotype become

z	0	1	2
$W(z)$	1.00	1.05	1.10

For example, *AAbb, Aabb, aaBb,* and *aaBB* all have a trait value of one, and (under linkage equilibrium), each of these genotypes has frequency of 4/81, for a total frequency of 16/81. The expected fitness for $z = 1$ is therefore

$$\frac{4/81\,(W_{AAbb} + W_{Aabb} + W_{aaBb} + W_{aaBB})}{16/81} = \frac{1.0 + 1.1 + 1.1 + 1.0}{4} = 1.05$$

If we measured the value of z and fitness in a random sample of individuals from this population, we would conclude that z is under directional selection and expect the trait mean μ_z to increase over time. However, applying two-locus theory (numerical iteration of Equations 5.13a–5.13d) shows that at equilibrium, $p_A = p_B = 1/2, \overline{W} = 1.1$, and $\mu_z = 1.50$. Hence, despite the initial positive correlation between z and W, selection causes μ_z to *decline* from its initial starting value of 1.78.

Gametic Equilibrium Frequencies, Linkage Disequilibrium, and Mean Fitness

Now let us consider the equilibrium behavior of two-locus systems. The equilibrium value $\widehat{D}$ represents the balance between recombination driving disequilibrium to zero and selection generating new disequilibrium. Bounds on $\widehat{D}$ for general two-locus systems were given by Hastings (1981b, 1986). A nonzero value of $\widehat{D}$ requires epistasis in fitness (nonadditive fitnesses *across* loci), and such a nonzero value has implications for the behavior of mean population fitness. To see this, first note that at equilibrium, the gamete frequencies remain unchanged ($\widehat{x}_i' = \widehat{x}_i$), and Equation 5.13a becomes

$$\widehat{x}_i = [\,\widehat{x}_i\,\widehat{W}_{g_1} - c\,\widehat{D}\,W_{g_1g_4}\,]/\overline{W}, \quad \text{or} \quad \overline{W} = \widehat{W}_{g_i} - c\,W_{g_1g_4}\frac{\widehat{D}}{\widehat{x}_i}$$

Similarly, Equations 5.13b–5.13d yield

$$\overline{W} = \widehat{W}_{g_i} + \eta_i c\,W_{g_1g_4}\frac{\widehat{D}}{\widehat{x}_i}, \quad \text{where} \quad \eta_i = \begin{cases} -1 & \text{for } i = 1,4 \\ \;\;1 & \text{for } i = 2,3 \end{cases} \tag{5.14}$$

If linkage is complete ($c = 0$), then all marginal fitnesses are equal at equilibrium and equilibrium mean fitness is at a local maximum. This second result follows because complete linkage causes the system to behave like a single locus with four alleles, meaning that Kingman's (1961a) result that equilibrium mean fitness is at a local maximum applies. However, when $c \neq 0$, then because in general $\widehat{D} \neq 0$ (there is gametic-phase disequilibrium at the equilibrium gamete frequencies), the equilibrium gametic fitnesses are a function of the recombination frequency c. What is most interesting is that when $\widehat{D} \neq 0$, the marginal gametic fitnesses W_{g_i} are *not* equal, and equilibrium mean fitness is not at a local maximum. Indeed, it can be shown that mean fitness often decreases as the equilibrium values are approached. Typically, this decrease is quite small, but it no longer holds that mean fitness always increases under constant-fitness viability selection (Kojima and Kelleher 1961).

Example 5.9. A simple example shows that mean fitness in a two-locus model can continuously decline toward an equilibrium value, as opposed to the continual increase to an equilibrium seen with one locus. Suppose the fitness of *AaBb* is 1 + *s* (where $s > 0$), while all other genotypes have a fitness of 1. If we form a population by crossing *AABB* and *aabb* parents, then all F_1 individuals will be *AaBb* and the mean population fitness is 1 + *s*. In each subsequent generation, mean population fitness decreases as the frequency of *AaBb* double heterozygotes will be reduced by recombination until equilibrium is reached (which takes several generations even if $c = 1/2$). For example, if $s = 0.1$, $\overline{W} = 1.1$ for the F_1, while iteration of Equations 5.13a–5.13d shows $\overline{W} = 1.025$ at equilibrium (under loose linkage).

Results for Particular Fitness Models

There are a number of ways to parameterize the general two-biallelic-locus fitness model (Table 5.2). Under the assumption of no *cis/trans* effects, there are eight free parameters (one of the nine fitnesses can always be normalized to one). When fitnesses are additive *across* loci (i.e., no epistasis but the possibility of dominance at each locus), two-locus systems (or multi-locus systems for that matter) will be well behaved in that there is at most one polymorphic equilibrium for any given set of segregating alleles, at which $\overline{W}$ is at a local maximum (Karlin and Liberman 1979). While D is zero at equilibrium in such cases, selection generates non-zero D during the sojourn to the equilibrium value for the gamete frequencies.

In contrast to this fairly simple equilibrium behavior under additive fitnesses, when epistasis in fitness exists, the behavior of gamete frequencies can be extremely complicated. For example, with sufficiently tight linkage and certain fitness values, there can be as many as nine polymorphic equilibria (many of which may be stable) for the symmetric viability model given in Table 5.2 (Hastings 1985). Hence, even with constant fitnesses, the final equilibrium state is potentially highly sensitive to initial conditions. Further, stable limit cycles can also exist, where equilibria are no longer point values (Akin 1979, 1982; Hastings

Table 5.2 Alternative parameterizations and specific models for viability selection on two loci.

	General Fitness		
	BB	*Bb*	*bb*
AA	W_{AABB}	W_{AABb}	W_{AAbb}
Aa	W_{AaBB}	W_{AaBb}	W_{Aabb}
aa	W_{aaBB}	W_{aaBb}	W_{aabb}
	Fitness Additive Between Loci		
	BB	*Bb*	*bb*
AA	$1-a-b$	$1-a$	$1-a-c$
Aa	$1-b$	1	$1-c$
aa	$1-d-b$	$1-d$	$1-d-c$
	Symmetric Viability		
	BB	*Bb*	*bb*
AA	$1-a$	$1-b$	$1-d$
Aa	$1-e$	1	$1-e$
aa	$1-d$	$1-b$	$1-a$

1981a), although point equilibria always exist if epistasis and/or selection are sufficiently weak (Nagylaki et al. 1999).

Example 5.10. The symmetric viability model (Table 5.2) arises in certain models of stabilizing selection on a character determined by additive loci. Suppose that two loci contribute in a completely additive fashion (e.g., no dominance or epistasis) to a character z under stabilizing selection, with $W(z) = 1 - s(z-2)^2$, which implies an optimal phenotypic value of two. Fitness functions of this general form were first introduced by Wright (1935a, 1935b) with his **quadratic optimum model**. Assuming that each capital letter allele adds one to z (and that there is no environmental variance), the resulting phenotypic and fitness values are

	Character value z			Fitness		
	AA	*Aa*	*aa*	*AA*	*Aa*	*aa*
BB	4	3	2	$1-4s$	$1-s$	1
Bb	3	2	1	$1-s$	1	$1-s$
bb	2	1	0	1	$1-s$	$1-4s$

Hence, while the trait has a completely additive genetic basis, the (nonlinear) mapping from phenotype to fitness introduces epistasis in fitness. This is an important point: simply showing that a *trait* under selection has an additive genetic basis is *not* sufficient to imply that *fitnesses* are also additive.

Phenotypic Stabilizing Selection and the Maintenance of Genetic Variation

We have seen that when the heterozygote is favored at a *single* locus, selection maintains both alleles (Example 5.4). However, Example 5.6 showed that when a number of loci underlie a trait with an additive genetic basis under stabilizing selection, underdominant selection occurs (encouraging the removal, rather than the maintenance, of genetic variation). All of this begs a simple question (with a very complex answer): under what conditions does stabilizing selection maintain genetic variation at a number of loci? This is of one of many questions that follow from one of the most perplexing observations in quantitative genetics—the maintenance of high levels of genetic variation for most traits under apparent stabilizing selection. We consider this in earnest in Chapter 28, confining our remarks here to the prospect that selection alone can maintain variation.

As detailed below in Example 5.11, even Wright's simple quadratic optimum model with additive gene action exhibits considerable complexity. Further, in the virtually certain event that the double heterozygote does not exactly correspond to the optimal phenotypic value, the fitness matrix immediately becomes asymmetric, leaving the general (and hence unsolved) two-locus model, with all of its potentially complex behavior. This is also true with epistasis in the trait under selection.

Where does all this modeling leave us? As we detail in Example 5.11, analysis of the highly symmetric Wright model (equal allelic effects, a quadratic fitness function, and a double heterozygote value equal to the optimal phenotypic value) shows that stabilizing selection on an additive trait cannot maintain variation. As we start to disrupt these symmetries (e.g., by allowing unequal allelic effects), we find conditions under which stabilizing selection can maintain variation at one or both loci. Indeed, superficially minor issues, such as subtle differences in fitness functions or noncorrespondence between the double heterozygote and optimal trait value, can result in qualitatively different behavior relative to the Wright model.

What happens when we move beyond two loci? Ironically, the situation may start to become simpler again. Bürger and Gimelfarb (1999) simulated stabilizing selection under the generalized Wright model and found, for randomly generated parameter sets (linkage, allelic effects, and strength of selection) that roughly 17% of two-locus systems maintained alleles at both loci at equilibrium. However, as they considered three-, four-, and five-locus systems, this probability (of two or more loci being polymorphic at equilibrium) fell dramatically, to $< 0.5\%$ in the five-locus models.

As one adds more loci (while keeping the range of phenotypic values constant), the effects of selection on any individual locus will be reduced and the behavior of many models will become much simpler. For example, Hastings and Hom (1989, 1990) showed that, under weak selection, the number of polymorphic loci that can be maintained by stabilizing selection is bounded above by the number of independent traits under selection. Thus, with sufficiently weak selection, stabilizing selection on k independent traits can maintain variation at no more than k loci, implying that if only one trait is under stabilizing selection under these conditions, *at most* only one underlying locus will be polymorphic. As Example 5.6 showed, it can easily be the case that *no* loci remain polymorphic (in the absence of new mutation). We examine these issues in more detail in Chapter 28.

In summary, analyses of strong-selection two-locus models instill caution about general statements of selection in multilocus systems. These concerns are quite valid when a major gene (or genes) accounts for most of the variation in the trait of interest. However, if selection tends to be weak relative to recombination (as might be expected in systems with a large number of loci with equal effects), the response under such genetic architectures may have a simpler and more predictable behavior. As reviewed in the next chapter, the assumption of weak selection on individual loci is the basis of several general statements about the behavior of fitness and trait evolution under weak selection (on the individual underlying loci), such as the breeder's equation. While any such general statements are not true in all settings (as the strong-selection two-locus results bear out), they may be largely true in many biological settings.

Example 5.11. We now consider the generalized two-locus version of Wright's quadratic optimum model, which provides significant insight into many of the issues concering the maintenance of variation strictly by selection. This model has been examined by numerous authors (e.g., Wright 1935a, 1935b; Hastings 1987a; Gavrilets and Hastings 1993, 1994a; Bürger and Gimelfarb 1999), and we follow the excellent treatment of Bürger (2000, pp. 204–210). Willensdorfer and Bürger (2003) presented a similar analysis for Gaussian selection.

The generalized model makes four key assumptions. First, fitness is a quadratic function of the phenotypic value z, $W(z) = 1 - sz^2$, with an implicit optimum at $z = 0$. Second, the genotypic value of the double heterozygote exactly corresponds with the phenotypic optimum. Third, there are no environmental effects. Fourth, the trait under stabilizing selection has a completely additive genetic basis (no dominance or epistasis). Thus $-a_1 : 0 : a_1$ are the genotypic values at the first locus (corresponding to $aa : Aa : AA$), while the values for the second locus are $-a_2 : 0 : a_2$ (corresponding to $bb : Bb : BB$).

While Wright's original analysis (and that of several other authors) assumed that allelic effects are identical ($a_1 = a_2$), the generalized version allows for $a_1 \geq a_2$. The resulting trait values become

	aa	Aa	AA
bb	$-(a_1 + a_2)$	$-a_2$	$a_1 - a_2$
Bb	$-a_1$	0	a_1
BB	$a_2 - a_1$	a_2	$a_1 + a_2$

Substituting these trait values into the quadratic fitness function shows that this model corresponds to the symmetric viability model given in Table 5.2, with

$$a = s(a_1 + a_2)^2, \quad b = sa_2^2, \quad d = s(a_1 - a_2)^2, \quad e = sa_1^2$$

Note that there is a relationship among these selection coefficients, $a + d = 2(b + e)$, which follows from the quadratic fitness function used.

Depending on parameter values, this model can have up to 11 equilibria, 7 of which are potentially stable (but not simultaneously so). There are always four trivial **corner equilibria** corresponding to each of the four gametes being fixed. The equilibria corresponding to either AB or ab being fixed are always unstable, but the other two corner equilibria, corresponding to Ab or aB being fixed, are potentially stable (as they yield more intermediate phenotypic values). There may also be two **edge equilibria**, corresponding to fixation at one locus and segregation at the other. For either of these edge equilibria to be admissible (the equilibrium x_i values lying on the simplex), a_1 must exceed $2a_2$ (given our ordering of $a_1 \geq a_2$), and hence the designation of **A** as the major locus. By definition, disequilibrium is zero at both the corner and edge equilibria.

Finally, there are three potential internal equilibria. The first is the so-called **symmetric equilibrium**, where both loci are segregating with all alleles at frequency 1/2, with gamete frequencies

$$\widehat{x}_{AB} = \widehat{x}_{ab} = \frac{1}{4} + \widehat{D}_{AB} \quad \text{and} \quad \widehat{x}_{Ab} = \widehat{x}_{aB} = \frac{1}{4} - \widehat{D}_{AB}$$

where $\widehat{D}_{AB} < 0$, i.e., the *AB* and *ab* gametes are underrepresented (see Bürger 2000 for the value of $\widehat{D}_{AB}$). This implies hidden genetic variation, with the additive variance for this trait increasing in the event of cessation of selection and the subsequent restoration of linkage equilibrium by recombination. (Chapter 16). While there is additive variance in the *trait* under selection at the equilibrium mean, as we will see in the next chapter (Example 6.6), there is no additive variance *in fitness* at equilibrium. Although the symmetric internal equilibrium seems straightforward, there can also be two other stable interior equilibria, the so-called **unsymmetric equilibria**. The expressions for these are complex (see Bürger 2000, p. 205), and their existence requires that the recombination fraction lie in a narrow range (see Figure 5.4).

For this model there are four mutually exclusive regions of the parameter space that correspond to different stable equilibria (Figure 5.4). In region 0, only the two corner equilibria (with either *Ab* or *aB* fixed) are stable. Note that if the two loci have equal effects ($a_2/a_1 = 1$, as Wright originally assumed), these are the only stable equilibria, and stabilizing selection leads to the removal of variation from both loci. In Region 1, the two edge equilibria (with the major locus being polymorphic and **B** fixed for either *bb* or *BB*), are the only stable equilibria. Note that this requires both rather uneven effects ($a_1 \geq 2a_2$) and recombination sufficiently large relative to selection. Finally, there are two regions where the internal equilibrium is stable. The very narrow Region 2b corresponds to the two stable unsymmetric internal equilibria, which requires a very specific relationship between selection and recombination. Region 2a corresponds to uneven allelic effects and recombination that is weak relative to selection, where a symmetric equilibrium exists.

Thus, provided allelic effects are uneven and selection is strong relative to recombination, selection can maintain both alleles at both loci (Kearsey 1968; Kearsey and Gale 1968; Nagylaki 1989a; Gavrilets and Hastings 1993, 1994a, 1994b). Gavrilets and Hastings showed

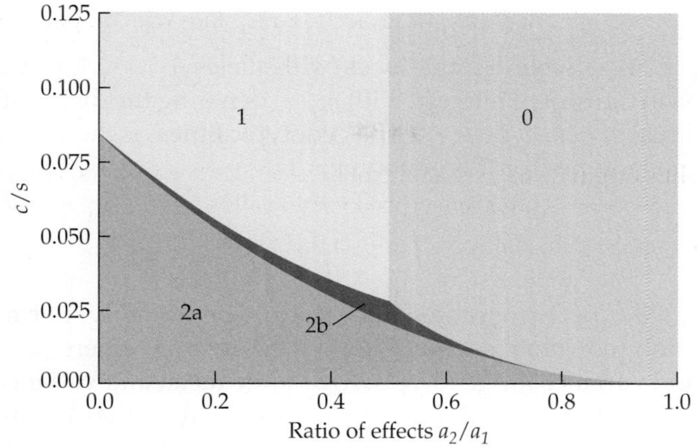

Figure 5.4 Different regions of equilibrium behavior under the generalized quadratic optimal model (Example 5.11). Here s is the strength of phenotypic selection, c is the recombination rate between loci, and a_1 and a_2 are the additive effects for alleles at the two loci (with $a_1 \geq a_2$). Region 0 corresponds to corner equilibria (Figure 5.3), where the population is fixed for a single gamete. Region 1 corresponds to edge equilibria, where the major locus is segregating and the minor locus is fixed. Region 2 corresponds to conditions where selection maintains polymorphisms at both loci (an internal equilibria). In region 2b, this equilibrium is unsymmetric (allele frequencies are unequal), while it is symmetric (all alleles have frequency 1/2) in region 2a. Wright's classic (1935a, 1935b) analysis, which assumes $a_2/a_1 = 1$, places the system in region 0, with the system fixed for a single gamete type (which varies depending on starting conditions). (After Bürger and Gimelfarb 1999.)

that, with strong selection, the mean trait value at equilibrium does not necessarily coincide with the optimum fitness value, so in general $\widehat{\mu}_z \neq 0$. Hence, at equilibrium, there can be the appearance of apparent directional selection.

The final subtlety is that very different results can arise from a simple change in the fitness function, for example from a quadratic to a Gaussian (Gimelfarb 1996b). Under a quadratic fitness function, $a + d = 2(b + e)$, which reduces what is normally a cubic equation for the equilibrium value of D to a quadratic equation (Gimelfarb 1996b). Under a Gaussian, this relationship no longer holds and the resulting equilibrium structure is potentially more complex. Indeed, Gimelfarb (1996b) showed that under sufficiently strong Gaussian selection, very unusual behavior can occur, such as the appearance of two internal symmetric equilibria with D values of opposite sign.

SELECTION ON A QUANTITATIVE TRAIT LOCUS

While population genetics is concerned with how the frequencies of specific *genotypes* change under selection, quantitative genetics is concerned with the evolution of composite features of these underlying genetic changes, i.e., the change in the mean or additive genetic variance of a *trait* under selection. Population-genetic models assume that we know the genotype-specific fitnesses and use these to generate expressions for the change in allele (one locus) or gamete (multiple-locus) frequencies. In contrast, quantitative genetics assumes fitness to be a function, $W(z)$, of the phenotypic value z of the focal trait. The connection between these two approaches starts by considering how selection on a particular trait maps into the average excess s_i in fitness for an allele at a locus underlying this trait.

Monogenic Traits

The simplest situation arises when a single locus (with alleles $A_1, \cdots, A_n$) entirely determines the genetic variation in the trait of interest, with $p_{ij}(z)$ denoting the distribution of character values for an individual of genotype A_iA_j. The genotypic fitness is the average of $W(z)$ over the distribution of phenotypes for this genotype,

$$W_{ij} = \int W(z)\, p_{ij}(z)\, dz \tag{5.15a}$$

In many situations, we expect environmental values to be (roughly) normally distributed about the mean genotypic value, so that $p_{ij}(z) \sim \mathrm{N}(\mu_{ij}, \sigma^2_{ij})$, where μ_{ij} and σ^2_{ij} are the phenotypic mean and variance for genotype A_iA_j. If the mean and variance are known for each genotype, and no other loci influence variation in z, then the W_{ij} are constant from one generation to the next (assuming no frequency-dependent selection or changes in the environment), and the values from Equation 5.15a can be substituted into Equation 5.1b or Equation 5.7c to directly compute the change in allele frequencies.

Likewise, if $p_i(z)$ denotes the phenotypic distribution for individuals carrying an A_i allele, the average fitness of individuals carrying an A_i allele is

$$W_i = \int W(z)\, p_i(z)\, dz \tag{5.15b}$$

Again, this can be directly substituted into Equation 5.7c to compute Δp_i. Note, however, that while $p_{ij}(z)$ can be independent of allele frequency, this is not the case for $p_i(z)$.

Many Loci of Small Effect Underlying the Character

When two or more loci underlie the character of interest, Equations 5.15a and 5.15b become problematic because the conditional densities $p_{ij}(z)$ and $p_i(z)$ are likely to change in each generation as selection changes the genotype frequencies at other loci. Ideally, we would like to have an approximation that uses only the unconditional phenotypic distribution $p(z)$ and some simple property of the locus being considered. Fortunately, in many situations, the average excess α_i of the trait (LW Chapter 4) provides such a connection for loci of small effect. It will prove slightly easier to work with relative fitnesses, so we will use $w(z) = W(z)/\overline{W}$, the expected relative fitness of an individual with phenotypic value z, throughout.

Following Bulmer (1971a) and Kimura and Crow (1978), assume that the average excess is small relative to the variance of z, as would occur if many loci of roughly equal effect underlie the character or if there are large environmental effects. Because having a copy of A_i increments the phenotype on average by α_i, as is shown in Figure 5.5, the conditional phenotypic distribution is, to a good approximation, the unconditional phenotypic distribution shifted by α_i, which can be written as

$$p_i(z) \simeq p(z - \alpha_i) \tag{5.16a}$$

Nagylaki (1984) showed that this approximation is correct only to linear order, e.g., to terms of order α_i (a point that we return to below). Alternatively, we could also consider the distribution given the *genotype* at this locus (rather than a specific *allele*), in which case

$$p_{ij}(z) \simeq p(z - a_{ij}) \tag{5.16b}$$

applies, where a_{ij} is the average deviation from the overall trait mean for an individual of genotype A_iA_j (again, this is correct only to linear order).

The approximation given by Equation 5.16a motivates two alternative expressions for the relative fitness, w_i. First, we have, directly that

$$w_i = \int w(z)\, p_i(z)\, dz \simeq \int w(z)\, p(z - \alpha_i)\, dz \tag{5.17a}$$

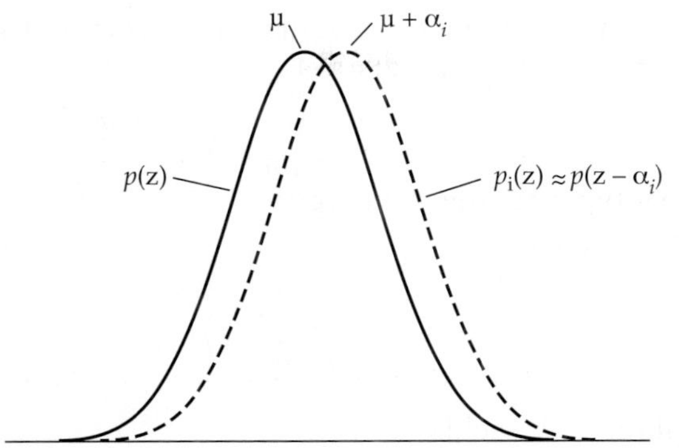

Figure 5.5 The unconditional phenotypic distribution $p(z)$ has a mean of μ, while the conditional phenotypic distribution $p_i(z)$ for an individual carrying a copy of allele A_i has a mean of $\mu+\alpha_i$, the mean plus the average excess for allele A_i. If α_i is small, then (to order α_i) we can approximate $p_i(z)$ by $p(z-\alpha_i)$, which shifts the phenotypic distribution to the right (for $\alpha_i > 0$) by α_i. This is only approximate, as knowing which allele is present at one locus decreases the genetic variance and results in the conditional phenotypic distribution having a (slightly) smaller variance.

(Bulmer 1971a; Kimura and Crow 1978). Alternatively, following Kimura and Crow (1978), a change of variables results in

$$w_i \simeq \int w(z+\alpha_i)\, p(z)\, dz \tag{5.17b}$$

For certain phenotypic distributions and fitness functions, these integrals can be evaluated exactly (Latter 1965a; Lynch 1984). However, even in these cases, the resulting w_i values are still only approximations because Equation 5.16a itself is only approximate. When the integral cannot be evaluated, a Taylor series expansion provides a useful approximation, often without having to completely specify the phenotypic distribution and/or fitness function. If the average excess α_i for the trait value is small,

$$p(z-\alpha_i) \simeq p(z) - \alpha_i \frac{d\,p(z)}{dz} \tag{5.18a}$$

$$w(z+\alpha_i) \simeq w(z) + \alpha_i \frac{d\,w(z)}{dz} \tag{5.18b}$$

Substituting into Equations 5.17a and 5.17b and recalling that $\int w(z)\, p(z)\, dz = 1$ yields the average excess in relative fitness (Equation 5.8b) as

$$s_i = w_i - 1 \simeq -\alpha_i \int w(z)\, \frac{d\,p(z)}{dz}\, dz \tag{5.19a}$$

and

$$s_i = w_i - 1 \simeq \alpha_i \int p(z)\, \frac{d\,w(z)}{dz}\, dz \tag{5.19b}$$

Equation 5.19a is applicable if the phenotypes are distributed continuously. For meristic (i.e., discrete) traits, Equation 5.19b applies provided $w(z)$ is differentiable (the integral is replaced by a weighted sum over the discrete trait values). Note that Equations 5.19a and 5.19b connect population genetics (the average excess s_i in fitness) with quantitative genetics ($w(z)$ and the average excess α_i in the trait value).

The integrals in Equations 5.19a and 5.19b represent the change in fitness associated with linear deviations of a character value from its mean (i.e., directional selection). To see this, consider the case in which phenotypic values are normally distributed, with a mean of μ and a variance of σ_z^2 at the population level,

$$p(z) = \left(2\pi\sigma^2\right)^{-1/2} \exp\left(\frac{-(z-\mu)^2}{2\sigma^2}\right) \tag{5.20a}$$

which upon differentiation results in

$$\frac{d\,p(z)}{dz} = -\left(\frac{z-\mu}{\sigma_z^2}\right) p(z) \tag{5.20b}$$

Substituting into Equation 5.19a yields

$$\begin{aligned} s_i &\simeq \frac{\alpha_i}{\sigma_z^2} \int w(z) \cdot (z-\mu)\, p(z)\, dz \\ &= \frac{\alpha_i}{\sigma_z^2} \left(\int z\, w(z) p(z)\, dz - \mu \int w(z)\, p(z)\, dz \right) = \frac{\alpha_i}{\sigma_z^2}\,(\mu^* - \mu) \end{aligned} \tag{5.20c}$$

The first integral is the mean μ^* after selection because $w(z)p(z)$ is the density of z following selection, while the second integral is one because $\overline{w} = 1$. Noting that the directional selection differential (the within-generation change in the mean from selection; Chapter 13) is $S = \mu^* - \mu$, we have

$$s_i \simeq \alpha_i \left(\frac{S}{\sigma_z^2}\right) = \overline{\imath}\left(\frac{\alpha_i}{\sigma_z}\right) \tag{5.21}$$

where $\overline{\imath} = S/\sigma_z$ is the standardized selection differential (or selection intensity; Chapter 13). Hence, to first order, the selection on an individual allele of small effect is approximately equal to its standardized average excess in the trait value (α_i/σ_z) multiplied by the selection intensity on the trait. As noted by Kimura and Crow (1978) and Milkman (1978), this approximation (Equation 5.21) is a well-known result for certain fitness functions, e.g., truncation selection (Haldane 1930; Griffing 1960a), and is a good approximation for arbitrary fitness functions when z is normally distributed.

One consequence of the first-order expression in s_i corresponding to the effects of directional selection is that for strictly stabilizing selection (i.e., no directional selection component), the first-order terms are zero, and we must consider second-order terms in order to have a proper approximation for s_i (to account for the fact that conditioning on the allelic state slightly reduces the variance). We will return to this point shortly.

A Population-genetic Derivation of the Breeder's Equation

The classic expression for the expected response R (the change in mean) of a single trait to selection is the **breeder's equation,** $R = h^2 S$ (Chapter 13). This expression is typically derived by assuming a linear midparent-offspring regression with slope h^2, although a few additional, and subtle, assumptions are required (Chapter 6). Here we show how the breeder's equation is obtained as an approximation of a population-genetic model of the response. As developed in Example 5.12 (below), the expected response, R_k, from a single locus k can be expressed in terms of the average effects, $\alpha_{k,i}$, on the trait and the average excesses on relative fitness, $s_{k,i}$, for all alleles at locus k, and the dominance deviations $\delta_{k,ij}$ for all pairs of alleles, with

$$R_k = 2\sum_i \alpha_{k,i} s_{k,i}\, p_{k,i} + \sum_{i,j} \delta_{k,ij}\, p_{k,i}\, s_{k,i}\, p_{k,j}\, s_{k,j} \tag{5.22}$$

where the sums are taken over all n_k alleles at the focal locus k. Recalling Equation 5.19, we can write $s_i \simeq \alpha_i\, I$, with I being the appropriate integral, which results in

$$R_k = I \sum_i 2\,\alpha_{k,i}^2\, p_{k,i} + I^2 \sum_{i,j} \delta_{k,ij}\, \alpha_{k,i}\, \alpha_{k,j}\, p_{k,i}\, p_{k,j} \tag{5.23a}$$

Assuming no epistasis, the total response, R, is simply the sum of all n single-locus responses,

$$R = \sum_{k=1}^{n} R_k = I \sum_{k=1}^{n} \sum_{i}^{n_k} 2\,\alpha_{k,i}^2\, p_{k,i} + I^2 \sum_{k=1}^{n} \sum_{i,j}^{n_k} \delta_{k,ij}\, \alpha_{k,i}\, \alpha_{k,j}\, p_{k,i}\, p_{k,j} \tag{5.23b}$$

This expression also holds when linkage disequilibrium is present, although the average excesses, average effects, and dominance deviations are expected to be different from their linkage-equilibrium values. For a random-mating population under linkage equilibrium, the first double sum is simply the additive genetic variance of this trait (LW Equation 4.23a). Assuming no epistasis and no linkage disequilibrium, summing over all loci yields a response of

$$R \simeq I\,\sigma_A^2 + I^2 \sum_{k=1}^{n} \sum_{i,j}^{n_k} \delta_{k,ij}\, \alpha_{k,i}\, \alpha_{k,j}\, p_{k,i}\, p_{k,j} \tag{5.23c}$$

As shown in Equation 5.21, if phenotypic values are normally distributed before selection, $I = S/\sigma_z^2$, and the response becomes

$$R = h^2 S + \frac{S^2}{\sigma_z^4} \sum_{k=1}^{n} \sum_{i,j}^{n_k} \delta_{k,ij}\, \alpha_{k,i}\, \alpha_{k,j}\, p_{k,i}\, p_{k,j} \tag{5.23d}$$

which recovers the breeder's equation, $R = h^2 S$ (response equals heritability times selection differential), plus a correction term. Still assuming there is no epistasis, in the absence of dominance (all $\delta_{k,ij} = 0$), the second term is zero. Even with dominance, the second term is of lower order than the first, and it vanishes as the number of underlying loci becomes large (Example 5.12). One way to view the correction term is to recall that when dominance is present, the parent-offspring regression is slightly nonlinear (LW Chapter 17), while the breeder's equation assumes linearity (among other things; see Chapter 6).

While the preceding derivation assumed normally distributed phenotypes, *exact* normality requires that the genotypic values at *each* locus be normally distributed (Nagylaki 1984). Because there are only a finite number of alleles, and hence a discrete number of genotypic values, this never holds exactly (see Chapter 24), but if the number of loci is large, the central limit theorem implies that the genotypic distribution is approximately normal. This points out one of the central assumptions of many quantitative-genetic selection models: *the number of loci is assumed to be sufficiently large that the amount of phenotypic variation attributable to any single locus is small, and hence the amount of selection on any locus is also small.* At its limit, we have the **infinitesimal model** (Chapter 24): an effectively infinite number of loci, each contributing an infinitesimal amount to the total phenotype. As the number of loci approaches infinity, the second sum in Equation 5.23d becomes vanishingly small (Example 5.12), and we recover the breeder's equation even when dominance is present.

Another class of models (Kimura 1965a; Lande 1975) allows for $n \geq 1$ loci by assuming there is a normal distribution of allelic effects at each locus underlying the character (effectively assuming an infinite number of alleles per locus). These two models (infinite number of loci versus infinite number of alleles at n loci) represent extreme approximations to the view that a moderate number of loci, each with a moderate number of alleles, underlie many quantitative characters. Chapter 24 explores these and other models in greater detail.

Example 5.12. Here we derive Equation 5.22, the expected response to selection associated with a single locus. Again, for ease of presentation, we suppress the subscripting indicating this locus. Assuming random mating, Equation 5.8b gives the single-generation allele-frequency dynamics as $p_i' = p_i(1 + s_i)$, where s_i is the average excess in relative fitness for allele A_i. To map these changes in allele frequencies into changes in mean genotypic values, we decompose the genotypic value of A_iA_j as $G_{ij} = \alpha_i + \alpha_j + \delta_{ij}$, where α_i is the average effect of A_i on the character value, δ_{ij} is the dominance deviation (LW Chapter 4), and the trait is scaled to initially have a mean of zero, with $E[G_{ij}] = E[\alpha_i] = E[\delta_{ij}] = 0$.

The contribution of this locus to the change in mean phenotype after a generation of selection and random mating is

$$\begin{aligned} R_k &\simeq \sum_{i,j} G_{ij}\, p_i'\, p_j' - \sum_{i,j} G_{ij}\, p_i\, p_j = \sum_{i,j} G_{ij}\, p_i(1+s_i)\, p_j(1+s_j) - \sum_{i,j} G_{ij}\, p_i\, p_j \\ &= \sum_{i,j} G_{ij}\, p_i p_j (1 + s_i + s_j + s_i s_j - 1) \\ &= \sum_{i,j} (\alpha_i + \alpha_j + \delta_{ij})\, p_i\, p_j\, (s_i + s_j) + \sum_{i,j} (\alpha_i + \alpha_j + \delta_{ij})\, p_i\, p_j\, s_i\, s_j \end{aligned} \tag{5.24a}$$

The $\sum G_{ij} p_i p_j$ term is the mean contribution before selection, which is zero, and hence vanishes. The careful reader will note that we made an approximation by using the decomposition of G_{ij} instead of decomposition of G'_{ij} in the very first sum, as we used the approximation

$$G'_{ij} \simeq \alpha_i + \alpha_j + \delta_{ij}$$

for

$$G'_{ij} = \alpha'_i + \alpha'_j + \delta'_{ij}$$

Because α_i and δ_{ij} are functions of the allele frequencies, they change as p_i changes, but we have assumed that these deviations are much smaller than the change in p_i itself (so that $\alpha'_i \simeq \alpha_i$ and $\delta'_{ij} \simeq \delta_{ij}$). To simplify Equation 5.24a further, recall (LW Chapter 4) that the additive and dominance effects are defined such that

$$\sum_i \alpha_i\, p_i = 0 \quad \text{and} \quad \sum_i \delta_{ij}\, p_i = 0 \tag{5.24b}$$

Separating the first term of Equation 5.24a into two parts based on s_i and s_j and applying Equation 5.24b yields

$$\begin{aligned} &\sum_{i,j} (\alpha_i + \alpha_j + \delta_{ij})\, p_i\, p_j\, s_i + \sum_{i,j} (\alpha_i + \alpha_j + \delta_{ij})\, p_i\, p_j\, s_j \\ &= 2 \sum_j s_j\, p_j \left(\alpha_j \sum_i p_i + \sum_i (\alpha_i + \delta_{ij})\, p_i \right) \\ &= 2 \sum_j s_j\, p_j\, (\alpha_j \cdot 1 + 0 + 0) = 2 \sum_j \alpha_j s_j\, p_j \end{aligned}$$

Likewise, a little more algebra (Nagylaki 1989b, 1991) simplifies the second sum in Equation 5.24a to yield a final expression for the expected contribution to response from locus k as

$$R_k = 2 \sum_j \alpha_j s_j\, p_j + \sum_{i,j} \delta_{ij}\, p_i\, s_i\, p_j\, s_j \tag{5.24c}$$

recovering Equation 5.22.

Note, however, that while the first term of Equation 5.24c recovers the breeder's equation, the second term

$$B_k = \sum_{i,j} \delta_{ij}\, p_i\, s_i\, p_j\, s_j$$

is a measure of departure from the breeder's equation at the focal locus (k). Nagylaki (1991) showed that the total departure over all loci $B = \sum_k B_k$ is bounded by

$$|\, B \,| \leq \left(\sum_{k=1}^{n} \sigma_{D(k)}(z) \right) \cdot \frac{\sigma_A^2(w)}{2}$$

where $\sigma_{D(k)}(z)$ is the square root of the dominance genetic variance in the character contributed by locus k, and $\sigma_A^2(w)$ is the additive variance in relative fitness. If all n loci underlying the character are identical (the **exchangeable model**), this bound reduces to

$$|\,B\,| \le \frac{\sigma_D(z) \cdot \sigma_A^2(w)}{2\,\sqrt{n}}$$

where $\sigma_D^2(z) = n\sigma_{D(k)}^2(z)$ is the total dominance genetic variance for the trait. Hence, assuming there is no epistatic genetic variance, even if dominance is present, as the number of loci increases, any departure from the breeder's equation will become increasingly small.

Correct Quadratic Terms for s_i

As mentioned earlier, the approximations given by Equations 5.16a and 5.16b are correct only to linear order, whereas quadratic (second-order) terms are required to properly account for selection acting directly on the trait variance. One source of error is that the conditional distribution of phenotypes for individuals carrying a particular allele has a lower variance than the unconditional phenotypic distribution. Partial knowledge of the genotype reduces the uncertainty in genotypic value, with the phenotypic variance of individuals with a fixed genotype at the kth locus being $\sigma^2 - \sigma_k^2$, where σ_k^2 is the contribution of the kth locus to the total phenotypic variance (Bulmer 1971a; Lynch 1984; Nagylaki 1984; Walsh 1990). In the absence of epistasis, gametic-phase disequilibrium, and genotype-environment interaction/correlation,

$$\sigma_k^2 = \sum_{i,j}^{n_k} a_{ij}^2\, p_i\, p_j$$

where $a_{ij} = G_{ij} - \mu_G$ is the deviation of the genotypic value from the genotypic mean and n_k is the number of alleles at locus k (Nagylaki 1984; Walsh 1990).

Using an expansion that accounts for this reduction in variance, Hastings (1990a) showed that s_i is approximated to quadratic order by

$$s_i \simeq -I_1\alpha_i + \frac{I_2}{2}\left[\sum_j a_{ij}^2\, p_j - \sigma_k^2\right] \tag{5.25a}$$

where

$$I_1 = \int w(z)\,\frac{d\,p(z)}{dz}\,dz \quad \text{and} \quad I_2 = \int w(z)\,\frac{d^2p(z)}{dz^2}\,dz \tag{5.25b}$$

Note that I_1 is simply the integral in Equation 5.19a and measures the effects of selection on the mean. For a normally distributed trait, substituting Equation 5.20b into the expression for I_1 yields

$$I_1 = \int w(z)\,\frac{d\,p(z)}{dz}\,dz = -\int w(z)\left(\frac{z-\mu}{\sigma_z^2}\right)p(z)\,dz = -\frac{\mu^* - \mu}{\sigma_z^2} = -S/\sigma_z^2 \tag{5.25c}$$

Hastings (1992) showed how this approach extends to a locus influencing n characters under selection.

The I_2 term measures the effect of selection on the variance. To see this, again suppose that the phenotypes are normally distributed. Differentiating Equation 5.20b a second time yields

$$\frac{d^2p(z)}{dz^2} = -\frac{p(z)}{\sigma_z^2} + \frac{(z-\mu)^2}{\sigma_z^4}\,p(z) \tag{5.26a}$$

and a bit of algebra (Example 5.13) yields

$$I_2 = \frac{\delta(\sigma_z^2) + S^2}{\sigma_z^4} \tag{5.26b}$$

where $\delta(\sigma_z^2)$ is the within-generation change in phenotypic variance due to selection (Chapters 16 and 24). While we use δ for both dominance and within-generation change, the distinction between δ_{ij} (the dominance associated with ij), and $\delta(\sigma_z^2)$, the within-generation change in σ_z^2, should be obvious from context. The reason for using $\delta(\sigma_z^2)$ is that it will be important in subsequent chapters to distinguish between a within-generation change $\delta(\sigma_z^2)$ and a between-generation change $\Delta(\sigma_z^2)$.

If phenotypes are normally distributed, $-I_1 = S/\sigma_z^2$ (Equation 5.25c), yielding

$$s_i \simeq \alpha_i \frac{S}{\sigma_z^2} + \frac{\delta(\sigma_z^2) + S^2}{2\sigma_z^4} \left[\sum_j a_{ij}^2 \, p_j - \sigma_k^2 \right] \tag{5.27a}$$

When alleles are completely additive, $a_{ij} = \alpha_i + \alpha_j$ and the term in brackets reduces to $\alpha_i^2 - \sum_j \alpha_j^2 \, p_j$. Substituting this improved value of s_i into Equation 5.22 and summing over all loci gives the response to selection (under a completely additive model) as

$$R = h^2 S + \frac{\delta(\sigma_z^2) + S^2}{2\sigma_z^4} \sum_{k=1}^{n} \sum_{i}^{n_k} (\alpha_{k,i})^3 \, p_{k,i} \tag{5.27b}$$

where the subscript k on the α_i and p_i terms reminds the reader that these can vary over loci.

As shown by Equation 29.16a, selection acting on the mean still results in a change in the variance, with $\delta(\sigma_z^2) = -S^2$. In this case, the second term of Equation 5.27b is zero and we recover the breeder's equation. More generally, however, if selection is also acting directly on the trait variance (such as, but not limited to, stabilizing or disruptive selection), $\delta(\sigma_z^2)$ departs from $-S^2$ (Chapter 29). Because the double sum in Equation 5.27b is the skewness in the genotypic distribution, if skew is present, changes due to selection on the variance also change the mean (Figure 29.7). Equation 5.27b raises several issues that will be examined in detail in Chapter 24. In particular, even if the distribution of *phenotypes* is normal, the response still depends on rather fine details (such as the third moment of allelic effects *at each locus*) of the distribution of *genotypic* values.

Example 5.13. To obtain Equation 5.26b, we first substitute Equation 5.26a into Equation 5.25b

$$I_2 = \int w(z) \left(-\frac{p(z)}{\sigma_z^2} + \frac{(z-\mu)^2}{\sigma_z^4} \, p(z) \right) dz \tag{5.28a}$$

Next, we use a standard trick by noting that

$$\begin{aligned} (z-\mu)^2 &= [(z-\mu^*) + (\mu^* - \mu)]^2 \\ &= (z-\mu^*)^2 + 2(z-\mu^*)(\mu^*-\mu) + (\mu^*-\mu)^2 \end{aligned} \tag{5.28b}$$

with Equation 5.28a becoming

$$\begin{aligned} I_2 = \frac{1}{\sigma_z^4} \Bigg(&-\sigma_z^2 \int w(z)p(z)dz + \int (z-\mu^*)^2 w(z)p(z)dz \\ &+ 2(\mu^*-\mu) \int (z-\mu^*) w(z)p(z)dz + (\mu^*-\mu)^2 \int w(z)p(z)dz \Bigg) \end{aligned} \tag{5.28c}$$

Noting that $\int w(z)p(z) = \overline{w} = 1$ and that $\int f(z)w(z)p(z) = E_s[f(z)]$, namely, the expected value of $f(z)$ in the selected individuals (as $w(z)p(z)$ is the distribution of z following selection), Equation 5.28c reduces to

$$\begin{aligned} I_2 &= \frac{1}{\sigma_z^4} \Big(-\sigma_z^2 + E_s[(z-\mu^*)^2] + 2(\mu^*-\mu)E_s(z-\mu^*) + (\mu^*-\mu)^2 \Big) \\ &= \frac{1}{\sigma_z^4} \Big(-\sigma_z^2 + \sigma_{z^*}^2 + 0 + S^2 \Big) \end{aligned}$$

Here σ^2_{z*} denotes the variance after selection, so that $-\sigma^2_z + \sigma^2_{z*} = \delta(\sigma^2_z)$, and we recover Equation 5.26b.

6

Theorems of Natural Selection: Results of Price, Fisher, and Robertson

Is there some reorientation for the expression of natural selection that may provide subtle perspective, from which we can understand our subject more deeply and analyze our problems with greater ease and greater insight? My answer is ... that the Price equation provides that sort of reorientation. Frank (2012, p. 1003)

One of the messages from Chapter 5 is that selection, even on just one or two loci, can have very complex dynamics. Indeed, outside of the special case of frequency-independent viability selection at a single locus, selection is not even guaranteed to increase mean fitness. Nonetheless, the search for general theorems (exact mathematical expressions) of selection response has motivated population and quantitative geneticists for over 80 years. By shifting attention away from trying to model individual allele-frequency change over a large number of loci to considering the dynamics of some composite feature of these loci, such as the mean of a trait, the hope was that some general statements might hold. Here we focus on three classical "theorems"—**Fisher's fundamental theorem of natural selection**, **Robertson's secondary theorem of natural selection**, and **Price's theorem**—as well as **Lush's breeder's equation**, summarized in Table 6.1. Ironically, these attempts to bring some order to population- and quantitative-genetic theory have instead resulted in a considerable amount of confusion and chaos in the literature.

As detailed below, the classical interpretation of Fisher's theorem, along with Robertson's theorem and the breeder's equation, are all approximations (albeit often good ones), and not formal mathematical theorems. In contrast, Price's general expression for *any* type of selection response *is* a formal theorem, as is a special case of it, the **Robertson-Price identity** for the within-generation response in a trait due to selection. A source of confusion is that the classical view assumed by most of the literature was apparently different from Fisher's own interpretation, which *is* a formal (but not very useful) theorem. As our discussion of these various "theorems," exact and otherwise, will highlight, it is the transmission of trait values from parent to offspring that generally induces complications and makes the theoretician's job challenging. This can be seen in Table 6.1 by simply comparing the exact Robertson-Price result for within-generation change (S) with results for cross-generational change ($S + E[w\delta_z]$). The surprising result is not that these "theorems" are wrong, but rather that they often are reasonable-to-excellent approximations for much of the dynamics of a trait under short-term selection.

Our treatment of this rather convoluted area is structured as follows. We start with Price's theorem, which is a *completely general* description of *any* selection response under *any* model of transmission. It does not rely on any explicit genetic model and thus serves as an ideal platform from which to examine the other "theorems." We next turn to Fisher's fundamental theorem, which has a rich and somewhat checkered history, in part due to Fisher's failure to be fully explicit with his definitions. An important corollary of Fisher's theorem is that, in the absence of mutation, selection will drive the additive variance in fitness toward zero, and we next examine some of the biological implications and misunderstandings of this result. We then turn to Robertson's secondary theorem, which focuses on the selection response of any arbitrary trait, not just fitness. Finally, because the breeder's equation is the basic workhorse result for much of selection theory in quantitative genetics (Chapters 13–20), we conclude by examining its robustness in some detail.

Evolution and Selection of Quantitative Traits. Bruce Walsh & Michael Lynch.
Published 2018 by Oxford University Press. © Bruce Walsh & Michael Lynch 2018.
DOI 10.1093/oso/9780198830870.001.0001

Table 6.1 General expressions for the response of a trait to selection, with fitness as an important special case. Here w denotes relative fitness (with mean value $E[w] = 1$); z, the value of an arbitrary trait; A_z, the breeding value of trait z (with A_w as the breeding value for relative fitness); R_z, the total response to selection (change in the mean across generations) of trait z (with R_w for the special case where the trait is mean fitness); ∂R_w, the partial response in mean fitness due exclusively to allele-frequency change (see the text for details); S_z, the within-generation change in z following selection but prior to gene transmission (the selection differential); and $\overline{\delta}_z$, the expected change between the mean value of a trait in selected parents and their progeny (also see text for details). Expressions denoted by † are true mathematical theorems, whereas the rest are approximations.

Fisher's Fundamental Theorem		Fisher (1930)
Classical interpretation	$R_w = \sigma^2(A_w)$	
Exact version†	$\partial R_w = \sigma^2(A_w)$	Price (1972b), Ewens (1989)
Breeder's Equation	$R_z = h_z^2 S_z$	Lush (1937)
Robertson-Price Identity†	$S_z = \sigma(w, z)$	Robertson (1966a), Price (1970)
Robertson's Secondary Theorem		
1966 version	$R_z = \sigma(w, A_z)$	Robertson (1966a)
1968 version	$R_z = \sigma(A_w, A_z)$ $= \sigma_A(w, z)$	Robertson (1968)
Price's Theorem†	$R_z = \sigma(w, z) + E(w\,\overline{\delta}_z)$ $= S + E(w\,\overline{\delta}_z)$	Price (1970)

PRICE'S GENERAL THEOREM OF SELECTION

The thoughtful reader might ask if there is a general, assumption-free statement about selection response under *any* situation. There is: namely, **Price's theorem** (Price 1970, 1972a), also referred to as the **Price equation**. Price's theorem provides a notationally elegant way to describe *any* evolutionary response. It makes *no assumptions* about the mechanism of transmission of a trait from some ancestral category (such as an individual or group) to its descendants. As such, it works for traits transmitted by standard genetics, epigenetics, and culture, and as such has been applied to everything from the evolution of languages to community structure in ecology. Recent reviews include Frank (1995, 1997, 1998, 2012), Rice (2004a), Boyd and Richerson (2005), Okasha (2006), Gardner (2008), Helanterä and Uller (2010), and Luque (2017), while van Veelen (2005; van Veelen et al. 2012) championed a more cautious use of Price's theorem.

The Life and Times of George Price

George Price was one of the most enigmatic figures in modern evolutionary biology (Frank 1995, Schwartz 2000, and Harman 2011 all reviewed Price's life and contributions). After obtaining a Ph.D. in chemistry from the University of Chicago, he worked on the Manhattan Project before joining IBM as an engineer. At age 44, Price quit his job and started working under Cedric Smith at University College London (from 1968 to 1974). In this brief tenure, he only published four solo papers and was the coauthor on two others, but in his roughly 25 total pages of publications, he made three fundamental contributions to modern evolutionary theory. In addition to Price's theorem, he introduced the power of game theory to evolutionary biology (Maynard Smith and Price 1973), and he was the first to fully grasp what Fisher had really meant by his enigmatic fundamental theorem (Price 1972b). Price left academia in 1974, working as a night janitor before giving away all his worldly possessions to homeless alcoholics and dying by his own hand in 1975 while a squatter in one of the poorest areas of London.

Price's Theorem, $R_z = \sigma(w_i, z_i) + E(w_i\,\overline{\delta}_i)$

Price's theorem expresses the expected selection response in a trait in general terms of covariances, rather than relying on any explicit model of transmission, and as such is a ***completely general description of any evolutionary response***. As succinctly stated by Rice (2004a, p. 170), "it is an exact characterization of a relationship that must hold between phenotype, fitness, selection, and evolution." The key to Price's equation is to first consider the effect of selection on *specific categories* (measured by how many descendants each leaves) and then consider how trait values may differ between an ancestral category and its descendants. This is an extremely subtle shift in focus, one that is easy to miss and misinterpret. However, this perspective nicely decomposes the evolutionary change (R_z) into a selection term and a remainder term due to all other forces, such as (but not limited to) imperfect transmission.

Consider selection first. Suppose there are N categories in the population, where q_i and z_i, respectively, denote the frequency of category i and the mean value for the trait of interest over all members of this category. Note that z can be *any* trait measure. For example, if x denotes the value of a trait, taking $z = (x - \mu_x)^2$ or $z = (x - \mu_x)^4$ gives the response in the variance and the fourth moment, respectively. Averaging over all categories, the mean trait value is

$$\overline{z} = \sum_{i=1}^{N} q_i\, z_i \tag{6.1}$$

Suppose that the members of category i leave a total of n_i descendants (the absolute fitness, W_i, for that category). The average number of descendants over categories (the mean fitness) is

$$\overline{W} = \sum_{i=1}^{N} q_i\, W_i \tag{6.2a}$$

The relative fitness of category i is $w_i = W_i/\overline{W}$.

Price's key idea was to define q_i' as the frequency of all *descendants* that have category i as their *ancestor*

$$q_i' = w_i q_i \tag{6.2b}$$

Note, in particular, that q_i' is *not* the frequency of descendants *in* category i, but rather the fraction of all descendants that are *from* ancestors in category i. The focus is entirely on the categories of ancestors, *not* on which categories the descendants are in. As an example of this shift in focus, suppose our three categories of interest are the genotypes *AA*, *Aa*, and *aa* at a diallelic locus. In a traditional population-genetics analysis (Chapter 5), we would write equations to describe how the frequency of each category changes. Price used a different focus, considering instead the frequency of the *descendants* that come from each category. Suppose category $i = 2$ corresponds to *Aa* and imagine an extreme case where only *Aa* individuals survive. Here $q_2' = 1$, as all offspring have *Aa* parents. However, in the next generation (before selection), segregation results in the genotypes *AA*, *Aa*, and *aa* at the frequencies 0.25, 0.5, and 0.25, i.e., all three categories are present *in the offspring*, but all have only *Aa parents*.

Now consider the transmission phase (which more generally includes everything other than selection). Let $\overline{z}_i$ denote the mean value of the descendants from category i, which we can decompose as

$$\overline{z}_i = z_i + \overline{\delta}_i \tag{6.3a}$$

namely, the mean value, z_i, of their ancestors plus a deviation, $\overline{\delta}_i$, due to imperfect transmission. Taking the average over all ancestral categories, the average trait value over all the descendants becomes

$$\overline{z}\,' = \sum_{i} q_i'\, \overline{z}_i \tag{6.3b}$$

Recalling Equations 6.1, 6.2b, and 6.3b, the response in trait value, $R_z = \overline{z}' - \overline{z}$, becomes

$$\begin{aligned} R_z &= \sum_i q_i' \, \overline{z}_i - \sum_i q_i z_i \\ &= \left(\sum_i q_i' z_i - \sum_i q_i z_i \right) + \left(\sum_i q_i' \, \overline{z}_i - \sum_i q_i' z_i \right) \\ &= \sum_i (q_i' - q_i) \, z_i + \sum_i q_i' \, (\overline{z}_i - z_i) \\ &= \sum_i \Delta q_i z_i + \sum_i q_i' \, \overline{\delta}_i \end{aligned} \tag{6.4}$$

The second line follows by adding and subtracting $\sum q_i' \, z_i$, and the third by suitably gathering terms. This version of the Price equation is based on Frank (1997, 2012). The first term (containing $\Delta q_i z_i$) represents the change due to *selection based entirely on ancestral values*, or the **partial evolutionary change caused by natural selection** (Price 1972b; Ewens 1989; Frank 2012). The second term (containing $q_i' \, \overline{\delta}_i$) is the part of total change caused by *imperfect transmission* of ancestral values to their descendants. Equation 6.4 is conceptually very powerful, as it decomposes the response into two separate components, one strictly based on the nature of selection (Δq_i) and the other on transmission ($\overline{\delta}_i$).

Both these terms can be expressed in a more transparent form. For the first term,

$$\begin{aligned} \sum_i \Delta q_i z_i = \sum_i (w_i q_i - q_i) \, z_i &= \sum_i w_i z_i q_i - \sum z_i q_i = E(w_i z_i) - E(z) \cdot 1 \\ &= E(w_i z_i) - E(z) \cdot E(w) \\ &= \sigma(w_i, z_i) \end{aligned} \tag{6.5a}$$

Note that to obtain this result, we use the identity $E(w) = 1$. For the second term,

$$\sum_i q_i' \, \overline{\delta}_i = \sum_i q_i w_i \, \overline{\delta}_i = E(w_i \overline{\delta}_i) \tag{6.5b}$$

Substituting Equations 6.5a and 6.5b into Equation 6.4 yields the more traditional form of the Price equation,

$$R_z = \overline{z}' - \overline{z} = \sigma(w_i, z_i) + E(w_i \, \overline{\delta}_i) \tag{6.6}$$

(Price 1970, 1972a). Note that all expectations are computed with respect to the *pre-selection* frequencies of the ancestral categories (q_i). The first term is the covariance between phenotype and fitness (the within-generation change, S, in the mean of ancestral values), while the second is the fitness-weighted transmission of any changes between the value of an ancestor and the mean of its descendants.

Several equivalent expressions for the Price equation are useful in that they emphasize different features. For example, using the definition of a covariance

$$E(w_i \, \overline{\delta}_i) = \sigma(\, w_i, \overline{\delta}_i) + E(w_i) \cdot E(\overline{\delta}_i) = \sigma(\, w_i, \overline{\delta}_i) + E(\overline{\delta}_i) \tag{6.7a}$$

and substituting this result into Equation 6.6 yields

$$R_z = \sigma(w_i, z_i) + \sigma(\, w_i, \overline{\delta}_i) + E(\overline{\delta}_i) \tag{6.7b}$$

Using Equation 6.3a to combine the first two covariances in Equation 6.7b gives

$$R_z = \sigma(w_i, \overline{z}_i) + E(\overline{\delta}_i) \tag{6.7c}$$

Equation 6.7c relates the selection response to the covariance between the fitness, w_i, of an ancestor and the mean value, $\overline{z}_i$, of its descendants. The second term, $E(\overline{\delta}_i)$, is often thought of as the expected change in mean value from ancestor to descendant in the absence of selection. As we will see at the end of the chapter, this interpretation is not quite correct, as this last term does contain a contribution from selection (Equation 6.38).

To summarize, if z_i denotes the average value in category i, which has a frequency of q_i before selection and a frequency of q_i' after selection, and whose offspring have average value $\overline{z}_i = z_i + \overline{\delta}_i$, then equivalent forms of Price's theorem are

$$R_z = \begin{cases} \sum_i \Delta q_i z_i + \sum_i q_i' \overline{\delta}_i \\ \sigma(w_i, z_i) + E(w_i\,\overline{\delta}_i) \\ \sigma(w_i, z_i) + \sigma(\,w_i, \overline{\delta}_i) + E(\overline{\delta}_i) \\ \sigma(w_i, \overline{z}_i) + E(\overline{\delta}_i) \end{cases} \tag{6.8}$$

As noted in Table 6.1, and as will be shown shortly (Equation 6.10), the selection differential, S, can be substituted for $\sigma(w_i, z_i)$ in the middle two expressions of Equation 6.8. As is further shown below, many of the standard approximations for evolutionary response (e.g., the breeder's equation, Fisher's fundamental theorem, Robertson's secondary theorem) follow directly from the $\sigma(w_i, z_i)$ selection term under the assumption that the residual term $E(w_i\,\overline{\delta}_i)$ is zero. Hence, these approximations fail when this term is significant.

While a discussion of Price's theorem often assumes the ancestor to be a parent or midparent and the descendants to be their offspring in the next generation, the theorem holds for *any* time interval and for *any* set of ancestors (such as a group of individuals; Chapter 22) that one wishes to consider. In this sense, Price's theorem is completely general and makes absolutely no assumptions about the mechanism of transmission of trait values from ancestors to their descendants, although it *does* make the assumption that all descendants have ancestors. This may seem trivial, but it is violated by migration, wherein an individual appears in the next generation from ancestors not considered. Kerr and Godfrey-Smith (2008) generalized the Price equation to accommodate missing ancestors and more general causal connections between ancestors and descendants.

Example 6.1. Let the ancestor (i) be the midparent (the average value of the two parents) with a phenotypic value of z_i and the descendants be the offspring in the next generation. If the average value of offspring, $\overline{z}_i$, is exactly the same as the value, z_i, of their ancestral midparent, then $\overline{\delta}_i = 0$ for all i and, from Equation 6.6, the response is simply $R = \sigma(w_i, z_i)$. From the Robertson-Price identity (Table 6.1, Equation 6.10), $\sigma(w_i, z_i) = S$, the selection differential, so that $R = S$ in this case of perfect transmission. However, the mean value of offspring generally differs from the average value of their parents, in which case the second term in Equation 6.6 is nonzero. The simplest transmission model is a linear midparent-offspring regression, $z_{ij} = \mu + b(z_i - \mu) + e_{ij}$. Here z_{ij} is the trait value for the jth offspring from midparent i, giving the mean value of offspring from i as $\overline{z}_i = \mu + b(z_i - \mu) + e_i$. The expected deviation then becomes

$$\overline{\delta}_i = \overline{z}_i - z_i = \mu + b(z_i - \mu) + e_i - z_i = (b-1)(z_i - \mu) + e_i \tag{6.9a}$$

Hence,

$$E(w_i\,\overline{\delta}_i) = E\{w_i\,[(b-1)(z_i - \mu) + e_i]\}$$
$$= (b-1)[E(w_i\,z_i) - \mu\,E(w_i)] + E(w_i e_i)$$

Recalling that $E(w_i) = 1$ gives

$$E(w_i\,\overline{\delta}_i) = (b-1)\,[\,E(w_i\,z_i) - \mu\,] + E(w_i e_i)$$

From the definition of a covariance

$$E(w_i\,z_i) = \sigma(w_i, z_i) + E(w_i)\,E(z_i) = S + 1\cdot\mu$$
$$E(w_i\,e_i) = \sigma(w_i, e_i) + E(w_i)\,E(e_i) = \sigma(w_i, e_i) + 1\cdot 0$$

Putting these results together into Equation 6.6 yields

$$R_z = S + E(w_i\,\overline{\delta}_i) = S + (b-1)(S + \mu - \mu) + \sigma(w_i, e_i) = bS + \sigma(w_i, e_i) \qquad (6.9b)$$

Provided that the residual, e_i, of the midparent-offspring regression and the fitness, w_i, of the midparent are uncorrelated, $R = bS$. When these are uncorrelated and the midparent-offspring slope equals the heritability, $b = h^2$, we recover the breeder's equation. While at first blush Equation 6.9b appears to be a rather general statement about the accuracy of the breeder's equation, we made a few subtle assumptions (besides the obvious one of linearity) about the parent-offspring relationship, although we will defer discussion of these until later sections of this chapter.

Example 6.2. Consider the change in allele frequency when a single diallelic locus (alleles *A* and *a*) determines fitness. Assume random mating among the survivors, with p = freq(A), and

Genotype	Frequency (before selection)	Fitness
AA	p^2	W_{AA}
Aa	$2p(1-p)$	W_{Aa}
aa	$(1-p)^2$	W_{aa}

To apply Price's theorem, we need to specify the categories to be followed, which we take as the alleles *A* and *a*. We index these by $i = 1$ and $i = 2$, respectively, and code their associated values as $z_1 = 1$ and $z_2 = 0$, which implies a mean value of $\overline{z} = (1 \cdot p) + (0 \cdot [1-p]) = p$, so R_z represents the change in p. In the absence of mutation, transmission is perfect, as the descendant allele from an *A* allele is always *A*, resulting in $\overline{\delta}_i = 0$. Putting these together, Equation 6.6 becomes

$$\Delta p = R_z = \sigma(w_i, z_i) + E(w_i \cdot 0) = \sigma(w_i, z_i)$$

Under random mating, the fitness W_1 of an *A* allele is simply its marginal fitness (Equation 5.7b), $W_1 = pW_{AA} + (1-p)W_{Aa}$. Similarly, $W_2 = pW_{Aa} + (1-p)W_{aa}$ and $E(W_i) = \overline{W} = pW_1 + (1-p)W_2$. Recalling that $w_i = W_i/\overline{W}$ and the definition of a covariance, we have

$$\Delta p = \sigma(w_i, z_i) = \frac{\sigma(W_i, z_i)}{\overline{W}} = \frac{1}{\overline{W}}\Big(E(W_i\,z_i) - E(W_i)E(z_i)\Big)$$

To show that this recovers the standard population-genetic equation for allele-frequency change, note that

$$E(W_i\,z_i) = \sum_{i=1}^{2} W_i\,z_i\,\text{freq(category } i) = [W_1 \cdot 1 \cdot p] + [W_2 \cdot 0 \cdot (1-p)] = p \cdot W_1$$

Using this result along with $E(W_i) = \overline{W}$ and $E(z_i) = p$

$$\Delta p = \frac{1}{\overline{W}}\left(p\,W_1 - \overline{W}\,p\right) = p\,\frac{(W_1 - \overline{W})}{\overline{W}}$$

which recovers Equation 5.7c.

The Robertson-Price Identity, $S = \sigma(w,z)$

When our concern is strictly on the *within-generation* change in trait value, then $\Delta\overline{z} = \mu_z^* - \mu_z$, the difference between the fitness-weighted mean after selection (but before reproduction),

μ_z^*, and the overall mean before selection, μ_z, which is the selection differential S (Chapter 13). Because the within-generation change is not influenced by cross-generation transmission, any terms involving $\overline{\delta}$ in Equation 6.6 are zero, and we recover the **Robertson-Price identity**

$$S = \sigma(w, z) \tag{6.10}$$

as first obtained by Robertson (1966a) and Price (1970). We derive this result by another route in Chapter 13, where we use this identity extensively in selection-response theory. The critical insight from Equation 6.10 is that no matter how complex the relationship between phenotype z and fitness w, the within-generation change in the mean only depends on the covariance between these measures.

The Breeder's Equation, $R = h^2S$

Our derivation of the breeder's equation via Price's theorem (Example 6.1) made three assumptions dealing with the midparent-offspring regression that need to be highlighted. The most obvious assumption was that the parent-offspring regression is linear. Two more subtle assumptions are that the parent and offspring means are unchanged in the absence of selection, and that the regression slope b is the same for selected and unselected parents. We address these assumptions separately, by considering the impact of changing means first, then nonlinearity, and finally changes in the regression slope. While each assumption is considered in isolation, one or more may occur simultaneously.

Example 6.1 used the linear regression $z_{ij} = \mu + b(z_i - \mu) + e_{ij}$ for the jth offspring from midparent combination i (Example 6.1), which assumes that the trait mean is the same in the offspring (o) and parental populations (p) in the absence of selection ($\mu_o = \mu_p = \mu$). More generally, the parent and offspring means can differ, making the regression $z_{ij} = \mu_o + b(z_i - \mu_p) + e_{ij}$, in which case

$$\begin{aligned}\delta_i = \overline{z}_i - z_i &= \mu_o + b(z_i - \mu_p) + e_i - z_i \\ &= (b-1)z_i - b\mu_p + \mu_o + e_i \\ &= (b-1)(z_i - \mu_p) + (\mu_o - \mu_p) + e_i\end{aligned} \tag{6.11}$$

Following the same logic leading to Equation 6.9b yields the response to selection

$$R = b\,S + \sigma(w_i, e_i) + (\mu_o - \mu_p) \tag{6.12}$$

where the last term $(\mu_o - \mu_p)$ accounts for changes in mean phenotype from parent to offspring (in the absence of selection), due, for example, to the decay of linkage disequilibrium (Chapter 15), nonrandom associations of environmental values (Chapter 15), or inbreeding (Chapter 23; LW Chapter 10).

Now suppose the parent-offspring regression is nonlinear. Assuming the simplest such departure, a quadratic, makes the key point. The mean value $\overline{z}_i$ of offspring from a midparent with phenotypic value z_i is given by

$$\overline{z}_i = a + b\,z_i + c\,z_i^2 + e_i \tag{6.13a}$$

Hence,

$$\sigma(w_i, \overline{z}_i) = b\,\sigma\left(w_i, z_i\right) + c\,\sigma\left(w_i, z_i^2\right) + \sigma\left(w_i, e_i\right) \tag{6.13b}$$

Assuming the residuals from the midparent-offspring regression are uncorrelated with the fitness of the midparent $\sigma\left(w_i, e_i\right) = 0$, Equation 6.13b becomes

$$\sigma(w_i, \overline{z}_i) = b\,S + c\,\sigma\left(w_i, z_i^2\right) \tag{6.13c}$$

From Equation 6.7c, the selection response is given by this expression plus $E(\overline{\delta}_i)$. Thus, with a nonlinear parent-offspring regression, selection on the variance z_i^2 (i.e., $\sigma\left(w_i, z_i^2\right) \neq 0$; see

Chapters 29 and 30) can enter into the response in the mean, in which case the strict linear version of the breeder's equation no longer holds.

A final source of error when using the breeder's equation is that the regression coefficient may differ between selected and unselected parents. To explicate this point, we follow Frank (1997) and consider any generalized linear predictor of some value z_i for individual i from k underlying predictor variables $x_{i1}, \cdots, x_{ik}$,

$$z_i = \sum_{j=1}^{k} b_j \, x_{ij} + e_i \tag{6.14}$$

For example, an individual's phenotype can be written as the sum of average allelic effects over all k loci plus a residual error. With a single diallelic locus, $z_i = b_1 x_{i1} + b_2 x_{i2} + e_i$, where b_j is the average effect of allele j and x_{ij} is the number of copies of allele j in individual i (values of 0, 1, or 2), so that (for a diploid) $\overline{x}_{\cdot j}/2 = p_j$, the frequency of allele j. The average of Equation 6.14 becomes

$$\overline{z} = \sum_j b_j \, \overline{x}_{\cdot j} = \sum_j 2b_j \, p_j \tag{6.15a}$$

At some time in the future, the values of both p_j (the allele frequency) and b_j (its average effect) may change to new values

$$p_j' = p_j + \Delta p_j \quad \text{and} \quad b_j' = b_j + \Delta b_j \tag{6.15b}$$

We write the new mean as

$$\overline{z}\,' = \sum_j 2b_j' \, p_j' \tag{6.15c}$$

and the same logic leading to Equation 6.4 allows us to decompose the total change as

$$\overline{z}\,' - \overline{z} = \sum_j 2b_j \Delta p_j + \sum_j 2p_j' \, \Delta b_j \tag{6.15d}$$

There are two sources for response: a change, $b_j \Delta p_j$, from the change in allele frequencies weighted by their original average effects, and a change, $p_j' \, \Delta b_j$, in the average effects weighted by the new allele frequencies. These two terms exactly correspond to the selection and transmission terms in Price's equation (compare this with Equation 6.4). The $b_j \Delta p_j$ terms correspond to partial evolutionary change caused by natural selection, while the transmission term represents the response due to changes, Δb_j, in the average effects themselves. These changes in turn influence the parent-offspring regression slope.

Example 6.3. The setting in which the breeder's equation is expected to be least accurate involves a trait entirely determined by a single dominant locus. This results in both a nonlinear parent-offspring regression and the potential for significant allele-frequency change following even a single generation of selection. Suppose that, at such a locus, genotypes *QQ* and *Qq* have a phenotypic value of 1, while *qq* has a value of 0 (there is no environmental variance), and that *qq* individuals have a survival rate twice as high as the survival rate of *QQ* and *Qq* individuals.

We now contrast the exact response from single-locus models (Chapter 5) with that predicted by the breeder's equation. Starting with the exact population-genetic model, and letting p be the frequency of allele q

Genotype	*QQ*	*Qq*	*qq*
Trait value z	1	1	0
Frequency (before selection)	$(1-p)^2$	$2p(1-p)$	p^2
Fitness	1	1	2
Frequency (after selection)	$(1-p)^2/\overline{W}$	$2p(1-p)/\overline{W}$	$p^2(2/\overline{W})$

The trait mean before selection is $\mu(p) = 1 - p^2$, with a mean fitness of $\overline{W}(p) = 1 + p^2$. The new frequency, p', of allele q after selection is half the frequency of Qq plus the frequency of qq (both after selection)

$$p' = \frac{(1/2)2p(1-p) + 2p^2}{1+p^2} = \frac{p(1+p)}{1+p^2} \tag{6.16a}$$

yielding an allele-frequency change of

$$\Delta p = p' - p = \frac{p(1+p)}{1+p^2} - p\,\frac{1+p^2}{1+p^2} = \frac{p^2(1-p)}{1+p^2}$$

which translates into a change in mean phenotype of

$$R_z = \mu(p') - \mu(p) = p^2 - (p')^2 = -2p^3\left(\frac{1 - p(1+p^2)/2}{(1+p^2)^2}\right) \tag{6.16b}$$

This exact single-generation response in the trait mean is plotted in Figure 6.4.

Now consider the response predicted from the breeder's equation. Applying the standard trait-value parameterization of $2a : a(1+k) : 0$ to the preceding trait values, we have $a = 1/2$ and $k = 1$, yielding (LW Chapter 4) the additive and dominance variances of the trait

$$\sigma_A^2 = 2p(1-p)a^2[1 + k(2p-1)]^2 = 2p(1-p)(1/4)(2p)^2 = 2p^3(1-p)$$

$$\sigma_D^2 = [2p(1-p)ak]^2 = p^2(1-p)^2$$

Because σ_E^2 is assumed to be zero, the heritability becomes

$$h^2 = \frac{\sigma_A^2}{\sigma_A^2 + \sigma_D^2} = \frac{2p^3(1-p)}{2p^3(1-p) + p^2(1-p)^2} = \frac{2p}{1+p} \tag{6.16c}$$

Following selection (but before reproduction), only QQ and Qq survive, each with a trait value of one, yielding the fitness-weighted mean

$$\mu^* = 1 \cdot \frac{1}{\overline{W}}\,(1-p^2) + 0 \cdot \frac{2}{\overline{W}}\,p^2 = \frac{1-p^2}{\overline{W}} = \frac{1-p^2}{1+p^2} \tag{6.16d}$$

yielding the selectional differential

$$S = \mu^* - \mu = \frac{1-p^2}{1+p^2} - (1-p^2) = -p^2\,\frac{1-p^2}{1+p^2} \tag{6.16e}$$

Equations 6.16c and 6.16e give the predicted response from the breeder's equation as

$$R_z = h^2 S = \left(\frac{2p}{1+p}\right)\left(-p^2\,\frac{1-p^2}{1+p^2}\right) = -\frac{2p^3(1-p)}{1+p^2} \tag{6.16f}$$

As shown in Figure 6.1, we see that the approximation given by the breeder's equation (Equation 6.16f) generally does well but slightly underestimates the exact response (Equation 6.16b), predicting (in the worst case) only about 90% of the actual response when $p \simeq 0.4$. For this simple one-locus model, two factors account for this discrepancy. First, owing to dominance, the parent-offspring regression is *not* linear. Second, the change in allele frequency in the selected parents results in changes in the parent-offspring covariance, and hence the parent-offspring regression slope.

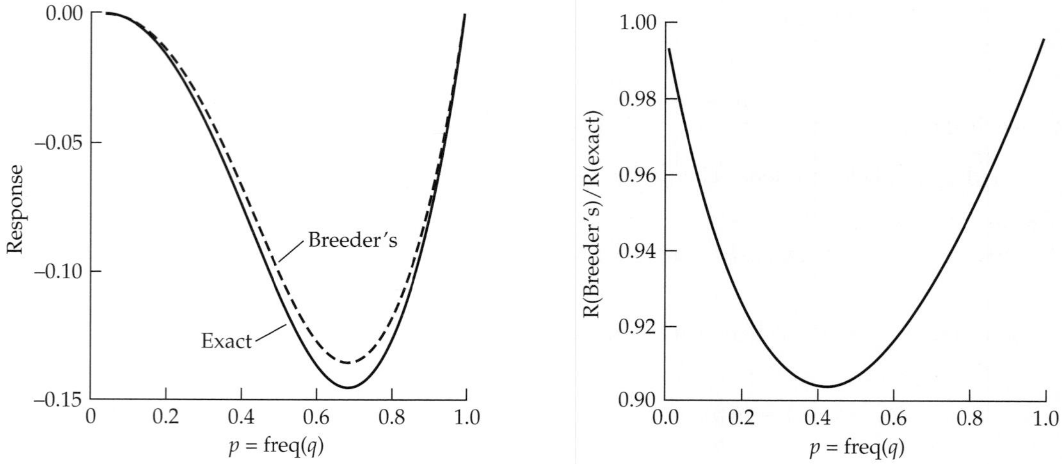

Figure 6.1 Analysis of the model from Example 6.3. (**Left**) Graph of the exact (using one-locus theory; Equation 6.16b) and predicted (via the breeder's equation; Equation 6.16f) response as a function of allele-frequency p, for a trait under selection detemined by a single dominant locus. (**Right**) The relative accuracies of the breeder's equation as a function of p.

FISHER'S FUNDAMENTAL THEOREM OF NATURAL SELECTION

The rate of increase in fitness of any organism at any time is equal to its genetic variance in fitness at that time. Fisher (1930, p. 35)

This simple statement from Fisher's (1930) book (which was dictated to his wife as he paced about their living room) has generated a tremendous amount of work, discussion, and sometimes heated arguments. Fisher claimed his result was exact, a true theorem. Historically, the **classical** (and seemingly obvious) interpretation of this quote is that the rate of increase in mean fitness equals the additive variance in fitness, $R_w = \sigma^2(A_w)$. Because variances are nonnegative, this interpretation implies that mean population fitness never decreases in a constant environment. However, we already know from Chapter 5 that this statement is incorrect, as mean fitness can decline even under simple models of selection. As a result, the mathematician Sam Karlin referred to this interpretation as "neither fundamental nor a theorem" as it requires rather special conditions to hold, especially when multiple loci influence fitness. As is discussed below, the classical interpretation of Fisher's theorem only holds *exactly* under restricted conditions, but is often a good approximate descriptor. We first review the classical interpretation, and then discuss what it appears that Fisher actually meant.

The Classical Interpretation of Fisher's Fundamental Theorem, $R_w = \sigma^2_A(w)$

One way to demystify the classical version of Fisher's theorem is to suppose that fitness is just a trait, and use the breeder's equation, $R = h^2 S$, to predict the response to selection on that trait. Letting $z = W$, and recalling that $w = W/\overline{W}$, the Robertson-Price identity (Equation 6.10) yields the selection differential

$$S_W = \sigma(z, w) = \sigma(W, w) = \frac{\sigma(W, W)}{\overline{W}} = \frac{\sigma^2(W)}{\overline{W}} \tag{6.17a}$$

Substituting into the breeder's equation shows

$$R_W = h^2_W S_W = \frac{\sigma^2_A(W)}{\sigma^2(W)} \frac{\sigma^2(W)}{\overline{W}} = \frac{\sigma^2_A(W)}{\overline{W}} \tag{6.17b}$$

Expressed in terms of *relative fitnesses*,

$$R_w = \Delta \overline{w} = \frac{R_W}{\overline{W}} = \frac{\sigma_A^2(W)}{\overline{W}^2} = \sigma_A^2(w) \tag{6.17c}$$

recovering the classical view of Fisher's theorem.

One can also use a population-genetics framework to motivate Fisher's theorem in terms of allele-frequency change. Consider a diallelic locus with constant fitnesses under random mating, and define $\overline{W}(p)$ to be the mean population fitness at allele frequency p (Equation 5.1a). The change in mean fitness is a function of the allele-frequency change, Δp,

$$R_W = \overline{W}(p + \Delta p) - \overline{W}(p) \tag{6.18a}$$

If the allele-frequency change is small, a first-order Taylor-series approximation yields

$$\overline{W}(p + \Delta p) \simeq \overline{W}(p) + \frac{d\overline{W}}{dp}\,\Delta p \tag{6.18b}$$

so

$$R_W = \overline{W}(p + \Delta p) - \overline{W}(p) \simeq \frac{d\overline{W}}{dp}\,\Delta p \tag{6.18c}$$

From Equation 5.5a

$$\begin{aligned}\frac{d\overline{W}}{dp} &= 2pW_{AA} + 2(1-2p)W_{Aa} + 2(p-1)W_{aa} \\ &= 2\left[\,p(W_{AA} - W_{Aa}) + (1-p)(W_{Aa} - W_{aa})\,\right] = 2(\alpha_A - \alpha_a)\end{aligned} \tag{6.18d}$$

where the last equality follows from the definition of the average effects (under random mating; see LW Chapter 4) of alleles A and a on fitness, namely,

$$\alpha_A = p\,W_{AA} + (1-p)\,W_{Aa} - \overline{W} \quad \text{and} \quad \alpha_a = p\,W_{Aa} + (1-p)\,W_{aa} - \overline{W} \tag{6.18e}$$

Recall that the quantity $\alpha = \alpha_A - \alpha_a$ is the average effect of an allelic substitution (LW Equation 4.6), as the difference in the average effects of these two alleles yields the mean effect on fitness from replacing a randomly chosen a allele with an A allele.

Applying Wright's formula (Equation 5.5b) together with Equation 6.18d returns

$$\Delta p = \frac{p(1-p)}{2\overline{W}}\,\frac{d\overline{W}}{dp} = \frac{p(1-p)}{2\overline{W}}\,(2\alpha) \tag{6.18f}$$

and substitution into Equation 6.18c then yields

$$R_W \simeq \frac{p(1-p)(2\alpha)^2}{2\overline{W}} = \frac{\sigma_A^2(W)}{\overline{W}} \tag{6.18g}$$

The last step follows from the fact that the additive genetic variance is related to α by $\sigma_A^2 = 2p(1-p)\alpha^2$ (LW Equation 4.12a). Thus, under this approximation of small allele-frequency change (in which terms of order $(\Delta p)^2$ can be ignored), the change in mean fitness is indeed proportional to the additive genetic variance in fitness.

As Example 6.4 (below) highlights, Equation 6.18g, and thus the classical view of Fisher's theorem, is only approximate. Under what conditions does this expression actually hold? While Equation 6.18g is correct for multiple additive loci (i.e., no dominance or epistasis) under both random and nonrandom mating (Kempthorne 1957; Ewens 1969), it is generally compromised by departures from additivity. Even when the theorem does not hold exactly, does it still remain a good approximation? Nagylaki (1976a, 1977a, 1977b, 1991, 1992b, 1993) examined ever more general models of fitness under the assumption of weak selection (i.e., the fitness of genotypes being approximately $1 + bs$ with s small and $|\,b\,| \ll 1$) and random mating. Selection is further assumed to be much weaker than the recombination frequency c_{min} for the closest pair of loci under selection ($s \ll c_{min}$). Under these conditions, the evolution of mean fitness falls into three distinct stages. During the first phase (roughly $t < 2\ln s/\ln[1-c_{min}]$ generations), any initial disequilibrium has a transient impact on the dynamics until the point where disequilibrium reaches a steady-state value. At this point, we enter the central phase, with the change in mean fitness becoming

$$R_W = \frac{\sigma_A^2(W)}{\overline{W}} + O(s^3) \tag{6.19}$$

where $O(s^3)$ means that terms on the order of s^3 have been ignored. Because additive genetic variance is expected to be of order s^2, Fisher's theorem is expected to hold to a good approximation during this period. The central phase of evolution lasts roughly $1/s$ generations. However, as gametic frequencies approach their equilibrium values, we reach the third phase, where the additive variance in fitness can be much smaller than order s^2, in which case terms of order s^3 can be important. During the first and third phases, mean fitness can decrease, but the fundamental theorem holds during the central phase of evolution. Because we expect the bulk of evolution (the majority of change in $\overline{W}$) to occur during this middle phase, Fisher's theorem approximately holds over the major part of evolutionary change. While Nagylaki's results are weak-selection approximations, we often expect weak selection to be the norm for quantitative traits, as even strong selection on a *trait* translates into weak selection on the *underlying loci* if each of these has a small effect (Equation 5.21).

Unlike Fisher's theorem, because Price's theorem is exact, we can go further and apply Price's results to make an exact statement about the evolution of mean fitness. Letting $z_i = A_i$ denote the breeding value for the fitness of the ith midparent, the mean breeding value in their offspring becomes

$$\overline{z}_i = A_i + \overline{\delta}_i \tag{6.20a}$$

where (as above) $\overline{\delta}_i$ is the difference between the trait value in the ancestor (the breeding value of the midparent) and the mean value in its offspring. The phenotypic value of fitness for this midparent (the average of the two parental fitnesses) can be written as $W_i = A_i + \epsilon_i$. Substituting these results into Equation 6.7c gives the between-generation change in the mean breeding value ($\Delta\overline{A}_W$) for fitness as

$$\begin{aligned} R_{A_W} &= \sigma(w_i, A_i + \overline{\delta}_i) + E(\overline{\delta}_i) \\ &= \sigma(w_i, A_i) + \sigma(w_i, \overline{\delta}_i) + E(\overline{\delta}_i) \\ &= \frac{1}{\overline{W}}\sigma(A_i + \epsilon_i, A_i) + \sigma(w_i, \overline{\delta}_i) + E(\overline{\delta}_i) \\ &= \frac{\sigma_A^2(W)}{\overline{W}} + \sigma(w_i, \overline{\delta}_i) + E(\overline{\delta}_i) \end{aligned} \tag{6.20b}$$

where the last expression follows because $\sigma(A_i, \epsilon_i) = 0$ by construction. When $E(\overline{\delta}_i) = \sigma(w_i, \overline{\delta}_i) = 0$, we recover the classic version of Fisher's fundamental theorem. If the mean breeding value for offspring is the average of their parent's breeding values, then $\overline{\delta}_i = 0$ and these conditions hold. Even if not *exactly* true, often $\overline{\delta}_i$ is very close to zero and the leading term (and hence Fisher's theorem) dominates.

Example 6.4. The accuracy of the first-order Taylor-series approximation used in Equation 6.18b was examined by Li (1967). Because $\overline{W}$ is a quadratic polynomial of p, the second-order Taylor series is exact,

$$\Delta\overline{W} = \frac{d\,\overline{W}}{dp}\,\Delta p + \frac{1}{2}\frac{d^2\,\overline{W}}{d\,p^2}\,(\Delta p)^2$$

As shown by Equations 6.18c and 6.18f, the first term recovers Fisher's theorem, while the second term is the error resulting from this approximation.

Taking the derivative of Equation 6.18d yields

$$\frac{d^2\,\overline{W}}{d\,p^2} = 2\,(W_{AA} - 2W_{Aa} + W_{aa})$$

and recalling Equation 6.18f, the residual term becomes

$$\frac{1}{2}\frac{d^2\,\overline{W}}{d\,p^2}\,\Delta p^2 = (W_{AA} - 2W_{Aa} + W_{aa})\left(\frac{p(1-p)}{\overline{W}}\,\alpha\right)^2$$

Thus, if fitnesses are additive, meaning that $[W_{AA} + W_{aa}]/2 = W_{Aa}$, the residual term is zero, and Fisher's theorem holds. However, when dominance in fitness is present, Fisher's theorem fails even for a single locus under random mating.

What Did Fisher Really Mean?

Fisher (1930, p. 35) warned that his theorem "requires that the terms employed should be used strictly as defined," and much of the confusion being referred to is concerned with what Fisher meant by "fitness." Crow (2002) noted that in stating his theorem, "Fisher was indulging in his usual elegant obscurity." Price (1972b) and Ewens (1989, 1992) argued that Fisher's theorem is always true because he had a very narrow interpretation of the change in mean fitness (also see Edwards 1990, 1994; Frank 1995; Lessard and Castilloux 1995; Lessard 1997; Plutynski 2006). They argued that rather than considering the *total* rate of change in fitness, Fisher was instead concerned only with the *partial* rate of change, that due only to changes in allele frequency, without considering any corresponding changes in the average excesses or effects of these alleles.

Placed in the framework of Price's theorem, this "partial increase" interpretation becomes clear. From Equation 6.15d (setting $b_j = \alpha_j$), the total response in fitness can be decomposed into two components

$$R_w = \sum_j 2\alpha_j \Delta p_j + \sum_j 2p'_j\, \Delta\alpha_j \tag{6.21a}$$

where α_j is the average effect of an allele on fitness. Recalling Equation 6.5a, the first sum is simply $\sigma(w_i, A_i) = \sigma(A_i + \epsilon_i, A_i) = \sigma^2_A(w)$, yielding

$$R_w = \sigma^2_A(w) + \sum_j 2p'_j\, \Delta\alpha_j \tag{6.21b}$$

Note that the first term in Equation 6.21a is the partial change due *solely* to changes in allele frequencies

$$\sum_j 2\alpha_j \Delta p_j$$

which we denote by ∂R_w to emphasize that only a *specific part* of the total change is being considered. Price argued that Fisher's interpretation of his theorem was that

$$\partial R_w = \sigma^2(A_w) \tag{6.21c}$$

Thus, the exact version of Fisher's theorem (Equation 6.21c) simply concerns the partial evolutionary response caused by natural selection, as Price argued that Fisher essentially regarded the second term as a change in the "environment" within which alleles find themselves after selection, with Fisher having a very broad interpretation of "environment," referring to both physical and genetic backgrounds. In the words of Price (1972b, p. 130), Fisher

> regarded the natural selection effect on fitness as being limited to the additive or linear effects of changes in gene (allele) frequencies, while everything else—dominance, epistasis, population pressure, climate, and interactions with other species—he regarded as a matter of the environment.

A nice discussion of this point was offered by Frank and Slatkin (1992), who pointed out that the change in mean fitness over a generation is also influenced by the change in "environment," E. Specifically,

$$R_W = (\,\overline{W}\,'\,|\,E\,'\,) - (\,\overline{W}\,|\,E\,) \tag{6.22a}$$

where the prime denotes the fitness or environment in the next generation. Expanding the preceding expression, we can partition the contributions from the change in fitness and the change in the environment by writing

$$R_W = [\,(\,\overline{W}\,'\,|\,E\,) - (\,\overline{W}\,|\,E\,)\,] + [\,(\,\overline{W}\,'\,|\,E\,'\,) - (\,\overline{W}\,'\,|\,E\,)\,] \tag{6.22b}$$

where the first term in brackets represents the change in mean fitness under the initial "environment," while the second represents the change in mean fitness due to changes in the environmental conditions. This decomposition is simply another way of stating Equation 6.21a. The exact version of Fisher's theorem relates solely to changes in the first component, $(\,\overline{W}\,'\,|\,E\,) - (\,\overline{W}\,|\,E\,)$, which he called the change in fitness due to natural selection. In considering this exact version of Fisher's theorem, Ewens (1994, p. 187) stated

> I believe that the often-made statement that the theorem concerns changes in mean fitness, assumes random-mating populations, is an approximation, and is not correct in the multi-locus setting, embodies four errors. The theorem relates the so-called partial increase in mean fitness, makes no assumption about random mating, is an exact statement containing no approximation, and finally is correct (as a theorem) no matter how many loci are involved.

Nagylaki (1993) suggested that the statement $R_W = \sigma^2_A(W)/\overline{W}$ be referred to as the **asymptotic fundamental theorem of natural selection**, while Fisher's more narrow (and correct) interpretation based on partial change should be referred to as the **Fisher-Price-Ewens theorem of natural selection**. A clear distinction between these two very different interpetations seems quite reasonable given their considerable past history of confusion. Warren Ewens (personal communication, 2005) said it best by noting that

> one should think of two totally different results, holding under totally different sets of circumstances, not intersecting with each other much, and which should not be put under the same umbrella.

Thus, the bold, sweeping classical interpretation of Fisher's statement is replaced in the exact version by an unfortunately all-too common outcome of mathematics—a result that is absolutely correct, but not really useful. While the nature of the partial response is certainly elegant, our interest is in the total response.

IMPLICATIONS OF FISHER'S THEOREM FOR TRAIT VARIATION

While the classical view of Fisher's theorem does not generally hold and the exact version has rather limited utility, Fisher's theorem has still had an enormous impact on how quantitative geneticists view trait variation. This is because an important corollary holds under very general conditions in a constant environment (Kimura 1965b; Ewens 1976; Nagylaki 1976a, 1977b; Ewens and Thomson 1977; Charlesworth 1987): in the absence of new variation from mutation or other sources such as migration, ***selection is expected to eventually remove all additive genetic variation in fitness***. This can be seen immediately for a single locus by considering Equation 5.8b—if the population is at equilibrium, all average excesses

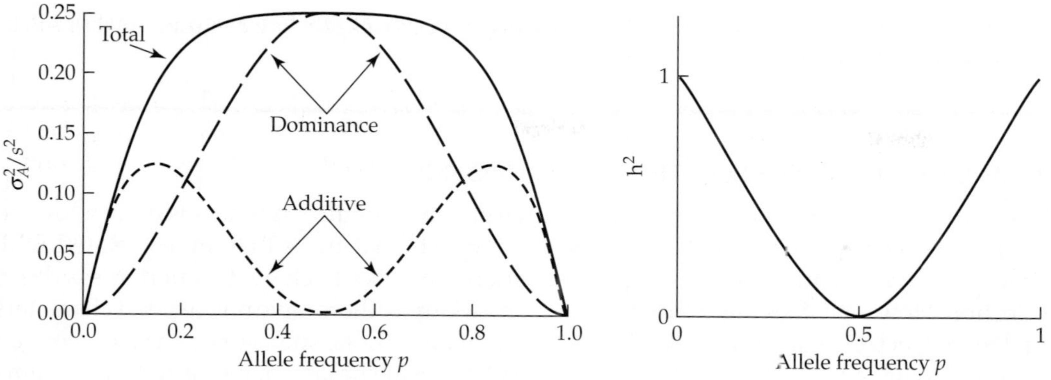

Figure 6.2 The behavior of genetic variance components and the heritability for fitness as a function of allele frequency p under the parameters given in Example 6.5. (**Left**) Note that total genetic variance is maximized at the value ($p = 1/2$), where the additive variance is zero. (**Right**) Assuming that there is no environmental variance, the heritability is simply the ratio of additive to total genetic variance, which is also zero when total variance is maximized.

are zero, as all segregating alleles have the same marginal fitness and hence no additive variation *in fitness* is present (Fisher 1941).

This corollary makes the general prediction that characters that are strongly genetically correlated with fitness should show reduced additive variance relative to characters that are less well correlated with fitness (Robertson 1955b), reflecting the removal of additive variance by selection (which is partly countered by new mutational input). As we now will review, there is indeed a loose trend for traits that are correlated with fitness to show reduced heritability, which is often interpreted as being due to a reduction in σ^2_A for fitness-related traits. However, note that we can express heritability as $h^2 = \sigma^2_A/(\sigma^2_A + \sigma^2_e)$, where $\sigma^2_e = \sigma^2_z - \sigma^2_A$ is the residual variance (the non-additive genetic and environmental variances). While a reduction in σ^2_A with σ^2_e held constant results in a smaller h^2, by sufficiently increasing σ^2_e, one can still have a reduced h^2, *even* when σ^2_A has *also* increased. Indeed, as we will show, a closer look at the data shows that the additive variance is often *greater* for traits that are correlated with fitness, but that this increase is overwhelmed by an increased residual variance, resulting in a lower heritability.

Example 6.5. Consider a locus with two alleles (A_1 and A_2) and overdominance in fitness,

$$W_{11} = 1 - s \quad W_{12} = 1 \quad W_{22} = 1 - s$$

Letting $p = \text{freq}(A_1)$, from Equation 6.18e and the definition of the average effect of a substitution under random mating, we have

$$\alpha = \alpha_A - \alpha_a = p[\,(1-s) - 1\,] + (1-p)[\,1 - (1-s)\,] = s(1-2p)$$

giving the additive genetic variance in fitness as

$$\sigma^2_A(W) = 2p(1-p)\alpha^2 = 2p(1-p)s^2(1-2p)^2$$

The dominance variance is computed from LW Equation 4.12b to be $\sigma^2_D(W) = [2p(1-p)s]^2$. For simplicity, the plot of h^2 for fitness (Figure 6.2) assumes no environmental effects, so $h^2 = \sigma^2_A/\sigma^2_G = \sigma^2_A/(\sigma^2_A + \sigma^2_D)$. As illustrated in Figure 6.2, these variance components change dramatically with p. The maximum total genetic variance in fitness occurs at $p = 1/2$, but in this particular example none of this variance is additive, so the heritability in fitness is zero, as the corollary of Fisher's theorem predicts. Even though *total* genetic variation in fitness is maximized at $p = 1/2$, no change in $\overline{W}$ occurs at this frequency, as none of the variance is additive.

Traits Correlated With Fitness Have Lower Heritabilities

The corollary to Fisher's theorem, that additive genetic variance in fitness is driven toward zero by selection in a constant environment, suggests that traits that are correlated with fitness will have reduced levels of additive variance relative to characters under less direct selection. How well does this prediction hold up? Many authors have noticed that characters that are expected to be under selection (e.g., life-history traits, such as clutch size) tend, on average, to have lower heritabilities than morphological characters measured in the same population or species (reviewed by Robertson 1955b; Charlesworth 1987; Mousseau and Roff 1987; Roff and Mousseau 1987; also see LW Figure 7.10). However, some notable exceptions are also apparent (see Charlesworth 1987).

The difficulty with these general surveys is knowing whether a character is highly *genetically* correlated with *lifetime* fitness. Clutch size, for example, would seem to be highly correlated with total fitness, but if birds with large clutch sizes have poorer survivorship, the correlation with lifetime fitness may be weak. Negative genetic correlations between components of fitness allow for significant additive variance in each component at equilibrium, even when additive variance in *total* fitness is zero (Robertson 1955b; Rose 1982).

Unfortunately, estimates of lifetime fitness in natural populations and their correlation with components of fitness (such as clutch size) are rare. One example was given by Gustafsson (1986; see also Merilä and Sheldon 2000), who measured lifetime reproductive success as well as the heritabilities of fitness and other characters in a closed natural population of collared flycatcher birds (*Ficedula albicollis*) in the Baltic Sea. Lifetime reproductive success had an estimated heritability that was not significantly different from zero, as expected from the corollary to Fisher's theorem. Clutch size had a rather high heritability, 0.32 ± 0.15, but the estimated phenotypic correlation between clutch size and total fitness was very low, $r^2 = 0.03$. In general, as the phenotypic correlation between a trait and total fitness decreased, its heritability increased (Figure 6.3).

McCleery et al. (2004) also found a negative relationship between trait heritability and trait-fitness phenotypic correlation in an English population of great tits (*Parus major*) that was followed for almost 40 years. Similar findings were seen by Teplitsky et al. (2009) in red-billed gulls (*Larus novaehollandiae*). Conversely, Schwaegerle and Levin (1991) found no significant association between the heritability of a character and its phenotypic correlation to fruit production (chosen as one measure of total fitness) in a wild population of the plant *Phlox dummondii* (Figure 6.3). While the evidence is mixed, these studies suggest a mild trend for characters that are phenotypically correlated with fitness to have reduced heritabilities relative to other characters. One important caveat is that this association is based on phenotypic, rather than genetic, correlations with fitness.

Under the classical view, if a trait is known to be under selection, one might be tempted to assume it is still far from its genotypic equilibrium value if it shows a modest to large heritability. This is false. As Example 6.6 highlights, a trait that is under selection can still have a high *trait* additive variance even when the additive genetic variance in *fitness* is near zero. This can happen if there is a nonlinear transformation of the trait value into fitness, such as occurs with stabilizing selection.

More generally, Price and Schluter (1991) noted that even with a simple linear relationship between a trait and fitness, it is quite likely that there may be a modest heritability for the trait but a very low heritability for fitness. The following simple model makes most of their main points. Assume fitness is entirely determined by a metric character, with fitness being a linear function of the phenotypic value z plus a residual deviation e, $W(z) = a + \beta z + e$, making the total variance in fitness $\sigma^2[\,W(z)\,] = \beta^2\sigma^2_z + \sigma^2_e$. Writing $z = A + E$, the additive genetic value A plus all other sources of variance (environmental and genetic), makes the

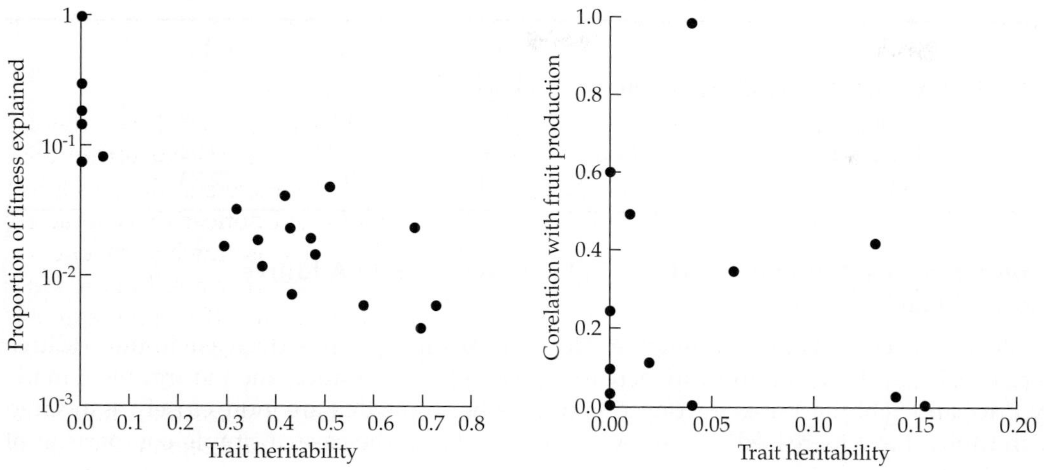

Figure 6.3 Two studies examining the association between a character's heritability and its total fitness, measured by r^2, the squared phenotypic correlation between the character and lifetime fitness. (**Left**) Gustafsson's (1986) work on the collared flycatcher *Ficedula albicollis* on the island of Gotland in the Baltic Sea. (**Right**) Schwaegerle and Levin's (1991) study of *Phlox drummondii*, with fruit production used as a measure of total fitness.

additive variance in fitness $\beta^2\sigma_A^2$. The heritability of fitness can be expressed in terms of the variance components for z as follows:

$$h_W^2 = \frac{\beta^2\,\sigma_A^2}{\beta^2\,(\sigma_A^2 + \sigma_E^2) + \sigma_e^2} = \frac{\sigma_A^2}{\sigma_A^2 + \sigma_E^2 + \sigma_e^2/\beta^2} < h_z^2 = \frac{\sigma_A^2}{\sigma_A^2 + \sigma_E^2} \tag{6.23}$$

Thus, even when fitness is entirely determined by a single trait, the heritability of fitness is less than the heritability of the character under selection, due to the variation, σ_e^2, in fitness about its expected value for a given trait value (which is expected to be considerable; see Chapter 29). If the heritability of *fitness* is found to be close to zero in this case, there still could be a significant heritability in the actual *character* under selection, implying that the trait mean can continue to change over time, albeit slowly.

Example 6.6. Even if Fisher's theorem holds exactly, its implication for character evolution can often be misinterpreted. Suppose that locus **A** in Example 6.5 completely determines a character under stabilizing selection. Let the genotypes AA, Aa, and aa have discrete phenotypic values of $z = -1$, 0, and 1, respectively (so that this locus is strictly additive with respect to the trait), and let the fitness function be $W(z) = 1 - sz^2$. If we assume no environmental variance, this generates the fitnesses for each genotype assumed in Example 6.5. The additive genetic variance for the *trait* z is $2(1)^2p(1-p)$, which is maximized at $p = 1/2$, precisely the allele frequency at which the additive genetic variance in fitness, $\sigma_A^2(W)$, equals zero. This difference emphasizes that Fisher's theorem concerns additive genetic variance in *fitness*, not in the *character*. In this example, the nonlinear mapping of the phenotypic character value, z, into the trait fitness, W, results in a situation where a character that is correlated with fitness retains additive variance in the trait but *not* in fitness at its equilibrium value.

Table 6.2 Heritabilities and coefficients of additive genetic (CV_A) and residual (CV_R) variation for representative traits in *Drosophilia melanogaster*. Both CV_A and CV_R values are multiplied by 100. Here n is the number of studies, and the median estimates are reported. (After Houle 1992.)

Trait	n	h^2	CV_A	CV_R
Sternopleural bristles	21	0.44	8.39	7.97
Wing length	31	0.36	1.56	2.09
Fecundity	12	0.06	11.90	39.02
Longevity	7	0.11	9.89	27.73

Traits Correlated With Fitness Have Higher Levels of Both Additive and Residual Variance

While a reduced h^2 value is often interpreted as resulting from a decrease in the additive variance, it can also result from an increase in the residual variance, and the argument made by Price and Schluter raises the question of whether traits that are more closely associated with fitness have increased residual variances. If this is the case, a simple comparison of heritabilities can be misleading (Houle 1992).

When comparing additive variances across traits, a standardized measure is required, and a common approach is to compare coefficients of variation, the ratio of the standard deviation to the mean. This led Houle (1992; see also Charlesworth 1984a) to suggest that the **coefficient of additive genetic variance**, $CV_A = \sigma_A/\mu$ (where μ is the trait mean) is the appropriate scale-free measure for comparing the amount of additive genetic variation across traits. To distinguish this measure from the heritability, Houle coined the term **evolvability** for CV_A (a term that has unfortunately since been co-opted in a number of ill-defined ways by developmental biologists). As a representative sample of *Drosophila* traits illustrates (Table 6.2), traits with low h^2 values can have very high CV_A values. Indeed, in a recent meta-analysis, Hansen et al. (2011) found essentially no correlation between heritability and evolvability.

However, in a survey of over 800 estimates for CV_A from a variety of traits, Houle (1992) found that characters assumed to be closely related to fitness (such as life-history traits) have higher evolvabilities (larger CV_A values) than do traits that are more loosely associated with fitness. Thus, the pattern of trait heritabilities decreasing with their correlation with fitness is not due to smaller additive variances, but instead is caused by larger residual (nonadditive plus environmental) variances, as quantified by the coefficient of residual variation (CV_R).

A study by Kruuk et al. (2000) on Scottish red deer (*Cervus elaphus*) offers some additional insight. The authors estimated components of additive genetic, maternal, and residual variances from pedigree data for this wild population (on the Isle of Rum in Scotland) for five life-history and three morphological traits in addition to lifetime fitness. As shown in Figure 6.4, they also found heritabilities to be negatively correlated with fitness, whereas the coefficient of additive genetic variance CV_A, was *positively* correlated with fitness in males (but negatively correlated in females). Moreover, CV_A values were higher for life-history traits than for morphological traits, and the coefficient of residual variation CV_R was also positively correlated with fitness. Similar patterns of both CV_A and CV_R being positively correlated with fitness have been seen in the seed beetle *Callosobruchus maculatus* (Messina 1993), in an Alberta population of bighorn sheep (Coltman et al. 2006), in the perennial herb *Ipomopsis aggregata* (Campbell 1997), and in natural bird populations (collared flycatchers by Merilä and Sheldon 2000; great tits by McCleery et al. 2004). For these studies, high residual variance, not low σ_A^2, accounts for the observed lower h^2 values for traits related to fitness. Conversely, while Teplitsky et al. (2009) also observed lower heritabilities for fitness-related traits in red-billed gulls, both higher residual variance *and* lower additive variance accounted for their trend.

What accounts for the higher coefficients of additive variance in traits associated with fitness? The prediction of lower additive variance in fitness-related traits is based on the notion that σ_A^2 is removed by selection. However, the loss of variation is partly countered

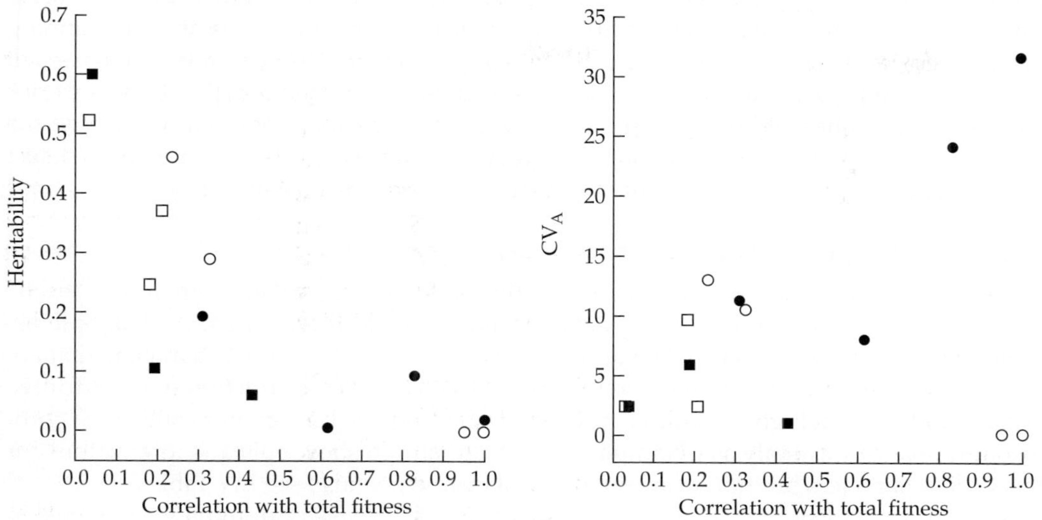

Figure 6.4 Kruuk et al.'s (2000) study of life-history and morphological traits in the Scottish red deer (*Cervus elaphus*). Circles denote life-history traits, squares morphological traits. Filled symbols are for males, open for females. (**Left**) The heritability of a trait is negatively associated with the correlation of that trait with fitness. (**Right**) The coefficient of additive genetic variation CV_A (the square root of the additive genetic variance of a trait divided by the trait mean, here multiplied by 100) is positively associated with fitness in males, and negatively associated with fitness in females.

by new mutational input, leaving some nonzero amount even in the face of strong selection (Chapter 28). If all traits have similar mutational variances, a faster removal of σ_A^2 by selection for the traits that are more closely related to fitness would indeed lead to lower equilibrium levels of additive genetic variance. However, it appears that traits that are more closely associated with fitness actually have *higher* mutational variances (Houle et al. 1996), most likely due to a larger number of loci influencing fitness (Houle 1992; Houle et al. 1996; Merilä and Sheldon 1999).

As with any summary metric, CV_A should be used with a little caution. Garcia-Gonzalez et al. (2012) found that roughly 45% of the studies they examined used incorrect methods in calculating CV_A (the most common being use of the sire variance $\sigma_A^2/4$ in place of σ_A^2, which results in a two-fold lower estimate). A second issue is that coefficients of variation should only be compared among items with the same dimensions (Lande 1977c). CVs for linear measures tend to be less than those for volumetric measures. Hence, something like body mass might intrinsically have a higher CV_A than some linear morphological trait. These and other issues are discussed by Garcia-Gonzalez et al. (2012), who provide a delta-method approximation (LW Appendix 1) for the standard error of a CV estimate.

A final caveat with these studies is that they examined the correlation between fitness (either total fitness or one of its components) and additive variance of a *trait*, while Fisher was concerned with the additive genetic variance of *fitness itself*. Estimates of the additive variance in total fitness in natural populations are very rare, as these require estimates of lifetime fitness (not a trivial task; see Chapter 29), further compounded by the difficulty of having to obtain such estimates over a set of relatives. To date, estimates of the additive variation associated with lifetime reproductive success in natural populations are from vertebrates with extensive known pedigrees. Kruuk et al. (2000) found no evidence for a significant heritability of this trait in red deer. McCleery et al. (2004) found positive, but not significant, estimates in great tits. Merilä and Sheldon (2000) found a significant additive variance for females and a positive (but not significant) variance for males in collared flycatchers. Teplitsky et al. (2009) found positive (but not significant) variance in females and no variance in males for red-billed gulls (*Larus novaehollandiae*). Blomquist (2010) obtained rather high ($\sim$0.4) estimates of heritability for lifetime reproductive success in a free-ranging population of rhesus macaques (*Macaca mulatta*) introduced in 1938 (from India) to a small island in Puerto Rico. Finally, Papaïx et al. (2010) obtained a very low estimate (a posterior mean of 0.02) for heritability of adult survival in the wild for a population of blue tits (*Cyanistes caeruleus*). One issue with all of these studies is the expected low power to detect small amounts of variances, so negative results should be viewed cautiously.

Nonadditive Genetic Variance for Traits Under Selection

As selection drives the additive variance in fitness toward zero, any remaining genetic variance is expected to be increasingly composed of nonadditive terms. As Example 6.5 highlights, this nonadditive genetic variance can be considerable. Thus, characters that are more closely associated with fitness are expected to have a higher fraction of nonadditive variance and hence a higher residual variance. This trend can be seen in results from chromosome substitution analysis (Example 6.7), which tend to show epistatic interactions for life-history characters but not for morphological characters (also see LW Table 5.1).

Crnokrak and Roff (1995) examined roughly 340 estimates of dominance variance in both life-history and morphological traits from 17 wild and 21 domestic species. In the wild species, traits assumed more closely connected with fitness (life-history traits) showed significantly higher dominance genetic variation (measured relative to total variance) than did morphological traits. In domesticated species, however, there were no significant differences in dominance genetic variance between life-history and morphological traits. The presumption is that many of the morphological traits examined in the domesticated species were themselves the result of strong recent selection during domestication (and thus, both groups of traits may have been under similar selection). This is supported by the observation that morphological traits in domesticated species showed significantly higher dominance variance than morphological traits in wild species. While certainly not conclusive, these results are consistent with the prediction of higher dominance genetic variance in traits more closely associated with fitness. Ideally, as above, the comparison of the amount of dominance variance should also be examined in terms of the coefficients of variation, as opposed to just the fraction of total variance.

Roff and Emerson (2006) presented a somewhat complementary analysis, using 90 estimates for life-history traits and over 140 estimates for morphological traits from line-cross data. Line-cross analysis examines the components (additive, dominance, etc.) contributing to the between-line divergence, rather than the variance segregating in any particular population (LW Chapter 9), and it is by no means clear if additive variance being driven to zero by selection within populations translates into significant nonadditive contributions to differences between line means. This important caveat aside, Roff and Emerson found that the magnitude of dominance effects (relative to additive effects) in line differences was much greater for life-history traits. Further, epistatic effects were more often detected for life-history traits, and the ratio of total nonadditive effects (dominance plus epistasis) relative to additive effects for life-history traits was roughly double that for morphological traits. Finally, DeRose and Roff (1999) showed that (in animals) inbreeding depression is greater for life-history than morphological traits, indicating higher amounts of directional dominance among segregating alleles for life-history traits (see also LW Chapter 10).

What accounts for the increase in residual variation for traits associated with fitness? One source, as suggested by Price and Schluter (1991), is higher environmental variance associated with fitness. As we have just seen, a second source is an increase in nonadditive variance. While both factors likely play a role, their relative importance is unknown (Merilä and Sheldon 1999). Due to the difficulty of estimating nonadditive genetic variance components without special mating designs (LW Chapter 20), resolving this question for natural populations is likely to prove quite difficult.

Example 6.7. Mackay (1985b) examined total fitness (measured by competition against a marked balancer stock) of 41 third chromosomes extracted from a natural population of *Drosophila melanogaster*. Using these chromosomes, lines with an otherwise common background could be made homozygous for a particular extracted third chromosome, and likewise the performance of that chromosome as a heterozygote could also be assayed. If there is significant additive genetic variance in fitness, a correlation between homozygote and heterozygote fitness is expected. Such a correlation was found for viability, suggesting some additive genetic variance in this character. However, when total fitness was examined, no correlation was found, suggesting no significant additive variation in total fitness. Mackay observed strong inbreeding depression, consistent with the total genetic variation in fitness being caused by segregation of rare deleterious recessive alleles (LW Chapter 10). A very similar experiment using segregating third chromosomes within a *Drosophila* population selected for domestication, Fowler et al. (1997) and Gardner et al. (2005) found high homozygote-heterozygote correlations for total fitness, and thus significant additive variance in fitness.

ROBERTSON'S SECONDARY THEOREM OF NATURAL SELECTION

In two separate papers, Robertson (1966a, 1968) suggested that the expected reponse to selection of any trait is a function of the covariance between the breeding value for that trait and relative fitness. Robertson (1968) called this relationship the **secondary theorem of natural selection**. While these two papers are usually jointly cited as the source for this theorem, a more careful reading shows that Robertson proposed two *different* formulae, based on whether the covariance with trait breeding value A_z involves the phenotype (w) or the breeding value (A_w) of relative fitness. As a result, there is confusion in the literature as to the exact form of Robertson's secondary theorem.

1968 Version: $R_z = \sigma_A(w,z)$

We first consider Robertson's (1968) version, as it is widely used in the population-genetics literature (e.g., Crow and Nagylaki 1976; Nagylaki 1992b). Robertson's (1968, p. 13) paper does not contain either a proof or a formal expression, just the clear statement that "The secondary theorem of natural selection states that the change in any character produced by a selection process is equal to the additive covariance between fitness and the character itself." In equation form,

$$R_z = \sigma_A(w, z) = \sigma(A_w, A_z) \tag{6.24a}$$

The expression $\sigma_A(w, z)$, which is shorthand for the covariance between the breeding values for the trait and relative fitness $\sigma(A_w, A_z)$, is often called the **Robertson covariance**. The term *secondary theorem* is reasonable for this version, as it directly follows from Fisher's fundamental theorem, using a simple regression argument (Falconer 1985). The expected change in a trait is given by its change in mean breeding value, ΔA_z, which in turn is the change in the breeding value of relative fitness, $R_w = \Delta A_w$ times the regression of the trait breeding value A_z given A_w, or

$$R_z = \Delta A_z = \beta_{A_z|A_w} \, \Delta A_w \tag{6.24b}$$

From standard regression theory (LW Chapter 3), the slope of the regression of A_z on A_w is simply

$$\beta_{A_z|A_w} = \frac{\sigma(A_w, A_z)}{\sigma^2(A_w)} \tag{6.24c}$$

Likewise, under the fundamental theorem (Equation 6.17c), $\Delta A_w = \sigma^2(A_w)$. Substituting these two results into Equation 6.24b recovers Robertson's 1968 version of his theorem,

$$R_z = \left(\frac{\sigma(A_w, A_z)}{\sigma^2(A_w)} \right) \sigma^2(A_w) = \sigma(A_w, A_z)$$

1966 Version: $R_z = \sigma(w, A_z)$

Although Robertson coined the term "secondary theorem" in his 1968 paper, a careful reading of Robertson's rather opaque 1966 paper suggests that his original result arose as a natural extension of the Robertson-Price identity, with

$$R_z = \sigma(w, A_z) \tag{6.25a}$$

Equation 6.10 shows that the within-generation change in the mean of a trait z is simply $\sigma(w, z) = S$. If one takes z as the breeding value of the trait, the Robertson-Price identity yields Equation 6.25a as the within-generation change in the breeding value of the trait caused by selection, and hence the expected response in the next generation when the average parental breeding value predicts the mean of their offspring. Equation 6.25a is widely used in evolutionary quantitative genetics (e.g., Lande 1976; Frank 1997), and is closely connected with the breeder's equation.

To see this connection, taking the regression of breeding value on phenotype as $A_z = h^2(z - \mu) + e_A$ and substituting into Equation 6.25a yields

$$R_z = \sigma\left[w, h^2(z - \mu) + e_A\right] = h^2\sigma(w, z) + \sigma(w, e_A) = h^2 S + \sigma(w, e_A) \tag{6.25b}$$

and we recover the breeder's equation when $\sigma(w, e_A) = 0$. Biologically, this last assumption implies that the relative fitness of an individual is uncorrelated with the residual error, e_A, when using their phenotype to predict their breeding value. The covariance, $\sigma(w, e_A)$, can be nonzero if an environmental factor influences both fitness and trait value, in which case the breeder's equation can fail, while Robertson's theorem (Equation 6.25a) may still hold. Indeed, as is detailed in Chapter 20, increased attention is being paid to Robertson's theorem in ecological genetics, as a discrepancy between the predicted response using the breeder's equation versus that using Robertson's theorem suggests that the focal trait is not the only target of selection.

Finally, we can connect the two versions of Robertson's theory by writing the phenotype, w, of relative fitness in terms of its breeding value, A_w, plus a residual, e_w (note that this is different from the residual e_A from Equation 6.25b, which is the error in predicting trait breeding value given trait phenotypic value). Substituting this expression for w in Equation 6.25a yields

$$R_z = \sigma(w, A_z) = \sigma(A_w + e_w, A_z) = \sigma(A_w, A_z) + \sigma(e_w, A_z) \tag{6.25c}$$

showing that while the 1966 version is more general, the two versions (Equations 6.24a and 6.25a) are equal when $\sigma(e_w, A_z) = 0$.

Accuracy of the Secondary Theorem

Formal population-genetic analysis of the secondary theorem is based on the 1968 version, and is very closely related to work on the accuracy of the classical interpretation of Fisher's theorem. Recall Equation 5.22, which showed the contribution to the selection response in trait z from the kth locus (in the absence of epistasis) as

$$R_{z,k} = 2\sum_j \alpha_j s_j\, p_j + \sum_{i,j} \delta_{ij}\, p_i\, s_i\, p_j\, s_j$$

where α_j is the average effect of allele i, δ_{ij} the dominance deviation associated with alleles i and j, and all terms relate to the values for locus k. The first sum is the expected product of the average effect, α_j, of an allele on character value and the average excess, s_i, of that allele on relative fitness. Recall that the definition of a covariance is $\sigma(x, y) = E(x\,y) - E(x)\,E(y)$, and note that the first term above is simply $2E(\alpha_j s_j)$. Because (by definition), $E(\alpha_j) = 0$, the first sum is thus the covariance between α_j and s_j, in other words, the additive genetic covariance between relative fitness and the focal trait. Summing over all loci, we can express Equation 5.22 as

$$R_z = \sigma_A(z, w) + B \tag{6.26}$$

If the character has no dominance (all $\delta_{ij} = 0$), the correction term, B, vanishes, recovering Robertson's original suggestion. Even if dominance is present, in the absence of epistatic variance, the error in the secondary theorem becomes increasingly small as the number of loci increases (Example 5.12).

The most general statement on the validity of Robertson's 1968 version, which is from Nagylaki (1992b, 1993), assumes weak selection on the underlying loci and random mating but allows for arbitrary epistasis and linkage disequilibrium. Similar to his weak selection analysis of Fisher's theorem discussed above, Nagylaki showed that after a sufficient time, the change in mean trait value is given by

$$R_z = \sigma_A(z, w) + O(s^2) \tag{6.27}$$

As with the fundamental theorem, when gametic frequencies approach their equilibrium values, terms of order s^2 may become significant and the mean response may differ significantly from Robertson's prediction, but again, the bulk of evolutionary change likely occurs before the equilibrium value is approached too closely. Consequently, the amount of change during the final approach to the equilibrium is generally expected to be quite small, so (as with the fundamental theorem), Robertson's theorem holds for the bulk of evolutionary change under weak selection on the underlying loci.

Connecting Robertson's Results With Those of Price, Fisher, and Lush

As we have shown, Robertson's 1968 version directly follows from the classical version of Fisher's theorem, while his 1966 version easily recovers the breeder's equation. Recall that we also showed that the breeder's equation (introduced by Lush 1937) recovers the classical version of Fisher's theorem (Equation 6.17b). All that remains is to consider how Robertson's results fit with Price's theorem.

If we let z in Equation 6.7b denote the breeding value of the trait of interest, the between-generation change in the mean breeding value for our focal trait is

$$\begin{aligned} \Delta\overline{A}_z &= \sigma(w, A_z + \overline{\delta}) + E(\overline{\delta}) \\ &= \sigma(w, A_z) + \sigma(w, \overline{\delta}) + E(\overline{\delta}) \end{aligned} \tag{6.28}$$

Since we are following breeding values, $\overline{\delta}$ represents the difference between parent and offspring breeding values. As with our analysis of Fisher's theorem within the Price equation framework, if parental breeding values exactly predict mean offspring breeding values ($\overline{\delta} = 0$), Equation 6.28 reduces to $R_z = \sigma(w, A_z)$. In this setting, Robertson's (1966a) version (Equation 6.25a) naturally follows from Price's theorem, and holds exactly. When parental breeding values are good predictors of mean offspring breeding value, meaning that $\overline{\delta}$ is not zero but is still small, then Roberston's (1966a) theorem is a good approximation for the response to selection. If we write $w = A_w + e_w$, Equation 6.28 becomes

$$\begin{aligned} \Delta\overline{A}_z &= \sigma(A_w + e_w, A_z) + \sigma(w, \overline{\delta}) + E(\overline{\delta}) \\ &= \sigma(A_w, A_z) + \sigma(e_w, A_z) + \sigma(w, \overline{\delta}) + E(\overline{\delta}) \end{aligned} \tag{6.29}$$

which recovers Robertson's 1968 version (Equation 6.25a), provided that the last three terms are zero (or else are very small relative to the first term).

THE BREEDER'S EQUATION FRAMED WITHIN THE PRICE EQUATION

The breeder's equation forms the backbone for much of the theory of selection response in quantitative genetics (Chapters 13–20). Almost all of the theory of breeding is framed around it, as is much of evolutionary quantitative genetics, although Robertson's secondary theorem has been gaining some recent traction in studies in natural populations (reviewed in Chapter 20). As these results show, the breeder's equation is an approximation, with a linear parent-offspring regression as a necessary, but not sufficient, condition (Example 6.1). When viewed in a population-genetic framework, Example 5.12 showed that the breeder's equation is generally a fairly good approximation under weak selection on each of the loci underlying a trait. Conversely, as highlighted in Example 6.3, the performance of the breeder's equation can be compromised if there are major genes, especially if they display dominance. Here we conclude our discussion on theorems of selection by expanding upon our earlier discussion (Equations 6.12 and 6.13) to fully place the breeder's equation into the exact Price equation framework.

Equation 6.12 shows that when the parent-offspring regression is linear, the expression for selection response reduces to the breeder's equation plus two correction terms. What can be said if we make no assumptions about the functional form of this regression? The most general solution is **Heywood's decomposition** (2005), which we build up to after introducing the partial covariance and the effect of the reference population on the parent-offspring regression. Based on the Price equation, Heywood's decomposition yields components with biologically meaningful interpretations that go beyond simply being mathematically convenient correction terms. A key point from this decomposition was foreshadowed in Example 6.3—it is the parent-offspring regression *after*, as opposed to before, selection that is more appropriate for predicting the response to selection.

Partial Covariance and the Spurious Response to Selection

To express the selection response in terms of the Price equation, as done above, let z_i denote the phenotypic value of the ith parent whose relative fitness is w_i and whose offspring have mean value $\overline{z}_i$ and, hence, a parent-offspring deviation $\overline{\delta}_i = \overline{z}_i - z_i$. Equation 6.7c shows the general expression for response as

$$R_z = \sigma(w_i, \overline{z}_i) + E(\,\overline{\delta}_i\,) \tag{6.30a}$$

Heywood's key insight is that Equation 6.30a can be decomposed into contributions from two biologically interpretable linear regressions and their corresponding residuals. The first regression involves offspring ($\overline{z}_i$) and parental (z_i) values

$$\overline{z}_i = \mu + \beta_{\overline{z}|z}\, z_i + e_{\overline{z},i}, \quad \text{where} \quad \beta_{\overline{z}|z} = \frac{\sigma(\overline{z}_i, z_i)}{\sigma_z^2} \tag{6.30b}$$

The second is the regression of relative fitness w_i of a parent i on its phenotypic value z_i

$$w_i = 1 + \beta_{w|z}\, z_i + e_{w,i}, \quad \text{where} \quad \beta_{w|z} = \frac{\sigma(w_i, z_i)}{\sigma_z^2} = \frac{S}{\sigma_z^2} \tag{6.30c}$$

It is important to stress that we are *not* assuming that the *true* relationship between either the mean offspring value $\overline{z}_i$ or paternal fitness w_i and paternal phenotype value z_i is linear, but rather simply considering the best linear regression for these relationships and the resulting residuals.

A key feature of Heywood's decomposition concerns the covariance $\sigma(e_w, e_{\overline{z}})$ between the residuals of these two regressions. As developed in Example 6.8, the covariance among the residuals for two variables regressed on a third is as known as the **partial covariance** between two variables (here w_i and $\overline{z}_i$) given a third (here z_i), which is denoted as

$$\sigma(e_w, e_{\overline{z}}) = \sigma(w, \overline{z} \,||\, z) \tag{6.31a}$$

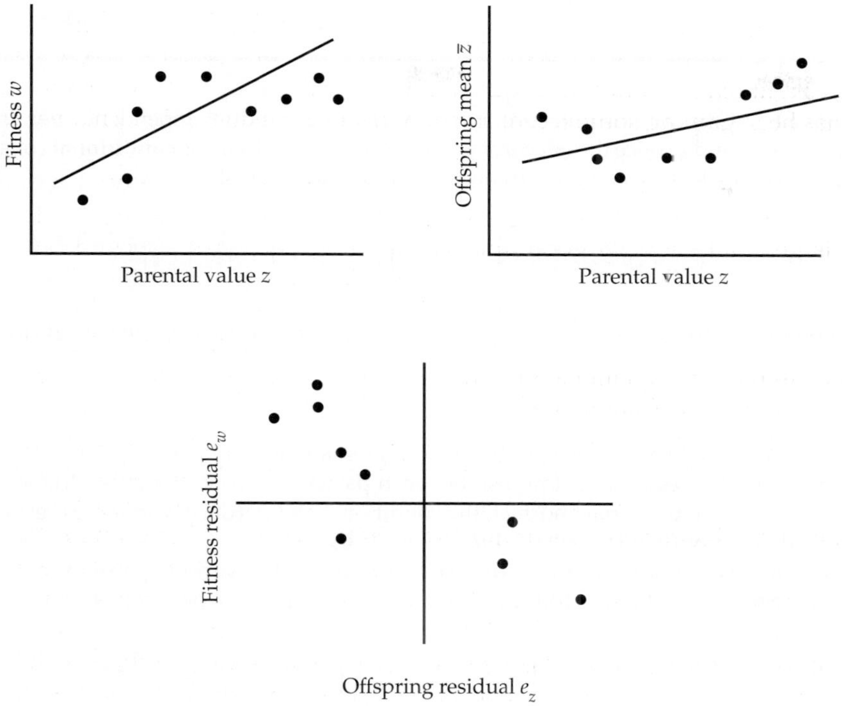

Figure 6.5 The partial covariance, $\sigma(w, \overline{z} \,||\, z)$, is the covariance $\sigma(e_w, e_{\overline{z}})$ between the residuals ($e_{\overline{z}}$) of the linear regression of $\overline{z}_i$ on z_i (the parent-offspring regression; Equation 6.30b) and the residuals (e_w) of the linear regression of w_i on z_i (fitness-phenotype regression; Equation 6.20c). In this example, both of these relationships are nonlinear, resulting in the distribution of residuals for the best fitting linear regression being nonrandomly distributed. A plot of these residuals against each other shows a strong negative covariance, so while $\sigma(w, z)$, $\sigma(\overline{z}, z)$, and $\sigma(w, \overline{z})$, are all positive, $\sigma(w, \overline{z} \,||\, z)$ is negative.

As shown in Example 6.8 (Equation 6.33f), we can write

$$\begin{aligned} \sigma(w, \overline{z}) &= \sigma(w, \overline{z} \,||\, z) + \beta_{\overline{z}|z}\, \sigma(w, z) \\ &= \sigma(w, \overline{z} \,||\, z) + \beta_{\overline{z}|z}\, S \end{aligned} \tag{6.31b}$$

Substitution of this expression into Equation 6.30a yields

$$R_z = \beta_{\overline{z}|z}\, S + \sigma(w, \overline{z} \,||\, z) + E(\,\overline{\delta}\,) \tag{6.32}$$

Thus, even when the parent-offspring regression is nonlinear, we can express the selection response in trait z as a linear response (breeder's equation) term $\beta_{\overline{z}|z}S$ plus a correction $\sigma(w, \overline{z} \,||\, z) = \sigma(e_w, e_{\overline{z}})$ accounting for (among other things) nonlinearity and an additional correction $E(\,\overline{\delta}\,)$ for transmission. Because $\sigma(w, \overline{z} \,||\, z)$ removes the (linear) effect of parental value z_i, on both its offspring mean $\overline{z}_i$ and its own relative fitness, w_i, any residual association between w_i and $\overline{z}_i$ is uncorrelated with z_i and hence uncorrelated with selection on the parent. Thus, the response from this component need not be adaptive, leading Heywood to denote $\sigma(w, \overline{z} \,||\, z)$ as the **spurious response to selection**.

There are two ways to generate a nonzero $\sigma(w, \overline{z} \,||\, z)$. First, both regressions (w_i on z_i and $\overline{z}_i$ on z_i) may be nonlinear, and, as a result, their residuals may be correlated, generating a spurious response (e.g., Figure 6.5). Second, even if one (or both) of these regressions is linear, if both w_i and $\overline{z}_i$ are correlated through an unmeasured variable (such as an environmental effect), their residuals, after being regressed on z_i, can still be correlated, again generating a potentially spurious response (Chapter 20 examines this process in more detail).

Example 6.8. One can imagine several measures of the residual dependency between two variables (say x and y) once the effects of a third (z) is removed. One is **conditional covariance**, $\sigma(x, y \mid z)$, defined as the covariance conditioned on the value of z:

$$\sigma(x, y \mid z) = E\left[\left(x - E(x|z)\right)\left(y - E(y|z)\right) \mid z\right] \tag{6.33a}$$

Heywood's decomposition uses an alternative measure of dependency, the **partial covariance**

$$\sigma(x, y||z) = \sigma(e_x, e_y) \tag{6.33b}$$

where e_x and e_y are the *residuals* for the linear regression of x on z and of y on z. We use the notation $||z$ in place of $|z$ to distinguish between partial and conditional covariances. If the three variables are multivariate normal, the conditional and partial covariances are equal, but generally they differ (Lawrance 1976; Baba et al. 2006).

The partial covariance can be expressed in terms of the various pairwise covariances between the three variables, as follows. The linear regression of x on z can be written as

$$x = \mu_x + \beta_{x|z} z + e_x = \mu_x + \left(\frac{\sigma(x, z)}{\sigma_z^2}\right) z + e_x$$

implying that

$$e_x = x - \left(\mu_x + \beta_{x|z}\, z\right) = x - \mu_x - \left(\frac{\sigma(x, z)}{\sigma_z^2}\right) z \tag{6.33c}$$

with a similar expression for e_y. Substituting these into Equation 6.33b and ignoring the mean terms (which, as constants, do not factor in the covariance) yields

$$\sigma(x, y||z) = \sigma(e_x, e_y) = \sigma\left[x - \left(\frac{\sigma(x, z)}{\sigma_z^2}\right) z, y - \left(\frac{\sigma(y, z)}{\sigma_z^2}\right) z\right]$$

which, by expansion, leads to

$$\begin{aligned}\sigma(x, y||z) &= \sigma(x, y) - \left(\frac{\sigma(x, z)}{\sigma_z^2}\right)\sigma(y, z) - \left(\frac{\sigma(y, z)}{\sigma_z^2}\right)\sigma(x, z) + \left(\frac{\sigma(y, z)}{\sigma_z^2}\right)\left(\frac{\sigma(x, z)}{\sigma_z^2}\right)\sigma_z^2 \\ &= \sigma(x, y) - \frac{\sigma(x, z)\,\sigma(y, z)}{\sigma_z^2}\end{aligned} \tag{6.33d}$$

The last term can be expressed as either $\beta_{x|z}\sigma(y, z)$ or $\beta_{y|z}\sigma(x, z)$. Applying Equation 6.33d by taking $x = w$ and $y = \overline{z}$ yields

$$\sigma(w, \overline{z}\,||\, z) = \sigma(w, \overline{z}) - \frac{\sigma(\overline{z}, z)}{\sigma_z^2}\,\sigma(w, z) = \sigma(w, \overline{z}) - \beta_{\overline{z}|z}\,\sigma(w, z) \tag{6.33e}$$

Rearranging this last expression yields

$$\begin{aligned}\sigma(w, \overline{z}) &= \sigma(w, \overline{z}\,||\, z) + \beta_{\overline{z}|z}\,\sigma(w, z) \\ &= \sigma(w, \overline{z}\,||\, z) + \beta_{\overline{z}|z}\, S\end{aligned} \tag{6.33f}$$

which is used to obtain Equation 6.32.

Parent-Offspring Regressions Before and After Selection

The second feature leading to Heywood's decomposition concerns the reference population for the parent-offspring regression. A subtle but important point to stress about Equation 6.32 is that $\beta_{\overline{z}|z}$ is the slope of the regression of offspring mean on parental phenotypes *following* selection. This slope may be different from the regression based upon unselected parents. In particular, when considering a *single* parent-offspring (as opposed to a midparent-offspring) regression, selection can change the mean of the offspring by changing the distribution of genotypes for the other parent with which the parent mates.

To see this point, suppose a single locus determines the trait of interest, and the selected parent chosen has genotype *Aa*. If the frequency of *A* before selection was 1/4, then the expected offspring frequencies when this parent is randomly mated to an unselected random parent (and hence a 1/4 chance of an *A* and a 3/4 chance of an *a*) are (1/8) *AA*, (1/2) *Aa*, and (3/8) *aa*, and these frequencies determine the trait value in the offspring from this parent. Now suppose that selection changes the frequency of *A* to 0.35. The resulting offspring frequencies when our *Aa* parent is now crossed to a random parent from the selected population become 0.175 *AA*, 0.5 *Aa*, and 0.325 *AA*, resulting in a different mean offspring value for this parent versus that when crossed to an unselected parent. When the allele-frequency change is very small (as would occur with weak selection on each underlying locus), these two regressions are very similar. However, when a gene of large effect is under selection, the regression using selected versus unselected parents can be different (Example 6.3).

Thus, the single-parent–offspring regression can change between selected and unselected populations, as the mate is not specified. This is *not* the case when we take the unit to be the midparent, because, given the genotypes of both parents, their distribution of offspring is fully determined, independent of the frequency of other genotypes in the population. The same is true for parents that asexually reproduce or that self.

Thus, when considering single parents, Equation 6.32 needs to be modified to account for the possibility that the mean offspring from this parent is influenced by the distribution of other parents after selection. To do so, let $\overline{z}_i$ and $\overline{z}_i^\circ$ denote the offspring mean from parent i when mated to parents from selected and unselected populations, respectively. Given that Equation 6.32 is expressed in terms of $\overline{z}_i$, following Heywood, we wish to translate this equation into an expression involving $\overline{z}_i^\circ$ (the offspring means of the parents in the absence of selection). To do so, we define $\delta_i^\circ = \overline{z}_i - \overline{z}_i^\circ$ as the difference in the offspring means for parent i when crossed to selected versus unselected parents. Because $\overline{z}_i = \overline{z}_i^\circ + \delta_i^\circ$

$$\sigma(\overline{z}_i, z_i) = \sigma(\overline{z}_i^\circ + \delta_i^\circ, z_i) = \sigma(\overline{z}_i^\circ, z_i) + \sigma(\delta_i^\circ, z_i) \tag{6.34}$$

Thus

$$\beta_{\overline{z}|z} = \frac{\sigma(\overline{z}, z)}{\sigma_z^2} = \frac{\sigma(\overline{z}^\circ, z) + \sigma(\delta^\circ, z)}{\sigma_z^2} = \beta_{\overline{z}^\circ|z} + \frac{\sigma(\delta^\circ, z)}{\sigma_z^2} \tag{6.35}$$

Equation 6.35 relates the parent-offspring regressions using unselected ($\beta_{\overline{z}^\circ|z}$) versus selected ($\beta_{\overline{z}|z}$) parents.

The same logic leading to Equation 6.33f (Example 6.8) can be used to obtain two useful identities:

$$\sigma(w, \overline{z}^\circ) = \sigma(w, \overline{z}^\circ \,||\, z) + \beta_{\overline{z}^\circ|z}\sigma(w, z) \tag{6.36a}$$

and

$$\sigma(w, \overline{z} \,||\, z) = \sigma(w, \overline{z}^\circ \,||\, z) + \sigma(w, \delta^\circ) - \frac{\sigma(\delta^\circ, z)\,\sigma(w, z)}{\sigma_z^2} \tag{6.36b}$$

Finally, while $\overline{\delta}_i$ is often considered to be the change from parent to offspring in the absence of selection, this is not strictly correct. Formally, it has two components,

$$\overline{\delta}_i = \overline{z}_i - z_i = (\overline{z}_i^\circ + \delta_i^\circ) - z_i = \delta_i^\circ + (\overline{z}_i^\circ - z_i) \tag{6.37}$$

Table 6.3 Terms in Heywood's (2005) decomposition of response and their biological interpretations. See the text for further details.

Term	Heywood's Interpretation
βS	Linear response to selection. Breeder's equation analog.
$\sigma(w, \overline{z}^\circ \,\|\|\, z)$	Spurious response to selection. Nonadaptive response to selection (i. e., uncorrelated with selection on the parent).
$E(\delta^\circ)$	General induced transmission bias. Difference between offspring mean when the selected parent is crossed to a selected vs. unselected individual.
$\sigma(w, \delta^\circ), \sigma(w, \delta^\circ \,\|\|\, z)$	Special induced transmission bias. Covariance between parental fitness and difference in offspring mean when a parent is crossed to a selected vs. unselected individual. Which version is used depends on whether regression is based on selected (Equation 6.40) or unselected (Equation 6.39) parents.
$E(\overline{z}^\circ - z)$	Constitutive transmission bias. Change in offspring mean in unselected individuals.

Hence

$$E(\overline{\delta}_i) = E(\delta_i^\circ) + E(\overline{z}_i^\circ - z_i) \tag{6.38}$$

where $E(\overline{z}_i^\circ - z_i)$, which Heywood calls the **constitutive transmission bias**, is the change from parent to offspring using *unselected* parents. Inbreeding, the decay of linkage disequilibrium, and randomization of correlated environmental effects are all examples of situations with the potential for a nonzero constitutive transmission bias (Chapters 15 and 23). The first component, $E(\delta_i^\circ)$, Heywood's **general induced transmission bias**, is the difference between the offspring mean when the parent is crossed to the selected population and the offspring mean when crossed to the unselected population. If the change in genotypic frequencies in the selected population is small relative to the unselected population, then the general induced transmission bias will be low. This is expected to be the case when the trait is determined by loosely linked loci, each with a small (to very small) effect.

Heywood's Decomposition of Response

We now have all of the components in place for Heywood's decomposition of the selection response. Expressed in terms of a regression using parents before selection ($\beta_{\overline{z}^\circ, z}$), substituting Equations 6.35–6.38 into Equation 6.32 yields

$$R = \beta_{\overline{z}^\circ|z}\, S + \sigma(w, \overline{z}^\circ \,||\, z) + \sigma(w, \delta^\circ) + E(\delta^\circ) + E(\overline{z}^\circ - z) \tag{6.39}$$

Conversely, when expressed in terms of the regression after selection ($\beta_{\overline{z}|z}$), the response is

$$R = \beta_{\overline{z},z} S + \sigma(w, \overline{z}^\circ \,||\, z) + \sigma(w, \delta^\circ \,||\, z) + E(\delta^\circ) + E(\overline{z}^\circ - z). \tag{6.40}$$

This is obtained from Equation 6.39, using Equation 6.35 and the identity

$$\sigma(w, \delta^\circ) - \frac{\sigma(\delta^\circ, z)\, \sigma(w, z)}{\sigma_z^2} = \sigma(w, \delta^\circ \,||\, z) \tag{6.41}$$

Equation 6.41 follows, using the same logic leading to Equation 6.31f. Beside the different regression coefficients ($\beta_{\overline{z}^\circ|z}$ vs. $\beta_{\overline{z}|z}$), Equations 6.39 and 6.40 differ in that $\sigma(w, \delta^\circ)$ in Equation 6.39 is replaced by $\sigma(w, \delta^\circ \,||\, z)$ when we use the regression on selected parents.

Heywood noted that the last four terms in Equations 6.39 and 6.40 arise because of imperfect transmission of parental to offspring phenotype. If transmission is perfect, then $z_i = \overline{z}_i^\circ = \overline{z}_i$ and $\delta_i^\circ = 0$, and these last four terms are all zero. Heywood pointed out that although all four terms are due to imperfect transmission, they have different causes and can be thought of as logically distinct processes, as is summarized in Table 6.3. In the decomposition given by Equation 6.39, the linear response to selection (βS) is exactly the breeder's equation (assuming that $\beta_{\overline{z}^\circ|z} = h^2$), as the regression coefficient is based on unselected parents. In the version given by Equation 6.32, β is the parent-offspring regression following selection. This regression may change as the fitness scheme changes, while the before-selection regression remains unchanged under any fitness scheme (assuming an unselected base population). Once selection has occurred, the genotypic frequencies in the offspring from selected parents will depart from those in the unselected base population. Hence, the regression based on unselected individuals starting from this new population will likely be different (although perhaps only trivially so) from that in the founding, unselected base population. These issues are examined in Chapters 15, 16, and especially 24–26. See Heywood (2005) for several worked examples using this decomposition.

7

Interaction of Selection, Mutation, and Drift

In recent years, there has been some tendency to revert to more or less mystical conceptions revolving about such phrases as "emergent evolution" and "creative evolution." The writer must confess to a certain sympathy with such viewpoints philosophically but feels that they can have no place in an attempt at scientific analysis of the problem. Wright (1931)

In the previous chapters, we treated the response to selection as an effectively deterministic process, making the assumption that the stochastic force of random genetic drift is negligible relative to the power of selection, and also ignoring the origin of new variation by mutation. Such an approach often works well when the focus is on short-term evolutionary issues. However, on longer time scales, interactions between selection, mutation, and drift can influence patterns of variation within and among populations in significant and sometimes counterintuitive ways. As all populations are finite in size and all genomes are subject to mutation, such factors must be incorporated into any general theory of evolution. Although the material in this chapter is confined to one- and two-locus systems, the resultant principles provide the basic building blocks for more complex models for the evolution of quantitative traits, which are presented in subsequent chapters.

Generally, mutation and drift, respectively, introduce and remove variation from populations, but selection can have either effect, depending on whether it is directional, stabilizing, or purifying in nature. Of special interest is the degree to which all three forces interact to define the distribution of allele frequencies in an equilibrium population (or more precisely, in a **quasi-equilibrium** population, as with drift there is always some stochastic wandering of allele frequencies around a long-term expectation). One of the key issues considered in the following pages concerns the amount of variation maintained in the face of opposing pressures. We initially address this matter by retaining the assumption of an effectively infinite population size, considering the issue of selection-mutation balance and the fitness load that recurrent mutation always imposes upon a population. We then evaluate the situation in which drift is sufficiently strong to compete with, or even overpower, the effects of selection. The latter issue is of special interest when we consider selection on a quantitative trait, as strong selection at the phenotypic level does not necessarily translate into strong selection on any particular underlying locus (Equation 5.21). We also show that even when they are completely penetrant, only a small fraction of advantageous mutations will be successfully fixed in a population owing to the overwhelming influence of stochastic forces when alleles are rare.

Because the ways in which genes evolve often depend on the background context, we also use this chapter to introduce some key issues regarding the evolution of multilocus systems. First, drawing on results outlined in Chapter 3 for the effects of linkage on the effective population size for a chromosomal region, we explore how this translates into a reduction in the efficiency of selection for advantageous alleles. Second, using compensatory mutations as an entrée into the matter of epistasis, we evaluate the extent to which such pairwise changes are promoted in small vs. large populations. Third, we evaluate the situation in which two or more key mutations are required for a new adaptation, showing that some relatively simple scalings apply to the time to establishment with respect to population sizes and mutation rates.

SELECTION AND MUTATION AT SINGLE LOCI

As discussed in Chapter 5, many of the central questions in population and quantitative

Evolution and Selection of Quantitative Traits. Bruce Walsh & Michael Lynch.
Published 2018 by Oxford University Press.
DOI 10.1093/oso/9780198830870.001.0001

genetics concern the mechanisms responsible for the maintenance of genetic variation in natural populations. Here, we introduce a few classical models for the balance between the opposing forces of mutation and directional selection. Our preliminary focus will be on the simple case of two alleles, as this serves as the foundation for more complex models for the maintenance of quantitative variation, which are covered in Chapter 28.

Consider a locus with an advantageous allele A and a deleterious allele a, which have respective frequencies of $1-p$ and p. Let μ be the mutation rate from A to a, let v be the rate of back mutation to A, and assume there is selection preceding mutation in each generation, followed by random mating in a population that is effectively infinite in size. From Chapter 5, the new frequency of a after a generation of viability selection is

$$p' = p\,\frac{W_{\rm a}}{\overline{W}} \tag{7.1}$$

where $W_{\rm a}$ is the marginal fitness of a (Equation 5.7b), and $\overline{W}$ is the mean fitness. Letting p'' be the allele frequency following mutation, we then have

$$p'' = (1-v)p' + \mu(1-p') = (1-\mu-v)p' + \mu \tag{7.2}$$

This follows because $1-v$ is the fraction of a that remains unchanged following mutation, while a fraction μ of all A alleles (with frequency $1-p'$) mutate to a. Thus, one generation of the joint action of selection and mutation leads to the new frequency of a

$$p'' = (1-\mu-v)p\,\frac{W_{\rm a}}{\overline{W}} + \mu \tag{7.3}$$

Haldane (1927) was the first to consider the equilibrium allele frequencies that are eventually reached under this model of opposing mutation and selection pressures. If we let the fitnesses of genotypes AA, Aa, and aa be 1, $1-hs$, and $1-s$, respectively, the equilibrium frequencies, $\widetilde{p}$, satisfying $\Delta p = p'' - p = 0$ are given by the solutions of the rather complicated cubic equation

$$\begin{aligned}(1-\widetilde{p})^3 s(2h-1) &+ (1-\widetilde{p})^2[2-3h+\mu h+v(1-h)] \\ &+ (1-\widetilde{p})[-s(1-h)+\mu(1-hs)+v(1-2s+hs)] - v(1-s) = 0\end{aligned} \tag{7.4}$$

(Bürger 2000). Provided $0 < s < 1$ and $h \le 0.5$, this expression has a single stable equilibrium and considerable simplification is possible in a number of biologically realistic cases. For example, for the case of neutrality ($s = 0$), the equilibrium is simply defined by the opposing forces of mutation

$$\widetilde{p} = \frac{\mu}{\mu+v} \tag{7.5}$$

A situation of special interest concerns the polymorphism maintained by a balance between selection and mutation when allele a is at a selective disadvantage. To simplify the solution, it is generally assumed that back mutation to the advantageous allele is a negligible force. There are several mathematical justifications and one biological justification for such an assumption. First, unless the selection coefficient is small relative to the mutation rate, the frequency of the mutant allele will generally be low enough that back mutation will be a second-order effect. Second, although functional genes may mutate to deleterious alleles through numerous mechanisms, precise back mutations to normal alleles will necessarily be much rarer events, i.e, we expect that $v \ll \mu$. Letting $v = 0$, Equation 7.4 can be reduced to a more manageable quadratic equation, with a solution of

$$\widetilde{p} = \frac{(1+\mu)hs - \sqrt{[hs(1+\mu)]^2 - 4(1-2h)\mu s}}{2(2h-1)s} \tag{7.6a}$$

assuming $s > \mu$. For the general case of intermediate dominance ($0 < h \leq 1$)

$$\widetilde{p} \simeq \frac{\mu}{hs}, \quad \text{provided} \quad h \gg \sqrt{\mu/s} \tag{7.6b}$$

For the extreme situation in which *a* is a completely dominant deleterious mutation ($h = 1$)

$$\widetilde{p} = \frac{\mu}{s} \tag{7.6c}$$

whereas if A is recessive ($h = 0$)

$$\widetilde{p} = \sqrt{\frac{\mu}{s}} \tag{7.6d}$$

A number of other special cases were presented in Nagylaki (1992a) and Bürger (2000); for example, $\widetilde{p} \simeq 3\mu/s$ for sex-linked recessives.

The multiple-allele version of this model can be obtained in a straightforward manner. Suppose there are k alleles ($A_1, \cdots, A_k$) and let μ_{ij} be the probability that allele A_i mutates to allele A_j. If we let $\mu_i = \sum_{j \neq i} \mu_{ij}$ be the total mutation rate from allele A_i to any other allele, and assuming constant viability selection followed by mutation and then random mating, the allele-frequency change equations become

$$p_i'' = \frac{1}{\overline{W}} \left((1 - \mu_i)\, W_i\, p_i + \sum_{j \neq i} \mu_{ji} W_j\, p_j \right) \tag{7.7}$$

where W_i is the marginal fitness of allele A_i (Equation 5.7b). The equilibrium behavior of this system can be quite complex, and with sufficiently strong mutation there will be the possibility that stable cycles exists (Bürger 2000).

Clark (1998) examined a special case of the multiple-allele model in which there is one optimal allele, and all heterozygotes for single mutations have a fitness of $1 - hs$, while heterozygotes for two different mutant alleles have a fitness of $1 - ks$, where k is a measure of complementation between two deleterious alleles (with $k = 0$ implying that each allele compensates for the other allele's deficiencies). Under this model, multiple deleterious alleles are maintained by mutation pressure, and provided $k < 1$, the sum of their frequencies was higher than expected under the two-allele model. The latter result arises as interallelic complementation reduces the magnitude of selection operating on mutant alleles when they are jointly present in the same genotype.

Example 7.1. How much variation can mutation maintain when a mutant allele is lethal ($s = 1$)? The equilibrium frequency of a dominant lethal allele is

$$\widetilde{p} = \mu$$

(Equation 7.6b), whereas for a recessive lethal

$$\widetilde{p} = \sqrt{\mu}$$

(Equation 7.6c). Thus, because $\mu \ll 1$ (Chapter 3), recessive lethals are expected to be much more common than dominant lethals, a pattern that is seen for numerous human genetic disorders (Cavalli-Sforza and Bodmer 1971). Drawing from a tradition starting with Haldane (reviewed in Nachman 2004), these expressions are often used to estimate the lethal mutation rate for monogenic human diseases under the assumption that the observed allele frequencies are at mutation-selection equilibrium (e.g., Kondrashov 2003).

For a dominant lethal, the frequency of selected individuals in the equilibrium population is

$$\text{freq}(aa) + \text{freq}(Aa) = \mu^2 + 2\mu(1-\mu) \simeq 2\mu$$

whereas for a recessive, the frequency of selected individuals is

$$\text{freq}(aa) = (\sqrt{\mu}\,)^2 = \mu$$

Thus, despite the great disparity in allele frequencies for dominant and recessive lethals, when there are low mutation rates, there will only be a two-fold difference in the expected frequencies of affected individuals.

What about the equilibrium mean fitness of the population? With a dominant lethal

$$\overline{W} = \text{freq}(AA) = (1-\mu)^2 \simeq 1 - 2\mu$$

while for a recessive lethal

$$\overline{W} = 1 - \text{freq}(aa) = 1 - \mu$$

owing to the two-fold lower incidence of affected individuals.

Example 7.2. Albinism in humans is caused by a recessive allele, with an estimated frequency of albinos of around 1/20,000 (Cavalli-Sforza and Bodmer 1971). If we assume that albinos are at a moderate selective disadvantage ($s = 0.1$) and at mutation-selection equilibrium, what is the estimated mutation rate to albino alleles? Assuming genotype frequencies in Hardy-Weinberg equilibrium, so that $\widetilde{p}^{\,2} = 1/20,000$, from Equation 7.6c,

$$\widetilde{p}^{\,2} = \frac{1}{20,000} = \left(\sqrt{\frac{\mu}{0.1}}\right)^2$$

which implies that $\mu = 5 \times 10^{-6}$. Conversely, if we were to assume a mutation rate of $\mu = 10^{-5}$, the strength of selection against albinism would be inferred from

$$\widetilde{p}^{\,2} = \frac{1}{20,000} = \left(\sqrt{\frac{10^{-5}}{s}}\right)^2$$

implying $s = 0.2$, i.e., a 20% reduction of fitness in albinos.

Example 7.3. Our treatment of mutation-selection balance has assumed that there is a single reoccurring allele. For many human diseases, however, a large number of different mutational events can have the same fitness effect, e.g., numerous kinds of mutations can inactivate a gene. In such cases, $\tilde{p}$ is the equilibrium frequency of the entire *set* of such deleterious alleles. An important question in human genetics is the diversity (or **spectrum**) of alleles within this set—is the disease largely dominated by a single allele or is it a diverse collection of rare alleles?

Hartl and Campbell (1982) found that the probability that two random alleles in the disease class are identical by descent is

$$\varphi = \frac{1}{1 + 4N_e\mu(1-\tilde{p})} = \frac{1}{1 + \theta(1-\tilde{p})}$$

where $\theta = 4N_e\mu$, with μ being defined here as the total mutation rate to deleterious alleles. This expression is closely connected to the results on mutation-drift balance presented in Chapter 2, where we showed that $\theta/(1+\theta)$ is the probability of a heterozygote, and hence $1/(1+\theta)$

is the probability of a homozygote. To obtain the diversity within the set of disease alleles, here we restrict ourselves to the population of normal alleles, $2N_e(1 - \tilde{p})$, as these are the source of mutations to new disease alleles. At mutation-selection equilibrium, this flux of new mutations is countered by the selective removal of existing alleles. The reciprocal of φ, which asymptotically approaches 1.0 for $\theta \ll 1$, provides an estimate of the effective number of alleles.

Reich and Lander (2001) considered the implication of this result for two equilibrium frequencies of a monogenic disease, rare ($\tilde{p} = 0.001$) and common ($\tilde{p} = 0.2$). Assuming an ancestral human N_e of 10^4 and a total deleterious mutation rate of $\mu = 3 \times 10^{-6}$ (meaning that $\theta = 0.12$), they obtained the effective number of alleles as $\varphi^{-1} = 1 + 0.12 \cdot (1 - 0.001) = 1.12$ and $\varphi^{-1} = 1 + 0.12 \cdot (1 - 0.2) = 1.10$ for the rare and common disease, respectively. Thus, both diseases have a very simple and similar expected allelic spectrum under the historical effective population size. However, as the human population rapidly expanded to 6×10^9 individuals, meaning that currently $\theta = 72,000$, the expected numbers of alleles *at equilibrium* now become $\varphi^{-1} = 1 + 72,000 \cdot (1 - 0.001) \simeq 72,000$, and $\varphi^{-1} \simeq 58,000$ for the rare and common diseases, respectively. These spectra are clearly very complex.

Reich and Lander noted that, owing to the rapidity of human population-size expansion, we are still far from having these much higher equilibrium values, and that the transient dynamics of the two disease classes are significantly different en route to their new equilibrium values. The number of new disease alleles arising each generation is $2N\mu(1 - \tilde{p})$, the number of normal alleles times the mutation rate. Because the mutation-selection equilibrium frequency is not changed by the increase in population size, the per-generation turnover within the disease allele class must then scale as $2N\mu(1 - \tilde{p})/(2N\tilde{p}) = \mu(1 - \tilde{p})/\tilde{p}$, which is the fraction of the standing pool of disease alleles that arises on a per-generation basis. With this mutation rate, the turnover amounts to 0.3% per generation for the rare disease and 0.0012% for the common disease.

The fraction of the ancestral disease alleles that remains after t generations following the population expansion is $\exp[-t\mu(1 - \tilde{p})/\tilde{p}]$, which implies a half-life of $\ln(2)\tilde{p}/[\mu(1 - \tilde{p})]$ generations. Assuming a human generation time of 25 years, the ancestral spectrum half-life is ~5,000 years for the rare disease and ~1.3 $\times 10^6$ for the common disease. Hence, under the expected pattern of demographic change in the human population, a common disease should largely retain the ancestral spectrum (very few alleles), while a rare disease should have an increasingly complex one. Consistent with this prediction, Reich and Lander (2001) found a strong negative correlation between the observed frequency of a monogenic disease and its associated number of alleles.

SELECTION AND DRIFT AT SINGLE LOCI

In the preceding section, we assumed that we were examining a situation in which the forces of selection and mutation are powerful enough to ignore the stochastic consequences of random genetic drift, at least in the short term. This deterministic approach to population genetics yields explicit equilibrium solutions for allele frequencies, usually with no oscillatory behavior. In reality, however, drift plays a significant role in all long-term population-genetic contexts. For example, even when selection against deleterious mutations is strong, the defective alleles segregating in a population today will generally be descendants of entirely different mutations than those that segregated many millenia in the past.

All mutations eventually experience one of two alternative fates, complete loss or fixation, and our focus now becomes the latter, specifically the probability that by the spread of its descendants, an allele expands to a total frequency of 1.0. In general, drift reduces the efficiency of selection because the sampling of gametes to form each consecutive generation results in random deviations of allele frequencies from their deterministic expectations based on selection alone. If drift is strong relative to selection, a favored allele may stochastically decrease in frequency, and sometimes, eventually become lost. Throughout the following subsections, we ignore the effects of recurrent mutation, focusing instead on a single specific event—the fate of a preexisting allele or newly arisen mutation.

Most of the theory of the interaction between selection and drift was developed for a single diallelic locus under viability selection, which allows the change in allele frequency per generation to be treated as the sum of changes resulting from selection and drift,

$$\Delta p = \Delta p_s + \Delta p_d$$

where Δp_s is given by Equation 5.1b and 5.1c, and Δp_d (the per-generation change due to drift) is a random variable. Drift causes no directional tendency in the change in allele frequency, and hence $E(\Delta p_d) = 0$. Thus, the simplest measure of the strength of drift is the expected variance in allele-frequency change due to gamete sampling, which, under the standard Wright-Fisher model (Chapter 2), is defined by the binomial distribution

$$\sigma^2(\Delta p_d) = \frac{p(1-p)}{2N_e} \tag{7.8}$$

where p is the allele frequency prior to sampling, N_e is the variance effective population size, and the 2 accounts for diploidy (and is replaced by a 1 for a haploid population; Chapter 3). If $\sigma^2(\Delta p_d)$ is small relative to Δp_s, allele-frequency changes will not be dramatically different from their expectations under selection in an infinite population, but if $\sigma^2(\Delta p_d) > \Delta p_s$, drift can substantially obscure the deterministic force of selection.

Consider the situation in which alleles have additive fitness effects, with the genotypes *AA*, *Aa*, and *aa* having respective fitnesses of $1,\ 1+s$, and $1+2s$. Letting p be the frequency of allele *a*, then, from Equation 5.2, $\Delta p_s \simeq s\,p(1-p)$, assuming weak selection ($|s| \ll 1$). If we compare this result with Equation 7.8, it becomes clear that directional selection dominates drift when $2N_e|s| \gg 1$, whereas drift dominates directional selection when $2N_e|s| \ll 1$.

Because the intensity of drift scales with $1/(2N_e)$, a useful heuristic is that $2N_e s$ approximates the ratio of the power of selection to drift. This argument is not quite precise because the variance of allele-frequency change is only a rough indicator of the sampling properties of the allele-frequency distribution. However, **diffusion theory**, which gives an essentially complete description of the dynamics of a diallelic locus under drift and selection, upholds this general conclusion (Appendix 1). We will frequently encounter the composite parameter, $2N_e s$, in the following paragraphs.

Probability of Fixation Under Additive Selection

There is no possibility of having a perfectly stable polymorphism when drift and selection interact. Indeed, even in the case of overdominant selection (where there is a stable equilibrium in an infinite population; Chapter 5), one allele will eventually drift to fixation unless both homozygotes are lethal. Under this scenario, all new mutations ultimately become either lost or fixed at the population level, and those that become fixed will themselves be subject to replacement by subsequently arising mutations. Thus, when finite populations are considered, we need to think in terms of fixation probabilities and sojourn times of mutations. Even highly favorable alleles have fixation probabilities of less than 1.0 to a degree that depends on the initial frequency p_0, the strength of selection, and the effective population size N_e.

Suppose we denote by $u_f(p_0)$ the probability that an allele starting at initial frequency p_0 will become fixed. As noted in Chapter 2, under neutrality, the probability of fixation depends only on an allele's initial frequency regardless of population size, so that

$$u_f(p_0) = p_0 \tag{7.9}$$

Depending on the magnitude and direction of selection, this probability will either increase or decrease. When allelic effects on fitness behave additively, such that each copy of allele *a* changes fitness by s (giving fitnesses of $1,\ 1+s$, and $1+2s$)

$$u_f(p_0) \simeq \frac{1-e^{-4N_e s p_0}}{1-e^{-4N_e s}} \tag{7.10a}$$

$$\simeq p_0 + 2N_e s p_0(1-p_0) \quad \text{when } 2N_e|s| \leq 1 \tag{7.10b}$$

Equation 7.10a, due to Kimura (1957) with a slightly improved version given by Cash (1977), was derived using diffusion theory in Appendix 1. The simplified version, Equation 7.10b, was developed by Robertson (1960a) using the Taylor series approximation $e^{-x} \simeq 1 - x + x^2/2$ for $|x| \ll 1$, and an alternative derivation is given below. Although these approximations apply to both beneficial ($s > 0$) and deleterious ($s < 0$) alleles, and work especially well with favorable alleles (Carr and Nassar 1970), they can significantly *overestimate* the fixation probabilities of highly deleterious alleles ($N_e\,s \leq -1$), an issue examined in detail by Bürger and Ewens (1995).

It is critical to note that even when an allele is under strong selection, drift still plays a powerful role when allele frequencies are near zero. Starting with a single copy of an advantageous allele (with frequency $p_0 = 1/(2N)$, where N is the number of reproductive adults in the population), Equation 7.10a implies that the probability of fixation of a new mutation is approximately $2s\,(N_e/N)$ when $4N_e s \gg 1$. As we expect N_e to generally be $\ll N$ (Chapter 3) and s is typically $\ll 1$, this implies that a newly arisen favorable mutation will usually be lost by drift, no matter how beneficial. However, once the frequency of a strongly beneficial allele becomes sufficiently high, fixation is almost certain. For example, if $N_e s p_0 > 0.5$, the probability of fixation exceeds 0.70, while if $N_e s p_0 > 1$, the probability of fixation exceeds 0.93.

For mutations with a weak effect, it is informative to consider the probability of fixation of a newly arisen mutation relative to the neutral expectation of $1/(2N)$. Returning to Equation 7.10a, and approximating the numerator as $4N_e s p_0$, with $p_0 = 1/(2N)$, the scaled probability of fixation

$$u'_f(p_0) = \frac{u_f(p_0)}{1/(2N)} \simeq \frac{4N_e s}{1 - e^{-4N_e s}} = \frac{S}{1 - e^{-S}} \tag{7.11}$$

is found to be entirely a function of the composite parameter $S = 4N_e s$, which, as noted above, is a measure of the strength of selection ($2s$ in favor of homozygotes) relative to that of drift, $1/(2N_e)$ (Figure 7.1). For positive selection with S = 0.01, 0.1, and 1.0, respectively, $u'_f(p_0) \simeq$ 1.005, 1.05, and 1.58, respectively, whereas with negative selection with the same absolute values, $u'_f(p_0) \simeq$ 0.995, 0.95, and 0.58, respectively. This shows that the fixation probability of a mutant allele will be very close to the neutral expectation of $1/(2N)$ provided $|S| \ll 1$. This domain of **effectively neutrality** is potentially significant in a number of different contexts. For example, populations of sufficiently small size are unable to purge deleterious mutations or promote beneficial mutations with $|s| < 1/(4N_e)$.

A number of other useful approximations for alleles with additive effects on fitness have been derived from diffusion theory. For example, Kimura (1969) found that the average cumulative contribution of a new mutation to the population-level heterozygosity (summed over all generations until it is lost or fixed) is equal to

$$H_T = \left(\frac{4N_e}{N}\right)\left(\frac{S - 1 + e^{-S}}{S[1 - e^{-S}]}\right) \tag{7.12}$$

Although this measure may seem somewhat abstract, the product of H_T times the number of new mutations arising in the population per generation, $2N\mu$, is equal to the expected heterozygosity under selection-mutation-drift equilibrium. For neutral mutations ($S \to 0$), $H_T \to 2N_e/N$, implying an expected heterozygosity of $4N_e\mu$ (which, assuming $4N_e\mu \ll 1$, is consistent with results in Chapter 2 that were obtained by a different method). For large positive values of S (strongly beneficial mutations), H_T approaches a limiting value of $4N_e/N$, implying that on a per-mutation basis, such mutations make twice the contribution to the heterozygosity as neutral mutations. Finally, for deleterious mutations with sufficiently strong effects to be eliminated by selection, $H_T \simeq 2/(N|\,s\,|)$.

As in the case of the fixation probability, the expected heterozygosity at a locus scaled to the neutral expectation (dividing $2N\mu H_T$ by $4N_e\mu$) is a simple function of S (Figure 7.1). Viewed in this way, it can be seen that although both the relative fixation rate and the

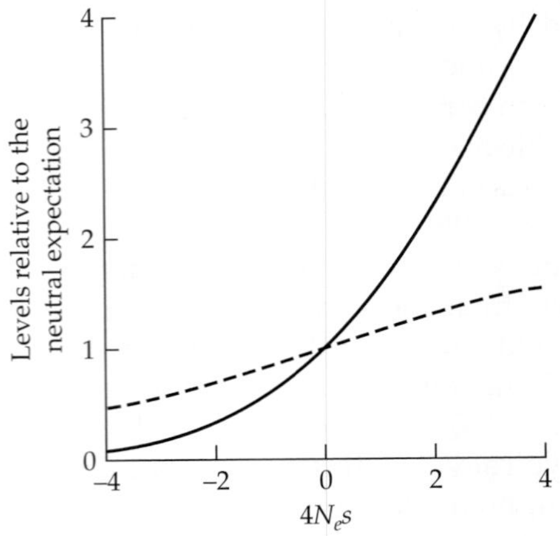

Figure 7.1 Probability of fixation (solid line) and lifetime contribution to heterozygosity (dashed line) of a new mutant allele with additive effects on fitness as a function of $4N_es$ (using Equations 7.11 and 7.12), both relative to the neutral expectation.

contribution to heterozygosity increase with S, the former responds much more rapidly. This is because deleterious mutations that essentially never fix in a population nevertheless make transient contributions to the heterozygosity prior to their elimination by selection, whereas positively selected mutations that are driven through the population relatively rapidly contribute to heterozygosity for only a relatively short period of time.

A useful approximation for newly arisen mutations with additive effects is that, conditional upon fixation, the expected number of generations spent at frequency x will be

$$\Phi_f(x) = \frac{2N_e(1 - e^{-Sx})(1 - e^{-S(1-x)})}{SNx(1 - x)(1 - e^{-S})} \tag{7.13a}$$

where $x = 1/(2N), \cdots, (2N - 1)/(2N)$ (from Equation 8.66 in Kimura 1983). There are two notable points with respect to this residence-time relationship (Figure 7.2). First, provided $|S| < 1.0$, conditional upon fixation, a new mutant allele will spend approximately $2N_e/N$ generations in each frequency class. Second, the residence-time features of a deleterious mutation en route to fixation are exactly the same as those for a beneficial mutation with the same absolute fitness effects, implying that both have the same mean time to fixation, even though the probability of fixation is lower in the former case. First pointed out by Maruyama and Kimura (1974), this counterintuitive behavior results from the fact that if a deleterious allele is to become fixed, it must do so as a consequence of some fortuitously rapid and extreme sampling errors.

It is also sometimes useful to know the expected residence times of mutations that eventually become lost, $\Phi_l(x)$. From Equation 8.70 in Kimura (1983), the unconditional mean residence time for mutations (regardless of being fixed or lost) is

$$\Phi(x) = \frac{2N_e(1 - e^{-S(1-x)})}{Nx(1 - x)(1 - e^{-S})} \tag{7.13b}$$

and using the fact that

$$\Phi(x) = u_f(1/2N) \cdot \Phi_f(x) + [1 - u_f(1/2N)] \cdot \Phi_l(x) \tag{7.13c}$$

yields the residual times conditional upon eventual loss

$$\Phi_l(x) = \frac{N_e e^{Sx}(e^{S(1-x)} - 1)^2}{N^2x(1 - x)(e^S - 1)(e^{S[1-(1/2N)]} - 1)} \tag{7.13d}$$

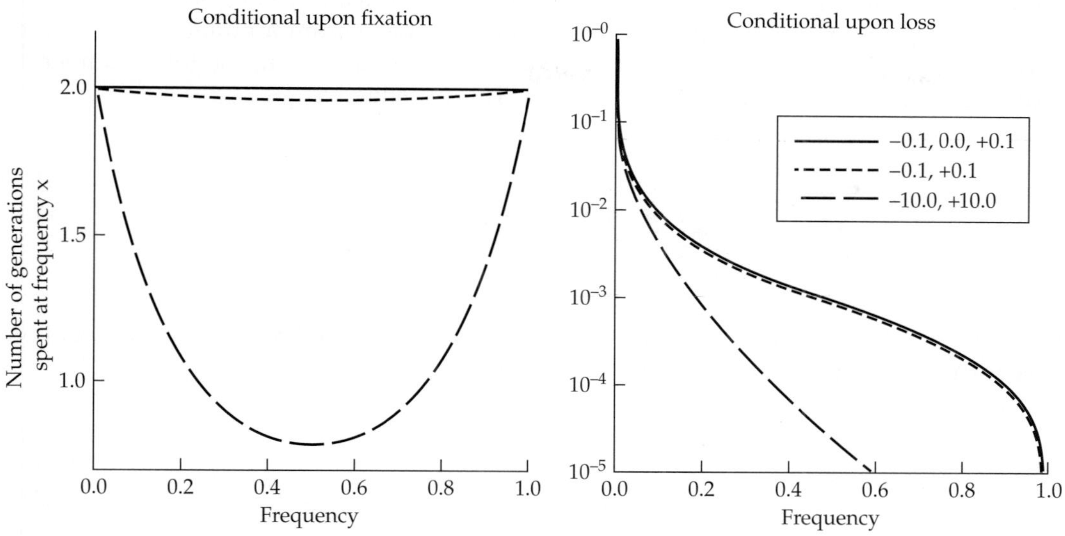

Figure 7.2 Average number of generations that a new mutation spends within different frequency classes, $x = 1/(2N), \cdots, (2N-1)/(2N)$, conditional on going to fixation (**Left**) or conditional on being lost (**Right**), given as a function of the scaled selection parameter $S = 4N_e s$ (**inset values**), obtained using Equations 7.13a and 7.13d, with $N = N_e = 1000$. Note that in each case, the results are identical for beneficial and deleterious mutations with the same absolute values of s. With $N_e \neq N$, the results must be multiplied by N_e/N.

Again, we see that the residence times conditional upon loss are essentially the same for positive and negative selection coefficients of the same absolute magnitude (Figure 7.2). This is not true for the unconditional residence times, $\Phi(x)$, which are functions of $\Phi_f(x)$ and $\Phi_l(x)$ weighted by the probabilities of fixation and loss (Equation 7.13c).

For effectively neutral mutations destined to loss, $|S| < 1.0$,

$$\Phi_l(x) \simeq \frac{N_e(1-x)}{N\lambda x} \tag{7.14a}$$

where $\lambda = 1 - [1/(2N)]$, whereas the unconditional residence time is

$$\Phi(x) \simeq \frac{N_e}{Nx} \tag{7.14b}$$

i.e., the average time spent in frequency class x is inversely proportional to x.

The preceding expressions are useful in a number of applications. For example, the mean numbers of generations to fixation, loss, or either (removal of either allele) can be obtained, respectively, by summing Equations 7.13a, 7.13d, and 7.13c over all frequency classes in the interval $[(1/(2N), 1 - 1/(2N)]$. Simplifications can be made possible in some cases. For example, as noted above, a neutral mutation that is destined for fixation spends an average of $2N_e/N$ generations in each frequency class, and because there are $2N - 1$ classes, the time to fixation of effectively neutral alleles is essentially $4N_e$ generations, an outcome obtained in Chapter 2 by different means. The conditional time to loss of a neutral mutation is

$$t_l = \frac{2N_e \ln(2N)}{N\lambda} \tag{7.15}$$

(derived in Example A1.8). The mean number of generations until the complete loss of a new mutation with a deleterious heterozygous effect of $s < 0$ is

$$t_l \simeq 2(N_e/N)[\ln(2N/|S|) + 0.423] \tag{7.16}$$

provided $|S| \gg 1$ (Kimura and Ohta 1969b; Nei 1971). More general expressions, which require some numerical integration, can be found in Kimura and Ohta (1969a).

Knowing the mean total number of copies descendent from a mutation prior to its loss or fixation is useful in a number of contexts, e.g., determining the total number of individuals affected by a deleterious mutation. This is defined as

$$\overline{n} = \sum_{y=1}^{2N-1} \Phi(y/2N) \cdot y \tag{7.17a}$$

with a shift of the function Φ to Φ_l or Φ_f, leading to the expected numbers conditional on loss or fixation, respectively. For the case of neutral mutations

$$\overline{n} = 4N_e\lambda \tag{7.17b}$$

$$\overline{n}_f = 4N_e N\lambda \tag{7.17c}$$

$$\overline{n}_l = 2N_e\lambda \tag{7.17d}$$

The mean frequency prior to absorption is simply $\overline{n}/(2N)$ divided by the average absorption time.

Example 7.4. Although it is generally thought that selection will increase the determinism of a system, this is not necessarily the case. Cohan (1984b) showed that, starting with identical allele frequencies, the probability of divergence between replicate populations can *increase* relative to the situation under pure drift if the initial frequency of the advantageous allele is sufficiently small. We refer to this phenomenon as the **Cohan effect.** This point can easily be seen as follows. Supposing two replicate populations are segregating alleles *A* and *a* at a locus, with the frequency of *A* being $p = 0.25$, then under pure drift, the probability that one replicate will become fixed for *A* and the other for *a* is $2 \cdot 0.25 \cdot (1 - 0.25) = 0.375$. Now suppose that *A* is favored by selection, with $N_e s = 0.5$. Again assuming $p_0 = 0.25$, Equation 7.10a gives the fixation probability of *A* as 0.46, implying that the probability of fixing alternative alleles is $2 \cdot 0.46 \cdot 0.54 = 0.496$. Thus, in this case, divergence is substantially *increased* by the interaction between selection and drift.

In general, the probability of fixing alternative alleles in two replicates is $2u_f(p)\,[\,1 - u_f(p)\,]$, which is maximized when $u_f(p) = 1/2$. Thus, the probability of divergence is increased by selection if $u_f(p)$ under selection is closer to $1/2$ than $u_f(p) = p$ under drift; and because $u_f(p) > p$ for a selectively favored allele, a minimum requirement for increased divergence under pan-selection is that the starting frequency of the advantageous allele be $< 1/2$. Figure 7.3 shows that under additive selection, the conditions for the probability of divergence under drift plus selection to exceed that under drift alone are not very restrictive.

The Cohan effect has a number of practical implications. For example, an elevated level of population subdivision for a quantitative trait relative to the neutral expectation is often taken to imply that there are divergent selective regimes across subpopulations (Chapter 12). However, here we see that under identical directional selection pressures, populations that initiate with low-frequency, advantageous alleles can exhibit levels of divergence that are conventionally interpreted as being associated with diversifying selection. Whether allele frequencies, selection coefficients, and drift intensities commonly have the right mixes for uniform selection to enhance the magnitude of phenotypic divergence remains to be seen, but a wide range of conditions appears to yield divergence levels that would be difficult to discriminate from the neutral expectation (Lynch 1986).

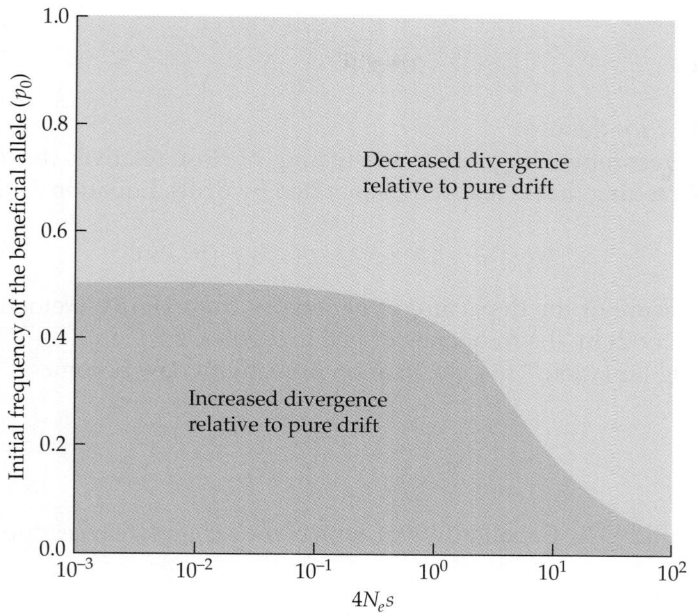

Figure 7.3 The influence of drift on the probability of fixation of alternative alleles in a pair of populations starting from an identical state. A diallelic locus under additive selection with fitnesses $1, 1+s$, and $1+2s$ is considered. The slightly darker shaded area on the lower left is the region of p_0 (the initial frequency of A) and $4N_e s$ space where the probability that isolated populations are eventually fixed for alternative alleles under selection and drift is higher than under drift alone. In this region, parallel selection *increases* the amount of evolutionary indeterminism relative to drift alone.

Probability of Fixation Under Arbitrary Selection

We now consider the more general model, allowing for dominance, with the genotypes aa, Aa, and AA having fitnesses of 1, $1+sh$, and $1+2s$, respectively. Diffusion theory (as developed in Appendix 1) then shows the fixation probability of allele A as

$$u_f(p_0 \mid s, h) \simeq \frac{\displaystyle\int_0^{p_0} e^{G(x)}\, dx}{\displaystyle\int_0^{1} e^{G(x)}\, dx} \tag{7.18a}$$

where

$$G(x) = -4N_e sx(h-x) \tag{7.18b}$$

For a new mutant introduced as a single copy, $p_0 = 1/(2N)$, under random mating and at least partial dominance,

$$u_f\left(\frac{1}{2N}\right) \simeq \frac{2N_e sh}{N[1-e^{-4N_e sh}]} \tag{7.19a}$$

This shows that the probability of fixation of a new mutation is largely determined by the heterozygous effect, as almost all copies of a mutation remain in this state until the allele frequency has achieved a moderately high level. For a complete recessive ($h = 0$), the approximation leading to Equation 7.19a breaks down, and higher-order terms in the approximation of Equation 7.18a are required. However, for strong positive selection on homozygotes of a completely recessive allele ($4N_es \gg 1$), a close approximation is given by

$$u_f\left(\frac{1}{2N}\right) \simeq \frac{\sqrt{4N_es/\pi}}{N} \tag{7.19b}$$

(see Example A1.7 for details).

If there is direct inbreeding due to the mating of close relatives (beyond the amount of long-term inbreeding that is naturally generated by drift), Equation 7.18a will still hold, but now with

$$G(x) = -4N_esx[2f + (1-f)(h-x)] \tag{7.20a}$$

where f is a measure of the departure of genotypes from Hardy-Weinberg expectations, defined (in Chapter 2) by the frequency of heterozygotes, $2p(1-p)(1-f)$ (Caballero and Hill 1992b). Using Equation 7.18a, the fixation probability now becomes

$$u_f\left(\frac{1}{2N}\right) \simeq \frac{2N_es[2f + (1-f)h]}{N} \tag{7.20b}$$

(Caballero and Hill 1992b; Caballero 1996), which, for a complete recessive ($h = 0$), reduces to

$$u_f\left(\frac{1}{2N}\right) \simeq \frac{4N_efs}{N} \tag{7.20c}$$

Thus, with even a small amount of inbreeding, the probability of fixation of a beneficial recessive allele is considerably higher than under random mating (Equation 7.19b) due to the elevated exposure in homozygotes (Caballero et al. 1991). In contrast, inbreeding has much more moderate effects on the fixation probabilities of alleles with additive ($h = 1$) or dominant ($h = 2$) fitness effects. Glémin (2012) showed that inbreeding also speeds up the loss and fixation times of a new allele relative to panmixia.

By indirectly causing localized inbreeding, population subdivision can also influence the probability of fixation. Whitlock (2003) found that, for a wide variety of population structures, the global probability of fixation of a new beneficial mutation is well approximated by

$$u_f\left(\frac{1}{2N}\right) = \frac{2N_esh(1-F_{ST})}{N} \tag{7.21}$$

where the effective and total population sizes (N_e and N) are defined at the metapopulation level and F_{ST} is an index of population subdivision (defined as the fraction of metapopulation variation for neutral alleles that is distributed among populations; see Chapter 2). Note that with complete population subdivision ($F_{ST} = 1$), fixation is impossible at the metapopulation level as mutations will be permanently confined to the demes in which they arise.

One cannot immediately infer from Equation 7.21 whether population subdivision will enhance or reduce the probability of fixation because subdivision influences both F_{ST} and N_e. Expressions for effective population sizes under a number of metapopulation structures were presented in Chapter 3, and parallel expressions for F_{ST} can be found in most of the literature cited there. In the case of the ideal island model with symmetric migration between demes and equal contributions of all demes to the entire metapopulation (Chapter 3), $N_e = N/(1-F_{ST})$, and Equation 7.21 reduces to $2hs$, showing that in this particular case the probability of fixation is independent of the magnitude of population subdivision and simply equal to twice the selective advantage in heterozygotes (Maruyama 1970). Analyses of more complex population structures (Slatkin 1981b; Barton 1993) are all special cases of Whitlock's (2003) expression provided the assumption of equal deme productivity is met; and the modifications that are necessary when this condition are violated were developed by Whitlock (2003) as well. The more complex situation in which the strength of selection varies among demes was taken up by Whitlock and Gomulkiewicz (2005).

Otto and Whitlock (1997) provided results for fixation probabilities in populations of changing size, and showed that selection is more effective in growing populations (increasing the probabilities that favorable alleles will be fixed and that deleterious alleles will be lost) than in declining populations. This result has obvious implications for managed populations. Fortuitously, the limiting expression for the fixation probability of alleles with additive effects (given above as $2sN_e/N$) applies to populations that are changing in size, provided appropriate modifications are made in the definition of N_e (Otto and Whitlock 1997). The much more complex issue of jointly varying population sizes and selection coefficients was taken up by Uecker and Hermisson (2011). Finally, a number of additional diffusion results are given for a diallelic locus in Appendix 1, but simple expressions are generally unavailable for multiple alleles.

Fixation of Overdominant and Underdominant Alleles

A case of special interest is the effect of drift on a locus experiencing selective overdominance, where the heterozygote has higher fitness than either homozygote. Whereas such balancing selection permanently maintains both alleles in an infinite population (Example 5.4), drift will ultimately fix one allele in a finite population provided that the homozygote has a nonzero fitness. Although it might seem that balancing selection will always magnify the longevity of a polymorphism, contrary to intuitive expectations, selection sometimes *increases* the rate of fixation at an overdominant locus in a finite population (Robertson 1962; Ewens and Thomson 1970; Chen et al. 2008).

If the equilibrium frequency expected in an infinite population is extreme (roughly $\widetilde{p} < 0.2$ or $\widetilde{p} > 0.8$), a polymorphism starting at $\widetilde{p}$ in a finite population will usually be lost *more rapidly* under balancing selection than under drift alone, thereby accelerating the removal of heterozygosity. Such behavior arises because selection keeps allele frequencies fairly close to their equilibrium values. If such values are near 0.0 or 1.0, the minor allele will be impeded from drifting to more protective states of moderate frequencies, thereby increasing the likelihood of loss by drift.

Nei and Roychoudhury (1973) evaluated this issue further with newly arisen overdominant alleles with an initial frequency of $1/(2N)$. In this case, the mutant allele is initially confined to the heterozygous state, so its early fate is largely independent of its own homozygous effect, but highly dependent on the magnitude of its heterozygous advantage over the ancestral homozygote. Fixation probabilities can only be obtained by numerical analysis in this case, but the results depend only on two parameters, $N_e(s_1 + s_2)$ and the infinite-population equilibrium frequency, $\widetilde{p} = s_2/(s_1 + s_2)$, where s_1 and s_2 are, respectively, the selection coefficients against the homozygotes associated with the mutant and resident alleles. If $\widetilde{p}$ for the derived allele under consideration is much less than 0.5, the fixation probability is less than the neutral expectation, for the reasons already noted. However, if $\widetilde{p} > 0.5$ (meaning that the fitness of the ancestral homozygote is lower than that of the mutant homozygote), the fixation probability will always be greater than the neutral expectation, even though fixation results in the loss of the optimal (heterozygous) genotype. Moreover, in this case, the fixation probability of the mutant allele is only slightly smaller than that predicted by Equation 7.10a when s_2 is used as a selection coefficient (Nei and Roychoudhury 1973). If $2N_e(s_1+s_2) \ll 1$, selection will be uniformly overpowered by drift, and the system will behave in an effectively neutral fashion.

The fixation times for newly arisen overdominant mutations parallel the patterns of loss of variation that Robertson (1962) first noted (Nei and Roychoudhury 1973). When the equilibrium frequency is outside of the range of (0.2, 0.8), the mean fixation time will be lower than the neutral expectation of $4N_e$ generations, whereas for $0.2 < \widetilde{p} < 0.8$, the time is elevated, with more extreme behaviors seen at high values of $N_e(s_1+s_2)$ (Figure 7.4). Particularly intriguing is the fact that the fixation time of an overdominant mutation will be symmetrical around $\widetilde{p} = 0.5$, i.e., for a given strength of selection $N_e(s_1+s_2)$, the time to fixation is the same at equilibrium frequencies $\widetilde{p}$ and $1-\widetilde{p}$. This is consistent with the situation for mutants with additive effects that was already noted, and indicates that when an overdominant

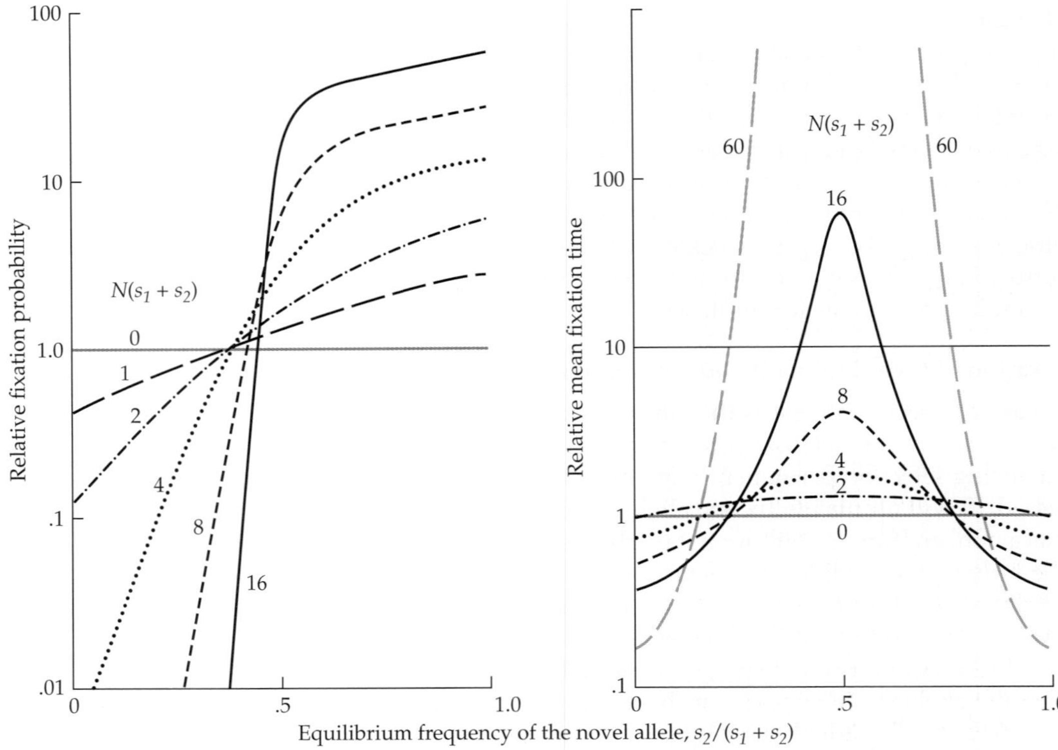

Figure 7.4 Ratios for the fixation probabilities and expected times to fixation for a newly arisen overdominant mutation relative to the expectation for a neutral mutation. These are given as a function of the equilibrium frequency expected in a population of infinite size, $\widetilde{p} = s_2/(s_1+s_2)$, where the fitnesses are $1-s_1$, 1, and $1-s_2$ (with the first value being the fitness for the mutant homozygote). Each curve gives results for a different value of $N_e(s_1 + s_2)$, a measure of the ratio of the overall power of selection to drift, where N_e is the effective population size. For any value of $N_e(s_1 + s_2)$, the probability of fixation increases with the magnitude of selection against the alternative homozygote, as this defines the selective advantage of the novel allele in the heterozygous state. (From Nei and Roychoudhury 1973.)

mutant allele is associated with the least fit homozygous type, for the rare occasions in which fixation occurs, it does so just as rapidly, on average, as when it is associated with the most fit homozygote (in which case it also fixes more frequently). Further considerations for the situation in which populations are subdivided were given in Nishino and Tajima (2004).

Important situations also exist in which a new mutation will be underdominant with respect to the ancestral allele, i.e., will have reduced fitness when in the heterozygous state, but equal or higher fitness as a homozygote. In an infinite population, such an allele will always be driven from the population if its marginal fitness at low frequency is less than that of the ancestral allele (Chapter 5). In a finite population, however, there is some chance that the mutant allele might drift to a high frequency, transiently taking the population through a reduction in mean fitness (during the period in which heterozygotes are common), but possibly eventually becoming fixed.

Such a scenario has generated considerable interest in the area of speciation biology, as the fixation of an underdominant mutation in a subpopulation will lead to a situation in which hybrids between subpopulations have reduced fitness. In principle, such a condition can constitute the first stage in the development of reproductive isolation.

For the situation in which the two homozygotes have equal fitness and heterozygotes experience a reduction in fitness of s, Lande (1979b) found that if $sN_e/N \ll 1$ (a condition likely to be met based on empirical information on N_e/N; Chapters 3 and 4), then

$$u_f(1/2N) \simeq \frac{\sqrt{N_e s/\pi}}{N \cdot e^{N_e s} \cdot \mathrm{erf}(\sqrt{N_e s})} \tag{7.22}$$

where the **error function**

$$\mathrm{erf}(x) = (2/\sqrt{\pi}) \int_0^x e^{-y^2} dy \tag{7.23}$$

is related to the cumulative frequency of a unit normal (Abramowitz and Stegun 1972). If the efficiency of selection is sufficiently low ($N_e s \ll 2$), then $u_f(1/2N) \simeq 1/(2N)$, as expected for an effectively neutral allele. However, if the efficiency of selection is high ($N_e s > 2$), so that $\mathrm{erf}(\sqrt{N_e s}) \simeq 1$, then

$$u_f(1/2N) \simeq \frac{\sqrt{N_e s/\pi}}{Ne^{N_e s}} \tag{7.24}$$

Of special interest in the study of speciation are chromosomal rearrangements that cause problems during meiosis in chromosomal heterozygotes, with values of s as large as 0.5 being quite plausible (Lande 1979b, 1984b). With $N_e s = 2$, 5, and 10, Equation 7.24 predicts fixation rates that are, respectively, 0.22, 0.017, and 0.00016 times the neutral expectation. Such results imply that if heterozygote fitness is greatly reduced, transitions to alternative allelic states (with equivalent homozygous fitness) will only be possible if N_e is extremely small. However, when such fixations do occur, they proceed much more rapidly than the neutral expectation of $4N_e$ generations (Lande 1979b).

Walsh (1982) generalized these results to the situation in which the fitness in the novel homozygote is elevated to $1+t$, such that after passage through a fitness bottleneck, fixation of the underdominant allele leads to an increase in mean population fitness. Letting $\theta = N_e s$, and $\omega = 1 + (t/2s)$

$$u_f(1/2N) = \frac{\mathrm{erf}\{[(1/2N) - (0.5/\omega)]\sqrt{4\theta\omega}\} + \mathrm{erf}\{\sqrt{\theta/\omega}\}}{\mathrm{erf}\{[1 - (0.5/\omega)]\sqrt{4\theta\omega}\} + \mathrm{erf}\{\sqrt{\theta/\omega}\}} \tag{7.25}$$

For $t < 2s$, the fixation probability is close to that predicted by Equation 7.22, whereas for very large values of t, $u_f(1/2N)$ can moderately exceed the neutral fixation probability provided $N_e s$ is not so strong that the allele is incapable of drifting to a high enough frequency to be favored by selection (Figure 7.5).

The latter case is of special interest, as one can identify a critical effective population size (N_e^*) above which the efficiency of selection is so strong that there is essentially no possibility of the population passing through the fitness bottleneck imposed by heterozygotes. With heterozygotes having a fitness reduction of s, derived homozygotes having an advantage of t, and p being the frequency of the mutant allele, the mean population fitness is $\overline{W} = 1 - 2p(1-p)s + p^2 t$, which reaches a minimum at $\widehat{p} = s/(t+2s) = 0.5\omega$, so that $p < \widehat{p}$ implies net selection against, and $p > \widehat{p}$ net selection in favor of, the mutant allele. Thus, the key issue is whether the mutant allele can drift from an initial frequency of $1/(2N)$ to $\widehat{p}$, at which point selection can pull it to fixation. When p is small, the frequency of mutant homozygotes is negligible, and the new allele effectively behaves like a deleterious mutation being removed from the population at rate s, and it can be shown that there is essentially no chance of the allele drifting to $\widehat{p}$ if

$$N_e^* > \frac{t+2s}{s^2} \tag{7.26}$$

(Lynch 2012a). For example, with a mutant allele with a disadvantage of $s = 0.01$ in the heterozygous state but an advantage of $t = 0.01$ in the homozygous state, an effective population size above 300 imposes a very strong barrier to its establishment. Lande (1979b, 1985)

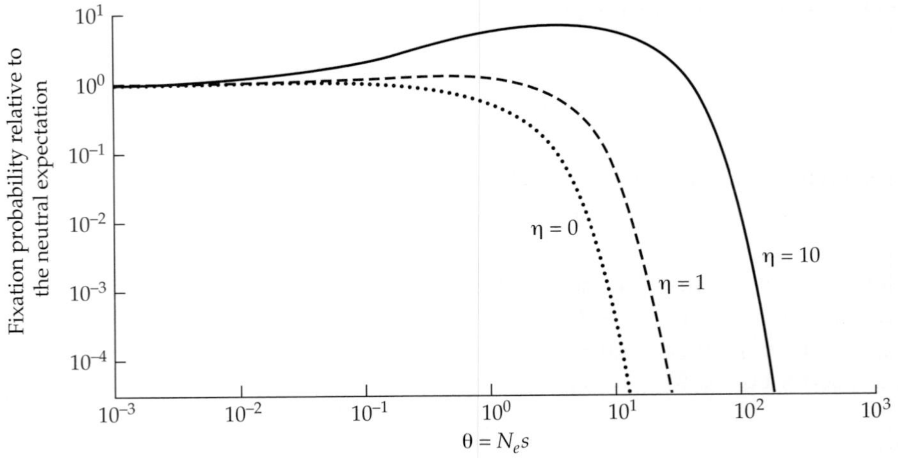

Figure 7.5 The probability of fixation of a newly arisen underdominant mutation, relative to the neutral expectation of $1/(2N)$, with a selective disadvantage of s in the heterozygous state and an advantage of t in the derived homozygous state, and $\eta = t/(2s)$. (After Walsh 1982.)

showed that such selective valleys are much more likely to be vaulted in subdivided populations, where local extinction and recolonization permit individual demes to make transitions to an alternative genotypic state and then export such a fixed change to a newly opened habitat.

Expected Allele Frequency in a Particular Generation

A number of applications, including attempts to predict the response to selection, arise for which it is useful to know the expected allele frequency at time t, $E(p_t)$. While exact results can be obtained from probability transition matrices (Hill 1969a; Carr and Nassar 1970) and good approximations can be derived from diffusion theory (Appendix 1; Maruyama 1977; Ewens 2004) and other approaches (Curnow and Baker 1968, 1969; Pike 1969), these methods tend to be numerically intensive. Fortunately, simple approximations have been developed for dealing with weak selection.

In a finite population, drift can reduce the selection response by progressively diminishing the expected heterozygosity in each succeeding generation. Consider a locus with additive selection, with the genotypes *aa*, *Aa*, and *AA* having fitnesses of 1, $1+s$, and $1+2s$, respectively. If there is weak selection, such that changes in allele frequencies associated with selection are relatively minor compared to those induced by drift, we can use Equation 5.1b to show that the expected per-generation frequency change for an allele in the jth generation of additive selection can be described as

$$E(\Delta p_j) \simeq sE[\,p_j(1-p_j)\,] \simeq sp_0(1-p_0)\left(1-\frac{1}{2N_e}\right)^j \tag{7.27}$$

where p_0 is the initial allele frequency. The last approximation follows directly from the expression for the expected heterozygosity for a neutral locus in a finite population after j generations with a starting allele frequency of p_0 (Equation 2.5). Summing over generations reveals that the expected frequency after t generations of selection and drift is

$$\begin{aligned} E(p_t) &= p_0 + \sum_{j=0}^{t} E(\Delta p_j) \simeq p_0 + sp_0(1-p_0)\sum_{j=0}^{t}\left(1-\frac{1}{2N_e}\right)^j \\ &\simeq p_0 + 2N_e s\, p_0(1-p_0)\left(1-e^{-t/2N_e}\right) \end{aligned} \tag{7.28a}$$

where the last step follows from the useful approximation for large values of N_e

$$\sum_{j=0}^{t} \left(1 - \frac{1}{2N_e}\right)^j \simeq 2N_e \left(1 - e^{-t/2N_e}\right) \tag{7.28b}$$

More generally, if the genotypes *aa*, *Aa*, and *AA* have fitnesses of 1, $1 + hs$, and $1 + 2s$, respectively, then for small values of $N_e|s|$ and $N_e|sh|$, the expected frequency of *A* is

$$E(p_t) \simeq p_0 + 2N_e s p_0(1 - p_0) \left[\left(1 - e^{-t/2N_e}\right) + \frac{(h-1)(1-2p_0)}{3} \left(1 - e^{-3t/2N_e}\right) \right] \tag{7.29}$$

These approximations provide a remarkably simple route to obtaining fixation probabilities under weak selection ($N_e s \ll 1$). Because an allele will ultimately be either fixed ($p_\infty = 1$) or lost ($p_\infty = 0$), the asymptotic mean frequency as $t \to \infty$ is equal to the fixation probability

$$E(p_\infty) = \{1 \cdot u_f(p_0)\} + \{0 \cdot [1 - u_f(p_0)]\} = u_f(p_0)$$

Thus, taking the limit of Equation 7.29 as $t \to \infty$ yields a useful expression for the probability of fixation under weak selection and arbitrary dominance

$$u_f(p_0) \simeq p_0 + 2N_e s p_0(1 - p_0) \left(1 + \frac{(h-1)(1-2p_0)}{3}\right) \tag{7.30}$$

For additive fitness effects ($h = 1$), this expression is identical to Equation 7.10b. Hill (1969a, 1969b) found this approximation to be reasonable provided $N_e|s| < 1$. The more general versions (Equations 7.29 and 7.30) were produced by Silvela (1980).

JOINT INTERACTION OF SELECTION, DRIFT, AND MUTATION

We now turn to the situation in which selection, drift, and mutation operate simultaneously. Under these conditions, alleles are not simply permanently lost or fixed. Rather, the allele frequencies in a population of constant size eventually reach a stochastic equilibrium (or **stationary distribution**), $\phi(x)$, where x denotes the allele frequency. Recall from Chapter 2 that we can interpret such an equilibrium in two different ways. First, given a conceptually large number of replicate populations, $\phi(x)$ closely approximates the frequency histogram of the numbers of populations with specific allele frequencies at the locus. Conversely, if we were to follow a single population temporally and construct a histogram of the historical record of allele frequencies at the locus over a very large number of widely separated time points, under constant population conditions, we would again recover $\phi(x)$.

Diffusion theory provides a general solution to this problem (Appendix 1). For the simple biallelic case in which mutations from allele *A* to *a* occur at a rate of μ, and v is the reciprocal rate, Wright (1949) found that the equilibrium distribution for the advantageous *A* allele is given by

$$\phi(x) = C\overline{W}^{2N_e} x^{4N_e v - 1} (1 - x)^{4N_e \mu - 1} \quad \text{for } 0 < x < 1 \tag{7.31a}$$

where C is a normalization constant such that Equation 7.31a integrates to one, and hence is a proper probability density (Example A1.4 provides a derivation of this expression). Here, $\overline{W}$ is the mean population fitness, which is itself a function of x and the selection coefficients associated with different gametic states. Note that when both mutation rates are substantially $< 1/(4N_e)$, conditions that may frequently be met for single nucleotide sites (Chapter 4), this simplifies to

$$\phi(x) \simeq \frac{C\overline{W}^{2N_e}}{x(1-x)} \tag{7.31b}$$

showing that with weak mutation pressure, the expected allele frequencies that are *conditioned on the population being polymorphic* are independent of both the mutation rate and the mutation bias.

This result, which represents yet another counterintuitive consequence of the influence of drift on gene frequencies, can be understood as follows. Suppose that allele A has a selective advantage s over allele a, and again let the rates of mutation from A to a and vice versa be μ and v, respectively. At a stationary state, the ratio of times that a population is completely fixed for optimal versus suboptimal alleles is

$$\frac{\widetilde{P}_A}{\widetilde{P}_a} = \left(\frac{v}{\mu}\right) e^S \tag{7.32}$$

where $S = 4N_e s$ (Wright 1931; Li 1987; Bulmer 1991; McVean and Charlesworth 1999). Note that (v/μ) and e^S are, respectively, the mutation and selection biases in favor of allele A, with the latter being equivalent to the ratio of fixation probabilities of newly arising beneficial and detrimental alleles with the same absolute s (obtainable from Equation 7.10a). Equation 7.32 demonstrates that although the distribution of allele frequencies that is conditional on polymorphism can be independent of mutational properties, the frequency of alternative fixed classes is not. In addition, it is apparent that the ratio at which the two monomorphic classes produce polymorphisms (μ/v) is perfectly compensated by the differential densities of the two classes, and provided the population is sufficiently small that each new mutation is either lost or fixed before another one is produced at the locus, this effect will not be influenced by secondary mutations. Equation 7.31b breaks down, however, when population sizes are large enough that the waiting times for new mutations are smaller than the sojourn times of mutant alleles.

Because Equation 7.31a treats allele frequencies as continuously distributed variables, they may behave aberrantly at the absorbing boundaries of the frequencies $x = 0$ and 1. However, an approximation for the absolute frequencies of the fixed classes can be obtained by noting that the equation

$$\widetilde{P}_p \simeq 2N(\widetilde{P}_A \mu \bar{t}_a + \widetilde{P}_a v \bar{t}_A) \tag{7.33}$$

is the equilibrium proportion of time for which the sites are polymorphic, with $\bar{t}_a$ and $\bar{t}_A$ being, respectively, the mean sojourn times of mutations to alleles a and A. Using Equation 7.32 and the fact that $\widetilde{P}_a + \widetilde{P}_A + \widetilde{P}_p = 1$, the solution can be obtained for all three components of this equation. By multiplying the values of Equation 7.31a by $\widetilde{P}_p$ over the range of $x = 1/(2N)$ to $1 - [1/(2N)]$, we then obtain the spectrum of alternative population states of polymorphism.

Figure 7.6 provides some examples of the form of the stationary distribution for biallelic loci experiencing bidirectional mutation. For neutral mutations, the distribution is highly U- or J-shaped (depending on the magnitude of mutation bias) at low population mutation rates ($4N\mu$ and $4Nv \ll 1$), as the population is almost always in a nearly fixed state, with the probability of the alternative fixed states being given by Equation 7.5. The distribution becomes flat with values of $4N\mu$ and $4Nv$ near 1.0, and then becomes more peaked as $4N\mu$ and $4Nv$ become progressively larger (with the mean centered on the infinite-population expectation given by Equation 7.5). Selection skews the distribution toward the more favorable allele, but even with an S as large as 10, a moderate frequency of the deleterious allele can be expected (even though fixation of the latter would essentially never occur).

Equation 7.31a is useful in a number of applications. Consider, for example, the case of a deleterious recessive allele maintained by mutation (with μ being the mutation rate to deleterious alleles, and s being the selective disadvantage of mutant homozygotes). If we let x be the frequency of the deleterious allele, the mean population fitness is $\overline{W} = 1 - sx^2$. Using the approximation $(1-y)^{2N_e} \simeq e^{-2N_e y}$ for small values of y, so that $\overline{W}^{2N_e} \simeq e^{-2N_e s x^2}$, and ignoring back mutation to the advantageous allele, yields the equilibrium distribution

$$\phi(x) = Ce^{-2N_e s x^2} x^{4N_e\mu - 1} (1-x)^{-1} \quad \text{for } 0 < x < 1 \tag{7.34}$$

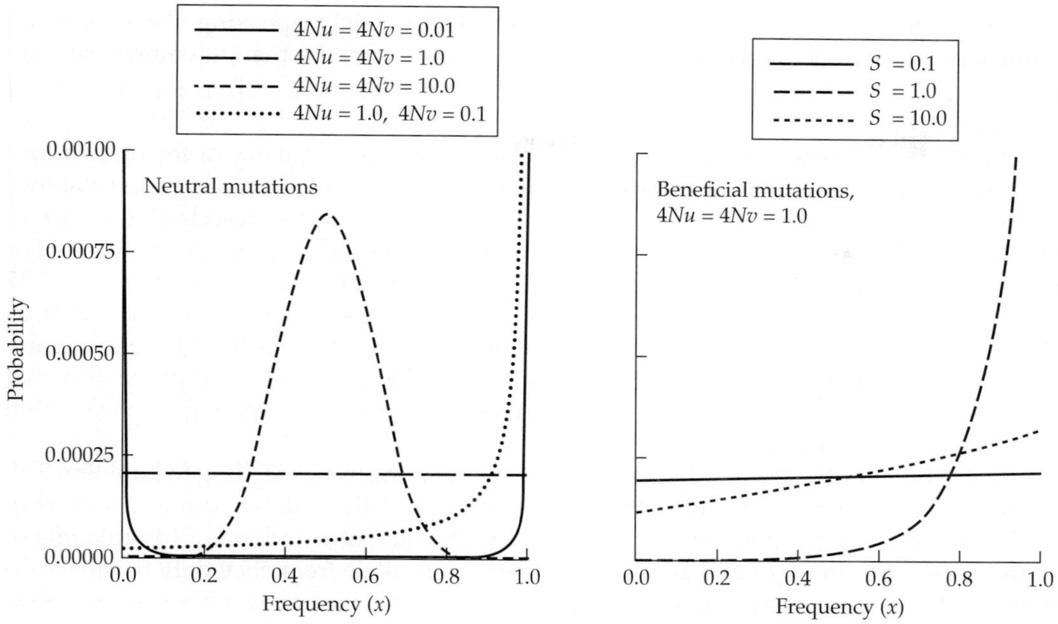

Figure 7.6 Stationary distributions of allele frequencies under the joint forces of mutation, selection, and random genetic drift (Equation 7.31a). An absolute population size of $N = 2000$ is assumed with $N_e = N$.

a result originally due to Wright (1938b).

Nei (1969) provided a broad overview of the allele-frequency spectrum for lethal mutations, including those that are entirely recessive or overdominant. As neither of these conditions are commonly observed (LW Chapter 10), we note only some of the results for partially recessive lethals. In this case, the average expected frequency at selection-mutation balance is given by Equation 7.6d, essentially independent of population size, and provided that $2N_e hs \gg 1$ (i.e., the power of selection against heterozygotes exceeds the power of drift), the variance in allele-frequency will be approximately

$$\sigma^2(p) = \widetilde{p}/(4N_e hs) \tag{7.35}$$

Nei (1971) and Li and Nei (1972) gave expressions for the expected total number of individuals affected by a newly arisen deleterious mutation prior to its elimination by selection.

An area of special interest is the behavior of the four possible nucleotides at a particular site. If we denote the four frequencies as x_i (where $i = 1, \cdots, 4$) and their selection coefficients as s_i (here assumed to be weak and additive), under the assumption that all nucleotides mutate to each other type at the same rate, μ, Equation 7.31a reduces to

$$\phi(x_1, x_2, x_3, x_4) = C\overline{W}^{2N_e} (x_1\, x_2\, x_3\, x_4)^{4N_e\mu - 1} \tag{7.36}$$

where $\overline{W} = 1 + 2\sum_{i=1}^{4} x_i s_i$ is the mean population fitness. It is not surprising that the solution to this trivariate expression (x_4 being defined as $1 - x_1 - x_2 - x_3$) is quite cumbersome (Li 1987; Zeng et al. 1989; Bulmer 1991; McVean and Charlesworth 1999).

Consider, however, the situation in which there is one optimal nucleotide, the frequency of which is denoted by x, with the three others having an equal selective disadvantage, s, in the heterozygous state. If we scale the less-fit alleles to have a fitness of 1, the mean population fitness will then be $\overline{W} = 1 + 2xs$, which is closely approximated by e^{2xs} under the assumption of a small s. If we let the mutation rate of all nucleotides to the optimal state be v and the total mutation rate of the optimal nucleotide to the other states be μ, it follows from Equation 7.32 that the expected frequency of the optimal nucleotide is

$$\widetilde{P}_{\text{opt}} \simeq \frac{(v/\mu)e^S}{1 + (v/\mu)e^S} \tag{7.37}$$

(Li 1987; Bulmer 1991; McVean and Charlesworth 1999). Strictly speaking, this expression applies to the weak-mutation limit (where $N(\mu + v) \ll 1$ ensures that polymorphisms are rare), so that $\widetilde{P}_{\text{opt}}$ denotes the fraction of time for which the site is fixed for the optimal nucleotide.

Equation 7.37 makes a simple, intuitive statement—the frequency of the optimal nucleotide at a site is a function of a single composite quantity, $(v/\mu)e^S$, which, as noted above, denotes the net pressure toward the optimal state. As $N_e s \to 0$, the expected frequency of the optimal allele approaches the expectation under pure mutation pressure, $v/(\mu + v)$. For populations that are sufficiently large to maintain substantial heterozygosity, Equation 7.37 will no longer serve as a strict definition of the probability of sampling an optimal allele, as prior to fixation the descendants of a new mutation will themselves have time to acquire secondary mutations. In this case, P_{opt} is more appropriately viewed as the probability that the most recent common ancestor of the alleles that are currently segregating in a population is an allele of the optimal type.

Sella and Hirsh (2005) and Lynch (2012b) expanded the model leading to Equation 7.37 to allow for multiple alleles with different fitness states. Both models assumed a stepwise-mutation model, with allele i mutating to $i - 1$ with a rate of μ and to $i + 1$ with a rate of v, and, again, are strictly valid as indicators of average allele frequency only in the weak-mutation limit, where the population is typically expected to be nearly monomorphic. Sella and Hirsh (2005) assigned a fitness of $W_i = 1 + s_i$ to allele i, and also assumed that there was symmetric mutation ($\mu = v$). If we let $S_i = 4N_e s_i$ (assuming diploidy), the equilibrium probability that i is the fixed (or nearly so) allele is completely independent of the mutation rate,

$$\widetilde{p}_i = \frac{e^{S_i}}{T}, \quad \text{where} \quad T = \sum_{i=1}^{n} e^{S_i} \tag{7.38}$$

and n is the number of alleles. Whereas the Sella-Hirsh model makes no assumptions about fitness ordering between alleles, Lynch's (2012b) model assumes there will be an ordered fitness increase in a series of alleles, such that $W_i = 1 - e^{-ki}$, with the constant k setting the granularity of fitness change between adjacent alleles, and a fitness of 1.0 being approached asymptotically as $i \to \infty$. In this case, the stationary distribution is

$$\widetilde{p}_i = \frac{(v/\mu)^i e^{-S_i}}{T}, \quad \text{where} \quad T = \sum_{i=1}^{\infty} (v/\mu)^i e^{-S_i} \tag{7.39}$$

and $S_i = 4N_e e^{-ki}$.

Formulae such as these can be readily modified into alternative fitness schemes. Among other things, they are useful for determining the extent to which drift limits the level of adaptation attainable by a population. For example, if we assume that there are higher mutation rates to unfavorable states ($\mu > v$), the advancement toward ever-higher (and fitter) allelic states will stall around a critical value in the allelic series, above which $s_i \simeq e^{-ki}$ is sufficiently small that drift (combined with mutation pressure) will overwhelm selection, thereby preventing any further adaptive progress (Lynch 2012b). Although alleles that are in a fitness state above this critical point can still arise by mutation, they will be unable to avoid being lowered by drift back down to the critical value. On the other hand, alleles with sufficiently large disadvantages will be incapable of proceeding to fixation and therefore will be purged by selection. Thus, as will be further discussed in the following section, under virtually all models of adaptation, a **drift barrier** will ultimately prevent a population from achieving a perfect state of adaptation, even in a constant environment.

HALDANE'S PRINCIPLE AND THE MUTATION LOAD

Having established the expected allele frequencies at a locus that is jointly influenced by mutation, selection, and drift, we will now consider in more detail the price that all organisms pay for the privilege of evolving. Because most mutations are deleterious, and many are unconditionally so, for every beneficial allele created by mutation, many more detrimental mutations will be introduced to a population. In populations of sufficiently large size, the majority of such mutations will be kept at a low frequency and eventually purged, but the relentless flux of new mutations will nevertheless result in an equilibrium load on the mean fitness in the population (Muller 1950; Crow 1993). What is remarkable is that, under reasonably general conditions, this load is often essentially independent of the effects of individual mutations.

In an elegant display of population-genetic reasoning, Haldane (1937b) proposed that the reduction in fitness resulting from recurrent deleterious mutations is a function of the deleterious mutation rate alone, an observation that has come to be known as **Haldane's principle**. To illustrate, consider a deleterious recessive allele *a*, with a selective disadvantage, s, in homozygotes. Recalling Equation 7.6d reveals that the mean population fitness when this locus is in selection-mutation balance is

$$\overline{W} = 1 - s \cdot \text{freq}(aa) = 1 - s\left(\sqrt{\frac{\mu}{s}}\right)^2 = 1 - \mu \tag{7.40a}$$

Because the expected frequency of recessive homozygotes is inversely proportional to the selective disadvantage, the reduction in mean fitness (the **mutation load**) is independent of the strength of selection and simply equal to the deleterious mutation rate per allele.

Now consider an allele with partial to complete dominance and with heterozygote a fitness of $1 - hs$. If we recall from Equation 7.6d that the equilibrium allele frequency is $\widetilde{p} = \mu/(hs)$, the mean population fitness is

$$\begin{aligned}\overline{W} &= 1 - 2hs\,\widetilde{p}(1-\widetilde{p}) - s\widetilde{p}^2 \\ &\simeq 1 - 2hs\,\widetilde{p} = 1 - 2hs\left(\frac{\mu}{hs}\right) = 1 - 2\mu\end{aligned} \tag{7.40b}$$

The approximation, which assumes weak mutation, meaning that that $\widetilde{p}^2 \ll 1$, shows that the expected mean fitness is independent of both h and s. Bürger (2000) explored these expressions in considerable detail and showed that the error in ignoring secondary terms in the preceding expressions is on the order of μ^2/s or smaller. With multiple deleterious alleles per locus, these same expressions apply if μ is interpreted as the total mutation rate of the most beneficial allele to all classes of deficient alleles at a locus (Crow and Kimura 1964; Clark 1998).

One potential caveat about these results is that the derivation assumes a situation in which there are negligible epistatic effects on fitness. To examine the robustness of this assumption, Kimura and Maruyama (1966) considered a quadratic fitness function of the form $w_i = 1 - h_1 i - h_2 i^2$, where i is the number of mutations carried by the individual. With $h_2 = 0$, the model of additive effects assumed above is closely approximated, and Haldane's principle continues to hold, with a mean fitness approximately equal to e^{-U}, where U is the deleterious mutation rate per diploid genome. However, at the opposite extreme, with $h_1 = 0$, fitness declines with the square of the number of mutations, and mean fitness is elevated to $\sim e^{-U/2}$ regardless of the magnitude of h_2. A more general analysis, which allows for nonzero values of both h_1 and h_2, was provided by Kimura and Maruyama (1966) and demonstrated that this type of **synergistic epistasis** always reduces the mutational load on a sexual population. In contrast, when there is **diminishing-returns epistasis**, where the decline in fitness with increasing numbers of deleterious mutations becomes progressively shallower, the mutation load will be elevated above the Haldane expectation.

Fitness functions involving epistasis have played a significant role in our attempt to understand the evolution of sexual reproduction, primarily because the behavior just noted does not extend to asexual genomes, as first shown by Kimura and Maruyama (1966) in a remarkably simple way. Consider an asexual population of mixed clones, with p_0 being the frequency of the clone with the minimum number of mutations in one generation and p_0' being its frequency in the next generation. Then, accounting for selection and mutation,

$$p_0' = \frac{p_0 W_0 e^{-U}}{\overline{W}} \tag{7.41}$$

where $\overline{W}$ is the mean population fitness, $W_0 = 1$ is the fitness of the optimal genotype, and e^{-U} is the fraction of the members of this class that do not acquire mutations. Note that no assumptions have been made here with respect to the mode of gene action or on the form of the fitness distribution, and yet at equilibrium ($p_0' = p_0$) we obtain the very general result that mean fitness, $\overline{W}$, equals e^{-U}. Thus, if synergistic epistasis among deleterious mutations is important, a matter on which there is little empirical consensus (Rice 2002b; Barton and Otto 2005; Kouyos et al. 2007; Keightley and Halligan 2009), a sexual population will have a long-term advantage in terms of mean fitness. Substantial additional work has been done on this subject (e.g., Kondrashov 1984, 1988; Charlesworth 1990; Agrawal and Chasnov 2001; Otto 2003; Haag and Roze 2007).

An additional issue with respect to Haldane's principle is that N_e must be several-fold greater than $1/(hs)$ for Haldane's principle to be closely approximated. If this is not the case, deleterious alleles will be capable of drifting to frequencies higher than what would be expected under selection-mutation balance alone. Although this observation led Kimura et al. (1963) to conclude that the mutational load due to segregating mutations will increase monotonically with decreasing N_e, their study invoked a relatively high level of back mutation in order to maintain a quasi-equilibrium allele frequency.

If, instead, one treats back mutation as a negligible force (for reasons stated above), it can be shown that the load associated with segregating mutations is nonmonotonic with respect to N_e. The segregational load actually reaches a maximum (in excess of the Haldane expectation) at the point where $1/(2N_e) \simeq hs$, as it is at this point that mutations have a maximum deleterious effect that is still consistent with being highly vulnerable to random genetic drift (Lynch et al. 1995a, 1995b). As N_e declines below this point, the segregational load approaches zero simply because drift becomes so strong that few segregating polymorphisms of any kind are maintained, and at this point permanent damage accrues via the fixation of deleterious alleles, i.e., there is a **fixation load** in addition to any **segregational load**. Indeed, once a population enters this small-population-size domain, the mutation load may no longer even be maintained at a quasi-equilibrium state as a continual flux of new rounds of weakly deleterious mutations will lead to further fixations. If unopposed for a sufficiently long time, such a condition can eventually reduce mean population fitness to the point at which the average individual will be incapable of replacing itself, leading to population extinction via a **mutational meltdown** (Lynch et al. 1995a, 1995b).

Even populations large enough to avoid extinction by a mutational meltdown must experience some fixation load, as they will often include mutationally derived alleles with small enough deleterious effects to be immune to selection (i.e., $|s| < 4N_e$). The issue has been explored with a variety of models for mutational passage between allelic classes (Hartl and Taubes 1998; Poon and Otto 2000; Sella and Hirsh 2005; Lynch 2012b). Although the exact results vary somewhat among studies, in every case the load resulting from the fixation of suboptimal alleles is inversely proportional to the effective population size, often with an upper bound on the order of $1/(4N_e)$.

One way to arrive at this result is to recall the additive two-allele model given above as Equation 7.37. Noting that the load associated with a fixed deleterious mutation is the homozygous effect, $2s$, multiplied by the expected fraction of time for which the deleterious allele is fixed, we then have

$$L = 2s(1 - \widetilde{P}_{\text{opt}}) = \frac{2s\mu/v}{e^S + (\mu/v)}$$

$$\simeq \frac{2s\mu/v}{1 + 4N_e s + (\mu/v)} \tag{7.42a}$$

The approximation applies when $S = 4N_e s < 1$, which must be the case for there to be a significant chance of fixation of a deleterious allele. Under the latter conditions, with symmetrical mutation rates ($\mu = v$),

$$L = \frac{1}{2N_e + (1/s)} < \frac{1}{4N_e} \tag{7.42b}$$

Mutational bias in the direction of deleterious alleles ($\mu/v > 1$) will elevate this load, but the point remains the same. Finite population size imposes an ultimate barrier to adaptive refinements that can be maintained in a population. Although this load may appear to be small, as noted in Chapter 4, in all known cases, $\mu < 1/(2N_e)$, suggesting that the drift load per locus is likely to be typically greater than Haldane's segregational load. In addition, the previous derivations apply to single loci, whereas the cumulative load over all n loci contributing to a trait will be roughly n times the single-locus load, assuming weak multiplicative fitness effects. Thus, drift appears to generally impose a nontrivial barrier to adaptive perfection.

There has been considerable debate about the meaning and consequences of the genetic load (Wallace 1991; Crow 1993; Kondrashov and Crow 1993; Reed and Aquadro 2006). As deleterious mutations impose differences in survival and/or reproduction, they must have some demographic consequences. Taken literally, if the deleterious mutation-free genotype is viewed as the standard ($W_0 = 1$), an equilibrium load, L, would imply approximately e^{-L} viability (not including mortality unassociated with genetic variation) provided its entire influence is on survivorship. This would then require an inflation of family sizes by a factor of e^L relative to the minimum value of two that is necessary to maintain population-size stability. Under this view, the load concept is paradoxical in that a low-fecundity organism such as a vertebrate would never be able to bear the demographic costs should the genome-wide deleterious mutation rate exceed $\sim$1.0, which is likely the case in vertebrates (Chapter 4). Lesecque et al. (2012) showed, however, that the magnitude of selective death is greatly diminished if the fitness of individuals is scaled relative to the actual mean fitness in the population rather than to the idealized $W_0 = 1$. Such a situation would be expected if selection operates mainly through competition of the actual members of the population, rather than by comparison to a nonexistent (idealized) genotype.

FIXATION ISSUES INVOLVING TWO LOCI

Populations and species diverge from each other through successive fixations of new mutations, which can be effectively neutral, advantageous, or even slightly deleterious. The relative contributions from these classes, especially the fraction of advantageous and hence adaptive substitutions, is of considerable interest (Kimura 1983; Gillespie 1994). Our goal here is to broaden the preceding outline of fixation theory by considering the influence of the genetic background on expected substitution rates.

There are a number of situations in which fixation probabilities of alleles are influenced by factors operating at other loci. For example, as discussed in Chapter 3, selection operating on any locus, either positive or negative, results in a reduction in the effective population size in the local chromosomal region, thereby reducing the efficiency of selection operating on all loci linked to the target of selection. Such effects will reduce the fixation probabilities for beneficial alleles, while enhancing the likelihood of fixation of deleterious alleles. In addition, for mutations with contextual (epistatic) effects, fixation probabilities depend critically on the genetic background, and hence on the frequencies of alternative alleles at interacting loci.

The Hill-Robertson Effect

We first consider the matter of selective interference associated with linked variation involving beneficial alleles. Suppose that the gamete with the highest fitness, *AB*, is initially absent and can only be generated by recombination in *Ab*/*aB* double heterozygotes. If we let x_2 and x_3 denote the frequencies of the *Ab* and *aB* gametes, and c be the recombination frequency between the two loci, then the probability of *AB* being generated in the population is related to the product of the expected frequency of *Ab*/*aB* heterozygotes and the probability that a random gamete from such individuals is *AB*, $(2x_2x_3)(c/2)$. Because $x_2x_3 \leq 1/4$ and a population with a stable size must produce $2N$ successful gametes, the upper bound to the expected number of *AB* gametes generated in any generation is $(2N)(c/4)$. Thus, if $Nc < 2$, fewer than one *AB* gametes will be produced in each generation by recombination, so unless there is a strong advantage to *AB*, one of the intermediate gamete types will most likely become fixed before *AB* can reach a sufficiently high enough frequency to be deterministically promoted by selection. Such fixation of one of the intermediate types will then leave new mutation as the only mechanism for the generation of *AB*. For this special case, where the optimal gamete is initially absent, Latter (1966b) developed approximate expressions for the mean time to the first appearance of the *AB* gamete by recombination and for its subsequent fixation probability.

Although there is no general expression for the probability of fixation when alleles at two or more loci are competing for fixation, a number of important results were developed by Hill and Robertson (1966). Most notably, they obtained a weak-selection approximation for the probability of fixation for the following case. Let two diallelic loci (with designated alleles of A/a and B/b) have a recombination frequency of c, p_0 be the initial frequency of A, and D_0 be the initial gametic-phase disequilibrium (as defined in Chapter 2). Assuming completely additive selection (no dominance or epistasis), with each copy of A adding s_1 and each copy of B adding s_2 to total fitness, the probability that A becomes fixed is

$$u_f(p_0) \simeq p_0 + 2N_e s_1 p_0(1 - p_0) + \frac{2N_e s_2}{2N_e c + 1} D_0 \tag{7.43}$$

provided that $2N_e|s_1|$ and $2N_e|s_2| < 1$. A comparison of this two-locus approximation to the single-locus result (Equation 7.10b) shows that the probability of fixation can be increased or decreased depending on the sign of the initial gametic-phase disequilibrium, D_0.

Computer simulations show that when selection is strong ($N_e|s_1|$ and/or $N_e|s_2| \gg 1$), linkage (i.e., $c < 0.5$) generally *decreases* the probability of fixation of an advantageous allele relative to the single-locus result (Hill and Robertson 1966). If *A* and *B* are favored alleles, linkage will have little effect on the probability of fixation of the *ab* gamete, but the probabilities of fixation of the *Ab* and *aB* gametes increase at the expense of the optimal *AB* gamete (Latter 1965b; Hill and Robertson 1966). This decrease is maximized when $N_e c$ is small and both loci have the same effect (e.g., $s_1 = s_2$), as then there is no selective distinction between the two intermediate gametes, rendering them neutral with respect to each other. This is a significant point, as most theoretical investigations on the effects of linkage on the selection response have assumed loci with equal effects (e.g., Fraser 1957; Latter 1965b, 1966a, 1966b; Gill 1965a, 1965b, 1965c; Qureshi 1968; Qureshi and Kempthorne 1968; Qureshi et al. 1968), thereby inflating the perceived importance of linkage.

The general phenomenon of selective interference between linked loci was subsequently nicknamed the **Hill-Robertson effect** by Felsenstein (1974). As discussed in Chapter 3, the primary implication of the Hill-Robertson effect is that selection renders the behavior of linked loci closer to that expected under neutrality by reducing the effective population size for the chromosomal region (Birky and Walsh 1988; Charlesworth 1994b; Peck 1994). This effect applies to the efficiency of selection on all nonneutral alleles, both advantageous and deleterious. For example, sometimes a moderately beneficial mutation will arise in tight linkage to a highly detrimental allele at another locus, which will result in the former's rapid elimination from the population if the net fitness of the chromosomal region remains lower than that of the population mean. In addition, the average substitution rate at a locus generating deleterious alleles will be *increased* if that locus is linked to another locus generating either deleterious or beneficial alleles (Birky and Walsh 1988). In other words, the net effect of linkage is to reduce the overall efficiency of selection for fitness-enhancing mutations, magnifying the accumulation of mildly deleterious mutations at the expense of fixing more advantageous alleles.

This realization, that the majority of Hill-Robertson effects have the functional consequence of reducing N_e, greatly facilitates the estimation of fixation probabilities of new mutations subject to background selection. Indeed, in most contexts that have been examined thus far, the standard fixation expressions given above still apply provided the appropriate modifications are made to the definition of N_e (Stephan et al. 1999), as was also found for subdivided and growing or declining populations. These redefinitions, which were already outlined at the end of Chapter 3, again point to the great technical utility of the concept of effective population size. Nonetheless, as detailed in the next chapter, interference among tightly linked loci can influence the dynamics beyond the expectation with a simple reduction in N_e.

Mutations with Contextual Effects

To this point, we have generally been assuming that the magnitude of selection operating directly on an allele is independent of the genetic background (other than effects associated with linkage disequilibrium) on which it resides. However, there are numerous situations in which this will not be the case. Most notable is the broad category of **compensatory mutations**, wherein specific single mutations at either of two loci cause a reduction in fitness, while their joint appearance restores fitness or even elevates it beyond the ancestral state. Such epistatic interactions play a prominent role in Wright's (1931, 1932) **shifting balance theory** for adaptive evolution, under which an adaptive valley between two fitness peaks is traversed in a local subpopulation, with the locally fixed advantageous genotype then being exported to surrounding demes by migration. Compensatory mutations appear to play a number of important roles in protein-sequence evolution and in the composition of nucleotides in the stems of RNA molecules (Stephan and Kirby 1993; Kondrashov et al. 2002; Kulathinal et al. 2004; Azevedo et al. 2006; Breen et al. 2012).

Ascertaining the conditions under which evolution by compensatory mutation is most likely to occur is challenging because unlike the situation in which a single mutation fixes at a rate depending only on its own initial frequency, the success of a mutation involved in an interlocus interaction depends on the frequency of alleles at the interacting locus, on the fitnesses associated with the nine possible two-locus genotypes, and on the recombination rate between the two loci. Consequently, no general theory for the long-term evolution of interacting loci has yet been developed, although considerable progress has been made in a number of special cases.

Because the matter of fixation probability becomes less clear in the case of adaptations involving more than one mutation, in this final section, we will shift our focus slightly to the rate and mean time to **establishment** of an adaptation. The latter is defined to be the expected arrival time of the final multisite adaptation destined to be fixed in the population, starting from a state in which all participating mutations are absent. This excludes the additional time required for fixation, which can generally be obtained from the expressions given above, and will often be considerably smaller than the first-arrival time. When considering the response to a long-term regular regime of selection, the steady-state rate of evolution is expected to be close to the rate of establishment, as the extra time to fixation simply elongates each individual event, leaving the intervals between events the same. Assuming a constant influx of adaptive mutations, the steady-state rate of adaptation is then simply the inverse of the time to establishment.

As a benchmark for the following theoretical results, we start with the rate of establishment of a single-site adaptation, with mutations having additive fitness effects. Given a per-site mutation rate of μ, $2N\mu$ new mutations are expected to arise in each generation, each at frequency $1/(2N)$. As noted above, if the population size is sufficiently large that $4N_e s \gg 1$, the fixation probability $u_f(1/2N) \simeq 2sN_e/N$, and the rate of establishment becomes

$$r_e = (2N\mu)(2sN_e/N) = 4N_e\mu s \tag{7.44}$$

which is directly proportional to the effective population size, the mutation rate to adaptive changes, and the selective advantage. This approach, of course, assumes that the response to selection is limited by the appearance of new adaptive alleles, and in subsequent chapters we will consider in detail the situation in which part or all of the selection response is a consequence of preexisting variation. It also ignores the point made in the previous section, that if $2N\mu > 1$ (more than one favorable mutation arises per generation), the simultaneous presence of multiple segregating mutations will reduce the effectiveness of selection, lowering the expected substitution rate (Chapters 8 and 10). The waiting time for establishment follows a geometric distribution with success parameter r_e, and hence the expected waiting time is $\bar{t}_e = 1/r_e$ generations, with a variance of $\sigma^2(t_e) = (1 - r_e)/r_e^2 \simeq 1/r_e^2$.

As the simplest possible model for the rate of adaptation by new mutations, Equation 7.44 relies on the assumption that fixations have no bearing on subsequent events. However, this assumption can be violated for at least two reasons. First, the fixation of a mutation can alter the selection coefficients of future mutations by, for example, moving the mean phenotype closer to the optimal state, and consequently reducing the magnitude of selection for further change. This point is implicit in the drift barrier to adaptation noted above, and it relates to the idea of Hartl et al. (1995) that the ultimate consequence of the relentless improvement of traits by natural selection is the evolution of effective neutrality among the remaining pool of segregating alleles. Second, when mutations have epistatic effects on fitness, i.e., when they depend on the genetic background, the possibility exists that neutral or even deleterious mutations may become beneficial in certain genetic contexts. We refer to multisite traits exhibiting the latter types of genetic behavior as **complex adaptations** because the paths for their evolution are much less obvious than under conditions of additive fitness effects.

Stochastic Tunneling

How do adaptations depending on the joint presence of more than one mutation become established? One possibility is simply that double mutations, while extremely rare, will still arise, with one eventually being carried to fixation by selection. If, however, the mutation rate at a nucleotide site is 10^{-9} (Chapter 4), a population size in excess of 10^{18} is required to routinely see such double mutations, making this route unlikely for all but enormous populations. On the other hand, in very small populations, the path toward adaptation must involve successive fixations via drift, which is also likely to be a very long process. In contrast, moderately large populations offer a dual problem in that the fixation of key intermediate mutations can be problematic if they are neutral (owing to the very long time to drift to fixation) and highly unlikely if they are deleterious.

Starting with Gillespie (1984b), it became clear that another pathway, often referred to as **stochastic tunneling** (Komarova et al. 2003; Iwasa et al. 2004), offers a route for the establishment of complex adaptations in large populations even when the intermediate states are deleterious. Under this scenario, secondary mutations arise within the small pool of segregating deleterious first-step mutations, resulting in fixation of the double mutant without either single mutation becoming common, and hence without a bottleneck in mean population fitness.

The power of stochastic tunneling is that it allows selection to explore (and exploit) the fitness surface more broadly than is possible by single-step mutations, and there is a growing, technical body of work on the subject (Carter and Wagner 2002; Komarova et al. 2003; Iwasa et al. 2004; Weinreich and Chao 2005; Gokhale et al. 2009; Weissman et al. 2009, 2010; Lynch 2010; Lynch and Abegg 2010). Drawing from this literature, our goal is to provide approximate answers to three basic questions regarding complex adaptations. First, what is the critical population size below which sequential fixation dominates tunneling as a mechanism for adaptation? Second, what is the expected rate of establishment of pairs of mutations? Third, how does recombination influence these processes?

To put the first question in context, we note that there must be a critical population size, N^*, below which adaptations are essentially only acquired via sequential fixations, owing to the extreme rarity of occasions in which multiple mutations simultaneously segregate at key sites. Below this threshold value, selection is restricted to exploring the fitness landscape by single mutational steps from the currently fixed genotype. While a single chance fixation can place a population one step closer to a distant adaptive peak, it can also move it even further away. Conversely, for population sizes exceeding N^*, stochastic tunneling allows selection to explore the consequences of genotypes that are two (and in that are large populations, even more) mutational steps away from the currently most common state. This argument suggests that adaptation in small populations will typically occur by simple, single-step hill climbing, occasionally supplemented by fortuitous drift across a sufficiently shallow adaptive valley (with a reduction in fitness incurred during such a phase). In contrast, large populations should experience episodes in which adaptive events involve the simultaneous fixation of two (or more) mutations, without any intervening period of fitness loss at the population level.

A simple statement on the critical population size, N^*, can be made for the situation in which first-step mutations are neutral (Walsh 1995; Lynch and Abegg 2010). Consider a complex adaptation requiring two mutations, with the two sites completely linked, and suppose that an *A* mutation that is destined to fix has arisen. How likely is it that a *B* mutation will arise within a member of this lineage on its way to fixation? Because the first-step mutation is neutral, on average, the second mutation will have a window of $4N_e$ generations (the mean fixation time for a neutral mutation) within which it can arise on an *A* background. During this period, the average frequency of *A* will be 0.5, so the expected number of *A*-bearing alleles acquiring the second-site mutation will be $4N_e \cdot (2N\mu) \cdot (1/2) = 4N_eN\mu$, where μ is the site-specific mutation rate. Hence, when $N_e \simeq N$, there is essentially no chance of a two-mutation haplotype even arising during the fixation of a one-step mutation if the population size is much smaller than $1/(2\sqrt{\mu})$. Obviously, if the first-step lineage is destined to become lost, even fewer copies of the double mutation will be produced. Now suppose that the double mutation has a selective advantage, s, so that the fixation probability of the *AB* haplotype is $\simeq 2s$. Again assuming $N_e \simeq N$, the adaptation will almost certainly arise by stochastic tunneling rather than by sequential fixation if the population size exceeds

$$N^* \simeq \frac{1}{2\sqrt{2\mu s}} \tag{7.45}$$

Note that this is not a terribly stringent condition, as with $\mu = 10^{-9}$ and $s = 0.01$, $N^* \simeq 112{,}000$. The critical population size will be larger by a factor of $1/\sqrt{x}$ if $N_e = xN$ (recall from Chapter 3 that x is usually $\ll 1$). When the intermediate step is strongly deleterious (with an effect of s_d), then provided $4N_es_d \gg 1$, first-step mutations will almost be never fixed, with tunneling dominating over sequential fixation.

We now turn to the matter of rates of establishment, focusing again on the situation in which two loci are fixed for alleles *A* and *B*, respectively, and inquiring as to the time to reach an alternative state of fixation at both loci, with respective alleles *a* and *b*. We will assume equivalent mutation rates of μ from *A* to *a* and *B* to *b*. The simplest selection scenario in this case, which was first explored by Kimura (1985), assumes that the gametes *Ab* and *aB* have equivalent fitnesses of $1 - s$ and the gametes *AB* and *ab* have equivalent fitnesses of 1.0. In this case, although transitions between pure population states of *AB* and *ab* may occur, nothing is gained in terms of fitness. Within the sequential fixation domain, the degree of linkage can be ignored (as only one locus is polymorphic at a time), and the mean time to establish the novel *ab* type from *AB* (or vice versa) is the sum of the waiting times for the two mutational steps,

$$\bar{t}_e = \frac{1}{2N\mu}\left(\frac{1}{2u_{fd}} + \frac{1}{u_{fb}}\right) \tag{7.46}$$

where u_{fd} and u_{fb} are, respectively, the probabilities of fixation of deleterious (first-step) and beneficial (second-step) alleles (obtained by applying the selection coefficients $-s$ and s to Equation 7.10a). Transitions to states *Ab* or *aB* (from the higher-fitness states *ab* or *AB*) occur at a rate of $(4N\mu)(u_{fd})$, which is the product of the population mutation rate and twice the rate of first-step fixation (because there are two ways to produce first-step mutations); and then, conditional on the first change, the second change occurs at a rate of $2N\mu u_{fb}$. Because the probability of fixation of a deleterious allele is e^{-4N_es} that of a beneficial allele (above), the establishment time in this case is expected to be primarily determined by the time required for the fixation of first-step alleles, so that the rate of establishment (the reciprocal of the time to establishment)

$$r_e \simeq 4N\mu u_{fd} \tag{7.47}$$

If, on the other hand, selection against the intermediate haplotypes is much stronger than drift so that fixation of the intermediate state is unlikely (the stochastic-tunneling domain), the most likely scenario for a transition from the *AB* type to *ab* type is a population that is initially residing in a state of selection-mutation balance at both loci. If we assume complete linkage, and a selection coefficient of $-s$ associated with the *a* and *b* alleles when not appearing alone, the *Ab* and *aB* gametes, each with initial frequency $\tilde{p} = \mu/s \ll 1$ (from Equation 7.6d), would then serve as staging grounds for mutations to the *ab* type. Mutant *ab* gametes arise at a rate of μ from each of the $4N\mu/s$ intermediate types ($2N$ times the frequency of each heterozygote, $\simeq 2\tilde{p}$), and fix in an essentially neutral fashion with a probability of $1/(2N)$ (as most resident gametes are of the type *AB*, with equivalent fitness to *ab* as noted above). Thus, the rate of establishment of the *ab* type by stochastic tunneling to an equivalent fitness state is

$$r_e \simeq (2\mu/s)(\mu) = \frac{2\mu^2}{s} \tag{7.48}$$

(Gillespie 1984b; Stephan 1996), which is essentially independent of population size.

When mutations are reversible, the question also arises as to the long-term stationary distribution of alternative states. Adhering to the reasoning that *Ab* and *aB* gametes will generally be maintained at low levels by selection-mutation balance, and assuming equal back and forward mutation rates, Higgs (1998) elegantly showed that the stationary distribution for the frequency (x_0) of the *AB* gamete is

$$\phi(x_0) = \frac{1}{(1-z)^{2\alpha-1}} \frac{\Gamma(2\alpha)}{\Gamma(\alpha)^2} \left[x_0(1-z-x_0)\right]^{\alpha-1} \tag{7.49}$$

where $\alpha = 8N\mu^2/s$ is the population rate of mutational production of *ab* gametes, $z = 2\mu/s$ is the summed frequency of the *Ab* and *aB* gametes, and Γ denotes the gamma function (Equation 2.25b). The frequency of the *ab* gamete is simply $1 - x_0 - z$. With $\alpha < 1$, the distribution of x_0 is highly U-shaped, with the probabilities of the population being fixed for alternative *AB* and *ab* states being nearly equal. A more general analysis, which allows for weaker efficiency of selection ($4N_e s < 1$) and for differential selection and mutation operating on the intermediate states was presented by Innan and Stephan (2001).

Now suppose that the secondary mutation has an advantage of s_b, and we denote the disadvantage of first-step mutations as s_d. The general Equation 7.46 still applies in the sequential-fixation domain, and we again expect the rate of establishment to be approximated closely by Equation 7.47 owing to the long waiting time for the fixation of a first-step mutation. For the stochastic-tunneling domain, however, Equation 7.48 must be modified to account for the fact that the fixation probability of the double mutant is $\sim 2s_b(N_e/N)$

$$r_e \simeq (2\mu/s_d)(2N\mu)[2s_b(N_e/N)] = \frac{8N_e\mu^2 s_b}{s_d} \tag{7.50}$$

The key observations here are that the rate of establishment now depends on the effective population size, while also scaling linearly with the square of the mutation rate and the ratio of selection coefficients associated with first- and second-step mutations.

Finally, we consider the special situation in which first-step mutations are effectively neutral. Again, Equation 7.46 provides an accurate description for the sequential-fixation domain, and with substitution of the appropriate fixation probabilities, it reduces to

$$\bar{t}_e = \frac{1}{2N\mu}\left(\frac{1}{2[1/2N]} + \frac{1}{u_{fb}}\right) \simeq \frac{1}{2\mu}\left(1 + \frac{1}{2N_e s}\right) \tag{7.51}$$

with the last approximation obtained by using $u_{fb} \simeq 2sN_e/N$. Thus, provided $2N_e s \gg 1$, when the intermediate mutation is effectively neutral, the expected rate of establishment will be $\simeq 2\mu$ and will only be weakly dependent on the size of the population in the sequential-fixation domain.

To obtain the expected rate of tunneling for the case of neutral intermediates, we require the probability that tunneling occurs within a descendant lineage of a first-step mutation before it becomes lost from the population. By various methods, and again assuming complete linkage, this probability has been found to be approximately $\sqrt{2\mu s N_e/N}$ in large populations (Komarova et al. 2003; Iwasa et al. 2004; Weissman et al. 2009, 2010; Lynch and Abegg 2010). With $4N\mu$ first-step mutations arising per generation, the rate of establishment via tunneling is then

$$r_e \simeq 4N\mu\sqrt{2\mu s N_e/N} = 4\mu\sqrt{2\mu s N_e N} \tag{7.52}$$

If the mutation rates at the two steps are different, μ inside and outside of the square-root expression should designate mutation rates at the first and second steps, respectively. The key observation here is that when the intermediate stages are neutral, tunneling occurs at a *higher rate* in larger populations, contrary to the situation with deleterious intermediates. Moreover, although two mutations are required for the final adaptation, the rate of establishment depends on the 3/2 power of the mutation rate, unlike the square scaling with deleterious intermediates.

While these analyses assume an evolutionary path to a final adaptation through just a single intermediate step, actual fitness surfaces are likely to be more complex, with a variety of potential pathways through any number of mutations. The rates of establishment of complex adaptations under these alternative scenarios have been examined by Gokhale et al. (2009), Weissman et al. (2009), and Lynch and Abegg (2010). Simple analytical expressions were found in only a few cases, two of which we now summarize. As complex adaptations involving more than two mutations are unlikely to evolve by sequential fixation, owing to the long time necessary for cumulative fixations, we restrict our attention to the stochastic-tunneling domain, focusing on how the establishment rate r_e scales with the underlying features of population size, mutation rate, and selection intensity.

For the case of neutral intermediates with increasing numbers (d) of mutations required for the final adaptation (and the order of events assumed to be irrelevant), the rate of establishment can be viewed as a series of nested tunneling events. For example, for the case of $d = 3$ (two neutral mutations required before the final adaptation is assembled with a third mutation), Equation 7.52 expands to

$$r_e = 6N\mu\sqrt{2\mu\sqrt{2\mu s N_e/N}} \tag{7.53a}$$

Note that the first term is now $6N\mu$ because first-step mutations can arise at three sites. The next step then starts at either of the two remaining sites, with the final stage being initiated at the one remaining site and involving tunneling within the sublineage containing the first two mutations. For an arbitrary d, this expression generalizes to

$$r_e = d\lambda\mu(2N\mu)^{1-0.5^{d-1}}S^{0.5^{d-1}} \tag{7.53b}$$

where $S = 4N_e s$, and

$$\lambda = \prod_{i=1}^{d-1}(d-i)^{0.5^i} \tag{7.53c}$$

This result shows that, with neutral intermediates, the rate of establishment by tunneling scales with no more than the square of the mutation rate and with no less than linearly with the absolute population size, these extremes being approached at high d. Thus, the rate of establishment of complex adaptations can be much more rapid than expected under the naive assumption that independently arising mutations would lead to a scaling with the dth power of the mutation rate.

For the case of deleterious intermediates, suppose that all haplotypes involving one to $d - 1$ mutations are equally deleterious (with a fitness of $1 - s_d$), with the final mutation conferring an advantage, s_b. First-step mutations then arise at rate $2Nd\mu$, but owing to selection, they have an expected survivorship time of $1/s_d$ generations, during which period $d - 2$ additional intermediate-step mutations must be acquired, followed by the appearance of a final-step mutation destined to fixation. This leads to a rate of establishment via tunneling of

$$r_e \simeq 4N_e d!(\mu/s_d)^d s_d s_b \tag{7.54}$$

which reduces to Equation 7.50 when $d = 2$. Here we see that r_e scales with the dth power of the mutation rate owing to the limited opportunities for mutation during the short sojourn times of deleterious mutations, whereas there is a linear scaling with the effective population size. One cautionary note with respect to all of these scaling features is that mutation rates appear to generally evolve to be inversely related to the effective size of a population, which will tend to reduce the dependence of rates of establishment on μ and measures of population size, as these two factors often enter as products of each other (Lynch 2010b).

Finally, we note that all of these analyses assume an absence of recombination. This is a matter of significance, as it is often surmised that recombination facilitates the evolution of complex adaptations. In the sequential-fixation regime, recombination can be ignored simply because multiple polymorphic sites are never present simultaneously. However, in the stochastic-tunneling domain, opportunities will exist for both the creation and breakdown of optimal haplotypes. For the case of deleterious intermediates but selectively equivalent end states (above), Higgs (1998) provided general expressions, allowing for arbitrary levels of recombination. Strong linkage substantially accelerates the rate of peak shifts with this fitness landscape because the frequencies of the low-fitness intermediates remain nearly unchanged during transitions to alternative high-fitness states, ensuring that the population does not pass through a phase of reduced mean fitness (Kimura 1985; Michalakis and Slatkin 1996; Stephan 1996; Innan and Stephan 2001). In contrast, recombination between the high-fitness *AB* and *ab* gametes during a peak shift produces low-fitness intermediates, imposing a bottleneck on mean population fitness, and thereby inhibiting the movement from one state to the other.

Lynch (2010b) and Weissman et al. (2010) examined this problem with a broad class of models, and reached the conclusion that recombination is most likely to have either a minor or an inhibitory effect on the establishment of a complex adaptation. Consider, for example, the case of a two-site adaptation, starting with a population fixed for the suboptimal *ab* haplotype. The overall influence of recombination on the rate of establishment of the *AB* haplotype is a function of two opposing effects—the rate of origin of *AB* gametes by recombination within doubly heterozygous (*aB*/*Ab*) parents is proportional to the rate of recombination between the sites (c), whereas the net selective advantage of the resultant *AB* haplotypes is discounted from s to $s - c$ by subsequent recombinational breakdown (as in the early stages, *ab* haplotypes still predominate and are the primary partners in recombination events with *AB*). Thus, because the product $c(s - c)$ is maximized at $c = s/2$, two-site adaptations are expected to emerge most rapidly in chromosomal settings where the recombination rate is half the selective advantage of the final adaptation.

For the case of neutral intermediates, details in Lynch (2010b) suggest that even at the optimal recombination rate, the rate of establishment is generally enhanced by much less than an order of magnitude relative to the situation with complete linkage, whereas $c > (s/2)$ is not greatly inhibitory. In contrast, when first-step mutations are deleterious, if the rate of recombination exceeds the selective advantage of the *AB* haplotype, recombination will present an extremely strong barrier to establishment of the *AB* haplotype (Lynch 2010b). The latter result arises because almost all recombinational events involving a newly arisen *AB* haplotype will involve an *ab* participant, thus generating the maladaptive *Ab* and *aB* products.

Taken together, these results suggest that only a narrow range of recombination rates (in the neighborhood of $s/2$) can enhance the rate of establishment of a complex adaptation from de novo mutations. Moreover, because the role that recombination plays in the origin of specific adaptations depends on both the selective advantage of the final product and the physical distance between the genomic sites of the underlying sites, the issue cannot be reduced to a simple generalization. With a highly context-dependent optimal recombination rate (per nucleotide site), it becomes unclear whether selection is likely to have any general influence on the promotion of recombination-rate modifiers (Chapter 4).

These kinds of observations, in which a two-locus system stochastically shifts from one semistable state to another through evolutionary time, appear to be closely related (albeit not transparently) to the features of a number of models of quantitative traits. For example, diallelic models of quantitative traits under stabilizing selection often exhibit multiple equilibria for allele frequencies (including alternative monomorphic and polymorphic states), depending on the effects of alleles and the ways of assembling a multilocus phenotype that is closest to the optimum (Bulmer 1972; Barton 1986, 1989; Bürger 1989; Gavrilets and Hastings 1994a). One can easily imagine that finite populations would wander from one local equilibrium to another through time, depending on the history of mutation and drift, although no formal theory on the rate of such internal shifts has been worked out.

8

Hitchhiking and Selective Sweeps

When a mutation B without much selective advantage occurs in the proximity of another mutant gene A with a high selective advantage, the survival chance of gene B is enhanced, and the degree of such enhancement is a function of the recombination fraction between the two loci. Gene B under this situation resembles a hitch–hiker riding along with a host driver.
Kojima and Schaffer (1967)

As first noted by Kojima and Schaffer (1967) and Maynard Smith and Haigh (1974), the dynamics of a neutral allele are strongly influenced by selection at a linked locus. Almost 50 years later, we are still trying to fully understand all of the ramifications of this idea. Chapter 3 provided a brief introduction to two rather different linked-selection scenarios: selective sweeps (the fixation of newly beneficial mutations) and background selection (the removal of deleterious new mutations). In this chapter we further unpack these concepts, presenting a much richer theoretical treatment and a more detailed account of some of their potential consequences. The results presented here also underpin many of the tests for detecting currently ongoing, or very recent, selection developed in Chapter 9.

Our treatment is structured as follows. We start with a review of the basic terminology for different scenarios, all loosely referred to as **sweeps**. Next, we review the population genetics of hard sweeps (those arising from a single mutation that is immediately favorable), detailing how neutral variation is perturbed by positive selection at linked sites. We then turn to soft sweeps, wherein a preexisting allele is suddenly placed under selection, generating a different pattern of background neutral variation relative to a hard sweep. This naturally leads to a discussion as to whether adaptation to a new challenge occurs via existing variation or by waiting for a new favorable mutation, as well as to the notion of a polygenic sweep (small allele-frequency changes at a number of loci). Another scenario for a soft sweep to occur involves multiple favorable mutations arising *during* a sweep, which may be more likely than intuition might suggest. Next, we examine the consequences of repeated bouts of selection at linked sites (be they recurrent sweeps or background selection) for levels of nucleotide diversity, substitution rates, and codon usage bias. We conclude with a discussion on whether the current data warrant a paradigm shift away from Kimura's (1983) classical neutral theory of molecular evolution (Chapter 2).

SWEEPS: A BRIEF OVERVIEW

Given that our treatment of the mathematics of sweeps will be rather technical in places, we first start with a brief overview of the basic terminology and key ideas about sweeps. Casual readers may find this section sufficient for their needs, while it will serve to orient more diligent readers before proceeding onward.

Hitchhiking, Sweeps, and Partial Sweeps

Although usually attributed to Maynard Smith and Haigh (1974), the term **hitchhiking** was introduced by Kojima and Schaffer (1967) to describe the increase in frequency of a neutral allele linked to an allele under directional selection. Plant breeders were also aware of this phenomenon, in the context of **linkage drag** (Brinkman and Frey 1977), wherein an introgressed favorable region may drag with it unfavorable linked genes. The term **selective sweep** (Berry et al. 1991), which is often treated as synonymous with hitchhiking, originally

Evolution and Selection of Quantitative Traits. Bruce Walsh & Michael Lynch.
Published 2018 by Oxford University Press.
DOI 10.1093/oso/9780198830870.001.0001

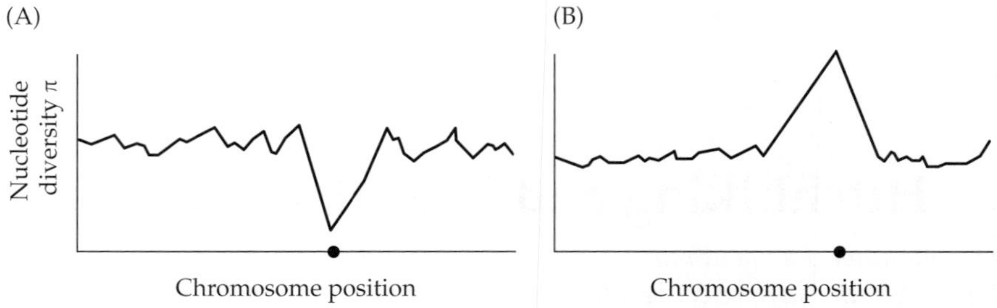

Figure 8.1 (**A**) The signature of positive directional selection (a selective sweep) around a selected site (the solid circle). The background levels of linked neutral variation (measured as the average nucleotide diversity, π, in a sliding window of markers) show a significant decrease around the selected site, reflecting the decreased effective population size (and hence a shorter time to the most recent common ancestor, TMRCA) for regions linked to this site. (**B**) By contrast, balancing selection generates an *increase* in nucleotide diversity at linked neutral markers, reflecting a longer TMRCA, and hence more opportunities for mutation to generate variation.

referred to the sweeping away of variation around a selected site following the fixation of a favorable allele (Figure 8.1A). This cleansing effect occurs because selection reduces the effective population size at linked regions, shortening the coalescence times for surviving neutral alleles relative to pure drift, and thus reducing the time for new variation to accumulate (Figure 8.3).

A **partial sweep** refers to the setting in which the favored allele is currently increasing in frequency, as would occur during an ongoing sweep. A partial sweep signal is also generated during **balancing selection** (Chapter 5), again when an allele is increasing in frequency toward some internal equilibrium value (as opposed to fixation). Indeed, any initially rapid increase in frequency by selection generates some type of transient sweep-related signal (albeit potentially quite small), even if the nature of selection changes before fixation (Coop and Ralph 2012).

Just as a specific signal for an ongoing selection event transitions into a different signal following the conclusion of the sweep, so too does the signal for a site under balancing selection. As shown in Figures 8.1 and 8.2, a region held under long-term balancing selection (meaning that the allele frequencies have been near their equilibrium values for a sufficient time) will show an *increase* in the amount of variation at linked neutral sites (Strobeck 1983; Hudson and Kaplan 1988; Kaplan et al. 1988). This occurs because selection holds alternate alleles at intermediate frequencies for a much longer time than excepted under drift, resulting in an older common ancestor relative to the neutral expectation (Figure 8.3), and hence more time for variation to accumulate. Over time, however, recombination will constantly shrink the size of the region around the selected site with an elevated coalescence time, continually shrinking this signal of long-term balancing selection (Wiuf et al. 2004; Pavlidis et al. 2012).

Selection Alters the Coalescent Structure at Linked Neutral Sites

The underpinning for many tests of selection using polymorphism data (Chapter 9) is that selection changes the coalescent structure at linked neutral sites. If we view the structure of genealogical relationships among the alleles in a sample as a **tree** (Figure 8.3), recent positive selection shortens the **total branch length** (the sum of the lengths of all the branches, representing the coalescent times), thus decreasing the amount of variation. Conversely, long-term balancing selection generates longer times to common ancestors (as alleles are retained in the population longer than expected under drift), thus increasing the amount of variation. This effect is equivalent to a change in the effective population size, with a sweep reducing the effective population size in a linked region (Chapter 2), generating shorter coalescence times, and balancing selection increasing N_e and coalescent times.

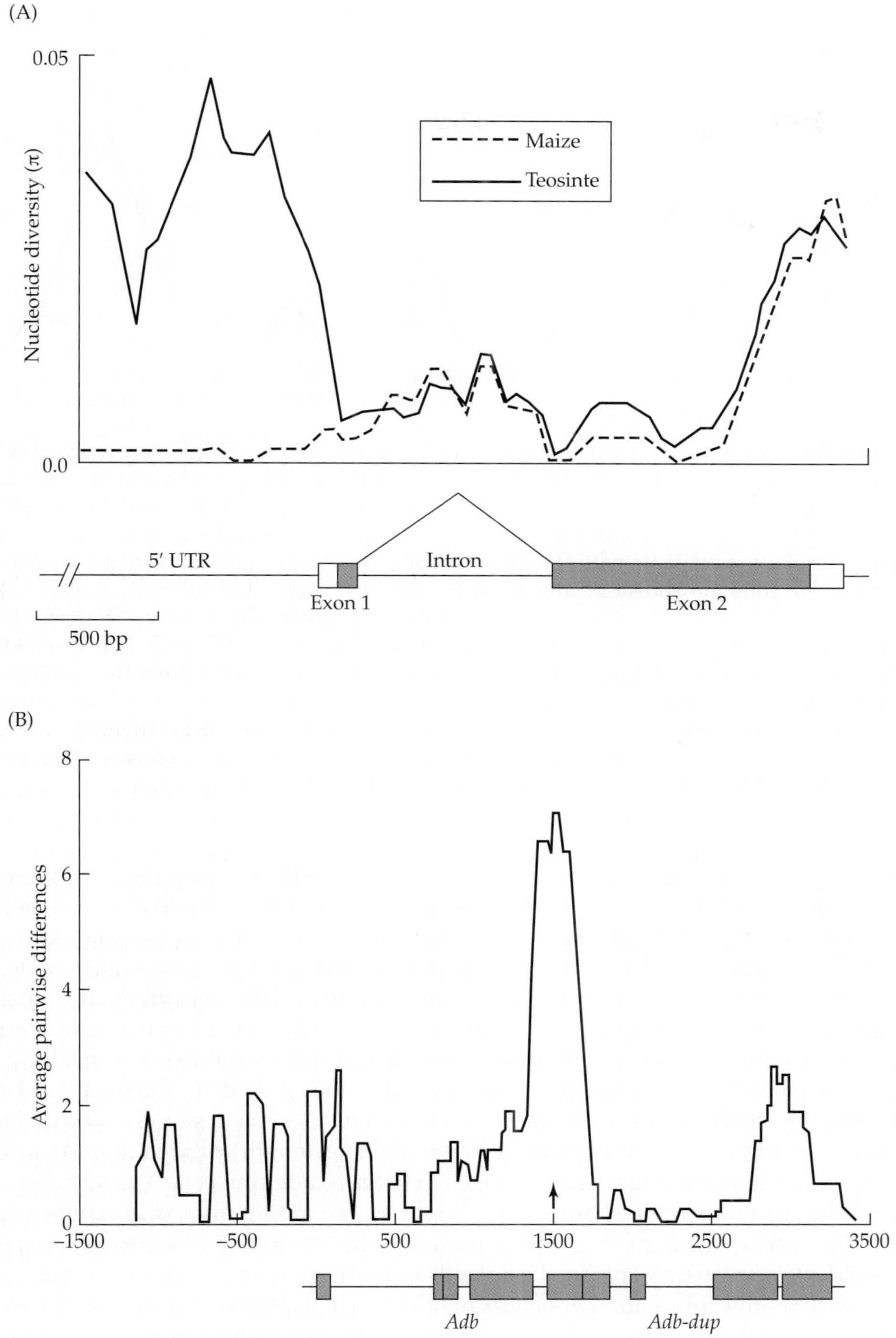

Figure 8.2 Examples of selection influencing levels of variation at linked neutral sites. (**A**) A sliding-window plot of levels of nucleotide diversity, π, around the *tb1* gene in maize (corn) and teosinte, a candidate gene for the domestication of teosinte into corn. Relative to teosinte, maize variation is dramatically reduced in the 5′ UTR region of *tb1*, suggesting a sweep linked to this region. (After Wang et al. 1999.) (**B**) Inflated levels of variation are seen around a site that results in an amino acid change (arrow), generating two allelic classes, *fast* and *slow*, at the *Adh* gene in *Drosophila melanogaster*. These alternative alleles have long been thought to be under balancing selection, which is consistent with the increased variation observed around this site. However, the nature of selection on this gene is still an open issue, as Begun et al. (1999) showed that the excess variation is *within slow* haplotypes, *not* between *fast* and *slow* haplotypes as would be expected if these two alleles were under balancing selection. (After Kreitman and Hudson 1991.)

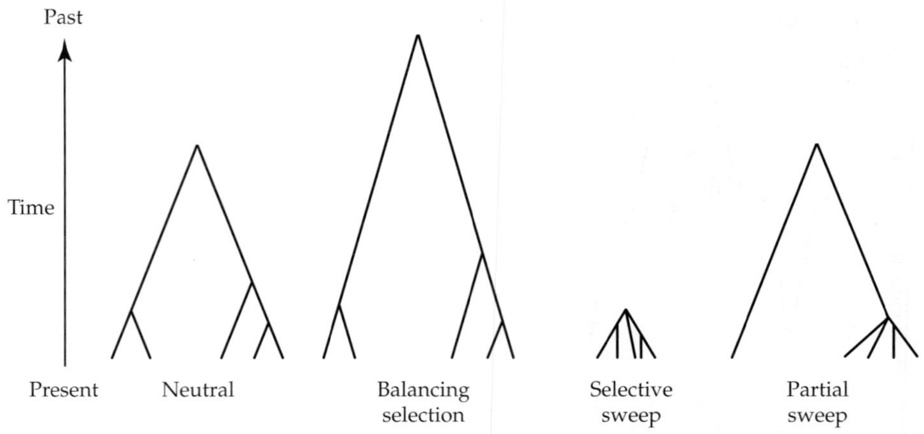

Figure 8.3 Example coalescent (or genealogical) structures (Chapter 2) for loci from the same population under pure drift, balancing selection, a selective sweep, and a partial sweep (ongoing selection in which an allele is still increasing in frequency). The tips of the tree at the bottom of the graph represent five sampled alleles, which eventually coalesce into a single lineage as we go back in time (the top of the graph). This final coalescent point represents the most recent common ancestor (MRCA) for the sampled alleles. For balancing selection, the time to the MRCA (TMRCA) is greater than for neutral genes, which in turn is greater than a region undergoing a sweep. The shape (measured by the relative distances between nodes) of the coalescent is also influenced by selection. Individual coalescent times for a sweep are much more compressed (the nodes are closer together) as one moves back in time, while under a neutral drift process, coalescent times *increase* as we approach the MRCA (Equation 2.40). A partial sweep represents a mixture, with a sweep-like structure in one part of the genealogy (here, the right branch) and a drift-like structure in the other (the left-branch).

Equally important, selection at a linked site does more than simply shorten or lengthen the coalescent structure. It ***alters its shape*** as well, namely, the expected distances between nodes conditioned on the total length of the tree (Figure 8.3). For a neutral coalescent, trees generated with different value of N_e have the same expected shape when scaled to the same total length (Chapter 2). Under a selective sweep, the **nodes** of the tree (the coalescent points of the genealogies in the sample) are compressed as one moves back in time. In the extreme, positive directional selection can be approximated by a **star** (or **palmetto**) **genealogy**, with all lineages coalescing at a single point. In contrast, under pure drift, the expected longest branch lengths are those that coalesce the final two lineages into a single ancestral lineage (Equation 2.40, Figure 2.10). While differences in the total length of the coalescent influence the total *amount* of neutral variation, changes in its *shape* alter the *pattern* of variation from that expected from a simple change in N_e (such as generating an excess number of rare alleles). The consequences of this change in tree shape are manifested through changes in the site and allele-frequency spectra (Chapter 2) and in the pattern of linkage disequilibrium, differences that underpin a number of tests of selection (Chapter 9). Unfortunately, recovery from a sharp population bottleneck (a crash in population size) also generates a very similar, although not quite identical, compression of the nodes (Barton 1998).

The precise coalescent structure generated by positive selection depends on when the process is being captured. The structure for a just-completed sweep is different from that generated during the initial phase of selection when a favorable allele is increasing in frequency (Figure 8.3). This latter setting is called a partial sweep and could represent either an allele ultimately on its way to fixation or an allele increasing to some equilibrium frequency under balancing selection. In either case, the resulting tree during the partial sweep phase can be rather unbalanced, with one branch displaying a sweep-like pattern (compressed nodal distances reflecting those lineages influenced by the selected

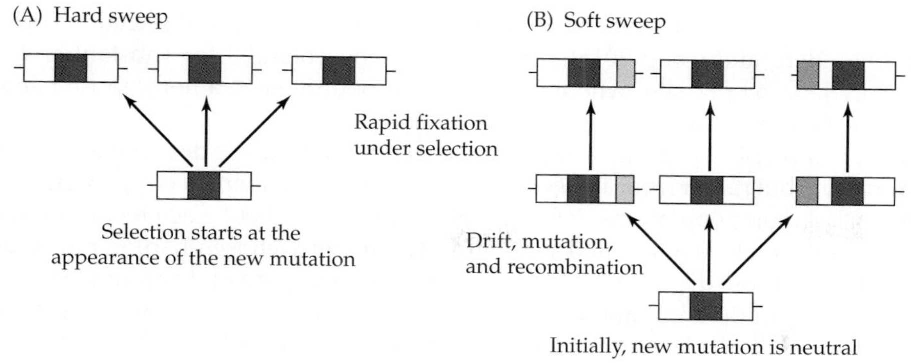

Figure 8.4 Types of sweeps and their corresponding collection of haplotypes. (**A**) A hard sweep. A new mutation is immediately favored, resulting in only a single haplotype sweeping to high frequency. (**B**) A single-origin soft sweep. Here a single mutation is initially neutral or even slightly deleterious. It drifts around the population, appearing in new haplotypes through recombination. At some point, an environmental change places this site under strong selection, which sweeps it to fixation, carrying along a sample of its existing collection of linked haplotypes.

allele) and the other a more drift-life pattern (those lineages that have yet to be affected). This partial-sweep coalescent structure is *transient*, and with time it may resolve to a sweep (if fixed by selection) or a balancing-selection structure.

Alternatively, the nature of selection may shift before fixation/equilibrium is reached. In such settings, an allele is driven to a certain frequency by selection, and then may subsequently be fixed by drift or by much weaker selection. Any initial rapid (relative to drift) change in allele frequency generates a signal, even if strong selection stops before fixation, as it modifies the coalescent structure of linked neutral alleles relative to that expected under pure drift (Coop and Ralph 2012).

Hard vs. Soft Sweeps

Not all sweeps, even those involving strong selection, are expected to leave a detectable signal. A **hard sweep** refers to a single favorable new mutation arising and immediately coming under selection (Perlitz and Stephan [1997] also refer to this as a **catastrophic sweep**). The fixation of this mutation drags the haplotype on which it arose to a high frequency, leaving a strong signal (Figure 8.4A). In contrast, under a **soft sweep** (Hermisson and Pennings 2005, 2017; Messer and Petrov 2013a) multiple haplotypes initially carry the favorable allele. The realization that there are different types of sweeps resolved one of the earlier criticisms of the Maynard Smith and Haigh model. Their analysis suggested a major impact on the amount of heterozygosity at linked neutral sites (detailed below), while Ohta and Kimura (1975) showed that recurrent hitchhiking had only a minor effect on heterozygosity. The resolution of these apparently contradictory results was that Maynard Smith and Haigh assumed a hard sweep, while Ohta and Kimura made what amounted to an assumption of a soft-sweep.

Soft sweeps can occur under two scenarios, which have different consequences for the nature and strength of the signal left by the sweep. Under a **single-origin soft sweep**, the eventually favorable mutation predates the start of selection, being either initially neutral or even slightly deleterious. It drifts around the population, potentially spreading to different haplotype backgrounds via recombination, until eventually a change in the environment results in it being favored. This has the effect of selection acting on a more diverse initial collection of haplotypes, giving off a much weaker signal than under a hard sweep. If p is the frequency of an allele at the start of selection, a soft sweep starts with $2pN$ copies of the favorable allele. Within this collection of haplotypes containing the eventually favored allele, the mean coalescence time for a completely linked site in two random individuals is $t = 2pN_e$, where N_e is the effective population size at the start of selection (Innan and Tajima 1997; Berg and Coop 2015). Thus, there is the potential for substantial variation ($2t\mu = 4pN_e\mu = p\theta$ per site, where $\theta = 4N_e\mu$) at neutral sites among those haplotypes carrying the favorable allele at the start of selection.

Under a **multiple-origin soft sweep**, the fixed favorable allele does not descend from a *single* mutation, but rather from a *collection* of multiple independent events, a scenario that is more likely in larger populations (Pennings and Hermisson 2006). Each recurrent mutation to the favorable allele is associated with an independently chosen haplotype, potentially resulting in even more diversity at fixation than a soft sweep involving a single preexisting mutation. Such multiple mutations may either be present at the start of selection or arise during the sweep itself. The term **standing sweep** is used to distinguish sweeps that start with preexisting variation, in contrast to settings where sweeps are driven by new mutations that arise following the selective change in the environment.

While the difference between hard and soft sweeps is often framed as being rather sharp, this distinction is largely a function of the population sample (Messer and Petrov 2013a). If the coalescence time for the haplotypes in a sample is more recent than the start of selection, we will see a hard sweep. If the coalescence time for sampled haplotypes carrying favorable alleles predates the start of selection, we will see a soft sweep. As we detail below, demography and population structure further influence how we perceive a given sweep. For example, a population bottleneck may **harden** a soft sweep by allowing only a single favorable haplotype to pass through.

THE BEHAVIOR OF A NEUTRAL LOCUS LINKED TO A SELECTED SITE

We start our discussion of the population-genetic theory of sweeps by first considering hard sweeps and their effects on linked neutral loci. Parts of this discussion are rather technical, but the main theoretical results are summarized in Table 8.1, with the expected signatures from a hard sweep summarized in Table 8.2. Throughout, we assume that there is strong selection on the favorable allele ($4N_e s \gg 1$, where, unless otherwise specified, s is the heterozygote fitness advantage). We also assume that no new mutations (either at neutral sites or for the favorable allele) occur during the sweep. This negligible mutation approximation, which reflects a rapid sweep through the population of a new favorable allele and hence reduced time for new mutations to arise, is relaxed later in this chapter.

Allele-frequency Change

To quantify the impact of a sweep, we need to determine how selection influences the frequency of a neutral allele, m, at a linked locus. The ultimate variable of interest is $q(t)$, the frequency of m in generation t of the sweep, with $q(\infty)$ denoting its frequency at the sweep's completion. In the absence of any recombination during the sweep, $q(\infty) = 1$, as the initial haplotype bearing the new favorable mutation (and m) is swept to fixation. The degree to which $q(\infty)$ actually approaches one depends on the relative strengths of recombination, c (between the selected site and the neutral marker), and the strength of selection, s, on the favored site. In Chapter 5 we showed that the time to fixation from selection scales with $1/s$, so it is reasonable to expect that $q(\infty)$ is some function of c/s.

In order to obtain $q(t)$, we need to follow the frequencies of m on the two different chromosomal backgrounds, those with selected allele A and those carrying the alternative allele, a. If we let $q_A(t)$ and $q_a(t)$ denote the frequencies of allele m on A- and a-bearing chromosomes, conditional probability gives the population frequency of m as

$$q(t) = q_A(t) \cdot p(t) + q_a(t) \cdot [1 - p(t)] \tag{8.1a}$$

where $p(t)$ is the frequency of A in generation t of the sweep. When A becomes fixed, $q(\infty) = q_A(\infty)$, the fraction of A-bearing chromosomes that still contain m.

A related quantity is the difference, δ_q, in the frequency of m on these two backgrounds

(A versus a), whose initial value is

$$\delta_q(0) = q_A(0) - q_a(0) \tag{8.1b}$$

Nonzero values of δ_q imply linkage disequilibrium between A and m, with the frequency of m on A-bearing chromosomes differing from its unconditional frequency in the general population. Hitchhiking is basically a race between the rate of recombination reducing the initial disequilibrium (and hence δ_q) and the rate of selection fixing A, eliminating any further recombination. Smaller values of c (slower rates of recombination) and/or larger values of s (faster time to fixation) increase the impact of a sweep. When A is introduced as just one or a few copies, then $q_a(0) \simeq q(0)$, as essentially all of the copies of m are on a chromosomes. If A arises as a single copy on an m-bearing chromosome, then $q_A(0) = 1$ (as the only A-bearing chromosome contains m), yielding $\delta_q(0) = 1 - q(0)$. Finally, let

$$\Delta_q = q(\infty) - q(0) \tag{8.1c}$$

denote the final change in the frequency of the linked neutral allele m after A has swept to fixation. Summarizing our notation to this point, we have

$q(t)$	Unconditional frequency of m in generation t
$q_A(t)$	Conditional frequency of m on A-bearing chromosomes
$q_a(t)$	Conditional frequency of m on a-bearing chromosomes
$\delta_q(t) = q_A(t) - q_a(t)$	Difference in frequency of m on A- versus a-bearing chromosomes
$\Delta_q = q(\infty) - q(0)$	Change in frequency of m after a sweep

In particular, note that $\delta_q(\infty)$ and Δ_q, while related, are still distinct.

To proceed, we define the critical measure of the strength of a hitchhiking event,

$$f_s = \frac{\Delta_q}{\delta_q(0)} = \frac{q(\infty) - q(0)}{q_A(0) - q_a(0)} \simeq \frac{q(\infty) - q(0)}{1 - q(0)} \quad \text{when } p(0) \ll q(0) \tag{8.1d}$$

which is the fraction of the initial excess association of m with A, $\delta_q(0)$, that persists when A is fixed (Δ_q). If the sweep is started with a single lineage, f_s is the probability of identity-by-descent at the m locus among fixed A chromosomes (Gillespie 2000; Kim and Nielsen 2004). In the absence of recombination, $f_s = 1$, resulting in an allele-frequency change of $\Delta_q = \delta_q(0)$, which is $1 - q(0)$ at the point where A arose as a single copy. With recombination, $f_s < 1$, and our task is to evaluate how the relative values of selection (s) and recombination (c) determine the magnitude of this effect. An important result in sweep theory is that the final allele frequency of m for the case of A arising as a single copy can be expressed as

$$q(\infty) = q(0) + \Delta_q = q(0) + f_s\delta_q(0) = q(0) + f_s[1 - q(0)] = f_s + q(0)(1 - f_s) \tag{8.1e}$$

The derivation of the standard deterministic approximation for Δ_q (given in Example 8.2) requires a few tricks, and the basic biology can get somewhat lost during its development. Hence, we first sketch a rough approximation of how selection and recombination compete before presenting more accurate results. As noted by Maynard Smith and Haigh (1974) and Barton (2002), because m is neutral, its frequency on either background only changes through recombination, with

$$\delta_q(t) = q_A(t) - q_a(t) = (1 - c)^t\,[\,q_A(0) - q_a(0)\,] \simeq \delta_q(0)\,e^{-ct} \tag{8.2a}$$

A rapid increase of A reduces the opportunity for recombination between A- and a-bearing chromosomes, which becomes nonexistent when A is fixed. Assuming A is under additive selection with fitness s, then as shown in Example 8.1, if A is introduced into the population as a single copy and is destined to become fixed, its approximate time to fixation is $\tau \simeq 2\ln(4N_es)/s$. Substituting this time into Equation 8.2a yields

$$\delta_q(\infty) = \delta_q(0)e^{-c\tau} \simeq \delta_q(0)\exp\left[-(2c/s)\ln(4N_es)\right] = \delta_q(0)\,(4N_es)^{-2c/s} \tag{8.2b}$$

so that

$$\frac{\delta_q(\infty)}{\delta_q(0)} = \left(\frac{1}{4N_e s}\right)^{2c/s} \tag{8.2c}$$

This shows that the fraction of initial association that persists following the sweep is a function of c/s, which is the key parameter for determining the strength of hitchhiking. When $c/s \ll 1$, most of the initial association remains at the conclusion of the sweep, while as ever-more-distant sites are considered (so that c/s increases), any initial association decay toward zero.

While Equation 8.2c conveys the general notion of competition between recombination and selection, this is a rather crude analysis, as it only considers the time to fixation for A (and hence the end of any opportunity for further recombination) and uses $\delta_q(\infty)$ in place of the real quantity of interest, Δ_q. As presented in Example 8.2, an improved analysis accounts for how the actual change in the frequency of A influences the opportunity for recombination. This problem has received considerable attention, starting with a strictly deterministic analysis by Maynard Smith and Haigh (1974; also see Stephan et al. 2006), followed by analyses allowing for finite population size by Kaplan et al. (1989), Stephan et al. (1992), Barton (1995a, 1998, 2000), Otto and Barton (1997), Durrett and Schweinsberg (2004), Etheridge et al. (2006), Pfaffelhuber et al. (2006), Pfaffelhuber and Studeny (2007), Ewing et al. (2011), and Coop and Ralph (2012).

The details for a deterministic analysis accounting for the change in A are given in Example 8.2. Under this analysis, if $p(0)$ is the starting frequency of A at the time of selection, then for $c/s \ll 1$, the change in q at the fixation of A is

$$\Delta_q \simeq \delta_q(0)\, p(0)^{c/s} \tag{8.3a}$$

so as a result, from Equation 8.1d,

$$f_s = \frac{\Delta_q}{\delta_q(0)} = p(0)^{c/s} \tag{8.3b}$$

This result shows that our approximate analysis leading to Equation 8.2c underestimated the effect of a sweep, as it scaled as the power $2c/s$, inflating the importance of recombination (Example 8.2 details why this occurs). Recalling that

$$x^a = e^{a\ln(x)} \simeq 1 + a\ln(x) \qquad \text{for} \quad |a\ln(x)| \ll 1 \tag{8.3c}$$

and applying this approximation to Equation 8.3b recovers the original result of Maynard Smith and Haigh (1974),

$$f_s \simeq 1 + \frac{c}{s}\,\ln\left[\,p(0)\right] \tag{8.3d}$$

$$\simeq 1 - \frac{c}{s}\,\ln(2N) \quad \text{for} \quad p(0) = \frac{1}{2N} \tag{8.3e}$$

This approximation can fail if $(c/s)\,|\ln\left[\,p(0)\right]|$ is too large, although the more exact Equation 8.3b will still hold. If the favorable mutation is initially linked to deleterious alleles at nearby loci, this will result in a lower net selective advantage during the sweep (diminishing the effective value of s), and hence a diminished value of f_s (Hartfield and Otto 2011; Good and Desai 2014).

As Equation 8.3e shows, the hitchhiking effect for a favorable mutation introduced as a single copy diminishes with increasing population size, reflecting the longer time to reach fixation in larger populations and hence a greater reduction of any initial association by recombination. This effect, however, is rather modest, scaling as the log of population size.

When dominance is present, so that the fitnesses are $1 : 1+2hs : 2s$, then c/s in Equations 8.3a–8.3d is replaced by $c/(2hs)$ for $h \neq 0$. For the case of a completely recessive allele ($h = 0$), Maynard Smith and Haigh (1974) found that

$$f_s \simeq 1 - \left(\frac{c}{2s}\,\frac{1}{p(0)}\right) \tag{8.3f}$$

For example, with $p(0) = 1/(2N)$, Equation 8.3f becomes $1 - (c/s)N$ versus $1 - (c/s)\ln(2N)$ (Equation 8.3d). The much weaker hitchhiking effect (smaller value of f_s) reflects the much longer fixation time for a recessive mutation and hence the greater opportunity for recombination to decay any initial disequilibrium. Conversely, the decreased fixation time for a favorable dominant allele ($h = 1$) effectively doubles the strength of selection (with $c/[2s]$ replacing c/s in Equation 8.3a), resulting in a larger region influenced by the sweep (also see Teshima and Przeworski 2006; Ewing et al. 2011).

These results are for a deterministic analysis. When an analysis allowing for drift is performed, using the initial frequency $1/(2N)$ for a single copy *underestimates* the effects of hitchhiking, as those alleles that become fixed leave the drift-dominated boundary region of the allele-frequency space *faster* than would be predicted by a deterministic analysis (Example 8.1). As a reasonable approximation, this can be accommodated by replacing $p(0) = 1/(2N)$ with $p(0) = 1/(4N_e s)$ in all of the previous expressions (as already seen in Equation 8.2c), yielding

$$f_s \simeq 1 - \frac{c}{2hs}\ln(4N_e s) \quad \text{for } p(0) = 1/(2N) \tag{8.3g}$$

For those who wish a more refined analysis, there is a growing body of very technical literature focusing on the genealogical structure of a sample from a hard sweep (Kaplan et al. 1989; Barton 1998; Etheridge et al. 2006; Pfaffelhuber et al. 2006; Pfaffelhuber and Studeny 2007; Ewing et al. 2011).

Example 8.1. What is the expected time to fixation for an allele with additive fitness effects under strong selection ($4N_e s \gg 1$)? In a strictly deterministic analysis, the fixation time is infinite, as the allele frequency gets arbitrarily close to, but never actually reaches, one. However, in a finite population, once the allele frequency is driven sufficiently close to one by selection, it is rapidly fixed by drift. If the scaled strength of selection is large relative to drift ($4N_e s \gg 1$), we can approximate the change in $p(t)$ by a deterministic process, provided p is not too close to zero or one. Near these boundary values, drift determines the dynamics. Hence, a standard approach is to treat $p(t)$ as a deterministic process when it is in the range $\epsilon < p < 1 - \epsilon$, for $\epsilon \ll 1$ (Kurtz 1971; Norman 1974; Kaplan et al. 1989; Stephan et al. 1992). Once the allele reaches a frequency of $1 - \epsilon$, it is assumed to be quickly fixed by drift, and this additional time is assumed to be small and thus is ignored.

Let $p(t)$ denote the frequency of the favored allele A at time t. If s is small (but $4N_e s$ large), the deterministic allele-frequency dynamics are well approximated by Equation 5.3b, which can alternatively be expressed as

$$\frac{p(t)}{1 - p(t)} = \frac{p(0)}{1 - p(0)} e^{st} \tag{8.4a}$$

The time, τ, for the frequency of A to change from $p(0) = \epsilon$ to $p(\tau) = 1 - \epsilon$ is obtained by substituting these values into Equation 8.4a and rearranging to yield

$$\tau = -2\ln(\epsilon)/s \tag{8.4b}$$

Taking $\epsilon = 1/(2N)$, the required time starting from a single copy to reach a frequency very close to one ($1 - 1/[2N]$) is

$$\tau \simeq -2\ln(1/[2N])/s = 2\ln(2N)/s \tag{8.4c}$$

While Equation 8.4c appears often in the literature, it actually *overestimates* the time to fixation in a finite population (and hence *underestimates* the strength of the sweep). A slightly better approximation follows if we first recall that only a fraction, $2sN_e/N$, of single introductions of A are expected to fix (Chapter 7). *Conditioned* upon those paths where A is fixed, its frequency must increase at a faster rate than would be predicted from the deterministic analysis. Barton (1995a, 2000; Otto and Barton 1997) showed that, in such cases, the rate of increase in allele frequency is initially inflated (multiplied) by an amount of $1/(2sN_e/N)$, as the allele frequency moves more quickly into the region where deterministic dynamics, not drift, dominate. As a result, they showed that a more accurate estimate of the time for an allele to reach a high frequency (essentially becoming fixed), given it started as a single copy, is given by replacing $\epsilon = 1/(2N)$ by

$$\epsilon = \frac{1}{2N}\frac{N}{2sN_e} = \frac{1}{4N_e s}$$

which results in

$$\tau = 2\ln(4N_e s)/s \tag{8.4d}$$

The standard finite-population-size correction for hitchhiking models starting with $p(0) = 1/(2N)$ is to replace $2N$ by $4N_e s$ (i.e., take $p[0] = 1/[4N_e s]$) to account for this effect.

Finally, the expected time, t_x, to reach a frequency of $x < 1$ appears in the analysis of partial sweeps (Coop and Ralph 2012). Again, starting at an initial frequency of $\epsilon \simeq 0$, Equation 8.4a yields

$$\frac{x}{1-x} = \epsilon\, e^{st_x}$$

which, upon rearranging (with $\epsilon = 1/[4N_e s]$), yields

$$t_x = \frac{1}{s}\ln\left(\frac{\epsilon^{-1}x}{1-x}\right) = \frac{1}{s}\ln\left(\frac{4N_e sx}{1-x}\right) \tag{8.4e}$$

Example 8.2. To obtain the final change, $\Delta_q = q(\infty) - q(0)$, in the frequency of a neutral linked allele under a deterministic model of hitchhiking, we follow Barton (2000). Letting $q'(t)$ and $p'(t)$ denote the frequencies of alleles m and A in generation t after selection, and recalling Equation 8.1a, we can express the change in q in generation t by selection (but before recombination). Letting $dp(t) = p'(t) - p(t)$, the change in the frequency of A resulting from selection in generation t,

$$\begin{aligned} dq(t) &= q'(t) - q(t) = \{p'(t)q_A(t) + [1-p'(t)]q_a(t)\} - \{p(t)q_A(t) + [1-p(t)]q_a(t)\} \\ &= [p(t) + dp(t)]q_A(t) + [1-p(t) - dp(t)]q_a(t) - \{p(t)q_A(t) + [1-p(t)]q_a(t)\} \\ &= dp(t)\,[q_A(t) - q_a(t)] \end{aligned}$$

Using Equation 8.2a to substitute for $[q_A(t) - q_a(t)]$, we have

$$dq(t) = dp(t)\delta_q(0)\,e^{-ct}$$

The final frequency is just the sum of all these single-generation changes, which we approximate by an integral,

$$\Delta_q = q(\infty) - q(0) = \sum_{i=0}^{\infty} dq(i) \simeq \int_0^\infty dq(t)\,dt = \delta_q(0)\int_0^\infty dp(t)\,e^{-ct}dt$$

A more accurate result follows if we note that $dp(t) = \Delta p/\Delta t \simeq dp/dt$, yielding

$$\Delta_q = \delta_q(0)\int_0^\infty e^{-ct}\frac{dp}{dt}dt = \delta_q(0)\int_{p(0)}^1 e^{-ct}dp$$

where the final integral follows by a change of variables with $p(\infty) = 1$. The trick to evaluating this last integral is to recall Equation 8.4a for the behavior of a selected allele, and note that $1 - p(0) \simeq 1$ (because $p(0) \ll 1$), yielding

$$\frac{p(t)}{1-p(t)} = \frac{p(0)}{1-p(0)}e^{st} \simeq p(0)\,e^{st}$$

Rearranging yields

$$p(0)\,\frac{1-p(t)}{p(t)} = e^{-st}$$

Noting that $e^{ab} = (e^a)^b$, we can write

$$e^{-ct} = \left(e^{-st}\right)^{c/s} = \left(p(0)\,\frac{1-p(t)}{p(t)}\right)^{c/s} = p(0)^{c/s}\left(\frac{1-p(t)}{p(t)}\right)^{c/s}$$

which, upon substitution into the previous integral, yields

$$\Delta_q = \delta_q(0)\int_{p_o}^{1} e^{-ct}dp = \delta_q(0)\,p(0)^{c/s}\int_{p_o}^{1}\left(\frac{1-p}{p}\right)^{c/s} dp$$

For $c/s < 0.1$, the integral is close to one, and we recover Equation 8.3a, $\Delta_q = \delta_q(0)\,p(0)^{c/s}$. Barton (Otto and Barton 1997; Barton 1998) showed that a more accurate result is given by

$$\Delta_q \simeq \delta_q(0)\,p(0)^{c/s}\,\{\Gamma\,[1+(c/s)]\}^2\,\Gamma\,[1-(c/s)] \tag{8.5a}$$

where Γ denotes the gamma function (Equation 2.25b). For $c/s < 0.1$, this reduces to

$$\Delta_q \simeq \delta_q(0)\,\left(1+\frac{c}{s}\,[\ln[p(0)]+0.5772]\right) \tag{8.5b}$$

which offers a slight improvement over Equation 8.3d, but only when $p(0)$ is not very small.

Reduction in Genetic Diversity Around a Sweep

By fixing any neutral alleles completely linked to the selected site (and usually on a faster time scale than the scale by which mutation can introduce new variation), a sweep removes any initial variation that is present over some region. Two central questions are how large a region is influenced by a recently completed sweep and how strong an effect is seen? To address these questions, we continue to assume that any effect from mutation occurring *during* the sweep can be ignored (a point that we will address shortly). The first treatment of this topic, and one of the more widely cited results on sweeps, is due to Kaplan et al. (1989). They showed that the expected coalescence time for two alleles at a neutral site that is linked (at recombination fraction c) to the site under selection differs significantly from $2N_e$ (the neutral value) when $c/s < 0.01$ and the sweep has been recent (fixation less than $0.2N_e$ generations ago, so that the effects of new mutations following fixation are negligible). This leads to their often-quoted approximation that ***neutral sites within 0.01 s/c of a selected site will be significantly influenced by a recent sweep.***

To express this notion in terms of physical distances, let c_0 be the recombination rate per unit distance (for example, per Mb), so that sites separated by ℓ units have a recombination rate of $c = c_0\ell$. The expected total length L (scaled in our chosen units) of depressed variation associated with a recent sweep then becomes

$$L = 0.02\,\frac{s}{c_0} \tag{8.6a}$$

where the factor of two arises because the influence extends on both sides of the sweep. If we suppose that c scales as 1 cM/Mb ($c_0 = 0.01$ for each 10^6 bases), then this approximation implies that a recent sweep with a selection coefficient of $s = 0.01$ is expected to influence variation in a region of size $0.02 \cdot (0.01/0.01) = 0.02$ Mb, or roughly 20 kb (Example 8.3 gives a more refined result). Likewise, a selection coefficient of $s = 0.1$ leaves an initial signature over a region of roughly 200 kb.

Rearrangement of Equation 8.6a can be used to obtain a rough estimate of s. Given a total length, L, of decreased variation and a value of c_0 for this interval,

$$s \simeq \frac{c_0 \cdot L}{0.02} \tag{8.6b}$$

For example, if a sweep roughly covers 50 kb (or $L = 0.05$ Mb) in a region where $c_0 = 0.02$ (2 cM/Mb), then an order-of-magnitude approximation for s is

$$s \simeq \frac{0.02 \cdot 0.05}{0.02} = 0.05$$

This is a crude approach, requiring a reasonable estimate of the size of the region influenced by a very recently completed sweep. Further, simulation studies have shown that ***sweeps can be asymmetric around the site under selection*** (Kim and Stephan 2002; Schrider et al. 2015), reflecting the random location of rare recombination events between markers and the selected site that occur early in a sweep. Simply taking the middle of a region of depressed variation can be a poor approach for localizing the site under selection.

A more accurate expression for the amount of variation remaining after a very recent sweep follows from the expected allele-frequency change (Equation 8.3a). Let $q(0)$ denote the initial frequency of allele m at a linked neutral marker, with $H_0 = 2q(0)[1 - q(0)]$ denoting the initial heterozygosity, which is typically measured as per-nucleotide diversity, π (Chapters 2 and 4). Hitchhiking during the fixation of a linked selected allele changes a marker allele frequency to $q(\infty) = q(0) + \Delta_q$, and hence the heterozygosity becomes

$$\begin{aligned} H &= 2q(\infty)[1 - q(\infty)] = 2[q(0) + \Delta_q][1 - (q(0) + \Delta_q)] \\ &= H_0 - 2[1 - 2q(0)]\Delta_q - 2\,(\Delta_q)^2 \end{aligned} \tag{8.7a}$$

The expected heterozygosity is the average of H over two scenarios. With a probability of $q(0)$, the favorable mutation arises on an m background, yielding $q_A(0) = 1$, $\delta_q(0) = 1 - q(0)$ and $\Delta_q \simeq [1 - q(0)]\,p(0)^{c/s}$ (Equation 8.3a). Conversely, with a probability of $1 - q(0)$, the favorable allele arises on a non-m background, yielding $q_A(0) = 0$, $\delta_q(0) = 0 - q(0) = -q(0)$ and $\Delta_q \simeq -q(0)\,p(0)^{c/s}$. Using these results, the expected allele-frequency change is

$$\text{E}(\Delta_q) = q(0) \cdot [1 - q(0)]\,p(0)^{c/s} + [1 - q(0)] \cdot \left[-q(0)\,p(0)^{c/s}\right] = 0 \tag{8.7b}$$

Taking the expectation of Equation 8.7a and using Equation 8.7b yields

$$H_h = \text{E}(H) = H_0 - 2\text{E}\left[\Delta_q^2\right] \tag{8.7c}$$

where

$$\begin{aligned} \text{E}\left[\Delta_q^2\right] &= q(0)\left[[1 - q(0)]p(0)^{c/s}\right]^2 + [1 - q(0)]\left[-q(0)p(0)^{c/s}\right]^2 \\ &= q(0)[1 - q(0)]p(0)^{2c/s} \end{aligned} \tag{8.7d}$$

Combining Equations 8.7c and 8.7d yields

$$H_h = H_0\left(1 - p(0)^{2c/s}\right) \tag{8.8a}$$

Recalling that this result is an approximation (as Equation 8.3a approximates the allele frequency change), our final result is

$$\frac{H_h}{H_0} \simeq 1 - p(0)^{2c/s} \simeq -\frac{2c}{s}\,\ln[p(0)] \quad \text{for} \quad c/s \ll 1 \tag{8.8b}$$

Following from our previous discussion, for a sweep starting from a single mutation, we can improve on Equation 8.8b by replacing $p(0) = 1/2N$ by $1/(4N_e s)$, yielding

$$\frac{H_h}{H_0} \simeq 1 - (4N_e s)^{-2c/s} \tag{8.8c}$$

Stephan et al. (1992) and Barton (1998) presented more accurate (and complex) expressions for the reduction in heterozygosity in a finite population.

An alternative way to obtain Equation 8.8b is to consider the fraction, f_s, of the initial associations that persist when A is fixed (Equation 8.1c), as with a probability of f_s^2, neutral alleles at our site for two randomly drawn chromosomes (under a catastrophic sweep) are identical-by-descent and hence (in the absence of mutation) homozygous. With a probability of $1 - f_s^2$, two randomly drawn alleles are not identical-by-descent (from the sweep), displaying the pre-sweep heterozygosity value. Hence, from Equation 8.3b, which assumes additive selection,

$$\frac{H_h}{H_0} = 1 - f_s^2 = 1 - p(0)^{2c/s} \tag{8.8d}$$

When dominance is present (implying a heterozygote fitness of $1+2hs$ instead of $1+s$), Equation 8.8b holds, with $2hs$ replacing s (for $h > 0$). For a complete recessive ($h = 0$, with fitnesses of $1 : 1 : 1 + 2s$), Ewing et al. (2011) showed that

$$\frac{H_h}{H_0} \simeq \frac{\eta}{1+\eta} \quad \text{where} \quad \eta = c\sqrt{4N_e/s} \tag{8.8e}$$

As expected, a recessive sweep produces a much weaker signal, reflecting the greater chance for recombination given the much longer time to fixation ($\sim\sqrt{N_e/s}$ generations; Ewing et al. 2011). It is important to stress that Equations 8.8d and 8.8e both refer to the reduction in heterozygosity *immediately following* a sweep. This is the maximal signature, as mutation will gradually rebuild neutral variation, an effect we will examine shortly.

Finally, we can examine the accuracy of Kaplan and Hudson's approximation (Equation 8.6a), which states that a sweep influences all sites with $c/s \leq 0.01$, using Equation 8.8b to find the reduction in heterozygosity for sites at this critical value. Assuming there is a single copy of the selected allele at the start of selection, $p(0) = 1/(2N)$, the simple approximation given by Equation 8.8b (with $c/s = 0.01$) yields

$$\frac{H_h}{H_0} = \frac{2c}{s}\ \ln(2N) = 0.02\ \ln(2N) \tag{8.9}$$

The dependence on N is very weak. For example, for $N = 10^4$, Equation 8.9 gives an 80% reduction ($H_h/H_0 = 0.20$), while for $N = 10^9$, it corresponds to a 57% reduction. A more refined analysis can be performed using the drift-corrected expression Equation 8.8c, with $4N_e s$ replacing $2N$.

Example 8.3. Assume a recombination rate of 1 cM/Mb (c_0 = 0.00001 per kb), and consider the expected reduction in heterozygosity at a site 10 kb away from a sweep ($c = 10 \cdot 0.00001 = 0.0001$). For an allele with additive fitness effects, with $s = 0.01$ and $N_e = 10^6$, Equation 8.8b yields $H_h/H_0 \simeq 0.19$, so that (ignoring any new mutation) only 19% of the initial heterozygosity is present immediately following a sweep. For a dominant allele, we replace $s = 0.01$ by $2s = 0.02$ in Equation 8.8b, giving $H_h/H_0 \simeq 0.10$. Finally, when the favored allele is recessive,

$$\eta = c\sqrt{4N_e/s} = 0.0001\sqrt{4 \cdot 10^6/0.01} = 2$$

and Equation 8.8c yields $H_h/H_0 \simeq 0.67$. Using the same parameters, the values for H_h/H_0 at different distances away from the selected site are as follows:

	1 kb	5 kb	10 kb	25 kb	50 kb	100 kb
Dominant	0.01	0.05	0.10	0.23	0.41	0.65
Additive	0.02	0.10	0.19	0.41	0.65	0.88
Recessive	0.17	0.50	0.67	0.83	0.91	0.95

The sweep from a dominant allele has the largest effect (roughly twice the reduction at small distances compared to additive selection), while the effect of a recessive allele is fairly weak except at very short distances from the site. For these three modes of gene action and $s = 0.01$, a 50% reduction ($H_h/H_0 = 0.5$) in heterozygosity occurs over a distance of 5 kb on either side of a selected recessive allele, the distance is 31 kb when additive, and the distance is 66 kb when dominant, giving the size of the sweep regions as 10, 62, and 132 kb, respectively.

The Messer-Neher Estimator of *s*

As Equation 8.6b illustrates, one can manipulate any of these expressions for the reduction in H to obtain an estimate of s (given H_0 and c). In Chapter 9, this approach is placed into a more sophisticated likelihood framework where the entire spatial pattern of genetic variation (as a function of the distance, c, from a putative site) is used to estimate s. These approaches all require knowledge of the recombination rate, c. How can we estimate s in situations with little or no recombination, such as in an organelle genome, on a very small chromosome, or in a highly inbred or asexual species? A creative solution to this problem was presented by Messer and Neher (2012).

As a favored allele exponentially increases in frequency during its initial sojourn to fixation, new mutations can appear on its initial haplotype. While rare, these mutations will mostly be at a higher frequency than those arising after fixation, as each hitchhikes up to some modest frequency. By considering a region around a putative selected site and counting the resulting haplotypes following a recently completed sweep, Messer and Neher obtained a very simple approximate relationship between the frequencies of the ranked haplotype classes. Letting n_0 denote the number of the most frequent haplotype, n_1 the next frequent, and so forth, they found that

$$\frac{n_i}{n_0} \simeq \left(\frac{\mu}{i\,s}\right)^{1-(\mu/s)} \simeq \frac{\mu}{i\,s} \tag{8.10}$$

namely, a power-law relationship as a function of the total mutation rate, μ, for the region being considered and the strength, s, of the favored allele. For example, for $\mu/s = 0.02$, n_0 is 50 and 100 times more frequent, respectively, then n_1 and n_2. Based on the relationship given by Equation 8.10, Messer and Neher developed a regression-based estimator of s.

While this approach works on genomic regions showing little to no recombination (the authors applied it to HIV), it also requires deep population sequencing, as the higher mutational classes (n_i for $i > 2$) are rare. For example, for $\mu/s = 0.1$, this approach requires accurate estimates of frequencies with an expected value around 2% (assuming a regression fitted with four values, and hence requiring the class n_4), implying a sample size of around 10^3 sequences. While c (the recombination rate over the chosen haplotype length) does not appear in Equation 8.10, Messer and Neher found that simply replacing μ by $\mu+c$ works well if the ancestral diversity is high but overestimates the rate of formation of new haplotypes if this diversity is low. They found that simply pruning the collection of haplotypes to ignore those that were clearly generated by recombination slightly underestimates s but otherwise performs rather consistently.

Recovery of Variation Following a Sweep

The signal left by even a strong sweep is transient, as new mutations will eventually restore heterozygosity at neutral sites back to their equilibrium value ($H_0 = 4N_e\mu$) before the sweep. Kim and Stephan (2000) found that the expected heterozygosity t generations after a completed sweep is approximately

$$E[\,H(t)\,] \simeq H_0\left[1-(4N_e s)^{-2c/s}\cdot e^{-t/(2N_e)}\right] \tag{8.11}$$

where $H_0(4N_e s)^{-2c/s} = H_0\, f_s^2$ is the reduction immediately following the conclusion of the sweep. Mutation following a cleansing sweep recovers variation, and this can be envisioned as a decay in the initial reduction, $H_0 f_s^2$, eventually driving this back to zero (and hence driving heterozygosity back to its original value, H_0). From Equation 8.11, the initial reduction decays by an amount of $1/(2N_e)$ in each generation, with the cumulative decay of this initial variation due to drift being $(1-1/2N_e)^t \simeq \exp(-t/2N_e)$. The expected time to recover half the variation lost during the sweep (its half-life) is $\exp(-t_{0.5}/2N_e) = 0.5$ or $t_{0.5} = -2\ln(0.5)N_e \simeq 1.4N_e$. Note the counterintuitive and important result that the ratio $E[\,H(t)\,]/H_0$ is *independent* of the actual mutation rate, μ. This is because a low (or high) mutation rate implies a slow (or fast) accumulation of new mutations following the sweep but also implies a low (or high) target heterozygosity to reach that point.

Effects of Sweeps on the Variance in Microsatellite Copy Number

These results on the behavior of nucleotide diversity (heterozygosity) during and after a sweep apply to SNP data. Because per-nucleotide mutation rates are very low (Chapter 4), the infinite-sites model offers a good approximation for such data, as recurrent neutral mutation during the sweep can largely be ignored because back mutations are unlikely and mutations are rare in general. Both of these assumptions are violated when microsatellite markers (simple tandem repeats [STRs]) are considered. These have high mutation rates (on the order of 10^{-2} to 10^{-4}), and recurrent mutation can regenerate the same allele (scored as the number of repeats at a site). A further complication is that a commonly used measure of STR variability is not heterozygosity, but rather the variance, V, in repeat number among alleles at the microsatellite marker.

Given the considerable uncertainty about the mutational dynamics of STRs, as well as a wealth of potential models (reviewed by Bhargava and Fuentes 2010), we only consider briefly results from the simple stepwise mutation model (an STR allele of length k has equal probability of changing to length $k+1$ or $k-1$). Under this model, Wiehe (1998) found that if V_0 denotes the initial variance in repeat number, its expected value, V_h, immediately following the sweep has a very similar form to Equation 8.8b

$$\frac{V_h}{V_0} = 1-\beta\cdot p(0)^{2c/s} \tag{8.12a}$$

the difference being a scaling factor of $\beta < 1$, which discounts the removal of variation by the sweep from the continual input from new mutation. Note that even at completely linked sites ($c=0$), the fraction of variance after a sweep is not zero, but rather $1-\beta$, due to mutational variation being generated during the sweep event itself. Under a simple model where the total mutation rate scales with allele length ($k\mu$ is the rate of an allele containing k repeats), Wiehe found that

$$\frac{V_h}{V_0} = 1-p(0)^{(4\mu+2c)/s} \tag{8.12b}$$

When $4\mu+2c > s$, little depression in the copy-number variance is expected. In this setting, mutation rates are such that new STR alleles are generated at a high rate even as the sweep is occurring, so that even if only the descendants of the single original haplotype are fixed, they will still show significant variation. For example, given an STR mutation rate of $\mu = 1/400$, s must exceed 0.01 for any significant sweep signal. Hence, STR-based scans are likely to miss all but the strongest of sweeps.

Using results from Slatkin (1995b), the rate of recovery in V following the sweep is a modification of Equation 8.11

$$V(t) = V_0\left(1-p(0)^{(4\mu+2c)/s}\cdot e^{-t/(2N_e)}\right) \tag{8.12c}$$

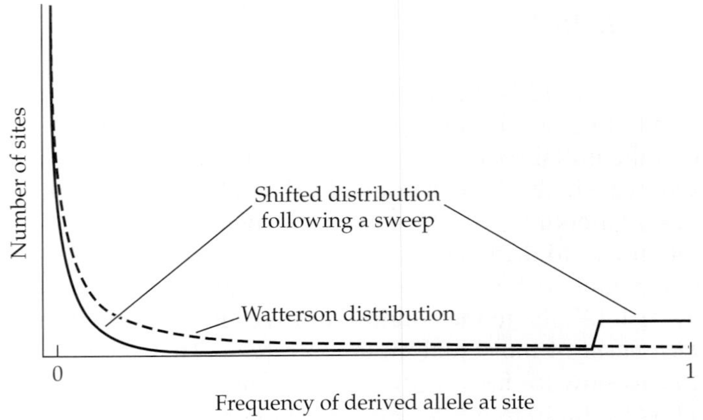

Figure 8.5 The effect of a hard sweep on the unfolded site-frequency spectrum of derived alleles. Under the equilibrium neutral model, this distribution is hyperbolic, an L-shaped curve that is monotonically declining, with most derived alleles being at low frequencies (the Watterson distribution; Equation 2.34a). The effect of a sweep is to shift some derived alleles to very high frequencies while shifting the others to frequencies near zero, resulting in a more U-shaped distribution (Equation 8.13).

As with Equation 8.11, $t_{0.5} \simeq 1.4N_e$ generations are required to recover half the decrease in V immediately following the sweep. It is often stated that microsatellites recover variation faster because they quickly generate new mutations *following* a sweep. This is incorrect, as the high level of variation is due to mutations arising *during* the sweep, not following it. Equation 8.12c again shows that the time to recover variation following the sweep (the time to decay the reduction present immediately following the sweep) is independent of the mutation rate.

The Site-Frequency Spectrum

Recall the concept of a frequency spectrum (Chapter 2), the expected distribution of the frequencies of different alleles or sites in a sample. In particular, the site-frequency spectrum (SFS), $\phi(x)$, gives the expected frequency of sites having a frequency of x for the derived (most recent) allele, and our focus is on how a sweep changes the SFS for linked, neutral alleles. Under the equilibrium neutral model, the SFS is given by the Watterson distribution of $\phi(x) = \theta/x$ (with $\theta = 4N_e\mu$ referring to the expected per-nucleotide heterozygosity value; Equation 2.34a), with most sites having very low derived-allele frequencies. As shown in Figure 8.5, a sweep transforms the (unfolded) SFS of derived alleles from the L-shaped Watterson distribution to a more U-shaped one (Fay and Wu 2000; Kim and Stephan 2002), with ***excesses of both low- and high-frequency sites***. When considered as a folded frequency spectrum (the spectrum over 0 to 0.5 based on the minor-allele frequency; Chapter 2), these excesses result in an increase in the fraction of sites with rare minor allele frequencies. Przeworski (2002) showed that both of these excesses are present immediately following a sweep, but that the excess of sites with high-frequency derived alleles rapidly dissipates (within $0.2N_e$ generations) as they become fixed. The excess of sites with rare alleles persists a bit longer (roughly $0.5N_e$ generations), as it is bolstered by new mutations following the sweep.

As detailed in Example 8.4, Fay and Wu (2000) found that a sweep transforms a pre-sweep Watterson distribution, $\phi(x)$, into a new post-sweep SFS, $\phi^*(y)$, which is a function of the critical parameter, f_s, for the impact of a sweep (Equation 8.1d), with

$$\phi^*(y) = \begin{cases} \theta\left(\dfrac{1}{y} - \dfrac{1}{1-f_s}\right) & \dfrac{1}{2N} \le y \le 1 - f_s \\ 0 & 1 - f_s < y < f_s \\ \dfrac{\theta}{1-f_s} & f_s \le y \le 1 - \dfrac{1}{2N} \end{cases} \tag{8.13}$$

These expressions show how a sweep distorts the frequency spectrum toward both higher and lower frequencies of derived alleles. Linked neutral alleles are swept toward fixation (inflating the high-frequency range) if they are on the original haplotype of the new favorable mutation (i.e., in coupling with the favorable allele). If x is the frequency of the derived neutral marker allele, then x is the probability it is coupled with the favorable allele and changed to a new (post-sweep) frequency of $y = f_s + x(1 - f_s)$ (Equation 8.1e). Otherwise, when not on the favorable haplotype (in repulsion), which occurs with probability $1 - x$, the linked neutral allele is swept toward loss (a new frequency of $y = x[1 - f_s]$).

If two *concurrent* sweeps are influencing the same region, the resulting site-frequency spectrum is rather different from the pattern for a single hard sweep. Simulations by Chevin et al. (2008) found an *excess of intermediate-frequency alleles* in such cases, mimicking the signature of balancing selection. However, concurrent sweeps also generate both an excess of high-frequency derived alleles and a deficiency of low-frequency alleles. The combination of these three features seems unique to concurrent sweeps.

Example 8.4. The derivation of Equation 8.13 follows in large part from Equation 8.1e, the expected change in marker allele frequency following a sweep for a given value of f_s. Let x denote the frequency of a linked neutral derived allele before the start of selection and y its frequency immediately following the sweep. We assume a pre-sweep Watterson SFS, $\phi(x) = \theta/x$ (Equation 2.34a). Given a single favorable mutation occurring at random on one of the existing chromosomes, then if x is the frequency of a linked neutral allele, with a probability of x, then the allele is in coupling phase with the new mutation; otherwise (with a probability of $1 - x$), it is in repulsion phase.

First consider a neutral derived allele (at a frequency of x before the sweep) that is in repulsion with the favorable allele. From Equation 8.1e, the post-sweep frequency is $y = x(1 - f_s)$, for which the range is obtained as follows. For derived alleles that are initially near fixation ($x \simeq 1$), $y = (1 - f_s)$, while rare alleles are driven toward zero. Because our focus is on segregating alleles, the lower limit becomes $1/(2N)$, giving the post-sweep frequency range for neutral markers in repulsion with the favorable allele as $1/(2N) \le y \le 1 - f_s$. The resulting SFS for derived neutral alleles in repulsion is given by

$$\phi^*(y)dy = (1 - x)\phi(x)dx = (1 - x)\frac{\theta}{x}dx$$

The first step follows because $(1 - x)$ is the probability that a neutral allele (initially at a frequency of x) is in repulsion with the favorable allele, and the second step follows from Equation 2.34a. Recalling (for alleles in repulsion) that $y = x(1 - f_s)$, we have both $x = y/(1 - f_s)$ and $dx = dy/(1 - f_s)$. Substituting for x and dx in this expression yields

$$(1 - x)\frac{\theta}{x}dx = \left(1 - \frac{y}{1 - f_s}\right)\frac{\theta(1 - f_s)}{y}\frac{dy}{1 - f_s}$$

which, after multiplication, yields the repulsion piece of the post-selection SFS as

$$\phi^*(y)dy = \theta\left(\frac{1}{y} - \frac{1}{1 - f_s}\right)dy \quad \text{for} \quad 1/(2N) \le y \le 1 - f_s$$

Conversely, with a probability of x, the derived allele is in coupling phase and is swept to a frequency of $y = f_s + x(1 - f_s)$. For $x \simeq 0$, $y \simeq f_s + 0(1 - f_s) = f_s$, while the upper limit follows from our restriction to segregating sites, yielding $f_s \le y \le 1 - 1/(2N)$. Following the same logic as before yields

$$\phi^*(y)dy = x\phi(x)dx = x\frac{\theta}{x}dx = \theta dx$$

As before, the first step follows because x is the probability that a neutral allele (initially at a frequency of x) is in coupling with the favorable allele. With an allele in coupling, $y = f_s + x(1 - f_s)$, yielding $x = (y - f_s)/(1 - f_s)$ and $dx = dy/(1 - f_s)$. Substituting for dx into the previous expression yields the coupling piece of the post-sweep SFS as

$$\phi^*(y)dy = \frac{\theta}{1 - f_s}dy \quad \text{for} \quad f_s \le y \le 1 - 1/(2N)$$

Putting these two pieces together recovers Equation 8.13.

The Pattern of Linkage Disequilibrium

The pattern of linkage disequilibrium (LD) generated by a sweep has been extensively studied (Thomson 1977; Gillespie 1997; Przeworski 2002; Kim and Nielsen 2004; Stephan et al. 2006; Jensen et al. 2007; McVean 2007; Jones and Wakeley 2008; Pfaffelhuber et al. 2008), and it turns out to be both complicated and surprising. The conventional wisdom was that a selective sweep increases LD across the site of selection (Thomson 1977; Przeworski 2002), with the increase in LD *during* a sweep providing a signal of selection (Chapter 9). Starting with Kim and Nielsen (2004), however, it has been realized that the spatial and temporal patterns in LD associated with a sweep are far more subtle.

The temporal dynamics of this structure are shown in Figure 8.6, which plots the dynamics of LD for two linked neutral sites separated by $c = 0.0002$. The midpoint between the sites is set at zero, with the value x representing the position of the selected locus, so that Figure 8.6 displays the expected LD between this pair of neutral sites at different distances from the selected site. For $|x| > 0.0001$, both markers are on the same side of the selected site, while for $|x| < 0.0001$, the selected site is between the markers. Vertical slices on the curve (the shape of the curve at a set time point) represent the expected LD (measured as D; Equation 2.18) between the two markers as a function of x for a given time point during the sweep. As shown, during the initial phase of the sweep (proceeding from the rear-most slices), strong LD accrues between sites that flank the selected site as the sweep progresses, with LD initially showing a peak at $x = 0$. As the sweep progresses (i.e., as slices move toward the front), the LD for a pair of markers flanking the selected site ($|x| < 0.0001$) approaches zero, while the LD between sites that are both on the same side of the selected site ($|x| > 0.0001$) largely remains intact. The plot in Figure 8.6 is based on a deterministic analysis of a three-locus model (one selected and two neutral) by Stephan et al. (2006). As such, it depicts a very smooth and symmetric view of the LD on either site of the selected site, representing the *average* behavior over a large number of identical sweeps. In reality, however, there is considerable variance in the amount of LD due to finite population size, the stochastic location of rare recombination events, and differences in allele frequencies across markers at the start of the sweep. Simulation studies (e.g., Kim and Nielsen 2004; Schrider et al. 2015) have often found a very asymmetric pattern of LD across a selected site, with a strong signal on one side and little to no signal on the other.

Thus, while LD does indeed increase during the early phase of the sweep of a favorable allele to fixation, it actually starts to *decrease* in the neighborhood of the site once the frequency of the favorable allele reaches roughly 0.5 (Stephen et al. 2006). Upon fixation, the result is a region flanking the selected site that has an LD level *lower* than the background level at unlinked neutral loci, and hence potentially reduced from its initial starting value. Conversely, for pairs of markers on either side of the selected site, LD significantly increases, so that strong LD can be found on the left and/or right sides of a selected site, but with

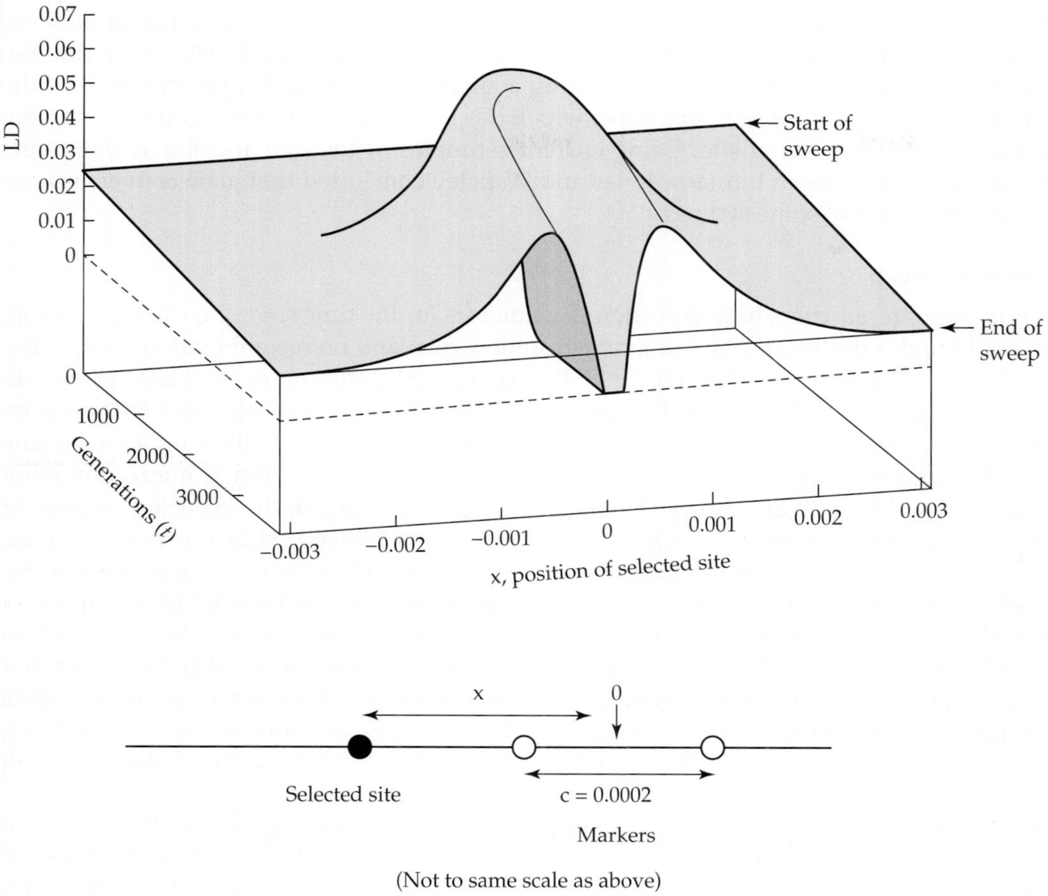

Figure 8.6 The dynamics of linkage disequilibrium (LD) between two neutral markers as a function of their location relative to the selected site. The neutral markers in this figure are separated by $c = 0.0002$, with x (measured in units of c) representing the location of the selected site. A value of $x = 0$ implies that the selected site is exactly between the two markers and it is flanked by the markers provided $|x| < 0.0001$, whereas both markers are on the same side of the site when $|x| > 0.0001$ (see the figure panel). The figure plots the LD dynamics (measured by D) for pairs of markers during the time course of a sweep, which starts at generation 0 (the rear slice along the generations axis) and runs until the conclusion of the sweep (the front slice). One sees a strong signal of LD *across* the site ($|x| < 0.0001$) during the early phase of the sweep (the partial sweep stage), with a sharp increase in LD for markers flanking the selected site. As the favorable allele reaches intermediate frequency, the LD flanking the site starts to decay (the region $|x| < 0.0001$), while the LD on either side largely remains intact ($|x| > 0.0001$). Upon fixation, the result is very little LD between markers that flank the site (often below that in the the starting background), while strong regions of LD exist for pairs of markers on the same side of the selected site. (After Stephan et al. 2006.)

no association *across* the site. These results have strong implications for tests of selection (Chapter 9), in that strong LD across the site is expected during a partial sweep (with the selected allele still increasing in frequency). Conversely, for a recently completed sweep, the LD pattern is very different, with LD between neutral sites on the same side of the selected site but little LD across sites. Thus, different LD-based tests are required for these two different situations.

Jones and Wakeley (2008) suggested some caution with these results, as even a very small amount of gene conversion involving the selected site can disrupt these LD patterns and obscure any LD-based hard-sweep signature. In particular, a gene conversion event involving just the favorable site can result in substantial LD *across* a selected site (i.e., between markers on opposite sides of the site), and hence has the opposite effect of recombination. This happens because a conversion tract that covers the selected site acts like a double-recombinant within that region, transferring a small section of the haplotype around the selected site onto a different haplotype, which has an effect on LD throughout the region, not just at the conversion site. Using current estimates of the rate and the average tract length of conversions in humans, Jones and Wakeley concluded that gene conversion can impact the signal of some sweeps.

Age of a Sweep

A number of researchers have considered estimators for the time since the start of a sweep, typically under the assumption of a catastrophic sweep and no recombination (Wiehe and Stephan 1993; Perlitz and Stephan 1997; Enard et al. 2002; Jensen et al. 2002; Przeworski 2003; Li and Stephan 2005, 2006; Ormond et al 2015). The simplest estimator follows from the infinite-sites model (Chapter 2) and assumes the sweep is essentially complete. Assume that S segregating sites are observed in a sample of n sequences from a nonrecombining region around the site of a sweep. Under the infinite-sites model, the expected number of segregating sites in a sample is $E(S) = \mu T_n$, where μ is the total mutation rate over the entire region of interest and T_n is the total branch length of the entire genealogy of the sample. Under a catastrophic sweep that started τ generations ago, the coalescent tree has its nodes sharply compressed, and can be approximated by a star phylogeny. In this case, the total branch length is $n\tau$ (as the length along each of the n branches is roughly τ), giving $\mu n\tau$ as the expected number of segregating sites, and leading to a simple method-of-moments estimator of the time τ,

$$\widehat{\tau} = \frac{S}{\mu\, n} \tag{8.14}$$

More sophisticated approaches for estimating τ are discussed in Chapter 9.

Example 8.5. Akey et al. (2004) found a 115-kb region on human chromosome 7 showing signatures of a sweep: excess rare alleles, excess high-frequency derived alleles, and a reduction in nucleotide diversity. Eleven segregating sites were found in a sample of 45 African- and European-American chromosomes. From Equation 8.14, using a mutation rate of 10^{-8} per site per generation, the total mutation rate over the entire region is $115{,}000 \cdot 10^{-8} = 0.00115$ per generation, yielding

$$\widehat{\tau} = \frac{11}{0.00115 \cdot 45} = 213 \text{ generations}$$

using the star phylogeny approximation. Assuming a generation time of 25 years for humans, this translates into ~ 5300 years. Example 9.15 shows how confidence intervals are estimated under this model.

Partial Sweeps

An ongoing selection event (i.e., the favored allele is still increasing in frequency), whether toward fixation or to some internal equilibrium frequency, generates a partial sweep signal. Coop and Ralph (2012) considered an additional scenario, wherein a newly arising allele is rapidly swept up to a frequency of x by selection, after which the environment changes, reducing the strength of selection (perhaps even to the point of making the allele effectively neutral). The key parameter is now t_x, the time to reach this frequency, which, from Equation 8.4e (for additive fitnesses and starting from $p(0) = 1/[2N]$), is

$$t_x = \frac{1}{s} \ln\left(\frac{4N_e s x}{1-x}\right)$$

Coop and Ralph found that the reduction in heterozygosity (for a neutral marker at recombination distance c) when the sweep pauses at frequency x is

$$\frac{H_h}{H_0} \simeq 1 - x^2 e^{-2ct_x} = 1 - x^2 \left(\frac{4N_e s x}{1-x} \right)^{-2c/s} \tag{8.15}$$

This follows from the probability of sampling a chromosome carrying the advantageous mutation being x and the probability that a site at distance c has not recombined over the t_x generations required to reach a frequency of x being $(1-c)^{t_x} \simeq e^{-ct_x}$, for a total probability of xe^{-ct_x}. The chance of drawing two such alleles is simply the square of this product, which is the expected homozygosity. Coop and Ralph concluded that the initial rapid rise in frequency of a selected allele during the partial sweep determines how much of a sweep signal is generated, even if the allele is ultimately fixed by drift. With recurrent partial sweeps, such shifts in the strength of selection can result in a reduction in heterozygosity (Equation 8.26) but will produce little shift in the frequency spectrum.

Impacts of Inbreeding

Three separate consequences of inbreeding enhance the signal of a sweep. First, under inbreeding, the effective population size is smaller, with $N_e = N/(1+F)$ (Chapter 3), where $F = \eta/(2-\eta)$ for a partially-selfing (with selfing rate η) population in equilibrium (Chapter 23). Equation 8.3g shows the impact of a sweep (measured by the critical parameter, f_s) is slightly larger in smaller populations, although the effect only scales as $\ln(N_e)$. The reason for this slightly greater impact is a faster time to fixation, and hence less opportunity for recombination. This point highlights the other two features of inbreeding that are more impactful on sweeps. First, selection is generally more efficient, especially when the favored allele is recessive or partly recessive. Indeed, the expected time to fixation is reduced at least by a factor of $(1+F)$ under inbreeding (Caballero and Hill 1992b, Glémin 2012). Inbreeding is also more democratic in that allele-frequency change trajectories under different amounts of dominance become increasingly similar under increased inbreeding (Hartfield et al 2017). Second, because of the reduction in the frequency of heterozygotes, recombination is correspondingly weaker, with an effective recombination rate of $c_e \simeq c(1-F)$ (Nordborg 2000). Together these features imply that the length of the sweep signal under inbreeding should be longer relative to the same selection and recombination parameters under panmixia. This results in easier detection of a sweep, but greater obfuscation in terms of localizing the causative site of the sweep (Hartfield et al 2017).

Summary: Signatures of a Hard Sweep

The key summary parameter for the potential impact of a sweep is the fraction of original haplotypes that stay intact following a sweep, $f_s = \Delta_q/\delta_q(0)$ (Equation 8.1d). If $f_s \simeq 1$, the sweep will have a major impact on the structure of variation at linked neutral sites, while if $f_s \simeq 0$, it will have essentially no impact. Table 8.1 summarizes expressions for f_s and also for the population-genetic impacts of a sweep on a linked neutral site. Table 8.2 summarizes more subtle signatures of a sweep beyond the simple reduction in variation. As detailed in the next chapter, all of the observations listed in Table 8.2, either singularly or in combination, have been used as the basis of tests of ongoing or recent selection. It is important to stress that the results in these two tables are *restricted to hard sweeps*, wherein the favorable allele is only present in (at most) a few copies at the start of selection. As shown below, many of these signals are either muted or washed out entirely under soft sweeps.

SOFT SWEEPS

Whereas a hard sweep starts with selection on a single haplotype, a soft sweep refers to situations where, at the start of the sweep, multiple haplotypes contain the favored allele

Table 8.1 Summary of various features associated with a hard sweep of a favorable allele A with fitnesses $1 : 1+2hs : 1+2s$ (for $h \neq 0$). Let $q(0)$ denote the frequency of a neutral marker allele at the start of selection at a distance (recombination fraction) of c from a strongly selected site ($4N_e s \gg 1$). Assume that the frequency of the favorable allele is $p(0)$ at the start of selection, and let $q(\infty)$ denote the final frequency for a neutral allele initially associated with the favorable allele, and H_h be the heterozygosity at the neutral site immediately following the sweep.

Fraction f_s of initial associations remaining at fixation:

$$f_s \simeq \begin{cases} p(0)^{-c/(2hs)} \simeq 1 - \dfrac{c}{2hs}\ln[p(0)] & \text{for } p(0) \gg 1/(2N_e s) \\ (4N_e s)^{-c/(2hs)} \simeq 1 - \dfrac{c}{2hs}\ln(4N_e s) & \text{for } p(0) = 1/(2N) \end{cases}$$

Total change in the frequency of a linked neutral marker: $\Delta_q \simeq [1 - q(0)]f_s$

Final frequency of a linked neutral marker: $q(\infty) = q(0) + \Delta_q = f_s + q(0)(1 - f_s)$

Reduction in heterozygosity immediately following the sweep: $\dfrac{H_h}{H_0} = 1 - f_s^2$

Heterozygosity t generations after sweep completed: $\dfrac{H(t)}{H_0} = 1 - f_s^2\, e^{-t/(2N_e)}$

Table 8.2 Expected population-genetic patterns associated with a hard sweep.

Signatures of spatial patterns of variation near a selected site:

(1) An excess of sites with rare alleles (in either the folded or unfolded frequency spectrum).

(2) An excess of sites with high-frequency *derived* alleles in the unfolded frequency spectrum.

(3) Depression of genetic variation, often *asymmetrically*, around the site of selection.

Signatures in the spatial pattern of LD differ during the sweep and after its completion:

(4a) When a favorable allele is at moderate frequency (a partial sweep), we see an excess in LD throughout the region surrounding the sweep.

(4b) As the favorable allele approaches fixation, we see an excess in LD on either side of the site, but a *depression* in LD between markers flanking the site.

Signatures of a sweep are *very fleeting*.

(5) Remaining on the order of $0.5N_e$ generations for signature (1), $0.4N_e$ generations for (2), $1.4N_e$ generations for (3), and $0.1N_e$ generations for (4b).

(Figure 8.4). Under a single-origin soft sweep, a single copy of the mutation arose in an environment that did not yet favor it, drifting around before an environmental change placed all of the haplotypes associated with it under positive selection. A response using existing variation, such as a soft sweep or a polygenic sweep, is also referred to as a **standing sweep** (or, more precisely, a **standing-variation sweep**). Under a multiple-origin soft sweep, the favored variant consists of a collection of independent mutational lineages, which arose in the standing variation before the allele became favored and/or *during* the sojourn to fixation for this allele.

A powerful way to understand many of the results presented below is to use time-scale arguments (Messer and Petrov 2013a). For example, the **establishment time** is the expected time for the appearance of a new mutation that is destined to become fixed. If this time is short relative to the time to fix such an allele once it appears (the **fixation time**), then multiple mutations (and hence a soft sweep) are expected. Likewise, in a geographically structured population, if the establishment time within a deme is short relative to the time of spread of favorable alleles between demes, a sweep through the entire population is likely to again involve the contributions of multiple favorable mutations.

Signatures of a Soft Sweep

The effect of a single-origin soft sweep is to diminish, and perhaps even erase, most of the signatures expected under a hard sweep (Table 8.2). If the favorable variant is at a very low frequency at the start of selection ($p[0] < 1/[4N_e s]$), a hard-sweep signature may be generated. However, hard-sweep signatures are very unlikely when the initial frequency exceeds this threshold. This erosion of hard-sweep signatures is even more dramatic for multiple-origin soft sweeps (Pennings and Hermisson 2006b). If $\theta_b = 4N_e\mu_b$ is the scaled mutation rate for beneficial alleles, the heterozygosity following such a sweep will become

$$\frac{H_h}{H_0} \simeq 1 - \frac{1}{1+\theta_b}(4N_e s)^{-2c/s} \tag{8.16a}$$

which contrasts with Equation 8.8c for a hard sweep. Thus, even with a completely linked site, the loss of variation is not complete, as

$$\frac{H_h}{H_0} \simeq 1 - \frac{1}{1+\theta_b} = \frac{\theta_b}{1+\theta_b} \tag{8.16b}$$

Some variation is preserved following a sweep because independent favorable mutations arose on random haplotypes during the sojourn of the favorable allele to fixation. For $c = 0$, if $\theta_b = 0.01$, then $H_h/H_0 \simeq 0.01$, while if $\theta_b = 0.5$, then $H_h/H_0 = 0.33$. Pennings and Hermisson found that in addition to reducing the heterozygosity signal, soft sweeps also significantly depress any sweep signal in the SFS. Indeed, even when $c = 0$, the folded frequency spectrum after a soft sweep can be very close to the neutral (Watterson) spectrum.

The feature where soft sweeps can leave a strong signature is in linkage disequilibrium (LD). A lower number of haplotypes and a higher level of association between sites relative to drift are expected, at least during a short window following the sweep ($\sim 0.1N_e$ generations). The resulting LD signature, however, is quite different from that for a hard sweep. Under the latter, LD is close to zero *across* the selected site following fixation (Figure 8.6), while under a soft sweep, LD extends *through* a site. This generates different LD-based tests for hard vs. soft sweeps, which are discussed in Chapter 9, with the ω^2 statistic (Equation 9.37) for hard sweeps, and the Z_{nS} test (Equation 9.36b) for soft and ongoing (i.e., partial) sweeps. Pennings and Hermisson found that the power of LD-based tests for detecting soft sweeps is significantly enhanced by ignoring new neutral mutations. They suggest that when a closely related population or sister species is available, using only sites that are shared polymorphisms (and hence not recent mutations) in both populations can improve the power of LD-based tests for detecting soft sweeps.

Under a soft sweep (especially when $\theta_b > 1$), there is no single dominant haplotype, as would be expected under a hard sweep. However, Garud et al. (2015) and Garud and Rosenberg (2015) noted that there may instead be a *few* dominant haplotypes (provided that $\theta_b < 10$, so that the sweep is not too soft), and suggested a simple modification of a standard hard-sweep test to detect soft sweeps. If p_i denotes the frequency of the ith haplotype in a sample (ranked, so that $i = 1$ is the most frequent), then under a hard sweep, the **haplotype homozygosity** $H_1 = \sum p_i^2$ should be excessive relative to its expectation under neutrality (Chapter 9). (As a notational aside, up to this point, we have used H to denote heterozygosity-based measures, while in the literature H is used for both heterozygosity and haplotype-homozygosity measures. Rather than define new notation, we will also use H_1 and H_{12} for haplotype homozygosity measures, as the distinction between them and heterozygosity-based measures should be quite clear from context.)

Garud et al. (2015) suggested a modified haplotype homozygosity statistic, H_{12}, which lumps the first two haplotypes into a single class

$$H_{12} = (p_1 + p_2)^2 + \sum_{i>3} p_i^2 = \sum_i p_i^2 + 2p_1p_2 = H_1 + 2p_1p_2 \tag{8.16c}$$

This test has reasonable power to detect *both* hard and soft sweeps. Under a hard sweep there is only one dominant haplotype, reducing this to the H_1 test as $H_1 \simeq H_{12}$. Under a soft sweep, the favorable site is on multiple haplotypes, and lumping the two most common haplotypes picks up at least part of this signal.

Garud et al. (2015) suggested a further test to distinguish between soft and hard sweeps for situations in which H_{12} is significant. Again, the logic is that under a soft sweep, we expect a few haplotypes to be common, whereas only a single common haplotype is expected under a hard sweep. Their statistic, $H_2 = \sum_{i>1} p_i^2$, is simply haplotype homozygosity with the most common haplotype removed, with the ratio of H_2/H_1 expected to be very small under a hard sweep, but modest under a soft sweep. Using this approach, they found much stronger support for soft, rather than hard, sweeps over a set of 50 regions (detected using H_{12} statistics; Chapter 9) in a North American sample of *Drosophila melanogaster*.

Sweeps Using Standing Variation

The hard-sweep model implies a lag in adaptation, with populations experiencing an environmental change having to wait for favorable mutations to appear in order to respond. In contrast to this situation, artificial selection for most traits in outbred populations generates an immediate response (Chapter 18), showing that a large reservoir of **standing** (or preexisting) **variation** typically exists for most traits. Thus, hard sweeps are expected to be more frequent when standing variation is low, such as in small or inbred populations, or for traits with a previous history of strong selection. On the other hand, while much of the initial selection response might occur from standing variation in outbred populations (a standing sweep), new mutations play a critical role in the continued response once this initial variation has been exhausted (Chapters 26 and 27). The molecular signature resulting from a sweep arising from standing variation has been examined by Innan and Kim (2004), Przeworski et al. (2005), and Berg and Coop (2015).

Innan and Kim (2004) were interested in domestication selection, clearly a radical change in the environment to a new selection regime. The reduction in neutral diversity at linked sites in this case can be much less than for a hard sweep, as much of the initial response may be based on alleles that were segregating before domestication started. Hence, the time to the most recent common ancestor for the favorable allele may significantly predate the start of selection. Innan and Kim found that if the frequency of the favored allele at the start of selection is greater than 0.05, at best only a weak reduction in background variation will be generated by a sweep. However, a factor not considered by these authors is that domestication usually involves a strong bottleneck, which can result in a preexisting allele being reduced to one (or a very few) lineages that survived the bottleneck before being selected, generating a more hard-sweep pattern. A potential example of this situation is the maize domestication gene *tb1*. While this locus shows a classic hard-sweep pattern (Figure 8.2), the domestication allele was created by an insertion of the retroposon *Hopscotch* that predated domestication by at least 20,000 years (Studer et al. 2011). Using DNA from ancient maize samples, Jaenicke-Després et al. (2003) suggested that domestication selection started to occur on *tb1* roughly 4,400 years ago.

Further insight into the signal from a standing sweep was provided by Przeworski et al. (2005), who found that as long as the initial frequency of a selected allele is $< 1/(4N_e s)$, the signal is the same as for a hard sweep. With higher initial allele frequencies, the situation is more complex. In some settings, the result is simply a weaker footprint, but with the normal features of a sweep (reduced diversity, excess of rare alleles, and excess of high-frequency derived alleles). However, in other cases, a sweep associated with a very weakly selected preexisting allele can result in an *excess of intermediate-frequency alleles*. In still other settings, essentially no detectable pattern is seen in the reduction of diversity, the frequency spectrum, or the distribution of LD. In particular, if the new environment favors an *ancestral* allele, especially one at high frequency, there will be no discernible change over the background pattern (Przeworski et al. 2005). The salient point is that selection on standing variation need not leave a hard-sweep signature, and significant ongoing or recent selection can easily be missed, even when strong selection is occurring.

Example 8.6. The myostatin gene (*MSTN*) is a negative regulator of skeletal muscle growth. Mutations in this gene underlie the excessive muscle development in double-muscled (DM) breeds of cattle, such as the Belgian Blue, Asturiana de los Valles, and Piedmontese. Wiener et al. (2003) compared variation at microsatellite loci as a function of their distance from *MSTN* in DM and non-DM breeds. For DM breeds, measures of variation decreased relative to non-DM breeds as the marker loci approached the *MSTN* locus. While this observation strongly suggests a genomic region under selection, the authors expressed skepticism about using this approach to fine-map the target of selection (i.e., localize it with high precision within this region), as the region under detected selection was only localized to a broad area around the known target (*MSTN*). At first glance, this seems surprising, given that *MSTN* variants have a major effect on the selected phenotype (beef production). However, the authors noted that the Belgian Blue was a dual-purpose (milk and beef) breed until the 1950s, and that in both the Belgian Blue and the Piedmontese there are records of *MSTN* mutations before World War I, predating the intensive selection on the double-muscled phenotype. By contrast, they found that the selective signal is stronger in the Asturiana, where the first definitive appearance of the mutation was significantly later. Thus, in both the Belgian Blue and the Piedmontese, selection on this gene likely resulted in a soft sweep (adaptation from preexisting mutations), while in the Asturiana, the time between the initial appearance of the mutation and strong selection on it was much shorter, resulting in a more traditional hard sweep (adaptation from a new mutation). O'Rourke et al. (2012) used haplotype homozygosity to estimate the age of the Belgian Blue mutation (*821dell11*) at ~200–400 years and that of the Piedmontese mutation (*C313Y*) at ~200 years.

Example 8.7. The threespine stickleback (*Gasterosteus aculeatus*) is a species complex of small fish widespread throughout the Northern Hemisphere in both freshwater and marine environments. The marine form is usually armored with a series of over 30 bony plates running the length of the body, while exclusively freshwater forms (which presumably arose from marine populations following the melting of the last glaciers) often lack some, or all, of these plates. Given the isolation of the freshwater lakes, it is likely that the reduced armor phenotype has independently evolved multiple times. What was the mechanism for this?

Colosimo et al. (2005) showed that this parallel evolution occurred by independent fixation of alleles derived from an ancient ($\sim 2 \cdot 10^6$ years) low-armor haplotype of the *Eda* locus (a component of the ectodysplasin signaling pathway). Surveying populations from Europe, North America, and Japan, they found that most nuclear genes showed a clear Atlantic-Pacific division. Conversely, at the *Eda* locus, low-armored populations shared a more recent history than full-armored populations, independent of their geographic origins, presumably reflecting more recent ancestry at the site due to the sharing of an ancient allele. In marine populations, low-armor alleles at *Eda* are present at low ($< 5\%$) frequency. Presumably, these existing alleles were repeatedly selected following the colonization of freshwater lakes from marine founder populations. Barrett and Schluter (2008), Messer and Petrov (2013a), and Jensen (2014) reviewed a number of other examples of adaptation from preexisting mutations.

How Likely is a Sweep Using Standing Variation?

Hermisson and Pennings (2005), Przeworski et al. (2005), and Berg and Coop (2015) used population-genetic models to examine the likelihood of a sweep arising from standing variation. To consider the probability of such an event over a series of replicate populations, let $\phi(x)$ denote the distribution of the frequency, x, for the soon-to-be favored allele A, and $u(x)$ its probability of fixation under the new environment given x. The probability, $\Pr_{sv}$, that a sweep occurs using standing variation at this locus is simply

$$\Pr_{sv} = E\left[u(x)\right] = \int_{1/(2N)}^{1-1/(2N)} u(x)\phi(x)dx \tag{8.17a}$$

The limits on the integral confine us to considering only segregating alleles. Przeworski et al. (2005) assumed that $\phi(x)$ is given by the neutral Watterson distribution (Equation 2.34a), while Hermisson and Pennings (2005) considered a more general setting, where the genotypes $aa : Aa : AA$ have fitnesses of $1 : 1 - 2h_d s_d : 1 - 2s_d$ in the old environment and $1 : 1 + 2hs : 1 + 2s$ in the new environment. This allows for the allele to be either neutral ($s_d = 0$) or deleterious ($s_d > 0$) before being favored.

Assuming selection-mutation-drift equilibrium on the allele prior to becoming favored, $\phi(x)$ is a function of N_e, the selection parameters (h_d, s_d), and the mutation rate μ_b of a to A, and can be obtained using diffusion machinery (Chapter 7; Appendix 1). Likewise, the fixation probability under the new fitnesses can also be obtained using diffusion results (Chapter 7). Putting these together, Hermisson and Pennings found that

$$\Pr_{sv} \approx 1 - e^{-\theta_b \ln(1+R)}, \quad \text{where} \quad R = \frac{2h\alpha_b}{2h_d\alpha_d + 1} \tag{8.17b}$$

with $\alpha_b = 4N_e s$ and $\alpha_d = 4N_e s_d$ being the scaled strengths of selection in the new and old environments, respectively, and $\theta_b = 4N_e\mu_b$ being the scaled mutation rate at which the (eventually beneficial) allele arises.

The alternate scenario to a standing-variation sweep is the need to wait for new favorable mutations to arise and subsequently become fixed (a classic hard sweep). Recall from Chapter 7 that the fixation probability of a single new mutation is $\sim 4hs(N_e/N)$, so roughly $N/(4N_e hs)$ such mutations must appear to have a reasonable chance of one becoming fixed. The expected number of such beneficial mutations arising in each generation is $2N\mu_b$, giving

$$\left[4hs(N_e/N)\right]\left[2N\mu_b\right] = 2hs(4N_e\mu_b) = 2hs\theta_b \tag{8.18a}$$

as the expected number of mutations that are destined to become fixed arising each in generation. The reciprocal of this quantity

$$t_e = \frac{1}{2hs\theta_b} \tag{8.18b}$$

is the mean establishment time for a new favorable mutation (Messer and Petrov 2013a).

Before proceeding, it is useful to rescale time from the number of generations τ to $T = \tau/(2N_e)$. Under this scheme, $\tau = 2N_e T$, with $T = 1$ corresponding to $2N_e$ generations, which is the natural scale for genetic drift (being the expected coalescence time for two neutral alleles; Chapter 2). The expected total number of beneficial mutations that will have appeared by time T is then

$$\tau \cdot 2hs\theta_b = Th(2N_e 2s)\theta_b = Th\alpha_b\theta_b$$

Hence, the probability that at least one favorable mutation that is destined to become fixed will appear by generation T is just one minus the probability that none have appeared, which, from the Poisson distribution, is

$$\Pr_{new}(T) = 1 - e^{-Th\alpha_b\theta_b} \tag{8.18c}$$

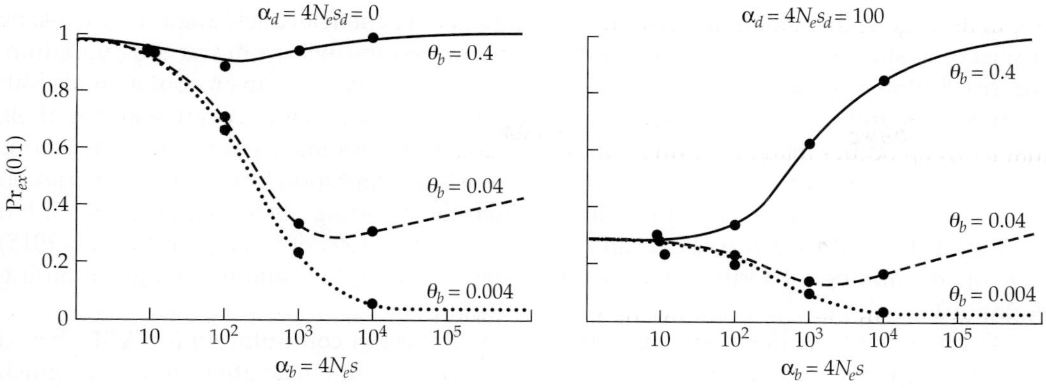

Figure 8.7 Plots of $\Pr_{ex}(0.1)$, the probability of a selected sweep from standing variation, *given* that a sweep has occurred within $0.1N_e$ generations following a change in the environment (Equation 8.20, with simulated data given by the points). This is a function of the beneficial mutation rate, $\theta_b = 4N_e\mu_b$ (separate curves within each graph), and the scaled strength of selection, $\alpha_b = 4N_e s$, during the sweep (horizontal axis). (**Left**) The allele is neutral in the old environment ($\alpha_d = 0$). (**Right**) The allele is deleterious in the old environment ($\alpha_d = 4N_e s_d = 100$) (After Hermisson and Pennings 2005.)

as obtained by Hermisson and Pennings (2005). When $\alpha_b\theta_b$ is small, the waiting time for a mutation that is destined to become fixed is quite long. In such cases, mutation is the rate-limiting step for adaptation, unless there is standing variation to exploit.

The total waiting time (in generations) until the *fixation* of a favorable allele with additive effects on fitness ($h = 0.5$) is approximately

$$t_{fix} = \frac{1}{s}\left(\frac{1}{\theta_b} + 2\ln(4N_e s)\right) \tag{8.19}$$

where the first term is the mean waiting time for the first appearance of a successful mutation (the establishment time; Equation 8.18b) and the second is its fixation time (Equation 8.4d). Karasov et al. (2010) and Messer and Petrov (2013a) developed similar expressions.

One might expect that a successful sweep (i.e., the selected allele is destined for fixation) starting early after selection is applied is likely due to standing variation, but as time proceeds, a successful sweep will more likely be due to new mutations. We can quantify this by conditioning on a sweep by time T (assuming that the fixation time of a successful mutation is very small relative to T), giving the probability that it is from an existing allele (i.e., a standing sweep) as

$$\begin{aligned} \Pr_{ex}(T) &= \Pr(\text{existing allele} \mid \text{sweep by generation } T) \\ &= \frac{\Pr_{sv}}{\Pr_{sv} + (1-\Pr_{sv})\Pr_{new}(T)} \\ &= \frac{1-\exp\left[-\theta_b \ln(1+R)\right]}{1-\exp\left\{-\theta_b[\ln(1+R) + Th\alpha_b]\right\}} \end{aligned} \tag{8.20}$$

This follows because $\Pr_{sv}$ is the probability that, in the absence of any new mutation, a variant segregating at the start of the new selection regime will be fixed, while the probability that the fixation will occur via a new mutation (arising by time T) is $(1-\Pr_{sv})\Pr_{new}(T)$. The denominator is the sum of the probabilities of these two different events, and hence the probability of a sweep occurring by time T. For a sufficiently large T, $\Pr_{new}(T) = 1$ (the probability of a new successful mutation approaches one), and Equation 8.20 reduces to $\Pr_{sv}$ (Equation 8.17b). This sets the *lower limit* on the probability that the favorable mutant fixed by the sweep was preexisting in the population before the start of selection. For shorter amounts of time, there is a reduced chance that the fixed beneficial allele arose after the start of selection, and hence a higher probability that a sweep resulted from standing variation. Figure 8.7 plots Equation 8.20 at $0.1N_e$ generations ($T = 0.05$) after an environmental shift. When both θ_b and α_b are high, most sweeps are from existing variation, even when the allele was deleterious before the shift. When θ_b is small, most sweeps are from new mutations unless both α_b and α_d are small. The reason is that while adaptation from standing variation is unlikely with a small α_b, it is more likely when the standing-allele frequency is not too small (which requires that α_d be small) before the start of selection. Berg and Coop (2015) developed a number of additional results for standing sweeps (assuming a single founding mutation).

Peter et al. (2012) developed an **approximate Bayesian computation** (or **ABC**)-based approach (Appendix 3) that combines several tests of selection in an attempt to distinguish between sweeps from de novo mutation versus those from standing variation (also see Ormond et al. 2015). Their key idea is that allelic trajectories under either setting are identical once selection starts, but that a preexisting allele starts at some frequency $> 1/(2N)$, while a de novo mutation is initially absent. The fit of several summary statistics of selection (Chapter 9) is then compared with various models assuming drift (i.e., no selection) at some point in the history of that allele. However, simulation studies showed that their test has little power unless selection is strong ($4N_es > 100$ under additivity). When their test was applied to seven of the strongest known sweeps in humans, two were found to be most likely from standing variation, three fit the hard-sweep model, and two were equivocal. While interesting, this small sample does not provide much insight into the relative frequency of the different types of sweeps, as these loci were generally detected using methods based on strong-sweep signals, creating an ascertainment bias in favor of hard sweeps.

Example 8.8. One measure of how rate-limiting mutation can be is the expected time at which there is a 50% chance that a mutation that is destined to be fixed has arisen, which follows from Equation 8.18c. Suppose that adaptation can only occur through mutation at one of five nucleotide sites, in each case generating an allele with an additive effect on fitness ($h = 1/2$) and a heterozygote advantage of 1% ($s = 0.01$). At each of these five sites, only one third of all new point mutations (assuming equal rates to all four nucleotides) generate the favorable allele, giving the beneficial mutation rate as $(5/3)\mu$, where μ is the per-site rate. In humans, assuming a historical value of $N_e = 10^4$ and a per-site mutation rate of 10^{-8}, we have $h\alpha_b = (1/2)4N_es = 2 \cdot 10^4 \cdot 0.01 = 200$, while $\theta_b = 4N_e\,\mu_b = 4 \cdot 10^4 \cdot [(5/3) \cdot 10^{-8}] = 0.00067$, giving $h\alpha_b\theta_b = 200 \cdot 0.00067 = 0.13$. Solving Equation 8.18c for the value of T giving a 50% probability,

$$0.5 = 1 - \exp(-T_{0.5}h\alpha_b\theta_b) = 1 - \exp(-0.13 \cdot T_{0.5})$$

which rearranges to $T_{0.5} = -\ln(0.5)/0.13 \simeq 5.33$, or $5.33(2N_e) = 10.66N_e = 106,600$ generations. Further, once such a mutation that is destined to be fixed arises, it still takes (on average) an additional $2\ln(4N_es)/s \sim 1060$ generations to become fixed (Equation 8.4d).

Example 8.9. As a sample calculation of how often sweeps occur from existing versus new mutations, suppose $N_e = 10^6$ and there is a genome average per-site mutation rate of $\mu = 2.5 \cdot 10^{-9}$, yielding $\theta = 4N_e\mu = 0.01$. For a beneficial mutation that can only occur by a change to a specific nucleotide at a specific site, one third of mutations at that site are beneficial, yielding $\theta_b = (1/3)\theta = 0.0033$. For an allele with additive fitness effects ($h = 1/2$) with $s = 10^{-4}$, we have $\alpha_b = 4N_es = 4 \cdot 10^6 \cdot 10^{-4} = 400$. If this mutation was neutral before being favored, $\alpha_d = 0$, and from Equation 8.17b, $R = 2h\alpha_b = 400$, yielding

$$\begin{aligned}\Pr_{sv} &\approx 1 - \exp\left(-\theta_b \ln[1+R]\right)\\ &= 1 - \exp\left(-0.0033\ln[1+400]\right) = 0.013\end{aligned}$$

Hence, there is only a 1.3% chance that a sweep will occur at this site in the absence of new mutation.

Now suppose that we examine this population at $T = 0.5$, namely, at N_e generations. From Equation 8.18c, the probability that at least one mutation that is destined to become fixed will arise by this time is

$$\begin{aligned}\Pr_{new}(T) &= 1 - \exp\left(-Th\alpha_b\theta_b\right) \\ &= 1 - \exp\left(-0.5 \cdot (1/2) \cdot 400 \cdot 0.0033\right) = 0.281\end{aligned}$$

Thus, *provided* we see a sweep at this locus by N_e generations, then from Equation 8.20, the probability that it was due to an existing allele present at the time of the environmental shift is

$$\Pr_{ex}(0.5) = \frac{\Pr_{sv}}{\Pr_{sv} + (1 - \Pr_{sv})\Pr_{new}(T)} = \frac{0.013}{0.013 + (1 - 0.013) \cdot 0.281} = 0.05$$

"Hardening" and "Softening" of a Sweep

Population sizes are expected to fluctuate over time, and such demographic complexity impacts whether an initially soft sweep (multiple adaptive haplotypes at the start of selection) is perceived as soft or hard (Messer and Petrov 2013a; Wilson et al. 2014; Hermisson and Pennings 2017). The reason, as seems to be the case for the maize *tb 1* gene (Figure 8.2), is that taking a population through a bottleneck (either just before or during a sweep) may leave only a single adaptive lineage, even when several were present before the bottleneck, thus **hardening** an initially soft sweep. Such hardening events result in undercounting the frequency of soft sweeps (and hence the frequency of adaptation from standing variation).

The strength of selection plays a critical role in determining the expected types of sweeps in a population that regularly contracts in size. If selection is sufficiently strong (and $\theta_b > 1$), multiple independent adaptive alleles may have their frequencies increased during the non-bottleneck phase to a level sufficiently high that they avoid being lost during the next bottleneck. For a population with a normal size of N_1 and bottleneck size of $N_2 \ll N_1$, whether a soft sweep is observed depends upon the establishment time, t_e, of new mutations plus the expected time, t_g, until their frequency is $\gg 1/(N_2)$, which means they will successfully pass through the bottleneck. Solving for $p(t) = 1/(N_2)$ in Equation 8.4a gives this latter time as $t_g \simeq (1/s)\ln(sN_1/N_2)$, as found by Messer and Petrov (2013a), while Equation 8.18b yields $t_e = 1/(\theta_1 s)$, where $\theta_1 = 4N_1\mu_b$. Hence, if the time, t_b, between bottlenecks exceeds $t_e + t_g$, a soft-sweep signature is expected; otherwise, a hard-sweep signature is seen, as at most one adaptive lineage is likely to both be present and to escape the bottleneck. In populations with both $\theta_b > 1$ and a history of large size fluctuations, strongly selected sites are thus more likely to produce soft sweeps than are weakly selected ones. Conversely, if most favorable mutations are under modest selection, there will be a bias toward hard sweeps, as the frequencies of most independent lineages are too small to escape the bottleneck.

While demography can harden an initially soft sweep (multiple haplotypes carrying the favorable allele at the start of selection) by allowing only a dominant haplotype to pass through a bottleneck, the converse is also true. That is, a hard sweep can give soft-sweep signals at sites close to, but not at, the selected target, or if one examines the sweep after some amount of time. Schrider et al. (2015) referred to the former as **soft shoulders** and the latter as **soft shadows**. As previously mentioned, the region of greatest decline in variation from a hard sweep is often asymmetric around the selected site. Hence, estimating the location of the sweep by the taking the midpoint of the region of greatest decline can result in an erroneous identification of the shoulder of a sweep as the actual target. Schrider et al. found that the shoulders of hard sweeps often give either soft-sweep or partial-sweep signatures. The same is true for an older hard sweep, whose decaying shadow can again show either partial- or soft-sweep signatures. This is especially problematic for tests utilizing a small window around some candidate region. As noted by Schrider et al., this leads to the ironic conclusion that spurious partial- and/or soft-sweep signatures become more common as the frequency of hard sweeps increases. Despite some early optimism, Berg and Coop (2015) also highlighted the difficulties in distinguishing between sweep types, thus supporting Jensen's (2014) earlier plea for caution when declaring there is a soft sweep.

Finally, it must be stressed that the distinction between a perceived hard vs. soft sweep is in part a function of the sample size (Messer and Petrov 2013a; Hermisson and Pennings 2017). If the TMRCA in our sample of alleles predates the start of selection, a soft-sweep signature may appear, while if it does not, the signal will be that of a hard-sweep, even when the actually biology was a soft sweep.

Recurrent Mutation of the Favorable Allele Cannot be Ignored

In their analysis of the effects of sweeps from standing variation, Innan and Kim (2004), Przeworski et al. (2005), and Berg and Coop (2015) all assumed a *single origin* for the favorable mutation. Likewise, while the analysis leading to Equation 8.20 does consider recurrent mutation, it simply allows new copies of the favorable allele to arise by mutation once selection starts and keeps track of how long one must wait until a copy that is destined to be fixed arises. It ignores any ongoing mutation either during the fixation of a preexisting copy of the favorable allele or following the introduction of a favorable allele that is destined to undergo fixation. However, even if a sweep *starts* as a single favorable allele on its way to fixation, *additional* new copies can arise by mutation (again in random backgrounds) during the sojourn of the original copy, potentially diffusing any pattern from the sweep over a set of multiple haplotypes. How likely is such a scenario?

Pennings and Hermisson (2006a, 2006b) approached this problem by considering the number of independent lineages of the favorable allele that are expected to be observed in a sample of n sequences following a sweep. Their rather remarkable result is that, to a first-order approximation, this is a function of θ_b, and *not* the strength of selection, α_b. In particular, an upper bound for the probability of a multiple-origin soft sweep (two or more independent lineages in our sample of size n) is

$$\Pr(\text{soft} \,|\, n) \leq \theta_b \left(\sum_{i=1}^{n-1} \frac{1}{i} \right) \simeq \theta_b[0.577 + \ln(n-1)] \tag{8.21}$$

Pennings and Hermisson also showed that the number of distinct lineages in the sample approximates the Ewens (1972) sampling distribution (Equation 2.30a) by using θ_b in place of θ. A more detailed analysis suggests the following general rules: multiple-origin soft sweeps are rare (even in a large sample) when $\theta_b < 0.01$, somewhat common when $0.01 \leq \theta_b \leq 1$, and almost certain when $\theta_b > 1$.

Orr and Betancourt (2001) also examined this problem, but from the perspective of standing variation alone, asking if **Haldane's sieve**, wherein dominant alleles are postulated to be more likely to contribute to selection response than recessive alleles (Turner 1981; Charlesworth 1992), is correct. They were also interested in the number of original copies of an allele that leave descendants in the fixed population. Assuming adaptation from standing variation alone, they found that the dominance of an allele has little effect on its probability of sweeping from standing variation when the dominance relationship is roughly the same under the old, deleterious and new, favorable environments. Recessive deleterious alleles are maintained at a higher frequency than dominant deleterious alleles, which compensates for their lower probability of fixation in the new environment (assuming roughly equal mutation rates for recessive and dominant alleles). Further, they showed that $\theta_b s_b / s_d$ is the critical determinant of the number of independent lineages leaving descendants in the fixed population. When $\theta_b s_b / s_d >\sim 1.3$, the fixed collection of favorable alleles is more likely to contain multiple lineages. If s_b and s_d are roughly of the same magnitude, their effects cancel each other out, again showing the strong dependence of a multiple-origins soft sweep on the value of θ_b.

Multiple-origin soft sweeps are therefore expected to occur under certain biologically realistic conditions. Pennings and Hermisson highlighted two scenarios where this might be expected: very large effective population sizes and large mutational target sizes. An example of the latter is the case where loss-of-function mutations are favored, as the presence of numerous pathways by which function can be lost significantly increases the value of μ_b.

A relevant question regarding multiple-origin sweeps concerns the number of recurrent favorable mutations expected to appear during the sojourn of the favored allele towards fixation. Recalling Equation 8.4d, the expected time for a single copy (a de novo mutation) of the favorable allele to sweep through a population is $\tau \approx 2\ln(4N_e s)/s$. If N is the population size, then the expected number of new favorable mutations arising in a generation is $2N(1-x)\mu_b$, where x is the current frequency of the favorable allele. Thus, we can approximate the expected number of new favorable mutations that will arise by noting that the average frequency of a favored additive allele over its sojourn from near zero to near fixation is 1/2, hence

$$\begin{aligned} E(\text{number of new favorable mutations that arise}) &\approx 2N(1/2)\mu_b\tau = (N/N_e)(N_e\mu_b)\tau \\ &= (N/N_e)(\theta_b/4)\cdot 2\ln(4N_e s)/s \\ &= 2N\theta_b \ln(\alpha_b)/\alpha_b \end{aligned} \tag{8.22a}$$

as obtained by Pennings and Hermisson (2006a). Although this is the *total* number of recurrent favorable mutations that arise, each has only a probability of $2s\,(N_e/N)$ of increasing (Chapter 7). Hence, the expected number of new mutations that arise *and* increase in frequency (i.e., are likely to become part of the fixed pool of the favorable allele after the sweep) is approximately $2s\,(N_e/N)$ times the result in Equation 8.22a, yielding

$$E(\text{number of new favorable mutations that increase}) \approx \theta_b \ln(4N_e s) \tag{8.22b}$$

Again, this is the number of favorable new mutations that increase in frequency during the sojourn of the initial allele to fixation, so that approximately $1 + \theta_b \ln(4N_e s)$ distinct lineages in the *population* are expected at fixation, placing an upper bound on the expected number of lineages in a *sample* of any size from such a population.

Example 8.10. *Caspase-12* (a cysteinyl asparate proteinase) is involved in the inflammatory and innate immune response to endotoxins (Wang et al. 2006). In humans, most alleles are nulls and nucleotide diversity around this locus is sharply reduced (relative to levels in the chimpanzee), suggesting a selective sweep. Using the current frequency of roughly 0.9 for null alleles, and assuming that the sweep favoring null alleles started shortly before the migration of modern humans out of Africa, the authors estimated $s \simeq 0.009$ (using Equation 5.3b). Previous work showed that the fully functional allele attenuates the response to endotoxins (Saleh et al. 2004), increasing the odds of severe sepsis (bacterial infection of the blood). Wang et al. hypothesized that null alleles were favored due to a change in the environment increasing the odds of severe sepsis when this gene is active. Consistent with this hypothesis, two other primate genes related to sepsis are also pseudogenes in humans. Similar findings were reported by Xue et al. (2006).

Example 8.11. To get a feel for the expected number of new favorable mutations that arise during a sweep, consider the values used in Example 8.9 ($N_e = 10^6, \theta_b = 0.0033, s = 10^{-4}, \alpha_b = 400$). From Equation 8.22a (and assuming $N \simeq N_e$), we expect

$$2N_e\theta_b \ln(\alpha_b)/\alpha_b = 2 \cdot 10^6 \cdot 0.0033 \ln(400)/400 \approx 90$$

new favorable mutations to arise, but the number expected to increase in frequency (and hence contribute to the pool of favorable alleles following the sweep) from Equation 8.22b is just

$$\theta_b \ln(4N_e s) = 0.0033 \ln(400) = 0.02$$

Hence, even though a large number of favorable mutations arise, none really contribute to the sweep. This is consistent with the general rule that multiple-origin soft sweeps are unlikely to occur when $\theta_b < 0.01$. Suppose we increase θ_b to 0.5, while keeping the other parameter values the same. Now roughly 15,000 recurrent favorable mutations will be expected, three of which are expected to increase (and hence result in a soft sweep).

While the reader may feel that the critical parameter for observing a soft sweep ($\theta_b = 4N_e\mu_b$) is generally expected to be very small, important exceptions can occur. A common view is that the target site for a beneficial mutation is small (a single change at only one or a few sites) and hence the small per-nucleotide mutation rates (10^{-9} to 10^{-8}; Chapter 4) suggest that such events are highly unlikely. This may indeed be true when adaptation requires a very specific amino acid substitution to be favorable. However, if a potentially large number of sites are targets, then the *total* beneficial mutation rate will be much higher. One example of the latter is the inactivation of a gene that has become deleterious due to a shift in the environment (Example 8.10). A second example is a beneficial regulatory change, which potentially could occur through changes at a number of nucleotide sites, or by other events (such as a mobile-element insertion in some rough neighborhood of the gene) that can influence regulation. We have already seen one example of the latter, the *Hopscotch* insertion underlying the maize domestication allele *tb1*. Another example is the famous peppered moth, *Biston betularia*, whose industrial melanism mutation was generated by a transposable element (van't Hof et al. 2016) that appeared to insert in the *cortex* gene around 1820, about 30 years before the first melanic form was reported (1848).

Mobile elements may thus be an under-appreciated source of adaptive mutations (Casacuberta and Gonzáez 2013; Villanueva-Cañas et al. 2017). Consistent with this idea, González et al. (2008) found that transposable genetic elements (TEs) may induce adaptation in *Drosophila melanogaster*. In a set of 909 TEs that inserted into new sites following the spread of this species out of Africa, at least 13 show signs of being adaptive (associated with signatures of partial sweeps). They suggested that the majority of these insertions likely induce regulatory changes. The much higher rate of TE mobilization (relative to nucleotide mutation rates), coupled with their much larger target of action (their insertion at a large number of sites can influence regulation over some distance), suggests that μ_b may often be much larger than one expects.

Even independent single-site beneficial mutations may be more likely than expected. A potential human example of this derives from the work of Enattah et al. (2007) on the lactase gene (*LCT*). Variants at this locus are correlated with lactase persistence (the ability to utilize milk as an adult) and hence are candidates for selection following the invention of dairy farming. Enattah et al. found that the T_{-12910} variant upstream of *LCT* appears to have at least two independent origins. In addition to the common northern European allele, an independent origin appears to have occurred in an isolated region in eastern Europe. Further, Tishkoff et al. (2007) found independent mutants at different sites in the *LCT* gene in African populations also leading to lactase persistence.

The second component to θ_b is N_e. While long-term values of N_e may be small (e.g., due to rare, but strong, bottlenecks), short-term values of N_e may be much larger (perhaps approaching the population census size). Current estimates of N_e are often based on levels of nucleotide diversity, which are generated by the cumulative joint action of mutation and drift over rather long periods of time. Conversely, when a favorable mutation appears, during the early phases of a sweep, its population size may still be much higher than its long-term harmonic average, offering the opportunity for multiple favorable mutations to arise during its sojourn.

The notion of independent mutants arising during a sweep is a special example of the more general question of how often the same genes are independently used for specific adaptations (reviewed by Arendt and Reznick 2008). Independent recurrent mutations in even distant taxa may be more common than intuition might suggest. Orr (2005c) considered the probability, P, of the occurrence of a parallel change when n different beneficial mutations can each provide an adaptive solution. Using the framework of extreme value theory (Chapter 27), even when these potential solutions have rather different selection coefficients, the probability that two independent taxa use the same solution (beneficial mutation) is just $P = 2/(n+1)$. Hence, if the number of paths leading to an adaptation is quite small (e.g., only a few amino acid changes will confer the adaptive change), then *provided* such an event has occurred in two independent taxa, there is a high probability of a parallel change.

Example 8.12. Karasov et al. (2010) examined *Drosophila melanogaster* mutations at the *Ace* locus, which codes for the neural signaling enzyme acetylcholinesterase, a target for many commonly used insecticides. Single nucleotide changes at four highly conserved sites confer partial insecticide resistance, with combinations of these changes conferring successively greater resistance. Single, double, and triple mutations are all found in natural populations. While one model is that these variants existed at the start of major insecticide use (the 1950s), the authors found that mutations in North America and Australia appear to have arisen de novo following the *D. melanogaster* migration out of Africa. Given that only 1000 to 1500 fly generations have elapsed since the start of the widespread use of insecticides that target the *Ace* product, estimates of $\theta \sim 0.01$ based on nucleotide diversity (and hence a θ_b of one third this value at each of the four sites) are not consistent with the independent origins of single new mutations (Equation 8.21) in this gene over this short time scale, and hence multiple new mutations (changes in two or three additional sites) are expected to be extremely unlikely. However, if the actual effective population size was 10^8 instead of the standard assumed value of 10^6 during the past 50 years, then $\theta_b \sim 1$ over this time window, and multiple independent origins by mutation are highly likely. The effective population size that matters for these mutations is that during their origin and spread, not that set by any history predating their appearance (Messer and Petrov 2013a).

Impact of Geographic Structure

Our analyses thus far have assumed a panmictic population, but the effects of population structure on sweep signals can be dramatic. One simple consequence is that hard sweeps from independent mutations in different subpopulations generate a soft-sweep signature when the data are combined (Ralph and Coop 2010; Messer and Petrov 2013a). Given the limitless model space for population structure, current models typically assume one of two highly idealistic forms, representing the ends of the continuum of possible population densities and gene flow in space. At the discrete end are simple two- (or multiple) deme models with weak migration (Slatkin and Wiehe 1998; Santiago and Caballero 2005; Kim and Maruki 2011; Messer and Petrov 2013a; Roesti et al. 2014). At the fully continuous end is a single population with uniform density and migration rates, spread out over some two-dimensional space (Ralph and Coop 2010; Novembre and Han 2012; Messer and Petrov 2013a). As with many complex issues in population and quantitative genetics, the hope is that these simple models capture most of the salient features of more complex ones. Whether this is true remains to be seen, and models incorporating sweeps in structured populations remain an important area for future research (Stephan 2010a).

Consider the simple two-deme models first. In ecology, it is of great interest to distinguish between uniform and heterogeneous selection pressures. A hard sweep initially confined to one deme may represent either a local adaptation to the unique environment

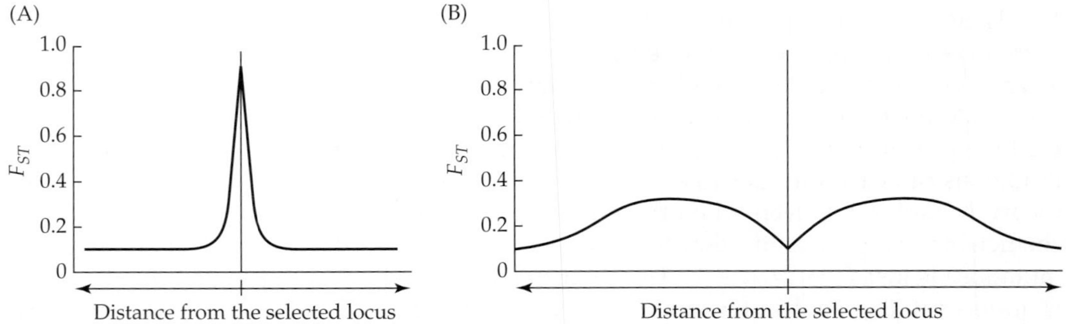

Figure 8.8 The expected pattern of among-population divergence (measured by F_{ST}; Chapter 2) around a selected site. (**A**) A localized hard sweep, with an allele that is favored in only a subset of the demes. (**B**) A global hard sweep, with the same favorable allele fixed over all demes. In both figures, the expected F_{ST} value for markers sufficiently far way from the selected site asymptotes out to the neutral mutation-migration-drift values for the given population structure. (After Bierne 2010.)

of that deme (**heterogeneous selection**), or simply an adaptive mutation with the same advantage in all demes (**uniform selection**) that has not yet spread out of its founding deme. As shown in Figure 8.8, a plot of F_{ST} (Chapter 2) around a selected site shows different patterns of divergence for a localized hard sweep (Figure 8.8A) versus a global hard sweep through all demes (Figure 8.8B). In the former, one sees a spike in population divergence around the selected site, as it is a local advantage, and hence the deme or demes in which it is favored have a very strong divergence at that site.

More striking is the **hill-valley-hill** pattern seen in Figure 8.8B for a global sweep (Bierne 2010; Roesti et al. 2014). For a hard sweep through all of the demes, there is little to no divergence at the selected site, but on either side there is an increase in population divergence, which then declines to the neutral mutation-migration-drift values at more distant sites. This F_{ST} pattern around the selected site arises because migration brings in the favored allele, which, coupled with recombination with local haplotypes, spreads the adaptive allele to specific backgrounds in each deme (for sites at some moderate recombinational distance from the selected site). The net result is that marker diversity among subpopulations (measured by F_{ST}) *increases* for a short distance as one moves away from the site (Figure 8.8B), and also shows an excess of sites with intermediate allele frequencies (mimicking signatures for balancing selection). Bierne (2010) observed such an F_{ST} signature for a set of mussel (*Mytilus edulis*) populations. A one-sided scan away from a length polymorphism in the third intron of the $EF1\alpha$ gene showed a valley-hill pattern, suggesting $EF1\alpha$ as a candidate for a global sweep. See Bierne (2010) for a detailed discussion on distinguishing local from global sweeps.

How often does one expect soft sweeps to occur in a structured population under uniform selection, where different favorable alleles arise by mutation in different demes? Again, a simple time-scale argument can be used in the situation where the population structure consists of a set of discrete demes (Messer and Petrov 2013a). Consider a given deme where the favorable mutation is currently absent, but that receives migrants from a deme (of size N) where it is fixed. If m is the per-individual migration rate between demes, then $2Nm$ favorable mutations are introduced in each generation, each with a probability $2s\,(N_e/N)$ of fixation (Chapter 7). This yields an approximate waiting time (in generations) for the appearance of a successful migrant from a fixed deme as $t_m \simeq 1/(4N_e ms)$. Conversely, the establishment time, $t_e = 1/(\theta_b s)$, yields the average waiting time for the appearance of new (destined to be fixed) favorable mutations in that deme (Equation 8.18b). When $t_e < t_m$, namely $\theta_b > 4N_e m$ (i.e., $\mu_b > m$), soft sweeps are expected, with the sweep through the metapopulation being the result of independent favorable mutations arising within different demes, as opposed to one mutation from just a single deme (Messer and

Petrov 2013a).

For populations continuously distributed over some area, again independent adaptive mutations can arise in different parts of the range, resulting in a patchwork pattern of alleles, which will eventually resolve into a soft-sweep signature as the process of migration mixes the favored haplotypes (Ralph and Coop 2010; Novembre and Han 2012; Messer and Petrov 2013a). In a key paper, Ralph and Coop started with Fisher's (1937) classic result that the speed ν, at which a favorable allele with additive effects on fitness spreads over a spatially continuous population is $\nu = \sigma\sqrt{2s}$, where σ is the standard deviation of dispersal distance. In a uniform two-dimensional space, a new favorable mutation will spread out from its origin in a radial pattern at a rate of ν. After t generations, the mutation will encompass a region of radius $t\nu$ from its origin, covering an area of roughly $\pi(t\nu)^2$. Turning to the rate at which favorable mutations arise, if ρ is the population density (the average number of individuals per unit area), then the (per unit area, per generation) rate, λ, at which new favorable mutations that are destined to be fixed arise is $2\rho\mu_b{\cdot}2s$, as $2\rho\mu_b$ favorable mutations arise per unit area, per generation, each with a chance of $2s$ of becoming fixed.

The final assumption made by Ralph and Coop is allelic exclusion—when two favorable alleles (of different origin) come into contact, they each behave neutrally with respect to each other, and hence the spreading process into each others' range effectively stops (becoming a drift-like process on a much longer time scale). Using these results for the rate of spread ν, and the rate of appearance of new successful mutations, λ, Ralph and Coop found that the key model parameter was a **characteristic dispersal length**

$$\chi = \left(\frac{\nu}{\pi\lambda}\right)^{1/3} = \left(\frac{\sigma}{2\pi\rho\mu_b\sqrt{2s}}\right)^{1/3} \tag{8.23}$$

This is the distance a new favorable mutation must travel before encountering another favorable mutation. If a species range is much less than the characteristic length (which varies over traits, as μ_b and s are trait dependent), then a hard sweep is seen, as independent mutations are unlikely. If χ is short relative to the range, independent mutations are likely, and these will interfere with each other as they come into contact.

In this latter case, populations appear as a patchwork of different alleles, a pattern that can be easily mistaken for local adaptation to selectively heterogeneous environments, when in fact this pattern is generated by independent mutations under uniform selection. Eventually this initial geographic patchwork of different adaptive alleles resolves into a soft-sweep pattern as these alleles, which are now neutral (with respect to each other), diffuse over the range. Even if one allele is superior to all the other advantageous mutations and eventually sweeps through, it will take longer to become fixed because its selective advantage with respect to the other favorable alleles is small. This increases the time for recombination to act, reducing any signature from the final sweep.

Ralph and Coop applied their model by considering the expected characteristic length relative to the range of humans over Eurasia (~8000 km). If population densities are low, χ can be sufficiently large that multiple-origin sweeps are unlikely. However, the combination of both a large mutational target and low density can allow for multiple-origin sweeps. Ralph and Coop noted that two complementary settings, moderate density ($\rho = 2$, namely two individuals per km^2) and low mutation ($\mu_b = 10^{-8}$) or low density and high mutation ($\rho = 0.002, \mu_b = 10^{-5}$), yield give $\chi \simeq 3000$ km. Hence, under these conditions, multiple mutations would be expected to be the norm over Eurasia (even under uniform selection).

In contrast with this expectation in humans, Andersen et al. (2012) detected several worldwide hard sweeps in *Caenorhabditis elegans*. They argued that the likely combination of both human-aided dispersal and common selection pressures from agriculture may account for the global nature of these sweeps in an organism that would otherwise be expected to have high population density.

A final issue relating to sweeps and geographic structure is the notion of **divergence hitchhiking** (Via and West 2008). Via and West noted excessive F_{ST} values localized around QTLs for ecological isolation in a pair of incipient aphid species. This observation fits with

Barton's (1979) classic work on gene flow in a cline, wherein regions involved in either partial reproductive isolation and/or divergence selection retain their identity, while the rest of the genome is homogenized by gene flow. When selection opposes recombinants between populations around isolation-inducing sites, the suppression of gene flow will result in such regions showing higher divergence than the background levels.

POLYGENIC ADAPTATION

The strength of a signal left by a hard sweep is a function of the magnitude of selection, with any signal being significantly diminished under soft-selection scenarios. This suggests that weak selection using standing variation at a number of loci is the worst-case scenario for detecting recent or ongoing selection solely from sequence data. Unfortunately, this appears to be *exactly* the situation for many, and perhaps most, quantitative traits (at least initially). As detailed in Chapter 18, almost all traits in outbred populations show some, usually rather significant, responses to artificial selection. The immediate nature of a selection response indicates that standing genetic variation underlies almost all initial responses to selection on complex traits, although contributions from new mutations become increasingly important over time (Chapters 26 and 27).

Recall that the strength of selection $s = \bar{\imath}(a/\sigma_z)$ on a QTL allele underlying a complex trait (Equation 5.21) is a function of the strength of selection on that trait ($\bar{\imath}$, the within-generation change in mean, expressed in standard deviations) and the fractional contribution of the allele to overall trait variation (a/σ_z, the additive effect for that allele, scaled in phenotypic standard deviations). With modest selection on the trait (say the mean is $0.1\sigma_z$ higher in the survivors following selection) and a modest contribution from an underlying QTL (an effect of 0.01 standard deviations), $s = 0.001$. Assuming a recombination fraction of 1 cM/Mb, Equation 8.6a suggests that a sweep at such a site should influence roughly $0.02\,(0.001/0.01) = 0.002$ Mb = 2000 bases. This is a small region, and yet it represents the *best-case* scenario, a hard sweep. Under a soft sweep (as when the selection response is derived from existing variation), this signal is further degraded. Moreover, for most complex traits, the situation is likely even worse. This is because the response to selection is distributed over a number of loci, allowing for substantial change in the trait mean with only modest allele-frequency change at any underlying loci (Chapter 24). In the case of changing the mean to some new optimum value, any resulting signal would likely be diffused by weak partial sweeps based on standing variation over a modest to large number of sites.

Given these concerns, Pritchard and Di Rienzo (2010) and Pritchard et al. (2010) have suggested that such polygenic adaptation is likely to leave little, if any, signal under traditional screens (also see Chevin and Hospital 2008; Pavlidis et al. 2012). This prediction was observed in a scan of selection on cattle by Kemper et al. (2014). Strong signals were seen around major loci that define breeds (e.g., the absence of horns and coat color loci), and five major loci with strong effects on production traits (stature, milk production, muscle mass) also showed signatures, but not as strong. However, no significant traditional signals were observed around a large number of small-effect QTLs known to be involved in strongly selected production traits.

How might such "**polygenic sweeps**" be detected? An interesting suggestion was offered by Hancock et al. (2010a, 2010b), who looked for subtle allele-frequency shifts that were concordant among human populations experiencing similar environments but in different geographic regions (also see Fumagalli et al. 2011; Fraser 2013; Berg and Coop 2014). Such approaches (examined in detail in Chapter 9) clearly have power issues (being a function of the number of independent replicates under the same environmental conditions) and also rely on the same alleles responding in the same environmental conditions. One interesting variant of this idea involved a comparison of height-QTL allele frequencies in northern versus southern Europeans (Turchin et al. 2012). Of 139 variants associated with height, the plus allele (increasing height) was more common in northern Europeans in 85 out of 139 cases ($p = 0.01$), with a highly significant average frequency increase of around

1.2%. The nature of this response was truly polygenic, as the effect for any single locus is quite small (average effects are $\leq 10^{-2}\sigma$, and involve small allele-frequency changes). Robinson et al. (2015) reported a similar story for both height and body mass index, while Mathieson et al. (2015) reported evidence for polygenic selection on height using ancient DNA (a set of 230 Western Eurasians who lived between 6000 and 300 BC).

The lack of power for sweep-based tests for detecting single sites under polygenic selection arises (in part) from low expected allele-frequency change at any particular site. In the extreme (the infinitesimal model), there can be a significant shift in the mean with essentially no detectable allele-frequency change at the underlying loci (Chapter 24). However, selection *will* generate LD, even among unlinked selected loci, which reduces the additive variance (Chapters 16 and 24). Chapter 12 discusses the $Q_{ST} - F_{ST}$ method of comparing the population structure of the genetic variance of a trait (Q_{ST}) with that for neutral molecular markers (F_{ST}), and the related Q_X test of Berg and Coop (2014). Q_{ST} is impacted not just by allele-frequency change, but also by LD, and thus can give a signal even when allele-frequency change is not detectable. Hence, Q_{ST}-based approaches (which suffer from their own host of problems; see Chapter 12) potentially offer one solution for detecting signals of polygenic sweeps (when gene action is additive).

Because the expected signal for any *single* site under polygenic selection is expected to be (at best) very hard to detect, most of the approaches for attempting to detect polygenic selection are in effect *trait-based*, relying on some composite measure, such as an estimated genotypic value obtained by summing GWAS (or QTL) effect-size estimates over a large number of sites (e.g., Berg and Coop 2014). Hence, these approachess can work for scans of *specific traits* using genomic information about those traits. However, at present, they are not applicable to a genomic scan of random markers. A number of such trait-assisted marker methods (candidate markers are chosen because they impact a known trait) for detecting selection are examined in Chapter 12, while Chapters 9 and 10 focus on detecting selection simply from genomic information without any regard to any specific trait.

Under what situations might one expect hard sweeps versus polygenic adaptation? Given the vast reservoir of standing variation for most traits, a shift to a new environment will likely have an initial polygenic response, but a major allele or major mutation could still occur and have very dramatic effects (Chapter 25). Thus, hard sweeps are expected in situations where little standing variation for the trait is present, as might occur for traits with a long history of consistent directional selection or those with a simple genetic basis (i.e., that are only influenced by a few loci). In such cases, any further response might be mutation limited. However, it can also be the case that low standing variation is in part due to one (or more) alleles of major effect initially residing at a low frequency because of deleterious fitness effects in the original environment (Lande 1983). A sudden shift in the environment may increase selection pressure to the point where the major effect on a trait now gives the allele a net selective advantage, which can result in a hard sweep (from standing variation) if the allele is sufficiently rare (we revisit this topic in Chapter 25).

GENOME-WIDE IMPACT OF REPEATED SELECTION AT LINKED SITES

Up to this point, our focus has been on the local impact of a single sweep. There is a much broader picture as well—*recurrent* hitchhiking events can have profound effects over the entire genome, a topic first introduced in Chapter 3. Indeed, Maynard Smith and Haigh (1974) proposed that **recurrent selective sweeps** could depress variation throughout a genome, providing insight into the observation that levels of polymorphism (having an expected value of $4N_e\mu$ under the equilibrium drift model) only weakly scale with N_e (as noted in Chapter 4, an evolutionary decrease in μ with increasing N_e is likely a contributing factor). Large-scale sequencing has led to the current view that recurrent selection at linked sites does indeed have a profound effect on many, and perhaps most, genomic regions, reducing standing levels of variation by lowering N_e (reviewed by Lynch 2007; Leffler et al. 2012; Cutter and Payseur 2013; Corbett-Detig et al. 2015). Such a reduction in N_e increases the fraction of new mutations that are effectively neutral, potentially increasing substitution rates in regions where weakly deleterious mutations are common. The current debate concerns how much of these genome-wide effects are due to recurrent sweeps (adaptive evolution) versus background selection against deleterious mutations (purifying selection). Building on our discussion on N_e in Chapter 3, our goal in this section is to introduce the basic theoretical results relevant to this issue, many of which follow from the above machinery for sweeps. In light of the theory, we also examine the empirical data. It is worth stressing that in the following discussion we are not considering the effects of *specific* sweeps at specific sites, but rather the expectation of their impact over all times and sites, where beneficial and deleterious mutations are expected to occur (roughly) uniformly over a chromosome.

Association Between Levels of Variation and Localized Recombination Rates

Early studies in *D. melanogaster* showed that regions with reduced recombination also have reduced genetic variation, which is typically measured by the amount, π, of nucleotide diversity at putatively neutral sites (Aguadé et al. 1989; Begun and Aquadro 1991; Berry et al. 1991). The fact that between-species divergence does not appear to be depressed in these regions suggests that a reduction in the mutation rate is not the culprit, and the initial interpretation was that this pattern is generated by recurrent sweeps of favorable mutations, which reduce linked neutral variation. While such positive correlations between levels of recombination and π are widespread, they are by no means universal. Such an association may be restricted to particular regions, such as chromosome arms, or it may be missing entirely in a species. For example, Flowers et al. (2012) and Cutter and Payseur (2013) found a positive correlation between π and the per-base-pair recombination rate in at least some genomic regions in 14 of 20 sampled organisms, showing that while common, this pattern is not universal. Many features, such as the strength and frequency of adaptive sweeps or background selection, likely vary over genomic regions, and could account for the lack of a consistent pattern. In particular, predicted patterns of variation should not be based on recombination rates alone, but rather on the ratio of the mutational target size for fitness effects relative to the recombination rate. For the same level of recombination, a stronger effect is expected in more gene-rich regions (under both background-selection and recurrent-sweep scenarios).

As whole-genome sequencing becomes widespread, broader surveys of the relationship between π and c over entire genomes (as opposed to randomly selected regions) should provide further insight into the underlying causes and constraints. In an initial survey of 40 obligately sexual eukaryotes, Corbett-Detig et al. (2015) found that the effects of linked selection are pervasive, with levels of variation positively correlated with local recombination rates. Further, the strength of this association is larger in species with larger census sizes, N (using adult body size and range as surrogate variables for N), as might be expected, given that larger populations allow selection to overpower drift over a wider range of selection coefficients. The authors fit likelihood-based models for background selection (BGS), recurrent sweeps (hitchhiking, HH), and BGS+HH, finding that the BGS model significantly improved the fit in all cases. Evidence for HH (based on a likelihood-model fit with either HH or BGS+HH) was stronger in species judged to have larger census sizes. The authors concluded that BGS was pervasive over all population sizes, suggesting (as did Cutter and Payseur 2013), that ***BGS, not neutrality, should be the default model for tests of positive selection***. However, while BGS was important at all population sizes, Corbett-Detig et al. (2015) found that recurrent sweeps become progressively more important as population size increases. One caveat with their analysis was raised by an earlier study by Romiguier et al. (2013), who found that body size and species range are not very predictive of π, and hence may be poor predictors of N_e.

Example 8.13. Some insight into the relative contributions of gene density versus recombination rates to patterns of variation can be gleaned by comparing species that show different associations between regional gene density and recombination rates. In rice, the highest gene density is found in high recombination regions (the chromosome arms), and Flowers et al. (2010) found that localized levels of variation were negatively correlated with gene density, but either uncorrelated or weakly *negatively* correlated with recombination rate. They suggested that the negative association of variation with gene density (less variation in more gene-dense regions) is consistent with background selection. Accounting for the observed negative association with recombination (higher variation in regions of low recombination) is more problematic. One possible explanation is Hill-Robertson interference (Chapter 7), as selection during the domestication of rice would be more efficient under higher recombination rates, resulting in stronger sweeps (and hence lower levels of π) in regions with elevated recombination.

The genome of the nematode *Caenorhabditis elegans* is structured in a very different fashion than rice. In *C. elegans*, high gene density occurs in the low-recombination regions around chromosome centers, lower gene density with higher recombination in chromosome arms, and intermediate gene density with no recombination at chromosome tips. Rockman et al. (2010) examined the genomic distribution of **local expression quantitative trait loci (eQTLs)** in *C. elegans*, defined as QTLs influencing the expression of genes close to the mapped eQTL location (within one LOD score; see LW Chapter 14). Simply based on gene density, one might expect an excess of eQTLs in chromosome centers, as more genes are located there. In fact, however, Rockman et al. observed the opposite, with 34% of transcripts in the low-density arms having local eQTLs, but with only 10% of transcripts from the high-density centers having local eQTLs. Rockman et al. suggest that background selection (BGS), which is expected to be higher in the centers (there is a higher ratio of gene density to recombination), could account for this reduction in variation. BGS could also account for the observation that only 20% of the transcripts from chromosome tips have eQTLs. Although there is little recombination, sites in the tips are only linked to deleterious mutations on one side, and hence experience somewhat weaker background selection. A striking feature of this analysis is that the authors were not judging variation by sequence diversity, but rather through a *functional* screen (measuring the presence or absence of eQTLs). In their words, "the propensity of traits to vary in *C. elegans* is explained by processes independent of the functions of the individual transcript," with those processes involving selection at linked sites.

Impact of Recurrent Hard Sweeps on Levels of Neutral Variation

As illustrated in the previous example, levels of recombination often vary widely over the genome. In a region with low recombination (such as telomeric and centromeric regions, very small chromosomes such as the fourth of *D. melanogaster*, and most of the heterogametic sex chromosome), even weak selection at a distant location can have an impact. In the extreme case where an entire genome has *no* recombination (such as in some bacteria or organelles), a single advantageous mutation can sweep a single genomic type to fixation. Laboratory populations of bacteria often show the phenomena of **periodic selection** (Atwood et al. 1951a, 1951b; Kock 1974; Dykhuizen 1990; Guttman and Dykhuizen 1994), wherein genetic diversity builds up slowly over time, only to be rapidly removed before starting to build all over again. Presumably, this is due to the periodic fixation of newly appearing favorable mutations, which generate sweeps that fix single chromosomal lineages. In this setting, the standing level of variation is a function of the frequency of sweeps. If it is sufficiently common, the population never has a chance to reach mutation-drift equilibrium before the next sweep, while if rare, the population may be at the mutation-drift equilibrium for much of the time. Thus, ***the rate of adaptation at least partly determines the amount of neutral variation***, a theme returned to throughout this section.

For a population of constant size undergoing periodic hard sweeps, Wiehe and Stephan (1993) found the equilibrium level of heterozygosity to be a decreasing function of the rate of adaptation, λ. Here, λ is defined as the number of beneficial mutations destined to become fixed that arise in each generation. This is given by the product of the number of new beneficial mutations arising in each generation ($2N\mu_b$) and their individual fixation probabilities $(2sN_e/N)$

$$\lambda = (2N\mu_b)(2sN_e/N) = \theta_b s \tag{8.24a}$$

where $\theta_b = 4N_e\mu_b$. Typically, we take μ_b as a per-nucleotide rate, which scales λ on a per-nucleotide basis, so that $L\lambda$ is the expected rate for a region of L nucleotides. We have assumed no epistasis in fitness, so that each new beneficial mutation adds a fitness contribution that is not influenced by the current genotype.

Wiehe and Stephan found that with recurrent sweeps, the expected heterozygosity, as measured by nucleotide diversity, π, at linked neutral sites, is

$$\frac{\pi}{\pi_0} \simeq \frac{c_0}{c_0 + \lambda\,\gamma\,\kappa} = \frac{1}{1 + \lambda\,\gamma\,\kappa / c_0} \tag{8.24b}$$

where $\pi_0 = 4N_e\mu$ is the average heterozygosity at a single site for an equilibrium neutral population under no sweeps, c_0 is the average per-nucleotide recombination rate over the region of interest (such as an entire chromosome), $\gamma = 2N_e s$ is the scaled strength of selection per adaptive nucleotide, λ is the per-nucleotide adaptive substitution rate over the region of interest, and there is a constant, $\kappa \simeq 0.075$. For $c_0 > \lambda\gamma\kappa$, because $1/(1+x) \simeq 1 - x \simeq e^{-x}$ for $|x| \ll 1$, Equation 8.24b rearranges to

$$\frac{\pi}{\pi_0} \simeq 1 - \frac{\lambda\gamma\kappa}{c_0} \simeq \exp\left(-\lambda\gamma\kappa/c_0\right) \tag{8.24c}$$

Rearranging the linear approximation gives the **Stephan regression** (Stephan 1995) for explaining the average variation, π, within a region as a function of the region's per-base recombination rate, c_0,

$$\pi = \pi_0 - (\pi_0\lambda\gamma\kappa)\,x, \quad \text{where } x = 1/c_0 \tag{8.24d}$$

Fitting Equation 8.24d, Wiehe and Stephan (1993) obtained an estimate of $\lambda\,\gamma \simeq 1.3 \cdot 10^{-8}$, based on 17 loci in medium to high recombinational backgrounds in *D. melanogaster*. For a modest recombination rate of 1 cM per megabase ($c_0 = 0.01/10^6 = 10^{-8}$), applying Equation 8.24b with this value of $\lambda\,\gamma$ yields

$$\frac{\pi}{\pi_0} \simeq \frac{10^{-8}}{10^{-8} + (1.3 \cdot 10^{-8} \cdot 0.075)} = 0.911$$

or roughly a 9% reduction in background heterozygosity. For smaller recombination rates, say 0.1 cM per megabase ($c_0 = 10^{-9}$), standing levels of variation are reduced by 49%. In a region of high recombination (2.5 cM/Mb, $c_0 = 2.5 \cdot 10^{-8}$), the reduction in π is only 3.7%. Hence, in regions of low recombination, recurrent selective sweeps can have a dramatic effect on standing levels of variation. Additional estimates of $\lambda\gamma$ are summarized in Table 8.3, and further developed in Chapter 10.

Example 8.14. Using machinery from Chapter 10 on a set of X-linked genes in *D. melanogaster*, Andolfatto (2007) and Jensen et al. (2008) obtained estimates of λ of $7.5 \cdot 10^{-10}$ and $4.2 \cdot 10^{-11}$ (respectively) for the per-nucleotide adaptation rate (summarized in Table 8.3). Consider a region with a length of 100 kb. Under Andolfatto's estimate, the per-generation rate of adaptive substitutions over a region of this size is $10^5 \cdot 7.5 \cdot 10^{-10} = 7.5 \cdot 10^{-5}$ or one sweep in roughly every 13,300 generations. Under Jensen's estimate, a sweep influencing this region occurs roughly every 238,000 generations. Potential reasons for such a vast difference in the estimates will be examined shortly.

A Few Large or Many Small Sweeps?

Because reduction in diversity from sweeps is a function of the product $\lambda\gamma$, the same average reduction in π could be caused by either a few large sweeps (where λ is small and γ large) or many smaller sweeps (where λ is large and γ small), as long as their product is held constant. With rare, strong sweeps, there would be dramatic reduction in variation over a fairly large region, but many regions would see little effect, as no recent sweep would have occurred in their vicinity. Conversely, with many weaker sweeps, most regions would be influenced, but each by a smaller amount. While the expected value of π is the same under both models, the variance in π is expected to be much greater under rare but strong sweeps (Jensen et al. 2008).

Distinguishing between the strong and weak selection scenarios requires an independent estimate of either λ or γ in addition to an estimate of $\lambda\gamma$. Methods to accomplish this are more fully developed in Chapter 10, but one approach is as follows. Suppose L sites within a region of interest are examined between two populations that separated t generations ago (for a total divergence time of $2t$). If a total of D sites show divergence, $d = D/L$ is the per-site divergence. Ignoring multiple mutations at the same site, if α denotes the fraction of all divergent sites that are adaptive, $d\alpha$ is the per-site number of adaptive fixations that occurred over $2t$ generations. This gives the rate as $\lambda = d\alpha/(2t)$. Estimates of t can be obtained from several sources (such as rates of neutral divergence), but estimates of the adaptive fraction, α, are more elusive. However, as detailed in Chapter 10, estimates of α for coding regions follow by noting that, under the assumption of neutrality, the ratio of the number of silent to replacement polymorphic sites should equal the ratio of the number of silent to replacement substitutions (Example 10.1). An excess of replacement substitutions presumably reflects the role of adaptive evolution, and the amount of excess provides an estimate of α (e.g., Example 10.1), and hence of λ. As detailed in Chapter 10, this approach for estimating α makes the strong assumption of long-term constancy in N_e.

Letting d_a denote the per-site rate of amino acid divergence, and substituting $\lambda = d_a\alpha/(2t)$ into Equation 8.24b, yields the **Andolfatto regression**

$$\pi \simeq \pi_0 \frac{c}{c + \lambda\gamma\kappa} = \pi_0 \frac{c}{c + [\alpha\gamma\kappa/(2t)]\, d_a} = \frac{\pi_0}{1 + \beta x} \tag{8.25}$$

where $x = d_a/c$ and $\beta = \alpha\gamma\kappa/(2t)$, and because λ is now measured on a per-amino acid basis, c is now the recombination rate per codon (Andolfatto 2007). In other words, scaling the per-site divergence, d_a, for each gene by its local rate of recombination, c, yields a nonlinear regression between observed nucleotide diversity, π, and the recombinational-scaled divergence, d_a/c. When one assumes a constant recombination rate, c, among the sampled sites, the regression is just on $x = d_a$ (which varies over regions), where now $\beta = \alpha\gamma\kappa/(2tc)$, which (with estimates of α, c, and t) returns γ (as is done in Figure 8.9). Chapter 10 reviews other approaches for estimating α and/or γ from joint polymorphism and divergence data at single loci.

A technically more demanding approach for gaining insight into the parameters of adaptation is to jointly estimate two of these three parameters (λ, γ, or $\lambda\gamma$) using the spatial pattern of genetic variation over a genomic region (Macpherson et al. 2007; Jensen et al. 2008). Figure 8.10 shows the motivation for this idea. Jensen et al. (2008) noted that strong selection should produce a higher variance in π and other measures of genetic variation that are impacted by a sweep (Table 8.2), for example, the number of segregating sites (S), an excess of high-frequency derived alleles, and pairwise linkage disequilibrium (LD). Using simulations, they examined the behavior of the coefficient of variation (CV) for summary statistics for these quantities as a function of the size, L, of the unit of analysis. Over small regions (L of 500 to 1000 bp), there was little difference in the CV between the rare and strong and the frequent and weak sweep models, but as the size of the analysis region increased, so did the CV for strong, but not weak, selection. Based on this observation, Jensen et al. (2008) developed an approximate Bayesian computation (ABC) approach that jointly considers

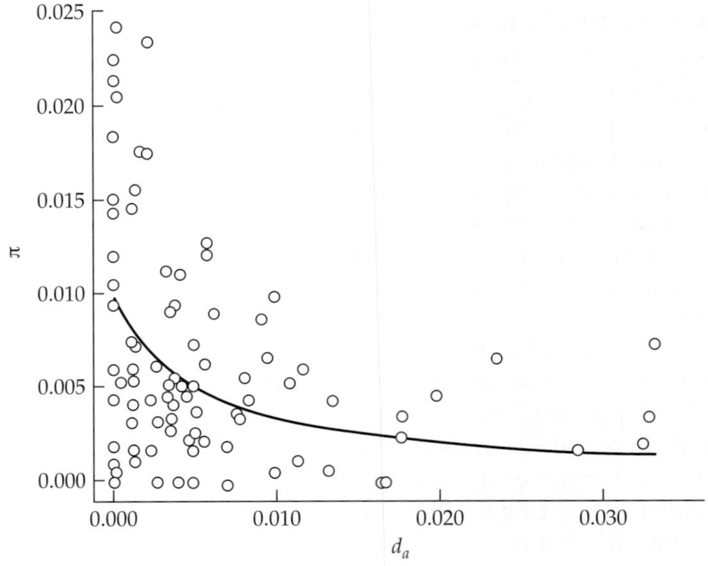

Figure 8.9 An example of the Andolfatto regression of nucleotide diversity, π, on the per-site amino acid divergence, d_a, in *Drosophila miranda,* under the assumption of the same recombination rate (c) among all sampled sites. In this setting, the regression given by Equation 8.25 becomes $\pi = \pi_0/(1+\beta x)$, where $x = d_a$ (the mean number of amino acid substitutions per codon in the region for the estimated π value) and $\beta = \alpha\gamma\kappa/(2tc)$. The solid curve is the least-squares fit of Equation 8.25, which gives estimates of π_0 and γ (as α, c, and t were independently estimated). (After Bachtrog 2008.)

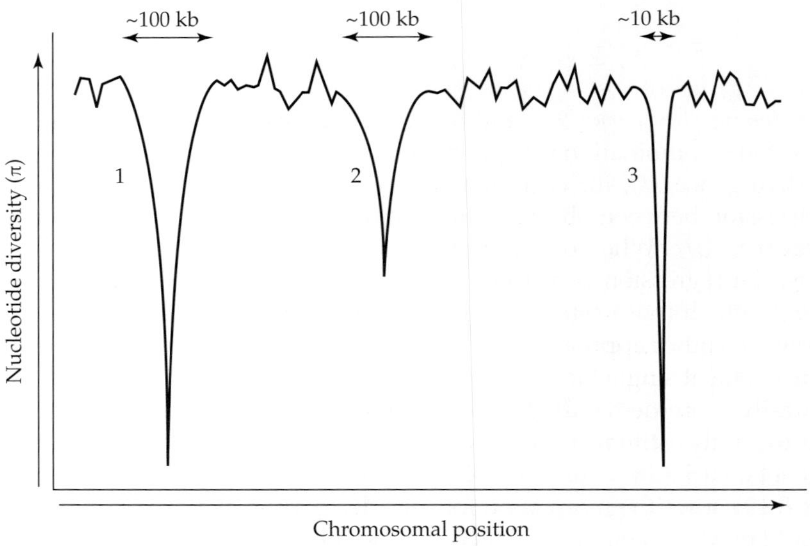

Figure 8.10 The pattern of nucleotide diversity over a large region may provide clues about the frequency and strength of past sweeps. Within this hypothetical region, three hard sweeps have occurred. Sweep 1 is a strong, recent sweep; sweep 2 is a strong older sweep; and sweep 3 a weak, recent sweep. Strong sweeps result in a depression in variation over a significant region (1). As the signal from a past sweep decays, its window of influence stays roughly the same size, but its impact within that window vanishes over time. An old, strong sweep leaves a weak signal of depressed variation over a fairly large region (2), while an old weak sweep leaves a signal similar to (2), but over a much smaller region. (After Macpherson et al. 2007.)

Table 8.3 Estimates of the per-nucleotide adaptive evolution rate, λ, and its components for several *Drosophila* species, aspen (*Populus tremula*), and humans. Depending on the study, the scale of analysis is either chromosomal or genome-wide. Methods for estimating individual components of the product $\lambda\gamma$ are developed in Chapter 10. Here $\gamma = 2N_e s$ is the scaled strength of selection, λ is the rate of adaptive substitutions per base pair, per generation, and s is the average strength of selection of a beneficial mutation. The species listed provided the polymorphism data used in these methods, while an outgroup was used for other estimates of λ (Equation 10.11a).

Organism	$\lambda\gamma$	γ	s	λ	Reference
D. melanogaster	$3.9 \cdot 10^{-7}$	34,400	$2.0 \cdot 10^{-3}$	$6.0 \cdot 10^{-11}$	Li and Stephan 2006
	$5.1 \cdot 10^{-8}$	74	$2.3 \cdot 10^{-5}$	$7.0 \cdot 10^{-10}$	Bachtrog 2008
	$2.6 \cdot 10^{-8}$	35	$1.2 \cdot 10^{-5}$	$7.5 \cdot 10^{-10}$	Andolfatto 2007
	$4.0 \cdot 10^{-7}$	10,000	$2.0 \cdot 10^{-3}$	$4.2 \cdot 10^{-11}$	Jensen et al. 2008
				$1.8 \cdot 10^{-11}$	Smith & Eyre-Walker 2002
				$3.6 \cdot 10^{-11}$	Andolfatto 2005
	$1.3 \cdot 10^{-8}$				Wiehe & Stephan 1993
		10			Schneider et al. 2011
D. simulans	$1.1 \cdot 10^{-7}$	30,000	$1.0 \cdot 10^{-2}$	$3.6 \cdot 10^{-12}$	Macpherson et al. 2007
D. miranda	$1.2 \cdot 10^{-6}$	3100	$2.7 \cdot 10^{-3}$	$4.0 \cdot 10^{-10}$	Bachtrog 2008
P. tremula	$1.5 \cdot 10^{-7}$				Ingvarsson 2010
Humans				$2.3 \cdot 10^{-12}$	Example 10.12

the means and variances of summary statistics measuring these various sweep features to obtain separate estimates of λ and s (for details see Example A3.9).

Macpherson et al. (2007) also used the chromosomal pattern of nucleotide diversity around putative sweeps to obtain separate estimates of λ and $\lambda\gamma$. As above, an estimate of the latter was obtained from the regression (Equation 8.24b) of π on c_0 (nucleotide diversity on local recombination rates). They then introduced a new statistic, Q_s, the ratio of a minimal estimate of heterozygosity within a region of fixed length (a window) to the average heterozygosity over the region scanned by the windows. They showed that the expected value of Q_s is a function of both λ and $\lambda\gamma$, so that the joint pair of statistics, Q_s and π, allows for separate estimates of λ and $\lambda\gamma$. One significant caveat, as stressed by the simulations of Schrider et al. (2015), is that the observed minimal region of heterozygosity may poorly correspond to the selected site, which can bias estimates using this method.

As summarized in Table 8.3, while estimates of the product, $\lambda\gamma$, for various studies in *Drosophila* are reasonably compatible, individual estimates of its components γ (or s) and λ differ by several orders of magnitude. There are several potential reasons for this. Different studies of even the same species may use different populations as well as different sets of genes, such as autosomal (Macpherson et al. 2007) versus X-linked (Andolfatto 2007; Bachtrog 2008; Jensen et al. 2008). However, the different methods employed may be the major contributor to the significant disparity between studies. Estimates based on short regions (single genes) as the unit of analysis, such as those by Andolfatto (2007) and Bachtrog (2008), found small estimates of γ and s in *D. melanogaster* (γ between 35 and 74, implying an s of around 10^{-5}). Estimates based on much longer regions (10–100 kb), such as Macpherson et al. (2007) and Jensen et al. (2008), found much larger estimates, with $\gamma \simeq$ 10,000 to 30,000 and $s \simeq 0.002$ to 0.01. Estimates obtained by Bachtrog (2008) for *D. miranda* using a number of small regions were intermediate, with $\gamma = 3000$, $s = 10^{-3}$.

Sella et al. (2009) suggested that these estimates of γ and λ may actually be more compatible than their spread suggests (Figure 8.10). Weak selection leaves a strong signal over only a very small region, while strong selection leaves a signal over a much larger region. For example, using a recombination rate of 1 cM/Mb ($c_0 = 10^{-8}$), Equation 8.6a suggests that weak sweeps ($\gamma = 35$, $s = 10^{-5}$) influence at most a few hundred bases, while strong sweeps ($\gamma = 10,000$, $s = 0.01$) can influence almost a hundred kilobases. Sella et al. suggested that methods using small regions (such as single genes) for their units of analysis are biased toward the detection of weak selection, while methods using much larger regions are biased toward strong selection. If the size of the region of analysis does does bias results, then weak selection accounts for most of the observed between-population divergence, while strong selection accounts for most of the reduction in diversity. Sattah et al. (2011) found strong support for this view by examining the pattern of nucleotide variation around substitutions in *Drosophila simulans*. As expected from the neutral theory, there was a slight elevation in standing diversity around silent-site substitutions (higher neutral substitution rates implying higher mutation rates, and thus higher levels of polymorphism). After adjusting for this effect, they found a trough in nucleotide diversity around sites that resulted in amino acid substitutions, as would be expected given a sweep. Using a composite-likelihood approach to test for a sweep given the local diversity pattern around a site (Chapter 9), they estimated that around 13% of the substitutions resulted in sweeps. A mixture model allowing for different strengths of selection fit the genomic data the best and suggested that about 30% of the sweep sites (4% of the total substitutions) were under strong selection, with a mean s value of $\sim$0.005, while the remainder had a mean s of $\sim 4 \cdot 10^{-5}$.

Finally, up to this point our analysis of the impact of recurrent sweeps has assumed that all sweeps were hard. At the other extreme are partial sweeps, where an allele increases rapidly in frequency (to a value of x) in one selective environment, which then changes. Equation 8.15 shows the impact of a *single* partial sweep on π. Coop and Ralph (2012) showed that the reduction in nucleotide diversity due to *recurrent* partial sweeps has a form very similar to Equation 8.24b, namely

$$\frac{\pi}{\pi_0} \simeq \frac{1}{1 + \lambda\zeta/c_0} \tag{8.26}$$

where ζ is a function of the time for the sweep to reach its target frequency, x (Equation 8.4e).

Selective Interference and the Hill-Robertson Effect

The above results for recurrent sweeps assume that *concurrent* sweeps influencing the same region are rare. Given that the expected number of new beneficial mutations introduced into the population in each generation scales as NU_b (with U_b being the total beneficial mutation rate for the region of interest), this assumption breaks down in sufficiently large populations. In this setting, multiple sites are segregating beneficial mutations, and sweeps can interfere with each other (Kim and Stephan 2003; Neher et al. 2010; Yu and Etheridge 2010; Neher and Shariman 2011; Weissman and Barton 2012; Neher 2013; Neher et al. 2013).

If sites with segregating beneficial mutations are loosely linked, their effect is to lower the effective population size that a particular site experiences, making selection slightly less efficient (the classic interpretation of a Hill-Robertson effect; Chapter 7). A more dramatic effect occurs when some of these sites are tightly linked, in which case there can be strong **selective interference** among them. One example is **clonal interference** (Gerrish and Lenski 1998), seen in very large populations with nearly complete linkage (i.e., bacteria). Competition occurs among the set of lineages (clones) carrying different beneficial mutations, as each such clone has a strong advantage against lineages lacking such mutations but (at best) only a weak advantage against other beneficial lineages. In a sexual population, recombination can shuffle genomes to combine beneficial mutations, but the dynamics are very different in large asexual populations. In the latter, NU_b is expected to be large, with multiple favorable mutations arising in each generation but with little (or no) recombination. In this setting, simply acquiring a new beneficial mutation is *not* sufficient for the fixation of a particular genome, as additional beneficial mutations continue to arise and an otherwise favorable genome may not acquire *new* additive favorable mutations as quickly as other genomes and thus become lost (Desai and Fisher 2007). Neher et al (2010) showed that while the appearance of favorable new mutations scales linearly with N (as NU_b), selective interference significantly *reduces* the probability of individual fixation as N increases, resulting in an eventual rate of adaptive evolution that scales as $\log(NU_b)$.

Weissman and Barton (2012) extended this analysis to sexual species. Let $\Lambda = L\lambda$ be the total rate of adaptation over a region of length L nucleotides with a total recombination rate of $C = Lc_0$, where c_0 is the average recombination rate between adjacent nucleotides. If $\Lambda_0 = L\lambda_0$ is the expected rate of adaptation ignoring the effects of interference, then

$$\Lambda \simeq \frac{\Lambda_0}{1+2\Lambda_0/C} \quad \text{or} \quad \lambda \simeq \frac{\lambda_0}{1+2\lambda_0/c_0} \tag{8.27a}$$

Expressing this as

$$\Lambda/C \simeq \frac{\Lambda_0/C}{1+2\Lambda_0/C} \tag{8.27b}$$

shows that for very large populations, where $\Lambda_0/C \gg 1$ (i.e., $\lambda_0/c_0 \gg 1$), the upper limit of adaptation is given by $\Lambda \simeq C/2$ (or $\lambda \simeq c_0/2$). For example, for a region of 1 centimorgan ($C = 0.01$), the total rate of adaptive substitutions over that region is bounded by $\Lambda \leq 0.005$, or the appearance of one successful beneficial mutation in every 200 generations. There is an additional effect from selection at unlinked sites, which reduces the rate further by $\exp(-4\Lambda s)$, which (for small s) is negligible relative to the effects of linkage. Weissman and Barton used their results to examine the impact of selective interference on the levels of variation at linked neutral sites. For small values of λ_0/c_0, there is not much interference among sweeps and Equation 8.24b is a good approximation for the loss of linked neutral variation. Conversely, when interference is strong, substantial neutral variation can still be maintained, as the density of sweeps (λ/c_0) approaches a limiting asymptotic value and the coalescent times for pairs of neutral alleles (and hence the amount of variation that is maintained) is a function of sweep density.

This notion of interference among linked selected sites is not restricted to beneficial alleles. Indeed, it may have a significant impact when tightly linked, weak *deleterious* alleles are involved. We have referred to situations where selection at one site influences the effectiveness of selection at a second as a Hill-Robertson (HR) effect (Chapter 7). As the previous discussion on selective interference illustrates, while the HR effect is usually regarded as a reduction in N_e, this is not the whole story (Felsenstein 1974; Comeron and Kreitman 2002; Neher and Shariman 2011; Schiffels et al. 2011; Neher 2013; Neher et al. 2013), as selective interference must also be considered. With free recombination between selected sites, gametes quickly arise that contain multiple favorable alleles, giving a wider range of fitness values and hence more efficient selection due to a higher additive variance in fitness. Conversely, with linkage, the production of gametes with multiple favored alleles is retarded (selection generates negative linkage disequilibrium among favorable alleles; Chapters 5 and 16), resulting in a smaller additive variance in fitness among individuals in the population and hence less efficient selection.

Background Selection: Reduction in Variation Under Low Recombination or Selfing

Charlesworth et al. (1993a) challenged the view that the reduction of variation in regions of low recombination was evidence for periodic selective sweeps (and hence the frequent substitution of adaptive alleles). They noted that the same pattern can be generated by selection *against* new deleterious mutations. Hence, purifying selection can potentially account for this pattern of reduced variation without the need to invoke adaptive selection. This occurs because the removal of new deleterious mutations also lowers the effective population size, and in a sufficiently long region of low recombination, the number of targets for mutation may be large enough to generate a high total deleterious mutation rate and therefore a significant reduction in variation. They referred to this process as **background selection, BGS**, which was introduced in Chapter 3. We review (and generalize) some of our results from that chapter here in order to more fully contrast BGS with recurrent sweeps.

Charlesworth et al. (1993a) estimated the potential impact of BGS as follows. First, consider a neutral site completely linked to a region in which new deleterious mutations arise at a rate of U (per diploid). A key assumption is that these new mutations are sufficiently deleterious to be removed rapidly, so that the population is at an equilibrium with the removal of mutation-bearing chromosomes through selection that is balanced by the creation of new chromosome types (haplotypes) by mutation. If we assume that the fitness of a new deleterious mutation (in the heterozygous state) is $1 - s$ and that fitness over the loci is multiplicative, the expected number of deleterious mutations per gamete at the mutation-selection equilibrium is $U/(2s)$ (Kimura and Maruyama 1966). Further, the number of mutations follows a Poisson distribution, so the probability of a mutation-free gamete is given by the zero term of a Poisson,

$$f_0 = \exp\left(-\frac{U}{2s}\right) \tag{8.28}$$

The effect of background selection is to reduce the effective population size from N_e to $f_o N_e$, giving an expected reduction in neutral variation of $\pi/\pi_0 = f_0$ as well. Because selfing reduces the opportunity for recombination, the effects of background selection can be quite significant in highly selfing populations, such as in many plants. Charlesworth et al. (1993a) noted that the reduction in π in strict selfers is given by Equation 8.28, with s replaced by $2s$, the selection against mutant homozygotes (assuming additivity).

Hudson and Kaplan (1995) extended these results by allowing for recombination. For a neutral locus in the middle of a region of length L with a total recombination frequency of C,

$$\frac{\pi}{\pi_0} \simeq \exp\left(-\frac{U}{2s + C}\right) \tag{8.29a}$$

where $U = L2\mu$ is the total diploid deleterious mutation rate within this region and $C = Lc_0$, and where (as above) μ and c_o denote the average rates of deleterious mutation and recombination per nucleotide. When the total amount of recombination within the region is large relative to s ($C \gg s$),

$$\frac{\pi}{\pi_0} \simeq \exp\left(-\frac{2\mu}{c_0}\right) \tag{8.29b}$$

and thus the decline in heterozygosity is approximately independent of the strength of selection. Note that this expression is of the same form as the exponential approximation for recurrent sweeps (Equation 8.24c). Likewise, because $e^{(-x)} \simeq 1 - x$ for $|x| \ll 1$, it follows for moderate to high recombination ($2\mu/c_0 \ll 1$) that

$$\frac{\pi}{\pi_0} \simeq 1 - \frac{2\mu}{c_0} \tag{8.29c}$$

This is the same form ($\pi/\pi_0 = 1 - b/c_0$) as our result for moderate to high recombination under recurrent selective sweeps ($b = 2\mu$ in Equation 8.29c, $b = \lambda\gamma\kappa$ in Equation 8.24c). As a consequence, in regions of moderate to high recombination, the regression of π on c_0 cannot distinguish between hitchhiking and background selection.

Conversely, in regions where recombination is weaker (so that the linear approximation given by Equation 8.29c no longer holds), the functional form of the relationship between π with c_0 differs between BGS and recurrent sweeps, potentially offering some hope of resolution between the two competing models in regions with very low recombination rates. Both Hudson and Kaplan (1995) and Charlesworth (1996) found that the BGS model gave a poor fit in *D. melanogaster*. Hudson and Kaplan were able to improve the fit of a BGS model to the third chromosome of *D. melanogaster*, but only after assuming much smaller selection coefficients for the low-recombination regions than used for the recombining regions on this same chromosome. The problem, as noted by Kaiser and Charlesworth (2009), is that the standard BGS model *overpredicts* the reduction in π in regions of very low recombination. They reasoned that this poor fit might occur in regions where U is sufficiently large that multiple deleterious alleles are segregating at any given time. As with beneficial mutations, multiple segregating deleterious mutations interfere with each other, reducing the efficiency of selection and resulting in less reduction in variation at linked sites. Incorporating this effect into their simulation results yielded reductions consistent with observed values in regions of very low recombination. We return shortly to the implications of such selective interference in regions of very low recombination.

A complicating factor when modeling background selection is **Muller's ratchet** (Muller 1964; Felsenstein 1974): in a region of very low recombination, the class of chromosomes carrying no deleterious mutations may become lost due to drift. If such an event occurs, then without recombination, there is no way (other than by extremely fortuitous back mutations) to recover mutation-free chromosomes. At this point, a new class (say, harboring just a single deleterious mutation) now becomes the most fit chromosomal type. This new class also may eventually be lost by drift, leading to chromosomes with two deleterious mutations now being the most fit class, turning the ratchet once again, and so on. The assumption leading to Equation 8.28 is that the zero class is at equilibrium (i.e., is unlikely to be lost in reasonable biological time).

Recall our discussions of effective neutrality in Chapter 7, where drift overpowers selection unless $N_e|s| \gg 1$. Under background selection, an unselected effective population size of N_e is reduced to $\sim f_0 N_e$, requiring $s f_0 N_e \gg 1$ for selection to overpower drift (and hence preventing the fixation of most deleterious alleles). Thus, weak selection and/or a small N_e are required for the ratchet to operate. Provided that $f_0 N_e s > 10$, selection is efficient enough to prevent operation of the ratchet and the reduction in effective population size from background selection is well approximated by Equation 8.28. When the ratchet is operating, in addition to reducing the background variation, an excess of rare alleles is generated, skewing the site-frequency spectrum (Gordo et al. 2002). This has significant implications if one is tying to distinguish between BGS and recurrent sweeps.

Of course, it is likely that *both* background selection and recurrent sweeps operate at some level. Kim and Stephan (2000) showed that Equation 8.24b can be modified to give the approximate diversity when both act, yielding

$$\frac{\pi}{\pi_0} \simeq \frac{f_0 c_0}{c_0 + \lambda_c\, f_0 \gamma\, \kappa} = \frac{1}{(1/f_0) + (\lambda_c\, \gamma\, \kappa / c_0)} \tag{8.30a}$$

where $f_0 N_e$ is the effective population size after correcting for BGS. This reduction in effective population size changes the scaled strength of selection from $\gamma = 2N_e s$ to the value appearing in Equation 8.30a, namely, $f_0\gamma = 2N_e f_0 s$. The more subtle correction is that the reduction in effective population size from BGS (which changes with c_0) also changes the fixation probabilities for new favorable mutations, so that λ, the product of the fixation probability and the number of new adaptive mutations arising per generation, is now a function of the recombination rate, c_0 (and it is indexed as λ_c in Equation 8.30a to remind the reader of this fact). Kim and Stephan (2000) suggested that recurrent sweeps are likely to be more important in regions of very low recombination, while BGS is more relevant in high-recombination regions. On a practical note, comparison of Equations 8.24b and 8.30a shows that ***ignoring background selection results in an inflated estimate of*** $\lambda\gamma$, and hence an inflated estimate of the rate of adaptation (Kim 2006).

Partial sweeps, wherein selection changes a selected allele to some moderate frequency, also change the coalescence rate relative to drift at linked neutral loci, although not as dramatically as for a completed hard sweep. Corbett-Detig et al. (2015) used the partial sweeps model of Coop and Ralph (Equation 8.26) and showed that Equation 8.30a can be generalized to

$$\frac{\pi}{\pi_0} \simeq \frac{1}{1/f_0 + \eta/c_0} \tag{8.30b}$$

where η is the number of new favorable mutations arising per generation times the pairwise coalescence rate (per generation) due to sweeps. The key feature is that the scaling with respect to c_0 is the same as in Equation 8.30a.

Our penultimate comment on BGS is to stress that it is not strictly a phenomenon of coding sequences. Indeed, the rather high rate of sequence conservation (and hence, functional constraints) seen for some noncoding DNA in *Drosophila* has important implications for background selection. Taking into account both its abundance and average level of constraint (Chapter 10), Andolfatto (2005) and Halligan and Keightley (2006) suggested that *Drosophila* noncoding DNA is likely a much larger deleterious-mutation target (by at least a factor of two) than coding DNA. This is still an open issue, as the actual fraction of noncoding DNA that may have a function is the subject of much debate (ENCODE Project Consortium 2012; Doolittle 2013; Fu and Akey 2013; Graur et al. 2013).

Finally, as first introduced by McVicker et al. (2009), one can use functional genomics data to construct a more granular accounting of background selection (Example 8.15). While this approach may seem overly technical, it leverages all available genomic information to more fully account for BGS. Because some level of BGS is always expected (deleterious mutations being ever-present), removing any of its potential influence is critical to understanding the roles of other forces in shaping the genome.

Example 8.15. Consider the impact of background selection on a neutral target site, i, due to k linked conserved sites (and hence, targets for background selection). Building on Hudson and Kaplan (1995), sum the impact over each potential site (assuming multiplicative fitnesses) and allow for a distribution of fitness effects over different classes of sites. The impact of background selection on site i is to reduce N_e to $B_i N_e$, where

$$B_i = \exp\left(-\sum_{j=1}^{k} \int_0^1 \left[\frac{\mu_i}{s(1+c_{ij}/s)^2} \right] \phi_j(s)\, ds \right) \tag{8.31a}$$

The term in the brackets is simply Hudson and Kaplan's (1995) expression for the impact of a deleterious mutation of effect s at distance c_{ij} from the target site j (an approximation of this leads to Equation 8.29a). The integral is taken over the distribution $\phi_j(s)$ of fitness effects for site j, and the results are summed for each potential site of background selection. While initially this expression seems unduly complex, it allows us to sum over sites with different effects, such as potential replacement sites and known conserved regions. For example, with v different classes of functional sites,

$$B_i = \exp\left(-\sum_{\ell=1}^{v} \sum_{j=1}^{k_\ell} \int_0^1 \frac{\mu_\ell}{s(1+c_{ij}/s)^2} \phi_\ell(s)\, ds \right) \tag{8.31b}$$

where we assume the same mutation rate, μ_ℓ, and distribution of deleterious fitness effects, ϕ_ℓ, for all linked sites in category ℓ. With a large amount of genomic data in hand (such as the site-frequency spectrum over a large number of sites), one can start to fit the distribution $\phi_\ell(s)$ of fitness effects within a class using the method of maximum likelihood (Chapter 9). See McVicker et al. (2009) for details.

Background Selection vs. Recurrent Selective Sweeps

While both BGS and recurrent sweeps reduce neutral variation in regions of low recombination, they represent the very different processes of purifying selection versus adaptive change. As such, evolutionary geneticists have spent considerable effort trying to distinguish between the two, but no clear answer has yet emerged (Hudson 1994; Andolfatto 2001; Sella et al. 2009; Charlesworth 2009, 2012; Stephan 2010b). As comparison of Equations 8.24c and 8.29c shows, for regions of moderate to high recombination, both processes predict a relationship of the form $\pi/\pi_0 \simeq 1-b/c \simeq e^{-b/c}$, where b is an unknown constant to be estimated. Hence, there is little resolution using the relationship between recombination and heterozygosity in moderate- to high-recombination genes. However, this is *not* the case for regions of low (but not too low) recombination. Innan and Stephan (2003) noted that in this region the regression of π on c is convex for recurrent sweeps and concave for BGS (compare Equations 8.24b and 8.29a). They applied this approach to a set of low-recombination X-linked genes in *D. melanogaster*, and found that recurrent sweeps gave a much better fit than BGS. However, when they examined two highly selfing species of tomatoes (*Lycopersicon*), BGS provided the better fit. In humans, Hellmann et al. (2008) found that recurrent sweeps gave a better fit than BGS but cautioned that this may simply be an artifact of the simplistic nature of the BGS model leading to Equation 8.29a (i.e., assuming no variation in s).

One distinct prediction between BGS and recurrent sweeps is the expected effect on the site-frequency spectrum (SFS). Under the "strong" version of BGS, deleterious mutations have *strong* effects ($4N_e|s| \gg 1$) and are quickly removed by selection. In this case, the effect is to simply lower N_e to $f_o N_e$, but not otherwise change the frequency spectrum away from its neutral shape (Charlesworth et al. 1993a, 1995). Conversely, under selective sweeps, an excess of sites with rare alleles is expected (Braverman et al. 1995; Kim 2006). As detailed in Chapter 9, a standard indicator of an excess of rare alleles (relative to the neutral Watterson SFS) is a negative value of Tajima's D statistic (Equation 9.24a), and such values are often (but not always) associated with genes showing reduced variation in regions of low recombination in *Drosophila* (e.g., Langley et al. 2000). A survey of 29 *D. melanogaster* genes by Andolfatto and Przeworski (2001) revealed a highly significant positive association ($r^2 = 0.31, p = 0.002$) between Tajima's D and recombination rate—as the recombination rate decreased, D became more negative. Such an observation is consistent with a recurrent sweep model, but not with the strong BGS model.

Nevertheless, while findings like these are suggestive of recurrent selection as opposed to BGS, they are not as conclusive as one might think. In fact, a model with *weakly* deleterious alleles can generate an excess of rare alleles relative to neutrality (Tachida 2000; Comeron and Kreitman 2002; Comeron et al. 2008). While a weak BGS model will not generate a significant reduction in variability relative to neutrality (Golding 1997; Neuhauser and Krone 1997; Przeworski et al. 1999), a process generating both strong and weak deleterious alleles could generate both a reduction in variability *and* a negative skew in the frequency spectrum (Gordo et al. 2002). Likewise, a more careful analysis of BGS under very low recombination shows that selective interference can also generate negative D (Kaiser and Charlesworth 2009). It is worth emphasizing that BGS and recurrent sweeps are but two of many possible models of selection. Equally realistic models of linkage to sites experiencing fluctuating selection coefficients can generate the same patterns (Gillespie 1997, 2000).

Sweeps, Background Selection, and Substitution Rates

Both recurrent sweeps and background selection are expected to lower the effective population size, N_e, and hence reduce neutral variation at tightly linked sites. Do these processes also influence the rate of divergence at such sites? For *strictly* neutral alleles ($s = 0$), changes in N_e have no effect on the substitution rate, which is simply equal to the rate of mutation, μ, to neutral alleles (Chapter 2). However, when alleles have a *distribution* of fitness effects (so that some values of s may be very small, but not zero), this is no longer true. Accepting the view that many mutations may be slightly deleterious (Ohta 1973, 1992, 2002), the same allele can be **effectively neutral** ($4N_e|s| < 1$) in smaller populations, while being deleterious in larger populations (when $4N_e|s| \gg 1$). In genomic regions where the effect of recurrent sweeps and/or background selection is expected to be strong (such as regions of low recombination), an *increase* in the divergence rate might be expected, as the fraction of new mutations that are effectively neutral increases. Likewise, in such regions, the rate of adaptive changes may decrease, as weakly favorable mutations are overpowered by the effects of drift, thus reducing their fixation rates. This alteration of the substitution pattern through the fixation of a greater fraction of weakly deleterious alleles and a decrease in the fixation rate of adaptive changes is an example of a Hill-Robertson effect.

As highlighted in Example 8.16, the direction of a potential change in the rate of replacement substitutions as recombination decreases is a function of whether there are more weakly positively selected alleles (the rate goes down) or weakly negative selected alleles (the rate goes up). Betancourt and Presgraves (2002) found, in a comparison of ~250 genes between *D. melanogaster* and *D. simulans*, that the nonsynonymous divergence rate is reduced in regions of low recombination, consistent with reduced fixation of weakly positive alleles in these regions due to a reduction in N_e. However, their gene set contained a large number of male accessory gland proteins (*Acps*), which are rapidly evolving and hence might have biased their results. When the *Acps* genes were removed from the analysis, there was no significant relationship between substitution and recombination rates. Among *Acps* genes, rapid protein evolution was largely confined to regions of high recombination, again consistent with a reduction in N_e retarding the rates of adaptive evolution for these genes.

Conversely, for genes in regions with essentially *no* crossovers in *D. melanogaster* and *D. yakuba* (such as on the tiny fourth chromosome), *elevated* rates of replacement substitution of amino acids are found, consistent with weakly deleterious alleles behaving as if effectively neutral due to a reduction in N_e in low-recombination regions (Haddrill et al. 2007). Similarly, in comparisons of the small (and largely nonrecombining) "dot" chromosome of *D. americana* with its other autosomes, Betancourt et al. (2009) found an increased rate of replacement substitutions on the dot. Further, estimates of the fraction, α, of adaptive substitutions (using methods discussed in Chapter 10) were significantly smaller for the dot than for the other autosomes, suggesting that the increase in replacement substitutions was largely due to the fixation of slightly deleterious alleles. Finally, Bullaughey et al. (2008) found no effect of recombination on rates of protein evolution over the human, chimpanzee, and rhesus macaque lineages. This lack of any consistent relationship between divergence and recombination rates is perhaps not surprising. The nature of any potential signal is population-specific, depending on the distribution of selection coefficients relative to the reduction in N_e in low-recombination regions. In particular, if only a very small fraction of alleles becomes effectively neutral due to a reduction in N_e, the impact will be negligible. Such can be the case for species whose effective population size is already small (such as some mammals), where further reduction in N_e by linked selection effects is unlikely to significantly change the frequency of effectively neutral mutations.

Because both BGS and recurrent sweeps can reduce diversity in regions of low recombination, we can also ask the related question of whether the amount of *divergence* at a site influences the amount of *linked neutral variation* within populations. This is a different question from that of whether linkage influences divergence. Under a strictly neutral model (Chapter 2), the amount of divergence and polymorphism is a function of the mutation rate, so sites with higher divergence should also display higher levels of polymorphism (when jointly compared in the same population so that N_e is approximately constant across genes). Under recurrent sweeps, if a gene shows a high rate of divergence due to more frequent sweeps, this is expected to lower diversity due to the local reduction in N_e that accompanies these sweeps. Such a negative correlation between synonymous nucleotide diversity and the substitution rate at replacement sites is seen in *Drosophila melanogaster* (Andolfatto 2007), *D. simulans* (Macpherson et al. 2007), *D. miranda* (Bachtrog 2008; Jensen and Bachtrog 2010), *D. pseudoobscura* (Jensen and Bachtrog 2010), the European rabbit (*Oryctolagus cuniculus*, Carneiro et al. 2012), the European aspen (*Populus tremula*, Ingvarsson 2010), and humans (Cai et al. 2009).

Cai et al. (2009) suggested that selection at linked sites in humans appears to reduce nucleotide diversity at putative neutral sites by 6% genome-wide and by 11% in the gene-rich half of the genome. McVicker et al. (2009) obtained even higher values: between 19% and 26% for autosomes and between 12% and 40% on the X. One reason for this apparent discrepancy between studies is that Cai et al. excluded regions immediately adjacent to genes, which are likely to experience some of the strongest linkage-related effects due to selection at nearby sites. Reduction in diversity at these various genomic locations could be due to recurrent sweeps, BGS, or (most likely) a combination of both. Several groups have suggested that the human data are better explained by BGS (Hernandez et al. 2011; Lohmueller et al. 2011; Alves et al. 2012). An analysis by Hernandez et al. suggested that classic sweeps have been rare in recent human history. Although local troughs in nucleotide diversity are found around amino acid substitutions (as expected under recurrent sweeps), this reduction in variation is essentially the same as seen around synonymous substitutions. This last observation is more consistent with BGS than with recurrent sweeps (if one assumes that synonymous sites are indeed neutral). This is in sharp contrast to the findings of Sattah et al. (2011) in *Drosophila*, where the data are more consistent with recurrent sweeps.

A confounding factor in these human studies was noted by Enard et al. (2014): sites showing nonsynonymous substitutions may simply reside in regions with *less* functional constraints and thus *weaker* BGS, resulting in an inflated level of neutral variation. Detecting reduced variation around substituted sites requires that the impact of sweeps in reducing variation will overpower the impact of reduced BGS in increasing the level of neutral variation. Enard et al. found that the correlation between the strength of BGS and amount of sequence conservation significantly reduces any signal from sweeps (i.e., reduced variation around substituted sites) in humans. They argue that the lack of an association between substitutions and lower diversity (which led Hernandez et al. 2011 and Lohmueller et al. 2011 to conclude that classic sweeps are rare in humans) is likely a byproduct of failing to account for variation in BGS levels. After being corrected for this effect, Enard et al. found numerous signatures of positive selection in humans.

Example 8.16. Modern rice was domesticated from *Oryza rufipogon* to form the indica (*Oryza sativa indica*) and japonica (*O. sativa japonica*) lineages (Huang et al. 2011). Lu et al. (2006) examined the ratio of the replacement- to silent-site substitution rates, K_a/K_s (Chapter 10), between both subspecies and an outgroup, *O. brachyantha*. In a comparison of over 15,000 genes, the K_a/K_s ratio for divergence between indica and japonica was 0.498. Conversely, in a comparison of roughly 5000 genes between japonica and the outgroup, $K_a/K_s = 0.259$, a highly significant difference. This increase in K_a/K_s between the domesticated lineages occurs throughout the genome, with most regions showing elevated values when comparing the two modern cultivars on the outgroup. Regions of lower recombination showed the largest K_a/K_s values, with a highly significant negative regression of K_a/K_s on the recombination rate. The authors interpreted these data to imply an increase in the fixation rate of deleterious alleles due to a decrease in N_e during the domestication of both cultivar lineages. If the increase in K_a/K_s ratios was due to the accelerated fixation of favorable alleles, this ratio should *increase* along with the recombination rate, as the effective population size is higher in regions of higher recombination, thus increasing the fixation rate of favorable alleles. Conversely, the fixation rate of (slightly) deleterious alleles should increase with decreasing recombination, as the smaller N_e in these regions allows more of these alleles to behave as if effectively neutral. The initial founding of lines during the early phases of domestication reduced N_e, a process that the authors suggest was exacerbated by strong selfing, and hence reduction of the effective amount of recombination throughout the genome. This reduced recombination resulted in the influence on larger regions of the genome by selective sweeps associated with the fixation of domestication genes, further reducing N_e.

To examine whether there was indeed an increased fixation rate for slightly deleterious alleles, the authors analyzed the nature of replacement substitutions, using a regression method developed by Tang et al. (2004). When comparing indica and japonica, a disproportionate amount of change involved radical amino acid replacements over conservative replacements, with the authors estimating that around a quarter of the replacement substitutions were likely deleterious. No such pattern was seen when comparing the divergence between the wild rices *rufipogon* and *brachyantha*.

Sweeps, Background Selection, and Codon Usage Bias

One of the most sensitive indicators of localized changes in N_e is provided by the behavior of sites under very weak selection ($N_e|s| \sim 1$). Under this setting, weakly favorable alleles are still selected for, while weakly deleterious alleles are selected against. However, a small decline in N_e in such settings (be it from recurrent sweeps and/or background selection), or in the effective strength of selection (e.g., from interference among multiple segregating selected alleles) can make a significant fraction of these weakly selected alleles behave in a neutral fashion. Hence, if one has a group of sites within a species where we expect an average of $N_e|s| \sim 1$ over the genome, *localized* reductions in N_e may leave a signature, as within this region the pattern shifts from weak selection to effective neutrality. One potential set of such markers involves synonymous codons, which in *some* species (such as certain *Drosophila*) appear to have a genome-wide average of $N_e|s| \sim 1$ (Lynch 2007).

Although synonymous codons are typically used as proxies for neutral sites, the observation of **codon usage bias** (the nonrandom use among the set of all synonymous codons for a given amino acid) in many organisms suggests that this is only approximately correct. In reality, synonymous sites often appear to be under very weak selection for **optimal** (or **preferred**) codons, which are found to be more frequent than expected (after correction for mutational bias, which can be substantial). As potential alleles under very weak selection, synonymous codons may be rather sensitive (at least in some species) to subtle changes in N_e. We first examine the evidence suggesting there is selection on synonymous codons and the genomic patterns of codon usage before considering what this might tell us about selection at linked sites. We stress that local changes in N_e are expected to generate *subtle* signals at weakly selected sites that can be detected only when one examines the average pattern over hundreds of genes.

The classical view of codon bias is that selection is likely to be stronger on more highly expressed genes, so that bias is expected to vary over genes. Further, the actual strength of selection, which is postulated to be generated by improved translational efficiency and accuracy due to the optimal codon matching the most abundant tRNA for that amino acid, is expected to be quite weak—so weak, in fact, that for an average gene, bias is expected to be significant only in organisms with sufficiently large effective population sizes. Surprisingly, however, bacteria, yeast, and *Drosophila* all have roughly similar levels of bias, despite their perceived great differences in effective population size (Powell and Moriyama 1997). This is tantalizingly reminiscent of Lewontin's (1974) observation that the level of average protein heterozygosity within a species (the surrogate for genetic variation at the time) is much narrower than expected given the range of census population sizes. As mentioned in Chapter 4, this trend is in part due to the evolution of lower mutation rates in populations with larger effective population sizes.

One of the first studies to suggest that *segregating* synonymous alleles may be under selection was performed by Akashi (1995) in *Drosophila*. By using an outgroup, Akashi polarized segregating alleles (Chapter 2), to determine the ancestral allele (that fixed in a sister species) and its new mutation. For a particular amino acid showing codon usage bias, **preferred codons** are those used more frequently than expected (correcting for any mutational bias), while **unpreferred codons** are those used less frequently than expected. Akashi placed segregating and fixed differences into two categories: those involving a preferred codon that mutated to an unpreferred one (denoted by $P \rightarrow U$), and those involving an unpreferred codon that mutated to a preferred one ($U \rightarrow P$). Under the expectation that $P \rightarrow U$ alleles are slightly selected against and $U \rightarrow P$ alleles are weakly selected for, Akashi compared the divergence to polymorphism ratio of $P \rightarrow U$ to that for $U \rightarrow P$. If unpreferred codons are selected against, we would expect a higher ratio of polymorphism (the ratio of segregating U to P alleles) to divergence (the ratio of fixed U to P mutations), as alleles under weakly deleterious selection can segregate (contributing to polymorphism levels) but are unlikely to be fixed (Figure 7.1). The polymorphism ratio was indeed greater than the divergence ratio in both *D. simulans* and *D. pseudoobscura*, while an excess of unpreferred fixations was seen in a sample of 28 *D. melanogaster* genes (Akashi and Schaeffer 1997). The authors attributed this latter observation to the three- to six-fold reduction in N_e in *D. melanogaster* relative to *D. simulans* reducing the effectiveness of selection on weakly selected sites. An important cautionary note for such studies is that mutational bias must be fully accounted for, otherwise pressure from this force may mimic selection. Historically, many early papers simply compared the frequency of third-base nucleotides to genome-wide nucleotide frequencies, but the latter can be a poor proxy for any underlying mutational bias.

Example 8.17. Maside et al. (2004) examined codon usage in *Drosophila americana*, a member of the *virilis* species group. Using *virilis* as an outgroup, they observed 84 synonymous substitutions (fixed differences or divergence) between the two species and 144 segregating synonymous sites within *americana*. Following Akashi, the authors classified these as either $P \to U$ or $U \to P$ changes, and observed the following pattern:

	Substitutions	Polymorphic (*americana*)	Polymorphism/Divergence
$P \to U$	52	124	2.38
$U \to P$	32	20	0.62

This roughly four-fold higher ratio of polymorphism to divergence for the putative deleterious mutations ($P \to U$) is highly significant (Fisher's exact test yields $p = 6.4 \cdot 10^{-5}$). Similarly, note that the divergence ratio for these two classes

$$\frac{P \to U}{U \to P} = \frac{52}{32} = 1.625$$

is substantially less than the polymorphism ratio (124/20 = 6.2).

Finally, if the class U is indeed deleterious, we would expect $P \to U$ mutations to be at lower frequencies within the population sample than $U \to P$ mutations, and such a significant difference was observed. This difference in the site-frequency spectrum was first noticed by Akashi (1999) for *D. simulans*, which was shifted toward lower frequencies for unpreferred mutations and toward higher frequencies for preferred mutations.

How strong is selection against unpreferred codons? Using the Poisson random field (PRF) method for analysis of the pattern of fixed differences and polymorphic sites (examined in detail in Chapter 10), estimates of $N_e|s| \sim 1$ were obtained for *simulans* and *pseudoobscura* (Akashi 1995; Akashi and Schaeffer 1997). An alternative approach to estimate $N_e|s|$ follows from Equation 7.37, which gives Li's (1987) expression for the expected frequency, $\widetilde{p}$, of a preferred codon at the mutation-selection-drift equilibrium,

$$\widetilde{p} \simeq \frac{\exp(2\gamma)}{\exp(2\gamma) + \zeta} \tag{8.32}$$

where $\gamma = 2N_e s$ is the scaled strength of selection for preferred codons and $\zeta = \mu_{P\to U}/\mu_{U\to P}$ measures any mutation bias (also see Bulmer 1991; McVean and Charlesworth 1999, 2000; Zeng and Charlesworth 2009, 2010; Zeng 2010). If ζ is known, Equation 8.32 can be used to directly estimate γ for a given synonymous codon set (averaged over genes). Maside et al. (2004) offered an alternative (but related) procedure that does not involve estimating ζ. They showed that the fraction, p_U, of segregating sites where the *derived* allele is the unpreferred synonymous codon (i.e., $P \to U$ mutations as opposed to the derived allele being a $U \to P$ mutation) in a sample of n alleles can be expressed as a function of γ alone. Using this approach, which measures contemporaneous selection coefficients (unlike PRF estimates which use divergence data, and hence are influenced by historical selection; see Chapter 10), Maside et al. (2004) obtained an estimate of $N_e|s| \simeq 0.65$ in *D. americana*.

Thus, for several *Drosophila* species, the strength of selection on synonymous codon usage is roughly $N_e|s| \simeq 1$, suggesting the possibility that small localized genomic changes in N_e can significantly impact codon bias. The prediction is that ***codon bias is reduced in regions where N_e is lowered***. Three observations offer support for this, with bias being less extreme: (1) in regions of low recombination, (2) for genes that are rapidly diverging, and (3) in the middle of long exons. We examine each of these observations in turn, again with the caveat that most of these observations come from *Drosophila*. Organisms where the scaled strength of selection is weaker (i.e., those with much smaller N_e) or much stronger (i.e., those with much larger N_e) might not show these trends, as in these cases an order of magnitude change in the N_e for a genomic region will still leave drift overpowering selection (small baseline N_e) or selection still overpowering drift (large baseline N_e).

There are mixed reports of codon bias depending to some extent on recombination rates in *Drosophila*. Kliman and Hey (1993) examined roughly 400 loci in *D. melanogaster*, finding reduced codon bias in regions of low recombination. The relationship was not linear, but rather was only apparent for genes in the regions of lowest recombination. A much larger study involving 14,000 *melanogaster* genes by Hey and Kliman (2002) attempted to control for concerns of a potential mutation bias toward G/C bases (commonly used in the optimal codon) in regions of high recombination (Marais et al. 2001; Duret and Galtier 2009). Again, a significant positive correlation between codon usage bias and recombination rate was seen for regions of low recombination, while genes in regions with a modest to high recombination rate ($c > 1.5$ cM/Mb) showed no association. Finally, Haddrill et al. (2007) found essentially no codon bias for genes in *D. melanogaster* and *D. yakuba* residing in regions with no recombination, and Betancourt et al. (2009) found that a significantly smaller fraction of genes on the small (dot) chromosome of *D. americana* used optimal codons relative to sites on larger chromosomes. Taken together, these observations suggest that in some *Drosophila* species, codon bias is less severe in regions of very low recombination, consistent with reduced efficiency of selection from a local decrease in N_e.

In addition to these regional effects over the scale of a small chromosomal segment, there are also reports of effects on much finer scales, namely gene-by-gene and even different regions within the same gene. For example, genes undergoing multiple sweeps (and hence higher rates of substitutions) might be expected to have lower effective population sizes, and hence less codon bias. In a study involving roughly 250 genes, Betancourt and Presgraves (2002) found that those with higher replacement rates tended to show less codon usage bias in both *D. melanogaster* and *D. simulans*. Maside et al. (2004) examined over 600 *melanogaster* genes, also finding a negative association between rates of replacement substitution and codon bias. However, they also noted that both codon bias and replacement rates are correlated with gene expression, so perhaps the latter is the driver for the correlation. Andolfatto (2007) found both reduced codon bias and reduced synonymous-site diversity, among the more rapidly evolving proteins in a survey of roughly 140 proteins on the high-recombination region of the X chromosome from *D. melanogaster*. Similar results are reported by Bachtrog (2008) for *D. miranda*. While most observations are restricted to *Drosophila*, Ingvarsson (2010) found a weakly negative (but not significant) relationship between codon bias and protein evolution rates in the European aspen (*Populus tremula*). One important caveat about these negative associations between the strength of codon bias and divergence rates is that relaxed selection on both the gene sequence and codon bias can also generate this pattern.

On an even finer scale are reports of correlations between codon bias and gene length in *Drosophila* (Comeron et al. 1999). For short genes (less than 750 bp), those residing in regions of lower recombination show reduced codon usage bias, but this effect is less for genes with longer coding regions. Moreover, the length of a coding region is negatively correlated with bias (longer genes have less bias) over all recombination values. Strikingly, Comeron and Kreitman (2002) found that codon bias decreases in the *middle* of long exons. A more detailed analysis by Qin et al. (2004) showed that codon bias decreases at the ends, as well as the middle, of long genes in *Drosophila*, while yeast and several species of bacteria showed no such pattern. Comeron and Guthrie (2005) estimated the strength of selection, γ, on synonymous codons and found this to be significantly reduced in long versus short genes. Consistent with a reduction in the efficiency of selection, longer exons also had higher rates of synonymous substitution, as would be expected if reduction in N_e made weakly deleterious synonymous mutations behave in a more neutral fashion. However, any factor resulting in a reduction in selective constraints could also generate such a pattern.

All of these signals of more neutral patterns of codon usage in specific genomic regions are consistent with the effects of selection at linked sites. Both recurrent sweeps and background selection could generate this reduction in bias in regions of low recombination, while the observation of lower codon bias for genes with high replacement-substitution rates is consistent with recurrent sweeps (Kim 2004). The most interesting observations, however, involve the very fine-scale differences, in particular the decrease in bias in the middle of long exons. Loewe and Charlesworth (2007) suggest that background selection could generate such a pattern, with sites in the middle of exons having deleterious mutations arise for some distance on both sides of them, increasing their local U value, and thus creating a local decrease in N_e. These very fine-scale effects are very sensitive to recombination. Hey and Kliman (2002) found that gene density (when measured by the number of genes per kilobase) had no effect on codon bias. However, very tightly spaced genes did exhibit bias, showing that potential linked effects of selection operate over very short distances.

An additional explanation for these very short range effects is *interference among selected sites*. Models of background selection and recurrent sweeps typically assume that alleles are under strong enough selection to have only a short persistence time in the population. Conversely, alleles under weak selection segregate for longer periods of time, allowing for multiple segregating mutations of weak effect within a gene, with interference among these alleles reducing the efficiency of selection. For example, if selected mutations are in negative LD, this reduces the additive variance in fitness (Chapter 16), lowering the efficiency of selection (Chapters 5 and 13). Likewise, within a set of alleles that are nearly selectively equivalent, drift can occur, reducing the efficiency of selection on any particular allele.

This reduction in selection efficiency due to interference among linked sites has been called a **small-scale Hill-Robertson effect** (Comeron et al. 1999), **weak selection Hill-Robertson interference** (McVean and Charlesworth 2000), and **interference selection** (Comeron and Kreitman 2002). If multiple weak, positively selected alleles are segregating in a tightly linked region (such as multiple preferred codons within an exon), they mutually interfere with each other, resulting in weaker section and a smaller codon usage bias. To see this, suppose that all favorable segregating sites have the same advantage. The frequency of this *entire set* of alleles increases over time, but within this set, the behavior of individual sites is essentially neutral. The same is true for a collection of weakly deleterious alleles. The key is extremely tight linkage. Simulation studies (Comeron and Kreitman 2002; Comeron et al. 2008) have shown that selective interference can indeed produce a decrease in codon bias in the middle of long exons, with bias decreasing with the number of selected sites. Its effect, however, is *extremely* local, except in regions of extremely low recombination.

McVean and Charlesworth (2000) showed that selective interference can also account for the puzzling observation of the relative insensitivity of codon usage bias to changes in N_e seen in cross-species comparisons (Powell and Moriyama 1997). When selective interference is present, it tends to moderate the effects of selection, making the expected bias relatively similar over several orders of magnitude in N_e. Selective interference can reduce bias even in genes in regions of moderate recombination because there is still tight linkage over very small regions segregating multiple sites under weak selection (McVean and Charlesworth 2000). As noted by Comeron and Kreitman (2002), exons and their adjacent control regions are prime candidates for selective interference, as the physical clustering of functional sites offers the possibility of weak selection over a number of tightly linked sites.

One final unexplained aspect of codon bias is the intriguing observation that X-linked genes tend (as a group) to have higher codon bias than autosomal genes in *D. melanogaster* (reviewed by Campos et al. 2012). Given that one would expect the effective population size of X-linked genes to be lower than those for autosomes ($\sim 3/4$; see Lynch 2007 for an extended discussion of the assumptions leading to this approximation), these arguments suggest that X-linked bias should be reduced. Campos et al. examined whether increased recombination on the X might counter this by reducing Hill-Robertson effects from selection at linked sites. The argument for higher average recombination on the X in *Drosophila* is that this chromosome spends two thirds of the time in females, which have normal recombination, and one third of the time in males, with no recombination. Conversely, autosomes spend equal amounts of time in both sexes, resulting in a one third higher expected effective recombination rate on the X. When comparing sets of X and autosomal genes with similar effective recombination rates (using regions of the X with lower rates and autosomal regions with higher rates, such that their scaled recombination values are roughly equal), Campos et al. still observed higher X-linked codon bias, suggesting that this pattern results from stronger selection for preferred codons for X-linked genes (relative to autosomes) in *D. melanogaster*, rather than from localized changes in effective population size (also see Zeng and Charlesworth 2010). One possible molecular basis for this apparently stronger selection for codon bias on X-linked genes is that *Drosophila* performs dosage-compensation by overexpressing X-linked genes in males (to twice the level for X-linked genes in females). This potentially results in X-linked and autosomal genes experiencing differing selection pressures for expression.

A Paradigm Shift Away From the Neutral Theory of Molecular Evolution?

As ably summarized by Charlesworth (2010), molecular population genetics has a rich and dynamic history. Its current focus traces back to the **neutral theory of molecular evolution**, which was born in the late 1960s in response to the higher than expected levels of protein polymorphism found in natural populations (Kimura 1968b; King and Jukes 1969), and gained strength through the 1980s as more molecular data became available (Kimura 1983). Under initial versions of the neutral theory, the vast majority of new mutations were assumed to be either neutral or rather strongly deleterious. As selection is expected to rapidly remove the latter, such mutations contribute little to the levels of polymorphism and even less to divergence (Figure 7.1). Under this view, advantageous mutations can indeed occur, but they are assumed to be extremely rare and rapidly fixed (or lost) when they arise, resulting again in little impact on polymorphisms and (given their rarity) at best a modest impact on divergence. Given these assumptions, it then follows that most fixed differences between populations or species and most segregating variation within a population or species are due to neutral variation.

Ohta (1973, 1992, 2002) presented an important modification, the **nearly-neutral** (or **slightly deleterious**) **theory**, allowing for slightly deleterious alleles, which can be close to effectively neutral and hence contribute significantly to polymorphisms but far less to divergence (reviewed by Akashi et al. 2012). A key prediction from both versions of the neutral theory is that regions with fewer functional constraints (and therefore a higher fraction of neutral or nearly neutral mutations) evolve faster. This prediction is strongly supported by the observation of slower substitution rates at replacement sites than at synonymous sites in protein-coding genes, and ever faster rates in pseudogenes.

The key assumption of all versions of the neutral theory is that while purifying selection can be very common, ***adaptive evolution at the molecular level is rare***, so that most segregating alleles and most substitutions involve alleles that are effectively neutral. Some researchers have questioned this key assumption, calling for a more nuanced view of the nature of stochastic changes in allele frequencies that can dominate weakly selected sites (Hahn 2008; Rockman 2012; but see Nei et al. 2010 for a contrary opinion). As discussed in Chapter 10, the estimates of high α values (the fraction of replacement substitutions that are adaptive) in some species are strongly at odds with the view of neutral or nearly neutral theories. A second potential problem relates to genomic effects from selection at linked sites, the most celebrated of which is the correlation between recombination rates and levels of variation already noted. If this is due to background selection, such a correlation is still consistent with the classical neutral theory, being generated by the removal of new deleterious mutations. However, if periodic selective sweeps generate this correlation, then much of the genome may be impacted by *positive selection*, either directly or indirectly through the effects of selection at linked sites, which change the local effective population size. Finally, observations consistent with selective constraints on silent sites and even noncoding DNA in some species (reviewed in Chapter 10) are also somewhat problematic for the neutral theory. While the removal of new deleterious mutations (BGS) falls under the umbrella of neutral theory, the converse, fixation of slightly favored sites (such as the fixation of a silent mutation to a preferred codon), is an example of positive selection. The inescapable conclusion from the *Drosophila* data is that weak selection is occurring throughout the genome, with patterns of variation being shaped by selection at linked sites. These effects can operate over quite small scales, on the level of differences between the ends and middle of a long exon, presumably due to interference among weakly selected sites. Whether these observations generalize beyond *Drosophila* remains an open question.

The great irony of gaining a deeper appreciation for how rampant selection (and especially weak selection) is throughout the genome (of at least some species) is that it likely makes *more* alleles behave as if they were effectively neutral. Kimura's original grand vision of the role of selection acting as a giant filter, through which only neutral and a very few advantageous alleles pass, now appears to have been replaced by the role of selection throughout the genome making weakly selected alleles behave in a more neutral fashion. Chapter 3 introduced Gillespie's (1997, 2000) concept of **genetic draft**, whereby the frequencies of neutral alleles fluctuate more from their random association with ongoing sweeps than they do from drift. While drift may be important in small populations, in sufficiently large populations, sweep-generated stochastic fluctuations of allele frequencies can still overpower weak selection. A combined analysis of drift and draft by Schiffels et al. (2011) found effective neutrality when the absolute value of a selection coefficient of an allele is less that $1/(2N) + \lambda$ (N being the census size, and λ the per-nucleotide rate of adaptive evolution at linked sites; Equation 8.24a), a feature they refer to as **emergent neutrality**. When N is small, the rate of adaptation, λ (which scales as NU_b for small to medium values of N), is likely small, and drift dominates. Conversely, when N is large, $1/N$ can be dominated by λ. For a very large N, interference among sweeps strongly constrains λ, causing a transition of the scaling of the adaptation rate from NU_b to $\ln[NU_b]$ (Neher 2013). Thus, at *every* population size, stochastic fluctuations in allele frequencies are important, but the forces underlying them differ. Drift dominates in small populations, and draft in large populations. Once the draft stage is reached, because λ increases with N (albeit very slowly), a *higher fraction* of new alleles becomes effectively neutral as population size increases. When coupled with selective interference among weakly selected sites, this leads to a view whereby much of the segregating variation in a large population may be effectively neutral. However, it is not due to a lack of positive selection, but rather a *consequence* of it.

9

Using Molecular Data to Detect Selection: Signatures from Recent Single Events

For the past 20 years, there has been a tendency on the part of journal editors and reviewers to assume that every case of alleged statistical evidence for positive selection is worthy of publication, even in the absence of a plausible biological mechanism underlying the alleged selection. Hughes (2007)

While the ubiquity of **purifying** (or **negative**) **selection** (the removal of deleterious alleles, i.e., background selection) at the molecular level is well established, the frequency of **positive** (or **adaptive** or **Darwinian**) **selection** remains unclear. Because of this, the development of methods to detect the latter is a major growth industry in evolutionary genetics. There is a massive population-genetics literature on this subject, and a partial (but not exhaustive) list of reviews includes Kreitman (2000), Nielsen (2001), Ford (2002), Bamshad and Wooding (2003), Schlötterer (2003), Guinand et al. (2004), Nielsen (2005), Storz (2005), Wright and Gaut (2005), Biswas and Akey (2006), Sabeti et al. (2006), Thornton et al. (2007), Holderegger et al. (2008), Pavlidis et al. (2008), Stinchcombe and Hoekstra (2008), Akey (2009), Nei et al. (2010), Oleksyk et al. (2010), Siol et al. (2010), Stephan (2010a), Crisci et al. (2012), Fu and Akey (2013), Vitti et al. (2013), Bank et al. (2014), Forester et al. (2016), Malaspinas (2016), Stephan (2016), Vatsiou et al. (2016), and Xiang-Yu et al. (2016).

As detailed in Chapter 8, a single recent event of positive selection can leave a transient signal in the pattern of linked neutral variation. The detection of such events is the subject of this chapter. Most approaches for detecting recent or ongoing selection use the segregating variation in a sample from a contemporaneous population, and we loosely refer to these as **polymorphism-based tests**. Because such signals are transient, these methods work only over ecological time scales, detecting events that are either ongoing or that concluded less than $\sim N_e$ generations ago. In contrast, a *history* of positive selection on a gene over evolutionary time can leave a *cumulative* signal in the pattern of substitutions. **Divergence-based tests** to detect these patterns, which require data on substitutions between species (or very distantly related populations), are developed in Chapter 10 (which also covers tests that jointly use polymorphism and divergence data). These different approaches are complementary, as an adaptive substitution could leave a strong (but fleeting) signature over an ecological time scale but essentially no signal in that gene over an evolutionary time scale (adding just one more substitution in a potential background of numerous neutral fixations). Likewise, the vast majority of adaptive events that have shaped a gene likely occurred in its distant past, leaving no currently detectable polymorphism pattern and only potential signals in divergence data (Chapter 10).

Because the search for sites under selection can be seductive, it is important to stress that the methods developed in this chapter have *limitations*. They *potentially* can be useful in detecting *some* events involving a single selective event at a single site, allowing for the prospect of studying individual (as opposed to cumulative) selective events. As discussed in Chapter 8, one source of these limitations is the nature of the signal left by the sweep itself. First, it is *very fleeting*, typically persisting for only $0.1N_e$ to N_e generations, depending on the feature being examined (Table 8.2). Second, *only a fraction of such events*, even if ongoing or very recent, can be detected. A weak, but nontrivial, selection event may be too small to leave a meaningful signal against a noisy molecular background. Finally, even very strongly selected sites may not be detectable, especially if they involve soft or polygenic sweeps (Chapter 8). Even with all these concerns, the most critical problem with polymorphism-

Evolution and Selection of Quantitative Traits. Bruce Walsh & Michael Lynch.
Published 2018 by Oxford University Press.
DOI 10.1093/oso/9780198830870.001.0001

based tests is the ***confounding effect from demography and population structure***. For example, rapid expansion following a population bottleneck leaves a sweep-like signal over the entire genome, while the presence of population subdivision can mimic balancing selection. One reason for the vast number of tests discussed below is that no single one is best in all settings, and the search to find strong signals *unique* to positive selection has, for the most part, been unsuccessful.

Besides being fundamentally important to our understanding of evolution, tests of selection can also be helpful to a breeder. Scans for ongoing and recently selected sites can provide a useful complement to QTL- and association-mapping studies. In these latter approaches, one specifies the traits of interest and then searches for marker-trait associations (LW Chapters 13 through 16). While this is a powerful approach, it is limited by having to specify the traits of interest. In a population undergoing selection for (say) water stress, one might miss pathways for adaptation that are not obvious and hence do not involve the traits chosen to be mapped. Conversely, one could perform a genomic scan on a population under relatively recent water stress to look for sites showing signatures of ongoing or recent selection. These, in turn, could suggest genome regions harboring genes for traits or pathways under selection, without having to specify particular traits. This approach has been termed **natural selection mapping** (Kohn et al. 2000), **hitchhiking mapping** (Schlötterer 2003), and **reverse ecology** (Li et al. 2008), and is widely used in the search for **domestication genes** responsible for the transition from wild relatives into domesticated lines. We review the results of several such scans at the conclusion of the chapter.

Finally, it is important to stress that our focus over these next two chapters is on using marker information *independent* of any specific traits. As mentioned in Chapter 8, polygenic response is likely to leave little, if any, detectable signal at *single sites*. However, a composite signal may be generated over a *collection* of such sites (i.e., a collection of unlinked, or loosely-linked, loci underlying the trait). Such trait-augmented marker approaches—wherein a set of markers is chosen because they impact a specific trait, with tests based on summary statistics over this ensemble of markers—are covered in Chapter 12.

AN OVERVIEW OF STRATEGIES BASED ON SEGREGATING VARIATION

For a multitude of reasons, there is no single omnibus test for selection. First, different scenarios (e.g., hard sweeps, soft sweeps, partial sweeps, balancing selection) leave different, and often conflicting, signatures (Chapter 8), so that tests for one type (e.g., hard sweeps) may easily miss signatures from another (e.g., soft sweeps). Second, different tests are designed to detect signals from *different time periods* during (and following) a sweep. Third, different sampling schemes are possible. Many tests assume there is only a single sample from a current population, but one might instead have contemporaneous samples from several related populations or a temporal series of samples from a single population.

Advances in genomics have enormously expanded our ability to score molecular variation, and this is reflected in the historical development of tests. The first test followed changes in a single allele at one locus through time (Examples 9.1 and 9.2), while later tests evolved to use data from **genomic scans**, where a very large number of sites are scored, and potentially phased (generating haplotype, as opposed to sequence, data). Haasl and Payseur (2016) suggested the terms **GWSS** (**genomewide scans for natural selection**) when a whole genome is scanned, and **ESS** (**exomic scans for natural selection**) when only coding or transciptome sequences are used. Approaches to detect recent selection can be classified into five categories, which loosely follow the historical development of the field:

1) **Excessive allele-frequency change**. The first formal test of selection was proposed by Fisher and Ford (1947), who used the machinery developed in Chapter 2 for the divergence under drift to test for excessive change in a time-series of allele frequencies from a single population. While perhaps the most unambiguous signature of selection, this approach requires long-term monitoring of a population and having some rea-

sonably independent estimate of N_e. The ever-increasing availability of **ancient DNA (aDNA)** samples opens up exciting new data sets for this type of analysis (Mathieson et al. 2015, Malaspinas 2016; Schraiber et al. 2016).

2) Excessive allele-frequency divergence. Lewontin and Krakauer (1973) proposed using the divergence between a series of contemporaneously sampled populations (presumably from a common ancestor) to test for selection. The machinery from Chapter 2 predicts the expected divergence under drift, as measured by Wright's F_{ST} statistic for population structure. Loci displaying excessive F_{ST} values relative to drift are selection candidates. Using an incorrect model of population structure can seriously compromise these tests.

The above two categories require samples from multiple populations (either temporally or spatially), which limits their widespread use. A less demanding design is a single population sample, as employed by the three remaining categories.

3) Chromosomal spatial patterns of variation. As detailed in Chapter 8, a sweep leaves a characteristic decrease in polymorphism around a selected site, and a number of formal likelihood tests are based on the expected pattern of the nucleotide diversity, π, as a function of the recombination distance, c, from the sweep (Equation 8.8a). Early versions of these tests assumed that the population was in mutation-drift equilibrium at the start of the sweep, while more recent versions have relaxed this strong assumption.

The final two categories divide tests by whether they assume an infinite-sites, or an infinite-alleles, framework, using the neutral equilibrium results for these models developed in Chapter 2. Recall that the infinite-sites framework considers a sequence as a series of separate sites (e.g., SNPs), while the infinite-alleles framework treats each different DNA sequence (haplotype) as a different allele (Figure 2.9). Both models assume that the region being considered is small enough that recombination within the sample can be ignored. Given the large (and diverse) number of tests in both of these categories, each section reviewing these different approaches concludes with a summary table of proposed tests (Table 9.1 for infinite-sites and Table 9.3 for haplotypes).

4) Changes in the site-frequency spectrum. Under the infinite-sites model, the frequency spectrum of neutral sites at mutation-drift equilibrium is given by the Watterson distribution (Equation 2.34). Starting with Tajima (1989), a number of tests have been proposed that search for shifts in this spectrum following a sweep, such as an excessive number of sites with rare alleles or with high-frequency derived alleles. The major complication with this class of tests is that changes in population demography (such as a recent expansion or contraction) or the presence of population structure (migration between partly isolated populations) can mimic signatures of selection.

5) Tests based on haplotype information. Under the infinite-alleles model, the number of alleles (haplotypes) in a sample at mutation-drift equilibrium is given by the Ewens sampling formula (Equation 2.30a) and their allele-frequency spectrum by Equation 2.33b. Starting with Ewens (1972) and Watterson (1977, 1978), a number of tests have used these expressions to detect departures from the neutral equilibrium model. As with tests based on the site-frequency spectrum, significant departures can occur for neutral alleles if the population is not in equilibrium or if population structure is present.

Two other strategies use haplotype information. The first searches for the distinct signatures in the pattern of pairwise linkage disequilibrium (LD) predicted around a hard or a soft sweep (Table 8.2). The second considers the frequency of a neutral allele as a function of its age (Equation 2.12). Under neutrality, a common allele is an old allele, with shorter blocks of LD, reflecting a longer history of recombination. The presence of high-frequency alleles with **long haplotypes** (large blocks of LD) offers a signature of selection (these are often called **LRH**, for **long-range haplotype**, tests). A key point is that haplotype structure provides ***signals that can be missed by site-frequency and hard-sweep tests***, and thus offers more power in some settings.

Attempts to Account for Departures From the Equilibrium Model

Most tests for selection are based on the null hypothesis of the neutral equilibrium (or standard neutral) model (Chapter 2). While rejection of this null can indeed imply a signature of selection, rejection can also occur if a neutral population is not in mutation-drift equilibrium. Cavalli-Sforza (1966) noted that demography and population structure should leave a common signal over all genes within a genome, and this observation has been used in attempts to correct for any genome-wide nonequilibrium features in the data. The simplest approach is the **outlier method**, whereby values of the test statistic are computed for a large number of genes, with outliers suggesting potential targets of selection. This is an *enrichment method*, not a formal test. The second approach is to use data from presumably neutral markers unlinked to a region of interest to infer the population history (e.g., bottlenecks, expansions, population structure). These histories can then be used to simulate the coalescent structure (Chapter 2) for neutral alleles under this nonequilibrium model, which in turn can be used to generate the distribution of the test statistic under this more appropriate null. A final approach is to use presumably neutral sites to generate an **empirical site-frequency spectrum** to use in place of the equilibrium Watterson distribution.

These approaches are based on information from a large number of loci obtained in a genomic scan, with the assumption that most sites are not under positive selection and hence provide information to better shape the null hypothesis. This critically relies on the validity of Cavalli-Sforza's assumption of a common demographic or population structure signal over all loci, upon which any additional signal from selection is placed. Unfortunately, this need not be the case, especially in a population that is expanding over space. **Allelic surfing** can occur, wherein random alleles (and new mutations) on the leading edge of a wave of population expansion can "surf " (this wave) rather quickly to high frequencies in newly founded parts of the population (Edmonds et al. 2004; Klopfstein et al. 2006; Hallatschek et al. 2007; Travis et al. 2007; Excoffier and Ray 2008; Hallatschek and Nelson 2008, 2009; Excoffier et al. 2009a; Hallatschek 2011). Because neutral alleles on the leading wave of expansion are largely random, surfing ***does not affect all genomic locations equally***, and as a result can mimic signatures of selection even after correcting for demography or structure based on others markers within the sample. This is especially troublesome as the model species most surveyed for recent selection—humans, cosmopolitan human commensal *Drosophila* (*melanogaster* and *simulans*), and *Arabidopsis*—all have undergone massive range expansions. Hofer et al. (2009) found that while a large fraction of the human single-nucleotide polymorphisms (SNPs), short tandem repeats (STRs), and indels show large (greater than 0.3) differences in frequency across world populations, this pattern is easily accounted by allelic surfing, suggesting that this phenomenon can be a considerable problem in the search for sites under recent selection in humans.

SNP Ascertainment Bias

Another (increasingly historical) concern is **SNP ascertainment bias**, which arises when molecular variation is scored using prechosen SNPs. In a typical SNP discovery setting, one sequences a relative small pool of individuals (the **SNP discovery panel**) to "discover" SNPs—polymorphic nucleotides whose minor allele is above some critical frequency in the panel. These are then used to score a much larger sample of individuals, thus creating a severe bias in favor of SNPs at intermediate frequencies and against rare SNPs. Likewise, if the SNP discovery panel is from a different population than the screened sample, this also creates bias in that important SNPs in the population of interest can be missed (e.g., Ptak and Przeworski 2002). When the frequencies of SNP minor alleles in the discovery panel are known, corrections for ascertainment can be straightforward (Nielsen et al. 2004). However,

SNP discovery is often a more complex process, creating biases that simple methods can reduce, but not remove (Clark et al. 2005). With the ever-increasing availability of whole-genome sequencing, this is rapidly becoming an issue of diminishing concern.

SNP Polarity Assignment Errors

A final source of bias can appear in tests requiring the **polarity** status (i.e., ancestral or derived) of a SNP allele. Recall from Chapter 2 the distinction between **unfolded** and **folded** frequency spectra. The former is based on the frequency of derived alleles (i.e., the Watterson distribution), whereas the latter is based on the frequency of the minor allele and thus is immune to polarity assignment errors. Typically, polarity is accessed using an outgroup, with the outgroup allele assumed to be the ancestral stage. This is a parsimony assumption, which requires that no back or parallel mutations occur, and that the site is monomorphic in the outgroup. Incorrect polarity assignments can result in mislabeling a low-frequency derived allele as a high-frequency ancestral one, and even a few such errors can significantly impact certain tests (Baudry and Depaulis 2003; Hernandez et al. 2007).

Background Selection as the More Appropriate Null?

Recently, it has been suggested that strict neutrality may not be the correct null hypothesis. Cutter and Payseur (2013) and Corbett-Detig et al. (2015) have both stressed that background selection (BGS) is a more appropriate null, given how widespread BGS appears to be (Chapter 8). If BGS is taken as the null, then test comparisons must accommodate differences in gene density per recombination unit, as the impact for BGS is expected to scale with this ratio (Chapter 8). If one is attempting to correct for nonequilibrium features by using a set of putatively neutral markers in a comparison with a possible region under selection, care must be taken to ensure that these markers are from regions with a similar gene-density to recombination value as the region of interest.

Structure of the Remainder of This Chapter

The rest of the chapter is structured into treatments based on our five categories of tests. These categories were largely constructed for convenience of presentation, and some tests draw upon ideas from several different approaches. Given the amount of information in this chapter, we have tried to make the discussion of each category largely autonomous of the others, thus allowing readers to skip directly to the section most appropriate for their needs. We conclude with a brief review of scans for recent positive selection in humans and domesticated organisms.

ALLELE-FREQUENCY CHANGE IN A SINGLE POPULATION

There are several settings where tests based on allele-frequency change may be appropriate. One is a population monitored over some reasonable period of time, which was the basis for the first formal test by Fisher and Ford (1947), of whether a specific gene is under selection (Examples 9.1 and 9.2). The second is a population under artificial selection, which has also been proposed as an approach for QTL mapping for a trait of interest (Nuzhdin and Pasyukova 1991; Keightley and Bulfield 1993; Nuzhdin et al. 1993; Ollivier et al. 1997). Most recent are studies where candidate allele frequencies are estimated from a small sample of ancient DNA and then compared with their frequencies in a more contemporary sample (e.g., Schraiber et al. 2016).

While excessive allele-frequency change is perhaps the most unambiguous signal of selection, there are power issues when the number of generations separating the first and last samples is modest (De Kovel 2006). Given that the time scale for significant allele-frequency change under selection is $\sim 1/s$ (Equation 5.3c), sampling based on a modest number of generations requires strong selection for a signal. In particular, to detect a significant change even in the absence of drift requires that the sample size, n, and number of generations, t, satisfy $tn \gg 1/s$.

Allele-frequency Change Over Two Sample Points: The Waples Adjusted Test

Chapter 4 considered the estimation of N_e from allele-frequency change, a setting where one typically averages over a number of loci to reduce the evolutionary sampling variance. Here our task is the complementary problem. *Given* some estimate of N_e, is the observed change in allele frequency at a candidate locus excessive? If so, this presumably reflects directional selection acting at, or close to, this region. In theory, one could also test for too little divergence (reflecting balancing selection), although this is rarely done, given the high sampling variance (and hence low power), unless sample sizes are extremely large.

In the early literature, a number of workers tested for excessive divergence by simply querying whether allele frequencies in two samples were significantly different. As noted by Gibson et al. (1979) and Waples (1989b), this is inappropriate, as it does not account for the evolutionary *drift variance* in allele frequencies

$$\sigma^2(p_t) = p_0(1-p_0)\left[1-\left(1-\frac{1}{2N_e}\right)^t\right] \simeq p_0(1-p_0)\frac{t}{2N_e} \quad \text{for } t \ll N_e$$

where p_0 is the initial allele frequency (Equation 2.14a). Consider a population sampled at two time points (0 and t), with sample sizes of n_0 and n_t, respectively. The estimated divergence is

$$\widehat{\delta}_t = \widehat{p}_t - \widehat{p}_0 \tag{9.1a}$$

This divergence has an expected value of zero (as $E[\,\widehat{p}_t] = p_0$), with a variance of

$$\sigma^2\left(\widehat{\delta}_t\right) = \sigma^2\left(\widehat{p}_t - \widehat{p}_0\right) = \sigma^2\left(\widehat{p}_t\right) + \sigma^2\left(\widehat{p}_0\right) - 2\sigma\left(\widehat{p}_t, \widehat{p}_0\right) \tag{9.1b}$$

where $\widehat{p}_i = p_i + e_i$, the true value plus an error due to finite sampling. Because these are draws from a binomial distribution, the sampling variance of the initial frequency is

$$\sigma^2\left(\widehat{p}_0\right) = \frac{p_0(1-p_0)}{2n_0} \tag{9.2a}$$

while the final allele frequency is influenced by both the drift and sampling variances (Waples 1989a, 1989b)

$$\begin{aligned}\sigma^2\left(\widehat{p}_t\right) &= p_0(1-p_0)\left[1-\left(1-\frac{1}{2n_t}\right)\left(1-\frac{1}{2N_e}\right)^t\right]\\ &\simeq p_0(1-p_0)\left[\frac{1}{2n_t}+\frac{t}{2N_e}\left(1-\frac{1}{2n_t}\right)\right]\end{aligned} \tag{9.2b}$$

If sampling is done (without replacement) before reproduction, then $\sigma\left(\widehat{p}_t, \widehat{p}_0\right) = 0$, and substitution of Equations 9.2a and 9.2b into Equation 9.1b yields

$$\sigma^2\left(\widehat{\delta}_t\right) \simeq p_0(1-p_0)\left[\frac{1}{2n_0}+\frac{1}{2n_t}+\frac{t}{2N_e}\left(1-\frac{1}{2n_t}\right)\right] \tag{9.2c}$$

If sampling is done either after reproduction or with replacement, this generates a covariance between the sample estimators $\widehat{p}_t$ and $\widehat{p}_0$; see Nei and Tajima (1981b) and especially Waples (1989a, 1989b) for details. Assuming this is not the case, so that $\sigma(\widehat{p}_0, \widehat{p}_t) = 0$, Equation 9.2c yields the correct variance for the null hypothesis of random genetic drift, giving the test statistic as

$$\frac{\widehat{\delta}_t^{\,2}}{\sigma^2\left(\widehat{\delta}_t\right)} \tag{9.2d}$$

which is approximately χ_1^2-distributed. This is the **Waples adjusted test** (Waples 1989b).

The application of this test requires an accurate estimate of p_0 to compute the sample variance (Equation 9.2c). While the sample estimate, $\widehat{p}_0$, can be used, this can be improved upon by noting that the expected allele frequency change is zero, meaning that $\widehat{p}_t$ also contributes information about p_0. A simple average of the two frequencies is not appropriate, as $\widehat{p}_0$ has a smaller drift variance and the two estimates may differ in informational value due to differences in sample size, n_i. Given these concerns, Schaffer et al. (1977) and Waples (1989b) proposed a generalized (i.e., weighted) least-squares (GLS) estimator (LW Chapter 8) for p_0. Let $\mathbf{p} = (\widehat{p}_0, \widehat{p}_t)^T$ denote the two sample estimates and denote its sampling variance-covariance matrix by

$$\mathbf{V} = \begin{pmatrix} \sigma^2(\widehat{p}_0) & \sigma(\widehat{p}_0, \widehat{p}_t) \\ \sigma(\widehat{p}_0, \widehat{p}_t) & \sigma^2(\widehat{p}_t) \end{pmatrix} \tag{9.3}$$

Finally, let $\mathbf{1} = (1,1)^T$ be a vector of ones. The underlying statistical model is $p_i = p_0 + e_i$, which can be written in general linear model form as $\mathbf{p} = p_0\mathbf{X} + \mathbf{e}$, where $\mathbf{V}$ is the covariance matrix for the vector, $\mathbf{e}$, of residuals and $\mathbf{X} = \mathbf{1}$. Recalling LW Equation 8.34 for GLS regression, the resulting estimate of p_0 is given by $\left(\mathbf{X}^T\mathbf{V}^{-1}\mathbf{X}\right)^{-1}\mathbf{X}^T\mathbf{V}^{-1}\mathbf{p}$, which reduces to

$$\text{GLS}\,(p_0) = \frac{\mathbf{1}^T\mathbf{V}^{-1}\mathbf{p}}{\mathbf{1}^T\mathbf{V}^{-1}\mathbf{1}} \tag{9.4}$$

because both quadratic products are scalars.

Example 9.1. One of the classic papers in evolutionary biology is Fisher and Ford's (1947) study of the *medionigra* gene in the scarlet tiger moth *Panaxia dominula*, a colorful day-flying species with one generation per year. A single diallelic locus has a major effect on the forewing pattern. Individuals that are homozygous for the *dominula* allele have multiple forewing spots, while individuals that are homozygous for the *medionigra* allele have a darkly suffused forewing with, typically, two small spots (the *bimacula* phenotype). Heterozygotes show an intermediate pattern, which is called the *medionigra* phenotype. In 1938, Ford began studying a small colony of this species in Cothill Fen, just southwest of Oxford, England. Starting in 1941, capture-recapture data were used to estimate the census population size, with the smallest estimated size between 1941 and 1947 being 1000. In 1939 ($t = 0$) the frequency of the *medionigra* allele was estimated (from a sample size of $n_0 = 223$) as $\widehat{p}_0 = 0.092$, while by 1947 ($t = 8$), its sample frequency had decreased to $\widehat{p}_8 = 0.037$ ($n_8 = 1341$). Taking $N_e = 1000$ (this being the smallest estimated census value over any of the generations, and hence most favorable to supporting drift), do these data show evidence of a departure from drift?

For simplicity, assume sampling without replacement, so that $\sigma(\widehat{p}_0, \widehat{p}_t) = 0$, with the variances are given by Equations 9.2a and 9.2b. The resulting covariance matrix, $\mathbf{V}$, becomes

$$\frac{\mathbf{V}}{p_0(1-p_0)} = \begin{pmatrix} \dfrac{1}{2\cdot 223} & 0 \\ 0 & \dfrac{1}{2\cdot 1341} + \dfrac{8}{2000}\left[1 - \dfrac{1}{2\cdot 1341}\right] \end{pmatrix} = \begin{pmatrix} 0.0022 & 0 \\ 0 & 0.0044 \end{pmatrix}$$

Because $\mathbf{V}^{-1}$ appears in both the numerator and the denominator of Equation 9.4, the unknown constant, $p_0(1-p_0)$, cancels out, allowing us to simply use the above right-hand matrix for $\mathbf{V}$, yielding

$$\text{GLS}\,(p_0) = \frac{\mathbf{1}^T\mathbf{V}^{-1}\mathbf{p}}{\mathbf{1}^T\mathbf{V}^{-1}\mathbf{1}} = \frac{49.496}{674.762} = 0.0734$$

Equation 9.2c yields the sampling variance for the difference in allele frequencies as

$$\sigma^2\left(\widehat{\delta}_t\right) \simeq p_0(1-p_0)\left[\frac{1}{2n_0} + \frac{1}{2n_t} + \frac{t}{2N_e}\left(1 - \frac{1}{2n_t}\right)\right]$$
$$= 0.0734 \cdot 0.9266\left[\frac{1}{446} + \frac{1}{2682} + \frac{8}{2000}\left(1 - \frac{1}{2682}\right)\right] = 0.0004495$$

The resulting Waples test statistic for fit to pure drift becomes

$$\frac{(0.037 - 0.092)^2}{0.0004495} = 6.729$$

The probability that a χ_1^2 random variable is this big or larger is 0.0095, implying strong rejection of neutrality. By using different values of N_e in the above calculation, we can find the largest effective population size that would still allow drift to account for these data. For $N_e = 500$, the test statistic becomes 4.19 (a p value of 0.040), while for $N_e = 250$, the statistic is 2.39 (a p value of 0.12). Hence, any effective population size slightly smaller than 500 would be compatible with a hypothesis of the observed allele-frequency change being driven by drift.

Except for a gap between 1979 and 1987, the Cothill Fen population has been surveyed yearly since 1939; see Jones (1989) and Cook and Jones (1996) for reviews (Jones provides a handy table of all data through 1988). O'Hara (2005) used a hierarchical Bayesian analysis to examine a 60-year time series of these data. He assigned genotypes fitness drawn from a lognormal prior, allowing them to vary yearly. While selection was found to significantly contribute to the change in allele frequency, most of the variance was attributable to drift. Cook and Jones (1996) and Mathieson and McVean (2013) estimated a selection coefficient against the *medionigra* allele of around 10% (assuming additivity in fitness), while Mathieson and McVean noted that a recessive model of selection provided a better fit, but required much stronger selection (essentially a lethal). A more recent analysis by Foll et al. (2015) found that both the weakly selected codominant and strongly selective recessive models are supported when N_e is fixed at 500. However, when N_e is jointly estimated from the data, there is stronger support for the lethal recessive model. While this is one of the best temporal data sets available, and selection appears to be strongly acting on a single gene, all of this uncertainty highlights the difficulty of dealing with natural populations.

While one might think that tests based solely on allele-frequency change are among the most convincing, this is not the case. As Example 9.1 shows, rejection of the neutral model can easily result from an overestimation of the true effective population size. Fisher and Ford took their results as evidence against Sewall Wright's notion of the importance of genetic drift. In his reply, Wright (1948a) noted that values of N_e simply based on census numbers can easily be contested by the widespread observation that the effective population size is generally (and often dramatically) lower than the observed number of individuals in the population (Chapter 3). In addition, tests of allele-frequency change suffer from low power. If selection is modest relative to $1/N_e$ or $1/n$ (with n being the sample size), the sample variance can obscure any selection signal. Waples (1989b) examined some of these design issues.

Although we have presented this test for a single locus with just two alleles, its extension to multiple alleles is straightforward (e.g., Waples 1989b; Goldringer and Bataillon 2004). A more subtle issue is the fit of a χ^2 distribution to the test statistic given by Equation 9.2d, which can be poor when alleles are rare, the number of alleles is large, or the number of generations is large (Goldringer and Bataillon 2004). While more sophisticated modifications (e.g., Sandoval-Castellanos 2010) can avoid some of these issues, the use of simulations that incorporate as much of the specific biology of the species as is known (e.g., Mueller et al. 1985) to model the change in the neutral alleles under drift is strongly preferred over parametric tests.

Allele-frequency Change Over a Times Series: The Fisher-Ford Test

The test given by Equation 9.2d assumes we have data from just two time points, but often one has **time-series** data for a number of generations. In such cases, the strong temptation to simply test the two most extreme values should be avoided, as such nonrandom sampling gives a highly biased result. Rather, specific tests have been developed that jointly consider all of the data. Indeed, the original test of Fisher and Ford involved such a temporal sequence of data. While one can use frequencies directly, Fisher and Ford used the arcsin-square-root transform to both stabilize the variance (making it independent of the initial frequencies) and improve the fit to normality, especially at extreme frequencies (note that the arcsin is measured in radians, rather than degrees). Such variance-stabilizing transformations were discussed in LW Chapter 11.

Let y_t denote the transformed frequency of the allele in generation t. For a t that is small relative to N_e, we find (approximately) that

$$y_t = 2\sin^{-1}\left(\sqrt{p_t}\right) \sim N\left(y_0, t/[2N_e]\right) \tag{9.5a}$$

where $y_0 = 2\sin^{-1}\left(\sqrt{p_0}\right)$ is the transformed value of the initial frequency. Estimates of allele frequencies are made at k time points, with no requirements about the temporal spacing between samples. Let $\mathbf{y}$ denote the vector of the transformed estimates of the k sampled allele-frequencies, and let $\mathbf{1}$ denote a vector of ones of the same length

$$\mathbf{y} = 2\begin{pmatrix} \sin^{-1}\left[\sqrt{p_1}\right] \\ \vdots \\ \sin^{-1}\left[\sqrt{p_k}\right] \end{pmatrix}, \qquad \mathbf{1} = \begin{pmatrix} 1 \\ \vdots \\ 1 \end{pmatrix} \tag{9.5b}$$

Finally, we need the covariance matrix, $\mathbf{V}$, whose elements are independent of the allele frequency (because of the variance-stabilizing transformation; Equation 9.5a). The sample indices denote the sequence of samples, not the actual sampled generation itself (see Example 9.2), with t_i the generation number associated with the ith sample. The diagonal terms of $\mathbf{V}$ are given from Equation 9.2c

$$V_{ii} = \frac{1}{2n_{t_i}} + \frac{t_i}{2N_e}\left(1 - \frac{1}{2n_{t_i}}\right) \simeq \frac{1}{2n_{t_i}} + \frac{t_i}{2N_e} \tag{9.5c}$$

Now consider the covariance between samples i and j, which correspond to generations t_i and t_j, respectively (where $i > j$ and $t_i > t_j$). The estimates for these two sample points have a shared history (from the base value, p_0) of drift up through generation t_j, yielding

$$V_{i,j} = V_{j,i} = \frac{t_j}{2N_e} \quad \text{where} \ \ t_j < t_i \tag{9.5d}$$

Note that the covariance with the base generation ($t = 0$) is always zero (which is why the off-diagonal covariances for $\mathbf{V}$ in Example 9.1 were set to zero). The $k \times k$ matrix, $\mathbf{V}$, contains only those rows and columns corresponding to the k specific generations sampled.

This is now a goodness-of-fit problem for a linear model. Using Equation 9.4, we obtain a generalized least-squares (GLS) estimate of the (transformed) initial frequency

$$\widehat{y}_0 = \frac{\mathbf{1}^T\mathbf{V}^{-1}\mathbf{y}}{\mathbf{1}^T\mathbf{V}^{-1}\mathbf{1}} \tag{9.6a}$$

Using this value, the vector of deviations is

$$\boldsymbol{\delta}_{\mathbf{y}} = \mathbf{y} - \widehat{y}_0 \cdot \mathbf{1} \tag{9.6b}$$

and the test statistic, the weighted sum of the squared (transformed) allele-frequency differences,

$$\boldsymbol{\delta}_{\mathbf{y}}^T\,\mathbf{V}^{-1}\boldsymbol{\delta}_{\mathbf{y}} \tag{9.6c}$$

is expected to be approximately χ^2_{k-1} distributed due to the normality assumption on y_i.

Example 9.2. We now revisit Fisher and Ford (Example 9.1), and consider a test based on the data from 1939, 1943, and 1947, where

Year	t	$\widehat{p}$	$y = 2\sin^{-1}\left(\sqrt{p}\right)$	n
1939	0	0.092	0.616	223
1943	4	0.056	0.478	269
1947	8	0.037	0.387	1341

Assuming $N_e = 1000$, the resulting covariance matrix, $\mathbf{V}$ (on the transformed scale), becomes

$$\mathbf{V} = \begin{pmatrix} V_{0,0} & V_{0,4} & V_{0,8} \\ V_{4,0} & V_{4,4} & V_{4,8} \\ V_{8,0} & V_{8,4} & V_{8,8} \end{pmatrix} = \frac{1}{2000}\begin{pmatrix} \frac{2000}{2\cdot 223}+0 & 0 & 0 \\ 0 & \frac{2000}{2\cdot 269}+4 & 4 \\ 0 & 4 & \frac{2000}{2\cdot 1341}+8 \end{pmatrix}$$

$$= \frac{1}{2000}\begin{pmatrix} 4.484 & 0 & 0 \\ 0 & 7.717 & 4 \\ 0 & 4 & 8.745 \end{pmatrix}$$

In addition,

$$\mathbf{y} = \begin{pmatrix} 0.616 \\ 0.478 \\ 0.387 \end{pmatrix}, \quad \mathbf{1} = \begin{pmatrix} 1 \\ 1 \\ 1 \end{pmatrix}, \quad \text{yielding } \widehat{y}_0 = \frac{\mathbf{1}^T\mathbf{V}^{-1}\mathbf{y}}{\mathbf{1}^T\mathbf{V}^{-1}\mathbf{1}} = \frac{418.851}{774.701} = 0.541$$

Using this estimate for y_0, the vector of deviations from the initial value becomes $\boldsymbol{\delta}_{\mathbf{y}} = \mathbf{y} - 0.541\cdot\mathbf{1}$, returning a test statistic value of $\boldsymbol{\delta}_{\mathbf{y}}^T\mathbf{V}^{-1}\boldsymbol{\delta}_{\mathbf{y}} = 7.964$, which when compared to a χ^2_2 distribution, returns a significance level of 0.0186. For $N_e = 500$, Equation 9.6c returns a value of 5.398, for a significance of 0.067, so the hypothesis that drift alone accounts for the observed pattern of change cannot be rejected under this smaller value of N_e.

A number of generalizations, as well as increasingly sophisticated tests building on the basic elements of the Fisher-Ford framework, have been proposed, including extending this methodology to handle data from high-throughput sequencing. A partial list includes Templeton (1974), Schaffer et al. (1977), Gibson et al. (1979), Wilson (1980), Watterson (1982), Waples (1989b), De Koeyer et al. (2001), Goldringer and Bataillon (2004), Bollback et al. (2008), Wisser et al. (2008), Sandoval-Castellanos (2010), Malaspinas et al. (2012), da Fonseca et al. (2013), Mathieson and McVean (2013), Feder et al. (2014), Lacerda and Seoighe (2014), Steinrücken et al (2014), Foll et al. (2015), Terhorst et al. (2015), Topa et al. (2015), Gompert (2016), and Schraiber et al. (2016).

In addition to its use in studying natural populations, this machinery can be applied to artificial selection experiments to detect regions of interest. In the pre-genomics era, this approach was pioneered by Stuber and Moll (1972) and Stuber et al. (1980), who looked for shifts in the frequencies of isozyme markers in lines of maize selected for yield. Other examples for maize include Labate et al. (1999) and Coque and Gallais (2006) for yield selection, and Wisser et al. (2008) for disease resistance, while De Koeyer et al. (2001) examined yield in oats. In the genomics era, extensions of this machinery have been used in microorganisms, such as the analysis by Foll et al. (2014) on the target (or targets) of selection for influenza A virus exposed to the drug oseltamivir, as well as with data from **evolve and resequence** experiments (**E&R**), such as those on *Drosophila* (Terhorst et al. 2015; Topa et al. 2015).

Schaffer's Linear Trend Test

A variation of the Fisher-Ford test was suggested by Schaffer et al. (1977), who noted that power might be improved by going beyond a simple lack of fit test against the model $y_t = \mu + e$ (where μ is the transformed initial allele frequency), by asking if a *significant linear trend* is present. The model now becomes

$$y_t = \mu + \beta t + e \tag{9.7a}$$

where a trend is indicated if β is significantly different from zero (the Fisher-Ford test assumes $\beta = 0$). Such a linear trend is not expected under drift but would be expected under directional selection, assuming that the direction of selection is not changing (migration from a population with a different allele frequency could also generate a linear trend). In general-linear-model form (LW Chapter 8), Equation 9.7a becomes $\mathbf{y} = \mathbf{X}\boldsymbol{\beta} + \mathbf{e}$, where

$$\mathbf{X} = \begin{pmatrix} 1 & t_1 \\ \vdots & \vdots \\ 1 & t_k \end{pmatrix}, \quad \boldsymbol{\beta} = \begin{pmatrix} \mu \\ \beta \end{pmatrix}, \quad \widehat{\boldsymbol{\beta}} = (\mathbf{X}^T\mathbf{V}^{-1}\mathbf{X})^{-1}\mathbf{X}^T\mathbf{V}^{-1}\mathbf{y} \tag{9.7b}$$

where the elements of $\mathbf{V}$ are given by Equations 9.5c and 9.5d. For the data in Example 9.2, the resulting $\mathbf{X}$ matrix and the GLS estimate, $\widehat{\boldsymbol{\beta}}$, of the vector of parameters becomes

$$\mathbf{X} = \begin{pmatrix} 1 & 0 \\ 1 & 4 \\ 1 & 8 \end{pmatrix}, \quad \widehat{\boldsymbol{\beta}} = \begin{pmatrix} 0.609 \\ -0.028 \end{pmatrix}$$

Applying LW Equation 8.35, the standard error on the slope is found to be 0.0086, showing that it is highly significant. Stuber et al. (1980) used this approach to infer selection at sites linked to several allozyme markers in a series of selected maize lines. One advantage of the linear-trend test is that it does not require highly accurate estimates of N_e.

Scans and Simulation-based Approaches

As presented, these tests for shifts in allele frequencies are performed one marker at a time, as they herald from the days of testing just one or a few unlinked candidate genes. With a few unlinked markers, Bonferroni corrections (or the slightly more powerful sequential methods; Appendix 4) can be applied to assign overall significance levels. Likewise, FDR approaches can use used to assign **false-discovery rates** among the set of markers declared to be significant (the fraction of tests declared to be significant that are actually from the null; Appendix 4). However, in the **genomic-scan** era, with the potential for thousands of linked markers on each chromosome, tests are no longer independent, thus compromising FDR approaches (Chen and Storey 2006). Even if tests are largely independent, their vast number makes Bonferroni-type corrections untenable for a test to have any power (Appendix 4). How then can these tests be extended to the dense marker maps used in genomic scans?

For starters, analyzing markers one at a time is rather inefficient, in that one potentially loses shared information from linked markers. A better approach is to compute the average allele-frequency change within a small sliding window. Starting with the initial allele frequencies within a given window and either their known or assumed recombination rates, simulations under pure drift (and recombination) can generate a null distribution of average change as a comparison point for the actual observed divergence (e.g., Example 9.3; Johansson et al. 2010). Other, less formal approaches can also be used to simply indicate regions of interest (as opposed to regions that are formally statistically supported); see Figure 9.1.

Birthdate Selection Mapping (*BDSM*)

A very interesting genomic-scanning approach for sites under selection becomes possible when one has extensive pedigree data, such as for cattle (Decker et al. 2012). With extensive pedigrees, one has information on the date of birth (DOB) of most individuals, whose value can be expressed as years since the start of the pedigree. Individuals with a later date of birth (i.e., more recent in the pedigree) have likely experienced more selection than earlier-born individuals (those deeper down in the pedigree). As such, one expects to find a positive relationship between a marker linked to a selected site and DOB. For example, if allele *A* was initially rare but favored, *AA* individuals are expected to be more common in animals with higher (i.e., more recent) DOB values. Decker et al. turned this

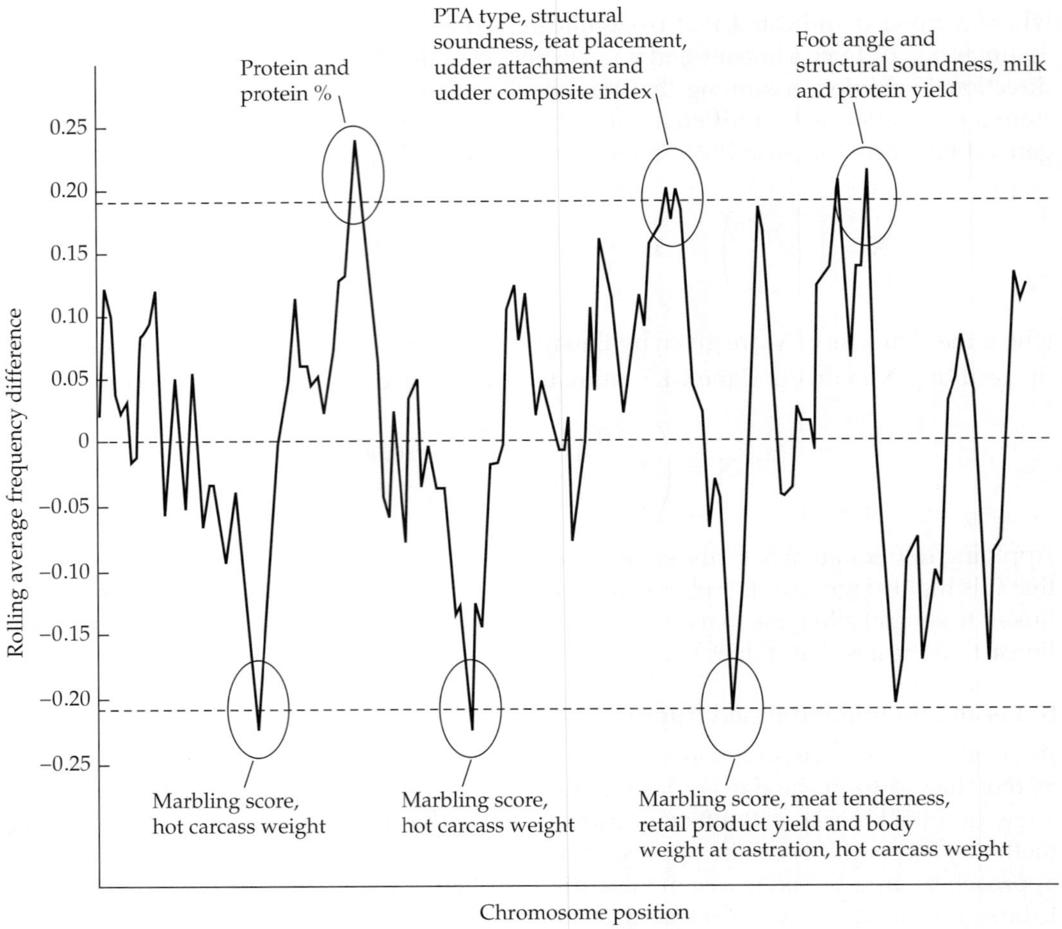

Figure 9.1 A scan of *Bos taurus* chromosome 19, contrasting differences in SNP allele frequencies between specialized dairy (Holstein) and meat (Angus) breeds. Positive values indicate alleles at higher frequencies in Holstein cattle (dairy-specific), and negative values indicate alleles that are more common in Angus cattle (meat-specific). Differences were based on a sliding window of five adjacent markers, using a set of 175 SNPs. The horizontal axis represents chromosomal position, and the vertical axis is the average between-breed difference in SNP allele frequencies over the five-SNP window. The upper and lower dashed lines indicate the 5% threshold levels as assessed via a permutation test (see the text for details). As annotated in the figure, the authors were able to associate these exceptional peaks and valleys with known QTLs for dairy and beef traits. Because QTL intervals tend to be rather vague (averaging around 20 megabases, or roughly 20 cM, for these traits), the significance of these associations with known QTLs, while suggestive, is unclear. (After Prasad et al. 2008.)

relationship around, realizing that by treating DOB values as a quantitative trait, one could use association mapping (LW Chapter 16), with sites showing an association with DOB (i.e., those that are under- or over-represented later in the pedigree) being candidates for selection. This approach is called **birthdate selection mapping** (or **BDSM**), and because association mapping is done in a mixed-model framework (Chapter 19), it accounts for biases introduced by family structure. The authors applied this approach to U. S. Angus cattle born over a 50-year period (roughly 10 generations). A standard random-effects mixed model of association mapping detected 11 loci significantly associated with DOB, while a Bayesian model found that $\sim$2% of the SNPs were strongly associated with DOB. The former model assumes an infinitesimal structure, while the latter allows for genes of larger effect that are embedded in a sea of smaller-effect genes. While **BDSM** requires large, deep pedigrees, it is an intriguing and potentially powerful approach.

DIVERGENCE BETWEEN POPULATIONS: TWO-POPULATION COMPARISONS

While most of the analysis of divergence data in structured populations is based on F_{ST} statistics (Chapter 2), we start with a few comments on the simple situation in which one is comparing a biallelic locus between two populations. As in the case of the divergence of a single population measured at starting and ending time points, divergence can be measured as the squared allele-frequency difference,

$$\widehat{\delta}_t = \left(\widehat{p}_{t,1} - \widehat{p}_{t,2}\right)^2 \tag{9.8}$$

namely, the squared difference between the frequency in the two populations at some sample time, t, following their isolation from a common ancestor in generation 0. Whether $\widehat{\delta}_t$ is too large, or too small, relative to drift can be evaluated using a simple modification of the Waples test, wherein the denominator in Equation 9.2d is replaced by $\sigma^2\left(\widehat{p}_{t,1}\right) + \sigma^2\left(\widehat{p}_{t,2}\right)$, the sum of the allele-frequency sampling variances for each population (defined as in Equation 9.2b). This expression requires estimates of the divergence time, t, as well as the average effective sizes for both populations. More generally, because $E\left[\widehat{p}_{t,i}\right] = p_0$, in theory one could sample the two populations at different time points (t_1 and t_2), but now using $\sigma^2\left(\widehat{p}_{t_1,1}\right) + \sigma^2\left(\widehat{p}_{t_2,2}\right)$ in the denominator of the test statistic.

A common scenario involves the comparison of two subpopulations, descendent from a shared ancestor, that presumably have experienced different selection pressures. The typical setting is either selection in one population (e.g., adaptation to elevation assessed by comparing a derived highland population to a lowland control) or lines selected in divergent directions (e.g., meat versus dairy cattle). If one has a fairly dense marker map, adjacent markers should be in LD and the joint use of multiple markers can enhance the signal of selection. A standard approach is a genomic scan using a sliding window of markers, contrasting the average frequency differences between populations for the alleles within windows (Figure 9.1). While the window size is arbitrary, it should be no smaller than the average size of an LD block for the populations being compared.

As with temporal data, the significance of divergence data can be assessed using simulations (Example 9.3), although other, less formal approaches are often used to highlight regions of interest. One example of the latter is the work of Prasad et al. (2008), presented in Figure 9.1, which depicts a scan for excess SNP-frequency differences in a sliding window analysis over *Bos taurus* (cattle) chromosome 19. Regions of interest were identified by a permutation test in which breed labels were randomized over SNPs to generate the null. Threshold values indicate the 95% limit for the range of maximal within-window differences in frequencies between the two populations in the randomized data sets. This permutation approach is only approximate, as it ignores LD among linked neutral markers within breeds (a more careful analysis, based on simulations, is detailed in Example 9.3). However, it still serves to indicate sites likely enriched for differentially selected genes.

Finally, a very simple statistic that often appears in comparisons of selected versus control populations is Grossman et al.'s (2010) ΔDAF statistic. This metric is a natural outgrowth of the type of comparisons shown in Figure 9.1, which focuses on the difference in the **derived allele frequency** (DAF) between a control and a selected population. For a candidate SNP, let $\overline{D}_{NS}$ denote the frequency of the derived allele in a nonselected control population (or its average frequency if multiple control populations are used) and its frequency, D_S, in the putatively selected population, with $\Delta DAF = D_S - \overline{D}_{NS}$. This statistic ranges between plus one and minus one, and standard outlier approaches are used to highlight SNPs with excessive values.

Example 9.3. Hayes et al. (2008) contrasted SNP frequencies between Australian populations of Holstein (dairy) and Angus (beef) cattle, specialized breeds developed by selection from a common ancestral stock. Whole-genome scans of both populations were performed using 7032 sliding windows of ten adjacent SNPs to measure average allele-frequency change between the corresponding windows for both populations (the difference at each SNP, averaged over all ten sites). Simulations were used to assess significance (extreme departures in average allele-frequency in homologous windows between the two populations) relative to the values expected under drift alone. The authors had to model two issues: linkage and breed formation (time in some initial early domesticated population, and then subsequent time in separate bottlenecks representing the formation of the two specialized breeds).

The authors simulated roughly 300 SNPs per chromosome (to account for LD in addition to allele-frequency change), while the population structure during breed formation was modeled as follows. The authors simulated 900 generations of drift and mutation in a base population of $N_e = 1000$ to generate a common domesticated stock population. Specialized breed formation from this common stock was then simulated by sampling from this stock population to form two subpopulations of size $N_e = 125$, each of which was simulated for an additional 100 generations (these values represented the best assumption regarding these parameters during domestication and breed formation). The observed genome-wide F_{ST} between the Holstein and Angus populations was 0.08, and only simulations whose genome-wide F_{ST} values matched this value were kept as the null.

Taking those windows in the data set that were in the upper 0.1% of excessive divergence (positive or negative) relative to the simulated data resulted in 15 significant regions (windows). Focusing on windows with the uppermost 0.5% of divergence (relative to simulations) yielded 84 candidate regions. To assess what fraction of these extreme windows might be false positives, the authors computed the false-discovery rate (FDR), the fraction of those tests declared to be significant that are likely to be false positives (Appendix 4). FDR provides a measure of how enriched a set of results declared to be significant is for true positives. At a 0.1% level of significance, one expects $7032 \cdot 0.001 = 7.03$ tests to be significant by chance alone, while for 0.5%, this increases to 35 false positives among the 7032 tests for each window. The FDRs are 7/15 = 47% for tests of 0.1%, and 35/84 = 42% for tests at the 0.05% significance level. Hence, the expectation is that slightly over half (53% and 58%, respectively) of the windows initially flagged as significant are true positives.

DIVERGENCE BETWEEN POPULATIONS: F_{ST}-BASED TESTS

When comparisons involve more than two populations or markers with more than two alleles, a more natural measure of divergence is Wright's F_{ST} statistic of population structure (Wright 1951). Recall from Chapter 2 that this statistic measures the fraction of total variation over a set of populations that is due to among-population differences in allele frequencies, and is easily extended to multiple alleles and multiple populations. For a biallelic locus, $F_{ST} = \sigma^2_B(p)/[\overline{p}\,(1-\overline{p}\,)]$, where $\sigma^2_B(p)$ is the variance in allele frequency, p, over the populations around its average value, $\overline{p}$.

There are important caveats when using F_{ST}. First, the analysis of F_{ST} statistics assumes that back mutations are sufficiently rare to be safely ignored. This is not the case for microsatellite (STR/SSR) markers, which have both high mutation rates and a high chance of convergent mutation (alleles of different origins having the same repeat copy number), and their use requires specific divergence metrics, such as R_{ST} (Slatkin 1995a; Goodman 1997). Excoffier et al. (2009b) showed that using F_{ST} in place of R_{ST} for the analysis of STR data can significantly inflate the false-positive rate. Second, the upper limit of F_{ST} is set by the expected heterozygosity, meaning that a highly variable locus has a maximal F_{ST} value smaller than a less variable one (Charlesworth 1998; Hedrick 1999). Jakobsson et al. (2013) and Edge and Rosenberg (2014) provided upper bounds on F_{ST} given the frequency of the most common allele (averaged over all populations), and found that when this is either small (i.e., very many alleles) or large (close to population-wide fixation), F_{ST} is restricted to values far below 1.0. Hence, the levels of diversity within a region constrain the possible F_{ST} values. Various standardization measures of F_{ST} (and related statistics) have been proposed, and their strengths and weaknesses were reviewed by Meirmans and Hedrick (2011).

Even using these standardizations, F_{ST}-based tests perform poorly when the expected genome-wide divergence due to drift is sufficiently large that many neutral loci are expected to have alternative alleles near fixation between populations. Against such a background, the effect of selection at a candidate region is hard to detect. This situation emphasizes that the tests considered in this chapter apply over *ecological time scales*, as they assume (for the null) that most neutral markers are segregating in most subpopulations, which puts the time scale for their use at no more than $\sim N_e$ generations.

Ecological geneticists have coined the term **landscape genetics** for the study of the distribution of genetic variation over spatial structures (Manel et al. 2003; Manel et al. 2010; Manel and Holderegger 2013), and F_{ST} and allele-environmental correlations (discussed in the next section) are central to this emerging field. As a result, the literature on these classes of tests is rapidly expanding.

Example 9.4. The effectiveness of F_{ST} to detect selection was examined by Taylor et al. (1995), using a putative target of selection in the tobacco budworm (*Heliothis virescens*), a noctuid moth and a major cotton pest in the United States. Pyrethroid insecticides have been used in control efforts, and these act on voltage-gated sodium channels in the nervous system. The historical usage patterns of these insecticides, and hence the putative selection pressures on sodium channel genes, differed over the sampled populations examined by the authors. As a result, they predicted that F_{ST} values at the sodium channel *Hpy* gene should be significantly higher than for background loci, reflecting this differential selection over the sampled subpopulations. Samples of adults from widely spaced locations in the United States revealed an F_{ST} value of 0.041 ± 0.005 at the *Hpy* marker, in contrast to values of 0.002 ± 0.001 at 14 other loci, with the latter result indicating fairly weak population structure in this species.

Outlier Approaches

The underlying premise for most F_{ST}-based tests of selection was the suggestion by Cavalli-Sforza (1966) that all neutral loci should have the same expected value of F_{ST}, reflecting the genome-wide impact of common demographic and population-structure forces. Thus, one can (in theory) use a large number of marker loci to estimate the baseline F_{ST} value for the set of populations being compared, and then search for outlier loci. This approach is easily modified to look for specific loci being outliers in specific populations (e.g., Akey et al. 2002; Kayser et al. 2003; Akey et al. 2010). Loci with excessively high values indicate more divergence than expected under drift, and the possibility that the marker is linked to a site that is under differential selection over the demes. Likewise, excessively low values indicate less divergence than expected under drift, and hence the potential for a site that is under balancing selection near the marker. While the historical interpretation of F_{ST} data follows from these last two statements, results from Chapter 8 on sweeps under *uniform* selection in structured populations suggest that a more nuanced view is needed. Recall from Figure 8.8 that uniform selection over the entire metapopulation can generate excessive divergence (Figure 8.8A) during a hard sweep of a single allele when it is still restricted to a subset of the demes. Similarly, a soft sweep under uniform selection can also generate excessive divergence. Conversely, a completed hard sweep through the sampled demes generates a reduction in divergence relative to background levels of F_{ST} (Figure 8.8B).

The outlier strategy makes two assumptions: the vast majority of scored loci are neutral, and all neutral sites reflect the same underlying population demography. As discussed in the introduction to this chapter, new alleles arising on the leading wave of a population expansion can "surf" to high frequencies, generating excessive values over the expected background. Likewise, differences in the ratio of gene density to recombination rate in different parts of the genome change the expected pattern of background selection, potentially creating outliers even among neutral markers.

A final complication is that when the population structure departs from the island model (equal divergence is expected between all demes; Chapter 2), the variance in F_{ST} is inflated, generating an excess of outliers. An interesting example of this phenomenon appears in the work of Fourcade et al. (2013), who found that river fishes showed an unusually high number of outlier loci. While such an observation might be taken as evidence that river species have higher rates of local adaptation, simulations by these authors showed that species with a **fractal** (highly branching) **population structure** have a greatly inflated variance in F_{ST} relative to the island model. This arises because migration on fractal structures (such as rivers or valleys) generates a complex pattern of correlated allele frequencies. Other types of population structures, such as hierarchical island models (Figure 2.11), population expansions from refugia, and allelic surfing, can all inflate the number of outliers (Excoffier et al. 2009a; Bierne et al. 2013).

Tests Based on F_{ST}-generated Branch-lengths

When migration and new mutation can be ignored, F_{ST} provides an estimate of the divergence time, T (scaled in $2N_e$ generations), between two populations. Rearranging Equation 2.43, taking the log of both sides, and recalling that $\ln(1-x) \simeq -x$ (for $|x| \ll 1$) yields

$$\ln(1 - F_{ST}) = t \ln\left(1 - \frac{1}{2N_e}\right) \simeq -t/2N_e \tag{9.9}$$

Hence $T = -\ln(1-F_{ST}) \simeq t/2N_e$, and one can recast an excessive F_{ST} value as an excessive separation time required for drift to account for the observed divergence. These estimated times are called **branch lengths** and (following the Cavalli-Sforza premise) should have the same expected value over all neutral genes. An excessive branch length for a candidate gene relative to some reference set of genes suggests excessive change relative to drift (Vitalis et al. 2001; Rockman et al. 2003), and is the basis of the **population branch statistics** (**PBS**) of Yi et al. (2010); see Figure 9.2.

The Lewontin-Krakauer Test: Basics

The above outlier methods (for either F_{ST} or branch lengths) are rather *ad hoc*, and best viewed as *enrichment* methods, distilling down a reduced set of markers that is likely enriched for selected sites. The critical missing element in these methods is the expected *distribution* of F_{ST} values for a random marker, allowing p values to be placed on outliers. Formal distribution-based tests were introduced by Lewontin and Krakauer (1973), who considered the distribution of F_{ST} values for a random biallelic locus sampled over n populations under an island model (Figure 2.11). If we assume that the distribution (over populations) of the frequency of an allele is roughly normal, the expected large-sample distribution of F_{ST} values approximately follows a $\lambda\chi^2_{n-1}$ distribution, with a scaling factor of $\lambda = E(F_{ST})/(n-1)$. Given Cavalli-Sforza's assumption that, on average, population structure influences all neutral loci equally, Lewontin and Krakauer estimated $E(F_{ST})$ from the average $\overline{F}_{ST}$ over all scored loci, giving the distribution for a random realization F_{ST} as

$$\frac{1}{\lambda} F_{ST} = \frac{(n-1)F_{ST}}{\overline{F}_{ST}} \sim \chi^2_{n-1} \tag{9.10a}$$

In other words, scaled individual F_{ST} values follow a chi-square distribution with $n-1$ degrees of freedom. This is a large-sample approximation, as the sampling error in estimating the true realization of the F_{ST} value for a given marker is ignored. Because the variance of a χ^2_n random variable is $2n$ (LW Equation A5.15b), the variance among realizations of F_{ST} values is approximately

$$\sigma^2(F_{ST}) \simeq \sigma^2\left(\lambda\chi^2_{n-1}\right) = 2(n-1)\lambda^2 = 2(n-1)\left(\frac{E[F_{ST}]}{n-1}\right)^2 \simeq 2\,\frac{\overline{F}^{\,2}_{ST}}{n-1} \tag{9.10b}$$

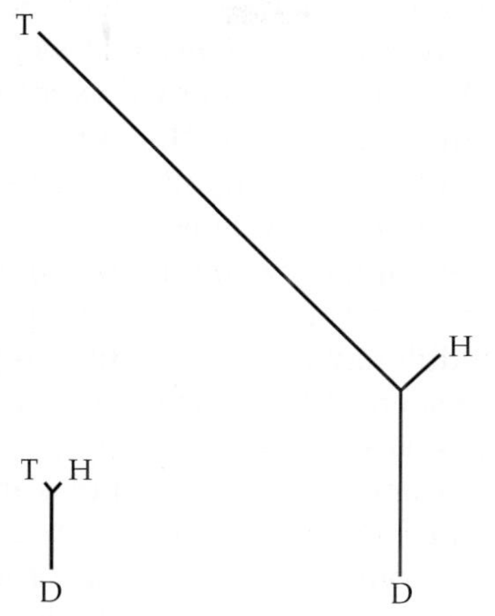

Figure 9.2 F_{ST}-based branch lengths for Tibetan (T), Han (H), and Danish (D) populations. (**Left**) Lengths based on the average F_{ST} values for all sampled markers. (**Right**) The tree for the *EPAS1* gene. While the D and H branches show increased divergence relative to the average F_{ST}, the divergence along the T lineage is far more dramatic. This is consistent with excessive allelic divergence due to selection for living at high altitude (or perhaps other features, such as allelic surfing). (After Yi et al. 2010.)

Baer (1999) empirically showed that the variance of a wide range of fish F_{ST} values is more accurately given by replacing the 2 in Equation 9.10b by a value between 5 and 7. As mentioned in the previous section, the fractal structure of many fish populations (Fourcade et al. 2013), and hence a significant departure from the assumed island-model underlying Equation 8.10b, likely account for at least part of this inflated variance.

There are a number of additional potential problems with this approach of using $\overline{F}_{ST}$ to provide an estimator of λ. First, this estimate can be biased by skew resulting from a few excessive F_{ST} values. Specifically, if $F \sim \lambda\chi^2_{n-1}$, estimating λ by comparing means yields the method-of-moments estimator, $\widehat{\lambda} = \overline{F}/(n-1)$, as $E[\chi^2_n] = n$. However, even just a few loci that are under selection—and hence with extreme large values of F_{ST}—inflate $\overline{F}$ and bias the estimate of λ under the null. A more robust approach is to replace the usage of the means with **medians**, the 50% values of the two distributions (Devlin and Roeder 1999). Specifically, $\text{med}(F) = \text{med}(\lambda\chi^2_{n-1})$, or

$$\widehat{\lambda} = \frac{\text{med}(F)}{\text{med}(\chi^2_{n-1})} \tag{9.10c}$$

For example, suppose the median for single-locus F_{ST} values among a collection of loci sampled over five populations is 0.127. Because $\Pr(\chi^2_4 \leq 3.357) = 0.5$, the median value of a χ^2_4 is 3.357, yielding

$$\widehat{\lambda} = \frac{\text{med}(F)}{\text{med}(\chi^2_{n-1})} = \frac{0.127}{3.357} = 0.038$$

as a more robust estimate of λ under the null (drift) relative to that based on the mean, $\overline{F}_{ST}$, because the median-based estimate is not biased by the presence of a modest number of loci under selection.

Second, Equation 9.10a depends on the validity of the χ^2 approximation for the distribution of F_{ST} values. This approximation fails when too many alleles (more than five) are present at a locus, the minor-allele frequency is small (< 0.1), or the divergence time is too large (Goldringer and Bataillon 2004). Indeed, Whitlock and Lotterhos (2015) recommended that loci with low heterozygosities be excluded in estimates of $\overline{F}_{ST}$.

A third, and deeper, problem is the implicit assumption of Lewontin and Krakauer that neutral allele frequencies are *independent among demes*. This is correct under the standard island model (Figure 2.11), which yields equal expected divergence among any pair of demes, and the same amount of variation within any deme (assuming no among-deme differences in N_e). However, this assumption fails under more complex population structures, such as unequal migration between demes (e.g., the **isolation by distance** model, wherein closer demes exchange migrants at higher rates) or hierarchical structure among demes generated by their founding. These population-structure issues create correlations among allele frequencies from different demes, inflating the variance of F_{ST} relative to the expectations under the island model, which impacts the χ^2 assumption (Nei and Maruyama 1975; Robertson 1975a, 1975b; Tsakas and Krimbas 1976).

As a result of these concerns (and others; see Nicholas and Robertson 1976), the original version of the Lewontin-Krakauer test quickly languished. However, its basic simplicity, coupled with its requirement of only the type of data routinely gathered by ecological geneticists (estimates of locus-specific F_{ST} values), fueled the search for ways to correct these initial flaws.

Whitlock and Lotterhos (2015) recently suggested a potentially simple work-around for many of these issues, going by the name of ***OutFLANK***. They noted through extensive simulations of very different population structures that the distribution for F_{ST} values (provided heterozygosity levels were not too small) was very close to χ^2, but with different degrees of freedom from the Lewontin-Krakauer value of $(n - 1)$. This difference in the degrees of freedom makes sense, given a lack of independence among demes, and they recommended a two-step approach for obtaining approximate p values. First, the upper and lower 5% of the empirical F_{ST} values are trimmed. The logic being that loci under uniform selection (generating excessive low values) and under divergent selection (generating expressive high values) are expected to be only a tiny fraction of all tested sites. The remaining trimmed distribution, representing the core 90% of the values, is then used in a ML setting to estimate the appropriate degrees of freedom for such a doubly truncated χ^2. (More generally, Table A2.1 shows that the χ^2 distribution is a special case of the gamma distribution, and fitting the latter allows for what amounts to fractional degrees of freedom, which might further improve the fit.) With the corresponding null density now estimated, appropriate p values for outliers can be obtained. Their simulations showed that this approach worked well for excessively high values (i.e., the right-hand tail of the distribution), but very poorly for the left-hand tail (those loci showing small F_{ST} values than expected).

Second-generation Lewontin-Krakauer Tests: Model-based Adjustments

One proposed strategy to resuscitate the Lewontin-Krakauer test was to use knowledge of the population structure as the basis for simulations of the distribution of F_{ST} under the null hypothesis of no selection. Such an analysis was first performed by Bowock et al. (1991), who had a rough idea of this structure for the five human populations they surveyed. Vigouroux et al. (2002) used coalescent simulations incorporating a founding bottleneck in a screen of 501 maize genes to find those with excessive F_{ST} values. Similarly, Ross-Ibarra et al. (2008) estimated the parameters of a complex model of the population structure of *Arabidopsis lyrata*, and then used simulations based on their estimated demography to detect outliers. The concern with any null distribution generated by simulations is robustness to assumptions about the population structure, as even the most careful simulations can be misleading (e.g., Carret et al. 2006; Excoffier et al. 2009a, 2009b). For example, most analyses of robustness to different demographic models fail to consider the effects of spatial expansion, and hence ignore concerns raised by allelic surfing.

An alternative strategy involves a more careful examination of outliers detected from simulation results. Beaumont and Nichols (1996) suggested that outliers on a two-dimensional plot of a gene's F_{ST} value versus its (population-wide) heterozygosity value (H) offered a more robust signal of selection. Their logic was that the sampling variance in F_{ST} becomes more sensitive when allele frequencies are skewed over populations, implying that sites with low-frequency alleles can generate an excess of extreme values simply by chance. Further, H constrains the possible range of F_{ST}. Given an estimate of the average neutral F_{ST}, coalescent simulations are performed to generate a joint distribution of F_{ST} versus H values under an island model (the Beaumont-Nichols ***FDIST*** and ***FDIST2*** tests). Excoffier et al. (2009b) used a similar approach but with simulations assuming a hierarchical island model (Figure 2.11).

A more robust approach is to move away from the island model as the basis for tests (Meirmans 2012). Because the potential underlying population structure is arbitrary and unknown, successively more complex models have been proposed in the attempt to capture at least some of the true covariance structure among demes. The most common model, from Beaumont and Balding (2004), is to assume that demes are formed as independent draws from the same common ancestral base population (potentially with different N_e values in each of the resulting demes). Once formed, no subsequent migration occurs among demes, resulting in a star phylogeny (all branches radiate from a single ancestral population). Under this model, the vector, $\mathbf{p}$, of the deme-specific allelic frequencies (under the null) is assumed to have been drawn from a Dirichlet distribution (Equation A2.37) with an ancestral allele frequency of p_0. Beaumont and Balding modeled the divergence associated with locus i in population j as

$$\ln\left(\frac{F_{ST,ij}}{1-F_{ST,ij}}\right) = 1 + \beta_j + \alpha_i + \epsilon \qquad (9.11)$$

where the locus-specific (α_i) and population-specific (β_j) values are estimated using a Bayesian hierarchical model (Appendix 3). Here, all loci in population j contribute to the estimation of β_j, while the values for locus i over all populations contribute to the estimation of its specific effect, α_i. Beaumont and Balding's ***BayesFST*** approach flags loci of interest when their α_i values fall significantly below zero (with F_{ST} below the population expectation, suggesting balancing selection) or significantly above zero (with F_{ST} above the population expectation, suggesting divergence selection). A number of investigators have refined this approach (Foll and Gaggiotti 2008; Riebler et al. 2008; Bazin et al. 2010; Gautier et al. 2010; Vitalis et al. 2014; de Villemereuil and Gaggiotti 2015).

The power of these model-based approaches has been examined by a number of authors (Pérez-Figueroa et al. 2010; Narum and Hess 2011; Vilas et al. 2012; de Mita et al. 2013; Lotterhos and Whitlock 2014, 2015; de Villemereuil et al. 2014), albeit usually under a modest range of simulated population structures. The conclusion is that these methods, and in particular Foll and Gaggiotti's (2008) ***Bayescan*** method, can perform well when the populations are either independent draws from a common ancestral population or are well described by an island model. In other settings, such as a spatial migration structure (e.g., alleles from closer demes are more correlated than those from more distant demes) or nonequilibrium conditions (e.g., expansion out of a refugium, and hence the potential for allelic surfing), these approaches have a high false-positive rate. This is true for both apparent signatures of divergent selection (with excessively high F_{ST} values) and balancing selection (with excessively low F_{ST} values). The presence of selfing within demes further exacerbates these issues (de Mita et al. 2013).

Third-generation Lewontin-Krakauer Tests: Correcting for Population Structure

If we refer to the above model-based approaches as **second-generation** Lewontin-Krakauer (LK) tests, **third-generation** LK tests attempt to estimate the covariance structure among the neutral alleles without using a formal model, and then use this to detect outliers. Their motivation traces back to Felsenstein's (2002) extension of his method for independent contrasts among taxa in a phylogeny (Felsenstein 1985) to within-population comparisons, and to the mixed-model approach of Yu et al. (2006) to correct for population structure under association mapping. These third-generation approaches were first used in methods (discussed below) that were intended to detect correlations between allele frequencies and environmental factors (Hancock et al. 2008; Coop et al. 2010; Eckert et al. 2010; Günther and Coop 2013), and then subsequently applied to Lewontin-Krakauer tests (i.e., F_{ST} data) by Bonhomme et al. (2010), Fariello et al. (2013), Günther and Coop (2013), Duforet-Frebourg et al. (2104), and Gautier (2015). Their basic structure (outlined in Example 9.5) is to first use all (or a part) of the data to estimate either a kinship or covariance matrix of correlations among neutral allele frequencies between demes, and then use this correlation structure to provide adjusted F_{ST} values.

The ***FLK* test** (for F-matrix LK test; see Example 9.5) of Bonhomme et al. (2010) starts with a set of assumed neutral alleles, together with an outgroup (to root the estimated phylogenic tree of relatedness between demes). Genetic distances are computed for all pairwise combinations of the sampled populations, and a standard neighbor-joining tree is constructed, whose pattern and branch lengths determine the between-deme correlations. This approach is a significant extension from either the star phylogeny (independent draws from an ancestral population) or equal branch length (island model) assumptions of second-generation approaches, allowing for arbitrary historical relationships among the demes (but no migration between them). The ***hapFLK* test** (Fariello et al. 2013) extends the Bonhomme approach to haplotypes. Fariello et al. (2014) applied both the *FLK* and *hapFLK* tests in a genome scan of worldwide sheep populations. The ***Bayenv/Bayenv2*** method of Coop et al. (2010) and Günther and Coop (2013) does not build the correlation matrix from a phylogeny (and hence does not require an outgroup), but rather constructs a covariance matrix for allelic correlations (within and between demes) directly from a set of assumed neutral alleles. As detailed in Example 9.5, it handles migration and also accommodates at least some local inbreeding, making it a slightly more robust approach. The ***PCAdapt*** approach of Duforet-Frebourg et al. (2014) accounts for population structure through the use of latent factors (Example 9.5). While the *Bayenv* approach is population-based (using average allele frequencies within a deme), *PCAdapt* is individual-based, using data from individuals, and hence does not require first structuring sampled individuals into groups.

Simulation studies find that these third-generation approaches perform significantly better than their second-generation counterparts in terms of controlling false positives and are much more robust to different population structures (Bonhomme et al. 2010; Günther and Coop 2013: de Mita et al. 2013; Lotterhos and Whitlock 2014, 2015; de Villemereuil et al. 2014). As succinctly noted by Lotterhos and Whitlock (2014), third-generation approaches "show great promise for accurately identifying loci under spatially divergent selection." However, they can still generate false positives when allelic surfing occurs, or when there is substantial genomic variance in the impact of background selection, as all of these methods assume all neutral markers in a genome experience a common population-structure effect. They are also biased when selectively influenced loci are included in the markers that are used to estimate the correlation structure.

Despite these numerous issues, F_{ST}-based methods remain popular and have strong supporters (e.g., Beaumont 2005; Novembre and Di Rienzo 2009) as well as detractors (Hermisson 2009). There is no question that they can provide a very useful tool for finding potential *regions of interest*, but great care should be exercised in making anything other than cautious statements about the *statistical significance* of such regions. Whenever possible, third-generation methods should be used.

ALLELE-FREQUENCY CORRELATIONS WITH ENVIRONMENTAL VARIABLES

A final approach for comparing allele frequencies over a set of populations was introduced in Chapter 8, namely to search for correlations between allele frequencies and environmental factors. This approach is often referred to as **environmental association analysis (EAA)** or **genetic-environmental analysis (GEA)**, although our preference is for the former to avoid confusion of the latter with the analysis of genotype × environment interactions. In such studies, typically, a large number of potential factors are initially considered, and then the method of principal components (Appendix 5) is used to extract a smaller set of environmental features. If polygenic adaptation is the norm, classic hard-sweep (Table 8.2) or even soft-sweep signals will be unlikely, as the response is driven by modest allele-frequency changes over a number of small-effect loci. Hancock et al. (2010a, 2010b) suggested that such polygenic sweeps might be detected through subtle allele-frequency shifts that are concordant in populations experiencing similar environments but in different geographic regions.

Searching for correlations between environmental factors (such as locations along some environmental gradient) and the frequency of alleles is a time-honored tradition in population genetics, tracing back to Dobzhansky's work on clines in *Drosophila* chromosome inversions (Lewontin et al. 1981). Historically, these approaches have assumed the presence of a candidate gene and some specific environmental factor or surrogate (such as latitude or altitude). A number of complications arise when moving from testing the association of a single gene with a single environmental variable to scanning a large number of genes and a set of environmental factors. First, all frequency-environmental correlation tests must deal with lack of power (the limiting feature is the number of sampled locations). This limited power is exacerbated by multiple comparisons (Appendix 4), as with n_g biallelic loci and n_e environmental factors, there are $n_g n_e$ comparisons. To accommodate this concern, results are often reported using Storey's q values (Equation A4.24), a measure of the false-discovery rate given the significance of a test. Appendix 4 details how a list of p values for a collection of tests is translated into a list of associated q values, whose interpretation is as follows. Suppose $q \leq 0.025$ for a specific comparison in a collection of tests. This implies that the false-discovery rate for that association is no greater than 2.5%. It further implies that the false-discovery rate for *any* other test in this collection with the same, or a smaller, p value (as our focal test) also has a false-discovery rate of no greater than 2.5%.

Many factors conspire to reduce the power of such association approaches by decreasing the correlation between present environmental values and current allele frequencies. Even when allele-frequency change has been shaped by the environment, the currently measured environment may be rather different from the historical values that generated the present frequency of an allele. There is also an assumption that the same allele is acted on by selection, and with the same LD pattern at nearby markers, over a majority of the demes where the selection pressure is present. If different alleles (perhaps at different loci) accomplish the same adaptation to an environmental feature or if the LD structure between markers and selected alleles varies over demes, this will further erode any signal. Finally, false-positives can be introduced when gene-flow correlates with environmental features. Consider a population expanding from a southerly glacial refugium. As individuals migrate out, the result is a north-south cline in neutral allele frequencies. Such a north-south gradient can also occur in environmental variables (e.g., temperature or hours of sunlight), creating a correlation between such features and allele frequencies. Despite these concerns, EAA approaches have the potential to detect effects that would be missed by standard F_{ST} outlier approaches that are "blind" with respect to any environmental information.

Joost's Spatial Analysis Method (*SAM*)

The extension of testing for an association between a specified candidate gene and a single environmental factor to a more general genome scan over a set of environmental features starts with Joost et al. (2007). Their **spatial analysis method** (*SAM*) computes separate logistic regressions for each allele-environment combination. As discussed in Chapter 14, logistic regressions are commonly used to model how the probability of an event varies with some other variable, in this case predicting allele frequency as function of the environmental value. As with second-generation LK tests, *SAM* has a critical limitation in assuming that neutral alleles from different populations are uncorrelated. Failing to account for the natural correlation in neutral allele frequencies shaped by shared migrations and/or history will yield incorrect sampling errors. Further, populations in geographic proximity are expected to have both correlated allele frequencies (due to migration) *and* correlated environmental values, generating many false positives. While Poncet et al. (2010) extended *SAM* by allowing for small-scale correlations in allele frequencies within spatially proximate demes, their approach does not adjust for larger-scale correlations.

Accounting for Population Structure: Coop's *Bayenv* and Frichot's *LFMM*

Coop et al. (2010; Eckert et al. 2010; Günther and Coop 2013; also see Gautier 2015) attacked the problem of adjusting for unknown population structure by using marker data to estimate the expected correlation pattern among neutral alleles for the sampled populations. This is akin to the kinship matrix approach used by Bonhomme et al. (2010) to adjust for correlations among allele-frequency values from different demes. Example 9.5 sketches the basic structure of their ***Bayenv*** approach, which uses Bayes factors (Appendix 2) to gauge the support for an allele-environmental correlation after the effects of population structure have been removed. Formally, however, this is still an outlier method, as it generates an empirical distribution of Bayes factors for each SNP and uses this to assess the strength of association for a given locus. An alternative implementation to adjust for population structure, which is very closely related to Coop's method (as well as to Duforet-Frebourg et al.'s previously mentioned *PCAdapt* approach), is the **latent factor mixed model** (***LFMM***) approach of Frichot et al. (2013), which is also outlined in Example 9.5.

Simulations by Frichot et al. (2013) and de Villemereuil et al. (2014) found that the *LFMM* approach, along with *Bayenv*, is more powerful and less prone to false positives than methods that do not account for allelic correlations. These authors further found that *LFMM* tends to be slightly less biased than *Bayenv*, perhaps because, under this approach, both the locus-specific environmental effects and the latent population structure are estimated simultaneously, while in *Bayenv* the latter is estimated first and then used to estimate environmental effects. One further advantage of *LFMM* is that the lower-dimensional representation of the covariance structure can result in less bias than using the full structure, whose minor components are generally estimated with error (we return to this issue in Volume 3 when examining the structure of the genetic variance-covariance matrix, **G**, associated with multivariate selection; Equation 13.23b)

De Mita et al. (2013), de Villemereuil et al. (2013), and Lotterhos and Whitlock (2015) compared the power of divergence-based (F_{ST}) and correlation-based (EAA) approaches for detecting selection under a number of population structures and assumed selection strengths. The results were somewhat mixed. For example, EAA can outperform divergence-based approaches under an island model, but tended to do poorly under an isolation-by-distance structure. Lotterhos and Whitlock (2015) made the critical point that the sampling design has a major impact. They contrasted three different designs: random sampling over a metapopulation, paired-sampling, and transect-sampling. Under paired sampling, one specifically chooses pairs of populations that are proximally close but that differ substantially in the target environmental variables, and Lotterhos and Whitlock found that this tended to be the most powerful design.

We close this section to noting that EAA is an extremely active research area, with new methods appearing frequently. Recent reviews of some of the evolving issues are given by Rellstab et al. (2015), François et al. (2016), and Hoban et al. (2016).

Example 9.5. Here we sketch out the basic structure of four extensions (*FLK, Bayenv, LFMM,* and *PCAdapt*) of LK and allele-environment correlation tests that attempt to account for among-deme correlations in allele frequencies. The details are fairly technical, but the basic idea is very similar to that for the Fisher-Ford test (Example 9.2). There, under the assumption of drift, a vector, $\mathbf{y}$, of allele-frequency changes is turned into a test statistic that is in the form of a quadratic product (Equation 9.6c), $\mathbf{y}^T\mathbf{V}^{-1}\mathbf{y}$, where $\mathbf{V}$ is the covariance for these expected changes under drift alone. This is simply the generalization of a test based on the squared difference between two allele frequencies to a vector of allele frequencies. If $\mathbf{y}$ is multivariate normal, this test statistic follows a χ^2 distribution and excessive values indicate a departure from the pure-drift model. This basic idea, and structure, also follows here. The extensions follow from an adjustment of the vector of changes to account for environmental influences on allele frequencies and more generalized covariance matrices, given the population structure.

The F_{ST} extension (*FLK*) of Bonhomme et al. (2010) uses a set of neutral loci together with an outgroup to construct a kinship matrix, $\mathcal{F}$, of populations, based on branch lengths of the estimated phylogenetic tree among the sampled populations. The assumption is that some pattern of evolution (described by $\mathcal{F}$) unfolds from an ancestral population with an allele frequency of p_0, but with no further migration between subpopulations. For n populations, the *FLK* test statistic is given by

$$T_{FLK} = \frac{(\mathbf{p} - \widehat{p}_0\mathbf{1})^T\mathcal{F}^{-1}(\mathbf{p} - \widehat{p}_0\mathbf{1})}{\widehat{p}_0(1-\widehat{p}_0)}, \quad \text{with} \quad \widehat{p}_0 = \frac{\mathbf{1}^T\mathcal{F}^{-1}\mathbf{p}}{\mathbf{1}^T\mathcal{F}^{-1}\mathbf{1}} \tag{9.12}$$

where $\mathbf{p}$ is a vector of the allele frequencies for one particular locus over the n sampled demes and $\mathbf{1}$ is a column vector of n ones. Bonhomme et al. showed that T_{FLK} follows a χ^2 distribution under the null model of no selection, provided allele frequencies are not too extreme, with outliers deemed to be candidates for loci under selection. Note that $\widehat{p}_0$ is of the same form as the GLS estimators for the initial frequency (Equations 9.4 and 9.6a), and that T_{FLK} has the same general form as the test statistic for the Fisher-Ford test for excessive allele-frequency change (Equation 9.6c).

The *Bayenv* test for allele-frequency and environmental correlation (Coop et al. 2010; Eckert et al. 2010; Günther and Coop 2013) starts with a clever latent-variable approach (from Nicholson et al. 2002) to model the allele frequencies, which allows us to work with an (approximately) multivariate-normal vector. The motivation for this approach is the constraint that allele frequencies be confined to the interval [0,1], while the multivariate normal can generate values outside this region. To avoid this issue, assume there is some underlying latent (unseen) variable, θ, that is normally distributed and maps into the deme allele frequency p as follows

$$p = \begin{cases} 0 & \text{if } \theta \le 0 \\ \theta & \text{for } 0 < \theta < 1 \\ 1 & \text{if } \theta \ge 1 \end{cases} \tag{9.13a}$$

This truncated normal transformation allows for nonzero probabilities that p is at zero or one, corresponding to loss or fixation of the allele, respectively. This transformation allows us to work with a normally distributed random variable, θ, which is mapped into allelic loss ($\theta < 0$) or fixation ($\theta > 1$) if it is too extreme, and otherwise maps into the frequency of a segregating allele. As with most F_{ST}-based methods, the assumption here is that there is very little probability mass at zero or one (i.e., there is modest divergence at most loci), so p is essentially θ.

Following Nicholson et al. (2002), Coop et al. (2010) assumed that the vector, $\boldsymbol{\Theta}$, of θ values for the n sampled populations is multivariate normal (approximating drift by Brownian motion; Appendix 1), with some ancestral frequency (p_0) and correlation structure given by $\boldsymbol{\Omega}$, so that

$$\boldsymbol{\Theta} \sim \text{MVN}_n\left[p_0\mathbf{1}, p_0(1-p_0)\boldsymbol{\Omega}\right] \tag{9.13b}$$

where $\boldsymbol{\Omega}$ is the empirical estimate of the covariance matrix of the underlying θ (and hence of the allele frequencies), based on a set of presumed neutral markers (for details see Coop et al. 2010 and Gautier 2015). This is their base model. Note that standardizing the observed vector of frequencies, $\mathbf{p}$, by centering it around the mean and adjusting for the covariance structure yields a χ^2_{n-1} random variable (as the quadratic product is the sum of squared unit normals). This recovers the T_{FLK} test statistic, but with $\boldsymbol{\Omega}$ and $\boldsymbol{\Theta}$ in place of $\mathcal{F}$ and $\mathbf{p}$. Operationally, if all elements of $\mathbf{p}$ fall within the range $(0,1)$, then $\mathbf{p}$ replaces $\boldsymbol{\Theta}$, yielding the test statistic as

$$\frac{(\mathbf{p} - \widehat{p}_0\mathbf{1})^T\boldsymbol{\Omega}^{-1}(\mathbf{p} - \widehat{p}_0\mathbf{1})}{\widehat{p}_0(1-\widehat{p}_0)} \quad \text{where} \quad \widehat{p}_0 = \frac{\mathbf{1}^T\boldsymbol{\Omega}^{-1}\mathbf{p}}{\mathbf{1}^T\boldsymbol{\Omega}^{-1}\mathbf{1}} \tag{9.13c}$$

which is identical in form to Equation 9.12, except that the covariance matrix $\boldsymbol{\Omega}$ replaces $\mathcal{F}$. Günther and Coop (2013) denoted this test statistic for extreme divergence at a specific locus by $\mathbf{X}^\mathbf{T}\mathbf{X}$, which some authors refer to as the ***XtX statistic***.

Some comment on $\boldsymbol{\Omega}$ versus $\mathcal{F}$ is in order. $\boldsymbol{\Omega}$ is the empirical estimate of the allelic-covariance matrix, while $\mathcal{F}$ is based on the estimated branching pattern of deme formation (assuming there is no between-deme migration). Hence, $\boldsymbol{\Omega}$ is expected to perform better in situations where the correlations induced by population structure are poorly approximated by a phylogenetic tree, as can be the case with migration between demes (Günther and Coop 2013). Further, because $\boldsymbol{\Omega}$ also contains the within-deme variance, it should be slightly more robust to within-deme inbreeding relative to $\mathcal{F}$, which only considers the covariances between demes. Finally, it should be stressed that the χ^2 distribution only formally arises when $\boldsymbol{\Omega}$ is known without error. When $\boldsymbol{\Omega}$ is replaced by some estimate, $\widehat{\boldsymbol{\Omega}}$, of its true value, this distribution no longer holds.

Coop's base model (Equation 9.13c) is extended to account for environmental factors that influence the allele frequencies as follows. Consider a vector, $\boldsymbol{\beta}$, of potential regression coefficients for the impact of environmental factors on allele frequencies, and a matrix, $\mathbf{X}$, whose values in row i correspond to the environmental parameters measured for the ith population (this is simply a GLS linear model; see LW Chapter 8). The null mean p_0 for an allele (Equation 9.13c) is augmented by the environmental effect to give

$$\boldsymbol{\Theta} \sim \text{MVN}_n\left[p_0 \mathbf{1}_n + \mathbf{X}\boldsymbol{\beta}, p_0(1-p_0)\boldsymbol{\Omega}\right] \tag{9.13d}$$

where $\mathbf{1}_n$ is an n-dimensional vectors of ones. This model assumes that any relationships between allele frequencies and the environmental variables have some linear component. The addition of the vector $\mathbf{X}\boldsymbol{\beta}$ to account for environmental effects is an example of a **factorial regression** (e.g., Baril et al. 1992), which is discussed at length in Volume 3 in the context of analyzing genotype-by-environment interactions.

Coop et al. (2010) couched their model in a Bayesian framework, which involves a different logic for hypothesis testing than the standard likelihood-ratio approach. As reviewed in Appendix 2, hypothesis testing in a Bayesian framework uses **Bayes factors**, which account for how much the data shift any prior belief in favor of a hypothesis (Equations A2.10a and A2.10b; Example A2.3). Coop et al. cautioned that hypothesis testing with their model (Equation 9.13d) is not as straightforward as simply asking whether the inclusion of a nonzero $\boldsymbol{\beta}$ (meaning that environmental factors influence at least some allele frequencies) significantly improves the fit (a large Bayes factor in favor of $\boldsymbol{\beta} \neq 0$ over the null $\boldsymbol{\beta} = 0$), as their model is only an approximation of the population structure. They noticed that the inclusion of environmental variables strongly influenced the distribution of Bayes factors among their set of control (and presumably neutral) loci used to estimate $\boldsymbol{\Omega}$. As a result, they recommend an outlier approach, in which the Bayes factor for a given locus is compared with the empirical distribution of Bayes factors for control loci that include that environmental variable. This is easiest to handle by using principal components to transform the environment variables to a new and independent set (e.g., Equation A5.15a), and then testing these one at a time. Test-statistic values are computed by adjusting the base-model quadratic product (Equation 9.13c) to account for environmental variables, which shifts the vector of means from $p_0\mathbf{1}$ to $p_0\mathbf{1} + \mathbf{X}\boldsymbol{\beta}$ (Equation 9.13d), yielding

$$\frac{(\mathbf{p} - \widehat{p}_0\mathbf{1} - \mathbf{X}\widehat{\boldsymbol{\beta}})^T \boldsymbol{\Omega}^{-1} (\mathbf{p} - \widehat{p}_0\mathbf{1} - \mathbf{X}\widehat{\boldsymbol{\beta}})}{\widehat{p}_0(1-\widehat{p}_0)} \tag{9.13e}$$

The *Bayenv* method of Coop et al. is a two-step approach: (i) $\boldsymbol{\Omega}$ is estimated from a presumed set of neutral markers, and (ii) the model is run with this matrix (or in a Bayesian framework, with draws of this matrix to generate a posterior accounting for the uncertainty in its estimation; Appendix 2).

Finally, the *LFMM* approach of Frichot et al. (2013) is a one-step model, which jointly fits $\boldsymbol{\beta}$ along with a series of random vectors (latent factors) to account for the population structure. In essence, this model approximates the allelic covariance structure by a matrix of lower rank (the rank given by the number of latent factors included in the model), by using what in essence are the first k principal components of $\boldsymbol{\Omega}$. These lower-rank approximations are discussed extensively in Volume 3, both in terms of estimating the rank of a multivariate selection covariance matrix, $\mathbf{G}$, and in computing additive main effects and multiplicative interaction (*AMMI*) models for genotype-by-environment interactions. Frichot et al. claimed slightly better results are obtained with *LFMM* than with *Bayenv* as a result of using a one-step approach. As mentioned, however, this could simply be due to a less-than-full rank estimate of the covariance matrix being better behaved, as the imprecise estimation of eigenvectors associated with minor eigenvalues may slightly bias the results. The *PCAdapt* approach of Duforet-Frebourg et al. (2014) is essentially the same as *LFMM*, except with $\boldsymbol{\beta} = 0$ (i.e., environmental associations are not considered).

CHANGES IN THE CHROMOSOMAL PATTERN OF NEUTRAL VARIATION

The classic signature of a recent hard sweep is a chromosomal region of depressed variation, while a site under long-term balancing selection displays enhanced variation, albeit over an ever-shrinking region (Chapter 8). This section develops methods based on the expected spatial pattern of variation on a chromosome around a selected site following a hard sweep. We start with simple graphical methods for suggesting interesting regions before considering a number of approaches based on maximum likelihood (ML) (LW Appendix 4). The details of some ML-based positional models can be challenging, but there is little question that using genomic positional information can significantly improve our ability to detect a recent hard sweep and provide estimates of its strength of selection. We present much of the technical detail to the examples, which the casual reader may prefer to skip.

Simple Visual Scans for Changes in Nucleotide and STR Diversity

The most basic approach is a simple plot of variation as a function of genomic location, looking for either peaks (long-term balancing selection) or valleys (a recent sweep); see, for example, Figures 8.1 and 8.2. With SNP data, variation is typically scored as average nucleotide diversity, π (Chapter 4), within a sliding window to smooth out the inherent noisiness from individual sites. With simple sequence repeats or microsatellite markers (also known as simple tandem repeats, or STRs, and simple sequence repeats, or SSRs), several different metrics of variation are available, such as copy-number variance, number of alleles, and probability of heterozygosity. With their large number of alleles per marker and high mutation rates, STRs provide a more consistent signal and are usually plotted on a per-marker basis (as opposed to a sliding-window analysis); see Figure 9.3 and Example 9.6. A point of caution is that mutation rates at STRs can be length dependent, with smaller arrays often expected to show less variation.

Example 9.6. Domesticated breeds of dogs are ideal candidates for the detection of regions influenced by selective sweeps (Schlamp et al. 2016). Most breeds are rather recent ($\sim$200 generations or less) and often exhibit large phenotypic effects (and hence harbor the potential of strong selection on just a few genes). Simulation studies by Pollinger et al. (2005) found that a few moderately linked, highly polymorphic loci can give a strong sweep signal under realistic conditions for the formation of dog breeds. They tested their idea using a scan of microsatellites around candidate genes in two different breeds. The Large Munsterlander is a recent breed (originating around 1910) and categorized by a black coat color, with the pigment gene *TYRP1* being suggested as a candidate for this trait. As Figure 9.3A shows, there is a roughly 50 Mb region of depressed microsatellite variation around *TYRP1* relative to the control (neither black or brown) and brown-coated populations. Note that the region under the sweep is rather large, and if indeed *TYRP1* was the actual target, this region is asymmetric, showing more reduced variation to the right of *TYRP1* than at the locus itself. (Recall from Chapter 8 that such asymmetries around a selected site are not uncommon.) A more striking example is offered by dachshunds (Figure 9.3B), which showed no variation at three microsatellites surrounding the *FGFR3* candidate gene that is involved in achondroplasia (limb-shortening).

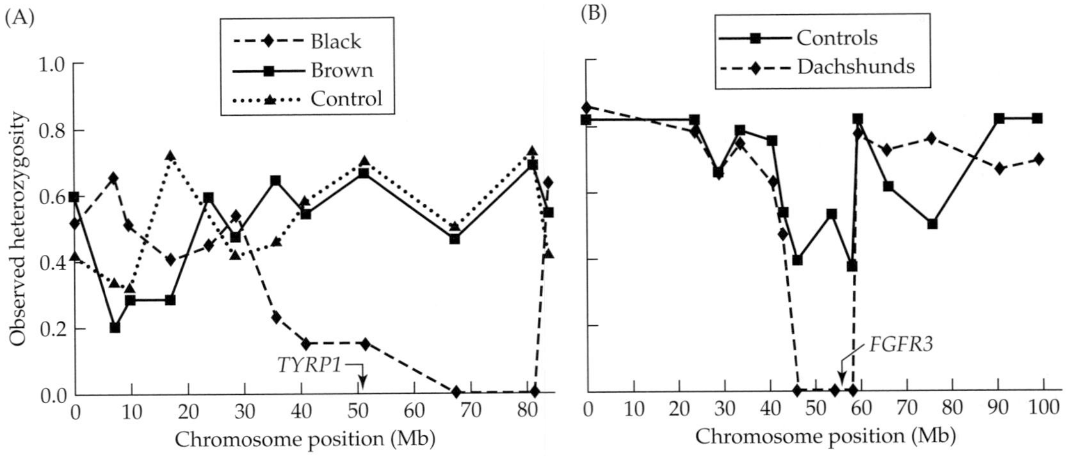

Figure 9.3 Using microsatellites in the search for dog domestication genes. (**A**) Large Munsterlanders have a black coat, suggesting the pigment gene *TYRP1* on chromosome 11 may be a possible domestication gene. A plot of variation for this breed (black) relative to both control (neither black or brown) and brown individuals shows depressed variation spanning this gene. (**B**) Dachshunds are characterized by shortened limbs, suggesting the *FGFR3* gene on chromosome 3 as a candidate. Dachshunds have an absence of variation at three microsatellites spanning this gene, while variation is present in controls (normal-limbed breeds). (After Pollinger et al. 2005.)

Both Munsterlanders and dachshunds went through strong bottlenecks during their formation, so sampling noise that was overlaid on the general reduction in variation from the bottleneck may have generated these depressions. To test for this, Pollinger et al. simulated the founding process (akin to what was done in Example 9.3). Assuming the loci are highly polymorphic (STR heterozygosity in excess of 30% prior to the bottleneck), simulations showed a less than 5% chance of finding three adjacent STR loci with no variation under a simple model of a genome-wide population bottleneck during founding (conditioned on the average levels of heterozygosity seen throughout the genome).

These examples in dogs show the power (strong signal with only modestly dense markers) but also the pitfalls (poor localization) when a strong sweep occurs. Because a sweep depresses variation throughout a region, additional sweep-based fine-mapping within this region would be futile. Ironically, localization is easier under a weak sweep than a strong one, as the region of depressed variation is smaller. Because the same traits often appear in independent dog breeds, one potential solution is that if the same gene (or genes) were the targets of selection in each breed there is the potential for improved resolution by searching for overlapping intervals in the detected sweep regions. Chan et al. (2012) used a variant of this approach by comparing the results of seven independent mouse lines selected for increased weight, finding ~70 **parallel selected regions** (**PSRs**), where most of the high-weight selected lines shared alleles rarely found in the controls.

While such plots of the spatial structure of genome variation are visually appealing, they are not formal tests of selection and can indeed be rather misleading. A change in the background level of variation can arise for several reasons besides selection, such as variation in the local mutation rate. This is especially true for STR markers, whose mutation rates are expected to vary considerably, reflecting differences in the composition (size and sequence) of their repeat units. Second, the inherent stochasticity of recombination and drift can generate considerable variation in the coalescent process across a genome, so that even with a sliding window analysis for a set of markers with the same mutation rate, strong peaks and valleys are routinely found in neutral simulations (Kim and Stephan 2002; Jensen at al. 2005).

Two different strategies can address these complications. The first is to use information from the same (and other) regions in several populations. If a region has a low (or high) mutation rate relative to the rest of the genome, presumably this will also be true in other populations or closely related species. For example, a large reduction in STR variation was seen around the *pfcrt* gene, which has been implicated in chloroquine resistance in the malaria parasite *Plasmodium falciparum* (Wootton et al. 2002). This reduction was seen in resistant lines from both South America and Asia and Africa, but was absent in sensitive lines from Asia and Africa, strongly suggesting that the reduction is not due to genomic differences in mutation rates.

However, if one compares *just* a narrow region of interest between two populations, one might (incorrectly) infer a sweep-like signal if one of the populations has experienced a bottleneck (as might occur in the domestication process). Ideally, one would compare k different regions in $n \geq 2$ populations. As detailed below, this can be done with a simple ANOVA approach or through more formal ML-based **bottleneck models**. These approaches examine if some loci appear to have experienced a more severe bottleneck than others in the same genome, which would be consistent with hard sweeps in these regions.

Second, in the absence of information from other populations, **explicit chromosomal spatial-structure models** can be applied. These use ML to compare the genomic positional pattern of variation as a function of recombination fraction over a region of interest, testing whether the fit is consistent with the pattern expected from a sweep (e.g., Equation 8.8a). This approach is both the most powerful (when modeling assumptions hold), and potentially the most fragile (when they do not), and we will address some of these concerns and their potential corrections.

Tests Based on STR Variation Across Populations

Schlötterer and colleagues (Schlötterer et al. 1997; Harr et al. 2002; Schlötterer et al. 2002; Kauer et al. 2003; Schöfl and Schlötterer 2004) proposed several straightforward tests using k unlinked markers sampled from n populations. While their tests used STR/SSR data, this same approach can be applied to SNP data in nonoverlapping windows. ANOVA is used to obtain site- and population-specific values to correct (respectively) for variation in mutation rates over loci and variation in N_e over populations, and then test if a specific locus-population combination is unusual.

Consider a set of STRs, and let V_{ij} denote the variance in repeat copy number at locus i in population j. Because variances typically have a skewed distribution, we work instead with $v_{ij} = \ln(V_{ij})$, as a log transform reduces the effects of skew. The effects of locus-specific mutation rates are accounted for by averaging locus i over all populations ($v_{i\cdot}$), while population-specific effects are accommodated by using the average ($v_{\cdot j}$) of population j over all loci. Under the assumption of no locus $\times$ population effects and no LD between scored markers, the expected log-variance can be written as a simple ANOVA model

$$v_{ij} = v_{i\cdot} + v_{\cdot j} - v_{\cdot\cdot} + e_{ij} \quad \text{for} \quad 1 \leq i \leq k, \ 1 \leq j \leq n \tag{9.14}$$

A sweep in population j near the ith STR is indicated by a significant deficiency relative to the predicted value, $\widehat{v}_{ij} = v_{i\cdot} + v_{\cdot j} - v_{\cdot\cdot}$, which can be tested using a t-like statistic (Schlötterer et al. 1997).

A related approach is the ***Log RV*** statistic of Schlötterer (2002). For a single-step mutation model (meaning that STR repeat length is equally likely to change by plus or minus one following a mutation), the expected variance is $\sim 4N_e\mu$ where μ is the locus-specific mutation rate (Slaktin 1995b). Thus, the locus-specific mutation rate cancels out in the ratio $RV = V_{ij}/V_{i\ell} = 4N_{j,e}\mu_i/4N_{\ell,e}\mu_i = N_{j,e}/N_{\ell,e}$ for the same locus (i) in two different populations (j and ℓ), leaving only the ratio of effective population sizes. (The related ***Log RH*** statistic of Kauer et al. 2003 uses the ratio of heterozygosities, as opposed to copy-number variance.) Simulation studies showed that the distribution for the logs of these two ratios is approximately a normal under many demographic scenarios, with the exception of extreme bottlenecks. While similar in spirit to Equation 9.14, the *Log RV* and *Log RH* statistics are outlier approaches, computing all pairwise values and using outliers as potential sites of selection. Modifications have been proposed for two linked STRs (Harr et al. 2002), while Wiehe et al. (2007) proposed a follow-up statistic as an additional test for an STR showing signs of selection.

Tests of Sweeps Using Bottleneck Models

In domesticated species, one can imagine a founding bottleneck that reduced variation across all loci relative to the ancestral source population. However, *in addition* to this common bottleneck, a *further* reduction will likely be associated with genes selected during domestication (assuming these generate hard sweeps), which can be thought of as an additional bottleneck beyond the genome-wide founding bottleneck. This idea leads to a more formal ML-based test. Data for multiple loci from two (or more) populations are first used to estimate a common bottleneck value for loci in one population (relative to another). One then tests whether the model fit is improved by allowing a subset of these loci to experience an additional bottleneck (as would happen with a sweep following the initial domestication bottleneck). One potential weakness of this approach is background selection (BGS). If there is significant variance over the genome in the ratio of gene density to recombination rate, a model assuming an additional bottleneck may provide an improved fit (even in the absence of sweeps), simply by capturing some of the genomic heterogeneity in loss of variation from BGS.

The bottleneck approach was first considered by Galtier et al. (2000). They assumed the existence of a population in mutation-drift equilibrium at some ancestral time, where $\theta_i = 4N_e\mu_i$ measures variation at the ith locus (Chapter 2). At some time, T, in the past (scaled in terms of $2N_e$ generations), a bottleneck occurred, and after some passage of time in the bottleneck, the population quickly expanded back to its previous size. The authors made the clever observation that while a bottleneck is a reduction in population size, the net result is that the number of distinct lineages going into a bottleneck is far greater than the number that escape it. Motivated by this observation, they introduced a measure of bottleneck strength, B, which is the expected amount of time required to lose the same number of lineages as in a model of constant population size. They then assumed for some time, 0, until T, a standard neutral coalescent process is occurring, with mutations and coalescent events (Chapter 2). From time $T < t < T + B$, only coalescences are allowed (mutation is effectively turned off), reducing the number of surviving lineages. Finally, from time $t > T + B$, mutation is turned back on again. Given a population sample of segregating sites over k loci, the method of Griffiths and Tavare (1994a, 1994b) can be used to obtain maximum likelihood estimates of T, B, and θ_i for $1 \leq i \leq k$. Galtier et al. then constructed a model where the bottleneck is potentially different for each locus, in which case T and B are now locus specific, and one estimates T_i, B_i, and θ_i for $1 \leq i \leq k$. A standard likelihood ratio test (LW Appendix 4) determines if the second model provides a better fit.

Example 9.7. Wright et al. (2005) used a multiple bottleneck model in their search for genes under selection in maize. The authors used SNP data on 774 genes from 14 maize and 16 teosinte inbred lines. Collectively, the sampled maize lines had roughly 60% of the heterozygosity found in teosinte lines, showing a strong bottleneck signal across the entire maize genome (as expected from the initial domestication process). The authors quantified the strength of this bottleneck using the parameter $b = N_b/d$, the size of the population in the bottleneck divided by its duration. Smaller values of b imply stronger bottlenecks. Using simulations with different values of b, the maximum likelihood given the numbers of segregating sites at a given locus in maize and teosinte occurred at a value of $\widehat{b} = 2.45$.

The authors then fit a second model that assumed two classes of loci: a fraction, $(1 - q)$, experiencing a bottleneck of strength b_1 and a fraction, q, experiencing a much stronger bottleneck of strength, $b_2 < b_1$, giving the resulting likelihood for locus i as

$$L(q, b_1, b_2|S_i) = (1-q)L(b_1|S_i) + q\,L(b_2|S_i) \tag{9.15a}$$

where $L(b_j|S_i)$ is the likelihood function for b_j ($j = 1, 2$), given the number, S_i, of segregating sites in maize at locus i. Assuming loci are independent, the full likelihood is the product over all loci. This is a mixture model (LW Chapter 13), with parameters of q, b_1, and b_2, and again, the model can be searched by simulation to locate the ML estimates (MLEs). The resulting MLEs using all 774 genes were $\widehat{b_1} = 2.45, \widehat{b_2} = 0.15$, and $\widehat{q} = 0.02$. However, many of the loci had low variation even in teosinte and offered little information on b. Using a set of 275 genes with high variation (10 or more segregating sites in teosinte) returned MLEs of $\widehat{b_1} = 2.45, \widehat{b_2} = 0.01$, and $\widehat{q} = 0.036$. This sample of $\sim$4% of maize genes potentially experienced a much greater bottleneck (smaller b) than the rest of the genome, and hence this set contains strong candidates for sites influenced by a sweep.

With these estimates in hand, one can use Bayes' theorem (Equation A2.2) to obtain the posterior probability of a locus being in this selected (stronger bottleneck) class, and hence localize genes that potentially were under past selection. Recall that Bayes' theorem allows one to "flip" the conditional, as we can easily compute $\Pr(S_i \mid b_j)$—indeed, this is just the likelihood, $L(b_j \mid S_i)$— but we are much more interested in $\Pr(b_j \mid S_i)$. Bayes' theorem connects these as

$$\Pr(b_2\,|S_i) = \frac{q\,L(b_2 \mid S_i)}{(1-q)\,L(b_1 \mid S_i) + q\,L(b_2 \mid S_i)} \tag{9.15b}$$

This gives a posterior probability for a particular locus (i) being in the strong bottleneck class (b_2), given that it has S_i segregating sites in maize. This same approach for posterior prediction reappears in Example 10.15 in the context of predicting which sites in a protein have been under positive selection.

Tests of Sweeps Using Genomic Positional Information: *CLRT-GOF*

While bottleneck approaches are elegant, they ignore any information from the expected spatial pattern of variation *across* a chromosomal region, for example, the functional form of the expected decrease in variation as one moves closer to the site of selection. Further, they require samples from two (or more) populations, which may be impractical. Kim and Stephan (2002) proposed a likelihood-based test using the positional pattern of variation from a single population, which not only detects a sweep, but to also localizes its position and estimates its strength of selection. A similar regression-based approach was suggested by Wiener and Pong-Wong (2011).

The basic structure of their test (and several extensions) is as follows. Suppose m linked segregating sites from a local chromosomal region of interest are scored in a sample of n chromosomes. Using an outgroup, we can polarize any segregating alleles, determining which are derived. The resulting data are the number, k_i, of copies of the derived alleles at a segregating site, i, where $1 \leq k_i \leq n-1$. Building on Equation 2.36c, the probability of observing k_i given the sample size n and the vector $\boldsymbol{\Theta}$ of model parameters is simply

$$\Pr(k_i \mid n, \boldsymbol{\Theta}) = \binom{n}{k_i} \int_{1/(2N)}^{1-1/(2N)} x^{k_i}(1-x)^{n-k_i}\,\phi_i(x \mid \boldsymbol{\Theta})\,dx \tag{9.16a}$$

where $\phi_i(x \mid \boldsymbol{\Theta})$ is the frequency spectrum for site i under the model (specified by the distribution parameter vector, $\boldsymbol{\Theta}$) of interest. Equation 9.16a is also the likelihood for $\boldsymbol{\Theta}$ given the data, $L(\boldsymbol{\Theta} \mid k_i, n)$. Under the equilibrium neutral model, the Watterson distribution (Equation 2.34a) is used for ϕ_i, where the vector of distribution parameters, $\boldsymbol{\Theta}$, is just a scalar, as $\boldsymbol{\Theta} = (\theta_i) = (4N_e\mu_i)$, the scaled mutation rate for site i. If one starts with a Watterson distribution and then has a sweep at a linked site, ϕ_i is now given by Equation 8.13, with parameters $\boldsymbol{\Theta} = (\theta_i, f_{s,i})$, where $f_{s,i}$ measures the strength of the sweep at site i. From Table 8.1

$$f_{s,i} = (4N_e s)^{-c_i/(2hs)} = e^{-c_i\,\lambda} \tag{9.16b}$$

with $\lambda = (1/2hs)\ln(4N_e s)$, and c_i representing the recombination fraction between site i and the site of the sweep. Usually one assumes that $\theta_i = \theta$ is the same for all sites, in which case, under the null model, all sites follow the same Watterson distribution (with no expected position-specific genomic pattern).

Because Equation 9.16a gives a proper likelihood for each site, one might imagine that the total likelihood $L(\boldsymbol{\Theta} \mid \mathbf{k}, n)$, given the vector $\mathbf{k}$ of the k_i values for each site, is simply the product of the site-specific likelihoods over all m sites

$$L(\boldsymbol{\Theta} \mid \mathbf{k}, n) = \prod_{i=1}^{m} L(\boldsymbol{\Theta} \mid k_i, n) \tag{9.16c}$$

with the MLE for $\boldsymbol{\Theta}$ being the value that maximizes $L(\boldsymbol{\Theta} \mid \mathbf{k}, n)$. However, this product is *not* a proper likelihood, as adjacent sites are *correlated* (due to shared history). The product of the site likelihoods (Equation 9.16c) is thus a **composite likelihood** (also called a **quasi-maximum likelihood** or **pseudo-likelihood**), which is an approximation of the total likelihood. Wiuf (2006) showed that the composite and true likelihoods are usually consistent for many population-genetic problems. Kim and Stephan (2002) contrasted the maximum value of this composite likelihood under the Watterson distribution (using Equation 2.34a for ϕ) with the maximum obtained under a sweep model (using Equation 8.13 for ϕ). Because f_s changes with distance c_i from any particular site (Equation 9.16b), the location giving the maximum value of f_s corresponds with the estimated position of the selected site. Kim and Stephan compared the ratio of the two composite likelihoods, $L_{sweep}/L_{neutral}$, corresponding to

$$\Lambda_{CLR} = \frac{\max L(\theta, f_s(\lambda, c_i) \mid \mathbf{k})}{\max L(\theta, f_s = 0 \mid \mathbf{k})} \tag{9.16d}$$

and called this approach the ***CLR*** test (or ***CLRT***), for **composite-likelihood ratio test**. Because this is *not* a strict likelihood ratio, large-sample approximations (LW Appendix 4) for its distribution are *not* valid, and the critical values must be obtained by simulation. Boitard et al. (2009; also see Kern and Haussler 2010) proposed the use of hidden Markov models to account for the correlations among markers that are ignored under a composite-likelihood framework. Li and Stephan (2005, 2006) developed a true maximum-likelihood method that uses only a subset of the frequency spectrum (the **compact frequency spectrum**, recording the number of singletons, doubletons, and sites greater than two) to estimate the position of the sweep.

Jensen et al. (2005) found that the *CLR* test is *not* robust to population structure or recent bottlenecks. While neither of these demographic features is expected to generate the same spatial pattern as seen under a sweep, they *can* generate an excess of sites segregating rare derived alleles, so that some aspects of the frequency spectrum can be similar to a sweep, and this in turn could improve the fit relative to a model assuming the equilibrium Watterson distribution. To distinguish sweeps from false signals generated by demography and population structure, Jensen et al. proposed that any significant *CLR* result be subjected to an additional goodness-of-fit (***GOF***) test to see how well it fits a sweep model (Example 9.8). Again, the key is that a sweep is expected to generate a specific *spatial pattern* of variation, which will be absent under neutrality even if the site-frequency spectrum is shifted because the population is not at equilibrium.

Kim and Nielsen (2004) extended the *CLRT* to include information on linkage disequilibrium using their ω statistic (Equation 9.37), which contrasts measures of LD among all pairs of loci on the same side of a putative sweep with those for pairs of loci on opposite sides of the sweep. While incorporation of this LD information did improve estimation, the gain was rather modest. Finally, Meiklejohn et al (2004) and Vy and Kim (2015) have extended the *CLRT* for detecting incomplete sweeps. The basic logic is as above, but with only a partial shift in the frequency of the selected allele (as it is still on its sojourn to fixation).

Example 9.8. The *GOF* test of Jensen et al. (2005) starts by comparing the maximum of the composite-likelihood function under the sweep model (as the null) against the maximum of the likelihood under a more general model, L_A, where the population frequency of the derived allele is unique for each site, but *uncorrelated between sites*. This latter assumption is the key, as a sweep is expected to generate a specific correlated pattern of variation (runs of extreme frequencies) near the selected site. The likelihood function for site i under the L_A model—namely, an unknown, but arbitrary, frequency p_i at site i— follows from the binomial distribution. Given the observation of k_i copies of the derived allele (out of n sampled)

$$L_A(p_i \,|\, k_i) = \binom{n}{k_i} p_i^{k_i} (1-p_i)^{n-k_i} \tag{9.17a}$$

The resulting maximum-likelihood estimate (MLE) for p_i is simply $\widehat{p}_i = k_i/n$ (LW Appendix 4), giving the value of the likelihood function evaluated at the MLE as

$$L_A(\widehat{p}_i \,|\, k_i) = \binom{n}{k_i} (\widehat{p}_i)^{k_i} (1-\widehat{p}_i)^{n-k_i} = \binom{n}{k_i} \left(\frac{k_i}{n}\right)^{k_i} \left(1-\frac{k_i}{n}\right)^{n-k_i} \tag{9.17b}$$

Assuming that the site-specific p values are uncorrelated, the total composite likelihood is the product of Equation 9.17a over all sites. If there are n_i sites, each with i copies of the derived allele, their contribution to the total likelihood is the n_ith power of Equation 9.17b, with the maximum of the composite likelihood being the product of the individual maximum likelihoods (Equation 9.17b) across all levels of polymorphisms

$$\max\left(CL_A\right) = \prod_{i=1}^{n-1} \left[\binom{n}{i} \left(\frac{i}{n}\right)^{i} \left(1-\frac{i}{n}\right)^{n-i} \right]^{n_i} \tag{9.17c}$$

The *GOF* test is the log of the ratio of the maxima of the two different likelihoods

$$\Lambda_{GOF} = \ln\left(\frac{\max\left(CL_A\right)}{\max\left(CL_0\right)}\right) \tag{9.17d}$$

where $\max\left(CL_0\right)$ is the value of the sweep composite-likelihood function evaluated at the MLEs for the sweep parameters. Again, because these are not true likelihoods, large-sample distribution theory cannot be used to assess their significance. Instead, Jensen et al. used the MLEs for the sweep parameters to generate a large number of data sets under the null (which is the sweep model under the GOF test), using these to compute $\max\left(CL_A\right)$, and hence a distribution of Λ_{GOF} under the null. Support for a sweep is indicated when: (i) the *CLR* test gives a significant result, *and* (ii) the *GOF* test is *not* significant. If both the *CLR* and *GOF* tests *are* significant, support for a sweep is questionable, as demographic features are likely the cause of departures from the neutral equilibrium model. Jensen et al. found that this two-step approach was much more robust to population structure and demography issues than the *CLR* test alone, giving false positives only when very severe bottlenecks occurred. However, the improved control over the false-positive rate comes at the cost of decreased power (Jensen et al. 2006; Boitard et al. 2009).

Tests of Sweeps Using Genomic Positional Information: "*SweepFinder*"

Nielsen et al. (2005b) proposed a modification of the *CLRT* approach, replacing the Watterson distribution by an empirical site-frequency spectrum, $\mathbf{p} = (\widehat{p}_1, \cdots, \widehat{p}_{n-1})$, where $\mathbf{p}$ is estimated by using a reference sample of m segregating, and presumed neutral, sites. Here $\widehat{p}_k = n_k/m$ is the fraction of these m sites that contain exactly k copies of the derived allele, where n_k is the number of such sites in the sample. Their idea is that the Watterson distribution assumes an equilibrium neutral population, while using the actual distribution observed in the population of interest (at presumed neutral sites) to a large extent can accommodate any demographically induced departures. The use of an empirical site-frequency spectrum should also at least partly correct for any SNP ascertainment bias. While an elegant approach, the delicate issue is being able to find an appropriate (and large) set of presumed neutral sites. This approach goes by the name ***SweepFinder***, and the resulting likelihood function is derived in Example 9.9. Simulations by Nielsen et al. (2005b) showed that while this approach is more robust than the *CLR* test, demography can still influence the test statistic. In particular, intermediate bottlenecks (of a size reduced to 5% to 10% of the original N_e) seem to be the most problematic in terms of false positives (Williamson et al. 2007; Stephan 2010a), while the test has little power under strong bottlenecks (Poh et al. 2014). Hence, significance values should be obtained by simulating this procedure under a set of assumed demographic models.

A cautionary tale on the importance of accounting for demography was given by Long et al (2013) and Huber et al. (2014). Long et al. sequenced 180 lines of *Arabidopsis thaliana* from Sweden, with 130 from southern Sweden (a population generally regarded as being continuous with the main European population) and 50 lines from an isolated population in northern Sweden. *Sweepfinder* was run separately on both populations, yielding 22 strong signals from the northern population and only one from the southern population. This disparity was surprising, especially given the much larger sample size for the southern population. Huber et al. reanalyzed these data by first constructing a demographic model for these populations from the sequence data. They found that the resulting CLR distribution under the null was left-shifted by demography, making the cutoffs used by Long et al. too conservative, while it was right-shifted for the northern population, making the cutoffs too liberal. Using these demographic-corrected cutoffs, Huber et al. found that only three of the original 22 northern signals were significant, while they found significant signatures for nine sweeps in the southern population.

While *SweepFinder* uses only sites that show variation within the sample, Huber et al. (2016) expanded the original concept by including sites that show fixed differences with an outgroup. This partly accounts for the variation in mutation rates, and their simulations showed that this improves both power and precision. Finally, if one assumes that background selection is a more appropriate null (as suggested in the introduction to this chapter), the presumed neutral sites must be chosen from genomic regions whose ratio of gene density to recombination rate is similar to that of the target region. Huber et al. (2016) discussed how to replace the empirical $\mathbf{p}$ (the observed site-frequency for reference, and presumably neutral, loci) with estimates corrected for background selection.

Example 9.9. While using the same basic logic as the *CLRT*, namely, constructing a likelihood model where the frequency spectrum is a function of distance from the site, there is a bit more bookkeeping required to obtain the likelihood function for *SweepFinder*. The task is to translate an empirical site-frequency spectrum, $\mathbf{p}$ (at a set of presumably neutral loci), before a sweep into a pattern, $\mathbf{p}_s$, after the sweep. Nielsen et al. (2005b) approached this problem by focusing on a site at distance (recombination frequency) c from the target of selection (with c being estimated by the model), and conditioning on the number of lineages (from the preselection population) that are now present in a sample of size n. Suppose ℓ of these represent lineages that have escaped the sweep (and hence will be reflective of the normal background distribution, $\mathbf{p}$), in which case the remainder, $n - \ell$, will be lineages that did not escape, and hence will either all contain the derived allele (if it was associated with the initial favorable mutation), or all lack it. From Chapter 8, the probability that any sample sequence escaped a sweep is $1 - f_s$, where f_s is a function of the distance, c, from the selected site (Equation 9.16b). The probability that ℓ out of n sample sequences are lineages that escaped the sweep is binomial, with a success probability of $1 - f_s$, yielding

$$P_e(\ell) = \binom{n}{\ell}(1-f_s)^\ell f_s^{n-\ell} = \binom{n}{\ell}\left(1-e^{-c\lambda}\right)^\ell \left(e^{-c\lambda}\right)^{n-\ell} \tag{9.18a}$$

where $\lambda = (1/2hs)\ln(4N_e s)$ measures the strength of selection (Equation 9.16b) and is constant over all linked sites, while c varies with the distance of our focal marker from the target of selection. There are $M = \min(\ell+1, n)$ lineages (the ℓ distinct lineages and the single lineage associated with the sweep). Conditioned on ℓ, we need to compute $p_{s,i}$, the probability of seeing i/n derived alleles in our sample. We then average this over $P_e(\ell)$ to obtain the likelihood for a site. The probability of finding i derived alleles in a sample of the M lineages following the sweep is

$$\Pr(i \mid M) = \sum_{j=i}^{n-1} p_j \frac{\binom{j}{i}\binom{n-j}{M-i}}{\binom{n}{M}}, \quad \text{for} \quad 0 \leq i \leq M \tag{9.18b}$$

where the combinatorial term in the sum is from the hypergeometric distribution. Assuming there are j copies of the derived allele in a sample of n initial lineages before the sweep, this term gives the probability that M draws (without replacement) will yield i copies in our sample. Averaging over the probability, p_j, that j copies of the derived allele (the jth element in **p**) were in this ancestral sample yields $\Pr(i \mid M)$. Given that there are i lineages of the M carrying the derived allele, the probability that the derived allele is in the lineage that did not escape is simply i/M.

For a derived allele to have i copies at a site following the sweep means that it either was linked to the favorable allele, and hence was present in $(i+\ell+1-n)$ of the $\ell+1$ lineages at the start of the sweep, leading to i copies in the sample, or it was not associated with the favorable mutation, so that i copies in the initial $\ell+1$ lineages results in i copies in the sample. Putting these together gives the likelihood as

$$\begin{aligned} p_{S,i} = P_e(n)\, p_i + \sum_{\ell=0}^{n-1} P_e(\ell) \Bigg[&\Pr\left(i+1+\ell-n \,\middle|\, \ell+1\right) \frac{i+1+\ell-n}{\ell+1} \\ &+ \Pr\left(i \,\middle|\, \ell+1\right) \frac{\ell+1-i}{\ell+1} \Bigg] \end{aligned} \tag{9.18c}$$

The first term is the probability that all of the sampled lineages escaped the sweep, simply recovering the background spectrum (p_i). The values given by Equation 9.18c replace those given by Equation 9.16a to construct the likelihood under a sweep. It is useful to note where parameters appear. The sweep strength (λ) and location (c) appear through f_s in the $P_e(\ell)$ term (Equation 9.18a), the number of lineages that escaped the sweep. The empirical background spectrum, **p**, appears in the $\Pr(i \mid M)$ terms (Equation 9.18b) that populate the original sample of lineages at the start of the sweep. Improvements to the original *SweepFinder* have been suggested, such as the computationally more efficient version ***SweeD*** (Pavlidis et al. 2013).

Tests of Sweeps Using Genomic Positional Information: *XP-CLR*

Chen et al. (2010) introduced a test that is similar in spirit to the *CLRT* but uses the genomic positional pattern of *allele-frequency differences* between two populations (a reference and a candidate), as opposed to heterozygosity data from a single population. Their **cross-population composite-likelihood ratio test**, or ***XP-CLR***, is constructed as follows. They assumed a biallelic marker, such as a SNP, and scored the frequency of an allele in the reference and candidate populations, with selection assumed to have occurred in the latter (but not the former). First, they modeled the neutral divergence in allele frequency between two isolated populations that originally shared a common ancestor. To simplify matters, they used a Brownian motion (or Wiener process) approximation for drift (Appendix 1), using a normal for the expected distribution of allele frequencies. This approximation assumes that alleles are segregating and hence frequencies are in the $(0, 1)$ range. For pure drift, the expected change, $m(x)$, in an allele at frequency x is zero, while the expected variance of the change is $v(x) = x(1-x)/(2N_e)$. If we build the Wiener process by using Equation A1.31a, the expected frequency, x, of an allele at generation t, given that it started at a frequency of p_0, is approximately normally distributed, with

$$x(t) \sim \mathrm{N}\left(p_0, t\,\frac{p_0(1-p_0)}{2N_e}\right) \tag{9.19a}$$

More generally, we can write the variance as $\beta\, p_0(1-p_0)$, where the β term accounts for the population history (allowing for variation in population size, etc.), so that if p_r is the allele frequency in some reference population, then under a pure drift model, the distribution of frequencies in the candidate population should follow

$$x \sim \mathrm{N}\left(p_r, \sigma^2\right) \quad \text{where} \quad \sigma^2 = \beta\, p_r(1-p_r) \ll 1 \tag{9.19b}$$

The last condition follows because this approximation only works well for σ^2 small. Due to the shared population history, all neutral genes should have (roughly) the same β value, which one can estimate directly from the data. If k out of n sampled chromosomes in the control population contain the allele, the likelihood (for β) for this site is given by using Equation 9.16a, but with $\phi(x)$ now replaced by the normal density function, with parameters given by Equation 9.19b. Chen et al. (2010) introduced an additional refinement. Instead of simply multiplying all individual likelihoods together, they formed their composite likelihood by multiplying the *weighted* product of the likelihoods, downweighting SNPs that are in LD in the reference population.

Now consider the effect of a sweep in the candidate population. Chen et al.'s argument follows the same logic leading to the shift in the frequency spectrum under a sweep (Equation 8.13), but now the focus is the shift in the distribution of the allele frequency in the candidate population. Suppose a single mutation immediately under selection arises that is linked to the SNP being followed. If the SNP allele frequency is x, then with a probability of x, the new favorable allele occurs on this background, and the resulting sweep changes the SNP allele frequency to $f_s + x[1-f_s]$ (Equation 8.1e). Conversely, with a probability of $1-x$, its frequency decreases to $x(1-f_s)$, where f_s is again given by Equation 9.16b. Using the same logic and changes-of-variables employed in Example 8.4, Chen et al. showed that the distribution of x in the candidate population shifts to a mixture of two normals, the first representing SNP alleles not initially associated with the new favorable mutation, and hence driven toward a frequency of 0, and the second, where they were associated with the mutation and driven toward a frequency of 1, yielding

$$x \sim \left(\frac{f_s - x}{f_s^2}\right) I_{[0,f_s]}\,\mathrm{N}\left(f_s p_r, f_s^2\sigma^2\right) + \left(\frac{x + f_s - 1}{f_s^2}\right) I_{[1-f_s,1]}\,\mathrm{N}\left(f_s p_r + 1 - f_s, f_s^2\sigma^2\right) \tag{9.20}$$

where $I_{[a,b]}$ denotes an indicator function that is 1 when x is in the interval (a, b) and zero otherwise. This distribution now replaces $\phi(x)$ in Equation 9.16a to give the likelihood under selection for this site. The formal test consists of the ratio of the maximum of the composite likelihood under the sweep model (Equation 9.20) to the maximum of the composite likelihood under the neutral model (Equation 9.19b).

What makes Chen et al.'s method compelling is that the empirical null distributions of their test statistic over a wide range of demographic models were essentially identical. Further, their approach had greater power than the *CLRT* in their simulation studies and showed robustness to variation in recombination rates. Peng et al. (2011) presented an interesting application of this model, locating candidate genes for the adaptation of humans to high altitudes by contrasting a Himalayan Tibetan population (as the candidate) with Han

Chinese from Beijing (as the reference). Finally, an alternative (nonparametric) test using between-population divergence and chromosomal positional information was suggested by Oleksyk et al. (2008).

Ascertainment Issues

Because many of these likelihood models exploiting genomic positional information are computationally demanding, they are typically employed *following* a general scan of a genome for some signature of selection, such as regions of depressed variation, or showing unusual site-frequency spectra (such as those with a negative Tajima's D or positive Fay and Wu's H values, which are discussed in the next section). Choosing the region or regions in which to perform the likelihood tests based on the appearance of these special features creates a strong ascertainment bias that dramatically shifts the null distribution. (Note that this is different from SNP ascertainment bias arising from the nonrandom choice of SNPs at the start of the analysis.) The coalescent process can be noisy, and regions with unusual underlying genealogies (such as strong compression of the nodes) can occur by chance even under the equilibrium neutral model. This is especially true when a large number of sites are sampled, presenting more draws from the same underlying process, some of which will be realizations that are extreme values.

Thornton and Jensen (2007) outlined an approach to adjust for both ascertainment and nonequilibrium population structure when simulating the neutral null distribution. First, genome-scan data are used to estimate the parameters for an appropriate demographic model (such as the time and duration of a bottleneck). Next, these are used in neutral coalescent simulations. Normally, this would be the null distribution, but we need to also model the ascertainment process itself as well. For example, suppose the lowermost 2% of regions of reduced variation are chosen for follow-up *CLR* tests. The appropriate null would be constructed by also sampling such low-variation regions from the unconditional null, and then using these as the appropriately ascertained null distribution. Thornton and Jensen noted that increasing the length of the region of analysis is a good general strategy for increasing power. Likewise, they and Teshima et al. (2006) both noted that measures of either diversity (such as reduced heterozygosity) or population differentiation (such as F_{ST} values) seem more reliable for identifying outliers than do frequency-spectrum based approaches.

Model Fragility: Demography, Mutation, Recombination, and Gene Conversion

A concern to always keep in mind when using a sophisticated model is its *robustness* to errors in the underlying assumptions. Simple approaches often have modest power but considerable robustness, while highly specialized models can be quite powerful *when the data fit the assumptions* but may be quite fragile when they do not. Given the constant concern about nonequilibrium and structured populations, and hence generation of the appropriate null model, the robustness to demography is critical (Akey et al. 2004).

While sensitivity to demographic assumptions is generally well appreciated, less appreciated are the effects of genomic assumptions, such as constant mutation and recombination rates across the region. We have already stressed that variation in mutation rates can generate peaks and valleys in the background patterns of neutral variation. These patterns can easily be declared as signals of selection by most of the previously mentioned likelihood tests. The exception is *XP-CLR*, which is based on between-population differences, and thus controls for this problem to some extent. Likewise, although the recombination rate, c, between a neutral site and the selected target appears in most of the likelihood models, operationally one assumes a constant rate, c_0, per nucleotide, with $c = c_0 L$, where L is the distance in nucleotides. However (as discussed in Chapter 4), recombination rates can vary dramatically over very fine scales (Coop and Przeworski 2007), and gene conversion also needs to be considered (Andolfatto and Nordborg 1998). As mentioned in Chapter 8, conversion events can disrupt the signal expected from a sweep (Jones and Wakeley 2008). For example, Glinka et al. (2006) observed a sharp peak of variation in the middle of a valley of depressed variation around the *unc-119* gene in a European population of *Drosophila*. They reasoned that a sweep plus two conversion events generated this unusual pattern.

Finally, background selection, which is expected to be a function of the ratio of gene density to recombination rate, also creates outliers if there is significant genomic variation in this ratio. To some extent, this can be controlled by examining whether significant tests disproportionately fall into such regions.

TESTS BASED ON SITE-FREQUENCY SPECTRUM STATISTICS

Under the infinite-sites model, a sequence is treated as a series of L sites, with each new mutation assumed to occur at a new site (Chapter 4). At mutation-drift equilibrium, most features of this model, including the site-frequency spectrum (SFS), are fully specified by the population-size-scaled mutation rate, $\theta = 4N_e\mu$. Depending on the nature of the data, an observed frequency spectrum is viewed as either folded or unfolded (Chapter 2). An unfolded spectrum considers the frequency of the derived allele (Equation 2.35a), and such data are said to be polarized (typically using an outgroup to distinguish between ancestral and derived, or mutant, alleles). The folded spectrum (Equation 2.35b) uses the minor-allele frequency, ignoring whether the rarer allele is ancestral or derived. To distinguish between these different spectra, we use the notation that ζ_i denotes the number of sites that contain exactly i derived alleles ($1 \le i \le n-1$), yielding the observed unfolded SFS as the vector $(\zeta_1, \cdots, \zeta_{n-1})$. Similarly, η_i denotes the number of sites with exactly i copies of the minor allele ($1 \le i \le [n/2]$), with $(\eta_1, \cdots, \eta_{[n/2]})$ being the observed folded SFS, where

$$[n/2] = \begin{cases} n/2 \text{ for } n \text{ even} \\ (n-1)/2 \text{ for } n \text{ odd} \end{cases}$$

The η_i and ζ_i are simply related by

$$\eta_i = \zeta_i + \zeta_{n-i} \quad \text{for} \quad 1 \le i \le [n/2]$$

For example, both ζ_1 and η_1 denote the number of sites that are **singletons**, with ζ_1 being the number of sites with a single copy of the derived allele and η_1 being the number of sites with a single copy of the minor allele. The latter could be due to *either* a single copy of the derived allele *or* a single copy of the ancestral allele, with $\eta_1 = \zeta_1 + \zeta_{n-1}$.

As detailed in Chapter 8, both a hard sweep and long-term balancing selection are expected to perturb a starting site-frequency spectrum into some new distribution. A hard sweep increases the frequency of sites with rare derived alleles and also sites with high-frequency derived alleles (Figure 8.5). In a folded frequency spectrum, these jointly appear as an increase in the frequency of sites with rare alleles. Conversely, long-term balancing selection is expected to increase the number of sites with intermediate-frequency alleles, albeit over a region that becomes ever-narrower over time (Chapter 8). While they are widely used, a problem with all site-frequency spectrum tests is that nonequilibrium conditions (e.g., during recovery following a bottleneck) or spatial population structure cause the frequency spectrum of neutral alleles to depart from the benchmark Watterson distribution. Thus, a significant amount of the following discussion deals with these concerns.

Summary Statistics Based on Infinite-sites Models

As introduced in Chapter 4, a variety of summary statistics can be used to estimate $\theta = 4N_e\mu$ under the infinite-sites model. For a sample of L sites (generally nucleotides), suppose that there are S segregating sites, with η_1 and ζ_1 sites harboring (folded and unfolded, respectively) singletons, and Π denoting the average number of pairwise differences between two random sequences. If our goal is to estimate θ on a *per-nucleotide* basis, so that μ is the per-nucleotide mutation rate, then (as in Chapter 4), we would

Table 9.1 Summary of the site-frequency tests presented in this chapter, which contrast estimates of θ based on different regions of the site-frequency spectrum. The estimators $\widehat{\theta}_S$, $\widehat{\theta}_\Pi$, $\widehat{\theta}_1$, and $\widehat{\theta}_{1*}$ are given by Equation 9.21a; $\widehat{\theta}_{S-\eta_1}$ and $\widehat{\theta}_{S-\zeta_1}$ are Achaz's (2008) estimators using S but removing unfolded and folded singletons; $\widehat{\theta}_{\Pi-\eta_1}$ and $\widehat{\theta}_{\Pi-\zeta_1}$ are Achaz's analogous estimators using Π; and $\widehat{\theta}_H$ (Equation 9.27a) and $\widehat{\theta}_L$ (Equation 9.28a) are developed below. The column labeled "Spectrum" indicates whether the test requires unfolded data, with alleles designated as ancestral or derived. Further details are given in the text.

Test	Contrast	Spectrum	Signal
Tajima's D	$\widehat{\theta}_S$ vs. $\widehat{\theta}_\Pi$	Folded	< 0: Excess of rare alleles Sweep or population bottleneck > 0: Excess of intermediate-frequency alleles Balancing selection or population structure
Achaz's Y^*	$\widehat{\theta}_{S-\eta_1}$ vs. $\widehat{\theta}_{\Pi-\eta_1}$	Folded	Same as for Tajima's D
Achaz's Y	$\widehat{\theta}_{S-\zeta_1}$ vs. $\widehat{\theta}_{\Pi-\zeta_1}$	Unfolded	Same as for Tajima's D
Fu and Li's D	$\widehat{\theta}_S$ vs. $\widehat{\theta}_1$	Unfolded	Same as for Tajima's D
Fu and Li's D^*	$\widehat{\theta}_S$ vs. $\widehat{\theta}_{1*}$	Folded	Same as for Tajima's D
Fu and Li's F	$\widehat{\theta}_\Pi$ vs. $\widehat{\theta}_1$	Unfolded	Same as for Tajima's D
Fu and Li's F^*	$\widehat{\theta}_\Pi$ vs. $\widehat{\theta}_{1*}$	Folded	Same as for Tajima's D
Fay and Wu's H	$\widehat{\theta}_\Pi$ vs. $\widehat{\theta}_H$	Unfolded	< 0: Excess of high-frequency derived alleles. Sweep or allelic surfing
Zeng et al.'s E	$\widehat{\theta}_\Pi$ vs. $\widehat{\theta}_L$	Unfolded	< 0: Excess of low- vs. high-frequency derived alleles. Signal of a recent *past* sweep

consider the fraction of segregating sites as S/L; the nucleotide diversity as $\pi = \Pi/L$; and the fraction of sites that are singletons as S_1/L. When searching for genomic regions under selection, our goal shifts to $\theta_L = 4N_e\mu_L$, the corresponding value for the *region*, where $\mu_L = L\mu$ is the total mutation rate over the L sites. Because the focus in this section is on specific regions, in an effort to keep the notation simple, we suppress the subscript and use θ and μ to denote the region-side values (θ_L, μ_L).

While S and Π have the same values for polarized and unpolarized data, the number of singletons can be slightly different. All of these summary statistics yield estimates of θ for a region of interest, with

$$\widehat{\theta}_S = \frac{S}{a_n} \qquad \widehat{\theta}_\Pi = \Pi \qquad \widehat{\theta}_1 = \zeta_1, \qquad \widehat{\theta}_{1*} = \frac{n}{n-1}\,\eta_1 \tag{9.21a}$$

where $a_n = \sum_{j=1}^{n-1} 1/j$ (Equation 4.3b). These four expressions correspond (respectively) to: the Watterson estimator (Equation 4.3a, which is also commonly denoted by θ_W); Tajima's estimator (Equation 4.1); our previous singleton estimator, $\widehat{\theta}_1$, using unfolded data (Equation 4.6a); and the corresponding singleton estimator, $\widehat{\theta}_{1*}$, using folded data. The sampling variances for these estimates are given by Equations 4.4a ($\widehat{\theta}_S$), 4.2 ($\widehat{\theta}_\Pi$), and 4.6b ($\widehat{\theta}_1$). These expressions for the variance are functions of both θ and θ^2, and are typically (e.g., Tajima 1989) computed by replacing

$$\theta \text{ by } S/a_n \quad \text{and} \quad \theta^2 \text{ by } \frac{S(S-1)}{a_n^2 + b_n} \tag{9.21b}$$

where $b_n = \sum_{j=1}^{n-1} 1/j^2$ (Equation 4.4b).

The idea behind site-frequency tests of neutrality is to compare two different estimates of θ based on information from *different regions* of the site-frequency spectrum. When the infinite-sites model holds and the population is at mutation-drift equilibrium, these estimates should be within the sampling error of each other, while they can be significantly different when the neutral equilibrium model does not hold. Table 9.1 summarizes the various site-frequency test statistics discussed here, all of which have the form

$$t = \frac{\widehat{\theta}_i - \widehat{\theta}_j}{\sigma(\widehat{\theta}_i - \widehat{\theta}_j)} \tag{9.21c}$$

When applying any of these tests, care must be taken to avoid SNP ascertainment bias. If the process by which SNPs are chosen is biased by their frequency (SNP discovery panels are generally biased in favor of intermediate-frequency sites), this results in a biased estimate of the frequency spectrum, potentially compromising the tests. Likewise, when tests are based on the unfolded SFS, errors introduced by incorrect polarity assignment (incorrectly assigning a derived allele ancestral status, and vice versa) can be very serious (Baudry and Depaulis 2003; Hernandez et al. 2007). Finally, sequencing errors result in an increase in singletons, which can bias tests away from neutrality (Achaz 2008; Johnson and Slatkin 2008), although maximum-likelihood approaches exist that deal with sequencing errors when estimating SNP frequencies (Lynch 2009a; Maruki and Lynch 2013).

Example 9.10. Suppose we sample ten alleles from a population and observe a total of 12 segregating sites ($S = 12$), an average of four differences between alleles ($\Pi = 4$), and three segregating sites that have only a single copy of the minor allele ($\eta_1 = 3$). What are the estimates of θ based on these three summary statistics? Using Equations 9.21a yields

$$\widehat{\theta}_S = \frac{S}{a_{10}}, \quad \text{with } a_{10} = \sum_{i=1}^{9} \frac{1}{i} = 2.83 \text{ yielding } \widehat{\theta}_S = \frac{12}{2.83} = 4.24$$

$$\widehat{\theta}_\Pi = \Pi = 4 \qquad \widehat{\theta}_{1^*} = \frac{n}{n-1}\eta_1 = \frac{10}{9} \cdot 3 = 3.33$$

Example 9.11. As we now illustrate, all of the tests summarized in Table 9.1 follow from a general family of estimators of θ based on the discrete Watterson distribution (Equation 2.35). For a sample of n sequences with L sites, the expected number of segregating sites with i copies of the derived (unfolded, ζ_i) or of the minor (folded, η_i) allele are

$$E(\zeta_i) = \frac{\theta}{i} \qquad \text{for} \qquad 1 \le i \le n-1$$

$$E(\eta_i) = E(\zeta_i) + E(\zeta_{n-i}) = \frac{\theta}{i} + \frac{\theta}{n-i} = \frac{\theta}{i}\frac{n}{n-i} \qquad \text{for} \qquad 1 \le i \le [n/2] \tag{9.22a}$$

where $\theta = 4N_e\mu$ is the scaled mutation rate for the entire region.

Hence, a method-of-moments estimator for θ using only the number in the ith class from either SFS is simply

$$\widehat{\theta}_i = \begin{cases} i \cdot \zeta_i & i \text{ copies of the derived allele} \quad 1 \le i \le n-1 \\ \dfrac{i \cdot (n-i)}{n}\eta_i & i \text{ copies of the minor allele} \quad 1 \le i \le [n/2] \end{cases} \tag{9.22b}$$

Nawa and Tajima (2008) suggested that a plot of $\widehat{\theta}_i$ versus i can be helpful for visualizing departures from the neutral SFS, although values for large i may be more problematic as the variance of $\widehat{\theta}_i$ dramatically increases with i.

Following Zeng et al. (2006), consider any summary statistic, g, of the unfolded site-frequency spectrum of the form

$$g = \sum_{i=1}^{n-1} c_i \, \zeta_i \tag{9.23a}$$

From Equation 9.22a

$$E(g) = \sum_{i=1}^{n-1} c_i \, \frac{\theta}{i} = \theta \, h(n) \quad \text{where} \quad h(n) = \sum_{i=1}^{n-1} \frac{c_i}{i} \tag{9.23b}$$

Thus, a family of estimators for θ based on an arbitrary vector $(c_1, \cdots, c_{n-1})$ of weights is given by

$$\widehat{\theta}_g = \frac{g}{h(n)} \tag{9.23c}$$

where $h(n)$ is a function of the sample size n and the chosen weights c_i, and g is the observed value of the statistic.

The choice of weights allows one to tailor statistics to use different parts of the frequency spectrum when estimating θ. Taking $c_i = 1$ yields $g = S$ and $h(n) = a_n$, recovering the Watterson estimator, $\widehat{\theta}_S = S/a_n$. Taking $c_i = i(n-i)$

$$h(n) = \sum_{i=1}^{n-1} i(n-i)/i = n(n-1)/2$$

yielding

$$\widehat{\theta} = \sum_{i=1}^{n-1} \frac{2i(n-i)}{n(n-1)}$$

which is simply the average pairwise difference, Π. As with S, Π is symmetric with respect to i and $n - i$, so that both folded and unfolded data return the same estimate. Taking $c_1 = 1, c_{i>1} = 0$ yields $g = \zeta_1$ (the number of derived singletons) and $h(n) = 1$, recovering the $\widehat{\theta}_1$ estimator.

Similarly, for a folded frequency spectrum,

$$g = \sum_{i=1}^{[n/2]} c_i \, \eta_i, \qquad \widehat{\theta}_g = \frac{g}{f(n)}, \qquad f(n) = \sum_{i=1}^{[n/2]} c_i \, \frac{n}{i(n-i)} \tag{9.23d}$$

Consider the estimator using only folded singletons, η_1. Here, $c_1 = 1, c_i = 0$ for $i > 1$, and hence $f(n) = n/(n-1)$, giving $\eta_1(n-1)/n$ as an estimator of θ, which recovers $\widehat{\theta}_{1^*}$. Achaz (2009) provided general expressions for the variance of any estimator of the form given by Equations 9.23c or 9.23d, providing all of the machinery to develop general tests in the form of Equation 9.21c using any feature of interest in the SFS.

Tajima's *D* Test

The first proposed, and most widely used, site-frequency spectrum test is **Tajima's *D*** (1989), which contrasts θ estimates based on the number of segregating sites (S) and average pairwise difference (Π),

$$D = \frac{\widehat{\theta}_\Pi - \widehat{\theta}_S}{\sqrt{\alpha_D S + \beta_D S^2}} \tag{9.24a}$$

where

$$\alpha_D = \frac{1}{a_n}\left(\frac{n+1}{3(n-1)} - \frac{1}{a_n}\right) - \beta_D \tag{9.24b}$$

$$\beta_D = \frac{1}{a_n^2 + b_n}\left(\frac{2(n^2+n+3)}{9n(n-1)} - \frac{n+2}{a_n n} + \frac{b_n}{a_n^2}\right) \tag{9.24c}$$

with a_n and b_n as above (Equations 4.3b and 4.4b). Being based on S and Π, this test does not require unfolded data. Tajima's motivation was that θ_S and θ_Π measure different features of the frequency spectrum. The number of segregating sites, S (and thus θ_S), counts polymorphic sites independent of their frequencies, making it more sensitive to changes in the frequencies of rare alleles (as small changes can cause sites segregating rare alleles to either enter or drop out of the sample). Conversely, the average pairwise difference, Π (and thus θ_Π), is a frequency-weighted measure and more sensitive to changes in intermediate-frequency alleles. A negative value of D indicates that there are too many low-frequency sites, while a positive value of D indicates that there are too many intermediate-frequency sites. Expressed another way, D is a test for whether the amount of heterozygosity per site is consistent with the number of polymorphic sites expected under the equilibrium neutral model. Under selective sweeps (and population expansion), heterozygosity should be significantly less than is predicted from the number of polymorphisms. As with all site-frequency spectrum tests, the distribution of D *critically* depends on adherence to the neutral equilibrium assumptions.

Tajima obtained upper (D_{max}) and lower (D_{min}) bounds on D, so that

$$(D - D_{min})/(D_{max} - D_{min})$$

lies in the range (0, 1). Under the equilibrium neutral assumption, Tajima showed that D is well approximated on this modified scale by a Beta distribution (Equation A2.38a), with distribution parameters

$$\alpha = -\frac{(1 + D_{min}D_{max})D_{max}}{D_{max} - D_{min}} \quad \text{and} \quad \beta = \frac{(1 + D_{min}D_{max})D_{min}}{D_{max} - D_{min}} \tag{9.25a}$$

Innan and Stephan (2000) and Ẑivković and Wiehe (2008) showed that the distribution of D in a population of changing size is significantly different from this equilibrium neutral result. Because the minimal value of D varies with the number of segregating sites, S, Schaeffer (2002) proposed a standardized $D' = D/D_{min}$ to adjust for this, allowing for more fair comparisons of D across loci. The minimum value of Π given S was obtained by Tajima (1989) as

$$\Pi_{min} = S\,\frac{2(n-1)}{n^2} \tag{9.25b}$$

D_{min} is computed from Equation 9.24a, with Π_{min} replacing $\widehat{\theta}_\Pi$.

Example 9.12. Two interesting applications of the D test were offered by Tajima (1989). First, he considered Aquadro and Greenberg's (1983) data for 900 base pairs in the mitochondrial DNA of seven humans. They observed 45 segregating sites and an average number of nucleotide differences between all pairs of 15.38. Hence,

$$a_7 = \sum_{i=1}^{6} \frac{1}{i} = 2.45, \qquad b_7 = \sum_{i=1}^{6} \frac{1}{i^2} = 1.49$$

$$\widehat{\theta}_S = \frac{S}{a_n} = \frac{45}{2.45} = 18.38, \qquad \widehat{\theta}_\Pi = \Pi = 15.38$$

$$\beta_D = \frac{1}{2.45^2 + 1.49}\left(\frac{2(7^2+7+3)}{9\cdot 7(7-1)} - \frac{7+2}{7\cdot 2.45} + \frac{1.49}{2.45^2}\right) = 0.00475$$

$$\alpha_D = \frac{1}{2.45}\left(\frac{7+1}{3(7-1)} - \frac{1}{2.45}\right) - 0.00475 = 0.01005$$

$$D = \frac{\widehat{\theta}_\Pi - \widehat{\theta}_S}{\sqrt{\alpha_D S + \beta_D S^2}} = \frac{15.38 - 18.38}{\sqrt{0.01005\cdot 45 + 0.00475\cdot 45^2}} = -0.945$$

Table 2 of Tajima (1989) gives the 95% confidence interval for D under strict neutrality for $n = 7$ as -1.608 to 1.932, so that this value is not significantly different from its neutral expectations.

Tajima also applied his test to the data of Miyashita and Langley (1988), who examined 64 samples of a 45-kb region of the *white* locus in *D. melanogaster*. Taking large insertions and deletions as the polymorphic sites, they found $S = 15$ and $\Pi = 0.94$, which yields a value of $D = -2.0709$. Given that the 95% confidence interval under neutrality is -1.795 to 2.055, the site-frequency spectrum associated with this locus shows evidence (a significantly negative D value) of either directional selection or a population bottleneck.

Achaz's *Y* and *Y** Tests

Achaz (2008) noted that estimates of both S and π can be biased by sequencing errors, which introduce an excess of singletons, thus skewing D toward more negative values. His ***Y*** and ***Y**** **tests** modify Tajima's D by computing $\widehat{\theta}_\Pi$ and $\widehat{\theta}_S$ after removing singletons. With unfolded data, his Y test replaces these estimates by $\widehat{\theta}_{\Pi-\zeta_1}$ and $\widehat{\theta}_{S-\zeta_1}$, while with folded data, his Y^* test uses $\widehat{\theta}_{\Pi-\eta_1}$ and $\widehat{\theta}_{S-\eta_1}$. Expressions for these estimators and the sampling variances of the tests can be found in his paper. While initially proposed as a method to deal with potential sequencing errors (which can now be accounted for by using ML approaches), these tests are still a useful metric, as a comparison of Y and Y^* with D providing information about the impact of singletons.

Fu and Li's *D** and *F** Tests

Fu and Li (1993b) introduced tests based on other contrasts among the infinite-sites θ estimators given by Equation 9.21a. Both proposed tests use the number of singleton sites, with variants using either folded (η_1 sites with a single copy of the minor allele) or unfolded (ζ_1 sites with a single copy of the derived allele) singletons. Using these statistics gives estimates of θ based on the rare-alleles region of the SFS, which are then contrasted with estimates based on either S or Π. Using folded data, this gives rise, respectively, to their D^* and F^* tests. Their exact counterparts for unfolded data (using ζ_1 in place of η_1) are, respectively, their D and F tests, which are not discussed further. Given the widespread use of Tajima's D, when we simply reference a "D test," this always refers to Tajima's test.

Fu and Li's ***D**** **test** compares the segregating sites (S) versus the folded-singleton (η_1) estimators of θ

$$D^* = \frac{\widehat{\theta}_S - \widehat{\theta}_{1^*}}{\sqrt{\alpha_* S + \beta_* S^2}} \tag{9.26a}$$

$$\alpha_* = \frac{1}{a_n}\left(\frac{n+1}{n} - \frac{1}{a_n}\right) - \beta_* \tag{9.26b}$$

$$\beta_* = \frac{1}{a_n^2 + b_n}\left[\frac{b_n}{a_n^2} - \frac{2}{n}\left(1 + \frac{1}{a_n} - a_n + \frac{a_n}{n}\right) - \frac{1}{n^2}\right] \tag{9.26c}$$

Their ***F**** **test** compares the average pairwise divergence (Π) with the folded-singletons (η_1) estimators

$$F^* = \frac{\widehat{\theta}_\Pi - \widehat{\theta}_{1^*}}{\sqrt{\alpha_F S + \beta_F S^2}} \tag{9.26d}$$

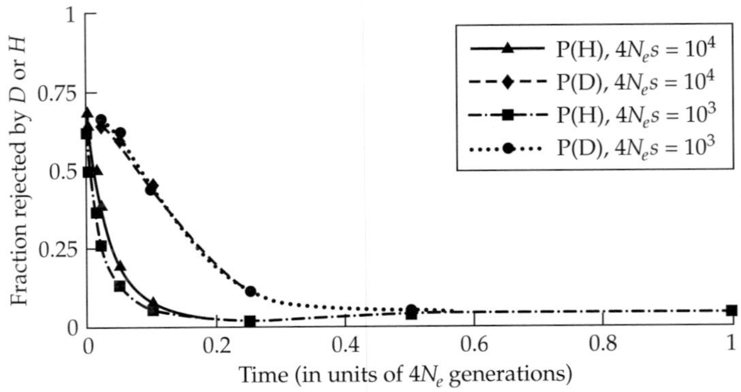

Figure 9.4 The power (P) of the H and D tests to detect signatures of a recent sweep is very fleeting. The power of H (which is based on high-frequency derived alleles) falls off especially rapidly after a sweep (as high-frequency alleles are fixed), essentially having power only within $0.2N_e$ generations following a sweep. D, which is based on an excess of rare alleles, can detect new mutations that enter following the sweep and has power over about $0.5N_e$ generations following a sweep. A value of $c/s = 0.01$ was used and power (for a sample size of 50) was graphed for two different values of $4N_es$. For D (the upper two curves in the figure, which are essentially superimposed on each other), the power is essentially identical for these two values of $4N_es$, while for H (the lower two curves), there is a very slight power increase for $4N_es = 10^4$ when $t < 0.1N_e$. (After Przeworski 2002.)

$$\alpha_F = \frac{1}{a_n}\left(\frac{4n^2 + 19n + 3 - 12(n+1)a_{n+1}}{3n(n-1)}\right) - \beta_F \tag{9.26e}$$

$$\beta_F = \frac{1}{a_n^2 + b_n}\left(\frac{2n^3 + 110n^2 - 255n + 153}{9n^2(n-1)} + \frac{2(n-1)a_n}{n^2} - \frac{8b_n}{n}\right) \tag{9.26f}$$

These expressions are from Simonsen et al. (1995), with Equation 9.26e correcting the original Fu and Li paper. Critical values (assuming no recombination in the region) were tabulated by Fu and Li (1993b). While these tests are fairly widely used, Simonsen et al. (1995) found that they are not as powerful as Tajima's test for detecting a selective sweep or demographic features (bottlenecks or population subdivision).

Fay and Wu's *H* Test

The first test to use the full power of the unfolded frequency spectrum was proposed by Fay and Wu (2000), who noted that a hard sweep results in an excess of sites with high-frequency derived alleles (Figure 8.5). Although the signature is rather fleeting (Figure 9.4), this excess forms the basis for their ***H* test**. Their idea is to disproportionately weight sites containing derived alleles at high frequencies, and they chose to do so using the weights $c_i = i^2$. From Equation 9.23b, these weights imply $h(n) = n(n-1)/2$, and Equation 9.23c yields

$$\widehat{\theta}_H = \frac{2}{n(n-1)}\sum_{i=1}^{n-1} i^2\,\zeta_i \tag{9.27a}$$

The H test is the scaled difference between Fay and Wu's estimator for θ and that based on average pairwise differences,

$$H = \frac{\widehat{\theta}_\Pi - \widehat{\theta}_H}{\sigma(H)} \tag{9.27b}$$

Zeng et al. (2006) obtain the sampling variance as

$$\sigma^2(H) = \left[\frac{n-2}{6(n-1)}\right]\theta + \left[\frac{18n^2(3n+2)b_{n+1} - (88n^3 + 9n^2 - 13n + 6)}{9n(n-1)^2}\right]\theta^2 \tag{9.27c}$$

As above, Equation 9.21b is used for θ and θ^2 when computing Equation 9.27c. Because Π is a measure of the intermediate-frequency sites, H is a contrast between high- and intermediate-frequency variation, with a negative H indicating an excess of sites with a high frequency of derived alleles. Jointly negative (and significant) values of D and H are consistent with a selective sweep, indicating both an excess of rare alleles and an excess of high-frequency derived alleles. One caution when applying the H test is its extreme sensitivity to polarity errors (Baudry and Depaulis 2003).

Przeworski (2002) showed that both the D and H tests have moderate power immediately after a sweep but the power of the H test rapidly dissipates (within $\sim 0.2N_e$ generations) as the high-frequency alleles become fixed (Figure 9.4). The D test retains power a bit longer ($\sim 0.5N_e$ generations), as it is sensitive to the generation of rare alleles by new mutations immediately after the sweep. As a result of using different signatures in the SFS, one can easily encounter situations where, following a sweep, one test is highly significant while the other is not. Even for a strong hard sweep, neither D nor H may be significantly negative, depending on the time since the sweep was completed (Figure 9.4).

Zeng et al.'s *E* Test

A variant of the H test was proposed by Zeng et al. (2006), who noted that the most powerful contrasts between regions of the unfolded frequency spectrum following selection should involve high- versus low-frequency sites. However, most contrasts involve a comparison with θ_Π, which is a measure of intermediate-frequency alleles. To rectify this, Zeng et al. introduced the estimator, θ_L, based on a weight, $c_i = i$, that places more emphasis on high-frequency sites than θ_S (but not as much as θ_H). For these weights, Equation 9.23b implies $h(n) = n - 1$, and hence Equation 9.23c yields

$$\widehat{\theta}_L = \frac{1}{n-1} \sum_{i=1}^{n-1} i\,\zeta_i \tag{9.28a}$$

Zeng et al.'s ***E* test** contrasts the high- and low-frequency regions of the frequency spectrum,

$$E = \frac{\widehat{\theta}_L - \widehat{\theta}_S}{\sigma(E)} \tag{9.28b}$$

where

$$\sigma^2(E) = \left[\frac{n}{2(n-1)} - \frac{1}{a_n}\right]\theta + \left[\frac{b_n}{a_n^2} + 2\left(\frac{n}{n-1}\right)^2 b_n - \frac{2(nb_n - n + 1)}{(n-1)a_n} - \frac{3n+1}{n-1}\right]\theta^2 \tag{9.28c}$$

Again the variance is approximated by replacing θ and θ^2 with their estimates from Equation 9.21b. A negative E indicates an excess of low- versus high-frequency sites relative to expectations of the equilibrium neutral model. This occurs *immediately after* a sweep, as the excess of high-frequency alleles is quickly lost by drift (due to their fixation), while at the same time mutation is generating an excess of low-frequency sites, which have yet to drift up to their neutral equilibrium values. The unique feature of E is its ability to yield a signal *after* a sweep, which can persist up to $2N_e$ generations (and hence much longer than H).

Adjusting the Null to Account for Nonequilibrium Populations

The site-frequency tests summarized in Table 9.1 critically depend on the Watterson distribution as the null model, as do the entire family of θ estimators given by Equation 9.23. As such, they are especially susceptible to false positives when samples come from a population not satisfying the underlying assumptions (a panmictic population in mutation-drift equilibrium). Four strategies have been proposed to address this concern. The first three

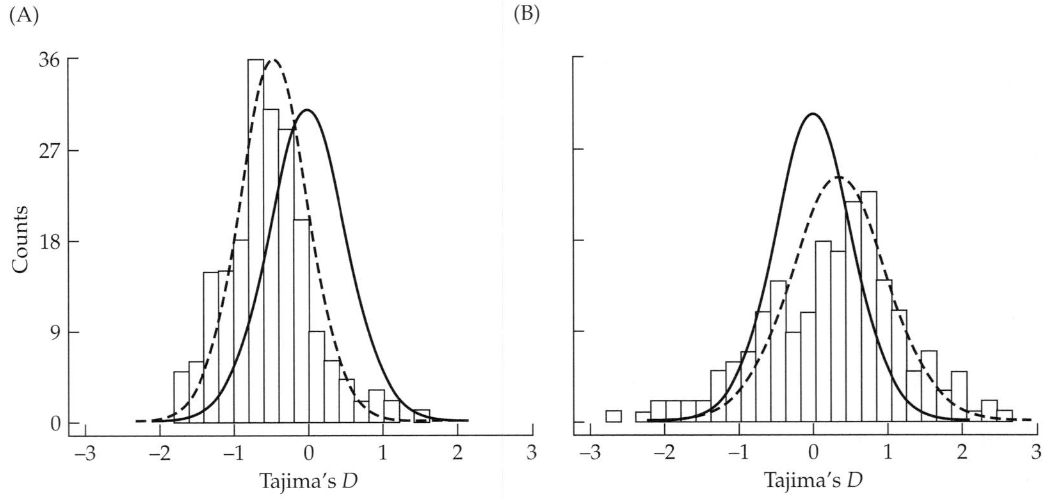

Figure 9.5 Distribution of Tajima's D for 201 genes in African-American (**A**) and European-American (**B**) samples. The empirical distribution is represented by the histogram, the solid line represents the simulated values under the equilibrium neutral model, and the dashed line the simulated distribution under the best-fitting demographic model. For (**A**), this is exponential growth starting 50,000 years ago, while for (**B**), this is a bottleneck starting 40,000 years ago. (After Ronald and Akey 2005.)

are standard approaches to refine the null to better suit the sampled population: (i) using the empirical distribution of test statistics from a set of genes in the sample (the outlier approach), (ii) using coalescent simulations with marker-based estimates of demographic parameters, and (iii) using the empirical site-frequency spectrum at reference locations as the null. The final strategy, **support via a preponderance of evidence**, considers the joint signatures from a number of different tests and will be discussed separately. While approaches attempting to account for nonequilibrium populations offer improvements over tests based on the standard (i.e., equilibrium) neutral model, they still do not guarantee that significant signatures represent true regions influenced by positive selection. Because of this, the current operational use for many of these tests (such as Tajima's D) are as *convenient summary statistics* for features in a region of interest, such as whether there is an excess of rare alleles.

The first approach assumes that the empirical distribution of a test statistic over a large number of genes sampled from the target population can provide useful information. Under the equilibrium neutral model, the test statistics reviewed in Table 9.1 should have a mean of zero, while the empirical distribution shows whether the tests trend away from this expectation in the target population. Figure 9.5 gives examples of the empirical distribution of D for 201 genes in two different human populations. For African-Americans, the mean D is negative, while it is positive for European-Americans. A gene whose negative D value is significant under the equilibrium neutral model is likely to be even more significant in this European-American population (given this population's trend toward a positive D), but is problematic in this sample of African-Americans (given that random tests in this group trend toward a negative D). While the mean of this distribution can be informative, one cannot simply use it to adjust test statistics for individual genes. This is because departures from the standard neutral model often *inflate the variance* of test statistics (Nielsen 2001). Thus, even when the mean of the empirical distribution is zero, the variance under the standard model may be too narrow, and hence significance can be overstated. Finally, the empirical distribution is largely shaped by common demographic features that influence all genes. Allelic surfing of neutral alleles does not leave a constant genome-wide signature, and sites experiencing surfing can easily create outliers, mimicking signatures of selection.

The second approach to account for a nonequilibrium population is to use genomic data to infer demographic parameters (such as the size and duration of any past bottleneck), which are then used as the basis for coalescent simulations (Chapter 2). This generates a more appropriate null distribution of the test statistic for the target population (e.g., Figure 9.5). An example of this approach was provided by Schaffner et al. (2005), who used human data to find the best-fitting model over a rather rich parameter space, including population structure, bottleneck times, and variation in recombination rates. Tenaillon et al. (2004) performed a similar analysis on the bottleneck during the formation of maize from teosinte. With estimated demographic parameters in hand, coalescent simulation programs such as MSMS (Ewing and Hermisson 2010), GENOME (Liang et al. 2007), cosi (Schaffner et al. 2005), or MS (Hudson 2002) can be used. Again, this approach only corrects for demographic features that leave a common signal over the entire genome, implying that sites experiencing allelic surfing can still generate false signals of selection even after this correction. As recently summarized by Li et al. (2012), the joint estimation of both demographic and selection parameters is still somewhat problematic.

The final approach is to use the empirical site-frequency spectrum vector, **p**, from a reference set—as opposed to the Watterson distribution—as the null (Nielsen et al. 2005b, 2009). Here p_i is the fraction of sites in the reference set with i copies of the allele (derived or minor, for the unfolded and folded spectra, respectively). A standard goodness-of-fit test (such as the G-test; LW Appendix 2) is then used to assess whether the spectrum $n_1, \cdots n_{n-1}$ in a candidate region is consistent with the multinominal probabilities given by **p**. One can also compare different parts of the spectrum, such as searching for an excess of low-frequency alleles, or high-frequency derived alleles, relative to this standard. Nielsen et al. (2009) used this approach for their ***MWU-low*** and ***MWU-high*** tests, respectively, where *MWU* stands for the Mann-Whitney U test (a common nonparametric test for comparing two groups, e.g., Conover 1999). One major reservation with these nonparametric approaches is the choice of the reference set of sites for the neutral background spectrum. Even if these site are neutral, local effects such as differences in the mutation rates (and hence in θ) and the background recombination rates that influence the levels of standing variation (Chapter 8) can result in the target sites (even if strictly neutral) differing from the distribution at reference sites. If one assumes background selection as the appropriate null, the sites used in constructing **p** should (at a minimum) come from genomic regions with very similar values of gene density to recombination rates as the tested region.

Support via a Preponderance of Evidence

A common strategy in the literature to support a claim for selection is to show that a number of different tests are all highly significant. To this aim, a number of authors have proposed compound tests based on the joint distribution of two (or more) summary statistics of selection. Zeng et al (2006) proposed their ***DH* test**, which combines signals from Tajima's D and Fay and Wu's H tests. Zeng et al. (2007b) extended this approach with their ***HEW*** and ***DHEW* tests**, which combine either Fay and Wu's H statistic or Zeng et al.'s DH test (respectively) with the Ewens-Watterson test (introduced shortly as Equation 9.30b). Pavlidis et al. (2010) and Lin et al. (2011) also suggested approaches that combine multiple signals. This approach was taken to the extreme by Grossman et al. (2010, 2013), who combined test statistics based on both population differentiation and haplotype information to construct a likelihood based on the values of multiple test statistics (their **composite of multiple signals**, or ***CMS***, approach). While composite tests likely do not return proper significance values (due to a lack of independence between tests), they can still have considerable utility. Grossman et al. noted that *CMS* often yields a substantially narrower region for a selected site, increasing resolution up to 100-fold. Further, given that different tests are optimal over different time scales during a sweep (Table 8.2), a composite test offers the possibility of having power over a larger time span.

Others have advocated **meta-analysis** approaches, combining the significance values over multiple tests (Appendix 4). This can be accomplished in several ways. Utsumomiya et al. (2103) proposed ***meta-SS***, using Stouffer's Z score (Equation A4.2) to combine p values for different tests applied in a particular region to obtain a single overall p value for that region. Randhawa et al. (2014) used a slightly different approach, their **composite selection signals** or ***CSS***. Here, for a given test, a standardized rank score, $R_k/(n+1)$, is computed for each of the n SNPs (R_k is the rank, from lowest to highest, of the p value of the test). The resulting scores (for a given test) for each SNP range from $1/(n+1)$ to $1-1/(n+1)$, which are then probit-transformed (Equation 14.2) and averaged over all of the tests to obtain a Z score for each particular SNP. Again, such meta-analysis p values are only approximations, as they assume the p values for different tests are uncorrelated, which is usually not true. Their utility is largely as a convenient summary statistic for evidence of selection in a particular region, rather than as a definitive probability statement.

Ma et al. (2015) proposed a simple measure to deal with test correlations, their **de-correlated composite of multiple signals**, or ***DCMS*** statistic. Let $p_{i,k}$ denote the p value for test k for site i, and let r_{kj} be the empirical correlation among the values of the test statistics for tests k and j over all of the scored sites, so that $r_{kk}=1$ and $r_{kj}=0$ when tests k and j are uncorrelated. Ma's *DCMS* statistic for site i is given by

$$DCMS_i = \left(\frac{1}{W}\right)\sum_{k=1}^{t} \ln\left(\frac{1-p_{i,k}}{p_{i,k}}\right), \quad \text{where} \quad W = \sum_{k=1}^{t} |\, r_{kt}| \tag{9.29a}$$

The terms in the sum are the odds ratio for each test (which Ma et al. used in place of Bayes factors with equal prior weight on the null and alternative; see Equation A2.10b). The weighting term (W) ranges from 1 (none of the tests are correlated, so that $W = r_{tt}+0=1$), to the case were all of the tests are perfectly correlated, so that $W=t$. In the former case, the composite measure is simply the sum of the odds ratios, while in the latter it is the average of the odds ratio. Ma at el. found in their simulations that *DCMS* had higher power than either *meta-SS* or *CSS* under most settings.

A final class of composite measures are **multivariate outlier metrics**. Just as the outlier approach is widely used to highlight sites that have exceptional values in a given single test statistics, one can also consider outliers from a *collection* of test statistics. Assuming all the tests have a mean of zero under the null, the total Euclidean distance of a vector of test statistics from the mean value under the null (**0**) would be one approach. However, different test statistics have different variances, and further they are correlated. One standard approach in such cases is to transform all of the tests statistics to have the same variance and to be uncorrelated, which leads to the **Mahalanobis distance** (Equation A5.19),

$$D_i^2 = \mathbf{t}_i^T\, \boldsymbol{\Sigma}_{\mathbf{t}}^{-1}\, \mathbf{t}_i \tag{9.29b}$$

where **t** is the vector of test statistics for site i and $\boldsymbol{\Sigma}_{\mathbf{t}}$ is the empirical variance-covariance matrix for the vector of test scores over all of the sites. Lotterhos et al. (2017) used this metric and a variant replacing the vector (**t**) of test statistics with a vector whose elements were based on the ranks of the p values for a given site (along the lines of Randhawa et al. 2014). They then took the negative log of these rank-based p values as the elements of **t** for the Mahalanobis distance. This approach goes by the compact name of **Mahalanobis distance based on negative-log rank-based p-values**, or ***Md-rank-P***. They found that this approach worked the best of the composite measures they tested, followed by *DCMS*.

While seemingly logical, there are a number of subtle concerns with these composite approaches. First, although the tests reviewed in Table 9.1 highlight different features of the site-frequency spectrum, they are generally still correlated (Fu 1997; Achaz 2009). Hence, when choosing a region because of an abnormal D value, one might expect to find other abnormal site-frequency values as well, even if the region is neutral. This also holds for other types of tests, such as parametric sweep–based approaches (*CLR, Sweepfinder*) and haplotype-based tests (discussed below). When a region is ascertained by having an unusual test statistic value, this skews the distribution of other tests as well. For example, a region of low recombination can amplify random departures from the neutral equilibrium model. This point was stressed by O'Reilly et al. (2008), who noted that loci with significant selection tests scores in humans are disproportionately found in regions of low recombination. On the one hand, this makes sense, as regions with low c values are expected to have stronger signals from sweeps (Chapter 8). However, on the other hand, regions of low recombination create longer correlations among adjacent sites as well, so that an extreme discordance of a local coalescence from the neutral equilibrium model extends over a larger region. This is one reason to be very cautious of tests that look for localized runs of a particular statistic. For example, consider the **continuous regions of Tajima's D reduction** (or ***CRTR***) test (Carlson et al. 2005). While a run of sites with negative D values is expected under a sweep, it is also expected around an unlikely—but not exceptional—neutral genealogy in a region of low recombination.

The strongest preponderance of evidentiary support comes from *completely independent* tests, such as site-frequency data from one population coupled with an abnormal F_{ST} value for that site between populations. Even in these cases, the skeptic can suggest that most of the signal is coming from an unusual event in a single population, but an event that could be an outlier from a neutral drift process. For example, if one catches a surfing allele in one population, it has the potential for generating a number of selection-like signals and will also give a large F_{ST} value relative to other populations where it has not surfed.

Recombination Makes Site-frequency Tests Conservative

A final comment on frequency-spectrum tests is that, ignoring demographic concerns, they are likely conservative in many settings. In particular, Wall (1999) noted that site-frequency spectrum tests all assume that there is no recombination within the region of interest. While recombination does not bias the expected values for various statistics, it does *reduce* their variances (Rozas et al. 1999: Wall 1999), as the observed values represent the average across several genealogies (Depaulis et al. 2003). As a result, when recombination *does* occur within a region, tests are *conservative*, with the true p value being smaller than the zero-recombination values tabulated by the original authors of the various tests. As a result of this conservative nature of SFS tests under recombination, they are often significantly *underpowered*, using more stringent critical values than necessary. Wall found this effect to be significant when the rate of recombination is on the order of the total regional mutation rate, as is often the case (Table 4.1). Coalescent simulations allowing for recombination can significantly improve the power of tests by obtaining more accurate p values. As discussed in Chapter 4, the four-gamete test (Hudson and Kaplan 1985) can be used to detect recombination in the coalescence history of the sample, and the R_M statistic suggested by these authors estimates the minimal number of recombinants in the sample, which can then be incorporated into an appropriate coalescent simulation (e.g., Depaulis et al. 2005).

HAPLOTYPE-BASED TESTS

While powerful in some settings, the site-frequency spectrum does not contain all the information in a sample of sequences, as it ignores their **haplotype structure**—the nature of the association (linkage disequilibrium, or LD), among segregating sites. Treating distinct haplotypes as distinct alleles moves us from an infinite-sites model of sequence analysis to an infinite-alleles framework (Chapter 2, especially Figure 2.9). It is important to note that tests based on haplotype information tend to be among the most powerful means of detecting an ongoing sweep. Before examining such tests, we first address the central question of just how one defines a haplotype.

Defining and Inferring Haplotypes

If one considers a sufficiently long stretch of DNA, every sequence is a unique haplotype, so just how are haplotypes defined? The answer depends on both the test being used and the features of LD that are of interest. If we are interested in number and diversity of haplotypes in an infinite-alleles framework, the unit of analysis is a sufficiently small region, ideally with no recombination observed in the sample. The four-gamete test of Hudson and Kaplan (1985) can be used to detect recombination in the sample (Chapter 4), helping to define the size of a region (for example, by setting the size of a sliding window moving through a larger region). Practically, one may be constrained to find regions with sufficient haplotype diversity for analysis given either the marker density or background levels of variation, so that small amounts of recombination within the defined region may appear in the sample. For tests based on the average pairwise disequilibrium among all sites within a region, one actually wants some (but not too much) recombination. Finally, tests based on long haplotypes require a **core haplotype** (either a single SNP or a set of a few tightly linked SNPs) to define distinct allelic classes, with the disequilibrium patterns within each class (i.e., as one moves away from the core) forming the basis of tests. Again, recombination (outside of the core) is critical to these tests.

Determining haplotypes requires *phased genotypes*, which are not required for analysis under an infinite-sites model. For example, an *AaBb* individual is segregating at both sites in an infinite-sites analysis and no data on phase are needed. Conversely, for a haplotype (infinite-alleles) analysis, we need to determine if this individual is composed of either *AB, ab* or *Ab, aB* chromosomes. How, then, are haplotypes inferred? In the simplest case, one has haploid sequence data, which can include X chromosome data from males (and Z chromosome data from species with heterogametic females), or mitochondrial or chloroplast sequences. One can also have effectively haploid data, such as sequences from fully inbred lines (Example 9.13). Often, however, haplotypes have to be *inferred* from sequence data. The ideal setting for phasing genotypes is **trio** data—both parents and their offspring—but such data are not independent, giving an effective number of sequences in the analysis that is less than the actual number of sequences. More generally, haplotypes for unrelated individuals are inferred from unphased data by a variety of methods (reviewed in Stephens and Scheet 2005), the most popular being PHASE and its descendants (Stephens et al. 2001; Stephens and Scheet 2005). Surprisingly little discussion or analysis appears in the literature as to whether these reconstructions are biased by selection or other demographic departures. Given this concern, it is reasonable for one to feel a little uneasy when using inferred haplotypes in tests of selection.

Overview of Haplotype-based Tests

As reviewed in Table 9.2, a number of haplotype features can be used as the basis for tests of ongoing selection. **Strong haplotype structure** occurs when there are fewer haplotypes than expected given the number, S, of segregating sites within a region. This *underdispersion* of haplotypes is a signature of excessive LD within a region. Strong haplotype structure also results in a deficiency in **haplotype diversity**, H (the probability that two random haplotypes from the sample are different, analogous to Π under the infinite-sites model), and an excess of high-frequency haplotypes (roughly analogous to Fay and Wu's H test; Equation 9.27b). Such signatures are created by any process generating a coalescent with long internal branches (relative to the equilibrium neutral model; see Figure 8.3), such as a partial sweep (the favorable allele is not yet fixed), recovery from a moderate bottleneck, balancing selection, or population structure. Conversely, we can have the opposite pattern (*overdispersion* of haplotypes), with an excess of haplotypes, excess haplotype diversity, and an excess of rare-frequency haplotypes. Such signals are generated by a star-like coalescent genealogy, as would occur near the conclusion of a hard sweep, or the recovery from an extreme population bottleneck. However, in these overdispersed settings, LD summary statistics typically have low power, as S is small (most of the variation is removed), so that while haplotype overdispersion occurs, its signal is often weak.

Table 9.2 Haplotype-based signals of positive selection under different types of sweeps.

Completed or Nearly Completed Hard Sweep		
	Overdispersion of haplotype structure relative to S	
		Excess number of haplotypes
		Excess haplotype diversity
		Excess of low-frequency haplotypes
	LD structure	
		High LD on either side of selected site, little across site
Partial Sweep or Recent Balancing Selection		
	Strong haplotype structure	
		Deficiency in number of haplotypes
		Deficiency in haplotype diversity
		Excess of high-frequency haplotypes
	LD structure	
		Alleles with long haplotypes at excessive frequencies
	Allele age	
		Alleles with long haplotypes at excessive frequencies
Soft Sweep		
	Moderate haplotype structure	
		A few dominant haplotypes
	LD structure	
		High pairwise LD across entire region

Another classic LD signature of ongoing selection involves **long haplotypes**, regions of LD far longer than expected given the observed frequency of an allele. High-frequency alleles are (on average) *older* alleles under neutrality (Figure 2.3), and hence have experienced more recombination, resulting in shorter haplotypes. Finally, as developed in Chapter 8, there is a characteristic LD structure around a selected site following a completed sweep. For a soft sweep, an excess of pairwise LD is expected *throughout* this region, even when the site-frequency spectrum shows little change (Pennings and Hermisson 2006b). For a hard sweep, a different pattern is seen with strong LD between sites on the same side of the sweep, but no LD *across* the site (Figure 8.6).

Based on these different possible signatures (summarized in Table 9.2), we place haplotype-based tests into three loose categories. The first are based on the infinite-alleles model, such as the number of unique haplotypes and their frequency distribution within a sample. These are the analogs of site-frequency tests but now under an infinite-alleles framework, focusing on haplotypes instead of sites. The second class of tests utilizes summary statistics of all pairwise linkage disequilibria over sites within a region. The final class essentially uses linkage information to determine the age of an allele, either by looking at sequence variation within a haplotype (such as variability at tightly linked STRs) or by the decay of LD as one moves away from a core sequence. Table 9.3 (at the end of this section) summarizes the rather large number of tests based on these different strategies.

It bears emphasizing that haplotype (and LD) structure can provide signals of selection that are missed by site-frequency and ML-based hard-sweep tests, and thus offer more power in some settings, especially for the detection of partial and soft sweeps (Zeng 2007a; Ferrer-Admetlla et al. 2014). Age-of-allele tests (particularly in the form of detecting long haplotypes) are perhaps the most powerful approach for detecting an *ongoing* sweep, but usually have little to no power once the sweep is close to completion. Conversely, tests based on pairwise LD summary statistics offer significant power (albeit over a very short time window) for the detection of just-completed sweeps.

Example 9.13. Hudson et al. (1994) used a sample of 41 homozygous lines (making haplotypes easy to infer) of *Drosophila melanogaster* from California and Spain to survey variation at the superoxide dismutase (*Sod*) gene. For these data, neither Tajima's D or Fu and Li's D^* (both defined in Table 9.1) were significant. However, the haplotype data told a very different story. The authors classified the 41 sampled chromosomes into two classes: 19 slow and 22 fast, as judged by a fast/slow polymorphism for isozyme mobility. They found that all 19 slow chromosomes were identical in sequence through a 1410-bp region surrounding the fast/slow site, while the 22 fast chromosomes consisted of 10 different haplotypes within this region. They used coalescence simulations (with no recombination), conditional on the observed number of segregating sites and the sample size, to show that this is a significant decrease in variation of slow haplotypes relative to their frequency. This suggested that the slow allele experienced a recent, and rapid, expansion, as might occur under positive selection.

Other *Drosophila* examples where haplotype-based tests gave a strong signal, but site-frequency tests were not significant, include the work of Kirby and Stephan (1995), who found very strong haplotype structure at the *white* locus, but nonsignificant D and D^* tests. Andolfatto et al. (1999) examined a 1.4 kb region spanning the breakpoint of a naturally occurring chromosome inversion, also finding highly significant haplotype structure but a nonsignificant D value. Finally, Rozas et al. (2001) examined a 1.3 kb region around the ribosomal protein 49 (*rp49*) gene in *D. simulans*. Tajima's D, Fu and Li's D and F, and Fay and Wu's H (Table 9.1) were all nonsignificant, while a number of measures of haplotype structure (diversity, number of haplotypes, and the frequency of the most common haplotype) were all significantly different from neutral expectations in most populations.

The Ewens-Watterson Test

The first formal tests of selection in the molecular era were based on haplotypes (i.e., number of alleles), and their development proceeds as follows. Assuming that the region of analysis is sufficiently small such that no recombinants are expected in the sample, the simplest approach is to treat each distinct haplotype as a distinct allele. Following the notation from Chapter 2, let k be the number of alleles (distinct halotypes) in a sample of size n sequences, and n_i be the number of alleles in the sample present in exactly i copies. For example, if one allele is present as five copies, three alleles are each present as two copies, and four alleles are present as singletons, then $k = 8$, and the allele-frequency spectrum becomes $n_5 = 1, n_2 = 3, n_1 = 4$ (with all other $n_i = 0$), and $n = \sum i \cdot n_i = 15$. Equation 2.33b gives $\Pr(n_1, n_2, \cdots, n_k \,|\, n, k)$, namely, the expected frequency spectrum under neutrality given n and k. Ewens (1972) and Watterson (1977, 1978) proposed comparing the fit of the observed allele-frequency spectrum to the conditional distribution given the observed number of alleles, k.

Ewens suggested using the following summary statistic of the frequency spectrum,

$$I = -\sum_{i=1}^{n} n_i \left(\frac{i}{n}\right) \ln\left(\frac{i}{n}\right) \tag{9.30a}$$

His motivation for this statistic was its use as a general measure of dispersion (information) in the data. Watterson (1977, 1978) showed that the sample homozygosity

$$h = \sum_{i=1}^{n} n_i \left(\frac{i}{n}\right)^2 \tag{9.30b}$$

was a better choice for improved power to detect departures under weak overdominance (the selection model du jour of the time). Comparing the statistic given by Equation 9.30b with its value under the equilibrium neutral model is known as the **Ewens-Watterson test** (also the **Watterson test** or **homozygosity test**). Watterson proposed to assess significance by taking a large number of draws from Equation 2.33b (using the observed number, k, of alleles in the sample) to generate a null distribution of h values to compare against its value in the original sample. The same approach can also be used for the Ewens statistic (Equation 9.30a).

Advances in computational speed led Slatkin (1994, 1996) to propose an **exact Watterson test**, wherein one computes all possible h values over the set for a given value of n, as opposed to randomly sampling some number of draws from Equation 2.33b. This same approach is the basis for Fisher's exact test for contingency tables, and hence the name. The resulting value P_h is computed as

$$P_h = \sum_{\boldsymbol{n}^* \text{ such that } h(\eta^*) \leq h} \Pr(n_1, n_2, \cdots, n_k \,|\, n,\, k) \tag{9.31a}$$

namely, the sum over all *configurations* $\boldsymbol{n}^* = (n_1, \cdots n_k)$, constrained by $\sum i\, n_i = n$ (Equation 2.32), that give a value of h (Equation 9.30a) that is the same, or smaller than, the observed sample value of h.

Slatkin also suggested a second exact test, wherein one computes the probability over all possible configurations *directly*, as opposed to using the less informative summary statistic h, with

$$P_e = \sum_{\substack{\boldsymbol{n}^* \text{ such that} \\ \Pr(\boldsymbol{n}^* \,|\, n, k) \leq \Pr(\boldsymbol{n} \,|\, n, k)}} \Pr(n_1, n_2, \cdots, n_k \,|\, n, k) \tag{9.31b}$$

where the vector, $\boldsymbol{n}$, is the observed allele-frequency spectrum. The difference between P_e and P_h is that the sum is over a different set of $\boldsymbol{n}^*$. In Equation 9.31a, the sum is over those $\boldsymbol{n}^*$ that give smaller h values than observed in the sample, while in Equation 9.31b the sum is over those $\boldsymbol{n}^*$ values that give a smaller probability of a particular configuration than in the original sample. Slatkin found that the resulting p values for both tests are very similar for small n, but can be rather different for large n.

Zeng et al. (2007a) found that the Ewens-Watterson (***EW***) test was among the most powerful for detecting selection during the sweep phase, but Zhai et al. (2009) found that its power quickly falls off near fixation. Because the classic infinite-alleles model assumes the sequence has no recombination (at least in the sample), a reasonable concern is how robust the *EW* test is to recombination. Zeng et al. found that it is remarkably so, in part because the number of distinct alleles, k, contains information on local recombination rates, so that conditioning on k partly accommodates the impact from recombination.

Other Infinite-alleles Tests: Conditioning on $\widehat{\theta}$

Watterson-type tests use the *conditional* allele-frequency spectrum, where the observed number of alleles, k, is used in Equation 2.33b to generate the null distribution. What about tests based on k itself, such as whether there are too many, or too few, alleles based on some other diversity measure? Such tests use the sampling distributions given by either Equation 2.30a or Equation 2.33a, and require an estimate of θ. Fu (1996, 1997) used this approach to test whether a sample contains too many, or too few, alleles (haplotypes) relative to the neutral equilibrium model. His ***W* test** (1996) used the Ewens sampling formula (Equation 2.30a) with θ replaced by the Watterson estimator, $\widehat{\theta}_S$ (Equation 4.3a), and it returns the probability of seeing k (or fewer) alleles in the sample as

$$W = \Pr(K \leq k) = \sum_{i=1}^{k} \Pr(K = i \,|\, \widehat{\theta}_S, n) = \sum_{i=1}^{k} \frac{S_n^i \cdot [\,\widehat{\theta}_S\,]^i}{S_n(\widehat{\theta}_S)} \tag{9.32}$$

where S_n^i is the coefficient on $(\widehat{\theta}_S)^i$ in the polynomial

$$S_n(\widehat{\theta}_S) = \widehat{\theta}_S\,(\widehat{\theta}_S + 1)(\widehat{\theta}_S + 2) \cdots (\widehat{\theta}_S + n - 1)$$

These coefficients are called **Stirling numbers**, hence the S_n^i notation. This is a test for a *deficiency of rare alleles/haplotypes*, and hence is *one-sided*. Fu showed that the W test is more powerful that Tajima's D or Fu and Li's D^* and F^* tests (Table 9.1) for detecting signals of balancing selection or a structured population. Indeed, Strobeck (1987) proposed essentially the same test (using $\widehat{\theta}_\Pi$ in place of $\widehat{\theta}_S$) as a method for detecting population structure, rather than selection.

Fu's $\boldsymbol{F_S}$ **test** (1997) is the compliment of W, as it tests for an *excess of rare alleles/haplotypes*. It starts by computing the probability of seeing k *or more* alleles/haplotypes in a sample,

$$S' = \Pr(K \geq k) = \sum_{i=k}^{n} \frac{S_n^i \cdot [\,\widehat{\theta}_\Pi\,]^i}{S_n(\widehat{\theta}_\Pi)} \tag{9.33a}$$

but now using $\widehat{\theta}_\Pi$, the estimator of θ based on average number of pairwise differences (which is more sensitive to sites with intermediate allele frequencies). Fu noted that S' is not an optimal test statistic because its critical values are often too close to zero. Because of this, the test uses the transformation

$$F_S = \ln\left(\frac{S'}{1 - S'}\right) \tag{9.33b}$$

As with W, this is also a one-sided test. F_S is negative when there is an excess of rare alleles/haplotypes (as would occur with a selective sweep or population expansion), with a sufficiently large negative value serving as evidence for selection or population expansion. Fu (1997) showed that F_S is more powerful that Tajima's D and the Fu-Li D^* and F^* tests (Table 9.1) for detecting selective sweeps or population expansion following a bottleneck.

Other Infinite-alleles Tests: Conditioning on S

While elegant in using exact results from the allele-frequency sampling distributions, these tests for excessive k values *do not* return exact p values, as using an estimate, $\widehat{\theta}$, in place of the true value, θ, makes both Equation 2.30a and Equation 2.33a only approximations. Hudson et al. (1994) and Depaulis and Veuille (1998) noted that while θ is unknown, the number of segregating sites, S, is directly observed. Hence, one can generate coalescence genealogies (via simulation) and then randomly place the S segregating sites over them (at a rate proportional to the branch lengths within the coalescent), generating a distribution of haplotypes in the final sample. This procedure generates draws under the neutral equilibrium model conditioned on the observed number of segregating sites. In effect, these tests examine the sequence data from *both* the infinite-alleles and infinite-sites perspectives. While the number of alleles, k, is a sufficient statistic under the infinite-alleles model, S *is not*. Conditioning on S, the distribution still has a dependency on θ, although this is often weak (Griffiths 1982; Depaulis et al. 2001, 2005; Markovtsova et al. 2001; Wall and Hudson 2001; Innan et al. 2005).

A detailed analysis by Ramos-Onsins et al. (2007) examined the distributions associated with nine haplotype-based tests and compared those generated by conditioning on an observed S with those using a known value of θ. All distributions based on conditioning on S departed significantly from their corresponding distributions using a known θ. More critically, the departure was usually in the tails, where p values are obtained. As a result, Ramos-Onsins et al. recommended against using infinite-alleles methods that condition on S. Further, Zeng et al. (2007a) found that tests that are conditional on S suffer from low power (being overly conservative) in the presence of recombination and concluded that tests conditioning on the number of alleles, k, are more powerful than those conditioning on the number of segregating sites, S. Tests conditioned on k are more robust to recombination, because k includes some information on recombination (being a function of haplotypes), while S does not (being a function of individual sites). Despite these concerns, infinite-alleles tests based on conditioning on S appear widely in the literature.

Hudson et al. (1995) pioneered this approach of conditioning on S with their **haplotype test**, also referred to as the **haplotype partition** (***HP***) or Hudson's haplotype test (***HHT***). Its initial form (motivated by the observations discussed in Example 9.13) was a rather open-ended test: given a sample of n sequences with S segregating sites, what is the probability of observing j sequences with i or fewer alleles (for example, what is the chance of observing $j = 3$ alleles present as $i = 2$ copies or less, i.e., $n_2 + n_1 = 3$). This is akin to Slatkin's exact test (Equation 9.31b), concerning the likelihood of a given allele-frequency configuration. The difference is that Slatkin's test conditions on the observed number of alleles, k, while Hudson's test conditions on S. Hudson's test is typically implemented by asking if there is an excess of the most frequent haplotype (Depaulis et al. 2005; Innan et al. 2005). Suppose there are n sequences and the highest-frequency haplotype occurs j times. Hudson's test is

$$\Pr(n_i \geq 1 \text{ for } i \geq j \mid S)$$

where, as above, n_i is the number of alleles present as exactly i copies in the sample. Andolfatto et al. (1999) extended Hudson's test using a sliding window of variable size to scan the region of interest and developed a correction for multiple tests (the different windows). Again, hypothesis testing is done using the null generated from a coalescent with S segregating sites.

Depaulis and Veuille (1998) also used conditioning on S and developed two tests. Their **haplotype number**, or ***K*, test** is essentially Fu's W test (Equation 9.32), but using $\widehat{\theta}_S$ (and hence conditioning on S) rather than $\widehat{\theta}_\Pi$. Their **haplotype diversity**, or ***H*, test**, uses the statistic

$$H = 1 - \sum_{i=1}^{k} p_i^2 \quad \text{with} \quad p_i = \text{frequency of the } i\text{th haplotype} \tag{9.34a}$$

namely, the haplotype heterozygosity, which is compared to its expected neutral equilibrium value given S. A comparison with Equation 9.30b shows that the H test is essentially the Ewens-Watterson test, but with its significance assessed by conditioning on S rather than k. Note that the range on H is

$$\frac{2(n-1)}{n^2} \leq H \leq 1 - \frac{1}{n} \tag{9.34b}$$

with the lower bound set by the sample consisting of just two haplotypes, one with $n - 1$ copies and the other a singleton ($n_{n-1} = 1, n_1 = 1$), while the upper range is set by all of the haplotypes being present as singletons ($n_1 = n$). Critical values for these statistics (conditioned on n and S) generated from coalescent simulations were tabulated by Depaulis and Veuille (1998).

Finally, Innan et al. (2005) proposed a **haplotype configuration test** (***HCT***), based on the configuration of the haplotype (allele) frequency spectrum. Again, this is a version of Slatkin's exact test (Equation 9.31b), but now the conditioning is on S (as opposed to k),

$$P_E = \sum_{\substack{\boldsymbol{n}^* \text{ such that} \\ \Pr(\eta^* \mid n, k) \leq \Pr(\eta \mid n, S)}} \Pr(n_1, n_2, \cdots, n_k \mid S) \tag{9.34c}$$

These probabilities can easily be generated using the constant-S coalescent simulation approach discussed above. Depaulis et al. (2005) and Innan et al. (2005) discussed haplotype-frequency spectrum approaches in greater detail, while power and bias issues were examined by Ramos-Onsins and Rozas (2002), Depaulis et al. (2003, 2005), and Ramos-Onsins et al. (2007).

Other approaches based on haplotype number have also been suggested, although more as heuristics and summary statistics than as formal tests. Przeworski (2002) suggested standardizing the number of haplotypes by the number of segregating sites, using $k' = k/(S+1)$. A smaller k' implies higher LD, as specific combinations of segregating sites are locked into a small number of haplotypes. Przeworski noted that while k' tends to decrease as one approaches a selected site *during* an ongoing hard sweep, k' can actually be greater than its expectation under the equilibrium neutral model *after* a sweep is completed. This occurs because high-frequency variants are fixed and new mutations arise, most of which are singletons and hence form a large collection of unique haplotypes. She noted that excluding singletons when computing both k and S gives k' much more stability, with the adjusted k' sharply decreased at the completion of a hard sweep and then increasing back to its neutral expectation in $\sim N_e$ generations after the sweep.

Garud et al.'s H_{12} and H_2 Tests

A number of tests are built around **haplotype homozygosity** (*HH*), the probability that two randomly chosen haplotypes are identical. This is given by the complement of the Depaulis-Veuille H (haplotype heterozygosity) statistic (Equation 9.34a),

$$H_1 = 1 - H = \sum_{i=1}^{k} p_i^2 \tag{9.35a}$$

where p_i is the frequency of the ith haplotype in the sample. To adjust for sampling, some variants of this statistic replace p_i^2 with $[p_i + (1/k)]^2$, where k is the number of haplotypes (e.g., Kemper et al. 2014). As mentioned in Chapter 8, Garud et al. (2015) showed that a simple modification of this statistic results in a test that can detect *both* hard and soft sweeps. Their $\boldsymbol{H_{12}}$ test statistic combines the two largest haplotype classes into a single one,

$$H_{12} = (p_1 + p_2)^2 + \sum_{i>3} p_i^2 = H_1 + 2p_1p_2 \tag{9.35b}$$

The logic is that a soft sweep results in not one, but several, dominant haplotypes. If the sweep is not too soft, then the first two haplotypes, both presumably harboring the favored allele, will together comprise most of the haplotype variation. In the case of a hard sweep, the second-most frequent haplotype will be sufficiently rare that $H_{12} \simeq H_1$. The authors applied this approach to *Drosophila*, looking at windows with a fixed number of SNPs and adjusting for the local recombination rate and then used coalescent simulations to generate values under the null of neutrality.

Garud et al. considered a second modified *HH* statistic, namely, the homozygosity with the largest class removed

$$H_2 = \sum_{i>1} p_i^2 \tag{9.35c}$$

Under a hard sweep with its single dominant haplotype, H_2 should be considerably smaller than H_1, while under a soft sweep the drop-off in value from H_1 to H_2 should be much less dramatic. Based on this observation, the ratio H_2/H_1 forms the basis of a test as to whether a detected sweep is hard or soft, with moderate values suggesting soft sweeps and very small values suggesting hard sweeps (Garud et al. 2015; Garud and Rosenberg 2015).

Recombination and Infinite-alleles-based Tests

What is the effect of recombination on these various infinite-alleles-based tests? Recall that recombination had a uniform effect on the interpretation of site-frequency spectrum tests, making all SFS tests conservative, and hence underpowered. This is *not* the case for haplotype-based tests. Recombination creates new alleles, which inflates the number of haplotypes, the diversity of haplotypes, and the number of rare haplotypes, while decreasing the frequencies of the most common haplotypes. Thus, depending on the test, recombination can either make a test conservative, with p values being too large, or anticonservative, with p values being too small (Depaulis et al. 2001; Rozas et al. 2001; Wall and Hudson 2001; Depaulis et al. 2005; Ramos-Onsins et al. 2007; Zeng et al. 2007a). Tests for a low number of haplotypes, low haplotype diversity, and excessive high frequency of the most common haplotype are all conservative under recombination. Tests for excess rare haplotypes, excess haplotype diversity, and excess haplotype number all have their p values reduced by recombination (making these tests anticonservative). Finally (as mentioned above), the Ewens-Watterson test (Equation 9.30b) is relatively robust to recombination, as conditioning on the number of alleles, k, partly accounts for recombination.

Incorporating recombination into coalescent simulations significantly improves the power for conservative tests and creates more accurate p values for anticonservative tests. However, using incorrect recombination values can significantly bias a test. Under the coalescent framework, recombination is measured by $4N_e c$, the population-scaled rate (Chapter 4), with c as the recombination frequency for the size of haplotypes being considered. Depaulis et al. (2005) offered the following suggestion. First, estimate Hudson's minimal number of recombinants, R_M, in the sample from the four-gamete test (Chapter 4), and then choose $4N_e c$ in the coalescent simulations as the value that gives 5% (or less) of the samples showing R_M or more recombinations.

Pairwise Disequilibrium Tests: Kelly's Z_{nS} and Kim and Nielsen's ω_{max}

Positive selection can produce two very different patterns of disequilibrium around a site. For soft and partial sweeps, there is an excess of LD *throughout* a region, while for completed hard sweeps, LD is found on *either side* of the selected region, but not *through it* (Chapter 8). To test for these patterns, two different averages of pairwise disequilibrium within a region have been proposed. Both start with Hill and Robertson's (1968) scaled measure of the disequilibrium, which is expressed as a correlation coefficient

$$r_{ij}^2 = \frac{D_{ij}^2}{p_i(1-p_i)q_i(1-q_i)} \tag{9.36a}$$

where D_{ij} is the disequilibrium between sites i and j, and p_i and q_i are the frequencies of leading alleles at the two sites. For a general measure of the average amount of LD throughout a region with S segregating sites, Kelly (1997) proposed using the average over all of the pairwise squared correlations

$$Z_{nS} = \frac{2}{S(S-1)} \sum_{i=1}^{S-1} \sum_{j=i+1}^{S} r_{ij}^2 \tag{9.36b}$$

This is often computed over windows of various sizes, so that (for example) Z_{n5} and Z_{n8} denote values for windows with five and eight segregating sites, respectively. Kelly showed that values of Z_{nS} are largely determined by the final coalescent time in the sample (the time for the last two lineages to coalesce into the ancestral lineage for the entire sample). The longer this time, the larger the value of Z_{nS}. This statistic is smallest under a star genealogy, as most of the coalescence events will have occurred at roughly the same time (i.e., the nodes of the genealogical tree are extremely compressed).

Thus, a small value of Z_{nS} is consistent with a hard sweep or an extreme bottleneck. In such cases, there is usually a reduced amount of site polymorphism, which in turn reduces the power of Z_{nS}. Conversely, with a partial or soft sweep, Z_{nS} increases. Critical values of Z_{nS} are determined by coalescent simulations that are conditional on S (as discussed above). When recombination is ignored, the one-sided test of excessive Z_{nS} values is conservative (as recombination lowers Z_{nS}), while the test of reduced Z_{nS} values is anticonservative. Pennings and Hermisson (2006b) suggested that LD-based tests, such as Z_{nS}, may have the most power to detect a *very recent* soft sweep.

Kim and Nielsen (2004) proposed a different measure of pairwise disequilibrium designed for the expected pattern of LD following a hard sweep (Figure 8.6). Here, there is LD on either side, but not *across* the site (this LD signal dissipates rather quickly, roughly on the order $\sim 0.1N_e$ generations immediately following a sweep; Przeworski 2002; Jensen et al. 2007). Based on this disjoint LD pattern, Kim and Nielsen proposed a test statistic, ω, comparing LD within, versus across, the left and right sides of a sliding window. Suppose there are ℓ sites $(1, \cdots, \ell)$ on the left (L) side of the putative selected region and $S-\ell$ (sites

$\ell+1,\cdots,S$) on the right (R). Define

$$\omega = C_{S,\ell}\,\frac{\sum_{i,j\in L} r_{ij}^2 + \sum_{i,j\in R} r_{ij}^2}{\sum_{i\in L,j\in R} r_{ij}^2}, \qquad C_{S,\ell} = \frac{\ell(S-\ell)}{\binom{\ell}{2} + \binom{S-\ell}{2}} \tag{9.37}$$

where the combinatorial term, $C_{S,\ell}$, is a function of the number of sites contrasted over the three comparisons (within L, within R, and between L and R). Under the distinct signal of LD from a completed hard sweep, one expects a large value for the numerator (strong LD within either side) and small denominator value (little LD across sides) around a sweep, giving a large value of ω. Because r^2 is sensitive to small allele frequencies, polymorphic sites showing singletons are best ignored when computing ω. Kim and Nielsen's test statistic ω_{max} (which gives rise to the name of their test) is the maximum value of Equation 9.37, a function of both window size and window position, with critical values determined via a coalescent simulation. Pavlidis et al. (2010) offered improvements that allow this test to be scaled up to scan an entire genome efficiently, which are implemented in their **OmegaPlus** package (Alachiotis et al. 2012).

Jensen et al. (2007) found that this approach had promise for a very vexing situation: detecting sweeps that start in nonequilibrium populations (i.e., those having experienced a recent change in population size). Indeed, Crisci et al. (2013) found that OmegaPlus outperformed SFS-based (*SweepFinder*, *SweepD*) and long-haplotype (iHS; see below) tests under both equilibrium and nonequilibrium settings in detecting an ongoing sweep.

Jensen et al. found that population structure does not cause the distinctive hard-sweep LD pattern detected by ω, but that a very strong bottleneck (1% of the population surviving) can give modest ω_{max} scores, and hence some chance of a false-sweep signal. The effects of recombination on this test (i.e., recombination within the test window) are a bit unclear. On the one hand, by reducing r_{ij}^2 values within each side, recombination should be conservative. Conversely, by also reducing r_{ij}^2 values between sides under the null, this can inflate p values, making the test anticonservative. Kim and Nielsen found that while assuming the incorrect value for the scaled recombination rate, $4N_ec$, in coalescent simulations for the null decreases power, ω_{max} is more robust to incorrect $4N_ec$ values than Z_{nS}.

To summarize these two LD-based test statistics, Z_{nS} has the power to detect partial sweeps and recently completed soft sweeps, but poor power to detect recently completed hard sweeps. Conversely, ω_{max} has good power to detect a recently completed hard sweep and also has robustness against many demographic concerns, although a severe bottleneck can also generate modest genome-wide ω_{max} scores. While most LD/haplotype tests have more power *during* a sweep, these two statistics have power only during a very short time window ($\sim 0.1N_e$ generations) immediately *following* a successful sweep. A final concern is that even rare amounts of gene conversion involving the selected site can significantly distort any disequilibrium-based signature (Andolfatto and Nordborg 1998; Jones and Wakeley 2008).

Contrasting Allele-frequency vs. Intra-allelic Variation Estimates of Haplotype Age

As discussed above, we define an allele by either a single SNP or a set of SNPs sufficiently close together that no recombination has occurred between them in our sample. Moving outward from this core, we can examine the structure of haplotypes for a given allele in progressively longer regions, which we loosely refer to as the haplotypes for a given allele. The pattern of variation among the haplotypes associated with a single allele provides information on its age.

Assuming that an allele arose as a single mutation, it initially was on a single haplotype background in complete LD with tightly linked markers. As the allele ages, the fraction of copies associated with the original background decays through recombination. Likewise, new mutations at tightly linked sites arise, with the number of segregating SNPs (and copy-number variation at STRs) both increasing with age. Hence, the diversity of haplotypes associated with that allele provides information about its age (reviewed by Slatkin and Rannala 1997, 2000). A common approach (especially in human genetics) is to contrast one (or more) of these **intra-allelic variation** estimates of age with the estimate of age based on allele frequency (e.g., Example 2.3). Figure 2.3 shows that, under the equilibrium neutral model, ***a common allele is an old allele***. As such, there is time for mutation and recombination to act, generating a more diverse collection of haplotypes associated with a particular SNP variant. While a few formal tests have been proposed to determine whether there is too little (directional selection) or too much (balancing selection) intra-allelic variation (Slatkin 2000, 2008; Slatkin and Bertorelle 2001), our goal here is to review the different age estimators, as discrepancies between them are often offered as evidence supporting selection. As with many tests, allelic surfing can result in neutral alleles of a young age that are common at the edges of an expanding population, generating false positives in allelic-age tests.

While Equation 2.12 gave a simple expression for the expected age of an allele as a function of its frequency, it is a bit misleading, due to its very large variance. A better estimator follows from Slatkin and Rannala (2000), based on their approximation for the probability distribution of age as a function of frequency under the equilibrium neutral model (also see Watterson 1976; Griffiths and Tavare 1998; Griffiths 2003). Letting T denote time scaled in $2N_e$ generations, then for an allele whose frequency is p in a sample of size n, the probability that the true age, T, is less than τ (namely, no older than $2N_e\tau$ generations) is given by

$$\Pr(T \leq \tau) \simeq (1-p)^{-1+n/(1+n\tau/2)} \tag{9.38a}$$

This is simply Equation 2.13, now expressed in units of $2N_e$ generations, which is a more natural time scale for drift (being the expected coalescent time for two neutral alleles). Taking the derivative with respect to τ recovers the probability density function, and hence the likelihood function (LW Appendix 4). The resulting distribution is very skewed, with a long heavy tail toward increased age. As a result, the mode of the distribution (the maximum likelihood estimator or MLE; LW Appendix 4) is less, and usually substantially so, than the mean. In particular, the MLE (in units of $2N_e$ generations) for T becomes

$$\text{MLE}(T) = -\ln(1-p) - \frac{2}{n} \tag{9.38b}$$

Unlike Equation 2.12, the MLE accounts for the sample size, n, used to estimate p. A $100(1-\alpha)\%$ confidence interval for age is given by $(\tau_{\alpha/2}, \tau_{1-\alpha/2})$, where τ_x is the solution in Equation 9.38a of

$$\Pr(T \leq \tau_x) = x \tag{9.38c}$$

For large n, $n/(1+n\tau_x/2) \simeq 2/\tau_x$, reducing Equation 9.38a to

$$x \simeq (1-p)^{-1+2/\tau_x}$$

which yields a solution to Equation 9.38c as

$$\tau_x \simeq \frac{2}{1+\ln(x)/\ln(1-p)} \tag{9.38d}$$

This equation can be used to approximate confidence intervals when n is large. When n or x are small, it serves as a starting value for numerically solving Equation 9.38c (for example, one can plot Equation 9.38a as a function of τ around the value give by Equation 9.38d to visually search where it equals x). The weakness of this estimator is that it is *extremely* sensitive to demography (Slaktin 2000; Slaktin and Rannala 2000).

Example 9.14. Recall Example 2.3, where we showed (following Stephens et al. 1998) that the estimated age of the HIV-resistant *CCR5-δ32* mutation (roughly 5100 generations from Equation 2.12) was incompatible with an independent estimate of its age based on intra-allelic variation (roughly 28 generations). Is this conclusion changed if Equation 9.38b and 9.38d are used instead of Equation 2.12 to estimate age from frequency? In Example 2.3, we used $p = 0.10$ and $N_e = 5000$. Taking $n = 4000$ as the approximate sample size of Europeans used to estimate the allele frequency, Equation 9.38b gives the MLE as

$$\text{MLE}(t) = -2N_e \left(\ln(1-p) + \frac{2}{n} \right) = -10,000 \left(\ln(0.9) + \frac{2}{4000} \right) = 1048 \text{ generations}$$

One reason for this discrepancy relative to the estimate of 5100 generations from Equation 2.12 is that the latter uses the mean, while Equation 9.38b uses the mode, of the allelic-age distribution. The long tail of the age distribution significantly skews the mean to much higher values, due to rare events associated with very large ages.

While the MLE-based estimator is much smaller than that given by Equation 2.12, does its 95% confidence interval still exclude the variation-based estimate of ~30 generations? Values of $\tau_{0.025}$ and $\tau_{0.975}$ are obtained by solving Equation 9.38c. Using the approximation offered by Equation 9.38d yields starting values (in units of $2N_e$ generations) of 0.055 and 1.61. Plotting Equation 9.38c around these values gives the exact answers, $\tau_{0.025} = 0.067$ and $\tau_{0.975} = 1.61$, which translates into 670 to 16,100 generations ($2N_e\tau_{0.025}$ and $2N_e\tau_{0.975}$, respectively). Despite this very wide range, the lower value still greatly exceeds 28 generations, meaning that the frequency of this allele is not consistent with an equilibrium neutral model.

Novembre et al. (2005) examined spatial models of the spread of this allele in Europe, showing that both strong selection and long-distance dispersal are required to account for its current geographic distribution. The agent of selection is thought to be smallpox (reviewed by de Silva and Stumpf 2004; Stumpf and Wilkinson-Herbots 2004; Galvani and Novembre 2005). However, as is often the case with claims of selection, the evidence may not be as definitive as the previous analysis suggests. Sabeti et al. (2005), using refined recombination estimates, obtained an allelic-age estimate of around 280 generations with a 95% confidence interval of 116 to 630. This is consistent with the observation that this allele is found in 2900-year-old skeletal remains (roughly 120 generations ago) from different locations in southern Europe. The lower end of our previous confidence interval (670) comes fairly close to Sabeti et al.'s (2005) upper estimate of 630 generations. Hence, the *CCR5-δ32* mutation may be significantly older than suggested from the Stephens et al. (1998) analysis. Further, Sabeti et al. found that this allele exhibited none of the classic long-haplotype, or other, features expected from a partial sweep (as would be generated by a selected allele increasing in frequency).

As Example 9.14 shows, potential evidence for selection is offered when a frequency-based estimator of age (which assumes a pure-drift model) is too large relative to an estimate based on within-haplotype variation. Because these comparisons involve *segregating* alleles, they are designed to detect sweeps that are currently underway (ongoing selection) as well as recent partial sweeps (where selection in the past has swept an allele up to near its current frequency, but is either now much weaker or no longer occurring). Intra-allelic age estimators are also important in their own right, as the age of an allele currently under selection, or the time since a sweep has completed, is of interest. We briefly consider three haplotype-based estimators of allelic age here: (i) the persistence of disequilibrium with a marker at a known recombination fraction, c; (ii) the number of segregating sites within a very closely linked region; and (iii) the variance at STRs. The last two measures are assumed to be scored in a sufficiently close region around the allele that recombination can be ignored. The major caveat with any of these approaches is that intra-allelic age estimates generally tend to have a downward bias, making alleles appear too young (Slatkin and Rannala 2000). Given that most contrasts of frequency and intra-allelic variation invoke selection when the frequency estimate is much older than the variation-based estimator, ***caution is always in order when using this approach***.

The technique of estimating age from the persistence of LD between an allele and a linked marker (Example 2.3) was first proposed by Serre et al. (1990). Again, we assume there is no selection. Let A denote a marker allele at a site closely linked to the target allele, B, that we are trying to age. The marker allele is typically chosen because the target allele is overrepresented on A-bearing haplotypes (relative to its population frequency), suggesting that the initial mutation arose on such a background (i.e., an AB haplotype). Let $x(t)$ denote the current frequency of B-bearing haplotypes that also contain the marker allele A. If we assume there is a single-origin mutation that occurred on an A background, then $x(0) = 1$. If we let y be the frequency of haplotypes carrying A in the ancestral population, we further assume that this ancestral value is close to the present-day value (i.e., we assume that $t \ll 1/[2N_e]$). The resulting decay in LD by recombination is

$$x(t) - y = (1-c)^t [\,x(0) - y] = (1-c)^t (1-y) \tag{9.39a}$$

This follows because $x(0) - y$ is the initial disequilibrium, which decays at a rate of $(1 - c)$ per generation (note that this is just a modification of Equation 8.2a). Solving for t yields

$$t = \frac{1}{\ln(1-c)} \ln\left(\frac{x-y}{1-y}\right) \tag{9.39b}$$

where x is the current frequency of AB haplotypes. This approach is very closely related to LD mapping (LW Chapter 14), where in the latter, one assumes that t is known and solves for c via Equation 9.39a. Equation 9.39b is a slightly more accurate approach for estimating age than was used in Example 2.3, as it accounts for the population-wide frequency, y, of the marker allele, A. Risch et al. (1995) and McPeek and Strahs (1999) discussed how information from multiple marker loci linked to the target site can be simultaneously used to estimate t. The delicate issue in applying Equation 9.39b is that it is sensitive to c. Using a value of c less than the true value overestimates the age, while the age will be underestimated if too large a c value is used.

A second approach to estimate the age of an allele using intra-allelic diversity is based on the number of segregating sites in a tightly linked region around the target allele (Example 8.5). This approach is typically used to estimate the age of a sweep (as opposed to, e.g., Equation 9.39b, which provides an age estimate for a contrast to a frequency-based age estimator). Under a hard selective sweep (and hence an approximate star genealogy for the sampled sequences), Equation 8.14 provides a rough estimate of age as $t = S/\mu n$, where n is the sample size, S is the number of segregating sites, and μ is the total mutation rate for the region being considered. Alternatively, one could use the fact that S follows a Poisson distribution to obtain both an ML estimator and confidence intervals (Example 9.15). Examples 8.5 and 9.15 both assume there is a very defined genealogy, which in reality is unknown, introducing considerable noise (and possible bias) into these estimators. In particular, the assumption of a star genealogy creates a process with too little variation, and hence confidence intervals that are too narrow. As emphasized by Slatkin and Rannala (2000), there is considerable variation in a particular genealogy *even* under an equilibrium neutral model.

Example 9.15. Meiklejohn et al. (2004) observed three segregating sites ($S = 3$) in a population sample of $n = 26$ sequences from a region spanning the *janB* locus in *Drosophila simulans*. They assumed that the total mutation rate for this region was $\mu = 1.73 \times 10^{-5}$ per year. For a mutation arising t years ago, the expected number of mutations on a single sampled allele is μt. Under the approximation that the coalescence has a star-shaped structure (very little shared coalescence time among sampled alleles, as would be expected in a sweep or partial sweep), each of the n lineages in the sample is essentially independent of the others, giving the expected number of segregating sites in a sample of n alleles as $\lambda = n\mu t$, with the distribution of observed S values within the sample following a Poisson distribution with a parameter of λ. The resulting MLE for t given these data is simply the value that maximizes the likelihood, given $S = 3$,

$$\Pr(S = 3 \,|\, t\,\mu) = \frac{\lambda^3}{3!} \exp(-\lambda) \quad \text{where} \quad \lambda = n\mu t = 0.0004498\, t$$

Plotting this expression as a function of t shows that the maximum of this expression (and hence the MLE) occurs at t = 6667 years (which is essentially identical to the simpler estimate given by Equation 8.14). The 95% confidence interval follows from the Poisson distribution, by finding those values of $t_{0.025}$ and $t_{0.975}$ such that $\Pr(S \leq 3 \mid t_{0.025}\, \mu) = 0.025$ and $\Pr(S \leq 3 \mid t_{0.975}\, \mu) = 0.975$, where (from the Poisson)

$$\Pr(S \leq 3 \mid \lambda) = \exp(-\lambda)\left(1 + \lambda + \frac{\lambda^2}{2!} + \frac{\lambda^3}{3!}\right)$$

Numerically solving this equation yields $\Pr(S \leq 3 \mid \lambda = 8.77) = 0.025$ and $\Pr(S \leq 3 \mid \lambda = 1.09) = 0.975$. Because $\lambda = 8.77 = 26 \cdot (1.73 \times 10^{-5}) \cdot t$, the upper limit of t becomes $8.77/[26 \cdot (1.73 \times 10^{-5})]$ ~19,500 years. Similarly, the lower limit is ~1400 years.

The copy-number variance at an STR also provides an estimator of age. As above, a new mutation arises on a haplotype carrying a specific allele (whose state is defined by repeat copy length) at a linked STR. As that haplotype spreads and ages, variants arise at the STR, generating variance in copy number at the STR for the haplotype carrying the mutation. As noted previously, using copy-number variance to estimate the age is very model dependent, especially given the unknown structure of STR mutation. Assuming a simple stepwise mutation scheme with a mutation rate that is constant over array size, Thomas et al. (1998) noted that the average squared difference between copy number over all sampled haplotypes and the ancestral haplotype has an expectation of μt, where μ is the microsatellite mutation rate. Much more sophisticated analyses can also be used (e.g., Wilson et al. 2003). Given their high mutation rates, STR variation can allow for more precise estimation of younger alleles than approaches using segregating sites or recombination with very tightly linked sites. The major caveat is that using the wrong mutational model can yield biased estimates (as rates can change with copy number and single mutations may result in copy-number changes greater than one).

Finally, we have seen examples in Chapter 8 where the insertion of a mobile element generated an adaptive mutation (most likely due to regulatory changes). Mobile elements provide a unique age-based test for adaptation, as one can date the insertion (and hence the age of the new mutation it creates). A typical element is several kilobases long, and its age of insertion can be estimated by looking at two features. Some elements generate direct or inverted flanking repeats of known length. While typically just a few bases long, divergence between these short repeat segments can be used to date an insertion. Much more information is provided by the sequence divergence of the inserted element itself, which has a mutational target size of several kb . One approach is to compare the insertion against a consensus sequence for active elements. While not fool-proof (there can be existing divergence in active elements and the insertion event itself might be error-prone, such as in retrotransposons), this does provide an approach to age the insertion. See Blumenstiel et al. (2014) for details.

Long-haplotype Tests Using Within-population Data: *rEHH, LDD, iHS, nS*$_L$*, SDS,* and *DIND*

The currently most popular class of allele-age-based tests are those that search for alleles that are at moderate to high frequency and with long haplotypes. Under the neutral model, a common allele is an old allele and hence has experienced many generations of recombination. Long haplotypes for strictly neutral alleles are only expected when either the allele is very young or when the recombination rate is greatly suppressed.

Recall our previous discussion on the definition of an allele, namely a core SNP or set of very tightly linked SNPs that define alternate classes. For alleles defined by a single biallelic SNP, this generates two classes (sequences carrying the alternative SNP alleles). The haplotype structure within each allelic class is examined by looking at shared variants

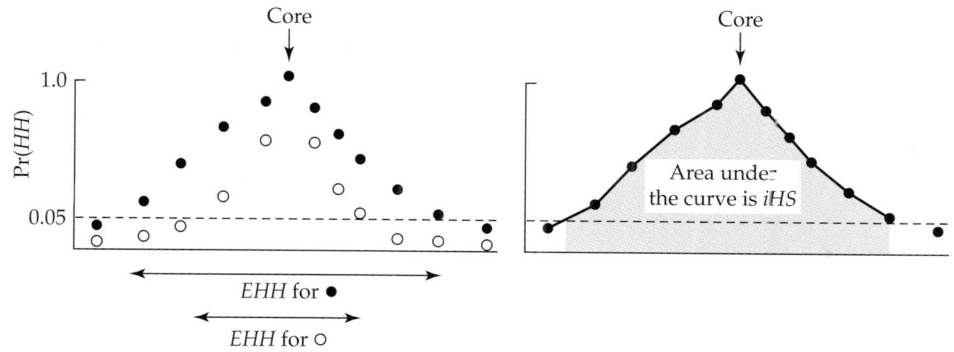

Figure 9.6 Haplotype homozygosity (*HH*) is defined as the probability that two randomly chosen chromosomes containing the same core SNP variant (used to define allelic classes) are identical (homozygous) at all markers within a defined window. In the figure, *HH* is computed at a series of SNP markers moving away from the core (allelic-defining) SNP. The open and filled circles correspond to the *HH* values at a given SNP in the two allelic classes, namely, the probability that random draws of chromosomes from the same allelic class are identical within the region between the core SNP and the marker SNP. The relationship between *HH* and distance from the core is usually summarized using one of two statistics. (**Left**) The extended haplotype homozygosity (*EHH*) for an allelic class is the length of the region around the core where the *HH* value is $\geq 5\%$ (above the dashed line). The allele corresponding to the filled circles has a larger *EHH* value, and thus a longer haplotype. (**Right**) A potentially more informative measure is given by the **integrated EHH score**, *iHS*, the total area under the *HH* curve over the region spanned by the *EHH* for that allele. For ease of presentation, only the values corresponding to the allele with the larger *EHH* value (filled circles) are plotted.

as one moves away from the core. The standard metric for the length of an allele is based on its haplotype homozygosity (*HH*), the probability that two randomly chosen chromosomes containing the same SNP variant (or core set of SNPs) are identical (homozygous) for *all markers* within a specified region. Sabeti et al. (2002) defined **extended haplotype homozygosity (*EHH*)** as the length of a region around the core allele (SNP) where *HH* has a value of 5% or greater, namely, the length around the core where there is a 5% or greater chance that any two random haplotypes of that allele are identical at all markers (Figure 9.6).

While alleles with excessive values of *EHH* are produced by partial sweeps, simply scanning for sites with large *EHH* values will not serve as a sufficient indicator of selection, as a localized decrease in the recombination rate inflates the *EHH* value. The formal use of *EHH* as a selection-detecting statistic thus requires an internal control. Sabeti et al. (2002) proposed considering the **relative extended haplotype homozygosity (*rEHH*)** of a particular allele (SNP variant), defined as the ratio of the *EHH* value for that allele divided by the average *EHH* value for all other core alleles at the focal locus. For allele i, this is given by

$$rEHH_i = \frac{EHH_i}{\text{ave}(EHH_j) \text{ for } j \neq i} \tag{9.40}$$

where ave(EHH_j) denotes the average *EHH* values for all SNPs at the allelic-defining site. For the biallelic case (an allele defined by a single SNP, as opposed to a collection of tightly linked SNPs), *rEHH* is simply the ratio of the *EHH* values for the two alleles. By contrasting different alleles at the same site, most concerns about local variation in the recombination rates are ameliorated. However, if there are haplotype-specific recombination rates (e.g., the insertion of a mobile element reducing local recombination rates; Macpherson et al. 2008), then this test may be compromised. One consequence of comparing different alleles at a site is that as one allele approaches fixation, the power of the test disappears, as there are too few individuals in the comparison class to produce a meaningful statistic. As a result, the *rEHH* test has a rather narrow time window for the detection of a sweep: a rough rule is

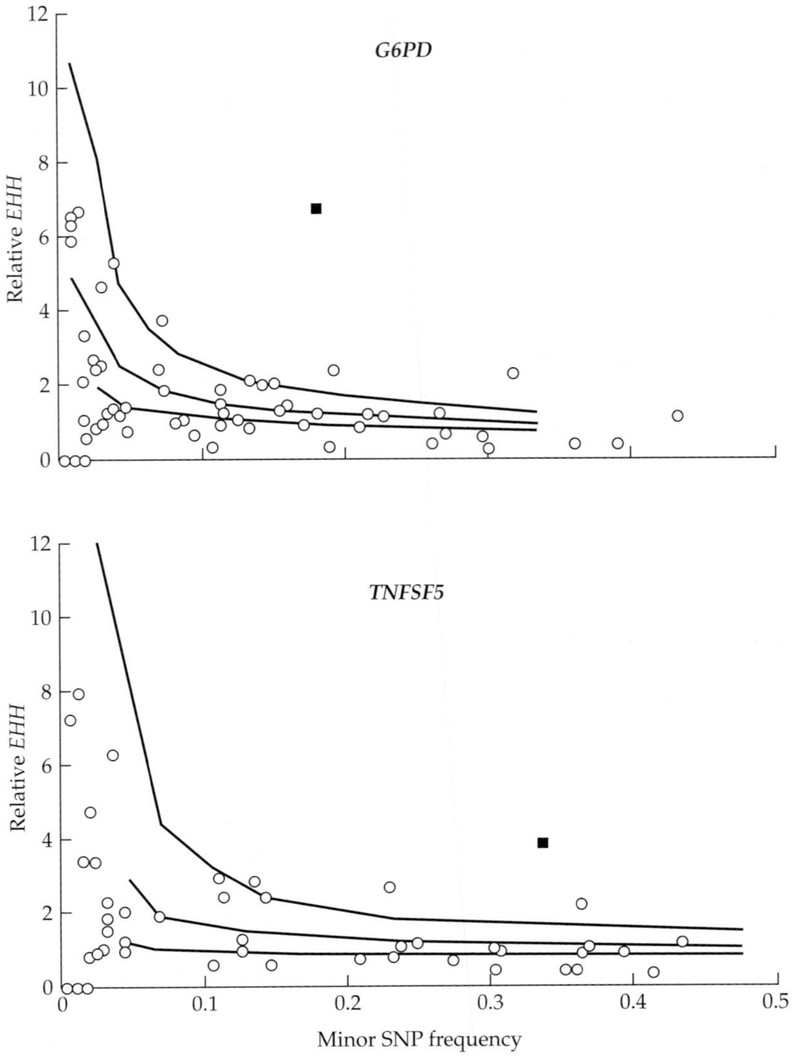

Figure 9.7 As a proof-of-concept of the *rEHH* method, Sabeti et al. (2002) looked for signatures of selection at two loci, *G6PD* and the *CD40* ligand gene (*TNFSF5*), that carry segregating alleles that are strongly suspected of increasing resistance to malaria. Standard site-frequency tests (Taijma's *D*, Fu and Li's D^*, and Fay and Wu's *H*; see Table 9.1) were all nonsignificant. However, recall from Chapter 8 that site-frequency spectrum signals are weak when the favored allele is at a modest frequency. The figure displays *rEHH* versus allele frequency for the candidate alleles (solid squares) along with values for alleles at other randomly chosen autosomal loci (open circles). The curves (from top to bottom) correspond to the empirical 95th, 75th, and 50th percentiles, respectively, of the cumulative distribution. (After Sabeti et al. 2002.)

that the frequency of the favored allele must be 0.7 or less. Within such a time window, this test is among the most powerful for detecting selection. Nonetheless, a large *rEHH* value is not sufficient for suggesting selection, as some rare alleles (potentially being very young, and hence with reduced time for recombination) are expected to have large *rEHH* values. To detect selection, Sabeti et al. (2002) plotted the *rEHH* value versus frequency for each allele and looked for outliers (Figure 9.7). Coalescent simulations were then performed under different neutral models (with different demographic assumptions) to assess significance. Wang et al.'s (2006) **linkage disequilibrium decay** (*LDD*) test is a modification of *EHH* that does not require phasing.

Variant tests based on the length of shared haplotypes have been proposed by a number of researchers (e.g., Toomajian et al. 2003, 2006; Hanchard et al. 2006; Wang et al. 2006); see Table 9.3. Perhaps the most powerful modification is from Voight et al. (2006), who extracted more LD information than simply the size of the *EHH* and corrected for differences in the local recombination rate and the target-allele frequency. One potential advantage of this approach is that while the *EHH* test has high power when the correct SNP is chosen to define alleles for the haplotype-length comparisons, its power falls off dramatically if the choice is off by even one polymorphic site (Zeng et al. 2007a). Voight et al.'s more comprehensive statistic may avoid this problem. Their approach used polarized data, with p denoting the frequency of the derived (D) SNP and $1-p$ denoting the frequency of the ancestral (A) SNP. To extract more information, they computed an **integrated EHH score** (***iHS***), the area under the curve drawn by connecting the adjacent values for the SNPs within the *EHH* (Figure 9.6). They defined the (unstandarized) integrated *EHH* score (iHS_{us}) as the log of the ratio of the iHS score for the ancestral allele to that for the derived allele

$$iHS_{us} = \ln\left(\frac{iHS_A}{iHS_D}\right) \tag{9.41a}$$

A negative score occurs when the iHS value for the derived allele exceeds that of the ancestral allele, whereas the converse occurs when the ancestral allele has the larger iHS score. One can either keep the sign or use the absolute iHS score, $|\,iHS\,|$. Voight et al. standardized the (signed) iHS_{us} score by defining the statistic

$$iHS = \frac{\ln\left(\frac{iHS_A}{iHS_D}\right) - E_p\left[\ln\left(\frac{iHS_A}{iHS_D}\right)\right]}{\mathrm{SD}_p\left[\ln\left(\frac{iHS_A}{iHS_D}\right)\right]} \tag{9.41b}$$

The expectation (E_p) and standard deviation (SD_p) are subscripted by p to highlight that these statistics are computed over all iHS_{us} values in the genome for SNPs whose derived allele frequency is p. Standardizing the score with respect to p automatically incorporates any relationship between the iHS_{us} score and the allele frequency (and hence the age for a neutral allele). The authors noted that this approach seems fairly robust to demographic departures from the equilibrium neutral model, especially at extreme values of the standardized score. Despite this, Voight et al. correctly did not assign significance values to individual iHS values, but rather used large (absolute) scores as a screening method for potential sites under selection.

Ferrer-Admetlla et al. (2014) proposed a statistic that is very similar in form to iHS but counts length variation differently. Their **number of segregating sites by length** (***nS***$_L$) statistic replaces the average area under the iHS curve by the average number of consecutive segregating sites shared by two randomly sampled chromosomes around a specific SNP variant. This average statistic for the ancestral and derived alleles replaces iHS_A and iHS_D (respectively) in Equation 9.41b. Ferrer-Admetlla et al. noted that this simple change in the metric results in a test that is significantly more robust to recombination and slightly more robust to nonequilibrium departures than the iHS statistic. When applied in a human genome scan, the method did not yield the large enrichment of significant scores in regions of low recombination typically seen when other (more recombination-sensitive) tests are used (e.g., O'Reilly et al. 2008). Further, their simulations found that nS_L has reasonable power to detect ongoing sweeps, even those from standing variation.

Another variant of this basic idea was recently suggested by Field et al (2016). Their **singleton density score** (***SDS***) measures the length of haplotypes by calculating the distance to the nearest singleton from a candidate site (looking on either side). This distance can be turned into an estimate of the mean branch length in the coalescent tree for that allele, and the estimates for the ancestral and deviate allele at a target site are contrasted. Specially, the test statistic is

$$SDS = \ln\left(\frac{\widehat{t}_A}{\widehat{t}_D}\right) \tag{9.42}$$

where $\widehat{t}$ are the estimated coalescent times from the singleton distance. As with several of the above test, the contrast the two alleles at a site controls for local variation in recombination and mutation rates. Under recent selection, the average branch lengths for an allelic class should be much shorter, resulting in longer distances to singletons. As with other haplotype-based approaches, comparisons are made over classes with the same derived allele frequencies. Field et al. found that their *SDS* test had power to detect very recent selective events (within the last $\sim$100 generations), a time scale usually too short for other haplotype-based methods (e.g., *iHS*) to show a strong signal. Further, they showed that with a sample size of 3000 individuals (and a derived allele frequency of 0.7), that they could detect ongoing events with a 2% selective advantage.

Finally, Barreiro et al. (2009) proposed a hybrid between long-range haplotype tests and diversity tests. Their **derived intra-allelic nucleotide diversity (*DIND*) test** compares the nucleotide diversities associated with the ancestral ($i\pi_A$) and derived ($i\pi_D$) alleles ($i\pi$ is used to denote that these are intra-allelic nucleotide diversities). A core SNP is used to define the two allelic classes, and the diversity, π, within some specified length around each core variant is computed in the same fashion as an estimate of pairwise diversity from a sample (Equation 4.1)). The logic is that derived alleles, being more recent, should show little diversity relative to their ancestral (and thus older) counterparts. The *DIND* test uses an approach very similar to that of the $rEHH$ test, plotting $i\pi_D/i\pi_A$ versus the frequency of the derived allele and then looking for outliers. Both the *SDS* and *DIND* tests were designed to fully leverage extensive resequencing data. Fagny et al. (2014) found that both the *DIND* and iHS tests were powerful in detecting recent selection, and also robust over demography. However, the *DIND* was more powerful than the iHS when the resequencing data had either poor coverage or low-quality genotype calling.

Long-haplotype Tests Using Between-population Data: *XP-EHH, Rsb, rHH, rMHH,* and χ_{MD}

Tests based on comparing the haplotype lengths of alternative alleles lose all power as a favorable allele approaches fixation. However, if the favored allele is only fixed in a single population, a *between*-population comparison of haplotype length still has power immediately following fixation. This approach was proposed by both Tang et al. (2007) and Sabeti et al. (2007), and it follows the same logic leading Equation 9.41a and 9.41b. However, instead of contrasting the *EHH* or *iHS* score for alternate alleles in the same population, they contrast values for the *same* allele in *different* populations. Sabeti et al. referred to this as the **cross-population extended haplotype homozygosity** (or ***XP-EHH***) test. A similar test (with a few subtle differences) was proposed by Tang et al., who defined the analog of Equation 9.41b as their ***ln(Rsb)*** statistic.

A second cross-population comparison test was suggested by Kimura et al. (2007), who considered two ratios of haplotype homozygosity in a control versus test population. The first, ***rHH***, is the ratio of haplotype homozygosities in the two populations. The second, ***rMHH***, is the ratio of haplotype homozygosities, but computed using just the most frequent haplotype in the control population (i.e., the probability that two random draws of sequences are both the most common haplotype). For a recently fixed allele, the expectation is a low *rMHH* value (a high population divergence), but a high *rHH* value (a population-specific decrease in haplotype diversity). Simulation studies with neutral models (under limited demographic conditions) suggested that the combination of low *rMHH* and high *rHH* values is rather unusual. Again, this is an outlier-based approach, with outliers on an *rMHH*-by-*rHH* plot of the genomic data suggesting potential targets of selection.

Table 9.3 Summary of tests reviewed in this chapter that use haplotype and LD information to indicate positive selection. AFS(k) denotes the equilibrium neutral distribution for the allele-frequency spectrum (the distribution of the number of haplotype classes within a sample), conditioned on the number of alleles/haplotypes, k (Equation 2.33b). S denotes the number of segregating sites, HH denotes the haplotype homozygosity (the probability that two randomly chosen sequences are identical over some defined region), and EHH is the extended haplotype homozygosity (the length of the region for a given allele over which $HH > 0.05$). See text for specific references.

Tests based on the allele-frequency spectrum AFS(k):

Ewens-Watterson Test: Observed allelic homozygosity vs. expected homozygosity under AFS(k)
Slatkin's Exact Test: Observed AFS(k) vs. expected AFS(k)
Innan et al.'s *HCT*: Observed AFS(k) vs. expected AFS(k) conditioned on observed S
Hudson's *HP* Test: Frequency of most common haplotype given S
Fu's *W*: Test for deficiency of rare haplotypes given S
Fu's F_s: Test for excess of rare haplotypes given $\widehat{\theta}_\Pi$ (average pairwise difference estimator)
Depaulis & Veuille's *K*: Observed number of haplotypes given S
Depaulis & Veuille's *H*: Observed haplotype diversity given S
Garud et al.'s H_{12}: Observed haplotype diversity combining the two most frequent classes
Garud et al.'s H_2: Observed haplotype diversity ignoring the most frequent class

Tests based on averages of pairwise disequilibria:

Kelly's Z_{nS}: Average of all pairwise disequilibria between all sites in a region
Kim and Nielson's ω_{max}: Pairwise LD among sites within vs. between sides of a region

Tests based on frequency estimates of age vs. allelic-diversity estimates of age:

Age estimated by decay of LD between allele and a linked marker
Age estimated by number of segregating sites S within an allelic haplotype class
Age estimated by copy-number variance at tightly linked STRs in the allelic class
Age of a mobile element insertion estimated by divergence from its consensus sequence

Tests contrasting haplotype lengths of alternative alleles in the same population:

Sabeti et al.'s *rEHH*: Ratio of the haplotype lengths (EHH) of two alternative alleles
Wang et al.'s *LDD*: Rate of linkage disequilibrium decay, modification of EHH
Hanchard et al.'s *nHS*: Haplotype diversity of the derived allele relative to the ancestral allele
Voight et al.'s *iHS*: Ratio of area under the EHH curve for ancestral vs. derived alleles
Ferrer-Admetlla et al.'s nS_L: Very similar to iHS, with the number of consecutive shared polymorphic sites replacing the area under the EHH curve
Field et al.'s *SDS*: Distance to nearest singleton, yielding an estimated mean allelic branch length
Barreiro et al.'s *DIND*: Ratio of nucleotide diversity in derived vs. ancestral allele

Tests contrasting haplotype lengths of the same allele in two populations:

Sabeti et al.'s *XP-EHH*, Tang et al.'s *ln(Rsb)*: Ratio of area under the EHH curve in different populations
Kimura et al.'s *rHH* vs. *rMHH* plot: Ratios of overall HH to HH based on most frequency haplotype
Lange and Poll's χ_{MD} test: Contrast of pairwise haplotype sharing between populations

---

A final cross-population test was proposed by Lange and Pool (2006). Their **comparative haplotype identity** method (χ_{MD}) examines whether the average length of pairwise sharing of haplotypes is large in a target population relative to a control (as would be expected if the target was under selection). For a given window around a target site, they first set a threshold value, then in each population compute the total lengths of haplotypes for a given pairwise comparison (i.e., between two members from the same population). Their χ statistic is this sum over all pairwise comparisons within a population, and then contrasts this value (for a given window) between the target and control populations. They found that this approach had good power and outperformed *XP-EHH* in many cases of either partial and/or soft sweeps.

Summary: Tests Based on Haplotype/LD Information

As summarized in Table 9.2, different kinds of sweeps (hard, partial, and soft) leave different haplotype signals. Given the diversity of such signals, it is not surprising that there are a number of haplotype-based tests to detect these different features (Table 9.3). LD-based tests are generally regarded as the *most powerful for sweeps that are currently underway*. Site-frequency spectrum tests often perform poorly under a partial sweep, as the distortion in the frequency spectrum is often not sufficiently powerful. Signatures from both a recently completed partial sweep, and a currently ongoing hard sweep, include long haplotypes at excessive frequencies, alleles that are at too high a frequency given other estimates of their age, an excess of one or a few haplotypes, and a reduction in haplotype diversity.

In addition to their unique role in detecting partial sweeps, LD summary statistics can also offer significant power to detect *just-completed sweeps*. Under a hard sweep, the unusual pattern of high LD on either side of, but not across, a selected site can be detected using Kim and Nielson's ω statistic (Equation 9.37). However, this statistic has no power to detect a soft sweep. Conversely, Kelly's Z_{ns} statistic (measuring average pairwise LD throughout a region; Equation 9.36b) can detect a recently completed soft sweep but has no power to detect a just-completed hard sweep.

As with almost all the tests discussed in earlier sections, haplotype-based tests can also generate false positives for neutral alleles in nonequilibrium populations. The standard approach of using outlier analysis to suggest regions of interest and coalescent simulations (using marker-based demographic estimates) can also be used here, with the same caveats. As mentioned, both outlier analysis and coalescent simulations use corrections based on genome-wide patterns and thus do not adjust for allelic surfing. This is especially troublesome, for as outlined below, the species most surveyed for recent selection—humans, the cosmopolitan human commensal *Drosophila* (*melanogaster* and *simulans*), and *Arabidopsis*—are all known to have undergone massive spatial expansions over the last 100,000 years, making them prime candidates for surfing.

SEARCHES FOR SELECTION: HUMANS

Given the huge array of tests introduced in this chapter, we end on a purely empirical note, examining what their applications to both natural and domesticated populations have told us about the nature of selection. Chapter 8 examined the theory and data for the impact of *recurrent* sweeps on a genome (a topic further examined in Chapter 10), while our focus here is on what is known about *particular* sweeps. The search for selection is motivated by both gene-specific and genome-wide questions. At the level of an individual gene, we would like to understand how an ecological challenge is met by a molecular solution: is it highly idiosyncratic or do some general patterns emerge? On a genomic scale, we are interested in general trends of adaptation. What is the relative importance of regulatory versus structural changes? Are genes of major effect more important than genes of minor effect? A growing consensus, at least in multicellular species, on the former question is that regulatory changes may be the predominant route for adaptation (e.g., Grossman et al. 2013), but whether their contribution is just roughly equal to, or significantly greater than, structural changes remains an open question. Our current understanding is certainly based on a nonrandom sample of loci, as methods for detecting sweeps are strongly biased toward genes of major effect that are under strong selection. We stress that our brief discussion is not meant to be comprehensive, as any review in this area will be out of date by the time it is published. Rather, we highlight a few case studies to illustrate the issues facing an investigator when trying to understand the results of a study for molecular signatures of selection.

Table 9.4 Overlap in sweep detection in three early scans (Carlson et al. 2005; Voight et al. 2006; Wang et al. 2006) that used different statistics to infer positive selection in humans. Diagonal elements represent the number of sites declared to be under positive selection in each given study, and off-diagonal elements represent the number shared between studies. See the text for further details. (After Biswas and Akey 2006.)

	Wang (*LDD*)	Voight (*iHS*)	Carlson (*D*)
Wang	1799	125	47
Voight		455	11
Carlson			176

At present, genome-wide scans for genes under recent, or ongoing, selection have been performed on only a modest (but growing) number of species. For natural populations, the most extensive work has been done on humans, *Drosophila*, and *Arabidopsis thaliana*. Given that we know a great deal about the genetics, genomics, and molecular biology of these species, this choice is not surprising. All three groups have undergone major expansions into a wide range of new habitats over the last 100,000 years, and hence harbor the potential for a significant response to evolutionary challenges. For humans, the movement out of Africa into more temperate climates, coupled with the transition from hunting and gathering to agriculture and the resulting increase in population density, generated novel environmental pressures. The commensal *D. melanogaster* and *D. simulans* followed humans into these new environments, while in the northern hemisphere, *Arabidopsis* underwent significant range expansion following the end of the ice age. The environmental challenges faced by these species, as well as demographic changes (such as massive population expansions), leads us to expect a history of recent selection.

Recent and Current Selection in Humans

Early searches in humans looked for molecular signals at candidate genes either believed, or very strongly suspected, to be under selection in particular environments. Examples include disease-resistance genes such as *Duffy* (*FY*) and *G6PD*, dietary genes such as lactase (*LCT*), and climate-related genes such as *MC1R* (influencing skin color). Ronald and Akey (2005) and Harris and Meyer (2006) reviewed these and other candidate genes, and found strong signals (such as skewed site-frequency spectra, long haplotypes, and/or excessive F_{ST}), adding support to the belief that they have experienced recent selection. Signals of recent hard sweeps should persist for no more than $2N_e$ generations (Table 8.1), or roughly covering no longer than the past 250,000 years (assuming 25 years per generation and $N_e \sim 5,000$).

Starting with the advent of dense-SNP maps and continuing as whole-genome sequencing became economically feasible, candidate-gene studies were replaced by **genomic scans**, searching the genome without any preconception of what sites might be under selection. Biswas and Akey (2006) reviewed six early scans that used different statistics to infer positive selection. In total, roughly 2300 genes were found with signals of selection in at least one of these studies, but the overlap between studies was quite small. For example, consider the overlap between the findings of Wang et al. (2006) and Voight et al. (2006), who used long-range haplotype tests (*LDD* and *iHS*, respectively; Table 9.3), and Carlson et al. (2005), who used outliers in Tajima's D (Table 9.1). As shown in Table 9.4, of the 455 sites detected by Voight, 125 (27%) were also seen by Wang. Conversely, of the 176 sites with outliers in D, only 6% (11) of these were also detected by Voight, while 27% (47) were detected by Wang.

More recent summaries (Akey et al 2009; Fu and Akey 2013) have echoed this general lack of cross-study replication. A number of factors can contribute to these discrepancies besides false positives. The first is clearly power. When power is low, many sites that are under positive selection are missed, and the overlap between studies is expected to be small. Second, as we have stressed throughout this chapter, *different tests detect selection over different time scales.* A partial sweep gives an LD signal but is not likely to give a site-frequency signal, while the converse is true for a recently completed sweep. Thus, LD-based tests (such as *rEHH* and *iHS*) are *expected* to detect different genes than site-frequency tests. Finally, selection can be *spatially localized*, being experienced in only a subset of populations. If a selected population is included in one study but not in another, the result is a lack of replication.

With this last concern in mind, several scans have searched for geographically localized selection by contrasting F_{ST} values among samples of different populations (and hence allowing for population-specific selection). Barreiro et al. (2008) examined the F_{ST} values associated with roughly 3 million SNPs over four populations (Nigerians, Europeans, Chinese, and Japanese). They used a modification of the outlier approach, binning SNPs by functional categories (e.g., synonymous, nonsynonymous, 5′ UTR, etc.). They observed an excess of higher F_{ST} values (relative to the genome-wide distribution) in both nonsynonymous and 5′ UTR SNPs, suggesting that there were around 600 sites under local selection. Further, the excess nonsynonymous SNPs were enriched for long haplotypes, as might be expected under a partial sweep. Pickrell et al. (2009) also found evidence of significant local adaptation (population-specific changes) in a survey of 53 populations, although Hofer et al. (2009) noted that the striking differences in allele frequencies between human populations could have easily arisen as a consequence of population expansion (and the accompanying allelic surfing). One additional concern with these studies is that (as mentioned earlier) F_{ST} values are constrained by the level of heterozygosity (which is influenced by background selection), with SNPs with higher minor-allele frequencies having higher maximal F_{ST} values.

Tempering these results was the declaration by some researchers that classic hard sweeps appear to be rare in humans (Hernandez et al. 2011; Lohmueller et al. 2011; Alves et al. 2012), or "have played a moderate, albeit significant, role" (Fagny et al. 2014). However, as discussed in Chapter 8, Enard et al. (2014) noted that a failure to account for background selection (BGS) can result in a distorted view of the importance of sweeps. After adjusting for this effect, they detected widespread signals for positive selection in humans, which were more correlated with regulatory sequences than amino acid changes. Others have stressed the importance of polygenic sweeps (Hancock et al. 2010a, 2010b; Amato et al. 2011; Fumagalli et al. 2011; Turchin et al. 2012; Daub et al. 2013; Zhang et al. 2013; Berg and Coop 2014; Mathieson et al. 2015, Robinson et al. 2015; Field et al. 2016).

The take-home message is that genomic scans in humans reveal numerous potential sites under selection but that sites are usually not replicated across studies. This could simply be a consequence of low power, but as is the case for QTL and association-mapping studies, an initial exciting finding is only the start, and not the conclusion, in the search for genes that are under selection. Unlike association studies, where support is offered by independent replication, the concept of independent replication in the search for genes under selection is more problematic. Finding similar support in two independent samples drawn from the same population is comforting, but it is not a formal validation, as demographic features could have generated the signal, and no amount of resampling will remove its effect. Similarly, a lack of replication between two different populations can easily be explained by differential selection pressures. Support for positive selection would be offered if we observe differences in fitness among alternative alleles at candidate genes, but failure to do so is also not damning. Fitness differences between alleles far below the level of detection (typically at least 1% in most studies of natural populations) can still be critically important (Chapter 5). Likewise, the nature of selection changes with the environment, and a signal could be the result of a past environmental effect that is no longer important. Finally, as discussed in Chapter 8, the geographic structure of human populations not only leads to allelic surfing, but also increases the likelihood that there will be multiple independent adaptive mutations (spread out over the geographic range) that address the same environmental challenge. Human geneticists thus face the very real possibility that only a biased set of events are being detected (along with a potentially large number of false positives), giving a very distorted view of the genetics of human adaptation.

Balancing Selection in Humans

As discussed in Chapter 8, the expected signal from balancing selection significantly changes over time. Initially, as an allele is sweeping from some low frequency up toward its equilibrium value (under either selective overdominance or frequency-dependent selection), a partial sweep signal is expected (i.e., long haplotypes). After equilibrium is reached, the site-frequency spectrum is expected to show an excess of intermediate-frequency alleles. If the equilibrium persists beyond the neutral coalescent time, a region of increased diversity is expected around the site that is under balancing selection (Figures 8.1B and 8.2B). As more time passes, the size of the region showing excessive polymorphism and intermediate-frequency alleles will continue to be shrunk by recombination, leaving an increasingly smaller window as the age of the selective polymorphism increases. As a result, a single site under very long-term balancing selection is likely to leave very little signal (Wiuf et al. 2004; Charlesworth 2006a; Pavlidis et al. 2012). In contrast, a larger, and more persistent, window of excessive diversity occurs when multiple linked loci are under balancing selection, especially if there are epistatic fitness interactions between them, resulting in selection against recombinants (Kelly and Wade 2000; Barton and Navarro 2002).

Given this expected change in signal over time, searches for balancing selection in humans have used different statistics, depending on the time frame of interest (nicely reviewed in Fijarczyk and Babik 2015). Andrés et al. (2009) were concerned with relatively recent events. They used the intersection of two test statistics: one for excessive diversity (the *HKA* test from Chapter 10) and a nonparametric test for excessive intermediate-frequency alleles in the site-frequency spectrum (the *MWU-high* test of Nielsen et al. 2009). They scored roughly 13,000 genes in two populations (African- and European-Americans) and found 60 candidate sites that were significant for both tests. Most of these appeared to be under balancing selection in both populations, but a few were population specific. Because approximately 12 significant results were expected by chance, the false-discovery rate (Appendix 4) for these data is roughly 12/60 (20%), suggesting that roughly 50 are true positives, comprising around 0.4% of the genes surveyed. Hence, this study suggests that intermediate-term balancing selection is not common.

In searching for sites that have been under balancing selection for longer than the neutral coalescent time, Bubb et al. (2006) used a simple approach based on two results from coalescent theory. First, the expected coalescent time for two random neutral alleles is $t_2 = 2N_e$ generations (Equation 2.38), while the expected coalescent time for *all* alleles is $t_A \simeq 4N_e$ generations (Equation 2.41). Hence, while the average diversity between two random alleles is $2t_2\mu = 4N_e\mu = \theta$ (Chapter 2), the maximal expected diversity between the two most distinct alleles is twice this value, or $2t_A\mu = 8N_e\mu = 2\theta$. This result is only approximate, as it is based on expected values and ignores the rather large variances in these coalescent times (Chapter 2). Using these results, Bubb et al. searched the genome for regions where the diversity between the most extreme alleles greatly exceeded the value of 2θ predicted from the neutral theory. Excluding *MHC* and *AB0* (the former is known to be under balancing selection, and the latter is strongly suspected to be), the authors found 16 such regions, but they concluded from simulations that these outliers could easily have been generated by chance variation in the coalescent times under a strictly neutral process.

Searches for balancing selection over an even longer time scale involve the search for **trans-species polymorphisms** (Klein 1980)—polymorphic alleles shared by species whose divergence time significantly predates the expected coalescent time for either species. Classic examples are the vertebrate *MHC* (major histocompatibility complex) genes and self-incompatibility alleles of flowering plants (Klein et al. 1998; Richman 2000; Charlesworth 2006a). Asthana et al. (2005) compared roughly 8000 chimpanzee transcripts with their human counterparts and found evidence of eight potentially shared polymorphic sites, three of which were nonsynonymous. In contrast, 12 trans-species polymorphisms were detected at the *MHC* region, leading to the conclusion that ancient polymorphisms, and hence very long-term balancing selection, were very rare in humans outside of the *MHC* complex. A more recent study by Leffler et al. (2013) found 125 candidate regions (based on shared haplotypes) outside of *MHC*. Surprisingly, all but two of these involved non-coding regions, suggesting that if long-term balancing selection occurs in humans, it usually involves regulatory changes.

Another approach in the search for loci under ancient balancing selection are extensions of the composite likelihood ratio (CLR) test (developed earlier for hard sweeps) by DeGiorgio et al. (2014). Recall that under the structure of CLR, one computes a likelihood for each site as a function of the putative distance (c) from a site under selection, and then (as an approximation of the full likelihood) treats each site as being independent (Equations 9.16–9.18). These authors developed two tests, T_1 based on divergence and polymorphism data, and T_2 based on the site-frequency spectrum and divergence pattern. Given that they found the T_2 test always performed at least at well as the T_1 test, this should be used. Their tests picked up strong signals for numerous *MHC* loci, but (surprisingly) did not find a signal for the *ABO* locus, which is widely speculated to be under long-term balancing selection. Rather few new candidate loci were detected by this approach. Perhaps the most interesting was the *FANK1* gene expressed during meiosis in males. They hypothesized (with some support) that any balancing selection on this locus operates by segregation distortion being balanced by negative selection, so a rather nontraditional setting for balancing selection.

As mentioned, the caveat about these studies is that the size of the region showing signals from balancing selection that is still ongoing (e.g., an excess of intermediate-frequency alleles) is progressively shrunk by recombination. Hence, most ancient events that still under selection are likely to be missed, especially if they involve only a single locus. If the region that is under balancing selection is generated by linked, epistatic loci, this can result in selection against recombination within the region separating them, allowing the signal to persist longer. A caveat with trans-species polymorphisms is that they can be generated by rare introgression events, rather than balancing selection over evolutionary time scales (Fijarczyk and Babik 2015).

Although for power reasons alone, the detection of balancing selection will almost always be more problematic than the detection of positive directional selection, a more subtle issue is the fact that balancing selection initially generated by overdominance is likely to be evolutionarily unstable, even if the selection pressures remain unchanged (Spofford 1969; Asthana et al. 2005; Bubb et al. 2006). If heterozygote advantage is generated by one of the alleles being deleterious as a homozygote, more fit alleles can arise by mutation and become fixed. A potential example is the *HbC* (hemoglobin C) allele, which provides resistance to malaria without being as deleterious as the essentially lethal sickle-cell *HbS* allele. As a result, we might expect the *C* allele to replace the *S* allele over evolutionary time (Walters and Lehmann 1956; Modiano et al. 2001). Even if both homozygotes have the same fitness, an overdominant polymorphism is evolutionarily unstable when gene duplication occurs. A gamete in which the tandem-duplicated copies are alternative alleles has a fitness advantage and can subsequently become fixed (Spofford 1969).

SEARCHES FOR SELECTION: DOMESTICATED ORGANISMS

Extensive searches of recent selection have been conducted in domesticated plants and animals. Domestication represents a major change in the environment, and hence the opportunity for significant selection to occur within a very recent time period (roughly within the past 10,000 years). Domestication thus offers model systems of adaptation (Darwin 1868; Ross-Ibarra et al. 2007), providing a "telescoped time frame in which both antecedent and descendant conditions remain extant and available for comparison" (Olsen and Wendel 2013).

The distinction between genes involved in the initial domestication events and those involved in the subsequent improvement of varieties can be rather subtle. Operationally, **domestication genes** are assumed to be present in all descendant varieties from the domestication event, while **improvement** (or **diversification**) genes are viewed as being further selected in only a subset of varieties. The latter could be the result of deliberate selection, such as for sticky rice, or the result of local selection conditions that lead to the formation of **landraces** (locally adapted varieties). A knowledge of improvement genes can give the breeder insight into achieving specific objectives. For example, genes whose selection signatures are limited to varieties in high drought conditions can suggest important target genes for drought tolerance in current elite germplasms. An example of this was revealed by Kane and Rieseberg (2007), who used the *log RH* and *log RV* tests (Equation 9.14) to search for sunflower (*Helianthus annuus*) genes with signatures of selection restricted to populations in drought and/or high salt environments. A second example was provided by Vielle-Calzada et al. (2009), who detected potentially selected genes for metal tolerance in the Mexican highland popcorn landrace *Palomero Toluqueño*.

The genetics of domestication is a rapidly growing field, which we only touch on briefly here. Reviews and perspectives were offered by Diamond (2002), Doebley et al. (2006), Purugganan and Fuller (2009), Gross and Olsen (2010), Larson and Burger (2013), Meyer and Purugganan (2013), Olsen and Wendel (2013), Larson and Fuller (2014), Larson et al. (2014), and Gaut et al. (2015).

The Process of Domestication

While domestication is often perceived to be the result of weak, or even unintentional, selection over some period of time, these settings need not be the norm. Some events are very sharp and deliberate, such as the creation of novel populations by intentional species crosses. A classic example is the mule, the sterile offspring of a male donkey and a female horse. Likewise, a number of allopolyploid crops resulted from crosses between two (or more) progenitor species. At the other extreme is the genetic modification of an ancient progenitor into a modern variety, such as the dramatic changes in plant architectural from teosinte to modern maize (LW Figure 5.2). As first noted by Darwin (1868), selection for domestication traits can be due to conscious action (**conscious selection**, or **methodical selection** in Darwin's terms) or it can be entirely **unconscious**. The latter are an indirect consequence of human-induced changes in the environment, where selection can reward certain strategies or traits that favor the domestication of wild species. For example, wolves that were less timid around human refuse piles may have gathered more food, leading to higher fitness while moving them closer to domestication.

The threshold beyond which a wild species is said to be domesticated can be challenging to assess. One operational definition is that domesticated varieties survive very poorly in a natural setting, due to the establishment of traits that increase fitness in the domesticated environment but decrease it in the wild. As best stated by Zeder et al. (2006), "domestication is a unique form of mutualism," leaving both genetic and archaeological signals (see Zeder et al. for several interesting examples). It is also worth emphasizing that domestication is not a uniquely human enterprise. For example, several species of insects cultivate fungal species, and the search for domestication genes in such systems (in both the domesticating insect and their cultivated fungus) remains an intriguing possibility.

Some domesticated species appear to have a single origin. Such seems to be the case for maize (Matsuoka et al. 2002), emmer and einkom wheats (*Triticum turgidum* and *T. monococcum*; Zohary 1999), potatoes (*Solanum tuberosum*; Spooner et al. 2005), and peanuts (*Arachis hypogaea*; Kochert 1996). The inference of a single origin is often based on the observation of a monophyletic clade when using neutral markers. A caveat with this approach is that simulations by Allaby et al. (2008) showed that such clades can be produced in crops with multiple origins, provided there is a rather protracted period of domestication. Other crops, such as barley (*Hordeum vulgare*; Zohary 1999) and *Phaseolus* beans (Gepts et al. 1986), show clear evidence of multiple domestication events.

Gene flow between lineages of independent origin, and between domesticated lines and their wild ancestors, further complicates the interpretation of any origins story. One such example is Asian rice (*Oryza sativa*), whose *indica* and *japonica* varieties have been regarded as a single domestication event (Molina et al. 2011), as a pair of distinct domestication events (Londo et al. 2006; Sang and Ge 2007), and as three independent domestication events (with a separate origin for the variety *aus*; Civáň et al. 2015). Huang et al. (2012) suggested an even more complicated story, with *japonica* first domesticated from its wild progenitor, *O. rufipogon*, in southern China and *indica* being subsequently developed by crossing *japonica* with *rufipogon* strains from South and Southeast Asia. Even with multiple origins, gene flow between *indica* and *japonica* was likely, however, as they share a number of key domestication alleles (such as *sh4*, which reduces grain shattering) that might otherwise suggest a single origin (Sang and Ge 2007; He et al. 2011). Introgression between nascent domesticated and wild populations also appears to have been widespread in animals (Larson and Burger 2013; Larson and Fuller 2014), obscuring both their center of origin and number of founding events.

Example 9.16. A complicating factor in both resolving the origins of any particular domestication event and detecting unambiguous signatures of selection is the **introgression** of wild genes into domesticated lineages. Such gene flow can be substantial, especially in the early stages of domestication (Larsen et al. 2014). The ***ABBA-BABA* test**, which was first introduced to infer the flow of Neandertal genes into modern humans, provides one approach for detecting such introgression (Green et al. 2010; Durand et al. 2011). Unfortunately, this approach is also called the ***D* test**, creating potential confusion with Tajima's D (Equation 9.24a), especially when both appear together in the same paper (e.g., da Fonseca et al. 2015). Our strong recommendation is to always retain the label *ABBA-BABA* test to avoid any such confusion.

Figure 9.8 outlines the basic structure of this test, which requires an outgroup, O. An important assumption is that the outgroup is sufficiently distant that no shared segregating alleles from it are present in the other taxa. Rosenberg (2003) showed that 99.9% of all loci have monophyletic genealogies when compared across lineages separated by more than $5.3N_e$ generations, which sets a lower bound on the ancestral time required for a proper outgroup to develop. Let 1, 2, and 3 denote three nested taxa, with 1 and 2 sharing a common ancestor with 3 (see Figure 9.8). In the original application by Green et al. (2010), 1 and 2 were African and non-African human populations, 3 was Neandertal, and the chimpanzee was the outgroup, O.

Define a string of length four and with elements A (for the ancestral allele present in O) and B (a derived allele present in taxon 3), with the positions in this sequence corresponding to species 1, 2, 3, and the outgroup. For example, the configuration given by I in Figure 9.8 is denoted ABBA. Suppose that at a given locus, the ancestral population of 1, 2, and 3 was segregating for A and B, and that in taxon 3, the lineage was sorted such that B was fixed. Conditioned on taxon 3 containing the derived allele, when 1 and 2 carry different alleles (one ancestral and the other derived), the direction should be entirely random (as the sorting would be random for neutral alleles), and hence both cases (*AB* vs. *BA* in these two species, translating into *ABBA* vs. *BABA* for the four-species comparison) should be equally likely. A systematic departure in one direction (i.e., far more *ABBA* than *BABA*) implies introgression from 3 into either 2 or 1 (respectively). Green et al. found a significant skew in favor of introgression from Neandertal into non-African humans. Their D statistic is given by

$$D_{ABBA-BABA} = \frac{N_{ABBA} - N_{BABA}}{N_{tot}} \tag{9.43}$$

where N_x is the number of events in class x and $N_{tot} = N_{ABBA} + N_{BABA}$ is the total number of the two events. Significance ($D \neq 0$) is assessed using a jackknife approach. See Durand et al. (2011) for a detailed discussion and development.

Finding Domestication and Improvement Genes in Crops

One standard approach for finding domestication and improvement genes is QTL mapping in a cross between the wild ancestor (provided it still exists) and the domesticated or

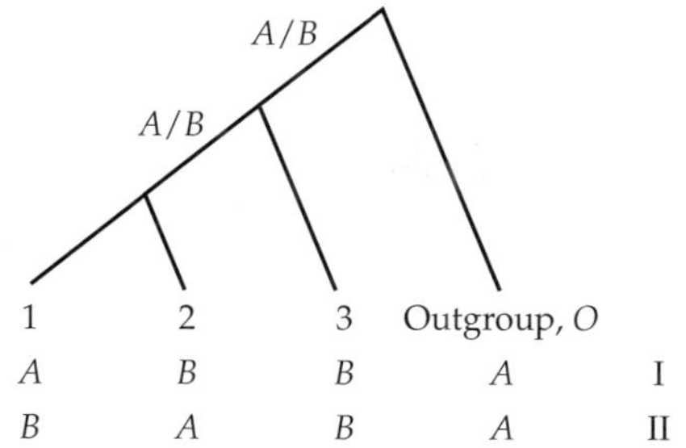

Figure 9.8 The ***ABBA-BABA*** test for detecting the introgression of genes from taxon 3 into either taxon 1 or 2; see Example 9.16 for details. Here *A* and *B* denote the ancestral and derived alleles, with the ancestral allele present in the outgroup, *O*. The test compares the distribution of *A* and *B* in taxa 1 and 2, conditioned on taxon 3 containing the derived allele, *B*. If there is simply neutral lineage sorting between the outgroup and the three resulting taxa, then configurations I (*ABBA*) and II (*BABA*) should be equally frequent. However, if there has been symmetric introgression of alleles from taxon 3 into one of these populations (but not the other), this pattern will be skewed, with one configuration being in excess of 50%.

improved variety. Such a strategy relies on knowing which *traits* are important. Classic examples of loci detected using this approach include *teosinte branched 1* (*tb1*) and *barren stalk 1* (*ba1*) for plant architecture in maize (Doebley et al. 1995; Gallavotti et al. 2004); *teosinte glume architecture 1* (*tga1*) for naked grains in maize (Wang et al. 2005); *fw2.2* for tomato fruit size (Frary et al. 2000); and *sh4*, *qSH1*, and *OsLG1* for reduced seed shattering in rice (Konishi et al. 2006; Li et al. 2006; Ishii et al. 2013). Given the obvious success of the QTL mapping approach, what role do signatures of selection play in the search for domestication and improvement genes? First, showing that QTL-detected regions were under selection provides independent support for their role in domestication. Second, one can estimate the average strength of selection on a given domestication allele, and hence obtain some indication of the required time to either fix, or substantially increase, its frequency during domestication. Finally, scans for selection are *trait-independent* searches. While some morphological features may be rather obvious candidates for domestication or improvement traits (and hence characters for QTL or association mapping), more subtle physiological changes may be less obvious. Notably, Hufford et al. (2012) found that the majority of scan-detected regions of selection in maize showed stronger signals than those for QTL regions associated with major morphological differences related to domestication.

Domestication can offer important insight into the genetics of adaptation, including the relative importance of loss-of-function versus gain-of-function mutations. For example, the *Q* gene in bread wheat (*Triticum aestivum*) is a critical domestication gene, allowing modern (nonhulled) wheat to be easily threshed (separation of the seed from chaff). A phylogenetic analysis by Simons et al. (2006) indicated that the *Q* gene had a single origin, and appears to involve a gain-of-function, rather than a loss-of-function, mutation. Likewise, Doebley et al. (2006) and Gross and Olsen (2010) noted that loss-of-function mutations are rare among known domestication genes in crops but not uncommon among improvement genes. Surveys of chickens and pigs (Rubin et al. 2010, 2012) found very little evidence that loss-of-function mutations are common in animal domestication and improvement genes.

A second general question concerns the role of regulatory changes in adaptation (Zuckerkandl 1968; King and Wilson 1975; Carroll 2008). As summarized by Doebley et al. (2006), Meyer and Purugganan (2013), and Olsen and Wendel (2013), regulatory changes underlie many domestication and improvement genes. One interesting example is the regulatory *BoCal* gene and its role in different varieties of domesticated cabbage (*Brassica oleracea*). Closely related to wild and domesticated cabbage (*B. oleracea oleracea*) are kale (*B. oleracea acephala*), cauliflower (*B. oleracea botrytis*), and broccoli (*B. oleracea italica*). The last two show significant modification of their inflorescence structures, while kale and cabbage have normal floral structures. Purugganan et al. (2000) showed that a nonfunctional allele (due to a premature termination codon) of *BoCal* is fixed in both cauliflower and broccoli, but segregating in wild cabbage and kale. Strong haplotype structure was seen, with a reduction in nucleotide diversity around this gene relative to other sites. A sample of cauliflower and broccoli alleles showed significantly negative Tajima's D and Fu and Li's D^* (Table 9.1), all consistent with recent positive selection. Neither D nor D^* was significant in a sample of kale and wild-cabbage alleles. This loss-of-function allele appears to have a single origin and is a regulatory mutation. However, the presence of this allele (at lower frequencies) in normal flowering populations of wild cabbage and kale shows that it is not sufficient by itself for the inflorescence modification that arose during the domestication.

Domestication and Improvement Genes in Rice

As perhaps the most important single staple in the world, rice has been widely searched for domestication and improvement genes. A key change during the domestication of Asian rice involved moving from a reasonably outcrossed species to a highly selfing one. Selfing reduces the effective recombination rate, causing the effects of a sweep to extend over a larger region of the genome. In particular, if η is the rate of selfing, the effective recombination rate, c^*, is well approximated by

$$c^* \simeq c\left(1 - \frac{\eta}{2-\eta}\right) = c(1 - \widetilde{F}) \tag{9.44}$$

where $\widetilde{F}$ is the equilibrium level of inbreeding under partial selection (Chapter 23; Nordborg 2000). This expression is reasonable given that $(1 - F)$ is the reduction in the frequency of heterozygotes (and hence opportunities for recombination) under inbreeding. For modern Asian rice, $\eta \simeq 0.99$, giving a roughly 50-fold decrease in the effective recombination rate. This reduction, when combined with small genome size (less than 400 Mb), implies that a significant impact on most of the rice genome is expected if even a modest number of sweeps occurred during domestication (Example 8.16). Caicedo et al. (2007) noted that domesticated rice shows a genome-wide excess of high-frequency derived alleles, which is not consistent with a simple founding bottleneck but is consistent with sweeps impacting much of the genome. Both He et al. (2011) and Huang et al. (2012) detected numerous regions of reduced diversity over a panel of domesticated lines relative to wild *O. rufipogon* populations, many of which exceeded 200 kb.

An example of a long region of depressed variation is seen around the *Waxy* gene, where a splice mutant results in low amylose levels and producing "Sticky" (glutinous) rice (reviewed by Olsen et al. 2006). This is an improvement trait, which is largely restricted to temperate *japonica* varieties. There is a massive sweep signature around this gene, with a 97% reduction in nucleotide diversity ($\pi = 0.0002$ versus normal levels of $\pi = 0.0064$ in wild accessions). The sweep signature spans 250 kb, encompassing ~40 genes. Further, there is a strong *EHH* signal (Table 9.3) around *Waxy*, and alleles from temperate *japonica* lines show a highly negative Tajima's D. Olson et al. assumed that $c = 3.7 \times 10^{-7}$ per bp (Inukai et al. 2000) and used Equation 8.6b to estimate the strength of selection as

$$s \simeq \frac{3.7 \times 10^{-7} \cdot 250,000}{0.02} = 4.6$$

This estimated value implies incredibly strong selection, with individuals carrying this allele leaving (on average) close to five times as many offspring as those without it. However, this estimate does not account for the reduction in recombination from selfing. Using the effective recombination rate (Equation 9.44) reduces the estimate to a more modest value of s ~0.1 (assuming a high selfing rate of $\eta = 0.99$).

Domestication and Improvement Genes in Maize

Maize is the king of crops when it comes to both genomic scans and searches for selection at specified candidate loci, and excellent overviews are given by Doebley (2004) and Tian et al.

(2009). The first demonstration in any species of selection on a putative domestication gene was for *teosinte branched 1*, (*tb1*) in maize (Wang et al. 1999; Clark et al. 2004, 2006). Given its obvious role from QTL studies as a candidate domestication gene, Wang et al. compared the levels of nucleotide diversity around this locus in maize with the corresponding region in teosinte. As shown in Figure 8.2, maize was found to have reduced levels of polymorphism (about 75%) throughout this region relative to teosinte, although this is consistent with a bottleneck during domestication influencing all loci in modern maize. More importantly, Wang et al. observed a significant *further* decrease in the amount of variation in the 5′ UTR region of maize (but not teosinte) *tb1*, suggesting that a selective sweep influenced this specific region (Figure 8.2). Surprisingly, the sweep did not influence the coding region, suggesting that the selected site was in the 5′ regulatory region, and not a change in the amino acid sequence of *tb1*. Clark et al. (2004) examined the 5′ region of *tb1* in more detail and found evidence for a sweep influencing a region of 60–90 kb in the 5′ UTR, with an average strength of selection of $s \sim 0.05$. This value implies an expected time to fixation of around 300 to 1000 years (using Equation 5.3b), indicating a fairly long period of domestication. Stuber et al. (2011) identified the likely selected site as an insertion of a *Hopscotch* retrotransposon roughly 64 kb upstream of *tb1*, which resulted in up-regulation in the amount of *tb1* transcripts. It is of note that this insertion predated domestication by at least 10,000 years, showing that standing variation was exploited during this domestication event, and yet a strong hard-sweep signal is seen (the sweep likely being hardened by the founding population passing through a bottleneck).

Other examples of selection signals on putative domestication genes in maize include the *c1* gene, which regulates anthocyanin production, and hence the transition from colorless to colored kernels in early maize (Hanson et al. 1996), genes in the starch pathway (Whitt et al. 2002), *Y1* for yellow kernels (Palaisa et al. 2004), *barren stalk 1* (Gallavotti et al. 2004), the *tga1* gene for naked seeds (Wang et al. 2005), and MADS-box regulators of plant floral development (Zhao et al. 2010).

Moving beyond tests for specific candidate genes, modest-scale genomic scans have been performed in maize by Vigouroux et al. (2002), Yamasaki et al. (2005), Wright et al. (2005), and Hufford et al. (2007). Based on the finding that 2% to 4% of 774 sampled genes showed signatures of selection, Wright et al. (2005) suggested that over 1200 maize genes have likely been influenced by artificial selection during domestication and subsequent improvement. Based on an analysis of 30 of Wright et al.'s candidates, Hufford et al. inferred that $\sim$40 % of these are domestication genes and the remainder are improvement genes (domestication genes showing sweep signatures in all lines, but improvement genes in only a subset of lines). Regulatory genes (such as transcription factors) were not overrepresented among these candidates. However, a more recent study by Zhao et al. (2010) sequenced 32 MADS-box genes (transcription factors) and 32 randomly chosen loci and found that eight MADS-box genes were targets for domestication and an additional one was a target for improvement, while two of the random genes were domestication targets and an additional four were improvement targets. Hufford et al. (2007) also noticed that candidate genes detected from scans were significantly overrepresented in expression in ear tissue relative to vegetative tissues, again suggesting an important regulatory component to the adaptive response.

A more comprehensive scan by Hufford et al. (2012) examined 35 improved lines, 23 landraces, and 17 wild relative lines with the *XP-CLR* test (Equation 9.20). Recall that this likelihood-based test compares the genomic spatial F_{ST} pattern in a selected line relative to an unselected control and returns an estimate of the strength of selection during the sweep. Domestication genes were detected by contrasting landraces (selected lines) with wild relatives (control), while improvement genes were located by contrasting improved lines against landraces (as the controls). The regions with the highest 10% of test scores included 484 potential domestication genes and 695 improvement genes. The average selection coefficients for these groups were $s = 0.015$ for domestication and $s = 0.003$ for improvement. Relative to random genes, domestication candidates showed greater changes in gene expression from their teosinte ancestor, tending to have higher levels of expression and more stability in expression over maize lines. Divergence in gene expression between teosinte and maize was further studied by Swanson-Wagner et al. (2012), who found that the regions detected by Hufford et al. were significantly enriched for both differences in expression, and altered coexpression profiles, relative to random genes from the maize genome.

An especially interesting study on maize domestication was performed by Jaenicke-Després et al. (2003), who used ancient maize ears as a "time machine" to look at the fixation of domestication alleles. Five maize cobs from the Ocampo Caves in Northeast Mexico were carbon dated, with two estimated at around 4300 years old, and the other three at between 2300 and 2800 years old. Six ancient cobs from Tularosa Cave in New Mexico were also examined, two of which dated to around 1900 years old, with the remaining four dating to around 650 to 900 years ago. DNA extracted from all cobs contained the modern maize allele at *tb1*. Examination of second domestication gene, *pbf* (which is involved in seed storage protein production), had the modern allele in all cobs as well. The final domestication gene examined was *sugary 1* (*su1*), which is involved in starch expression in the kernels. Here the pattern was mixed. The alleles $M1$ and $M2$ at this locus are found in 30% and 62% (respectively) of modern maize lines, whereas both are around 7% in teosinte. All the cobs from Mexico were homozygous for $M2$, while the four younger cobs from New Mexico were homozygous for $M1$. However, the two older cobs from New Mexico were heterozygotes, $M1/M2$ and $M1/T1$, where the $T1$ allele is not seen in modern maize and found in only ~4% of current teosinte populations. Thus, it appears that while much of the initial domestication was completed by 4000 years ago, allelic selection (at least in the New Mexico populations) was still ongoing as of ~2000 years ago. See da Fonseca et al. (2015) for additional analysis of maize domestication using ancient DNA samples spanning ~6000 years.

Finally, a cautionary tale in the search for domestication genes is offered by observations on *Shrunken2* (*Sh2*; Whitt et al. 2002; Manicacci et al. 2006). This gene is involved in endosperm starch biosynthesis, and it was suggested as a target domestication gene from QTL studies that showed a seed-weight QTL in a maize-teosinte cross in the *Sh2* region. However, a more careful analysis by Whitt et al. and Manicacci et al. showed similar reduced levels of nucleotide diversity in both maize and teosinte at *Sh2*. A comparison with two sister species suggested that a sweep in the 3′ region of *Sh2* occurred in teosinte prior to domestication. Because the wild ancestors of our current crops were themselves subject to selection, caution is in order when declaring selection by contrasting diversity in a domesticated variety with that in a sister species of the progenitor, rather than the progenitor itself.

Relative Strengths of Selection on Domestication vs. Improvement Genes

An unresolved question concerns the relative strength of selection on domestication versus improvement genes, a contrast first discussed by Olsen et al. (2006). Based on local estimates of recombination rates and the length of depressed variation around the candidate genes, two domestication genes in maize had estimated selection coefficients of s between 0.02 (*tga1*; Wang et al. 2005) and 0.05 (*tb1*). However, the improvement gene *Y1* has a 600 kb sweep, giving an estimated strength of selection of $s = 1.2$ (Palaisa et al. 2004), and (as previously discussed) the strength of selection on the rice improvement gene *Waxy* (correcting for the effective recombination rate) is $s = 0.1$. This small initial sample suggested that there was stronger selection on improvement genes. Conversely, in the Hufford et al. (2012) survey of the maize genome, regions involved in domestication had an average estimated value of $s = 0.015$, while regions associated with improvement had an average estimated $s = 0.003$. One potential reason for the significant decrease in s for improvement genes is that the authors lumped together both tropical and temperate landraces, which could reduce the average estimated strength of selection that would be seen if landraces were differentially selected.

If selection is too intense (especially when selfing can occur), considerable linkage drag can allow deleterious alleles to accumulate and potentially favorable alleles to become lost. Because wild species that are subjected to very strong selection may not retain sufficient variation for subsequent improvement, it would have benefited our first breeders if selection during domestication was relatively weak.

Silkmoths and Flies

When one envisions domesticated animals, pets or farm animals usually come to mind. However, insect populations have been domesticated as well, most notably honey bees and silkmoths. Xia et al. (2009) sequenced the genomes of 29 lines of domesticated silkmoths (*Bombyx mori*) and of 11 lines from the wild progenitor species (*B. mandarina*). Their analysis clearly showed that a single domestication event gave rise to *B. mori*, with only a mild bottleneck (90% of the ancestral diversity is maintained). Using a joint statistic based on reduction in diversity (π_{mori}/π_{mand}) within a region, coupled with a low Tajima's D score, they identified slightly over 1000 regions of interest, spanning 3% of the genome. This suggested around 350 protein-coding regions as candidates for domestication genes (given the study's focus on structural, as opposed to regulatory, changes). Of these, 159 showed differential expression between *mori* and its wild relative, 90 of which are expressed in the silk gland, midgut, or testis. Two of the candidate genes in the silk gland were related to counterparts in *Drosophila* involved in transcriptional regulation of the glue genes (whose product is used to glue pupae to a substrate).

The selection pressures during *Bombyx* domestication were likely both deliberate (increased silk production) and unintentional (easier handling and better survival under cultivation). Such domestication selection pressures for growth and survival under laboratory conditions can potentially occur in any organism under long-term captivity, and *Drosophila* is no exception. Montgomery et al. (2010) examined the behavior of eight STRs over time in a series of replicate populations of different sizes with very tight control over the effective population sizes through complete knowledge of the population pedigree (meticulously accomplished by recording every set of parents in the population). The loss of heterozygosity at all STRs was about 12% faster than predicted from the pedigree-generated inbreeding coefficients, while the between-population divergence (F_{ST}) and temporal within-population changes in allele frequencies were significantly greater (by 25% and 33%, respectively) than predicted under drift. The authors interpreted these results as support for multiple ongoing partial sweeps throughout the genome influencing the dynamics at linked STRs, suggesting that there was ongoing selection to adapt to the domestication conditions.

Constraints on Finding Domestication and Improvement Genes Through Selective Signals

While numerous putative domestication genes have been located in a variety of species, there are reasons to suspect that many more have been missed. Given that domestication represents a sudden change in the environment, at least a fraction of any initial response likely results from standing variation. Thus, many domestication genes likely experienced soft or polygenic sweeps, potentially giving a very reduced signal (Chapter 8). For example, comparisons of dairy versus beef cattle largely rely on differences in allele frequencies, with few fixed differences (Example 9.3; Figure 9.1). Indeed, one might imagine that there is a bias in the current set of domesticated species, in that ancient farmers used the species that they could most easily exploit. This likely favored species with at least some standing variation for traits of interest, with these surpassing competing species for which farmers had to wait for new mutations to show improvement.

Other genomic features can complicate the detection and localization of domestication genes. If the ancestral species had low levels of polymorphism at the start of selection (perhaps from passing through bottlenecks and/or being under selection themselves), the additional reduction in polymorphism around the selected site will leave a much weaker signal. Thus, for some species it may be very difficult to detect signatures of selection, even for sites under strong selection. For example, Hamblin et al. (2006) found that the genome-wide background variation in sorghum (*Sorghum bicolor*) was too low to reliably detect signatures of selection, given the markers and density they used. Wild accessions of sorghum had levels of nucleotide diversity of around $\pi = 0.0027$, far lower than teosinte, while domesticated varieties had even lower levels ($\pi = 0.0008$). The presence of low background levels of variation in the progenitor coupled with an obvious strong bottleneck during domestication makes detection of regions under selection challenging, but not impossible (Casa et al. 2006).

Finally, the average size of a domestication sweep has important evolutionary implications. Signals of a sweep arise because of a reduction in the effective population size around the selected site, resulting in decreased efficiency of selection at linked genes (Chapter 8). Within a sweep region, linked deleterious alleles are more likely, and linked favorable alleles are less likely, to become fixed, compared to sites outside of the sweep. In species with high effective recombination rates, only small genomic regions (and hence few nontarget genes) are influenced by sweeps. However, in a highly selfing species, sweeps can influence the behavior of numerous genes well beyond the target site (as we saw with the *Waxy* gene in rice). Thus, in a species where a high fraction of the genome has been influenced by domestication sweeps, numerous deleterious mutations may have become fixed as a consequence of domestication. There is at least some suggestive evidence of this occurring in rice (Example 8.16), and it is expected to be more of a concern in selfing species. This reduction in fitness caused by domestication has been called the **cost of domestication** or the **domestication load** (Gaut et al. 2015).

With the advent of high-precision **gene-editing** methods, any potential domestication load now represents an opportunity. If deleterious alleles were indeed dragged along during domestication or improvement, in many cases they can be localized through whole-genome sequencing. For example, if elite lines contain nonsynonymous substitutions that are absent (or at very low frequency) in wild lines, some fraction of these may be deleterious. By using editing to revert them to their wild-type versions, the performance of an elite line may be improved, perhaps substantially so, especially if editing is done over a number of such sites.

10

Using Molecular Data to Detect Selection: Signatures from Multiple Historical Events

Model selection is a process of seeking the least inadequate model from a predefined set, all of which may be grossly inadequate as a representation of reality. Welch (2006)

Chapter 9 reviewed tests for detecting an ongoing, or very recently completed, *single episode* of positive selection. Here we examine the complementary issue, the nature of the cumulative signature left in divergence data by *multiple* historical selective events. In contrast to the large variety of tests for detecting ongoing or recent selection, only two basic approaches use divergence data to detect the signal from multiple episodes of positive selection. The first contrasts the levels of polymorphism within a reference population with the level of divergence between populations or species, using either different classes of sites within the same gene (the **McDonald-Kreitman**, or **MK**, **test**) or different genes (the **Hudson-Kreitman-Aguadé**, or **HKA**, test). Because these tests require a population sample to determine the amount of polymorphism, we refer to them as **population-based divergence tests**.

The second category of tests, which uses **phylogeny-based divergence** data, contrasts the rates of evolution at different sites within a gene over a number of species in a phylogenetic context. These tests do not require a population sample, as the signal comes entirely from the pattern of divergence, and not from polymorphisms. Specifically, for protein-coding sequences in the absence of positive selection, the rate of replacement substitution is generally expected to be less than the rate of silent substitution. A replacement rate *exceeding* the silent substitution rate provides a very robust signal of positive selection. However, when applied over an entire gene, this approach has almost no power, as any signal within a gene from a few positively selected codons is swamped by a much larger signal from the majority of codons that are likely to be under purifying (negative) selection. As a result, most current phylogeny-based tests scan a gene of interest for an excess of substitutions at *single codons* when examined over a phylogeny. Given their focus on *a very special class of events*—repeated positive selection on the same codon over a number of species—phylogeny-based divergence tests likely detect only a small fraction of actual selection episodes. By contrast, the HKA and MK tests are less stringent, as they simply require multiple substitutions over the *entire gene*. Finally, hybrid population genetic-phylogenetic tests are starting to appear (e.g., Wilson et al. 2011), but these tests will not be considered further here.

The tests outlined in the previous chapter are complementary to the approaches examined here, with each detecting signals of selection that would be missed by the other. Tests from Chapter 9 cast a wide net, in that many ongoing or recent events leave some signal, albeit perhaps a very weak one. However, this signal decays very quickly, so that most sites experiencing positive selection prior to some very modest amount of evolutionary time in the past ($>N_e$ generations) will leave essentially no signal for these tests (Table 8.2). Conversely, the HKA and MK tests entirely miss genes with one or two relatively recent adaptive substitutions, as such a small number of additional substitutions will not leave a sufficient divergence signal to be detected. They can, however, detect situations in which numerous adaptive substitutions have occurred across a gene during the divergence of two populations. Phylogeny-based tests are even more restrictive, showing a signal only in very special cases: those in which the same codons are repeated targets of selection over the species in a phylogeny.

The time scale over which positive selection can be detected varies between the two

Evolution and Selection of Quantitative Traits. Bruce Walsh & Michael Lynch.
Published 2018 by Oxford University Press.
DOI 10.1093/oso/9780198830870.001.0001

divergence-based approaches, as there must be a sufficient number of adaptive substitutions to give some signal above the expected number of neutral substitutions (Example 10.1). Given that the entire gene is the unit of analysis, HKA and MK tests require at least a relatively modest amount of evolutionary time to acquire a sufficient signal. Phylogeny-based tests, which have a much smaller unit of analysis (individual codons), typically require much longer time scales for a sufficient number of adaptive substitutions to accrue. As such, phylogeny-based methods can work over deep time scales, provided that the number of silent substitutions is not excessive (i.e., the sites have not become too saturated). They can also be applied over short time scales in rapidly evolving viruses (such as HIV), whose high mutation rates and large population sizes can introduce sufficient variation for multiple rounds of adaptive evolution to occur, even over short time spans.

Finally, it is worth noting that most applications of phylogeny-based methods use only a single sequence from each species, thus inflating the divergence, as a chance polymorphism in the sampled sequence may be recorded as a fixation. If only a few true fixations have occurred, this error can be significant. This is also a potential problem for population-based tests that use only a single reference sequence for making divergence estimates.

BRIEF OVERVIEW OF DIVERGENCE-BASED TESTS

We start with a short overview of population- versus phylogeny-based approaches before considering each approach in considerable detail. As was done in Chapter 9, the next few pages introduce the key ideas without most of the technical details.

A History of Selection Alters the Ratio of Polymorphic to Divergent Sites

Population-based tests contrast the patterns of within-species polymorphism and between-species divergence to see if they are in concordance with their neutral expectations. Under the equilibrium neutral model, two standard measures of polymorphism under the infinite-sites model are functions of $4N_e\mu$ (where μ is the per-site mutation rate): the nucleotide diversity, π, and the number of segregating sites, S. These have expected values of $E[\pi] = 4N_e\mu$ and $E[S] = 4N_e\mu\, a_n$, where a_n is a constant that depends only on the sample size, n (Equation 9.21a). Under the assumptions of the equilibrium neutral model, the relationship between polymorphism (measured by nucleotide diversity, π) and the between-population divergence (D) for the ith gene being considered is

$$\pi_i = 4N_e\mu_i, \qquad D_i = 2t\mu_i \tag{10.1a}$$

where N_e is the effective population size, and t is the divergence time in generations. Hence,

$$\frac{\pi_i}{D_i} = \frac{4N_e\mu_i}{2t\mu_i} = \frac{2N_e}{t} \tag{10.1b}$$

Because the gene-specific mutation rates cancel, under the equilibrium neutral model, the π/D ratio at all loci should be roughly the same, namely $2N_e/t$ (subject to random sampling). When polymorphism is instead scored as the number of segregating sites, S, then

$$\frac{S_i}{D_i} = \frac{2N_e a_n}{t} \tag{10.1c}$$

Again, this ratio is expected to be roughly constant over neutral genes.

Example 10.1. McDonald and Kreitman (1991a) examined the *Adh* (alcohol dehydrogenase) locus in the sibling species *Drosophila melanogaster* and *D. simulans*. Within this gene, they contrasted **replacement (nonsynonymous)** and **silent (synonymous)** sites. At the DNA level, a replacement-site mutation results in an amino acid change, while a silent-site mutation still codes for the ancestral amino acid. Equation 10.1c indicates that, under neutrality, the ratio of the number of segregating sites to the number of fixed differences should be the same for both categories of sites. This results in a simple association test, and significance can be assessed using either a χ^2 approximation or the (much better) Fisher's exact test, which accommodates small numbers in the observed table entries. Of the 24 fixed differences between the two species seen by McDonald and Kreitman, 7 were replacement-site mutations and 17 were silent-site mutations. The total number of polymorphic sites segregating in either species was 44, 2 of which were replacement and 42 of which were silent. The resulting association table becomes

	Fixed	Polymorphic
Silent	17	42
Replacement	7	2

Fisher's exact test gives a p value of 0.0073, indicating a highly significant lack of fit to the neutral equilibrium model. Based on the ratio of 42:2 silent/replacement polymorphisms, the expected number, x, of replacement fixations is $17/x = 42/2$, or $x = 0.81$, i.e., $\sim$1 replacement substitution is expected under neutrality. Because 7 were seen, this suggests roughly 6 adaptive substitutions, or that 86% (6/7) of the *Adh* amino acid substitutions between these species are adaptive.

A History of Positive Selection Alters the Ratio of Silent- to Replacement-site Substitution Rates

Phylogeny-based divergence tests do not require polymorphism data, but rather simply contrast the divergence rates at silent versus replacement sites. Silent sites are treated as proxies for neutral sites, although we have seen that they may be under (at least) weak selection (Chapter 8). Mutations at replacement sites are generally viewed as being under much stronger selection, most of it purifying. The primary evidence that such *negative* selection (removal of new deleterious mutations) is widespread is the observation that silent-site substitution rates are almost always much higher than those for replacement sites, when averaged over an entire gene. This pattern is expected if a higher fraction of mutations in replacement sites is deleterious relative to that in silent sites. However, there are cases where, for a limited region within a gene, the replacement-site substitution rate exceeds that for silent sites, suggesting the presence of adaptive fixation (i.e., positive selection).

While there are several variant notations in the literature, we use K_s to denote the per-site silent substitution rate and K_a to denote the per-site replacement rate between taxa (the subscript a indicating a change in an amino acid); K_{ns} and K_n are also used in the literature to denote replacement-site (i.e., nonsynonymous) substitution rates. A value of $K_a/K_s > 1$ indicates a long-term pattern of positive selection at replacement sites. As Example 10.2 illustrates, even if this is occurring at *specific regions* within a gene, when averaged over an *entire* gene, K_a/K_s is usually < 1. Thus, while an observation of $K_a/K_s > 1$ is almost universally accepted as a signature of a long-term pattern of multiple episodes of positive selection, such inflation is almost never seen if the entire gene is taken as the unit of analysis. Phylogeny-based methods (examined below) accommodate this concern by taking the codon as the unit of analysis, first placing genes within a phylogeny and then using codon-evolution models to test whether $K_a/K_s > 1$ for some subset of codons.

Example 10.2. One of the classic examples of using sequence data to detect signatures of positive selection is the work of Hughes and Nei (1988, 1989). They examined the major histocompatibility complex (MHC) Class I and Class II loci of mice and humans, highly polymorphic genes involved in antigen recognition. A large number of prior studies on other genes had found that an excess of silent substitutions is almost always the norm, implying that most replacement changes are selected against. Indeed, when one looks over an entire Class I (or II)

MHC gene, this pattern is also seen. The insight of Hughes and Nei was to use data on protein structure to specifically focus on the putative antigen-binding site and to compare this region with the rest of the gene as an internal control.

Hughes and Nei compared the ratio of silent- to replacement-site nucleotide substitution rates in the putative antigen recognition sites versus the rest of the gene. For both Class I and Class II loci, they found a significant excess of replacement substitutions in the recognition sites and a significant deficiency of such substitutions elsewhere. If both types of substitutions were neutral, the per-site rates should be roughly equal. If negative selection is acting, the expectation is that the silent-site substitution rate would be significantly higher (reflecting the removal of deleterious replacement mutations). However, if positive selection is sufficiently common among new mutations, one expects to find an excess of replacement substitutions. The observed patterns for both Class I and II loci were consistent with positive selection within the part of the gene coding for the antigen recognition site and purifying selection on the rest of the gene.

Divergence-based Tests are Biased Toward Conservative Sites

A major (but subtle) distinction between most methods in this chapter and those in Chapter 9 are that the latter usually have very little restrictions on the kinds of sequences being scanned for selection. In contrast, most divergence-based tests were built (at least initially) around analyses of protein-coding sequences (HKA is an exception), such as contrasts between silent and replacement sites or the substitution patterns at a codon (or set of codons) over a phylogeny. In such settings, these methods focus almost exclusively on detecting *structural* adaptations, namely, adaptive changes in the amino acid sequence. As we saw in Chapter 9, regulatory changes are thought to be at least as important as structural changes for short-term adaptation.

One reason for the focus on protein-coding regions in divergence-based tests is that one must be able to ***align homologous sequences***. Because they accept relatively few insertion or deletion mutations, long open-reading frames allow one to align homologous coding sequences, even over fairly substantial periods of evolutionary time. By contrast, this is often *not* the case for regulatory sequences, especially when considering that we still have a limited (albeit improving) ability to detect the full universe of such sequences. As shown in several examples below, divergence-based approaches have been applied to *highly conserved* regulatory regions, which offer a better opportunity for comparing homologous sequences over evolutionary time. However, this also biases these tests toward regions under strong functional constraints. Thus, the very interesting question of whether structural changes may be more important than regulatory changes for long-term adaptation cannot be fully addressed by divergence-based data alone, as these have a bias toward detection in highly conserved regions, whether structural or regulatory. Extensive regulatory changes in less-conserved regions may be entirely missed by most divergence-based tests. Despite these isses, there are hints starting to emerge of at least as many adaptive substitutions in noncoding regions as there are in coding regions (as we detail below).

What Fraction of the Genome is Under Functional Constraints?

The amount of metazoan DNA that codes for proteins and structural RNAs (the so-called **coding DNA**) is usually just a fraction of their total genome. The role of the remaining (and usually majority) component of the genome, the **noncoding DNA**, has been the subject of numerous debates as to its evolutionary role and function. This raises a central question of just what fraction of the genome is under some sort of functional constraint (and therefore, selection). Chiaromonte et al. (2003) denoted this fraction by α_{sel}, which is somewhat unfortunate notation given the widespread use of α for the fraction of *adaptive* substitutions (to be covered in detail shortly). One obvious approach for estimating α_{sel} is from the amount shared conserved sequences between two divergent taxa. For example, early studies searched for regions first shared between mice, humans, and dogs, and later over a wider range of mammals, arriving at the result that around 6% of the human genome is conserved over such time scales (Lindblad-Toh et al. 2005, 2011). This is six-fold more than the 1% of the human genome that codes for proteins (~33 MB out of a total of ~3100 MB; Church et al. 2009). Andolfatto (2005) estimated a much higher value of α_{sel}, between 40% and 70%, for *Drosophila melanogaster*, with about twice as many constrained sites in noncoding, as opposed to coding, regions. Such comparisons, especially when based on widely-divergent taxa, are simply *lower bounds*, as sequences under functional constraints can still turnover through time, escaping detection (Dermitzakis and Clark 2002). Indeed, Pheasant and Mattick (2007) suggested that the functional portion of the human genome may exceed 20%, basing their argument on the fact that rapidly evolving regions will not be detected through sequence conservation studies.

Further insight into α_{sel} can be gained by examining how the amount of conserved sequences shared between species pairs changes with their divergence times. This approach was used by Meader et al. (2010), who found that the fraction of shared conserved sequences among mammals decreased over time, and used the rate of this decrease to estimate that between 200 and 300 MB (6.5% to 10%) of the human genome is under functional constraints. A more refined estimate arrived at a value of around 8% (Rands et al. 2014). Hence, roughly 88% (7/8) of human constrained sites are found in noncoding regions. Meader et al. also used their approach on *Drosophila melanogaster*, finding an α_{sel} value of between 47% and 55%. Given around 22 MB for coding DNA and their estimate of 35–45 MB of constrained noncoding DNA, roughly two-thirds of the constrained sites are in noncoding regions.

These estimates of the amount of constrained noncoding DNA raise a number of important evolutionary questions (beyond the obvious one of their functional role). How strong is selection in noncoding regions? How often do adaptive mutations arise from these noncoding regions? What fraction of segregating deleterious mutations are attributable to these regions? While unbiased answer to these questions remain elusive, preliminary estimates based on conserved noncoding regions and on transcription factor binding sites suggest that noncoding DNA is likely a rich source of adaptive substitutions.

THE HKA AND MCDONALD-KREITMAN TESTS

Building on the basic ideas just introduced, we now develop the HKA test and present a much more in-depth discussion of the McDonald-Kreitman test, focusing on important caveats in its application.

The Hudson-Kreitman-Aguadé (HKA) Test

Hudson, Kreitman, and Aguadé (1987) proposed the first approach to jointly use polymorphism and divergence data. Unlike many of the other divergence-based tests, their's can be applied to any type of sequence data (not just a contrast between replacement and silent sites). Their **HKA test** is formulated as follows. Consider two species (or very distantly related populations) A and B, which are both at mutation-drift equilibrium with effective population sizes of $N_A = N_e$ and $N_B = \delta N_e$. Further assume that they separated $\tau = t/(2N_e)$ generations ago from a common population of size $N_e^* = (N_A + N_B)/2 = N_e(1+\delta)/2$, the average of the two current population sizes. Suppose $i = 1, \cdots, L$ unlinked loci are examined in both species. We allow the neutral mutation rate, μ_i, to vary over loci, but assume (for a given locus) that it has been the same in both species, and hence unchanged during divergence. The expected number of neutral segregating sites at locus i is a function of $\theta_i = 4N_e\mu_i$ in species A, and $4N_B\mu_i = 4(\delta N_e)\mu_i = \delta\theta_i$ in species B. The expected divergence between A and B is $2t\mu_i$, which we can express as

$$2t\mu_i = 2\,\frac{t}{2N_e}\,2N_e\mu_i = \tau\theta_i$$

Under this model, the levels of polymorphism (measured by the number of segregating sites) and divergence at the L loci are a function of $L + 2$ parameters: L gene-specific θ_i

values, and two demographic parameters (δ and τ) shared by all loci. To estimate these parameters, we have $3L$ observations: the numbers S_i^A and S_i^B of segregating sites at each of the L loci in each species or population, and the number, D_i, of substitutions between each pair of L loci. Under the HKA test, these data are first used to estimate the model parameters, and then a goodness-of-fit test is performed on the observed data. If the model provides a sufficiently poor fit, the equilibrium neutral model is rejected.

More formally, the HKA test statistic, X^2, is given by

$$X^2 = \sum_{i=1}^{L} X_i^2 \tag{10.2a}$$

where

$$X_i^2 = \frac{\left(S_i^A - \widehat{E}[S_i^A]\right)^2}{\text{Var}(S_i^A)} + \frac{\left(S_i^B - \widehat{E}[S_i^B]\right)^2}{\text{Var}(S_i^B)} + \frac{\left(D_i - \widehat{E}[D_i]\right)^2}{\text{Var}(D_i)} \tag{10.2b}$$

is the contribution to overall lack-of-fit from gene i. We use the notation $\widehat{E}[\,]$ to denote the estimate of an expectation, which is obtained by using estimates (also denoted by carets) of the parameters (Equations 10.3a through 10.3d). For n_A samples (haploid sequences from each of the L loci) from species A and n_B samples from species B,

$$\widehat{E}[S_i^A] = \widehat{\theta}_i a_{n_A}, \quad \widehat{E}[S_i^B] = \widehat{\delta}\,\widehat{\theta}_i a_{n_B}, \quad \text{where} \quad a_{n_x} = \sum_{i=1}^{n_x-1} \frac{1}{i} \tag{10.3a}$$

$$\text{Var}(S_i^A) = \widehat{\theta}_i a_{n_A} + \widehat{\theta}_i^{\,2} b_{n_A}, \quad \text{Var}(S_i^B) = \widehat{\delta}\,\widehat{\theta}_i a_{n_A} + \widehat{\delta}^{\,2}\,\widehat{\theta}_i^{\,2} b_{n_B}, \quad b_{n_x} = \sum_{i=1}^{n_x-1} \frac{1}{i^2} \tag{10.3b}$$

$$\widehat{E}[D_i] = \widehat{\theta}_i \left(\widehat{\tau} + \frac{1+\widehat{\delta}}{2}\right) \tag{10.3c}$$

$$\text{Var}(D_i) = \widehat{\theta}_i \left(\widehat{\tau} + \frac{1+\widehat{\delta}}{2}\right) + \left(\frac{\widehat{\theta}_i(1+\widehat{\delta})}{2}\right)^2 \tag{10.3d}$$

Equations 10.3a and 10.3b follow from the infinite-sites model (Equations 4.3a and 4.4a, respectively). Equation 10.3c follows if we rewrite

$$\theta_i \left(\tau + \frac{1+\delta}{2}\right) = 4N_e\mu_i \left(\frac{t}{2N_e} + \frac{1+\delta}{2}\right) = 2\mu_i t + 4\mu_i \frac{N_e(1+\delta)}{2} = 2\mu_i t + 4N_e^*\mu_i$$

The first term in the right-most expression is the between-population divergence due to new mutations, while the second term is the divergence from the partitioning of any initial polymorphism, $4N_e^*\mu_i$, present in the ancestral population. The HKA test statistic X^2 is approximately χ^2-distributed with $3L - (L+2) = 2L - 2$ degrees of freedom, given the $3L$ observations and $L+2$ parameters to estimate. Hudson et al. suggested the following system of equations for estimating the unknown parameters $(\theta_1, \cdots, \theta_l, \delta, \tau)$, given the $1 \le i \le L$ observed values of S_i^A, S_i^B, and D_i

$$\sum_{i=1}^{L} S_i^A = a_{n_A} \sum_{i=1}^{L} \widehat{\theta}_i \tag{10.4a}$$

$$\sum_{i=1}^{L} S_i^B = \widehat{\delta}\, a_{n_B} \sum_{i=1}^{L} \widehat{\theta}_i \tag{10.4b}$$

$$\sum_{i=1}^{L} D_i = \left(\widehat{\tau} + \frac{1+\widehat{\delta}}{2}\right) \sum_{i=1}^{L} \widehat{\theta}_i \tag{10.4c}$$

$$S_i^A + S_i^B + D_i = \widehat{\theta}_i \left(\widehat{\tau} + \frac{1+\widehat{\delta}}{2} + a_{n_A} + \widehat{\delta} \cdot a_{n_B} \right) \quad \text{for } i = 1, \cdots, L-1 \tag{10.4d}$$

Equations 10.4a through 10.4c are each single equations, while Equation 10.4d is a set of $L-1$ equations (Equation 10.4d is automatically satisfied for $i = L$ when Equations 10.4a–10.4c hold). This set of $L+2$ equations can be solved numerically for the L $\widehat{\theta}_i$ values unique to each locus and the common demographic values $\widehat{\delta}$ and $\widehat{\tau}$, thus generating the estimated values for the X^2 statistic (Equations 10.3a–10.3d). The HKA model assumes that there is no recombination within a gene but that there is free recombination between genes, thus treating distinct genes as independent. If a significant HKA value is found, the gene-specific X_i^2 values (Equation 10.2b) indicate which loci contributed the most to the lack of fit.

Modifications of the HKA test were proposed by Wright and Charlesworth (2004), who presented a maximum-likelihood version, and Innan (2006), who framed the test in terms of the polymorphism-divergence ratio, r. This formulation allowed Innan to consider a joint test involving r and a site-frequency measure (such as Tajima's D) to provide more support for selection at a site (Innan's **two-dimensional test**). An interesting application of this class of tests is the work of Ochola et al. (2010) in searching for vaccine targets in the malaria parasite *Plasmodium falciparum*. Their reasoning was that proteins in the parasite that are the target of naturally acquired host immunity (i.e., arms-race genes) are often under balancing selection. Hence, searching for loci with balancing-selection signals (a high HKA ratio coupled with high positive value of Tajima's D) can suggest potential candidates.

Example 10.3. Hudson et al. (1987) partitioned the *Adh* gene into two regions, silent sites and 4-kb of the 5′ flanking region, corresponding to a test using $L = 2$ loci. (The careful reader might be concerned that these loci are linked, while the HKA test assumes independence across loci. The high recombination rates in *Drosophila* result in LD generally being over only very small distances.) A sample of 81 *Drosophila melanogaster* alleles was examined, along with a single allele from its sibling species *D. sechellia*. Based on sequencing data, the divergence was 210 differences in the 4052-bp flanking region and 18 differences in the 324 silent sites, amounting to roughly equal levels of divergence per base pair between the two "loci." Based on restriction-enzyme data, within *melanogaster*, 9 of the 414 5′ flanking sites were variable, while 8 of 79 *Adh* silent sites were variable. Thus, while the per-site divergence was roughly equal, there was a four-fold greater polymorphism level at silent sites.

Hudson et al. modified their test to accommodate the use of polymorphism data from only a single population, as with no polymorphism data available from *D. sechellia*, there is no Equation 10.4b, and thus can be no S_i^B or a_{n_B} terms in Equation 10.4d. In this setting, δ cannot be estimated, so the authors assumed $\delta = 1$ (i.e., that both species have the same effective population size; an alternative approach would be to use the value of δ giving the smallest X^2 value). Given that there are different numbers of sites between the polymorphism and divergence data (which are based on restriction sites and sequence data, respectively), let θ_i be the population-scaled per-nucleotide mutation rate (for locus i), so that we have to weight the θ_i value for each term by the number of sites compared, giving Equations 10.4a, 10.4c, and 10.4d as, respectively,

$$S_1^A + S_2^A = 9 + 8 = a_{81}(414 \cdot \widehat{\theta}_1 + 79 \cdot \widehat{\theta}_2)$$

$$D_1 + D_2 = 210 + 18 = (\widehat{\tau} + 1)\left(4052 \cdot \widehat{\theta}_1 + 324 \cdot \widehat{\theta}_2\right)$$

$$D_1 + S_1^A = 210 + 9 = 4052 \cdot \widehat{\theta}_1 \,(\widehat{\tau} + 1 + a_{81})$$

where $a_{81} = \sum_{i=1}^{80} 1/i = 4.965$. The solutions to this system were found to be

$$\widehat{\tau} = 6.73, \quad \widehat{\theta}_1 = 6.6 \cdot 10^{-3}, \quad \text{and} \quad \widehat{\theta}_2 = 9.0 \cdot 10^{-3}$$

yielding the resulting modified X^2 statistic (Equation 10.2a but with the terms involving S_i^B in Equation 10.2b excluded) as 6.09. There are four observations (S_1^A, S_2^A, D_1, D_2) and three parameters to fit $(\theta_1, \theta_2, \tau)$, which results in a test with one degree of freedom. Because $\Pr(\chi_1^2 > 6.09) = 0.014$, the test indicates a significant departure from the equilibrium neutral model.

Although Equations 10.3 and 10.4 assume that all loci are autosomal, *with care*, sex-linked and organelle genes can also be incorporated. If all compared loci are X-linked, Equations 10.3 and 10.4 apply. However, if the loci are a mixture of autosomal and sex-linked, the θ_i terms for sex-linked loci must be multiplied by 3/4 (under equal sex ratios), as their expected levels of neutral polymorphism are $3N_e\mu_i$ (Begun and Aquadro 1991; see Lynch 2007 for more general results). Finally, while the HKA test can accommodate mitochondrial or chloroplast genes, they introduce three concerns. First, all sequences from a given organelle are generally completely linked (because such genomes typically are nonrecombining), and thus must be treated as a single locus. Second, organelle loci have a different effective population size from autosomal genes, which also requires a scaling of their θ value (typically by 1/4 to 1/2, but other values may be justified). The third issue is a bit more subtle. Given that most organelle genomes are only transmitted through females, the population structure and demographic history of nuclear genes (which are an average of the two parents) can be significantly different from that of organelle genes (females only). This raises special concerns in HKA comparisons of genes between nuclear and organelle genomes.

Example 10.4. Ingvarsson (2004) examined chloroplast (cpDNA) diversity in two plants in the genus *Silene* (family Caryophyllaceae). A standard HKA test contrasting four noncoding regions of the chloroplast (treated as a single locus) and two unlinked autosomal genes between *S. vulgaris* and *S. latifolira* gave a highly significant value, with most of the signal (using Equation 10.2b) coming from the cpDNA region. However, the estimated F_{ST} value (Chapter 2) for cpDNA was 0.546 versus 0.056 for nuclear genes, showing strong population structure at the organelle-gene level but only modest structure for nuclear genes. Ingvarsson attempted to correct for these between-gene differences in the amount of structure as follows. Under an island model of migration (Chapter 2), to a first approximation, population structure increases the amount of segregating sites and decreases the divergence, both by a factor of $1 - F_{ST}$. Ingvarsson thus corrected the observed number, S, of segregating sites by using $S_c = (1 - F_{ST})S$ and the divergence by $D_c = D/(1 - F_{ST})$. Applying these corrections to both the cpDNA and nuclear genes and using the S_c and D_c values in the HKA test yielded a nonsignificant result. Thus, the apparently strong signal of selection appears to simply be an artifact generated by nuclear and organelle genes having different population structures.

The McDonald-Kreitman (MK) Test: Basics

One of the most straightforward, and widely used, tests of selection was proposed by McDonald and Kreitman (1991a), who contrasted the amounts of polymorphism and divergence between two categories of sites within a single gene (Example 10.1). Typically, these categories are silent versus replacement sites, but the basic logic can be extended to other comparisons. Under the neutral theory, deleterious mutations are assumed to occur, but to then be quickly removed by selection, thus not contributing to either polymorphism or divergence (Figure 7.1). In the standard neutral-theory expressions for the amount of polymorphism $(4N_e\mu)$ and divergence $(2t\mu)$, μ is the *effectively neutral* mutation rate, which is the rate at which effectively neutral $(4N_e|s| \ll 1)$ mutations arise. While most mutations at silent sites may often be effectively neutral, a much smaller fraction, f, of new mutations at replacement sites are neutral, resulting in a lower effectively neutral mutation rate, $f\mu$. Given that f is the fraction of replacement mutations that is effectively neutral, $1 - f$ is a measure of *functional constraints*, with values of $1 - f$ near one ($f \simeq 0$) implying that most new mutations are not effectively neutral (i.e., they are deleterious). A minor bookkeeping detail is that the silent and replacement mutation rates in the MK test refer to the sum over all sites, so that $\mu_s = \mu n_s$ and $\mu_a = \mu f n_a$ are the total neutral mutation rates over the collection of n_s silent and n_a replacement sites in the gene of interest (generally $n_a > 2n_s$, as all second-base and many third-base positions within codons are replacement sites).

As before, under the equilibrium neutral model, the expected number of substitutions (D_i) in site class i is $2t\mu_i$, while the expected number of segregating sites (S_i) in a sample of n sequences is $a_n\theta_i$ (Equation 9.21a). Because S_i is a measure of the amount of polymorphism, we denote it by P_i to conform to the standard notation for MK tests. Thus, under neutrality,

$$\frac{D_a}{D_s} = \frac{2t\mu_a}{2t\mu_s} = \frac{2t\mu f n_a}{2t\mu n_s} = f\,\frac{n_a}{n_s}, \qquad \frac{P_a}{P_s} = \frac{S_a}{S_s} = \frac{a_n\theta_a}{a_n\theta_s} = \frac{4N_e\mu f n_a}{4N_e\mu n_s} = f\,\frac{n_a}{n_s} \tag{10.5a}$$

where the subscript a denotes replacement (amino-acid changing) sites, and s denotes silent sites. Hence, under the equilibrium neutral model, we expect that, on average,

$$D_a/D_s = P_a/P_s \tag{10.5b}$$

If some replacement sites are under positive selection, because of their rapid sojourn times relative to drift, these will generally contribute very little to the within-species polymorphism (Kimura 1969; Smith and Eyre-Walker 2002; Figure 7.1), but they will result in an excess of replacement substitutions, so that $D_a/D_s > P_a/P_s$. Similarly, note that

$$\frac{P_a}{D_a} = \frac{a_n\theta_a}{2t\mu_a} = \frac{a_n 4N_e\mu f n_a}{2t\mu f n_a} = \frac{a_n 2N_e}{t}, \qquad \frac{P_s}{D_s} = \frac{a_n\theta_s}{2t\mu_s} = \frac{a_n 2N_e}{t} \tag{10.5c}$$

and thus, under neutrality, we also have

$$P_a/D_a = P_s/D_s \tag{10.5d}$$

which is just a simple rearrangement of Equation 10.5b. It is worth noting that a very similar approach to the MK test was proposed by Templeton (1987, 1996), based on contrasting patterns in the tips versus interiors of estimated gene-tree topologies, and predates the MK test.

McDonald and Kreitman provided a more general derivation of the polymorphism ratio in Equation 10.5a, replacing $4N_e$ (the equilibrium value) by T_{tot}, the total time on all of the within-species coalescent branches (Chapter 2). By considering the ratio of the number of polymorphic sites in the two categories, the common term T_{tot} cancels, so that any effects of demography also cancel. Hence, ***provided the effectively neutral mutation rates remain unchanged***, the MK test is ***unaffected by population demography*** (Hudson 1993; Nielsen 2001). Because the coalescent structure that determines the amount of polymorphism is explicitly removed by using the P_a/P_s ratio, there is no assumption that the allele frequencies are in mutation-drift equilibrium nor any assumption about constant population size. This is a very robust feature not shared by most other tests of selection.

Thus, while Zhai et al. (2008) found that the HKA test was more powerful than the MK test when the equilibrium assumptions hold, the robustness of the MK test (and lack of robustness of the HKA test) when demographic issues are present favors the use of the former. However, as we will see shortly, the MK test is by no means foolproof, as changes in the effective population size can influence the *effectively neutral* mutation rates (the rate at which alleles with $4N_e|s| < 1$ arise), which can bias some of the comparisons used by the test. Another complication is that mildly deleterious alleles can contribute to within-species polymorphisms, but not to between-species divergence, and thus their presence inflates the polymorphism ratio over the divergence ratio, reducing the power to detect positive selection.

The MK test is performed by contrasting polymorphism and divergence data at silent and replacement sites for the gene in question. Given that these two ratios are expected to be equal under neutrality, the test uses a simple 2×2 contingency table (Example 10.1). The presentation of the data required for the MK test is often referred to as either an **MK table** or a **DPRS table**, the latter based on the (clockwise order) of the table's four categories: **D**ivergence (number of substitutions), **P**olymorphism (number of segregating sites), **R**eplacement, and **S**ilent (or **S**ynonymous):

	Divergence	Polymorphism
Silent	D_s	P_s
Replacement	D_a	P_a

Example 10.1 presented the original data used by McDonald and Kreitman, while Example 10.5 shows how their test can be modified to examine different regions within the same gene.

Example 10.5. Le Corre et al. (2002) examined the *FRIGIDA* (*FRI*) gene in *Arabidopsis thaliana*, a key regulator of flowering time. European populations show significant variation in flowering time, with potentially strong selection for earlier flowering having arisen following the end of the last ice age. For the data below, fixed differences (divergence) were obtained by comparing *A. thaliana* with *A. lyrata*, while data on numbers of segregating sites are based on *A. thaliana* populations.

	Fixed	Polymorphic	
Entire coding region	Fixed	Polymorphic	
Silent	59	7	
Replacement	68	21	Fisher test $p = 0.056$
Exon 1	Fixed	Polymorphic	
Silent	30	2	
Replacement	38	16	Fisher test $p = 0.013$
Exons 2 and 3	Fixed	Polymorphic	
Silent	29	5	
Replacement	30	5	Fisher test $p = 1.000$

The *FRI* locus clearly shows heterogeneity in patterns of selection when contrasting exon 1 with the remaining exons, and detecting such within-gene heterogeneity may provide important clues for a putative region under functional selection.

These data could be interpreted simply as a reduction on functional constraints in exon 1, resulting in a smaller fraction of segregating replacement mutations being deleterious. In principle, this could occur because of a shift in the selection pressures or for purely demographic reasons, such as a recent reduction in the effective population size increasing the effectively neutral mutation rate. However, there is a nice internal control in that exons 2 and 3 do not display a decrease in the ratio of fixed to polymorphic replacement sites relative to silent sites, which appears to rule out a reduction in effective population size in *thaliana* accounting for the reduction in constraints. The authors noted that roughly half of the replacement polymorphisms in exon 1 are loss-of-function mutations, which result in early flowering. Hence, it appears that the excess number of replacement polymorphisms in exon 1 likely results from selection for early flowering in some populations. Further, because a nonfunctional copy of *FRI* results in early flowering, there are a large number of mutational targets to achieve this phenotype (and hence a high effective mutation rate), which likely explains the large number of replacement polymorphisms. In effect, these data appear to show an ongoing multiple-origins soft sweep (Chapter 8).

This example introduces two important statistical issues in the analysis of MK data. First, one should always use Fisher's exact test for the goodness-of-fit (which can be found in standard statistical packages, such as R). The χ^2 and G tests (LW Appendix 4) for contingency tables are large-sample approximations, and tend to perform poorly when any table entry has an expected value of less than 5 (note the value of 2 in one of the MK data cells for exon 1). Second, many tests are often performed in a single analysis, raising the thorny issue of multiple comparisons (Appendix 4). If one desires a false-positive rate, q, over an entire collection of n independent tests, then the Bonferroni correction requires a critical value of $p = q/n$ for each test (Equation A4.4). Under this criterion, there is a probability of q that none of the tests are false positives when each is declared significant only when $p \leq q/n$.

Here there were three comparisons (entire, exon 1, and exons 2 and 3), suggesting a critical value of $p = q/3$ to give an experiment-wide value of q that none of the tests are false-positives. However, these three comparisons use overlapping data (e.g., the category "entire" contains all three exons). For the sake of discussion, assume that there are only $n = 2$ independent tests in this example. In that case, a significance threshold of $p = 0.01/2 = 0.005$ is required for each test to give a false-positive rate of 1% over the entire set of comparisons. Likewise, using $p = 0.05/2 = 0.025$ gives a false-positive rate of 5% over the entire collection. Hence, the experiment-wide significance level is closer to 5% than the 1.3% reported for exon 1. As detailed in Appendix 4, Bonferroni corrections are rather strict, and they can be improved by use of sequential Bonferroni methods, or (where appropriate) using control of the *false-discovery* rate. The latter gives the estimated fraction of false positives among a set of tests declared to be significant ("discoveries"); see Example 9.3 for an application, and Appendix 4 for much more details.

While initially presented as a contrast between silent and replacement sites within a single gene, the basic logic of the MK test is not limited to this specific type of comparison. Other types of sites can be contrasted (e.g., noncoding versus silent), and one can easily construct tests involving more than two categories with a simple extension of the contingency table logic underlying MK tests (e.g., Hudson 1993; Templeton 1996; Podlaha et al. 2005; Egea et al. 2008; Chen et al. 2009). Further, as Example 10.5 highlights, one often performs separate MK tests in different regions of the same gene. The general issue of how to detect selection heterogeneity based on a scan of a region is examined by McDonald (1996, 1998) and Goss and Lewontin (1996).

A McDonald-Kreitman test will be significant when P_a/D_a is significantly different from P_s/D_s (Equation 10.5d). Because it is assumed that the silent-site ratio is unchanged by selection, a significant MK test can occur either through an excess of replacement polymorphisms (P_a too large relative to D_a and P_s/D_s) or through an excess of replacement substitutions (D_a too large relative to P_a and P_s/D_s). The **neutrality index** of Rand and Kann (1996),

$$NI = \frac{P_a/D_a}{P_s/D_s} = \frac{P_a\, D_s}{P_s\, D_a} \tag{10.6a}$$

indicates which of these two scenarios occurs. Note that *NI* is simply the odds ratio for the MK contingency table (Jewell 1986). A value greater than one indicates more polymorphic replacement sites than expected, while a value less than one indicates an excess of replacement substitutions. Values less than one suggest that some of the substitutions are adaptive, while values greater than one are suggestive of weakly deleterious segregating alleles.

Note that *NI* is not defined if either P_s or D_a are zero, and it is biased if either is small (Stoletzki and Eyre-Walker 2011). Hence, its use is problematic when the gene being considered shows little divergence. When the observed cell numbers in any MK table are small (less than 5), a number of corrections are possible, which basically start by adding an extra count to D_a and P_s (Haldane 1956; Jewel 1986). Stoletzki and Eyre-Walker (2011) noted that these corrections are still biased, and they proposed a **direction of selection** (***DoS***) statistic,

$$DoS = \frac{D_a}{D_a + D_s} - \frac{P_a}{P_a + P_s} \tag{10.6b}$$

Positive values indicate an excess of replacement substitutions (suggesting adaptive evolution), while negative values imply an excess of replacement polymorphisms (suggesting that slightly deleterious alleles are segregating).

While the *DoS* statistic is appropriate when comparing divergence to polymorphism features of genes as a function of some other variable (such as recombination rate or GC content), other approaches have been used when the aim is to return a single summary statistic for the entire genome. A simple average of the *NI* values over all sampled genes is biased, as genes for which *NI* is not defined (P_s or D_a are zero) are excluded, and those genes with small values for either P_s or D_a return biased estimates. Example 10.8 illustrates one commonly used approach to avoid these issues, namely, summing over all sites to create a grand MK table for the entire collection of sampled genes.

Example 10.6. Andolfatto (2005) examined 35 coding and 153 noncoding fragments from a Zimbabwe sample of 12 *D. melanogaster* X chromosomes, with a single *D. simulans* X as an outgroup. The numbers of observed polymorphic and divergent sites were then lumped into various classes as follows:

		Polymorphisms		Fisher Test p value	
Mutational Class	Fixed	All sites	Minus singletons	All sites	Minus singletons
Silent	604	502	323	——	——
Replacement	260	115	52	$4.7 \cdot 10^{-7}$	$4.3 \cdot 10^{-10}$
Noncoding	3168	2386	1295	$1.4 \cdot 10^{-2}$	$5.2 \cdot 10^{-3}$
5′ UTRs	328	160	71	$2.7 \cdot 10^{-6}$	$1.7 \cdot 10^{-10}$
3′ UTRs	143	86	36	$3.3 \cdot 10^{-2}$	$8.2 \cdot 10^{-5}$

Given the small sample size ($n = 12$ chromosomes), polymorphism data are reported both as the total number of segregating sites (all sites) and the total number of segregating sites minus the singletons. The logic for removing singletons is the concern that slightly deleterious alleles can contribute to segregating sites (although they will be rare) but are unlikely to become fixed, and if retained in the analysis, will result in the polymorphism ratio overpredicting the number of fixed sites. Using the silent class as the neutral reference, McDonald-Krietman tests were performed against each of the four remaining categories (replacement, noncoding, 5′ UTR, and 3′ UTR), and computed separately using either all polymorphisms or only polymorphisms that were not singletons. The exclusion of singletons ("Minus singletons" column above) decreases the p values (increasing significance) in all cases. Even after correcting for multiple tests, all of the comparisons based on polymorphisms minus singletons were highly significant.

Andolfatto also observed that the average nucleotide diversity, π, was higher for silent sites than for any of the other categories displayed above. This suggests that there are stronger constraints on the sampled noncoding regions than on silent sites, and hence stronger purifying selection on these noncoding sites. Conversely, these test values all show excessive substitutions relative to the amount of within-population variation, suggesting that many of the differences were likely fixed by positive selection. Both of these results (stronger purifying selection on polymorphisms and stronger positive selection for substitutions) for noncoding DNA relative to silent sites were very surprising, and they suggested that part of what is called noncoding DNA may have some functional role (the same appears to be at least partly true for humans: ENCODE Project Consortium 2012). A similar study using polymorphism data from *D. simulans* (with *D. melanogaster* as the outgroup), which has a larger effective population size than *D. melanogaster*, found an even stronger signature of purifying selection against *D. simulans* noncoding polymorphisms (Haddrill et al. 2008).

One concern when dealing with noncoding DNA is obtaining the correct alignment to ensure that homologous sites are being compared. This can be problematic for even moderately divergent species, as insertions and deletions run rampant, making correct alignment nearly impossible. Care must then be taken, as one may discard much of the noncoding sequence because of alignment issues, which could enrich the sequences remaining in the analysis with those sites under stronger functional constraints (which are more conserved and thus more easily aligned). Conversely, with coding regions, strong historical selection to keep the sequence in frame usually results in few insertions and deletions.

Example 10.7. Consider Le Corre et al.'s data on the *FRI* gene (Example 10.5). For exon 1, the neutrality index is

$$NI = \frac{P_a/D_a}{P_s/D_s} = \frac{16/38}{2/30} = 6.42$$

showing that the significant result is due to an excess of segregating replacement sites. Conversely, for exons 2 and 3

$$NI = \frac{5/30}{5/29} = 0.97$$

suggesting a good fit to the neutral model, with neither an excess of polymorphic site nor of fixed replacement sites.

Our interpretation of the signal in exon 1 was as a sign of ongoing selection of alleles for earlier flowering (Example 10.5). However, the *NI* value is also consistent with an excess of slightly deleterious alleles in this region, thus inflating the levels of replacement polymorphisms. The lack of such a signal in exons 2 and 3 argues against this, but it remains a formal possibility that slightly weaker selection in exon 1 (relative to exons 2 and 3), coupled with a genomewide reduction in N_e, could account for the excess polymorphism in exon 1. However, evidence for a recent population expansion argues against this.

This discussion raises the more general question of how often we can safely use the MK test to detect signatures of ongoing *positive* selection. For example, what is the impact of an ongoing hard sweep? This would greatly reduce the number of all segregating sites (both silent and replacement), and hence give the MK test little, if any, power. Generally speaking, the safest interpretation of excess replacement polymorphisms is that they are slightly deleterious. Example 10.6 shows that additional information is required to make a case that these segregating sites are beneficial. Further, such an excess is expected only in rare settings, such as a multiple-origins soft sweep, in the case of *FRI* likely fueled by a large mutational target size (simple deactivation of a function) to produce the putatively favored phenotype (early flowering).

Example 10.8. Bustamante et al. (2005) sequenced roughly 11,600 genes in 39 humans and contrasted the results with human-chimp divergence at these same loci. Summing over all sites, the resulting DPRS table (where SNPs denote polymorphic sites) was

	Divergence	SNPs
Silent	34,099	15,750
Replacement	20,467	14,311

As in Example 10.6, this analysis differs from a standard MK test, as the values for a large number of loci are aggregated into a single table. The resulting p value, $< 10^{-16}$, was highly significant, meaning that the neutral model is rejected.

What is the source of the discrepancy? Equation 10.6a gives the neutrality index as

$$NI = \frac{P_a/D_a}{P_s/D_s} = \frac{14,311/20,467}{15,750/34,099} = 1.514$$

showing that the lack-of-fit to the neutral model is driven by an excess of replacement polymorphisms (SNPs). The authors suggest that these polymorphisms are mainly deleterious, a view echoed by Hughes et al. (2003). Consistent with this conclusion, in an analysis of ~47,500 replacement SNPs in a sample of 35 humans, Boyko et al. (2008) used the site-frequency spectrum to estimate that 27–29% of these SNPs were effectively neutral, 30–42% were moderately deleterious, and nearly all of the rest were highly deleterious (we will discuss how such values are obtained shortly). This large fraction of segregating deleterious alleles significantly lowers the power of MK tests. Indeed, Charlesworth and Eyre-Walker (2008) noted that because of excessive replacement polymorphisms, MK tests in humans are very underpowered.

While commonly used, the approach in Examples 10.6 and 10.8 of summing the MK tables for single genes to create a single grand MK table for the entire genome is potentially problematic because of the **Yule-Simpson effect** (Yule 1903; Simpson 1951), also known as **Simpson's paradox** (Blyth 1972). This is a well-known statistical phenomena wherein the results of individual 2×2 contingency tables suggest a trend in one direction, whereas their amalgamated table suggests a trend in the opposite direction (reviewed by Good and Mittal 1987). More generally, the (unweighted) average of the odds ratios over individual tables is different from the odds ratio in the amalgamated table. This discrepancy commonly arises when there are large disparities in the sample sizes over individual tables. For MK data, this is expected when there are large between-gene differences in the number of replacement substitutions (D_a). To avoid this issue, Stoletzki and Eyre-Walker (2011) suggested using a weighted approach proposed by Tarone (1981) and Greenland (1982) for combining the odds ratio over general 2×2 contingency tables,

$$NI_{TG} = \frac{\sum_i D_{si} P_{ai} / (P_{si} + D_{si})}{\sum_i P_{si} D_{ai} / (P_{si} + D_{si})} \tag{10.6c}$$

where i denotes the ith gene. This index is defined for all genes that show any silent-site variation (either P_s or D_s is nonzero), and also weights each gene by its total silent-site sample variation ($P_s + D_s$).

The McDonald-Kreitman Test: Caveats

One of the initial criticisms of the McDonald-Kreitman test was that estimates of the number of segregating sites are rather sensitive to sampling, especially when the sample size is small (Graur and Li 1991; Whittam and Nei 1991). McDonald and Kreitman (1991b) countered that this problem is not serious, as these effects would equally influence estimates of the number of polymorphic silent and replacement sites. While largely correct, this is not always true, however, as there are generally two- to three-fold more potential replacement sites than silent sites, giving the former a slightly smaller sampling error. However, this difference in variances has more to do with power, and is unlikely to lead to false positives.

One potentially significant advantage of the MK test is that it does not assume constant population size or that mutation-drift equilibrium has been reached, and hence is rather robust against many of the demographic concerns that plague other tests. Balancing this advantage are two subtle (but serious) problems, both relating to how the distribution of fitness values for new alleles impacts the observed data (polymorphisms and substitutions).

First, the MK framework assumes that *deleterious mutations are strongly deleterious and make essentially no contribution to either the number of segregating or fixed sites.* In fact, however, weakly deleterious mutations (i.e., $-10 < 4N_e s < -1$) can contribute to segregating polymorphisms (especially because the MK test uses the number of polymorphic sites, not their frequencies), but they are highly unlikely to become fixed (Figure 7.1). Such mutations are overrepresented in polymorphic sites relative to fixed sites, which reduces the power of the MK test to detect an excess of replacement substitutions (and hence a signature of positive selection). We assume that the impact from any overrepresentation of selected polymorphisms at silent sites (our neutral proxy) is small, as these are either neutral or under very weak purifying selection. Conversely, overrepresentation is potentially a significant problem at polymorphic replacement sites. One proposed correction for this problem is to drop "rare" polymorphisms, but this is a rather subjective endeavor. Dropping singletons (Templeton 1996) as in Example 10.5 provides one simple correction, while other authors (e.g., Fay et al. 2002; Smith and Eyre-Walker 2002; Gojobori et al. 2007) have suggested including only "common" polymorphisms in the analysis, such as those with minor-allele frequencies above 0.10. We return to this issue shortly.

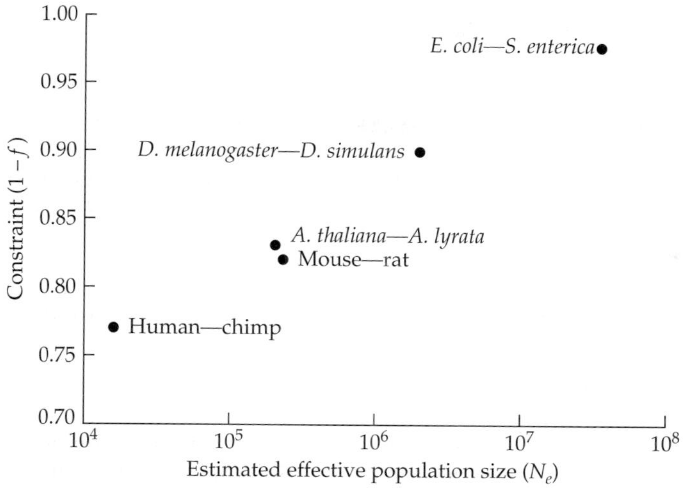

Figure 10.1 The estimated constraint, $1 - f$, on replacement sites as a function of effective population size, where f is the ratio of effectively neutral mutation rates (the fraction of new mutations that efficiently behave as neutral alleles) at replacement versus silent sites. As N_e increases, more deleterious mutations move from the effectively neutral class into the strongly deleterious class (f decreases), reducing the effectively neutral mutation rate and increasing the amount of constraint on a gene. (After Wright and Andolfatto 2008.)

The second concern is even more problematic. At the heart of the MK test is Equation 10.5a. Under the neutral hypothesis, the ratio of polymorphic sites and the ratio of substitutions both estimate the same quantity, f (scaled by the sample-size correction factor n_a/n_s), the ratio of effectively neutral mutation rates for the two categories. Recalling (Chapter 7) that any mutation for which $4N_e|s| \ll 1$ behaves as if it were effectively neutral, the caveat is that the *effectively neutral mutation rate, $f\mu$, changes with N_e*. It is important to stress that the *total* mutation rate, μ, remains unchanged, but the fraction, f, of these mutations that are effectively neutral can decline with increasing N_e, resulting in a decline in $f\mu$. Figure 10.1 shows that estimates of f do indeed decrease as the effective population size, N_e, increases, as the amount of constraint, $1 - f$, increases with N_e. For the same distribution of selection coefficients, one can raise (or lower) f (and hence the effectively neutral substitution rate) by decreasing (or increasing) the effective population size. If the effective population size is significantly different during the divergence phase (when substitutions were fixed) than in the current phase (which generates the observed number of polymorphisms), then these two phases could have different fractions of mutations that are effectively neutral. Because the ratios D_a/D_s and P_a/P_s estimate the f values for these two different phases, they can have different expected values.

McDonald and Kreitman (1991a) were aware that an increase in the effective population size could create a situation where slightly deleterious mutations that were fixed during divergence under a smaller population size do not even contribute to within-species polymorphisms. Such an increase in N_e following the bulk of divergence time would result in an inflated value of D_a and a deflated value of P_a, resulting in an inflated D_a/P_a ratio, and hence a false signal of positive selection. Eyre-Walker (2002) showed that even a modest increase in N_e can generate such false signals, and that the problem is exacerbated by culling rare polymorphisms, which (as discussed above) is common practice. In the words of Hughes (2007), this feature implies that the MK test "cannot distinguish between positive Darwinian selection and any factor that causes purifying selection to become relaxed or to become less efficient." Phrased in terms of the neutrality index (Equation 10.6a), an *NI* value > 1 can be generated by either segregating deleterious alleles or by a relaxation in the functional constraints during the polymorphism phase. The latter could occur in response to a change in the environment (Example 10.9) or a change in N_e. Conversely, a value of

$NI < 1$ (which is normally taken as support for adaptive evolution) could similarly be generated by a relaxation of functional constraint during the divergence phase, so that more mutations (relative to those currently segregating in the population) were effectively neutral, and hence fixed. This effect can also occur between populations of the same species. For example, Lohmueller et al. (2008) observed a higher fraction of segregating deleterious mutations in human populations from Europe than from Africa, which they attributed to the bottleneck in the founding European population (and hence a reduction in N_e) during the migration out of Africa.

Example 10.9. An example of some of the potential difficulties in interpreting the results of a McDonald-Kreitman test was seen in a study of the human melanocortin 1 receptor (*MC1R*), a key regulatory gene in pigmentation (Harding et al. 2000). In comparing the canonical *MC1R* haplotype in humans with a sequence from chimpanzees, these authors found 10 replacement and 6 silent substitutions. An African population sample revealed no replacement and 4 silent polymorphisms, giving the MK table as

	Fixed (Human-Chimp)	Polymorphic (African)
Silent	6	4
Replacement	10	0

Fisher's exact test gives a p value of 0.087, close to significance. Taken at face value, one might assume that these data imply that the majority of the replacement substitutions between human and chimp were selectively driven. However, the authors also had data from populations in Europe and East Asia, which showed 10 replacement and 3 silent polymorphisms, resulting in a new MK table:

	Fixed (Human-Chimp)	Polymorphic (Europe/East Asia)
Silent	6	3
Replacement	10	10

with a corresponding p value of 0.453. The authors suggested that the correct interpretation of these data is as very stringent purifying selection due to increased functional constraints in African populations (due to selection for protection against high levels of UV exposure), with a release of constraints in Europe and East Asia. Asians in Papua New Guinea and India (populations living in high-UV environments) also showed very strong functional constraints (few replacement polymorphisms), consistent with a model of selection for UV protection.

The key point is that the ***population chosen as the reference standard for the polymorphism ratio is critical***. The two tests above used the same divergence data, but the significance (or lack thereof) of the MK test critically depended on whether the population sample was African or European and East Asian.

Example 10.10. The effect of slightly deleterious alleles on the expected value of the neutrality index was examined by Welch et al. (2008). Assume that the scaled selection coefficient values, $\gamma = 4N_e s$, of new mutations are drawn from a reflected gamma distribution (Equation A2.25a) over the range of $-\infty < \gamma < 0$, with a shape parameter of $\beta > 0$ (the coefficient of variation for γ is given by $1/\sqrt{\beta}$, with $\beta = 1$ corresponding to the exponential distribution). Under these conditions, Welch showed that the expected value of the neutrality index is

$$NI \simeq 1 + \beta K$$

where $K > 0$ is a function of the sample size (n).

Welch further cautioned that even the usual interpretation of $NI < 1$ as positive selection is not generally true. Assume we have the same model as above, with new mutations only being deleterious, but now suppose that the population size has changed over time. In particular, suppose that the population had a constant size, N_e, for some fraction, q, of the total divergence time, after which it increased by a factor of $\delta > 1$ to $\delta N_e > N_e$. In this case, the expected value of the neutrality index becomes

$$NI \simeq \frac{1 + \beta K}{1 + q(\delta^{\beta} - 1)}$$

Welch noted that if the population expansion is recent or substantial (q is near one and/or δ is large), NI can easily be less than one, giving a false signature of positive selection.

Finally, silent sites may be a rather poor proxy for neutral sites, especially in species with large effective population sizes. Chapter 8 reviewed codon usage bias, wherein some synonymous codons are preferentially used over others. Selection is thought to be weak on such sites, but it can still have an impact (Hartl et al. 1994; Akashi 1995). For example, DuMont et al. (2004) found that "preferred" synonymous codons are substituted significantly faster than unpreferred synonymous changes at the *Notch* locus in *D. simulans*, while *D. melanogaster* (with a smaller N_e) has a significantly higher substitution rate for unpreferred changes. The consensus on codon bias is that the strength of selection is very weak ($s < 10^{-5}$), making synonymous changes effectively neutral in small populations but subject to selection pressures in populations where $4N_es$ is sufficiently large.

The nature of selection on third-base positions is further complicated by the observation that weak selection for preferred codons is not the only constraint on synonymous sites. Lawrie et al. (2013) estimated that roughly 20% of synonymous sites in *Drosophila melanogaster* are under very strong functional constraints (with an estimated $-4N_es$ on the order of 10^3 based on the distribution of rare alleles in the site-frequency spectrum). It is striking that they found that this strong constraint was *independent* of codon bias, being driven by some feature other than selection on preferred codons.

In some settings, selection on the neutral proxy sites may actually provide some robustness to MK tests. As mentioned, false positives under an MK test can be generated by an increase in the effective population size. Eyre-Walker (2002) showed that selection on the neutral proxy sites (synonymous codons) restricts the conditions under which a false positive signal can arise via a change in N_e. Presumably this occurs because changes in N_e influence both the replacement and the neutral proxy sites, thus creating somewhat of an internal control.

Dominance in Fitness and the MK Test

One might be concerned that dominance alters the ratio of polymorphism to divergence at replacement sites, as both the frequency spectrum and the probability of fixation for a selected site are influenced by dominance (Chapter 7). While Weinreich and Rand (2000) and Williamson et al. (2004) showed that most types of dominance have little impact on this ratio, an important exception concerns weak-to-moderate overdominance. Williamson et al. showed that overdominance can increase the substitution rate at replacement sites relative to that predicted from the amount of polymorphism, giving a signal of positive directional selection in an MK test (a neutrality index less than one). The reason for this behavior follows from Robertson's (1962) classic result examined in Chapter 7, wherein overdominance can *increase*, rather than retard, the rate of fixation when the equilibrium allele-frequency values are extreme (a minor equilibrium allele-frequency of 0.2 or less; see Figure 7.4). The idea is that selection rapidly moves allele frequencies to these equilibrium values, at which point drift can cause alleles to become fixed if selection is relatively weak.

Fluctuating Selection Coefficients and MK Tests

While we have been assuming that the selection coefficient, s, on a new mutation remains constant over its sojourn in the population, this is likely not the case. The impact of fluctuating selection on MK tests was examined by Huerta-Sanchez et al. (2008) and Gossmann et al. (2014), both of which assumed selection coefficients randomly sampled over time from a distribution with a mean value of zero. Huerta-Sanchez et al. found that fluctuating values of s result in an increase in the probability of fixation (relative to a neutral allele) and a decrease in the amount of polymorphism. This can generate false positives for positive selection in an MK test. Gossmann et al., however, noted that the results are more subtle. Because selection coefficients are randomly sampled over time, some alleles will, by chance, end up with a net positive value of s over their entire sojourn, and such mutations contribute disproportionately to levels of polymorphism and divergence. They concluded that MK signals under fluctuating selection are therefore genuine, as fixed mutations are those that, by chance, end up with a net positive s value over their entire history on their way to fixation. Further, they found that the real impact of fluctuating selection is that MK methods tend to *underestimate* the fraction of adaptive sites, as those alleles with $E[s] > 0$ during their sojourn to fixation tend to be undercounted.

Recombinational Bias in Extended MK Tests

The standard MK test, contrasting silent and replacement sites *within a single gene*, is very robust to recombination. As noted by Andolfatto (2008), this occurs because the comparison sites are fully interdigitated, with silent sites interspersed among replacement sites. If we denote these two classes of sites by a and b, standard MK tests have comparisons of the form *abababab*, with adjacent sites sharing the same coalescent structure. In this setting, recombination (or lack thereof) has little effect. Conversely, recombination *can* impact extensions of the MK test that compare classes of *non-interspersed* sites, such as contrasting the silent sites in a gene with 3′ or 5′ UTR sites adjacent to that gene. Again denoting classes as a or b, these comparisons are now of the form *aaaa–bbbb*. Hence, they are *not* fully interdigitated, and potentially have different coalescent structures when the distance between these comparison blocks is sufficiently large. Andolfatto (2008) found that in such noninterspersed settings, recombination can indeed bias the MK test, generating an increased number of false positives. The bias is most severe for noninterdigitated comparisons when the ratio of recombination to mutation rates is around 1.0, whereas for very small (no recombination), and very large (unlinked sites), values of this ratio, there is little bias.

ESTIMATING PARAMETERS OF ADAPTIVE EVOLUTION

As shown in Example 10.1, DPRS tables lead to a simple prediction about the expected number of replacement substitutions, given the ratio of silent to replacement polymorphisms. Under certain assumptions, this allows us to directly ask how many (if any) excess substitutions at replacement sites have occurred within a target gene. While straightforward, one issue is power: at any particular gene, the true excess has to be fairly substantial in order for the MK test to be significant. However, when we sum up such excesses over a large number of genes, we have the power to detect even a small average increase. This ability to look at the cumulative evidence over a large number of genes in order to detect a small average individual effect is one of the advantages of genomewide studies. A second approach to estimating the number of adaptive substitutions places this idea into a more formal statistical framework, called the **Poisson random field model**, which allows us to estimate the average selection coefficients of sites under positive selection. We will examine this latter approach shortly.

Estimating the Fraction, α, of Substitutions That are Adaptive

It was quickly realized that DPRS tables offer much more than simply an opportunity to test for selection (Sawyer and Hartl 1992; Charlesworth 1994b; Fay et al. 2001, 2002; Smith and Eyre-Walker 2002). A neutrality index < 1 indicates that the observed number of replacement substitutions is greater than expected from the ratio of the number of replacement to silent polymorphic sites. Assuming that the P_a/P_s ratio does indeed reflect the ratio of effectively neutral mutation rates at these two classes of sites, then, when coupled with the observed number of silent substitutions, it predicts the expected number of effectively neutral replacement substitutions. Any statistically significant excess over this predicted value either reflects sites fixed under positive selection or is the result of changes in the effectively neutral mutation rate between the population (or populations) generating the polymorphism data and those responsible for the divergence data. As mentioned, excessive divergence can occur if the effective population size was much smaller during the divergence phase, allowing more slightly deleterious mutations to escape selection and become fixed.

As above, let μ and $f\mu$ denote the per-site rates at which effectively neutral mutations arise at silent and replacement sites, so that $\mu_a = f\mu n_a$ and $\mu_s = \mu n_s$ are the total rates for the replacement and silent sites in our sample (where n_a and n_s are, respectively, the number of replacement and silent sites). Under neutrality, the expected numbers of effectively neutral substitutions for each class are $D_s = 2\mu_s t$ and $D_{a,n} = 2\mu_a t$. Now suppose there are η_a additional replacement substitutions fixed by positive selection, giving the total number of replacement substitutions as $D_a = D_{a,n} + \eta_a = 2\mu_a t + \eta_a$. Ideally, we would like to estimate both the number, η_a, and the fraction, $\alpha = \eta_a / D_a$, of replacement substitutions that are adaptive. To estimate η_a, note that the expected number of segregating sites for category x is given by $\theta_x a_n$ (Equation 9.21a), yielding $P_s = 4\mu_s N_e a_n$ and $P_a = 4\mu_a N_e a_n$, where the latter assumes that the vast bulk of segregating sites are neutral (adaptive mutations are assumed to be both rare and also fixed quickly, and hence make little contribution to P_a). First note that

$$D_s \frac{P_a}{P_s} = 2\mu_s t \frac{\mu_a}{\mu_s} = 2\mu_a t \tag{10.7a}$$

From above, this last expression is simply the expected number of neutral replacement substitutions, $D_{a,n}$, and because $\eta_a = D_a - D_{a,n}$, our estimate of the number of adaptive replacement substitutions becomes

$$\widehat{\eta}_a = D_a - D_s \frac{P_a}{P_s} \tag{10.7b}$$

as obtained by Charlesworth (1994b), Fay et al. (2001, 2002), and Smith and Eyre-Walker (2002). This immediately suggests an estimator for the fraction, α, of replacement substitutions that are adaptive,

$$\widehat{\alpha} = \frac{\widehat{\eta}_a}{D_a} = 1 - \frac{D_s P_a}{D_a P_s} = 1 - NI \tag{10.7c}$$

Note that a positive estimate of α requires a neutrality index <1. Using the data from Example 10.6 for noncoding regions on the X chromosome in *D. melanogaster*, $\widehat{\alpha} = 1 - 0.906 = 0.094$ using all polymorphic sites, and $\widehat{\alpha} = 1 - 0.764 = 0.236$ if singletons are ignored. Hence, between roughly 10% and 25% of all substitutions in these noncoding regions might be adaptive. Similarly, Kousathanas et al. (2010) obtained estimates of around 10% adaptive substitutions in the immediate up- and downstream regions around protein-coding genes in the house mouse (*Mus musculus castaneus*).

Finally, note that we can estimate the fraction, f, of replacement mutations that are effectively neutral by rearranging Equation 10.5a to

$$\widehat{f} = \frac{P_a}{P_s}\frac{n_s}{n_a} = \frac{P_a/n_a}{P_s/n_s} \tag{10.7d}$$

This is simply the ratio of the fraction of replacement sites that are polymorphic divided by the fraction of silent sites that are polymorphic. Recall that $1-f$ is a measure of the amount of constraint relative to a silent site, as f is the fraction of replacement-site mutations (relative to those at silent sites) that are effectively neutral. For *Drosophila*, estimated $1 - f$ values are 0.94 for replacement sites, 0.81 for UTRs, 0.61 for intergenic regions, and 0.56 for intron sequences (summarized by Sella et al. 2009). Further, Halligan and Keightley (2006) showed that silent sites are not the fastest evolving sequences in *Drosophila*: rather, this distinction belongs to **FEI (fastest evolving intronic) sites**. In comparison to these sites, the constraint on silent sites is 0.09, suggesting that 9% of new silent mutations are deleterious. Lawrie et al. (2013), used site-frequency spectrum data to obtain a more extreme value, finding that close to 20% of *D. melanogaster* silent sites are under strong functional constraints.

While Equations 10.7b and 10.7c can be applied to single genes, individual-gene estimates of α are expected to have a large sampling variance and low power. If the actual fraction of adaptive substitutions is small, the modest increase in the number of substitutions will often not be large enough to be significantly different from its neutral expectation, and the resulting estimate of α will not be significantly different from zero. For example, if five substitutions are expected at our focal gene given the ratio of silent to replacement polymorphisms, an observed value of eight substitutions is unlikely to be excessive enough to be declared significantly different from five. However, if three of the eight substitutions were indeed driven to fixation by positive selection, then $\alpha = 0.375$, which is quite substantial.

Despite low power for estimating α at any *single* locus, considerable power can be obtained by estimating the expected value, $E[\alpha] = \overline{\alpha}$, over a *number of loci*. To accomplish this task, Fay et al. (2001, 2002) suggested the estimator

$$\widehat{\overline{\alpha}}_{Fay} = 1 - \frac{\overline{D}_s}{\overline{D}_a}\left(\frac{\overline{P}_a}{\overline{P}_s}\right) \tag{10.8a}$$

where the bar implies the average of that quantity over all sampled genes, e.g., $\overline{D}_s$ is the average number of silent substitutions over all the sampled genes. Note that we use α when referring to a single gene, $\overline{\alpha}$ for its expected value over a set of genes, and $\widehat{\overline{\alpha}}$ as an estimate of $\overline{\alpha}$.

The estimator given by Equation 10.8a has two potential sources of bias, both of which can lead to an overestimation of $\overline{\alpha}$ (Smith and Eyre-Walker 2002; Welch 2006). Let μ and $f\mu$ denote the effectively neutral per-site substitution rates for silent and replacement sites within a gene, where f is allowed to vary over genes. Following Welch (2006), one can show that

$$E\left[\frac{\overline{D}_s}{\overline{D}_a}\right] = \frac{\overline{n}_s}{\overline{n}_a}\frac{1}{E[f]}\left(E\left[\frac{1}{1-\alpha}\right]\right)^{-1} \simeq \frac{\overline{n}_s}{\overline{n}_a}\frac{1}{E[f]}\left[1-\overline{\alpha}-\sigma^2(\alpha)\right] \tag{10.8b}$$

where n_x is the average number of sites of type x over all genes, $E[\cdot]$ is the expectation over all sampled genes, and $\sigma^2(\alpha) = E[\alpha^2] - (E[\alpha])^2$ is the among-gene variance in the fraction of adaptive substitutions (α), with the last approximation following from the delta method (LW Equation A1.3). Equation 10.8b shows that when there is among-locus variation in α (so that $\sigma^2(\alpha) > 0$), $\overline{\alpha}$ is overestimated by Equation 10.8a.

A more subtle bias occurs if f and $4N_e\mu$ are *negatively correlated* over genes, as

$$E\left[\frac{\overline{P}_a}{\overline{P}_s}\right] = \frac{\overline{n}_a}{\overline{n}_s}\left(E[f] + \frac{\sigma(4N_e\mu, f)}{4E[N_e\mu]}\right) \tag{10.8c}$$

as obtained by Smith and Eyre-Walker (2002) and Welch (2006). Hence, Equation 10.7d *underestimates* f, and therefore results in an overestimation of $\overline{\alpha}$, if $4N_eu$ and f are negatively correlated (and underestimates $\overline{\alpha}$ if they are positively correlated). Smith and Eyre-Walker (2002) noted that a negative correlation is biologically reasonable, as the effective population size can vary over the genome (Chapters 3 and 8), and regions with smaller N_e are likely have higher f values (Figure 10.1), as more mutations become effectively neutral.

To reduce bias from correlations between f and N_e, Smith and Eyre-Walker (2002) suggested the estimator

$$\widehat{\overline{\alpha}}_{SEW} = 1 - \frac{\overline{D}_s}{\overline{D}_a}\overline{\left(\frac{P_a}{P_s+1}\right)} \tag{10.9a}$$

where the second term is the average of the quantity $P_a/(P_s+1)$ over the sampled genes. Provided that the number of polymorphic silent sites in the sample is modest (five or greater), this adjusted polymorphism ratio is unbiased by correlations between f and N_e, with

$$E\left[\widehat{\overline{\alpha}}_{SEW}\right] \simeq \overline{\alpha} + \sigma^2(\alpha) \tag{10.9b}$$

as shown by Smith and Eyre-Walker (2002) and Welch (2006). While this correction removes concern over correlations between f and N_e, overestimation of $\overline{\alpha}$ remains when among-locus variation in α is present.

Example 10.11. A simple model provides some insight into the amount of bias possible when using Equation 10.9b. Suppose there are just two types of genes: a fraction, q, having $\alpha = \alpha_* > 0$, and the rest having only neutral substitutions ($\alpha = 0$). Under this model, $\overline{\alpha} = q\alpha_*$, while

$$\sigma^2(\alpha) = E[\alpha^2] - \overline{\alpha}^2 = q\alpha_*^2 - q^2\alpha_*^2 = \alpha_*^2 q(1-q)$$

Suppose that $\alpha_* = 0.2$ and $q = 0.5$, so that $\overline{\alpha} = 0.1$. From Equation 10.9b, the expected value of the Smith-Eyre-Walker estimate is

$$\overline{\alpha} + \sigma^2(\alpha) = 0.1 + \left[0.2^2 \cdot 0.5(1-0.5)\right] = 0.11$$

or a 10% overestimation. Conversely, consider a more extreme case where at 10% of the genes, all substitutions are adaptive, so that $\alpha_* = 1$ and $q = 0.1$. Again, $\overline{\alpha} = 0.1$, while the expected value from the Smith-Eyre-Walker estimate is $0.1 + 0.1 \cdot 1^2(1-0.1) = 0.19$, so in this extreme case, $\overline{\alpha}_{SEW}$ returns only a two-fold overestimate of $\overline{\alpha}$.

A potential concern with Equations 10.8a and 10.9a is bias due to the Yule-Simpson effect. Recalling Equations 10.7c and 10.6c suggests that the estimator

$$\widehat{\overline{\alpha}}_{TG} = 1 - NI_{TG} = 1 - \frac{\sum_i D_{si}P_{ai}/(P_{si}+D_{si})}{\sum_i P_{si}D_{ai}/(P_{si}+D_{si})} \tag{10.9c}$$

is perhaps the most robust approach to this problem. While Stoletzki and Eyre-Walker (2011) found very close agreement between $\widehat{\overline{\alpha}}_{TG}$ and $\widehat{\overline{\alpha}}_{Fay}$ over the data sets they examined, all of the above considerations suggest that the most prudent estimator is $\widehat{\overline{\alpha}}_{TG}$. We will refer to estimators of α that use departures from the expectation under neutrality in a DPRS table collectively as **MK estimators** (Equations 10.7c, 10.8a, 10.9a, and 10.9c).

Confidence intervals for $\overline{\alpha}$ using any of these estimators can be obtained using bootstrap resampling. One generates a sample of genes by drawing *with replacement* from the original list of all genes and estimates $\overline{\alpha}$ for this sample. This process is repeated a large number of times to generate a distribution of the estimate under resampling. Taking the lower 2.5% and upper 97.5% in this distribution yields the 95% bootstrap confidence interval.

While the above sources of bias (among-locus variation in α and correlations between f and $4N_e\mu$; Equations 10.8b and 10.8c) are generally modest and in a predictable direction (overestimation of $\overline{\alpha}$), the presence of mildly deleterious alleles provides a major bias, which can be either positive or negative (Eyre-Walker 2002; Bieren and Eyre-Walker 2004; Welch 2006; Charlesworth and Eyre-Waker 2008; Eyre-Walker and Keightley 2009; Halligan et al. 2010; Schneider et al. 2011; Keightley and Eyre-Waker 2012; Messer and Petrov 2013b). Estimates of α are downwardly biased by the presence of low-frequency deleterious alleles that contribute to P_a but not D_a, thus inflating the polymorphism ratio relative to the divergence ratio (Eyre-Walker 2006; Eyre-Walker and Keightley 2009). As with MK tests, one approach is to count only "common" polymorphisms for P_a and P_s. However, Charlesworth and Eyre-Walker (2008) noted that while this approach is "better than doing nothing," estimates of α still tend to be downwardly biased even after making this correction unless the true α is fairly substantial. Further, the bias is a function of the complex distribution of fitness effects (Charlesworth and Eyre-Walker 2008; Welch et al. 2008; Eyre-Walker and Keightley 2009; Schneider et al. 2011; Keightley and Eyre-Waker 2012).

Messer and Petrov (2013b) suggested that one simple solution is to estimate $\overline{\alpha}$ using different cutoff levels for rare polymorphisms, with $\overline{\alpha}(x)$ denoting the estimate that ignores polymorphisms whose *derived* allele frequency is below x. Note that $\overline{\alpha}(x)$ could be based on any of our previous MK estimators (e.g., Equations 10.8a, 10.9a, and 10.9c) simply by ignoring polymorphisms below this threshold. Recalculating this statistic for increasing values of x, an exponential regression of the form $\alpha(x) = a + b\exp(-cx)$ is fit to the data, and the asymptotic value (the projected value at $x = 1$) is given by the **Messer-Petrov asymptotic estimate** of $\overline{\alpha}$

$$\overline{\alpha}_{MP} = a + b\exp(-c) \tag{10.9d}$$

The presence of mildly deleterious alleles also biases estimates of α if the population size differed during the divergence and polymorphism phases. If the population has recently undergone an expansion, this can upwardly bias estimates of α. In such cases, slightly deleterious alleles may have previously been fixed (contributing to divergence), but would be quickly removed in the new, larger population, thus not contributing to P_a. Conversely, if the population has recently undergone a contraction, this inflates P_a as more deleterious alleles are segregating, downwardly biasing estimates of α. Eyre-Walker and Keightley (2009) and Halligan et al. (2010) obtained a simple expression for the bias in α when the recent population size, N_P, generating the polymorphism data differs from the ancestral size, N_D, generating the divergence data. Assuming beneficial mutations are sufficiently strong that α is invariant under the two population sizes, and that deleterious new mutations have their fitness effects drawn from a gamma distribution with a shape parameter of $\beta > 0$, then the connection between the expected value, α_{est}, of an estimated α and its true value is

$$\alpha = 1 + (\alpha_{est} - 1)\left(\frac{N_P}{N_D}\right)^{\beta} \tag{10.10}$$

A contraction in N_e ($N_P < N_D$) leads to an underestimation of α, while an expansion ($N_P > N_D$) results in an overestimation. The same approach leading to Equation 10.10 was used in Example 10.10 to examine the behavior of the neutrality index (which is closely related to α; see Equation 10.7c) under changes in N_e.

Maximum-likelihood (ML) estimators of α have been proposed that attempt to account for segregating deleterious mutations (Bierne and Eyre-Walker 2004; Welch 2006; Boyko et al. 2008; Eyre-Walker and Keightley 2009; Schneider et al. 2011; Keightley and Eyre-Waker 2012). This is done by assuming a standard form (such as a gamma) for the distribution of deleterious fitness effects, and then using site-frequency spectrum data to estimate the parameters of this distribution. We sketch the basic outline of this approach in the next section (in the context of Poisson random field models). While it is elegant and powerful when the model assumptions are correct, the concern is that this approach is highly dependent on the assumed functional form (e.g., gamma, normal, or other) of the unknown distribution of fitness effects for the slightly deleterious mutations. Indeed, Kousanthanas and Keightley (2013) found that these models perform poorly when the distribution of fitness effects is multimodal, and they suggested using nonparametric approaches for such cases.

Another potential source of bias, which was first noted by Akashi (1995), is codon usage. Strong codon usage bias results in the synonymous substitution rate underestimating the neutral divergence rate, which in turn inflates estimates of α. If strong bias occurs on just a few genes, this will have a minor impact on $\overline{\alpha}$. However, chromosome-wide biases can cause problems. Recall that the *D. melanogaster* X chromosome has a higher codon usage bias than the autosomes (Chapter 8). Because this appears to be due to stronger selection on X-specific genes, it can result in different biases in α between estimates based on X versus those based on autosomal loci (Campos et al. 2012).

Given these competing sources of bias (overestimation of $\overline{\alpha}$ when $\sigma^2(\alpha) > 0$ and underestimation of α when deleterious alleles are segregating), are MK estimators more likely to over- or underestimate the true $\overline{\alpha}$? Indeed, are they reliable at all? As Example 10.11 highlights, the overestimation of $\overline{\alpha}$ when α varies over genes, while not trivial, is often modest, especially if $\overline{\alpha}$ is moderate to large. Conversely, as mentioned above, the presence of segregating weakly deleterious replacement alleles (which are unlikely to become fixed) inflates the polymorphism ratio, P_a/P_s, leading to an underestimation of the actual excess number of adaptive substitutions. This effect can be quite dramatic. In particular, if deleterious alleles are common, a neutrality index value greater than 1.0 can occur, which results in a negative estimate of α (Equation 10.7c). Upon putting these two sources of bias together, $\overline{\alpha}$ is generally likely to be underestimated unless the population has undergone a recent size expansion.

A final complication, noted by Fay (2011), is the assumption that each site evolves independently (also see Messer and Petrov 2012). Two possible sources of overestimation of α are possible when this assumption fails. First, slightly deleterious mutations may be fixed by hitchhiking to a favorable substitution, potentially inflating D_a and hence estimates of α. Second, epistasis in fitness between sites may occur such that fixation at one site changes the constraints on other sites within the gene, which again can potentially result in an inflation in D_a.

How Common Are Adaptive Substitutions?

There has been an explosion of genomewide estimates of $\overline{\alpha}$ (Eyre-Walker 2006) that will likely continue, as the required data (the amount of divergence between a set of genes in two species, and polymorphism data, or the number of segregating sites, for the same genes from one, or both, species) are becoming increasingly easy to obtain. Table 10.1 summarizes some of these studies, and Figure 10.2 shows an analysis from ten species-pairs in plants. The quest for $\overline{\alpha}$ values is very reminiscent of the mad "find them and grind them" dash in the 1970s to estimate levels of protein variation in a menagerie of species (e.g., Lewontin 1974).

The general observation for *Drosophila* is that estimates of $\overline{\alpha}$ for amino acid substitutions are high, averaging around 50%, with estimates of the fraction of adaptive changes in noncoding regions also approaching 30% in some cases. High $\overline{\alpha}$ values for replacement sites are also observed for the mouse, bacteria, and three plants (*Populus, Helianthus*, and *Capsella*), while very low levels are seen in other plants (Table 10.1 and Figure 10.2). Low levels in *Arabidopsis thaliana* were originally attributed to the high levels of selfing in this species (Bustamante et al. 2002), but a close outcrossing relative (*A. lyrata*) similarly shows very low levels of $\overline{\alpha}$ (Foxe et al. 2008). The case receiving the most interest is humans, where an initially rather high estimate of 0.35 by Fay et al. (2001) for a small set of genes was followed by several studies showing much lower values (Table 10.1).

One trend that has been suggested is that $\overline{\alpha}$ increases with effective population size (Eyre-Walker 2006). While intriguing, there are also apparent counterexamples. For example, Bachtrog (2008) found that *D. miranda*, which is thought to have a low effective population size, has a similar value of $\overline{\alpha}$ as *Drosophila* species thought to have a significantly larger values for N_e.

Drawing a clear conclusion from these initial data is problematic for several reasons. First, even in the same species, different genes may be used or different populations may be chosen as the polymorphism benchmark. The effect of the latter is especially prominent in Figure 10.2, with the same divergence data between two sunflower species (*Helianthus annuus* versus *H. petiolaris*) showing a significantly positive estimate of mean α when using *Helianthus petiolaris* as the polymorphism reference population, but a negative (but not significant) estimate when using *H. annuus* as the reference population (reminiscent of Example 10.9). Differences in N_e values between the two species being considered can inflate or deflate estimates of α (Equation 10.10). Second, different studies used different methods, ranging from simple MK-type estimators (Equations 10.8 and 10.9) to much more sophisticated, ML-based estimators that attempt to account for both changes in N_e and the presence of segregating deleterious alleles (Bierne and Eyre-Walker 2004; Welch 2006; Eyre-Walker and Keightley 2009). While they are certainly powerful when the modeling assumptions are correct, the robustness of these ML approaches against model misspecification is unclear.

Despite these potential misgivings, the pattern of estimates of $\overline{\alpha}$ over species, and even

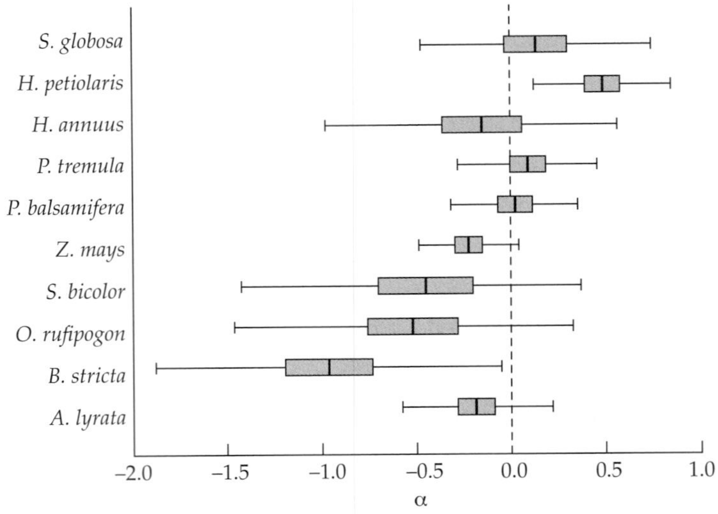

Figure 10.2 Estimated $\overline{\alpha}$ values for ten plant species, where the listed species supplied the polymorphism data. Boxes and whiskers indicate, respectively, the 50% and 95% confidence intervals for the estimates of $\overline{\alpha}$, obtained using Eyre-Walker and Keightley's (2009) ML method, which allows for a distribution of deleterious fitness effects and potentially different effective population sizes in the divergence and polymorphism phases. Only the comparison involving sunflowers (polymorphism data from *Helianthus petiolaris*, divergence between *petiolaris* and *annuus*) had an estimated average α that was significantly positive. Surprisingly, the comparison using polymorphism data from *H. annuus* and the same divergence (*petiolaris* versus *annuus*) gave a negative estimate of average α (but was not significantly different from zero). Note that most estimates are negative (although only one was significantly so,). Recall that negative estimates of α occur when the neutrality index (Equation 10.6a) is greater than one, namely, when there is an excessive number of segregating replacement sites. As mentioned throughout this chapter, this can occur if weakly deleterious alleles are common, which inflate the number of polymorphisms but not fixations. (After Gossmann et al. 2010.)

within genomes, is of fundamental importance to evolutionary biologists. If, indeed, some species have very low $\overline{\alpha}$ values, does this automatically imply that they have lower rates of adaptation? One surprising taxon that shows a very low estimated $\overline{\alpha}$ is the Hawaiian silversword plant genus *Schiedea* (family Caryophyllaceae), a group with rapid (and dramatic) morphological evolution over a very recent time window (Gossmann et al. 2010). One possible resolution to this apparent disconnect is that most current studies have focused on the estimation of α in coding sequences, whereas considerable adaptation (especially over short time scales) may occur at the level of gene regulation. Based upon the estimated α values in noncoding regions, Andolfatto (2005, Wright and Andolfatto 2008) suggested that the number of adaptive substitutions in noncoding regions in *Drosophila* could be far greater than the number of adaptive replacement substitutions. Given that *Drosophila* has a compact genome relative to humans and many other metazoans and land plants, the bulk of adaptive variation may not reside in the coding regions that are the focus of most current estimates of $\overline{\alpha}$. An alternative, and not necessarily exclusive, explanation for the *Schiedea* data is that only a few key genes underlie most of the morphological change, resulting in very little change in the genomewide value of $\overline{\alpha}$.

One observation consistent with the importance of regulatory changes derives from the work of Torgerson et al. (2009), who compared polymorphism and divergence levels in roughly 15,000 **conserved noncoding** (**CNC**) regions flanking human genes. CNCs are

Table 10.1 Partial list of estimates of the fraction, $\overline{\alpha}$, of replacement substitutions that are adaptive. The organism listed is the species that provided the polymorphism data. MK refers to a MacDonald-Kreitman estimator (Equations 10.8 or 10.9), ML refers to maximum-likelihood extensions of MK estimators (Bierne and Eyre-Walker 2004; Welch 2006; Eyre-Walker and Keightley 2009; Schneider et al. 2011), PRF refers to Poisson random field estimators, and IN to the *INSIGHT* estimator (details in the text). Estimates of zero indicate a neutrality-index score exceeding one (and hence a negative estimate of $\overline{\alpha}$).

Organism	$\overline{\alpha}$	Method	Reference
Mus musculus castaneus (mouse)	0.57	ML	Halligan et al. 2010
Oryctolagus cuniculus (rabbit)	0.60	MK, ML	Carneiro et al. 2012
Gallus gallus (chicken)	0.20	MK	Axelsson and Ellegren 2009
Drosophila simulans	0.45	MK	Smith and Eyre-Walker 2002
	0.43	ML	Bierne and Eyre-Walker 2004
	0.41	ML	Welch 2006
D. melanogaster	0.44	ML	Bierne and Eyre-Walker 2004
	0.95	PRF	Sawyer et al. 2007
	0.85	ML	Schneider et al. 2011
D. miranda			
Total	0.48	ML	Bachtrog 2008
X chromosome	0.33	MK	Haddrill et. al. 2010
	0.14	ML	
autosomal	0.00	MK	
	0.00	ML	
D. pseudoobscura			
X chromosome	0.44	MK	Haddrill et. al. 2010
	0.70	ML	
autosomal	0.59	MK	
	0.87	ML	
Escherichia coli	0.56	MK	Charlesworth and Eyre-Walker 2006
Arabidopsis thaliana	0.00	PRF	Bustamante et al. 2002
A. lyrata	0.00	PRF	Foxe et al. 2008
Capsella grandiflora (crucifer)	0.40	ML	Slotte et al. 2010
Populus tremula (aspen)	0.30	ML	Ingvarsson 2010
Helianthus annuus (sunflower)	0.75	MK	Strasburg et al. 2009
Humans	0.35	MK	Fay et al. 2001
	0.00	MK	Zhang and Li 2005
	0.06	PRF	Bustamante et al. 2005
	0.12	MK	Gojobori et al. 2007
	0.2	IN	Arbiza et al. 2013

operationally defined as noncoding sequences of at least 100 nucleotides in length that show at least 70% conserved sequence identity between mice and humans. The idea is that these sequences are putative regulatory control regions, and hence under purifying selection. When comparing human-chimpanzee divergence, the authors estimated an overall $\overline{\alpha} \sim 0.05$ for all CNC, $\sim$0.15 and $\sim$0.23 for 5′ and 3′ UTRs regions of known genes, respectively, and $\sim$0.12 for upstream and downstream regions from known genes. Their most interesting finding was of a lack of correlation between the estimated α values for the CNC regions flanking a gene versus that for replacement substitutions within that gene, namely, an apparent disconnect between regulatory (CNC) and structural (amino acid) substitutions.

Estimating the Rate, λ, of Adaptive Substitutions

A quantity that prominently appeared in expressions in Chapter 8 on the effects of recurrent sweeps was λ, the per-generation rate at which adaptive substitutions occur. While it might seem that estimates of λ (the number of adaptive substitutions per site divided by the total time of divergence, $2t$) would be very difficult to obtain, fortunately this is not the case, as they follow almost directly from estimates of α (Smith and Eyre-Walker 2002; Andolfatto 2007). If $d_a = D_a/n_a$ denotes the per-site number of replacement substitutions between two species that separated t generations ago, then an upper bound for λ is simply $d_a/(2t)$. The use of D_a to compute d_a involves the assumption that all substitutions have been observed, so that no corrections for multiple substitutions at the same site are needed, which is not unreasonable when comparing two closely related species. With an estimate of α, the number of adaptive replacement substitutions is just αD_a, yielding **Andolfatto's estimator** (2007),

$$\widehat{\lambda} = \frac{\alpha\, d_a}{2t} \tag{10.11a}$$

for the per-site, per-generation rate of adaptive substitutions.

In order to apply Equation 10.11a, one must have an estimate of the divergence time, t. This can be estimated (scaled as $\tau = t/(2N_e)$ generations) from the ratio of D_s/P_s, as follows. From Equations 10.12a and 10.12b,

$$\frac{E[D_s]}{E[P_s]} = \frac{1}{a_m + a_n}\left(\tau + \frac{1}{m} + \frac{1}{n}\right) \tag{10.11b}$$

where m and n are the sample sizes for the two populations and the sample size feature, a_x, is given by Equation 4.3b. Substituting the observed values of D_s and P_s for their expected values and rearranging provides a simple method-of-moments estimator for the scaled divergence time

$$\widehat{\tau} = (a_m + a_n)\,\frac{D_s}{P_s} - \left(\frac{1}{m} + \frac{1}{n}\right) \tag{10.11c}$$

Using this estimate yields $\widehat{t} = 2N_e\widehat{\tau}$, and substituting into Equation 10.11a yields

$$\widehat{\lambda} = \frac{\alpha\, d_a}{2N_e\widehat{\tau}} \tag{10.11d}$$

Note that the estimate offered by Equations 10.11a and 10.11d for the rate, λ, is typically based on *structural changes*, namely, the adaptive rate of amino acid replacement substitutions in protein-coding genes. A more inclusive estimate would also account for regulatory adaptations, which are expected to be *at least* on par with protein structural adaptations (Chapter 9).

Example 10.12. The estimated amino acid divergence between human and chimpanzee proteins is $d_a = 0.008$ (Chimpanzee Sequencing and Analysis Consortium 2005), with a divergence time of roughly 7 million years. If we take $\alpha = 0.10$ (10% of replacement substitutions are adaptive, the rough average for human studies in Table 10.1), then from Equation 10.11a, our estimate of the rate of adaptive replacement substitutions per site, per generation is

$$\lambda = \frac{0.10 \cdot 0.008}{14 \cdot 10^6} = 5.7 \cdot 10^{-11} \text{ per site, per year}$$

Assuming a generation time of 25 years, this corresponds to a rate of $2.3 \cdot 10^{-12}$ per site, per generation.

As a point of comparison, Andolfatto (2007) contrasted X chromosome genes in *Drosophila melanogaster* (for polymorphism data) and *D. simulans* (as the outgroup for divergence). The estimated α was 0.5, while $d_a = 0.028$, and $t = 10^7$ generations, yielding

$$\lambda = \frac{0.50 \cdot 0.028}{2 \cdot 10^7} = 7.0 \cdot 10^{-10} \text{ per site, per generation}$$

Hence (for these data), *Drosophila* have a 12-fold higher per-site adapation rate than humans.

THE SAWYER-HARTL POISSON RANDOM FIELD MODEL

Another approach for extracting information from DPRS tables on the nature and amount of selection is the **Poisson random field (PRF) model** of Sawyer and Hartl (1992). Their initial version assumed that all sites within a region evolve independently and that the strength of selection on all replacement sites was the same. Strongly deleterious mutations were allowed to occur, but the assumption is that these do not contribute to either polymorphism (observed segregating sites) or divergence, and they are accounted for by simply reducing the mutation rate to exclude such mutations. Under this model, the observed counts (P_s, D_s, P_a, and D_a) in a DPRS table follow independent Poisson distributions, whose expected values are functions of four parameters (θ_a, θ_s, τ, and γ). With four observations (the DPRS entries) and four unknowns, we can estimate these parameters, but we cannot assess how well the model fits the data. Two of the parameters are the scaled total mutation rates, $\theta_a = 4N_e\mu_a$ and $\theta_s = 4N_e\mu_s$, while the third parameter is the scaled divergence time, $\tau = t/(2N_e)$. Of most interest is the fourth parameter, the scaled strength of selection, $\gamma = 2N_e s$. Sawyer and Hartl assumed there was additive fitness, so that a new mutation has a fitness of $1 + s$ as a heterozygote and $1 + 2s$ as a homozygote. In contrast to MK approaches, the PRF model does not estimate the fraction, α, of adaptive substitutions directly, but knowledge of γ can allow one to do so indirectly (Example 10.13).

Basic Structure of the Model

The PRF model assumes that each site evolves independently, and hence there are no effects from selection at linked sites—the assumption is that selection only influences a site by directly acting on it. To obtain the expected values for the entries in a DPRS table, Sawyer and Hartl used results from diffusion theory (Appendix 1) on the equilibrium distributions (under mutation-selection-drift balance) for polymorphisms at neutral and selected sites, as well as the expected divergence between sites. The PRF model is an infinite-sites model (Chapter 2), with each new mutation assumed to be unique and at a different site from previous ones. For a sample of m and n sequences from the two species, the expected values for the DPRS entries are

$$E[D_s] = \theta_s\left(\tau + \frac{1}{m} + \frac{1}{n}\right) \tag{10.12a}$$

$$E[P_s] = \theta_s\left(\sum_{j=1}^{m-1}\frac{1}{j} + \sum_{j=1}^{n-1}\frac{1}{j}\right) = \theta_s\,(a_m + a_n) \tag{10.12b}$$

$$E[D_a] = \theta_a\left(\frac{2\gamma}{1 - e^{-2\gamma}}\right)\Big(\tau + G(m,\gamma) + G(n,\gamma)\Big) \tag{10.12c}$$

$$E[P_a] = \theta_a\Big(F(m,\gamma) + F(n,\gamma)\Big) \tag{10.12d}$$

where

$$F(n,\gamma) = \int_0^1 \left(\frac{1 - x^n - (1-x)^n}{x(1-x)}\right)\left(\frac{1 - e^{-2\gamma(1-x)}}{1 - e^{-2\gamma}}\right) dx \tag{10.13a}$$

$$G(n,\gamma) = \int_0^1 x^{n-1}\left(\frac{1 - e^{-2\gamma(1-x)}}{2\gamma(1-x)}\right) dx \tag{10.13b}$$

The full derivation is given by Sawyer and Hartl, but a brief sketch of the underlying ideas is as follows.

Recall (Equation 7.13b) Wright's (1938b) classic result for the amount of time that a new segregating mutation (with selection coefficient s) spends in the interval $(x, x + dx)$,

$$\phi(x \mid N_e s) = \frac{1 - e^{-2\gamma(1-x)}}{1 - e^{-2\gamma}} \frac{1}{x(1-x)} dx \tag{10.14a}$$

where x is the frequency of the derived allele. In the limit as $\gamma \to 0$, this reduces to dx/x, recovering Watterson's expression for the (unfolded) site-frequency spectrum (SFS) for neutral alleles (Equation 2.34a). Equation 10.14a is the expected equilibrium unfolded frequency spectrum for segregating sites under selection, and it is valid for both positive and negative values of s.

As a brief aside, we mentioned above that certain maximum-likelihood versions of the basic MK test use a **distribution of fitness effects** (often denoted as **DFE** in the literature), $\varphi(s \mid \boldsymbol{\Delta})$, where $\boldsymbol{\Delta}$ denotes the vector of distribution parameters (such as the α and β parameters for a gamma distribution; see Equation A2.25a) (Bierne and Eyre-Walker 2004; Eyre-Walker et al. 2006; Welch 2006; Boyko et al. 2008; Eyre-Walker and Keightley 2009; Keightley and Eyre-Walker 2012). Under this model, the expected SFS becomes

$$\phi(x \mid N_e, \boldsymbol{\Delta}) = \int \phi(x \mid N_e s)\, \varphi(s \mid \boldsymbol{\Delta})\, ds \tag{10.14b}$$

This is simply the average of the frequency spectrum, $\phi(x \mid N_e, s)$, for alleles with a set value of s over the assumed distribution, $\varphi(s \mid \boldsymbol{\Delta})$, of s. Equation 10.14b is then used to obtain a maximum-likelihood estimate of the vector of distribution parameters, $\boldsymbol{\Delta}$, given the SFS data, with the resulting DFE used to adjust for the effects of segregating deleterious alleles.

Returning to the PRF model, we do not use the site-frequency spectrum, but rather translate the SFS into the four cell counts in the DPRS table. If x is the population frequency of a segregating allele, the probability that we score it as a polymorphic site in a sample of size n is $1 - x^n - (1-x)^n$, where the last two terms account for all n draws either being only the derived allele or only the ancestral allele (Equation 2.36b). Hence, the probability that we will score a truly segregating site as polymorphic in a sample of size n becomes

$$\int_0^1 [1 - x^n - (1-x)^n]\ \phi(x \mid N_e, s) \tag{10.14c}$$

This follows by averaging the probability of scoring a site as polymorphic given a derived allele frequency of x over the distribution $\phi(x \mid N_e, s)$ of x given the selection model. The function $F(n, \gamma)$ given by Equation 10.13a follows upon substitution of Equation 10.14a in Equation 10.14c (a similar approach was used in Chapter 9 for ML-based detection of hard sweeps; see Equation 9.16a).

The Sawyer-Hartl model also correctly accounts for the possibility that segregating mutations are scored as substitutions because the sample size is insufficient to contain both alleles. If the derived-allele frequency is x, the probability that we score a polymorphic site as a substitution event (for the derived allele) is x^n, giving the additional increment to the probability of an observed substitution as

$$\int_0^1 x^n\ \phi(x \mid N_e, s) \tag{10.14d}$$

This term is added to the probability of a true substitution to give a full accounting of the number of sites in the sample that are scored as substitutions (Equation 10.12c), with $G(n, \gamma)$ following from Equations 10.14a and 10.14d.

The basic similarities, and fundamental differences, between MK estimators (e.g., Equations 10.7–10.9) and the PRF approach can be easily obscured by the imposing nature of the PRF equations. The similarity is that both approaches use the same data, the four values in a DPRS table. However, the two approaches estimate different quantities and have different underlying model assumptions. MK estimators make no assumption about the nature or strength of selection on replacement sites, but instead estimate f, the reduction in the effectively neutral substitution rate at replacement sites, and α, the fraction of replacement substitutions at a gene that are adaptive. The effect of purifying selection enters only through f, while the effects of positive selection enter only through α.

In contrast, the PRF equations estimate θ_a and θ_s, the scaled total mutation rates over all sites of the two categories within the gene. The ratio of θ_a/θ_s (suitably corrected for the number of sites within each category; see Equation 10.7d) is *not* an estimate of f, as the PRF model *does* allow for slightly deleterious alleles to be segregating (i.e., the estimate of $\gamma = 2N_e s$ might be negative). The original Sawyer-Hartl model was very restrictive, with only a single fitness class for replacement sites (which is approximately treated as an average selection coefficient over mutations). Extensions (discussed shortly) remove this restriction, allowing for neutral, deleterious, and advantageous classes, with either constant values of γ within each class, or (more generally) class-specific *distributions* of γ values. Thus, the PRF model does *not* estimate α directly, but given estimates of γ, we can compute the expected fraction of substitutions that are fixed by positive selection (Example 10.13 and Equation 10.16c).

The original Sawyer-Hartl analysis equated the observed entries in a DPRS table with their corresponding expected values (Equations 10.12a–10.12d), and then solved for the unknowns of interest (the ratio $\theta_a/\theta_s = \mu_a/\mu_s$, the scaled average strength of selection $\gamma = 2N_e s$, and the scaled time of divergence $\tau = t/[2N_e]$). A value of γ significantly different from zero implies selection on replacement sites, with $\gamma > 0$ implying positive selection and $\gamma < 0$ implying negative selection (the latter applies only to mildly deleterious alleles, as the PRF model treats very deleterious alleles by lowering the mutation rate: these are assumed to be not seen as either polymorphisms or divergences). This original model, which only assumes a single selective class with silent sites being neutral, can be placed in a likelihood framework by recalling that each observed entry in a DPRS table is an independent Poisson random variable. The probability that the count X in a specific category is x, given its expected value ζ, follows from the Poisson distribution,

$$\text{Prob}(X = x \mid \zeta) = \zeta^x \exp(-\zeta)/x!, \quad \text{where} \quad \zeta = E[X]$$

The likelihood of the data in the DPRS table for gene i is thus given by

$$L_i = \prod_{j=1}^{4} \left(\frac{\zeta_{i,j}^{x_{i,j}} \exp(-\zeta_{i,j})}{(x_{i,j})!} \right) \tag{10.15}$$

where $x_{i,j}$ denotes the observed DPRS table values for category j in gene i, with

$$x_{i,1} = P_{s,i}, \quad x_{i,2} = P_{a,i}, \quad x_{i,3} = D_{s,i}, \quad x_{i,4} = D_{s,i}$$

and $\zeta_{i,j}$ are the corresponding gene-specific expected values,

$$\zeta_{i,1} = E[P_{s,i}], \quad \zeta_{i,2} = E[P_{a,i}], \quad \zeta_{i,3} = E[D_{s,i}], \quad \zeta_{i,4} = E[D_{a,i}]$$

Note from Equations 10.12a–10.12d that $\zeta_{i,1}$ through $\zeta_{i,4}$ are functions of the unknown parameters ($\theta_{a,i}, \theta_{S,i}, \gamma_i, \tau$) that we wish to estimate by ML. A numerical search over all possible values of these parameters for the combination that maximizes Equation 10.15 given the data (treating the $x_{i,j}$ as fixed constants) yields the ML solutions (LW Appendix 4). Under the assumption of independence across genes, the combined likelihood over k genes becomes

$$L = \prod_{i=1}^{k} L_i$$

where θ_a, θ_S, and γ can potentially vary over the genes, while the divergence time, τ, is shared by all. Hence, for M genes, there are $3M + 1$ unknown parameters.

As noted, this basic model can be expanded by considering more realistic fitness models. For example, Nielsen et al. (2005a) allowed three fitness classes for replacement sites: neutral, deleterious, and beneficial (advantageous). While fitness is assumed to be the same within each class, this is a significant improvement over the original Sawyer-Hartl model. The resulting likelihood now has four parameters for selection (as opposed to one, γ). These are p_b, p_0, and p_d, the frequencies of beneficial, neutral, and deleterious mutations (where $p_b = 1 - p_0 - p_d$), and γ_b and γ_d, the scaled selection coefficients for the beneficial and deleterious alleles (which are assumed to be the same over all genes). Nielsen et al. applied their method to a set of 50 human genes with prior evidence for possible positive selection. The resulting ML estimates were $p_d = 0.748, p_0 = 0.172$, and $p_b = 0.080$ as the fraction of deleterious, neutral, and advantageous mutations, and $\gamma_d = -34.96$ and $\gamma_b = 267.11$ as the scaled strengths of selection of deleterious and advantageous mutations. Note that even in this case where genes were ascertained as likely to be under positive selection, most mutations were still deleterious. A similar analysis of two *Drosophila melanogaster* data sets by Schneider et al. (2011) found that ~1.5% of all replacement mutations were adaptive (i.e., $p_b \sim 0.015$), but with a much smaller scaled strength of selection, $\gamma_b \sim 10$.

While the PRF model does not directly estimate the fraction of adaptive replacements (α), this can be obtained from the estimates of γ and the fraction, p_b, of advantageous mutations as follows. The expected rate of effectively neutral substitutions at replacement sites is μp_0 (the neutral mutation rate), whereas the expected number of favorable mutations arising in each generation is $2N\mu p_b$, where μp_b is the favorable mutation rate. For large γ, each favorable mutation has a fixation probability of $2sN_e/N$ (Chapter 7), for an expected per-generation substitution rate of favorable alleles of

$$\lambda \simeq (2N\mu p_b)(2sN_e/N) = \mu p_b(2\gamma) \tag{10.16a}$$

The fraction of adaptive substitutions is the rate of adaptive substitutions divided by the total rate of substitutions (adaptive plus neutral),

$$\alpha = \frac{\lambda}{\lambda + \mu p_0} \tag{10.16b}$$

Substituting Equation 10.16a yields

$$\alpha = \frac{2\gamma\mu p_b}{2\gamma\mu p_b + \mu p_0} = \frac{2\gamma}{2\gamma + (p_0/p_b)} \tag{10.16c}$$

Equation 10.16c relates the selection estimates p_b and γ from a PRF model with the selection estimate α from an MK approach. Inspection shows that small p_b (or more precisely a small value of p_b/p_0) does not mean that α is small, as $\alpha > 0.5$ when $2\gamma > p_0/p_b$. One final result emerges from Equation 10.16a. Because μp_b is the rate of beneficial mutation, which (in keeping with our notation from Chapter 8) we denote by μ_b, Equation 10.16a becomes

$$\lambda = 2\gamma\mu_b \tag{10.16d}$$

which immediately suggests the **Bachtrog estimator** (2008),

$$\mu_b = \frac{\lambda}{2\gamma} \tag{10.16e}$$

Example 10.13. What is the estimate of α for the subset of human genes considered by Nielsen et al. (2005a) that was previously discussed (immediately proceeding Equation 10.16a)? Here $p_b = 0.08, p_0 = 0.172$, and $\gamma_b = 267.11$. While only 8% of all new replacement mutations were deemed to be advantageous, α is considerably larger than 0.08, as Equation 10.16c yields

$$\alpha = \frac{2 \cdot 267.11 \cdot 0.08}{(2 \cdot 267.11 \cdot 0.08) + 0.172} = 0.996$$

The reason for this high value is that the estimated advantageous mutation rate (0.08μ) is just slightly below half of the estimated neutral rate (0.172μ), while the fixation probabilities for advantageous mutations are over 500 times greater. If we lumped the neutral and deleterious mutations rates together and assumed these were all effectively neutral (i.e., replacing 0.172 by $1 - p_b = 0.920$), our estimate of α would still be very high (0.980). It is also important to recall that Nielsen et al. focused on a highly biased set of genes, which were chosen to be enriched for positive selection. It is thus likely that the p_b, γ, and α estimates based on this set of loci are larger than those for typical human genes.

Now consider the Schneider et al. (2011) values for *Drosophila melanogaster* ($p_b \sim 0.015$, $\gamma_b \sim 10$). If we assume that all of the remaining mutations are neutral ($p_0 = 1 - p_b = 0.985$), Equation 10.16c yields

$$\alpha = \frac{2\gamma}{2\gamma + (p_0/p_b)} = \frac{20}{20 + 0.985/0.015} = 0.23$$

If we assume that 50% of all new mutation are deleterious ($p_0 = 1 - p_b - 0.5 = 0.485$), then $\alpha = 0.38$. A key point of this example is that α can be quite substantial even when p_b is very small.

The robustness of PRF estimates to violations of model assumptions has been examined by several authors. While the model assumes additive selection, estimates of γ are relatively insensitive to dominance (Williamson et al. 2004). Wakeley (2003) examined the effects of population structure (assuming an island model; Chapter 2). While estimates of the divergence time, τ, are significantly affected, estimates of γ are only weakly affected and tend to be conservative (closer to neutrality). Desai and Plotkin (2008) noted that the infinite-sites assumption (that mutations never reoccur at the same site) breaks down under high scaled mutation rates ($\theta > 0.05$). In such cases, recurring mutations at the same site can result in genes under weak negative selection giving a signal of strong positive selection. This may be especially problematic for viral populations with high mutation rates and large population sizes.

One critical difference between PRF and MK analyses is the contribution of information from silent sites (e.g., P_s, D_s), a point stressed by Li et al. (2008). Estimates of selection under an MK analysis are in the form of estimates of α, which are critically dependent upon P_s and D_s (e.g., Equations 10.8a and 10.9a), in addition to D_a and P_a. Conversely, under the PRF model, positive selection is estimated only through γ. An examination of Equations 10.12c and 10.12d shows that estimates of γ depend *only* on D_a and P_a, and that information from silent sites (P_s and D_s) does not enter into them. As a consequence, the control for demographic effects on P_a provided by P_s does not enter, and over- or under-inflated estimates of P_a from population structure can significantly bias estimates of γ. Further, Equation 10.14a (from which the PRF equations follow) is an *equilibrium* model, which assumes that the population size has been stable for sufficient time to reach the mutation-selection-drift equilibrium. Chapter 9 was littered with the bodies of tests that critically depend on this same assumption.

In contrast, because MK estimates involve the ratio of P_a/P_s, recent demographic effects influencing polymorphism levels are accounted for, and there is no assumption about the population being at an equilibrium value for the current amount of genetic variation (see the discussion following Equation 10.5d). Thus, while both MK and PRF approaches face bias from differences in population size between the divergence and polymorphism phases, PRF approaches have additional bias introduced by any nonequilibrium patterns in the polymorphism data. As noted by Li et al. (2008), tests of selection using PRF theory (i.e., γ significantly greater than zero) are closer to an HKA than an MK test, as the former compares the P/D ratio over different genes and lacks the internal control of comparing polymorphism levels from two different classes *within* the same gene.

Finally, while we have framed the PRF approach in terms of simple DPRS count data, it can be modified to directly estimate γ using the site-frequency spectrum from a single population (Hartl et al. 1994; Bustamante et al. 2001; Williamson et al. 2004; Huerta-Sanchez et al. 2008). DPRS data are very granular, collapsing all of the polymorphism and divergence information into just four data points. In contrast, the site-frequency spectrum is a very rich source of additional information on the structure of the polymorphism data (Chapters 2 and 9). Using the PRF model to estimate γ directly from the frequency spectrum is done in a fashion analogous to estimating sweep parameters using the frequency spectrum discussed in Chapter 9. In particular, Equation 10.14a is substituted into Equation 9.16a to form the (composite) likelihood, from which an MLE for γ can be obtained by standard approaches (LW Appendix 4). While elegant, this approach is not generally recommended due to the very delicate dependence of the frequency spectrum on demographic structure, which is not accounted for by the current models. Likewise, Equation 10.14a assumes additive fitnesses, whereas even small amounts of dominance can alter the site-frequency spectrum (Williamson et al. 2004).

Bayesian Extensions

More fined-grain variation in the fitness of replacement mutations was allowed by Bustamante et al. (2002) and Sawyer et al. (2003) in the form of Bayesian models (an approach discussed more fully in Chapter 19 and in great detail in Appendices 2 and 3). Instead of returning a point estimate, $\widehat{\theta}$, for an unknown parameter, θ (or vector of parameters, $\boldsymbol{\Theta}$), a Bayesian analysis returns the full distribution (the **posterior**), $\varphi(\theta \,|\, \mathbf{x})$, for that parameter, given any previous information (the **prior** for $\boldsymbol{\Theta}$) and the likelihood given the data, $\mathbf{x}$.

Bayesian analysis of PRF data typically uses a **hierarchical model**, the motivation for which comes from random-effects models (Chapter 19). Suppose we have p parameters of interest. Treating the parameters as fixed effects requires p degrees of freedom, but often there are more parameters than observations ($p \gg n$). In some settings, we can treat these p quantities as random effects: draws from some unknown distribution, such as a normal, with unknown mean and variance. Because all draws (realizations) are assumed to come from this common distribution, we can borrow information across observations to estimate the distribution parameters, using (for the case of a normal) only two degrees of freedom (estimation of the unknown mean and variance).

Bayesian hierarchical models take this idea a step further. Consider data structured into a number of categories (say, genes), with multiple observations (draws) from each category (say, new mutations in a particular gene). Assuming that the draws from a given category are all from the same distribution (say, a normal with a category-specific mean and variance), then when the number of categories is large, so too is the parameter set (all of the category-specific means and variances). A hierarchical model reduces the number of parameters to estimate by assuming that the mean (and/or variance) for each category-specific distribution is *itself* a draw from a second distribution. Once each draw is made, these parameter values are fixed for that category. This reduces the estimation problem to one of simply estimating the parameters in the second distribution.

An example of this approach was presented by Bustamante et al. (2002), who assumed that all new replacement mutations at gene i have the same selection value, γ_i, but allowed these gene-specific values to vary among loci. This was done by assuming each γ_i to be a random variable drawn from a normal distribution with a mean of μ_γ and a variance of σ^2_γ, both estimated from the data. In other words, this model allows selection to vary over loci (but not between replacement mutations in the same gene) as a function of just two parameters ($\mu_\gamma, \sigma^2_\gamma$). Formally, the selection coefficient associated with the jth new replacement mutation at locus i is

$$\gamma_{i,j} = \gamma_i, \quad \text{where } \gamma_i \text{ is a single draw from a } N(\mu_\gamma, \sigma^2_\gamma) \tag{10.17a}$$

Because the divergence time, τ, is a common factor over all genes, this allows information to be borrowed across loci (i.e., all loci contribute to the estimation of τ), improving power, while only loci with sufficient polymorphism and divergence information (a rough rule of thumb is $P_a + D_a \geq 4$) are likely to be informative about γ. Figure 10.3 shows an example of

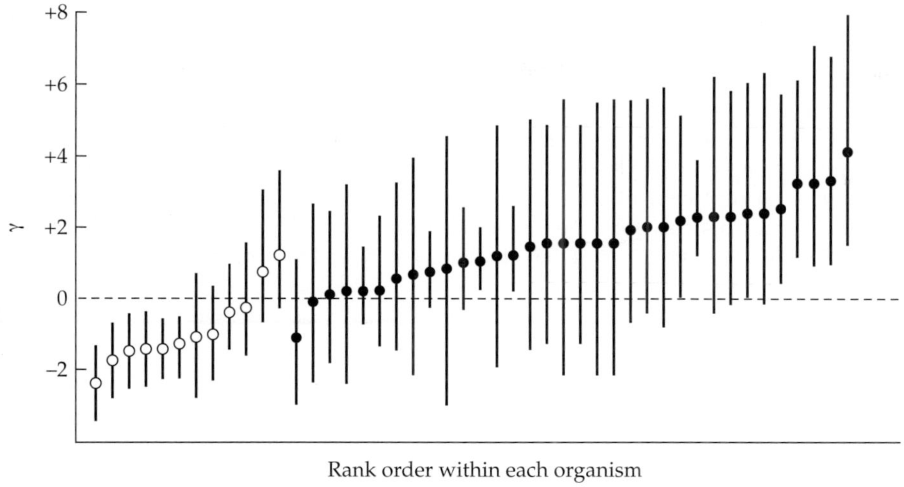

Figure 10.3 Bustamante et al. (2002) examined 12 genes from *Arabidopsis thaliana* (using a single allele from *A. lyrata* to compute divergence) and 34 genes from *D. melanogaster* (with a single allele for *D. simulans*). This figure plots the resulting posterior distribution for γ for each gene (i.e., the locus-specific value, γ_i from Equation 10.17a). The circle represents the mean, and the vertical lines denote the 95% credible intervals (the shortest span of the posterior distribution containing 95% of the probability; Appendix 2). These are plotted by rank order within the two species, with *Arabidopsis* plotted first (open circles) and *D. melanogaster* second (filled circles). If the vertical line is entirely below zero, selection on mutations at this locus is significantly negative (i.e., purifying selection). For lines entirely above zero, selection on new variants is significantly positive. Half (6 of 12) of the *Arabidopsis* genes are significantly negative, while none are significantly positive. Conversely, no *Drosophila* genes are significantly negative, while 9/34 are significantly positive.

the output from such an analysis. The analysis uses estimates of the distribution parameters (μ_γ and σ^2_γ; obtained from using all of the loci) to return estimates (formally, these are random-model predictors) of the selective value, γ_i, associated with each gene, as plotted in the figure.

Sawyer et al. (2003) extended the Bustamante et al. (2002) approach by allowing each new mutation (j) at locus i to potentially have a different fitness value, $\gamma_{i,j}$, with $\gamma_{i,j} \sim N(\mu_{\gamma,i}, \sigma^2_w)$. Hence, each new mutation has a fitness value drawn from a distribution with a locus-specific mean, $\mu_{\gamma,i}$, and a variance, σ^2_w, that is common over all loci (allowing us to share information over genes). This is a two-stage hierarchical model, where (as in Equation 10.17a) the mean fitness effect, $\mu_{\gamma,i}$, for locus i is drawn from a normal distribution with a mean of μ_γ and a variance of σ^2_γ. Once the locus-specific mean fitness effects are assigned, the fitness of a new replacement mutation at gene i is drawn from a *second* normal, with this locus-specific mean, $\mu_{\gamma,i}$, and a variance, σ^2_w, assumed to be common over all loci (again allowing us to share information over genes). This model can be more compactly written as

$$\gamma_{i,j} \sim N(\mu_{\gamma,i}, \sigma^2_w), \quad \text{where} \quad \mu_{\gamma,i} \sim N(\mu_\gamma, \sigma^2_\gamma) \tag{10.17b}$$

which has three distribution parameters to estimate: $\mu_\gamma, \sigma^2_\gamma$, and σ^2_w. Comparison with Equation 10.17a shows that *each* replacement mutation at a given locus is now a random draw (as opposed to all having the same value), and that (as before) the locus-specific mean also varies. This increased flexibility comes at the cost of only a single additional parameter, σ^2_w, the variance in gene-specific γ values about their mean (under the assumption of homoscedasticity).

Example 10.14. Sawyer et al. (2007) applied their 2003 model (Equation 10.17b) to a sample of 91 genes from an African population of *D. melanogaster*, using a *D. simulans* sequence to assess divergence. After ignoring very strong deleterious mutations that are unlikely to contribute to polymorphisms, they found that approximately 95% of all new replacement mutations are deleterious (estimates of $\gamma_{i,j} < 0$), with an estimated 70% of all replacement polymorphisms observed in the sample being deleterious. Conversely, they estimated that over 95% of the fixed differences at replacement sites are due to positive selection ($\gamma_{i,j} > 0$), albeit it was fairly weak. Within this class of replacement substitutions with estimated positive values, 46% were estimated to have $\gamma_{i,j} < 4$, 85% have $\gamma_{i,j} < 8$, and 99% have $\gamma_{i,j} < 14$.

While Bayesian models allowing fitness to vary over new mutations are powerful and potentially offer a solution to the vexing problem of segregating deleterious mutations that plagued MK tests and estimates of α, their robustness remains unclear. Current versions (as applied to PRF data) all assume normal distributions of fitness effects, but this is clearly not a realistic model (Eyre-Walker et al. 2006; Eyre-Walker and Keightley 2007; Boyko et al. 2008; Welch et al. 2008). The normal distribution has symmetry about the mean, while asymmetric (e.g., the gamma) and/or more heavy-tailed (e.g., t) distributions appear to be a better reflection of biology based on these references.

A second concern is somewhat technical. In a Bayesian analysis, one has to specify a prior (some statement about our uncertainty as to the true parameter values), and a key measure of this uncertainty is the prior's variance (Appendix 2). For example, our initial uncertainty in σ^2_γ would be set by the variance chosen in the prior for this parameter. As developed in Appendix 2, the usual prior for an unknown variance in a normal distribution is a scaled inverse chi-square distribution (Equations A2.30a and A2.32a), which is a function of two parameters (the prior **hyperparameters**). Ideally, we would like the choice of these hyperparameters to have little impact on the posterior, with the signal from the data overwhelming any impact of the prior. Instead, however, Li et al. (2008) found that the prior can have a very strong effect in PRF models. Specifically, the number of genes with γ values declared to be significantly different from zero increased with the assumed uncertainty in the variance of σ^2_γ (i.e., the variance hyperparameter in the scaled-inverse chi-square prior). This makes sense, in that restricting σ^2_γ to be small constrains most realized values of γ to be close to the mean value, while increasing it allows estimates to deviate substantially from the mean (and hence have their credible intervals avoid overlapping zero). Strong dependency of the posterior on the prior is always problematic in a Bayesian analysis, and it is good practice to run the model over several, rather different, sets of prior hyperparameters (such as using a range for the variance of the prior) to assess the stability of the posterior under different priors. Li et al. (2008) noted that a plot of a number of positively selected sites (genes with γ values whose credible intervals are all greater than zero) increases with the assumed variance in γ, but appeared to show signs of approaching an asymptote in humans and *Drosophila simulans* over the values for σ^2_γ used in the analysis. However, the same curve for yeast (*Saccharomyces cerevisiae*) showed no signs of approaching an asymptote over this range.

INSIGHT Analysis of Human Trascription Factors

As mentioned above, the use of conserved noncoding sequences as a proxy for all regulatory sequences likely represents a rather biased set, and global inferences on regulatory selection from this collection may be distorted. A different (albeit also potentially biased) group of noncoding DNAs are transcription factor binding sites. Through the use of **ChIP-seq (chromatin immnoperception and sequencing)** technology, nucleotide-level resolution can be achieved in delimiting binding sites, which while short, are abundant throughout the genome. Hence, issues of alignment are somewhat avoided by having a direct functional assay.

To analyze such data, Gronau et al. (2013; Arbiza et al. 2013) proposed a method that falls between the MK and PRF approaches, in that it contrasts polymorphism and divergence data between a target and control sequence, but also uses (in a coarse granular fashion) information on allele frequencies. Their approach (***INSIGHT*****: Inference of Natural Selection from Interspersed Genomically coHerent elemenTs**) is essentially a likelihood model, where the target sites are the very short sequences used by transcription factors, which were contrasted over humans and chimp. As a control sequence, they examined a 20K flanking block around each target sequence, which was then trimmed to consider only sites regarded as neutral. This left a contrast of around 10 bases for a typical binding site and around 7000 bases for the neutral control region for each comparison. The logic of *INSIGHT* was that only three types of events are presented in the divergence and polymorphism data. Strongly deleterious mutations never appear as either polymorphisms or as substitutions. Neutral mutations can contribute to both polymorphisms and divergence, while strongly selected mutations only contribute as substitutions (of derived alleles). Finally, weakly deleterious mutations appear as polymorphisms with low-frequency derived alleles, but such derived alleles are never fixed. Sites were binned as substitutions or either low or high-frequency derived-allele polymorphisms. A probability model was then fit using maximum likelihood to estimate the fraction α of adaptive substitutions (among other parameters).

Using this approach, Arbiza et al. (2013) estimated a value of $\alpha = 0.05$ for transcription-factor binding sites. As an internal control, they also ran the same analysis on second-base positions in codons, which returned an α value of essentially zero. When applied to a filtered set of roughly 15,000 proteins, their method estimated $\alpha \simeq 0.2$ over all coding-region substitutions. Given that their analysis examined roughly 5% of all transcription factors, Arbiza et al. extrapolated that adaptive substitutions were equally likely to occur in binding sites as in protein-coding genes. When coupled with the fact that binding sites represent a fraction of all regulatory DNA (assuming that most conserved sequences are regulatory in some sense), this suggests that adaptive substitutions are far more likely to arise in regulatory, as opposed to structural, sites. Obviously, these are only preliminary analyses, but they make an intriguing case for the importance of regulatory substitutions.

PHYLOGENY-BASED DIVERGENCE TESTS

Finally, we briefly consider divergence tests that examine the pattern of substitutions over a known phylogeny. These tests are designed to detect a rather different pattern of selection than was assumed in Chapter 9 (single events) or earlier in this chapter (multiple substitutions over an *entire gene* between two populations or species). While multiple substitutions are also required for a signal in phylogeny-based divergence tests, these must be at the same *site* (typically a codon) within a gene. Single substitutions over a number of different codons across a gene may leave very little signal for these tests (unless very few silent substitutions have occurred). As such, phylogenetic tests are biased toward detecting sites that undergo repeated evolution, and are likely to miss many, indeed perhaps *most*, adaptive substitutions (Hughes 2007). Given this restriction, these methods may work well in so-called "**arms race**" scenarios, in which trait values between two interacting species escalate to increasingly extreme values (Bergelson et al. 2001), such as the interactions between hosts and parasites.

The required input for phylogeny-based tests is a set of aligned DNA sequences and a predetermined phylogenetic tree for the sampled species. The assumption is that all sequence differences are the result of fixation events. Thus, if a site is segregating in one (or more) of the taxa from which a single sequence is drawn, one may incorrectly infer that it is a substitution event. The taxa must also have the correct amount of divergence, as either too little or too much, will result in very low power. With too little divergence, there are not enough substitutions, and hence there is little power to detect small percentage differences in silent versus replacement changes at particular sites. Further, if little true divergence has occurred, even a few segregating sites incorrectly called as substitutions can significantly inflate the divergence. Conversely, with too much divergence, multiple substitutions at single sites may occur between lineages, and adjustments for such multiple hits can introduce substantial bias if an incorrect statistical model is used to account for these.

A few comments are in order on the phylogeny for the sampled taxa, as this determines the covariance structure of the data. For proper analysis, we require, not only the topology (the pattern of common ancestry), but also the *branch lengths*, or the distances (divergence times) between taxa. Errors in either type of information obviously compromise these tests. For example, one expected pattern of repeated selection is independent mutations of the same key amino acid at a particular critical site (e.g., Example 10.15). The topology of a phylogeny can inform us as to whether a cluster of taxa sharing this key amino acid are all descendants from a single fixation event or comprise a collection of independent events. An erroneous topology for the species tree can thus introduce serious errors. Likewise, errors in some of the branch lengths relative to the rest of the phylogeny can bias rate estimates, which in turn can generate false positives.

There is a very rich literature on molecular evolution, and our purpose here is only to provide a brief overview of divergence-based tests at the phylogenetic level. As such, our treatment of phylogeny-based methods is more superficial than the more detailed treatment of tests for recent or ongoing selection (Chapter 9) and polymorphism-based tests (covered above). Readers seeking a fuller treatment of many of the important side issues (such as tree construction) not addressed here should consult any number of excellent texts on the subject (Kimura 1983; Page and Holmes 1998; Hughes 1999; Graur and Li 2000; Nei and Kumar 2000; Felsenstein 2004; Yang 2006, 2014; Li 2006).

The K_a to K_s ratio, ω

The basis for divergence-based tests is $\omega = K_a/K_s$, the (per-site) ratio of replacement (nonsynonymous) to silent (synonymous) substitution rates, which Miyata and Yasunaga (1980) referred to as the **acceptance rate** and which also appears in the literature as the **width of the selective sieve**. For sites under the standard neutral model (deleterious mutations can arise, but are quickly removed), the expected value of ω at a site (or gene) is $\omega = \mu f/\mu = f \leq 1$, where f is the ratio of the effectively neutral mutation rates. Thus, in the absence of positive selection, we expect $\omega < 1$. Moreover, if adaptive mutations are absent (or very rare), then $1-\omega$ is a direct measure of the amount of constraint $(1-f)$ on a site. Conversely, $\omega > 1$ is usually taken as an unmistakable signature of selection (Kimura 1983). Even if a demographic change results in a lowering of the effective population size (increasing the effectively neutral mutation rate at replacement sites), such a change (in the absence of positive selection) only brings K_a/K_s closer to, but still likely leaves it smaller than, 1.0.

There are cases where $\omega > 1$ is *not* a signal for positive selection. Ratnakumar et al. (2010) noted that a resolution of heteroduplex DNA during gene-conversion events often results in a bias toward G and C bases (also see Galtier et al. 2001; Webster and Smith 2004; Lassalle et al. 2015). Given that replacement-codon positions often have lower GC content than synonymous sites, there can be more opportunities for A/T at these sites to be changed to G/C, resulting in replacement substitutions and potentially inflating the K_a/K_s ratio (Berglund et al. 2009; Galtier et al. 2009). Ratnakumar et al. analyzed a dataset of roughly 18,000 human genes compared against their orthologs in at least two other mammalian genomes. They found that genes giving divergence-based signals of selection had a significant tendency to also display genomic signals of GC conversion bias. They estimated that >20% of elevated ω values in this dataset could be the result of biased gene conversion. A second factor is mutational bias. McVean and Charlesworth (1998) and Lawrie et al. (2011) found the counterintuitive result that weak selective constraints that oppose a mutational bias can actually *accelerate* the rate of evolution over that of a neutral site. In the words of Lawrie et al., this occurs because

> Common mutations drive substitutions away from the fitter states despite purifying selection, whereas selection favors fixation of uncommon mutations resulting in faster back substitutions

to the fitter states. This allows for greater overall flux between states and thus a higher rate of substitution at the constrained sites compared with the neutrally evolving sites.

A final factor that can upwardly bias estimates of ω is the presence of strong selective constraints on *silent* sites, was found in *Drosophila* by Lawrie et al. (2013). Chamary et al. (2006) reviewed some of the evidence that silent sites may still be subjected to constraints (beyond any weak ones from codon usage bias; Chapter 8) because they affect mRNA stability, splicing, or microRNA binding. A cautionary tale is offered by the work of Hurst and Pál (2001), who examined constraints on the breast cancer *BRCA1* gene. A sliding window of roughly 300 nucleotides, allowing for average regional estimates of K_a and K_s, was used to scan across this gene in two pairs of comparisons, human-dog and mouse-rat. The window around position 200–300 showed a relatively normal level of K_a (relative to the rest of the gene), while K_s plummeted dramatically, especially in the human-dog comparison. The result was an ω value significantly greater than one, not due to an elevation in the replacement-substitution rate, but rather to a decrease in the silent-substitution rate. Wolf et al. (2009) found that an upward bias in ω from reduced K_s values can be especially problematic when using closely related taxa, as a small value of D_s causes excessive stochastic variation in the denominator of a K_a/K_s ratio. Pond and Muse (2005) noted that if variation in K_s occurs over the gene, failure to include this heterogeneity in the model can easily result in false positives (estimated $\omega > 1$ for particular codons). Thus, while $\omega > 1$ is usually taken as a gold standard for positive selection, a little more humility in its use may be in order.

While conceptually straightforward, the operational problem in using ω is that while one or a few sites may be under *repeatedly* strong directional selection ($\omega > 1$ at these residues), most sites in a protein are expected to be under some selective constraints ($\omega < 1$), so that the average over all sites yields $\omega < 1$. Indeed, a meta-analysis by Endo et al. (1996) found that only 17 out of 3595 proteins (from a wide range of species comparisons) showed $\omega > 1$. There were, however, a few early success stories. Example 10.2 discussed the work of Hughes and Nei (1988), who used the three-dimensional (3-D) protein structure of the major histocompatibility complex to suggest potential sites to examine (those amino acids on the surface in critical positions). Within this set of residues, $\omega > 1$, while $\omega < 1$ when averaged over the entire gene. Unfortunately, most proteins lack this amount of detailed biological knowledge for an investigator to draw upon. Because amino acid residues in close proximity on the 3-D structure of a protein can be scattered all over the primary (i.e., linear) sequence, grouping sites for analysis by their position in the primary sequence can be very ineffective, and even misleading. The key is to base tests of ω values on a *codon-by-codon* basis, so that codons, rather than genes, become the unit of analysis. The limitation for this approach is the need for sequences from a sufficiently dense and well-supported phylogeny.

Two general approaches have been suggested to estimate ω. Both require a phylogeny, and issues such as the correct multiple-sequence alignment as well as errors in the assumed tree potentially loom in the background. **Parsimony-based** approaches reconstruct the ancestral sequence at each node in the tree, and then use these to count up the number of silent and replacement substitutions for each codon. **Likelihood approaches** (LW Appendix 4) are on a much firmer statistical footing, but they are computationally intense and can be rather model-specific. Both approaches allow for tests of whether a protein is under positive selection and, more specifically, tests of positive selection at specific *sites* in that protein. As with extensions to PRF models, more recent tests are being built around **Bayesian approaches** that extend the ML models (Appendix 2), which allow for the management of uncertainty in very complex statistical models.

Parsimony-based Ancestral Reconstruction Tests

Fitch et al. (1997) and Suzuki and Gojobori (1999) proposed similar parsimony-based approaches for detecting selection at single sites. Both start with a phylogeny and then use **parsimony** (i.e., choosing the solution requiring the fewest number of changes) to reconstruct the ancestral sequences at all of the nodes in the tree. With these estimated sequences in hand, one can then simply count the number of silent and replacement substitutions on the tree. The method of Fitch et al. computes an average, ω, rate for the entire gene and then looks for excessive variation at particular codons, while the method of Suzuki and Gojobori performs the analysis by considering each codon separately. The false-positive rate of both methods is generally small (Suzuki and Gojobori 1999; Suzuki and Nei 2002), but they suffer from low power (Wong et al. 2004). Further, both are subject to several rather delicate issues of sequence evolution that, if not correctly accounted for, can introduce rather significant artifacts.

First, it is well known that **transitions** (A $\leftrightarrow$ G, C $\leftrightarrow$ T) can occur at different rates than **transversions** (e.g., A $\leftrightarrow$ T, etc.), and (at third-base positions) transitions are more likely to give synonymous changes. Failure to incorporate these rate differences can result in an overestimation of the number of replacement substitutions (Yang and Nielsen 2002). Second, any codon usage bias (Chapter 8) must be accommodated. Third, when divergence times are modest to large, to avoid undercounting the number of the actual substitution events one must correct for the possibility of multiple substitutions between lineages at a site. All of these issues can have a highly significant effect on estimates of ω (Yang and Bielawski 2000). Finally, given that the ancestral states are likely estimated with error, parsimony analysis has no formal procedure to take this uncertainty into account. Bayesian posterior distributions can account for these errors, but this requires moving from a parsimony to a likelihood framework. For these reasons, most analyses use likelihood-based approaches (and their Bayesian extensions), wherein one explicitly allows the model to account for transitions vs. transversions, codon usage bias, and multiple substitutions.

Maximum-likelihood-based Codon Tests

Maximum-likelihood (ML) methods following the evolution of a codon over a phylogenetic tree were introduced by Goldman and Yang (1994) and Muse and Gaut (1994). While conceptually straightforward, they involve a fair bit of bookkeeping. They assume that each site is evolving independently, which can be compromised by two rather different factors. First, a substitution at one site can change the nature of selection at other sites (epistasis). Second, high levels of recombination can lead to false signals of selection, as recombination results different parts of the gene having different coalescent structures (Anisimova et al. 2003). While generally not an issue, this can be problematic with viral populations that have high sequence diversity and frequent recombination.

ML methods require a specific probability model for the movement among the 64 different codons. They start with a vector representing the 61 different nonstop codon states (stop codons are assumed lethal). At any point in time, a codon can mutate to one of nine other codons following a single base change (Figure 10.4). The model given by Goldman and Yang (1994) defines the following relative rates for movement between codons i and j,

$$Q_{ij} = \begin{cases} 0 & \text{if } i \text{ and } j \text{ differ at more than one position} \\ \pi_j & \text{for a silent transversion} \\ \kappa\pi_j & \text{for a silent transition} \\ \omega\pi_j & \text{for a replacement transversion} \\ \omega\kappa\pi_j & \text{for a replacement transition} \end{cases} \qquad \text{for } 1 \le i, j \le 61 \qquad (10.18)$$

The 61×61 **Q** matrix is specified by Equation 10.18. The π_j are the equilibrium frequencies of codon j (often calculated from the nucleotide frequencies at the three codon positions), while κ and ω are estimated parameters intended to account for biases in codon changes. Potential differences in transition versus transversion rates are accounted for by κ. One takes the currently observed codons over the phylogeny, and then runs the model backward in time by considering all possible ancestral codons at each of the internal nodes (ancestors) in the tree. The model thus corrects for multiple hits. The key parameter of interest is ω, the strength of selection on replacement substitutions. In the early models using

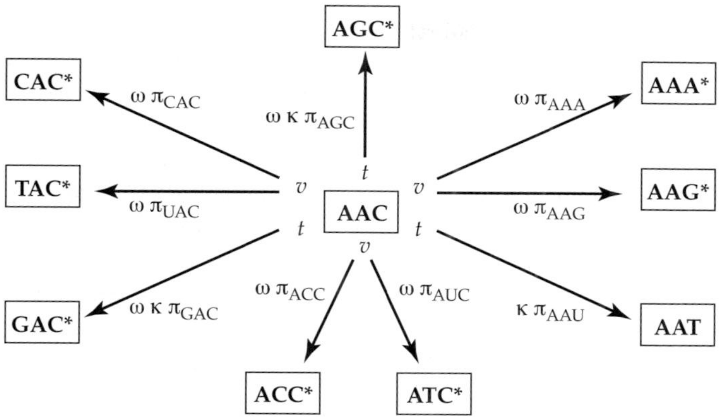

Figure 10.4 The various possible state changes and their rates under the codon evolution model (Equation 10.18) for the nine new codons that are within a single nucleotide change from the target codon (here AAC). Asterisks denote a replacement change, where the rate is a function of selection, and hence ω. Because transitions (denoted in the figure by t) and transversions (v) may occur at different rates, setting the transversion rate as the baseline, κ denotes any transition rate correction (with $\kappa = 1$ if the two rates are equal). All changes are a function of π_j, the equilibrium frequency of the mutant codon, j. Performing these same calculations over all 60 other nonstop codons generates the full transition matrix , $\mathbf{Q}$.

this approach, ω was a fixed constant over all genes, but later models allowed ω to vary over sites (nicely paralleling the development of extensions of the Poisson random field model to allow γ to vary over genes/alleles), a point upon which we will expand shortly. Figure 10.4 shows the basic structure of these various state changes for a particular codon. The mutational structure in Equation 10.18 (where mutation rates are functions of κ and π_j) is known in the molecular evolution literature as the **Hasegawa-Kishino-Yano**, or **HKY, model** (Hasegawa et al. 1985).

Tests for directional selection on a gene are accomplished by using this codon model superimposed on the phylogenetic tree to run the likelihood calculation (over all codons) to find the ML solutions for the $\mathbf{Q}$ matrix parameters. This allows for a direct test that $\omega > 1$ using the likelihood-ratio approach (LW Appendix 4). The key to these likelihood calculations is that $\mathbf{P}(t)$, the codon state matrix at time t, is related to the instantaneous rate matrix, $\mathbf{Q}$, by

$$\mathbf{P}(t) = \exp(\mathbf{Q}t) \tag{10.19}$$

The corresponding elements of the 61×61 matrix $\mathbf{P}$ are

$$P_{ij}(t) = \Pr(\text{codon } = i \text{ at time } t \,|\, \text{codon is } j \text{ at time } t = 0) \tag{10.20}$$

The matrix exponential, $\exp(\mathbf{Q}t)$, is computed by diagonalizing the $\mathbf{Q}$ matrix by writing $\mathbf{Q} = \mathbf{U}\boldsymbol{\Lambda}\mathbf{U}^T$, where $\boldsymbol{\Lambda}$ is a diagonal matrix, whose ith diagonal element is the eigenvalue λ_i of $\mathbf{Q}$ (Equation A5.10a), and $\mathbf{U}$ is a matrix of the eigenvectors of $\mathbf{Q}$ (Equation A5.10c). With this transformation, Equation 10.19 now becomes

$$\exp(\mathbf{Q}t) = \mathbf{U}\exp(t\boldsymbol{\Lambda})\mathbf{U}^T$$

where

$$\exp(t\boldsymbol{\Lambda}) = \begin{pmatrix} e^{t\lambda_1} & 0 & \cdots & 0 \\ 0 & e^{t\lambda_2} & \cdots & 0 \\ \vdots & & \ddots & \vdots \\ 0 & \cdots & \cdots & e^{t\lambda_n} \end{pmatrix} \tag{10.21}$$

A variety of likelihood models based on Equation 10.19 are typically tested (much in the same way that one tests subsets of complex segregation analysis models; see LW Chapter

13), adding additional factors (i.e., $\kappa \neq 1$, etc.) if they improve the model fit (i.e., give a significantly better likelihood ratio). Evidence for positive selection on a gene is indicated if the likelihood-ratio test for $\omega > 1$ is significant.

The base model (Equation 10.18) assumes that all codons within a given gene have the same ω value, which is not only unreasonable but also destroys most of the power of this approach, as it returns an estimate of ω based on a gene-wide average. Given that $\omega < 1$ for most codons, the signal from the majority of codons then masks the signal from any small fraction of codons where indeed $\omega > 1$ (e.g., Example 10.2). Nielsen and Yang (1998) and Yang et al. (2000) extended the base model by assuming a mixture-model (LW Chapter 13), with the codons in a sequence being drawn from one of several selection categories, each with different ω values. For codons from selection category k, Equation 10.18 becomes

$$Q_{ij}^{(k)} = \begin{cases} 0 & \text{if } i \text{ and } j \text{ differ at more than one position} \\ \pi_j & \text{for a silent transversion} \\ \kappa\pi_j & \text{for a silent transition} \\ \omega^{(k)}\pi_j & \text{for a replacement transversion} \\ \omega^{(k)}\kappa\pi_j & \text{for a replacement transition} \end{cases} \tag{10.22a}$$

The simplest version of biological interest has three selection classes, with codons either being neutral (with probability p_0), deleterious (with probability p_d), or advantageous (with probability $p_b = 1 - p_n - p_d$). Within each class there is a fixed selective value, with

$$\omega^{(k)} = \begin{cases} 0 & \text{deleterious class} \\ 1 & \text{neutral class} \\ \omega > 1 & \text{positively selected class} \end{cases} \tag{10.22b}$$

The parameters p_0, p_d, and ω are estimated from the data by maximum likelihood (LW Chapter 13 examines ML on mixture models). The idea is that one fits a base model (allowing only neutral and deleterious classes), and then fits the full model (Equation 10.22b or other extensions), using a likelihood-ratio test to see if the fit is significantly improved. If so, this is taken as support for a history of repeated positive selection on a subset of codons in the gene of interest.

While Equation 10.22b is clearly an improvement over models assuming a single value of ω for *all* replacement mutations, assigning all codons in the deleterious class an ω value of 0 (i.e., no substitutions) is clearly too restrictive, as is assigning all codons in the advantageous class the same ω value. Nielsen and Yang (1998) and Yang et al. (2000) further expanded Equation 10.22b by taking

$$\omega^{(k)} = \begin{cases} w^{(d)} \sim (0,1) & \text{deleterious class} \\ 1 & \text{neutral class} \\ w^{(a)} \sim (1,\infty) & \text{positively selected class} \end{cases} \tag{10.23}$$

where now the fitness values, $\omega^{(k)}$, for any particular codon in class k are *random draws* from some specified distribution (as opposed to Equation 10.22b, which assumed they are unknown constants) whose parameters are again estimated by maximum likelihood. This is exactly the approach used previously to allow γ to vary over genes in the PRF model (e.g., Equations 10.17a and 10.17b). A number of candidate distributions for ω are possible, depending on whether we wish to restrict values to between $(0, 1)$ or to $(1, \infty)$, for codons in the deleterious and positively selected classes (respectively). For example, Nielsen and Yang (1998) and Yang et al. (2000) used either a beta or truncated gamma distribution (restricted to returning values of $0 < \omega < 1$) for the deleterious class and a truncated gamma (restricted to returning values of $\omega > 1$) for the positively selected class (Appendix 2 reviews the beta and gamma distributions). Again, a model-fitting approach is used where one first fits a lower-order model, and then progressively adds in additional parameters to see if the fit is significantly improved.

The power of the basic ML approach was examined by Anisimova et al. (2001, 2002, 2003) and Wong et al. (2004), and is a function of two different sample sizes: the number of codons in the sequence, and the number of actual sequences (number of taxa in the phylogeny). The more codons in a gene the better, although 100 seems to give reasonable power. Power is more efficiently increased by adding more sequences (species), as opposed to looking at longer genes. For moderately long sequences ($\sim$100 codons) with a modest phylogeny (10–20 species), power can be quite reasonable, at least under the parameters simulated (typically 5–10% adaptive codons, each with ω around 5). They also found that the χ^2 test used to compute the significance of likelihood ratios is conservative, and hence can be safely used, albeit suffering some reduction in power.

As might be expected, this basic framework can be modified in a number of ways, for example by letting some branches of the phylogeny be under selection and others not (Yang and Nielsen 2002; Zhang et al. 2005). **Branch models** assume the same value of ω over all sites, but allow ω to vary over branches; **site models** (our main focus here) allow ω to vary over sites, but not branches; while **branch-site** models allow ω to vary over both (e.g., Kosakovsky Pond and Frost 2005; Anisimova and Yang 2007; Kosiol et al. 2008; Kosakovsky Pond et al. 2011). These ML approaches can also be used to estimate the strength of selection on particular substitutions (Halpern and Bruno 1998; Yang and Nielsen 2008; Rodrigue et al. 2010; Tamuri et al. 2011).

ML methods are not foolproof, and their robustness to the underlying distributional assumptions (for ω) remains unclear (Suzuki and Nei 2004; Nozawa et al. 2009). For example, Zhang (2004) found 20–70% false positives in a branch model by Yang and Nielsen (2002) that allowed selection to operate on some branches but not others. By simply replacing the assumption of $\omega = 0$ for the deleterious class with ω being an unknown to be estimated that lies within the interval (0,1), Zhang et al. (2005) obtained much better branch-model behavior (also see Yang and dos Reis 2011). These results point out how fragile some of these models can be, with essentially no internal controls to check for model consistency.

Finally, while our discussion of phylogeny-based divergence tests has focused exclusively on coding sequences, such a restriction is not essential. Wong and Nielsen (2004) extended the logic of codon-based models to noncoding regions. Here the test is the substitution rate in noncoding regions versus the rate at nearby silent sites. Using this approach, Wong and Nielsen found little signal of selection on noncoding regions of the sequences from 13 viral data sets, but strong signals of positive selection in protein-coding regions in five of these data sets. As noted above, the major complication with using noncoding sequences is alignment, as homologous positions need to be compared over a phylogeny. Given that insertions and deletions are common in such regions, the time window for unambiguous alignment tends to be rather short.

Bayesian Estimators of Sites Under Positive Selection

Provided that one has the correct model, likelihood methods can be used to infer which actual *sites* have likely been under repeated positive selection. This powerful idea, which due to Nielsen and Yang (1998), first tests to see if a model allowing for a subset of codons to be positively selected significantly improves the fit. If so, this provides evidence of positive selection *somewhere* in the gene of interest, but it does not specify *which* particular codons are the actual targets. To find these, Nielsen and Yang used Bayes' theorem (Equation A2.2). Equations 10.18, 10.22, and 10.23 can be used to generate the conditional probability $\Pr(D \,|\, \omega_i)$ of the observed states at a particular codon over the sampled tree (the data D), given that the codon came from fitness class i (typically there are three classes: neutral, deleterious, and advantageous; each with a different ω value). However, it is more desirable to flip this conditional and obtain $\Pr(\omega_i \,|\, \text{data})$, i.e., $\Pr(\text{ in class } i \,|\, \text{data})$. Our particular interest is the posterior probability of a codon being in the advantageous class given the observed data. Bayes' theorem allows us to do this.

Suppose there are k classes, with each class having a different associated ω. The posterior probability that a specific codon is in fitness class i is

$$\Pr(\text{class } i \mid D) = \frac{\Pr(D \mid \text{class } i)\Pr(\text{class } i)}{\Pr(D)} = \frac{\Pr(D \mid \omega_i)\Pr(\omega_i)}{\sum_{i=1}^{k}\Pr(D \mid \omega_i)\Pr(\omega_i)} \tag{10.24a}$$

where D is the pattern of codons for that site in the tree, and the prior Pr(class i)—the values for p_0, p_b, and p_d —is estimated by maximum likelihood. The case of interest is whether the codon belongs to the class of advantageous sites, $\Pr(\omega > 1 \mid D)$,

$$\Pr(\text{advantageous} \mid D) = \frac{\Pr(D \mid \omega > 1)p_b}{\Pr(D \mid \omega < 1)p_d + \Pr(D \mid \omega = 1)p_0 + \Pr(D \mid \omega > 1)p_b} \tag{10.24b}$$

This approach allows us to directly assign probabilities of selective classes to any particular site. Anisimova et al. (2002) found that large ω values and a modest to large number of sequences are required for this approach to have reasonable power. A number of technical issues that arise when applying Equation 10.24a were examined by Huelsenbeck and Dyer (2004), Newton et al. (2004), Scheffler and Seoighe (2005), Yang et al. (2005), Aris-Brosou (2006), Guindon et al. (2006), and Anisimova and Liberles (2007).

Example 10.15. Bishop et al. (2000) examined the class I chitinase genes from 13 species of mainly North American *Arabis* (tower mustards), crucifers closely related to *Arabidopsis*. Chitinase genes are thought to be involved in pathogen defense, as they destroy the chitin in cell walls of fungi. Many fungi have evolved resistance to certain chitinases, so these genes are excellent candidates for repeated cycles of selection (i.e., an "arms race" scenario). Codon-evolution models estimated that between 64 and 77% of replacement substitutions are deleterious, with 5–14% being advantageous (analyses using phylogenies estimated by different methods all yielded similar results). These favored sites had an estimated value of $\omega = 6.8$. Using the criterion of a posterior probability of membership in the advantageous class in excess of 0.95 (i.e., Pr(advantageous class | D) > 0.95), 15 putative sites were located (using Equation 10.24b). Seven of these sites involved only one substitution type, which evolved multiple times over the phylogeny. The authors had access to the 3-D structure of chitinase, which shows a distinctive cleft thought to be the active site. Mapping putative sites of positive selection showed a significant excess of these sites clustered at the cleft.

Balancing this apparently successful application of these methods to detect selected sites is the work of Yokoyama et al. (2008). These authors examined the evolution of dim-light vision in vertebrates, which is determined by the wavelength of maximal absorption of rhodopsin. This can be directly measured in the lab, allowing the authors to experimentally determine the role of particular substitutions in dim-light adaptation using 11 engineered ancestral rhodopsin sequences. They found that most of the change in maximal absorption can be accounted for by 12 sites. In contrast, Bayesian methods predicted a total of 8 positively selected sites, none of which corresponded to sites shown by mutagenesis to have adaptive roles.

CONNECTING THE PARAMETERS OF ADAPTIVE EVOLUTION

As summarized in Table 10.2, a number of different parameters of adaptive evolution have been introduced in this chapter (as well as in Chapter 8), along with various machinery for estimating them. We have examined the connections between some of these parameters (e.g., Equations 10.16a–10.16e). However, we have yet to develop a connection between the two key parameters: the scaled strength of selection, $\gamma = 2N_e s$, at a site (Poisson random field models) and the ratio of substitution rates, $\omega = K_a/K_s$ (codon models).

We can connect these parameters as follows. Assume that silent sites are taken as the neutral benchmark, so that (as a first approximation) their per-site mutation rate, μ_s is also

Table 10.2 Summary of the key parameters of adaptive evolution and their connections. Chapter 8 first introduced several of these (α, γ, and μ_b), while ω and f were introduced in this chapter.

α	The fraction of substitutions that are adaptive
γ	The scaled strength of selection, $2N_e s$
μ	The total per-site mutation rate
μ_s	The effectively neutral per-site mutation rate at silent sites (usually assume $\mu_s \simeq \mu$)
μ_b	The adaptive (benefical) mutation rate
p_b	The fraction of new mutations at a site that are advantageous, $\mu_b = p_b \mu$
λ	The rate of adaptive fixations, $\lambda = 2\gamma\mu_b$
$f = p_0$	The fraction of neutral mutations
$1 - f$	The amount of constraint on a site (relative to some standard, typically silent sites)
ω	The ratio of the replacement- to silent-site substitution rates

$$\omega = f + 2\gamma p_b = \frac{2\gamma p_b}{\alpha} \quad \text{(Equations 10.25a and 10.25c)}$$

$$\gamma = \frac{\omega - f}{2p_b} = \frac{\omega - p_0}{2p_b} \quad \text{(Equation 10.25b)}$$

$$\alpha = \frac{\lambda}{\lambda + \mu p_0} = \frac{2\gamma}{2\gamma + p_0/p_b} = \frac{2\gamma p_b}{\omega} \quad \text{(Equations 10.16b, 10.16c, and 10.25c)}$$

the actual mutation rate, μ. Two types of mutations contribute to the rate of replacement substitutions: a fraction f (notationally interchangeable with p_0, as $f = p_0$) that is effectively neutral and a much smaller (perhaps zero) fraction p_b that are favored. Effectively neutral substitutions accrue at a rate of $f\mu_s$, while (Equation 8.24a) beneficial substitutions accrue at rate $\lambda = (2N\mu_b)(2sN_e/N) = 2(2N_e s)\mu_b = 2\gamma\mu_b = 2\gamma p_b \mu_s$. Hence

$$\omega = \frac{K_a}{K_s} = \frac{f\mu_s + 2\gamma p_b \mu_s}{\mu_s} = f + 2\gamma p_b = p_0 + 2\gamma p_b \tag{10.25a}$$

so that very strong, or frequent, selection ($\gamma\, p_b > 1$) is required for $\omega > 1$. Similarly, we can rearrange this equation to solve for γ,

$$\gamma = \frac{\omega - f}{2p_b} = \frac{\omega - p_0}{2p_b} \tag{10.25b}$$

If $f = 0.5$ and $p_b = 0.01$, so that half of the mutations are effectively neutral and 1% are favored, $\gamma = 25$ is required for $\omega = 1$, while $\omega = 3$ requires $\gamma = 125$. If p_b is 0.001, a value of $\gamma = 400$ only gives $\omega = 1.3$, which is a sufficiently small deviation to avoid detection in many cases. Finally, to connect α and ω, from Equations 10.16b and 10.25a, we have

$$\alpha = \frac{2\gamma p_b}{2\gamma p_b + p_0} = \frac{2\gamma p_b}{\omega} \tag{10.25c}$$

which can alternately be expressed as

$$\alpha\,\omega = 2\gamma p_b, \qquad \text{and} \qquad \omega = \frac{2\gamma p_b}{\alpha} \tag{10.25d}$$

THE SEARCH FOR SELECTION: CLOSING COMMENTS

Detecting selection using molecular data is a major growth industry and will continue to accelerate as whole-genome sequencing becomes increasingly faster and cheaper. As the last two chapters indicate, there is an enormous amount of statistical machinery proposed to carry out this task, but every method has major limitations. As detailed in Chapters 29 and 30, ecologists and evolutionary biologists search for selection using complementary trait-based approaches, which require specifying potential traits under selection, and measuring the association between these and individual fitness. While such trait-based approaches allow us to consider particular traits of interest, molecular data have several advantages.

Two advantages are fairly obvious, in that traits need not be specified and measurement of individual fitness is not needed. The greatest advantage of the molecular approach, however, is that molecular data are a time machine. We cannot go back in the past to measure traits and fitness, but past selection *may* leave a number of different signals in the genome. Very recent events *may* leave sweep-like signatures (Chapter 8); and Chapter 9 reviews the myriad of tests for detecting these. If polygenic adaptation is the rule, major changes in trait values can occur through only cumulative changes at multiple loci, each of small effect. In this case, very little molecular signal is expected, whereas ongoing selection can be detected using trait-based methods, *provided* one knows the correct traits! Over a longer time scale, repeated selection events *may* leave molecular patterns. Population-based divergence tests (HKA, MK) can detect patterns of repeated positive selection over an entire gene during the divergence of two populations or species, while phylogeny-based divergence tests (codon models) can detect repeated positive selection at the same *codon* over a phylogeny.

Finally, there is an intermediate between this marker-based vs. trait-based dichotomy, namely **trait-augmented marker methods**, which are examined in Chapter 12. Here, one focuses on the QTL and GWAS hits associated with a particular trait and queries whether something about them is nonrandom (such as a set of correlated allele-frequency changes).

Caution is in Order When Declaring Positive Selection

Because nearly every test can give a false positive for reasons other than positive selection, any detected region should *always* be viewed as no more than a candidate to be followed up by direct work assessing any functional impact and, if so, the nature of selection. In particular, investigators should be extremely wary of "just-so" stories, wherein once a region is detected, some clever story is proposed as to the cause of selection. One must resist the notion that functional differences can automatically be equated to adaptive changes (Gould and Lewontin 1979; Storz and Wheat 2010; Barrett and Hoekstra 2011). In the words of Nielsen (2009), "evidence of selection, and knowledge of the function of a gene, does not constitute evidence for adaptation," as the following cautionary tale illustrates.

Example 10.16. Humans show a dramatic expansion of brain size with respect to most mammals, with this increase in (relative) size usually being assumed to be correlated with increased cognitive abilities. Primary microcephaly is a pathological condition in humans resulting in small heads, but other normal features. Nonfunctional alleles at the genes *microcephalin* and *ASPM* (abnormal spindle-like microcephaly associated) both display microcephaly phenotypes, with a typical individual having a brain size of around 400 cm^3, comparable to that in early hominids (versus the normal 1400 cm^3 in modern humans). Not surprisingly, several studies have searched for selection on these genes within the primate lineage.

Zhang (2003) inferred a K_a/K_s ratio of 1.03 for *ASPM* on the branch from the human-chimpanze common ancestor to humans, but a ratio of 0.66 on the branch from this ancestor to chimpanzees. Values of 0.43 to 0.29 were found along other branches in mammals, suggesting positive selection along the human lineage. Evans et al. (2004a) also examined *ASPM* over a larger phylogeny ranging from New World monkeys through humans. Accelerated rates of evolution ($K_a/K_s > 1$) were seen between gibbons and the ancestor of the great apes, and a large acceleration ($K_a/K_s = 1.44$) was seen on the lineage from the human and chimpanze ancestor to humans. Evans et al. also performed a McDonald-Kreitman test, comparing the polymorphisms within humans to the divergence between the human-chimpanze common ancestor and finding

	Fixed	Polymorphic
Synonymous	7	10
Replacement	19	6

Fisher's exact test gives a p value of 0.01, with these data showing an excess of around 15 replacement substitutions over what is expected from the replacement-to-silent ratio seen in the polymorphism data ($\alpha = 15/19$ ~80%).

Similar results were seen for *microcephalin*. Upon examining different parts of the phylogeny of this gene, Evans et al. (2004b) found K_a/K_s = 1.05 in the simian lineages leading to humans, and ratios of 0.4 to 0.6 along other mammalian lineages. A further breakdown showed that most of the excess in K_a/K_s occurred from prosimians to the branching of the great apes, with values < 1 within the great apes. They also found a significant McDonald-Kreitman result, with an estimated 45 adaptive replacement substitutions occurring between prosimians and humans. Thus, the data suggest that *microcephalin* is associated with expansion of brain size leading to the great apes, while *ASPM* is inferred to be further associated with the increase in brain size specifically along the lineage leading to humans.

Building on these strong observations of selection leading to the human lineage, Mekel-Bobrov et al. (2005) and Evans et al. (2005) searched for *ongoing* selection in these two genes, and found strong signals in each. Evans et al. (2005) found that the *microcephalin* gene had one haplotype in humans (associated with a replacement substitution) at much higher frequencies than the others, with extended linkage disequilibrium and small intra-allelic variation. Using intra-allelic variation, the age of this haplotype was estimated at 37,000 years (with a range of 14,000 to 60,000). Young alleles at high frequencies are hallmark indicators of positive selection (Chapter 9). Extensive coalescent simulations using a variety of population structures all gave high levels of significance to these results. This same pattern was seen by Mekel-Bobrov et al. (2005) with *ASPM*: a common haplotype with long LD and a very recent estimated origin (5800 years). Again, coalescent simulations of neutral drift under a variety of proposed models of human population growth and expansion showed these results to be highly significant.

Given two functional genes that both influence brain size, a presumed correlate of intelligence, coupled with a history of past and ongoing selection, these data do indeed seem to suggest a case for selection on intelligence. As such, they gathered a significant amount of attention, in part from the finding that the putative adaptive haplotypes were in higher frequencies in Europe and Asia relative to Africa.

This view, however, was quickly dispelled. Timpson et al. (2007) and Mekel-Bobrov et al. (2007) showed in large sample sizes (900 and 2400, respectively) that there was no correlation between the putative adaptive haplotypes and increased intelligence. Thus, any ongoing selection on these genes does not appear to correlate with selection for increased cognition. Currant et al. (2006) further noted that *spatial* models of population growth were not considered, and the above patterns for neutral mutations can passively arise along the leading edge of a recent population expansion (through allelic surfing; see Chapter 9). If not for the concern among many geneticists at drawing social implications from the initial selection findings, this saga might have become a textbook standard in the search for selection. This case provides as especially strong lesson, as in most studies, few loci with signatures of selection would have received this level of scrutiny.

Curbing Our Enthusiasm

We started this set of chapters with a plea for caution and will do so again to draw our discussion to a conclusion. As in the great electrophoresis hunt in the 1970s (grinding up every species or population in sight to measure segregating protein variation) and the great QTL hunt in the 1990s (trying to find QTLs for traits in favorite organisms), we are now experiencing the great selection hunt phase of evolutionary genomics. The obvious excitement of detecting either ongoing selection or targets with a history of repeated past selection must also be tempered with caution. There are a huge variety of different tests, but no one best test even for a particular situation (much less over all settings). Simple methods may lack power, but very sophisticated, highly parametric tests may not be very robust to modeling assumptions. As mentioned on numerous occasions, complications involving demography (changes in population size) and population structure can cripple most tests. More sophisticated methods developed to circumvent some of these issues are not yet fully vetted, and must be used with caution. Finally, there is the **Beavis effect** (LW Chapter 15), also known as the **winner's curse** (Kraft 2008), in which a parameter (such as a strength of selection) that is declared significant often has its value overestimated. This problem is especially acute when the power for detection is low, as often occurs with using more stringent individual p values to control false positives under multiple comparisons. When a selection signal is detected (likely out of a sea of candidates, with each test having moderate to low power), the actual effect is likely overestimated, and potentially by a very large amount. These comments are not meant to discourage the use of these molecular-marker methods, but rather to ensure that the enthusiasm with which they are applied is tempered by the cold reality of their limitations.

Finally, as stressed throughout the last few chapters, even when successful, these tests give us an insight into *just a tiny fraction* of all selective substitutions. How representative this subsample is of adaptive evolution in general is unclear, but it is certainly biased toward large-effect mutations. Thus, significant caution is in order in extrapolating results to general statements about adaptation. In closing, what is clear is that multiple selection events (be they recurrent sweeps or background selection) leave an impact on linked neutral sites, and most genomes show ample signals that this a very common phenomena (Chapter 8).

III

Drift and Quantitative Traits

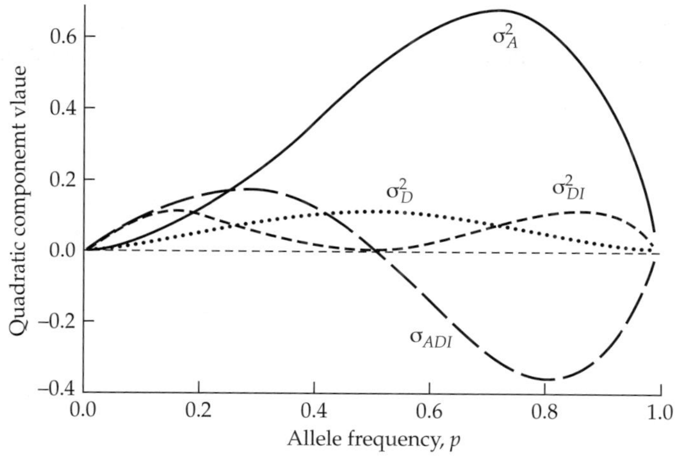

11

Changes in Genetic Variation Induced by Drift

You want to shave with Occam's razor, not cut your throat. J. B. Walsh

We noted in Chapter 2 that, when operating as the sole evolutionary force, random genetic drift leads inevitably to the loss of alleles within populations as well as to the fixation of alternative alleles in different populations. These conclusions extend logically to quantitative characters. Following a reduction in population size, for example, we expect the genetic variance within populations to decline and the mean phenotypes of isolated populations to diverge. There are some interesting surprises, however, particularly when the mode of gene action has a nonadditive component. In this case, the genetic variance for a trait is not a simple function of the underlying heterozygosity (LW Chapter 4), so we cannot expect the temporal dynamics of the genetic variance to strictly reflect changes in heterozygosity. Indeed, as we will show, under certain conditions, the genetic variance for a quantitative trait is expected to transiently *increase* during the early phase of a population bottleneck.

The goal of the following two chapters is to develop a null (neutral) hypothesis for quantitative-trait evolution under the assumption that selection is a negligible evolutionary force. For the most part, we will continue to adhere to an ideal Wright-Fisher population structure, with random mating and discrete generations. In this vein, our conceptual approach will be to consider a series of replicate populations, all isolated at the same time from a large base population, which is generally assumed to be in Hardy-Weinberg and gametic-phase equilibrium, and all subsequently kept indefinitely at an identical population size. The current chapter focuses on the expected dynamics of the genetic variance within populations, whereas Chapter 12 focuses on interpopulation divergence.

In both chapters, we initially assume that the dynamics of evolutionary change are due entirely to genetic properties of the base population, which is essentially the case with short-term population bottlenecks. Then, the role of mutation will be taken up. In this chapter, for example, we will end by considering the levels of genetic variance expected in the absence of selection, when a stochastic equilibrium has been reached between the input of variation by mutation and the loss by genetic drift. We also consider the statistical underpinnings of the covariance between relatives in inbred populations, as this has special relevance in attempts to derive inferences about the mode of gene action from phenotypic observations.

The subject material of this chapter is rather technical in places, as it involves the expected temporal dynamics of higher-order gene-frequency moments, including, at the minimum, fourth-order moments, as well as a number of quadratic components of gene action not previously encountered with outbred populations. With quantitative traits, we must also worry about the joint distribution of allele frequencies at different loci, so issues of linkage disequilibrium come in as well. These complexities quickly get out of control in considering the sampling variances of genetic variances and covariances, and we will keep our treatment of these issues to a minimum (which is not to deny their substantial practical significance). Finally, we note that most of our coverage will be concerned with genetic *variance*, although all moments of the phenotype distribution are subject to change in the presence of drift and mutation.

RESPONSE OF WITHIN-POPULATION GENETIC VARIANCE TO DRIFT

Complete Additivity

Consider a diallelic locus (indexed by i) with a strictly additive genetic basis, such that

Evolution and Selection of Quantitative Traits. Bruce Walsh & Michael Lynch.
Published 2018 by Oxford University Press.
DOI 10.1093/oso/9780198830870.001.0001

the three genotypic values contributing to a quantitative trait are scaled to be 0, a_i, and $2a_i$. Assuming Hardy-Weinberg equilibrium, at any particular generation, t, the total (and additive) genetic variance associated with this locus is $2a_i^2 p_i(t)[1 - p_i(t)]$, where $p_i(t)$ is the allele frequency at time t (LW Chapter 4). Assuming gametic-phase equilibrium, this expression is readily extended to a multilocus trait with a purely additive genetic basis. If we sum over all n loci contributing to the trait, the expected within-population genetic variance is

$$\sigma_A^2(t) = 2\sum_{i=1}^{n} E\left\{a_i^2 p_i(t)[1 - p_i(t)]\right\} \tag{11.1}$$

From Chapter 2, we know that the expected heterozygosity after t generations at an effective population size of N_e is simply $[1 - 1/(2N_e)]^t$ times the initial value. Moreover, under the assumption of neutrality, there should be no correlation between allele frequency and effect, so, substituting from Equation 2.5,

$$\begin{aligned} \sigma_A^2(t) &= 2\sum_{i=1}^{n} a_i^2 p_i(0)[1 - p_i(0)]\left(1 - \frac{1}{2N_e}\right)^t \\ &= \sigma_A^2(0)\left(1 - \frac{1}{2N_e}\right)^t \simeq \sigma_A^2(0)\exp\left(-\frac{t}{2N_e}\right) \end{aligned} \tag{11.2}$$

as obtained by Wright (1951).

Equation 11.2 illustrates the simplest possible behavior that can be expected for the genetic variance within a finite population, starting with a baseline level of $\sigma_A^2(0)$. For a character with a purely additive genetic basis, in the absence of any significant replenishing forces for variation (mutation or migration), the additive genetic variance within populations is expected to decline exponentially at the rate of $1/(2N_e)$ per generation. When linkage disequilibrium is present, the additive *genic* (as opposed to the additive *genetic*) variance declines at this rate (Chapters 16 and 24). A key point worth stressing is that Equation 11.2 describes the *expected* behavior of the genetic variance, as averaged over a very large number of replicate populations. As discussed below, as a consequence of the stochastic sampling of gene frequencies, any single replicate population can deviate substantially from its expected trajectory.

The Effects of Dominance

Robertson (1952) extended the preceding theory to loci with dominance and obtained the surprising result that rare recessive alleles can sometimes cause an initial *increase* in both the additive and dominance components of variance in an inbreeding population. A rare neutral allele will usually be lost from a small population, in which case the variance will decline, but if the frequency of a rare recessive allele stochastically increases, the frequency of the extreme genotype will also increase. For completely recessive alleles, a temporary inflation of the expected within-population variance will occur, provided the initial frequency of the recessive genotype is less than 0.17 (Robertson 1952). Although an inflation of the expected variance can also occur with partial dominance, the critical initial frequency for the recessive allele becomes progressively smaller as additivity is approached. Regardless of the degree of dominance, however, the within-population variance eventually declines to zero as loci move toward fixation, as in the case of pure additivity.

Robertson (1952), and thereafter Willis and Orr (1993), considered only a single diallelic locus, whereas the analysis is much more complex with multiple loci or with more than two alleles per locus. Fairly general results have been obtained for the case in which all of the genetic variance can be partitioned into additive, dominance, and additive $\times$ additive epistatic components (Cockerham 1984a, 1984b; Cockerham and Tachida 1988; Tachida and Cockerham 1989). Even for this case, however, and assuming initial conditions of Hardy-Weinberg and gametic-phase equilibrium, the temporal dynamics of genetic

Table 11.1 Factors contributing to the additive, dominance, and additive $\times$ additive components of genetic variance in finite populations. Here n is the number of loci, n_k is the number of alleles at the kth locus, p_{ki} is the frequency of the ith allele at locus k, α_{ki} is the additive effect of the ith allele at locus k, δ_{kij} is the dominance effect at locus k associated with genotype ij, and $(\alpha\alpha)_{ki,mj}$ is the additive $\times$ additive effect of alleles i and j from different loci (k and m) (LW Chapters 4 and 5). The inbreeding depression is defined for individual loci (ι_k for locus k) as well as for the sum over all loci (ι). The α_{ki}, δ_{kij}, and $(\alpha\alpha)_{ki,mj}$ are defined from the standpoint of a randomly mating base population (LW Chapter 4).

Additive variance	$\sigma^2_A = 2\sum_{k=1}^{n}\sum_{i=1}^{n_k} p_{ki}\alpha^2_{ki} = 2\sum_{k=1}^{n} E[\alpha^2_{k\cdot}]$
Dominance variance	$\sigma^2_D = \sum_{k=1}^{n}\sum_{i=1}^{n_k}\sum_{j=1}^{n_k} p_{ki}p_{kj}\delta^2_{kij} = \sum_{k=1}^{n} E[\delta^2_{k\cdot\cdot}]$
Epistatic variance	$\sigma^2_{AA} = 4\sum_{k,m=1}^{n}\sum_{i=1}^{n_k}\sum_{j=1}^{n_m} p_{ki}p_{mj}(\alpha\alpha)^2_{ki,mj} = 4\sum_{k,m=1}^{n} E[(\alpha\alpha)^2_{k\cdot,m\cdot}]$
Inbreeding depression	$\iota_k = \sum_{i=1}^{n_k} p_{ki}\delta_{kii} = E[\delta_{kii}] \qquad \iota = \sum_{k=1}^{n} \iota_k$
Sum of squared locus-specific inbreeding depressions	$\iota^* = \sum_{k=1}^{n} \iota^2_k$
Variance of dominance effects in inbred individuals	$\sigma^2_{DI} = \sum_{k=1}^{n}\sum_{i=1}^{n_k} (p_{ki}\delta^2_{kii} - \iota^2_k) = \sum_{k=1}^{n} (E[\delta^2_{kii}] - \iota^2_k)$
Covariance of additive and dominance effects in inbred individuals	$\sigma_{ADI} = 2\sum_{k=1}^{n}\sum_{i=1}^{n_k} p_{ki}\alpha_{ki}\delta_{kii} = 2\sum_{k=1}^{n} E[\alpha_{ki}\delta_{kii}]$

variance depend on seven quadratic properties of the base population (Table 11.1), as well as on several expectations for the higher-order moments of allele and gamete frequencies. We first present the general model, and then consider some illuminating results that arise under special conditions.

Quadratic Components for Inbred Populations

When dominance is present, the covariance between relatives (and hence the trait variance) under inbreeding is no longer fully described by just σ^2_A and σ^2_D. Rather, additional **quadratic components of covariance** are required, as outlined in Table 11.1. For the case of dominance and additive $\times$ additive variance, these include the familiar parameters σ^2_A, σ^2_D, and σ^2_{AA} (i.e., the additive, dominance, and additive $\times$ additive components of genetic variance; LW Chapter 5); the inbreeding depression, ι, here defined to be the difference between the mean phenotypes of outbred and completely inbred individuals (by construction, $E[G] = 0$ for an outbred population, where G denotes the genotypic value, measured as a deviation from the overall mean); the sum, ι^*, of squared locus-specific inbreeding depressions; the variance of dominance effects among inbred individuals, σ^2_{DI}; and the covariance of additive and dominance effects in inbred individuals, σ_{ADI}. Simplification is possible under certain circumstances. Most notably, with only two alleles per locus, $\iota^* = \sigma^2_D$, and if all alleles have a frequency of 0.5, as in a cross between two pure (i.e., fully inbred) lines, then $\sigma^2_{DI} = \sigma_{ADI} = 0$.

Table 11.2 Some of the alternative notations used for the genetic components required under inbreeding: Harris = Harris (1964); Gallais = Goldringer et al. (1996); Jacquard = Jacquard (1974); Cornelius = Cornelius (1975); Cockerham = Cockerham (1984a, 1984b); de Boer = de Boer and Hoeschele (1993); Smith = Smith and Mäiki-Tanila (1990); Abney = Abney et al. (2000).

Ours	Harris	Gallais	Jacquard	Cornelius	Cockerham	de Boer	Smith	Abney
σ^2_A	σ^2_{Ar}	σ^2_A	V_A	σ^2_A	σ^2_A	σ^2_{Ar}	σ^2_a	V_a
σ^2_D	σ^2_{Dr}	σ^2_D	V_D	σ^2_D	σ^2_D	σ^2_{Dr}	σ^2_d	V_d
ι	–	–	–	–	H	Δ_I	μ_δ	–
ι^*	D^2_I	$\sum Do^2$	D^2_H	μ_∞	H^*	Δ^2_I	–	SS_{μ_h}
σ^2_{DI}	σ^2_{DI}	σ^2_{Do}	$V_h - D^2_H$	$\sigma^2_\infty - 2C + 2\sigma^2_A$	D^*_2	σ^2_{DI}	σ^2_δ	V_h
σ_{ADI}	σ_{ADI}	σ_{ADO}	$2\text{Cov}_H(\text{A}, \text{D})$	$C - 2\sigma^2_A$	$2D_1$	σ_{ADI}	$\sigma_{a\delta}$	$\text{Cov}_h(a, d)$

As shown in Table 11.2, numerous alternative notations for these quadratic components appear in the literature, with almost every new paper seeming to invent its own terminology, which is often a hybrid of several previous papers. Additionally, there are also reparameterizations of these components, such as the **Q model** of Cornelius and Van Sanford (1988). As Example 11.1 shows, the dominance-related quadratic components are easily computed if one knows the genotypic frequencies in the (randomly mating) base population. Because these components depend on α and δ, which themselves depend on allele frequencies, their behavior under a change in allele frequency can be complex.

Example 11.1. Consider a population with a single locus with genotypic values of $A_1A_1 = 0$, $A_1A_2 = 1.67$, and $A_2A_2 = 2$. What are the quadratic components when $p = \text{freq}(A_1) = 0.8$? Using standard expressions (LW Chapter 4), the random-mating parameters are

α_1	α_2	δ_{11}	δ_{12}	δ_{22}	σ^2_A	σ^2_D
-0.2804	1.1216	-0.0536	0.02144	-0.8576	0.628993	0.045967

Note that the mean value of $G = 2E[\alpha] + E[\delta] = 0$ under random mating, as

$$(2\alpha_1 + \delta_{11})p^2 + (\alpha_1 + \alpha_2 + \delta_{12})2p(1-p) + (2\alpha_2 + \delta_{22})(1-p)^2 = 0$$

The mean value (ι) of G under complete inbreeding (which is the inbreeding depression change in the mean as $E[G] = 0$ under random mating), follows upon recalling that a fraction, p_1, are A_1A_1 and a fraction, p_2, are A_2A_2 and that $p_1\alpha_1 + p_2\alpha_2 = 0$, yielding

$$\begin{aligned}\iota &= p_1(2\alpha_1 + \delta_{11}) + p_2(2\alpha_2 + \delta_{22}) \\ &= p_1\delta_{11} + p_2\delta_{22} = 0.8 \cdot (-0.0536) + 0.2 \cdot (-0.8576) = -0.2144\end{aligned}$$

Because there are only two alleles, $\iota^* = \sigma^2_D$ (Cockerham and Matzinger 1985), and this is confirmed as

$$\iota^* = (p_1\,\delta_{11} + p_2\,\delta_{22})^2 = (-0.2144)^2 = 0.045967 = \sigma^2_D$$

As for the other two quadratic components,

$$\begin{aligned}\sigma^2_{DI} &= p_1\,\delta^2_{11} + p_2\,\delta^2_{22} - \iota^* \\ &= 0.8\,(-0.0536)^2 + 0.2\,(-0.8576)^2 - 0.045967 = 0.103427 \\ \sigma_{ADI} &= 2(p_1\,\alpha_1\,\delta_{11} + p_2\,\alpha_2\,\delta_{22}) \\ &= 2\big[\,0.8\,(-0.2804)\,(-0.0536) + 0.2\,(1.1216)\,(-0.8576)\,\big] = -0.360707\end{aligned}$$

These quadratic components for other allele frequencies are graphed below.

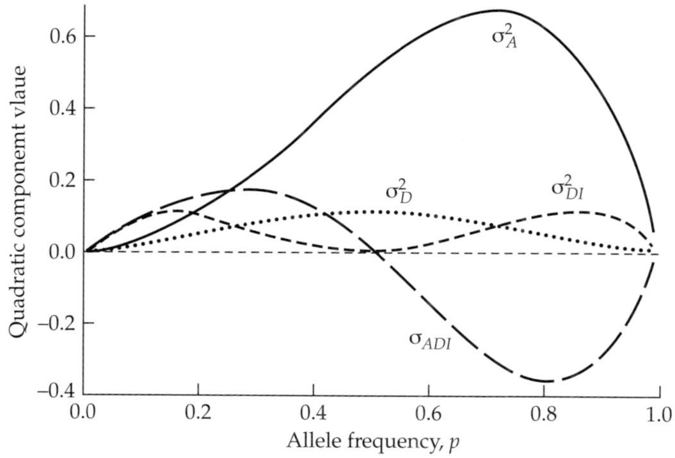

One- and Two-locus Identity Coefficients

The contributions of the factors in Table 11.1 to the traditional components of genetic variance (σ^2_A, σ^2_D, and σ^2_{AA}) in a finite population depend upon several one- and two-locus identity coefficients. As shown in Figure 11.1, these give the probabilities that randomly drawn combinations (from the population) of two, three, or four alleles at a given locus are identical by descent (both IBD and ibd are used in the literature; LW Chapter 7), with extensions to the two-locus case for randomly drawn combinations of two, three, or four two-locus gametes. Of the one-locus coefficients, f is the familiar inbreeding coefficient, i.e., the probability that two alleles are identical by descent at a particular locus (Chapter 2). The probabilities that the members of random groups of three and four alleles are all identical by descent are denoted by γ and δ (the latter is not to be confused with the dominance effects, which are subscripted in Table 11.1).

The coefficient Δ also involves four gametes (Figure 11.1), and is the probability of IBD within two pairs of gametes (including the possibility that all four genes are IBD). Under the random union of gametes, this corresponds to the probability that two (diploid) individuals have genotypes that are IBD. With a probability of δ, both individuals are inbred, and thus all four alleles are IBD, with the resulting genotypes being homozygotes. The more interesting scenario, which occurs with a probability of $\Delta - \delta$, is that where the two genotypes are IBD but the alleles within each individual are not (i.e., neither individual is inbred). Hence, the shared genotypes could be either heterozygotes or homozygotes. For example, suppose both of diploid genotypes are AA and that the maternal copy of A in both individuals is IBD, as is the paternal copy of A, but the maternal and paternal copies are not IBD. This generates two homozygotes that are not inbred (the two copies of A *within* each individual are not IBD), yet both genotypes are IBD to each other. Similarly, if both genotypes are Aa, neither individual is inbred, but if the A is IBD in both individuals, and so too are the a alleles, then the genotypes are IBD, but not inbred.

For randomly mating monoecious populations under the classical Wright-Fisher model, the transition equations for these coefficients are functions of N_e and t, with

$$f_t = 1 - \lambda_1^t \tag{11.3a}$$

$$\gamma_t = 1 - \frac{3\lambda_1^t}{2} + \frac{\lambda_2^t}{2} \tag{11.3b}$$

$$\Delta_t = 1 - \frac{24\lambda_1^t - 10\lambda_2^t + \lambda_3^t}{15} + \frac{\lambda_1^t - \lambda_3^t}{5(5N_e - 3)} \tag{11.3c}$$

$$\delta_t = 1 - \frac{9\lambda_1^t - 5\lambda_2^t + \lambda_3^t}{5} - \frac{3\lambda_1^t}{20(5N_e - 3)} + \frac{\lambda_2^t}{12(N_e - 1)} - \frac{(8N_e - 3)\lambda_3^t}{30(5N_e - 3)(N_e - 1)} \tag{11.3d}$$

where $\lambda_j = 1 - (j/2N_e)$ for $j = 1, 2, 3$ (Cockerham and Weir 1983).

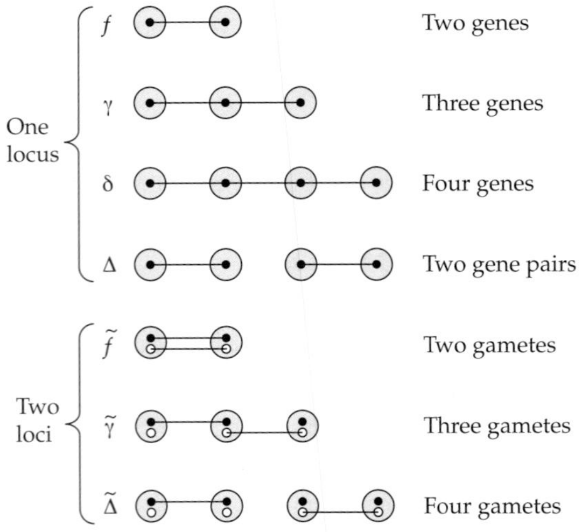

Figure 11.1 Measures of identity by descent (IBD) for single loci (f, γ, δ, Δ) and pairs of loci ($\widetilde{f}$, $\widetilde{\gamma}$, $\widetilde{\Delta}$). The large circles denote gametes (alleles when restricted to a single locus), and the open and closed dots within them represent alleles from one (top four relationships) or two (bottom three) loci. Identity by descent is indicated by a horizontal line. For example, δ is the probability that four randomly chosen alleles are all IBD, while Δ is the probability that, in any two pairs of gametes, each pair has an IBD allele (i.e., two diploid genotypes are IBD). As discussed in the text, Δ includes δ as a special case.

The three two-locus coefficients (denoted by tildes in Figure 11.1) refer to joint identities by descent at two loci. First, $\widetilde{f}$ refers to pairs of alleles on two gametes and is the joint probability of IBD at locus 1 and IBD at locus 2. Under neutrality, random mating, and linkage equilibrium, $\widetilde{f}$ cannot be less than the product of the separate identity probabilities for each locus, f^2. Second, $\widetilde{\gamma}$ refers to the situation in which each member of a pair of genes in one gamete is identical by descent with a gene in a separate gamete (IBD at locus 1 for gametes one and two, IBD at locus 2 for gametes two and three). Finally, $\widetilde{\Delta}$ is the joint identity by descent of genes (two at each locus) in two pairs of different gametes (IBD at locus 1 for gametes one and two, IBD at locus 2 for gametes three and four). The transition equations for these double identity-by-descent measures, which were derived by Weir and Cockerham (1969) for ideal monoecious populations, depend upon N_e, t, and the linkage parameter $\rho = 1 - 2c$ (where $c \leq 0.5$ is the recombination frequency between loci). Letting $\widetilde{f}_t = \widetilde{f}_t^* + 2f_t - 1$, $\widetilde{\gamma}_t = \widetilde{\gamma}_t^* + 2f_t - 1$, and $\widetilde{\Delta}_t = \widetilde{\Delta}_t^* + 2f_t - 1$, the coefficients are obtained by use of Equation 11.3a and the following matrix expression,

$$\begin{pmatrix} \widetilde{f}_{t+1}^* \\ \widetilde{\gamma}_{t+1}^* \\ \widetilde{\Delta}_{t+1}^* \end{pmatrix} = \mathbf{M} \begin{pmatrix} \widetilde{f}_t^* \\ \widetilde{\gamma}_t^* \\ \widetilde{\Delta}_t^* \end{pmatrix} \tag{11.4a}$$

starting with $\widetilde{f}_0^* = \widetilde{\gamma}_0^* = \widetilde{\Delta}_0^* = 1$, where

$$\mathbf{M} = \begin{pmatrix} \dfrac{(1+\rho)^2}{4} - \dfrac{\rho}{2N_e} & \dfrac{(N_e-1)(1-\rho^2)}{2N_e} & \dfrac{(N_e-1)(1-\rho)^2}{4N_e} \\ \dfrac{1+\rho}{4N_e} - \dfrac{\rho}{4N_e^2} & \dfrac{(N_e-1)[N_e+1+\rho(N_e-2)]}{2N_e^2} & \dfrac{(N_e-1)(2N_e-3)(1-\rho)}{4N_e^2} \\ \dfrac{2N_e-1}{4N_e^3} & \dfrac{(N_e-1)(2N_e-1)}{N_e^3} & \dfrac{(N_e-1)(2N_e-1)(2N_e-3)}{4N_e^3} \end{pmatrix} \tag{11.4b}$$

Table 11.3 Coefficients for the quadratic properties defined in Table 11.1 (necessary for the definition of the variance components noted in the Source column) for lines derived from a base population. For example, the total within-population genetic variance (first row) is equal to the weighted sum over all the quadratic components. The among-population variance follows similarly, with the total genetic variance being the sum of the within- and among-population components. Likewise, we can decompose the within-population variance into the contributions from the additive, dominance, and additive-by-additive components under inbreeding (the sum of these equal the within-population variance). The numerical values of the coefficients must be computed with Equations 11.3 and 11.4, and in practice, the two-locus identity coefficients need to be averaged over all pairs of loci, with each two-locus estimate depending on the recombination fraction.

Source	σ_A^2	σ_D^2	σ_{ADI}	σ_{DI}^2
Within	$1-f$	$1-f-2(\Delta-\delta)$	$2(f-\gamma)$	$f-\delta$
A	$1-f$	$2[f-\gamma-2(\Delta-\delta)]$	$2(f-\gamma)$	$2(\gamma-\delta)$
D	0	$1-3f+2(\Delta+\gamma-\delta)$	0	$f+\delta-2\gamma$
AA	0	0	0	0
Among	$2f$	$2(\Delta-\delta)$	2γ	δ
Total	$1+f$	$1-f$	$2f$	f

Source	ι^*	$\iota^2-\iota^*$	σ_{AA}^2
Within	$f-\Delta$	$\widetilde{f}-\widetilde{\Delta}$	$1+2f-2\widetilde{\gamma}-\widetilde{\Delta}$
A	$2(\gamma-\Delta)$	$2(\widetilde{\gamma}-\widetilde{\Delta})$	$4f-\widetilde{f}-2\widetilde{\gamma}-\widetilde{\Delta}$
D	$f+\Delta-2\gamma$	$\widetilde{f}-2\widetilde{\gamma}+\widetilde{\Delta}$	0
AA	0	0	$1-2f+\widetilde{f}$
Among	$\Delta-f^2$	$\widetilde{\Delta}-f^2$	$\widetilde{f}+2\widetilde{\gamma}+\widetilde{\Delta}$
Total	$f(1-f)$	$\widetilde{f}-f^2$	$1+2f+\widetilde{f}$

Impact of Drift Under Nonadditive Variance

With definitions in hand for the quadratic expressions in the base population (Table 11.1) and the temporal dynamics of the identity coefficients (Equations 11.3 and 11.4), we are now in a position to explore the impact of finite population size on the components of variance for a quantitative trait with a nonadditive genetic basis. The expected dynamics are determined by summing the products of the seven quadratic terms listed across the top of Table 11.3 with their associated tabulated identity coefficients in the table. For example, the expected within-population dominance variance is

$$[1-3f+2(\Delta+\gamma-\delta)]\sigma_D^2+(f+\delta-2\gamma)\sigma_{DI}^2+(f+\Delta-2\gamma)\iota^*+(\widetilde{f}-2\widetilde{\gamma}+\widetilde{\Delta})(\iota^2-\iota^*)$$

Here, σ_D^2, σ_{DI}^2, etc. are the base-population values of these components under random mating.

To gain a more intuitive feel for the source of the expressions in Table 11.3, we first consider the total genetic variance in a collection of lines, each inbred to level f, while ignoring epistasis until the following section. Subscripting loci with k and the two alleles at a locus as i and j, the genotypic value, G (expressed as a deviation from the mean), of an individual can be written as

$$G=\sum_{k=1}^{n}[(1-\phi_{kij})(\alpha_{ki}+\alpha_{kj}+\delta_{kij})+\phi_{kij}(2\alpha_{ki}+\delta_{kii})] \qquad (11.5)$$

where ϕ_{kij} is equal to one if the two alleles at a locus are identical by descent and equal to zero otherwise.

To compute the mean genotypic value, $\mu_G = E[G]$, note that the expected value of ϕ_{kij} is f and that we can write the expected value of G as two components: if not inbred ($\phi_{kij} = 0$), then $E[\alpha_{ki} + \alpha_{kj} + \delta_{kij}] = 0$, as these are deviations from the mean, leaving

$$E[G] = E\left[\sum_{k=1}^{n} \phi_{kij}(2\alpha_{ki} + \delta_{kii})\right] = f\left[\sum_{k=1}^{n}\sum_{i=1}^{n_k} p_{ki}(2\alpha_{ki} + \delta_{kii})\right]$$

If inbred, the two alleles at locus k are identical, giving the frequency of genotype $A_{ki}A_{ki}$ as $fp_{k,i}$, yielding the last equality. Because $\sum_i \alpha_{ki}p_{ki} = 0$ (by construction), we are left with

$$E[G] = \mu_G = f\sum_{k=1}^{n}\sum_{i=1}^{n_k} \delta_{kii}p_{ki} = f \cdot \iota$$

the last step following from the definition of ι (Table 11.1). Under complete inbreeding, $\mu_G = \iota$, the change in the mean from inbreeding depression (as $\mu_G = 0$ under random mating). In addition,

$$E[G^2] = \sum_{k=1}^{n}\left\{(1-f)E[(\alpha_{ki} + \alpha_{kj} + \delta_{kij})^2] + fE[(2\alpha_{ki} + \delta_{kii})^2]\right\} + \widetilde{f}(\iota^2 - \iota^*) \qquad (11.6a)$$

The final term in Equation 11.6a summarizes the consequences of joint inbreeding at pairs of loci, with $\widetilde{f}$ being the probability that two loci in the same individual are inbred (Figure 11.1) and $\iota^2 - \iota^* = 2\sum_{k<m}^{n} \iota_k\iota_m$ being the sum of cross-products of the locus-specific inbreeding depressions. In obtaining Equation 11.6a, note that all other products across loci have expectations equal to zero because $E[\alpha_{ki}]$ is always equal to zero and $E[\delta_{kij}] = 0$ at non-inbred loci (LW Chapter 4).

Recalling that the genetic variance is defined to be $\sigma^2_G = E[G^2] - \mu^2_G$, we obtain

$$\sigma^2_G = \sum_{k=1}^{n}\{(1-f)(2E[\alpha^2_{ki}] + E[\delta^2_{kij}]) + f(4E[\alpha^2_{ki}] + 4E[\alpha_{ki}\delta_{kii}] + E[\delta^2_{kii}])\}$$
$$+ \widetilde{f}(\iota^2 - \iota^*) - f^2\iota^2 \qquad (11.6b)$$

Further simplification is achieved by adding and subtracting $f(1-f)\iota^*$ on the right side of this expression, which, after using the expressions in Table 11.1, leads to

$$\sigma^2_G = (1+f)\sigma^2_A + (1-f)\sigma^2_D + 2f\sigma_{ADI} + f\sigma^2_{DI} + f(1-f)\iota^* + (\widetilde{f} - f^2)(\iota^2 - \iota^*) \qquad (11.6c)$$

in agreement with the final row in Table 11.3.

Although the preceding results apply to the genetic variance summed within and among a set of hypothetical isolated subpopulations, an expression for the average within-population variance can be obtained by removing the among-population component. The simplest route to this result is to recall the general rule that the variance among groups is equivalent to the covariance between individuals within groups (e.g., LW Chapter 18). Using this principle, the contribution of each quadratic component in Table 11.1 to the among-population variance can be obtained in the following way.

First, in the context of the entire collection of populations, f is equivalent to the probability that single alleles in the same population are identical by descent (the average coefficient of coancestry), so the additive genetic covariance for members of the same population is equal to $2f\sigma^2_A$ (LW Chapter 7). Second, two individuals within a population may also share both genes at a locus, in which case they will exhibit dominance genetic covariance, the magnitude of which will depend on whether the locus is inbred or not. From Figure 11.1, we see that the probability that both individuals are inbred and share the same genotype by descent is equal to the identity measure δ, so the genetic covariance by this route is $\delta\sigma^2_{DI}$.

The probability that the two individuals are not inbred but share identical genotypes by descent is $2(\Delta - \delta)$ (the 2 accounts for paternal-paternal and maternal-maternal versus cross paternal-maternal sources of identity by descent), so the covariance by this route becomes $2(\Delta - \delta)\sigma_D^2$. Third, the probability that three alleles in two members of a population are identical by descent is equal to γ, and there are two ways in which this can arise, so the covariance between additive and dominance effects is $2\gamma\sigma_{ADI}$. Finally, the covariance resulting from shared inbreeding depression is $(\Delta - f^2)\iota^*$ because Δ is the probability that two members of the same population are jointly inbred at the same locus, while the average fraction of individuals that are inbred over all populations is f per locus, and ι^* is the sum of squared per-locus inbreeding depressions. Similarly, the covariance due to joint inbreeding depression at different loci is equal to $(\widetilde{\Delta} - f^2)(\iota^2 - \iota^*)$, with the latter term being the sum of cross-products of per-locus inbreeding depressions.

Upon summing all six of these contributions, we obtain the genetic variance among populations given in the second row from the bottom of Table 11.3. The within-population genetic variance is then obtained by subtracting the among-population component from the total genetic variance. Results such as these provide a mechanistic explanation for the changes in components of genetic variance that can be induced by small population size. For example, it can be seen from the second line of Table 11.3 that inbreeding always converts some initial dominance genetic variance into additive genetic variance. This does not necessarily imply that there will be a net increase in additive genetic variance in a bottlenecked population, as the total dynamics depend critically on the relative magnitudes and temporal dynamics of all five of the quadratic components involving dominance in the base population (Figure 11.2). However, it does imply that Equation 11.2 cannot be strictly correct in the presence of dominance. While the contribution to the additive genetic variance from the base population, σ_A^2, declines in each generation, Figure 11.2 shows that all other contributions first increase before eventually decreasing to zero. Thus, whether a population bottleneck will induce an increase in additive genetic variance depends critically on the magnitude of σ_A^2 relative to the other quadratic components in the base population.

Note that the two-locus identity coefficients ($\widetilde{f}, \widetilde{\gamma}$, and $\widetilde{\Delta}$) appear only in association with quadratic terms involving pairs of loci, in this case $(\iota^2 - \iota^*)$. Two-locus identity by descent is of relevance in finite populations because the gametic-phase disequilibrium that inevitably develops by chance causes **identity disequilibrium** between loci (Weir and Cockerham 1968)—individuals that are inbred at one locus are likely to be so at other loci, causing a transient inflation of the genetic variance through the production of extreme phenotypes.

Although it may not be immediately apparent, the coefficients in the final two (bottom) columns in Table 11.3 are equivalent to measures of identity disequilibrium (Cockerham 1984a). For example, $\widetilde{f} - f^2$ is the deviation of the double identity-by-descent within gametes in the same population from that based on the assumption of independence between loci. This difference is simply the logical extension of the notion of disequilibrium, where we defined $D_{AB} = p_{AB} - p_A p_B$, with $\widetilde{f}$ being the analog of p_{AB} (the frequency of the AB gamete) and f being the analog of p_A and p_B. Although $\widetilde{f}$ depends upon the average linkage relationships between all relevant pairs of loci (Weir and Cockerham 1968, 1969), in most cases, if most pairs of loci are on different chromosomes, or if the population is randomly mated and expanded following the bottleneck, $\widetilde{f}$ will be approximately equal to f^2. Under such conditions, $(\widetilde{f} - f^2)$, as well as the other coefficients of $(\iota^2 - \iota^*)$ in Table 11.3, will be very close to zero, removing at least this one term from the variance expressions (Figure 11.2).

The Effects of Epistasis

The fundamental point in the preceding section is that because dominance is a function of a two-allele interaction, the variance in dominance effects can be altered in unexpected ways when inbreeding alters the average background on which an allele appears. This same issue applies to epistatic effects, although on a potentially larger scale because the

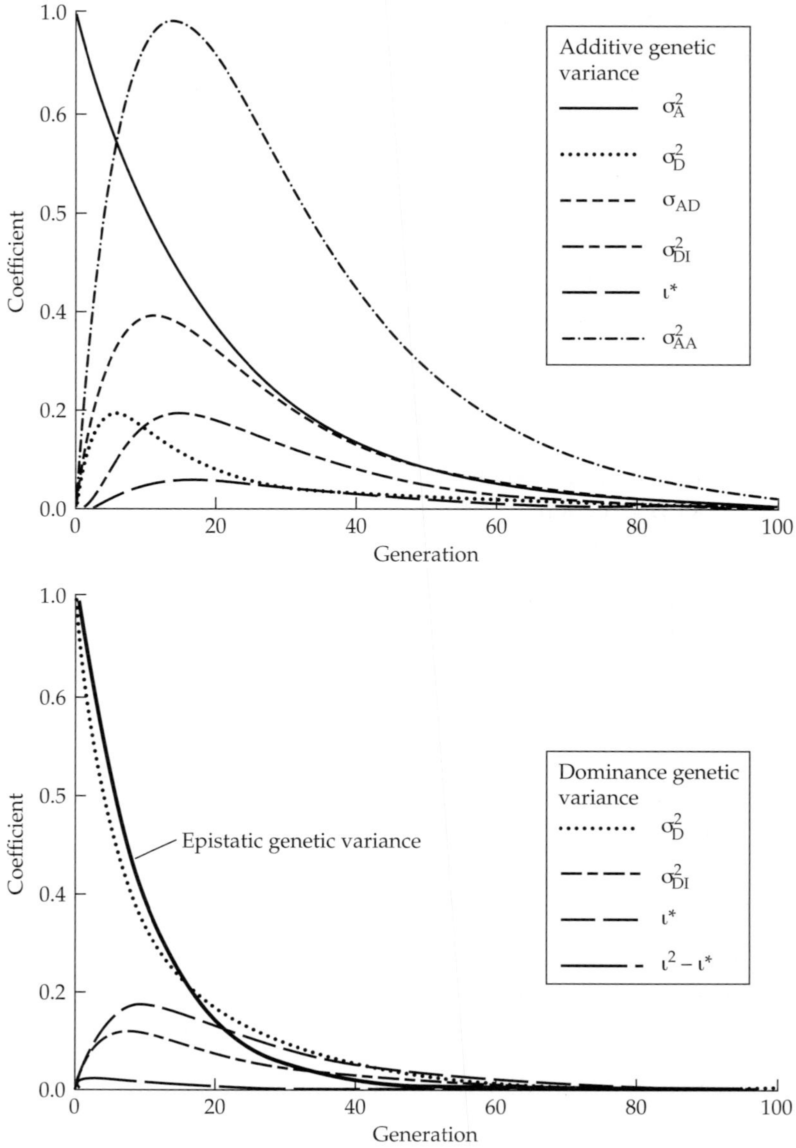

Figure 11.2 Dynamics of the coefficients for the terms contributing to the additive, dominance, and additive × additive genetic variance within populations for an effective population size of 10 and freely recombining loci ($c = 0.5$), obtained by use of Equations 11.3 and 11.4, along with Table 11.1. The top panel gives the contributions of each term to the within-population additive variance, while the bottom panel gives the same for the within-population dominance (and $A \times A$) variance. The coefficient for the contribution of $(\iota^2 - \iota^*)$ to the additive genetic variance is barely visible on the scale in the bottom graph. These results apply approximately to any other population size, N_e, if the time scale is transformed by multiplying by $N_e/10$. To obtain the actual dynamics of the variance components, the coefficients need to be multiplied by the base-population properties. For example, the additive genetic variance in generation 50 is approximately $0.08(\sigma_A^2 + \sigma_{ADI}) + 0.04\sigma_{DI}^2 + 0.01(\sigma_D^2 + \iota^*) + 0.28\sigma_{AA}^2$, while the additive × additive genetic variance is $\simeq 0$, and the dominance genetic variance is $\simeq 0.04(\iota^* + \sigma_D^2)$.

additive × additive epistatic variance (σ_{AA}^2) is a function of n^2 terms, while all of the other quadratic components in Table 11.1 (except $\iota^2 - \iota^*$, which seems to be of little significance) are functions of only n terms. If we assume unlinked loci, the coefficient of the σ_{AA}^2

Table 11.4 A simple two-locus system with epistasis. Elements in the table are the expected genotypic values for the two-locus genotypes.

	A_1A_1	A_1A_2	A_2A_2
B_1B_1	$4a+i$	$3a$	$2a-i$
B_1B_2	$3a$	$2a$	a
B_2B_2	$2a-i$	a	i

contribution to the additive genetic variance rises to nearly 1.0 in a little over N_e generations, i.e., the equivalent of all of the base-population σ^2_{AA} is added to the (otherwise declining) additive genetic variance at this point (Figure 11.2).

Thus, the potential exists for substantial additive $\times$ additive epistatic variance in the base population to spawn a prolonged increase in the additive genetic variance following a reduction in population size, or at least to slow the erosion relative to the expectation given by Equation 11.2. The matter is of considerable interest because whereas inflations in the additive genetic variance induced by dominance effects are accompanied by the maladaptive effects of inbreeding depression (i.e., a deleterious change in the mean phenotype), those caused by a conversion of epistatic additive effects have no side effects on the mean (unless there is linkage disequilibrium, in which case the Griffing effect, discussed in Chapter 15, can be important), and simply increase the range of variation upon which natural selection can act.

To see how this might happen, consider the following. From the standpoint of any locus, variation in epistatic interactions with genes at other loci amounts to a reduction in the efficiency with which allelic effects are transmitted from generation to generation—segregation and recombination ensure that interlocus interactions in parents are not transmitted faithfully through gametes. However, as genetic drift moves alleles toward fixation at one or both loci or as identity disequilibria increase, this variation in the genetic environment is reduced. In Table 11.4, for example, the A_1 allele is present in genetic backgrounds that lead to five distinct genotypic values in a randomly mating population. If, however, the B_2 allele becomes fixed, then an A_1 allele can only be in two backgrounds ($A_1A_1B_2B_2$ and $A_1A_2B_2B_2$). In this case, the epistatic interactions are still present, but they are transmitted reliably as additive effects (the difference between adjacent pairs of A-locus genotypes being a constant, $a-i$).

Some simple insight into the role of additive $\times$ additive epistatic variance in the dynamics of genetic variance of finite populations can be achieved if one is willing to assume that the loci involved are unlinked ($c = 0.5$), meaning that identity disequilibria are of negligible significance. Returning to Table 11.3, it can be seen that the coefficient for the contribution of base-population additive $\times$ additive variance to future additive genetic variance is $(4f - \widetilde{f} - 2\widetilde{\gamma} - \widetilde{\Delta})$, which reduces to $4f(1-f)$ under the assumption that all of the two-locus identities $\simeq f^2$. If we ignore any contributions from dominance, the expression for the dynamics of the additive genetic variance then simplifies to

$$\sigma^2_A(t) \simeq (1-f_t)\sigma^2_A(0) + 4f_t(1-f_t)\sigma^2_{AA}(0) \tag{11.7a}$$

and the expression for the additive $\times$ additive variance simplifies to

$$\sigma^2_{AA}(t) \simeq (1-f_t)^2\sigma^2_{AA}(0) \tag{11.7b}$$

assuming the absence of any higher-order epistatic variance (Cockerham and Tachida 1988; Goodnight 1988; López-Fanjul et al. 1999). Equation 11.7a shows that the conversion of additive $\times$ additive to additive genetic variance, which scales as $(1-f_t)\cdot f_t$, is maximized at $f_t = 0.5$. Because $f_t \simeq 1 - \exp(-t/2N_e)$, this translates to $t \simeq 1.4N_e$ generations in accordance with Figure 11.2.

Based on results from Barton and Turelli (2004), van Buskirk and Willi (2006) proposed to approximate the additive variance (with epistasis and dominance) under inbreeding by

$$\sigma_A^2(t) \simeq (1 - f_t)\left[\sigma_A^2(0) + 2f_t\sigma_D^2 + 4f_t\sigma_{AA}^2(0)\right] \tag{11.7c}$$

This is simply Equation 11.7a with a dominance term (σ_D^2) added. While this expression appears in the literature (e.g., Taft and Roff 2012), comparison with the exact result (given from the A row in Table 11.3) shows that Equation 11.7c ignores other quadratic components involving dominance (σ_{ADI} and σ_{DI}^2) and the contributions from other identity coefficients besides f, and thus may be a poor approximation in some settings.

Limited attention has been given to the role of two-locus epistasis involving dominance effects in finite populations (Weir and Cockerham 1977; Cheverud and Routman 1996; López-Fanjul et al. 1999; Barton and Turelli 2004), and no general formulation exists for the dynamics of genetic variance resulting from higher-order epistatic interactions. We can anticipate that the necessary algebra for such a solution would be extremely tedious, as it would involve descent measures involving three and more loci, and as will be discussed below, the existing data do not support the need for such a theory. For heuristic purposes, however, we will consider the approximate case for higher-order epistasis involving only additive effects, again assuming freely recombining loci and ignoring identity disequilibrium.

As a simple entrée into this matter, recall that in the absence of dominance, the expected covariance between the relatives x and y is $\sigma_G(x,y) = 2\theta_{xy}\sigma_A^2 + (2\theta_{xy})^2\sigma_{AA}^2 + \cdots + (2\theta_{xy})^n\sigma_{A^n}^2$, where θ_{xy} is the coefficient of coancestry (LW Chapter 7), and $\sigma_{A^n}^2$ refers to epistatic variance involving the additive effects of n loci. The total genetic variance (summed over the within- and among-population components) is equivalent to the covariance of individuals with themselves, which is obtained by letting $\theta_{xy} = (1+f)/2$ (LW Chapter 7), whereas the variance among isolated subpopulations is equivalent to the covariance of random members from the same subpopulation, which is obtained by letting $\theta_{xy} = f$. Thus, for any n-locus epistatic interaction, the contribution to the total genetic variance is $(1+f)^n\sigma_{A^n}^2$, to the among-population component of variance is $(2f)^n\sigma_{A^n}^2$, and to the within-population component is the difference, $[(1+f)^n - (2f)^n]\sigma_{A^n}^2$. This implies that the base-population additive and additive $\times$ additive genetic variances contribute $(1-f)\sigma_A^2$ and $(1+2f-3f^2)\sigma_{AA}^2$, respectively, to the within-population genetic variance, a result that can also be obtained directly from Equations 11.7a and 11.7b. The contribution from additive $\times$ additive $\times$ additive epistatic variance is $(1+3f+3f^2-7f^3)\sigma_{AAA}^2$, etc.

Each of these terms (except those involving σ_A^2) reaches a maximum at an intermediate level of inbreeding and then declines to zero as $f \to 1$. For additive epistatic effects involving $n = 2$, 3, and 4 loci, the peak contributions to the within-population genetic variance occur when f is approximately 0.33, 0.55, and 0.66, respectively. For randomly mating populations, these maxima occur at $0.8N_e$, $1.6N_e$, and $2.2N_e$ generations, with the peak contributions to the total within-population genetic variance being equal to $1.33\sigma_A^2$, $2.39\sigma_{AA}^2$, and $4.56\sigma_{AAA}^2$, respectively. Thus, even if levels of higher-order epistatic genetic variance are relatively low in a base population, they may have a significant influence on the within-population variance under inbreeding, with the full impact not being revealed for many generations. One potential consequence of this, as noted by Naciri-Graven and Goudet (2003), is that as the number of loci increases, epistasis becomes more important than dominance. However, López-Fanjul et al. (2002) reached the opposite conclusion for two loci, namely, that dominance is more important than epistasis, thus showing that intricacies of the genetic architecture impact our predictions.

Under the model of only additive epistasis, the components of within-population genetic variance are described by the following general expression,

$$\sigma_{A^n}^2 = (1-f)^n \sum_{i=0}^{x-n} \binom{n+i}{n} (2f)^i \sigma_{A^{n+i}}^2 \tag{11.8}$$

where x denotes the highest level of epistasis involving additive effects influencing the trait (Barton and Turelli 2004; Hill et al. 2006). When $x = 2$, this expression recovers Equations 11.7a and 11.7b, and with x as high as 3, we obtain

$$\sigma_A^2(t) = (1 - f_t)[\,\sigma_A^2(0) + 2(2f_t)\sigma_{AA}^2(0) + 3(2f_t)^2\sigma_{AAA}^2(0) + \cdots\,] \tag{11.9a}$$

$$\sigma_{AA}^2(t) = (1 - f_t)^2[\,\sigma_{AA}^2(0) + 3(2f_t)\sigma_{AAA}^2(0) + \cdots\,] \tag{11.9b}$$

$$\sigma_{AAA}^2(t) = (1 - f_t)^3[\,\sigma_{AAA}^2(0) + \cdots\,] \tag{11.9c}$$

with the dots denoting potential contributions from higher-order effects.

These expressions show that, under progressive inbreeding, the expected values for each variance component depend on all higher-order epistatic variances, and that the erosion of the higher-order components proceeds most rapidly. Most notably, Equation 11.9a shows that the presence of any epistatic variance will inflate the additive genetic variance above the simple expectation $(1 - f)\sigma_A^2(0)$, but whether $\sigma_A^2(t)$ rises beyond the base-population level, $\sigma_A^2(0)$, depends on the magnitude of the base-population epistatic variance components. From Equation 11.8, it can be seen that for any $n > 1$, the peak contribution of $\sim(2^{n-1}/2.72)\sigma_{A^n}^2$ to the additive genetic variance occurs at $f = 1 - (1/n)$ (Turelli and Barton 2006).

A practical way of evaluating the conditions necessary for a net increase in the additive genetic variance is to consider the nature of empirical estimates of additive genetic variance. As noted in Lynch and Walsh (1998), clean estimates of the causal components of genetic variance are generally unachievable. For example, although twice the parent-offspring covariance is often used as an estimate of the additive genetic variance, the true expectation is actually $\sigma_A^2 + (\sigma_{AA}^2/2) + (\sigma_{AAA}^2/4) + \cdots$. Ignoring all but the additive $\times$ additive genetic variance, Equations 11.7a and 11.7b can be used to show that the parent-offspring covariance after inbreeding to level f will exceed that in the base population if $\sigma_{AA}^2 > 2\sigma_A^2/(6 - 7f)$, which reduces to $\sigma_{AA}^2 > \sigma_A^2/3$ as $f \to 0$.

Although it is exceedingly difficult to obtain perfectly isolated estimates of σ_A^2 and σ_{AA}^2, a survey of the existing data combined with a number of indirect arguments suggests that the condition of $\sigma_{AA}^2 > \sigma_A^2/3$ is hardly ever met in natural populations (Hill et al. 2008; Mäki-Tanila and Hill 2014). As reviewed in Lynch and Walsh (1998) and reemphasized by Hill et al. (2008), this situation is not likely to be a consequence of limited epistatic interactions among genes. Rather the very nature of variance-component partitioning, with higher-order effects being defined as residual deviations from expectations based on lower-order effects, largely ensures that epistatic components of variance will be small relative to σ_A^2, especially when most alleles have frequencies far from 0.5.

Finally, we emphasize that although all of the previous results strictly apply to ideal monoecious populations that become inbred via random genetic drift, the general approach applies to any mating system, provided appropriate modifications are made to the recursion formulae for the identity coefficients. For monoecy with the avoidance of selfing and for separate sexes, the appropriate expressions were given by Weir et al. (1980) and Weir and Hill (1980), and explicit formulae for obligate self-fertilization, full-sib mating, and other special systems of mating are developed in Cockerham and Weir (1968, 1973) and Weir and Cockerham (1968), with a useful review provided in Cockerham and Weir (1977).

Sampling Error

It cannot be emphasized too strongly that the preceding expressions give only the *expected* change of the within-population variance for a neutral quantitative character. Due to the stochastic nature of random genetic drift, departures from this expectation will arise in any individual population, so a central concern is the degree to which the average behavior of a small number of populations (e.g., a typical replicated experiment) will represent the expected pattern.

In the following discourse, we denote the realized additive genetic variance for any particular population by $\widehat{\sigma}_A^2(t)$. Estimation error on the part of the investigator aside, three sources of error contribute to the variation in $\widehat{\sigma}^2_A(t)$ among replicate populations: (1) variation in the genetic variance among founder populations caused by sampling; (2) subsequent departures of the within-population heterozygosity from its expectation caused by drift; and (3) deviations from Hardy-Weinberg and gametic-phase equilibrium.

Quantification of these sources of variation is difficult, but some general results have been obtained for characters with a purely additive genetic basis. The additive genetic variance within a particular population can be written as

$$\widehat{\sigma}^2_A(t) = \sigma^2_a(t) + \widehat{\sigma}_{HW}(t) + \widehat{\sigma}_L(t) \tag{11.10}$$

where $\sigma^2_a(t)$ is the variance due to the true gene effects expected if the line were expanded into an infinitely large, randomly mating population with global Hardy-Weinberg and gametic-phase disequilibrium (the **genic variance**; Chapters 16 and 24), while $\widehat{\sigma}_{HW}(t)$ and $\widehat{\sigma}_L(t)$ are transient covariances of genic effects within and among loci caused by disequilibria within and between loci. The expected value of $\sigma^2_a(t)$, given by Equation 11.2, is $\sigma^2_A(t)$, as the disequilibria are equally likely to occur in positive and negative directions in the absence of selection. Thus, the expected value of $\widehat{\sigma}^2_A(t)$ is also equal to $\sigma^2_A(t)$.

Each of the terms on the right side of Equation 11.10 has a variance associated with it, meaning that the expected variance of the within-population additive genetic variance among hypothetical replicate populations can be expressed as

$$\sigma^2\left[\widehat{\sigma}^2_A(t)\right] = \sigma^2\left[\sigma^2_a(t)\right] + \sigma^2\left[\widehat{\sigma}_{HW}(t)\right] + \sigma^2\left[\widehat{\sigma}_L(t)\right] \tag{11.11}$$

The variance of the "true" additive genetic variance is

$$\sigma^2\left[\sigma^2_a(t)\right] = \sum_{i=1}^{n} a_i^4 \sigma^2_{H_i}(t) \tag{11.12}$$

where $\sigma^2_{H_i}(t)$ is the expected variance of heterozygosity, $H_i(t) = 2p_i(t)[1 - p_i(t)]$, at a locus i among replicate populations t generations after divergence. Bulmer (1980) obtained an expression for $\sigma^2_{H_i}(t)$ for a locus with two alleles, and a very close approximation to this is given in Example 2.5. While the exact dynamics of $\sigma^2[\sigma^2_a(t)]$ will depend on the initial allele frequencies at all loci, which are generally unknown, a useful qualitative statement can be made. For fixed initial genetic variance in the base population, the average value of a_i^2 must scale inversely with the number of loci. Thus, because $\sigma^2[\sigma^2_a(t)]$ is the sum of n terms, each of which is a function of $a_i^4 \propto n^{-2}$, then $\sigma^2[\sigma^2_a(t)]$ must be inversely proportional to n. Therefore, for characters with large effective numbers of loci, deviations from the true additive genetic variance caused by variance in heterozygosity are likely to be of negligible importance.

The expected variance of the within-population variance from Hardy-Weinberg deviations is $\sigma^2[\widehat{\sigma}_{HW}(t)] \simeq \sigma^4_A(t)/N_e$ (Bulmer 1976, 1980), but the variation due to gametic-phase disequilibrium is more substantial, and the details of the rather tedious derivations appear in Avery and Hill (1977) and Bulmer (1980). Regardless of the degree of linkage, $\sigma^2[\widehat{\sigma}_L(1)] \simeq \sigma^4_A(0)/N_e$ in the first generation of inbreeding, and thereafter for the special case of unlinked loci, $\sigma^2[\widehat{\sigma}_L(t)] \simeq 5\sigma^4_A(t)/(3N_e)$. With linkage $\sigma^2[\widehat{\sigma}_L(t)]$ is necessarily larger, but for most cases it will not be substantially so (Avery and Hill 1977), and regardless of the state of disequilibrium in the base population, the expected value of $\sigma^2[\widehat{\sigma}_L(t)]$ is almost always attained within five generations.

An advantage of the preceding expressions for the variance of the components of the within-population genetic variance is that they are defined in terms of measurable quantities. However, to achieve this useful property, several assumptions (ideal population structure, no association between map distances and effects of genes, additivity of gene effects) had to be made, violations of which will tend to inflate the variance of $\widehat{\sigma}^2_A(t)$. Thus, summing over the two disequilibrium sources, we find that $\sigma^2[\widehat{\sigma}^2_A(t)]$ must be at least $8\sigma_A^4(t)/3N_e$. A similar conclusion was reached by Zeng and Cockerham (1991), who presented a more thorough and highly technical analysis.

These theoretical results have significant implications for the interpretation of observed changes of genetic variance in small populations, in particular in the use of such observations to infer any significant conversion of nonadditive to additive genetic variance. Clearly, estimates of $\widehat{\sigma}_A^2(t)$ derived from a small number of replicate populations, even over several generations, provide unreliable assessments of the expected dynamics of $\sigma_A^2(t)$. If we average over L independent lines, the sampling variance of the average of the within-line additive variances is at least $8\sigma_A^4(t)/(3LN_e)$. Therefore, if it is desirable to keep the standard error of an estimate of the additive genetic variance at a level of 10% of the expectation, $\sigma_A^2(t)$, the design must be such that $N_eL \simeq 270$, i.e., approximately 70 lines of $N_e = 4$, or 17 of $N_e = 16$. For self-fertilizing lines, the sampling variance is closer to $7\sigma_A^4(t)/L$ over the first five generations of inbreeding (Lynch 1988a), so on the order of 700 lines would have to be monitored to achieve a similar level of precision. In practice, one would need to set the target number of lines even higher than these estimates, because the additional variation due to parameter estimation, i.e., the deviation of the observation Var(A,t) from the realized parameter $\widehat{\sigma}_A^2(t)$, which may be considerable, has been ignored in the preceding arguments.

One final problem that bears mentioning is that the values of $\widehat{\sigma}_A^2(t)$ observed in successive generations are not independent, as the minimum correlation between adjacent generations equals one-half for unlinked loci. Thus, if the genetic variation within a particular population exceeds the expectation due to chance in one generation, it is likely to remain in excess for several consecutive generations. When this problem is confounded with the sampling variance described above, there is a substantial possibility that $\widehat{\sigma}_A^2(t)$ for a particular replicate population may on occasion increase for several generations, contrary to the expected trend, and even for characters with a purely additive genetic basis (Avery and Hill 1977; Bulmer 1980).

In summary, even in the case of purely additive gene action, obtaining a reliable empirical view of the expected dynamics of the additive genetic variance requires a very large number of replicate populations. There are three levels at which sampling error plays a role. First, in each replicate, the variance observed by the investigator, Var(A), is likely to deviate substantially from the parametric value $\widehat{\sigma}_A^2$ for the replicate, due simply to the finite number of individuals monitored. Second, the true realized variance, $\widehat{\sigma}_A^2$, in each line may deviate considerably from the actual equilibrium value, σ_a^2, expected in the absence of Hardy-Weinberg and gametic-phase disequilibria. Finally, random genetic drift will cause σ_a^2 to deviate from the global expectation, σ_A^2. One can expect the situation to get even more complex in the presence of nonadditive gene action, but mastering the details of the sampling theory remains a formidable challenge.

Empirical Data

The influence of small population size on components of genetic variance is of substantial relevance to several areas of inquiry. For example, an underlying assumption of much of conservation genetics is that the loss of heterozygosity from small populations translates immediately into a loss of variation for adaptive traits. As noted above, however, this need not be the case in the presence of nonadditive gene action. A key additional question is whether increases in the additive genetic variance following a population bottleneck, if they do indeed occur, are accompanied by changes in the mean phenotype that are contrary to the maintenance of high fitness. Nothing is gained from a population bottleneck if the extreme phenotypes that are produced are simply low-fitness individuals resulting from inbreeding depression.

The preceding theory is also of potential relevance to the field of speciation. Substantial uncertainty exists over the importance of population bottlenecks for the speciation process (Mayr 1954; Templeton 1980; Barton and Charlesworth 1984; Carson and Templeton 1984), and much of the debate revolves around verbal arguments regarding additive and epistatic

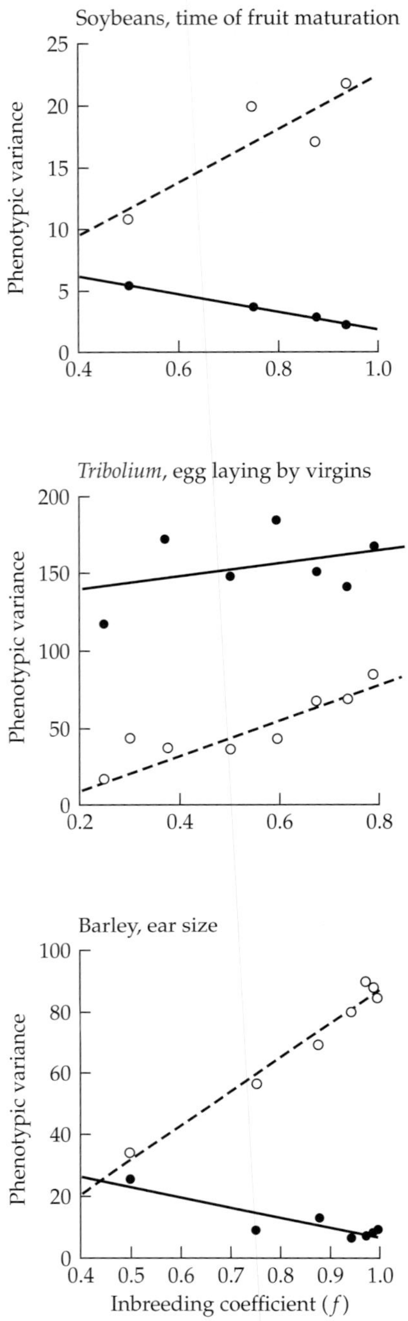

Figure 11.3 Response of the average within-line and among-line phenotypic variance to inbreeding in experimental lines. References and system of mating: **top:** Horner and Weber (1956), selfing; **middle:** López-Fanjul and Jódar (1977), full-sib mating, control-corrected; **bottom:** Bateman and Mather (1951), selfing. Solid and open points denote the within- and among-population components of phenotypic variance.

gene action. In Carson's (1968, 1975) **founder-flush** theory, for example, it is assumed that a period of population expansion following a bottleneck will often result in a conversion of various types of epistatic interactions into additive genetic variance. Similar issues were raised by Templeton (1980) in his hypothesis of speciation via **genetic transilience**. Although such arguments sometimes appear intuitive on the surface, the preceding theoretical examples amply illustrate that intuition can be quite misleading with respect to the dynamics of genetic variance in small populations. The consequences of a population bottleneck are highly sensitive to the nature of gene action and the frequency distribution of alleles, and establishing whether increases in bottleneck-induced variance are common is ultimately an empirical question.

Unfortunately, only a few well-designed empirical studies have addressed the influence of inbreeding on the genetic variance within populations. Studies that strictly focus on *phenotypic* variance often reveal essentially linear declines in the phenotypic variance with f, as expected for a character with a purely additive genetic basis, but in other cases the response has been so noisy that no general conclusion could be drawn, and sometimes the within-population variance steadily increases over time (Figure 11.3). A substantial limitation of studies of this sort is that the environmental component of variance often increases with inbreeding as a consequence of reduced developmental stability (Chapter 17; LW Chapter 6; Whitlock and Fowler 1999; Kelly and Arathi 2003), thereby obscuring the relationship between phenotypic and genetic variance.

A study by Cheverud et al. (1999) provided a clear example of the creation of additive genetic variance by a population bottleneck. By crossing two long-established mouse lines, one selected for large body size and the other for small body size, an F_2 base population with high genetic variance for adult weight was constructed. Thirty-nine replicate inbred lines were then initiated from the F_3 generation, each maintained as two pairs of males and females through four generations of inbreeding to yield an average $f = 0.39$. Two contemporary control strains were maintained by randomly mating 60 pairs of individuals derived from the base (hybrid) population. Using a full-sib analysis, the authors found that the average additive genetic variance for adult weight after inbreeding was about 1.75-fold greater (and significantly so) than expected under the additive model (a fraction, $1 - f = 0.61$, of the additive variation in the base population) and slightly greater than that in the controls. Two lines of evidence suggest that this inflation in σ_A^2 was largely, if not entirely, due to the conversion of additive $\times$ additive epistatic variance. First, the absence of any significant change in mean adult weight throughout the period of inbreeding implies that directional dominance is negligible for this trait. Second, previous QTL analysis of this experimental population had revealed pervasive epistatic interactions between loci influencing body size (Routman and Cheverud 1997; Kramer et al. 1998). As a caveat, however, it must be emphasized that this study is quite artificial, in that by constructing a base population with intermediate gene frequencies, the epistatic genetic variance was maximized at the outset. We now consider the few results that have emerged for more naturally derived populations.

Bryant et al. (1986b) put populations of houseflies (*Musca domestica*) through single-generation bottlenecks of 1, 4, and 16 pairs, and then rapidly expanded them for several generations prior to the measurement of the additive genetic variance (to reduce the variation in the within-line variance caused by gametic-phase disequilibrium). Analyses of several morphological characters suggested an increase in σ_A^2 in the bottlenecked lines relative to a control (Figure 11.4), which the authors surmised to be a consequence of the conversion of epistatic to additive genetic variance. Although this study has become something of a flagship example of bottleneck-induced increases in genetic variance, it also serves to highlight the extreme difficulties that exist in interpreting the dynamics of genetic variance in inbred populations.

First, only four replicate populations were maintained at each population density in this study, so there is a substantial chance that the average within-line variance may have increased entirely by chance, even in the absence of nonadditive genetic variance. Second, some characters exhibited a five-fold inflation in the additive genetic variance over the control, and based on the considerations outlined above, this is hard to accept as a real consequence of inbreeding in the essentially non-inbred ($f \simeq 0.03$) 16-pair lines (traits IE and SW in Figure 11.4). In contrast, although the evidence that inbreeding created a real increase in additive genetic variance in these lines is not very compelling, it is equally true that there

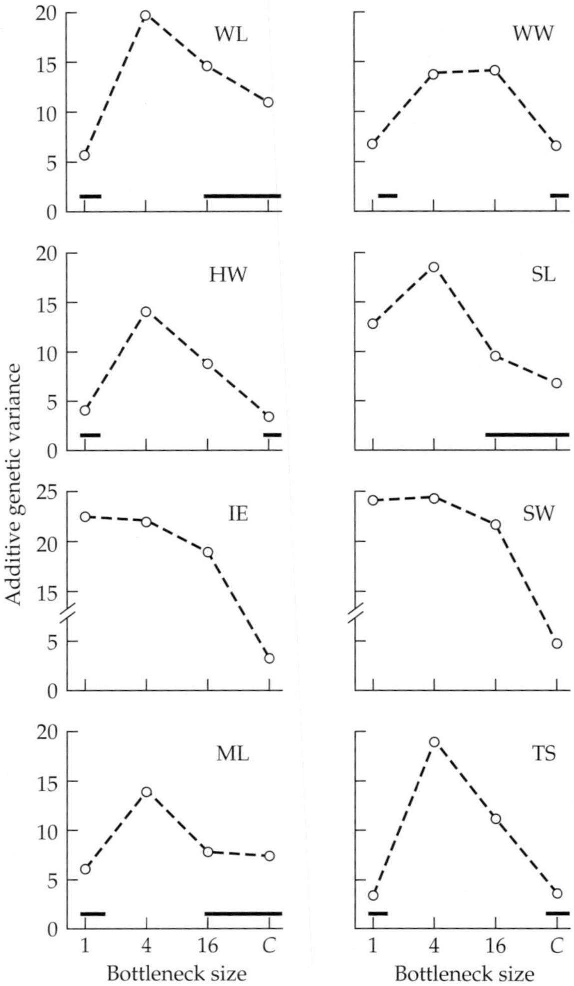

Figure 11.4 Additive genetic variances for eight morphometric traits averaged over four replicate lines of bottlenecked housefly (*Musca domestica*) populations. Horizontal lines (along the bottom axes) connect variances that were not significantly different at the 0.05 level. C denotes a large randomly mating control population, whereas the remaining populations were propagated through single-generation bottlenecks of 1, 4, and 16 pairs. WL denotes wing length; WW, wing width; HW, head width; SL, scutellum length; IE, inner-eye separation; SW, scutellum width; ML, metafemur length; and TS, thoracic-suture length. (From Bryant et al. 1986b.)

is no evidence of a substantial erosion in additive genetic variance following inbreeding. In subsequent studies involving single-generation bottlenecks of four individuals, Meffert (1995) did not detect any overall change in the additive genetic variance for various aspects of courtship behavior (six replicate populations), and Bryant and Meffert (1996) observed increases in σ_A^2 for two morphological characters but decreases for two others (two replicate populations) thought to have had relatively high levels of additive × additive epistatic variance in the base population.

Replication is less of a problem in a few other recent studies. For example, starting from a large base population of *D. melanogaster,* Whitlock and Fowler (1999) subjected 52 lines to a single generation of full-sib mating ($f = 0.25$) and then expanded them to a large size. Within each line, the additive genetic variance for various aspects of wing structure was estimated by parent-offspring regressions involving 90 families, and a similar procedure was applied to a large control population; see also Example 11.2. No evidence for an increase in the additive genetic variance emerged from this well-designed study, and a similar conclusion was reached in a study on sternopleural bristles in bottlenecked populations of *D. bunnanda* (van Heerwaarden et al. 2008). Although a few individual lines exhibited moderate increases in σ_A^2, these increases were always compatible with the expectations of additive genetic theory (i.e., consistent with the predicted sampling variance), as was the average reduction in σ_A^2 across all lines.

In a smaller study with the flour beetle *(Tribolium castaneum)*, involving three inbreeding levels ($f = 0, 0.375$, and 0.672, with five replicates each), Wade et al. (1996) also observed average changes in the additive genetic variance for pupal weight that were entirely compatible with expectations of the additive model. Likewise, changes in the additive genetic variance of wing pigmentation patterns in bottlenecked populations of the butterfly *Bicyclus anynana* (Saccheri et al. 2001), sternopleural bristle number in *Drosophila* (Kristensen et al. 2005), and flower size in *Nigella degenii* (Andersson et al. 2010) were all consistent with expectations of the additive model. However, a meta-analysis dominated by morphological traits led to a slightly different conclusion (Taft and Roff 2012). In order to compare the results of a number of different studies, Taft and Roff considered the log of the ratio of the estimated additive variance in a bottlenecked population to the estimated variance in the control. Under the additive model (Equation 11.7a), one expects that

$$R = \ln\left(\frac{V_A(bott)}{V_A(cont)}\right) = \ln(1-f)$$

in which case the regression of R on $\ln(1-f)$ would have a slope of one and an intercept of zero. More generally, one could fit $R = a + b\ln(1-f)$ to look for departures from the predictions of the additive model. While the slope of this regression was not significantly different from its expected value ($b = 1$), Taft and Roff observed a significant nonzero intercept. They interpreted this as arising from trying to force a linear relationship onto a collection of response curves, some of which were nonlinear (i.e., with the additive variance increasing over some initial range in f). A simple alternative explanation is that estimation error could generate a nonzero intercept (especially considering that variance estimators have highly asymmetric confidence intervals; see Figure 12.1). Taken together, these diverse studies provide little justification for the view that the *expected* additive genetic variance for *morphological* traits commonly increases during early phases of inbreeding, although transient increases associated with sampling are certainly expected in some replicates.

These results are in striking contrast to those from studies on fitness-related traits. For example, in a parallel study of offspring production in *Tribolium*, Wade et al. (1996) observed no significant decline in additive genetic variance at inbreeding levels up to $f = 0.672$. Likewise, López-Fanjul and Villaverde (1989) took 16 replicate populations of *D. melanogaster* through single generations of full-sib mating and assayed them for egg-to-pupa viability. The average additive genetic variance in the control lines was not significantly different from zero, whereas that in lines inbred to $f = 0.25$ was five-fold (and significantly) higher. In a study involving 32 lines of *D. melanogaster*, again inbred to $f = 0.25$, Fernández et al. (1995) observed a ten-fold increase in the additive genetic variance for viability, whereas that for fecundity remained approximately equal to that of the control; and a similar increase in additive genetic variance for viability following inbreeding was seen in still another study by García et al. (1994). In each of these studies, the characters of interest exhibited significant inbreeding depression.

This dichotomy between the behavior of additive variance under inbreeding for metric traits versus those more directly related to fitness is predicted from theoretical results. As discussed in Chapter 6, we expect a higher fraction of nonadditive variance in fitness-related traits. Assuming dominance (but no epistasis), Zhang et al. (2004b) noted that the equilibrium distribution of allele frequencies for a nearly neutral trait is very different from that for a fitness-related trait under mainly purifying selection. In the latter, deleterious alleles tend to be rare and at least partly recessive (Chapter 7), exactly the condition that facilitates an increase in additive variance under inbreeding. In contrast, no such association in the joint distribution of allelic frequencies and effect sizes is expected for a nearly neutral trait, and thus no increase in σ_A^2 is expected.

If any general message can be taken from these limited results, it is that increases in additive genetic variance following a population bottleneck are largely restricted to fitness characters harboring substantial dominance genetic variance, with morphological and behavioral traits exhibiting genetic-variance dynamics that are not greatly different from expectations based on the additive model (Wang et al. 1998; Van Buskirk and Willi 2006). Moreover, there is, as yet, no firm empirical evidence that population bottlenecks create significant levels of *adaptive* variation. With the exception of the intentionally artificial setting used by Cheverud et al. (1999), all observed increases in additive variation following inbreeding have been accompanied by substantial inbreeding depression—while the variance increased, the mean changed in a direction contrary to high fitness.

Might the creation of new additive genetic variance nevertheless compensate for slippage in the mean via inbreeding depression? In the two *D. melanogaster* studies in which selection for increased fitness was imposed on inbred lines, a substantial increase in the response to selection (relative to the controls) was observed, but this was more than offset by the loss of fitness due to inbreeding depression (López-Fanjul and Villaverde 1989; García et al. 1994), i.e., there was an overall reduction in viability even after selection utilized the released genetic variance. In another study, involving bottlenecked populations of *Drosophila bunnanda*, van Heerwaarden et al. (2008) found that although inbreeding resulted in an inflation of the additive genetic variance for desiccation resistance, there was no increase in the response to selection relative to control populations. The same pattern was seen in selected populations of the mustard plant *Brassica rapa*—bottlenecking led to a significant increase in additive genetic variance for cotyledon size, apparently via a release from the dominance component, but a reduction in the long-term response to selection (Briggs and Goldman 2006).

Example 11.2. The significance of the problem of the variance of the within-population variance is highlighted by a massive experiment performed by López-Fanjul et al. (1989). Starting from a large random-bred base population of *D. melanogaster,* 304 non-inbred lines were constructed, and another 300 inbred lines were produced by four generations of full-sib mating followed by population expansion for six generations. The components of variance for abdominal bristle number were evaluated for the initial 304 lines ($f = 0$) and for the fourth and tenth generations after the bottleneck/expansion treatment (both $f = 0.5$) by several techniques including sib analysis. Consistent with the view that this character has a largely additive genetic basis (LW, pp. 171–172), the mean ($\overline{z}$) was unaffected by inbreeding (table below). Moreover, averaging over all of the inbred lines, there was an approximately 50% reduction in the additive genetic variance, as predicted by additive theory.

The data from this experiment are in excellent accord with the sampling theory for the additive genetic variance presented above. Summing the expected variances contributed by Hardy-Weinberg and gametic-phase disequilibria, $\sigma_A^4(0)/N_e + \sigma_A^4(0)/N_e$, the expected coefficient of variation for the additive genetic variance in the non-inbred lines (random populations with $N_e \simeq 8$ and $t = 0$) is $(2/N_e)^{1/2} = 0.50$, which is reasonably close to the observed value 0.35 (table below).

Generation	f	$\overline{z}$	Var(A)	CV[Var(A)]
1	0.0	41.4	5.2	0.35
4	0.5	41.4	2.5	1.05
10	0.5	41.4	1.8	1.15

In addition, for both of the inbred generations, the observed values of the coefficient of variation for Var(A), CV[Var(A)], are close to the theoretical minimum $\sqrt{8/(3N_e)} = 1.04$ (using $N_e = 2.5$ for full-sib mating). In principle, several generations of random mating would be expected to cause a reduction in CV[Var(A)] through the elimination of gametic-phase disequilibrium if significant numbers of linked loci contributed to the genetic variance, but a comparison of results for $t = 4$ and 10 shows that this was not observed.

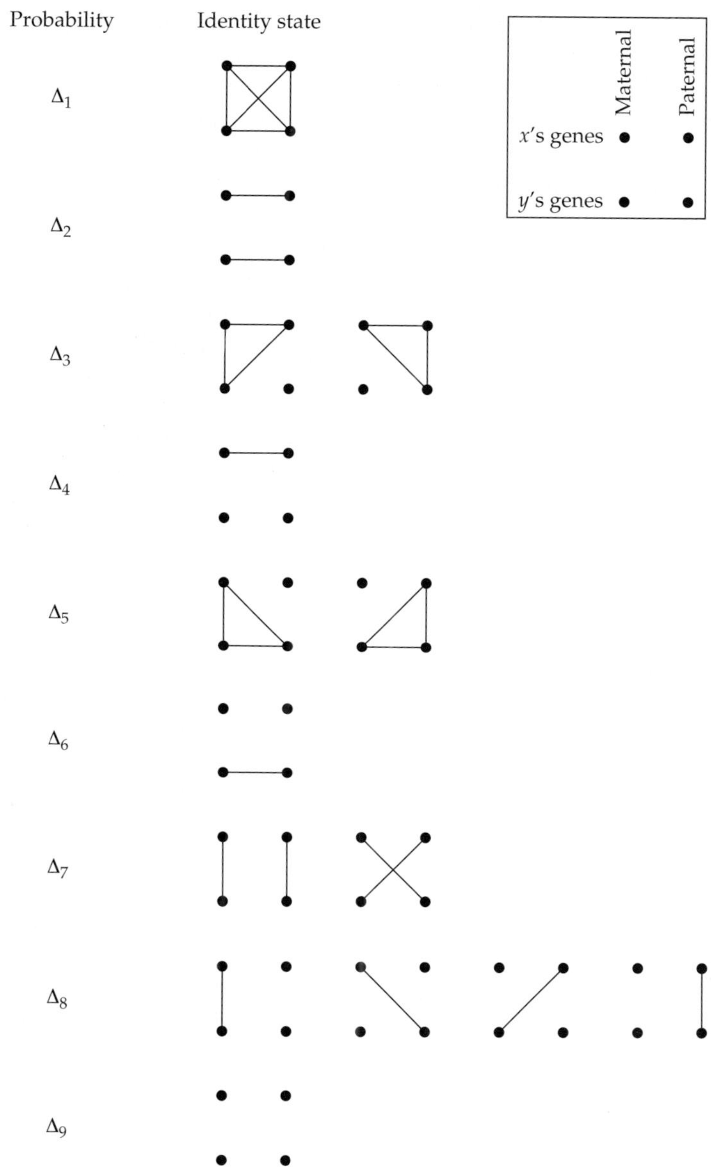

Figure 11.5 The 15 possible states of identity by descent for a locus in individuals x and y; condensed into nine classes. Alleles that are identical by descent are connected by lines, with horizontal lines indicating an inbred individual (Δ_1 through Δ_6). Note that Δ_4, Δ_6, and Δ_9 involve unrelated individuals (there are no lines between any gene of x to any gene of y).

COVARIANCE BETWEEN INBRED RELATIVES

In the preceding sections, we assumed there was a parallel series of small populations, each being propagated across generations as progeny derived from randomly mating populations of size N_e. Even in the simplest case of no epistasis, we found that the dynamics of the genetic variance within populations is a potentially complex function of six quadratic parameters of gene effects in the base population (Table 11.1). What remains to be considered is how these contributions can be estimated in a practical sense. Not surprisingly, the key strategy is the usual one in quantitative genetics—the resemblance between relatives (LW Chapter 7).

When individuals are inbred with respect to the base population, the expressions for the genetic covariance between relatives become functions of all of the parameters outlined in Table 11.1, not just the usual σ_A^2 and σ_D^2. On the other hand, with inbreeding there are also many more potential types of relationships than in the conventional case, as the latter are supplemented by inbred relatives. One can imagine, for example, a multigenerational series of individuals resulting from continuous selfing, full-sib mating, or both. Given phenotypic information on these additional types of relatives, it should be possible to estimate several different factors contributing to phenotypic covariance (as many as the number of observed relationships). The little experience we have gained in this area, however, indicates that the statistical difficulties in achieving accurate estimates are still quite formidable, even in the absence of epistasis, which we will assume in the following.

Three new issues arise in considering the sources of phenotypic resemblance between inbred relatives. First, inbreeding causes a statistical dependence between alleles within individuals (generating an excess of homozygotes), and this creates a covariance between the additive effects, α_i, in one relative and the dominance effects, δ_{ii} in the other, as represented by $\sigma(\alpha_i, \delta_{ii}) = \sigma_{ADI}$ (Table 11.1). Second, if two individuals have identical genotypes by descent, their dominance covariance will differ depending on whether they are inbred or outbred (because inbred individuals cannot be heterozygous), and this will generally vary from locus to locus. Third, with dominance, the mean phenotype of inbred individuals will generally differ from that of non-inbred individuals, and this can inflate the covariance between certain types of relatives by breaking the population up into classes of inbred vs. non-inbred individuals.

The one- and two-locus identity measures (Figure 11.1) used to examine the within- and among-genetic variances in a random-mating population undergoing drift are insufficient to describe all of the potential relationships among two diploids when generalized inbreeding is occurring. To do so, we need the nine **condensed coefficients of identity**, Δ_i, between two individuals (Figure 11.5). Introduced in LW Chapter 7, these coefficients sum to one and have a natural connection to the quadratic components in Table 11.1. As shown in the figure, three situations (Δ_4, Δ_6, and Δ_9) correspond to comparisons between unrelated individuals (although one or both may be inbred), and hence do not contribute to the genetic covariance between relatives.

Harris (1964) and Gillois (1965) first derived an expression for the covariance between inbred relatives, assuming gametic-phase equilibrium and an absence of epistasis, and Cockerham (1984a) extended their analyses to allow for gametic-phase disequilibrium, showing that the genetic covariance between individuals x and y is

$$\begin{aligned}\sigma_G(x,y) =& 2\Theta_{xy}\sigma_A^2 + \Delta_{7xy}\sigma_D^2 + \Delta_{1xy}\sigma_{DI}^2 + (2\Delta_{1xy} + \Delta_{3xy} + \Delta_{5xy})\sigma_{ADI} \\ &+ (\Delta_{2xy} - f_x f_y)\iota^* + (\widetilde{\Delta}_{xy} - f_x f_y)(\iota^2 - \iota^*)\end{aligned} \tag{11.13}$$

where $\widetilde{\Delta}_{xy}$ indicates our previous two-locus identity measure (Figure 11.1), applied to individuals x and y (as opposed to randomly drawn gametes).

To see how the Δ_i enter into the covariance expressions, consider the coefficient of coancestry Θ_{xy}, the probability that two genes, one drawn from x and the other from y are identical by descent. In terms of the condensed coefficients of identity,

$$\Theta_{xy} = \Delta_{1xy} + \frac{1}{2}(\Delta_{3xy} + \Delta_{5xy} + \Delta_{7xy}) + \frac{1}{4}\Delta_{8xy} \tag{11.14}$$

(LW Chapter 7), where each condensed coefficient of identity is weighted by the conditional probability that a gene that is randomly drawn from x is identical by descent with a gene that is randomly drawn from y. There are four different ways to randomly choose an allele from

Table 11.5 The expected genetic covariance generated by each Δ_i relationship.

Δ_1	Δ_2	Δ_3, Δ_5	Δ_7	Δ_8	$\Delta_4, \Delta_6, \Delta_9$
$2\sigma_A^2 + \sigma_{DI}^2 + 2\sigma_{ADI}$	$(1 - f_x f_y/\Delta_2)\iota^*$	$\sigma_A^2 + \sigma_{ADI}$	$\sigma_A^2 + \sigma_D^2$	$\sigma_A^2/2$	0

each of two diploids. In each case, with a probability of Θ_{xy} the two chosen alleles are IBD, in which case their contribution to the genetic covariance is $\sigma_A^2/2$. Hence, the contribution to the additive variance resulting from shared additive effects is $4\Theta_{xy}\sigma_A^2/2 = 2\Theta_{xy}\sigma_A^2$.

Similarly, Δ_{7xy} and Δ_{1xy} account for the probabilities that the two individuals share identical genotypes by descent, in the absence or presence of inbreeding, respectively, so the dominance genetic covariance is $(\Delta_{7xy}\sigma_D^2 + \Delta_{1xy}\sigma_{DI}^2)$. The term $(2\Delta_{1xy} + \Delta_{3xy} + \Delta_{5xy})$ is a measure of the expected number of ways in which three alleles in the two individuals are identical by descent, and when multiplied by σ_{ADI}, it yields the expected covariance between individuals resulting from the covariance between additive and homozygous dominance effects. Finally, $(\Delta_{2xy} - f_x f_y)$ is the probability that the two individuals are inbred at the same locus in excess of that expected for random members of the population ($f_x f_y$), whereas $(\widetilde{\Delta}_{xy} - f_x f_y)$ is the excess joint inbreeding at one locus in x and another in y. These latter two coefficients are multiplied, respectively, by the quadratic terms describing inbreeding depression at the same and at different loci. Table 11.5 summarizes the expected genetic covariance contributed by the various relationships.

One conclusion that can be drawn immediately from Equation 11.13 is that with inbreeding, dominance can contribute to the covariance between many types of inbred relatives. All of the quadratic components in this equation are necessarily positive except σ_{ADI}, which can be positive or negative (Example 11.1). Thus, while it is likely that inbreeding will inflate the covariance between relatives, this cannot be stated with certainty.

Example 11.3. Consider the situation in which fathers are mated to their daughters. What is the genetic covariance between the offspring (y) of such matings and their fathers (x)? Assuming the father is not inbred (there is no line connecting the maternal and paternal alleles of x in Figure 11.5), $\Delta_{1xy} = \Delta_{2xy} = \Delta_{3xy} = f_x = \widetilde{\Delta}_{2xy} = 0$, so to complete the solution of Equation 11.13, we only require values for the coefficients Θ_{xy}, Δ_{7xy}, and Δ_{5xy}. The inbreeding coefficient of y is the same as the coefficient of coancestry between the parents (the father and his daughter), $f_y = 1/4$. Moreover, because y inherits only one gene from x directly, if y is inbred, then identity relationship 5 must hold, so $\Delta_{5xy} = f_y \cdot 1 = 1/4$. A gene in x can be identical with one in y by direct descent from the father or by indirect descent from the father through his first daughter (the mother of y), so $\Theta_{xy} = (1/4) + (1/8) = 3/8$. Finally, given that y has inherited one gene directly from x, the probability that x's other gene has been transmitted through his first daughter is $1/4$. Thus, $\Delta_7 = 1/4$. Substituting into Equation 11.13,

$$\sigma_G(x, y) = \frac{3}{4}\sigma_A^2 + \frac{1}{4}\sigma_D^2 + \frac{1}{4}\sigma_{ADI}$$

This may be contrasted with $\sigma_G(x, y) = \sigma_A^2/2$, the expectation for the parent-offspring covariance under random mating (the mother of y and its father, x, are unrelated).

Some attention has been given to the contribution of additive $\times$ additive genetic variance to the resemblance between inbred relatives (Cockerham 1984b; Cockerham and Tachida 1988; Tachida and Cockerham 1989). In this case, Equation 11.13 requires an additional term,

$$\left(\widetilde{f}_{xy} + \widetilde{\gamma}_{\bar{x}y} + \widetilde{\gamma}_{x\bar{y}} + \widetilde{\Delta}_{\bar{x}\bar{y}}\right) \sigma_{AA}^2$$

Table 11.6 Coefficients for the components of genetic covariance for an equilibrium population undergoing mixed selfing and random mating (in proportions of β and $1-\beta$, respectively). The equilibrium variance in the inbreeding coeffcient among individuals is $\sigma_f^2 = f(1-f^2)/(2+f)$, with $f = \beta/(2-\beta)$. (From Cockerham and Weir 1984.)

Relationship	σ_A^2	σ_D^2	σ_{ADI}	σ_{DI}^2	ι^*	$\iota^2 - \iota^*$
Parent and outcrossed offspring	$\frac{1+f}{2}$	0	$\frac{f}{2}$	0	0	0
Parent and selfed offspring	$1+f$	$\frac{1-f}{2}$	$\frac{1+7f}{4}$	f	$\frac{f(1-f)}{2}$	$\frac{\sigma_f^2}{2}$
Parent and mixed offspring	$\frac{1+3f}{2}$	$\frac{2f(1-f)}{2(1+f)}$	$\frac{f(1+3f)}{1+f}$	$\frac{2f^2}{1+f}$	$\frac{f^2(1-f)}{1+f}$	$\frac{f\sigma_f^2}{1+f}$
Selfed sibs	$1+f$	$\frac{1-f}{2}$	$\frac{1+3f}{2}$	$\frac{1+7f}{8}$	$\frac{f(1-f)}{4}$	$\frac{\sigma_f^2}{4}$
Selfed sib and outcrossed sib	$\frac{1+f}{2}$	0	$\frac{1+3f}{8}$	0	0	0
Full sibs	$\frac{1+f}{2}$	$\frac{(1+f)^2}{4}$	0	0	0	0
Half sibs	$\frac{1+f}{4}$	0	0	0	0	0

Here, the double identity measures are analogous to those described in Figure 11.1, with the overbars (on gamete subscripts) denoting that the two gametes contributing to that individual are involved. These coefficients depend upon the previous inbreeding in the population and the amount of recombination that occurs between individuals x and y. The algebraic details may be found in the references given above.

Equation 11.13 provides a practical way to obtain estimates of the quadratic components described in Table 11.1 from estimates of the phenotypic covariances between various types of inbred relatives and solution of the resultant set of equations (the usual method-of-moments approach). An optimal design for such an analysis employs a number of very small populations in order to maximize the temporal change in the identity coefficients and to allow a high degree of replication. For systems of selfing and full-sib mating, there is an added advantage of simplicity in formulating the identity coefficients, as we will now show, while Table 11.6 gives the genetic covariances in populations undergoing mixed selfing and random mating.

If we assume there is negligible linkage, all identity coefficients under obligate self-fertilization can be expressed in terms of the inbreeding coefficient (Cockerham 1983; Wright and Cockerham 1986a; Wright 1988), a point that will be quite useful in Chapter 23, when we examine the response to selection under selfing. For a set of selfed lines derived from a random-mating base population existing in generation 0, the covariance of relatives in generations (of selfing) i and j whose last common ancestor occurred in generation t is

$$\sigma_G(x_i, y_j, t) = (1+f_t)\sigma_A^2 + \left(\frac{(1-f_i)(1-f_j)}{1-f_t}\right)(\sigma_D^2 + f_t\iota^*) + \left(\frac{f_i+f_j+2f_t}{2}\right)\sigma_{ADI}$$
$$+ \left(f_t + \frac{(f_i-f_t)(f_j-f_t)}{2(1-f_t)}\right)\sigma_{DI}^2 + (1+f_t)^2\sigma_{AA}^2 \tag{11.15}$$

where $f_k = 1-(1/2)^k$. For example, the covariance of a parent in generation t and a descendant in generation j is

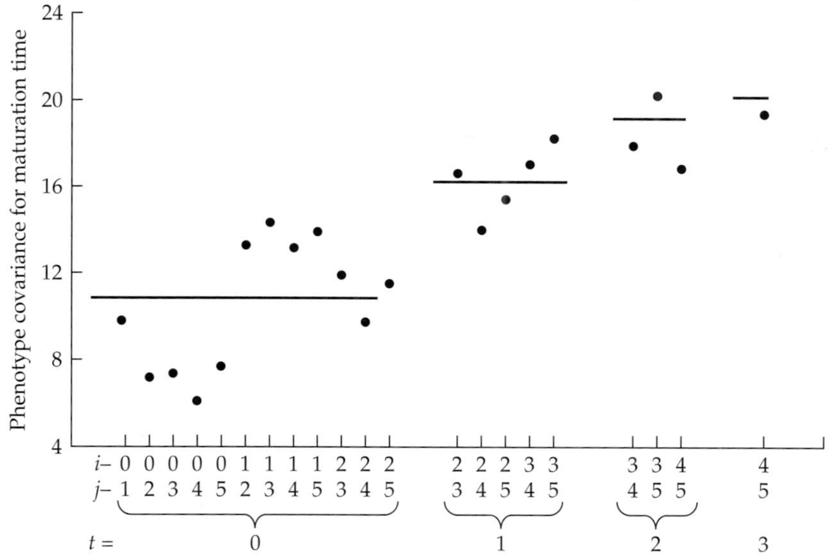

Figure 11.6 Observed covariances between relatives in a selfing series starting from a highly heterozygous F_2 synthetic population (i.e., that formed by all pairwise crosses among a set of lines) of soybeans ($t = 0$). Here i and j denote the generations of the individuals under consideration, and t is the generation of their last common ancestor. For example, the covariance between individuals in generations 2 and 3 with a last common ancestor at generation 0 is indicated by $i = 2,\ j = 3,\ t = 0$. The lines represent the expectations (for a given value of t) under the assumption of an additive model, $(1+f_t)\sigma_A^2$ with $f_t = 1-(1/2)^t$ and $\sigma_A^2 = 10.9$. (Data are from Horner and Weber 1956.)

$$\sigma_G(x_t, y_j, t) = (1 + f_t)\sigma_A^2 + (1 - f_j)(\sigma_D^2 + f_t\iota^*) + \frac{f_j + 3f_t}{2}\sigma_{ADI} + f_t\sigma_{DI}^2 + (1 + f_t)^2\sigma_{AA}^2 \tag{11.16}$$

For a parent-offspring analysis, $j = t + 1$. Additional terms involving σ_{AA}^2 and $(\iota^2 - \iota^*)$ are required if there are pairs of linked loci with major effects (Cockerham 1983, 1984b).

Although Equation 11.15 applies to an *entire collection* of selfed lines, *within* a single selfed line (from the F_1 of pure-line cross), there are two equally frequent alleles per polymorphic locus, which leads to $\sigma_{ADI} = \sigma_{DI}^2 = 0$ and $\iota^* = \sigma_D^2$. The expected covariance between relatives *within lines* then becomes

$$\sigma_G(x_i, y_j, t) = (1/2)^t\sigma_A^2 + (1/2)^{i+j-t}\sigma_D^2 + (1/2)^{2t}\sigma_{AA}^2 \tag{11.17}$$

(Wright and Cockerham 1986a), which will also prove very useful in Chapter 23 when examining within-line selection. Note that for $t > 5$, the within- and among-population components of variance are very close to 0 and $2\sigma_A^2 + 2\sigma_{ADI} + \sigma_{DI}^2 + 4\sigma_{AA}^2$, respectively. Wright (1987) extended Equation 11.17 to include additive $\times$ dominance and dominance $\times$ dominance epistasis, but even in the absence of linkage, 12 terms are necessary to define the genetic covariance in this case.

An example of the utility of the selfing theory is provided by a study with soybeans, a predominantly self-fertilizing species (Horner and Weber 1956). Two inbred varieties were crossed to produce a uniform F_1 population, which was then selfed to produce a segregating F_2 population. Random F_2 plants were then selfed to produce F_3 plants, and so on down to the F_7. The covariances between many possible types of relatives for the timing of seed maturation were then assessed. Under a simple additive genetic model, Equation 11.15 reduces to

$$\sigma_G(x_i, y_j, t) = (1 + f_t)\sigma_A^2$$

which indicates that the genetic covariances of all types of direct descendants from generation t plants should be independent of i and j. The observed covariances are in fair accord with these expectations with $\sigma_A^2 = 10.9$ (Figure 11.6). Although there is a certain amount of noise in the data, the inclusion of other base-population properties does not significantly improve the fit, and it is likely that some of the scatter in the data is caused by year-to-year differences in growth conditions.

For the special case of full-sib mating, Cornelius and Dudley (1975) provided a general solution (ignoring epistasis and linkage) for the covariance between parents and descendants, full-sibs, and uncle (or aunt) and niece (or nephew). They presented tables of the coefficients needed for Equation 11.13 for the first eight generations of consanguineous mating. Cockerham (1971) derived a transition matrix that allows the computation of all of the coefficients for the covariance between full-sibs,

$$\begin{pmatrix} 1-\Delta_1 \\ 1-\Delta_3 \\ 1-\Delta_7 \\ 1-\Delta_2 \\ 1-f \\ 1-\Theta \end{pmatrix}_{t+1} = \begin{pmatrix} 1/4 & 1/2 & 0 & 0 & 0 & 1/4 \\ 0 & 1/2 & 0 & 0 & 0 & 1/2 \\ 0 & 1/4 & 1/8 & 1/8 & 1/4 & 1/8 \\ 0 & 1/2 & 1/4 & 0 & 0 & 1/4 \\ 0 & 0 & 0 & 0 & 0 & 1 \\ 0 & 0 & 0 & 0 & 1/4 & 1/2 \end{pmatrix} \begin{pmatrix} 1-\Delta_1 \\ 1-\Delta_3 \\ 1-\Delta_7 \\ 1-\Delta_2 \\ 1-f \\ 1-\Theta \end{pmatrix}_t \tag{11.18}$$

where in this case $\Delta_3 = \Delta_5$.

In closing, it needs to be emphasized that all of the expressions developed above have been written in terms of the quadratic components for the random-mating base population. Provided that mating remains random in a small population, there is no reason why the simpler and more familiar expressions of LW Chapter 5 cannot be relied upon, provided it is understood that the variance and covariance components apply to the *current population*. For example, the expected genetic covariance between half-sibs in generation t may be written either as $\sigma_A^2(t)/4$ or in terms of base-population properties with Equation 11.13. The advantage of interpreting the covariance between relatives in terms of the base population properties is that it provides a mechanistic explanation for the temporal changes in the usual components of variance, $\sigma_A^2(t)$ and $\sigma_D^2(t)$.

REML Estimates

As with all approaches to variance-component estimation, comparisons among appropriate sets of relatives (to provide the correct independent contrasts) can be used to estimate σ_{ADI}, σ_{DI}^2, and ι^*. One can use specific crossing designs to ensure that modest to large numbers of the correct types of relatives are included. More generally, one can potentially use pedigree data, provided there is sufficient inbreeding. Both of these settings can be handled under the very flexible mixed-model framework offered by REML variance estimation (Chapters 19, 20, and 22; LW Chapter 27). The basic idea is that the covariance matrix, $\mathbf{V}$, for the vector, $\mathbf{y}$, of observations (sets of relatives, where $V_{ij} = \sigma[y_i, y_j]$) can be decomposed into the sum of the products of appropriate relationship matrices times their associated quadratic components. From Equation 11.13, and ignoring epistasis,

$$\mathbf{V} = \text{Cov}(\mathbf{y}) = \sigma_A^2 \mathbf{A} + \sigma_D^2 \mathbf{D} + \sigma_{DI}^2 \mathbf{M}_1 + \sigma_{ADI} \mathbf{M}_2 + \iota^* \mathbf{M}_3 + \sigma_e^2 \mathbf{V}_e \tag{11.19a}$$

where $\mathbf{A}_{ij} = 2\Theta_{ij}$ is the (additive) relationship matrix (Chapters 19, 20, and 22; LW Chapters 26 and 27), $\mathbf{D}_{ij} = \Delta_{7ij}$ is the dominance relationship matrix, $\mathbf{V}_e$ is the residual covariance structure (often assumed to be $\mathbf{V}_e = \mathbf{I}$, implying independent and homoscedastic residuals), and the elements of the other three relationship matrices are given by

$$\mathbf{M}_{1ij} = \Delta_{1ij}, \quad \mathbf{M}_{2ij} = 2\Delta_{1ij} + \Delta_{3ij} + \Delta_{5ij}, \quad \mathbf{M}_{3ij} = \Delta_{2ij} - f_i f_j \tag{11.19b}$$

as obtained by Smith and Mäiki-Tanila (1990), de Boer and Hoeschele (1993), Shaw and Woolliams (1999), and Abney et al. (2000). This approach is computationally demanding,

Table 11.7 Average values for all nine Δ_i coefficients for all 324,415 pairwise combinations of a measured set of 806 individuals from a 13-generation pedigree of South Dakota Hutterities. A religious sect originally from the Tyrolean Alps, Hutterities are well-studied by human geneticists as all extant individuals can be traced by genealogical records back to less than 90 ancestors from the 1700s to the early 1800s. This table shows the power (or lack thereof) for computing the various genetic components that appear in Equation 11.13. For example, information to estimate σ^2_{DI} comes from the roughly 70 pairs (324,415 · 0.000217) that are expected to involve Δ_1. (From Abney et al. 2000.)

Δ_1	Δ_2	Δ_3, Δ_5	Δ_4, Δ_6	Δ_7	Δ_8	Δ_9
0.000217	0.000993	0.00411	0.0283	0.00444	0.141	0.788

as the standard iterative approach used to obtain the REML estimates must invert the **V** matrix following each update of its component variances (LW Chapter 27).

Two historical approaches have been used to obtain the simplest relationship matrix, **A**, given a pedigree. Wright (1921a) used path analysis (counting all paths connecting two individuals within a pedigree), while Emik and Terrill (1949) and Crudent (1949) introduced the **tabular method**, a quicker approach that starts with the founding relatives and tabulates the relationships forward in each generation. Extensions of both approaches have been proposed to calculate the more general relationships, Δ_i, needed for the matrices in Equation 11.19: path analysis approaches were proposed by Jacquard (1966) and Nadot and Vaysseix (1973), and tabular methods by Smith and Mäiki-Tanila (1990). Issues with pedigree-free (i.e., marker based) estimates of the Δ_i are examined by García-Cortés et al. (2014) and Ackerman et al. (2017).

The central issue of whether a pedigree/designed cross is able to uniquely estimate each quadratic component is the concept of **identifiability** (Rothenberg 1971; Jiang 1996), discussed in Chapter 22. This is the random-effects counterpart of the concept of estimability for fixed effects (LW Chapter 26). Basically, variance components can be uniquely estimated, *provided* the relationship matrices are sufficiently independent of each other. For example, from Equation 11.16, the among-(selfed)-line variation for $t > 5$ is essentially $2\sigma^2_A + 2\sigma_{ADI} + \sigma^2_{DI}$, so that these three components are fully confounded and not separable if the data consist of only the between-line variance for a series of lines with (say) 5, 10, and 15 generations of selfing. Power is another issue. If a relationship matrix is sparse (contains mostly zeros), then even a large dataset may not have much power, as the numbers in the required identity classes may be quite small (Table 11.7).

Empirical Observations

Unfortunately, data on the parameters σ^2_{DI}, σ_{ADI}, and ι^* are scant, although some progress has been made with annual plants. Starting from a random-mating base population of maize, Cornelius (1988) produced a series of selfed and full-sib mated lines, all of which were assayed in a common-garden experiment. The parameter estimates given in Table 11.8 best describe the overall set of observed covariances. Except for yield, all of the characters exhibit significant additive genetic variance, and four of the six traits exhibit significant squared inbreeding effects, ι^*. However, nearly all of the estimates for σ^2_D, σ^2_{DI}, and σ_{ADI} are nonsignificant. In part, this is clearly a power issue, as the significant inbreeding depression implies the presence of dominance.

The few other attempts to estimate quadratic components involving inbreeding have yielded mixed results. For example, starting with 300 inbred lines of the monkeyflower (*Mimulus guttatus*), Kelly and Arathi (2003) crossed triplets of lines to create outbred full-sib and half-sib families, allowing a joint analysis of σ^2_A, σ^2_D, σ^2_{DI}, and σ_{ADI} for six floral traits. Although each character exhibited significant inbreeding depression, and σ^2_D was significant for two traits and σ^2_{DI} was significant for six traits, σ_{ADI} was significant in only one case. In a study of another flowering plant, *Nemophila menziesii*, >1000 plants with f up to 0.75 were

Table 11.8 Estimates of the quadratic components of Equation 11.13 from phenotypic data on selfed and full-sib mated lines derived from a panmictic base population of maize. The parameter estimates, which were obtained by a maximum-likelihood procedure, are those that give the overall best fit to a large number of observed relationships. (From Cornelius 1988.)

Character	Var(A)	Var(D)	Var(DI)	Cov(ADI)	ι^*
Plant height (cm)	370 ± 99	-57 ± 139	225 ± 220	-258 ± 178	1045 ± 341
Ear height (cm)	382 ± 83	-103 ± 98	383 ± 179	-450 ± 152	430 ± 238
Grain yield (g/plant)	-125 ± 231	1403 ± 436	-129 ± 552	330 ± 468	3286 ± 886
% Moisture of seed	5.9 ± 1.7	-1.0 ± 2.5	-5.0 ± 3.6	3.2 ± 2.8	15.3 ± 6.9
% Oil of seed	0.14 ± 0.05	0.05 ± 0.08	-0.02 ± 0.12	0.02 ± 0.10	0.31 ± 0.19
Kernel wt. (g/100)	14.7 ± 4.2	-2.3 ± 6.3	4.5 ± 9.9	-1.6 ± 7.6	15.3 ± 16.0

evaluated in a common garden for two morphological and two floral traits (Shaw et al. 1998). A REML analysis revealed significant inbreeding depression for all traits, but σ^2_{DI} was significant only for the floral traits, and σ_{ADI} was uniformly nonsignificant.

Approaches based entirely on pedigree data (as opposed to defined crossed populations) have been much less successful. Hoeschele and Vollema (1993) obtained inconsistent estimates of the inbreeding quadratic terms for milk and fat yield in Holsteins, using least-squares on subsets of the data. Shaw and Woolliams (1999) used REML to study a pedigree of 2000 sheep with variable inbreeding up to a high of $f = 0.6$, but found no evidence for significant σ^2_{DI} or σ_{ADI} for body weight or fleece quality, despite the presence of significant inbreeding depression for both. The application of REML to humans by Abney et al. (2000) found no significant inbreeding quadratic components for a cholesterol measure (HDL) in a population of Hutterites, but this may have been a consequence of lack of power. Table 11.7 makes the key point that the effective numbers of individuals in these relationship classes needed to estimate σ^2_{DI}, σ_{ADI}, and ι^* are expected to be small in most pedigrees, even those that are rather inbred. By contrast, a controlled crossing design can generate large numbers of the appropriate sets of relatives.

Kelly (1999c) suggested a selection-based approach to estimate the ratio of σ_{ADI}/σ^2_A as a test for rare recessives. His motivation can be seen in the figure for Example 11.1, where $\sigma_{ADI}/\sigma^2_A > 1$ for a rare recessive (allele frequency, p, is small), while for intermediate allele frequencies, this ratio is close to zero, or even negative. Kelly's idea (examined more fully in Chapter 28) is to compare the change in mean, ΔM, and the change in the inbreeding depression coefficient, ΔB, following selection, as he noted that $\Delta B/\Delta M$ tracks σ_{ADI}/σ^2_A. Significant, and negative, estimates of this ratio were seen following artificial selection experiments on flower size in *Mimulis guttatus* (Kelly and Willis 2001), and in female fecundity in *Drosophila melanogaster* (Charlesworth et al. 2007), suggesting significant, and negative, estimates of σ_{ADI}, as well as the presence of recessive alleles at intermediate frequencies in these traits.

Although these limited surveys do not rule out important contributions from quadratic inbreeding components in some cases, combined with the observations on the dynamics of genetic variance described above, they do raise questions about the general necessity of incorporating such complexities into expressions for genetic variances and covariances. Again, this is not to deny an important role for dominance and epistasis in the expression of complex traits, for which the evidence is substantial (Wolf et al. 1980; Lynch and Walsh 1998). However, despite arguments to the contrary (Templeton 1980; Nelsen et al. 2013), there is little compelling evidence that we need to abandon the existing theoretical framework for quantitative-trait evolution, even in the presence of substantial physiological epistasis. Although substantial progress has been made in incorporating the complexities of nonadditive gene action into the theory of quantitative traits in finite populations, the limited empirical evidence to date implies a second-order nature of such effects.

DRIFT-MUTATION EQUILIBRIUM

The models introduced in the previous sections predict that finite populations eventually lose all of their genetic variation, at which point the genotypic means of isolated populations will have attained a maximum level of divergence. These results arose because we assumed there was an absence of significant evolutionary forces countering the loss of variance caused by random genetic drift. In reality, however, there is one such force that cannot be prevented—the continual input of new variation by polygenic mutation. When this is accounted for, we can expect neutral quantitative traits to approach an equilibrium level of within-population variance as a balance is struck between the opposing forces of drift and mutation. The means of such characters should also continue to diverge as isolated populations become fixed for unique mutations, a subject that will be dealt with in detail in Chapter 12. We also explore the within-population drift-mutation variance in more detail in Chapter 28, as part of a larger discussion on what forces account for the observed levels of quantitative-genetic variation.

Consider a character with a purely additive genetic basis in a population with a constant effective size. In each generation, a fraction $1/(2N_e)$ of the genetic variation is lost by drift, while new variation in the amount σ_m^2 is introduced by mutation. In mechanistic terms, σ_m^2 is defined as $2\sum_{i=1}^{n} \mu_i E(a_i^2)$, where μ_i denotes the per-locus mutation rate for gene i, n denotes the number of loci contributing to the trait, and $E(a_i^2)$ is the average squared heterozygous effect of a new mutation on the phenotypic value (LW Chapter 12). This leads to the simple recursion equation,

$$\sigma_A^2(t) = \left(1 - \frac{1}{2N_e}\right)\sigma_A^2(t-1) + \sigma_m^2 \tag{11.20a}$$

(Clayton and Robertson 1955), which has the approximate solution

$$\sigma_A^2(t) = 2N_e\sigma_m^2 + [\sigma_A^2(0) - 2N_e\sigma_m^2]\exp(-t/2N_e) \tag{11.20b}$$

Thus, the equilibrium (obtained as $t \to \infty$) genetic variance for a neutral quantitative trait with an additive genetic basis is simply

$$\sigma_A^2(\infty) = \widetilde{\sigma}_A^{\,2} = 2N_e\sigma_m^2 \tag{11.20c}$$

(Lande 1976; Chakraborty and Nei 1982; Lynch and Hill 1986). Starting from a completely homozygous base population, the times to 50% and 95% of the equilibrium variance are approximately $1.4N_e$ and $6.0N_e$ generations, respectively (Lynch and Hill 1986). Note that this time to converge to the equilibrium is *independent* of the mutational variance, σ_m^2. Because $\sigma_A^2(0)$ will typically be greater than zero, small isolated populations can be expected to reach the equilibrium quite rapidly. On the other hand, if a population is suddenly reduced to an unusually small N_e, such that $\sigma_A^2(0) \gg 2N_e\sigma_m^2$, for the several generations immediately following the bottleneck, $\sigma_A^2(t) \simeq \sigma_A^2(0)e^{-t/2N_e}$, justifying the use of Equation 11.2 for the short term.

If we let $h_m^2 = \sigma_m^2/\sigma_E^2$, where σ_E^2 is the environmental variance of the trait, be the **mutational heritability** and again assuming additivity of genetic effects, the expected equilibrium heritability for a neutral character under the model given by Equation 11.20b is

$$\widetilde{h}^{\,2} = \frac{\widetilde{\sigma}_A^{\,2}}{\widetilde{\sigma}_A^{\,2} + \sigma_E^2} = \frac{2N_e\sigma_m}{2N_e\sigma_m + \sigma_E^2} = \frac{2N_e h_m^2}{2N_e h_m^2 + 1} \tag{11.21a}$$

Almost all estimates of h_m^2 are in the range of 0.01 to 0.0001 with a median value near 0.001 (LW Chapter 12). Thus, populations with $N_e \simeq 100$ are expected to have small to moderate levels of heritability for neutral characters, but nearly all of the phenotypic variation for neutral characters is expected to have a genetic basis if $N_e > 10^4$ (Figure 11.7).

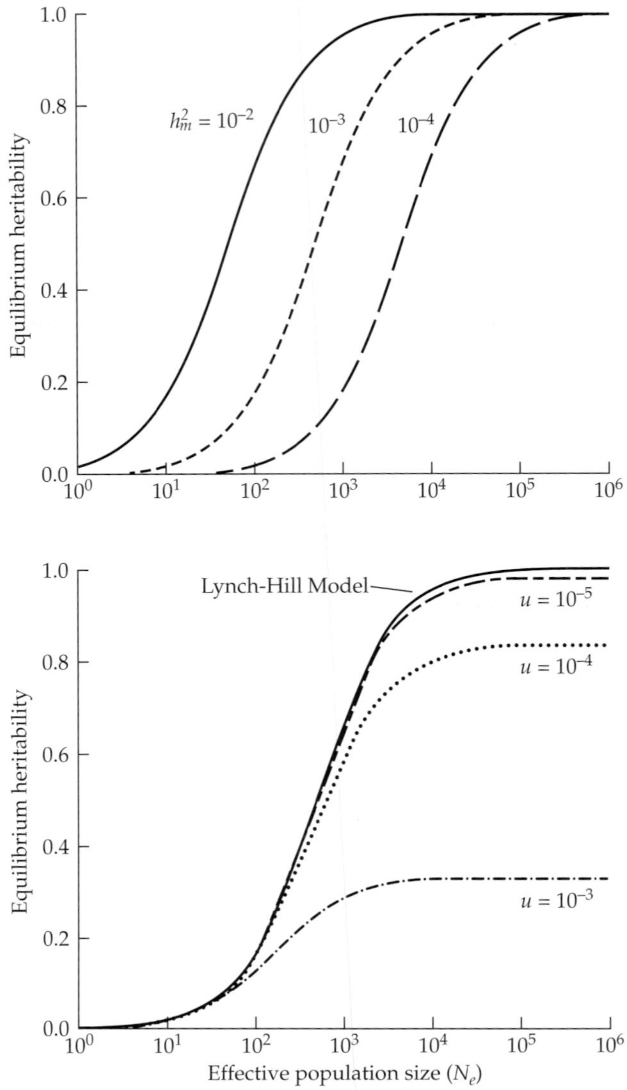

Figure 11.7 (**Top**) Levels of heritability expected for neutral characters with an additive genetic basis under drift-mutation equilibrium, assuming the Lynch-Hill (1986) incremental (i.e., Brownian-motion) model. The three levels of mutational heritability, σ^2_m/σ^2_E, span the range of observed values. (**Bottom**) Comparison of the predictions of the Lynch-Hill model (solid line) with that of the Cockerham-Tachida model for three different gametic mutation rates for the trait (dotted and dashed lines), with $h^2_m = 0.001$ in both cases.

It is informative to note the similarity of Equation 11.21a with the expected heterozygosity for sites in drift-mutation equilibrium

$$E(H) = \frac{4N_e\mu_0}{1 + 4N_e\mu_0} \tag{11.21b}$$

where μ_0 is the per nucleotide mutation rate (derived in Chapter 2). A comparison of these two expressions shows that because μ_0 is on the order of 10^{-9} to 10^{-8} (Chapter 4) while $h^2_m \simeq 10^{-3}$, substantial heritability can exist for quantitative traits in populations with low or undetectable levels of molecular heterozygosity.

Lynch and Hill (1986) generalized the preceding results to allow for dominance and linkage. Let $a(1 + k)$ denote the value of a heterozygote. If we assume a population size small enough that no more than two alleles are likely to be segregating simultaneously per locus, and letting $\bar{k}$ and σ_k^2 be the mean and variance of dominance effects, with $k = 0$ implying additivity, the equilibrium levels of additive and dominance genetic variance are

$$\widetilde{\sigma}_A^2 \simeq \frac{2N_e\sigma_m^2(3+2\bar{k})}{3} \tag{11.22a}$$

$$\widetilde{\sigma}_D^2 \simeq \frac{2N_e\sigma_m^2(\bar{k}^2+\sigma_k^2)}{3} \tag{11.22b}$$

Thus, unless new mutations tend to be highly dominant, highly recessive, or highly variable in their dominance effects, most of the genetic variation will be additive in nature. If we assume that there is no overdominance, the bounds on $\widetilde{\sigma}_A^2$ are $(2/3)N_e\sigma_m^2$ and $(10/3)N_e\sigma_m^2$. Although these formulae ignore the fact, discussed above, that dominant mutations in gametic-phase disequilibrium can inflate the genetic variance, this effect only magnifies the variance by a factor of approximately $0.02\overline{ak}/\overline{a^2}$, which is unlikely to be very large (Lynch and Hill 1986).

The preceding results were obtained by use of Kimura and Crow's (1964) **infinite-alleles model**, which postulates that although a large number of alleles need not be segregating at a particular locus at any point in time, each new mutation gives rise to a novel allele. From a quantitative-genetic perspective, the additive effect of each new mutant allele is assumed to equal to that of the ancestral allele plus a random deviate with a mean of zero. Under this **Brownian-motion** (or **incremental**) model of mutational effects, there is no directional change of the mean but also no upper or lower bound on the range of mutational effects. This implies that $\sigma_A^2 \to \infty$ as $N_e \to \infty$.

Taking exception to this assumption, Cockerham and Tachida (1987) assumed a finite number of possible additive allelic states, with new (mutant) allelic effects following the so-called **house-of-cards model** of Kingman (1977, 1978). Under this model, each new mutant allele has a new effect drawn randomly from the distribution of possible effects and independent of the prior state (so that new mutations collapse the previous allelic values like a house of cards). Hence, unlike under the Brownian motion model, allelic effects remain bounded, rather than wandering off to infinity. Under these conditions, the equilibrium genetic variance within a finite population becomes $\widetilde{\sigma}_A^2 = E(H)\sigma_A^2(\infty)$, where $\sigma_A^2(\infty)$ is the equilibrium level of genetic variance expected in a hypothetical population of infinite size, and $E(H)$ is the equilibrium heterozygosity for the loci underlying the trait. Under the Cockerham-Tachida model, $\sigma_A^2(\infty) = nE(a^2)$, and $E(H)$ is defined by Equation 11.21b. Thus, when $4N_e\mu \ll 1$ (with μ being the per-locus rate at which new alleles appear, with $\mu = L\mu_0$ for a locus of length L nucleotides, where μ_0 is the per-nucleotide mutation rate), the Cockerham-Tachida expression for the equilibrium variance is very close to $4N_e\mu nE(a^2) = 2N_e\sigma_m^2$, which is identical to the Lynch-Hill expression.

As it is not likely that mutant alleles will have effects that are entirely independent of their ancestral alleles, nor that mutational effects can grow without bounds, reality must lie between these two extremes. Zeng and Cockerham (1993) presented a **regression model** approach that joins these two limiting cases. They imagined a situation in which the effect of a mutant allele (a_m) is that of a random deviate around a linear regression on the ancestral state (a_0), i.e.,

$$a_m = \tau a_0 + e_a \tag{11.23}$$

where e_a denotes the deviation around the expectation. When $\tau = 1$, Equation 11.23 is equivalent to the Lynch-Hill (incremental) model, whereas when $\tau = 0$, it is equivalent to the Cockerham-Tachida (house-of-cards) model. The general solution to the equilibrium additive genetic variance under this model is

$$\widetilde{\sigma}_A^2 = \frac{2N_e\sigma_m^2}{1+4N_e\mu(1-\tau)} \simeq \begin{cases} 2N_e\sigma_m^2 & \text{for} \quad 4N_e\mu \ll 1 \\[2ex] \dfrac{\sigma_m^2}{2\mu(1-\tau)} & \text{for} \quad 4N_e\mu \gg 1 \end{cases} \tag{11.24}$$

Thus, both the house-of-cards and regression models predict a linear increase in the equilibrium genetic variance with population size so long as N_e is smaller than the reciprocal of the mutation rate to alleles affecting quantitative-trait expression ($4N_e\mu \ll 1$). As noted in Chapter 4, when defined at the nucleotide level, $4N_e\mu$ is generally in the range of 0.001 to 0.05 in eukaryotes. Because a typical protein-coding locus contains $\sim$1000 sites where a nucleotide substitution can result in an amino acid replacement, and regulatory sequences may comprise another 100 to 2000 sites per gene (Lynch 2007), then if we assume that a moderate fraction of such sites yield mutations with phenotypic effects, a typical mutational target size per locus will be on the order of $L \sim 10^3$ sites. Recalling the survey in Chapter 4, the mutation rate per site per generation is $\mu_0 \sim 10^{-9}$ in microbes and a high of 10^{-8} in humans: thus $\mu = L\mu_0$ for a quantitative-trait locus is expected to be in the range of 10^{-6} to 10^{-5}. We revisit these mutational models in detail in Chapter 28.

Finally, due to the randomness of both the drift and mutation processes, the within-population genetic variance is expected to vary considerably around its expectation both among populations of the same size and from generation to generation in the same population. Assuming that there is a large number of unlinked loci, the coefficient of variation of the average within-population genetic variance under the infinite-alleles model is

$$\text{CV}(\widetilde{\sigma}^2_A) \simeq \left[\frac{1}{L}\left(\frac{E(a^4)}{12N_eU[E(a^2)]^2} + \frac{2}{3N_e} + \frac{2}{s}\right)\right]^{1/2} \tag{11.25}$$

where $U = n\mu$ is the gametic mutation rate for the trait, L is the number of lines examined, and s is the sample size per line (Lynch and Hill 1986; Keightley and Hill 1989; Zeng and Cockerham 1991). If the effects of new mutations are approximately normal, with an average of zero, then $E(a^4) = 3[E(a^2)]^2$. Further considering only the true evolutionary variance, and assuming $\mu \ll 1$, the CV for a single line reduces to $\simeq(4N_eU)^{-1/2}$, or the inverse of the square root of twice the effective number of new mutations entering the population per generation. Bürger and Lande (1994) further considered the temporal correlation in σ^2_A over consecutive generations.

Subdivided Populations

In closing, we emphasize that the results given in the previous section apply to the ideal situation in which individual demes are completely isolated from each other. In nature, however, it is common for a total metapopulation to be fragmented into multiple demes held together in a genealogical sense by restricted gene flow. Borrowing from results presented in Chapter 3, we now explore the quantitative-genetic consequences of population subdivision. Throughout, it will be assumed that there is some possible migratory route, either direct or indirect, between all pairs of demes under consideration. In other words, even if two particular demes are incapable of directly exchanging genes, they are assumed to be connected by a corridor through other subpopulations. In this case, at least for characters with an additive genetic basis following the Lynch-Hill model, the average within-deme genetic variance exhibits some remarkably general behavior, although the results for traits with a nonadditive genetic basis remain to be worked out.

Recall Wright's (1951) ideal island model, discussed in Chapter 2, in which the metapopulation consists of d demes, each comprised of an equivalent number (N) of (ideally) randomly mating individuals, with each deme contributing an identical fraction, m, of its genes to a pool of migratory genes. Under this model, with equal exchange rates between all deme pairs, the migration rate from any subpopulation to any other is $m/(d-1)$. In Chapter 2, we noted the geographic invariance principle for this model under neutrality, which indicates that the mean coalescence time between random alleles within a deme is simply equal to $2dN$, independent of the migration rate (Li 1976; Slatkin 1987b; Strobeck 1987; Nagylaki 2000). If we let μ be the per-locus mutation rate, it then follows that the mean number of mutations separating two random alleles is $4dN\mu$, or more generally $4N_T\mu$ with unequal deme sizes (Slatkin 1987b; Strobeck 1987), where $N_T = \sum_{i=1}^{d} N_i$ is the sum of effective sizes of the individual demes. If we note that the contribution of each haploid mutational change to the genetic variance is $E(a^2)/2$, with $2n$ genes contributing to the character, the average within-deme additive genetic variance is

$$\sigma_A^2 = 4N_T\mu \cdot 2n \cdot E(a^2)/2 = 2N_T\sigma_m^2 \tag{11.26}$$

which is identical in form to the expectation for a single isolated deme, $2N_e\sigma_m^2$, but with N_T being substituted for the N_e of a single deme. Depending on the exact population structure, individual demes may have higher or lower equilibrium variances than this quantity, but Equation 11.26 gives the expectation over all demes.

That the preceding result was obtained by a much more detailed route by Lynch (1988b) for two demes and Lande (1992) for an arbitrary number of demes illustrates the substantial utility of results from coalescence theory (Chapter 2) for problems in quantitative genetics involving traits with an additive-genetic basis. As anticipated from the coalescent, with an ideal island structure, the equilibrium additive-genetic variance is not only completely independent of the migration rate (provided that it is nonzero), but most remarkably, it behaves as though the average deme were panmictic, with an effective size of N_T. Although lower migration rates imply a lower rate of replenishment of alleles lost locally by random genetic drift, a greater degree of isolation also increases the level of the interpopulation divergence of alleles, so that a rare immigration event will likely introduce a more substantial allelic variant. Under the ideal island model, these two opposing effects perfectly compensate for each other.

Because the expected coalescence time, $2Nd$, applies to all types of population structures, so long as they allow for migratory routes between all pairs of subpopulations, this result generalizes to situations well beyond the ideal island model, including the **stepping-stone model**, in which migration events are restricted to adjacent demes. Slatkin and Voelm (1991) evaluated the genealogical properties of a population with a hierarchical metapopulation where there are k neighborhoods, each containing d demes, and even this structure yields a result analogous to that presented above. Again, provided that there are potential migratory routes between demes within neighborhoods as well as between neighborhoods, the expected genetic variance within a deme can be shown to be $\sigma_A^2 = 2Nkd\sigma_m^2$, where Nkd is the sum of demic effective population sizes (over all neighborhoods). Thus, we again see that provided the trait has an additive genetic basis, the expected within-population additive variance under neutrality (and assuming the Lynch-Hill model) is $2N_T\sigma_m^2$. This result does assume that gametic-phase disequilibria do not substantially influence the expected standing level of variation, but this is reasonable for a neutral trait, as there will be no tendency for disequilbria to favor coupling over repulsion effects.

These results apply to the variation within single demes, and it is of additional interest to determine the equilibrium features of the entire metapopulation. This requires a measure of among-deme divergence in addition to the within-deme variance, as the total genetic variation in the metapopulation is the sum of the two. This matter is also readily resolved using results from coalescence theory, again assuming a neutral character with an additive genetic basis. Consider, for example, Wright's island model. If two genes are randomly drawn from an entire metapopulation, they will be derived from the same subpopulation with a probability of $1/d$, in which case they will have an average coalescence time of $2Nd$ generations, and from different subpopulations with a probability of $(d-1)/d$, in which case they will have an elevated average coalescence time of $2Nd+[(d-1)/(2m)]$ as a consequence of divergence during isolation (Li 1976). Weighting these two coalescence times by their respective probabilities yields the average coalescence time given, as Equation 2.45b. Again noting that the expected number of mutations separating two alleles is 2μ times the average coalescence time, and that there are $2n$ genes involved, each with respective contributions to the variance of $E(a^2)/2$, the total additive genetic variance for the metapopulation is

$$\sigma_{A,T}^2 = \left(2Nd + \frac{(d-1)^2}{2dm}\right)\sigma_m^2 \tag{11.27}$$

Again, essentially the same result was obtained by Lande (1992) by a more circuitous route.

Thus, for the island model, the within- and among-deme components of additive-genetic variance are equal to $2Nd\sigma_m^2$ and $[(d-1)^2/(2dm)]\sigma_m^2$, respectively. Recalling that $m/(d-1)$ is the genic migration rate per deme, this shows that the among-deme component of genetic variance is inversely proportional to the exchange rate among demes. It is also notable that the among-deme component of genetic variance is completely independent of the sizes of the individual demes. If we assume that there is a large d, the fraction of the total genetic variance associated with the interdemic component is $1/(1+4Nm)$, showing that under this type of population structure, the relative contribution from interdemic variance is low unless the expected number of migrants per deme per generation (mN) is less than one.

Note that Equation 11.27 can be obtained directly from our general expression for the equilibrium additive-genetic variance, $2N_e\sigma_m^2$, using the coalescent-based definition of N_e for the entire metapopulation under the island model, Equation 2.45b. Using this general strategy, the expressions for N_e given in Chapter 3 can be used to obtain results for a variety of other types of population structures. Moreover, as summarized in Lynch (1994), coalescent results can also be used to estimate the genetic variances and covariances for pairs of populations separated by various distances for assessing situations in which migration is spatially restricted. Provided there are possible migratory routes between demes, the expected excess variance between any pair of demes is simply the product of the coalescence time (in excess of the within-deme expectation) and the rate of polygenic mutation, σ_m^2.

12

The Neutral Divergence of Quantitative Traits

From the perspective of the evolutionary biologist, missing or incomplete null or neutral models for many omics data (for example, transcriptome and metabolome data) limit our understanding of how selection has shaped their evolution. Leinonen et al. (2013)

In the preceding chapter, we learned how the opposing forces of random genetic drift and mutation lead to an equilibrium level of within-population genetic variance. In contrast, the phenotypic variance among isolated populations may continue to increase nearly indefinitely for neutral characters, as isolated demes or species recurrently acquire and become fixed for independent mutations. Here, we explore neutral factors that can drive the evolutionary dynamics of the among-population variance. As in Chapter 11, we will start with the situation in which the time span is short enough that most of the change in population-mean phenotypes is driven by drift acting on existing variation rather than by new alleles introduced by mutation. We then explore the consequences of longer-term divergence, with mutation playing an increasingly dominant role, and show that eventually, the rate of divergence for neutral characters may become essentially independent of local effective population sizes. We conclude by using this theory to develop statistical tests of whether an observed pattern of phenotypic divergence is consistent with a model of strict neutral drift and mutation.

Although few quantitative traits may actually evolve in a purely neutral fashion, a more compelling case for selection can always be made if the hypothesis of neutrality can be formally rejected. For example, an observed divergence of isolated lines that is significantly less than the neutral expectation provides evidence of stabilizing selection, whereas the reverse supports a role for diversifying selection. In addition, as populations become diminishingly small in size, drift begins to overwhelm selection, promoting nearly neutral patterns of evolution.

Tests for departures from neutral-trait divergence come in several different forms. First, one can compare the observed rate of divergence in the trait mean with the rate that would be expected given estimates of the effective population size, time, and genetic variation. Second, one may have a time series of data (such as a fossil record sequence) and test whether the observed pattern is more consistent with a neutral random walk, a biased random walk, or stasis. Third, one can compare the within- and between-population structure of the genetic variance of a candidate trait (measured by Q_{ST}) against a genome-wide estimate based on markers that are presumed neutral (F_{ST}). Finally, a number of **trait-augmented marker-based approaches** have been proposed. These use markers from either a QTL or a GWAS study (and hence require a candidate trait to chose the set of markers being examined). Examples include tests based on the distribution of QTL effects detected by crossing two divergent lines and tests using GWAS marker information (such as comparing marker-allele frequencies between two populations).

This highly diverse collection of neutral divergence tests has been applied to an equally diverse collection of traits, ranging from studies of morphological changes in the fossil record to evaluations of evolutionary forces shaping omics data. To highlight the latter, we conclude the chapter by applying several of these approaches to examine if (and if so, how) divergence in gene expression departs from neutrality. Chapters 8–10 considered the complementary topic of tests for departures from neutrality at specific *loci*, as opposed to our focus here, which is specific *traits*.

Evolution and Selection of Quantitative Traits. Bruce Walsh & Michael Lynch.
Published 2018 by Oxford University Press.
DOI 10.1093/oso/9780198830870.001.0001

SHORT-TERM DIVERGENCE

We start with the special case in which all gene action is additive and random genetic drift is the only evolutionary force. Most of the predictions of this model can be expressed in terms of two observable quantities: the additive genetic variance in the base population, $\sigma_A^2(0)$, and the effective population sizes, N_e, of the isolated lineages. The expected among-population genetic variance, $\sigma_B^2(t)$, under neutrality is obtained by noting that the mean genotypic value at a diallelic locus i is $2a_ip_i$ (there being two genes per locus, each with additive effect a_i with probability of p_i, and effect 0 with probability $1-p_i$). The variance among populations for this locus is (from the definition of the variance), $E\{[2a_ip_i(t)]^2\} - \{E[2a_ip_i(t)]\}^2$, which simplifies to

$$4a_i^2\left(E\{[p_i(t)]^2\} - \{E[p_i(t)]\}^2\right) = 4a_i^2\sigma_{p_i}^2(t),$$

where $\sigma_{p_i}^2(t)$ is the expected among-population variance in allele frequency. Summing over all loci, assuming negligible gametic-phase disequilibrium, and substituting from Equation 2.14a yields

$$\sigma_B^2(t) = 4\sum_{i=1}^{n} a_i^2 p_i(0)[1-p_i(0)]\left\{\frac{1}{N_{fo}} + \left[1-\left(1-\frac{1}{2N_e}\right)^t\right]\right\} \tag{12.1a}$$

$$= \left(\frac{1}{N_{fo}} + 2f_t\right)\sigma_A^2(0) \tag{12.1b}$$

where N_{fo} is the effective number of founders per line, the inbreeding coefficient (f_t) follows from Equation 2.4c, and the time index is defined such that $t=0$ denotes the final generation of the base population and $t=1$ denotes the founding generation for the isolated lines. Equation 12.1 shows that, under the assumptions of this ideal model, the expected variance among genotypic means of isolated populations increases linearly with the inbreeding coefficient, asymptotically approaching a limit (as $f_t \to 1$) that is very close to twice the additive genetic variance in the base population (Wright 1951). Under the assumption of additivity, Equation 12.1b holds regardless of the number of alleles at the underlying loci.

If we ignore the generally minor contribution (N_{fo}^{-1}) from the baseline founder effect, this limiting result may be obtained in a simpler manner. Because the probability of fixation of a neutral allele is equal to its initial frequency, when the process of random drift is completed, a proportion $p_i(0)$ of the populations will have genotypic value $2a_i$, while the remaining proportion, $1-p_i(0)$, will have genotypic value 0. The mean genotypic value is, therefore, $2a_ip_i(0)$ and the mean squared value is $(2a_i)^2p_i(0)$, which yields the among-population variance, $4a_i^2p_i(0)[1-p_i(0)] = 2\sigma_{A_i}^2(0)$.

The expression for $\sigma_B^2(t)$ given by Equation 12.1 only considers the true genetic divergence among lines (the **evolutionary variance**), which can, in principle, be obtained by an analysis of variance of phenotypic variation within and among lines. If, however, one simply focuses on the raw variance of the observed means, additional sources of variation associated with finite sample sizes also contribute to the observed divergence (Hill 1972a; Lynch 1988a). For example, when the mean phenotype of each line is estimated using n progeny from $N/2$ matings (involving $N/2$ males and females, for a total parental sample size of N), there can be three additional sources of variance to add to Equation 12.1:

(i) The **segregational variance**, $(1-f_{t-1})\sigma_A^2(0)/(Nn)$, of the mean offspring value about the mean breeding value of their parents resulting from the sampling of $Nn/2$ individuals. This follows because the segregational variance (in the absence of linkage disequilibrium) equals half the additive variance (Chapters 16 and 24);

(ii) The **maternal sampling variance**, $\sigma_{E_m}^2/(N/2)$, associated with any maternal effects resulting from the sampling of $N/2$ mothers;

(iii) The **residual variance**, $\sigma_{E_s}^2/(Nn/2)$, associated with special environmental effects averaged over the entire progeny pool.

Finally, the among-line variances in consecutive generations will be correlated as a consequence of shared ancestry,

$$\sigma_B(t,t') = \left(\frac{1}{N_{fo}} + 2f_t\right)\sigma_A^2(0) \qquad \text{for } 0 < t < t' \tag{12.2}$$

Equation 12.2 assumes no transmission of maternal effects across generations, which, if present, would further inflate this covariance.

A few words should also be said about the potential importance of nonadditive gene action. From Table 11.3, it can be seen that in the presence of dominance, the among-population variance (in the absence of any new mutation) eventually asymptotes at $\sigma_B^2 = 2\sigma_A^2 + 2\sigma_{ADI} + \sigma_{DI}^2$. Thus, dominance can magnify or reduce the among-population variance depending upon the magnitudes of σ_{DI}^2 and σ_{ADI} and on the sign of the latter. In addition, the asymptotic contribution from epistatic interactions involving additive effects is equal to $2^n\sigma_{A^n}^2$ for n-locus epistasis, i.e., $4\sigma_{AA}^2$ for additive $\times$ additive epistasis, and $8\sigma_{AAA}^2$ for additive $\times$ additive $\times$ additive epistasis (again, all based on the base-population values). Thus, epistasis involving large numbers of loci can, in principle, greatly magnify the among-population variance, even if it appears to be of relatively minor importance within the populations.

Sampling Error

We now consider the sampling properties of the among-population genetic variance by reference to a particular experimental design, again assuming a character with a strictly additive basis (Hill 1972a; Lynch 1988a). Starting from a base population with additive genetic variance, $\sigma_A^2(0)$, L replicate lines are isolated and subsequently maintained in each generation with $N/2$ random monogamous matings. Due to the fact that only a finite number of lines is studied, the among-population variance that actually develops in any particular experiment (the **realized variance**), $\widehat{\sigma}_B^2(t)$, will deviate from the expectation (the evolutionary variance), $\sigma_B^2(t)$, given by Equation 12.1b. Moreover, due to finite sample sizes within populations, the among-population variance estimated by the investigator, $V_B(t)$, will further deviate from $\widehat{\sigma}_B^2(t)$. This first source of variation, $\sigma^2[\,\widehat{\sigma}_B^2(t) - \sigma_B^2(t)]$, is a function of population-genetic structure and, for a fixed system of mating, is largely beyond the control of the investigator. The second source of variation, the **sampling variance**, $\sigma^2[V_B(t) - \widehat{\sigma}_B^2(t)]$, arises in estimating $\widehat{\sigma}_B^2(t)$ from the among-line sample variance $V_B(t)$. Its contribution can be minimized by the use of large sample sizes.

Because our concern here is variation in divergence due to genetic changes generated by random drift, we focus on the situation in which the among-line divergence has been measured in such a way as to eliminate nongenetic causes (such as a common-garden experiment designed to remove any environmental trends). Suppose that the same experiment has been repeated many times, on each occasion starting with L lines from the same base population. Due to the variation in the drift process and the finite number of observed lines, each set of experimental lines will develop its own temporal pattern of realized among-population variance. The expected variation in the realized variance among these hypothetical replicate experiments provides a measure of confidence that one can have in the results of any single experiment. If we let $\widehat{\sigma}_B^2(t)$ be the realized among-population variance at generation t for a particular experiment, the expected variance of this quantity among replicate experiments is

$$\sigma^2[\,\widehat{\sigma}_B^2(t)\,] \simeq \frac{4\sigma_A^4(0)}{L-1}\left[\frac{1}{2N_{fo}^2} + 2\left(1 + \frac{1}{N_{fo}}\right)f_t^2 + \sigma_f^2(t)\right] \tag{12.3}$$

Although, in practice, one generally performs a divergence experiment only once, the utility of Equation 12.3 is that it is entirely expressed in terms of observable parameters, so that some idea of the reliability of estimates of $\sigma_B^2(t)$ can be determined in advance. In most situations, the terms in Equation 12.3 involving the founder number (N_{fo}) will be of second or third order and can be ignored.

The variance, $\sigma_f^2(t)$, in the amount of actual inbreeding between individuals in the population requires additional comments. This has been examined in detail in Lynch (1988a), drawing heavily from the results of Weir et al. (1980) and Cockerham and Weir (1983). The theoretical value of $\sigma_f^2(t)$ under different systems of mating is of special interest because empirical studies usually do not record the essential pedigree information for its computation. For freely recombining loci, σ_f^2 is zero when the pedigree structure is fixed, e.g., for obligate selfing, full-sib mating, the maximum avoidance systems of Wright (1921b), and the circular systems of Kimura and Crow (1963a); and even with fairly tightly linked loci, $\sigma_f^2(t)$ is generally negligible in any generation under selfing or full-sib mating. However, under most natural mating schemes, some individuals mate by chance with closer relatives than do others. This results in variation in f among members of the same population, which, because of sampling, accumulates as the among-population variance in f (different lines stochastically accumulate different amounts of inbreeding). For larger population sizes, even with unlinked loci, if the sexes are separate and matings are monogamous, the squared coefficient of variation of $f(t)$, $[\text{CV}(f_t)]^2 = \sigma_f^2(t)/f_t^2$, can attain values of 0.1 to 1.0 in the first two to four generations of isolation, which is enough to contribute significantly to $\sigma^2[\widehat{\sigma}_B^2(t)]$. However, after six or so generations have passed, $\sigma_f^2(t)$ can be safely ignored, regardless of the population size, even with tightly linked loci.

Ignoring the initial founder effect, these results indicate that the coefficient of variation of the among-population variance is $\sqrt{2\{1+[\text{CV}(f_t)]^2\}/(L-1)}$, which is generally on the order of $\sqrt{2/L}$, although in some cases may be as high as $2/\sqrt{L}$. Thus, studies of phenotypic divergence ***need to have very large number of replicates to be statistically reliable***. For example, if it is desirable to reduce the standard error of the among-line variance to 10% of the expectation under the null hypothesis of neutrality and additivity ($\sqrt{2/L} = 0.1$), a minimum of 200 lines should be studied.

One can assess the fit of the additive theory to actual data under two different settings. In the first, we have a single estimate of the among-line variance, and we compare this result to the value expected from theory (as later illustrated in Example 12.2). In the second, we have a series of among-line estimates at different time points, allowing us to consider the temporal pattern of increase in σ_B^2, which, as noted above, should eventually reach a constant (in the absence of mutational input) as $f \to 1$. When a temporal series of $V_B(t)$ is available, these may be regressed on f_t. Under the null hypothesis of neutral additive genes, from Equation 12.1b, the expected slope of such a regression is $2\sigma_A^2(0)$. However, because of shared ancestry, consecutive estimates of mean phenotypes obtained from the same lines are nonindependent (Equation 12.2). This violates a fundamental assumption of ordinary least-squares (OLS) regression analysis, and generalized least-squares (GLS) must be used instead (Chapter 18; LW Chapter 8). For example, once the lines have become completely inbred (and ignoring mutation), all future values of $\widehat{\sigma}_B^2(t)$ must be fixed, and therefore they should not be given equal weight with earlier time points in the regression analysis. The expected covariance of $\widehat{\sigma}_B^2$ between generations with inbreeding levels of f_t and $f_{t'}$ is

$$\sigma[\,\widehat{\sigma}_B^2(t),\, \widehat{\sigma}_B^2(t')\,] \simeq \frac{4\sigma_A^4(0)}{L-1}\left[\frac{1}{2N_{fo}^2} + 2\left(1+\frac{1}{N_{fo}}\right) f_t f_{t'} + \lambda_1^{t'-t}\sigma_f^2(t)\right] \quad \text{for } t < t' \qquad (12.4)$$

where $\lambda_1 = 1 - 1/(2N)$. Lynch (1988a) provided approximate expressions for the standard errors of the slope and intercept that account for the intrinsic correlations in the data, assuming measurements of $V_B(t)$ in progressive generations. Chapter 18 also considers this problem, but in the context of the response in a selection experiment. The variance of the regression coefficient increases with the duration of the experiment, but it is essentially constant after the fourth generation of inbreeding. At that point, the standard error ranges from approximately $4\sigma_A^2(0)/\sqrt{L}$ under obligate self-fertilization to $3\sigma_A^2(0)/\sqrt{L}$ with larger N_e, implying coefficients of variation in the range of $1.5/(f\sqrt{L})$ to $2/(f\sqrt{L})$. For large values of f, these are not greatly different from the sampling variances of single-point estimates noted above.

Sample Variance Confidence Intervals

Given the critical role played by the sample variance in empirical tests of the additive-drift model, we digress here to briefly consider a few statistical issues (construction of confidence intervals and power) related to estimating a variance from a sample, which are used throughout the rest of the chapter. This section may be skipped by both the statistically sophisticated and the casual reader.

Let z_i be the trait value in individual i, with population mean μ and variance σ^2. Provided that individual observations used to estimate a sample variance are uncorrelated and normally distributed with $z_i \sim N(\mu, \sigma^2)$, then (LW Equation A5.14c) for a sample of size n, for the sample estimator $\text{Var} = \sum(z_i - \overline{z})^2/(n-1)$, we have that

$$(n-1)\text{Var} \sim \sigma^2 \chi^2_{n-1} \tag{12.5a}$$

namely, that the variance estimator, Var, is distributed as a scaled chi-square random variable with $n-1$ degrees of freedom. As a result, confidence intervals for the true variance, σ^2, based on the observed sample variance, Var, follow from critical values for a χ^2 distribution. Letting $X_{p,n}$ satisfy $\Pr(\chi^2_n \leq X_{p,n}) = p$, then

$$\Pr(X_{\alpha/2,n} \leq \chi^2_n \leq X_{1-\alpha/2,n}) = 1-\alpha \tag{12.5b}$$

From Equation 12.5a, substituting $(n-1)\text{Var}/\sigma^2$ for χ^2_{n-1}, we have

$$\Pr\left(X_{\alpha/2,n-1} \leq \frac{(n-1)\text{Var}}{\sigma^2} \leq X_{1-\alpha/2,n-1}\right) \tag{12.6a}$$

$$= \Pr\left(\frac{1}{X_{\alpha/2,n-1}} \geq \frac{\sigma^2}{(n-1)\text{Var}} \geq \frac{1}{X_{1-\alpha/2,n-1}}\right) = 1-\alpha \tag{12.6b}$$

yielding

$$\Pr\left[\left(\frac{n-1}{X_{1-\alpha/2,n-1}}\right)\text{Var} \leq \sigma^2 \leq \left(\frac{n-1}{X_{\alpha/2,n-1}}\right)\text{Var}\right] = 1-\alpha \tag{12.6c}$$

This motivates a $(1-\alpha)100\%$ confidence interval for the true variance, σ^2, given the observed sample variance, Var. As shown in Figure 12.1, confidence intervals for σ^2 are asymmetrical about the estimated value, Var, and tend to be quite large, even for modest sample sizes.

A second important issue is power: how likely are we to declare the variance to be statistically distinct from some hypothesized null value, σ_0^2? Intuitively, power increases with both the sample size (n) and the difference between the true and null values. To formally develop this relationship, note from Equation 12.6a that the upper and lower critical values for a $100(1-\alpha)\%$ confidence interval on the value of Var when the true variance is σ_0^2 are

$$\left(\frac{\sigma_0^2}{n-1}\right) X_{\alpha/2,n-1} \leq \text{Var} \leq \left(\frac{\sigma_0^2}{n-1}\right) X_{1-\alpha/2,n-1} \tag{12.7a}$$

If Var falls outside this interval, we declare that the true variance is statistically different (at level α) from the assumed value, σ_0^2. If β denotes the probability that a sample value of Var falls *within* this internal, namely the probability of a type-II error (failing to declare a test significant when the null is false, i.e., when the true variance is σ_1^2), then the power is $1-\beta$. We can obtain β by noting that the true distribution is given by $[(n-1)/\sigma_1^2]\,\text{Var} \sim \chi^2_{n-1}$. Multiplying all terms of Equation 12.7a by $(n-1)/\sigma_1^2$ gives the probability β of a sample variance failing to be declared significant as

$$\beta = \Pr\left[\left(\frac{\sigma_0^2}{n-1}\right)\left(\frac{n-1}{\sigma_1^2}\right) X_{\alpha/2,n-1} \leq \chi^2_{n-1} \leq \left(\frac{\sigma_0^2}{n-1}\right)\left(\frac{n-1}{\sigma_1^2}\right) X_{1-\alpha/2,n-1}\right]$$

$$= \Pr\left[\left(\frac{\sigma_0^2}{\sigma_1^2}\right) X_{\alpha/2,n-1} \leq \chi^2_{n-1} \leq \left(\frac{\sigma_0^2}{\sigma_1^2}\right) X_{1-\alpha/2,n-1}\right] \tag{12.7b}$$

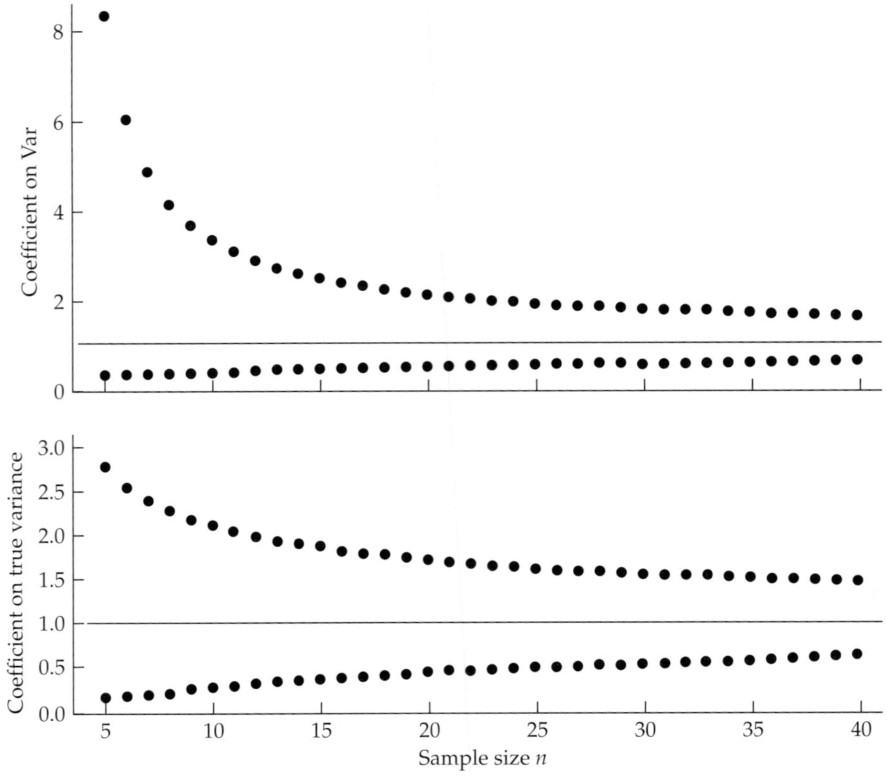

Figure 12.1 Confidence limits and critical values for σ^2 estimated from a sample of n observations. Define $X_{p,n}$ as satisfying $\Pr(\chi^2_n \leq X_{p,n}) = p$. (**Top**) Upper and low values correspond to $[(n-1)/X_{0.025,n-1}]$ and $[(n-1)/X_{0.975,n-1}]$, respectively, the coefficients that multiply the estimated sample variance, Var, to yield a 95% confidence interval for σ^2 (Equation 12.6c). For example, for $n = 10$, the 95% confidence interval for σ^2 is 0.44·Var to 3.33·Var. (**Bottom**) Upper and low values correspond to $X_{0.975,n-1}/(n-1)$ and $X_{0.025,n-1}/(n-1)$, respectively, the coefficients that multiply the assumed variance, σ_0^2, to yield upper and lower 2.5% critical values for an observed sample variance (Equation 12.7a). For example, for $n = 10$, 95% of the values of Var are expected to fall within the interval of $0.30 \cdot \sigma_0^2$ to $2.11 \cdot \sigma_0^2$.

The complement of this event (namely, either less than the lower critical value or greater than the upper critical value) is the power,

$$1 - \beta = \Pr\left[\chi^2_{n-1} \leq \left(\frac{\sigma_0^2}{\sigma_1^2}\right) X_{\alpha/2,n-1}\right] + \Pr\left[\chi^2_{n-1} \geq \left(\frac{\sigma_0^2}{\sigma_1^2}\right) X_{1-\alpha/2,n-1}\right] \tag{12.7c}$$

Example 12.1. Consider a sample variance estimated from $n = 10$ observations. Because $\Pr(\chi^2_9 \leq 2.700) = 0.025$ and $\Pr(\chi^2_9 \leq 19.023) = 0.975$, Equation 12.6c gives the 95% confidence interval ($\alpha = 0.05$) for the true variance σ^2 as between (9/19.023)Var and (9/2.7)Var, or 0.473·Var to 3.333·Var, implying an uncertainty in σ^2 spanning almost a full order of magnitude.

What observed values of the sample variance are unlikely given an assumed true variance of σ_0^2? From Equation 12.7a, the upper and lower critical values (for a two-sided test with α = 0.05) are $(2.700/9)\sigma_0^2 = 0.3 \cdot \sigma_0^2$ and $(19.023/9)\sigma_0^2 = 2.11 \cdot \sigma_0^2$.

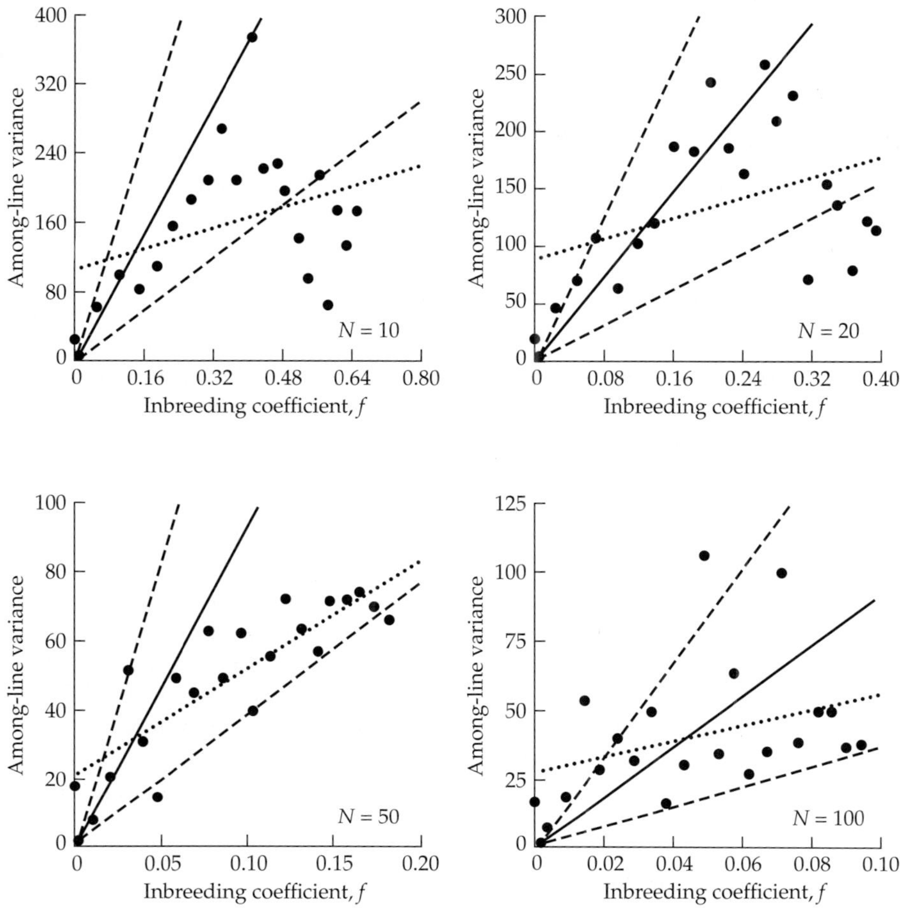

Figure 12.2 Observed and expected levels of the among-population variance for pupal weight in a divergence experiment with the flour beetle *Tribolium*. Solid lines are the expected divergence ($2\sigma_A^2(0)f_t = 920f_t$), dotted lines are the least-squares regressions of the observations, and the paired dashed lines give the approximate 95% confidence interval. (Data from Rich et al. 1984.)

Finally, what is the power of this design to declare the null model (the true variance is σ_0^2) false when the correct variance is $\sigma_1^2 = 2\sigma_0^2$? Taking $\alpha = 0.05$, Equation 12.7b (with $\sigma_0^2/\sigma_1^2 = 1/2$) gives the power as

$$\Pr\left(\chi_9^2 \leq \frac{2.700}{2}\right) + \Pr\left(\chi_9^2 \geq \frac{19.023}{2}\right) = 0.39$$

and hence a type II error rate of 61% when the true variance is twice the assumed variance. A similar calculation assuming $\sigma_1^2 = \sigma_0^2/2$ gives a power of 0.20, or a type II error rate of 80%. Useful `R` commands are (i) `pchisq(x,n)`, which returns $\Pr(\chi_n^2 \leq x)$, and hence `1-pchisq(x,n)`, which returns $\Pr(\chi_n^2 \geq x)$; and (ii) `qchisq(p,n)`, which returns $X_{p,n}$.

Empirical Observations

As an example of the application of the preceding theory, consider the results from a large drift experiment with laboratory cultures of the flour beetle *Tribolium castaneum* (Rich et al. 1984). The authors followed 12 replicate populations at four population sizes (1:1 sex ratio, random mating) over 20 consecutive generations. In each generation, the mean pupal weight (in μg) of each population was obtained from a bulk sample of 100 random individuals.

The additive genetic variance was estimated to be 460 in the base population. The observed values of $V_B(t)$ are plotted as a function of f_t in Figure 12.2, along with the expected divergence, $2\sigma_A^2(0)f_t = 920f_t$ (solid lines). The dashed lines, which were obtained by using Equation 12.3 for the expected variance and substituting this into Equation 12.6c (using $\alpha = 0.05$ and $n = 12$), give the limits of the among-population variance beyond which there is less than a 5% chance for the realization of the drift process under the null to generate these values. Because these bounds ignore measurement error and hence are too narrow, they may be regarded as conservative confidence limits. Nevertheless, almost all of the observations, with the exception of the clusters of the late generations (which are expected to have the largest sampling variances) at $N = 10$ and 20, lie within these limits. The least-squares regressions of the data are given by the dotted lines (more formally, a GLS regression would be used to account for correlated and heteroscedastic residuals; see Chapter 18). The slope of each regression is less than the expected 920, but all slope estimates are within two standard errors of the expectation. Thus, the observed patterns are fairly consistent with a hypothesis of random drift of neutral additive genes. The observed declines in $V_B(t)$ late in the experiment at the two smallest population sizes may have simply arisen by chance and remained there due to intergenerational correlations (Equation 12.2). The results of some other short-term divergence experiments previously given in Figure 11.3 show no evidence for nonlinear increases in the among-population variance with inbreeding.

Eisen and Hanrahan (1974) have argued that the divergence of growth and reproductive rates in inbred lines of mice is more rapid than can be accounted for by the additive genetic variance in the base population, and Bryant et al. (1986a) suggested the same for morphological traits in bottlenecked housefly lines. The implication of these authors is that some nonadditive variance is converted by inbreeding into σ_A^2 (Chapter 11), leading to a faster among-line divergence. However, their designs have low power and in neither case was it verified that the departures from expectations were significant.

Estimating the Among-group Variance

With L replicate populations, a common estimate in the literature for $\sigma_B^2(t)$ is

$$V_B(t) = \frac{1}{L-1}\sum_{i=1}^{L}\left[\,\overline{z}_i(t) - \overline{z}.(t)\right]^2 \tag{12.8a}$$

the sample variance among the sample means, $\overline{z}_1, \cdots, \overline{z}_L$, of the replicate population. When just two populations are being considered (as in some of the tests developed below), their squared difference

$$d^2(t) = \left[\,\overline{z}_1(t) - \overline{z}_2(t)\right]^2 \tag{12.8b}$$

is often used. This is easily related to Equation 12.8a by noting for $L = 2$ that

$$V_B = \frac{1}{2-1}\sum_{i=1}^{2}\left(\overline{z}_i - \frac{\overline{z}_1+\overline{z}_2}{2}\right)^2 = \frac{(\overline{z}_1-\overline{z}_2)^2}{4} + \frac{(\overline{z}_2-\overline{z}_1)^2}{4} = \frac{d^2}{2} \tag{12.8c}$$

These expressions for V_B overestimate the true among-line variance σ_B^2, as the sample means are measured with error. In particular, $\overline{z}_i = \mu_i + e_i$, so that

$$\sigma^2(\overline{z}_i) = \sigma^2(\mu_i) + \sigma^2(e_i) = \sigma_B^2 + \frac{\sigma_z^2}{n} \tag{12.8d}$$

where σ_z^2 is the trait variance and n is the sample sized used to estimate μ_i. When $\sigma_B^2 = 2f_t\sigma_A^2 \gg \sigma_z^2/n$ (which is equivalent to $2f_th^2 \gg 1/n$), the difference between $\sigma^2(\overline{z}_i)$ and σ_B^2 is small.

As suggested by a number of authors (Lynch 1988a, 1990; Turelli et al. 1988; Bjöklund 1991; Savalli 1993), a simple way to avoid this issue is to estimate the among-group variance from a standard one-way ANOVA, with

$$V_B(t) = \frac{MS_B - MS_W}{n_0} \tag{12.8e}$$

using the among- and within-group mean squares (MS_B and MS_W, respectively), with n_0 being a measure of the average sample size per group (see LW Table 18.1).

Lande's Constant Variance Test, F_{CV}

Is an observed divergence over a modest amount of time significantly different than that expected by drift? For the case in which one has only a single estimate of the among-population divergence, Lande (1977b) suggested the statistic

$$F_{CV} = \frac{V_B(t)}{t \cdot \mathrm{V}_A(0)/N_e} \tag{12.9a}$$

as a test for neutrality, where $\mathrm{V}_A(0)$ is an estimate of the base-population additive variance, and t is the number of generations of divergence among the replications. This is the **constant variance** version of Lande's test, as it assumes that the additive variance remains unchanged over the time period being followed (Turelli et al. 1988). As noted by Lande, under approximate assumptions, his test statistic F_{CV} follows an F distribution (LW Appendix 5), which can be shown as follows. If we assume the trait is normally distributed, then so is the sample mean $\overline{z}_i \sim N[\mu(0), \sigma^2_B(t)]$, where we have assumed that terms associated with the sampling variance of the mean are small enough to be ignored (i.e., $\sigma^2_B \gg \sigma^2_z/n$). With L independent lines drawn from a common population at the same time (i.e., a star phylogeny), Equation 12.5a yields

$$V_B(t) = \frac{1}{L-1}\sum_{i=1}^{L}(\,\overline{z}_i - \overline{z}.\,)^2 \sim \frac{\sigma^2_B(t)}{L-1}\chi^2_{L-1} \tag{12.9b}$$

If we ignore the (usually) small founder effect, Equation 12.1b returns

$$\sigma^2_B(t) \simeq 2f_t\sigma^2_A(0) = 2\left[1 - \left(1 - \frac{1}{2N_e}\right)^t\right]\sigma^2_A(0) \simeq t\sigma^2_A(0)/N_e \quad \text{for } t \ll N_e \tag{12.9c}$$

and hence

$$V_B(t) \sim \left(\frac{t\sigma^2_A(0)}{N_e[L-1]}\right)\chi^2_{L-1} \tag{12.9d}$$

Assuming that $\mathrm{Var}_A(0)$ is a good estimate of $\sigma^2_A(0)$, substitution of Equation 12.9d into Equation 12.9a yields

$$F_{CV} \sim \frac{\chi^2_{L-1}}{L-1} \sim F_{L-1,\infty} \tag{12.9e}$$

The last step follows from the definition of an F distribution (LW Appendix 5). Hence, Lande's F_{CV} statistic follows an F distribution with $L-1$ numerator and infinite denominator degrees of freedom. More generally, because $\sigma^2_A(0)$ is estimated by $\mathrm{Var}_A(0)$, the denominator degrees of freedom are those associated with this estimate (e.g., $F_{CV} \sim F_{L-1,df}$, where df is the degrees of freedom associated with the estimate of $\sigma^2_A[0]$). As noted in the previous section, Lande's original test statistic can be improved by using Equation 12.8e to estimate $V_B(t)$.

A couple of approximations were required to reach Equation 12.9e. One check of their validity is that if the distribution of some summary statistic x is given by a scaled-χ^2, with $x \sim \sigma^2\chi^2_n/n$, then the variance of x should equal $2\sigma^4/n$ (as $\sigma^2(\chi^2_n) = 2n$; LW Equation A5.15b). Hence, the numerator of Equation 12.9a should have a variance approximately equal to $2[2f_t\sigma^2_A(0)]^2/(L-1)$. Ignoring the added contribution from sampling error, note that this last result matches Equation 12.3 when N_{fo} is large and $\sigma^2_f(t)$ is small. Thus, Lande's approach should be restricted to lines with at least moderate effective population size. Moreover, as we will see below, all of the preceding formulae for σ^2_B become questionable for $t > N_e$ because they ignore the contribution from new mutations. Hence, Lande's F_{CV} test is best thought of as one for *short-term divergence*, such as would be seen in a laboratory experiment or, at most, a modest amount of time in a set of natural populations.

Example 12.2. Lande (1977b) used Equation 12.9a to evaluate the results of a 12-year divergence experiment involving five populations of *Drosophila pseudoobscura* (Anderson 1973). Two of the populations had been maintained at 25°C, two at 27°C, and one at 16°C. They were then raised in two common environments (16 and 25°C) and measured for wing length. Estimates of the additive genetic variance for these two environments were 0.88 and 0.77, respectively, while the among-population variances were approximately 6.62 and 4.37, respectively. An approximate upper bound for the number of generations of divergence is $t = 150$, whereas the effective population size probably always exceeded $N_e = 1000$. The use of these extreme bounds gives conservative estimates of F_{CV}, making it more difficult to demonstrate diversifying selection on wing length. The resulting values (for the two environments) were

$$F_{CV,1} = \frac{6.62}{150 \cdot 0.88/1000} = 50.15, \quad \text{and} \quad F_{CV,2} = \frac{4.37}{150 \cdot 0.77/1000} = 37.84$$

Given that $\Pr(F_{4,\infty} > 4.6) = 0.001$, both values are highly significant. Thus, the hypothesis that the observed line divergence is solely attributable to random genetic drift can be rejected confidently. More likely, the different thermal conditions resulted in selection for different wing lengths.

LONG-TERM DIVERGENCE

Our previous results were simply concerned with how any initial variation is partitioned among lines during drift/inbreeding. While this is occurring, new variation is constantly being generated by mutation, further driving divergence (Haldane 1949). Polygenic mutation was first incorporated into the theory of population divergence by Dempster (in the appendix to Bailey 1959) and was subsequently studied by Lande (1976), Chakraborty and Nei (1982), and Lynch and Hill (1986). As noted in Chapter 11, the assumed mutational model generating the new mutational effect (a_m) given the ancestral allelic effect (a_0) is critical. The standard model is a version of the infinite-alleles approach, where each new mutation gives rise to a new allele, with $a_m = a_0 + e_a$, namely, the ancestral value plus a random increment, e_a. To distinguish this from the standard infinite-alleles model of Chapter 2 (where the allelic effect of a mutation were not a concern), this approach is also known as the **incremental** or **Brownian-motion** mutational model in the literature; it is examined more fully in Chapter 28.

Again focusing on a character with a purely additive genetic basis, starting with an ancestral-population genetic variance of $\sigma_A^2(0)$, and assuming the infinite-alleles model, the expected variance of genotypic means for replicate populations isolated t generations in the past is

$$\sigma_B^2(t) = 2\sigma_m^2 t + 2\big(\sigma_A^2(0) - 2N_e\sigma_m^2\big)\left(1 - e^{-t/(2N_e)}\right) \tag{12.10}$$

where σ_m^2 is the per-generation mutational rate of input of genetic variance, as described in Chapter 11. This expression shows that as t becomes large, the expected rate of increase of the among-population variance for a neutral quantitative trait becomes a constant $2\sigma_m^2$ per generation. The same formulation applies to the among-species genetic covariance for a pair of traits, if the mutational rate of production of covariance between the traits is substituted for σ_m^2 (Lande 1979a).

Thus, under the infinite-alleles model, the asymptotic divergence rate is independent of the population size, just as it is in the neutral theory of molecular evolution (Chapter 7; Kimura 1983). Although the expected number of new mutations entering a population in each generation is $2Nu$ per locus, the probability of fixation of a new mutation is its initial frequency $1/(2N)$, so (at equilibrium), the expected number of mutations that is fixed per locus,

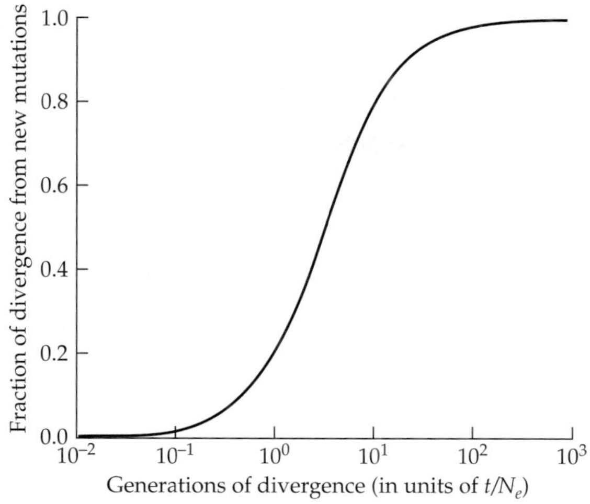

Figure 12.3 The expected fraction of neutral among-population variance attributable to mutations arising subsequent to the isolation event. It is assumed that the base population is in drift-mutation equilibrium, $\sigma_A^2(0) = 2N_e\sigma_m^2$, with the same effective size as the daughter species, so that from Equation 12.1b, the divergence due to base-population variance is $4N_e\sigma_m^2[1 - e^{-t/(2N_e)}]$. To obtain the actual number of generations of divergence for any population size, the horizontal axis is multiplied by N_e.

per population, per generation is simply the mutation rate, u. With each fixed mutation causing an increase in expected among-population variance of $\sim E[(2a)^2]$, and n loci contributing, the asymptotic divergence rate is $nuE[(2a)^2] = 2\sigma_m^2$.

Under the assumptions of the infinite-alleles model, the asymptotic divergence rate of $2\sigma_m^2$ is a fairly general result. It is independent of the degree of dominance of new mutations, the linkage relationships of the constituent loci, and the mating system (Lynch and Hill 1986). This is because both dominance and gametic-phase disequilibrium are transient properties of alleles en route to loss or fixation, and not cumulative phenomena, and because the probability of fixation of a new neutral mutation is equal to its initial frequency regardless of the breeding system.

How long should populations be isolated before one should start to worry about the contribution of new mutations to divergence? From Equation 12.10, it can be seen that this depends on the initial level of genetic variance and on the effective sizes of the derived isolates. In Figure 12.3, it is assumed that the initial base population is in drift-mutation equilibrium, so that $\sigma_A^2(0) = 2N_e\sigma_m^2$, and that the isolated lineages have rapidly attained the same effective sizes (N_e). Under these circumstances, by the time N_e generations have elapsed, polygenic mutation subsequent to the isolation event has caused about 20% of the divergence, whereas for $t > 3N_e$ generations, the majority of the divergence is due to new mutations.

Note that $\sigma_B^2(t)$ is unbounded in t under the infinite-alleles model. However, as emphasized in the preceding chapter (and in Chapter 28), alternatives exist to this mutational model, raising questions about the appropriate structure of a neutral null model. For example, Cockerham and Tachida's (1987) model, which assumes that there are a finite number of alleles with each new allelic state being independent of the prior allelic state (the house-of-cards model), yields a bounded equilibrium among-population variance

$$\widetilde{\sigma}_B^2 = 2[1 - E(H)]\sigma_A^2(\infty) \tag{12.11}$$

where $E(H)$ is the expected heterozygosity per locus, and $\sigma_A^2(\infty) = 2nE(a^2)$ is the expected additive genetic variance in a population of infinite size (Chapter 11). Note that under this model, not only does the among-population variance not build up indefinitely, but because $E[H] = 4N_eu/(1 + 4N_eu) \to 1$ as $N_e \to \infty$, the among-population component of variance asymptotically approaches zero as populations become progressively larger. This is because under the house-of-cards model, replicate populations that are each effectively infinite in size eventually will harbor the same alleles with the same frequency spectrum defined by the mutational interconversion rates.

If nothing else, these dichotomous results indicate that although neutral models are essential to demonstrating the necessity of invoking natural selection to explain an observed pattern of divergence, the actual *construction* of the null model depends on unresolved biological issues. Recall from Chapter 11 that Zeng and Cockerham (1993) proposed a model bridging the infinite-alleles (incremental) and house-of-cards approaches, wherein the effect of a mutant allele is given by $a_m = \tau a_o + e_a$. Taking $\tau = 1$ recovers the incremental (Lynch-Hill) model, while $\tau = 0$ recovers the house-of-cards (Cockerham-Tachida) model. With τ being a measure of the dependency of the mutational effect on the current allelic value, this **regression model** (see Chapter 28 for more details) yields an equilibrium among-population variance of

$$\widetilde{\sigma}_B^2 = \frac{4E(a^2)}{(1-\tau)^2[1 + 4N_e u(1-\tau)]} \tag{12.12}$$

For $\tau < 1$, the temporal approach to the equilibrium level of divergence is defined by the mutation rate (u), assuming an identical N_e in the base and descendant populations,

$$\sigma_B^2(t) = [1 - (1-u)^{2t}]\,\widetilde{\sigma}_B^2 \tag{12.13}$$

and hence is quite slow (approximately $2u$ per generation).

Finally, we note that the expression for the variance of the among-population variance (i.e., the variance of $\sigma_B^2(t)$ among replicate experiments with mutational input) is algebraically complex, and it has only been worked out for the infinite-alleles model (Lynch and Hill 1986). However, if it is assumed that the number of loci is large and the distribution of mutational effects is normal with a mean of zero, the variance of the realized among-population variance approaches $2(2\sigma_m^2 t)^2/L$ for large values of t. This is simply twice the square of the expected among-population variance. Thus, for large t, the coefficient of variation of a realized among-population variance based on L lines is expected to be on the order of $\sqrt{2/L}$, and, we have noted before, unless L is quite large, estimates of $\sigma_B^2(t)$ can deviate quite far from their expectation.

Effectively Neutral Divergence and the Estimation of Rates of Mutational Variance

As discussed in detail in LW Chapter 12, the theoretical expectations of the neutral model provide the basis for estimating the rates of polygenic mutation. Starting from an inbred base population, experimental lines with known times of divergence can be used to estimate the amount of polygenic mutation necessary to account for the distribution of the resultant mean phenotypes. In one of the earliest applications of this approach, Russell et al. (1963) started with several lines of maize that had been maintained by prolonged self-fertilization. They then performed a dichotomous branching experiment for five generations in which each plant was self-fertilized to produce two new daughter sublines. Seed was saved from each generation, so that at the end of the experiment members of all generations could be assayed simultaneously in a common environment, and then sib analysis was used to estimate the additive genetic variance for the *total* population in each generation. Assuming the within-population variance to be in drift-mutation equilibrium, this type of population expansion should give rise to an average rate of increase in the total genetic variance of $2\sigma_m^2$ per generation. In accordance with this prediction, the regressions of the genetic variance on time were positive for every character investigated (Figure 12.4). The rate of polygenic mutation for each of the traits is estimated by one-half the slopes of these regressions.

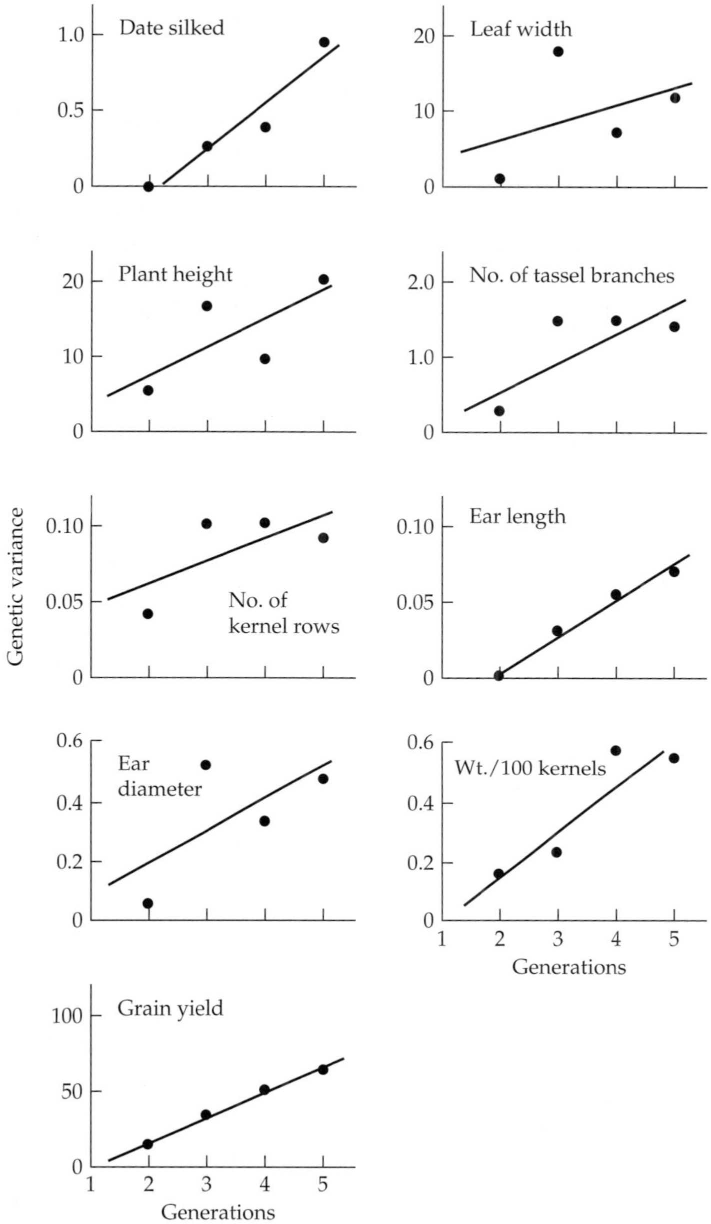

Figure 12.4 The increase in total additive genetic variance (within- plus among-population components) from new mutational variance in an expanding set of lines of corn. See text for details. (After Russell et al. 1963.)

Results from many other experiments of this sort were reviewed in LW Chapter 12. Although a number of additional results have emerged since then, most of these are confined to a small number of model systems, and the conclusions reached in our earlier review remain unaltered. Here, we simply give a brief update, providing references only to post-1998 papers. Most estimates are framed in terms of the mutational heritability, $h_m^2 = \sigma_m^2/\sigma_e^2$. Estimates of h_m^2 for a diversity of morphological, physiological, and life-history traits in *D. melanogaster* are consistently in the range of 0.001 to 0.005. Mutational heritabilities for body size and life-history traits in nematodes fall in the range of 0.001 to 0.008 (Vassilieva et al. 2000; Baer et al. 2006; Ostrow et al. 2007), and the same is true for life-history traits in the microcrustacean *Daphnia pulex* (Lynch et al. 1998; Latta et al. 2013) and in the grape phylloxera insect *Daktulosphaira vitifoliae* (Downie 2003). Thus, essentially all studies with invertebrates imply $0.001 < h_m^2 < 0.01$ for complex traits.

Although the numbers of studies are still rather limited, estimates of h_m^2 for some land plants and vertebrates appear to be several-fold higher than those noted above. Mutational heritabilities for growth and reproductive traits in *Arabidopsis thaliana* are in the range of 0.001 to 0.008 (Schultz et al. 1999; Shaw et al. 2000; Chang and Shaw 2003; Kavanaugh and Shaw 2005), but the average h_m^2 for maize, from the study of Russell et al. (1963), is 0.0092. In addition, mutational heritabilities for morphological and reproductive traits in mice fall in the range of 0.003 to 0.023 (Casellas and Medrano 2008). Thus, there is at least a rough indication that mutational heritabilities are increased in organisms with longer life spans, which might in principle be a consequence of elevated rates of mutation per generation (Chapter 4).

Finally, it should be emphasized that in all mutation-accumulation experiments, fitness declines in the vast majority of lines, indicating that mutations are, on average, deleterious, although the fraction of mutations that are beneficial remains unclear (Shaw et al. 2002; Keightley and Lynch 2003; Charlesworth and Eyre-Walker 2007; Eyre-Walker and Keightley 2007; Dickinson 2008; Hall et al. 2008). Equally important, for characters that influence fitness only indirectly (e.g., morphology), the fraction of new mutations with negative pleiotropic effects on fitness remains unclear (Chapter 28). Hence, estimates of h_m^2 from mutation-accumulation experiments with their very small effective population sizes may *overestimate*, ***perhaps significantly***, the actual *usable* amount of h_m^2 for most populations. Even if the focal trait is neutral, if some of its underlying mutants have deleterious pleiotropic effects on fitness (i.e., fitness impacts *independent* of the actual value of the focal trait), then the effective number of neutral mutations in the trait changes with effective population size. As N_e increases, this fraction decreases, so that the effective value of h_m^2 is likely a decreasing function of N_e. What is unclear is whether this decline plateaus out fairly quickly or continues to decrease over a large range of N_e. Resolving these issues is critical to any attempts to utilize estimates of mutational heritability to infer long-term mechanisms of evolution, as illustrated in the following sections.

TESTING THE NULL HYPOTHESIS OF NEUTRAL PHENOTYPIC DIVERGENCE: RATE-BASED TESTS

One of the enduring problems in evolutionary biology is the struggle to demonstrate that various aspects of biodiversity are products of diversifying selection. It is one thing to concoct plausible adaptive scenarios to explain patterns of morphological, physiological, or behavioral divergence, but quite another to formally demonstrate that an observed level of divergence cannot simply be explained by a null model of random genetic drift. This is especially the case with the massive volume of omics data flooding out of large-scale functional genomics studies.

Four broad classes of tests for departures from neutral divergence have been proposed, each requiring successively more information. The first (**rate tests**) simply require the means of divergent populations, and ask (given N_e, t, and either h^2 or h_m^2) whether the rate of divergence is consistent with drift. A second approach is based entirely on the internal information in a temporal sequence of trait values, typically from the fossil record. A third class requires information on the additive variation of a trait and the frequencies of a set of markers in a subdivided population. Such tests compare the within- and among-population structure of the additive variation of a trait (measured by Q_{ST}) with that from some presumably neutral set of markers (measured by F_{ST}; Chapter 2). The final class, trait-augmented marker-based tests, examine patterns in the set of markers associated with a specified trait detected using either a QTL (fixed differences) or GWAS (segregating alleles) study.

A genomics-focused researcher may (incorrectly) get the impression that despite the use of marker information in some of these tests, none of these four approaches are relevant to their work. Nothing could be further from the truth. For example, methods initially developed to test for neutrality in changes in morphology (often over a fossil record) are now widely applied to omics data. To highlight this transition from morphology to molecules, we conclude this chapter by applying these tests to gene-expression data to examine if such divergence is largely governed by neutral drift, stabilizing selection, or directional selection.

Rate Tests

Lande's F_{CV} (Equation 12.9a) is one example of a rate test, comparing the amount of divergence across a set of lines or populations with the expected among-population variance (σ_B^2). As mentioned, this test is best applied over short time scales ($t \ll N_e$), as in a selection experiment or over a short to modest amount of time in nature.

Here we focus on tests over potentially much longer time scales and ask whether an observed amount of total divergence, $d = |\mu(t) - \mu(0)|$, within a single lineage is excessively large (or small) relative to drift. Tests for unusual amounts of divergence (either too much, or too little) are framed by asking what critical values for an effective population size (N_e) or mutation variance (σ_m^2) are consistent with the amount of divergence and whether these values are biologically reasonable. For example, a smaller effective population size and/or higher mutation variance results in increased divergence. Based on the observed rate of divergence, rate tests return either a critical effective population size $N_e(d)$, above which the divergence is unlikely due to drift, or a critical mutational variance, $\sigma_m^2(d)$, below which divergence is unlikely from drift. Likewise, either large population size or small mutation variance will constrain neutral divergence, yielding a critical value of $N_e(s)$ below which the lack of divergence observed is unlikely, and conversely a critical mutation variance, $\sigma_m^2(s)$, above which lack of divergence is unlikely. Hence, if $N_e > N_e(d)$ or $\sigma_m^2 < \sigma_m^2(d)$, the amount of divergence is excessive relative to drift, consistent with directional selection; while if $N_e < N_e(s)$ or $\sigma_m^2 > \sigma_m^2(s)$, there is too little divergence relative to the drift prediction, consistent with stabilizing selection. If none of these conditions is satisfied, the hypothesis of drift alone accounting for the pattern is not rejected. It should be stressed that situations exist in which considerable selection shapes the observed pattern, and yet we can still fail to reject the drift model.

While these tests are widely used, they have several important caveats. For any analysis of this sort to be meaningful, one must be confident that the majority of population divergence is genetic, and not inflated by environmental effects on phenotypes. This is clearly problematical when populations cannot be assayed in a common-garden environment, as is the case with data from the fossil record. A further complication is that expressions for critical effective population sizes or mutational variances ignore the sampling error of all other terms. Finally, as we have discussed, the infinite-alleles model for neutral-trait evolution (where σ_B^2 is an ever-increasing function of time; Equation 12.10) may be too extreme, as the usable amount of σ_m^2 likely decreases with increasing N_e. The among-population divergence under mutation might be better described by either the House-of-cards (Equation 12.11) or the regression (Equation 12.12) models, that by the incremental model. Given all these considerations, it should be clear that the following tests for neutrality cannot be regarded as being very rigorous in a statistical sense. As is the case for most of the tests for selection on specific genes covered in Chapters 9 and 10, tests of neutral phenotypic evolution are best employed as diagnostic guides to prioritize traits for further study.

Lande's Brownian-motion Model of Neutral Trait Evolution

The basic structure of tests for neutral trait divergence have the form of $\mu_t \sim (\mu_0, \sigma_B^2[t])$, namely, the mean phenotype at time t has an expected value equal to the initial mean

μ_0 and variance $\sigma_B^2(t)$. To proceed further, we need additional assumptions about the actual *distribution* from which the means are sampled, which is generally assumed to be Gaussian (normal). Support for this assumption traces back to Lande (1976), who framed mean divergence in terms of a **Brownian-motion process** (Appendix 1). Under the simplest Brownian-motion model, the expected change in value over a small time interval is zero with a constant variance of b. Under this model, the distribution of values at time t is normal with mean x_0 (the initial value) and variance $\sigma_t^2 = bt$ (Equation A1.31b). Assuming a strictly additive model with no environmental trends, Lande noted that if we sample N_e individuals, their mean breeding value (i.e., the trait mean) will have a sampling variance in each generation of σ_A^2/N_e, which is used for b. Hence, at generation t, the distribution of phenotypic means is approximately normal with mean μ_0 (the initial mean) and variance

$$\sigma_t^2 = t\sigma_A^2/N_e \tag{12.14a}$$

This approach assumes a constant additive genetic variance as well as a constant effective population size during the period of divergence being considered. Because drift can also change σ_A^2, the assumption of a constant σ_A^2 is reasonable only for $t \ll N_e$, unless the initial variance is close to its mutation-drift equilibrium value ($2N_e\sigma_m^2$; see Equation 11.20b). More generally, if the additive variance and N_e are both changing in each generation, then under the Brownian-motion model (taking the first time point at $i = 0$ and the last at $i = t - 1$, bringing us up to generation t),

$$\sigma_t^2 = \sum_{i=0}^{t-1} \left(\frac{\sigma_A^2(i)}{N_e(i)} \right) \tag{12.14b}$$

For example, assuming a constant effective population size, the additive genetic variance at time t under drift and mutation is given by Equation 11.20b. Substituting this into Equation 12.14b yields

$$\sigma_t^2 = \frac{1}{N_e} \sum_{i=0}^{t-1} \left[2N_e\sigma_m^2 + \left(\sigma_A^2(0) - 2N_e\sigma_m^2\right) e^{-i/2N_e} \right]$$

$$= 2\sigma_m^2 t + \left(\sigma_A^2(0) - 2N_e\sigma_m^2\right) \left(\frac{1}{N_e} \sum_{0=1}^{t-1} e^{-i/2N_e} \right) \tag{12.14c}$$

By noting that

$$\frac{1}{N_e} \sum_{i=0}^{t-1} e^{-i/2N_e} \simeq 2\left(1 - e^{-t/2N_e}\right) \tag{12.15a}$$

recovers the previous expression (Equation 12.10) for σ_B^2 under drift and mutation. This useful identity follows by recalling that the partial sum of a geometric series is

$$\sum_{i=0}^{k-1} x^i = \frac{1 - x^k}{1 - x}. \tag{12.15b}$$

Taking $x = e^{-1/2N_e}$ and noting from a first-order Taylor series that

$$1 - e^{-1/2N_e} \simeq 1 - \left(1 - \frac{1}{2N_e}\right) = \frac{1}{2N_e} \tag{12.15c}$$

returns Equation 12.15a.

Thus, as expected, the variance (σ_t^2) of the Brownian-motion process corresponds to the among-group drift variance $\sigma_B^2(t)$ in population means. The notion that (under a pure drift model) the additive variance within populations reaches a drift-mutation equilibrium value has resulted in different parameterizations of σ_t^2 in tests of drift (Lande 1976; Turelli et al. 1988). Lande assumed a constant variance, $\sigma_t^2 = t\sigma_A^2/N_e$, but because part of his concern was evolution in the fossil record, he replaced σ_A^2 by $h^2\sigma_z^2$, yielding

$$\sigma_t^2 = h^2\sigma_z^2 t/N_e \tag{12.16a}$$

His logic was that σ_z^2 could be estimated directly from a sample in the fossil record, while h^2 values for many morphological traits fall within a relatively narrow range. Hence, either a representative value for h^2 could be used, or different values of h^2 could be explored to examine the robustness of any conclusions. This results in a test based on joint considerations of N_e and h^2.

Conversely, Turelli et al. (1988) noted that if the population has been at its current size sufficiently long enough that the additive genetic variance is at its mutation-drift equilibrium value, then (assuming the infinite-alleles model) $\sigma_A^2 = 2N_e\sigma_m^2$, yielding

$$\sigma_t^2 = 2tN_e\sigma_m^2/N_e = 2t\sigma_m^2, \tag{12.16b}$$

Under this setting, N_e does not appear in σ_t^2 and tests are based on whether the values of σ_m^2 required to be consistent with drift are plausible. This is the quantitative-trait analog of the divergence ($d = tu$) between two neutral sequences separated by t total generations (Chapter 7), with the *allelic* mutation rate (u) replaced by twice the *trait* mutational variance, $2\sigma_m^2$. It needs to be stressed, however, that this apparent independence of N_e is a bit misleading. As mentioned previously, some fraction of its underlying variation may be due to alleles with deleterious pleiotropic effects on fitness (Chapter 28). Any such alleles are expected to have ever-decreasing frequencies (and hence less impact on h_m^2) as N_e increases. Hence, the evolutionary relevant fraction of h_m^2 very likely declines with N_e, *even for a neutral trait*, resulting in less neutral divergence than expected.

Tests Based on the Brownian-motion Model

Under the Brownian-motion model, the mean phenotype $\mu_t \sim N(\mu_0, \sigma_t^2)$, providing the basis for tests of either too much, or too little, divergence based on simple normal theory. Suppose an absolute divergence of $d = |\,\mu(t) - \mu(0)\,|$ is observed, where $\mu(t)$ and $\mu(0)$ are the means from two samples from the same population taken t generations apart. The probability of this level of divergence under drift alone is given by

$$\Pr\left(\,|\,\mu(t) - \mu(0)\,| \leq d\right) = \Pr\left(\frac{|\,\mu(t) - \mu(0)\,|}{\sigma_t} \leq \frac{d}{\sigma_t}\right) = \Pr\left(|\,U\,| \leq \frac{d}{\sigma_t}\right) \tag{12.17}$$

where U is a unit normal random variable. Lande's (1976) original test (distinct from his 1977 F_{CV} test; Equation 12.9a) was based on the constant variance assumption, $\sigma_t^2 = th^2\sigma_z^2/N_e$. Recalling that $\Pr(|\,U\,| \leq 1.96) = 0.95$, Lande's critical effective population size below which there is a $< 5\%$ probability of an absolute deviation as large as d satisfies

$$1.96 = \frac{d}{\sqrt{th^2\sigma_z^2/N_e}}, \quad \text{implying} \quad (1.96)^2 th^2\sigma_z^2 = N_e d^2 \tag{12.18a}$$

Equation 12.18a allows one to determine critical values for either divergence time, t, heritability, h^2, or N_e that are consistent with drift. For example, solving for the upper bound, $N_{e,u}$ on the effective population size that is compatible with drift yields

$$N_{e,u} = \frac{t \cdot h^2 \cdot 1.96^2}{d_*^2} = 3.84 \cdot \frac{t\,h^2}{d_*^2} \tag{12.18b}$$

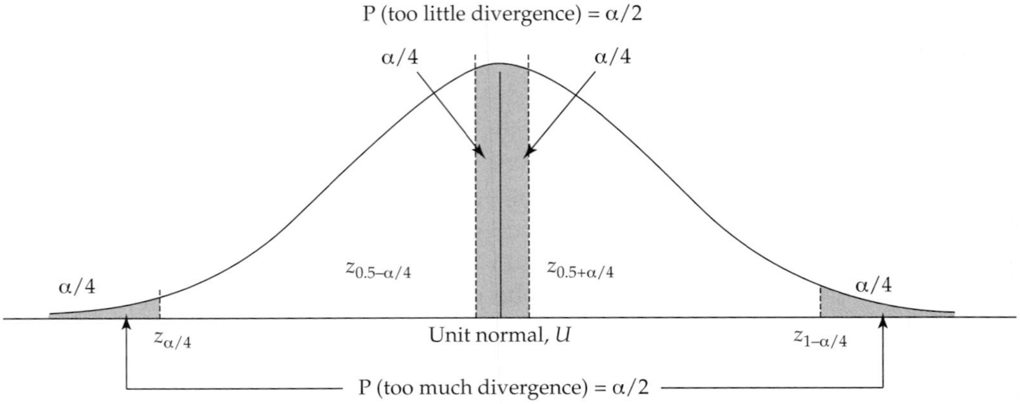

Figure 12.5 Critical values for an α-level test of a departure from drift having either too little, or too much, absolute divergence. Too much absolute divergence occurs when the unit-normal scaled test score is either in the lower $\alpha/4$ or upper $\alpha/4$ tail (for a total probability of $\alpha/2$). Too little absolute divergence occurs when the unit–normal scaled test score is too close to zero, namely, a region of probability $\alpha/4$ below zero and a region of probability $\alpha/4$ above zero (for a total probability of $\alpha/2$). Here, z_p satisfies $\Pr(U \leq z_p) = p$, where U is a unit-normal random variable. See the text for further details.

where $d_* = d/\sigma_z$ is the divergence scaled in phenotypic standard deviations. Drift with $N_e > N_{e,u}$ is unlikely to generate the observed amount of divergence. For a test with an arbitrary α, one replaces 1.96 by $z_{1-\alpha/2}$, where z_p satisfies $\Pr(U \leq z_p) = p$ for a unit normal U ($\alpha/2$ is used, as we are considering an absolute difference). Likewise, if one is comparing the means of two species with a common ancestor τ generations ago, then $t = 2\tau$ and d is the absolute difference between their means. For historical reasons, the above discussion has been framed in terms of d^2. Recalling Equation 12.8e, more generally, the ANOVA-based estimate V_B of the between-group variance can be used in place of $d^2/2$.

As noted by Turelli et al. (1988), the population-size test for departures from drift given by Equation 12.18b is really *two-sided*. Lande's original test examines whether N_e is too large to account for the observed divergence (as might occur if directional selection was changing the mean). However, any formal test of departures from neutral trait drift must also inquire as to whether the *stability* of population means is too great to be compatible with neutrality (too little divergence). For a two-tailed test of neutrality with an $\alpha = 5\%$ overall significance level, we use a 2.5% probability cutoff for the observed divergence being too small to be consistent with drift and a 2.5% cutoff for excessively high divergence. Because $\Pr(|U| \geq 2.24) = 0.025$ for a unit–normal random variable U, the critical upper bound, $N_{e,u}$, on population size in a test that evolution has been too fast for drift is

$$N_{e,u} \leq \frac{t \cdot h^2 \cdot 2.24^2}{d_*^2} = 5.02 \cdot \frac{t\,h^2}{d_*^2} \tag{12.19a}$$

Because populations with smaller N_e should show more drift (and divergence), Equation 12.19a gives the largest value of N_e that is consistent with drift generating the observed amount of divergence. If the assumed N_e exceeds $N_{e,u}$, we reject the hypothesis that drift can account for this fast a divergence. Likewise, because $\Pr(\,|U|\, < 0.03) = 0.025$, the critical lower-bound population size $N_{e,l}$ in a test that evolution has been too slow (support for stabilizing selection) is

$$N_{e,l} \geq \frac{t \cdot h^2 \cdot 0.03^2}{d_*^2} = 0.0009 \cdot \frac{t\,h^2}{d_*^2} \tag{12.19b}$$

If our assumed N_e is *less* than $N_{e,l}$, we reject the hypothesis that drift can account for this slow a divergence.

More generally, for a two-sided test at overall significance level α ($\alpha/2$ for too much absolute divergence and $\alpha/2$ for too little absolute divergence), the above values of 2.24 and 0.03 are replaced by $z_{1-(\alpha/4)}$ and $z_{0.5+\alpha/4}$ (Figure 12.5). To see how these critical values arise, first consider the probability of excessive absolute divergence, which means that either the difference between means is too negative or too positive. Because $\alpha/2$ is the critical value for either of these two events occurring, we set the negative lower limit (the difference between means is too negative) to occur with probability $(\alpha/2)/2 = \alpha/4$ (i.e., the probability of divergence in the lower tail is less than $\alpha/4$), and likewise set the upper positive limit (the difference between means is too positive) also at $\alpha/4$. Given the symmetry of the normal, we can compactly express the total probability for excessive divergence in either direction as

$$\Pr\left(|U| \geq z_{1-(\alpha/4)}\right) = \alpha/2$$

Turning to tests of too little divergence, instead of focusing on the tails of the normal, we focus around its mode (its mean, which corresponds to $z_{0.5}$), with a section of probability $\alpha/4$ below the mean and a corresponding section above the mean, where the total area in the region of too little divergence corresponds to $\alpha/2$. Putting these together,

$$\Pr\left(|U| \leq z_{0.5+\alpha/4}\right) = \alpha/2$$

Figure 12.5 illustrates this logic.

Example 12.3. Reyment (1982) observed a change of $1.49\sigma_z$ in the size of a Cretaceous foraminifer over roughly $5 \cdot 10^5$ generations. Taking a typical heritability value of 0.3, Equation 12.18b (i.e., assuming a one-sided test, namely, a test only for excessive divergence) gives the upper bound, $N_{e,u}$, on population size consistent with this amount of divergence as

$$N_{e,u} = 3.84 \cdot \frac{t\,h^2}{d_*^2} = 3.84 \cdot \frac{5 \cdot 10^5 \cdot 0.3}{1.49^2} \simeq 260{,}000$$

However, paleontological data suggest that the census population size $N \gg 10^6$, implying that drift was unlikely to account for such a rapid divergence (of course, caveats from Chapter 3 apply in that usually $N_e \ll N$). Assuming h^2 values of 0.5, 0.7, and 1.0 yields $N_{e,u}$ values of 433,000, 607,000, and 867,000 respectively, so that only for assumed h^2 values close to one does the critical N_e under drift approach the assumed size of $N_e > 10^6$.

Using the two-sided test (strict departure from drift, either too little or too much absolute divergence), the value of the 3.84 used above is replaced by 5.02 (Equation 12.19a), resulting in an ~31% (5.02/3.84 = 1.307) increase in $N_{e,u}$ value, with critical values of $\sim$ 340,000, 566,000, 794,000, and 1,133,000 for h^2 values of 0.3, 0.5, 0.7, and 1.0, respectively. Similarly, the lower critical $N_{e,l}$ (the size below which the lack of divergence is too improbable under drift) is 61 (using Equation 12.19b with $h^2 = 0.3$).

The structure of the tests given by Equations 12.17 through 12.19 depends on N_e and h^2. A second approach is to instead base tests on the mutational variance, σ_m^2, alone. The idea is that if N_e has been roughly constant for a sufficient amount of time, then the additive-genetic variance for a neutral trait approaches its mutation-drift equilibrium value, $2N_e\sigma_m^2$ (Equation 11.20c). Under this condition, Equation 12.16b shows that the among-group variance becomes $\sigma_B^2 = 2t\sigma_m^2$, giving the **MDE** (mutation-drift equilibrium) version of Lande's F test (Equation 12.9a) as

$$F_{MDE} = \frac{V_B(t)}{2t\sigma_m^2} \tag{12.20a}$$

We can also arrive at this test by substituting $2N_e\sigma_m^2$ for $V_A(0)$ in Equation 12.9a. As above, V_B is best estimated from the among-group variance in a one-way ANOVA (Equation 12.9f).

When V_B is based on more than two lineages, Equation 12.20a assumes a star phylogeny (Chapter 8). If the phylogeny is more complex, one has to place the lineage relationships into a phylogenetic framework to account for the covariance structure imparted by shared common ancestry (Felsenstein 1985, 2004, 2008; Lynch 1991; Gu 2004). Using the same logic leading to Equation 12.9e with L lineages (under a star phylogeny), Equation 12.5a gives $(L-1)V_B(t) \sim 2t\sigma_m^2 \chi^2_{L-1}$, and Equation 12.6c yields

$$\Pr\left[\left(\frac{L-1}{X_{1-\alpha/2,L-1}}\right) V_B(t) \leq 2t\sigma_m^2 \leq \left(\frac{L-1}{X_{\alpha/2,L-1}}\right) V_B(t)\right] = 1-\alpha \tag{12.20b}$$

where $X_{p,n}$ satisfies $\Pr(\chi_n^2 \leq X_{p,n}) = p$. Thus, if an estimate of t is available, one can test the neutral hypothesis without an estimate of N_e by inquiring whether there has been too little or too much divergence given some assumed value of σ_m^2 (Turelli et al. 1988). We can also frame the test in terms of the mutational heritability $h_m^2 = \sigma_m^2/\sigma_e^2$ by dividing all terms in Equation 12.20b by $2t\sigma_e^2$, yielding

$$\Pr\left[\left(\frac{L-1}{X_{1-\alpha/2,L-1}}\right) \frac{V_B(t)}{2t\sigma_e^2} \leq h_m^2 \leq \left(\frac{L-1}{X_{\alpha/2,L-1}}\right) \frac{V_B(t)}{2t\sigma_e^2}\right] = 1-\alpha \tag{12.20c}$$

A slightly different formulation of this test is based in terms of the observed rate of divergence (Lynch 1990). Letting $\Delta = (V_B/t)/\sigma_e^2$ be the estimated rate of divergence scaled in units of the environmental variance, Equation 12.20c becomes

$$\Pr\left[\left(\frac{(L-1)/2}{X_{1-\alpha/2,L-1}}\right) \Delta \leq h_m^2 \leq \left(\frac{(L-1)/2}{X_{\alpha/2,L-1}}\right) \Delta\right] = 1-\alpha \tag{12.20d}$$

yielding the upper and lower bounds on the mutational heritability h_m^2 consistent with drift. For $\alpha = 0.05$ and $L = 2$, Equation 12.20b becomes

$$\Pr\left(0.10 \cdot \Delta \leq h_m^2 \leq 509 \cdot \Delta\right) = 0.95 \tag{12.20e}$$

Thus, the hypothesis of drift is rejected (at $\alpha = 0.05$) if the mutational heritability is too small to account for the observed divergence rate, namely

$$h_m^2 < 0.10 \cdot \Delta \simeq 0.10 \cdot \frac{d_*^2}{t} \tag{12.21a}$$

Just as a smaller N_e allows for more divergence (and hence we set a critical *upper* value for N_e in Equation 12.18a), so does a larger mutational heritability, h_m^2, and we set a critical *lower* value, above which drift can account for the observed divergence. Given that a typical upper-range value is $h_m^2 = 0.05$ (LW Table 12.1), if $0.05 < 0.10 \cdot \Delta$ (i.e., $\Delta > 0.5$), the rate of divergence is too high to reasonably be accounted for by drift.

Conversely, the divergence is too slow to be accounted for by drift if the assumed mutational heritability is too high to account for the observed divergence rate, or when

$$h_m^2 \geq 509 \cdot \Delta \simeq 509 \cdot \frac{d_*^2}{t} \tag{12.21b}$$

A mutational heritability above this value would lead to significantly more divergence than observed. There is one minor bookkeeping detail with both Equations 12.21a and 12.21b. The careful reader might recall Equation 12.8c, where we showed that $V_B = d^2/2$. So why did we assume that $\Delta \simeq d_*^2/t$ in these two equations? Recalling that $\sigma_e^2 = (1-h^2)\sigma_z^2$, we have

$$\Delta = \frac{V_B}{t\sigma_e^2} = \frac{d^2}{2t(1-h^2)\sigma_z^2} = \frac{1}{2(1-h^2)}\frac{d_*^2}{t} \simeq \frac{d_*^2}{t} \tag{12.21c}$$

with the last step following when $h^2 \simeq 1/2$.

One important caveat for σ_m^2-based tests of stabilizing selection is that estimates of h_m^2 are obtained using very small effective population sizes, and hence most mutations are likely effectively neutral in these settings (Chapter 7). Because even a completely neutral trait may have underlying loci with deleterious pleiotropic fitness effects, the evolutionarily relevant fraction of h_m^2 may be far less than that suggested from laboratory experiments. Hence, using laboratory estimates for the polygenic mutation rate may generate a considerable number of false positives for stabilizing selection (the mutation rate is too large relative to the small amount of divergence), so this test should be used with considerable caution.

Example 12.4. We now return to Reyment's foraminifer data from Example 12.3. Using the original Lande model (Equation 12.18b), we rejected the hypothesis that drift could have accounted for the divergence. Applying Equation 12.21a, the hypothesis of drift accounting for excessive divergence is not rejected when

$$h_m^2 > 0.10 \cdot \frac{d_*^2}{t} = 0.10 \cdot \frac{1.49^2}{5 \cdot 10^5} = 4.4 \cdot 10^{-7}$$

This critical value of the mutational heritability is several orders of magnitude lower than typical values of this parameter, implying that this pattern of divergence is not too excessive for drift.

Thus, we reach two very different conclusions depending on whether the constant variance (Example 12.3) or equilibrium variance (Example 12.4) assumption is used. Which is the better choice? In our view, the constant variance assumption (Equation 12.16a) is less problematic, as the usable amount of σ_m^2 and h_m^2 may *decrease* with increasing N_e. In such cases, $2N_e\sigma_m^2/N_e$ may *not* be a constant over N_e, complicating tests based on critical mutational variances. Conversely, most trait heritabilities typically fall within a modest window of values, and one can vary the assumed value of h^2 in Equation 12.16a to examine its consequences.

Ornstein-Uhlenbeck Models

As developed in Appendix 1, the **Ornstein-Uhlenbeck (OU) process** provides a model of Brownian motion drift coupled with a restoring force back to some optimal value (θ), as might be expected with drift and stabilizing selection. This process has been used to model the divergence of traits over a phylogeny (Felsenstein 1988, 2004, 2008; Garland et al. 1993; Martins 1994; Hansen and Martins 1996; Hansen 1997; Martins and Hansen 1997; Butler and King 2004; Beaulieu et al. 2012), including gene-expression data (Bedford and Hartl 2009; Kalinka et al. 2010; Brawand et al. 2011; Rohlfs et al. 2014).

Under the OU model, the expected change in the mean value of a process at a value of x is $a(\theta - x)$ with $a > 0$, so that if $x < \theta$, it increases, while it decreases for $x > \theta$. The parameter a, which measures the strength of the restoring force, is a measure of the strength of stabilizing selection. Under the standard model of Gaussian stabilizing selection (Example 5.6; Equation 16.17), where ω^2 measures the strength of selection (smaller ω^2 implies stronger selection), Example A1.13 shows that

$$a = \frac{\sigma_A^2}{\sigma_z^2 + \omega^2} \tag{12.22a}$$

As with Brownian motion, the value of the process at time t is normally distributed (Equation A1.33b), but now with mean and variance

$$\mu_t = \theta + [x_o - \theta]e^{-at} \tag{12.22b}$$

$$\sigma_t^2 = \frac{b}{2a}[1 - e^{-2at}] \tag{12.22c}$$

where $b = \sigma_A^2/N_e$ under the constant-variance model. For large t, the mean value approaches the optimal value (θ), while the divergence variance approaches an asymptotic value of

$$\frac{b}{2a} = \frac{\sigma_z^2 + \omega^2}{2N_e} \tag{12.22d}$$

Initially, the among-population divergence increases linearly with divergence time t under both the pure drift (e.g., Equation 12.10) and OU models. However, unlike pure drift (which retains its linear divergence over all time), the between-lineage variance under an OU eventually asymptotes at a fixed level of divergence after sufficient time. Thus, while initially both models have very similar behavior, they become increasingly distinct as time progresses. It should be noted that divergence approaching an asymptotic variance can also occur under the purely neutral house-of-cards (Equation 12.11) and regression (Equation 12.12) mutational models, mimicking a process under stabilizing selection.

Divergence in Morphological Traits

Numerous attempts have been made to apply the above procedures, or variants of them, to data from the fossil record to test the hypothesis that levels of morphological divergence over geological time scales have been driven by directional selection. In the first such study, Lande (1976) showed that changes in tooth-size dimensions over a 42 million year period in early horse evolution are consistent with the hypothesis of random genetic drift if the heritabilities of the traits had been near 0.5 and the long-term effective population size was smaller than 60,000 or so individuals. Given the generally high levels of heritability observed for mammalian morphological traits (Lynch and Walsh 1998), an assumption of $h^2 = 0.5$ is not unreasonable, and the argument that the long-term N_e in such lineages could be smaller than the critical value of $N_e^* = 60,000$ is also plausible (Chapter 4). Analyses of tooth morphometrics in two additional lineages of extinct mammals (condylarths and oreodonts) suggested critical effective sizes of 80,000 to 120,000, below which the observed changes would be compatible with a neutral hypothesis (Lande 1976). Thus, only if the effective sizes of these ancient mammalian taxa were actually in excess of 10^5, a matter that remains unclear, would the observed changes require some mechanism of directional selection.

Several other studies of this nature have been applied to aspects of mammalian skull evolution. For example, by setting the upper and lower limits to mutational heritability, σ^2_m/σ^2_e, at 10^{-2} and 10^{-4}, Lynch (1990) found that the rates of evolution of cranial morphology in a wide array of placental mammalian lineages are one to two orders of magnitude *below* the minimum neutral rate, and Lemos et al. (2001) observed a similar pattern in marsupials. The only exception to this general trend concerns the races of modern man, which appear to have diverged at a rate slightly above the minimum neutral expectation (Lynch 1990; Ackermann and Cheverud 2004; Roseman 2004). Although they leave many questions unanswered, these kinds of results put in perspective previous arguments that rates of morphological evolution are exceptionally high in mammals, and especially so in the great apes (e.g., Cherry et al. 1982; Wyles et al. 1983; Van Valen 1985). Clearly, the predominant mode of evolution in mammalian skeletal morphology has been one of stabilizing selection, not of strong diversifying selection. Similarly, Spicer (1993) found widespread evidence of stabilizing selection in a variety of morphological traits in *Drosophila*, but some caution is in order here as the tests were based on critical mutation variances (Equation 12.21b). As mentioned, this approach likely generates many spurious calls of too little divergence (because h^2_m estimates are likely inflated by inclusion of deleterious mutations), hence biasing tests toward inferring stabilizing selection.

TIME SERIES DIVERGENCE TESTS

The methods discussed above are based on *external comparisons*—one has information from two time points and examines whether the observed amount of divergence is consistent with some external information (either estimated or assumed), such as effective population size (N_e), time of divergence (t), and a measure of genetic variation (h^2 or σ^2_m). Conversely, starting with Raup (1977; Raup and Crick 1981), a number of methods have been proposed (largely from paleontology) to detect departures from drift entirely from observations on the *internal* characteristics of a series of trait values over time (no external information, such as an estimate of N_e or h^2, is required by the test). The sequence of trait data from a set of subsamples within a stratigraphic column has been called a **stratophenetic series** (Gingerich 1979). This is not a new enterprise. Almost a century ago, Ronald Brinkmann (1929) published morphological information on close to 3,000 ammonites in the genus *Kosmoceras* from a 14-meter stratigraphic section from the Middle Jurassic. The number of available stratophenetic series is considerable, and growing. A recent review by Hunt et al. (2015) examined 709 such series from roughly 200 lineages. The number of samples per sequence (populations measured from different locations within the same stratigraphic column) ranged from 7 to 114 (with a median of 14), covering approximate time ranges from 5000 to more than 50 million years, with most between 10^5 and 10^7 years in duration. Such a temporal series of data makes it possible to look for statistical trends in mean phenotypes or for correlations in rates of change in adjacent intervals, neither of which are expected in a strictly neutral model (under the strong assumption of no environmentally influenced change in mean phenotype). The motivation for many of these tests is to provide a statistical framework to examine fossil data in the context of the **punctuated equilibrium** debate, which postulates directional selection to be rare in the fossil record, with a pattern of **stasis** (very little change) being predominant (Eldredge and Gould 1972; Gould and Eldredge 1977; Eldredge et al. 2005).

There are two important caveats with random-walk models. First, any observed phenotypic trend could be entirely environmental, with changes in the mean being independent of any underlying genetic change. Second, as highlighted by Raup (1977), a pattern indistinguishable from a random walk can mask significant underlying selection, such as short, episodic bursts of directional selection in shifting directions or stabilizing selection with drift occurring in the optimal value. These are examples of **hierarchical models** of random change, wherein selection is driving the generational change, but the focus of selection (either directional or stabilizing) is randomly changing, generating an random walk.

Tests for Departures From Symmetric Random Walks

A number of tests of departure from a symmetric random walk have been proposed. All assume uncorrelated changes over time increments and an equal chance of positive and negative increments (with the mean incremental change equaling zero). Raup (1977) and Raup and Crick (1981) proposed using the **Wald–Wolfowitz runs test** (under the null hypothesis of an equal number of positive and negative changes). Here the test statistic, R_n, for the number of runs (changes in the direction of the walk) in a sample of size n is approximately normally distributed with mean $n/2+1$ and variance $n/4$. A sequence showing excessive runs of the same sign is consistent with directional selection, while a sequence with an excessive number of sign reversals is consistent with stabilizing selection (as might be expected for a population mean fluctuating around an optimum). Similarly, one could simply test the number of positive increments against the value expected from a binomial with success parameter $1/2$ and sample size n. Both the runs and binomial tests do not use any information on the *size* of any jump, but rather simply test against a null of equally likely up versus down change over any given time point.

A more sophisticated approach was taken by Bookstein (1987, 1988), who used results from the theory on the maximal excursion of a symmetric random walk. The sequence of measured phenotypes is scaled so that the initial mean is zero, with x_i denoting the mean of the ith sample. The standard error for the expected divergence over the entire sequence (n steps) of a symmetric random walk is $\sigma\sqrt{n}$, where σ^2 is the variance in change per increment. Assuming roughly equal time intervals between samples, this standard error can be estimated as

$$\widehat{\sigma^2} = \frac{1}{n}\sum_{i}^{n}(x_i - x_{i-1})^2 \tag{12.23a}$$

Further, define S_k as the sum of the first k increments (the displacement of the mean from its original value after k steps). Bookstein obtained a large-n expression for the distribution of the largest scaled excursion,

$$\gamma = \frac{\max_k |\, S_k \,|}{\widehat{\sigma}\sqrt{n}} \tag{12.23b}$$

Namely, the largest absolute value of the walk ($\max_k |\, S_k \,|$) over any of the sampled times, expressed in terms of the expected standard error of the walk value at the final sample time ($\sigma\sqrt{n}$). For $\gamma > 1$, critical values (the upper p in the tail of the null distribution) are very closely given by the corresponding $p/4$ critical values for a unit normal. For example, the upper 5% tail corresponds to $\gamma = 2.25$, consistent with the value of 2.24 for a normal with $p/4 = 0.0125$. The upper 1% and 0.1% upper tail probabilities correspond to γ values of 2.8 and 3.5, respectively. Series with values exceeding these critical values are said to be **improbably directional**, consistent with directional selection (or an environmental trend). Conversely, a series that does not vary enough is said to be **improbably constrained**, consistent with some sort of stabilizing selection or other cause of stasis. The lower 5%, 1%, and 0.1% values correspond to γ values of 0.62, 0.49, and 0.41, respectively. Failure to reject the null of a random walk still allows for considerable selection, either due to a lack or power or randomness in the direction of selection over time. Multivariate random-walk tests are discussed by Bookstein (2013).

Another widely used test for departures against the null of a symmetric random walk (closely related to Bookstein's approach), called **scaled range analysis**, is based on **Hurst exponents** (Hurst 1951). The idea behind this approach is that the absolute difference of a symmetric random walk $|\, x_t - x_0 \,|$ scales as $\sigma\sqrt{t}$ (which can be estimated from Equation 12.23a). Defining the **standardized range**, $R(\tau)$, for a time interval (τ) as

$$R(\tau) = \frac{|\, x_\tau - x_0 \,|}{\sigma} \tag{12.24a}$$

one then regresses $R(\tau)$ on ever-increasing values of τ, fitting the log-log regression

$$\ln[R(\tau)] = H \ln(\tau) + \epsilon \tag{12.24b}$$

where the slope (H) is the Hurst exponent (i.e., $R \propto \tau^H$). Under a symmetric random walk with uncorrelated increments, absolute trait divergence is expected to scale with the square root of time, giving $H = 0.5$. As increments become more positively correlated, H increases to 1.0 (**directional persistence**), consistent with directional selection. As adjacent increments become increasingly negatively correlated, H decreases to zero (**anti-persistence**), consistent with stabilizing selection or some other form of stasis. Roopnarine (2001) discussed permutation tests for the significance of $H \neq 0.5$. Gingerich's (1993) ***LRI*** (log rate versus log interval) method is a version of this test, where the slope (G) of his *LRI* regression is simply $G = H - 1$ (Roopnarine et al. 1999).

While straightforward and widely applied in the early literature, these methods typically have low power, meaning that the null hypothesis of a symmetric random walk is hard to reject (Roopnarine et al. 1999; Roopnarine 2001; Sheets and Mitchell 2001). This is especially the case with stratophenetic series, with their usual incompleteness and sporadic coverage due to the vagaries of the fossilization process. Further, as noted by Sheets and Mitchell (2001), there is an asymmetry of detection in that stabilizing selection is easier to detect than directional selection. They showed that the Hurst exponent (and, by extension, the *LRI* method) has the highest power to detect stabilizing selection, followed by Bookstein's approach, and then the runs test. Conversely, for detecting directional selection, the runs test is often the most powerful, followed by the Hurst exponent, and then Bookstein's approach.

Hunt's Approach for Comparing Different Models

Hunt (2006, 2007, 2008a, 2008b; Hunt and Carrano 2010) noted that the low power for tests of departures from symmetric random walks creates a "tyranny of the null hypothesis," potentially overinflating the role of drift. He suggested that instead of testing against

Table 12.1 Summary of the 251 fossil sequences examined by Hunt (2007), each fit using three models of divergence: random walk, directional selection, and stasis. Counts given under the Trait and Fossil group categories are the numbers of times a model had the highest Akaike weight (Example 12.5) for a fossil sequence. For example, 13 of the 251 sequences (0.052) had directional selection as the model with the highest support, while 5 of 114 (0.044) size-related traits had directional selection as the most-supported model. Values under the Median column correspond to the median fraction of support over all sequences for a given model. For example, half of all sequences had a support for directional change model of 0.06 or less, while 95% of all sequences have a fractional support for directional selection in the 0.04 to 0.08 range. The fossil groups are planktonic and benthic microfossils (Plank and Benth) and macrofossils (Macro).

		Trait				Fossil group		
Model	Median, 95% CI	All	Size	Shape	Other	Plank	Benth	Macro
Directional	0.06 (0.04, 0.08)	13	5	4	4	5	3	5
Random	0.47 (0.39, 0.56)	123	67	43	13	24	57	42
Stasis	0.34 (0.20, 0.50)	115	42	68	5	12	37	66
		251	114	115	22	41	97	113

the random null, one should examine a set of candidate models, using Akaike weights (Anderson et al. 2000) to indicate support for each (see Example 12.5 for details). The Akaike weights for a set of competing models sum to one, providing a useful indicator of their relative support.

Hunt initially considered three basic models: a symmetric random walk (with an incremental mean value of zero); a directional (or generalized) random walk (mean increment $\neq 0$); and stasis. For the two random walks, he modeled the incremental δ (the change over an interval) with normal random variables. For the symmetric random walk, $\delta \sim N(0, \sigma_\delta^2)$, while for the general random walk, $\delta \sim N(\mu_\delta, \sigma_\delta^2)$, with parameters fit by maximum likelihood. For stasis, he assumed a simple model initially suggested by Sheets and Mitchell (2001). Instead of constructing likelihood models around increments (so that $z_t = z_0 + \sum \delta_i$), they simply took the trait value at time t to be $z_t \sim N(\mu, \sigma^2)$, namely a constant variance (σ^2) over all time (rather than the linear increased under Brownian motion). While at first glance this appears to be an OU process at stationarity ($t \to \infty$), this is not the case, as an Ornstein-Uhlenbeck process has correlated means (reflecting the shared history on a common evolutionary path). Akaike weights allow for a more nuanced interpretation of model fits (see Figure 12.6), and are based on the relative likelihoods among a set of candidate models, summing to one over all the models considered (Example 12.5). For example, suppose the Akaike weights for a fossil sequence are 0.50, 0.48, and 0.02 for the random, directional, and stasis models, respectively. Clearly, stasis and a random walk are almost equally likely explanations in this case, but the support for a directional model is very slim.

It should be stressed that the power of this approach is not the initial small set of candidate models, but rather that it serves as a much more general framework for examining an ever-richer set of models. Indeed, Hunt et al. (2015) examined more complex models that allow for the sequence of mean phenotypes to shift between modes (e.g., random vs. stasis) over the time sampled. One could also use the Akaike weights strategy to contrast the simple stasis model used by Hunt with the Ornstein-Uhlenbeck or other competing models of stabilizing selection (Estes and Arnold 2006; Uyeda et al. 2011), as well as considering other models of directional evolution (e.g., Charlesworth 1984b).

Using this approach, Hunt (2007), Hopkins and Lidgard (2012), and Hunt et al. (2015) examined an evergrowing number of stratophenetic series (251, 635, and 709 studies, respectively). The basic conclusions from Hunt et al. (2007), which hold for these larger (and more recent) studies as well, are given in Table 12.1. For each fossil series, the fraction of support for the three models was computed using Akaike weights. In only 13/251 (5.2%) of the sequences was directional selection (a generalized random walk with $\mu_\delta \neq 0$) the most

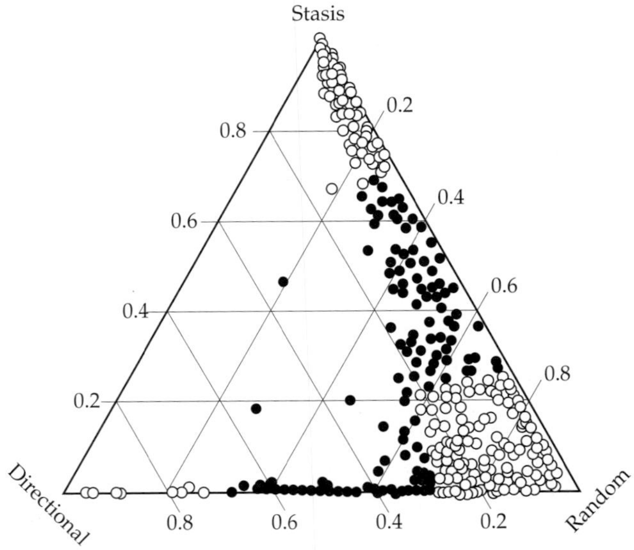

Figure 12.6 A De Finetti diagram of the support for the random walk, directional walk, and stasis models. Each point corresponds to the coordinates of the Akaike weights for these three models (which sum to one) for a single stratophenetic series. Points near vertices corresponds to almost 100% support for a particular model, hence the lables at the vertices. Points along an edge of the triangle indicate very little support for the model perpendicular to that edge. Unfilled points indicate strong support (weight for most supported model at least 2.7 times the weight of any other model). (After Hopkins and Lidgard 2012.)

strongly supported model (had the largest Akaike weight). The random walk model was the most supported overall (49% of the time), while 46% of the fossil series had stasis as the most supported model. Figure 12.6 presents the relative support for all three models from the analysis of Hopkins and Lidgard (2012), presented as a **De Finetti diagram** (or **De Finetti triangle**). Values in the middle of the triangle have roughly equal support for all three models (which was rarely seen). Values near the edges of the triangle have very weak support for at least one model, and values near the vertices correspond to very high support for a single model. Note that there is little support along the directional selection axis for any sequence, with most of the support lying along the stasis-random walk axes.

An interesting perspective on the rates of macroevolution was offered by Uyeda et al. (2011), who examined a vast data set of traits followed over time, with time-span ranging from fractions of years to over 350 million years. For periods of a million years or less, rapid evolution was seen to occur, but it is constrained, and does not accumulate over time. This matches the earlier observation by Estes and Arnold (2007) that the expected magnitude of divergence over samples is largely time-independent up to about a million generations. However, as Figure 12.7 illustrates, Uyeda et al. observed an accumulation of cases of rapid divergence starting at $\sim 10^6$ year intervals, generating what they called a **blunderbuss pattern**, as the spread of values resembles the flared muzzle of the seventeenth-century firearm of the same name.

While Estes and Arnold (2007) were able to account for their observed pattern in terms of stabilizing selection with a fluctuating optimum, this model did not fit the data of Uyeda et al. (2011) presented in Figure 12.7. Rather, the best fit was a model of essentially stasis (the Sheets-Mitchell model), coupled with rare random bursts of significant change (with an average waiting time of $\sim 10^7$ years). The model allowing for multiple (as opposed to single) bursts fit the data best. While this pattern is striking and reproducible over the several different taxonomic data sets used by the authors, the underlying mechanism is unclear. Uyeda et al. suggested that this pattern might be correlated with the opening of new niches following species turnover (as species life spans are typically in the million-year range).

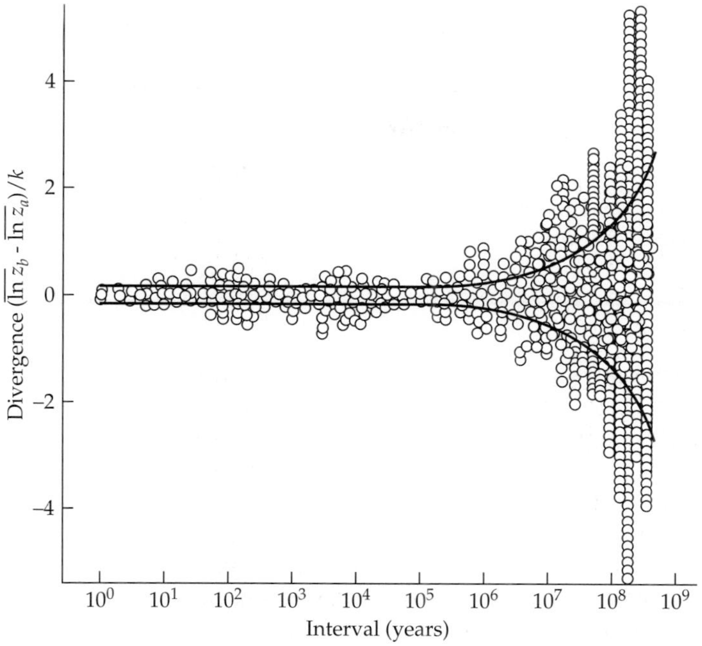

Figure 12.7. The blunderbuss pattern of divergence observed by Uyeda et al. (2011). Bounded variation is seen over the first 10^6 years, after which considerable divergence can occur. Divergence is scored as the log difference between means (at time points a and b, for an interval of $b - a$), scaled by the dimension k of the data ($k = 2$ for area, $k = 3$ for mass). (After Uyeda et al. 2011.)

Example 12.5. Anderson et al. (2000) proposed that a series of candidate models can be compared via their **Akaike weights**, an approach used by Hunt (2007) to assess the relative fit of a series of candidate models for long-term evolution (Figure 12.6). Suppose one has a series of models that were fit using maximum likelihood. If these models are not nested (so that likelihood-ratio tests for comparing fit are not available; LW Appendix 4), then comparison statistics, such as the **Akaike information criterion** (**AIC**; Akaike 1974) can be used to rank them. AIC rewards goodness of fit (higher log-likelihood, L), while penalizing for the number of parameters (k) with smaller AIC values implying a better model. With $n < 40k$ observations, Anderson et al. suggested that a corrected version of AIC should be used,

$$AIC_c = -2\ln(L) + 2k + \frac{2k(k+1)}{n-k-1} \tag{12.25a}$$

which differs from the standard AIC measure in the addition of the last term. Suppose that one has a set of m candidate models and computes the support for model i relative to the best fitting of all m models,

$$\Delta_i = AIC_{c,i} - \min(AIC_c) \tag{12.25b}$$

The resulting Akaike weights for each of the candidate models are given by

$$w_i = \frac{e^{-(\Delta_i/2)}}{\sum_{j=1}^{m} e^{-(\Delta_i/2)}} \tag{12.25c}$$

By construction, these weights sum to one, with each giving the relative support for model i, given the collection of proposed models.

Note that Equation 12.25c gives each model equal prior weight (for m models, a prior weight of $1/m$ on each cancels in the denominator and numerator of Equation 12.25c). More generally, if one has prior weights π_i for each model,

$$w_i = \frac{\pi_i e^{-(\Delta_i/2)}}{\sum_{j=1}^{m} \pi_j e^{-(\Delta_i/2)}} \tag{12.25d}$$

For example, one might, a priori, assume that 60% of the traits are under drift, 30% are under stabilizing selection, and the rest are under directional selection. This would yield different Akaike weights than when our prior belief is equal for all three cases.

POPULATION STRUCTURE BASED–TESTS: Q_{ST} VERSUS F_{ST}

Species are often distributed in space as a series of populations isolated by semipermeable migration barriers. In such settings, the variation at a neutral locus represents a balance between input of new variation by mutation and migration, countered by its removal by drift (Chapter 2). One measure of the resulting genetic population structure is Wright's (1951) F_{ST} statistic, reviewed by Holsinger and Weir (2009). As discussed in Chapter 2, F_{ST} partitions the total genetic variance of the entire metapopulation (measured as heterozygosity under the assumption of panmixia) into the fractions within $(1 - F_{ST})$ and among (F_{ST}) populations (Cockerham 1973; Nei 1987; Weir 1996). As developed in Chapter 9, one signature of selection at a candidate locus is whether it shows too much or too little population structure with respect to the F_{ST} value estimated from a set of (presumably) neutral markers (the Lewontin-Krakauer test). An excessive F_{ST} value at the candidate locus suggests too much divergence relative to the drift-migration expectation, consistent with directional selection varying over populations. Too little divergence (F_{ST} is too small) is consistent with stabilizing selection over populations retarding the divergent effects of drift. Although there are a number of problems with such tests (Chapter 9), they still have wide appeal, as the required data are reasonably straightforward to collect.

The *trait*-based analog of this test is based on comparing Q_{ST} to F_{ST}. Q_{ST} (developed below) is a measure of the population structure of the genetic variance underlying a quantitative trait, which is then compared with a neutral-marker based estimate of F_{ST} for the same population. While the candidate-gene tests in Chapter 9 compare candidate-*locus* F_{ST} values with the neutral standard, the *trait*-based test compares Q_{ST} for the candidate *trait* against this standard. This is a rather active area of research, and reviews can be found in Merilä and Crnokrak (2001), McKay and Latta (2002), Whitlock (2008), and Leinonen et al. (2008, 2013). Standard Q_{ST} analysis assumes only a single level of structure, but extensions to more hierarchically structured populations were proposed by Whitlock and Gilbert (2012). Likewise, multivariate extensions (properly accounting for the genetic correlations among a series of measured traits) were developed by Chenoweth and Blows (2008), Martin et al. (2008), Chapuis et al. (2008), Ovaskainen et al. (2011), and Karhunen et al. (2013, 2014).

Q_{ST}: Partitioning Additive Variance Over Populations

Consider a quantitative trait in a diploid with a purely additive-genetic basis, and denote its genetic variance over the entire metapopulation by σ^2_G. From Table 11.3 (setting $f = Q_{ST}$), the within- and among-population components of variance can be represented as $\sigma^2_{GW} = (1 - Q_{ST})\sigma^2_G$ and $\sigma^2_{GB} = 2Q_{ST}\sigma^2_G$, respectively, for a total variance in a structured population of $(1 + Q_{ST})\sigma^2_G$. Rearranging yields

$$Q_{ST} = \frac{\sigma^2_{GB}}{\sigma^2_{GB} + 2\sigma^2_{GW}} \tag{12.26a}$$

While the term Q_{ST} was introduced by Spitze (1993), this metric was proposed earlier by Prout and Barker (1989, 1993) and Lande (1992), and strongly hinted at by Rogers and

Table 12.2 Interpretation of Q_{ST} versus F_{ST} comparisons.

Observation	Interpretation
$Q_{ST} > F_{ST}$	**Divergent selection**: spatial variation in trait values in excess of neutral expectation.
$Q_{ST} = F_{ST}$	Consistent with divergence expected under drift. Does not rule out selection, but does not support it either.
$Q_{ST} < F_{ST}$	**Convergent selection**: spatial variation in trait values less than neutral expectation. Similar trait values are favored over populations.

Harpending (1983). Equation 12.26a is a very general result, applicable to a wide range of population structures and migration patterns provided the character does indeed have an entirely additive genetic basis (Whitlock 1999). When dealing with haploid populations (Whitlock 2008) or collections of entirely selfed lines (Bonnin et al. 1996; Le Corre 2005; Rhoné et al. 2010), σ^2_{GW} in Equation 12.26a is weighted by one, rather than two. More generally, when f is the amount of inbreeding within each population, then following Bonnin et al. (1996),

$$Q_{ST} = \frac{(1+f)\sigma^2_{GB}}{(1+f)\sigma^2_{GB} + 2\sigma^2_{GW}} \tag{12.26b}$$

Equations 12.26a and 12.26b provide a potential approach for testing the hypothesis of neutral divergence among population means. Provided that a sufficient number of families from multiple populations can be grown in a common environment, appropriate statistical methods can be used to estimate σ^2_{GW} and σ^2_{GB} (e.g., by ANOVA; Lynch and Walsh 1998). The resultant estimate of Q_{ST} can then be compared to a parallel measure of subdivision (F_{ST}) derived from putatively neutral molecular markers. Under the assumption of neutrality, Q_{ST} should not be significantly different from F_{ST}. On the other hand, $Q_{ST} > F_{ST}$ is expected if population differentiation has been primarily driven by adaptive divergence, whereas the opposite relationship is expected if the mean phenotypes of all or most populations are kept relatively uniform by stabilizing selection for the same optima (Table 12.2). It is important to stress that any comparison of this sort *must* be performed using the ***same set of populations*** to obtain both Q_{ST} and F_{ST}. An analysis using an estimate of F_{ST} from one set of populations and Q_{ST} from another is not trustworthy.

The first (of many) caveats with respect to this strategy is that, even under neutrality, the expected value of Q_{ST} will not necessarily equal F_{ST} if the trait of interest is influenced by nonadditive genetic effects. As outlined in Chapter 11, with nonadditive gene action, the within- and among-population components of genetic variation for neutral characters under short-term divergence are no longer equal to $\sigma^2_{GW} = (1-f)\sigma^2_G$ and $\sigma^2_{GB} = 2f\sigma^2_G$ (where f is the parameter estimated by F_{ST}), but instead are influenced by a number of higher-order terms (see Table 11.3). In general, because the within-population genetic variance declines less rapidly with inbreeding under nonadditivity (and sometimes even increases; Chapter 11), Q_{ST}, as defined by Equation 12.26b, will tend to be smaller than F_{ST} under neutrality. In particular, Whitlock (1999) showed that additive $\times$ additive variance always results in $Q_{ST} < F_{ST}$ under neutrality. Dominance also causes Q_{ST} and F_{ST} to deviate under neutrality, with the direction of the inequality depending on the details of the population structure. There is disagreement as to the practical importance of these departures, especially given the large variances associated with Q_{ST} estimates (López-Fanjul et al. 2003, 2006, 2007; Goudet and Büchi 2006; Goudet and Martin 2007; Whitlock 2008; Santure and Wang 2009). However, because these violations of assumptions often (but not always) result in $Q_{ST} < F_{ST}$, this general behavior makes conclusions regarding adaptive divergence based on elevated Q_{ST} conservative, while rendering observations of $Q_{ST} < F_{ST}$ ambiguous. Violations of the assumption of additivity may not be a serious issue for most morphological traits, but given that life-history traits often show considerable nonadditive variance (Chapter 6), these may be more vulnerable to false impressions under a comparison of Q_{ST} and F_{ST}.

A second caveat is that the choice of markers used to estimate F_{ST} can introduce bias. The strong assumption is that the markers chosen are neutral, such that any structure associated with the markers reflects the neutral population structure. Historically, allozyme markers were commonly used to estimate F_{ST}, and because these represent variant protein products, some may not be neutral. Another problematic (but widely used) marker class is microsatellites. For F_{ST} to serve as a neutral proxy for the behavior of alleles underlying a focal trait, the mutational structure of the markers and QTLs must be compatible. Microsatellite alleles can easily back-mutate, resulting in underestimation of F_{ST} (Hendry 2002; Kronholm et al. 2010). While microsatellite-specific distance metrics (such as R_{ST}; Slatkin 1995a; Goodman 1997) have been proposed, these should *not* be used in place of F_{ST} for comparison with Q_{ST}. These modified metrics adjust for high rates of back-mutations, something not expected at QTL alleles, potentially resulting in different adjusted measures of allelic divergence at the markers versus QTLs. These issues are of special concern given that many early studies used microsatellites (Edelaar and Björklund 2011; Edelaar et al. 2011). The ever-increasing use of SNPs to estimate F_{ST} avoids these concerns.

P_{ST}: Approximating Q_{ST} with Phenotypic Data

Because of the requirement for assays in a common-garden arena, true joint studies of Q_{ST} and F_{ST} are not common. Pujol et al. (2008) noted that roughly half of the wild population studies they reviewed were not based on estimated additive variances. Instead, a phenotypic-based proxy for Q_{ST} was used, where within- and/or among-population *phenotypic* variances replace the more challenging estimates of additive variation. The former can easily be obtained via a standard ANOVA (e.g., Holand et al. 2011), while the latter require a series of parent-offspring or sib rearings in a common environment. A modification of this purely phenotypic approach is to use

$$\widehat{\sigma}^2_{GB} = c\,\widehat{\sigma}^2_{PB}, \qquad \widehat{\sigma}^2_{GW} = h^2\,\widehat{\sigma}^2_{PW} \tag{12.27a}$$

where c reflects the fact that only part of an observed phenotypic difference in means may be genetic (Merilä 1997; Leinonen et al. 2006; Sæther et al. 2007; Brommer 2011). Substitution into Equation 12.26 yields the P_{ST} statistic of Leinonen et al. (2006),

$$\widehat{P}_{ST,L} = \frac{\widehat{\sigma}^2_{PB}}{2(h^2/c)\,\widehat{\sigma}^2_{PW} + \widehat{\sigma}^2_{PB}} \tag{12.27b}$$

When $c = h^2$, this reduces to Equation 12.26a, with phenotypic variances replacing their genetic counterparts. Holand et al. (2011) suggested doing a sensitivity analysis by varying the value of c (for a fixed h^2 value), and using simulations to find critical upper and lower c values for which Q_{ST} is significantly above and significantly below F_{ST}. While enticing because of their simplicity and relative ease of application (only phenotypic data are required), strong caution is advised when replacing Q_{ST} by a phenotypic surrogate (Pujol et al. 2008; Brommer 2011). At a minimum, such estimates should *always* be denoted as P_{ST} whenever *any* variance component is based on a purely phenotypic measure. Although the biology or ecology of a species might be such that only P_{ST} estimates are possible, in such cases the investigator needs to seriously consider if such a resulting study can give truly worthwhile results.

Even when genetic data (information from crosses) are used, bias can still be introduced into Q_{ST} estimates. Ideally, additive variances should be estimated from the covariance among paternal half-sibs. When covariance among full sibs is used, additive variance estimates can be inflated by the presence of dominance or maternal effects. Likewise, if among-group differences are not measured in a common garden, shared environmental effects can inflate this estimate. Conversely, a common garden may obscure any evolved plastic response that is part of the adaptive response to specific environments. See Whitlock (2008) for further discussion of such sources of bias.

A slightly optimistic note was struck by Pujol et al. (2008), who noted that the onerous requirement of a common garden may be circumvented through the use of BLUP-based genetic-group mixed models (e.g., Westell et al. 1988; Quaas 1988), which allow for the estimation of both within- and among-group additive variance (a variant of this approach was used by Roberge et al. 2007). BLUP uses the genetic relationships among all measured individuals to separate genetic from environmental contributions (Chapters 19 and 20). This requires good estimates of these genetic relationships, which (in the absence of pedigree data) requires a rather dense set of markers (a very large number of SNPs) for the accuracy needed, given the expected rather distant connections among groups.

Testing Q_{ST} Versus F_{ST}

The construction of rigorous statistical tests for comparing Q_{ST} with F_{ST} is problematic on several levels. First, both are ratios of variances, so that estimates obtained by directly substituting variance estimates into Equation 12.26 are biased, as the expectation of a ratio is not the same as a ratio of expectations (LW Equation A1.19a). Second, the sampling distribution of Q_{ST} is complex, as one must use a crossing design to estimate the variance components. Hence, the correct construction of dispersion intervals (such as standard errors in a frequentist setting or credible intervals in a Bayesian setting) is not trivial. Finally, there is the issue of formally comparing a somewhat noisy estimate (F_{ST}) with a very noisy estimate (Q_{ST}), which were obtained using very different designs. Some of these issues were addressed by O'Hara and Merilä (2005), Whitlock (2008), and Whitlock and Guillaume (2009). As noted by O'Hara and Merilä, one significant problem is simply power. The among-group variance is a function of the number of groups, with at least 20 needed for any substantial power. Unfortunately, the typical group number is around 7 for the studies reviewed by Merilä and Crnokrak (2001).

An important advance was the observation by Whitlock (2008) that the distribution of realized Q_{ST} values (ignoring, for now, the additional error introduced by using the sample estimate, $\widehat{Q}_{ST}$, for the true value of the realization for a particular trait) can often be approximated using the Lewontin-Krakauer distribution for F_{ST} values (Equation 9.10a). Simulations by Whitlock confirmed the suggestion by Rogers and Harpending (1983) that, provided F_{ST} is small, the amount of information on population structure derived from the variance components of a quantitative trait is equivalent to that from a single-marker F_{ST}. Provided that the average F_{ST} is small, then under the null that $Q_{ST} = F_{ST}$, to a very good approximation, we have

$$\frac{n_d - 1}{\overline{F}_{ST}} Q_{ST} \sim \chi^2_{n_d-1}, \quad \text{implying} \quad Q_{ST} \sim \frac{\overline{F}_{ST}}{n_d - 1} \chi^2_{n_d-1} \tag{12.28a}$$

where $\overline{F}_{ST}$ is the average F_{ST} over the scored molecular marker loci, and n_d is the number of demes. This expression assumes that Q_{ST} is estimated without error, a point addressed shortly.

The requirement that $\overline{F}_{ST}$ is small arises (in part) from χ^2 being defined over $(0, \infty)$, while Q_{ST} is restricted to $(0, 1)$. Hence, the approximation given by Equation 12.28a assumes that there is essentially no probability in the upper tail of a χ^2 above a critical value,

$$\Pr\left(\frac{\overline{F}_{ST}}{n_d - 1} \chi^2_{n_d-1} > 1\right) = \Pr\left(\chi^2_{n_d-1} > \frac{n_d - 1}{\overline{F}_{ST}}\right) \simeq 0 \tag{12.28b}$$

To achieve this condition, Whitlock (2008) recommended an upper limit of $\overline{F}_{ST} < 0.1$. For example, with $n_d = 2$, 5, and 10, the probabilities in Equation 12.28b (with $\overline{F}_{ST} = 0.1$) become 0.002, $4{\cdot}10^{-8}$, and $2{\cdot}10^{-15}$, respectively.

Insight into power is obtained by asking, under the null, how often the ratio Q_{ST}/F_{ST} exceeds some value, δ. Rearranging Equation 12.28a yields

$$\Pr\left(\frac{Q_{ST}}{\overline{F}_{ST}} > \delta\right) = \Pr\left(\frac{(n_d - 1)Q_{ST}}{\overline{F}_{ST}} > \delta(n_d - 1)\right) = \Pr\left(\chi^2_{n_d-1} > \delta(n_d - 1)\right) \tag{12.28c}$$

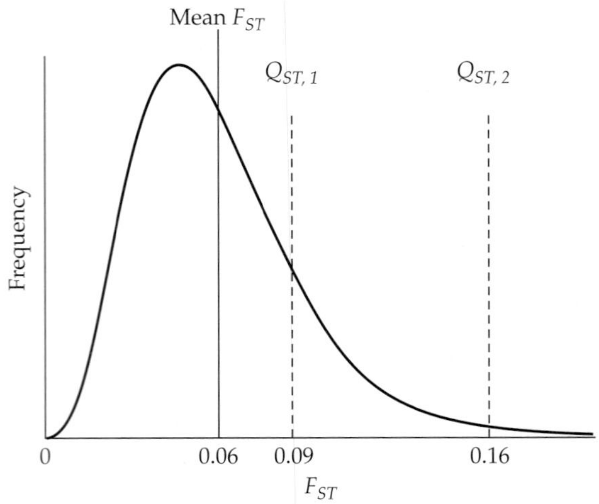

Figure 12.8 When $\overline{F}_{ST}$ is small, the Q_{ST} distribution for a neutral, completely additive trait should approximately follow the Lewontin-Krakauer distribution (Equations 9.10a and 12.28a). In this example, two traits, one with Q_{ST} = 0.09, and a second with Q_{ST} = 0.16 are both larger than $\overline{F}_{ST}$ = 0.06, but only trait 2 is significant. (After Whitlock 2008.)

Consider $n_d = 2$, as occurs when comparing two populations. How much larger must the true value of Q_{ST} be than the true value F_{ST} for this difference to be significant at the $\alpha = 0.05$ level? Because tests involving Q_{ST} are two-sided (either too large or too small being of interest), and $\Pr(\chi_1^2 > 5.02) = 0.025$, Equation 12.28c gives the critical value as $\delta = 5.02$. Hence, Q_{ST} must be in excess of 5 times $\overline{F}_{ST}$ to be significant at the 5% level. For $n = 10$, $\Pr(\chi_9^2 > 19.03) = 0.025$, or $\delta = 19.03/3 = 2.1$, and hence only a two-fold difference is required for significance. The same logic can be used to obtain the critical value when $Q_{ST} < F_{ST}$. For example, because $\Pr(\chi_9^2 < 2.7) = 0.025$, a value of Q_{ST} less than one third of $\overline{F}_{ST}$ (2.7/9 = 0.3) is significant at the 5% level when $n_d = 10$.

Figure 12.8 shows the basic structure of tests based on this simple approach: compute Q_{ST} and compare this value with the distribution of realized values for single-locus F_{ST}, where the mean of this latter distribution as taken is $\overline{F}_{ST}$, the mean F_{ST} value over all loci in the sample. This approach assumes that just a *single* trait is of interest and that Q_{ST} is measured without error (again, we return to this below). In the typical study setting, however, one has k Q_{ST} values (one for each of the k traits in the study), but uses the same set of markers (and hence the same $\overline{F}_{ST}$ value) for all traits. This is now a multiple-comparisons setting (Appendix 4). One approach to accommodate this concern is to use the first k order statistics from the Lewontin-Krakauer distribution (Equation 12.28a), which can be obtained as follows. A large number of samples are generated by randomly drawing k χ^2 random variables and scaling each using Equation 12.28a to generate an empirical distribution of the k order statistics (i.e., the values of the k realizations, ranked from largest—the first order statistic—to smallest; Chapter 14). The largest Q_{ST} value is assessed by comparing it against critical values for the empirical distribution of the largest value from each of the simulated samples. If this Q_{ST} value is significant (for example, only 2% of the simulated samples of k draws each have a greater value for their largest order statistic), one can then turn to the second largest Q_{ST} value and compare it with the simulated distribution of the second largest order statistic, and so on until a Q_{ST} value is no longer significant relative to its corresponding order statistic.

The major flaw with using Equation 12.28a is that it ignores the very important sampling variances of both our estimates, $\widehat{Q}_{ST}$ and $\overline{F}_{ST}$. Whitlock and Guillaume (2009), building

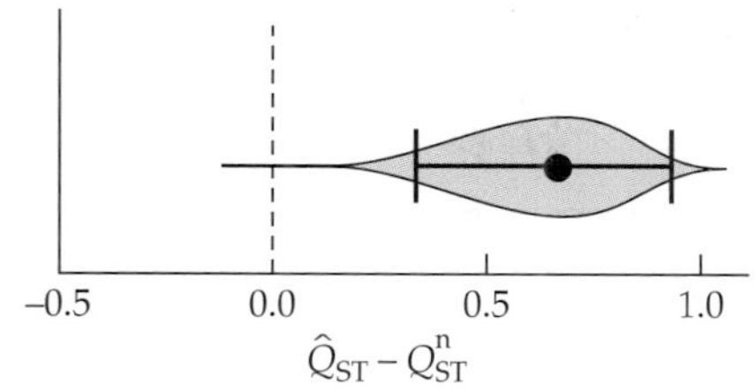

Figure 12.9 A violin plot for the distribution of the difference $(\widehat{Q}_{ST} - Q^n_{ST})$ for body length in the sea-run brown trout (*Salmo trutta*), using the resampling scheme suggested by Whitlock and Guillaume (2009), and detailed in Example 12.9. The width of the "violin" indicates the probability mass in that interval, the dot denotes the highest posterior probability, and the error bars the 95% credibility interval. Here this interval is completely above zero, demonstrating that $\widehat{Q}_{ST}$ is significantly in excess of its predicted neutral value given $\overline{F}_{ST}$. (After Rogell et al. 2012.)

on Equation 12.28a, showed how to incorporate such uncertainty to construct the distribution (which they denoted by Q^n_{ST}) of the estimated $\widehat{Q}_{ST}$ values under the assumption of an additive and neutral trait, which is also specific for the design used to estimate variance components (Example 12.6). A plot of the resulting distribution of the difference $(\widehat{Q}_{ST} - Q^n_{ST})$ provides a formal statistical test of whether the observed value, $\widehat{Q}_{ST}$, is excessively large (the 95% credible interval is entirely above zero) or small (the 95% credible interval exclusively below zero). Figure 12.9 shows how **violin plots** provide a useful way to display these results.

Example 12.6. To generate draws from the distribution Q^n_{ST} accounting for the sampling of an estimated $\widehat{Q}_{ST}$ value, Whitlock and Guillaume (2009) started with the clever observation that because $Q_{ST} = F_{ST}$ for a neutral additive trait, Equation 12.26a can be rearranged to yield

$$\sigma^2_{GB} = \frac{2\,F_{ST}\,\sigma^2_{GW}}{1 - F_{ST}} \tag{12.29}$$

This trick allowed them to generate a random value for σ^2_{GB}, given values of σ^2_{GW} and F_{ST}. Their scheme to generate the ith sample of Q^n_{ST} was as follows:

1) Draw a bootstrap sample of the n markers by sampling with replacement. Compute the mean F_{ST} value for this (the ith) sample, $\overline{F}_{ST,i}$.

2) Draw a sample value of $\sigma^2_{GW,i}$ as follows:

   a) If σ^2_{GW} is computed by a standard ANOVA, recall that $(MS/df) \sim \chi^2_{df}$, where MS is the observed mean sum of squares and *df* its degrees of freedom (LW Chapter 18). Generate X_i, a draw from a χ^2_{df} and take the sample i value for MS as $MS/df \cdot X_i$, using this value to compute the variance estimates (O'Hara and Merilä 2005; Whitlock and Guillaume 2009). When σ^2_{GW} is estimated as a function of several different sums of squares, compute each in the fashion above, and then use these realizations to compute $\sigma^2_{GW,i}$.

   b) If σ^2_{GW} is estimated by Bayesian methods, then $\sigma^2_{GW,i}$ is a sample drawn from the resulting posterior (i.e., Rogell et al. 2012).

4) Use the sample values of $\overline{F}_{ST,i}$ and $\sigma^2_{GW,i}$ with Equation 12.29 to generate a draw of the among-group variance, $\sigma^2_{GB,i} = 2\overline{F}_{ST,i}\,\sigma^2_{GW,i}/(1 - \overline{F}_{ST,i})$.

4) Take $Q^n_{ST,i} = \sigma^2_{GB,i}/(\sigma^2_{GB,i} + 2\sigma^2_{GW,i})$.

5) Repeat steps (1) through (4) until a sufficient number of draws (say m) have been computed to generate a sufficiently dense empirical distribution under the null. Figure 12.9 was com-

puted using the m values $(\widehat{Q}_{ST} - Q^n_{ST})$ obtained by subtracting the estimated Q_{ST} from each sample drawn from the underlying null distribution Q^n_{ST}.

Empirical Data

Results from the large number of Q_{ST} vs. F_{ST} comparisons from natural populations were summarized by Merilä and Crnokrak (2001), McKay and Latta (2002), and Leinonen et al. (2008, 2013). Values of Q_{ST} and F_{ST} are positively correlated, with $r = 0.24$ (Leinonen et al. 2013). Thus, there is a modest tendency for the structure of quantitative-trait variation to parallel the population structure for neutral alleles. The striking finding is that $Q_{ST} > F_{ST}$ for $\sim 70\%$ of all traits, which, taken at face value, suggested that diversifying selection was very widespread (Figure 12.10). Conversely, values of $Q_{ST} < F_{ST}$ are rare, despite the bias in this direction for neutral traits under a variety of conditions (discussed above), suggesting that persistent stabilizing or uniform selection is far less common.

One potential explanation for this trend of $Q_{ST} > F_{ST}$ was offered by Miller et al. (2008). They found that the variance of Q_{ST} is significantly larger than that for F_{ST} and noted a strong positive correlation in the data between Q_{ST} and the difference $(Q_{ST} - F_{ST})$. Hence, populations with larger Q_{ST} values tend to also have greater departures from F_{ST}. In particular, they noted that if more variable traits are overrepresented in the sampling process, this generates outliers of Q_{ST}, given the latter's larger variance, which in turn generates excessive $(Q_{ST} - F_{ST})$ values, even under neutrality. Whitlock (2008) further stressed this concern:

> It will always be possible to choose a set of traits that have higher than average Q_{ST} values. Traits chosen in this way cannot reliably be used to infer the extent of spatially heterogeneous selection. Examination of the traits chosen for many Q_{ST} studies makes one wonder whether traits are in fact always chosen with previous knowledge of the likely results.

A second source of bias in comparisons of Q_{ST} and F_{ST} was noted by Edelaar and Björklund (2011) and Edelaar et al. (2011). Markers with high mutation rates underestimate F_{ST}, and the most widely used markers in early Q_{ST}/F_{ST} studies, microsatellites, have high mutation rates. As shown in Figure 12.11, there is a strong positive relationship between the polymorphism level of a marker (with highly polymorphic markers having higher mutation rates) and the excess values of Q_{ST} over F_{ST}. Note that most of this trend is driven by studies employing microsatellites, with allozyme studies showing an excess of Q_{ST} largely independent of their polymorphism level.

Thus, the striking trend of $Q_{ST} > F_{ST}$ is certainly inflated by ascertainment bias, and somewhat inflated by the use of highly polymorphic markers (which is a more recent trend), making it difficult to make any general statement about how commonly diversifying selection structures quantitative traits in subdivided populations. As noted by Whitlock (2008), "While useful, Q_{ST} is a crude measure of the genetic differentiation of a trait caused by local adaptation."

One check of theory is to compare Q_{ST} and F_{ST} values between control and artificially selected groups. Morgan et al. (2005) examined the results of a 14-generation replicated selection experiment for increased wheel-running activity in mice. A base population was split into a control and a selected group, each with four replicate lines. The average selection intensity per generation was close to one ($\bar{\imath} \simeq 1$: Chapter 14), and significant response was seen in both the target trait, and (as a correlated response) in body mass. Theory predicts that Q_{ST} contrasting the control versus selected group should exceed F_{ST}, and this was observed for both the directly selected (wheel running) and correlated (body mass) traits. Q_{ST} and F_{ST} should be similar *among* the replicate lines of the selected group, and this was indeed seen among the wheel-running treatments, where Q_{ST} was actually below F_{ST} for body mass, but not significantly so.

Support for Q_{ST} as a method for detecting selection was also offered by Rhoné et al.

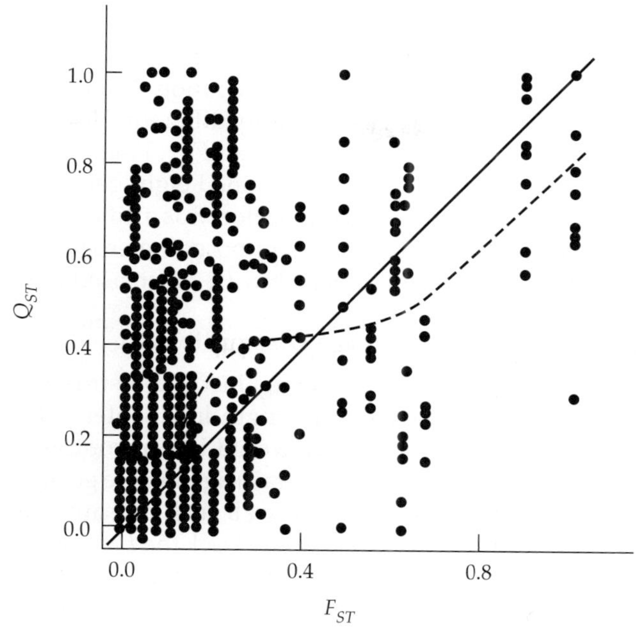

Figure 12.10 The joint distribution of Q_{ST} vs. F_{ST} seen in the meta-analysis of Leinonen et al. (2008). The solid line represents the neutral expectation, $Q_{ST} = F_{ST}$, while the dashed line is their smoothed nonlinear regression. There is a very strong tendency for $Q_{ST} > F_{ST}$. While consistent with widespread diversifying selection, as discussed in the text, such a pattern can also arise from ascertainment bias or the use of highly polymorphic markers (which underestimate F_{ST}).

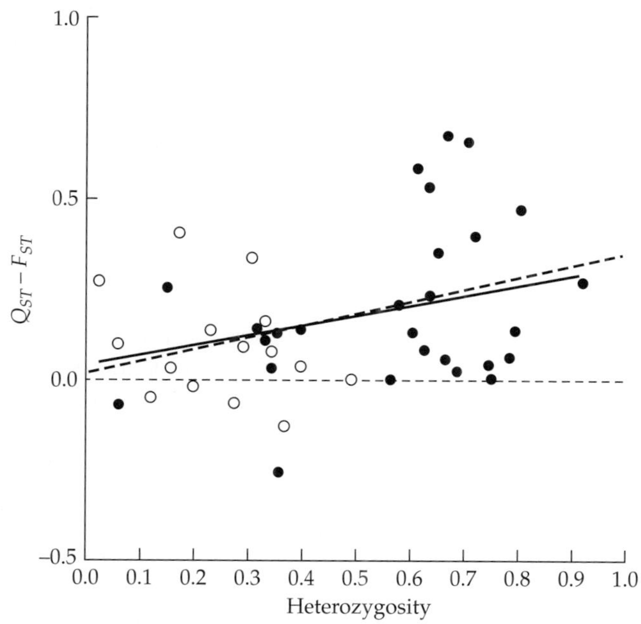

Figure 12.11 Correlation between the difference $(Q_{ST} - F_{ST})$ and heterozygosity at the marker loci, with each point representing one trait comparison. Filled circles involve microsatellites, and open circles denote allozyme markers. The solid line is the regression slope of $(Q_{ST} - F_{ST})$ on heterozygosity; the dashed line is the same regression, but correcting the previously discussed decline in F_{ST} with heterozygosity. (After Edelaar et al. 2011.)

(2010). They examined the response to 12 generations of natural selection on flowering times in a synthetic population of wheat grown in three locations in France (which experienced rather different environmental conditions). For generation 2 remnant seed assessed in a common garden, Q_{ST} and F_{ST} were not significantly different. However, individuals from generations 7 and 12 had Q_{ST} significantly larger than F_{ST}. Finally, agreement with theory was mixed in Porcher et al. (2004, 2006), who examined eight generations of selection in a series of structured *Arabidopsis* populations (migration was artificially controlled over a set of demes). Larger Q_{ST} values were seen under imposed heterogeneous selection among the experimental demes, consistent with theory, but F_{ST} increased as well.

Closing Comments: Q_{ST}, F_{STQ}, and Linkage Disequilibrium

Tests comparing F_{ST} values at candidate loci against the distribution of F_{ST} values at putatively neutral markers were discussed at length in Chapter 9. Comparisons of Q_{ST} to F_{ST} are a step removed, in that, ideally, we would like to contrast the **F_{STQ}** value (the average F_{ST} value for loci underlying our focal trait) against the genome-wide F_{ST} neutral standard. Given the near impossibility of locating all such causative loci, we have instead been using Q_{ST}, as with an additive trait, this should track the F_{STQ} values at the underlying causative loci. However, as is detailed in Chapters 16 and 24, allele-frequency changes are not the only route through which genetic variances (and hence the components of Q_{ST}) can change. Selection-generated gametic-phase disequilibrium (LD)—*even among unlinked loci*—can have a dramatic effect, even in situations where little allele-frequency change occurs. This impact of LD on Q_{ST} was stressed first by Latta (1998, 2005), and later by Le Corre and Kremer (2003, 2012; Kremer and Le Corre 2012). Because Q_{ST} is based on variance components, it can be influenced by linkage disequilibrium, which generates covariances between alleles at different loci, either inflating or deflating the resulting variances. When this happens, the values of Q_{ST} and F_{STQ} can become decoupled, and (as we will see) Q_{ST} can have more power to detect selection than F_{STQ} (even presuming we could locate all the underlying loci).

Thus, while a significant departure of Q_{ST} from the background value of F_{ST} is usually taken as indicating a shift in the F_{STQ} values at the underlying trait loci, this is only strictly correct when linkage disequilibrium is absent. Even in cases where selection induces little allele-frequency change (and hence little shift in F_{STQ} relative to the background F_{ST}), selection-induced disequilibrium (i.e., shifts in gamete, as opposed to allele, frequencies) can still generate a significant Q_{ST} signal. In particular, under the infinitesimal model, there is essentially no shift in the allele frequencies at underlying loci ($F_{STQ} \simeq F_{ST}$), but there can be a substantial change in the genetic variances due to selection-induced LD (Chapters 16 and 24), and hence a perturbation of Q_{ST} away from F_{STQ}. In such a setting, a direct comparison of F_{STQ} to the genome-wide F_{ST} standard would not reveal any evidence of selection, *but* a comparison of Q_{ST} (with its LD-shifted variance components) against F_{ST} might. Hence, under polygenic sweep conditions (Chapter 8), an appropriately performed Q_{ST} test might detect selection signatures missed by allele-frequency based tests.

To expand on this point, we need to consider how the within- and among-population LD (Ohta 1982) impact Q_{ST}. Letting the subscript x denote either within- or among-population values ($x = w$ and $x = a$, respectively), we can express the variances comprising Q_{ST} as

$$\sigma_x^2 = \sigma_{x,0}^2 + d_x = (1+\phi_x)\sigma_{x,0}^2, \quad \text{where} \quad \phi_x = \frac{d_x}{\sigma_{x,0}^2} \tag{12.30a}$$

where $\sigma_{x,0}^2$ is the linkage equilibrium value, d_x is the disequilibrium contribution generated by covariance among alleles at different loci (Equations 16.1 and 16.2), and ϕ_x is the ratio of the disequilibrium contribution to the linkage-equilibrium (i.e., genic) variance (note that ϕ_x is negative when d_x is negative). As discussed in Chapter 16, stabilizing or directional selection within a population generates negative d, so we often expect negative within-population LD (negative values of d_w and ϕ_w).

Turning to the among-population LD, Latta (1998) noted that if each population is under stabilizing selection for a different optimum value (θ), then for an additive trait where the population means have reached their optimal values,

$$d_a = \sigma_\theta^2 - 2F_{STQ}\,\sigma_A^2 \tag{12.30b}$$

where σ_θ^2 is the variance in the optimum value over populations, and σ_A^2 is the expected additive genetic variation if the populations were to be randomly mated to form a single, panmictic, population (in linkage equilibrium). With nearly uniform selection (the variance in θ values over demes is small) and reduced migration (so that F_{STQ} is large), Equation 12.30b gives a negative covariance ($d_a, \phi_a < 0$) between trait-increasing alleles at different loci across demes, reducing the among-group variance σ_{GB}^2 below its linkage-equilibrium value. Conversely, if diversifying selection is strong (σ_θ^2 is large) and gene flow is high (F_{STQ} is small), a positive covariance is expected ($d_a, \phi_a > 0$), and σ_{GB}^2 is inflated relative to its value in the absence of LD. Thus, Q_{ST} often magnifies the effect of selection over what is expected from changes in F_{STQ} alone, with significant changes in Q_{ST} (relative to F_{ST}) possible even when little differentiation has occurred at the underlying QTLs ($F_{STQ} \simeq F_{ST}$).

For a completely additive trait, Le Corre and Kremer (2003) quantified the influence of LD on Q_{ST} by noting that the relationship between Q_{ST} (based on variance components) and F_{STQ} (based on the underlying loci) is given by

$$Q_{ST} = \frac{(1+\phi_a)F_{STQ}}{(\phi_a - \phi_w)F_{STQ} + 1 + \phi_w} \tag{12.30c}$$

where ϕ_x is given by Equation 12.30a. Note that Q_{ST} equals F_{STQ} only when the among- and within-population LD values are equal ($\phi_a = \phi_w$). Using Equation 12.30c, Kremer and Le Corre (2012) showed that $Q_{ST} > F_{STQ}$ when $\phi_a > \phi_w$. Given that stabilizing selection within populations generates negative values of ϕ_w, while diversifying selection (variation in the optimum over populations) generates positive values of ϕ_a (Equation 12.30b), this combination amplifies the signal in Q_{ST} over that generated from F_{STQ}. As $Q_{ST} > F_{ST}$ is the signal for divergent selection (Table 12.2), while our last result implies that $Q_{ST} > F_{STQ} > F_{ST}$, the impact of LD is to *magnify* the impact of divergent selection over that expected from allele-frequency change alone (F_{STQ}). Again, the salient point is that even if the difference between F_{STQ} and F_{ST} is small, the difference between Q_{ST} and F_{ST} can still be large.

Hence, while Q_{ST}-based tests are fraught with complications, if properly performed (which is no small feat), they may actually be more powerful than a scan for F_{ST} outliers at known candidate genes for the trait of interest (Chapter 9). While F_{ST}-based scans are trait independent, knowledge of the potential target trait or traits allows Q_{ST}, and thus further information from LD, to be exploited. We return to this point below when considering certain trait-augmented marker-based tests.

Example 12.7. Using candidate genes in the photoperiod pathway (detected in *Arabidiopsis thaliana*), Ma et al. (2010) explored whether variation in these loci is involved in growth cessation in populations of European aspen (*Populus tremula*) across a latitudinal gradient in Sweden. Their population sample consisted of 10 trees from each of 12 sites (spanning roughly ten degrees of latitude), scoring 113 SNPs from 23 photoperiod genes and 93 SNPs from 21 random control genes. Six of the photoperiod SNPs showed significant associations with growth cessation (with no evidence of epistatic interactions between these detected loci). While four of the photoperiod SNPs showed a significant correlation with latitude, the F_{ST} values for the photoperiod and control groups of SNPs were not significantly different (0.018 vs. 0.016, corresponding to F_{STQ} and F_{ST}, respectively), although photoperiod SNPs showed a significantly greater variance in F_{ST} values relative to the control SNPs. None of the individual SNPs showed significant F_{ST} departures from the control loci, so that even when using candidate genes in a known pathway that is likely to be under selection, no signature of selection was observed in individual F_{STQ}

values. However, as a group, the photoperiod SNPs showed a significant excess of pairs of alleles from different loci correlated with each other (i.e., showed LD), while no such pattern was seen with the control SNPs. Further, the highest five of the allelic pairs correlated between loci also involved either one (or both) alleles (SNPs) that showed significant clines with latitude. Thus, while selection in this study did not seem to generate a significant departure between F_{STQ} and F_{ST}, it did generate among-population covariances.

TRAIT-AUGMENTED MARKER-BASED APPROACHES: TESTS USING QTL INFORMATION

Our last class of tests for neutral trait evolution exploit marker information from either a QTL mapping experiment or a GWAS study. We refer to these as trait-augmented marker-based approaches, as tests are not based on a set of *random* markers (as was the case for genome scans; Chapter 9), but rather on a very *specific* set of markers, namely those chosen because they are associated with a target trait (as either markers linked to QTLs or GWAS hits). We first examine approaches using QTLs, which focus fixed sites, and then consider GWAS information, which focuses on segregating sites (i.e., changes in allele frequencies).

Leveraging QTL Studies

In theory, one could take localized QTL regions detected from such a cross (LW Chapters 14–16) as candidate regions for tests of selection using the machinery in Chapters 9 and 10. Here we examine a different class of tests, based not on a signature from a *single* QTL, but rather on the signature from an *entire collection* of QTLs for a given trait. We assume that the lines have been fixed (or nearly so) for alternative alleles at the underlying causative QTLs, and the pattern of fixation (i.e., which alleles were fixed in which line) provides information on whether this pattern was neutral.

The basic idea traces back to three papers, all coincidentally examining crosses for male secondary traits involving *Drosophila simulans* (Coyne 1996; Laurie et al. 1997; True et al. 1997). Under the neutral hypothesis, the relative abundances of "plus" and "minus" QTL alleles (associated with larger versus smaller trait values, respectively), are expected to be randomly distributed over lines and thus should not differ over the crossed lines. Intuitively, this might suggest a simple **sign test**: is there an excessive number of plus alleles in one line? If so, this is not consistent with neutral drift (which is agnostic with respect to the direction and effect sizes of QTLs being fixed). This general strategy will be biased if the parental lines are intentionally chosen (ascertained) to have extreme phenotypes, as the high line would naturally be expected to be enriched with "plus" alleles. Orr (1998a) suggested several approaches to correct for any such bias.

Orr's *QTLST* and *QTLST-EE* Sign Tests

Assume that n detected QTL differences (alternative fixed alleles at n loci) are found via a standard QTL mapping experiment involving a cross between two lines (LW Chapter 15). Under neutrality, there should be no systematic directionality as to whether a line is fixed for increasing (plus) alleles over decreasing (minus) alleles at any particular QTL. This simple idea forms the basis of sign tests, but it requires modifications to account for the actual biology. For example, when the line means differ, the high (larger trait value) line is expected to contain more plus alleles (assuming equal effects; with a distribution of allelic effects, this need not be the case, as is discussed below). Orr noted that by choosing the larger line, we have introduced an ascertainment bias, as this line is *expected* to contain an excess of plus alleles. To proceed, we need some appropriate conditioning on this fact to obtain an unbiased statistic representing the value that constitutes an excess of plus alleles. The simplest approach is Orr's **equal-effects model**, where all n QTLs have close to equal effects. Here, the large line must contain at least $[n/2]$ high (plus) QTLs, where

$$[n/2] = \begin{cases} (n/2)+1 & \text{for } n \text{ even} \\ (n+1)/2 & \text{for } n \text{ odd} \end{cases}$$

In other words, the high line must contain *at least* one more high allele than the low line (because all have equal effects). Determining whether an observed number, n_{high}, of plus alleles in the high line constitutes an excess now becomes a simple combinatorial problem. The probability of k high alleles in one line (under neutrality) follows from the binomial, where there is an equal chance that a random line gets a plus or a minus allele at any particular QTL, yielding

$$\Pr(n_+ = k) = \binom{n}{k}(1/2)^k(1/2)^{n-k} = \binom{n}{k}(1/2)^n$$

Note that all values of k contain a $(1/2)^n$ term. We now condition this probability of k alleles in the high line on the fact that this line *must* contain at least $[n/2]$ plus alleles, yielding

$$\Pr(n_+ \geq n_{high} \,|\, n_+ \geq [n/2]) = \frac{\Pr(n_+ \geq n_{high})}{\Pr(n_+ \geq [n/2])} = \sum_{i \geq n_{high}}^{n} \binom{n}{i} \Big/ \sum_{j \geq [n/2]}^{n} \binom{n}{j} \tag{12.31a}$$

where the common term of $(1/2)^n$ in both the numerator and denominator cancels. This is Orr's **QTL sign test for equal effects**, or ***QTLST-EE***. Orr noted that a *minimum* of $n = 6$ detected QTLs is required for this test to be applied. To see this, note for $n = 6$ that $[n/2] = 4$, and the most extreme value, $n_{high} = 6$, gives a p value of $1/22 \sim 0.05$, while for $n = n_{high} = 5$, the smallest p is $1/16 \sim 0.0625$. For large values of n, Orr noted that Equation 12.31a can be approximated by a normal, with

$$\Pr\left(n_+ \geq n_{high} \,|\, n_+ \geq [n/2]\right) \simeq 2\left[1 - \Phi\left(\frac{n_{high} - [n/2]}{\sqrt{n/4}}\right)\right] \tag{12.31b}$$

where $\Phi(x) = \Pr(U \leq x)$ for $U \sim N(0,1)$.

Example 12.8. True et al. (1997) found that none of the eight detected QTLs for the posterior lobe in the male genitalia in a *Drosophila* cross were **antagonistic**, i.e., with effects in the opposite direction of the line value, such as low (minus) alleles in the high line or high (plus) alleles in the low line. Orr suggested that the equal-effects model may be reasonable for this trait. Assuming that this model holds, we have $n = 8$, $[n/2] = 5$, $n_{high} = 8$, and Equation 12.31a yields the probability of having all eight detected QTLs from the high line being plus alleles as

$$\sum_{i=8}^{8} \binom{8}{i} \Big/ \sum_{j \geq 5}^{8} \binom{8}{j} = \frac{\binom{8}{8}}{\binom{8}{5} + \binom{8}{6} + \binom{8}{7} + \binom{8}{8}} = \frac{1}{56 + 28 + 8 + 1} = 0.011$$

showing that this is a highly significant excess. Orr's normal approximation (Equation 12.31b) yields

$$p \simeq 2\left[1 - \Phi\left(\frac{8-5}{\sqrt{8/4}}\right)\right] = 2\left[1 - \Phi\left(2.1213\right)\right] = 0.034$$

The latter approximation is rather conservative, but not surprising, as this is a large-sample approximation and the number of detected QTLs here is very modest.

Under this strong assumption of equal effects, *QTLST-EE* is a nonparametric test, making no other assumptions, and not using any information on the actual difference between the high and low lines. Anderson and Slatkin (2003) noted that this test can be highly biased by trait choice (whereby the investigator, often unconsciously, chooses traits showing excessive divergence). While Orr's approach corrects for ascertainment bias *within* any specific trait comparison, it assumes that the traits were chosen at random. To examine the impact from nonrandom trait sampling (which, as previously discussed, also biases Q_{ST} tests), Anderson and Slatkin simulated T identical and independently distributed traits, each with 10 QTLs of equal effects, and then chose the most divergent single trait from this set for the subject of a *QTLST-EE* test. They found that this process of trait ascertainment introduces a significant bias. For $n = 10$ QTLs, Equation 12.31a gives the probability of 9 or more plus alleles in the high line as 0.0285. However, when the trait was not randomly chosen, but rather the high line from the *most divergent* trait in a set of 25 traits was used, then over 50% of the time it contained at least 9 high alleles. This lack of robustness with respect to the trait ascertainment scheme means that significant *QTLST-EE* results must be interpreted with caution.

A second class of tests proposed by Orr avoids this problem, and indeed, simulations show that it is conservative (Anderson and Slatkin 2003; Rice and Townsend 2012). For these tests, let D be the difference between the high and low lines. This may be either the actual observed difference, or the difference based on summing the effects over all detected QTLs. With a distribution of QTL effects in hand, one can then condition on the number of plus alleles, *given* the observed difference, D. This is Orr's **QTL sign test** (***QTLST***). The seemingly problematic issue of the distribution of QTL effects can be easily handled via a bootstrap approach in one of two ways. First, one could use the observed distribution of absolute QTL effects ($|a|$), and then fit this using some standard distribution. Orr used the gamma distribution (Equation A2.25a; Figure A2.2) because of its flexibility and the fact that the exponential, a commonly assumed distribution of effect sizes (Chapter 27), is a special case. Note that estimating the distribution parameters that give the best fit is done using a truncated distribution, as QTL effects below a critical absolute size would not be detected. One can then generate the p value for the observed number of plus alleles through parametric bootstrapping. To do so, we generate n draws from this distribution, randomly assign each a sign, and only keep those samples for which the total (G) of signed QTL effects equals or exceeds D. The resulting distribution of plus alleles in the high line is now conditioned on neutrality (signs drawn at random), the assumed distribution of QTL effects, and the actual divergence D, yielding

$$p = \Pr(n_+ \geq n_{high} \,|\, G \geq D) = \sum_{i \geq n_{high}}^{n} \Pr(n_+ = i \,|\, G \geq D) \tag{12.32}$$

where $\Pr(n_+ = i \,|\, G \geq D)$ is simply the fraction of times that exactly i plus alleles were found in the retained bootstrap samples (i.e., those showing a divergence of at least D). Alternatively, instead of sampling from the fitted distribution, one could use standard bootstrapping and directly sample (with replacement, and with draws randomized with respect to sign) from the observed distribution of QTL effects (e.g., Rice and Townsend 2012).

While the *QTLST* adjusts for ascertainment bias, it does so the expense of power. As noted by Rice and Townsend (2012), the difference (D) provides some information on the amount of any previous selection, and by conditioning on its value, we are removing this evidence. Consider the extreme case where a line fixes plus alleles at all ten QTLs, and all have equal effect. In order to obtain the observed value of D, we must condition on only those cases where all ten are fixed, giving this test zero power (Griswold and Whitlock [2003] also noted the low power of this test). Rice and Townsend found that both the power and the false-positive rate increase with the variance of QTL effects.

Orr's tests have a simple appeal, but (as we have shown) there are a number a caveats

Table 12.3 Summary of the analysis of Rieseberg et al. (2002) on the signs of QTLs in traits from wild species, analyzed by trait categories. Within a category, number of antagonistic (opposite sign) and total QTLs are given, along with their QTL ratio (the the fraction of antagonistic QTLs). Under the equal-effects assumption, this ratio should be close to 0.5. As indicated by **, all ratios are significant at $p < 0.001$ (using *QTLST-EE*, with p values adjusted using a sequential Bonferroni correction; Appendix 4), except for *, which denotes $p < 0.01$. A clearer comparison of the category effects is offered by the LS means estimate of the QTL ratio, which uses a linear model to estimate the direct effect of a category. For example, 0.139 is the average fraction of antagonistic QTLs for life history traits, after removing effects of taxon type, species comparison, and mating system. For the LS mean column, † denotes a mean in excess of two standard deviations from zero. Note that the presence of either a smaller QTL ratio or a smaller LS mean implies a stronger effect (a smaller fraction antagonistic QTLs, and hence greater departure from the neutral expectation of close to 0.5).

Trait Category	Antagonistic QTLs	Total QTLs	QTL ratio	LS Means
Animals	73	312	0.234**	$0.185 \pm 0.039^{\dagger}$
Plants	128	439	0.292**	$0.202 \pm 0.025^{\dagger}$
Interspecific	47	245	0.192**	0.137 ± 0.154
Intraspecific	154	506	0.304**	0.250 ± 0.243
Outcross	98	425	0.231**	0.170 ± 0.174
Self	103	326	0.316**	0.217 ± 0.262
Life history	111	540	0.206**	0.139 ± 0.175
Morphology	138	508	0.272**	0.266 ± 0.255
Physiology	8	40	0.200*	0.176 ± 0.125
Phenology	37	124	0.298**	0.236 ± 0.219
Total	201	751	0.268	

concerning their application. Another restriction is that no significant epistasis is allowed, as both tests sum over single-locus effects. There are additional biological reasons for Orr's tests to be conservative. A trait with a smaller fraction of antagonistic QTLs could imply either stronger past selection or more temporally consistent selection. If the sign of directional selection changes on a trait over evolutionary time, then different bouts of selection may fix alleles of opposite signs, lowering the observed excess. If the trait is under stabilizing selection and at its optimal value, random fixations of small-effect alleles are equally likely to be positive or negative. Likewise, if a trait fixes a major allele that causes it to overshoot an optimal value, subsequent fixations of smaller effects in the corrective direction will also reduce any signature from this test. Despite these concerns, Orr's tests can still be useful, even if one treats them as nothing more than an exploratory device, rather than as a formal test (which was our suggestion for most of the locus-specific tests of selection discussed in Chapters 9 and 10).

Applications of QTL Sign Tests

Using *QTLST-EE*, Rieseberg et al. (2002) performed a meta-analysis of over 2600 QTL effects from 572 traits in 86 studies. Their summary statistic was the QTL ratio: the fraction of antagonistic QTLs for the comparison of interest (Table 12.3). Roughly half of the studies involve wild × domesticated crosses, where strong directional selection is suspected for domestication traits. Upon restricting analysis to those examples with six or more QTLs per trait (Orr's condition for such tests to have any power), 35 of the 54 qualifying traits (65%) believed to be involved in domestication showed significant departures from neutrality (i.e., too few antagonistic QTLs). By contrast, only 14 of 84 nondomestication traits (15.6%) in crosses involving domesticated species showed significant departures. Treating this latter class of traits as a control demonstrates that *QTLST-EE* behaved in the direction predicted for these crosses (revealing signatures for domestication traits and a lack of signatures for nondomestication traits).

Given that most studies in this survey involved just four or five QTLs per trait, the restriction of *QTLST-EE* to traits with six or more QTLs discards much potential information. To utilize these additional data, Rieseberg et al. amalgamated traits into a series of categories to accrue a sufficient number of QTLs for the *QTLST-EE* test to be applicable. As shown in Table 12.3, two different analyses were performed on these amalgamated data. First, the table simply reports the unadjusted QTL ratio (the fraction of antagonistic QTLs) for each category. Given that the same study can appear in multiple categories, the unadjusted QTL ratio is potentially influenced by these other categories. The second analysis in Table 12.3 (LS means) uses a linear model to estimate the impact from each category on the fraction of antagonistic QTLs, once the effects of other categories have been removed. This latter analysis more accurately reflects the category contribution.

As shown in the table, all (unadjusted) categories showed significant departures from neutrality, while only two of the adjusted categories (the LS means) did. Rieseberg et al. noted that interspecific crosses had a smaller fraction of antagonistic QTLs than did within-species (intraspecific) crosses (LS mean ratios of 0.137 versus 0.250), which they interpreted as a stronger role for directional selection in generating among-species differences. They also noted that life-history traits (LS ratio 0.139) were (by this criterion) more strongly selected than morphological traits (LS 0.266). While all of these differences are suggestive, it needs to be stressed that none are significant. Further, this pattern of more strongly selected life-history traits is at odds with meta-analyses based on the strength of selection as directly measured in the field (Figure 30.10). Rieseberg noted that the widespread departures from neutrality over all traits could result from an ascertainment bias due to investigators focusing on divergent traits in their crosses.

Ironically, a contrary example to Rieseberg's pattern of life-history and physiological traits tending to have fewer antagonistic QTLs than morphological traits comes from Rieseberg's own group (Lexer et al. 2005). A cross of two wild sunflowers (*Helianthus annuus* and *H. petiolaris*) revealed QTL ratios for life-history, physiological, and morphological traits of 0.345, 0.388, and 0.287, respectively. However, the reduction in the fraction of antagonistic QTLs for morphological traits appears to be due entirely to flower morphology (QTL ratio of 0.214; significant under *QTLST*), with the other measured morphological features (root/shoot and leaf traits) having (nonsignificant) QTL ratios (0.333 and 0.322, respectively) similar to those for life-history and physiological traits.

A second QTL meta-analysis worth mentioning is that of Louthan and Kay (2011), who focused on traits that were expected to be under biotic rather than abiotic selection. While larger QTL effects were more common for biotically selected traits, the fraction of antagonistic QTLs did not differ between the two groups. Both trait groups showed significant departures from neutrality using *QTLST-EE*.

We conclude by briefly highlighting the utility of two applications of sign tests to specific biological problems (as compared to the broad generalizations explored above). Albertson et al. (2003) examined traits in the massive species radiation occurring among cichlid fishes in the East African rift lakes. One striking feature of this radiation is extensive convergent evolution across lakes in feeding morphology, suggesting parallel directional selection. The authors used *QTLST* (with effect sizes drawn from a gamma distribution) to examine the genetic basis of feeding morphology through crossing two wild species from Lake Malawi. Because most individual traits had less than six QTLs, they grouped the traits, finding that only 4 of the 46 QTLs were antagonistic for jaw and teeth features. The highly significant p value supports directional selection on these feeding traits. Muir et al. (2014) examined QTLs in tomatoes (*Lycopersicon*) to explore leaf-related traits in wild species thought to be associated with adaptation to precipitation. They found no significant departure from neutrality for two leaf and two trichome (leaf hair) traits, but a significant departure from neutrality for two stomatal traits. They computed p values using both *QTLST* and *QTLST-EE*, and they found (in agreement with Anderson and Slatkin 2003) that *QTLST* was more conservative, yielding p values about twice as large as those obtained from *QTLST-EE*.

TRAIT-AUGMENTED MARKER-BASED APPROACHES: TESTS USING GWAS INFORMATION

While QTL data usually involves fixed differences that are revealed by crossing two divergent lines, genome-wide associaton studies (GWAS) provide information on currently segregating alleles within a target population (or set of populations). As such, GWAS data potentially offer inroads into the vexing problem of detecting a polygenic sweep (Chapter 8). In this setting, allele-frequency shifts at the underling loci are expected to be small, and hence missed by most standard tests that look for single-locus signatures of selection (Chapter 9). The power of trait-augmented marker tests is that one chooses a set of markers given a trait, and then pools information across all of these markers, potentially generating a much stronger signal than could be found by considering any single marker in isolation.

Approaches Based on Combining Signals

The basic idea of combining signals over a set of GWAS markers has been exploited in several different ways. The initial suggestion was **gene set enrichment analysis (GSEA)** from genomics (Subramanian et al. 2005), wherein one considers clusters of genes on the basis of membership in some functional group (i.e., the same gene ontology, GO, class). This tactic was used by Daub et al. (2013), who computed the average F_{ST} value over a set of pathway-connected genes and contrasted this with the average F_{ST} value over a same-sized set of putatively neutral markers. Using this approach, they found evidence for selection on several human pathways, many connected with pathogen response. They also noted that long-distance LD was detected, which they attributed to epistatic interactions. While this could be correct, a confounding factor is that selection is also expected to generate such long-distance (i.e., between loosely-linked sites) LD with strictly additive genes (Chapters 16 and 24).

In the Daub et al. analysis, the "trait" was a specific pathway, while other analyses have consider more classical human traits, in particular, height. A simple, but robust, approach was used by Turchin et al. (2012). They examined allele frequency differences for 139 GWAS markers for height between Northern- and Southern-European populations. Under neutrality, allele-frequency increases in plus alleles should be randomly distributed between two populations (i.e., the sign test introduced above). Instead, what they found was that 85 of the 139 markers (sign test $p = 0.01$) showed an increase for high alleles in the Northern-European population. Note that one advantage of GWAS data is that typically a reasonable number of hits (marker-trait associations) are found, while QTL-based studies often fail to have more that five detected QTLs for a focal trait (and hence Orr's test is not applicable).

Tests Based on *tSDS* Scores

Recall Field et al.'s (2016) singleton density score, *SDS* (Equation 9.42), for detecting very recent selection on a given single site. In that paper, they also showed how to extend this approach to search for polygenic selection on a given candidate trait, given a set of associated GWAS marker scores. This requires that both the *SDS* values and GWAS test statistics (such as a z value under a normality test) for a set of markers were generated using the same population. The *SDS* score for a given marker is first translated into a ***tSDS*** (**trait-*SDS***) score, where the sign of the *SDS* score is changed so that trait-increasing markers receive positive scores. Their simplest approach was to combine the *tSDS* scores associated with all the significant GWAS markers for a target trait, using this mean as the test statistic.

Field et al. noted that most of the trait variance is usually explained by markers whose GWAS test statistics do not pass the genome-wide significance threshold (Chapter 24), and hence are not included in the test set. To incorporate information from these nonsignificant (but potentially biologically important) markers, they used a regression-based approach. Data points for the regression were generated by first binning SNPs with very similar GWAS scores, taking the bin average GWAS score as the predictor variable and bin average *tSDS* score as the associated response variable. A significant regression (or correlation) is expected under selection, but not under drift. Both the average *tSDS* score and regression approaches detected clear signals of selection for increased height in Britain over the past 2000 to 3000 years. Several other traits (infant head size, body mass index [BMI], and female hip size, to name a few) also showed evidence of recent polygenic selection.

The Berg and Coop Q_x Test

The final approach leveraging GWAS-estimated marker effects for a target trait is due to Berg and Coop (2014), building on previous work by them (Coop et al. 2010; Günther and Coop 2013) as well as by Ovaskainen et al. (2011). Let $\mathbf{p}_i^T = (p_{i,1}, p_{i,2}, \cdots, p_{i,m})$ denote the vector of allele frequencies for the ith marker over m subpopulations, where $p_{i,j}$ denotes the allele frequency in population j. Example 9.5 showed that the expected distribution of $\mathbf{p}_i$ under neutrality is approximately given by

$$\mathbf{p}_i \sim \text{MVN}_m \left[p_{i,0}\mathbf{1}, p_{i,0}(1 - p_{i,0})\boldsymbol{\Omega} \right] \tag{12.33a}$$

where $\boldsymbol{\Omega}$ is a (marker-estimated) matrix of expected covariances in allele frequencies over the subpopulations and $p_{i,0}$ is the allele frequency in the ancestral population. As detailed in Chapter 9, this formed the basis of Coop's *Bayenv* test for excessive divergence at a *specific locus*. Berg and Coop (2014) extended this result to a *trait-based* test as follows. Suppose n GWAS hits are discovered for the focal trait, where the trait-increasing allele for the ith marker changes the trait by a value of g_i, with $p_{i,j}$ denoting the frequency of this allele in population j. The GWAS-predicted mean genetic value for the trait in population j thus becomes

$$a_j = 2\sum_{i=1}^{m} g_i\, p_{ij} \tag{12.33b}$$

Letting $\mathbf{a}^T = (a_1, a_2, \cdots, a_m)$ be the vector of mean trait genetic values over the m populations, then combining Equations 12.33b and 12.33a gives the expected distribution of trait means under drift as

$$\mathbf{a} \sim \text{MVN}_m \left[\mu\mathbf{1}, 2V_A\boldsymbol{\Omega} \right] \tag{13.33c}$$

where

$$\mu = \frac{2}{m}\sum_{i=1}^{n} g_i\, p_{i,0} \qquad \text{and} \qquad V_A = 2\sum_{i=1}^{n} g_i^2\, p_{i,0}(1 - p_{i,0}) \tag{13.33d}$$

represent the expected genetic value and additive variance in the ancestral population.

To proceed, Berg and Coop expressed all of the a_j values as deviations from the grand mean, yielding $a_j^* = a_j - \overline{a}$. This uses one degree of freedom, and returns the vector $(\mathbf{a}^*)^T = (a_1^*, a_2^*, \cdots, a_{m-1}^*)$, where one population is dropped. As Berg and Coop note, information from the dropped population is fully retained by the vector $\mathbf{a}^*$, so that the choice of which population to drop has no impact on the resulting analysis. The resulting vector is now distributed as

$$(\mathbf{a}^*)^T \sim \text{MVN}_{m-1} \left[\mathbf{0}, 2V_A\boldsymbol{\Omega} \right] \tag{13.33e}$$

As discussed in Appendix 5, a standard trick with a vector of correlated variables is to use a transformation to return a vector of uncorrelated variables of unit variance. Berg and Coop did this by using the Cholskey decomposition (Appendix 5) of $\boldsymbol{\Omega} = \mathbf{C}\mathbf{C}^T$, using the transformation

$$\mathbf{x} = \frac{1}{\sqrt{2V_A}}\mathbf{C}^{-1}\mathbf{a}^* \tag{13.33f}$$

which returns

$$\mathbf{x} \sim \text{MVN}_{m-1}(\mathbf{0}, \mathbf{I}) \tag{13.33g}$$

This is the basis for the **Berg-Coop Q_x test**, whose statistic is given by

$$Q_x = \mathbf{x}^T\mathbf{x} = \frac{(\mathbf{a}^*)^T\boldsymbol{\Omega}^{-1}\mathbf{a}^*}{2V_A} \tag{12.33h}$$

Under neutrality, $Q_x \sim \chi^2_{m-1}$, as $\mathbf{x}^T\mathbf{x}$ is the sum of $(m-1)$ squared unit-normal random variables. Note by comparing this result to Equation 9.13c, that the Q_x test is very similar in form to the Günther-Coop (2013) $\mathbf{X^TX}$ test for selection on a single site, but with estimated trait genetic values replacing allele frequencies.

Robinson et al. (2015) applied this test to height and BMI based on ~9400 individuals from 14 European countries, finding evidence that selection favored increased height and reduced BMI. Mathieson et al. (2015) also applied this test to Europeans, but used ancient DNA from 230 individuals (who lived between 6400 to 300BC), and reported evidence for two independent episodes of selection for height.

DIVERGENCE IN GENE EXPRESSION

The power of quantitative genetics is that its machinery can be applied to *any* character of interest, including omics traits (e.g., amounts of specific transcripts, proteins, and metabolites). Application of quantitative-genetic machinery to such omics traits has been coined **genetical genomics** by Jansen and Nap (2001), and it traces back to Damerval et al. (1994), who mapped QTLs controlling the spot volume of anonymous maize proteins detected by two-dimensional gel electrophoresis (LW Figure 15.10).

Level of Gene Expression as a Quantitative Trait

Much of the current work in genetical genomics has focused on the transcriptome, treating the level of expression of a specific gene as a quantitative trait and then attempting to map **eQTLs** (expression QTLs) or **eSNPs** (in GWAS studies) that influence this trait. Modern transcriptomic tools (initially using **microarray analyses**, and more recently, **RNA-Seq**) allow one to measure the level of expression for essentially the full repertoire of an individual's genes (Schena et al. 1995; Brown and Botstein 1999; Duggan et al. 1999; Wang et al. 2009). The amount of mRNA present (either measured by the intensity of hybridization against probes for a gene or from the amount present in massive sequencing of an RNA pool) is a typical quantitative trait, showing both genetic and environmental sources of variation, and further confounded by measurement error. This transcriptomics approach yields thousands of traits, as expression levels of each gene are separate, although potentially highly correlated, characters.

Our treatment of the evolutionary analysis of gene expression glosses over a number of very important concerns in the actual generation and processing of the raw data. Gene expression is both highly environmentally dependent and tissue specific, and formally, it should be viewed as a function-valued trait (Volume 3)—a character whose value is indexed by time and potentially other features (such as tissue type or developmental stage). In the following discussion, we simply refer to "the" expression level of a gene, but this is highly context-specific, with many transcripts showing considerable variation within an individual (over both time and tissues). There is also **technical variation** due to sampling and hybridization/amplification, so that an otherwise identical sample might still show considerable variation. Much of the early work used whole organisms (and hence all tissues at the sampled developmental stage), and it often pooled multiple individuals. RNA-seq relaxes many of these restrictions, as it requires much smaller amounts of material. Whichever method is used to assess expression levels, significant care is required to ensure that traits being compared are indeed the same character (i.e., expression levels at the same development time in the same set of tissues). As highlighted by Lynch and Walsh (1998), accurate quantitative-genetic studies, especially with noisy traits, require very significant replication, something lacking in many published studies of expression variation. All of these concerns highlight the fact that careful experimental design is absolutely critical in expression studies (Kerr and Churchill 2001a, 2001b; Churchill 2002; Yang and Speed 2002; Kerr 2003; Rosa et al. 2005).

While microbes have historically had only a relatively minor role in classical quantitative genetics (largely due to their limited number of easily scored traits), they have flourished in the genetical genomics era, in part because they allow many of these design issues to be addressed. Starting with Brem and Kruglyak (Brem et al. 2002, 2005; Brem and Kruglyak 2005), yeast was quickly adapted as a model system for the quantitative genetics of gene expression, allowing investigators to examine the heritability of expression and map eQTLs for thousands of transcripts. This work was closely followed by similar analyses in mice, humans, and maize (Schadt et al. 2003), and has subsequently been extended to an ever-growing number of species. Such genome-wide transcription studies offer very high phenotypic throughput, allowing thousands of traits to be scored in a single experiment. While there is some modest bias against weakly expressed genes, the sample of expression levels over thousands of loci offers a largely unbiased view of the quantitative genetics of this class of traits. This is an extremely active area, with the early work reviewed by (among others) Stamatoyannopoulos (2004), de Koning and Haley (2005), Gibson and Weir (2005), Ranz and Machado (2006), Rockman and Kruglyak (2006), Whitehead and Crawford (2006), Fay and Wittkopp (2007), Gilad et al. (2008), Skelly et al. (2009), Emerson and Li (2010), Romero et al. (2012), and Albert and Kruglyak (2015). The conclusion from this early work is that the control of gene expression often has considerable heritability, involves both *cis* and *trans* factors, and can be very polygenic.

In order to understand the nature of evolutionary forces shaping the complex webs of gene regulation, appropriate null models are needed. That our initial impression of the origins of a given network structure can be highly misleading was stressed in Kauffman's (1969) classic paper. He showed that randomly constructed gene networks appear to be highly coordinated and hence give the appearance of being highly evolved (i.e., highly structured by natural selection). Thus, tests of whether certain features of gene regulation (or any omics data) are largely neutral are critical to understanding which evolutionary forces might shape these features. Much of the machinery developed in this chapter for detecting departures from neutral drift has been applied to the amount of divergence in gene expression.

Finally, before proceeding, a subtle, but important, clarification is in order. Our focus here is on detecting selection on particular *traits*, while Chapters 9 and 10 discuss machinery for detecting selection on particular *genes* (or, more correctly, specific *sequences*). Selection on the *expression level* at a specific gene is different from selection acting on the *sequence* of that gene. To see this, suppose only *trans*-acting factors influenced the expression levels at our target gene. In this case, *trans*-factor alleles at the eQTLs for that transcript would be under selection, *not* the alleles at the target gene itself. A test of a sequence-specific signature of selection would not pick up the target gene, but might detect genes coding for these *trans*-acting factors (subject to all the caveats discussed in Chapters 9 and 10). Likewise, a genome-wide association study would highlight sequence variation in *trans*-acting genes, *not* sequence variation in the gene whose expression is under selection. This distinction becomes murky when a gene also has *cis*-acting sites, as, while the expression level is the target, sequence variation near the gene (*cis*-acting sites), as well as at more distant (often unlinked) sites, would be the genetic targets of selection. An example of such a *cis*-acting site is in the *tb1* gene, which is involved in the domestication of maize (Chapter 9). This site is roughly 60 kilobases upstream of the *tb1* gene itself, and it is influenced by the insertion of a *Hopscotch* retrotransposon that increases the amount of *tb1* transcripts (Studer et al. 2011).

Rate-based Tests for Neutrality in Divergence of Gene Expression

Early attempts to detect departures of gene expression evolution from neutral trait predictions used rate-based approaches, based on Lande's F_{MDE} test (Equation 12.20a). While the standard version of this test uses the ratio of the observed among-group variance, V_B, to the expected among-group variance (expressed as $2t\sigma_m^2$, where t is the separation time in generations), a modified version uses an estimate of V_A (the additive-genetic variance of expression in the reference population or species) in place of $2t\sigma_m^2$. An important caveat is that these early studies were based on the phenotypic variance, such as the among-line variance, as a surrogate in place of the additive-genetic variance (Hsieh et al. 2003; Rifkin et al. 2003; and Nuzhdin et al. 2004). This approach raises some of the issues discussed previously when using P_{ST} as a surrogate of Q_{ST}; see Example 12.9.

With this concern in mind, the version of Lande's test used by these investigators starts by noting that $E[V_A] = 2N_e\sigma^2_m$ for an additive trait at mutation-drift equilibrium (Equation 11.20c). Assuming expression values are drawn from a normal distribution under the null model, then when L lineages are used to estimate the among-group variance and k individuals per line were used to estimate V_A, Equation 12.5a shows that both estimators approximately follow χ^2 distributions, with

$$V_B \sim (2\,t\,\sigma^2_m) \cdot \chi^2_{L-1}/(L-1) \qquad \text{and} \qquad V_A \sim (2N_e\,\sigma^2_m) \cdot \chi^2_{k-1}/(k-1) \tag{12.34a}$$

where t is the time of divergence since the common ancestor. These expressions suggest a modified version of Lande's F_{MDE} test statistic,

$$F^*_{MDE} = \frac{V_B/(2t\sigma^2_m)}{V_A/(2N_e\sigma^2_m)} = \frac{V_B}{V_A} \cdot \left(\frac{N_e}{t}\right) \sim \frac{\chi^2_{L-1}/(L-1)}{\chi^2_{k-1}/(k-1)} \tag{12.34b}$$

where this statistic follows an F distribution, with $F^*_{MDE} \sim F_{L-1,k-1}$ (as it is the ratio of two χ^2 random variables, scaled by their degrees of freedom; see LW Appendix 5). A scaled ratio less than a critical value of $F_{\alpha/2}$ is suggestive (at level α) of too little divergence, and hence suggestive of stabilizing selection, while a scaled ratio in excess of $F_{1-\alpha/2}$ implies too much divergence, suggestive of directional selection. These critical values are given by

$$\frac{V_B}{V_A} \le F_{\alpha/2,L-1,k-1}\left(\frac{t}{N_e}\right) \qquad \text{and} \qquad \frac{V_B}{V_A} \ge F_{1-\alpha/2,L-1,k-1}\left(\frac{t}{N_e}\right) \tag{12.34c}$$

where $F_{\alpha,M,N}$ denotes critical values for an F distribution and satisfies

$$\Pr\left(F_{M,N} \le F_{\alpha,M,N}\right) = \alpha$$

In the case where just two populations ($L = 2$) are compared by using their squared difference, d^2, then recalling that $V_B = d^2/2$ (Equation 12.8c), the conditions given by Equation 12.34c become

$$\frac{d^2}{V_A} \le F_{\alpha/2,1,k-1}\left(\frac{2t}{N_e}\right) \qquad \text{or} \qquad \frac{d^2}{V_A} \ge F_{1-\alpha/2,1,k-1}\left(\frac{2t}{N_e}\right) \tag{12.34d}$$

Example 12.9. Rifkin et al. (2003) examined variation in gene expression at the start of metamorphosis in six inbred lines of *Drosophila*: four *melanogaster*, one *simulans*, and one *yakuba*. Of the roughly 12,900 genes whose transcripts were examined, 52% (~6700 genes) showed expression changes in at least one lineage (either between species or within the *melanogaster* lines). For ~4500 of these genes, the authors could not reject the hypothesis that all six lineage-specific samples came from the same distribution, and these were deemed to be evolutionarily stable and potentially under stabilizing selection. Of the remainder, ~1700 genes showed no significant variation across the sampled *melanogaster* lines, but a significant difference between *melanogaster* and one of the other species. These were deemed to be under lineage-specific selection.

The evolutionary forces acting on the remaining 527 genes were examined using Equation 12.34b. Divergence was scored separately between *melanogaster* and each of the other two species ($L = 2$) using d^2, with V_A estimated from the among-group variance in four fully inbred *D. melanogaster* lines ($k = 4$). Because d^2 is used, critical values are given by Equation 12.34d, with one correction. The expected among-group variance (for an additive trait) between a set of fully inbred lines is twice the additive variance (from Table 11.3, with $2f = 2$), so that $2N_e$ replaces N_e in the critical values. The resulting upper and lower 2.5% critical values follow first by noting that $\Pr(F_{1,3} \geq 17.4) = 0.025$ and $\Pr(F_{1,3} \leq 0.001) = 0.025$. These authors used an estimated effective population size for *D. melanogaster* of $N_e = 3 \cdot 10^6$, while the total divergence times (twice the separation time, in generations) were estimated as $2t = 4.6 \cdot 10^7$ (*melanogaster-simulans*) and $2t = 10.2 \cdot 10^7$ (*melanogaster-yakuba*). Hence, the critical values for excessive divergence were

$$F_{c,mel\text{-}sim} = 17.4 \cdot \frac{4.6 \cdot 10^7}{6 \cdot 10^6} = 133.4 \quad \text{and} \quad F_{c,mel\text{-}yak} = 17.4 \cdot \frac{10.2 \cdot 10^7}{6 \cdot 10^6} = 195.8$$

Transcripts whose ratio of d^2/V_A exceeded these values are unusually divergent. Using this criterion (as well as the lower threshold for too little divergence), of these remaining 527 genes, 464 were consistent with drift, while 63 were consistent with excessive divergence between at least one species pair.

One caveat with using half the among-group variance over a set of fully inbred lines as an estimate of σ^2_A is that, when dominance is present, the among-line genetic variance becomes $2\sigma^2_A + 2\sigma_{ADI} + \sigma^2_{DI}$ (Equation 11.6a). Depending on the sign in the middle covariance and the magnitude of the last two terms, this can be considerably larger, or smaller, than $2\sigma^2_A$, and hence can bias the results.

While straightforward to apply, a concern with Equation 12.34b is the estimation of t and N_e. One approach is that, under neutrality, the ratio of the expected divergence D_s at silent sites divided by their expected polymorphism (P_s) is $D_s/P_s \simeq t/(2N_e)$ (Equation 10.1b). While this marker-based estimate could be substituted into Equation 12.34b, this is an ad hoc approach, as the sampling error of this estimator of t/N_e is ignored. The idea of combining silent-site information with the within- and between-population expression variances was also considered by Warnefors and Eyre-Walker (2012), who proposed a MacDonald-Kreitman-type approach (Chapter 10); see their paper for details. Again, these are useful metrics, but not formal statistical tests.

The rate-based test given by Equation 12.34b was scaled to be independent of σ^2_m, at the cost of assuming or estimating an effective population size. Conversely, Equation 12.20d can be used to compare the rate divergence to that of candidate values of σ^2_m (or h^2_m), which circumvents the potentially problematic issue of estimating N_e, although (as with Equation 12.34b) one must still estimate t. This was the approach taken by Lemos et al. (2005). Using a diverse series of lineage comparisons (within mice strains, populations of *Drosophila*, and between species in primates and flies), and assuming h^2_m is in the range of 10^{-4} to 10^{-2}, they found that the majority of gene-expression differences were consistent with stabilizing selection (an average of 85% of all transcripts), with drift comprising the next largest category (an average 11.5%), and directional selection comprising the smallest (an average of <4%).

These approaches either ignore σ^2_m or estimate its required value under drift given the observed divergence. A more arduous approach is to directly estimate σ^2_m from a mutation-accumulation experiment (LW Chapter 12). This estimate is then used to predict either the within-population variation ($2N_e\sigma^2_m$) or the among-population divergence ($2t\sigma^2_m$) and to assess if these are consistent or too extreme. The general conclusion from such studies is that stabilizing selection plays a prominent role in reducing the amount of variation in gene expression below the neutral expectation—both within and among species, levels of variation are much lower than expected based on the estimated mutational variance. For example, using lines of the nematode *C. elegans* from a long-term (280 generation) mutation-accumulation experiment, Denver et al. (2005) estimated values of σ^2_m for several thousand genes. By comparing levels of variation among a global collection of natural isolates, they found that the gene-specific ratios of the standing level of genetic variance to the estimated σ^2_m value were generally no greater than a few hundred. Given that this ratio provides an estimate of $4N_e$ under the assumption of neutrality in a selfing organism (as opposed to $2N_e$ in an outcrosser, as σ^2_A is inflated by $2f = 2$ with complete inbreeding), these observations provide a firm rejection of the hypothesis that gene-expression levels evolve in a neutral fashion. Rifkin et al. (2005) were able to estimate mutational heritabilities for expression in mutation-accumulation lines of *D. melanogaster* by factoring out the variance at the level of the individual fly to obtain an estimate of σ_e^2. They found a median of $h_m^2 \simeq 2.4 \cdot 10^{-5}$ across all genes, and they showed that although interspecific variance in the expression of a gene was correlated with its mutational variance (in qualitative accordance with the neutral theory), the absolute level of divergence was too low to be compatible with neutrality, consistent with the results of Denver et al. (2015).

An especially interesting analysis of expression levels was offered by Hodgins-Davis et al. (2015), who used the machinery developed in Chapter 28 on the expected level of variation under a balance between mutation and stabilizing selection. Using data sets for yeast and *Drosophila*, they found that the model of mutation-selection balance that best explains the observed pattern of variation is one of large mutational effects and weak stabilizing selection (details are given in Example 28.3).

Largely Neutral Evolution of Expression Levels in Primates?

While these results strongly suggest a leading role for stabilizing selection, there has been considerable discussion regarding the evolution of gene expression among the great apes. One key focus has been the expected pattern of divergence under pure drift. However, as with many of the early applications of rate-based tests to expression data, much of the analysis is largely phenotypic in nature, and therefore does not utilize the much stronger comparisons based on estimates of the additive variances of these traits.

Under a Brownian motion model, the expected divergence (measured by the among-group variance) scales linearly with divergence time, t (Equation 12.10, under the infinite-alleles assumption). In contrast, under an Ornstein-Uhlenbeck (OU) process (drift countered by stabilizing selection), the total divergence approaches an asymptotic value (Equation 12.22c). Bedford and Hartl (2009) used an OU process to fit the pattern of expression divergence within a clade of seven species of *Drosophila*. In accordance with the OU model (and consistent with stabilizing selection), they found that the divergence variance does not increase linearly with time but, rather, quickly approaches an asymptotic value.

In contrast, Khaitovich et al. (2004, 2005) argued that gene expression can evolve in a mostly neutral fashion, based in large part on an observed linear increase in the divergence of among-species expression with time within the clade of great apes. They also noted the observation of Riftkin et al. (2005), namely, a positive correlation between levels of within- and among-species variation for the expression of different genes. Such a pattern is expected under neutrality, as both divergence and standing variation are functions of σ_m^2. However, this is not strong support for neutrality, as a number of other features can create such a correlation. For example, genes whose expression is strongly influenced by the environment may naturally exhibit higher levels of variation, both within and among samples. Likewise, linearity in divergence, by itself, is suggestive, but not sufficient. Unless the actual rate of divergence is consistent with the rate of polygenic mutation, linear patterns of evolutionary diversification need not imply neutrality.

An important complication is that Khaitovich et al. (2004, 2005) used human probes to measure differences in expression among species. Because sequence divergence between the probes and target sites accumulates over time, generating reduced levels of hybridization, this technique will result in an artifactual increase in apparent expression divergence over time. Broadley et al. (2008) reported a similar linear divergence of expression variance with time in a series of 14 taxa in the Brassicaceae but, again, the probes were based on a single species (*Arabidopsis*). In evaluating primates more broadly (human, chimpanzee, orangutan, and rhesus macaque), Gilad et al. (2006) found that the between-species variance in expression of most genes did not increase with divergence time, contrary to the neutral expectation. This study was well designed in that it employed only probes for which the sequences were identical across all four species, thereby removing any species-specific bias in hybridization. Thus, the conclusion that primate gene expression is evolving in a neutral fashion is questionable, and it has, in fact, been essentially retracted in a more recent analysis (Chaix et al. 2008), which suggested a rate elevation specific to the human lineage.

Conversely, drift was suggested by an analysis by Perry et al. (2012), who used RNA-seq data (very high-coverage sequencing of an RNA pool) to examine liver transcript levels for roughly 5700 genes over humans and 11 other primates. Expression levels were modeled by a Brownian motion model over the entire phylogeny. The base model assumed a constant rate of change (per unit time), and this was contrasted with a model in which the rate of change was allowed to vary over the phylogeny. The latter model provided a significantly better fit for slightly less than 10% of the genes, with specific branches showing accelerated rates of evolution (and hence being candidates for directional selection). Taken as a whole, this analysis suggested that most of the expression changes over this large phylogeny were at least partly consistent with neutral drift. This conclusion, however, is somewhat tempered, as the contrast between two different Brownian motion models is not nearly as conclusive as a contrast between a Brownian and an Ornstein-Uhlenbeck model to compare drift versus stabilizing selection. Further, there were no quantitative genetics involved in this analysis—the more powerful comparison of the expected rate of change (given an estimated additive variance) with the observed rate is lacking.

Transcriptional Q_{ST}, tQ_{ST}

Gibson and Weir (2005) suggested that comparisons of Q_{ST} and F_{ST} can be applied to gene expression data, and proposed the term $\boldsymbol{tQ_{ST}}$ for such a transcriptome scan. By scoring a very large number of traits at once, the vexing issue of ascertainment bias (wherein researchers are naturally drawn to the most variable traits) that plagues Q_{ST} tests can be largely avoided. However, other problems with this approach persist. One involves obtaining the estimated genetic variances of within- and among-population components, as opposed to their phenotypic proxies (P_{ST}). The second problem is low power, especially when only two populations are compared. Perhaps because of these concerns, this approach has not been not widely applied to expression data. A Q_{ST} approach was hinted at by Whitehead and Crawford (2006), but these authors ultimately resorted to rate-based tests for comparing transcripts.

A formal application of this approach was performed by Roberge et al. (2007) to study a very recent population divergence, which was created by the installation of a fish ladder in 1981 on the Sainte-Marguerite River in Quebec. Upstream and downstream populations of Atlantic salmon (*Salmo salar*) showed an F_{ST} divergence of just over 0.03 after roughly six generations of presumed differential natural selection. The authors used a mixed-model framework (Chapters 19 and 20) to estimate the among-group genetic variance for transcripts from these two populations, and then searched for transcripts showing up as Q_{ST} outliers. They found 16 such transcripts, with an average Q_{ST} roughly three times the F_{ST} value between these two populations, leading them to suggest that the expression levels for these genes were under extensive directional selection following the population subdivision. However, Equation 12.28c shows that Q_{ST} must be roughly five times as large as F_{ST} for significance when $L = 2$. Thus, while there are hints of selection, low power prevents a definitive conclusion.

Cis Versus *Trans*, Local Versus Distant, and Allele-Specific Expression (ASE)

Finally, before discussing applications of sign-based tests to gene-expression data, we need to review a few additional features of transcriptional regulation. Historically, the term *cis* refers to a control element that only acts on a gene residing on the same DNA molecule. *Cis*-acting control elements are thought to be binding sites for diffusible factors (e.g., transcription factors, small RNAs, etc.), target sequences for gene-processing features (intron splice sites, poly-A sites, etc.), or sites that exert some local control over chromatin structure. By contrast, *trans*-acting factors are diffusible and exert their influence throughout the genome, presumably by coding for proteins or RNAs that interact with specific *cis* sites to control regulation. Formally, the terms *cis* and *trans* refer to this difference in functionality. However, the early eQTL mappers co-opted them to refer (respectively) to sites that closely mapped to the gene coding for the target transcript (whose expression is being followed) and those that mapped further away (often on different chromosomes). Rockman and Kruglyak (2006) suggested that the terms **local** and **distant** are more appropriate for describing eQTL location. Given that the uncertainty region for a typical QTL spans tens of centimorgans (and hence tens of megabases), what is called a *cis* eQTL could actually be several genes away and thus could act in *trans*.

There are formal genetic procedures for determining whether a region truly does act in *cis*, and these are exploited by a few of the sign-based tests. These employ a more nuanced view of the expression from a single gene in an individual, as not simply its total amount of mRNA, but rather at the levels of expression from different *alleles* at that gene—**allele-specific expression** (**ASE**; Wright and Moyer 1966; Knight 2004). When the gene products from the different alleles at a locus can be distinguished (either by hybridization or sequencing), then the expression of each product can be followed (e.g., Yan et al. 2002). Cowles et al. (2002) and Wittkopp et al. (2004) both proposed that the use of hybrids (crosses of different, often inbred, lines) can distinguish *cis* from *trans* control of expression. If allele-specific differences seen in the parental lines persist in an F_1 hybrid, these are (at least in part) due to *cis*-acting factors, as the two alternative alleles in the hybrid both experience the same environment, and hence the same set of *trans*-acting factors. Wittkopp et al. (2008) defined the amount, C, of *cis*-acting differences by the ratio of expression of the two alleles in the hybrid (C = allele 1/allele 2), while P = strain 1/strain 2 is the ratio of expression in the parental homozygous lines. The amount of *trans*-acting differences is estimated by the difference between these two sets of ratios, $T = P - C$, although this approach can be complicated by *cis* $\times$ *trans* interactions. Because parent-of-origin (i.e., imprinting) effects can also create ASE, reciprocal crosses are used to rule out such effects. Emerson et al. (2010) offered an alternative approach for determining the amount of *cis* and *trans* effects.

Applications of Sign-based Tests to Expression Data

While Orr's tests were framed in the increasingly dated technology of QTL mapping, their central underlying idea (effects are randomly distributed among lines under neutrality) fits very nicely with genomics-era data. We already mentioned a GWAS application of sign-based tests, and there is an increasing use of sign-based approaches to explore the nature of selection on gene expression. The standard QTL-based tests discussed above are not directly applicable, as most genes have very few detected eQTLs, and thus do not qualify for testing based on Orr's requirement of at least six QTLs per trait. However, as reviewed by Fraser (2011), two rather different approaches have been used to circumvent this limitation.

The first approach is simply to shift focus from the expression levels at *single genes* to the pattern of expression over a *set of genes*, pooling these to create a setting with more than six eQTLs for the trait. Bullard et al. (2010) used this approach in a cross of two closely related yeast species, *Saccharomyces cerevisiae* and *S. bayanus*. One key requirement in the statistical analysis is that each eQTL is independent, as a single eQTL that simultaneously influences k genes should be weighted as one change, not k changes, in the same direction. The use of *cis*-regulatory alleles ensures independence over a set of loosely linked genes. Bullard et al. accomplished this by only considering alleles showing ASE. An excessive number of up-regulated ASEs over a specific gene set from one species indicates the presence of lineage-specific selection, and a number of pathways were detected showing this feature. Fraser et al. (2011) used a similar approach in a cross of two subspecies of the mouse (*Mus musculus*). They chose gene sets defined by shared GO (Gene Ontology Consortium) membership, and found over 100 genes with evidence of lineage-specific selection. These studies are important, as (at least for these two crosses) they suggest that adaptation via gene-expression changes may be widespread, highly polygenic, and involves *cis*-regulatory sites.

As noted by Fraser (2011; Fraser et al. 2011), a significant result (an excess in one direction) is not necessarily indicative of directional selection, as regulatory mutations are biased toward down-regulation. A pattern that is seen could simply be a relaxation of purifying selection in one lineage, resulting in a series of neutral, but down-regulated, substitutions, given this inherent mutational bias. In essence, this is the same limitation seen in the McDonald-Kreitman test (Chapter 10). Fraser et al. (2011) noted that a simple solution to this problem is to examine expression levels in an outgroup and assess whether a directional change was due to up-regulation (relative to the outgroup) in one lineage, which is likely due to selection, or down-regulation, which could simply be due to relaxed selection. However, selection for reduced expression in a pathway cannot formally be ruled out in the latter case, nor can overexpression be interpreted with certainty as being adaptive.

A second modification of a sign-based test for expression data, which was offered by Fraser et al. (2010), applies to genes whose expression levels are influenced by both *cis* and *trans* eQTLs. The central premise of sign-based tests is that ***directionality is random under the null***, so that in a cross of lines A $\times$ B, if an eQTL from A is a *cis* up-regulator, this should provide no information as to whether a *trans*-acting factor from A (acting on the expression level at the same target gene) is an up- or down-regulatory allele. *Cis*- and *trans*-acting alleles whose influence is in the same direction (up and up, down and down) are called **reinforcing**, and those acting in opposite directions are called **opposing**. With a collection of genes whose expression is influenced by both *cis* and *trans* eQTLs, a simple 2×2 contingency table can be constructed and tested for departures from randomness. If a significant departure is seen, it is a straightforward process to estimate the amount of excess in a particular class (e.g., Example 10.1). Fraser et al. applied this approach in a cross of two yeast (*Saccharomyces cerevisiae*) strains that diverged roughly 10^7 generations ago and found an excess of roughly 242 genes showing reinforcing levels of *cis* and *trans*. While this approach suggests significant regulatory evolution over the genome, it does not indicate *which* specific transcripts are involved. This result is reminiscent of some of the approaches for detecting genome-wide signatures of selection examined in Chapter 10: evidence of a genome-wide pattern is seen, but no particular gene can be singled out with confidence as being a target of the selection process generating the observed pattern.

Artieri and Fraser (2014) used this statistical machinery to examine the nature of selection on the *translational* profiles of mRNA, using **ribosome profiling**—extracting mRNA bound to ribosomes to create a sample of the mRNA pool actually undergoing translation. Using this enriched pool, Artieri and Fraser examined the translation rates of specific transcripts within and between two species of yeast, and found both *cis*- and *trans*-acting regulatory divergence. They reported that the majority of translational divergence appears to buffer the amounts of mRNA, consistent with stabilizing selection on expression levels acting at both the transcriptional and translational stages of gene regulation.

Example 12.10. Using *cis* and *trans* expression data, Emerson et al. (2010) suggested a test for neutral expression evolution that is related in spirit to sign-based tests. They combined their polymorphism data with divergence data from Tirosh et al. (2009) to examine the within- and among-species expression control in the yeasts *Saccharomyces cerevisiae* and *S. paradoxus*. They used a MacDonald-Kreitman approach (Chapter 10) by examining the fraction of *cis*- and *trans*-controlled transcripts measured within and between species. Their resulting contingency table,

	Polymorphism	Divergence
Cis	396	1270
Trans	412	541

was highly significant, with *trans* polymorphisms being slightly more common than *cis*, but over twice as many *cis* regions were fixed. Such a pattern could arise from either an excessive number of *cis* fixations between species, an excessive amount of *trans* polymorphism within a species, or a combination of both. Analogous to arguments with interpretations of MacDonald-

Kreitman data discussed in Chapter 10, an excessive amount of polymorphism could arise from a high mutation rate for slightly deleterious alleles. This generates an excess amount of within-species polymorphism that does not transfer to among–species differences (as they are not fixed). Wittkopp et al. (2008), working with *Drosophila*, also noted an excess of *cis* regions being fixed over *trans* regions and suggested that the fixation of some *cis* mutations by directional selection, coupled with a larger number of slightly deleterious *trans* alleles, likely underlies this pattern.

Example 12.11. Fraser (2013) suggested yet another approach for detecting selection on the expression level of specific sets of transcripts. First, the genes comprising a specific functional set are chosen (e.g., UV protection), and then an expression score for a population is calculated from the mean frequency of all eSNPS that up-regulate members of this set. This was done over a series of roughly 60 human populations that have lived and evolved under different environmental conditions (such as summer UV flux), computing the correlation between the environment variable and expression score. The significance of this correlation was tested using a randomization approach, wherein the correlation between expression score and environmental variable is computed using a random set of genes. This was repeated $\sim 10^6$ times to generate a distribution for each gene-environment correlation under the null. This approach is very similar to methods examined in Chapter 9 to search for individual SNP frequency–environmental associations (Hancock et al. 2010a, 2010b, 2011; Fumagalli et al. 2011). However, the latter is performed on a SNP-by-SNP basis, whereas the focus here is on the *entire set* of regulatory actors in some network. Using Fraser's approach, significant signals were detected for transcript sets involved in UV response, immune cell proliferation, and diabetes. Further, using a catalogue of putative locally adaptive human SNPs, Hancock et al. (2011) found a roughly ten-fold enrichment of eSNPS and SNPs in *cis*-regulatory regions over amino-acid replacement SNPs in the same genes. This suggests a more important role in local adaptation for regulatory, as opposed to structural, changes.

Evolution of Expression Levels: Drift, Directional, or Stabilizing Selection?

The study of gene regulatory evolution is still in a rather embryonic state. Historically, the field has moved from an early era of speculation that regulatory changes may be at least as, if not more, important than structural changes (Britten and Davidson 1969, 1971; King and Wilson 1975), to a broader acceptance favoring regulatory evolution, at least in some groups or traits (Carroll 2005, 2008; Wray 2007); but see Hoekstra and Coyne (2007) for a counterperspective. As our previous discussion suggests, the initial view provided from transcriptome-wide studies is that stabilizing selection is important, but that some drift in expression level can occur, which is more constrained as phylogenetic distance increases. There is also evidence that directional selection on some *cis*-acting sites has occurred between closely related species.

Taken together, these results (along with those described above for more general traits) suggest, perhaps not surprisingly, that at both the phenotypic and gene-regulatory levels, evolution is primarily characterized by periods of stabilizing selection, although episodes of directional selection certainly occur. However, it must also be emphasized that the interpretation of conservative rates of evolution is far from clear-cut. In principle, evolutionary divergence rates that are below the expectation of the infinite-alleles model of mutation may be a consequence of the general opposition of selection to all allelic changes associated with the trait. Alternatively, they might reflect a situation in which a fraction of mutations is truly neutral, with the rest having negative pleiotropic effects on fitness (Chapter 28). In that case, an observed level of divergence could actually be entirely based on neutral mutations, but with the appropriate measure of mutational variance being lower (and likely *much* lower) than the actual value observed in mutation-accumulation experiments (where even highly deleterious mutations can accumulate). In addition, if the house-of-cards is a more appropriate model for mutation effect sizes, then one would expect cumulative levels of divergence to plateau in time rather than to increase indefinitely, not because of direct selective constraints, but rather due to the limited availability of alternative allelic states.

IV

Short-Term Response on a Single Character

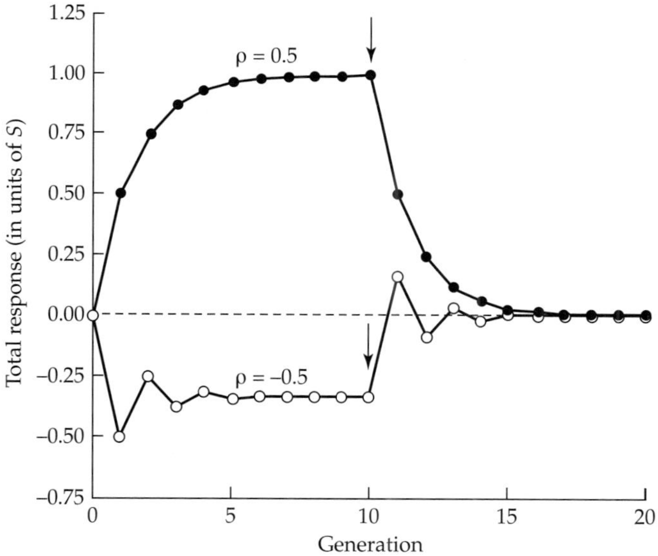

13

Short-term Changes in the Mean: 1. The Breeder's Equation

Prediction is very difficult, especially if it's about the future. Niels Bohr

The topic of selection on quantitative traits, and its consequences, comprises the remainder of this book. We start by discussing the simplest situation—the expected change in the mean of a single character following a single generation of selection from an unselected base population. This response is reasonably predictable in a wide variety of settings, using a regression framework and the appropriate covariances between relatives. By contrast, the response after a number of generations is much less predictable, as allele- and gamete-frequency change alter genetic variances (and hence the resemblance between relatives) from their initial values. Provided that each locus has only a small effect on the trait, only small allele-frequency changes are expected over the first several generations. In the extreme under the infinitesimal model (the limit of a very large number of loci, each with a vanishingly small effect), the additive *genic* variance (that part of the additive genetic variance that is independent of any disequilibrium effects) remains essentially unchanged during selection. **Short-term** response refers to these early generations, where allele-frequency change has a negligible effect on the initial additive variance. As discussed in Chapters 16 and 24, gametic-phase disequilibrium is generated by even a single generation of selection, changing gametic frequencies (and hence genetic variances) even in the absence of allele-frequency change. As detailed in Chapter 16, such short-term changes in the additive variance from disequilibrium are easily computed under the infinitesimal model. Over longer time scales, allele-frequency evolution results in substantial changes in the variance that are extremely difficult to predict; this is the setting for **long-term response** (Chapters 25–28).

Selection can occur in a myriad of ways. Our focus in this chapter is **individual** (or **mass**) **selection** under random mating, wherein individuals are chosen solely on the basis of their phenotypic value (i.e., information from relatives and other such factor are ignored). **Family selection**, whereby individuals are chosen based on their family mean and/or ranking within a family, is discussed in Chapter 21. Chapter 22 discusses the setting in which individuals interact in groups (**kin selection** if they are related) and selection may operate at the individual and/or group level (**group selection**), while Chapter 23 examines response in inbred populations. Using additional information, such as the trait value in relatives and/or the values of other traits in the focal individual, can improve the accuracy in predicting an individual's breeding value and hence increase the expected selection response relative to individual selection. One way to accomplish this is by **index selection**, which generalizes to BLUP-based selection (Chapters 19, 20, and 22; LW Chapter 26), both of which, along with a number of other important selection schemes (such as multivariate selection, marker-assisted and genomic selection, selection for outcross performance, pure-line selection, and selection in age-structured populations) are largely deferred until Volume 3.

There is a huge literature on breeding schemes that exploit specific features of the reproductive biology of a target organism (such as artificial insemination in animals and complex crossing schemes in selfing plants). See Lush (1945), Turner and Young (1969), Pirchner (1983), Ollivier (1988), Weller (1994), Cameron (1997), Simm (1998), and Kinghorn et al. (2000) for applications in animal breeding, and Namkoong (1979), Wricke and Weber (1986), Mayo (1987), Stoskopf et al. (1993), Bos and Caligari (1995), Gallais (2003), Hallauer et al. (2010), and Bernardo (2010) for applications in plant breeding.

Evolution and Selection of Quantitative Traits. Bruce Walsh & Michael Lynch.
Published 2018 by Oxford University Press.
DOI 10.1093/oso/9780198830870.001.0001

SINGLE-GENERATION RESPONSE: THE BREEDER'S EQUATION

The Breeder's Equation: A General Approximation for Response

Previous chapters developed explicit expressions for a single generation of response in the mean of a trait, based on either specific population-genetic models (Equations 5.23c and 5.27b) or completely general covariance-based expressions using Price's theorem (Equations 6.12, 6.39, and 6.40). These results show that either a large number of underlying loci of small effect and/or a linear parent-offspring regression generally will recover the simple **breeder's equation**

$$R = h^2\, S \tag{13.1}$$

plus correction terms that are often small. This approximation is perhaps the most well-known expression in quantitative genetics, and its myriad of extensions form the backbone of the quantitative-genetic theory of short-term response. Although the actual origin of the breeder's equation is somewhat unclear, elements of it (in multivariate form) appear in the early writings of Pearson (1903), and it was popularized by Lush (1937). Indeed, Ollivier (2008) made the quite reasonable suggestion that it be called the **Lush equation**. Its simplicity is compelling, as it relates the change in mean *across* a generation (the **response**, R) to the product of the *within*-generation change (the **selection differential**, S) and a measure of how the character value is passed across generations (the heritability, h^2).

As discussed in Chapter 6, a necessary (but not sufficient) condition to recover the breeder's equation is a linear parent-offspring regression, with the phenotypic value, z_o, of an offspring whose parents have the mean phenotypic value, z_{mp}, given by

$$z_o = \mu + b_{o|mp}(z_{mp} - \mu) + e$$

where $b_{o|mp}$ is the slope of the midparent-offspring regression, which in this chapter is usually assumed to be equivalent to the narrow-sense heritability, h^2 (but is generalized later). If we take the average over all selected parents, then $E_s[z_{mp} - \mu] = S$, while the difference between the expected value of the offspring from such parents and the overall mean is the selection response, R, which yields

$$E_s[z_o - \mu] = R = b_{o|mp}\, E_s[z_{mp} - \mu] = b_{o|mp}\, S = h^2 S$$

Recall (Equation 6.12) that two other technical restrictions are also required to formally obtain $R = b_{o|mp}\, S$. First, it is assumed that the residuals of the linear parent-offspring and fitness-phenotype regressions are uncorrelated with each other. Second, it is assumed that the mean does not change in the absence of selection. As will be discussed below, cases do exist in which the mean and/or variance can change under random mating as disequilibrium induced by prior selection decays (Chapters 15 and 16). In our treatment we either assume that these potential complications introduce only very small errors, or we explicitly model their effects (e.g., Chapters 15, 20, 22, and 23).

The Importance of Linearity

A variety of factors, such as a major gene with dominance, can result in a nonlinear parent-offspring regression (Chapter 6; LW Chapter 17). In such cases, the mean of the selected parents (and hence the selection differential, S) is *not* sufficient to predict the offspring mean. As Figure 13.1 shows, two selected parental populations with the same mean, but different variances, can have different expected responses when this regression is nonlinear. Even if phenotypes are normally distributed and the character is completely determined by additive loci (no dominance or epistasis), if the underlying distribution of genotypic values is skewed, selection on the variance (e.g., selection for, or against, extreme phenotypes) also results in a change in the mean (see Equation 5.27b). In this case, S is again not sufficient to describe the expected response to selection. While

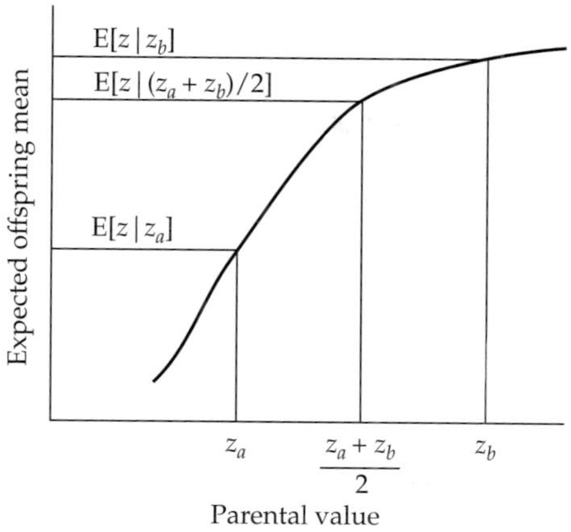

Figure 13.1 The importance of linearity in the parent-offspring regression. If this regression is nonlinear, different subsets of the population with the same mean can have different offspring means. Suppose equal numbers of parents with values z_a and z_b are chosen. If we denote the expected value of offspring from parents with value z_x by $E[\,z \mid z_x\,]$, the offspring mean in this case is given by $(E[\,z \mid z_a\,] + E[\,z \mid z_b\,])/2$. In contrast, choosing parents all with value $(z_a + z_b)/2$ gives the same parental mean as in the case of mixed parents, and hence the same S, but the expected offspring mean is now $E[\,z \mid (z_a + z_b)/2\,]$, which, as shown above, can deviate considerably from $E[\,z \mid z_a\,] + E[\,z \mid z_b\,])/2$ when nonlinearity is significant.

a sufficient condition for linearity is that the joint distribution of breeding and phenotypic values be bivariate normal (LW Chapter 8), selection generally causes the distribution of genotypic values to depart from normality (Chapters 16 and 24), creating at least slight departures from linear parent-offspring regressions. The selection response under strongly nonnormal distributions can be very complicated, depending on summary statistics of the underlying genetic architecture, which do not easily translate into standard (and measurable) variance components (Chapter 24).

Response is the Change in Mean Breeding Value

Under the infinitesimal model and a linear parent-offspring regression, a key concept is that the ***response equals the mean breeding value of the selected parents***. Recall that (noninbred, sexually reproducing) parents pass along only a fraction of their total genotypic value, namely, their breeding value, A, to their offspring. Under the infinitesimal model, the expected offspring value is simply the average breeding values of its parents (LW Chapter 4).

Trait improvement by artificial selection is achieved by choosing parents with the most favorable breeding values. The problem is that we cannot completely predict the breeding value of an individual from its phenotype alone (unless $h^2 = 1$). Because the phenotype of an individual is an imperfect indicator of its breeding value, the offspring of phenotypically exceptional parents are generally not themselves as exceptional. From standard regression theory (LW Chapter 3), the predicted breeding value, $\widehat{A}$, for an individual with a phenotypic value of z (given no other information) is

$$\widehat{A} - \mu_A = \frac{\sigma(A,z)}{\sigma_z^2}(z - \mu_z) \quad \text{or} \quad \widehat{A} = h^2(z - \mu_z)$$

where μ_z and μ_A are the mean phenotype and breeding value, respectively, and $\sigma(A,z)$ is the covariance between the breeding value and the phenotype. This expression follows because the regression $y = a + bx$ can be expressed as $y - \mu_y = b(x - \mu_x)$, where $b = \sigma(x,y)/\sigma_x^2$.

For the regression of A on z, the means are $\mu_A = 0$ and μ_z, respectively, while the slope is $\sigma(z, A)/\sigma_z^2 = \sigma_A^2/\sigma_z^2 = h^2$, which follows because

$$\sigma(z, A) = \sigma(G + E, A) = \sigma(A + D + E, A) = \sigma(A, A) = \sigma_A^2$$

The expected breeding value for a set of selected parents thus becomes

$$E_s[\widehat{A}] = E_s[h^2(z - \mu_z)] = h^2 E_s[z - \mu_z] = h^2 S$$

The change in the mean value of their offspring (relative to the base population) is simply the mean breeding value of the selected parents, because (by definition) $\mu_A = 0$ in the base population. Thus, the response equals $h^2 S$, and we recover the breeder's equation. A key assumption is $E_s[h^2] = h^2$, namely, that the regression using the selected parents is the same as (or extremely close to) the regression in the absence of selection, an assumption discussed at length in Chapter 6.

Response Under Sex-Dependent Parent-Offspring Regressions

It is not uncommon for a trait to show different variances between the sexes or to have a less than perfect correlation across the sexes. In such cases, the coefficients for parent-offspring regressions can vary with the sex of both the parents and of the offspring. We denote the phenotypic values of the *fa*ther and *mo*ther by z_{fa} and z_{mo} and an offspring by z_o (if its sex is unimportant) or by z_{so} and z_{da} for *so*ns and *da*ughters (respectively) if sex is important. Let $E[z_o \,|\, z_{fa}, z_{mo}]$ be the expected phenotypic value of an offspring whose parents have phenotypic values z_{mo} and z_{fa}. The importance of this conditional expectation (the **biparental regression**) is that the expected character value in the next generation (assuming there are no fertility differences) is the average of this expectation over all selected parents. Taking expectations is straightforward when the biparental regression is linear, i.e.,

$$E[z_o \,|\, z_{fa}, z_{mo}] = \mu_o + b_{o,fa}\,(z_{fa} - \mu_{fa}) + b_{o,mo}\,(z_{mo} - \mu_{mo}) \tag{13.2}$$

where μ_{fa} and μ_{mo} are the mean character values of males and females before selection, and μ_o is the mean for the offspring sex being considered. Taking the expectation over all selected parents, the expected offspring mean after selection is

$$\begin{aligned} E_s\big[\, E(z_o \,|\, z_{fa}, z_{mo})\,\big] &= \mu_o + b_{o,fa}\, E_s\big[\,(z_{fa} - \mu_{fa})\,\big] + b_{o,mo}\, E_s\big[\,(z_{mo} - \mu_{mo})\,\big] \\ &= \mu_o + b_{o,fa}\, S_{fa} + b_{o,mo}\, S_{mo} \end{aligned} \tag{13.3}$$

where S_{fa} and S_{mo} are the directional selection differentials for fathers and mothers.

Given that Equations 13.2 and 13.3 acknowledge the presence of differences between sexes in regression coefficients, separate equations for sons and daughters are required (e.g., Example 13.1). For example, the expected change in the mean character value of daughters, R_{da}, equals the expected mean of daughters of selected parents minus the mean of females before selection. Applying Equation 13.3,

$$E_s\big[\, E(z_{da} \,|\, z_{fa}, z_{mo})\,\big] = \mu_{mo} + b_{da,fa}\, S_{fa} + b_{da,mo}\, S_{mo}$$

implying

$$R_{da} = b_{da,fa}\, S_{fa} + b_{da,mo}\, S_{mo} \tag{13.4a}$$

where $b_{da,fa}$ is the regression coefficient of daughters on fathers and $b_{da,mo}$ is the mother-daughter regression coefficient. Likewise, for sons

$$R_{so} = b_{so,fa}\, S_{fa} + b_{so,mo}\, S_{mo} \tag{13.4b}$$

Equations 13.4a and 13.4b require that the biparental regression be linear, in which case $b_{o,fa}$ and $b_{o,mo}$ are partial regression coefficients and can be obtained from the sex-specific covariances between relatives. Again, linearity is ensured if the joint distribution of breeding values in both parents and their offspring is multivariate normal. If there is no correlation between the phenotypes of the parents (which is guaranteed under random mating), the partial regression coefficients are standard univariate regression coefficients (LW Chapter 8) and applying LW Equation 3.14b yields

$$b_{o,fa} = \frac{\sigma(z_o, z_{fa})}{\sigma^2(z_{fa})} \quad \text{and} \quad b_{o,mo} = \frac{\sigma(z_o, z_{mo})}{\sigma^2(z_{mo})}$$

If mating is random, and genotype $\times$ environmental interactions, shared environmental effects, epistasis, and sex-specific effects (i.e., the need for separate regression coefficients) can all be neglected, the regression slope (for each parent-offspring combination) is $b_{o,p} = h^2/2$ (LW Chapters 7 and 17). If we define the total selection differential as the average of both parental values, $S = (S_{fa} + S_{mo})/2$, we again will recover the breeder's equation

$$R = \frac{h^2}{2}\, S_{fa} + \frac{h^2}{2}\, S_{mo} = h^2\, S \tag{13.5}$$

Equation 13.5 shows how differential selection on parents is incorporated into the breeder's equation. For example, consider selection on dioecious plants. If plants that form the next generation are chosen *after* pollination, fathers (pollen donors) are chosen at random with respect to the character under selection ($S_{fa} = 0$), yielding $R = (h^2/2)S_{mo}$. If parents are selected before pollination with equal amounts of selection (S) on both sexes, $R = h^2 S$. Chapter 21 examines family-based breeding schemes that ensure equal selection on both pollen and seed parents.

Example 13.1. Coyne and Beecham (1987) estimated the following parent-offspring regression coefficients for abdominal bristle number in laboratory populations of *Drosophila melanogaster*:

Mother-son	$b_{so,mo} = 0.39 \pm 0.08$
Mother-daughter	$b_{da,mo} = 0.32 \pm 0.08$
Father-son	$b_{so,fa} = 0.13 \pm 0.10$
Father-daughter	$b_{da,fa} = 0.40 \pm 0.08$

Note that the father-son regression has a significantly smaller slope than the three other parent-offspring sex combinations. Other *Drosophila* examples where the regressions differ significantly between sons and daughters were given by Gimelfarb and Willis (1994).

Suppose that different amounts of selection are applied to fathers and mothers, with selected fathers showing an increase of two bristles, while selected mothers show a decrease of one bristle. What is the expected change in mean bristle number in the male and female offspring using these estimated regression coefficients, assuming all parent-offspring regressions are linear? Here $S_{mo} = -1$ and $S_{fa} = 2$, and from Equation 13.4a, the expected change in bristle number in females becomes

$$R_{da} = b_{da,fa}\, S_{fa} + b_{da,mo}\, S_{mo} = 0.40\,(2) + 0.32\,(-1) = 0.48$$

Likewise, from Equation 13.4b, the expected change in males is

$$R_{so} = b_{so,fa}\, S_{fa} + b_{so,mo}\, S_{mo} = 0.13\,(2) + 0.39\,(-1) = -0.13$$

This expected response of a decrease in males and an increase in females is the exact opposite of the pattern of selection on the sexes.

The Selection Intensity, $\bar{\imath}$

While the selection differential (S) is a convenient and simple measure of selection on the mean, it does not tell us the actual *strength* of selection. Consider selection acting on the same character in two different populations. In one, the largest 5% of measured individuals are allowed to reproduce, while in the second, the largest 25% reproduce. Clearly, selection is more intense in the first population. However, under truncation selection on a normally distributed trait, the selection differentials for these two populations are $S_1 = 2.06\,\sigma_1$ and $S_2 = 1.27\,\sigma_2$, respectively, where σ_k^2 is the phenotypic variance in population k (Equation 14.3a). Thus, if the second population is more variable than the first, it may have the larger selection differential even though it clearly experiences less intense selection.

For this reason, in many applications, a more informative measure of the strength of selection is the **selection intensity** (or **standardized selection differential**), which is the selection differential expressed in phenotypic standard deviations

$$\bar{\imath} = S/\sigma_z \tag{13.6a}$$

and also denoted by i or $\imath$ in the literature (we will use $\bar{\imath}$ throughout to avoid any confusion with i as an index variable). The selection intensity accounts for differences in the phenotypic variances, in the same way that a correlation coefficient is a better measure of the strength of association than a covariance (LW Chapter 3). Substituting $\bar{\imath}\,\sigma_z$ for S gives the **selection-intensity version** of the breeder's equation

$$R = h^2\,\bar{\imath}\,\sigma_z = \bar{\imath}\,h\,\sigma_A = \sigma_A^2\,\bar{\imath}/\sigma_z \tag{13.6b}$$

which follows from

$$h^2\sigma_z = \frac{\sigma_A^2}{\sigma_z^2}\sigma_z = \frac{\sigma_A}{\sigma_z}\sigma_A = h\,\sigma_A$$

The various forms of Equation 13.6b will prove to be useful starting points for generalizations (developed below) of the breeder's equation to accommodate more general types of selection. A second reason why breeders and experimentalists generally work with $\bar{\imath}$ is that specifying the fraction (p) of adults saved to form the next generation determines the expected value of $\bar{\imath}$, and hence $S = \bar{\imath}\sigma_z$, in some future selection experiment.

The Robertson-Price Identity, $S = \sigma(w, z)$

As introduced in Chapter 6, the selection differential can be written as the covariance between relative fitness and trait value

$$S = \sigma(w, z) \tag{13.7a}$$

This is the Robertson-Price identity (Equation 6.10), which was first noted by Robertson (1966a) and later elaborated on by Price (1970, 1972a). Chapter 6 showed how this expression directly follows from Price's theorem. For an alternative derivation, let z_i, p_i, and w_i be the trait value, frequency before selection, and relative fitness, respectively, of class i. The selection differential is simply the mean after selection minus the mean before

$$S = \mu_s - \mu = \sum z_i\,w_i\,p_i - \sum z_i\,p_i = E[\,z\,w\,] - E[\,z\,]$$

Because $E[\,w\,] = 1$ by definition, we can write this as

$$S = E[\,z\,w\,] - E[\,z\,]\,E[\,w\,] = \sigma(w, z)$$

thus recovering the Robertson-Price identity. The last step follows from the standard definition of a covariance (LW Equation 3.8).

If, instead of considering the phenotypic value (z) of a trait, we consider its breeding value, A_z, then from the Robertson-Price identity, the expected change in breeding value following selection is

$$\Delta\mu_{A_z} = R_{A_z} = \sigma(A_z, w) \tag{13.7b}$$

Under the conditions of the breeder's equation, this change in the breeding value in the selected parents equals the change in the offspring mean (Chapter 6), thereby equating this covariance with the response (R) in the phenotypic mean of the trait. Equation 13.7b is the 1966 version of Robertson's secondary theorem of natural selection (Equation 6.25a). The more restricted 1968 version, $R = \sigma(A_z, A_w)$, based on the *breeding value* of relative fitness (A_w versus w, Equation 6.24a), appears in the literature as well (Chapters 6 and 20). These covariance-based identities for S and R play important roles in evolutionary quantitative genetics. Chapter 20 examines applications of Equation 13.7b in predicting the selection response in natural populations, while Equation 13.7a routinely appears in selection theory (Chapters 15, 20–23, 29, and 30).

An important application of the Robertson-Price identity follows if we consider the slope of the least-squares linear regression of relative fitness (w) on phenotypic value, z

$$w = a + \beta z + e \tag{13.8a}$$

The interpretation of the slope is that a unit change in z results in a change in relative fitness of β (Chapter 29 examines this regression in detail). From the theory of least-squares regression (LW Chapter 3)

$$\beta = \frac{\sigma(z, w)}{\sigma_z^2} = \frac{S}{\sigma_z^2} \tag{13.8b}$$

Substituting $S = \sigma_z^2\,\beta$ into Equation 13.1 yields

$$R = \sigma_A^2\,\beta \tag{13.8c}$$

which relates the strength of association, β, between trait value and fitness with the response. This is the univariate version of the multivariate Lande equation ($\mathbf{R} = \mathbf{G}\beta$), to be introduced shortly (Equation 13.26a).

Correcting for Reproductive Differences: Effective Selection Differentials

In artificial selection experiments, S is usually estimated as the difference between the mean of the selected adults and the sample mean of the population before selection. However, selection need not stop at this stage. For example, strong artificial selection to increase a character might be countered by natural selection associated with a decrease in the fertility of individuals with extreme trait values. This is the simplest example of a partitioning of **episodes of selection** (multiple rounds of selection within the same generation), in this case a single episode of viability selection followed by fertility selection, which will be explored more broadly in Chapter 29.

Biases introduced by such differential fertility in experimental or breeding settings can be removed by randomly choosing the same number of offspring from each selected parent, thus ensuring equal fertility. Alternatively, differential fertility can be accounted for by using the **effective** (or **realized**) **selection differential**, S_e,

$$S_e = \frac{1}{n_p}\sum_{i=1}^{n_p}\left(\frac{n_i}{\overline{n}}\right)(z_i - \mu_z) \tag{13.9}$$

where z_i and n_i are the phenotypic value and total number of offspring of the ith parent, n_p is the number of parents selected to reproduce, $\overline{n}$ is the average number of offspring from the selected parents, and μ_z is the mean before selection. If all selected parents have the same number of offspring ($n_i = \overline{n}$ for all i), then S_e reduces to S. If there is variation in the number of offspring among selected parents, S_e can be considerably different from S.

The derivation of Equation 13.9 follows directly from the Robertson-Price identity. If we examine a total of N individuals, n_p of which are selected as parents, then

$$S = \sigma(z, w) = E[\, w\, z\,] - E[\, z\,]\, E[\, w\,] = \frac{1}{N} \sum_{i=1}^{N} \left(\frac{W_i}{\overline{W}} \right) z_i - \mu_z \cdot 1$$

where the fitness of individual i is $W_i = n_i$ (with $n_i = 0$ for individuals not chosen as parents). The mean fitness becomes

$$\overline{W} = \frac{1}{N} \sum_{i=1}^{N} n_i = \frac{\overline{n}\, n_p}{N} \qquad \text{where} \qquad \overline{n} = \sum_{i=1}^{N} n_i / n_p$$

and therefore $\overline{n}$ is the mean number of offspring left by the adults that were selected to reproduce. Hence

$$w_i = \frac{W_i}{\overline{W}} = \frac{n_i\, N}{\overline{n}\, n_p}, \quad \text{yielding} \quad \sigma(z, w) = \frac{1}{n_p} \sum_{i=1}^{n} z_i \frac{n_i}{\overline{n}} - \mu_z$$

Rearranging recovers Equation 13.9.

Example 13.2. Consider a trait with heritability 0.3 and a before-selection mean of 30. Suppose five parents are selected, with the following trait values and offspring numbers:

Parent	Phenotypic value	Number of offspring
1	45	1
2	40	2
3	35	3
4	33	5
5	32	5

The unweighted phenotypic mean of the selected parents is 37, yielding S = 37 − 30 = 7 and an expected response of $R = 0.3 \cdot 7 = 2.1$. Is the predicted response altered when differential fertility is taken into account? Computing the effective selective differential by weighting the selected parents by the number of offspring they leave yields

i	z_i	n_i	$n_i/\overline{n}$
1	45	1	0.313
2	40	2	0.625
3	35	3	0.938
4	33	5	1.563
5	32	5	1.563

$$\frac{1}{n_p} \sum_{i=1}^{n_p} \left(\frac{n_i}{\overline{n}} \right) z_i = 34.70$$

Hence, $S_e = 4.70$, yielding an expected response of $R = 0.3 \cdot 4.70 = 1.41$. In this case, not using the effective differential results in a 50% overestimation of the expected response.

EXPANDING THE BASIC BREEDER'S EQUATION

The basic breeder's equation predicts the mean breeding value of the set of parents chosen to form the next generation because of their exceptional phenotypic values. However,

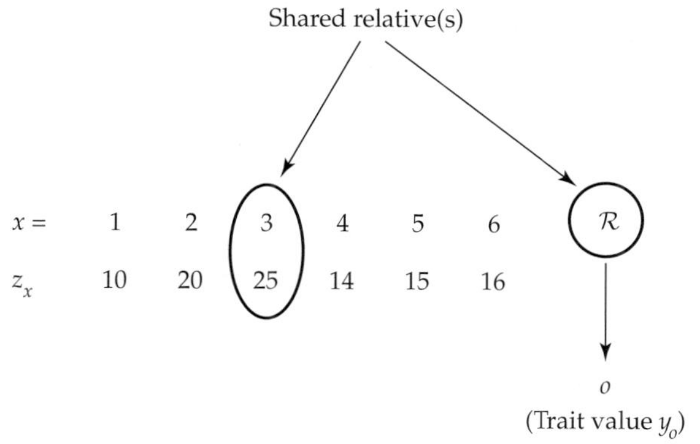

Figure 13.2 The general selection problem: The ultimate goal is to predict the selection response in some response trait (whose values are denoted by y), based on the values of a potentially different—but genetically correlated—selected trait (whose values are denoted by z). The values of the selected trait are measured on one set of individuals (indexed by x_i). For example, the value (z_x) of the selected trait in individual x may be an index that weights x's value for the response trait, as well as the values of the response trait in several of x's relatives. In the figure, x_3 has the highest value of the selected trait, but instead of using x_3 as a parent for the next generation (which would correspond to individual selection), we instead use a relative, $\mathcal{R}$, of x_3, with o denoting an offspring from $\mathcal{R}$. The covariance required for predicting the mean change in the response trait is $\sigma(z_x, y_o)$, namely, the covariance between the selection trait value (z_x) in individual x and the response trait value (y_o) in the offspring of parent $\mathcal{R}$. Under our infinitesimal assumption that the expected value of an offspring is the average of the two parental breeding values, this covariance is also $\sigma[z_x, A_y(\mathcal{R})/2]$, half the covariance between the phenotypic value of the selection trait in x and the breeding value for the response trait in x's relative, $\mathcal{R}$.

breeders, experimentalists, and natural selection all may use additional information in determining the fitness of individuals. For example, one may measure traits in one set of individuals and then use this information to predict the mean breeding value for a second set of related individuals that will be the actual parents of the next generation. One such setting is family selection, wherein one measures the values of a number of family members (for example, by growing seed from a family over a number of environments) and, based on their means, selects the exceptional families. Remnant seeds from these families (i.e., seeds whose phenotypes are not scored) are then used to form the next generation. The prediction of the selection response now involves predicting the breeding value of a family member given the mean of other family members. The breeder's equation is easily extended to these more complex settings, as we now demonstrate. The structure of the general selection problem is given by Figure 13.2.

Accuracy

Suppose our goal is increased milk production. The top females are easy to score, but with no selection on males (who do not display the trait), Equation 13.5 gives the selection response as $h^2(S/2)$. Selection on males is made possible, however, by choosing brothers of the top-scoring females as the sires for the next generation, as breeding values are correlated between relatives. Predicting selection response in this case depends upon the genetic covariance between the phenotypic value (z_x) of individual x and the phenotypic value (z_o) in a relative (o) of x. Here, z_x is milk-yield in female x, whose relative, $\mathcal{R}$ (her brother), is then used as a parent for the next generation and whose resulting offspring are denoted by o (Figure 13.2).

Figure 13.2 shows the most general setting, where selection is based on one trait (the selected trait), while our interest is the resulting response in another, genetically correlated, trait (the response trait). Let z_x and y_o denote, respectively, the value of the selected trait (z) in a measured individual x (upon which selection decisions are based), and the value of the response trait (y) in the offspring, o. Assuming that the regression of y_o on z_x is linear (i.e., the same assumption as for the breeder's equation), standard regression theory (LW Chapter 3) yields

$$E[\, y_o - \mu_y \mid z_x \,] = \frac{\sigma(z_x, y_o)}{\sigma^2(z)} \, (z_x - \mu_z) \tag{13.10a}$$

where μ_y and μ_x are the preselection means of the selected and response traits, respectively. Taking expectations over the selected parents gives the expected change in y from selection on x as

$$R_y = \mu_y^{**} - \mu_y = \frac{\sigma(z_x, y_o)}{\sigma^2(z)} \, (\, \mu_z^* - \mu_z \,) = \frac{\sigma(z_x, y_o)}{\sigma^2(z)} \, S_z = \frac{\sigma(z_x, y_o)}{\sigma(z)} \, \bar{\imath}_z \tag{13.10b}$$

where μ^* denotes the within-generation change in the mean following selection (but before reproduction), μ^{**} denotes the offspring mean, and $\bar{\imath}_z$ is the selection intensity used when choosing parents. The latter is a function of the fraction, p, of measured individuals chosen to have a relative as a parent of the next generation. For example, if brothers from only the top 5% of females are used, then $p = 0.05$, yielding $\bar{\imath}_z \simeq 2.06$ (Example 14.1).

The selection-intensity version of Equation 13.10b can alternatively be expressed as

$$R_y = \frac{\sigma(z_x, y_o)}{\sigma(z)} \, \bar{\imath}_z = \frac{\sigma(z_x, y_o)}{\sigma(z)\, \sigma(y)} \, \bar{\imath}_z \, \sigma(y) = \bar{\imath}_z \, \rho(z_x, y_o) \, \sigma(y) \tag{13.11a}$$

where $\rho(z_x, y_o)$ is the correlation between z_x and y_o. This correlation is referred to as the **accuracy** in predicting the value of the response trait measured in o (y_o) from the selected trait measured in x (z_x). One immediately sees that by improving the accuracy of our selection scheme, we improve the response. Expressing Equation 13.11a in terms of the **relative response**, the change in y in phenotypic standard deviations, gives the expected change in the response trait when using the selected trait measure in a relative as

$$\frac{R_y}{\sigma(y)} = \bar{\imath}_z \, \rho(z_x, y_o) \tag{13.11b}$$

A more powerful way of viewing Equation 13.11a is in terms of the *breeding values*, A_y, for the trait of interest. Under the assumption (used throughout this chapter) that the expected mean of the offspring equals the average breeding values of its parents (Chapter 6 examined this assumption in detail), the mean of the response trait in the offspring is simply the mean breeding value of this trait in the parents. Hence, $R_y = R_{A_y}$, namely, the difference between the mean breeding value for the response trait in the selected parents versus that for the overall population. By taking the response trait to be the breeding value, A_y, of the trait of interest, Equation 13.11a becomes

$$R_{A_y} = \bar{\imath}_z \, \rho(z_x, A_y) \, \sigma(A_y) \tag{13.11c}$$

Hence, the breeder's equation can be considered as a special case of the more general expression

$$\text{Response} = (\text{Intensity}) \cdot (\text{Accuracy in predicting breeding value using } z_x) \cdot (\text{Usable variation}) \tag{13.11d}$$

An accuracy of special interest that arises during individual selection is the correlation between an individual's phenotype (z) and breeding value (A), where

$$\rho(z, A) = \sigma(z_x, A_x)/(\sigma_z \, \sigma_A) = \sigma_A^2/(\sigma_z \, \sigma_A) = \sigma_A/\sigma_z = h \tag{13.11e}$$

This accuracy corresponds to individual selection (x is the parent, $x = \mathcal{R}$; and the selected and response traits are the same, $y = z$). When substituted into Equation 13.11c, this recovers Equation 13.6b. As illustrated in Example 13.4, this is a key result, as whether some proposed selection scheme is more efficient than individual selection depends on the correlation between the breeding value of a chosen parent and the selection variable used for that scheme (e.g., z_x could be x's family mean of the response trait). If this correlation exceeds h, then from Equation 13.11c (because $\bar{\imath}$ and $\sigma^2[A]$ are the same), the response is larger. Equations 13.11a through 13.11c form the foundation for most of Chapter 21, which deals with various selection schemes using family information (e.g., family means and within-family deviations).

The greatest selection response occurs if we take the selection variable (z_x) with the largest correlation with the breeding value for the response trait (assuming $\bar{\imath}$ and σ_A are the same over all comparisons). This idea forms the foundation of **index selection**, whereby individuals are chosen based on some index, $z_x = \sum a_i z_i$, a linear combination of trait values in the relatives and/or correlated traits in the focal individual (Volume 3).

Example 13.3. Consider selection on clones or other pure lines, where parents pass on their entire genome to their offspring. The phenotypic value (z_o) of an offspring from a parent with a genotypic value G_p can be written as $z_o = G_p + E_o$, so that the parent-offspring covariance (in the absence of any genotype × environment covariance and/or interactions) equals

$$\sigma(z_o, z_p) = \sigma(G_p + E_o, G_p + E_p) = \sigma(G_p, G_p) = \sigma(G, G)$$

namely, the total genetic variance, σ^2_G. The resulting parent-offspring regression has a slope $b_{op} = \sigma^2_G/\sigma^2_z = H^2$, the broad-sense heritability (LW Chapter 20), yielding

$$R = H^2\, S$$

Because $H^2 \geq h^2$ (as $\sigma^2_G \geq \sigma^2_A$), the single-generation response to selection is at least as large for clones as for a sexual population with the same variance components. However, when selection continues for several generations, using clones is expected to be far less effective, as selection among clones very rapidly removes any genetic variation from the population without any mechanism (other than mutation) to regenerate it. Likewise, the assumption of a normal distribution of genotypic values quickly breaks down as only a few genotypes remain. By contrast, with selection among sexually reproducing individuals, segregation and recombination will generate an almost endless supply of new variation if a large number of segregating loci underlie the trait. Special issues with regard to the selection and development of pure lines are examined in Volume 3.

Example 13.4. **Progeny testing** uses the mean of an individual's offspring (here, all are assumed to be half-sibs) to predict its breeding value. In order to predict the selection response using this scheme, we first need the correlation between the mean ($\overline{z}_o$) of n half-sib offspring and the breeding value (A) of the common parent. This is given by

$$\rho(\overline{z}_o, A) = \frac{\sigma(\overline{z}_o, A)}{\sigma_A \cdot \sigma(\overline{z}_o)}$$

To obtain this, first note that the covariance between the parent and its offspring is $\sigma^2_A/2$

$$\sigma(\overline{z}_o, A) = \frac{1}{n}\sum_{i=1}^{n} \sigma\left(z_i, A\right) = \frac{1}{n} n\sigma(z_i, A) = \sigma^2_A/2$$

The expression for $\sigma^2(\overline{z}_o)$ requires a bit more bookkeeping, as sibs are correlated. Assuming half-sibs and that there are no shared environmental effects, $\sigma(z_i, z_j) = \sigma^2_A/4$ (for $i \neq j$), yielding

$$\sigma^2(\overline{z}_o) = \sigma\left(\frac{1}{n}\sum_{i=1}^{n} z_i, \frac{1}{n}\sum_{j=1}^{n} z_j\right) = \frac{1}{n^2} n\sigma(z_i, z_i) + \frac{n(n-1)}{n^2}\sigma(z_i, z_j)$$

$$= \frac{\sigma^2_z}{n} + \left(1 - \frac{1}{n}\right)\frac{\sigma^2_A}{4} = \frac{\sigma^2_z}{4n}\left[4 + (n-1)h^2\right]$$

Combining these results yields

$$\rho(\overline{z}_o, A) = \frac{\sigma_A^2/2}{\sigma_A \sigma_z \sqrt{\frac{4+(n-1)h^2}{4n}}} = \frac{\sigma_A}{\sigma_z}\sqrt{\frac{4n/4}{4+(n-1)h^2}} = \sqrt{\frac{h^2 n}{4+(n-1)h^2}} = \sqrt{\frac{n}{n+\gamma}}$$

where $\gamma = (4-h^2)/h^2$. For large n, note that the accuracy approaches one. Substituting this result into Equation 13.11c, the response to selection becomes

$$R = \overline{\imath}\,\rho(\overline{z}_o, A)\,\sigma_A = \overline{\imath}\sigma_A\sqrt{\frac{n}{n+\gamma}} = \overline{\imath}\sigma_A\sqrt{\frac{h^2 n}{4+h^2(n-1)}}$$

Recalling Equation 13.6b, the ratio of response for progeny testing (R_{pt}) to mass selection (R_{ms}) becomes

$$\frac{R_{pt}}{R_{ms}} = \frac{\overline{\imath}_{pt}\,\rho(\overline{z}_o, A)\,\sigma_A}{\overline{\imath}_{ms}\,\rho(z, A)\,\sigma_A} = \left(\frac{\overline{\imath}_{pt}}{\overline{\imath}_{ms}}\right)\frac{1}{h}\sqrt{\frac{h^2 n}{4+h^2(n-1)}}$$

$$= \left(\frac{\overline{\imath}_{pt}}{\overline{\imath}_{ms}}\right)\sqrt{\frac{n}{4+h^2(n-1)}}$$

The selection intensity under progeny testing is likely to be lower, as it is easier (and cheaper) to score a phenotype than to progeny test. When the intensities are equal, the ratio of responses approaches $1/h$ for large n. Assuming $\overline{\imath}_{pt} = \overline{\imath}_{ms}$, progeny testing gives a larger response when $\rho(\overline{z}_o, A) > \rho(z, A)$, or when

$$\sqrt{\frac{n}{4+h^2(n-1)}} > 1 \quad \text{or} \quad n > \frac{4-h^2}{1-h^2}$$

In particular, $n > 4$, 5, and 7 is required for h^2 = 0.1, 0.25, and 0.5, respectively, for progeny testing to give a larger response. Hence, when the heritability of a trait is high, more offspring must be scored for the accuracy of progeny selection to exceed that of mass selection. A high heritability implies that an individual's phenotype is a good predictor of its breeding value, which requires increasingly more observations of an indirect measure (offspring values) to obtain higher accuracy than the simple direct measure of the individual phenotype.

Example 13.5. Suppose a character of interest is extremely hard to measure in live individuals. For example, one cannot directly select on individuals that have to be killed to measure a trait such as meat quality. Similarly, consider traits expressed in only one sex, such as milk production. How can we select on males if they do not express the trait themselves? One solution to both of these problems is **sib selection** (Chapter 21), or using sibs of exceptional individuals as the parents for the next generation (e.g., choosing males based on the milk production of their sisters).

Under sib selection, the selection unit (x) is the trait value in sib s_1, with the correlation between its phenotypic value (z_{s_1}) and the breeding value (A_{s_2}) of sib s_2 being (LW Table 7.3)

$$\sigma(z_{s_1}, A_{s_2}) = \begin{cases} \sigma_A^2/2 & \text{for full sibs} \\ \sigma_A^2/4 & \text{for half-sibs} \end{cases} \quad \text{yielding} \quad \rho(z_{s_1}, A_{s_2}) = \begin{cases} h/2 & \text{for full sibs} \\ h/4 & \text{for half-sibs} \end{cases}$$

For example, when s_1 and s_2 denote full sibs

$$\rho(z_{s_1}, A_{s_2}) = \frac{\sigma(z_{s_1}, A_{s_2})}{\sigma(z_{s_1})\cdot\sigma(A_{s_2})} = \frac{\sigma_A^2/2}{\sigma_z\sigma_A} = \frac{\sigma_A}{2\sigma_z} = \frac{h}{2}$$

with the correlation between half-sibs obtained similarly. Applying Equation 13.11c, the resulting response to selection (assuming equal selection intensity on both sexes) based on the performance of a sib is

$$R = \begin{cases} \bar{\imath}\,(h/2)\,\sigma_A & \text{for full sibs} \\ \bar{\imath}\,(h/4)\,\sigma_A & \text{for half-sibs} \end{cases}$$

where the selection intensity, $\bar{\imath}$, is a function of the fraction of measured sibs chosen in order to have a relative as a parent of the next generation. For example, if only 5% of the measured sibs are chosen to have a sib as a parent in the next generation, then $\bar{\imath} \simeq 2.06$ (Example 14.1). Comparison with Equation 13.6b shows that using a single full sib (in place of the measured individual) gives a response that is half that of mass selection, while using a single half-sib gives only a quarter of the response.

One can also have a mixture of direct and sib selection, as can occur when selection is based on milk production. Here, there is direct selection on females (based on their trait value) and sib selection on males (based on the trait value of a sister). Summing the separate responses from selection on females and males, the expected response becomes

$$R = (1/2)\,\bar{\imath}\,h\,\sigma_A + (1/2)\,\bar{\imath}\,(h/2)\,\sigma_A = (3/4)\,\bar{\imath}\,h\,\sigma_A$$

where the first term is the response from using superior females (Equation 13.11a with selection on only one sex, giving $\bar{\imath}/2$), and the second term is the response using a brother of a superior sister (full sib selection, or this term divided by two for half-sib selection). Here, we have assumed the same intensity, $\bar{\imath}$, in both sexes, while with different amounts of selection

$$R = (1/2)\,\bar{\imath}_d\,h\,\sigma_A + (1/2)\,\bar{\imath}_s\,(h/2)\,\sigma_A = [(2\,\bar{\imath}_d + \bar{\imath}_s)/4]\,h\,\sigma_A$$

While one could develop additional extensions (for example, by using progeny testing for males), in practice, information from all relatives is handled using BLUP selection, a robust and general approach for predicting response under very diverse schemes of selection that will be introduced shortly.

Reducing Environmental Noise: Stratified Mass Selection

Accuracy (and hence response) can also be increased by using designs that reduce environmental noise. One approach is Gardner's (1961) method of **stratified mass selection**: a population is stratified into a number of blocks (potentially representing different microenvironments) and selection occurs *within* each block. The motivation behind Gardner's method was to improve individual selection for yield in maize. At the time of his paper, selection based solely on the observed yield of individual plants resulted in a very poor response, largely because environmental effects overwhelm genetic differences, resulting in very small h^2 values. Simply by selecting for plants within blocks of presumably similar environments, Gardner was able to use mass selection to obtain fairly significant gains (about 4% per year). Stratified mass selection is an important component in Burton's (1974, 1982) method of **recurrent restricted phenotypic selection (RRPS)** for turf grass breeding.

To obtain the expected response under stratified mass selection, we need to compute the accuracy, which first requires the covariance between within-block deviations and an individual's breeding value. Suppose n individuals are measured within each block, and selection occurs on the deviation from the block mean, e.g., on $z_{ij} - \bar{z}_i$ where z_{ij} is the jth individual from block i and $\bar{z}_i$ is the block mean. An individual's phenotypic value can be expressed as its genotypic value, G_{ij} (indicating the jth individual from block i), plus an environmental value consisting of a block effect, B_i, and the residual environmental value, e_{ij},

$$z_{ij} = \mu + G_{ij} + B_i + e_{ij} \tag{13.12a}$$

The total environmental variance equals the variance among blocks, σ^2_B, plus the within-block variance, σ^2_e (the variance of the residuals e_{ij}), resulting in a total variance of

$$\sigma^2_z = \sigma^2_G + \sigma^2_E = \sigma^2_G + \sigma^2_B + \sigma^2_e \tag{13.12b}$$

For the jth individual in block i, the covariance between individual breeding value and within-block deviation is

$$\sigma(z_{ij} - \overline{z}_i, A_{ij}) = \sigma(z_{ij}, A_{ij}) - \frac{1}{n}\sum_{k=1}^{n} \sigma(\, z_{ik}, A_{ij}) = \sigma_A^2 \left(1 - \frac{1}{n}\right) \tag{13.13}$$

as the assumption is that individuals within blocks are unrelated. The variance of deviations within a block is $\sigma_G^2 + \sigma_e^2$, making the accuracy

$$\rho(z_{ij} - \overline{z}_i, A_{ij}) = \frac{\sigma_A^2(1 - 1/n)}{\sigma_A\sqrt{\sigma_G^2 + \sigma_e^2}} = \frac{\sigma_A(1 - 1/n)}{\sqrt{\sigma_G^2 + \sigma_e^2}} \tag{13.14a}$$

Applying Equation 13.11c yields a the resulting response of

$$R = \overline{\imath}\,\rho(z_{ij} - \overline{z}_i, A_{ij})\,\sigma_A = \frac{\overline{\imath}\,\sigma_A^2(1 - 1/n)}{\sqrt{\sigma_G^2 + \sigma_e^2}} \simeq \frac{\overline{\imath}\,\sigma_A^2}{\sqrt{\sigma_G^2 + \sigma_e^2}} \tag{13.14b}$$

where $\overline{\imath}$ is the selection intensity *within blocks*, and the final approximation assumes a large n.

In contrast, if the effects of blocks are ignored and individuals are simply selected from the entire population, the between-block variance is incorporated into the variance of z, and from Equation 13.6b the response becomes

$$R = \frac{\overline{\imath}\,\sigma_A^2}{\sqrt{\sigma_G^2 + \sigma_B^2 + \sigma_e^2}} \tag{13.14c}$$

The relative advantage of stratification (assuming the block size is modest to large, so that $1 - 1/n \simeq 1$) is

$$\sqrt{\frac{\sigma_G^2 + \sigma_B^2 + \sigma_e^2}{\sigma_G^2 + \sigma_e^2}} = \sqrt{1 + \frac{\sigma_B^2}{\sigma_G^2 + \sigma_e^2}} \tag{13.15}$$

Thus, within-block selection can significantly improve the selection response when the among-block variance accounts for a significant fraction of the total variation. Schutz and Cockerham (1966) extend this idea of selecting within blocks to a number of other designs.

Reducing Environmental Noise: Repeated-Measures Selection

The **repeated-measures** design is a second example of increasing accuracy (and response) by providing some control over environmental noise. Here the character of interest is measured n different times on each individual, and selection occurs on $\overline{z}_i$, the mean value for individual i. For example, if we are considering the number of days to ripening, a better approach is to use a collection of fruit, rather than a single one, to assign a value to an individual tree. Repeated-measures selection is a common design in behavioral experiments, wherein a single measure (such as wheel-running speed) may vary greatly within an individual over time.

Our analysis depends on the **repeatability** (LW Chapter 6) of the trait. The character value for the jth measure of individual i is decomposed as

$$z_{ij} = G_i + E_i + e_{ij} \tag{13.16a}$$

where G_i and E_i are the genotypic and (permanent) environmental values common to all measures of i, and e_{ij} is the special environmental value restricted to the jth measure of i, with the repeatability, r, being defined as

$$r = \frac{\sigma_G^2 + \sigma_E^2}{\sigma_z^2} = 1 - \frac{\sigma_e^2}{\sigma_z^2} \tag{13.16b}$$

yielding

$$r\sigma_z^2 = \sigma_G^2 + \sigma_E^2 \quad \text{and} \quad (1-r)\sigma_z^2 = \sigma_e^2 \tag{13.16c}$$

To obtain the accuracy in using $\overline{z}_i$ to predict A_i, we need both the covariance between $\overline{z}_i$ and A_i, and the variance of $\overline{z}_i$. The former is simply

$$\sigma(A_i, \overline{z}_i) = \sigma\left(A_i, \frac{1}{n}\sum_{j=1}^{n} z_{ij}\right) = \frac{1}{n}\sum_{j=1}^{n}\sigma(A_i, z_{ij}) = \frac{1}{n}\, n\,\sigma(A_i, A_i) = \sigma_A^2 \tag{13.17a}$$

To obtain the variance of $\overline{z}_i$, starting with

$$\overline{z}_i = \frac{1}{n}\sum_{j=1}^{n} z_{ij} = G_i + E_i + \frac{1}{n}\sum_{j=1}^{n} e_{ij} \tag{13.17b}$$

it immediately follows from Equation 13.16c that

$$\begin{aligned}\sigma^2(\overline{z}_i) &= \sigma_G^2 + \sigma_E^2 + \sigma_e^2/n \\ &= \sigma_z^2\, r + \sigma_z^2\,\frac{1-r}{n} = \sigma_z^2\left(\frac{1+(n-1)r}{n}\right)\end{aligned} \tag{13.17c}$$

The resulting accuracy in using $\overline{z}_i$ to predict A_i becomes

$$\rho(\overline{z}_i, A_i) = \frac{\sigma(A_i, \overline{z}_i)}{\sigma_A\,\sigma(\overline{z}_i)} = \frac{\sigma_A^2}{\sigma_A\sqrt{\sigma_z^2\left(\frac{1+(n-1)r}{n}\right)}} = h\sqrt{\frac{n}{1+(n-1)r}} \tag{13.18a}$$

giving the response as

$$R = \overline{\imath}\,\rho(\overline{z}_i, A_i)\,\sigma_A = \overline{\imath}\,h\sqrt{\frac{n}{1+(n-1)r}}\;\sigma_A \tag{13.18b}$$

The ratio of accuracies under repeated-measures versus (single-measure) mass selection becomes

$$\frac{\rho(\overline{z}_i, A_i)}{\rho(z_i, A_i)} = \sqrt{\frac{n}{1+(n-1)r}} \tag{13.19a}$$

which approaches $1/\sqrt{r}$ for large values of n. Hence, when repeatability is low ($\sigma_e^2 \gg \sigma_G^2 + \sigma_E^2$), repeated-measures selection can result in a considerable improvement in response.

This comparison assumes the same selection intensity under single- versus repeated-measures selection, but one might imagine that they could differ. For example, if one has the time and resources to only score 500 individuals, and (for breeding reasons) must keep (at least) 50 parents, then the fraction to be saved can be as small as 50/500 = 10% under single-measure selection. However, with five replicate measures per individual, we can only score 100 individuals, if we are to save a fraction no smaller than 50/100 = 50%. Applying Equation 14.4c, these translate into selection intensities of $\overline{\imath}_{\overline{z}_i} = 0.79$ for repeated-measures selection and $\overline{\imath}_{z_i} = 1.75$ for single-measures selection, respectively. Such differences in the potential selection intensity are easily incorporated into comparisons of different selection strategies, with the comparison now being

$$\frac{R_{\overline{z}_i}}{R_{z_i}} = \frac{\overline{\imath}_{\overline{z}_i}}{\overline{\imath}_{z_i}}\sqrt{\frac{n}{1+(n-1)r}} \tag{13.19b}$$

Finally, we note that our analysis of repeated-measures selection assumes that the additive-genetic correlation across individual measurements is 1.0, which is expected for many traits. However, if measures are sufficiently separated in time that age effects are important, or if they represent significantly distinct events (such as litter size at different **parities**, i.e., distinct litters), these correlations can be less than one. In other words, the traits measured at different ages may not be the same genetically. In such cases, one should treat these measurements as a set of correlated traits and use index-selection theory (Volume 3). With measurements at arbitrary time points that may differ over individuals, the method of random regression (Volume 3) is used.

Example 13.6. As an example of the consequences of basing selection decisions on single versus multiple measurements, consider the following data set, which was simulated by assuming a character with $h^2 = 0.1$, $\sigma_z^2 = 100$, $\mu = 50$, and $r = 0.2$. The simulated values for 20 individuals for either a single measurement, $z(1)$, or the average of five measurements, $\overline{z}(5)$, are

j	$z_j(1)$	$\overline{z}_j(5)$	j	$z_j(1)$	$\overline{z}_j(5)$
1	54.97	56.80	11	49.81	48.76
2	64.01	54.51	12	51.92	46.76
3	42.64	52.61	13	43.56	51.79
4	42.70	38.69	14	41.60	47.23
5	61.62	56.42	15	51.80	48.90
6	39.86	47.70	16	52.88	47.21
7	56.54	48.63	17	63.86	54.03
8	35.88	47.26	18	39.76	49.62
9	54.32	53.93	19	36.45	47.78
10	57.85	45.10	20	59.07	51.16

where j indexes the individuals. Suppose the uppermost 25% (top 5 of the 20) are chosen for selection. Based on single measures, individuals 2, 5, 10, 17, and 20 would be chosen, while based on five measures, individuals 1, 2, 5, 9, and 17 would be selected. Using the single (initial) measurement, the overall mean is 50.05, while the mean of selected individuals is 61.28, yielding an S of 11.23.

If we use repeated-measures selection, the overall mean of the five-sample averages for each individual is 49.74, while the mean of selected individuals is 55.14, resulting in an S of 5.39. The smaller value of S under repeated measures under the same selection intensity, $\bar{\imath}$, is a simple consequence of the reduced variance associated with using the mean (repeated measures) versus a single observation. To see this, note from Equation 13.6a that $S_{\overline{z}}/S_z = (\bar{\imath}\sigma_{\overline{z}})/(\bar{\imath}\sigma_z) < 1$, as $\sigma_{\overline{z}} < \sigma_z$ (Equation 13.17c). From the breeder's equation, the expected response based on single measures is

$$R = h^2 S = 0.1 \cdot 11.23 = 1.12$$

To express the response given by Equation 13.18b (the repeated-measures expression) in terms of S, note that here $\rho = \sigma_A^2/(\sigma_A\,\sigma_{\overline{z}}) = \sigma_A/\sigma_{\overline{z}}$, and hence $R = \bar{\imath}\,\rho\,\sigma_A = \bar{\imath}\,\sigma_A^2/\sigma_{\overline{z}} = (\bar{\imath}\,\sigma_{\overline{z}})\,\sigma_A^2/\sigma_{\overline{z}}^2 = S\sigma_A^2/\sigma_{\overline{z}}^2$, which, using Equation 13.17c, yields

$$R = h^2 \left(\frac{n}{1 + (n-1)r} \right) S = 0.1 \cdot 2.78 \cdot 5.39 = 1.50$$

Thus, the reduction in S under repeated measures (5.39 versus 11.23) is more than made up for by increased accuracy, yielding a larger expected response relative to mass selection (i.e., selection based on single measures). From Equation 13.19a, the expected ratio of the accuracies (and hence responses, assuming the same selection intensities) of the five-measure to single-measure schemes is 1.67, which approaches $1/\sqrt{r} \simeq 2.24$ for large values of n. The value of the response ratio for our simulated data (1.50/1.12 = 1.34) deviated from the expected value due to the small number (20) of randomly sampled individuals.

Adjustments for Overlapping Generations

Thus far, we have been assuming that we are examining nonoverlapping generations, with all parents reproducing in a discrete single generation. However, domesticated animals, perennial plants, and many species in nature can have offspring over multiple years and for varying life spans. In such cases, generations overlap and the selection response should be considered on an absolute time scale (typically years) rather than a per-generation scale. To express the breeder's equation in terms of a yearly rate of response, we first need to compute the **generation intervals**, L_x (the average age of parents when progeny are born), for both sexes.

Assuming that the variance components are independent of age and sex, the yearly rate of response, r_y, can be expressed as

$$r_y = \left(\frac{\bar{\imath}_s + \bar{\imath}_d}{L_s + L_d} \right) h^2 \sigma_z = \left(\frac{\bar{\imath}_s + \bar{\imath}_d}{L_s + L_d} \right) h \sigma_A \tag{13.20}$$

where $\bar{\imath}_s$ and $\bar{\imath}_d$ denote the selection intensities of the sire (father) and dam (mother). This result (in a slightly different form) is from Rendel and Robertson (1950), although the basic idea traces back to Dickerson and Hazel (1944). Thus, one way to increase the rate of response is to reduce the generation intervals, for example, by using younger parents. However, the problem here is that there is a tradeoff between generation interval and selection intensity. In species that are reproductively limited (with few offspring per dam), using younger dams means that a higher fraction of the dams must be kept to replace the population. As a consequence, the selection intensity on the parents is reduced. Equation 13.20 is an *asymptotic result*, as it takes time for the selection response to propagate through an age-structured population. Volume 3 examines the effects of age structure on selection response in greater detail.

Example 13.7. Compute the sire, L_s, and dam, L_d, generation intervals for the following age structure:

Parental Age at Birth of Progeny

Sires	Year 2	Year 3	Year 4	Year 5	Total
Number	760	380	0	0	1140
Dams	Year 2	Year 3	Year 4	Year 5	Total
Number	400	600	100	40	1140

The resulting sire generation interval is the average age of sires when offspring are born. Here, 760/1140 = 2/3 of the sires are age two, while 1/3 are age three, yielding

$$L_s = 2 \cdot \frac{760}{1140} + 3 \cdot \frac{380}{1140} = \frac{2 \cdot 760 + 3 \cdot 380}{1140} = 2.33$$

Similarly, the dam generation interval is

$$L_d = \frac{2 \cdot 400 + 3 \cdot 600 + 4 \cdot 100 + 5 \cdot 40}{1140} = 2.81$$

Because each offspring has a single mother and father, the population-level average generation interval is just the average of the two parental intervals, or 2.57 years.

Maximizing Response Under the Breeder's Equation

We can combine both the selection accuracy (Equations 13.11c) and generation-interval (Equation 13.20) versions of the breeder's equation to give a more general expression, with the expected rate of response being

$$r_y = \left(\frac{\bar{\imath}_s + \bar{\imath}_d}{L_s + L_d} \right) \rho(A, x)\, \sigma_A \tag{13.21a}$$

where x is the measure used to choose the parents to form the next generation. Even more generally, if the accuracies vary over sex

$$r_y = \left(\frac{\bar{\imath}_s\, \rho_s(A, x) + \bar{\imath}_d\, \rho_d(A, x)}{L_s + L_d} \right) \sigma_A \tag{13.21b}$$

Beyond importing new genetic material, there is not much a breeder can do to increase σ_A^2, which leaves three selection features that the breeder has some control over (Dickerson and Hazel 1944; Kinghorn et al. 2000):

(i) selection intensity, $\bar{\imath}$
(ii) generation interval, L
(iii) selection accuracy, ρ

The response rate increases with ρ and $\bar{\imath}$, and it decreases with increasing values of L. We have already discussed tradeoffs between L and $\bar{\imath}$, and there are similar tradeoffs between L and ρ. Clearly, the longer we wait to allow a parent to reproduce, the more accurately we can predict its breeding value, as information from other relatives and from progeny testing accumulates over time. However, these increases in ρ also result in increases in L. An optimal selection program must balance all of these competing interests.

Equation 13.21 highlights the importance to animal breeding of advances in reproductive technologies such as **artificial insemination** (**AI**) and **multiple ovulation embryo transfer (MOET)** schemes (e.g., Woolliams 1989). The more offspring a parent can produce, the stronger is the selection intensity that can be applied while still keeping a fixed number of animals in a population. AI has resulted in the potential for far greater sire selection intensities (but as a side effect, far more inbreeding) than would be possible under natural insemination. Likewise, MOET schemes that increase the number of offspring from females allow for increases in the selection intensity on dams as well as decreases in the generation interval.

Equation 13.21a is also highly relevant to **genomic selection** (wherein high-density marker information is used to predict the values of offspring; Volume 3). The gain from genomic selection is generally *not* the result of an increased accuracy, ρ, when using marker information, but rather from much quicker and earlier scoring of phenotypes, which lowers L and increases $\bar{\imath}$.

Example 13.8. As an example of the tradeoff between accuracy and generation intervals, consider a trait with $h^2 = 0.25$ and selection only on sires (fathers). One scheme is to simply select on the sire's phenotype, which results in a sire generation interval of (say) 1.5 years. Alternatively, one might perform progeny testing to improve the accuracy of the selected sires, which results in an increase of the sire generation interval to (say) 2.5 years. Suppose that in both cases, the dam (mother) interval is constant at 1.5 years. Because the additive-genetic variation is the same in both schemes, the ratio of response under mass selection to response under progeny testing becomes

$$\frac{r_y(\text{sire phenotype})}{r_y(\text{progeny mean})} = \left(\frac{\bar{\imath}_{sp}}{\bar{\imath}_{pt}} \right) \frac{\rho(A, \text{sire phenotype})/(L_s + L_d)}{\rho(A, \text{progeny mean})/(L_s + L_d)}$$

where $\bar{\imath}_{sp}$ and $\bar{\imath}_{pt}$ are the selection intensities under individual selection (sire phenotype) and progeny testing, which can differ due to costs in scoring. From Equation 13.11e, $\rho(A, \text{sire phenotype}) = h = \sqrt{0.25} = 0.5$, while the generation interval becomes $L_s + L_d = 1.5 + 1.5 = 3$. With progeny testing, Example 13.4 yields

$$\rho(A, \text{progeny mean}) = \sqrt{\frac{n}{n+\gamma}} = \sqrt{\frac{n}{n+15}}$$

as $\gamma = (4 - h^2)/h^2 = 15$, with a total generation interval of $L_s + L_d = 2.5 + 1.5 = 4$. Rearrangement of this expression yields

$$\frac{r_y(\text{sire phenotype})}{r_y(\text{progeny mean})}\left(\frac{\bar{\imath}_{pt}}{\bar{\imath}_{sp}}\right) = \frac{0.5/3}{(1/4)\sqrt{\frac{n}{n+15}}} = \frac{2}{3}\cdot\sqrt{\frac{n+15}{n}}$$

For $n = 2$ progeny tested per sire, this ratio is 1.95, resulting in a much larger rate of response under sire-only selection. For $n = 12$, the ratio is exactly one, while for a very large number of offspring tested per sire, the ratio approaches $2/3$, or a 1.5-fold increase in the rate of response under progeny testing, despite the increase in the sire generation interval. Thus, taking into account the ratio of selection intensities, mass selection always gives a higher per-year rate of response (for the values of the other parameters assumed in this example) when $\bar{\imath}_{pt}/\bar{\imath}_{sp} < 3/2$. By contrast, from Example 13.4, when $n > 4$, progeny testing yielded a larger response (for $h^2 = 0.25$), but that example did not discount for the effect of the longer sire generation intervals required for progeny testing.

Maximizing the Economic Rate of Response

Example 13.8 hints at another important feature of the selection response, economics. Notice that by scoring more than 12 offspring, we can obtain a larger expected rate of response using progeny testing (assuming equal selection intensities). Why not simply score 30 progeny, giving a 122% rate of response relative to simple mass selection? The economic reality relates to the cost of a raising and scoring a large number of progeny. Much of applied breeding is concerned with the *economic* rate of response—trying to maximize the rate of response per unit of capital, although this point is often underappreciated, even by some breeders. Along these same lines, much of current selection in animal breeding is for increased efficiency (conversion of resources into desirable traits), and hence greater economic gain per unit of input at the production level. Weller (1994) presented a nice development of how to incorporate economics into breeding.

Mean- Versus Variance-Standardized Response

As was the case with the selection differential, S, in order to assess the relative strength of response one needs some sort of standardization. One obvious approach is to express the response in units of phenotypic standard deviations (**variance-standardization**). From Equation 13.6b,

$$\frac{R}{\sigma_z} = h^2\,\bar{\imath} \tag{13.22a}$$

implying that a (scaled) strength of selection of $\bar{\imath} = 1/h^2$ is required for a standard deviation of response. For example, with $h^2 = 0.25$, a total selection intensity of $\bar{\imath} = 4$ is required to achieve a total response of one phenotypic standard deviation. Example 14.1 shows that $\bar{\imath} = 2.06$ for truncation selection saving the upper 5%, so that only two generations of such selection are required for a one standard deviation change in the trait mean.

While scaling traits in units of phenotypic standard deviations is extremely common, it can potentially be rather misleading (Houle 1992; Houle et al. 2011). To see this, imagine two traits, both with a standard deviation of 2.0. Trait one has a mean of 10 and trait two has a mean of 100. A response of one standard deviation increases trait one by 20%, but trait two by only 2%. From a variance-standardized viewpoint, the response is equal, but as a proportional response of the total mean, trait one has clearly experienced a stronger response.

Houle and colleagues (Houle 1992; Hansen et al. 2003; Hansen et al. 2011; Houle et al. 2011) argued that using **mean-standardization**, R/μ_z, namely the *proportional amount* of response, is often more appropriate. Again using Equation 13.6b,

$$\frac{R}{\mu_z} = \bar{\imath}\, h \frac{\sigma_A}{\mu_z} = \bar{\imath}\, h\, CV_A \tag{13.22b}$$

where $CV_A = \sigma_A/\mu_z$, the coefficient of additive genetic variation, is Houle's (1992) **evolvability** index, which he argued was a better measure of evolutionary potential than h^2 (Chapter 6). Houle (1992) and Hansen et al. (2011) found that h^2 is essentially uncorrelated with evolvability, so that a trait with a lower h^2 could still have high evolvability (i.e., potential for a significant proportional change in the mean), and vice versa.

BLUP SELECTION

LW Chapter 26 introduced the basic **mixed model** for estimating a vector, $\mathbf{a}$, of breeding values for a set of individuals given some vector, $\mathbf{y}$, of *records* (observations), a relationship matrix, $\mathbf{A}$, connecting individuals with records with individuals whose breeding values are of interest, and a set, $\boldsymbol{\beta}$, of fixed effects to estimate

$$\mathbf{y} = \mathbf{X}\boldsymbol{\beta} + \mathbf{Z}\mathbf{a} + \mathbf{e}, \quad \mathbf{a} \sim MVN(\mathbf{0}, \sigma^2_A \cdot \mathbf{A}), \quad \mathbf{e} \sim MVN(\mathbf{0}, \sigma^2_e \cdot \mathbf{I})$$

where (as detailed in Chapters 19 and 20) the matrices $\mathbf{X}$ and $\mathbf{Z}$ are given from the data (relating which observations contribute information to which fixed and random effects), and $\mathbf{A}$ is obtained from the pedigree or from sufficiently dense genetic markers. Solving the model returns a vector, $\widehat{\mathbf{a}}$, of BLUP (best linear unbiased prediction) breeding values.

Much of modern animal breeding (and a growing amount of plant breeding) is based on using BLUP estimates to select individuals with the highest breeding values for the trait to form the next generation. This is called **BLUP selection**. The expected response is simply given by the difference between the mean breeding value of selected parents and the population from which they were chosen. This is an *extremely* flexible methodology, with all of the examples in this chapter being special cases of this general approach. Indeed, provided relatives are measured, BLUP can be used to predict the breeding value of individuals with no phenotypic values (such as milk production in sires). Relatedness information for all measured individuals enters through $\mathbf{A}$, and multiple records (repeated measurements) from the same individual are easily incorporated, as are additional fixed (and random) factors such as plot, location, and herd effects. Further, the effects of age structure are fully accounted for by the relationship matrix $\mathbf{A}$. Chapters 19, 20, and 22 discuss features of BLUP estimation of breeding values, while more technical details (such as maximal avoidance of inbreeding) are deferred until Volume 3 (also see Henderson 1984; Simm 1998; Bernardo 2010; and Mrode 2014).

THE MULTIVARIATE BREEDER'S EQUATION

Expressing the heritability in terms of additive-genetic and phenotypic variances, the breeder's equation can be written as

$$R = \sigma^2_A\, \sigma_z^{-2}\, S \tag{13.23a}$$

While this decomposition seems rather trivial, it suggests (as we formally show in Volume 3) that its multivariate version (under appropriate linearity assumptions) is given by

$$\mathbf{R} = \mathbf{G}\mathbf{P}^{-1}\mathbf{S} \tag{13.23b}$$

where $\mathbf{R}$ and $\mathbf{S}$ are the vectors of responses and selection differentials for each character, and $\mathbf{G}$ and $\mathbf{P}$ are the additive-genetic and phenotypic covariance matrices (LW Chapter 21), with

$$P_{ij} = \sigma(z_i, z_j) \qquad \text{and} \qquad G_{ij} = \sigma(A_i, A_j) \tag{13.23c}$$

Here, we briefly consider a few features of Equation 13.23b; we examine its full range of consequences and applications at length in Volume 3. As an aside, note that Equation 13.23b breaks the standard convention that vectors (here $\mathbf{R}$ and $\mathbf{S}$) are usually written as lower case bold letters. This is to conform with the standard notation for these two vectors in the literature.

Response With Two Traits

One expects that selection is always acting on more than a single trait, as even with strong artificial selection on a single character, natural selection is likely operating on other traits as well. What risks do we run by ignoring this expectation and treating selection as a univariate problem? While this is examined much more fully in Volume 3, we can gain significant insight by considering the simple case of two traits, both of which are potentially under selection. Equation 13.23b gives the expected vector of responses as

$$\mathbf{R} = \begin{pmatrix} R_1 \\ R_2 \end{pmatrix} = \mathbf{GP}^{-1}\mathbf{S} = \begin{pmatrix} G_{11} & G_{12} \\ G_{12} & G_{22} \end{pmatrix} \begin{pmatrix} P_{11} & P_{12} \\ P_{12} & P_{22} \end{pmatrix}^{-1} \begin{pmatrix} S_1 \\ S_2 \end{pmatrix} \tag{13.24a}$$

Using LW Equation 8.11 to compute the inverse of $\mathbf{P}$, and, recalling for a covariance that $P_{12} = \rho_z\sqrt{P_{11}P_{22}}$, where ρ_z is the phenotypic correlation between the two traits, yields

$$\begin{aligned} \mathbf{P}^{-1}\mathbf{S} &= \frac{1}{P_{11}P_{22} - P_{12}^2} \begin{pmatrix} P_{22} & -P_{12} \\ -P_{12} & P_{11} \end{pmatrix} \begin{pmatrix} S_1 \\ S_2 \end{pmatrix} \\ &= \frac{1}{P_{11}P_{22}(1-\rho_z^2)} \begin{pmatrix} S_1P_{22} - S_2P_{12} \\ -S_1P_{12} + S_2P_{11} \end{pmatrix} \end{aligned} \tag{13.24b}$$

Substituting into Equation 13.24a and recalling that $h_i^2 = G_{ii}/P_{ii}$, the response in trait one becomes

$$\begin{aligned} R_1 &= \frac{1}{P_{11}P_{22}(1-\rho_z^2)} \begin{pmatrix} G_{11} & G_{12} \end{pmatrix} \begin{pmatrix} S_1P_{22} - S_2P_{12} \\ -S_1P_{12} + S_2P_{11} \end{pmatrix} \\ &= \frac{G_{11}\,(S_1P_{22} - S_2P_{12}) + G_{12}\,(-S_1P_{12} + S_2P_{11})}{P_{11}P_{22}(1-\rho_z^2)} \\ &= \frac{h_1^2}{(1-\rho_z^2)} \left(S_1 - S_2\frac{P_{12}}{P_{22}} \right) + \frac{G_{12}\,(-S_1P_{12} + S_2P_{11})}{P_{11}P_{22}(1-\rho_z^2)} \end{aligned} \tag{13.24c}$$

with an analogous expression for R_2. The breeder's equation is recovered only when trait one is phenotypically ($P_{12} = \rho_z = 0$) *and* genetically ($G_{12} = 0$) uncorrelated with trait two. As we now demonstrate, this complicated expression masks the rather different roles played by phenotypic and genetic correlations, impacting (respectively) the within- and between-generation changes. Volume 3 examines these points in some detail.

Accounting for Phenotypic Correlations: The Selection Gradient

Recall from Equation 13.8b that the univariate **directional selection gradient**, $\beta = S/\sigma_z^2$, is the slope of the linear regression of relative fitness, w, as a function of the phenotypic value, z, of the trait (Equation 13.8a). The multivariate extension is given by

$$\boldsymbol{\beta} = \mathbf{P}^{-1}\mathbf{S} \tag{13.25a}$$

where the vector $\boldsymbol{\beta} = (\beta_1, \cdots, \beta_n)^T$ contains the coefficients for the multiple linear regression

$$w = a + \sum_i \beta_i z_i + e = a + \boldsymbol{\beta}^T\,\mathbf{z} + e \tag{13.25b}$$

of relative fitness, w, on the vector, $\mathbf{z}$, of trait values (LW Equation 8.10c). The interpretation of β_i is that it represents the change in relative fitness given a one unit change in trait i while holding all other trait values constant. In other words, β_i measures the amount of direct selection on trait i, removing any indirect effects from selection on phenotypically correlated traits *included in the analysis*, i.e., all traits in the vector $\mathbf{z}$ (Chapter 30). Because $\boldsymbol{\beta} = \mathbf{P}^{-1}\mathbf{S}$, then $\mathbf{S} = \mathbf{P}\boldsymbol{\beta}$, resulting in an observed selection differential of

$$S_i = P_{ii}\beta_i + \sum_{j\neq i} P_{ij}\beta_j \tag{13.25c}$$

The *within-generation* change, S_i, in the mean of trait i following selection thus consists of an effect from direct selection on that trait ($P_{ii}\beta_i$) plus the effects of selection on all other phenotypically correlated traits ($P_{ij}\beta_j \neq 0$). Hence, the sign of S_i tells us nothing about the sign of β_i (the amount of direct selection on that trait), as **correlated selection** ($P_{ij}\beta_j$ terms) can easily overpower the direct effect. The selection gradient, $\boldsymbol{\beta}$, presents the correct picture of which traits are under selection, *provided* there are no additional traits under direct selection that are phenotypically correlated with our focal vector, $\mathbf{z}$, of traits (Chapter 30).

Accounting for Genetic Correlations: The Lande Equation

To see the effects of genetic correlations, substituting $\mathbf{P}^{-1}\mathbf{S} = \boldsymbol{\beta}$ into Equation 13.23b gives the **Lande equation** (Lande 1979a),

$$\mathbf{R} = \mathbf{G}\boldsymbol{\beta} \tag{13.26a}$$

which is the multivariate version of Equation 13.8c, $R = \sigma^2_A\beta$. Considering two traits, using the Lande equation

$$\mathbf{R} = \begin{pmatrix} R_1 \\ R_2 \end{pmatrix} = \mathbf{G}\boldsymbol{\beta} = \begin{pmatrix} G_{11} & G_{12} \\ G_{12} & G_{22} \end{pmatrix} \begin{pmatrix} \beta_1 \\ \beta_2 \end{pmatrix} \tag{13.26b}$$

yields greatly simplified expressions (relative to Equation 13.24c) for the responses

$$\begin{aligned} R_1 &= G_{11}\beta_1 + G_{12}\beta_2 \\ R_2 &= G_{12}\beta_1 + G_{22}\beta_2 \end{aligned} \tag{13.26c}$$

The role of genetic correlations (G_{12}) is now obvious, in that direct selection on trait two ($\beta_2 \neq 0$) influences the response (between-generation change) in trait one *only* when the two traits have a nonzero genetic correlation ($G_{12} \neq 0$). The contribution, $G_{ii}\beta_i$, from direct selection is called the **direct response**, and the contribution to response on trait i from direct selection on other *genetically* correlated traits ($G_{ij}\beta_j \neq 0$) is called the **correlated response**. If two traits are genetically uncorrelated, selection on one has no impact on the *response* of the other, *even if they are phenotypically correlated*. More generally, with n potentially correlated traits

$$R_i = G_{ii}\beta_i + \sum_{j\neq i} G_{ij}\beta_j \tag{13.26d}$$

Comparison with Equation 13.25c shows that P_{ij} for the within-generation changes (S_i) are replaced by G_{ij} for the between-generation changes (R_i).

The Lande equation shows that when the multivariate breeder's equation holds, we can distinguish between **phenotypic selection**, which is the change in a phenotypic distribution *within* a generation (measured by $\mathbf{S}$, with the nature of selection, i.e., the direct effects, summarized by $\boldsymbol{\beta}$), and the **evolutionary response to selection**, which is the transmission of these within-generation changes to the next generation (given by $\mathbf{R}$). Lande and Arnold (1983) and Arnold and Wade (1984a, 1984b), following Fisher (1930, 1958) and Haldane (1954), have stressed the utility of this approach. Attempts to measure selection by comparing phenotypic distributions across generations are confounded by inheritance, as $\mathbf{R}$ depends on $\boldsymbol{\beta}$ through $\mathbf{G}$. Chapters 29 and 30 examine in detail methods for estimating the nature of phenotypic selection in natural populations. When the breeder's equation fails, this separation of selection from inheritance may no longer be possible, leading to the recent use of Robertson's secondary theorem, $\mathbf{R} = \sigma(w, \mathbf{A}_z)$, for examining response in natural populations (Chapter 20).

Selection Gradients and Mean Population Fitness

Under appropriate conditions, the selection gradient, $\boldsymbol{\beta}$, demonstrates how a within-generation change in the vector of trait means maps into a change in the mean fitness of a population. If $W(z)$ denotes the expected fitness of an individual with a character value of z, then when phenotypes are normally distributed and fitness is frequency-independent (individual fitnesses are not a function of the means of the characters), the directional selection gradient satisfies $\beta = \partial \ln \overline{W} / \partial \mu$ (Lande 1976; Example A6.3 gives the full multivariate derivation). Hence we can express the breeder's equation as

$$R = \sigma^2_A \left(\frac{\partial \ln \overline{W}}{\partial \mu} \right) \tag{13.27a}$$

The multivariate version of this partial derivative is the **gradient of mean fitness** with respect to the vector of character means, which is the vector of partials of the log of mean fitness with respect to each trait mean under consideration

$$\boldsymbol{\beta} = \frac{\partial \ln \overline{W}}{\partial \boldsymbol{\mu}} \tag{13.27b}$$

with $\beta_i = \partial \ln \overline{W} / \partial \mu_i$ (the change in log mean fitness from a change in the mean of trait i). The resulting gradient version of the Lande equation becomes

$$\mathbf{R} = \mathbf{G}\, \frac{\partial \ln \overline{W}}{\partial \boldsymbol{\mu}} \tag{13.27c}$$

The vector $\boldsymbol{\beta}$ represents the direction for the joint change in the means that maximizes the local increase in mean fitness. In contrast, the *actual* response involves the product (or **projection**) of this vector with the genetic covariance matrix **G**. The resulting response vector is generally not parallel to $\boldsymbol{\beta}$, as the genetic covariance structure causes the character means to change in a direction that does not necessarily result in the optimal change in population fitness. We examine the implications of genetic constraints imposed by the structure of **G** in detail in Volume 3.

We can connect the somewhat abstract notion of variance in fitness with a measurable quantity, the amount of selection, $\boldsymbol{\beta}$, on a vector, **z**, of traits, as follows. Walsh and Blows (2009) showed that the additive variance in relative fitness w accounted for by selection on **z** is

$$\sigma_A(\mathbf{z}^T, w)\, \mathbf{G}^{-1} \sigma_A(\mathbf{z}, w) \tag{13.28a}$$

where the notation $\sigma_A(x, y)$ denotes the covariance between the breeding values of x and y. Recalling Robertson's secondary theorem (1968 version; Equation 6.24a), $\sigma_A(\mathbf{z}, w) = \mathbf{R} = \mathbf{G}\boldsymbol{\beta}$ (assuming the conditions for the multivariate breeder's equation hold), yielding

$$(\mathbf{G}\boldsymbol{\beta})^T \mathbf{G}^{-1} (\mathbf{G}\boldsymbol{\beta}) = \boldsymbol{\beta}^T \mathbf{G} \boldsymbol{\beta} \tag{13.28b}$$

The additive variance in fitness that remains unaccounted for after the effects of **z** are removed becomes

$$\sigma^2_A(w) - \boldsymbol{\beta}^T \mathbf{G} \boldsymbol{\beta} \tag{13.28c}$$

In theory, if one had an estimate of $\sigma^2_A(w)$ in hand (Chapter 20), the significance of a set, **z**, of traits can be determined. If these account for most of the variation, there is little need to consider additional traits. If these account for only a small fraction, important traits are missing.

Table 13.1 Alternate versions and extensions of the basic breeder's equation. Refer to the specific equation number for discussion and explanation of the symbols.

Version	Expression	Equation Number
Basic breeder's equation	$R = h^2 S$	13.1
Sex-specific response (sex s)	$R_s = b_{s,fa}\, S_{fa} + b_{s,mo}\, S_{mo}$	13.4
Selection intensity	$R = h^2\, \bar{\imath}\, \sigma_z = \bar{\imath}\, h\, \sigma_A = \sigma_A^2\, \bar{\imath}/\sigma_z$	13.6b
Response (in trait y, selection using x)	$R_y = \dfrac{\sigma(x,y)}{\sigma_x^2}\, S_x = \dfrac{\sigma(x,y)}{\sigma_x}\, \bar{\imath}_x$	13.10b
Accuracy (in trait y, selection using x)	$R_y = \bar{\imath}_x\, \sigma_y\, \rho(x,y)$	13.11a
Accuracy (breeding values)	$R = \bar{\imath}_x\, \rho(x, A)\, \sigma_A$	13.11c
Rate of response (per year)	$r_{year} = \left(\dfrac{\bar{\imath}_s + \bar{\imath}_d}{L_s + L_d} \right) h\sigma_A$	13.20
Rate of response using accuracy	$r_{year} = \left(\dfrac{\bar{\imath}_s + \bar{\imath}_d}{L_s + L_d} \right) \rho(A, x)\, \sigma_A$	13.21
Variance-standardized response	$R/\sigma_z = h^2\, \bar{\imath}$	13.22a
Mean-standardized response	$R/\mu_z = \bar{\imath}\, h\, CV_A$	13.22b
Robertson's secondary theorem		
1966 version	$R = \sigma(w, A_z)$	13.7b
1968 version	$R = \sigma(A_w, A_z)$	6.24a
Univariate Lande equation	$R = \sigma_A^2\, \beta$	13.8c
	$R = \sigma_A^2\, \dfrac{\partial \ln \overline{W}}{\partial \mu}$	13.27a
Multivariate breeder's equation	$\mathbf{R} = \mathbf{G}\mathbf{P}^{-1}\mathbf{S}$	13.23b
Multivariate Lande equation	$\mathbf{R} = \mathbf{G}\boldsymbol{\beta}$	13.26a
	$\mathbf{R} = \mathbf{G}\, \dfrac{\partial \ln \overline{W}}{\partial \boldsymbol{\mu}}$	13.27c

Finally, we can use the multivariate breeder's equation to make a connection with the classical interpretation of Fisher's fundamental theorem (Chapter 6). If the fitness determined by the vector of traits, $\mathbf{z}$, can be expressed as a linear regression (Equation 13.25b), then the expected change in fitness from selection response on these traits is

$$\Delta \overline{w} = \boldsymbol{\beta}^T\, \Delta \mathbf{z} = \boldsymbol{\beta}^T\, \mathbf{R} = \boldsymbol{\beta}^T \mathbf{G} \boldsymbol{\beta} \tag{13.29}$$

From Equation 13.28b, this is just the additive variance in fitness associated with these traits. Thus, we recover Fisher's theorem that the change in mean fitness equals the additive variance in relative fitness (Equation 6.17c).

LIMITATIONS OF THE BREEDER'S EQUATION

As we have seen, the basic breeder's equation has many alternative expressions and extensions (summarized in Table 13.1). All are based on Equation 13.1, which assumes a linear midparent-offspring regression with slope h^2. This single-generation prediction is a good approximation over multiple generations provided that selection does not result in a significant change in the base-population heritability, a region we call short-term response. Chapter 16 shows that even a single generation of selection will change h^2 through the generation of linkage disequilibrium, but because this is straightforward to correct for when

Table 13.2 Summary of various factors complicating the prediction of short-term selection response in the phenotypic mean, even assuming all regressions are linear and considering just a single generation of selection from an unselected base population. Short-term response specifically refers to conditions where the effects of any allele-frequency change on the additive variance are negligible. Models of long-term response (Chapters 25–28) relax this restriction.

Major gene with dominance (Chapter 6; LW Chapter 17)	Can generate a nonlinear parent-offspring regression.
Epistasis (Chapter 15)	The component of response due to epistasis is transient. Parent-offspring covariance overestimates permanent response.
Correlated environmental effects (Chapter 15)	The component of response due to correlated environmental effects is transient.
Maternal effects (Chapters 15, 22)	The potential for complicated lags in response—the mean changes unpredictably after selection is relaxed. Possibility of reversed response.
Gametic-phase disequilibrium (Chapter 16)	Changes the additive genetic variance. Directional selection generates negative gametic-phase disequilibrium, reducing h^2 and slowing response.
Assortative mating (Chapter 16)	Generates gametic-phase disequilibrium, which either enhances (positive correlation between mates) or retards (negative correlation between mates) response.
Environmental change (Chapters 18 - 20)	A significant change in the environment can obscure the true amount of genetic change.
Drift (Chapters 18, 19)	Generates variance in the short-term response.
Environmental correlations (Chapter 20)	Environmental factors can influence both the trait and fitness, confounding both the nature of selection and the true amount of genetic change.
Associative effects (Chapter 22)	Trait influenced by both direct and social components from group members. A decline in the mean social value can swamp an increase in mean direct value. Possibility of reversed response.
Inbreeding (Chapter 23)	Response depends on additional variance components that are difficult to estimate (σ^2_{DI}, σ_{ADI}, etc). Response has permanent and transient components.
Age-structure (Volume 3)	Several generations are required to propagate genetic change uniformly through the population.
Selection on correlated characters (Volume 3)	Response completely unpredictable unless selection on correlated characters accounted for. Possibility of reversed response.
G × E interactions (Volume 3)	Possibility of nonlinear parent-offspring regressions. Often treated as a correlated characters problem, with traits measured in different environments treated as correlated traits. Possibility of reversed response.

allele-frequency change is infinitesimal, we treat this as a special case of short-term response (details in Chapter 16). The more serious problem is that eventually allele-frequency change significantly alters the genetic variance (long-term response), and these variances changes cannot be predicted without extensive (and essentially unavailable) knowledge about the distribution of allelic effects and their frequencies (Chapters 25 and 26).

However, even over the short-term response time frame, there are a number of complications that compromise the basic breeder's equation (Table 13.2). One particularly important (and usually unstated) assumption is that we start from an *unselected* base population. If the base population itself has been under selection, decay of transient response components from previous selection compromises the predicted single-generation response (Chapter 15). In the Price equation setting (Equation 6.12), this occurs because the mean of the population changes in the *absence* of selection, as the population regains Hardy-Weinberg proportions and linkage equilibrium following a perturbation from past selection.

Another troublesome feature of the breeder's equation is the assumption that all of the selection on the character of interest is accounted for. This is especially problematic as selection on any character correlated with the one of interest can introduce significant bias to the expected selection response. This problem is examined in Chapter 20, but generally there is no easy solution, or even any indication of a problem before an experiment or field study begins. Thus, even in the best of situations (linearity and no selection-induced changes in allele and gamete frequencies), there are still pitfalls in predicting even a single generation of response from the slope of the parent-offspring regression. The situation gets worse if the parent-offspring regression is nonlinear, as the single-generation change in the mean can then depend on higher-order moments of the genotypic distribution, and hence is not predictable from simple variance components. See Equations 5.23c and 5.27b for population-genetic expressions, Equations 6.12, 6.39, 6.40 for expressions based on the Price equation, and Chapter 24 for a detailed discussion.

Table 13.2 summarizes some of these factors compromising the breeder's equation, giving the chapters in which these complications are examined in detail. Provided one can assume linearity of the regressions of relatives, we can account for many of these concerns, as when the regression of an individual on all of its direct relatives selected in previous generations (back to the original unselected base population) remains linear, the selection response is entirely determined by the covariances between a current individual and these previous relatives (Chapter 15).

As mentioned at the start of the chapter, even if we have corrected for all of the potential complications listed in Table 13.2, the breeder's equation (using the base population h^2) is expected to become an increasingly poor predictor as selection proceeds. If there are segregating alleles of large effect, even a single generation of selection can significantly change the underlying variance components, which in turn changes the regression coefficients. Further, selection can introduce nonlinearities into an initially linear regression by perturbing the starting distribution of breeding values away from normality, although these departures are usually small (Chapter 24). In the absence of major genes, allele-frequency changes over the first few generations of selection are expected to be rather small, but *genotype* frequencies can change dramatically due to selection generating gametic-phase disequilibrium (Chapters 16 and 24). Directional selection generates negative disequilibrium, decreasing heritabilities and hence reducing the selection response, and this reduction can be significant if heritability is high. Likewise, selection on the variance itself (through disruptive or stabilizing selection) also creates disequilibrium effects on the expressed genetic variance. Chapter 16 examines the effect of such short-term changes in disequilibrium on the additive genetic variation. An additional complication occurs when there is genetic variance for the amount of environmental variability that a genotype displays, and this is discussed in Chapter 17. As selection continues over several generations, even if all loci have very small effects, allele frequencies themselves start to appreciably change (Chapters 25 and 26). Drift and mutation also become increasingly important, and these complications are examined in Chapters 26–28.

14

Short-term Changes in the Mean: 2. Truncation and Threshold Selection

Far better an approximate answer to the right question, which is often vague, than an exact answer to the wrong question, which can always be made precise. Tukey (1962)

This brief chapter first considers the general theory of truncation selection on the mean and then examines a number of more specialized (but related) topics, which may be skipped by the casual reader. Truncation selection (Figure 14.1), which occurs when all individuals with trait values on one side of a threshold are chosen, is by far the commonest form of artificial selection. One key result to be presented is that, for a normally distributed trait, the expected selection intensity, $\bar{\imath}$, is fully determined by the fraction, p, saved, provided that the chosen number of adults is large. This allows a breeder or experimentalist to predict the expected selection response, given the choice of p and knowledge of h^2. When small numbers of adults are chosen to form the next generation, one should apply a small-sample correction, otherwise $\bar{\imath}$ will be overestimated. However, as will be shown, this correction is generally relatively minor unless only a few individuals form the next generation.

We then turn our attention to the selection response for discrete traits, which (as we will show) has close connections with certain aspects of truncation-selection theory. We start with a binary (present or absent) trait, such that, when some continuous underlying **liability value** exceeds a particular (usually unknown) threshold, the trait takes on one value (or is displayed), while it takes on another (or is absent) when its liability is below the threshold (LW Chapter 25). The key assumption is that the liability value is a normally distributed trait, with the breeder's equation holding on this scale. We also examine binary-trait response in a logistic regression framework (estimating the probability that an individual exhibits the trait, given some underlying liability score).

We conclude with a few brief comments on selection response when a trait is better modeled as Poisson, rather than normally, distributed, as can occur with certain types of count data (such as number of offspring). Again, the key idea is that there is some underlying (and unobserved, or **latent**) liability scale that maps into the discrete character space.

TRUNCATION SELECTION

In addition to being the commonest form of artificial selection, truncation selection is also the most efficient, as it provides the largest selection intensity of any scheme culling the same fraction of individuals (Kimura and Crow 1978; Crow and Kimura 1979). Truncation selection is described by either the fraction, p, of the population saved or the threshold phenotypic value, T, below (or above) which individuals are **culled**. The investigator usually sets one of these (usually p) in advance of the actual selection. Hence, while S is trivially computed *after* the parents are chosen, we would like to *predict* the expected selection differential, given either p or T. Specifically, given p or T, what is the expected mean of the selected parents? In our discussion of this topic, we first assume that a large number of individuals is saved, before turning to complications introduced by finite sample size.

Selection Intensities and Differentials Under Truncation Selection

Given a threshold cutoff, T, the expected mean of the selected adults is given by the conditional mean, $E[\, z \mid z \geq T\,]$. Generally, it is assumed that phenotypes are normally distributed,

Evolution and Selection of Quantitative Traits. Bruce Walsh & Michael Lynch.
Published 2018 by Oxford University Press.
DOI 10.1093/oso/9780198830870.001.0001

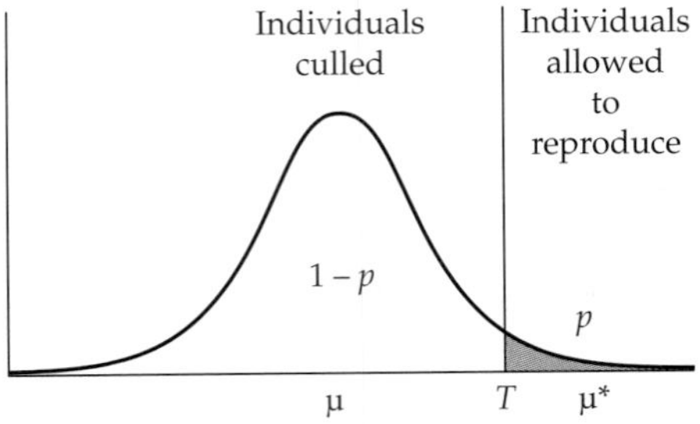

Figure 14.1 Under truncation selection, the uppermost (or lowermost) fraction, p, of a population is selected to reproduce. Alternatively, one could set a threshold level, T, in advance. To predict the selection response, given either p or T, we need to know the expected mean of the selected tail (μ^*), from which we can compute either $S = \mu^* - \mu$ or $\bar{\imath} = S/\sigma$, and then apply the breeder's equation.

and we use this assumption throughout (unless stated otherwise). With an initial mean of μ and variance of σ^2, and $p = \Pr(z \geq T)$ being the fraction saved, this conditional mean is given by LW Equation 2.14, which yields the expected selection differential as

$$S = \varphi\left(\frac{T-\mu}{\sigma}\right)\frac{\sigma}{p} \tag{14.1}$$

where $\varphi(x) = (2\pi)^{-1/2}\exp(-x^2/2)$ is the unit normal density function evaluated at x.

Usually the fraction saved p (rather than T) is preset by the investigator. Given p, to apply Equation 14.1, we must first find the threshold value T_p satisfying $\Pr(z \geq T_p) = p$. Notice that T in Equation 14.1 enters only as $(T-\mu)/\sigma$, which transforms T_p to a scale with a mean of zero and unit variance. Hence,

$$\Pr(z \geq T_p) = \Pr\left(\frac{z-\mu}{\sigma} > \frac{T_p-\mu}{\sigma}\right) = \Pr\left(U > \frac{T_p-\mu}{\sigma}\right) = p$$

where $U \sim N(0,1)$ denotes a unit normal random variable. Define $x_{[p]}$, the **probit transformation** of p (LW Chapter 11), as satisfying

$$\Pr(U \leq x_{[p]}) = p \tag{14.2a}$$

so that

$$\Pr(U > x_{[1-p]}) = p \tag{14.2b}$$

It immediately follows that $x_{[1-p]} = (T_p - \mu)/\sigma$, and Equation 14.1 yields the expected selection intensity as

$$\bar{\imath} = \frac{S}{\sigma} = \frac{\varphi(x_{[1-p]})}{p} \tag{14.3a}$$

Note that $\bar{\imath}$ is ***entirely a function of p*** provided z is normal. Equation 14.3a can be approximated by

$$\bar{\imath} \simeq 0.8 + 0.41\,\ln\left(\frac{1}{p} - 1\right) \tag{14.3b}$$

a result due to Smith (1969). Simmonds (1977) found that this approximation is generally quite good for $0.004 \leq p \leq 0.75$ and offered alternative approximations for p values outside this range, as did Saxton (1988). Montaldo (1997) gives an approximation for the standardized truncation value $z = (T-\mu)/\sigma$ in terms of $\bar{\imath}$. Finally, when one selects for the *lowest* fraction p of the population, Equation 14.3a still holds, provided we take its negative value.

Example 14.1. Consider selection on a normally distributed trait for which the upper 5% of the population is saved (p = 0.05). Here $x_{[1-0.05]} = x_{[0.95]}$ is obtained by the R command `qnorm(0.95)`, which returns 1.645, as Pr[U > 1.645] = 0.05. The R command `dnorm(x)` returns the value of $\varphi(x)$, with `dnorm(1.645)` returning a value of 0.103. Hence,

$$\bar{\imath} = \frac{\varphi(1.645)}{0.05} = \frac{0.103}{0.05} \simeq 2.06$$

More compactly, the R command for any value of p is `dnorm(qnorm(1-p))/p`. In contrast, Smith's approximation calculates the selection intensity as

$$\bar{\imath} \simeq 0.8 + 0.41 \ln\left(\frac{1}{0.05} - 1\right) \simeq 2.01$$

Finally, if we selected for the lowermost 5%, $\bar{\imath} = -2.06$.

Correcting the Selection Intensity for Finite Sample Sizes

If the number of individuals saved is small, Equation 14.1 *overestimates* the selection differential because of sampling effects (Nordskog and Wyatt 1952; Burrows 1972). To see this, suppose 100 observations are put randomly into ten groups of size 10 and the largest value is selected from each group. These values will be, on average, not as extreme as when selecting the best 10 from the entire 100, as the best observation within a random group of ten can be the 11th largest (or even smaller) for the entire group.

To more formally treat the effects of finite sample size, assume M adults are sampled at random from the population, with the largest N of these being used to form the next generation, yielding $p = N/M$. The expected selection coefficient is computed from the distribution of **order statistics**. We rank the M observed phenotypes as $z_{1,M} \geq z_{2,M} \cdots \geq z_{M,M}$, where $z_{k,M}$ denotes the kth-order statistic (the kth largest value) when M observations are sampled. The expected selection intensity is given by the expected mean of the N selected parents, which is the average of the first N order statistics

$$E\left[\,\bar{\imath}_{(N,M)}\,\right] = \frac{1}{\sigma}\left(\frac{1}{N}\sum_{k=1}^{N} E\left[z_{k,M}\right] - \mu\right) = \frac{1}{N}\sum_{k=1}^{N} E\left[z'_{k,M}\right]$$

where $z'_{k,M} = (z_{k,M} - \mu)/\sigma$ are the **standardized order statistics**. While the properties of order statistics have been worked out for many special cases (Harter 1961; Sarhan and Greenberg 1962; Harter 1970a, 1970b; Kendall and Stuart 1977; David 1981), values for any distribution are easily be obtained via simulation. For example, Figure 14.2 plots 10,000 random draws of the largest order statistic in a sample of ten unit normals. Note that the distribution of realized differentials is asymmetric about its mean, implying that the variance alone is not sufficient for computing confidence intervals. Figure 14.3 plots the expected selection intensity for small values of N (assuming normality), showing that Equation 14.3a overestimates the intensity, although the difference will be small unless N is small.

Burrows (1972) developed a finite-sample approximation for the expected selection intensity for any reasonably well-behaved continuous distribution. Using the standardized variable $y = (z-\mu)/\sigma$ simplifies matters considerably. Letting $\phi(y)$ be the probability density

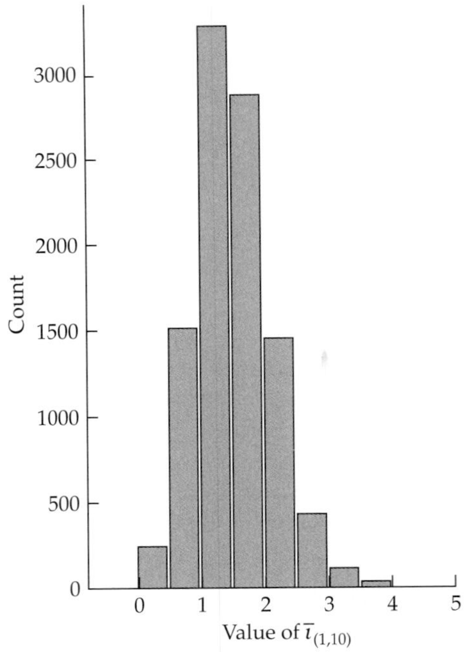

Figure 14.2 The distribution of 10,000 random draws of $\bar{\imath}_{(1,10)}$, the largest order statistic in a sample of ten unit normal random variables. The mean value is 1.54, as opposed to the expected value of $\bar{\imath} = 1.75$ for $p = 0.1$ in an infinite population (Equation 14.3a). Notice that there is a considerable spread about the mean, and that the distribution is not symmetric, but rather is skewed toward higher values.

function of the phenotypic distribution, and y_p be the truncation point (i.e., $\Pr(\, y \geq y_p \,) = p$), Burrows's approximation is

$$E\left[\, \bar{\imath}_{(N,M)} \,\right] \simeq \mu_{y_p} - \frac{(M-N)\,p}{2N(M+1)\,\phi(\, y_p \,)} \tag{14.4a}$$

where

$$\mu_{y_p} = E\left[\, y \,|\, y \geq y_p \,\right] = \frac{1}{p} \int_{y_p}^{\infty} x\,\varphi(x)\,dx$$

is the truncated mean (which can be obtained by numerical integration), and $\phi(\, y_p \,)$ is the height of the density function at the truncation point. Because the second term of Equation 14.4a is positive, if M is finite, the expected truncated mean overestimates the expected standardized selection differential. For a unit normal distribution, $\mu_{y_p} = \varphi(y_p)/p = \bar{\imath}$, yielding

$$E\left[\, \bar{\imath}_{(N,M)} \,\right] \simeq \bar{\imath} - \left[\frac{M-N}{2N(M+1)} \right] \frac{1}{\bar{\imath}} = \bar{\imath} - \left[\frac{1-p}{2p(M+1)} \right] \frac{1}{\bar{\imath}} \tag{14.4b}$$

where $\bar{\imath}$ is given by Equation 14.3a, using $p = N/M$. Lindgren and Nilsson (1985) found this approximation to be quite accurate for $N \geq 5$. Bulmer (1980) suggested an alternative approximation under normality, using Equation 14.3a with p replaced by

$$\widetilde{p} = \frac{N + 1/2}{M + N/(2M)} \tag{14.4c}$$

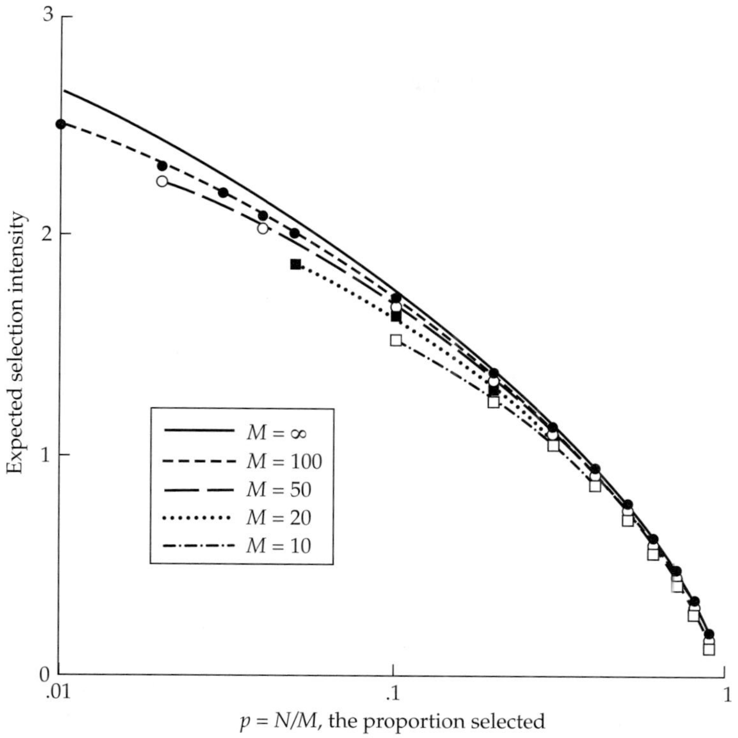

Figure 14.3 The expected selection intensity, $E[\,\bar{\imath}_{(N,M)}]$, under truncation selection on a normally distributed phenotype, as a function of the total number of individuals measured, M, and the fraction of these saved, $p = N/M$. The curve $M = \infty$ is given by Equation 14.3a, while the curves for M = 10, 20, 50, and 100 were obtained from the average of the expected values of the $N = pM$ largest unit normal order statistics (Harter 1961). Note that Equation 14.3a is generally a good approximation, even when N is fairly small.

Burrows (1975) provided expressions for the variance of $\bar{\imath}_{(N,M)}$.

A final correction for finite population size was noted by Rawlings (1976) and Hill (1976, 1977b). If families are sampled, such that n individuals are chosen per family, then the selection intensity is further reduced because there are positive correlations between family members. This effectively lowers the sample size below n—in the extreme case where all individuals are clones with little environmental variance, all have essentially the same value, and hence $n \sim 1$. If a total of M individuals is sampled, with n individuals per family, then Burrows's correction (Equation 14.4b) is modified to become

$$E\left[\,\bar{\imath}_{(N,M)}\right] \simeq \bar{\imath} - \left[\frac{1-p}{2p(M+1)(1-\tau+\tau/n)}\right]\frac{1}{\bar{\imath}} \tag{14.5}$$

where τ is the intraclass correlation of family members. This important result for certain types of family selection is revisited in more detail in Chapter 21.

Example 14.2. Consider the expected selection intensity on males when the upper 5% are used to form the next generation and phenotypes are normally distributed. If the sampled number is large, $\bar{\imath} \simeq 2.06$ (Example 14.1). Suppose, however, that just 20 males are scored (phenotyped), with only the largest allowed to reproduce, in order to yield $p = 0.05$. The expected value for this individual is the expected value of the largest order statistic for a sample of size 20. For the unit normal, this is $\simeq$1.87 (Harter 1961), and hence $E[\,\bar{\imath}_{(1,20)}\,] \simeq 1.87$. There is considerable spread about this expected value, as the standard deviation of this order statistic is 0.525 (Sarhan and Greenberg 1962). How well do the approximations of $E[\,\bar{\imath}_{(1,20)}\,]$ perform? Burrows's approximation (Equation 14.4b) yields

$$E\left[\,\bar{\imath}_{(1,20)}\,\right] \simeq 2.06 - \frac{(20-1)}{2\,(20+1)\,2.06} = 2.06 - 0.22 = 1.84$$

Bulmer's approximation (Equation 14.4c) uses

$$\widetilde{p} = \frac{1+1/2}{20+1/40} \simeq 0.075$$

which returns $x_{[1-0.075]} \simeq 1.44$. Because $\varphi(1.44) = 0.1415$, $E[\,\bar{\imath}_{(1,20)}\,] \simeq 0.1415/0.075 \simeq 1.89$.

RESPONSE IN DISCRETE TRAITS: BINARY CHARACTERS

The Threshold/Liability Model

One application of truncation-selection theory is in the response to selection of **binary traits**, which are characterized simply by presence or absence (such as normal or diseased). The basic trait model to this point assumed a continuous character, which initially seems at odds with a binary trait. However, as discussed in LW Chapters 11 and 25, discrete characters can often be modeled by mapping an underlying (unobserved, or **latent**) continuous character, the **liability**, z, onto the observed discrete character states, $y = 0$ or $y = 1$ (Figure 14.4). The assumption is that the breeder's equation holds on the liability scale, and our goal is to predict how changes on this scale map onto changes in the frequency of a binary trait. The simplest assumption is a **threshold model**, wherein the character is present if liability exceeds some threshold value T ($z \geq T$), and otherwise is absent ($z < T$). Roff (1996) reviewed a number of examples of such threshold-determined morphological traits in animals. Our analysis is restricted to a single threshold, but extension to multiple thresholds is straightforward (Lande 1978; Korsgaard et al. 2002).

To predict the selection response, let μ_t be the mean liability and q_t be the frequency of individuals displaying the character in generation t, i.e., $q_t = \Pr(y_t = 1)$. If liability is well enough behaved to satisfy the assumptions of the breeder's equation, then $\mu_{t+1} = \mu_t + h^2 S_t$. As shown in Figure 14.4, three tasks must be performed to predict the selection response: (i) estimate the mean liability, μ_t, from the observed frequency, q_t, of the trait; (ii) estimate S on the liability scale, given the change in the frequency of the binary trait following selection; and (iii) translate μ_{t+1} into q_{t+1}. We assume liability to be normally distributed on some appropriate scale, in which case we can also choose a scale that sets the threshold value at $T = 0$ and assigns z a variance of 1.0. Because $z - \mu_t$ is a unit normal, $\Pr(\,z \geq 0\,) = \Pr(\,z - \mu_t \geq -\mu_t\,) = \Pr(\,U \geq -\mu_t\,) = q_t$, and by analogy with Equation 14.2b, where $\Pr(U \geq x_{[1-p]}) = p$, we have

$$\mu_t = -x_{[1-q_t]} \tag{14.6}$$

where $x_{[p]}$ is the probit transformation of p (Equation 14.2a), as suggested by Wright (1934). For example, if 5% of the population displays the trait, $\Pr(U \leq 1.65) = 0.95$, implying $x_{[0.95]} = 1.65$, and yielding the mean on the underlying liability scale as $\mu = -x_{[0.95]} = -1.65$. When the mean liability is at this value, only 5% of the population has a value exceeding the threshold ($T = 0$), and hence displays the trait.

The response to selection, as measured by the new frequency, q_{t+1}, of the trait in the next generation, is given by

$$\begin{aligned} q_{t+1} &= \Pr(U \geq -\mu_{t+1}) \\ &= \Pr(U \geq -\mu_t - h^2\, S_t\,) \\ &= \Pr(U \geq x_{[1-q_t]} - h^2 S_t\,) \end{aligned} \tag{14.7}$$

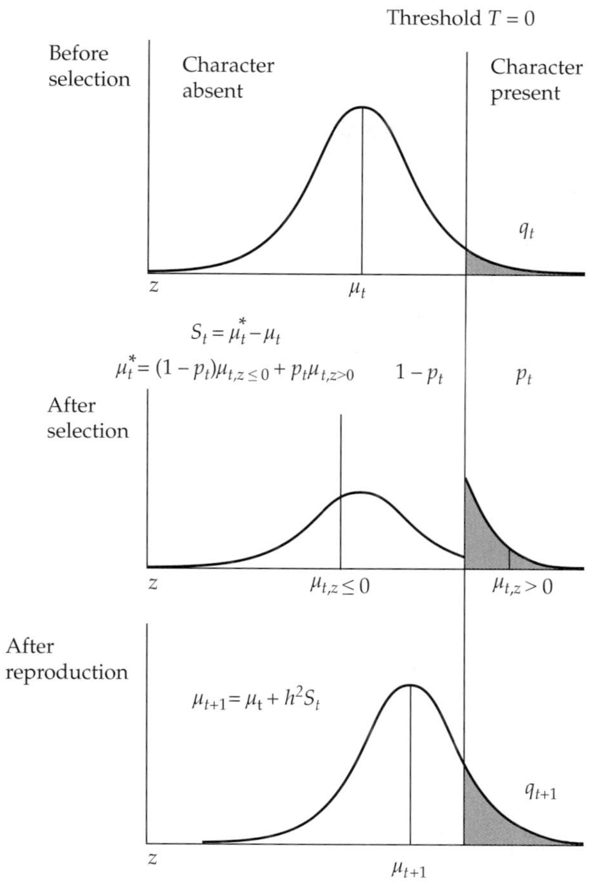

Figure 14.4 Selection response for a binary trait when the underlying liability, z, exceeds some threshold value, T. We assume that an appropriate scale can be found such that $z \sim N(\mu_t, 1)$, where μ_t is the current mean and $T = 0$. Under this scaling for T, a mean liability of zero ($\mu = 0$) implies that 50% of the population shows the trait, while <50% display the trait when $\mu < 0$ and >50% do when $\mu > 0$. Because z is normally distributed, the probit transform estimates μ_t from the frequency, q_t, of individuals displaying the character (Equation 14.6). We assume that the breeder's equation holds on the liability scale, so that $\mu_{t+1} = \mu_t + S_t h^2$, where $S_t = \mu_t^* - \mu_t$. Using properties of the unit normal allows translation of the mean liability following selection, μ_{t+1}, into the new frequency, q_{t+1}, of the trait (Equation 14.7). Note that after selection, where a fraction, p_t, of the selected parents display the trait, the mean liability value is now the weighted average of the means of two truncated normal distributions (Equation 14.8a).

It remains to obtain $S_t = \mu_t^* - \mu_t$, where μ_t^* is the mean liability value in the selected parents in generation t. While the selected population may consist entirely of adults displaying the trait, more individuals than this may be required to keep the population at a constant size, especially if q_t is small (i.e., the trait is rare). In this case, the selected adults consist of two groups of individuals: those displaying the trait (hence having $z \geq 0$) and those not displaying it ($z < 0$). Letting p_t be the fraction of selected adults displaying the character,

$$\mu_t^* = (1 - p_t)\, E\,[z|z < 0, \mu_t] + p_t\, E\,[z|z \geq 0, \mu_t] \tag{14.8a}$$

Applying LW Equation 2.14, and noting that the unit normal density function satisfies $\varphi(x) = \varphi(-x)$, yields

$$E\,[z|z \geq 0, \mu_t] = \mu_t + \frac{\varphi(\mu_t)}{q_t} \quad \text{and} \quad E\,[z|z < 0, \mu_t] = \mu_t - \frac{\varphi(\mu_t)}{1 - q_t} \tag{14.8b}$$

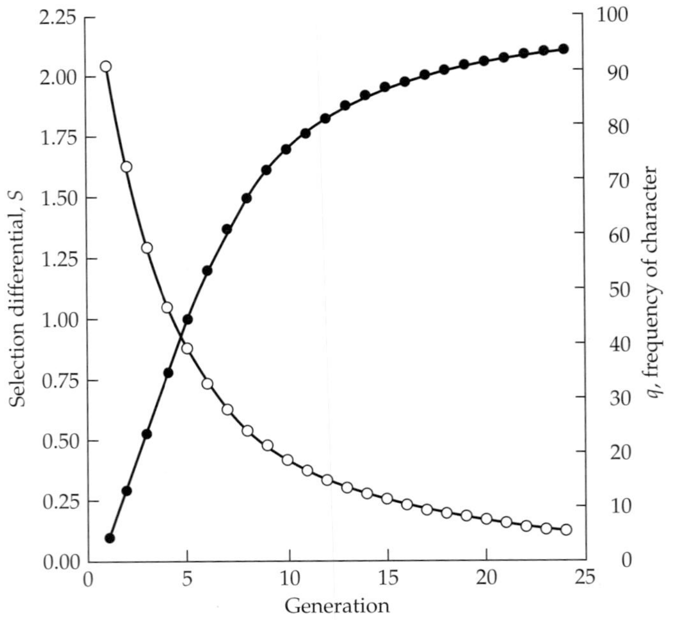

Figure 14.5 The response to selection on a threshold trait. Changes in q and S are plotted below, where solid circles denote q_t, and open circles denote S_t. See Example 14.3 for details.

where $\varphi(x)$ is the unit normal density function evaluated at x. Substituting into Equation 14.8a yields

$$S_t = \mu_t^* - \mu_t = \frac{\varphi(\mu_t)}{q_t}\,\frac{p_t - q_t}{1 - q_t} = \frac{\varphi(-x_{[1-q_t]})}{q_t}\,\frac{p_t - q_t}{1 - q_t} \tag{14.9}$$

As expected, if $p_t > q_t$, then $S_t > 0$. Maximal selection occurs if only individuals displaying the trait are saved ($p_t = 1$), in which case Equation 14.9 reduces to $S_t = \varphi(-x_{[1-q_t]})/q_t$.

Why did we not simply estimate μ_t^* using $x_{[1-q_t^*]}$, i.e., using the frequency, q^*, of the trait in the selected parents? The reason is that the distribution of z values in selected parents is a weighted average of two truncated normal density functions (Equation 14.8a), and this distribution is *not* normal (see Figure 14.4). However, we assume that normality is restored in the liability distribution at the start of the next generation due to segregation plus the addition of the environmental value. We examine the validity of this assumption in Chapter 24. Finally, a diligent reader might be concerned that this lack of normality violates the breeder's equation, and hence our assumption that $\mu_{t+1} = \mu_t + h^2 S$. However, this expression is a weighted form of truncation selection on a normally distributed trait, with a fraction $1 - p$ with $z < T$, and a fraction p with $z \geq T$, with the change in breeding value given by the weighted average of these two pools.

One important feature about selection on threshold traits is that the *response to selection is not necessarily symmetric*—a selected 5% increase in the trait may not yield the same response as a selected 5% decrease. The reason for this is that the mapping between phenotypes and their underlying liability is highly nonlinear. Even though the parent-offspring regression on the liability scale is assumed to be linear (and hence liability response is symmetric), the parent-offspring regression on the *phenotypic* level is not linear, resulting in an asymmetric response.

Example 14.3. Consider a threshold trait whose liability has a heritability of $h^2 = 0.25$ (Example 14.4 and especially LW Chapter 25 discuss how h^2 can be estimated on this scale). What is the expected response to selection if the initial frequency of individuals displaying the character is 5% and selection is practiced by choosing only adults displaying the character? Only six generations are required to increase the frequency of the trait to 50% ($\mu = 0$). Note that even though all selected parents exhibit the trait, the selection differential rapidly declines in a nonlinear fashion (Figure 14.5).

The values plotted in Figure 14.5 were obtained as follows. As was calculated earlier, $q_0 = 0.05$, which implies $\mu_0 = -1.645$ (the mean liability is 1.65 standard deviations below the threshold). Only individuals displaying the trait are allowed to reproduce, yielding (from Equation 14.9) the resulting selection differential on the liability scale:

$$S_0 = \varphi(-1.645)/0.05 \simeq 0.106/0.05 \simeq 2.062$$

Applying the breeder's equation returns the new mean value of liability:

$$\mu_1 = \mu_0 + 0.25 \cdot S_0 = -1.645 + 0.25 \cdot 2.062 = -1.129$$

Equation 14.7 translates this new mean into the fraction of the population now above the threshold

$$q_1 = \Pr(U \geq -\mu_1) = \Pr(U \geq 1.129) = 0.129$$

Thus, after one generation of selection, the character frequency is expected to increase from 5% to 12.9%. Further iteration in this fashion recovers the rest of the values in Figure 14.5.

Example 14.4. The effectiveness of selection on wing morphs in females of the whitebacked planthopper (*Sogatella furcifera*) was examined by Matsumura (1996). While this hemipteran is a serious rice pest in Japan, it is unable to overwinter. Rather, each year it migrates from southern China to recolonize Japan. Females exhibit two wing morphs: *macropterous* females are fully winged, while *brachypterous* females have reduced wings and cannot fly. Further, increasing nymphal population density increases the frequency of macropterous females (leading to increased dispersal). Using three replicate experiments at each of three densities, Matsumura selected for increased macroptery in one replicate, decreased macroptery in another, and a control (no selection) in the third. For the replicates with a density of one nymph per container, roughly 40–90 adults were scored, and 20 were chosen to form the next generation. The resulting data for the first five generations in the up-selected line was as follows (Matsumura, pers. comm.):

Generation	q	μ	p	S	R
1	0.224	-0.76	1.00	1.34	0.35
2	0.340	-0.41	0.80	0.75	0.54
3	0.551	0.13	1.00	0.72	0.33
4	0.675	0.45	1.00	0.53	-0.07
5	0.651	0.39	1.00	0.57	0.16
6	0.708	0.55			

Here q is the frequency of macroptery before selection and p is the frequency of macroptery in the selected parents. Translation from q into the mean liability, μ, follows from Equation 14.6. The response (on the liability scale) to selection on generation 1 is

$$R(1) = \mu_2 - \mu_1 = -x_{[1-0.340]} - (-x_{[1-0.224]}) = -0.41 - (-0.76) = 0.35$$

Likewise, the total response was

$$\mu_6 - \mu_1 = 0.55 - (-0.76) = 1.31$$

Selection differentials were calculated from q and p using Equation 14.9. For example, for generation 2,

$$S_2 = \frac{\varphi(\mu_2)}{q_2}\,\frac{p_2 - q_2}{1 - q_2} = \frac{\varphi(-0.41)}{0.34}\,\frac{(0.80 - 0.34)}{1 - 0.34} = 0.75$$

The total selection differential is $\sum_i S_i = 3.91$. One key summary statistic for any selection experiment is the *realized heritability*, the ratio of response to selection differential. As detailed in Chapter 18, there are several ways to compute this for a multigeneration selection experiment. One simple estimate is the ratio of the total response to the total differential,

$$\widehat{h^2} = \frac{\sum R_i}{\sum_i S_i} = \frac{1.31}{3.91} = 0.33$$

giving an estimated heritability of the underlying liability for macroptery of around 30%.

Direct Selection on the Threshold, T

It is biologically quite reasonable to imagine that there is variation in T itself (Hazel et al. 1990). Suppose the trait of interest appears when the size of an organism exceeds some critical value, which itself varies over individuals, with certain genotypes and/or environments lowering the value of T, thus allowing individuals with a lower liability score to display the trait. Decomposing both the liability and threshold in terms of genetic and environmental factors gives $z = g_z + e_z$ and $T = g_T + e_T$. The trait appears when $z \geq T$, or

$$g_z + e_z - (g_T + e_T) = (g_z - g_T) + (e_z - e_T) = g + e \geq 0$$

Thus, even though both the liability and threshold values are variable, we can simply consider a single new **risk liability**, the difference between the liability and threshold values, and the analysis proceeds as above. If interest is simply on presence or absence of the binary trait, it does not matter whether the liability or threshold, or both, show variation. However, as Example 14.5 (below) shows, there are situations where we can directly measure the threshold value itself, in which case we can estimate the realized heritability of the threshold level by a selection experiment.

The Logistic Regression Model for Binary Traits

The threshold approach offers one model for mapping an underlying continuous liability, z, into a discrete character space, y (which is either zero or one, corresponding to trait absence or presence). This is a deterministic model, with all individuals with $z \geq T$ displaying the trait ($y = 1$), while all those with $z < T$ do not display it ($y = 0$). A potentially more realistic model is that trait presence is stochastic, with the underlying liability, z, mapping onto a *probability* of displaying the trait, e.g., $p(z) = \text{Prob}(y = 1 \mid z)$. Under the threshold model, this probability is 1.0 for $z \geq T$, and 0 otherwise. From a biological standpoint, one imagines that $p(z)$ is a monotonically increasing function of z, approaching 0 for low values and 1 for high values. One reasonable candidate that satisfies these requirements is the **logistic function**,

$$\ell(z) = \frac{\exp(z)}{1 + \exp(z)} = \frac{1}{1 + \exp(-z)} \tag{14.10a}$$

with $\ell(z) \simeq 0$ for $z \ll -1$, $\simeq 1$ for $z \gg 1$, and $\ell(0) = 1/2$. A more general version is

$$\ell[\alpha\,(z - m)] = \frac{1}{1 + \exp[-\alpha\,(z - m)]} \tag{14.10b}$$

which has a value of 0.5 at $z = m$ and a scaling factor, α, that sets the abruptness of the transition from low to high probability. The larger the value of α, the more abrupt is the transition, approaching the threshold model for sufficiently large values (Figure 14.6). Equation 14.10b is often called a **logistic regression**.

Biologically speaking, the logistic regression and threshold models may be viewed as essentially identical. To see this, recall that the threshold model very easily extends to the

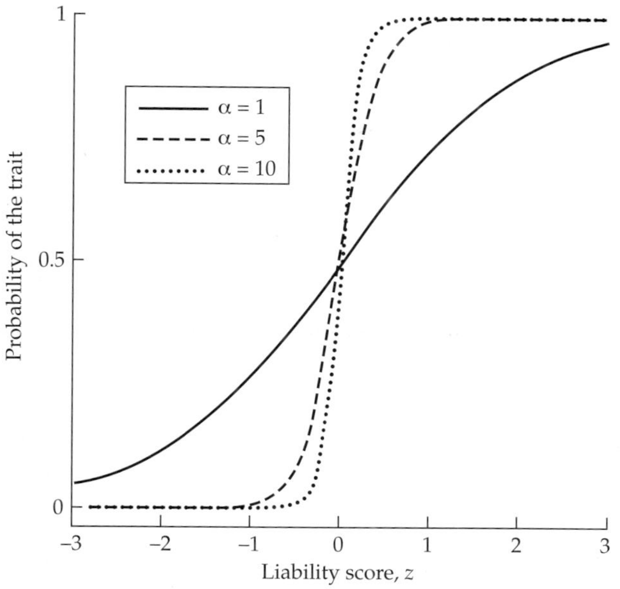

Figure 14.6 A more realistic model of threshold traits is that the liability, z (horizontal axis), determines the probability, $p(z)$, of displaying the trait (vertical axis). One flexible model is to assume that $p(z)$ follows a logistic function (Equation 14.10b) with a scale parameter of α, plotted here for values of α = 1, 5, and 10. For α values in excess of 5, the logistic function essentially recovers the discrete threshold model.

case where T varies over individuals. In such cases, if the liability value of an individual is z, the trait will only be displayed if $T \leq z$. Now consider the logistic regression model where $p(z)$ denotes the probability that an individual with a liability value of z displays the trait. One source of this stochasticity could simply be population variation in T, so that $p(z)$ can be viewed as the *cumulative distribution function* (cdf; LW Chapter 2) for the threshold value T, e.g., $p(z) = \Pr(T \leq z)$. In this case, a fraction, $p(z)$, of individuals with a liability of z are above the threshold, and hence display the trait.

If the logistic gives the cdf of random threshold values, then the **logistic distribution**, $\phi(x, \alpha, m)$, gives the actual distribution of T. From the definition of a cumulative distribution function,

$$\int_{-\infty}^{z} \phi(x, \alpha, m)dx = \frac{1}{1 + \exp[-\alpha\,(z - m)]} \tag{14.11a}$$

Taking derivatives of both sides yields

$$\phi(x, \alpha, m) = \frac{\alpha\,\exp[-\alpha\,(z - m)]}{(1 + \exp[-\alpha\,(z - m)])^2} \tag{14.11b}$$

Johnson and Kotz (1970b) gave the first three moments of this distribution as

$$\mu = m, \quad \sigma^2 = \frac{1}{3}\left(\frac{\pi}{\alpha}\right)^2, \quad \text{and} \quad \mu_3 = 0 \tag{14.11c}$$

As shown in Figure 14.7, the normal and logistic distributions have very similar cumulative distribution functions. Indeed, for a unit normal random variable, U,

$$\Pr(U \leq x) \simeq \frac{1}{1 + \exp(-\alpha\, x)}, \quad \text{where} \quad \alpha = \frac{\pi}{\sqrt{3}} \tag{14.12}$$

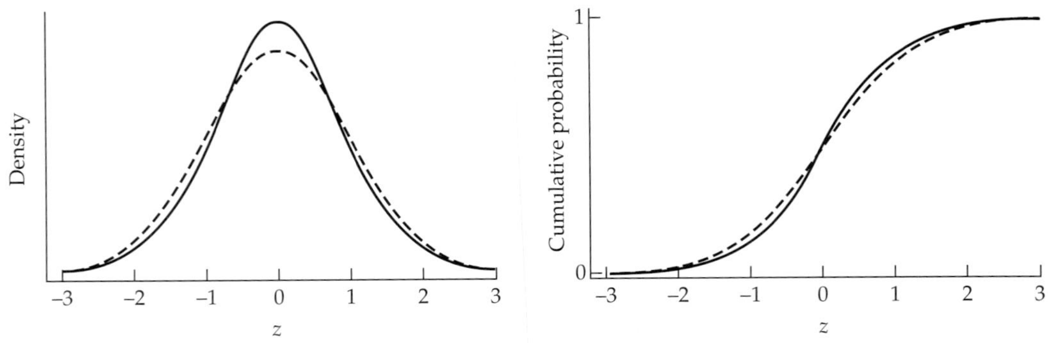

Figure 14.7 A comparison of the unit normal and unit logistic ($\mu = 0, \sigma^2 = 1$) distributions (dashed and solid curves, respectively), with the horizontal axis denoting the value of z. (**Left**) Probability density functions: the logistic is more peaked, with positive kurtosis. (**Right**) The cumulative distribution functions are extremely similar.

which is the logistic distribution with a variance of 1 (see Equation 14.11c).

An interesting biological interpretation of the scale parameter, α (that sets the abruptness of the transition of the logistic regression), is as a measure of the strength of developmental canalization. When α is small, small changes in the liability usually map into small changes in the probability of the trait being displayed ($\alpha = 1$ in Figure 14.6). When α is large, one sees robustness when the liability is away from the mean, m, as small changes in the liability have very little impact on the probability of the trait being displayed. However, when near m, small liability changes can result in dramatically different probabilities of displaying the trait. Chevin and Lande (2013) essentially used this idea to explore the conditions under which a relatively continuous norm of reaction changes into a very discrete step-function (i.e., the conditions leading to the evolution of a large α value).

Thus, we have two approaches for mapping liability values into binary traits: the strict threshold approach (a deterministic mapping of liability onto the discrete trait) and the logistic regression approach (a stochastic mapping translating a liability value into a probability of observing the trait). Given the very close connection between the threshold and logistic regression models, for most purposes using the simple threshold model is a reasonable approach, even if the underlying mapping is stochastic, and as illustrated above, it can easily be used to predict selection response. One setting where the logistic regression is more appropriate is in the actual analysis of the behavior of the threshold when one either knows the liability value or has at least a strong proxy (such as size).

Example 14.5. An interesting analysis of selection on a threshold trait using logistic regressions was provided by Wesselingh and de Jong (1995), who studied the connection between plant size (measured by the proxy of dry weight) and flowering in hound's-tongue (*Cynoglossum officinale*). This species is a facultative biennial, which means that, like an annual plant, it flowers only once, but unlike an annual, it may live several years before flowering. This strategy represents a tradeoff between reduced survival and a larger seed set from a larger size at flowering. For *Cynoglossum*, it has been shown that vernalization (cold treatment) followed by an appropriate photoperiod is required for flowering. However, unless plants are at (or above) a certain threshold size, they are unresponsive to vernalization, and will grow without flowering through the next growing cycle. The authors were interested in the threshold size that triggers the binary trait (vernalization sensitivity), and in particular whether this size is both variable and heritable. To examine this, they grew plants for different numbers of days (ranging from 31 to 86) to generate individuals of different sizes before the vernalization treatment. This generated two selection groups based on threshold size: the smallest plants that flowered following vernalization were chosen as the low-line (low-threshold) parents, while those plants that did

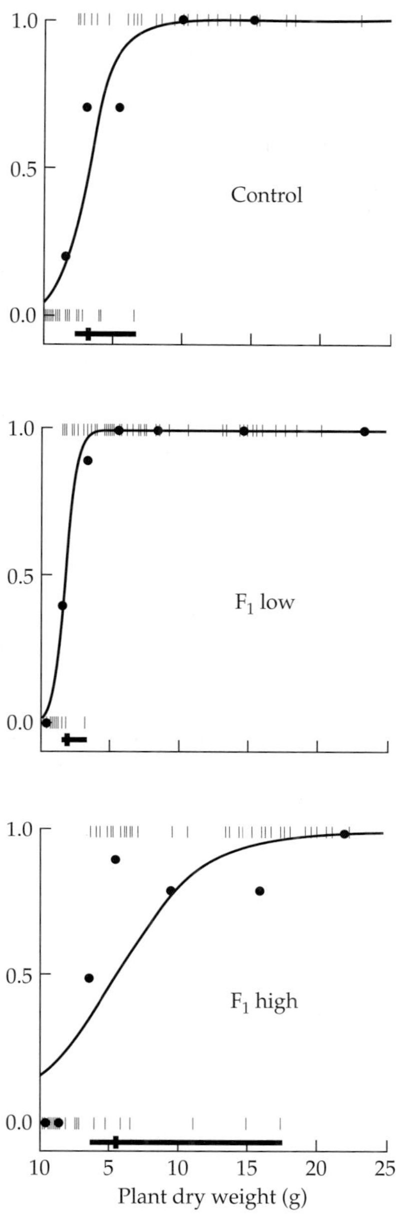

Figure 14.8 The logistic regressions for the relationship between dry weight and flowering in hound's-tongue (*Cynoglossum officinale*). Data were obtained from Wesselingh and de Jong (1995). Regressions are given for control, high, and low lines, grown contemporaneously. See Example 14.5 for further details.

not respond to the first vernalization treatment were allowed to grow a second cycle and were chosen as the parents for the high lines. The response to a single generation of selection can be assessed by comparing the offspring from the low (or high) selected parents against those from an unselected control, as plotted in Figure 14.8.

The data available to the authors were 0 or 1 (insensitive or sensitive to vernalization) values as a function of size. To estimate the distribution of threshold sizes, they performed a logistic regression on these data, using maximum likelihood (LW Appendix 4) to fit the α and mean (m) terms of Equation 14.10b. Data for the high and low lines are plotted in Figure 14.8 along with the ML solution for the logistic regression. Each individual has a 0 or 1 data point (individual ticks), while the circles represent the average value for weight classes with more than ten individuals.

Logistic regressions were estimated for progeny from the low- and high-line parents and for a control line grown contemporaneously with these progeny. The ML estimates of the mean, m (which corresponds to the weight yielding 50% flowering), and value of α for these regressions were

Line	m	α
Control	3.30	0.97
Low	1.85	2.58
High	5.41	0.31

Note that the low line not only had a smaller mean size for vernalization (1.85), but also a much larger α value (2.58), and therefore a more abrupt transition between insensitivity and sensitivity. Using these estimates, Equation 14.10b yields the expected percent of vernalization sensitivity (flowering) for a given weight. For a 3 gram plant, this is 0.43 in the control line and 0.32 in the high lines, but 0.95 in the low line.

By estimating both the response, R, to selection as well as the within-generation change, S, we can estimate the heritability of this trait by calculating R/S. First, the response to selection (change in m) can be estimated by using the contemporaneously grown control line as a standard. From the previous table, the response in the high line is 5.41 − 3.30 = 2.11, while the response in the low line is 1.85 − 3.30 = −1.45. Turning to estimating S, the selection truncation point for the low line is the largest low parent (2.74 grams, or 25.5% of the left tail of the founding source population), while the smallest flowering high parent was 9.95 grams (corresponding to the upper 12.2% of the founding source population). From Equation 14.3a, these translate into selection intensities of $\bar{\imath} = -1.26$ and 1.66, respectively. To obtain the selection differentials, S, for each line, recall that $S = \bar{\imath}\sigma_p$. To estimate σ_p, the authors note that the 0.25 quartile for a normal distribution is 0.674σ from the mean. Although the assumption is that the threshold values follow a logistic distribution, the cumulative probability functions are rather similar for both the normal and the logistic (Figure 14.7). Hence, taking the observed 0.25 quartile (in the founding lines) of 2.68, and its mean of 5.12, suggests

$$\sigma_p = \frac{5.12 - 2.68}{0.674} = 3.63$$

The response (change in m), selection intensity, and estimated heritability, $\widehat{h^2} = R/S$, for the high and low lines are

Line	$\bar{\imath}$	S	R	$\widehat{h^2} = R/S$
Low	−1.26	−4.58	−1.45	0.32
High	1.66	6.02	2.11	0.35

Thus, there is heritable variation in threshold size, as there was response to selection for both larger and smaller threshold sizes. Further, the estimated heritability (based on the single-generation response to selection) was around 0.3.

BLUP Selection With Binary Data: Generalized Linear Mixed Models

Animal breeders routinely use the **general linear mixed model** to obtain BLUP estimates of breeding values for normally distributed traits, where the expected value of the phenotype, y_i, of individual i can be written as

$$E[y_i] = \mu_i = \mu + \sum \beta_k x_{k,i} + A_i \tag{14.13a}$$

Here A_i is an individual's breeding value and the β_k are fixed effects (such as adjustments for age and sex, whose values in individual i are given by the $x_{k,i}$). A particular realization (i.e., a scored phenotype) from this individual can be written as

$$y_i = \mu_i + e_i \tag{14.13b}$$

where the residual, e_i, is normally distributed with a mean of zero, implying that y_i is also normally distributed. Putting these together, an observation from individual i can be written as

$$y_i = \mu + \sum \beta_k x_{k,i} + A_i + e_i \tag{14.13c}$$

As described in Chapters 13, 19, 20, and LW Chapters 26 and 27, in addition to its observed phenotype, y_i, additional information to estimate A_i is borrowed from the y values of relatives through the relationship matrix $\mathbf{A}$. In a selection scheme, those individuals with the largest estimated A values are then chosen to form the next generation.

This same strategy can be extended to cases where the mean μ_i is *not* a simple linear function of A_i and where the residual error is not normally distributed. In particular, the phenotype, y_i, of a binary trait takes on a value of either 0 or 1. The expected value of y_i is just the probability of displaying the trait, i.e., the trait is **Bernoulli-distributed** (a binomial with a single draw), so that y_i is either 0 or 1 given this mean μ_i (its success parameter in the Bernoulli). Equation 14.13a is generally not applicable in this setting as μ_i is constrained (being a probability) to the 0 to 1 range, while no such constraints occur on Equation 14.13a. A second issue is that the residual error structure is not normal.

The solution to both satisfying this constraint and using the correct residual distribution relies on the use of **generalized** (as opposed to general) **linear mixed models**. These extend Equation 14.13c to cases where the expected value of y conditioned on the variables of interest is *not* a linear function of breeding value, and the residuals about this expected value are not necessarily normal (Bolker et al. 2009; de Villemereuil et al. 2016). The basic structure of a generalized linear model can be thought of as an extension of the liability model to more general functions.

Let the liability value (z_i) for individual i be some linear function of the breeding value, A_i, an environmental value, E_i, plus (potentially) fixed effects,

$$z_i = \mu + \sum \beta_k x_{k,i} + A_i + E_i \tag{14.14a}$$

which is mapped onto a mean trait value, y, by a monotonic function, g. Specifically, the conditional expectation of y given z is

$$E[y \,|\, z] = g(z) \tag{14.14b}$$

where

$$g^{-1}\left(E[y \,|\, z]\right) = z = \mu + \sum \beta_k x_{k,i} + A_i + E_i \tag{14.14c}$$

The inverse g^{-1} is called the **link function**, as it transforms the conditional expectation into a linear model. The function g imposes the desired constraint on the mean of y, mapping some unconstrained value, z, into the desired constraint space (such as 0 to 1).

For binary data, Equation 14.14b becomes $E[y\,|z] = g(z) = p(z)$, where $p(z) = \Pr(y = 1\,|\,z)$, the success parameter, with y following a Bernoulli distribution. Thus, we desire some function that maps the underlying liability, z, onto a value, $p(z)$, constrained to the space $[0,1]$. As we have seen, one candidate function is the simple logistic function, $\ell(z)$ (Equation 14.10a). The corresponding link function (Equation 14.14c), which is the inverse of the logistic function, is given by the **logit function**, defined as

$$\text{logit}(p) = \ln\left(\frac{p}{1-p}\right) \tag{14.15a}$$

namely, the log of the odds ratio (probability of the trait being present divided by probability that the trait is absent). If $\ell(z) = p$, then $\text{logit}(p) = z$, so that a logit-transformed p value recovers the liability value,

$$\text{logit}(p|z_i) = z_i = \mu + \sum \beta_k x_{k,i} + A_i + E_i \tag{14.15b}$$

Under this framework, BLUP selection for individuals with the highest breeding values for a binary trait proceeds by taking the 0/1 binary data from a set of individuals (along with other fixed, and possibly random, effects of interest) and using either maximum likelihood or Bayesian approaches to estimate the breeding values on the liability scale given by Equation

14.15b (Foulley et al. 1983; Foulley 1992; Vazques et al. 2009). This approach can be extended to $k \geq 2$ thresholds in the mapping of liability into k character states (Korsgaard et al. 2002).

RESPONSE IN DISCRETE TRAITS: POISSON-DISTRIBUTED CHARACTERS

Discrete characters with a large number of possible states, such as numbers of leaves on a tree, can be treated as a continuous trait. However, what about a discrete trait with a rather compact distribution? A common example is the number of offspring, such as the clutch size of a bird, which might range from 0 to 10 eggs in an observed sample. This discreteness is of special concern when the trait has a significant probability mass at a particular value (especially zero), as often happens with offspring number (corresponding to failure to reproduce).

A natural way to model such traits is to use the **Poisson distribution**, where the probability of observing a trait value of k is given by

$$\Pr(y = k) = e^{-\lambda}\,\frac{\lambda^k}{k!} \tag{14.16}$$

where $\lambda = E[y]$ is the expected value of the trait, which is constrained to $\lambda > 0$. Motivated by the previous treatment of binary traits, one might imagine some underlying liability value, z, with additive effects (Equation 14.14a) mapped onto a mean value, λ, constrained to be positive. For example, we can use

$$\lambda = \exp(z) \tag{14.17a}$$

ensuring that $\lambda > 0$ for all z and hence is a proper expectation for a Poisson. In the context of generalized linear models, $g(z) = \exp(z)$, so that the link function $g^{-1}(z)$ is simply $\ln(z)$, as $g^{-1}[g(z)] = \ln[\exp(z)] = z$, yielding

$$\ln(\lambda) = z = \mu + \sum \beta_k x_{k,i} + A_i + e_i \tag{14.17b}$$

This is called a **log-linear model**, as the log of the distribution parameter, λ, is a linear function of the variables of interest (in particular, the breeding value).

On this log scale, both the breeding and environmental values are assumed to be normal, with a mean of zero and variances of σ^2_A and σ^2_e. As with binary traits, BLUP selection based on this generalized linear model framework can be used to estimate the A_i values (Foulley 1993; Foulley and Im 1993; Korsgaard et al. 2002; Vazques et al. 2009; Morrissey 2015; de Villemereuil et al. 2016). Other models are also possible, such as a **zero-inflated Poisson**, which has extra probability mass at zero relative to a standard Poisson (e.g., Rodrigues-Motta et al. 2007; see Chapter 29). A trait can also be underdispersed relative to the Poisson (as seems to consistently be the case for bird clutch size; J. Hadfield, pers. comm.), in which case other distributions, such as a **zero-truncated Poisson** (the nonzero data follow a Poisson distribution) can be used (Chapter 29). A nice discussion of estimation issues under generalized linear models (beyond the binary and log-linear models presented here) is given by de Villemereuil et al. (2016), while Morrissey (2015) presented the theory of selection response under more general settings (which we will examine in Volume 3).

Under the log-linear model, the liability, z, of an individual yields an expected value of $\lambda = \exp(z)$, with a realization (e.g., the number of offspring for an individual with trait value z) drawn from a Poisson to return the observed trait value. For example, if the latent value, λ_i, for individual i has a value of 0.2, then (from Equation 14.16) the probability this individual has a trait of value zero is $\exp(-0.2) \simeq 0.82$, the probability they have value 1 is $0.2 \cdot \exp(-0.2) \simeq 0.16$, value 2 is $0.2^2 \cdot \exp((-0.2)/2! \simeq 0.02$, and so on.

The resulting mean trait value in a population under the log-linear model becomes

$$\begin{aligned} E[y] &= E[\lambda] = E[\exp(z)] \\ &= E[\,\exp(\mu) \cdot \exp(A) \cdot \exp(e)\,] \\ &= \exp(\mu) \cdot E[\,\exp(A)\,] \cdot E[\,\exp(e)\,] \end{aligned} \tag{14.18}$$

where the last step follows because (by construction) A and e are uncorrelated (and thus independent if they are bivariate normal, as is usually assumed), while μ is a constant. To compute these expectations, recall that the expression $E[e^{tx}]$ is the **moment-generating function** of the random variable, x (Johnson and Kotz 1970a). For $x \sim N(\mu, \sigma^2)$,

$$E\left[e^{tx}\right] = \exp\left(\mu t + \frac{\sigma^2}{2} t\right) \tag{14.19a}$$

In this case, with a mean, μ, equal to zero and variance of σ^2, setting $t = 1$ yields

$$E\left[e^{x}\right] = \exp\left(\frac{\sigma^2}{2}\right) \tag{14.19b}$$

Substituting into Equation 14.18 shows that the expected mean trait value is a function of both the mean, μ, and variance, σ_z^2, of the underlying liability value, with

$$E[y] = \exp(\mu) \cdot \exp\left(\frac{\sigma_A^2 + \sigma_e^2}{2}\right) = \exp(\mu) \cdot \exp(\sigma_z^2/2) \tag{14.20a}$$

One might initially expect that if A is the breeding value for liability, then the mean phenotype of an individual would simply be $\exp(\mu + A)$. However, Equation 14.20a shows that if we condition on the breeding value being A, the conditional mean now becomes $\mu + A$ and the variance, σ_A^2, vanishes, leaving

$$E[y \mid A] = \exp(\mu + A) \cdot \exp\left(\frac{0 + \sigma_e^2}{2}\right) = \exp(\mu + A) \cdot \exp(\sigma_e^2/2) \tag{14.20b}$$

Following a single generation of selection, the distribution of liability values has approximately the same variance, but now the mean is shifted to $\mu + h^2 S$ (where S is the selection differential *on the liability scale*). Applying Equation 14.20a, the response on the phenotypic scale becomes

$$\begin{aligned} R &= E[y_{t+1}] - E[y_t] \\ &= \left[\exp(\mu + h^2 S) - \exp(\mu)\right] \cdot \exp(\sigma_z^2/2) \\ &= \left[\exp(h^2 S) - 1\right] \cdot \exp(\mu) \cdot \exp(\sigma_z^2/2) \\ &= \left[\exp(h^2 S) - 1\right] \cdot E(y_t) \end{aligned} \tag{14.21}$$

Notice that, as was the case for selection on a binary trait, the response to selection is not symmetric, as $S = +\delta$ does not give the same increment of response as $S = -\delta$.

15

Short-term Changes in the Mean: 3. Permanent Versus Transient Response

The phenotypic gains from selecting for epistatic differences come from distorting the gametic array and soon disappear after selection is relaxed, as the gametic array returns to random. By contrast, the gains from changes in gene frequency are permanent. Lush (1948)

While the basic breeder's equation is the foundation for much of the theory of selection response in quantitative traits, its elegant form of $R = h^2 S$ arises because of certain simplifying assumptions (Table 13.2). One interesting complication is that the selection response can have both transient and permanent components, and this chapter examines common situations where this can occur (also see Chapter 23). Selection can change allele and genotype frequencies, generate linkage disequilibrium, and create nonrandom associations among environmental values, all of which can contribute to the response. Of these, only allele-frequency changes are permanent, as any such changes remain after selection stops. In contrast, random mating eventually ensures there will be Hardy-Weinberg frequencies for genotypes, linkage equilibrium, and random associations of environmental effects. The contributions to the selection response generated by departures from these equilibrium values are therefore **transient**, decaying to zero following the cessation of selection.

Transient components of response can be both positive and negative, so the selection response may decay or increase following the cessation of selection. In the extreme situation, a **reversed response** occurs, wherein a trait selected to increase actually *decreases* due to a negative transient component overwhelming a positive permanent response. The potential of transient contributions from previous generations of selection is one reason why the breeder's equation requires an unselected base population.

Our treatment starts with a brief discussion of why the breeder's equation focuses on h^2, even though other factors can also contribute to the selection response. We then turn to two genetic situations that generate transient components to response: additive epistasis in diploids and dominance in autotetraploids. This leads to a discussion of the method of using ancestral regressions to determine the conditions under which a component of response is transient. The implications of correlated environmental effects between relatives are considered next, as we examine the selection response when the parent-offspring covariance is entirely due to shared environmental effects (which also serves as model for epigenetic effects). We conclude by considering the complications introduced by the presence of heritable maternal effects, which show share properties seen with shared environmental values.

WHY ALL THE FOCUS ON h^2?

As discussed in LW Chapters 7 and 17, epistasis or environmental effects shared between parents and their offspring can cause the parent-offspring regression to deviate significantly from $h^2/2$, thus altering the response from that predicted by the breeder's equation. If we allow for epistasis (among unlinked loci) and correlation between parental and offspring environmental values, the single parent-offspring slope will be

$$b_{op} = \frac{\sigma(z_p, z_o)}{\sigma_z^2} = \frac{1}{\sigma_z^2}\left(\frac{\sigma_A^2}{2} + \frac{\sigma_{AA}^2}{4} + \frac{\sigma_{AAA}^2}{8} + \frac{\sigma_{AAAA}^2}{16} + \cdots + \sigma(E_p, E_o) \right) \tag{15.1a}$$

If we assume there is a linear biparental regression with identical slopes for both parents,

Evolution and Selection of Quantitative Traits. Bruce Walsh & Michael Lynch.
Published 2018 by Oxford University Press.
DOI 10.1093/oso/9780198830870.001.0001

the response to a single generation of selection becomes

$$R = 2b_{op}S = h^2S + \frac{S}{\sigma_z^2}\left(\frac{\sigma^2_{AA}}{2} + \frac{\sigma^2_{AAA}}{4} + \frac{\sigma^2_{AAAA}}{8} + \cdots + \sigma(E_{fa}, E_o) + \sigma(E_{mo}, E_o)\right) \quad (15.1b)$$

which can deviate significantly from h^2S. Why, then, is so much attention focused on h^2?

The reason is that we are generally interested in the **permanent response** to selection. Selection changes allele frequencies, which in turn changes the mean genotypic value. Under the infinitesimal model, very small changes in allele frequencies over a large number of loci can generate a significant change in the mean without much change in the additive variance (Chapter 24), and these changes are reflected in the response component, h^2S. Any additional response from epistasis or shared environmental effects arises because of gametic-phase disequilibrium or nonrandom associations of environmental values. Recombination and the randomization of environmental effects cause these associations to decay to zero under random mating. Conversely, changes in allele frequencies are permanent, as once selection stops, the new allele frequencies will be stable (ignoring the longer-term effects of drift). Hence, as will be shown shortly, the permanent response under the conditions leading to Equation 15.1b is $h^2\,S$. One exception, discussed in Chapter 23, is when significant inbreeding occurs. In this case, σ^2_{AA}, and other nonadditive variance components specific to inbred populations (σ^2_{DI} and σ_{ADI}; introduced in Chapter 11), can contribute to permanent response.

GENETIC SOURCES OF TRANSIENT RESPONSE

Diploid parents undergoing sexual reproduction pass along single alleles at each locus to their offspring, with a parent's breeding value being the sum of its allelic effects over all loci (LW Chapter 4). Genotypic values are generally different from breeding values, with deviations representing interaction terms, such as dominance (between alleles at the same loci) or epistasis (between alleles at different loci). Such interactions are often not passed to an offspring of a randomly mating diploid parent, as their transmission requires that the parent pass along all of the component parts of the interaction (such as both alleles at a locus for dominance). As such, only part of an exceptional genotypic value of a selected parent is passed on to its offspring. However, there are two common settings where at least some of these interaction terms can be passed to an offspring under random mating—additive epistasis in a diploid (and higher ploidy) and dominance in an autotetraploid. When nonrandom mating occurs, such as inbreeding, the possibility exists that other interactions can be passed on to an offspring (such as dominance in a diploid); see Chapter 23.

Additive Epistasis

Pairwise epistasis involves interactions between either single alleles at different loci (additive-by-additive epistasis, $A \times A$), between entire genotypes at different loci (dominance-by-dominance epistasis, $D \times D$), or between an allele at one locus and the genotype at a second (additive-by-dominance epistasis, $A \times D$). Because a randomly mating diploid passes along a single allele at each locus to its offspring, additive ($A \times A$, $A \times A \times A$, etc.) interactions between alleles at different loci can be passed from a parent to its offspring. Under random mating in a diploid, any epistatic term involving dominance ($A \times D$, $D \times D$, etc.) cannot be passed along, as a parent must contribute *both* alleles at a given locus to its offspring to transmit a D component.

Before proceeding, it is important to realize the important distinction between epistatic *effects* (the expected multilocus value does not equal the sum of the single-locus values) and epistatic *variances*. The former can be quite large and yet the latter can still remain small. In large part this is due to the additive genetic variance being the variance accounted for by the best linear fit of a potentially highly nonlinear system, so that (as is the case for dominance) epistatic effects are partly accounted through σ^2_A (Crow 2010; LW Chapter 5). Under fairly general patterns of multilocus interactions, extreme allele frequencies (near zero or one), as would be expected under drift (Chapter 2), result in most of the *genetic variance* being additive even when the *genotypic values* have a highly nonadditive basis (Hill et al. 2008; Mäki-Tanila and Hill 2014). Our concern here is how the residual nonadditive effects (e.g., σ^2_{AA}) that are not absorbed into additive variance influence the selection response.

Although Lush (1948) and Kempthorne (1957) clearly grasped the key idea that the component of selection response from additive epistasis is transient, the first quantitative analysis was by Griffing (1960a, 1960b), who assumed the infinitesimal model. Under the assumptions that (i) phenotypes are normally distributed, (ii) effects at any particular locus are very small relative to the total phenotypic variation, and (iii) no third- (or higher-) order additive epistasis is present, the response to one generation of selection is

$$R = S\left(h^2 + \frac{\sigma^2_{AA}}{2\sigma^2_z}\right) \tag{15.2}$$

One might expect that $R(t)$, the cumulative response after t generations of selection, is simply t times the result given by Equation 15.2. However, any increased response (relative to Sh^2) due to epistasis is only temporary, reflecting gametic-phase disequilibrium generated by selection. As disequilibrium decays under recombination, so does the component of response due to epistasis. This occurs because the $A \times A$ contribution comes from favorable *combinations* of alleles at different loci, above and beyond their individual contributions (which are accounted for by changes in the mean breeding value of the selected parents, h^2S). Recombination breaks down these combinations, removing the epistatic contribution. Griffing showed that for two linked loci separated by a recombination fraction of c, the response when a generation of selection is followed by τ generations of random mating is

$$S\left(h^2 + (1-c)^\tau\,\frac{\sigma^2_{AA}}{2\sigma^2_z}\right) \tag{15.3}$$

which converges to h^2S as $\tau \to \infty$. Equation 15.3 follows by noting that the probability of a gamete containing specific alleles from both loci remaining intact following one generation of recombination is $1-c$. Thus, after τ generations, only $(1-c)^\tau$ of the favorable two-locus combinations selected at $\tau = 0$ remain unaltered by recombination. This transient component of response from epistasis is often called the **Griffing effect**.

Summing Equation 15.3 over t yields the cumulative response after t generations with a constant selection differential, S, as

$$R(t) = t\,h^2S + R_{AA}(t) \tag{15.4a}$$

where

$$R_{AA}(t) = S\,\frac{\sigma^2_{AA}}{2\sigma^2_z}\left(\sum_{i=1}^{t}(1-c)^{i-1}\right) = S\left(\frac{1-(1-c)^t}{c}\right)\left(\frac{\sigma^2_{AA}}{2\,\sigma^2_z}\right) \tag{15.4b}$$

denotes the cumulative additive $\times$ additive epistatic contribution. The last equality follows using the partial sum of a geometric series

$$\sum_{i=1}^{n} x^{i-1} = \sum_{i=0}^{n-1} x^i = \frac{1-x^n}{1-x} \tag{15.5a}$$

A related result (used later in the chapter) is that

$$\sum_{i=1}^{n} x^i = \frac{1-x^{n+1}}{1-x} - 1 = \frac{x(1-x^n)}{1-x} \tag{15.5b}$$

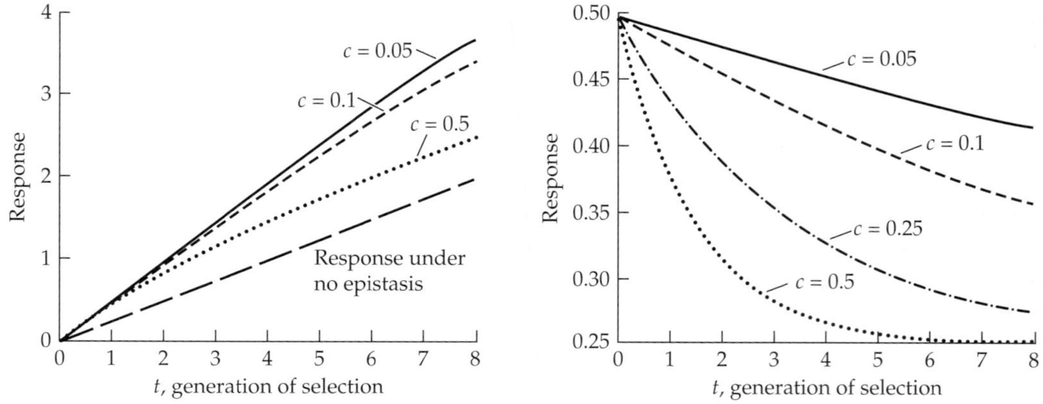

Figure 15.1 The permanent and transient response to selection (scaled in units of S) assuming pairwise epistasis in a diploid, with $h^2 = 1/4$ and $\sigma^2_{AA}/\sigma^2_z = 1/2$. **Left**: The cumulative response assuming a constant amount of selection for various values of c. Note that even with this large amount of epistasis (σ^2_{AA} accounts for half the total variance), it is difficult to distinguish the curvilinear response with epistasis from a linear response. **Right**: The decay of response following a single generation of selection due to the decay of the contribution from epistasis. Provided $c > 0$, the cumulative response eventually decays to $h^2 S = S/4$, the expectation under no epistasis.

If loci are completely linked ($c = 0$), $R_{AA}(t) = t\,S\,\sigma^2_{AA}/(2\sigma^2_z)$, but provided $c > 0$, the total epistatic contribution approaches a limiting value of

$$\widetilde{R}_{AA} = \lim_{t \to \infty} R_{AA}(t) = \frac{1}{c}\left(S\,\frac{\sigma^2_{AA}}{2\,\sigma^2_z} \right) \tag{15.6a}$$

$\widetilde{R}_{AA}$ represents the balance between selection generating disequilibrium and its removal by recombination. Equation 15.6a shows that the limiting contribution is $1/c$ times the epistatic response in the first generation. For unlinked loci ($c = 0.5$), this is simply twice the initial response. With tight linkage, the total response can be significantly larger.

The expected time, $t_{1/2}$, for half the total epistatic response to accumulate, $R_{AA}(t_{1/2}) = \widetilde{R}_{AA}/2$, is obtained by noting from Equations 15.4b and 15.6a that this occurs when $1 - (1 - c)^{t_{1/2}} = 1/2$, or

$$t_{1/2} = \frac{-\ln(2)}{\ln(1-c)} \tag{15.6b}$$

For small value of c, this is approximately $\simeq 0.68/c$. As linkage becomes tighter, the total cumulative epistatic response increases, as does the time for half of this total response to occur (Figure 15.1).

Once selection stops, the epistatic contribution decays to zero. With t generations of selection followed by τ generations of random mating, the cumulative response under the infinitesimal model is

$$t\,h^2\,S + (1-c)^{\tau}\,R_{AA}(t) \tag{15.7}$$

which (for large τ) converges to $R = t\,h^2\,S$, the value predicted from the breeder's equation. The half-life for decay of the epistatic response, R_{AA}, is also given by Equation 15.6b.

The presence of epistasis can result in a curvilinear selection response if σ^2_{AA}/σ^2_z is sufficiently large. However, as Figure 15.1 shows, any such curvilinearity is usually difficult to distinguish from a linear response, especially given the very noisy nature of selection response data (Chapter 18). Unless there is tight linkage, most of the nonlinearity occurs in the first few generations (Equation 15.6b). With a constant selection differential, the additional increment to response from epistasis decreases in each generation as R_{AA} approaches its limiting value, $\widetilde{R}_{AA}$, at which point the per-generation response is simply h^2S and linear over future generations.

Under the infinitesimal model, once selection is relaxed, the total response decays back to that predicted from the breeder's equation. It is interestingly that this situation mimics the effects of natural selection countering artificial selection, which also result in a decay of the cumulative response once artificial selection stops. Thus, in order to predict the *permanent* response correctly, we must know h^2. If only the parent-offspring slope is estimated, this may overestimate the final amount of response due to the inclusion of σ^2_{AA} and higher-order (additive) epistatic variances, although the bias is generally likely to be small, as the values of the higher-order variances are expected to be small relative to σ^2_A.

Griffing's analysis is restricted to two loci, and hence limited to only pairwise (additive $\times$ additive) epistasis. In contrast, Equation 15.1 gives the single-generation response for arbitrary levels of additive epistasis, provided the biparental offspring regression is linear. These higher-order contributions also rapidly decay under random mating. Again assuming the infinitesimal model (and unlinked loci), Bulmer (1980) found that the response due to a single generation of selection (Equation 15.1) decays after one generation of random mating to

$$R = S\left(h^2 + \frac{1}{4}\frac{\sigma^2_{AA}}{\sigma^2_z} + \frac{1}{16}\frac{\sigma^2_{AAA}}{\sigma^2_z} + \frac{1}{64}\frac{\sigma^2_{AAAA}}{\sigma^2_z} + \cdots\right) \tag{15.8}$$

which again rapidly converges to $R = h^2\,S$ after several generations of random mating. For n-locus additive epistasis (e.g., $\sigma^2_{A\cdots A}$, where there are n A's), the per-generation decay rate for unlinked loci is $(1/2)^{n-1}$, which is the probability that a parental gamete containing specific alleles at n unlinked loci is passed on intact to an offspring. The probability that such a gamete will remain unchanged after t generations is $2^{-t(n-1)}$, which rapidly converges to zero. Example 15.2 (below) uses the method of ancestral regressions to more fully examine the response under n-locus additive epistasis. A final caveat is that these results apply to infinite populations. As shown in Chapter 11, in finite populations, some of the additive epistatic contribution can be permanent due to some of σ^2_{AA} being transformed into simple additive variation by drift.

Dominance in Autotetraploids

Polyploidy, which is very common in plants and occurs in some animals (e.g., salmonid fishes), can introduce complications in predicting selection response (Gallais 2003). In particular, the dynamics of selection response for autotetraploids with dominance is very similar to the dynamics of diploids with epistasis. From LW Equation 7.22 and LW Table 7.5, the tetraploid parent-offspring covariance when dominance (but no epistasis) is present is

$$\sigma(z_p, z_o) = \frac{\sigma^2_A}{2} + \frac{\sigma^2_D}{6}$$

The inflation of the parent-offspring covariance over that for a diploid is due to dominance interactions between the two alleles per locus that each tetraploid parent passes on to its offspring. Thus, like additive epistasis in diploids, favorable *combinations* of alleles can be passed down from parent to offspring in polyploids. With equal amounts of selection on both sexes (e.g., selection occurs before pollination), the resulting response (assuming linearity of the parent-offspring regression) is

$$R = S\left(h^2 + \frac{\sigma^2_D}{3\sigma^2_z}\right) \tag{15.9}$$

If selection occurs after pollination, S is replaced by $S/2$. Gallais (1975) extended Griffing's (1960a) method (and hence assumed normally distributed phenotypes, with each gene having very small effects on the character) to obtain the response after t generations of selection with a constant differential of S,

$$R(t) = th^2S + R_D(t) \tag{15.10a}$$

where

$$R_D(t) = S\,\frac{3}{2}\left[1 - \left(\frac{1}{3}\right)^t\right]\frac{\sigma_D^2}{3\sigma_z^2} \tag{15.10b}$$

which converges to a limiting value of

$$\widetilde{R}_D = S\,(\sigma_D^2/2\sigma_z^2) \tag{15.10c}$$

This is just a 50% increase over the first-generation dominance response. Thus, as with the contribution from epistasis in diploids, the total contribution from dominance in polyploids approaches a limiting value representing the balance between selection favoring specific combinations of alleles and reproduction reshuffling those combinations.

Again, as with epistasis, the contribution from dominance decays upon the cessation of selection as genotype frequencies return to their Hardy-Weinberg values. As discussed in LW Chapter 4, the rate of approach to tetraploid Hardy-Weinberg expectations depends on the probability, c, of a double reduction. This is the chance that a gamete contains two copies of one of the four alleles in the parent, an event that requires a crossover (see LW Figure 4.3). In the absence of double reduction ($c = 0$), as would occur for a locus completely linked to the centromere, the difference in the frequency of pairs of alleles from Hardy-Weinberg expectation decays by one third in each generation (and this is assumed in obtaining Equation 15.10b). In such a setting, the response to t generations of selection followed by τ generations of random mating is

$$t\,h^2\,S + (1/3)^\tau\,R_D(t) \tag{15.10d}$$

which again rapidly converges to $t\,h^2\,S$. More generally, if c is the per-generation probability of a double reduction, the decay rate of $(1/3)^t$ is replaced in Equations 15.10b and 15.10d by $[(1-c)/3]^t$. Swanson et al. (1974) found that if some double reductions occur ($c > 0$), the additive genetic variance is slightly inflated over the value expected with no double reductions ($c = 0$), which permanently increases selection response. This results from the slight excess of homozygotes at equilibrium over the Hardy-Weinberg expectation (see LW Chapter 4).

Wricke and Weber (1986) discussed additional topics on tetraploid selection, while selection under single-locus tetraploid models was examined by R. Hill (1971; Hagg and Hill 1974; Hill and Hagg 1974; Rowe and Hill 1984). By far the most complete treatment of selection with polyploids was presented by Gallais (2003).

ANCESTRAL REGRESSIONS

A general approach for examining which components of response are transient is to consider the expected value of an offspring as a function of all of its direct relatives that have been under selection (Pearson 1898; Bulmer 1971b, 1980). If this **ancestral regression** is linear (as would occur if the joint distribution of the phenotypic values of all relatives is multivariate normal), the selection response can be described by specifying the regression coefficients via an obvious extension of the biparental regression, to now include all selected relatives back to the original unselected base population. For example, if selection starts in generation 0, the response in the first generation is $R(1) = 2\,\beta_{1,0}\,S_0$, where $\beta_{1,0}$ is the regression of offspring at generation 1 on a parent from generation 0 (this assumes that both parents have the same regression coefficients and selection differentials, an assumption that will be relaxed shortly). Likewise, the total response after two generations, $R(2) = 4\,\beta_{2,0}\,S_0 + 2\,\beta_{2,1}\,S_1$, depends on the nature of selection on the four grandparents (S_0) and both parents (S_1) as well as on the transmission from grandparent to grandchild ($\beta_{2,0}$) and that from parent to offspring ($\beta_{2,1}$). Note that this formulation allows the parent-offspring regression to change through time (e.g., $\beta_{2,1}$ need not equal $\beta_{1,0}$), as can happen with inbreeding (Chapter 23). Similarly, the response following three generations of selection depends upon the nature of selection on that individual's eight great-grandparents, four grandparents, and two parents

$$R(3) = 8\,\beta_{3,0}\,S_0 + 4\,\beta_{3,1}\,S_1 + 2\,\beta_{3,2}\,S_2$$

Proceeding in this fashion gives the total response by generation T as

$$R(T) = \sum_{t=0}^{T-1} 2^{T-t}\,\beta_{T,t}\,S_t \tag{15.11a}$$

where $\beta_{T,t}$ is the partial regression coefficient for the phenotype of an individual in generation T on one (out of 2^{T-t}) of its ancestors in generation $t < T$ (with selection starting at generation 0). Because the total response is simply the sum of the independent contributions from each generation, the regression coefficients are simply standard univariate regression coefficients (i.e., $\beta_i = \sigma(y, x_i)/\sigma^2_{x_i}$), yielding

$$R(T) = \sum_{t=0}^{T-1} 2^{T-t}\,\frac{\sigma_G(T,t)}{\sigma^2(z_t)}\,S_t \tag{15.11b}$$

where $\sigma_G(T,t) = \sigma(z_T, z_t)$ is the **cross-generation covariance**, the phenotypic covariance between an individual in generation t and its descendant in generation $T > t$. We usually assume that $\sigma(z_T, z_t)$ is entirely genetic (e.g., Examples 15.1 and 15.2), and hence denote it by $\sigma_G(T,t)$, although environmental factors shared across generations can also contribute to $\sigma(z_T, z_t)$.

With pure selfing, each individual has only a single relative, so the 2^{T-t} term in Equation 15.11a is absent, which makes the ancestral regression

$$R(T) = \sum_{t=0}^{T-1} \beta_{T,t}\,S_t \tag{15.11c}$$

As we will see in Chapter 23, ancestral regression offers a powerful approach for the analysis of selection under inbreeding, where the cross-generation covariances depend on *both* T and t, rather than simply on the number of generations $(T-t)$ between the ancestor and the descendant. For example, the parent-offspring covariance changes over generations as inbreeding proceeds.

If different relatives in the same generation experience different amounts of selection, with $S_{k,i}$ being the selection differential on relative i in generation k, then

$$R(T) = \sum_{t=0}^{T-1} \left[\beta_{T,t}\left(\sum_{i=1}^{n(t,T)} S_{t,i}\right)\right] \tag{15.12a}$$

where $n(t,T)$ is the number of relatives in generation t that contribute to response in generation T. Note for the case of pure selfing, $n(t,T) = 1$, which reduces Equation 15.12a to Equation 15.11c. Finally, we can also allow for different regression coefficients on each relative to completely generalize this approach

$$R(T) = \sum_{t=0}^{T-1} \left(\sum_{i=1}^{n(t,T)} \beta_{T,t,i}\,S_{t,i}\right) \tag{15.12b}$$

where $\beta_{T,t,i}$ is the regression coefficient of the phenotype of an individual in generation T on its ith relative in generation t.

To apply ancestral regression for predicting response, we require that the regression remain linear and that *selection-induced* changes in the variances and covariances be negligible. Thus, while we allow changes in $\beta_{T,t}$ due to the particular genetic system being considered (e.g., selfing, wherein the additive genetic variance decreases by a predictable amount in each generation in the absence of selection), we assume that selection does not confound these changes. Bulmer (1980) showed that the joint distribution of an offspring and all its direct ancestors is multivariate normal, and hence the ancestral regression is linear (LW Chapter 8), under the infinitesimal model. Because selection does not change allele frequencies under the infinitesimal model, this might suggest that the regression coefficients, $\beta_{T,t}$ are unaffected by selection. The problem, however, is that selection generates gametic-phase disequilibrium, which can significantly alter the genotypic moments (Chapters 16 and 24). For now, we assume that these changes are small enough to be neglected. In Chapter 19 we show that BLUP estimates of breeding values are a type of ancestral regression, with the relationship matrix, **A**, accounting for drift and disequilibrium.

The behavior of the regression coefficients over time informs us about the permanency of the response. Consider the contribution from generation τ at t generations later, so that in the notation of our previous expressions, $T = t + \tau$. Equation 15.11a shows that unless $2^t\,\beta_{\tau+t,\tau}$ remains constant as t increases, the contribution to cumulative response from selection on adults in generation τ changes over time. For strictly additive loci (under random mating), $\sigma_G(\tau+t,\tau) = 2^{-t}\,\sigma^2_A(\tau)$ and thus $2^t\,\beta_{\tau+t,\tau} = h^2_\tau$, the standard result from the breeder's equation. Conversely, any term of $\sigma_G(\tau + t, \tau)$ that decreases by more than one half in each generation contributes only to the transient response, as $2^t\,\sigma_G(\tau+t,\tau) \to 0$ as $t \to \infty$. An exception to this condition occurs under pure selfing. Here, an individual has only a single ancestor t generations ago (as opposed to 2^t ancestors under random mating), resulting in the total contribution in generation $t + \tau$ from an ancestor in generation τ of $\sigma_G(\tau + t, \tau)$. Hence, any term contributing to $\sigma_G(\tau + t, \tau)$ that decreases to 0 as $t \to \infty$ contributes only to the transient response.

Example 15.1. As an application of ancestral regressions, consider the situation that arises with additive-by-additive epistasis. In this case, Cockerham (1984b) found (under the infinitesimal model) that for two linked loci, the cross-generation covariance is

$$\sigma_G(\tau+t,\tau) = \frac{\sigma^2_A(\tau)}{2^t} + \frac{\sigma^2_{AA}(\tau)}{2}\left(\frac{1-c}{2}\right)^t$$

yielding

$$2^t\,\sigma_G(\tau+t,\tau) = \sigma^2_A(\tau) + (1-c)^t\,\frac{\sigma^2_{AA}(\tau)}{2}$$

Provided the genetic variances remain constant, applying Equation 15.11a recovers Equation 15.3.

Example 15.2. Consider the expected response under arbitrary levels of additive epistasis (under the infinitesimal model with unlinked loci). LW Equation 7.12 shows the genetic covariance between relatives x and y as

$$\begin{aligned}\sigma^2_G(x,y) &= (2\Theta_{x,y})\,\sigma^2(A) + (2\Theta_{x,y})^2\,\sigma^2(AA) + \cdots + (2\Theta_{x,y})^i\,\sigma^2(A^i)\\ &= \sum_{i=1}^{n}(2\Theta_{x,y})^i\,\sigma^2(A^i)\end{aligned}$$

where $\sigma^2(A^i)$ denotes the genetic variance due to ith-order additive epistasis. The coefficient of coancestry $\Theta_{t,t+\tau}$ between a parent in generation t and a direct descendant in generation $t+\tau$ under random mating is

$$\Theta_{t,t+\tau} = \left(\frac{1}{2}\right)^{\tau+1}$$

Using this result, the contribution, $\sigma_{G,i}(t+\tau,t)$, to the total genetic covariance due to ith-order additive epistasis (A^i) becomes

$$\sigma_{G,i}(t+\tau,t) = (2\Theta_{t+\tau,t})^i\,\sigma^2(A^i) = 2^i\left(\frac{1}{2}\right)^{(\tau+1)i}\sigma^2(A^i) = \left(\frac{1}{2}\right)^{\tau i}\sigma^2(A^i)$$

and the ancestral regression terms involving $\sigma^2(A^i)$ become

$$2^\tau\frac{\sigma_{G,i}(t+\tau,t)}{\sigma^2(A^i)} = 2^\tau\frac{(1/2)^{\tau i}\sigma^2(A^i)}{\sigma^2(A^i)} = 2^\tau 2^{-\tau i} = \left(\frac{1}{2}\right)^{\tau(i-1)} = \left(\frac{1}{2^{i-1}}\right)^\tau$$

With constant selection, S, the contribution to total response from ith-order additive epistasis follows from Equation 15.11b:

$$R_{A^i}(t) = S\,\frac{\sigma^2(A^i)}{\sigma_z^2}\sum_{\tau=1}^{t}\left(\frac{1}{2^{i-1}}\right)^\tau \tag{15.13a}$$

Recalling Equation 15.5b,

$$\sum_{\tau=1}^{t}\left(\frac{1}{2^{i-1}}\right)^\tau = \frac{x(1-x^t)}{1-x}, \quad \text{where} \quad x = (1/2)^{i-1} \tag{15.13b}$$

The limit of this sum (as $t \to \infty$) is

$$\frac{x}{1-x} = \frac{(1/2)^{i-1}}{1-(1/2)^{i-1}} = \frac{1}{2^{i-1}-1} \tag{15.13c}$$

Because the scaled initial contribution, $R_{A^i}(1)/[S\sigma^2(A^i)/\sigma_z^2]$, equals $(1/2)^{i-1}$, the additional total increment to response beyond that seen in the first generation follows from Equations 15.13a and 15.13c:

$$\frac{\widetilde{R}_{A^i} - R_{A^i}(1)}{S\sigma^2(A^i)/\sigma_z^2} = \frac{1}{2^{i-1}-1} - \frac{1}{2^{i-1}} = \frac{1}{(2^{i-1}-1)2^{i-1}}$$

The response in generation one, $R(1)$, and at the limit, $\widetilde{R}$ (both scaled in units of $S\sigma^2(A^i)/\sigma_z^2$), and the fraction of total response occurring in the first generation, $R(1)/\widetilde{R}$, are as follows:

	AA	AAA	AAAA	AAAAA
$R(1)/[S\sigma^2(A^i)/\sigma_z^2]$	0.500	0.250	0.125	0.063
$\widetilde{R}/[S\sigma^2(A^i)/\sigma_z^2]$	1.000	0.333	0.143	0.067
$R(1)/\widetilde{R}$	0.500	0.750	0.875	0.938

As shown in the second line of this table, the limiting contribution, $\widetilde{R}$, is small for higher-order additive epistasis among unlinked loci. For example, for four-way additive epistasis, the total response is ~14% (0.143/1.000) of the expected response under two-way additive epistasis when the two variance components are equal, $\sigma^2(AA) = \sigma^2(AAAA)$ Further, as the final line in the table shows, ~88% of the total response from four-way additive epistasis occurs in the first generation of selection.

RESPONSE DUE TO ENVIRONMENTAL CORRELATIONS

Imagine a situation where a bird is large by chance, enabling it to defend a larger and more productive breeding location. As a result, its offspring are better provisioned and are larger themselves, above and beyond any genetic effects on size. This is an example of a **shared environmental effect** and is a special case of the more general setting where at least part of the environment experienced by an individual is a function of the phenotypes of other individuals with which they interact (Chapter 22). Rossiter (1996), Bonduriansky and Day (2009), and Bonduriansky et al. (2012) reviewed a number of such shared environmental effects, while Day and Bonduriansky (2011) modeled the expected response in a number of different settings using the Price equation framework (Chapter 6).

As Equation 15.1a indicates, shared parent-offspring environmental effects—a nonzero value of $\sigma(E_p, E_o)$—can influence the selection response. This contribution is also transient. Consider a character whose variation is *entirely* environmental, in which case the phenotypic value can be decomposed as

$$z = \mu + E = \mu + e_{fa} + e_{mo} + e$$

where μ is the mean value of the character when environmental effects are randomly distributed, and the environmental value, E, has been decomposed into the maternal and paternal contributions to the offspring due to shared environmental effects (e_{mo} and e_{fa}) and a residual due to special environmental effects (e), where $E[e] = 0$. In order to predict the shared environmental contribution from a parent, we assume the simplest model, wherein a fraction, b, of the total environmental value of a parent is passed on to its offspring.

This model serves as a useful introduction to some of the dynamics that can occur for certain models of maternal effects (examined in the next section). It also serves as a model for selection response under **epigenetics**, wherein a nongenetic modification is passed through generations (e.g., factors bound to the DNA that alter the expression level of a gene). While much excitement is made by some on the role of such epigenetic inheritance in evolution, as we will see, its impact is transient if there is even the smallest amount of imperfection in transmission.

To simplify matters further, assume that this regression coefficient is independent of the sexes of the parent and offspring, although this condition can easily be relaxed. Here the expected contribution from a father to his offspring is $e_{fa} = b\,E_{fa} = b\,(z_{fa} - \mu)$, where E_{fa} is the father's total environmental value. Further assuming that the parents and the offspring have the same phenotypic variance, then $b = \rho/2$, where ρ is the slope of the midparent-offspring regression. In principal, b can be negative under some environmental conditions, e.g., if parents and offspring compete for a limited common resource, and larger parents gather a disproportionate share of resources, resulting in smaller offspring.

Provided E_{fa} and E_{mo} are uncorrelated, the expected value of an offspring from parents with phenotypic values of z_{fa} and z_{mo} is

$$E\,[\,z_o \,|\, z_{mo}, z_{fa}\,] = \mu + \frac{\rho}{2}\,(\,z_{fa} - \mu\,) + \frac{\rho}{2}\,(\,z_{mo} - \mu\,) \tag{15.14a}$$

where μ is the population baseline mean under randomized environmental effects. Denoting the mean of adults selected in generation t by μ_t^*, the mean at generation $t+1$ is given by

$$\mu_{t+1} = \mu + \rho\,(\,\mu_t^* - \mu\,) \tag{15.14b}$$

where $(\,\mu_t^* - \mu\,)$ is the average environmental deviation of the selected parents at generation t, a fraction, ρ, of which is passed on to their offspring. If shared environmental effects are only passed on through one parent, $\rho/2$ replaces ρ in Equation 15.14b. Rewriting the mean after selection as $\mu_t^* = \mu_t + S_t$,

$$\mu_{t+1} = \mu + \rho\,(\,\mu_t + S_t - \mu\,) \tag{15.15}$$

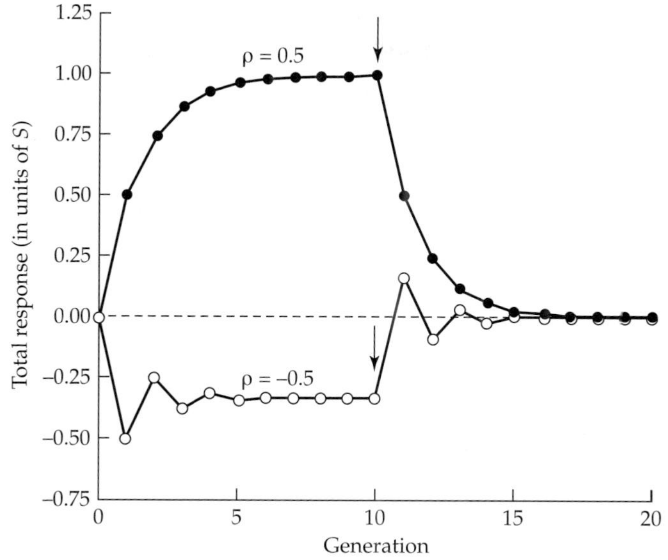

Figure 15.2 The response when resemblance between relatives is due entirely to correlation between environmental values in parents and offspring. Selection with a constant value of S starts at $t = 0$ and continues until generation 10 (indicated by the arrow), at which point selection is stopped. Note the interesting dynamics that occur if environmental values are negatively correlated, wherein the response to selection is reversed with respect to the selection differential. In this case, selection for *increased* character value results in a *decreased* mean value, with the total response in this example eventually converging to $-S/3$ (for $\rho = -0.5$). Once selection is relaxed (denoted by the arrows), there is an initial positive response (generation 11), which then quickly decays to zero.

and defining the change in mean as $\Delta\mu_t = \mu_{t+1} - \mu_t$ yields

$$\begin{aligned}\Delta\mu_t &= [\,\mu + \rho\,(\,\mu_t + S_t - \mu)\,] - [\,\mu + \rho\,(\mu_{t-1} + S_{t-1} - \mu\,)\,] \\ &= \rho\,[\,(\,\mu_t - \mu_{t-1}) + (S_t - S_{t-1}\,)\,] \\ &= \rho\,[\,\Delta\mu_{t-1} + (S_t - S_{t-1})\,] \qquad (15.16a)\end{aligned}$$

Suppose that constant selection (with a differential of S) is applied starting at generation 1. Here, $\Delta\mu_0 = 0$, $S_0 = 0$, and $S_t = S$ for $t \geq 1$. Equation 15.16a then gives $\Delta\mu_1 = \rho S$. Further iterations yield the increment from selection in generation t to the total response as

$$\Delta\mu_t = \rho^t\,S \qquad (15.16b)$$

which (like epistasis) decreases in each generation, and approaches zero for large values of t. Hence, even under continued selection, the total response to selection eventually reaches a limit. Note that if $\rho < 0$, the sign of the increment-specific response switches in each generation (Figure 15.2).

The reason for this decline in the per-generation rate of response, $\Delta\mu_t$, can be seen from Equation 15.16a. The change in the character mean due to previous selection decays, countering the gain from selection in the current generation. Only a fraction, ρ, of the change from generation $t-1$ is passed on, and in general, only ρ^k of the response from generation $t-k$ persists by generation t. Summing over Equation 15.16b, the total response to selection after t generations is

$$R(t) = \mu_t - \mu_0 = \sum_{i=1}^{t} \Delta\mu_i = S \sum_{i=1}^{t} \rho^i \qquad (15.16c)$$

Recalling the partial sum of a geometric series (Equation 15.5b), this reduces to

$$R(t) = S \, \frac{\rho}{1-\rho} \, (\, 1 - \rho^t \,) \tag{15.17a}$$

Note from Equation 15.17a that the sign of the total response does not change over t (as $|\rho| < 1$) and equals the sign of ρ. While the per-generation response, $\mu_{t+1} - \mu_t = \rho^t S$ (Equation 15.16b), may change sign each generation (provided $\rho < 0$), the absolute magnitude of this per-generation change rapidly decreases with t, resulting in no change of sign in the total response.

As was the case for epistasis, the total cumulative response reaches an equilibrium value representing the balance between the generation of correlations by selection and their removal by reproduction, with

$$\widetilde{R} = \lim_{t \to \infty} R(t) = S \, \frac{\rho}{1-\rho} \tag{15.17b}$$

Thus, no matter how long selection is applied, the mean can never change by more than $S\rho/(1-\rho)$. Further, none of this response is permanent. Suppose selection is stopped after t generations, resulting in $S_t = S$, $S_{t+\tau} = 0$ for $\tau \geq 1$. If we substitute these values of S into Equation 15.16a and use Equation 15.16b, the expected change in generation $t + \tau$ is

$$\Delta\mu_{t+\tau} = \rho^\tau \, (\, \Delta\mu_t - S \,) = \rho^\tau \, (\, \rho^t S - S \,) = -S \, \rho^\tau (\, 1 - \rho^t \,) \tag{15.18}$$

Using this result, by generation $t + \tau$, the cumulative response is

$$R(t+\tau) = R(t) + \sum_{i=1}^{\tau} \Delta\mu_{t+i} = R(t) - S \, (\, 1 - \rho^t \,) \sum_{i=1}^{\tau} \rho^i = \rho^\tau R(t) \tag{15.19a}$$

The last step follows if we first note from Equation 15.17a that $S\,(1-\rho^t) = [(1-\rho)/\rho]R(t)$. Using this identity, and applying Equation 15.15b with $x = \rho$ and $n = \tau$, yields

$$-S \, (\, 1 - \rho^t \,) \sum_{i=1}^{\tau} \rho^i = -\left[R(t) \frac{1-\rho}{\rho} \right] \left[\frac{\rho(1-\rho^\tau)}{1-\rho} \right] = -(1-\rho^\tau) R(t) \tag{15.19b}$$

Observe that Equation 15.19a converges to zero, with the rate of decay set by ρ (Figure 15.2). The half-life, $t_{1/2}$, of the decay in response follows from solving $\rho^{t_{1/2}} = 1/2$, or

$$t_{1/2} = -\frac{\ln(2)}{\ln(\rho)} \tag{15.19c}$$

Hence, while there can be some selection response when the resemblance between relatives is entirely environmental, any response is transient and decays away once selection stops. Further, no matter how long selection proceeds, the total response asymptotes to a limit beyond which no further response is excepted (Equation 15.17b).

As mentioned, these results bear on the role of epigenetic inheritance. The correlation, ρ, is a measure of the **persistence of an epigenetic modification** (be it cultural, hormonal, or from proteins bound to DNA). When the persistence of an effect is high (ρ is close to one), a trait can achieve a significant selection response under purely epigenetic inheritance (Equation 15.17b), and the response can persist for several generations following the cessation of selection (Equation 15.19a). For example if $\rho = 0.95$ (a 5% loss of the effect each generation), then a total response of $S \cdot 0.95/0.05 = 19\,S$ can be achieved. After selection stops, half this expected gain is lost in $-\ln(2)/\ln(0.95) \simeq 14$ generations (Equation 15.19c). While impressive, from an evolutionary standpoint, none of this short-term gain translates into a long-term, permanent, selection response.

SELECTION IN THE PRESENCE OF HERITABLE MATERNAL EFFECTS

A mother can influence the phenotype of her offspring in two ways. The usual assumption simply involves gene transmission (including those genes on organelle genomes). Together with paternally derived genes, these maternally transmitted genes create the genotype of the offspring, and the expression of this genotype, together with the environment it experiences, determines the phenotype. The second route is through **maternal effects**, which are traits or genes *expressed in the mother* that influence offspring phenotypes (Wolf and Wade 2009). These can be thought of as influencing the *environment* that a mother's offspring experiences, which can, in turn, influence its phenotype. Examples include **maternal performance characters** such as her body size, amount of care invested in offspring, milk yield, and endosperm production, all of which can influence a variety of progeny traits. These maternal performance traits can themselves have a genetic basis, and hence can also respond to selection, thus creating potentially very complex dynamics even in the simplest settings.

Paternal effects are also possible, especially in situations where the father plays a role in caring for the offspring. While not considered here, they can be treated in exactly the same fashion as maternal effects. There is an extensive literature on maternal effects and their evolutionary implications (reviewed by Roach and Wulff 1987; Bernardo 1996; Rossiter 1996; Mousseau and Fox 1998; Reinhold 2002; Räsänen and Kruuk 2007; Mousseau et al. 2009). Here we introduce some of the basic ideas, with additional material developed in Chapters 22, 29, and 30.

Example 15.3. Maternal effects include genes expressed in the mother that influence the very early development of her offspring, with the maternal phenotype of these genes being extremely difficult to measure in the mother. For many metazoans, much of the gene expression during the first few rounds of cell division in a newly fertilized zygote is due to the translation of maternal mRNAs deposited in the egg. The resulting maternal gene products provide information for early key development steps in the nascent embryo, with the impact of these maternal genes seen in the trait values of the offspring, but not in the mother's own phenotype. A classic example involves the work of Sturtevant (1923) on right- versus left-handed shell coiling of the snail *Limnaea*, where the genotype of the mother, not the offspring, determines offspring phenotype. Offspring from a mother homozygous for the *sinistral* allele are all left-handed, independent of their genotype, whereas the presence of at least one *dextral* allele in the mother yields all right-handed offspring. While the maternal trait is easy to assay from observations on her offspring, it cannot be assayed using just the mother alone, as the products from the *sinistral/dextral* locus are still unknown, although Kuroda et al. (2009) have showed that the direction of coiling is determined at the eight-cell stage of the embryo.

Decomposing Maternal Effects

Assuming a maternal effect, the environmental value, E, for the focal trait in an individual can be decomposed as $E = M + e$, comprised of a **maternal performance** component, M, contributed by the mother plus an environmental deviation, e, resulting in a trait value of

$$z = G + E = G + M + e \tag{15.20a}$$

The genotypic value, G, of the offspring is often referred to as the **direct** (or **intrinsic**) **effect** on the trait, reflecting the inherent genetics of the individual. If M is entirely environmental in basis, the results of the previous section (on correlated environmental values) apply, and any contribution from maternal effects is transient. However, there is often a *genetic* component to M, with the result that selection can change the population means of *both* the direct and the maternal values (Dickerson 1947; Willham 1963, 1972; Cheverud 1984a;

Riska et al. 1985; Kirkpatrick and Lande 1989). In this case, $M = G_m + E_m$, where G_m is the contribution to the offspring value resulting from the mother's genotypic value for the maternal performance character, while E_m is the contribution resulting from the environmental value of the maternal performance character (reviewed in LW Chapter 23). Thus, although M is treated as an environmental effect from the offspring's standpoint, it can have both a genetic and an environmental basis in the mother.

Letting G_d denote the genotypic value of the direct effect (previously G), the phenotypic value of an individual can be written in terms of direct and maternal contributions

$$z = G + M + e = G_d + G_m + E_m + e \tag{15.20b}$$

This significantly complicates the selection response, as if the correlation between G_d and G_m is negative, selection to increase a trait might result in an improved direct value ($\Delta\mu_{G_d} > 0$) but a decreased maternal value ($\Delta\mu_{G_m} < 0$). In the extreme, this can result in a *reversed selection response*, with the decline in maternal environment overwhelming any improvement in direct effects. Because trait value, z, is a function of two potentially correlated genetic components, G_d and G_m, this is formally a multiple-trait problem and can be attacked with the machinery of multivariate selection (Cheverud 1984a; Kirkpatrick and Lande 1989, 1992; Lande and Kirkpatrick 1990). However, we first consider the simplest case, where M is a function of the value of the focal trait in the mother, which collapses the model to a single-trait problem, but with much more complicated dynamics than suggested by the simple breeder's equation. We then very briefly discuss a two-trait model, while Volume 3 examines maternal effects in the full multivariate framework.

An important point to note is that maternal-effects models are a special case of **associative effects** (or **social**, or **indirect genetic effects**) models, where the phenotype of an individual within a group is a function of both its direct value plus the associative values from members of its group. Here the group is an offspring and its mother. Both kin and group selection, which are special cases of this more general formulation, are examined in detail in Chapter 22.

Selection Response Under Falconer's Dilution Model

One of the simplest models of maternal effects (motivated by the inheritance of litter size in mice) is that of Falconer (1965): the value of the focal trait in the mother determines M (reviewed in LW Chapter 23). Falconer reasoned that offspring from large litters receive less maternal resources per individual than those in smaller litters, and that this can have carryover effects when they become mothers themselves.

Assume that the maternal contribution is a linear function of the maternal phenotype, z_{mo} (the size of her litter), so that $M = m z_{mo}$ and, upon substitution of this result into Equation 15.20a, the phenotypic decomposition becomes

$$z = G + M + e = G + m\, z_{mo} + e \tag{15.21}$$

Conceivably, M could be a nonlinear function of z_{mo}, but linearity is assumed for tractability. Equation 15.21 is called the **dilution model**, because the effect of the maternal phenotype becomes diluted over successive generations (for $|m| < 1$). The parameter m can be regarded as the partial regression coefficient (i.e., holding the offspring's genotypic value, G, constant) of offspring phenotype on maternal phenotype and can be estimated as the difference between slopes of the maternal- and paternal-offspring regressions (LW Equation 23.13).

Negative estimates of m have been reported: values of -0.15 for litter size in mice (Falconer 1965); of -0.58 and -0.40 for age of maturity in two replicate lines of the springtail insect, *Orchesella cincta* (Janssen et al. 1988); of -0.25 for clutch size in the collared flycatcher, *Ficedula albicollis* (Schluter and Gustafsson 1993); and of -0.29 for juvenile growth rate in the red squirrel *Tamiasciurus hudsonicus* (McAdam and Boutin 2003).

If we further assume that the joint distribution of phenotypes and breeding values in parents and offspring is multivariate normal and there is an absence of epistasis, the expected phenotypic value of an offspring whose mother has a phenotypic value of z_{mo} is

$$E\left[z_o \mid A_{mo}, A_{fa}, z_{mo}\right] = \frac{A_{mo}}{2} + \frac{A_{fa}}{2} + m\, z_{mo} \tag{15.22a}$$

where A_{mo} and A_{fa} are the maternal and paternal breeding values. Averaging over the selected parents, the mean in generation t becomes

$$\mu_z(t+1) = \frac{A^*_{fa}(t) + A^*_{mo}(t)}{2} + m\, \mu^*_{mo}(t) \tag{15.22b}$$

where $A^*_{fa}(t)$ and $A^*_{mo}(t)$ are the mean breeding values of the selected parents and $\mu^*_{mo}(t)$ is the mean phenotypic value of selected mothers in generation t. The first term is the expected breeding value in the offspring, $\mu_A(t)$, and the second term is the additional maternal contribution.

To proceed in using this model to predict response, we first use the regression of breeding value on phenotype

$$A = \mu_A + b_{A|z}\,(\, z - \mu_z\,) + e$$

In particular, we can rewrite $A^*_{mo}(t) = E_s[A_{mo}]$, the expected breeding values in selected mothers, as

$$\begin{aligned} A^*_{mo}(t) = E_s[A_{mo}(t)] &= E_s\left[\, \mu_A(t) + b_{A|z}\left[\, z_{mo} - \mu_z(t)\,\right] + e \,\right] \\ &= \mu_A(t) + b_{A|z} S_{mo}(t) \end{aligned} \tag{15.22c}$$

where $E_s[\;]$ denotes the expected value over the selected parents and S_{mo} denotes the selection differential on the trait in mothers in generation t. A similar expression for $A^*_{fa}(t)$ includes the selection differential, S_{fa}, on fathers. In the absence of maternal effects, $b_{A|z} = h^2$. However, the dilution model generates an equilibrium covariance between M and A, specifically $\sigma_{A,M} = m\,\sigma^2_A \,/\, (\, 2 - m\,)$, which in turn alters the covariance between z and A (Falconer 1965; Kirkpatrick and Lande 1989; see LW Equation 23.12a). The resulting regression slope (at equilibrium) is

$$b_{A|z} = h^2 \frac{2}{2-m} \tag{15.23a}$$

The expected breeding value in the offspring is the average of the breeding values of the selected parents, which (from Equations 15.22c and 15.23a) yields

$$\mu_A(t+1) = \frac{A^*_{mo}(t) + A^*_{fa}(t)}{2} = \mu_A(t) + \frac{h^2}{2-m}\Big(S_{mo}(t) + S_{fa}(t) \Big) \tag{15.23b}$$

Substituting this result for the average of the breeding values in Equation 15.22b and using $\mu^*_{mo}(t) = \mu_z(t) + S_{mo}(t)$ yields

$$\mu_z(t+1) = \mu_A(t) + \frac{h^2}{2-m}\Big(S_{mo}(t) + S_{fa}(t) \Big) + m\Big(\mu_z(t) + S_{mo}(t) \Big) \tag{15.24}$$

The change in population mean over one generation, $\Delta\mu_z(t)$, is thus

$$\begin{aligned} \Delta\mu_z(t) = \mu_z(t+1) - \mu_z(t) = &\left[\, \mu_A(t) + \frac{h^2}{2-m}\Big(S_{mo}(t) + S_{fa}(t) \Big) + m\Big(\mu_z(t) + S_{mo}(t) \Big) \right] \\ &- \left[\, \mu_A(t-1) + \frac{h^2}{2-m}\Big(S_{mo}(t-1) + S_{fa}(t-1) \Big) + m\Big(\mu_z(t-1) + S_{mo}(t-1) \Big) \right] \end{aligned}$$

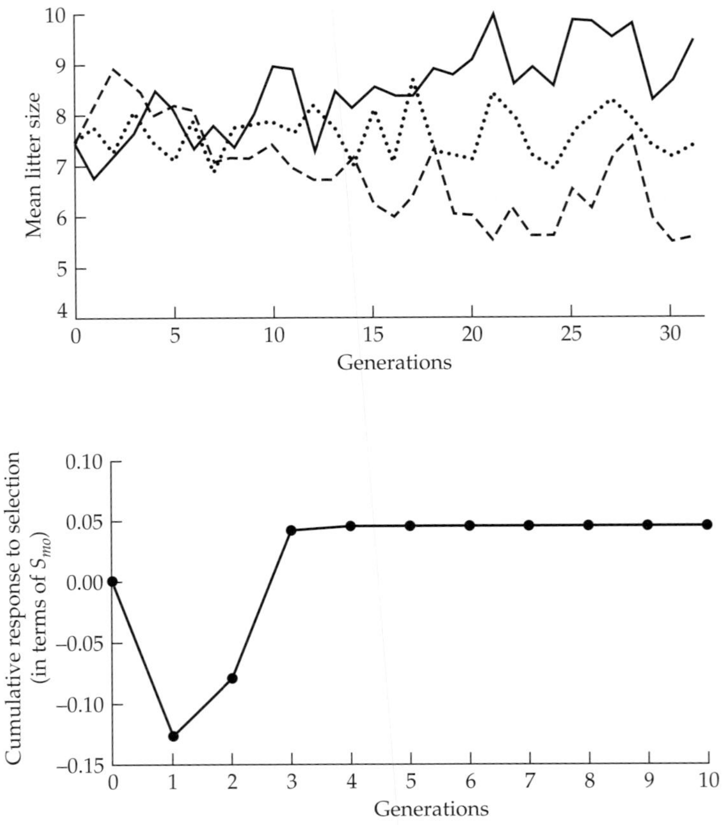

Figure 15.3 **Top:** Falconer's experiments on selection response for litter size in mice. The dashed line is the response to selection for smaller litters, the thick line is the response to selection for larger litters, and the dotted line is the control. Note the reversed response in the first generation in both the up- and down-selected lines. (After Falconer 1960b.) **Bottom:** The predicted change in population mean following a single generation of selection on females with $S_{mo} > 0$ (using Falconer's estimated values of $h^2 = 0.11$ and $m = -0.13$). There is a reversed response in the first generation, even though the net genetic change is to increase the character. By generation 3, the net nongenetic change in phenotypic mean has largely decayed away, revealing the net genetic change of $S_{mo}\, h^2/[(1-m)(2-m)] = 0.044 \cdot S_{mo}$ (Equation 15.33).

which, using Equation 15.23b, simplifies to

$$\Delta\mu_z(t) = \frac{h^2}{2-m}\Big(S_{mo}(t) + S_{fa}(t)\Big) + m\, S_{mo}(t) + m\Big(\Delta\mu_z(t-1) - S_{mo}(t-1)\Big) \tag{15.25}$$

This result can be interpreted as follows: the first two terms are the change in character value resulting from selection in generation t due to genetic ($h^2\,/\,[\,2-m\,]$) and maternal (m) contributions. The final term, which can be expressed as $m\,[\,\mu_z(t) - \mu_z^*(t-1)\,]$, represents the decay in the maternal contribution from the previous generation.

Assuming that an equilibrium for maternal effects has been reached such that Equation 15.23a holds and then starting with an unselected base population, the response to a single generation of selection is

$$\Delta\mu_z(1) = \frac{h^2}{2-m}\Big(S_{mo}(1) + S_{fa}(1)\Big) + mS_{mo}(1) \tag{15.26}$$

An interesting implication of Equation 15.26 is that if $m < 0$, there is some possibility of a reversed response, with $\Delta\mu_z$ being opposite in sign from S. If $S_{fa} = S_{mo} = S$, a reversed response is expected if

$$m < 1 - \sqrt{1 + 2h^2} \tag{15.27a}$$

If selection is only occurring on females, this condition is

$$m < 1 - \sqrt{1 + h^2} \tag{15.27b}$$

An example of an apparent maternally induced reversed response was seen by Falconer (1960b, 1965) in his selection experiments on litter size in mice. This character shows a negative maternal effect, with m and h^2 estimated to be -0.13 and 0.11, respectively. Because selection for litter size occurs only in females, Equation 15.27b implies that a reversed response in the first generation is expected (as $1 - \sqrt{1 + 0.11} \simeq -0.05 > m = -0.13$). As Figure 15.3 shows, a reversed response was indeed observed.

An observed reversed response can be misleading because the *permanent* response is expected to have the same sign as S, while the initial observed response also includes a transient component that (in this case) is of the opposite sign and of a larger magnitude than the permanent response component. It may take several generations for this transient component to decay to the point that the actual genetic changes are revealed (Figure 15.3). For a single generation of selection, the expected genetic change is $2S\,h^2/\,[\,(2-m)(1-m)\,]$, as shown by Equation 15.33 (below).

The possibility of a reversed selection response hints at some of the complicated dynamics that can appear when maternal effects are present. To examine these dynamics in more detail, consider the dilution model with constant directional selection occurring equally on both sexes, i.e., $S_{fa}(t) = S_{mo}(t) = S$ for $t \geq 1$. Iteration of Equation 15.25 yields

$$\Delta\mu_z(t) = S\left[\frac{2h^2}{(1-m)(2-m)}\left(1 - m^t\right) + m^t\right] \tag{15.28a}$$

which (for $|m| < 1$) converges to

$$\Delta\mu_z = S\,\frac{2h^2}{(1-m)(2-m)} \tag{15.28b}$$

Hence, after a sufficient number of generations, the per-generation change is constant. If $|m|$ is near zero, the per-generation response rapidly converges to this asymptotic value, while if $|m|$ is near one, the rate of convergence is considerably slower. Summing over the single-generation changes (Equation 15.28a) and recalling Equation 15.5b, the cumulative response to t generations of selection is

$$R(t) = \sum_{i=1}^{t} \Delta\mu_z(i) = \frac{S}{1-m}\left[t\,\frac{2h^2}{2-m} + m\,(1-m^t)\left(1 - \frac{2h^2}{(1-m)(2-m)}\right)\right] \tag{15.29a}$$

which converges (for $|m| < 1$) to

$$\frac{S}{1-m}\left[\frac{2h^2}{2-m}\left(t - \frac{m}{1-m}\right) + m\right] \tag{15.29b}$$

How much of this response is permanent? Suppose selection ends at generation t, and suppose we denote by τ the number of generations since selection was stopped. Iterating Equation 15.25 with $S(t) = S$, $S(t+\tau) = 0$ for $\tau \geq 1$ yields

$$\Delta\mu_z(t+\tau) = m^\tau\,[\Delta\mu_z(t) - S] \tag{15.30}$$

where $\Delta\mu_z(t)$ is given by Equation 15.28a. Thus, *the selection response continues even after the cessation of selection* (Figure 15.3), a feature that Kirkpatrick and Lande (1989) called **evolutionary momentum**. Using Equation 15.5b to sum Equation 15.30 over τ yields the cumulative response *following* the last generation of selection

$$R^*(\tau) = \frac{m(1-m^\tau)}{1-m}\,[\,\Delta\mu_z(t) - S\,] \tag{15.31}$$

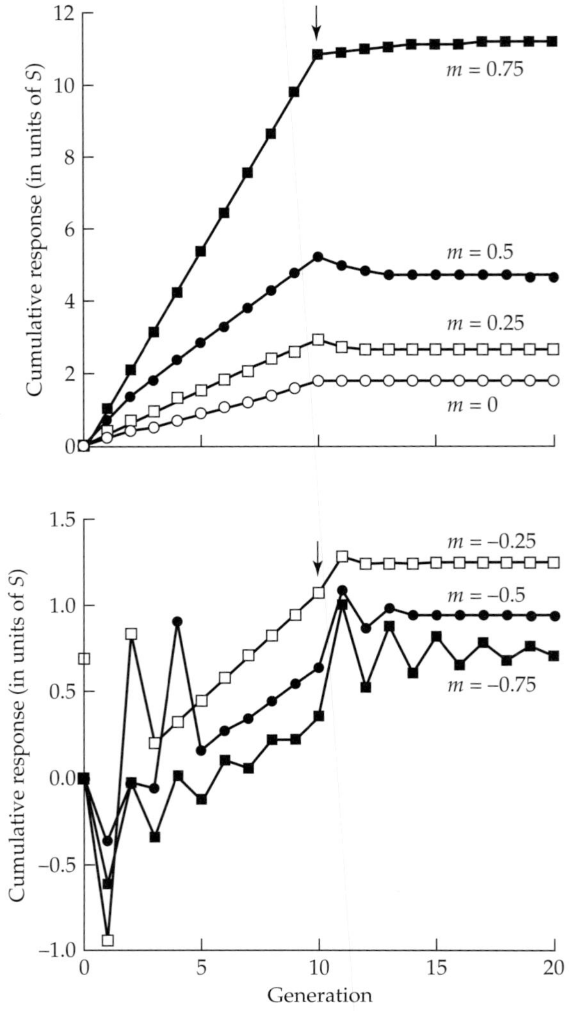

Figure 15.4 Examples of the predicted selection response with maternal effects under Falconer's dilution model. Selection starts at generation 0, with $S_{fa} = S_{mo} = S$ until generation 10 (arrows), at which point selection stops. We assume that $h^2 = 0.35$ throughout, with the different curves corresponding to different maternal effect coefficients, m. **Top:** Positive maternal effects ($m > 0$). For this value of h^2, Equation 15.34a gives the critical m value (below which some erosion of response occurs upon cessation of selection) as 0.52, meaning that the selection response continues (for a few generations) after selection is relaxed for $m = 0.75$, while response decays for $m = 0.5$ and 0.25. **Bottom:** Negative maternal effects ($m < 0$). The dynamics here are considerably more interesting, with additional response following the cessation of selection for all values of $m < 0$.

which, as $\tau \to \infty$, converges to

$$R^* = S\,\frac{m\,(1-m^t)}{1-m}\left[\frac{2h^2}{(1-m)(2-m)} - 1\right] \tag{15.32}$$

Summing Equations 15.29a and 15.32, the *permanent* response to t generations of selection is

$$R(t) + R^* = t\,h^2\,S\,\frac{2}{(1-m)(2-m)} \tag{15.33}$$

which is just t times the asymptotic response (Equation 15.28b). If R^* is opposite in sign to S, there is some erosion of the cumulative response upon relaxation of selection (we have already seen a special case of this with reversed response). For $|m| < 1$, erosion in response occurs if

$$0 < m < \frac{3 - \sqrt{1 + 8h^2}}{2} \tag{15.34a}$$

On the other hand, if maternal effects are either negative ($m < 0$) or sufficiently large

$$m > \frac{3 - \sqrt{1 + 8h^2}}{2} \tag{15.34b}$$

the response *will continue* for a few generations following the relaxation of selection. This occurs by the transient component of response decaying away to reveal the actual permanent response due to changes in breeding values. Figure 15.4 plots some sample trajectories.

In summary, the presence of maternal effects introduces several complications even under this simplest of models (in which the direct and maternal trait are the same). First, predicting the response to selection in a given generation requires, not only knowledge of the inheritance parameters (m, h^2) and current selection differential, but also knowledge of *previous selection*, namely, $\Delta\mu_z(t-1)$ and $S_{mo}(t-1)$. Second, after selection is stopped, the mean is likely to continue to change due to lag effects (e.g., Figure 15.4). If $m < 0$, the response will continue, while if $m > 0$, the response may either continue or partly decay. This clearly causes problems if we are trying to estimate the nature of selection acting on a character by comparing changes in means between generations. For example, an observed cross-generation decrease in a character could be due to four very different causes: (i) $S < 0$, (ii) $S > 0$ and there is a reversed response due to maternal effects, (iii) there is no selection in the observed generation but a previous history of $S > 0$, with the decrease in mean due to a positive maternal effect (reflecting a decay in response), or (iv) there is no current selection but a previous history of $S < 0$, with the decrease in mean due to a negative (or sufficiently large positive) maternal effect (reflecting a continuation of response).

Separate Direct and Maternal Traits

In contrast to Falconer's simple model of maternal effects, where the direct trait has a dual role as the maternal trait, a more general model allows us to separate the direct and maternal traits, e.g., weight in an offspring and milk production in its mother. In this case, Equation 15.21 becomes

$$z = G_d + m\, z_{m,mo} + e \tag{15.35a}$$

where z is the value of the focal (direct) trait in an offspring (e.g., its weight), and $z_{m,mo}$ is the value of the maternal trait in its mother (e.g., her milk production, which influences offspring weight). Clearly, this expression generalizes to a vector, $\mathbf{z}_d$, of direct traits that depends on a vector, $\mathbf{z}_m$, of maternal traits, as modeled by Kirkpatrick and Lande (1989, 1992) and Lande and Kirkpatrick (1990). We examine their full multivariate treatment in detail Volume 3. Here, we briefly comment on the bivariate case (a single direct trait, $z = z_d$, and a single, and distinct, maternal trait, z_m). Note that the selection response in z depends on two components, a direct response, R_d, due to changes in G_d and a response, R_m, due to changes in the maternal trait (which influences the phenotype of the direct trait), hence

$$R_z(t) = \Delta\mu_z(t) = R_d(t) + mR_m(t-1) \tag{15.35b}$$

where the generation index value of $(t-1)$ in the maternal response occurs because its contribution to the change in z for the population in generation t depends on the values in their mothers (which are themselves from generation $t-1$). Additionally, there can be a component of response due to the effects of previous selection. Putting all these together, the joint dynamics obtained by Kirkpatrick and Lande (1989, 1992) are

$$R_d(t) = \left(G_{dd} + \frac{m}{2}G_{md}\right)\beta_z(t) + G_{md}\beta_m(t) \tag{15.36a}$$

$$R_m(t) = \left(G_{md} + \frac{m}{2}G_{mm}\right)\beta_z(t) + G_{mm}\beta_m(t) \tag{15.36b}$$

$$R_z(t) = R_d(t) + mR_m(t-1) + m\left[P_{mm}\Delta\beta_m(t-1) + P_{mz}\Delta\beta_z(t-1)\right] \tag{15.36c}$$

Here, β_z and β_m are the selection gradients on the direct and maternal traits, respectively, and $G_{xy} = \sigma(A_x, A_y)$ and $P_{xy} = \sigma(z_x, z_y)$ are, respectively, the additive-genetic and phenotypic covariances between traits x and y. Note that the expression given for Equation 15.36c is based on the corrected version of Kirkpatrick and Lande (1992). As with the simple Falconer model, there are time lags in the response, R_z, on the focal trait due to dependency of the response on selection in the previous, as well as the current, generation (Equation 15.36c). By contrast, no such lags are seen for the maternal response, R_m.

A few other features of this model are worth noting. In the absence of additive-genetic variance for the direct trait ($G_{dd} = G_{md} = 0$), there may still be a permanent response in the direct trait due to response in the maternal trait, which is the $mR_m(t-1)$ term in Equation 15.36c. Likewise, in the absence of selection on the maternal effect ($\beta_m = 0$), Equation 15.36b shows that the maternal trait value, z_m, can still evolve when the direct and maternal trait are genetically correlated ($G_{md} \neq 0$). As reviewed by Wilson and Réale (2006), this covariance is often both significant and negative, so selection to increase the direct trait value can result in a declining maternal value (Cheverud 1984a). Chapter 22 greatly expands on this idea of an evolving environmental background (the overall social effects, and not just the maternal value) that influences the phenotype of a focal individual.

Example 15.4. Assume that there are constant selection differentials following the start of selection, so that (after the first generation) terms involving $\Delta\beta$ (changes in β) are zero. Equations 15.36b and 15.36c then simplify to

$$R_m(t) = \left(G_{md} + \frac{m}{2}G_{mm}\right)\beta_z + G_{mm}\beta_m \tag{15.37a}$$

$$\begin{aligned} R_z(t) &= \left(G_{dd} + \frac{m}{2}G_{md}\right)\beta_z + G_{md}\beta_m + mR_m(t-1) \\ &= \left(G_{dd} + \frac{m}{2}G_{md}\right)\beta_z + G_{md}\beta_m + m\left[\left(G_{md} + \frac{m}{2}G_{mm}\right)\beta_z + G_{mm}\beta_m\right] \\ &= \left(G_{dd} + \frac{3m}{2}G_{md} + \frac{m^2}{2}G_{mm}\right)\beta_z + (G_{md} + mG_{mm})\,\beta_m \end{aligned} \tag{15.37b}$$

One central question in these (and more general) models concerns which trait (or traits) in the mother determine the value of the maternal effect in her offspring. Falconer made the simplification that the value, z_{mo}, of the direct trait in a mother is scaled (by m) to give the maternal effect in her offspring ($M = mz_{mo}$). The classical bivariate counterpart to this one-trait model is to assume a general measure z_m of maternal performance, where $M = z_{m,mo}$, namely its value in the mother (Dickerson 1947; Willham 1963, 1972; Cheverud 1984a; Riska et al. 1985; Kirkpatrick and Lande 1989). Under this model, the trait—or, more likely, some composite index of traits—comprising z_m is not specified (and hence not directly scored in mothers), although BLUP allows us to estimate its variance components (e.g., G_{mm} and G_{md}), given an appropriate pedigree design (Chapter 22). Under this model, $m = 1$ (as $M = z_{m,mo}$), which simplifies Equations 15.37a and 15.37b. If we assume that there is no direct selection on the maternal performance trait, $\beta_m = 0$, the response equations further simplify to

$$R_m = \left(G_{md} + \frac{1}{2}G_{mm}\right)\beta_z \tag{15.38a}$$

$$R_z = \left(G_{dd} + \frac{3}{2}G_{md} + \frac{1}{2}G_{mm}\right)\beta_z \tag{15.38b}$$

The assumption of selection only on the direct trait implies that $\beta_z = S_z/\sigma_z^2$, yielding

$$R_z = \left(h_d^2 + \frac{3}{2}\frac{G_{md}}{\sigma_z^2} + \frac{G_{mm}}{2\sigma_z^2} \right) S_z \qquad (15.38c)$$

as $h_d^2 = G_{dd}/\sigma_z^2$. This is the **Dickerson-Willham model** of the selection response in the presence of heritable maternal effects (Chapter 22), which can also be expressed in terms of a **total heritability**

$$h_T^2 = \frac{G_{dd} + 1.5G_{md} + 0.5G_m}{\sigma_z^2} = \frac{\sigma^2(A_d) + 1.5\,\sigma(A_d, A_m) + 0.5\sigma^2(A_m)}{\sigma_z^2} \qquad (15.38d)$$

yielding the response as

$$R_z = h_T^2 S_z \qquad (15.38e)$$

If $G_{md} > 0$, then $h_T^2 > h_d^2$, and the direct heritability underestimates the total response (e.g., McAdam et al. 2002). Conversely, if G_{md} is sufficiently negative, then $h_d^2 > h_T^2$, and the direct heritability overestimates the response (e.g., Wilson et al. 2005a). Bijma (2011) more fully detailed the connections between the Dickerson-Willham and Falconer models (also see Chapter 22).

Maternal Selection vs. Maternal Inheritance

Finally, we note that, in addition to influencing offspring *phenotypes*, parents can potentially influence offspring *fitness*, which would then be a function of both the offspring's phenotype and that of its parents. Indeed, such a parental effect on fitness is simply a special case where the trait experiencing the maternal effect is fitness itself. Kirkpatrick and Lande (1989) called this process **maternal selection**, as opposed **maternal inheritance** (what we have been calling maternal effects, wherein the trait value is influenced by the mother). Hence, the latter is a maternal influence on the value of a specific trait (other than fitness itself), while the former is a maternal influence directly on fitness. One can have maternal selection (the mother's phenotype directly influences offspring fitness) without maternal inheritance (the mother's phenotype influences offspring trait value), and vice versa, or a trait could jointly experience both maternal selection and maternal inheritance (see Figure 15.5). A further complication is that the maternal trait that influences offspring fitness may itself be under selection in the mother. As detailed in Volume 3, accounting for all of these different interactions leads to more complex expressions for the inheritance of these traits than presented by Equations 15.36a–15.36c.

To see how the presence of maternal selection changes the selection-response equations, consider its simplest formulation (analogous to the Falconer dilution model of maternal inheritance), where the expected fitness of individual i is a function of its current phenotype, z_i, and the phenotype of the same trait, z_{mo_i}, in its mother. The linear regression predicting i's relative fitness (e.g., Equation 13.25b) becomes

$$w_i = 1 + (\beta_z \cdot z_i) + (\beta_{ms} \cdot z_{mo_i}) + e_i \qquad (15.39a)$$

Here, β_{ms} is expected change in relative offspring fitness from a one-unit change in the maternal phenotype while holding the offspring phenotype constant (the impact from maternal selection). Likewise, β_z is the impact on relative fitness of a unit change in offspring phenotype, while holding the maternal phenotype constant (the impact from direct selection on the offspring trait). Using the Robertson-Price identity (Equation 6.10), the selection differential, S, on this trait becomes

$$\begin{aligned} S &= \sigma\,(z_i, w_i) = \sigma\,(z_i, 1 + \beta_z z_i + \beta_{ms} z_{mo_i} + e_i) \\ &= \beta_z \sigma(z_i, z_i) + \beta_{ms} \sigma(z_i, z_{mo_i}) \end{aligned} \qquad (15.39b)$$

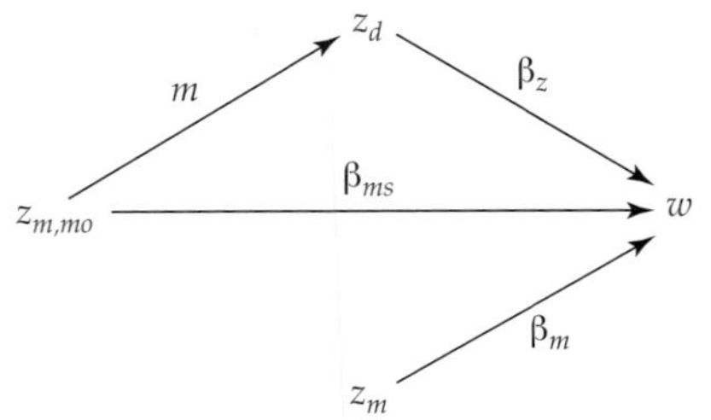

Figure 15.5 Some of the complications for predicting selection response that arise when maternal selection or maternal inheritance occur. The figure displays the simplest bivariate model, where the direct and maternal effects are separate traits. Here z_d and z_m denote the direct trait and maternal trait values, respectively, in the focal offspring, whose fitness is w. Similarly, $z_{m,mo}$ denotes the value of the maternal trait in the offspring's mother. For ease of presentation, paths involving potential genetic correlations (e.g., $z_m \leftrightarrow z_d$; $z_{m,mo} \rightarrow z_m$) are not shown. Maternal inheritance ($m \neq 0$) occurs when the value of $z_{m,mo}$ influences the value of z_d in her offspring. In the figure, this is indicated by the path whose coefficient is given by m, as we assume the maternal inheritance model given by Equation 15.35a. If β_z denotes the direct selection gradient on the focal trait, under maternal inheritance, the mother's maternal trait value indirectly influences the fitness of her offspring through the path $z_{m,mo} \rightarrow z_d \rightarrow w$, or an effect of $m\beta_z$. Maternal selection ($\beta_{ms} \neq 0$) occurs when there is a *direct* path from the mother's trait value to the fitness of her offspring (the single-step path $z_{m,mo} \rightarrow w$, whose coefficient is β_{ms}). The figure also allows for the possibility that there is direct selection on the value of the maternal trait itself (the gradient β_m, shown in the figure operating on the offspring), if, e.g., there is a direct fitness cost for high maternal performance, $\beta_m < 0$. Equations 15.35 through 15.37 give the dynamics for the case when $m \neq 0$ and $\beta_{ms} = 0$, while Equation 15.40 gives the relative fitness when $m = 0$ and $\beta_{ms} \neq 0$.

where $\sigma(z_i, z_{mo_i})$ is the phenotypic covariance between the trait in the offspring and its mother. If we assume that there are no maternal effects acting on the trait, then the covariance between an offspring's phenotypic value and that of its mother is simply half the additive genetic variance in the trait (LW Chapter 7), yielding $\sigma(z_i, z_{mo_i}) = \sigma_A^2/2$. Substituting this into Equation 15.39b yields the selection response in z as

$$R = h^2 S = h^2 \left(\beta_z \sigma_z^2 + \beta_{ms} \sigma_A^2/2\right) = \sigma_A^2 \left(\beta_z + \beta_{ms} h^2/2\right) \tag{15.39c}$$

The first term, $\sigma_A^2\beta_z$, corresponds to the standard breeder's equation (Equation 13.8c), while the second term, $(h^2/2)\sigma_A^2\beta_{ms}$, is the correlated change from selection on the mother's phenotype.

Because the value of the trait in the mother, z_{mo_i}, influences the fitness of her offspring, selection on this trait in the prereproductive females, which potentially alters the population distribution of trait values, impacts the fitness of the offspring (Kirkpatrick and Lande 1989). For example, suppose that larger mothers protect their young offspring better than smaller mothers ($\beta_{ms} > 0$), but that larger individuals are also selected against (for example, because they are more obvious to predators), so $\beta_z < 0$. In this case, there is selection for smaller individuals before reproduction, but the offspring of surviving larger mothers have increased fitness. Whether the response is positive or negative depends on the relative magnitudes of β_z and β_{ms}. From Equation 15.39c, $R > 0$ when $\beta_z + \beta_{ms}h^2/2 > 0$ or when $-\beta_z/\beta_{ms} < h^2/2$.

We can extend this simple model by assuming that the trait also experiences maternal inheritance. Under the dilution model, the maternal effect is a fraction, m, of the mother's trait value, $M = m \cdot z_{mo_i}$ (Equation 15.21), resulting in an equilibrium mother-offspring phenotypic covariance of $\sigma(z_i, z_{mo_i}) = m\sigma_A^2/[2-m]$ (LW Chapter 23). From Equation 15.39b, the resulting selection differential becomes

$$S(t) = \beta_z(t)\, \sigma_z^2 + \beta_{ms}(t)\, \sigma_A^2 \left(\frac{m}{2-m}\right) \tag{15.39d}$$

which is then used in the response dynamics under the dilution model (Kirkpatrick and Lande 1989). The multivariate extension of maternal selection (following a vector of traits influencing offspring fitness and, potentially, phenotype) was given by Kirkpatrick and Lande, and is examined in Volume 3.

Our treatment (Equations 15.39a through 15.39d) of maternal selection collapsed an inherently multivariate problem (separate traits in the mother influence offspring trait values and directly impact offspring fitness) into a univariate problem by assuming that a single trait does both (a dilution model for both maternal selection and maternal inheritance). Such a single-trait analysis obscures the important distinction between the gradient, β_{ms}, of *offspring* fitness on the maternal value in Equation 15.39a (maternal selection, the direct impact of a *mother's* phenotype on the fitness of her offspring), and the gradient, β_m, on maternal effects (selection on the maternal *trait* itself), see Figure 15.5.

Consider the simplest bivariate model, where z_d and z_m denote the values of the direct and maternal traits in a focal individual, and $z_{m,mo}$ denotes the value of the maternal trait in its mother. For example, suppose that fitness is a function of the height, z_d, and weight, z_m, of an individual and the weight, $z_{m,mo}$, of its mother. The regression for relative fitness now becomes

$$w_i = 1 + (\beta_z \cdot z_{d,i}) + (\beta_m \cdot z_{m,i}) + (\beta_{ms} \cdot z_{m,mo_i}) + e_i \tag{15.40}$$

Here, β_m represents the expected change in relative fitness given a one-unit increase in an individual's maternal trait value when holding its direct value ($z_{d,i}$) and the maternal value (z_{m,mo_i}) of the mother constant. This distinction between the fitness consequences of the value of the maternal trait to herself (β_m) and to her offspring (β_{ms}) is critical—a maternal-effect trait value may be harmful to an individual but helpful to its offspring, generating **antagonistic selection**. For example, having a large litter size is beneficial to a female but harmful to her offspring, as individuals from larger litters may have lower survivorship (Wilson et al. 2005b). Determining when a fitness component is assigned to an offspring versus a parent can be problematic (Wolf and Wade 2001), a point we consider further in Chapter 30.

16

Short-term Changes in the Variance: 1. Changes in the Additive Variance

In artificial selection experiments it seems likely that the effects of linkage disequilibrium will be more important than the slower and less dramatic effects due to changes in gene frequencies. Michael Bulmer (1976a)

Up to this point, we have been assuming that selection does not significantly change the variance of a trait, at least over the short term (a few generations). This arises from our focus on **directional selection** (direct selection on the mean) and the assumption that a large number of loci, each with a small effect, underlie a trait. Under such a genetic architecture, allele-frequency changes over the short term can be cumulatively large enough to have a significant effect on the mean while having little effect on the variance (Chapter 24). This constancy of variance assumption ignores the fact that selection also generates gametic-phase disequilibrium (LD), even among unlinked loci, which can swiftly and dramatically change the variance even in the absence of *any* allele-frequency change. Further, natural and artificial selection can act directly on the variance of a trait itself, as in the case of **stabilizing selection** for more uniformity or **disruptive selection** for more extreme phenotypes on either side of the mean. The breeder's equation only considers the change in mean and hence is uninformative in these latter two cases.

As we will show below, both directional and stabilizing selection generate negative disequilibrium (as alleles that increase trait values become negatively correlated within a gamete, even if unlinked), thus reducing σ^2_A. Conversely, disruptive selection increases the phenotypic variance, and this generates positive disequilibrium, inflating the additive variance. This chapter develops the **Bulmer equation**, an analog of the breeder's equation for the change in the additive genetic variance from selection-induced LD. This equation predicts how changes in LD change h^2, allowing updated values to be substituted into the breeder's equation for more accurate prediction of the selection response in the mean. It also predicts the short-term change in the variance under stabilizing and disruptive selection. Phenotypic assortative mating also generates disequilibrium, and we can use extensions of the Bulmer equation to simultaneously consider the effects of assortative mating and selection. Throughout this chapter, we assume (unless otherwise stated) the infinitesimal model holds and there is an infinite population size. Chapter 24 continues this discussion, relaxing many of the infinitesimal model assumptions (such as allowing for linkage and a finite number of loci) and more fully considering the impact from drift.

Changes in the genetic variance (though either LD or allele-frequency change) are not the only route by which selection can alter the phenotypic variance, σ^2_z. If there is heritable variation in the environmental variance (i.e., σ^2_E varies over genotypes), σ^2_E itself can respond to selection, also resulting in a change in σ^2_z, as discussed in Chapter 17.

CHANGES IN VARIANCE DUE TO GAMETIC-PHASE DISEQUILIBRIUM

In the absence of epistasis, gametic-phase disequilibrium does not change the population mean (Chapter 15). However, as first pointed out by Lush (1945), it affects the response to selection by introducing correlations between alleles at different loci (even if unlinked), thus altering the additive genetic variance, even in the absence of any allele-frequency change.

Evolution and Selection of Quantitative Traits. Bruce Walsh & Michael Lynch.
Published 2018 by Oxford University Press.
DOI 10.1093/oso/9780198830870.001.0001

To see this, let $a_1^{(k)}$ and $a_2^{(k)}$ be the average effects of the two alleles at locus k from a random individual, where the subscripts 1 and 2 denote the maternally and paternally derived alleles, respectively. Assuming (for now) random mating, there is no covariance between alleles of maternal and paternal origin, so that $\sigma(a_1^{(k)}, a_2^{(j)}) = 0$ for all k and j. However, when gametic-phase disequilibrium is present, there *can* be covariances between alleles at different loci from the *same* parent, so that $\sigma(a_1^{(k)}, a_1^{(j)})$ and $\sigma(a_2^{(k)}, a_2^{(j)})$ can be nonzero. That is, there can be correlations between alleles in any particular gamete. Because σ_A^2 is the variance of the sum of average effects over all loci

$$\sigma^2\left(\sum_{k=1}^{n}\left[a_1^{(k)} + a_2^{(k)}\right]\right) = 2\sum_{k=1}^{n}\sigma^2\left(a^{(k)}\right) + 4\sum_{k=1}^{n}\sum_{k<j}^{n}\sigma\left(a^{(j)}, a^{(k)}\right) \tag{16.1a}$$

$$= 2\sum_{k=1}^{n} C_{kk} + 4\sum_{k=1}^{n}\sum_{k<j}^{n} C_{jk} \tag{16.1b}$$

where n is the number of loci and C_{jk} is the covariance between allelic effects at loci j and k (when contributed by the same parent, and hence on the same gamete). This decomposes the additive variance as

$$\sigma_A^2 = \sigma_a^2 + d \tag{16.2}$$

where $\sigma_a^2 = 2\sum C_{kk}$ is the additive variance in the absence of gametic-phase disequilibrium, while the disequilibrium contribution $d = 4\sum_{j<k} C_{kj}$ is the covariance between allelic effects at different loci (in terms of the notation used in LW Equation 7.14, $d = \sigma_{A,A}$).

The component of the additive genetic variance that is unaltered by changes in gametic-phase disequilibrium, σ_a^2, is often referred to as the **additive genic variance** (or simply the **genic variance**) to distinguish it from the additive *genetic* variance, σ_A^2. In the absence of disequilibrium, the genic and genetic variances are equivalent. Negative disequilibrium ($d < 0$) implies the presence of hidden additive variance ($\sigma_A^2 < \sigma_a^2$), with σ_A^2 increasing toward σ_a^2 as the disequilibrium decays. If $d > 0$, the additive variance is inflated relative to a random-mating population ($\sigma_A^2 > \sigma_a^2$), with σ_A^2 decreasing toward σ_a^2 as disequilibrium decays. Because $n(n-1)$ terms contribute to d, whereas n terms contribute to σ_a^2, gametic-phase disequilibrium can generate large changes in the additive genetic variance even when changes in the individual covariances, C_{jk}, are all very small (Chapter 24).

The allelic effects, $a^{(k)}$ (and hence the genic variance, σ_a^2), are altered as allele frequencies change, resulting in a permanent change in σ_A^2. Changes in σ_a^2 due to selection strongly depend on the initial distribution of allelic effects and frequencies (Chapters 5 and 24–26), both of which are extremely difficult to estimate. Changes in d, however, are generally less sensitive to the initial distribution of allelic effects (Sorensen and Hill 1982). Any changes in σ_A^2 due to changes in d are *transient*—in the absence of selection, recombination removes disequilibrium and the additive genetic variance, σ_A^2, returns to the additive genic variance, σ_a^2, as d decays to zero.

Thus, under our (short-term response) assumption that allele frequencies remain effectively constant, changes in σ_A^2 are due entirely to changes in d, as the C_{kk} terms in Equation 16.1b (corresponding to σ_a^2) are only altered by allele-frequency change. Hence, the additive genetic variance in generation t is calculated by $\sigma_A^2(t) = \sigma_a^2 + d(t)$, yielding a change in variance of $\Delta\sigma_A^2(t) = \Delta d(t)$. Under random mating in the absence of selection, the disequilibrium between pairs of unlinked loci is halved in each generation (LW Equation 5.12), thus halving the covariance between allelic effects

$$d(t+1) = \frac{d(t)}{2} \tag{16.3}$$

Countering this process, selection tends to generate gametic-phase disequilibrium. For example, directional selection to change the mean usually reduces the variance of a character, thus generating negative values of d anew in each generation (Felsenstein 1965).

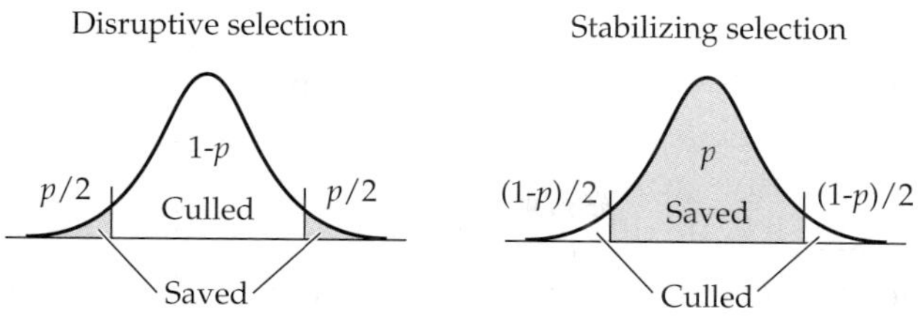

Figure 16.1 Artificial stabilizing and disruptive selection using **double truncation**. In both cases, a fraction, p, of the population is allowed to reproduce. Under stabilizing selection, the central p of the distribution is saved, while under disruptive selection, the uppermost and lowermost $p/2$ are saved.

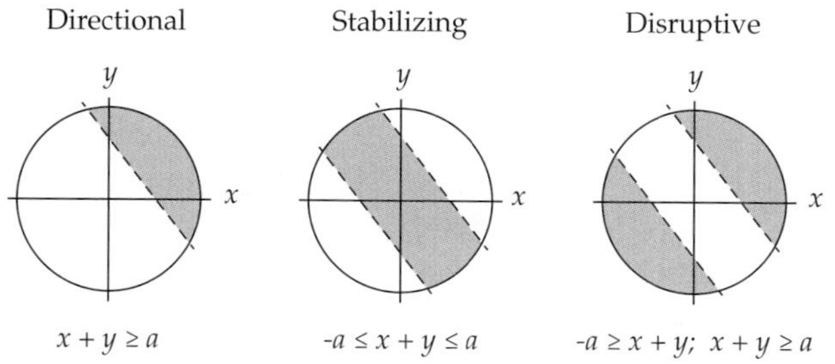

Figure 16.2 The generation of covariances (gametic-phase disequilibrium, LD) by the various type of truncation selection. The variables x and y (e.g., allelic effects at two different loci) are uncorrelated before selection, with their distribution indicated by the open circle. **Left:** Under directional selection, only those values of $x+y$ above some threshold (say, a) are retained. The resulting distribution (the shaded area above the line for $x+y=a$) now displays a negative covariance between the remaining x and y values. **Middle:** Under stabilizing selection, only those values in the range of $-a \leq x+y \leq a$ are retained, also generating a negative covariance. **Right:** Under disruptive selection, only values of $x+y \geq a$ or $x+y \leq -a$ are retained, now resulting in a positive covariance between the remaining x and y values.

As shown in Figure 16.1, stabilizing selection reduces the phenotypic variance and, in the process, creates negative values of d (as we will shortly demonstrate), while disruptive selection increases the phenotypic variance, generating positive d. Stabilizing and disruptive selection are occasionally referred to as **centripetal selection** and **centrifugal selection**, respectively (Simpson 1944). Figure 16.2 shows how directional and stabilizing selection generate negative covariances (and hence negative values of d) between loci under selection, while disruptive selection generates positive covariances (and hence positive values of d).

More generally, for values of z that are normally distributed, selection reduces the phenotypic variance when $\partial^2 \ln w(z)/\partial z^2 < 0$ for all z (Shnol and Kondrashov 1993), generating negative values of d. If this partial differential is > 0 for all values of z, selection increases the variance, generating positive values of d. One function that does not change the variance (when $z \sim$ normal) is the exponential fitness function, $w(z) = \exp(-az)$, as

$$\frac{\partial^2 \ln w(z)}{\partial z^2} = -\frac{\partial^2 az}{\partial z^2} = 0$$

for all z, and hence the variance following selection neither increases or decreases (Charlesworth 1990).

Assuming the validity of the infinitesimal model, Bulmer (1971b, 1974a, 1976a, 1980) solved the question of how these within-generation changes in the variance translate into between-generation changes (response in the variance). Chapter 24 moves beyond the infinitesimal model by considering the impact of linkage with a finite number of loci and finite population size. Estimation of the nature and amount of selection acting on the mean and the variance of a character is examined in Chapters 29 and 30.

CHANGES IN VARIANCE UNDER THE INFINITESIMAL MODEL

Because allele frequencies remain essentially constant under the assumptions of the infinitesimal model, the additive genic variance, σ_a^2, remains constant and all changes in the additive genetic variance, σ_A^2, are due to changes in d. Assuming the population is initially in gametic-phase equilibrium, so that $d(0) = 0$, then $\sigma_A^2(0) = \sigma_a^2$, yielding

$$\sigma_A^2(t) = \sigma_a^2 + d(t) = \sigma_A^2(0) + d(t)$$

Let $h^2(t)$ and $\sigma_z^2(t)$ denote the heritability and phenotypic variance before selection in generation t, and h^2 and σ_z^2 be the values of these quantities in the absence of gametic-phase disequilibrium.

While LD can alter the additive variance, what is its impact on the dominance genetic variance? Under the infinitesimal model, Bulmer (1971b) suggested that gametic-phase disequilibrium does not change σ_D^2. To see Bulmer's argument, first note from LW Equation 5.16b, that with a finite number of loci (n), the disequilibrium contribution to dominance genetic variance is of the order of $n^2\,\overline{D}^{\,2}$, where $\overline{D}$ is the average pairwise disequilibrium. Under the infinitesimal model, the total disequilibrium (summing over all pairs of loci) remains bounded as the number of loci increases, implying that $\overline{D}$ is of the order of n^{-2} as there are $n(n-1)/2 \approx n^2/2$ pairs of loci contributing to $\overline{D}$. The contribution to dominance variance from disequilibrium is thus of the order of $n^2(n^{-2})^2 = n^{-2}$, which converges to zero in the infinitesimal-model limit (as the number of loci $n \to \infty$). There is, however, some delicacy involved in this argument. In Chapter 24, we show that with strong directional dominance, the amount of inbreeding depression becomes unbounded as one increases the number of loci to the infinitesimal limit. These arguments on the lack of impact from LD on σ_D^2 assume the presence of a large number of small-effect loci. By contrast, Jorjani et al. (1998) used simulations to show that with a modest number of loci and small population size, changes in disequilibrium (in their case, from assortative mating) can significantly change the dominance variance. While we make the Bulmer assumption that there is no effect of dominance throughout this chapter, further exploration of the impact of disequilibrium on σ_D^2 is warranted.

With this assumption in mind, in the absence of any epistatic variance, genotype × environment interactions or correlations, the phenotypic variance and heritability at generation t become

$$\sigma_z^2(t) = \sigma_E^2 + \sigma_D^2 + \sigma_A^2(t) = \sigma_z^2 + d(t) \tag{16.4a}$$

$$h^2(t) = \frac{\sigma_A^2(t)}{\sigma_z^2(t)} = \frac{\sigma_a^2 + d(t)}{\sigma_z^2 + d(t)} \tag{16.4b}$$

where $\sigma_z^2 = \sigma_z^2(0)$ is the phenotypic variance before selection in the initial (unselected) population ($d(0) = 0$). Thus, knowledge of the value of $d(t)$ is sufficient to predict the variances in generation t, and hence the heritability and the response in the mean, using

$$R(t) = h^2(t)\,S(t) = h^2(t)\,\bar{\imath}(t)\,\sigma_z(t) \tag{16.4c}$$

Under the infinitesimal model, genotypic values are normally distributed before selection (Bulmer 1971b, 1976b). Recalling that $z = G + E$, we then see that if the environmental values, E, are also normally distributed, the joint distribution of phenotypic and genotypic values is multivariate normal. Hence, from standard statistical theory (e.g., LW Chapter 8), the regression of offspring phenotypic value, z_o, on parental phenotypes, z_m and z_f, is linear and homoscedastic, with

$$z_o = \mu + \frac{h^2}{2}(z_m - \mu) + \frac{h^2}{2}(z_f - \mu) + e \tag{16.5a}$$

where

$$\sigma_e^2 = \left(1 - \frac{h^4}{2}\right)\sigma_z^2 \tag{16.5b}$$

The derivation of Equation 16.5 follows from standard multiple-regression theory and the correlations between relatives (see Example 6 in LW Chapter 8 for details).

We denote the within-generation change in variance by $\delta(\sigma_z^2) = \sigma_{z^*}^2 - \sigma_z^2$, where z^* refers to a phenotypic value from the selected population. (More generally, $\sigma_{z^*}^2$ is the fitness-weighted trait variance; see Chapter 29). Throughout this chapter we use the notation δ to distinguish the *within-generation* change of a variable from its *between-generation* change, Δ, as the latter incorporates the effects of genetic transmission across a generation. If we take variances of both sides of Equation 16.5a and assume random mating (so that $\sigma(z_f, z_m) = 0$) and identical selection on both sexes, the phenotypic variance among the offspring from selected parents becomes

$$\begin{aligned}\sigma^2(z_o) &= \frac{h^4}{4}\left[\sigma^2(z_m^*) + \sigma^2(z_f^*)\right] + \sigma_e^2 \\ &= \frac{h^4}{2}\left[\sigma_z^2 + \delta(\sigma_z^2)\right] + \left(1 - \frac{h^4}{2}\right)\sigma_z^2 \\ &= \sigma_z^2 + \frac{h^4}{2}\,\delta\left(\sigma_z^2\right) \qquad (16.6)\end{aligned}$$

The change in phenotypic variance in the offspring due to selection on their parents generating disequilibrium is thus $(h^4/2)\,\delta(\sigma_z^2)$. Because there is no change in the environmental, dominance, or genic variances, this change is all in the disequilibrium component, d, of the additive genetic variance, σ_A^2. Combining Equations 16.6 and 16.3 yields a general recursion for changes in the variance under the infinitesimal model with unlinked loci of

$$d(t+1) = \frac{d(t)}{2} + \frac{h^4(t)}{2}\,\delta\left(\sigma_{z(t)}^2\right) \qquad (16.7a)$$

implying that the between-generation change in the disequilibrium contribution is

$$\begin{aligned}\Delta d(t) = \Delta\sigma_{z(t)}^2 &= \Delta\sigma_A^2(t) \\ &= -\frac{d(t)}{2} + \frac{h^4(t)}{2}\,\delta\left(\sigma_{z(t)}^2\right) \qquad (16.7b)\end{aligned}$$

This **Bulmer equation** (1971a) is the variance analog of the breeder's equation, which predicts short-term changes in the variance as opposed to the mean. The first term is the decay due to recombination in the disequilibrium contribution from the previous generation (assuming loci are unlinked), while the second term is the amount of new disequilibrium generated by selection that is passed onto the offspring generation. As with the breeder's equation, this second term is a function of the within-generation change (δ) in the phenotypic variance and the fraction ($h^4/2$) transmitted to the next generation. When loci are linked with a recombination fraction of c, a larger fraction $(1-c)$ of any previous d remains, yielding the change (from recombination alone) of $\Delta d(t) = (1-c)d(t) - d(t) = -cd(t)$. Chapter 24 examines the effect of linked loci in more detail.

Note from Equation 16.7b that if we start from a base population in linkage equilibrium, $d(0) = 0$, the sign of the within-generation change in the variance, $\delta(\sigma_{z(t)}^2)$, equals the sign of d. Selection that *decreases* the phenotypic variance generates negative values of d, while selection that *inflates* the variance generates positive values of d. This change in the variance (typically a reduction) due to selection generating disequilibrium is called the **Bulmer effect**. Provided the joint distribution of phenotypic and genotypic values remains multivariate normal, under the infinitesimal model, the complete dynamics of the phenotypic distribution are jointly described by the change in the variance (Equations 16.7a and 16.7b), while the change in the mean is given by updating the breeder's equation, $R(t) = h^2(t)\,S(t)$, using Equation 16.4b.

Equation 16.7a makes the further point that if we wish to use variance components to predict the response to selection, we need to start from an *unselected base population*. If a population has experienced recent prior selection, then $d(0) \neq 0$, and hence the change in σ_A^2 (and, in turn, the response to selection) cannot be predicted without knowing the value of d in the starting population. Finally, setting $\Delta d = 0$ in Equation 16.7b shows that, at equilibrium,

$$\widetilde{d} = \widetilde{h}^4\, \widetilde{\delta}(\sigma_z^2) \tag{16.7c}$$

where the tilde denotes an equilibrium value.

One very important implication for evolution follows from Equation 16.7a, which shows that loci underlying traits whose variance is reduced following selection tend to be in negative disequilibrium. Specifically, the frequency of gametes containing two positive (or two negative) alleles are underrepresented compared to the situation in random mating. Fitness itself is also a quantitative trait, which is influenced by numerous loci and environmental effects. Following selection, the variance in fitness is decreased, and hence Equation 16.7a implies that favorable alleles will be in negative disequilibrium. This reduces any additive variance in fitness, which in turn reduces the efficiency of selection.

Example 16.1. Data from Rendel (1943) suggest that stabilizing selection occurs on egg weight in ducks. Of 960 eggs followed, 64.5% hatched. The change in mean egg weight (in grams) after selection was negligible, but the variance showed a significant decrease. The variance before selection was 52.7 (using all 960 eggs), and declined to 43.9 after selection (in those eggs that hatched), yielding $\delta(\sigma_z^2) = -8.8$. Assuming that the reduction in variance due to gametic-phase disequilibrium is at equilibrium and setting $\widetilde{h}_z^2$ at 0.60 (the heritability for egg weight in poultry) gives from Equation 16.7c,

$$\widetilde{d} = \widetilde{h}^4\, \widetilde{\delta}(\sigma_z^2) = (0.6)^2(-8.8) = -3.2 \quad \text{and} \quad \widetilde{\sigma}_A^2 = \widetilde{h}^2\, \widetilde{\sigma}_z^2 = 0.6 \cdot 52.7 = 31.6$$

Assuming the infinitesimal model holds, if selection stops, the additive variance is expected to eventually increase to

$$\sigma_A^2 = \sigma_a^2 = \widetilde{\sigma}_A^2 - \widetilde{d} = 31.6 + 3.2 = 34.8$$

with half of this change occurring in one generation (assuming all underlying loci are unlinked). Similarly, $\sigma_z^2 = 52.7 + 3.2 = 55.9$ and $h^2 = 34.8/55.9 = 0.62$. (Example from Bulmer 1971b.)

Within- and Among-Family Variance Under the Infinitesimal Model

An alternative, instructive approach to the phenotypic regression argument leading to Equation 16.7a is to consider the regression of offspring breeding value (A_o) on the breeding values of its parents (A_f, A_m). Under the infinitesimal model, the joint distribution of parental and offspring breeding values before selection is multivariate normal (Bulmer 1971b), and Example 7 in Chapter 8 of LW shows that the distribution of breeding values in the offspring of parents with breeding values of A_f and A_m is given by the regression

$$A_o = \frac{1}{2}A_m + \frac{1}{2}A_f + e \tag{16.8a}$$

The residual e is the contribution due to segregation, which is normally distributed with a mean of zero and variance of $\sigma_a^2/2$, half the current genic variance (Bulmer 1971b; Felsenstein 1981; Tallis 1987), see Example 16.2.

Because e is the residual of a regression, it is uncorrelated with both A_f and A_m (LW Chapters 3 and 8). Computing variances and assuming random mating (so that A_f and A_m

are uncorrelated),

$$\sigma^2_A(t+1) = \sigma^2_{A_o}(t+1) = \sigma^2\left(\frac{A_m(t)}{2} + \frac{A_f(t)}{2}\right) + \sigma^2_e$$
$$= \frac{1}{4}\left(\sigma^2_{A_m}(t) + \sigma^2_{A_f}(t)\right) + \frac{1}{2}\sigma^2_A(0)$$
$$= \frac{1}{2}\sigma^2_{A^*}(t) + \frac{1}{2}\sigma^2_a \tag{16.8b}$$

where $\sigma^2_{A^*}(t)$ is the variance of the breeding values of the selected parents (with assortative mating, Equation 16.8b has an additional term, $\sigma(A^*_m, A^*_f)/2$; see Equation 16.21b). Equation 16.8b shows that additive variance can be decomposed into an *among-family* component (half the additive *genetic* variance, $\sigma^2_{A^*}(t)/2$), that measures the differences between the mean breeding values of families, and a *within-family* component (half the additive *genic* variance, $\sigma^2_a/2$) due to segregation that measures the variation within families. Equations 16.8a and 16.8b imply that under the infinitesimal model (and in an infinite population), ***the within-family additive variance remains constant***. The change in the additive-genetic variance is thus entirely due to changes in the variance of the mean values of different families. Positive disequilibrium ($d > 0$) increases the among-family component while negative disequilibrium ($d < 0$) decreases it (Reeve 1953). For example, under directional selection, selected parents (being chosen for exceptional trait values) are more similar to each other than are two random individuals from the unselected base population.

The within-family variance, $\sigma^2_a/2$, deserves additional comment. This is often called the **Mendelian sampling variance** or the **segregation variance**. Notice that this variance (under the infinitesimal model) is *not* affected by selection, as we assume there is only negligible change in allele frequencies. As we will see shortly, however, it can be decreased by drift or inbreeding. Likewise, with a finite (but large) number of loci, σ^2_a can indeed be affected by selection, but the change per generation is typically very small (Chapter 24). An especially important implication of this constant within-family segregation variance is that it tends to largely restore a normal distribution of breeding values following selection. As Equation 16.8a demonstrates, the distribution of breeding values in the offspring is the sum of two components: the breeding values of the selected parents plus the contribution due to segregation. Even if the distribution of breeding values in the selected parents departs significantly from normality, segregation tends to reduce this departure for a Gaussian. Interestingly, Smith and Hammond (1987) found that the short-term deviation from normality is largest when selection is moderate, with deviations becoming smaller as selection increases. This can be seen from Equation 16.8a by writing $A_o = A_{mp} + e$, where A_{mp} is the midparental breeding value and e is the contribution due to segregation (offspring receiving alternative alleles from heterozygous loci). As selection intensity increases, the variation due to A_{mp} decreases (as the selected individuals fall into an ever-decreasing range of phenotypes), and the majority the variation of A_o is accounted for by the normally distributed random variable, e, thus decreasing any departure from normality induced by the distribution of A_{mp}.

The derivation of Equations 16.7a, 16.7b, and 16.8a assumes that breeding values remain normally distributed. If selection changes the distribution of breeding values sufficiently away from normality, the parent-offspring regression may cease to be linear and homoscedastic. Consequences of departures from linearity were briefly discussed in Chapter 13 and are explored more fully in Chapter 24. The heteroscedasticity of the residuals implies that σ^2_e in Equation 16.8a may depend on the actual parental values chosen, which greatly complicates matters. In all discussions that follow, we assume that these departures from normality can be ignored. Chapter 24 works at relaxing these assumptions.

Example 16.2. To show why the residual variance in Equation 16.8a depends on the genic variance, σ^2_a, we assume the presence of random mating and unlinked loci, and that allele-frequency changes from selection can be ignored. Focusing on a single locus, suppose a parent has a genotype of A_iA_j, with allelic effects of a_i and a_j, where i and j index random alleles. The expected contribution from this locus to the breeding value of its offspring is $(a_i + a_j)/2$, as each allele is transmitted with a probability of 0.5. The resulting deviation between the actual contribution and expected contribution when A_i is transmitted is $a_i - (a_i + a_j)/2 = (a_i - a_j)/2$. Likewise, this deviation is $(a_j - a_i)/2$ when A_j is transmitted. Hence, the offspring variance in the contribution to its breeding value for this particular locus is simply the average of the squares of these two deviations, or

$$\frac{1}{2}\left[\frac{(a_j - a_i)^2}{4} + \frac{(a_i - a_i)^2}{4}\right] = \frac{(a_j - a_i)^2}{4}$$

Expanding the quadratic product and taking the expected value yields

$$E\left[\frac{(a_j - a_i)^2}{4}\right] = \frac{E[a_j^2]}{4} + \frac{E[a_i^2]}{4} - \frac{E[a_i \cdot a_j]}{2}$$

Under random mating, the alleles at a locus are independent, and the last term is simply $E[a] \cdot E[a] = 0$, which yields a contribution to the segregation variance from a single locus in a single parent of $E[a^2]/2$. Summing over both parents gives the variance in the breeding value from this locus in their offspring as $E[a^2]$, which is half the random-mating, linkage-equilibrium additive variance at this locus, $\sigma_a^2 = 2E[a^2]$ (Equation 16.1b). Further, because of independent assortment, the contributions from each unlinked loci from a parent are uncorrelated, and summing over all loci yields a residual variance of $\sigma_a^2/2$.

Conversely, if a parent is inbred (with f being the probability that both alleles at a randomly chosen locus are identical by descent, IBD), then the alleles A_i and A_j are no longer independent. With a probability of f they are IBD, and hence the same allele; otherwise they are independent alleles, yielding

$$E[a_i \cdot a_j] = fE[a^2] + (1 - f)E[a] \cdot E[a] = fE[a^2]$$

and now

$$E\left[\frac{(a_j - a_i)^2}{4}\right] = \frac{E[a^2]}{4} + \frac{E[a^2]}{4} - f\frac{E[a^2]}{2} = (1 - f)\frac{E[a^2]}{2}$$

Summing over all loci results in a segregation variance of $\sigma_a^2(1 - \overline{f})/2$, where $\overline{f}$ is the average of the inbreeding levels of the two parents.

Accounting for Inbreeding and Drift

As shown in Example 16.2, the effects of drift and regular inbreeding are easily accommodated under the infinitesimal model (Verrier et al. 1989). Recall that the segregation variation is simply half the additive genic variance of the parental population. When genetic drift is present, Equation 11.2 yields a genic variance in generation t of

$$\sigma_a^2(t) = \sigma_a^2(0)\left(1 - \frac{1}{2N_e}\right)^t \tag{16.9a}$$

resulting in a segregation variance in generation t of $\sigma_a^2(t)/2$. As shown by Keightley and Hill (1987), drift has only a small effect on the disequilibrium

$$\Delta d(t) = -\frac{d(t)}{2}\left(1 + \frac{1}{N_e}\right) - \frac{1}{2}\left(1 - \frac{1}{N_e}\right)\kappa h^2(t)\sigma_A^2(t) \tag{16.9b}$$

where $\kappa = 1 - \sigma_{z^*}^2/\sigma_z^2$ is the fractional reduction in phenotypic variance following selection (Equation 16.10a). When population size is finite, the additive variance in any particular generation, $\sigma_A^2(t) = \sigma_a^2(t) + d(t)$, can be computed by jointly iterating Equations 16.9a and 16.9b.

Similarly, when the parents are inbred, the segregation variance is also correspondingly reduced. This variance arises from the segregation of alleles in heterozygotes in the parents (and hence the term Mendelian sampling variance). As parents become more inbred, the heterozygosity, and hence the segregation variance, decreases. Assuming there is no correlation between the parents, the within-family segregation variance under inbreeding is

$$\frac{\sigma_a^2(t)}{2} = \frac{\sigma_a^2(0)}{2} \left[1 - \frac{f_m(t) + f_f(t)}{2} \right] \tag{16.9c}$$

where f_m and f_f denote the average amount of inbreeding in the selected male and female parents (measured by their respected inbreeding coefficients, f; Chapter 2; LW Chapter 10). The additive variance recursion (Equation 16.8b), under the assumptions of the infinitesimal model, becomes

$$\sigma_A^2(t+1) = \frac{1}{4} \left[\sigma^2_{A^*_m}(t) + \sigma^2_{A^*_f}(t) \right] + \frac{\sigma_a^2(0)}{2} \left[1 - \frac{f_m(t) + f_f(t)}{2} \right] \tag{16.9d}$$

These results for the reduction in σ_a^2 under inbreeding apply to the case of only additive variance. When nonadditive variance is present, the changes in additive variance under inbreeding are potentially much more complex (Chapter 11).

CHANGES IN VARIANCE UNDER TRUNCATION SELECTION

Provided the normality assumptions of the infinitesimal model hold, the changes in variance under any selection model can be computed by obtaining the within-generation change in the phenotype variance, $\delta(\,\sigma^2_{z(t)}\,)$, and applying Equation 16.7a or 16.7b. In the general case, this requires numerical iteration to obtain the equilibrium heritability and genetic variance. However, in many cases, the phenotypic variance after selection can be written as

$$\sigma^2_{z^*} = (1-\kappa)\,\sigma_z^2 \tag{16.10a}$$

where κ is a constant independent of the current value of the variance. In such settings,

$$\delta\left(\sigma_z^2\right) = \sigma^2_{z^*} - \sigma_z^2 = -\kappa\,\sigma_z^2 \tag{16.10b}$$

When Equation 16.10a holds (implying that selection generates a constant proportional reduction in variance), simple analytic solutions for the equilibrium variances and heritability can be obtained (again, assuming the validity of the infinitesimal model). Truncation selection—both as we have defined it for directional selection (Chapter 14) and **double truncation** giving disruptive or stabilizing selection (Figure 16.1)—satisfies Equation 16.10. As shown in Table 16.1, for truncation selection on a normally distributed phenotype, κ is strictly a function of the fraction, p, of the population saved and the type of truncation selection used. Figure 16.3 plots values of κ as a function of p for these three different truncation selection schemes.

Suppose selection is such that Equation 16.10a is satisfied. We allow for differential selection on the sexes by letting the variance after selection in males and females be $\sigma^2(z_m^*) = (1-\kappa_m)\sigma_z^2$ and $\sigma^2(z_f^*) = (1-\kappa_f)\sigma_z^2$, respectively. If parental phenotypes are uncorrelated (i.e., there is random mating),

Table 16.1 Changes in the phenotypic variance under the various schemes of single and double truncation given in Figure 16.1. Assuming the character is normally distributed before selection, the phenotypic variance after selection is calculated as $\sigma^2_{z^*} = (1-\kappa)\,\sigma^2_z$, where κ (as shown in the table) is a function of the fraction, p, of individuals saved. Here φ denotes the unit normal density function and $x_{[p]}$ satisfies $\Pr(U \le x_{[p]}) = p$ (equivalently, $\Pr[U > x_{[1-p]}] = p$), where U is a unit normal random variable. Finally, $\bar{\imath}$ is the selection intensity and is also a function of p (Equation 14.3a). While first presented in the quantitative genetics literature by Bulmer (1976a), these expressions can be found in Johnson and Kotz (1970a).

Directional Truncation Selection: Uppermost p saved

$$\kappa = \frac{\varphi\left(x_{[1-p]}\right)}{p}\left(\frac{\varphi\left(x_{[1-p]}\right)}{p} - x_{[1-p]}\right) = \bar{\imath}\left(\bar{\imath} - x_{[1-p]}\right) \tag{16.11a}$$

Stabilizing Truncation Selection: Middle fraction p of the distribution saved

$$\kappa = \frac{2\,\varphi\left(x_{[1/2+p/2]}\right)\,x_{[1/2+p/2]}}{p} \tag{16.11b}$$

Disruptive Truncation Selection: Uppermost and lowermost $p/2$ saved

$$\kappa = -\frac{2\,\varphi\left(x_{[1-p/2]}\right)\,x_{[1-p/2]}}{p} \tag{16.11c}$$

$$\sigma^2_{z^*(t)} = \frac{\sigma^2\left[\,z^*_f(t)\,\right]}{2} + \frac{\sigma^2\left[\,z^*_m(t)\,\right]}{2} = (1-\kappa)\,\sigma^2_z(t) \tag{16.12a}$$

where

$$\kappa = \frac{\kappa_f + \kappa_m}{2} \tag{16.12b}$$

The within-generation change in the variance due to selection becomes

$$\delta(\sigma^2_{z(t)}) = -\kappa\,\sigma^2_z(t) = -\kappa\,\frac{\sigma^2_A(t)}{h^2(t)} \tag{16.12c}$$

where we have used the identity $\sigma^2_z = \sigma^2_A/h^2$. Substituting Equation 16.12c into Equation 16.7a recovers the result of Bulmer (1974a),

$$d(t+1) = \frac{d(t)}{2} - \frac{\kappa}{2}\,h^2(t)\,\sigma^2_A(t) = \frac{d(t)}{2} - \frac{\kappa}{2}\,\frac{[\,\sigma^2_a + d(t)\,]^2}{\sigma^2_z + d(t)} \tag{16.12d}$$

with last step following from $h^2\sigma^2_A = (\sigma^2_A/\sigma^2_z)\sigma^2_A = \sigma^4_A/\sigma^2_z$.

At equilibrium, $\widetilde{d} = -\kappa\,\widetilde{h}^2\,\widetilde{\sigma}^2_A$, and because $\widetilde{\sigma}^2_A = \sigma^2_a + \widetilde{d}$ and $\widetilde{h}^2 = (\sigma^2_a + \widetilde{d})/(\sigma^2_z + \widetilde{d})$, we have

$$\widetilde{d} = -\kappa\,\frac{(\,\sigma^2_a + \widetilde{d}\,)^2}{\sigma^2_z + \widetilde{d}} \tag{16.12e}$$

This quadratic equation in $\widetilde{d}$ has one admissible solution (the constraint being that $\widetilde{\sigma}^2_A = \widetilde{d} + \sigma^2_a \ge 0$). Solving yields

$$\widetilde{\sigma}^2_A = \sigma^2_z\,\gamma, \quad \text{where} \quad \gamma = \frac{2h^2 - 1 + \sqrt{1 + 4h^2(1-h^2)\kappa}}{2(1+\kappa)} \tag{16.13a}$$

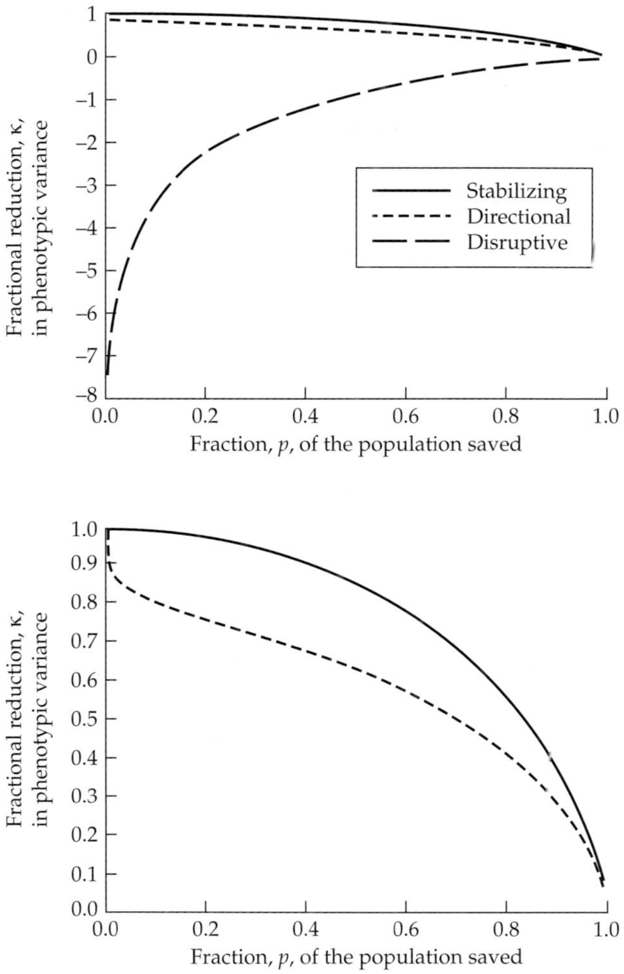

Figure 16.3 The fractional reduction, κ, of phenotypic variance removed by truncation selection (Figure 16.1) as a function of the fraction, p, of individuals saved. Following selection, the new variance is $(1-\kappa)\sigma_z^2$. **Top:** The lower-most curve (values of $\kappa < 0$) corresponds to disruptive selection (and hence an *increase* in the variance, $\sigma_{z^*}^2 > \sigma_z^2$), while the upper two curves ($\kappa > 0$) correspond to directional (middle curve) and stabilizing selection (upper curve), and hence a decrease in the variance, $\sigma_{z^*}^2 < \sigma_z^2$. **Bottom:** Close-up for directional (lower curve) and stabilizing selection (upper curve).

Because $\widetilde{\sigma}_A^2 - \sigma_A^2 = \widetilde{d}$, we can write

$$\widetilde{\sigma}_z^2 = \sigma_z^2 + (\widetilde{\sigma}_A^2 - \sigma_A^2) = \sigma_z^2(1+\gamma-h^2) \tag{16.13b}$$

yielding an equilibrium heritability of

$$\widetilde{h}^2 = \frac{\widetilde{\sigma}_A^2}{\widetilde{\sigma}_z^2} = \frac{\gamma}{1+\gamma-h^2} \tag{16.13c}$$

Following Gomez-Raya and Burnside (1990), we can also express the equilibrium heritability as

$$\widetilde{h}^2 = \frac{-1+\sqrt{1+4h^2(1-h^2)\kappa}}{2\,\kappa\,(1-h^2\,)} \tag{16.13d}$$

Figure 16.4 plots the equilibrium heritability as a function of κ and the initial heritability in the absence of any disequilibrium.

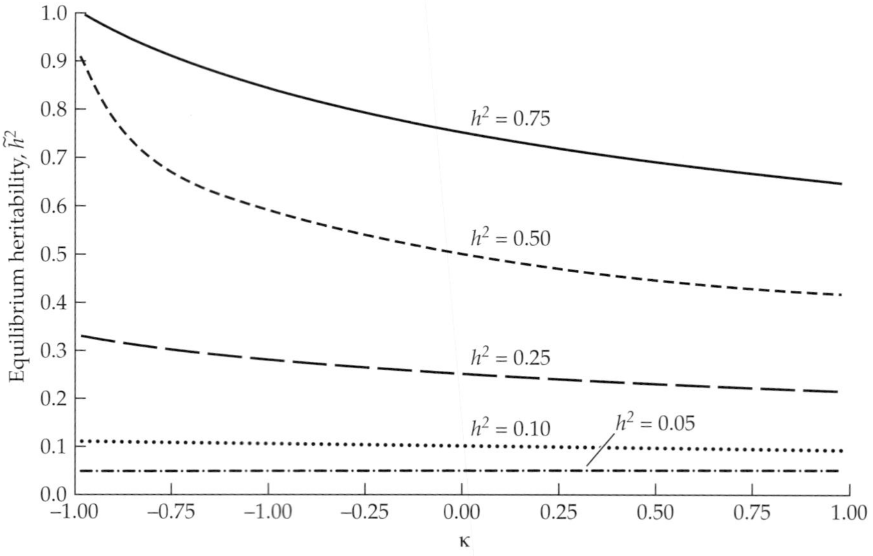

Figure 16.4 Equilibrium h^2 values as a function of κ and the initial heritability, h^2. Note that for $\kappa < 0$, the variance is *increased* by selection ($\sigma^2_{z^*} > \sigma^2_z$, as occurs with disruptive selection) and the equilibrium h^2 is greater than its initial value.

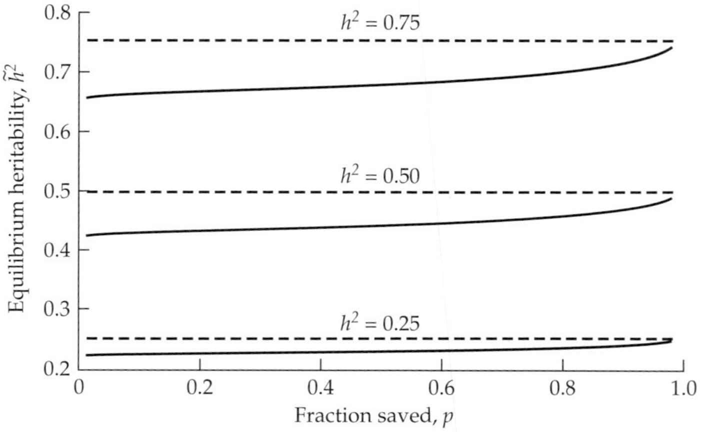

Figure 16.5 Equilibrium heritability values under directional (truncation) selection as a function of the fraction, p, saved and the initial heritability, h^2. The three curves correspond to initial heritability values of 0.75, 0.5, and 0.25, with the dashed lines displaying the constant heritability values and the solid line displaying the value at equilibrium.

Changes in Correlated Characters

Suppose the joint distribution of phenotypic values for our trait under selection, z, and two other phenotypically correlated traits, x and y, is multivariate normal. If the within-generation change in the phenotypic values of z is given by Equation 16.10a, then classical results (Pearson 1903) for the multivariate distribution imply that the variance in x following selection on (only) z is calculated by

$$\sigma^2(x^*) = (1 - \kappa\,\rho^2_{x,z})\sigma^2(x) \tag{16.14a}$$

implying that

$$\delta\left[\sigma^2(x)\right] = -\kappa\,\rho^2_{x,z}\,\sigma^2(x) \tag{16.14b}$$

where $\rho_{x,z}$ is the phenotypic correlation between traits x and z. If selection reduces the variance in z ($0 < \kappa < 1$), then the variance in any correlated character is also reduced, *independent* of the sign of the correlation (as change is a function of ρ^2). Likewise, the covariance between x and y following selection on z is calculated by

$$\sigma(x^*, y^*) = \sigma(x, y) - \kappa \, \frac{\sigma(x, z)\, \sigma(y, z)}{\sigma_z^2} \tag{16.14c}$$

yielding a within-generation change in this covariance of

$$\delta\,[\,\sigma(x, y)\,] = -\kappa \, \frac{\sigma(x, z)\sigma(y, z)}{\sigma_z^2} \tag{16.14d}$$

These results will prove useful in Volume 3 when we consider the Bulmer effect for multivariate traits, such as selection on an index or using BLUP.

Directional Truncation Selection: Theory

A fractional change, $\sigma^2_{z^*} = (1-\kappa)\,\sigma_z^2$, in the phenotypic variance occurs under various forms of truncation selection. Directional truncation selection results in a reduction in the phenotypic variance following selection ($\kappa > 0$), generating negative values of d and a corresponding reduction in both the additive variance and heritability. When the trait is normally distributed, recalling LW Equation 2.15 yields

$$\sigma^2_{z^*} = \left[\,1 - \bar{\imath}\,\left(\bar{\imath} - x_{[1-p]}\right)\,\right]\,\sigma_z^2 \tag{16.15a}$$

and hence, as given in Table 16.1,

$$\kappa = \bar{\imath}\,\left(\bar{\imath} - x_{[1-p]}\right) \tag{16.15b}$$

where $\bar{\imath}$ is the selection intensity (Equation 14.3).

The stronger the selection (i.e., the smaller the value of p and hence the larger the value of $\bar{\imath}$), the larger is the disequilibrium generated and the greater is the reduction in additive variance (Figure 16.5). Because the response to selection depends on the additive genetic variance in the selected *parents*, the response to selection in the first generation is unaffected (assuming the parents from the base population are in gametic-phase equilibrium). However, over the next two or three generations, essentially all of the reduction in h^2 due to buildup of negative d occurs, after which heritability remains essentially constant (see Example 16.3). Equations 16.13a through 16.13d provide the equilibrium (or **asymptotic**) variances and heritabilities. The ratio of the asymptotic to initial (assuming $d = 0$) rates of response is given by

$$\frac{\widetilde{R}}{R(0)} = \frac{\bar{\imath}\,\widetilde{h}\,\widetilde{\sigma}_A}{\bar{\imath}h(0)\,\sigma_A(0)} = \sqrt{\frac{\widetilde{h}^2}{h^2(0)\,[\,1 + \kappa\,\widetilde{h}^2\,]}} \tag{16.16}$$

as obtained by Gomez-Raya and Burnside (1990). As shown in Figure 16.5, the reduction in heritability is greatest when selection is strong (i.e., when the fraction saved, p, is small) and heritability is high.

Example 16.3. Suppose directional truncation selection is performed (equally on both sexes) on a normally distributed character with $\sigma_z^2 = 100$, $h^2 = 0.5$, and $p = 0.20$ (the upper 20% of the population is saved). To examine the impact of LD on the selection response, we first need to compute the fraction, κ, of phenotypic variance removed by selection. To apply Equation 16.15b, we need to compute both $x_{[0.8]}$ and $\bar{\imath}$. To do so, note that for a unit normal random variable, U, that $\Pr(U \leq 0.842) = 0.8$, yielding $x_{[0.8]} = 0.842$, and from Equation 14.3a

$$\bar{\imath} = \varphi(x_{[0.8]})/0.2 = \varphi(0.842)/p = 0.280/0.20 = 1.400$$

From Equation 16.15b, the fraction of variance removed by selection is

$$\kappa = \bar{\imath}\left(\bar{\imath} - x_{[0.8]}\right) = 1.400\,(1.400 - 0.842) = 0.781$$

With κ in hand, Equation 16.12d becomes

$$d(t+1) = \frac{d(t)}{2} - 0.391\,\frac{[\,50 + d(t)\,]^2}{100 + d(t)}$$

Starting selection in generation 0 on a base population in gametic-phase equilibrium (and hence $d(0) = 0$), iteration of this expression yields

Generation	0	1	2	3	4	5	∞
$d(t)$	0.00	−9.78	−11.90	−12.39	−12.51	−12.54	−12.54
$\sigma^2_A(t)$	50.00	40.22	38.10	37.61	37.49	37.46	37.46
$h^2(t)$	0.50	0.45	0.43	0.43	0.43	0.43	0.43

For example, in generation 2

$$h^2(2) = \frac{\sigma^2_A(2)}{\sigma^2_z(2)} = \frac{\sigma^2_A(0) + d(2)}{\sigma^2_z(0) + d(2)} = \frac{50 - 11.90}{100 - 11.90} = 0.43$$

Note that essentially all of the decline in additive variance occurs in the first three generations.

An important point to note is that the within-generation reduction in the variance is close to 80% ($\kappa = 0.781$), but the resulting decrease in the phenotypic variance (at equilibrium) is 100 −12.54 = 87.46, only a 13% decrease. As was found for the response in the mean using the breeder's equation, only a fraction of the within-generation change in the variance is transmitted between generations. As with the change in mean (Chapter 13), this arises because only a fraction of the variation is due to additive-genetic variance. Another contributing factor is the additional variation generated by Mendelian sampling in each generation, which partly mitigates the decrease in variance from selection.

We can also obtain the equilibrium additive variance directly from Equation 16.13a. Here $\widetilde{\sigma}^2_A = \sigma^2_z\,\gamma = 100\,\gamma$, with

$$\gamma = \frac{2 \cdot 0.5 - 1 + \sqrt{1 + 4 \cdot 0.5 \cdot (1 - 0.5) \cdot 0.781}}{2(1 + 0.781)} = 0.3746$$

and hence $\widetilde{\sigma}^2_A = 37.46$, as found by iteration. Likewise, Equation 16.13c returns the equilibrium heritability as

$$\widetilde{h}^2 = \frac{\gamma}{1 + \gamma - h^2} = \frac{0.3746}{1 + 0.3746 - 0.5} = 0.43$$

Again, this matches the value found by iteration.

How does this reduction in σ^2_A influence the per-generation change in mean, $R(t)$? Because $\bar{\imath}$ is unchanged (being entirely a function of the fraction, p, of adults saved), but h^2 and σ^2_z change over time, substituting $\sigma_A(t) = h(t)\sigma_z(t)$ into Equation 13.6b yields the single-generation response in generation t as

$$R(t) = h^2(t)\,\bar{\imath}\,\sigma_z(t) = 1.40\,h^2(t)\,\sqrt{\sigma^2_z + d(t)} = 1.40\,h^2(t)\,\sqrt{100 + d(t)}$$

Thus, the selection response declines from an initial value of $R = 1.4 \cdot 0.5 \cdot 10 = 7$ to an asymptotic per-generation value of $\widetilde{R} = 1.4 \cdot 0.43 \cdot \sqrt{87.46} = 5.6$. Using the unadjusted breeder's equation to predict change in mean over several generations without accounting for the Bulmer effect would have overestimated the expected response by 25%.

Table 16.2 Heritability and additive genetic variance in an experimental population undergoing directional selection on abdominal bristle number in *Drosophila melanogaster*. The base population is denoted by B. At the third generation of selection (H3), and following four generations of selection plus three generations of no selection (C7, in generation 7), h^2 was estimated from the response to divergent selection (Chapter 18) and σ_A^2 was subsequently estimated by $\widehat{h}^2\sigma_z^2$. The standard error for $\widehat{h}^2$ in all cases was 0.04. (After Sorensen and Hill 1982.)

	$\widehat{h}^2(t)$			$\widehat{\sigma}_A^2(t)$		
	B	H3	C7	B	H3	C7
Replicate 1	0.42	0.45	0.59	3.63	5.83	7.66
Replicate 2	0.38	0.26	0.26	2.96	2.28	2.08

Directional Truncation Selection: Experimental Results

How well do these predictions, which make a number of assumptions (additivity, infinitesimal model, normality), hold up for directional selection? Somewhat surprisingly, not many experiments have directly examined these issues. One reason is that the predicted change in h^2 under directional selection is usually expected to be small (Figure 16.5) and hence laborious to detect (requiring very large sample sizes, even when h^2 is large and p is small). One indirect study is that of Atkins and Thompson (1986), who subjected Blackface sheep to selection for increased bone length. Following 18 years of selection, the realized heritability (the ratio of observed response to selection differential; see Equation 18.10) was estimated to be 0.52 ± 0.02. Using the infinitesimal model, they predicted the expected base population heritability to be 0.57, in agreement with the estimated base population heritability of 0.56 ± 0.04. Further, the infinitesimal model predicts a 10% decrease in phenotypic variance, and the authors observed a 9% decrease in the upwardly selected line and an 11% decrease in the downwardly selected line.

A more direct study is that of Sorensen and Hill (1982), who subjected two replicate lines of *Drosophila melanogaster* to directional truncation selection on abdominal bristle number for four generations and then relaxed selection (Table 16.2). They interpreted their data as being consistent with the presence of a major allele (or alleles) at low frequency in the base population. These alleles are lost by sampling accidents in some lines (e.g., replicate 2, which shows no net increase in additive variance). If not lost, they are expected to increase rapidly in frequency due to selection, thus increasing additive variance (replicate 1), with this increase being partly masked by the generation of negative disequilibrium with other loci. Once selection stops, disequilibrium breaks down, resulting in a further increase in additive variance (compare the additive variance in lines H3 and C7 in replicate 1). Hence, even when major alleles are present, generation of gametic-phase disequilibrium reduces the rate of selection response.

Effects of Epistasis: Does the Griffing Effect Overpower the Bulmer Effect?

As discussed in Chapter 15, Griffing (1960a, 1960b) showed that when additive epistasis is present, gametic-phase disequilibrium increases the response to directional selection, with the change in mean augmented by $S\sigma_{AA}^2/2\sigma_z^2$. This (transient) increase in the rate of response has been termed the Griffing effect. Thus, in the presence of additive epistasis, disequilibrium is, on one hand, expected to increase the rate of response, while on the other hand it is also expected to decrease the rate of response by decreasing the expressed additive genetic variance (the Bulmer effect). Which change is more important?

Based on a small simulation study, Mueller and James (1983) concluded that if epistatic variance is small relative to additive variance and the proportion of pairs of loci showing epistasis is also small, the Bulmer effect dominates the Griffing effect, and disequilibrium reduces the response to selection. More generally, as Chapter 15 stresses, the Griffing effect only transiently inflates the response. It has no effect on the permanent component of response, while the Bulmer effect does. Specifically, while the change in variance under the

Bulmer effect and change in the mean from additive-by-additive genetic variance under the Griffing effect both decay to zero under random mating once selection stops, the change in the mean from h^2S is permanent. By lowering the additive variance during selection, the Bulmer effect results in a reduced permanent response. Hence, under the infinitesimal model, final response is lowered by the Bulmer effect and not influenced by the Griffing effect.

Double-Truncation Selection: Theory

Table 16.1 and Figure 16.3 show that $\kappa > 0$ under stabilizing double-truncation selection, so that selection reduces the within-generation phenotypic variance and generates negative disequilibrium. Conversely, $\kappa < 0$ for disruptive selection, with selection increasing the within-generation variance and generating positive disequilibrium. Hence, when the assumptions of the infinitesimal model hold, heritability is expected to decrease under stabilizing selection and increase under disruptive selection (Figure 16.4), although all of this response in the variance is transient. Upon the cessation of selection, the additive genetic variance decays back to its base-population value.

Consideration of Equation 16.13a shows that under stabilizing selection ($\kappa > 0$), the value $\gamma = \widetilde{\sigma}_A^2/\sigma_z^2$ (which measures the fraction of the initial phenotypic variance that is additive genetic variance at equilibrium) satisfies $0 < \gamma < h^2$. Similarly, under disruptive selection, $\gamma > h^2$, with one twist. If disruptive selection is sufficiently strong, $\kappa < -[4h^2(1-h^2)]^{-1}$, there is no positive real root for γ, and the variance increases without limit in the infinitesimal model (Bulmer 1976a). This is a consequence of the infinite number of loci in the infinitesimal limit. What happens with a finite number of loci is suggested from simulation studies of Bulmer (1976a), who examined the behavior when disruptive selection generated sufficiently negative κ values to ensure that there is no positive real root of Equation 16.13a. Bulmer assumed 12 identical additive diallelic loci (alternative alleles contributing 1 and 0, respectively, to the genotypic value). After a few generations, this population showed essentially complete disequilibrium, with most individuals having values of 0, 12, and 24 (with frequencies of $1/4$, $1/2$, $1/4$). At equilibrium, the population behaved as though there were a single locus segregating two alleles (contributing 0 and 12), each with a frequency of $1/2$. Thus, the expectation when there is no positive real solution for $\widetilde{\sigma}_A^2$ is that the population approaches a state of essentially complete disequilibrium while under selection.

The approach to the equilibrium value, $\widetilde{d}$, also behaves differently under disruptive selection. Under directional and stabilizing selection, the majority of reduction in the additive variance occurs in the first few generations. However, the increase in the variance toward its equilibrium value under disruptive selection requires many more generations, as Example 16.4 illustrates.

Example 16.4. Consider a normally distributed character with $\sigma_z^2 = 100$ and $h^2 = 0.5$ in a random-mating population. To compare the effects of stabilizing and disruptive selection, suppose that in one replicate, disruptive selection is practiced via double truncation with $p = 0.5$ (the upper and lower 25% of the population are saved), while stabilizing selection with $p = 0.5$ (the upper and lower 25% are culled) is practiced in the other. To obtain κ (from Table 16.1), first note that for stabilizing selection, $x_{[1/2+p/2]} = x_{[0.75]} = 0.674$, while for disruptive selection, $x_{[1-p/2]} = x_{[0.75]}$, with $\varphi(0.674) = 0.318$. Using these values, Equations 16.11b and 16.11c yield

$$\kappa = \pm(2 \cdot 0.318 \cdot 0.674)/0.5 = \pm 0.857$$

where the plus sign is used for stabilizing selection, and the minus sign is used for disruptive selection. Equation 16.12d becomes

$$d(t+1) = \frac{d(t)}{2} \mp 0.429\,\frac{[\,50 + d(t)\,]^2}{100 + d(t)}$$

where, because the equation is in terms of $-\kappa$, the minus sign in this expression corresponds to stabilizing selection and the plus sign to disruptive selection. Starting selection on a base population with $d = 0$ and iterating the above equation yields

Disruptive selection

Generation	0	1	2	3	4	5	20	∞
$d(t)$	0.00	10.73	19.65	27.22	33.72	39.34	74.83	82.69
$\sigma^2_A(t)$	50.00	60.37	69.65	77.22	83.72	89.34	124.83	132.69
$h^2(t)$	0.50	0.55	0.58	0.61	0.63	0.64	0.71	0.73

Stabilizing selection

Generation	0	1	2	3	4	5	∞
$d(t)$	0.00	−10.73	−12.77	−13.20	−13.29	−13.31	−13.32
$\sigma^2_A(t)$	50.00	39.27	37.23	36.80	36.71	36.69	36.68
$h^2(t)$	0.50	0.44	0.43	0.42	0.42	0.42	0.42

Note that with disruptive selection, the absolute change in d is much greater and the time to approach equilibrium considerably longer than with stabilizing selection.

Now suppose that after the equilibrium value of d has been reached, both of these two populations are then subjected to directional selection. In particular, assume directional truncation selection, with the upper 20 percent of the population being saved. The recursion equation for the disequilibrium contribution is given in Example 16.3, but now the initial disequilibrium values are $d(0) = -13.32$ for the population with a previous history of stabilizing selection and $d(0) = 82.69$ for the population with a previous history of disruptive selection. From Example 16.3, the per-generation response is $R(t) = h^2(t)\, 1.4\, \sigma_z(t)$, which is plotted in Figure 16.6 for both populations. The resulting changes in d and the heritability under these two treatments are also plotted in Figure 16.6. If stabilizing or disruptive selection are stopped for k generations before truncation selection, $d(0)$ is replaced by $d(0)/2^k$.

Initially there is a large difference in d (and hence in h^2 and in the response) between the replicates, but after three generations, both have essentially the same value, converging to $\widetilde{d} = -12.54$ and hence a per-generation response of $\widetilde{R} = 5.6$, the equilibrium value under directional truncation selection (see Example 16.3). As plotted in Figure 16.6, the replicate that was originally subjected to disruptive selection shows a higher cumulative response, due to larger responses resulting from higher heritabilities in the first few generations. This difference in cumulative response is permanent—it does not decay away once selection stops.

Hence, we expect that if an unselected base population is divided into three replicates, one first subjected to disruptive selection, another subjected to stabilizing selection, and the third to no selection, and if directional selection is then applied, the largest response should occur in the disruptively selected replicate and the smallest in the replicate that underwent stabilizing selection. This pattern is indeed seen in artificial selection experiments on *Drosophila* sternopleural bristles (Thoday 1959) and wing length (Tantawy and Tayel 1970).

Double Truncation Selection: Experimental Results

Experiments examining the effects of selection on the variance were reviewed by Prout (1962a), Thoday (1972), Soliman (1982), and Hohenboken (1985). One complication with many of these results is that only phenotypic variances are examined, making it problematic to distinguish between changes in genetic and environmental contributions (Chapter 17).

Nonetheless, as expected under the infinitesimal model, several experiments using stabilizing artificial selection (typically by double truncation) have revealed a reduction in the phenotypic variance that is largely due to reduction in the additive variance. Examples include sternopleural bristle number (Gibson and Bradley 1974), developmental time (Prout 1962a), wing venation (Scharloo 1964; Scharloo et al. 1967), and wing length (Tantawy

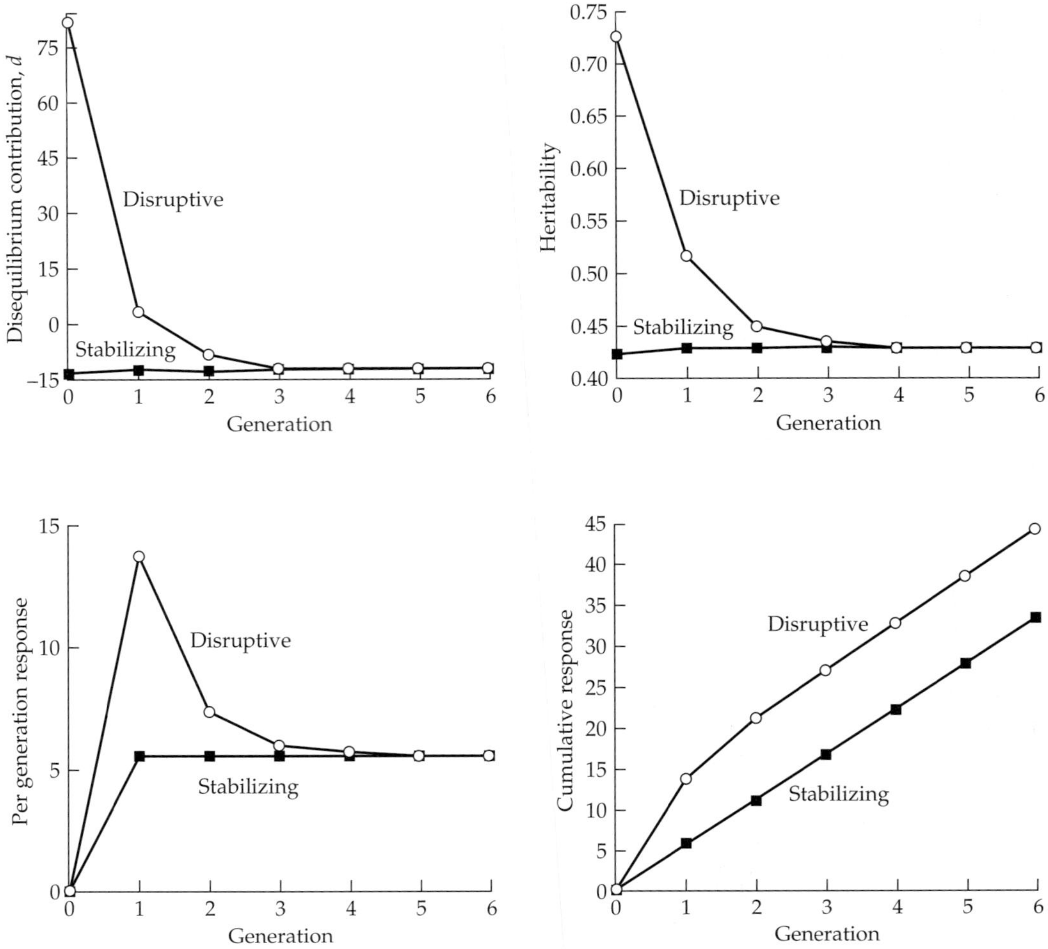

Figure 16.6 The response to the same amount of direction selection for two different starting populations, one initially subjected to stabilizing selection and the other to disruptive selection. See Example 16.4 for details.

and Tayel 1970) in *Drosophila melangaster*, and developmental time in *Tribolium castaneum* (Soliman 1982). Gibson and Bradley (1974) found that some decrease in the phenotypic variance of bristle number was also due to a decrease in the environmental variance.

Other experiments obtained different results. For example, selection on sternopleural bristle number done by Gibson and Thoday (1963) resulted in no change in the phenotypic variance because the decrease in additive variance was apparently countered by an increase in the environmental variance (strictly speaking, the increase was in the residual variance, which could include nonadditive genetic variance as well as environmental effects). Likewise, 95 generations of stabilizing selection on pupal weight in *T. castaneum* done by Kaufman et al. (1977) resulted in a decrease in the additive variance, but only a slight decrease in the heritability, reflecting a corresponding decrease in the residual variance as well. Bos and Scharloo (1973a, 1973b) observed no decrease in the phenotypic variance following stabilizing selection on *Drosophila* body size. Grant and Mettler (1969) observed a significant increase in variance in one replicate and a significant decrease in the other for two lines subjected to stabilizing selection for a *Drosophila* behavioral trait (escape behavior). Falconer (1957) reported finding no decrease in additive variance when abdominal bristle number in *Drosophila melanogaster* was subjected to stabilizing selection. However, a reanalysis by Bulmer (1976a) suggested that a reduction in variance had indeed occurred, close to the value predicted from the infinitesimal model.

The conclusion from this collection of studies is that while reductions in the environmental variance itself sometimes occur, the reduction in the additive variance is often the main source for the observed decrease in phenotypic variance. We will return to the expected response in the environmental variation in Chapter 17.

In contrast, disruptive selection experiments generally show rather large increases in the phenotypic variance. Increases in the heritability and/or additive variance in response to disruptive selection were observed in *Drosophila* for sternopleural bristle number (Thoday 1959; Millicient and Thoday 1961; Barker and Cummins 1969) and wing venation traits (Scharloo 1964; Scharloo et al. 1967), and for pupal weight in *Tribolium* (Halliburton and Gall 1981). Increases in the residual variance were also seen in many of these studies, reflecting changes in either the environmental or nonadditive genetic variances, or both. On the other hand, for *Drosophila* development time, Prout (1962a) observed that the heritability actually decreased relative to the base population, indicating that the large increase observed in phenotypic variance was due to changes in the residual variance. Robertson (1970c) observed an increase in the phenotypic variance following disruptive selection on *Drosophila melangaster* sternopleural bristles, but no significant corresponding increase in heritability.

While a change in variance is one prediction from the infinitesimal model, a more striking prediction is the behavior of the variance upon relaxation of selection, as any gametic-phase disequilibrium generated by selection quickly decays (for unlinked loci). Thus, more solid support for the predictions of the infinitesimal can come from experiments that also follow the variance upon relaxation of selection. This was done by Sorensen (1980) and Sorensen and Hill (1982), who disruptively selected on abdominal bristle number in *Drosophila melanogaster*. They observed large changes in the phenotypic variance, with realized heritability increasing from 0.37 to 0.69 in two generations of selection. Following four generations of no selection, heritability decreased to 0.44 (the standard error for all heritability estimates was 0.04). This pattern is consistent with the expected decay in response due to the decay of gametic-phase disequilibrium (which here is expected to be positive, thus inflating σ^2_A).

RESPONSE UNDER NORMALIZING SELECTION

While double truncation is the common mode of artificial stabilizing selection, one standard model for approximating stabilizing selection in natural populations is **normalizing** (or **nor-optimal**) selection (Weldon 1895; Haldane 1954),

$$W(z) = \exp\left(-\frac{(z-\theta)^2}{2\omega^2}\right) \tag{16.17}$$

Here, the optimal value is $z = \theta$, and the strength of selection is given by the width, ω^2, of the fitness function, which corresponds to the variance term in a normal distribution. When ω^2 is large (corresponding to a large variance in the fitness function), selection is weak, as the fitness function falls off slowly around the optimal value, θ. Conversely, a small value of ω^2 corresponds to a small variance in the fitness function and strong selection, with fitness quickly declining away from θ. Formally, it is useful to compare ω^2 to the phenotypic variance, with $\omega^2 \gg \sigma^2_z$ corresponding to weak selection and $\omega^2 \ll \sigma^2_z$ corresponding to strong selection.

If phenotypes are normally distributed before selection, with a mean of μ and variance of σ^2_z, then after selection, phenotypes remain normally distributed, with a new mean and variance

$$\mu^* = \mu + \frac{\sigma^2_z}{\sigma^2_z+\omega^2}(\theta-\mu) \quad \text{and} \quad \sigma^2_{z^*} = \sigma^2_z\left(1 - \frac{\sigma^2_z}{\sigma^2_z+\omega^2}\right) \tag{16.18a}$$

Writing the change in variance as $\sigma^2_{z^*} = (1-\kappa)\,\sigma^2_z$, Equation 16.18a shows that $\kappa = \sigma^2_z/(\sigma^2_z + \omega^2)$ is no longer a constant, but rather a function of the changing variance, σ^2_z. Thus, previous results assuming a constant, κ, no longer apply. However, under normalizing selection, the distribution of genotypes remains normal after selection, and hence parent-offspring regressions remain linear throughout. Thus, we can apply the breeder's equation to predict changes in the mean and Equation 16.7a to predict changes in the variance (under the infinitesimal model). Here,

$$S = \frac{\sigma_z^2}{\sigma_z^2 + \omega^2}(\theta - \mu) \quad \text{and} \quad \delta\left(\sigma_{z^*}^2\right) = -\frac{\sigma_z^4}{\sigma_z^2 + \omega^2} \tag{16.18b}$$

Note that both directional and stabilizing selection can occur simultaneously with normalizing selection, as when $\mu \neq \theta$, the mean changes under selection. Substituting Equation 16.18b into the breeder's equation yields

$$R(t) = h^2(t)\, S(t) = h^2(t)\, \frac{\sigma_z^2(t)\, [\,\theta - \mu(t)\,]}{\sigma_z^2(t) + \omega^2} \tag{16.19}$$

This shows that the mean converges to θ, as the sign of R is given by the sign of $\theta - \mu(t)$, which is positive when $\mu(t)$ is below θ and negative when above it. However, Equation 16.18a shows that the change in variance is independent of the current mean value, μ. From Equation 16.7a, the change in the disequilibrium contribution is given by

$$d(t+1) = \frac{d(t)}{2} - \frac{h^4(t)}{2}\, \frac{\sigma_z^4(t)}{\sigma_z^2(t) + \omega^2} = \frac{d(t)}{2} - \frac{1}{2}\, \frac{[\,\sigma_a^2 + d(t)\,]^2}{\sigma_z^2 + d(t) + \omega^2} \tag{16.20a}$$

implying that the equilibrium value, $\widetilde{d}$, satisfies

$$\widetilde{d} = \frac{\widetilde{d}}{2} - \frac{1}{2}\, \frac{[\,\sigma_a^2 + \widetilde{d}\,]^2}{\sigma_z^2 + \widetilde{d} + \omega^2} \tag{16.20b}$$

This rearranges to the quadratic equation

$$2\widetilde{d}^{\,2} + \left(\sigma_z^2 + \omega^2 + 2\sigma_a^2\right)\widetilde{d} + \sigma_a^4 = 0 \tag{16.20c}$$

which has one admissible solution (again, the constraint being that additive genetic variance must be nonnegative, hence $\widetilde{d} + \sigma_a^2 \geq 0$),

$$\widetilde{d} = \frac{-b + \sqrt{b^2 - 8\sigma_a^4}}{4} \quad \text{with} \quad b = \sigma_z^2 + \omega^2 + 2\sigma_a^2 \tag{16.20d}$$

Example 16.5. Suppose that normalizing selection occurs on a normally distributed character with $\sigma_z^2 = 100$, $h^2 = 0.5$, and $\omega^2 = 200$. From Equation 18.20a, with $\sigma_a^2 = 0.5 \cdot 100 = 50$, the dynamics of the disequilibrium contribution are given by

$$d(t+1) = \frac{d(t)}{2} - \frac{1}{2}\, \frac{[50 + d(t)]^2}{300 + d(t)}$$

Equation 16.20d gives $\widetilde{d} = -6.46$, and hence $\widetilde{h}^2 = (50 - 6.46)/(100 - 6.46) = 0.47$. Most of the reduction in heritability occurs in the first few generations, as iteration of Equation 16.20a yields

Generation	0	1	2	3	4	5	∞
d_t	0	-4.17	-5.64	-6.16	-6.35	-6.42	-6.46
$\sigma_A^2(t)$	50.00	45.83	44.37	43.84	43.65	43.58	43.54
$h^2(t)$	0.50	0.48	0.47	0.47	0.47	0.47	0.47

Thus, under the infinitesimal model, the distribution reaches an equilibrium, with the phenotypes (before selection) normally distributed with a mean of θ and variance of $\widetilde{\sigma}_z^2 = 100 - 6.46 = 93.54$.

SELECTION WITH ASSORTATIVE MATING

Assortative mating changes the additive genetic variance relative to expectations in a randomly mating population, mainly by generating gametic-phase disequilibrium (LW Chapter 7). Assortative mating also results in some inbreeding (measured by a slight increase in homozygosity), but if the number of loci is large, the deviation of genotypes from Hardy-Weinberg frequencies is expected to be small. In the limiting infinitesimal model, no changes in genotypic frequencies occur at single loci, although large changes in variance can occur due to gametic-phase disequilibrium. Positive assortative mating (in which the phenotypic correlation, ρ, between mates is positive) generates positive d, thus increasing σ_A^2, while negative assortative mating ($\rho < 0$, also called **disassortative mating**) generates negative d, decreasing σ_A^2. As with selection, these changes in the variance are temporary and dissipate after a few generations of random mating (for unlinked loci).

Results Using the Infinitesimal Model

Assortative mating is easily incorporated into the infinitesimal model (Fisher 1918; Bulmer 1980). Previously, we assumed that selected parents are randomly mated to form the next generation, independent of their phenotype. However, under assortative mating, the selected parents no longer randomly mate, as mating is based on the phenotypes. We assume that assortative mating follows selection, and so the selected parental phenotypic values, z_f^* and z_m^*, are correlated. Returning to Equation 16.5a, the offspring variance is given by

$$\sigma^2(z_o) = \frac{h^4}{4}\sigma^2\left(z_m^* + z_f^*\right) + \sigma_e^2 \tag{16.21a}$$

If we write the variance of a sum as $\sigma^2(x+y) = \sigma_x^2 + \sigma_y^2 + 2\rho_{xy}\sigma_x\sigma_y$, this becomes

$$\sigma^2(z_o) = \frac{h^4}{4}\left(\sigma^2(z_m^*) + \sigma^2(z_f^*) + 2\rho\,\sigma(z_f^*)\,\sigma(z_m^*)\right) + \sigma_e^2 \tag{16.21b}$$

Assuming selection is such that $\sigma^2(z_x^*) = (1-\kappa_x)\sigma_z^2$ for $x = f$ or m, Equation 16.21b becomes

$$\sigma^2(z_o) = \frac{h^4}{2}\,\sigma_z^2\left(1 - \frac{\kappa_f + \kappa_m}{2} + \rho\sqrt{(1-\kappa_f)\,(1-\kappa_m)}\right) + \sigma_e^2 \tag{16.21c}$$

If we compare this expression with Equation 16.12a, we see that under assortative mating, Equation 16.12d holds, with

$$\kappa = \frac{\kappa_f + \kappa_m}{2} - \rho\sqrt{(1-\kappa_f)\,(1-\kappa_m)} \tag{16.21d}$$

Likewise, Equations 16.13a through 16.13d hold with κ now given by Equation 16.21d. This generalization is from Tallis (1987; Tallis and Leppard 1988a), and was extended to multiple traits by Tallis and Leppard (1988b).

If there is no selection ($\kappa_f = \kappa_m = 0$), $\kappa = -\rho$ and previous results for assortative mating (LW Equations 7.18 through 7.20) follow immediately from Equations 16.12, 16.7c, and 16.13, respectively. More generally, when the amount of selection and assortative mating change in each generation,

$$d(t+1) = \frac{d(t)}{2} - \frac{\kappa(t)}{2}\,h^2(t)\,\sigma_A^2(t) \tag{16.22}$$

where $\kappa(t)$ is given by Equation 16.21d, with κ_f, κ_m, and ρ taking on values for generation t.

Under the infinitesimal model, analyzing the joint effects of assortative mating and selection is straightforward. When selection is the same in both sexes, the effect of assortative mating is to change a value of κ in the absence of assortative mating to a new value of $\kappa - \rho(1-\kappa)$. Negative gametic-phase disequilibrium is generated when this quantity is positive (indicating a reduction in variance), while positive disequilibrium is generated when it is negative. Note that if $\kappa > 0.5$, then $\kappa - \rho(1-\kappa) > 0$ and no amount of positive assortative mating can generate positive disequilibrium. However, for all values of κ, there is some amount of negative assortative mating such that $\kappa - \rho(1-\kappa) > 0$. Even if selection generates positive disequilibrium ($\kappa < 0$, such as with disruptive selection), sufficiently strong negative assortative mating ($\rho < \kappa/[\,1-\kappa\,]$) generates negative disequilibrium, thus reducing the additive genetic variance.

Assortative Mating and Enhanced Response

Given that positive assortative mating increases the additive genetic variance, Breese (1956) and James and McBride (1958) suggested that the response to selection could be increased by employing assortative mating among the selected parents. However, experimental support for such an increase is mixed. Studies in both *Drosophila melanogaster* (McBride and Robertson 1963) and *Tribolium castaneum* (Wilson et al. 1965; Campo and Garcia Gil 1993, 1994) exhibited slight (but not statistically significant) increases in response when parents were assortatively mated. Conversely, Sutherland et al. (1968) and Garcia and Sanchez (1992) found no effect of assortative mating when selecting on body weight in mice and pupal weight in *Drosophila*.

One reason for an apparent absence of an impact from assortative mating is simply a lack of power due to a small expected effect (Example 16.6). Biological reasons may further obscure any such difference. Wright (1921c) first noticed that apparently random mating in small populations can still stochastically generate correlations between mates, creating what he termed **unconscious assortative mating**. Simulation studies showed that rather large population sizes ($N_e > 400$) are required to avoid unconscious assortative mating (Jorjani et al. 1997a, 1997b, 1997c), and most selection experiments employ much smaller effective population sizes (Chapter 26). Jorjani (1995) suggested that unconscious assortative mating in the presumed random-mating controls (diluting any expected differences), when coupled with low power, may account for this lack of experimental consistency with the theory.

The effect of coupling assortative mating with truncation selection was examined in detail by Baker (1973), DeLange (1974), Fernando and Gianola (1986), Smith and Hammond (1987), and Tallis and Leppard (1988a). Shepherd and Kinghorn (1994) found a much larger effective gain with assortative mating when selection (and mating) is based on estimated breeding values (BLUP selection) as opposed to simple individual selection (i.e., individual phenotypes). The general conclusion is that the relative increase in selection response is greatest when h^2 is large and selection is weak. However, unless the population is subjected to multiple generations of assortative mating before selection, the increase (for individual selection under the infinitesimal model) is at most 6%, consistent with the very small increases seen in experiments. When the number of loci is small, assortative mating can have a larger effect, due to faster allele-frequency change, as opposed to generation of positive disequilibrium (Fernando and Gianola 1986).

Example 16.6. Starting with a base population that is initially in gametic-phase equilibrium with $h^2 = 0.5$ and $\sigma_z^2 = 100$, individuals are positively assortatively mated (with $\rho = 0.5$) for k generations before being subjected to directional truncation selection, with $p = 0.20$. What is the difference in response to one generation of selection in the assortatively mated population relative to a random-mating base population?

To answer this question, we first need to compute the expected disequilibrium contribution following k generations of assortative mating in the population before selection. Here, $\kappa_f = \kappa_m = 0$ and Equation 16.21d yields $\kappa = -\rho = -0.5$, which is used to iterate Equation 16.22,

$$d(t+1) = \frac{d(t)}{2} + 0.25\, h^2(t)\sigma_A^2(t) \tag{16.23a}$$

where $d(0) = 0$, $h^2(0) = 0.5$, and $\sigma_A^2(0) = 50$. As shown in the table below, assortative mating, by itself, generates $d > 0$, reaching an upper limit (for these genetic and mating parameters) of $\widetilde{d} = 20.71$. From Example 16.3, the single-generation response using a population with $d(k)$ is given by

$$R(k) = 1.4\, h^2(k)\, \sigma_z(k) = 1.4 \left(\frac{\sigma_A^2(0) + d(k)}{\sigma_z^2(0) + d(k)} \right) \sqrt{\sigma_z^2(0) + d(k)}$$

$$= 1.4 \left(\frac{50 + d(k)}{100 + d(k)} \right) \sqrt{100 + d(k)} \tag{16.23b}$$

Iterating Equation 16.23a for k generations and substituting the resulting $d(k)$ value into Equation 16.23b gives the following values for a single generation of response following k generations of assortative mating:

k	0	1	2	3	5	10	∞
$d(t)$	0.00	6.25	10.57	13.58	17.17	20.09	20.71
$h^2(t)$	0.50	0.53	0.55	0.56	0.57	0.58	0.59
$R(t)$	7.00	7.64	8.06	8.35	8.69	8.95	9.01
$R(t)/R_{rm}$	1.00	1.09	1.15	1.19	1.24	1.28	1.29

R_{rm} = 7.00 is the first-generation response under random mating and $d(0)$ = 0. With k = 3 generations of assortative mating in the base population, at the start of selection $d(0)$ = 13.58 and h^2 = 0.56, giving a response of 8.35.

While there can be up to a 29% increase in the rate of response to a single generation of selection when starting from an assortatively mated base population, it is extremely inefficient to assortatively mate a population for several generations before applying selection. For example, with a single generation of assortative mating followed by a single generation of selection, the total response is 7.64. If instead one just selected both generations (from a randomly mated base population), the responses are 7.00 in the first generation and (from Example 16.3) 5.93 in the second, for a total response almost twice as large (12.93/7.64). Further, once selection starts, the initially positive d from assortative mating (which inflates σ_A^2, and hence h^2 and the selection responses) decays very rapidly (Figure 16.6). Only half the previous value of d is passed onto the next generation, where it is supplemented by negative d values generated by selection (Equation 16.7b), which quickly drives d to negative values and greatly reduces R/R_{rm}. Assuming $d(0)$ = 20.71, after a single generation of selection, $d(1)$ = −3.55 and h^2 = 0.48, and the selection response in the next generation drops to 6.62 (compared to a response of 5.93 under random mating of the selected parents). In the second generation of selection, $d(2)$ = −9.25, with a response of 5.98.

Example 16.7. Consider the same population and selection parameters as in the previous example, but with selection and assortative mating now occurring simultaneously, starting at generation 0. What is the difference in response if random mating occurs in one replicate and assortative mating with $\rho = 0.5$ occurs in the other? From Example 16.3, directional truncation selection with $p = 0.2$ on both sexes yields $\kappa_f = \kappa_m = 0.781$, with the variance reduction from Equation 16.21d becoming

$$\kappa = \kappa_f - \rho(1 - \kappa_f) = 0.781 - 0.5(1 - 0.781) = 0.672$$

and (from Equation 16.22), the change in d in the assortatively mated population becomes

$$d(t+1) = \frac{d(t)}{2} - \frac{0.672}{2}\, h^2(t)\, \sigma_A^2(t)$$

Assuming the initial population is in gametic-phase equilibrium $d(0) = 0$, iteration yields

Generation	0	1	2	3	4	5	∞
$d(t)$	0.00	−8.40	−10.55	−11.12	−11.27	−11.31	−11.33
$h^2(t)$	0.50	0.45	0.44	0.44	0.44	0.44	0.44
$R(t)$	7.00	6.09	5.84	5.77	5.76	5.75	5.75
$R_{rm}(t)$	7.00	5.93	5.68	5.63	5.61	5.61	5.61
$R(t)/R_{rm}(t)$	1.00	1.03	1.03	1.02	1.02	1.02	1.02

where $R_{rm}(t) = \bar{\imath}\, h^2_{rm}(t) \sqrt{\sigma_z^2 + d_{rm}(t)}$ is the selection response in generation t of random mating, and the values of $h^2_{rm}(t)$ and $d_{rm}(t)$ come from Example 16.3.

Note that the response in the first generation (generation 0) is the same in both populations: the response to selection depends on the additive variance of the parents, and in the first generation, both populations have the same variance (as $d(0) = 0$ in both). Thereafter, there is at most a 3% increase in the rate of response under assortative mating. Even with perfect positive assortative mating ($\rho = 1$), $\kappa - \rho(1-\kappa) = 2\kappa - 1 = 0.564$ gives a maximum value of $R(t)/R_{rm}(t) \simeq 1.05$.

Disruptive Selection, Assortative Mating, and Reproductive Isolation

We would be remiss if we did not mention the historical interest in the connection between disruptive selection and assortative mating as a mechanism for reproductive isolation. In the early 1960s, the general view was that speciation (reproductive isolation between populations) required geographic (or other) isolation, a view strongly championed by Mayr (1963). However, the idea that **sympatric speciation** (Maynard-Smith 1962a, 1966) could develop without the need for such isolation was bolstered by an experimental observation by Gibson and Thoday (1962). They observed that disruptive selection on sternopleutral bristle number in *D. melanogaster* seemed to generate two distinct groups (high vs. low flies), which appeared to assortatively mate (individuals with an intermediate phenotype were absent from the population, whereas they would be expected under random mating). Their explanation was that crosses between high and low parents generate less fit offspring (having intermediate values), and that selection generated preferential mating over the short time course of this experiment; i.e., it appeared that only 12 generations of disruptive selection had generated partial reproductive isolation.

However, this striking observation was not reproducible (Scharloo et al. 1967; Barker and Cumming 1969; Charbora 1968; Thoday and Gibson 1970). Indeed, Scharloo (1971) suggested that the base population for selection used by Thoday and Gibson might have been composed of flies from different geographic origins, and hence already possessing partial isolation that was uncovered, rather than evolved, by their experiment. While Thoday and Gibson's interpretation of their experiments is now largely discounted, the notion of **reinforcement** (the evolution of mating preferences to reduce the production of less fit hybrids when diverged populations come back into contact) remains a concept of interest (Noor 1999; Servedio and Noor 2003; Ortiz-Barrientos et al. 2009).

17

Short-term Changes in the Variance: 2. Changes in the Environmental Variance

It is the purpose of this short communication to suggest that recent views on the nature of the developmental process make it easier to understand how the genotypes of evolving organisms can respond to the environment in a more co-ordinated fashion. Waddington (1942)

In our discussion of the response to selection, one assumption thus far has been that the environmental variation is **homoscedastic**—constant across genotypes—and hence not subject to modification by selection. However, a fairly universal, and very striking, observation is that most traits show at least some genetic variation in an outbred population. One can imagine that sensitivity to the environment, as measured by the environmental variance, is such a trait (Waddington 1957; Hill 2007), and thus can potentially respond to selection. If true, selection for (or against) extreme individuals, such as directional and disruptive selection for the former, and stabilizing selection for the latter, may also result in selection for increased (or decreased) values of the environmental variance, σ_E^2. There are also settings that favor *direct* selection on σ_E^2, such as breeding for more uniformity in an agricultural or laboratory trait (Hohenboken 1985). There can also be fitness consequences for uniformity in domesticated populations. For example, preweaning survival increases as the within-litter variance (a function of σ_E^2) in weight decreases in both pigs (Milligan et al. 2002) and rabbits (Garreau et al. 2008). Selection on σ_E^2 likely occurs in natural populations as well, such as selection on the within-plant variation in flowering time (Devaux and Lande 2009). Finally, Gibson (2009) and Feinberg and Irizarry (2010) argued that selection on the inherent stochasticity of developmental systems may play an important role in our understanding of human diseases. All of these considerations have spurred an interest in selection response in σ_E^2 (reviewed by Hill and Mulder 2010).

One technical comment before proceeding is that simple scale effects can also result in a change in the variance—if the coefficient of variation of a trait remains constant as its mean changes, then its variance must also change as well. As discussed in LW Chapter 11, a suitable transformation (such as the log of the trait value) often removes these scale effects, and we assume this has been done previous to any analysis.

BACKGROUND: HERITABLE VARIATION IN σ_E^2

Scales of Environmental Sensitivity

The environment an organism experiences can be partitioned into many different scales of resolution, but operationally we are usually concerned with just two: those features shared by all individuals in some common setting (**macroenvironments**), and those features unique to each individual (**microenvironments**). Differential sensitivity (i.e., performance) of genotypes over any of these scales indicates genotype × environmental (G × E) interactions (LW Chapter 22). Volume 3 examines the selection response in the presence of G × E over macro-scale differences (such as different growing regions for a crop or different host plants for an insect), by treating the trait value in each macroenvironment as a correlated character (Falconer 1952). A related topic concerns **norms of reactions**, which are performance curves over a gradient for a particular environmental feature (such as temperature). The analysis of response for such **function-valued traits** is also deferred until our final volume.

Our focus here is on sensitivity to microenvironmental variation, which itself can oc-

Evolution and Selection of Quantitative Traits. Bruce Walsh & Michael Lynch.
Published 2018 by Oxford University Press. © Bruce Walsh & Michael Lynch 2018.
DOI 10.1093/oso/9780198830870.001.0001

cur over several different scales. The most fundamental consideration is **developmental noise**, which can sometimes be measured by differences in the trait values of homologous structures within an individual, such as the amount of **fluctuating asymmetry** (differences in trait values on the left and right side of bilaterally symmetric organisms; LW Chapter 11; see also Leamy and Klingenberg 2005; Dongen 2006; Hansen et al. 2006; Graham et al. 2010). Presumably such within-individual variation reflects "noise" in the developmental process—variation in the end product of the same genotype in the same macroenvironment. A related measure of microenvironmental variation is the variance in the repeated performance (records) of an individual over time, such as offspring size in different litters from a single mother. While any source of environmental variation contributes to σ^2_E, different pathways may be involved in environmental sensitivity at these different scales (Pélabon et al. 2010).

Environmental vs. Genetic Canalization

The idea that genotypes may vary in their microenvironmental sensitivity has a rich history, dating back to Waddington's (1942) notion of **canalization** and Schmalhausen's (1949) **autoregulation**—developmental buffering against small perturbations (be they environmental or genetic). Under these views, a wide range of genotypes and environments yield essentially the same developmental end product. Waddington (1957, 1959) also stressed that canalization is an *evolved system*, and hence to some extent a selectable trait. Part of Waddington's concern was sensitivity to the environment, with genotypes showing **environmental canalization** (or **environmental robustness**) having lower environmental variances. However, he was also concerned with the fact that a particular genotype may find itself in a variety of different *genetic* environments, and that genotypes may also differ in their sensitivities to these different backgrounds. **Genetic canalization** (or **genetic robustness**) is the stability of a particular genotype when placed in a variety of different genetic backgrounds, and it is a function of the epistasis between a genotype of interest and the universe of genetic backgrounds in which it may find itself. These two measures of sensitivity can be easily confounded, yet they are fundamentally different. Environmental robustness (which requires G x E if genotypes vary in environmental sensitivity) does not necessarily imply genetic robustness (which requires epistasis), and vice versa. As reviewed in Flatt (2005) and Hansen (2006), the conditions for the evolution of genetic canalization (an overall reduction in the sensitivity of a genotype to its genetic background) are more restrictive than for environmental canalization, in part because the target background under the former is itself continually evolving.

Using an appropriate design, the genetic and environmental sensitivities for a particular genotype can be separated. Under a repeated-measures design, the genetic background remains constant and the residual variance is due entirely to environmental sensitivity (plus measurement error). A second (but obviously more restrictive) design is the use of a series of inbred lines or clones (Fraser and Schadt 2010; Geiler-Samerotte et al. 2013). For a particular genotype of interest (such as a marker locus tagging a quantitative trait location [QTL]), the *among-line* trait variance of a genotype across a series of lines (and hence different genetic backgrounds) is a measure of its genetic sensitivity, while the *within-line* variance of the target genotype is a measure of its environmental sensitivity.

Example 17.1. Fraser and Schadt (2010) considered mRNA expression levels for thousands of genes over a series of 19 mouse inbred lines. Within each line, roughly 20 individuals were scored at ~160,000 markers. For a given trait (the expression level of a specified gene), the within-line variation was contrasted between the two alternative homozygous genotypes at each marker. Because there is essentially no genetic variation within an inbred line, significant differences in the within-line variance over marker genotypes indicate linkage to a QTL influencing environmental robustness (differences in σ^2_E). Conversely, a significant difference in the dispersion of the mean values of the marker genotypes *across* inbred lines (increased among-line variance) indicates that the marker is linked to a QTL influencing genetic robustness. Using this approach, these authors found QTLs for both types of robustness. QTLs for environmental robustness were largely *trans*-acting and sex-specific (different QTLs in the two sexes). In contrast, QTLs for genetic robustness were often *cis*-acting and were not sex-specific. There was no detected overlap between the two classes of QTLs. In reporting their results, the authors used the convention that an eQTL mapping close to its target site was regarded as *cis*, while an eQTL mapping further away, or on a different chromosome, was regarded as *trans* (Chapter 12).

One caveat about this study concerns the among-line variance. Given the small number (19) of genotypic backgrounds (inbred lines), we expect that by chance some markers will be nonrandomly distributed with respect to QTLs that influence the line means. This could result in alternative genotypes at these markers showing different patterns of among-line variance, not because of any epistatic interactions, but rather because they were not sufficiently randomized with respect to background QTLs influencing the mean of the target trait.

Evidence for Heritable Variation in the Environmental Variance

The suggestion that different genotypes may have different environmental variances is not new. For example, Robertson and Reeve (1952) and Lerner (1954) noted that inbred lines often have larger environmental variances than their outbred counterparts (see Whitlock and Fowler [1999] for a more recent example). This led Lerner to propose that **genetic homeostasis** (developmental buffering across environments) is facilitated by heterozygosity, while environment sensitivity (σ^2_E) increases with homozygosity. Consistent with this suggestion is the observation that developmental noise (measured by the amount of fluctuating asymmetry) often decreases with increasing levels of protein (i.e., isozyme) heterozygosity (reviewed in Mitton and Grant 1984; Livshits and Kobyliansky 1985; Chakraborty 1987; Zouros and Foltz 1987; Britten 1996; Vøllestad et al. 1999).

Direct evidence for genetic variation in σ^2_E is provided by comparing inbred lines. Mackay and Lyman (2005) observed different amounts of environmental variation for bristle number across chromosomal substitution lines of *Drosophila* from a common source population. Similar findings using inbred lines were seen by Ordas et al. (2008) in maize, by Hall et al. (2007) in *Arabidopsis*, by Ansel et al. (2008) in yeast, and by a number of workers using the *Drosophila melanogaster* Genetic Reference Panel (DGRP) lines (Harbison et al. 2013; Ayroles et al. 2015; Huang et al. 2015; Morgante et al. 2015; Sørensen et al. 2015). Characters showing among-line variation in σ^2_E included morphological, physiological, and behavioral traits, as well as expression-level traits (i.e., mRNA levels). The DGRP lines show an exception to the inverse relationship between σ^2_E and levels of heterozygosity. These lines contain varying amounts of residual heterozygosity (which is substantial in several cases due to segregating chromosomal inversions), with the amount of residual heterozygosity being uncorrelated with σ^2_E.

While these studies provide direct evidence for *genetic* variation in σ^2_E, our concern is with *heritable* variation—additive genetic variation in the trait that can respond to selection. Direct evidence for heritable variation for the level of developmental noise derives from traits that usually respond to selection for either increased or decreased fluctuating asymmetry (LW Chapter 11). However, this is only one potential component of the microenvironmental variance, so what evidence is there for a heritable component of σ^2_E in general?

Indirect support comes from observations of heritable variation in the within-family variance in livestock traits. Van Vleck (1968) and Clay et al. (1979) observed significant sire differences in the variation in milk yield in dairy cattle across half-sib families, while Rowe et al. (2006) found significant sire variation in the within-family residual variance for 35-day body weight in broiler chickens. While variation among sires is consistent with a heritable component for within-family variances, it can also arise simply from genetic segregation. In particular, heteroscadisticity of family variances is a classic (but weak) test for the presence of a major gene, with parents heterozygous for the major allele having a larger within-family variance than homozygous parents (LW Chapter 13).

A more recent line of evidence for heritable variation in σ_E^2 comes from a significantly improved fit of statistical models assuming a heritable component of the residual variance (and hence a correlation in σ_E^2 among relatives) over those that assume no such heritable variation. Such an improved fit was seen for fecundity in sheep (SanCristobal-Gaudy et al. 2001), body weight in the snail *Helix aspersa* (Ros et al. 2004), and litter size in pigs (Sorensen and Waagepetersen 2003), with additional examples listed in Table 17.2. The caveat about these results is the concern that violations of the underlying statistical models may lead to an incorrect suggestion that such genetic variation exists when in fact it is absent. Indeed, E. Yang et al. (2011) showed that these analyses are strongly biased by the presence of intrinsic skew in the data, as the presence of heritable variation in σ_E^2 is also manifested as skew (Ros et al. 2004). Yang et al. simultaneously fitted a model along with a general Box-Cox transformation (LW Chapter 11) of their data to remove any intrinsic skewness. After accounting for skew, evidence for genetic variance in σ_E^2 was reduced in some cases, while it was strengthened in others (Table 17.2). The bottom line from the analysis of these models is that there does appear to be real evidence for heritable variation in σ_E^2, but estimating some of the genetic parameters associated with this variance (in particular, the correlation between breeding values for trait means and residual variances) can be very delicate.

Another line of evidence derives from the mapping of major genes involved in trait variances. The classic example involves the heat shock protein HSP90, which has been shown to buffer both genetic and environmental effects (reviewed in Sangster et al. 2008). In a second potential example, J. Yang et al. (2012) showed that different genotypes at the *FTO* (fat mass and obesity-associated protein) locus display different residual variances for body mass in humans. However, because *FTO* effects were scored in segregating populations, it is unclear whether its variance effect is due to genetic or environmental canalization.

Finally, QTLs can be associated with trait variances. Denoted as **vQTLs** (variance QTLs) by Rönnegård and Valdar (2011) and **veQTLS** (variance in expression-level traits) by Huang et al. (2015), such QTLs denote sites where the trait variance differs over marker genotypes. While early QTL mapping projects noted that some marker genotypes differ in trait variances (e.g., Edwards et al. 1987), the formal development of specific methods to map such QTLs is more recent (Ordas et al. 2008; Paré et al. 2010; Struchlain et al. 2010; Visscher and Posthuma 2010; Jimenez-Gomes et al. 2011; Rönnegård and Valdar 2011, 2012; Hothorn et al. 2012; Shen et al. 2012). While these studies have found a number of candidate regions, as with among-sire differences in family variances, they reflect differences in the *residual* (as opposed to strictly the *environmental*) variance for marker genotypes, which can arise from differences in sensitivity to genetic background when alternative vQTL genotypes are assessed in segregating populations. Indeed, Paré et al. (2010) and Deng and Paré (2011) suggested using variance heterogeneity across markers as a preliminary scan for potentially epistatic loci. Nonetheless, a number of recent studies have mapped vQTLs using variation in σ_E^2 over inbred lines, directly showing at least some genetic control on σ_E^2 (Hall et al. 2007; Ansel et al. 2008; Harbison et al. 2013; Huang et al. 2015; Sørensen et al. 2015).

Collectively, these observations suggest that heritable variation in the environmental variation likely exists for many traits, and that selection for changes in phenotypic variance can result in a response that in part derives from changes in the overall environmental variance of the population. Consistent with this view, recall that changes in σ_E^2 were seen in several of the stabilizing- and disruptive-selection experiments reviewed in Chapter 16.

MODELING GENETIC VARIATION IN σ_E^2

A variety of statistical models have been proposed to account for the heritable transmission of at least part of the environmental variance. The starting point for each model is that the phenotypic value of an individual of genotype i can be written as

$$z_i = \mu + G_i + E \quad \text{where} \quad E \sim (0, \sigma_i^2) \tag{17.1a}$$

The notation $x \sim (\mu, \sigma^2)$ denotes that x comes from a distribution with a mean of μ and variance of σ^2. For ease of development, we generally assume that the trait has an entirely additive-genetic basis, meaning that $G = A$, namely the standard breeding value for the trait. Normally, this breeding value is either unsubscripted or is denoted as A_z to connect it with z. However, for models with a heritable (i.e., a breeding-value) component to σ_E^2, we need to keep track of two *separate* breeding values, *both* of which influence z. One, A_m, influences the mean of z (so that $G = A_m$ in Equation 17.1a), while another breeding value, A_v, influences the environmental variance of z. As we will detail shortly, several different models have been proposed to connect A_v and σ_E^2 (reviewed in Table 17.1). These two breeding values, A_m and A_v, can be correlated, which further complicates the dynamics of selection response. Changes in the mean of A_m change the mean of z (the standard response to selection as measured by a change in the mean), while changes in the mean of A_v change the average value of σ_E^2 (yielding a response in the environmental variance).

Because we allow σ_E^2 to vary over genotypes, its population value is not the usual constant that was assumed in previous chapters, but rather an average value that may change over time. Taking the expectation (to avoid confusion, in this chapter we will use roman E for expectation and italic E for environmental values), the population environmental variance is the average of the σ_i^2,

$$\sigma_E^2 = \mathrm{E}[\,\sigma_i^2\,] \tag{17.1b}$$

When working with a series of pure lines, one can estimate σ_i^2 directly. The more interesting (and difficult) problem arises when considering an outbred population. In this case, we have to deal with both estimation and the vexing issue of modeling transmission. Models allowing for heterogeneity of environmental variance were introduced by breeders in the 1990s (e.g., Foulley et al. 1992; Foulley and Quaas 1995; Cullis et al. 1996), but these models ignored the question of selection (and evolution) of the environmental variance itself. As outlined below, the first formal analyses of the evolution of the environmental variance were population-genetic models presented by Gavrilets and Hastings (1994c) and Wagner et al. (1997), and breeding-value-based models presented by SanCristobal-Gaudy et al. (1998).

The Multiplicative Model

Gavrilets and Hastings (1994c) assumed some underlying environmental factor, e (such as temperature), with different genotypes having different sensitivity, γ_i, to this factor, hence

$$E = \gamma_i\, e \quad \text{where} \quad e \sim (0, \sigma_e^2) \tag{17.2a}$$

This **multiplicative model** is simply the joint-regression model for genotype×environment interactions (LW Equation 22.13b; Volume 3), and was also used by Wagner et al. (1997). Under Equation 17.2a, the conditional environmental variance (given the genotypic value and its environmental sensitivity) is

$$\sigma^2[E\,|\,G, \gamma_i] = \gamma_i^2 \sigma_e^2 \tag{17.2b}$$

As shown in Example 17.2 (below), taking the expected value over the population distribution of sensitivity values, $\gamma \sim (\mu_\gamma, \sigma_\gamma^2)$, yields the unconditional environmental variance in the population

$$\sigma_E^2 = \left(\mu_\gamma^2 + \sigma_\gamma^2\right) \sigma_e^2 \tag{17.2c}$$

Note that while γ_i can be negative, Equation 17.2b shows that it influences the environmental variance through its square, γ_i^2. Thus, the magnitude, rather than the sign, of γ_i determines its impact on σ_E^2. Under the multiplicative model, the environmental variance for the population decreases by selecting μ_γ to zero and/or by decreasing the variance, σ_γ^2, of environmental sensitivities. The problematic issue here is modeling the change in the distribution of the genotypic-specific sensitivities, γ. The simplest approach is to assume that the environmental sensitivity, γ, is an entirely additive quantitative trait, so that $\gamma = A_v$, namely, the breeding value for the environmental variance.

Analysis of Equation 17.2c led Gavrilets and Hasting to comment on the relationship between developmental noise and heterozygosity mentioned previously. Lerner (1954) assumed this relationship to be causative, with higher levels of heterozygosity directly causing decreased environmental variance. However, Gavrilets and Hastings noted that when $\mu_\gamma^2 = 0$, as might occur with selection to decrease σ_E^2, then under the simple additive model ($\gamma = A_v$), the environmental variance for a given trait is proportional to the additive genetic variance in environmental sensitivity, $\sigma_\gamma^2 = \sigma^2(A_v)$.

Gavrilets and Hastings noted that a correlation between heterozygosity and σ_E^2 simply falls out as a consequence of their model, rather than from any functional relationship between the two. They reach this conclusion by using the result that, for an additive trait (in our case, $\gamma = A_v$), the genetic variance is a *decreasing* function of the number of heterozygous loci (Chakraborty 1987), so that when the average heterozygosity in a population increases, $\sigma^2(A_v)$ decreases, and hence (from Equation 17.2c), so does σ_E^2. While their theoretical point is valid, it is actually addressing a slightly different issue than Lerner's argument. The Gavrilets and Hastings model suggests that *populations* with less heterozygosity are expected to show increased levels of σ_E^2, while Lerner was suggesting that *individuals* with reduced heterozygosity show increased σ_E^2.

If we allow for dominance in the quantitative-trait formulation of γ, the result is $\gamma = A_v + D_v$, where the dominance value, D_v, is *not* transmitted from parent to offspring. Further, by construction, D_v has a mean value of zero and under the infinitesimal model, the dominance variance is not changed by selection (Chapter 16). Under this extension, the mean environmental variance becomes

$$\sigma_E^2 = \left(\mu_{A_v}^2 + \sigma_{A_v}^2\right) \sigma_e^2 + \sigma_{D_v}^2 \, \sigma_e^2 \tag{17.2d}$$

This same argument applies if we replace $\sigma_{D_v}^2$ by the total nonadditive genetic variance. While selection can reduce the first component in Equation 17.2d (either by driving the mean breeding value to zero and/or reducing $\sigma_{A_v}^2$), the component involving nonadditive variance remains unchanged. Hence, implicit in assuming a breeding value for this model (or any of the others discussed below) is that any nontransmissible genetic variation in σ_E^2 remains unchanged by selection. Genetic variation in σ_E^2, by itself, is *not sufficient* for a selection response, as the latter requires that *at least part of this variation must be transmissible* under the breeding scheme being used.

The Exponential Model

While we have presented the multiplicative model within a breeding-value framework, this was not explicitly done by Gavrilets and Hastings (1994c), who (coming from a population-genetics background) were more concerned with the evolution of the environmental variance than with estimating A_v. Conversely, SanCristobal-Gaudy et al. (1998), coming from an animal breeding background, were more concerned with the estimation of A_v. They did so by modeling E using an **exponential model**

$$E = \exp\left(\frac{A_v}{2}\right) \cdot e \quad \text{where} \quad e \sim N(0, \sigma_e^2) \quad \text{and} \quad A_v \sim N(\mu_{A_v}, \sigma_{A_v}^2) \tag{17.3a}$$

Why the breeding value appears as $A_v/2$ (rather than simply A_v) will become apparent shortly.

The connection with the multiplicative model follows if we note that for small $|x|$, $e^x \simeq 1 + x$, so that $E \simeq [1 + (A_v/2)] \cdot e$ for $|A_v| \ll 2$, implying $\gamma_i \simeq 1 + (A_v/2)$. By assuming normality and independence (of e, A_v, and A_m), SanCristobal-Gaudy et al. (1998) obtained likelihood estimators for the breeding values for the environmental variance (A_v) and trait

Table 17.1 Models for heritable variation in the environmental value, E, involve two separate breeding values, A_m and A_v, underlying the phenotype, z, of a focal trait. The former breeding value is associated with the mean of the focal trait and the latter influences the environmental variance. The table reviews three different models for translating A_v into a value of σ^2_E, assuming some intrinsic environmental value, $e \sim N(0, \sigma^2_e)$. All three models start with the usual decomposition of $z = \mu + A_m + E$, where $A_m \sim N(\mu_{A_m}, \sigma^2_{A_m})$ is the breeding value for z (the subscript m is a mnemonic for mean). The departure from this standard model is that rather than assuming σ^2_E to be constant over all genotypes, we assume it has a heritable component (a breeding value, A_v) that influences the emergent value of σ^2_E. We assume $A_v \sim N(\mu_{A_v}, \sigma^2_{A_v})$, and let U denote a unit normal random variable. See the text for full details on each model.

Model	E	$\sigma^2(E \mid A_v)$	$\sigma^2(E) = \mathrm{E}\left[\sigma^2(E \mid A_v)\right]$
Multiplicative	$A_v \cdot e$	$A_v^2 \, \sigma^2_e$	$\left(\mu^2_{A_v} + \sigma^2_{A_v}\right) \sigma^2_e$
Exponential (or log-additive)	$\exp(A_v/2) \cdot e$	$\exp(A_v)\, \sigma^2_e$	$\exp\left(\mu_{A_v} + \sigma^2_{A_v}/2\right) \sigma^2_e$
Additive	$\sqrt{A_v + \sigma^2_e} \cdot U$	$A_v + \sigma^2_e$	$\mu_{A_v} + \sigma^2_e$

mean (A_m). They explicitly considered estimation under either a sire design (using half-sib values to estimate the values of A_v and A_m in the common parent) or under a model where repeated measurements on a single individual and its relatives are used to estimate breeding values for σ^2_E (Chapters 13 and 19). SanCristobal-Gaudy et al. (2001) extended this approach to threshold traits (in particular, litter size). Bayesian estimators under this model were developed by Sorensen and Waagepetersen (2003) and Ros et al. (2004).

Given A_v, the conditional distribution of the environmental variance becomes

$$\sigma^2(E \mid A_v) = \sigma^2\left[\exp(A_v/2) \cdot e \mid A_v\right] = \left[\exp(A_v/2)\right]^2 \sigma^2_e = \sigma^2_e \exp(A_v) \tag{17.3b}$$

where the last step follows by recalling that $[\exp(a)]^2 = \exp(2a)$. Hence, the environmental variance is a constant (σ^2_e) multiplied by a scaling factor that is an exponential function of the breeding value A_v for the environmental variance (which motivates our use of $A_v/2$ in Equation 17.3a). Decreasing A_v results in an individual with reduced environmental sensitivity (reduced σ^2_E). The constant, σ^2_e, can be interpreted as the environmental variance for an individual with an environmental breeding value of zero, $A_v = 0$.

The exponential model is also called the **log-additive model**, as the breeding value is additive on a log scale

$$\ln\left[\sigma^2(E \mid A_v)\right] = \ln\left(\sigma^2_e\right) + A_v \tag{17.3c}$$

As detailed in Example 17.2, the expectation of Equation 17.3b (over the population distribution of A_v values) yields a mean environmental variance of

$$\sigma^2_E = \sigma^2_e \exp\left(\mu_{A_v} + \sigma^2_{A_v}/2\right) \tag{17.3d}$$

Equation 17.3d shows that either decreasing the mean breeding value, μ_{A_v}, or its additive variance, $\sigma^2_{A_v}$, decreases the environmental variance. Comparison of Equations 17.2c and 17.3d shows one subtle difference between the multiplicative and exponential models. Under the former, the contribution to the environmental variance is a function of $\mu^2_{A_v}$, meaning that the minimal population environmental variance occurs when $\mu_{A_v} = 0$, and any deviation away from zero increases the average environmental variance. By contrast, under the exponential model, decreasing μ_{A_v} *always* decreases the average value of σ^2_E in the population. Thus, under the exponential model, σ^2_E can be selected to be arbitrarily small, while under the multiplicative model, it has a lower bound set by $\sigma^2_{A_v}$ (and more generally, by the dominance variance as well; see Equation 17.2d).

The Additive Model

Our last formulation for modeling genetic variation in E was suggested by Hill and Zhang (2004) and Mulder et al. (2007), where

$$E = U \cdot \sqrt{\sigma_e^2 + A_v} \quad \text{where} \quad U \sim N(0,1) \quad \text{and} \quad A_v \sim N(\mu_{A_v}, \sigma^2_{A_v}) \tag{17.4a}$$

This is the **additive model**, as the environmental variance for an individual with breeding value A_v is simply

$$\sigma^2(E \,|\, A_v) = \sigma_e^2 + A_v \tag{17.4b}$$

with the constraint on the breeding value being that $\sigma_e^2 + A_v > 0$. The additive model is a *local* analysis around the current mean, as selection to decrease A_v can eventually result in this constraint being violated, which generates a negative variance. Under the additive model, the mean population value for the environmental variance is simply

$$\sigma_E^2 = \text{E}\left(\sigma_e^2 + A_v\right) = \sigma_e^2 + \mu_{A_v} \tag{17.4c}$$

Unlike in the multiplicative and exponential models, changes in σ_E^2 under the additive model depend only on changes in the mean breeding value, and not on its variance (Table 17.1).

The additive model has the advantage of being much more tractable, but it has the disadvantage that it breaks down when the breeding value becomes sufficiently negative ($A_v < -\sigma_e^2$). In contrast, the exponential model has additivity on the log of the variance scale, which is a nice statistical feature, as log variances are approximately normally distributed (Box 1953; Layard 1973). Mulder et al. (2007) discussed additional connections between the additive and exponential models, while Hill and Mulder (2010) reviewed different estimation methods under these models.

Example 17.2. Here we derive the unconditional variances for the models summarized in Table 17.1. Consider the multiplicative model first, where

$$\sigma_E^2 = \text{E}[\,\gamma^2\sigma_e^2\,] = \sigma_e^2\,\text{E}[\,\gamma^2\,]$$

Recalling that $\text{E}[\,x^2\,] = \mu_x^2 + \sigma_x^2$ yields

$$\sigma_E^2 = \sigma_e^2\,\text{E}[\,\gamma^2\,] = \sigma_e^2\left(\sigma_\gamma^2 + \mu_\gamma^2\right)$$

Now consider the exponential model (Equation 17.3a). By construction, both E and e have expected values equal to zero, and the variances of E and e are simply the expected values of E^2 and e^2. Taking expectations

$$\sigma_E^2 = \text{E}\left[\left(e \cdot \exp\{A_v/2\}\right)^2\right] = \sigma_e^2\,\text{E}\left[\left(\exp\{A_v/2\}\right)^2\right] = \sigma_e^2\,\text{E}\left[\exp\left(A_v\right)\right]$$

This follows by again recalling that $[\exp(x/2)]^2 = \exp(2x/2) = \exp(x)$. The rightmost expectation in this expression is calculated with respect to the distribution of breeding values, A_v, by recalling (Equation 14.19a) that for a normally distributed random variable, x, with a mean of μ and variance of σ^2, $\text{E}[e^x] = \exp(\mu + \sigma^2/2)$. Because we assumed that $A_v \sim N(\mu_{A_v}, \sigma^2_{A_v})$, the average environmental variance for the population becomes

$$\sigma_E^2 = \sigma_e^2\,\exp\left(\mu_{A_v} + \frac{\sigma^2_{A_v}}{2}\right)$$

The Heritability of the Environmental Variance, h_v^2

Estimates of $\sigma^2(A_v)$ under any of the models for σ_E^2 reviewed in Table 17.1 are obtained using fairly complicated likelihood functions on data from sets of relatives; see SanCristobal-Gaudy et al. (1998, 2001), Sorensen and Waagepetersen (2003), Ros et al. (2004), or any of the other references in Table 17.2 for details. Table 17.2 presents these estimates scaled as heritabilities and evolvabilities (Equation 13.22b) to facilitate comparison over traits, which raises the issue how the heritability of the environmental variance is defined.

Mulder et al. (2007) suggested that one definition is as the slope of the regression of the breeding value, A_v, of an individual on some appropriate function of the phenotype value, z. Under the additive-model framework (Equation 17.4a), they show that the appropriate transformation is the square, z^2, of phenotypic value. To see this, recall from Equation 17.4a that under this model

$$z = \mu + A_m + E = \mu + A_m + U\sqrt{A_v + \sigma_e^2}$$

If we assume that A_m and A_v are uncorrelated and recall for a unit normal random variable, U, that $E(U^2) = 1$ (as U^2 is a χ_1^2 random variable, which has a mean of 1; LW Equation A5.15b), then

$$\begin{aligned}\sigma(A_v, z^2) &= \sigma(A_v, [\mu + A_m + E]^2) = \sigma(A_v, E^2)\\ &= \sigma(A_v, U^2[A_v + \sigma_e^2]) = \sigma(A_v, A_v) = \sigma^2(A_v)\end{aligned} \tag{17.5a}$$

From regression theory (LW Chapter 3), the slope of the regression of A_v on z^2 is simply this covariance divided by the variance of the predictor variable

$$h_v^2 = \frac{\sigma(A_v, z^2)}{\sigma^2\left(z^2\right)} \tag{17.5b}$$

If z is normally distributed, then $\sigma^2(z^2) = 2\sigma_z^4 + 3\sigma^2(A_v)$ (see Mulder et al. 2007 for details), yielding a heritability of

$$h_v^2 = \frac{\sigma^2(A_v)}{2\sigma_z^4 + 3\sigma^2(A_v)} \tag{17.5c}$$

Hence, estimates of h_v^2 are obtained by substituting an estimate of $\sigma^2(A_v)$—which can be estimated within a likelihood framework (see the previously mentioned references)—along with the value of σ_z^4, into Equation 17.5c.

Table 17.2 reviews estimated h_v^2 values from a number of studies. Note from this table that values of h_v^2 are low (typically less that 0.05), while the evolvabilities, i.e., $\sigma(A_v)/\sigma_E^2$, the coefficient of variation of the environmental variance (Equation 13.22b), are large. Although the selection response may be slow (given the low heritability), there is much variation to exploit, as a high evolvability implies that significant proportional change in the trait value can be achieved (Chapter 13). Recall from Equation 13.22b that the expected response, scaled in terms of the mean value, can be expressed as

$$R(\sigma_E^2)/\sigma_E^2 = \bar{\imath} h_v CV_{A_v} = \bar{\imath}\sqrt{0.038} \cdot 0.41 = 0.08\,\bar{\imath}$$

where we have used the average values for h_v^2 and CV_{A_v} from Table 17.2. Recalling that $\bar{\imath} \simeq 2$ when we save the upper 5% of the population (Example 14.1), the expected scaled response per generation is 0.16 in this setting, implying that slightly more than six generations are required to double the mean value of the variance under this strength of selection.

The table also shows that some caution is in order when using these likelihood-based estimates, which can be very model-specific. In particular, the fragility of these estimates can be seen by comparing the estimated additive-genetic correlation, $\rho(A_m, A_v)$, in the litter-size studies (Yang et al. 2011). Data are often transformed before an analysis for any number of reasons (LW Chapter 11), and one of the more flexible approaches is the Box-Cox

Table 17.2 Estimates of the heritability, h_v^2, and evolvability, $CV_{A_v} = \sigma(A_v)/\sigma_E^2$, of the environmental variance (Equation 13.22b), as well as bivariate-model estimates of the additive-genetic correlation, ρ, between A_m and A_v. For the Yang et al. (2011) results, BC denotes that a Box-Cox transformation was fitted simultaneously with the model, while results without this notation indicate that this transformation was not used. (Based, in part, on Mulder et al. 2007 and Hill and Mulder 2010.)

Species	Trait	h_v^2	CV_{A_v}	ρ	Reference
Pig (*Sus*)	Meat pH	0.039	0.40	0.79	SanCristobal-Gaudy et al. (1998)
	Litter size	0.026	0.31	−0.62	Sorensen & Waagepetersen (2003)
		0.021	0.27	−0.64	Yang et al. (2011)
		0.012	0.19	0.70	Yang et al. (2011), BC
	Weight	0.011	0.34	−0.07	Ibáñez-Escriche et al. (2008c)
Sheep (*Ovis*)	Litter size	0.048	0.51	0.19	SanCristobal-Gaudy et al. (2001)
Snail (*Helix*)	Body weight	0.017	0.58	−0.81	Ros et al. (2004)
Chicken (*Gallus*)	Body weight (male)	0.029	0.30	−0.17	Rowe et al. (2006)
		0.046	0.49	−0.45	Mulder et al. (2009)
		0.030	0.32	−0.23	Wolc et al. (2009)
	Body weight (female)	0.031	0.32	−0.11	Rowe et al. (2006)
		0.047	0.57	−0.41	Mulder et al. (2009)
		0.038	0.37	−0.27	Wolc et al. (2009)
Rabbit (*Oryctolagus*)	Litter Size	0.045	0.42	−0.74	Ibáñez-Escriche et al. (2008b)
		0.041	0.37	−0.73	E. Yang et al. (2011)
		0.017	0.24	0.28	E. Yang et al. (2011), BC
	Birth weight	0.013	0.25	—	Garreau et al. (2008)
Mouse (*Mus*)	Litter size	0.048	0.44	−0.93	Gutierrez et al. (2006)
	Litter weight	0.039	0.37	−0.81	Gutierrez et al. (2006)
	Birth weight	0.208	1.21	0.97	Gutierrez et al. (2006)
	Body weight	0.006	0.36	−0.31	Ibáñez-Escriche et al. (2008a)
	Weight gain	0.018	0.47	−0.19	Ibáñez-Escriche et al. (2008a)
Average		0.038	0.41	−0.24	

transformation (LW Equation 11.4), which includes the standard log transform as a special case. For pigs, using untransformed data resulted in $\rho(A_m, A_v) = -0.64$, which changed to 0.70 when a Box-Cox transformation was first applied to the data. For rabbits, $\rho(A_m, A_v)$ changes from -0.73 to 0.28. Likewise, estimates of the heritabilities and evolvabilities were also lower when the likelihood model included a Box-Cox transformation.

Finally, our discussion thus far has focused on narrow-sense heritabilities. Broad-sense heritability estimates, H_v^2, for the genetic variance of σ_E^2, based on among-line variation in σ_E^2, are often an order of magnitude higher than the narrow-sense values shown in Table 17.2. For example, Morgante et al (2015) observed H_v^2 values for σ_E^2 of 0.75, 0.54, and 0.36 for three behavioral and physiological traits in *Drosophila*, which were of comparable magnitude with the broad-sense estimates for the traits themselves (values of 0.37, 0.56, and 0.58, respectively). This apparent difference between the broad- and narrow-sense estimates for the genetic variance of σ_E^2 may reflect something deep, such as a significant amount (indeed, the majority) of genetic variance for σ_E^2 being nonadditive (and hence the numerator of H_v^2, the total genetic variance, being much larger than the numerator of h_v^2, the additive genetic variance). Or it may simply reflect the fact that direct estimates for H_v^2 by comparing variances over inbred lines may be much more powerful than the more complex likelihood-based methods used in the estimation of variance components of σ_E^2 in outbred populations.

SELECTION ON σ_E^2

As with the breeder's equation for the response of the mean to selection, the response of σ_E^2 is a function of two features: the nature of transmission and the nature of selection. Our following discussion is thus partitioned into these two features. We begin with a discussion of transmission, examining how a change in the mean value of A_v (the response R_{A_v}) translates into a change in σ_E^2 in the next generation. As might be expected, the results are highly dependent on which of the models given in Table 17.1 is assumed.

We then examine the nature of selection on σ_E^2. This can occur via three different pathways. The first route is through direct selection on A_v generated by selection on the phenotypic value, z, of a trait. A second route is that selection can be based on direct expression of σ_E^2 in an individual through repeated measurements, selecting for individuals with a larger (or smaller) range in these records. The final route is as a correlated response (Equation 13.26c), with selection on z resulting in selection on the breeding value, A_m, for the trait, which in turn may be correlated with A_v. The machinery of multivariate selection is needed to consider the totality of response in this latter case, so we focus solely here on the direct response (i.e., assuming $\rho[A_m, A_v] = 0$), and defer a full discussion of this general case until Volume 3.

However, a few brief comments on the nature of this potential genetic correlation, $\rho(A_m, A_v)$, are still in order. If the coefficient of variation, σ_z/μ_z, remains roughly constant under selection, this implies that as the mean increases, so does the variance (and, thus, presumably, so does σ_E^2). When A_m and A_v are positively correlated, simple selection to increase the mean (to increase μ_{A_m}) results in a correlated response that also increases μ_{A_v}, and therefore σ_E^2. While most estimated $\rho(A_m, A_v)$ values are negative (Table 17.2), there are reasons these should be viewed with caution. Current statistical models for estimating A_v assume no intrinsic skew in the data other than that generated by any correlation between A_m and A_v. If skew is present for other reasons, this can significantly bias estimates of this correlation (Ros et al. 2004), as shown in the previous discussion on the results of Yang et al. (2011).

Translating the Response in A_v Into Response in σ_E^2

A number of authors have modeled the selection response of the phenotypic variance (from either direct or indirect selection on the variance) when there are heritable differences in environmental sensitivity (Gavrilets and Hastings 1994c; Wagner et al. 1997; SanCristobal-Gaudy et al. 1998, 2001; Sorensen and Waagepetersen 2003; Hill and Zhang 2004; Ros et al. 2004; Mulder et al. 2007, 2008). A critical step in modeling this selection response is treating the phenotypic value and residual variance as two separate (and potentially correlated) traits, both with heritable (i.e., additive-genetic) variation. While some models (Gavrilets and Hastings 1994c; Wagner et al. 1997; Hill and Zhang 2004) are based on strict population-genetic analysis (following the change in individual allele frequencies), most are based on schemes that assign breeding values to the heritable component of σ_E^2 (Table 17.1). Under the infinitesimal model, the expected breeding value in the offspring (for either A_m or A_v) is simply the mean breeding values of its parents (Chapter 13). Using the expressions given in Table 17.1 allows us to map changes in μ_{A_v}, the mean breeding value for environmental sensitivity, onto changes in σ_E^2.

The simplest case is the additive model (Equation 17.4b). Suppose we let the selection response, R_{A_v}, denote the change in the mean breeding value (for the environmental variance) of the selected parents from the mean breeding value of the entire population. The resulting change in σ_E^2 becomes

$$\begin{aligned}\Delta\sigma_E^2(t) &= \sigma_E^2(t+1) - \sigma_E^2(t)\\ &= \left[\,\mu_{A_v}(t+1) + \sigma_e^2\,\right] - \left[\,\mu_{A_v}(t) + \sigma_e^2\,\right]\\ &= \left[\mu_{A_v}(t) + R_{A_v}(t)\right] - \mu_{A_v}(t) = R_{A_v}(t)\end{aligned} \tag{17.6a}$$

The response is a bit more complex under the multiplicative and exponential models, as the mean population value, σ_E^2, for the environmental variance is a nonlinear function of both the mean and the variance of A_v (Table 17.1). For the multiplicative model

$$\sigma_E^2 = \left[\mu_{A_v}^2 + \sigma^2(A_v)\right] \sigma_e^2$$

Because the mean breeding value in generation $t+1$ can be expressed as the previous mean plus the response, $\mu_{A_v}(t+1) = \mu_{A_v}(t) + R_{A_v}(t)$, if we assume that there is no change in the additive variance of environmental sensitivities following selection ($\Delta\sigma^2(A_v) = 0$), we have

$$\begin{aligned}
\Delta\sigma_E^2(t) &= \sigma_E^2(t+1) - \sigma_E^2(t) \\
&= \left(\left[\mu_{A_v}(t) + R_{A_v}(t) \right]^2 + \sigma_{A_v}^2 \right) \sigma_e^2 - \left[\mu_{A_v}^2(t) + \sigma_{A_v}^2 \right] \sigma_e^2 \\
&= \left[\, 2\mu_{A_v}(t) R_{A_v}(t) + R_{A_v}^2(t) \, \right] \sigma_e^2
\end{aligned} \tag{17.6b}$$

Change in $\sigma^2(A_v)$ due to selection-generated linkage disequilibrium can be similarly accounted for by using Equation 16.8b.

Under the exponential model and again assuming that there is no change in $\sigma_{A_v}^2$, Equation 17.3d implies that

$$\begin{aligned}
\Delta\sigma_E^2(t) &= \sigma_E^2(t+1) - \sigma_E^2(t) \\
&= \sigma_e^2 \, \exp\left[\mu_{A_v}(t) + R_{A_v}(t) + \sigma_{A_v}^2/2 \right] - \sigma_e^2 \, \exp\left[\mu_{A_v}(t) + \sigma_{A_v}^2/2 \right] \\
&= \sigma_e^2 \, \exp\left[\mu_{A_v}(t) + \sigma_{A_v}^2/2 \right] \left[\exp\left(R_{A_v}\{t\} \right) - 1 \right] \\
&= \sigma_E^2(t) \cdot \left[\exp(R_{A_v}\{t\}) - 1 \right]
\end{aligned} \tag{17.6c}$$

These expressions translate a response, R_{A_v}, in the mean breeding value for the environmental variance into the expected change in σ_E^2 for the different transmission models reviewed in Table 17.1. We now consider two different settings by which such a response can occur: as a consequence of direct selection on phenotypic value, z, and as the result of direct selection on σ_E^2 itself. A third possibility—a correlated response in A_v (i.e., in the environmental variance of the trait) due to direct selection on A_m (i.e., the value z of the trait itself) when $\rho(A_v, A_m) \neq 0$—is examined in Volume 3.

Response From Stabilizing Selection on Phenotypic Value, z

We have previously suggested that selection either for, or against, extreme individuals may also result in some selection for genotypes with higher, or lower, environmental variances. We formalize this by considering how selection on a phenotypic value, z, maps onto selection on A_m and A_v. First consider a quadratic fitness model of stabilizing selection. Here, the expected fitness of an individual with a phenotypic value of z is

$$W(z) = 1 - s(z - \theta)^2 \tag{17.7a}$$

where θ is the optimal trait value and s is the strength of stabilizing selection. This is a weak selection model, as $W \geq 0$ only for sufficiently small s relative to the total variance of z. Note that if we set $s < 0$, Equation 17.7a becomes a model of (weak) disruptive selection. Gavrilets and Hastings (1994c) examined how this phenotypic fitness function translates into selection on (A_m, A_v) under the multiplicative model. To do so, we replace z by $A_m + A_v e$ and take the expectation over e. Noting that $\text{E}[e] = 0$, $\text{E}[e^2] = \sigma_e^2$, the expected fitness as a function of A_m and A_v becomes

$$\begin{aligned}
W(A_m, A_v) &= 1 - s \, \text{E}\left[\, (A_m + A_v \, e - \theta)^2 \mid A_m, A_v \right] \\
&= 1 - s \left[\, (A_m - \theta)^2 + 2(A_m - \theta) A_v \text{E}\left[\, e \, \right] + A_v^2 \, \text{E}\left(e^2 \right) \right] \\
&= 1 - s \left[\, (A_m - \theta)^2 + A_v^2 \, \sigma_e^2 \right]
\end{aligned} \tag{17.7b}$$

Similar fitnesses arise under the Gaussian model of weak stabilizing selection (Equation 16.17); see Hill and Mulder (2010), and Example 17.3 for a more exact analysis. Equation 17.7b shows that phenotypic stabilizing selection favors A_v values near zero, thus decreasing σ_E^2 (Hansen et al. 2006), which has two important consequences. First, the reduction in phenotypic variance can be significantly greater than predicted from the simple reduction in the additive variance from the Bulmer effect (Chapter 16). And second, there can be cases where the trait heritability, h_z^2, will *increase* under stabilizing selection. Because both additive and environmental variances are decreased, if the decrease in environmental variance is sufficiently greater, then h_z^2 increases. Results for quadratic disruptive selection follow by changing the sign on s, which results in selection to increase A_v.

Example 17.3. The quadratic fitness function (Equation 17.7a) is a model for weak stabilizing selection, as it can generate negative (and hence undefined) fitness values when selection is sufficiently strong. An alternative model of stabilizing selection without this constraint is normalizing selection (Equation 16.17), where θ denotes the optimal phenotypic value and ω^2 denotes the strength of selection around this optimum. Devaux and Lande (2009) used this fitness function in their study of selection on the flower-timing variance within an individual. They assumed that the additive model for genetic variation in σ_e^2 (Equation 17.4c) holds and that repeated expressions z of the trait from an individual with breeding values A_m (for the trait) and A_v (for σ_E^2) are drawn from a normal, so that

$$p(z \,|\, A_m, A_v) = \frac{1}{\sqrt{2\pi(\sigma_e^2 + A_v)}} \exp\left[-\frac{(z - A_m)^2}{2(\sigma_e^2 + A_v)} \right]$$

Integration of $W(A_m, A_v) = \int W(z)\, p(z \,|A_m, A_v)\, dz$ yields

$$W(A_m, A_v) = \sqrt{\frac{\omega^2}{\omega^2 + \sigma_e^2 + A_v}} \exp\left[-\frac{(A_m - \theta)^2}{2(\omega^2 + \sigma_e^2 + A_v)} \right] \tag{17.8a}$$

When $A_m \simeq \theta$, the exponential term is near one (as its numerator is near zero), and so fitness is largely driven by the square root term. For weak selection ($\omega^2 \gg \sigma_e^2 + A_v$) when $A_m = \theta$, a first-order Taylor series approximation yields

$$W(A_m, A_v) \simeq 1 - \frac{\sigma_e^2 + A_v}{2\,\omega^2} \tag{17.8b}$$

As was the case for weak quadratic selection, fitness increases as A_v decreases.

The more interesting case is that where the population is far from the equilibrium, so $|\, A_m - \theta \,| \gg 1$, and the numerator in the exponential term in Equation 17.8a is large. In this case, fitness can be improved by *increasing* the value of A_v (i.e., moving it closer to θ), which reduces the magnitude of the exponential term. Thus, as also noted by Lande (1980b) and Bull (1987), stabilizing selection can actually favor an *increase* in σ_E^2 when the population is far from its optimum, as the larger variance increases the chance that some phenotypes will be near θ. Svardal et al. (2011) similarly found that selection favors increased values of σ_E^2 when there are strong fluctuations in θ over time.

Response From Directional Selection on *z*

Now consider directional selection on the trait phenotype, z. We first assume that the multiplicative model holds and that we have a simple linear fitness function

$$W(z) = 1 + sz \tag{17.9a}$$

Taking the expectation over environmental values, Gavrilets and Hastings (1994c) found that

$$W(A_m, A_v) = 1 - s\text{E}(A_m + eA_v \mid A_m, A_v) = 1 - sA_m + A_v\text{E}(e) = 1 - sA_m \tag{17.9b}$$

Under this setting, there is no direct selection on A_v.

A rather different outcome was noted by both Hill and Zhang (2004) and Mulder et al. (2007) for truncation selection on a normally distributed trait. As a measure of fitness, Hill and Zhang considered the probability, $P(a, b)$, that a genotype with a mean of $\mu + a$ and a variance of $\sigma^2 + b$ is selected by using a multidimensional Taylor series approximation (Equation A6.6a) for the probability that such a genotype exceeds the truncation threshold when a fraction, p, is saved and $z \sim N(\mu, \sigma^2)$. Keeping only first-order terms in a and b yields

$$P(a, b) \simeq p\left(1 + a\frac{\bar{\imath}}{\sigma_z} + \frac{b}{2}\frac{\bar{\imath}\, x_{[1-p]}}{\sigma_z^2}\right) \tag{17.10a}$$

Here $\bar{\imath}$ is the selection intensity (Equation 14.3a), and $x_{[1-p]}$ satisfies $\Pr(U \geq x_{[1-p]}) = p$, where $U \sim N(0, 1)$. Thus, truncation selection generates selection pressure, $\bar{\imath}$, on A_m and $\bar{\imath}\, x_{[1-p]}$ on A_v. When A_m and A_v are uncorrelated, the expected response in the trait mean is simply our standard result from Chapter 13 (Equation 13.6b),

$$R_{A_m} = h_m^2\, \bar{\imath}\, \sigma_z \tag{17.10b}$$

Under the additive model for the environmental variance (Equation 17.4a), Hill and Zhang found that the response in the mean breeding value for the environmental variance is

$$R_{A_v} = h_v^2\, \bar{\imath}\, x_{[1-p]}\, \sigma_z^2 \tag{17.10c}$$

which is also the response in σ_E^2 (Equation 17.6a). Equation 17.10a assumes that the population distribution of the trait value, z, is approximately normal, which breaks down at extreme trait values when there is heritable variation in σ_E^2 (as z is now no longer normally distributed, but rather has become a weighted mixture of normals). Hence, for strong selection these results are potentially biased.

Example 17.4. Consider a trait with $\sigma_z^2 = 100$, $h_m^2 = 0.3$, and $h_v^2 = 0.03$ (the latter a typical value from Table 17.2). Assume that the additive model for the environmental variance (Equation 17.4b) holds. What is the expected response in the mean and σ_E^2 following a single generation of truncation selection with $p = 0.1$? Because $h_m^2 \sigma_z = h_v^2 \sigma_z^2 = 3$, Equations 17.10b and 17.10c show that any difference in response is due entirely to differences in the strength of selection ($\bar{\imath}$ vs. $\bar{\imath}\, x_{1-p]}$), not the genetic variances of these traits. Noting that $\Pr(U > 1.282) = 0.1$, we have $x_{[1-0.1]} = 1.282$ and (recalling Equation 14.3a) $\bar{\imath} = \varphi(1.282)/0.1 = 1.755$, with Equations 17.10b and 17.10c yielding

$$R_{A_m} = 0.3 \cdot 1.755 \cdot 10 = 5.265 \quad \text{and} \quad R_{A_v} = 0.03 \cdot 1.755 \cdot 1.282 \cdot 100 = 6.750$$

meaning that a single generation of selection increases the mean by 5.3 and the environmental variance increases by 6.75.

Using these same parameter values, the Bulmer equation (Equation 16.12d) yields the change in the additive genetic variance of the trait, $\sigma^2(A_m)$, after one generation of selection as $d = -3.74$, for (ignoring changes in σ_E^2) a phenotypic variance of $100 - 3.74 = 96.26$ and a heritability of $(30 - 3.74)/(100 - 3.74) = 0.27$. Accounting for changes in σ_E^2 yields a phenotypic variance after one generation of $100 - 3.74 + 6.75 = 103.1$ and heritability of $(30 - 3.74)/103.1 = 0.25$. Because the response in the trait mean is given by $R(t) = h^2(t)\, \bar{\imath}\, \sigma_z(t)$, the decrease in h^2 (from increased σ_E^2) is somewhat offset by the increase in the phenotypic variance. The response in the trait mean in generation 2 becomes $0.27 \cdot 1.77 \cdot \sqrt{96.26} = 4.69$ when ignoring the change in σ_E^2 and $0.25 \cdot 1.77 \cdot \sqrt{103.1} = 4.49$ when including it.

Now consider stronger selection, $p = 0.01$. Here $x_{[1-p]} = 2.326$ and $\bar{\imath} = 2.666$, yielding

$$R_{A_m} = 0.3 \cdot 2.666 \cdot 10 = 7.998 \quad \text{and} \quad R_{A_v} = 0.03 \cdot 2.666 \cdot 2.326 \cdot 100 = 18.603$$

Relative to $p = 0.1$, this is roughly a 50% increase in the response in the mean, but a 275% increase in the response in the environmental variance. The Bulmer equation yields $d = -4.06$ for one generation of selection and a resulting heritability (ignoring any changes in σ_E^2) of 0.27. If we include the change in environmental variance, the new phenotypic variance is 100 - 4.06 + 18.60 = 114.54, resulting in a heritability of 25.94/114.54 = 0.23. As above, the actual trait heritability is less than predicted from the Bulmer equation, but the resulting impact on the response in the mean is again partly offset by the increase in the phenotypic variance, with the expected response in generation two of 7.06 (Bulmer) and 6.78 (Bulmer plus changes in σ_E^2).

Example 17.4 illustrates the fact that as truncation selection becomes stronger, there is a disproportionate change in the variance relative to the mean, as selection is favoring outliers, and hence is more strongly influenced by genotypes associated with larger environmental variances. The effect on σ_E^2 from directional selection on the trait value, z, is thus expected to be greatest under strong selection (Hill and Zhang 2004). We can quantify this assertion by using Equation 17.10a. As shown in Figure 17.1, from moderate to large values of p (i.e., close to 1 and implying weak selection, as most of the population is saved), selection on the mean ($\bar{\imath}$) dominates. The two strengths of selection are equal around $p = 0.16$ (as $x_{[1-0.16]} = 1$), below which ($p < 0.16$) selection on the variance ($\bar{\imath}\, x_{[1-p]}$) is stronger (as $x_{[1-p]} > 1$). For $p > 0.5$ (more than half the population is saved), $x_{[1-p]} < 0$, implying that weak directional selection results in a slight *decrease* in σ_E^2 (Hill and Zhang 2004). The effect is largest around $p = 0.80$ (only 20% of the population is culled), but even here the strength of selection on σ_E^2 is fairly small, with $\bar{\imath}\, x = -0.3$. This slight decrease in σ_E^2 under weak directional selection occurs because only outliers with very low values of z are selected *against*, and such phenotypes arise in genotypes that have higher variances.

As we have seen, there are two very different pathways, through either $\sigma^2(A_m)$ or σ_E^2, for short-term change in the phenotypic variance ,σ_z^2. Generation of gametic-phase disequilibrium by selection changes $\sigma^2(A_m)$ without requiring significant allele-frequency change (Chapters 16 and 24). Likewise, the presence of heritable variation in σ_E^2 can also generate a short-term response in the total variance. As noted by Bull (1987), "environmental and genetic factors may thus compete to produce a given selected level of phenotypic variance."

What insights do these results offer on which factor is more important? The general conclusion is that while direct selection pressure on σ_E^2 *often* has the same sign as the selection on $\sigma^2(A_m)$, this is not *always* the case. Under disruptive selection, there is direct selection for positive disequilibrium, and hence an increase in $\sigma^2(A_m)$, along with direct selection to increase σ_E^2, so the Bulmer equation is expected to underpredict the increase in phenotypic variance. With stabilizing selection (when the population mean is close to the optimum value), the direct selection pressures on $\sigma^2(A_m)$ and σ_E^2 also align, favoring a decrease in each, and again resulting in an underprediction of the total change in σ_z^2 if only the Bulmer equation is used. However, if the current population mean is far from the optimum, there can be selection pressure to *increase* σ_E^2 (Example 17.3). The most direct conflict between these two potential components of change in the phenotypic variance occurs under directional selection. This always generates a negative disequilibrium value d (Equation 16.2), and hence a reduction in the additive variance, $\sigma^2(A_m)$, of the trait. However, under modest to strong selection, it also favors an *increase* in σ_E^2 which often results in both an increase in the phenotypic variance and a further decrease in the heritability. The net result is that the Bulmer equation underpredicts the expected change in the variance (Example 17.4). With very modest selection (over 50% of the population saved), there is weak selection pressure for a slight *decrease* in σ_E^2. It is important to stress that all of these

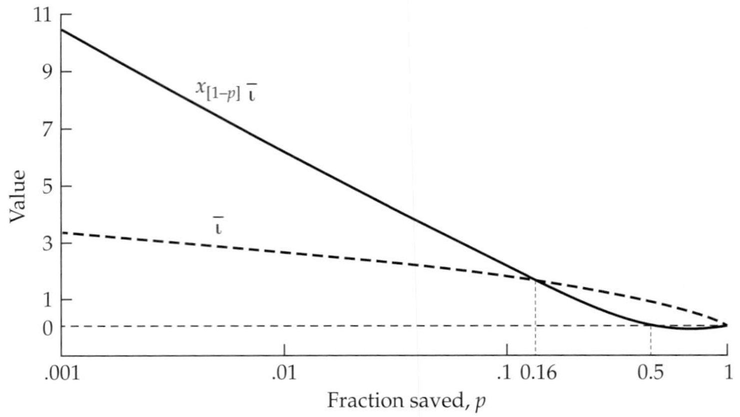

Figure 17.1 The relative strengths of selection on the mean ($\bar{\imath}$) and variance ($x_{[1-p]}\,\bar{\imath}$) under truncation selection as a function of the fraction, p, saved. The two strengths of selection are equal around $p = 0.16$ (as $x_{[1-0.16]} = x_{[0.84]} = 1$). Note that for $p > 0.5$, there is (weak) selection to *decrease* the variance, as the curve for $x_{[1-p]}\,\bar{\imath}$ dips below the horizontal dashed line that indicates a value of zero.

results only consider the *direct* response in σ^2_E, as it is assumed that $\rho(A_m, A_v) = 0$. When the breeding values for the trait value and its environmental variance are negatively correlated, the sign of response on σ^2_E can depart from these predictions (examined in Volume 3).

A major distinction in these two potential routes for changing σ^2_z is that (under the infinitesimal model) changes in $\sigma^2(A_m)$ are transient, and decay away after selection stops. However, change in σ^2_E are due to changes in the mean breeding value, μ_{A_v}, and this response (being due to changes in allele frequencies) is stable after selection stops. Hence, the genetic variance in the trait value, $\sigma^2(A_m)$, returns to its base population value, while the environmental variance stays at its new population value.

Direct Selection on σ^2_E Using Repeated Records

While σ^2_E can change as a consequence of simple selection on the trait value, z, a breeder may wish to target σ^2_E directly. While simple selection on z can result in direct selection on A_v, it also targets A_m (and hence changes the mean). Through the use of an appropriate selection index, one can directly select on A_v alone (and hence directly target σ^2_E), even with only a single observation per individual. This is possible because A_m is linearly associated with z, while A_v is associated with z^2. If we rescale z to have a mean of zero, an index of the form $I_i = az_i + bz_i^2$ can be constructed to specifically target individuals with high (or low) A_v values. We examine this index, and the component responses, in Volume 3.

Another approach involves selecting those individuals with the smallest residual variances under a repeated-measures design. We considered this design in Chapter 13 as an approach to reduce environmental noise when selecting on the *mean* trait value of an individual, but here the target is the actual *variation* among the records themselves. Individuals are chosen based on the index

$$V_i = \frac{1}{n-1}\sum_{j=1}^{n}(z_{ij} - \overline{z}_i)^2 \tag{17.11}$$

where z_{ij} denotes the jth record (observation) from individual i, with selection for uniformity favoring individuals with smaller V_i values. Natural selection can also act in a repeated-records setting, such as on the within-individual variation in flowering time. Depending on the ecological setting, selection can favor individuals with either larger or smaller values of within-individual variances (Example 17.3; Devaux and Lande 2009).

Assuming the exponential model holds for σ^2_E, San Cristobal-Gaudy et al. (1998) and

Ibáñez-Escriche et al. (2008b) approximated the expected response in the mean breeding value of A_v, given selection intensity, $\bar{\imath}$, over the index, V, as

$$R_{A_v} \simeq \bar{\imath}\,\frac{\sigma^2(A_v)}{\sqrt{\exp\left[\sigma^2(A_v)\right]\left[(n+1)/(n-1)\right]-1}}, \tag{17.12}$$

where n is the number of repeated records per individual. Substitution into Equation 17.6c provides the expected response in σ_E^2. More exact expressions were given in Ibáñez-Escriche et al., who also examined the power and required sample sizes when using repeated-measures selection experiments to detect heritable environmental variation.

Under a strict repeated-records design, all observations in V_i have the same genotype. A related design is to select based on variation in trait value among the *offspring* of an individual. For example, Garreau et al. (2008) selected rabbit dams based upon Equation 17.11, using the weights of offspring (suitably corrected for fixed effects such as litter size and parity). As mentioned at the start of the chapter, offspring mortality is lower within litters with more similar individual weights (Milligan et al. 2002; Garreau et al. 2008), and such selection conceivably occurs in natural populations as well. Under an offspring-based index, the multiple records are based on individuals with potentially different genotypes, and hence a large V score could arise from a high breeding value for σ_E^2, segregation of a major gene, nonadditive variance, or (most likely) some combination of these. Garreau et al. observed significant responses in the first generation for both the increased- and decreased-variance selected lines. The selection pressure was weaker in subsequent generations in their experiment, and the response was largely flat. While these data are consistent with a response in σ_E^2, the majority of the initial response may simply have arisen from selection for, or against, females that are heterozygous for major genes influencing weight.

18

Analysis of Short-term Selection Experiments: 1. Least-squares Approaches

To consult the statistician after an experiment is finished is often merely to ask him to conduct a postmortem examination. He can perhaps say what the experiment died of.
Fisher (1938)

This chapter examines the analysis of short-term selection experiments, whose duration is such that changes in allele frequency (from either selection or drift) are assumed to be relatively negligible. We restrict our attention here to the change in the mean under directional selection. The bulk of the chapter consists of developing least-squares (LS) estimates (and their corresponding standard errors) of **realized heritabilities** from the observed response. We also examine the empirical evidence for the goodness-of-fit of the breeder's equation to short-term artificial selection experiments. This is followed by a discussion of experimental and optimal designs for selection experiments, and we conclude with a detailed discussion of **generation means analysis**, the analysis of response when individuals from different cycles of selection are crossed (e.g., by using remnant seed).

Under the LS framework (LW Chapter 8), the only data required are the sample means (before and after selection) over generations, which are usually straightforward to obtain. Conversely, when one has access to the phenotypic values of essentially *all* of the individuals in the experiment, as well as their complete pedigree, more powerful mixed-model (MM) methods can be used. These are discussed in Chapter 19, while Chapter 20 considers the analysis of response in the much more complex setting of natural populations, where there is little control over either the nature of selection or the environment.

Finally, while our focus here is on phenotypic changes, we would be remiss if we did not point out that an area of very considerable interest is the genetic basis behind any such response. We previously discussed (Chapter 9; LW Chapter 15) a number of tools to aid in this dissection (e.g., QTL mapping using crosses between selected lines; tests of selection on specific loci/markers). The most recent extension of this quest are **evolve & resequence (E&R) experiments** (a term coined by Turner et al. [2011]), wherein a population is allowed to evolve under laboratory conditions and then genomic sequencing is applied in the search for target loci. See Long et al. (2015) for a recent review, and Kofler and Schlötterer (2013) and Bladwin-Brown et al. (2014) for design and power issues. Chapters 25 and 26 discuss the findings of such experiments.

The literature on selection experiments is truly massive, and our goal here is to introduce the important concepts rather than to exhaustively review all experiments. Reviews of selection experiments (mainly on metazoans) include Wilson (1977), Wright (1977), Robertson (1980), Mather (1983), Hill (1984), Sheridan (1988), Eisen (1989), Hill and Mackay (1989), Falconer (1992), Hill and Caballero (1992), Garland and Rose (2009), and Hill (2011). Fueled by the rapid advances in next-generation DNA sequencing, artificial selection experiments have undergone a renaissance, both in the form of **experimental evolution studies** (Kawecki et al. 2012) and in a growing number of experiments involving viral and microbial systems (Chapter 26). Our focus here is on goodness-of-fit to the breeder's equation, while in Chapters 25 and 26 we review empirical trends from long-term selection experiments.

VARIANCE IN SHORT-TERM RESPONSE

As we saw in Chapter 12, the means for a series of initially identical lines will inevitably diverge over time through the action of genetic drift, generating an among-line variance. The same is true for lines under selection, with this among-line variance being manifested

Evolution and Selection of Quantitative Traits. Bruce Walsh & Michael Lynch.
Published 2018 by Oxford University Press.
DOI 10.1093/oso/9780198830870.001.0001

as **variation in response** (Figure 18.1). While the grand mean over a series of replicate lines under pure drift should remain essentially unchanged, the similar mean over a series of replicate selection lines should show a response over time, meaning that any particular realized response equals this expected response plus noise due to the evolutionary process of drift. Because artificial selection usually involves choosing a small number of parents to form the next generation, this among-line variance can be considerable. As we saw in Chapter 12, drift also changes the additive variance, σ_A^2, within any given line. Our short-term assumption implies that $t/N_e \ll 1$, in which case drift can generate a significant among-line variance in means but with little within-line change in σ_A^2 (Chapters 11 and 12). Our focus on short-term response assumes that selection has only a minor role in changing the genetic variance over the course of the experiment. This is the standard infinitesimal model assumption used throughout Chapters 13–20 (and relaxed in Chapters 24–28). One complication that we initially ignore (addressing it later in the chapter) is that directional selection decreases the additive genetic variance by creating negative gametic-phase disequilibrium (Chapter 16).

Expected Variance in Response Generated by Drift

The expected variation among the means of replicate lines subjected to the same amount of selection was first considered by Prout (1962b) and then examined in detail by Hill (1971, 1972c, 1972d, 1974b, 1977a, 1980, 1986). The variation between the sample means, $\overline{z}_{t,i}$, of a series of replicate lines (in generation t) has two components: an **evolutionary variance** due to differences between the true means, $\mu_{t,i}$, generated by drift and selection, and a sampling (or residual) variance from estimating the true mean, $\mu_{t,i}$, by a sample mean, $\overline{z}_{t,i}$, based on $M_{t,i}$ individuals. Chapter 12 examined these components under pure drift, while here we consider the joint action of drift and selection.

With selection, the M individuals are not chosen at random with respect to their phenotypes, which *decreases* the among-line variance in population means relative to pure drift. This is partly countered by the fact that selection generates a lower effective population size than expected from drift alone (Chapters 3 and 26). Further, selection can increase the among-line variance relative to drift by changing allele frequencies more rapidly than expected from drift alone. It is these opposing (and potentially offsetting) changes in variance that led to suggestions that many of these effects will cancel out, leaving the pure-drift variance (with a suitably adjusted N_e) as an adequate approximation (Hill 1977a, 1980, 1986; Robertson 1977a; Nicholas 1980). Simulations (Robertson 1977a) and experimental results (Falconer 1973; López-Fanjul 1982) suggest that the pure-drift variance is a reasonable approximation.

Using the pure-drift results (Chapter 12), the sample mean in generation t for a particular realization (line) can be decomposed as

$$\overline{z}_t = \mu_t + e_t \tag{18.1a}$$

where μ_t is the true mean of this line and e_t is the residual error in estimating this true mean from a population sample. The true mean can be further decomposed as

$$\mu_t = \mu + g_t + b_t \tag{18.1b}$$

where g_t is the mean breeding value (relative to a base population value of $g_0 = 0$) and b_t is the mean environmental deviation in generation t, yielding an expected value for the sample mean of a random line in generation t of

$$E(\,\overline{z}_t\,) = \mu + E(\,g_t\,) + b_t \tag{18.2}$$

Under pure drift, $E(g_t) = 0$, while under selection, $E(g_t)$ is given by the breeder's equation, $\sum_i^t h_i^2 S_i = th^2 S$, if the heritability and selection differential remain constant (given the usual

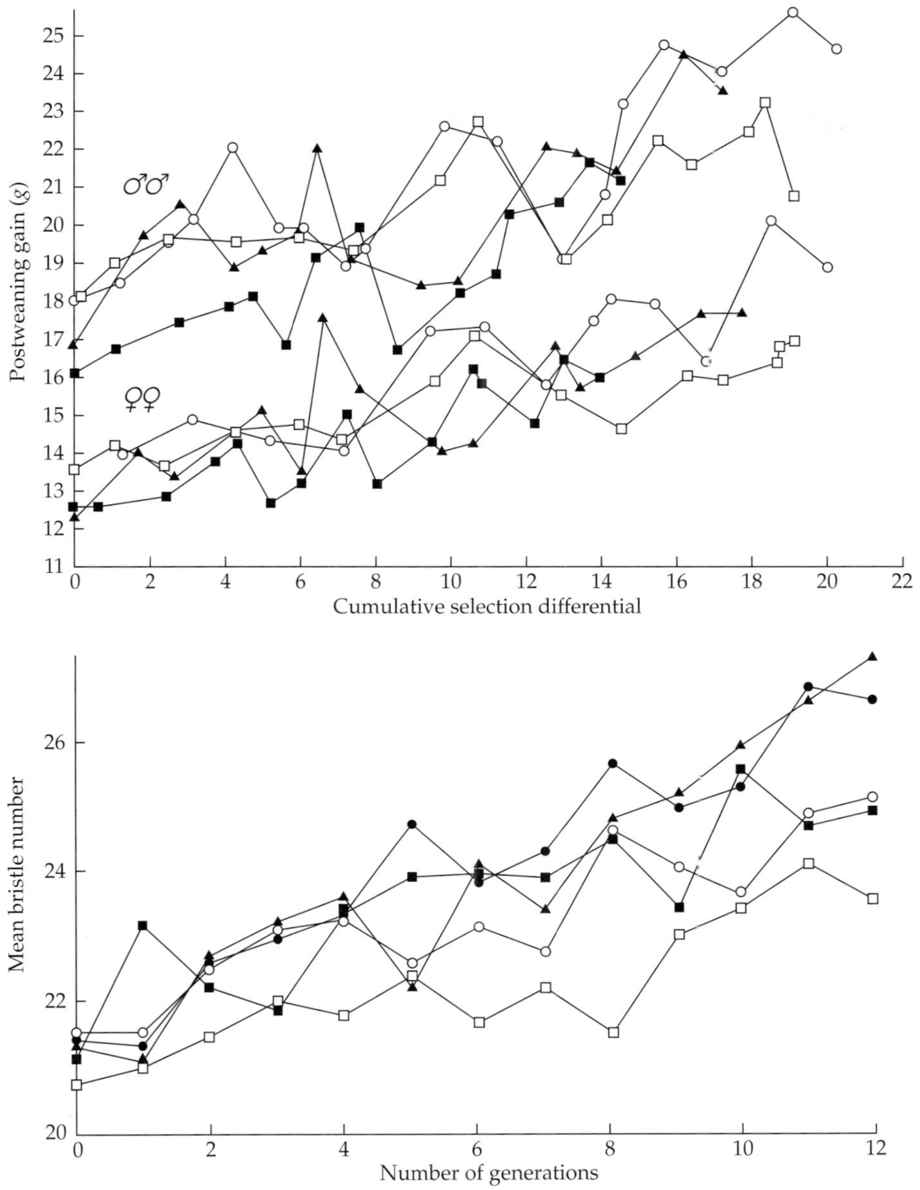

Figure 18.1 Examples of the variance in response to selection among replicate lines. **Top:** Postweaning weight gain in male and female mice (Hanrahan et al. 1973). Each replicate consists of a full-sib single family, which was propagated by selecting the largest pair within each family. **Bottom:** Abdominal bristle number in *Drosophila melanogaster* (Frankham et al. 1968a). Here, 50 pairs of parents were scored and the largest 10 of each sex were used to form the next generation. Notice that these graphs differ in the way the response is plotted. In the upper graph, response is given as a function of the **cumulative selection differential**, which has an expected slope of h^2 under the breeder's equation. On the lower graph, response is a function of the **number of generations**, which can depart from linearity if S varies over time.

caveats summarized in Table 13.2). Assuming $b_t \sim (0, \sigma_b^2)$, the variance about the expected mean value is

$$\sigma^2_{\bar{z}}(t) = \sigma^2_{\bar{g}}(t) + \sigma^2_{\bar{e}}(t) + \sigma^2_b \tag{18.3}$$

As discussed later in the chapter, comparison between contemporaneous populations, such as a selected and control line in the same generation or an upselected versus downselected line, can remove the shared environmental effect, and hence the σ_b^2 term.

Given our previous comments, we assume the evolutionary sampling variance, $\sigma_g^2(t)$, about the mean breeding value for selected lines is, to a good approximation, the same as that for lines under drift alone. If M_0 individuals are initially sampled to form each line, then from Equation 12.1b

$$\sigma_g^2(t) = \left(\frac{1}{M_0} + 2f_t \right) h^2 \sigma_z^2 \tag{18.4}$$

The M_0 term accounts for variation in mean breeding value among lines in the founding generation, while f_t (the amount of inbreeding at generation t) accounts for variation generated by subsequent drift. If population size remains constant

$$2f_t = 2 \left[1 - \left(1 - \frac{1}{2N_e} \right)^t \right] \simeq t/N_e \qquad \text{for } t/N_e \ll 1 \tag{18.5}$$

If different numbers of males (N_m) and females (N_f) are sampled or if N varies over time

$$2f_t \simeq \sum_{k=0}^{t-1} \left[\frac{1}{4N_m(k)} + \frac{1}{4N_f(k)} \right] \tag{18.6}$$

The variance, $\sigma_e^2(t)$, in Equation 18.3, which is associated with estimating the true mean from a sample mean, depends on the relatedness among the M_t individuals chosen to estimate the mean. If these individuals are unrelated, then $\sigma_e^2(t) = \sigma_z^2(t)/M_t$. If some selected individuals are related (such as from the same family), the positive covariance between them reduces the effective sample variance, with the exact form of σ_e^2 depending on the distribution of family sizes within the sample. Hill (1971, 1980) shows that it is bounded by

$$\frac{\sigma_z^2 - \sigma_A^2/2}{M_t} \leq \sigma_e^2(t) \leq \frac{\sigma_z^2}{M_t} \tag{18.7}$$

which corresponds to the range where all individuals are from the same family (lower bound) to all being unrelated (upper bound). Conservatively choosing the upper bound, Equation 18.3 becomes

$$\sigma_{\bar{z}}^2(t) = \left(\frac{1}{M_0} + 2f_t \right) h^2 \sigma_z^2 + \sigma_b^2 + \sigma_z^2/M_t \tag{18.8}$$

Because drift variance accumulates in each generation (via f_t increasing each generation) while the other variance coefficients do not change, when $h^2\sigma_z^2 > \sigma_b^2$, the drift term in Equation 18.8 is expected to eventually dominate, usually after a few generations. Whereas Equation 18.8 describes the divergence *between* lines due to drift, drift also introduces a positive covariance between the means at different generations *within* a line (Equation 12.2)

$$\sigma(g_t, g_{t'}) = \left(\frac{1}{M_0} + 2f_t \right) h^2 \sigma_z^2 \quad \text{for } t < t' \tag{18.9}$$

Assuming that cross-generational environmental and residual effects are uncorrelated, so that $\sigma(b_t, b_{t'}) = \sigma(e_t, e_{t'}) = 0$, then $\sigma(\bar{z}_t, \bar{z}_{t'}) = \sigma(g_t, g_{t'})$.

The careful reader will notice that neither h^2 or σ_z^2 are indexed by generation, and hence they are assumed to be constant. This treatment is in keeping with our assumption that these parameters remain essentially unchanged over the short time course of the experiment. This is a major assumption and can be violated in several ways. First, changes in the underlying allele frequencies can change these variances (Chapters 5 and 24–28). If alleles of major effect are segregating, large changes in the variance can occur within a few generations. Further, when major alleles are segregating at low frequencies, the sampling to create lines from a base population can result in a substantial increase in the among-line variance over that predicted by Equation 18.4, as these alleles are absent in some lines and overrepresented (relative to the base population) in others. This results in lines having a larger range of starting additive variances, thus increasing the variance in response (James 1970).

Second, directional selection generates negative gametic-phase disequilibrium, reducing the additive genetic variance within a line over the first few generations before approaching an equilibrium value (Chapter 16). We will deal with this shortly. Third, inbreeding due to finite population size reduces the additive genetic variance within lines. For a completely additive locus, the expected additive genetic variance (in the absence of mutational input) within a line in generation t is expressed by Equation 11.2

$$E\left[\sigma_A^2(t)\right] = \left(1 - \frac{1}{2N_e}\right)^t \sigma_A^2(0) \simeq \left(1 - \frac{t}{2N_e}\right) \sigma_A^2(0) \quad \text{for} \quad t \ll N_e$$

where $\sigma_A^2(0)$ is the variance in the base population (the last expression follows from $(1-x)^t \simeq 1 - xt$ for $|\,xt\,| \ll 1$). If dominance or epistasis is present, the within-line variance can actually increase (for a time) under drift (Chapter 11), further inflating the among-line variance. Provided $t/2N_e \ll 1$, the error introduced in Equations 18.8 and 18.9 by ignoring the reduction in the within-line variance due to inbreeding is small. Finally, when there is heritable variation in the environmental variance, σ_E^2 is expected to increase under moderate to strong directional selection (Chapter 17), increasing σ_z^2 and thus decreasing h^2.

Despite all of these potential complications, those few experimental tests of the pure-drift approximation to artificial-selection data have found it to be fairly reasonable (Falconer 1973; López-Fanjul 1982). Mixed-model (REML/BLUP) methods developed in LW Chapters 26 and 28 (and reviewed here in Chapter 19), account for some of these concerns, but they require significantly more information (the phenotypic values of all individuals and their complete pedigree), which may be difficult to obtain.

Variance in Predicted Response Versus Variance in Actual Response

It is important not to confuse the previous expressions for variance in response (the variation about the mean response when the genetic parameters and selection differentials are *known without error*) with variance in the *predicted* response, which has (at least) two *additional* sources of variation. The first is the uncertainty produced by using estimates in place of the true values for the genetic and phenotypic variances. The second is uncertainty in the realized selection intensity. While selecting a preset fraction, p, of the population specifies the *expected* selection intensity, there is variation about this mean value for any particular realization (Chapter 14). For example, *before* selection is performed, we expect that truncation selection saving the uppermost 5% will have an *average* selection intensity of 2.06 (Example 14.1). However, when selecting the largest 5% from a small population (so that the number saved is small, e.g., $N \leq 25$ or so), its realized value can be significantly larger or smaller than the expectation, thus generating additional variation in the predicted response.

Several papers (Tai 1979; Knapp et al. 1989; Bridges et al. 1991) have presented confidence intervals for the expected selection response, but these simply focus on the variance generated by the uncertainty in the initial estimates of additive and phenotypic variances. They do not consider the additional evolutionary variance in response (discussed above), nor do they consider the variance in the particular realization of a prespecified selection intensity (although, retrospectively, we can deal with the latter by using the observed values).

REALIZED HERITABILITIES

The breeder's equation, $R = h^2 S$, immediately suggests that heritability can be estimated as the ratio of the observed response to the observed selection differential

$$\widehat{h}_r^2 = \frac{R}{S} \tag{18.10}$$

Falconer (1954) referred to Equation 18.10 as the **realized heritability**, a term now more broadly defined to include any estimate of heritability based on the observed response to selection.

While one can use this approach to estimate h^2, any complication in predicting response using the breeder's equation (Table 13.2) will usually make $\widehat{h}_r^2$ a biased estimator. Turning this point around suggests that one test for the success of the breeder's equation is to compare how close realized heritabilities are to estimates based on resemblance between relatives in the unselected base population. If the breeder's equation generally provides an accurate model of selection response, we expect these two different estimates to be similar (i.e., within sampling error).

Example 18.1. An interesting example of applying artifical selection in a natural setting is the work of Flux and Flux (1982), who examined response on clutch size in starlings (*Sturnus vulgaris*) in New Zealand. Eggs were laid almost exclusively in nest boxes, allowing for identification of the mothers and careful monitoring of their offspring. Increased clutch size was selected for by removing all eggs from clutches below a specified brood size (artificial truncation selection on clutch size). Pooling all of the single-generation responses over the entire study, the 516 clutches from the offspring of selected females had an average size of 5.60 $\pm$ 0.04, while the average size of 2050 clutches from offspring of unselected females was 5.48 $\pm$ 0.02, yielding an estimated response of R = 0.12 $\pm$ 0.04. The mean clutch size of selected female parents was 6.20 versus 5.48 for unselected (control) females, yielding S_f = 0.72. Because there is no selection on fathers under this design, $S = S_f/2$ (Chapter 13), yielding an expected response in daughters of $R = h^2\, S_f/2$, and a realized heritability of

$$\widehat{h}_r^2 = 2R/S_f = 2 \cdot 0.12/0.72 = 0.33$$

This value was in good agreement with the estimated heritability based on mother-daughter regressions of $\widehat{h}^2$ = 0.34 $\pm$ 0.08.

Estimators for Several Generations of Selection

While the estimator given by Equation 18.10 is unambiguous for a single generation of selection, two different estimation approaches have been proposed for an experiment of a length of $T > 1$ generations. Both approaches are based on the **cumulative selection response**, $R_C(t)$, and **cumulative selection differential**, $S_C(t)$

$$S_C(t) = \sum_{i=1}^{t} S_i \qquad \text{and} \qquad R_C(t) = \sum_{i=1}^{t} R_i \tag{18.11}$$

where $S_i = \overline{z}_i^* - \overline{z}_i$ and $R_i = \overline{z}_{i+1} - \overline{z}_i$ are the selection differential and single-generation response (respectively) for generation i, and $\overline{z}_i$ and $\overline{z}_i^*$ denoting the preselection and post-selection means in generation i.

The first approach, the simple **ratio estimator** of the realized heritability, is to use the total response divided by the total differential

$$\widehat{b}_T = \frac{R_C(T)}{S_C(T)} \tag{18.12}$$

We use the notation throughout the chapter of $\widehat{b}_x$ for a particular estimator, as the realized heritability is some multiple of this, $\widehat{h}_r^2 = c\cdot\widehat{b}_x$, where c depends on the experimental design. For individual selection with only one sex selected, $c = 2$ (Example 18.1), while $c = 1$ when both sexes are selected.

The ratio estimator makes no assumption about a constant heritability, and instead returns an average value over the experiment. When h^2 remains constant, the breeder's equation predicts a linear cumulative response in R as a function of S. This observation leads to the second (and much more widely used) approach of estimating the realized heritability from the slope of the regression of cumulative response on cumulative selection differential

$$R_C(t) = b_C S_C(t) + e_t \qquad \text{for } t = 1, \cdots, T \tag{18.13}$$

with $\widehat{h}_r^2 = b_C$ for individual selection on both sexes (Falconer 1954). Modifications of Equations 18.12 and 18.13 can be used for family selection (Chapter 21) and other designs, such as divergent selection (discussed at the end of this chapter). Because the expected response is zero if there is no selection, the regression line is constrained to pass through the origin and hence lacks an intercept term. Standard linear model approaches (LW Chapter 8) can be used to test for goodness-of-fit, such as testing whether a quadratic regression gives an improved fit (which suggests a changing heritability).

Recall from LW Chapter 8 that a regression estimator depends on the assumed covariance structure of the residuals. Most applications of the regression estimator (Equation 18.13) in the literature have assumed ordinary least-squares (OLS), which requires that the residuals, e_t, be homoscedastic and uncorrelated (LW Chapter 8). However, the careful reader will recall from Equations 18.4 and 18.9 that drift generates heteroscedastic and correlated residuals, thus requiring a generalized least-squares (GLS) solution (LW Chapter 8). We return to this issue after first examining the OLS estimator.

Recalling LW Equation 8.33a, the OLS estimator for the slope is $(\mathbf{X}^T\mathbf{X})^{-1}\mathbf{X}^T\mathbf{y}$. Because the design matrix, $\mathbf{X}$, for the regression given by Equation 18.13 is simply the vector of cumulative selection differentials, $\mathbf{S}$, and $\mathbf{y}$ is the vector of cumulative responses, $\mathbf{R}$, it follows that the OLS estimate of the slope is given by

$$\widehat{b}_C(\text{OLS}) = \left(\mathbf{S}^T\mathbf{S}\right)^{-1}\mathbf{S}^T\mathbf{R} = \frac{\sum_{i=1}^T S_C(i)\cdot R_C(i)}{\sum_{i=1}^T S_C^2(i)} \tag{18.14}$$

We will refer to this as the **OLS regression estimator** of the realized heritability. While this is the most common approach used in the literature, by assuming the wrong residual error structure, the OLS approach substantially *underestimates* the standard error for the slope. Richardson et al. (1968) and Irgang et al. (1985) noted that when the number of individuals, M_t, used to obtain the sample mean varies over generations, sample means based on more individuals contain more information, and thus have smaller residual errors. In this setting, the residual variances (roughly σ_z^2/M_t) are no longer homoscedastic, prompting these authors to suggest that a weighted least-squares method (e.g., LW Example 8.11) be used to correct for this. While it is an improvement over OLS, however, this approach still ignores the very significant impact from additional heteroscedasticity and correlations in the residual errors generated by drift.

The theory for estimating realized heritabilities (via a LS analysis) in populations with overlapping generations is less well-developed. As will be discussed in Volume 3, approaches assuming an asymptotic selection response are flawed in that many generations are required to reach a stable genetic structure when starting from an unselected base population. For nonasymptotic response, see Hill (1974a) and Johnson (1977a) for the relevant theory, and Atkins and Thompson (1986) for an example with Scottish Blackface sheep. Mixed-model approaches (Chapter 19) easily accommodate overlapping generations, but they require the full pedigree of all measured individuals used in making selection decisions. This is often quite feasible with selection on large domesticated animals, and it is becoming increasingly feasible in more general settings using dense-marker estimates of relationships (Chapter 20).

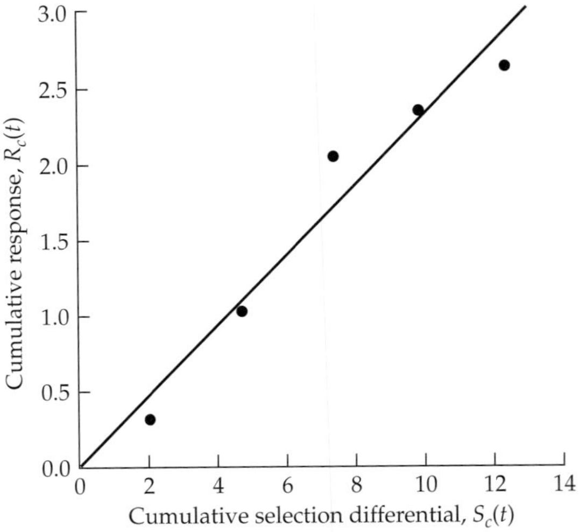

Figure 18.2 Response seen in Mackay's (1985a) divergent selection experiment on abdominal bristle number in *Drosophila melanogaster*. Details are provided in Example 18.2.

Example 18.2. Consider the following data from Mackay (1985a), who performed a divergent selection experiment on abdominal bristle number in replicate lines of *Drosophila melanogaster*. Fifty males and 50 females were measured in each line, with 10 of each sex selected to form the next generation. Her data for the High (up-selected) line from replicate pair 2 for the first five generations of selection are

t	$\overline{z}$	$\overline{z}^*$	$S(t)$	$R(t)$	$S_C(t)$	$R_C(t)$
1	18.02	20.10	20.10 − 18.02 = 2.08	18.34 − 18.02 = 0.32	2.08	0.32
2	18.34	21.00	21.00 − 18.34 = 2.66	19.05 − 18.34 = 0.71	4.74	1.03
3	19.05	21.75	21.75 − 19.05 = 2.70	20.07 − 19.05 = 1.02	7.44	2.05
4	20.07	22.55	22.55 − 20.07 = 2.48	20.36 − 20.07 = 0.29	9.92	2.34
5	20.36	22.95	22.95 − 20.36 = 2.59	20.65 − 20.36 = 0.29	12.51	2.63
6	20.65					

The ratio estimate of the realized heritability (Equation 18.12) is $\widehat{h}_r^2 = 2.63/12.51 = 0.2102$. The regression, forced through the origin, of cumulative response (R_C) on cumulative selection differential (S_C) is plotted in Figure 18.2. Equation 18.14 yields an OLS regression estimator of the realized heritability of

$$\widehat{h}_r^2 = \widehat{b}_C(OLS) = \frac{\sum_{i=1}^5 S_C(i) \cdot R_C(i)}{\sum_{i=1}^5 S_C^2(i)} = \frac{78.96}{350.45} = 0.2245$$

Weighted Least-squares Estimates of Realized Heritability

Ordinary least-squares regression assumes that the residuals are homoscedastic and uncorrelated, yielding a covariance matrix for the vector of residuals, **e**, as Var(**e**) = $\sigma_e^2 \cdot \mathbf{I}$ (LW Chapter 8). Genetic drift causes the residual covariance structure of the regression given by Equation 18.13 to depart significantly from this simple form. In particular, the residual variance increases with time (Equation 18.8) and residuals from different generations within a given line are correlated (Equation 18.9). Thus, Var(**e**) is not simply $\sigma_e^2 \cdot \mathbf{I}$, and generalized least-squares (GLS) must be used to account for the overall covariance structure. From LW Equation 8.34, the GLS estimator of the regression slope is given by

$$\widehat{b}_C(\text{GLS}) = (\mathbf{S}^T\mathbf{V}^{-1}\mathbf{S})^{-1}\mathbf{S}^T\mathbf{V}^{-1}\mathbf{R} \tag{18.15a}$$

where $\mathbf{V}$ is the variance-covariance matrix associated with selection response, with elements

$$V_{ij} = \sigma_e(i,j) = \sigma\,[\,R_C(i), R_C(j)\,] \tag{18.15b}$$

The elements of $\mathbf{V}$ can be obtained from the pure-drift approximation, with the variances, V_{ii}, given by Equation 18.8 and the covariances, V_{ij}, by Equation 18.9, yielding

$$V_{ij} = \begin{cases} \left(\frac{1}{M_0} + 2f_i\right) h^2\sigma_z^2 + \sigma_z^2/M_i & \text{for } i = j \\ \left(\frac{1}{M_0} + 2f_i\right) h^2\sigma_z^2 & \text{for } i < j \end{cases} \tag{18.15c}$$

where $2f_i \simeq i/N_e$ (Equation 18.5). If differences in environmental values across generations are not accommodated for by the design, then V_{ii} has an additional term, σ_b^2, reflecting random changes in the environment that influence the mean trait value (as discussed below, the use of control lines can often remove this term). Even though OLS assumes an incorrect residual structure, it still provides an unbiased estimate of b_C. However, *OLS significantly underestimates the standard error* of b_C, and it is for this reason that GLS estimators should be used whenever possible (Hill 1971, 1972c, 1972d, 1974b, 1977a, 1980, 1986).

Example 18.3. Computing the GLS regression using the data from Example 18.2 requires the variance-covariance matrix, $\mathbf{V}$, of the residuals, which is obtained using the pure-drift approximation. From Example 18.2, $M = 100$ and $N = 20$, while the estimated phenotypic variance is Var(z) = 3.293 (Mackay, personal communication). Assuming $N_e = N$, Equation 18.5 yields $f_i \simeq i/20$. Further assuming that both initial sampling and among-generation environmental effects can be ignored (i.e., $M_0 \gg 1$ and $\sigma_b^2 \simeq 0$), Equation 18.8 gives the variance associated with the response in generation i as

$$V_{ii} = 2f_i h^2\sigma_z^2 + \sigma_z^2/M_i = \left(\frac{i}{20}\right) h^2 \cdot 3.293 + \frac{3.293}{100} = 0.1647 \cdot (i \cdot h^2 + 0.2)$$

Similarly, Equations 18.9 and 18.5 give the covariance between generations as

$$V_{ij} = \left(\frac{i}{N}\right) h^2\sigma_z^2 = i \cdot h^2 \cdot 0.1647 \qquad \text{for } i < j$$

The resulting covariance matrix becomes

$$\mathbf{V} = 0.1647 \cdot \begin{pmatrix} h^2+0.2 & h^2 & h^2 & h^2 & h^2 \\ h^2 & 2h^2+0.2 & 2h^2 & 2h^2 & 2h^2 \\ h^2 & 2h^2 & 3h^2+0.2 & 3h^2 & 3h^2 \\ h^2 & 2h^2 & 3h^2 & 4h^2+0.2 & 4h^2 \\ h^2 & 2h^2 & 3h^2 & 4h^2 & 5h^2+0.2 \end{pmatrix}$$

Because $\mathbf{V}$ is a function of the unknown heritability, estimation is an iterative process. Starting with some initial estimate of h^2, each new h_r^2 estimate is used to update $\mathbf{V}$ in subsequent iterations until convergence occurs. Using the ratio estimator $h^2 = 0.21$ (Equation 8.2) as the starting value, Equation 18.15 gives a first estimate as

$$\widehat{b}_C(GLS)^{(1)} = (\mathbf{S}^T\mathbf{V}^{-1}\mathbf{S})^{-1}\mathbf{S}^T\mathbf{V}^{-1}\mathbf{R} = 0.2222$$

Substituting this new estimate of h^2 into $\mathbf{V}$ gives, upon a second iteration, $\widehat{b}_C(GLS)^{(2)} = 0.2221$, which remains unchanged in subsequent iterations.

Standard Errors of Realized Heritability Estimates

The final pieces of statistical machinery necessary for assessing the success of the breeder's equation are the standard errors associated with the different realized heritability estimators. Consider first the realized heritability estimated from the unweighted (OLS) regression (Equation 18.14). Recalling LW Equation 8.33b for the variance for an OLS estimator

$$\text{Var}\left[\widehat{b}_C(\text{OLS})\right] = \sigma_e^2 \left(\mathbf{X}^T\mathbf{X}\right)^{-1} = \sigma_e^2 \left(\mathbf{S}^T\mathbf{S}\right)^{-1} = \sigma_e^2 \Big/ \sum_{i=1}^{T} S_C^2(i) \tag{18.16a}$$

The residual variance, σ_e^2, can be estimated from the residual sums of squares divided by the degrees of freedom (see LW Chapter 8). For an experiment lasting for T generations

$$\widehat{\sigma}_e^2 = \frac{1}{T-1}\sum_{i=1}^{T} \widehat{e}_i^2 = \frac{1}{T-1}\sum_{i=1}^{T}\left(R_C(i) - \widehat{h}_r^2\, S_C(i)\right)^2 \tag{18.16b}$$

As mentioned, because the OLS estimator assumes that residuals are uncorrelated and have equal variances (both of which are incorrect under drift), it significantly *underestimates* the correct variance (Example 18.4). The GLS regression estimator (Equation 18.15) avoids these problems by properly accounting for the residual variance-covariance structure generated by drift. From standard GLS theory (LW Equation 8.35)

$$\text{Var}\left[\widehat{b}_C(\text{GLS})\right] = (\mathbf{S}^T\mathbf{V}^{-1}\mathbf{S})^{-1} \tag{18.17}$$

As was done above, the pure-drift approximation is used to obtain the elements of **V**, with $\widehat{h}_r^2$ used in place of h^2.

Finally, consider the variance for the ratio estimator b_T, the total response to total selection (Equation 18.12). Because Var(y/c) = Var(y)/c^2 for a constant, c, we have

$$\text{Var}(\,\widehat{b}_T) = \frac{\text{Var}\left[R_C(T)\right]}{S_C^2(T)} \simeq \frac{(T/N)\,\widehat{h}_r^2\,\sigma_z^2 + \sigma_z^2/M_T}{S_C^2(T)} \tag{18.18}$$

The numerator (the variance in response in generation T) follows from the pure-drift approximation (Equation 18.8), assuming that initial sampling can be ignored (e.g., $M_0 \gg 1$) and there is no significant between-generation environmental variance ($\sigma_b^2 = 0$). Hill (1972c, 1972d) noted that $\widehat{b}_C$(GLS) is generally a slightly better estimator (i.e., returning a smaller standard error) than $\widehat{b}_T$ when h^2 is small, while $\widehat{b}_T$ is a slightly better estimator when h^2 or the number of generations is large.

Example 18.4. Using the data from Examples 18.2 and 18.3, let us compare the standard errors associated with the three different realized heritability estimates. Consider the unweighted regression estimator, $\widehat{b}_C$(OLS), first. The residual sum of squares is

$$\sum_{i=1}^{T}\left(R_C(i) - \widehat{h}_r^2 S_C(i)\right)^2 = 0.091$$

yielding an estimated residual variance of $\widehat{\sigma}_e^2$ = 0.091/4 = 0.0228. Equation 18.16a then yields

$$\text{Var}\left[\,\widehat{b}_C(\text{OLS})\,\right] = \sigma_e^2 \Big/ \sum_{i=1}^{T} S_C^2(i) = \frac{0.0228}{350.45} = 0.0000649$$

Taking the square root yields a standard error of 0.0081.

Turning to the estimate, $\widehat{b}_T$, based on the total response to total selection, Equation 18.18 returns

$$\text{Var}(\,\widehat{b}_T\,) = \frac{(5/20)\cdot 0.21 \cdot 3.292 + 0.03292}{12.51^2} = 0.00132$$

for a standard error of 0.0363. Finally, if we substitute the GLS estimate of $\widehat{h}_r^2 = 0.222135$ (Example 18.3) into $\mathbf{V}$, Equation 18.17 will yield a variance for this estimate of $(\mathbf{S}^T\mathbf{V}^{-1}\mathbf{S})^{-1} = 1/790.4$, for a standard error of 0.0356.

To summarize, the three approaches give extremely similar estimates

Unweighted least-squares regression, $\widehat{b}_C$(OLS)	$\widehat{h}_r^2 = 0.2245 \pm 0.0081$
Total response/total differential, $\widehat{b}_T$	$\widehat{h}_r^2 = 0.2102 \pm 0.0363$
Weighted least-squares regression, $\widehat{b}_C$(GLS)	$\widehat{h}_r^2 = 0.2221 \pm 0.0356$

Note that the difference in the standard error between $\widehat{b}_C$(GLS) and $\widehat{b}_T$ is very small, with $\widehat{b}_T$ considerably more straightforward to compute. Also note that the unweighted regression greatly underestimates the standard error, making it four-fold smaller than the other two standard errors.

Power: Estimation of h^2 from Relatives or Selection Response?

Estimates of heritability from relatives are based on variance components (e.g., LW Chapters 17 and 18), while estimates from selection response are based on ratios of means. Thus, one might expect that realized heritability estimates may have greater power than those based on relatives. Is this indeed the case? Consider the sampling variance for the simple ratio estimator (Equation 18.18)

$$\text{Var}(\widehat{h}_r^2) = \frac{\text{Var}\,[R_C(T)]}{S_C^2(T)} \simeq \frac{(T/N)\,\widehat{h}_r^2\,\sigma_z^2 + \sigma_z^2/M}{S_C^2(T)} = \frac{(T/N)\,\widehat{h}_r^2 + 1/M}{S_C^2(T)/\sigma_z^2}$$

where T is the number of generations and M individuals are measured per generation, with the most extreme N of these allowed to reproduce. Suppose that the strength of selection is the same in each generation ($S_i = S$), in which case we have that

$$S_C^2(T)/\sigma_z^2 = T^2\,S^2/\sigma_z^2 = T^2\cdot\bar{\imath}^2$$

and hence

$$\text{Var}(\widehat{h}_r^2) \simeq \frac{(T/N)\,\widehat{h}_r^2 + 1/M}{T^2\cdot\bar{\imath}^2} = \frac{(1/TN)\,\widehat{h}_r^2 + 1/(T^2M)}{\bar{\imath}^2} \simeq \frac{\widehat{h}_r^2}{TN\,\bar{\imath}^2} \tag{18.19}$$

where the last step follows by ignoring the $1/(T^2M)$ term.

Turning to relative-based estimators of h^2, the optimal midparent-offspring design (N_{mp} sets of parents each with a single offspring) has (LW Equation 17.11b) a standard error of

$$\text{SE}(h_{mp}^2) = \left(\frac{2-h^4}{N_{mp}}\right)^{1/2}$$

This gives a ratio of sampling variances for the two estimators of

$$\frac{\text{Var}(h_{mp}^2)}{\text{Var}(\widehat{h}_r^2)} \simeq \frac{(2-h^4)/N_{mp}}{h^2/(TN\,\bar{\imath}^2\,)} = \left(\frac{TN}{N_{mp}}\right)\left[\bar{\imath}^2\left(\frac{2-h^4}{h^2}\right)\right] \tag{18.20}$$

In applying Equation 18.20 to determine which method (N_{mp} midparent-offspring pairs versus TN total adults over the course of a selection experiment) is more accurate, a few points are in order. In a selection experiment, one measures M adults in each generation, where N of these reproduce. Hence, if the limiting factor for an investigator is in raising offspring, then the comparison of N_{mp} midparents with TN adults that are allowed to reproduce is a fair one. However, if the limiting factor is in the actual *measurement* of individuals, then the fair comparison should be designs based on TM measured adults versus $3N_{mp}$ midparents (as the latter involves two adults and one offspring measure per pair). As Example 18.5 illustrates, even when T and M are fixed, the choice of N is important, as the standard error (Equation 18.19) is minimized when the product, $N\,\bar{\imath}^2$, is maximized. There is a tradeoff in this maximization, as increasing selection decreases N but increases $\bar{\imath}$ (which is a nonlinear function of N/M).

There are a few final caveats in using a realized heritability estimator in place of a more traditional relative-based one. Clearly, if any of the assumptions of the breeder's equation (Table 13.2) are violated, we obtain a potentially biased estimate. A more subtle issue is the target population for the heritability estimate—is this the entire population or some subset? An example of why this distinction may matter derives from the work of Skibinski and Shereif (1989). These authors initiated lines for selection on sternopleural bristle number in *Drosophila melanogaster* by taking parents from the central part of the distribution versus taking parents with extremely high or extremely low bristle number. The lines founded from the central part of the distribution had higher heritabilities, and larger responses, than lines founded from the extremes of the distribution. This is not expected under the infinitesimal model (Chapter 24), but it *is* expected if there are alleles of modest (or greater) effect segregating in the population. Thus, the subtle point is that realized heritability estimates may actually be sampling a different part of the population than relative-based estimators (this was also stressed in Chapter 6; see Equations 6.39 and 6.40). For example, if we sample rare alleles of large effect, these will rapidly increase in selected populations, thus inflating the heritability.

Example 18.5. Consider a trait with a true heritability of $h^2 = 0.30$. This trait is expensive to measure, while offspring are inexpensive to rear. With the resources to measure 3000 individuals, two possible designs are to measure 1000 midparent-offspring trios or to measure a total of 3000 individuals over the course of a selection experiment. The expected standard error for the midparent-offspring regression (LW Equation 17.11b) is

$$\mathrm{SE}(h^2_{mp}) = \sqrt{\frac{2-h^4}{N_{mp}}} = \sqrt{\frac{2-0.30^2}{1000}} = 0.044$$

For a T-generation selection experiment, we have $T \cdot M = 3000$ as the design constraint. Thus, if we measure M initial parents to select the most extreme N, and do this for six generations (the parents = generation 0, followed by selected generations 1, 2, 3, 4, 5), with 3000/6 = 500 individuals being measured in each generation.

Consider three different designs, with N = 250 (50% selected), 100 (20%), or 50 (10%). Applying Equation 14.3a, these correspond to selection intensities (not correcting for finite population size) of $\bar{\imath}$ = 0.80, 1.40, and 1.75, respectively. From Equation 18.19, the standard errors for h^2_r are given by

$$\sqrt{\frac{h^2}{TN\,\bar{\imath}^2}} = \sqrt{\frac{0.3}{5N\,\bar{\imath}^2}}$$

or 0.0194, 0.0175, and 0.0197 (for $N = 250$, 100, and 50, respectively). Thus, while all three selection-experiment designs have a smaller standard error than the midparent-offspring estimate, the design with $N = 100$ is the best among the three candidates. Trial and error yields an optimal N of 135, resulting in a standard error of 0.0172. Similar calculations can also be performed assuming experiments of different length (with T varying).

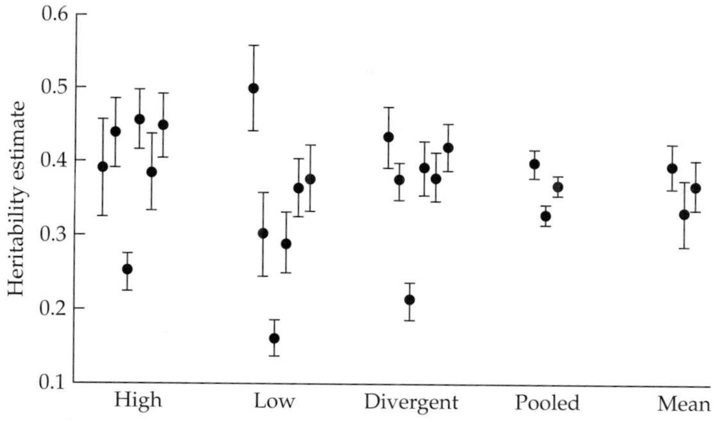

Figure 18.3 Realized heritability estimates (and their associated standard errors) from Falconer's (1973) selection experiment in mice. Six replicate experiments were performed and realized heritabilities within each replicate were estimated by comparing up- and down-selection lines to a control (High and Low, respectively) as well as to each other (Divergent). For each comparison, the estimated heritability plus or minus one standard error (using the OLS regression) is plotted, resulting in six estimates per class. Two combined estimates were also used with each yielding a single estimator for the three classes. The three Pooled estimates (for High, Low, and Divergent) used an OLS regression that took the average value over the six replicates at the data point for any given generation. The three Mean estimates correspond to the mean value of the estimates over the six replicates, with the error bars corresponding to the empirical standard error among the heritability estimates for these six different realizations. While these observed variances were larger than predicted from an OLS regression on the pooled data, this is not unexpected, as the OLS method underestimates the true standard errors.

Empirical Versus Predicted Standard Errors

While the standard error in response can be directly estimated from the among-line variance over a series of replicate lines subjected to identical selection, this approach is generally not cost-effective given the large standard error on such estimates (Chapter 12). As a result, the standard errors reported for most experiments use the pure-drift approximation discussed previously. How closely do the empirical (observed) standard errors match those predicted by the drift approximation? Unfortunately, very few experiments have addressed this issue, and those few that do often use OLS-generated standard errors (which are too small) to compare the fit with the observed among-line variance.

Perhaps the most extensive study is that of Falconer (1973), who performed two-way selection for 6-week weight in mice (Figure 18.3). Six replicate sets of lines were used. Each set consisted of a line selected for larger size, a line selected for smaller size, and an unselected control, for a total of 18 lines in the entire experiment. Realized heritabilities were estimated (by OLS regression) using the three different contrasts available within each replicate set: large versus control (High), small versus control (Low), and large versus small (Divergent).

In addition to these three estimates of $\widehat{h}_r^2$ for each of the six replicates, Falconer also considered two combined estimates, based on all of the lines. First, a regression-based *Pooled estimate* was obtained by using the average of the six means for a given contrast (High, Low, or Divergence) as the data points in a regression. For example, the Pooled High estimate was obtained by regressing the cumulative difference between the mean of all high lines and the mean of all control lines on generations. Second, Falconer also considered a simple *Mean estimate* for the High, Low, and Divergent comparisons, which was computed as the average of the realized heritability estimates for each comparison over all six replications.

The standard error for the Mean estimate measures the variation seen over the replicates. This is the empirical estimate of the standard error used to compare against the theoretical predictions. The average realized heritability estimates under each of the three different contrasts were very similar: 0.395, 0.331, and 0.369 for High, Low, and Divergent lines (respectively). The three Pooled estimates give very similar estimates. As expected (because OLS estimated standard errors are expected to be downwardly biased), the standard errors for the Mean estimates (based on variation between the replicate lines) were 1.6, 3.3, and 2.3 times larger than the estimated standard error computed using OLS regression of the pooled data.

A second experiment (López-Fanjul and Domínguez 1982; summarized by López-Fanjul 1982) followed 20 replicate *Drosophila* lines selected for sternopleural bristle number. They found that the empirical standard errors were less than OLS-estimated SEs, which in turn were less than GLS-estimated SEs. This is contrary to the expectation that the OLS-estimated SEs are smaller than the true variances, while the empirical and GLS-estimated SEs are expected to be roughly equal. The authors suggested that this may be at least partly due to a scale effect (LW Chapter 11), as the phenotypic variance greatly decreased during selection.

Finally, (Bohn et al. (1983) examined the variance in response to selection on pupal weight in *Trioblium castaneum*, examining three replicates from each of six different base populations. The authors found a poor fit when using Hill's (1971) variance formula, which attempts to correct for the effects of selection. Interestingly, a strong correlation was observed between departures in the predicted variance and departures from normality in the base population. The variance among replicate lines drawn from base populations with increasing (absolute) amounts of kurtosis showed larger departures from the predicted value.

Realized Heritability With Rank Data

Realized heritability estimates are not limited to normally distributed traits. Indeed, the heritability of the liability underlying a threshold trait (Chapter 14) can be computed using a realized heritability estimator, which compared performance of the offspring from different parental classes (Example 14.5; LW Chapter 25). In a similar fashion, Schwartz and Wearden (1959) considered a realized heritability estimator when the data are reduced to ranks, e.g., pecking order. They considered the case where individuals were divided into high- and low-dominance groups, by first ranking the dominance of individuals (in their case, chickens) within a flock, and then assigning those in the upper half to the high group and those in the lower half to the low group. The selection differential was computed as the difference in the mean rank of high and low parents, $a_h - a_l$. Conversely, the selection response was measured as the difference in the mean rank of progeny from high parents versus low parents, $b_h - b_l$, with

$$\widehat{h}_r^2 = \frac{b_h - b_l}{a_h - a_l}$$

The authors cleverly noted how this estimator can be related to the Mann-Whitney U statistic (for comparing two sample means based on ranks; Conover 1999), and they used this to develop large-sample confidence intervals for the resulting heritability estimate.

Infinitesimal-model Corrections for Disequilibrium

Our final comment on estimation is that directional selection is expected to decrease the additive variance (and hence the heritability), due to the generation of negative gametic-phase disequilibrium (Chapters 16 and 24). Hence, realized heritability is depressed relative to the (unselected) base population value (Figure 18.4). Under the infinitesimal model, the

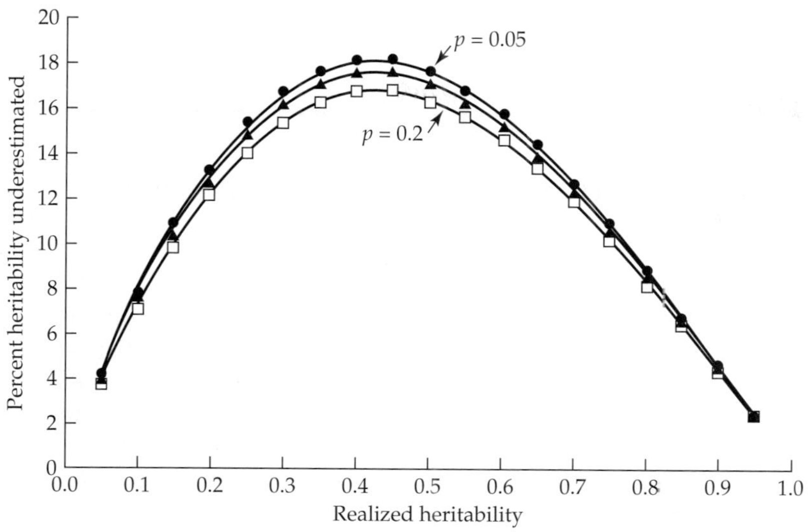

Figure 18.4 The relative percentage by which the base population heritability is underestimated by the realized heritability due to reduction in the additive variance by gametic-phase disequilibrium. The three curves correspond to different levels of directional truncation selection on a normally distributed trait, with 5% (upper curve), 10% (middle curve), and 20% (lower curve) of the population saved (corresponding to $\bar{\imath}$ values of 2.06, 1.75, and 1.40, respectively). Values were obtained by numerically solving Equation 18.21 and assuming the infinitesimal model (no significant selection-induced changes in allele frequencies) holds.

majority of this decrease occurs over the first few generations, after which the heritability is essentially at its equilibrium value, $\widetilde{h}^2$, where (from Equations 16.13a and 16.13c)

$$\widetilde{h}^2 = \frac{\gamma}{1+\gamma-h^2} \quad \text{with} \quad \gamma = \frac{2h^2 - 1 + \sqrt{1+4h^2(1-h^2)\kappa}}{2(1+\kappa)} \tag{18.21}$$

Here h^2 is the (unselected) base population heritability and κ is a measure of the reduction in phenotypic variance due to selection (the variance following selection being $(1-\kappa)\,\sigma_z^2$; Equation 16.10a). In particular, $\kappa = \bar{\imath}\,(\bar{\imath} - z_{[1-p]})$ for truncation selection that saves a fraction, p, of the population for a normal trait (Table 16.1). One can thus approximately treat the realized heritability as an estimate of the equilibrium heritability ($\widehat{h}_r^2 \simeq \widetilde{h}^2$), using Equation 18.21 to numerically solve for the base population heritability (h^2), an approach first used by Atkins and Thompson (1986). Figure 18.4 shows the relative percentage by which the base population heritability is underestimated by the uncorrected realized heritability.

Example 18.6. What are the disequilibrium-corrected estimates for the realized heritability values obtained in Examples 18.2 and 18.3? Recall that for this experiment, $p = 20/100 = 0.2$. From Example 16.3, this value of p yields $\kappa = 0.781$. Substituting this value into Equation 18.21 and using each of the three previous estimates of h_r^2 gives infinitesimal-model corrected realized heritabilities as:

	Estimate of h_r^2	
Estimator	Uncorrected	Corrected
Unweighted least-squares regression, $\widehat{b}_C$(OLS)	0.22	0.25
Total response/total differential, $\widehat{b}_T$	0.21	0.24
Weighted least-squares regression, $\widehat{b}_C$(GLS)	0.22	0.25

If we assume that the infinitesimal model is a reasonable approximation over the course of this experiment, all three methods underestimated the base population heritability by about 13% (see Figure 18.4).

EXPERIMENTAL EVALUATION OF THE BREEDER'S EQUATION

Although the breeder's equation requires a number of assumptions (Table 13.2) and, strictly speaking, holds only for a single generation of selection from an unselected base population, the general claim is that it is usually satisfactory (using the base population h^2 value) over a few (5–10) generations (e.g., Falconer 1981). A slightly more refined statement is that the number of generations should not exceed $N_e/2$ (Hill 1977a), as drift significantly changes the genetic variance within a line after this amount of time has elapsed (Equation 11.2). After a sufficient number of generations, drift and selection change the genetic variances substantially from their base population values, and the breeder's equation (using the initial heritability value) fails (Chapters 24–26).

Our concern here is the adequacy of the breeder's equation for short-term artificial selection experiments, with other empirical trends from selection experiments (both short- and long-term) reviewed in Chapters 25 and 26 and with response in natural populations examined in Chapter 20. Starting with the first formal comparisons by Reeve and Robertson (1953) and Clayton and Robertson (1957), a number of authors have compared their observed short-term responses with those predicted from the breeder's equation. As summarized below, the results are mixed. One complication is that most analyses use ordinary (unweighted) least squares, resulting in significantly underestimated standard errors. Likewise, almost all realized heritability estimates have not been adjusted for the expected decline due to gametic-phase disequilibrium.

Most Traits Respond to Selection

Before delving into the goodness-of-fit of the breeder's equation, it is important to emphasize a fundamental observation in quantitative genetics: ***Most traits in outbred populations respond to selection***. This is such a fundamental observation that the few exceptions are classic and well-known. The first such exception is sex ratio in chickens. Despite enormous economic incentive to create female-biased lines, to our knowledge this has not happened.

The second exception is selection for **directional asymmetries, DA**, (i.e., "handedness") in traits in *Drosophila* (Lewontin 1974). While selection to change the *variance* in asymmetry is usually successful (Chapter 17), attempts to select for a consistently higher trait value on one *specific* side of an organism (such as more bristles on the right side) when DA are absent in the base population have historically been unsuccessful (Maynard Smith and Sondhi 1960; Purnell and Thompson 1973; Coyne 1987; Tuinstra et al. 1990; Carter et al. 2009). Because there are numerous examples, such as handedness of the larger claw in fiddler crabs and wing features in many insects (reviewed by Pélabon and Hansen 2008), of such directional asymmetries in nature, this lack of response is intriguing. Two studies offer some potential insight. First, Pélabon et al. (2005) observed changes in DA as a correlated response to selection on an index of wing traits in *Drosophila*. The key difference from previous unsuccessful experiments is that this index itself showed DA in the control population, while the previously cited failures of response in DA occurred in traits that initially were symmetric. Hence, when DA are already present, it may be possible to amplify them via selection. Second, Rego et al. (2006) found significant genetic variation in DA following hybridization of two closely related *Drosophila* species, showing that there is at least between-population genetic variation in DA for some traits.

When we move from univariate to multivariate traits, the notion that each trait usually responds to selection is less clear. There often appear to be serious constraints (i.e., little

Table 18.1 Comparison of realized heritabilities ($\widehat{h}_r^2$) and heritability estimates based on resemblances between relatives ($\widehat{h}^2$). Within each group, the table gives the distribution of the percent absolute disagreement ($|\,\widehat{h}^2 - \widehat{h}_r^2\,|/\widehat{h}_r^2$) between the two estimates, where n is the number of experiments considered for each species group. For example, 8% of the 60 *Drosophila* experiments had a percent absolute disagreement between estimates of 30% to 50%. (After Sheridan 1988.)

	Distribution of absolute disagreement (relative to $\widehat{h}_r^2$)					
Species	0–10%	10–20%	20–30%	30–50%	>50%	n
Drosophila	0.48	0.12	0.05	0.08	0.27	60
Tribolium	0.31	0.08	0.12	0.38	0.12	26
Mice and rats	0.23	0.00	0.13	0.23	0.41	39
Poultry and quail	0.20	0.05	0.17	0.05	0.54	41
Swine and sheep	0.20	0.00	0.10	0.20	0.50	10
Summary over all groups						
Laboratory species	0.37	0.07	0.09	0.19	0.28	125
Commercial species	0.19	0.05	0.18	0.11	0.47	62
All species	0.31	0.06	0.12	0.17	0.34	187

genetic variation) for specific *combinations* of traits, despite significant heritabilities in each *component*. In such settings, selection on *only* a specific trait results in a response, while selection on that trait *as part of* a multivariate selection index may not yield any response, or even may yield a response in the opposite direction of selection on that trait (Volume 3).

Sheridan's Analysis

One of the most extensive reviews of the fit of the breeder's equation is that of Sheridan (1988), who examined 198 experiments involving laboratory and domesticated animals and compared realized heritabilities with estimates of heritability based on resemblances between relatives. Sheridan first considered those experiments in which the base populations already had an extensive past history of artificial selection on the target trait. In these populations, the response was very poorly predicted by the breeder's equation, with all 11 experiments (6 in *Drosophila*, 2 in *Tribolium* and 3 in mice) showing greater than 50% disagreement between the realized and estimated (i.e., base population) heritabilities.

Table 18.1 shows the fit for the remaining 187 experiments, namely, for those traits without an extensive past history of artificial selection. As the data show, the fit was rather poor in many experiments—roughly half have a disagreement of at least 30%, and one in three exceeds 50%. In addition to the biological reasons listed in Table 13.2, there are also design issues that could account for the apparently poor fit. First, none of the experiments reviewed by Sheridan corrected for the expected decline in the realized heritability due to gametic-phase disequilibrium. Second, small absolute disagreements can translate into large relative percentages of disagreement for traits with low heritabilities (Hill and Caballero 1992). For example, $|\,\widehat{h}_r^2 - \widehat{h}^2\,| = 0.04$ is a 20% relative disagreement if $\widehat{h}_r^2 = 0.2$, but an 80% disagreement if $\widehat{h}_r^2 = 0.05$. Third, variance in response will also generate some level of disagreement. This raises the questions as to which differences in the table are significant. As a rough approximation, Sheridan considered the disagreement to be significant if it exceeded two standard errors (which amounts to $p < 0.05$ if the data are normally distributed). Assuming the two estimates are uncorrelated, the variance for their difference is the sum of the variance for each estimate, giving a standard error of

$$\text{SE}\left(\widehat{h}^2 - \widehat{h}_r^2\right) = \sqrt{\left(\text{SE}[\,\widehat{h}^2\,]\right)^2 + \left(\text{SE}[\,\widehat{h}_r^2\,]\right)^2} \tag{18.22}$$

Table 18.2 Tests of significance between relative-based estimates of h^2 and realized heritabilities. Differences of more than two standard errors (computed from Equation 18.22) were regarded as significant. Only those experiments (from Table 18.1) with estimated standard errors for both heritability estimates are included. As above, n denotes the number of such experiments. Given that most studies use OLS (resulting in SEs that are too small), the fraction of significant differences is best viewed as an upper bound. (After Sheridan 1988.)

Species	Proportion of Significant Differences	Total
Drosophila	0.23	61
Tribolium	0.27	26
Mice and rats	0.18	34
Poultry and quail	0.45	11
Swine and sheep	0.53	15
Summary over all groups		
Laboratory species	0.21	131
Commercial species	0.37	30
All species	0.25	151

Table 18.3 Agreement between realized and estimated base population heritability as a function of the duration of the experiment. (After Sheridan 1988.)

	Percent absolute disagreement (relative to $\widehat{h}_r^2$)			
Generations	0–10%	10–30%	> 30 %	n
1–5	0.18	0.27	0.55	44
6–10	0.24	0.17	0.59	98
10–15	0.52	0.10	0.38	90

Taking those 131 experiments with standard errors for both $\widehat{h}_r^2$ and $\widehat{h}^2$, Table 18.2 shows that 25% had realized heritabilities significantly different from $\widehat{h}^2$. One problem with this approach is that most of the reported standard errors for $\widehat{h}_r^2$ used by Sheridan were based on unweighted least-squares (OLS), which underestimates the true standard error, giving confidence intervals that are too narrow and thereby inflating the level of significance. The overall 25% of experiments reported by Sheridan as showing a significant departure from the predicted response is thus an *upper bound*, suggesting that the breeder's equation may not be doing such a poor job after all.

Finally, Sheridan looked at the goodness-of-fit as a function of the duration of the experiment (Table 18.3). Surprisingly, longer experiments tended to have a better fit. While this is contrary to expectations, it could simply be a design artifact. In many cases, longer experiments employ larger population sizes than experiments of shorter duration, thus reducing the effects of drift.

Realized Heritabilities, Selection Intensity, and Inbreeding

In addition to quantitative differences in the predictions of the breeder's equation discussed above, there are also reported cases of major *qualitative* departures. For example, the breeder's equation predicts that while selection response should increase with selection intensity, the ratio of response to selection differential, R/S, should be constant. Some studies have reported a dependence of realized heritability on the selection intensity, although a survey of selection experiments found no consistent pattern (Table 18.4).

Table 18.4 Summary of experiments examining the effects of selection intensity on $\widehat{h}_r^2$.

Abdominal bristles in *D. melanogaster* Clayton and Robertson (1957)	$\widehat{h}_r^2$ decreases with increasing selection intensity.
Abdominal bristles in *D. melanogaster* Frankham et al. (1986)	Agreement between base population estimate of h^2 and $\widehat{h}_r^2$ is best at highest selection intensity, becoming worse as selection intensity decreases.
Postweaning weight gain in mice Hanrahan et al. (1973)	No consistent effect of selection intensity on $\widehat{h}_r^2$.
Pupal weight in *Tribolium castaneum* Meyer and Enfield (1975)	$\widehat{h}_r^2$ decreases with selection intensity in down-selected lines; no effect in up-selected lines.
Kernal oil content in maize Silvela et al. (1989)	No effect of selection intensity on $\widehat{h}_r^2$.

Table 18.5 Results of experiments examining the effects of finite population size and inbreeding on short-term response. Chapter 26 examines drift and long-term response in more detail.

Wing length in *D. melanogaster* Tantawy and Reeve (1956)	Short-term response for outbreds > double first cousins > sib mating.
Gonadotrophic hormone level in rats Chung and Chapman (1958)	Outbred and crossbred lines had larger short-term responses than inbred lines.
Body size in mice Lewis and Warwick (1953)	The outbred line had a larger short-term response than an inbred line.
Abdominal bristles in *D. melanogaster* Frankham et al. (1968a)	Weak trend for larger population sizes to have a greater short-term response.
Postweaning weight gain in mice Hanrahan et al. (1973)	Short-term response increased with increasing population size.
Kernal oil content in maize (Silvela et al. 1989)	Short-term response increased with increasing population size.

There are several reasons why a dependency may arise between realized heritability and selection intensity. First, increasing the selection intensity increases gametic-phase disequilibrium, thus reducing σ_A^2 and hence $\widehat{h}_r^2$ (Chapter 16). As Figure 18.4 shows, the prediction is that (uncorrected) realized heritabilities should decrease with increasing selection intensity (albeit only slightly in many cases). Again, essentially none of the reported selection experiments correct for this expected reduction. Second, allele frequencies are expected to change more rapidly as the strength of selection increases. Whether this results in an increased or decreased response depends on the initial distribution of allele frequencies and effects (see the discussion on genetic asymmetries below).

Finally, N_e decreases as selection intensity increases (Chapter 3 and 26), thus increasing the amount of inbreeding and (generally) reducing the additive variance (Chapter 11). When inbreeding occurs, the short-term response is generally found to be less than that predicted from the breeder's equation when using variance components estimated from the base population (Table 18.5). This is expected, as inbreeding (generally) decreases the additive genetic variance within a line (Chapter 11), thus reducing response. Exceptions can occur if nonadditive genetic variance is present, in which case inbreeding may actually result in an *increase* in the within-line additive variance (Chapter 11), increasing response. The effects of drift on long-term experiments are examined in detail in Chapter 25. Inbreeding depression is another complication, which we will address shortly.

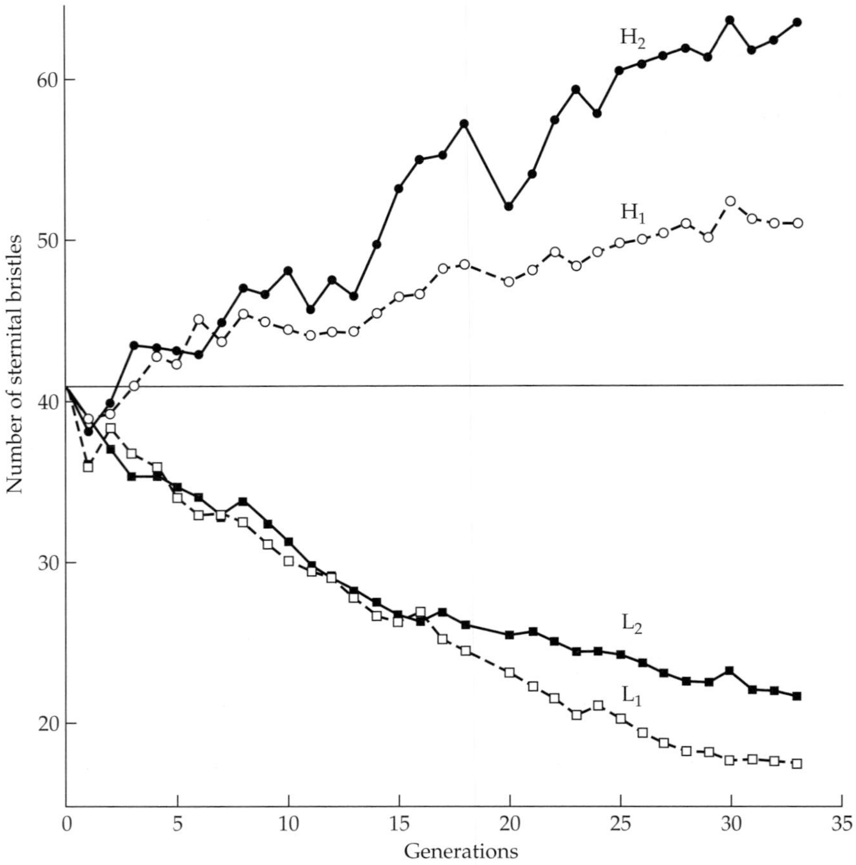

Figure 18.5 Selection response on abdominal bristle number in male *Drosophila melanogaster*. Two replicates in each direction (High and Low) were performed, all with the same selection differential. Contrary to the symmetric response expected under the breeder's equation, an asymmetric selection response was observed. (After Sheldon 1963.)

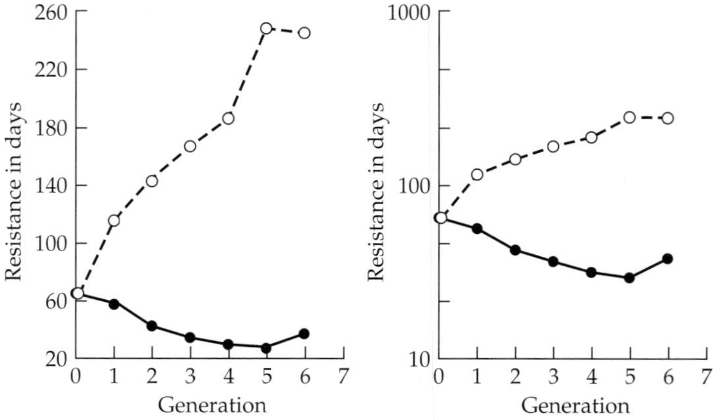

Figure 18.6 An example of a scale effect generating an asymmetric selection response. **Left:** Selection for resistance to dental caries (cavities) in albino rats (*Rattus norvegicus*), response measured as the expected number of days to develop caries on a standard diet. **Right:** The same data presented on a log scale show a symmetric response. (Based on Falconer's 1954 analysis of data from Hunt et al. 1944.)

Table 18.6 Possible explanations for asymmetric response (including reversed response).

Design artifacts	Drift (Chapter 12) Scale effects (LW Chapter 11) Different effective selection differentials (Chapter 13) Undetected environmental trends (Chapter 20) Transient effects from previous selection in the base population, e.g., decay of response from $A \times A$ or genetic maternal effects (Chapter 15) Undetected selection on correlated characters (Chapter 20)
Nonlinear parent-offspring regressions	Major genes with dominance (LW Chapter 17) Genotype $\times$ environment interactions (Volume 3) Departures from normality (Chapter 24)
Other sources	Genetic asymmetries (this chapter) Inbreeding depression (Chapter 23) Maternal effects (Chapters 15 and 22) Associative effects (Chapter 22)

Asymmetric Selection Response

A common design is to perform a **divergent selection experiment**, wherein replicate lines are selected in opposite directions. Many such experiments (e.g., Figures 18.5–18.8) show different amounts of response in the up versus down direction, a phenomenon referred to as an **asymmetric selection response** (Falconer 1954). This is in sharp contrast with the expectation from the breeder's equation, which predicts that the absolute magnitude of response should depend only the absolute value of S, and not on its sign.

There are a variety of possible explanations for asymmetric responses (Table 18.6). They may simply be artifacts of experimental design or analysis. In particular, the prediction of equal positive and negative slopes holds only for plots of cumulative responses versus cumulative *selection differentials*, $R_C(t)$ versus $S_C(t)$. Asymmetry in response based on differences in slope of cumulative response versus *generations* of selection, $R_C(t)$ versus t, can thus be misleading, as the different lines may have experienced different amounts of selection. Likewise, differences in response could simply be scale effects (Figure 18.6; see LW Chapter 11). For example, if the genetic variance increases with the mean, the heritability can also increase with the mean, resulting in a faster response in the upwardly selected lines. Asymmetric differences in selection response can also be due to biological features that were not considered by the design. For example, even if the amount of *artificial* selection is the same in both directions, there may be major differences in *natural* selection (e.g., up-selected lines might experience lower fertility). Using effective selection differentials (Equation 13.9) corrects for this source of bias, but the investigator often lacks the data (e.g., fertilities for each parent) necessary to compute them. Likewise, we have seen that if a population has been under previous selection, its mean can change even after selection has stopped (Chapter 15). If we start a divergent selection experiment using such a nonequilibrium population as our base, we can bias the response in at least one direction, thus generating an asymmetric response even when the true genetic trend from the experiment is symmetric.

Even though lines may have quite different values of $\widehat{h}_r^2$, there remains the issue of whether these differences are statistically significant. Genetic drift can generate considerable variation between replicate lines, and it is important to distinguish between real differences in the expected response and the variation among realizations of a process with the same (absolute) expected value. Directional trends in the environment can also produce asymmetries, inflating the apparent response in the direction of the trend and retarding it in the opposite direction. An example of this can be seen in the bottom graph in Figure 18.7. If one simply compares the High and Low lines, a clear asymmetric response is seen.

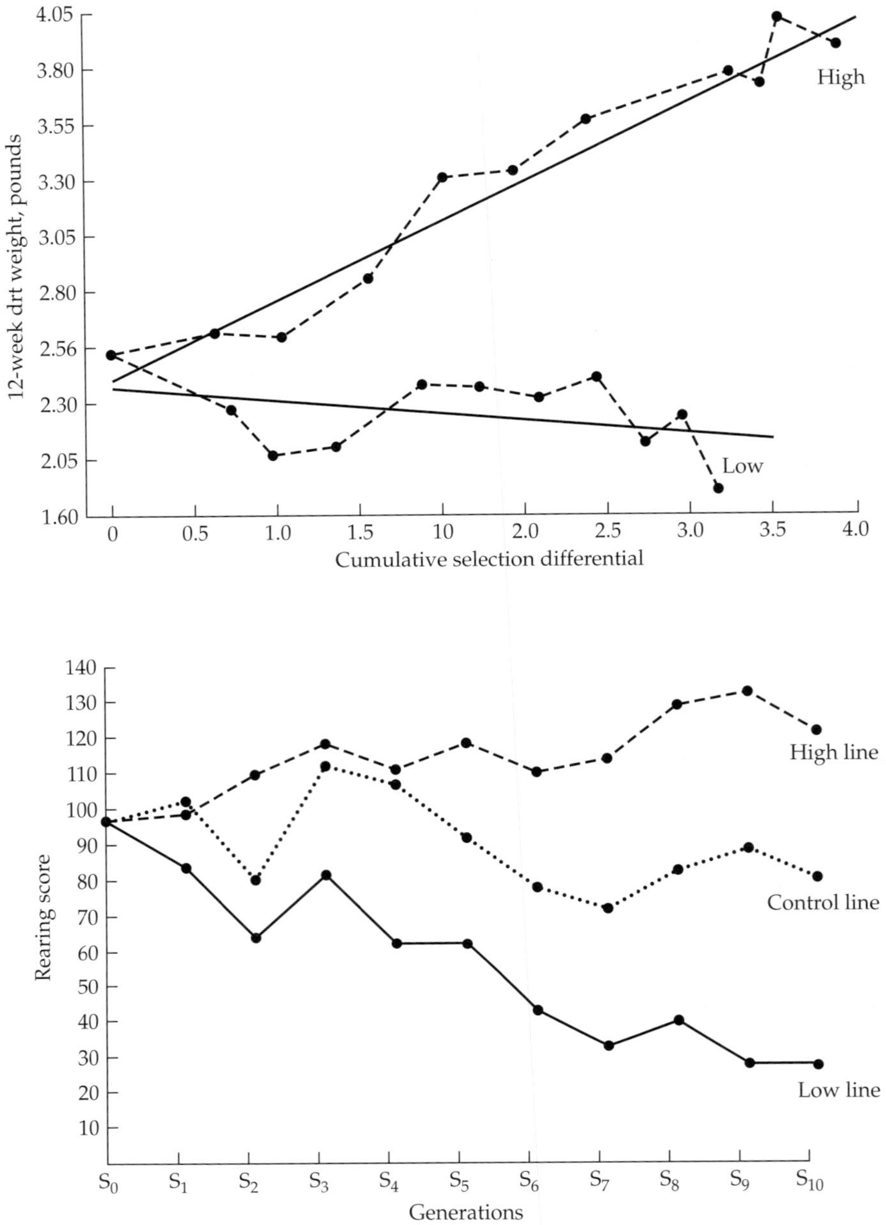

Figure 18.7. Further examples of asymmetric response to selection. **Top:** Twelve-week body weight in chickens (Maloney et al. 1963). **Bottom:** Rearing activity in rats (Sanders 1981).

However, when comparing each line as measured as a deviation from the Control line, the response is much more symmetric.

Although spurious asymmetric responses can result from these defects in design and analysis, true asymmetric responses can be generated by a variety of genetic situations. For example, if the parent-offspring regression is nonlinear, S will not be sufficient to predict the response (Gimelfarb and Willis 1994; Chapters 6, 13, and 24), and it is not surprising that asymmetric responses can be generated in such cases. Failure to initially detect departures from linearity in the base population does not rule out nonlinearity as an explanation for an observed asymmetric response, as the range of variation in the base population may not be sufficient to detect any such departures at the extreme ends of the initial range of phenotypes. As the selected lines diverge, however, differences in the tails of the initial phenotypic distribution can be revealed. Likewise, as previously mentioned, heritability can be a function of the part of the phenotypic distribution from which individuals are drawn (e.g., Skibinski and Shereif 1989).

Characters displaying inbreeding depression (LW Chapter 10) show an asymmetric selection response, with the change in mean from inbreeding accentuating the response in one direction and retarding it in the other. A simple test for inbreeding as an explanation of an asymmetric response is to see whether the mean of an unselected control population (when inbred) changes in the direction of greater response. The effects of inbreeding depression can be corrected by inbreeding a control population to the same level as the divergent selection lines and using the contrast between selected and control lines to estimate response. However, this process is not necessarily as straightforward as it appears, as selection generally increases the amount of inbreeding within a line by decreasing its effective population size (Chapters 3 and 26). Simply keeping the control line at the same size as the selected lines thus underestimates the amount of inbreeding (and hence any correction for inbreeding depression), especially when selection is intense. An alternative tests is to cross two replicate lines, with heterosis in the resulting F_1 indicating inbreeding depression.

The assumption of negligible allele-frequency change over a few generations of selection can be violated when alleles of major effect are segregating. In this setting, asymmetries can arise as a consequence of allele frequencies changing in different directions in the differentially selected populations. An increase in an allele from its initial frequency nearly always results in a different additive variance from that produced by an equal decrease in frequency. Falconer (1954) refers to this feature as **genetic asymmetry**. This is most easily seen by considering LW Figure 4.6, which shows additive genetic variation as a function of allele frequency for a single diallelic locus. If alleles are completely additive, then σ_A^2 as a function of allele frequency is symmetric about $p = 1/2$. Suppose that the frequency of an allele that increases the character value is initially below $1/2$. In upwardly selected lines, σ_A^2 (and h^2) increase as this allele increases to a frequency of 0.5 and decrease when it exceeds this value. Conversely, in downwardly selected lines, the contribution to σ_A^2 from this locus always decreases. If dominance is present, then σ_A^2 will no longer be a symmetric function of allelic frequencies (LW Figure 4.8) and asymmetric changes in the contribution to σ_A^2 from a single locus are almost always expected.

Frankham and Nurthen (1981) provided an interesting example of this phenomena. They started with a base population with the major recessive allele sm^{lab} (which greatly reduces abdominal bristle number in *Drosophila melanogaster*) at low frequency. As shown in Figure 18.8, in two of three down-selected lines, this allele increased in frequency, resulting in a large increase in h^2 as the sm^{lab} allele reached intermediate frequencies. Presumably, this allele was either lost or not sampled when Line 1 was formed, as it does not show this pattern. Heritability returned to the base population value as this allele approached fixation, suggesting much smaller allele-frequency change in the residual polygenic genetic variation that remains once the major allele has been removed.

While we have focused on the effects of a single major gene, the effects of unequal allele frequencies apply to any QTL. Alleles of small effect are expected to have slower changes in allele frequencies (and thus are expected to have smaller effects on asymmetry) over the time scales of most short-term experiments. Hence, if many loci of small effect underlie a character, the effects of genetic asymmetry on selection response are expected to be slight unless the number of generations is large. Further, at least some cancellation is expected between the asymmetries in σ_A^2 generated by allele-frequency changes at different loci. Analogous to directional dominance being required for inbreeding depression (a systematic trend in the direction of dominance over loci; LW Chapter 10), there must also be some systematic trend in the allele frequencies at loci underlying the trait for genetic asymmetries to occur (in the absence of major genes). For example, if most alleles that decrease the trait value tend to be rare, then up-selected versus down-selected lines are expected to show an asymmetric response as allele-frequency changes become significant, with the rare alleles increasing (and increasing variance) in the down-selected, but not the up-selected, lines.

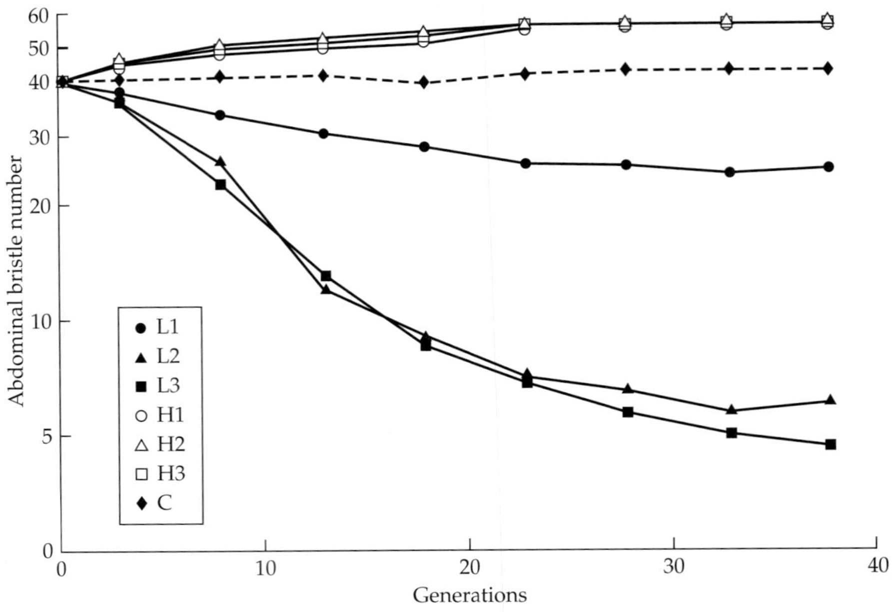

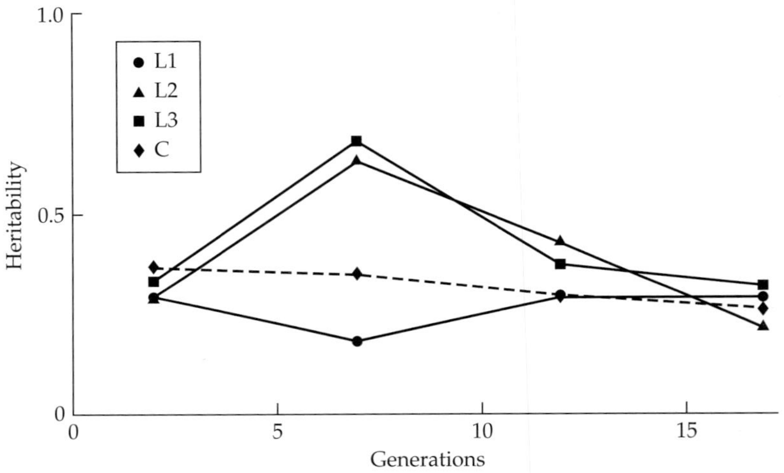

Figure 18.8 **Top:** Asymmetric selection response for abdominal bristle number in *Drosophila melanogaster* from a base population containing a major allele (sm^{lab}) that was initially at low frequency. L1–L3 are three replicate low lines, H1–H3 are three replicate high lines, and C is the control line. Data were log-transformed to remove scale effects, and on this new scale, low line L1 has a symmetric response with high lines H1–H3, while lines L2 and L3 do not. Further, there is also a greater variance in response among the low lines. **Bottom:** Changes in heritabilities in the control (dashed curve) and lines L1–L3. Heritability was estimated from phenotypic correlations between bristle numbers of adjacent segments of the same fly (see Frankham and Nurthen [1981] for details). Note the large increases in heritability for lines L2 and L3, the lines that show an asymmetric selection response (relative to H1–H3), while the heritability is roughly constant in L1, which does not show an asymmetric response. The increase in h^2 reflects an increase in the major allele sm^{lab} due to selection. This allele increased rapidly in frequency after generation 5 and was essentially fixed by generation 10, which is reflected by a rapid increase in h^2 after generation 5, with h^2 returning to normal (the level in the unselected control) as the allele becomes fixed. (After Frankham and Nurthen 1981.)

When might such an asymmetric distribution of allele frequencies be present? One situation is that where there has been a recent history of selection on the trait before the start of the

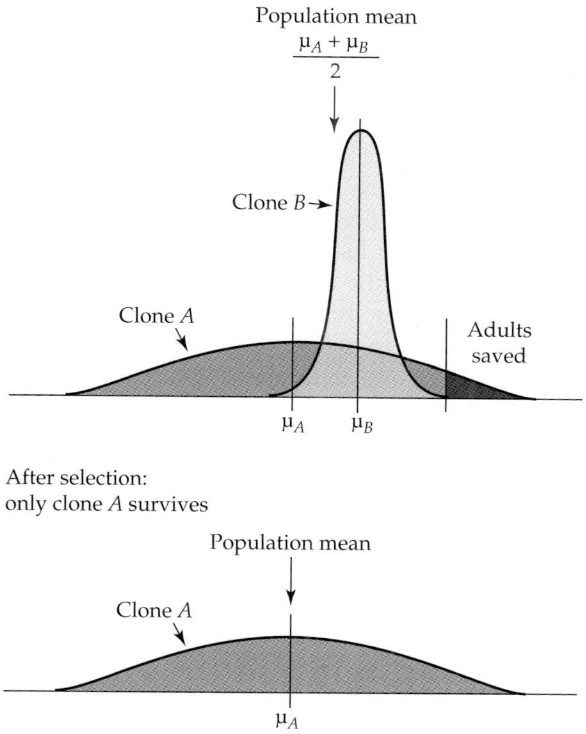

Figure 18.9 Haldane's (1931) example of a reversed response generated by genotype $\times$ environment interaction. The population consists of equal numbers of two asexual clones, A and B. The mean phenotypic value of A (μ_A) is less than the mean phenotypic value of B (μ_B). However, clone A has a larger environmental variance than B. Strong truncation selection culls out all members of clone B in the population, leaving only A. The new mean in the generation following selection, μ_A, is less than the mean before selection, $(\mu_A + \mu_B)/2$. In this case, selection for *increased* character value z, resulted in a *lowering* of the population mean.

artificial selection experiment.

Components of reproductive fitness, such as fecundity and development time, are expected to have asymmetric allele frequencies, as natural selection increases the frequency of alleles increasing fitness. This condition suggests that asymmetric responses in the selection for reproductive traits are expected, and it further predicts that response should be larger in lines selected for a decrease in reproductive fitness. This was seen in a review by Frankham (1990), who found a larger response in the direction of reduced reproductive fitness in 24 of 30 experiments. Frankham suggested that the presence of rare recessives decreasing reproductive fitness was the most likely cause for this trend. As Figure 25.2 will show, very marked asymmetric responses are expected in such situations.

Reversed Response

The most extreme departure from the breeder's equation is a **reversed response**, a response in the *opposite* direction of selection. Negative maternal effects can result in such a response (Figures 15.3 and 15.4). Likewise, they can arise from sufficiently large genotypic-dependent environmental variances (Haldane 1931; Chapter 17). Haldane's intuition was based on selection with two asexual clones, the first with a higher mean but smaller environmental variance than the second. As shown in Figure 18.9, if the most extreme individuals are selected, there will be an excess of the clone with the smaller mean but higher variance, resulting in a decrease in mean value in the next generation. Wright (1969) expanded Haldane's model to a single diallelic locus. Gimelfarb (1986b) presented a particularly interesting analysis when there is a multiplicative genotype $\times$ environment interaction, showing for this model that while *phenotypes* are subjected to directional selection, the nature of the interaction between genotype and environment is such that *genotypic* values can actually be under either stabilizing or disruptive selection. The interesting feature of both Haldane's and Gimelfarb's models is that reversed response is most probable when selection is very intense, as this is the setting most favorable to genotypes with high variances (Chapter 17).

Important, and not yet widely appreciated, sources of reversed response are **associative effects** (also called **social effects** or **indirect genetic effects**), of which maternal effects are a special case. As introduced by our discussions on maternal effects in Chapter 15, and more fully developed in Chapter 22, the environment that a focal individual experiences can be strongly influenced by its neighbors. In such cases, an individual's phenotype is the result of both **direct effects**, due to its own intrinsic genes, and **social effects**, from its neighbors. If the latter have a heritable component, these effects can also evolve. If direct and social effects are genetically negatively correlated, a positive response in the direct trait can be overcome by a negative social response, resulting in a net *decrease* in trait value. Consider selection for egg production in caged chickens. Aggressive individuals gather more food in the cage thus producing more eggs. With selection based solely on individual egg production, more aggressive chickens are chosen (i.e., those with negative associative effects), creating a deleterious cage environment that can overshadow any direct genetic advance in terms of number of eggs, resulting in a *reduction* in the number of eggs per cage in the next generation.

CONTROL POPULATIONS AND EXPERIMENTAL DESIGNS

We conclude this chapter by considering several specialized issues in the analysis of selection experiments, topics that may be skipped by the casual reader. We first examine the strengths and weaknesses of different types of designs (such as the use of a control population), then consider optimal designs, and finish with a discussion of a modified line-cross analysis (LW Chapter 9) to examine the genetic components of a selection response.

One complication with estimating realized heritabilities is distinguishing between genetic versus environmental trends. For example, Newman et al. (1973) found that 60% of the increase in yearling weight in a selected line of Shorthorn cattle was due to environmental, rather than genetic, improvement. Using a large unselected **control population** reared under the same environmental conditions as the selected line or lines allows for correction of such between-generation environmental changes (e.g., the bottom graph in Figure 18.7). Hill (1972a, 1972b) reviewed this approach and some of the features seen in control populations from a number of different species. The use of control populations is not a foolproof approach for removing environmental trends, as the control and selected lines can evolve different genotype $\times$ environment interactions. Likewise, extensive genetic drift in the mean of the control population can also result in biases, resulting in the corrected (selection minus control) values underestimating or overestimating the true selection trend. When full pedigree data are available, the powerful methods of mixed-model analysis (Chapter 19) can be used to remove genetic and environmental trends, even in the absence of control populations (although incorporating such controls is preferred). Undetected environmental trends are especially problematic in the analysis of response in natural populations (Chapter 20).

Basic Theory of Control Populations

Assuming no genotype $\times$ environment interaction, the true mean of a line in generation t can be decomposed as $\mu + g_t + b_t$, where μ is the base population mean, g_t is the change in mean breeding value due to selection and drift, and b_t is the change in mean due to environmental change. Under this model, observed means, $\overline{z}$, for a selected (s) and control (c) population reared in a common environment can be decomposed as

$$\overline{z}_{s,t} = \mu + g_{s,t} + b_{s,t} + e_{s,t} \tag{18.23a}$$

$$\overline{z}_{c,t} = \mu + g_{c,t} + b_{c,t} + e_{c,t} \tag{18.23b}$$

where e_t is the error in estimating the mean breeding value $(\mu + g_t)$ from the observed mean, corrected for the change in the environment $(\overline{z}_t - b_t)$, and has an expected value of 0.

Assuming that the breeder's equation holds, the expected total (i.e., cumulative) response at generation t is $E[\,R_C(t)\,] = E[\,g_{s,t}\,] = h^2 S_C(t)$. Under drift alone, there is no expected directional change in the mean breeding value of the control population, $E[\,g_{c,t}\,] = 0$. If we assume that there are no genotype $\times$ environment interactions (i.e., $b_{s,t} = b_{c,t}$), the contrast between the selected and control populations has an expected value of

$$E[\,\overline{z}_{s,t} - \overline{z}_{c,t}\,] = E[\,g_{s,t}\,] - E[\,g_{c,t}\,] = h^2 S_C(t) \tag{18.24a}$$

If the control population is small, $g_{c,t}$ can drift significantly away from zero, resulting in an overestimation (or underestimation) of the true selection response. If the goal is simply to remove an environmental trend, control populations should be kept as large as possible so that $g_{c,t}$ is close to zero. However, as previously mentioned, control populations are also used in an attempt to correct for any effects of inbreeding depression. In such cases, the size of the control population is usually on par with that for the selected population, meaning that significant drift has likely occurred. In this setting, the use of a control population may introduce additional uncertainty in the estimate of the response (which may be more than compensated for by the control accounting for inbreeding depression and environmental trends).

If genotype $\times$ environmental interactions are present, then

$$E[\,\overline{z}_{s,t} - \overline{z}_{c,t}\,] = h^2 S_C(t) + (b_{s,t} - b_{c,t}) \tag{18.24b}$$

resulting in $\overline{z}_{s,t} - \overline{z}_{c,t}$ being a potentially biased estimator of $h^2 S_C(t)$. In principle, if the environmental trends are positively correlated between populations, then the use of a control will still improve the estimate of the genetic trend. If they are negatively correlated, however, the use of a control can lead to a less accurate estimate than simply using a selected population without a control. As discussed in Chapter 19, mixed-model analyses may be able to provide some insights (as they estimate the $b_{x,t}$), but they require extensive information (full pedigrees) and that the model assumptions hold.

If a control population is used, the (per-generation, R_t, and cumulative, $R_C[t]$) responses and differentials are estimated by

$$R_t = (\,\overline{z}_{s,t} - \overline{z}_{c,t}) - (\,\overline{z}_{s,t-1} - \overline{z}_{c,t-1}) \tag{18.25a}$$

$$R_C(t) = \overline{z}_{s,t} - \overline{z}_{c,t} \tag{18.25b}$$

$$S_t = \overline{z}^{\,*}_{s,t-1} - \overline{z}_{s,t-1} \tag{18.25c}$$

$$S_C(t) = \sum_{i=1}^{t} S_i \tag{18.25d}$$

where $\overline{z}^{\,*}$ denotes the mean of the selected individuals. If selection differs between the sexes

$$S_t = \frac{S_t(m) + S_t(f)}{2}$$

where $S_t(m)$ and $S_t(f)$ are, respectively, the observed selection differentials on males and females.

Muir (1986a, 1986b) has suggested that an analysis of covariance approach (along the lines of LW Equation 10.12) be used when extensive G $\times$ E is expected. In its simplest form, Muir's idea is to adjust the mean of the selected population in generation i by using the deviation of $\overline{z}_{c,i}$ from its grand mean over the course of the experiment, $\overline{z}_{c,\cdot}$, namely

$$\overline{z}^{\,\prime}_{s,i} = \overline{z}_{s,i} - \beta(\,\overline{z}_{c,i} - \overline{z}_{c,\cdot})$$

If there is no genotype $\times$ environment interaction (so that the environmental effect at time i has the same influence on both the selected and control lines), then $\beta = 1$ and we recover our previous results. However, if there is G × E, such that the environmental deviation influences the control differently from the selected population, then this expression provides a less biased correction than simply subtracting off the mean of the control. Note that this covariate approach at least partly corrects for the most extreme case, namely that in which the environmental effect results in different signs in the control and selective populations ($\beta < 0$).

Divergent Selection Designs

A related approach is the **divergent** (or **bidirectional**) selection design, wherein one compares lines that were selected in opposite directions. Again assuming there are no significant genotype × environment interactions between lines, the basic statistical model for this design is

$$\overline{z}_{u,t} = \mu + g_{u,t} + b_{u,t} + e_{u,t} \tag{18.26a}$$

$$\overline{z}_{d,t} = \mu + g_{d,t} + b_{d,t} + e_{d,t} \tag{18.26b}$$

where the subscripts u and d refer to the upwardly and downwardly selected lines, respectively. Using this design (and assuming that $b_{u,t} = b_{d,t}$), the responses and differentials are estimated by

$$R_t = (\,\overline{z}_{u,t} - \overline{z}_{u,t-1}) - (\,\overline{z}_{d,t} - \overline{z}_{d,t-1}) \tag{18.27a}$$

$$R_C(t) = \overline{z}_{u,t} - \overline{z}_{d,t} \tag{18.27b}$$

$$S_t = (\,\overline{z}^{\,*}_{u,t-1} - \overline{z}_{u,t-1}) - (\,\overline{z}^{\,*}_{d,t-1} - \overline{z}_{d,t-1}) \tag{18.27c}$$

Again, the expected response (using Equations 18.27a and 18.27c for the response and selection differentials) is simply $R = h^2 S$. More generally, if $S_{x,i}$ denotes the selection differential in line x at generation i, the total (cumulative) expected divergence between lines is

$$R_C(t) = h^2 \sum_{i=1}^{t} (S_{u,i} - S_{d,i}) = h^2\,(S_{C,u} - S_{C,d}) \tag{18.27d}$$

Variance in Response

Recall the pure-drift approximation for the variance (Equation 18.8) and the within-line covariance across generations (Equation 18.9) for the design of a single selected line. Here we present corresponding expressions for the selection plus control and divergence selection designs (assuming there are no genotype × environmental interactions between lines). For unidirectional selection plus a control population

$$\begin{aligned} R_C(t) = \overline{z}_{s,t} - \overline{z}_{c,t} &= (\,\mu + g_{s,t} + b_t + e_{s,t}) - (\,\mu + g_{c,t} + b_t + e_{c,t}\,) \\ &= (g_{s,t} - g_{c,t}) + (e_{s,t} - e_{c,t}) \end{aligned} \tag{18.28}$$

Similarly, for divergent selection

$$R_C(t) = \overline{z}_{su,t} - \overline{z}_{sd,t} = (g_{su,t} - g_{sd,t}) + (e_{su,t} - e_{sd,t}) \tag{18.29}$$

Because each term in Equations 18.28 and 18.29 is independent, applying Equations 18.4 and 18.5 yields a pure-drift approximation for the variance in response in generation t of

$$\sigma^2\,[\,R_C(t)\,] = (2f_t + B_0)\;h^2\sigma^2_z + B_t\,\sigma^2_z \simeq (t\,A + B_0)\,h^2\sigma^2_z + B_t\,\sigma^2_z \tag{18.30a}$$

and the covariance between generations within the same line

$$\begin{aligned} \sigma\,[\,R_C(t), R_C(t')\,] &= (2f_t + B_0)\;h^2\sigma^2_z \\ &\simeq (t\,A + B_0)\,\sigma^2_z\,h^2 \qquad \text{for } t < t' \end{aligned} \tag{18.30b}$$

Table 18.7 Coefficients for the pure-drift variances and covariances in response (Equations 18.30a and 18.30b). $M_{x,t}$ individuals are sampled in population x at generation t, of which N_x are allowed to reproduce. The subscripts $x = s$ and c refer to the selected and control populations, $x = u$ and d refer to the up-selected and down-selected lines, and $f_{x,t}$ refers to the amount of inbreeding in population x at time t.

Selection in a single direction without a control line. Equation 18.30a has an extra term, σ_b^2, accounting for the between-generation variation in environmental effects.

$$f_t = f_{s,t}, \qquad A = \frac{1}{N_s}, \qquad B_t = \frac{1}{M_{s,t}} \quad \text{for } t \geq 0$$

Selection in a single direction with a control line.

$$f_t = f_{s,t} + f_{c,t}, \qquad A = \frac{1}{N_s} + \frac{1}{N_c}, \qquad B_t = \frac{1}{M_{s,t}} + \frac{1}{M_{c,t}} \quad \text{for } t \geq 0$$

Divergent selection without a control line.

$$f_t = f_{u,t} + f_{d,t}, \qquad A = \frac{1}{N_u} + \frac{1}{N_d}, \qquad B_t = \frac{1}{M_{u,t}} + \frac{1}{M_{d,t}} \quad \text{for } t \geq 0$$

---

where the (design-specific) coefficients, A and B_t, are given in Table 18.7. If the number of reproducing individuals varies over time, then the tA term is replaced by the sum over time of the components in A, for example by $\sum_{i=1}^{t} (1/N_i)$, or related expressions. Finally, recall (Equation 18.3) that for unidirectional selection without a control, the variance in response has an additional term, σ_b^2, accounting for the between-generation environmental variation.

Control Populations and Variance in Response

When does using a control population in a undirectional selection experiment reduce the variance in response? Equation 18.30a, along with the coefficients from Table 18.7, gives the expected variance with and without the use of a control. Assuming $M = M_s = M_c$ and $N = N_s = N_c$, this equation yields

$$\sigma^2 \left[R_C(t) \right]_{control} - \sigma^2 \left[R_C(t) \right]_{no\ control} = \left(\frac{t}{N} + \frac{1}{M_0} \right) h^2 \sigma_z^2 + \frac{1}{M} \sigma_z^2 - \sigma_b^2 \tag{18.31a}$$

Assuming the between-line drift variance dominates (terms involving M can be ignored), the condition for the variance in response with a control to be larger than the response without one is approximately

$$t\sigma_z^2 h^2/N > \sigma_b^2 \tag{18.31b}$$

Hence, regardless of the value of σ_b^2, if sufficient generations are used, the optimal design (in terms of giving the smallest expected variance in response) is not to use a control. However, this approach runs the risk that an undetected directional environmental trend will

Table 18.8 Coefficients of variation (CV) for various designs, assuming the pure-drift approximation and that $\sigma^2\left[R_C(t)\right] \simeq t\,Ah^2\sigma_z^2$. The latter assumes that the selection experiment is of sufficient duration that the among-line drift dominates (i.e., $t\,A \gg B_0$ and $t\,Ah^2 \gg B_t$, with these parameters defined in Table 18.7). We assume that the absolute selection intensity in all selected lines is $\bar{\imath}$.

Selection in a single direction with a control line.

$$E\left[R_C(t)\right] = t\,h^2\,\bar{\imath}\,\sigma_z, \qquad \text{CV}\left[R_C(t)\right] \simeq \frac{1}{h\bar{\imath}}\sqrt{\frac{2}{Nt}} \tag{18.32a}$$

Selection in a single direction without a control line.

$$E\left[R_C(t)\right] = t\,h^2\,\bar{\imath}\,\sigma_z, \qquad \text{CV}\left[R_C(t)\right] \simeq \frac{1}{h\bar{\imath}}\sqrt{\frac{1}{Nt}} + \frac{1}{t\,h^2\,\bar{\imath}}\left(\frac{\sigma_d}{\sigma_z}\right) \tag{18.32b}$$

Divergent selection without a control line.

$$E\left[R_C(t)\right] = 2t\,h^2\,\bar{\imath}\,\sigma_z, \qquad \text{CV}\left[R_C(t)\right] \simeq \frac{1}{h\bar{\imath}}\sqrt{\frac{1}{2Nt}} \tag{18.32c}$$

compromise the estimated heritability. Thus, the cost from to an increase in SE when using a control seems very worth enduring in many cases, given the more general benefits of having one or more control lines.

OPTIMAL EXPERIMENTAL DESIGNS

As Equation 18.31b illustrates, it is not entirely obvious which design is optimal. What in general can we say? The coefficient of variation (CV) of the selection response

$$\text{CV}\left[R_C(t)\right] = \frac{\sigma\left[R_C(t)\right]}{E[R_C(t)]} \tag{18.33}$$

is especially useful in comparing efficiencies of different designs, as it is independent of σ_z^2. Further, it provides an appropriate measure of comparative efficiencies when the expected response differs between designs. Table 18.8 gives expressions for the CV under some simplifying assumptions. The coefficient of variation is a function of tN, the total number of adults selected during the course of the experiment, provided drift variance dominates the error variance. A short experiment with many selected adults per generation thus gives the same expected CV as a longer experiment with fewer adults per generation (provided the total numbers, tN, are the same). However, if the error variance is nontrivial relative to the drift variance (as would be expected if h^2 is small), increasing the duration of the experiment results in some improvement in precision (Hill 1980).

As an example of using the CV of response, consider unidirectional selection without a control population versus divergent selection. Which is more efficient if the same total number of adults are selected (e.g., N under unidirectional selection and $N_d = N_u = N/2$ under divergent selection)? If there is no between-generation environmental variance, then both designs are equally efficient, while divergent selection is more efficient if $\sigma_b^2 > 0$.

Example 18.7. Suppose we plan to select the upper 5% of a population for a normally distributed character with $h^2 = 0.25$. What value of Nt is needed for the expected CV of response to be no greater than 0.01 if no control population is used? From Example 14.1, $E[\,\bar{\imath}\,] = 2.06$ if the population is large, and slightly less in small populations (for simplicity we assume the large population value applies). To solve for the value of Nt, we use the expression for the CV in response without a control line (Equation 18.32b), and further assume that the drift variance dominates, so that we ignore the contribution from σ_b^2. This yields

$$CV = 0.01 = \frac{1}{0.5 \times 2.06} \times \sqrt{\frac{1}{Nt}}$$

which has a solution of $Nt \simeq 9426$. Hence, during the entire course of the experiment using a total of at least 9426 selected parents returns an approximate expected CV of less than 1%. If the desired CV is 0.05 or 0.10, then $Nt \simeq 377$ and $Nt \simeq 94$, respectively. It is important to stress that these numbers are somewhat misleading, in that N is the number of *saved* individuals, while $M = N/p$ is the number of *measured* individuals. For $p = 0.05$, the value of Mt is 1/0.05 = 20 times the value of Nt, so one must measure roughly 18,900, 7540, and 1880 individuals to obtain these desired CVs.

Nicholas' Criterion

An alternative criterion for choosing Nt was suggested by Nicholas (1980). Often the investigator is interested in ensuring that at least a certain response will occur with a preset probability. To a reasonable approximation, the expected mean value in any given replicate selection line after t generations of selection is normally distributed, with a mean of $E[R_C(t)]$ and a variance of $\sigma^2\,[R_C(t)]$. Consider the probability that the observed response is at least β of the expected response

$$\Pr\Big(R_C(t) > \beta E[R_C(t)]\Big) = \Pr\left(\frac{R_C(t) - E[R_C(t)]}{\sigma\,[R_C(t)]} > \frac{(\beta - 1)E[R_C(t)]}{\sigma\,[R_C(t)]}\right)$$

$$= \Pr\left(U > \frac{\beta - 1}{\text{CV}\,[\,R_C(t)\,]}\right) \tag{18.34}$$

where U is a unit normal random variable. Note that the probability that the observed response *exceeds* the expected response ($\beta = 1$) is one half, as $\Pr[U > 0] = 1/2$.

Example 18.8. Again suppose that $\bar{\imath} = 2.06$, $h^2 = 0.25$, and the design is unidirectional selection without a control population. What value of Nt is required in order to obtain a 95% probability that the observed response is at least 90% of its expected response? Here, $\beta = 0.9$ and $\Pr[U > -1.65] = 0.95$. Hence,

$$\frac{\beta - 1}{\text{CV}\,[\,R_C(t)\,]} = \frac{-0.1}{\text{CV}\,[\,R_C(t)\,]} = -1.65$$

Rearranging yields

$$\text{CV}\,[\,R_C(t)\,] = \frac{1}{0.5 \times 2.06}\sqrt{\frac{1}{Nt}} = \frac{0.1}{1.65}$$

implying that $Nt \simeq 257$. As in Example 18.7, the required total number of *measured* individuals is $1/p$ times this value, yielding $Mt = 257/0.05 \simeq 5140$.

Replicate Lines

There is no loss of efficiency when replicate lines are used (Hill 1980). To see this, let $\bar{z}_{i,t}$ for $1 \leq i \leq r$ be the sample mean for replicate population i at time t. The overall mean is

$$\bar{z}_t = \frac{1}{r}\sum_{i=1}^{r} \bar{z}_{i,t} \tag{18.35a}$$

If we take the variances and assume each line is independent

$$\sigma^2_{\bar{z}}(t) = \frac{1}{r^2}\sum_{i=1}^{r}\sigma^2_{\bar{z}_i}(t) = \frac{1}{r}\,\sigma^2_{\bar{z}_1}(t) \tag{18.35b}$$

If the number sampled and number used as parents within a replicate are $M^* = M/r$ and $N^* = N/r$, respectively, then it is easily seen from Table 18.8 that the variance of a replicate line is simply r times the variance of a population with N and M. Hence, variance in the sample mean from r replicate lines with N^* and M^* is the same as the variance with a single line with N and M, provided that the number of individuals within each replicate line is sufficiently large to avoid significant inbreeding. Richardson et al. (1968), Irgang et al. (1985), and Muir (1986a, 1986b) developed regression approaches that correct for between-generation environmental changes when replicate lines are used.

LINE-CROSS ANALYSIS OF SELECTION EXPERIMENTS

Plant-breeding schemes often involve two (or more) generations for each **cycle** of selection (Chapter 23; Volume 3). The amount of selective progress can be examined by growing seeds from different cycles together in a common garden thus offering a direct assessment of the genetic response. For example, one could regress the mean of a line on its selection-cycle number (e.g., Burton et al. 1971). An even more powerful design involves growing *crosses* between remnant seed from different cycles of selection in a common garden. Such a **line-cross analysis** yields summary statistics about the genetic nature of the divergence between lines, such as the role of additive versus nonadditive effects (LW Chapters 9 and 20). As we saw in LW Chapter 20, a large number of parameters are required when one has a collection of *unrelated* lines (i.e., the general and specific combining abilities of each line and each pair, respectively). However, in the analysis of a selection experiment, lines are not unrelated but rather are genetically connected, requiring far fewer parameters to model. The key to connecting lines from different cycles is to assume a constant rate of allele-frequency change over the course of the experiment (Hammond and Gardner 1974; Smith 1979a, 1979b, 1983; Melchinger and Flachenecker 2006), which is referred to as a **generation means analysis** (**GMA**). With a large number of loci, each of which has a small effect, this assumption may be reasonable. If there are major alleles present or selection runs long enough such that substantial allele-frequency change occurs, then this model becomes more problematic. While commonly used in plant breeding, GMA is applicable to any organism for which an equivalent of "remnant seed" is available, such as frozen breeding stock or lines extracted from different generations of selection.

The Simple Additive Model

To see the logic behind a GMA, consider the simplest case: two alleles at each locus, with only additive effects and no G × E present. Focusing on a particular locus, the genotypes contribute values of $0 : a : 2a$ to the overall mean, where we sum contributions across all loci (because we assumed there is no epistasis). If p denotes the frequency of the favorable allele, then the contribution from this locus is $2ap$. If we further assume a roughly constant rate of change in allele frequency (at least over a few generations), then the frequency, $p(k)$, of the favorable allele in cycle k of selection is

$$p(k) \simeq p + k\Delta p$$

where p is the initial frequency and Δp is the per-generation rate of change for this locus. Summing over all loci, the mean following k cycles of selection is

$$C_k = \mu + \sum_i 2a_i\,p_i(k) \simeq \mu + \sum_i 2a_i(p_i + k\Delta p_i) = \mu + \sum_i 2a_ip_i + k\sum_i 2a_i\Delta p_i$$

Next, we define

$$m = \mu + 2\sum p_i\, a_i \qquad \text{and} \qquad A = 2\sum \Delta p_i\, a_i \tag{18.36}$$

where m is the line mean before selection and A is a weighted measure of change in the mean given constant allele-frequency change, Δp_i, at a locus (which can vary over loci). The expected line mean from selection cycle k becomes

$$C_k = m + A \cdot k \tag{18.37a}$$

Now suppose individuals from cycles k and j are crossed. With the frequencies of the favorable allele in these lines being $p(k)$ and $p(j)$, $[\,1-p(k)\,]\,[\,1-p(j)\,]$ is the chance of getting the 0-valued homozygote, while $p(k)\,[\,1-p(j)\,] + p(j)\,[\,1-p(k)\,]$ is the chance of getting an a-value heterozygote, and $p(k)\,p(j)$ is the chance of getting a $2a$ homozygote. The resulting contribution from a particular locus to this cross becomes

$$0 \cdot \Big\{[\,1-p(k)\,]\,[\,1-p(j)\,]\Big\} + a \cdot \Big\{p(k)\,[\,1-p(j)\,] + p(j)\,[\,1-p(k)\,]\Big\} + 2a \cdot \Big\{p(k)\,p(j)\Big\}$$

$$= a \cdot [\,p(k) + p(j)\,] = a \cdot [\,2p + (k+j)\Delta p\,]$$

Summing over all loci, and recalling Equation 18.36, we get an expected mean for this cross of

$$C_{k\times j} = m + A \cdot \frac{k+j}{2} \tag{18.37b}$$

Under this simple additive model, the mean of any particular population (a line, C_k, or the progeny from a line cross, $C_{k\times j}$) is a function of the constants m and A and the selection cycle number. The goodness-of-fit can be assessed by comparing the predicted and actual mean values (LW Chapter 9).

Equations 18.37a and 18.37b have a simple form because of the assumption of strict additivity (no dominance or epistasis). However, when loci show dominance, the expressions for line and line-cross means are more complex. Line means may then change under selfing (assuming that there is inbreeding depression), although any such change allows us to estimate certain dominance components. Further, when dominance is present, care is required when considering the *progeny* of line crosses. We can regard the progeny from a particular cross (say $C_{k\times j}$) as an F_1, and can generate an F_2 by either selfing individuals (when possible) or by letting the F_1 randomly mate. Under strict additivity, the means of both types of F_2 offspring (from either selfing or random mating) are still given by Equation 18.37b, the mean value of the F_1. When loci display dominance, the F_1 mean and the two different types of F_2 all differ from each other. Again, these differences offer opportunities to estimate various dominance components, as we now will describe.

The Hammond-Gardner Model

Motivated by the general diallel analysis of Gardner and Eberhart (1966), Hammond and Gardner (1974) extended the simple additive model to allow for dominance, which requires three additional parameters (D_0, D, and D_q). Assuming all lines start from the same base population, as derived below in Example 18.10, the **Hammond-Gardner (HG) model** returns means for various lines and line crosses of:

$$C_k = m + A \cdot k + D_0 + D \cdot k + D_q \cdot k^2 \tag{18.38a}$$

$$C_k(s) = m + A \cdot k + \frac{1}{2}\left(D_0 + D \cdot k + D_q \cdot k^2\right) \tag{18.38b}$$

$$C_{k\times j} = m + \frac{A}{2}(k+j) + D_0 + \frac{1}{2}\,[\,D \cdot (k+j) + 2D_q \cdot kj\,] \tag{18.38c}$$

$$C_{k\times j}(s) = m + \frac{A}{2}(k+j) + \frac{D_0}{2} + \frac{1}{4}\,[\,D \cdot (k+j) + 2D_q \cdot kj\,] \tag{18.38d}$$

$$C_{k\times j}(rm) = C_{k\times j} + \frac{D_q}{4}(k-j)^2 \tag{18.38e}$$

where (s) and (rm), respectively, denote the selfed and randomly mated next generations given the cross or line. For example, $C_4(s)$ is the mean among the selfed progeny of individuals from cycle four. Likewise, for $C_{2\times 6}(rm)$, we cross cycle two and cycle six individuals, let their offspring (the F_1, $C_{2\times 6}[rm]$) randomly mate, and then measure their resulting progeny (the F_2). As expected, when all of the dominance terms are zero, Equations 18.38a–18.38e reduce to Equations 18.37a and 18.37b. With no dominance, there is no change in the line mean following inbreeding, so $C_k = C_k(s)$, and the F_1 progeny mean from a cross has the same mean as the progeny from either a selfed or randomly mated F_2, $C_{k\times j} = C_{k\times j}(s) = C_{k\times j}(rm)$.

Five parameters (m, A, D, D_0, and D_q) are required under the full model, and closer inspection of the coefficients on these model parameters in Equations 18.38a–18.38e reveals what types of crosses are required to estimate parameters of interest. If the investigator only has the unselfed line means, C_k, and their F_1 crosses, $C_{k\times j}$, A and D are not separately estimable, as they appear as $k(A + D)$ in expressions for these means (see Equations 18.38a and 18.38c). Obtaining separate estimates requires having different coefficients on the A and D terms, as occurs with either selfed lines or selfed F_2 crosses (Equations 18.38b and 18.38d). These particular crosses provide additional information on the genetic nature of the between-generation divergence that is not given by the simple line means, or their F_1s, alone. Likewise, one cannot obtain separate estimates of m and D_0 unless there is at least one selfed line in the study. These terms enter as $m+D_0$ in all randomly mating populations, but they enter as $m + D_0/2$ in selfed lines.

As with the strictly additive model, we can express the parameters of the Hammond-Gardner model in terms of locus-specific parameters, with m and A as defined by Equation 18.36. Letting d_i be the dominance for locus i, which now makes the genotypic values $0 : a_i + d_i : 2a_i$, the dominance-related parameters, D_0, D, and D_q, in Equations 18.38a–18.38e are as follows. D_0 is the contribution (to the mean) from dominance in the initial (cycle-zero) population

$$D_0 = 2\sum p_i(1 - p_i)d_i \tag{18.39a}$$

The role of dominance in response in the line means enters through D and D_q, which are the linear and quadratic regression coefficients on cycle number. D is calculated by

$$D = 2\sum \Delta p_i(1 - 2p_i)d_i \tag{18.39b}$$

and it appears in Equations 18.38a–18.38e as terms of the form $D \cdot k$, namely, D times the appropriate selection cycle number, k. Recalling (LW Equation 4.10b) that the average effect, α, of an allelic substitution (replacing an unfavorable allele by a favorable one) is given by $\alpha = a + d(1 - 2p)$, we have

$$\begin{aligned} A + D &= 2\left(\sum \Delta p_i a_i + \sum \Delta p_i(1 - 2p_i)d_i\right) \\ &= 2\left(\sum \Delta p_i\left[a_i + (1 - 2p_i)d_i\right]\right) = 2\sum \Delta p_i \alpha_i \end{aligned} \tag{18.39c}$$

with $A+D$ representing the per-generation change in the trait mean as a linear regression on cycle number (e.g., $k[A+D]$ for line k). This is is also twice the average effect of substituting the unfavorable allele with the favorable one weighted by the allele-frequency change.

The final dominance term

$$D_q = -2\sum (\Delta p_i)^2\, d_i \tag{18.39d}$$

appears as a quadratic function of cycle number, with terms of the form $k^2 D_q$ for cycle k and kjD_q for crosses between lines from cycles j and k. This is perhaps the most interesting dominance term, as it measures both inbreeding depression and the expected heterosis when two lines are crossed. As discussed below, D_q has also been interpreted as the component of change in the mean due to the effects of drift, with $A + D$ representing the response from selection (the expected deterministic change). Recall (LW Chapter 10) that the heterosis, H, between two lines

$$H = \mu_{1\times 2} - \frac{\mu_1 + \mu_2}{2}$$

is the difference between the mean of the F_1 and the average of its parental lines. In the absence of epistasis

$$H = \sum (\delta p_i)^2 d_i \tag{18.39e}$$

where δp_i is the between-line difference in allele frequency for the favorable allele at the ith locus and d_i is the associated dominance term. Because the expected difference, δp_i, in allele frequency between lines separated by a single cycle of selection is Δp_i, comparing Equations 18.39d and 18.39e shows that D_q is the negative of twice the heterosis generated by a single generation of selection, namely, $D_q = -2H$.

The composite parameters (m, A, D, D_0, and D_q) are typically estimated by OLS, with the vector of estimates given by $\left(\mathbf{X}^T\mathbf{X}\right)^{-1}\mathbf{X}^T\mathbf{y}$, with $\mathbf{y}$ being the vector of population means and the design matrix, $\mathbf{X}$, containing the appropriate coefficients for the parameters of interest (Example 18.9). The machinery introduced in LW Chapter 9 for standard line-cross analysis, such as hypothesis testing, testing goodness-of-fit, and estimation via GLS when the standard errors vary over means, can all be applied in a straightforward fashion. As with a standard line-cross analysis, one starts a GMA with the simple additive model and then tests whether adding additional model terms significantly improves the fit (LW Chapter 9). Typically one ends up with a reduced model, as some of the parameters turn out to be nonsignificant.

GMA can be extended to situations where some of the crosses are between lines from different base populations (Smith 1979b, 1983; Helms et al. 1989; Butruille et al. 2004; Melchinger and Flachenecker 2006). The same logic holds as before, with the frequency of the favorable allele in cycle k from base population j being $p_j(k) \simeq p_j + k\Delta p_j$, and with associated parameters $m_j, A_j, D_j, D_{0,j}$, and $D_{q.j}$. Additional terms are required to account for any initial heterosis (cycle-zero crosses between base populations j and i) which, from Equation 18.39e, is a function of the square of the difference in their allele frequencies, $\delta_{i,j}^2 = (p_i - p_j)^2$. Finally, terms for any additional heterosis that has accrued during selection are required, and again these are functions of the squared allele-frequency differences. For a cycle-k line from base population i crossed to a cycle-ℓ line from base population j, this is

$$[\delta_{i,j}(k,\ell)]^2 = [p_j(k) - p_i(\ell)]^2 = [(p_j + k\Delta p_j) - (p_i - \ell\Delta p_i)]^2$$

The motivation for this type of analysis is examined in our final volume, where we discuss the breeding of superior hybrids between lines. Basically, one tries to improve the performance of both selected lines, as well as that of their resulting hybrid. Generation means analysis can help partition any observed improvement into components of within-line and between-line (hybrid) improvement.

Example 18.9. As a simple worked example of the Hammond-Gardner model, consider eight populations (cycles 0, 4, 7 and their crosses) from a much larger sample generated by a selection experiment for yield in maize analyzed by Smith (1983). Suppose we wish to estimate the parameters in the Hammond-Gardner model, fitting a reduced model where D_q is excluded. The crosses (which include two selfed populations), their means, and the resulting coefficients (in Equation 18.38) for the parameters to estimate are as follows:

		Coefficients			
Line/cross	Mean	m	A	D_0	D
C_0	59.7	1	0	1	0
C_4	63.4	1	4	1	4

		Coefficients			
Line/cross	Mean	m	A	D_0	D
C_7	68.9	1	7	1	7
$C_7(s)$	42.3	1	7	$\frac{1}{2}$	7/2
$C_{0\times4}$	61.2	1	$\frac{0+4}{2}$	1	$\frac{0+4}{2}$
$C_{0\times7}$	79.6	1	$\frac{0+7}{2}$	1	$\frac{0+7}{2}$
$C_{4\times7}$	72.3	1	$\frac{4+7}{2}$	1	$\frac{4+7}{2}$
$C_{4\times7}(s)$	40.3	1	$\frac{4+7}{2}$	$\frac{1}{2}$	$\frac{4+7}{4}$

The resulting vector, $\mathbf{y}$, of means and the design matrix, $\mathbf{X}$, become

$$\mathbf{y} = \begin{pmatrix} 59.7 \\ 63.4 \\ 68.9 \\ 42.3 \\ 61.2 \\ 79.6 \\ 72.3 \\ 40.3 \end{pmatrix}, \qquad \mathbf{X} = \begin{pmatrix} 1 & 0 & 1 & 0 \\ 1 & 4 & 1 & 4 \\ 1 & 7 & 1 & 7 \\ 1 & 7 & 0.5 & 3.5 \\ 1 & 2 & 1 & 2 \\ 1 & 3.5 & 1 & 3.5 \\ 1 & 5.5 & 1 & 5.5 \\ 1 & 5.5 & 0.5 & 2.75 \end{pmatrix}$$

Solving for the vector of parameter estimates yields

$$\begin{pmatrix} \widehat{m} \\ \widehat{A} \\ \widehat{D}_0 \\ \widehat{D} \end{pmatrix} = \left(\mathbf{X}^T\mathbf{X}\right)^{-1}\mathbf{X}^T\mathbf{y} = \begin{pmatrix} 4.266 \\ 1.072 \\ 57.402 \\ 0.523 \end{pmatrix}$$

Based on this subset of data, we note that there is considerable initial dominance ($\widehat{D}_0 = 57.402$) influencing the initial value of the trait. Of the per-generation change of $A + D$ = 1.595 per generation, roughly two-thirds are from A (1.072), and one-third is from D (0.5223). Standard errors for these estimates follow using standard OLS machinery (LW Chapter 9).

Example 18.10. Here we derive the line and cross means when dominance is present. Because the focus is on the contribution from a specific locus, for ease of presentation we suppress the subscript (on a, d, and p) denoting the locus being followed. Assuming no epistasis, the mean is derived by summing the contributions over all loci. To allow for dominance, the genotypic values at our target locus become $0 : a + d$: $2a$. With p being the frequency of the favorable allele, the mean contribution from this locus is

$$0 \cdot (1 - p^2) + (a + d) \cdot 2p(1 - p) + (2a) \cdot p^2 = 2ap + 2dp(1 - p)$$

Notice that the contribution from a is a linear function of the allele frequency (and hence a linear function of allele frequency change), while the contribution from d has both linear (p) and quadratic (p^2) components. The contribution from this locus to the line mean in cycle k follows by replacing p with $p(k) \simeq p + k\Delta p$ in Equation 18.36b to yield

$$\begin{aligned} C_k &= \mu + 2ap(k) + 2dp(k)[\,1 - p(k)\,] \\ &= \mu + 2a[\,p + k\Delta p\,] + 2d[\,p + k\Delta p\,][\,1 - (p - k\Delta p)\,] \\ &= \left(\mu + 2ap\right) + k \cdot 2a\Delta p + 2dp(1 - p) + k \cdot 2d\Delta p \cdot (1 - 2p) - k^2 \cdot 2d\,(\Delta p)^2 \\ &= m + k \cdot A + D_0 + k \cdot D + k^2 \cdot D_q \end{aligned}$$

Now suppose that this line is selfed. Homozygotes replicate themselves, while segregation of genotypes occurs when heterozygotes are selfed, with half of the offspring being $a + d$ heterozygotes, one-quarter of the offspring being $2a$ homozygotes, and the final one-quarter being 0-valued homozygotes. The resulting mean value following a generation of selfing is the sum of these contributions,

$$p^2\,[2a] + 2p(1-p)\left[(1/4)(2a) + (1/2)(a+d) + (1/4)\cdot 0\right] + (1-p^2)\cdot 0 = 2ap + dp(1-p)$$

Note that all terms involving d are exactly half of their random-mating values, yielding

$$C_k(s) = m + k\cdot A + \frac{1}{2}\left(D_0 + k\cdot D + k^2\cdot D_q\right)$$

Now consider the mean of the cross of lines from cycles k and j. As above, the chance of a 0-valued homozygote is $[\,1-p(k)\,]\,[1-p(j)\,]$, whereas it is $p(k)\,[\,1-p(j)\,]+p(j)\,[\,1-p(k)\,]$ for an $a+d$ heterozygote, and $p(k)\,p(j)$ for a $2a$ homozygote, yielding a resulting mean of

$$\begin{aligned}
C_{k\times j} &= \mu + (a+d)\cdot\Big(p(k)\,[\,1-p(j)\,] + p(j)\,[\,1-p(k)\,]\Big) + 2a\cdot p(k)p(j)\\
&= \mu + a\Big(p(k)+p(j)\Big) + d\cdot\Big(p(k)+p(j)-2p(k)p(j)\Big)\\
&= (\mu+2ap) + (k+j)a\Delta p + d\Big(2p(1-p) + (k+j)\cdot\Delta p(1-2p) - k\,j\cdot 2\,(\Delta p)^2\Big)\\
&= m + \frac{k+j}{2}\cdot A + D_0 + \frac{k+j}{2}\cdot D + k\,j\cdot D_q
\end{aligned}$$

If we self these individuals to form an F_2, by the same argument used above for selfing a line, each of the dominance terms is reduced by one-half, recovering Equation 18.38d.

Finally, when we form an F_2 by randomly mating the F_1 progeny from a line cross, the new allele frequency is

$$\frac{p(k)+p(j)}{2} \simeq p + \frac{k+j}{2}\Delta p$$

yielding a mean of

$$\begin{aligned}
C_{k\times j}(rm) &= \mu + 2a\left(p + \frac{k+j}{2}\Delta p\right) + 2d\left(p + \frac{k+j}{2}\Delta p\right)\left(1 - p - \frac{k+j}{2}\Delta p\right)\\
&= (\mu+2ap) + \frac{k+j}{2}2a\Delta p + 2dp(1-p) + \frac{k+j}{2}2d\Delta p(1-2p) - \left(\frac{k+j}{2}\right)^2 2d(\Delta p)^2\\
&= m + \frac{k+j}{2}\cdot A + D_0 + \frac{k+j}{2}\cdot D + \left(\frac{k+j}{2}\right)^2\cdot D_q
\end{aligned}$$

Recalling Equation 18.38c, we can express this as

$$C_{k\times j}(rm) = C_{k\times j} - kj\cdot D_q + \left(\frac{k+j}{2}\right)^2\cdot D_q = C_{k\times j} + \frac{(k-j)^2}{4}\cdot D_q$$

The additional term is due to the decay in heterosis following random mating, as the heterosis in the F_1 cross between lines from cycles j and k is

$$\begin{aligned}
H &= \sum (p_i(k) - p_i(j))^2 d_i = \sum (p + k\Delta p - p - j\Delta p)^2 d_i\\
&= (k-j)^2 \sum (\Delta p)^2 d_i = -(k-j)^2 D_q/2
\end{aligned}$$

Recall (LW Chapter 9) that the amount, H, of heterosis from the F_1 is reduced to $H/2$ in the (random-mating generated) F_2. Hence

$$C_{k\times j}(rm) - C_{k\times j} = -\frac{H}{2} = (k-j)^2\frac{D_q}{4}$$

as found above.

Accounting for Inbreeding Depression and Drift

The deterministic assumption of a constant rate of allele-frequency change in the Hammond-Gardner model requires large population sizes. Because a typical selection experiment often involves small populations, genetic drift can also be important and thus needs to be taken into account. Smith (1979a, 1979b, 1983) was the first to incorporate drift and inbreeding depression in a GMA, while the most general treatment is from Melchinger and Flachenecker (2006). The latter authors' key idea was to separate allele frequency at cycle k into deterministic and drift components

$$p(k) = p(0) + k\Delta p + \delta p_k \tag{18.40}$$

where the Δp represents the expected large-population change (i.e., the Hammond-Gardner model), while δp_k accounts for the additional change due to drift. Recall from Chapter 2 that under drift $E[\,\delta p_k\,] = 0$, so that $E\left[\,(\delta p_k)^2\,\right] = \sigma^2(\delta p_k)$. Under pure drift, Equation 2.14a yields an allele-frequency variance after k generations of

$$\sigma^2(\delta p_k) = p(0)[1 - p(0)]f_k \tag{18.41a}$$

where $f_k \simeq k/(2N_e)$ is the amount of inbreeding in cycle k. Melchinger and Flachenecker suggested that, with selection, Equation 18.41a can be approximated by using the average frequency over the k cycles

$$E[\,(\delta p_k)^2\,] \simeq \left(\frac{p(0)[\,1-p(0)\,] + p(k)[\,1-p(k)\,]}{2}\right) f_k \tag{18.41b}$$

Using the deterministic approximation $p(k) = p(0) + k\Delta p$, this reduces to

$$E[\,(\delta p_k)^2\,] \simeq p(0)[\,1-p(0)\,]\,f_k + \frac{k}{2}\cdot(1-2p)\Delta p\, f_k - \frac{k^2}{2}\,(\Delta p)^2\; f_k \tag{18.41c}$$

Notice the close similarity of these three terms to D_0, D, and D_q (c.f. Equations 18.39a–18.39d). Using Equation 18.40 as the model for allele frequency and following the logic used in Example 18.10, the adjustment in the mean for inbreeding due to drift becomes

$$I_k = \left(D_0 + \frac{k\cdot D}{2} - \frac{k^2\cdot D_q}{2}\right) f_k \tag{18.42a}$$

Assuming $f_k = k\Delta f$ (typically, with $\Delta f = 1/[2N_e]$), this simplifies to

$$I_k = k\left(D_0 + \frac{k\cdot D}{2} - \frac{k^2\cdot D_q}{2}\right)\Delta f \tag{18.42b}$$

The resulting line means are

$$C_k = C_k(HG) - I_k \qquad \text{and} \qquad C_k(s) = C_k(s, HG) - I_k/2 \tag{18.43a}$$

while the line-cross means become

$$C_{k\times j} = C_{k\times j}(HG) - I_k \quad \text{for } j > k \tag{18.43b}$$

where HG denotes the value under the Hammond-Gardner model (Equations 18.38a–18.38e).

Equations 18.42 and 18.43 define the **Melchinger-Flachenecker model**. The presence of drift ($f_k > 0$) modifies the contributions from the three dominance terms, resulting in a reduction in mean (relative to the HG model) when both D_0 and D are positive (as D_q is ≤ 0 when directional dominance is present). Smith (1979a, 1983) equated this reduction to inbreeding depression, which arises from drift causing some favorable alleles to decrease in frequency. On average, drift is equally likely to increase or decrease the change in frequency of an allele over its deterministic value, Δp. Thus, under drift, some alleles have a greater-than-deterministic increase in frequency, but this is countered by (on average) others having a less-than-expected increase. For additive genes, the expected value of these changes exactly cancel (as $E[\delta p_k] = 0$). However, with dominance, this variance in allele-frequency change around Δp does not cancel out, as the mean is a quadratic function of d.

While Melchinger and Flachenecker's analysis is exact, the approximate results of the work by Smith (1979a, 1983) had a significant influence on the GMA literature, so a few comments on his approximations are in order. Smith's (1979a) model was

$$C_k(s, S) = C_k(s, HG) + k \cdot I \tag{18.44}$$

namely, it added a term, kI, to the Hammond-Gardner model expression for $C_k(s)$ (Equation 18.38b) to account for inbreeding, with $I = (D_0 + D)$. Note by comparison to the exact result (Equation 18.42b) that this is only correct when $D_0 \gg D$, and D_q is negligible (as Smith assumed). Further, Smith neglected to include I in his expressions for the line-cross means, instead using the HG expression for $C_{k \times j}(s)$, Equation 13.38d. Equation 18.43b shows this to be incorrect. Smith (1979a) ignored the D_q term, as he felt that Δp was expected to be small, and hence its square was likely to be negligible (or in any case, potentially difficult to estimate with precision).

In a later paper, Smith (1983) instead ignored the inbreeding correction, I, and suggested that for experiments with very small effective population size, $E\left[\,(\delta p_k)^2\,\right] \gg (\Delta p)^2$, and thus D_q was a surrogate measure for drift, as most of its value was due to the drift variance. Equations 18.42b and 18.43 show that both of these ideas, while correct in spirit, are approximations of the more general solution (Equation 18.42b). However, they can often yield a fairly good approximation. For example, Flachenecker et al. (2006) found that both the Smith (1983) and Melchinger-Flachenecker models gave rather similar parameter estimates. The critical insight offered by Smith's analysis was that by adjusting for the effects of drift, one could estimate the potential response if a larger population size was used, as the following two examples illustrate.

Example 18.11. Helms et al. (1989) obtained an estimate of $\widehat{D}_q = -0.024$ (which was significant at the 1% level) in an analysis of a maize line selected for yield. Assuming small effective population size, how much was the population mean at cycle ten reduced due to drift-generated inbreeding depression? Recalling from Equation 18.38a that the contribution from D_q in cycle k is $D_q \cdot k^2 = -0.024 \cdot 10^2 = -2.4$, then the observed mean (provided the assumptions of the Smith [1983] model hold) is expected to be reduced by 2.4 units (relative to a randomly mating population) by the effects of drift generating inbreeding depression.

Example 18.12. Smith (1979b) examined the relative effectiveness of two different selection schemes (using half sib [HT] versus S_1 [S] family selection; both procedures are examined in Chapters 21 and 23) for yield in maize using the cultivar BSK. Both the Hammond-Gardner and Smith (1979a) models were fit to the data, yielding the following values

		Hammond-Gardner	Smith (1979a)
BSK(HT)	A	0.7 $\pm$ 0.15	1.22 $\pm$ 0.15
	I	—	-1.83 $\pm$ 0.28
	D, D_q	not significant	—

		Hammond-Gardner	Smith (1979a)
BSK(S)	A	1.42 ± 0.15	1.97 ± 0.15
	I	—	-1.83 ± 0.28
	D, D_q	not significant	—

By ignoring inbreeding, the Hammond-Gardner model resulted in a significant *underestimation* of A in both experiments. Smith's GMA shows that both selection schemes result in significant amounts of inbreeding depression (-1.83 in both cases), but also significant gains (1.22 and 1.97). Increasing the effective population size is expected to decrease $\Delta f = 1/(2N_e)$, and thus the magnitude of I, in turn enhancing response. Both the Smith and Hammond-Gardner analyses showed that S_1 selection was more effective than half-sib selection, as measured in terms of average genetic gain ($A + D$) per cycle (1.42/0.7 = 203% and 1.97/1.22 = 160% , respectively).

Finally, a rough calculation suggests the conditions under which $\sigma^2(\delta p) \gg (\Delta p)^2$, i.e., when the variance in allele-frequency change from drift exceeds the square of the deterministic change from selection. Equation 5.2b shows that $\Delta p \simeq sp(1-p)$ for a very weakly selected additive locus, while Equation 5.21 yields $s \simeq \bar{\imath}\, a/\sigma_z$, where $\bar{\imath}$ is the selection intensity on the trait, yielding $\Delta p_i \simeq (\bar{\imath}\, a_i/\sigma_z)p_i(1-p_i)$. Recalling that $\sigma^2(A_i) = 2a_i^2 p_i(1-p_i)$ is the additive genetic variance associated with locus i

$$(\Delta p_i)^2 \simeq (\bar{\imath}\, a_i/\sigma_z)^2 p_i^2(1-p_i)^2 = \bar{\imath}^2 p_i(1-p_i)\left(\frac{a_i^2 p_i(1-p_i)}{\sigma_z^2}\right)$$

$$= \bar{\imath}^2 p_i(1-p_i)\left(\frac{\sigma^2(A_i)/2}{\sigma_z^2}\right) = \frac{\bar{\imath}^2 p_i(1-p_i)}{2} h_i^2$$

where h_i^2 is the heritability due to locus i. Assuming that n loci underlying the trait contribute roughly the same to the overall heritability, h^2, then

$$(\Delta p_i)^2 \simeq \frac{p_i(1-p_i)}{2}\frac{\bar{\imath}^2 h^2}{n} \tag{18.45a}$$

Drift dominates the expected quadratic change when $\sigma^2(\delta p) \gg (\Delta p)^2$, or

$$\frac{p(1-p)}{2N_e} \gg \frac{p(1-p)}{2}\frac{\bar{\imath}^2 h^2}{n}$$

which reduces to

$$n \gg N_e\,(\bar{\imath}^2 h^2) \tag{18.45b}$$

For example, for a trait with $h^2 = 0.25$ under truncation selection to save the largest 5% ($\bar{\imath} = 2.06$; Example 14.1), $\bar{\imath}^2 h^2 \simeq 1$, so drift dominates the quadratic term when the number of loci is much greater than the effective population size (assuming loci have roughly equal effects). Expressed in another way, the fractional contribution to the expected square in allele-frequency change from drift is

$$\frac{\sigma^2(\delta p)}{\sigma^2(\delta p) + (\Delta p)^2} \simeq \frac{1/N_e}{1/N_e + \bar{\imath}^2 h^2/n} = \frac{1}{1+\bar{\imath}^2 h^2 N_e/n} \tag{18.45c}$$

or

$$\frac{\sigma^2(\delta p)}{\sigma^2(\delta p) + (\Delta p)^2} \simeq 1 - \bar{\imath}^2 h^2 N_e/n, \quad \text{when} \quad n \gg Ne \tag{18.45d}$$

For $h^2 = 0.25, \bar{\imath} = 2.06, n = 100$, and $N_e = 25$, Equation 18.45c returns a value of 0.79, while the approximation given by Equation 18.45d yields 0.74. Hence, in this setting almost 80% of the expected squared allele-frequency change is due to drift.

19

Analysis of Short-term Selection Experiments: 2. Mixed-model and Bayesian Approaches

Unnecessarily complex analysis should not be used as a foil to disguise lower quality datasets: estimates of genetic parameters are only as good as the data on which they are based.
Kruuk (2004)

While a least-squares (LS) analysis of a selection experiment distills the data down to the trait mean and variance for each generation, one often has more information. In the extreme, one has measurements (or **records**) for all individuals throughout the course of the experiment (or breeding program) as well as their complete pedigree. When such additional data are available, an LS analysis simply ignores them. A **mixed-model** (**MM**) **analysis** (LW Chapters 26 and 27), on the other hand, fully considers the covariances between *all* observations. By virtue of using this additional information, an MM is potentially far more powerful. It is also more flexible, easily incorporating complicated fixed effects and highly unbalanced designs. Finally, breeders and evolutionary biologists are especially concerned with the realized genetic gain from selection, which, due to shifting environments, may be different from the change in mean phenotype. LS estimates the latter, while MM, by estimating the mean breeding value in each generation, estimates the former. An important application of the analysis of selection experiments is in the **evaluation of breeding programs**, and the success of almost all animal, and increasingly many plant, breeding programs is gauged by examining the response using a mixed-model framework.

There are two different frameworks for a mixed-model analysis. The first is the **two-step approach**, wherein one first employs **REML** (LW Chapter 27) or some other method to estimate the appropriate variance components, and then uses these with **BLUP** (best linear unbiased predictor) to estimate breeding values (LW Chapter 26). With BLUPs for individual breeding values in hand, one can estimate the mean breeding value for any particular generation and directly follow *genetic*, as opposed to *phenotypic*, change. This allows for the separation of genetic versus environmental change, even in the absence of a control population. While straightforward, the two-step approach does not account for the uncertainty in BLUP estimates that arises from using estimates of the variances (as opposed to their true values). In contrast, **Bayesian approaches** provide for an exact accounting of the uncertainty in the estimation of the breeding values by integrating over the uncertainty from the confounding effects of nuisance parameters, such as the variance components.

By building around applications to the analysis of selection experiments (including applied breeding), we use this chapter to review some of the basic statistical machinery behind mixed models (which is also used extensively in Chapters 20 and 22) and to more formally introduce the Bayesian framework (briefly touched upon in Chapter 10, and more fully developed in Appendices 2 and 3). We start with a brief review of the theory of mixed models and then consider various applications of the **animal** (or **individual**) **model** to the analysis of selection experiments. We conclude by briefly examining mixed models under a Bayesian framework.

MIXED-MODEL VERSUS LEAST-SQUARES ANALYSIS

Figure 19.1 illustrates the result of a mixed-model analysis of selection response. Note that instead of measuring response from the *observed phenotypic means* (the LS approach), response is measured from the *estimated mean breeding values* obtained from BLUP. Further, instead of estimating a realized heritability, a mixed-model (REML) analysis estimates the

Evolution and Selection of Quantitative Traits. Bruce Walsh & Michael Lynch.
Published 2018 by Oxford University Press. © Bruce Walsh & Michael Lynch 2018.
DOI 10.1093/oso/9780198830870.001.0001

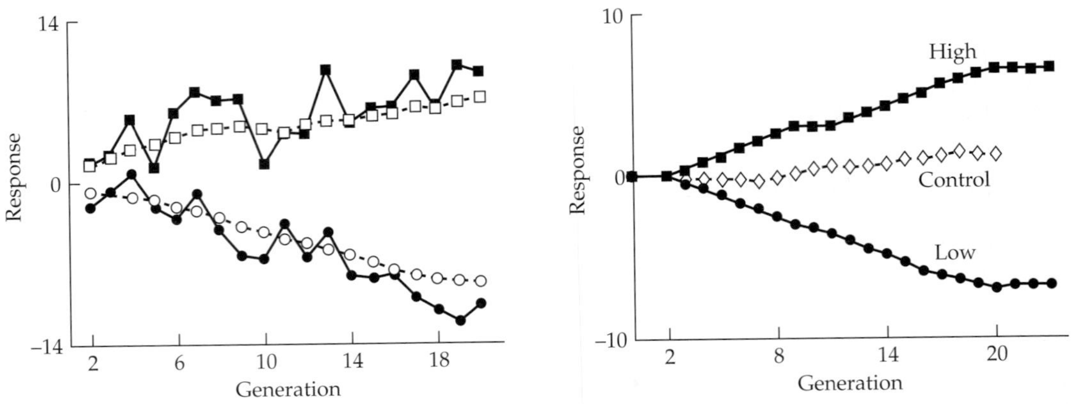

Figure 19.1 Results from high and low selection on 6-week weight in mice. **Left:** Observed (filled boxes and circles) and predicted (open boxes and circles) mean phenotypic values in the up- and down-selected lines, expressed as deviations from the control population. The predicted mean for a specific generation is given by the estimated mean breeding value plus the estimated environmental value. **Right:** Estimated mean breeding values for both selected populations and the control. See Examples 19.2 and 19.7 for more details on this experiment. (After Meyer and Hill 1991.)

additive genetic variance in the *base population*.

Mixed-models readily allow records to be adjusted for any number of **fixed effects** (differences between the expected value of individuals from different categories). For example, one might correct for trait differences between sexes, trait differences between individuals from different litter sizes, age effects, known environmental factors, etc. Such adjustments result in a more accurate prediction of an individual's genetic value, and thus a more accurate estimate of the population's genetic response. Further, a properly formatted MM analysis can separate phenotypic changes into genetic and environmental components without using a control population. In contrast, an LS analysis cannot separate genetic from environmental trends when only a single line is considered. Under a mixed model, such a separation is possible because the covariance structure associated with the pedigree of all individuals in the experiment allows information to be borrowed from relatives across generations.

Another advantage of a mixed-model analysis is its enormous flexibility in handling almost any selection design. For example, an MM analysis easily allows for overlapping generations (when a parent contributes offspring over several different years of selection), while analysis of response in overlapping generations under LS can be difficult to formulate correctly. Further, as detailed below, a properly designed MM analysis can also account for assortative mating, drift, and selection-induced gametic-phase disequilibrium (provided we can assume that the infinitesimal model holds).

Nonetheless, despite its power, an MM analysis has tradeoffs relative to a simpler LS analysis. First, an MM analysis requires far greater record keeping (e.g., following all individuals and their relatives) and is more computationally demanding. Second, an MM analysis can be rather model-sensitive. In particular, assumption of the infinitesimal model is critical. If selection-induced changes in allele frequencies are significant during the course of the experiment, the assumptions of an MM analysis will be violated. Finally, MM analysis *critically* depends on the covariance structure of the random effects. If this is incorrectly specified, an MM analysis can lead to highly biased results.

BLUP Selection

A previously introduced MM application is **BLUP selection** (Chapter 13), wherein a mixed model is used to find those individuals with the highest estimated breeding values, which are then used as parents to form the next generation. This is the main route of selection used by animal and tree breeders, and to a growing extent, by plant breeders working with outcrossing species. BLUP selection uses all of the information up to a given generation to choose the parents for the next cycle of selection.

Conversely, an MM analysis of selection response is a *retrospective* analysis of the genetic gain of a population, wherein we start at some final time point and infer the trajectory of past genetic gain during the experiment or breeding program. The actual scheme used to choose parents (such as mass, index, or BLUP selection) is irrelevant to the MM analysis of the final genetic gain, which is solely based on the values of all individuals, and their relationships (pedigree), over the course of the experiment.

BASICS OF MIXED-MODEL ANALYSIS

Here, we start with a quick review of some of the key theoretical results of mixed-model analysis and then examine specific applications. We encourage the reader to review LW Chapters 26 and 27 (either before, or after, reading this short introduction). These chapters provide additional worked examples to give a better feel for mixed models and also consider advanced topics in MM analysis in greater detail.

Mixed models are so named because they consider both fixed and random effects. Recall that fixed effects are unknown constants, while random effects have values that are drawn from some underlying distribution (LW Chapters 8 and 26). Hence, any particular value for a random effect represents just one possible realization from this underlying distribution, which is usually assumed to be normal with a mean of zero and an unknown (yet to be estimated) variance.

A brief example will remind the reader of a powerful feature of random-effects models. Suppose we have T time points and include an environmental value, E_i, for each time point in our model. Treating these values as fixed effects makes no assumption as to how the E_i from different generations are related to each other, but the cost is T degrees of freedom. Conversely, if we make a random-effects assumption that the E_i values are drawn from some underlying distribution, we use far fewer degrees of freedom. If we assume the E_i are independent draws from an underlying normal, then we only need a single degree of freedom (the variance, σ^2_E, of this distribution, as the mean value of a random effect is zero), no matter the value of T. More generally, one could make additional assumptions about the distribution of the E_i values, at the cost of additional degrees of freedom. For example, one could assume that all the E_i values for adjacent generations are autocorrelated by the same amount, ρ, which introduces an additional parameter to be estimated.

Typically, statisticians speak of *estimating* fixed effects and *predicting* the realized values of random effects. Both LS and MM analyses estimate the fixed effects in a model, while MM analysis also predicts the values of the random effects by using the covariances between observations (after adjusting for fixed effects). (Note that under a Bayesian analysis, as will be discussed later, every effect is assumed to be random, and this distinction between fixed and random effects is more subtle.) The standard mixed model for a vector, $\mathbf{y}$, of n observations is

$$\mathbf{y} = \mathbf{X}\boldsymbol{\beta} + \mathbf{Z}\mathbf{a} + \mathbf{e} \tag{19.1}$$

where $\boldsymbol{\beta}$ is a $q \times 1$ vector of q fixed effects, $\mathbf{a}$ is a $p \times 1$ vector of p random effects (in our case, the breeding values of the individuals in our experiment), and $\mathbf{e}$ is the $n \times 1$ vector of n residuals (which is also random). The matrices $\mathbf{X}$ and $\mathbf{Z}$ are, respectively, the $n \times q$ **design matrix** and the $n \times p$ **incidence matrix** associated with the fixed and random effects, respectively.

In the absence of the vector, $\mathbf{a}$, of random effects, Equation 19.1 reduces to a least-squares model, $\mathbf{y} = \mathbf{X}\boldsymbol{\beta} + \mathbf{e}$, and whether ordinary (unweighted) least-squares (OLS) or generalized (weighted) least-squares (GLS) analysis is used to estimate $\boldsymbol{\beta}$ depends on our assumption about the covariance structure of the vector, $\mathbf{e}$, of the residuals. OLS assumes that residuals are uncorrelated and homoscedastic, yielding $\mathbf{Var(e)} = \sigma_e^2\,\mathbf{I}$. More generally, if $\mathbf{Var(e)} = \mathbf{V}$, where the only constraint is that $\mathbf{V}$ be symmetric and positive-definite, then generalized GLS is used (Equation 19.3a).

In order to solve Equation 19.1, we need to specify the covariance structure for the vectors of random effects, $\mathbf{a}$ and $\mathbf{e}$. As just noted, it is generally assumed that the residuals are uncorrelated and homoscedastic, so $\mathbf{Var(e)} = \sigma_e^2\,\mathbf{I}$. The covariance of $\mathbf{a}$ (the vector of breeding values) has a more complicated structure, which is given by the pedigree, $\mathbf{Var(a)} = \sigma_A^2\,\mathbf{A}$, with $\sigma(a_i, a_j) = \sigma_A^2 A_{ij}$. Here $\mathbf{A}$ is a matrix of known constants (the **numerator relationship matrix**, often abbreviated as simply the **relationship matrix**), whose elements are given by the pedigree structure (or, more recently, from sufficiently dense marker information). The resulting $n \times n$ covariance matrix, $\mathbf{V}$, for the vector of observations, $\mathbf{y}$, becomes

$$\mathbf{V} = \sigma_A^2\,\mathbf{ZAZ}^T + \sigma_e^2\,\mathbf{I} \tag{19.2a}$$

For the case of a single phenotypic measurement for each individual with no missing phenotypic data, $\mathbf{Z} = \mathbf{I}$ and Equation 19.2a reduces to $\mathbf{V} = \sigma_A^2\,\mathbf{A} + \sigma_e^2\,\mathbf{I}$. More generally, when there are multiple observations per individual, or individuals with missing phenotypic data, then $\mathbf{Z}$ departs from an identity matrix (e.g., Example 19.5).

The covariance matrix, $\mathbf{V}$, is thus a function of the (usually unknown) variance components (σ_A^2 and σ_e^2) and matrices of known constants ($\mathbf{Z}$, $\mathbf{A}$, and $\mathbf{I}$). Because $\mathbf{a}$ is a vector of breeding values, we can alternatively express $\mathbf{V}$ as a function of the heritability (h^2) and the (fixed-effect-adjusted) phenotypic variance ($\sigma_z^2 = \sigma_A^2 + \sigma_e^2$) of the trait of interest

$$\mathbf{V} = \sigma_z^2 \cdot \left(h^2\,\mathbf{ZAZ}^T + (1-h^2)\,\mathbf{I} \right) \tag{19.2b}$$

as $\sigma_e^2 = \sigma_z^2 - \sigma_A^2 = \sigma_z^2(1-h^2)$. Because our focus is generally on breeding values, any dominance variance gets swept into σ_e^2. This potentially results in e_i values within families being correlated (due to full sibs sharing $\sigma_D^2/4$), and we will discuss corrections for this shortly.

Assuming $\mathbf{V}$ is known exactly, estimation of the vector, $\boldsymbol{\beta}$, of fixed effects follows from GLS (LW Chapter 8)

$$\widehat{\boldsymbol{\beta}} = \left(\mathbf{X}^T\mathbf{V}^{-1}\mathbf{X} \right)^{-1} \mathbf{X}^T\mathbf{V}^{-1}\mathbf{y} \tag{19.3a}$$

Equation 19.3a is called the **best linear unbiased estimator** (**BLUE**) of the vector, $\boldsymbol{\beta}$, of fixed effects. The **estimability** of the fixed effects can be an issue, as the structure of the data (indicated by the number of independent columns—the **column rank**—of $\mathbf{X}$) may not allow for unique estimates of all fixed effects. In such cases, generalized inverses can be used to obtain unique estimates of certain linear combinations of the fixed effects (LW Appendix 2). If $\mathbf{X}$ has a column rank of $\ell \leq q$, then exactly ℓ combinations of the q fixed effects can be estimated; see LW Chapter 26, and LW Appendix 2, for further details. Finally, note that although the BLUEs are a function of $\mathbf{V}$ (and hence of σ_A^2 and σ_e^2), applying Equation 19.2b shows that the phenotypic variance, σ_z^2, in $\mathbf{V}$ cancels out in Equation 19.3a ($\mathbf{V}^{-1}$ scales as $1/\sigma_z^2$, while the inverse of $\mathbf{V}^{-1}$ scales as σ_z^2), leaving the BLUE estimate (for Equation 19.1) as a function of the heritability alone.

If we again assume that $\mathbf{V}$ is exactly known, the **best linear unbiased predictor** (**BLUP**) of the vector of random effects is given by

$$\widehat{\mathbf{a}} = \mathbf{Var(a)}\mathbf{Z}^T\mathbf{V}^{-1}\left(\mathbf{y} - \mathbf{X}\widehat{\boldsymbol{\beta}} \right) = \sigma_A^2\mathbf{AZ}^T\mathbf{V}^{-1}\left(\mathbf{y} - \mathbf{X}\widehat{\boldsymbol{\beta}} \right) \tag{19.3b}$$

The BLUPs for breeding values are often called **PBVs** or **EBVs**, for **predicted** or **estimated breeding values**. Equation 19.3b is the regression of $\mathbf{a}$ on $(\mathbf{y} - \mathbf{X}\widehat{\boldsymbol{\beta}})$, the vector of observations, $\mathbf{y}$, adjusted for their expected mean values, $\mathbf{X}\widehat{\boldsymbol{\beta}}$. Recall that for the univariate regression predicting a from y, the scaling of $(y - \widehat{y})$ is given by the regression slope, $\sigma(a,y)/\sigma^2(y)$. In Equation 19.3b, this scaling is of the form of the covariance matrix, $\sigma(\mathbf{a},\mathbf{y}) = \sigma(\mathbf{a},\mathbf{Za}) =$

$\sigma_A^2 \mathbf{A}\mathbf{Z}^T$ times the inverse of $\mathbf{V} = \sigma(\mathbf{y}, \mathbf{y})$. Even if the number of random effects exceeds the number of actual observations (i.e., $p > n$), Equation 19.3b still provides unique estimates of each (provided $\mathbf{V}^{-1}$ exists). This occurs because the $p \times p$ covariance structure ($\mathbf{A}$) of the vector, $\mathbf{a}$, is incorporated in the model. As with the BLUEs, BLUPs are simply functions of h^2 as $\sigma_A^2 = h^2\sigma_z^2$, while $\mathbf{V}^{-1}$ scales as $1/\sigma_z^2$. An alternative expression, due to Kennedy and Trus (1993), is

$$\widehat{\mathbf{a}} = \left(\mathbf{Z}^T\mathbf{M}\mathbf{Z} + \lambda\mathbf{A}^{-1}\right)^{-1}\mathbf{Z}^T\mathbf{M}\mathbf{y} \tag{19.3c}$$

where $\lambda = \sigma_e^2/\sigma_A^2 = (1 - h^2)/h^2$, and

$$\mathbf{M} = \mathbf{I} - \mathbf{X}\left(\mathbf{X}^T\mathbf{X}\right)^{-1}\mathbf{X}^T \tag{19.3d}$$

is the **absorption matrix** for the fixed effects, and the design matrix, $\mathbf{X}$, assigns the fixed effects associated with any particular observation (Equation 19.1). As an aside

$$\mathbf{H} = \mathbf{X}\left(\mathbf{X}^T\mathbf{X}\right)^{-1}\mathbf{X}^T \tag{19.3e}$$

is referred to as the **hat matrix** (Hoaglin and Welsch 1978), because for an OLS estimate

$$\widehat{\mathbf{y}} = \mathbf{X}\widehat{\boldsymbol{\beta}} = \mathbf{X}\left(\mathbf{X}^T\mathbf{X}\right)^{-1}\mathbf{X}^T\mathbf{y} = \mathbf{H}\mathbf{y} \tag{19.3f}$$

and $\mathbf{H}$ maps the observed $\mathbf{y}$ values onto those predicted, $\widehat{\mathbf{y}}$, by the fixed effects. Hence

$$\mathbf{M}\mathbf{y} = (\mathbf{I} - \mathbf{H})\mathbf{y} = \mathbf{y} - \widehat{\mathbf{y}} = \mathbf{y} - \mathbf{X}\widehat{\boldsymbol{\beta}} \tag{19.3g}$$

are the adjusted values for $\mathbf{y}$ after the fixed effects have been removed, namely, the vector of the deviations of the observations, $\mathbf{y}$, from their expected values, $\widehat{\mathbf{y}}$ (based on fixed effects only).

In practice, Equations 19.3a–19.3c are often not used, as they require the inversion of the potentially very large ($n \times n$) matrix, $\mathbf{V}$. As an alternative, $\widehat{\boldsymbol{\beta}}$ and $\widehat{\mathbf{a}}$ can be obtained without computing an inverse by numerically solving (e.g., by applying Gaussian elimination to) **Henderson's mixed-model equations**, (derived in Example A6.5)

$$\begin{pmatrix} \mathbf{X}^T\mathbf{X} & \mathbf{X}^T\mathbf{Z} \\ \mathbf{Z}^T\mathbf{X} & \mathbf{Z}^T\mathbf{Z} + \lambda\,\mathbf{A}^{-1} \end{pmatrix} \begin{pmatrix} \widehat{\boldsymbol{\beta}} \\ \widehat{\mathbf{a}} \end{pmatrix} = \begin{pmatrix} \mathbf{X}^T\mathbf{y} \\ \mathbf{Z}^T\mathbf{y} \end{pmatrix} \tag{19.4}$$

The careful reader might wonder why we worried so much about avoiding computing the inverse of $\mathbf{V}$, given that the mixed-model equations contain $\mathbf{A}$, which at first blush looks just as complicated to invert as $\mathbf{V}$. In fact, however, $\mathbf{A}$ turns out to be very easy to invert, as a slight modification of the recursive approach used to compute $\mathbf{A}$ from a pedigree can be used to directly compute $\mathbf{A}^{-1}$ (Henderson 1976; Quaas 1976).

The sampling variance-covariance matrices for $\widehat{\mathbf{a}}$ and $\widehat{\boldsymbol{\beta}}$ also follow from the mixed-model equations. First we partition the inverse of the $(p+q) \times (p-q)$ matrix in Equation 19.4 as

$$\begin{pmatrix} \mathbf{X}^T\mathbf{X} & \mathbf{X}^T\mathbf{Z} \\ \mathbf{Z}^T\mathbf{X} & \mathbf{Z}^T\mathbf{Z} + \lambda\,\mathbf{A}^{-1} \end{pmatrix}^{-1} = \begin{pmatrix} \mathbf{C}_{11} & \mathbf{C}_{12} \\ \mathbf{C}_{12}^T & \mathbf{C}_{22} \end{pmatrix} \tag{19.5a}$$

where $\mathbf{C}_{11}$, $\mathbf{C}_{12}$, and $\mathbf{C}_{22}$ are, respectively, $q \times q$, $q \times p$, and $p \times p$ submatrices. Henderson (1975) showed that the $q \times q$ covariance matrix for the BLUE vector of fixed effects, $\boldsymbol{\beta}$, is given by

$$\mathbf{Var}(\widehat{\boldsymbol{\beta}}) = \sigma_e^2\,\mathbf{C}_{11} \tag{19.5b}$$

The variance of the predicted breeding values, $\widehat{\mathbf{a}}$, is a bit more subtle, as our real interest is not $\mathbf{Var}(\,\widehat{\mathbf{a}})$, but rather the **prediction error variances (PEVs)**, $\mathbf{Var}(\,\widehat{\mathbf{a}}-\mathbf{a}\,)$. These are the variances and covariances among the vector of prediction errors $(\,\widehat{\mathbf{a}}-\mathbf{a})$, and they are given by the $p \times p$ matrix

$$\mathbf{Var}(\,\widehat{\mathbf{a}}-\mathbf{a}\,) = \sigma_e^2\,\mathbf{C}_{22} \tag{19.5c}$$

Note that

$$\mathbf{Var}(\,\widehat{\mathbf{a}}\,) = \mathbf{Var}(\,\widehat{\mathbf{a}}-\mathbf{a}\,) + \mathbf{Var}(\,\mathbf{a}\,) = \sigma_e^2\,\mathbf{C}_{22} + \sigma_A^2\mathbf{A} \tag{19.5d}$$

Following Equation 19.3c, Kennedy and Trus (1993) showed that we can also express Equation 19.5c as

$$\mathbf{Var}(\,\widehat{\mathbf{a}}-\mathbf{a}\,) = \sigma_e^2\left(\mathbf{Z}^T\mathbf{M}\mathbf{Z} + \lambda\mathbf{A}^{-1}\right)^{-1} \tag{19.5e}$$

The prediction error variances for *individual* EBVs are not of serious interest in this chapter, as we measure selection response by taking the average of the EBVs over all individuals within each generation. However, individual EBVs can be important when we try to disentangle selection in natural populations, and we will revisit PEVs in Chapter 20.

Finally, the covariances between estimated fixed effects and prediction errors are

$$\boldsymbol{\sigma}(\widehat{\boldsymbol{\beta}}, \widehat{\mathbf{a}}-\mathbf{a}\,) = \sigma_e^2\,\mathbf{C}_{12} \tag{19.5f}$$

Equations 19.5b–19.5f assume that both σ_e^2 and $\lambda = \sigma_e^2/\sigma_A^2$ are known without error. When BLUP estimates are obtained using *estimated* values (e.g., $\widehat{\sigma}_e^{\,2}$), this additional source of error further inflates the sampling variances. Bayesian methods (discussed below) accommodate this additional uncertainty.

One concern is that the BLUPs and BLUEs obtained using these procedures may be biased by selection. However, Henderson (1975) showed that, provided variances are known, the estimates will be unbiased by selection when two conditions hold. First, selection decisions must be based on linear combinations of data (such as truncation selection based on individual phenotypes or a linear index based on the phenotypes of an individual and its relatives). Second, that the model must use estimates of fixed effects that are unbiased when selection is absent. These conditions hold under many reasonable forms of artificial selection.

REML Estimation of Unknown Variance Components

The variance components (σ_A^2 and σ_e^2), or, at a minimum, the heritability $h^2 = \sigma_A^2/(\sigma_A^2+\sigma_e^2)$, must be specified to obtain $\widehat{\boldsymbol{\beta}}$ and $\widehat{\mathbf{a}}$. Although these variances are generally unknown, they can be estimated, for example, by using **restricted maximum likelihood (REML)**. REML is closely related to BLUP, with (roughly speaking) REML estimates obtained from iterating and updating BLUP estimates until there is suitable convergence (LW Chapter 27). REML maximizes that part of the likelihood function that is unaffected by fixed effects (Patterson and Thompson 1971). Harville (1977) coined the term **restricted ML**, but Thompson (2008) noted that REML maximizes a *residual* likelihood, and hence preferred the term **residual maximum likelihood**. One advantage of REML estimates (over those obtained by other estimation procedures) is that they are unbiased by the estimates of fixed effects (Patterson and Thompson 1971). For further details, we refer the reader to the extensive discussion of REML variance estimation in LW Chapter 27, and to Hofer (1998), Thompson and Mäntysaari (2004), Thompson et al. (2005), Misztal (2008), Thompson (2008), and Gianola and Rosa (2015) for a review of more recent developments, including computational issues.

For much of this chapter, we assume that BLUP variance components are first obtained by REML (although we relax this assumption when discussing Bayesian mixed-model analysis). This two-stage approach of BLUP using estimated variance components (in place of their true values) is called **empirical BLUP** or **REML/BLUP** (Sorensen and Kennedy 1986; Kennedy and Sorensen 1988; Harville 1990). Kackar and Harville (1981) and Gianola et al.

(1986, 1988) showed that using REML estimation does not result in biased values for BLUPs, but that the resulting predictors may not be "best" (there may be other linear predictors with smaller mean-squared errors).

The matrix of sample variances and covariances for the REML variance component estimates can be approximated by using the best quadratic fit of the restricted likelihood surface, centered at the REML estimates (Smith and Graser 1986; Graser et al. 1987). If $\boldsymbol{\sigma}$ is a vector of m estimated variance components (σ_A^2 and σ_e^2 in the models thus far considered), we compute the restricted likelihood, $L(\boldsymbol{\sigma})$, for a grid of values close to the REML solution and then fit the best quadratic surface to the data, using

$$L(\boldsymbol{\sigma}) = \mathbf{b}_0 + \boldsymbol{\sigma}^T \mathbf{b}_1 + \boldsymbol{\sigma}^T \mathbf{Q}\, \boldsymbol{\sigma} \tag{19.6a}$$

With v variance components, $\mathbf{b}_0$ and $\mathbf{b}_1$ are $v \times 1$ vectors, and $\mathbf{Q}$ is an $v \times v$ symmetric matrix of quadratic regression terms (in Chapter 30 we discuss the fitting of such quadratic surfaces in the context of fitness surface estimation). The approximate covariance matrix for the vector of REMLs, $\widehat{\boldsymbol{\sigma}}$, is calculated by

$$\mathbf{Var}(\widehat{\boldsymbol{\sigma}}) \simeq (-2\mathbf{Q})^{-1} \tag{19.6b}$$

The rationale for this approach is that, for large samples, the inverse of the matrix of second-order partial derivatives of the likelihood surface (evaluated at the maximum likelihood estimate, MLE) approaches the covariance matrix of these estimates (LW Appendix 4). Equation 19.6a is a (second-order) multidimensional Taylor series (Equation A6.7b), with $2\mathbf{Q}$ corresponding to the matrix of second-order partial derivatives (the **Hessian matrix**) of the likelihood function. Alternatively, Meyer (2008) suggested that the use of the likelihood profile function can often return more appropriate estimated confidence intervals than those based on these large-sample variances (see her paper for details).

Although Equation 19.6b returns a large-sample approximation for the uncertainty in REML variance estimates, this cannot be easily translated into how much additional uncertainty is introduced into BLUP estimates by using REML estimates of variance components in place of their true values. These concerns can be addressed in a Bayesian framework (see below), as Bayesian approaches are exact for any sample size (rather than large-sample approximations). Further, they fully incorporate any uncertainty in the variance estimates into the uncertainty in the BLUP estimates.

REML Often Returns Variance Estimates Unbiased by Selection

As with BLUPs and BLUEs (which assume the presence of known variances), REML variance estimates are often unbiased by selection. As we will show, under the infinitesimal model, the relationship matrix, $\mathbf{A}$, fully accounts for any change in σ_A^2 from disequilibrium, inbreeding, and drift. As a result, if the base population consists of unselected and noninbred individuals in linkage equilibrium and phenotypic data are available for all selected and unselected individuals, then under the infinitesimal model, REML yields essentially unbiased estimates of the additive genetic variance in the base population (Henderson 1949; Henderson et al. 1959; Curnow 1961; Thompson 1973; Rothschild et al. 1979; Sorensen and Kennedy 1984b; Gianola and Fernando 1986; Gianola et al. 1988; Juga and Thompson 1989; Gianola et al. 1989; Im et al. 1989; Fernando and Gianola 1990; Piepho and Möhring 2006). Simulations by van der Werf and de Boer (1990) showed that if the model includes the pedigree information for all individuals but is missing records (trait values) for some, then REML does not necessarily yield unbiased estimates of σ_A^2, and in this case bias increases with heritability (Jeyaruban and Gibson 1996).

When the base population consists of previously selected individuals, REML no longer provides protection from biased estimates of the additive genetic variance in the population prior to selection, even if the entire pedigree of individuals back to the base population is included (van der Werf 1990; van der Werf and de Boer 1990; van der Werf and Thompson 1992). This arises because of disequilibrium, $d(0) \neq 0$, in the base population (recall from

Equation 16.2 that d is the difference between the additive genetic and genic variances). While Equation 16.7a allows us to predict the dynamics of $d(t)$, it requires the value of $d(t-1)$, so without knowledge of the actual value of the base population, $d(0)$, the relationship matrix cannot fully account for the dynamics of d. Finally, if selection acts on a suite of unmeasured characters whose breeding values are correlated with characters included in the model, REML can generate biased estimates of the variances and covariances of the measured characters (Schaeffer and Song 1978).

ANIMAL-MODEL ANALYSIS OF SELECTION EXPERIMENTS

The basic building block of a mixed-model analysis of selection experiments is the **animal model** (Quaas and Pollak 1980), which estimates the breeding (or additive-genetic) values of all individuals measured during the course of the experiment. We examine its simplest version first, and consider various elaborations in later sections (also see Chapter 22 and LW Chapter 26). While this model has its origin in the animal-breeding literature, it has very widespread applicability. We trust that plant scientists will not be greatly offended, as the "animal" (or better yet, the "individual") model can be used to analyze plant-selection experiments as well (e.g., Piepho et al. 2008; Bernardo 2010).

The Basic Animal Model

Mixed-models easily allow for overlapping generations by simply predicting breeding values at discrete time points (e.g., every year or at each natural cohort) rather than in each generation. Hence, in the discussion that follows, one can easily replace "generation" by "year" or some other time measure.

To apply the animal model to a selection experiment, one first vectorizes the observations from the entire experiment by letting y_{ij} denote the trait measurement on the jth individual from generation i, where $0 \leq i \leq t$ (generation 0 represents the unselected base population) and $1 \leq j \leq n_i$, where n_i is the number of measured individuals in generation i. Let the vector $\mathbf{y}$ denote the observations on all measured individuals from the entire experiment

$$\mathbf{y} = \begin{pmatrix} \mathbf{y}_0 \\ \mathbf{y}_1 \\ \vdots \\ \mathbf{y}_t \end{pmatrix} \qquad \text{where} \qquad \mathbf{y}_i = \begin{pmatrix} \mathbf{y}_{i1} \\ \vdots \\ \mathbf{y}_{in_i} \end{pmatrix}$$

The vector $\mathbf{y}_i$ includes the values for all measured individuals from generation i, *including* those culled as well as those that were allowed to reproduce. The simplest animal model for these data is

$$y_{ij} = \mu + a_{ij} + e_{ij} \tag{19.7a}$$

where μ is an overall mean, a_{ij} is the breeding value of the jth measured individual from generation i, and e_{ij}, the deviation between breeding and phenotypic values. Equation 19.1 gives the mixed model as $\mathbf{y} = \mathbf{X}\boldsymbol{\beta} + \mathbf{Za} + \mathbf{e}$. With exactly one record per individual, $\mathbf{Z} = \mathbf{I}$. In this simple model, the only fixed effect is the mean, returning $\boldsymbol{\beta} = (\mu)$ and $\mathbf{X} = \mathbf{1}$ (a vector of ones) thus reducing Equation 19.1 to

$$\mathbf{y} = \mu \cdot \mathbf{1} + \mathbf{a} + \mathbf{e} \tag{19.7b}$$

where $\mathbf{a}$ is the vector of breeding values for all individuals measured during the course of the experiment, with $\mathbf{Var(a)} = \sigma_A^2\mathbf{A}$.

The relationship matrix, $\mathbf{A}$, is the key to the animal model, as it includes all of the pedigree information. Because of this information, the animal model is easily extended to allow breeding values to be estimated for individuals without records, *provided* they have measured *relatives* in the analysis (see Example 19.5). The diagonal elements of $\mathbf{A}$ describe the amount of inbreeding, with $A_{ii} = (1 + f_i) = 2\,\Theta_{ii}$, while the off-diagonal elements given by $A_{ij} = 2\Theta_{ij}$ (twice the coefficient of coancestry; see LW Chapters 7 and 26) describe the relatedness of individuals i and j. Recursive methods for obtaining the elements of $\mathbf{A}$ (and $\mathbf{A}^{-1}$), given a pedigree, are discussed in LW Chapter 26. The simple animal model assumes that all genetic variance is additive, so there is no genetic covariance between residuals. In this case, it is generally assumed that $\mathbf{Var}(\mathbf{e}) = \sigma_e^2\,\mathbf{I}$, and the mixed-model equations (Equation 19.4) simplify to

$$\begin{pmatrix} \widehat{\mu} \\ \widehat{\mathbf{a}} \end{pmatrix} = \begin{pmatrix} n & \mathbf{1}^T \\ \mathbf{1} & \mathbf{I} + \lambda\,\mathbf{A}^{-1} \end{pmatrix}^{-1} \begin{pmatrix} n\,\overline{y} \\ \mathbf{y} \end{pmatrix} \tag{19.7c}$$

where n is the total number of individuals in the experiment, $\lambda = \sigma_e^2/\sigma_A^2 = (1-h^2)/h^2$, $\widehat{\mathbf{a}}$ is the n-dimensional vector of the predicted breeding values of all measured individuals, and $\mathbf{1}$ is an n-dimensional vector of ones. Likewise, the covariance matrices for the fixed effects estimates, $\mathbf{C}_{11}$, the predictor errors for the BLUPs of breeding values, $\mathbf{C}_{22}$, and the covariances, $\mathbf{C}_{12}$, between these estimates are given by

$$\begin{pmatrix} \mathbf{C}_{11} & \mathbf{C}_{12} \\ \mathbf{C}_{12}^T & \mathbf{C}_{22} \end{pmatrix} = \begin{pmatrix} n & \mathbf{1}^T \\ \mathbf{1} & \mathbf{I} + \lambda\,\mathbf{A}^{-1} \end{pmatrix}^{-1} \tag{19.7d}$$

Response Is Measured by Change in Mean Breeding Values

Under a mixed-model analysis, response is measured by the change in the mean breeding value of a selected population over time. The estimated mean breeding value in generation k (or, alternatively, at time point k) is simply obtained by calculating the average of individual breeding-value estimates for that generation

$$\widehat{\overline{a}}_k = \frac{1}{n_k} \sum_{j=1}^{n_k} \widehat{a}_{kj} \tag{19.8a}$$

Because the predicted mean breeding value in generation 0 (the unselected base population) is zero by construction ($\overline{a}_0 = 0$), the total (cumulative) response at generation k is $\overline{a}_k - \overline{a}_0 = \overline{a}_k$. In matrix notation, the vector, $\overline{\mathbf{a}}$, of mean breeding values is estimated by

$$\widehat{\overline{\mathbf{a}}} = \begin{pmatrix} \widehat{\overline{a}}_0 \\ \vdots \\ \widehat{\overline{a}}_{t-1} \end{pmatrix} = \mathbf{K}^T \widehat{\mathbf{a}} \tag{19.8b}$$

where elements in the kth row of $\mathbf{K}$ are $1/n_k$ when the column corresponds to an individual from generation k and otherwise are zero (see Example 19.1). Thus, for t generations of data (corresponding to $t-1$ generations of selection, as the analysis includes the unselected base population, generation 0), $\mathbf{K}$ is an $n \times t$ matrix that satisfies $\mathbf{K}^T \mathbf{1}_n = \mathbf{1}_t$ (a $t \times 1$ vector of ones).

From Equation 19.5a, and recalling that $\mathbf{Var}(\mathbf{Bx}) = \mathbf{B}\,\mathbf{Var}(\mathbf{x})\mathbf{B}^T$ (LW Equation 8.21b), the sampling covariance matrix for the vector of estimated genotypic means becomes

$$\mathbf{Var}\left(\widehat{\overline{\mathbf{a}}}\right) = \sigma_e^2\,\mathbf{K}^T \mathbf{C}_{22} \mathbf{K} \tag{19.8c}$$

where the $n \times n$ matrix, $\mathbf{C}_{22}$, is the solution to Equation 19.7d (under the simple animal model), or more generally, Equation 19.5a. Again, these expressions assume that the residual variance is known without error, and using an estimate for σ_e^2 adds an additional uncertainty by increasing the sample variance.

Example 19.1. As an example of how one performs a mixed-model analysis of a selection experiment, consider the following, very simple, situation. From a base population of unrelated and noninbred individuals, four (indexed by 1–4) are measured and have trait values of 3, 6, 5, and 2, respectively. The two largest individuals are mated, and their resulting offspring (individuals 5–8) have values of 4, 5, 6, and 5. Now suppose that we have either a REML-based estimate of the heritability (which is not advisable here given the very small sample size) or have previous knowledge of its value. Our goal is to estimate the size of the genetic response. Assuming the only fixed effect is the mean, the resulting animal model is $\mathbf{y} = \mathbf{1}\boldsymbol{\beta} + \mathbf{a} + \mathbf{e}$, where

$$\mathbf{y} = \begin{pmatrix} y_1 \\ y_2 \\ y_3 \\ y_4 \\ y_5 \\ y_6 \\ y_7 \\ y_8 \end{pmatrix} = \begin{pmatrix} 3 \\ 6 \\ 5 \\ 2 \\ 4 \\ 5 \\ 6 \\ 5 \end{pmatrix}, \quad \mathbf{a} = \begin{pmatrix} a_1 \\ a_2 \\ a_3 \\ a_4 \\ a_5 \\ a_6 \\ a_7 \\ a_8 \end{pmatrix}, \quad \boldsymbol{\beta} = (\,\mu\,), \quad \mathbf{X} = \mathbf{1} = \begin{pmatrix} 1 \\ 1 \\ 1 \\ 1 \\ 1 \\ 1 \\ 1 \\ 1 \end{pmatrix}$$

What is the relationship matrix, $\mathbf{A}$? Because individuals 2 and 3 are the parents and all offspring are full sibs, all related individuals have values of $A_{ij} = 1/2$, as $2\theta_{ij} = 1/2$ for both parent-offspring pairs and full sibs. The resulting numerator relationship matrix becomes

$$\mathbf{A} = \begin{pmatrix} 1 & 0 & 0 & 0 & 0 & 0 & 0 & 0 \\ 0 & 1 & 0 & 0 & 1/2 & 1/2 & 1/2 & 1/2 \\ 0 & 0 & 1 & 0 & 1/2 & 1/2 & 1/2 & 1/2 \\ 0 & 0 & 0 & 1 & 0 & 0 & 0 & 0 \\ 0 & 1/2 & 1/2 & 0 & 1 & 1/2 & 1/2 & 1/2 \\ 0 & 1/2 & 1/2 & 0 & 1/2 & 1 & 1/2 & 1/2 \\ 0 & 1/2 & 1/2 & 0 & 1/2 & 1/2 & 1 & 1/2 \\ 0 & 1/2 & 1/2 & 0 & 1/2 & 1/2 & 1/2 & 1 \end{pmatrix}$$

For example, individuals 2 and 5 are parent and offspring, so $A_{5,2} = A_{2,5} = 1/2$, as the parent-offspring covariance $\sigma^2_A/2$. Similarly, individuals 7 and 8 are full-sibs, so $A_{7,8} = A_{8,7} = 1/2$. Note that the relationship matrix for the founders (base population members) is given by the 4×4 identity submatrix in the upper left of $\mathbf{A}$. This identity matrix implies that noninbred individuals (diagonal elements are one) and unrelated individuals (off-diagonal elements are zero) formed the base population.

Turning to the covariance matrix of the residuals, we make the standard assumption that $\mathbf{Var}(\mathbf{e}) = \sigma^2_e \mathbf{I}$, or in other words, that all residuals are uncorrelated with common variance, σ^2_e. However, if there is dominance, the residuals among full-sibs will be inflated by $\sigma^2_D/4$. Likewise, if there are common-family effects (e.g., maternal effects or other shared environmental effects), the residuals are inflated by σ^2_c, where c is the common-family effect. For now we will ignore these possible complications, which are easily accommodated by adding additional vectors of random effects to the model (see below).

Suppose (from REML or prior knowledge) that the heritability of the trait is $h^2 = 0.3$. Applying Equation 19.3a yields

$$\widehat{\mu} = \left(\mathbf{1}^T \mathbf{V}^{-1} \mathbf{1}\right)^{-1} \mathbf{1}^T \mathbf{V}^{-1} \mathbf{y} = 4.22$$

where we have computed $\mathbf{V}$ using Equation 19.2b as scaled to remove the phenotypic variance σ^2_z, for example, $\mathbf{V} = 0.3\,\mathbf{A} + 0.7\,\mathbf{I}$ (note that σ^2_z cancels in the above expression, as it appears in both $\mathbf{V}$ and $\mathbf{V}^{-1}$). Substituting into Equation 19.3b yields the 8×1 vector, $\widehat{\mathbf{a}}$, of BLUPs for the individual genetic values, and the resulting 2 × 1 vector, $\mathbf{K}^T\widehat{\mathbf{a}}$, of genetic means

$$\widehat{\mathbf{a}} = \begin{pmatrix} -0.366 \\ 0.666 \\ 0.366 \\ -0.666 \\ 0.386 \\ 0.562 \\ 0.739 \\ 0.562 \end{pmatrix} \quad \mathbf{K} = \frac{1}{4}\begin{pmatrix} 1 & 0 \\ 1 & 0 \\ 1 & 0 \\ 1 & 0 \\ 0 & 1 \\ 0 & 1 \\ 0 & 1 \\ 0 & 1 \end{pmatrix} \quad \mathbf{K}^T\widehat{\mathbf{a}} = \begin{pmatrix} 0 \\ 0.562 \end{pmatrix} \quad \mathbf{K}^T\mathbf{1}_8 = \begin{pmatrix} 1 \\ 1 \end{pmatrix}$$

Note that (by construction) the mean breeding value in the base population is zero. Hence, the estimated response (for $h^2 = 0.3$) is 0.562. The estimated genetic gain (response) for different assumed heritabilities is found to be as follows

h^2	Estimated response	h^2	Estimated response
0.0	0	0.6	0.940
0.1	0.211	0.7	1.026
0.2	0.398	0.8	1.083
0.4	0.707	0.9	1.095
0.5	0.833	1.0	1.000

The estimated gain increases with the assumed h^2 until it reaches a maximum of $\simeq$1.098 for $h^2 = 0.86$, after which it decreases as the assumed h^2 increases. A Bayesian analysis removes this dependency of the estimated response on h^2 by computing a weighted average of response over all possible h^2 values (weighted by their posterior values), yielding a marginal posterior distribution for the response that fully accounts for any uncertainty introduced from estimating the heritability.

Turning to a more standard analysis, the selection differential is the mean of the selected parents minus the mean of all parents, $S = 5.5 - 4 = 1.5$. Likewise, the selection response is the mean of the offspring minus the mean in the previous generation, $5 - 4 = 1$, which yields a realized heritability of $R/S = 1.0/1.5 = 0.67$. Using this value for h^2 in the MM analysis returns a genetic gain of 1.000.

As is apparent from the above example, a mixed-model analysis of a selection experiment has a very different character than an LS analysis. In the latter, one estimates the realized heritability from a suitable regression of phenotypic means on selection differentials. Under a REML/BLUP analysis, however, one first estimates the heritability in the base population (using REML), and then uses (empirical) BLUP to estimate breeding values for all individuals. The genetic response in a given generation is given by the mean of the elements in the estimated vector, $\mathbf{a}$, of breeding values in that generation, allowing for the separation of genetic from environmental response, even in the absence of a control population. Thompson and Atkins (1994) noted a fundamental difference between the two approaches in separating genetic from environmental change: an LS analysis typically uses *between-population* information (e.g., contrasts of the means of selection vs. control, or up- vs. down-selected lines), while a REML/BLUP analysis uses *within-population* information (the connections between relatives across generations contained within $\mathbf{A}$).

Under an MM analysis, the estimate of heritability should be based on REML estimates of the base population variance components, $\widehat{h^2} = \widehat{\sigma}^2_A/(\widehat{\sigma}^2_A + \widehat{\sigma}^2_e)$. One must avoid the temptation to estimate realized heritabilities using estimated mean breeding values. For example, Blair and Pollak (1984) regressed the BLUP mean breeding values on cumulative selection differentials to obtain a realized heritability estimate. The problem with this approach is that these mean estimates depend on the *assumed*, rather than the *actual*, heritability (Thompson 1986; Sorensen and Johannsson 1992; Ollivier 1999). In contrast, under a Bayesian analysis, the marginal posterior estimate of mean breeding value in any particular generation averages over all possible h^2 values and is independent of the heritability (Example 19.10).

Hence, a regression of Bayesian-derived mean breeding values on generations can return an unbiased estimate of realized heritability.

Fixed Effects Alter Heritabilities

Variance components in a mixed model are estimated *after* any variation introduced by fixed effects is removed, and this has an impact on the estimated heritability. Under the mixed-model framework, $h^2 = \sigma_A^2/(\sigma_A^2 + \sigma_e^2)$, where σ_e^2 is the *residual* variance. Here, the denominator is the **fixed-effects-adjusted** phenotypic variance (the sum of the variance components). This is potentially different from the more standard $h^2 = \sigma_A^2/\sigma_z^2$, which is a function of the (total) phenotypic variance, σ_z^2. In the absence of fixed effects that differ over individuals (such as sex- or age-specific means), $\sigma_z^2 = \sigma_A^2 + \sigma_e^2$, meaning that the total and fixed-effects-adjusted phenotypic variances are equivalent, as are the two definitions of h^2. However, if such class-specific differences exist and are explicitly modeled in the analysis, then the variation they contribute to the overall phenotype is *removed*, and the resulting residual variance reduced, so $\sigma_z^2 > \sigma_A^2 + \sigma_e^2$. For example, if a trait mean varies over the sexes, the variance of the trait in the entire population is greater than the variance of the trait within each sex. Consequently, for a model with sex added as a fixed effect, this source of variation is removed, resulting in a smaller residual variance and a larger heritability.

As a result, mixed-model heritability estimates should be larger than estimates that ignore fixed effects. A related issue is that mixed models using the same data but assuming different fixed effects can differ significantly in their estimated residual variances, σ_e^2, and thus, their resulting heritabilities (Wilson 2008). The implication is that comparison of heritabilities for different traits or populations estimated under a mixed-model framework can be somewhat problematic due to differences in the incorporated fixed effects.

Model Validation

Given the sensitivity of a mixed-model analysis to the validity of the assumptions (in particular, the infinitesimal model), some form of model validation is required to be able to apply these methods with confidence. One approach is to test the infinitesimal-model prediction that estimates of the base population, σ_A^2, should remain stable as additional generations of selection are considered. If the infinitesimal model holds, **A** completely accounts for changes in the additive variance in these later generations from both drift and selection-generated LD (a point that will be more fully developed shortly). If, on the other hand, σ_A^2 changes during selection in ways that are not predictable from the infinitesimal model (e.g., significant allele-frequency change occurs), using data from additional generations of selection may result in rather different estimates of the base-population additive variance.

Example 19.2. One of the first REML/BLUP analyses of a selection experiment was performed by Meyer and Hill (1991), who examined the response to selection for adjusted food intake (AFI) in mice (Figure 19.1). AFI is defined as food intake between 4 and 6 weeks, corrected for 4-week weight. Meyer and Hill had three replicate sets, each consisting of high, low, and control lines, for a total of almost 11,000 mice over the course of the experiment. Within-family selection (Chapter 21) on AFI was followed for 23 generations. Meyer and Hill included a number of fixed effects in their model, as well as adding a random effect to control for common litter (i.e., family) effects (see Example 19.7 for details).

As a check of the validity of the MM assumptions (in particular, the infinitesimal model), Meyer and Hill compared variance estimates based on data from generations 5–7 with estimates based on generations 14–23. In both cases, the full pedigree structure was incorporated into the relationship matrix. While incorporation of the complete pedigree information reduces the bias in estimates of the base-population additive variance, some bias will remain if records from some of these individuals are missing (van der Werf and de Boer 1990). In other words, knowledge of the pedigree is not sufficient; the phenotypic values of these individuals (even if not selected) are also needed to remove bias. Meyer and Hill observed a dramatic decline in the estimated additive variances (from 19.2 based on generations 5–7 to 2.5 based on generations 14–23). Under the infinitesimal model, both estimates should infer the base-population variance, so this large decrease suggests that the infinitesimal model may not be appropriate for assessing this trait. It is interesting to note that this decrease in the estimate of σ_A^2 occurred even as the total variance increased dramatically (from 23.88 to 33.93). This increase resulted mainly from an increase in the estimated environmental variance (from 12.9 to 25.5), although there was also a slight increase in the litter-effects variance (from 4.78 to 5.96).

Several other REML/BLUP analyses of selection experiments in mice also found differences between estimates of base-population additive variance when comparing data from early and late generations. Beniwal et al. (1992a and 1992b) observed decreases in the estimated additive variance (in body weight, litter size, and lean mass), while Heath et al. (1995) observed an increase in the additive variance in body weight.

In contrast, Martinez et al. (2000) found no changes in estimates of σ_A^2 over 20 generations of selection for body composition (fat pad to body weight ratio) in mice. These authors examined REML estimates of the additive variance (and heritability) using various subsets of the full 20 generation data. Consistent estimates of the additive variance were obtained (i) using the set of all records and the complete pedigree from generations 0–20; (ii) using records from generations 9–20, but with pedigree information from generation 0; and (iii) using only three-generation blocks of the phenotypic data as the entire dataset. They concluded that the selection response, while resulting in a roughly four-fold change in mean, was still well-fit by the infinitesimal model. Finally, Holt et al. (2005) observed no changes in base-population additive variance estimates over time for a high line of mice selected for litter size, but did see a variance reduction in the low line.

Separating Genetic and Environmental Trends

Observed improvement in a trait over time (such as milk yield) may be due entirely to improvement in the environment (better husbandry and nutrition), entirely from genetic changes (response from selected breeding), or (most likely) a combination of both. Thus, it is critical to partition an observed phenotypic change into genetic and environmental components. For example, Southwood and Kennedy (1991) showed that the improvement in several litter-size-related traits in pigs over a 10-year period in Quebec was entirely due to environmental, rather than genetic, changes. Recent plant examples are given by Laidig et al. (2014, 2017) and Piepho et al. (2014).

In a least-squares analysis, any underlying environmental trend is assumed to be removed by contrasting selected and control populations (or contrasting populations selected in opposite directions). The rationale is that the kth individual from population j in generation t can be described as

$$y_{j,tk} = \mu + b_{j,t} + a_{j,tk} + e_{j,tk} \tag{19.9}$$

where $b_{j,t}$ is the environmental trend in population j. If the common environmental value is the same in both the selected and the control populations ($b_{s,t} = b_{c,t}$), then the difference in phenotypic means in generation t between these populations is

$$\overline{y}_{s,t} - \overline{y}_{c,t} = (\overline{a}_{s,t} - \overline{a}_{c,t}) + (e_{s,t} - e_{c,t}) \tag{19.10}$$

Because the residuals, e, have an expected value of zero, this contrast provides an unbiased estimate of $\overline{a}_{s,t}$ *provided* there is no significant drift in the mean breeding value of the control population (meaning that $\overline{a}_{c,t} \simeq 0$). However, if genotype-environment interactions are present, the environmental values can differ between populations, in which case Equation 19.10 has an additional term, $(b_{s,t} - b_{c,t})$. Hence, even when a control population is used, a least-squares analysis can still give biased results if there is significant drift in the mean of the control population ($|\,\overline{a}_{c,t}\,| \gg 0$) or significant $G \times E$.

Insight into another way in which the use of a control in an LS analysis can be misleading was offered by Su et al. (1997), who examined response in body weight in chickens starting from a base population with a known previous history of selection on this trait. Because of prior selection, the base population showed a slippage of the mean back to the original (unselected) value. If this population was used as a control, this slippage would be taken as a decay in the environment over time, resulting in an overestimation of the selection response when using an LS analysis. Because of this concern, Su et al. used a mixed model to estimate the genetic response.

A mixed-model analysis estimates the mean *breeding value*, rather than the *phenotypic mean*, of the population, which allows for this separation of the genetic change from any environmental change (Henderson et al. 1959; Blair and Pollak 1984; Sorensen and Kennedy 1984a; Kennedy 1990). Such a separation is possible because **A** tracks the flow of genes through the population, which allows us to make estimates of breeding values independent of environmental effects by borrowing information from relatives across generations. This is dependent on the model assumptions holding, but when they do, a mixed-model analysis does not require a control population. This being said, Sorensen et al. (2003) showed that the inclusion of a control in an MM analysis generally improves the efficiency of estimates (resulting in smaller standard errors).

Common-environment effects are incorporated into the basic animal model by simply adding a fixed effect, b_t, for the common environmental effect in generation t

$$y_{tk} = \mu + b_t + a_{tk} + e_{tk} \tag{19.11}$$

For T generations, there are T estimable fixed effects (μ and all but one b_t). Typically, one either constrains the b_t to sum to zero or arbitrarily sets one of the b_t equal to zero. We do the latter with b_1, so the vector of fixed effects becomes

$$\boldsymbol{\beta} = (\mu, b_2, \cdots, b_T)^T$$

and the corresponding design matrix, **X**, has zeros or ones in columns 2 though T, corresponding to the generation in which the individual was scored

$$\mathbf{X} = \begin{pmatrix} 1 & 0 & \cdots & 0 \\ 1 & 0 & \cdots & 0 \\ \vdots & & & \vdots \\ 1 & 1 & \cdots & 0 \\ 1 & 1 & \cdots & 0 \\ \vdots & & & \vdots \\ 1 & 0 & \cdots & 1 \\ 1 & 0 & \cdots & 1 \end{pmatrix}$$

To remind the reader how **X** is obtained, recall that $\mathbf{X}\boldsymbol{\beta}$ returns a vector that adjusts each observation for the fixed effects. From the rules of matrix multiplication, the adjustment for observation i is the inner product, $\mathbf{x}_{i\cdot}^T\boldsymbol{\beta}$, where the row vector $\mathbf{x}_{i\cdot}^T$ corresponds to the ith row of **X**. Likewise, the jth column of **X** corresponds to the vector of weights (over all observations) for the jth fixed effect. Here, the first column is all ones, as all observations contain the mean, μ. Entries in column 2 (corresponding to b_2) are zero, unless an observation is from generation 2 (and hence has a mean of $\mu + b_2$), in which case the entry is 1, and so forth. Hence, the first two displayed rows of the above **X** correspond to observations from generation 1, which have means of μ, the second pair of displayed rows to generation 2 (with means of $\mu + b_2$), and the last pair to generation T (wth means of $\mu + b_T$).

An alternative approach is to treat the b_i as random (as opposed to fixed) effects, drawn independently (i.e., uncorrelated across generations) from a normal distribution with a mean of zero and an unknown variance, σ_b^2. The assumption of no environmental correlation between adjacent generations can be questionable (at best), and treating the b_i as fixed, rather than randon, removes these concerns (although at a cost of absorbing more degrees of freedom than with a random-effects analysis).

Example 19.3. To examine the potential bias from $G \times E$ and drift in the control population, Blair and Pollak (1984) examined a seven-generation selection experiment on 14-month greasy fleece weight in sheep. The model they assumed was that the mth individual in generation t, with fixed sex effect (male/female), v_i, fixed age of dam effect (mature/immature), d_j, rearing rank effect (single/twin), r_k, and year effect, b_t, had a phenotypic value of

$$y_{ijktm} = v_i + d_j + r_k + b_t + a_{tm} + e_{tm}$$

In matrix form, $\mathbf{y} = \mathbf{X}\boldsymbol{\beta} + \mathbf{Za} + \mathbf{e}$, where the vector, $\boldsymbol{\beta}$, contains the fixed effects for sex (v), dam age (d), and rearing rank (r), in addition to the effects for years (b). Both the selected and control lines were subjected to separate BLUP analyses using this model, and three different estimates of selection response were considered:

(i) $\widehat{\overline{y}}_{s,t} - \widehat{\overline{y}}_{c,t}$, the estimated phenotypic means following adjustment for the fixed effects (v, d, and r), obtained by $\widehat{\overline{y}}_{x,t} = \widehat{\overline{a}}_{x,t} + \widehat{b}_{x,t}$, where $x = c$ or s, corresponding to the control and selection lines, respectively. This is an unbiased estimate of the response if there is no drift in the control population ($\overline{a}_{c,t} \simeq 0$) and no $G \times E$, so that $b_{c,t} = b_{s,t}$.

(ii) $\widehat{\overline{y}}_{s,t} - \widehat{b}_{c,t}$, the (fixed-effects-adjusted) phenotypic mean of the selected population minus the common environmental effect, as estimated from the control population.

(iii) $\widehat{\overline{a}}_{s,t}$, the BLUP estimate of the mean breeding value in the selected population.

Estimate (i) mimics the estimate used in a least-squares analysis, and Blair and Pollak showed that it is independent of the assumed heritability. Estimates (ii) and (iii) are highly dependent on the assumed (or estimated) heritabilities in the control and selected populations. As Figure 9.2 reveals, all three estimates show a positive genetic trend (following a reversed response over the first few generations). As expected, the estimated response using only the predicted mean breeding value is smoother than the other two estimates.

The potential biases in a least-squares analysis of these data based on the contrast between the control and selected phenotypic means are seen in the two lower graphs in Figure 19.2. The lower left graph plots the difference in the estimated common environmental effects between selected and control populations ($\widehat{b}_{ct} - \widehat{b}_{st}$). Ignoring the inherent variance in estimating the $\widehat{b}$, the average difference (via a paired t test) is not significantly different from zero. The right-hand graph plots the predicted mean breeding value of the control population, which is assumed to be zero under the least-squares analysis. As can be seen, there is a slight, but positive, trend in the mean. When the selected mean is adjusted by subtracting from the control mean, the net result is that the LS analysis slightly underestimates the true response, $\overline{a}_{s,t}$. Thus, there is no evidence of error being introduced by $G \times E$ differences between the control and selected lines, but error is introduced if the mean breeding value of the control population departs significantly from zero.

Validation That a Trend Is Indeed Genetic

The estimated common-environmental (b) and additive-genetic (a) effects are highly dependent on the estimated (or assumed) base population heritability, h^2. Hence, using BLUP to separate genetic from environmental values is highly dependent on the heritability used being close to its true value. Other departures from mixed-model assumptions (e.g., the infinitesimal model and assuming that BLUP and REML estimates are unaffected by selection) can also result in incorrect assignment of the relative importance of environmental versus genetic values. Thus, achieving some sort of validation of a detected trend is critical.

We have already discussed validation of the general animal model, namely, by examining the consistency of the estimated additive variance over different subsets of the data. Similar validation that a trend is indeed genetic can be performed by again examining the consistency across an analysis. For example, if a control population is used, a BLUP

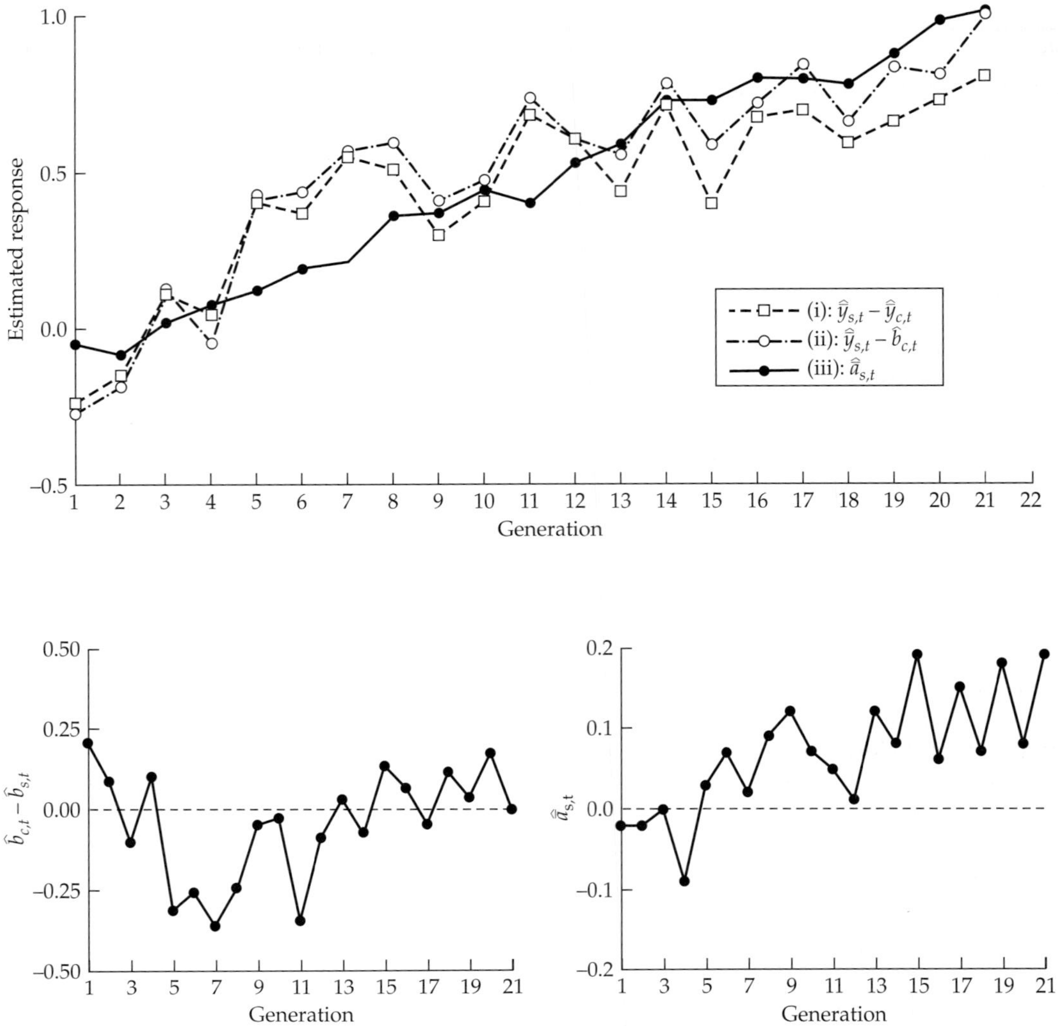

Figure 19.2 Trends for the selection experiment in Example 19.3. See the example for details.

analysis can estimate the amount of drift in the mean breeding value from its expected value of zero. Likewise, estimates of the selection and control environmental effects can be compared (Figure 19.2), and if these are reasonably consistent, then a joint analysis (assuming the same environmental values in both populations) may yield more precise estimates of the generational common-environmental values. Likewise, if the common-environmental estimates are significantly different between selected and control populations, the possibility of $G \times E$ needs to be seriously considered. Even when using a control population, there is still much to be gained by subjecting each group to a mixed-model (e.g., BLUP) analysis (Sorensen et al. 2003).

Even in the absence of a control population, one can still attempt to validate a trend. For example, Boichard et al. (1995) examined several different methods to attempt to validate genetic trends in dairy cattle, all of which involved the comparison of predicted trend values using different subsets of the data. The authors were interested in comparing the performance of a mixed model (AM90) used for French Holsteins from 1990 to 1992 with a more recent model (AM93), whose use started in 1993 (the models differed in their choice of included fixed and random effects). For a variety of reasons, there was concern that the older AM90 model yielded biased trend estimates. One check was based on the fact that milk yield data are in the form of multiple records per individual, so a repeatability model was appropriate (Example 19.6). The authors compared EBVs based only on the first lactations with estimates using all records. They found that the trends estimated from the first versus all lactations agreed well under the AM93 model, but differed dramatically under the AM90 model. Other measures of consistency (for example, looking at the stability of estimated breeding values of individuals as more information is added) also showed that the newer AM93 model seemed relatively robust, while the older AM90 model seemed to produce biased estimates of the trend.

Replicate Lines

It is straightforward to jointly analyze multiple lines simultaneously. For n lines, we write the total vector of observations as $\mathbf{y}^T = (\mathbf{y}_1^T, \mathbf{y}_2^T, \cdots, \mathbf{y}_n^T)$, where $\mathbf{y}_k$ is the vector of all observations from line k. If the generational environmental effects are assumed to be the same across the lines, the model for the ith individual in generation t from line k is

$$y_{k,ti} = \mu + b_t + a_{k,ti} + e_{k,ti} \tag{19.12a}$$

Alternatively, if the environmental effects are potentially different in each line

$$y_{k,ti} = \mu + b_{k,t} + a_{k,ti} + e_{k,ti} \tag{19.12b}$$

The power of combining multiple lines arises when environmental effects can be assumed to be the same across the lines. In this case, the effective sample size for estimating each effect is increased, and (presumably) the resulting sampling variance is decreased, improving the precision of the estimates.

The assumed covariance matrix for the vector of joint breeding values over a set of replicates can take several forms. If the founding members for each line are drawn from the same base population (but otherwise unrelated), then the covariance matrix for the vector of breeding values $\mathbf{a}$ has a block-diagonal form, with the ith block corresponding to $\sigma_A^2\mathbf{A}(i)$, the relationship matrix for line i times the base population additive variance. If the founding members of some lines are related, then $\mathbf{A}$ is more complex, reflecting these relationships. Further modifications for joint analysis were proposed by Visscher and Thompson (1990), and extended by Beniwal et al. (1992a, 1992b) and Heath et al. (1995) by allowing the additive *genic* variance to change over time (discussed below; recall from Chapter 16 that the additive genic variance is the value for σ_A^2 in the absence of LD). For example, one might assume that the additive genic variance remains constant for the first few generations of selection, after which it assumes a different value. This is a reasonable, but still ad hoc, approach for attempting to deal with potential departures from the infinitesimal model.

Estimating the Additive Variance at Generation *t*

Even under the infinitesimal model, the additive variance changes over time by the generation of disequilibrium (Chapter 16) and inbreeding (Chapters 3 and 24), a point more fully explicated in the next section. While REML provides an estimate of the base population additive genic variance, $\sigma_a^2(0)$ (which is unbiased provided the model assumptions hold), it does not immediately provide estimates of the actual additive variance, $\sigma_A^2(t) = \sigma_a^2(t) + d(t)$, in any particular generation of selection. The most straightforward approach is to use the parent-offspring regression for each generation of selection to estimate the additive genetic variance in the parents. With parents from generation t and offspring in generation $t+1$, the regression estimates the heritability of the parents, $h_A^2(t)$ (Robertson 1977b). The drawback with this approach is the typically small sample size associated with each generation (resulting in large standard errors for each heritability or variance estimate). Ideally, one would like to be able to combine information across generations in such a way as to improve the variance estimates.

Sorensen and Kennedy (1984b) suggested one approach, which is to use a mixed-model analysis treating generation t as the base population. In particular, one considers only the data from generation t onward (say, to generation T), and the relationship matrix is adjusted to assume that generation t is the base population. The resulting covariance matrix for the breeding values becomes

$$\mathbf{Var}\begin{pmatrix} \mathbf{a}_t \\ \mathbf{a}_{t+1} \\ \vdots \\ \mathbf{a}_T \end{pmatrix} = \sigma_A^2(t) \begin{pmatrix} \mathbf{I} & \mathbf{A}_{t,t+1} & \cdots & \mathbf{A}_{t,T} \\ \mathbf{A}_{t,t+1} & \mathbf{A}_{t+1,t+1} & \cdots & \mathbf{A}_{t+1,T} \\ \vdots & \ddots & \vdots & \vdots \\ \mathbf{A}_{t,T} & \mathbf{A}_{t+1,T} & \cdots & \mathbf{A}_{T,T} \end{pmatrix} \tag{19.13}$$

where $\mathbf{a}_k$ is the vector of breeding values in generation k, and $\mathbf{A}_{j,k}$ is the relationship matrix of associations between individuals in generations j and k. By taking $\mathbf{Var}(\mathbf{a}_t) = \sigma_A^2(t)\mathbf{I}$, we are assuming that all individuals in generation t are unrelated and noninbred, as this is now our base population. All measured individuals from generation t (including those not leaving offspring) are included in the base population. While this approach seems logical, it is still somewhat ad hoc and not exact. Simulation studies by van der Werf and de Boer (1990) showed that Sorensen and Kennedy's approach tends to overestimate the true additive-genetic variance.

Another potential approach to estimating the additive variance in generation t would be to use the variance among the predicted breeding values within a generation

$$\text{Var}(A_t) = \frac{1}{n-1}\sum_{i=1}^{n_t}\left(\widehat{a}_{ti} - \overline{\widehat{a}}_t\right)^2 \tag{19.14}$$

Again, however, there are complications. One is that the assumed genetic variance used to obtain the BLUP estimates has a strong influence on the values of the estimated $\widehat{a}_{ti}$. Another is that Equation 19.14 estimates $\sigma^2(\widehat{a})$, which is different from σ_A^2, as the former estimates $\rho^2\sigma_A^2$, where $\rho^2 < 1$ is the accuracy of the predicted breeding values (Equation 20.23c). Hence, Equation 19.14 is expected to underestimate the true variance, σ_A^2. Sorensen et al. (2001) noted that the use of Equation 19.14 in a *Bayesian* framework (wherein the uncertainty in variance estimates is naturally incorporated into the analysis) avoids both of these problems.

THE RELATIONSHIP MATRIX, A, ACCOUNTS FOR DRIFT AND DISEQUILIBRIUM

There are three potential sources of change in the additive-genetic variance, σ_A^2, during a selection experiment: allele-frequency change from selection; allele-frequency change from drift; and gametic-phase disequilibrium generated by selection (Chapters 5 and 16). Under the infinitesimal model, selection does not change allele frequencies, leaving only drift and disequilibrium as potential agents of change. If the average effect for any allele is small, allele-frequency changes (from selection) are also small, at least for a modest number of generations (Chapters 5 and 25), and the infinitesimal assumption can be approximately correct. In these settings, the relationship matrix, $\mathbf{A}$, in an MM analysis accounts for the effects of gametic-phase disequilibrium as well as genetic drift (Sorensen and Kennedy 1983, 1984a), and hence for all changes in σ_A^2, *provided* the base population consists of unrelated, noninbred individuals in linkage equilibrium. As a result, as long as selection-induced allele-frequency change is negligible, the variance-covariance matrix of the vector of breeding values remains the product of the base population additive-genetic variance and the relationship matrix, $\mathbf{Var}(\mathbf{a}) = \sigma_A^2\mathbf{A}$.

To both see this point and provide connections among these three potential agents of change, recall two important concepts from Chapter 16. First, the additive-genetic variance can be decomposed as $\sigma_A^2 = \sigma_a^2 + d$, the sum of the additive *genic* variance, σ_a^2, and the disequilibrium contribution, d (Equation 16.2). The former is the additive variance under linkage equilibrium and Hardy-Weinberg. Under drift and assuming only additive effects, the expected genic variance at generation t is simply $\sigma_a^2(t) = \sigma_a^2(0)(1 - f_t)$, where f_t is the inbreeding coefficient in generation t (Equation 16.9c). Second (Equation 16.8a), the regression of the breeding value, A_i, of an individual on the breeding values, A_{m_i} and A_{f_i}, of its parents is

$$A_i = \frac{1}{2}A_{f_i} + \frac{1}{2}A_{m_i} + s_i \tag{19.15a}$$

where the **segregation residual**, s, results from Mendelian sampling due to the segregation of alleles at heterozygous loci in the parents (Example 16.2).

The critical idea in this section—that the conditional independence of the covariance relationships of the vector, **a**, from selection and drift (given **A**)—follows as a consequence of the behavior of the segregation residuals under the infinitesimal model. To see this, note that if the joint distribution of breeding values for parent and offspring is multivariate normal (as occurs under the standard infinitesimal model), the regression given by Equation 19.15a is linear and homoscedastic, meaning that s is *independent* of the parental breeding values. Its variance, $\sigma^2_{s_i}$, for individual i is also independent of parental breeding values, but it does depend on the average inbreeding, $\overline{f}_i$, of its parents, such that

$$\sigma^2_{s_i} = (1 - \overline{f}_i)\,\sigma^2_a/2 \tag{19.15b}$$

Provided the vector, s, of Mendelian sampling residuals remains multivariate normal, then s $\sim$ MVN($\mathbf{0}, [\sigma^2_a/2]\,\mathbf{F}$), with $\mathbf{F}$ representing a diagonal matrix whose ith element is $(1 - \overline{f}_i)$, namely, one minus the average inbreeding of the parents of i. If k and j are the parents of i, then

$$F_{ii} = (1 - \overline{f}_i) = \left(1 - \frac{f_k + f_j}{2}\right) = \left(2 - \frac{A_{kk} + A_{jj}}{2}\right) \tag{19.16}$$

where $A_{kk} = (1 + f_k)$ denotes the kth diagonal element of the relationship matrix, **A**. Thus, the distribution of the Mendelian sampling terms, s, is unaffected by the breeding values of the parents (and hence by selection or assortative mating), while the effects of drift are fully accounted for by **F**.

The final key concept is that when we have the complete pedigree of all individuals in the selection experiment, any breeding value can be expressed as a linear function of the base population breeding values and Mendelian sampling terms (Sorensen and Kennedy 1984a; Kennedy et al. 1988). This is an extension of the idea of ancestral regressions discussed in Chapter 15. Example 19.4 shows a critical consequence of this: because the mean breeding value in the base population is zero (by construction), ***most of the response in a selection experiment comes, not from base individuals with exceptional breeding values, but rather from Mendelian sampling (i.e., segregation) generating new variation.***

Example 19.4. Suppose individuals 1 through 4 are from the base population (and assumed unrelated), while individuals 5 and 6 are the offspring from 1 and 2, and 3 and 4, respectively. The offspring breeding values can be written as

$$A_5 = \frac{1}{2}A_1 + \frac{1}{2}A_2 + s_5 \qquad \text{and} \qquad A_6 = \frac{1}{2}A_3 + \frac{1}{2}A_4 + s_6$$

If individual 7 is the offspring of parents 5 and 6, its breeding value is

$$\begin{aligned} A_7 &= \frac{1}{2}A_5 + \frac{1}{2}A_6 + s_7 = \frac{1}{2}\left(\frac{1}{2}A_1 + \frac{1}{2}A_2 + s_5\right) + \frac{1}{2}\left(\frac{1}{2}A_3 + \frac{1}{2}A_4 + s_6\right) + s_7 \\ &= \frac{1}{4}(A_1 + A_2 + A_3 + A_4) + \frac{1}{2}(s_5 + s_6) + s_7 \end{aligned}$$

Note that as individuals become increasingly more distant from the base population, the majority of their breeding value is determined by Mendelian sampling, rather than the base population breeding values of their ancestors.

This linear relationship, highlighted in the last example, between the $n \times 1$ vector, $\mathbf{a}$, of all breeding values and the $m \times 1$ vector, $\mathbf{a}_b$ (breeding values for the m base population founders), and the $(n-m) \times 1$ vector, $\mathbf{s}$ (Mendelian segregation values), can be formally expressed as follows. Defining $\mathbf{w}^T = (\mathbf{a}_b^T, \mathbf{s}^T)$, we can write

$$\mathbf{a} = \mathbf{T}\mathbf{w} = \mathbf{T}\begin{pmatrix} \mathbf{a}_b \\ \mathbf{s} \end{pmatrix} \tag{19.17a}$$

where the $n \times n$ matrix, $\mathbf{T}$, describes the flow of genes from ancestors to relatives. Indexing the elements in $\mathbf{a}$ so that ancestors precede their relatives, $\mathbf{T}$ is a lower-triangular matrix (all above-diagonal elements are zero) with diagonal values of one. Further, the upper-left $m \times m$ submatrix of $\mathbf{T}$, corresponding to the base population (and thus, no known ancestors), has all below-diagonal elements with value zero. For descendants of the base population founders, the off-diagonal element of $\mathbf{T}$ associated with (descendant) individuals i, j is

$$T_{i,j} = \frac{1}{2}\left(T_{f,j} + T_{m,j}\right) \quad \text{for} \quad j < i \tag{19.17b}$$

where f and m correspond to the index for the father and mother of i. Equations 19.17a and 19.17b formalize the decomposition (highlighted in Example 19.4) of a current breeding value into base population and segregation contributions.

From Equation 19.17a

$$\mathbf{Var}(\mathbf{a}) = \mathbf{Var}(\mathbf{T}\mathbf{w}) = \mathbf{T}\,\mathbf{Var}(\mathbf{w})\mathbf{T}^T \tag{19.18a}$$

Because we assume that the base population consists of unrelated individuals ($\mathbf{A}_b = \mathbf{I}$) in gametic-phase equilibrium (so that $\sigma_A^2 = \sigma_a^2$), $\mathbf{Var}(\mathbf{a}_b) = \sigma_a^2\,\mathbf{I}$, while $\mathbf{Var}(\mathbf{s}) = (\sigma_a^2/2)\,\mathbf{F}$. Under multivariate normality, $\mathbf{a}_b$ and $\mathbf{s}$ are independent, and hence uncorrelated, yielding

$$\mathbf{Var}(\mathbf{w}) = \mathbf{Var}\begin{pmatrix} \mathbf{a}_b \\ \mathbf{s} \end{pmatrix} = \sigma_a^2\begin{pmatrix} \mathbf{I} & \mathbf{0} \\ \mathbf{0} & \mathbf{F}/2 \end{pmatrix} = \sigma_a^2\,\boldsymbol{\Lambda} \tag{19.18b}$$

where $\boldsymbol{\Lambda}$ is a diagonal matrix with elements of one when both parents are unknown (base population) and $0.5(1-\overline{f})$ when both parents are known. If, due to missing data, only a single parent is known, the diagonal element becomes $(3-f)/4$, where f is the inbreeding in the known parent (Kennedy et al. 1988). Putting Equations 19.18a and 19.18b together yields

$$\mathbf{Var}(\mathbf{a}) = \sigma_A^2\mathbf{A} = \mathbf{T}\,\mathbf{Var}(\mathbf{w})\mathbf{T}^T = \sigma_a^2\,\mathbf{T}\,\boldsymbol{\Lambda}\mathbf{T}^T$$

Assuming $\sigma_A^2 = \sigma_a^2$ (no disequilibrium in the base population), then

$$\mathbf{A} = \mathbf{T}\boldsymbol{\Lambda}\mathbf{T}^T \tag{19.18c}$$

as obtained by Henderson (1976) and Thompson (1977). As an aside, this expression, coupled with the simple form for $\mathbf{T}$, is what allowed Henderson (1976) to obtain a very rapid way of computing $\mathbf{A}^{-1}$. The critical feature of Equation 19.18c is that $\boldsymbol{\Lambda}$, which describes the Mendelian sampling residuals, is ***independent of selection and assortative mating*** (under the infinitesimal model), and also accounts for the reduction in additive variance from genetic drift.

If the infinitesimal model does not hold, then residual values may vary with parental breeding values, and hence selection can influence the distribution of residuals. Provided that the change in allele frequencies over the course of the experiment is small, this bias may not be too serious (e.g., Maki-Tanila and Kennedy 1986). Another key assumption from the infinitesimal model is that the distribution of residuals does not significantly deviate from normality. Chapter 24 examines this rather technical issue in some detail.

When the genic variance, σ_a^2, itself is changing due to selection, knowledge of $\mathbf{A}$ is no longer sufficient to account for all changes in σ_A^2. Beniwal et al. (1992a) suggested that this complication might be (somewhat) accommodated by a modification of the additive-genetic covariance matrix. Consider the simple case of two time blocks with different (unknown) genic variances: the breeding values, $\mathbf{a}_1$, in the first block have a base genic variance, $\sigma_a^2(1)$, while the vector of breeding values, $\mathbf{a}_2$, from the second block start with a genic variance of $\sigma_a^2(2)$. First we decompose the diagonal matrix $\boldsymbol{\Lambda}$ into two blocks

$$\boldsymbol{\Lambda} = \begin{pmatrix} \boldsymbol{\Lambda}_1 & \mathbf{0} \\ \mathbf{0} & \boldsymbol{\Lambda}_2 \end{pmatrix}$$

where $\boldsymbol{\Lambda}_i$ represents block i. The covariance matrix for the breeding values can be written as

$$\mathbf{Var}(\mathbf{a}) = \mathbf{Var}\begin{pmatrix} \mathbf{a}_1 \\ \mathbf{a}_2 \end{pmatrix} = \sigma_a^2(1)\mathbf{T}\begin{pmatrix} \boldsymbol{\Lambda}_1 & \mathbf{0} \\ \mathbf{0} & \mathbf{0} \end{pmatrix}\mathbf{T}^T + \sigma_a^2(2)\mathbf{T}\begin{pmatrix} \mathbf{0} & \mathbf{0} \\ \mathbf{0} & \boldsymbol{\Lambda}_2 \end{pmatrix}\mathbf{T}^T \tag{19.19}$$

The variances $\sigma_a^2(1)$ and $\sigma_a^2(2)$ can be estimated separately using REML, and they correspond to the genic variances at the start of each block, which may be reduced by drift within the block. Heath et al. (1995) generalized this approach to allow the genic variance to change in each generation.

MODIFICATIONS OF THE BASIC ANIMAL MODEL

In many cases, it is prudent to modify the simple animal model by considering additional fixed and random effects to more fully accommodate any relevant biological features. For example, genetic and environmental effects can be separated without a control population by adding fixed effects to account for environmental trends. Likewise, as we will discuss shortly, it is often reasonable to include additional random effects, such as maternal or litter effects, to reduce potential correlations between the residuals.

Another modification of the basic model occurs when the phenotypic scores (records) of the parents are unknown. Example 19.5 shows how we can estimate these as random effects, but likely with bias (when selection occurs). As we will cover shortly, an alternative is to treat these breeding values of unmeasured parents as fixed, rather than random.

Example 19.5. A feature of mixed models is that they can often predict the values of random effects (in this case, a breeding value) for an individual with no records (but whose relatives have records). Let 0 (unmeasured) and 1 (measured) index parents that each have a measured offspring (indexed as 2 and 3, respectively), via unrelated and unmeasured individuals. If we assume that there is a single fixed effect (the mean, μ), the resulting mixed model becomes

$$\begin{pmatrix} y_1 \\ y_2 \\ y_3 \end{pmatrix} = \begin{pmatrix} 1 \\ 1 \\ 1 \end{pmatrix}\mu + \begin{pmatrix} 0 & 1 & 0 & 0 \\ 0 & 0 & 1 & 0 \\ 0 & 0 & 0 & 1 \end{pmatrix}\begin{pmatrix} a_0 \\ a_1 \\ a_2 \\ a_3 \end{pmatrix} + \begin{pmatrix} e_1 \\ e_2 \\ e_3 \end{pmatrix}$$

where y_1 through y_3 denote the values of the three measured individuals. Notice that there is no y_0 in this model (as there is no observed value for individual 0), but that parent 0's breeding value, a_0, *is* included. To complete the mixed-model analysis, we need the relationship matrix, $\mathbf{A}$. Because all parents are assumed to be unrelated, so are the two measured offspring, yielding

$$\mathbf{A} = \begin{pmatrix} 1 & 0 & 1/2 & 0 \\ 0 & 1 & 0 & 1/2 \\ 1/2 & 0 & 1 & 0 \\ 0 & 1/2 & 0 & 1 \end{pmatrix}$$

Provided unmeasured individuals have measured relatives, $\mathbf{A}$ allows us to estimate their breeding values. In this case, we can estimate the breeding value, a_0, for the unmeasured parent (for example, by using Equation 19.3b to estimate the vector, $\mathbf{a}$, of the four breeding values), as $\mathbf{A}$ shows that information on a_o is provided from its observed offspring value, y_2.

Connections between relatives are often referred to as **links**, and the number and strengths of links (or the **connectiveness**) in a relationship matrix is one measure of its precision in estimating breeding values (i.e., the prediction error variance; see Kennedy and Trus 1993). The breeding value of an unmeasured individual with few measured relatives will have much less precision than that of an individual with a large number of measured, and close, relatives. Despite this flexibility in predicting breeding values for individuals with missing records, it is important to again stress that the simple inclusion of all pedigree relationships appears *not* to be sufficient to yield unbiased BLUP/REML estimates when selection occurs. The records (measured values) of all individuals upon which selection decisions have been based must also be included (van der Werf 1990; van der Werf and de Boer 1990; van der Werf and Thompson 1992).

Models with Additional Random Effects

We have assumed that residuals of a mixed model are uncorrelated and homoscedastic, making their covariance matrix $\mathbf{Var}(\mathbf{e}) = \sigma^2_e\,\mathbf{I}$. When additional random effects are present but ignored by the model, they are inadvertently subsumed into the residuals, introducing correlations and heteroscedasticity, which, in turn, may bias the BLUP estimates. For example, if sibs share a common maternal environment, this introduces correlations between them beyond those accounted for by $\mathbf{A}$. If the model only includes $\mathbf{a}$ and $\mathbf{e}$, this additional covariance appears between the residuals of sibs, and the true covariance matrix for $\mathbf{e}$ is no longer the assumed diagonal, which leads to biased estimates of the BLUEs and BLUPs. By suitably incorporating additional random effects, we can develop a new model, in which the residuals again have the simple OLS covariance structure, $\mathbf{Var}(\mathbf{e}) = \sigma^2_e\,\mathbf{I}$.

Suppose there is a second vector, $\mathbf{u}$, of m random effects (e.g., the common effects from the m families in our sample) in addition to the vector, $\mathbf{a}$, of p breeding values, and vector of residuals, $\mathbf{e}$. Equation 19.1 becomes

$$\mathbf{y} = \mathbf{X}\boldsymbol{\beta} + \mathbf{Za} + \mathbf{Wu} + \mathbf{e} \tag{19.20a}$$

where $\mathbf{X}$, $\mathbf{Z}$, and $\mathbf{W}$ are $n \times q$ (n observations and q fixed effects), $n \times p$, and $n \times m$ incidence matrices. Assuming that common-family effects are *not* shared across different families (even if the families are related), the assumed covariance structures for the three vectors of random effects become $\mathbf{Var}(\mathbf{a}) = \sigma^2_A\,\mathbf{A}$, $\mathbf{Var}(\mathbf{u}) = \sigma^2_u\,\mathbf{I}$, and $\mathbf{Var}(\mathbf{e}) = \sigma^2_e\,\mathbf{I}$, where $\mathbf{a}$, $\mathbf{u}$, and $\mathbf{e}$ are uncorrelated. This can be compactly written as

$$\mathbf{Var}\begin{pmatrix} \mathbf{a} \\ \mathbf{u} \\ \mathbf{e} \end{pmatrix} = \begin{pmatrix} \sigma^2_A\,\mathbf{A} & \mathbf{0} & \mathbf{0} \\ \mathbf{0} & \sigma^2_u\,\mathbf{I} & \mathbf{0} \\ \mathbf{0} & \mathbf{0} & \sigma^2_e\,\mathbf{I} \end{pmatrix} \tag{19.20b}$$

yielding a covariance matrix of $\mathbf{y}$ as

$$\mathbf{Var}(\mathbf{y}) = \mathbf{V} = \sigma^2_A\,\mathbf{ZAZ}^T + \sigma^2_u\,\mathbf{WW}^T + \sigma^2_e\,\mathbf{I} \tag{19.20c}$$

If we had assumed the validity of the model $\mathbf{y} = \mathbf{X}\boldsymbol{\beta} + \mathbf{Za} + \mathbf{e}$, the resulting covariance matrix for the residuals would be $\mathbf{Var}(\mathbf{e}) = \sigma^2_u\,\mathbf{WW}^T + \sigma^2_e\,\mathbf{I}$, which shows how the additional random effects alter the covariance structure, which is no longer of OLS form, $\sigma^2\mathbf{I}$. Analysis of this model assuming an OLS covariance structure would thus be erroneous.

The resulting set of Henderson's mixed-model equations (Equation 19.4) for the model specified by Equations 19.20a and 19.20b becomes

$$\begin{pmatrix} \mathbf{X}^T\mathbf{X} & \mathbf{X}^T\mathbf{Z} & \mathbf{X}^T\mathbf{W} \\ \mathbf{Z}^T\mathbf{X} & \mathbf{Z}^T\mathbf{Z} + \lambda_A\mathbf{A}^{-1} & \mathbf{Z}^T\mathbf{W} \\ \mathbf{W}^T\mathbf{X} & \mathbf{W}^T\mathbf{Z} & \mathbf{W}^T\mathbf{W} + \lambda_u\mathbf{I} \end{pmatrix} \begin{pmatrix} \widehat{\boldsymbol{\beta}} \\ \widehat{\mathbf{a}} \\ \widehat{\mathbf{u}} \end{pmatrix} = \begin{pmatrix} \mathbf{X}^T\mathbf{y} \\ \mathbf{Z}^T\mathbf{y} \\ \mathbf{W}^T\mathbf{y} \end{pmatrix} \tag{19.21a}$$

where

$$\lambda_A = \frac{\sigma_e^2}{\sigma_A^2} \quad \text{and} \quad \lambda_u = \frac{\sigma_e^2}{\sigma_u^2} \tag{19.21b}$$

Additional vectors of random effects can be incorporated in a similar manner; see LW Chapters 26 and 27 for details. The mixed-model equations again form the basis for iterative REML estimates of the unknown variance components (σ_A^2, σ_u^2, and σ_e^2), as discussed in detail in LW Chapter 27.

Example 19.6. Often the same trait is measured multiple times in the same individual, for example, the sizes of different litters from a single female. When multiple records are present for at least some individuals, a **repeatability model** is appropriate (Chapter 13; LW Chapter 26). Repeated measures from the same individual have three components: a breeding value, a_k, and a common (permanent) environmental value, p_k, which are the same in each measurement of individual k, and the residual environmental value, e, which varies between measurements, yielding the ith measurement of the kth individual as $a_k + p_k + e_{ki}$. The **repeatability** of the trait is $r = (\sigma_A^2 + \sigma_p^2)/\sigma_z^2$, making the variance of the residuals $\sigma_e^2 = (1-r)\sigma_z^2$ and the variance of permanent environmental effects as $\sigma_p^2 = (r - h^2)\sigma_z^2$, where σ_z^2 is the trait variance after accounting for fixed effects. We remind the reader that permanent "environmental" effects can also include nonadditive genetic components, as these are not passed along to offspring under random mating.

The repeatability model was used by Estany et al. (1989) to examine the selection response for litter size in rabbits. Their model assumed that there are two groups of fixed effects: the year-season (environmental) effects, b_t which had 22 levels over the course of this experiment (with b_0 set to zero), and the reproductive state, l_i, of the doe (three levels; because only two of these factors are estimable, l_1 was assigned a value of zero). Their model had three random effects: a_k and p_k, for the additive genetic and permanent environmental effect of the kth doe, respectively, and the residual e, resulting in an overall model of

$$y_{tk\ell i} = \mu + l_i + b_t + a_k + p_k + e_{tk\ell i}$$

where $y_{tk\ell i}$ denotes the size of the ℓth litter of doe k in reproductive state i in season-year t. In matrix form, the mixed model becomes

$$\mathbf{y} = \mathbf{X}\boldsymbol{\beta} + \mathbf{Za} + \mathbf{Zp} + \mathbf{e}$$

where $\mathbf{a}$ and $\mathbf{p}$ are $n \times 1$ vectors corresponding to the n does, $\mathbf{Var(a)} = \sigma_A^2\,\mathbf{A}$, $\mathbf{Var(p)} = \sigma_p^2\,\mathbf{I}$, and $\mathbf{Var(e)} = \sigma_e^2\,\mathbf{I}$. $\mathbf{X}$ and $\mathbf{Z}$ are incidence matrices, and the vector of fixed effects is

$$\boldsymbol{\beta} = \begin{pmatrix} \mu \\ l_2 \\ l_3 \\ b_1 \\ \vdots \\ b_{22} \end{pmatrix}$$

The mixed-model equations are given by Equation 19.21a, with

$$\lambda_A = \frac{\sigma_e^2}{\sigma_A^2} = \frac{1-r}{h^2} \quad \text{and} \quad \lambda_u = \frac{\sigma_e^2}{\sigma_p^2} = \frac{1-r}{r-h^2}$$

The careful reader might notice that the two vectors of random effects, the breeding values, $\mathbf{a}$, and permanent environment effects, $\mathbf{p}$, enter the model as $\mathbf{Za}$ and $\mathbf{Zp}$, respectively. Why, then, do we simply not combine these as, for example, $\mathbf{Zu}$ where $\mathbf{u} = \mathbf{a} + \mathbf{p}$? The reason we do not do this (and indeed, the reason we can estimate $\mathbf{a}$ and $\mathbf{p}$ separately) is that $\mathbf{a}$ and $\mathbf{p}$ have *different covariance structures*, $\sigma_A^2\,\mathbf{A}$ versus $\sigma_p^2\,\mathbf{I}$. While estimates of $\mathbf{a}$ borrow information from relatives (correlated observations provide information to supplement a direct observation), estimates of $\mathbf{p}$ depend only on the focal individual (as we assume these are uncorrelated between individuals).

Common Family and Maternal Effects

When sibs are present, any common-family environmental effects (if not included directly in the model) are subsumed into the residuals, creating a correlation between the residuals of sibs. For example, if two sibs (i and j) share a common environmental value, c, then $\sigma(e_i, e_j) = \sigma_c^2$. Hence, there are off-diagonal elements in the covariance matrix of residuals, and we no longer have the standard OLS assumption of $\mathbf{Var}(\mathbf{e}) = \sigma_e^2\,\mathbf{I}$. Such correlations can easily be accommodated by adding an additional vector of random effects.

Example 19.7. Meyer and Hill (1991) examined the response to selection on adjusted food intake (AFI) in mice (Example 19.2) and formulated a model incorporating shared family values, c, as random effects. In addition, their model accounts for fixed effects due to common environmental effects associated with generations (b, 22 levels), lines (L, 3 levels), sex (v, male/female), and litter size (l, 7 levels for litters of size 6 to 12 individuals). Under their model, the observed value for AFI from the ith individual from generation t, line ℓ and full-sib family k is given by

$$y_{\ell,tki} = \mu + b_t + L_\ell + v_j + l_m + a_{\ell,tki} + c_{\ell,tk} + e_{\ell,tki}$$

where this individual has sex j and was reared in a litter of size m. In matrix form, $\mathbf{y} = \mathbf{X}\boldsymbol{\beta} + \mathbf{Za} + \mathbf{Wc} + \mathbf{e}$. The vector of fixed effects, $\boldsymbol{\beta}$, contains the b, L, v, and l values, while the random effects are the vector of common family effects, $\mathbf{c}$, the vector of additive genetic values, $\mathbf{a}$, and the vector of residuals, $\mathbf{e}$.

The $\mathbf{W}$ incidence matrix has 1 as its ikth element if individual i is from family k, or else the element is 0. Note that Meyer and Hill have two model variables to account for litter effects—a fixed effect, l, common to all litters of the same size, and a random effect, $c_{\ell,tk}$, that varies between families but is the same for all individuals within a particular family. The resulting REML estimate for heritability was 0.15, while the fraction of the total variation (after removal of the fixed effects) accounted for by random family effects was estimated to be $\gamma^2 = \sigma_c^2/\sigma_z^2 = 0.22$, where $\sigma_z^2 = \sigma_A^2 + \sigma_c^2 + \sigma_e^2$ is the fixed-effects adjusted trait variance. Because the intraclass correlation between sibs is $(h^2/2) + \gamma^2$ (Chapter 21), a larger fraction of the resemblance between sibs is due to shared family environments (potentially including maternal effects), rather than due to shared genes. One caveat is that the model assumes no dominance variance. If this are present, sibs also share dominance variance ($\sigma_D^2/4$), but under this model, it would be absorbed into σ_c^2.

Sib correlations (beyond those accounted for by their correlations in breeding value for the focal trait) can arise for multiple reasons. The first is dominance, which we will address shortly. The second and third reasons are common family effects, which can arise from two different sources: shared *environmental* effects and shared maternal effects with a genetic component (i.e., maternal performance itself has a *genetic* component; see Chapters 15 and 22). The distinction between the genetic and the environmental components of maternal effects (both of which may be included in a model) was foreshadowed in Example 19.6, in that they result in different covariance structures. Environmental correlations are unique to a given family, while maternal-genetic effects are correlated over relatives, even those in different generations. In Example 19.7, the common environment, c, was assumed to be uncorrelated between sibships, implying that its covariance structure is $\mathbf{Var}(\mathbf{c}) = \sigma_c^2\,\mathbf{I}$, which is a diagonal matrix whose dimension is set by the number of families. There is no correlation across different sibships and thus no shared information.

When a mother has several litters (especially when each may contain a modest to large number of offspring), modifications beyond assigning a single (random) common-family environment value for all her litters may be considered. One approach is to assume that each particular litter has a unique common-family effect, that is uncorrrelated across litters from the same mother. Here, the covariance structure remains $\mathbf{Var}(\mathbf{c}) = \sigma_c^2\,\mathbf{I}$, but now each litter (rather than each mother) has its own c value.

A second, and perhaps biologically more realistic, approach is to consider something akin to a repeatability model (Example 19.6), wherein the contribution from the mother to a particular litter has two components, one shared over all litters, and a second unique to each litter. Under this model, the common-family environmental effect for all sibs in the ith litter from mother k is given by $c_{ki} + p_k$, a permanent environmental effect for this mother, p_k (a common effect shared by all her litters, as in our initial model), plus a unique environment shared by all members of her ith litter, c_{ki}. For this model, we need to estimate the variances σ_p^2 and σ_c^2, which correspond to the contributions from p_k and c_{ki}, respectively. We assume that at least one litter has two or more individuals, otherwise c_{ki} and p_k cannot be estimated separately. More generally, this model is not expected to be very efficient unless the expected number of offspring per litter is modest to large. If the common-litter effect under this model is environmental, namely, that a mother does not pass along any of her performance genetically to her daughters, then the covariance structure for these two vectors of random effects will be $\mathbf{Var}(\mathbf{c}) = \sigma_c^2\,\mathbf{I}$ and $\mathbf{Var}(\mathbf{p}) = \sigma_p^2\,\mathbf{I}$, square matrices whose dimensions are the total number of separate litters in the experiment and the number of mothers, respectively.

However, maternal performance could also have a *genetic* component, such that female relatives have correlated maternal performances. In this case, we would add a third random effect, a_m (the breeding value of this maternal effect), to the model. This vector of breeding values for maternal performance has a covariance matrix of $\mathbf{Var}(\mathbf{a}_m) = \sigma_A^2(a_m)\mathbf{A}$. Distinguishing between maternal and direct effects requires that there are paternal, as well as maternal, links in the pedigree (Clément et al. 2001; Kruuk 2004; Chapter 22). A complication is that the breeding values for the focal trait and maternal performance can be correlated. This raises a multiple-trait problem (LW Chapter 27). We discuss maternal effects in greater detail in Chapter 22 in the context of more general models of associative effects.

In summary, from an operational standpoint, if maternal effects are suspected, at a minimum, a common-family effect should be included, and it should be in the form of a repeatability model if the female has multiple litters (provided each has several offspring). More generally, if there are many links between female relatives (with litters) in the data set, one should seriously consider a genetic maternal-effects model. Failure to do so may result in contributions from shared genetic maternal performance being regarded as breeding values for the direct trait, giving a biased picture of the nature of the selection response (Milner et al. 2000; Clément et al. 2001; Kruuk 2004). Chapter 22 will examine in more detail when to include, and when to exclude, random effects when constructing more complex models.

Treating Certain Breeding Values as Fixed Effects

How should one proceed if the base population has, itself, been under selection? Graser et al. (1987) suggested that if this is known, or suspected, to be the case, the base population breeding values should be treated as fixed, rather than random, effects. The motivation for this suggestion is that if the parents are selected, they are not a random sample from the base population. Because REML variance estimates are unbiased by fixed effects, any bias in the variance of the initial sample is ignored by treating the original parental breeding values as fixed. However, simulation studies show that even if initial bias is reduced by treating the parents as fixed, selection on the resulting offspring (or future generations) introduces additional bias (van der Werf 1990).

Despite this reservation, we briefly review the approach here, as parents (beyond the base population) whose records are missing are also often treated as fixed, which requires some modifications of the mixed-model equations. Let $\mathbf{a_b}$ be the vector of breeding values for the base population and $\mathbf{a_r}$ be breeding values of the remaining individuals that descend from the base population. Following Graser et al. (1987), we can express the dependence of $\mathbf{a_r}$ on the base population breeding values, $\mathbf{a_b}$, as follows

$$\begin{pmatrix} \mathbf{a_b} \\ \mathbf{a_r} \end{pmatrix} = \begin{pmatrix} \mathbf{I} & \mathbf{0} \\ \mathbf{P}_1 & \mathbf{P}_2 \end{pmatrix} \begin{pmatrix} \mathbf{a_b} \\ \mathbf{a_r} \end{pmatrix} + \begin{pmatrix} \mathbf{0} \\ \mathbf{s} \end{pmatrix} \tag{19.22}$$

where $\mathbf{s}$ is the vector of segregational residuals (Equation 19.15a) and $\mathbf{P}_1$ and $\mathbf{P}_2$ are matrices with values of 1/2 in the parent's column in each row. Here, $\mathbf{a_r}$ is treated as a random effect because it is a function of a fixed effect ($\mathbf{a_b}$) and a random effect ($\mathbf{s}$). The resulting mixed-model is

$$\mathbf{y} = \mathbf{X}\boldsymbol{\beta} + \mathbf{Z}_1\mathbf{a_b} + \mathbf{Z}_2\mathbf{a_r} + \mathbf{e} \tag{19.23}$$

Graser et al. showed that the associated mixed-model equations are

$$\begin{pmatrix} \mathbf{X}^T\mathbf{X} & \mathbf{X}^T\mathbf{Z}_1 & \mathbf{X}^T\mathbf{Z}_2 \\ \mathbf{Z}_1^T\mathbf{X} & \mathbf{Z}_1^T\mathbf{Z}_1 + \lambda\mathbf{Q}^T\mathbf{G}^{-1}\mathbf{Q} & -\lambda\mathbf{Q}^T\mathbf{G}^{-1} \\ \mathbf{Z}_2^T\mathbf{X} & -\lambda\mathbf{G}^{-1}\mathbf{Q} & \mathbf{Z}_2^T\mathbf{Z}_2 + \lambda\mathbf{G}^{-1} \end{pmatrix} \begin{pmatrix} \widehat{\boldsymbol{\beta}} \\ \widehat{\mathbf{a}}_\mathbf{b} \\ \widehat{\mathbf{a}}_\mathbf{r} \end{pmatrix} = \begin{pmatrix} \mathbf{X}^T\mathbf{y} \\ \mathbf{Z}_1^T\mathbf{y} \\ \mathbf{Z}_2^T\mathbf{y} \end{pmatrix} \tag{19.24a}$$

where $\lambda = (2\sigma_e^2/\sigma_A^2)$, and

$$\mathbf{Q} = (\mathbf{I} - \mathbf{P}_2)^{-1}\mathbf{P}_1 \quad \text{and} \quad \mathbf{G} = (\mathbf{I} - \mathbf{P}_2)^{-1}\mathbf{F}\left[(\mathbf{I} - \mathbf{P}_2)^{-1}\right]^T \tag{19.24b}$$

with the elements of the diagonal matrix, $\mathbf{F}$, given by Equation 19.16. Note that the information used to estimate the additive-genetic variance comes from the vector, s, of segregation values, and thus estimates the additive genic variance, σ_a^2. If no linkage disequilibrium is present in the base population, this segregation-based estimate is also, then, an estimate of the additive genetic variance, σ_A^2; otherwise it is biased (as $\sigma_A^2 \neq \sigma_a^2$ when $d \neq 0$; Chapter 15).

Dominance

Up to this point, we have been assuming that all genetic variation is additive, requiring us to only consider the vector, $\mathbf{a}$, of breeding values and its numerator relationship matrix, $\mathbf{A}$. When nonadditive genetic variance is present, it creates additional genetic correlations between certain relatives beyond those accounted for by $\mathbf{A}$. The simplest setting is that in which dominance occurs, which inflates the covariance among (noninbred) full sibs by $\sigma_D^2/4$. As demonstrated previously, sibs can also have their covariance inflated by common-family effects, and separating the contribution of dominance from common-family environment is nontrivial, as it requires very specific types of links in the pedigree.

If the goal is simply to reduce the bias in predicted breeding values when dominance is present, an animal model with an additional random factor for common-family effects (e.g., Example 19.7) will often be satisfactory. This model simply estimates the common sib variance, σ_c^2, which may include contributions from both dominance and shared-family environments.

The goal of estimating the dominance variance directly, and thus predicting dominance values, is considerably more difficult. Misztal (1997) found that roughly 20-fold more data are required for dominance estimates to match the precision of their additive counterparts. This is because information on dominance only arises from relatives with a nonzero coefficient of fraternity (Equations 11.13, 11.19a, and 11.19b; LW Equation 7.7), which requires that each parent from one individual be related to at least one parent of the other individual (see Equation 19.26a). Even in these cases, the coefficient of fraternity (which yields the corresponding weight on σ_D^2 in the covariance between relatives; LW Equation 7.7) is small (e.g., 0.25 for full sibs, 0.0625 for double first-cousins, and much less for more distant relatives). If the only such links in a pedigree are between full sibs, then common-family environmental effects and dominance are fully confounded and cannot be separated (meaning that one must have a sufficient number of relatives with different, but nonzero, coefficients of fraternity).

Assuming no common-family environmental effects (which is a *major* assumption), we can attempt to estimate dominance as follows. Letting the vector $\mathbf{d}$ denote the dominance effects, the mixed model becomes

$$\mathbf{y} = \mathbf{X}\boldsymbol{\beta} + \mathbf{Za} + \mathbf{Zd} + \mathbf{e} \tag{19.25a}$$

The overall genetic merit of an individual is estimated by $\widehat{\mathbf{g}} = \widehat{\mathbf{a}} + \widehat{\mathbf{d}}$. Turning to the covariance structure of this model, as before, $\mathbf{Var}(\mathbf{a}) = \sigma_A^2\,\mathbf{A}$ and $\mathbf{Var}(\mathbf{e}) = c_E^2\,\mathbf{I}$, while the covariance matrix for dominance effects is $\mathbf{Var}(\mathbf{d}) = \sigma_D^2\,\mathbf{D}$, yielding

$$\mathbf{Var}(\mathbf{y}) = \mathbf{V} = \sigma_A^2\,\mathbf{ZAZ}^T + \sigma_D^2\,\mathbf{ZDZ}^T + \sigma_e^2\,\mathbf{I} \tag{19.25b}$$

where $\mathbf{D}$ is the **dominance genetic relationship matrix**, which will be detailed shortly. Equation 19.25b shows that if we assumed the validity of the model

$$\mathbf{y} = \mathbf{X}\boldsymbol{\beta} + \mathbf{Za} + \mathbf{e}^* \tag{19.25c}$$

then the covariance structure is given by

$$\mathbf{Var}(\mathbf{e}^*) = \sigma_D^2\,\mathbf{ZDZ}^T + \sigma_e^2\,\mathbf{I} \tag{19.25d}$$

If we assumed the validity of Equation 19.25c and the absence of dominance, we would (incorrectly) use $\sigma_e^2\,\mathbf{I}$, instead of the correct error structure, which potentially would bias our estimates of $\mathbf{a}$.

The elements of $\mathbf{D}$ are obtained as follows. The covariance between dominance effects for (noninbred) individuals *i* and *j* is the product of the dominance genetic variance and the coefficient of fraternity, $\sigma_D^2\Delta_{ij}$. From LW Equation 7.7, this is given by

$$\Delta_{ij} = \Theta_{gk}\,\Theta_{hl} + \Theta_{gl}\,\Theta_{hk} \tag{19.26a}$$

where *i*'s parents are indexed by *g* and *h* and *j*'s parents are indexed by *k* and *l*, and where (as above), Θ is the coefficient of coancestry. Recalling that the elements of the numerator relationship matrix, $\mathbf{A}$, are $2\Theta_{ij}$, the off-diagonal elements of $\mathbf{D}$ can be computed from the elements of $\mathbf{A}$ by

$$D_{ij} = \frac{A_{gk}\,A_{hl} + A_{gl}\,A_{hk}}{4} \tag{19.26b}$$

whereas the diagonal elements are all $D_{ii} = 1$ (when i is not inbred). Note that $\mathbf{D}$ is expected to be considerably more **sparse** (most of its off-diagonal elements are zero) than $\mathbf{A}$, and hence may not contribute information for most individuals. Ovaskainen et al. (2008) noted that Equation 19.26a is an approximation, requiring that the four probabilities (i.e., the Θ_{ij}) determining Δ_{ij} be independent. This is usually not a serious problem unless the pedigree is highly inbred. The resulting mixed-model equations for Equation 19.25a become

$$\begin{pmatrix} \mathbf{X}^T\mathbf{X} & \mathbf{X}^T\mathbf{Z} & \mathbf{X}^T\mathbf{Z} \\ \mathbf{Z}^T\mathbf{X} & \mathbf{Z}^T\mathbf{Z} + \lambda_A\mathbf{A}^{-1} & \mathbf{Z}^T\mathbf{Z} \\ \mathbf{Z}^T\mathbf{X} & \mathbf{Z}^T\mathbf{Z} & \mathbf{Z}^T\mathbf{Z} + \lambda_D\mathbf{D}^{-1} \end{pmatrix} \begin{pmatrix} \widehat{\boldsymbol{\beta}} \\ \widehat{\mathbf{a}} \\ \widehat{\mathbf{d}} \end{pmatrix} = \begin{pmatrix} \mathbf{X}^T\mathbf{y} \\ \mathbf{Z}^T\mathbf{y} \\ \mathbf{Z}^T\mathbf{y} \end{pmatrix} \tag{19.27}$$

where $\lambda_A = \sigma_e^2/\sigma_A^2$ and $\lambda_D = \sigma_e^2/\sigma_D^2$. Hoeschele and Van Raden (1991) presented a rapid method for computing $\mathbf{D}^{-1}$ for a noninbred population.

In addition to these concerns, inbreeding (which occurs in all selection experiments) introduces major complications. First, there may be inbreeding depression. In some situations, this can be dealt with by simply including the level of inbreeding (f) as a covariate, for example, by using a model such as

$$y_{ti} = \mu + (B \cdot f_{ti}) + a_{ti} + b_{ti} + e_{ti} \tag{19.28}$$

where f_{ti} is the inbreeding coefficient for the ith individual in generation t and B is the inbreeding depression under complete inbreeding, which is a fixed factor to be estimated. Because $A_{ii} = (1 + f_i)$, the value for f_{ti} immediately follows from the diagonal element of $\mathbf{A}$ corresponding to individual ti, namely $f_{ti} = A_{ti,ti} - 1$.

When the only nonadditive genetic interaction is dominance, inbreeding depression is a linear function of the inbreeding level, f, but this relationship is nonlinear when certain types of epistasis are present (LW Chapter 10). Thus, if there is significant epistasis, Equation 19.28 may not appropriately correct for inbreeding depression, especially at very high values of f (those approaching one). If the levels of inbreeding are very similar between a selected and a control population, the use of a control can account for nonlinear inbreeding depression. If f is expected to be small or modest (say $f \leq 0.3$), then f^2 will be 0.01 or less and the weighting on any epistatic term will be quite small. In such cases, a simple linear model for inbreeding should be sufficient.

A second, and more subtle, complication is that the covariance between inbred relatives with dominance is no longer a function of just σ^2_A and σ^2_D. As discussed in Chapter 11, these covariances now depend upon four other quadratic components (σ^2_{DI}, σ_{ADI}, $\imath^*$, $\imath^2$); see Equation 11.13 and Table 11.2. While one could formulate a mixed model incorporating all six quadratic components (Equations 11.19a and 11.19b present the required covariance structures), the resulting model is extremely complex, numerically very demanding, and expected to yield very low precision estimates of nonadditive genetic effects. A start toward including these other variance components was developed by Smith and Mäki-Tanila (1990), who should be consulted for more details. A simpler, but only approximate, approach is to use Equation 19.27, with a $\mathbf{D}$ matrix that approximates the elements under inbreeding (Smith and Mäki-Tanila 1990). One could also combine the use of a modified $\mathbf{D}$ with a covariate for inbreeding depression (Equation 19.28), but this is still a largely ad hoc approach to a complex problem.

Some guidance on dealing with dominance is offered from simulations by de Boer and van Arendonk (1992), who examined the consequences of ignoring a cofactor for inbreeding depression and not using the full (i.e., correct) covariance structure under inbreeding. When a standard dominance model not accounting for inbreeding was used, estimates of both additive and dominance effects were biased. However, when a simple cofactor for inbreeding depression was included, but the full covariance structure under inbreeding was ignored, estimates of $\mathbf{a}$ were generally unbiased, at least up to the level of inbreeding used in the simulations ($f = 0.35$). Thus, simply including a fixed effect for inbreeding depression (Equation 19.28) and a random effect for common-family effects (Example 19.7) appears to be a relatively robust approach for handling dominance, at least for modest levels of inbreeding (those typically seen in most animal-breeding programs). With highly inbred populations (such as selfed lines), other approaches are required (Chapter 23).

The situation with epistasis is even more complex than that for dominance. The encouraging news is that the weighting of the nonadditive-variance component terms for the covariances between even modestly distant relatives is very small (e.g., LW Equation 7.12), so even if nonadditive components are significant, their *actual* contribution to the genetic covariance of most relatives, especially those separated by more than one generation, are very minor and can usually be ignored. In theory, epistatic terms can be included in the mixed-model equations in a similar fashion as with dominance; see LW Chapter 26 for details, and LW Chapter 27 for modifications of the REML equations to estimate nonadditive variances. In practice this is almost never done, as the effects are generally small and the precision of estimates is quite poor at best.

BAYESIAN MIXED-MODEL ANALYSIS

As mentioned throughout this chapter, a standard mixed-model analysis does not fully account for the uncertainty introduced into EBVs by using estimates of the model variance components (instead of their true values). While there are large-sample approximations for the sample variance of a REML variance estimator, it is never fully clear what constitutes "large." Further, quantities of interest, such as the heritability, are often functions of the estimated quantities. The sample variances and distributions of such functions are very complicated, and they are typically obtained through either simulations or approximations (e.g., the delta method; LW Appendix 1). **Bayesian approaches** offer solutions to both of these problems. While Bayesian statistics (as opposed to more standard, or **frequentist**, statistics) are often touted for their ability to incorporate prior information, their key utility is in providing a more complete description of the uncertainty of an estimate.

Frequentists assume that the true value of a parameter is (typically) a constant and the samples are variable. Statistics (such as confidence intervals) are computed by conceptually drawing an infinite number of samples. For example, the frequentist's expectation is that the true value of a parameter is contained within constructed 95% confidence intervals in all but 5% of such draws. In contrast, a Bayesian assumes that the sample is fixed while the parameter is random, the focus being on how the data changes our prior assumptions for the probability distribution for possible locations of the parameter. Thus, the term "Bayesian mixed model" is (formally) inappropriate, as *all* terms in a Bayesian analysis are assumed to be random, and hence never "mixed." However, we use this term to emphasize that many of the basic foundations (such as the model formulation) of an MM analysis of selection experiments remain unchanged in a Bayesian framework. What does change is how we analyze the data.

Appendix 2 introduces some of the basic ideas in a Bayesian analysis (beyond our short introduction here). Computational issues are extremely important, and they are covered in Appendix 3. Indeed, the recent explosion in the application of Bayesian approaches largely follows from relatively recent computational approaches, such as Markov Chain Monte Carlo (MCMC) methods that allow complex distributions to be handled through straightforward (but computationally intensive) procedures (Appendix 3).

Introduction to Bayesian Statistics

While very deep (and subtle) differences in philosophy separate hard-core Bayesians from hard-core frequentists (Glymour 1981; Efron 1986), our treatment here of Bayesian methods is motivated simply by their use as a powerful statistical tool. Their introduction into quantitative genetics can be largely credited to the influential paper of Gianola and Fernando (1986), which reviewed Bayesian applications to animal breeding. Blasco (2001, 2017) provided nice overviews of Bayesian versus frequentist approaches in quantitative genetics and both are highly recommended, while a detailed treatment of applications to quantitative genetics was provided by Sorensen and Gianola (2002). Gianola and Rosa (2015) presented an excellent review of the history of statistical methods in animal breeding, culminating with the wide acceptance of Bayesian approaches.

The foundation of Bayesian statistics is **Bayes' theorem** (Appendix 2), which provides the relationship between $\Pr(x \mid y)$ and $\Pr(y \mid x)$, namely, the flipped conditional probabilities. The continuous, vector-valued version of this theorem is

$$p(\boldsymbol{\Theta} \mid \mathbf{y}) = \frac{p(\mathbf{y} \mid \boldsymbol{\Theta})\, p(\boldsymbol{\Theta})}{p(\mathbf{y})} = \frac{p(\mathbf{y} \mid \boldsymbol{\Theta})\, p(\boldsymbol{\Theta})}{\int p(\mathbf{y}, \boldsymbol{\Theta})\, d\boldsymbol{\Theta}} \tag{19.29a}$$

where $\boldsymbol{\Theta}^T = (\theta^{(1)}, \theta^{(2)}, \cdots, \theta^{(k)})$ is a vector of k random variables. Here $p(\boldsymbol{\Theta})$ is our prior belief (**prior** for short) about the distribution of the unknown values, $\boldsymbol{\Theta}$, while $p(\mathbf{y} \mid \boldsymbol{\Theta})$ is simply a standard likelihood function for the probability density of the observed vector of data, $\mathbf{y}$, given that the unknown parameters have a specified value of $\boldsymbol{\Theta}$ (LW Appendix 4). The product of these two variables, normalized by $p(\mathbf{y})$ to form a proper probability distribution (i.e., that integrates to 1), is our posterior belief (the **posterior**), $p(\boldsymbol{\Theta} \mid \mathbf{y})$, for the distribution of the unknown parameters, given both the data, $\mathbf{y}$, and the prior information, $p(\boldsymbol{\Theta})$. Because $p(\mathbf{y})$, the probability of the data vector $\mathbf{y}$, is a constant independent of $\boldsymbol{\Theta}$, it is typically ignored, and the posterior is often simply written as

$$p(\boldsymbol{\Theta} \mid \mathbf{y}) \propto p(\mathbf{y} \mid \boldsymbol{\Theta})\, p(\boldsymbol{\Theta}) \tag{19.29b}$$

In words, the posterior is the product of the likelihood and the prior multiplied by a normalization constant to return a proper probability distribution.

Often, only a subset of the unknown variables is of concern, with the rest being regarded as **nuisance variables** (or **nuisance parameters**) that we wish to remove (or at least ignore). To see how this can be accomplished, partition the vector of unknown parameters as $\boldsymbol{\Theta}^T = (\boldsymbol{\Theta}_1^T, \boldsymbol{\Theta}_{nu}^T)$, where $\boldsymbol{\Theta}_{nu}$ is the column vector of nuisance variables. Integrating the full posterior over $\boldsymbol{\Theta}_{nu}$ yields the **marginal posterior distribution** for the variables of interest, $\boldsymbol{\Theta}_1$, as

$$\begin{aligned} p(\boldsymbol{\Theta}_1 \mid \mathbf{y}) &= \int p(\boldsymbol{\Theta}_1, \boldsymbol{\Theta}_{nu} \mid \mathbf{y})\, d\boldsymbol{\Theta}_{nu} \\ &= \int p(\boldsymbol{\Theta}_1 \mid \boldsymbol{\Theta}_{nu}, \mathbf{y})\, p(\boldsymbol{\Theta}_{nu} \mid \mathbf{y}) d\boldsymbol{\Theta}_{nu} \\ &= E_{\boldsymbol{\Theta}_{nu}} \left[p(\boldsymbol{\Theta}_1 \mid \boldsymbol{\Theta}_{nu}, \mathbf{y}) \right] \end{aligned} \tag{19.30}$$

The last line highlights the fact that the marginal posterior is the conditional distribution of the parameters of interest, averaged over the distribution of nuisance parameters.

The marginal posterior is a very powerful tool, as it accounts for how uncertainty in the nuisance variables influences the level of uncertainty in the variable or variables of interest. This marginal probability calculation illustrates both the strength and the weakness of a Bayesian analysis before the advent of MCMC approaches. The strength is that obtaining such a marginal is very powerful for inference. The weakness is that the integration to obtain this marginal can be daunting (at best). Fortunately, MCMC techniques allow one to easily simulate draws from most marginal distributions (Appendix 3).

The dependence of the posterior on the prior (which can be partly assessed by examining the posterior's stability over different priors) provides an indication of how much information on one or more unknown variables is contained in the data. If the posterior is highly dependent on the prior, the data likely will have little signal, while if the posterior is largely unaffected by the shape of the assumed prior, the data will be highly informative. Such explorations of the effects of a prior under a careful Bayesian analysis offer some protection from incorrect conclusions based on weak likelihoods (a flat likelihood surface) or an overly strong prior. It is not uncommon to find, in the same analyses, that the marginal posterior for some variables is very robust to the choice of priors (with the data containing a strong signal generating a highly peaked likelihood surface for these variables), while for other variables, the marginal is highly dependent on the prior (with very little information contained in the data on these variables, generating a flat likelihood surface).

Example 19.8. As an example of a Bayesian analysis, consider the simple case of n observations from a normal distribution with an unknown mean, μ, but known variance, σ^2. The details for this analysis (and more realistic cases, such as those in which both the mean and the variance are unknown) are developed in Appendix 2. Assuming the data $\mathbf{y} = (y_1, \cdots, y_n)^T$ are n independent draws from this distribution, the resulting likelihood function, which corresponds to $p(\mathbf{y} \mid \mu)$, is

$$p(\mathbf{y} \mid \mu) = \frac{1}{\sqrt{2\pi\sigma^2}} \exp\left(-\sum_{i=1}^{n} \frac{(y_i - \mu)^2}{2\sigma^2} \right)$$

Suppose we assume a Gaussian prior for the location of the mean, $\mu \sim \mathrm{N}(\mu_0, \sigma_0^2)$, so that

$$p(\mu) = \frac{1}{\sqrt{2\pi\sigma_0^2}} \exp\left(-\frac{(\mu-\mu_0)^2}{2\sigma_0^2} \right)$$

While μ is treated as a fixed (but unknown) value in a standard (frequentist) analysis, it is treated as a *random variable* in a Bayesian analysis. There are no fixed effects in a Bayesian analysis, as everything is treated as a random variable.

The mean, μ_0, and variance, σ_0^2, are referred to as the prior **hyperparameters**, where μ_0 specifies a prior location for the mean and σ_0^2 specifies the uncertainty in this prior location. The larger is this variance, the greater is our uncertainty. In the limit (as $\sigma_0^2 \to \infty$), this corresponds to $p(\mu) = c$, a constant, which is a **uniform** or **flat** prior, where all values of μ between some upper and lower value are assumed to be equally likely. (Keynes 1921 called this the **principle of indifference**—all possible outcomes are equally probable.) A little algebra (Appendix 2) yields

$$p(\mu \mid \mathbf{y}) \propto p(\mathbf{y} \mid \mu)\, p(\mu) \propto \exp\left\{ -(\mu-\mu_*)^2/[2\sigma_*^2] \right\}$$

where the expressions for μ_* and σ_*^2 are given in Appendix 2 (Equation A2.22b). Thus, the posterior density function for μ is a normal with a mean of μ_* and variance of σ_*^2, which can be formally expressed as

$$\mu \mid (\mathbf{y}, \sigma^2, \mu_0, \sigma_0^2) \sim \mathrm{N}\left(\mu_*, \sigma_*^2\right)$$

Notice that our Gaussian prior for μ **conjugated** with the likelihood function, with the product of the prior and likelihood returning a distribution in the same family as the prior (but whose distribution parameters, which here are the mean and the variance, have been changed from their values in the prior by the data). The use of such **conjugate priors** (for a given likelihood), which is a key tool in Bayesian analysis, is explored in detail in Appendix 2. For example, with normally distributed data and an unknown variance, using a **scaled inverse-χ^2** prior for the variance also conjugates the likelihood, with the posterior distribution for the variance also following a scaled inverse-χ^2, but (as above) also with different distribution parameters (i.e., the shape and scale parameters are changed; Appendix 2).

A Bayesian analysis returns *distributions*, rather than *point estimates*. A number of summary statistics can be reported for the posterior, such as its mean, median (50% value), and mode (most frequent value, which is equivalent to the MLE in a likelihood analysis). More generally, when closed-form analytical expressions are not available, one can simply plot the posterior distribution via a histogram from values generated through MCMC methods (as discussed in Appendix 3).

What is the relative importance of the prior information, $p(\mu)$, compared to the actual data, $\mathbf{y}$? Equation A2.22b gives the mean of the posterior distribution as

$$\mu_* = \mu_0 \, \frac{\sigma_*^2}{\sigma_0^2} + \overline{y} \, \frac{\sigma_*^2}{\sigma^2/n}$$

With a very diffuse prior on μ (i.e., $\sigma_0^2 \gg \sigma^2$), Equation A2.22b shows that $\sigma_*^2 \to \sigma^2/n$ and thus $\mu_* \to \overline{y}$, so with very weak prior information, the mean of the posterior distribution is close to the usual estimate of μ, which is the sample mean, $\overline{y}$. Conversely, as we collect enough data (i.e., large n), $\sigma_*^2 \to \sigma^2/n$ and again $\mu_* \to \overline{y}$. Thus, even with a very strong prior belief about the location of the mean (σ_0^2 is small), as our sample size becomes sufficiently large, the mean of the posterior approaches the sample mean. With a modest prior and data (meaning that σ_0^2 and n are modest), this expression shows how these two contributions are weighted.

Example 19.9. A simulation study by Sorensen et al. (1994) examined the effects on the posterior estimate of heritability by using different priors on the variance components. To study the impact of sample size, the data were analyzed in two sets: the entire data set (Large) and a partial subset (Small). The simulated heritability was 0.5. Both uniform and scaled inverse-χ^2 priors for the additive and residual variance were used. The uniform spreads belief evenly over all possible values within a defined range, while the inverse-χ^2 places more weight on specific values (Appendix 2). The mean and variance of the marginal posterior distribution for h^2 (denoted by $E[\,h^2\,|\,\mathbf{y}]$, and $\sigma^2[\,h^2\,|\,\mathbf{y}]$, respectively) in these four cases were as follows

Data Set	Prior	$E[\,h^2\,\|\,\mathbf{y}]$	$\sigma^2[\,h^2\,\|\,\mathbf{y}]$
Small	Uniform	0.737	0.0226
Small	Inverse-χ^2	0.501	0.0029
Large	Uniform	0.550	0.0163
Large	Inverse-χ^2	0.529	0.0024

Note the disparity of the estimates under the two priors in the partial data case (Small) and their agreement in the full data case (Large). In the partial data case, the effect of the prior had a strong influence, indicating a weak signal (likelihood) for h^2 in this particular data set. With the full data set, the signal greatly increased, mitigating the effects of the prior. As expected, the posterior variance, $\sigma^2[\,h^2\,|\,\mathbf{y}]$, decreased under the larger sample size (Small versus Large) for both priors. Also note that the posterior variance was smaller under the assumed inverse-χ^2 priors. Thus, the choice of a prior influences not only the mean of the estimate, but also its variance. In this case, while the different priors yielded essentially the same mean heritability in the full data set, their variances differed by an order of magnitude.

This is an example of a **sensitivity analysis** using different priors to probe the stability of the posterior. With complex posteriors, one can observe broad stability for many of the variables (insensitivity to changes in the priors), but extreme dependence in the others. The use of different priors provides one means to explore the amount of signal along the different directions (variables) of the likelihood surface.

Example 19.10. In the context of analyzing a selection response experiment, the vector of breeding values, $\mathbf{a}$, is of interest, while the q fixed effects ($\boldsymbol{\beta}$) and variances (σ_A^2, σ_e^2) are often regarded as nuisance parameters. In this case, Equation 19.30 gives the marginal distribution of the breeding values $\mathbf{a}$, given the data $\mathbf{y}$, as

$$p(\mathbf{a}\,|\,\mathbf{y}) = \int p(\mathbf{a}, \boldsymbol{\beta}, \sigma_A^2, \sigma_e^2\,|\,\mathbf{y})\; d\boldsymbol{\beta}\; d\sigma_A^2\; d\sigma_e^2$$

The integration is over the $q + 2$ dimensional space given by the q elements in $\boldsymbol{\beta}$ and the two variances. This conditioning removes any dependencies of estimates of the response on estimates of the variance components and fixed effects. Uncertainties introduced by estimating these nuisance parameters are automatically accommodated when considering the marginal distribution. While solving this multidimensional integral is extremely challenging, a Gibbs sampler (below) can often be constructed to obtain draws from this marginal distribution.

With the marginal density, $p(\mathbf{a}\,|\,\mathbf{y})$, in hand, and recalling Equation 19.8b, one can obtain estimates of the response to selection, $\mathbf{K}^T\mathbf{a}$, that are *independent* of the assumed (or estimated) additive-genetic variance, σ_A^2. The error due to estimation of the additive variance from the data is directly incorporated when the marginal is computed, as we integrate over possible values of σ_A^2 and their support, given the data. This independence of the estimated selection response from the estimate of additive variance and the subsequent incorporation of the error in estimating σ_A^2 in the estimate of the response are two very compelling reasons for performing a Bayesian analysis of response.

This example hints at a key feature noted by Gianola and Fernando (1986). If all the data on which selection was based are included in the analysis, then (by integrating over all nuisance parameters) the Bayesian approach accounts for any potential bias introduced by selection (*provided* that the model assumptions, such as multivariate normality, hold).

Computing Posteriors and Marginals: MCMC and the Gibbs Sampler

Historically, the widespread implementation of Bayesian approaches was limited by the difficulty in obtaining marginal posterior distributions, which typically requires the integration of complex, high-dimensional functions (e.g., Equations 19.29a and 19.30). **Markov Chain Monte Carlo (MCMC)** approaches (Appendix 3) provide a solution by offering straightforward (although computationally demanding) procedures for generating random draws from very complex distributions. One such distribution is the posterior (given the data and the prior) in a Bayesian analysis, and it often turns out to be fairly easy to implement a **sampler** that allows draws from this distribution.

Simulating random vectors directly drawn from some complex **target distribution** (such as the posterior for a given model and particular data set) can be a very difficult task. The idea behind MCMC approaches is to successively draw samples from far simpler distributions in such a way that the distribution of the samples converges to the target distribution. These approaches are so named because one uses the previous sample value to randomly generate the next sample value, thus generating a **Markov chain** (Appendix 3). While there are a wide range of MCMC methods, two of the most commonly encountered in the quantitative-genetics literature are the **Metropolis-Hastings algorithm** (Metropolis and Ulam 1949; Metropolis et al. 1953; Hastings 1970), and the **Gibbs sampler** (Geman and Geman 1984).

Under Metropolis-Hastings, one simulates draws from a complex target distribution by first drawing a random variable from a specified (and simpler) distribution and then using a probability-based decision rule to decide whether to keep that realization or reject it (details are in Appendix 3). The strength of Metropolis-Hastings is that is can be applied to a very wide range of problems, such as priors that do not conjugate with the likelihood (and hence do not have a simple form). Its weakness is that candidate values can end up being rejected with a very high probability, making the sampler very inefficient (requiring very long runs to produce a reasonably sized trimmed sequence with low correlation between elements), especially in a multivariate setting, where each simulated draw is a vector of random variables.

The Gibbs sampler is a special case of Metropolis-Hastings sampling wherein the random value is always accepted. The key to this sampler is that one only considers univariate conditional distributions—the distribution that results when all of the random variables but one are assigned fixed values. Typically, one uses conjugate priors to form a Gibbs sampler (see below). More generally, one can also use a **block implementation** of the sampler, generating draws using conditional multivariate distributions

To introduce the Gibbs sampler, consider a bivariate random vector (x, y), and suppose we wish to compute one or both marginals, $p(x)$ and $p(y)$. The idea behind the sampler is that it is far easier to consider a sequence of conditional distributions, $p(x \mid y)$ and $p(y \mid x)$, than it is to obtain the marginal by integration of the joint density $p(x, y)$, e.g., $p(x) = \int p(x, y)dy$. The sampler starts with some initial value y_0 for y and obtains x_0 by generating a random variable from the conditional distribution, $p(x \mid y = y_0)$. The sampler then uses x_0 to generate a new value, y_1, drawing from the conditional distribution based on the value x_0, $p(y \mid x = x_0)$. The sampler proceeds as follows:

$$x_i \sim p(x \mid y = y_{i-1}) \tag{19.31a}$$

$$y_i \sim p(y \mid x = x_i) \tag{19.31b}$$

Repeating this process k times generates a **Gibbs sequence** of length k, where a subset of points (x_j, y_j) is taken to represent our simulated draws from the full joint distribution. To obtain the desired total of m sample points, we first sample the chain after a sufficient **burn-in** period to remove the effects of the initial starting values and then at set time points (say, every n samples) following the burn-in (**trimming** or **thinning** the sequence). For example, Wang et al. (1994b) created a Gibbs sampler for an animal model that generated a total of 1,205,000 sample vectors. The first 5000 were discarded (corresponding to the burn-in), and then every tenth subsequent iteration was saved (to reduce correlations between sample vectors), to yield a total sample of 120,000 vectors. The burn-in period, and sampling interval following the burn-in can be delicate, and careful analysis of the resulting Gibbs sequence using convergence diagnostic tools is critical (Appendix 3).

When more than two variables are involved, the sampler is extended in the obvious fashion. For example, if there are four variables, (w, x, y, z), the sampler becomes

$$\begin{aligned} w_i &\sim p(w \,|\, x = x_{i-1}, y = y_{i-1}, z = z_{i-1}) \\ x_i &\sim p(x \,|\, w = w_i, y = y_{i-1}, z = z_{i-1}) \\ y_i &\sim p(y \,|\, w = w_i, x = x_i, z = z_{i-1}) \\ z_i &\sim p(z \,|\, w = w_i, x = x_i, y = y_i) \end{aligned}$$

Any feature of interest for the marginals can be computed from the m realizations of the Gibbs sequence. For example, the expectation of any function, f, of the random variable, x, is approximated by

$$E[f(x)]_m = \frac{1}{m}\sum_{i=1}^{m} f(x_i) \tag{19.32a}$$

which is simply the average of the function evaluated over the points in the sampler. This is the **Monte Carlo** (MC) **estimate** of $f(x)$, as $E[f(x)]_m \to E[f(x)]$ as $m \to \infty$. Likewise, the MC estimate for any function of n variables $(\theta^{(1)}, \cdots, \theta^{(n)})$ is given by

$$E[f(\theta^{(1)}, \cdots, \theta^{(n)})]_m = \frac{1}{m}\sum_{i=1}^{m} f(\theta_i^{(1)}, \cdots, \theta_i^{(n)}) \tag{19.32b}$$

Example 19.11. As a toy example of how to use the output of a Gibbs sampler, suppose we are interested in the distribution of breeding values in a particular generation (measured by four individuals in the analysis), as well as in the base population heritability. A Gibbs sampler has been implemented and the realizations at three different iterations (say 100, 200, and 300) after a sufficient burn-in period are as follows:

Factor	Sample 100	Sample 200	Sample 300
$a(1)$	1.5	1.8	2.2
$a(2)$	2.1	3.4	1.4
$a(3)$	3.1	2.9	4.4
$a(4)$	3.3	4.3	3.6
σ_A^2	0.55	0.64	0.46
σ_e^2	1.10	0.98	1.20

Here $a(1)$ through $a(4)$ correspond to the values (realizations of the posterior distribution) of the four breeding values in that particular iteration of the sampler, and σ_A^2 and σ_e^2 are, similarly, the realizations for the variances in that iteration. The mean and variance for the breeding value of these four individuals, in sample i, are

$$\overline{a}_i = \frac{1}{4}\sum_{j=1}^{4} a_i(j) \quad \text{and} \quad \text{Var}(a)_i = \frac{1}{4-1}\sum_{j=1}^{4}\left[\, a_i(j) - \overline{a}_i \,\right]^2$$

and the base population heritability

$$h_i^2 = \sigma_{A,i}^2 / (\sigma_{A,i}^2 + \sigma_{e,i}^2)$$

Using these three realizations

	Sample 100	Sample 200	Sample 300
$\overline{a}$	2.5	3.1	2.9
Var(a)	0.72	1.09	1.83
h^2	0.33	0.40	0.28

Thus, the sampler has returned three values for each of the quantities of interest. Of course, a full sampler consists of thousands to hundreds of thousands of such realization, allowing us to empirically generate the full distribution of any for these functions. For example, the mean of the marginal posterior for the mean breeding value over these four individuals is simply the mean of $\overline{a}$ over the entire sample from the Gibbs sequence. Likewise, the empirical histogram of h_i^2 values is the marginal posterior for the heritability, and thus accounts for any source of variation in estimating h^2 (given that the assumed model is correct).

Bayesian Analysis of the Animal Model

The use of Bayesian approaches for the analysis of selection experiments was first suggested by Sorensen and Johansson (1992). Starting with the standard animal model

$$\mathbf{y} = \mathbf{X}\boldsymbol{\beta} + \mathbf{Za} + \mathbf{e}, \quad \text{where} \quad \mathbf{y} \sim \text{MVN}(\mathbf{X}\boldsymbol{\beta}, \sigma_A^2\,\mathbf{ZAZ}^T + \sigma_e^2\,\mathbf{I})$$

The (unconditional) mean and variance for this model follow from $E[\mathbf{y}] = \mathbf{X}\boldsymbol{\beta}$, and $\text{Var}(\mathbf{y}) = \text{Var}(\mathbf{Za} + \mathbf{e}) = \sigma_A^2\mathbf{ZAZ}^T + \sigma_e^2\,\mathbf{I}$. Wang et al. (1993, 1994a, 1994b), Sorensen et al. (1994), and Jensen et al. (1994) developed Gibbs samplers for this model and its extensions. As before, the conditional distribution of the data, given the vectors of fixed effects, $\boldsymbol{\beta}$, breeding values, $\mathbf{a}$, and the environmental variance, σ_e^2, is multivariate normal

$$\mathbf{y} \,|\, \boldsymbol{\beta}, \mathbf{a}, \sigma_e^2 \sim \text{MVN}(\mathbf{X}\boldsymbol{\beta} + \mathbf{Za}, \sigma_e^2\,\mathbf{I}) \tag{19.33a}$$

The infinitesimal model is assumed, with the distribution of breeding values, given the relationship matrix, $\mathbf{A}$, and additive genetic variance, σ_A^2, being multivariate normal

$$\mathbf{a} \,|\, \mathbf{A}, \sigma_A^2 \sim \text{MVN}(\mathbf{0}, \sigma_A^2\,\mathbf{A}) \tag{19.33b}$$

Sorensen et al. (1994) assumed a uniform prior for $\boldsymbol{\beta}$, a normal prior for $\mathbf{a}$, and both uniform and inverse-χ^2 priors for the variances. For example, the joint prior for $\mathbf{a}$ and σ_A^2 is the product of $p(\mathbf{a} \,|\, \sigma_A^2, \mathbf{A}) \cdot p(\sigma_A^2)$, where the first distribution is a multivariate normal (Equation 19.33b) and the second is either a uniform or an inverse-χ^2 (Appendix 2). These distributions are chosen because they are conjugate priors (Appendix 2) for the multivariate normal, resulting in analytical expressions for the $p + q + 2$ univariate conditional distributions for each factor in the model (p breeding values, a_k, q fixed effects, β_j, and the variances, σ_A^2 and σ_e^2). Using these univariate conditionals, a Gibbs sampler can be constructed. The outline for the sampler is as follows:

1. Set initial values for $\mathbf{a}$, $\boldsymbol{\beta}$, σ_A^2, and σ_e^2.
2. Using the current values of $\mathbf{a}$, $\boldsymbol{\beta}$, σ_A^2, and σ_e^2 and the conditional distributions (see Sorensen et al. 1994 for exact expressions):

   (i) Update the fixed effects by sequentially drawing (for $j = 1, \cdots, q$) from the conditionals (which are univariate normals)

   $$\beta_{j,i} \sim p(\beta_j \,|\, \beta_{1,i}, \cdot\cdot, \beta_{j-1,i}, \beta_{j+1,i-1}, \cdot\cdot, \beta_{q,i-1}, \mathbf{a}_{i-1}, \sigma_{A,i-1}^2, \sigma_{e,i-1}^2)$$

   where $\beta_{j,i}$ is the value of β_j during the ith iteration of the sample. The current values of these parameters define the mean and variance for a normal, from which a random value is drawn. For factor j, we take the values for $\mathbf{a}$ and the variances from the last iteration ($i - 1$), the values of β_1 to β_{j-1} from the current iteration (i), and the values of β_{j+1} to β_q from the last iteration ($i - 1$). These values are inserted to give the parameters (here, the conditional mean and variance) for the univariate normal that corresponds to the conditional distribution for β_j and a random variable is drawn from this distribution to yield $\beta_{j,i}$.

(ii) Update the breeding values by sequentially drawing (for $i = 1, \cdots, p$) from the conditionals (again univariate normals), where the vector, $\boldsymbol{\beta}$, of fixed effects is taken as the updated version

$$a_{j,i} \sim p(a_j \mid \boldsymbol{\beta}_i, a_{1,i}, \cdot\cdot, a_{j-1,i}, a_{j+1,i-1}, \cdot\cdot, a_{p,i-1}, \sigma^2_{A,i-1}, \sigma^2_{e,i-1})$$

(iii) Update the additive variance by drawing from the conditional (a scaled inverse-χ^2 distribution)

$$\sigma^2_{A,i} \sim p(\sigma^2_A \mid \boldsymbol{\beta}_i, \mathbf{a}_i, \sigma^2_{e,i-1})$$

(iv) Update the error variance by drawing from the conditional (again, a scaled inverse-χ^2 distribution)

$$\sigma^2_{e,i} \sim p(\sigma^2_e \mid \boldsymbol{\beta}_i, \mathbf{a}_i, \sigma^2_{A,i})$$

3. Using the updated values, repeat (**2**) until k samples have been obtained, from which m are extracted (following the burn-in and trimming) for the Gibbs-sampler chain.

This Bayesian analysis makes most of the standard animal-model assumptions, in particular that the infinitesimal model and multivariate normality, hold (as with an MM analysis, a Bayesian approach is potentially biased by selection-induced changes in allele frequencies). A Bayesian analysis has all the advantages of an MM analysis (over an LS analysis) and, in addition, the posterior marginals correctly give the distribution of any parameter of interest, independent of the values assumed for other parameters. Any uncertainty introduced by estimating these additional parameters is fully captured by the marginal posteriors. The Bayesian approach yields the correct distribution (assuming that the model assumptions hold and the prior is reasonable) for the estimated selection response, independent of the additive genetic variance. By contrast, an MM analysis is highly dependent on the assumed (or estimated) additive variance, and the standard error of a REML/BLUP estimate for the response (Equation 19.8c) does not account for the additional uncertainty introduced by using a REML estimate of σ^2_A. One standard package for Bayesian analysis is Winbugs, and Damgaard (2007) outlined how to apply this software to animal models.

Application: Estimating Selection Response in Pig Litter-size Components

Blasco et al. (1998) used the method of Sorensen et al. (1994) to estimate the response to selection on ovulation rate and prenatal survival in French Large White pigs. Three lines were followed, with single lines selected separately on each trait (ovulation rate and survival) and a control line. The relevant selection and control lines were jointly analyzed to estimate the selection response. Ovulation rate was examined using the standard animal model

$$\mathbf{y} \mid \boldsymbol{\beta}, \mathbf{a}, \sigma^2_e \sim \text{MVN}(\mathbf{X}\boldsymbol{\beta} + \mathbf{Z}\mathbf{a}, \mathbf{I}\sigma^2_e) \quad \text{and} \quad \mathbf{a} \mid \mathbf{A}, \sigma^2_A \sim \text{MVN}(\mathbf{0}, \mathbf{A}\,\sigma^2_A)$$

Prenatal survival (as a function of the mother) was examined using the repeatability model, where c is the permanent environmental effect of a mother over multiple litters

$$\mathbf{y} \mid \boldsymbol{\beta}, \mathbf{a}, \mathbf{c}, \sigma^2_e \sim \text{MVN}(\mathbf{X}\boldsymbol{\beta} + \mathbf{Z}\mathbf{a} + \mathbf{W}\mathbf{c}, \mathbf{I}\,\sigma^2_e)$$

$$\mathbf{a} \mid \mathbf{A}, \sigma^2_A \sim \text{MVN}(\mathbf{0}, \mathbf{A}\,\sigma^2_A) \quad \text{and} \quad \mathbf{c} \mid \sigma^2_c \sim \text{MVN}(\mathbf{0}, \mathbf{I}\,\sigma^2_c)$$

Among the fixed effects in $\boldsymbol{\beta}$ are terms for the **parity** of the mother (first litter, second litter, and so on). The marginal posterior distribution for breeding values (and hence, recalling Equation 19.8b, for the response via $\mathbf{K}^T\mathbf{a}$) was obtained by using the Gibbs-sampler approach of Sorensen et al. (1994) outlined previously. For each trait, two independent chains

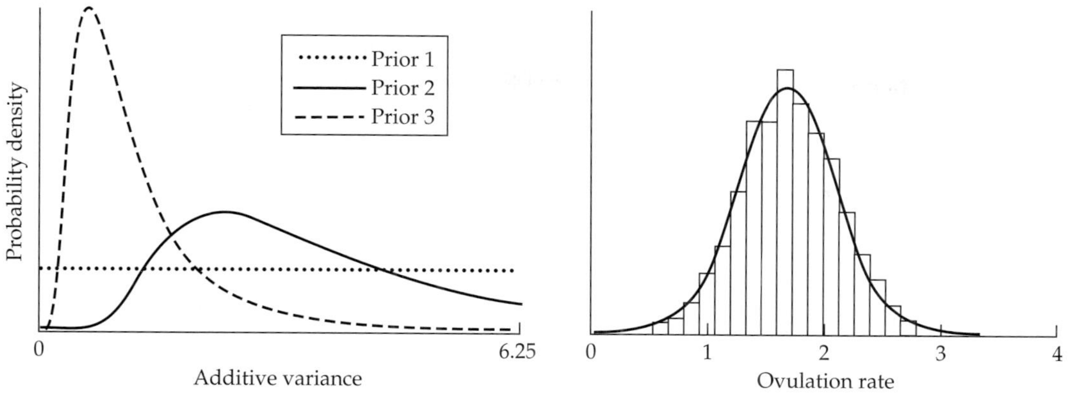

Figure 19.3 Analysis of ovulation rate at puberty in French Large White pigs. **Left:** Assumed priors for σ_A^2 (see text for details). **Right:** The Bayesian estimate of selection response is shown by presenting the marginal posterior density for the mean breeding value in ovulation rate in the last generation of selection as a histogram (all three priors gave very similar results). This distribution is approximately normal (the solid curve). (After Blasco et al. 1998.)

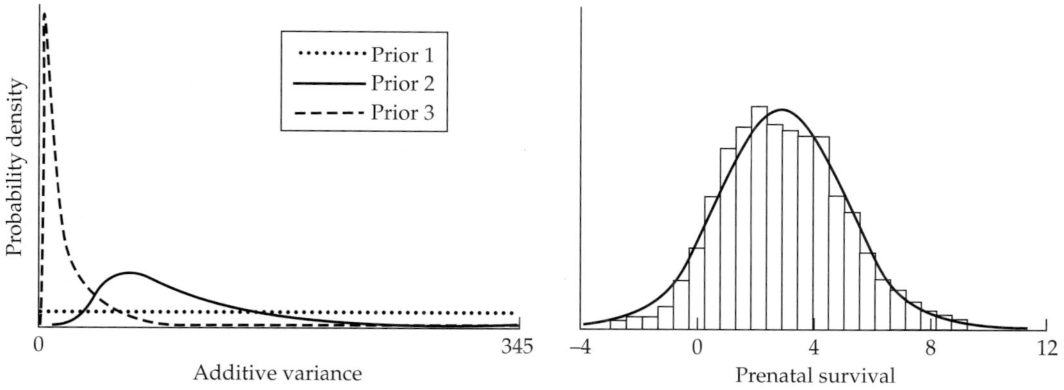

Figure 19.4 Analysis of French Large White pig prenatal survival. **Left:** Priors for the additive variance (see text for details). **Right:** Posterior distribution of mean breeding values (at generation 4) in prenatal survival (based on prior 1), which deviates from the best-fitting normal. (After Blasco et al. 1998.)

of length 100,000 were computed, with the first 10,000 samples discarded (to remove burn-in effects) and sampling at every 30 iterations thereafter, which generated a trimmed sampler of length 3000. The authors obtained these burn-in and resampling values after several initial runs, using the diagnostics suggested by Raftery and Lewis (1992a) for level of precision and Geyer (1992) for autocorrelation between samples. A uniform prior was taken for the fixed effects, while different priors were used for the variances (discussed below).

We will consider the results for ovulation rate at puberty first. Figure 19.3 shows the three priors assumed for the additive variance in this trait. The phenotypic variance of this trait is 6.25, which sets an upper limit on σ_A^2. Prior 1 is a uniform distribution that weights all values in the parameter space equally. Priors 2 and 3 (which are scaled inverse-χ^2

Table 19.1 Estimated response to selection for ovulation rate at puberty and prenatal survival in French Large White pigs. Bayesian analysis with three different priors (Figure 19.3 for ovulation and Figure 19.4 for prenatal survival) were used to obtain Monte-Carlo estimates of the mean response and their associated standard deviations (the latter incorporating the additional error from estimating the additive variance and other parameters). For comparison, least-squares (LS) estimates ($\overline{z}_{i+1} - \overline{z}_i$) and mixed-model (REML/BLUP) estimates are also included. (After Blasco et al. 1998.)

Total Response in Ovulation Rate at Puberty

Method	Gen. 1	Gen. 2	Gen. 3	Gen. 4
Bayesian, Prior 1	0.30 ± 0.31	0.51 ± 0.35	1.03 ± 0.39	1.58 ± 0.43
Bayesian, Prior 2	0.31 ± 0.30	0.51 ± 0.34	1.05 ± 0.38	1.55 ± 0.42
Bayesian, Prior 3	0.31 ± 0.31	0.51 ± 0.35	1.01 ± 0.35	1.53 ± 0.38
LS	-0.09	0.35	1.98	1.87
REML/BLUP	0.27	0.45	1.00	1.54

Total Response in Prenatal Survival

Method	Gen. 1	Gen. 2	Gen. 3	Gen. 4
Bayesian, Prior 1	-0.53 ± 1.44	1.23 ± 1.61	2.83 ± 1.94	2.89 ± 2.12
Bayesian, Prior 2	-0.64 ± 1.70	1.50 ± 1.87	3.46 ± 2.05	3.49 ± 2.30
Bayesian, Prior 3	-0.46 ± 1.45	1.22 ± 1.60	2.84 ± 1.82	2.90 ± 2.01
LS	-5.71	2.11	4.13	-2.82
REML/BLUP	-0.54	1.49	3.26	3.42

distributions) reflect additional information. Published heritabilities for this trait in pigs and rabbits range from 0.1 to 0.6, and prior 2 assumes a broad distribution around the approximate median value ($\sigma^2_A = 0.4 \cdot 6.25 = 2.5$). A study specifically in French Large Whites gave an estimate of $h^2 = 0.11 \pm 0.02$, and the tight distribution around this value is reflected in prior 3. Blasco et al. obtained Monte Carlo estimates of the (base population) heritability under these three priors of $h^2 = 0.39 \pm 0.07, 0.39 \pm 0.06$, and 0.32 ± 0.06. Table 19.1 shows the estimated cumulative response after each of the four generations of selection, in comparison with the least-squares (differences between generation means) and mixed-model (REML/BLUP) estimates. Note that the three different priors give very consistent estimates of response, implying that the data contained sufficient information to overpower most of the signal coming from the assumed prior. The Bayesian and MM analyses yielded very similar results, while LS analysis yields a substantially different estimates of response.

While the results for ovulation rate were consistent across the three priors and with the MM analysis, those for prenatal survival were more problematic (Table 19.1). Figure 19.4 shows the assumed different priors. In this case, as with ovulation rate, prior 1 is uninformative, weighting all potential additive variances equally. Prior 2 (as with prior 2 for ovulation rate) assumed a broad distribution around the mean heritability ($h^2 \simeq 0.2$) of this trait in a number of studies, while prior 3 used the estimate of $h^2 = 0.03 \pm 0.03$ found using French Large Whites. The three priors returned Monte Carlo estimates of heritability (and its standard deviation) of $h^2 = 0.12 \pm 0.06, 0.16 \pm 0.04$, and 0.11 ± 0.04. Likewise, these priors give Monte Carlo estimates of the repeatability (r) of $0.23 \pm 0.05, 0.23 \pm 0.04$, and 0.19 ± 0.04. As Table 19.1 shows, the standard deviations for the Monte Carlo estimates of mean response were very large, but the three priors and the MM analysis gave consistent results, while the LS results were quite different. Clearly, the information on prenatal survival in the experiment was modest, as the posterior is significantly influenced by the prior. Additional early examples of Bayesian analysis of selection experiments involved lean growth in pigs (Rodriguez et al. 1996) and body weight in chickens (Su et al. 1997).

LS, MM, OR BAYES?

Just what analysis should an investigator use for a selection experiment? Obviously, in the absence of any pedigree information, a least-squares analysis is the only option, although this could also be placed in a Bayesian framework. With the pedigree in hand (either observed or inferred; see Chapter 20), a mixed-model analysis is much more powerful and is strongly preferred over LS, unless there is strong evidence that model assumptions have been violated. If a mixed-model approach is appropriate and chosen, should the analysis be frequentist or Bayesian? As mentioned, the Bayesian approach does a much better job of treating uncertainty, but this comes at a higher computational cost, especially when one does a proper analysis using several different priors to assess sensitivity. Perhaps the best advice is that offered by Blasco (2001):

> "The choice of one school or the other should be related to whether there are solutions in one school that the other does not offer, to how easily the problems are solved, and to how comfortable scientists feel with the way they convey their results."

Blasco's last point is especially important: it is much more important for investigators to use a method with which they are comfortable, in the sense of knowing its limitations and having some intuition into the approach, than to simply use a method because it is new and trendy.

Generally speaking, simpler methods (such as LS) tend to be more robust to model fragility than more complex approaches (e.g., mixed-models). While the latter can be considerably more powerful when the model assumptions *do hold*, they can also be significantly more biased when they fail. Ideally, one should use several different approaches in the analysis of any dataset. If the results are consistent, one can have additional confidence that the model assumptions may be holding or that the methods are immune to violations. If the results are rather different, this provides a critical indicator to the investigator that a much more careful examination of model assumptions is in order.

20

Selection Response in Natural Populations

Associations between phenotype and fitness, however appealing, will give a misleading impression of the potential for evolution in a trait if the true target of selection is unmeasured or immeasurable. Kruuk et al. (2002)

Under artificial selection (animal and plant breeding and laboratory selection experiments), the breeder's equation machinery developed in Chapters 13 through 19 for the prediction and analysis of response generally works well. However, there is considerable angst as to whether this is also true for natural populations (e.g., Merilä et al. 2001c; Morrissey et al. 2010, 2012; Pemberton 2010; Timothée et al. 2017). There are two principal reasons for this. First, under artificial selection, individuals are strictly chosen by the phenotypic value of their focal trait or traits. By contrast, in natural populations, one *infers* the target of selection, typically by looking for changes in the mean and/or variance of certain candidate traits within and/or across generations. The problem is that the phenotypic moments of an unselected trait can change if it is correlated with another trait under selection. A within-generation change (a nonzero selection differential) occurs if an unselected trait is phenotypically correlated with a selected one (Equation 13.25c), while a between-generation change (a response) occurs if the traits are genetically correlated (Equation 13.26c).

The second complication is lack of control over the environment. With artificial selection, there is generally considerable environmental control, in part due to husbandry and cultivation methods designed to standardize rearing and growing conditions and to mitigate extreme environmental events. This is certainly not the case when attempting to track selection in natural populations. Indeed, we are largely unable to determine which environmental factors may be important, let alone be able to control them. Further, artificial selection experiments generally impose considerable control over the *biotic* (in addition to the physical) environment in which an organism finds itself (e.g., the collection of species interacting with the focal population). In natural populations, the biotic environment is both absolutely critical and largely uncontrollable. One potential consequence of lack of environmental control arises when unmeasured environmental factors jointly influence the trait and fitness. Changes in the environment can also mask underlying genetic changes and can lead to significant changes in the nature of selection from one generation to the next (such as favoring larger trait values in wet years and smaller values in dry years). Finally, changes in the environment can modify genetic and environmental variances, thus altering the heritability.

This chapter addresses these concerns in two parts. The first is largely theoretical, centering on the bias caused by selection on unmeasured features. We initially frame this concern within the context of correlated characters, and then focus on the special case where an unmeasured variable is entirely environmental, which can generate a nonzero selection differential but no response. Next, we extend the univariate breeder's equation to account for all of the possible biases that arise from any unmeasured traits influencing the focal trait. We conclude by recasting the selection response under both versions of Robertson's secondary theorem, $R = \sigma(A_z, A_w)$ and $R = \sigma(A_z, w)$ (Chapter 6), as opposed to the breeder's equation ($R = h^2 S$) framework. Here, w is relative fitness, and A_z and A_w are, respectively, the breeding values for the trait and relative fitness. Contrasts between these two predictions suggest tests for assessing whether a focal trait is the sole target of selection.

The second part of our treatment is largely empirical, examining the advantages and pitfalls of applying mixed models (Chapter 19) in natural populations. During the first decade of the 2000s, there was much excitement that BLUP predictions of breeding values

Evolution and Selection of Quantitative Traits. Bruce Walsh & Michael Lynch.
Published 2018 by Oxford University Press.
DOI 10.1093/oso/9780198830870.001.0001

would offer powerful insight into the nature of selection response in natural populations. A rash of results, many initially viewed as classic, quickly appeared from the analysis of pedigrees from several vertebrate populations under long-term observation (reviewed in Kruuk et al. 2008). However, problems with BLUP estimates of breeding values in natural settings, initially noticed by Postma (2006), were shown by Hadfield (2008) and Hadfield et al. (2010) to be extremely serious. Hence, many of these initial results need to be seriously reconsidered.

Nonetheless, certain aspects of the animal model, in particular REML estimates of specific covariances (such as that between the breeding values of a trait and fitness), remain powerful approaches. We first review how a BLUP analysis in a natural population proceeds (with a specific focus on estimating the relationship matrix, **A**), then show where flaws can appear, and finally examine how certain features of the animal model can still prove useful. The rapid rise, even quicker demise, and then phoenix-like resurrection (and reorientation) of animal-model applications in natural populations can be quite confusing to the novice reading the historical literature, so we try to carefully navigate through the shoals of confusion. We conclude by reviewing a number of examples of selection response (or lack thereof) in natural populations, using the developed theoretical and statistical machinery to highlight problems that can arise when attempting to predict response in a natural population.

EVOLUTION IN NATURAL POPULATIONS: WHAT IS THE TARGET OF SELECTION?

While there are many assumptions underlying the breeder's equation (Chapter 6; Table 13.2), the one that is most likely to fail in natural populations, and the one that is most challenging to test, involves *causality*—our assumption that the phenotype of the focal trait is the sole target of selection (in the sense that it is genetically and phenotypically uncorrelated with any other traits under selection). Under the breeder's equation, the trait value (z) entirely governs fitness, and transmission of the resultant change in the mean of z to the next generation is entirely described by h^2. The conceptual beauty of the breeder's equation is that it partitions evolution into separate, and distinct, ecological (S) and genetical (h^2) processes, allowing ecologists to focus on the former (the nature of selection) and geneticists on the latter (the inheritance of the trait). If we incorrectly assign the target of selection, the breeder's equation will give misleading results.

Direct and Correlated Responses

Bias from correlated traits can be removed by using the multivariate breeder's equation, *provided* all relevant traits are included. This equation expresses the vector, **R**, of responses (changes in means) as a function of the genetic (breeding value), **G**, and phenotypic, **P**, covariance matrices for the traits of interest, and the vector, **S**, of their selection differentials. From Equations 13.23b and 13.26a,

$$\mathbf{R} = \mathbf{G}\mathbf{P}^{-1}\mathbf{S} = \mathbf{G}\boldsymbol{\beta}$$

where the selection gradient, $\boldsymbol{\beta} = \mathbf{P}^{-1}\mathbf{S}$, controls for any phenotypic correlations among the *measured* traits, returning the amount of direct selection acting on each particular character (LW Chapter 8; Chapter 30).

Focusing on the bivariate version of this equation provides insight into some of the complications that can arise by ignoring selection on correlated traits. Suppose we are following trait 1, which is influenced by a second (and unmeasured) feature, which we denote as trait 2. Noting that $\mathbf{S} = \mathbf{P}\boldsymbol{\beta}$, the selection differential on trait 1 becomes

$$S_1 = P_{11}\beta_1 + P_{12}\beta_2 = \sigma^2(z_1)\beta_1 + \sigma(z_1, z_2)\beta_2 \tag{20.1a}$$

A within-generation change ($S_1 \neq 0$) in trait 1 occurs from: (i) direct selection on trait 1 ($\beta_1 \neq 0$), and/or (ii) indirect selection from a *phenotypically* correlated one (trait 2) under directional selection ($\beta_2 \neq 0$ and $\sigma(z_1, z_2) \neq 0$). As a result, the signs of S_1 and β_1 can differ, and even strong direct selection ($\beta_1 \neq 0$) on a trait can still be associated with a net selection differential of nearly zero (or worse, of opposite sign). Turning to the expected response in trait 1,

$$R_1 = G_{11}\beta_1 + G_{12}\beta_2 = \sigma^2(A_1)\beta_1 + \sigma(A_1, A_2)\beta_2 \tag{20.1b}$$

Trait 1 can evolve as a consequence of direct selection (if it has additive variation) or as a correlated response from direct selection on another *genetically* correlated trait (with the breeding values of the two traits being correlated within individuals, $\sigma(A_1, A_2) \neq 0$).

As the following example highlights, some of the best studies of the response to selection in natural populations come from birds (reviewed by Merilä et al. 2001b; Merilä and Sheldon 2001; Gienapp et al. 2008; Kruuk et al. 2008; Clutton-Brock and Sheldon 2010; Charmantier et al. 2014). In certain settings (such as isolated islands), the entire population can be banded and all nests located (often through the use of nest boxes), allowing for accurate measurement of individual fitness (Chapter 29).

Example 20.1. Alatalo et al. (1990) examined tarsus length in a population of collared flycatchers (*Ficedula albicollis*) residing in the southern part of the island of Gotland in the Baltic Sea. Measurements of lifetime fitnesses in this isolated bird population were possible because most surviving offspring (which are tagged before **fledging**, i.e., before leaving the nest) return to breed in the area where they were reared as offspring. In addition to tarsus length, fledgling weight was also measured (with both traits scaled in standard-deviation units) and Pearson-Lande-Arnold regressions (Chapter 30; LW Chapter 8) were performed to compute the amounts of direct selection (the estimated selection gradients) on both characters, yielding

Year	Observed $\bar{\imath}$ on tarsus length	Estimated selection gradients, $\widehat{\beta}$: Tarsus length	Fledgling weight
1981	0.19**	0.01	0.25*
1983	0.08	-0.01	0.21*
1984	0.20**	0.12	0.33***
1985	0.02	-0.06	0.27***
pooled	0.12**	0.03	0.27***

$^*p < 0.05,\quad ^{**}p < 0.01,\quad ^{***}p < 0.001$

Although there was a significant selection differential (which is presented as the selection intensity, $\bar{\imath}$, because the trait was scaled in standard deviations; Equation 13.6a) on tarsus length in two of the years (and in the pooled data), there was no significant *direct* selection on tarsus length itself (none of the estimated selection gradients for this trait were significant). Rather, direct selection was on fledgling weight. While there is a significant phenotypic correlation between tarsus length and fledgling weight ($r = 0.32; p < 0.001$), it appears to be entirely due to within-individual correlations of environmental effects, as there is no correlation between the fledging weight of an offspring and the tarsus length of its parent ($r = -0.01; p > 0.1$). The latter observation implies an absence of a genetic correlation between tarsus length and fledgling weight, and hence (from Equation 20.1b) no response in tarsus length is expected.

Example 20.2. As reviewed in Grant and Grant (1995, 2002; and references therein), one of the best documented cases of natural selection is on body size and bill morphology in Darwin's finches (*Geospiza fortis*) on the Galápagos island of Daphne Major. Two strong episodes of selection were observed during their long-term study, due to serious droughts in 1976–1977 (when the population crashed from 634 birds down to 95, a 15% survival rate), and in 1984–1986 (556 birds reduced to 180, a 32% survival rate). Six (log-transformed) morphological traits were followed through both episodes, and (after rescaling all traits to have unit variances) the selection differentials, $S = \bar{\imath}$ (as traits are scaled to unit variance), and gradients, β, for the two episodes were as follows (where * denotes $p < 0.05$):

Trait	1976–1977 $\bar{\imath}$	1976–1977 β	1984–1986 $\bar{\imath}$	1984–1986 β
Weight	0.74*	0.477*	−0.11	−0.040
Wing length	0.72*	0.436*	−0.08	−0.015
Tarsus length	0.43*	0.005	−0.09	−0.047
Bill length	0.54*	−0.144	−0.03	0.245*
Bill depth	0.63*	0.528*	−0.16*	−0.135
Bill width	0.53*	−0.450*	−0.17*	−0.152

Two striking features are apparent. First, the observed (within-generation) change in the mean, $\bar{\imath}$, was not a good predictor of the actual amount of direct selection, β, on a trait, and can even have a different sign (e.g., bill length). Second, the nature of selection changed over the two drought periods. During the 1976–1977 drought, larger individuals were favored, and there was selection on bill shape (increased bill depth, decreased bill width). A change in the dominant food supply during a subsequent drought in 1984–1986 resulted in selection favoring smaller birds. Hence, the two episodes of selection were in opposite directions (at least in terms of body size).

Grant and Grant had an estimate of the genetic variance matrix, $\mathbf{G}$, for these traits in hand, allowing them to substitute these estimated βs into the multivariate breeder's equation (13.26a) to examine how well responses were predicted. Response was well predicted in 1976, but overpredicted in three of the six traits in the 1984 episode. They suggested that the main reason for these discrepancies was a change in the biotic environment. Higher population densities for offspring in 1984 retarded growth, resulting in an overprediction of size-related traits.

Character	1976–1977 Predicted	1976–1977 Observed	1984–1986 Predicted	1984–1986 Observed
Weight	17.39 ± 0.22	17.52 ± 0.25	16.82 ± 0.13	$15.48 \pm 0.08^{*}$
Wing length	69.98 ± 0.39	69.65 ± 0.35	67.93 ± 0.17	$67.21 \pm 0.11^{***}$
Tarsus length	19.45 ± 0.09	19.32 ± 0.14	19.02 ± 0.04	19.02 ± 0.04
Bill length	11.14 ± 0.10	11.06 ± 0.11	10.86 ± 0.05	10.96 ± 0.03
Bill depth	9.83 ± 0.12	9.94 ± 0.09	9.51 ± 0.06	$9.32 \pm 0.03^{**}$
Bill width	8.96 ± 0.08	8.97 ± 0.08	8.77 ± 0.04	8.70 ± 0.03

$^{*}p < 0.05$, $^{**}p < 0.01$, $^{***}p < 0.001$

Environmentally Generated Correlations Between Fitness and Traits

In natural populations, one or more environmental factors can influence both an individual's trait value and its fitness. This generates a correlation between the trait and fitness, and hence (from the Robertson-Price identity; Equation 6.10) a selection differential, $S = \sigma(z, w)$, on the trait, even if there is no direct selection on the trait itself. Consider the following example, which was suggested by Rausher (1992).

Suppose, for example, that soil nitrate concentration influences both fitness (seed production) and the amount of alkaloids (secondary plant chemicals) in the foliage of a plant. As Figure 20.1 shows, if we were able to partition individuals from a population into high- and low-nitrate environments, within each group we might find no association between alkaloid concentration and fitness. However, if we ignored this partition and simply lumped all individuals into a single group, we would find a positive covariance between alkaloid concentration and fitness. An investigator unaware of this difference in soil nitrates might conclude that there was a fitness effect from the presence of alkaloids (for example, as an insect deterrent), when in fact the correlation between trait and fitness arises solely because both are influenced by a third, and unmeasured, variable. In this setting, the breeder's equation would lead to an erroneous prediction of a selection response in alkaloid levels because it assumes that alkaloids are the causative agent of fitness differences. (As an aside, this particular observation is an example of Simpson's paradox, introduced in Chapter 10, wherein individual association tables give a different picture from their amalgamated table.)

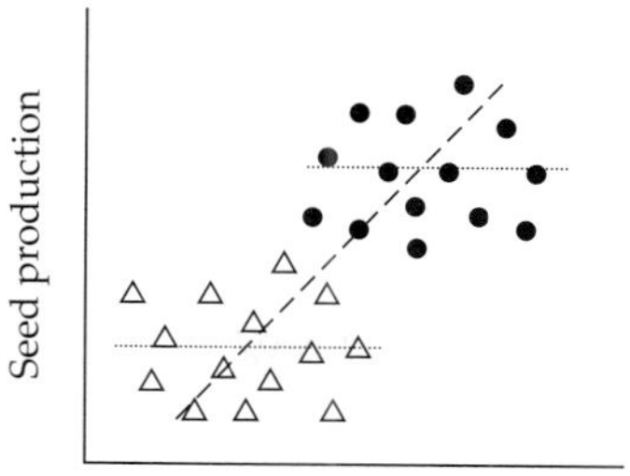

Figure 20.1 An environmental variable (soil nitrate) influences both fitness and trait value (alkaloid levels), creating a covariance between the trait and fitness (measured by seed production), when in fact the trait value is not a causal determinant of fitness. In low-nitrate soils (open triangles), plants have low fitness and low levels of alkaloids. In high-nitrate soils (filled circles), plants have high fitness and high levels of alkaloids. Within each of the two environments, there is no association between the trait and fitness (dotted regressions). If one ignores the environmental effects and simply lumps all the individuals together, there is a strong association between fitness and the trait value (dashed regression). (Based on Rausher 1992 and Mauricio and Mojonnier 1997.)

Rausher's example (in the previous paragraph) introduces the concept of selection on a nonheritable feature that also influences the focal trait. While the feature, soil nitrate, is a primary determinant of plant fitness, it is an environmental factor that also influences alkaloid levels. From Equations 20.1a and 20.1b, the selection differential and response are

$$S_{\text{alkaloid level}} = \sigma(\text{alkaloid level, soil nitrate})\beta_{\text{soil nitrate}} \neq 0$$

$$R_{\text{alkaloid level}} = \sigma(A_{\text{alkaloid level}}, A_{\text{soil nitrate}})\beta_{\text{soil nitrate}} = 0 \cdot \beta_{\text{soil nitrate}} = 0$$

While plants with high alkaloid levels (which are heritable) appear to have higher fitness ($S > 0$), alkaloid levels do not increase in the next generation ($R = 0$). This is because plants with high *environmental* values for alkaloids (due to high soil nitrate values) were selected, not plants with high *genetic* values, and hence there is no selection response. Put another way, there is selection on a component of the *environmental* value of alkaloid level (that due to soil nitrate value), but *not* on its *breeding* value, and hence no response (Chapter 13).

Example 20.3. Considering the evolution of **clutch size** (z_1) in birds (number of eggs laid in a particular episode), Price and Liou (1989) suggested that fitness is largely determined by the nutritional state, z_2, of a mother, which also influences her own clutch size, $\sigma(z_1, z_2) \neq 0$. They assumed that nutritional state is entirely environmental, $\sigma^2(A_2) = 0$. Equation 20.1a implies that, even if there is no direct selection on clutch size per se ($\beta_1 = 0$), we would still observe a selection differential on clutch size if it is phenotypically correlated with nutritional state and the latter is itself under selection ($\beta_2 \neq 0$), as $S_1 = \beta_2\,\sigma(z_1, z_2) \neq 0$. The resulting selection response in clutch size, R_1, is $\beta_2\,\sigma(A_1, A_2) = 0$, because nutritional state is assumed to have no heritable variance, which implies $\sigma(A_1, A_2) = 0$. As a result, there is apparent directional selection on clutch size ($S_1 \neq 0$), but no response ($R_1 = 0$).

The notion of nutritional status, or some other measure of well-being, of an organism is often referred to as **condition** by ecologists (Le Cren 1951). Although this term is often used fairly loosely, without any formal definition, one common operational measure is the residual from a regression of weight on some measure of body size (i.e., size-adjusted weight). The motivation for this metric is that individuals in good condition will be heavier than expected given their size, while individuals in poor condition will be underweight. Jakob et al. (1996), Green (2001), and Schulte-Hostedde et al. (2005) discussed the merits of this metric. While condition is often treated entirely as a product of the environment, as with any standard quantitative trait, it is reasonable to assume that it may have some genetic component as well (e.g., Gosler and Harper 2000; Merilä et al. 2001a).

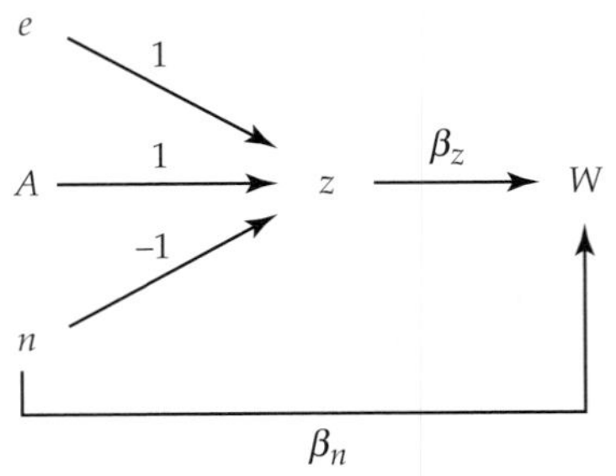

Figure 20.2 A path diagram (LW Appendix 2) of the components in the Fisher-Price-Kirkpatrick-Arnold model, showing the connections between breeding date, z, nutritional state, n, and fitness, W. The breeding value, A, general environmental value, e, and nutritional state, n, all influence the breeding date, z, which itself influences fitness (path coefficient β_z). Further, there is a second path to fitness directly from the nutritional state (β_n), which represents the direct contribution to W of n after its indirect contribution through breeding date is removed. We assume that A, e, and n are all uncorrelated, and hence not connected by any paths. (After Price et al. 1988.)

THE FISHER-PRICE-KIRKPATRICK-ARNOLD MODEL FOR EVOLUTION OF BREEDING DATE

The idea of an environmental feature being the target of selection dates back to Darwin and Fisher. Fisher (1958), based on observations by Darwin (1871), suggested that the condition of a bird influences both her clutch size and the date at which she breeds, with healthier females breeding earlier and having larger clutch sizes. Price, Kirkpatrick, and Arnold (1988) used Fisher's idea as an explanation for the apparent lack of selection response for breeding date in many birds in the temperate zone. Birds that reproduce early have higher fitness than those that breed later in the season, and hence S for breeding date is negative (selection to move the breeding date earlier). Further, when examined, the breeding date typically has moderate to high heritability. Because both h^2 and S are nonzero, the breeder's equation predicts a response to selection resulting in a decrease in breeding date, but this is not seen.

The model of Price et al. (1988) to explain this lack of response is shown in Figure 20.2. For brevity, we refer to the **Fisher-Price-Kirkpatrick-Arnold model** as simply **Fisher's model** (we will resist the temptation of referring to this as the Fisher-Price toy model). The model is as follows: assume that the breeding date, z, of a female has three components

$$z = A - n + e \tag{20.2}$$

A is the breeding value for breeding date, e is the environmental value, and n is the nutritional state (or condition) of the female. Equation 20.2 shows that females with a higher value of n (higher nutritional status) breed earlier (i.e., z declines with increasing values of n). Price et al. (following Fisher) treated n as a nonheritable environmental factor, but one could also model n as a heritable trait, thus changing this to a multivariate selection problem (Chapter 13; Volume 3). The three components of Equation 20.2 are assumed to be uncorrelated and normally distributed, with variances of σ_A^2, σ_n^2, and σ_e^2. Let μ be the current mean breeding value and assume that the means of n and e are zero.

Price et al. (1988) modeled the process of selection by considering two separate components of fitness. First, they assumed there is an optimum breeding date, θ, so that z is under stabilizing selection. Recall from Equation 16.17 that a standard model for stabilizing selection in natural populations is nor-optimal (or normalizing) selection (Weldon 1895; Haldane 1954), where

$$W(z) = \exp\left(-\frac{(z-\theta)^2}{2\,\omega^2} \right) \tag{20.3a}$$

This function, giving the expected fitness, $W(z)$, of an individual with a phenotypic value of z, has the same form as a normal distribution, with the highest fitness at the optimal phenotypic value ($z = \theta$) and declining as one moves away from θ. The strength of selection is described by ω^2, the "width" of the fitness function. The larger the value of ω^2, the more slowly fitness declines as one moves away from the optimum. If $\omega^2 \gg \sigma_z^2$, selection is weak (most of the population has roughly the same fitness), while selection is strong when $\omega^2 \ll \sigma_z^2$. One advantage of this fitness function is that if z is normally distributed before selection, it remains normally distributed following selection, and expressions for the new mean and variance are easily obtained (Equation 16.18a).

Second, Price et al. assumed that fitness increases with the nutritional status, n. One way to express this is to assume that

$$W(n) = \exp(\alpha n) \qquad \text{for } \alpha > 0 \tag{20.3b}$$

Note that if $|\alpha\, n| \ll 1$, then $W(n)$ is approximately $1 + \alpha\, n$. As was the case with nor-optimal fitness, under the fitness function given by Equation 20.3b, traits that were normally distributed before selection remain normal following selection.

Conditioned on breeding date, z, and nutritional value, n, the resulting fitness is

$$W(n, z) = W(z) \cdot W(n) = \exp\left(\alpha n - \frac{(z-\theta)^2}{2\,\omega^2}\right) \tag{20.4a}$$

Recalling Equation 20.2, fitness can be expressed in terms of the components of the model

$$W(n, e, A) = \exp\left(\alpha n - \frac{(A - n + e - \theta)^2}{2\,\omega^2}\right) \tag{20.4b}$$

Under the assumption that A is normally distributed with a mean of μ, while n and e are normally distributed with a mean of zero, Heywood (2005) found the change in mean to be

$$R = \sigma_A^2 \left(\frac{\theta - \mu + \alpha\sigma_n^2}{\omega^2 + \sigma_z^2}\right) \tag{20.5}$$

where σ_n^2 is the variance in nutritional value. From Equation 20.5, at equilibrium ($R = 0$), the mean breeding date is

$$\widehat{\mu} = \theta + \alpha\sigma_n^2 \tag{20.6}$$

which is later than the optimal breeding date, θ. Price et al. (1988) commented that this displacement of the equilibrium mean above θ occurs because females that are in good nutritional condition ($n > 0$) breed *earlier* than the mean breeding value at equilibrium, as $z = A - n + e$ (Equation 20.2), and hence

$$E[z|n > 0] = E[A - n + e|n > 0] = E[A] + E[e] - E[n|n > 0] = \widehat{\mu} + 0 - E[n|n > 0] < \widehat{\mu}$$

Because $\widehat{\mu} > \theta$, females that are in good nutritional condition ($n > 0$) have a mean breeding date below $\widehat{\mu}$, and therefore closer to the optimal value, θ. Price et al. noted that this model may also apply to clutch size in birds and might be a reasonable model for seed germination time (especially for the latter if there is a significant nonheritable nutritional contribution from the maternal endosperm).

MODIFYING THE BREEDER'S EQUATION FOR NATURAL POPULATIONS

As these examples show, one of the most serious limitations in applying the breeder's equation to natural populations is that fitness can be influenced by unmeasured traits and environmental features. Additionally, genotype-environment *correlations* can be a concern,

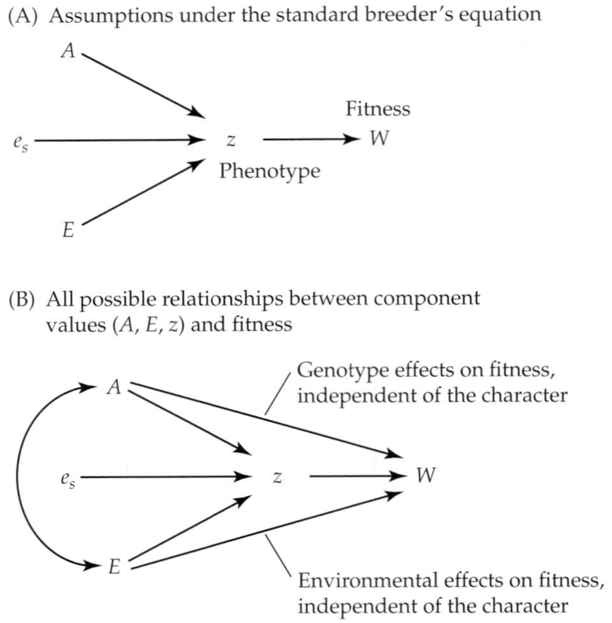

Figure 20.3 The pathways by which the components of a character (phenotype, z, additive genetic value, A, common environmental effect, E, and special environmental effect, e_s) influence fitness, W. **A:** The breeder's equation assumes that only the phenotype (z) of a character causally influences fitness. This is not an unreasonable starting assumption for artificial selection, wherein the breeder directly chooses individuals on the basis of phenotypes and randomizes environments with respect to phenotypes. **B:** Other pathways by which the components of a character can influence fitness. Either (or both) of the additive genetic and environmental values can influence fitness *independently* of their influence on phenotype. For example, an environmental value can both influence the character of interest and independently influence fitness. The influence of other traits that are also under selection, whose breeding values are correlated with our focal trait, appears through A and not through z. Similarly, the effect of shared environmental factors on phenotypic correlations appears through E. Finally, genotypic and environmental values may be correlated ($\sigma(A, E) \neq 0$), which is indicated by the double-headed arrows connecting A and E.

as (for example) larger individuals may be able to occupy better environments. In artificial selection and breeding situations, such a correlation is less of a concern because there is usually some attempt to randomize individuals over environments. Here we develop a general expression for the selection response of a single trait when all of these factors are in play. We do so by first assuming that there is a static environment (no environmental change), namely, that the distribution of environmental effects within the population remains constant over the generations of response being predicted.

Complications in the Absence of Environmental Change

How these complications bias the breeder's equation was examined by van Tienderen and de Jong (1994), with similar analysis (under a multivariate breeder's equation framework) by Hadfield (2008). van Tienderen and de Jong assumed complete additivity (no dominance or epistasis), multivariate normality, and linear parent-offspring regressions. As shown in Figure 20.3, they used a path analysis argument (LW Appendix 2) to explore the relationship between the selection response (R) and the selection differential (S) when complications such as selection on correlated characters and genotype-environment correlations exist.

To proceed, we decompose the phenotype, z, as

$$z = A + E + e_s$$

where A is the additive genetic value, E is the general environmental effect (for example, the average impact of a particular macrohabitat), and e_s is the special (or residual) environmental effect unique to each individual (LW Chapter 6). By construction, e_s is independent of the other variables (meaning that the total environmental variance is $\sigma^2_E + \sigma^2_{e_s}$), but A and E may be correlated. Consider Figure 20.3, which shows possible paths of how the environmental value, E, the genotypic value, A, and the phenotypic value, z, can influence fitness. Figure 20.3A shows the breeder's equation assumption that E and A influence fitness only through the phenotypic value, z.

Figure 20.3B shows the general situation where E and A can influence fitness independent of (or, in addition to) their effects on z, as can occur if the focal trait is phenotypically and/or genetically correlated with other characters that are under selection. If fitness is entirely determined by the phenotypic value of the focal trait, there should be no expected differences in the fitness of individuals with the same phenotypic value, z, but different underlying genetic (A) or environmental values (E). Suppose two individuals both have $z = 100$ and $e_s = 0$, but individual 1 has $A = 80$ and $E = 20$, while individual 2 arrives at this phenotypic value by $A = 10$ and $E = 90$. If selection is entirely on the phenotype, both individuals have the same expected fitness, but their expected fitnesses may differ if there is additional selection on A and/or E beyond that based on direct selection on z. For example, if correlated characters are under selection, then individuals with the same z value can have different fitnesses due to correlations between their A and/or E values with the genetic and/or environmental values at other traits that influence fitness.

To quantify the effects from these different paths influencing fitness, van Tienderen and de Jong considered the multiple regression of relative fitness, w, as a function of z, A, and E,

$$w = \alpha + \beta_z\, z + \beta_A\, A + \beta_E\, E + \epsilon \tag{20.7a}$$

The partial regression coefficients (β) represent the expected change in fitness when holding the other variables constant (LW Chapter 8). For example, β_z is the expected change in fitness from a unit change in the phenotype, z, holding the other variables (A and E) constant. In particular, if selection is *entirely* based on phenotypic value, then $\beta_A = \beta_E = 0$, because once we have controlled for z, A and E will have no effect on fitness. As shown by Queller (1992a), the condition for this to occur is that the partial covariances (given z) of breeding value and fitness, and of environmental value and fitness, are both zero,

$$\sigma(A, w \,||\, z) = \sigma(E, w \,||\, z) = 0 \tag{20.7b}$$

We return to this observation in the next section. The notation $\sigma(x, y \,||\, z)$ is used to remind the reader that the partial covariance of x and y (given z) can be *different* from $\sigma(x, y \mid z)$, which is the covariance of x and y conditional on z (see Example 6.8).

A notational aside is that we use β for the *partial* regression coefficients with respect to fitness (a special case of which is β_i, the selection gradient on trait i in the multivariate breeder's equation, e.g., Equation 20.1a), and we use b in the next section for the slopes of the *univariate* fitness regressions based on either A, E, or z separately. For example, for the univariate regression $w = 1 + b_z z + \epsilon$, the regression slope is $b_z = \sigma(z, w)/\sigma^2_z$, while β_z denotes the partial regression slope on fitness when A and E are also included (Equation 20.7a). As developed in the next section, comparing the appropriate univariate (b) and partial (β) regression slopes provides information on whether unmeasured variables are potentially influencing response.

From multiple regression theory (LW Chapter 8), the partial regression coefficients in Equation 20.7a satisfy the matrix equation

$$\begin{pmatrix} \sigma(w,z) \\ \sigma(w,A) \\ \sigma(w,E) \end{pmatrix} = \begin{pmatrix} S \\ R \\ \sigma(w,E) \end{pmatrix} = \begin{pmatrix} \sigma^2_z & \sigma(z,A) & \sigma(z,E) \\ \sigma(z,A) & \sigma^2_A & \sigma(E,A) \\ \sigma(z,E) & \sigma(E,A) & \sigma^2_E \end{pmatrix} \begin{pmatrix} \beta_z \\ \beta_A \\ \beta_E \end{pmatrix} \tag{20.8a}$$

The left-most vector contains the covariances between relative fitness, w, and the predictor variables (z, A, and E), with $S = \sigma(w, z)$ following from the Robertson-Price identity

(Equation 6.10) and $R = \sigma(w, A)$ from Robertson's secondary theorem (his 1966 version; Equation 6.25a). The 3×3 matrix on the right-hand side of Equation 20.8a is the variance-covariance matrix for the three predictor variables, where the covariance

$$\sigma(z, A) = \sigma(A + E + e_s, A) = \sigma^2_A + \sigma(E, A) \tag{20.8b}$$

In a similar fashion, one can show that $\sigma(z, E) = \sigma^2_E + \sigma(E, A)$. Using these identities, and considering the first two rows of Equation 20.8a after multiplying out the matrix product, returns the within-generation change as

$$\begin{aligned} S &= \sigma^2_z\,\beta_z + \sigma(z, A)\beta_A + \sigma(z, E)\beta_E \\ &= \sigma^2_z\,\beta_z + \left[\,\sigma^2_A + \sigma(E, A)\,\right]\beta_A + \left[\,\sigma^2_E + \sigma(E, A)\,\right]\beta_E \\ &= \sigma^2_z\,\beta_z + \sigma^2_A\,\beta_A + \sigma^2_E\,\beta_E + \sigma(E, A)\,(\beta_A + \beta_E) \end{aligned} \tag{20.9a}$$

and the response as

$$\begin{aligned} R &= \sigma(A, z)\,\beta_z + \sigma^2_A\beta_A + \sigma(E, A)\beta_E \\ &= \left[\,\sigma^2_A + \sigma(E, A)\,\right]\beta_z + \sigma^2_A\,\beta_A + \sigma(E, A)\,\beta_E \\ &= \sigma^2_A\,\beta_z + \sigma^2_A\,\beta_A + \sigma(E, A)\,(\,\beta_z + \beta_E) \end{aligned} \tag{20.9b}$$

If there are no genotype-environment correlations [$\sigma(E, A) = 0$], then

$$S = \sigma^2_z\,\beta_z + \sigma^2_A\,\beta_A + \sigma^2_E\,\beta_E \tag{20.10a}$$

and

$$R = \sigma^2_A\,(\beta_z + \beta_A) \tag{20.10b}$$

It is worth noting the connection between the expression for S from Equation 20.10a (based on following a single focal trait) and Equation 20.1a, where the selection differential on a focal trait is expressed in terms of direct selection on that trait (β_1) plus indirect selection from a correlated trait under selection ($\beta_2 \neq 0$). If we equating the two equations, we obtain

$$S = \sigma^2(z_1)\beta_1 + \sigma(z_1, z_2)\beta_2 = \sigma^2_z\,\beta_z + \sigma^2_A\,\beta_A + \sigma^2_E\,\beta_E \tag{20.11a}$$

where trait 2 (with a value of z_2) is the unmeasured correlated trait and trait 1 (with a value of $z_1 = z$) is the focal trait. In the notation of Equation 20.1a, $\sigma^2(z_1)\beta_1 = \sigma^2_z\,\beta_z$, giving

$$\sigma(z_1, z_2)\beta_2 = \sigma^2_A\,\beta_A + \sigma^2_E\,\beta_E \tag{20.11b}$$

Writing $z_1 = A + E$, we have

$$\sigma(z_1, z_2)\beta_2 = \sigma(A + E, z_2)\beta_2 = \sigma(A, z_2)\beta_2 + \sigma(E, z_2)\beta_2 \tag{20.12a}$$

If we match the terms in Equations 20.11b and 20.12a,

$$\sigma(A, z_2)\beta_2 = \sigma^2_A\,\beta_A \qquad \text{and} \qquad \sigma(E, z_2)\beta_2 = \sigma^2_E\,\beta_E \tag{20.12b}$$

Under the formulation for S given by Equation 20.9a, selection on any phenotypically correlated traits appears as a nonzero value of β_A (if the phenotypic correlations are, at least in part, due to correlated breeding values) and/or β_E (if the correlations are due, at least in part, to shared environmental values). Turning to the response, Equation 20.9b is much more general than the single correlated-trait expression given by Equation 20.1b, as β_A and β_E encompass *all* of the effects of *any* genetically and/or phenotypically correlated traits.

Further insight into the generalized response when no genotype-environment correlations are present follows if we first rearrange Equation 20.10a to isolate $\sigma^2_z\,\beta_z$ and then multiply both sides of Equation 20.10a by h^2, which yields

$$h^2\sigma^2_z\,\beta_z = \sigma^2_A\,\beta_z = h^2\left[\,S - (\,\sigma^2_A\,\beta_A + \sigma^2_E\,\beta_E\,)\,\right] \tag{20.13a}$$

Substituting this expression for $\sigma_A^2\,\beta_z$ into Equation 20.10b yields

$$R = h^2\,S + \sigma_A^2(1-h^2)\,\beta_A - h^2\,\sigma_E^2\,\beta_E \tag{20.13b}$$

Any extra (positive) selection on additive genetic values, $\beta_A > 0$ (due to selection on genetically correlated traits), inflates the selection response over the value predicted by the breeder's equation, while extra (positive) selection on environmental values ($\beta_E > 0$) decreases the response. The response is similarly decreased when $\beta_A < 0$.

Finally, following this same approach, Equations 20.9a and 20.9b yield a more general response (when $\sigma(E,A) \neq 0$) of

$$R = h^2\,S + \sigma_A^2(1-h^2)\,\beta_A - h^2\sigma_E^2\,\beta_E + \sigma(E,A)\,\left[\beta_z - h^2\,\beta_A + (1-h^2)\,\beta_E\right] \tag{20.14}$$

These expressions can be further simplified if selection acts only on the phenotype of the character being considered. In this case, $\beta_A = \beta_E = 0$, and Equation 20.9a reduces to $S = \sigma_z^2\,\beta_z$, implying that $\beta_z = S/\sigma_z^2$. Substituting these values into Equation 20.9b gives the response as

$$R = \beta_z\left[\sigma_z^2 + \sigma(E,A)\right] = \left(h^2 + \frac{\sigma(E,A)}{\sigma_z^2}\right)\,S \tag{20.15a}$$

which (as expected) reduces to the breeder's equation when there is no genotype-environment correlation. As discussed in Chapter 15, unless the correlation between E and A is perfect, the component of response from $\sigma(E,A)$ is transient, decaying to zero once selection stops.

Finally, these expressions provide insight into a key difference between artificial and natural selection. Under artificial selection, it is generally assumed that individual fitness is entirely based on the phenotype of the character of interest, specifically those phenotypes chosen by the breeder. In this case, the partial regression coefficients of fitness on genotype and environmental values are zero, as the phenotype entirely determines fitness. In natural populations, however, we do not have this luxury, and another possibility is that there is no natural selection on the character of interest (i.e., its phenotype, by itself, has no effect on fitness, meaning that $\beta_z = 0$), but rather that selection occurs on characters correlated with the one we are following. If these traits under selection are only connected to the focal trait through its breeding value (i.e., there is no environmental correlation between characters), then $\beta_A \neq 0$ while $\beta_z = \beta_E = 0$. In this case, using Equation 20.9a to express β_A in terms of S returns the response as

$$R = \beta_A\,\sigma_A^2 = S\left(\frac{\sigma_A^2}{\sigma_A^2 + \sigma(E,A)}\right) \tag{20.15b}$$

which reduces to $R = S$ in the absence of a genotype-environment correlation for the focal trait. The reason for this strong response is that (in this extreme setting) all of the selection is on the breeding value of the trait. With selection on phenotypes (the breeder's equation), only a fraction (h^2) translates into selection on breeding values.

A final possibility is that the only correlation between features influencing fitness and our focal character is through shared environmental effects, giving $\beta_E \neq 0$, while $\beta_A = \beta_z = 0$. In this case, using Equation 20.9b, the response becomes

$$R = \beta_E\,\sigma(E,A) = S\left(\frac{\sigma(E,A)}{\sigma_E^2 + \sigma(E,A)}\right) \tag{20.15c}$$

which equals zero unless a genotype-environment correlation exists. Again, in the absence of a perfect correlation between E and A, this response is transient (Chapter 15).

Additional Complications From Environmental Change

The above analysis considered the complications from uncontrolled, but static, environmental effects. A further layer of complexity arises when the environment (more formally, the distribution of possible environments) changes from year to year. First, the target of selection may radically change from one year to the next (e.g., Example 20.2). The significance of such temporal variation in selection remains an unresolved question. Siepielski et al. (2009) claimed that it is rather common and that changes in sign can be expected. Conversely, a reanalysis of the Siepielski et al. dataset by Morrissey and Hadfield (2012) concluded that the strength of directional selection in these studies is actually remarkably consistent after accounting for sampling variation. A related question is whether evolution is largely shaped by relatively rare, but major, events, (e.g., Example 20.2; Marrot et al. 2017), or rather by more gradual, but constant, pressures with less temporal variation.

A second complication is that a major shift in the environment can result in a shift in the trait mean even in the absence of any genetic change. As we will see later, a deterioration in the environment can mask significant underlying genetic change, leading to the appearance of stasis if the response and environmental change are in opposite directions. Finally, changes in the environment can result in changes in components of genetic (and environmental) variance and hence in h^2.

IS A FOCAL TRAIT THE DIRECT TARGET OF SELECTION?

Causality—wherein the phenotypic value of a focal trait is the sole target of selection—is a critical assumption when applying the breeder's equation to natural populations. With multiple traits, causality means that the covariance of the focal traits with fitness is entirely due to the phenotypic values of that set of focal traits. As we have seen, however, an observed selection differential can be generated by direct selection on a trait, direct selection on phenotypically correlated traits, an environmental covariance between the focal trait and fitness, or a combination of all of these (Equations 20.1a and 20.9a). One approach to control for phenotypically correlated traits is to include them in the analysis and then compute the vector, $\boldsymbol{\beta} = \mathbf{P}^{-1}\mathbf{S}$, of selection gradients (Equation 13.25a; Chapter 30). However, how does one ascertain if all relevant traits are included in the analysis? Many missing factors that are assumed to be traits could in fact be environmental features that influence both the focal trait and fitness, some (or all) of which could easily be overlooked in even the most careful analysis. One approach to assess causality (initially suggested by Rausher and Simms 1989; Queller 1992a; and Rausher 1992) is intimately connected with Robertson's secondary theorem of natural selection (Chapter 6). If the predicted selection response using Robertson's theorem is consistent with that from the breeder's equation, meaning that $\sigma(A_z, w) \simeq h^2 S$, one can have significantly increased confidence that the phenotypic value of the focal trait is indeed the target of selection. We refer to this basic strategy, and its variants, as **Robertson consistency tests**. This strategy can also involve a comparison of the fraction of the selection differential that is associated with a trait's breeding versus environmental values (more formally, the latter is the residual trait value following the removal of its breeding value) and whether either of these differentials is inconsistent with selection based solely on the phenotypic value, z, of the focal trait.

What is the advantage of a consistency test versus simply comparing the realized response in a natural population with its prediction from the breeder's equation? A lack of fit between observed and predicted response, by itself, is not informative as to which assumptions underlying the breeder's equation failed. By contrast, if a consistency test fails, this strongly suggests that selection is acting on more than just the phenotype of the focal trait or traits.

Robertson's Theorem: Response Prediction Without Regard to the Target of Selection

Recall from Chapter 6 that the breeder's equation is not the only expression for predicting the selection response (Table 6.1). Exact (but largely unusable) expressions follow from Price's theorem (Equations 6.8, 6.39, and 6.40). Under the assumption that parental breeding values are good predictors of the mean trait value of their offspring, Robertson's (two) secondary theorems of natural selection (Equations 6.24a and 6.25a) provide alternative expressions for the selection response.

As discussed in Chapter 6, there is some confusion in the literature on the secondary theorem, as Robertson actually suggested two slightly different versions. Robertson (1966a) suggested that $R = \sigma(A_z, w)$, namely, that response in a specific trait is equal to the covariance between the breeding value of that trait (A_z) and relative fitness (w), while later (Robertson 1968) he suggested that $R = \sigma(A_z, A_w)$, where relative fitness, w, is replaced by its breeding value, A_w. The relationship between the 1966 and 1968 versions follows (Equation 6.25c) by noting

$$\sigma(A_z, w) = \sigma(A_z, A_w + e_w) = \sigma(A_z, A_w) + \sigma(A_z, e_w)$$

showing that while the 1966 version is more general, the two are equal when $\sigma(A_z, e_w) = 0$. There is no reason to suggest that this covariance should generally be zero, as it simply states that there is a covariance between the residual component of fitness (once the effect of the breeding value of fitness has been removed) and the breeding value of the trait itself.

Example 20.4. One of the earliest applications of Robertson's theorem to natural populations examined mean nestling weight and offspring survival in great tits (*Parus major*) (van Noordwijk 1988). The key idea exploited by van Noordwijk was to consider two different covariances involving an individual's nestling weight, z. The first, $\sigma(z, w)$, was with its own survival (standardized by mean survival of the population to obtain a relative fitness, w). From the Robertson-Price identity (Equation 6.10), this is simply the selection differential, S, on nestling weight. With an estimate of h^2 for nestling weight, the expected responses of $R = h^2 S$ under the breeder's equation are given in the table below.

In the same study van Noordwijk also considered the covariance of parental nestling weight with the survival of its *offspring*. In the absence of any shared environmental effects, the covariance between one trait in a parent (nestling weight) and a second trait in its offspring (offspring survival) is through the breeding value of the parental trait. He examined the difference between the mean weight of all parents and the mean weight of parents who had surviving offspring. While not stated as such by van Noordwijk, this selection differential (on the parents) conditioned on the survival of their offspring is an estimator of $\sigma(A_z, w)$, and hence this is the predicted response under Robertson's theorem. The results for 1975 to 1978 are in the final column in the following table.

Year	S	h^2	$R = h^2 S$	$R = \sigma(A_z, w)$
1975	0.24	0.38	0.07	0.00
1976	0.68	0.47	0.32	0.03
1977	0.16	0.26	0.04	0.06
1978	0.53	0.29	0.15	0.05
mean	0.40	0.35	0.14	0.035

The breeder's equation significantly overpredicts the selection response relative to Robertson's theorem, suggesting that factors correlated with nestling weight also influence fitness.

Robertson Consistency Tests

Queller (1992a), Rausher (1992), and Morrissey et al. (2010, 2012) all suggested that an analysis that estimates the required parameters for both the breeder's equation and Robertson's theorem can provide insight on whether the breeder's equation applies to a focal trait in a natural population (Example 20.4). If the two estimates of response agree, this suggests that the phenotypic value, z, of the focal trait is largely causative as the target of selection.

If the predictions are significantly different, other forces beside direct selection on z are involved. Note that this analysis checks the prediction under the *static environment assumption*. Generational changes in E (e.g., shifting selection targets, shifts in trait mean from entirely environmental factors, or changes in variance components due to $G \times E$) can all cause Robertson's theorem (as well as the breeder's equation) to fail. Likewise, if the standard breeding-value model is not a good approximation of the genetics of transmission, both Robertson's theorem and the breeder's equation can fail (Chapter 6).

Under what conditions should the breeder's equation and Robertson's theorem agree? If we use the more general, 1966, version of Robertson's theorem, the two predicted responses are equal when

$$R = h^2 S = \frac{\sigma^2(A_z)}{\sigma^2(z)}\,\sigma(z,w) = \sigma(A_z,w) \tag{20.16a}$$

Rearranging the last equality yields the result of Queller (1992a; also Hadfield 2008)

$$\frac{\sigma(z,w)}{\sigma^2(z)} = \frac{\sigma(A_z,w)}{\sigma^2(A_z)} \tag{20.16b}$$

The left-hand side of Equation 20.16b is the slope (b_z) of the univariate linear regression of w on trait phenotypic value (z), while the right-hand side is the slope (b_A) of the univariate regression of w on the breeding value (A_z) of the trait. If $b_z \simeq b_A$, then the breeder's equation is likely to hold (subject to the assumptions of a static environment and the infinitesimal model for inheritance). However, if b_z and b_A are significantly different, additional targets of selection besides the phenotypic value of the focal trait influence the selection response of the focal trait. This test hinges on the ability to obtain an unbiased estimate of $\sigma(A_z, w)$, a subject discussed in detail below. Finally, note that when Equation 20.16b is satisfied, this expression can be rearranged to yield

$$h^2\sigma(z,w) = \sigma(A_z,w) \tag{20.16c}$$

Rausher's Consistency Criteria

Rausher (1992) obtained a multivariate version of Equation 20.16b by equating the vector of responses, $\mathbf{R}$, under Robertson's theorem with that predicted from the multivariate breeder's equation (Equation 13.23b)

$$\mathbf{R} = \sigma(A_{\mathbf{z}},w) = \mathbf{G}\mathbf{P}^{-1}\mathbf{S} \tag{20.17a}$$

where the ith component of the vector $\sigma(A_{\mathbf{z}}, w)$ is $\sigma(A_i, w)$, the covariance between the breeding value of trait i and relative fitness (Equation 6.25a; the 1966 version of Robertson's theorem). If we multiply both sides by $\mathbf{G}^{-1}$ and recall the multivariate version of the Robertson-Price identity ($\mathbf{S} = \sigma(\mathbf{z}, w)$; Equation 6.10), Equation 20.17a can be restated as

$$\mathbf{G}^{-1}\sigma(A_{\mathbf{z}},w) = \mathbf{P}^{-1}\sigma(\mathbf{z},w) \tag{20.17b}$$

which is **Rausher's consistency condition** and the multivariate version of Equation 20.16b. This is a slight generalization of Rausher's (1992) original result, which assumed the 1968 version of Robertson's theorem, with $\sigma(A_{\mathbf{z}}, A_w)$ replacing $\sigma(A_{\mathbf{z}}, w)$. Notice in Equation 20.17b that $\mathbf{G}^{-1}\sigma(A_{\mathbf{z}}, w)$ is the vector of coefficients for the regression of relative fitness on the vector of trait breeding values, while $\mathbf{P}^{-1}\sigma(\mathbf{z}, w)$ is the vector of coefficients for the phenotype-fitness regression. Rausher's consistency condition is that the coefficients for the regression of fitness are the same (for a given trait) when one uses breeding values in place of phenotypic values.

For a univariate trait, Equation 20.17b reduces to Equation 20.16b, in which case Equation 20.16c immediately yields

$$h^2 S_z = S_A \tag{20.18}$$

namely, the selection differential, $S_A = \sigma(A, w)$, based on the breeding value of a trait is simply h^2 times the phenotypic selection differential, $S_z = \sigma(z, w)$. Although Rausher's condition (Equation 20.17b) directly leads to Equation 20.18, the formal regression test he proposed (Rausher and Simms 1989; Rausher 1992) is slightly different (Equation 20.20), and thus Equation 20.18 is referred to as **Postma's test** (Postma 2006).

Rausher framed his consistency test in terms of the relative strengths of selection on the *components* of the focal trait's phenotypic value, namely, its breeding (A) and environmental (E) values (more formally, the latter is the residual value in z after A is removed, meaning that E can also include nonadditive genetic terms). Consider the slopes, b_z and b_A, of the *univariate* regressions of relative fitness on phenotype, z, and breeding value, A. When Equation 20.16b holds, then

$$b_A = b_z \tag{20.19a}$$

which is simply the univariate version of the condition given by Equation 20.17b. Similarly, for the univariate regression of fitness on E, $b_E = \sigma(E, w)/\sigma^2_E$. If we note that $\sigma^2(A)/h^2 = \sigma^2(z)$ and $\sigma^2_E/(1-h^2) = \sigma^2(z)$, we can relate the slope, b_z, of the univariate regression of fitness on z with the corresponding univariate regression slopes b_A and b_E (based on A and E, respectively) as follows:

$$\begin{aligned} b_z = \frac{\sigma(w,z)}{\sigma^2(z)} &= \frac{\sigma(w,A)+\sigma(w,E)}{\sigma^2(z)} \\ &= \frac{\sigma(w,A)}{\sigma^2(A)}h^2 + \frac{\sigma(w,E)}{\sigma^2(E)}(1-h^2) \\ &= b_A h^2 + b_E(1-h^2) \end{aligned} \tag{20.19b}$$

When the identity given by Equation 20.16b is satisfied, then $b_z = b_A$ (Equation 20.19a), and Equation 20.19b becomes

$$b_z = b_A = b_A h^2 + b_E(1-h^2), \quad \text{or} \quad b_A(1-h^2) = b_E(1-h^2)$$

and hence $b_z = b_E = b_A$ (provided $\sigma[A, E] = 0$).

This observation suggests the **Rausher-Simms equality test** (Rausher and Simms 1989; Rausher 1992). Here, one computes the multiple regression of fitness on both the trait breeding value, A, and the residual deviation, E, namely,

$$w = 1 + \beta_A A + \beta_E E + e \tag{20.20}$$

If z is the sole target of selection, then $\beta_A = \beta_E$, which can be tested in a straightforward fashion using standard results from regression theory (LW Chapter 8), *assuming* that A and E are known. A closely related test queries whether β_A is significantly different from zero, as this indicates that at least *some* of the selection is translated onto the breeding value of the focal trait. The equality test is more stringent, asking whether selection is *strictly* a function of the phenotypic value of the focal trait, no matter how that phenotype is obtained (e.g., individuals with high breeding value versus high environmental deviation, but the same phenotype, experience the same amount of selection).

The reader might be inclined to assume that when $\beta_A \neq \beta_E$, the component with the larger partial regression coefficient experienced stronger selection. To see that this reasoning can be misleading, suppose that phenotypic selection influences both β_A and β_E by 0.4 (a unit change in z changes w by 0.4). Assume also that additional selection on A (over and above that through selection on z, such as on a genetically correlated, but unmeasured, trait) adds -0.35, and additional selection on E (for example, a specific environmental factor improving fitness beyond that achieved through direct selection on trait value z) adds 0.05, resulting in $\beta_A = 0.05$ and $\beta_E = 0.45$. While this superficially suggests that there is more selection on E, the additional direct component of selection (beyond that due to z) is actually much stronger on A.

The $\beta_A = \beta_E$ and $\beta_A \neq 0$ tests were suggested before the application of mixed models to natural populations, leaving the critical issue of how to estimate breeding values unresolved. Focusing on plants, Rausher and Simms (1989) and Stinchcombe et al. (2002) replicated genotypes (when clones were available) or sibs (half, full, or selfed) over environments, estimating the genotypic value of a clone by its average over the sampled environments, and likewise assigning all sibs the same breeding value, namely, their family mean. Because sib (or clone) means replicated over environments are used as estimates of the breeding or genotypic values, there are different sample sizes associated with β_A (number of families) and β_E (number of individuals). Stinchcombe et al. (2002) discussed how to deal with this issue. Using the regression approach given by Equation 20.20, Stinchcombe et al. and Scheiner et al. (2002) compared estimates of selection for six plant species grown on experimental plots (and hence under stricter environmental control than expected for populations in nature). Even in these settings, these authors found that a significant fraction (around 25%) of the traits they measured appeared to show selection on factors other than z (β_A was significantly different from β_E). While this bias rarely resulted in a change in sign, it often significantly impacted the estimated strength of selection directly on z.

While tests based on sibs or replicated genotypes were an important conceptual advance, their actual utility was rather limited. In addition to the logistical issues involved in implementing such a design, this approach *critically* depends upon the randomization of genotypes over environments. The estimated breeding value assigned to all members of a sibship is their family effect, which is a function of the mean breeding value of their parents but also of maternal effects, dominance (for full sibs), and common-family environmental values. If environments are not randomized, a common-family environment could influence both the trait and fitness, and this would appear in the family effect. In this case, β_A could be significantly different from zero, but as a reflection of selection on common-family environmental values rather than on breeding values. A further complication with full sibs is that they potentially share an additional covariance of $\sigma^2_D/4$, which could be loaded into β_A even when environments are randomized.

The realization in the early 2000s that mixed models (Chapter 19; LW Chapters 26 and 27) could return estimates of trait breeding values for *individuals* led to a brief period (with a flury of publications) during which BLUP-estimated breeding values were used to test for associations between trait breeding value and fitness (e.g., Kruuk 2004). While potentially much more powerful than clone or family studies (using individual, rather than group, breeding values), as we detail below, given the structure of most natural pedigrees, BLUP-estimated breeding values have a strong environmental bias, and were eventually realized to be highly unreliable for these sort of studies (Postma 2006; Postma and Charmantier 2007; Hadfield 2008; Hadfield et al. 2010; Wilson et al. 2010). However, while estimates of *individual* breeding values may be suspect, the power of a mixed-model analysis can still be used through direct REML estimates of *population*-level parameters, such as either $\sigma(A_z, A_w)$ or $\sigma(A_z, w)$. By directly estimating such covariances under a bivariate animal model, the pitfalls of using predicted breeding values of individuals can be avoided (Hadfield et al. 2010). We will examine all of these issues in detail shortly.

Morrissey et al.'s Consistency Criteria

An alternative expression for consistency can be obtained as follows. Writing

$$\sigma(z, w) = \sigma(A_z + E_z, A_w + e_w) = \sigma(A_z, A_w) + \sigma(E_z, e_w) + \sigma(A_z, e_w) + \sigma(E_z, A_w)$$

where $E_z = z - A_z$ is the residual trait value after the removal of the breeding value, with $e_w = w - A_w$ similarly defined. Likewise noting that $\sigma(A_z, w) = \sigma(A_z, A_w) + \sigma(A_z, e_w)$, the consistency condition given by Equation 20.16b becomes

$$\frac{\sigma(A_z, A_w) + \sigma(E_z, e_w) + \sigma(A_z, e_w) + \sigma(E_z, A_w)}{\sigma^2(A_z) + \sigma^2(E_z)} = \frac{\sigma(A_z, A_w) + \sigma(A_z, e_w)}{\sigma^2(A_z)} \tag{20.21a}$$

which can be rearranged to

$$1 + \frac{\sigma(E_z, e_w) + \sigma(E_z, A_w)}{\sigma(A_z, A_w) + \sigma(A_z, e_w)} = 1 + \frac{\sigma^2(E_z)}{\sigma^2(A_z)}$$

implying

$$\frac{\sigma(E_z, e_w) + \sigma(E_z, A_w)}{\sigma(A_z, A_w) + \sigma(A_z, e_w)} = \frac{\sigma^2(E_z)}{\sigma^2(A_z)}$$

Finally, this rearranges to yield an alternative consistency condition

$$\frac{\sigma(E_z, e_w) + \sigma(E_z, A_w)}{\sigma^2(E_z)} = \frac{\sigma(A_z, A_w) + \sigma(A_z, e_w)}{\sigma^2(A_z)} \tag{20.21b}$$

By assuming Robertson's 1968 version, with $\sigma(A_z, e_w) = 0$, and also that $\sigma(E_z, A_w) = 0$, this reduces to

$$\frac{\sigma(E_z, e_w)}{\sigma^2(E_z)} = \frac{\sigma(A_z, A_w)}{\sigma^2(A_z)} \tag{20.21c}$$

which is the **Morrissey consistency condition** (Morrissey et al. 2010, 2012).

Example 20.5. As detailed shortly, mixed models can be used to estimate the variance components required for Equation 20.21c. This was done by Morrissey el al. (2012), who used a bivariate animal model (the focal trait plus fitness as the second trait). They examined four morphological traits in Soay sheep (*Ovis aries*) on the island of St. Kilda. Body size was of special interest, because estimates of S and h^2 suggested a positive response using the breeder's equation, yet the sheep were, if anything, getting smaller. By contrast, the expected response under the secondary theorem (1968 version), $R = \sigma(A_z, A_w)$, was slightly negative (but not significantly different from zero). Using estimates of the components of Equation 20.21c showed that the two sides of this consistency condition were significantly different ($p = 0.048$). Thus, the failure of the selection response to match that predicted by the breeder's equation is likely a result of unmeasured factors that do not influence selection on the breeding value, but upwardly bias estimates of the amount of selection on the phenotype.

The Breeder's Equation Versus the Secondary Theorem

The elegance of the breeder's equation is that it fully separates ecology (S) from genetics (h^2). Queller (1992a) noted that when Equation 20.16b is satisfied, this separation occurs. More formally, **Queller's separation condition** is that the partial covariance (Example 6.8) of A and w given z is zero, $\sigma(A, w \,||\, z) = 0$ (Equation 20.7b; which also implies $\sigma(E, w \,||\, z) = 0$; see Queller 1992a). This is simply an alternative way of interpreting Equation 20.16b: that the residual values of A and w (following their separate regression on z) are uncorrelated (e.g., Equation 6.31a). Thus, after accounting for the phenotypic value, there is no residual correlation between breeding value and fitness. (As a technical aside, note that when the separation condition holds, Heywood's spurious response term, Equation 6.31a, is zero.)

In contrast, the secondary theorem fully *confounds* (rather than separates) selection and genetics, as the covariance of A (genetics) with w (ecology) is a combined, rather than a separable, function of these two features. Further, the secondary theorem says absolutely nothing about the nature of selection on the *phenotype*. Rather, it simply does the accounting and asks what fraction of selection translates into direct selection on the breeding value. The secondary theorem is thus largely about genetics (van Tienderen and de Jong 1994; Morrissey et al. 2012) and rather devoid of ecology. As such, it is generally expected to be more predictive than the breeder's equation, as it ignores the actual target of selection (but, as mentioned, can still fail). When the two predicted responses (from the breeder's equation and Robertson's theorem) agree, we can have some confidence that we have found a causal target of selection (z), implying that the breeder's equation is not significantly compromised by selection on unmeasured variables.

Which approach, the breeder's equation, $R = h^2 S = \sigma^2(A)\boldsymbol{\beta}$, or the secondary theorem, $R = \sigma(A_z, w)$, should be used by an investigator? In large part, this depends on the question being asked. In a conservation biology setting, such as when trying to predict if a species has sufficient genetic variation to withstand a major environmental change, selection response is the major issue of concern, as opposed to the actual targets of selection. An example of this was provided by Etterson and Shaw (2001), who used Robertson's theorem to show that there were significant constraints in response to selection from climate change in a native annual legume from the Great Plains region. Antagonistic genetic correlations among the traits under selection reduced the amount of usable additive variation in the direction favored by selection. While knowing the targets of selection (i.e., those particular traits favored by selection) is always of interest, the more pressing concern for Etterson and Shaw was whether the population could mount a successful selection response to the pressures generated by climate change. Robertson's theorem can address this question without any bias from unmeasured characters influencing the focal traits by examining if there is a sufficiently high covariance between the trait breeding value and relative fitness to generate some required amount of response. Conversely, the targets of selection are generally of great interest to ecologists and evolutionary biologists, and the joint use of the breeder's equation and Robertson's theorem can help clarify the importance of candidate traits.

APPLYING MIXED MODELS TO NATURAL POPULATIONS: BASICS

Recall from Chapter 19 that mixed models offer a very flexible platform for genetic analysis in the presence of multiple fixed effects and multigenerational relatives. In particular, the general **animal model**, so called because of its initial focus on estimating the breeding values of a single trait in a collection of individual animals (originally in dairy cattle),

$$\mathbf{y} = \mathbf{X}\boldsymbol{\beta} + \mathbf{Z}_a\mathbf{a} + \sum_{i=1}^{k} \mathbf{Z}_i\mathbf{u}_i + \mathbf{e} \tag{20.22a}$$

has been widely used in animal breeding since the 1970s. As detailed in Chapter 19, $\boldsymbol{\beta}$ is the vector of unknown fixed effects, $\mathbf{a}$ is the random vector of breeding values, $\mathbf{e}$ is the random vector of residuals, and the $\mathbf{u}_i$ denote k other possible vectors of random effects. These additional random effects can accommodate permanent environmental effects under a repeated-records design, common family or maternal effects, and other factors that can complicate the residual error structure (Chapters 19 and 22).

In Equation 20.22a, $\mathbf{y}$ is an observed vector of trait values, $\mathbf{X}$ is the design matrix for the fixed effects, and $\mathbf{Z}_a$ and the $\mathbf{Z}_i$ are incidence matrices for the random effects. The power of a mixed model is its ability to borrow information on random effects from correlated observations. This is done through their covariance structure, which determines the strength of additional information provided by correlated observations. The vector, $\mathbf{a}$, of breeding values has a covariance structure determined by the (assumed known) relationship matrix, $\mathbf{A}$, and it is assumed that $\mathbf{a}$ and $\mathbf{e}$ are uncorrelated, so

$$\begin{pmatrix} \mathbf{a} \\ \mathbf{e} \end{pmatrix} \sim \begin{pmatrix} \mathbf{0} \\ \mathbf{0} \end{pmatrix}, \begin{pmatrix} \sigma^2(A)\,\mathbf{A} & \mathbf{0} \\ \mathbf{0} & \sigma_e^2\,\mathbf{I} \end{pmatrix} \tag{20.22b}$$

where $\sigma^2(A)$ denotes the additive-genetic variance of the trait. Similar assumptions are made about the covariance structures for any additional random effects (Chapters 19 and 22), and the assumed covariance structure in combination with Equation 20.22a fully specify the model.

Variance components are estimated by REML (Chapter 19; LW Chapter 27), which are then used to estimate the vector, $\widehat{\mathbf{a}}$, of **predicted breeding values** (**PBVs**) using BLUP (Chapter 19). These are also called **estimated breeding values** (**EBVs**) in the literature, but our preference is to use *predicted* for random effects and *estimated* for fixed effects. As we will show, while REML estimates of variances and covariances are appropriate when using animal models in wild populations, using individual PBVs is generally not appropriate unless performed within an appropriate Bayesian framework. Indeed, Hadfield et al. (2010) said they would "discourage future use of BLUP as an inferential tool in the fields of ecology and evolutionary biology." The reasoning here, and solutions to some of the issues, are examined in the next major section. First, however, we consider the limitations of constructing an animal model for a wild population, with analysis issues examined later.

Animal-model Analysis in Natural Populations: Overview

Given sufficiently large, complete, and accurate pedigrees, animal models can help separate genetic from environmental trends (Chapter 19), and their ability to estimate *individual* breeding values seemed to offer the possibility of conducting more accurate Robertson consistency tests. Given these features, it is surprising that the application of mixed models to natural populations was rather recent, starting with suggestions by Shaw (1987) and then by Konigsberg and Cheverud (1992) and Cheverud and Dittus (1992), who applied them to free-living primate populations. These papers went somewhat unnoticed, and a second wave of applications to ungulate mammals and nesting birds started in 1999 (Réale et al. 1999), and has been a rapid growth industry ever since (Kruuk 2004; Kruuk and Hadfield 2007; Postma and Charmantier 2007; Kruuk and Hill 2008; Clutton-Brock and Sheldon 2010; Hadfield et al. 2010; Wilson et al. 2010; Postma 2014).

The animal model has generally been quite successful in the analysis of artificial selection experiments and breeding programs (Chapter 19). However, natural populations differ in fundamental ways from these more controlled settings, leading to a number of design issues (Table 20.1). First, in natural populations, the relationship matrix must be ascertained indirectly, and this is usually done with a bias toward finding mother-offspring (maternal) links, while missing (or misspecifying) father-offspring (paternal) connections.

Second, artificial selection experiments, and many breeding programs, involve **closed populations**, with little to no immigration from outside sources once selection has started. Further, most (if not all) organisms in the population are included in the analysis. When immigration occurs, it is usually controlled, and hence immigrants can be identified in the pedigree. However, in most natural populations, immigration from outside the study area is generally the norm, which causes serious ascertainment problems. Immigrants potentially bring in a different distribution of breeding values, and sufficiently high immigration rates can remove any signal of local genetic change or falsely create such a signal when none is present. Further, analysis in natural populations is usually based on a somewhat haphazard sample of individuals, rather than information from the entire population (which is often available in artificial selection experiments or commercial-breeding populations).

Finally, an important consequence of the more open structure of natural populations is that the connectedness (number of relatives) is lower, often substantially so, than for breeding programs. The relationship matrices, **A**, from breeding programs tend to be denser than those for natural populations (the former having more nonzero and larger off-diagonal elements than the latter), as most individuals in a breeding-program sample have measured relatives in previous generations. The presence of such measured relatives is *not* ensured for a sample of individuals from a natural population.

To see the significance of a sparse relationship structure, consider an individual that is unconnected to any others in the pedigree. In this setting, that individual's PBV is simply the estimated heritability times its phenotypic value (adjusted for fixed effects). In the simplest case of a single fixed effect (the mean μ), its PBV is simply $\widehat{a} = h^2(z - \mu)$. When an individual has links (via **A**) to other members in the sample, BLUP uses this covariance information to obtain an improved estimate of its PBV, making the latter less dependent on just the individual's own phenotypic value (which has some environmental influence). In large pedigrees with many links (such as in most breeding programs), this substantial additional information can significantly improve the accuracy of PBVs. In natural populations, the number of links may be far smaller, in which case an individual's predicted breeding value may be largely determined by its own phenotype alone. In such cases, PBVs can

Table 20.1 Design limitations when applying animal models to natural populations.

The relationship matrix, **A**, must be estimated. Pedigree errors result in bias and lower power.
Open population structure. Immigration from outside of the study area complicates the interpretation of model results.
Lack of sufficient size or depth of the sampled pedigree (low connectedness). Low variation in sample relatedness results in low power, potentially confounding parameters of interest.
Lack of power complicates model selection. Low power to detect variance components can complicate results. Interpretations of the key parameters can substantially change over alternative models incorporating different sets of random effects. Such additional effects are often only included when their associated variance components are significant.

be strongly influenced by the environment, a point to which we will return to below. Because of this fragility of individual PBV estimates when pedigree links are sparse, their use is now strongly discouraged in such settings (Postma 2006; Hadfield 2008; Hadfield et al. 2010).

One important observation is that, to date, most estimates of heritability based on mixed-model analyses of wild populations are *lower* than more traditional h^2 estimates based on parent-offspring regressions (Kruuk 2004; Postma 2014). As mentioned in Chapter 19, the *opposite* pattern is expected. This is because heritability under a mixed model is defined as the ratio of the additive variance to the sum of all variance components, with the latter being computed *after* fixed effects (and thus a significant source of variation) have been removed (Wilson 2008). Phrased another way, once the fixed effects have been removed, the sum of all variance components is less than the trait variance, $\sigma^2(z)$, when individuals differ in their fixed effects. Thus, parent-offspring regressions and mixed-model h^2 estimates are looking at slightly different quantities, with the latter estimate expected to be larger. Why does this not seem to be the case?

Åkesson et al. (2008) and de Villemereuil et al. (2013) suggested that one reason for this apparent discrepancy might involve subtle differences in the datasets, with parent-offspring regressions (PO) often using the average values of offspring, while mixed-models (MM) use individual measures, and hence have a higher intrinsic variance. These authors found that when mixed models were run using the offspring mean, parent-offspring and animal-model estimates of heritabilities were very similar, while when they were run using a repeated-measures mixed model (i.e., keeping the individual measures; Chapter 19), heritability estimates were lower.

Biological factors could also be involved. Given that the vast majority of mixed-model estimates are for species with extensive parental care (birds and mammals), part of the difference between MM and PO estimates may arise from a lack of control over maternal effects. Most parent-offspring regressions in the wild involve mother-offspring relationships, confounding direct and maternal effects, which in turn can inflate regression-based estimates of h^2. Mixed models that include maternal effects remove this bias.

An extension of this general idea is that parents and offspring often share similar environments in natural populations, inflating their phenotypic similarity (Magnussen 1993). Indeed, Stopher et al. (2012) noted that when home-range overlap is included in mixed models, the heritability estimates for several traits in red deer decreased, consistent with the notion that relatives in natural populations share environmental effects. While the potential presence of shared environmental effects is poorly controlled in parent-offspring regressions, they may still not be fully accounted for in mixed models, as the chain of relatives used to estimate the breeding value of a focal individual may still (albeit more weakly) share environmental features due to living in relatively close proximity to each other.

Obtaining the Relationship Matrix: Direct Observation of the Pedigree

The central difficulty in applying the animal model to free-living populations is obtaining the relationship matrix, **A**, for the measured sample of individuals. One source of information is the **social pedigree** based on field observations, especially for birds and mammals (which to date comprise the majority of BLUP applications to wild populations). If we observe a mother nursing an offspring, we have fairly high confidence that the offspring is from that mother. Accessing paternity is more difficult, however. Again, field observations may be useful, for example, observations of which male visits the nest or appears to be the dominant male in other social settings.

Of course, none of these social observations is foolproof. For example, **intraspecific brood parasitism** occurs when a female lays an egg in the nest of another female. Likewise, even with (apparently) pair-bonded couples, **extra-pair paternities** can occur. The frequency of such extra-pair events is $\sim$15% in the collared flycatcher (*Ficedula albicollis*) population discussed in Example 20.1 (Sheldon and Ellegren 1999), similar to the values for other socially monogamous birds (Firth et al. 2015). Hence, the simple observation of a male helping at the nest does not imply that he is the father. Similarly, it is not guaranteed that the dominant male in a harem sired all of the offspring.

Pedigree errors can be high even in systems with apparently strong control over matings. Visscher at al. (2002) estimated a sire error rate of $\sim$10% for UK dairy cattle, despite the very widespread use of artificial insemination, while Leroy et al. (2011) found rates of 1–9% for dogs, 1–10% for sheep, and 4% for a French cattle population. Recording errors, as well as the ingenuity of organisms searching for mates, should never be underestimated! Because of this intrinsic bias toward determining the mother, pedigrees from wild populations often show an excess of **maternal linkages**. This has implications when maternal effects are present, as the pedigree must contain a sufficient number of **paternal linkages** to disentangle direct effects from maternal effects (Clément et al. 2001; Kruuk 2004).

Obtaining the Relationship Matrix: Marker Data

A second source of information on relatedness is provided by polymorphic molecular markers. Methods estimating ancestry from marker data can be grouped into two categories: those that are hypothesis-driven (e.g., tests for paternity from a pool of candidate males, or for individuals being half- or full-sibs) and those that make no prior assumptions about relatedness. These two approaches can be restated as a focus on categorical **relationships** (assigning pairs of individuals into discrete classes such as parent-offspring, or full- or half-sibs) versus continuous measures of **relatedness** (estimates of the coefficient of coancestry; LW Chapter 7). A number of methods to estimate pairwise relatedness have been proposed (reviewed by Ritland 2000; van de Casteele et al. 2001; Blouin 2003; Garant and Kruuk 2005; Thomas 2005; Csilléry et al. 2006; Oliehoek et al. 2006; Weir et al. 2006; Frentiu et al. 2008; Pemberton 2008; Powell et al. 2010; Sillanpää 2011; Gay et al. 2013; Bérénos et al. 2014; Jensen et al. 2014; Speed and Balding 2015; Conomos et al. 2016; Ackerman et al. 2017; Wang et al. 2017).

At first blush, one might think to simply use one of these methods to estimate the pairwise relatedness between all sampled individuals, substituting these as the elements of **A**. However, there are numerous problems with this approach. First, there are high sampling variances for these estimates (see the reviews mentioned above). Second, such procedures often result in the molecular-based relationship matrix used to estimate **A** not being positive-definite (not having all positive eigenvalues; see Appendix 5) and hence it is not a proper covariance matrix (Frentiu et al. 2008). There is also the issue that some pairwise methods may return negative estimates of relatedness for nonrelatives. Although such estimates are often set to zero, doing so introduces a bias akin to that introduced by setting negative variance estimates to zero (Ackerman et al. 2017).

One approach for constructing a marker-based relationship matrix is to ignore more distant relationships that must be inferred solely from molecular markers and instead use markers to confirm (or find) sets of close relatives, such as parent-offspring (Blouin 2003;

Jones and Ardren 2003; Jones et al. 2010; Walling et al. 2010) and sibs (Thomas and Hill 2000; Wang 2004; Wang and Santure 2009; Huisman 2017). Much of the power in a mixed model comes from data on the closest relatives, which justifies an initial focus on detecting and confirming close pedigree linkages. Further, for distant relatives, the sampling (and segregation) variances for relatedness measures can be considerable (Speed and Balding 2015). Early studies using markers to infer paternity or to assign individuals to sibships typically used no more than one or two dozen microsatellite loci. Although these are highly polymorphic markers, and hence have significant power for verifying very recent ancestry (such as first-degree relatives, which share half their alleles IBD), the expected fraction of alleles shared between two relatives with a common ancestor k generations in the past is $(1/2)^{2k-1}$. Relatives with a common ancestor two generations in the past thus share (on average, but with considerable variance) only 1/8 of their alleles IBD, which greatly reduces the power for a relatively small number of markers to detect this degree of ancestry, yet alone more distant relationships.

In cases where there are a modest number of markers, the most powerful approach for reconstructing a natural pedigree is to combine marker data with additional information, such as ranges of specific individuals and their behavior (e.g., apparent position in a dominance hierarchy, and hence the likelihood of being a sire). Hadfield et al. (2006) presented such an analysis, which they set within a Bayesian framework, so that uncertainty in relationship estimates is fully captured in the posterior uncertainty of parameter estimates. In a full Bayesian analysis, information at different levels can inform each other (O'Hara et al. 2008). For example, consider a setting where, based on marker information, individual A is ever so slightly more likely than B to be the father of C. Phenotypic data provide additional information on whether C is closer to A or B, but is ignored in a sequential likelihood analysis (which estimates relationships first and then uses these to estimate genetic parameters), which would use A in this case. Under a Bayesian analysis, this additional phenotypic information will influence the posterior paternity estimates.

In the few cases where (for a natural population) the estimates of a relationship matrix from an explicit pedigree plus marker-information study have been compared to an entirely marker-inferred relatedness matrix, erratic behavior in the estimated variance components has been seen (e.g., Thomas et al. 2002; Coltman 2005; Frentiu et al. 2008; Pemberton 2008). In part, this is likely due to the very low resolution offered by the limited number of markers used in these early studies to estimate relationships. Indeed, using a much larger set of markers ($\sim$800 SNPs), Santure et al. (2010) obtained much better behavior over a 20-generation pedigree of a captive zebra finch population. They still, however, recommended using markers to estimate one- or two-generation pedigree links (i.e., sibs, father-offspring), which are then assembled into a pedigree matrix (e.g., connecting grandsons to grandfathers through separately estimated grandfather-to-father and father-to-son linkages), as opposed to simply using the marker relationship matrix directly. As Example 20.6 suggests, this may be less of a limitation when one scores thousands of markers. Lopes et al. (2013) found that roughly 2,000 SNPs worked well when comparing marker- and pedigree-based estimates in pigs, while Rolf et al. (2010) suggested that 2,000 to 10,000 markers would be required to construct reasonable molecularly based pedigrees in cattle. Indeed, Bérénos et al. (2014) obtained very similar heritability estimates using pedigrees that estimated only parent-offspring relationships compared to those based on whole-genome relatedness at $\sim$40,000 SNPs in a wild population of Soay sheep.

Given the potential of very dense marker information to more accurately infer relationships, it has been suggested that most wild populations will soon have the potential to permit an animal-model style analysis, with **A** directly estimated from marker information alone (Moore and Kukuk 2002; Gienapp et al. 2017). Does this mean that, in the near future, animal-model analyses will be practical for many, or even most, natural populations? The answer is likely no, as even if **A** is estimated with complete accuracy, any analysis is still limited by the *variance among relationships* in the sample (Visscher and Goddard 2015). If the sample lacks sufficient diversity in links between relatives (**low variance in relatedness**), it will contain little information for an animal-model analysis (Thomas and Hill 2000; Thomas et al. 2002; Csilléry et al. 2006). For example, if one randomly samples a very large population over multiple generations, there is a reasonable expectation that very few, if any, relatives will be found. Although a very large sample might suggest significant power, if the true relationship matrix for the population sample is of the form $\mathbf{A} = \sigma^2(A)\,(\mathbf{I} + \epsilon\,\mathbf{B})$, where $\mathbf{B}$ is a matrix of off-diagonal elements (indicating sets of relatives in the sample) and $\epsilon \ll 1$ (meaning that any off-diagonal elements are very small), then practically speaking, the sample consists of unrelated individuals (**A** is essentially a diagonal matrix). With little information from relatives, the power of a mixed-model analysis vanishes, as most breeding values are estimated solely from an individual's own phenotype. Balancing this pessimistic view are two studies on free-living fish that spend part of their time in the open ocean, which found sampled individuals to be enriched for close relatives (Thériault et al. 2007; DiBattista et al. 2009).

Example 20.6. We now possess the ability—either through dense SNP chips or by whole-genome sequencing—to score thousands to millions of SNPs, which offers a very simple approach for obtaining the relationship matrix, **A**. Given their very low mutation rates, two SNP alleles that are **alike in state** (**AIS**), or, equivalently, show **identity by state** (**IBS**), can be viewed as also being identical by descent (IBD) with respect to some ancient base population (Speed and Balding 2015), allowing us to compute the coefficient of coancestry, θ_{ij} (LW Chapter 7), directly from the SNP data, and hence the entry $A_{ij} = 2\theta_{ij}$ in the relationship matrix.

The use of dense marker data highlights the important distinction between **pedigree kinship** and **realized kinship** (Wang et al. 2017). The value of θ calculated using a known pedigree (e.g., LW Chapter 7) is the *expected* value of the kinship, given the relationship between two individuals. However, with the exception of clones and parent-offspring pairs (which always shared *exactly* one allele IBD), all other relationships have some *variation* in the fraction of alleles shared about their expected kinship value due to Mendelian sampling (Risch and Lange 1979; Suarez et al. 1979; Stam 1980; Guo 1996; Visscher et al. 2006; VanRaden 2007, 2008; Hill and Weir 2011). Consider outbred full sibs. At any given locus, the probability that a pair shares 0, 1, or 2 IBD alleles is 1/4, 1/2, and 1/4, respectively. In a pedigree approach, all pairwise A_{ij} values among full sibs would be the same (1/2), which is the expectation of 2θ for noninbred full sibs (LW Chapter 7). This is the pedigree kinship. However, there is variation in the actual fraction of shared IBD alleles, so that (for example) sibs 1 and 2 may have a realized value of $2\theta_{12} = 0.55$, while 1 and 3 have a realized value of $2\theta_{13} = 0.42$. Dense SNP data capture this variation in relatedness, giving more accurate weights when using information from relatives (replacing expected values by their actual realizations). This is the basis of the genomic selection method known as **G-BLUP** (genomic-BLUP), wherein a marker-estimated (**genomic relationship**) matrix is used in place of a pedigree relationship matrix for **A** to improve the BLUP estimates (e.g., VanRaden 2007, 2008; Hayes et al. 2009; Volume 3 examines this method in detail).

There are a large number of proposed methods that translate SNP data into estimates of θ. The basic approach is as follows. Consider two individuals, x and y. We denote the two alleles in x by a and b (which may be alike in state), and similarly in y by c and d. The **molecular similarity** at locus ℓ between x and y is defined by

$$S_{xy,\ell} = \frac{I_{ac} + I_{ad} + I_{bc} + I_{bd}}{4} \tag{20.23a}$$

where I_{ad} is an indicator function that equals one if a and d are AIS, and otherwise is zero. For diallelic loci (such as most SNPs), $S_{xy,\ell}$ takes on values of 0, 1/2, or 1. A value of 1/4 requires at least three distinct alleles, and values of 3/4 do not occur as, if the first three combinations are one, so is the last (Oliehoek al. 2006). Toro et al. (2002) referred to Equation 20.23a as **molecular coancestry**, as when AIS equals IBD, then $E[S_{xy,\ell}] = \theta_{xy}$, with the average over all loci giving an estimate of the elements of the relationship matrix

$$\widehat{A}_{xy} = 2\widehat{\theta}_{xy} = \frac{2}{L}\sum_{\ell=1}^{L} S_{xy,\ell} \tag{20.23b}$$

While simple, the issue with this estimator is the equating of AIS with IBD. In order to adjust for AIS status, one needs to assign a base (or reference) population and use the expected genotype frequencies in this base population as the correction for AIS.

Specifically, suppose we let s_ℓ denote the probability that two randomly drawn alleles in the base population are AIS. Obviously, s_ℓ is (at a minimum) a function of the allele frequencies at ℓ. As shown by Lynch (1988c), the expected value for $S_{xy,\ell}$ is given by

$$E[\,S_{xy,\ell}\,] = \theta_{xy} + (1-\theta_{xy})s_\ell \tag{20.23c}$$

For a diallelic locus in Hardy-Weinberg,

$$s_\ell = p_\ell^2 + (1-p_\ell)^2 = 1 - 2p_\ell(1-p_\ell)$$

Rearranging Equation 20.23c suggests a more general estimator

$$\widehat{\theta}_{xy} = \frac{1}{L}\sum_{\ell=1}^{L} \frac{S_{xy,\ell} - s_\ell}{1-s_\ell} \tag{20.23d}$$

where L is the number of SNPs for which x and y contain no missing data. Negative estimates of θ can arise when $S_{xy,\ell} < s_\ell$ over a large number of loci, implying that these individuals are less related than expected by chance. Assuming $s_\ell = 0$ eliminates this problem (the assumption behind Equation 20.23b), but also introduces bias (Speed and Balding 2015; Ackerman et al. 2017). Oliehoek al. (2006) obtained an adjusted value for s_ℓ to ensure that all the $\widehat{\theta}_{xy}$ are nonnegative, but again this likely introduces some slight bias.

Alternatively, one can base a kinship estimator on the total number of shared AIS alleles over L loci (Day-Williams et al. 2011). First, we define

$$S_{xy} = \sum_{\ell=1}^{L} S_{xy,\ell} \tag{20.23e}$$

Summing Equation 20.23c over all loci yields

$$E[\,S_{xy}\,] = \sum_{\ell=1}^{L} E[\,S_{xy,\ell}\,] = \sum_{\ell=1}^{L} [\theta_{xy} + (1-\theta_{xy})s_\ell] = L\theta_{xy} + (1-\theta_{xy})\sum_{\ell=1}^{L} s_\ell \tag{20.23f}$$

Rearranging yields the **Day-Williams** estimator

$$\widehat{\theta}_{xy,DW} = \frac{S_{xy} - \sum_{\ell=1}^{L} s_\ell}{L - \sum_{\ell=1}^{L} s_\ell} \tag{20.23g}$$

Wang et al. (2017) showed that Equations 20.23d and 20.23g are special cases of a general weighting scheme of per-locus information, which ignore any linkage disequilibrium among SNPs. They also examined estimators that included information from SNP LD patterns.

Closely related to these approaches are estimators of θ based upon the average correlation among alleles (Cotterman 1940; Malécot 1948). This equivalence of θ with a correlation immediately suggests why some estimates can be negative, as, unlike a probability, a correlation can be negative, and $\theta < 0$ simply suggests that the two individuals are more *dissimilar* than expected by chance. To proceed, code the two alleles at a given SNP locus (ℓ) as **0**/**1** and let the random variable b_ℓ denote the value of a randomly drawn allele from this SNP, where $E[b_\ell] = p_\ell$ is the frequency of allele **1**, and $\sigma^2(b_\ell) = p_\ell(1-p_\ell)$. Analogous to Equation 20.23c, the probability that a randomly drawn allele from x and y are both **1** is

$$\Pr(b_{x,\ell} = b_{y,\ell} = 1) = \theta_{xy}p_\ell + (1-\theta_{xy})p_\ell^2 \tag{20.23h}$$

which rearranges to

$$\theta_{xy} = \frac{\Pr(b_{x,\ell} = b_{y,\ell} = 1) - p_\ell^2}{p_\ell(1-p_\ell)} = \frac{E[b_{x,\ell}\, b_{y,\ell}] - (E[b_\ell])^2}{p_\ell(1-p_\ell)} = \frac{\sigma(b_{x,\ell}, b_{y,\ell})}{\sigma(b_{x,\ell})\sigma(b_{y,\ell})} = \text{corr}(b_{x,\ell}, b_{y,\ell}) \tag{20.23i}$$

namely, the correlation between a random allele in x and a random allele in y.

Similarly, we can consider the correlation in allelic copy number between x and y. Let $T_{x,\ell}$ denote the number of copies of allele **1** (from SNP ℓ) that x carries, where $T_{x,\ell} = 0, 1,$ or 2. Hence, $E[T_\ell] = 2p_\ell$, yielding a contribution to the covariance in T_ℓ between two relatives of

$$(T_{x,\ell} - 2p_\ell)\,(T_{y,\ell} - 2p_\ell)$$

The variance in T_ℓ is $\sigma^2(T_\ell) = E[T_\ell^2] - (E[T_\ell])^2$, which under, Hardy-Weinberg, becomes

$$\left[0^2 \cdot (1-p_\ell)^2 + 1^2 \cdot 2 \cdot p_\ell(1-p_\ell) + 2^2 \cdot p_\ell^2\right] - (2p_\ell)^2 = 2p_\ell(1-p_\ell)$$

yielding a correlation in T between x and y at SNP ℓ of

$$\text{corr}(T_{x,\ell}, T_{y,\ell}) = \frac{\sigma(T_{x,\ell}, T_{y,\ell})}{\sigma(T_{x,\ell}) \cdot \sigma(T_{y,\ell})} = \frac{(T_{x,\ell} - 2p_\ell)\,(T_{y,\ell} - 2p_\ell)}{2p_\ell(1-p_\ell)} \tag{20.23j}$$

To relate $\text{corr}(T_{x,\ell}, T_{y,\ell})$ with θ_{xy}, we write $T_{x,\ell} = b_{x,\ell} + b'_{x,\ell}$, where b and b' represent the two SNP alleles in x, to yield that

$$\sigma(T_{x,\ell}, T_{y,\ell}) = \sigma(b_{x,\ell} + b'_{x,\ell}, b_{y,\ell} + b'_{y,\ell}) = 4\sigma(b_{x,\ell}, b_{y,\ell})$$

Substituting this result into Equation 20.23j and recalling Equation 20.23i yields

$$\text{corr}(T_{x,\ell}, T_{y,\ell}) = \frac{4\sigma(b_{x,\ell}, b_{y,\ell})}{2p_\ell(1-p_\ell)} = 2\left[\frac{\sigma(b_{x,\ell}, b_{y,\ell})}{p_\ell(1-p_\ell)}\right] = 2\text{corr}(b_{x,\ell}, b_{y,\ell}) = 2\theta_{xy} \tag{20.23k}$$

Summing over uncorrelated SNPs returns

$$\widehat{A}_{xy} = 2\widehat{\theta}_{xy} = \frac{1}{L}\sum_{\ell=1}^{L} \frac{(T_{x,\ell} - 2p_\ell)\,(T_{y,\ell} - 2p_\ell)}{2p_\ell(1-p_\ell)} \quad \text{for } x \neq y \tag{20.23l}$$

Yang et al (2010) showed that a slight modification is required when considering the coancestry of x with itself,

$$\widehat{A}_{xx} = 1 + \frac{1}{L}\sum_{\ell=1}^{L} \frac{T_{x,\ell}^2 - (1+2p_\ell)T_{x,\ell} + 2p_\ell^2}{2p_\ell(1-p_\ell)} \tag{20.23m}$$

where the term following the one is the estimate of inbreeding. Note that this term could be negative, implying that the individual shows less homozygosity than expected (Ackerman et al. 2017). Two technical comments are in order concerning these estimators. First, formally speaking, the correlations are between deviations of allelic counts with their expected values under the *assumed base population*, for example, $\text{corr}(b_{x,\ell} - p_\ell, b_{y,\ell} - p_\ell)$ and $\text{corr}(T_{x,\ell} - 2p_\ell, T_{y,\ell} - 2p_\ell)$, as p_ℓ may differ from the average allele frequency, $\widehat{p}_\ell$, in the sample. Second, Yang et al. (2010) showed that Equations 20.23l and 20.23m are slightly biased due to sampling error and suggested a regression-based correction (Yang et al. 2011a).

There are a number of subtleties in connecting AIS/IBS with IBD, largely dealing with how one defines a base population, whose descendants have alleles that are IBD. As detailed in Powell et al. (2010), IBD estimators that adjust AIS data for marker-allele frequencies assume a particular base population (set by the assumed allele frequencies, p_ℓ) as the point of reference for IBD estimates. Using the currently observed allele frequencies sets the current population as the base. Powell et al. noted that the objective of most IBD estimators is to predict the AIS status at unobserved loci. In our case, these unobserved loci are the QTL underlying the traits of interest. If these causative alleles have a different allele-frequency spectrum than the marker alleles being used to estimate IBD (for example, the trait has been under selection, with most sites having rare alleles), this introduces some error. IBD estimates from pedigree data (which estimates IBD independent of the underlying genetic structure) are robust to this effect.

Consequences of Pedigree Errors

Because the strength of a mixed-model analysis arises from accurately borrowing information from relatives, pedigree errors result in both bias and a loss of power. Generally speaking, there are two types of errors: (i) missing a link (setting relatives to unrelated); and (ii) incorrectly linking unrelated individuals together. In natural populations, there is a strong bias toward maternal connections, in that (depending on the biology of the focal species) many mother-offspring connections will be found and the incorrect assignment of maternity is unlikely (but certainly possible). In contrast, assigning fathers is much more problematic, with many offspring potentially having unassigned fathers, even when their true fathers are in the sample. Missed or incorrect paternal assignments have the effect of making what the model assumes to be unrelated individuals more similar (when the father is in the sample but not assigned to its offspring) and related individuals less similar (an incorrect father is assigned). This generally reduces heritability estimates. More importantly, there are significant implications for the detection of maternal effects. If most pedigree errors are paternal, then offspring will tend to resemble their mothers more than their fathers, resulting in a false signal of maternal effects (Postma and Charmantier 2007). Proper estimation of maternal effects requires a significant number of correct paternal links, as these allow direct and maternal effects to be disentangled (Chapter 22).

The consequences of misassigned paternities are a function of the pedigree structure and the trait heritability. In animal-breeding designs, where there is a great excess of mothers (dams) over fathers (sires), the effects can be substantial. In beef cattle, simulations by Lee and Pollak (1997) revealed a significant reduction in the estimated heritability (0.1 versus the true value of 0.3) when 20% of the sires were misidentified. Their pedigree structure had roughly 2% sires and 22% dams, and the rest were nonparents. In contrast, simulations by Charmantier and Réale (2005), assuming roughly equal percentages of sires and dams, found that if the rate of misassigned paternity was under 20%, then the bias in h^2 was modest (a relative error of less than 15%). Similar findings were reported by Firth et al. (2015). For Soay sheep (*Ovis aries*), Milner at al. (2000) found that estimates of h^2 were higher when using a pedigree with 95% confidence on a paternity than when based on a pedigree with 80% paternity confidence. Working with morphological traits in Darwin's finches (*Geospiza fortis*), Keller et al. (2001) found that not accounting for maternal effects introduced a much greater bias than did extra-pair matings (i.e., misassigned paternities).

Estimates of individual predicted breeding values (PBVs) are more impacted by incorrect pedigree links than are estimates of h^2, which is critical as PBVs are the key in assessing direct selection on breeding values. As seen with the prediction of a breeding value from a single phenotype, $\widehat{a} = h^2(z - \mu)$, the effect of the heritability is to shrink an estimate back toward the mean breeding value (zero). If h^2 is large, most of the phenotypic deviation is retained by the estimate, while if h^2 is close to zero, all estimates are shrunk back to very near zero. Because pedigree errors typically result in underestimated heritabilities, this results in excessive shrinkage of PBVs (Geldermann et al. 1986; Israel and Weller 2000). As a result, the magnitudes of true extreme breeding values are underestimated (see Figure 20.4).

How does an investigator deal with all of this pedigree uncertainty? Henderson (1988) suggested that if a father is not known with certainty, it may be more efficient to include all possible sires (weighted by their paternal probabilities, e.g., weighting all possible sires equally) than to simply not include any sire-offspring linkages in the pedigree. He introduced the idea of an **average numerator relationship matrix** to accomplish this. Konigsberg and Cheverud (1992) applied this approach to estimate heritabilities of several craniometric traits in a macaque (*Macaca sinica*) colony on Cayo Santiagos. Here, mothers were known with certainty but sires were unknown, although field and social data could be used to exclude many possible sires for each offspring. If there were k possible sires, Konigsberg and Cheverud simply weighted them all with equal probability ($1/k$) and applied Henderson's method. The natural (and more sophisticated) extension of this idea is a fully Bayesian approach, wherein uncertainty in the pedigree estimates is directly built into the model, with the resulting marginal posteriors fully incorporating all of this uncertainty (e.g., Hadfield et al. 2006).

With all these potential uncertainties in the pedigree, performing a sensitivity analysis given its basic structure in a study is critical before applying an animal model. Following Morrissey et al. (2007), the investigator first assumes a rough pedigree framework for the study population and then randomly incorporates the types of pedigree errors suspected, given the biological system in question. Simulation studies can then be used to examine the power (the ability to estimate parameters) and sensitivity (how robust these estimates are in the face of pedigree errors). Software for such an analysis was developed by Morrissey and Wilson (2010). Using a framework pedigree for Soay sheep (*Ovis aries*), Morrissey et al. (2007) found that the simple animal model (where breeding values are the only random effect) was relatively robust to pedigree errors, but that when maternal effects were present, the results were more fragile (i.e., much more sensitive to pedigree errors). This finding is not surprising, as separating maternal and direct effects is fairly sensitive to the types of links in the pedigree. Quinn et al. (2006), using the pedigrees for two bird species, suggested a rough rule of thumb that at least three years and 100 individuals per year are required to estimate heritability with confidence. These numbers should be treated as lower bounds, as other sampled natural populations may contain fewer relatives than do the samples for Quinn et al.'s species, significantly reducing the power.

In summary, although there appears to be a wealth of tools for using molecular markers to assign relationships, using them as the sole means to reconstruct **A** is suspect unless the marker density is high. Rather, using field observations to first suggest potential relationships and then molecular markers to confirm them should provide fairly reliable **A** matrices (albeit culled for more distant relatives). In a multigenerational study, this approach can provide links across generations, and connecting these links over several generations can largely fill out the important entries in **A**. A further caveat is that, as mentioned in Chapter 19, BLUP and REML methods can be compromised by previous selection, which is exactly what is expected in natural populations. Given the generally smaller size and shallower depth (connectiveness) of wild pedigrees relative to those from much larger, designed breeding programs, a full and formal accounting of all uncertainty is critical, and our recommendation is that Bayesian approaches be used whenever possible.

Model Selection

A final concern when using wild pedigrees is the delicate issue of model selection—determining which additional random effects among a set of candidates should be incorporated (e.g., whether a common family effect should be added to the model). Typically one incorporates additional random effects when their associated variance component is significant or when the model fit is improved by some criterion (e.g., the likelihood ratio test or model selection statistics such as AIC or BIC; see Example 12.5). One interesting consequence of a Bayesian analysis is that unless a variance prior has some point mass at zero, the resulting posterior confidence interval will always exclude zero, resulting in all variance components (for which the MCMC converges) being significant in the sense of a value of zero being excluded. In such settings, goodness-of-fit measures such as BIC should be the criteria used for model selection.

The incorporation of additional random effects into a model may significantly change the interpretation of a key feature (such as the additive variance). The problem faced by an investigator is determining when to include such additional random effects, given that the power to declare a variance component significant is likely small. Further, it is not uncommon for some variance components in natural populations to be almost fully confounded (e.g., Kruuk and Hadfield 2007; Ovaskainen et al. 2008). Given the expected low power for a mixed model using a pedigree sampled from a wild population, the sensibilities of the investigator come into play. If the a priori feeling is that the biology of the trait dictates the inclusion of a specific random effect (such as a maternal effect), this is best left in the model. Its inclusion will not result in bias estimates of the breeding values, but it may result in a

Table 20.2 Summary of best practices for examining common evolutionary questions on response to selection within the animal-model framework. BV is the true breeding value (A) of the focal trait, PBV is the predicted BV from an animal model ($\widehat{a}$), and w denotes relative fitness. Full details are included in the text.

Task: **Conducting Robertson consistency tests based on BV-fitness associations**

Problem: The variance of PBVs is less than the additive variance of the trait (Postma 2006). This biases Rausher's test (Equation 20.20) for direct selection on the focal trait, as the slope, $b_{\widehat{a}}$, of a fitness-PBV regression overestimates the slope, b_A, of the fitness-BV regressions (Equation 20.25b).

Solution: Use a bivariate (trait, relative fitness) animal model and frame tests in terms of REML variance components, for example, Morrissey's test (Equation 20.21c).

Problem: If trait BVs are estimated with a univariate animal model, $\sigma(\widehat{a}, w)$ is a biased estimate of $\sigma(A, w)$, as when A and w are correlated, these must be estimated *jointly*, which is not the case when using a univariate model (Hadfield 2008). This biases Postma's test for direct selection on the focal trait (Equation 20.18).

Solution: Use a bivariate animal model (focal trait, fitness) to estimate S_z and S_A directly from REML variance components (Equations 20.28a, 20.28b, 20.29c, and 20.29d).

Task: **Detecting genetic trends using temporal regressions based on PBVs**

Problem: The error structure associated with PBVs is of GLS, not OLS, form, with heteroscedastic and correlated residuals. This results in strongly anticonservative tests, with p values highly biased toward smaller values (Hadfield et al. 2010).

Solution: Use a Bayesian posterior for the regression slopes (Figure 20.6; Example 20.7).

reduction of power. At a minimum, one strategy is to present analyses under a series of models (even if the resulting variance components may not be significant) if they result in substantially different interpretations of key results.

APPLYING MIXED MODELS TO NATURAL POPULATIONS

Animal models have been used to address two important questions regarding the nature of the response to selection on a focal trait. First, is there evidence of selection on the breeding values of that trait, and if so, is it consistent with selection largely on the phenotype of the focal trait (as opposed to some other target)? Second, is there any genetic trend (change in mean breeding values) in the data? Historically (during the first decade of the 2000s), these tasks were accomplished by regressing relative fitness on individual PBVs to address selection on the breeding value of a trait and by regressing population mean PBVs on year (or generation number) to detect any genetic trend. As detailed below, both of these approaches are flawed when simply using the PBVs directly, but both questions can be safely addressed within an animal-model framework with the appropriate adjustments (Table 20.2).

Consistency Tests: Accuracy, Reliability, and Caveats when Using PBVs

Many of the initial applications of animal models in natural populations used individual PBVs in tests of selection on breeding, versus environmental, values (Equations 20.18 and 20.20), for example, fitting the Rausher-Simms regression using PBVs, $w = 1 + \beta_A\,\widehat{a} + \beta_E\,\widehat{e} + \epsilon$. However, Postma (2006) noted that there are significant problems with using an estimated value, $\widehat{a}$, in place of the true breeding value, A (to conform with the literature, we use an italicized A for the true breeding value, $\widehat{a}$ for the PBV, and a bold roman $\mathbf{A}$ for the relationship matrix). His central point was that when the pedigree adds little additional information, the PBV for an individual is almost entirely determined by its phenotype. The resulting

PBVs are biased by environmental values influencing the phenotypes of individuals, which confounds their ability to separate A from E. Further, as shown by Hadfield (2008, Hadfield et al. 2010), PBVs are correlated with each other, and this correlation structure must be taken into account for proper inference.

To quantify these concerns, we first need to consider several related measures of the uncertainty of predicted breeding values. Their **accuracy**, ρ, is the correlation between the predicted, $\widehat{a}$, and actual, A, breeding values (Chapter 13), while the **reliability**, ρ^2, is the fraction of variation in A accounted for by the PBVs. When using only the phenotype of an individual to obtain its PBV, $\rho = h$ (Chapter 13). The difference, $\rho^2 - h^2$, between the reliability of a particular PBV and the heritability is a measure of how much additional information is provided from relatives (i.e., from the pedigree). Expressing a PBV as its expected value (the true breeding value) plus an uncorrelated residual, $\widehat{a} = A + \epsilon$, then $\sigma^2(\widehat{a} - A) = \sigma^2(\epsilon)$ is the **prediction-error variance** (**PEV**; Chapter 19). When the PEVs are small, the PBVs will be very close to the true breeding values.

These measures of uncertainty (ρ and PEV) in the PBVs are connected as follows. By construction, A and the residual ϵ are uncorrelated, implying $\sigma(\widehat{a}, A) = \sigma(\widehat{a}, \widehat{a}+\epsilon) = \sigma^2(\widehat{a})$. This allows us to express the PEV as

$$\sigma^2(\epsilon) = \sigma^2(\widehat{a} - A) = \sigma^2(\widehat{a}) - 2\sigma(\widehat{a}, A) + \sigma^2(A) = \sigma^2(A) - \sigma^2(\widehat{a}) \tag{20.24a}$$

Second, the definition of a correlation yields the accuracy as

$$\rho = \frac{\sigma(\widehat{a}, A)}{\sqrt{\sigma^2(\widehat{a})\,\sigma^2(A)}} = \sqrt{\frac{\sigma^2(\widehat{a})}{\sigma^2(A)}} \tag{20.24b}$$

Hence,

$$\sigma^2(\widehat{a}) = \rho^2\sigma^2(A) \tag{20.24c}$$

Because $\rho^2 \leq 1$, it immediately follows that $\sigma^2(\widehat{a}) \leq \sigma^2(A)$.

We have already seen a hint of this result, in that individual PBVs are shrunk back toward their expected values, A, reducing their variance relative to the variance of their true values (Figure 20.4). These results imply that the correlation between the PBV and the phenotypic value of the trait, z, is greater than the correlation between A and z, as $\sigma(\widehat{a}, z) = \sigma(A - \epsilon, z) = \sigma(A, z)$, hence

$$\begin{aligned} |\,\mathrm{corr}(\widehat{a}, z)\,| &= \frac{|\,\sigma(\widehat{a}, z)\,|}{\sqrt{\sigma^2(\widehat{a})\,\sigma^2(z)}} = \frac{|\,\sigma(A, z)\,|}{\sqrt{\sigma^2(\widehat{a})\,\sigma^2(z)}} \\ &= \frac{|\,\sigma(A, z)\,|}{\rho\sqrt{\sigma^2(A)\,\sigma^2(z)}} = \frac{|\,\mathrm{corr}(A, z)\,|}{\rho} \geq |\,\mathrm{corr}(A, z)\,| \end{aligned} \tag{20.24d}$$

where we use corr to denote a correlation, as ρ here is restricted to a very specific correlation, namely, that between $\widehat{a}$ and A (the accuracy). Equations 20.24a and 20.24c show the connection between ρ and the PEV, with

$$\sigma^2(\epsilon) = \sigma^2(\widehat{a} - A) = \left(1 - \rho^2\right)\sigma^2(A) \tag{20.24e}$$

Finally, for $z = A + E$, Postma (2006) showed that the prediction-error variance equals the covariance between the PBV and the environmental (or more generally, residual) deviation, E, so that we can also write

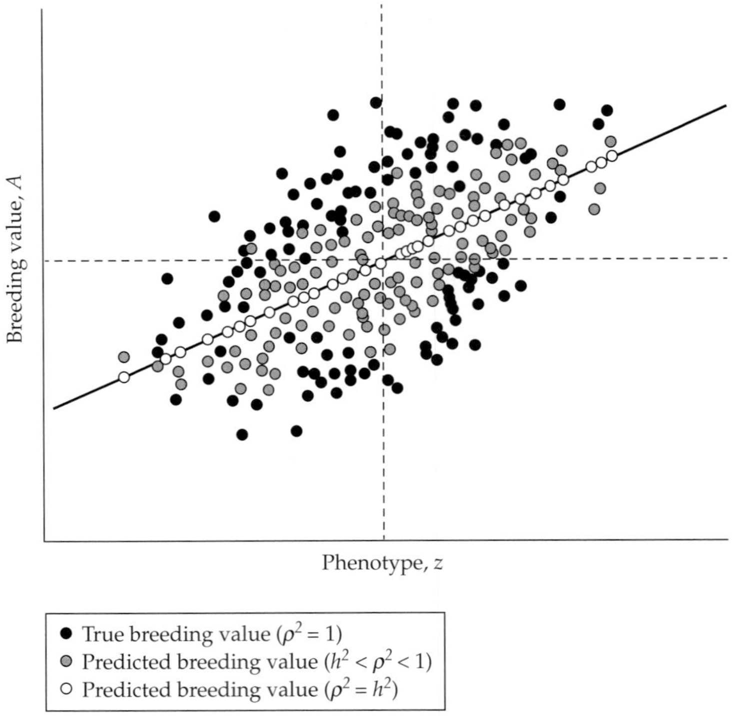

Figure 20.4 Comparison of predicted breeding values (PBVs) as a function of phenotypic value, z, based on different amounts of information from relatives. With only a single observation (i.e., no relatives), PBVs (open circles) show no variation about their predicted value of $\widehat{a} = h^2(z - \mu)$, and $\rho^2 = h^2$. As more information from relatives accrues, $\rho^2 > h^2$, and the PBVs (gray-filled circles) become less dependent on an individual's phenotype, showing greater spread about the regression. Now, for a given phenotypic value, z, the residual variance of a PBV around its mean predicted value of $h^2(z - \mu)$ is $(\rho^2 - h^2)\sigma^2(A)$. When $\rho = 1$ (PBV = true BV, shown in the figure as black circles) this spread around z is $(1 - h^2)\sigma^2(A)$, which can be substantial when heritability is low, as z is a poor predictor of A. (After Postma and Charmantier 2007.)

$$\sigma^2(\epsilon) = \sigma(\widehat{a}, E) = \left(1 - \rho^2\right)\sigma^2(A) \tag{20.24f}$$

When $\rho^2 < 1/2$, the implication is that PBVs resemble individual *phenotypes* more than they do the *true breeding values* (Figure 20.4), and hence PBVs are ***biased by environmental influences on the phenotype***. In particular, Equation 20.24f shows that the prediction error is the covariance of an individual's environmental value and its PBV. When $\rho^2 \sim 1$, the environmental value has no impact on the estimate, $\widehat{a}$, but when ρ^2 is modest to small, the PBV is a mixture of the true breeding value, A, plus an error reflecting environmental effects.

Figure 20.4 illustrates this phenomenon in a plot of predicted breeding values versus phenotypic values. When the predicted breeding value is based solely on the phenotypic value, z, of a single individual (open circles), there is no variation about its predicted value of $h^2(z-\mu)$, as all individuals with the same z value have the same predicted breeding value. In this case, while there is no variance in the PBVs for a given z (as all PBVs are the same), the residual variance between z and the *true* breeding value can be considerable, namely $\sigma^2(\epsilon) = \left(1 - h^2\right)\sigma^2(A)$. This follows from Equation 20.24e and the fact that $\rho^2 = h^2$ when only z is used to predict A. With a standard heritability of around 0.3, this amounts to 70% of the additive variance remaining unaccounted for by the PBVs. When PBVs are based on information from the phenotypes of relatives (in addition to an individual's phenotype), the reliability ρ^2 exceeds h^2, and influenced by this information from relatives, the predicted values start to vary about their mean value of $h^2(z - \mu)$ for a given z, giving a more accurate picture of the true values. The influence of relatives is to make an individual's predicted breeding value less dependent on its own phenotypic (and hence, environmental) value. The residual variance for the true breeding value for a given phenotypic value, z, is $(\rho^2 - h^2)\sigma^2(A)$, which is maximized at $(1-h^2)\sigma^2(A)$, when the reliability is complete ($\rho = 1$). Postma and Charmantier (2007) noted that ρ^2 is often around 0.5 for wild pedigrees, so that roughly half of the estimate of a typical PBV is influenced by the environment (Equation 20.24f).

The prediction-error variance for any specific PBV can be obtained from Equation 19.5c, which returns the covariance matrix of the PEVs for each of the estimated breeding values (also see Meyer 1989; Tosh and Wilton 1994). The ith diagonal element of this covariance matrix yields a prediction-error variance for individual i of

$$PEV_{ii} = (1 - \rho_i^2)\sigma^2(A) \tag{20.24g}$$

and hence a prediction reliability of

$$\rho_i^2 = 1 - \left(\frac{PEV_{ii}}{\sigma^2(A)} \right) \tag{20.24h}$$

The degree to which ρ_i^2 exceeds h^2 is a measure of the amount of additional information on that individual's breeding value provided by the pedigree (i.e., the additional information beyond that provided by the phenotype of the focal individual). A more subtle, but equally important, point is that the PEV matrix is not a diagonal, but rather contains numerous nonzero off-diagonal elements, as the PBVs for relatives are correlated (because they are based on shared information). Hence, the residual-error structure for the Rausher-Simms regression (Equation 20.20) is complex, requiring GLS, not OLS, regression (Chapter 18; LW Chapter 8). We return to this point below, as it has consequences for tests of genetic trends.

These results also have important implications for populations under selection. If an individual is lost before it leaves offspring (i.e., its fitness is zero), it will have fewer links in the pedigree than higher-fitness individuals who survive to leave offspring. The breeding-value predictions for individuals of low fitness are thus expected to have lower reliabilities than for individuals with higher fitness, and are thus more influenced by environmental values than those for higher-fitness individuals (Postma 2006). This can bias estimates of the amount of selection on breeding value.

Postma (2006) noted that besides the prospect of differential bias in PBVs for individuals with different fitnesses, a critical component of Rausher's consistency condition (Equation 20.17b) fails to hold when this regression is based on $\widehat{a}$. Recall that when selection is entirely on phenotypic values (i.e., when Equation 20.16b is satisfied), then

$$b_A = \frac{\sigma(w, A)}{\sigma^2(A)} = \frac{h^2 \sigma(w, z)}{h^2 \sigma^2(z)} = b_z \tag{20.25a}$$

namely, the univariate regression slope is the same when relative fitness is regressed on either breeding (A) or phenotypic (z) value. However, when using predicted breeding values ($\widehat{a}$) in place of A, recalling Equation 20.24c yields

$$b_{\widehat{a}} = \frac{\sigma(w, \widehat{a})}{\sigma^2(\widehat{a})} = \frac{h^2 \sigma(w, z)}{\rho^2 h^2 \sigma^2(z)} = \frac{b_z}{\rho^2} \tag{20.25b}$$

Hence, the slope of the regression of relative fitness on PBV ($\widehat{a}$) differs from that using phenotypic value (z). Thus, when $\rho^2 < 1$ (as would always be expected in wild pedigrees), $b_{\widehat{a}} > b_A$, showing that a selection gradient based on predicted breeding values ($b_{\widehat{a}}$) overestimates the gradient expected for true breeding values (b_A), thus compromising the Rausher-Simms equality test (Equation 20.20) for selection entirely on phenotypes.

To circumvent this problem, Postma (2006) suggested that the consistency test $h^2 S_z = S_A$ (Equation 20.18) be used instead, as when selection is based entirely on the phenotype of the focal trait, then

$$S_A = \sigma(A, w) = S_{\widehat{a}} = \sigma(\widehat{a}, w)$$

A caveat with this approach is that $\sigma(\widehat{a}, w)$ is a biased estimate of $\sigma(A, w)$ in a univariate animal model because the breeding values are estimated *separately* from fitnesses (Hadfield 2008). When the components are estimated separately, $\sigma(\widehat{a}, w)$ is unbiased only when Equation 20.16b is satisfied (i.e., exactly one of the Robertson consistency conditions is tested). Fortunately, a solution to both these of problems is achieved by *jointly* modeling the trait and fitness in a bivariate animal model, and then using REML-estimated variance components for consistency tests, as detailed in the following section.

Bivariate Animal Models: REML Estimates of $\sigma(A_z, A_w)$, S_A, and S_z

LW Chapter 26 introduced the multivariate animal model, in which a set of potentially correlated traits is measured in a pedigree of individuals. For trait j, a standard animal model is fit,

$$\mathbf{y}_j = \mathbf{X}_j\mathbf{b}_j + \mathbf{Z}_j\mathbf{a}_j + \mathbf{e}_j \tag{20.26a}$$

where $\mathbf{X}_j$ and $\mathbf{Z}_j$ are, respectively, the design and incidence matrices (Chapter 19; LW Chapter 26) associated with trait j, $\mathbf{a}_j$ is the vector of breeding values for trait j, and the covariance model for the random effects associated with j is

$$\begin{pmatrix} \mathbf{a}_j \\ \mathbf{e}_j \end{pmatrix} \sim \begin{pmatrix} \mathbf{0} \\ \mathbf{0} \end{pmatrix}, \begin{pmatrix} \sigma^2(A_j)\,\mathbf{A} & \mathbf{0} \\ \mathbf{0} & \sigma^2(e_j)\,\mathbf{I} \end{pmatrix} \tag{20.26b}$$

Consider the bivariate case, where $\mathbf{y}_1 = \mathbf{z}$ is the vector of phenotypes for the trait of interest and $\mathbf{y}_2 = \mathbf{w}$ is the vector of corresponding relative fitnesses, so (z_i, w_i) are the values of the focal trait and relative fitness for individual i. The resulting bivariate mixed model becomes

$$\begin{pmatrix} \mathbf{y}_1 \\ \mathbf{y}_2 \end{pmatrix} = \begin{pmatrix} \mathbf{z} \\ \mathbf{w} \end{pmatrix} = \begin{pmatrix} \mathbf{X}_z & \mathbf{0} \\ \mathbf{0} & \mathbf{X}_w \end{pmatrix} \begin{pmatrix} \mathbf{b}_z \\ \mathbf{b}_w \end{pmatrix} + \begin{pmatrix} \mathbf{Z}_z & \mathbf{0} \\ \mathbf{0} & \mathbf{Z}_w \end{pmatrix} \begin{pmatrix} \mathbf{a}_z \\ \mathbf{a}_w \end{pmatrix} + \begin{pmatrix} \mathbf{e}_z \\ \mathbf{e}_w \end{pmatrix} \tag{20.27a}$$

Note that the structure of this model allows the trait and fitness to have different fixed effects. The model can logically be extended to include more complex designs, such as maternal effects or repeated measures (e.g., Morrissey et al. 2012). The resulting covariance structure for the stacked vector of breeding values is

$$\boldsymbol{\sigma}(\mathbf{a}) = \boldsymbol{\sigma}\begin{pmatrix} \mathbf{a}_z \\ \mathbf{a}_w \end{pmatrix} = \begin{pmatrix} \sigma^2(A_z)\mathbf{A} & \sigma(A_z, A_w)\mathbf{A} \\ \sigma(A_z, A_w)\mathbf{A} & \sigma^2(A_w)\mathbf{A} \end{pmatrix} = \mathbf{G} \otimes \mathbf{A} \tag{20.27b}$$

where $\otimes$ denotes the Kronecker (or direct) product (LW Chapter 26) and

$$\mathbf{G} = \begin{pmatrix} \sigma^2(A_z) & \sigma(A_z, A_w) \\ \sigma(A_z, A_w) & \sigma^2(A_w) \end{pmatrix} \tag{20.27c}$$

is the matrix of genetic covariances of interest. Similarly, the covariance structure for the stacked vectors of residuals is

$$\boldsymbol{\sigma}\begin{pmatrix} \mathbf{e}_z \\ \mathbf{e}_w \end{pmatrix} = \mathbf{E} \otimes \mathbf{I}, \quad \text{where} \quad \mathbf{E} = \begin{pmatrix} \sigma^2(e_z) & \sigma(e_z, e_w) \\ \sigma(e_z, e_w) & \sigma^2(e_w) \end{pmatrix} \tag{20.27d}$$

Finally, we need to specify any covariances between $\mathbf{a}$ and $\mathbf{e}$. By construction $\sigma(A_z, e_z) = \sigma(A_w, e_w) = 0$, while the standard assumption is $\sigma(A_z, e_w) = \sigma(A_w, e_z) = 0$, yielding an assumed covariance structure of

$$\boldsymbol{\sigma}\begin{pmatrix} \mathbf{a}_z \\ \mathbf{a}_w \\ \mathbf{e}_z \\ \mathbf{e}_w \end{pmatrix} = \begin{pmatrix} \mathbf{G} \otimes \mathbf{A} & \mathbf{0} \\ \mathbf{0} & \mathbf{E} \otimes \mathbf{I} \end{pmatrix} \tag{20.27e}$$

The resulting six variance components for **G** (Equation 20.27c) and **E** (Equation 20.27d) are estimated by REML, and these components are our main interest. For example, $\sigma(A_z, A_w)$ is the expected selection response under the 1968 version of Robertson's theorem (Equation 6.24a). From the assumption that $\sigma(A_z, e_w) = \sigma(A_w, e_z) = 0$,

$$S_z = \sigma(z, w) = \sigma(A_z + e_z, A_w + e_w) = \sigma(A_z, A_w) + \sigma(e_z, e_w) \tag{20.28a}$$

is a direct estimate of the selection differential on the phenotypic value (z) of the focal trait, while

$$S_A = \sigma(A_z, w) = \sigma(A_z, A_w + e_w) = \sigma(A_z, A_w) \tag{20.28b}$$

is an estimate of the selection differential on the breeding value (A_z) of the focal trait. By using these variance components directly, we avoid the pitfalls associated with working with individual PBVs, and Postma's (Equation 20.18) and Morrissey's (Equation 20.21c) consistency conditions are easily computed from these variance components. Ideally, this is done within a Bayesian setting, so that the posterior distribution reflects all of the model uncertainty (from pedigree estimation on down). For example, draws from the MCMC sampler for such an analysis (Appendix 3) can be used to compute the posterior distribution of (say) $S_A - h^2 S_z$, and thus test whether there is sufficient support to include (consistent with $S_A = h^2 S_z$) or exclude (consistent with $S_A \neq h^2 S_z$) a zero value for the difference, $S_A - h^2 S_z$.

Removing the assumption that $\sigma(A_z, e_w) = \sigma(A_w, e_z) = 0$, the covariance structure given by Equation 20.27e is replaced by

$$\boldsymbol{\sigma}\begin{pmatrix} \mathbf{a}_z \\ \mathbf{a}_w \\ \mathbf{e}_z \\ \mathbf{e}_w \end{pmatrix} = \begin{pmatrix} \mathbf{G} \otimes \mathbf{A} & \mathbf{C} \\ \mathbf{C}^T & \mathbf{E} \otimes \mathbf{I} \end{pmatrix} \tag{20.29a}$$

where

$$\mathbf{C} = \begin{pmatrix} \boldsymbol{\sigma}(\mathbf{a}_z, \mathbf{e}_z) & \boldsymbol{\sigma}(\mathbf{a}_z, \mathbf{e}_w) \\ \boldsymbol{\sigma}(\mathbf{a}_w, \mathbf{e}_z) & \boldsymbol{\sigma}(\mathbf{a}_w, \mathbf{e}_w) \end{pmatrix} = \begin{pmatrix} \mathbf{0} & \sigma(A_z, e_w)\mathbf{A} \\ \sigma(A_w, e_z)\mathbf{A} & \mathbf{0} \end{pmatrix} \tag{20.29b}$$

These two additional variance components allow us to use the more general consistency condition given by Equation 20.21b. Likewise, to apply Postma's test (Equation 20.18), the phenotypic and breeding-value selection differentials now become

$$S_z = \sigma(A_z, A_w) + \sigma(e_z, e_w) + \sigma(A_z, e_w) + \sigma(A_w, e_z) \tag{20.29c}$$

and

$$S_A = \sigma(A_z, A_w) + \sigma(A_z, e_w) \tag{20.29d}$$

Next-generation Analysis: Generalized Mixed Models

This bivariate analysis of a trait and fitness makes two critical assumptions that are starting to be relaxed by using the more sophisticated analyses that are possible within a Bayesian framework. The first is that the standard mixed model assumes normally distributed random variables. Such is clearly *not* the case with fitness, in which random variables are often expected to be **zero-inflated**, with a point mass at zero fitness (which can be substantial). One way to handle such data is to use a zero-inflated Poisson distribution (a Poisson with an additional point mass at zero; Chapters 14 and 29). As discussed in Chapter 29, zero-inflated models are one attempt to deal with **overdispersion**. If count data follow a Poisson, then the mean and variance should be equal. Overdispersion occurs when the variance exceeds the mean, which can occur if there are an excess of zeros. Overdispersion can also occur if there are an excessive of other values, and Chapter 29 examines models that allow for both zero inflation *and* any additional overdispersion that is present after accounting for any excess of zeros. When fitness does not follow a normal distribution, the standard general linear model is replaced by generalized linear models (Chapters 14 and 29), which extend to **generalized mixed models** (Bolker et al. 2009; de Villemereuil et al. 2016) when the distribution of random effects is nonnormal.

The basic strategy in applying a generalized linear model was seen in our discussion of threshold models in Chapter 14. Here, on some underlying (latent) scale, an individual's breeding and environmental values determine its liability value, which is then translated into a realization of the trait value. For the threshold-trait model, if the liability exceeds a critical value, a particular trait value is seen. More generally, the liability value on this latent scale can be the parameter for some distribution from which a realization is drawn. In a fitness setting, suppose that an individual's liability value is 1.5; then the number of offspring it has is (say) a random draw from a Poisson with a mean of 1.5. As with a standard mixed model, we can borrow information on (liability) breeding values from the relatives of a focal individual to estimate the additive variance on this latent scale (see Example 20.7).

As detailed in Chapter 29, while a zero-inflated Poisson may be an improvement over assuming normality, in reality fitness is expected to have a more complex distribution. The Aster model approach detailed in Chapter 29 allows for a more realistic fitness distribution to be constructed by assuming a set of appropriate distributions over each fitness episode (e.g., Bernoulli for survival data, zero-inflated and/or overdispersed discrete distributions for number of mates, etc.). Bayesian approaches (using MCMC samplers) can deal with these more complex and realistic distributions.

Next-generation Analysis: Modeling Missing Data

By their very nature, fitness data have missing values. Individuals may not be recaptured (even when still alive), the number of mates and offspring may be undercounted, and selection on a trait may occur before we measure it for the first time in a cohort. The nature of the process causing certain data to be absent is critical in determining whether ignoring these missing values yields biased estimates for parameters of interest (Little and Rubin 2002; Hadfield 2008; Nakagawa and Freckleton 2008). Consider a simple case, where egg production, y, is a linear function of body size, x, and where we have body-size measures for all individuals in a sample but some of the egg-production values are missing (Figure 20.5). Following the terminology of Little and Rubin (2002), the missing y values are said to be **missing completely at random** (**MCAR**) when the process generating the missing values is entirely independent of either the x or y values (i.e., the missing values are simply a random sample of the population; Figure 20.5B). In such cases, the standard process of simply ignoring data with missing values (i.e., discarding those x observations with missing y) does not bias estimates of the relationship between x and y.

In contrast, the y values are said to be **missing at random** (**MAR**) when the process generating the missing data depends on the observed (x) values, but *not* on the associated y values. For example, suppose individuals above a certain body size have missing y values. The regression (of y as a function of x) ignoring these missing values recovers the same regression (within sampling error) as the full data set (Figure 20.5C). In other words, we can potentially take the observed data and use them to impute the missing data. Finally, suppose that the missing data are generated by a process that depends instead on y; for example, we do not score individuals whose egg production is outside of a certain range. Removing these data points (here based on truncated y values) results in biased estimates of the regression parameters (Figure 20.5D). Such data are said to be **missing not at random** (**MNAR**). MCAR and MAR data are said to be **ignorable**, in that estimates are not biased by the missing data. Conversely, MNAR data are **nonignorable**, and one requires a model of the process generating the missing data to obtain unbiased parameter estimates.

Quantitative-genetic considerations of missing data trace back to Im et al. (1989) in animal breeding and Hadfield (2008) in evolutionary genetics. In particular, Hadfield found that viability selection data are usually MNAR, implying that one must model the censoring process to yield unbiased estimates. He showed that modeling the survival process and

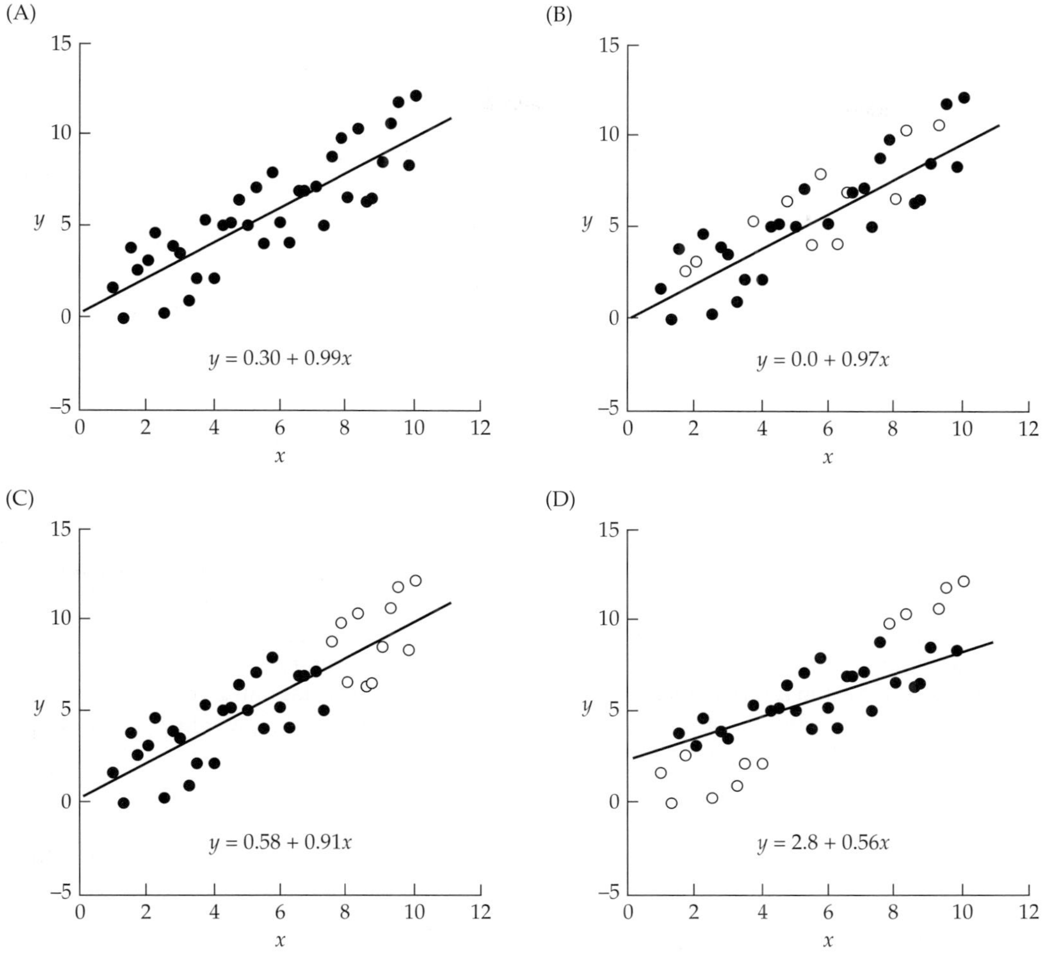

Figure 20.5 The different categories of missing data, as illustrated through a simple example where the missing data, y (vertical axes), are a linear function of x (horizontal axes). Regression lines using the observed data (filled circles) are illustrated. **A:** The uncensored full dataset. In the remaining three panels, roughly 20% of the original data is missing (unfilled circles). **B:** Here, y values were removed entirely at random. Such data are said to be missing completely at random (MCAR). **C:** The data are not missing at random, but rather missing as a function of x (the y values for large x values are missing). Note, however, that in this setting (because y is a function of x), our estimate of the relationship between x and y (as in **B**) is not biased. Such data are said to be missing at random (MAR). In the settings (**B**) and (**C**), namely MCAR and MAR data, the missing data are said to be ignorable, as estimates of the regression parameters are not biased by the missing data. **D:** Here the data are missing as a function of y (culled for extreme values). The resulting estimated regression is seriously biased in both its slope and its intercept. Such data are said to be missing not at random (MNAR), and said to be nonignorable, as estimates of the regression parameters *are indeed* biased by the missing data.

using an animal-model framework (wherein information can be borrowed from measured relatives) improves estimation. Papaïx et al. (2010) and Steinsland et al. (2014) developed missing-data extensions for capture-recapture analysis and for the invisible fraction (selection before traits are measured; Chapter 29), respectively. The coupling of non-Gaussian distributions for fitness with the modeling of the missing-data process represent an important extension towards more realistic models for the analysis of trait-fitness data from natural populations.

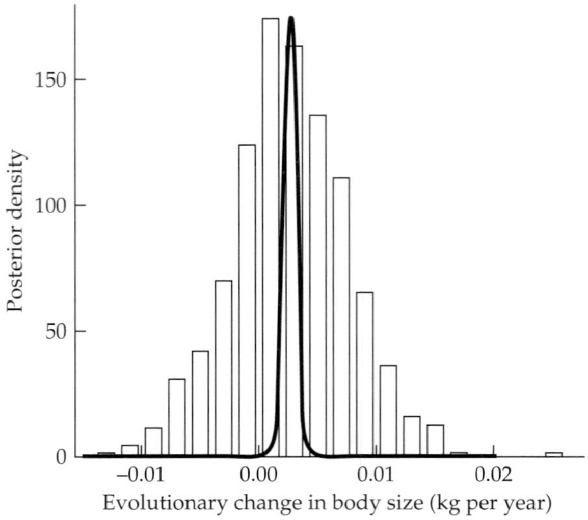

Figure 20.6 The anticonservative nature of using mean PBVs regressed on time to detect a genetic trend, with support for a trend given by a significantly nonzero slope for this regression. Here, the trait is body weight in Soay sheep. The peak of the thick smooth curve centered slightly to the right of zero represents the OLS estimate of the slope for the regression of mean BLUP-PBVs on generation number (i.e., time), and the spread of the curve represents the associated sampling error, which is almost entirely to the right of zero. This shows that the OLS slope estimate is significantly positive. Conversely, the histogram gives samples from the posterior distribution of slopes from a Bayesian analysis (details in the text). While this distribution has the same mean as the BLUP-based slope estimates, its variance is significantly greater. Indeed, 28% of the probability mass is less than zero, showing that the Bayesian estimate of the slope (which more fully accounts for the uncertainty and correlations among individual estimates) is not significantly different from zero. (After Hadfield et al. 2010.)

Detecting Genetic Trends

The gold standard for detecting a genetic change is to grow two populations in a common-garden experiment, ensuring that any change is genetic, rather than environmental. With the exception of the use of remnant seed (or, for other organisms where we can freeze or otherwise immortalize individuals), making contemporaneous comparisons of the genetic composition of different generations is usually not possible. However, as we saw in Chapter 19, it is often possible to make this comparison statistically through the animal model, *provided the population is sufficiently connected across the generations by sampled relatives*. Due to their pedigree depth and connectedness, this is reasonable for most breeding programs and artificially selected populations. The reliability of PBVs in such settings is fairly high, as these conditions reduce the environmental influence on PBVs. Further, PBVs are unbiased (provided individuals are randomized over environments), and taking their average for a sample (such as a generation) smooths out some of the environmental noise associated with individual estimates, thus allowing a plot of mean PBVs over time to show a trend (or lack thereof). In theory, this same approach can be applied to natural populations, provided the genotype-environment correlations are largely ignorable.

The more delicate issue is that of *inference*. As we saw in Chapter 18, both OLS and GLS regressions of the selection response on the selection differential gave unbiased *estimates* of the realized heritability, but because the residuals are heteroscedastic and correlated, the sampling variances under OLS are too small relative to the correct values under GLS (see Example 18.4). As a result, OLS-based tests are **anticonservative**; namely, the p values (testing for a potential genetic trend by regressing mean breeding value on time) are heavily biased toward smaller, more significant, values. The same issue arises with the error structure of PBVs (Hadfield et al. 2010), as Figure 20.6 illustrates.

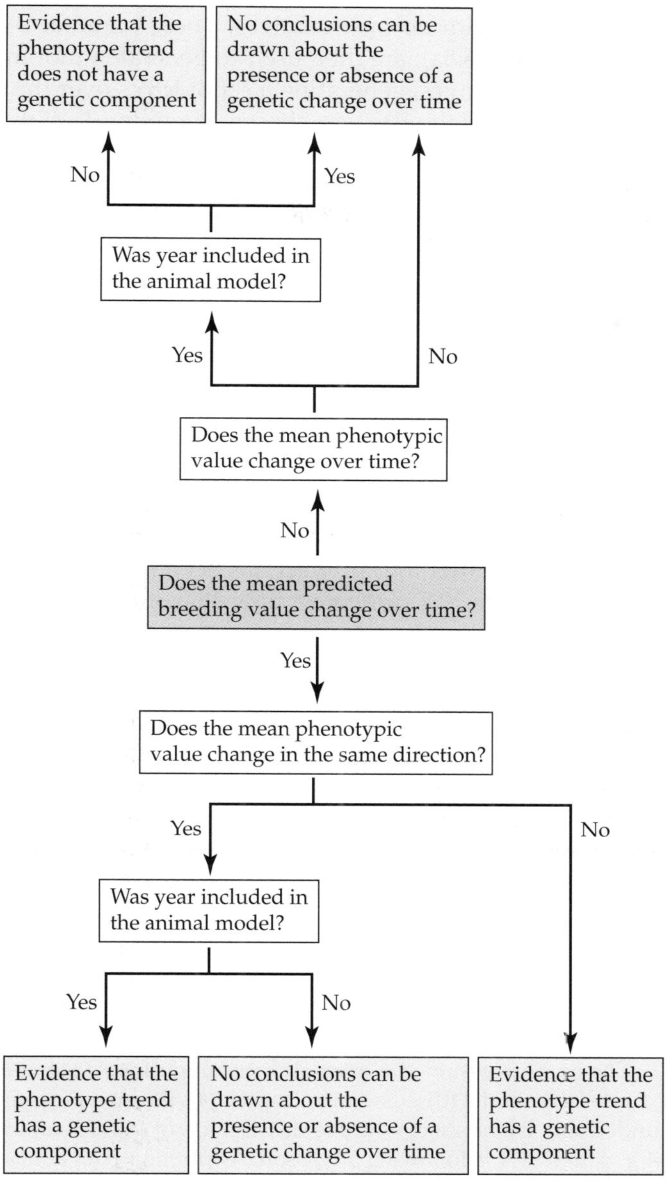

Figure 20.7 Postma and Charmantier's (2007) recommendations for the interpretation of genetic trends.

Fortunately, there is a simple solution (Hadfield al. 2010), again within the powerful Bayesian framework that (additionally) accounts for all levels of uncertainty in our analysis. A vector, $\widehat{\mathbf{a}}$, from the posterior PBV distribution is drawn from a given iteration of the MCMC sampler (Appendix 3). Then the resulting series of PBVs is used to regress (using OLS) mean PBV on generation number (i.e., time) and the resulting slope is recorded. This sampling process is repeated thousands of times to construct an empirical histogram of the posterior slope distribution, which is then used for inference (see Figure 20.6). The resulting empirical distribution of slopes fully accounts for the correlated structure among the PBVs.

Detection of a significant genetic trend is not, by itself, evidence of a selection response, as this could arise simply from drift. A powerful feature of the MCMC approach is that it allows for a straightforward test of drift versus selection (Hadfield et al. 2010). First, a value for $\sigma^2(A)$ is sampled from its marginal posterior. Breeding values are then drawn from a normal distribution (with a mean of zero, and the sampled value of $\sigma^2[A]$) and assigned to individuals in the pedigree with no past relatives. These parental BVs, coupled with Mendelian sampling (Equation 16.8a), are then used to generate random breeding values (RBVs) for their downstream relatives, generating a set of RBVs over the specific pedigree under the assumption of drift (as the BVs are chosen at random). With a set of RBVs in hand, a genetic trend is computed as above (regression of mean RBVs over generations), and the slope is recorded. Extracting many such samples (redrawing $\sigma^2(A)$ at the start of each run) generates an empirical histogram of the posterior slope distribution under drift alone, which can be compared with those generated from the MCMC using the full data.

An issue debated in the literature is whether one should also include a year (or generation) effect in models for trends. When PBVs have low reliability, they are strongly influenced by the environment, with the PBV trend partly reflecting any underlying environmental trend (Postma 2006). Incorporation of a year effect (a fixed value for each year or generation) can account for such an environmental trend, but it will also partly absorb any true genetic trend, thus reducing the power. Postma and Charmantier (2007) offered some guidelines as to how to proceed (Figure 20.7). Given that PBVs are biased toward an environmental trend when their reliability is low, the finding of genetic and environmental trends in opposite directions provides evidence of a genetic trend. If the change is in the same direction and still persists after a year effect has been incorporated, this is also supportive of a genetic trend. Figure 20.7 presents a flow chart for the interpretation of other combinations of trends.

A final issue is that if one has individuals throughout the pedigree with missing parents, these are treated as de facto founders and their BVs come from a common distribution of those older individuals at the top of the pedigree that really are biologically founders. This effect will generally dampen any inferred trend in breeding values (M. Morrissey, pers. comm.)

Example 20.7. Milot et al. (2011) examined the evolution of age at first reproduction (AFR) of human settlers in the isolated island of île aux Coudres in Québec. This island was populated by thirty families between 1720 and 1772. Because of careful church records, it has a very detailed chronicle of births, marriages, and deaths, which allowed the authors to construct a bivariate animal model for AFR and lifetime reproductive success (LRS). The latter is a proxy for total fitness and was defined in this study as the total number of a woman's offspring reaching age 15. The authors assumed a normal distribution for AFR but a latent Poisson model for LRS. Under this model (discussed in Chapter 14), the breeding value for AFR was defined on an underlying latent scale, with the distribution of LRS for an individual with a liability score of $y = \mu + A + e$ following a Poisson distribution with a mean of e^y. MCMC methods were used in a Bayesian analysis of this bivariate model, which found significant heritability in both AFR and LRS (the latter as measured on the latent scale). Further, these two traits showed a significant negative genetic correlation (posterior mode of -0.81, 95% credible interval of -0.97 to -0.48). Thus, Robertson's theorem (Equation 6.24a) suggests direct selection to reduce the age of first reproduction, which declined from $\sim$26 to $\sim$22 years over a 140-year period. The regression of predicted breeding value over time (measured as eight 20-year cohorts) is given in Figure 20.8, with the diamonds representing the average of 1,000 MCMC samples from the marginal posterior ($\pm$ their standard errors).

As also shown in Figure 20.8, the authors tested whether this trend could be due to drift, using the approach suggested by Hadfield et al. (2010) discussed above. Given the posterior distribution of the additive variance for AFR, RBVs were generated over the known pedigree, and the regression of the RBVs over time was compared to that for PBVs for each run of the sampler. The proportion of times during which the absolute regression slope of the RBVs exceeded the slope based on PBVs was taken as the posterior probability that the response is due to drift, and was found to be less than 0.01. The average slope for RBVs is given by the solid line in Figure 20.8.

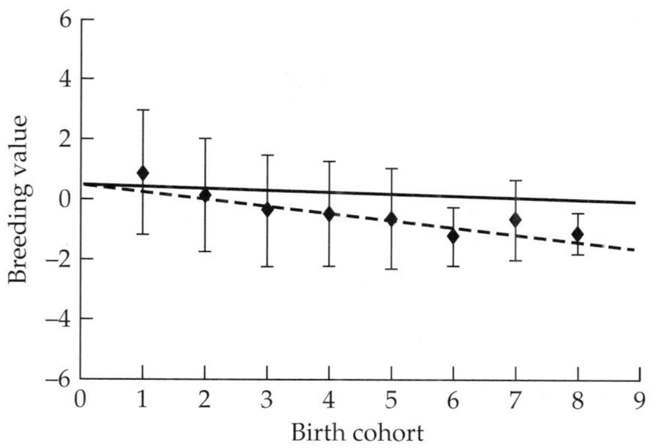

Figure 20.8 An application of Hadfield et al.'s (2010) approach for testing whether a genetic trend is significantly different from that expected under drift alone. Data are for age at first reproduction (AFR) for human settlers on an isolated island in Québec. Diamonds represent the average of 1,000 MCMC samples from the marginal posterior (± their standard errors) of the mean predicted breeding value for AFR for each of the eight 20-year cohorts. The regression of these mean PBVs on cohort number is given by the dashed line. As a test for drift, a posterior-sampled value for the additive variance is drawn, and then used to generate random breeding values (RBVs), given the pedigree structure. The average slope for mean RBVs is given by the solid line, showing that the observed PBV trend (dashed line) is more excessive than that expected by drift. This was confirmed by a comparison of the posterior distribution of regression slopes based on observed PBVs and RBVs (Example 20.7).

CAUSES OF APPARENT FAILURES OF RESPONSE IN NATURAL POPULATIONS

Given the above concerns on the suitability of the breeder's equation in natural populations, what do the data say? A detailed review of well-studied mammal and bird populations by Merilä et al. (2001c) noted a number of cases where: (i) there was a consistent selection differential on a particular trait; (ii) the trait was heritable; and (iii) yet no selection response (or worse, response in the opposite direction) was observed over a lengthy period (many generations) (Table 20.3). While there are several classic examples of natural populations responding either to imposed artificial selection (Example 18.1; Semlitsch and Wilbur 1989), a new environmental challenge (such as a habitat shift, major weather event, or introduction of a novel selective agent, e.g., Example 20.2; Losos et al. 1997; Reznick et al. 1997), or even to the introduction of a new competitor species (Grant and Grant 2006), Merilä et al. (2001c) lamented the apparent lack of response outside of these situations, raising the central question as to the basis for apparent **stasis** in response in the face of apparent selection.

Table 20.4 summarizes possible (and not mutually exclusive) explanations for this apparent failure, most of which have been discussed previously. The most obvious is that the phenotype of the focal trait is not the sole target of selection. However, an apparent lack of response could be as trivial as a lack of sufficient power to detect a small expected change (e.g., Gienapp et al. 2006; Postma et al. 2007). A related design issue is that most studies only sampled a small part of an open population, meaning that immigration and differential dispersal could either mask or enhance any local selection response (Garant et al. 2005).

Likewise, changing environments can result in multiple effects that can cause apparent stasis. First, the selection differential could be changing sign over generations, resulting in a net long-term differential close to zero. Second, deterioration in the environment can mask underlying genetic change. Third, when G × E interactions are present, heritabilities can

Table 20.3 Examples of natural populations of mammals and birds in which apparent strong directional selection on a heritable trait fails to show response. Duration is the length of the study (in years), and $\bar{\imath} = S/\sigma$ is the selection intensity. (After Merilä et al. 2001c and Gienapp et al. 2008.)

Species/Trait	h^2	$\lvert\bar{\imath}\rvert$	Response	Duration	Reference
Mammals					
Cervus elaphus (red deer)					
Antler mass	0.33	0.44	Opposite	29	Kruuk et al. (2000, 2002)
Birth mass (male)	0.11	0.40	No change		
Birth mass (female)	0.25	0.22	No change		
Ovis aries (soay sheep)					
Body mass (Male)	0.12	0.11	No change	12	Milner et al. (1999, 2000)
Body mass (Female)	0.24	0.07	No change		
Ovis canadensis (big horn sheep)					
Body weight	0.23	0.30	As expected	29	Coltman et al. (2003, 2005)
Horn length	0.39	0.33	As expected	26	
Tamiascurus hudsonicus (red squirrel)					
Parturition date	0.16	0.17	As expected	10	Réale et al. (2003)
Chiononmys nivalis (snow vole)					
Body Mass	0.17	0.21	Opposite	10	Timothée et al. (2017)
Birds					
Branta leucopsis (barnacle goose)					
Tarsus length (male)	0.53	0.03	Opposite	13	Larsson et al. (1998)
Tarsus length (female)		0.09	Opposite		
Anser caerulescens (snow goose)					
Clutch size	0.20	0.30	Opposite	20	Cooke et al. (1990)
Cygnus olor (mute swan)					
Clutch size	0.20	0.66	As expected	25	Charmantier et al. (2006)
Ficedula albicollis (collared flycatcher)					
Relative mass	0.30	0.23	Opposite	17	Merilä et al. (2001a, 2001b)
Tarsus length	0.52	0.12	No change	4	Alatalo et al. (1990)
	0.35	0.18	No change	17	Kruuk et al. (2001)
Breeding time	0.19	0.22	No change	19	Sheldon et al. (2003)
Cyanistes caeruleus (blue tit)					
Body mass	0.27	0.31	No change	14	Charmantier et al. (2004)
	0.35	0.42	No change	12	
Tarus length	0.47	0.27	No change	13	
	0.48	0.21	No change	12	
Parus major (great tit)					
Breeding time	0.17	0.21	No change	30	Perrins and Jones (1974) Gienapp et al. (2006)
Egg size	0.80	0.38	No change	7	Hõrak et al. (1997)
Fledging mass	0.24	0.21	Opposite	36	Garant et al. (2004)
	0.20	0.14	Opposite	36	Garant et al. (2005)
	0.29	0.18	No change	36	

change with the environment, raising the possibility of low heritabilities when selection is most intense. The tools developed in this chapter may help an investigator sort through these possible explanations, as the following case studies illustrate.

Table 20.4 Possible causes for an observed stasis in response, despite heritable variation and a significant selection differential.

Genetic response has occurred, but not is detected.
Low power to detect a genetic trend.
Genetic gain countered by environmental deterioration.
The focal trait is not the target of selection.
Trait and fitness are correlated through an environmental variable.
Selection on a phenotypically, but not genetically, correlated trait (i.e., $S \neq 0, \beta = 0$).
Consequence of open population structure.
Immigration from populations outside of the study area.
Consequence of fluctuating environmental conditions.
Fluctuating selection differential, with little net selection.
Fluctuating h^2, with smallest h^2 when selection is strongest.
Constraints and tradeoffs.
Direct response on a trait countered by correlated responses from other traits.
Measured fitness component is an incomplete measure of total fitness.

Cryptic Evolution: Genetic Change Masked by Environmental Change

One explanation for stasis is that change in the environment can dilute, and indeed even swamp, any underlying genetic gain. In the extreme, one can have **cryptic evolution**: significant genetic change that does not show up as phenotypic change because it is countered by environmental change. Levins (1968) and Conover and Schultz (1995) coined the phrase **countergradient variation** for situations in which the environmental trend is opposite to the direction of selection. Such situations can increase the strength of selection on a trait, as the population struggles to keep pace with the declining environment. In the extreme, a population faces the risk of extinction if the environment is deteriorating at a faster rate than the rate at which compensating trait values can evolve. Obviously, this is an important issue for populations when attempting to track climate change.

One striking example of apparent cryptic evolution is the study by Merilä et al. (2001b) on the Gotland population of collared flycatchers. These authors examined body condition (a measure of relative body weight) at fledging. They defined condition as the residual from the regression of body mass on tarsus (leg) length, and found that this trait has substantial heritable genetic variation (estimated h^2 = 0.30). Further, it appears to be under constant positive selection, with an average selection intensity of 0.23 (i.e., survivors, on average, were 0.23 standard deviations above the mean before selection). Despite the heritable nature of this trait, which was coupled with strong positive selection, mean condition *declined* over time (Figure 20.9), with the regression of mean condition from 1981 to 1999 showing a significant negative slope ($b = -0.036$ per year). Merilä et al. found that the covariance between fitness, w, and the breeding value of condition, A_z, was nonzero, meaning that selection for condition indeed occured directly on the breeding values. Because $\sigma(A_z, w) \neq 0$, Robertson's secondary theorem predicts a nonzero response (Equation 6.25a). Why, then, is there an the apparent lack of response? As shown in Figure 20.9, the regression of predicted breeding values on time has a positive slope (b = 0.0022), with the population showing genetic improvement, despite the mean phenotype declining over time. The environmental component of condition has been declining over time, and at a rate faster than the genetic improvement, resulting in a net phenotypic decline. Merilä et al. (2001b) suggested this is likely attributable to reductions in the caterpillar food supply due to climatic trends.

Apparent stasis can also actually mask a *decrease* in mean genetic value. Such an example was seen by Timothée et al. (2017), who examined body mass in snow voles (*Chiononmys nivalis*). While the predicted response (based upon the observed values of S and h^2)

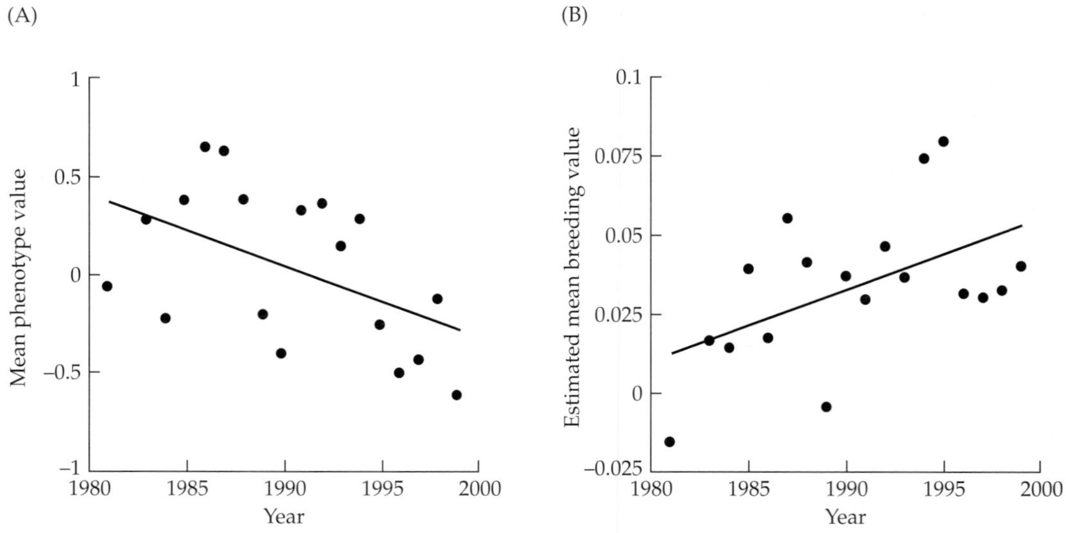

Figure 20.9 Body condition at fledging for a Gotland population of collared flycatchers between 1980 and 1999. **A:** Trend for the mean phenotype. **B:** Trend for the mean breeding value (the average of the individual PBVs in a given year). (After Merilä et al. 2001b.)

was for an increase of ~0.17 grams per year, the actual change over a nine year window was slightly positive, but not significant. A mixed-model analysis showed that mean breeding value had in reality significantly *decreased* over this period, with the apparent status in body mass the result of phenotypic plasticity. Selection appeared to be largely on rapidly maturity in juveniles, which leads to low potential adult mass.

Example 20.8. A second example of a negative environmental trend was offered by Larsson et al. (1998), who examined body size in the barnacle goose (*Branta leucopsis*). The natural colonization of the Baltic area of Sweden by this normally Arctic species started with a single breeding pair in 1971, followed by subsequent rapid increase in the population size. The authors studied the two largest Baltic colonies from 1984 to 1996. Head size and tarsus length were measured to extract a structural body-size index, and larger females were found to have larger, and earlier, clutches (with larger eggs, resulting in more and heavier young). Further, juvenile body mass was significantly positively correlated with post-fledging survival. Both size measures were highly heritable, but average body size declined over the 13-year study period (by 0.7 and 0.5 standard deviations for head and tarsus length, respectively). The authors concluded (from a variety of evidence) that the environment had declined due to growth of the colony. For example, the values of birds from the earlier (and smaller) colony were initially around a standard deviation larger than the values of birds from the older (and larger) colony. Thus, the declining environment seems to involve density-dependent effects on individual growth brought on by the overall success of the colony itself.

As Example 20.8 highlights, as organisms evolve, they can change at least part of their environment. Indeed, van Valen's (1973) **Red Queen hypothesis** states that organisms have to evolve just to stay where they are relative to the evolving biosphere around them. (The Red Queen, a character from Lewis Carroll's *Through the Looking Glass*, had to run simply to stay in place.) Cooke et al. (1990) suggested that the lack of response to selection on clutch size in birds may be explained by the Red Queen hypothesis. While Price and Liou (1989) suggested that apparent selection for clutch size is largely based on selection for nutritional state (Example 20.3), Cooke countered that birds with better-quality territories have larger clutches and higher fitness, so a component of selection for clutch size involves the ability to compete for such territories. Although competitive ability may increase over time, average territory quality, and hence average clutch size, remains relatively constant, and hence no response is seen (this is an example of selection of associative effects, examined in detail in Chapter 22). Under the Cooke et al. model, if one could hypothetically let ancestral and current populations compete for territories, current individuals, which possess higher breeding values for competitiveness, would win.

Building on this theme, Hadfield et al. (2011) noted that a number of distinct biological processes may present the appearance of cryptic evolution (no apparent change in mean phenotype, while the mean breeding value changes), such as sib competition and selection on a trait before it is scored (the so-called invisible fraction discussed in Chapter 29). In particular, they stress the importance of **evolutionary environmental deterioration**, namely, an adaptive response resulting in a harsher biotic environment (following Cooke et al. 1990 and Fisher 1958). We consider this process in detail in Chapter 22, in the context of selection under the joint action of both direct and associative genetic effects.

Antler Size in Red Deer: The Focal Trait Is Not the Target of Selection

Free-living red deer (*Cervus elaphus*) on the Isle of Rum in Scotland are another well-studied natural population with a largely complete pedigree spanning several decades. Males fight to compete for mates, suggesting antler size as a potential trait under selection. Males shed antlers in the early spring, and given that antler shape is very individual-specific, cast antlers found in the field can easily be assigned to a specific stag. Kruuk et al. (2002, 2014) found that males with larger antlers (measured as the mass of the annually shed antlers) had increased lifetime breeding success (total number of offspring), with a selection differential of $S/\sigma_z = \bar{\imath} = 0.445 \pm 0.094$. Although body size (measured by leg length) also had an effect on lifetime breeding success, antler size still had a significant effect on fitness even after accounting for body size, with a standardized (scaled to unit variance) selection gradient of $\beta = 0.44 \pm 0.18$. Antler size was heritable ($h^2 = 0.329 \pm 0.12$), and the breeder's equation would suggest a response of $R/\sigma_z = h^2 S/\sigma_z = 0.329 \cdot 0.445 = 0.146$ standard deviations per generation. Given a generation time of roughly 8 years and a standard deviation of 163 grams for antler mass, this suggests an expected change of roughly 2.3 grams per year. However, the average mass of antlers actually *declined* by 6.7 grams per year over the study period. One apparent reason for the decline was an environmental change due to increased population density over the study period, with antler size decreasing with increasing density.

Was this also a case of genetic change being masked by this environmental change? Apparently it was not. The REML estimate of the genetic correlation between lifetime breeding success and antler size was not significant, -0.254 ± 0.289. Thus the significant selection differential appears to be generated through selection on some feature that is phenotypically, but not genetically, correlated with antler size. The authors suggested that male fighting ability is, at least in part, a function of the nutritional condition of a male, and males with better nutritional value may be both better fighters and also more likely to grow larger antlers. Being better fighters, they have a greater lifetime reproductive success, which generates a correlation between antler mass and fitness.

Lower Heritabilities in Environments with Stronger Selection?

A more subtle implication of environmental change arises when genotype $\times$ environment interactions are present. As the environment changes, so can heritabilities, due to either changes in the environmental or in the genetic components of variance (Hoffmann and Parsons 1991, 1997a, 1997b; Hoffmann and Merilä 1999; Merilä and Sheldon 2001; Sgró and Hoffmann 2004; Charmantier and Garant 2005). There are some suggestions of a weak trend in the direction of change in h^2. Data from wild vertebrate populations show increased heritabilities for morphological traits in more favorable environments, while traits more closely associated with fitness show no pattern (Merilä and Sheldon 2001; Charmantier and

Garant 2005). Charmantier and Garant examined 46 traits, 38 of which showed no significant difference in heritabilities in good versus poor environments, but of the remaining 8 traits that were significant, all were higher in the more favorable environment. Roughly 65% of the traits showed decreased additive variation in less favorable environments, but most differences were not significant. In contrast, environmental variances tended to increase under poor conditions. Lower values of h^2 in more unfavorable environments suggests that there is less response in more stressful environments, exactly those that are likely to impose more selective pressures (Example 20.9).

These surveys of wild populations are at odds with older laboratory experiments in *Drosophila*, which found higher additive genetic variance (and heritabilities) in stressed environments (Hoffmann and Merilä 1999). While this observation may simply suggest that there are no general trends, it may also be a reflection of conditions in the lab versus the field. This observation is also consistent with **Holloway's conjecture**, which states that adaptive traits may show higher additive variance and reduced genetic correlations in novel environments, as an environmental change might disrupt the genetic architecture that evolved in response to a different set of environments (Holloway et al. 1990). One could view the laboratory *Drosophila* experiments in this light, in that many of the artificially imposed stresses create novel environments, especially for laboratory-adapted strains. Holloway's conjecture is also consistent with some results for Soay sheep, which exhibit smaller genetic correlations in more favorable environments (measured by first year population-wide survival) than in poorer environments (Robinson et al. 2009).

Example 20.9. Charmantier et al. (2004) examined chick tarsus length and body mass at fledging in three French populations of blue tits (*Cyanistes caeruleus*), two on the island of Corsica, and a third on the mainland. Their study followed ∼8000 banded chicks from ∼1200 individual broods, representing three different habitats, which the authors were able to rank in quality. They found that poorer habitats showed weak selection to increase tarsus length and strong selection to increase body mass, while in good habitats there was no significant selection on either trait. It is interesting that heritability for body mass increased with habitat quality, with the lowest heritability occurring in the poor habitats that experienced the strongest selection for increased body weight. In such low-quality environments, strong selection would be at least partly countered by lower heritabilities, leading to a weaker response.

A similar situation was observed by Wilson et al. (2006) for birthweight in Soay sheep (*Ovis aries*). These authors used a random-regression animal model (Volume 3), which allowed for the estimation of maternal performance over a continuous environmental variable (here the population-wide neonatal mortality for a given year). Harsh environmental conditions generated strong selection for higher birthweights but also resulted in a lower genetic variance in this trait. More benign environments resulted in weaker selection but higher birthweight heritabilities. A counterexample to these two studies was provided by Husby et al. (2011), who found, in a population of great tits (*Parus major*), that both the selection differential on earlier breeding and the heritability in this trait increased with increasing spring temperatures.

Fitness Tradeoffs and Multivariate Constraints

Finally, we note that our discussion of the selection response in natural populations has neglected two extremely important issues that will be addressed later. The first is the estimation of fitness, which is examined in detail in Chapters 29 and 30. Because lifetime (or total) fitness is very difficult to measure, a component of fitness (such as viability or fecundity) is often measured and assumed to be a faithful proxy of the total fitness. This, however, raises concerns about fitness **tradeoffs**, wherein a trait has a positive effect on one fitness component but a negative effect on another, so that its *net* effect on fitness is far less than expected when considering the effect of either component separately. Similarly, one might imagine sex-specific tradeoffs, wherein a trait has a positive fitness effect in one sex and a negative effect in the other. Because selection tends to remove additive variation in fitness (Chapters 5, 6, and 28), there is a widespread view that such tradeoffs likely occur. The logic behind this assumption is that selected alleles lacking tradeoffs are either quickly lost or fixed, so that alleles with roughly constant total-fitness effects segregate in a population for a longer amount of time, and hence might be expected to comprise some of the standing genetic variation. However, despite an obsession among ecologists and evolutionary biologists with such tradeoffs, they have proven rather elusive to detect (Chapters 29 and 30; Volume 3). A full discussion of this topic touches on the very vexing question of what maintains quantitative-genetic variation, which we return to in Chapter 28.

The second issue, **multivariate constraints**, is a potential concern because all selection response is inherently multivariate. A critical observation is that each component in a vector of traits under selection can have genetic variation ($h_i^2 \neq 0$), but the multivariate direction favored by response (the selection gradient vector, $\boldsymbol{\beta}$) may contain little, or no, usable additive variation. To see this, suppose that only two traits are under selection, yielding a response in trait 1 of $\sigma^2(A_1)\beta_1 + \sigma(A_1, A_2)\beta_2$ (Equation 20.1b). This shows that direct selection on the phenotype of trait 1 ($\beta_1 \neq 0$), coupled with heritable variation ($\sigma^2(A_1) \neq 0$), is *not* sufficient to ensure that $R_1 \neq 0$. If trait 1 is genetically correlated with trait 2, then for certain combinations of selection, namely, $\sigma^2(A_1)\beta_1 \simeq -\sigma(A_1, A_2)\beta_2$, the selection response in trait 1 is effectively zero. In this setting, even if the phenotype of trait 1 is a focal target of selection and the trait is heritable, there will still be essentially no response. It is worth noting that in this case, Robertson's theorem would predict a small to absent response, with $\sigma(A_z, w)$ being very close to zero. Although it is a univariate treatment, the covariance of the breeding value of trait 1 with fitness would be nearly zero in this case, as Robertson's theorem fully accounts for all of the genetically correlated traits that impact the focal trait-fitness covariance.

The general multivariate-constraint problem is examined in some detail in Volume 3, expressed in terms of matrix geometry (Appendix 5). A lack of response implies that the **G** matrix (of the genetic variances and covariances in the multivariate breeder's equation: Equation 13.23b) has eigenvectors (axes of variation) whose associated eigenvalues are close to (or at) zero, implying essentially no variation in these directions. If the angle between these nearly null eigenvectors of **G** and the direction, $\boldsymbol{\beta}$, favored by selection is very small, there is essentially no response in *any* of the component traits.

V

Selection in Structured Populations

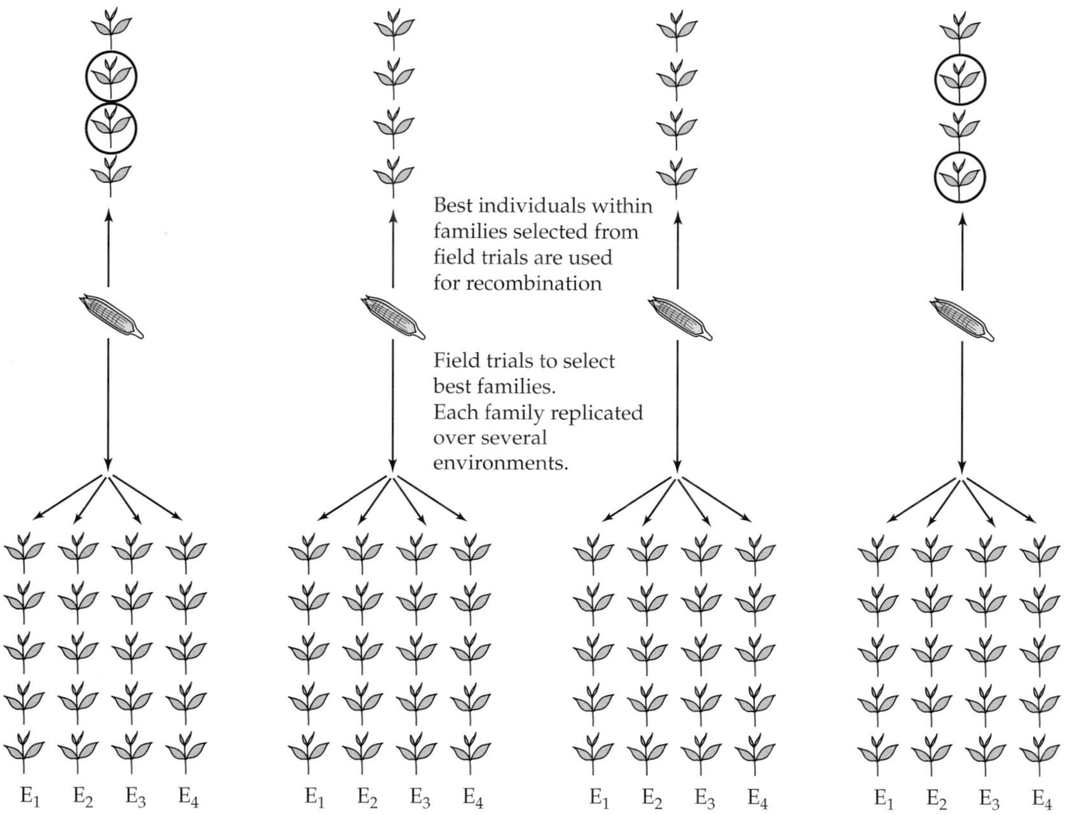

21

Family-Based Selection

Practical breeding programs must be commercially optimal,
not theoretically maximal. Fairfull and Muir (1996)

Up to now, we have focused on **individual selection**, wherein selection decisions are based solely on the phenotypes of single individuals (this is also referred to as **mass selection** or **phenotypic selection**, and we use all three terms interchangeably). Selection decisions can also incorporate the phenotypic values of an individual's measured relatives, and in fact, most plant- and animal-breeding schemes do so. The focus of this chapter is on **family-based** selection—using family information to select individuals. While we restrict discussion in this chapter to using sibs, the culmination of this approach is BLUP selection using an index based on the entire known pedigree of an individual, which is the major route for artificial selection in most domesticated animals (Chapters 13 and 19). While our focus here is on short-term selection response (formally, the single-generation response), certain family-based schemes can give a greater long-term response than individual selection, even when their initial response is less. This long-term advantage arises because of larger effective population sizes associated with selection schemes that down-weight among-family differences, a point examined in detail in Chapter 26.

There are a variety of reasons for using family-based schemes. Employing mass selection may be impractical in many settings due to difficulties in measuring trait values in single individuals (e.g., most forage crops and cereals). Family-based designs can also provide greater accuracy in predicting an individual's breeding value, and hence can give a larger (short-term) response. In particular, an appropriately weighted index of an individual's family mean and phenotypic value has an expected response at least as large as mass selection. When significant environmental heterogeneity exists (e.g., crops planted across a broad climatic range), the replication of families over environments provides a more efficient method than mass selection for choosing higher-performing genotypes. This is one major factor leading crop breeders to favor family-based schemes over individual selection.

The structure of this chapter is as follows. We start with a brief overview of the nature and types of family-based selection schemes before considering extensions of the generalized breeder's equation to accommodate these approaches. Next, we develop the variances and covariances required to apply these equations, and then consider a number of these schemes in detail. The relative efficiencies of within- and among-family selection compared to mass selection are then examined, followed by a consideration of designs in which families are replicated over environments, as is usually the case in plant breeding. We conclude by examining the properties of family-index selection. While most of the concepts in this chapter are straightforward, the bookkeeping can be tedious at times. Thus, we summarize key results at the end of various sections to allow the casual reader to more easily navigate through this material.

INTRODUCTION TO FAMILY-BASED SELECTION SCHEMES

Family-based designs are based on two approaches: **among-family** schemes, which choose entire families on the basis of their mean performance, and **within-family** schemes, which choose individuals based on their relative performance within their families. While many designs are based on just one of these components, the most general approach is **family-index selection**, wherein individuals are chosen based on a weighted index of among- and within-family components. Mass selection is a special case of a family index, where the within- and among-family components are weighted equally. (Although the phrase

Evolution and Selection of Quantitative Traits. Bruce Walsh & Michael Lynch.
Published 2018 by Oxford University Press.
DOI 10.1093/oso/9780198830870.001.0001

"between-family selection" is widely used in the literature, "between" refers to a comparison of two items, while "among" refers to comparisons of two or more, which is the general setting here. Hence, while between- and among- are used interchangeably in the literature, we will use "among" throughout our discussion here.)

While we assume that the parents in any particular family are from the same population, in some breeding settings the parents are from different populations. Examples of this are **interpopulation improvement** schemes, where the goal is to improve the performance of *hybrids* among populations (Volume 3). The focus in this chapter is family-based schemes for **intrapopulation improvement** (increasing the performance of the population under selection).

Overview of the Different Types of Family-based Selection

The key to making sense out of the bewilderingly large number of family-based designs in the literature is to consider the individual components that together define any particular scheme. The first component is the ***type of sib family providing information for selection decisions***. A family may consist of half-sibs, full-sibs, or full-sibs nested within half-sibs (e.g., the NC Design I; see LW Chapter 18). Sibs can also be generated by one (or more) generations of selfing (e.g., S_1, S_2), and we examine such families in Chapter 23. While the family-based schemes developed in this chapter are generally used with **allogamous** species (outcrossers, cross-pollination), they can also be applied to facultatively **autogamous** species (facultative selfers, self-pollination) through the use of controlled pollination and/or the introduction of male-sterile genes under open pollination (e.g., Gilmore 1964; Doggett and Eberhart 1968; Brim and Stuber 1973; Burton and Brim 1981; Sorrells and Fritz 1982).

Once a particular family type has been chosen, the second component is ***how sib data are used for selection decisions***. One could use among-family selection, choosing the best families (i.e., those with the largest family means). Alternatively, one could use within-family selection, choosing either the best individuals within each family (**strict within-family [WF] selection**), or the individuals with the largest *deviations* from their family means (**family-deviations [FD] selection**). While WF and FD selection are very similar, there are subtle differences between the two schemes, as they do not necessarily select the same individuals. One could also consider an index weighting both family mean and family deviations.

The final design component is the ***relationship between the measured sibs and the individuals serving as parents for the next generation***. Under either within-family or family-index selection, the selected individuals are used to form the next generation. However, in among-family selection, we can use any number of relatives of the chosen (selected) families to form the next generation. The most straightforward approach is to use measured sibs from the chosen families (**family selection**). However, some characters cannot be scored on living organisms, such as carcass traits in production animals, or can only be scored after reproduction. In such cases, one can use unmeasured sibs from the best families as the parents of the next generation (**sib selection**), which is often used to improve selection on sex-specific traits. For example, milk production can be selected for in males by choosing sires from families whose sisters show high levels of production. An important variant of sib selection is the use of **remnant seeds** from the best families, which are planted and subsequently crossed to form the next generation. In perennial species and in annual species that can be asexually propagated (cloned), one can select the best parents by the performance of their offspring (**parental selection** or **progeny testing**). Finally, an option available for facultatively autogamous species is to both self an individual to generate S_1 progeny (S_1 seeds) and likewise outcross it to one or more individuals to generate a family for testing. For such species, one can grow and intercross the remnant S_1 seed from the chosen families to form the next generation (the **S_1 seed design**).

Plant Versus Animal Breeding

While animal breeders typically employ only a few standard sib-based designs (Turner and

Young 1969), plant breeders can choose from a vast array of options (e.g., Hallauer and Miranda 1981; Schnell 1982; Nguyen and Sleper 1983; Wricke and Weber 1986; Hallauer et al. 1988; Aastveit and Aastveit 1990; Nyquist 1991; Vogel and Pedersen 1993; Holland et al. 2003; Hallauer et al. 2010). Furthermore, the final product desired by a plant breeder can vary considerably: it could be an open-pollinated population, an F_1 hybrid, a pure (i.e., fully inbred) line, or a synthetic line. Thus, it is not surprising that the literatures on family-based selection in the two fields are rather divergent. Much of the animal-breeding literature is expressed in terms of the phenotypic (t) and additive-genetic (r) correlations among sibs, while much of the plant-breeding literature is expressed in terms of variance components. As our discussion attempts to interweave both approaches, we will typically present selection-response equations in both forms.

Reproductive differences between plants and animals underlie many of the differences in the designs that are available to breeders. Historically, plant breeders have had more options than animal breeders because of the reproductive flexibility of many plants (i.e., selfing, stored seed, vegetation propagation; see Fehr and Hadley 1980). With the cloning of several domesticated animals, animal breeders now have the option of exploiting some of these classical plant-breeding schemes.

One obvious difference between plants and many animals is the ability to easily store progeny for many generations in the form of seed. Generally speaking, plants also produce far more offspring than domesticated animals, providing more offspring per family, and thus allowing for more extensive replication of families across environments. Another reproductive advantage of plants is that asexual propagation (cloning) is straightforward in many species, allowing individual genotypes to be preserved over generations.

Yet another key difference is in the control of crosses. While simple isolation will prevent most undesirable crosses in animals, either complete isolation or extensive manual control may be required to prevent pollination vectors from generating undesirable crosses in plants. When studying facultatively autogamous species, the investigator may be faced with either trying to prevent selfing or trying to prevent outcrossing, or to allow for both while identifying which seed came from which type of cross. Options for controlled crosses range from complete manual control over pollination at one extreme to open pollination at the other. Given that most plants have multiple flowers (which are often both numerous and small), large-scale controlled crosses can be much more labor intensive than similar crosses among animals, as hand pollination and the control of external and/or self-pollinators may be required. Even under **open pollination** (allowing seed plants to be pollinated at random), the investigator still has different levels of control over the pollen spectrum that a seed plant experiences. In a **test cross** or **topcross** design, the population of plants supplying the pollen is controlled. For example, individual maize plants can be **detasseled** by hand (removing the pollen-producing tassels) or have their tassels bagged to prevent the plants from either selfing or pollinating other plants. Such plants serve only as seed plants and are intergrown with rows of the **tester** strain, which provide the pollen. Under true open pollination, seed parents are randomly pollinated from the population, with no control of the pollen parent. A consequence of open pollination is that while most half-sib families in animal breeding are paternal, most half-sib families in plant breeding are maternal (they share a common seed parent).

There are also more subtle biological differences between plants and animals that drive differences in designs. While one can usually score many traits in individual animals, this is often not done in plants. For example, many traits of forage grasses, grains, and legumes are scored as **plot totals**, which involves measuring the mean performance of an entire family (or line) instead of each separate individual. When individuals cannot be directly scored, among-family selection is possible, but within-family and family-index selection are not. Similarly, many selected traits in plants can be scored only after reproduction (seed or fruit yield being prime examples), and this influences the types of relatives that can be used to form the next generation.

Among- Versus Within-family Selection

When the heritability of a trait is high, an individual's phenotype is an excellent predictor of its breeding value, and mass selection is more efficient than either strict within- or among-family selection. When heritability is low, individual phenotypic value is a poor predictor of breeding value, in which case an individual's family mean or its relative performance within its family may be better predictors.

The relative efficiencies of among- versus within-family selection depend on the relative magnitudes of the common-family (E_c) and individual-specific (E_s) environmental variances. A large common-family effect severely compromises the phenotype as a predictor of breeding value. However, *within* each family, all members share the same environmental effect, and differences between individuals more accurately reflect differences in breeding value. In this case, selection within families (for example, by choosing the largest individuals from each family) can yield a larger response than individual selection. Many mouse selection experiments use within-family selection, especially for traits with suspected maternal effects, such as body weight (Falconer and Latyszewski 1952; Falconer 1953, 1960a; Eisen and Hanrahan 1972; von Butler and Pirchner 1984; Nielsen and Anderson 1987; Siewerdt et al. 1999), litter size (Falconer 1960b), and nesting behavior (Lynch 1980).

Conversely, suppose that environmental effects unique to each individual account for a large fraction of the phenotypic variance ($\sigma^2_{E_s} \gg \sigma^2_{E_c}$). In this case, selecting whole families as units can give a larger response than individual selection, as the family mean averages out differences based on environmental values, revealing those families with the most extreme breeding values. An important example of this family-averaging of environmental effects is the use of among-family selection to improve performance across multiple environments. Under mass selection, a genotype is represented by a single individual in a single environment, while family-based approaches allow the performance of different families to be compared over multiple environments. Such studies are by no means restricted to plant breeding, as animal selection experiments examining phenotypic plasticity (norms of reaction), in which genotypes must also be assessed over multiple environments, almost exclusively use among-family selection (e.g., Waddington 1960; Kindred 1965; Waddington and Robertson 1966; Druger 1967; Scharloo et al. 1972; Brumpton et al. 1977; Minawa and Birley 1978; Scheiner and Lyman 1991).

DETAILS OF FAMILY-BASED SELECTION SCHEMES

Selection and Recombination Units

Under mass selection, individuals are scored and those with the best phenotypic values are used as parents to form the next generation. Here groups of individuals upon which selection decisions are based and those used for **recombination** (gamete production to form the next generation) are one and the same, and a single **cycle of selection** takes a single generation. In family-based selection schemes, the individuals used for selection decisions may be entirely separate from those used as parents to form the next generation. Further, a single cycle of selection may take two (or more) generations, as one must generate, score, and recombine families. For perennial species (such as forage crops), traits may be scored over several years before selection decisions are made, such as selecting for winter hardiness (Vogel and Pedersen 1993).

Following the convention of plant breeders, we distinguish between an individual, x_i, in the **selection unit** (those measured individuals upon which selection decisions are made, which throughout this chapter are we assume are sibs) and an individual, $\mathcal{R}_i$ (a relative of x_i, potentially including x_i itself), from the **recombination unit** (individuals serving as parents for the next generation) whose resulting offspring are y. Even though we may not *directly* select on the parents ($\mathcal{R}_1, \mathcal{R}_2$) of y, we expect some response in y due to the genetic correlation between x_i and $\mathcal{R}_i$ caused by their sharing of (at least) one common relative, P_i (Figure 21.1). An equivalent way to think about this distinction is that selection response occurs due to

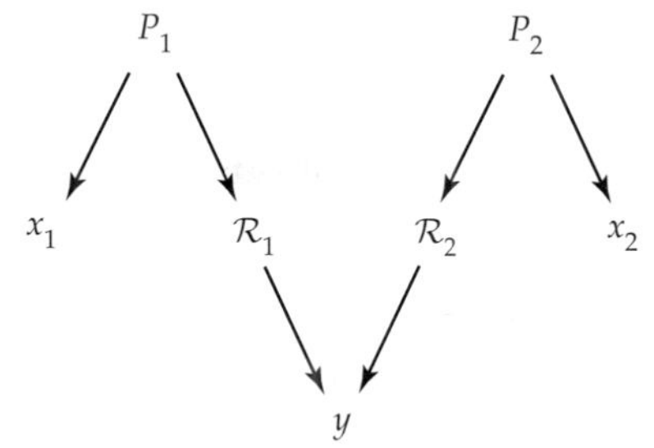

Figure 21.1 Under family-based schemes, selection decisions are based on some function of the values of measured sibs (x_i) in the selection unit. An offspring, y, in the next generation has parents, $\mathcal{R}_1$ and $\mathcal{R}_2$, that are chosen on the basis of the selection unit. Members of the selection (x_i) and recombination ($\mathcal{R}_i$) units are related as they both share a common relative, P_i, which in this case is the parent of sib x_i. Under within-family or family-index selection, $\mathcal{R}$ is simply one of the measured sibs, while under among-family selection, $\mathcal{R}$ is often an *unmeasured* relative. See Figure 21.2 for specific examples.

observations on the selection unit, x, providing information to predict the breeding value of $\mathcal{R}$.

As mentioned in the introduction, the variety of family-based schemes appearing in the literature arises from the combination of four specific components:

1. **Type of sib family** comprising the selection unit. Sibs can be half- or full-sibs, full-sibs nested within half-sibs (NC Design I), or selfed sibs (which are considered in Chapter 23).
2. **Nature of the selection decisions** based on the sib information. Selection can be based on sib-family means, the deviations of individuals within families, an index of both, or strict rank within families.
3. **Selection on one versus both parents.** Often selection decisions involve only one sex, with the parents of the opposite sex chosen at random (and hence being unselected). For example, a trait may not be scorable until after pollination, resulting in selection on seed parents (females) but not on pollen parents (males). In such cases, we are only concerned with one side of the pedigree, for example, involving $\mathcal{R}_1$ but not $\mathcal{R}_2$ (Figure 21.1). More generally, the two parents ($\mathcal{R}_1$ and $\mathcal{R}_2$) of the offspring, y, may be chosen using different schemes, which generates a variety of family-based schemes.
4. **Nature of the relationship between a measured sib, x_i, in the selection unit and a parent, $\mathcal{R}_i$, of the next generation.** Under within-family or family index selection, $\mathcal{R}$ is one of the measured sibs ($\mathcal{R}_i = x_i$), while under among-family selection, $\mathcal{R}$ is often an unmeasured relative. For example, $\mathcal{R}_i$ could be the parent of the sibs ($\mathcal{R}_i = P_i$), meaning that the relationship between x and $\mathcal{R}$ is that of parent-offspring, or it could be an unmeasured sib, meaning that the relationship between x and $\mathcal{R}$ is that of either half- or full-sib (depending on the type of family).

While the variety of family-based selection schemes may seem a bit overwhelming at first (especially in the plant-breeding literature), considering each design in terms of these four components greatly simplifies matters.

Variations of the Selection Unit

Once the type of family (half-sib, full-sib, nested, or inbred S_i) has been specified, there is still the issue of how to incorporate sib information when making selection decisions. To distinguish between a particular sib and the *trait value* of that sib, we use x_i to denote the ith sib and z_i to denote its trait value, and more generally, x_{ij} and z_{ij} for the jth individual from the ith family. We select the uppermost fraction, p, of the relevant population, with m families each with n sibs, for a total of $M = mn$ scored individuals, which we use to choose N parents. Four different approaches for weighting sib information are commonly used:

1. **Among-family selection**: Individuals are selected solely on the basis of their family means, $\overline{z}_i$, with the result that all individuals from the same family have the same selective rank. Here, the best $N = pm$ families are chosen.

2. **Strict within-family (WF) selection**. The best pn individuals from *each* family are chosen ($N = pnm = pM$), so that individuals are ranked within each family. WF selection increases the effective population size because the among-family variance in offspring number is zero (Chapters 3 and 26).

3. **Selection on within-family deviations (FD)**: Individuals are ranked solely on the basis of their within-family deviation, $z_{ij} - \overline{z}_i$. The $N = pM$ individuals with the largest deviations (regardless of family) are chosen.

4. **Family-index selection**: Individuals are ranked using an index weighting within- and among-family components

$$I = b_1\,(z_{ij} - \overline{z}_i\,) + b_2\,\overline{z}_i = b_1\,z_{ij} + (b_2 - b_1)\,\overline{z}_i$$

   The pM individuals with the best index scores are chosen. Note that the index with weights (cb_1, cb_2) chooses the same individuals as an index with weights (b_1, b_2). Thus, one of the index weights is often set to one, as the indices with weights (b_1, b_2), $(1, b_2/b_1)$, and $(b_1/b_2, 1)$ are all equivalent (in that they all choose the same individuals). Individual selection, among-family selection, and selection on family deviations (FD) are special cases, being indices with weights (b_1, b_2) = (1,1), (0,1), and (1,0), respectively. Note, however, that strict within-family (WF) selection cannot be expressed in terms of an index. Family-index selection is also referred to as **combined selection**, which is unfortunate, as the same term is also used by breeders to refer to approaches that combine *different types* of selection schemes in a single cycle (such as modified ear-to-row selection, discussed below).

The choice of the particular scheme has implications for the selection intensity (Example 21.1). When the fraction saved (p) is fixed, among-family and strict within-family selection have lower selection intensities than family-deviations, index, or mass selection. The former selects the best pm of m families and pn of n sibs, while the latter three select the best pM of M individuals. Because M is greater than either n or m, the finite-sample value for $\overline{\imath}$ is larger when sampling from M than from n or m (Chapter 14).

Example 21.1. Suppose that a total of 100 sibs are measured and the fraction that is selected is $p = 0.2$. As a benchmark, for this level of selection, the infinite-population value for the selection intensity is $\overline{\imath} = 1.40$ (Equation 14.3a). Suppose that the $M = 100$ total measured sibs are distributed into 20 families of five sibs each ($m = 20$, $n = 5$). Under within-family selection, the top 1 of 5 within each family is selected. Under among-family selection, the top 4 of the 20 families are selected. Finally, under family-deviations or index selection, the top 20 of the 100 measured individuals are selected. Using the finite-size correction approximation offered by Equation 14.4b yields the following selection intensities:

Individual selection (infinite population)	Best 20%	$\overline{\imath}_\infty = 1.40$
Individual selection, index selection, family-deviations (FDs) selection	Best 20 of 100	$\overline{\imath}_{(20,100)} = 1.39$
Among-family selection	Best 4 of 20	$\overline{\imath}_{(4,20)} = 1.33$
Strict within-family selection (WF)	Best 1 of 5	$\overline{\imath}_{(1,5)} = 1.16$

As shown later (Equations 21.40 and 21.57), additional corrections to the selection intensity are required in some cases, as family members are correlated, which changes the variance

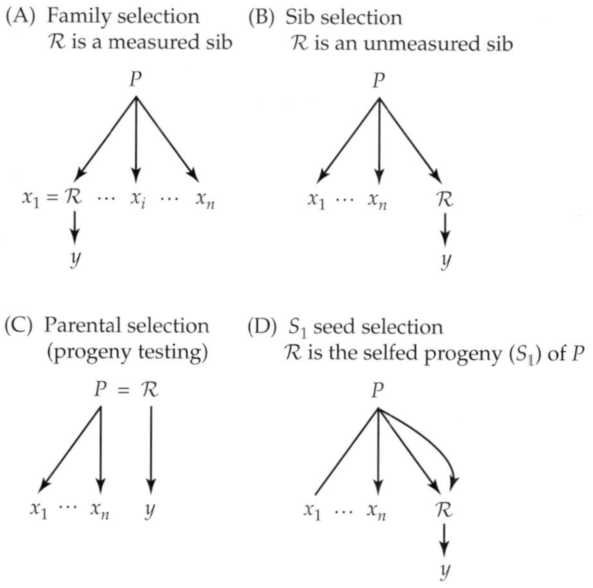

Figure 21.2 Under among-family selection, decisions as to which families to choose are made on the basis of observations from sibs, while the next generation is formed by crossing relatives ($\mathcal{R}$) of sibs from the chosen families. The measured sibs upon which selection decisions are based are denoted by $x_1, \cdots, x_n$, while y denotes a random offspring from a random member, $\mathcal{R}$, from the recombination unit. Different types of relatives can be used for $\mathcal{R}$, with a few of the most common types illustrated here. Let P denote the shared parent(s) of x_i and $\mathcal{R}$. The pedigrees illustrated here all focus on just one parent of y, with a corresponding pedigree for the other parent. **A:** Family selection: $\mathcal{R}$ is one of the measured sibs ($x_1 = \mathcal{R}$). **B:** Sib selection: $\mathcal{R}$ is an unmeasured sib. **C:** Parental selection (also known as progeny testing): $\mathcal{R}$ is the parent of the sibs ($\mathcal{R} = P$). **D:** S_1 seed selection: $\mathcal{R}$ is the selfed progeny of the parent of the sibs, but $\mathcal{R}$ is then outcrossed to generate the offspring, y. In this chapter, we assume offspring are generated by outcrossing ($\mathcal{R}_1$ and $\mathcal{R}_2$ are unrelated), whereas in Chapter 23 we examine the setting wherein y is obtained by selfing $\mathcal{R}$, as well as more general inbreeding schemes (such as the tested sibs being the result of selfing).

relative to n unrelated individuals. Ignoring this correction (for now), note that strict within-family (WF) selection has only 83% of the selection intensity (for this example) as family-deviations (FDs) selection.

Finally, the choice of the selection scheme also influences the long-term effective population size (and hence the long-term response; see Chapter 26), with schemes that place more weight on among-family components resulting in smaller effective population sizes (due to larger among-family offspring variances) than those that place more weight on within-family components (Chapter 3).

Variations of the Recombination Unit

Under either within-family or index selection, measured individuals are selected as the parents for the next generation, which forms the recombination unit. By contrast, with among-family selection there are a variety of options for the nature of the relatives that comprise the recombination unit (Table 21.1; Figure 21.2). The most straightforward situation is **family selection**, using measured sibs from each chosen family as the parents for the next generation (Figure 21.2A). Under **sib selection**, unmeasured sibs from the chosen families are used to form the next generation (Figure 21.2B).

Table 21.1 Family-based selection schemes using outbred sibs. Families are selected based on the sib values $z_{i1}, \cdots, z_{in}$. $\mathcal{R}_i$ denotes a relative of the ith selected family used to form the next generation. The variables $\overline{z}_{HS}$ and $\overline{z}_{FS}$ denote the sample means, and μ_{HS} and μ_{FS} denote the true means, of half- and full-sib families, respectively, while P is the parent of the measured sibs, and z_{ij} denotes the jth measured sib from family i.

Among-family Selection	Recombination Unit $\mathcal{R}$	Selection Unit x
Family selection	Measured sib	
Half-sib family selection		$\overline{z}_{HS}$
Full-sib family selection		$\overline{z}_{FS}$
Sib selection / Remnant seed	Unmeasured sib	
Half-sib sib selection		$\overline{z}_{HS}$
Full-sib sib selection		$\overline{z}_{FS}$
Parental selection / Progeny testing	Parent P	$\overline{z}_{HS}$
S_1 Seed Selection	S_1 Seed of P	
Half-sib S_1 seed selection		$\overline{z}_{HS}$
Full-sib S_1 seed selection		$\overline{z}_{FS}$
Within-family Selection		
Family deviations (FD) selection	Measured Sib	
Half-sib family deviations selection		$z_{ij} - \overline{z}_{HS}$
Full-sib family deviations selection		$z_{ij} - \overline{z}_{FS}$
Strict within-family (WF) selection	Measured Sib	
Half-sib strict within-family selection		$z_{ij} - \mu_{HS}$
Full-sib strict within-family selection		$z_{ij} - \mu_{FS}$

In animal breeding, sib selection is often used for traits that are sex-limited or that cannot be scored without sacrificing the individual. Plants breeders routinely use sib selection in the form of **remnant seeds**. Here, seeds from a cross are split into two batches and one is planted and used to assess families while the other is held in reserve. Seeds from the choosen families are then grown and crossed to form the next generation. Under this design, a single **cycle of selection** takes (at least) two generations—(at least) one to assess the families and a second to grow and cross the remnant seeds. Given this extra generation, what is the advantage of crossing plants from remnant seeds to form the next generation? For annual plants, any traits that are expressed during or after flowering can only be directly selected in already pollinated females, with seeds from the best-performing plants forming the next generation. Because these plants were pollinated at random, selection has occurred for the seed, but not the pollen, parents. By using remnant seeds, one can chose the best families, grow their remnant seeds, and allow the resulting plants to randomly intercross. Because both seed and pollen parents have now been selected (through their families), a single cycle of selection using remnant seed has double the response of family selection on seed from open pollinated plants. This doubling of response per cycle exactly counters the extra generation in each cycle, so open-pollinated family selection and sib selection using remnant seed have the same expected response per generation. One potential advantage with the use of remnant seed is that the extra generation to grow the seeds to mature plants for crossing can be used for selection on other characters, for example, culling those otherwise elite families that show poor disease or insect resistance.

Another common among-family design is **parental selection** (or **progeny testing**), where $\mathcal{R} = P$, the parent of the measured sibs (Figure 21.2C). This design typically involves evaluation of half-sib families with selection on just one sex. In animal breeding, these are typically sires, elite males chosen by the performance of their half-sib families, which is greatly facilitated by the use of artificial insemination and frozen semen. The ability to clone domesticated animals (e.g., sheep, Campbell et al. 1996; goats, Baguisi et al. 1999;

and cattle, Wells et al. 1999a, 1999b) is likely to further increase the importance of progeny testing in animal-breeding settings. (The most elaborate, and widely used, extension of progeny testing is BLUP selection wherein the entire pedigree is used for information on selection decisions; Chapter 19). Plant breeders typically perform progeny testing using maternal half-sib families (seed from the common parent). Vegetative propagation (cloning) allows even some annual plants to be used as parents in future generations. Depending on reproductive timing, if the species being selected is **monoecious** (single individuals produce both seed and pollen), one potentially may be able to obtain elite plants for both seed and pollen on the basis of female (seed) performance, and hence select on both sexes.

Finally, with self-compatible species, an alternative to vegetative propagation is the **S_1 seed design** (Figure 21.2D). For each parent, a subset of flowers is selfed to produce S_1 seed and the remainder are outcrossed. The outcrossed seed is then grown to produce the sibs in which the trait of interest is assessed. Following selection of the best families, their S_1 seed grown and the adults from different families are crossed to form the next generation. As with remnant seed, a single cycle takes two generations. In maize, the S_1 seed design requires the use of **prolific plants** (those with more than one ear), as one ear is selfed, and the other(s) outcrossed. Hallauer and Mirana (1981) noted that the use of such plants also results in selection for prolificacy, which by itself can increase yield. An advantage of designs using remnant seed is that traits can be scored over several years before selection, providing the opportunity to select over temporal variation in the environment. As presented in this chapter, the S_1 seed design has a random-mated family as the selection unit. Obviously, one could collect only S_1 seed from a plant and use some for selection decisions (i.e., the selection unit is an S_1 family) and the rest for future breeding. Such designs, where the selection unit is a selfed family, are examined in Chapter 23.

THEORY OF EXPECTED SINGLE-CYCLE RESPONSE

Response is typically given on a *per-cycle*, rather than per-generation, basis. A cycle begins with choosing the parents, P, to form the sib families and ends with the creation of offspring, y, formed by crossing members, $\mathcal{R}$, from the recombination unit. The expected response is the difference in the means of these two populations (P vs. y). When comparing the efficiencies of different schemes, response per cycle should be converted to a response per generation (for discrete generations) or per unit time (for overlapping generations).

Our treatment of the theory of response starts by developing several equivalent modifications of the breeder's equation (Chapter 13) to accommodate family-based selection. To apply these expressions, we require the **selection unit-offspring covariance**, $\sigma(x, y)$, and the **variance of the selection unit**, σ_x^2, for various family-based designs. The full development of these variances and covariances is straightforward but involves a fair amount of bookkeeping. The reader wishing to skip the details can find the results summarized below in Tables 21.3 and 21.4.

Modifications of the Breeder's Equation for Predicting Family-based Response

Response is a function of how selection decisions based on the sib families (x_1 and x_2) translate into selection on the corresponding parents ($\mathcal{R}_1$ and $\mathcal{R}_2$) of the offspring, y. Phrased in terms of breeding values, we predict response by using the sib information to predict the breeding values of the parents, $\mathcal{R}$, for the next generation. Under the infinitesimal model, the expected mean of the offspring equals the mean breeding value of the chosen parents (Chapters 6 and 13).

Making the standard assumption that all appropriate regressions are linear (which follows under the infinitesimal model assumptions; Chapters 6 and 24), the expected response is given by the general form of the breeder's equation (Equations 13.4a and 13.4b),

$$R_y = \frac{\sigma(x_m, y)}{\sigma_{x_m}^2} S_{x_m} + \frac{\sigma(x_f, y)}{\sigma_{x_f}^2} S_{x_f} \tag{21.1a}$$

Here x_m and x_f correspond to individuals from the selection units associated with the male (sire/pollen) and female (dam/seed) parents ($\mathcal{R}_m$ and $\mathcal{R}_f$) of the offspring, y. Equation 21.1a allows the male and female parents to be chosen by completely different schemes. For example, sib selection could be used on males and individual selection on females when selecting for a female-limited character (Example 13.5). The selection unit-offspring covariance, $\sigma(x, y)$, can be directly computed from the pedigree connecting P, a sib in x, and $\mathcal{R}$ through the use of path analysis (LW Appendix 2). The path (or correlation) between selection on the unit, x_f, through the female parent, $\mathcal{R}_f$, and its offspring, y, is

$$x_f \leftarrow P \rightarrow \mathcal{R}_f \rightarrow y$$

Because the path connecting x_f and y is through $\mathcal{R}_f$, we often write $\sigma(x, y \,|\, \mathcal{R}_f)$ in place of $\sigma(x_f, y)$ to remind the reader of this fact. Path(s) connecting x_m and y through $\mathcal{R}_m$ are similarly defined. If P consists of multiple relatives, each path connecting x_i and $\mathcal{R}_i$ (and hence y) needs to be counted. For example, if x_i and $\mathcal{R}_i$ are full-sibs, we must compute the paths through each of the common parents (e.g., Figure 21.3D). If the selection unit-offspring covariances are the same for both parents, Equation 21.1a simplifies to

$$R_y = \frac{\sigma(x, y)}{\sigma_x^2}\, S_x \tag{21.1b}$$

where $S_x = (S_{x_m} + S_{x_f})/2$ is the average selection differential on the unit(s) leading to the parents and

$$\sigma(x, y) = \sigma(x, y \,|\, \mathcal{R}_f) + \sigma(x, y \,|\, \mathcal{R}_m) = 2\sigma(x, y \,|\, \mathcal{R}) \tag{21.1c}$$

is the covariance between the value of selection unit, x, and the offspring, y, counting the paths through both parents ($\mathcal{R}_m$ and $\mathcal{R}_f$). When covariances are equal, this is twice the single parent-covariance, $\sigma(x, y \,|\, \mathcal{R}_1)$. By analogy with the breeder's equation, Equation 21.1b is often written as

$$R_y = h^2_{x,y} S_x \tag{21.2a}$$

where the **generalized heritability** of y given x,

$$h^2_{x,y} = \frac{\sigma(x, y)}{\sigma_x^2} = 2\left[\frac{\sigma(x, y \,|\, \mathcal{R})}{\sigma_x^2}\right] \tag{21.2b}$$

is twice the slope of the regression of y on x (LW Chapter 3). Just as the individual heritability, h^2, is the accuracy in using an individual's phenotypic value to predict the breeding value (Chapter 13), the generalized heritability is the accuracy of using the sib data, x, to predict the breeding value of $\mathcal{R}$.

Example 21.2. Consider family selection, wherein the selection unit is the family mean, $\overline{z}_i$, and the recombination units are measured sibs (those whose trait values have been scored) from this family. Assuming the covariance between the sib mean and an individual sib is independent of sex, Equations 21.1b and 21.1c yield a response of

$$R_b = \frac{2\,\sigma(\,\overline{z}_i, y \,|\, \mathcal{R}_i\,)}{\sigma^2(\,\overline{z}_i\,)}\, S_b$$

Recall (Equation 21.1c) that the numerator is twice the covariance between the value of the family mean, $\overline{z}_i$, and the offspring, y, from a parent chosen from this family, $\mathcal{R}_i$, which (in this case) is one of the measured sibs. The preceding expression can be more compactly written as $R_b = h_b^2\, S_b$, where the **among-family heritability** is

$$h_b^2 = \frac{2\,\sigma(\,\overline{z}_i, y \,|\, \mathcal{R}_i\,)}{\sigma^2(\,\overline{z}_i\,)}$$

We used the notation R_b and h_b^2 for "between family" in keeping with the literature (although this is formally among families, as it generally involves more than two families). Similarly, for selection on within-family deviations, the value of the selection unit is $z_{ij} - \overline{z}_i$, which yields

$$R_{FD} = \frac{2\,\sigma(\,z_{ij} - \overline{z}_i, y\,|\,\mathcal{R}_i\,)}{\sigma^2(\,z_{ij} - \overline{z}_i\,)}\,S_{FD}$$

where $\mathcal{R}_i = x_{ij}$. Response can also be expressed in terms of the **family-deviations heritability**, with $R_{FD} = h^2_{FD}\,S_{FD}$, where

$$h^2_{FD} = \frac{2\,\sigma(\,z_{ij} - \overline{z}_i, y\,|\,\mathcal{R}_i\,)}{\sigma^2(\,z_{ij} - \overline{z}_i)}$$

Tables 21.3 and 21.4 (below) give expressions for these variances and covariances.

Other (equivalent) versions of Equations 21.1a and 21.2a appear in the literature. The **selection-intensity version** allows for standardized comparisons of different selection schemes. Defining the selection intensity on x by $\overline{\imath}_x = S_x/\sigma_x$, Equation 21.1a becomes

$$R_y = \frac{\sigma(x_m, y\,)}{\sigma_{x\,m}}\,\overline{\imath}_{x\,m} + \frac{\sigma(x_f, y\,)}{\sigma_{x\,f}}\,\overline{\imath}_{x\,f} \tag{21.3a}$$

If the regressions are the same for both parents,

$$R_y = \frac{\sigma(x, y)}{\sigma_x}\,\overline{\imath}_x \tag{21.3b}$$

where $\overline{\imath}_x = (\overline{\imath}_{x_m} + \overline{\imath}_{x_f})/2$ is the average selection intensity. This expression is frequently written in terms of the selection unit-offspring correlation, $\rho(x, y)$,

$$R_y = \sigma_z\,\overline{\imath}_x\,\rho(x, y) \tag{21.4a}$$

where (counting both parents) $\rho(x, y) = 2\rho(x, y\,|\,\mathcal{R}\,)$. Equation 21.4a follows immediately from Equation 21.3b by recalling that $\rho(x, y) = \sigma(x, y)/(\sigma_x\,\sigma_y\,)$ and that the trait variance in the offspring, y, is simply the phenotypic variance of the character ($\sigma^2_y = \sigma^2_z$). A variant of Equation 21.4a commonly seen in the literature is

$$R_y = \sigma_A\,\overline{\imath}_x\,\rho(x, A_{\mathcal{R}}\,) \tag{21.4b}$$

where $\rho(x, A_{\mathcal{R}})$, the correlation between the value of the selection unit, x, and the breeding value of a parent, $\mathcal{R}$, of y, is the accuracy of selection (Equation 13.11a). Equation 21.4b holds in the absence of epistasis, while Equations 21.1–21.3 hold for arbitrary epistasis. Recall that the accuracy of individual selection (the correlation between an individual's phenotypic and breeding values) is $\rho(z_{\mathcal{R}}, A_{\mathcal{R}}) = h$. A particular family-based approach is favored over individual selection if x is a more accurate predictor of the breeding value of $\mathcal{R}$ than is $\mathcal{R}$'s phenotypic value, that is when $\rho(x, A_{\mathcal{R}}) > h$.

Equation 21.4b follows by first recalling that the mean value of an offspring is the average of its parental breeding values, $y = \mu + (A_{\mathcal{R}_m}/2) + (A_{\mathcal{R}_f}/2) + e_y$. Hence,

$$\sigma(x, y) = \frac{1}{2}\sigma(x, A_{\mathcal{R}_m}) + \frac{1}{2}\sigma(x, A_{\mathcal{R}_f}) + \sigma(x, e_y)$$

In the absence of epistasis, inbreeding, and shared environmental effects, $\sigma(x, e) = 0$. If the regression is the same for both sexes, then $\sigma(x, y) = \sigma(x_1, A_{\mathcal{R}_1})$. Recalling that $\sigma_y = \sigma_z$,

$$\rho(x, y) = \frac{\sigma(x, y)}{\sigma_x\,\sigma_z} = \left(\frac{\sigma_A}{\sigma_z}\right)\frac{\sigma(x_1, A_{\mathcal{R}_1})}{\sigma_x\,\sigma_A} = h\,\rho(x, A_{\mathcal{R}}) \tag{21.5}$$

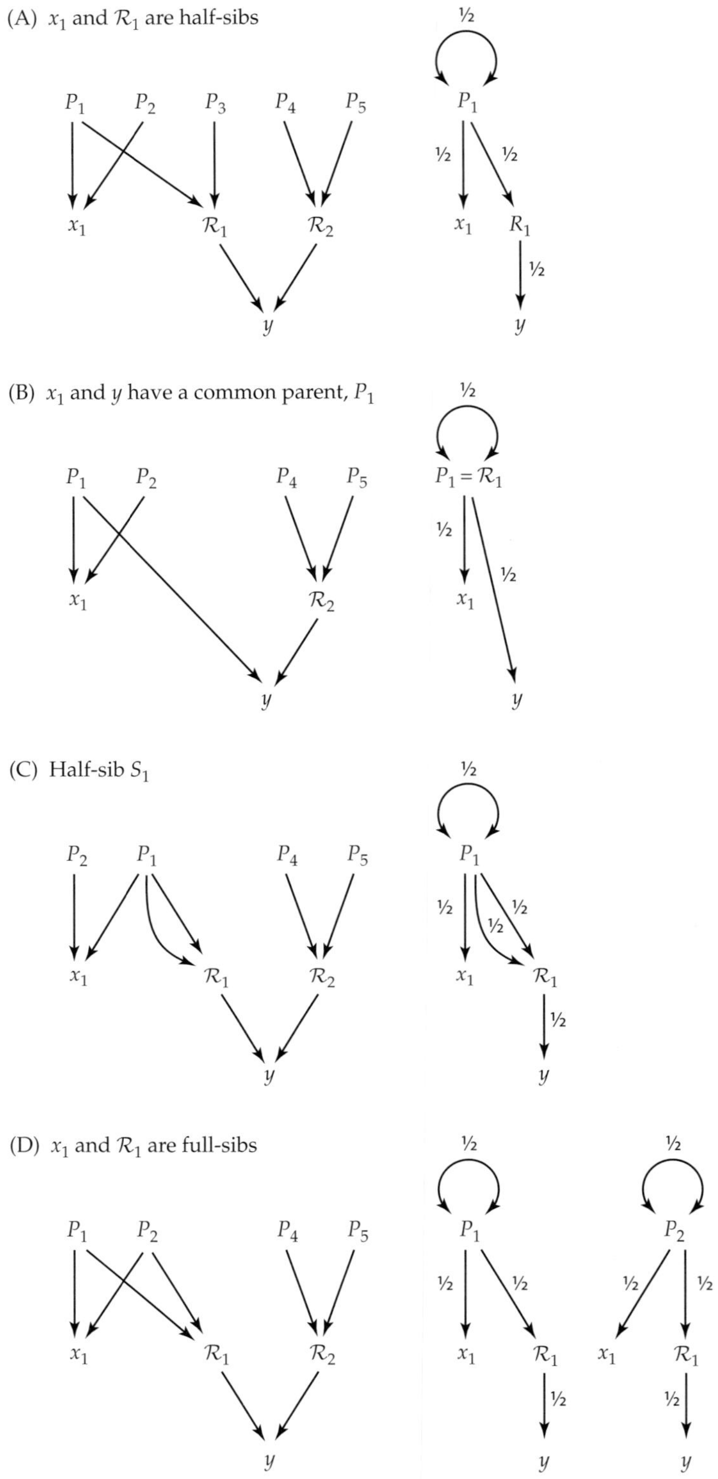

Figure 21.3 Derivation of the coefficient of coancestry, Θ, values in Table 21.2, showing pedigrees (left) and associated path diagrams (right) for computing Θ between a measured sib, x_1, and an offspring, y, from the parent, $\mathcal{R}_1$. P_1 to P_5 are assumed to be unrelated and noninbred. **A:** x_1 and $\mathcal{R}_1$ are half-sibs. The product of the path coefficients yields $\Theta_{x_1 y} = (1/2)^4 = 1/16$. **B:** x_1 and y are half-sibs, with $\Theta_{x_1 y} = (1/2)^3 = 1/8$. **C:** $\mathcal{R}_1$ is a selfed progeny from the common parent, P_1. There are two separate paths between x_1 and y (two different routes through P_1), yielding $\Theta_{x_1 y} = 2{\cdot}(1/2)^4 = 1/8$. **D:** x_1 and $\mathcal{R}_1$ are full-sibs. Again there are two paths between x_1 and y (one through each parent), each being $(1/2)^4$, giving a total of $\Theta_{x_1 y} = 2{\cdot}\,(1/2)^4 = 1/8$.

Table 21.2 Coefficients of coancestry, Θ, between an offspring, y (of parent $\mathcal{R}_1$), and a member of the selection unit, x_1. Genetic covariances, $\sigma_G(x_1, y)$, are computed with the assumption of no epistasis. Derviations are given in Figure 21.3.

Relationship between x_1 and $\mathcal{R}_1$	$\Theta_{x_1 y}$	$\sigma_G(x_1, y) = 2\Theta_{x_1 y}\,\sigma_A^2$
$x_1 = \mathcal{R}_1$ (the sib is also the parent of y)	1/4	$\sigma_A^2/2$
x_1 and $\mathcal{R}_1$ are half-sibs (Figure 21.3A)	1/16	$\sigma_A^2/8$
x_1 and $\mathcal{R}_1$ are full-sibs (Figure 21.3D)	1/8	$\sigma_A^2/4$
$\mathcal{R}_1$ is the parent of both x_1 and y (Figure 21.3B)	1/8	$\sigma_A^2/4$
$\mathcal{R}_1$ is an S_1 offspring of the parent of x_1 (Figure 21.3C)	1/8	$\sigma_A^2/4$

Substitution into Equation 21.4a recovers Equation 21.4b (as $\sigma_z\, h = \sigma_z\,(\sigma_A/\sigma_z) = \sigma_A$). Equations 21.1–21.4 provide equivalent expressions for computing the expected selection response. To apply these expressions to a particular selection scheme, we need to compute the selection unit-offspring covariance, $\sigma(x, y)$, and the variance of the selection unit, σ_x^2.

The Selection Unit-offspring Covariance, $\sigma(x, y)$

Recall that the genetic covariance between two (noninbred) relatives is a function of their **coefficients of coancestry,** Θ, and **fraternity**, Δ, (LW Chapter 7). If we ignore epistasis (for now), the genetic covariance between a particular sib, x_i, and y is $\sigma_G(x_i, y) = 2\Theta_{x_i y}\,\sigma_A^2 + \Delta_{x_i y}\,\sigma_D^2$ (LW Equation 7.12). In the absence of inbreeding in y (the parents $\mathcal{R}_1$ and $\mathcal{R}_2$ are from different, unrelated families; $\Theta_{\mathcal{R}_1,\mathcal{R}_2} = 0$), Δ_{xy} is zero. Note that $\Delta = 0$ even when $\mathcal{R}_1$ and/or $\mathcal{R}_2$ are themselves inbred, provided that they are unrelated. For dominance effects to be shared by relatives, there must be paths wherein *both* alleles from an individual, x, in the selection unit are passed onto the offspring, y, which cannot occur if the parents of y ($\mathcal{R}_1$ and $\mathcal{R}_2$) are unrelated.

The coefficient of coancestry between x_1 and y depends upon the relationship between $\mathcal{R}_1$ and x_1. The designs covered in Table 21.1 involve four different relationships (Figure 21.2): (i) $x_1 = \mathcal{R}_1$ (a measured sib is a parent of y), (ii) x_1 and $\mathcal{R}_1$ are sibs, (iii) $\mathcal{R}_1 = P_1$ (the parent of x_1), and (iv) $\mathcal{R}_1$ is the selfed-progeny of the parent of x_1. The path diagrams for computing $\Theta_{x_1 y}$ for these four relationships are given in Figure 21.3, and Table 21.2 summarizes the resulting genetic covariances. The parents, P_i, are assumed to be non-inbred (i.e., $\Theta_{P_i P_i} = 1/2$). If they are inbred, then $\Theta_{P_i P_i} = (1 + f_i)/2$, where i is the inbreeding coefficient on that parent, and the expressions in Table 21.2 are multiplied by this additional factor (for each inbred parent).

As an example of how the coefficients of coancestry given in Table 21.2 are used, consider family selection. Ignoring epistasis,

$$\sigma(\,\overline{z}_i, y\,|\,\mathcal{R}_1 = x_{ij}\,) = \frac{1}{n}\sum_k \sigma(z_{ik}, y\,|\,\mathcal{R}_1 = x_{ij}) = \frac{1}{n}\sigma(z_{ij}, y) + \left(1 - \frac{1}{n}\right)\sigma(z_{ik}, y)$$

$$= \sigma_A^2\left[\frac{1/2}{n} + \left(1 - \frac{1}{n}\right)2\Theta_{z_{ik}y}\right] \tag{21.6a}$$

This follows because the first covariance, $\sigma(z_{ij}, y)$, is for parent and offspring ($\sigma_A^2/2$), while the second covariance, $\sigma(z_{ik}, y)$, follows using the appropriate value of 2Θ from Table 21.2 (1/8 for half-sibs and 1/4 for full-sibs). Using the results from Table 21.2, expressions for the sib selection, parental selection (progeny testing), and S_1 seed designs follow in similar fashion. These are summarized in Table 21.3.

In much of the animal-breeding literature, **Wright's coefficient of relationship**, r, is used in place of 2Θ. Assuming no inbreeding, $r = 1/4$ for half-sibs and 1/2 for full-sibs.

Table 21.3 Summary of the covariances between the selection unit and one parent ($\mathcal{R}_1$) from the recombination unit. As given by Equation 21.6b, $r_n = r + (1-r)/n$, where (for non-inbred sibs), $r = 1/2$ and $1/4$, for full-sibs and half-sibs, respectively.

Among-family Selection:

Family selection ($\mathcal{R}_1$ is a measured sib from family i)

$$\sigma(\,\overline{z}_i, y \,|\, \mathcal{R}_1\,) = r_n\,(\sigma_A^2/2) = \begin{cases} (1+3/n)\,(\sigma_A^2/8) & \text{half-sibs} \\ (1+1/n)\,(\sigma_A^2/4) & \text{full-sibs} \end{cases}$$

Sib selection / Remnant seed ($\mathcal{R}_1$ is an unmeasured sib from family i)

$$\sigma(\,\overline{z}_i, y \,|\, \mathcal{R}_1\,) = r\,(\sigma_A^2/2) = \begin{cases} \sigma_A^2/8 & \text{half-sibs} \\ \sigma_A^2/4 & \text{full-sibs} \end{cases}$$

Parental selection / Progeny testing ($\mathcal{R}_1$ is a parent of the measured sibs)

$$\sigma(\,\overline{z}_i, y \,|\, \mathcal{R}_1\,) = \sigma_A^2/4$$

S_1 seed design ($\mathcal{R}_1$ is a selfed progeny of a parent of the measured sibs)

$$\sigma(\,\overline{z}_i, y \,|\, \mathcal{R}_1\,) = \sigma_A^2/4$$

Within-family Selection:

Selection on family deviations (FD)

$$\sigma(\,z_{ij} - \overline{z}_i, y \,|\, \mathcal{R}_1\,) = (1-r_n)\,(\sigma_A^2/2) = \begin{cases} (1-1/n)\,(3/8)\,\sigma_A^2 & \text{half-sibs} \\ (1-1/n)\,(\sigma_A^2/4) & \text{full-sibs} \end{cases}$$

Strict within-family selection (FW)

$$\sigma(\,z_{ij} - \mu_i, y \,|\, \mathcal{R}_1\,) = (1-r)\,(\sigma_A^2/2) = \begin{cases} (3/8)\,\sigma_A^2 & \text{half-sibs} \\ \sigma_A^2/4 & \text{full-sibs} \end{cases}$$

Using Wright's coefficient, Equation 21.6a simplifies to

$$\sigma(\,\overline{z}_i, y \,|\, \mathcal{R}_1 = x_{ij}\,) = r_n \frac{\sigma_A^2}{2} \qquad \text{where} \qquad r_n = r + \frac{1-r}{n} \tag{21.6b}$$

Considering the paths through both parents ($\mathcal{R}_1$ and $\mathcal{R}_2$) of y,

$$\sigma(\,\overline{z}_i, y) = 2\,\sigma(\,\overline{z}_i, y \,|\, \mathcal{R}_1\,) = r_n\,\sigma_A^2 \tag{21.6c}$$

Likewise, the covariance between an individual's family deviation and its offspring's phenotypic value is

$$\sigma(\,z_{ij} - \overline{z}_i, y \,|\, \mathcal{R}_1 = x_{ij}\,) = \sigma(\,z_{ij}, y \,|\, \mathcal{R}_1\,) - \sigma(\,\overline{z}_i, y \,|\, \mathcal{R}_1\,) = (1-r_n)\frac{\sigma_A^2}{2} \tag{21.7a}$$

which follows because $\sigma(\,z_{ij}, y \,|\, \mathcal{R}_1\,)$ is the parent-offspring covariance, $\sigma_A^2/2$. Doubling the single-parent contribution yields a total contribution (considering both parents of y) of

$$\sigma(\,z_{ij} - \overline{z}_i, y\,) = (1-r_n)\,\sigma_A^2 \tag{21.7b}$$

The covariance for strict within-family (WF) selection is slightly different (with r replacing r_n; see Table 21.3), as the appropriate covariance here is $\sigma(\,z_{ij} - \mu_i, y\,)$, with μ_i in place of $\overline{z}_i$. The rankings of individuals under WF selection is simply their ranking within each family, while their ranking under FD selection further depends on how much an individual actually deviates from its family mean. Thus, the top-ranked individuals in two families are always chosen under WF selection, but may not be chosen under FD selection. As a consequence, FD selection is influenced by the observed family mean, $\overline{z}_i$, while WF selection is a function of the true mean, μ_i (Dempfle 1975, 1990; Hill et al. 1996).

A few simple rules emerge from Table 21.3. The number, n, of measured sibs only influences the covariance for family selection and family-deviations selection. Even in these cases, its effect is small unless the number of sibs is small. Under sib selection (and family selection ignoring terms of order $1/n$), the selection unit-offspring covariance contributed through one parent ($\mathcal{R}_i$) is $\sigma^2_A/8$ when the selection unit consists of half-sibs and $\sigma^2_A/4$ when the selection unit consists of full-sibs. For parental selection and S_1 seed designs, this covariance is $\sigma^2_A/4$ (independent of whether full-sibs or half-sibs are used in the selection unit). The covariance under WF selection (and FD selection when ignoring terms of order $1/n$) is $3\sigma^2_A/8$ for half-sibs and $\sigma^2_A/4$ for full-sibs.

Variance of the Selection Unit, σ^2_x

The variance, σ^2_x, of the selection unit is a function of the within- and among-family variances, and obtaining it requires a bit of bookkeeping. We start by assuming that the total environmental value can be partitioned as $E = E_c + E_s$, a common-family effect (E_c) plus an individual-specific effect (E_s). This decomposes the total environmental variances into among- and within-family components, $\sigma^2_E = \sigma^2_{E_c} + \sigma^2_{E_s}$. When families are replicated over plots and environments, the environmental variance contains additional structure and is usually partitioned into further components (Equations 21.41 and 21.42).

The among-family variance σ^2_b (the variance among the expected family means, μ_i) is

$$\sigma^2_b = \sigma^2(\mu_i) = \sigma^2_{GF} + \sigma^2_{Ec} \tag{21.8a}$$

where σ^2_{GF}, the among-family genetic variance (the variance in the expected mean genotypic value among families), is developed below (Equations 21.11a and 21.26a). Likewise, the within-family variance about the expected family mean is

$$\sigma^2_w = \sigma^2(z_{ij} - \mu_i) = \sigma^2_{Gw} + \sigma^2_{Es} \tag{21.8b}$$

where σ^2_{Gw} is the within-family genetic variance (Equations 21.11b and 21.26b). Note that σ^2_b and σ^2_w are functions of the true family mean, μ_i, while the variance of the selection unit usually relies upon the variances about the *observed* mean, $\overline{z}_i$, of each family. Replacing μ_i with $\overline{z}_i$ results in a slight inflation of the among-family variance and a slight reduction in the within-family variance (this is formally shown below in Example 21.3). With n sibs in each family, the among-family variance based on the observed means becomes

$$\sigma^2(\,\overline{z}_i\,) = \sigma^2(\mu_i + \overline{e}_i\,) = \sigma^2_b + \sigma^2_w/n \tag{21.8c}$$

namely, the among-family variance, $\sigma^2(\mu_i) = \sigma^2_b$, plus the variance, $\sigma^2(\,\overline{e}_i\,) = \sigma^2_w/n$, in the error from estimating μ_i from $\overline{z}_i$. Because the total variance is the sum of the among- and within-family variances ($\sigma^2_b + \sigma^2_w$), the within-family variance (about the observed, rather than expected, mean) is correspondingly reduced to

$$\sigma^2(z_{ij} - \overline{z}_i) = (1 - 1/n)\sigma^2_w \tag{21.8d}$$

Equation 21.8c thus implies

$$\sigma^2(\,\overline{z}_i\,) = \sigma^2_{GF} + \sigma^2_{E_c} + \frac{\sigma^2_{Gw} + \sigma^2_{E_s}}{n} \tag{21.9a}$$

In the animal-breeding literature, this equation is often more compactly written in terms of t, the phenotypic correlation between sibs (the **intraclass correlation coefficient**; see

Table 21.4 Within- and among-family variances as functions of the genetic and environmental variance components. Epistasis is assumed to be absent and the environmental value is partitioned as $E = E_c + E_s$, a common-family value plus an individual-specific value; n is the number of measured sibs.

Half-sib among-family variance

$$\sigma^2(\,\overline{z}_{HS}\,) = \frac{\sigma_A^2}{4} + \frac{(3/4)\sigma_A^2 + \sigma_D^2 + \sigma_{E_s}^2}{n} + \sigma_{E_c(HS)}^2$$

Full-sib among-family variance

$$\sigma^2(\,\overline{z}_{FS}\,) = \frac{\sigma_A^2}{2} + \frac{\sigma_D^2}{4} + \frac{(1/2)\sigma_A^2 + (3/4)\sigma_D^2 + \sigma_{E_s}^2}{n} + \sigma_{E_c(FS)}^2$$

Half-sib with nested full-sibs (nested sibs) among-family variance
(n_f females per male, n_s offspring per female, $n = n_f n_s$ offspring per male)

$$\sigma^2(\,\overline{z}_{HS(FS)}\,) = \frac{\sigma_A^2}{4}\left(1 + \frac{1}{n_f} + \frac{2}{n}\right) + \frac{\sigma_D^2}{4n_f}\left(1 + \frac{3}{n_s}\right) + \frac{\sigma_{E_s}^2}{n} + \frac{\sigma_{E_c(FS)}^2}{n_f} + \sigma_{E_c(HS)}^2$$

Half-sib within-family variance

$$\sigma^2(\,z_{ij} - \overline{z}_i \,|\, HS\,) = \left(1 - \frac{1}{n}\right)\left(\frac{3}{4}\sigma_A^2 + \sigma_D^2 + \sigma_{E_s}^2\right)$$

Full-sib within-family variance

$$\sigma^2(\,z_{ij} - \overline{z}_i \,|\, FS\,) = \left(1 - \frac{1}{n}\right)\left(\frac{1}{2}\sigma_A^2 + \frac{3}{4}\sigma_D^2 + \sigma_{E_s}^2\right)$$

LW Chapter 7). The phenotypic covariance between sibs can be expressed as $t\sigma_z^2 = \sigma_b^2 = \sigma_{GF}^2 + \sigma_{E_c}^2$ (Example 21.3), implying that

$$\sigma^2(\,\overline{z}_i\,) = t_n\,\sigma_z^2 \tag{21.9b}$$

where, akin to our use of r_n (Equation 21.6b),

$$t_n = t + \frac{1-t}{n}. \tag{21.9c}$$

Likewise, the within-family variance (about the observed mean) is

$$\sigma^2(\,z_{ij} - \overline{z}_i\,) = \left(1 - \frac{1}{n}\right)\left(\sigma_{Gw}^2 + \sigma_{E_s}^2\right) \tag{21.10a}$$

which is usually written as

$$\sigma^2(\,z_{ij} - \overline{z}_i\,) = (1 - t_n)\,\sigma_z^2 \tag{21.10b}$$

Table 21.4 gives these family variances in terms of genetic and environmental variance components, which follow upon expressing the within- and among-family genetic variances in terms of additive and dominance variance components. Recalling from ANOVA theory that the among-group variance equals the within-group covariance (LW Chapter 18), the among-family component, σ_{GF}^2, equals the genetic covariances between sibs. If, for now, we ignore epistasis,

$$\sigma_{GF}^2 = \begin{cases} \frac{1}{4}\sigma_A^2 & \text{half-sibs} \\ \frac{1}{2}\sigma_A^2 + \frac{1}{4}\sigma_D^2 & \text{full-sibs} \end{cases} \tag{21.11a}$$

Because the total genetic variance (σ^2_G) equals the among-family genetic variance plus the within-family variance,

$$\sigma^2_{Gw} = \sigma^2_G - \sigma^2_{GF} = \begin{cases} \frac{3}{4}\sigma^2_A + \sigma^2_D & \text{half-sibs} \\ \\ \frac{1}{2}\sigma^2_A + \frac{3}{4}\sigma^2_D & \text{full-sibs} \end{cases} \tag{21.11b}$$

Finally, under a **nested-sib design** (the **North Carolina Design I** of Comstock and Robinson 1948), one sex (typically a male or a pollen plant) is mated to each of n_f (unrelated) females (or seed parents), each of which produces n_s sibs, for a total of $n = n_f n_s$ sibs per male. The expression in Table 21.4 for the among-family variance under the nested-sib design follows, with similar logic as in Example 21.3, and with

$$\sigma^2(\,\overline{z}_{HS(FS)}\,) = \sigma^2_{GF(HS)} + \frac{\sigma^2_{G(f|m)}}{n_f} + \frac{\sigma^2_{Gw(FS)}}{n_s n_f} + \frac{\sigma^2_{E_s}}{n} + \frac{\sigma^2_{E_c(FS)}}{n_f} + \sigma^2_{E_c(HS)} \tag{21.12a}$$

where $\sigma^2_{G(f|m)}$, the genetic variances of females nested within males, is

$$\sigma^2_{G(f|m)} = \sigma^2_{GF(FS)} - \sigma^2_{GF(HS)} = \frac{\sigma^2_A + \sigma^2_D}{4} \tag{21.12b}$$

When epistasis is present, Equation 21.26a (below) provides the appropriate additional genetic variance terms in $\sigma^2_{G(f|m)}$. The among-family variance under a nested design is bounded below by the half-sib variance ($n_f = n$ and $n_s = 1$) and above by the full-sib variance ($n_f = 1$ and $n_s = n$).

Example 21.3. To obtain the within- and among-family variances for families with n sibs, decompose the phenotypic value of the jth individual from family i as

$$z_{ij} = G_{ij} + E_{ij} = \mu + GF_i + Gw_{ij} + Ec_i + Es_{ij}$$

where the genotypic value, $G_{ij} = \mu + GF_i + Gw_{ij}$, has both a family genotypic effect, GF_i (the expected genotypic value of a random sib from that family), and a deviation, Gw_{ij}, the departure of jth individual's genotypic value from its family average. The environmental value is similarly decomposed, with $E_{ij} = Ec_i + Es_{ij}$, an environmental effect, Ec_i, common to family i, and a special environmental effect, Es_{ij}, unique to the jth individual from this family. Because $GF_i + Ec_i = b_i$ are the effects common to a family, the among-family variance becomes

$$\sigma^2_b = t\,\sigma^2_z = \sigma^2_{GF} + \sigma^2_{Ec}$$

The equality $\sigma^2_b = t\sigma^2_z$ follows from the ANOVA identity that the among-group variance equals the covariance among group members (LW Chapter 18).

Similarly, $Gw_{ij} + Es_{ij} = w_{ij}$ are the within-family effects, yielding a within-family variance (around the expected family mean) of

$$\sigma^2_w = (1-t)\sigma^2_z = \sigma^2_{Gw} + \sigma^2_{Es}$$

The equality $\sigma^2_w = (1-t)\sigma^2_z$ again follows from ANOVA theory, as the total variance equals the among-group variances plus the within-group variances, $\sigma^2_z = \sigma^2_b + \sigma^2_w = t\sigma^2_z + \sigma^2_w$.

Using these results, we can decompose the observed mean of a family of size n as

$$\overline{z}_i = \frac{1}{n}\sum_{j=1}^{n} z_{ij} = \frac{1}{n}\sum_{j=1}^{n} (\mu + GF_i + Gw_{ij} + Ec_i + Es_{ij})$$

$$= \mu + GF_i + Ec_i + \sum_{j=1}^{n} \frac{(Gw_{ij} + Es_{ij})}{n}$$

Because they are deviations from the mean, Es_{ij} and Gw_{ij} are uncorrelated with each other, yielding

$$\sigma^2(\,\overline{z}_i\,) = (\sigma^2_{GF} + \sigma^2_{Ec}) + \frac{1}{n^2}\sum_{j=1}^{n}(\sigma^2_{Gw} + \sigma^2_{Es}) = \sigma^2_b + \frac{n\,\sigma^2_w}{n^2}$$

$$= \left(t + \frac{1-t}{n}\right)\sigma^2_z = t_n\,\sigma^2_z$$

which recovers Equation 21.9b.

Now consider the variance of the within-family deviations from the observed means. Recalling the expression for the variance of a sum (LW Equation 3.11a), we have

$$\sigma^2(\,z_{ij} - \overline{z}_i\,) = \sigma^2_z + \sigma^2(\,\overline{z}_i\,) - 2\sigma(z_{ij}, \overline{z}_i\,)$$

To refine this further, first note (Equation 21.9b) that $\sigma^2(\,\overline{z}_i\,) = t_n\,\sigma^2_z$, and that the covariance term simplifies to

$$\sigma(z_{ij}, \overline{z}_i\,) = \frac{1}{n}\left[\sigma(z_{ij}, z_{ij}) + \sum_{k \neq j}^{n}\sigma(z_{ij}, z_{ik})\right] = \frac{\sigma^2_z}{n} + \frac{n-1}{n}\,t\,\sigma^2_z = t_n\,\sigma^2_z$$

as $\sigma(z_{ij}, z_{ik}) = t\sigma^2_z$ (for $j \neq k$). Thus, the variance of within-family deviations reduces to

$$\sigma^2(\,z_{ij} - \overline{z}_i\,) = \sigma^2_z + t_n\,\sigma^2_z - 2\,t_n\,\sigma^2_z = (1 - t_n)\,\sigma^2_z$$

which recovers Equation 21.10b.

RESPONSE FOR PARTICULAR DESIGNS

The formal development of the response equations for any particular design follows from the generalized breeder's equation (Equations 21.1 through 21.4), using the appropriate selection-unit variance (Table 21.4) and selection unit-offspring covariance (Table 21.3). Results for a number of standard among- and within-family designs are developed below, with family-index selection examined at the end of the chapter.

Overview of Among- and Within-family Response

The selection response for a particular family-based scheme depends on how the additive-genetic (breeding value) and total (phenotypic) variances are partitioned within and among families. When the number, n, of sibs per family is large (meaning that the observed mean will be very close to the true mean), these variances are partitioned as

	Within-family	Among-family
Breeding values	$(1-r)\,\sigma^2_A$	$r\,\sigma^2_A$
Phenotypic values	$(1-t)\,\sigma^2_z$	$t\,\sigma^2_z$

where t and r are, respectively, the phenotypic and additive-genetic correlations between sibs (r = 1/4 for noninbred half-sibs and 1/2 for noninbred full-sibs). When the number of measured sibs within each family is small, $t_n = t + (1-t)/n$ replaces t, and r_n (similarly defined) replaces r. Because the response to selection depends on the ratio of the available additive-genetic variance to the phenotypic variance, the response, R_b, to among-family selection is of the form

$$R_b = \frac{r_n\,\sigma^2_A}{t_n\,\sigma^2_z}\,S = \sigma_A\left(\frac{\sigma_A}{\sigma_z}\right)\left(\frac{r_n}{\sqrt{t_n}}\right)\left(\frac{S}{\sqrt{t_n}\,\sigma_z}\right) = \sigma_A\,h\,\frac{r_n}{\sqrt{t_n}}\,\overline{\imath} \tag{21.13a}$$

Equation 21.13a is the exact expression for family selection and is due to Lush (1947). Expressions for the predicted selection response under other among-family designs (e.g., sib, parental, or S_1 seed selection) are very similar (see below).

Similarly, the response to within-family selection is a function of the within-family additive-genetic and phenotypic variances, leading us to expect that the response will be in the form of

$$R_{FD} = \frac{(1-r_n)\,\sigma_A^2}{(1-t_n)\,\sigma_z^2}\,S = \sigma_A\, h\,\frac{1-r_n}{\sqrt{1-t_n}}\,\bar{\imath} \tag{21.13b}$$

Indeed, this is the exact expression for selection on family deviations (FD), while the response under strict within-family (WF) selection is given by replacing r_n and t_n with r and t.

Equations 21.13a and 21.13b are the standard response equations that appear in much of the elementary animal-breeding literature, as the use of r and t allows these results to be presented in a very compact fashion. When the design is more complicated, such as when it involves the replication of families over environments or the use of nested-sib families, expressions are given in terms of variance components, as shown below.

Among-family Selection

Here the selection unit is $\overline{z}$, the mean of a half-, full-, or nested-sib family. The type of sib family, together with the relatives used to produce the next generation, specifies the particular among-family design (Table 21.1). Tables 21.3 and 21.4 and Equation 21.13a yields the selection response, R_b, to a single cycle of among-family selection as

$$R_b = \frac{\gamma}{\sqrt{t_n}}\,\frac{\sigma_A}{2}\,h\left(\bar{\imath}_{x_m} + \bar{\imath}_{x_f}\right) = \frac{\gamma}{\sigma(\,\overline{z}\,)}\,\frac{\sigma_A^2}{2}\left(\bar{\imath}_{x_m} + \bar{\imath}_{x_f}\right) \tag{21.14a}$$

The left equality holds when the sib families are not nested and the families are not replicated, while the rightmost expression is completely general (using $\sigma^2(\,\overline{z}\,)$ in place of $t_n\sigma_z^2$). The selection unit-offspring covariance is $\gamma\,\sigma_A^2/2$, where

$$\gamma = \begin{cases} r_n = r + (1-r)/n & \text{family selection} \\ r & \text{sib selection} \\ 1/2 & \text{parental or } S_1 \text{ seed selection} \end{cases} \tag{21.14b}$$

Recall that these different values arise because r is the genetic correlation among sibs (1/4 and 1/2, respectively, for half- and full-sibs), and that parental and S_1 selection correspond to the case where $r = 1/2$. Under strict sib selection, no measured individual is a parent of the next generation and hence all the correlations between an individual in the selection unit and a parent of the next generation are the same (namely, $r\sigma_A^2$). Under family selection, one of the n measured individual sibs is also the parent of the next generation, and hence has a genetic covariance of $\sigma_A^2/2$, while the other $n-1$ individuals are sibs of this parent, each with a genetic covariance of $r\sigma_A^2/2$ with the offspring (a covariance of $r\sigma_A^2$ between sibs times 1/2 for that between parent and offspring).

The variance of the selection unit, $\sigma^2(\,\overline{z}\,) = t_n\sigma_z^2$, depends only on the types of sibs that are measured and is independent of the types of relatives used to form the next generation. The theory of expected response to among-family selection traces back to Lush's classic 1947 paper, and Equation 21.14a is a generalization of his results. Table 21.5 expresses the response in terms of variance components.

Several variants of Equation 21.14a appear in the literature. Noting that $\sigma_A\, h = \sigma_z h^2$, the response can be expressed as

$$R_b = \frac{\gamma}{\sqrt{t_n}}\,\sigma_z\, h^2\,\bar{\imath} \tag{21.15a}$$

where $\bar{\imath} = (\bar{\imath}_{x_f} + \bar{\imath}_{x_m})/2$. Similarly, the response can be expressed in terms of the among-

Table 21.5 Variance-component expressions of the expected response to among-family selection schemes using outbred sibs. Here $\bar{\imath}_{xm}$ is the selection intensity on individuals in the selection unit used to choose the male parent of the offspring, y (similarly, $\bar{\imath}_{xf}$ for the female parent). The number, n, of measured sibs is assumed to be sufficiently large that terms of order $1/n$ can be ignored (i.e., $r_n \simeq r$ and $t_n \simeq t$). We also assume no epistasis and a simple structure, $E = E_c + E_s$, for environmental values. We allow the within-family common environmental factor (E_c) to vary over the type of family, with $\sigma^2_{E_c(HS)}$ and $\sigma^2_{E_c(FS)}$, respectively, as the corresponding variances for half- and full-sib families.

	Half-sibs	Full-sibs
Family or sib selection	$\dfrac{(\sigma^2_A/8)\,(\bar{\imath}_{xm}+\bar{\imath}_{xf})}{\sqrt{\sigma^2_A/4+\sigma^2_{E_c(HS)}}}$	$\dfrac{(\sigma^2_A/4)\,(\bar{\imath}_{xm}+\bar{\imath}_{xf})}{\sqrt{\sigma^2_A/2+\sigma^2_D/4+\sigma^2_{E_c(FS)}}}$
Parental or S_1-seed selection	$\dfrac{(\sigma^2_A/4)\,(\bar{\imath}_{xm}+\bar{\imath}_{xf})}{\sqrt{\sigma^2_A/4+\sigma^2_{E_c(HS)}}}$	$\dfrac{(\sigma^2_A/4)\,(\bar{\imath}_{xm}+\bar{\imath}_{xf})}{\sqrt{\sigma^2_A/2+\sigma^2_D/4+\sigma^2_{E_c(FS)}}}$

family heritability, with

$$R_b = h^2_{b,\gamma}\, S, \qquad \text{where} \qquad h^2_{b,\gamma} = \frac{\gamma}{t_n}\, h^2 \tag{21.15b}$$

and with $S = (S_f + S_m)/2$ being the average selection differential on the parents.

Turning now to particular among-family designs, we start with family selection. Here, measured sibs (either all or a random subset) from the chosen families form the parents for the next generation. To reduce the effects of inbreeding, crosses between sibs from the same family are typically avoided (Chapter 23 examines the response when sibs are crossed, resulting in offspring that are inbred). With family selection, Equation 21.14a becomes

$$R_b = \begin{cases} \dfrac{(1+3/n)}{\sqrt{t_n(HS)}}\,\dfrac{\sigma_A}{8}\,h\,\left(\bar{\imath}_{xm}+\bar{\imath}_{xf}\right) & \text{half-sibs} \\[2ex] \dfrac{(1+1/n)}{\sqrt{t_n(FS)}}\,\dfrac{\sigma_A}{4}\,h\,\left(\bar{\imath}_{xm}+\bar{\imath}_{xf}\right) & \text{full-sibs} \end{cases} \tag{21.15c}$$

as first obtained by Lush (1947). While full-sibs have twice as much usable among-family additive variance relative to half-sibs ($\sigma^2_A/2$ vs. $\sigma^2_A/4$), this advantage is reduced somewhat because half-sibs have a smaller among-family phenotypic variance, with $t_{HS}/t_{FS} \leq 1$. This inequality follows by recalling that $\sigma^2(\,\bar{z}\,) = t\sigma^2_z$ and noting that $(t_{FS} - t_{HS})\,\sigma^2_z = \sigma^2(\,\bar{z}_{FS}) - \sigma^2(\,\bar{z}_{HS})$, where

$$\sigma^2(\,\bar{z}_{FS}) - \sigma^2(\,\bar{z}_{HS}) = \frac{\sigma^2_A + \sigma^2_D}{4} + \left(\sigma^2_{Ec(FS)} - \sigma^2_{Ec(HS)}\right) \tag{21.16a}$$

Given that full-sibs share a common mother (and hence potentially share maternal effects), we expect $\sigma^2_{Ec(FS)} \geq \sigma^2_{Ec(HS)}$ and, hence $\sigma^2(\,\bar{z}_{FS}) > \sigma^2(\,\bar{z}_{HS})$. Assuming the same selection intensity, Equation 21.15c yields the ratio of response for full- to half-sib family selection as

$$\frac{R_b(FS)}{R_b(HS)} = \left(\frac{1+1/n}{1+3/n}\right)\left(\frac{8\sqrt{t_n(HS)}}{4\sqrt{t_n(FS)}}\right) < 2\sqrt{\frac{t_n(HS)}{t_n(FS)}} < 2 \tag{21.16b}$$

with the last equality following from Equation 21.16a.

If the character can only be measured after reproduction, females (or seed parents) from the chosen families have already been fertilized, and hence selection has occurred on only one sex ($S_m = \bar{\imath}_{xm} = 0$). Planting these seeds (or, in animals, examining the offspring from fertilized females) and evaluating the resulting families allows for half-sib selection (under random pollination). Full-sib selection can also be accomplished, but each cycle takes an additional generation. Here seeds from open-pollinated selected females are grown and controlled crosses are made between the offspring from different seed parents to create full-sib families for the next cycle of selection.

Example 21.4. Clayton et al. (1957) examined family selection on abdominal bristle number in *Drosophila melanogaster* (LW Figure 14.1). Their estimated intraclass correlations for half- and full-sibs were 0.121 and 0.265, respectively, while the estimated additive variance and heritability were 5.59 and 0.52, respectively. Hence,

$$t_{HS} = 0.121, \quad t_{FS} = 0.265, \quad \text{and} \quad \sigma_A\, h = \sqrt{5.59 \cdot 0.52} = 1.70$$

Clayton et al. performed selection in two different settings: (i) the top 2 of 10 half-sib families were saved; and (ii) the top four of 20 full-sib families were saved. The expected selection intensities under these two schemes were, respectively, $\bar{\imath}_{HS} = \bar{\imath}_{(2,10)} = 1.27$, and $\bar{\imath}_{FS} = \bar{\imath}_{(4,20)} = 1.33$ (Equation 14.4b). The family sizes, n, used were 20 half-sibs and 12 full-sibs. Because of the laboratory mating design used by the authors, there was a 1 in 10 chance that the half-sibs are actually full-sibs, resulting in a slight inflation of r from 0.25 to 0.275 (= $0.25 \cdot [9/10] + 0.5 \cdot [1/10]$). To summarize:

	Half-sibs	Full-sibs
r	0.275	0.5
n	20	12
t_n	0.165	0.326
r_n	0.311	0.542

Equation 21.13a gives an expected response to half-sib family selection of

$$R_b(HS) = (\sigma_A\, h)\, \frac{r_n}{\sqrt{t_n}} \cdot \bar{\imath}_{HS} = 1.70\, \frac{0.311}{\sqrt{0.135}} \cdot 1.27 = 1.67$$

while the expected response to full-sib family selection is

$$R_b(FS) = 1.70\, \frac{0.542}{\sqrt{0.326}} \cdot 1.33 = 2.15$$

Clayton et al. obtained slightly different estimated responses (1.33 and 2.02 for half- and full-sibs, respectively). This occurred because they used $R = h_b^2\, S_b$, with $S_b = \sigma_b\, \bar{\imath}$ computed by taking the observed among-family variance, σ_b^2 (in place of the estimates σ_A^2, t, and h^2). The observed responses (averaged over the first five generations) were, respectively, 1.38 and 0.94 for up- and down-selected half-sibs, and 1.62 and 1.36 for up- and down-selected full-sibs. The authors noticed a fairly sizable reduction in the estimated additive variance during generations two through four, which (in addition to sampling error; Chapter 18) likely accounts for the discrepancy between observed and predicted response.

Under sib selection, unmeasured sibs from each chosen family are used to form the next generation. The most common response equation for sib selection in the literature, which is due to Robertson (1955a), is

$$R_{sib} = \bar{\imath}\, \sigma_A\, h\, \frac{n\, r}{\sqrt{n(1 + [n-1]\, t)}} \tag{21.17}$$

where $\bar{\imath}$ denotes the average selection intensity used to choose both parents. Equation 21.17 follows from Equation 21.15a because $\gamma = r$ for sib selection and we use $\sigma_A h$ in place of

$\sigma_z h^2$. The use of remnant seed is a variant of sib selection. Forming offspring for the next cycle of selection by randomly crossing plants grown from the remnant seeds of the selected families allows these offspring to be the product of selection on both sexes of parents, but at the cost of an extra generation.

Under parental selection (progeny testing), parents are chosen based on the performance of a trial set of their offspring. Historically (until it was replaced by BLUP selection), this was the approach used to select the top bulls for dairy production. Typically, half-sib families are used and selection is on only one sex. In this case, the expected response is

$$R_{pt} = \frac{\sigma_A/4}{\sqrt{t_n(HS)}}\, h\,\bar{\imath} \tag{21.18a}$$

where $\bar{\imath}$ is the intensity on the selected sex. In monoecious species, the expected response is double that given by Equation 21.18a if one uses the selected parents for both seed and pollen. The use of maternal half-sib families (as commonly occurs in plant breeding) is expected to inflate $t(HS)$ relative to paternal half-sibs (and hence reduce response), as common-family environmental effects can be rather significant due to shared maternal effects.

If males (sires or pollen plants) are progeny tested using a nested-sib design, wherein each male is crossed to n_f (unrelated) females (dams or seed plants), each of which has n_s sibs, the among-family variance is given in Table 21.4, and the response becomes

$$\begin{aligned} R_{pt} &= \frac{h\,\bar{\imath}\,\sigma_A^2/4}{\sqrt{\sigma^2_{GF(HS)} + \dfrac{\sigma^2_{G(f|m)}}{n_f} + \dfrac{\sigma^2_{Gw(FS)}}{n_f n_s} + \dfrac{\sigma^2_{E_s}}{n_f n_s} + \dfrac{\sigma^2_{E_c(FS)}}{n_f} + \sigma^2_{E_c(HS)}}} \\ &= \frac{h\,\bar{\imath}\,\sigma_A^2/4}{\sqrt{\dfrac{\sigma_A^2}{4}\left(1+\dfrac{1}{n_f}+\dfrac{2}{n_f n_s}\right) + \dfrac{\sigma_D^2}{4n_f}\left(1+\dfrac{3}{n_s}\right) + \dfrac{\sigma^2_{E_s}}{n_f n_s} + \dfrac{\sigma^2_{E_c(FS)}}{n_f} + \sigma^2_{E_c(HS)}}} \end{aligned} \tag{21.18b}$$

For progeny testing of females using a nested design, the roles of males and females are reversed in the above expression. Because $\sigma^2(\,\bar{z}_{HS}\,) \leq \sigma^2(\,\bar{z}_{HS(FS)}\,) \leq \sigma^2(\,\bar{z}_{FS}\,)$, the response using a nested progeny test is intermediate to that for schemes using half- or full-sibs. All these comments for parental selection equally apply to the S_1 seed design, as the expected response is the same.

Among-family Selection: Which Scheme Is Best?

Given the number of among-family selection designs, which scheme should be used? Biological and economic restriction may preclude the use of certain designs and make others more feasible. These logistical considerations aside, there are three issues that must be weighed: (i) cycle time versus selection on one or both sexes, (ii) performance evaluation using half- versus full-sib families (the value of t_n, and more generally, $\sigma^2[\,\bar{z}\,]$), and (iii) choice of relatives for the recombination unit (the value of γ in Equation 21.14a). As mentioned above, a common reason for using a two-generation cycle (e.g., crossing plants grown from remnant seed from superior families) is the inability to select on both sexes. In such cases, the doubling of the cycle time is countered by selection on both sexes doubling the response per cycle, which yields the same expected rate of progress on a per-generation basis. In many cases, a multigenerational method is used because selection on other characters beside the primary one of interest is also performed during one (or both) generations of the cycle.

The second choice is the type of family comprising the selection unit. While the type of sibs changes the value of γ under family- and sib-selection, it does not influence γ under parental or S_1 selection (Equation 21.14b). Indeed, for these last two designs, it is more efficient to use half-sib families, as (from Equation 21.14a) the ratio of response of a parental half-sib to a parental full-sib scheme is $\sqrt{t(FS)/t(HS)} > 1$.

Provided that the same type of families (half-, full-, or nested-sibs) are measured, choosing relatives that increase the recombination unit-offspring covariance (by increasing γ) increases the expected response. For half-sib families, both parental and S_1 selection yield twice the response per cycle as sib or family selection (assuming the same number of sexes are under selection in the comparison). With full-sibs, Table 21.5 shows that, given the same selection intensity, the response per cycle under all four methods (family, sib, parental, and S_1 selection) is the same. While the response to selection using full-sib families is greater than that of family or sib selection using half-sibs, the use of full-sibs does not result in a doubling of the response (Equation 21.16b). This less-than-twofold increase in response per cycle using full-sibs is thus not sufficient to cover the cost of the extra generation that is often required to create full-sib families.

Once one has chosen a particular design, there is also the issue of allocation of the number of sibs (n) per each of the m families, given constraints on the total number of sibs, $N = mn$, measured over each cycle of selection. One increases the accuracy by increasing the number of sibs per family, but one does so by decreasing the selection intensity (for fixed N, increasing n decreases m, and hence $\bar{\imath}$; see Example 21.1). Robertson (1957, 1960b), Rendel (1959), and Lindgren et al. (1997) examined this problem of optimal family size. To maximize response, the breeder usually has two fixed constraints: the total number of sibs, N, examined and the number, n_p, of families used to form the next generation. A low value of n_p increases inbreeding, and thus not only invites inbreeding depression, but also reduces the eventual long-term response (Chapter 26). For fixed n_p and N, the goal is to find the number of sibs, n, per family that maximizes response. Noting that $\sigma_z\, h^2$ is fixed, while $n_p = mp$ (with p being the fraction saved) and $m = N/n$, Equation 21.15a shows that the single-generation response is maximized by maximizing the quantity $\gamma[\,\bar{\imath}_{(n_p,N/n)}/\sqrt{t_n}\,]$ with respect to n. With the exception of family selection (where $\gamma = r_n$), γ is a fixed constant with respect to n. Maximizing of the long-term response (or more generally, the expected response after $k > 1$ generations) also needs to consider the differences in the effective populations sizes. This is examined in Chapter 26.

Within-family Selection

Within-family selection chooses individuals based on their relative performance within families. Under family-deviations (FD) selection, individuals with the largest family deviations are chosen, independent of which family they come from. In contrast, strict within-family (WF) selection chooses the largest individuals from each family, independent of how much they actually deviate from their family means. Suppose that in family one the deviations for three measured sibs are 4, 3, and -7, while the deviations in family two are 1, 0, and -1. If we select the upper one-third, then under WF selection, the top individual from each family is chosen, while under FD selection, two individuals from family one and none from family two are chosen. The result of this rather subtle distinction is that FD selection is influenced by the observed mean, $\bar{z}_i$, while WF selection is not. Family deviations and strict within-family selection have been confused in the literature, and the correct expression for WF selection was provided by Dempfle (1975, 1990) and Hill et al. (1996). Because WF selection ensures an equal representation of families, while FD selection does not, WF selection has a larger effective population size (Equation 3.4), and hence an expected larger long-term response (Chapter 26).

Under family-deviations (FD) selection, the selection unit is the value of an individual's within-family deviation, $z_{ij} - \bar{z}_i$. Using the results from Tables 21.3 and 21.4, Equation 21.1a yields an expected response of

$$\begin{aligned} R_{FD} &= \frac{\sigma(\,z_{ij} - \bar{z}_i, y\,|\,\mathcal{R}_m)}{\sigma(\,z_{ij} - \bar{z}_i\,)}\,\bar{\imath}_{x_m} + \frac{\sigma(\,z_{ij} - \bar{z}_i, y\,|\,\mathcal{R}_f)}{\sigma(\,z_{ij} - \bar{z}_i\,)}\,\bar{\imath}_{x_f} \\ &= \frac{1 - r_n}{\sqrt{1 - t_n}}\,\sigma_A\, h \left(\frac{\bar{\imath}_{x_m} + \bar{\imath}_{x_f}}{2}\right) \end{aligned} \qquad (21.19)$$

with the last equality following from $\sigma_A^2/\sigma_z = \sigma_A\, h$.

Under strict within-family (WF) selection, individuals are chosen entirely on their rank within each family, resulting in the observed mean, $\overline{z}_i$, being replaced by the true (and unobserved) mean, μ_i (Dempfle 1975, 1990; Hill et al. 1996). The response becomes

$$\begin{aligned} R_{WF} &= \frac{\sigma(\, z_{ij} - \mu_i, y \,|\, \mathcal{R}_m)}{\sigma(\, z_{ij} - \mu_i\,)}\, \overline{\imath}_{x_m} + \frac{\sigma(\, z_{ij} - \mu_i, y \,|\, \mathcal{R}_f)}{\sigma(\, z_{ij} - \mu_i\,)}\, \overline{\imath}_{x_f} \\ &= \frac{1-r}{\sqrt{1-t}}\, \sigma_A\, h \left(\frac{\overline{\imath}_{x_m} + \overline{\imath}_{x_f}}{2} \right) \end{aligned} \tag{21.20}$$

Noting that

$$\frac{1-r_n}{\sqrt{1-t_n}} = \frac{(1-1/n)(1-r)}{\sqrt{(1-1/n)(1-t)}} = \frac{1-r}{\sqrt{1-t}} \sqrt{1-\frac{1}{n}}$$

it follows that

$$R_{FD} = R_{WF}\, \frac{\overline{\imath}_{FD}}{\overline{\imath}_{WF}} \sqrt{1-\frac{1}{n}} \tag{21.21}$$

Thus, when the number of measured sibs in each family is modest to large (meaning that the selection intensities are essentially equal, $\overline{\imath}_{FD} \simeq \overline{\imath}_{WF}$; see Example 24.1), the difference between the expected responses under WF versus FD selection is very small. For large values of n, Equation 21.20 yields a resulting response for strict within-family (WF) selection using half- and full-sib families of

$$R_{WF} = \begin{cases} \dfrac{(3/8)\,\sigma_A}{\sqrt{1-t(HS)}}\, h\, (\overline{\imath}_{x_m} + \overline{\imath}_{x_f}) & \text{half-sibs} \\[2ex] \dfrac{(1/4)\sigma_A}{\sqrt{1-t(FS)}}\, h\, (\overline{\imath}_{x_m} + \overline{\imath}_{x_f}) & \text{full-sibs} \end{cases} \tag{21.22a}$$

When expressed in terms of variance components,

$$R_{WF} = \begin{cases} \dfrac{(3/8)\sigma^2_A}{\sqrt{(3/4)\sigma^2_A + \sigma^2_D + \sigma^2_{E_s}}}\, (\overline{\imath}_{x_m} + \overline{\imath}_{x_f}) & \text{half-sibs} \\[2ex] \dfrac{(1/4)\sigma^2_A}{\sqrt{\sigma^2_A/2 + (3/4)\sigma^2_D + \sigma^2_{E_s}}}\, (\overline{\imath}_{x_m} + \overline{\imath}_{x_f}) & \text{full-sibs} \end{cases} \tag{21.22b}$$

With their smaller amounts of among-family genetic variance, there is more usable within-family variance among half-sibs, namely, a within-family additive variance of $(3/4)\,\sigma^2_A$. Only half of this variance is passed from parent to offspring, giving the $(3/8)\,\sigma^2_A$ term in Equations 21.22a and 21.22b. For full-sibs, the within-family additive variance is $(1/2)\,\sigma^2_A$, again only half of which is passed onto offspring, which results in the $\sigma^2_A/4$ term in these equations.

The **within-family heritability**, h^2_w, is the same under both FD and WF within-family selection because

$$\frac{1-r_n}{1-t_n} = \frac{(1-1/n)(1-r)}{(1-1/n)(1-t)} = \frac{(1-r)}{(1-t)}$$

yielding

$$h^2_w = \frac{2\,\sigma(\, z_{ij} - \overline{z}_i, y \,|\, \mathcal{R}_1\,)}{\sigma^2(\, z_{ij} - \overline{z}_i)} = \frac{(1-r_n)\,\sigma^2_A}{(1-t_n)\,\sigma^2_z} = \frac{(1-r)}{(1-t)}\, h^2 \tag{21.23}$$

Example 21.5. Using the data of Clayton et al. (1957) from Example 21.4, what is the expected response under the two within-family selection schemes (WF and FD)? Suppose we assume that there is a full-sib family design, with 10 families of 20 sibs each, and that we perform strict within-family selection, with the upper 20% chosen from each family (the top 4 of the 20 measured sibs). Correcting for finite population size (Equation 14.4b), the expected selection intensity is $\bar{\imath}_{(4,20)} = 1.33$, and from Equation 21.20, the predicted response is

$$R_{WF} = \bar{\imath} \cdot (\sigma_A\, h)\, \frac{1 - r}{\sqrt{1 - t}} = 1.33 \cdot 1.70\, \frac{1 - 0.275}{\sqrt{1 - 0.121}} = 1.75$$

If we use within-family deviations (FD), selecting the uppermost 20% of all 200 individuals gives a corrected selection intensity of $\bar{\imath}_{(40,200)} = 1.39$, and Equation 21.19 returns a predicted response of

$$R_{FD} = \bar{\imath} \cdot (\sigma_A\, h)\, \frac{1 - r_n}{\sqrt{1 - t_n}} = 1.39 \cdot 1.70\, \frac{1 - 0.311}{\sqrt{1 - 0.165}} = 1.78$$

The selection intensity values used here can be further corrected to account for correlations among sibs, and we do so later in the chapter (Equation 21.57b).

Realized Heritabilities

By analogy with individual selection, one can estimate the realized heritability (Chapter 18) associated with a particular family-based scheme from the ratio of observed response to selection differential, namely,

$$\widehat{h}^2_{r,x} = \frac{R_x}{S_x} \tag{21.24a}$$

Falconer and Latyszewski (1952) used this approach to estimate a realized within-family heritability for response to selection on body size in mice. These authors computed the standard error of this estimate by noting that

$$\sigma^2\left(\widehat{h}^2_{r,wf}\right) = \sigma^2\left(\frac{R_{wf}}{S_{wf}}\right) = \frac{\sigma^2\left(R_{wf}\right)}{S^2_{wf}} \tag{21.24b}$$

The last equality follows by assuming that the error variance in measuring S can be ignored.

Because the among- and within-family heritabilities can be expressed as a function of the individual heritability, h^2 (Equations 21.15b and 21.23), we can similarly translate a realized heritability estimate for a particular family-based design into a realized individual heritability, $\widehat{h}^2_r$. With among-family selection,

$$\widehat{h}^2_r = \left(\frac{t_n}{\gamma}\right) \widehat{h}^2_{r,b} \tag{21.25a}$$

where γ is given by Equation 21.14b. For within-family selection, these two heritabilities are connected by

$$\widehat{h}^2_r = \left(\frac{1 - t}{1 - r}\right) \widehat{h}^2_{r,wf} \tag{21.25b}$$

These expressions apply to a single generation of selection. Additional uncertainty is introduced into the estimate if the sib phenotypic correlation (t) is unknown and must itself be estimated. Equations 21.25a and 21.25b should be used with caution when multiple cycles of selection have occurred, as the sib additive-genetic correlation (r) increases in each successive generation due to inbreeding, which in turn changes the phenotypic correlation, t (Chapter 26).

Accounting for Epistasis

The response to within- and among-family selection in the presence of epistasis was briefly examined by Nyquist (1991), and we expand upon his results here. As with individual selection, additive epistasis contributes to the initial response under family-based selection, but its contribution to the ultimate response rapidly decays over time as recombination breaks up favorable combinations of alleles at different loci (Chapter 15). We first consider the single-generation response and then briefly examine the transient dynamics.

Recalling that the among-group variance equals the within-group covariance (LW Chapter 18), the among-family genetic variance, σ^2_{GF}, with arbitrary epistasis immediately follows from the genetic covariance between sibs (LW Table 7.2),

$$\sigma^2_{GF} = \begin{cases} \frac{1}{4}\sigma^2_A + \frac{1}{16}\sigma^2_{AA} + \frac{1}{64}\sigma^2_{AAA} + \cdots & \text{half-sibs} \\ \\ \frac{1}{2}\sigma^2_A + \frac{1}{4}\sigma^2_D + \frac{1}{4}\sigma^2_{AA} + \frac{1}{8}\sigma^2_{AD} + \frac{1}{16}\sigma^2_{DD} + \frac{1}{8}\sigma^2_{AAA} + \cdots & \text{full sibs} \end{cases} \tag{21.26a}$$

Likewise, the within-family genetic variance, $\sigma^2_{Gw} = \sigma^2_G - \sigma^2_{GF}$, becomes

$$\sigma^2_{Gw} = \begin{cases} \frac{3}{4}\sigma^2_A + \sigma^2_D + \frac{15}{16}\sigma^2_{AA} + \sigma^2_{AD} + \sigma^2_{DD} + \frac{63}{64}\sigma^2_{AAA} + \cdots & \text{half-sibs} \\ \\ \frac{1}{2}\sigma^2_A + \frac{3}{4}\sigma^2_D + \frac{3}{4}\sigma^2_{AA} + \frac{7}{8}\sigma^2_{AD} + \frac{15}{16}\sigma^2_{DD} + \frac{7}{8}\sigma^2_{AAA} + \cdots & \text{full-sibs} \end{cases} \tag{21.26b}$$

The among- and within-family variances, $\sigma^2(\,\overline{z}_i\,)$ and $\sigma^2(\,z_{ij} - \overline{z}_i\,)$, immediately follow if we substitute Equations 21.26a and 21.26b into Equations 21.9a and 21.10a, respectively.

The genetic covariance between an individual (x) from the selection unit and the offspring (y) under epistasis follows if we use LW Equation 7.12,

$$\sigma_G(x,y) = (2\,\Theta_{xy})\,\sigma^2_A + (2\,\Theta_{xy})^2\,\sigma^2_{AA} + \cdots = \sum_{u=1} (2\,\Theta_{xy})^u\,\sigma^2_{A^u}$$

The previous expression assumed that $\Delta_{xy} = 0$, meaning that terms involving dominance are not included. Using the values of Θ_{xy} from Table 21.2, the parent-offspring covariance is

$$\sigma(\mathcal{R}_1, y) = \frac{\sigma^2_A}{2} + \frac{\sigma^2_{AA}}{4} + \frac{\sigma^2_{AAA}}{8} + \cdots = \sum_{u=1} \left(\frac{1}{2^u}\right)\sigma^2_{A^u} \tag{21.27a}$$

Table 21.2 shows that Θ_{xy} = 1/16 and 1/8 when x is a half- or full-sib, respectively, of $\mathcal{R}$. Expressed in terms of Wright's coefficient of relationship, r,

$$\sigma(x_1, y\,|\,\mathcal{R}_1\,) = (r/2)\sigma^2_A + (r/2)^2\sigma^2_{AA} + (r/2)^3\sigma^2_{AAA} + \cdots = \sum_{u=1} \left(\frac{r}{2}\right)^u \sigma^2_{A^u} \tag{21.27b}$$

Substituting Equation 21.27a and 21.27b into Equation 21.6a yields a covariance for family selection of

$$\begin{aligned} \sigma(\,\overline{z}, y\,|\,\mathcal{R}_1\,) &= \frac{1}{n}\sum_{u=1}\left(\frac{1}{2^u}\right)\sigma^2_{A^u} + \left(1 - \frac{1}{n}\right)\sum_{u=1}\left(\frac{r}{2}\right)^u \sigma^2_{A^u} \\ &= \sum_{u=1}\left(\frac{1}{2^u}\right) r^u_n \sigma^2_{A^u} \end{aligned} \tag{21.28}$$

where $r^u_n = r^u + (1 - r^u)/n$. For a large family size, the coefficient for u-fold additive epistasis approaches $r^u/2^u$, which is the value under sib selection. Taking $r = 1/2$ returns the coefficients for parental and S_1 seed selection. Applying Equation 21.28, the single-parent covariance for half-sib family selection ($r = 1/4$) becomes

$$\sigma(\,\overline{z}_{HS}, y\,|\,\mathcal{R}_1\,) = \left(1 + \frac{3}{n}\right)\frac{\sigma^2_A}{8} + \left(1 + \frac{15}{n}\right)\frac{\sigma^2_{AA}}{64} + \left(1 + \frac{63}{n}\right)\frac{\sigma^2_{AAA}}{512} + \cdots \tag{21.29a}$$

Likewise, the single-parent covariance for full-sib family selection ($r = 1/2$) is

$$\sigma(\,\overline{z}_{FS}, y\,|\,\mathcal{R}_1\,) = \left(1+\frac{1}{n}\right)\frac{\sigma^2_A}{4} + \left(1+\frac{3}{n}\right)\frac{\sigma^2_{AA}}{16} + \left(1+\frac{7}{n}\right)\frac{\sigma^2_{AAA}}{64} + \cdots \tag{21.29b}$$

For sib-selection, $\sigma(\,\overline{z}_1, y\,|\,\mathcal{R}_1)$ is directly provided by Equation 21.27b, and Equations 21.29a and 21.29b apply if terms of order $1/n$ are ignored. For among-family selection using parental selection or S_1 seed, the covariance is the same as that for full-sibs under sib selection, namely, Equation 21.29b (as all three have the same value of Θ_{xy}).

The covariance for within-family deviations (again considering the contribution through a single parent of y) becomes

$$\begin{aligned}\sigma(\,z_{ij} - \overline{z}_i, y\,|\,\mathcal{R}_1\,) &= \sigma(\,\mathcal{R}_1, y\,) - \sigma(\,\overline{z}_i, y\,|\,\mathcal{R}_1\,)\\ &= \sum_{u=1}\left(\frac{1}{2}\right)^u\left(1 - r_n^u\right)\sigma^2_{A^u}\\ &= \left(1-\frac{1}{n}\right)\sum_{u=1}\left(\frac{1}{2}\right)^u(1-r^u)\,\sigma^2_{A^u}\end{aligned} \tag{21.30}$$

where we have used the identity $(1 - r_n) = (1 - 1/n)(1 - r)$. Ignoring the common $(1 - 1/n)$ factor found in all terms, for half-sibs we have

$$\sigma(\,z_{ij} - \overline{z}_{HS}, y\,|\,\mathcal{R}_1\,) = \left(\frac{3}{8}\right)\sigma^2_A + \left(\frac{15}{64}\right)\sigma^2_{AA} + \left(\frac{63}{512}\right)\sigma^2_{AAA} + \cdots \tag{21.31a}$$

while for full-sibs

$$\sigma(\,z_{ij} - \overline{z}_{FS}, y\,|\,\mathcal{R}_1\,) = \left(\frac{1}{4}\right)\sigma^2_A + \left(\frac{3}{16}\right)\sigma^2_{AA} + \left(\frac{7}{64}\right)\sigma^2_{AAA} + \cdots \tag{21.31b}$$

Equations 21.29 and 21.31 show that additive epistasis contributes to the short-term response. However, as with individual selection, this contribution is transient and decays over time as recombination breaks up linkage groups of favorable alleles (Chapter 15). For u-locus additive epistasis ($\sigma^2_{A^u}$), the per-generation decay rate for unlinked loci is $[1 - (1/2)^{u-1}]$, or one minus the probability that a parental gamete containing specific alleles at u unlinked loci will be passed onto an offspring. The probability that such a gamete remains unchanged after τ generations is $2^{-\tau(u-1)}$, which rapidly converges to zero. Thus, if R_{A^u} is the contribution due to u-locus additive epistasis, after τ generations, the contribution from a single generation of selection becomes $2^{-\tau(u-1)}R_{A^u}$.

Response with Autotetraploids

Recall from Chapter 15 that selection response with autotetraploids (which are common among crop plants) has similar features to selection in the presence of additive epistasis—there is a transient component to the response contributed by nonadditive gene action. In the case of autotetraploids, this is the dominance variance, which occurs because autotetraploid parents pass along two alleles at each locus to their offspring. As with epistasis, the contribution to the selection response from nonadditive variance arises because the genotypes are not in Hardy-Weinberg equilibrium. After several generations of random mating, the selection-induced allele frequencies remain unchanged (and hence, any additive contribution is permanent), but any nonadditive contributions decay away as the population approaches Hardy-Weinberg.

This section is a bit technical, with some of the details developed in Examples 21.6 and 21.7, so we will first review the key results. Except in the case of selfing (using S_1 seed), the permanent response to selection is the same as with a diploid. The transient contribution from dominance is generally small (indeed, it is smaller than its contribution under individual selection; Equation 15.9) and is only significant when the dominance variance is substantially larger than the additive variance. Further, this (generally small) transient contribution quickly decays under random mating. An additional complication involving autotetraploids is deferred until Chapter 23, namely, that the offspring from a cross of two (unrelated) autotetraploid parents from S_1 seed are inbred, as the two alleles from each parent can be identical by descent.

Using the results from Example 21.6, we find that

$$\sigma_G(x,y) = \begin{cases} \frac{1}{2}\sigma_A^2 + \frac{1}{6}\sigma_D^2 & \text{parent, offspring} \\ \frac{1}{4}\sigma_A^2 + \frac{1}{36}\sigma_D^2 & \text{half-sibs} \\ \frac{1}{2}\sigma_A^2 + \frac{2}{9}\sigma_D^2 + \frac{1}{12}\sigma_T^2 + \frac{1}{36}\sigma_Q^2 & \text{full-sibs} \end{cases} \tag{21.32a}$$

where σ_T^2 and σ_Q^2 are the variances of third- and fourth-order interactions within loci (see Example 21.6; LW Chapters 5 and 7, for details). Using these covariances and following the same logic leading to Equations 21.26a and 21.26b yields an among-family genetic variance of

$$\sigma_{GF}^2 = \begin{cases} \frac{1}{4}\sigma_A^2 + \frac{1}{36}\sigma_D^2 & \text{half-sibs} \\ \\ \frac{1}{2}\sigma_A^2 + \frac{2}{9}\sigma_D^2 + \frac{1}{12}\sigma_T^2 + \frac{1}{36}\sigma_Q^2 & \text{full-sibs} \end{cases} \tag{21.32b}$$

The within-family genetic variances follow from $\sigma_{Gw}^2 = \sigma_G^2 - \sigma_{GF}^2$, which yields

$$\sigma_{Gw}^2 = \begin{cases} \frac{3}{4}\sigma_A^2 + \frac{35}{36}\sigma_D^2 + \sigma_T^2 + \sigma_Q^2 & \text{half-sibs} \\ \\ \frac{1}{2}\sigma_A^2 + \frac{7}{9}\sigma_D^2 + \frac{11}{12}\sigma_T^2 + \frac{35}{36}\sigma_Q^2 & \text{full-sibs} \end{cases} \tag{21.32c}$$

The among- and within-family variances, $\sigma^2(\overline{z}_i)$ and $\sigma^2(z_{ij} - \overline{z}_i)$, immediately follow, respectively, if we substitute Equations 21.32b and 21.32c into Equation 21.9a, and Equation 21.32c into Equation 21.10a. One of the few attempts to measure variance components in a tetraploid (alfalfa) was done by Dudley et al. (1969), who found that only σ_A^2 was significant for the five yield-related traits that they measured. While estimates of σ_D^2 were negative, estimates of σ_T^2 and σ_Q^2 were of the same order as σ_A^2, but they had standard errors an order of magnitude higher than those of σ_A^2, and hence were not significant.

To proceed further, we need to compute the expected genetic covariance between y (an offspring of $\mathcal{R}$) and x, a sib upon which selection decisions are made. This requires us to compute two additional genetic covariances for autotetraploids, namely, for a half-uncle or half-aunt and a nephew (when x and $\mathcal{R}$ are half-sibs, x and y are related as half-uncle or half-aunt and nephew) and for an uncle or aunt and nephew (when x and $\mathcal{R}$ are full-sibs, x and y are related as uncle or aunt and a nephew). Example 21.7 carries out the bookkeeping.

Using the results from Example 21.7, the covariance between the family mean and an offspring generated using family selection again has two terms: a parent-offspring contribution $1/n$ (from the measured sib serving as a parent, $\mathcal{R}$, of y), and the covariance between x and y when $\mathcal{R}$ is an unmeasured sib of x, yielding

$$\begin{aligned} \sigma(\,\overline{z}, y \,|\, \mathcal{R}_1) &= \frac{1}{n}\sigma(y \,|\, \mathcal{R}_1) + \left(1 - \frac{1}{n}\right)\sigma(x, y \,|\, \mathcal{R}_1) \\ &= \frac{1}{n}\left(\frac{1}{2}\sigma_A^2 + \frac{1}{6}\sigma_D^2\right) + \left(1 - \frac{1}{n}\right)\sigma(x, y \,|\, \mathcal{R}_1) \end{aligned} \tag{21.33a}$$

Using the results from Example 21.7, when x and $\mathcal{R}$ are half-sibs, we have

$$\begin{aligned} \sigma(\,\overline{z}, y \,|\, \mathcal{R}_1) &= \frac{1}{n}\left(\frac{1}{2}\sigma_A^2 + \frac{1}{6}\sigma_D^2\right) + \left(1 - \frac{1}{n}\right)\left(\frac{1}{8}\sigma_A^2 + \frac{1}{216}\sigma_D^2\right) \\ &= \frac{1}{8}\sigma_A^2\left(1 + \frac{3}{n}\right) + \frac{1}{216}\sigma_D^2\left(1 + \frac{35}{n}\right) \end{aligned} \tag{21.33b}$$

Turning to full-sibs,

$$\sigma(\bar{z}, y \,|\, \mathcal{R}_1) = \frac{1}{n}\left(\frac{1}{2}\sigma_A^2 + \frac{1}{6}\sigma_D^2\right) + \left(1 - \frac{1}{n}\right)\left(\frac{1}{4}\sigma_A^2 + \frac{1}{27}\sigma_D^2\right)$$
$$= \frac{1}{4}\sigma_A^2\left(1 + \frac{1}{n}\right) + \frac{1}{27}\sigma_D^2\left(1 + \frac{7}{n}\right) \tag{21.33c}$$

In both cases, the additive-genetic contribution to the genetic covariance is the same as for diploids, while the extra contribution from dominance is small and transient, decaying under random mating by two-thirds each generation (Equation 15.10d). Similar expressions can be obtained for within-family selection. Again, the contribution from additive variance is the same as for diploids, while the contribution from dominance is small and decays by two-thirds each generation of random mating.

Example 21.6. For the response under family-based selection schemes involving tetraploids, we will need a bit more detailed treatment of resemblance between polyploid relatives than was given in LW Chapter 7. To begin, we label the four alleles in a tetraploid by B_1, B_2, B_3, and B_4. There are six possible gametes from this parent, (B_1, B_2), (B_1, B_3), (B_1, B_4), (B_2, B_3), (B_2, B_4), and (B_3, B_4). Allowing for nonadditive interaction between alleles, the genotypic value can be decomposed into four additive (single-allele) terms, six dominance (two-allele) terms, four three-way interactions, and one four-way interaction:

$$G_{1234} = a_1 + a_2 + a_3 + a_4 + d_{12} + d_{13} + d_{14} + d_{23} + d_{24} + d_{34} +$$
$$t_{123} + t_{124} + t_{134} + t_{234} + q_{1234} \tag{21.34a}$$

The resulting total genetic variation can be partitioned as

$$\sigma_G^2 = 4\sigma_a^2 + 6\sigma_d^2 + 4\sigma_t^2 + \sigma_q^2 = \sigma_A^2 + \sigma_D^2 + \sigma_T^2 + \sigma_Q^2 \tag{21.34b}$$

If two relatives share only one allele IBD, then their genetic covariance is $\sigma_a^2 = (1/4)\sigma_A^2$. If they share exactly two IBD alleles, the genetic covariance is $2\sigma_a^2 + \sigma_d^2 = (1/2)\sigma_A^2 + (1/6)\sigma_D$. If we fill out the rest of these covariances and let π_i denote the probability that two relatives share exactly i IBD alleles, we will have

IBD alleles	Prob.	σ_A^2	σ_D^2	σ_T^2	σ_Q^2
1	π_1	1/4	0	0	0
2	π_2	1/2	1/6	0	0
3	π_3	3/4	1/2	1/4	0
4	π_4	1	1	1	1

Using these results, observe that the genetic covariance between any two relatives can be expressed as a function of their π_i values, namely,

$$\left(\frac{\pi_1 + 2\pi_2 + 3\pi_3 + 4\pi_4}{4}\right)\sigma_A^2 + \left(\frac{\pi_2 + 3\pi_3 + 6\pi_4}{6}\right)\sigma_D^2 + \left(\frac{\pi_3 + 4\pi_4}{4}\right)\sigma_T^2 + \pi_4\sigma_Q^2 \tag{21.34c}$$

With a parent-offspring relationship, exactly two alleles are IBD, so that $\pi_2 = 1$. With half-sibs, by looking at the 36 entries in the 6×6 table of pairs of gametes from the same parent, we see that 6 share two alleles, 24 share one, and 6 share zero. Hence, $\pi_1 = 24/36 = 2/3$, $\pi_2 = 6/36 = 1/6$. Results for full-sibs follow from these half-sib results. Let P_1 and P_2 denote the shared parents. The probability that four alleles are shared is the probability the sibs share two alleles from P_1 times the probability they share two alleles from P_2, or $\pi_4 = (1/6)(1/6)$, assuming the parents are unrelated and not inbred. Now consider the case of sharing exactly two alleles IBD. This can happen in three different ways: sharing one IBD allele from each parent (probability [2/3][2/3]), sharing two IBD alleles from P_1 and zero from P_2 (probability [1/6][1/6]), or sharing two alleles from P_2 and zero from P_1 (probability [1/6][1/6]), yielding

$$\pi_2 = (2/3)^2 + (1/6)^2 + (1/6)^2 = 18/36 = 1/2$$

Similar logic yields $\pi_1 = 2/9$ and $\pi_3 = 2/9$. To summarize,

Relative pair	π_0	π_1	π_2	π_3	π_4
Parent-offspring	0	0	1	0	0
Half-sibs	1/6	2/3	1/6	0	0
Full-sibs	1/36	2/9	1/2	2/9	1/36

Substitution of the above results for the π_i into the general expression for the covariance yields Equation 21.32a.

Example 21.7. Example 21.6 computed the π_1, π_2, π_3, and π_4 values for half- and full-sibs. As we saw above, expressions for the selection response under various family-based selection schemes requires the covariance between a member of the selection unit (x) and an offspring (y) of $\mathcal{R}$, a relative of x. We can obtain these covariances by conditioning on the number of IBD alleles shared by x and $\mathcal{R}$, and then computing the probability that $\mathcal{R}$ passes along one or two of these IBD alleles to its offspring, y. For example, if x and $\mathcal{R}$ share exactly one IBD allele, then with a probability of one-half, that allele is also transmitted to y, in which case x and y share one IBD allele. The 1/2 comes from considering the six possible gametes that $\mathcal{R}$ can generate. Let B_1 denote the IBD allele for x and $\mathcal{R}$. When enumerating all six possible biallelic gametes, we see that three contain B_1, while the other three do not. Similar enumeration fills out the table below. For example, suppose x and $\mathcal{R}$ share two IBD alleles, B_1 and B_2. Again counting the six possible gametes of $\mathcal{R}$, only one contains both alleles (1/6 have two IBD), while four of six contain either B_1 or B_2 (but not both). The values when x and $\mathcal{R}$ share three and four IBD alleles are given below:

	Prob(IBD shared by x and y)	
IBD shared by x and $\mathcal{R}$	1	2
1	1/2	0
2	2/3	1/6
3	1/2	1/2
4	0	1

When x and $\mathcal{R}$ are half-sibs, Example 21.6 shows that 2/3 of sibs share one IBD allele, while 1/6 share two. We let the notation (I$[x, y] = 1$) and (I$[x, y] = 2$) denote that the pair (x, y) share, respectively, exactly one or two IBD alleles. Using the above table, the probability that y and x share one or two IBD alleles becomes

$$\pi_1 = \Pr(\mathrm{I}[x,y] = 1|\mathrm{I}[x,\mathcal{R}] = 1)\Pr(\mathrm{I}[x,\mathcal{R}] = 1) + \Pr(\mathrm{I}[x,y] = 1|\mathrm{I}[x,\mathcal{R}] = 2)\Pr(\mathrm{I}[x,\mathcal{R}] = 2)$$
$$= (1/2)(2/3) + (2/3)(1/6) = 4/9$$
$$\pi_2 = \Pr(\mathrm{I}[x,y] = 2\,|\,\mathrm{I}[x,\mathcal{R}] = 2)\ \Pr(\mathrm{I}[x,\mathcal{R}] = 2) = (1/6)(1/6) = 1/36$$

Substituting these into Equation 21.34c returns

$$\sigma_G(x,y) = \left(\frac{1(4/9) + 2(1/36)}{4}\right)\sigma_A^2 + \frac{1/36}{6}\sigma_D^2 = \frac{1}{8}\sigma_A^2 + \frac{1}{216}\sigma_D^2 \qquad (21.34d)$$

as the genetic covariance between x and y when the relationship is that of half-uncle (x and $\mathcal{R}$ are half-sibs) and nephew (y, an offspring of $\mathcal{R}$).

When x and $\mathcal{R}$ are full-sibs, Example 21.6 showed that the probability they share one, two, three, and four IBD alleles is 2/9, 1/2, 2/9, and 1/36, respectively. Following the same logic and using the above table,

$$\pi_1 = (1/2)(2/9) + (2/3)(1/2) + (1/2)(2/9) + (0)(1/36) = 20/36 = 5/9$$

$$\pi_2 = (0)(2/9) + (1/6)(1/2) + (1/2)(2/9) + (1)(1/36) = 2/9$$

yielding

$$\sigma_G(x,y) = \left(\frac{1(5/9) + 2(2/9)}{4}\right)\sigma_A^2 + \frac{2/9}{6}\sigma_D^2 = \frac{1}{4}\sigma_A^2 + \frac{1}{27}\sigma_D^2 \tag{21.34e}$$

as the genetic covariance between x and y when the relationship is that of uncle (x and $\mathcal{R}$ are full-sibs) and nephew (y, an offspring of $\mathcal{R}$).

EFFICIENCY OF FAMILY-BASED VS. INDIVIDUAL SELECTION

Intuition suggests that individual selection is better than either within- or among-family selection when h^2 is modest to large, as in this case, individual phenotypes, z, are good predictors of individual breeding values, A. When h^2 is small, we expect within-family selection to be more efficient if there is a large common family environmental effect ($\sigma_{E_c} \simeq \sigma_z^2$) and among-family selection to be more efficient if the individual-specific environmental effects are large ($\sigma_{E_s} \simeq \sigma_z^2$).

To more formally develop these points, recall that the expected response under mass (individual) selection is $R_m = \bar{\imath}_m\, \sigma_A\, h$ (Equation 13.6b). Applying Equation 21.14a, the ratio of response of among-family selection to individual selection becomes

$$\frac{R_b}{R_m} = \left(\frac{\bar{\imath}_b}{\bar{\imath}_m}\right)\left(\frac{\gamma}{\sqrt{t_n}}\right) \tag{21.35a}$$

where γ is a function of the type of among-family selection (Equation 21.14b), t is the intraclass correlation among sibs, with t_n given by Equation 21.9c, and $\bar{\imath}$ is the average selection intensity on the two sexes. Likewise, for family-deviations selection, Equation 21.19 yields

$$\frac{R_{FD}}{R_m} = \left(\frac{\bar{\imath}_{FD}}{\bar{\imath}_m}\right)\left(\frac{1-r_n}{\sqrt{1-t_n}}\right) \tag{21.35b}$$

where r_n is calculated by Equation 21.6b. Finally, Equation 21.20 yields a response ratio for strict within-family to mass selection of

$$\frac{R_{WF}}{R_m} = \left(\frac{\bar{\imath}_{WF}}{\bar{\imath}_m}\right)\left(\frac{1-r}{\sqrt{1-t}}\right) \tag{21.35c}$$

Equations 21.35a–21.35c show that the relative efficiency of any particular family-based scheme is the product of the ratio of selection intensities (the first term) and the **accuracy of selection** relative to individual selection (the second term). This accuracy ratio measures how well (relative to individual selection) the selection criterion predicts the breeding values of the parents. We focus first on the accuracy ratio, as the selection-intensity ratio is generally close to one unless sample sizes are very small (Example 21.1; Table 21.6).

The Relative Accuracies of Family-based vs. Individual Selection

Relative accuracies are typically expressed in terms of the phenotypic correlation, t, between sibs and their coefficient of relatedness, r. Under the simple environmental model ($E = E_c + E_s$), the variance of family means is $\sigma^2(\mu_i) = t\sigma_z^2 = \sigma_{GF}^2 + \sigma_{Ec}^2$. Hence,

$$t = \frac{\sigma_{GF}^2}{\sigma_z^2} + \frac{\sigma_{Ec}^2}{\sigma_z^2} = \frac{r\sigma_A^2}{\sigma_z^2} + \frac{(\sigma_{GF}^2 - r\sigma_A^2) + \sigma_{Ec}^2}{\sigma_z^2} = rh^2 + \frac{(\sigma_{GF}^2 - r\sigma_A^2) + \sigma_{Ec}^2}{\sigma_z^2} \tag{21.36a}$$

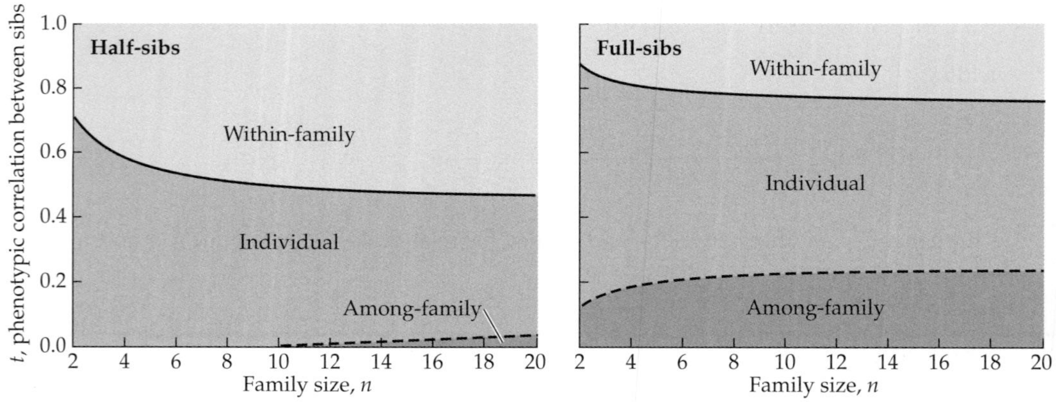

Figure 21.4 Regions of the family size (n)-sib correlation (t) space where individual, among-family (family selection) and within-family (selection of family deviations, [FD]) are the most accurate (based on Equations 21.35a and 21.35b). If t is sufficiently large, within-family selection yields the largest response (for large values of n; $t > 7/16 = 0.4375$ for half-sibs and $t > 3/4$ for full-sibs). Among-family selection is best when t is sufficiently small (for large values of n; $t < 1/16 = 0.0625$ for half-sibs, and $t < 1/4$ for full-sibs). Individual selection yields the largest response for intermediate values of t. For large n (t_n, r_n approaching t, r), among-family selection equals sib selection, as does parental selection (using the curve for full-sibs), while family-deviations selection approaches strict within-family (WF) selection.

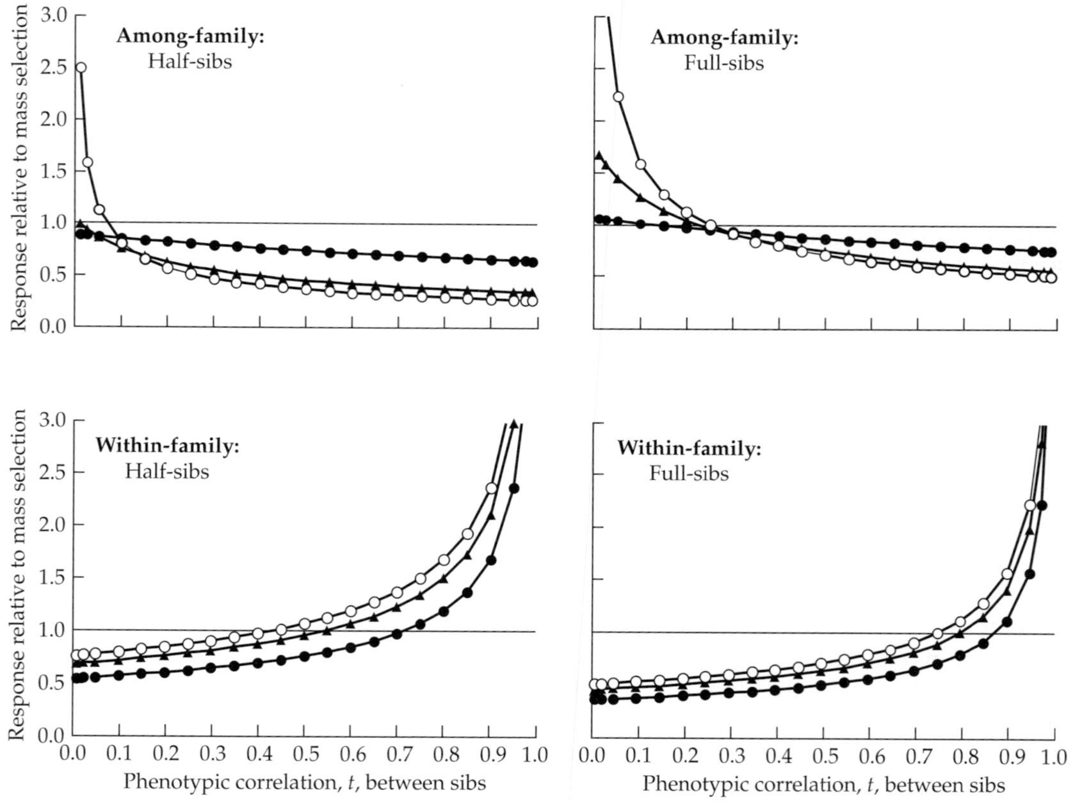

Figure 21.5 Accuracies of among-family selection (**top row**) and selection on family deviations (FD) (**bottom row**) relative to individual selection. In all graphs, filled circles correspond to $n = 2$ and open circles to large n. In the upper two graphs, the filled triangles correspond to $n = 10$ and correspond to $n = 5$ in the lower two graphs. Strict within-family selection (WF) corresponds to the large-n values for FD selection (open circles). Assuming equal selection intensities, values exceeding one indicate an increased single-generation response relative to individual selection.

In the absence of epistasis, Equation 21.11a yields

$$t = rh^2 + c^2, \quad \text{with} \quad c^2\sigma_z^2 = \begin{cases} \sigma^2_{Ec(HS)} & \text{half-sibs} \\ \\ \frac{1}{4}\sigma_D^2 + \sigma^2_{Ec(FS)} & \text{full-sibs} \end{cases} \tag{21.36b}$$

where c^2 scales the residual among-family variance (upon removal of any shared additive variance). Figures 21.4 and 21.5 plot the relative accuracies and responses under among-family (family selection) and within-family (family-deviations) selection. Note that if $c = 0$, then $t = rh^2$, which is bounded above by 1/4 in half-sibs and 1/2 in full-sibs (because $h^2 \leq 1$, it follows that $t \leq r$). For t to exceed r requires that $c > 0$.

What are the exact conditions for a particular method to be more accurate than individual selection? Equation 21.35a shows that among-family selection is more accurate when $\gamma^2 > t_n$, or

$$t_n = t + \frac{1-t}{n} = (rh^2 + c^2)\left(1 - \frac{1}{n}\right) + \frac{1}{n} < \gamma^2 \tag{21.37a}$$

For values of n that are moderate to large (such that $t_n \simeq t$), among-family selection is more accurate than mass selection when c^2, the fraction of the total due to residual among-family effects, is sufficiently small. Substituting the value of γ associated with a particular selection scheme (Equation 21.14b) into Equation 21.37a (for moderate to large values of n) yields the condition for among-family selection to be more accurate than individual selection as

$$c^2 < \begin{cases} \dfrac{1}{16}\left(1 - 4\,h^2\right) & \text{half-sibs (for family and sib selection; } \gamma = 1/4) \\ \dfrac{1}{4}\left(1 - h^2\right) & \text{half-sibs (for parental and S}_1\text{ seed selection; } \gamma = 1/2) \\ \dfrac{1}{4}\left(1 - 2h^2\right) & \text{full-sibs } (\gamma = 1/2) \end{cases} \tag{21.37b}$$

If $h^2 > 1/2$, the condition given by Equation 21.37b when full-sibs comprise the selection unit becomes $c^2 < 0$, and among-family selection is always less efficient than individual selection. With half-sibs comprising the selection unit, family- and sib-selection are always less efficient than individual selection when $h^2 > 1/4$. Among-family selection is, therefore, only more effective than mass selection when heritability is small *and* the fraction, c^2, of total variation due to common-family residual variation is also small.

Turning to within-family selection, Equation 21.35b shows that family deviations (FD) selection yields a larger response than individual selection when $(1 - r_n)/\sqrt{1 - t_n} > 1$. When families are large ($n \gg 1$, such that $t_n \simeq t$ and $r_n \simeq r$), this condition reduces to

$$t = rh^2 + c^2 > 1 - (1 - r)^2 \tag{21.38a}$$

or

$$c^2 > 1 - (1-r)^2 - rh^2 = \begin{cases} \dfrac{7}{16} - \dfrac{h^2}{4} & \text{half-sibs } (r = 1/4) \\ \\ \dfrac{3}{4} - \dfrac{h^2}{2} & \text{full-sibs } (r = 1/2) \end{cases} \tag{21.38b}$$

Because $h^2 + c^2 \leq 1$ (both being fractions of the total variance due to different sources), there is an additional constraint that $1 - h^2 \geq c^2$. When $h^2 > 0.75$, within-family (half-sib) selection is always less efficient than individual selection, as here $c^2 \leq 1 - h^2 = 0.25$, while the critical c^2 value that must be exceeded is 0.25 $[(7/16) - (3/4)/4] = 1/4$. By the same logic, for full-sibs, individual selection is more efficient than within-family selection whenever $h^2 > 0.5$. Because we assumed that n is large, r and t replace r_n and t_n, respectively, and hence Equations 21.38a and 21.38b are also the conditions for strict within-family (WF) selection. Within-family selection is thus more efficient than individual selection only when the heritability is low and the residual among-family variance ($c^2\sigma_z^2$) accounts for a very significant fraction of the total variance; in other words, common-family effects account for much of the phenotypic variance.

Willeke (1982) suggested that an excellent candidate trait for within-family selection would be litter size in pigs. Recall (Chapter 15) that mice from large litters tend to have a negative environmental value for litter size. Given that the heritability estimates for pig litter size from a grandmother-granddaughter regression are higher than those based on a mother-daughter regression, a similar situation likely occurs in pigs. Thus, there is a large family contribution that obscures prediction of the breeding value from the phenotypic value, resulting in the female's ranking within a family being a more informative predictor of her breeding value than her phenotypic value.

Example 21.8. Wilson (1974) examined family selection (using full-sibs) on larval and pupal weight in *Tribolium castaneum*. Correlations among full-sibs were estimated to be $t = 0.16$ for larval weight and $t = 0.20$ for pupal weight. Family size was $n = 12$. Under family selection, Equation 21.14b implies that $\gamma = r_n$. With noninbred full-sibs, $r = 1/2$, yielding (from Equation 21.35a) a ratio of the accuracies of family to mass selection on larval weight of

$$\frac{\gamma}{\sqrt{t_n}} = \frac{r_n}{\sqrt{t_n}} = \frac{r + (1-r)/n}{\sqrt{t + (1-t)/n}} = \frac{0.5 + 0.5/12}{\sqrt{0.16 + (0.84/12)}} = 1.13$$

Likewise, the relative accuracy for pupal weight is

$$\frac{0.5 + 0.5/12}{\sqrt{0.20 + (0.80/12)}} = 1.05$$

showing that both characters are expected to show a slightly larger response under family selection than under mass selection. Note from Equation 21.14b that the expected response for sib-selection using full-sibs is the same as for parental selection (as $\gamma = 1/2$ in both cases). The relative accuracy of these two methods (sib and parental selection) on larval weight is

$$\frac{\gamma}{\sqrt{t_n}} = \frac{0.5}{\sqrt{0.16 + (0.84/12)}} = 1.04$$

while their relative accuracy for pupal weight is

$$\frac{\gamma}{\sqrt{t_n}} = \frac{0.5}{\sqrt{0.20 + (0.80/12)}} = 0.97$$

Thus, for pupal weight, family selection is slightly more accurate than mass selection, while sib selection and parental selection are slightly less accurate.

Several other studies have compared family and individual selection. Campo and Tagarro (1977) compared full-sib family and individual selection on *Tribolium* pupal weight, using experiments with family sizes of 4 and 10. In both experiments, family selection gave the larger single-generation response, while mass selection had the larger response after six generations. None of these differences were significant. Two other studies compared individual and among-family selection, both using half-sib family selection in chickens. Garwood et al. (1980) examined laying rate ($h^2 = 0.22$) and egg weight ($h^2 = 0.55$) and found that individual selection yielded a greater single-generation response for both characters, but the difference for egg weight was not significant. Kinney et al. (1970) examined several characters, and found that the response under individual selection exceeded that from family selection, although again none of the differences were significant. Lack of significance is not surprising given the (often) small expected differences between methods, coupled with the large evolutionary sampling variance in selection experiments with small to modest populations (Chapter 18).

Finally, it is informative to compare methods using the heritability version of response, $R_x = h_x^2 S_x$. From Equation 21.23, the within-family heritability exceeds the individual heritability when $1 - r > 1 - t$. Hence,

$$h_w^2 > h^2 \quad \text{when} \quad t > r \tag{21.39a}$$

Likewise, from Equation 21.15b, the among-family heritability satisfies

$$h_b^2 > h^2 \quad \text{when} \quad \gamma > t_n \tag{21.39b}$$

The careful reader may have noticed that these conditions are rather different from those given by Equations 21.37a and 21.38a. For example, Equation 21.37a implies that among-family selection yields a larger response than individual selection when $\gamma > \sqrt{t_n}$, whereas Equation 21.39b implies that the among-family heritability is greater than h^2 when $\gamma > t_n$. What is the discrepancy between these two approaches (accuracies versus heritabilities)?

The key is that the variances of the groups being selected differ. Because $\sigma_z^2 = \sigma_b^2 + \sigma_w^2$, the among-family and within-family variances are each less than the phenotypic variance of a random individual. Because $S_x = \bar{\imath}_x \sigma_x$, larger selection intensities are required to give a family-based approach the same selection differential as individual selection. Because the within- and among-family variances are $(1 - t_n)\sigma_z^2$ and $t_n \sigma_z^2$, respectively, it follows that

$$\frac{S_b}{S_m} = \frac{\bar{\imath}_b\, \sigma_b}{\bar{\imath}_m\, \sigma_z} = \frac{\bar{\imath}_b}{\bar{\imath}_m} \sqrt{t_n} \quad \text{and} \quad \frac{S_w}{S_m} = \frac{\bar{\imath}_w\, \sigma_w}{\bar{\imath}_m\, \sigma_z} = \frac{\bar{\imath}_w}{\bar{\imath}_m} \sqrt{1 - t_n} \tag{21.39c}$$

Under identical selection intensities, the differentials for among- and within-family selection are $\sqrt{t_n}$ and $\sqrt{1 - t_n}$, respectively, of the differential under mass selection. Thus, even when h_w^2 or h_b^2 exceeds h^2, this advantage is partially countered by smaller selection differentials due to smaller variances. The contrast of heritabilities as a comparison of expected response assumes that there is the same selection differential, and thus has a hidden assumption of more selection under family-based selection, as the presence of identical S values implies a larger value of $\bar{\imath}$ for family-based selection (Equation 21.39c).

Comparing Selection Intensities: Finite Size Corrections

While not nearly as dramatic as the above differences in the selection differentials (Equation 21.39c), the selection intensities can differ across methods even if the same fraction, p, is saved (Example 21.1). These differences arise from the finite sample-size correction of $\bar{\imath}$ (Chapter 14). Suppose nine individuals are measured, three from each of three families. If we select for the upper 1/3, we keep the best one of three families under among-family selection, and the best of the three individuals within each family under WF selection, resulting in an expected selection intensity of $\bar{\imath}_{(1,3)} = 0.846$ (the expected value of the largest of the first three order statistics). Under family deviations (FD) and mass selection, we chose the largest three of nine values, resulting in an expected selection intensity of $\bar{\imath}_{(3,9)} = 0.996$. Table 21.6 summarizes the selection intensities for the different methods, and shows that $\bar{\imath}_b = \bar{\imath}_{WF} \leq \bar{\imath}_m \leq \bar{\imath}_{FD}$.

An additional subtlety in adjusting the selection intensity was pointed out by Hill (1976, 1977b). The expected selection intensity is computed by taking the expected value of the largest standardized order statistics (Chapter 14), under the assumption that the order statistics are *uncorrelated*. However, with family deviations (FD), family index, and even mass selection, there is the potential for correlations between order statistics. This arises if families contribute different numbers of individuals, resulting in correlations between those measures from the same family, and hence correlations between some of the order

Table 21.6 Selection intensities for various forms of family-based selection schemes corrected for finite sample size. The upper p of the population is saved and the population consists of m families each with n members, for a total of $M = mn$ measured individuals. Tables of exact values for $\bar{\imath}_{(K,M)}$ (the average value of the top K of the M standardized order statistics; see Chapter 14) are given by Becker (1992), and can also be easily obtained via simulations. Approximations for $\bar{\imath}_{(K,M)}$ are given by Equations 14.4a–14.4c.

Selection Type	Corrected Selection Intensity
Individual	$\bar{\imath}_m = \bar{\imath}_{(pM,M)}$
Among-family	$\bar{\imath}_b = \bar{\imath}_{(pm,m)}$
Family-deviations	$\bar{\imath}_{FD} = \bar{\imath}_{(pM,M)}\sqrt{1 + \dfrac{1}{M-1}}$
Within-family	$\bar{\imath}_{WF} = \bar{\imath}_{(pn,n)}$

statistics. The correction for mass selection is generally very small and will be ignored here (see Equation 21.57b). Within-family deviations are negatively correlated within a family, ($\rho = -1/[n-1]$ for a family of size n), as they are deviations from a common family mean. As a result, Dempfle (1990) and Hill et al. (1996) found that the resulting selection intensity for within-family deviations is thus slightly larger than the intensity for mass selection $\bar{\imath}_m$, with

$$\bar{\imath}_{FD} = \bar{\imath}_m\sqrt{1 + \frac{1}{M-1}} \tag{21.40}$$

where M is the total number of measured sibs. On the other hand, with selection on a family index, the correlations between index scores are positive and can be considerable even for large n (Equation 21.58). We will consider the appropriate correction for $\bar{\imath}$ in our treatment of family index selection at the end of this chapter.

Within-family Selection Has Additional Long-term Advantages

The above discussion of the relative efficiencies of different methods focused on a single generation of response from an unselected base population. As we saw in Chapter 16, after one generation of selection, gametic-phase disequilibrium (LD for short) is generated (even among unlinked loci), which (for directional selection) results in a reduction in the additive variance. This reduction, and the resulting decrease in selection response, arises entirely from among-family effects, and thus impacts both individual selection and among-family selection. For unlinked loci, LD does not, however, impact the amount of additive variation *within* a family (Example 16.2), meaning that under strict within-family (WF) selection, there is no decrease in the additive variance from negative LD. Specifically, the amount of within-family additive variance (in the absence of drift or inbreeding) remains at $\sigma_a^2/2$ (half the genic variance, σ_a^2, where $\sigma_a^2 = \sigma_A^2$ in the absence of LD), while the amount of among-family variance is $\sigma_a^2/2 + d$, where $d < 0$ (Chapter 16). Hence, the above comparisons for a single generation undervalue the relative short-term gains from WF selection (with respect to either mass or among-family selection). As we saw in Chapter 16, the reduction in the amount of additive variance due to among-family differences can be substantial ($|d| \gg 0$), especially for a trait that has a moderate to high heritability and is under strong selection (Figure 16.4).

A further advantage of within-family selection appears over the longer term. As mentioned previously, under strict within-family selection (WF), all families contribute the same number of offspring to the next generation, which results in a doubling of the effective population size relative to other schemes that weight among-family information (the latter generates an among-family variance in offspring number, which reduces N_e; Equation 3.4).

As developed in Chapter 26, the long-term response to selection is a function of the effective population size, which results once again in the single-generation comparison of WF to mass or among-family selection that underestimates its relative long-term importance. We examine these issues further in Chapter 26.

RESPONSE WHEN FAMILIES ARE REPLICATED OVER ENVIRONMENTS

Family members are often raised in multiple plots and/or environments, and carefully designed family replication in such a setting offers two potential advantages. First, it allows for the selection of families that perform best over a range of environments, even when extensive genotype $\times$ environment interactions (G$\times$E) are present. Second, replication within an environment reduces the effects of microenvironmental differences, thus increasing the predictability of a family's breeding value and resulting in a larger selection response.

Because family replication is a hallmark of plant breeding, we will examine several schemes used by breeders in detail in this section (in Chapter 23 we examine related designs under inbreeding, while both line crossing and selection in the presence of G$\times$E are more fully examined in Volume 3). Detailed reviews of plant-breeding methodology are given by Namkoong (1979), Hallauer (1981, 1985), Hallauer and Miranda (1981), Nguyen and Sleper (1983), Wricke and Weber (1986), Mayo (1987), Hallauer et al. (1988), Gallais (1990, 2003), Nyquist (1991), Stoskopf et al. (1993), Bos and Caligari (1995), Allard (1999), Holland et al. (2003), Sleper and Poehlman (2006), Acquaah (2007), Bernardo (2010), and Hallauer et al. (2010).

Among-family Variance Under Replication

The expected response to among-family selection under replication follows from Equation 21.14a, using the appropriate among-family variance, $\sigma^2(\,\overline{z}\,)$, given the replication design used by the breeder. In the simplest case, only a single macroenvironment (such as a growing region) is considered, and the family is replicated by raising n_s sibs in each of n_p separate plots (for a total of $N = n_p n_s$ sibs per family). Under this replication scheme, the total environmental value can be partitioned as $E = E_c + E_p + E_s$, representing a common-family effect (E_c), a plot-specific effect (E_p), and individual-specific effects (E_s). Following similar logic to that in Example 21.3, the resulting variance becomes

$$\sigma^2(\,\overline{z}\,) = \sigma^2_F + \frac{\sigma^2_{E_p}}{n_p} + \frac{\sigma^2_w}{N} \tag{21.41a}$$

where $\sigma^2_{E_p}$ is the plot-to-plot variance (the environmental variance among plots in the same macroenvironment), $\sigma^2_F = \sigma^2_{GF} + \sigma^2_{E_c}$ is the among-family variance, and $\sigma^2_w = \sigma^2_{Gw} + \sigma^2_{E_s}$ is the within-plot variance of individuals about their family averages. Recall that $\sigma^2_{Gw} = \sigma^2_G - \sigma^2_{GF}$, and values for the among- and within-family genetic variances are given by Equations 21.11a and 21.11b, respectively, when epistasis is absent, and more generally by Equations 21.26a and 21.26b.

An alternative way to express the variance of family means is

$$\sigma^2(\,\overline{z}\,) = \sigma^2_{GF} + \sigma^2_{E_c} + \sigma^2(\epsilon) \tag{21.41b}$$

where for the design given by Equation 21.41a, the residual variance is

$$\sigma^2(\epsilon) = \frac{\sigma^2_{Gw}}{N} + \frac{\sigma^2_{E_s}}{N} + \frac{\sigma^2_{E_p}}{n_p} = \left(\frac{1}{N}\right)\left(\sigma^2_{Gw} + \sigma^2_{E_s} + n_s\sigma^2_{E_p}\right) \tag{21.41c}$$

The critical observation is that the contribution from $\sigma^2(\epsilon)$ can be largely controlled by the experimental design (here, the choice of n_s and n_p).

More generally, if the family is replicated over n_e distinct macroenvironments, each with n_p plots and n_s sibs per plot, for a total of $N = n_p n_s n_e$ sibs, Equation 21.41b holds, with the residual variance now being

$$\sigma^2(\epsilon) = \frac{\sigma^2_{Gw}}{N} + \frac{\sigma^2_{GF\times E}}{n_e} + \frac{\sigma^2_{E_p}}{n_e n_p} + \frac{\sigma^2_{E_s}}{N}$$
$$= \left(\frac{1}{N}\right)\left(\sigma^2_{Gw} + \sigma^2_{E_s} + n_s\sigma^2_{E_p} + n_p n_s\,\sigma^2_{GF\times E}\right) \tag{21.42a}$$

where σ^2_{GF} is the genetic variance among family means over this set of environments, and $\sigma^2_{GF\times E}$ is the variance from the family-environment interaction (LW Chapter 22).

Plant breeders often use an alternative partition of the environment into location (L) and year (Y) effects. Suppose a family is replicated over n_ℓ locations over n_y years, where each of the $n_\ell\, n_y$ year-location combinations is replicated as n_p plots of n_s sibs each, for a total of $N = n_\ell\, n_y\, n_p\, n_s$ sibs per family. Again, Equation 21.41b holds, with a residual variance of

$$\sigma^2(\epsilon) = \frac{\sigma^2_{GF\times L}}{n_\ell} + \frac{\sigma^2_{GF\times Y}}{n_y} + \frac{\sigma^2_{GF\times L\times Y}}{n_\ell\, n_y} + \frac{\sigma^2_{E_p}}{n_\ell\, n_y\, n_p} + \frac{\sigma^2_{Gw} + \sigma^2_{E_s}}{n_\ell\, n_y\, n_p\, n_s}$$
$$= \left(\frac{1}{N}\right)\left[\sigma^2_{Gw} + \sigma^2_{E_s} + n_s\sigma^2_{E_p} + n_p n_s\left(n_y\sigma^2_{GF\times L} + n_\ell\sigma^2_{GF\times Y} + \sigma^2_{GF\times L\times Y}\right)\right] \tag{21.42b}$$

where $\sigma^2_{GF\times L}$, $\sigma^2_{GF\times Y}$, and $\sigma^2_{GF\times L\times Y}$ are the family by environment (year, location, and year-location) interactions (Lonnquist 1964; Comstock and Moll 1973; Patterson et al. 1977; Brennan and Byth 1979; Thompson and Cunningham 1979).

The above expressions for $\sigma^2(\epsilon)$ show the importance of replication and provide some guidance as to how one should allocate resources. For a fixed number of sibs per family (N), how should one choose n_e, n_p, and n_s to minimize $\sigma^2(\overline{z})$? If N is fixed, then the relative weightings on the within-family genetic variance and individual-specific environmental variance are fixed. When the genotype×environment interaction variance ($\sigma^2_{GF\times E}$) is large, its effect on the selection response can be reduced by replicating families across more environments (increasing n_e). More generally, when viewing environments as locations-years (Equation 21.42b), the total number of environments is $n_e = n_\ell\, n_y$, and preliminary estimates of the variation components ($\sigma^2_{GF\times L}$, $\sigma^2_{GF\times Y}$, and $\sigma^2_{GF\times L\times Y}$) can suggest the appropriate allocation over locations versus years for a fixed value of n_e. When the among-plot variance ($\sigma^2_{E_p}$) is large, its effect is reduced by increasing n_p or n_e. With preliminary estimates of the variance components in hand, one can numerically search for the optimal values of n_e, n_p, and n_s that give the smallest $\sigma^2(\overline{z})$ for a fixed value of $N = n_e\, n_p\, n_s$. Using replication can result in a considerable improvement over mass selection. For example, using variance components estimated for maize lines grown in several locations in India, Sanghi (1983) estimated that full-sib selection with replication would be three to sixfold times more efficient than mass selection.

One consequence of replication is that the among-family heritability, $h^2_b = \gamma\,\sigma^2_A/\sigma^2(\overline{z})$ (Equation 21.15b), is now a complex function of the design, namely, the values of n_e and n_p, in addition to the total number of sibs, which enter through $\sigma^2(\overline{z})$, via $\sigma^2(\epsilon)$. Thus, with replication, an among-family heritability does not directly translate into an individual heritability (Hanson 1963; Nyquist 1991; Holland et al. 2003). Even with the same variance components, h^2_b changes as a function of the replication design. Hanson suggested that, when replication is present, the among-family heritability needs to be defined with respect to a particular standard design, such as his proposal in soybeans of a design with two years over two locations, with two replications in each location-year combination.

Finally, consider the among-family variance under a nested-sib design with replication. Suppose (as before) that there are n_f females per male, but now that each full-sib family is replicated as n_s sibs over n_e environments. The resulting variance becomes

$$\sigma^2(\,\overline{z}\,) = \sigma^2_{GF(HS)} + \frac{\sigma^2_{G(f|m)}}{n_f} + \frac{\sigma^2_{GF(HS)\times E}}{n_e} + \frac{\sigma^2_{G(f|m)\times E}}{n_f\, n_e} + \frac{\sigma^2_{Gw(FS)} + \sigma^2_{E_s}}{N} \tag{21.43a}$$

where $N = n_f n_e n_s$ is the total number of half-sibs per male (Robertson et al. 1955; Webel and Lonnquist 1967; da Silva and Lonnquist 1968). Assuming no epistasis or genotype by environment interaction (G×E), we can express this among-family variance as

$$\frac{\sigma^2_A}{4} + \frac{\sigma^2_A + \sigma^2_D}{4\,n_f} + \frac{\sigma^2_{A\times E}}{4\,n_e} + \frac{\sigma^2_{A\times E} + \sigma^2_{D\times E}}{4\,n_f\,n_e} + \frac{(1/2)\sigma^2_A + (3/4)\sigma^2_D + \sigma^2_{E_s}}{N} \tag{21.43b}$$

The extension of this result to multiple plots per location when G×E is present follows in a similar fashion from our development above for Equation 21.42b.

An example of family selection with replication was provided by selection for increased grain yield in maize by the International Maize and Wheat Improvement Center (CIMMYT), summarized by Pandey et al. (1986, 1987) and Crossa and Gardner (1989). The goal of the CIMMYT selection schemes was to develop varieties of maize that yield well over a wide range of environments. Starting in 1974, 250 full-sib families, along with six local **checks** (control lines to allow for standardized comparisons), were evaluated at six lowland tropical locations (with two replications per location) in the northern and southern hemispheres. A total of 28 countries were used during the course of five cycles of selection. Selection (initially) was strictly among families with the international field trials conducted on full-sib families, while the recombination unit consisted of S_1 seed from the superior families. The selection scheme was later modified to allow for within-family selection as well. Roughly 50% of the families were selected based on the international trials, about 20% of which were subsequently rejected given their poor performance in disease- and insect-resistance trials in separate nurseries. The average gain in yield per cycle was around 2%.

Example 21.9. Eberhart et al. (1966) estimated genetic variance components for seven characters in two open-pollinated maize varieties. Using individuals grown in two locations in North Carolina, they obtained the following estimates for yield in the variety Jarvis:

$$\sigma^2_A = 120, \quad \sigma^2_{A\times L} = 114, \quad \sigma^2_D = 270, \quad \sigma^2_{D\times L} = 98, \quad \sigma^2_{E_s} = 508$$

Estimates of epistatic variances were not significantly different from zero. Consider the expected response under a design with 25 half-sib families, each with a total of 50 offspring scored over five environments ($n_e = 5$). The top five families were selected, using S_1 seed to form the next generation (allowing for selection on both sexes). Recalling Equation 21.14a (with $\gamma = 1/2$ for S_1 seed; Equation 21.14b), the expected response will be

$$R = \frac{2\,\bar{\imath}_{(5,25)}\,(\sigma^2_A/4)}{\sigma(\,\overline{z}\,)} = \frac{2 \cdot 1.345 \cdot 30}{\sigma(\,\overline{z}\,)} = \frac{80.7}{\sigma(\,\overline{z}\,)}$$

using Equation 14.4b to obtain $\bar{\imath}_{(5,25)}$. If we use the above variance estimates, then $\sigma^2_{GF\times E} = \sigma^2_{A\times L}/4 = 28.5$, while Equation 21.26a and 21.26b yield, respectively, $\sigma^2_{GF} = \sigma^2_A/4 = 30$ and $\sigma^2_{Gw} = (3/4)\sigma^2_A + \sigma^2_D = 360$. If the families being scored are strict half-sibs (meaning that all offspring from a pollen parent each have a different seed parent, $n_f = N = 50$), then Equation 21.42a returns

$$\sigma^2(\,\overline{z}_{HS}\,) = \sigma^2_{GF} + \frac{\sigma^2_{Gw} + \sigma^2_{E_s}}{N} + \frac{\sigma^2_{GF\times L}}{n_e}$$
$$= 30 + \frac{360 + 508}{50} + \frac{28.5}{5} = 53.06$$

and the expected response becomes $80.7/\sqrt{53.06} = 11.08$.

Now suppose that the sibs are from a nested design with each male pollinating five seed parents, and with each cross producing 10 offspring ($n_f = 5$, $N = 50$). Using the above variance components, Equation 21.43b yields

$$\sigma^2(\,\overline{z}\,) = \frac{\sigma_A^2}{4} + \frac{\sigma_A^2 + \sigma_D^2}{4\,n_f} + \frac{\sigma_{A\times L}^2}{4\,n_e} + \frac{\sigma_{A\times L}^2 + \sigma_{D\times L}^2}{4\,n_f\,n_e} + \frac{(1/2)\sigma_A^2 + (3/4)\sigma_D^2 + \sigma_{E_s}^2}{N}$$
$$= \frac{120}{4} + \frac{120+270}{20} + \frac{114}{20} + \frac{114+98}{100} + \frac{(1/2)120 + (3/4)270 + 508}{50}$$
$$= 72.73$$

resulting in an expected response of $80.7/\sqrt{72.73} = 9.47$. Hence, the strict half-sib design has a smaller among-family variance, and thus a 117% larger expected response than expected under a nested design.

Ear-to-Row Selection

One of the earliest examples of family-based selection was the **ear-to-row** selection method in maize, first used by Hopkins (1899) to start his classic long-term selection experiment (Chapter 26). Here the seeds from each maize ear are planted in a single row (so that a row corresponds to a family), with individuals from the best rows chosen as seed parents for the next generation. Plants in the rows to be scored are either detassled or have their tassles (pollen-producing structures) bagged, removing their ability to contribute pollen. As a result, these plants can neither self nor pollinate. Pollen is provided by rows planted with bulk of all seeds (a **polycross** mating design). Assuming open pollination, the seeds on a single ear are half-sibs (with a common mother), which means that the ear-to-row method is an example of half-sib family selection, with selection on only one sex (the seed parent). In rice, **panicle-to-row** selection has been used (e.g., Ntanos and Roupakias 2001), where the **panicle** is essentially the equivalent of the maize ear, and again a row equals a family.

Suppose a total of $n = n_e\,n_p\,n_s$ sibs per family are scored, by growing n_p rows of n_s sibs, replicated over n_e distinct environments. From Equation 21.15c, the expected response under ear-to-row selection, when choosing the top K of M families ($p = K/M$), is

$$R_{ER} = \bar{\imath}_{(pM,M)} \frac{(1+3/n)\,(\sigma_A^2/8)}{\sigma(\,\overline{z}_{HS})} \simeq \bar{\imath}_{(pM,M)} \frac{\sigma_A^2/8}{\sigma(\,\overline{z}_{HS})} \tag{21.44}$$

where $\sigma^2(\,\overline{z}_{HS}\,)$ is calculated by Equation 21.42a. For large values of n (in the absence of epistasis),

$$R_{ER} = \bar{\imath}_{(pM,M)} \frac{\sigma_A^2/8}{\sqrt{\dfrac{\sigma_A^2}{2} + \dfrac{\sigma_{GF\times E}^2}{n_e} + \dfrac{\sigma_{E_p}^2}{n_e\,n_p} + \dfrac{\sigma_{E_s}^2}{n}}} \tag{21.45}$$

Modified Ear-to-Row Selection

The ear-to-row method has the advantage of being fairly easy to implement for testing a family (with replication reducing the effects from the environmental variance), coupled with the same cycle time as mass selection (one generation). As a result, this method was commonly used by early maize breeders, for example, Hopkins (1899), Smith (1908, 1909), Montgomery (1909), Williams and Walton (1915), Kiesselbach (1916), and Hume (1919). While it proved effective at modifying highly heritable traits (such as kernel protein and oil content), ear-to-row selection was generally not successful in improving yield (Kiesselbach 1922; Richey 1922; Smith and Bruson 1925), and it was not regarded as a practical scheme for yield improvement. Sprague (1955) suggested that the failure for yield improvement

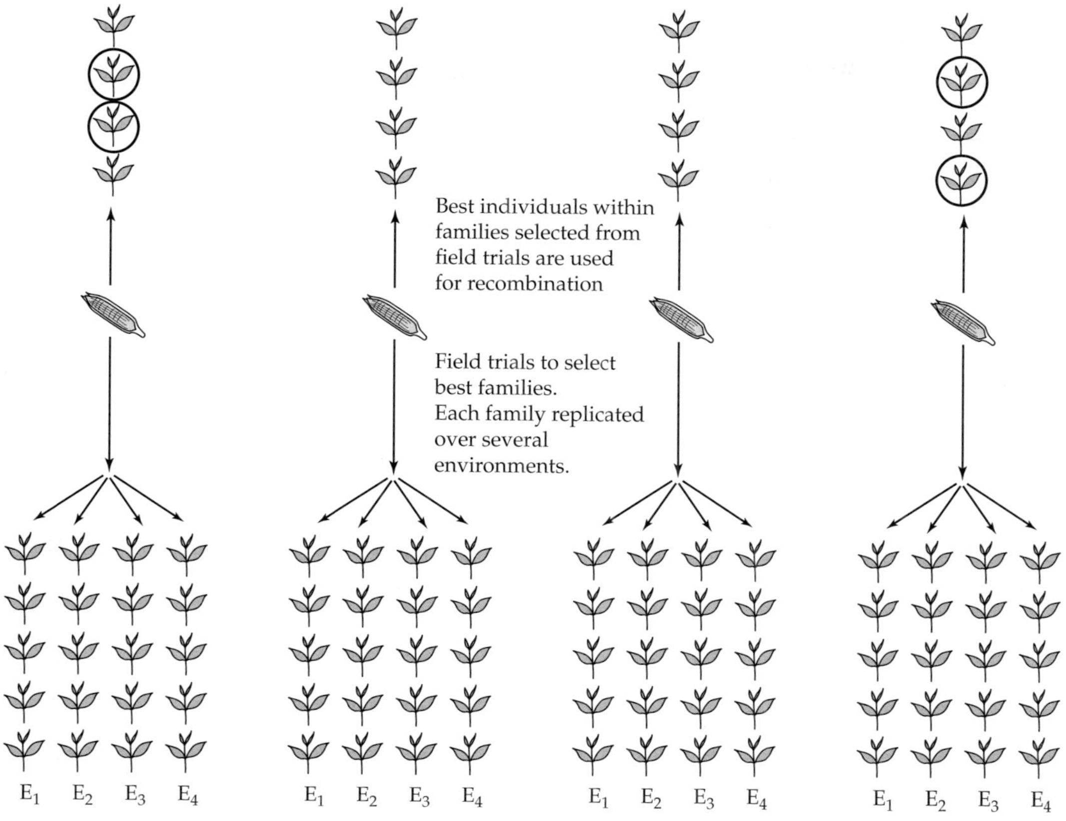

Figure 21.6 Lonnquist's (1964) modified ear-to-row selection scheme. Half-sib families (represented here by the maize ears in the middle of the figure) are planted both as rows in multiple environments (the yield trials over environments E_1 through E_4 at the bottom of the figure) and as a single additional row in yet another location, the so-called crossing block (the rows at the top of the figure). From the best families in the yield trials (the first and fourth in the above figure) one then chooses the best individuals (indicated by the circled plants) from their sibs in the crossing block (within-family selection) to form the next generation.

was largely the result of insufficient control over environmental variance, which resulted in σ_E^2 largely obscuring the additive variance. (For this same reason, mass selection was also regarded as being impractical for improving maize yield.) An alternative hypothesis was suggested by Hull (1945, 1952), who thought that the lack of response in yield was a result of most of the genetic variance being nonadditive. The finding of considerable additive variance in yield by a number of maize geneticists motivated Lonnquist's (1964) development of the **modified ear-to-row** scheme, a combined selection approach involving both among-family (ear-to-row) and within-family (within-row) selection (Figure 21.6).

Under Lonnquist's design, seed from each family is planted as rows in several environments. These form the **yield** or **performance trials** for selecting the best-performing families averaged over these environments. On a separate plot (the **crossing block**), additional seeds from each family are planted as a single row. Within the crossing block, the best individuals from the rows corresponding to the families with the best performance in the yield trials are used as the seed parents for the next generation. Selection is only on one parent in the crossing block, as plants are detassled and open pollinated from a random bulk of all the initially planted families. One advantage of this scheme is that one can use bulk measures over rows in the yield trials and more detailed (and labor-intensive) individual plant measures in the smaller crossing block.

Under Lonnquist's original design, the replicated field trials and the crossing block are grown contemporaneously (planting of the crossing block may be delayed slightly to ensure that all field information from the yield trials can be gathered). Thus, one cycle of modified ear-to-row selection can be carried out in a single generation. The expected total response is the sum of the expected gains at each step in the cycle, $R_{ER(m)} = R_{ER} + R_{ER(w)}$. The response, R_{ER}, under the first step (choosing the best families) is the same as for standard ear-to-row selection (Equations 21.44 and 21.45). Because plants in the crossing block are open pollinated using a bulk of all families, selection is only on females within each row. If one chooses the best $k = qn_s$ of n_s plants within each selected row (i.e., strict within-family [FW] selection saving the upper fraction, q), the expected response to within-row selection becomes

$$R_{ER(w)} = \bar{\imath}_{(qn_s,n_s)} \frac{(3/8)\,\sigma_A^2}{\sigma_{w(HS)}} \tag{21.46}$$

Because families are not replicated within the crossing block, then

$$\sigma^2_{w(HS)} = \sigma^2_{Gw(HS)} + \sigma^2_{E_s}$$

Hence, in the absence of epistasis, the component of response from within-row selection becomes (Equation 21.22b)

$$R_{ER(w)} = \bar{\imath}_{(qn_s,n_s)} \frac{(3/8)\,\sigma_A^2}{\sqrt{(3/4)\sigma_A^2 + \sigma_D^2 + \sigma_{E_s}^2}} \tag{21.47}$$

Ignoring any potential changes in σ_A^2 due to the first step of selection (ear-to-row), the expected response becomes

$$\begin{aligned} R_{ER(m)} &= R_{ER} + R_{ER(w)} \\ &= \bar{\imath}_{(pM,M)} \frac{\sigma_A^2/8}{\sigma(\,\overline{z}_{HS})} + \bar{\imath}_{(qn_s,n_s)} \frac{(3/8)\sigma_A^2}{\sigma_{w(HS)}} \end{aligned} \tag{21.48a}$$

where we have chosen the best $K = pM$ of M families in the yield trials and the best $k = qn_s$ of n_s within each selected family in the crossing block. With a large number of sibs per row (n_s is large) and a roughly equal selection within and among rows ($\bar{\imath}_{(pM,M)} \simeq \bar{\imath}_{(qn_s,n_s)} = \bar{\imath}$), the expected response to modified ear-to-row selection is

$$R_{ER(m)} = \frac{\bar{\imath}\,\sigma_A^2/8}{\sqrt{\dfrac{\sigma_A^2}{2} + \dfrac{\sigma^2_{F\times E}}{n_e} + \dfrac{\sigma^2_{E_p}}{n_e\,n_p} + \dfrac{\sigma^2_{E_s}}{N}}} + \frac{\bar{\imath}\,(3/8)\sigma_A^2}{\sqrt{\dfrac{3\sigma_A^2}{4} + \sigma_D^2 + \sigma_{E_s}^2}} \tag{21.48b}$$

Inspection of Equation 21.48b shows that it is not obvious which component (within- vs. among-family) contributes more to the total selection response. The threefold increase in usable additive variance in the within-family component in the numerator can be partly or fully offset by the fact that $\sigma^2_{Gw} > \sigma^2_{GF}$ (the within-family genetic variance is greater than the among-family variance; see Equations 21.26a and 21.26b). Likewise, it is not clear whether the among- or the within-family environmental variance is expected to be larger. Some fine-tuning is possible on the among-family component, as, if estimates of the appropriate environmental variances are available, changing the experimental design (the values of n_p, n_s, and n_e) can reduce $\sigma^2(\epsilon)$.

Example 21.10. Webel and Lonnquist (1967) used modified ear-to-row selection for yield in the Hays Golden open-pollinated variety of maize. Performance of each family was evaluated using single rows grown in three different locations. Based on these yield trials, the best 44 of roughly 220 families were identified. In the crossing block, the best 5 of the 25 (or so) plants were chosen in each of the 44 rows corresponding to the selected families. The resulting expected selection intensities for the among- and within-family components were $\bar{\imath}_{(44,220)} = 1.40$ and $\bar{\imath}_{(5,25)} = 1.35$, respectively (Equation 14.4b). Over the first four cycles of selection, Webel and Lonnquist observed a 9.4% increase in yield per cycle, compared with the 3% increase per cycle observed under mass selection (Gardner 1973). The predicted response was 8.4%, with expected contributions of 4.6% from among-families (55% of predicted response) and 3.8% from within-families. The results for 10 cycles of selection were summarized by Compton and Bahadur (1977). Paterniani (1967) also used modified ear-to-row selection for yield for three cycles in Brazilian maize populations. The average yield increased by 42% over the course of the experiment.

Compton and Comstock (1976) suggested a variant of Lonnquist's design. This approach is also referred to as **among-and-within-family selection (AWF)** or **between-and-within-family selection (B&WFS)** by forage breeders (Aastveit and Aastveit 1990; Vogel and Pedersen 1993; Casler and Brummer 2008). Families are again planted ear-to-row in performance trials, but remnant seed from each family is stored. The best families are chosen and the remnant seed for these families is planted to form the crossing block. The pollen plants in the crossing block are a bulk of the selected families. Hence, both parents in the crossing block are subjected to half-sib selection, which doubles the response from the among-family component, and yields

$$R_{ER(m)} = \bar{\imath}_{(pM,M)} \frac{(1/4)\sigma_A^2}{\sigma(\,\overline{z}_{HS})} + \bar{\imath}_{(qn_s,n_s)} \frac{(3/8)\sigma_A^2}{\sigma_{W(HS)}} \tag{21.49}$$

The Compton-Comstock modified ear-to-row scheme requires two generations per cycle, but it offers increased response (per cycle) as the pollen is also from selected parents. Using the predicted values of Webel and Lonnquist (Example 21.10), the expected response per cycle under the Compton-Comstock design would be 2 · 4.6 + 3.8 = 13, for an expected 155% increase per cycle over the Lonnquist design (which had a predicted response 8.4). However, the Compton-Comstock design also requires two generations per cycle, with the result that the response per generation is 6.5, 77% of that expected under the Lonnquist design. The use of **off-season** (or winter) **nurseries**, where seeds are grown in either the opposite hemisphere or in the tropics (such as the Hawaiian island of Moloka'i), can allow for two generations in the same calendar year, but this may require more resources than the breeder has available.

SELECTION ON A FAMILY INDEX

While our focus to this point has been on schemes that use either within- or among-family selection, the modified ear-to-row approach points out the advantage of using selection schemes containing both within- and among-family components. The modified ear-to-row approach is an example of **combined selection**, where the components are sequentially selected in different generations (and/or plots), and several such schemes are used by plant breeders. Alternatively, one can use both within- and among-family information to select individuals within a single generation. The most general way to do this is to select on a **family index**,

$$I_{ij} = b_1 \,(z_{ij} - \overline{z}_i) + b_2\, \overline{z}_i \tag{21.50a}$$

where the index value, I_{ij}, is for individual j from family i. Individuals with the largest index scores are mated (avoiding within-family crosses) to form the next generation. Note that individual ($I_{ij} = z_{ij}$), family ($I_{ij} = \overline{z}_i$), and family-deviations ($I_{ij} = z_{ij} - \overline{z}_i$) selection are all special cases of this general family index, which correspond to weights of $b_1 = b_2$, $b_1 = 0$, and $b_2 = 0$, respectively.

An important point is that the *relative values* of the index weights, not their absolute values, define the choice of individuals—if both weights are multiplied by the same constant, the same individuals are chosen by the new index. As a result, the family index is often written as

$$I_{ij} = z_{ij} + B\,\overline{z}_i \tag{21.50b}$$

where B is the relative weight on family mean compared to an individual's phenotype. As the reader can easily verify with a little algebra, this is equivalent to the index given by Equation 21.50a, with

$$B = \frac{b_2}{b_1} - 1 \tag{21.50c}$$

Response to Selection on a Family Index

Once again, either Equations 21.1a or 21.4a can be used to predict the single-generation response to selection. Taking $x = I$ returns

$$R_I = \frac{\sigma(I, y\,|\,\mathcal{R}_1)}{\sigma_I^2}\left(S_{I_m} + S_{I_f}\right) = \overline{\imath}_I\,\sigma_z\,\rho(I, y) \tag{21.51}$$

where $\sigma(I, y\,|\,\mathcal{R}_1)$ is the covariance between the index value, I, of a parent and the phenotype of its offspring, y. The variances and covariances required for Equation 21.51 are obtained as follows. Using the covariances summarized in Table 21.3,

$$\begin{aligned}\sigma(I, y\,|\,\mathcal{R}_1) &= b_1\,\sigma(z_{ij} - \overline{z}_i, y\,|\,\mathcal{R}_1 = x_{ij}) + b_2\,\sigma(\,\overline{z}_i, y\,|\,\mathcal{R}_1 = x_{ij})\\ &= b_1\,(1 - r_n)(\sigma_A^2/2) + b_2\,r_n(\sigma_A^2/2)\\ &= [\,b_1 + r_n(b_2 - b_1)\,]\;(\sigma_A^2/2)\end{aligned} \tag{21.52a}$$

Likewise, if we recall that $\sigma^2(x + y) = \sigma_x^2 + \sigma_y^2 + 2\sigma_{x,y}$, the variances summarized in Table 21.4 yield

$$\begin{aligned}\sigma^2(I) &= b_1^2\,\sigma^2(z_{ij} - \overline{z}_i) + b_2^2\,\sigma^2(\,\overline{z}_i) + 2\,b_1\,b_2\,\sigma(z_{ij} - \overline{z}_i, \overline{z}_i)\\ &= b_1^2\,(1 - t_n)\,\sigma_z^2 + b_2^2\,t_n\,\sigma_z^2 + 2\,b_1\,b_2\,\sigma(z_{ij}, \overline{z}_i) - 2\,b_1\,b_2\,\sigma^2(\,\overline{z}_i)\\ &= \left(b_1^2\,(1 - t_n) + b_2^2\,t_n + 2\,b_1\,b_2\,t_n - 2\,b_1\,b_2\,t_n\right)\sigma_z^2\\ &= \left[\,b_1^2 + t_n(b_2^2 - b_1^2)\,\right]\,\sigma_z^2\end{aligned} \tag{21.52b}$$

The resulting heritability of the index becomes

$$h_I^2 = \frac{2\,\sigma(I, y\,|\,\mathcal{R}_1)}{\sigma^2(I)} = h^2\left[\frac{b_1 + r_n(b_2 - b_1)}{b_1^2 + t_n(b_2^2 - b_1^2)}\right] \tag{21.53a}$$

Finally, because parents only pass along half their breeding value to an offspring (Chapters 6 and 16), it follows that $\sigma(I, y\,|\,\mathcal{R}_1) = \sigma(I, A)/2$, namely, half the covariance between the parent's index and breeding values. Hence, from Equations 21.52a and 21.52b, the correlation between an individual's index score (I) and breeding value (A) is

$$\rho(I, A) = \frac{\sigma(I, A)}{\sigma(I)\,\sigma(A)} = \frac{2\sigma(I, y\,|\,\mathcal{R}_1)}{\sigma(I)\,\sigma(A)} = h\left[\frac{b_1 + r_n(b_2 - b_1)}{\sqrt{b_1^2 + t_n(b_2^2 - b_1^2)}}\right] \tag{21.53b}$$

Given that $\rho(z, A) = h$ (Equation 13.11e), the term in the brackets represents the accuracy of the index relative to mass selection. Substituting Equation 21.53a into Equation 21.51 (and recalling that $\sigma_z h^2 = h\sigma_A$) yields an expected response of

$$R_I = \overline{\imath}_I\,h\,\sigma_A\,\frac{b_1 + r_n(b_2 - b_1)}{\sqrt{b_1^2 + t_n(b_2^2 - b_1^2)}} \tag{21.53c}$$

where $\bar{\imath}_I = (\bar{\imath}_{I_m} + \bar{\imath}_{I_f})/2 = (S_{I_m} + S s_{I_f})/(2\sigma_I)$ is the average selection intensity on both sexes. Observe from Equation 21.53c that if we create a new index with weights of ab_1 and ab_2, that the constant a cancels, and (as noted above) yields the same response.

Example 21.11. Again consider the work of Clayton et al. (1957) on abdominal bristle number in *Drosophila* (Examples 21.4 and 21.5). Here $r_n = 0.542$, $t_n = 0.326$, and $\sigma_A\, h = 1.70$. Suppose individuals with index scores in the upper 20% are chosen. What is the expected response if we place three times the weight on within-family deviations as we do on family means ($b_1 = 3, b_2 = 1$)? Because 20 families each with 12 sibs are scored, the expected selection intensity is $\bar{\imath}_{(48,240)} = 1.39$ (as 48 is the upper 20% of 20 · 12 = 240), and Equation 21.53c yields an expected response of

$$R_I = 1.39 \cdot 1.70 \left(\frac{3 + 0.542(1-3)}{\sqrt{3^2 + 0.326(1^2 - 3^2)}} \right) = 1.79$$

This is not as efficient as strict among-family selection (where R_b = 2.15; see Example 21.4). Likewise, the response under individual (i.e., mass) selection is $R_m = \bar{\imath}_m\, \sigma_A\, h = 2.36$. Because individual selection is a special case of the general index, we can always choose the index weights to give at least as large an expected response as individual selection. For example, placing twice the weight on family means relative to within-family deviations ($b_1 = 1, b_2 = 2$), returns an expected response of $R = 2.59$, which is 110% of the expected response under individual selection.

Lush's Optimal Index

As the previous example shows, by making the appropriate choice of index weights, we can always obtain a response at least as large as that expected under mass selection. Note from Equation 21.51 that σ_A and $\bar{\imath}_I$ remain constant under different index weights, implying that the maximal response occurs if we choose the weights that maximize the correlation, $\rho(I, y)$, between the index and offspring value (Equation 21.5 shows that this is equivalent to maximizing the correlation, $\rho(I, A)$, between the index and breeding values of an individual). Lush (1947) showed that the resulting optimal index weights are

$$b_1 = \frac{1-r}{1-t} \quad \text{and} \quad b_2 = \frac{1+(n-1)\,r}{1+(n-1)\,t} \tag{21.54}$$

The formal derivation (which follows from a Smith-Hazel index; Example A6.8) is given in our general treatment of index selection in Volume 3. We refer to the family index using these weights as the **Lush index**. Note that the weight (b_1) on family deviations is independent of the family size (n), while the weight on the family mean (b_2) depends on n, approaching r/t for large families. Figure 21.7 plots the ratio of among- to within-family weights (b_2/b_1) for the Lush index as a function of t and n. For small between-sib correlations (t), more weight is placed on family mean, while more weight is placed on within-family deviation when the sib correlation is large.

We can rearrange the Lush index as $I_L = z_{ij} + B_L \bar{z}_i$, where substituting Equation 21.54 into Equation 21.50c returns

$$B_L = \frac{(r-t)n}{(1-r)[\,1+(n-1)t\,]} \tag{21.55}$$

Using the optimal weights, Equation 21.53c simplifies to yield the response under Lush's index as

$$R_{LI} = \bar{\imath}\,\sigma_A\, h\, \sqrt{1 + \frac{(r-t)^2(n-1)}{(1-t)[1+(n-1)t]}} \tag{21.56a}$$

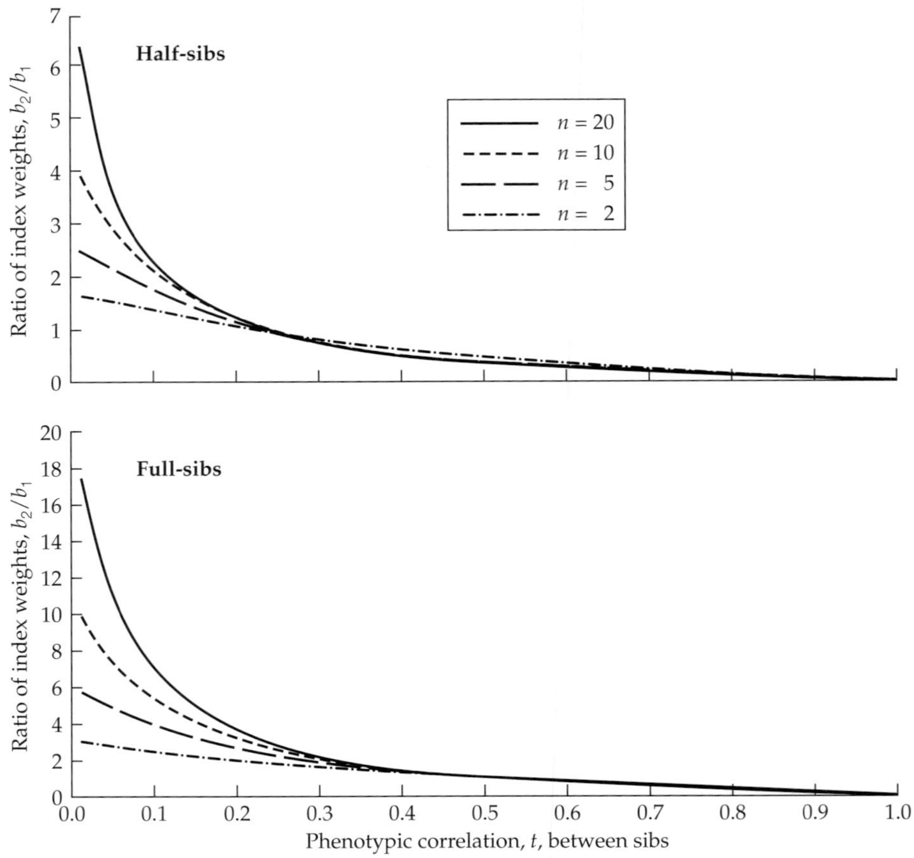

Figure 21.7 The ratio, b_2/b_1, of the weights placed on the among- (b_2) relative to within- (b_1) family weights under the optimal Lush index (Equation 21.54). Individual selection corresponds to $b_2/b_1 = 1$. These optimal weights are a function of the phenotypic correlation, t, between sibs and the number, n, of sibs per family.

The resulting increase in response over that expected under individual selection ($R_m = \bar{\imath}\,\sigma_A\,h$; Equation 13.6b) is thus

$$\frac{R_{LI}}{R_m} = \sqrt{1 + \frac{(r-t)^2(n-1)}{(1-t)[1+(n-1)t]}} \geq 1 \tag{21.56b}$$

Figure 21.8 plots Equation 21.56b as a function of t and n for half- and full-sibs. Because the quantity in the square root exceeds one, the expected response under Lush's index exceeds the response under individual selection, except at $r = t$ (i.e., $t = 0.25$ for half-sibs, $t = 0.5$ for full-sibs), in which case the expected responses are equal. For large values of n, Equation 21.56b converges to

$$\frac{R_{LI}}{R_m} = \sqrt{1 + \frac{(r-t)^2}{(1-t)t}} \tag{21.56c}$$

which can take on large values for t near zero or one, as seen by the roughly U-shaped plots in Figure 21.8.

Example 21.12. Recalling (Example 21.4) that $t = 0.265$ and $r = 0.5$ for full-sibs in Clayton et al.'s (1957) bristle experiments, the resulting Lush weight on family deviations becomes

$$b_1 = \frac{1-r}{1-t} = \frac{1-0.5}{1-0.265} = 0.680$$

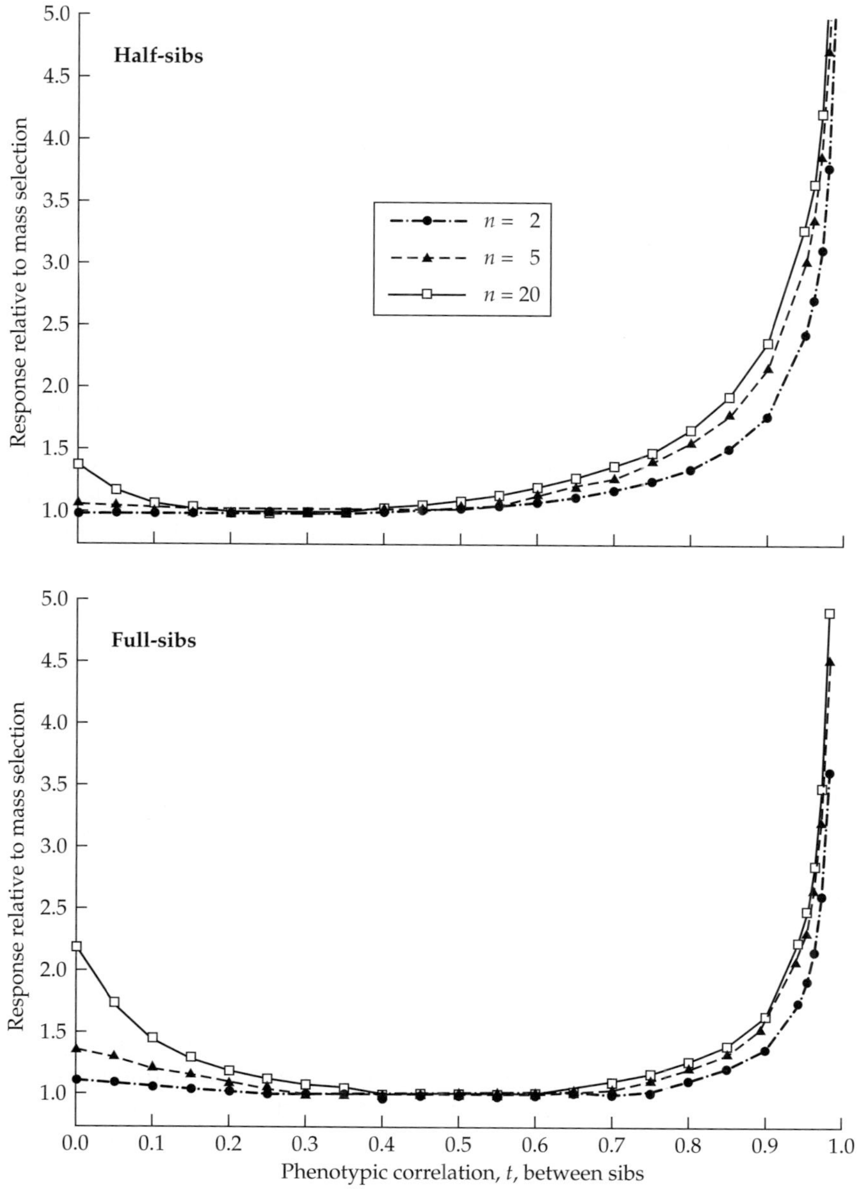

Figure 21.8 Response of Lush's index relative to individual selection, as a function of the number of sibs, n, for full-sibs ($r = 1/2$) and half-sibs ($r = 1/4$). Except at $r = t$ (where the expected responses are equal), Lush's index results in a larger expected response than individual selection.

If we further recall that the family size was $n = 12$, the optimal weight on family means becomes

$$b_2 = \frac{1 + (n-1)\,r}{1 + (n-1)\,t} = \frac{1 + (12-1)\,0.5}{1 + (12-1)\,0.265} = 1.66$$

We can rescale the weights as $b_1 = 1$ and $b_2 = 1.66/0.680 = 2.44$, with Equation 21.53c returning an expected response of

$$R_{LI} = \bar{\imath}_{LI}\,\sigma_A\,h\,\rho(I,y) = 1.39 \cdot 1.70 \cdot \left(\frac{1 + 0.542(2.44 - 1)}{\sqrt{1^2 + 0.326(2.44^2 - 1^2)}} \right) = 2.60$$

The expected response under individual selection was $R_m = 2.36$ (Example 21.11), so the expected response under the Lush index is 10% greater than that of mass selection.

The Lush index weights change with t and r, and so may have to be periodically updated as changes in the genetic variance change t and as inbreeding changes r (Chapters 16, 23, and 24). In particular, both drift and gametic-phase disequilibrium can be important when several generations of selection are considered (Chapter 16). As selection proceeds, both these forces increase the relative importance of within-family selection over among-family selection (Chapters 16 and 24). This results in individual values given increased weight and family means given decreased weight.

Specifically, the amount of within-family additive variance (in the absence of drift or inbreeding) remains at $\sigma_a^2/2$ (half the genic variance, the value of σ_A^2 in the absence of LD), while the amount of among-family variance is $\sigma_a^2/2 + d$, where $d < 0$ (Chapter 16). Hence, LD has no impact on the within-family component of additive variance, but it decreases the among-family component. Wray and Hill (1989) noted that while the relative efficiency of index selection over individual selection may be greatly diminished by gametic-phase disequilibrium, the relative rankings of the methods still hold.

A concern with any index is that the population parameters have to be correctly estimated, otherwise the index constructed from these estimates will have incorrect weights and be less than optimal (Volume 3). Fortunately, only the intraclass correlation, t, must be estimated for the Lush index, and Sales and Hill (1976) showed that the efficiency of index selection is quite robust to estimation errors in t (as initially suggested by Lush 1947).

Nonetheless, given some of these concerns, it is not surprising that experimental verification of the advantage of the Lush index over individual or family selection is mixed. Further, the common problem of low statistical power in most selection experiments due to small sample sizes makes negative results difficult to interpret (Chapter 18). McBride and Robertson (1963) and Avalos and Hill (1981) found that index selection resulted in a larger response than individual selection for abdominal bristles in *Drosophila melanogaster*. More conclusive results, also on *Drosophila* bristle number, were those of James (cited in Frankham 1982), who found that the observed increase in response under index selection (relative to mass selection) was 133% $\pm$ 9.7% and 111% $\pm$ 7% in two replicates, consistent with the expected increase of 121%. Results for selection for egg production in poultry were less conclusive, and although Kinney et al. (1970) found that individual selection gave a larger (but not significant) response than family index selection, while Garwood and Lowe (1981) found that index selection gave a larger response (again not significant) than family selection. Work on larval and pupal weight in *Tribolium* showed similar mixed results, as Wilson (1974) found that individual selection gave the largest response, while Campo and Tagarro (1977) did not find any significant differences (index selection gave a larger response in a replicate with large family size, while individual selection showed the larger response in a replicate with small family size).

We note in passing that a more general family index was considered by Osborne (1957a, 1957b) for the nested-sib design, which separately weights information from full- and half-sib families. If z_{ijk} denotes the kth full-sib from dam j and sire i, an index weighting both half- and full-sib information is

$$I = b_1\,(z_{ijk} - \overline{z}_{ij}) + b_2\,(\overline{z}_{ij} - \overline{z}_{i\cdot}) + b_3\,\overline{z}_{i\cdot}$$

where b_1 is the weight on the deviation within a full-sib family, b_2 is the weight on the deviation among dam-family means within a sire, and b_3 is the sire weight (half-sib means). Volume 3 examines this, and more general indices, in much greater detail.

Correcting the Selection Intensity for Correlated Variables

As mentioned previously, expressions for the selection intensity in finite populations make the assumption that the order statistics are uncorrelated. However, the selection of multiple individuals from the same family results in correlations among the order statistics due to the correlation between sibs. Our treatment of this issue follows that of Hill (1976, 1977b).

Suppose the population from which individuals are drawn consists of m families, each with n sibs, for a total of $M = m\,n$ measured individuals. If phenotypic values are uncorrelated among all members of the sample (the sib correlation, t, is zero), Burrow's correction (Equation 14.4b) yields a finite population size-adjusted selection intensity of

$$\bar{\imath}_{(K,M)} = \bar{\imath}_p - \frac{1-p}{2\,\bar{\imath}_p\, p(M+1)}$$

where a fraction, $p = K/M$, of the population is saved and $\bar{\imath}_p$ is the infinite-population selection intensity associated with the fraction p saved (Equation 14.3a). When some members are correlated, this reduces the effective number of independent variables to some value below M. This value ranges from $m\,n = M$, with no correlation between sibs ($t = 0$), to m, with a perfect correlation between sibs ($t = 1$). Using this observation, Hill (1976) suggested a linear approximation for the effective number, M_e, of independent variables of

$$M_e = M(1-t) + mt \tag{21.57a}$$

Substituting into Burrow's correction gives an expected selection intensity adjusted for correlations of approximately

$$\bar{\imath}_{(K,M)}(t) = \bar{\imath}_p - \frac{1-p}{2\,\bar{\imath}_p\, p[\,M(1-t) + mt - 1\,]} \tag{21.57b}$$

Note that $\bar{\imath}$ decreases as t increases. Simulation studies by Hill showed that this is a reasonable approximation, and Hill (1976) provided tables of exact values (over a limited set of n and t values). An alternative approximation was offered by Rawlings (1976), while Tong (1982) and Meuwissen (1991), respectively, considered contributions from unequal family size and under a nested full-sib-half-sib design.

The effect of sib-correlations on the selection intensity for individual selection is generally small, as t is typically less that 0.5, and has only a modest effect on reducing $\bar{\imath}$. In contrast, the presence of the family mean, $\bar{z}_i$, in the index scores greatly inflates the correlation between the sib index values, I, over the correlation among phenotypic values, z. Hill (1976) showed that if selection occurs on the index $I = z_{ij} + B\,\bar{z}_i$ (Equation 21.50b), the intraclass correlation, τ, among the *index values* of sibs is given by

$$\tau = 1 - \frac{n\,(1-t)}{n + B(2+B)[\,1 + (n-1)t\,]} \tag{21.58a}$$

where t is the intraclass correlation of individual *phenotypic values* among sibs. Note that for large B, τ approaches 1.0. Hence, for schemes that place considerable weight on family means, the index scores for sibs within a family are almost perfectly corrected, and the effective number of independent order statistics approaches the number of families chosen. This is very reasonable, as I approaches $\bar{z}_i$ for large B, which is equivalent to among-family selection, giving the number of independent order statistics as the number of families, m.

Using the value of B (from Equation 21.55) under Lush index weights, Hill (1976) showed (for large n) that

$$\tau \simeq \begin{cases} 1-t & \text{full-sibs} \\[2ex] \dfrac{1-t}{1+8t} \simeq \dfrac{1}{1+2h^2} & \text{half-sibs} \end{cases} \tag{21.58b}$$

Example 21.13. Once again, consider Clayton et al's (1957) experiment on *Drosophila* bristle number. From Example 21.12, the Lush index weights are $b_2/b_1 = 2.44$, with Equation 21.50c yielding $B = b_2/b_1 - 1 = 1.44$. Recalling that $t = 0.265$ and $n = 12$, Equation 21.58a returns the correlation, τ, among the index values of sibs as

$$\tau = 1 - \frac{12\,(1 - 0.265)}{12 + 1.44(2 + 1.44)[\,1 + (12 - 1)0.265\,]} = 0.72$$

which is 2.7 times the correlation, t, among sib phenotypic values. Note that under strict family selection ($\tau = 1$), the correlation among the index value increases to 3.8 times the sib phenotypic correlation.

Suppose we select on a Lush index using four families ($m = 4$). The resulting total number of individuals becomes $N = 12 \cdot 4 = 48$, while Equation 21.57a gives the effective number of independent variables as

$$M_e = 48(1 - 0.72) + 4 \cdot 0.72 = 16.3$$

which is just 34% of the actual number of total individuals. Because $p = 0.2$ (implying $\bar{\imath}_p = 1.40$), Equation 21.57b yields a corrected selection intensity of

$$\bar{\imath} = 1.40 - \frac{1 - 0.2}{2 \cdot 1.40 \cdot 0.2 \cdot (16.3 + 1)} = 1.32$$

or a reduction of ~6%.

22

Associative Effects: Competition, Social Interactions, Group and Kin Selection

These findings . . . support the writer's view that competitive ability should be accepted as it stands as a genetic character, simple or aggregate, a view of great importance in the discussion to follow. Sakai (1955)

This chapter weaves together several seemingly unrelated, but nevertheless important, topics: competition; altruism and other social behaviors; traits defined by group, rather than individual, attributes; maternal effects; and group and kin selection. The connection between all of these topics is the notion that the genotype (and hence phenotype) of one individual may influence the trait value of another. In this sense, the "environmental" component of the phenotype of a focal individual may itself have some heritable component (based on the contribution from some other individual), allowing for some part of the environmental component to evolve along with the focal trait. In such settings, the phenotype of a focal individual consists of two components: **direct effects** from the focal individual and **associative effects** contributed from other individuals within the group. A critical implication of this distinction is that the breeding value of an individual contains a component for direct effects that appear in its own phenotype (and hence can be influenced by individual selection) and a component for associative effects that only appears in the phenotypes of *other* group members. The exploitation of associative effects by selection generally requires either interactions among kin (**kin selection**) or selection based on some combination of both individual and group values (**multilevel selection**). In the extreme, **group selection** occurs when all of the weight is placed on among-group differences. Note that multilevel selection is an extension of family-index selection (Chapter 21) to more general groups.

The framework for dealing with these issues was laid out in a series of classical, but largely ignored, papers by Griffing (1967, 1968a, 1968b, 1969, 1976a, 1976b, 1977), who introduced the notion of associative effects. There are also roots extending to classical work on maternal effects based on trait phenotype (Falconer 1965) or on an unmeasured maternal value (Willham 1963), as well as to the foundations of the study of social evolution (Hamilton 1963, 1964a, 1964b). There are two modeling approaches for dealing with associative effects: **trait-based** and **variance component-based**. Trait-based approaches (Moore et al. 1997) have their roots in univariate (Falconer 1965) and multivariate (Kirkpatrick and Lande 1989; Lande and Kirkpatrick 1990) models of selection response under maternal effects. As their name implies, trait-based approaches assume that we know the particular traits in group members that influence the phenotype of the focal individual. This approach is best handled in a multivariate framework, so we will delay its full discussion until Volume 3.

The variance-component approach also has roots in maternal-effects models (Willham 1963), wherein a general (but unmeasured, i.e., latent) **maternal performance value** influences the phenotype of the focal individual. Using BLUP, we can estimate the genetic variance of the associative effects (as well as its covariance with the direct effects). Somewhat counterintuitively, variance component-based methods (where the actual traits that generate the associative effects are unspecified) are empirically more powerful than trait-based methods. The reason is that we can estimate this unspecified total contribution directly, while if characters that influence associative effects are left out of a trait-based model, this can introduce errors. McGlothlin and Brodie (2009) and Bijma (2014) show the congruence between these methods, which is also examined in detail in Volume 3.

Evolution and Selection of Quantitative Traits. Bruce Walsh & Michael Lynch.
Published 2018 by Oxford University Press.
DOI 10.1093/oso/9780198830870.001.0001

Traits whose phenotypes are determined, in part, by interactions with other individuals have important roles in both breeding and evolution. In breeding, we are often more interested in the performance of a group rather than that of an individual. For example, standard poultry husbandry is to keep several females together in a cage, with total egg production per cage being the key quantity of interest. In the extreme, an aggressive female may kill all her cage-mates, and, in less extreme cases, may largely dominate feeding, resulting in an individual benefit at the expense of the group. Hence, individual selection may result in a *decrease* in group performance, in which case the number of eggs per cage would decline.

The issue here is that ***individual selection cannot effectively utilize the genetic variation in associative effects*** to guarantee the response of the mean associative value in the direction favored by the breeder. The same concerns have long been raised in evolutionary biology, in particular to account for the evolution of **altruistic traits** (such as alarm calls in birds) that are expected to decrease individual fitness, yet still have evolved. There is a very rich, and stormy, evolutionary literature on the importance (or lack thereof) of selection based on group attributes. The general view in evolutionary biology has often been to invoke group selection arguments only as a last, desperate resort when all individual selection arguments fail (e.g., Williams 1966). As we will see, much of the debate regarding group versus kin selection is misplaced, as they are essentially manifestations of the same general process.

Our treatment starts with a formal definition of direct and associative effects, including the powerful concept of the total breeding value, A_T, of a trait (which requires measurements of group members). Next, we show how the presence of associative effects influences selection. One key result is that when the breeding values for direct and associative effects are negatively correlated, individual selection can result in a reversed response. Conversely, group selection (even when group members are unrelated) always results in an expected positive response, but it can be very ineffective when associative effects are small. We then examine selection based on an index of both individual and group information, including the optimal weighting for maximal response. A key innovation that we examine in detail is the use of BLUP/REML methodology (Chapters 19 and 20) to estimate the direct and associative effects of individuals, along with their variance components. We conclude by applying these results to some of the debates on group and kin selection in evolutionary biology. Our goal in this last section is not to extensively review this literature, which is often contradictory and, at times, was driven more by verbal models than detailed analyses. Rather, it is to show how the problem of selection based on group attributes can be easily placed in a quantitative-genetics framework.

For many readers, this may be one of the most important chapters in the book, as associative effects are potentially game-changing in the analysis of many traits. Evolutionary biologists, breeders, behavioral ecologists, and human geneticists all need to be aware of their importance and implications. They reshape many classic problems in evolutionary biology, such as Fisher's fundamental theorem (Chapter 6), inclusive fitness, and kin and group selection. Their presence fundamentally changes breeding strategies, as individual selection potentially leaves much of the usable genetic variation in a trait untapped and can result in reversed responses (Chapter 15). Most behavioral traits arise from interactions between individuals, which is exactly the framework for associative effects. Finally, their presence radically changes the way in which we analyze traits. An important example is disease resistance. As this is both a function of the susceptibility of an individual and the infectiousness of those around it, a full consideration requires a model with associative effects (Lipschutz-Powell et al. 2012a, 2012b; Costa e Silva et al. 2013). Partial reviews of some of the implications of associative effects are given by Griffing (1977), Moore et al. (1997), Wolf et al. (1998), Bijma and Wade (2008), McGlothlin et al. (2010), Wade et al. (2010), Wolf and Moore (2010), Bijma (2011, 2014), and Bailey (2012).

DIRECT VERSUS ASSOCIATIVE EFFECTS

All organisms interact with their external environment, and a very significant fraction of that environment is biological. In particular, interactions with conspecifics through competition, cooperation, parental care, or other social interactions can constitute an important part of the environment that an individual experiences, which, in turn, can influence trait values. Further, this "environment" may contain heritable components and coevolve with the trait of interest. The classic example of this is competition, which we briefly consider first.

Early Models of Competition

It has long been appreciated by breeders that competition among plants within a plot has a significant impact on important agricultural traits such as yield. While a particular genotype may have high yield when grown in isolation, when grown in a group, its competitive effects on other members within its group could result in a lower plot yield. Yield (and other traits) of a particular plant in a plot is therefore a function of two components. First, an individual's genotype has a direct influence on its ability to garner resources such as light, water, and nutrients. Second, that genotype influences others around it by competing for limiting resources. Other plants in the plot also compete, and these in turn influence the yield of the focal individual. One might expect that plants that are very successful at garnering resources have positive direct effects, but negative associative effects on nearby individuals. Thus, a plot of high-competing genotypes can have a low yield, as the positive direct effects for any particular plant are more than countered by negative associative effects from being surrounded by superior competitors.

A historically important paper on the evolution of competition is that by Sakai (1955), who noted that competition, like yield or height, is a genetic trait and hence can potentially evolve. Following Sakai, a number of workers developed single-locus population genetic models to examine the evolution of competition (Schutz et al. 1968; Allard and Adams 1969; Schutz and Usanis 1969; Cockerham and Burrows 1971; Cockerham et al. 1972). These studies all used simple ecological models of competition among a series of fixed types (here, all possible single-locus genotypes). While interesting, this class of models does not easily generalize beyond one locus. Griffing (1967) made the important extension of Sakai's idea by replacing a single-locus genotype with direct and associative values that are quantitative traits, consisting of breeding and residual values. Placed in this framework, such traits can potentially evolve and can also have their variance components estimated, allowing associative effects to be exploited by using appropriate selection designs.

Example 22.1. The point that high-competing genotypes can reduce yield was made in a classic paper by Wiebe et al. (1963), who examined yield in mixed- versus single-genotype plots of barley. They observed that genotypes that yielded well in mixed stands had poorer yield when grown as pure stands, while those genotypes that did poorly in mixed stands had the highest yield in pure stands. In our framework, we could imagine that lines that do well in mixed stands have both high positive direct effects and high negative associative effects, suppressing the phenotypes of their neighbors. When grown in a pure stand, the high negative associative effects suppress plot yield. Conversely, lines that perform poorly in mixed stands might have low direct effects but high positive associative effects, and so the phenotypes of their neighbors are enhanced (or at least not hindered). When grown as a pure stand, these high positive associative effects more than compensate for the low direct effects, increasing plot yield.

Direct and Associative Effects

A simple example will introduce Griffing's idea. As shown in Figure 22.1A, consider a group of four individuals. Our focal individual is 1, and its phenotype, z_1 (for the trait of interest), is determined by its own intrinsic value, $P_{d,1}$ (the subscript d indicating the direct effect), plus the associative effects, $P_{s,2}$, $P_{s,3}$, and $P_{s,4}$, contributed by other group members.

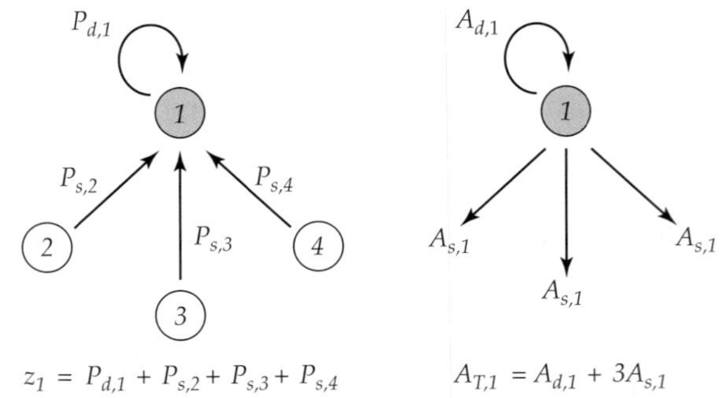

Figure 22.1 **Left:** The phenotypic value, z_1, of the focal individual is the sum of its direct phenotypic effect, $P_{d,1}$, plus the associative effects, $P_{s,2}$, $P_{s,3}$, $P_{s,4}$, of the three other members in its group. **Right:** The total breeding value ($A_{T,1}$) of individual 1 is its direct breeding value, $A_{d,1}$, plus the total contribution of the associative-effect breeding value, $A_{s,1}$, to the three members of its group. A key concept is that *only part of* A_T *(namely* A_d*) is embedded within its own phenotypic value.* The remaining part of A_T, namely its associative component, $3A_s$, is only expressed in the phenotypes of *other* group members.

Associative effects are also referred to in the literature as **indirect genetic effects (IGEs)** (Moore et al. 1997; Wolf et al. 1998; McGlothlin et al. 2010), or **social effects** (Bijma et al. 2007a, 2007b), and we use the subscript s (indicating social effects) to denote them. In our discussion, we will use the terms *associative* and *social effects* interchangeably. Note that the values of $P_{s,i}$ do not necessarily correspond to the phenotypes for the trait of interest in the other group members, but rather represent the contribution from these members to the phenotype of the focal individual. This contribution from fellow group members is part of the environment experienced by the focal individual.

More generally, for a group containing n equally interacting individuals, the resulting phenotype (z_i) for individual i becomes

$$z_i = P_{d_i} + \sum_{j \neq i}^{n} P_{s_j} \tag{22.1a}$$

where the sum has $n - 1$ terms. Each of these components can be decomposed into a breeding value, A, plus a residual component, E (containing environmental effects plus any nonadditive genetic effects), yielding

$$z_i = \mu + (A_{d_i} + E_{d_i}) + \sum_{j \neq i}^{n} (A_{s_j} + E_{s_j}) \tag{22.1b}$$

We can write this more compactly as

$$z_i = \mu + A_{d_i} + \sum_{j \neq i}^{n} A_{s_j} + e_i, \quad \text{with} \quad e_i = E_{d_i} + \sum_{j \neq i}^{n} E_{s_j} \tag{22.1c}$$

Because the environmental values have expected value of zero, the mean phenotypic value in the group is simply

$$\mu_z = \mu + \mu_{A_d} + (n-1)\mu_{A_s} \tag{22.1d}$$

Further, the change in the mean trait value within a group following selection is

$$\Delta\mu_z = \Delta\mu_{A_d} + (n-1)\Delta\mu_{A_s} = R_d + (n-1)R_s \tag{22.1e}$$

which decomposes the change in trait value into contributions from responses, R_d and R_s, respectively, in the direct and social values.

This equation foreshadows individual versus group selection. Individual selection targets the direct effect and results in a favorable change in μ_{A_d}. If the direct and social breeding values are correlated within an individual, namely, $\sigma(A_d, A_s) \neq 0$, then individual selection can also change μ_{A_s}, but not necessarily in a favorable direction. Indeed, as Example 22.4 will show, an increase in μ_{A_d} under individual selection can be more than countered by an unfavorable change in μ_{A_s}, resulting in the mean phenotype changing in an unfavorable direction. Direct selection on μ_{A_s} requires either undergoing group selection or having relatives within the group. All of these points will be expanded upon below. Our focus is entirely on additive genetic effects, as most of the theory has been developed under this assumption. Attempts to include nonadditive variance were developed by Gallais (1976) and Wright (1986). Finally, one way to make to concept of associative effects a bit more concrete is to note that one can map associative-effect QTLs; see Mutic and Wolf (2007) and Wolf et al. (2011) for examples.

Animal Well-being and the Improvement of the Heritable Social Environment

In high-intensity agricultural systems, competition has a strong effect on yield and other traits. Animals in such environments face significant stress, which impacts both their production and their well-being. As reviewed by Muir and Craig (1998), animal well-being is becoming an increasingly important aspect of animal production. Muir suggests that social aspects such as aggression, fighting, and sharing of common resources are all potential targets of selection, and responses in these traits (for less aggression and more sharing) improves both animal welfare and production. Further, for a number of species (such as certain fishes), domestication has proved somewhat problematic due to the tendency for cannibalism (and lesser forms of aggression), when individuals are grown under production conditions. Muir suggested improving welfare by selecting for an improved mean social environment through selection of individuals with favorable A_s values for the traits of interest. Again, these are aspects of the group environment and can respond favorably to an appropriate selection design, provided there is a heritable component of P_s, namely, $\sigma^2(A_s) > 0$.

What Do We Mean by Group?

Given that we use the term *group* extensively in this chapter, a more formal definition is required. Our focus here is on traits whose values are influenced by interactions with others. The set of individuals that interacts with the focal individual constitutes the unit we will call a group. This may be straightforward in some breeding settings, such as the specific animals in a pen or cage. However, in other settings, such as cattle in a very large feedlot, only some subset of all the individuals likely interact with the focal individual. Hence, group size may be much smaller than the number of individuals physically confined to some space. Likewise, individuals may be part of different groups for different traits, especially if those traits are expressed at different times during development. The same is true on a grander scale in natural populations. The key issue with traits influenced by interactions is that phenotypes of the group members provide some information on part of the breeding value of the focal individual—the part dealing with its associative effect—that is simply not provided by the phenotype of the focal individual. To exploit this additional heritable variation (when it exists), interactions with relatives or selection that puts at least some weight on group value is essential.

The second feature about groups is their formation and reproduction, an issue that is especially important under differential propagation of groups (i.e., group selection). Here, we are assuming a situation akin to our analysis of family selection (Chapter 21), in that, while group information may be used to select individuals to form the next generation, these individuals are then mated at random. In the group-selection literature, this is referred to as a **migrant pool model** (Levins 1970; Wade 1978). Such a structure only allows changes in breeding values (as opposed to genotypic values) to propagate to the next generation. In settings where entire groups are propagated as a unit (the **propagule pool model**; Wade

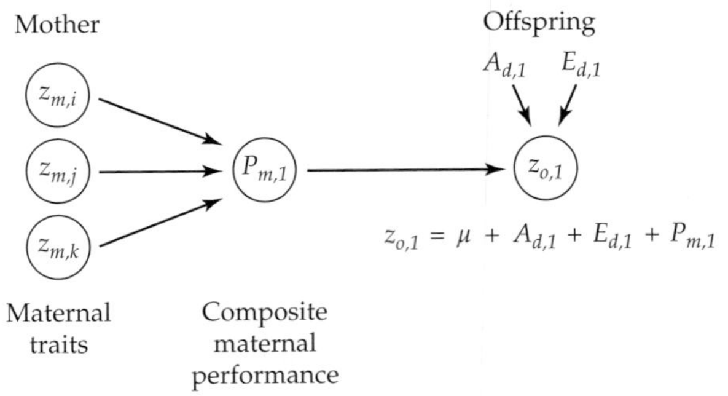

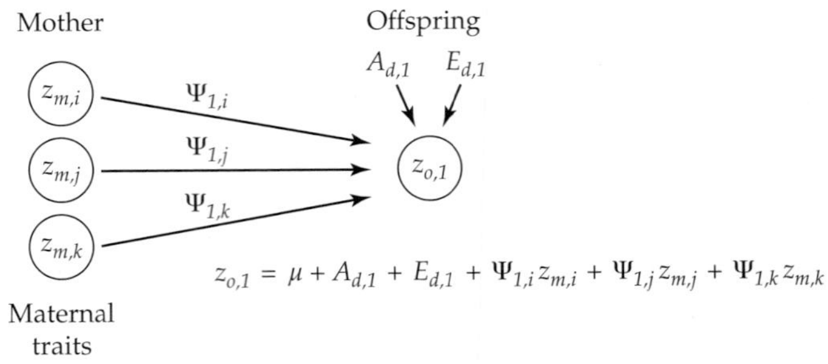

Figure 22.2 The difference between trait-based and variance-component based models. Here, the phenotypic value, $z_{o,1}$, of a trait (which we label 1) in an offspring is a function of maternal phenotype. We suppose that there are three maternal traits (i, j, k) whose phenotypes influence the offspring value. **Top:** Under a variance-component based approach, we ignore all the maternal trait values and simply estimate a single maternal performance value, $P_{m,1}$, that directly influences the trait value in the offspring. The resulting model becomes $z_{o,1} = \mu + A_{d,1} + E_{d,1} + P_{m,1}$, where $A_{d,1}$ is the trait breeding value in the offspring, $E_{d,1}$, its environmental value, and $P_{m,1} = A_{s,1} + E_{s,1}$ can be decomposed into the social breeding value on trait 1 plus a residual. **Bottom:** Under a trait-based model, *provided* we know all of the maternal traits whose phenotypes influence trait 1 in the offspring, then we directly incorporate these, along with their regression coefficients, $\Psi_{1,i}$, showing how these maternal phenotypes translate into offspring trait value. Here, $z_{o,1} = \mu + A_{d,1} + E_{d,1} + \Psi_{1,i} z_{m,i} + \Psi_{1,j} z_{m,j} + \Psi_{1,k} z_{m,k}$, where the last three terms together comprise $P_{m,1}$. Trait-based models are required if one wishes to consider the joint evolution of traits 1 and i, j, k. Their drawback is that one has to specifiy to all of the relevant maternal traits. Conversely, under a variance-component method, all of the maternal phenotypes are conveniently collapsed into a single value, whose breeding value can be estimated from an appropriate design (detailed below).

1978), the potential exists for nonadditive variance to contribute to the among-group variance.

Trait- vs. Variance Component-based Models

A brief comment is in order, expanding upon our earlier remarks on trait vs. variance-component based modeling (see Bijma 2014 for an extended discussion). The original trait-based model of associative effects was Falconer's (1965) model for litter size in mice (Equation 15.21), namely

$$z_i = G_i + e_i + (m \cdot z_{mo,i})$$

where G_i is the direct breeding value for litter size, while the associative effect is a function of the litter size of its mother ($z_{mo,i}$). Building on this idea, Moore et al. (1997) and Wolf et al. (1998) suggested a model wherein the value for trait i also depends upon the value, z'_j, of trait j (which may be a different trait from i) in an interacting individual,

$$z_i = A_i + e_i + \Psi_{ij} z'_j = A_i + e_i + \Psi_{ij} A'_j + \Psi_{ij} E'_j \tag{22.2}$$

where Ψ_{ij} (following Kirkpatrick and Lande 1989) is the multivariate extension of Falconer's m. This class of models can lead to some very interesting behavior, such as feedback loops that significantly modify Equation 22.2. Figure 22.2 illustrates this difference in modeling, while Volume 3 explores trait-based models in some detail.

Bijma (2014) noted that variance-component approaches are akin to using Robertson's secondary theorem, $R = \sigma(w, A)$ (Equation 6.25a), to predict response, which ignores any specific traits and simply considers the covariance between breeding value (which we generalize by calculating total breeding value, A_T, below) and relative fitness. In contrast, trait-based approaches are akin to using the multivariate Lande equation, $\mathbf{R} = \mathbf{G}\boldsymbol{\beta}$ (Equation 13.26a), to predict response. The Lande equation returns the response in all traits of interest, but it requires that all relevant traits be included in the analysis and is potentially erroneous if they are not (Volume 3).

The Total Breeding Value (TBV) and T^2

Given that an individual contains breeding values for both direct and social effects, what is its contribution to the next generation? We can directly see this from Equation 22.1d, where it is shown as the contribution to the population mean from individual 1 from its direct breeding value (A_{d_1}) plus its contribution to the $(n-1)$ other individuals in its group through its associative-effects breeding value, A_{s_1} (Figure 22.1b). Based on this observation, Bijma et al. (2007a) defined the **total breeding value (TBV)**, A_T, of a trait from an individual measured in a group of size n as the sum of its direct effect plus the total associative effects over all group members, or

$$A_{T_i} = A_{d_i} + (n-1)A_{s_i} \tag{22.3}$$

Moore et al. (1997) introduced a similar measure for trait-based models. Noting that the mean of the population is simply the mean breeding value allows Equation 22.3 to recover Equation 22.1d. The critical observation is that when associative effects are present, the total breeding value of an individual contains components that are ***not expressed in its own phenotype***, but rather, only in the phenotypes of other individuals with which it interacts.

Example 22.2. Consider a trait in a group of four (unrelated) individuals, where we assume there are no environmental effects, which means that $P_d = A_d$ and $P_s = A_s$. The population mean is 20, and the four group members have the following breeding values for direct, associative, and total effects:

Individual	A_d	A_s	A_T	$\sum_{j \neq i} A_{s_j}$	z
1	9	-4	-3	4	33
2	5	-1	2	1	26
3	-6	2	0	-2	12
4	-8	3	1	-3	9

Because $n = 4$, $A_T = A_d + 3\,A_s$. The contribution of the associative effects of the other three group members to i's phenotypic value is given by $\sum_{j \neq i} A_{s_j}$. For example, for individual 1, the contributions from individuals 2 through 4 is -1 + 2 + 3 = 4. Thus, from Equation 22.1c, the phenotypic value we would observe is

$$z_i = 20 + A_{d_i} + \sum_{j \neq i}^{4} A_{s_j}$$

Individual 1 has the largest direct effect (9) and the largest observed trait value (33). This individual also has the most unfavorable associative value (-4), and the smallest total breeding value (-3). Conversely, it has the largest contribution (4) to its trait value from the associative effects of the other group members. Its high trait value is due to this combination of a high direct effect and a high contribution from the associative effects of the other group members.

Its unfavorable associative effects do not appear in its own phenotype, but rather are expressed in the trait values of the other group members. As a result, its own phenotypic value is a poor predictor of A_T.

If the next generation is formed by crossing the two individuals (1 and 2) with the largest trait values, the expected offspring mean will be 20 + (-3+2)/2 = 19.5, which is the mean plus the average of the two individuals' total breeding values. Although the two largest individuals were chosen, the population mean *decreases*. Conversely, crossing the two smallest individuals gives an expected offspring mean of 20 + (0+1)/2 = 20.5, increasing the mean. While the two smallest individuals have the smallest direct effects, they also have the most favorable associative effects, and hence result in a more favorable response. The greatest expected response occurs by crossing the two individuals with the largest total breeding values (2 and 4), for an expected mean of 20 + (2+1)/2 = 21.5.

The covariance between an individual's phenotype and total breeding value is

$$\sigma(z_i, A_{T_i}) = \sigma\left[\mu + A_{d_i} + \sum_{j\neq i}^{n} A_{s_j} + e_i, A_{d_i} + (n-1)A_{s_i}\right]$$

$$= \sigma\left[A_{d_i}, A_{d_i} + (n-1)A_{s_i}\right] + \sum_{j\neq i}^{n} \sigma\left[A_{s_j}, A_{d_i} + (n-1)A_{s_i}\right] \tag{22.4a}$$

For now, we assume group members are unrelated, in which case the covariances in the summation are all zero,

$$\sigma(z, A_T) = \sigma^2(A_d) + (n-1)\sigma(A_d, A_s) \tag{22.4b}$$

If the direct and associative effects are uncorrelated, this reduces to the standard result that the covariance between an individual's phenotype and breeding value is simply the additive genetic variance (in this case, of direct effects). By contrast, the variance of the total breeding value becomes

$$\sigma^2(A_T) = \sigma^2\left[A_d + (n-1)A_s\right]$$

$$= \sigma^2(A_d) + 2(n-1)\sigma(A_d, A_s) + (n-1)^2\sigma^2(A_s) \tag{22.4c}$$

$$= \sigma(z, A_T) + (n-1)\left[\sigma(A_d, A_s) + (n-1)\sigma^2(A_s)\right] \tag{22.4d}$$

Equation 22.4d shows that the covariance between total breeding value and phenotype is different from the variance in total breeding value. This reflects the fact that the associative effects of an individual do not influence its own phenotype. Note from Equation 22.4c that $\sigma(A_d, A_s)$ and $\sigma^2(A_s)$ are scaled by $(n-1)$ and $(n-1)^2$, respectively, in $\sigma^2(A_T)$. Hence, with even modest group sizes, small values of $\sigma(A_d, A_s)$ and $\sigma^2(A_s)$ can still have a very significant impact. Some of the early papers reporting estimates of these two quantities ignored this scaling with n, and hence tended to downplay the importance of social interactions (Chen et al. 2006; Van Vleck et al. 2007).

Now consider the phenotypic variance,

$$\sigma_z^2 = \sigma^2\left(P_{d_i} + \sum_{j\neq i}^{n} P_{s_j}\right) \tag{22.5a}$$

If we assume (for now) that the group members are unrelated, then $\sigma(P_{d_i}, P_{s_j}) = 0$. For a group of size n, Equation 22.5a reduces to

$$\sigma_z^2 = \sigma^2(P_d) + (n-1)\sigma^2(P_s) \tag{22.5b}$$

$$= \sigma^2(A_d) + (n-1)\sigma^2(A_s) + \sigma^2(E_d) + (n-1)\sigma^2(E_s) \tag{22.5c}$$

$$= \sigma^2(A_d) + (n-1)\sigma^2(A_s) + \sigma^2(e) \tag{22.5d}$$

where e is given by Equation 22.1c. With the phenotypic variance in hand, we can define the heritability of the direct and associative effects, h_d^2 and h_s^2, respectively, as

$$h_d^2 = \frac{\sigma^2(A_d)}{\sigma_z^2} \quad \text{and} \quad h_s^2 = \frac{\sigma^2(A_s)}{\sigma_z^2} \tag{22.6a}$$

The careful reader will note that there is a different, but perhaps more natural, definition of these two heritabilities. Equation 22.6a standardizes the genetic variances with respect to the total trait variance, but one could also standardize them with respect to the variance of direct and associative effects, for example,

$$h_{d'}^2 = \frac{\sigma^2(A_d)}{\sigma^2(P_d)} \quad \text{and} \quad h_{s'}^2 = \frac{\sigma^2(A_s)}{\sigma^2(P_s)} \tag{22.6b}$$

We use a prime to distinguish these from the hertiabilities scaled to total trait variance ($\sigma^2(P_x)$ vs. σ_z^2). While heritabilities scaled by σ_z^2 (Equation 22.6a) are the most widespread in the literature, there are some advantages to scaling hertiabilities by $\sigma^2(P_x)$ (where $x = d$ or s). On this scale, the heritabilities measure the fraction of additive genetic variation in the actual effect (direct or associative) itself, rather than in the trait value. Further, $h_{x'}^2$ is independent of the group size (provided that A_s does not change with group size), as $\sigma^2(P_x)$ is independent of n, while σ_z^2 is a function of n (Equation 22.5b).

In keeping with Equation 22.6a, we can similarly define the "heritability" of the total breeding value as

$$T^2 = \frac{\sigma^2(A_T)}{\sigma_z^2} \tag{22.7a}$$

as suggested by Bijma et al. (2007a). The reason we have used T^2 rather than h_T^2 is that, unlike heritabilities, T^2 can exceed one. This can happen because $\sigma^2(A_T)$ contains additional terms not found in σ_z^2, as ***the associative effect of an individual influences others in the group, rather than the individual in which it resides***.

To see this, first assume that the environmental effects are all zero ($\sigma^2(e) = 0$), so that we can focus on differences in the genetic variance components. From Equations 22.4c and 22.5c,

$$\begin{aligned} \sigma^2(A_T) - \sigma_z^2 &= 2(n-1)\sigma(A_d, A_s) + (n-1)^2\sigma^2(A_s) - (n-1)\sigma^2(A_s) \\ &= (n-1)\left[2\sigma(A_d, A_s) + (n-2)\sigma^2(A_s)\right] \end{aligned} \tag{22.7b}$$

If this difference exceeds the contribution (σ_e^2) from environmental effects, then $T^2 > 1$.

Bijma (2011, 2014) noted that $\sigma^2(A_T)$ provides the appropriate (and general) definition for the amount of heritable variation underlying the potential for response. Recalling Equations 22.1e and 22.3, the Robertson-Price identity (Equation 6.10) yields the expected response (change in mean breeding value) to selection as

$$R = \sigma(w, A_T) \tag{22.8a}$$

Because the linear regression of w on A_T has a slope of

$$\beta_{w|A_T} = \frac{\sigma(w, A_T)}{\sigma^2(A_T)} \tag{22.8b}$$

(LW Equation 3.14b), the general expression for response can be written as

$$R = \beta_{w|A_T}\, \sigma^2(A_T) \tag{22.8c}$$

The apparent simplicity of this expression is somewhat misleading, as $\beta_{w|A_T}$ can be a very complex function of the relationship among group members (see Bijma 2011 for examples).

Example 22.3. Consider a trait in a group of 10 unrelated individuals, with $\sigma^2(P_d) = 10$, $\sigma^2(P_s) = 1$, and both direct and associative effects having modest heritabilities on the scale of the effects themselves ($h^2_{d'} = 0.4, h^2_{s'} = 0.3$). To simplify matters, assume $\sigma(A_d, A_s) = 0$. Applying Equation 22.5b, the resulting phenotypic variance is

$$\sigma_z^2 = \sigma^2(P_d) + 9 \cdot \sigma^2(P_s) = 10 + 9 \cdot 1 = 19$$

From Equation 22.4c, the variance in total breeding value becomes

$$\sigma^2(A_T) = \sigma^2(A_d) + 9^2 \cdot \sigma^2(A_s) = h^2_{d'}\,\sigma^2(P_d) + \left[9^2 \cdot h^2_{s'}\,\sigma^2(P_s)\right] = 4 + (81 \cdot 0.3) = 28.3$$

yielding (from Equation 22.7a) T^2 = 28.3/19 = 1.49.

A real-world example of large potential differences in h^2_d versus T^2 involves survival days in chickens (Bijma et al. 2007b). Ignoring associative effects yields a direct heritability of h^2_d= 0.07, while a mixed model incorporating associative effects (detailed later in the chapter) yielded an estimate of T^2= 0.20, a threefold increase. Hence, under the conditions of this study, roughly two-thirds of the heritable variation in the trait arises from interactions between individuals and is thus hidden from standard analyses that ignore them. As discussed below, this component is only fully accessible under individual selection if the group includes relatives.

A_s as a Function of Group Size

As the careful reader will have noted, the direct effect, A_d, is independent of group size, while the social effect, A_s, potentially changes with group size. Suppose a genotype has a breeding value for social effects of 10 when measured in groups of size four. Does this change with group size and, if so, how? This is an empirical issue, and one can frame it in a G × E setting. The environments here are different group size, and if A_s shows G × E, the value of A_s changes over n.

Two simple scenarios bracket the possible changes. First, suppose that an individual eats 500 grams of food daily. In a group with a fixed food supply, the associative effect of this individual is to remove 500 grams from the total food supply each day. Hence, in a group of size n, $P_{s_i} = -500/(n-1)$, while its *total* associative effect is the sum over all group members, $(n-1)P_{s_i} = -500$. Here, the total associative effect remains unchanged over group size, while the *individual* associative effect on any group member shows a **dilution** with increasing group size. Alternatively, consider a large tree whose associative effect results from shading individuals under its canopy. In such a case, its associative effect shows no dilution with group size. Similarly, Bijma et al. (2007a) noted that alarm calls are also expected to show no dilution with group size.

More generally, we have been assuming that all group members experience the same social effect from a conspecific (i.e., they all experience $P_{s,i}$ from individual i). However, one can imagine settings where $P_{s,i}$ is some base value, but its effect on specific individuals depends on their physical distance from individual i (e.g., Muir 2005; Cappa and Cantet 2008), or the total amount of time that they interact with each other (Cantet and Cappa 2008) (e.g., Example 22.11). Bijma (2014) presented a more general treatment of this issue. A second complication, wherein interactions may differ between kin and nonkin (e.g., Sherman 1977), was examined by Alemu et al. (2014).

A developing research area involves the further characterization of social effects and the degree to which they change over group size. Some initial insight was provided by Hadfield and Wilson (2007) and Bijma (2010b). Hadfield and Wilson assumed a simple regression model

$$P_{s_i,n} = P_{sb_i} + \frac{1}{n-1} P_{sr_i} \tag{22.9}$$

with the value for social effect in a group of size n being a function of two components: a base (or intercept) value, P_{sb_i}, and a linear dilution rate, P_{sr_i}. Note that the resulting total sum of associative effects from i over the $(n-1)$ group members becomes $(n-1)P_{sb_i} + P_{sr_i}$ meaning that P_{sr} is the constant contribution, while that from P_{sb} scales with group size.

Bijma (2010b) suggested a related model

$$P_{s_i,n} = \frac{1}{(n-1)^d} P_{s_i,2} \tag{22.10a}$$

which expresses all group social values as a function of the value for a group of size two ($P_{s_i,2}$) weighted by a power function of the **dilution fraction**, d (assumed to be the same over all genotypes). As we will see in the models below, Bijma's model is a bit more tractable, while the Hadfield-Wilson model is more general. When $d = 1$ and $P_{sp_i} = 0$, the two models are equivalent. Under the Bijma model, substituting Equation 22.10a into Equation 22.5a gives the total phenotypic variance as

$$\sigma^2_{z,n} = \sigma^2(P_d) + (n-1)^{1-2d}\sigma^2(P_{s,2}) \tag{22.10b}$$

Phenotypic variance increases with n for $d < 1/2$, remains constant for $d = 1/2$, and decreases with n for $d > 1/2$. Assuming that breeding values are diluted in the same fashion as phenotypic effects, then under the Bijma model

$$A_{s,n} = \frac{A_{s,2}}{(n-1)^d} \quad \text{and} \quad \sigma^2(A_{s,n}) = \frac{\sigma^2(A_{s,2})}{(n-1)^{2d}} \tag{22.10c}$$

Hence, $\sigma(A_d, A_s) = \sigma(A_d, A_{s,2})/(n-1)^d$, and substituting into Equation 22.4c gives the total additive-genetic variance for a group of size n as

$$\sigma^2(A_{T,n}) = \sigma^2(A_d) + (n-1)^{1-d}\left[2\,\sigma(A_d, A_{s,2}) + (n-1)^{1-d}\sigma^2(A_{s,2})\right] \tag{22.10d}$$

Hence, provided that $d < 1$, the additive total variance increases with n. Both Hadfield and Wilson (2007) and Bijma (2010b) have suggested methods to estimate the quantities in Equations 22.9 and 22.10a, respectively.

SELECTION IN THE PRESENCE OF ASSOCIATIVE EFFECTS

One of the key results when associative effects are present is that individual selection can result in a reversed response, while group selection always results in a positive response (although it may be far from optimal). These points were clearly made by Griffing (1967) for the simple case of two interacting, and unrelated, individuals within each group. For selection on individual phenotypes, the response becomes

$$R = \frac{\bar{\imath}}{\sigma(z)}\left[\sigma^2(A_d) + \sigma(A_d, A_s)\right] \tag{22.11a}$$

A negative covariance between direct and associative effects reduces the efficiency of selection, and if it is sufficiently negative, it gives a reversed response. This loss of efficiency occurs because the only information an individual's phenotype contains about its breeding value for associative effects is that provided by the covariance between the direct and associative breeding values (which can be negative). Conversely, if we select based on the mean of a group, we are selecting on both direct and associative effects to improve trait value. For the case of $n = 2$, Griffing obtained the expected response under group selection as

$$R = \frac{\bar{\imath}}{2\sigma(\overline{z})}\left[\sigma^2(A_d) + 2\sigma(A_d, A_s) + \sigma^2(A_s)\right] = \frac{\bar{\imath}}{2\sigma(\overline{z})}\sigma^2(A_T) \tag{22.11b}$$

While group selection always yields a nonnegative response, if the associative effects are weak, this approach will prove very inefficient relative to individual selection. For example, in the absence of associative effects, $\sigma^2(\overline{z}) = \sigma^2(z)/2$, and Equation 22.11b reduces to $\overline{\imath}\,\sigma(A_d)/[\sqrt{2}\,\sigma(z)]$, or $1/\sqrt{2} = 0.701$ of the response under individual selection.

Individual Selection: Theory

Consider individual selection in a group of size n, whose members are potentially related. Recalling Equation 22.4a, the correlation between phenotype and total breeding value is

$$\sigma(z_i, A_{T_i}) = \sigma^2(A_d) + (n-1)\sigma(A_d, A_s) + \sum_{j \neq i} \sigma\left[A_{s_j}, A_{d_i} + (n-1)A_{s_i}\right]$$

Let r_{ij} denotes the relationship between individuals i and j. When individuals within the group are related, then

$$\sigma(A_{s_j}, A_{s_i}) = r_{ij}\sigma^2(A_s)$$

Likewise if A_d and A_s are correlated, then for relatives we also have

$$\sigma(A_{d_j}, A_{s_i}) = r_{ij}\sigma(A_d, A_s)$$

Hence,

$$\begin{aligned}
\sum_{j \neq i} \sigma\left[A_{s_j}, A_{d_i} + (n-1)A_{s_i}\right] &= \sum_{j \neq i} \sigma\left(A_{s_j}, A_{d_i}\right) + (n-1)\sum_{j \neq i} \sigma\left(A_{s_j}, A_{s_i}\right) \\
&= \sigma(A_d, A_s)\sum_{j \neq i} r_{ij} + (n-1)\sigma^2(A_s)\sum_{j \neq i} r_{ij} \\
&= \left[\sigma(A_d, A_s) + (n-1)\sigma^2(A_s)\right]\left[\sum_{j \neq i} r_{ij}\right] \qquad (22.12a)
\end{aligned}$$

When all of the group members have the same relatedness ($r_{ij} = r$), the sum becomes $(n-1)r$, returning the result of Bijma et al. (2007a),

$$\begin{aligned}
\sigma(z, A_T) &= \sigma^2(A_d) + (n-1)\left[\sigma(A_d, A_s) + r\sigma\left(A_s, A_d\right) + r(n-1)\sigma^2\left(A_s\right)\right] \qquad (22.12b) \\
&= \sigma^2(A_d) + (n-1)(1+r)\sigma(A_d, A_s) + r(n-1)^2\sigma^2\left(A_s\right) \qquad (22.12c)
\end{aligned}$$

Equation 22.12c shows the impact of having relatives within the group, which is to shift some of the variance in social effects, $\sigma^2\left(A_s\right)$, into the covariance, $\sigma(z, A_T)$, between individual phenotype and total breeding value. The use of relatives in the group thus allows individual selection to access some of this otherwise untapped variance. This occurs because the breeding values for social effects of group members (which impacts the phenotypic value of the focal individual) are now correlated with an individual's own breeding value for social effects (where the latter has no direct impact on its phenotype).

A useful alternative expression is to partition $\sigma(z, A_T)$ into the contribution expected in unrelated groups (Equation 22.4b) plus the additional contribution due to individuals in the group being related, which yields

$$\sigma(z, A_T) = \sigma(z, A_T \,|\, r = 0) + (n-1)r\left[\sigma\left(A_s, A_d\right) + (n-1)\sigma^2\left(A_s\right)\right] \qquad (22.12d)$$

Alternatively, this can be expressed as

$$\sigma(z, A_T) = r\sigma^2(A_T) + (1-r)\left[\sigma^2(A_d) + (n-1)\sigma(A_d, A_s)\right] \qquad (22.12e)$$

showing that the more closely related group members are, the more weight individual selection puts on A_T. In the extreme, when groups are composed of clones, then $\sigma(z, A_T) =$

$\sigma^2(A_T)$. Plant breeding often selects among groups comprised of genetically identical individuals (i.e., inbred lines and clonally propagated lines), with such settings exploiting all of the heritable variation in both direct and associative effects without requiring any special design.

Similarly, when all members in the group have the same relatedness, r, the phenotypic variance becomes

$$\sigma^2(z) = \sigma^2(A_d) + \sigma^2(E_d) + (n-1)\left[\sigma^2(A_s) + \sigma^2(E_s)\right] + (n-1)r\left[2\sigma(A_s, A_d) + (n-2)\sigma^2(A_d)\right] \tag{22.13a}$$

$$= \sigma^2(z \,|\, r=0) + (n-1)r\left[2\sigma(A_s, A_d) + (n-2)\sigma^2(A_d)\right] \tag{22.13b}$$

where the phenotypic variance when all group members are unrelated, $\sigma^2(z \,|\, r = 0)$, is given by Equation 22.5c.

The response to selection is simply the change in the mean total breeding value, which (from Chapter 13) is the within-generation change in the phenotypic mean after selection (the selection differential, S) times the slope of the regression of A_T on phenotype z, yielding

$$R = \frac{\sigma(z, A_T)}{\sigma_z^2}\, S = \frac{\sigma(z, A_T)}{\sigma_z}\, \overline{\imath} \tag{22.14}$$

The second expression follows from the standard identity that $S = \sigma_z \, \overline{\imath}$ (Equation 13.6a). Substituting Equation 22.12c, with $n = 2$ and $r = 0$, into Equation 22.13 recovers Griffing's result (Equation 22.11a).

Example 22.4. Muir (2005) estimated variance components for six-week body weight in Japanese quail (*Coturnix coturnix japonica*) housed in groups of $n = 16$ per cage. REML estimates of the genetic variances were $\sigma^2(A_d) = 33.7$ and $\sigma^2(A_s) = 2.87$, while $\sigma(A_d, A_s) = -5.5$. Under these values, the predicted response to individual selection in a group of 16 unrelated individuals is

$$R = \frac{\overline{\imath}}{\sigma_z}\left[\sigma^2(A_d) + (n-1)\sigma(A_d, A_s)\right] = \frac{\overline{\imath}}{\sigma_z}\,[33.7 + 15 \cdot (-5.5)] = -48.8\,\frac{\overline{\imath}}{\sigma_z}$$

The strong negative covariance between direct and social (competitive) effects results in an expected reversed selection response if individual selection is used, as the positive gain from the improvement of direct effects (33.7) is swamped by the negative effects from the correlated response in social values (-82.5).

The presence of relatives within the group results in some fraction of $\sigma^2(A_s)$ being incorporated into the response under individual selection. Suppose the group of 16 consists of two half-sib families. In this case, the average relationship is 0.125, and from Equation 22.12d, the resulting covariance between phenotype and total breeding values becomes

$$\sigma(z, A_T) = \sigma(z, A_T \,|\, r = 0) + (n-1)r\left[\sigma\left(A_s, A_d\right) + (n-1)\sigma^2\left(A_s\right)\right]$$
$$= -48.4 + 15 \cdot 0.125\,(-5.5 + 15 \cdot 2.87) = 21.6$$

and from Equation 22.14,

$$R = \frac{\overline{\imath}}{\sigma_z}\,\sigma(z, A_T) = 21.6\,\frac{\overline{\imath}}{\sigma_z}$$

Hence, simply using groups of relatives (as opposed to groups of unrelated individuals) allows individual selection to give an expected positive response.

Example 22.5. Consider a trait with $\sigma^2(A_d) = 500$, $\sigma^2(A_s) = 50$, $\rho(A_d, A_s) = -0.25$, and $\sigma^2(E_d) = 400$. For ease of presentation, we assume that there are no social environmental effects ($E_s = 0$) and that E_d is uncorrelated across family members (i.e., no common family environment and no dominance). Consider a group size of $n = 6$. Given a selection intensity of $\bar{\imath}$, what response is expected under individual selection when group members are unrelated? Here $\sigma(A_d, A_s) = -0.25\sqrt{500 \cdot 50} = -39.5$. Substituting into Equation 22.4b gives the covariance as

$$\sigma(z, A_T) = \sigma^2(A_d) + (n-1)\sigma(A_d, A_s) = 500 - (5 \cdot 39.5) = 302.5$$

Likewise, from Equation 22.5c, the phenotypic variance is

$$\begin{aligned}\sigma^2(z) &= \sigma^2(A_d) + \sigma^2(E_d) + (n-1)\left[\sigma^2(A_s) + \sigma^2(E_s)\right]\\ &= 500 + 400 + 5 \cdot 50 = 1150\end{aligned}$$

Applying Equation 22.14, the resulting response becomes

$$R = \frac{\sigma(z, A_T)}{\sigma_z}\,\bar{\imath} = \frac{302.5}{\sqrt{1150}}\,\bar{\imath} = 8.92\,\bar{\imath}$$

Now suppose that group members are half-sibs ($r = 0.25$). What is the expected response? Applying Equation 22.12d yields

$$\begin{aligned}\sigma(z, A_T) &= \sigma(z, A_T \,|\, r = 0) + (n-1)r\left[\sigma\left(A_s, A_d\right) + (n-1)\sigma^2\left(A_s\right)\right]\\ &= 302.5 + 5 \cdot 0.25(-39.5 + 5 \cdot 50) = 565.5\end{aligned}$$

while Equation 22.13b yields a phenotypic variance of

$$\begin{aligned}\sigma^2(z) &= \sigma^2(z \,|\, r = 0) + (n-1)r\left[\sigma(A_s, A_d) + (n-2)\sigma^2(A_d)\right]\\ &= 1150 + \left[5 \cdot 0.25\left(-39.5 + 4 \cdot 50\right)\right] = 1350.6\end{aligned}$$

resulting in a response of $R = 15.39\,\bar{\imath}$. Likewise, if the group consists of full sibs ($r = 0.5$), the resulting covariance, variance, and response are, respectively, 828.5, 1551.2, and $21.04\,\bar{\imath}$. Thus, the response to selection increases with the relatedness of group members, with a 1.7- and 2.4-fold increase when using groups of half- and full-sibs (respectively) relative to groups of unrelated individuals.

Individual Selection: Direct vs. Social Response

Recalling Equation 22.1e, the response in the trait has two components: that from direct effects, $R_d = \Delta\mu_{A_d}$, and that from social effects, $R_s = \Delta\mu_{A_s}$. The relative contribution of each to the total response easily follows by considering the covariance of an individual's phenotype value, z, with either its direct, A_d, or social, A_s, breeding values. Specifically,

$$R_z = R_d + (n-1)R_s, \quad \text{where} \quad R_d = \frac{\sigma(A_d, z)}{\sigma_z}\,\bar{\imath} \quad \text{and} \quad R_s = \frac{\sigma(A_s, z)}{\sigma_z}\,\bar{\imath} \tag{22.15a}$$

Here

$$\sigma(A_d, z) = \sigma\left(A_d, A_d + \sum_{i \neq j} A_{s,i} + e\right) = \sigma^2(A_d) + r(n-1)\sigma(A_d, A_s) \tag{22.15b}$$

while

$$\sigma(A_s, z) = \sigma\left(A_s, A_d + \sum_{i \neq j} A_{s,i} + e\right) = \sigma(A_d, A_s) + r(n-1)\sigma^2(A_s) \tag{22.15c}$$

Equation 22.15b shows that the group must contain relatives ($r \neq 0$) in order for the covariance between direct and social values to impact the response in the direct value. Likewise, under individual selection, response in the social value only occurs if the direct and social values are correlated within individuals ($\sigma(A_d, A_s) \neq 0$) or if group members are related ($r \neq 0$), in which case the social value of the focal individual is correlated with the social values of those within its group.

Example 22.6. Consider the response in a family of half-sibs from Example 22.5, where the expected total response was 15.39 $\bar{\imath}$. What were the contributions from the direct and social responses? For the values used in that example,

$$\sigma(A_d, z) = \sigma^2(A_d) + r(n-1)\sigma(A_d, A_s) = 500 + [0.25 \cdot 5 \cdot (-39.5)] = 450.63$$
$$\sigma(A_s, z) = \sigma(A_d, A_s) + r(n-1)\sigma^2(A_s) = -39.5 + [0.24 \cdot 5 \cdot 50] = 23.0$$

Recalling from Example 22.5 that $\sigma_z^2 = 1350.6$ for half-sibs, Equations 22.15a and 22.15b return the two components of response as

$$R_d = \frac{450.63}{\sqrt{1350.6}}\,\bar{\imath} = 12.26\,\bar{\imath} \quad \text{and} \quad R_s = \frac{23}{\sqrt{1350.6}}\,\bar{\imath} = 0.63\,\bar{\imath}$$

Hence, 80% (12.26/15.39) of the total response was due to response in direct effects, while 20% was from the response in social effects (5·0.63/15.39). Under individual selection with half-sib families, both the mean direct and mean social values improved. By contrast, if group members are unrelated, then (Example 22.5) $\sigma_z^2 = 1150$, while

$$\sigma(A_d, z) = \sigma^2(A_d) = 500 \quad \text{and} \quad \sigma(A_s, z) = \sigma(A_d, A_s) = -39.5$$

resulting in responses of

$$R_d = \frac{500}{\sqrt{1150}}\,\bar{\imath} = 14.74\,\bar{\imath} \quad \text{and} \quad R_s = \frac{-39.5}{\sqrt{1150}}\,\bar{\imath} = -1.165\,\bar{\imath}$$

In this case, while the total response was positive, the large direct response (14.74) was partly offset by a decrease in the mean social environment ($[n-1]R_s = 5 \cdot [-1.165] = -5.83$), yielding a total response of (14.74 − 5.83) $\bar{\imath}$ = 8.91 $\bar{\imath}$. The lack of relatedness implies no direct selection involving $\sigma^2(A_s)$, and hence the social breeding values only change through their correlation with the direct values, which in this example was negative.

Individual Selection: Maternal Effects

An important special case, and indeed the forerunners of more general models of associative effects, are models of direct and **maternal** effects (Dickerson 1947; Willham 1963, 1972; Cheverud 1984a). Here, the trait value of an individual is a function of its direct effect, P_d, and a **maternal performance trait**, P_m, contributed by its mother, meaning that if j is the mother of i, then

$$z_i = P_{d_i} + P_{m,j} \tag{22.16a}$$

In the absence of inbreeding, $r = 1/2$ for this group (mother-offspring) with $n = 2$. From Equation 22.12c, the covariance between phenotype and total breeding value ($A_T = A_d + A_m$, with $A_s = A_m$) is

$$\sigma(z, A_T) = \sigma^2(A_d) + (3/2)\sigma(A_d, A_m) + (1/2)\sigma^2(A_m) \tag{22.16b}$$

while Equation 22.13a yields a phenotypic variance of

$$\sigma^2(z) = \sigma^2(A_d) + \sigma(A_d, A_m) + \sigma^2(A_m) + \sigma^2(e) \tag{22.16c}$$

making the resulting response to selection

$$R = \frac{\sigma(z, A_T)}{\sigma_z}\,\bar{\imath} = \frac{\sigma^2(A_d) + (3/2)\sigma(A_d, A_m) + (1/2)\sigma^2\,(A_m)}{\sqrt{\sigma^2(A_d) + \sigma(A_d, A_m) + \sigma^2(A_m) + \sigma^2(e)}}\,\bar{\imath} \tag{22.16d}$$

The total response can also be expressed in terms of the direct and maternal-effect response. From Equation 22.15,

$$R_d = \frac{\sigma(A_d, z)}{\sigma_z}\,\bar{\imath} = \frac{\sigma^2(A_d) + (1/2)\sigma(A_d, A_m)}{\sigma_z}\,\bar{\imath} \tag{22.17a}$$

and

$$R_m = \frac{\sigma(A_m, z)}{\sigma_z}\,\bar{\imath} = \frac{\sigma(A_d, A_m) + (1/2)\sigma^2(A_m)}{\sigma_z}\,\bar{\imath} \tag{22.17b}$$

with the response, R, in the trait mean being

$$R = R_d + (2-1)R_m = R_d + R_m \tag{22.17c}$$

Substitution of Equations 22.17a and 22.17b into Equation 22.17c recovers Equation 22.16d. As reviewed by Cheverud (1984a), most estimates of the direct-maternal covariance are negative. This raises the possibility of a reversed response due to a greater reduction in the maternal environment than improvement in the direct effect. It also allows for the trait to improve (via its direct value) at the expense of a declining maternal value.

The careful reader might recall from Chapter 15 that Falconer's trait-based model of a single maternal effect results in more complicated dynamics (such as time lags). Why do these not appear in this analysis? As noted by Bijma (2011), variance-component models essentially focus on the permanent component of response, ignoring transient contributions that can appear in a trait-based analysis. He showed that Equation 22.16d and Falconer's model both give the same value for the permanent response.

Group Selection: Theory

Under individual selection with unrelated group members, there is no contribution from $\sigma^2(A_s)$ to the response, and changes in A_s only enter as a correlated response to changes in A_d, which can be in an unfavorable direction when $\sigma(A_d, A_s) < 0$. As we will see, $\sigma^2(A_s)$ enters into the response under group selection even when there are no relatives in the group. The reason is that the group phenotype is a function of the distribution of A_s values.

Under strict group selection, selection is based on the group mean, $\bar{z}$, or equivalently the total value of the group, $n\bar{z} = \sum z$, and we will usually work with the latter. To obtain the covariance between the total value of a group and the total breeding value of one of its members, first note that

$$\begin{aligned}\sum_{j=1}^{n} z_j &= \sum_{j=1}^{n}\left[A_{d_j} + E_{d_j} + \sum_{k\neq j}^{n}(A_{s_k} + E_{s_k})\right] = \sum_{j=1}^{n} A_{d_j} + \sum_{j=1}^{n}\sum_{k\neq j}^{n} A_{s_k} + \sum_{j=1}^{n} e_j \\ &= \sum_{j=1}^{n} A_{d_j} + (n-1)\sum_{j=1}^{n} A_{s_j} + \sum_{j=1}^{n} e_j \\ &= \sum_{j=1}^{n} A_{T_j} + \sum_{j=1}^{n} e_j \end{aligned} \tag{22.18}$$

where the residual values, e_i, sweep up a variety of environmental terms, and are given by Equation 22.1c. The residuals are assumed to be uncorrelated with any breeding values, but of course residuals can be (and usually are) correlated within a group (e.g., Equation

22.23a). If r_{ij} is the relationship between individuals i and j, the covariance between the group total and the total breeding value of a group member, i, is

$$\sigma\left(A_{T_i}, \sum_{j=1}^{n} z_j\right) = \sigma\left(A_{T_i}, \sum_{j=1}^{n}[A_{T_j} + e_j]\right) = \sum_{j=1}^{n} \sigma\left(A_{T_i}, A_{T_j}\right) = \sigma^2(A_T) \sum_{j=1}^{n} r_{ij}$$

$$= \sigma^2(A_T)\left(1 + \sum_{j \neq i} r_{ij}\right) \tag{22.19a}$$

If the group members are unrelated, then

$$\sigma\left(A_{T_i}, \sum_{j=1}^{n} z_j\right) = \sigma^2(A_T) \tag{22.19b}$$

which implies that $\sigma(A_{T_i}, \overline{z}) = \sigma^2(A_T)/n$. Hence, group selection acts on the total breeding value of an individual, rather than on only part of A_T, as was the case with individual selection (e.g., Equation 22.12e). The contribution of associative effects to the total breeding value does not influence the phenotype of the focal individual, but *does* influence the phenotypes of other group members, and hence, $\overline{z}$. Group selection directly targets these effects. If all members have the same degree of relationship (r), then

$$\sigma\left(A_{T_i}, \sum_{j=1}^{n} z_j\right) = \sigma^2(A_T)\,[\,1 + (n-1)r\,] \tag{22.19c}$$

Selection can act on associative effects even when none of the individuals in the group are related, but its efficiency is amplified when using relatives (compare Equations 22.19b and 22.19c). From Equation 22.19c, the covariance of the total breeding value, A_T, of a group member with its group mean, $\overline{z}$, is

$$\sigma\left(A_{T_i}, \overline{z}\,\right) = \frac{1}{n}\,\sigma^2(A_T)\,[\,1 + (n-1)r\,] = \sigma^2(A_T)\left(r + \frac{1-r}{n}\right) \tag{22.19d}$$

Turning to the phenotypic variance of the group total, $n\,\overline{z}$, a little bit of algebra is required. From Equation 22.18, we can decompose this group variance into additive-genetic and environmental components

$$\sigma^2\left(\sum_{j=1}^{n} A_{T_j} + \sum_{j=1}^{n} e_j\right) = \sigma\left(\sum_{j=1}^{n} A_{T_j}, \sum_{k=1}^{n} A_{T_k}\right) + \sigma\left(\sum_{j=1}^{n} e_j, \sum_{k=1}^{n} e_k\right) \tag{22.20}$$

Tackling the genetic component first yields

$$\sigma\left(\sum_{j=1}^{n} A_{T_j}, \sum_{k=1}^{n} A_{T_k}\right) = \sigma^2(A_T) \cdot \sum_{j=1}^{n}\sum_{k=1}^{n} r_{ij} \tag{22.21a}$$

When all group members have the same degree of relationship, r, this reduces to

$$\sigma\left(\sum_{j=1}^{n} A_{T_j}, \sum_{k=1}^{n} A_{T_k}\right) = \sigma^2(A_T)\,n\,[1 + (n-1)\,r\,] \tag{22.21b}$$

Turning our attention to the residual terms, recall (Equation 22.1c) that the residual is a function of both direct and social environmental effects,

$$e_i = E_{d_i} + \sum_{k \neq i} E_{s_k}$$

Clearly, individuals within the same group are correlated because they share the E_s values from the other group members. Recalling that $\sigma(E_{d_i}, E_{s_k}) = 0$ for $i \neq k$, the residual variance becomes

$$\begin{aligned}\sigma^2(e) = \sigma(e_i, e_i) &= \sigma\left(E_{d_i} + \sum_{k\neq i} E_{s_k}, E_{d_i} + \sum_{k\neq i} E_{s_k} \right) = \sigma\left(E_{d_i}, E_{d_i}\right) + \sum_{k\neq i} \sigma\left(E_{s_k}, E_{s_k}\right) \\ &= \sigma^2(E_d) + (n-1)\sigma^2(E_s) \end{aligned} \tag{22.22a}$$

For $i \neq j$ in the same group, the covariance among residuals is

$$\begin{aligned}\sigma(e_i, e_j) &= \sigma\left(E_{d_i} + E_{s_j} + \sum_{k\neq i,j} E_{s_k}, E_{d_j} + E_{s_i} + \sum_{k\neq i,j} E_{s_k} \right) \\ &= \sigma\left(E_{d_i}, E_{d_j}\right) + \sigma\left(E_{d_i}, E_{s_i}\right) + \sigma\left(E_{d_j}, E_{s_j}\right) + \sigma\left(\sum_{k\neq i,j} E_{s_k}, \sum_{k\neq i,j} E_{s_k} \right) \\ &= 0 + 2\sigma(E_d, E_s) + \sum_{k\neq i,j} \sigma\left(E_{s_k}, E_{s_k} \right) \\ &= 2\sigma(E_d, E_s) + (n-2)\sigma^2(E_s) \end{aligned} \tag{22.22b}$$

The first term accounts for the fact that the direct and social environmental values can be correlated within the same individual, while the second term accounts for the shared environmental values contributed by the other $n-2$ group members. Putting these together yields

$$\sigma(e_i, e_j) = \begin{cases} \sigma^2(e) & i = j \\ \rho\,\sigma^2(e) & i \neq j, \ i \text{ and } j \text{ in the same group} \\ 0 & i \neq j, \ i \text{ and } j \text{ in different groups} \end{cases} \tag{22.23a}$$

where

$$\sigma^2(e) = \sigma^2(E_d) + (n-1)\sigma^2(E_s) \quad \text{and} \quad \rho = \frac{2\sigma(E_d, E_s) + (n-2)\sigma^2(E_s)}{\sigma^2(e)} \tag{22.23b}$$

Here ρ is the correlation among environmental values within a group, and can be either positive or negative. For large values of n, we expect $\sigma^2(E_s)$ to dominate the covariance term, yielding $\rho > 0$. Equations 22.23a and 22.23b were first obtained by Bijma et al. (2007b). Correlations among environmental residuals are also generated by shared maternal effects and (for full-sibs) dominance. If all group members are the same type of relative, this is simply incorporated into ρ. However, when a group consists of two (or more) families, the additional residual covariance among sibs needs to be accounted for (Example 22.14, below, shows how this is accomplished in a BLUP framework).

Using these results, and following the same logic as with additive-genetic values, yields

$$\sigma\left(\sum_{j=1}^{n} e_j, \sum_{k=1}^{n} e_k \right) = n\sigma^2(e) + \sum_{j\neq k} \sigma\left(e_j, e_k\right) = n\sigma^2(e)\left[1 + (n-1)\rho\right] \tag{22.24}$$

Substituting Equations 22.21b and 22.24 into Equation 22.20 returns the variance of the group total as

$$\sigma^2\left(\sum_{j=1}^{n} z_j \right) = n\sigma^2(A_T)\left[1 + (n-1)r\right] + n\sigma^2(e)\left[1 + (n-1)\rho\right] \tag{22.25a}$$

The variance of the group mean is simply $1/n^2$ of this value, or

$$\begin{aligned}\sigma^2(\,\overline{z}\,) &= \sigma^2(A_T)\left(\frac{1 + (n-1)r}{n}\right) + \sigma^2(e)\left(\frac{1 + (n-1)\rho}{n}\right) \\ &= \sigma^2(A_T)\left(r + \frac{1-r}{n}\right) + \sigma^2(e)\left(\rho + \frac{1-\rho}{n}\right) \end{aligned} \tag{22.25b}$$

Note the symmetric roles of the relatedness, r, of group members and the within-group correlation, ρ, of residuals with respect to, respectively, the variance in total breeding values and the residual variance.

Using the covariance between total breeding value and group mean (Equation 22.19d) and the variance of the group mean (Equation 22.25b), the resulting response to selection (i.e., the change in trait mean) follows from our general response expression (Equation 13.10b), and is

$$R = \frac{\sigma(A_T, \overline{z})}{\sigma^2(\overline{z})} S = \frac{\sigma^2(A_T) r_n}{\sigma^2(A_T) r_n + \sigma^2(e)\rho_n} S \tag{22.26a}$$

$$= \frac{\sigma(A_T, \overline{z})}{\sigma(\overline{z})} \bar{\imath} = \frac{\sigma^2(A_T) r_n}{\sqrt{\sigma^2(A_T) r_n + \sigma^2(e)\rho_n}} \bar{\imath} \tag{22.26b}$$

where

$$r_n = r + \frac{1-r}{n} \quad \text{and} \quad \rho_n = \rho + \frac{1-\rho}{n}$$

For $n = 2$ and $r = 0$, applying Equations 22.19b and 22.25a recovers Griffing's result (Equation 22.11b). As expected, in cases where there are only direct effects, Equations 22.26a and 22.26b reduce to our expressions for family selection (Chapter 21).

Example 22.7. Consider group selection using Muir's quail data from Example 22.4. Here $\sigma^2(A_d) = 33.7, \sigma^2(A_s) = 2.87, \sigma(A_d, A_s) = -5.5$, and $n = 16$. Muir estimated the residual variance as $\sigma^2(e) = 69.0$, while his model assumed $\rho = 0$, resulting $\rho_n = 1/n$, and hence $\sigma^2(e)\rho_n = 69.0/16 = 4.32$. Applying Equation 22.4e yields a total additive variance of

$$\sigma^2(A_T) = \sigma^2(A_d) + 2(n-1)\sigma(A_d, A_s) + (n-1)^2\sigma^2(A_s)$$
$$= 33.7 + [30 \cdot (-5.5)] + \left[30^2 \cdot 2.87\right] = 2451.7$$

while Equation 22.26b yields a response of

$$R = \frac{\sigma^2(A_T) r_n}{\sqrt{\sigma^2(A_T) r_n + \sigma^2(e)\rho_n}} \bar{\imath} = \frac{2451.7 \cdot r_n}{\sqrt{2451.7 \cdot r_n + 4.32}} \bar{\imath}$$

For groups of unrelated individuals, $r = 0$ and $r_n = 1/16 = 0.0625$, respectively, and the response becomes R= 12.2 $\bar{\imath}$. For half- and full-sibs (r = 0.25 and 0.5), r_n = 0.297 and 0.531, respectively, with responses of 26.9 $\bar{\imath}$ and 36.0 $\bar{\imath}$, twofold and threefold increases relative to that for a group of unrelated individuals.

While Equation 22.26a shows that group selection always results in an expected non-negative response (as $\sigma^2(A_T) \geq 0$), it may be less than optimal. If direct effects account for the majority of variance, group selection can be very inefficient relative to individual selection. To see this, consider groups of unrelated individuals and suppose the trait of interest has no associative effects, $\sigma^2(A_s) = 0$, so that $\sigma^2(A_T) = \sigma^2(A_d)$. Under individual (or mass) selection, the response is $R_m = h\sigma(A_d)\bar{\imath}$ (Equation 13.6b). Now consider the response, R_G, in the mean of trait z under group selection, where $\sigma(\overline{z}, A_T) = (1/n)\sigma^2(A_d)$ and $\sigma^2(\overline{z}) = \sigma_z^2/n$, giving the response (from Equation 22.26b) as

$$R_G = \frac{\sigma(\overline{z}, A_T)}{\sigma(\overline{z})} \bar{\imath} = \frac{(1/n)\sigma^2(A_d)}{\sigma_z/\sqrt{n}} \bar{\imath} = \frac{1}{\sqrt{n}} \frac{\sigma(A_d)}{\sigma_z} \sigma(A_d)\bar{\imath} = \frac{1}{\sqrt{n}} h\sigma(A_d)\bar{\imath} = \frac{1}{\sqrt{n}} R_m$$

Under these conditions, individual selection is always superior to group selection, with the superiority increasing with group size. For groups of 5, 10, and 25, group selection has only 44.7%, 31.6%, and 20% (respectively) of the expected response of individual selection.

Group Selection: Direct vs. Social Response

As was the case for individual selection, we can decompose the response under group selection into the responses from direct and social effects, $R_z = R_d + (n-1)R_s$. Under group selection, these response components are given by

$$R_d = \frac{\sigma(A_d, \sum z)}{\sigma(\sum z)}\,\bar{\imath} \quad \text{and} \quad R_s = \frac{\sigma(A_s, \sum z)}{\sigma(\sum z)}\,\bar{\imath} \tag{22.27a}$$

The covariance between the direct breeding value of a group member and the group total becomes

$$\begin{aligned}\sigma\left(A_{d_i}, \sum_{j=1}^{n} z_j\right) &= \sigma\left(A_{d_i}, \sum_{j=1}^{n} A_{d_j} + (n-1)\sum_{j=1}^{n} A_{s_j} + \sum_{j=1}^{n} e_j\right) \\ &= \sigma^2(A_d)\sum_{j=1}^{n} r_{ij} + (n-1)\sigma(A_d, A_s)\sum_{j=1}^{n} r_{ij} \\ &= \left[\sigma^2(A_d) + (n-1)\sigma(A_d, A_s)\right]\left[1 + (n-1)\overline{r}\,\right] \end{aligned} \tag{22.27b}$$

Where $\overline{r} = \sum_{j\neq i}^{n} r_{ij}/(n-1)$ is the average degree of relationship (for i) among group members (assuming that $r_{ii} = 1$, i.e., i is not inbred), resulting in $\sum_{j}^{n} r_{ij} = 1 + (n-1)\overline{r}$. Similarly, for the social breeding value

$$\sigma\left(A_{s_i}, \sum_{j=1}^{n} z_j\right) = \left[\sigma(A_d, A_s) + (n-1)\sigma^2(A_s)\right]\left[1 + (n-1)\overline{r}\,\right] \tag{22.27c}$$

Increasing the relatedness, r, of group members increases the contributions from $\sigma(A_d, A_s)$ and $\sigma^2(A_s)$ by the same proportional amount, $[1+(n-1)\overline{r}\,]$. Hence, the relative contribution of these two components is independent of the degree of relatedness within the group. By contrast, recall that under individual selection, the relative contributions of these two components changes (and potentially can change rather dramatically) with r (Equations 22.15b and 22.15c).

Group Selection: Experimental Evidence

How effective is group selection? As reviewed in Chapter 21, the special case of the group being a single family has a fairly robust experimental literature. What is seen in more general settings? Experiments in laboratory settings generally have proved effective in generating a positive response (Goodnight and Stevens 1997; Goodnight 2005). Especially telling are several reports of group selection yielding a positive response when individual selection either failed to do so or generated a negative response.

One of the first group-selection experiments was by Wade (1976, 1977), who found a rapid response to group selection for the (group-level) trait of population size in the flour beetle *Tribolium castaneum*. A series of 48 populations was founded, each with 16 unrelated individuals, and population size was measured at 37 days postfounding. Under the control (allowing for individual selection during the growout to 37 days), a group of 16 individuals was chosen at random from the 48 populations and used to found a new population, repeatedly (with the possibility of resampling from the same population), until 48 new populations were formed. Under group selection for increased population size, sets of 16 individuals were drawn from the largest population and used to found a new population, which was continued until the largest population was exhausted. When this happened, individuals were similarly used from the second largest population, and so forth, to fill out the new array of 48 populations. The group-selected populations showed significantly larger population sizes relative to the control, and they also showed reduced levels of cannibalism. Laboratory populations of *Tribolium* were also used by Craig (1982), who found that group selection was very efficient in increasing (and decreasing) emigration rates. In both studies, some degree of relationship might be expected within groups, which would be small at first, with $\overline{r}$ increasing under inbreeding as selection proceeds (albeit likely still remaining somewhat small at the end of the experiment).

Response under group selection is not limited to animals. Goodnight (1985) contrasted individual and group selection for leaf area in the mustard *Arabodopsis thaliana*. Plants were grown in groups of 16 unrelated individuals. Individual selection for increased leaf area actually resulted in a reversed response, with offspring showing smaller leaf area. In contrast, average leaf area per plant (i.e., a larger total leaf area for the group) increased under group selection.

Finally, dramatic responses with significant economic impact have occurred when using group selection in animal production settings. In chickens, high egg-production systems typically house several hens per cage. Aggressive behavior and mortality are common in such settings. Selection for improved individual production could result in increased aggression within the cage, and hence lower cage production (reviewed in Muir 1985). To assess whether group selection could improve performance, Muir (1996) made selections based on the mean value of nine-bird cages ($n = 9$). Eggs per hen per day, eggs per hen, and egg mass all increased dramatically. What was even more striking, was that annual percentage mortality declined from 68% to just under 9% at the end of generation 6, which is similar to the mortality in single-bird cages. Muir called the resulting selected strain KGB chickens (for Kinder, Gentler Birds). Selection based on the group (here, cage) mean improved total performance of the cage, in part by reducing the amount of aggression within the cage, as Craig and Muir (1996) found that KGB birds showed a significant reduction (relative to controls) in beak-inflicted injuries.

The benefits of group selection have often been framed in terms of exploiting non-additive variation that is not accessible by traditional individual selection (e.g., Goodnight and Stevens 1997). While we have focused here on genetic variation that is not directly accessible under individual selection when groups are unrelated ($\sigma^2[A_s]$), this variation is entirely *additive*. Specifically, when heritable associative effects are present, they can only be directly accessible through either group selection (with either related or unrelated members) or individual selection when interactions occur in groups of related individuals (as the appropriate covariances for response in either setting places nonzero weight on $\sigma[A_s]$). This is not to ignore the possibility of exploiting additional nonadditive variation under group selection, but rather to highlight the importance of associative effects.

INCORPORATING BOTH INDIVIDUAL AND GROUP INFORMATION

Given that group selection always results in an expected positive response, while individual selection can range from (at best) being far more efficient than group selection to (at worst) generating an expected reversed response, clearly the optimal approach is some combination of selection on both individual and group components. This is simply an extension of the concept of a family index introduced in Chapter 21, that combines both individual and family (now group) information.

Response on a Weighted Index

To combine both individual and group selection, consider the index, I, where the value of the index for the ith individual is given by

$$I_i = z_i + g \sum_{j \neq i} z_j \tag{22.28a}$$

This is a modification of the initial proposal by Griffing (1977). Letting $\overline{z}_i$ denote the mean of the group containing individual i, this index can also be written as

$$I_i = (1-g)z_i + g\sum_{j=1}^{n} z_j = (1-g)z_i + g\,n\,\overline{z}_i \tag{22.28b}$$

showing that individual selection corresponds to $g = 0$ and group selection to $g = 1$. Thus, the index given by Equation 22.28b includes both individual and group selection as special cases. Selection of individuals based on within-group deviations is also a special case of Equation 22.28a, as setting $g = -1/n$ yields

$$I_i = z_i - \frac{1}{n}\sum_{j=1}^{n} z_j = z_i - \overline{z} \tag{22.28c}$$

The response in the trait mean (μ_z) from selection on this index is

$$R = \frac{\sigma(I, A_T)}{\sigma(I)}\,\overline{\imath}_I \tag{22.28d}$$

This can also be written in terms of the **accuracy** of selection, a concept first introduced in Chapter 13 (Equation 13.11a), which is the correlation between the target of selection (here I) and the breeding objective (here the total breeding value, A_T). We can express the ratio in Equation 22.28d as

$$\frac{\sigma(I, A_T)}{\sigma(I)} = \frac{\sigma(I, A_T)}{\sigma(I)}\frac{\sigma(A_T)}{\sigma(A_T)} = \sigma(A_T)\frac{\sigma(I, A_T)}{\sigma(A_T)\,\sigma(I)} = \sigma(A_T)\,\rho(A_T, I) \tag{22.28e}$$

where the accuracy

$$\rho(A_T, I) = \frac{\sigma(I, A_T)}{\sigma(A_T)\,\sigma(I)} \tag{22.28f}$$

is the correlation between the index value of an individual and its breeding value. Using this result, Equation 22.28d becomes

$$R = \rho(A_T, I)\,\sigma(A_T)\,\overline{\imath}_I \tag{22.28g}$$

which is simply Equation 13.11c for the selection criteria $x = I$. This is a very useful expression for comparing different selection schemes, as $\sigma(A_T)$ remains unchanged (provided group size remains fixed), so the maximal response occurs by maximizing $\rho(A_T, I)\,\overline{\imath}_I$. Given that the fraction saved largely sets the selection intensity $\overline{\imath}_I$ (subject to minor variation due to finite populations; see Equation 14.4b), the optimal scheme (i.e., the optimal weight, g) is that which maximizes the accuracy, $\rho(A_T, I)$.

To obtain a general expression for response for any combination of group selection fraction (g) and average relatedness within groups (r), we first need the covariance of I and A_T within an individual. This is obtained as follows. First, note that

$$\sigma\left(A_T, I\right) = (1-g)\sigma\left(A_T, z\right) + g\sigma\left(A_T, \sum_{j=1}^{n} z_j\right) \tag{22.29a}$$

When group members are unrelated, Equations 22.4b and 22.19b give

$$\sigma\left(A_T, I\right) = (1-g)\left[\,\sigma^2(A_d) + (n-1)\sigma(A_d, A_s)\,\right] + g\,\sigma^2(A_T) \tag{22.29b}$$

When group members all have the same relationship, Equations 22.12e and 22.19c yield

$$\sigma\left(A_T, I\right) = (1-g)\Big(r\sigma^2(A_T) + (1-r)\left[\,\sigma^2(A_d) + (n-1)\sigma(A_d, A_s)\,\right]\Big) + g\left[1 + (n-1)r\right]\sigma^2(A_T) \tag{22.29c}$$

Collecting terms, Equation 22.29c reduces to

$$\sigma(A_T, I) = [g + r + (n-2)gr]\,\sigma^2(A_T) + (1-g)(1-r)\left[\sigma^2(A_d) + (n-1)\sigma(A_s, A_d)\right] \quad (22.29d)$$

While parts of this result (in a bit more cryptic form) appear in Griffing (1977), this, more general, version is due to Bijma et al. (2007a). Note that g and r have symmetric roles in the covariance between the index and the total breeding value. Thus, from the standpoint of this covariance, relatedness and group selection (r and g) are interchangeable. However, as we will soon demonstrate, g and r do not play symmetric roles in the variance, σ_I^2, of the index, so interchanging r and g values results in a different variance, and hence a different selection response (see Equation 22.28d and Example 22.8).

Now consider the variance of the index, I. From Equation 22.28a,

$$\begin{aligned}\sigma_I^2 &= \sigma\left(z_i + g\sum_{j\neq i} z_j,\, z_i + g\sum_{j\neq i} z_j\right)\\ &= \sigma_z^2 + 2g\sigma\left(z_i, \sum_{j\neq i} z_j\right) + g^2\sigma^2\left(\sum_{j\neq i} z_j\right)\end{aligned} \quad (22.30a)$$

If all group members have the same relationship, then

$$\sigma\left(z_i, \sum_{j\neq i} z_j\right) = (n-1)\sigma(z_i, z_j) \quad (22.30b)$$

and

$$\sigma^2\left(\sum_{j\neq i} z_j\right) = (n-1)\left[\sigma_z^2 + (n-2)\sigma(z_i, z_j)\right] \quad (22.30c)$$

Substituting these last two expressions into Equation 22.30a and collecting terms gives

$$\sigma_I^2 = \sigma_z^2\left[1 + g^2(n-1)\right] + \sigma(z_i, z_j)\big[g(n-1)\left(2 + g\{n-2\}\right)\big] \quad (22.30d)$$

As a check of Equation 22.30d, note that (as expected) this reduces to σ_z^2 when $g = 0$ and to $n\sigma_z^2 + n(n-1)\sigma(z_i, z_j)$ when $g = 1$. Equation 22.13a gives the expression for σ_z^2 when all relatives within the group are related by r. It remains to obtain $\sigma(z_i, z_j)$, the phenotypic covariance of group members, in order to apply Equation 22.30d. From Equation 22.1c, and ignoring the constant, μ,

$$\sigma(z_i, z_j) = \sigma\left(A_{d_i} + \sum_{k\neq i} A_{s_k} + e_i,\, A_{d_j} + \sum_{k\neq j} A_{s_k} + e_j\right) \quad (22.31a)$$

$$= \sigma\left(A_{d_i}, A_{d_j}\right) + 2\sigma\left(A_{d_i}, \sum_{k\neq i} A_{s_k}\right) + \sigma\left(\sum_{k\neq i} A_{s_k}, \sum_{k\neq j} A_{s_k}\right) + \sigma(e_i, e_j) \quad (22.31b)$$

If we expand and evaluate these covariance terms and collect the common terms, Equation 22.31b ultimately reduces to

$$\begin{aligned}\sigma(z_i, z_j) &= 2\sigma(A_d, A_s) + (n-2)\sigma^2(A_s) + \rho\,\sigma^2(e)\\ &\quad + r\left[\sigma^2(A_d) + 2(n-2)\sigma(A_d, A_s) + \left\{(n-1) + (n-2)^2\right\}\sigma^2(A_s)\right]\end{aligned} \quad (22.31c)$$

Notice, by comparison to Equation 22.4c, that the term appearing when group members are related ($r \neq 0$) is the variance of A_T when the group size is $(n-1)$ plus the correction $(n-1)\sigma^2(A_s)$. Equations 22.29d and 22.30d are substituted into Equation 22.28d to obtain the response. The interplay of $\sigma(A_T, I)$ and σ_I^2 (as functions of g and r) determine the accuracy of any particular index (Figure 22.3).

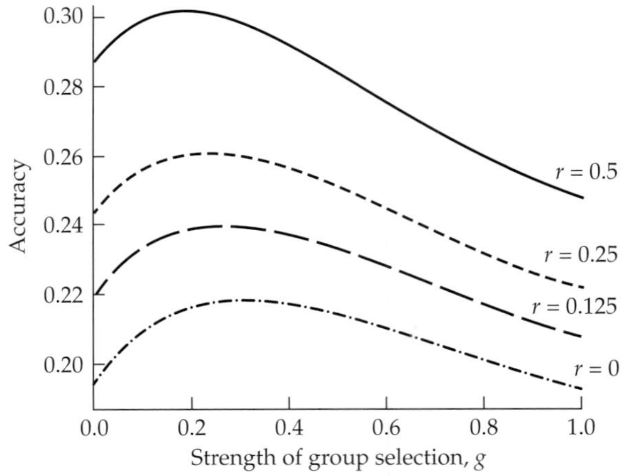

Figure 22.3 Accuracy of the index, I, as a function of the group weight, g, for groups of different types of relatives (the curves corresponding to different values of r). Accuracy was computed using Equation 22.28f, whose components are given by Equations 22.29d, 22.7b, and 22.30d. The variance components used were from Ellen et al. (2008) for survival days for chickens, and are given in Example 22.8, which also works through the calculations. Accuracy increases with r and is maximized at some intermediate strength of group selection, g.

Example 22.8. Ellen et al. (2008) estimated the following values for survival days in chickens raised in groups of $n = 4$: $\sigma^2(A_d) = 915, \sigma(A_d, A_d) = 62, \sigma^2(A_s) = 134, \sigma^2(e) = 11,500$, and $\rho = 0.08$. Applying Equation 22.4c yields

$$\sigma^2(A_T) = \sigma^2(A_d) + 2(n-1)\sigma(A_d, A_s) + (n-1)^2\sigma^2(A_s)$$
$$= 915 + (2 \cdot 3 \cdot 62) + (3^2 \cdot 134) = 2493$$

while

$$\sigma^2(A_d) + (n-1)\sigma(A_s, A_d) = 915 + (3 \cdot 62) = 1101$$

Substituting these results into Equation 22.29d returns a covariance between I and total breeding value of

$$\sigma(A_T, I) = [(g + r + 2gr) \cdot 2493] + [(1-g)(1-r) \cdot 1101] \tag{22.32a}$$

To obtain the variance in I using Equation 22.30d, we first need expressions for $\sigma^2(z)$ and $\sigma(z_i, z_j)$. From Equation 22.5d, the phenotypic variance when the group contains unrelated individuals is

$$\sigma^2(z \,|\, r = 0) = \sigma^2(A_d) + (n-1)\sigma^2(A_s) + \sigma^2(e) = 915 + (3 \cdot 134) + 11,500 = 12,817$$

Noting that

$$(n-1)\left[2\sigma(A_s, A_d) + (n-2)\sigma^2(A_d)\right] = 3 \cdot [(2 \cdot 62) + (2 \cdot 915)] = 5862$$

Equation 22.13b shows the phenotypic variance for a group with relationship r as

$$\sigma^2(z) = \sigma^2(z \,|\, r = 0) + (n-1)r\left[2\sigma(A_s, A_d) + (n-2)\sigma^2(A_d)\right]$$
$$= 12,817 + (r \cdot 5862) \tag{22.32b}$$

To obtain $\sigma(z_i, z_j)$, we first find that

$$2\sigma(A_d, A_s) + (n-2)\sigma^2(A_s) + \rho\,\sigma^2(e) = (2 \cdot 62) + (2 \cdot 134) + (0.08 \cdot 11,550) = 1312$$

and

$$\sigma^2(A_d) + 2(n-2)\sigma(A_d, A_s) + \left[(n-1) + (n-2)^2\right]\sigma^2(A_s)$$
$$= 915 + (2 \cdot 2 \cdot 62) + \left[(3 + 2^2) \cdot 134\right] = 2101$$

Substituting into Equation 22.31c gives the general covariance between group members when all members are related by an amount, r, as

$$\sigma(z_i, z_j) = 1312 + r \cdot 2101 \tag{22.32c}$$

Finally, substituting Equations 22.32b and 22.32c into Equation 22.30d gives the variance in I as

$$\sigma_I^2 = \sigma_z^2\left[1 + g^2(n-1)\right] + \sigma(z_i, z_j)g(n-1)\left[2 + g(n-2)\right]$$
$$= \left[12{,}817 + (r \cdot 5862)\right]\left(1 + 3g^2\right) + \left[1312 + (r \cdot 2101)\right]6g\left(1 + g\right) \tag{22.32d}$$

Note from Equation 22.32a that the roles of relatedness (r) and amount of group selection (g) are fully interchangeable in the covariance between I and A_T. However, Equation 22.32d shows that this is *not* the case for σ_I^2, and hence the expected responses when the values of r and g are swapped are not expected to be equal. For example, for $r = 0.5, g = 0$,

$$\sigma(I, A_T) = 1797, \quad \sigma_I^2 = 15{,}748, \quad \rho(A_T, I) = 0.287, \quad R = 14.32\,\bar{\imath}$$

where $\rho(A_T, I)$ is given by Equation 22.28f. For $r = 0, g = 0.5$,

$$\sigma(I, A_T) = 1797, \quad \sigma_I^2 = 28{,}334, \quad \rho(A_T, I) = 0.214, \qquad R = 10.68\,\bar{\imath}$$

As points of reference, the accuracy and response under individual selection with no relatives in the group ($g = r = 0$) are $\rho(A_T, I) = 0.195$ and $R = 9.73\,\bar{\imath}$, while for group selection with unrelated individuals within the group ($g = 1, r = 0$), the accuracy and response become $\rho(A_T, I) = 0.193$ and $R = 9.63\,\bar{\imath}$.

Optimal Response

In the index shown by Equation 22.28a, g is the fraction of weight placed on a random individual from the group that interacts with the focal individual. If this weight is zero, the index reduces to individual selection, while if this weight is 1, all individuals in the group are weighted equally in the selection decision and there is group selection An obvious question is to determine the optimal value for g that maximizes the selection response. From Equation 22.28g, we see that the optimal response occurs by using those weights in I that maximizes the correlation, $\rho(A_T, I)$, between I and A_T. To find these optimal weights, we start with the index

$$I = b_1 z + b_2 \sum_{j \neq i} z_j \tag{22.33a}$$

with no restrictions placed on the ranges of b_1 and b_2. Selection on this index is equivalent to selection using the index

$$I = z + \frac{b_2}{b_1} \sum_{j \neq i} z_j \tag{22.33b}$$

Hence, the connection between Equations 22.28a and 22.33a is that $g = b_2/b_1$. The difference is that we no longer restrict consideration of g to between zero and one. All of the previous results for selection response on Equation 22.28a hold for any value of g, but we focused on the range of zero to one given the transition from individual to group selection. More generally, we could have negative weights, or a g value exceeding one. In the former case, negative g values correspond to a weighted within-group deviation (e.g., Equation 22.28c). In the latter case ($g > 1$), we place more weight on a random individual from the interacting group than on the focal individual. This might occur when associative effects are much larger than direct effects, and hence group members provide more information than the focal individual about the value of A_T for that focal individual.

In Chapter 21, we were able to obtain straightforward expressions for optimal weights in a family index (Equation 21.54). Index selection theory (Volume 3) gives the optimal index weights in the form of a matrix expression (Equation 22.35a), which is greatly simplified under simple family selection (i.e., with no associative effects). Unfortunately, such is not the case here, and so we (very briefly) introduce the machinery for obtaining an optimal index, deferring the full development of the theory to Volume 3. The idea is that there are two potentially different indices: the index I, used for selection (i.e., to choose individuals to form the next generation), and the index H, whose response we wish to maximize. Specifically, we select on some index $I = \mathbf{b}^T\mathbf{x}$ where x_i is the value of trait i used to make selection decisions and b_i is the weight placed on that trait in the index. In keeping with Equation 22.33a, the vector of phenotypes for individual i is

$$\mathbf{x} = \begin{pmatrix} z_i \\ \sum_{j\neq i} z_j \end{pmatrix} \tag{22.34a}$$

Using this index to make selection decisions, we wish to find the weights, $\mathbf{b}$, that maximize the selection response for some weighted combination of variables, $H = \mathbf{c}^T\mathbf{a}$. Here the elements of $\mathbf{c}$ are the weights and $\mathbf{a}$ is the vector of breeding values for the traits of interest. In our case, we wish to maximize response in the total breeding value, which means that

$$H = A_T = A_d + (n-1)A_s = \mathbf{c}^T\mathbf{a} \tag{22.34b}$$

where

$$\mathbf{a} = \begin{pmatrix} A_d \\ A_s \end{pmatrix} \quad \text{and} \quad \mathbf{c} = \begin{pmatrix} 1 \\ n-1 \end{pmatrix} \tag{22.34c}$$

The optimal weights $\mathbf{b}_s$ in I for maximizing response in H (i.e., to obtain the highest correlation between A_T and I) are given by the **Smith-Hazel index** (Smith 1936; Hazel 1943), which is derived in Example A6.8, where

$$\mathbf{b}_s = \mathbf{P}^{-1}\mathbf{G}^T\mathbf{c} \tag{22.35a}$$

$\mathbf{P}$ is the phenotypic covariance matrix for the elements in $\mathbf{x}$, which in our case becomes

$$\mathbf{P} = \begin{pmatrix} \sigma^2(z) & \sigma\left(z_i, \sum_{j\neq i} z_j\right) \\ \sigma\left(z_i, \sum_{j\neq i} z_j\right) & \sigma\left(\sum_{j\neq i} z_j, \sum_{j\neq i} z_j\right) \end{pmatrix} \tag{22.36a}$$

$$= \begin{pmatrix} \sigma^2(z) & (n-1)\sigma(z_i, z_j) \\ (n-1)\sigma(z_i, z_j) & (n-1)\left[\sigma_z^2 + (n-2)\sigma(z_i, z_j)\right] \end{pmatrix} \tag{22.36b}$$

where we have used Equations 22.30b and 22.30c. $\mathbf{G}$ is the matrix of covariances between the breeding values in the index H and the trait values in the index I, with $G_{ij} = \sigma(a_i, x_j)$. Because different traits can be involved in the two indices, $\mathbf{G}$ need not be symmetric. For our case,

$$\mathbf{G}^T = \begin{pmatrix} \sigma\left(A_{d_i}, z_i\right) & \sigma\left(A_{s_i}, z_i\right) \\ \sigma\left(A_{d_i}, \sum_{j\neq i} z_j\right) & \sigma\left(A_{s_i}, \sum_{j\neq i} z_j\right) \end{pmatrix} \tag{22.37a}$$

where

$$\sigma\left(A_{d_i}, z_i\right) = \sigma^2(A_d) + r(n-1)\sigma(A_d, A_s) \tag{22.37b}$$

$$\sigma\left(A_{s_i}, z_i\right) = \sigma(A_d, A_s) + r(n-1)\sigma^2(A_s) \tag{22.37c}$$

$$\sigma\left(A_{d_i}, \sum_{j\neq i} z_j\right) = (n-1)\sigma(A_d, A_s) + r(n-1)\left[\sigma^2(A_d) + (n-2)\sigma(A_d, A_s)\right] \tag{22.37d}$$

$$\sigma\left(A_{s_i}, \sum_{j\neq i} z_j\right) = (n-1)\sigma^2(A_s) + r(n-1)\left[\sigma(A_d, A_s) + (n-2)\sigma^2(A_s)\right] \tag{22.37e}$$

Equations 22.37b through 22.37e follow from the approach used throughout this chapter of a term-by-term evaluation of the covariance. The use of index selection machinery to find the optimal value of g was initially outlined by Ellen et al. (2007).

Example 22.9. As an application of the previous theory, consider a trait where $\sigma(A_d, A_s) = 0$, and there are no correlations between environmental values within the group ($\rho = 0$) and no relatives in the group ($r = 0$). Equation 22.5d gives $\sigma^2(z) = \sigma^2(A_d) + (n-1)\sigma^2(A_s) + \sigma^2(e)$, while (with $\sigma(A_d, A_s) = r = \rho = 0$), Equation 22.31c reduces to $\sigma(z_i, z_j) = (n-2)\sigma^2(A_s)$. Hence, from Equation 22.36b,

$$\mathbf{P} = \begin{pmatrix} \sigma^2(z) & (n-1)(n-2)\sigma^2(A_s) \\ (n-1)(n-2)\sigma^2(A_s) & (n-1)\left[\sigma^2(z) + (n-2)^2\sigma^2(A_s)\right] \end{pmatrix}$$

Likewise, Equations 22.37b through 22.37e imply

$$\mathbf{G} = \begin{pmatrix} \sigma^2(A_d) & 0 \\ 0 & (n-1)\sigma^2(A_s) \end{pmatrix}$$

What are the optimal weight (g) for a trait with $\sigma^2(A_d) = 100, \sigma^2(A_s) = 9, \sigma^2(e) = 100$, and $n = 10$? For these values, Equation 22.35a yields

$$\mathbf{b}_s = \mathbf{P}^{-1}\mathbf{G}^T\mathbf{c} = \begin{pmatrix} 281 & 648 \\ 648 & 9090 \end{pmatrix}^{-1} \begin{pmatrix} 100 & 0 \\ 0 & 81 \end{pmatrix} \begin{pmatrix} 1 \\ 9 \end{pmatrix} = \begin{pmatrix} 0.2046 \\ 0.0656 \end{pmatrix}$$

The resulting index weight on group information is 0.0656/0.2046 = 0.32, giving the index as

$$I_i = z_i + 0.32 \cdot \sum_{j\neq i} z_j$$

If we increase $\sigma^2(A_s)$ to 15, redoing the above calculations for the elements of $\mathbf{P}$ and $\mathbf{G}$ gives the optimal weights as

$$\mathbf{b}_s = \mathbf{P}^{-1}\mathbf{G}^T\mathbf{c} = \begin{pmatrix} 335 & 1080 \\ 1080 & 13,950 \end{pmatrix}^{-1} \begin{pmatrix} 100 & 0 \\ 0 & 136 \end{pmatrix} \begin{pmatrix} 1 \\ 9 \end{pmatrix} = \begin{pmatrix} 0.0236 \\ 0.0852 \end{pmatrix}$$

yielding a weight on the group sum of 0.0852/0.0236 = 3.6, and an optimal index of

$$I_i = z_i + 3.6 \cdot \sum_{j\neq i} z_j$$

Hence, on average, the phenotypes of individuals with which a focal individual interacts are given almost four time the weight as the focal individual's own phenotype. Finally, suppose $\sigma^2(A_s)$ = 20. In this case

$$\mathbf{b}_s = \mathbf{P}^{-1}\mathbf{G}^T\mathbf{c} = \begin{pmatrix} 380 & 1440 \\ 1440 & 18,000 \end{pmatrix}^{-1} \begin{pmatrix} 100 & 0 \\ 0 & 180 \end{pmatrix} \begin{pmatrix} 1 \\ 9 \end{pmatrix} = \begin{pmatrix} -0.1120 \\ 0.0989 \end{pmatrix}$$

making the optimal index

$$I_i = 0.0989 \cdot \left(\sum_{j \neq i} z_j \right) - 0.112 \cdot z_i$$

Expressing the summation as $n\,\overline{z} - z_i$, this is equivalent to selection on the index

$$I_i = 0.989\,\overline{z} - (0.0989 + 0.112)\, z_i = 0.989\,\overline{z} - 0.2109\, z_i$$

We can also rewrite this index as

$$I_i = (0.989 - 0.2109)\,\overline{z} - 0.2109\,(z_i - \overline{z}) = 0.7781\,\overline{z} - 0.2109\,(z_i - \overline{z})$$

which is equivalent to selecting using the index

$$I_i = \overline{z} - \frac{0.2109}{0.7781}\,(z_i - \overline{z}) = \overline{z} - 0.2710\,(z_i - \overline{z})$$

Hence, the optimal index in this case is the group mean minus a weighted within-group deviation.

BLUP ESTIMATION OF DIRECT AND ASSOCIATIVE EFFECTS

While Griffing developed many of the basic equations for selection response with associative effects, one reason for the initially low impact of his important work was that, at the time, there was no reliable way to estimate the key variance components, $\sigma^2(A_d)$, $\sigma^2(A_s)$, and $\sigma(A_d, A_s)$. These are required to compare h_d^2 with T^2, and hence to judge the potential amount of additional genetic variation that cannot be exploited under individual selection. Further, reasonable estimates of these variance components are required to obtain the optimal index weights. Finally, without some tangible values, Griffing's work was, for some, a bit too abstract: the observed phenotype was decomposed as the sum of two unmeasured components, whose estimation was entirely unclear. The solution to these problems was suggested by Muir and Schinckel (2002) and detailed in the seminal paper of Muir (2005), who put these estimation problems into a standard BLUP/REML mixed-model framework (Chapters 19 and 20; LW Chapters 26 and 27).

Mixed-Model Estimation of Direct and Associative Effects

The general approach follows if we consider a standard animal model with additional random effects (Equations 19.20 and 19.21). Equation 22.1b shows how the phenotype of individual i is the sum of its direct breeding value, the social breeding values of its group members, and the environmental effects,

$$z_i = \mu + (A_{d_i} + E_{d_i}) + \sum_{j \neq 1} (A_{s_j} + E_{s_j}) \tag{22.38a}$$

To start, we assume a very simple residual structure

$$z_i = \mu + A_{d_i} + \sum_{j \neq 1} A_{s_j} + e_i \tag{22.38b}$$

where the e_i are uncorrelated and homoscedastic, so that $\mathbf{e} \sim (\mathbf{0}, \sigma^2(e)\,\mathbf{I})$. Letting $\mathbf{a}_d$ denote the vector of **direct breeding values (DBVs)**, and $\mathbf{a}_s$ be the vector of **social breeding values (SBVs)**, the resulting mixed model becomes

$$\mathbf{z} = \mathbf{X}\boldsymbol{\beta} + \mathbf{Z}_d\,\mathbf{a}_d + \mathbf{Z}_s\,\mathbf{a}_s + \mathbf{e}, \quad \text{with} \quad \mathbf{e} \sim (\mathbf{0}, \sigma^2(e)\,\mathbf{I}) \tag{22.38c}$$

Here $\boldsymbol{\beta}$ is the vector of fixed effects (which will be just the mean for our simple example) and $\mathbf{X}$ is the design matrix associated with these fixed effects. Likewise, $\mathbf{Z}_d$ and $\mathbf{Z}_s$ are the corresponding incidence matrices for the direct and social effects, which follow logically upon considering the group members (Examples 22.10 and 22.11).

To complete the model, we need to specify the covariance structures of the three vectors of random effects. Our initial assumption on the residual errors implies that the covariance matrix for the residuals is $\sigma^2(e)\,\mathbf{I}$. The covariance structure for the two vectors of random effects is a function of the relationship matrix $\mathbf{A}$ (Chapter 19) of the individuals in the study, which has block-matrix form

$$\mathbf{Var}\begin{pmatrix} \mathbf{a}_d \\ \mathbf{a}_s \end{pmatrix} = \begin{pmatrix} \sigma^2(A_d)\mathbf{A} & \sigma(A_d, A_s)\mathbf{A} \\ \sigma(A_d, A_s)\mathbf{A} & \sigma^2(A_s)\mathbf{A} \end{pmatrix} \tag{22.39a}$$

This is often written more compactly using the **Kronecker** or **direct product** notation as $\mathbf{G} \otimes \mathbf{A}$, where

$$\mathbf{G} = \begin{pmatrix} \sigma^2(A_d) & \sigma(A_d, A_s) \\ \sigma(A_d, A_s) & \sigma^2(A_s) \end{pmatrix} \tag{22.39b}$$

Because the residuals are assumed to be uncorrelated with the other random effects, the full covariance structure for this model is

$$\mathbf{Var}\begin{pmatrix} \mathbf{a}_d \\ \mathbf{a}_s \\ \mathbf{e} \end{pmatrix} = \begin{pmatrix} \sigma^2(A_d)\mathbf{A} & \sigma(A_d, A_s)\mathbf{A} & \mathbf{0} \\ \sigma(A_d, A_s)\mathbf{A} & \sigma^2(A_s)\mathbf{A} & \mathbf{0} \\ \mathbf{0} & \mathbf{0} & \sigma^2(e)\,\mathbf{I} \end{pmatrix} \tag{22.39c}$$

Example 22.10. To introduce how a mixed-model with direct and social effects is constructed, consider the following toy example where eight individuals are measured. Individuals 1 through 4 are (noninbred) half-sibs, as are 5 through 8, but they are unrelated to the first family. The relationship matrix $\mathbf{A}$ becomes

$$\mathbf{A} = \begin{pmatrix} 1 & 0.25 & 0.25 & 0.25 & 0 & 0 & 0 & 0 \\ 0.25 & 1 & 0.25 & 0.25 & 0 & 0 & 0 & 0 \\ 0.25 & 0.25 & 1 & 0.25 & 0 & 0 & 0 & 0 \\ 0.25 & 0.25 & 0.25 & 1 & 0 & 0 & 0 & 0 \\ 0 & 0 & 0 & 0 & 1 & 0.25 & 0.25 & 0.25 \\ 0 & 0 & 0 & 0 & 0.25 & 1 & 0.25 & 0.25 \\ 0 & 0 & 0 & 0 & 0.25 & 0.25 & 1 & 0.25 \\ 0 & 0 & 0 & 0 & 0.25 & 0.25 & 0.25 & 1 \end{pmatrix}$$

These eight individuals are placed into two groups of size four. Group 1 contains individuals 1, 2, 5, and 6; while group 2 contains 3, 4, 7, and 8. For simplicity, the only assumed fixed effect is the mean, μ. The resulting matrices for the mixed-model equations become

$$\mathbf{z} = \begin{pmatrix} z_1 \\ z_2 \\ z_3 \\ z_4 \\ z_5 \\ z_6 \\ z_7 \\ z_8 \end{pmatrix} \quad \mathbf{X} = \begin{pmatrix} 1 \\ 1 \\ 1 \\ 1 \\ 1 \\ 1 \\ 1 \\ 1 \end{pmatrix} \quad \mathbf{a}_d = \begin{pmatrix} A_{d,1} \\ A_{d,2} \\ A_{d,3} \\ A_{d,4} \\ A_{d,5} \\ A_{d,6} \\ A_{d,7} \\ A_{d,8} \end{pmatrix} \quad \mathbf{Z}_d = \begin{pmatrix} 1 & 0 & 0 & 0 & 0 & 0 & 0 & 0 \\ 0 & 1 & 0 & 0 & 0 & 0 & 0 & 0 \\ 0 & 0 & 1 & 0 & 0 & 0 & 0 & 0 \\ 0 & 0 & 0 & 1 & 0 & 0 & 0 & 0 \\ 0 & 0 & 0 & 0 & 1 & 0 & 0 & 0 \\ 0 & 0 & 0 & 0 & 0 & 1 & 0 & 0 \\ 0 & 0 & 0 & 0 & 0 & 0 & 1 & 0 \\ 0 & 0 & 0 & 0 & 0 & 0 & 0 & 1 \end{pmatrix} = \mathbf{I}_8$$

where $\boldsymbol{\beta} = (\,\mu\,)$.

Turning now to the incidence matrix for social effects, $\mathbf{Z}_s$, note that its ith row has a 1 for each member in i's group, and a 0 otherwise. For example, individual 1 is influenced by the social breeding values of individuals 2, 5, and 6, which are assigned values of one in the first row of $\mathbf{Z}_s$, while all other elements in row one are zero. Likewise, individual 3 is influenced by the social breeding values of its group (individuals 4, 7 and 8), and so forth. Filling in the rest of the matrix yields

$$\mathbf{Z}_s = \begin{pmatrix} 0&1&0&0&1&1&0&0\\ 1&0&0&0&1&1&0&0\\ 0&0&0&1&0&0&1&1\\ 0&0&1&0&0&0&1&1\\ 1&1&0&0&0&1&0&0\\ 1&1&0&0&1&0&0&0\\ 0&0&1&1&0&0&0&1\\ 0&0&1&1&0&0&1&0 \end{pmatrix} \qquad \mathbf{a}_s = \begin{pmatrix} A_{s,1}\\ A_{s,2}\\ A_{s,3}\\ A_{s,4}\\ A_{s,5}\\ A_{s,6}\\ A_{s,7}\\ A_{s,8} \end{pmatrix}$$

The group effects are made more apparent (but $\mathbf{A}$ is made more confusing) by reordering the individuals as $\mathbf{z}^T = (z_1, z_2, z_5, z_6, z_3, z_4, z_7, z_8)$, so that group members are clustered together. This gives

$$\mathbf{Z}_s = \begin{pmatrix} 0&1&1&1&0&0&0&0\\ 1&0&1&1&0&0&0&0\\ 1&1&0&1&0&0&0&0\\ 1&1&1&0&0&0&0&0\\ 0&0&0&0&0&1&1&1\\ 0&0&0&0&1&0&1&1\\ 0&0&0&0&1&1&0&1\\ 0&0&0&0&1&1&1&0 \end{pmatrix}, \qquad \mathbf{A} = 0.25 \cdot \begin{pmatrix} 4&1&0&0&1&1&0&0\\ 1&4&0&0&1&1&0&0\\ 0&0&4&1&0&0&1&1\\ 0&0&1&4&0&0&1&1\\ 1&1&0&0&4&1&0&0\\ 1&1&0&0&1&4&0&0\\ 0&0&1&1&0&0&4&1\\ 0&0&1&1&0&0&1&4 \end{pmatrix}$$

Groups of different sizes are easily incorporated through the use of $\mathbf{Z}_s$.

Example 22.11. Assigning an element in $\mathbf{Z}_s$ a value of one for each individual within a group weights all interactions equally. This need not be the case, however. For example, Muir (2005) noted that with plants (or other sessile organisms), the distance between two individuals likely influences their effects. In particular, he suggested that if trees are a distance d apart, a reasonable model for their associative interactions would be A_s/d^2, where a tree has some intrinsic social breeding value (A_s) whose effect is diminished by distance. In a case of three trees, where d_{ij} is the distance between trees i and j, the resulting incidence matrix for social breeding values would be

$$\mathbf{Z}_s = \begin{pmatrix} 0 & 1/d_{12}^2 & 1/d_{13}^2 \\ 1/d_{21}^2 & 0 & 1/d_{23}^2 \\ 1/d_{31}^2 & 1/d_{32}^2 & 0 \end{pmatrix}$$

Cantet and Cappa (2008) suggested similar "intensity of competition" weights for individuals within groups of animals, such as the total contact time between two individuals. Other measures of interactions could be used and easily incorporated into $\mathbf{Z}_s$ (e.g., Wey et al. 2007).

Because we allow for the possibility that the direct and social breeding values are correlated, the standard mixed-model equations for two vectors of random effects (Equation 19.21; LW Equations 26.19b and 26.30) must be slightly modified. They become

$$\begin{pmatrix} \mathbf{X}^T\mathbf{X} & \mathbf{X}^T\mathbf{Z}_d & \mathbf{X}^T\mathbf{Z}_s \\ \mathbf{Z}_d\mathbf{X}^T & \mathbf{Z}_d^T\mathbf{Z}_d + \lambda_1\mathbf{A}^{-1} & \mathbf{Z}_d^T\mathbf{Z}_s + \lambda_2\mathbf{A}^{-1} \\ \mathbf{Z}_s\mathbf{X}^T & \mathbf{Z}_s^T\mathbf{Z}_d + \lambda_2\mathbf{A}^{-1} & \mathbf{Z}_s^T\mathbf{Z}_s + \lambda_3\mathbf{A}^{-1} \end{pmatrix} \begin{pmatrix} \boldsymbol{\beta} \\ \mathbf{a}_d \\ \mathbf{a}_s \end{pmatrix} = \begin{pmatrix} \mathbf{X}^T\mathbf{X} \\ \mathbf{X}^T\mathbf{Z}_d \\ \mathbf{X}^T\mathbf{Z}_s \end{pmatrix} \tag{22.40a}$$

where the weights (λ_i) are related to elements in the inverse of $\mathbf{G}$, namely,

$$\begin{pmatrix} \lambda_1 & \lambda_2 \\ \lambda_2 & \lambda_3 \end{pmatrix} = \sigma^2(e)\,\mathbf{G}^{-1} = \sigma^2(e) \begin{pmatrix} \sigma^2(A_d) & \sigma(A_d, A_s) \\ \sigma(A_d, A_s) & \sigma^2(A_s) \end{pmatrix}^{-1} \tag{22.40b}$$

as obtained by Muir (2005) and Van Vleck and Cassady (2005).

In order to solve these equations, estimates of the variance components—$\sigma^2(e), \sigma^2(A_d)$, $\sigma^2(A_s)$, and $\sigma(A_d, A_s)$—are required, and within the mixed-model framework, these are obtained by REML (LW Chapter 27). Van Vleck and Cassady (2005) used simulated data to show that, under the appropriate design, REML does indeed provide separable estimates of the genetic variance components. However, two early applications to real data sets, weight gain in pigs within pens by Arango et al. (2005) and weight gain in Hereford cattle in feedlots by Van Vleck et al. (2007), found that the likelihood surface for $\sigma^2(A_s)$ was very flat, making model fitting challenging. We will examine such issues of identifiability shortly. While mixed-model methodology is very robust (for example, it easily handles missing data and variable group numbers), it can easily fail if the model is not correctly specified or the experimental design is such that random effects are not separable, points that we will address shortly.

Example 22.12. Using mixed-model methods, Bergsma et al. (2008) examined four traits (growth rate, feed intake, back-fat thickness, and muscle depth) in a sample of over 14,000 pigs with a known pedigree (from roughly 400 sires and 600 dams). The pigs were placed in pens of 6–12 animals and several different mixed models were fitted.

First, a mixed model allowing for only direct effects plus a separate vector for common litter effects, $\mathbf{c}$ (which is fairly standard; Chapters 19 and 20), was fit. Here, the model was

$$\mathbf{z} = \mathbf{X}\boldsymbol{\beta} + \mathbf{Z}_a\mathbf{a} + \mathbf{Z}_c\mathbf{c} + \mathbf{e}$$

where

$$\mathbf{a} \sim N(\mathbf{0}, \sigma^2(A)\cdot\mathbf{A}), \quad \mathbf{c} \sim N(\mathbf{0}, \sigma^2(c)\cdot\mathbf{I}), \quad \mathbf{e} \sim N(\mathbf{0}, \sigma^2(e)\cdot\mathbf{I})$$

The resulting estimates of additive variation and heritability for these traits were found to be

	Growth	Back fat	Muscle	Intake
$\sigma^2(A)$	2,583	2.83	7.94	41,275
h^2	0.37	0.36	0.25	0.41

Next, a model was fit that also included a random pen (group) effect, g_p, common to all members within the same group (but differing across groups). The model now becomes

$$\mathbf{z} = \mathbf{X}\boldsymbol{\beta} + \mathbf{Z}_a\mathbf{a} + \mathbf{Z}_g\mathbf{g}_p + \mathbf{Z}_c\mathbf{c} + \mathbf{e}, \quad \text{where} \quad \mathbf{g}_p \sim N(0, \sigma^2(g_p)\cdot\mathbf{I})$$

Use of this model did not change the heritabilty estimates for back fat and muscle depth, but decreased the estimates for growth and feed intake

	Growth	Back fat	Muscle	Intake
$\sigma^2(A)$	1,780	2.79	7.69	17,678
h^2	0.25	0.36	0.24	0.18

Comparison with the previous table shows that failure to include a group effect (here, assumed to be entirely nonheritable, i.e., all environmental), resulted in some traits (growth and intake) having their heritabilities overestimated. Finally, a model was fit allowing for heritable social effects

$$\mathbf{z} = \mathbf{X}\boldsymbol{\beta} + \mathbf{Z}_d\mathbf{a}_d + \mathbf{Z}_s\mathbf{a}_s + \mathbf{Z}_g\mathbf{g}_p + \mathbf{Z}_c\mathbf{c} + \mathbf{e}$$

which returned estimates of

	Growth	Back fat	Muscle	Intake
$\sigma^2(A_d)$	1,522	2.75	6.68	16,950
h_d^2	0.21	0.35	0.21	0.17
$\sigma^2(A_s)$	51	0.01	0.03	596
$\sigma^2(A_T)$	5,208	3.19	10.35	68,687
T^2	0.71	0.41	0.32	0.70

While both body-composition traits (back fat and muscle thickness) were largely unaffected by social effects, growth-related traits (growth and intake) were. Failure to incorporate group effects (either shared environmental, g_p, or genetic, A_s) resulted in an overestimation of the (direct) heritability for growth traits. The exploitable genetic variance (the total breeding value) for the two growth traits was about three times higher than suggested by the individual (direct) breeding values, and T^2 was about triple the value of h_d^2. Hence, significant potential for improvement would remain untapped when using individual selection with groups of unrelated individuals for growth traits. Conversely, the incorporation of some group-selection would have little additional effect on the response of the two body-composition traits, as their h_d^2 and T^2 values are very close.

The results in the above example are fairly typical of the published results from the animal-breeding literature. Often the estimates of $\sigma(A_d, A_s)$ and $\sigma^2(A_s)$ are quite small relative to $\sigma^2(A_d)$, but because terms involving social effects are scaled by roughly n or n^2 (for the covariance and variance, respectively), their contributions can be considerable. For example, a series of eight (mostly growth) traits in cattle, pigs, and chicken, $(n-1)\sigma(A_d, A_s)$ was between 5 and 40% of $\sigma^2(A_d)$, with an average value of 24% (Van Vleck et al. 2007; Chen et al. 2008, 2009; Hsu et al. 2010).

As discussed in Chapter 19, one could use a Bayesian analysis of a mixed model instead of BLUP estimates of the random effects and REML estimates of the variance. Recall that a BLUP/REML analysis returns point estimates and associated confidence intervals for variables of interest, while a Bayesian analysis returns the whole posterior distribution of potential values given the data (Chapter 19; Appendices 2 and 3). Arora and Lahiri (1997) showed for mixed models that "empirical BLUP," namely using REML estimates of variance components to solve the mixed-model equations, generally gives the same average value as a Bayesian analysis, but that the latter returns a smaller mean-squared error and hence offers more precision. Cappa and Cantet (2006, 2008) developed a Gibbs sampler (Appendix 3) for the mixed model with associative effects.

One of the strengths of mixed models is their flexibility. The basic model shown by Equation 22.38c, which allows for direct and associative effects, can easily be extended. For example, Bouwmann et al. (2010) included a separate maternal genetic effect, distinct from social effects, while Alemu et al. (2014) modified associative effects to allow kin and nonkin interactions to differ.

Muir's Experiment: BLUP Selection for Quail Weight

In his classic paper, Muir (2005) not only laid out the approach for incorporating social effects into a mixed-model framework, but also directly tested this method by examining the response to selection based entirely on the estimated breeding values (EBVs) obtained from the model. Muir selected on six-week weight in Japanese quail (*Coturnix coturnix japonica*), which are aggressive and cannibalistic. Groups were formed with 16 birds per cage, with each group consisting of several half-sib families. Banding of the birds allowed the pedigree of individuals to be followed through the 23 hatches of the experiment. As Example 22.4 showed, due to a negative covariance between associative and direct effects, individual selection is expected to produce a reversed response when using a group of unrelated individuals.

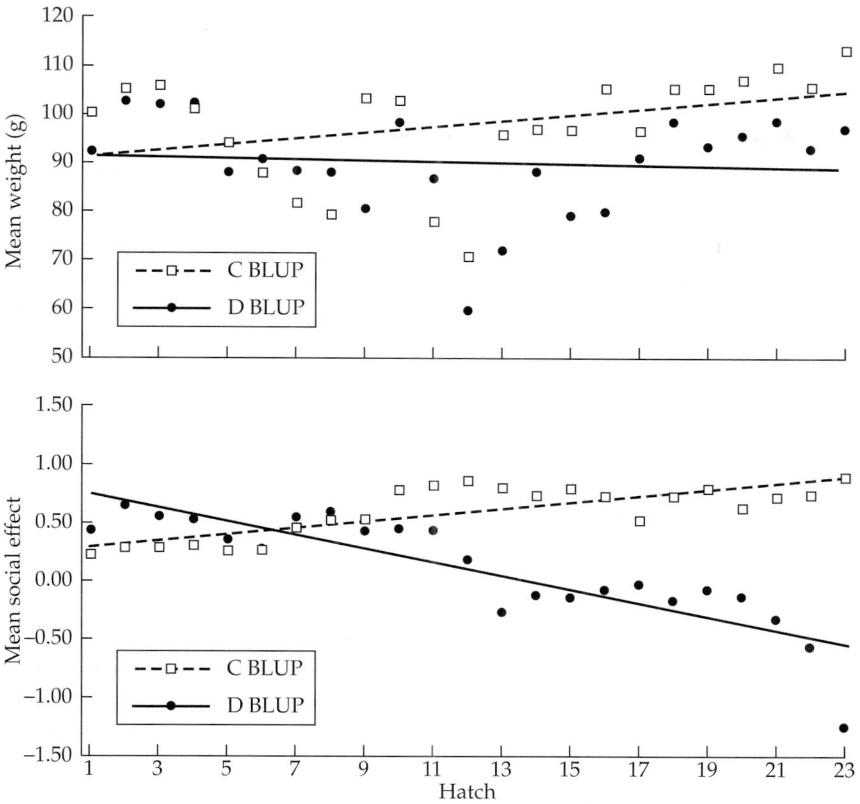

Figure 22.4 Selection response for two differentially selected lines of Japanese quail (Muir 2005). Both lines were selected for six-week weight using BLUP. Line D-BLUP selected individuals with the largest estimated direct breeding values, while line C-BLUP selected individuals with the largest estimated total breeding values. **A:** (Top) Mean response in six-week weight over 23 cycles of selection. The C-BLUP line showed a significant improvement, while the D-BLUP line showed a slight (but not significant) negative trend. (**B:** (Bottom) The trend in mean social values showed an increase in the C-BLUP lines, and a decrease in D-BLUP lines. Hence, competition increased in lines strictly selected for direct breeding value, while it decreased in lines selected on an index of direct and associative effects.

Rather than select using individual phenotype or group means, Muir used BLUP selection (Chapters 13 and 19), wherein a mixed model is used to estimate the breeding values, and those individuals with the largest EBVs are chosen. Starting with the same base population, two lines were selected using different BLUP criteria. For both lines, the mixed model allowing for both direct and social effects was fitted, using REML estimates of the variances to obtain BLUPs for the desired breeding values. In the D-BLUP line, individuals with the largest EBVs of A_d (direct effects) were selected. In the C-BLUP line, those individuals with the largest EBVs of A_T, namely $\text{EBV}(A_d) + (16-1)\text{EBV}(A_s)$, were selected. Figure 22.4A shows the results through 23 hatches (cycles of selection). Under BLUP-D selection, the mean six-week weight decreased (slightly, but not significantly), while it significantly increased under C-BLUP. Both D-BLUP and C-BLUP increased the mean of direct effects, although the response under D-BLUP was about twice as great.

As further shown in Figure 22.4B, the reason for the decrease in mean weight in the D-BLUP line was that the mean associative effect increased under C-BLUP (i.e., became more favorable toward others in the group), but as expected given the negative correlation between A_d and A_s) it decreased under D-BLUP (became less favorable). Two other improvements were observed in the C-BLUP line. Mortality increased significantly in the D-BLUP line, while it decreased slightly (but not significantly) in the C-BLUP line. Feed conversion was also better in the C-BLUP line, requiring 6.65 grams of feed per gram of gain, versus 7.26 in the D-BLUP line. Clearly, selection based on the mixed-model estimates of total breeding value resulted in significantly better results than lines selected by a more conventional (i.e., D-BLUP) approach.

Details: Environmental Group Effects and the Covariance Structure of e

Our simplifying assumption (Equation 22.28c), that the residuals, e_i, are homoscedastic and uncorrelated (meaning that $\boldsymbol{\sigma}(\mathbf{e}) = \sigma^2(e)\,\mathbf{I}$), is generally incorrect. As Equation 22.23a shows, individuals within the same group are correlated because they share the E_s values from the other group members, and not correctly accounting for these shared environmental values results in an overestimation of the variance of the social breeding values (Van Vleck and Cassady 2005; Bijma et al. 2007b; Bergsma et al. 2008; Chen et al. 2009). Equation 22.23a returns the correct covariance matrix for the residuals as

$$\boldsymbol{\sigma}(\mathbf{e}) = \sigma^2(e)\,\mathbf{R}, \quad \text{where} \quad R_{ij} = \begin{cases} 0 & i \text{ and } j \text{ in different groups} \\ \rho & i \text{ and } j \text{ in the same group} \\ 1 & i = k \end{cases} \tag{22.41}$$

where $\sigma^2(e)$ and ρ are given by Equation 22.23b.

Example 22.13. For the design used in Example 22.10 with group members clustered, so that $\mathbf{z} = (z_1, z_2, z_5, z_6, z_3, z_4, z_7, z_8)^T$, the corresponding covariance matrix for the residuals becomes

$$\boldsymbol{\sigma}(\mathbf{e}) = \sigma^2(e)\,\mathbf{R} = \sigma^2(e) \cdot \begin{pmatrix} 1 & \rho & \rho & \rho & 0 & 0 & 0 & 0 \\ \rho & 1 & \rho & \rho & 0 & 0 & 0 & 0 \\ \rho & \rho & 1 & \rho & 0 & 0 & 0 & 0 \\ \rho & \rho & \rho & 1 & 0 & 0 & 0 & 0 \\ 0 & 0 & 0 & 0 & 1 & \rho & \rho & \rho \\ 0 & 0 & 0 & 0 & \rho & 1 & \rho & \rho \\ 0 & 0 & 0 & 0 & \rho & \rho & 1 & \rho \\ 0 & 0 & 0 & 0 & \rho & \rho & \rho & 1 \end{pmatrix}$$

With the same number of individuals in all groups, the only two estimable parameters in the environmental covariance matrix are ρ and $\sigma^2(e)$. With groups of variable size (either by design or simply through the loss of data), the residual variances and covariances change with n (Equation 22.23b). In this case, the residual covariance matrix would be specified in terms of the three environmental variance/covariance terms, $\sigma^2(E_d), \sigma^2(E_s)$, and $\sigma(E_d, E_s)$.

Provided $\rho > 0$, an equivalent approach is simply to fit a random group effect (Bergsma et al. 2008; Ellen et al. 2008). Example 22.14 works through an example. This approach is computationally less demanding than jointly estimating $\sigma^2(e)$ and ρ in an $\mathbf{R}$ matrix. However, if the covariance, $\sigma(E_d, E_s)$, between environmental direct and social effects is sufficiently negative, ρ can be negative (Equation 22.23b) and the simple random group-effects model fails, as the group variance $\sigma^2(g_p)$ must be positive. As Equation 22.23b suggests, as group size increases, the contribution from $\sigma^2(E_s)$ eventually dominates ρ, making it positive. Thus, for a design with large group size, fitting a random group effect will often suffice.

Example 22.14. Suppose that instead of fully specifying the matrix $\mathbf{R}$ (Equation 22.41), we instead simply fit a random group effect. Here, all individuals in group i share the common random effect g_i, where we assume $g_i \sim N[0, \sigma^2(g_p)]$. The resulting mixed model becomes

$$\mathbf{z} = \mathbf{X}\boldsymbol{\beta} + \mathbf{Z}_d\mathbf{a}_d + \mathbf{Z}_s\mathbf{a}_s + \mathbf{Z}_g\mathbf{g}_p + \mathbf{e}$$

where, for k groups, $\mathbf{g}_p^T = (g_1, g_2, \cdots, g_k)$ is the vector of random group effects. The incidence matrix $\mathbf{Z}_g$ has k columns, the ith of which (corresponding to membership in group i) has a one for each individual in group i and a zero elsewhere. For example, for the design in Example 22.13,

$$\mathbf{g} = \begin{pmatrix} g_1 \\ g_2 \end{pmatrix} \quad \text{and} \quad \mathbf{Z}_g = \begin{pmatrix} 1 & 0 \\ 1 & 0 \\ 1 & 0 \\ 1 & 0 \\ 0 & 1 \\ 0 & 1 \\ 0 & 1 \\ 0 & 1 \end{pmatrix}$$

If we assume a simple covariance for the residuals, so that $\boldsymbol{\sigma}(\mathbf{e}) = \sigma^2(\epsilon)\,\mathbf{I}$, the resulting contribution to the covariance matrix of $\mathbf{z}$ from the group and residual terms becomes $\sigma^2(g_p)\,\mathbf{Z}_g\mathbf{Z}_g^T + \sigma^2(\epsilon)\,\mathbf{I}$, or

$$\sigma^2(g_p)\begin{pmatrix} 1 & 1 & 1 & 1 & 0 & 0 & 0 & 0 \\ 1 & 1 & 1 & 1 & 0 & 0 & 0 & 0 \\ 1 & 1 & 1 & 1 & 0 & 0 & 0 & 0 \\ 1 & 1 & 1 & 1 & 0 & 0 & 0 & 0 \\ 0 & 0 & 0 & 0 & 1 & 1 & 1 & 1 \\ 0 & 0 & 0 & 0 & 1 & 1 & 1 & 1 \\ 0 & 0 & 0 & 0 & 1 & 1 & 1 & 1 \\ 0 & 0 & 0 & 0 & 1 & 1 & 1 & 1 \end{pmatrix} + \sigma^2(\epsilon)\,\mathbf{I} = \sigma^2(e)\begin{pmatrix} 1 & \alpha^2 & \alpha^2 & \alpha^2 & 0 & 0 & 0 & 0 \\ \alpha^2 & 1 & \alpha^2 & \alpha^2 & 0 & 0 & 0 & 0 \\ \alpha^2 & \alpha^2 & 1 & \alpha^2 & 0 & 0 & 0 & 0 \\ \alpha^2 & \alpha^2 & \alpha^2 & 1 & 0 & 0 & 0 & 0 \\ 0 & 0 & 0 & 0 & 1 & \alpha^2 & \alpha^2 & \alpha^2 \\ 0 & 0 & 0 & 0 & \alpha^2 & 1 & \alpha^2 & \alpha^2 \\ 0 & 0 & 0 & 0 & \alpha^2 & \alpha^2 & 1 & \alpha^2 \\ 0 & 0 & 0 & 0 & \alpha^2 & \alpha^2 & \alpha^2 & 1 \end{pmatrix}$$

where

$$\sigma^2(e) = \sigma^2(\epsilon) + \sigma^2(g_p) \quad \text{and} \quad \alpha^2 = \frac{\sigma^2(g_p)}{\sigma^2(\epsilon) + \sigma^2(g_p)}$$

We use α^2 to remind the reader that (under this model), this is the ratio of two variances, and hence is always nonnegative. Comparison with Example 22.13 shows that adding a random effect for group corresponds to the more fully specified covariance residual model (Equation 22.41), with $\rho = \alpha^2$. When $\rho > 0$, both models are identical. However, if the within-group environmental correlations are negative ($\rho < 0$), then the simple group random-effects model fails.

Further, note that we can write the covariance matrix of group plus residual effects under this model as

$$\sigma^2(e) \cdot \left(\mathbf{I} + \alpha^2\left[\mathbf{Z}_g\mathbf{Z}_g^T - \mathbf{I}\right]\right)$$

showing that the matrix $\mathbf{R}$ in Equation 22.41 (provided $\rho > 0$) is given by

$$\mathbf{R} = \mathbf{I} + \alpha^2\left[\mathbf{Z}_g\mathbf{Z}_g^T - \mathbf{I}\right]$$

Finally, a standard approach when families are in the analysis is to include a common-family effect, c, that is due to shared maternal effects and dominance (if full sibs are present). This is simply done by adding an additional vector of random effects, $\mathbf{c}$, for the family effects. Using this approach, the model starting this example now becomes

$$\mathbf{z} = \mathbf{X}\boldsymbol{\beta} + \mathbf{Z}_d\mathbf{a}_d + \mathbf{Z}_s\mathbf{a}_s + \mathbf{Z}_g\mathbf{g}_p + \mathbf{Z}_c\mathbf{c} + \mathbf{e}$$

For example, suppose individuals 1, 2, 5, and 6 (in the ordering used in this example) are from one family, while the others are from a second then

$$\mathbf{c} = \begin{pmatrix} c_1 \\ c_2 \end{pmatrix} \quad \text{and} \quad \mathbf{Z}_c = \begin{pmatrix} 1 & 0 \\ 1 & 0 \\ 0 & 1 \\ 0 & 1 \\ 1 & 0 \\ 1 & 0 \\ 0 & 1 \\ 0 & 1 \end{pmatrix}$$

Note that if $\mathbf{Z}_g = \mathbf{Z}_c$, the group and family effects are *fully confounded* and cannot be separated (this point will be discussed in more detail shortly). This addition of a common family effect is easily incorporated into any of the above models.

Details: Ignoring Additive Social Values Introduces Bias

Before models directly accounting for social effects were developed, it was not unusual to add a fixed or random group effect to the standard animal model to account for common environments due to individuals being raised in the same pen, cage, or group. For example, if group effects are random, the corresponding animal model becomes

$$\mathbf{z} = \mathbf{X}\boldsymbol{\beta} + \mathbf{Z}\,\mathbf{a} + \mathbf{Z}_g\,\mathbf{g}_p + \mathbf{e} \tag{22.42}$$

where we (initially) assume $\sigma(\mathbf{g}_p) = \sigma^2(g_p)\,\mathbf{I}$. In this model, $\mathbf{a}$ would be the estimated vector of (direct) breeding values. As detailed above, $\mathbf{g}_p$ can often account for any shared environmental social values (i.e., E_s). However, if *heritable* associative effects are present, simply adding a group effect is *insufficient*, as it results in overestimation of $\sigma^2(g_p)$ and often an overestimation of the (direct) additive variance (Example 22.12). Hence, an analysis that simply includes a group effect (but no $\mathbf{a}_s$ vector) results in biased estimates of the direct breeding values when heritable associative effects are present.

Van Vleck and Cassady (2005) showed how the presence of additive associative effects inflates the estimate of group variance. Consider two members in the same group (with a common group effect, g_p),

$$z_1 = A_{d_1} + A_{s_2} + \sum_{k=3}^{n} A_{s_k} + g_p + e_1$$

$$z_2 = A_{d_2} + A_{s_1} + \sum_{k=3}^{n} A_{s_k} + g_p + e_2$$

Using the standard ANOVA identity that the covariance within a group equals the variance among groups (LW Chapter 18), for unrelated individuals, $\sigma^2(g_p)$, is estimated from the within-group covariance, which reduces to

$$\begin{aligned}\sigma(z_1, z_2) &= \sigma(A_{d_1}, A_{s_1}) + \sigma(A_{d_2}, A_{s_2}) + (n-2)\sigma^2(A_s) + \sigma^2(g_p) + \sigma(e_1, e_2) \\ &= 2\sigma(A_d, A_s) + (n-2)\sigma^2(A_s) + \sigma^2(g_p) + \sigma(e_1, e_2)\end{aligned} \tag{22.43}$$

If the residuals are uncorrelated, the bias in the within-group covariance-based estimate of $\sigma^2(g_p)$ is $2\sigma(A_d, A_s) + (n-2)\sigma^2(A_s)$, which can be considerable. Hence, when additive (i.e., heritable) associative effects are present, the simple model given by Equation 22.42 is inappropriate. This model, however, can be useful in a preliminary analysis. Van Vleck and Cassady suggested that obtaining a large estimated group variance when using Equation

22.42 indicates that a more detailed model including additive associative effects should be fit to the data. Hence, one approach is to do a quick fit to Equation 22.42. If the group variance is sufficiently small, it is unlikely that additive associative effects are present. However, this approach is not always foolproof. Inspection of Equation 22.43 shows that a sufficiently negative covariance between direct and social breeding values may result in a small estimated group variance.

Details: Identifiability of Variance Components

Due to potential confounding of effects, any particular design might not allow for all variables of interest to be uniquely estimated. For the vector β of fixed effects, the uniqueness of an estimated variable is indicated the concept of **estimability** (LW Chapter 26). For $\mathbf{z} \sim (\mathbf{X}\beta, \mathbf{V})$, the vector of fixed effects is estimable (β has a unique value) if $(\mathbf{X}^T\mathbf{V}^{-1}\mathbf{X})^{-1}$ exists. Otherwise, some of the fixed effects are confounded and cannot be separated by the design ($\mathbf{X}$) being used. A similar concept, **identifiability**, exists for random effects and is based on whether variance components (often called the **dispersal parameters**) can be uniquely estimated. If variance components are not identifiable in the design, then BLUPs for their associated vectors of random effects do not exist, and the model will fail.

The lack of identifiability has been a problem in some attempts to estimate associative effects, with lack of convergence of REML estimates, convergence to multiple peaks in the likelihood surface (depending on starting conditions), and very flat likelihood surfaces all being seen (Arango et al. 2005; Van Vleck et al. 2007; Chen et al. 2008). Cantet and Cappa (2008) formally showed that using a *fixed* group effect results in a lack of identifiability when the design matrix, $\mathbf{Z}_g$, contains equal weights for all group members. Thus, treating group effects as fixed is not recommended, while treating them as random can often account for environmental correlations (as discussed above). Another common reason for lack of identifiability is the *composition of the group*. If all group members are from a single half-sib or full-sib family, the covariance of group members equals the covariance among family members within a group, confounding variance components and leading to a lack of identifiability (Cheng et al. 2009). Bijma et al. (2007b) noted that this problem plagued one of the early attempts to estimate social variance components (Wolf 2003). The important caveat is that lack of identifiability can easily arise in attempts to estimate social effects even when using seemingly innocent designs (such as a fixed group effect or having each group be a single family). One key is that family members must be spread over at least two groups, and each group should contain at least two different families. This avoids confounding within groups and allows A_s to be estimated by borrowing information (via relatives) across groups.

Conditions for identifiability of REML estimates of (co)variance components were given by Rothenberg (1971), Jiang (1996), and Cantet and Cappa (2008). Before presenting these conditions, we first review a few details about REML. Recall (LW Chapter 27) that REML estimates are those that maximize that part of the likelihood function independent of the fixed effects (this is often stated as being the **translation invariant** part of the likelihood). Let $\mathbf{V}$ be the covariance matrix of $\mathbf{z}$, which is a function of its variance components. As detailed in LW Chapter 27, Harville (1977) showed that (if it exists) the transformation provided by the matrix

$$\mathbf{P} = \mathbf{V}^{-1} - \mathbf{V}^{-1}\mathbf{X}(\mathbf{X}^T\mathbf{V}^{-1}\mathbf{X})^{-1}\mathbf{X}^T\mathbf{V}^{-1} \tag{22.44a}$$

plays a critical role in REML estimates. (To be consistent with the literature, we use $\mathbf{P}$ for this transformation matrix, despite our previous use of $\mathbf{P}$ to indicate the phenotype variance-covariance matrix. The distinction between these two usages should be obvious given the context of the issue being discussed.)

That the matrix given by Equation 22.44a can remove fixed effects can be seen by recalling that (under GLS), $\widehat{\beta} = (\mathbf{X}^T\mathbf{V}^{-1}\mathbf{X})^{-1}\mathbf{X}^T\mathbf{V}^{-1}\mathbf{z}$, and hence Equation 22.44a implies that

$$\mathbf{P}\mathbf{z} = \mathbf{V}^{-1}\left(\mathbf{z} - \mathbf{X}\widehat{\beta}\right) \tag{22.44b}$$

where the vector $\mathbf{Pz}$ is a function of the data $\mathbf{z}$ adjusted by the estimated fixed effects, $\mathbf{X}\widehat{\boldsymbol{\beta}}$ (i.e., centered to have a mean of zero). Now consider covariance structures of the form

$$\mathbf{V} = \sum_{i=1}^{n} \mathbf{V}_i \theta_i \tag{22.45a}$$

where $\mathbf{V}_i$ is a matrix of known constants and the θ_i are unknown variances and covariances to be estimated. This is the structure for all of the $\mathbf{V}$ matrices presented in this chapter. The equations to maximize the likelihood over the restricted space (the REML estimates) are given by LW Equations 27.18 and 27.19, and are solved iteratively. These equations involve the **trace** (the sum of the diagonal elements) of matrix products involving $\mathbf{P}$ and the $\mathbf{V}_i$. Recall (LW Appendix 4) that for a vector $\boldsymbol{\Theta}$ of n unknowns, the Fisher information matrix, $\mathbf{F}$ (the matrix of second partial derivatives of the likelihood with respect to the parameters), can be used to provide large-sample standard errors. The resulting $n \times n$ information matrix for REML estimates of the unknown θ_i in Equation 22.45a has as its ijth element

$$F_{ij} = \text{trace}\,(\mathbf{PV}_i\mathbf{PV}_j) \tag{22.45b}$$

Much in the same fashion that the existence of $(\mathbf{X}^T\mathbf{V}^{-1}\mathbf{X})^{-1}$ informs us that all fixed effects are estimable in a given design, all variance components, θ_i, are identifiable if all of the eigenvalues of the matrix $\mathbf{F}$ are positive, that is if $\mathbf{F}$ is positive-definite (Rothenberg 1971; Jiang 1996). For the simplest associative-effects mixed model (Equation 22.38c), Equation 22.45a becomes

$$\mathbf{V} = \mathbf{V}_1\,\sigma^2(A_d) + \mathbf{V}_2\,\sigma(A_d, A_s) + \mathbf{V}_3\,\sigma^2(A_s) + \mathbf{V}_4\,\sigma^2(e) \tag{22.46a}$$

where

$$\mathbf{V}_1 = \mathbf{Z}_d\mathbf{A}\mathbf{Z}_d^T, \quad \mathbf{V}_2 = \left(\mathbf{Z}_d\mathbf{A}\mathbf{Z}_s^T + \mathbf{Z}_s\mathbf{A}\mathbf{Z}_d^T\right), \quad \mathbf{V}_3 = \mathbf{Z}_s\mathbf{A}\mathbf{Z}_s^T, \quad \mathbf{V}_4 = \mathbf{I} \tag{22.46b}$$

Substituting Equations 22.44a and 22.46b into Equation 22.45b fills out the $\mathbf{F}$ matrix (which is only 4×4 in this case, given the four unknown variance components). For any particular design (the values of $\mathbf{A}, \mathbf{Z}_d$, and $\mathbf{Z}_s$), the eigenvalues of this matrix can be computed to determine if the variance components are all identifiable. Cheng et al. (2009) used this approach to show that two of the eigenvalues of their information matrix were zero for a design where groups consist entirely of single full-sib families, showing the lack of identifiability in such settings.

Appropriate Designs for Estimating Direct and Associative Effects

While most of the statistical power for detecting associative effects arises from the number of groups, not numbers of individuals (Bijma 2010c), the relationship structure within groups is also critical. In contrast to selection response, where there is a benefit from having all group members from the same family (and hence an increased r value), in a design to estimate direct and associative values and variance components, groups should be composed of at least two different families. If there is no within-group variation in relationships, direct and associative effects cannot be separated. Groups can also consists of unrelated individuals, but Bijma (2010c) showed that, in general, using groups with two (or more) different families offers more power than using unrelated individuals (also see Ødegård and Olesen 2011).

Using the appropriate mixed model is also critical. Initially, one might think that associative effects could be accommodated by simply adding a random effect for group to an otherwise standard animal model. As previously shown (Equation 22.43), however, this approach typically overestimates the direct effects, as well as inflating the group variance (which is a measure of the environmental social effects), when heritable associative effects are present, namely, $\sigma^2(A_s) > 0$. Conversely, ignoring any environmental associative effects also introduces bias. For example, a model fitting just $\mathbf{a}_d$ and $\mathbf{a}_s$ using the simple error structure $\mathbf{e} \sim (0, \sigma^2(e) \cdot \mathbf{I})$ also introduces bias by ignoring the correlation among environmental associative effects within a group. As mentioned above, the correct residual covariance structure can be accounted for by incorporating a random group effect into the model (which assumes a positive correlation between social environmental effects within a group), or by using a model with $e \sim (0, \sigma^2(e) \cdot \mathbf{R})$ where the elements of $\mathbf{R}$ are given by 22.41, which allows for the within-group environmental correlations, ρ, to be negative.

Using Kin Groups: A Quick-and-dirty Way Around Associative Effects?

As the proceeding sections demonstrate, performing a Muir (2005)-style BLUP selection on total breeding value A_T (Figure 22.4) requires an appropriate design and care to ensure that groups contain a mixture of relatives and nonrelatives in order to provide separate estimates of A_d and A_s. Given this background, it may be counterintuitive that Muir et al. (2013) suggested that a quick-and-dirty way around dealing with associative effects is to ensure that groups are made up *entirely* of relatives. Their logic follows from Equation 22.12e, which shows that when the average relatedness within a group is r, selection based *entirely* on individual values still captures a fraction, r, of A_T. They suggested that in settings where individuals naturally interact in groups (such as caged birds), simply assigning relatives to groups provides a path for direct selection on A_T. As our above analysis suggests, such a setting may not allow for separate estimates of A_s and A_d (and hence a direct estimate of A_T), but it can provide a much simpler way to ensure at least some selection on A_T.

Their idea is that if relatives are assigned to the same groups, then standard BLUP selection based on individual phenotype and relatedness (Chapters 13 and 19) will capture part of A_T. While the accuracy will admittedly be lower than for a direct estimate from an appropriate design, it will still be far greater than when interacting groups are entirely comprised of nonrelatives (Equation 22.12e). To test this idea, Muir et al. (2013) essentially replicated Muir's classic (2005) experiment on weight gain in Japanese quail (Figure 22.4), but now using standard BLUP selection that completely ignores associative effects. They compared the response under two otherwise identical settings: one in which groups non-randomly consisted of half-sibs, and the second where groups were formed at random (and hence members were unrelated). The response using kin-groups was an order of magnitude greater than for the random groups. The beauty of this approach is that one simple action, ensuring interacting groups contain mostly relatives, allows individual selection to partially capture some of the variation in A_T without using all of the above extra machinery. However, one downside is that it may lead to increased inbreeding.

ASSOCIATIVE-EFFECTS, INCLUSIVE FITNESS, AND FISHER'S THEOREM

We conclude by examining some of the important implications for evolution when heritable associative effects ($\sigma^2(A_s) > 0$) are present. First and foremost, their presence has significant implications on the evolution of mean population fitness (Bijma 2010a), which forms the subject of this section.

Change in Mean Fitness When Associative Effects are Present

The most important trait in evolution is fitness, W (Chapters 6 and 29). Clearly, the fitness of any particular genotype is partly a function of the environment in which it finds itself. While we normally treat this environment as static, when conspecifics influence fitness (as is generally expected to be the case), part of this environment may also be evolving in response to selection (namely, conspecifics are constantly improving). In these settings, the use of models with associative effects is appropriate. Here, the individual fitness of a focal individual results from a direct fitness effect from its own genotype plus the associative effects on its fitness from the other genotypes with which it interacts. Competition, a detrimental fitness effect from other individuals, is one such associative effect, where the contribution from conspecifics is to lower fitness (e.g., Wilson et al. 2009, 2011). Conversely, with cooperation or mutalism, associative effects increase the fitness of the focal individual.

Examining the expected change in mean fitness is straightforward. Using previous results, we simply take the trait being followed as individual fitness ($z = W$). From Equation 22.1c, the fitness of individual i becomes

$$W_i = \mu + A_{d_i} + \sum_{j \neq i} A_{s_j} + e_i \tag{22.47a}$$

A_d is the direct breeding value of fitness, while A_s is the social breeding value (how a focal individual influences the fitness of others in its group). As above, A_{s_i} does not contribute to W_i, while A_{s_j} for $j \neq i$ does. Likewise, the total breeding value for the fitness of an individual is simply

$$A_{T_i} = A_{d_i} + (n-1)A_{s_i} \tag{22.47b}$$

with a variance of

$$\sigma^2(A_T) = \sigma^2(A_d) + 2(n-1)\sigma(A_d, A_s) + (n-1)^2\sigma^2(A_s) \tag{22.47c}$$

The first term, $\sigma^2(A_d)$, is the classical additive genetic variance in fitness in the absence of associative effects. When interactions are present, there is the potential for substantially more heritable variation in fitness. Indeed, the total genetic variance in fitness has the potential to exceed the actual variance in individual fitness, $\sigma^2(A_T) > \sigma^2_W$, as much of the variation is hidden in interactions with others, which do not appear in one's individual fitness.

When the trait is fitness itself, the response equation for individual selection (Equation 22.14) simplifies somewhat. Recall the Robertson-Price identity (Equation 6.10), $S = \sigma(z, w)$, where $w = W/\overline{W}$ is relative fitness. When $z = W$, the selection differential becomes

$$S_W = \sigma(W, w) = \frac{\sigma(W, W)}{\overline{W}} = \frac{\sigma^2(W)}{\overline{W}} \tag{22.48a}$$

Equation 22.14 expresses the response in terms of $\bar{\imath}/\sigma$. When the trait is fitness itself, Equation 22.48a shows that this simplifies to

$$\frac{\bar{\imath}_W}{\sigma(W)} = \frac{S_W/\sigma(W)}{\sigma(W)} = \frac{\sigma^2(W)/\overline{W}}{\sigma^2(W)} = \frac{1}{\overline{W}} \tag{22.48b}$$

Substituting Equation 22.48b into Equation 22.14 gives the response (the change in mean population fitness) as

$$R_W = \frac{\sigma(W, A_T)}{\sigma(W)}\,\bar{\imath}_W = \frac{1}{\overline{W}}\sigma(W, A_T) = \sigma(w, A_T) \tag{22.48c}$$

This is simply Price's equation (Equation 6.6), where we have restricted our discussion to cases where the transmission is such that mean breeding values of offspring equals the mean breeding values of their parents (and hence the correction term in Equation 6.6 for changes induced solely by transmission disappears). Such is expected to be the case for the infinitesimal model under random mating.

Applying Equation 22.12c gives the response in terms of the variance components as

$$R_W = \frac{1}{\overline{W}}\left[\sigma^2(A_d) + (n-1)(1+r)\sigma(A_d, A_s) + r(n-1)^2\sigma^2(A_s)\right] \tag{22.48d}$$

As we saw for other traits, when $r = 0$, the possibility of a reversed response occurs if the breeding values for direct and social effects on fitness are sufficiently negatively correlated. Hence, under rather realistic conditions, individual selection can result in a *decrease* (and a potentially rather significant one) in mean fitness.

Ironically, even though a negative response can occur in the presence of associative effects, there is actually more total additive-genetic variance in fitness available when such effects are present, as $\sigma^2(A_T) > \sigma^2(A_d)$. However, only a fraction of this variance may be accessible to *individual* selection, and this fraction (being a covariance rather than a variance) can be negative (Equation 22.4b).

The key for exploiting the available variance is either selection among groups or the presence of relatives in one's group of interacting individuals. To see this last point, note from Equation 22.12e that we can express Equation 22.48d as

$$R_W = \frac{1}{\overline{W}} \left(r\sigma^2(A_T) + (1-r)\left[\sigma^2(A_d) + (n-1)\sigma(A_d, A_s)\right] \right) \tag{22.48e}$$

The term in square brackets represents the response in a group of nonrelatives. When interactions occur among kin ($r > 0$), then for sufficiently close relatives, the response becomes positive (mean fitness increases), even if the response is negative when $r = 0$. At the extreme, when $r = 1$ (all interactions are among clonemates), the response in mean fitness is simply $\sigma^2(A_T)/\overline{W}$ and all of the heritable variance in fitness is utilized. Conversely, when interactions occur among unrelated individuals, only a fraction of this additive-genetic potential for fitness, $\sigma^2(A_T)$, is exploited. This observation led Bijma (2010a) to suggest that when heritable fitness interactions are present, the key to evolutionary success is interacting with relatives. The reason for this is clear from our previous discussion. With interactions among unrelated individuals, one's phenotype (here fitness) provides very little information about one's own social breeding value for fitness. With interacting kin, the breeding values of the kin's social effects influence one's fitness, and these effects are positively correlated (via kinship) with one's own breeding value for social effects, as $\sigma(A_{s_i}, A_{s_j}) = r_{ij}\sigma^2(A_s)$. Hence, even under individual selection, there is correlated selection on an individual's social value when some of its group members are relatives ($r_{ij} > 0$).

Finally, we can decompose the total response in fitness into response from changes in the mean of the direct effects and response from changes in the mean of the social effects. Equation 22.15a gives

$$R_W = R_{W,d} + (n-1)R_{W,s} \tag{22.49a}$$

Recalling Equation 22.48b, Equations 22.15a through 22.15c give these response components as

$$R_{W,d} = \frac{\sigma^2(A_d) + r(n-1)\sigma(A_d, A_s)}{\overline{W}} \tag{22.49b}$$

and

$$R_{W,s} = \frac{\sigma(A_d, A_s) + r(n-1)\sigma^2(A_s)}{\overline{W}} \tag{22.49c}$$

Example 22.15. As Equation 22.48d shows, a reversed response in mean population fitness can occur, in the extreme driving a population to extinction. A potential example of this was provided by Dawson (1969), who studied competition between two species of flour beetles (*Tribolium castaneum* and *T. confusum*). He found that *castaneum* won (driving the other species to extinction) in nine of ten replicates. In the remaining replicate, *castaneum* again appeared to be winning (with a frequency of over 90% by generation 4 from a starting frequency of 50%) when an eye color mutation allelic to *chestnut* spontaneously appeared. From that generation onward, the frequency of this allele increased while the *castaneum* population itself was driven to extinction. One explanation for such a **Trojan gene**—a gene driving the population to extinction (Muir and Howard 1999)—is a strong positive direct fitness effect (increasing the frequency of the allele), but with a strong negative associative value on conspecifics (decreasing mean population fitness).

Example 22.16. Haldane (1932a) coined the term **altruistic trait** to denote a behavior (or trait) that harms an individual, but benefits others. The classic example is an alarm call: others in a group are warned (increasing their fitness), but at some expense to the individual making the call (a direct effect decreasing fitness). Note that the increase in an altruistic trait is an example of a reversed response, as the trait lowers the fitness of the individual that bears it. What are the conditions for such traits to spread? In terms of our fitness model with associative effects (Equation 22.47a), we can rephrase this as the conditions for the mean value of A_s to increase, which are given by Equation 22.49c. From the definition of altruism, $\sigma(A_d, A_s) < 0$, as performing an altruistic act decreases direct fitness while increasing the fitness of those in the group. Equation 22.49c shows that a necessary (but not sufficient) condition for altruism to evolve ($R_{W,s} > 0$) under individual selection is that $r > 0$, namely, that individuals interact in groups of relatives.

As pointed out by Bijma and Wade (2008), we can view $\sigma(A_d, A_s)$ as the **cost** ($-c < 0$) for performing an altruistic act for others in a group. Conversely, the altruistic contribution from others in one's group is $(n-1)\sigma^2(A_s) \geq 0$, which we denote as the **benefit**, b. With these definitions, from Equation 22.49c, the condition for altruism to evolve under individual selection is

$$-c + rb > 0, \quad \text{or} \quad r > b/c \tag{22.50a}$$

This is the classic **Hamilton's rule** (Hamilton 1963, 1964a, 1964b), which is discussed shortly. For an altruistic trait to evolve under individual selection, individuals must interact in groups of relatives. If groups consist of unrelated individuals ($r = 0$), individual selection is not sufficient for altruism, and some component of group selection is required. Note from Equation 22.49c that a more general version of Hamilton's rule is

$$\sigma(A_d, A_s) + r(n-1)\sigma^2(A_s) > 0 \tag{22.50b}$$

which shows that the benefit scales with group size, provided $\sigma^2(A_s)$ is independent of n. In contrast, under Bijma's dilution model for social effects (Equation 22.10a), the variance in social effects is a function of n, and (applying Equation 22.10c) Hamilton's rule becomes

$$\sigma(A_d, A_s) + r(n-1)^{1-2d}\sigma^2(A_{s,2}) > 0 \tag{22.50c}$$

with benefit increasing with group size only when the dilution fraction $d < 1/2$.

The same logic extends to the evolution of altruism under group selection. From Equation 22.27a, the response in the mean social value to group selection depends on the covariance of A_s and the group total (here, the sum of the fitnesses of all group members). From Equation 22.27c,

$$\begin{aligned} \sigma\left(A_{s_i}, \sum_{j=1}^{n} W_j\right) &= \left[\sigma(A_d, A_s) + (n-1)\sigma^2(A_s)\right]\left[1 + (n-1)r\right] \\ &= (-c + b)\left[1 + (n-1)r\right] \end{aligned} \tag{22.50d}$$

As long as $b > c$ (the benefit exceeds the cost), altruism can evolve under group selection, even in groups of *unrelated* individuals. When individuals within the group are related ($r > 0$), this covariance is significantly larger, and hence the response to selection for altruism is greater.

Lynch (1987) showed that this simple expression for Hamilton's rule breaks down under more complex settings, such as multiple relatives with different levels of interactions between them and allowing for ontological changes in behavior, for example, an individual may act as an offspring early in life and as a parent later in life

---

Inclusive Fitness

As Equation 22.47a illustrates, when heritable interactions are present, the fitness of an individual depends on both its own genes as well as the genes in others. Hamilton (1964a, 1964b) suggested that evolutionary focus should shift from individual fitness to what he called **inclusive fitness**: that component of fitness influenced only by the alleles carried by the focal individual. Hamilton argued that evolution strives to increase inclusive, as opposed to individual, fitness (also see Michod and Abugov 1980; Grafen 2006; Akçay and Van Cleve 2016).

Formally, the inclusive fitness of an individual is context specific, and is defined as individual fitness plus the effect of that individual on the fitness of others (weighted by the relatedness to these others) minus any contribution to that fitness from the group environment. While sounding rather abstract, when placed in an associative-effect framework, this definition is quite clear. From Equation 22.47a, for individual i, A_{d_i} is the heritable component of individual fitness (W_i) remaining when the social contributions from others have been removed. The focal individual's social breeding value (A_{s_i}) does not influence its own fitness, but the social effects of other group members do, with the (heritable) contribution to individual i's fitness from individual j being A_{s_j}. The correlation between the breeding value A_{s_i} carried by i and the contribution to i's fitness from j is the relatedness r_{ij}, so that $r_{ij} A_{s_i}$ is the predicted value of A_{s_j} given A_{s_i}. Putting these together gives the heritable component (i.e., breeding value) of i's inclusive fitness as

$$A_{incf,i} = A_{d_i} + A_{s_i} \sum_{j \neq i}^{n} r_{ij} = A_{d_i} + r(n-1)A_{s_i} \tag{22.51a}$$

where the last equality makes our standard assumption that all group members are equally related (which is easily relaxed). Note that the presence of r makes a genotype's inclusive fitness context specific: if the same genotype interacts with two different groups (even when both have identical allele frequencies), it may have different inclusive fitnesses.

The resulting variance in the breeding value for inclusive fitness becomes

$$\sigma^2(A_{incf}) = \sigma^2(A_d) + 2r(n-1)\sigma(A_d, A_s) + r^2(n-1)^2\sigma^2(A_s) \tag{22.51b}$$

In the absence of heritable associative effects, $\sigma^2(A_s) = 0$, this reduces to the additive variance in direct fitness, as is also the case when $r = 0$. It is important to note that the heritable component of inclusive fitness is not the same as the total breeding value, A_T, for fitness, as a comparison of Equation 22.51a to Equation 22.3 shows that

$$A_{T_i} = A_{incf,i} + (1-r)(n-1)A_{s_i} \tag{22.51c}$$

Just as Equation 22.49a decomposed the total response into components from direct and associative effects, we can similarly decompose the change in mean individual fitness, R_W, into change in mean inclusive fitness, $R_{W,incf}$, plus a residual response. From Equation 22.51c

$$R_W = R_{W,incf} + (1-r)(n-1)R_{W,s} \tag{22.52a}$$

showing that the total response in fitness is the change in inclusive fitness plus any response in the residual of the mean social value (after the effects of group relatives are absorbed into inclusive fitness). Using the same logic leading to Equation 22.48c, the response in the mean inclusive fitness is given by

$$R_{W,incf} = \frac{1}{\overline{\overline{W}}}\sigma(W, A_{incf}) \tag{22.52b}$$

where, from Equations 22.47a and 22.51a, we have

$$\begin{aligned}\sigma(W, A_{incf}) &= \sigma\left(\mu + A_{d_i} + \sum_{j\neq i} A_{s_j} + e_i,\, A_{d_i} + r(n-1)A_{s_i}\right) \\ &= \sigma^2(A_d) + 2r(n-1)\sigma(A_d, A_s) + r^2(n-1)^2\sigma^2(A_s)\end{aligned} \tag{22.52c}$$

The last line follows by evaluating the covariance in a similar fashion as done throughout this chapter. Note by comparison with Equation 22.51b that Equation 22.52c is simply $\sigma^2(A_{incf})$, yielding

$$R_{W,incf} = \frac{\sigma^2(A_{incf})}{\overline{\overline{W}}} \tag{22.52d}$$

Hence (under our simple model), the response in mean inclusive fitness is proportional to the additive variance in inclusive fitness, so that mean inclusive fitness is nondecreasing.

Why, then, can the mean of *individual* fitness decline despite the continual increase in mean inclusive fitness? The reason is a decline in the mean (residual) social value. Recalling Equation 22.49c, Equation 22.52a becomes

$$R_W - R_{W,incf} = \frac{(1-r)(n-1)}{\overline{W}}\left(\sigma(A_d, A_s) + r(n-1)\sigma^2(A_s)\right) \tag{22.52e}$$

Hence, if the covariance between direct and associative effects is sufficiently negative, any increase in inclusive fitness is more than countered by the decline in the mean social environment. When $r = 0$,

$$R_W - R_{W,incf} = \frac{(n-1)\sigma(A_d, A_s)}{\overline{W}} \tag{22.52f}$$

which can be positive or negative, depending on the sign of the covariance. Note, from Equation 22.52e, that increasing the relatedness, r, of group members decreases the residual response between mean individual and inclusive fitness, which in turn increases the chances that individual mean fitness increases.

Bijma's Theorem: Inclusive Fitness and Fisher's Fundamental Theorem

As we have seen, when heritable associative effects are present, individual selection can cause mean individual fitness to decrease. Further, while mean *individual* fitness can decrease, mean *inclusive fitness* is nondecreasing, as the response in inclusive fitness is proportional to its additive variance (Equation 22.52d). Comparing these results with those from Chapter 6, we have an apparent conflict. For the simple model of additive fitness effects and random mating, the classical interpretation of Fisher's fundamental theorem (FFT) holds, with the change in mean individual fitness being proportional to the additive variance in individual fitness, so that (under these conditions) mean individual fitness is nondecreasing. Further, the Price-Ewens exact version of the FFT (Chapter 6) states that the partial increase in mean individual fitness (the change that occurs when set in a constant environment) is nondecreasing. Bijma (2010a) reconciled these results, showing that the Price-Ewens FFT corresponds to statements about inclusive fitness (which reduces to individual fitness in the absence of associative effects).

Recall from Chapter 6 that Price (1972b) and Ewens (1989, 1992) showed that Fisher appeared not to be concerned about the *total* change in individual fitness, but rather about only *one component* of that change, namely that caused by changes in the allele frequencies of genes under selection, when all other factors (such as change in the environment) are held constant. Bijma (2010a) made the important distinction between what we have been calling the total breeding value, A_T, and the traditional breeding value, which we here denote by BV, computed by considering only the effects of the alleles in the focal individual on its own phenotype. The later (BV) is the more traditional definition of breeding value and (LW Chapter 4) is simply the regression of fitness on the direct and social breeding values of individual i,

$$W_i = BV_i + e = \beta_{i,d}A_{d_i} + \beta_{i,s}A_{s_i} + e \tag{22.53a}$$

To compute these regression slopes, first note that the individual fitness of i is a function of its direct value plus the associative effects for every group member except i,

$$W_i = \mu + A_{d_i} + \sum_{j\neq i} A_{s_j} + e_i$$

Recalling that $\sigma(A_{s_i}, A_{s_j}) = r_{ij}\sigma^2(A_s)$, the traditional breeding value for individual fitness for i reduces to

$$BV_i = A_{d_i} + A_{s_i}\sum_{j\neq i} r_{ij} \tag{22.53b}$$

as obtained by Bijma (2010a). For the case of all relatives being equally related, this reduces to

$$BV_i = A_{d_i} + r(n-1)A_{s_i} \tag{22.53c}$$

Comparing this to Equation 22.51a shows that the traditional breeding value (BV) equals the breeding value for inclusive fitness. Thus, we have **Bijma's theorem:**

> *The traditional breeding value for individual fitness is equivalent to the breeding value for inclusive fitness.*

From Equation 22.52d, it immediately follows that the response in inclusive fitness is proportional to the traditional additive variance in fitness (which, in this case, is the variance in inclusive fitness). Hence, inclusive fitness provides the bridge between the FFT and response when associative effects are present.

HAMILTON'S RULE

Example 22.16 briefly introduced Hamilton's classic result: the condition for an altruistic trait to spread is $br - c > 0$, where c is the fitness cost to the individual (or **actor**) performing the altruistic act, b is the benefit to individuals with which it interacts (the **recipients**), and r is the relationship between the actor and the recipient (Hamilton 1963, 1964a, 1964b). Hamilton's original result followed upon consideration of an altruistic trait determined by a single locus with additive fitness effects in an outbred population. In this setting, $-c + rb$ is the inclusive fitness of the altruistic allele, whose frequency increases when its inclusive fitness is positive. He further showed (again under this simple model) that inclusive, rather than individual, fitness is maximized by selection. The roots of Hamilton's result, that for altruism to spread under individual selection requires interactions occur among sufficiently related individuals, dates back to a widely attributed quote by Haldane: "*Would I lay down my life to save my brother? No, but I would to save two brothers or eight cousins.*" (As an aside, this is often cited as Haldane 1955, but this quote does not appear in that paper). Hamilton's contribution was to generalize Haldane's intuition and to introduce the important concept of inclusive fitness.

How General Is Hamilton's Rule?

Hamilton's rule is a bit like the breeder's equation: it provides a simple expression that conveys the nature of interactions between the key quantities of interest, but it can fail (at least as an exact expression) under a variety of conditions. Given that Hamilton assumed a single additive locus under weak selection in an outbred population, just how general is his result? Assuming weak selection, a number of studies have shown that altruistic traits with a polygenic basis also generally satisfy Hamilton's rule (e.g., Yokoyama and Felsenstein 1978; Boyd and Richerson 1980; Aoki 1982; Engels 1983). Conversely, Cavalli-Sforza and Feldman (1978) found that it can fail for even a single locus. When fitnesses are no longer additive (i.e., the fitness of the heterozygote is no longer the average of the two homozygotes), then comparing r to c/b may not be sufficient to predict spread or loss, as the actual value of b (in addition to the ratio c/b) can also matter.

Starting with Hamilton (1970), attempts for a more general version of Hamilton's rule were built around the Price Equation. Recall from Equation 6.6 that Price's (1970, 1972a) theorem states that the response in any quantity, G, can be expressed as

$$\Delta G = \sigma(G, w) + E[w\,\delta_G] \tag{22.54}$$

where w is relative fitness and δ_G is any transmission bias, namely, the average deviation between the value of G in an ancestor and its mean value in their descendants. Typically, we can treat G as either the frequency of an allele (for a single-locus analysis) or as the breeding value of a trait (for a quantitative-genetic analysis). In the absence of drift and under normal Mendelian segregation (i.e., no meiotic drive), the allele frequencies in the offspring match the allele frequencies in their parents. Likewise, under the infinitesimal model (in an outbred population) the expected breeding value of an offspring equals the mean breeding value of its parents, and hence the second term in Price's equation is usually ignored (see Chapter 6 for a more detailed discussion).

One way to obtain Hamilton's result is to assume what is typically called **neighbor-modulated fitness** (Hamilton 1964a, 1964b), wherein the phenotypes of neighbors influences one's fitness (this has also been referred to as a **direct fitnesses** model, e.g., Taylor and Frank 1996; Taylor et al. 2007). Following Queller (1992a), we let the relative fitness of some focal individual, i, interacting with its neighbor, j, be

$$w_i = w_0 + w_{d,i} + w_{s,j} \tag{22.55a}$$

w_0 is a component independent of social interactions, $w_{d,i}$ is the direct effect of i on its own fitness, and $w_{s,j}$ is the (social) effect of the neighbor j on i's fitness. Now, use a linear regression to quantify the effects of phenotypes z_i on fitness, with

$$w_i = a + \beta_{w_d|z} z_i + \beta_{w_s|z} z_j + e_i \tag{22.55b}$$

Ignoring the transmission bias term in Equation 22.54, Price's theorem yields an expected response in the mean breeding value, μ_A, of the trait as

$$\begin{aligned}\Delta\mu_A = \sigma(w_i, A_i) &= \sigma(\beta_{w_d|z} z_i + \beta_{w_s|z} z_j + e_i, A_i) \\ &= \beta_{w_d|z}\, \sigma(z_i, A_i) + \beta_{w_s|z}\, \sigma(z_j, A_i) + \sigma(e_i, A_i)\end{aligned} \tag{22.55c}$$

Here, we have used the fact that $\sigma(a, A_i) = 0$ for the constant a. Note that, to this point, Equation 22.55c is exact. If the residual, e_i, from the regression of fitness on phenotype (Equation 22.55b) is uncorrelated with the breeding value, A_i, the result is exactly

$$\Delta\mu_A = \beta_{w_d|z}\sigma(z_i, A_i) + \beta_{w_s|z}\sigma(z_j, A_i) \tag{22.56a}$$

The residual condition, $\sigma(e_i, A_i) = 0$, is satisfied when w is entirely determined by the phenotypes z_i and z_j plus any additional components, *provided* the latter are uncorrelated with A_i (see Chapter 6). The mean value of the trait increases when $\Delta\mu_A > 0$. Dividing both sides of Equation 22.56a by $\sigma(z_i, A_i)$, this occurs when

$$\beta_{w_d|z} + \beta_{w_s|z} \frac{\sigma(z_j, A_i)}{\sigma(z_i, A_i)} > 0 \tag{22.56b}$$

where $\beta_{w_d|z}$ is the cost (c) of the trait, $\beta_{w_s|z}$ is the benefit (b), and the ratio is a generalized measure of relatedness. Given that the phenotypic fitness regression residuals are uncorrelated with breeding values (and that we can safely ignore the transmission bias term), then Equation 22.56b is a general version of Hamilton's rule.

Under the infinitesimal model assumptions in an noninbred population, the regression term (a covariance divided by a variance) quantifying relatedness reduces to

$$\frac{\sigma(z_j, A_i)}{\sigma(z_i, A_i)} = \frac{\sigma(A_j, A_i)}{\sigma(A_i, A_i)} = \frac{r_{ij}\, \sigma(A_i, A_i)}{\sigma(A_i, A_i)} = r_{ij}$$

and we recover the standard version of Hamilton's rule. However, when inbreeding is present or the infinitesimal model assumptions are not valid (i.e., significant allele-frequency change can occur within a generation), Price's equation shows that a more general definition of relatedness may be required. There is a detailed literature on the appropriate measure of relatedness to use; see Hamilton (1970), Orlove and Wood (1978), Michod and Hamilton (1980), Aoki (1981), Seger (1981), Uyenoyama and Feldman (1981), Pepper (2000), Goodnight (2005), Smith et al. (2015), and references therein. For single-locus models, relatedness measures attempt to account for the difference in the frequency of an altruistic allele in recipients versus the general population, and hence can depend on genotypic frequencies and other details of the assumed model. See Toro et al. (1982) and Michod (1982) for a more detailed discussion of these population-genetic models. There are a very large number of models for altruism and cooperation built around Hamilton's rule; see Lehmann and Keller (2006a, 2006b and references therein) for an overview, and McGlothlin et al. (2014) for a recent review of quantitative-genetic versions of Hamilton's rule.

Queller's Generalization of Hamilton's Rule

Equation 22.56b follows by considering fitness regressed on *phenotype*. This makes sense, as quantitative genetics tries to work with measurable quantities, such as the effect of phenotype on fitness and the correlation between breeding and phenotypic values. However, Queller (1992b) noted that by considering the regression of fitness directly on the *breeding values* of the selected traits, an exact expression for Hamilton's rule can be obtained. The key is that, by construction (e.g., LW Chapter 3), the residuals are uncorrrelated with the predictor variables in the regression. Hence, writing fitness as a multiple regression on the breeding values of the actor (i) and recipient (j),

$$w_i = a + \beta_{w_d|A}\, A_i + \beta_{w_s|A}\, A_j + e_i \tag{22.57a}$$

we always have $\sigma(e_i, A_i) = 0$, and (assuming no transmission bias so that we can ignore the second term in 22.54), Price's equation yields

$$\begin{aligned} \Delta\mu_A = \sigma(w_i, A_i) &= \sigma(a + \beta_{w_d|A}\, A_i + \beta_{w_s|A}\, A_j + e_i, A_i) \\ &= \beta_{w_d|A}\, \sigma(A_i, A_i) + \beta_{w_s|A}\, \sigma(A_j, A_i) \end{aligned} \tag{22.57b}$$

Dividing both sides by $\sigma(A_i, A_i)$ gives Queller's generalization of Hamilton's rule as

$$\beta_{w_d|A} + \beta_{w_s|A}\, \frac{\sigma(A_j, A_i)}{\sigma(A_i, A_i)} > 0 \tag{22.58a}$$

Recalling the definition of a regression slope (the covariance divided by the variance of the predictor, e.g., LW Chapter 3), Queller's exact result reduces to the very satisfying form of

$$\beta_{w_d|A} + \beta_{w_s|A} \cdot \beta_{A_j|A_i} > 0 \tag{22.58b}$$

where $\beta_{w_d|A}$ is the cost, $\beta_{w_s|A}$ the benefit, and $\beta_{A_j|A_i}$ is a generalized measure of relatedness between i and j. Gardner et al. (2007) cautioned that while Equation is 22.56b is rather general and Equation 22.58b completely general (both under the assumption of no biased transmission term), that "the cost of this generality is that it hides a lot of detail, and so naive application of Hamilton's rule may lead to mistakes." See Frank (1998) for a detailed discussion of potential pitfalls.

GROUP SELECTION, KIN SELECTION, AND ASSOCIATIVE EFFECTS

Kin, Group, and Multilevel Selection

There is a vast (and often heated) literature in evolutionary genetics and social evolution dealing with kin selection, group selection, and inclusive fitness (selected works include Wynne-Edwards 1962, 1986; Maynard Smith 1964, 1976; Williams 1966; Wade 1978; Wilson 1983; Frank 1998; Lehmann and Keller 2006a; Lehmann et al. 2007; West et al. 2006, 2008; Wilson and Wilson 2007; Bijma and Wade 2008; Nowak et al. 2010; Abbot et al. 2011; Van Veelen et al. 2012; Goodnight 2013, 2015; West and Gardner 2013; Birch 2014; Birch and Okasha 2015; Gardner 2015; Lehtonen 2016). Much of the debate has revolved around the evolutionary mechanism(s) needed to account for traits that reduce the fitness of an individual but increase the fitness of a group. As we have seen, Hamilton's rule gives the condition (a sufficiently high degree of relatedness among the interacting individuals) for such an allele to spread under individual selection. This is a **kin selection** model (Maynard-Smith 1964), whereby interactions among kin (i.e., $r > 0$) generate an inclusive fitness that can allow an allele, potentially harmful to the individual, but helpful to a group, to spread.

An alternative school of thought, **group selection**, which was proposed by Wynne-Edwards (1962, 1963), states that traits favoring a group spread via selection at the level of groups—those carrying alleles for beneficial group behavior tend to leave more groups that those that lack them (the roots of this idea trace back to Darwin 1871). In animal and plant breeding this is not at all controversial, as family selection (choosing only those individuals from the best families) is widely practiced (Chapter 21). In its most extreme form (e.g., Wynne-Edwards 1962, 1963), group selection is envisioned to occur through isolated demes that undergo differential extinction and propagation. Quantitative-genetic models of response to group selection due to population structure have been examined by Slaktin and Wade (1978), Slaktin (1981a), Crow and Aoki (1982), and Tanaka (1996a).

A less restricted form of group selection is the **levels-of-selection** (or **multilevel selection**) argument, where the fitness of an individual is influenced by those individuals with which it interacts, so that fitness is a function of the collection of interacting individuals, rather than a single individual. Family index selection (Chapter 21), wherein both within- and among-family information is used, is an example of a levels-of-selection approach. Levels-of-selection does not require isolated units, and groups (here, simply sets of interacting individuals) can form anew each generation even in a panmixtic population.

In evolutionary biology, the debate over the relative importance of kin vs. group selection has, at times, had the feeling of a holy war. One argument against group selection involves concerns about the formation and subsequent propagation of groups, as well as the generation of among-group variation upon which selection can act. The heavy hand of parsimony (running the risk of getting oneself cut with Occam's razor) is also raised against group selection—why invoke it if individual selection will do (Williams 1966)? Of course, one could argue this is entirely the wrong prior. Most biologists would not disagree with the idea that an individual's fitness is influenced by those with whom it interacts, in which case levels-of-selection would be the more reasonable default, especially because it includes individual selection as a special case. The issue then becomes an empirical one, namely, the relative fitness weights on group versus individual components. Estimation of levels-of-selection components is examined in Chapter 30.

Much Ado About Nothing?

The reader who is unfamiliar with the evolutionary literature might be a bit perplexed about this controversy, as when placed in a framework of associative effects, both group- and kin-selection components arise and have symmetric roles (e.g., Equation 22.29d). Indeed, a number of workers have suggested group and kin selection are simply two extremes on the continuum of potential interactions and hence closely related (Wade 1980; Queller 1991; Lehmann et al. 2007). Bijma and Wade (2008) succinctly make the point that

> the ongoing debate on equivalence of kin and levels-of-selection models is partly caused by the fact that levels-of-selection models tend to hide the relatedness component of response to selection, whereas kin selection models tend to hide the multilevel selection component of response to selection . . . the response to selection is naturally described by the combination of relatedness and the degree of multilevel selection, rather than by focusing on one or the other of the two factors.

As we will see, in the absence of associative effects influencing a trait value ($\sigma^2(A_s) = 0$), *both* kin ($r \neq 0$) *and* multilevel (group weight $g \neq 0$, see Equation 22.28a) selection are required for social selection to have a response that differs from the breeder's equation. When associative effects influence trait values, kin and multilevel selection appear as symmetric roles, and even in the absence of these latter two factors (namely, $r = g = 0$), the selection response can still differ from the breeder's equation. These results were first clearly stated by Bijma and Wade (2008), and we closely follow their development below.

Group and Kin Selection: Models Without Trait Associative Effects

Most models of kin and group selection assume that the trait of interest is not influenced by associative effects, so that we can decompose the phenotypic value of individual i as

$$z_i = A_i + E_i$$

where A_i denotes i's breeding value (we use A rather than A_d to stress that this model assumes *no* associative effects, so that no A_s terms appear). While no associative effects influence the *phenotypic value* of the trait of interest, we do assume that the fitness of an individual is influenced by the phenotypes of its neighbors. This results in *fitness* showing both direct and associative effects. This distinction between models where the trait does not show associative effects while fitnesses do, versus models where *both* the trait and fitness show associative effects is subtle, but rather important, as the resulting model behavior is quite different (Bijma and Wade 2008).

In particular, Bijma and Wade show that when traits lack associative effects, the response under either kin or group selection deviates from the breeder's equation only when *both* relatedness and multilevel selection occur. A slightly more general development of their result proceeds as follows.

First, consider a classical kin selection model, where the fitness of individual i is a function of its phenotypic value plus contributions that depend on the phenotypic values (for the same trait) of the $n-1$ individuals with which it interacts. Generalizing Equation 22.55b, we can express i's fitness as the multiple regression

$$w_i = a + \beta_{w_d|z} z_i + \beta_{w_s|z} \left(\sum_{j \neq i}^{n} z_j \right) + \epsilon_i \tag{22.59a}$$

where $\beta_{w_d|z}$ is the direct effect on fitness and $\beta_{w_s|z}$ the indirect (or social) effect on i's fitness given the phenotypes of its conspecifics. Assuming the residual (ϵ_i) is uncorrelated with i's breeding value (A_i) for the trait under selection, Equation 22.56a generalizes to give the selection response in the trait mean as

$$R = \Delta\mu_A = \beta_{w_d|z} \sigma(z_i, A_i) + \beta_{w_s|z} \left(\sum_{j \neq i}^{n} \sigma(z_j, A_i) \right) \tag{22.59b}$$

As mentioned previously, we can think of $\beta_{w_d|z} = c$ as the cost and $\beta_{w_s|z} = b$ as the benefit, so that for altruistic traits $\beta_{w_d|z} < 0$ and $\beta_{w_s|z} > 0$. For the infinitesimal model under random mating,

$$\sigma(z_i, A_i) = \sigma(A_i, A_i) = \sigma_A^2 \quad \text{and} \quad \sigma(z_j, A_i) = \sigma(A_j, A_i) = r_{ij}\sigma_A^2$$

If we assume that all interacting pairs have the same relationship (so that $r_{ij} = r$), Equation 22.59b reduces to

$$R = \Delta\mu_A = \sigma_A^2 \left(\beta_{w_d|z} + r(n-1)\beta_{w_s|z} \right) \tag{22.59c}$$

Hence, the trait increases when $\beta_{w_d|z} + r(n-1)\beta_{w_s|z} > 0$ (Hamilton's rule).

The multilevel selection connection that appears in kin selection models is seen by defining

$$g = \frac{\beta_{w_s|z}}{\beta_{w_d|z}} \tag{22.60a}$$

Using this definition, we can rewrite Equation 22.59a as

$$w_i = a + \beta_{w_d|z} \left(z_i + g \sum_{j \neq i} z_j \right) + \epsilon_i = a + \beta_{w_d|z}\, I_i + \epsilon_i \tag{22.60b}$$

where

$$I_i = z_i + g \sum_{j \neq i} z_j$$

is the index given by Equation 22.28a. Individual selection corresponds to $g = 0$, as $I_i = z_i$. Likewise, $g = 1$ corresponds to group selection, as

$$I_i = \sum_{j=1}^{n} z_j = n\overline{z}$$

so that fitness is entirely a function of group mean. As above, g need not be restricted to between zero and one. For example, negative values of g correspond to selection based on deviation within a group (see Example 22.9).

Expressed in terms of g, the expected response under kin selection (Equation 22.59c) becomes

$$\Delta\mu_A = \beta_{w_d|z}\, \sigma_A^2 \Big(1 + (n-1)gr \Big) \tag{22.61a}$$

This equation makes Bijma and Wade's main point: The response is a function of the *product* of relatedness (r) and levels-of-selection (g). If either is zero, Equation 22.61a reduces to

$$\Delta\mu_A = \beta_{w_d|z}\, \sigma_A^2 = \frac{\sigma(w_d, z_i)}{\sigma_z^2}\, \sigma_A^2 = S\, \frac{\sigma_A^2}{\sigma_z^2} = h^2 S \tag{22.61b}$$

where we have used the Robertson-Price identity (Equation 6.10), $\sigma(w_d, z_i) = S$. Hence, for the selection response to differ from the standard breeder's equation requires *both* relatedness ($r > 0$) *and* multilevel selection ($g \neq 0$). While the relatedness is obvious in kin-selection formulations, the levels-of-selection component historically has been a bit less transparent, being "hidden" in the costs and benefits, $\beta_{w_s|z}$ and $\beta_{w_d|z}$ (Bijma and Wade 2008).

Now consider the response under multilevel selection. Here, fitness is a function of both individual and group value, usually expressed as the components of fitness due to within-group deviation ($\Delta z_i = z_i - \overline{z}$) and to the group mean ($\overline{z} = \sum z_i/n$). As above, the fitness of i can be expressed as the regression

$$w_i = a + \beta_{w|\,\overline{z}}\, \overline{z} + \beta_{w|\Delta z}\, \Delta z_i + e_i \tag{22.62a}$$

Individual selection occurs when $\beta_{w|\,\overline{z}} = \beta_{w|\Delta z} = \beta$, as Equation 22.62a reduces to $w_i = a + \beta z_i + e_i$. Again assuming that e_i is uncorrelated with A_i, Price's theorem gives the response as $\Delta\mu_A = \sigma(w_i, A_i)$, which from Equation 22.62a becomes

$$R = \Delta\mu_A = \beta_{w|\,\overline{z}}\, \sigma(\,\overline{z}, A_i) + \beta_{w|\Delta z}\, \sigma(\Delta z_i, A_i) \tag{22.62b}$$

The first covariance term is given by

$$\sigma(\,\overline{z}, A_i) = \frac{1}{n}\, \sigma \left(\sum_{j=1}^{n} z_j, A_i \right) = \frac{\sigma(z_i, A_i) + \sum_{j \neq i} \sigma(z_j, A_i)}{n} = r_n \sigma_A^2 \tag{22.62c}$$

where $r_n = r + (1-r)/n$ (Equation 21.6b). This is just the among-group genetic variance (Chapter 21). Likewise,

$$\sigma(\Delta z_i, A_i) = \sigma\,(z_i - \overline{z}, A_i) = \sigma_A^2\,(1 - r_n) \tag{22.62d}$$

which is the within-group genetic variance (Chapter 21). Substitution of Equations 22.62c and 22.62d into Equation 22.62b gives the expected response to multilevel selection as

$$\Delta\mu_A = \sigma_A^2 \Big(r_n \beta_{w|\,\overline{z}} + (1 - r_n)\beta_{w|\Delta z} \Big) \tag{22.62e}$$

As noted by Wade (1980) and Cheverud (1985), total response is the sum of the among-group response, $r_n\,\sigma^2_A\,\beta_{w|\,\overline{z}}$, plus the within-group response, $(1-r_n)\,\sigma^2_A\,\beta_{w|\Delta z}$. Relatedness enters into multilevel selection models because r influences the relative amounts of within- versus among-group variance. Increasing r increases the among-group variation, $r_n\sigma^2_A$, while decreasing the within-group variation, $(1-r_n)\sigma^2_A$. With increasing relatedness, more of the response comes from among-group (as opposed to within-group) selection. In the absence of relatives within the group ($r=0$, so that $r_n=1/n$), Equation 22.62e becomes

$$\Delta\mu_A = \sigma^2_A\left(\frac{1}{n}\beta_{w|\,\overline{z}} + \left[1-\frac{1}{n}\right]\beta_{w|\Delta z}\right) \tag{22.62f}$$

Example 22.17 shows that the term in the parentheses reduces to $\beta_{w_d|z}$, and hence in the absence of relatives the response reduces to Equation 22.61b, namely the breeder's equation. Relatedness of group members is thus required for the response under the multilevel fitness model (Equation 22.62a) to depart from the breeder's equation.

As mentioned, although these models have no associative effects when *trait* values are considered, their fitness functions (Equation 22.59a and 22.62a) generate direct and associative effects in *fitness*. It will prove useful (especially when contrasting the above results with models that *do* allow traits to have associative effects) to consider the direct and associative components of fitness as they relate to the breeding value, A_i, of the focal individual. Write the index I_i as

$$I_i = z_i + g\sum_{j\neq i}^{n} z_j = A_i + g\sum_{j\neq i}^{n} A_j + e^*_i$$

where we have swept all of the terms not involving breeding values into the residual, e^*_i. Substitution into Equation 22.60b gives the fitness of i in terms of the breeding values A_i and A_j as

$$w_i = \beta_{w_d|z}A_{d,i} + g\,\beta_{w_d|z}\sum_{j\neq i}^{n} A_{d,j} + \epsilon^*_i = w_{d,i} + \sum_{j\neq i}^{n} w_{s,j} + \epsilon^*_i \tag{22.63a}$$

where we have used ϵ^*_i as the residual in the fitness regression to distinguish it from the residual (e^*_i) in our expression for the index I_i. We have now reverted to the A_d notation for the breeding values (of direct effects), as we will shortly expand this result to allow for breeding values of associative effects (A_s). The right side of Equation 22.63a decomposes the fitness into direct and associative components, with

$$w_{d,i} = \beta_{w_d|z}A_{d,i} \quad \text{and} \quad w_{s,j} = g\beta_{w_d|z}A_{d,j} \tag{22.63b}$$

The direct component ($w_{d,i}$) is the contribution from genes in the focal individual i toward its fitness, while the associative component ($w_{s,j}$) is the contribution from genes in j toward i's fitness. From Price's equation, we have the response as the sum of direct and associative contributions, where

$$R_{i,d} = \sigma(w_{d,i}, A_i) = \beta_{w_d|z}\sigma(A_{d,i}, A_i) = \beta_{w_d|z}\sigma^2(A_d) \tag{22.63c}$$

which is the breeder's equation, while any additional contribution from i due to genes in j is given by

$$R_{i,j} = \sigma(w_{s,j}, A_i) = g\beta_{w_d|z}\sigma(A_{d,j}, A_i) = rg\beta_{w_d|z}\sigma^2(A_d) \tag{22.63d}$$

Thus, two factors are required for genes in j to influence the response based on selecting i. First, multilevel selection ($g\neq 0$) is required in order for there to be an effect of genes in j on the *fitness* of i ($g\,\beta_{w_d|z}A_{d,j}$). Second, the genes in i and j must be correlated ($r>0$). Only when *both* $g\neq 0$ and $r\neq 0$ is there an additional increment in the selection response from the genes in j, and only in this case do we see departures from the breeder's equation. Summing over all $n-1$ of i's interacting neighbors recovers Equation 22.61a.

Example 22.17. Following Bijma and Wade, we can formally show the connections between the fitness models given by Equation 22.59a (kin selection) and Equation 22.62a (multilevel selection). Expanding $\overline{z}$ and Δz_i to express them in terms of z_i and $\sum_{j\neq i} z_j$, we have that

$$\begin{aligned}\beta_{w|\overline{z}}\,\overline{z} + \beta_{w|\Delta z}(z_i - \overline{z}) &= \beta_{w|\Delta z} z_i + (\beta_{w|\overline{z}} - \beta_{w|\Delta z})\,\overline{z} \\ &= \beta_{w|\Delta z} z_i + \frac{1}{n}\Big(\beta_{w|\overline{z}} - \beta_{w|\Delta z}\Big)\Big(z_i + \sum_{j\neq i} z_j\Big) \\ &= \frac{1}{n}\Big(\beta_{w|\overline{z}} + [n-1]\beta_{w|\Delta z}\Big) z_i + \frac{1}{n}\Big(\beta_{w|\overline{z}} - \beta_{w|\Delta z}\Big)\sum_{j\neq i} z_j \end{aligned} \tag{22.64a}$$

Matching terms with Equation 22.59a, the regression slopes for fitness in these two model are related as follows:

$$\beta_{w_d|z} = \frac{\beta_{w|\overline{z}} + (n-1)\beta_{w|\Delta z}}{n} \quad \text{and} \quad \beta_{w_s|z} = \frac{\beta_{w|\overline{z}} - \beta_{w|\Delta z}}{n} \tag{22.64b}$$

Similarly, we can show that

$$\beta_{w_d|z} z_i + \beta_{w_s|z} \sum_{j\neq i}^{n} z_j = \left(\beta_{w_d|z} - \beta_{w_s|z}\right)\Delta z_i + \left(\beta_{w_d|z} + (n-1)\beta_{w_s|z}\right)\overline{z} \tag{22.65a}$$

implying

$$\beta_{w|\Delta z} = \beta_{w_d|z} - \beta_{w_s|z} \quad \text{and} \quad \beta_{w|\overline{z}} = \beta_{w_d|z} + (n-1)\beta_{w_s|z} \tag{22.65b}$$

Thus, in the absence of *trait* associative effects ($\sigma^2(A_s) = 0$, and hence $\beta_{w_s|z} = 0$), $\beta_{w|\Delta z}$ and $\beta_{w|\overline{z}}$ are equivalent and simply differ by shifting focus over individual versus group values.

Finally, we can rearrange the left identity in Equation 22.64b to

$$\frac{1}{n}\beta_{w|\overline{z}} + \left(\frac{n-1}{n}\right)\beta_{w|\Delta z} = \beta_{w_d|z}$$

Using Equation 22.65b to substitute for $\beta_{w|\overline{z}}$ and $\beta_{w|\Delta z}$ yields

$$\frac{\beta_{w_d|z} + (n-1)\beta_{w_s|z}}{n} + \left(\frac{n-1}{n}\right)\left(\beta_{w_d|z} - \beta_{w_s|z}\right) = \beta_{w_d|z}$$

showing that the response to multilevel selection when group members are unrelated ($r = 0$) reduces to the breeder's equation,

$$\Delta\mu_A = \sigma^2_A\left(\frac{1}{n}\beta_{w|\overline{z}} + \left(\frac{n-1}{n}\right)\beta_{w|\Delta z}\right) = \sigma^2_A\,\beta_{w_d|z} = h^2 S$$

Hence, relatedness is required for the response under multilevel selection to deviate from the breeder's equation (Bijma and Wade 2008).

Group and Kin Selection in the Associative-effects Framework

The class of models just considered assumes that trait values are not influenced by associative effects (i.e., no $A_{s,j}$ terms appear in expression for z_i), but does allow for fitnesses to be influenced by the trait values of group members (Equations 22.59a and 22.62a). Under this assumption, the response to social selection only deviates from the breeder's equation when the product gr is nonzero. When associative effects are present in the trait ($\sigma^2(A_s) > 0$), *both* the trait value and fitness of the focal individual may be functions of the genotypes in the group. In this case, kin and group selection (r and g) play symmetric roles, and when *either* is nonzero, response can deviate from the breeder's equation. Indeed, even when $r = g = 0$, response can still deviate from the breeder's equation (e.g., Equation 22.11a). Hence, the addition of trait associative effects introduces a profound change in the behavior of the selection-response model. We first formally present the expected response to selection and then explore the source of this rather different behavior in the presence of trait associative effects by considering the contributions to response from direct and associative fitnesses.

When associative effects are present (A_s terms appear), from Equation 22.1b the index I becomes

$$I_i = z_i + g\sum_{j\neq i}^{n} z_j = \left(A_{d_i} + \sum_{j\neq i} A_{s_j}\right) + g\sum_{j\neq i}^{n}\left(A_{d_j} + \sum_{k\neq j} A_{s_k}\right) + e_i$$

$$= \left(A_{d_i} + g(n-1)A_{s_i}\right) + \sum_{j\neq i}\left(A_{s_j} + g\left[A_{d_j} + (n-2)A_{s_j}\right]\right) + e_i \qquad (22.66)$$

When associative effects are present, Equation 22.1e shows that the expected response is the change in the mean of total breeding value, $A_T = A_d + (n-1)A_s$, where A_d and A_s are the direct and social breeding values. Recalling Equation 22.60b ($w_i = \beta_{w_d|z} I_i + \epsilon_i$), applying the Price Equation, assuming that $\sigma(\epsilon_i, A_{T_i}) = 0$ and no transmission bias, we obtain

$$R = \Delta\mu_{A_T} = \sigma(w_i, A_{T_i}) = \beta_{w_d|z}\sigma(I_i, A_{T_i}) + \sigma(\epsilon_i, A_{T_i}) = \beta_{w_d|z}\sigma(I_i, A_{T_i}) \qquad (22.67a)$$

Substituting the expression for $\sigma(A_{T_i}, I_i)$ given by Equation 22.29d into Equation 22.67a yields

$$R = \beta_{w_d|z}\left(\,[g + r + (n-2)gr]\,\sigma^2(A_T) + (1-g)(1-r)\left[\sigma^2(A_d) + (n-1)\sigma(A_d, A_s)\right]\right) \qquad (22.67b)$$

Equivalently, we could have used a breeder's equation framework (Equation 22.28d) by recalling that for $w_i = \beta_{w_d|z} I_i + \epsilon_i$ (Equation 22.60b), the regression slope can be expressed as

$$\beta_{w_d|z} = \frac{\sigma(w, I)}{\sigma^2(I)} = \frac{S_I}{\sigma(I)}\frac{1}{\sigma(I)} = \frac{\bar{\imath}_I}{\sigma(I)}$$

recovering Equation 22.28d.

Note the completely symmetric roles of relatedness (r) and levels-of-selection (g) in Equation 22.67b. The term in the second set of square brackets can be negative, resulting in R and β potentially having opposite signs (and hence a maladaptive response). Increasing either relatedness, r, or the amount of weight, g, on the other individuals within the group results in increased weight on the $\sigma^2(A_T)$ term, which is always nonnegative, increasing the chance of congruence between the signs of R and β. Finally, there is a synergistic effect between r and g in groups of size greater than two, in that the product $(n-2)gr$ weights $\sigma^2(A_T)$. Bijma and Wade note that this occurs because $n-2$ is the number of group members that two individuals have in common.

Let's now examine Equation 22.67b for a couple of special cases. First (as expected) in the absence of heritable social effects ($\sigma^2(A_s) = 0$), then $\sigma^2(A_T) = \sigma^2(A_d)$ and $\sigma(A_d, A_s) = 0$, with Equation 22.67b reducing to

$$R = \beta_{w_d|z}\sigma^2(A_d)\left[1 + (n-1)gr\right]$$

which recovers Equation 22.61a, showing that (in this case) *both* relatedness *and* multilevel selection are required for the response to deviate from the breeder's equation.

Now suppose that heritable social effects are present, but interacting group members are unrelated. Here $r = 0$ and Equation 22.67b reduces to

$$R = \beta_{w_d|z} \left(g\sigma^2(A_T) + (1-g)\left[\sigma^2(A_d) + (n-1)\sigma(A_d, A_s)\right] \right) \tag{22.67c}$$

Hence, when associative effects on the phenotype occur ($\sigma^2(A_s) > 0$), relatedness is not required for traits with social effects to show a deviation in response from that predicted from the breeder's equation. Likewise, if $g = 0$ and $r > 0$, we see from Equation 22.67b that Equation 22.67c applies, provided that g is now replaced by r. Hence, when relatedness among group members is present ($r \neq 0$), multilevel selection is not needed ($g = 0$) for deviation from the breeder's equation.

Finally, in the absence of both kin and group selection ($g = r = 0$), Equation 22.67b reduces to

$$R = \beta_{w_d|z} \left(\sigma^2(A_d) + (n-1)\sigma(A_d, A_s) \right) = \beta_{w_d|z}\sigma^2(A_d) + (n-1)\beta_{w_d|z}\sigma(A_d, A_s) \tag{22.67d}$$

where the first term in the last equality is the standard breeder's equation (in Lande equation form; Equation 13.8c). Hence, the presence of associative effects, by themselves, are sufficient to produce deviations from the breeder's equation even in the absence of kin or group selection, provided the direct and social breeding values are correlated. This point was highlighted earlier in the chapter.

Another point stressed earlier is that if $\sigma(A_d, A_s)$ is sufficiently negative, the direction of response, R, in the trait may be different from the direction of direct selection, $\beta_{w_d|z}$, on that trait, producing a reversed response. Bijma and Wade (2008) took this point further, and noted that (by definition), selection for an altruistic trait results in a decrease in individual fitness when the mean trait value increases. Hence, spread of an altruistic trait is an example of a reversed response, and we see that if the direct and social breeding values are sufficiently negative correlated within an individual that this can happen *even in the absence of* kin or group selection. The careful reader might be concerned that this result appears to be at odds with Example 22.16, wherein we showed that relatedness was required for individual selection ($g = 0$) to spread an altruistic trait. The difference is in the models. Example 22.16 assumed associative effects only in fitness, but Equation 22.67d assumes associative effects on the *trait itself*. This means that the social breeding value in individual i influences not just the fitness of a group member j, but also j's *trait value*. Under this setting, altruism can evolve in the absence of either group selection or relatedness.

What is the basis of these rather dramatic differences in the behavior between models with and without associative *trait* effects? The key is to consider the direct and associative components of individual fitness, w_i, as we did in Equation 22.63b. From Equations 22.60b and 22.66, these are given by

$$w_{d,i} = \beta_{w_d|z}\big(A_{d_i} + g(n-1)A_{s_i}\big) \quad \text{and} \quad w_{s,j} = \beta_{w_d|z}\big(A_{s_j} + g\big[A_{d_j} + (n-2)A_{s_j}\big]\big) \tag{22.68}$$

as obtained by Bijma and Wade (2008). These equations generalize Equation 22.63b, reducing to it when associative trait effects are absent ($A_s = 0$). Careful inspection of these components show the sources and targets of selection, and the implications for response. For example, i's social breeding value (A_{s_i}) only enters the direct component of individual fitness ($w_{d,i}$) when $g \neq 0$ (i.e., multilevel selection of occurring). The same is true for the direct breeding value (A_{d_i}) to influence the associative component of individual fitness ($w_{s,i}$)

Likewise, to see the role of relatedness, the generalization of the predicted response given by Equation 22.67a to include the direct and associative components of fitness becomes

$$\Delta\mu_{A_T} = \sigma(w_i, A_{T_i}) = \sigma(w_{d,i}, A_{T_i}) + \sum_{j \neq i} \sigma(w_{s,j}, A_{T_i}) \tag{22.69a}$$

No relatedness is required to have $\sigma(w_{d,i}, A_{T_i}) \neq 0$, and hence contribute to the selection response. Conversely, i and j must be related ($r_{ij} > 0$) for $\sigma(w_{s,j}, A_{T_i})$ to be nonzero. This last point follows because $w_{s,j}$ is a function of j's direct and associative breeding values (A_{d_j} and A_{s_j}; Equation 22.68), and

$$\sigma(A_{d_j}, A_{T_i}) = r_{ij}\sigma(A_{d_i}, A_{T_i}) \quad \text{and} \quad \sigma(A_{s_j}, A_{T_i}) = r_{ij}\sigma(A_{s_i}, A_{T_i}) \tag{22.69b}$$

both of which are zero if i and j are unrelated.

Equations 22.68 and 22.69b show the roles played multilevel selection (g) and relatedness (r) when associative effects are present for the trait. Multilevel selection determines how the associative and direct breeding values of i and j are distributed over the direct and associative components of fitness (Equation 22.68), while relatedness allows the associative component of fitness to contribute to response (Equation 22.69b). For example, in the absence of multilevel selection ($g = 0$), Equation 22.68 reduces to

$$w_{d,i} = \beta_{w_d|z} A_{d_i} \quad \text{and} \quad w_{s,j} = \beta_{w_d|z} A_{s_j} \tag{22.70a}$$

In this setting, selection response can only utilize $\sigma^2(A_s)$ when $\sigma(w_{s,j}, A_{T_i}) \neq 0$, which requires that i and j are related (Equation 22.69b).

Conversely, in the absence of any relatives within a group, $w_{s,j}$ does not contribute to the selection response, which from Equation 22.68 is entirely determined by

$$w_{d,i} = \beta_{w_d|z}\left[A_{d_i} + g(n-1)A_{s_i}\right] \tag{22.70b}$$

Multilevel selection ($g \neq 0$) allows for A_{s_i} to be included in $w_{d,i}$, and hence $\sigma^2(A_s)$ is incorporated into the response.

Finally, if both r and g are zero, then only $w_{d,i} = \beta_{w_d|z} A_{d_i}$ enters into the response, giving (from Equation 22.69d)

$$R = \Delta\mu_{A_T} = \beta_{w_d|z}\sigma(A_{d_i}, A_{T_i}) = \beta_{w_d|z}\left[\sigma^2(A_d) + (n-1)\sigma(A_d, A_s)\right] \tag{22.70c}$$

which shows a departure from the breeder's equation when direct and social breeding values are correlated. Because A_s only enters through is covariance with A_d, we are not directly accessing $\sigma^2(A_s)$, so that no direct selection on social values occurs.

Closing Comments

In the absence of associative effects influencing the trait value of interest, both relatedness *and* multilevel selection are required for the expected selection response to deviate from that predicted from the breeder's equation. Hence, when $\sigma^2(A_s) = 0$ and r or g is zero, the standard breeder's equation holds. When $\sigma^2(A_s) > 0$ (the trait value of a group member depends in part on the genotypes of other group members), the selection response typically deviates from the breeder's equation. The only general setting where this is not true is when the direct and social breeding values are uncorrelated and there is no group or kin selection, $\sigma(A_s, A_d) = g = r = 0$.

It must be emphasized that any departure of the response from that predicted from the breeder's equation does not (by itself) imply that social (i.e., group-level) trait mean will evolve in a favorable direction. When the departure is entirely due to a correlation between social and direct breeding values, $\sigma(A_s, A_d) \neq 0$, the result can be a social response in an unfavorable direction when this correlation is negative. In order for selection to directly access social breeding values, $\sigma^2(A_s)$ must appear in the response equation, and when trait associative effects are present this only occurs when either r or g is nonzero. Hence, in the evolution of social traits, three different components are important to consider: the presence or absence of (i) associative effects influencing *trait* value ($\sigma^2(A_s) > 0$); (ii) kin selection ($r \neq 0$); and (iii) multilevel selection ($g \neq 0$). In the presence of associative effects on the trait value, beneficial changes in the mean social value typically requires either kin *or* multilevel selection. In the absence of associative effects ($\sigma^2(A_s) = 0$), *both* kin and multilevel selection are required for beneficial changes in the mean social value.

The associative effects framework is very powerful, as it brings the full machinery of quantitative genetics to bear to the evolution of group-level traits. One immediate advantage is conceptual, in that quantitative-trait models provide a more realistic description of complex traits (be they behavioral or morphological) than do the single-locus models upon which much of the earlier work on social selection is built. The other advantage is empirical: BLUP can be used to estimate the breeding values, and REML used to estimated the associated variances and covariances, of the direct and associative effects and hence model the transmission of a particular trait. The complementary empirical issue of estimating selection on different levels (e.g., individuals vs. groups) is developed in Chapter 30.

23

Selection Under Inbreeding

Either inbreeding or selection, never both at the same time. R. A. Fisher

There are several reasons for jointly considering inbreeding and selection. First, one may have little choice. For many species, such as the autogamous crops that provide much of our food, the extra work required to ensure outcrossing is often considerable. Second, in many cases, the creative use of inbreeding can increase selection response. Third, the development of elite pure lines starts with a cross between two (often inbred) lines, with the breeder then selfing the resulting progeny to fixation. Under such development schemes, selection can occur in the early generations of selfing (this chapter) and/or among the completely selfed lines (Volume 3). Finally, many wild populations undergoing natural selection have mating systems that result in offspring that are partly to highly inbred.

As we have seen, selection changes genotypic frequencies by changing allelic frequencies and creating gametic-phase disequilibrium (Chapters 5, 16, and 24). Under random mating, departures from generalized Hardy-Weinberg (such as those induced by gametic-phase disequilibrium) are reduced each generation. This erodes any transient contributions from nonadditive variances to the selection response, leaving only allele-frequency change (accomplished via σ_A^2) as the permanent component of response (Chapter 15). When inbreeding occurs, departures from generalized Hardy-Weinberg are *created*, rather than *destroyed*, by the mating system. By increasing the frequency of homozygotes, inbreeding redistributes the genetic variance in a population, reducing or removing it within a line undergoing inbreeding and increasing it among a collection of such lines (Chapter 11). Inbreeding also generally increases the covariance between relatives, as they become increasingly more genetically similar (Chapter 11). Finally, the reduction in the frequency of heterozygotes under inbreeding results in a corresponding reduction in the effective recombination rate, retarding the decay of multilocus departures from generalized Hardy-Weinberg. As we will see, all these of factors have important consequences for selection response.

When inbreeding occurs and nonadditive genetic action (dominance and/or epistasis) is present, the standard genetic variance components (σ_A^2, σ_D^2, and σ_{AA}^2) are no longer sufficient to predict response. As was discussed in Chapter 11, when dominance is present, at least three additional components (ι^*, σ_{DI}^2, and σ_{ADI}; see Table 11.1) are required to describe the covariance between inbred relatives, and hence to predict the short-term selection response. A further complication is inbreeding depression (LW Chapter 10), which can change the mean even in the absence of selection. Unless otherwise mentioned, we assume throughout this chapter that there is gametic-phase equilibrium and no epistasis. The complications these conditions introduce for selection response with inbreeding remain largely unexplored, but are potentially quite important. For example, directional selection creates negative disequilibrium (Chapter 16), which is reduced in each generation by recombination between heterozygotes. By reducing (and ultimately removing) heterozygotes, inbreeding can significantly retard the decay of any selection-induced disequilibrium, thus magnifying its impact relative to random mating (Allard 1975; Hayashi and Ukai 1994; Shaw et al. 1998; Kelly 1999a; Kelly and Williamson 2000).

Our examination of the selection response under inbreeding begins with a general overview of the machinery and concepts for treating the joint action of inbreeding and selection. This is followed by a discussion of family selection when the parents and/or scored progeny are inbred, which extends the results of Chapter 21. These first two sections

Evolution and Selection of Quantitative Traits. Bruce Walsh & Michael Lynch.
Published 2018 by Oxford University Press.
DOI 10.1093/oso/9780198830870.001.0001

form the basics of the selection response under inbreeding. The remainder of the chapter examines a number of special (but important) cases, such as selfing and partial selfing, in more detail. Additional aspects of selection and inbreeding are covered elsewhere, with the interaction between selection and drift examined in Chapter 26, and the generation and selection among pure lines examined more extensively in Volume 3. We conclude by examining the evolution of the selfing rate. We caution the reader that while most of the topics here are not conceptually challenging, the bookkeeping can be demanding.

BASIC ISSUES IN SELECTION RESPONSE UNDER INBREEDING

Animal and plant breeders generally treat inbreeding very differently and inbreed to very different levels. For animal breeders (and others dealing with species that mainly outcross, such as many trees), inbreeding is an undesirable complication that is a by-product of selection tending to promote related individuals. This inflates the level of inbreeding beyond that expected under drift, thus increasing any inbreeding depression (LW Chapter 10), which can offset gains from selection. In addition to any effect on the artificially selected trait, inbreeding depression also lowers overall fitness, with more inbred individuals displaying poorer performance, especially in survival and reproductive traits. Because levels of inbreeding tend to be modest in their breeding schemes, animal breeders generally approximate the selection response under inbreeding by using the breeder's equation as adjusted for inbreeding depression (Uimari and Kennedy 1990; de Boer and van Arendonk 1992; Shaw et al. 1998).

Conversely, plant breeders (and others dealing with species that mainly self) often *exploit* inbreeding through selection schemes whose endproducts are pure lines. As a result, selected individuals can be highly inbred, and a more proper accounting of their covariances requires the machinery of Chapter 11. A rough rule of thumb (examined in greater detail at the end of the chapter) is that outcrossing species can display significant inbreeding depression (so that even a modest change in the level of inbreeding, f, can have substantial consequences for the wellbeing of the organism), while species with high levels of natural selfing often show little inbreeding depression (LW Chapter 10). As a consequence, we deal with two settings in this chapter. The first (which derives largely from animal breeding) involves the impact of small to modest amounts of inbreeding, which mainly occurs through inbreeding depression. The second (which derives largely from plant breeding) concerns selection on lines undergoing selfing or some other regular system of inbreeding.

Accounting for Inbreeding Depression

Even in the absence of selection, changes in the population level of inbreeding, f, can induce changes in the mean due to inbreeding depression (LW Chapter 10). Let ΔI denote the change in mean from inbreeding depression (the difference between the mean of a randomly mated population versus its value under the current level of inbreeding). If dominance is the only nonadditive genetic effect, the change from inbreeding in generation t, $\Delta I_t = Bf_t$, is a linear function of the inbreeding coefficient. The parameter B is the difference in trait mean between a completely inbred ($f = 1$) and fully outbred ($f = 0$) population, and it can be estimated by regressing the trait mean on f under the assumption that allele frequencies have not dramatically changed during the inbreeding process, namely, that lines are not lost under inbreeding (LW Chapter 10). If higher-order epistasis is present, then $\Delta I = Bf + Cf^2 + \cdots$, where the order of polynomial in f depends on the type of dominant epistatic interactions present, e.g., order two for $D \times D$ (LW Chapter 10).

The genetic underpinnings of the inbreeding parameter B are as follows: if n diallelic loci underlie the trait, the ith of which has genotypic values of $0 : a_i + d_i : 2a_i$ with allele frequency p_i, then LW Equation 10.3 yields

$$B = -2\sum_{i=1}^{n} d_i p_i (1 - p_i) \tag{23.1a}$$

demonstrating that B changes as allele frequencies change. Inbreeding depression occurs when the values of d_i tend to be positive (**directional dominance**), implying that an average heterozygote has a genotypic value closer to that of the larger-valued homozygote. More generally (Table 11.1), with n_k alleles at locus k,

$$B = \sum_{k=1}^{n} \sum_{i=1}^{n_k} \delta_{kii} \, p_{ki} \tag{23.1b}$$

where δ_{kii} (which is a function of the base-population allele frequencies; LW Chapter 4) is the dominance deviation for the homozygote of allele i at locus k. Equation 23.1b follows because the expected value of the dominance deviations, $E[\delta_{ij}]$, is zero under random mating (LW Chapter 4), but nonzero under inbreeding. With probability f, the two alleles at a locus are IBD, with p_i being the frequency of A_iA_i homozygotes in such cases. Hence, the expected value of $E[\delta_{ij}] = (1-f)E[\delta_{ij} \,|\, \text{not inbred}] + fE[\delta_{ii}] = fE[\delta_{ii}]$. Summing over all alleles recovers Equation 23.1b. A bit of algebra (using results from LW Table 4.1) shows that Equation 23.1b reduces to Equation 23.1a for diallelic loci.

To distinguish between the change due to inbreeding depression and the response due to selection, we decompose the total change in the population mean after t generations as

$$\Delta_\mu(t) = \mu_t - \mu_0 = R(t) + \Delta I_t \tag{23.1c}$$

where typically, $\Delta I_t < 0$. When computing the response to selection, we ignore the change from inbreeding depression, so

$$R(t) = \Delta_\mu(t) - \Delta I_t \tag{23.1d}$$

If the selected population is subsequently randomly mated, the inbreeding depression term disappears, revealing the true genetic response, $R(t)$.

Equations 23.1a and 23.1b show that the importance of our standard assumption of very little allele-frequency change (i.e., the infinitesimal model) is magnified when predicting short-term selection response under inbreeding. In addition to concerns about changes in the genetic variances as allele frequencies change, when inbreeding is present, we are additionally concerned with changes in the composite inbreeding depression parameter, B, which is also a function of the allele frequencies in the base population. If these frequencies significantly change under selection, the impact from inbreeding depression will become increasingly unpredictable, reflecting (in part) the unpredictability in the change in B. Note from Equation 23.1a that as one drives alleles toward fixation, B decreases in magnitude, as $p(1-p)$, which is maximized at $p = 1/2$, approaches zero.

Response Under Small Amounts of Inbreeding

When the amount of inbreeding is small enough that changes in the covariances between relatives are negligible relative to their random-mating counterparts, its main effect is through inbreeding depression. Consider a population of modest size undergoing random mating, where the amount of inbreeding generated by genetic drift at generation t is $f_t \simeq t/(2N_e)$, provided $t \ll N_e$ and $f_0 = 0$. If no epistasis is present, then

$$\Delta I_t = -Bf_t \simeq -\frac{Bt}{2N_e} \tag{23.2a}$$

An important, but subtle, correction is that the act of selection on a heritable trait further increases the level of inbreeding relative to a finite population of the same size (due to a tendency to select related individuals; Chapters 3 and 26). Hence, the appropriate value of N_e for Equation 23.2a starts with the pure-drift value, given the sample size and sex ratio (Chapter 3), which serves as the base value of N_e for the corrections given by Equations 26.6–26.8, which account for this further reduction in N_e according to the strength of selection and heritability of the trait. The expected selection response with a small amount of inbreeding is the response from the breeder's equation (13.6b) plus a correction for any inbreeding depression,

$$\Delta_\mu(t) = R(t) + \Delta I_t \simeq t\,\bar{\imath}\,h^2\,\sigma_z - \frac{B\,t}{2N_e} = t\,\bar{\imath}\left(h^2\sigma_z - \frac{B}{2N_e\,\bar{\imath}}\right) \tag{23.2b}$$

(Nordskog and Hardiman 1980; Hill 1986). This not an unreasonable approximation for $f = t/(2N_e) < 0.1$. For $f > 0.1$, inbreeding significantly alters the genetic variances from their base population values, and this must be taken into account (Chapter 26).

Using these corrections, one can then apply the standard breeder's equation or its extensions (Chapter 13), adjusted for inbreeding depression, to predict the net selection response (Equation 23.2b). De Boer and van Arendonk (1992) found that this approach (over a few generations) is accurate even at intermediate levels of inbreeding ($f \leq 0.35$), while simulations by Shaw et al. (1998) suggested that this approach may generally be fairly accurate, provided the magnitude of σ_{ADI} is small. When this covariance is negative, alleles with positive average effects tend to have negative dominant deviations as homozoygotes, which accentuates the effects of inbreeding depression.

In a BLUP-selection framework (choosing individuals based on their mixed-model predicted breeding values; Chapters 13, 19, and 20) with inbreeding, it is often sufficient to simply include inbreeding depression as a cofactor, for example, the phenotypic value, y_i, of individual i can be decomposed as

$$y_i = \mu + A_i + Bf_i + e_i \tag{23.2c}$$

Otherwise, it suffices to use a standard additive model (Uimari and Kennedy 1990; de Boer and van Arendonk 1992). Recall that the amount of inbreeding for individual i is calculated by $f_i = A_{ii} - 1$, where A_{ii} is the associated diagonal element for that individual in the relationship matrix, $\mathbf{A}$ (Chapter 19). As detailed in Chapter 19, the use of mixed models accounts (through $\mathbf{A}$) for both gametic-phase disequilibrium (among unlinked loci) and the reduction in values of N_e from choosing relatives.

All of these approaches assume that B is relatively constant, but as mentioned, this parameter can change under selection. Indeed, as discussed in Chapter 28, Kelly (1999c) proposed a test for the presence of rare recessives by contrasting the relative change in two population parameters following selection: the trait mean and the value of B for that trait. Thus, the standard caveat for the breeder's equation, that it applies to a single generation of selection, also applies here, with the accuracy of this approximation breaking down as allele frequencies change, due to changes in either σ^2_A or B.

A key point in species showing inbreeding depression is that even if the target of artificial selection is not significantly affected ($B \simeq 0$), the overall fitness (performance) of the individual generally declines as f increases (LW Chapter 10). As a result, significant attention in the animal breeding literature has focused on maximizing selection response under either constrained levels of inbreeding or under the minimization of inbreeding, especially for BLUP selection (Toro and Pérez-Enciso 1990; Quinton et al. 1992; Grundy et al. 1994, 1998, 2000; Meuwissen and Woolliams 1994; Villanueva et al. 1994; Brisbane and Gibson 1995; Luo et al. 1995; Quniton and Smith 1995; Meuwissen 1997; Meuwissen and Sonesson 1998; Meszaros et al. 1999; Sonesson and Meuwissen 2000; Sonesson et al. 2000). We return to this topic in Volume 3.

Using Ancestral Regressions to Predict Response

The simplicity of Equation 23.2b follows from the assumption that a small amount of inbreeding does not greatly change genetic variances. With larger amounts of inbreeding, variances and covariances can change in each generation, which can depend on additional terms such as σ^2_{DI} and σ_{ADI}. Fortunately, the expected covariances between relatives under regular systems of inbreeding in the absence of selection are predictable (Chapter 11).

Motivated by this predictability, we make the key approximation throughout much of this chapter that selection does not substantially alter these covariances from their expected values in the absence of selection. *Provided* this assumption holds and all regressions are linear and homoscedastic, the method of ancestral regressions (Chapter 15) offers a powerful approach for predicting short-term selection response when inbreeding occurs.

Recall that under ancestral regression, the cumulative response can be expressed as a series of regression coefficients (a covariance divided by a variance) of the contribution to the current total response from selection in a previous generation, t, yielding an expected response after T generations of selection and inbreeding of

$$R(T) = \sum_{t=0}^{T-1} S_t \, \frac{\sigma_G(T,t)}{\sigma^2(z_t)} = \sum_{t=0}^{T-1} \bar{\imath}_t \, \frac{\sigma_G(T,t)}{\sigma(z_t)} \tag{23.3}$$

Here $\sigma_G(T,t)$ is the covariance (in genotypic values) between a relative in generation t and one in the current generation ($T > t$), while $\sigma^2(z_t)$ is the phenotypic variance in generation t. Note that the coefficients in Equation 23.3 are simply the regressions of z_T on z_t, which has a slope of $\sigma_z(T,t)/\sigma^2(z_t) = \sigma_G(T,t)/\sigma^2(z_t)$ in the absence of environmental correlations between generations T and t. The tth term in the sum in Equation 23.3 corresponds to the response from selection in generation t that remains in generaton $T > t$. As we will see later in this chapter, under complicated systems of inbreeding, a number of relatives with different degrees of inbreeding must be simultaneously followed, leading to additional indices in the covariance, such as $\sigma_G(T,\tau,t)$ or $\sigma_G(T,\tau,t,k)$, where the additional indices τ and k indicate the generation of founding relatives (see Figure 23.2). Equation 23.3 is a generalization of the breeder's equation to settings where the genetic variances change in predicable ways under the mating system, making the assumption that these inbreeding-induced changes are much more significant than any changes in these parameters from selection.

Equation 23.3 is based on the infinitesimal model, it assumes that selection-induced changes in allele frequencies are negligible. While *genotypic* frequencies change due to inbreeding (with homozygotes increasing and heterozygotes decreasing), we assume that there is no significant change in *allele* frequencies. Hence, if p_i is the frequency of allele $\mathbf{A}_i$ in the base population, the frequency of lines eventually fixed for the $\mathbf{A}_i\mathbf{A}_i$ genotype is assumed to essentially remain as p_i, despite selection. Formally, it is changed to $p_i + \epsilon_i$, where ϵ_i is a very small amount. With a very large number of loci, all of these very small values of ϵ_i can add up to a considerable change in the mean, while still resulting in little change in genetic variances (Chapter 24). While allele-frequency change is negligible under the infinitesimal assumption, gametic-phase disequilibrium can be considerable (the Bulmer effect; Chapters 16 and 24), which can significantly alter the genetic variance without any allele-frequency change. For now, we make the approximation of ignoring any such disequilibrium-based change, a point addressed later in this chapter.

Because the covariance function also gives the genetic variance in generation t, as $\sigma^2_G(t) = \sigma_G(t,t)$, with the covariance function for our particular system of inbreeding in hand (a point we address shortly), we can immediately write the selection response as

$$R(T) = \sum_{t=0}^{T-1} S_t \, \frac{\sigma_G(T,t)}{\sigma_G(t,t) + \sigma^2_e} = \sum_{t=0}^{T-1} \bar{\imath}_t \, \frac{\sigma_G(T,t)}{\sqrt{\sigma_G(t,t) + \sigma^2_e}} \tag{23.4}$$

For example, the response after two generations of inbreeding and selection is

$$R(2) = \bar{\imath}_0 \, \frac{\sigma_G(2,0)}{\sigma(z_0)} + \bar{\imath}_1 \, \frac{\sigma_G(2,1)}{\sigma(z_1)} \tag{23.5a}$$

The first term represents the response from selection in generation zero that carries over to the second generation, while the second term is the response to selection from generation one. If we stop selection after two generations but continue to inbreed the population to complete homozygosity, the permanent response (after correcting for any inbreeding depression) is

$$R_\infty(2) = \bar{\imath}_0 \, \frac{\sigma_G(\infty, 0)}{\sigma(z_0)} + \bar{\imath}_1 \, \frac{\sigma_G(\infty, 1)}{\sigma(z_1)} \tag{23.5b}$$

Inspection of Equation 23.5a and 23.5b highlights a key feature of response with inbreeding. In most cases, these covariances change, so that it is generally the case that $\sigma_G(i,t) \neq \sigma_G(j,t)$ for $i \neq j$. The relative contribution to response from selection in any particular generation t can thus change over time, meaning that there is both a transient and permanent component to response (Chapter 15).

The Covariance Between Inbred Relatives

To apply ancestral regressions, we must obtain the covariance between relatives under the particular inbreeding scheme of interest. Such covariances were discussed in detail in Chapter 11, and here we remind the reader of a few key concepts, which are summarized in Figure 23.1. Equation 11.13 gives the genetic covariance between individuals x and y under general inbreeding, but it assumes that there are no linkage effects or epistasis, as

$$\begin{aligned} \sigma_G(x,y) = {} & 2\Theta_{xy}\sigma^2_A + \Delta_{xy,7}\sigma^2_D + \Delta_{xy,1}\sigma^2_{DI} \\ & + (2\Delta_{xy,1} + \Delta_{xy,3} + \Delta_{xy,5})\sigma_{ADI} + (\Delta_{xy,2} - f_x f_y)\iota^* \end{aligned} \tag{23.6a}$$

where

$$2\Theta_{xy} = 2\Delta_{xy,1} + \Delta_{xy,3} + \Delta_{xy,5} + \Delta_{xy,7} + \frac{1}{2}\Delta_{xy,8} \tag{23.6b}$$

To aid in reading, we suppress the xy subscript on Δ for those cases where the two relatives being considered are obvious.

The nine possible Δ_i coefficients of relatedness between two (diploid) individuals are defined in Figure 11.5 (and summarized in Figure 23.1), while the composite genetic parameters (the familiar additive and dominance variances, σ^2_A and σ^2_D, and the less-familiar quadratic components, σ^2_{DI} and ι^*, and the covariance, σ_{ADI}) are defined from the standpoint of the noninbred base population (Table 11.1). While σ^2_{DI} and ι^* are nonnegative (by construction), σ_{ADI} is a covariance and hence can be either positive or negative.

The nature of the identity-by-descent (IBD) measures, Δ_i, provides some insight into which components contribute to the transient, as opposed to the permanent, component of response. If we inbreed to complete homozygosity, then both alleles in an individual y are identical by descent (indicated in Figure 23.1 by a horizontal line between its two alleles). Only three of the Δ_i measures in Figure 23.1 correspond to this condition (Δ_1, Δ_2, and Δ_5). Note that under IBD state 2, although x and y are both inbred (both have horizontal lines), they are also *unrelated* (there are no vertical lines between them), and hence this condition generally does not enter into discussions of selection response. When either Δ_1 or Δ_5 are nonzero, σ^2_{DI} and σ_{ADI} can contribute to the permanent response, while σ^2_D cannot (as it enters only through Δ_7). When $\Delta_1 = \Delta_3 = \Delta_5 = 0$, then σ^2_A and σ^2_D are sufficient to describe the covariance between relatives. As mentioned in Chapter 11, the literature on covariances under general inbreeding can be a bit daunting and requires care when reading, as there is no consistent notation for these additional genetic components (Table 11.2).

Limitations

The major limitation with the ancestral regression approach is the assumption that selection does not significantly alter the covariances between relatives over what is expected under the system of inbreeding (in the absence of selection). Clearly, if there are favorable major alleles, selection will favor individuals carrying them, thus further increasing the amount of inbreeding in the population. As a result, this general approach is best thought of as a weak selection approximation, that is, it assumes that selection is weak on any underlying *locus*,

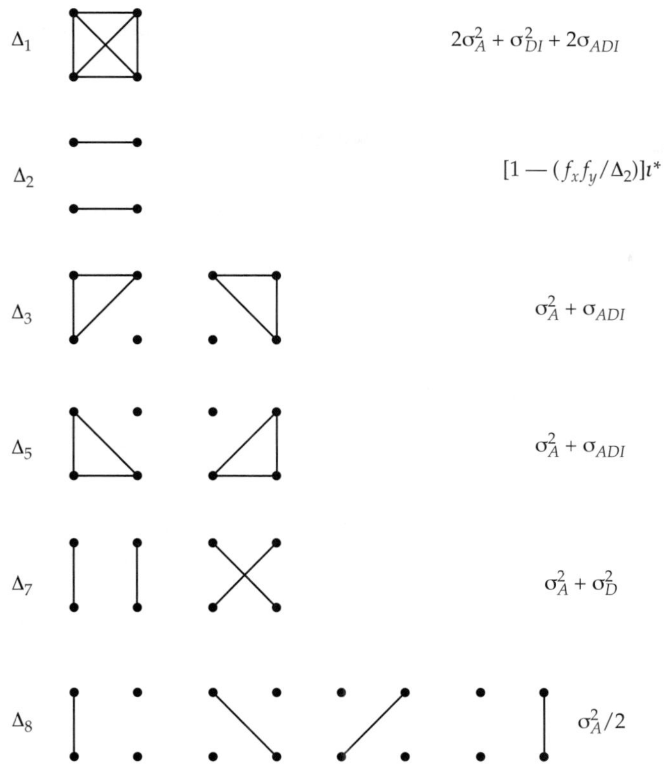

Figure 23.1 The Δ_i coefficients of relatedness and their impact on the genetic covariance between relatives. Following Figure 11.5, the upper two dots correspond to the two alleles in (diploid) relative x, and the bottom two to those in relative y. A horizontal line indicates inbreeding in that relative (the two alleles are identical by descent, IBD), while a line connecting alleles from different relatives indicates that these alleles are IBD. See Figure 11.5 and Chapter 11 for details.

although selection on the *trait* itself may still be strong. Even in the absence of major alleles, the effect of selection is to generally make individuals more inbred than expected by the particular system of inbreeding (Chapters 3 and 26). In such cases, the covariances between relatives are also affected. The other significant caveat is that, even under the infinitesimal-model, selection-induced disequilibrium between alleles at different loci can significantly alter the variances (the Bulmer effect, described in Chapter 16). By reducing the frequency of heterozygotes, inbreeding reduces the opportunities for recombination to remove linkage disequilibrium, which magnifies the Bulmer effect over its role (which is significant) under random mating. We return to this concern later in the chapter.

FAMILY SELECTION WITH INBREEDING AND RANDOM MATING

As detailed in Chapter 21, the motivation for family-based selection is to use family means to provide better estimates of the breeding values of the individuals that will be chosen to form the next cycle of selection. This section extends these results to inbreeding by allowing the sibs and/or their parents to be inbred. Here, we assume that relatives chosen on the basis of the best families are *outcrossed* to form the next generation, with the selection response when chosen relatives are *selfed* discussed later in the chapter. In the terminology of Chapter 21, the selection unit is the mean of the measured sibs, while the recombination unit is either a sib (measured or unmeasured) or the parent of a measured sib. Specifically, let $x_1, \cdots, x_n$ denote the n measured sibs in a family, with z_i denoting the value for sib i. Families are chosen based on their mean values, $\overline{z}$, with relatives, $\mathcal{R}$, of the chosen families (either one of the measured sibs, an unmeasured sib, or one of the parents of the sibs) outcrossed to form the next generation, and with y denoting an offspring from $\mathcal{R}$. The expected selection response becomes

$$R = \frac{\sigma(\overline{z}, y)}{\sigma^2(\overline{z})}\, \overline{\imath} \tag{23.7}$$

where the numerator is the selection-unit offspring covariance and the denominator the selection-unit variance (Equation 21.3b). When the offspring, parents, or both, are inbred, the variances of these expressions differ from their random-mating counterparts (given in Chapter 21). Using inbred sibs and/or sibs with inbred parents can increase the selection response (R) by increasing the selection unit–offspring covariance. Conversely, by increasing the among-family variance, the use of inbred sibs also increases the selection-unit variance, $\sigma^2(\overline{z})$, which can reduce the selection response. Hence, using inbred sibs or inbred parents can, in some cases, increase the selection response, while in other situations the response is less than with family selection using randomly mated sibs from outbred parents.

Two bookkeeping issues commonly arising in family-based selection account for some of the variety of selection-response equations found in the literature. First, under strict family selection, a measured sib is used as a parent for the next generation. In this case, the covariance between the family mean, $\overline{z}$, and an offspring, y, starting the next cycle of selection has two components. If z_1 denotes the value of a measured sib (x_1) used as a parent of y (in the notation of Chapter 21, $x_1 = \mathcal{R}$; see Figure 21.1), then with n measured sibs in a family

$$\sigma(\,\overline{z}, y) = \frac{1}{n}\sum_{i=1}^{n} \sigma(z_i, y) = \left(\frac{1}{n}\right)\sigma(z_1, y) + \left(1 - \frac{1}{n}\right)\sigma(z_2, y) \tag{23.8a}$$

The first covariance, $\sigma(z_1, y)$, is that between a parent and its offspring, while the second, $\sigma(z_2, y)$, is that between an individual, x_2, and the offspring, y, of its sib, x_1. Alternatively, when sib selection occurs (such as through the use of remnant seed), the sib used in the recombination unit (i.e., as a parent of the next generation) is *not* one of the sibs measured in the selection unit, and $\sigma(\overline{z}, y) = \sigma(z_2, y)$. To simplify our results, we assume only sib or parental selection (progeny testing, where parents are chosen based on the performance of their offspring). For a moderate to large number of measured sibs (n), the difference between sib selection and family selection is expected to be very small.

The second issue relates to the variance of the selection unit, $\sigma^2(\,\overline{z}\,)$. From Chapter 21, the variance in observed family means is the among-group variance, σ_b^2, plus the error in estimating their true mean, μ, from $\overline{z}$, which is σ_w^2/n. Recalling (LW Chapter 18) that the variance among groups equals the covariance within groups, then $\sigma_b^2 = \sigma_z(\text{sibs})$, the covariance between sibs, which is the sum of their genetic covariance, $\sigma_G(\text{sibs})$, plus the variance of any common-family environmental effects, $\sigma^2_{E_c}$ (such as maternal effects when sibs share a common mother). Likewise, the within-family variance can be decomposed into genetic and environmental components, $\sigma_w^2 = \sigma^2_{G_w} + \sigma^2_{E_s}$ (Equation 21.8b), where $\sigma^2_{G_w} = \sigma^2_G - \sigma_G(\text{sibs})$. Hence,

$$\begin{aligned}\sigma^2(\,\overline{z}\,) &= \sigma_b^2 + \sigma_w^2/n \\ &= \sigma_G(\text{sibs}) + \sigma^2_{E_c} + [\sigma^2_{G_w} + \sigma^2_{E_s}]/n \\ &= \sigma_G(\text{sibs}) + \sigma_e^2\end{aligned} \tag{23.8b}$$

Because a goal of this chapter is comparing the impact of inbreeding on $\sigma_G(\text{sibs})$, we combine the common-family variance ($\sigma^2_{E_c}$; which can be considerable if the sibs share a common mother for a trait with strong maternal effects) and the error in estimating the true family mean into a single error term,

$$\sigma_e^2 = \sigma^2_{E_c} + [\sigma^2_{G_w} + \sigma^2_{E_s}]/n \tag{23.8c}$$

This is mainly for ease of bookkeeping, as Equations 21.41–21.43 illustrated some of the complex expressions for σ_e^2 under different family replication designs. A caveat with this notational brevity is that different designs can have rather different σ_e^2 values. A paternal half-sib design avoids any shared maternal effects, which (for some traits) can be considerable. Likewise, because of the reproductive biology, some types of families may result in significantly more offspring, and hence a reduced sampling error in Equation 23.8c. Finally, different investigators using the same material can use very different structures for the error variances, depending on the family replication scheme chosen (Equations 21.41–21.43).

Family Selection Using Inbred Parents

One scheme for increasing the response to family selection is to cross inbred, but unrelated, parents, and then score the resulting half- or full-sib progeny as the family unit. This has two effects on selection response, one positive (increasing the covariance between relatives), and one negative (increasing the variance of the selection unit). The genetic covariance among half-sibs where the common parent is inbred (to an amount f) is

$$\sigma_G(HS) = \left(\frac{1+f}{4}\right)\sigma_A^2 + \left(\frac{1+f}{4}\right)^2\sigma_{AA}^2 + \cdots \left(\frac{1+f}{4}\right)^k\sigma_{A^k}^2 \tag{23.9a}$$

For full sibs, if $\overline{f} = (f_1 + f_2)/2$ is the average inbreeding coefficient for the parents, then

$$\sigma_G(FS) = \left(\frac{1+\overline{f}}{2}\right)\sigma_A^2 + \left(\frac{(1+f_1)(1+f_2)}{4}\right)\sigma_D^2 + \left(\frac{1+\overline{f}}{2}\right)^2\sigma_{AA}^2 + \cdots \tag{23.9b}$$

This inflation of the between-sib covariances relative to random mating also increases the variance of the selection unit (Equation 23.8b). For the reader wondering why the inbreeding variance components ($\sigma_{ADI}, \sigma_{DI}^2$, etc.) do not appear in Equation 23.9, it is because the parents of the sibs, while being inbred, are *unrelated* (their coefficient of coancestry, Θ, is zero). Hence, alleles within the resulting sibs are not identical by descent (because their parents are unrelated). This implies that the Δ_1 to Δ_5 coefficients (those associated with at least one relative being inbred) for such sibs are zero, and hence the contributions from their associated variance components (Figure 23.1 and Equation 23.6) are also zero. This also applies to the selection unit–offspring covariances (Equation 23.10).

Turning to the selection unit–offspring covariances, we will ignore the effects of additive epistasis, as this contributes to the transient, rather than permanent, component of response (because random mating among the parents, $\mathcal{R}$, in the recombination unit eventually decays linkage disequilibrium). The resulting covariances between a measured individual, x, and the offspring y (through a *single* parent, $\mathcal{R}$, of y) are

$$\sigma_G(x, y \,|\, \mathcal{R} = P \,\text{of}\, x\,) = \left(\frac{1+f}{4}\right)\sigma_A^2 \tag{23.10a}$$

$$\sigma_G(x, y \,|\, \mathcal{R} = HS \,\text{of}\, x\,) = \left(\frac{1+f}{8}\right)\sigma_A^2 \tag{23.10b}$$

$$\sigma_G(x, y \,|\, \mathcal{R} = FS \,\text{of}\, x\,) = \left(\frac{1+f}{4}\right)\sigma_A^2 \tag{23.10c}$$

with P, HS, and FS implying that the parent, $\mathcal{R}$, of y is related to the measured sibs as either a parent, a half-sib, or a full-sib (respectively). When the parents are inbred ($f > 0$), all of these covariances exceed their random-mating counterparts ($f = 0$). As mentioned, this increase is offset to some degree by the corresponding increase in the selection-unit variance that also occurs with inbreeding, as $\sigma^2(\overline{z}\,|\,f > 0) > \sigma^2(\overline{z}\,|\,f = 0)$.

Substitution of these results into Equation 21.1 yields the response to a single cycle of selection under various schemes, which are summarized in Table 23.1. As a comparison of Table 23.1 with its random-mating counterpart (Table 21.5) shows, for half-sibs, that the selection response when using inbred parents ($f > 0$) is greater than when using outbred

Table 23.1 The response to family selection when both parents are inbred (to a level of f). Depending on the trait, the common-family variance can be considerably less for paternal half sibs than for either maternal half-sibs or full-sibs, so we index our general expression for σ_e^2 (Equation 23.8c) to remind the reader of this. Half-sibs versus full-sibs refer to the family unit being measured, while the parents for the next generation are either remnant seed (sib selection) or the parent of the selection unit itself (progeny testing). For comparison purposes, selection on both parents is assumed. Response is halved if only a single parent has been chosen by family selection. The effects of epistatis are ignored. Additive × additive epistasis inflates the immediate response, but its contribution decays with recombination (as offspring in the next generation are formed by random mating). By inflating the selection-unit variance over the values given here, the presence of epistasis reduces the permanent response.

Selection Scheme	$R/(\sigma_A^2\,\bar{\imath})$
Half-sibs, remnant seed	$\dfrac{(1+f)/4}{\sigma(\,\overline{z}_{HS,f}\,)} = \dfrac{(1/2)\sqrt{1+f}}{\sqrt{\sigma_A^2 + [4\sigma_e^2(HS)/(1+f)]}}$
Half-sibs, parental	$\dfrac{(1+f)/2}{\sigma(\,\overline{z}_{HS,f}\,)} = \dfrac{\sqrt{1+f}}{\sqrt{\sigma_A^2 + [4\sigma_e^2(HS)/(1+f)]}}$
Full-sibs, remnant seed	$\dfrac{(1+f)/2}{\sigma(\,\overline{z}_{FS,f}\,)} = \dfrac{\sqrt{(1+f)/2}}{\sqrt{\sigma_A^2 + [(1+f)\sigma_D^2/2] + [2\sigma_e^2(FS)/(1+f)]}}$

parents ($f = 0$). This is also true for full-sibs when σ_D^2 is small. However, random mating can yield a larger response if σ_D^2 is sufficiently large.

Example 23.1. To get some feel for the impact of inbreeding, consider the selection response for different types of family selection as f increases. Assume we have a trait with $\sigma_A^2 = 50$ and $\sigma_D^2 = 25$. To simply the discussion, we assume the same value of the residual variance (Equation 23.8c) over all designs, namely, $\sigma_e^2 = 50$. Using the expression summarized in Table 23.1, the response to half-sib, full-sib, and parental selection for various values of parent inbreeding (f) are given below as a fraction of the response for $f = 0$ (random mating). The values in the $R/\bar{\imath}$ column are the expected responses under $f = 0$, while the values in the columns for $f > 0$ represent the ratio of response for that f value relative to the response under random mating. For example, the $R/\bar{\imath}$ ratio for full-sibs with f = 1/4 is 2.774 · 1.181 = 3.276.

		Fraction of random-mating response				
Selection	$R/\bar{\imath}$	f = 1/8	f=1/4	f=1/2	f = 3/4	f = 1
Half-sib						
Reminant seed	1.581	1.111	1.220	1.430	1.632	1.826
Parental	3.162	1.111	1.220	1.430	1.632	1.826
Full-sib	2.774	1.093	1.181	1.342	1.485	1.612

Finally, our assumption that σ_e^2 is a constant (over both f and the designs) deserves additional comment. Under the simplest design of n sibs scored in a single plot in a single environment, then from Equation 23.8c,

$$\sigma_e^2 = \sigma_{E_c}^2 + [\sigma_{G_w}^2 + \sigma_{E_s}^2]/n$$

Clearly, in the case of paternal half-sibs, $\sigma_{E_c}^2$ is expected to be smaller (and can be much smaller) relative to designs with a common mother (and hence, there is the prospect of a much larger common-family effect). Further, different types of families might logically be expected to vary in n, for example, one can usually generate more half-sibs than full sibs. For designs with more structure in the replication scheme, we can write $\sigma_e^2 = \sigma_{E_c}^2 + \sigma^2(\epsilon)$, where Equations 21.41c, 21.42a, and 21.42b provide examples of the structure of $\sigma^2(\epsilon)$. Finally, $\sigma_{G_w}^2$ also changes with inbreeding. Our assumption of a constant value of σ_e^2 is an approximation, assuming that n is relatively large, meaning that that changes in $\sigma_{G_w}^2$ with f do not significantly impact σ_e^2.

Progeny Testing Using Inbred Offspring

Building on an earlier suggestion by Mostagee (1971), Toro (1993) proposed that sire progeny testing be performed using inbred offspring (by crossing the sire to full-sib or half-sib sisters to generate the family), with the chosen superior sires then outcrossed. In animal breeding, such a scheme can be used in species for which artificial insemination and long-term sperm storage are feasible. This suggestion takes advantage of improved accuracy for testing using inbred sibs while still having an outcrossed population (and hence no inbreeding depression in the next generation). It is important to note that the inbred sibs upon which parental selection will be made could themselves experience inbreeding depression, so the assumption is that all such tested sibs have the same f value (or else they are all corrected to adjust for this possibility).

To quantify the advantage of testing inbred progeny, consider a sire crossed to a full-sib sister. Let $\mathcal{R}$ denote a sire and x denote one of the resulting offspring (the mean values of which are used to choose among sires). Assuming that $\mathcal{R}$ is chosen (selected), let y denote an offspring of $\mathcal{R}$ when now crossed (likely using stored sperm) to an unrelated dam (these sex roles can easily be reversed in designs involving plants). The selection unit–offspring covariance is that of an inbred sib (x) from a sire ($\mathcal{R}$) and of an outcrossed sib (y) from the same sire. The probability that the sire allele in the inbred and outcrossed offspring are identical by descent (IBD) is 1/2. Because the sire and dam are related, an IBD copy of this same allele can also be transmitted through the dam to the inbred offspring, generating a Δ_3 IBD state (both copies of an allele in x are IBD, and these are also IBD with the sire allele transmitted to its offspring, y). If the sire and dam are full-sibs, the probability of the dam transmitting this allele is 1/4, while if the sire and dam are half-sibs, this probability is 1/8. Hence, when the sire and dams are full-sibs (denoted as **SDFS**), then

$$\Delta_3 = (1/2)(1/4) = 1/8 \quad \text{and} \quad \Delta_8 = (1/2)(1-1/4) = 3/8 \tag{23.11a}$$

State Δ_8 corresponds to a single allele in x and y being IBD. For a half-sib sire and dam family (**SDHS**)

$$\Delta_3 = (1/2)(1/8) = 1/16 \quad \text{and} \quad \Delta_8 = (1/2)(1-1/8) = 7/16 \tag{23.11b}$$

Note that under either scheme $\Delta_1 = \Delta_2 = \Delta_5 = \Delta_7 = 0$. Substituting into Equations 23.6a and 23.6b gives the resulting covariance between inbred (I) and outcrossed (O) sibs for a full-sib sire and dam design as

$$\sigma_G(I, O\,|\,SDFS) = (5/16)\,\sigma_A^2 + (1/8)\sigma_{ADI} \tag{23.11c}$$

while for a half-sib sire and dam design

$$\sigma_G(I, O\,|\,SDHS) = (9/32)\,\sigma_A^2 + (1/16)\sigma_{ADI} \tag{23.11d}$$

By comparison, if the dam and sire are unrelated, the above two covariances are just that between outbred half-sibs, namely $(1/4)\sigma_A^2$. Thus, in the absence of dominance (and hence $\sigma_{ADI} = 0$), the sib covariance under SDFS (Equation 23.11c) is 125% that of outbred half-sibs ([5/16]/[1/4] = 5/4), and similarly, the sib covariance for SDHS (Equation 23.11c) is 112%

([9/32]/[1/4] = 9/8) of outbred half-sibs. When $\sigma_{ADI} < 0$, the possibility exists that these covariances are smaller than when the tested sibs are not inbred.

S_1, S_2, and $S_{i,j}$ Family Selection

Another scheme for family selection using inbreeding is **S_1 family selection**, wherein an (outbred) individual is selfed, and the family mean of the selfed progeny is used for selection decisions (i.e., individuals are chosen based on the mean trait value of their S_1 family). This scheme takes two generations. In the first generation, the selfed seed must be grown for scoring families. An additional generation is then required for remnant S_1 seeds from superior families to be grown and outcrossed to form the start of the next cycle of selection. Note that S_1 *family* selection is different from S_1 *seed* selection (the latter is discussed in Chapter 21). While seed selection also uses remnant S_1 seeds as the recombination unit, the tested family under S_1 seed selection consists of outbred half-sibs, rather than the S_1-sibs used in family selection.

Selection can also be based on S_2 families. Under classical **S_2 family selection**, an outbred individual is selfed to form an S_1, with a single S_1 plant selfed again to form the S_2 family (upon which selection decisions are made, i.e., the selection unit is an S_2 family). Remnant seed from the S_1 is used as the recombination unit, with seed from superior families grown and crossed at random to start the next cycle of selection. There is the potential for ambiguity with S_2 selection, as either: (i) the family to be tested could be (as above), the progeny from a *single* S_1 individual; or (ii) they could be a *collection* of progeny from a *set* of S_1 individuals. Because of this ambiguity, we use a modification of the notation suggested by Wricke and Weber (1986) and consider **$S_{i,j}$ family selection** (Wricke and Weber used $I_{i,j}$). Here j denotes the generations of inbreeding in the tested family and i denotes the generations of inbreeding for the founding individual for that family. Hence, S_1 family selection becomes $S_{0,1}$ selection (the selfed progeny from a noninbred plant) and classical S_2 family selection becomes $S_{1,2}$ (the selfed progeny from a single S_1), while $S_{0,2}$ corresponds to **bulk S_2 family selection**, wherein the tested family are the selfed-progeny from a set of S_1 lines from a single outbred individual (and hence $i = 0$).

Expressions for the response to S_1 selection in the literature (e.g., Choo and Kannenberg 1981; Hallauer and Miranda 1981; Bradshaw 1983) are based on the result of Empig et al. (1972). Let p_i and q_i denote the frequencies of alleles $A_{i,1}$ and $A_{i,2}$ at the ith underlying locus for the trait, whose genotypic values are given by $-a_i : d_i : a_i$. Assuming linkage equilibrium, Empig et al. found that the covariance between an individual, x, in the selection unit and the offspring, y, of its selfed sib is given by

$$\sigma(x,y) = \sigma_A^2 + \beta \qquad \text{where} \qquad \beta = \sum_{i=1}^{n} 2p_i q_i (p_i - 1/2) d_i [a_i + (q_i - p_i) d_i] \tag{23.12}$$

Similar expressions exist for the response to $S_{1,2}$ selection (Hallauer and Miranda 1981) and for $S_{0,j}$ selection (Wricke and Weber 1986).

However, it is fairly easily to obtain a variance component-based expression (and hence the ability of estimate the required quantities) for response under general $S_{i,j}$ family selection. Because a member of the recombination unit ($\mathcal{R}$) is outcrossed, it passes on only single alleles to each offspring, y. This situation excludes all of the identity states except for Δ_3 (both alleles in x are IBD, and one is passed on to the offspring y through $\mathcal{R}$), Δ_8 (the alleles in x are unrelated and one is passed onto y via $\mathcal{R}$), and Δ_9 (the alleles in x are unrelated to those in y). As a result, Equation 23.6a implies that the selection unit–offspring covariance only depends on σ_A^2 and σ_{ADI}. As shown in Example 23.2 (which can be skipped by the casual reader), the required values of Δ_i can be obtained by some simple bookkeeping as

$$\Delta_3 = f_i + (1 - f_i)\left(\frac{1 - 2^{-(j-i)}}{2}\right) = 1 - \frac{1}{2}\left(\frac{1}{2^i} + \frac{1}{2^j}\right) \tag{23.13a}$$

$$\Delta_8 = (1 - f_i)2^{-(j-i)} = 2^{-j} \tag{23.13b}$$

Table 23.2 Coefficients for Equation 23.14, the selection unit–offspring covariance under $S_{i,j}$ family selection. The column under σ_A^2 gives the coefficient for the additive variance (which is a function of only i), while the σ_{ADI} coefficient is also a function of j and is shown in the remaining columns.

		Coefficent on σ_{ADI} for $j =$					
i	σ_A^2	$i+1$	$i+2$	$i+3$	$i+4$	$i+5$	∞
0	1.00	0.50	0.75	0.88	0.94	0.97	1.00
1	1.50	1.25	1.38	1.44	1.47	1.48	1.50
2	1.75	1.63	1.69	1.72	1.73	1.74	1.75
3	1.88	1.81	1.84	1.86	1.87	1.87	1.88
4	1.94	1.91	1.92	1.93	1.93	1.94	1.94
5	1.97	1.95	1.96	1.96	1.97	1.97	1.97
6	1.98	1.98	1.98	1.98	1.98	1.98	1.98
7	1.99	1.99	1.99	1.99	1.99	1.99	1.99
8	2.00	1.99	2.00	2.00	2.00	2.00	2.00

Substituting these results into Equation 23.6b yields

$$2\Theta_{xy} = \Delta_3 + \Delta_8/2 = 1 - \frac{1}{2^{i+1}} \tag{23.13c}$$

The covariance $\sigma_G(x, y)$ immediately follows from Equations 23.6a and 23.6b, where only Δ_3 and Δ_8 are nonzero. Because both of the parents of y come from superior (i.e., selected) families, we double the covariance to give the total (i.e., accounting for both parents of y) selection unit–offspring covariance under $S_{i,j}$ family selection as

$$\begin{aligned} 2\sigma_G(x, y) &= 4\,\theta_{x,y}\sigma_A^2 + 2\,\Delta_3\sigma_{ADI} \\ &= 2\sigma_A^2\left(1 - \frac{1}{2^{i+1}}\right) + 2\sigma_{ADI}\left(1 - \frac{1}{2^{i+1}} - \frac{1}{2^{j+1}}\right) \end{aligned} \tag{23.14}$$

Numerical values for these coefficients are presented in Table 23.2. Using the results in Table 11.1, a little algebra shows that $\beta = \sigma_{ADI}/2$, which connects Equations 23.12 and 23.14.

Finally, the genetic variance among $S_{i,j}$ families is

$$\begin{aligned} \sigma_G^2(S_{i,j}) &= (2 - 2^i)\sigma_A^2 + 2^{-(2j-i)}\sigma_D^2 + (2 - 2^{-i} - 2^{-j})\sigma_{ADI}^2 \\ &+ \left(1 + 2^{-(2j+1-i)} - 2^{-j} - 2^{-(i+1)}\right)\sigma_{DI}^2 + 2^{-(2j-i)}\left(1 - 2^{-i}\right)\iota^* \end{aligned} \tag{23.15}$$

This expression is derived below in Example 23.7. Substituting Equation 23.14 and 23.15 into Equation 23.7 yields the general expression for a single generation of response to $S_{i,j}$ family selection as

$$R_{S_{i,j}} = \bar{\imath}\,\frac{2\sigma_A^2(1 - 2^{-(i+1)}) + 2\sigma_{ADI}(1 - 2^{-(i+1)} - 2^{-(j+1)})}{\sqrt{\sigma_G^2(S_{i,j}) + \sigma_e^2(S_{i,j})}} \tag{23.16}$$

In particular, the response to S_1 family selection ($i = 0, j = 1$) is

$$R_{S_{0,1}} = \bar{\imath}\,\frac{\sigma_A^2 + (1/2)\sigma_{ADI}}{\sqrt{\sigma_A^2 + (1/4)\sigma_D^2 + (1/2)\sigma_{ADI} + (1/8)\sigma_{DI}^2 + \sigma_e^2(S_{0,1})}} \tag{23.17a}$$

The response to "classic" S_2 family selection ($i = 1, j = 2$) is

$$R_{S_{1,2}} = \bar{\imath}\,\frac{(3/2)\sigma_A^2 + (5/4)\sigma_{ADI}}{\sqrt{(3/2)\sigma_A^2 + (1/8)\sigma_D^2 + (5/4)\sigma_{ADI} + (9/16)\sigma_{DI}^2 + (1/16)\iota^* + \sigma_e^2(S_{1,2})}} \tag{23.17b}$$

while the response to bulk S_2 family selection ($i = 0, j = 2$) is

$$R_{S_{0,2}} = \bar{\imath}\,\frac{\sigma_A^2 + (3/4)\sigma_{ADI}}{\sqrt{\sigma_A^2 + (1/16)\sigma_D^2 + (3/4)\sigma_{ADI} + (9/32)\sigma_{DI}^2 + \sigma_e^2(S_{0,2})}} \tag{23.17c}$$

Table 23.3 Comparison of the different types of family-based selection, under the assumption of no dominance. $R^* = R/[\,\bar{\imath}\,\sigma_A^2\,\sigma(\,\bar{z}\,)\,]$ is the scaled selection response *per cycle* per selected parent (using the contribution to the selection unit–offspring covariance from a *single* parent), g is the number of generations per cycle , and c (1 or 2) is the number of parents under selection. The response per generation is shown in the final column, $c\,R^*/g$.

Type	R^*	g	c	$c\,R^*/g$
S_1	1/2	2	2	1/2
$S_{1,2}$	3/4	3	2	1/2
$S_{0,2}$	1/2	3	2	1/3
Full-sibs	1/4	2	2	1/4
HS, S_1 seed	1/4	2	2	1/4
HS, remnant seed	1/8	2	2	1/8
HS, Parent	1/4	2	1	1/8

Starting with F_1s from a pure-line cross (and hence $\sigma^2_{DI} = \sigma_{ADI} = 0$ and $\iota^* = \sigma^2_D$, as $p = q = 1/2$; see Chapter 11), Equation 23.16 reduces to

$$R_{S_{i,j}} = \bar{\imath}\,\frac{2\sigma_A^2(1-2^{-(i+1)})}{\sqrt{(2-2^i\,)\sigma_A^2 + 2^{-(2j-i-1)}(1-2^{-(i+1)})\sigma_D^2 + \sigma_e^2(S_{i,j})}} \tag{23.18}$$

The simplifications for Equations 23.17a–23.17c in this setting follow in a similar fashion.

How does the use of selfed families compare with other types of among-family selection? Equation 23.16 shows that the response depends on $\sigma_G(\,x,y\,)$ in the numerator and $\sigma^2(\bar{z}) = \sigma^2_G(S_{i,j}) + \sigma^2_e(S_{i,j})$ in the denominator, making formal comparisons between methods a bit tedious. If we assume that the denominators are roughly similar (i.e., $\sigma^2(\bar{z})$ is roughly equal over different schemes), then we can simply compare the numerators. Because different schemes take different number of generations, the scaled response ratio, $R/[\,\bar{\imath}\,\sigma_A^2\,\sigma(\,\bar{z}\,)\,]$, should be expressed in terms of response per generation. We also need to adjust for whether one or both parents have been chosen from superior families. Table 23.3 presents the response per cycle after accounting for all these factors under the assumption of no dominance (i.e., $\sigma_{ADI} = 0$).

Table 23.3 shows that S_1 and $S_{1,2}$ selection are superior to the other approaches that were listed (under the assumption of no dominance and roughly equal family variances among the different approaches). While we have not included comparisons with methods using inbred parents, these are easily obtained by multiplying the scaled response per generation by $1+f$ (see Table 23.1). While $S_{1,2}$ selection yields a larger response per cycle (R^* = 3/4), this is countered by increased cycle time ($g = 3$). Note that S_2 bulk-family selection ($S_{0,2}$) is not as efficient as S_1 or $S_{1,2}$. For other types of $S_{i,j}$ selection, the tradeoff between an increase in additive variance, scaling as $2(1-2^{-(i+1)})\sigma_A^2$, versus the increase in cycle time (g increasing with i) is such that the scaled response per generation, $R/(\sigma_A^2\bar{\imath})$, is under 1/2 for $i > 2$ and hence not as efficient as either S_1 or $S_{1,2}$ selection.

Choo and Kannenberg (1979a, 1979b) compared (via computer simulations allowing for dominance) the relative efficiencies of S_1 family, mass, and modified (Lonnquist) ear-to-row selection (Chapter 21). S_1 selection was found to be overall superior, with the largest advantage occurring at low heritabilities. Similar conclusions were reached by Eberhart (1972), who found that favorable allele frequencies change most rapidly under S_1, but that the loss of genetic variance is also the fastest under this method. Both of these features are likely consequences of the smaller effective population size associated with S_1 selection (which is about 1/3 of a comparable mass or ear-to-row scheme). Indeed, Choo and Kannenberg (1979b) and Bradshaw (1984) observed that S_1 selection had the highest loss of favorable alleles, although this was not seen by Wright (1980), who assumed higher initial allele frequencies. Choo and Kannenberg also noted that linkage can slow S_1 response, as recombination occurs only every other generation (as opposed to every generation under mass and ear-to-row selection).

Consistent with these theoretical predictions, several researchers found that S_1 recurrent selection was better than testcross (half-sib) selection for increasing yield in maize (Duclos and Crane 1968; Burton et al. 1971; Carangal et al. 1971; Geneter 1973; Moll and Smith 1981; Tanner and Smith 1987) and sorghum (Doggett 1972). Likewise, Moll and Smith (1981) reported that S_1 selection for yield in maize resulted in a roughly 50% greater response than full-sib selection. S_1 lines, however, did show an increased loss of genetic variation (Mulamba et al. 1983; Tanner and Smith 1987). A caveat is that S_1 lines can show greater genotype × environment interaction (Lonnquist and Lindsey 1964; Wricke 1976; Jan-orn et al. 1976). Caution is thus in order for declaring the general superior of S_1 or $S_{1,2}$ selection over other family-based approaches.

The results in Table 23.3 rely on two assumptions: no dominance and equal among-family variance. Both among-family genetic differences over the different schemes, as well as G × E and other replication-dependent error terms in $\sigma^2(\overline{z})$, can cause this latter assumption to be incorrect (e.g., Equations 21.42a and 21.42b). One can easily imagine situations where the difference in error variance more than compensates for the difference in the selection unit–offspring covariances. For example, half-sib selection may generate far more family members for testing than an S_1, thus greatly reducing the error variance (by increasing n in Equation 23.8c). Likewise, σ_{ADI} can be negative, thus reducing the expected advantage of S_1 and $S_{1,2}$ selection. Indeed, Jan-orn et al. (1976) estimated that σ_{ADI}/σ_A^2 was around -0.5 for many traits in sorghum, suggesting that σ_{ADI} can be both negative and substantial.

Example 23.2. To compute the probabilities of the IBD states Δ_3, Δ_8, and Δ_9 required to predict the response under S_{ij} family selection, first recall the various relatives involved:

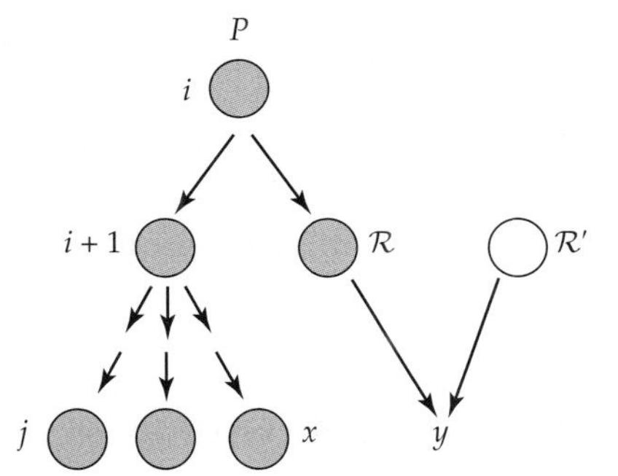

P is the parent that has undergone i generations of selfing, and it generates both a remnant seed, $\mathcal{R}$ (that will be grown and crossed with an unrelated individual, $\mathcal{R}'$, to form the offspring, y), and a collection of S_j families that will be scored. As in Chapter 21, x denotes a measured sib from the family that is used for selection decisions.

Consider a random locus. If the alleles at this locus are IBD in P (which occurs with a probability of f_i) , then the only IBD state between x and y is Δ_3, as both alleles in x are IBD and this allele is also passed onto $\mathcal{R}$, and hence to y. Otherwise (with a probability of $1 - f_i$), a locus in P is not IBD, and we denote P's two alleles as **A** and **a** (these designations are only for the purpose of following IBD alleles, and in fact, the two alleles could be identical in state). There are then three possible genotypes for x, **AA** **Aa**, and **aa**, while y can receive either **A** or **a** from $\mathcal{R}$ (and an unrelated allele from $\mathcal{R}'$). The IBD states for these different combinations are as follows:

	Allele in y from $\mathcal{R}$	
Genotype of x	**A**	**a**
AA	Δ_3	Δ_9
Aa	Δ_8	Δ_8
aa	Δ_9	Δ_3

Hence, Δ_8 equals the probability that a locus in P is not IBD, $(1 - f_i)$, times the probability that the locus is still not IBD by generation j (state **Aa**). Because the probability that a locus is not converted to an IBD state is 1/2 for each generation of selfing,

$$\Delta_8 = (1 - f_i)\, 2^{-(j-i)}$$

recovering Equation 23.13b.

If the two alleles in P are not IBD (probability $1 - f_i$), Δ_3 can still arise as twice the probability that x is the genotype **AA** and y receives allele **A**. The factor of two arises because of symmetry, as the case of **aa** and **a** have equal probability. Turning to the transmission for $\mathcal{R}$, if the focal locus in P is not IBD, then after one generation of selfing, 1/2 of the time $\mathcal{R}$ is IBD, and in half of these cases it will be **AA**. Otherwise, $\mathcal{R}$'s alleles are not IBD, in which case 1/2 are **A**, giving the chance of transmitting **A** to its selfed offspring as (1/2)(1/2) + (1/2)(1/2) = 1/2. Hence, the probability that y gets **A** from $\mathcal{R}$ is 1/2, while the probability x is **AA** (given that $\mathcal{R}$ was not IBD in generation i) is $(1-2^{-(j-i)})(1/2)$, which is the probability of the locus becoming IBD (i.e., not staying as non-IBD) times (1/2) for randomly fixing allele **A**. Hence,

$$\begin{aligned}\Delta_3 &= f_i + 2\,[\,(1 - f_i)(1/2)(1 - 2^{-(j-i)})\,]\,(1/2)\\ &= f_i + (1 - f_i)(1/2)(1 - 2^{-(j-i)})\end{aligned}$$

recovering Equation 23.13a.

Other Inbreeding-based Family-selection Schemes

Other family-selection schemes involving inbreeding have been proposed, such as the **selfed half-sib (SHS)** and **selfed full-sib family (SFS)** methods of Burton and Carver (1993). Here, progeny from either a half- or full-sib family are selfed, and these selfed individuals are then used as the family mean for selection decisions. The advantage of this approach is a large increase in the amount of seed (and hence the ability to more fully replicate a family, reducing the error variance)—in other words, if there are M initial sibs, each of which is selfed to obtain N selfed offspring, there are MN offspring per family. Burton and Carver suggested that this approach can be at least as efficient as S_1 family selection, largely due to the decreased sampling variance in the selection unit, $\sigma^2(\overline{z})$, compared to S_1 families.

Another variant is joint **half-sib, S_1 family selection**, which was proposed by Goulas and Lonnquist (1976) for maize. On prolific (multieared) plants, the lower ear is selfed, while the upper ear is outcrossed. Both the HS (upper ear) and S_1 (lower ear) progenies are jointly evaluated and the best families are chosen, with the remnant HS seed from the best families used as the parents for the next generation. Dhillon (1991b) proposed the **alternate recurrent selection** of S_1 and half-sib families, involving alternate cycles of S_1 selection and either ear-to-row or half-sib selection. The idea here is to take advantage of breeding situations that involve a trial field season and a winter nursery (and hence an extra generation per year) for creating and recombining new families. Under the right settings, this approach can exceed the per-year response of S_1 selection.

Cycles of Inbreeding and Outcrossing

Dickerson (1973) and Dickerson and Lindhé (1977) have suggested that, in some cases, the response to selection under a scheme of random mating alternating every other generation with full-sib mating enhances short-term response. Their logic is that a generation (or two) of inbreeding increases the among-group variance, and this increase can be exploited by selection. However, given the extra generation(s) used for inbreeding (instead of selection), the conditions for such a cyclic inbreeding-selection system to give a larger response than mass selection are rather stringent. Dickerson and Lindhé showed that the ratio of response under cyclic inbreeding (R_I) to response under mass selection (R_m) is approximately

$$\frac{R_I}{R_m} \simeq \left(\frac{\overline{\imath}_m\, g_I}{\overline{\imath}_I\, g_m}\right)\sqrt{\frac{(1+f)r_f}{h^2}} \tag{23.19a}$$

where g is the generation time per cycle (typically $g_m = 1$, $g_I = 2$) and r_f is the genetic correlation among the inbred line members. For example, if one crosses full-sibs and then crosses and selects on inbred families in alternate years, $f = 0.25$, $r_f = 0.6$, and $g_I/g_m = 1/2$, which implies that

$$\frac{R_I}{R_m} \simeq \left(\frac{\bar{\imath}_m}{\bar{\imath}_I}\right)\sqrt{\frac{0.1875}{h^2}} \tag{23.19b}$$

Under equal selection intensities ($\bar{\imath}_m = \bar{\imath}_I$), Equation 23.19b implies that $h^2 < 0.1875$ for cyclic inbreeding to exceed mass selection (Dickerson and Lindhé 1977). Further, Equation 23.19a assumes that there is no significant impact from inbreeding depression.

Given these stringent conditions, it is not surprising that experimental support for the advantage of cyclic inbreeding is lacking. Dion and Minvielle (1985) used 15 generations of cyclic full-sib versus random mating to select for increased pupal weight in *Tribolium castaneum*, and found no differences in the response or realized heritabilities relative to random mating. Similar results were observed in Japanese quail (Example 23.3). While López-Fanjul and Villaverde (1989) observed that one generation of full-sib mating resulted in a fourfold increase in the realized heritability of egg to pupal viability in *Drosophila melanogaster*, this was more than offset by the reduction in the mean from inbreeding depression.

Another cyclic scheme, **S_1 mass selection**, was proposed by Dhillon (1991a). Here, individuals are crossed and the resulting offspring are selfed. The S_1 are then evaluated by individual selection, with the superior individuals outcrossed to start the cycle again. To evaluate the expected selection response under such a scheme, we need to compute the covariance between an S_1 and its outbred offspring, which is obtained as follows. With probability of 1, an S_1 individual passes on a single allele to its outbred offspring, so $\Delta_3 + \Delta_8 = 1$. With probability 1/2, the S_1 individual has both alleles IBD at a locus (due to the generation of selfing), giving $\Delta_3 = \Delta_8 = 1/2$. More generally, if k generations of selfing are used before random mating, then (using the same logic as in Example 23.2) $\Delta_3 = f_k = 1 - 2^{-k}$, $\Delta_8 = 1 - f_k = 2^{-k}$, and $2\theta = \Delta_3 + \Delta_8/2 = (1/2)(2 - 2^{-k})$, returning the S_k-parent and offspring covariance as

$$\sigma(S_k, y) = (1/2)(2 - 2^{-k})\sigma_A^2 + (1 - 2^{-k})\sigma_{ADI} \tag{23.20a}$$

Assuming selection on both parents, the response per generation is then

$$R_{S_K} = \left(\frac{\bar{\imath}}{k+1}\right)\frac{(2 - 2^{-k})\sigma_A^2 + 2(1 - 2^{-k})\sigma_{ADI}}{\sqrt{\sigma_g^2(S_k) + \sigma_e^2(S_k)}} \tag{23.20b}$$

The factor of $1/(k+1)$ arises because there are k generations of selfing for each single generation of selection. The genetic variance, $\sigma_g^2(S_k)$, among S_k individuals is given (below) by Equation 23.23, and $\sigma_e^2(S_k)$ is the error variance for single S_k individuals, which is expected to be considerably larger than the error variance for families, as no replication is involved. For strict additivity, the ratio of per-generation response under S_k mass selection relative to standard mass selection is

$$\frac{R_{S_k}}{R_m} = \left(\frac{2 - 2^{-k}}{k+1}\right)\sqrt{\frac{\sigma_A^2 + \sigma_e^2(m)}{(2 - 2^{-k})\sigma_A^2 + \sigma_e^2(S_k)}} \tag{23.20c}$$

Note that $R_{S_k} < R_m$ for all values of k (assuming the error variances are roughly equal).

Dhillon assumed that a greenhouse could be used for the S_1, giving one cycle (i.e., two generations) per field generation. In this case, there is no generational cost for inbreeding, and the ratio in the parentheses of Equation 23.20c becomes $(2 - 2^{-1}) = 3/2$. However, a major biological limitation in the assumptions behind Equation 23.20c is that the selected traits must be expressed before reproduction. For traits expressed during or after reproduction, only a single sex will be selected upon (as presumably the S_1 is outcrossed to random individuals). In such cases, the response ratio (in this greenhouse setting) is reduced to (1/2)(3/2) = 3/4 of mass selection.

Example 23.3. MacNeil et al. (1984) subjected two populations of Japanese quail (*Coturnix coturnix japonica*) for an index of total egg mass divided by female weight. Standard mass selection with random mating was used for population two (filled circles), while alternative cycles of full-sib and random mating were used for population one (open squares).

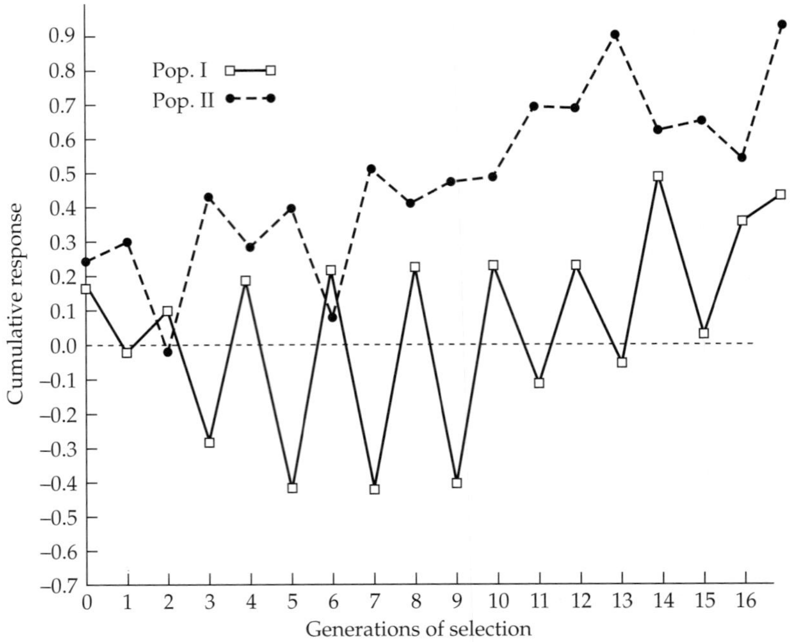

While both populations ultimately responded to selection, significant inbreeding depression was seen in population one. The cyclic mating scheme produced inbred individuals every odd generation, resulting in population one showing dramatic shifts between positive selection response and significant inbreeding depression, which countered any positive response.

INDIVIDUAL SELECTION UNDER PURE SELFING

Under pure selfing, one starts with a collection of individuals and continually selfs each to form a series of inbred lines. Let S_k denote such a line after k generations of selfing, with the S_0 being the collection of outbred individuals that are initially selfed to start the lines, and the S_∞ the collection of the completely inbred lines. A variety of options exist for generating the initial collection of lines. The simplest is to use a random sample of individuals from an outbred population. Another common situation is the **pure-line cross**, wherein one crosses two completely inbred (pure) lines, and continually selfs starting with the F_1. In this case, the initial cross produces a number of F_1 individuals, and even though these are selfed to create a series of F_2 lines, the first generation of selfing is formally defined as the F_3. The reason is that all of the F_1s are genetically identical, being heterozygous at every locus at which the two lines differ. Such a population consisting of only heterozygotes is not in Hardy-Weinberg equilibrium, but the F_2 population is (for diploid autosomal loci). Thus, the F_2 defines the base population, and this is the population from which variance components are extracted. The F_2 also sets the initial value for counting generations of selfing, so that $S_0 = F_2$, $S_1 = F_3$, and so on. If the loci are unlinked, then linkage disequilibrium (which is maximal in the F_1) is zero in the F_2. The instant achievement of linkage equilibrium in the F_2 from a pure-line cross arises because all F_1 individuals are genetically identical and heterozygous at all segregating loci, which is not the situation for crosses of three (or more) lines. If the loci are linked, it may take several rounds of random mating to mitigate the effects of the

F_1 disequilibrium between tightly linked loci. Finally, recall that $\sigma_{ADI} = \sigma^2_{DI} = 0$ in this line-cross setting (Chapter 11).

Several other line-cross designs can form the foundational population from which individuals are drawn for selfing. If one intermates a collection of lines, the first generation will not be in Hardy-Weinberg equilibrium unless the allele frequencies are the same in each line. However (for diploid autosomal loci), Hardy-Weinberg will be reached with an additional generation of random mating (sex-linked loci and polyploids take several additional generations; see LW Chapter 4). Linkage disequilibrium is also created in such a cross, due to differences in the gamete frequencies across lines. Unlike the case for crossing two pure lines, the F_2 from a multiple-line cross will not necessarily be in linkage equilibrium, even for unlinked loci. In this case, the disequilibrium decays as $(1/2)^t$, where t is the number of generations of random mating that the starting F_1 population experiences. For linked loci, the disequilibrium decays as $(1-c)^t$, where c is the recombination frequency. Other common types of line crosses are **three-way hybrids**, $(L_1 \times L_2) \times L_3$ (the F_1 from an $L_1 \times L_2$ crossed to L_3), and **double-crosses** (or **four-way hybrids**) $(L_1 \times L_2) \times (L_3 \times L_4)$, which commonly arose in maize breeding. Again, it is often advisable to take such crosses through at least one additional round of random mating to achieve Hardy-Weinberg (meaning that our expressions for response are valid) and approach linkage equilibrium before starting inbreeding.

Response Under Pure Selfing

Suppose we commence selfing from a collection of individuals that are in Hardy-Weinberg and linkage equilibrium. After all lines have become completely inbred, there is no response to selection *within* a line as there is no within-line genetic variation (in the absence of mutation). However, the response *among* lines involves the entire genotypic variance, as selection is among clones.

We first consider one extreme, inbreeding each line entirely to fixation and then selecting among the lines. At this point, a parent and its (selfed) offspring are genetically identical, and the resulting parent-offspring covariance equals the total genetic variance in the population (i.e., the variation over all sets of inbred lines). Inbreeding alters the total genetic variation from its random-mating value of $\sigma^2_A + \sigma^2_D$ to a new value of $\widetilde{\sigma}^2_G$ over the entire collection of pure lines. The resulting parent-offspring covariance among fully inbred lines is

$$\sigma(z_p, z_o) = \widetilde{\sigma}^2_G = 2\sigma^2_A + 2\sigma_{ADI} + \sigma^2_{DI} \tag{23.21}$$

When kth-order additive epistasis is present, $2^k\sigma^2_{A^k}$ is added in Equation 23.21 (e.g., $4\sigma^2_{AA}$, $8\sigma^2_{AAA}$, etc.). Assuming linearity of the parent-offspring regression, the response to a generation of selection among these inbred lines produces an expected response of

$$R = S\,\frac{\widetilde{\sigma}^2_G}{\widetilde{\sigma}^2_G + \sigma^2_e} \tag{23.22}$$

Even if selection is moderate, a single generation is likely to significantly alter the distribution of remaining genotypes (and hence change the genetic variance), and thus the validity of Equation 23.22 over more than a few generations is, at best, questionable. There are a number of subtleties when attempting to select the best pure line from a collection, which are examined in Volume 3.

Instead of waiting for inbreeding to be complete, suppose instead that we select among individuals while inbreeding is still ongoing (i.e., $f < 1$). This entails choosing individuals and then following their selfed offspring. The response in generation T from selection in generation $t < T$ is then a function of the **cross-generational covariance**, $\sigma_G(T, t)$. For strict selfing, Equation 11.16c yields the covariance between a relative from generation T and its ancestor in generation $t < T$ as

$$\sigma_G(T, t) = (1 + f_t)\sigma^2_A + (1 - f_T)(\sigma^2_D + f_t\iota^*) + \frac{f_T + 3f_t}{2}\,\sigma_{ADI} + f_t\,\sigma^2_{DI} \tag{23.23}$$

where $f_t = 1-\left(\frac{1}{2}\right)^t$ is the amount of inbreeding in generation t. The phenotypic variance in generation t is $\sigma^2(z_t) = \sigma_G(t,t) + \sigma_e^2$, where $\sigma_G(t,t) = \sigma_G^2(t)$ is obtained from Equation 23.23 by setting $T = t$. Equation 23.21 follows if we note that $f_\infty = 1$. Recall (Chapter 17; LW Chapter 6) that the environmental variance, σ_e^2, may increase with inbreeding, and thus we may need to account for this factor as well. Cockerham and Matzinger (1985) extend Equation 23.23 to include additive by additive epistasis (but still assuming gametic-phase equilibrium). If additive epistasis up to order k is present, extra terms are added to the covariance provided by Equation 23.23, namely,

$$(1+f_t)^2\sigma_{AA}^2 + \cdots + (1+f_t)^k\sigma_{A^k}^2$$

When all possible types of pairwise epistasis (e.g., $A \times A$, $A \times D$, $D \times D$) occur, 12 variance components are required to describe $\sigma_G(T,t)$ under selfing (Wright 1987, 1988), but we will ignore this level of complication.

Substitution of Equation 23.23 into Equation 23.4 gives the response to selection while the line is being inbred. For complete additivity, $\sigma_G(T,t) = (1+f_t)\sigma_A^2 = (2-2^{-t})\sigma_A^2$, yielding a response of

$$R(T) = \sum_{t=0}^{T-1} S_t \, \frac{(2-2^{-t})\sigma_A^2}{(2-2^{-t})\sigma_A^2 + \sigma_e^2} \tag{23.24}$$

as obtained by Brim and Cockerham (1961) and, under much more general conditions, by Pederson (1969a). If dominance is present, the selection response under selfing has both a transient and a permanent component (Chapter 15). When selection is relaxed (short of complete inbreeding), the mean potentially changes as the transient component decays. The expected total change in the mean after n generations, the first T of which were under selection (generations 0 to $T-1$), is thus given by

$$R(n\,|\,T) = \sum_{t=0}^{T-1} S_t \, \frac{\sigma_G(n,t)}{\sigma_G(t,t)+\sigma_e^2} = \sum_{t=0}^{T-1} \overline{\imath}_t \, \frac{\sigma_G(n,t)}{\sqrt{\sigma_G(t,t)+\sigma_e^2}} \tag{23.25}$$

The permanent response to T generations of selection, $\widetilde{R}(T)$, is given by

$$\widetilde{R}(T) = R(\infty\,|\,T) = \sum_{t=0}^{T-1} S_t \, \frac{\sigma_G(\infty,t)}{\sigma_G(t,t)+\sigma_e^2} = \sum_{t=0}^{T-1} \overline{\imath}_t \, \frac{\sigma_G(\infty,t)}{\sqrt{\sigma_G(t,t)+\sigma_e^2}} \tag{23.26}$$

Because $f_\infty = 1$, Equation 23.23 calculates the covariance between an individual in generation t and a completely inbred (F_∞) line descended from it as

$$\sigma_G(\infty,t) = \left(2 - \frac{1}{2^t}\right)\sigma_A^2 + \left(2 - \frac{3}{2^{t+1}}\right)\sigma_{ADI} + \left(1 - \frac{1}{2^t}\right)\sigma_{DI}^2 \tag{23.27}$$

which is essentially $\widetilde{\sigma}_G^2$ (Equation 23.21) for $t > 5$. Thus, additive variance contributes to the permanent response, while σ_D^2 and ι^* contribute to the transient, but not the permanent, response. Dominance does, however, make a contribution to the permanent response, through σ_{DI}^2 and σ_{ADI}. To see why, consider the case when inbreeding is complete. Here the only genotypes are of the form $\mathbf{A}_i\mathbf{A}_i$ and have a genotypic decomposition of $2\alpha_i + \delta_{ii}$. The frequency of such genotypes (in the collection of completely inbred lines) is $\simeq p_i$, assuming that there is no change in the population allele frequencies (i.e., our assumption of weak selection on the underlying loci). Recalling the definitions of σ_{DI}^2 and σ_{ADI} (Table 11.1), the resulting genetic variance among lines is

$$\widetilde{\sigma}_G^2 = \sigma^2(2\alpha_i + \delta_{ii}) = 4\sigma^2(\alpha_i) + 2\cdot 2\sigma(\alpha_i,\delta_{ii}) + \sigma^2(\delta_{ii}) = 2\sigma_A^2 + 2\sigma_{ADI} + \sigma_{DI}^2$$

The contribution from standard dominance variance, $\sigma_D^2 = \sigma^2(\delta_{ij})$, decays as $\mathbf{A}_i\mathbf{A}_j$ heterozygotes are lost due to inbreeding, leaving only the contribution from homozygotes, $\sigma_{DI}^2 = \sigma^2(\delta_{ii})$.

Example 23.4. As an example of what fraction of the response from various generations of selection is ultimately passed on to the completely inbred line, assume the genotypic values and frequencies from Example 11.1, which computed the components required to predict the selection response under inbreeding as

$$\sigma_A^2 = 0.629, \quad \sigma_D^2 = \iota^* = 0.046, \quad \sigma_{DI}^2 = 0.103, \quad \sigma_{ADI} = -0.361$$

Assume $\sigma_e^2 = 1$ (which corresponds to narrow- and broad-sense heritabilities of $h^2 = 0.38$ and $H^2 = 0.41$, respectively). If we substitute these values into Equation 23.23, and recall that $\sigma_z^2(t) = \sigma_G^2(t,t) + \sigma_e^2$, we get the response in various generations (t) of inbreeding. The fifth column in the table below shows the expected single-generation response (the response in $t+1$ from selection in t) and the fourth column shows the eventual contribution ($T = \infty$), while the final column shows the percentage of the response from that particular generation that is translated into the final response.

t	$\sigma_G(\infty,t)$	$\sigma_G^2(t,t)$	$\dfrac{\sigma_G(\infty,t)}{\sigma_z^2(t)}$	$\dfrac{\sigma_G(t+1,t)}{\sigma_z^2(t)}$	%
0	0.449	0.675	0.268	0.335	79.86
1	0.544	0.669	0.326	0.363	89.73
2	0.592	0.657	0.357	0.377	94.78
3	0.616	0.649	0.374	0.384	97.37
4	0.628	0.645	0.382	0.387	98.68
5	0.634	0.642	0.386	0.389	99.34
7	0.639	0.641	0.389	0.390	99.83

For example, if selection is among individuals that have experienced three generations of inbreeding (t = 3), then 97.4% of the response seen in generation 4 is seem among the fully inbred lines. This calculation essentially assumes that this is only a single generation of selection, as multiple generations of selection (especially among largely inbred lines) will quickly remove most of the existing among-line variation. Note that the largest reduction between the initial and permanent responses occurs in the first generation of selection, where selected individuals are selfed for the first time. As selfing progresses, the genotypic values between ancestors and descendants become increasingly more similar, and hence they retain almost all of their initial response. While more dramatic changes are expected under different combinations of σ_{ADI}, σ_{DI}^2, and ι^*, inbreeding will largely be complete after the first six generations, and the offspring will be almost genetically identical to their parents.

This example highlights the dilemma faced by plant breeders. Selection decisions based on early cycles of the inbreeding (**early-generations selection**) may be poor predictors of the ultimate inbred product (especially when dominance is present), but carrying out additional generations of inbreeding (to increase the genetic covariance between the tested and fully inbred genotypes) takes additional time. For the values assumed here, only 80% of the first-generation ($t = 0$) response is retained in fully inbred lines, which increases to almost 95% for the third-generation ($t = 2$) response. The use of doubled-haploids, which can make a plant fully inbred in a single generation, can have significant advantages in an inbred line breeding program (Gallais 1989; Volume 3).

It is important to again stress that these results for the expected selection response are based on infinitesimal-model approximations. The covariances between relatives under inbreeding that drive these equations are based on the assumption (as was the case for the breeder's equation) that selection does not significantly change the variance components (the values of σ_A^2, σ_D^2, σ_{ADI}, etc. remain largely unchanged). Clearly, selection with a small number of loci can substantially change allele frequencies, thus violating the assumptions leading to Equations 23.24 through 23.27. Likewise, with a small number of lines and/or strong selection, these results will also be biased. Another, more subtle, violation of this basic model will occur if some lines are disproportionately chosen over others (as one might expect might happen). In such cases, the covariances that are now appropriate are not those for the population as a whole, but rather those for individuals *within* particular sublines. The unstated assumption of Equation 23.23 is that when individuals are being compared for selection, their most recent ancestors are those that are drawn from the base population. If their most recent ancestor is more current, however, the covariances will be incorrect and the estimated response biased. Finally, as developed shortly, selection-generated gametic-phase disequilibrium can significantly reduce response.

Response When Inbreeding Pure-line Crosses

Considerable simplification occurs when two pure lines are crossed. In the resulting F_1, each locus will have only two segregating alleles (one from each line; both at frequency 1/2), and as a result, $\iota^* = \sigma_D^2$ and $\sigma_{DI}^2 = \sigma_{ADI} = 0$ (Chapter 11). Equation 23.23 then becomes

$$\sigma_G(T,t) = c_t\,\sigma_A^2 + 2^{-T} c_t\,\sigma_D^2 + c_t^2\,\sigma_{AA}^2 + \cdots c_t^k \sigma_{A^k}^2, \quad \text{where} \quad c_t = 2 - \frac{1}{2^t} \tag{23.28}$$

If we start selection on the F_2 population and denote this generation as generation 0, Equation 23.25 simplifies (Pederson 1969a) to

$$R(\,n\,|\,T\,) = \sum_{t=0}^{T-1} \bar{\imath}_t \; \frac{(\,2 - 2^{-t}\,)\,(\,\sigma_A^2 + 2^{-n}\sigma_D^2\,)}{\sqrt{(\,2 - 2^{-t}\,)\,(\,\sigma_A^2 + 2^{-t}\sigma_D^2\,) + \sigma_e^2}} \tag{23.29}$$

However, Equation 23.29 carries the unstated assumption that the environmental variance, σ_e^2, is unchanged over levels of inbreeding, which may not be correct (Chapter 17). More generally, σ_e^2 can be replaced by $\sigma_e^2(t)$ to accommodate this concern. Note that any initial contribution to the selection response from dominance will quickly decay (as 2^{-n}).

Example 23.5. Suppose a cross between two inbred lines is subjected to truncation selection for the uppermost 20% in the first two generations of selfing ($t = 0, 1$). What is the cumulative response to selection at the nth generation of selfing? Here $\bar{\imath} = 1.402$, and we assume random-mating variance components (i.e., values in the F_2 population) of $\sigma_A^2 = 50$, $\sigma_D^2 = 25$, and $\sigma_e^2 = 50$, yielding random-mating heritabilities of h^2 = 0.4 and H^2 = 0.6. The total phenotypic variances in the first two generations are $\sigma_z^2(0) = \sigma_G^2(0) + \sigma_e = 125$ and $\sigma_z^2(1) = \sigma_G^2(1) + \sigma_e = 143.75$. If we let $r(n,t)$ denote the response to selection in generation t that is present at generation n and let $R(n) = \sum_t r(n,t)$ denote the cumulative response from selection that is still present at generation n, Equation 23.39 returns the following values:

n	t	$\sigma_G(n,t)$	$r(n,t)$	$R(n)$
1	0	62.5	7.837	7.837
2	0	56.3	7.054	
2	1	84.4	9.866	16.920
3	0	53.1	6.662	
3	1	79.7	9.318	15.980
4	0	51.6	6.466	
4	1	77.3	9.044	15.510
5	0	50.8	6.368	
5	1	76.2	8.907	15.275
10	0	50.0	6.273	
10	1	75.0	8.774	15.047
∞	0	50	6.270	
∞	1	75	8.770	15.040

For example, by generation 3 ($n = 3$), the remaining amount of response from selection in generations 0 and 1 are 6.662 and 9.318, respectively, for a total response of 15.98. Of this response, only 15.04 persists when the lines are fully inbred, 6.27 from generation 0 and 8.77 from generation 1. Hence, for these starting variance-component values, only 80% (6.27/7.837) and 89% (8.77/9.866), respectively, of the initial first- and second-generation response is permanent.

The Bulmer Effect Under Selfing

Recall from Equation 16.2 that we can decompose the additive-genetic variance as $\sigma_A^2 = \sigma_a^2 + d$, the sum of the genic variance, σ_a^2, plus the impact, d, from any gametic-phase disequilibrium. In Chapter 16, we showed that directional selection generates $d < 0$, decreasing σ_A^2 from its linkage-equilibrium value of σ_a^2 (the Bulmer effect). Under random mating, recombination (for unlinked loci) removes half of the existing d in each generation, which quickly balances any new d introduced by selection (Chapter 16). Under inbreeding, the frequency of heterozygotes (where such recombination takes place) quickly declines, significantly enhancing the impact of linkage disequilibrium (LD) by slowing its decay from recombination.

The first attempts to study the impact of LD when inbreeding is present were in small-scale simulation studies by Bliss and Gates (1968) and Stam (1977), who assumed a finite number of loci in a completely additive model (with no dominance or epistasis). As expected, they found that linkage reduces the rate of selection response, while (for fixed σ_A^2) the per-generation response increases as the number of loci decreases. This latter observation is also to be expected, as increasing the number of loci (with σ_A^2 fixed) reduces the allelic effect at any single locus, reducing the amount of selection on that locus (Chapters 24 and 25).

The first analytical investigations on the magnitude of Bulmer effect during selfing were done by Silvela and Diez-Barra (1985) and Cornish (1990a, 1990b). Assume an F_2 population of lines is continually selfed until it is fully inbred, yielding an F_∞ collection of lines. In Cornish's model, a single generation of selection occurs in the F_2, and its effect on the final (F_∞) lines was examined. Cornish assumed directional truncation selection, whereby the upper fraction (p) of the population is saved. Recall (Equation 16.11a) that in this setting, the phenotypic variance following selection is reduced by $\bar{\imath}\,(\bar{\imath} - x_{[1-p]})\sigma_z^2$, where the selection intensity, $\bar{\imath}$, is a function of p (Equation 14.3a) and $x_{[1-p]}$ satisfies $\Pr(U > x_{[1-p]}) = p$, where U is a unit normal. Cornish found that the genetic variance in the offspring of the selected parents is reduced by $h^2\,\bar{\imath}\,(\bar{\imath} - x_{[1-p]})\sigma_A^2$, as only a fraction ($h^2$) of the phenotypic change is passed on to the offspring. In a randomly mating population, the reduction in variance from selection-induced negative LD rapidly approaches an equilibrium value (Equation 16.12e). Upon relaxation of selection under random mating, the variance eventually returns to its preselection value under the infinitesimal model as d decays to zero (Chapter 16). By contrast, because of the dramatic reduction in the fraction of heterozygotes in a selfing population, part of this reduction in variance from LD is permanent. Thus, while LD in a randomly mating population reduces the *rate* of response, with selection in a selfing population, it also impacts the final selection *limit*.

A far more detailed investigation of the Bulmer effect under the infinitesimal model was performed by Hayashi and Ukai (1994) and Kelly (1999a). We examine the results of Hayashi and Ukai here and return to Kelly's more general treatment (under partial selfing) at the end of the chapter. Hayashi and Ukai obtained recursion equations for the changes in variance and covariance for a pure-line cross. They assumed that truncation selection starts in the F_2 generation and continues for t generations. If only additive variance is present,

$$\sigma_A^2(t+1) = \sigma_{A_o}^2(t+1) - \sigma_{A_o}^2(t) + \left[1 - \bar{\imath}(\bar{\imath} - x_{[1-p]})\frac{\sigma_A^2(t)}{\sigma_A^2(t) + \sigma_e^2}\right]\sigma_A^2(t) \tag{23.30}$$

where $\sigma_{A_o}^2(t) = (2 - 1/2^t)\,\sigma_A^2(0)$ is the total additive variance in an unselected population of lines following t generations of selfing (this is akin to the genic variance, σ_a^2, in Equations

16.2 and 16.8b, which remains unchanged by selection under the infinitesimal when under random mating). Hence, $\sigma^2_{A_o}(t+1) - \sigma^2_{A_o}(t)$ can be regarded as the within-line segregation variance, which (for unlinked loci) is unaffected by selection (Equation 16.8b), but (unlike the random mating case) is strongly influenced by inbreeding, as

$$\sigma^2_{A_o}(t+1) - \sigma^2_{A_o}(t) = \left[(2 - 1/2^{t+1}) - (2 - 1/2^t)\right] \sigma^2_A(0) = \sigma^2_A(0)/2^{t+1}$$

which shrinks to zero as lines become progressively inbred. This fraction represents the opportunity for recombination in any remaining heterozygotes to reduce selection-induced disequilibrium. The remaining component in the brackets of Equation 23.30 represents the change in the among-line variance (the variance in progeny means), which is reduced by selection, and represents the impact of disequilibrium (akin to the generation of d in Chapter 16). Note that for the first generation of selection, this reduces to $[1 - h^2\bar{\imath}(\bar{\imath} - x_{[1-p]})]\sigma^2_A$, namely, Cornish's result. Note, however, that Equation 23.30 shows that Cornish's result neglects the partial decay of this disequilibrium due to recombination in any remaining heterozygotes (the $\sigma^2_{A_o}(t+1) - \sigma^2_{A_o}(t)$ term), and thus is an overestimation of $\sigma^2_A(t)$.

If both additive and dominance effects are present, they will have correlated changes and the recursion equations will be more complex. Some simplifications occur because of the pure-line cross assumption (as $\sigma_{ADI} = \sigma^2_{DI} = 0$ and $\iota^* = \sigma^2_D$). By letting $\sigma_{G_o}(T,t)$ denote the cross-generational covariance under pure selfing (starting with a pure-line cross), Hayashi and Ukai showed (for unlinked loci) that

$$\sigma_G(T,t) = \sigma_{G_o}(T,t) - \bar{\imath}(\bar{\imath} - x_{[1-p]}) \sum_{k=0}^{t-1} \frac{\sigma_G(t,k)\,\sigma_G(T,k)}{\sigma_G(k,k) + \sigma^2_e} \tag{23.31a}$$

where

$$\sigma_{G_o}(T,t) = \left(2 - 2^{-t}\right)\left(\sigma^2_A + 2^{-T}\sigma^2_D\right) \tag{23.31b}$$

is simply the covariance between relatives in generations t and T in the absence of selection (when $\sigma_{ADI} = \sigma^2_{DI} = 0$ and $\iota^* = \sigma^2_D$, Equation 23.23 reduces to Equation 23.31b). Equation 23.31a is solved by iteration, starting with

$$\sigma_G(T,0) = \sigma_{G_o}(T,0) = \sigma^2_A + 2^{-T}\sigma^2_D \tag{23.31c}$$

As mentioned, part of the relative simplicity of these expressions arises from the assumption of a pure-line cross. Kelly (1999a) considered more general cases, which require iterative expressions for the changes in σ^2_{DI} and σ_{ADI} from disequilibrium (see Equations 23.67–23.68).

Example 23.6. As an application of the Hayashi-Ukai variance correction, assume the values used in Example 23.5 ($\sigma^2_A = 50, \sigma^2_D = 25$, and $\sigma^2_e = 50$). The fraction saved was $p = 0.2$, and hence $\bar{\imath} = 1.402$. Likewise, $x_{[1-p]} = x_{[0.8]} = 0.84$ (Example 16.3), so $\bar{\imath}(\bar{\imath} - x_{[1-p]}) = 0.788$. From Equation 23.31c, the genetic covariance between a selected F_2 individual (generation $t = 0$) and a relative after T generations of selfing is

$$\sigma_G(T,0) = \sigma_{G_o}(T,0) = \sigma^2_A + 2^{-T}\sigma^2_D = 50 + 25/2^T$$

Note that this covariance is the same as with pure selfing. The covariance between an individual selected in the next generation (1) and its relative in generation T of selfing does, however, show a reduction, with Equation 23.31a yielding

$$\sigma_G(T,1) = \sigma_{G_o}(T,1) - \bar{\imath}\,(\bar{\imath} - x_{[1-p]})\,\frac{\sigma_G(1,0)\,\sigma_G(T,0)}{\sigma_G(0,0) + \sigma^2_e}$$

The first term is the pure-selfing covariance, and the second is the reduction due to selection. To obtain the value of the latter, first note that $\sigma_G(1,0) = 50 + (25/2) = 62.5$, while $\sigma_G(0,0) + \sigma_e^2 = 125$ and (Equation 23.31b) $\sigma_{G_o}(T,1) = (3/2)\left[50 - (25/2^T)\right]$, yielding

$$\begin{aligned}\sigma_G(T,1) &= (3/2)\left[50 + (25/2^T)\right] - 0.788\,\frac{62.5 \cdot \left[50 + (25/2^T)\right]}{125} \\ &= \left[50 + (25/2^T)\right]\left(3/2 - 0.788\,\frac{62.5}{125}\right)\end{aligned}$$

Because the quantity in the square brackets is proportional to $\sigma_{G_o}(T,1)$, it follows that $\sigma_G(T,1)/\sigma_{G_o}(T,1) = 0.737$ is a constant independent of T, which shows a significant (27%) reduction (over all T) in the predicted response from generation 1 when disequilibrium is ignored. Similarly, applying Equation 23.31c, the genetic variance in generation 1 becomes

$$\begin{aligned}\sigma_G(1,1) &= \sigma_{G_o}(1,1) - \bar{\imath}(\bar{\imath} - x_{[1-p]})\,\frac{\sigma_G(1,0)\,\sigma_G(1,0)}{\sigma_G(0,0) + \sigma_e^2} \\ &= (3/2)\,(50 + 25/2) - 0.788\frac{(62.5)^2}{125} = 69.125\end{aligned}$$

Again, the first quantity is the pure-inbreeding value and the second is the correction for disequilibrium. In this case, the correct genetic variance is only 0.737 of the pure-inbreeding value (which Equation 23.31b gives as 93.75). Because neither the genetic variance nor the covariance for generation 0 are affected by selection ($\sigma_G(0,0) = \sigma_{G_o}(0)$, and $\sigma_G(T,0) = \sigma_{G_o}(T,0)$), the response to selection in generation 0 is unaffected by disequilibrium (this is also the case for a random mating population; see Chapter 16). The response of σ_G^2 in generation 2, however, is influenced by selection. Because both covariance and the genetic variance are reduced by the same fraction (0.737) relative to strict inbreeding, applying Equation 23.3 with the above covariances shows that the ratio of response at generation T from selection in generation 1, $R(T,1)$, to its predicted value, ignoring disequilibrium, $R_0(T,1)$, is

$$\frac{R(T,1)}{R_0(T,1)} = \left(\frac{\sigma_G(T,1)}{\sigma_{G_o}(T,1)}\right)\left(\frac{\sqrt{\sigma_{G_o}(1,1)}}{\sqrt{\sigma_G(1,1)}}\right) = \frac{0.737}{\sqrt{0.737}} = 0.859$$

The presence of gametic disequilibrium thus reduces the selection response by 14%.

FAMILY SELECTION UNDER PURE SELFING

Predicting response to family selection—using the phenotypic means of the selfed offspring (perhaps several generations' worth) to choose lines—requires a consideration of the hierarchical structure among the selfed lines in a population (Figure 23.2). The collection of lines descended from a parent at time t (which we can think of as this individual's extended family) are expected to show less within-line variation than a collection of lines from an earlier ancestor at time $k < t$. For family selection, the goal is to predict the selection response, given that we select individuals from generation t on the basis of the performance of their offspring in generation $\tau > t$. We may then wish to know what fraction of this response still persists in some future generation, $T > \tau$ (such as the permanent response, i.e., $T = \infty$). For example, we may select the best lines in generation t based on the performance of their selfed offspring and using remnant seed from the selected families to form the next generation. In this case, $\tau = t + 1$. If individual plants do not produce sufficient seed for family testing, two generations of selfing may be needed to generate a sufficiently large family, in which case $\tau = t + 2$.

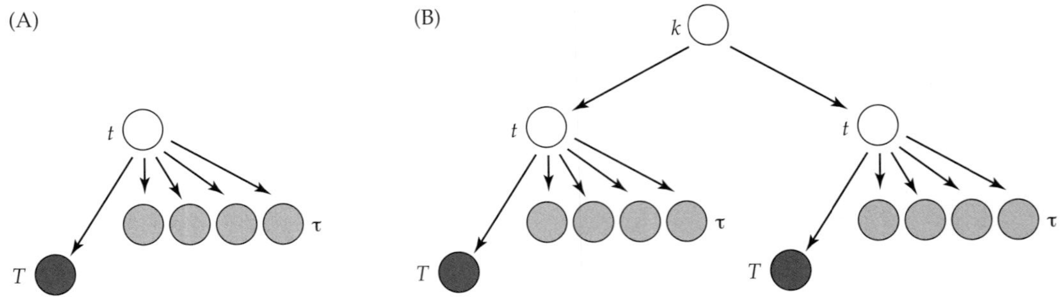

Figure 23.2 The hierarchical structuring of selfed populations. The gray circles denote sibs in the selection unit, the solid circles denote relatives of interest, and k, t, T, and τ denote the generations of selfing experienced by an individual. The arrows denote lines of descent through selfing and may be longer than one generation. **A:** Often we select using a parent in generation t of selfing by scoring its (selfed) offspring in generation τ, and we need the covariance between τ and some future generation, T, where the common relative to both is from generation t. **B:** Another level of hierarchical structuring of selfed populations: When selecting within a substructure of the selfing pedigree, we may be interested in the response using parents in generation t whose offspring are scored in generation τ and for which the response is across those families in the pedigree sharing an earlier common parent in generation $k < t$.

The Covariance Between Relatives in a Structured Selfing Population

Given the need to account for the structure in a selfing population, Cockerham (1983) and Cockerham and Martzinger (1985), built on concepts from Horner (1952) and Gates et al. (1957) and generalized the covariance given by Equation 23.23 to $\sigma_G(\,T,\tau,t\,)$, the covariance between a relative in generation T and another relative in generation $\tau \le T$, when the last common relative of both is from generation $t \le \tau$ (Figure 23.2A). This covariance was given by Equation 11.15, namely,

$$\begin{aligned}\sigma_G(\,T,\tau,t\,) &= (1+f_t)\sigma_A^2 + \left(\frac{(1-f_T)(1-f_\tau)}{1-f_t}\right)\sigma_D^2 + \left(f_t + \frac{f_T+f_\tau}{2}\right)\sigma_{ADI} \\ &+ \left(f_t + \frac{(f_T-f_t)(f_\tau-f_t)}{2(1-f_t)}\right)\sigma_{DI}^2 + \left(\frac{f_t(1-f_T)(1-f_\tau)}{1-f_t}\right)\iota^* \\ &+ (1+f_t)^2\sigma_{AA}^2 + (1+f_t)^3\sigma_{AAA}^2 + \cdots (1+f_t)^k\sigma_{A^k}^2 \end{aligned} \tag{23.32}$$

where $t \le \tau \le T$. Notice that Equation 23.32 reduces to Equation 23.23 when $\tau = t$ (parents are the selection unit), as $\sigma_G(\,T,t,t\,) = \sigma_G(\,T,t\,)$. The epistatic terms are often ignored, and Equation 23.32 does not account for nonadditive epistatic terms and assumes linkage equilibrium. For cross-generational covariances indexed by two or more relatives, such as $\sigma_G(\,T,t\,)$ and $\sigma_G(\,T,\tau,t\,)$, we use the convention that the rightmost index (t in this case) references the oldest (earliest-generation) individual, while the leftmost (T in this case) references the youngest (latest-generation). Thus, as one proceeds right-to-left in the index, more recent relatives are being considered; see Figure 23.2. Again, to use these covariances in predicting the selection response, we must make the strong assumption that selection does not significantly modify this covariance from the unselected version (Equation 23.32). The covariance between a parent in generation t and an offspring in generation T follows by noting that here $t = \tau$ and Equation 23.32 reduces to Equation 23.23.

For the special case of a pure-line cross ($\iota^* = \sigma_D^2$, $\sigma_{DI}^2 = \sigma_{ADI} = 0$), Equation 23.32 simplifies considerably, to

$$\begin{aligned}\sigma_G(\,T,\tau,t\,) &= (1+f_t)\sigma_A^2 + \left(\frac{(1+f_t)(1-f_T)(1-f_\tau)}{1-f_t}\right)\sigma_D^2 + (1+f_t)^2\sigma_{AA}^2 + \cdots \\ &= \left(2-\frac{1}{2^t}\right)\left(\sigma_A^2 + \frac{\sigma_D^2}{2^{T+\tau-t}} + \left(2-\frac{1}{2^t}\right)\sigma_{AA}^2 + \cdots\right) \end{aligned} \tag{23.33}$$

The permanent selection response is given by the covariance between a completely inbred F_∞ line ($f_\infty = 1$) and a relative (for example, from the selection unit) in generation τ who last both shared a relative in generation t (as would occur if remnant seed from t was used to form the new lines). Here, Equation 23.32 reduces to

$$\sigma_G(\infty, \tau, t) = (1 + f_t)\sigma_A^2 + \frac{1 + 2f_t + f_\tau}{2}\,\sigma_{ADI} + \frac{f_t + f_\tau}{2}\,\sigma_{DI}^2$$
$$+ (1 + f_t)^2\sigma_{AA}^2 + (1 + f_t)^3\sigma_{AAA}^2 + \cdots (1 + f_t)^k\sigma_{A^k}^2 \qquad (23.34a)$$

For offspring resulting from selfing a pure-line cross, $\sigma_{ADI} = \sigma_{DI}^2 = 0$, leaving the permanent response as only a function of the additive (and additive epistatic) effects,

$$\sigma_G(\infty, \tau, t) = (1 + f_t)\sigma_A^2 + (1 + f_t)^2\sigma_{AA}^2 + \cdots (1 + f_t)^k\sigma_{A^k}^2 \qquad (23.34b)$$

Equation 23.33 also provides the genetic variance for any particular generation. For the sake of a clearer exposition, we will ignore additive epistasis in what follows (although their inclusion is trivial). First, the total genetic variance in generation T is given by

$$\sigma_G(T, T, T) = (1 + f_T)\sigma_A^2 + (1 - f_T)\sigma_D^2 + 2f_T\sigma_{ADI} + f_T\sigma_{DI}^2 + f_T(1 - f_T)\iota^* \qquad (23.35a)$$

This is the genetic variance across the entire population (across *all* the lines present in generation T). For a pure-line cross, this reduces to

$$\sigma_G(T, T, T) = (1 + f_T)\sigma_A^2 + (1 - f_T)(1 + f_T)\sigma_D^2 \qquad (23.35b)$$

We also require the genetic variance in generation T among the subset of lines that descend from a single individual in generation t. Here, $\tau = T$ and the variance becomes

$$\sigma_G(T, T, t) = (1 + f_t)\sigma_A^2 + \frac{(1 - f_T)^2}{1 - f_t}\,\sigma_D^2 + (f_t + f_T)\,\sigma_{ADI}$$
$$+ \left(f_t + \frac{(f_T - f_t)^2}{2(1 - f_t)}\right)\sigma_{DI}^2 + \frac{f_t(1 - f_T)^2}{1 - f_t}\,\iota^* \qquad (23.36a)$$

An example of this would be the genetic variance across the collection of F_3 or F_4 **bulk families** from a single F_2 parent. For an F_3 family this is $\sigma_G(1, 1, 0)$, as the F_2 represents generation 0 of selfing, while the entire collection of F_4 families that trace back to this F_2 individual has a variance of $\sigma_G(2, 2, 0)$. For the selfed offspring from a pure-line cross, Equation 23.36a simplifies to

$$\sigma_G(T, T, t) = (1 + f_t)\sigma_A^2 + \frac{(1 - f_T)^2}{1 - f_t}\,(1 + f_t)\,\sigma_D^2 \qquad (23.36b)$$

Example 23.7. Suppose we wish to know the among-family genetic variance for $S_{i,j}$ families, namely, the bulk collection of S_j families from a single individual in generation i? From Equation 23.36a, $\sigma_G^2(S_{i,j}) = \sigma_G(j, j, i)$. Substitution of $f_t = 1 - 2^{-t}$ and simplifying yields

$$\sigma_G^2(S_{i,j}) = (2 - 2^i)\sigma_A^2 + 2^{-(2j-i)}\sigma_D^2 + (2 - 2^{-i} - 2^{-j})\sigma_{ADI}^2$$
$$+ \left(1 + 2^{-(2j+1-i)} - 2^{-j} - 2^{-(i+1)}\right)\sigma_{DI}^2 + 2^{-(2j-i)}\left(1 - 2^{-i}\right)\iota^*$$

Some particular results of interest are

$$\sigma_G(S_{0,1}) = \sigma_A^2 + (1/4)\sigma_D^2 + \sigma_{ADI}/2 + (1/8)\sigma_{DI}^2$$

$$\sigma_G(\mathrm{S}_{0,2}) = \sigma_A^2 + (1/16)\sigma_D^2 + (3/4)\sigma_{ADI} + (9/32)\sigma_{DI}^2$$

$$\sigma_G(\mathrm{S}_{1,2}) = (3/2)\sigma_A^2 + (1/8)\sigma_D^2 + (5/4)\sigma_{ADI} + (9/16)\sigma_{DI}^2 + (1/16)\iota^*$$

Again, under the common setting of a pure-line cross, these simplify as $\iota^* = \sigma_D^2$ and $\sigma_{DI}^2 = \sigma_{ADI} = 0$.

Finally, it will prove useful to decompose the total genetic covariance, $\sigma_G(T,t)$, into within- and among-family covariances, $\sigma_{Gw}(T,t)$ and $\sigma_{Gb}(T,t)$, respectively, where

$$\sigma_G(T,t) = \sigma_{Gw}(T,t) + \sigma_{Gb}(T,t) \tag{23.37a}$$

The among-family covariance in generation t is the covariance between sibs from a parent in generation $t-1$

$$\sigma_{Gb}(T,t) = \sigma_G(T,t,t-1) \tag{23.37b}$$

The within-family genetic covariance follows as

$$\begin{aligned}\sigma_{Gw}(T,t) &= \sigma_G(T,t) - \sigma_{Gb}(T,t)\\ &= \sigma_G(T,t,t) - \sigma_G(T,t,t-1)\end{aligned} \tag{23.37c}$$

For more general families, $t-1$ is replaced by $t-j$ when the last common ancestor to the family occured j generations before the collection of families was scored. Note that the within- and among-family genetic variances in generation t are given by $\sigma_{Gw}(t,t)$ and $\sigma_{Gb}(t,t)$, respectively. Recalling Equation 23.8b, this implies a phenotypic variance for the among-family means of $\sigma_{\bar{z}}(t,t) = \sigma_{Gb}(t,t) + \sigma_e^2$.

Example 23.8. Consider the within- and among-family genetic variances for an $\mathrm{S}_{j-1,j}$ family (the offspring from a single S_{j-1} individual). Here $T = \tau = j$ and $t = j-1$, and from Equation 23.37b, the among-family genetic variance is

$$\sigma_{Gb}(j,j) = \sigma_G(j,j,j-1)$$

For a pure-line cross, Equation 23.33 (ignoring epistasis) returns

$$\sigma_{Gb}(j,j) = \sigma_G(j,j,j-1) = \left(2 - \frac{1}{2^{j-1}}\right)\left(\sigma_A^2 + \frac{\sigma_D^2}{2^{j+1}}\right)$$

while from Equation 23.37a, the within-family variance becomes

$$\sigma_{Gw}(j,j) = \sigma_G(j,j,j) - \sigma_G(j,j,j-1)$$

For a pure-line cross, the genetic variation (ignoring epistasis) in the population is

$$\sigma_G(j,j,j) = \left(2 - \frac{1}{2^j}\right)\left(\sigma_A^2 + \frac{\sigma_D^2}{2^j}\right)$$

giving the within-family variance as

$$\sigma_{Gw}(\mathrm{S}_j) = \sigma_G(j,j,j) - \sigma_G(j,j,j-1) = \left(\frac{1}{2^j}\right)\sigma_A^2 + \left(\frac{3-2^{j+1}}{4^j}\right)\sigma_D^2$$

As Figure 23.2B illustrates, we can consider ever-deeper hierarchical levels of population structure with selfing. Suppose we are interested in the response in generation T due to selection among parents in generation t chosen on the basis of their relatives in generation τ, but that we are considering only the response among the subpopulation that descended from a common ancestor in generation k. For example, among all the descendants from a particular S_3 individual, what is the response to selection on their S_4 offspring if we base selection on the S_6 family means? Here T is the generation of interest for the response ($T = \infty$ for the permanent response), $k = 3$, $t = 4$, and $\tau = 6$.

Defining $\sigma_G(T, \tau, t, k)$ as the covariance between T and τ, and given that they shared an ancestor in generation t from subpopulation k, Wright and Cockerham (1986b), following Gates (1954), showed that

$$\sigma_G(T, \tau, t, k) = \sigma_G(T, \tau, t) - \sigma_G(T, \tau, k) \quad \text{for } t > k \tag{23.38}$$

Hence, Equation 23.32 can be used to compute these covariances.

Example 23.9. As an example of the difference between $\sigma_G(T, \tau, t, k)$ and $\sigma_G(T, \tau, t)$, consider the following situation. Suppose we randomly choose F_2s from a pure-line cross and self a large collection of them, generating a total of 1000 lines. While there are a number of options for advancing these lines, two extremes are to advance single lines from each of the original F_2s or to advance 1000 lines from a single F_2. Lacking any other information on the lines, it is obvious that keeping all of the original lines is likely the better strategy (indeed, this was proposed by Compton 1968), but just how much better is it? For both situations, consider the contribution to response in a completely inbred individual ($T = \infty$) from an individual in generation t chosen by the evaluation of relatives in generation $\tau \geq t$.

Because we assumed a pure-line cross, $\sigma_{ADI} = \sigma^2_{DI} = 0$, and Equation 23.34 yields, for the first strategy,

$$\sigma_G(\infty, \tau, t) = \left(1 - \frac{1}{2^{t+1}}\right) 2\sigma^2_A$$

If selection is delayed until at least the fifth generation of selfing, this is essentially $2\sigma^2_A$. Conversely, the covariance for the second strategy (advancing all the lines from a single F_2) is given by $\sigma_G(\infty, \tau, t, 0)$, as all individuals trace back to a single individual in generation 0 (the F_2). Here Equation 23.38 yields

$$\begin{aligned}\sigma_G(\infty, \tau, t, 0) &= \sigma_G(\infty, \tau, t) - \sigma_G(\infty, \tau, 0) \\ &= \left(1 - \frac{1}{2^{t+1}}\right) 2\sigma^2_A - \left(1 - \frac{1}{2}\right) 2\sigma^2_A = \left(1 - \frac{1}{2^{t+1}}\right) \sigma^2_A\end{aligned}$$

This is one half of the covariance when the entire collection of lines is used. This makes sense, as under the infinitesimal model, in an outbred population (such as the collection of F_2s), half the additive variance is between individuals and half is generated by segregation within individuals (Chapter 16). The first covariance, $\sigma_G(\infty, \tau, t)$, considers both the variance within a particular line and also the variance among the initial line founders, while the second, $\sigma_G(\infty, \tau, 0)$, focuses solely on the variance within a particular line.

Response to Family Selection

Our earlier discussions of selection with selfing (Equations 23.23–23.28) assumed that the selection unit was the parent (individual selection), meaning that $\tau = t$. More generally, consider a parent in generation t where we save selfed seed from this individual for the recombination unit and test the parent using the mean of its bulked selfed offspring in generation τ, namely, an $S_{t,\tau}$ family (Figure 23.3). For such cases, the response in generation T from

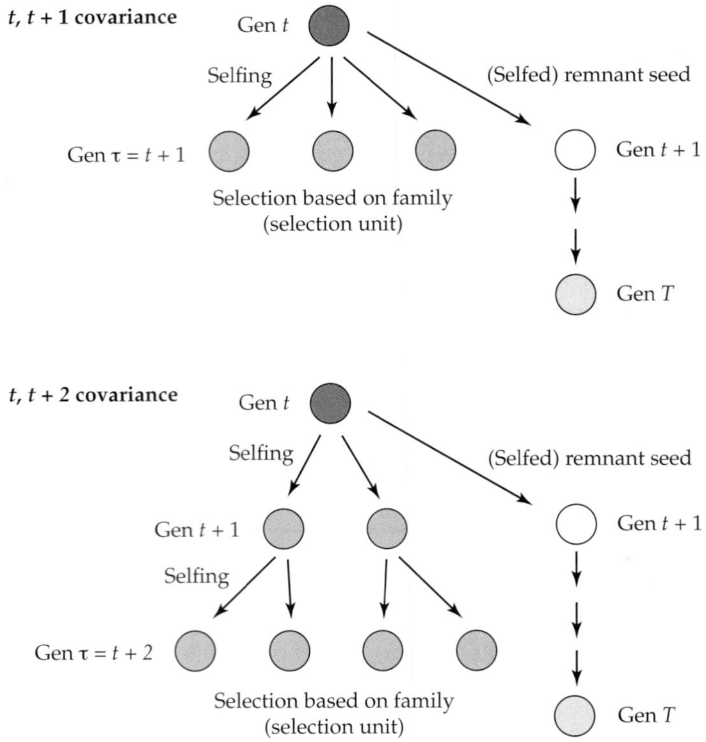

Figure 23.3 Examples of family selection in selfed lines. **Top:** The selection unit is the selfed offspring of a parent, and the recombination unit is a (selfed) remnant seed from this parent, so that selection is based on $S_{t,t+1}$ families. The covariance of interest is between an individual in the selection unit (generation $\tau = t + 1$) and a descendant of the recombination unit (measured in generation T), which had a common parent in generation t. **Bottom:** In species with a low seed set, a single plant may not generate sufficient seed for family testing. In this case, additional seed can be generated by a second round of selfing. If selection is based upon such $S_{t,t+2}$ families, and we use a remnant (selfed) seed from the parent as the recombination unit, the common parent is two generations removed from the selection unit ($\tau = t + 2$).

selection among parents in generation t is

$$R(T,\tau,t) = \bar{\imath}\,\frac{\sigma_G(T,\tau,t)}{\sqrt{\sigma_G(\tau,\tau,t) + \sigma_e^2}} \tag{23.39}$$

where σ_e^2 is the error variance for the mean of the $S_{t,\tau}$ family. Equation 23.39 follows because the genetic variance of the selection unit is $\sigma_G(\tau,\tau,t)$, while the covariance between the selection unit (τ) and an offspring of the recombination unit measured in generation T have their last common parent in generation t, and hence the appropriate covariance is given by $\sigma_G(T,\tau,t)$ (Figure 23.3).

For the cases in Figure 23.3, where families are selected and remnant seed from those families is used to form the next (selfed) generation, then for selection based on such $S_{t,t+1}$ families, Equation 23.34 returns the permanent response to selection (that remaining in the fully inbred lines) as

$$\sigma_G(\infty, t+1, t) = (1 + f_t)\sigma_A^2 + \frac{3 + 5f_t}{4}\,\sigma_{ADI} + \frac{1 + 3f_t}{4}\,\sigma_{DI}^2 + (1 + f_t)^2\sigma_{AA}^2 \tag{23.40a}$$

while for selection based on $S_{t,t+2}$ families,

$$\sigma_G(\infty, t+2, t) = (1 + f_t)\sigma_A^2 + \frac{7 + 9f_t}{8}\,\sigma_{ADI} + \frac{3 + 5f_t}{8}\,\sigma_{DI}^2 + (1 + f_t)^2\sigma_{AA}^2 \tag{23.40b}$$

As pointed out by Cockerham and Matzinger (1985), the long-term response under these two schemes differs only in the dominance-related terms, σ^2_{DI} and σ_{ADI}. Because both of these terms are zero in lines derived from the cross of two pure lines, the permanent response in such cases is entirely a function of additive and additive-epistatic effects.

If instead of using remnant seed, a measured sib from the selected families is selfed (strict family selection, as opposed to sib—remnant seed—selection), the expressions become a little more complex. If n family members are scored, the appropriate covariance for the permanent response selection based upon families from one generation of selfing is

$$\frac{1}{n}\,\sigma_G(\infty,t+1,t+1)+\frac{n-1}{n}\,\sigma_G(\infty,t+1,t) \tag{23.41a}$$

This arises if we follow the logic leading to Equation 23.8b. Because selection is based on the family mean, $1/n$ is the weight on the individual in the family that is also the recombination unit (hence, the common ancestor is simply this individual, which is in generation $t+1$) and the remainder $(1-1/n)$ are sibs of this individual (and the common ancestor is in the previous generation). If the number, n, of family members tested is large, the correction given by Equation 23.41a can be ignored and the simpler version (Equation 23.40a) can be used instead. Similarly, if family selection is based on two generations of selfing, then the appropriate covariance becomes

$$\frac{1}{n^2}\,\sigma_G(\infty,t+2,t+2)+\frac{n-1}{n^2}\,\sigma_G(\infty,t+2,t+1)+\frac{n-1}{n}\,\sigma_G(\infty,t+2,t) \tag{23.41b}$$

For large values of n, this reduces to Equation 23.40b.

Within-family Selection Under Selfing

Our results for selfing thus far have only been concerned with selection among lines. If selection is practiced entirely *within* a single selfed lineage (e.g., within the progeny of a single selfed individual), genetic variation is quickly removed and the selection response rapidly ceases. Pederson (1969b) obtained the response remaining in generation T from within-family selection in generation t as

$$r_w(T,t)=\bar{\imath}_t\,\frac{\sigma_{Gw}(T,t)}{\sqrt{\sigma_{Gw}(t,t)+\sigma^2_{E_w}}} \tag{23.42a}$$

where $\sigma^2_{E_w}$ is the within-family environmental variance among individuals. If we apply Equation 23.37c, this becomes

$$r_w(T,t)=\bar{\imath}_t\,\frac{\sigma_G(T,t,t)-\sigma_G(T,t,t-1)}{\sqrt{\sigma_G(t,t,t)-\sigma_G(t,t,t-1)+\sigma^2_{E_w}}} \tag{23.42b}$$

Recalling Equation 23.34, for a pure-line cross, we have

$$\sigma_G(T,t,t)-\sigma_G(T,t,t-1)=$$

$$2^{-t}\,\sigma^2_A+2^{-T}\,\sigma^2_D+\sum_{\ell=2}\left[\left(2-\frac{1}{2^t}\right)^{\ell}-\left(2-\frac{1}{2^t}\right)^{\ell-1}\right]\sigma^2_{A^\ell} \tag{23.42c}$$

where the sum is over the degree of A-epistasis. Ignoring epistasis, the cumulative response to k generations of selection at generation $T>k$ is

$$R(T\,|\,k)=\sum_{t=1}^{k}\,\bar{\imath}_t\,\frac{2^{-t}\,\sigma^2_A+2^{-T}\,\sigma^2_D}{\sqrt{2^{-t}\,(\sigma^2_A+\sigma^2_D)+\sigma^2_{E_w}}} \tag{23.42d}$$

(Pederson 1969b). Note that the sum here begins at selfing generation 1, as selection starts within the S_1 inbred family. Because any within-family genetic variation rapidly decays, after only a few generations the response essentially stops, with the amount of permanent response due to selection in generation t depending on $\sigma_A^2 \,/\, 2^t$. Because

$$\sigma^2_{E_w} \leq 2^{-t}\,(\sigma^2_A + \sigma^2_D) + \sigma^2_{E_w} \leq \sigma^2_z \tag{23.43a}$$

it follows that the permanent response ($T \to \infty$) can be bounded by noting

$$\bar{\imath}\,\frac{\sigma_A^2}{\sigma_{E_w}} \sum_{t=1}^{\infty} 2^{-t} \geq \bar{\imath}\,\sigma_A^2 \sum_{t=1}^{\infty} \frac{2^{-t}}{\sqrt{2^{-t}\,(\sigma_A^2 + \sigma_D^2) + \sigma^2_{E_w}}} \geq \bar{\imath}\,\frac{\sigma_A^2}{\sigma_z} \sum_{t=1}^{\infty} 2^{-t} \tag{23.43b}$$

Because the leftmost and rightmost power series sum to one, the total permanent response $R(\,\infty\,|\,\infty\,)$ under continued within-family selection (ignoring new mutation) is

$$\bar{\imath}\,\sigma^2_A/\sigma_{E_w} \geq R(\,\infty\,|\,\infty\,) \geq \bar{\imath}\,h\,\sigma_A \tag{23.43c}$$

showing that the total response is at least twice the response of the first generation ($\bar{\imath}\,h\,\sigma_A/2$).

Combined Selection

Suppose n_2 F_2 individuals are collected and selfed to create F_3 families of size n_3 for each of the F_2 founding lines. In trying to choose the best F_3 line, we might consider **combined selection**, basing the choice of which individuals to save on both their individual values (z_{ij}) and the average value of the entire F_2 family ($\overline{z}_{i\cdot}$) from which they derive, with

$$\overline{z}_{i\cdot} = \frac{1}{n_3} \sum_{j=1}^{n_3} z_{ij}$$

where n_3 is the number of measured sibs in the family. By analogy with the family index (Chapter 21), we can select individuals using the index

$$I_{ij} = b\,\overline{z}_{i\cdot} + z_{ij} \tag{23.44}$$

As with the standard family index, a large value of b places more weight on the family average, while a small value of b places more weight on the individual value. Individuals chosen of the basis of a superior I value are then selfed to complete fixation in order to generate a fully inbred line. The expected response in generation T of selfing follows from a slight modification to the generalized breeder's equation (Equation 13.10b), and is

$$R_y(T) = \bar{\imath}\,\frac{\sigma(I_{ij}, y_T)}{\sigma(I_{ij})} \tag{23.45}$$

where y_T is a direct descendant of the individual z_{ij} in generation T of selfing. The permanent response is given by considering $T = \infty$.

It will be useful at this point to introduce two variance components that will (somewhat) simplify our results. First define

$$\sigma^2_c(T) = \sigma_G(T, 1, 0) \tag{23.46a}$$

as the variance between crosses (the different F_2s) and

$$\sigma^2_{\ell|c}(T) = \sigma_G(T, 1, 1) - \sigma_G(T, 1, 0) \tag{23.46b}$$

as the variance in lines within crosses (the variance in deviations from the mean cross effect). For a pure-line cross, these variances are

$$\sigma^2_c(T) = \sigma^2_A + 2^{-(T+1)}\sigma^2_D + \sigma^2_{AA} + \cdots \sigma^2_{A^k} \tag{23.47a}$$

$$\sigma^2_{\ell|c}(T) = (1/2)\,\sigma^2_A + 2^{-T}\,\sigma^2_D + (5/4)\,\sigma^2_{AA} + \cdots + \left[\left(\frac{3}{2}\right)^k - 1\right]\sigma^2_{A^k} \tag{23.47b}$$

With these variance components in hand, consider the numerator of Equation 23.45. First note that

$$\sigma(I_{ij}, y_T) = b\sigma(\,\overline{z}_{i\cdot}, y_T) + \sigma(z_{ij}, y_T)$$

Because y_T is a direct descendant of line ij, $\tau = t = 1$, hence

$$\sigma(\,z_{ij}, y_T) = \sigma_G(T,1,1) \tag{23.48a}$$

Likewise,

$$\sigma(\,\overline{z}_{i\cdot}, y_T) = \frac{1}{n_3}\sum_{k=1}^{n_3}\sigma(\,z_{ik}, y_T) = \frac{\sigma(\,z_{ij}, y_T)}{n_3} + \left(1 - \frac{1}{n_3}\right)\sigma(\,z_{ik}, y_T)$$

The first term is given by Equation 23.48a, while the second is the covariance between a sib, z_{ik}, of individual ij. Here, the common parent between ik and y is the F_2, meaning that $t = 0$, $\tau = 1$, and this covariance is $\sigma_G(T,1,0)$. Putting these together,

$$\sigma(\,\overline{z}_{i\cdot}, y_T) = \sigma_G(T,1,0) + \frac{\sigma_G(T,1,1) - \sigma_G(T,1,0)}{n_3} = \sigma^2_c(T) + \frac{\sigma^2_{\ell|c}(T)}{n_3} \tag{23.48b}$$

The numerator covariance in Equation 23.45 thus becomes

$$\sigma(I_{ij}, y_T) = b\left(\sigma^2_c(T) + \frac{\sigma^2_{\ell|c}(T)}{n_3}\right) + \sigma^2_c(T) + \sigma^2_{\ell|c}(T) \tag{23.49}$$

Now turning to the variance of the selection index,

$$\sigma^2(I_{ij}) = \sigma^2\left(b\,\overline{z}_{i\cdot} + z_{ij}\right) = b^2\sigma^2\left(\,\overline{z}_{i\cdot}\,\right) + \sigma^2\left(\,z_{ij}\right) + 2b\sigma\left(\,\overline{z}_{i\cdot}, z_{ij}\right) \tag{23.50a}$$

First, consider the variance of a random F_3 line,

$$\sigma^2\left(\,z_{ij}\right) = \sigma_G(1,1,1) + \sigma^2_\epsilon = \sigma^2_c(1) + \sigma^2_{\ell|c}(1) + \sigma^2_\epsilon \tag{23.50b}$$

where σ^2_ϵ is the within-line error variance. Next, note that

$$\sigma^2\left(\,\overline{z}_{i\cdot}\,\right) = \frac{n_3\,\sigma^2(\,z_{ij}\,)}{n_3^2} + \frac{n_3(n_3-1)\,\sigma(\,z_{ij}, z_{ik}\,)}{n_3^2} = \frac{\sigma^2(\,z_{ij}\,)}{n_3} + \left(1 - \frac{1}{n_3}\right)\sigma(\,z_{ij}, z_{ik}\,)$$

The first covariance is given by Equation 23.50b, while the second is the covariance between sibs, $\sigma(1,1,0)$, which yields

$$\sigma^2\left(\,\overline{z}_{i\cdot}\,\right) = \frac{\sigma(1,1,1) + \sigma^2_\epsilon}{n_3} + \left(1 - \frac{1}{n_3}\right)\sigma(1,1,0) = \sigma^2_c(1) + \frac{\sigma^2_{\ell|c}(1) + \sigma^2_\epsilon}{n_3} \tag{23.50c}$$

Finally, we can show that $\sigma\left(\,\overline{z}_{i\cdot}, z_{ij}\right) = \sigma^2\left(\,\overline{z}_{i\cdot}\,\right)$. Putting all these together gives

$$\sigma^2(I_{ij}) = b(b+2)\left(\sigma^2_c(1) + \frac{\sigma^2_{\ell|c}(1) + \sigma^2_\epsilon}{n_3}\right) + \left(\sigma^2_c(1) + \sigma^2_{\ell|c}(1) + \sigma^2_\epsilon\right) \tag{23.51}$$

Substitution of Equations 23.49 and 23.51 into 23.45 shows the expected response for arbitrary b and $T > 1$.

An obvious question is what value of b maximizes response? Taking the derivative of Equation 23.45 with respect to b and solving for zero returns the optimal value as

$$b = \frac{n_3 \left[\sigma_c^2(T)\, \sigma_\epsilon^2 - \sigma_c^2(1)\, \sigma_{\ell|c}^2(1) + \sigma_c^2(T)\, \sigma_{\ell|c}^2(1) \right]}{\sigma_{\ell|c}^2(T) \left[n_3\, \sigma_c^2(1) + \sigma_{\ell|c}^2(1) + \sigma_\epsilon^2 \right]} \tag{23.52}$$

As with family-index selection (Chapter 21), the selection intensity is reduced by the fact that F_3 lines from the same F_2 are related and hence correlated, and we must correct for this correlation. Letting ρ be the correlation between the index scores from lines in the same cross, if we choose the best K from $N = n_2 n_3$ total lines (n_3 lines from each of n_2 F_2's), then Equation 21.57b returns the adjusted selection intensity as

$$\bar{\imath}_{n_2 n_3, K} = \bar{\imath}_q - \frac{1-q}{2\, \bar{\imath}_q\, q \left[n_2 n_3 (1-\rho) + n_2 \rho + 1 \right]} \tag{23.53}$$

where $q = K/N$ is the fraction saved (we use q in place of our normal usage of p to avoid confusion here with ρ) and $\bar{\imath}_q$ the infinite-population size selection intensity (Equation 14.3a). To compute the correlation, ρ, between index scores, first note that

$$\begin{aligned}
\sigma(I_{ij}, I_{ik}) &= \sigma\left(b\, \overline{z}_{i\cdot} + z_{ij},\, b\, \overline{z}_{i\cdot} + z_{ik}\right) \\
&= b^2 \sigma^2(\overline{z}_{i\cdot}) + 2b\, \sigma(\overline{z}_{i\cdot}, z_{ij}) + \sigma(z_{ij}, z_{ik}) \\
&= \left[b^2 \sigma^2(\overline{z}_{i\cdot}) + 2b\, \sigma(\overline{z}_{i\cdot}, z_{ij}) + \sigma^2(z_{ij}) \right] - \sigma^2(z_{ij}) + \sigma(z_{ij}, z_{ik}) \\
&= \sigma^2(I_{ij}) - \left(\sigma_c^2 + \sigma_{\ell|c}^2 + \sigma_\epsilon^2 \right) + \sigma_c^2 \\
&= \sigma^2(I_{ij}) - \left(\sigma_{\ell|c}^2 + \sigma_\epsilon^2 \right)
\end{aligned} \tag{23.54a}$$

The third step follows upon recalling Equation 23.50a. The correlation between index scores among individuals from the same line thus becomes

$$\rho = \frac{\sigma(I_{ij}, I_{ik})}{\sigma^2(I)} = \frac{\sigma^2(I_{ij}) - \left(\sigma_{\ell|c}^2 + \sigma_\epsilon^2 \right)}{\sigma^2(I_{ij})} = 1 - \frac{\sigma_{\ell|c}^2 + \sigma_\epsilon^2}{\sigma^2(I_{ij})} \tag{23.54b}$$

Substitution of this value into Equation 23.53 yields the appropriate selection intensity, which is corrected for both finite population size and correlation between lines.

When there is a fixed total number of lines to examine, there is a tradeoff between selection intensity and accuracy. The optimal design for known (or estimated) genetic variances can be obtained by numerically evaluating different combinations of the n_2/n_3 ratio (for fixed values of N) in Equation 23.45 (and its associated components, Equations 23.49 and 23.51–23.53). Weber (1982, 1984; Wricke and Weber 1986) examines index selection under selfing, including both optimal design (in the absence of dominance) and its extension to additional generations of selfing. For example, Weber (1982) considered the more general situation where each F_2 family consists of n_3 F_3 families, with each F_3 family consisting of n_4 F_4 families and so on to F_j families. In this case, the full index is

$$I = b_2\left(\overline{z}_2 - \overline{z}\right) + b_3\left(\overline{z}_3 - \overline{z}_2\right) + \cdots + b_j\left(\overline{z}_j - \overline{z}_{j-1}\right) \tag{23.55a}$$

The weights, b_k, are chosen to maximize the correlation between the index, I, and the final genetic value of the completely inbred lines (which we denote as g_∞), which (Weber 1982) implies

$$b_k = \frac{\sigma\left(\overline{z}_k - \overline{z}_{k-1}, g_\infty\right)}{\sigma^2\left(\overline{z}_k - \overline{z}_{k-1}\right)} \tag{23.55b}$$

Table 23.4 The covariances required to predict response under partial selfing using Equation 23.56. The probability of selfing is η, with an equilibrium level of inbreeding of $f = \eta/(2-\eta)$. Let σ_{G_O}(P,O) and σ_{G_O}(P,O) denote the parent-offspring covariance under outcrossing and selfing (respectively) at equilibrium, while $\sigma_{G(BR)}$ and $\sigma_{G(AR)}$ denote the population-averaged (i.e., over outcrossing and selfing) parent-offspring regression for selection before and after reproduction. (After Wright and Cockerham 1985.)

Cov	σ^2_A	σ^2_D	σ_{ADI}	σ^2_{DI}	ι^*	$\iota^2 - \iota^*$
σ_{G_O}(P,O)	$\frac{1+f}{2}$	0	$\frac{f}{2}$	0	0	0
σ_{G_S}(P,O)	$1+f$	$\frac{1-f}{2}$	$\frac{1+7f}{4}$	f	$\frac{f(1-f)}{2(2+f)}$	$\frac{f(1-f^2)}{2(2+f)}$
$\sigma_{G(AR)}$	$\frac{1+3f}{2}$	$\frac{f(1-f)}{1+f}$	$\frac{2f(1+3f)}{2(1+f)}$	$\frac{2f^2}{1+f}$	$\frac{f^2(1-f)}{(1+f)(2+f)}$	$\frac{f^2(1-f^2)}{(1+f)(2+f)}$
$\sigma_{G(BR)}$	$1+f$	$\frac{f(1-f)}{1+f}$	$\frac{f(3+5f)}{2(1+f)}$	$\frac{2f^2}{1+f}$	$\frac{f^2(1-f)}{(1+f)(2+f)}$	$\frac{f^2(1-f^2)}{(1+f)(2+f)}$
σ^2_G	$1+f$	$1-f$	$2f$	f	$\frac{f(1-f)}{2+f}$	$\frac{f(1-f^2)}{2+f}$

Consult Weber (1982) for further details. Finally, an important extension of combined selection schemes is the work of Cowling et al. (2015), who incorporated multiple-generation family information into a BLUP-based index (i.e., they used the animal model to predict the breeding values of focal individuals.

RESPONSE UNDER PARTIAL SELFING

One of the most widespread natural systems of inbreeding is **partial selfing**, wherein each individual can either self or outcross. If η is the probability of selfing, then an unselected population approaches a *mean* inbreeding value of $f = \eta/(2-\eta)$, but there is a *distribution* of inbreeding values among individuals within the population. In particular, the probability that an individual is inbred to a level of $f_i = 1 - 2^{-i}$ follows a geometric distribution with a success parameter, η, yielding $\Pr(f_i) = (1-\eta)\eta^i$. This lack of uniformity in f greatly complicates the prediction of selection response. In particular, nonlinear parent-offspring regressions can occur and selection can change the distribution of inbreeding-value classes away from a simple geometric form (Wright and Cockerham 1985; Kelly 1999a). Our treatment first presents approximate results using covariances by treating the entire population as a single unit. We then consider a more careful treatment due to Kelly (1999a, 1999b) that considers the selection response *within* each group at a given level of inbreeding (i.e., all group members have the same f value), which also examines the impact of selection-induced disequilibrium.

An Approximate Treatment Using Covariances

Partial selfing has been examined by Wright and Cockerham (1985, 1986a) and Wright (1988), who obtained the appropriate cross-generational covariances and predicted response when using the method on ancestral regression. They assumed that the population is at the equilibrium mean inbreeding value and that each individual has the same probability (η) of selfing (i.e., there is no genetic variation in selfing vs. outcrossing rates). Because a parent can either self or outcross (or both, e.g., on different flowers on the same plant), the parent-offspring covariance must take this into account. Further, because of selfing, the covariances for selection before and after reproduction are slightly different. For selection before reproduction (BR), the single-generation response is given by

$$R_{BR} = \bar{\imath}\,\frac{\sigma_{G(BR)}(t+1,t)}{\sqrt{\sigma^2_G + \sigma^2_e}} \tag{23.56a}$$

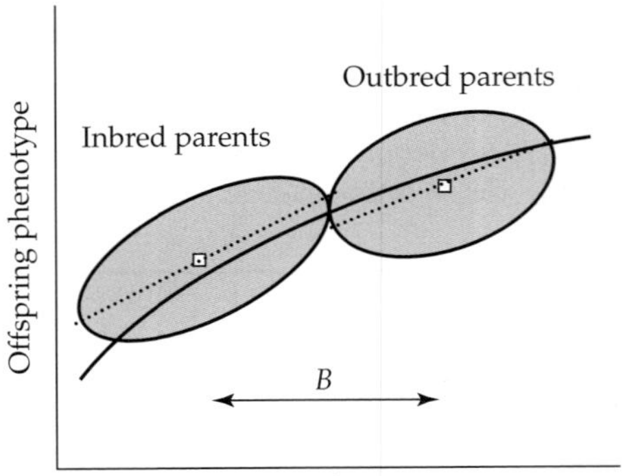

Figure 23.4 When inbreeding depression is present, the parent-offspring regression under partial selfing can be nonlinear. To see this, suppose we have just two groups: $f = 0$ (outcrossed) and $f = 1$ (fully inbred). Within each group, parent-offspring regressions are linear (as indicated by the dashed lines). However, for the population as a whole (i.e., examining individuals without knowledge of which group they belong to), the regression is nonlinear. Here B denotes the inbreeding depression (the change in mean, the open squares, between the outcrossed and fully inbred populations). (After Kelly 1999a.)

where the genetic covariance for selection before reproduction is

$$\sigma_{G(BR)}(t+1,t) = 2(1-\eta)\sigma_{G_O}(P,O) + \eta\,\sigma_{G_S}(P,O) \tag{23.56b}$$

where $\sigma_{G_0}(P,O)$ is the parent-offspring covariance under outcrossing and $\sigma_{G_S}(P,O)$ is the parent-offspring covariance under selfing. Because the population is assumed to be at its inbreeding equilibrium, the parent-offspring covariances and total genetic variation (σ^2_G) are assumed to be constants, independent of t. Table 23.4 shows the coefficients for these covariances, as well as those for σ^2_G. Note that an additional quadratic component, $\iota^2 - \iota^*$, also appears in the covariance between relatives (which arises because there is a distribution of inbreeding values in the population). As we will see, there is a significant transient component to the immediate response. For selection after reproduction (AR), an outcrossed individual has only one parent under selection (the pollen parent not being under selection), and the covariance in the numerator of Equation 23.56a is replaced by

$$\sigma_{G(AR)}(t+1,t) = (1-\eta)\sigma_{G_O}(P,O) + \eta\,\sigma_{G_S}(P,O) \tag{23.56c}$$

The permanent response to selection is given by replacing the numerator covariance in Equation 23.56a by the appropriate version of $\sigma_G(\infty,t)$. For selection before reproduction, Wright and Cockerham showed that this equals

$$\sigma_{G(BR)}(\infty,t) = (1+f)\,\sigma^2_A + \left[\frac{\eta(3-\eta)}{(2-\eta)^2}\right]\sigma_{ADI} + \left[\frac{\eta^2}{(2-\eta)^2}\right]\sigma^2_{DI} \tag{23.57a}$$

Further, the before- and after-selection covariances are related by

$$\sigma_{G(AR)}(\infty,t) = \left(\frac{1+\eta}{2}\right)\sigma_{G(BR)}(\infty,t) \tag{23.57b}$$

While the coefficient on σ^2_A is unchanged in $\sigma_G(\infty,t)$ ($1+f$ for any value of t), the limiting contributions ($T=\infty$) from all the other quadratic components have rather different coefficients relative to their single-generation values. Thus, the transient component of response is expected to be considerable. Wright and Cockerham found that

$$\sigma_{G(BR)}(t+T,t) = \sigma_{G(BR)}(\infty,t) + \left[\sigma^2_G - \sigma_{G(BR)}(\infty,t)\right]\left(\frac{\eta}{2}\right)^T \tag{23.58}$$

with a similar expression for the covariance for selection after reproduction. The transient component decays rather rapidly, as the rate of decay is no slower than 2^{-T}. Wright and Cockerham also showed that the final change in the mean from a generation of selection can be predicted from the response in the first two generations after selection, with

$$\overline{z}_\infty = \frac{2\,\overline{z}_2 - \eta\,\overline{z}_1}{2-\eta} \tag{23.59}$$

where $\overline{z}_i$ is the mean i generations after a generation of selection. Family selection was also examined by Wright and Cockerham (1986a) and Edwards (2008). With partial-selfing, there are a number of potential families that one can consider for the selection unit: outcrossed half- and full-sibs, selfed individuals, or naturally pollinated individuals (a mixture of selfed and outcrossed progeny). Consult Wright and Cockerham for details on response under these different systems.

While straightforward (beyond some tedious bookkeeping), there are several potential problems with this covariance approach for accommodating partial selfing, which arise from individuals varying in the amount of selfing. In particular, selected individuals may not be a random sample of the inbreeding classes. If the focal trait shows inbreeding depression, the less-inbred individuals are expected to be chosen by selection more often. This has the effect of creating a nonlinear parent-offspring regression (Figure 23.4). If the nonlinearity is significant, the covariance approach (which makes predictions based on a *linear* regression) is inappropriate. Even in the absence of inbreeding depression on the focal trait, more-inbred individuals are often expected to be generally less fit. Together with any inbreeding depression on the focal trait, this skews the distribution of selected individuals that have selfed for i generations away from the geometric value $(1-\eta)\eta^i$ predicted under no selection (Kelley 1999a).

A More Careful Treatment: Kelly's Structured Linear Model

Kelly (1999a, 1999b; Kelly and Williamson 2000) proposed an alternative approach for examining the selection response with partial selfing. By analogy with a geographically structured population, he suggested that a more careful approach is to consider the population as a set of cohorts, within which all individuals have the same level of inbreeding. Cohort 0 contains all individuals that arose through random mating in the previous generation, cohort 1 contains those individuals that arose from selfing in the previous generation, and cohort i contains those individuals that have undergone i rounds of selfing before they last enjoyed random mating, with $f_i = 1 - 1/2^i$. Kelly terms this approach a **structured linear model (SLM)**, as within each cohort, it is assumed that the parent-offspring regressions are linear. The response under selfing is then predicted by specifying both the within-cohort variances and covariances and the among-cohort transitions. Besides properly accounting for the structured nature of partly selfing populations, Kelly's treatment also accommodates the effects of linkage disequilibrium (under the infinitesimal-model assumptions).

Before examining the SLM approach, we discuss several important observations from simulations reported by Kelly (1999a; Kelly and Williamson 2000). If directional dominance is present (which implies that inbreeding depression occurs; LW Chapter 10), then the distribution of the cohorts will not follow the geometric distribution predicted from the strict (i.e, no selection) partial-selfing model. Instead, selection greatly skews this distribution. For example, when advantageous alleles are recessive, inbred individuals will be overrepresented, while when advantageous alleles are dominant, inbreds are underrepresented. In either setting, the Wright-Cockerham covariance approach provides incorrect results, as it assumes that there is a geometric distribution of inbreeding values. The deviation from this distribution caused selection can be quite striking. For example, after 10 generations of selection when beneficial recessives are present, the cohort distribution is strongly bimodal, with peaks at 0 (outbreeding) and 10 (selfing for all 10 generations). A second feature seen in Kelly's simulations was that linkage disequilibrium generated significant changes in the variances and covariances. Changes in the value of f for the population directly impact the usable level of additive variance, and selfing reduces the chance for recombination in heterozygotes to break down nonrandom associations generated by selection. Such selection-induced changes in the covariances from their strict (i.e., no-selection) partial-selfing values also compromises the results from the covariance approach.

Turning to the details of Kelly's approach, define cohort i as the collection of individuals that have undergone i generations of selfing since their last outcrossing. Let $\pi_i(t)$ denote the frequency of cohort i in generation t, where $\pi_i(t) = (1-\eta)\eta^i$ in the absence of selection. If we assume that selection acts before reproduction, the transition probabilities between cohorts are as follows: with probability $1-\eta$, an individual outcrosses, so

$$\pi_0(t+1) = 1-\eta \tag{23.60a}$$

For an individual to enter cohort $i+1$, it must have been a member of cohort i in the previous generation and it must self. Further, Kelly allows for selection against cohorts (for example, highly inbred individuals may have lower fitness). If $W(t)$ and $W_i(t)$ denote the average fitness of a random individual from the population and from cohort i (respectively) in generation t, then the fraction of the population that is in cohort i after selection (but before reproduction) is $\pi_i(t)\,W_i(t)/W(t)$. Putting all these together, the fraction of the population in cohort $i+1$ in the next generation is

$$\pi_{i+1}(t+1) = \eta\left(\frac{W_i(t)}{W(t)}\right)\pi_i \tag{23.60b}$$

In the absence of selection, namely, $W_i(t) = W(t)$ for all i, Equation 23.60b generates a geometric distribution, with $\pi_i(t) = (1-\eta)\eta^i$.

Now consider the dynamics of the mean. Because a member of cohort i has an inbreeding level of $f_i = 1-2^{-i}$, the mean value of inbreeding for the population is

$$f(t) = \sum \pi_i(t)\,f_i = \sum \pi_i(t)\left(1-2^{-i}\right) = 1-\sum \pi_i(t)/2^i \tag{23.61}$$

Denote the allele frequency-weighted sum of average effects by A (the average breeding value) and the weighted sum of homozygous dominance deviations by B, so that

$$A = 2\sum_k\sum_j \alpha_{kj}\,p_{kj}, \qquad B = \sum_k\sum_j \delta_{kjj}\,p_{kj} \tag{23.62}$$

where α_{kj} is the average effect for allele j from locus k, p_{kj} is the allele frequency, and δ_{kjj} represents the dominance deviations for homozygotes (where α_{kj} and δ_{kjj} are the values for a base population in Hardy-Weinberg equilibrium with the same allele frequencies, p_{kj}). If $A(t)$ and $B(t)$ denote the average values of A and B at generation t, then the mean population phenotype is

$$z(t) = \mu + A(t) + f(t)B(t) \tag{23.63a}$$

This is simply a restatement of Equation 23.1c, as $A(t)$ is the response in the absence of inbreeding and $f(t)B(t)$ is the amount of inbreeding depression (Equation 23.1b). If we let $A_i(t)$ and $B_i(t)$ denote the average value of A and B in cohort i in generation t, it follows that

$$A(t) = \sum \pi_i(t)\,A_i(t), \qquad \text{and} \qquad B(t) = \sum \pi_i(t)\,B_i(t) \tag{23.63b}$$

Kelly assumed that *within* each cohort, the parent-offspring regressions are linear, which means that the value, A', of A following selection in cohort i is

$$A_i'(t) = A_i(t) + \frac{\sigma_{Az,i}(t)}{\sigma^2_{z,i}(t)}\,S_i(t) \tag{23.64a}$$

where $\sigma_{Az,i}$ is the covariance between A and z (the breeding value and phenotype, respectively) prior to selection at an inbreeding level of f_i. Because the mean additive value of selfed offspring (now in cohort $i+1$) equals that of their parent (from cohort i),

$$A_{i+1}(t+1) = A_i'(t) = A_i(t) + \frac{\sigma_{Az,i}(t)}{\sigma^2_{z,i}(t)}\, S_i(t) \tag{23.64b}$$

Likewise, the mean additive value of outcrossed offspring equals the average additive values of their parents,

$$A_0(t+1) = \sum_i \pi_i(t)\, A_i'(t) = A(t) + \sum_i \pi_i(t)\, \frac{\sigma_{Az,i}(t)}{\sigma^2_{z,i}(t)}\, S_i(t) \tag{23.64c}$$

Under linkage equilibrium, Kelly showed that these covariance functions are given by

$$\sigma_{Az,i}(t) = (1+f_i)\sigma^2_A + f_i \sigma_{ADI} \tag{23.65a}$$

Similar expressions exist for $B_i(t)$, with B replacing A in Equations 23.64a–23.64c (under linkage equilibrium), and with the cohort-i covariance between B and z given by

$$\sigma_{Bz,i}(t) = \sigma_{ADI}(1+f_i)/2 + f_i \sigma^2_{DI} \tag{23.65b}$$

Finally, again under linkage equilbrium, the phenotypic variance in cohort i is given by

$$\sigma^2_{z,i}(t) = \sigma^2_G(i,i) + \sigma^2_{e,i} \tag{23.65c}$$

where $\sigma^2_G(i,i)$, the genetic variance after i generations of selfing, is shown by Equation 23.23. Iteration of these equations calculates the evolution of the $A_i(t)$ and $B_i(t)$ values, from which the population mean in generation t can be obtained from Equations 23.63b and 23.63c.

As mentioned throughout this chapter, selection-induced gametic-phase disequilibrium (LD) can have a dramatic effect on the covariances (even for unlinked loci), and indeed this was observed in Kelly's above-mentioned simulations. To accommodate this concern, Kelly suggested an iterative approach (along the lines of our results for random mating from Chapter 16 and of Hayashi and Ukai's (1994) results for pure selfing; Equation 23.30) for computing all the required covariances when disequilibrium is present. As in Chapter 16, let

$$\kappa_i(t) = 1 - \frac{\sigma^2_{z^*,i}(t)}{\sigma^2_{z,i}(t)} \tag{23.66}$$

denote the reduction in the phenotypic variance (in this case, in cohort i) caused by selection, where $\sigma^2_{z^*,i}(t)$ is the phenotypic variance (in inbreeding cohort i) after selection. Under the infinitesimal-model assumptions, Kelly showed that the recursion equations for the variances and covariances in the inbred cohorts ($i \geq 1$) become

$$\sigma^2_{A,i+1}(t+1) = \sigma^2_{A,i}(t) - \kappa_i(t)\, \frac{[\sigma_{Az^*,i}(t)]^2}{\sigma^2_{z,i}(t)} + 2^{-(i+1)}\sigma^2_A \tag{23.67a}$$

$$\sigma_{AB,i+1}(t+1) = \sigma_{AB,i}(t) - \kappa_i(t)\, \frac{\sigma_{Az^*,i}(t)\, \sigma_{Bz^*,i}(t)}{\sigma^2_{z,i}(t)} + 2^{-(i+2)}\sigma_{ADI} \tag{23.67b}$$

$$\sigma^2_{B,i+1}(t+1) = \sigma^2_{B,i}(t) - \kappa_i(t)\, \frac{[\sigma_{Bz^*,i}(t)]^2}{\sigma^2_{z,i}(t)} + 2^{-(i+2)}\sigma^2_{DI} \tag{23.67c}$$

where

$$\sigma_{Az^*,i}(t) = \sigma^2_{A,i}(t) + (1-2^{-i})\, \sigma_{AB,i}(t) \tag{23.67d}$$

$$\sigma_{Bz^*,i}(t) = \sigma_{AB,i}(t) + (1 - 2^{-i})\, \sigma^2_{B,i}(t) \tag{23.67e}$$

For Equations 23.67a–23.67c, the first term is the variance (covariance) before selection, the second term is the reduction from selection, and the final term is the contribution from segregation (these are the extensions of Equation 16.8b to selfing). These generalize the previous results of Hayashi and Ukai (1994; Equation 23.30). If we note that $B = 0$ when only additive gene action is present, then recalling Equation 16.11a recovers Equation 23.30.

The recursion equations for the outbred cohort are more involved, as parents are inbred to differing degrees. Letting $f'(t)$ denote the average value of inbreeding among selected parents, then

$$\sigma^2_{A,o}(t+1) = \frac{1}{2}\,\sigma^2_{A'}(t) + \frac{1}{2}\,[\,1 - f'(t)\,]\,\sigma^2_A \tag{23.68a}$$

$$\sigma_{AB,o}(t+1) = \frac{1}{2}\,\sigma_{AB'}(t) + \frac{1}{4}[\,1 - f'(t)\,]\,\sigma_{ADI} \tag{23.68b}$$

$$\sigma^2_{B,o}(t+1) = \frac{1}{2}\,\sigma^2_{B'}(t) + \frac{1}{4}[\,1 - f'(t)\,]\,\sigma^2_{DI} \tag{23.68c}$$

Variances and covariances denoted by a prime are the population averages in the selected parents, and f' denotes the average inbreeding over all selected parents. Unfortunately, these cannot be simply expressed as base-population genetic components; see Kelly (1999a) for expressions.

CONCLUSIONS: AN INCOMPLETE THEORY FOR SHORT-TERM RESPONSE

For all of the detailed formulae presented in this chapter, it must be stressed that predicting even the short-term selection response under inbreeding is based on a very incomplete theory. Most of the covariance expressions assume linear parent-offspring regressions and make the strong assumption that selection does not significantly change covariances from their neutral expectations under the mating system. While this may not be an unreasonable assumption in regard to allele-frequency change when many genes of small effect underlie a trait, it ignores the important impact of gametic-phase disequilibrium (LD), which can be dramatic (i.e., a 27% reduction after one generation was seen in Example 23.6). One effect of inbreeding is suppressing the impact of recombination, as the frequency of heterozygotes (which facilitate the breakdown of gametic phase disequilibrium via recombination) quickly becomes very small under inbreeding before vanishing entirely. Thus, gametic-phase disequilibrium is significantly more impactful than under random mating, where its effects are not trivial (Chapter 16).

Further, a practical concern is that, even if the above expressions are reasonable approximations despite these concerns, the inbreeding variance-components (ι^*, σ^2_{DI}, σ_{ADI}) are challenging, at best, to estimate. However, Chapter 11 shows that in those few cases where attempts have been made to estimate these, they are often very small, suggesting that standard additive and dominance terms dominate the covariances in many settings. If correct, this substantially reduces the complexity of predicting short-term selection response. Conversely, the impact from inbreeding depression, especially when starting from largely outbred populations, is typically not trivial (LW Chapter 10). Fortunately, it is straightforward to both estimate its impact and include its effect in standard models of short-term response (Equation 23.2b). It needs to be stressed, however, that the assumption that there is only a small amount of allele-frequency change is critical here, as the composite parameter B (Equations 23.1a, 23.1b, and 23.62) for the impact of inbreeding depression is a function of allele frequencies.

The robust results from all of this theory are that: (i) inbreeding can facilitate among-family selection while hampering within-family selection, (ii) when dominance is present, gains from early generations of inbreeding may significantly erode by the time inbreeding is complete, and (iii) inbreeding depression can have a significant impact on selection response in outbred species (depressing both the apparent response and the overall fitness of the population). From an applied standpoint, progeny testing and family selection can be improved with the judicious use of either inbred offspring or parents, but the cost in terms of additional time and resources can more than offset any gain.

THE EVOLUTION OF SELFING RATES

Finally, as we saw in Chapter 4, selection shapes the parameters of fundamental genetic processes, and the selfing rate is no exception. Monoecy (the lack of separate sexes, namely, **hermaphroditism**) is not uncommon among multicellular organisms, occurring in $\sim$94% of angiosperms (Renner and Ricklefs 1995) and $\sim$6% of animals (Jarne and Auld 2006). Hermaphroditism is common in 14 animal phyla, including the major phyla Porifera, Cnidaria, Platyhelminthes, Mollusca, and Annelida. When arthropods are excluded, the fraction of hermaphroditic animals increases to about 1/3 (Jarne and Auld 2006). Hence, the *opportunity* for selfing is widespread throughout multicellular life in that hermaphroditic species are widespread. However, many of these species have either genetic (self-incompatibility systems) or behavioral and morphological features that encourage outcrossing. Examples of the latter include flower morphology in the angiosperms (Barrett 2002), such as the distance between the anthers and stigma and their placement. Similarly, the timing of pollen release and the receptivity of a stigma may be completely or largely nonoverlapping (**dichogamy**), while sexually mature hermaphroditic animals often have a significant **waiting time** before they will engage in selfing.

Morphological, phenological, and behavioral features that influence the selfing rate are standard quantitative traits and hence have the potential to respond to selection. For example, selfing rates could potentially be changed by altering the timing of pollen release and sigma receptivity or the distance between anthers and stigma (Example 23.10). It could also be modified by selection for other features, such as reduced floral displays, an extreme example being **cleistogamy** (flowers whose petals do not open, facilitating self-pollination). These observations raise the widely debated question of the nature of the forces behind the evolution of various mating systems. While this is a very active and diverse field, we limit our discussion here to a brief overview of the forces influencing the selfing rate among hermaphrodites and mainly focusing on monoecious angiosperms.

Automatic Selection, Inbreeding Depression, and Reproductive Assurance

As first noticed by Fisher (1941), the foundational question of mating-system evolution is, "Why are not all hermaphrodites selfers?" This is because of **transmission bias**, which is often referred to as **automatic selection** or the **cost of outcrossing**. Consider a hermaphrodite that selfs. It contributes both eggs and pollen or sperm to its resulting selfed offspring, and it can potentially contribute sperm or pollen to other conspecifics that outcross. Conversely, an outcrossing hermaphrodite contributes an egg that combines with the pollen or sperm of another individual and pollen or sperm that fertilizes the egg of another conspecific. Hence, there is a three to two transmission bias in favor of selfers, generating a 50% selective advantage of an allele that promotes selfing (Fisher 1941; Nagylaki 1976b; Lloyd 1979).

As noticed by Nagylaki (1976b), such a selfing gene only spreads if it is also contributes to the pollen or sperm pool for outcrossers (otherwise, the allele never spreads beyond its selfed lineage). The generalization of this latter observation is the notion of **pollen discounting**, wherein the 50% selection advantage due to increased transmission of the selfing allele becomes discounted if the allele also results in reduced pollen for outcrossing, with any transmission advantage vanishing completely in the limit when no pollen is available for outcrossing.

Given this intrinsic selective advantage deriving solely from the genetics of transmission, coupled with the potential ecological advantage of not having to require pollinators or nearby conspecifics in order to produce offspring (**reproductive assurance**), why, then, are not fully selfing species more frequent? Especially in plants, the evolutionary transition from outcrossing to selfing is not uncommon. Yet despite these factors, only a small fraction (~10-15%) of plants are strictly selfing (Wright et al. 2013), with roughly half of all angiosperms having outcrossing enforced by either self-incompatibility (SI) systems or by dioecy (separate sexes) (Igic and Kohn 2006). Indeed, SI systems have evolved independently at least 35 times, and are found in at least 100 families of angiosperms, occurring in ~40% of all species (Igic et al. 2008). Hence, there must be strong selective pressure to maintain outcrossing. Conversely, SI is also frequently lost, for example, on at least 60 different occasions in the Solanaceae alone (Igic et al. 2008). Despite this loss, it is important to note that the transition from SI to self-compatibility is not sufficient for selfing, as other physical or temporal barriers may still have to be overcome. Evolutionary explanations for the relative lack of species with high rates of selfing, given all its apparent advantages, have been proposed at both the individual and group levels of selection, and we consider each in turn.

Because mating systems are products of individuals, any initial focus on evolutionary mechanisms should be on individual selection, namely, on the immediate fitness consequences to individuals that self (Charlesworth 2006b). The most obvious fitness cost is inbreeding depression (Darwin 1876, 1877; Charlesworth and Charlesworth 1995). Following Lande and Schemske (1985), we let $\overline{w}_0$ and $\overline{w}_1$ denote the fitnesses of random offspring produced by outcrossing and selfing, respectively. Let η denote the selfing rate, and assume identity equilibrium and linkage equilibrium, namely, random associations between loci involved in selfing and those involved in fitness. (Recall from Chapter 11 that inbreeding levels are correlated among loci when identity-disequilibrium is present.) Assuming no gamete discounting, the expected fitness of a genotype with selfing rate η becomes

$$w(\eta) = \eta\,\overline{w}_1 + \frac{1}{2}\,(1-\eta)\,\overline{w}_0 + \frac{1}{2}\,(1-\overline{\eta})\,\overline{w}_0 \tag{23.69a}$$

The first term corresponds to a selfing event, the second term to the focal individual producing outcrossed seed, and the last term to pollen or sperm from the focal individual fertilizing others (where $\overline{\eta}$ is the selfing rate for the population). The selection coefficient associated with a small change in the selfing rate becomes

$$\frac{\partial w(\eta)}{\partial \eta} = \overline{w}_1 - \frac{\overline{w}_0}{2} \tag{23.69b}$$

Hence, if the reduction in the fitness of selfed offspring is 50% or greater relative to outcrossed offspring ($\overline{w}_1/\overline{w}_0 < 0.5$), the intrinsic transmission bias of selfing is overcome, and other evolutionary forces would be required for a selfing allele to spread. With pollen discounting, this differences in fitness due to inbreeding depression to halt the spread of selfing can be much smaller than 50%.

Selfers can also have an immediate advantage when reproductive assurance is important, such as in low-density populations or when pollinators are scarce or absent (Darwin 1876). Hence, at least three different contributions to individual fitness potentially impact the evolution of selfing rates: transmission bias (and potential pollen discounting), the amount of inbreeding depression, and issues of reproductive assurance.

Stebbins (1957) suggested that the relative scarcity of predominately selfing species was the result of another level of selection, namely, at the species level. Stebbins envisioned obligatory selfing species as being an evolutionary dead end, as such populations have reduced genetic variation and hence are less likely to adapt to changing environments. Further, they can also accumulate higher levels of deleterious mutations (Muller's ratchet; Chapter 7). Both features suggest that the rate of taxon expansion (species formation minus loss from extinction) of fully selfing species may be less than that of outcrossers. Hence, even if selfing might be favored by individual selection, its long-term evolutionary footprint over a phylogeny would be reduced.

Beyond these simple verbal descriptions of potential sources of individual and group selection, what do theory and data have to say about the relative importance of the above features for the evolution of selfing rates?

The Lande-Schemske Model: Theory

A watershed paper on the evolution of selfing was published by Lande and Schemske (1985), who made the critical observation (as did Ohta and Cockerham 1974) that the amount of inbreeding depression may itself evolve during selfing. If inbreeding depression is largely caused by the presence of deleterious recessive or partly recessive alleles (as appears to be consistent with much of the data; Charlesworth and Willis 2009; LW Chapter 10), then the increased frequency of homozygotes under inbreeding increases the efficiency of selection for removing such alleles, which in turn can reduce the amount of inbreeding depression. Using the machinery presented in Chapter 7, Lande and Schemske showed that selfing results in lower equilibrium frequencies of recessive or partly recessive deleterious alleles under mutation-selection balance. For example, with 5000 loci, each with a mutation rate to lethal recessives of 2×10^{-6}, the expected fitness reduction in first-generation inbreds relative to outcrossed offspring is 97% based on the equilibrium allele-frequencies under outcrossing. This fitness reduction decreases to 61%, 15%, and 10% when based on the expected equilibrium frequencies for selfing rates of 0.01, 0.05, and 0.10, respectively. Hence, as the selfing rate increases, the fitness cost imposed by inbreeding depression can decrease over time, facilitating the spread of alleles that increase the selfing rate by reducing the fitness differences between outcrossed and selfed offspring.

Based on this observation, the simple Lande-Schemske model (with no pollen discounting or other selection from factors such as reproductive assurance, and ignoring any linkage or identity disequilibrium) suggests that, at equilibrium, one might expect a largely bimodal distribution of selfing rates. In other words, some species are predominately selfing (with reduced levels of inbreeding depression) and others are predominately outcrossing (with high levels of inbreeding depression), and there are relatively few species with intermediate levels of outcrossing. Their augment further suggests that once a species becomes predominately selfing, the (expected) reduced levels of inbreeding depression make it difficult for modifiers that increase the outcrossing rate to evolve. Hence, while the transition from outcrosser to selfing can occur under individual selection, the reverse transition, from selfer to outcrosser, is generally expected to be opposed by individual selection.

The simple equilibrium prediction of a bimodal selfing-rate distribution is compromised when there is selection for reproductive assurance. Even in settings where inbreeding depression cannot be purged, in the absence of pollinators (or under low population densities), even lower-fitness offspring produced under selfing are better than no offspring at all, potentially leading to a **mixed-mating** strategy, namely, an intermediate selfing rate. Lande and Schemske (1985) noted that pollinator failure or population bottlenecks (resulting in sufficiently low density to greatly reduce the potential for outcrossing) can favor the evolution of selfing. They noted that the approximate time scale for perturbed allele frequencies to return to their (random-mating) mutation-selection equilibrium frequency is $\sim 1/\sqrt{\mu}$ generations for lethal recessives (roughly a thousand generations for $\mu = 2 \times 10^{-6}$), while it is $\sim 1/(hs)$ generations for partial recessives, where hs is the selection against heterozygotes. For $h = 0.02$ for lethals, this is on the order of a hundred generations. Hence, they suggested that pollinator or population bottlenecks occurring roughly every hundred generations or so may result in selection favoring predominately selfers.

Schoen et al. (1996) argued that selection for selfing from reproduction assurance is more likely to involve multiple population bottlenecks than is automatic selection. They suggested looking for specific patterns in the reduction of genetic diversity (akin to the polymorphism-based tests for selection examined in Chapter 9) to examine whether such bottlenecks occurred during the initial transition to selfing. Unfortunately, as shown by Barrett et al. (2014), a number of confounding factors (such as increased background selection from the reduction in the effective recombination rate) lead to, at best, an ambiguous signal.

The Lande-Schemske Model: Data

The Lande-Schemske model makes two empirical predictions. First, levels of inbreeding depression should be less in organisms that have an evolutionary history of predominately selfing. This trend is supported by the data. A survey of 54 species of vascular plants from 23 different families examined by Husband and Schemske (1996) found a negative correlation between the amount of inbreeding depression and the selfing rate. The average fitness reduction when comparing offspring produced by selfing with those produced by outcrossing was 23% in predominately selfing species, while it was 53% for species that predominately outcross. However, a more detailed survey by Winn et al. (2011) gave a more nuanced view, in that species with intermediate levels of outcrossing displayed as much inbreeding depression as predominately outcrossing species. Such an outcome was suggested earlier by Lande et al. (1994), who showed that if inbreeding depression is sufficiently strong, no purging will occur until a threshold level of selfing is approached. If the amount of selfing is less than this threshold, little reduction in inbreeding depression is expected.

Turning to animals, a survey of 17 species of hermaphroditic gastropods by Escobar et al. (2011) found that selfing rates were negatively correlated with both inbreeding depression and the waiting time until an individual chooses to self. Further, they found that species showing higher levels of inbreeding depression had longer waiting times, presumably balancing the cost of delayed reproduction against the decrease in fitness when offspring are produced by selfing.

The second, and more direct, prediction of the Lande-Schemske model is that of a U-shaped distribution of selfing rates, namely, a bimodal distribution, with most species being either predominately selfing or predominately outcrossing. Here the data are less clear. The distribution of selfing rates is consistent with aspects of both the Lande-Schemske model and fitness advantages from reproductive assurance by at least some selfing. The original analysis of 55 species of vascular plants (both angiosperms and gymnosperms) by Schemske and Lande (1985) placed the estimated outcrossing rates into five bins (0–0.2, 0.2–0.4, etc.). In accordance with the predictions of the Lande-Schemske model, they observed a U-shaped distribution, with significant excesses in the upper (0.8 to 1.0; predominately outcrossing) and lower (0.0 to 0.2; predominately selfing) bins and far fewer species (31%) in the intermediate (0.2 to 0.8) range. Subsequent analyses of larger datasets of vascular plants also showed excesses of the upper and lower values, but also a higher fraction of species in the intermediate range than observed in the original Schemske and Lande dataset (Voglert and Kalisz 2001; Goodwillie et al. 2005). For example, 42% of the 345 vascular plant species considered by Goodwillie et al. were in the intermediate range of outcrossing (0.2 to 0.8). Estimates of the distribution of selfing rates for 142 species of hermaphroditic animals were also similar to the distribution seen in vascular plants, with modes at the high and low ends, but also with a substantial fraction (47%) in the intermediate range (Jarne and Auld 2006).

Given that the bimodal prediction is for populations that have reached equilibrium, intermediate selfing values might represent populations in transition (from outcrossing to selfing). The Lande-Schemske model predicts that such transient populations would show reduced levels of inbreeding depression, as they are in the process of purging deleterious alleles on their way to a predominately selfing existence. However, the data of Winn et al. (2011) showing that the levels of inbreeding depression in mixed-mating species are as high as those of predominately outcrossers suggests that these are unlikely to be transient populations and may instead be relatively stable states. One potential factor could be pollen discounting, which can result in stable, intermediate levels of selfing (Holsinger 1991; Porcher and Lande 2005, 2013). Pollinator behavior itself could also result in some fraction of selfing, leading to stable mixed-mating systems (Devaux et al. 2014).

Another force behind relatively stable mixed-mating systems could be selection for reproductive assurance. One suggestion from the data that reproductive assurance may be partially driving the intermediate selfing values is that when species are partitioned into wind- versus animal-pollinated, the former showed a more U-shape distribution (fewer species in the intermediate range) than the latter. In particular, Goodwillie et al. (2005) found that 46.5% of their 267 animal-pollinated species were in the immediate range, while only 26.9% were for their 78 wind-pollinated species. Although still impacted by population density, wind-pollinated species are not vulnerable to a lack of pollinators, and hence their selfing rates may be less impacted by reproductive assurance issues than those species that require pollinators.

Example 23.10. Two artificial selection experiments when outcrossing opportunities were reduced resulted in fairly rapid evolution of selfing and reduction in inbreeding depression. Bodbyl Roels and Kelly (2011) subjected yellow monkey flower (*Mimulus guttatus*) lines to either five generations with abundant bumblebees pollinators or five generations with no pollinators. The no-pollinator lines showed reduced seed set over the first few generations, but largely rebounded by generation five. Anther-stigma separation distances decreased in the no-pollinator lines, and this distance was negative correlated with seed set (smaller differences resulting in larger seed set).

Noël et al. (2016) subjected populations of a hermaphroditic freshwater gastropod (*Physa acuta*) to either outcrossing each generation or to a constrained setting where mates were often scarce, forcing selfing. The starting populations initially had high levels of inbreeding depression. After approximately 20 generations, the constrained lines showed a roughly 60% decrease in their waiting time (i.e., initiating selfing earlier) and a decrease in inbreeding depression (indicated by higher survival of offspring from selfing).

Baker's Law and the Demographic Advantages of Selfing

There are some significant ecological and demographic correlates with selfing among the flowering plants. Many annuals are predominately selfing, as are many species associated with frequent colonization events (such as weedy species) and those living in ephemeral habitats. Conversely, long-lived woody species tend to be predominately outcrossing (Vogler and Kalisz 2001; Barrett et al. 2014). Annuals and species in higher-risk ecological settings periodically run the risk of lack of pollinators or reduced population density in any given year, while these risks are reduced by averaging for species whose individuals have the capacity for reproduction over a number of years. Selfing is also more common in the marginal populations of a species range, again suggestive of some selection to mitigate a lack of outcrossing from either low population density and/or insufficient pollinators. Indeed, in a study involving approximately 200 species from 20 genera spanning 15 plant families, selfers tended to have larger geographic ranges (by about twofold) relative to outcrossing sister taxa (Grossenbacher et al. 2015), which was likely due to the ability of selfers to reproduce in more marginal environments.

The above observations are all loose correlates of **Baker's Law**, so named by Stebbins (1957), which is based on the suggestion that island species tend to be selfers (Baker 1955). For example, island species are much less likely to be self-incompatible (Igic et al. 2008). However, it was fairly quickly noted that dioecious species are common among native Hawaiian plants, suggesting that Baker's original observation is more of a trend than a law. Much has been written about Baker's Law and how much it may generalize to the above correlations between selfing and demography. We agree with the suggestion by Pannell et al. (2015) that species colonizing by long-distance dispersal are much more likely to be enriched for the *potential* to self (as opposed to being predominately selfing). Obviously, any low-density colonizer faces potentially serious reproductive assurance issues, and the ability (even if only sporadically) to self creates a demographic sieve enriching for such species among successful colonizing taxa.

Group-level Selection Against Selfing

The above individual selection pressures (transmission, reproductive assurance, and inbreeding depression) suggest how selfing can evolve, and also what individual-selection forces can enforce outcrossing. However, once a population undergoes a sufficient amount of selfing to reduce its level of inbreeding depression, any transition back towards increased outcrossing is expected to be opposed by individual selection. Hence, much of the historical discussion on the modest frequency of predominately selfing species is based on the Stebbins (1957) notion of the **SEDE** (Selfing as an Evolutionary Dead End) hypothesis.

The two tenets of SEDE are that: (i) selfing is largely irreversible; and (ii) selfed lineages have lower rates of species diversification than outcrossers, presumably because of higher extinction rates. The assumption of a very low reversion rate from selfers to outcrossers is not controversial, as the theory from individual selection suggests that such back transitions are rarely favored. Further, phylogenetic evidence suggests that such reversions are very rare (Takebayashi and Morrell 2001).

Identifying direct evidence for increased extinction rates in predominately selfing species is more problematic. Such evidence would be based on species comparisons over a phylogeny, which are limited by the precision of the reconstruction of the ancestral states (here, selfing or outcrossing) at the appropriate nodes of the phylogenetic tree. Given this difficulty, it is not surprising that while the current phylogenetic data appear to be consistent with the SEDE model, there is still uncertainty about its veracity (Takebayashi and Morrell 2001; Igic and Busch 2013).

The main arguments advanced in favor of increased extinction risk for predominately selfing populations are genetic: reduced genetic variation, excess accumulation of deleterious alleles, and slower rates of adaptation (Hartfield et al. 2017). The arguments for all three of these factors are based on the decreased effective population size of a selfed population, coupled with the reduction in recombination (due to a deficiency of heterozygotes). The latter magnifies both the role of sweeps and background selection in reducing standing variation (Chapters 3 and 8) and amplifies Hill-Robertson effects, reducing the efficiency of selection (Chapters 3 and 8). These factors are expected to result in an increased accumulation of deleterious alleles in lineages with reduced recombination (Muller's ratchet; Chapter 7) and reduction in the rate of fixation of favorable alleles.

A more careful examination of the theory shows that there are a number of subtleties with these initial predictions. First, consider the reduction in genetic variance from a reduction in the effective population size. The effect of inbreeding is the reduce the effective population size to $N_e = N/(1+f)$, which follows from Equation 3.20 with $f = \eta/(2-\eta)$. For a strictly neutral (and additive) trait, the expected mutation-drift equilibrium additive variance is $\widetilde{\sigma}^2_A = 2N_e\sigma^2_m$ (Chapter 11), showing that inbreeding does indeed reduce the additive variance, yielding

$$\widetilde{\sigma}^2_A = \frac{2N\sigma^2_m}{1+f} \tag{23.70a}$$

However, recall that selfing also *redistributes* the genetic variance, as the population is a collection of lineages, can still contain significant among-lineage variation. In particular, for an additive trait, the population-level variance is $(1+f)\sigma^2_A$ (Chapter 11). Putting these two expressions together, the expected equilibrium additive variance in the presence of inbreeding becomes

$$(1+f)\widetilde{\sigma}^2_A = (1+f)\frac{2N\sigma^2_m}{1+f} = 2N\sigma^2_m \tag{23.70b}$$

and hence the same over all values of f. Hence, for a neutral, additive trait, the reduction N_e from selfing alone is not expected to reduce the standing variation. The subtlety is that the second key feature of inbreeding—reduction in recombination—*further* reduces N_e due to a more important role for background selection when recombination is rare. Hence, it is *selection* that reduces the standing amount of genetic variation for a neutral trait in inbreds relative to outcrossers.

The strictly neutral model is a bit of a caricature, in that many traits are expected to be under stabilizing selection in natural populations (Chapters 28 and 29), so a more relevant comparison of the differences in the levels of genetic variation between inbred and outcrossed populations might be that expected under mutation-stabilizing-selection balance. This is the subject of Chapter 28. Initially, Lande (1977a) showed that the mating system has little effect on the amount of additive-genetic variation at equilibrium under stabilizing selection-mutation balance. However, as we detail in Chapter 28, this is a very model-dependent result (Table 28.2). Charlesworth and Charlesworth (1995) found that the amount of variation maintained under stabilizing selection-mutation balance is much lower in selfers than in outcrossers. These apparent discrepancies results were reconciled by Lande and Porcher (2015), who showed that outcrossers and selfers have very similar levels of variation under stabilizing-selection-mutation balance up to some threshold selfing value. With selfing rates higher than this threshold value, purging of inbreeding depression occurs, resulting in much less variation in selfed lineages. Hence, while predominately selfing populations *may* harbor less standing genetic variation than outcrossed populations, this may not always be true.

The second subtlety is *dominance*. Reduced effective population size and enhanced Hill-Robertson (HR) effects due to reduction in recombination weaken the efficiency of selection (Chapter 26), increasing the fixation of deleterious new mutations and decreasing the fixation of advantageous mutations (relative to an outcrossing population). However, when such alleles show partial or complete dominance, the efficiency of selection under inbreeding can be *increased* relative to outbreds (Charlesworth et al. 1993b; Glémin 2007, 2012; Hartfield et al. 2017). In particular, the fixation probabilities of partly to completely recessive beneficial alleles can increase, and their fixation times decreased, under inbreeding (Glémin 2012). Hence, if deleterious alleles tend to be recessive, the decreased efficiency from reduced N_e and HR effects may be partly countered by increased selection efficiency against recessives in inbreds. The same is true for recessive to partly recessive favorable alleles. Thus, while the initial effects of inbreeding appear to be rather unfavorable, if most selected alleles tend to be somewhat recessive, these negative effects may be somewhat offset.

Given that theory offers a bit of mixed message, what do the data say? The empirical data on reduced levels of genetic variation (and hence reduced ability to respond to a selective challenges) in predominately selfing species are mixed. Charlesworth and Charlesworth (1995) found little evidence of a reduction in either heritability or evolvability (Chapter 13) in highly selfed species. They noted the quality of data in their meta-analysis was rather mixed, and that a slight trend for reduced evolvability in predominately selfing species appeared when only the higher-quality data were used. Hence, the empirical evidence at present does not strongly support reduced standing genetic variation in predominately selfing populations.

Perhaps the more important evolutionary constraint faced by a predominately selfing population under adaptive pressure is not the amount of starting variation at the beginning of a new ecological challenge, but rather how it exploits this variation following the challenge. Outcrossing populations continue to generate new variation by segregation. In a strictly selfing population, this avenue is closed, and among-line selection will eventually result in only a few genotypes being present. This bottleneck in generating new variation to keep pace with environmental change may be the real impediment to the persistence of many selfing populations.

Example 23.11. Two recent animal-based studies examined the impact of selfing rates on the evolutionary fate of a population. Morran et al. (2009) used a very clever genetic approach in the nematode *Caenorhabditis elegans*, introducing either the *xol-1* or *fog-2* mutations into a wild-type line (with an outcrossing rate of roughly 5%). These mutations generate obligately selfing and obligately outcrossing lines, respectively, in otherwise identical genetic backgrounds. The authors applied this system to address two possible costs of selfing: a reduced rate of elimination of new deleterious mutations and a reduced ability to mount an adaptive response in a changing environment. They first artificially inflated the mutation rate (by roughly four-fold) by exposing individuals to the chemical mutagen EMS. The obligately selfing lines showed a significant decline in fitness, while the obligately outcrossing lines did not. Further, the wild-type lines showed an increased level of outcrossing after several generations of exposure. These observations are consistent with outcrossing (and hence recombination) facilitating the removal of deleterious mutations, suggesting that their accumulation could be a potential cost of selfing. Next, they subjected the three distinct mating-type lines (now under natural mutation rates) to 50 generations of selection in the presence of a novel bacterial pathogen. The obligately outcrossing and wild-type lines were able to adapt (evolve resistance) while the obligately selfing lines were not. Further, as with the EMS experiment, the outcrossing rate in the wild-type line had increased by the end of the selection experiment.

A second experiment was conducted by Noël et al. (2017), who examined the selection response in the gastropod *Physa acuta*. They contrasted response in populations with different starting histories of outcrossing and selfing, and also under different outcrossing and selfing regimes during the actual artificial selection itself. Their C lines had outcrossing each generation, while their S lines had self-fertilization imposed every other generation. Selection response with lines of approximately 30 generations of either the S or the C treatment as the base population were compared, as were settings where 100% selfing or 100% outcrossing occurred once selection started. Selection response was larger when the base population had a history of outcrossing (C) as opposed to selfing (S). Further, while the selection response (in this case, for shell morphology) was initially more rapid in lines undergoing 100% selfing during the course of selection, this response only lasted a few generations, after which it rapidly diminished. In contrast, the fully outcrossed populations continued to show a relatively constant response over the course of the experiment. Again, a history of either past or ongoing selfing resulted in a reduced selection response, and hence the potential (in some settings) for selfed populations being unable to mount an adaptive response given a sufficient environmental challenge.

VI

Population-Genetic Models of Trait Response

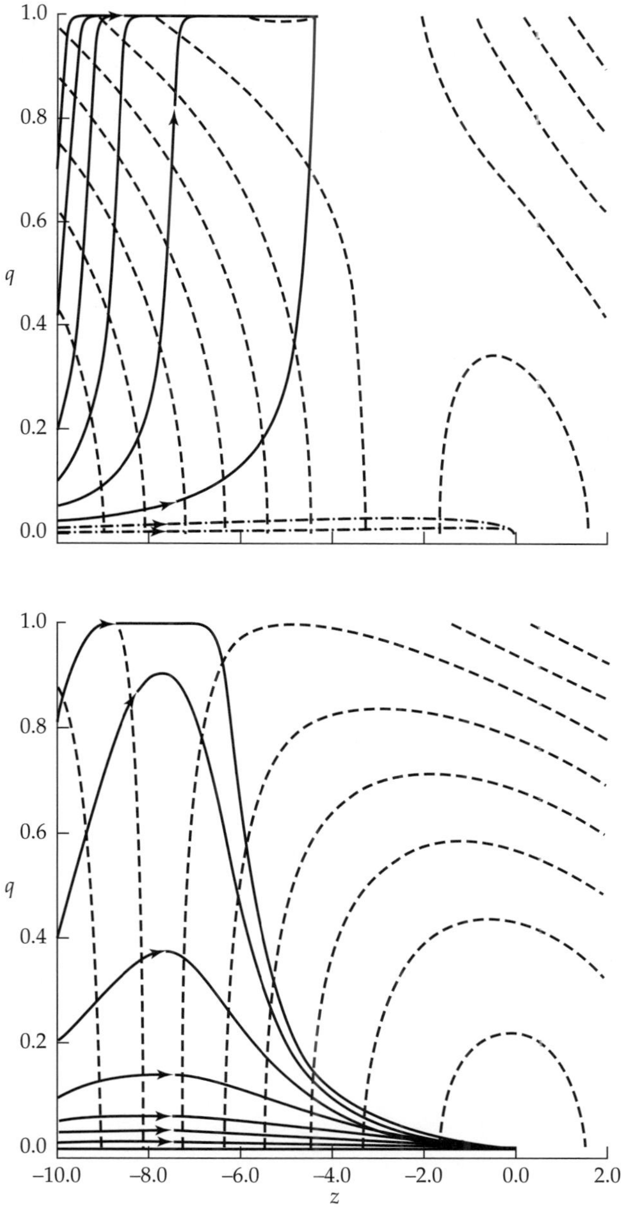

24

The Infinitesimal Model and Its Extensions

Normal theory is clearly the most powerful and problematic hypothesis in the present analysis. Chevalet (1988)

What, me normal? Turelli and Barton (1994)

As detailed in Chapters 6, 13, and 16, the response to a single generation of selection in a quantitative trait can often be closely predicted from **macroscopic** (observable) features of that trait, namely, its variance components. This is in sharp contrast to predicting selection response under one- and two-locus models, which requires detailed knowledge of the underlying genotype frequencies and effects (Chapter 5). Hence, population-genetic predictions typically require knowledge of **microscopic** (largely unobservable) genetic parameters. The breeder's equation (13.1) and the Bulmer equation (16.7b) are examples of macroscopic-based predictors. In contrast, the Price equation (6.8) is a microscopic-based predictor, given its composite transmission parameter, $\sigma(w, \delta)$, which is, at best, very challenging to measure (and is even harder to iterate over generations).

These variance-component predictors, which underlie much of the quantitative-genetic theory of response, are, of course, approximations, but often very good ones, especially over short time scales. Their single-generation versions follow from the assumption of a linear and homoscedastic parent-offspring regression, and their extension to multiple generations requires not only that linearity and homoscedasticity hold following selection, but also that any selection-induced changes in the variance components (which determine the slopes of these regressions) are largely predicable from macroscopic parameters (for example, via the Bulmer equation).

The justification for these approaches is usually framed by invoking a set of assumptions that we have been loosely calling the infinitesimal model. At various times, we have used either one, or more, of the assumptions of: (i) a very large (effectively infinite) number of loci, each with very small (effectively infinitesimal) effects; (ii) a normally distributed genotypic distribution in a randomly mating population under no selection, and, hence parent-offspring regressions that are linear and homoscedastic; and (iii) that the genotypic distributions stay at least close to **Gaussian** (normal) following selection such that the linear and homoscedastic regressions approximately hold.

The goal of this chapter, which is rather technical in places, is to examine in more detail the various infinitesimal approximations that we have made and to examine the consequences for selection response when they fail. We start by deconstructing what we have called the **standard infinitesimal model** (the joint assumptions of very large numbers of loci and parent-offspring regressions that are linear and homoscedastic) into its components and examine how these are connected, and also when they fail. As we will see, despite the use of the term *standard*, there is actually a nested family of infinitesimal models that make increasingly stringent assumptions, which correspond to assumptions (i) through (iii), which were just mentioned (Table 24.1).

We then introduce the **Gaussian approximation**, one of a broad group of **continuum-of-alleles** (or **COA**) **models**, assuming a *finite* number of loci, with a *distribution* of allelic effects at a locus (as opposed to a fixed number of possible alleles), with mutation likely generating alleles not previously segregating in the population. The Gaussian approximation assumes that the distribution of allelic effects at *each locus* is normal. With a large number of loci, the results from Gaussian COA models approach those obtained under the infinitesimal

Evolution and Selection of Quantitative Traits. Bruce Walsh & Michael Lynch.
Published 2018 by Oxford University Press.
DOI 10.1093/oso/9780198830870.001.0001

Table 24.1 Classification of the different versions of the infinitesimal model, based on Turelli (2017). Note that these models are nested, such that a model makes all of the assumptions of any model that proceeded it in the table.

Infinitesimal genetics A large number of loci, each with vanishingly small effects.	Fisher (1918)
Gaussian descendants (Fisher-Bulmer Infinitesimal) Within-family segregation variance independent of parental phenotypes, depending only on the relatedness of parents.	Bossert (1963)
In the limit, results in a Gaussian distribution of breeding values in their unselected descendants. Parent-offspring regressions are linear and homoscedastic.	Fisher (1918), Bulmer (1971b), Barton et al. (2017)
Gaussian populations The distribution of breeding values in a population is Gaussian.	

model, thus serving as a bridge between models assuming effectively an infinite number of loci (each with a small number of possible alleles, typically biallelic) and models assuming a very large (effectively infinite) potential number of alleles at a finite number of loci. We then examine the effects of linkage on the behavior of these models, and conclude by examining the selection response when the distribution of breeding values is no longer Gaussian.

The goal of this chapter is to start to bridge short-term predictors of response based on macroscopic parameters with predictors of long-term response based on microscopic parameters. After sufficient allele-frequency change accrues, these bridging models break down, thus imposing the need for explicit population-genetic models (Chapter 5), which are examined in Chapters 25–28.

THE INFINITESIMAL: A FAMILY OF MODELS

We have been somewhat cavalier in our use of the term *infinitesimal model*, as there is considerable ambiguity as to what this term formally means. Further, what is required from an infinitesimal assumption depends the on problem being considered: Is one making the assumption of very small allele-frequency change? The assumption of a linear, and homoscedastic, parent-offspring regression? The assumption that the distribution of breeding values in a population is Gaussian (normally distributed) and is largely unaltered by selection? As shown in Table 24.1, there is a nested hierarchy of models, which impacts the assumptions one is actually making when invoking the infinitesimal.

Following Turelli (2017), one can consider three nested versions of the model. The simplest, **infinitesimal genetics**, matches Fisher's (1918) approximation of describing the trait architecture as consisting of a very large number of loci, each of small effect. A direct consequence of this model is that allele-frequency changes are very small, as the time scale for frequency change scales with the inverse of allelic-effect size (Equation 5.21).

The next-level model, which Turelli (2017) named **Gaussian descendants**, starts with the assumption (for unrelated and outbred parents) of a constant within-family variance that is independent of the phenotypes of this parents. Such a constant **segregation kernel** variance was first considered in Bossert's (1963) unpublished thesis, and this is simply half of the genic variance (Equation 16.2). As we saw in Example 16.2, the within-family segregation variance is reduced when parents are inbred, so a slightly more generalized assumption is a segregation variance that only depends on the relatedness of the parents (Bulmer 1971b, Barton et al. 2017). As suggested by Fisher (1918), and most generally shown by Barton et al. (2017), the assumption of a large number of loci (the infinitesimal genetics model) in the limit (provided linkage is not too extreme) leads to a Gaussian (normally) distribution of breeding values among the offspring of a set of parents, whose variance only depends on the parental relationship. Hence, the oft-used infinitesimal assumption of linear regressions and homoscedastic variances follows from this limit.

The Gaussian descendants version of the infinitesimal model is the starting point for most of our previous discussions about the impact of a single generation of selection from an unselected base population, and it forms the foundation for BLUP. We take this version—an efficiently infinite number of loci, a segregation variance that only depends upon relatedness, and a Gaussian distribution of breeding values in the absence of selection (and hence linear and homoscedastic parent-offspring regression)—as the standard infinitesimal model. We will also refer to this particular set of assumptions as the **Fisher-Bulmer infinitesimal model**.

The most restrictive of the infinitesimal family of models is what Turelli (2017) called the **Gaussian populations** assumption. Here, the *population* (as opposed to the *within-family*) distribution of breeding values is Gaussian. This is the infinitesimal assumption which is used to predict selection response over multiple generations. As we will show later in the chapter, this generally does not hold, even when starting with the Fisher-Bulmer infinitesimal model, as selection (through generating disequilibrium) will drive any initially Gaussian distribution away from normality. The key observation is that while the distribution of breeding values *within* a family may stay Gaussian following selection, a population is a *collection* of families. If the distribution of *family means* is highly non-Gaussian (as might be expected to be generated following selection on a trait), then the population distribution will depart from normality. Gaussian within-family segregation will recover some of the normality in the offspring generation, and the critical question is whether this is usually enough to approximately recover the Gaussian populations assumption.

The key points we will make below are as follows: First, allele-frequency change from selection on a trait with a finite number of loci tends to drive the distribution of breeding values away from a Gaussian, but the degree of subsequent departure quickly diminishes as the number of loci increases. Second, selection-induced correlations over alleles from different loci (linkage disequilibrium) will drive even a trait with infinitesimal genetics away from a Gaussian. And third, when the within-family segregation remains Gaussian, this will partly restore a population-level Gaussian distribution, but departures away from a Gaussian will remain in the population until selection has ceased and recombination removes among-locus correlations.

THE INFINITESIMAL GENETICS MODEL: EMPIRICAL DATA

Fisher's (1918) motivation for the infinitesimal genetics model, which assumed that many genes of small effect underlie a typical trait, was part mathematical convenience and part biological approximation. Early on, many geneticists, trained in dissecting the action of single genes of modest to large effect, questioned the validity of this biological motivation. This objection was reinforced in the mid-1980s when QTL mapping experiments seemed to detect abundant major-effect QTLs, each accounting for 10% or more of the phenotypic variation (LW Chapter 15). However, with the advent of more powerful genomic tools, attempts to isolate the underlying nucleotide sites behind these major QTLs grew increasingly frustrating. A major QTL (indicated by a large peak in the likelihood surface spanning some genomic region; LW Chapter 15) seen in the cross of two inbred lines often fractionated into several minor peaks upon finer mapping, and each of these minor peaks further fractionated as attention was turned to them. While genes of large effect have been found, what frequently appeared to be a single major gene turned out to be a number of tightly linked regions of much smaller effect (reviewed by Flint and Mackay 2009; Mackay et al. 2009).

This trend continued during the GWAS (genome-wide association study) phase of quantitative genetics (starting in the early 2000s). Association mapping (LW Chapter 16) uses population-level disequilibrium and therefore allows for mapping on a kilobase scale, as opposed to the tens of megabase scale for QTL resolution (the typical confidence interval in a QTL mapping experiment; LW Chapter 15). The broad conclusions from a large number of studies on human traits (mainly, but not exclusively, diseases) with enormous sample sizes (in the tens of thousands or greater) are twofold (Visscher et al. 2012a).

First, the effects of detected sites (measured by the marker-associated variance) tend to be very small. For example, over 600 variants associated with human height variation have been detected, most of which typically account for only a minuscule fraction of the additive variance (Lango Allen et al. 2010; Yang et al. 2010; Wood et al. 2014). Further, estimates suggest that the actual number of genes influencing human height is significantly greater than 600, running from over 1600 (Kemper et al. 2012) to "a very large but finite number (thousands) of causal variants" (Wood et al. 2014). Gene knockout experiments (Reed et al. 2008) have suggested roughly 6000 genes with the potential to influence body weight in mice, while the response to artificial selection on body size in *Drosophilia* involves between 300 and 1200 regions (Turner et al. 2011). Finally, in a number of studied traits the fraction of additive-genetic variation accounted for by SNPs on a per-chromosome basis is proportional to chromosomal length (Visscher et al. 2007; Yang et al. 2011b; Yang et al. 2013). This observation is consistent with an infinitesimal-like model of a very large number of small-effect loci uniformly distributed over the genome.

Second, the total additive variance accounted for by *all* detected sites is only a small fraction (around 10%) of the total additive variance for the same trait estimated from the resemblance between relatives (in humans, typically twin studies). The latter observation led to concerns about "**missing heritability**" (Mather 2008; Manolio et al. 2009) and a large number of subsequent papers attempting to account for this apparent paradox. In reality, this observation of missing heritability provides strong support for the infinitesimal genetics model. In testing up to millions of SNPs for association in a GWAS, stringent thresholds are set to control for multiple comparisons (Appendix 4). This, in turn, requires larger effect sizes in order for a given marker to be declared significant, which excludes many biologically relevant SNPs from the model using all of the *detected* sites.

By using mixed-model approaches that allow *all* SNPs to be incorporated (by shrinking the effects of most SNPs toward zero), Yang et al. (2010) could account for $\sim$45% of the additive variance in human height. Similar findings were seen for schizophrenia (Purcell et al. 2009) and a growing number of other traits (e.g., Hill 2010; Vinkhuyzen et al. 2013; Yang et al. 2013; Robinson 2014). Example 24.1 illustrates how incomplete linkage disequilibrium (in which the marker and causative alleles are less than completely associated) can easily account for the remaining fraction of the "missing" heritability. Lee et al. (2011) reviewed approaches for estimating this **hidden** (not missing) **heritability** in GWAS studies using mixed models (and hence essentially assuming the Gaussian descendants version of the infinitesimal model). Thus, we have come full-circle from the early QTL days in that current genomic data for many traits are most consistent with a very large number of loci, each with a small effect. This is not to say that major alleles do not exist, but rather that they tend to be rare. Indeed, the effect detected in a GWAS study is the additive variance associated with a marker SNP (LW Chapter 16), so finding that the vast majority of sites have a low effect *variance* does not imply that they involve alleles of *small effects*. If alleles of large effect tend to be rare, they will display small variances, although the presence of such alleles has consequences for the prediction of selection response (as we detail below). Notably, Kemper et al. (2012) found a negative correlation between frequency and effect size of alleles influencing human height (see Figure 28.5).

Example 24.1. As an example of one source of "missing heritability," consider a site at which a new QTL allele, Q, with an additive effect of a, arises on a SNP marker background, with M and m being the two marker alleles. The strongest marker association occurs when Q is completely restricted to the background on which it arose, so we assume this here, with Q only found on M-bearing haplotypes. Moreover, only a fraction, ξ, of M haplotypes will harbor Q, and (as we now demonstrate), this results in hidden heritability. Summarizing these assumptions, we have:

Gamete	Frequency	Effect
MQ	$p\xi$	a
Mq	$p(1-\xi)$	0
mq	$1-p$	0
mQ	0	a

Hence, the frequencies of M and Q are, respectively, p and $p\xi$. The resulting additive variance due to the causal site becomes

$$\sigma_A^2(QTL) = 2a^2(p\xi)(1-p\xi) \simeq 2a^2 p\xi$$

with the approximation following because $p\xi \ll 1$. Conversely, the average effect of marker allele M is $a\xi + 0(1-\xi) = \xi a$, while the value for m is zero, making the additive variance associated with this marker

$$\sigma_A^2(M) = 2(\xi a)^2(p)(1-p) = 2a^2 p(1-p)\xi^2$$

The resulting ratio of the marker to causal additive variances is

$$\sigma_A^2(M)/\sigma_A^2(QTL) = \xi(1-p)/(1-\xi p) \simeq \xi$$

Thus, if Q is somewhat rare on M backgrounds, only of a fraction, ξ, of the actual variance is accounted for by the linked SNP. This situation would occur if the site-frequency spectrum (SFS) for QTLs alleles is shifted toward smaller values relative to the SFS for the marker alleles. Both the ascertainment for common SNPs and weak selection against QTL alleles could generate such a shift.

While the value of ξ is unknown, suppose that, on average, $\xi \simeq 0.5$: then, at most 50% of the total variation from the causative sites can be captured by markers. This value is very close to the value of 45% of the variance accounted for by Yang et al. (2010) for human height. Further, this calculation is biased in favor of marker variances as it assumes complete disequilibrium (i.e., Q is *only* found on the M background). If some of the m-bearing chromosomes also carry Q, the fraction of the causal variance accounted for by the marker variance will be even less.

THEORETICAL IMPLICATIONS OF THE INFINITESIMAL GENETICS MODEL

Under the classic infinitesimal genetics model introduced by Fisher (1918), a character is determined by an infinite number of unlinked and nonepistatic loci, each with an infinitesimal effect. It is occasionally assumed that each locus has two alleles and that the effects and frequencies are the same (or very similar) across all loci, although these constraints can be relaxed. Here we examine some of the properties resulting from the various infinitesimal assumptions (Table 24.1), providing a starting point for evaluating the consequences on the short-term selection response when they fail.

Selection Does Not Change Allele Frequencies

Recall from Chapter 16 that we can express the additive *genetic* variance, σ_A^2, as the sum of the *genic* variance, σ_a^2, and the disequilibrium contribution, d, with $\sigma_A^2 = \sigma_a^2 + d$. This partition decouples the effect of allele-frequency change (changes in σ_a^2) from the effect of changes from linkage disequilibrium (d).

Under the infinitesimal genetics model, *allele frequencies are essentially unchanged by selection*, and thus σ_a^2 is assumed to be constant over time (in an infinite population). Large changes in the mean can nonetheless occur via summation of infinitesimal allele-frequency changes over a large number of loci. To see this, consider a character determined by n completely additive biallelic loci. Suppose that all loci are **interchangeable**, with each having the same effects and frequencies (the **exchangeable model**). Each locus has two alleles, Q and q, with the genotypes QQ, Qq, and qq contributing $2a$, a, and 0 (respectively) to the genotypic value, so that allele Q has effect a, and let p denote the frequency of Q. The resulting mean is then $2nap$, with the change in mean due to a single generation of selection being $\Delta\mu = 2na\Delta p$.

Ignoring any contribution from gametic-phase disequilibrium (i.e., $d = 0$), the additive variance is $\sigma_A^2 = \sigma_a^2 = 2na^2 p(1-p)$. For σ_A^2 to remained bounded as the number of loci increase, a must then scale as $n^{-1/2}$. Assuming the frequency of Q changes by the same amount at each locus, then $\Delta p = \Delta\mu/(2na)$. Because a is of order $n^{-1/2}$, Δp is of order $1/(n \cdot n^{-1/2}) = n^{-1/2}$, and approaches zero as the number of loci becomes very large. The infinitesimal genetics model thus allows for arbitrary changes in the mean with essentially no change in the allele frequencies at underlying loci. Biologically (i.e., with a finite number of loci), the infinitesimal genetics model implies that large changes in the mean of a trait can occur with only small changes in allele frequencies if all loci each make only a small contribution to the trait variance.

What effect do small amounts of allele-frequency change have on the genic variance, σ_a^2? Letting $p' = p + \Delta p$ denote the frequency after selection, the change in genic variance is

$$\begin{aligned}\Delta\sigma_a^2 &= 2na^2 p'(1-p') - 2na^2 p(1-p)\\ &= 2na^2\Delta p(1-2p-\Delta p) = a(1-2p-\Delta p)\,[2na\Delta p]\\ &\simeq a\,(1-2p)\,\Delta\mu\end{aligned}$$

Because a is of order $n^{-1/2}$, the change in variance due to changes in allele frequencies scales as $(1/\sqrt{n}\,)$ times the change in the mean (assuming Δp is small). Thus, with a large number of loci, very large changes in the mean can occur without any significant change in the genic variance. The more loci of equal effect underlying a trait, the slower will be the change in σ_a^2, and hence the longer the selection response will be predictable (Chapter 25). In the limit of an infinite number of loci, there is no selection-induced change in the genic variance ($\Delta\sigma_a^2 = 0$), while arbitrary changes in the mean can occur.

Accounting for Dominance

While our previous focus has been exclusively on alleles with additive effects, dominance is not excluded under an infinitesimal model. The incorporation of dominance, however, requires very delicate conditions for the scaling of allelic effects, so as to bound both the dominance variance and any inbreeding depression. To see this, suppose we have n biallelic loci with no epistasis (the total genotypic value is simply the sum of the individual locus genotypic values), and let the genotypic values at locus i be $0 \,:\, a_i + \delta_i \,:\, 2a_i$, where the frequency of the increasing allele is p_i. The resulting dominance variance becomes

$$\sigma_D^2 = \sum_{i=1}^{n} [2p_i(1-p_i)\delta_i]^2$$

For n exchangeable loci, this simplifies to

$$\sigma_D^2 = 4np^2(1-p)^2\delta^2$$

For σ_D^2 to remain bounded as $n \to \infty$, δ must scale as $n^{-1/2}$, just as we found for a. Thus, if both a and δ scale as $1/\sqrt{n}$, the additive and dominance variances remain bounded as the number of locus goes to infinity.

Now consider the behavior of inbreeding depression, the difference between the mean trait value, μ_f, when population-level inbreeding is f versus that under random mating, μ_0 (LW Chapter 10). Again, assuming that there is no epistasis, from LW Equation 10.3, the inbreeding depression is given by

$$\mu_f - \mu_0 = -2f\sum_{i=1}^{n} p_i(1-p_i)\delta_i$$

Assuming n loci of equal effect gives

$$\mu_f - \mu_0 = -2nfp(1-p)\delta$$

Because δ scales as $n^{-1/2}$ under the infinitesimal genetics assumption, the amount of inbreeding depression scales as $n \cdot n^{-1/2} = n^{1/2}$ and hence goes to infinity with increasing values of n. Conversely, if we scale δ as order $1/n$, we have bounded inbreeding depression, but the dominance variance is now of order $n/n^2 = 1/n$ and hence is zero in the infinitesimal limit.

Thus, under the exchangeable infinitesimal model, one cannot have both bounded dominance variance and inbreeding depression, a point first made by Robertson and Hill (1983). Of course, a limitation of this argument is our assumption of equal effects over all loci, with all of the δ having the same sign. If we assume that $E[\,\delta\,] = 0$, namely, that there is no directional dominance, then we can have bounded dominance variance but no inbreeding depression. To have both dominance variance and inbreeding depression in the infinitesimal limit requires a great deal of delicacy, in that individual allelic effects have to be scaled so that $0 < n\,E[\,\delta\,] < \infty$ as $n \to \infty$ (there must be finite directional dominance). Related to this point, Wellmann and Bennewitz (2011) showed that the ratio of the squared inbreeding depression to the dominance variance sets a lower bound on the number of underlying loci.

Disequilibrium

While σ_a^2 remains unchanged by selection under the infinitesimal genetics model, selection-induced changes in d can significantly alter the additive genetic variance, σ_A^2 (Chapter 16). We can show this using the same scaling arguments we just employed. Changes in the covariances, C_{ij}, of allelic effects between loci i and j (for $i \neq j$) are roughly of order n^{-2} (Bulmer 1980; Turelli and Barton 1990). Because there are $\sim n^2$ terms contributing to d (Equation 16.1a), the total disequilibrium is of order one ($n^2 \cdot n^{-2}$), and does not necessarily approach zero as the number of loci becomes infinite. The same reasoning holds for changes in the higher-order moments, which are caused by higher-order associations between groups of loci. For the kth-order moment, there are $\sim n^k$ terms in the sum, each scaling as n^{-k} to potentially give a nonzero value in the limit (Turelli and Barton 1990).

THEORETICAL IMPLICATIONS OF THE STANDARD INFINITESIMAL MODEL

The standard (Fisher-Bulmer) infinitesimal model assumes both the infinitesimal *genetics* model *and* that parent-offspring regressions are linear and homoscedastic. As a result, the *unselected* descendants from a set of parents have a normal distribution of breeding values (and hence, Turelli's use of *Gaussian descendants* for this version of the infinitesimal).

What Generates a Gaussian Distribution Within a Family?

The **central limit theorem** from probability theory—that sums of random variables typically converge to a Gaussian distribution—implies that the distribution of breeding values is Gaussian under the infinitesimal genetics model (Bulmer 1971b; Barton et al. 2017). This Gaussian limit requires the key assumptions that loci are not too tightly linked, mating is random, and there has been no previous selection or any other force generating disequilibrium. If alleles are sufficiently correlated across loci, or if their effects are significantly different (for example, some remain at finite values while the rest become vanishingly small), then convergence to a normal is by no means assured (Lange 1978, 1997; Matthysse et al. 1979; Dawson 1997).

In particular, as we will demonstrate, selection can generate dependencies (linkage disequilibrium [LD]) among sets of three (or more) loci, which can transform a normal distribution into a non-normal one as selection proceeds. However, under the infinitesimal genetics model, there are no changes in allele frequencies, implying that once selection stops and random mating occurs, any departures from normality (being due to LD) quickly decay. Indeed, Bulmer (1980) showed that the kth-order departure from normality (measured by cumulants, which are introduced later in the chapter) decays by $(1/2)^{k-1}$ in each generation (assuming unlinked loci), so that following t generations of random mating, the kth-order initial departure from normality is scaled by $(1/2)^{t(k-1)}$, which quickly approaches zero. The issue remains as to how much of a departure from normality LD generates and whether this biases infinitesimal-model-based predictions of the selection response. We return to this issue later in the chapter.

A widely used feature of the Fisher-Bulmer infinitesimal model is (for unlinked loci in noninbred parents) a **constant within-family genetic variance**. Specifically, the distribution of breeding values, A_o, in the offspring, conditioned on the breeding values, A_f and A_m, of the parents, is assumed to be normally distributed, with a mean of $(A_f + A_m)/2$ and a variance of $\sigma_a^2/2$ (half the *genic* variance). Thus, we have homoscedasticity with the predictor error variance, $\sigma_a^2/2$, being a constant, independent of the parental values. This constant **Mendelian sampling variance** (or **segregation variance**) is caused by segregation of heterozygous loci in the parents. Provided the loci are unlinked, this feature holds even when gametic-phase disequilibrium is present, as with the independent assortment of heteroyzotes, the linkage phase of a parent does not influence gamete frequencies for unlinked loci (Example 16.2). More generally, the segregation variance is given by $\sigma_a^2(1-\overline{f})/2$, where $\overline{f}$ is the average amount of inbreeding in the two parents (reflecting the decrease in heterozygosity in the parents; Example 16.2). As detailed in Chapter 19, the additive relationship matrix, A, fully accounts for relatedness-based changes in σ_a^2 among a collection of individuals in a BLUP analysis.

Under the infinitesimal genetics model and its assumption of no selection-induced allele-frequency change, the genic variance (corrected for parental inbreeding) remains constant. However, if the infinitesimal genetic assumptions are violated, then even alleles with very small effects eventually will have their frequencies changed by selection. This alters the genic variance, and hence changes the segregation variance. When major genes are present, the segregation variance depends on parental genotypes, which compromises the constant-variance (i.e., genotype-independent) assumption made under the Fisher-Bulmer infinitesimal. Indeed, recall from LW Chapter 8, that such variation of the within-family variance over families is one (albeit weak) test for the presence of a major gene.

Modifications of the Fisher-Bulmer Infinitesimal Model

The rest of this chapter starts to move beyond the standard (Fisher-Bulmer) infinitesimal model. First, by assuming a Gaussian distribution of allelic effects *at each locus*, we can partly account for changes in allele frequencies, and hence changes in σ_a^2, caused by a finite number of loci and/or genetic drift. This approximation breaks down over time (Chapter 25) and is best regarded as an intermediate-term predictor of response. Next, we allow for linkage. Finally, we examine the selection response when the distribution of genotypic values is no longer normal. None of these approaches fully accounts for allele-frequency change, and they are best considered as predictors for intermediate-term response. The prediction of long-term response requires explicit population-genetic models (Chapters 5, and 25–28).

Finally, a number of authors have suggested finite-locus modifications of infinitesimal-like models, largely in the context of fitting the segregation term in a mixed model when major alleles are also segregating (Cannings et al. 1978; Fernando et al. 1994; Stricker et al. 1995; Lange 1997; Du et al. 1999; Pong-Wong et al. 1999; Goddard 2001). These methods foreshadowed certain aspects of genomic selection, which assigns weights to different chromosomal segments in order to use marker data to predict breeding values (Volume 3). Diffusion approximations (Appendix 1) of finite-locus models were proposed by Miller et al. (2006).

GAUSSIAN CONTINUUM-OF-ALLELES MODELS

Simulation studies (e.g., Bulmer 1974a, 1976; Mueller and James 1983; Sorensen and Hill 1983; Chevalet 1988) have shown that the infinitesimal model gives a reasonably good fit to the change in variance over a few generations of selection when the number of loci is large but finite (provided all alleles have small effects). With a finite number of loci (and hence nonvanishing individual locus effects), some (potentially very small) selection-induced allele-frequency change occurs in each generation. After a sufficient number of generations, the cumulative effect of these changes becomes so large that they cannot be ignored (Chapter 25). Likewise, if the population is finite, genetic drift also changes allele frequencies (Chapter 2). Thus, when either the number of loci, n, or the effective population size, N_e, is finite, we must incorporate changes in the *genic* variance, σ_a^2, into our model.

Is there an intermediate step between the short-term predictions from the breeder's equation/infinitesimal model and the unpredictable long-term behavior when significant allele-frequency changes have occurred? In many cases, an affirmative answer is provided by approximations using **continuum-of-alleles models** (**COA**). Our focus here is on the Gaussian COA model, with other COA models (such as the house-of-cards, rare alleles, or house-of-Gauss models) examined more fully in Chapter 28. For brevity, for the rest of this chapter, *continuum-of-alleles* refers strictly to the Gaussian COA. This model allows us to partly account for modest changes in allele frequencies due to selection (given a finite number of loci) and genetic drift (due to finite population size). The nice feature about these intermediate-term approximations for response is that they are based entirely on macroscopic parameters, for which there is some hope of estimation.

Infinite-alleles and Continuum-of-alleles Models

The historical roots of the continuum-of-alleles model trace back to the classic paper of Kimura and Crow (1964), which introduced the **infinite-alleles model** (Chapter 2). Before its publication, most population-genetic models typically assumed two (or at most a few) alleles per locus. Kimura and Crow, in the first serious treatment of molecular evolution, noted that with an allele being represented by a long DNA sequence, each new mutation likely creates a new sequence, implying an effectively infinite number of possible alleles. Kimura and Crow's original paper simply dealt with how much variation (measured in terms of heterozygosity) could be maintained by the balance between drift and mutation (Chapter 2), and it was not concerned with allelic effects. Crow and Kimura (1964) and Kimura (1965a) quickly applied this notion of a very large number of alleles per locus to quantitative genetics by considering the distribution of allelic effects at each locus. These Gaussian continuum-of-alleles models were further developed by Latter (1970), Lande (1975, 1977a), and Felsenstein (1977) to consider the amount of variation maintained by mutation-selection balance (Chapter 28). Kimura's (1965a) original analysis revealed that if new mutations have small effects relative to the existing variation at the locus, the distribution of effects (in an infinite population) converges to a normal.

Hence, Gaussian COA models make the assumption that the distribution of additive genetic values *at each locus* is Gaussian (and jointly multivariate normal over a vector of loci). This assumption is strictly correct only if the number of alleles at each locus is infinite, which further implies that there is an infinite population size. This assumption of a *Gaussian distribution of effects at each locus* is much more restrictive than the assumption that the distribution of the *total* genotypic values is normal. The latter follows from the central limit theorem, as the sum of non-normal distributions of single-locus effects converges to a Gaussian, provided that the loci are not overly correlated.

While Gaussian COA models are a very restrictive subset of all possible models that can lead (at their limit) to the infinitesimal, their advantage is that we can assume a *finite number of loci*, and hence partly accommodate allele-frequency change. As discussed in Chapter 28, Gaussian COA models assume that mutational input in any generation is much smaller than the standing additive variation at a locus, or equivalently, that mutation is much stronger

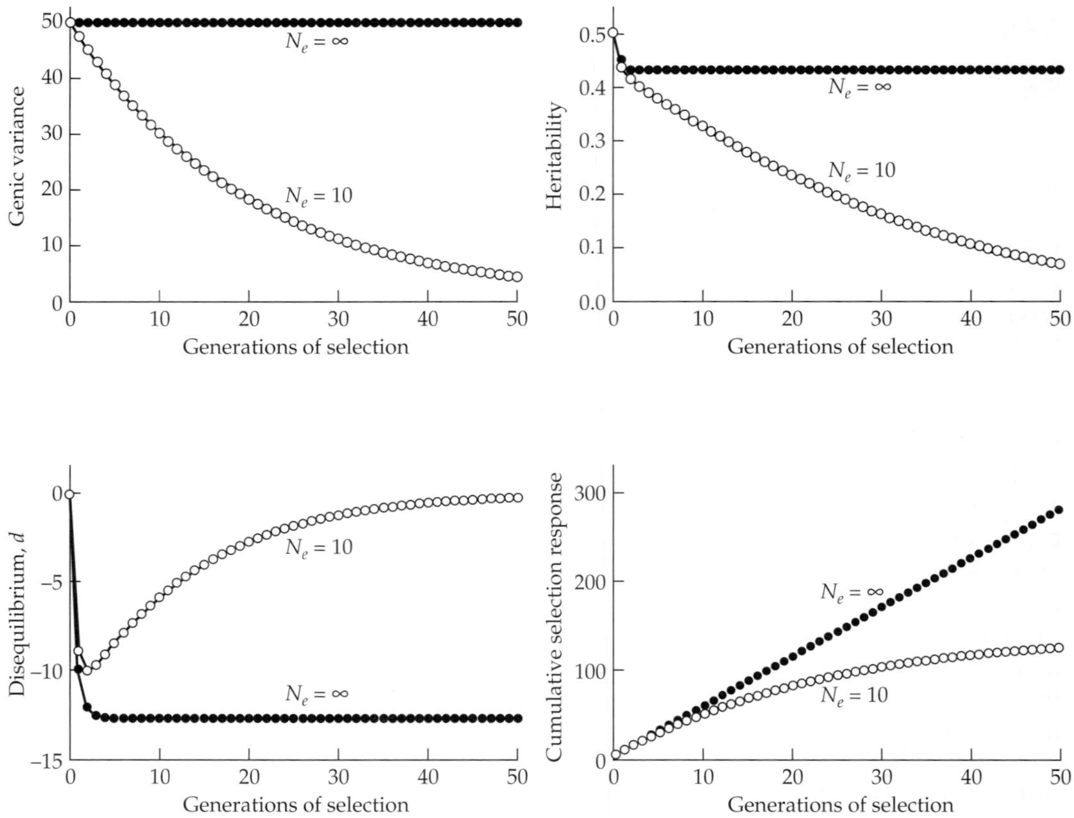

Figure 24.1 The impact of a finite population size, on the genic variance, σ_a^2 (upper left panel), heritability (upper right panel), disequilibrium d (lower left panel), and cumulative selection response (lower right panel). Open circles correspond to $N_e = 10$, and filled circles to the infinitesimal model in the absence of drift ($N_e = \infty$). Details in Example 24.2.

than selection (which allows significant amounts of polymorphism to accumulate at a given locus; Chapter 7). One scenario where biology suggests that the Gaussian model might be approached is in asexual species, where the entire genome is essentially transmitted as a single gene, and hence can have a very large number of possible alleles.

Gaussian COA models attempt to bridge short-term predictors (such as the breeder's and Bulmer equations) that rely on estimable qualities (σ_A^2, h^2) with the long-term predictors of response (Chapters 25 and 26) based on population-genetic models containing quantities that are essentially unestimable. COA models attempt, using estimable quantities, to capture the change in variance from changes in both allele frequencies (and hence changes in σ_a^2) and σ_A^2 from selection-generated disequilibrium (changes in d). Gaussian approximations of the Bulmer equation for the change in variance (Equation 16.7b) under a finite number of loci (n) were introduced by Lande (1975) and Felsenstein (1977, 1979), and Keightley and Hill (1987) further allowed for a finite effective population size (N_e). We consider the effects of drift first. The behavior of these models with mutation is discussed extensively in Chapter 28.

Drift

Assuming that the phenotypic variance after selection has the form $\sigma_{z^*}^2 = (1-\kappa)\,\sigma_z^2$ (Equation 16.10a), Keightley and Hill (1987) obtained equations for the change in the additive genic variance, σ_a^2, and the gametic-phase disequilibrium, d, when the population size is finite:

$$\Delta\, \sigma_a^2(t) = -\frac{\sigma_a^2(t)}{2N_e} \tag{24.1a}$$

$$\Delta\, d(t) = -\frac{1}{2}\left[\left(1+\frac{1}{N_e}\right) d(t) + \left(1-\frac{1}{N_e}\right) \kappa\, h^2(t)\, \sigma_A^2(t)\right] \tag{24.1b}$$

Note that these expressions were shown previously (Equations 16.9a and 16.9b). As before, $\sigma_A^2(t) = \sigma_a^2(t) + d(t)$ and $h^2(t) = \sigma_A^2(t)/\sigma_z^2(t)$, with $\sigma_z^2(t) = \sigma_A^2(t) + \sigma_E^2$, where $\sigma_E^2 = \sigma_z^2(0) - \sigma_A^2(0)$. As in Chapter 16, the selection response in the mean is given by the breeder's equation, $R(t) = h^2(t)S(t)$. If the population size is at least modest, the correction for drift effects on $d(t)$ is small, as $1 \pm 1/N_e \simeq 1$, and Equation 24.1b essentially becomes the Bulmer equation. Drift effects on the genic variance, however, are quite substantial, as they remove all of the initial genic variance (σ_a^2, and hence σ_A^2) after sufficient time ($t \simeq 2N_e$ generations, ignoring new mutation). Solving Equation 24.1a yields

$$\sigma_a^2(t) = \left(1-\frac{1}{2N_e}\right)^t \sigma_a^2(0) \simeq \sigma_a^2(0)\, \exp\left(-\frac{t}{2N_e}\right) \tag{24.1c}$$

This is simply the standard loss of genetic variation under drift (Chapter 2). When dominance or epistasis is present, the additive variance can actually *increase* (for a while) under inbreeding (Chapter 11), so the assumption of only additive gene action is critical. In the absence of mutation, the response runs out of standing variation in finite populations, as σ_a^2 is driven to zero by drift.

Example 24.2. To see the effects of drift on the infinitesimal model, reconsider Example 16.2 under finite population size. This example assumed that $h^2(0) = 0.5$ and $\sigma_z^2(0) = 100$, implying that $\sigma_A^2(0) = \sigma_a^2 = \sigma_E^2 = 50.00$ (assuming $d(0) = 0$). Truncation selection was applied, with the upper 20% saved (yielding $p = 0.2$, $\kappa = 0.781$, and $\bar{\imath} = 1.40$). Under the infinitesimal model, the genic variance, σ_a^2, remains unchanged at its original value of 50, while the additive genetic variance decreases to its equilibrium value of $\widetilde{\sigma}_A^2 = 37.46$, and hence $\widetilde{h}^2 = 37.46/(37.46 + 50.00) = 0.43$, yielding an asymptotic value of response of $\widetilde{R} = \bar{\imath}\widetilde{h}^2\widetilde{\sigma}_z = 5.6$ per generation.

Now consider a finite population size with $N_e = 10$, which is close to the effective population sizes of many artificial selection experiments (Chapter 26). Iteration of Equation 24.1 yields the dynamics depicted in Figure 24.1. Drift erodes away the genic variance, decreasing the heritability (and hence response) over time. The population (in the absence of mutation) will eventually run out of variation, reaching a **selection limit** (Chapters 25 and 26). Note the unusual behavior of the disequilibrium, d, which (following an initial drop) increases toward zero over time. This occurs because the genic variance is declining, which limits the amount of disequilibrium that is possible (Equation 24.4).

Drift and a Finite Number of Loci

Under the infinitesimal model, there is no selection-induced change in allele frequencies, leaving the genic variance unchanged by selection. However, when the population size is finite, alleles are subjected to drift, changing allelic frequencies and eventually reducing the genic variance to zero. A second route for allele-frequency change arises when the number of loci, n, is finite. In this case, there are nonzero selective effects on each locus and allele frequencies change (although potentially very slowly; recall from Equation 5.21 that s scales as a/σ_z). Assuming that the potential distribution of genotypic values at *each* locus is Gaussian, continuum-of-alleles (COA) models can account for both finite N_e and n. The most general result is due to Chevalet (1988, 1994; also see Verrier et al. 1991), where for loci of equal effect and assuming selection of the form such that the phenotypic variance after selection is $\sigma^2_{z^*} = (1-\kappa)\,\sigma^2_z$, we have

$$\Delta\,\sigma^2_a(t) = -\left[\frac{\sigma^2_a(t)}{2N_e} + \left(1 - \frac{1}{N_e}\right)\frac{\kappa\,h^2(t)\,\sigma^2_A(t)}{2n}\right] \tag{24.2a}$$

$$\Delta\,d(t) = -\frac{1}{2}\left[\left(1 + \frac{1}{N_e}\right)d(t) + \left(1 - \frac{1}{n}\right)\left(1 - \frac{1}{N_e}\right)\kappa\,h^2(t)\,\sigma^2_A(t)\right] \tag{24.2b}$$

Provided we are willing to accept the assumption that the distribution of effects at each locus remains normally distributed (a point addressed later), these expressions are iterated to obtain the current values of σ^2_a and d. Starting from an unselected base population (and hence assuming that $d(0) = 0$), the only genetic parameters required to iterate Equations 24.2a and 24.2b are $\sigma^2_A(0)$, h^2, n, and N_e, all of which are potentially estimable.

Equations 24.2a and 24.2b highlight the changes that occur when we assume a finite number of loci ($n < \infty$) and/or finite population size ($N_e < \infty$). When both are infinite, we recover the Bulmer equation (16.7b),

$$\Delta\,\sigma^2_a(t) = 0 \qquad \text{and} \qquad \Delta\,d(t) = -\frac{d(t) + \kappa\,h^2(t)\,\sigma^2_A(t)}{2} \tag{24.2c}$$

Notice that the additive genic variance, σ^2_a, remains unchanged (as allele frequencies remain unchanged), while disequilibrium (nonzero d) is generated by selection but decays to zero once selection stops (i.e., when $\kappa = 0$).

While finite n or N_e both result in modifications of the simple Bulmer equation for the dynamics of d, Equation 24.2b shows that these corrections are generally small. However, this is not the case for changes in σ^2_a. With either finite population size or a finite number of loci, the genic variance decreases in each generation, eventually decaying to zero (in the absence of mutation). The relative importance of drift in comparison to a finite number of loci for changes in σ^2_a can be compared using Equation 24.2a, which yields

$$\frac{\Delta\sigma^2_a}{2} = \begin{cases} -\dfrac{\sigma^2_a}{N_e} & \text{for } n = \infty \\[2ex] -\dfrac{\kappa h^2\sigma^2_A}{n} & \text{for } N_e = \infty \end{cases} \tag{24.2d}$$

Because either directional or stabilizing selection generates negative values of d, we have that $0 < \kappa < 1$ (and hence $\kappa h^2 < 1$), implying that $\sigma^2_A = \sigma^2_a + d < \sigma^2_a$, so that $\sigma^2_a > \sigma^2_A > \kappa\,h^2\,\sigma^2_A$. Using this inequality in Equation 24.2d shows that for comparable values of N_e and n, drift results in a greater per-generation reduction in σ^2_a (also see Figure 24.2).

Example 24.3. Suppose the number, n, of loci underlying a trait is finite. Assume the same model as in Example 24.2, but now let $n = 10$ and $N_e = \infty$ (i.e., reversing the values of n and N_e). We will contrast the behavior of this system with that in Example 24.2 ($N_e = 10$, $n = \infty$), and with the standard infinitesimal model ($N_e = n = \infty$). As Figure 24.2 shows, both h^2 and the selection response decrease over time with a finite number of loci, and eventually a selection limit is reached when all of the initial variation is lost. However, these decreases are not nearly as dramatic as those in Example 24.2 (see Figure 24.1). Figure 24.2 shows that the cumulative response for a model with $N_e = n = 10$ is only very slightly less than under drift alone ($N_e = 10$).

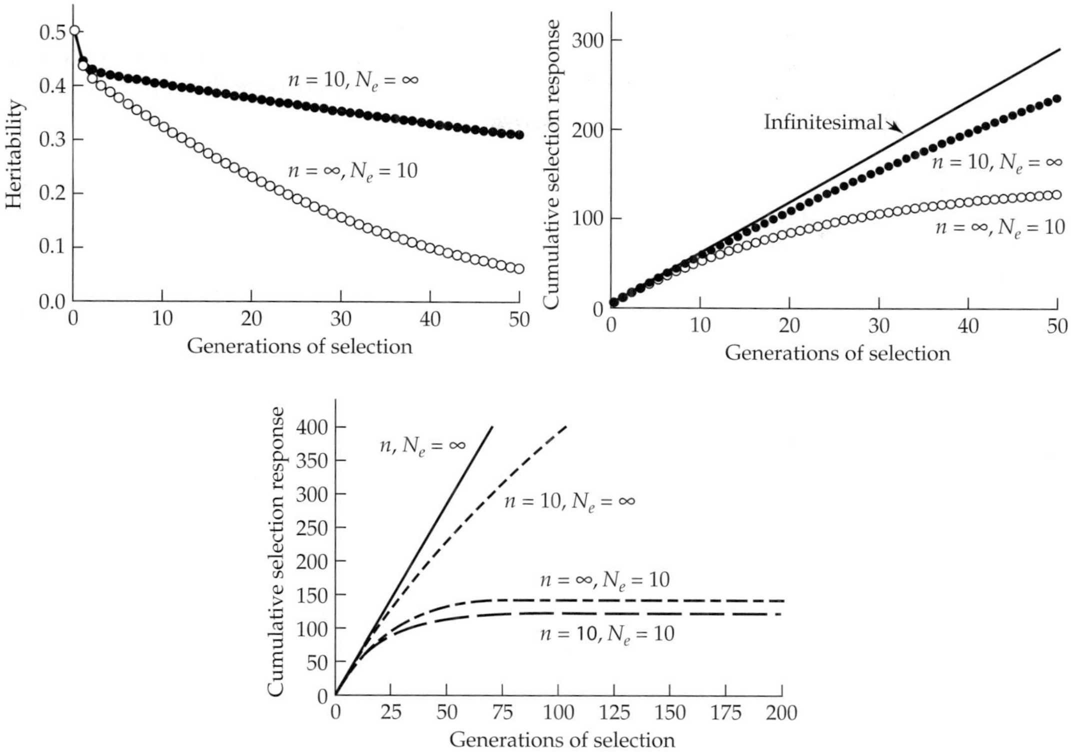

Figure 24.2 The impact of a finite number, n, of loci on heritability and selection response. In both upper panels, the filled circles correspond to the response, with $n = 10$ and $N_e = \infty$, while the open circles correspond to $N_e = 10$ and $n = \infty$. The lower panel displays the cumulative response under four different combinations of N_e and n. Note that the values of $N_e = n = \infty$ corresponds to the standard infinitesimal model.

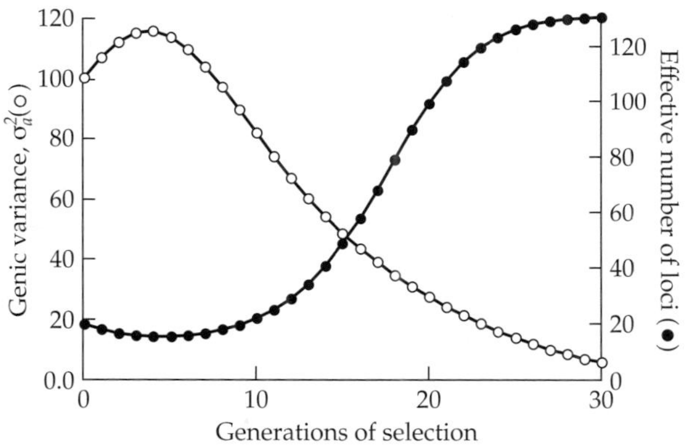

Figure 24.3 The change in the genic variance, σ_a^2 (open circles), and the effective number of loci, n_e (filled circles; see Equation 24.3), for a model with 5 major and 125 minor loci. See Example 24.4 for details.

The Effective Number of Loci, n_e

Chevalet (1994) allowed loci to differ in the amount of genetic variance they contribute, replacing the acutal number of loci, n, in Equations 24.2a and 24.2b by the **effective number of loci**, n_e, defined as

$$n_e = \frac{n}{1 + c_v^2} \tag{24.3}$$

where c_v is the coefficient of variation in the genic variance contributed by each locus,

$\sigma\left(\sigma_{ai}^2\right)/E[\,\sigma_{ai}^2\,]$, with σ_{ai}^2 being the genic variance contributed by locus i. (Note that Equation 24.3 is closely related to the Castle-Wright estimator for the effective number of segregating genes in a line-cross F_2 population; LW Chapter 9.) If all loci contribute the same variance, then $c_v = 0$ and $n_e = n$, while $n_e \ll n$ when $c_v \gg 1$.

Note that n_e changes over time, as allele-frequency changes alter the genic variance contributed by any particular locus. Indeed, loci with the largest genetic variance should show the largest initial response to selection, and also the fastest depletion of locus-specific variance (Chapter 25). In such cases, one can move from a situation where the effective number of loci is quite small (a few loci with large effects, and hence a high value of c_v) to a situation where n_e can be quite large (with the remaining loci all having roughly equal effects, so that c_v is small). Hence, n_e can increase over time, but we also correspondingly expect the total genic variance, σ_a^2, for the remaining loci to decrease (Figure 24.3).

Example 24.4. As an illustration of how the effective number of loci, n_e, can change over the course of selection, consider an additive model with both major and minor loci underlying a trait under selection. There are five major loci, each with frequency $p = 0.25$ and effect size $a = 5.16$, and 125 minor loci, each with $p = 0.5$ and $a = 0.89$. The resulting initial genic variance is $\sigma_a^2 = 100$ (half of which is from the major genes and half from the minor loci), and we assume an initial heritability of $h^2 = 0.5$ (by setting $\sigma_E^2 = 100$). Finally, assume truncation selection with the uppermost 20% saved (further details for this model are given in Example 25.2). We ignore any effects of disequilibrium, focusing instead on how the genic variance (the open circles in Figure 24.3) and the effective number of loci, n_e (the filled circles) change over time due to allele-frequency changes.

As shown in Figure 24.3, while there are 130 loci in this system, initially the effective number is around 20, due to the large coefficient of variation in the locus-specific genic variances. As selection proceeds, the genic variance initially *increases* as the major alleles increase their frequencies toward 0.5 (where they have maximal additive variance when dominance is absent; LW Figure 4.8). Such an increase in variance is not predicted by Gaussian COA models (which assume a continuous *distribution* of allelic effects at each locus, rather than simply two alleles). Notice that the effective number of loci decreases slightly during this increase in variance, as the c_v increases. As alleles at these major loci become fixed, the total genic variance decreases while the effective number of loci increases (to approximately 125, the number of minor loci), reflecting a decrease in the coefficient of variation (as the remaining loci all have very similar variances).

As Example 24.4 illustrates, a selection response can result in an increase in the additive variance at some point during selection (typically fairly early), as favored alleles at low frequencies increase to intermediate frequencies. Hospital and Chevalet (1996) observed a similar phenomenon in their simulation of linkage, namely, that finite-locus models can also show an increase in additive variation. The distinction is that this increase may come, not rather quickly (as was the case for rare alleles), but only many generations after selection is initiated, reflecting recombination finally generating favorable gametes, which then increase in frequency. They also found that linked systems are vulnerable to lower selection responses due to hitchhiking fixing less favorable alleles, an issue examined further in Chapter 26.

Dynamics: σ_a^2 and d Change on Different Time Scales

Chevalet (1988, 1994) and Gavrilets and Hastings (1994a, 1995) noted that the dynamics of the genic variance and the disequilibrium operate on rather different time scales. The change in d is rather rapid, quickly approaching a **quasi-equilibrium** value (Kimura 1965b),

$$d = -\kappa\left(1 - \frac{1}{n_e}\right) h^2 \sigma_A^2 \tag{24.4}$$

Note from Equation 24.2b that this is simply the amount of new disequilibrium generated by selection (taking $N_e = \infty$). This is not strictly an equilibrium value, as changes in σ_a^2, which occur over much longer time scales, also change σ_A^2, albeit much more slowly. As $n_e \to \infty$, we recover Equation 16.7c (as $\delta(\sigma_z^2) = -\kappa\sigma_z^2$), the equilibrium d value found by Bulmer (which is a true equilibrium, as σ_a^2 does not change under the infinitesimal model). Thus, for a given value of σ_a^2, there is a quick approach to the equilibrium value of d, which we can very closely approximate by treating the value of σ_a^2 as fixed and applying Equations 16.13a–16.13c.

One direct application of Equation 24.4 involves the distribution of the additive genetic variance within the population. This can be decomposed into that held between families (the difference in family means) and that generated by segregation within each family (Chapter 16). When no disequilibrium is present ($d = 0$), under random mating, each component is $\sigma_A^2/2 = \sigma_a^2/2$. However, with selection, the genetic variance *among* full-sib families is simply half the additive variance following selection, $\sigma_A^2 + d$, which (Equation 24.4) yields

$$\sigma^2(FS) = \frac{\sigma_A^2}{2}\left[1 - \left(1 - \frac{1}{n_e}\right)\kappa h^2\right] \tag{24.5a}$$

while the within-family (additive genetic) variance is

$$\sigma_A^2(\text{within-family}) = \frac{\sigma_a^2}{2} = \frac{\sigma_A^2}{2}\left[1 + \left(1 - \frac{1}{n_e}\right)\kappa h^2\right] \tag{24.5b}$$

Equation 24.5b follows by noting that $\sigma_A^2 = \sigma_a^2 + d$, and substituting Equation 24.4 for the value of d.

Gavrilets and Hastings (1994a, 1995) noted that this difference in time scales for changes in d versus σ_a^2 has important implications for the interpretation of experiments using artificial stabilizing selection, where response is measured by a decrease in the phenotypic variance. Simulations, as well as their analysis of two- and n-locus models, showed that while a rapid approach to the quasi-equilibrium d value (Equation 24.4) occurs, the rate of change of allele frequencies can be very slow—on the order of 100 (or more) generations even with strong selection under a two-locus model. As mentioned in Chapter 16, we expect an immediate decrease in the phenotypic variance due to selection generating negative disequilibrium, which reduces σ_A^2. Selection will also operate to eventually reduce the genic variance, σ_a^2, resulting in a further decrease in σ_A^2 over that predicted from disequilibrium alone. Gavrilets and Hastings suggested that, given the short time scales of most experiments, if decreases in σ_a^2 occur, these are more likely due to drift than selection. A related point is that an observed reduction in the phenotypic variance is usually assumed to arise solely from a reduction in σ_A^2. However, as mentioned in Chapter 17, if there is heritable variation in σ_E^2, the environmental variance is also reduced by stabilizing selection. Thus, selection-driven allele-frequency changes might be involved in an observed reduction in the phenotypic variance, but through changes in σ_E^2, rather than through σ_a^2 for the trait.

How Robust Is the Gaussian Continuum-of-alleles Model?

As will be fully discussed in Chapter 28, a key determinant of which continuum-of-alleles model is appropriate is the relative strength of selection and mutation at a locus (Table 28.2). The Gaussian model is appropriate when the per-locus mutation rate is far greater than the average strength of selection on a locus (in a large population). In such settings, a given locus is likely to be segregating a large number of alleles, and any new allele adds little to the standing variation. Conversely, other approximations (such as the rare-allele or the house-of-cards models) are more appropriate when selection is much stronger than mutation (i.e., $s \gg \mu$), in which case any given locus will have one dominant allele and a few very rare alleles. Our discussion here assumes that mutation is much stronger than selection.

Under this assumption, if the trait is determined by a modest to large number of loci, all of roughly equal effect and with alleles at intermediate frequencies, then Gaussian COA models can perform reasonably well, at least over intermediate time scales (Chevalet 1988, 1994). They generally tend to overestimate the cumulative response as generations increase, so that while the decrease in σ_a^2 from selection is partially captured (an improvement over the infinitesimal model), after sufficient time, the COA approximation breaks down.

What are the possible causes for the breakdown of the approximation? A subtle feature of the Gaussian assumption for each locus is the key. The COA approximation generally works well at the start of selection, provided the distributions of genotypic values at each locus are initially *close* to normal. However, as allele-frequency changes drive the individual locus distributions further from normality, the COA approximation breaks down, as loci with alleles at extreme frequencies (near zero or one) can show large departures from normality. A recurrent theme of this chapter is a focus on the skewness and kurtosis of distributions, as these provide very convenient measures of the departure from a normal. Skewness measures departures from symmetry, while kurtosis measures whether the tails of the distribution decline more or less rapidly than under a normal. Both of these statistics (with kurtosis appropriately defined) are zero under a normal, and hence they provide a simple quantification of the departure of a distribution from normality (LW Chapter 2).

What can we say about the skewness and kurtosis for an n-locus biallelic model? Assume n exchangeable loci, each with frequency p of the favorable allele. Zeng (1987) showed that the scaled (to unit variance) coefficients of skewness, k_3, and kurtosis, k_4, for the resulting distribution of genotypic values are

$$k_3 = \frac{2p-1}{\sqrt{2np(1-p)}} \qquad \text{and} \qquad k_4 = \frac{1-2p(1-p)}{2np(1-p)} \tag{24.6}$$

Note that skewness is zero and kurtosis is minimized at intermediate allele frequencies ($p = 1/2$). As allele frequencies become more extreme, so do the skew and kurtosis.

Rare alleles of large effect are especially problematic. Not only do these generate skewness and kurtosis, but as their frequencies increase, so does the genic variance (Figure 24.3). Lande (1983) and Zhang and Hill (2005a) both noted that natural selection tends to generate a correlation between allelic effect size and frequency, so that alleles of large effect may tend to be rare in natural populations due to pleiotropic deleterious fitness effects. If these rare alleles are captured when a population is sampled to form a laboratory stock for artificial selection, an increase in additive variance is expected during selection. However, if the founding population is under strong drift for a few generations (as with a founding bottleneck), rare alleles can be lost and the COA approximation may be a good predictor of short-term response. This theme of initially rare favorable alleles, and hence an early accelerated response to selection, will be revisited in Chapter 28, as it is central to certain predictions about response in a population under mutation-selection balance.

THE BULMER EFFECT UNDER LINKAGE

Both the standard infinitesimal and Gaussian COA models assume *unlinked* loci. When loci are linked, the contribution from gametic-phase disequilibrium, d, decays by less than half in each generation. This allows higher values to accrue, yielding a larger value of $|d|$ at equilibrium. We examine the impact of linkage under two different settings. First, we consider the Gaussian COA approximation (requiring multivariate normality of the locus-specific distributions of effects), and second, allowing for departures from normality serves as an entry point for our final section on treating non-Gaussian distributions of genetic values.

An Approximate Treatment

An approximate solution for the dynamics of σ_A^2 incorporating linkage was offered by Bulmer (1974a, 1980), whose approach we follow (a more general solution by Turelli and Barton will be considered shortly). Recall from Chapter 16 that C_{ij} denotes the covariance between allelic effects at loci i and j, meaning that C_{ii} is the genic variance for locus i, while C_{ij} for $i \neq j$ measures the contribution from disequilibrium between i and j. Thus, from Equation 16.1b, $d(t) = 4\sum_{i<j} C_{ij}(t)$, with the changes in the pairwise covariances describing the change in d. If r_{ij} is the recombination fraction between two loci (we use r_{ij} in place of our more standard c_{ij} used in previous chapters to avoid confusion with C_{ij}), then $(1 - r_{ij})\, C_{ij}(t)$ is the contribution passed on to $C_{ij}(t+1)$. Recalling Equation 16.6, the change in $d(t)$ due to selection when genotypic and phenotypic values are normally distributed is

$$\frac{h^4(t)}{2}\, \delta(\sigma^2_{z(t)}) = 4 \sum_{i=1}^{n} \sum_{i<j}^{n} \delta C_{ij}(t) \tag{24.7a}$$

where we use both δX and $\delta(X)$ to denote the *within-generation* change in the variable X. In order to approximate δC_{ij} (the change in disequilibrium generated by selection), Bulmer assumed that these changes are the same for each pair of loci (an exchangeable model). For n loci, there are $n(n-1)/2$ unique pairs, giving the contribution from each pair as

$$4\delta C_{ij}(t) \simeq \frac{h^4(t)\, \delta(\sigma^2_{z(t)})}{n(n-1)} \tag{24.7b}$$

Because the new disequilibrium equals the fraction of the current disequilibrium after recombination plus the fresh disequilibrium generated by selection

$$C_{ij}(t+1) = (1 - r_{ij})\, C_{ij}(t) + \delta C_{ij}(t) \tag{24.7c}$$

This equation is approximate, as the covariance *between* gametes, $C_{i,j}$, is ignored here. Equation 24.11c provides a more exact treatment (as does Example 28.7). Ignoring $C_{i,j}$, Equation 24.7c implies that, at equilibrium, $r_{ij}\, \widetilde{C}_{ij} = \widetilde{\delta C}_{ij}$, where the tildes denote equilibrium values. Using Equation 24.7b yields an equilibrium covariance of

$$\widetilde{C}_{ij} = \frac{\widetilde{h}^4\, \widetilde{\delta}(\sigma^2_z)}{4n\,(n-1)}\, \frac{1}{r_{ij}} \tag{24.7d}$$

Thus,

$$\widetilde{d} = 4 \sum_{i=1}^{n} \sum_{i<j}^{n} \widetilde{C}_{ij}(t) = 4\, \frac{\widetilde{h}^4\, \widetilde{\delta}(\sigma^2_z)}{4n\,(n-1)} \sum_{i=1}^{n} \sum_{i<j}^{n} \frac{1}{r_{ij}} = \frac{1}{2}\, \widetilde{h}^4\, \widetilde{\delta}(\sigma^2_z)\, \frac{1}{r_H} \tag{24.8a}$$

where r_H is the harmonic mean of all pairwise recombination distances between loci,

$$r_H = \left(\frac{1}{n\,(n-1)/2} \sum_{i=1}^{n} \sum_{i<j}^{n} \frac{1}{r_{ij}} \right)^{-1} \tag{24.8b}$$

The value of r_H varies with both the number of loci and chromosomes, decreasing as the number of loci per chromosome increases. Using simulations of randomly distributed loci, Bulmer (1974a) found that if the haploid chromosome number exceeds 10, r_H will likely be no smaller than 0.4, while in *Drosophila melanogaster*, with its three main chromosomes and lack of recombination in males, r_H is around 0.1 if there are many loci. However, even if only a few loci occur as tightly linked pairs, r_H can be considerably below 0.5, as the harmonic mean disproportionately weights very small values.

Assuming again that the phenotypic variance after selection can be written as $\sigma^2_{z^*} = (1-\kappa)\, \sigma^2_z$, Equation 16.13a can be modified for linkage to give the equilibrium additive genetic variance as $\widetilde{\sigma}^2_A = \sigma^2_z\, \gamma$. Letting h^2 and σ^2_z denote values in the absence of LD (i.e., $d = 0$, implying $h^2 = \sigma^2_a/\sigma^2_z$), then

$$\gamma = r_H \left(\frac{2h^2 - 1 + \sqrt{1 + 2h^2(1-h^2)\kappa/r_H}}{2r_H + \kappa} \right) \tag{24.9}$$

and the equilibrium heritability, $\widetilde{h}^2 = \widetilde{\sigma}^2_A/\widetilde{\sigma}^2_z = (\sigma^2_a + \widetilde{d})/(\sigma^2_z + \widetilde{d})$, is given by Equation 16.13c using the value of γ for Equation 24.9. Note that we can express this result in terms of $\widetilde{d}$ by noting that $\widetilde{\sigma}^2_A - \sigma^2_a = \widetilde{\sigma}^2_A - h^2\sigma^2_z = \widetilde{d}$, or that $\widetilde{d} = (\gamma - h^2)\sigma^2_z$. The general conclusion from Equation 24.9 is that, for a fixed value of κ, increasing the amount of linkage (e.g., decreasing r_H) increases the absolute value of $\widetilde{d}$ (Bulmer 1974a, 1976, 1980).

Finally, turning briefly to disruptive selection (where $\kappa < 0$, and hence $d > 0$), the standard infinitesimal model predicts that $\widetilde{d}$ should increase as linkage tightens. As we saw in Chapter 16, with sufficiently strong disruptive selection, namely $\kappa < -r_H/[2h^2(1-h^2)]$, there is no real positive root for γ in Equation 24.9, and the standard infinitesimal model predicts that d, and hence σ^2_A, increases without limit (Bulmer 1976). With a finite number of loci, this condition implies that selection creates almost complete disequilibrium, so that only a few of the possible gamete types are actually present, namely, most gametes are either $abcd\cdots$ or $ABCD\cdots$ (Chapter 16).

While the prediction that $\widetilde{d}$ increases under disruptive selection as linkage tightens holds for an infinite population, simulations by Sorensen and Hill (1983) found exactly the opposite with small values of N_e, with $|\widetilde{d}|$ actually *decreasing* as linkage tightens. They reasoned that this discrepancy arises due to the interaction between a finite number of loci and the finite population sizes used in their simulations. To see this, consider complete linkage. In a finite population, the most extreme gamete observed is affected by sampling, as selection can generate no gamete more extreme than those found in the initial sample (in the absence of recombination). If the number of loci is small, the probability of sampling the most extreme possible gamete is high, but this probability decreases as the number of loci increases. Countering this, as recombination (measured by r_H) or the population size increases, the probability increases that recombination can regenerate more extreme gametes before the relevant loci are fixed by drift or selection. When population size becomes large enough that drift effects are no longer important, $\widetilde{d}$ increases with increasing linkage. Interactions of this sort between drift, selection, and recombination are considered in detail in Chapter 26.

Example 24.5. As an example of the consequences of increased linkage, reconsider our analysis of the response under directional selection used in Examples 24.2 and 24.3. Here we assume an infinite number of loci and infinite population size. Substituting into Equation 24.9 to obtain γ and recalling Equations 16.13a–16.13c yields the following metrics of response for different r_H values:

r_H	γ	$\widetilde{d}$	$\widetilde{\sigma}^2_A$	$\widetilde{h}^2$	$\widetilde{R}$
0.5	0.37	−12.60	37.40	0.43	5.60
0.4	0.35	−14.50	35.50	0.42	5.37
0.3	0.33	−17.11	32.89	0.40	5.06
0.2	0.29	−20.97	29.03	0.37	4.57
0.1	0.23	−27.49	22.51	0.31	3.70

A value of $r_H = 0.5$ corresponds to free recombination, while $r_H = 0.1$ might be expected in *Drosophila melanogaster*. As expected, decreasing the average amount of recombination between loci increases the effect of linkage disequilibrium, generating more extreme d values, and hence (for directional and stabilizing selection) smaller additive variances, heritabilities, and selection responses. With strong linkage ($r_H = 0.1$), the response is only 66% of that for unlinked loci (3.70 versus 5.60).

A More Careful Treatment

A more rigorous treatment of how selection changes the within-gamete covariances, C_{ij}, requires consideration of the (pairwise) **between-gamete covariance,** $C_{i,j}$, as well as higher-order covariance terms that measure the amount of gametic-phase disequilibrium between groups of more than two loci. Here, we introduce some of the notation needed for non-normal distributions of genotypic and phenotypic values, returning to the consequences of relaxing the normality assumption in the next section.

We start by defining the between-gamete covariance,

$$C_{i,j} = \sigma\left(a_{fa}^{(i)}, a_{mo}^{(j)}\right) \tag{24.10}$$

which is the covariance between the effect of an allele at the ith locus in the paternal (fa) gamete and an allele at the jth locus in the maternal (mo) gamete. Under random mating, gametes unite at random, and hence $C_{i,j} = 0$ at the start of each generation. However, selection generates correlations *between* gametes in much the same way that it generates correlations among loci *within* gametes. For example, consider a particular chromosome containing multiple loci influencing a character under stabilizing selection. Initially, there is no correlation between the genetic values of the two copies of this chromosome in an offspring from randomly mated parents. Stabilizing selection changes this initial distribution, favoring adults with intermediate genotypic values. Thus, surviving adults with a large genetic value on one chromosome are expected to have a small value on the other and vice versa, generating negative $C_{i,j}$ (Figure 16.2; Example 28.7). Likewise, positive assortative mating generates positive $C_{i,j}$, while disassortative mating generates negative $C_{i,j}$.

We assume random mating, so that $C_{i,j}(t) = 0$ at the start of each generation. Letting C^* denote the covariance after selection, where

$$C_{ij}^*(t) = C_{ij}(t) + \delta C_{ij}(t) \quad \text{and} \quad C_{i,j}^*(t) = C_{i,j}(t) + \delta C_{i,j}(t) = \delta C_{i,j}(t) \tag{21.11a}$$

Assuming recombination follows selection, then with probability $1 - r_{ij}$, no recombination occurs between i and j and the within-gamete covariance is unchanged. Conversely, with a probability of r_{ij}, recombination occurs and the new covariance depends on the covariance *between* gametes, yielding the result of Lande (1975) and Bulmer (1980) that

$$C_{ij}(t+1) = (1 - r_{ij})C_{ij}^*(t) + r_{ij}C_{i,j}^*(t) \tag{24.11b}$$

Substituting for C^* from Equation 24.11a yields

$$\begin{aligned} C_{ij}(t+1) &= (1 - r_{ij})\,[\,\delta C_{ij}(t) + C_{ij}(t)\,] + r_{ij}\,\delta C_{i,j}(t) \\ &= (1 - r_{ij})C_{ij}(t) + \delta C_{ij}(t) - r_{ij}\,[\,\delta C_{ij}(t) - \delta C_{i,j}(t)\,] \end{aligned} \tag{24.11c}$$

Note that we recover Equation 24.7c only if $\delta C_{ij}(t) = \delta C_{i,j}(t)$, meaning that selection changes the within-gamete and between-gamete covariances by the same amount. Turelli and Barton (1990) showed that this occurs if there is either global gametic-phase equilibrium before selection (all groups of loci are in gametic-phase equilibrium) or the distribution of allelic effects over loci is multivariate normal (see Equations 24.13a and 24.13b). Thus, Equation 24.7c follows under Gaussian COA assumptions. However, selection can drive a distribution away from normality, in which case Equation 24.7c may no longer hold.

General expressions for $\delta C_{ij}(t)$ and $\delta C_{i,j}(t)$ were obtained by Turelli and Barton (1990) for the case of no dominance or epistasis. Their expressions involve generalizations of (i) measures of selection to higher moments of a distribution, and (ii) disequilibrium measures to groups of k loci. Starting with (i), recall (Equation 13.27b) that we defined the directional selection gradient, which measures how mean fitness varies with the phenotypic mean, as $\partial \ln \overline{w} \,/\, \partial \mu_z$. We can extend this notion to higher moments by considering $\partial \ln \overline{w} \,/\, \partial \mu_{k,z}$, where $\mu_{k,z} = E[\,(z - \mu_z)^k\,]$ is the kth central moment of the phenotypic distribution (for $k \geq 2$), with $\mu_{2,z} = \sigma_z^2$. If selection is primarily on the mean and variance of the phenotypic distribution, gradients for the skew and higher moments ($k \geq 3$) will generally be negligible. When phenotypes are normally distributed, the first two gradients are given by

$$\frac{\partial \ln \overline{w}}{\partial \mu_z} = \frac{S}{\sigma_z^2} \tag{24.12a}$$

$$\frac{\partial \ln \overline{w}}{\partial \sigma_z^2} = \frac{\delta(\sigma_z^2) + S^2}{2\,\sigma_z^4} \tag{24.12b}$$

(Lande 1976; Lande and Arnold 1983). As will be shown in Chapters 29 and 30, when selection acts only on the mean (such that $\partial \ln \overline{w}/\partial \mu_{k,z} = 0$ for $k \geq 2$), the within-generation change in the phenotype variance will be $\delta(\sigma_z^2) = -S^2$. Hence, $\delta(\sigma_z^2) + S^2$ is the change in variance over that expected due to selection simply on the mean.

Using these extended selection gradients, and ignoring selection acting on the skew and higher moments (i.e., $\partial \ln \overline{w}/\partial \mu_{k,z} = 0$ for $k \geq 3$), Turelli and Barton (1990) found that

$$\delta C_{ij} = \frac{\partial \ln \overline{w}}{\partial \mu_z} \sum_{h=1}^{n} C_{ijh} + \frac{\partial \ln \overline{w}}{\partial \sigma_z^2} \sum_{h=1}^{n} \sum_{\ell=1}^{n} (\, C_{ijh\ell} - C_{ij} C_{h\ell}\,) + \cdots \tag{24.13a}$$

$$\delta C_{i,j} = \frac{\partial \ln \overline{w}}{\partial \sigma_z^2}\, 2 \sum_{h=1}^{n} C_{ih} \sum_{\ell=1}^{n} C_{j\ell} + \cdots \tag{24.13b}$$

where C_{ijh} refers to the third-order covariance between the effects of alleles at loci i, j, and h. If X_i is the additive value of a randomly chosen allele at locus i, and $\mu_i = E(X_i)$ is the average value for this locus, then $C_{ijh} = E[\,(X_i - \mu_i)(X_j - \mu_j)(X_h - \mu_h)\,]$. Higher-order covariances are similarly defined. The covariances in Equation 24.13a measure the amount of third-order (C_{ijh}) and fourth-order (C_{ijhl}) gametic-phase disequilibrium (the departures from random assortment for triplets and quadruplets of loci). If selection on the third (skew) or higher-order moments is significant, then Equations 24.13a and 24.13b need to include covariance terms of order five and higher.

The key point about these equations is that ***changes in covariances depend critically on very fine details of the genotypic distribution***, details that are essentially impossible to estimate empirically in realistic situations. Thus, simplifying assumptions are required to proceed further. For example, if the distribution of the vector of individual-locus genotypic values is multivariate normal (which, as previously mentioned, involves the rather strong assumption that allelic effects *at each locus* are normally distributed), Equation 24.13a simplifies greatly, as $C_{ijk} = 0$ and C_{ijkl} can be expressed in terms of second-order covariances ($C_{ijlk} = C_{ij}\, C_{kl} + C_{ik}\, C_{jl} + C_{il}\, C_{jk}$). In this case, $\delta C_{ij} = \delta C_{i,j}$, and combining Equations 24.12b and 24.13b yields

$$\delta C_{i,j}(t) = \delta C_{ij}(t) \simeq \frac{\delta(\sigma_z^2) + S^2}{\sigma_z^4}\, C_i(t)\, C_j(t) \tag{24.14a}$$

where $C_i(t) = \sum_j C_{ij}(t)$. Thus, when allelic effects are multivariate normal (normal at each locus and multivariate normal for any subset of loci), Equation 24.11c yields a between-generation change (denoted by Δ, as opposed to a within-generation change, which is denoted by δ) in covariance of

$$\Delta C_{ij}(t+1) = C_{ij}(t+1) - C_{ij}(t) = \frac{\delta(\sigma_z^2) + S^2}{\sigma_z^4}\, C_i(t)\, C_j(t) - r_{ij} C_{ij}(t) \tag{24.14b}$$

a result due to Lande (1975, 1977a). Because $2\sum_i C_i = 2\sum_{ij} C_{ij} = \sigma_A^2$, and assuming all of the C_i are equivalent, it follows for n loci that $C_i = \sigma_A^2/(2n)$, and Equation 24.14a reduces to

$$\delta C_{ij} \simeq \frac{\sigma_A^4}{4n^2 \sigma_z^4} \left(\delta(\sigma_z^2) + S^2 \right) = \frac{h^4}{4n^2} \left(\delta(\sigma_z^2) + S^2 \right) \tag{24.14c}$$

When $S^2 \ll |\,\delta(\sigma_z^2)\,|$ (selection is mainly on the variance), we recover Bulmer's approximation (Equation 24.7b) when the number of loci, n, is large.

RESPONSE UNDER NON-GAUSSIAN DISTRIBUTIONS

The assumption that genotypic values of offspring are described by a linear and homoscedastic regression of the genotypic values of their parents most easily follows if the joint distribution of parental and offspring values is multivariate normal. This assumption, which is the basis for much of the theory of selection response, provides a simple solution to the vexing problem of modeling the transmission of a quantitative trait. The other option is Price's theorem (Equation 6.8), but its composite transmission parameter is very difficult to estimate and depends on the fine details of the underlying genetic architecture.

While changing allele frequencies and the generation of gametic-phase disequilibrium compromise the prediction of response by altering the genetic variance, a more subtle, but no less important, issue is that these changes also compromise predictions by driving the genotypic distribution away from a Gaussian. An active area of research is to both describe how selection can alter a distribution and extend selection theory to arbitrary distributions of genotypic values. While good progress has been made, we warn the reader that this can be a rather intimidating area of the literature. Our purpose here is to simply introduce some of the basic ideas and machinery used, as well as to summarize the major findings.

We start by considering how the distribution of effects at each of the individual loci translates into a distribution of genotypic values. In particular, we examine how within-locus moments of allelic effects translate into moments of the full distribution of genotypic values. While moments are more intuitive measures of the shape of a distribution, the cumulants of the distribution (to be described shortly) are more natural to work with when describing deviations from normality.

Once this basic machinery has been inroduced, we will then consider two types of models for the genetics of a trait: (i) a small to modest number of segregating loci, and (ii) a very large number of loci of small effect. With a small number of loci, to an initial approximation, one can ignore effects of gametic-phase disequilibrium and instead focus on the changes in the higher genotypic moments caused by allele-frequency changes. The key result from the analysis of such models is that even single-generation predictions require extensive information about the underlying genetics. In contrast, with a very large number of loci, the (short-term) effects of allele-frequency changes can be essentially ignored, and changes from gametic-phase disequilibrium become critical. The nice (and somewhat surprising) result for this latter class of models is that both the breeder's and Bulmer equations are quite accurate for both directional and strong disruptive selection (Turelli and Barton 1994).

Describing the Genotypic Distribution: Moments

Under our assumption that genotypic and environmental values are additive and independent, $z = G + E$. When environmental values, E, are normally distributed, phenotypes, z, are normally distributed if and only if the genotypic values are Gaussian (i.e., follow a normal distribution). However, the converse is not true—an approximately normal distribution of phenotypes *does not necessarily* imply that genotypic values are Gaussian. While we can test if *phenotypes* are normally distributed, this tells us little about the distribution of *genotypes*. In theory, we can estimate the genotypic distribution by estimating the breeding values for a sample of individuals (LW Chapter 26), but this is generally impractical in most studies (however, see Chapter 20). Further, methods used to estimate breeding values typically *start* with the assumption of normality (Chapters 19 and 20; LW Chapters 26 and 27), and hence bias the distribution of estimated values toward a Gaussian.

Because we assume that there are no genotype×environment interactions, if the environment remains constant over time, any changes in the phenotypic distribution are entirely due to changes in the genotypic distribution. The moments of a distribution provide a convenient measure to describe its shape, and hence changes in the moments provide descriptions of changes in the shape of the distribution. To see the connection between the moments of the phenotypic and genotypic distributions, note that the phenotypic mean, variance, and skew can be decomposed as $\mu_z = \mu_G$, $\sigma_z^2 = \sigma_G^2 + \sigma_E^2$, and $\mu_{3,z} = \mu_{3,G} + \mu_{3,E}$. Thus, assuming no environmentally induced changes in the moments, changes in any of the first three phenotypic moments will exactly equal the change in the corresponding genotypic moments. Example 24.6 (below) uses cumulants to derive the fourth phenotypic moment, yielding

$$\mu_{4,z} = \mu_{4,G} + \mu_{4,E} + 6\,\sigma_G^2\,\sigma_E^2 \tag{24.15}$$

This shows that any changes here can result from either changes in the second (variance) or fourth moments of the genotypic distribution. When E is normal, $\mu_{3,E} = 0$ and $\mu_{4,E} = 3\sigma_E^4$, simplifying these expressions.

How do the moments of G depend on the distribution of allelic effects at individual loci? If n loci control the character, our assumption of complete additivity implies

$$G = \sum_{i=1}^{n} (X_{fa,i} + X_{mo,i}) \tag{24.16}$$

where $X_{fa,i}$ ($X_{mo,i}$) is the value of the paternal (maternal) allele at the ith locus. Assuming both sexes have the same distribution of allelic effects, the moments of G can be related to moments of the distribution of allelic effects at individual loci by expanding

$$\begin{aligned}\mu_{k,G} &= E\left([G - \mu_G]^k \right) \\ &= E\left(\left[\sum_{i=1}^{n} \{ X_{fa,i} + X_{mo,i} - 2E(X_i) \} \right]^k \right) \qquad \text{for } k \geq 2\end{aligned} \tag{24.17}$$

Finally, we assume random mating, so that $X_{fa,i}$ and $X_{mo,i}$ are independent at the start of each generation. Because we assume that the distribution of allelic effects is the same in both sexes, we drop the subscript referring to parental origin.

When considering a particular moment of G, it will be important to distinguish between contributions to that moment from individual loci (**within-locus moments**) and contributions from gametic-phase disequilibrium (**between-locus covariances**). This partitioning, which was used earlier with the additive genetic variance (e.g., Equation 16.2), is extended here to the third and higher genotypic moments. To describe the distribution of effects at locus i, let $\mu_{1,i} = E(X_i) = m_i$ denote the average value of an allele at locus i, and define the kth moment for this locus by $\mu_{k,i} = E([X_i - m_i]^k)$ for $k \geq 2$. After summing over all n loci, we define

$$M_1 = 2\sum_{i=1}^{n} \mu_{1,i} \tag{24.18a}$$

$$M_2 = 2\sum_{i=1}^{n} \mu_{2,i} \tag{24.18b}$$

$$M_3 = 2\sum_{i=1}^{n} \mu_{3,i} \tag{24.18c}$$

as the mean, variance, and skewness, respectively, of the genotypic distribution in terms of the mean, variance, and skew at individual loci (the within-locus moments). Finally, we define the within-locus kurtosis as

$$M_4 = 2\sum_{i=1}^{n} \left(\mu_{4,i} - 3\,\mu_{2,i}^2 \right) \tag{24.18d}$$

While this may seem odd at first, recall that the fourth and second moments of a normal distribution are related by $\mu_4 = 3\mu_2^2$ (LW Chapter 2). Hence, if the distribution of allelic effects *at each locus* is normal, $M_4 = 0$ and, likewise, $M_3 = 0$ ($\mu_{3,i} = 0$ because a normal random variable does not display skew). On the other hand, nonzero values of M_3 or M_4 imply that G is non-Gaussian and provide a quantitative measure of the departure from normality.

The between-locus contributions from gametic-phase disequilibrium are described by C_{ij}, C_{ijk}, and C_{ijkl}, the covariances between, respectively, groups of two, three, and four loci, as defined previously. Note that with this notation, $C_{ii} = \mu_{2,i}$, $C_{iii} = \mu_{3,i}$ and $C_{iiii} = \mu_{4,i}$, referring to the moments at locus i. If loci are independent (i.e., in gametic-phase equilibrium), then all other combinations involving four (or fewer) loci are zero except C_{iijj}, which equals $C_{ii} \cdot C_{jj} = \mu_{2,i} \cdot \mu_{2,j}$.

Following Turelli and Barton (1990), the genotypic moments can be decomposed into within-locus effects (the M_i from Equation 24.18) due to the moments at individual loci and between-locus effects due to covariances generated by gametic-phase disequilibrium. Remember that we are assuming the simplest case, complete additivity (no dominance or epistasis), so the genotypic distribution G is the distribution of additive genetic (breeding) values (A), namely, the sum of allelic effects over all loci. Expanding Equation 24.17 and taking expectations yields the familiar expressions for the mean and variance

$$\mu_G = 2\sum_{i=1}^{n} \mu_{1,i} = M_1 \tag{24.19a}$$

$$\sigma_G^2 = \sigma_A^2 = 2\sum_{i,j=1}^{n} C_{ij} = M_2 + 2\sum_{i=1}^{n}\sum_{j\neq i}^{n} C_{ij} \tag{24.19b}$$

where $M_2 = 2\sum C_{ii}$ corresponds to the genic variance σ_a^2, and the double sum corresponds to the disequilibrium, d (Equation 16.1b). Similarly, the skew can be partitioned as

$$\mu_{3,G} = 2\sum_{i,j,k=1}^{n} C_{ijk} = M_3 + 2\sum_{i=1}^{n}\sum_{j,k\neq i}^{n} C_{ijk} \tag{24.19c}$$

All terms in the second sums of Equations 24.19b and 24.19c are zero when all groups of two and three loci (respectively) are in gametic-phase equilibrium ($C_{ij} = C_{ijk} = 0$). Partitioning the kurtosis requires a little more care. After some simplification (Turelli and Barton 1990), we obtain

$$\mu_{4,G} = 3\sigma_A^4 + M_4 + 2\sum_{i=1}^{n}\sum_{j,k,\ell\neq i}^{n} (C_{ij\ell k} - C_{ij}\, C_{k\ell} - C_{ik}\, C_{j\ell} - C_{i\ell}\, C_{jk}) \tag{24.19d}$$

Again, these fourth- and second-order covariance terms ($C_{ij\ell k}$ and C_{ij}, respectively) are zero when all groups of four loci are in gametic-phase equilibrium. Because $3\sigma_A^4$ is the kurtosis value that is expected when genotypic values are Gaussian-distributed, the last two terms partition any kurtosis in G into the contribution from individual locus kurtosis (M_4) and the contribution generated by gametic-phase disequilibrium between groups of four loci. If the distribution of allelic effects is multivariate normal, then $M_4 = 0$, and each term within the covariance sum is zero as $C_{ijlk} = C_{ij}\, C_{kl} + C_{ik}\, C_{jl} + C_{il}\, C_{jk}$.

Analogous to allele frequencies changing the genic variance, σ_a^2, and disequilibrium changing the covariances (and hence d), changes in M_3 and M_4 reflect *allele-frequency change*, while changes in the third- and fourth-order covariances reflect *changes from disequilibrium*. These higher-order moments can depart from their expectations under normality by the presence of skewness or kurtosis of allelic effects at the individual loci (generating nonzero M_3 and/or M_4), which can result from allele-frequency changes. Alternatively, even if the within-locus moments are normal ($M_3 = M_4 = 0$), gametic-phase disequilibrium (nonzero

C_{ijk} and/or $C_{ijk\ell}$) can introduce skewness and/or kurtosis. When the number of loci is small, the impact of skew or kurtosis at the individual loci can be significant (Equation 24.6), yielding nonzero M_3 and/or M_4, with the resulting genotypic distribution deviating from normality.

To see these points, first consider the changes due to within-locus moments. If n is the number of loci, then as we saw earlier the effects (a) of alleles at individual loci must scale as $1/\sqrt{n}$ in order for the genetic variance to remain bounded, hence C_{ii} terms scale as a^2 or n^{-1}. Summing over all n loci, M_2 will be of order $n \cdot n^{-1} = 1$ and, as required, remains bounded as the number of loci increases. What happens to the skew and kurtosis as n increases? Assuming a is of order $n^{-1/2}$, then $\mu_{3,i}$ will be of order a^3 or $n^{-3/2}$, implying that M_3 is of order $n \cdot n^{-3/2} = n^{-1/2}$ (also see Equation 24.6). Hence, as the number of loci becomes very large, the contribution from skew at individual loci becomes negligible. Likewise, $\mu_{4,i}$ is of order $n^{-4/2}$, implying M_4 is of order $n \cdot n^{-2} = n^{-1}$ (Equation 24.6). Changes in kurtosis generated by within-locus (i.e., allele-frequency) changes become negligible as the number of loci becomes sufficiently large, but this occurs even more rapidly than skew with increasing n.

The behavior of the between-locus contributions (correlations from disequilibrium) as n increases is quite different from that due to within-locus contributions (Turelli and Barton 1990). Under weak selection, Turelli and Barton showed that C_{ijk} is proportional to $C_{ii}C_{jj}C_{kk}$, and of order n^{-3}. However, there are $n(n-1)(n-2) \simeq n^3$ terms involving C_{ijk} in the covariance contribution to skew, so the total contribution is of order one and does not necessarily converge to zero as the number of loci approaches infinity. The same argument holds for the kurtosis and higher-order moments (Turelli and Barton 1990). If the number of loci is very large, the distribution of genotypic values can depart from a Gaussian due to selection generating third- and higher-order covariances between loci, which in turn creates skew and kurtosis in the genotypic distribution. Even if the distribution of genotypes is initially Gaussian, selection generates these higher-order disequilibria, driving it away from normality (Bulmer 1980; Zeng 1987; Turelli and Barton 1990, 1994). Conversely, when selection stops, this disequilibrium quickly decays to zero, restoring the Gaussian.

Describing the Genotypic Distribution: Cumulants and Gram-Charlier Series

While most readers are familiar with moments, an alternate approach to describing the shape of a distribution, and in particular its departures from a Gaussian, is to examine its **cumulants**. These quantities, which arise from the moment-generation function of a probability distribution, offer some advantages over moments, as we will discuss shortly. The first uses of cumulants in examining selection response appears in O'Donald (1972) and Bulmer (1980). Sophisticated (and highly technical) treatments were developed by Bürger (1991a, 1993) and Turelli and Barton (1994). Our aim here is both to give the fearless reader sufficient background to this literature and to show the connection between results derived using moments and those derived using cumulants.

Cumulants (the nth of which we denote by K_n) arise naturally in series approximations of probability distributions, and they can be related to the central moments (μ_n). For example, the first five central moments can be expressed as functions of the cumulants as follows:

$$\mu_1 = K_1, \quad \mu_2 = K_2, \quad \mu_3 = K_3, \quad \mu_4 = K_4 + 3K_2^2, \quad \mu_5 = K_5 + 10\,K_2 K_3$$

(Kendall and Stuart 1977). Hence, the first three cumulants are equal to the mean, variance, and skew, respectively, while the fourth and fifth cumulants are

$$K_4 = \mu_4 - 3\mu_2^2, \quad K_5 = \mu_5 - 10\,\mu_2\,\mu_3 \tag{24.20}$$

The major advantage of cumulants over moments is that they are *additive*, so that the nth cumulant of a sum of random variables is simply the sum of the cumulants for each, namely, $K_n(x+y) = K_n(x)+K_n(y)$. This linearity property does not hold for higher-order moments, which are highly nonlinear functions of the moments of the component distributions. For a normal distribution, cumulants of order three and higher are zero, so nonzero values for these higher-order cumulants provide a convenient measure of departures from normality.

The major disadvantage of using cumulants in place of moments arises when dealing with recombination (Turelli and Barton 1994; Bürger 2000). In such cases, one works with cumulants to compute within-generation changes, converts these to moments for recombination, and then converts the recombinant products back into cumulants.

Finally, cumulants appear in series approximations of arbitrary probability distributions. Consider a standardized random variable, $y = (z - \mu)/\sigma$, which has mean zero and variance one. If the true density function for y is $\phi(y)$, we can approximate it as a unit-normal density function, $\varphi(y)$, plus correction terms. In particular (Johnson and Kotz 1970a; Kendall and Stuart 1977), the **Gram-Charlier series** approximation (here shown to order five) is given by

$$\phi(y) \simeq \varphi(y)\left[1 + \frac{K_3}{6}\,H_3(y) + \frac{K_4}{24}\,H_4(y) + \frac{K_5}{120}\,H_5(y)\right] \tag{24.21a}$$

where $H_k(y)$ denotes the **Chebsyshev-Hermite polynominal** of order k, with

$$\begin{aligned} H_3(y) &= y^3 - 3y \\ H_4(y) &= y^4 - 6y^2 + 3 \\ H_5(y) &= y^5 - 10y^3 + 15y \end{aligned} \tag{24.21b}$$

Equation 24.21a shows how the higher-order cumulants (K_3 and above) quantify departures from normality. If all of these are zero, the distribution is Gaussian.

Bulmer (1980), Zeng (1987), and Turelli and Barton (1994) used Gram-Charlier series to examine departures from normality under selection. Further properties of cumulants and Gram-Charlier (and other) series approximations are discussed in Johnson and Kotz (1970a) and Kendall and Stuart (1977). One potentially troublesome issue with Equation 24.21a is that the term in the square brackets can be negative for some y values (and hence not a proper probability distribution) if too low an order of approximation is used.

Example 24.6. Cumulants can be used to easily compute the fourth and fifth central moments of the phenotypic distribution. Here, $z = G + E$, so (Equation 24.20) the fourth moment is

$$\begin{aligned} \mu_{4,z} &= K_{4,z} + 3K_{2,z}^2 \\ &= [\,K_{4,G} + K_{4,E}\,] + 3\,(K_{2,G} + K_{2,E})^2 \\ &= [(\,\mu_{4,G} - 3\mu_{2,G}^2\,) + (\,\mu_{4,E} - 3\mu_{2,E}^2\,)] + 3\,(\,\mu_{2,G} + \mu_{2,E}\,)^2 \\ &= \mu_{4,G} + \mu_{4,E} + 6\sigma_G^2\sigma_E^2 \end{aligned}$$

where the second and third steps, respectively, follow from the additivity property of cumulants ($K_{n,z} = K_{n,G} + K_{n,E}$) and from Equation 24.20, while the final step recovers Equation 24.15. Likewise

$$\begin{aligned} \mu_{5,z} &= K_{5,z} + 10K_{2,z}K_{3,z} \\ &= (\,K_{5,G} + K_{5,E}\,) + 10(K_{2,G} + K_{2,E})(K_{3,G} + K_{3,E}) \\ &= (\,\mu_{5,G} - 10\mu_{2,G}\mu_{3,G}\,) + (\,\mu_{5,E} - 10\mu_{2,E}\mu_{3,E}\,) \\ &\qquad + 10(\mu_{2,G} + \mu_{2,E})(\mu_{3,G} + \mu_{3,E}) \\ &= \mu_{5,G} + \mu_{5,E} + 10\,(\mu_{2,G}\mu_{3,E} + \mu_{2,E}\mu_{3,G}) \end{aligned}$$

These nonlinear expressions for the higher-order moments of a sum of variables are in sharp contrast to the expressions for cumulants, in which $K_{n,z} = K_{n,G} + K_{n,E}$.

Example 24.7. To see the advantage of working with cumulants, consider the fourth cumulant of the genotypic distribution. Equation 24.19d presented a rather complex expression for the fourth *moment*, but we can use cumulants to easily obtain this result. If the underlying genes are additive across loci (no epistasis), the nth cumulant of the genotypic distribution is the sum of the appropriate cumulants for each of the underlying loci. Following Turelli and Barton (1994)

$$K_{4,G} = \sum_{i,j,k,\ell=1}^{n} K_{ijk\ell} = \sum_{i=1}^{n} K_{iiii} + \sum_{i=1}^{n} \sum_{j,k,\ell \neq i}^{n} K_{ijk\ell}$$

the sum over K_{iiii} represents the within-locus contributions to the fourth cumulant, while the sums over the other indices are the contributions to K_4 from fourth-order disequilibria between loci. We recover Equation 24.19d by noting that $\mu_{4,G} = K_{4,G} + 3\sigma_A^4$ and substituting

$$M_4 = \sum_{i=1}^{n} K_{iiii} \quad \text{and} \quad K_{ijk\ell} = C_{ijk\ell} - C_{ij}\,C_{k\ell} - C_{ik}\,C_{j\ell} - C_{i\ell}\,C_{jk}$$

Application: Departure from Normality Under Truncation Selection

One application of the preceding machinery is to compute the distribution of breeding values following a single generation of truncation selection, assuming that the initial joint distribution of phenotypic and breeding values is multivariate normal. This was examined by Bulmer (1980) and Zeng (1987), and Turelli and Barton (1994) presented a very elegant (and elaborate) analysis for multiple generations. As before, we consider only additive models, so the distribution of genotypic values is also the distribution of breeding values.

First, we assume that there is initial (i.e., existing before selection) normality in phenotypic values, $z \sim N(\mu, \sigma_z^2)$, and compute the cumulants for the resulting distribution of phenotypic values after truncation selection. As before (Chapters 14 and 16), truncation selection saves the uppermost fraction, p, of the population, giving a selection intensity of

$$\bar{\imath} = \frac{\varphi(z_p)}{p}$$

where $\varphi(z_p)$ is the value of a unit-normal density function evaluated at z_p, and z_p satisfies $\Pr(U > z_p) = p$, with U denoting a unit-normal random variable (Equation 14.2b). From Chapters 14 and 16, we already have the first two cumulants following selection as

$$K_{1,z}^* = \mu^* = \mu + \bar{\imath}\,\sigma_z \quad \text{and} \quad K_{2,z}^* = \sigma_z^{2\,*} = [1 - \bar{\imath}(\bar{\imath} - z_p)]\;\sigma_z^2$$

while the next two cumulants are (from Bulmer 1980)

$$K_{3,z}^* = [(\bar{\imath} - z_p)(2\,\bar{\imath} - z_p) - 1]\;\bar{\imath}\,\sigma_z^3 \tag{24.22a}$$

$$K_{4,z}^* = \left[-6\,\bar{\imath}\,(\bar{\imath} - z_p)^2 + (3 - z_p^2)(\bar{\imath} - z_p) + \bar{\imath}\right]\,\bar{\imath}\,\sigma_z^4 \tag{24.22b}$$

Next, we translate these within-generation changes in the *phenotypic* distribution into the within-generation change in the distribution of *breeding values* and then examine how this breeding-value distribution changes (under random mating) during transmission to the next generation. Both steps rely critically on assumptions of normality. If the distribution of breeding and phenotypic values is bivariate normal before selection (as might occur in the initial round of selection, but not necessarily in subsequent rounds), then the regression of breeding values on phenotypic values is linear

$$A = \mu_z + h^2(z - \mu_z) + \epsilon$$

Rao et al. (1968) showed that a single generation of truncation selection does not alter this regression, which leads to our standard results (Chapter 16) for the mean and variance of the breeding values following selection

$$\mu_A^* = \mu_z + h^2\,\bar{\imath}\,\sigma_z \tag{24.23a}$$

and

$$\sigma^2_{A*} = \sigma^2_A\left[1 - h^2\,\bar{\imath}\,(\bar{\imath} - z_p)\right] \tag{24.23b}$$

Bulmer (1980) showed that when the joint distribution of breeding values and phenotypes is multivariate normal before selection, all higher cumulants follow a very simple relationship

$$K^*_{i,A} = \left(h^2\right)^i K^*_{i,z} \qquad \text{for} \quad i \geq 3 \tag{24.24a}$$

Assuming unlinked loci, the cumulants for the distribution of breeding values in the next generation become

$$K_{i,A}(t+1) = \left(\frac{1}{2}\right)^{i-1} K^*_{i,A}(t) \tag{24.24b}$$

Hence, the cumulants for the distribution of breeding values at the start of the next generation are related to the cumulants of the postselection phenotypic distribution by

$$K_{i,A}(t+1) = 2\left(\frac{h^2}{2}\right)^i K^*_{i,z}(t) \tag{24.24c}$$

Notice from Equation 24.22 that after a single generation of selection, $K_{3,A}$ and $K_{4,A}$ will be nonzero, and hence the distribution of breeding values will no longer be normal. At this point, the assumption of bivariate normality no longer holds, and there is no longer a simple relationship between $K^*_{i,A}$ and $K^*_{i,z}$. Thus, we cannot simply iterate this procedure over more than one generation of selection. See Turelli and Barton (1994) for a detailed analysis over multiple generations.

Example 24.8. Suppose truncation selection occurs on a normally distributed trait with an initial mean of $\mu_z = 0$ and variance of $\sigma_z^2 = 100$. Individuals whose phenotypes are in the upper 5% of the distribution are saved, so that $\bar{\imath} = 2.063$ and $z_p = 1.645$ (Example 14.1). To demonstrate an extreme case, assume that $h^2 = 1$, so that all variance is additive genetic. Applying Equation 24.22a, the resulting third-order cumulant in the phenotypic distribution following selection is

$$\begin{aligned} K^*_{3,z} &= [(\bar{\imath} - z_p)(2\,\bar{\imath} - z_p) - 1]\,\bar{\imath}\sigma_z^3 \\ &= [(2.063 - 1.645)(2 \cdot 2.063 - 1.645) - 1] \cdot 2.063 \cdot 100^{3/2} \\ &= 76.45 \end{aligned}$$

Applying Equation 24.24c translates this into the third cumulant in the genotypic distribution in the next generation, yielding

$$K_{3,A}(t+1) = 2\left(\frac{h^2}{2}\right)^3 K^*_{3,z}(t) = 2\left(\frac{1}{2}\right)^3 76.45 = 19.11$$

Using the machinery from Chapter 16, for $p = 0.2$, Equation 16.11a yields $\kappa = 0.862$ (i.e., selected indviduals have only a fraction, $[1 - 0.862] = 0.138$, of the variance of the preselection population), and Equation 16.12d returns $d(1) = (-\kappa/2)100 = -43.1$. Hence, the phenotypic variance in the first generation is $\sigma_z^2 = \sigma_A^2(1) = 100 + d(1) = 56.9$. Thus, the scaled skew becomes

$$\gamma_3 = \frac{K_3}{\sigma_A^3} = \frac{19.11}{56.9^{3/2}} = 0.045$$

A similar calculation using Equations 24.22b and 24.24c yields $K_4 = 59.7$ and $\gamma_4 = K_4/56.9^2 = 0.018$. Applying Equation 24.21a, the resulting (fourth-order) Gram-Charlier series approximation for the distribution, $\phi(A)$, of breeding values in generation 1 is

$$\begin{aligned}\phi(A') &\simeq \varphi(A')\left[1 + \frac{0.045}{6}\,H_3(A') + \frac{0.018}{24}\,H_4(A')\right]\\ &= \varphi(A')\,[1 + 0.0075\,H_3(A') + 0.00075\,H_4(A')]\end{aligned}$$

where $\varphi(x)$ is the normal distribution, the functions $H_i(x)$ are defined by Equation 24.21b, and $A' = A/\sigma_A$ is the standardized breeding value (which, initially, has a mean of zero). For example, consider $A' = 2$, a value two standard deviations away from the mean. Applying Equation 24.21b, $H_3(2) = 2$ and $H_4(2) = -5$, yielding the correction factor for the unit normal density function as

$$1 + 0.0075\,H_3(2) + 0.00075\,H_4(2) = 1.01125$$

The key point is that the resulting distribution is only very weakly perturbed away from a Gaussian (assuming that $h^2 = 1$ is the most extreme case). For a more typically heritability, say $h^2 = 0.3$, similar calculations yield $\gamma_3 = 0.0039$ and $\gamma_4 = 0.0007$, and

$$\phi(A) \simeq \varphi(A)\,[1 + 0.00065\,H_3(A') + 0.00003\,H_4(A')]$$

yielding an even smaller departure from normality (an adjustment factor of 1.00115 for $A' = 2$).

Thus, under the infinitesimal model, the generation of linkage disequilibrium by truncation selection has very little impact on driving the distribution of breeding values away from a Gaussian. While the disequilibrium introduced by truncation selection can indeed drive a distribution of breeding values away from a strict Gaussian, the error in assuming this remains Gaussian is generally small. This point was initially made by Bulmer (1980). The much more extensive analysis by Turelli and Barton (1994) showed that, even in the presence of strong truncation or disruptive selection, the Bulmer equation (Equation 16.7b) can be used with little error. Turelli and Barton's analysis assumed a sufficiently large number of loci so that changes in both the genic variance and locus-specific cumulants of order three or higher can be ignored.

Short-term Response Ignoring Linkage Disequilibrium

With this machinery in hand, we are now ready to examine the response to selection under non-Gaussian genotypic distributions. We first consider the situation in which a small to modest number of loci underlie the character, wherein most of the changes in the higher-order moments are due to changes in allele frequencies, rather than through generation of gametic-phase disequilibrium. Our treatment follows that of Barton and Turelli (1987).

If we are willing to assume additivity across loci and gametic-phase equilibrium, genetic changes in the character will be completely described by the dynamics of allele-frequency changes at each locus. The complete dynamics for a locus with k alleles are described by the $k-1$ allele-frequency change equations. Alternatively, we could fully describe the dynamics by using equations based on any set of $k-1$ independent new variables that can be expressed as functions of allele frequencies (this is the standard multivariate transformation problem of vector calculus and requires that the determinant of the Jacobian transformation matrix be nonzero). One such set of new variables involves the first $k-1$ moments of the allelic distribution. This is the motivation behind Barton and Turelli's (1987) approach, which focuses on *allelic moments*, rather than *allelic frequencies*.

If we ignore gametic-phase disequilibrium, then for n loci with k alleles each, we can completely describe the dynamics by using the first $n(k-1)$ moments of the genotypic distribution. This same approach can be used when linkage is considered, but in this case the number of equations increases dramatically (scaling as the number of distinct gametic types). While the process of using a new set of variables is exact, it is also just as challenging to solve as the original allele-frequency change equations. The hope, however, is that by considering the first few moments, we can gain considerable insight into the actual dynamics and a better feel for the conditions under which certain approximations work and those under which they break down.

To briefly sketch the approach used by Barton and Turelli, recall Wright's formula for frequency changes with multiple alleles (Equation 5.11a)

$$\Delta p_i = \sum_{j=1}^{k} G_{ij} \frac{\partial \ln \overline{w}}{\partial p_j} \tag{24.25a}$$

where $G_{ii} = p_i(1-p_i)/2$ and $G_{ij} = -p_i\, p_j/2$ (for $i \neq j$). The assumption of linkage equilibrium is needed here, as Wright's formula fails when single-locus fitnesses are background-dependent, which can occur even with constant genotypic fitnesses when linkage disequilibrium is present (see Example 5.7 for details). Now consider a function, $f(p_1, p_2, \ldots p_{k-1})$, that depends on the allele frequencies at this locus, such as a particular moment of the allelic distribution. The change in f due to changes in allele frequencies can be approximated by a second-order Taylor series to yield

$$\Delta f = \sum_{i=1}^{k} \frac{\partial f}{\partial p_i} \Delta p_i + \frac{1}{2} \sum_{i=1}^{k} \sum_{j=1}^{k} \frac{\partial^2 f}{\partial p_j\, \partial p_i} \Delta p_i\, \Delta p_j + \cdots \tag{24.25b}$$

where this expression ignores higher-order terms of Δp_i. Substituting for Δp_i via Equation 24.25a yields (to first order)

$$\Delta f \simeq \sum_{i=1}^{k} \frac{\partial f}{\partial p_i} \left(\sum_{j=1}^{k} G_{ij} \frac{\partial \ln \overline{w}}{\partial p_j} \right) \tag{24.25c}$$

Recall from the chain rule of differentiation that

$$\frac{\partial \ln \overline{w}}{\partial p_j} = \frac{\partial \ln \overline{w}}{\partial f} \frac{\partial f}{\partial p_j}$$

which, upon substitution in Equation 24.52c, yields

$$\Delta f = \frac{\partial \ln \overline{w}}{\partial f} \sum_{i=1}^{k} \sum_{j=1}^{k} \frac{\partial f}{\partial p_i} G_{ij} \frac{\partial f}{\partial p_j} \tag{24.25d}$$

This is a *weak-selection approximation*, as it assumes that terms of second order, $(\Delta p_i\, \Delta p_j)$, and higher can be ignored (if drift is considered, these second-order terms must be included even if selection is weak; see Turelli 1988).

Using this expression yields a set of equations in which changes in a certain moment depend on higher-order moments. After considerable algebra (for details, see Barton and Turelli 1987), the changes in genotypic moments (under the assumptions of complete additivity and gametic-phase equilibrium) can be expressed in matrix form as

$$\boldsymbol{\Delta}_{\mu_G} \simeq \mathbf{M}\, \nabla \ln \overline{w} \tag{24.26a}$$

where

$$\boldsymbol{\Delta}_{\mu_G} = \begin{bmatrix} \Delta\mu_{1,G} \\ \Delta\mu_{2,G} \\ \Delta\mu_{3,G} \\ \vdots \end{bmatrix} \quad \text{and} \quad \nabla \ln \overline{w} = \begin{bmatrix} \dfrac{\partial \ln \overline{w}}{\partial \mu_{1,z}} \\ \dfrac{\partial \ln \overline{w}}{\partial \mu_{2,z}} \\ \dfrac{\partial \ln \overline{w}}{\partial \mu_{3,z}} \\ \vdots \end{bmatrix} \tag{24.26b}$$

are, respectively, the vectors of changes in the genotypic moments ($\boldsymbol{\Delta}_{\mu_G}$) and the vector of partial derivatives of log mean fitness with respect to each moment ($\nabla \ln \overline{w}$), and a matrix of genotypic moments,

$$\mathbf{M} = 2\sum_{i=1}^{n} \begin{bmatrix} \mu_{2,i} & \mu_{3,i} & (\mu_{4,i} - 3\mu_{2,i}^2) & \cdots \\ \mu_{3,i} & (\mu_{4,i} - \mu_{2,i}^2) & (\mu_{5,i} - 4\mu_{3,i}\,\mu_{2,i}) & \cdots \\ (\mu_{4,i} - 3\mu_{2,i}^2) & (\mu_{5,i} - 4\mu_{3,i}\,\mu_{2,i}) & (\mu_{6,i} - \mu_{3,i}^2 - 6\mu_{2,i}\,\mu_{4,i} + 9\mu_{2,i}^2) & \cdots \\ \vdots & \vdots & \vdots & \ddots \end{bmatrix} \tag{24.26c}$$

The summation symbol used Equation 24.26c indicates that each element on $\mathbf{M}$ is twice the sum of a (locus-specific) quantity over all loci. The additional elements of $\mathbf{M}$ that are not displayed correspond to selection on the fourth and higher moments (M_{ij} for i and/or $j \geq 4$). The expressions for these terms are more complicated than may be suggested by the simple dots in the matrix due to the nonadditive nature of higher moments. Expressions based on $\partial \ln \overline{w} \,/\, \partial K_{i,z}$ (the partial derivative of fitness with respect to the ith *cumulant* of the phenotypic distribution) have a simpler form due to the additive nature of cumulants (Bürger 1991a, 1993; Turelli and Barton 1994), but retain the undesirable feature that the response of the ith cumulant depends on cumulants of higher order.

In order to close the set of equations given by 24.26a–24.26c, we must impose restrictions on the number of columns in $\mathbf{M}$, which requires assuming that selection mainly occurs on the first few moments. Likewise, the number of rows of $\mathbf{M}$ must also be restricted, which implies additional assumptions on the genetic moments. Examples 24.9 and 24.11 discuss the two most common assumptions to accomplish these goals, namely the Gaussian and rare-alleles models.

In cases where the first three phenotypic moments account for the majority of selection, the expected single-generation change in mean becomes

$$\Delta\mu_z \simeq \sigma_A^2\, \frac{\partial \ln \overline{w}}{\partial \mu_z} + \mu_{3,G}\, \frac{\partial \ln \overline{w}}{\partial \mu_{2,z}} + \kappa_4\, \sigma_A^4\, \frac{\partial \ln \overline{w}}{\partial \mu_{3,z}} \tag{24.27}$$

where $\kappa_4 = (\mu_{4,G} - 3\sigma_A^4)/\sigma_A^4$ is the scaled coefficient of kurtosis (Frank and Slaktin 1990 obtained the same result, for the special case of stabilizing selection using the Price equation; Chapter 6). If the distribution of G is Gaussian, then $\mu_{3,z} = \kappa_4 = 0$, and we recover the selection gradient version of the breeder's equation (Equation 13.27a). Under more general distributions, predicting changes in even the simplest genotypic moment, the mean, requires a detailed knowledge of both higher-order allelic moments ($\mu_{k,i}$) and the nature of selection on these higher-order moments ($\partial \ln \overline{w} \,/\, \partial \mu_{k,z}$). In order to proceed further, we have to make additional assumptions about the distribution of allelic effects at *individual loci.*

If the phenotypes are approximately normally distributed (but allelic effects at individual loci are not necessarily Gaussian), the mean and variance terms of the selection gradient vector generally will dominate. Recalling Equations 24.12 and 24.18, Equation 24.26a reduces to

$$\begin{bmatrix} \Delta\mu_{1,G} \\ \Delta\mu_{2,G} \\ \Delta\mu_{3,G} \\ \vdots \end{bmatrix} = \begin{bmatrix} \sigma^2_A & M_3 \\ M_3 & M_4 \\ M_4 & 2\sum_i(\mu_{5,i} - 4\mu_{3,i}\,\mu_{2,i}) \\ \vdots & \vdots \end{bmatrix} \begin{bmatrix} \dfrac{S}{\sigma^2_z} \\ \dfrac{\delta(\sigma^2_z) + S^2}{2\,\sigma^4_z} \end{bmatrix}$$

If we consider only the first three genotypic moments, then

$$\Delta\mu_G \simeq h^2 S + \left(\frac{\delta(\sigma^2_z) + S^2}{2\sigma^4_z} \right) M_3 \tag{24.28a}$$

$$\Delta\sigma^2_A \simeq \frac{S}{\sigma^2_z}\, M_3 + \left(\frac{\delta(\sigma^2_z) + S^2}{2\sigma^4_z} \right) M_4 \tag{24.28b}$$

$$\Delta\mu_{3,G} \simeq \frac{S}{\sigma^2_z}\, M_4 + \left(\frac{\delta(\sigma^2_z) + S^2}{\sigma^4_z} \right) \sum_{i=1}^{n} (\, \mu_{5,i} - 4\,\mu_{3,i}\,\mu_{2,i}\,) \tag{24.28c}$$

where M_3 and M_4 are as defined by Equations 24.18c and 24.18d. As is discussed in Chapters 29 and 30, when selection acts only on the mean ($\partial \ln \overline{w} \,/\, \partial\, \mu_{k,z} = 0$ for $k \geq 2$), $\delta(\sigma^2_z) = -S^2$, meaning that the first term in each of these three equations accounts for the effect of selection to change the mean and the second term accounts for the effect of selection acting directly on the variance (selection is based, in part, on an individual's squared deviation from the mean, $[z - \mu]^2$). We previously obtained Equation 24.28a by an alternative approach (Equation 5.27b). When the genotypic distribution is skewed ($M_3 \neq 0$), the single-generation change in the mean also depends on the nature of selection on the variance (O'Donald 1968, 1972; Bulmer 1980; Gillespie 1984a; Barton and Turelli 1987; Mitchell-Olds and Shaw 1987). Further, even if skew is initially absent, Equation 24.28c shows that if the kurtosis of the genotypic distribution differs from that expected for a Gaussian ($M_4 \neq 0$), selection that is strictly on the mean generates skew. Thus, even ignoring the effects of gametic-phase disequilibrium, selection on the mean generates skew when the genotypic distribution displays kurtosis.

Whether allele-frequency change at the individual loci or gametic-phase disequilibrium among loci is more important for producing departures from normality depends on whether there are alleles of modest to large effects. When alleles of major effect are present, locus-specific selection coefficients (e.g., Equation 5.21) can be sufficiently large that significant allele-frequency change can quickly occur, which has a far great effect on departures from normality than does selection-generated disequilibrium (Turelli and Barton 1990). Conversely, when all alleles have small effects (and hence selection is extremely weak on any single locus), then over modest time scales, departure from normality is largely due to selection generating third- (and higher-) order disequilibrium. Thus, when the number of loci is small, the error created by using Equation 24.26a (which assumes gametic-phase equilibrium) should be small. As the number of (equivalent) loci increases, within-locus effects make a smaller and smaller contribution, with departures from normality caused by disequilibrium eventually dominating any small departures caused by allele frequency change.

Gaussian Versus Rare-alleles Approximations

In order to have Equation 24.26a form a closed set of equations, additional assumptions are needed, both on the nature of selection (columns of **M**) and on the underlying genetics (rows of **M**). If phenotypes are (and remain) normally distributed, then one needs to only consider selection on the mean and variance, as these define all other moments. More generally, if selection is sufficiently weak, then any selection function can be well approximated by a quadratic Taylor series, which again only involves the first two moments.

Two different assumptions about the genotypic distribution have been used to reduce Equation 24.26a to the first three moments. The first is to assume that distributions start (and remain) Gaussian (Example 24.9). The second is the **rare-alleles model**, which assumes loci are very near fixation (Example 24.11). This model is very closely related to an important approximation (Turelli's 1984 house-of-cards) which appears in Chapter 28 and which assumes that the effect of a new mutation is large relative to the amount of standing variation at a locus, which is equivalent to assuming that selection is much stronger than mutation.

Example 24.9. If we assume that the Gaussian approximation holds, the distribution of allelic effects at each locus will be normal. In this case, all odd central moments at each locus are zero ($\mu_{2k+1} = 0$) and all even moments are related to the second moment by $\mu_{2k} = \mu_2^k\,(2k)!/(2^k k!)$ (Kendall and Stewart 1977). For example, $\mu_4 = 3\mu_2^2$, implying that $\mu_4 - \mu_2^2 = 2\mu_2^2$. Assuming that most of selection is on the mean and variance, we can neglect the third- and higher-order selection gradients. In this case, $\mathbf{M}$ (Equation 24.26c) reduces to a 2×2 matrix

$$\mathbf{M} = \begin{pmatrix} 2\sum_{i=1}^{n} \mu_{2,i} & 0 \\ 0 & 4\sum_{i=1}^{n} \mu_{2,i}^2 \end{pmatrix} = \begin{pmatrix} \sigma_A^2 & 0 \\ 0 & \dfrac{\sigma_A^4}{n_e} \end{pmatrix} \tag{24.29a}$$

where

$$n_e = \frac{\sigma_A^4}{4\sum_i \mu_{2,i}^2} \tag{24.29b}$$

is equivalent to Chevalet's (1994) effective number of loci (Equation 24.3); see Example 24.10. The expected response in the genotypic mean and variance then becomes

$$\begin{pmatrix} \Delta\mu \\ \Delta\sigma_A^2 \end{pmatrix} \simeq \begin{pmatrix} \sigma_A^2 & 0 \\ 0 & \sigma_A^4/n_e \end{pmatrix} \begin{pmatrix} \dfrac{\partial \ln \overline{w}}{\partial \mu_z} \\ \dfrac{\partial \ln \overline{w}}{\partial \sigma_z^2} \end{pmatrix} = \begin{pmatrix} \sigma_A^2 \dfrac{\partial \ln \overline{w}}{\partial \mu_z} \\ \dfrac{\sigma_A^4}{n_e} \dfrac{\partial \ln \overline{w}}{\partial \sigma_z^2} \end{pmatrix} \tag{24.30a}$$

If the phenotypic distribution is *exactly* normal, all moments can be expressed in terms of the mean and variance, only gradients measuring selection on the mean and variance will appear, and these equations will be exact. Recalling Equations 24.12a and 24.12b yields

$$\Delta\mu \simeq h^2 S \tag{24.30b}$$

and

$$\Delta\sigma_A^2 \simeq \frac{h^4}{2n_e}\left(\delta(\sigma_z^2) + S^2\right) \tag{24.30c}$$

Thus, the expected change in the mean follows the breeder's equation and short-term changes in variance (from allele-frequency change) are expected to be small when the value of n_e is modest to large. We remind the reader that this analysis ignores the effects of gametic-phase disequilibrium.

Because the locus-specific variances, $\mu_{2,i}$, change as allele frequencies change, predicting changes in variance over several generations, even under these simplifying assumptions, still requires a detailed knowledge about the distribution of allelic effects *at individual loci*. Thus, while short-term changes in the mean can be predicted without detailed knowledge of the underlying genetics (only σ_A^2 is required), changes in variance cannot (unless an estimate of $\sum \mu_{2,i}^2$ or n_e can be obtained). Further, as allele frequencies change, so does n_e, and Example 24.4 showed just how unpredictable such changes can be.

Finally, let's attempt to connect these results for the change in the genic variance, $\Delta\sigma_A^2 = \Delta\sigma_a^2$ (as we ignore any disequilibrium), with those obtained under the continuum-of-alleles (COA) approximation (Equation 24.2a). If the within-generation change in the phenotypic variance is $\delta(\sigma_z^2) = -\kappa\sigma_z^2$, then the Gaussian COA approximation for the change in genic variance (Equation 24.2a, ignoring drift by taking $N_e = \infty$) becomes

$$\Delta\sigma_a^2 = -\frac{\kappa\, h^2\, \sigma_A^2}{2n_e} \tag{24.30d}$$

By contrast, because $\kappa h^4\sigma_z^2 = \kappa h^2\sigma_A^2$, Equation 24.30c yields an allelic-moment approximation of

$$\Delta\sigma_a^2 \simeq \frac{h^4}{2n_e}\left(\delta(\sigma_z^2) + S^2\right) = -\frac{\kappa h^2\sigma_A^2}{2n_e} + \frac{h^4 S^2}{2n_e} \tag{24.30e}$$

Thus, the allelic-moment approximation has a positive term lacking in the COA approximation, and hence predicts a smaller change in σ_a^2 when $\kappa > 0$ (i.e., when selection reduces the phenotypic variance; Chapter 16). Nick Barton (pers. comm., 2014) suggested that this discrepancy between approximations arises because the selection-gradient approach relies on a weak selection assumption, so terms of order S^2 are not accurately predicted.

Example 24.10. Here we show that n_e, as defined in the previous example, is equivalent to Chevalet's (1994) n_e (Equation 24.3). This simply clears up a technical detail, and it can be skipped by the casual reader. Specifically, we need to show that

$$n_e = \frac{n}{1+c_v^2} = \frac{\sigma_A^4}{4\sum_i \mu_{2,i}^2}$$

where c_v is the coefficient of variation in the genic variance contributed by each locus. Because $\mu_{2,i}$ is the variance of allelic effects at locus i, the genic variance contributed by locus i is $2\mu_{2,i}$ (as there are two alleles per locus). If we recall that the coefficient of variation is defined as the standard deviation divided by the mean

$$1 + c_v^2 = 1 + \left(\frac{\sigma(2\mu_{2,i})}{E[\,2\mu_{2,i}\,]}\right)^2 = \frac{E[\,2\mu_{2,i}\,]^2 + \sigma^2(2\mu_{2,i})}{E[\,2\mu_{2,i}\,]^2} \tag{24.31a}$$

Recalling that $\sigma^2(x) = E[\,x^2\,] - E[\,x\,]^2$, we have

$$\sigma^2(2\mu_{2,i}) = E\left[(2\mu_{2,i})^2\right] - E[2\mu_{2,i}]^2 = \left[\frac{1}{n}\sum_{i=1}^{n}(2\mu_{2,i})^2\right] - E[2\mu_{2,i}]^2 \tag{24.31b}$$

which rearranges to

$$E[2\mu_{2,i}]^2 + \sigma^2(2\mu_{2,i}) = \frac{1}{n}\sum_{i=1}^{n}(2\mu_{2,i})^2 \tag{24.31c}$$

If we sum the genic variances, $2\mu_{2,i} = \sigma_{a_i}^2$, at each locus we get the total additive genic variance, σ_a^2. Because we are ignoring disequilibrium (i.e., assuming that $d = 0$), we have the result that $\sigma_a^2 = \sigma_A^2$, or that

$$\sigma_A^2 = \sum_{i=1}^{n} 2\mu_{2,i} = n\,E[2\mu_{2,i}], \quad \text{or that} \quad E[2\mu_{2,i}]^2 = \frac{\sigma_A^4}{n^2} \tag{24.31d}$$

First substituting Equation 24.31b into 24.31a and then using Equation 24.31d yields

$$1 + c_v^2 = \frac{E[\,2\mu_{2,i}\,]^2 + \sigma^2(2\mu_{2,i})}{E[\,2\mu_{2,i}\,]^2} = \frac{(1/n)\sum_{i=1}^{n}(2\mu_{2,i})^2}{\sigma_A^4/n^2} = \frac{4n\sum_{i=1}^{n}\mu_{2,i}^2}{\sigma_A^4}$$

Thus,

$$n_e = \frac{n}{1+c_v^2} = \frac{n\sigma_A^4}{4n\sum_{i=1}^{n}\mu_{2,i}^2} = \frac{\sigma_A^4}{4\sum_{i=1}^{n}\mu_{2,i}^2}$$

demonstrating the equivalence of n_e as defined by Equation 24.29b with that given by Equation 24.3.

Example 24.11. An important construct used in the analysis of population-genetic models for the maintenance of quantitative-trait variation (Chapter 28) is the **rare-alleles model** of Barton and Turelli (1987). This assumes that loci are very near fixation, which occurs when the strength of selection is much greater than the strength of mutation. Here we show that under this assumption, allelic moments are proportional to the rare-allele frequencies, meaning that products of moments can be ignored and higher-order moments can be expressed in terms of lower-order ones.

To see how this approximation arises, consider the simplest case of a biallelic locus (i), where the common allele has an effect of 0, while the rare allele ($p_i \simeq 0$) has an additive effect of a_i. The resulting mean is $\mu_{1,i} \simeq 2a_i p_i$ and because p_i is assumed small, quadratic and higher terms in p_i are ignored in higher moments. For example, the $(2k)$th moment becomes

$$\mu_{2k,i} = (a_i - \mu_{1,i})^{2k}\, p_i + (0 - \mu_{1,i})^{2k}\,(1 - p_i) \simeq a_i^{2k} p_i \tag{24.32a}$$

The last step follows because $\mu_{1,i}^k$ is of order p_i^k and is ignored (for $k \geq 2$). Thus, higher-order moments are related,

$$\mu_{4,i} \simeq a_i^4\, p_i = a_i^2\,(a_i^2 p_i) = \xi_i\, \mu_{2,i}$$

For now, we set $\xi_i = a_i^2$ as this, or a closely related term, appears in the expression for all higher-order moments. Likewise, products of moments are of quadratic- or higher-order in p_i, and thus are ignored. For example, $\mu_{4,i} - \mu_{2,i}^2 \simeq \mu_{4,i} = \xi_i\, \mu_{2,i}$, as $\mu_{2,i}^2$ is of order p_i^2. Turelli (1984) showed that these moment relationships also hold under his house-of-cards assumption that selection at a locus is much stronger than mutation (and hence most alleles are rare and deleterious), but now with $\xi_i = \sigma_{\alpha_i}^2$, the variance in the effects of new mutations at locus i, replacing a_i^2 (which follows from the fact that $E[\alpha_i^2] = \sigma_{\alpha_i}^2 + (E[\alpha_i])^2 = \sigma_{\alpha_i}^2$, as $E[\alpha_i] = 0$). Assuming n equivalent loci (hence, $\xi = \xi_i$), under the rare-alleles or house-of-cards assumption, the moments matrix, $\mathbf{M}$, from Equation 24.26c simplifies to

$$\begin{pmatrix} \Delta\mu \\ \Delta\sigma_A^2 \\ \Delta\mu_{3,G} \end{pmatrix} \simeq \begin{pmatrix} \sigma_A^2 & M_3 & \xi\sigma_A^2 \\ M_3 & \xi\sigma_A^2 & \xi M_3 \\ \xi\sigma_A^2 & \xi M_3 & \xi^2\sigma_A^2 \end{pmatrix} \begin{pmatrix} \dfrac{\partial \ln \overline{w}}{\partial \mu} \\ \dfrac{\partial \ln \overline{w}}{\partial \sigma_z^2} \\ \dfrac{\partial \ln \overline{w}}{\partial \mu_{3,z}} \end{pmatrix} \tag{24.32b}$$

Notice that $\Delta\mu_{3,G} = \xi\Delta\mu_{1,G}$, thus directly coupling changes in the mean and skew. If we assume no initial skew (and no selection on skew), Equation 24.32b further reduces to

$$\begin{pmatrix} \Delta\mu \\ \Delta\sigma_A^2 \end{pmatrix} \simeq \begin{pmatrix} \sigma_A^2 & 0 \\ 0 & \xi\sigma_A^2 \end{pmatrix} \begin{pmatrix} \dfrac{\partial \ln \overline{w}}{\partial \mu_z} \\ \dfrac{\partial \ln \overline{w}}{\partial \sigma_z^2} \end{pmatrix} = \begin{pmatrix} \sigma_A^2\, \dfrac{\partial \ln \overline{w}}{\partial \mu_z} \\ \xi\sigma_A^2\, \dfrac{\partial \ln \overline{w}}{\partial \sigma_z^2} \end{pmatrix} \tag{24.32c}$$

Comparing these results with Example 24.9 (the Gaussian approximation) shows that the expected change in the mean (in the absence of skew) is identical in both the rare alleles and Gaussian-approximation models. Under the Gaussian, the change in the variance is given by σ_A^4/n_e times the fitness gradient with respect to the phenotypic variance, while under the rare alleles model, the quantity $\xi\sigma_A^2$ is multiplied by the gradient. Under the rare alleles (and house-of-cards) models, the assumption is that the variance at a locus is small relative to the input from new mutation, implying that $\xi = \sigma_{\alpha_i}^2 \gg \sigma_A^2/n_e$, and thus $\xi\sigma_A^2 \gg \sigma_A^4/n_e$, predicting a much larger change in the variance than under the Gaussian. We will return to this important point in Chapter 28.

A second critical difference between the rare-alleles and Gaussian-approximation models can be seen from Equation 24.28a. This shows that selection on the variance only influences the response in the mean when there is skew in the breeding value distribution ($M_3 \neq 0$), which does not occur under the Gaussian approximation (although selection-induced third-order LD can create it from an initially Gaussian model). Conversely, the rare-alleles model can easily have skew, thus coupling changes in the mean with selection on the variance.

A final important point is that if the rare-alleles model is a good approximation of reality, then most genetic variation is additive. Even when significant interactions (dominance and epistasis) are present, most genetic variation loads onto the additive component in cases where all but a few of the multilocus genotypes under consideration are rare (Crow 2008; Hill et al. 2008; Mäki-Tanila and Hill 2014). One simple way to see this point is to consider a fully dominant, but rare, allele. In this case, the frequency of dominant homozygotes is extremely small, so that the additive effect of the allele is given almost entirely by the genotypic values of the recessive homozygote and the heterozygote, thus loading most of the effects into the additive component. With multilocus genotypes, most combinations of genotypes are so rare that they have little impact on the least-squares regressions that determine additive effects. For example, consider a highly nonlinear relationship between genotypes and trait value. If only a few of the genotypes are common, then most of the regression is determined by just a few points, meaning that a linear regression (and hence additive effects) is likely to account for a significant fraction of the variance.

Short-term Response Ignoring Allele-frequency Change

The last section considered one class of approximations for the short-term selection response for non-Gaussian distributions of genotypic values, focusing solely on allele-frequency changes. Here we consider the converse approximation: a large enough number of loci (all of small effect) that allele-frequency change (over the time span of interest) can be ignored, with the change in genotypic moments thus attributable entirely to selection-generated disequilibrium. Our discussion departs from the standard infinitesimal model in that we no longer make any Gaussian assumptions.

Turelli and Barton (1990, 1994) extended basic moments analysis (Equation 24.26a) to allow for gametic-phase disequilibrium, by considering both within-locus moment changes due to allele-frequency changes (M_{ii}) and between-locus contributions generated by disequilibrium ($M_{ij}, i \neq j$). Their 1994 paper is the more general of the two, with the analysis based on the cumulants of the distribution. While the mean, variance, and skew are equivalent to the first three cumulants, cumulants of order four and higher provide much more compact expressions than using moments, due to the additivity of cumulants versus the nonlinear nature of higher-order moments.

In parallel with their moments analysis, Turelli and Barton defined the gradients of selection associated with the ith cumulant of the phenotypic distribution $K_{z,i}$ by

$$L_i = \frac{\partial \ln(\overline{W})}{\partial K_{z,i}} \tag{24.33a}$$

L_1 and L_2 correspond to selection on the mean and variance, while L_i for $i \geq 3$ represents selection that drives the distribution away from normality (as cumulants of order three and higher are zero for a Gaussian). Turelli and Barton presented general expressions for the change in all cumulants of the distribution. In particular, for a large number of loci, they show that if the majority of selection is on the first four cumulants of the distribution, the changes in the mean and variance are given by

$$\Delta\mu = \sigma_A^2\, L_1 + K_{G,3}\, L_2 + K_{G,4}\, L_3 + K_{G,5}\, L_4 \tag{24.33b}$$

$$\begin{aligned}\Delta\sigma_A^2 &= \frac{\sigma_a^2 - \sigma_A^2}{2} - \frac{(\Delta\mu)^2}{2} + \frac{K_{G,3}}{2} L_1 + \left(\sigma_A^4 + \frac{K_{G,4}}{2}\right) L_2 \\ &+ \left(3\sigma_A^2 K_{G,3} + \frac{K_{G,5}}{2}\right) L_3 + \left(3K_{G,3}^2 + 4\sigma_A^2 K_{G,4} + \frac{K_{G,6}}{2}\right) L_4\end{aligned} \tag{24.33c}$$

where $K_{G,i}$ denotes the ith cumulant of the genotypic distribution. Note that for Equation 24.33b if some cumulants of order three or higher are nonzero, selection on higher-order cumulants of the phentoypic distribution (i.e., L_3 or $L_4 \neq 0$) also results in a change in the mean. Further, note the appearance of the *genic* variance σ_a^2 in Equation 24.33c. We are assuming (at least over our time scale) that allele-frequency change can be ignored and hence σ_a^2 is a constant. All changes in the variance (and higher-order moments or cumulants) are thus assumed to arise entirely from selection generated-disequilibrium.

Example 24.12. As an application of these results, when phenotypes are normally distributed, Equations 24.12a and 24.12b yield

$$L_1 = \frac{S}{\sigma_z^2} \quad \text{and} \quad L_2 = \frac{\delta(\sigma_z^2) + S^2}{2\,\sigma_z^4} \quad \text{with} \quad L_i = 0 \quad \text{for } i \geq 3$$

If the genotypic values also follow a normal distribution, then $K_{G,i} = 0$ for $i \geq 3$. In this case, Equation 24.33b reduces to

$$\Delta\mu = \sigma_A^2 \,\frac{S}{\sigma_z^2} = h^2 S$$

which recovers the breeder's equation. If we recall that $\sigma_A^2 = \sigma_a^2 + d$, using the preceding expressions reduces Equation 24.33c to

$$\begin{aligned}
\Delta\sigma_A^2 &= \frac{\sigma_a^2 - \sigma_A^2}{2} - \frac{(\Delta\mu)^2}{2} + \frac{0}{2} L_1 + \left(\sigma_A^4 + \frac{0}{2} \right) L_2 \\
&\quad + \left(3\sigma_A^2 \cdot 0 + \frac{0}{2} \right) \cdot 0 + \left(0^2 + 4\sigma_A^2 \cdot 0 + \frac{0}{2} \right) \cdot 0 \\
&= \frac{\sigma_a^2 - \sigma_A^2}{2} - \frac{(h^2 S)^2}{2} + \sigma_A^4 \left(\frac{\delta(\sigma_z^2) + S^2}{2\,\sigma_z^4} \right) \\
&= -\frac{d}{2} + \frac{h^4}{2}\,\delta(\sigma_z^2)
\end{aligned}$$

which recovers Bulmer's equation. Notice that there is no change in the genic variance, as we assume there are very large number of loci of small effect.

Turelli and Barton (1994) examined the effects of both strong truncation (directional) selection and strong disruptive selection on Gaussian (infinitesimal and COA) models when the number of loci is large. They found that while strong truncation selection does indeed generate nonzero cumulants of order three and higher (and hence departures from normality), these departures are generally quite small (e.g., Example 24.8). As a result, the breeder's equation with the variance changes predicted from the Bulmer equation (Equation 16.7b) gives quite accurate results for the predicted change in the mean and variance. Hence, the effects of disequilibrium in this case are essentially accounted for by considering only the second-order disequilibrium, which is done in the basic Bulmer model. Barton and Turelli found that the distribution of genotypic values is highly non-normal under strong disruptive selection, with a significant fourth cumulant (kurtosis) being generated by significant fourth-order disequilibrium (generating correlations between groups of four loci). Surprisingly, even in this case the change in variance is still well predicted by the Bulmer equation.

Effects of Linkage

As might be expected, when these results are generalized to allow for arbitrary linkage (as opposed to the previous expressions, which assume unlinked loci), they become rather complex, even when we assume that there is no allele-frequency change (Turelli and Barton 1990, 1994; Bürger 2000). However, when selection is weak, we can include linkage into an approximation for the asymptotic response for a generalized infinitesimal model that makes no assumptions about the distribution of genotypic values (Turelli and Barton 1990). In particular, Turelli and Barton showed under weak selection that higher-order genotypic moments can be expressed in terms of the initial additive variance in the absence of gametic-phase disequilibrium (the genic variance, σ_a^2, which is assumed to be constant). Using this result, the asymptotic response to selection can be found to be approximately

$$\Delta\mu_z \simeq \sigma_a^2 \left(\frac{\partial \ln \overline{W}}{\partial \mu_z} \right) + \frac{\sigma_a^4}{r_{H_2}} \left(\frac{\partial \ln \overline{w}}{\partial \mu_z} \cdot \frac{\partial \ln \overline{w}}{\partial \sigma_z^2} \right) + \frac{3\,\sigma_a^6}{2\,r_{H_3}} \left(\frac{\partial \ln \overline{w}}{\partial \sigma_z^2} \cdot \frac{\partial \ln \overline{w}}{\partial \mu_{z,3}} \right) \tag{24.34}$$

where r_{H_2} and r_{H_3} are the harmonic mean recombination rates (weighted by the allelic contributions at each locus) between pairs and triplets of loci (Turelli and Barton 1990). At equilibrium, the higher-order genotypic moments are constant, as allele frequencies do not change and, with constant selection, covariances between loci approach equilibrium values. Under these conditions, the expected change in the mean following t generations of selection is just t times Equation 24.34.

Recalling Equations 24.12a and 24.12b (the fitness gradients for the mean and variance of a normally distributed trait), if phenotypes are approximately normal, the asymptotic rate of response further reduces to

$$\Delta\mu_z \simeq \frac{\sigma_a^2}{\widetilde{\sigma}_z^2} \left[S + \sigma_a^2\, \frac{\widetilde{\delta}(\sigma_z^2) + S^2}{2\,\widetilde{\sigma}_z^2} \left(\frac{S}{r_{H_2}} + \frac{3\,\sigma_a^2}{2\,r_{H_3}} \frac{\partial \ln \overline{w}}{\partial \mu_{3,z}} \right) \right] \tag{24.35}$$

where $\widetilde{\sigma}_z^2$ is the equilibrium phenotypic variance and $\widetilde{\delta}(\sigma_z^2)$ is the equilibrium within-generation change in phenotypic variance due to selection. This generalizes Bulmer's results (Chapter 16), which correct the breeder's equation for changes in the variance due to pair-wise disequilibrium. Equation 24.35 demonstrates that further corrections are required to account for the third- (and higher-order) disequilibrium generated by selection.

SUMMARY: WHERE DOES ALL THIS MODELING LEAVE US?

Predicting selection response is complicated. Even in the ideal setting where the breeder's equation holds exactly, drift and segregation generate a *variance* in response about the expected value (Chapter 18). Thus, any specific *realization* of the selection response will be randomly distributed about its expected value, and hence be less predictable than suggested by the deterministic result for the expected response. Further, even when the initial parent-offspring regression is linear and homoscedastic, there is still a large number of confounding factors for even the single-generation response (Table 13.2). Despite these concerns, short-term prediction of the response to artificial selection is reasonable for many traits (Chapter 18), but far less so for natural selection, in part due to uncertainty as to the target(s) of selection (Chapter 20). In contrast, the prediction of long-term response is an unobtainable goal unless one essentially knows all of the very fine (microscopic) genetic details of a trait, *including* the distribution of allelic effects and frequencies.

As we have seen here and elsewhere (Chapters 5 and 16), selection compromises the simple breeder's equation, $R = h^2 S$, prediction in two different ways. First, when some of the underlying loci harbor alleles of modest to large effect, this can generate locus-specific selection coefficients sufficiently large to significantly change allele frequencies over short time scales (Equations 5.3 and 5.21). Such changes alter the base-population heritability in ways that are not predictable from observable macroscopic features (such as the initial additive variance). Rather, the dynamics of selection response depend on very fine (microscopic) details of the trait's genetic architecture. A more subtle consequence of allele-frequency change is that it can drive a genotypic distribution away from normality by generating locus-specific skewness and kurtosis. This results in nonlinear and heteroscedastic parent-offspring regressions, and hence potential failure of the breeder's equation, even when correctly updated values of h^2 are used.

Dealing with the second consequence of selection, generation of linkage (or, more correctly, gametic-phase) disequilibrium, is often much more manageable than accounting for changes in allele frequencies. Indeed, in the absence of allele-frequency change, changes in LD are temporary, and decay away under random mating following the cessation of selection. Further, unless linkage is very tight or there are genes of very large effect, changes in LD occur on a much faster time scale than do allele-frequency changes. When the trait is controlled by a large number of loci, each of small effect, allele-frequency change is negligible over short time scales. However, selection-induced correlations (even among unlinked loci) change not only the genetic variance (Chapter 16), but can also generate skewness and kurtosis via the creation of third- and fourth-order disequilibrium. While these two latter effects drive a genotypic distribution away from normality, this effect is often modest and does not greatly compromise predictions of response. Further, as we saw in Chapter 16, the Bulmer equation accounts for changes in variances from disequilibrium using easily observed parameters. In the words of Turelli and Barton (1990), "Though our work shows that the distribution of breeding values for an additive polygenic character is unlikely to be precisely Gaussian, we expect that the Gaussian approximation suffices for predicting short-term selection response in all but the most extreme cases."

Allele-frequency change is the more pernicious feature of selection (relative to disequilibrium), but (for short time scales) it is restricted to traits whose underlying genetic architectures harbor one or more alleles of large effect. Chapters 25 and 26 examine the long-term consequences of allele-frequency change, while the additional role of mutation when directional and stabilizing selection is occurring is examined in Chapters 26 and 28, respectively.

25

Long-term Response:
1. Deterministic Aspects

The depletion of the variance by fixation of favored alleles is compensated by bringing previously rare alleles into the range where they contribute substantially to the response. Crow (2010)

Previous chapters assumed that the genetic variances for traits under selection either remain constant or can be predicted solely from their base-population values. Under the infinitesimal model, selection does not alter allele frequencies (Chapters 13, 16, and 24), and hence the genic variance, σ_a^2, remains unchanged, while the additive variance, $\sigma_A^2 = \sigma_a^2 + d$, changes (from selection-induced disequilibrium, d) in a predictable way (Chapter 16). With finite population size, allele frequencies are changed by drift, but again in predictable ways under an additive model (Chapters 16 and 24). If allele-frequency change is entirely due to drift, the expectation is that the amount of change is *independent* of allelic-effect size. However, with selection-induced changes, we expect alleles with larger effects to experience greater allele-frequency change. In such cases, while short-term response can be reasonably predicted from the base-population variance components alone, long-term response depends on the underlying, and generally unknown, genetic architecture (number of genes, allelic effects and frequencies). Our discussion of long-term response, which spans the next four chapters, is divided into three major topics: (i) deterministic changes in very large populations (the focus of this chapter), (ii) the special features that emerge due to drift (Chapter 26), and (iii) the long-term consequences of mutational input (Chapters 26–28). Our focus over the next two chapters is directional selection. The long-term consequences of stabilizing selection are considered in Chapter 28.

We start by examining an idealized model where an initially linear response declines smoothly to an asymptotic **selection limit** as the additive-genetic variation from the initial (or **base**) population becomes exhausted. We will show more formally that while populations with the same variance components show essentially the same short-term response, their long-term responses can be very different, depending on their underlying genetic architectures. We then develop the deterministic theory for allele-frequency change under long-term response in order to quantify the expected time for a certain amount of response and the ultimate selection limit (using only the initial variation). While these models cannot be applied to most real populations (as they require detailed information on the joint distribution of allele frequencies and effects at each locus underlying the trait), they provide an important framework for examining empirical results. Next, we examine response with a major gene and background polygenes. We conclude by reviewing a few generalizations that emerge from long-term artificial selection experiments and examine the nature of the (apparent) selection limit, if present, in these experiments.

IDEALIZED LONG-TERM RESPONSE IN A LARGE POPULATION

The general pattern expected in the long-term response to directional selection is roughly as follows: in the absence of segregating major alleles, additive variance (and hence the selection response) is roughly constant over the first few generations, yielding a nearly linear response (Figure 25.1). As discussed in Chapters 16 and 24, there is an initial reduction

Evolution and Selection of Quantitative Traits. Bruce Walsh & Michael Lynch.
Published 2018 by Oxford University Press.
DOI 10.1093/oso/9780198830870.001.0001

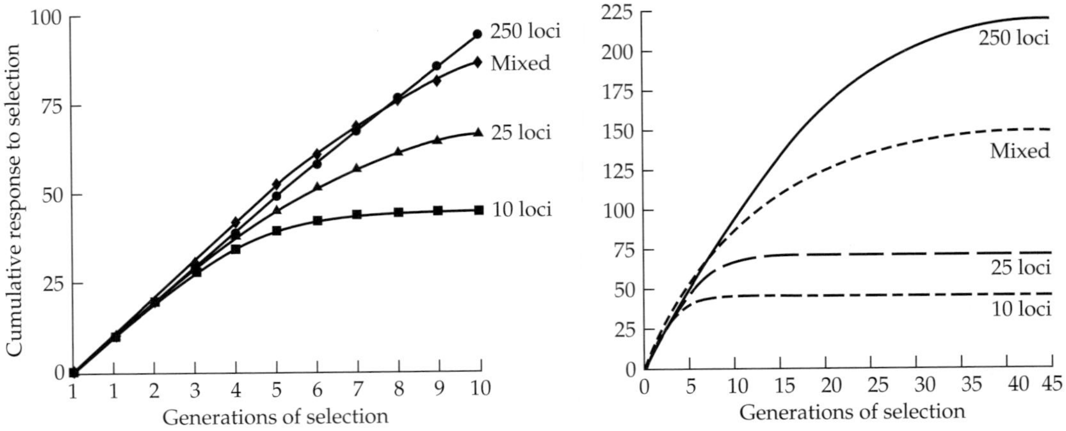

Figure 25.1 Examples of the expected response to selection, here assuming truncation selection (with the upper 20% saved) and n identical diallelic loci (each with genotypic values of $0:a:2a$, and a favorable allele frequency of p). We further assume that there is no epistasis and ignore any effects of gametic-phase disequilibrium. All populations start with $\sigma_A^2(0) = 100$ and $\sigma_E^2 = 100$, so $h^2(0) = 0.5$. Curves are plotted for models with 10, 25, and 250 equivalent loci, each with initial allele frequency $p = 0.5$ and a values of 4.47, 2.82, and 0.89, respectively. A mixed genetic model is also shown, which consists of 5 identical major loci ($p = 0.25$ and $a = 5.16$) and 125 identical minor loci ($p = 0.5$ and $a = 0.89$); as a consequence of these starting values, the major and minor loci contribute equally to the initial additive-genetic variance. **Left:** Short-term response over the first 10 generations. **Right**: Response over the first 40 generations. Note that the total response increases with the number of loci. In the infinitesimal-model limit, the response remains linear over all generations (after correcting for the slight decrease over the first few generations from linkage disequilibrium; see Example 16.3).

in the additive variance due to the generation of gametic-phase disequilibrium, but this is generally small unless directional selection on the trait is strong, heritability is high, and the number of underlying loci is large. As generations proceed, sufficient allele-frequency change accrues to significantly alter genetic variances, and in particular the genic variance, σ_a^2. At this point, the additive-genetic variance can either increase or decrease, depending on the starting distribution of allelic frequencies and effects. However, assuming no input of new variation (from mutation or migration), the additive variance generated from the initial variation in the base population eventually declines. Ultimately, a **selection limit** or **plateau** is reached, potentially reflecting the removal of all additive-genetic variance at the underlying loci, either by fixation of all segregating alleles at a locus or the absence of additive variance at those loci that are still segregating (e.g., the presence of overdominant alleles for the character under selection). This expectation follows from an important corollary of Fisher's fundamental theorem (Chapter 6), namely that, in the absence of new inputs, selection is expected to eventually remove all additive genetic variation in fitness.

If loci with both major and minor alleles influence the character under selection, an initial rapid response can result from large changes in allele frequencies at major loci, provided these alleles are initially not rare. This burst of response is followed by a much longer period of slower response due to allele-frequency changes at loci having smaller effects and major alleles that were initially very rare or partly recessive. Such differences in rates of response can make it difficult to determine whether a selection limit has actually been reached. As the genetic variation in the base population becomes fully exhausted, the effects of new mutations will drive any continued response, which is examined in Chapter 26.

Figure 25.1 illustrates differences in the long-term selection response for four hypothetical populations with the same initial heritability but different underlying genetic architectures. All show essentially the same response over the first few generations. By generation 5,

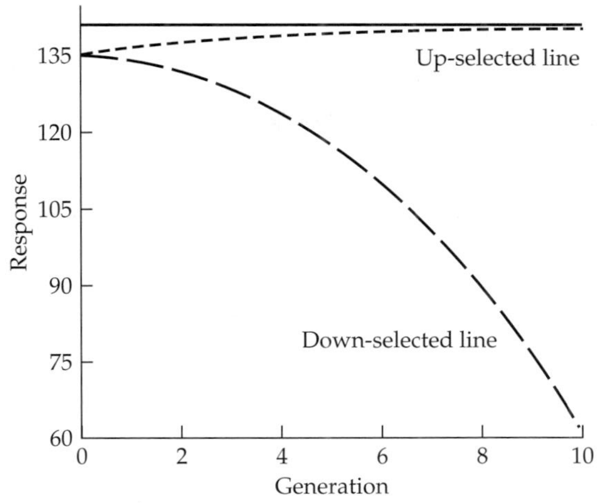

Figure 25.2 With strong directional dominance, an apparent selection limit can result when favored alleles are dominant, because selection is only acting against the increasingly rare recessive homozygotes. Here the genotypes have values of $0 : 2a : 2a$, and we ignore epistasis and gametic-phase disequilibrium. The population consists of 25 identical loci, with $a = 2.82$ and initial dominant-allele frequency of $p = 0.8$. We assume truncation selection with the upper (or lower) 20% of the population being saved. If all loci are fixed for the favored allele, the selection limit is $2 \cdot 2.82 \cdot 25 = 141$ (indicated by the horizontal line). There is little response to upward selection and the population appears at a selection limit, even though there was considerable genetic variation in the base population, as shown by the rapid response of the down-selected line.

however, selection has changed allele frequencies in the 10- and 25-locus populations enough to reduce the response, while the 250-locus population shows a roughly constant response through 20–25 generations. The mixed population (5 major loci, each with an initial frequency of the favored allele of $p = 0.25$, and 125 minor loci, each with $p = 0.5$) shows an enhanced response relative to the others in generations 3–7. This results from an increase in heritability, as the frequencies of alleles with large effects increase from 0.25 to 0.50, increasing the additive variance contributed by these loci (LW Figure 4.8). If rare recessives are present, there can be a considerable time lag until the enhanced response appears (Figure 25.10B). For all models, the time for allele-frequency change scales as $1/s$ (Equations 5.3d–5.3f), with s scaling with the allelic-effect size (Equation 5.21). Hence, as a rough approximation, if the effect size is halved, the same amount of allele-frequency change will take twice as long.

If alleles favored by selection are dominant, response slows considerably as they become common, reflecting the rarity of homozygous recessives. In such cases, the response can be so slow that the population appears to be at a limit. However, as Figure 25.2 demonstrates, reverse selection can result in a fairly rapid response, indicating the presence of substantial additive-genetic variation. As was mentioned in Chapter 18, divergent selection in this case generates a significant asymmetric response. This apparent limit due to the very slow removal of recessives can be partly overcome by inbreeding. By increasing the frequency of homozygotes relative to a random-mating population, inbreeding greatly improves the efficiency of selection, allowing favorable dominant alleles to be more rapidly fixed.

Example 25.1. Falconer (1971) examined an apparent limit in a mouse line selected for increased litter size. Four sublines were created from this plateaued line and subjected to inbreeding and selection. Selection on a new line formed by crossing these inbred-selected lines gave an improvement of 1.5 mice per litter over the original limit. Falconer's interpretation was that many recessive alleles decreasing litter size were segregating in the apparently plateaued line, some of which were lost during inbreeding within sublines. Crossing the inbred-selected lines generated a population segregating fewer recessives (i.e., fixed for more of the favorable dominant alleles), thus facilitating the total response. Several other selection experiments in mice also found segregating recessives in populations near apparent selection limits. For litter size, Eklund and Bradford (1977) found that inbreeding and selection increased the response. However, Al-Murrani and Roberts (1974) found that while a population that had plateaued for increased body weight was also segregating a number of recessives, their loss was expected to yield only a trivial increase in body weight, and no increase was detected using Falconer's inbred-selection method.

DETERMINISTIC SINGLE-LOCUS THEORY

The contribution to the selection limit from a single locus and the half-life associated with this contribution depend on a variety of genetic parameters: initial allele frequencies, the dominance relationship among alleles, and allelic-effect sizes. This section quantifies how these factors influence long-term response for a diallelic locus in the absence of drift, mutation, and epistasis. This basic model provides insight into the dynamics of response and serves as the foundation for theories incorporating drift and mutation (Chapters 26 and 28).

Expected Contribution From a Single Locus

We start with the expected total contribution from a given diallelic locus. Let *B* be the allele favored by directional selection, where the genotypes *bb* : *Bb* : *BB* have genotypic values of $0 : a(1+k) : 2a$. Assuming genotypes are in Hardy-Weinberg equilibrium, the contribution to the mean character value from this locus is a function of p (the frequency of *B*), namely,

$$m(p) = 2ap\,[\,1 + (\,1 - p\,)k\,] \tag{25.1a}$$

Provided there is no epistasis, gametic-phase disequilibrium will have no influence on this contribution to the mean. The total contribution from this locus if *B* is fixed, given that it starts at an initial frequency of p_0, is

$$m(1) - m(p_0) = 2a - 2ap_0\,[\,1 + (\,1 - p_0)k\,] = 2a\,(\,1 - p_0)(\,1 - p_0 k\,) \tag{25.1b}$$

Figure 25.3 plots the total contribution when allele *B* is either additive ($k = 0$), dominant ($k = 1$), or recessive ($k = -1$). The total response from this locus is largest when *B* is recessive and rare, and smallest when *B* is dominant and common. With overdominance ($k > 1$), the maximum value for $m(p)$ occurs at at $p = \widehat{p}$, where

$$\widehat{p} = \frac{1+k}{2k} \tag{25.1c}$$

which is obtained by taking the derivative of $m(p)$ with respect to p and solving for zero. With an overdominant locus underlying the trait, when $p_0 > \widehat{p}$, directional selection on the trait results in p decreasing to $\widehat{p}$, while if $p_0 < \widehat{p}$, then p increases to $\widehat{p}$. In either case, the final contribution from this locus is

$$m(\,\widehat{p}\,) - m(p_0) \tag{25.1d}$$

Finally, the allele frequency, p_β, at which a preset fraction, β, of the total contribution occurs is also of interest. This is determined by solving the quadratic equation

$$m(p_\beta) - m(p_0) = \beta\,[\,m(1) - m(p_0)\,] \tag{25.1e}$$

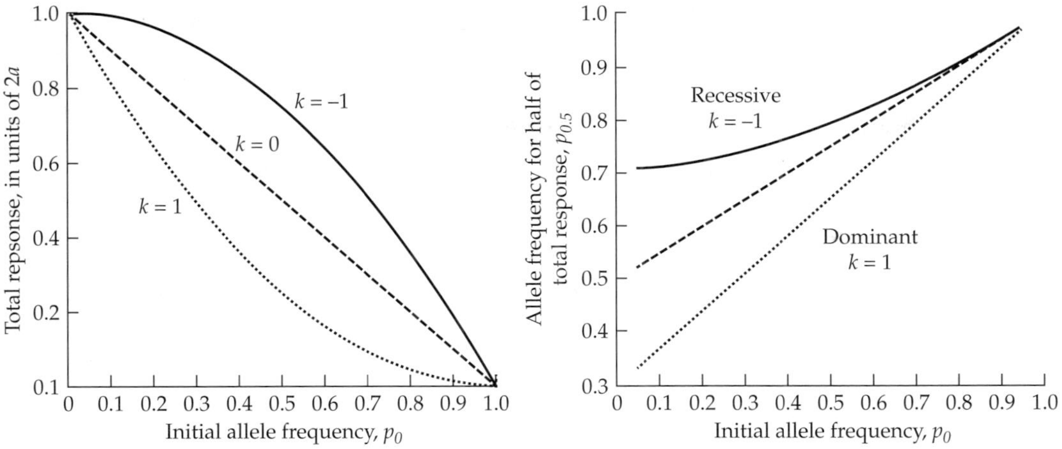

Figure 25.3 **Left:** The contribution to total response from a diallelic locus assuming allele B, starting at frequency p_0, is eventually fixed. The genotypes $BB : Bb : bb$ have values of $2a : a(1+k) : 0$. The three curves correspond to B being additive ($k = 0$), dominant ($k = 1$), and recessive ($k = -1$). The smallest contribution is made by dominant alleles at high frequencies, and the largest from recessive alleles at low frequencies. **Right:** The allele frequency, $p_{0.5}$, at which half the total response contributed by a locus occurs, as a function of its initial frequency, p_0.

Table 25.1 Total contribution to the selection limit and the allele frequency, $p_{0.5}$, at which half this response occurs for a diallelic locus when allele B has an initial frequency of p_0.

	Total Contribution	$p_{0.5}$
B additive ($k = 0$)	$2a(1 - p_0)$	$(1 + p_0)/2$
B dominant ($k = 1$)	$2a(1 - p_0)^2$	$1 - \sqrt{[1 - p_0(2 - p_0)\,]/2}$
B recessive ($k = -1$)	$2a(1 - p_0^2\,)$	$\sqrt{(1 + p_0^2\,)/2}$

for p_β, with $m(1)$ is replaced by $m(\widehat{p})$ when $k > 1$. A case of particular interest is $p_{0.5}$, the frequency at which half of the expected response occurs ($\beta = 1/2$). Expressions for $p_{0.5}$ as a function of initial allele frequency are given in Table 25.1 and plotted in Figure 25.3. Observe that rare recessives have to increase substantially in frequency to yield half their ultimate response (e.g., for $k = -1$, $p_{0.5} \simeq 0.71$ when $p_0 = 0.1$).

Dudley's Estimators of a, n, and p_0

In a similar fashion to the Wright-Castle estimator for the number of loci (LW Chapter 9), if we are willing to make the assumption of the exchangeable model (all loci are additive with the same effects and initial frequencies; Chapter 24), we can estimate a, n, and p_0 from data on selection limits. Under this model, Equation 25.1b gives the expected total response ($R_c(\infty)$, which we will simply write here as R) of $2na(1 - p_0)$, while the starting additive variance is $2na^2p_0(1-p_0)$. Using these expressions yields Robertson's (1970a) result for the total response, scaled by the square root of the initial additive variance, as

$$\frac{R}{\sigma_A} = \frac{2na(1 - p_0)}{\sqrt{2na^2p_0(1 - p_0)}} = \sqrt{\frac{2n(1 - p_0)}{p_0}} \tag{25.2a}$$

Dudley (1977) noted that, in a divergence selection experiment (Chapter 18), Equation 25.2a yields the limit, R_H, for response in the high direction, while

$$\frac{R_L}{\sigma_A} = \frac{2nap_0}{\sqrt{2na^2p_0(1 - p_0)}} = \sqrt{\frac{2np_0}{1 - p_0}} \tag{25.2b}$$

yields the limit to response in the low direction. Taking the ratio of these two limits yields

$$\frac{R_H}{R_L} = \frac{\sqrt{2n(1-p_0)/p_0}}{\sqrt{2np_0/(1-p_0)}} = \frac{1-p_0}{p_0} \tag{25.2c}$$

Equation 25.2c rearranges to suggest an estimate of p_0, with

$$\widehat{p}_0 = \frac{1}{R_H/R_L + 1} \tag{25.2d}$$

Next, because $R_H R_L = 4n^2 a^2 p_0(1-p_0)$, it follows that $R_H R_L/\sigma_A^2 = 2n$, which suggests an estimator for the number of loci

$$\widehat{n} = \frac{1}{2}\frac{R_H R_L}{\sigma_A^2} \tag{25.2e}$$

Finally, a little algebra recovers an estimate of the allelic effect, namely,

$$\widehat{a} = \sigma_A^2 \left(\frac{1}{R_H} + \frac{1}{R_L} \right) \tag{25.2f}$$

These estimates of the genetic architecture of a trait should be regarded as extremely crude (at best), but they nonetheless provide potential insight. One of the many caveats in applying these results is that the selected populations must be at their respective limits, which can be hard to access. Operationally, the limits could be estimated from curve-fitting of the data, a topic that will be discussed shortly.

Dynamics of Allele-frequency Change

To obtain approximate expressions for the actual dynamics of the selection response in the mean, we need to follow allele-frequency changes over time. Recall from Equation 5.21 that if a character is normally distributed, then the change in the frequency of allele B is $\Delta p \simeq \bar{\imath}\,(\alpha/\sigma_z)\,p$, where p and α are, respectively, the frequency and average excess of B. This is a weak-selection approximation, as it assumes that $|\,\bar{\imath}\,\alpha/\sigma_z| \ll 1$. It also assumes that the effects of epistasis, gametic-phase disequilibrium, and genotype × environment interactions are negligible. Assuming random mating, the average effect of an allele equals its average excess, and LW Equation 4.15a gives $\alpha = (1-p)a[1+k(1-2p)]$. Substituting yields

$$\Delta p \simeq \frac{a\,\bar{\imath}}{\sigma_z} p(1-p)[\,1+k(1-2p)\,] \tag{25.3}$$

Recall that this is correct only to linear order (terms of a^2 and higher order are ignored; see Equation 5.27a). Thus, there are potential pitfalls in applying Equation 25.3 when $\bar{\imath} \simeq 0$. One important example of this latter situtation is strict stabilizing selection, where $\bar{\imath} = 0$ but allele frequencies can still change due to selection on the phenotypic variance of the character, which enters as quadratic terms, a^2 (e.g., Equation 5.6f).

Example 25.2. The idealized response curves in Figure 25.1 were generated using Equation 25.3 to compute the expected allele-frequency change at each locus, assuming there is no gametic-phase disequilibrium. We assumed complete additivity ($k = 0$ and no epistasis), that $\sigma_E^2 = 100$, and n identical loci underlying the character. Equation 25.3 yields

$$\Delta p_t = \frac{a\,\bar{\imath}\,p_t(1-p_t)}{\sigma_z(t)} = \frac{a\,\bar{\imath}\,p_t(1-p_t)}{\sqrt{\sigma_A^2(t)+\sigma_E^2}} \simeq \frac{a\,\bar{\imath}\,p_t(1-p_t)}{\sqrt{2na^2p_t(1-p_t)+100}}$$

Strictly speaking, the last expression is an approximation, albeit a close one, as $2na^2p_t(1-p_t)$ is the *genic* variance, $\sigma_a^2(t)$, at generation t, while the *additive genetic* variance equals the genic variance plus the disequilibrium contribution, $\sigma_A^2(t) = \sigma_a^2(t) + d(t)$, as discussed in Chapters 16 and 24. Iteration generates the response curves shown in the figure.

Recall that the results for the single-locus selection response in Chapter 5 used the fitness parameterization where the genotypes *bb* : *Bb* : *BB* have fitnesses $1 : 1 + s(1+h) : 1 + 2s$. For weak selection (e.g., $|s|$, $|sh| \ll 1$), this model gives the change in the frequency of *B* as

$$\Delta p \simeq s\,p\,(1-p)\,[\,1 + h\,(1-2p)\,]$$

which follows from Equation 5.1c upon noting that $1/\overline{W} = 1 + O(s, sh)$, namely, one plus terms of order s and sh. Upon matching terms with Equation 25.3, we find that a QTL under directional selection has approximate selection parameters of

$$s = \frac{a}{\sigma_z}\,\bar{\imath} \qquad \text{and} \qquad h = k \tag{25.4}$$

Thus, as an initial approximation, the dynamics for a QTL with a small effect on a character under directional selection follow those of a locus under these constant fitnesses. If there is gametic-phase disequilibrium or epistasis, single-locus fitnesses change as the background genotype changes (Example 5.7). However, in the absence of these complications, fitnesses still change as the phenotypic variance, σ_z^2, changes. As other loci become fixed due to selection and drift, σ_z^2 generally decreases as the genetic variance decreases, which in turn increases s. Unless heritability is large, this effect is usually small. For example, assuming that all of the genetic variance is additive, then if $h^2 = 0.1$, the phenotypic standard deviation when all loci are fixed is 95% ($\sqrt{0.9}$) of its initial value (inflating s by 5%), while for $h^2 = 0.25$ and 0.5, s can be inflated by 15% and 43%, respectively. This decrease in the phenotypic variance can be countered if σ_E^2 increases as genotypes become more homozygous (LW Chapter 6) or if there has been selection to increase σ_E^2 (Chapter 17). It is worth stressing that these results are for a trait under directional selection. The dynamics for a locus for a trait under stabilizing selection are quite different (Example 5.6; Chapter 28).

We can use these results to compute the expected time to achieve a fraction of the response contributed by a locus in an infinite population. When selection is weak ($|s|$, $|hs| \ll 1$), Equation 5.3c gives the expected time for a favorable allele, *B*, to reach a frequency of p, given it starts at a frequency of p_0, for the general fitness model $1 : 1+s(1+h) : 1+2s$. Equation 5.3d gives the expected time (in generations) when *B* is additive ($h = 0$), Equation 5.3e when *B* is recessive ($h = -1$), and Equation 5.3f when *B* is dominant ($h = 1$). These expressions, together with Equations 25.1e and 25.4, allow us to obtain approximate expressions for the expected time until a fraction, β, of the total contribution from a single locus occurs (namely, the expected time to reach frequency p_β). Note that the rate of allele-frequency change scales as $s^{-1} = (\bar{\imath}\,a/\sigma_z)^{-1}$, meaning that the smaller is the allelic effect, the slower is the expected response time. Substituting $p_{0.5}$ for p_β gives the expected half-life of response associated with the locus under consideration (Figure 25.4). The half-life for rare recessives can be quite long, while the half-life of response for dominant loci *increases* with allele frequency when *B* is common (although in such cases, the additional gain made by fixing *B* is typically very small; see Figure 25.3).

These results ignore the effects of gametic-phase disequilibrium. Negative disequilibrium generated by directional selection reduces the average effect of an allele (plus alleles are associated with an excess of minus alleles at other loci, and vice versa, reducing allelic effects relative to a population in gametic-phase equilibrium). This results in a slower change in allele frequency. Hence, the half-lives plotted in Figure 25.4 are slight underestimates. In addition, for major alleles, our assumption that $|a|/\sigma_z$ and $|ak|/\sigma_z$ are small no longer holds, and the previous expressions for change in allele frequency and expected time to reach a given frequency can be poor approximations. More accurate versions for cases where major alleles are present were given by Latter (1965a) and Frankham and Nurthen (1981).

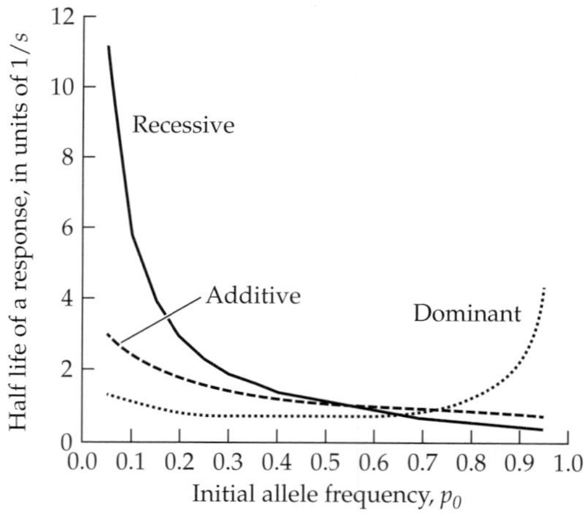

Figure 25.4 The expected times for a diallelic locus to contribute half of its total response, assuming the favored allele, *B*, is eventually fixed. These curves are obtained by substituting $p_{0.5}$ from Table 25.1 into the appropriate version of Equation 5.3. Note that times for half-life scale as $s^{-1} = (\bar{\imath}\, a/\sigma_z)^{-1}$ generations.

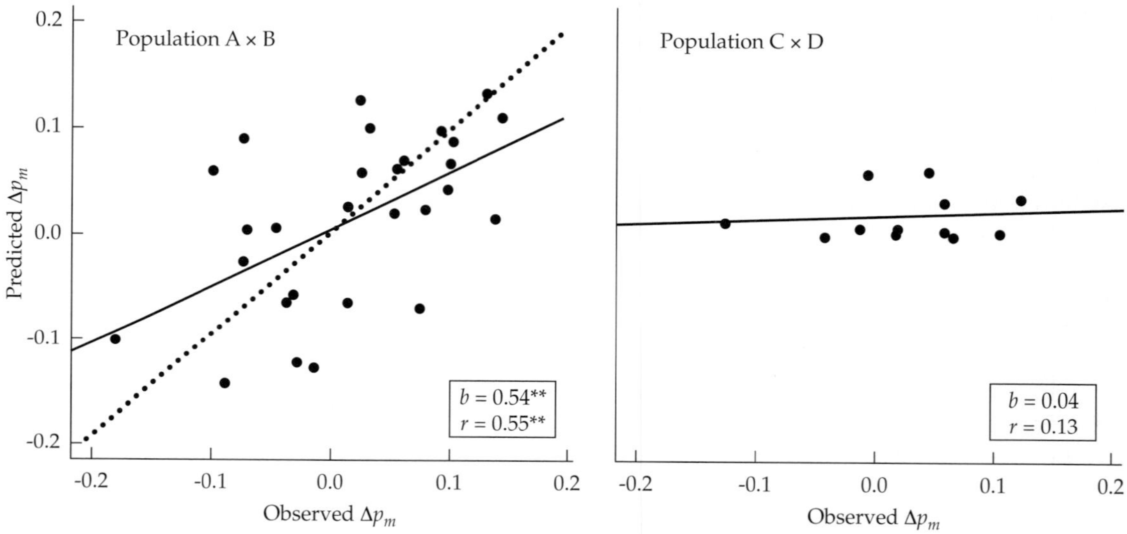

Figure 25.5 Correlation between observed and predicted (from estimated QTL effect size) changes in marker allele frequencies in two inbred-line crosses (A $\times$ B and C $\times$ D) of maize subsequently subjected to seven cycles of selection. **Left:** QTL effect size was a fairly good predictor in the A $\times$ B selected lines (r = 0.55). The expectation is a line with a slope of $b = 1$ (observed = predicted; the dotted line), while the solid line is the best-fitting regression, with a slope of $b = 0.54$. While significantly different from 0 (showing a positive association between observed and predicted values), the predicted allele-frequency change is *less* than the observed allele-frequency change. **Right:** No significant association between the observed and predicted allele-frequency change was seen in the C $\times$ D selected lines (r = 0.13, ns). Further details are provided in Example 25.2. (After Falke et al. 2007.)

Example 25.3. An ingenious experiment examining the fit between estimated QTL effects and their projected allele-frequency changes under selection was performed by Falke et al. (2007). By making crosses between inbred lines, the frequency of all segregating alleles in the F_1 is $1/2$. Assuming only additive effects, for $p = 1/2$, Equation 25.3 reduces to $\Delta p = a\,\bar{\imath}\,/(4\sigma_z)$. Further, the expected frequency change at a marker allele linked to n QTLs is

$$\Delta p_m = \frac{\overline{\imath}}{4\sigma_z} \sum_{i=1}^{n} a_i(1 - 2c_i) \tag{25.5}$$

where c_i is the recombination frequency between the ith-marker and the QTL (this expression is only approximate, as it ignores LD among the linked QTLs and quickly breaks down if more than a couple of QTL are linked to the marker).

Falke et al. examined two sets of crosses involving European flint maize. The A $\times$ B cross used roughly 270 $F_{2:3}$ lines (selfed F_2 lines) for QTL mapping and was subjected to four cycles of selection, while the C $\times$ D cross used roughly 130 $F_{3:4}$ (selfed F_3 lines) for QTL mapping and was subjected to seven cycles of selection. As Figure 25.5 shows, while the QTL effect was a modest predictor of change in marker allele frequency in the A $\times$ B cross ($r = 0.55$), the slope of the regression of predicted on observed change was roughly 0.5, implying that the observed marker allele frequency change exceeded the predicted value by roughly two-fold. The association was nonsignificant in the C $\times$ D cross ($r = 0.13$).

While the lack of fit should not be surprising given that Equation 25.5 is an approximation, the direction was unexpected. One would expect Equation 25.5 to *overpredict* (predict a greater change than was seen), rather than underpredict, the allele-frequency change. Overprediction is expected from either the Beavis effects (overestimation of the effect sizes of detected QTLs when power is low; LW Chapter 15; also see Göring et al. 2001; Xu 2003; Goddard et al. 2009) or from the generation of negative linkage disequilibrium among selected loci (Chapter 16), which reduces the selection response below the value predicted by Equation 25.5.

Example 25.4. As an example of the consequences for the limit, R, and half-life, $t_{0.5}$, as the number of loci increases, consider the exchangeable model with n completely identical additive loci (in the absence of mutational input). Suppose populations with different numbers of loci underlying the character all start with the same initial variances ($\sigma_A^2(0) = 100$ and $\sigma_z^2(0) = 200$) and with an initial frequency of $p_0 = 0.5$. To hold initial additive-genetic variance constant as n increases, the allelic effect, a, must decrease as the number of loci increases. If we ignore gametic-phase disequilibrium, then $\sigma_A^2(0) = 2na^2 p_0(1 - p_0) = na^2/2 = 100$, implying $a = 10\sqrt{2/n}$. From Table 25.1, the selection limit becomes $2na(1 - 1/2) = na = 10\sqrt{2n}$. With $p_0 = 1/2$, Table 25.1 gives $p_{0.5} = [1 + (1/2)]/2 = 3/4$. Substituting these values into Equation 5.3d yields the expected time, $t_{0.5}$, for this amount of allele-frequency change (from 0.5 to 0.75, or $t_{0.75,0.5}$ in the notation of Equation 5.3d) as

$$t_{0.5} = t_{0.75,0.5} \simeq \frac{1}{s} \ln\left(\frac{(3/4)\,(1 - [1/2])}{[1/2]\,(1 - [3/4])}\right) = \left(\frac{\sigma_z}{a\,\overline{\imath}}\right) \ln(3) = \frac{\sqrt{n}}{\overline{\imath}} \ln(3)$$

The resulting values of these various quantities for 5 to 500 loci become

n	a	R	$R/\sigma_z(0)$	$t_{0.5} \cdot \overline{\imath}$
5	6.32	31.6	2.2	2.5
10	4.47	44.7	3.2	3.5
25	2.82	70.7	5.0	5.5
50	2.00	100.0	7.1	7.8
100	1.41	141.4	10.0	11.0
250	0.89	223.6	15.8	17.4
500	0.63	316.2	22.4	24.6

At the selection limit, the mean phenotype is usually more extreme than any phenotype observed in the initial base population ($R > 3\sigma_z$). For example, when $n = 25$, the total response is 5 phenotypic standard deviations. For $U \sim N(0,1)$, $\Pr(U > 5) = 2.87 \cdot 10^{-7}$. Hence, on average, $2.87 \cdot 10^{-7} \cdot 10^6 = 0.287$ such extreme individuals are expected in a base population sample of size 10^6. From the zero term of the Poisson, the probability that no such individuals are seen in such a sample is $e^{-0.287}$ = 0.75. Hence, the limiting mean exceeds any phenotype likely to be found in the initial population. This is not surprising, as the probability of observing the most extreme genotype (*BB* at all loci) in the base population is $(1/4)^{25} \simeq 10^{-15}$

MAJOR GENES VERSUS POLYGENIC RESPONSE: THEORY

As highlighted by Example 24.4 and Figure 25.1, the presence of a major gene or genes can change the dynamics of response. A hotly debated issue in quantitative genetics and evolutionary biology is whether selection response is largely due to major genes or polygenes. At present, the data are still murky and likely biased. Before reviewing the evidence, we first consider Lande's theoretical work on conditions for major gene versus polygenic response (Lande 1983). Apparently unaware of Lande's work, plant and animal breeders have also conducted small-scale simulations on this issue (Sehested and Mao 1992; Cox 1995). A related topic, selection when a known major gene is included in an index of selection (e.g., Pong-Wong and Woolliams 1998), is a special case of marker-assisted selection and is covered in Volume 3.

Lande's Model: Response With a Major Gene in an Infinitesimal Background

Lande (1983) assumed a single major gene and an infinitesimal background of polygenes, and his concern was how often a selection response (such as an adaptation by natural selection) is primarily due to a single (or a very few) major genes versus being polygenic. Because genes with major effects on a trait often have deleterious effects on overall fitness (Wright 1977; Lande 1983; Kemper et al. 2012), Lande allowed for such pleiotropic fitness effects acting on the major locus in addition to its influence on the trait under phenotypic selection. The basic parameters of his model are given in Table 25.2. For each of the three major-locus genotypes, the distribution of phenotypic values is assumed to be normal with a mean of $\mu+\alpha_i$ and a variance of σ^2. Part of this variance, $h^2\sigma^2$, is from the additive-genetic variance of background polygenes, where h^2 is the polygenic heritability. As the expression for the trait mean in Table 25.2 suggests, the dynamics of the mean jointly depend on the change in frequency of the major allele (Δp) and the change in the mean ($\Delta\mu$) of the background polygenes (also see Equation 25.8e). While both p and μ change over time, to avoid excessive notation we suppress the subscript for generation on each.

Consider the change in the major allele-frequency, p, first. Using the mean and marginal fitnesses from Table 25.2, Wright's formula (Equation 5.5b) yields the expected change as

$$\Delta p = \frac{p(1-p)}{2\overline{W}}\frac{\partial \overline{W}}{\partial p}$$

$$= \frac{p(1-p)}{\overline{W}}\left[\,(p-1)\overline{W}_0 + (1-2p)(1-s_1)\overline{W}_1 + p(1-s_2)\overline{W}_2\,\right] \tag{25.6}$$

with the second step following upon differentiation of the mean fitness. Note that the $\overline{W}_i$ are not constants, but rather change with the polygenic mean, μ, whose expected change follows from the Lande equation (13.27a)

$$\Delta\mu = h^2\sigma^2\frac{\partial \ln(\overline{W})}{\partial \mu} = \frac{h^2\sigma^2}{\overline{W}}\frac{\partial \overline{W}}{\partial \mu} \tag{25.7a}$$

Taking the derivative of $\overline{W}$ with respect to μ returns

$$\Delta\mu = \frac{h^2\sigma^2}{\overline{W}}\left[(1-p)^2\frac{\partial \overline{W}_0}{\partial \mu} + 2p(1-p)(1-s_1)\frac{\partial \overline{W}_1}{\partial \mu} + p^2(1-s_2)\frac{\partial \overline{W}_2}{\partial \mu}\right] \tag{25.7b}$$

To evaluate the derivatives of the marginal fitnesses of the major locus, first note that the normal density function is given by

$$\varphi(z,\mu,\sigma^2) = \frac{1}{\sqrt{2\pi\sigma^2}}\cdot\exp\left[-\frac{(z-\mu)^2}{2\sigma^2}\right]$$

Table 25.2 Lande's (1983) model for simultaneous selection on a major locus and background polygenes. The distribution of phenotypic values for each major-locus genotype is assumed to be normal, with variance σ^2. The distribution of background polygenic values around each major genotype is normal with variance $h^2\sigma^2$. Here $\varphi(z, \mu, \sigma^2)$ denotes the density function for a normal random variable with mean μ and variance σ^2, and $w(z)$ is the expected fitness associated with phenotype z.

	Major-locus genotype		
	bb	Bb	BB
Frequency	$(1-p)^2$	$2p(1-p)$	p^2
Mean phenotype	μ	$\mu+\alpha_1$	$\mu+\alpha_2$
Natural selection	1	$1-s_1$	$1-s_2$
Mean fitness	$\overline{W}_0$	$(1-s_1)\overline{W}_1$	$(1-s_2)\overline{W}_2$

$$\overline{W}_i = \int w(z)\,\varphi(z, \mu+\alpha_i, \sigma^2)\,dz \quad \text{for } i = 0, 1, 2$$

Mean fitness:

$$\begin{aligned}\overline{W} &= (1-p)^2\, W_{bb} + 2p(1-p)W_{Bb} + p^2 W_{BB} \\ &= (1-p)^2\,\overline{W}_0 + 2p(1-p)(1-s_1)\overline{W}_1 + p^2(1-s_2)\overline{W}_2\end{aligned}$$

Mean phenotype:

$$\overline{z} = \mu\left(1 + 2\alpha_1 p(1-p) + \alpha_2\, p^2\right)$$

To obtain the required derivatives of $\varphi(z, \mu, \sigma^2)$, recall from the chain rule of differentiation that

$$\frac{\partial \exp[f(x)]}{\partial x} = \frac{\partial[f(x)]}{\partial x} \cdot \exp[f(x)]$$

yielding

$$\frac{\partial \varphi(z, \mu+\alpha_i, \sigma^2)}{\partial \mu} = \frac{z-(\mu+\alpha_i)}{\sigma^2}\,\varphi(z, \mu+\alpha_i, \sigma^2) \tag{25.8a}$$

Hence,

$$\begin{aligned}\frac{\partial \overline{W}_i}{\partial \mu} &= \int w(z)\,\frac{\partial \varphi(z, \mu+\alpha_i, \sigma^2)}{\partial \mu}\,dz \\ &= \frac{1}{\sigma^2}\left[\int z\,w(z)\,\varphi(z, \mu+\alpha_i, \sigma^2)dz - (\mu+\alpha_i)\int w(z)\,\varphi(z, \mu+\alpha_i, \sigma^2)dz\right] \\ &= \frac{1}{\sigma^2}\left[\int z\,w(z)\,p_i(z)\,dz - (\mu+\alpha_i)\,\overline{W}_i\right] = \frac{\overline{W}_i}{\sigma^2}\,S_i\end{aligned} \tag{25.8b}$$

where

$$S_i = \int z\,w(z)\,\frac{p_i(z)}{\overline{W}_i}\,dz - (\mu+\alpha_i) \tag{25.8c}$$

is the selection differential acting on the major-locus genotype, i. This equivalence follows from the fact that the integral represents the mean value for major-locus genotype i following selection (μ_{s_i}), the second term is the mean for i before selection (μ_i), and $S_i = \mu_{s_i} - \mu_i$. From Equations 25.7b and 25.8b, the expected change in the polygenic mean becomes

$$\Delta\mu = h^2\left[(1-p)^2\,\frac{\overline{W}_0}{\overline{W}}\,S_0 + 2p(1-p)(1-s_1)\,\frac{\overline{W}_1}{\overline{W}}\,S_1 + p^2(1-s_2)\,\frac{\overline{W}_2}{\overline{W}}\,S_2\right] \tag{25.8d}$$

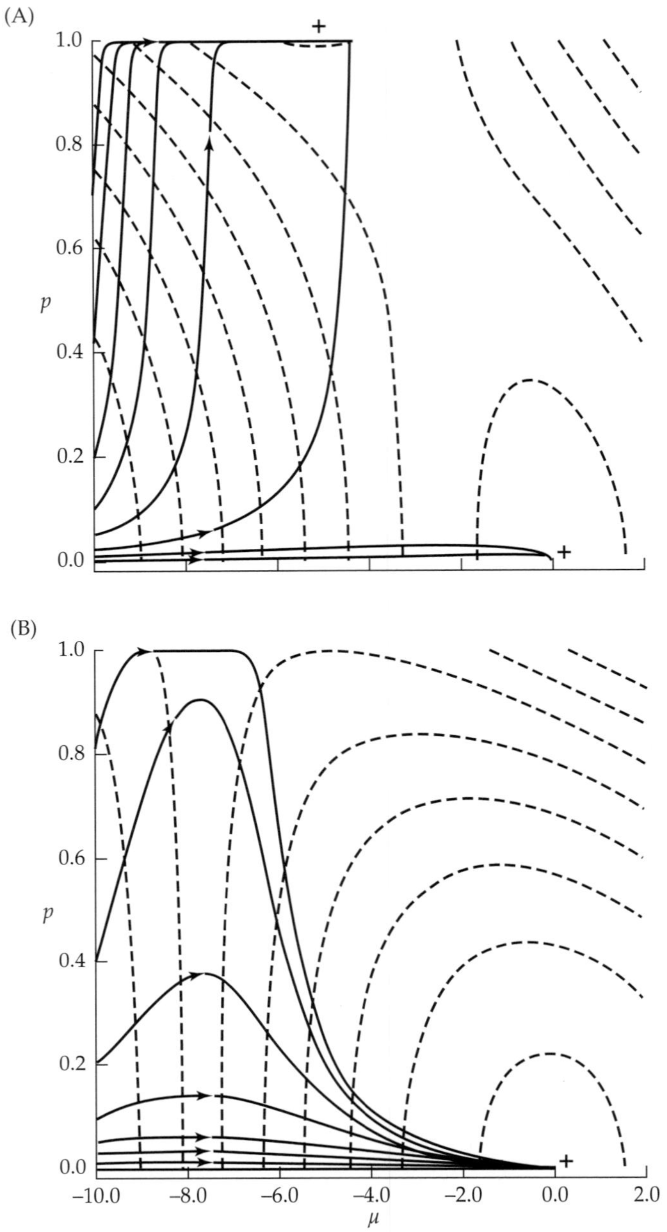

Figure 25.6 Lande's analysis of selection toward a new optimum when a major gene and polygenes are present. In both examples, the population initially starts 10 units below the new optimum value (zero), and the favored major-gene homozygote adds a value of 5. Dashed lines represent contours of equal fitness in the (p, μ) space, while the arrowed solid lines represent the allele-frequency trajectories of the major gene. **A:** The favored homozygote has a pleiotropic disadvantage of $s = 0.02$. Here there are two peaks on the (p, μ) fitness surface, on the lower right ($p = 0$, $\mu = 0$) and in the upper middle ($p = 1$, $\mu = -5$). If the initial frequency of the major allele is above 0.025, the major allele is fixed, while it is lost if the initial frequency is below this critical frequency. **B:** The favored homozygote has a pleiotropic disadvantage of $s = 0.40$. There is a single peak on the fitness surface ($p = 0$, $\mu = 0$), and although p may initially increase in frequency, it will always be lost, with the ultimate response being entirely polygenic. A significant fraction of the initial response can be through the major gene, but eventually this is replaced by the polygenic component. As polygenic response drives the population mean toward the optimum, the pleiotropic disadvantage of the major allele eventually exceeds its fitness advantage from selection on the trait, resulting in its eventual loss. (After Lande 1983.)

From Table 25.2, the new mean becomes

$$\overline{z} = (\mu + \Delta\mu)\left[1 + 2\alpha_1(p + \Delta p)(1 - p - \Delta p) + \alpha_2\,(p + \Delta p)^2\right] \tag{25.8e}$$

Note that while changes in p influence changes in μ, and vice-versa, Lande assumed that the infinitesimal variance, σ^2, and heritability, h^2, remain unchanged over time (ignoring the Bulmer effect). By using the machinery from Chapter 16, we could modify these expressions to allow for the changes caused by selection generating gametic-phase disequilibrium.

Lande provides extensive analysis of his model in several settings, two of which we will consider. The first is the evolution toward a new optimum. Suppose the optimal phenotype suddenly shifts (for example, due to a major environmental change). In such cases, if the frequency of the major allele is sufficiently rare relative to the strength of selection on the trait (and also relative to the strength of pleiotropic selection against the major allele), it will be lost and the response will be entirely polygenic. As Figure 25.6 illustrates, the dynamics can be complex in this case. The key to understanding this frequency-dependent behavior is to recall the dynamics of an underdominant locus (Figure 5.1). Here, there is an unstable internal equilibrium, with the allele becoming fixed provided it starts above this value, and otherwise being lost.

Two additional remarks concerning Lande's examples displayed in Figure 25.6 are in order. First, because *B* is strictly deleterious before the shift in optimum, its initial frequency is expected to be low. For example, for an additive allele with mutation rate ν, $p \simeq 2\nu/s$ (Equation 7.6b). While most of the trajectories in Figure 25.6A show fixations, almost all populations would be expected to start with the frequency of *B* below the 0.025 initial allele-frequency threshold for fixation (the initial frequency is below this value when $2\nu/0.02 < 0.025$, or when $\nu < 0.00025$). Second, this is a deterministic analysis, which has implications for Figure 25.6B. Notice that the trajectory for an allele starting at frequency 0.8 approaches one before eventually declining to zero. As polygenic response moves the trait mean toward the optimum, the deleterious pleiotropic effects of this allele become greater than its favorable effect on the trait, resulting in the major allele becoming selected against, and eventually removed. In a finite population, selection may initially drive the major-allele frequency sufficiently close to one for drift to fix *B* before polygenic response negates its favorable effect on the trait. In such cases, a major gene response will be seen.

Lande's analysis of directional selection used an exponential model of trait fitness, $w(z) \propto \exp(\beta z)$, which reduces to a simple linear fitness function, $w(z) \simeq 1 + \beta z$, for weak selection ($|\beta z| \ll 1$). Lande showed that under exponential fitnesses, his model is nicely behaved, with the polygenic mean evolving at a constant rate

$$\Delta\mu = \sigma_A^2\,\beta \tag{25.9a}$$

while the relationships between the major-locus genotypic fitnesses remain constant, with

$$\frac{\overline{W}_i}{\overline{W}_0} = e^{\beta\alpha_i} \quad \text{for } i = 1, 2 \tag{25.9b}$$

The resulting relative fitnesses of the three major locus genotypes become

$$W_{bb} = 1 \quad W_{Bb} = (1 - s_1)\,e^{\beta\alpha_1} \quad W_{BB} = (1 - s_2)\,e^{\beta\alpha_2} \tag{25.9c}$$

Because these are constants, the machinery of Chapter 5 quickly informs us as to the fate of the major gene. Selection maintains both alleles as a stable polymorphism when $W_{Bb} > W_{BB}, W_{bb}$. There is an unstable internal equilibrium when $W_{Bb} < W_{BB}, W_{bb}$, with *B* being lost if sufficiently rare (frequency below the equilibrium value), and otherwise being fixed. Finally, if $W_{bb} < W_{Bb} \leq W_{BB}$ or $W_{bb} \leq W_{Bb} < W_{BB}$, then the major allele is fixed (under our assumption, in this chapter, of no drift).

The simple fate of the major allele is not a complete analysis of the full dynamics of this system, as even if the allele is fixed, its contribution could be far outstripped by the polygenic response. Lande examined this possibility by letting α represent the difference between the two major-locus homozygotes and assuming that B is initially rare. He then compared the expected amount of time for a B allele (starting at frequency $p_0 \ll 0.5$) to increase to the point where the response from the locus is half its potential response, $\alpha/2$. This is accomplished by using Equation 25.1e to find the critical frequency, and then applying the appropriate version of Equation 5.3 to obtain the required time for this amount of allele-frequency change. If the polygenic response over this amount of time exceeds $\alpha/2$, the response is primarily polygenic, even if B is fixed. For weak exponential selection (Equation 25.9c), the resulting initial frequencies above which the major gene exceeds the polygenic response are approximately

$$p_0 > \begin{cases} 2/b & \text{recessive} \\ e^{-b/4} & \text{additive} \\ e^{-b/2} & \text{dominant} \end{cases} \qquad \text{where} \quad b = \left(1 - \frac{s}{\beta\,\alpha}\right)\frac{\alpha^2}{\sigma_A^2} \tag{25.9d}$$

where, before, β is the strength of directional selection on the trait.

Much of this discussion is framed by the assumption that genes that have a large effect on a trait generally have deleterious effects in natural populations. However, as Orr and Coyne (1992) pointed out, if genes with a small effect on the character also have similarly small (and negative) effects on fitness, their advantage over a major locus largely (or completely) disappears. However, if the potential pool of alleles of small effects is large, it will become enriched by natural selection for those alleles with nearly neutral pleiotropic fitness effects.

Gomulkiewicz et al. (2010) reached a somewhat different conclusion from Lande's suggestion that major genes generally require very strong selection on a trait in order to account for the majority of response. They focused on the time required for a population that has invaded a harsh new environment (equivalent to a major environmental shift) to evolve persistence. This is a different scenario from that envisioned by Lande, who focused on the genetic composition once an adaptation had occurred (the equilibrium value following a shift). One of the major points from Gomulkiewicz can be seen in Figure 25.6B. Here, a major allele can substantially increase in frequency, but at some point it is overtaken by polygenic response, and ultimately lost. Gomulkiewicz noted that such a situation may be critical for a population to evolve persistence in a harsh new environment. Even though the major gene is eventually lost, the population runs a risk of extinction if this allele is not initially present. Thus, by shifting the focus toward the dynamics during adaptation, rather than on the equilibrium values, a potentially different view of the relative importance of major and minor genes arises. Important roles of major genes may therefore be overlooked if one only focuses on the alleles that are ultimately fixed.

Finally, while the importance of deleterious pleiotropic fitness effects has been framed as a constraining force for major alleles, Otto (2004) showed that the presence of pleiotropy may impact small-effect alleles as well, which returns us to the point made earlier by Orr and Coyne (1992).

Example 25.5. Lande's method of analysis can be used with other fitness functions. For example, suppose the trait of interest is subjected to truncation selection, with only individuals above the threshold value, T, being allowed to reproduce. In this case

$$W(z) = \begin{cases} 1 & \text{for } z \geq T \\ 0 & \text{for } z < T \end{cases}$$

The marginal fitnesses become

$$\overline{W}_i = \int_T^\infty \varphi(z, \mu + \alpha_i, \sigma^2)\, dz = \Pr\left[U > \left(T^* - \frac{\alpha_i}{\sigma}\right)\right]$$

where $T^* = (T - \mu)/\sigma$ is the current (standardized) truncation value given μ, and U is a unit normal random variable. We usually analyze truncation selection in terms of the fraction, q, of individuals that are allowed to reproduce, rather than the threshold value, T, especially because T changes as the population mean increases. In our case, these are connected by

$$\overline{W} = q = (1-p)^2\Pr(U > T^*) + 2p(1-p)(1-s_1)\Pr\left[U > \left(T^* - \frac{\alpha_1}{\sigma}\right)\right] + p^2(1-s_2)\Pr\left[U > \left(T^* - \frac{\alpha_2}{\sigma}\right)\right]$$

This expresion for mean fitness is simply the probability of being above the threshold given a particular major-locus genotype, weighted by the frequencies of these genotypes. For a particular q value and the current μ and p values, one can numerically solve the these equation for T^*. Likewise, from LW Equation 2.14, the mean of genotype i following selection becomes

$$\mu_{s_i} = \mu_i + \sigma\,\frac{\varphi(T, \mu + \alpha_i, \sigma^2)}{\Pr[U > (T^* - \alpha_i/\sigma)]}$$

implying a directional selection differential of

$$S_i = \mu_{s_i} - \mu_i = \sigma\,\frac{\varphi(T, \mu + \alpha_i, \sigma^2)}{\Pr[U > (T^* - \alpha_i/\sigma)]}$$

To proceed with the analysis of the model dynamics, for a given (p, μ) vector, we first find T^* to obtain the specified strength of truncation selection, q, and then compute the $\overline{W}_i$ for Equation 25.6 and S_i for Equation 25.8d to update the p and μ values. Example 25.10 uses these results to obtain the equilibrium frequency of a major gene that is lethal as a homozygote, but improves the trait as a heterozygote.

MAJOR GENES VERSUS POLYGENIC RESPONSE: DATA

A long-running debate in evolutionary biology, dating back to the rediscovery of Mendel, is whether the majority of adaptations are due to a few alleles with large effects or to the accumulation of small changes over a large number of loci. Before the modern evolutionary synthesis, geneticists (the Mendelians) felt that macromutations drove evolution, while supporters of Darwin (the Biometricians) felt that evolution was driven by selection acting on numerous factors of small effect. These differing views (and more importantly, their vocal supporters and opponents) delayed the merging of modern genetics with Darwin's theory of evolution. Fisher's (1918) paper, founding quantitative genetics, was a watershed event in helping to fuse these two schools (see Provine 1971 for a historical overview of the Mendelian-Biometrician debate). This same debate, in slightly different forms, resurfaced in the 1940s with Goldschmidt's (1940) idea of **hopeful monsters** (single mutations with a large effect driving major evolutionary changes) and also in the late 1970s to early 1980s with the debate surrounding **punctuated equilibrium** (the causes of long periods of evolutionary stasis, punctuated by rapid change, in the fossil record; see Eldredge and Gould 1972; Charlesworth et al. 1982).

Major Genes Appear to Be Important in Response to Anthropogenically Induced Selection

One of Lande's (1983) conclusions was that sufficiently strong selection is required for a major-gene response when polygenic variation is available. One situation where strong selection is often assumed to occur involves the response of wild populations to **anthropogenically induced selection**, namely, a major (and sudden) environmental change induced by human activity. This could be in the form of toxins (pesticides, herbicides, pollutants) or other side effects of human activity, such as industrial melanism (Lees 1981).

In the pesticide and herbicide literature, a commonly expressed theme echoing Lande's theoretical predictions is that relatively weak selection (as might be expected to occur in laboratory settings, where at least a small percentage of the population is allowed to reproduce) leads to a polygenic response, whereas very strong selection (e.g., in a newly sprayed field where nearly everything is killed) leads to major-gene resistance (Greaves et al 1977; Clarke and McKenzie 1987; Macnair 1991; McKenzie et al. 1992; McKenzie 2000). However, a survey by Groeters and Tabashnik (2000) found that the strength of selection on insecticide resistance varies greatly in the field and overlaps the intensities used in laboratory experiments. Further, major-gene responses are not uncommon in the laboratory.

A relevant example involves resistance to Bt toxin (*Bacillus thuringiensis* Cry1Ac toxin), an organic insecticide widely used in both sprayed fields and as the foundation for some transgenic crops (e.g., Bt corn). Bt resistance is often due to recessive mutations in the same gene independently arising over different pest species. For example, independent mutations in a 12-cadherin-domain protein gene confer resistance in laboratory-selected strains from three very distantly related moth species, as well as in field populations of a fourth species (Zhang et al. 2012). Likewise, Baxter et al. (2011) found that independent recessive mutations in a different gene (the membrane transporter *ABCC2*) confers resistance in two very distant moth species.

In contrast to these finding of mainly recessive Bt-resistance alleles, Zhang et al. detected nonrecessive cadherin alleles in a Chinese population of cotton bollworms (*Helicoverpa armigera*). This observation has important biocontrol implications, because the strategy used to retard the evolution of Bt resistance is to plant refuge rows of non-Bt crops (Gould 1988; Tabashnik et al. 2008). Under the assumption that resistance is recessive, crosses of resistant homozygotes and susceptibles are expected to result in susceptible heterozygous offspring, which are killed when their larvae feed on Bt crops. This strategy fails if the response is either polygenic or due to nonrecessive major genes, as resistant alleles can then spread.

If the strength of selection is not the key factor explaining the difference between field (usually major genes) and lab (mainly polygenic) responses, then what is? One likely explanation is simply population size. Major alleles, especially those involved in detoxification, likely have deleterious side effects in toxin-free environments, and are thus expected to occur at very low frequencies in a population. As rare alleles are mainly present as heterozygotes (with frequency $\sim 2p$), the probability that a random sample of n individuals chosen to create a laboratory stock for selection does not contain the allele is $(1 - 2p)^n$. For $n = 1000$ and $p = 0.001$, this is 0.14. Using a more realistic founder stock of 100, this increases to 0.82. Even if such a mutation is present, it will likely be in just a few copies and can easily be lost early in an experiment by drift, even with strong selection. These arguments illustrate that the interaction of drift and mutation is often critical in determining the nature of the selection response, especially in smaller laboratory populations (Chapter 26).

Finally, the finding that major genes appear to be commonly involved in Bt resistance in laboratory populations might be explained by the fact that many of these mutations appear to be knockouts. One might expect a rather large mutational target size for loss-of-function mutations, meaning they might appear at a modest rate in laboratory populations.

What is the Genetic Architecture of Response in Long-term Selection Experiments?

With the advent of dense molecular markers and subsequent whole-genome sequencing, we have a new set of tools to examine the genetic makeup of long-term response (Stapley et al. 2010; Burke 2012). One approach is to use QTL mapping with large sample sizes to both detect alleles of small effect and avoid the overestimation of effects when power is low (the Beavis effect; LW Chapter 15). While a large number of studies crossing divergent lines have been performed (LW Chapter 15), we restrict attention to those crosses between lines generated by persistent selection in opposite directions. For these crosses, the general picture emerging is that much, if not most, of the selection response is often due to QTLs of small effect.

Perhaps the most careful studies involve the Illinois long-term selection experiment, which (as we will detail shortly) has been going on for over a century (Figure 25.9). The F_1 progeny crosses between 70th-generation high vs. low oil lines and high vs. low oil protein lines were randomly mated for 7–10 generations before QTL mapping (the advanced intercross, or AIC, design; LW Chapter 15). Such a design allows recombination to randomize even closely linked QTLs (the effect is a 7- to 10-fold map expansion relative to the F_2). Over 50 QTLs were detected, each with small, and additive, effects (Laurie et al. 2004; Clark et al 2006; Dudley et al. 2007). A similar finding is seen in chicken lines divergently selected for 50 generations (resulting in a nine-fold difference in body weight), which revealed mainly small-effect QTLs underlying the response (Jacobsson et al. 2005; Wahlberg et al. 2008).

Results from mouse lines have been more mixed. The majority of the roughly 40 QTLs detected in a cross from a 27-generation line selected for weight gain with a random control had effects of 1–3%, although a few had effects of around 5% (Allan et al. 2005). Moody et al. (1999) found QTLs with effects of ~3–4% in an analysis of lines divergently selected for energy balance for 16 generations. In contrast, Hovat et al. (2000) found that just four QTLs could account for most of the response in obesity in lines subjected to 53 generations of divergent selection. One caveat about these mouse results is that typically F_2, rather than AIC, designs were used. This can result in significant overestimates of QTL effects due to linkage of multiple QTLs. Typically, such large QTL peaks fractionate upon finer mapping (reviewed by Flint and Mackay 2009; Mackay et al. 2009).

A second class of approaches is to search for signatures of selection in the genomes of individuals sampled near the end of a long-term selection experiment, either from the fixation of alternate alleles in divergent lines (essentially a localized F_{ST} measure; e.g., Johansson et al. 2010) or by more classic hard-sweep tests (e.g., Chan et al. 2012). In addition to estimating the number of genomic regions under apparent selection, the machinery of Chapters 8 and 9 could also be used to estimate selection coefficients (and hence effect sizes). However, as discussed in Chapter 8, these approaches are strongly biased toward detecting alleles of large effect, and hence involve hard sweeps as opposed to either soft sweeps (existing variation) or polygenic sweeps (small changes at a number of loci). As mentioned previously, this bias toward detecting major genes is somehat countered by a bias *against* them by the founding of most experimental populations. The expectation is that alleles of large effect are at low frequencies in natural populations, and thus are unlikely to be routinely captured in the small to modest population samples used to found most laboratory populations.

What is clear from the existing data is that massive responses in long-term experiments can be entirely due to genes of small effect (Teotónio et al. 2009; Burke et al. 2010; Johansson et al. 2010; Parts et al. 2011; Turner et al. 2011; Zhou et al. 2011; Chan et al. 2012; Beissinger et al. 2014). What is unclear is the extent to which these results from long-term laboratory experiments with strong and constant artificial selection on a single trait translate to natural or domesticated populations undergoing mild (and likely constantly shifting) selection.

Finally, the results from examining adaptations (often inferred from species differences) in natural populations are also mixed. Hilu (1983) and Gottlieb (1984, 1985) suggested that major genes have played very important roles in species differences between plants (many of which are, presumably, adaptive), but Coyne and Lande (1985) disputed this view. A literature review by Orr and Coyne (1992) found that support for the polygenic model (i.e., that most adaptations are due to many genes of small effect) is also inconclusive. There may also be a publication bias as many of these example are color traits, which are often controlled by just a few genes and lead to differences that are visually obvious, and thus more easily detected (and therefore studied). Clearly, this is an area of active ongoing research, and we will return to a different aspect of the question, **adaptive walks** (the successive fixation of multiple mutations during adaptation), in Chapter 27.

AN OVERVIEW OF LONG-TERM SELECTION EXPERIMENTS

The above theory suggests that populations under selection should show a reasonably

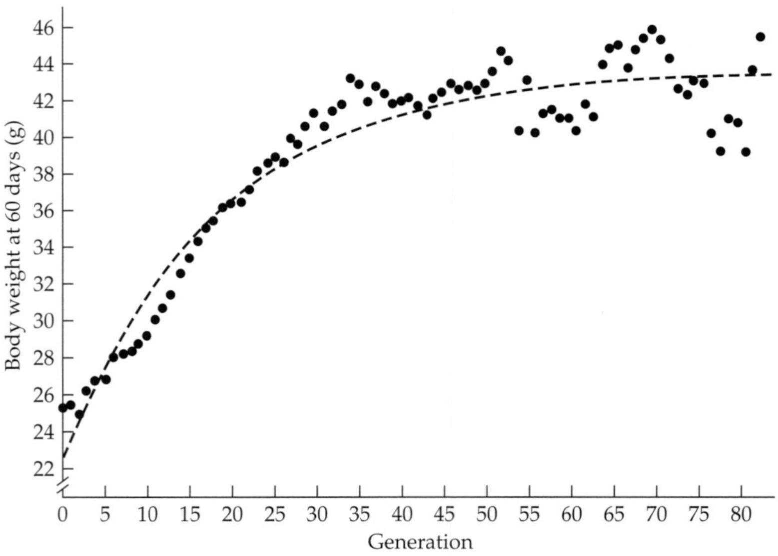

Figure 25.7 Bünger and Herrendörfer's (1994) fit of an exponential regression to the long-term selection experiment of Goodale on mouse weight at 60 days (Goodale 1938; Wilson et al. 1971). The estimated selection limit was 43.5 grams (for a total response of 21.3 grams), with a half-life of 12 generations. The plotted curve regresses the cumulative response as a function of generation number. When the regression is instead plotted as a function of the cumulative selection differential (as opposed to generations), the estimated total response was 17.5 grams.

smooth response (albeit often with considerable sampling noise; Chapter 18), which is initially linear, but eventually (in the absence of new mutations) asymptotes to a selection limit as base-population genetic variance becomes exhausted. Unfortunately, this simple picture is very often wrong. Selection response can be rather erratic, showing periods of acceleration even after many generations of selection, and limits often occur in spite of significant additive variance in the character under artificial selection. Before reviewing the experimental data, a few remarks on estimating the actual limit and duration of response are in order.

Estimating Selection Limits and Half-lives

Because the selection limit is approached asymptotically, the typical measure of duration is the **half-life** of response—the time for half the response to occur. As was the case for short-term response (Chapters 18 and 19), this parameter is generally estimated by curve-fitting, and in doing so accommodating the inherent sampling noise in selection response data (Chapter 18). Given that the response curve is nonlinear, a number of authors have used quadratic regressions, taking the maximum of the regression as the limit (James 1965; Eisen 1972; Rutledge et al. 1973). Grassini et al. (2013) used a piecewise approach, breaking the response into either two distinct linear regressions or a quadratic regression segment followed by a constant value (the latter is used to represent the plateau).

A more natural approach is to use **exponential regressions**, where the cumulative response at generation t is given by $R_c(t) = a + ce^{-bt}$ (more formally this is a **negative-exponential regression**). A number of variant expressions based on this approach appear in the literature (James 1965; Frahm and Kojima 1966; Harris 1982; Herrendörfer and Bünger 1988; Bünger and Herrendörfer 1994; Árnason 2001). The motivation for using exponential regressions is two-fold. First, these curves naturally generate an asymptotically maximum value (a), while quadratic regressions do not. Second, if the fraction of additive variance that is retained is just a constant amount, $(1-\alpha)$, of the previous generation (as would occur under drift with the infinitesimal model; e.g., Equation 26.15a), this leads to a cumulative response of the form

$$R_c(t) = \beta\,(1 - [1-\alpha]^t) \simeq \beta\,\left(1 - e^{-t\,\alpha}\right) \tag{25.10a}$$

This motivated Herrendörfer and Bünger (1988) and Bünger and Herrendörfer (1994) to suggest using an exponential regression of the form

$$R_c(t) = a\big[1 - \exp\left(-bt/a\right)\big] \tag{25.10b}$$

where a is the selection limit and b is the maximum rate of response (i.e., the response in the first generation). Figure 25.7 shows an application of this method, while optimal experimental design issues were examined by Rudolph and Herrendörfer (1995).

An interesting application of such estimated limits was presented by Árnason (2001). A fitted exponential regression for the racing speed of standardbred trotters in Sweden suggested that the limiting trotting time is around 68 sec/km. In the 1950s, the fastest times were just under 80 sec/km, reaching around 73 sec/km in the mid-1990s, which is slightly more than half of the expected total response of a limiting decrease of 12 sec/km (80–68, starting from the 1950 benchmark).

Using the fitted curve given by Equation 25.10b, the resulting half-life of the cumulative response becomes

$$t_{0.5} = -a \cdot \ln(0.5)/b \tag{25.10c}$$

Assuming constant selection, the tangent of the response curve, $R_c(t)$, at a particular generation provides an estimate of the rate of response in that generation (Frahm and Kojima 1966; Herrendörfer and Bünger 1988). Assuming that $R_c(t)$ is well approximated by an exponential regression, this is given by the derivative of Equation 25.10b, evaluated at the generation of interest, yielding

$$\frac{\partial R_c(t)}{\partial t} = b \cdot \exp(-b\,t/a) \tag{25.10d}$$

As was the case for linear regressions of short-term response, one decision is whether to regress cumulative response on number of generations or on the cumulative selection differential (Chapter 18). Under the breeder's equation, the short-term response is linear with respect to the cumulative selection differential, $R_c(t) = h^2 S_c(t)$ (see Chapter 18). For long-term response, if the expectation is a constant decline in additive variation each generation, then regression on number of generations is more logical, *provided* that the assumptions of a constant selection differential and a constant decline in variance are appropriate.

A second, more subtle issue, which also arose in Chapter 18, is the nature of the residual variance structure. One standard approach for curve-fitting is ordinary least-squares (OLS), which involves finding the parameter values for Equation 25.10b that minimize the sum of squared residuals, $\sum e_i^2 = \mathbf{e}^T\mathbf{e}$, where the ith residual, $e_i = R_c(t_i) - \widehat{R}_c(t_i)$, is the difference between the ith observed and predicted values for a given candidate regression model. OLS regression assumes homoscedasticty, with $\sigma^2(e_i) = \sigma^2_e$ for all i, and that residuals are uncorrelated, $\sigma(e_i, e_j) = 0$ for $i \neq j$. As with short-term response, the presence of drift compromises both of these assumptions. As was done in Chapter 18 to accommodate this concern, curves should be fitted using generalized least-squares, where parameters are chosen to minimize the quadratic product $\mathbf{e}^T\mathbf{V}^{-1}\mathbf{e}$, where $\mathbf{V}$ is the variance-covariance matrix for the residuals (Equation 18.15c).

Other candidate response curves have also been proposed, which also attempt to capture the asymptotic approach to a limit expected for an idealized long-term response. Wiser et al. (2013) suggested a **hyperbolic function**, with

$$R_c(t) = \frac{a\,t}{t+b} \tag{25.11a}$$

Here, the cumulative response approaches a limiting value of a, with a half-life of $t_{0.5} = b$.

Table 25.3 Estimates of the selection limit and half-life based on 22 generations of selection for increased 12-day litter weight in mice. *Selection limit* refers to response in grams as a deviation from the control, and half-life is given in generations. The quadratic and exponential models explain the same amount of variation ($r^2 = 0.81$ for both models) and cannot be discriminated on this basis. (Data for line W_3 from Eisen 1972.)

Estimate	Model	
Selection limit	Quadratic	5.79 ± 0.84
	Exponential	8.19 ± 0.29
Half-life	Quadratic	8.58
	Exponential	12.48

A much more intriguing function for very long-term selection data is a **power curve**, where

$$R_c(t) = b\,t^a \quad \text{for } a < 1 \tag{25.11b}$$

which has the feature that while the rate of response decelerates over time ($\partial^2\mu(t)/\partial t^2 < 0$ for $a < 1$), there is no upper limit. The power curve provided a better fit than the hyperbolic for data on the response over 50,000 generations of selection for fitness in *Escherichia coli* (Wiser et al. 2013), suggesting that slowly diminishing returns, rather than an approach to a true selection limit, is a better description of their data. A concern with all of these models is that the selection limit is *extrapolated* from the data. As Table 25.3 shows, different models can yields essentially the same fit of the data but very different estimates of the limit and half-life.

Some final cautionary notes are in order. First, scale effects (LW Chapter 11) can be important. Many continuously distributed characters have zero as a lower limit, and hence on a linear scale they always have a lower limit. This is not true on a log scale. Similarly, if we model a binary trait as resulting from the transformation of an underlying continuous variable (the liability), we should work with response as measured on the liability scale (see Chapter 14 and LW Chapters 11 and 25).

Finally, the entire issue of selection limits due to the exhaustion of additive-genetic variation is complicated by mutation. Most "long-term" experiments are long-term only from the viewpoint of the experimenter, rarely spanning more than 50 generations. As is discussed in Chapters 26 and 28, over longer time scales, mutational input becomes very important and observed limits can be artifacts of the relatively short time scales being used.

General Features of Long-term Selection Experiments

As Figure 25.8 illustrates, long-term selection experiments display a wide range of behaviors. Fortunately, a few generalizations do emerge:

1. *Selection routinely results in mean phenotypes that are far outside the range seen in the base population.* At the selection limit, the mean phenotype is usually many standard deviations away from the initial mean.
2. *Response can be very uneven.* Bursts of accelerated response after many generations of selection can be seen seen, and the additive-genetic and phenotypic variances can increase during part of the response.
3. *Reproductive fitness usually declines as selection proceeds.*
4. *Most laboratory populations approach a selection limit.* As discussed in Chapter 26, an apparent selection limit may simply be an artifact of the short time scale (and hence insufficient time for significant mutational input) and small population sizes of most experiments. However, this is not always the case (Figure 25.9).

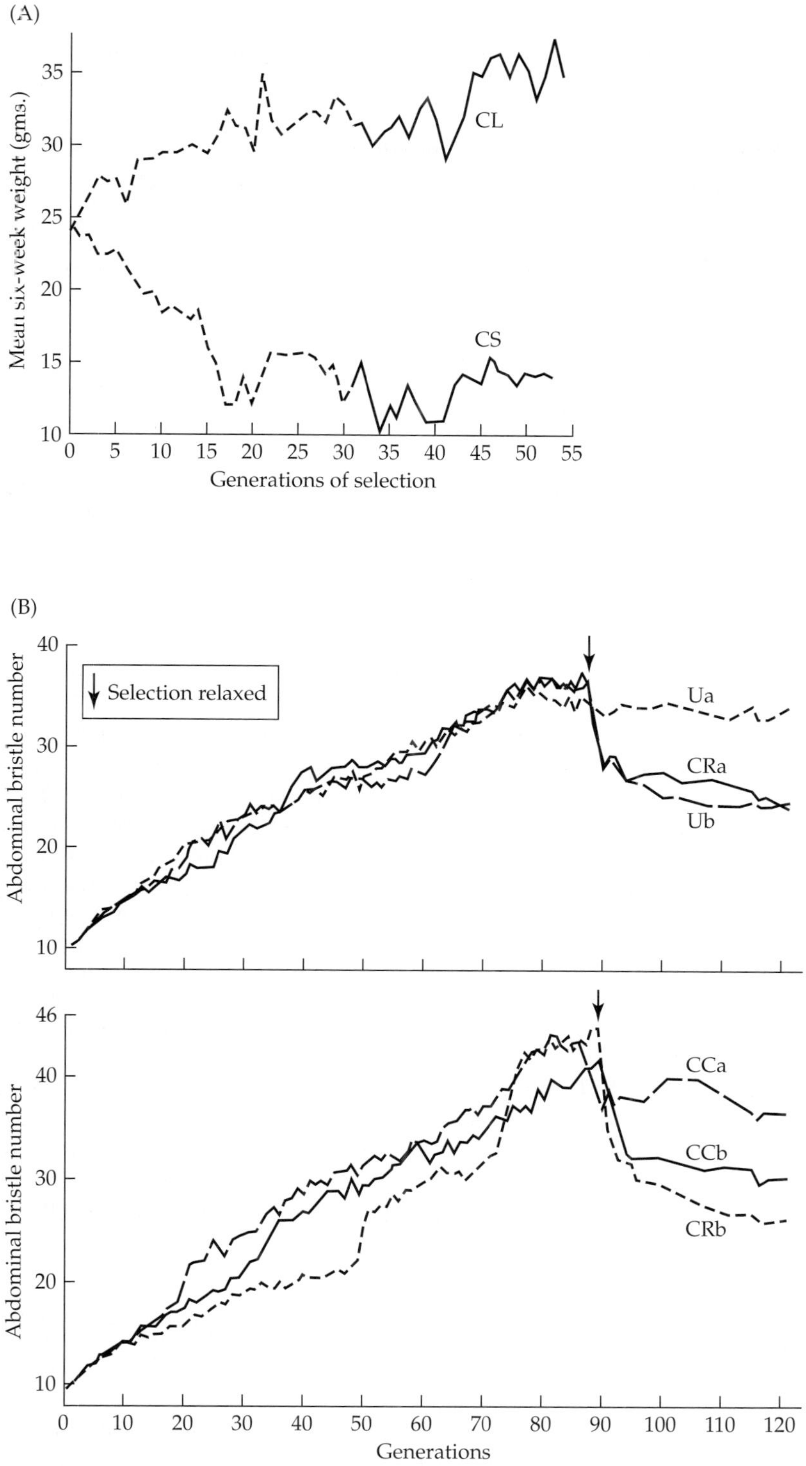

Figure 25.8 A few of the nonstandard behaviors observed in long-term selection experiments. **A:** Delayed accelerated response during selection for increased six-week body weight in mice. An apparent limit of 31 grams had been reached in the up-selected line (CL) by generation 15. A second burst of response occured around generation 43, with the mean weight increasing to around 35 grams (Roberts 1966). **B:** Selection for increased abdominal bristle number in *Drosophila*. At generation 90, selection was relaxed and most lines showed a considerable (but not complete) erosion of response. The presence of segregating lethals accounts for some of this erosion. Also note the bursts of response for line CRb (the short-dashed curve in the lower panel) around generations 50 and 75 (Yoo 1980a).

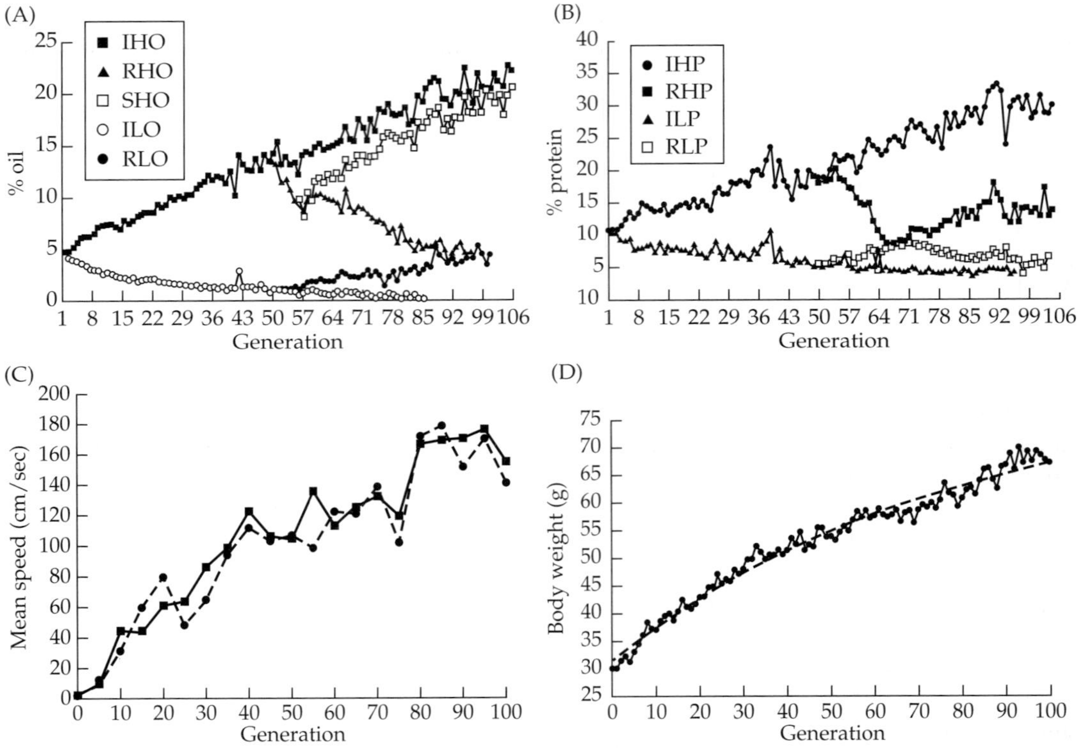

Figure 25.9 Example of long-term selection experiments showing no apparent selection limits. The top two panels are from the Illinois long-term selection experiment on oil and protein content in maize. **A:** Response over 106 generations of selection for increased or decreased oil percentage. Lines IHO and ILO (Illinois high and low oil) are up- and down-selected, while lines RHO and RIL are lines of IHO and ILO subjected to reversed selection around generation 50. Line SHO (switchover high oil) is an up-selected line using RHO. The responses in RHO, RLO, and SHO indicate significant additive variance present in the population when these new lines were formed. (After Dudley 2007.) **B:** Response over 106 generations of selection for changes in the percentage of protein. Lines IHP and ILP were up- and down-selected, while lines RHP and RLP are the result of reverse selection starting around generation 50. Again, the responses of RHP and RLP indicates significant additive variance. (After Dudley 2007.) **C:** One hundred generations of response for increased flight speed in *Drosophila melanogaster*. Two replicate lines showed very similar response. (After Weber 1996.) **D:** Response in the Dummerstorf long-term mouse lines, which were subjected to continuous selection for 42-day weight. Conducted over 160 generations, this is the longest continuous selection experiment in mammals (Renne et al. 2013). (After Bünger et al. 2001.)

5. *Considerable additive variance in the trait under artificial selection often exists at an apparent selection limit* (Table 25.4).

It is important to recognize that long-term selection experiments are a biased sample of organisms and traits. Controlled selection experiments in multicelluar organisms exceeding 20 generations are largely restricted to *Drosophila*, *Tribolium*, mice, and maize. Whether the genetic architectures of these organisms (and the easy-to-score traits used in experiments) are representative of typical characters in natural populations is unclear, although there is no serious reason to suspect that they are not. Another caveat in extrapolating from these model experimental systems to natural and domesticated populations is that the strength of continuous selection on a single character is likely much higher in artificial selection experiments. In addition, under natural selection, and in most breeding programs, selection operates on a suite of characters, so that for a given overall strength of selection, the fraction imparted on a particular trait likely decreases as the number of traits under selection increases. Further, the amount of natural selection on any particular character likely fluctuates over time. Conversely, laboratory experiments on artificial selection generally focus on a single character and involve strong and consistent selection, usually in highly controlled environments.

The Nature of Selection Limits

What is the nature of selection limits observed in artificial selection experiments? In particular, is there any genetic variation present at an apparent limit, and if so, is any of it additive? Correlations between relatives can be used to characterize the nature of residual variation. One caveat to this approach is that selection can generate strong gametic-phase disequilibrium, complicating standard methods for estimating components of variance (Robertson 1977b). However, changing selection schemes and inbreeding offer two simple approaches for characterizing the nature of any residual genetic variation. If additive variance is present, the line should respond to **reversed selection** (subjecting the line to selection in the opposite direction, e.g., Figures 25.9A and 25.9B). A decay in the mean of a plateaued line after selection is relaxed also indicates the possibility of additive variance in the selected trait (Figure 25.8B), although epistasis or maternal effects can also result in slippage of the mean (see Chapter 15). If nonadditive variance is present, the line can show inbreeding depression, with the mean changing as the line is inbred. The *absence* of inbreeding depression, however, does not imply a lack of nonadditive genetic variation, as directional dominance, and not simply the presence of σ_D^2, is required for inbreeding depression (LW Chapter 10).

Table 25.4 highlights some of the causes of selection limits observed in long-term artificial selection experiments. This is by no means a comprehensive listing. The general conclusion is that significant additive-genetic variance in the selected character is often present at an apparent limit. This is rather surprising given that most experiments have such small effective population sizes that drift is expected to remove most variation (Chapters 3 and 26).

One celebrated apparent selection limit is racing performance in thoroughbred horses, where the winning times in classic English races have not fallen substantially over the past 50 years (Gaffney and Cunningham 1998; Hill 1998). To be fair, this is a nonstandard trait, namely, the best single performance within a set of individuals and not their mean performance. However, if mean times have responded to selection, we also expect these outlier times (the best of a dozen or so) to fall as well. One possibility is that while means have fallen, so too has the genetic variance, potentially keeping low outliers (i.e., faster speeds) at a roughly constant value. Gaffney and Cunningham (1998) found ample additive variance in a strong correlate of performance, handicap weight, which Hill predicted should result in a mean improvement of roughly 0.1% per year. A more recent analysis by Wilson and Rambaut (2008) on the genetics of lifetime earning of racing horses (again correlated with speed) found that while roughly 90% of the variation was environmental (diet, trainer, etc.), there was a significant heritable component that should respond to selection.

Some long-term experiments have yet to reach their limit (Figure 25.9). The most iconic of these is the **Illinois long-term corn selection experiment**, started by the agricultural chemist Cyril Hopkins in 1896 and currently ongoing (Hopkins 1899; Smith 1908). The results after 76, 90, 100, and 106 generations of selection were summarized by Dudley (1977), Dudley and Lambert (1992, 2004), Moose et al. (2004), and Dudley (2007). As shown in Figure 25.9A, a fairly constant response for increased oil content is seen over 90 generations with no apparent selection limit, with a total cumulative response of $22\sigma_A$. Selection for low oil was stopped after 87 generations due to the difficulty of selecting among individuals with close to 0% oil. While a limit appears to have been reached, this is due to a scale effect, as oil percentage is bounded below by zero. Selection for protein shows a similar pattern to that for oil (Figure 25.9B), with the up-selection line (IHO) currently showing a cumulative increase of $26\sigma_A$ after generation 90 with no apparent limit, and the down-selected lines showing an apparent plateau, again likely due to scale effects. The interesting, and far-reaching, intellectual legacy of this experiment was nicely summarized by Goldman (2004).

Table 25.4 Nature of the selection limit observed in various laboratory selection experiments.

Reduced thorax length in *D. melanogaster* F. W. Robertson 1955	Apparent exhaustion of all genetic variation: no further change under inbreeding, no response to reversed selection.
Increased body weight in mice Falconer and King 1953 Roberts 1966	Exhaustion of σ_A^2: no response to reversed selection.
Egg production in *D. melanogaster* Brown and Bell 1961, 1980	Exhaustion of σ_A^2: significant nonadditive genetic variance present at selection limit. Lethals and sterility factors negligible.
Wing length in *D. melanogaster* Reeve and Robertson 1953	Significant σ_A^2 at limit: complicated interaction due to segregating lethals and an overdominant gene influencing wing length.
Reduced body weight in mice Falconer 1955 Roberts 1966	Opposing natural selection: response to reversed selection, mean slippage upon relaxation of selection. Likely due to reduction in viability.
Abdominal bristles in *D. melanogaster* Clayton and Robertson 1957 Yoo 1980b	Segregating lethals: major gene increases bristle number as a heterozygote, lethal as a homozygote.
Pupal weight in *Tribolium castaneum* Enfield 1980	Opposing natural selection: significant σ_A^2 at limit, large decay in response with relaxed selection. Sterility reduced and fertility improved in relaxed lines.
Shank length in chickens Lerner and Dempster 1951	Opposing natural selection: shank length negatively correlated with hatchability.
Litter weight in mice Eisen 1972	Negative genetic correlation between direct and maternal effects.
Increased body weight in mice Wilson et al. 1971	Negative correlation between weight and litter size.
Increased litter size in mice Falconer 1971	Apparent limit due to slow changes in in the frequency of dominant alleles.

A few other classical experiments continue to show selection response without an apparent limit. The **Dummerstorf long-term** selection lines (started in Dummerstorf, Germany) are the mammalian counterpart to the Illinois maize lines. Selecting for mouse weight at 42 days, this experiment has run for over 160 generations, making it the longest continuous selection experiment in mammals (Bünger et al. 2001). As shown in Figure 25.9D, it has not yet reached a limit. Another long-term mouse experiment is the work of Holt et al. (2005), who selected for litter size for over 122 generations, but with the selected population experiencing at least two waves of migration of new genetic material to break the selection limit. **Weber's long-term selection experiment for flight speed** in *Drosophila* involves over 600 generations of continuous selection (Weber 1996, 2004). Figure 25.9C shows no apparent limit after 100 generations, while by generation 300, the response was slowing down, but still continuing (Figure 26.4). Unpublished results by Weber (pers. comm.) indicate that the line was still responding after 640 generations of selection. The champion of continuous long-term studies is **Lenski's long-term evolution experiment** (**LTEE**) using the bacterium *Escherichia coli*, which continues to respond to selection for increased fitness after over 50,000 generations (Lenski et al. 1991; Lenski and Travisano 1994; Barrick et al. 2009; Wiser et al. 2013). The roles of both finite population size (larger populations have larger limits) and new mutation input are critical to the ongoing selection response in any very long-term experiments, and we examine this further in Chapter 26.

Fortunately, limits appear to be rare in many selection programs for important commercial traits in domesticated animals (Fredeen 1984; Hunton 1984; Kennedy 1984). This is perhaps not surprising given that breeders are constantly shifting the suite of characters under artificial selection, as well as searching out new sources of genetic material. A more dubious possibility is that, while not currently at a limit, breeders are quickly approaching one (e.g., Grassini et al. 2013).

Several strategies can be used to break an apparent selection limit and allow for further response. Relaxing selection for several generations followed by directional selection can break a limit caused by strong gametic-phase disequilibrium between segregating loci. Likewise, if the limit results from a balance between natural and artificial selection, increasing the amount of artificial selection can result in further response. If the limit is caused by a lack of genetic variation, crossing different lines can introduce additional variation. Over longer time scales, a limit can be exceeded simply by waiting for new mutational input, either to increase additive variance or by generating alleles that improve the artificially selected trait but with less deleterious pleiotropic effects on fitness (Chapters 26 and 28). A final approach is selection in a new environment, which can often exploit genetic variation that is not usable in the current environment. For example, Abplanalp (1962) was able to improve a chicken line, which was apparently at a plateau for increased egg number, by selecting in a different environment (females being subjected to one day without food every two weeks).

A final comment on the nature of limits, which is discussed in much greater detail in Volume 3, is essentially an extension of the concept of a lack of further selection response because of a trade-off between natural and artificial selection. In natural populations, where the target of selection is some multivariate phenotype (say an index of trait values), one can easily have very little or no additive variation in the index (and hence, selection response at a limit), yet still have ample additive variation in each of the component traits. If the focus is on a single character in nature or even some smaller subset of the traits making up the index upon which natural selection is operating, one could easily observe ample selection on the trait and the presence of ample additive variation for that trait, yet no selection response (Blows and Hoffmann 2005; Blows and Walsh 2009; Walsh and Blows 2009; Chapter 20). As stressed in Volume 3, treating multivariate selection as a series of univariate responses is extremely misleading.

INCREASES IN VARIANCES AND ACCELERATED RESPONSES

Contrary to the expectations of idealized long-term response, phenotypic and additive genetic variance can *increase* during the course of selection, often resulting in bursts of response (Figure 25.8B). As we detail here, a variety of different conditions can lead to such a burst in response, emphasizing just how unpredictable long-term responses can be.

Rare Alleles

One obvious source for increases in variance is the increase of favorable rare alleles under selection. For an additive locus, σ_A^2 is maximized at $p = 1/2$ (LW Figure 4.8). Thus, additive loci with favorable alleles below 50% show an increase in additive variance as the allele frequency approaches one-half, after which σ_A^2 starts to decline to zero as the alleles becomes fixed. If the allele is rare and also of large effect, the result can be an increase in response many generations after the start of selection (Figure 25.10). The magnitude of this effect depends on both the initial frequency of the allele and its effect size. Alleles of large effect are subjected to stronger selection, and hence show more rapid increases in allele frequencies and larger effects on response. As Figure 25.11 illustrates, if there is a distribution of allele frequencies (and allelic-effect sizes) at the underlying loci, then different alleles are increasing at different rates, which can result in a very erratic pattern of response. This is especially true when the allele frequencies follow the Watterson distribution (Chapter 2), the distribution of allele

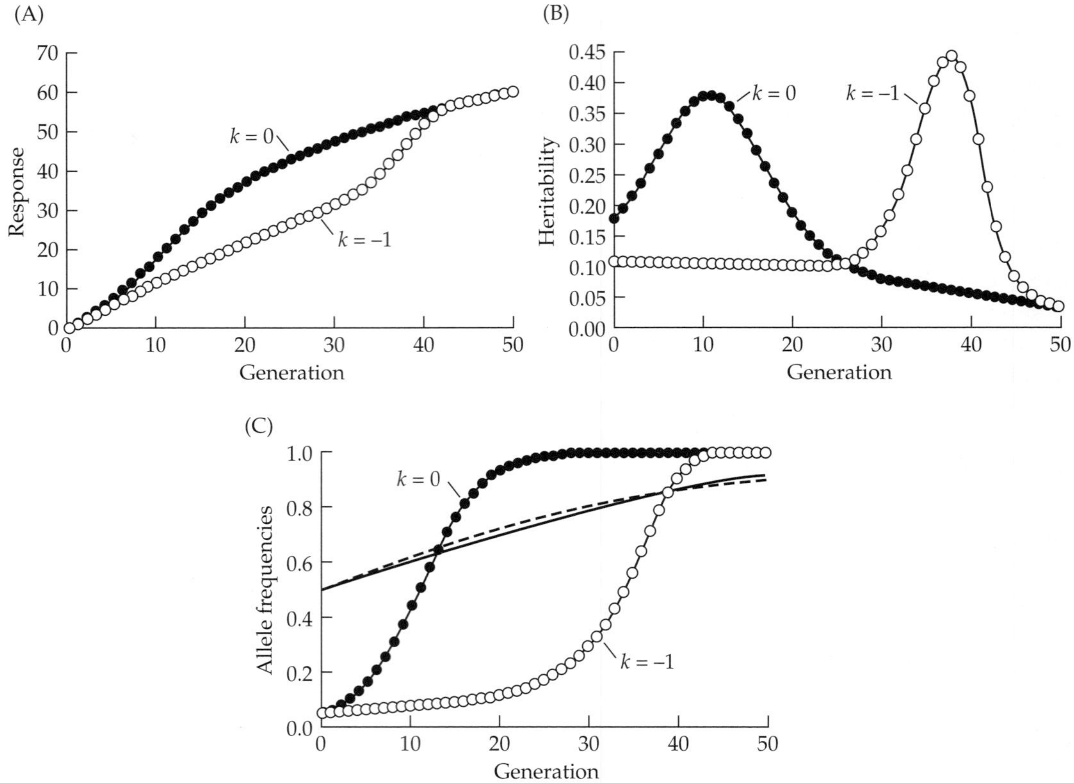

Figure 25.10 Examples of a delayed accelerated response due to the increase of an initially rare allele of major effect. The character is determined by a polygenic background (100 completely additive dialleic loci, with $a = 0.5$ and $p = 0.5$, so that the initial additive variance contributed by the polygenic background is $\sigma_A^2 = 12.5$) plus a major allele that is initially at low frequency ($a = 10$ and $p = 0.05$). We assume that the favored allele at the major locus is either additive ($k = 0$; contributing an initial additive variance of 9.5) or recessive ($k = -1$; contributing an initial additive variance of 0.095). **A:** Response under the recessive model shows an accelerated response around generation 35, while the additive major gene results in an acceleration around generation 10. **B:** Heritabilities clearly show the basis for this acceleration. **C:** Changes in the major-allele frequency show the much longer time required for the recessive major allele to increase in frequency. Note that the change in the polygenic frequencies (the middle two curves; solid for $k = 0$ and dashed for $k = -1$) are almost the same under the two different major-locus dominance values.

frequencies expected under drift-mutation balance, wherein most minor alleles are rather rare. The effects of natural selection may further exaggerate this distribution. If alleles of large effect also tend to be slightly deleterious, then these frequencies may be even lower than expected under the Watterson distribution, with a negative correlation between frequency and effect size (e.g., Figure 28.5).

Major Mutations

Major alleles can be arise by mutation while selection is ongoing, creating bursts of response throughout the course of the experiment. An example of this appeared in an experiment by Yoo (1980a), who selected for increased abdominal bristles in *Drosophila* for over 80 generations (Figure 25.8B). Five of the six replicate lines showed various periods of accelerated response after 20 generations of selection. Yoo was able to correlate many of these bursts with the appearance of new alleles that had major effects on bristle number as heterozygotes but were lethal as homozygotes.

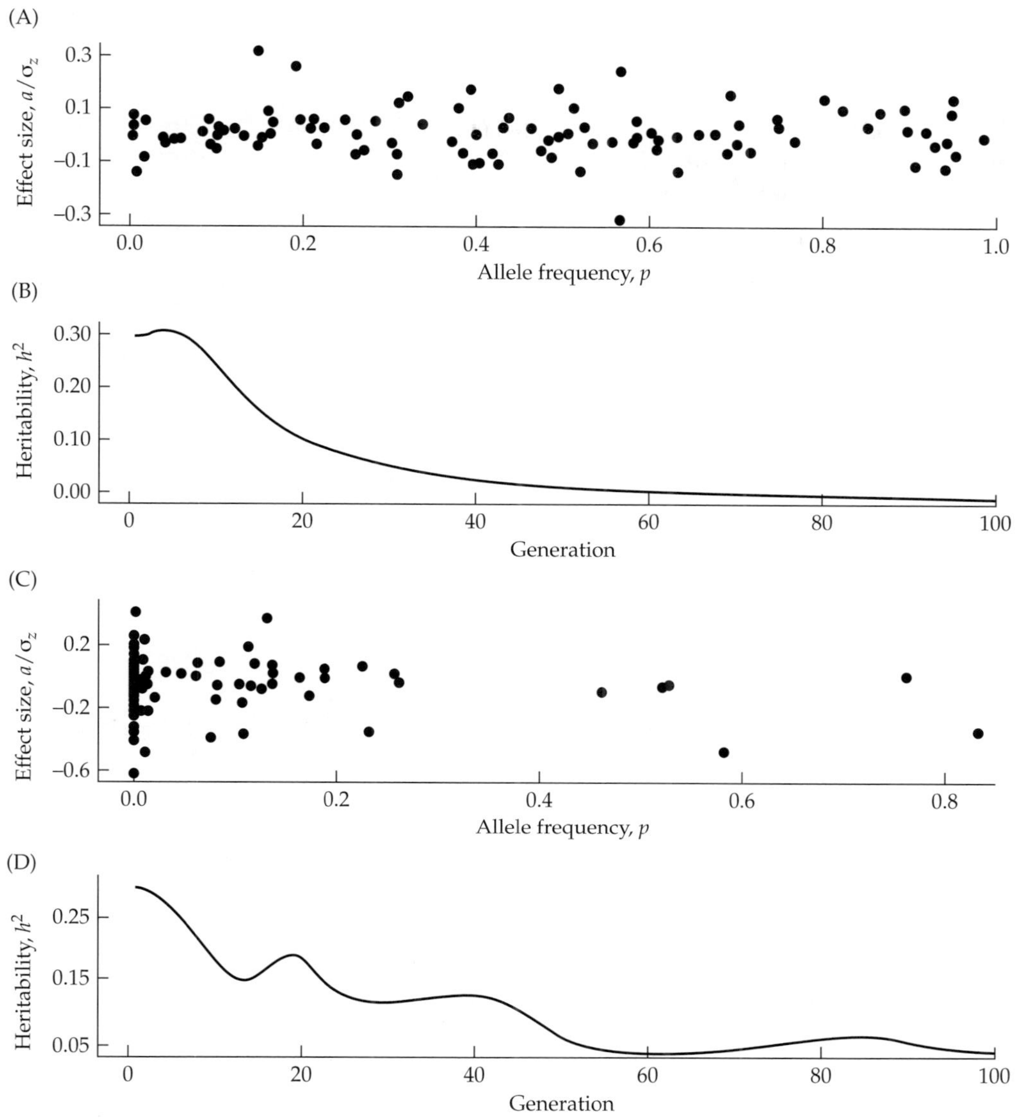

Figure 25.11 The impact of a distribution of initial allele frequencies and effects on the long-term response to directional truncation selection on a trait in an infinite population. The model simulated here assumes 100 underlying loci, that are unlinked and with completely additive effects within and among loci (no dominance or epistasis). Linkage disequilibrium is ignored, with Equation 25.3 iterated to generate the dynamics (see Example 25.2). Allelic effects were randomly sampled from an exponential distribution reflected about zero (effects were equally likely to be positive or negative). Initial allele frequencies were either sampled from uniform or Watterson (Equation 2.34a) distributions, with randomly assigned allelic effects. Given the initial distribution of allele frequencies and effects, a base-population additive variance, $\sigma_A^2(0)$, was computed, and σ_E^2 was set at $(7/3) \cdot \sigma_A^2(0)$ to give the trait a starting heritability of 30% (a typical value for many traits). Results from two realizations are presented here. The joint distribution of initial frequencies and their standardized effects, a/σ_z, is plotted for a given realization from the uniform (**A**) and Watterson (**C**) distributions. As (**B**) shows, under a uniform distribution of starting allele frequencies at the underlying loci, the temporal change in the heritability was generally well-behaved over the course of selection (here, a slight initial increase, followed by a nearly monotonic decrease). Conversely, (**D**) shows that change in the heritability under an initial Watterson distribution was highly erratic. While the specific realization shown here for the Watterson distribution was typical for a number of simulations, even more erratic patterns (i.e., heritabilities rapidly increasing after many generations of selection) were seen in some realizations. In experimental populations, drift and founder-sampling would obscure these patterns.

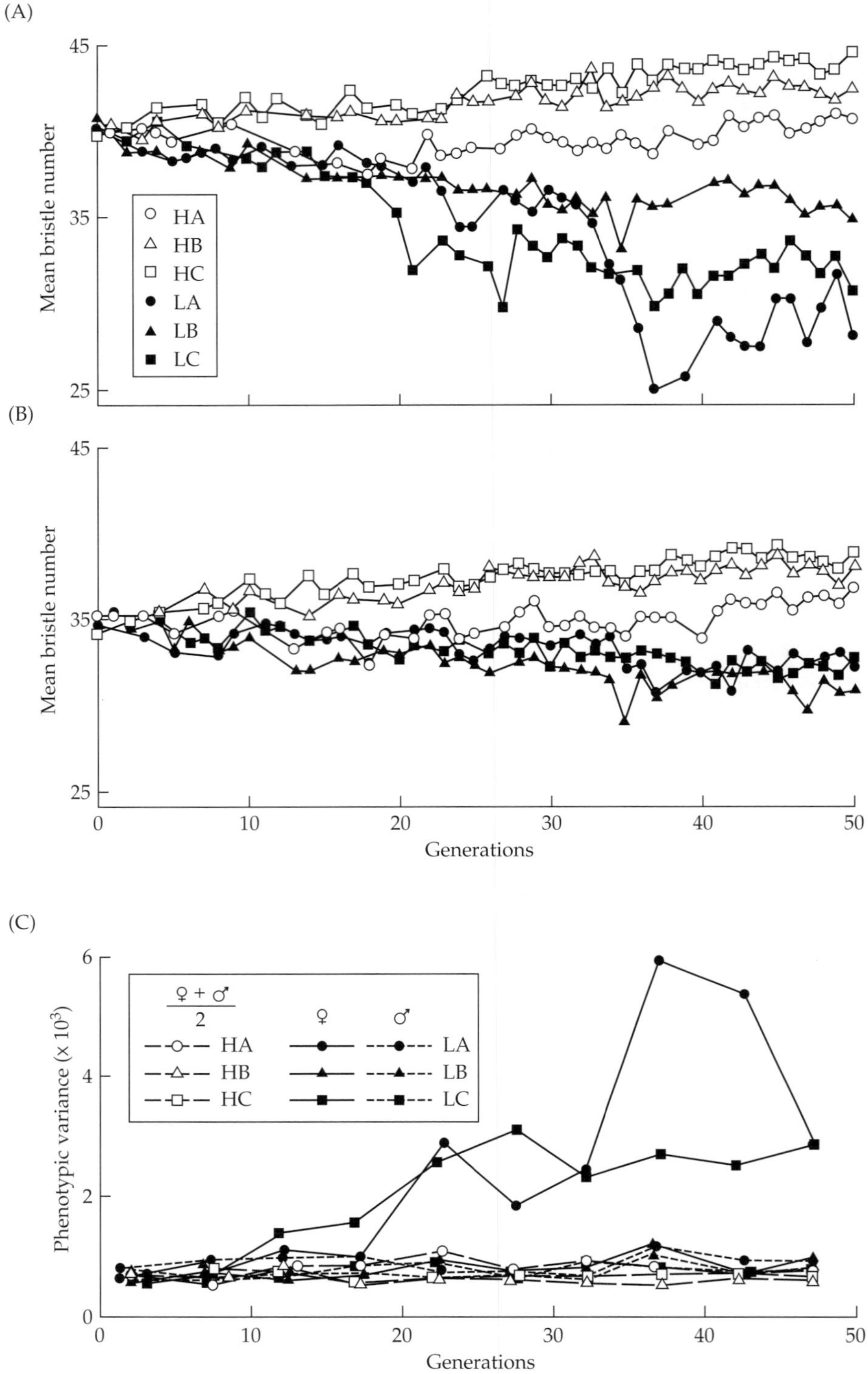

Figure 25.12 The response to selection in *Drosophila melanogaster* for high and low abdominal bristle numbers. While equal amounts of selection was applied to both sexes, results are separated by sex-specific response. **A:** Response in females. **B:** Response in males. While two of the down-selected lines (LA and LC) show bursts of response in females, no such response is seen in the males from these lines. **C:** The phenotypic variances for these lines. The figure plots the average of the male- and female-specific variances for all three high-selected lines (which showed no sex-specific differences over the response), while sex-specific differences were seen in two of the low-selected lines. Note that the variance increased only in females from the two lines showing a burst of response. (After Frankham et al. 1980.)

A second example of a mutation-induced burst of response in *Drosophila* bristle number was seen by Frankham's group (Frankham et al. 1978, 1980; Frankham 1988), which examined selection responses in lines initially containing very little variation. In two of their down-selected lines, females (but not males) showed a burst of response (Figure 25.12). This burst was accompanied by an increase in the phenotypic variance and heritability in females, but not in males. Females also showed reduced fitness, as indicated by a male-biased sex ratio in these lines. These effects were attributable to the appearance of *bobbed* mutants at the ribosomal gene cluster, a deficiency in the number of rRNA genes. The *bobbed* mutants arose on the X-chromosome rRNA cluster, while the Y-chromosome rRNA cluster remained normal, accounting for the sex-limited nature of the response. These mutants were generated by unequal crossingover within the rRNA gene cluster during the course of the selection experiment.

These examples involve mutations of major effects, with an almost immediate impact. The implications of ongoing mutations of minor effects are considered in Chapter 26.

Scale and Environmental Impacts on Variances

Scale effects can also result in increases in genetic variances and selection responses, for example, when the variance increases with the mean (LW Chapter 11). A possible example of this is Enfield's (1972) selection experiments for increased pupal weight in *Tribolium*. Both the additive-genetic and total phenotypic variance increased over time, while heritability remained roughly constant (meaning that response was fairly constant). Comstock and Enfield (1981) suggested that a multiplicative model of gene action was more appropriate in this case than an additive model, and could account for the observed increases in variance. As was discussed in Chapter 14, scale effects can be especially important in threshold characters (also see LW Chapters 11 and 25).

Variances can also increase due to environmental effects. For example, the environmental variance can increase as genotypes become more homozygous, although this is not inevitable (see LW Table 6.1). Likewise, we showed in Chapter 17 that directional selection on a trait can result in an increase in σ_E^2 if the environmental variance has heritable variation. Finally, changes in the environment during the course of selection can sometimes increase the additive variance. A possible example of this effect derives from observations on long-term selection on milk yield in North American dairy cows, where the additive variance in yield has been increasing rather than decreasing (Kennedy 1984). One explanation for such behavior is environmental change, as improved management techniques likely allow for greater discrimination between genotypes, although scale effects may also play a role.

Linkage Effects

Recombinational breakdown of preexisting gametic-phase disequilibrium can also generate an accelerated response. Why might such disequilibrium be present? Mather (1941, 1942, 1943) suggested that QTLs are often in negative disequilibrium as a result of previous natural selection (he considered mainly stabilizing selection), referring to this genetic architecture as **polygenic balance**. More generally, selection tends to build up such negative associations based on fitness (as, by definition, fitness is always under directional selection, which generates negative values of d; see Chapter 16). As a result, alleles influencing fitness tend to be in negative gametic-phase disequilibrium, with gametes containing two favorable alleles occuring at a lower frequency than expected under linkage equilibrium (Chapters 16 and 24). As a result, alleles favored by artificial selection on a character and alleles at linked loci that improve other components of fitness (Sved 1977) have a tendency to become negatively correlated. A character with extensive negative disequilibrium (either between QTLs controlling the character or between QTLs for the character and other fitness loci) can show accelerated selection response in the mean as this disequilibrium decays (Figure 25.13).

An accelerated response can also occur when recombination generates coupling

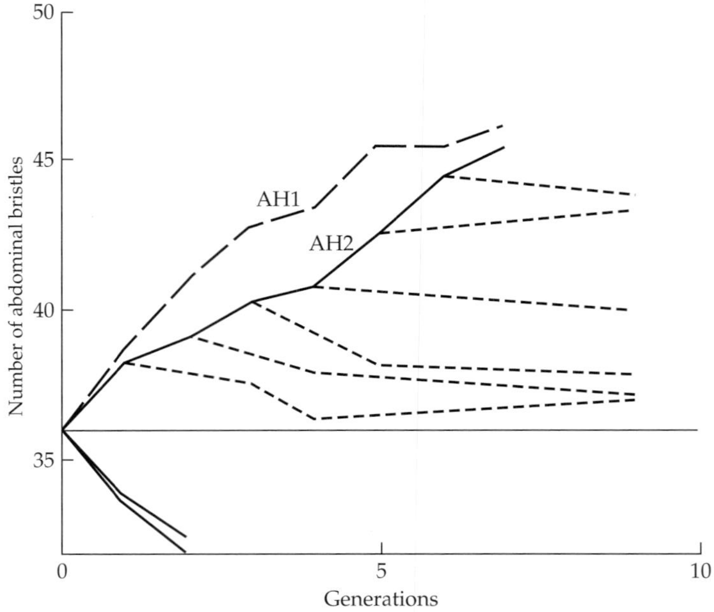

Figure 25.13 An apparent example of linkage between QTLs and deleterious fitness loci. Latter and Robertson (1962) selected for increased abdominal bristle number in *Drosophila melanogaster*, creating sublines (indicated by the dashed lines) from the selected lines at various generations and subjecting these sublines to relaxed selection. Sublines of line AH2 extracted in the first three generations of selection showed significant erosion of response upon relaxation of selection, while sublines extracted in later generations show little erosion. Note also that line AH2, which has a depressed response relative to line AH1 over generations 1–4, shows an accelerated response following generation 4. One explanation for this pattern is that alleles increasing the character were initially in gametic-phase disequilibrium with alleles having deleterious effects on fitness in line AH2. By generation 4, this disequilibrium had largely broken down, allowing the frequencies of alleles increasing the character value to remain stable following relaxation of selection and allowing a faster response to selection. (After Latter and Robertson 1962.)

gametes for alleles that increase character value. A classic example is Thoday's selection experiments for increased sternopleural bristle number in *Drosophila* (Thoday and Boam 1961; Thoday et al. 1964). As shown in Figure 25.14, a burst of response was seen after about 20 generations of selection. Using polygenic mapping, Thoday et al. (1964) were able to show that the initial population consisted mainly of $--$ chromosomes with only a few $+-$ and $-+$ chromosomes (each $+$ indicates a major allele increasing bristle number). Selection reduced the frequency of $--$ chromosomes, increasing the frequency of $+-/-+$ heterozygotes, which in turn increased the frequency at which $++$ chromosomes were generated by recombination. The selection response accelerated as these newly created gametes became sufficiently common to increase additive variance.

While recombination removes gametic-phase disequilibrium, selection generates it (Chapters 5, 16, and 24). It follows that if linkage effects are important, a relaxation of selection should facilitate long-term response by allowing negative gametic-phase disequilibrium to decay, which increases the additive variance (Chapter 16). Thoday and Boam (1961) observed a large increase in *Drosophila* sternopleural bristle number after reselecting a line in which selection was relaxed for several generations following an apparent selective plateau. Similar patterns were seen by Mather and Harrison (1949) in some of their lines that were selected for increased abdominal bristle number. On the other hand, Rathie and Barker (1968) compared the effects of cycles of selection followed by no selection versus continuous selection on abdominal bristles and found no differences in response. However, the continuously

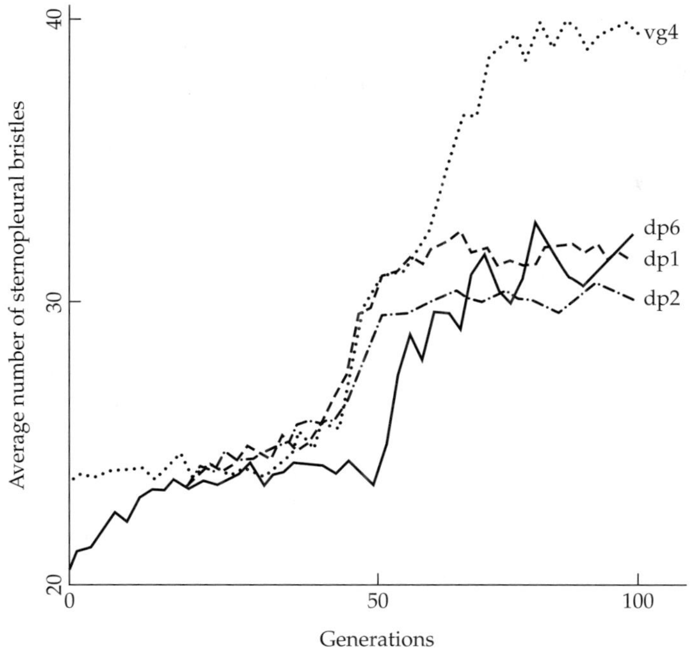

Figure 25.14 Accelerated response in sternopleural bristle number in *Drosophila melanogaster* lines selected by Thoday and Boam (1961). All lines showed acceleration in response, but the acceleration in line *vg 4* was especially dramatic.

selected lines showed larger erosions of response upon relaxation of selection and had greater decreases in reproductive fitness, suggesting that disequilibrium between QTLs and fitness loci was greater in these lines.

Our discussion of the effects of linkage has been restricted to largely deterministic considerations. Here, in the absence of epistasis, linkage influences the rate, but not the ultimate limit, of response. When population size is finite, linkage can have an important impact on the ultimate selection limit as well. For example, in a small population, selection and drift could have fixed $+-$ chromosomes in Thoday's experiment before $++$ chromosomes reached frequencies sufficiently high to overcome the effects of drift. We will defer further discussion of these complex interactions until Chapter 26.

Epistasis

While the permanent component of response in any particular generation is a function of the additive variance (Chapter 15), σ_A^2 itself changes as the frequencies of underlying genotypes change. While it is generally true that most genetic variance is additive, even when strong epistasis is present (Hill et al. 2008; Mäki-Tanila and Hill 2014), it is also true that changes in genotype frequencies can result in some of the epistatic variance being converted into additive variance (Goodnight 2004; also see Chapter 11). Eitan and Soller (2004) proposed the idea of **selection-induced genetic variation**, wherein the process of selection generates additional variation, as opposed to strictly removing it. An example of this is the study by Carlborg et al. (2006), who found a strongly epistatic locus in chicken lines that were divergently selected for growth. Alleles at three growth-specific loci (*Growth4, Growth6,* and *Growth12*) had much higher effects in a (high-growth) homozygous *Growth9* genotypic background. While Eitan and Soller were concerned with new variation being generated via epistasis, recall the very erratic pattern of response seen in Figure 25.11D for a *purely additive* model. If such a pattern were observed in an experiment, many researchers would declare that it could only be due to epistasis. This is based on the expectation that the additive variance is continually declining, which requires the assumption that the majority of loci are at, or above, their frequencies for maximal additive variation. If this is not correct, such low-frequency additive alleles can (for at least part of the response) generate new additive variance as they increase toward the frequency that yields a maximal value of σ_A^2, after which further allele-frequency change causes σ_A^2 to decline. As Figure 25.11 shows, bursts of response for rare additive alleles can occur > 50 generations after the start of selection.

There is a healthy debate on the impact of epistasis on the response to directional selection (Carlborg and Haley 2004; Malmberg and Mauricio 2005; Le Rouzic and Carlborg 2007; Crow 2008, 2010; Hansen 2013), with some researchers even claiming that models that do not incorporate epistasis are fundamentally flawed (Nelson et al. 2013). However, from the standpoint of predicting long-term response, epistasis is, in some sense, the least of our concerns. First, there is the important distinction between underlying epistasis being common versus epistatic *genetic variances* being large. By construction, even when the underlying gene action of a trait is highly nonadditive, the majority of the genetic variance tends to be additive, especially when contributing loci have alleles at extreme frequencies (LW Chapter 7; Hill et al. 2008). As the various mechanisms discussed previously illustrate, a number of different genetic conditions other than epistasis can result in bursts of selection response. Determining the underlying cause for a burst (or some other feature attributed to epistasis) is far from trivial, implying that epistasis should only be invoked when it is fully justified by the data.

CONFLICTS BETWEEN NATURAL AND ARTIFICIAL SELECTION

It is frequently seen that components of fitness (such as viability and fertility) decline rather dramatically during artificial selection experiments. Selected lines can even die out due to extreme declines in fitness. There are several (not mutually exclusive) reasons for these declines, which have rather different implications for long-term response.

1. *Selection increases the amount of inbreeding* relative to control populations of the same size, a point developed in Chapters 3 and 26. Drift effects associated with inbreeding can increase the frequency (and exposure) of deleterious recessives, as well as moving overdominant fitness loci away from their equilibrium frequencies. If inbreeding is sufficiently strong, deleterious alleles can be fixed.
2. *Loci favored by artificial selection can be in gametic-phase disequilibrium with loci having deleterious effects on fitness.* Fitness declines as these deleterious alleles increase in frequency due to hitchhiking with alleles favored under artificial selection. This disequilibrium need not be present initially—rather, it can be generated during artificial selection (Chapter 16). In infinite populations, the gametic-phase disequilibrium between QTL and fitness loci eventually decays, and deleterious alleles are not fixed. In small populations, deleterious alleles can be dragged along to fixation by linked major alleles.
3. *Alleles favored by artificial selection can have deleterious effects on fitness.* There are two different routes by which this can occur: the artificially selected character may itself be under natural selection, or loci controlling this character can have pleiotropic effects on other characters that are under natural selection (see Chapter 28). The impact on long-term response has been examined in some detail for two particular models: the **optimum model**, wherein the character under artificial selection is also subjected to natural stabilizing selection (Latter 1960; James 1962; Zeng and Hill 1986; Hill and Mbaga 1998); and the **homeostatic model**, wherein heterozygotes have the highest fitness under natural selection (Lerner 1954; Robertson 1956). While the genetic basis for these models is very different, Nicholas and Robertson (1980) noted that "despite the profound differences between the two models, the practical implications of each are essentially the same *in the context of artificial selection*. Consequently, there seems to be no aspect of observable response which would enable a distinction to be made between the two."

What are the implications of these different fitness-decreasing mechanisms for long-term response? The inbreeding effect of selection is a consequence of finite population size being further exaggerated by selection: these effects should largely disappear if the effective population size is kept sufficiently large during selection.

If loci influencing the character also influence fitness (either directly or because of gametic-phase disequilibrium with other fitness loci), response is expected to decay upon relaxation of selection, provided alleles decreasing fitness are not fixed (e.g., Figure 25.8B). Erosion of response, however, does not automatically imply that fitness effects are important. For example, some erosion is expected when additive epistasis or maternal effects have contributed to the response (Chapter 15). If erosion is largely due to fitness effects, it should be correlated with increases in fitness.

Example 25.6. Frankham et al. (1988) selected *Drosophila melanogaster* for increased ethanol tolerance. Following the suggestion of Gowe (1983), they attempted to reduce the expected decline in reproductive fitness by further culling those artificially selected pairs showing reduced number of offspring. Their logic was that if deleterious fitness effects during selection were largely caused by rare recessives (the exposure of which increases by inbreeding during selection), then removing a very small fraction of the lowest-fitness individuals would cull individuals that are homozygous for deleterious recessives. Following the selection of parents based on increased tolerance, Frankham et al. placed single mated pairs in vials that were subsequently ranked according to the number of pupae produced. Vials with the lowest number of pupae were culled. The HS line, which was subjected to both selection for tolerance and subsequent culling on reproductive fitness, had the same tolerance response as the HO line (which was selected only for increased tolerance). The unselected control line and the HS line had the same fitness, as measured by Knight and Robertson's (1957) very general competitive index measure. Conversely, the HO line had significantly reduced fitness. If alleles increasing tolerance had either pleiotropic or linkage effects on fitness, the HS line should have reduced the tolerance response relative to the HO line. Given that the responses were identical, however, Frankham et al. suggested that the reduction in fitness in the HO line was mainly due to the effects of inbreeding, rather than linkage or pleiotropy.

A similar study was reported by Gowe et al. (1993), who examined 30 years of selection on laying hens. Again, a two-stage selection approach was used: following truncation selection for increased egg production, the lowest 10% of hens chosen as parents by truncation selection were then culled again on the basis of hatchability. Using this selection scheme, the increased production lines retained the same levels of hatchability as an unselected control. Typically, selection only on increased egg production alone reduces hatchability.

A final example of an experiment attempting to control for deleterious fitness effects was provided by Imasheva et al. (1991), who combined directional selection for increased *radius incompletus* expression in the wing venation of *Drosophila melanogaster* with stabilizing selection on a suite of wing morphological characters. After 16 generations of selection, the control and directional plus stabilized selected lines had similar population sizes, both of which were higher than the population that was subjected to strict directional selection. The three lines, however, did not differ when fitness was measured by looking at competitive ability.

Example 25.7. Enfield (1980) subjected the flour beetle (*Tribolium castaneum*) to selection for increased pupal weight. As mean pupal weight increased, components of reproductive fitness (percent sterility and mean number of progeny per fertile mating) declined. Upon the relaxation of selection, pupal weight decreased and fitness increased. When relaxed lines were again subjected to selection, fitness components again decreased as pupal weight increased. However, Enfield reported evidence that increased pupal weight, by itself, does not necessarily decrease fitness, having found that lines can be created with rather large mean pupal weight, that remain stable upon the relaxation of selection. Thus, it appears that reproductive fitness declines as a result of a correlated selection response with pupal weight, rather than natural selection acting directly on pupal weight itself. At least some of the alleles for increased pupal weight thus appear to be associated with alleles that decrease reproductive fitness. If this is due to linkage disequilibrium, recombination will reduce this effect. If it is due to pleiotropy, however, one must select for modifiers of these deleterious effects.

Example 25.8. An interesting potential example of a decay in selection response upon relaxation of selection in a natural population was provided by Cruz and Wiley (1989), who examined egg-rejection behavior in the village weaver bird (*Ploceus cucullatus*) in Hispaniola. This species was introduced onto the island from western Africa around 200 years ago. Studies in western Africa by Victoria (1972) showed that female weavers can recognize their own eggs and eject foreign eggs from their nest, with the rate of rejection proportional to the amount of difference between eggs. Victoria postulated that this rejection behavior evolved in response to selective pressure from the Didric cuckoo (*Chrysococcyx caprius*), which is a brood parasite that lays its eggs in the nests of other species. The rejection of eggs that appear sufficiently different could lead to increased fitness when brood parasites are present but decreased fitness when they are absent (as any discarded eggs would be from the mother).

Victoria found an average rejection rate in Africa of eggs with a different appearance from their mothers of around 40–55%, while Cruz and Wiley found a rejection rate on Hispaniola of 12%. Because Hispaniola was free of brood parasites until the mid-1970s, Cruz and Wiley suggested that this difference in rejection rates amounts to a slippage in the selection gain (in Africa) following the relaxation of selection (in Hispaniola). Such a slippage shows that the 40–55% value in Africa does not represent a selection limit due to lack of additive variance in this behavioral trait. If a selection limit did exist in Africa, it would likely represent a tradeoff between the fitness costs of rejecting all eggs that appear to be different. This natural experiment continues today, as in the mid-1970s the shiny cowbird (*Molothrus bonariensis minimus*), another brood parasite, was introduced into Hispaniola.

Accumulation of Lethals in Selected Lines

Lethal alleles are often detected in lines that are subjected to long-term selection. If these alleles also influence the character under selection, they can result in increases in the additive variance during a period of the selection response, the presence of significant additive variance in the trait at an apparent selection limit, and some erosion of both the response and the additive variance upon relaxation of selection. In *Drosophila* experiments, lethals have been observed in lines that were subjected to directional selection on sternopleural bristles (Madalena and Robertson 1975; García-Dorado and López-Fanjul 1983), abdominal bristles (Clayton and Robertson 1957; Frankham et al. 1968b; Hollingdale 1971; Yoo 1980b), dorsocentral bristles (Domínguez et al. 1987), and wing length (Reeve and Robertson 1953). Skibinski (1986) also found that lethals accumulated during stabilizing selection on sternopleural bristle number. Yoo (1980b) and Skibinski (1986) found that most lethals arose during the course of the selection experiment, rather than being initially present in the base population. A similar example in mice involves the homozygous sterile allele *pygmy*, which reduces body size when heterozygous (Warwick and Lewis 1954; King 1955). This mutant arose during MacArthur's (1949) long-term selection experiments for decreased body size.

Newly arising lethals could be due to new mutation (such as the insertion of a mobile element; see Mackay 1988) or could be generated by recombination between strongly epistatic genes creating **synthetic lethals** (LW Chapter 10; Phillips and Johnson 1998). Once a lethal with a strong effect on the character appears, it partly shelters closely linked sites from further selection, creating linked clusters of lethals (Madalena and Robertson 1975; García-Dorado and López-Fanjul 1983).

Example 25.9. Consider the following estimated variance components from a selection experiment by Reeve and Robertson (1953) for increased wing length in *Drosophila melanogaster*:

Population	σ_z^2	σ_A^2	σ_E^2	h^2
Selected	4.65	2.50	1.72	0.54
Relaxed	4.50	1.80	1.72	0.40
Base	3.20	1.02	1.72	0.32

The selected line shows large increases in additive variance and heritability relative to the base population, while upon the relaxation of selection, both the additive variance and the heritability decline to values that are intermediate between those in the base and the selected lines. Reeve and Robertson attributed this behavior to the presence of at least two major alleles that are lethal as homozygotes. As these alleles increase in frequency, they increase additive variance. Because they are never fixed (as is discussed below, their maximum frequency is 1/3), the genetic variance attributable to these alleles does not subsequently decline as selection proceeds. However, upon relaxation of selection, the component of response due to these alleles decays as their frequency is reduced by natural selection. The additive variance is also expected to decline as these alleles are eventually lost due to natural selection following the relaxation of artificial selection.

Why do lethal alleles persist in some selected populations? In many cases, this seems to be a result of fitness overdominance resulting from a balance between natural and artificial selection—the allele increases character value as a heterozygote, but it is lethal (or sterile) as a homozygote. Such alleles can be maintained at rather high frequencies if artificial selection acting on the locus is strong (i.e., the allele has a major effect on the character that is under strong directional selection). We can show this informally as follows: let the *BB* homozygote be lethal, while the *Bb* heterozygote increases the genotypic value of the trait under selection (relative to *bb*). Under directional selection, the fitness of *Bb* relative to *bb* can be written as $1 + s$, where s increases as the effect of *B* on the character under artificial selection increases, yielding total fitnesses of $1 : 1 + s : 0$ for the genotypes *bb* : *Bb*: *BB*. The resulting mean population fitness becomes $\overline{W} = 1+2sp(1-p)-p^2$, where p is the frequency of *B*. Applying Wright's formula (Equation 5.5b) and solving $\Delta\widehat{p} = 0$ yields an equilibrium frequency of *B* in newly formed zygotes (before natural selection) of

$$\widehat{p} = \frac{s}{1+2\,s} \tag{25.12a}$$

Following the loss of lethal homozygotes in the surviving offspring, $\widehat{p}$ decreases to

$$\widetilde{p} = \frac{(1/2)\text{freq}(Bb)}{\text{freq}(Bb) + \text{freq}(bb)} = \frac{\widehat{p}\,(1-\widehat{p}\,)}{2\,\widehat{p}\,(1-\widehat{p}\,) + (1-\widehat{p}\,)^2} = \frac{\widehat{p}}{1+\widehat{p}} \tag{25.12b}$$

Substitution of Equation 25.12a into Equation 25.12b yields $\widetilde{p} = s/(1+3s)$. Thus, for large values of s, the equilibrium frequency of the allele approaches 1/3 at the start of each generation before artificial selection and increases to 1/2 after artificial selection. A more formal treatment of this problem is given in Example 25.10 (also see Figure 25.15).

While many lethal alleles have a demonstrated major effect on the character under selection, in some cases their frequencies are not consistent with this theory. Skibinski (1986) found no evidence that artificial selection accounts for the maintenance of lethals observed in his *Drosophila* lines. Instead, one lethal showed evidence of segregation distortion (Lyttle 1991, 1993; Taylor and Ingvarsson 2003), which could account for its observed frequency. Likewise, none of the *Drosophila* lethals isolated by Domínguez et al. (1987) had a significant effect on the character under selection. They also found evidence of segregation distortion, with at least one lethal allele being preferentially transmitted by males. Thus, in some experiments, lethal alleles may persist for reasons other than artificial selection and therefore should persist upon the relaxation of selection, whereas lethals maintained by artificial selection will not. The increased drift generated by artificial selection can increase the frequency of even strongly deleterious alleles, and this, especially when interacting with other factors such as segregation distortion, might account for the increase in lethals that do not affect the character under artificial selection.

Example 25.10. For a more formal treatment of the expected equilibrium value, consider a major gene that is lethal as a recessive (*BB*), but increases character value as a heterozygote (*Bb*). What are the dynamics of this locus when truncation selection is used to increase character value? Suppose that the distribution of phenotypes for the two viable genotypes is normal, with $z_{Bb} \sim N(\mu + a, \sigma^2)$ and $z_{bb} \sim N(\mu, \sigma^2)$, and let p be the frequency of *B*. Following random mating, the expected zygotic frequencies will be in Hardy-Weinberg frequencies, with freq(*Bb*) = $2p(1-p)$, freq(*bb*) = $(1-p)^2$, and freq(*BB*) = p^2. After natural selection, only the genotypes *Bb* and *bb* remain, and these now have frequencies of

$$\text{freq}'(Bb) = \frac{2p(1-p)}{1-p^2} = \frac{2p}{1+p} \quad \text{and} \quad \text{freq}'(bb) = \frac{(1-p)^2}{1-p^2} = \frac{1-p}{1+p}$$

Truncation selection occurs on the survivors of natural selection, generating a mixture distribution for the trait value (LW Chapter 13), with

$$\begin{aligned} z &= \text{freq}'(Bb)\, p_{Bb}(z) + \text{freq}'(bb)\, p_{bb}(z) \\ &= \left(\frac{2p}{1+p}\right) p_{Bb}(z) + \left(\frac{1-p}{1+p}\right) p_{bb}(z) \end{aligned}$$

Here $p_{Bb}(z)$ and $p_{bb}(z)$ denote the density functions for the normal distributions associated with these two genotypes.

Recall from Chapter 14 that truncation selection is usually framed in terms of the fraction, q, of individuals that are allowed to reproduce. However, it will prove easier to initially formulate this problem in reverse, assuming some trait threshold value, T, above which individuals are allowed to reproduce. Given the current mean and variance, we can obtain the value of T for a given value of q. For a fixed value of q, we expect T to increase in each generation as the trait mean increases. Hence, we first solve this problem for T, and then express the final result in terms of the fraction saved, q.

If the trait threshold value above which individuals are allowed to reproduce is T, then the fraction of individuals allowed to reproduce is given by

$$q = \left(\frac{2p}{1+p}\right) \Pr(z_{Bb} > T) + \left(\frac{1-p}{1+p}\right) \Pr(z_{bb} > T)$$

Because $(z_{Bb} - \mu - a)/\sigma \sim U$ and $(z_{bb} - \mu)/\sigma \sim U$, where U denotes a unit normal, this rearranges to yield

$$q(1+p) = 2p \Pr\left(U > T^* - \frac{a}{\sigma}\right) + (1-p)\Pr(U > T^*) \qquad (25.13)$$

where $T^* = (T - \mu)/\sigma$. The frequency of *Bb* following artificial selection becomes

$$\text{freq}''(Bb) = \left(\frac{2p}{1+p}\right) \frac{\Pr(U > T^* - a/\sigma)}{q}$$

yielding the frequency p'' of *B* after a single round of both natural and artificial selection as

$$p'' = \frac{1}{2}\,\text{freq}''(Bb) = \left(\frac{p}{1+p}\right) \frac{\Pr(U > T^* - a/\sigma)}{q} \qquad (25.14)$$

If we let $\widehat{p}$ denote the equilibrium frequency of *B* in the zygotes at the start of the next generation (before natural selection), by rearranging Equation 25.14, it follows that

$$q(1+p)p'' = p\Pr(U > T^* - a/\sigma)$$

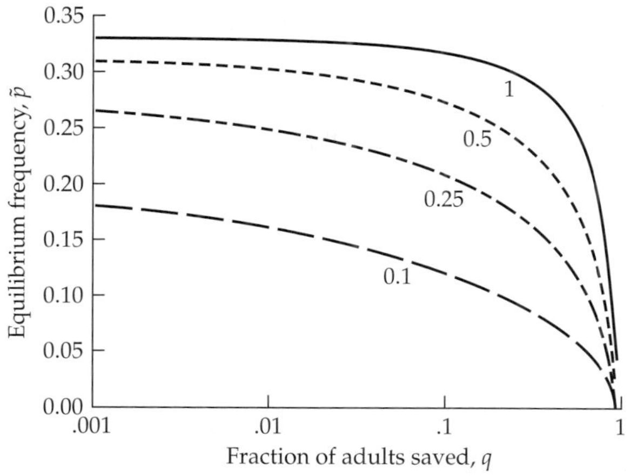

Figure 25.15 The equilibrium frequency, $\widetilde{p}$, of a lethal allele that also increases the value of a trait under artificial selection. Selection is assumed to act in the zygote, so that $\widetilde{p}$ is the frequency of the allele in surviving individuals before artifical selection is performed. The equilibrium frequency is a function of the strength of artificial selection (measure by the fraction, q, of adults saved under truncation selection) and the contribution, a/σ, of the allele to the trait under artificial selection. The four curves are for values of $a/\sigma = 1, 0.50, 0.25$, and 0.10, respectively. See Example 25.10 for further details.

Because $p'' = p = \widehat{p}$ at equilibrium, this reduces to

$$q\,(1+\widehat{p}) = \Pr(U > T^* - a/\sigma)$$

Combining this result with Equation 25.13 yields

$$2\,\widehat{p}\,\Pr\,(U > T^* - a/\sigma) + (1-\widehat{p})\,\Pr(U > T^*) = \Pr(U > T^* - a/\sigma)$$

Solving for $\widehat{p}$ returns

$$\widehat{p} = \frac{\Pr(U > T^*) - \Pr(U > T^* - a/\sigma)}{\Pr(U > T^*) - 2\,\Pr(U > T^* - a/\sigma)} \tag{25.15a}$$

Likewise, the equilibrium frequency, $\widetilde{p}$, following the removal of lethals (*BB* homozgyotes) is

$$\widetilde{p} = \frac{(1/2)\text{freq}(Bb)}{1 - \text{freq}(BB)} = \frac{\widehat{p}\,(1-\widehat{p})}{1-\widehat{p}^2} = \frac{\widehat{p}}{1+\widehat{p}} \tag{25.15b}$$

Figure 25.15 plots $\widetilde{p}$ as a function of q and a/σ. The figure was generated by applying Equations 25.15a and 25.15b for a given value of T^*, and then using Equation 25.13 to obtain the value of q, given the T^*, a/σ, and $\widehat{p}$ values.

Lerner's Model of Genetic Homeostasis

A second class of models assuming pleiotropic fitness effects is based on Lerner's (1954) theory of **genetic homeostasis**, which was motivated by the notion that natural selection tends to favor heterozygotes, a view that is still controversial and has weak support at best. Under Lerner's model, alleles segregating at a QTL are favored as heterozygotes by natural selection. The simplest case is that in which the QTL is additive for the character under selection. Let the genotypes *bb* : *Bb* : *BB* have fitnesses (under natural selection) of $1 - s_2 : 1 : 1 - s_1$. Further, suppose the expected trait values for these offspring are $\mu - a : \mu : \mu + a$, where the trait is normally distributed and a is small. Then, from Equation 5.21, the fitnesses under directional selection are approximately $1-\overline{\imath}\,a/\sigma_z : 1 : 1+\overline{\imath}\,a/\sigma_z$, resulting in (assuming weak selection; e.g. $s_1, s_2, \overline{\imath}\,a/\sigma \ll 1$) approximate total fitnesses of

$$(1-s_2)(1-\overline{\imath}\,a/\sigma_z) \,:\, 1 \,:\, (1-s_1)(1+\overline{\imath}a/\sigma_z)$$

If artificial selection is sufficiently strong relative to natural selection ($\overline{\imath}\,a/\sigma_z > s_1$), B is fixed. However, if $s_1 > \overline{\imath}\,(a/\sigma_z)$, the total fitness shows overdominance, and there is an internally stable equilibrium of

$$\widehat{p} = \frac{s_2 + \overline{\imath}\,(a/\sigma_z)\,(1-s_2)}{s_1 + s_2 + \overline{\imath}\,(a/\sigma_z)\,(s_1 - s_2)} \simeq \frac{s_2 + \overline{\imath}\,(a/\sigma_z)}{s_1 + s_2} \quad \text{for} \quad s_1, s_2 \ll 0 \tag{25.16}$$

This weak-selection result is due to Verghese (1974) and Nicholas and Robertson (1980), while Minvielle (1980) gave a more general equilibrium condition for alleles of major effect. The additive genetic variance for the trait contributed by this locus at equilibrium is $2a^2\,\widehat{p}\,(1-\widehat{p})$, which can be considerable.

Changes in reproductive fitness in divergent selection lines (Chapter 18) are often asymmetric, with lines selected in one direction showing a much larger decrease in fitness than lines selected in the opposite direction. Such asymmetries are not necessarily inconsistent with genetic homeostasis, as they can be accounted for by directional dominance in fitness (e.g., if $s_1 < s_2$, namely, that alleles increasing the character under artificial selection also tend to be more fit as homozygotes, holds for most loci).

Artificial Selection Countered by Natural Stabilizing Selection

Lerner's model is an example in which the QTL influencing a character under artificial selection also influences fitness under natural selection through paths that are independent of the phenotypic value of the focal trait. Under his model, extreme phenotypes are less fit because they are more homozygous than intermediate phenotypes. Alternatively, the phenotypic value, z, itself could be under natural selection. Here, extreme phenotypes are intrinsically less fit, independent of their genotypes. For example, z could be under natural selection for an intermediate optimum, with directional artificial selection being opposed by stabilizing natural selection. This can also lead initially to an apparent selection limit in the presence of additive variance in the artificially selected trait (Latter 1960; James 1962; Zeng and Hill 1986; Hill and Mbaga 1998).

Suppose we assume that stabilizing selection is occurring according to Equation 16.17, with the width of the selection function given by ω^2 (with smaller values implying stronger stabilizing selection) and the optimum set at zero. From Equation 16.18b, the selection differential imparted on a population whose mean is at a distance of μ from the optimum is

$$S_{st} = -\mu\,\frac{\sigma_z^2}{\sigma_z^2 + \omega^2} \tag{25.17a}$$

and, from Equation 16.18a, the new phenotypic variance after stabilizing selection becomes

$$\sigma_{z^*}^2 = \frac{\sigma_z^4 + \sigma_z^2\,\omega^2 - \sigma_z^4}{\sigma_z^2 + \omega^2} = \frac{\sigma_z^2\,\omega^2}{\sigma_z^2 + \omega^2} \tag{25.17b}$$

Now suppose directional selection operates with a selection intensity of $\overline{\imath}$, yielding a selection differential from artificial selection of

$$S_a = \overline{\imath}\sigma_{z^*} = \frac{\overline{\imath}\,\sigma_z\,\omega}{\sqrt{\sigma_z^2 + \omega^2}} \tag{25.17c}$$

for a total selection differential of

$$S = S_a + S_{st} = \frac{\overline{\imath}\,\sigma_z\,\omega}{\sqrt{\sigma_z^2 + \omega^2}} - \mu\,\frac{\sigma_z^2}{\sigma_z^2 + \omega^2} \tag{25.17d}$$

Response stops (a limit is reached) when $h^2 S = 0$, which can occur for $h^2 \neq 0$ when $S = 0$. Setting Equation 25.17d equal to zero and solving for the limiting mean yields

$$\mu_\infty = \bar{\imath}\,\omega\,\frac{\sqrt{\sigma_z^2 + \omega^2}}{\sigma_z} \tag{25.18a}$$

or a total response of $\mu_\infty - \mu_0$, a result first obtained by James (1962). Note that this limit does not appear to depend on h^2 (provided it is nonzero at the limit). However, this expected response is an approximation, as disequilibrium and allele-frequency change alter h^2, changing the value of σ_z^2 and altering the ultimate limit.

Hill and Mbaga (1998) noted that for standard $\bar{\imath}$ values of 1 to 2 (corresponding to saving between 5% to 35% of the population; Equation 14.3a) and the typically assumed value of ω^2/σ_z^2 of 5 to 20 (Chapter 28), that a total response of 5 to 40 phenotypic standard deviations can occur (assuming the limit is caused by opposing selection, not lack of variation). If the initial mean is zero ($\mu_0 = 0$, i.e., the population mean starts at the optimum), then recalling Equation 25.17d yields a response after an initial generation of selection of

$$R_1 = h^2 S = h^2 \left(\frac{\bar{\imath}\,\sigma_z\,\omega}{\sqrt{\sigma_z^2 + \omega^2}} \right) \tag{25.18b}$$

returning the ratio of the total to initial response as

$$\frac{R_c(\infty)}{R_1} = \frac{\sigma_z^2 + \omega^2}{\sigma_z^2\,h^2} = \frac{\sigma_z^2 + \omega^2}{\sigma_A^2} \tag{25.18c}$$

The half-life of response (in generations) is

$$t_{0.5} = \ln(2)\,\frac{\sigma_z^2 + \omega^2}{\sigma_z^2\,h^2} \tag{25.18d}$$

as found by James (1962) and Zeng and Hill (1986).

While the population will display additive variation in the selected trait when these forces of directional and stabilizing selection balance, this situation is similar to strict stabilizing selection. As introduced in Example 5.6 (and discussed at length in Chapter 28), in the absence of mutation, strict stabilizing selection eventually results in the loss of essentially all genetic variation (Robertson 1956). Hence, the balance between artificial directional and natural stabilizing selection is not, by itself, sufficient to maintain variation. However, the loss of additive variation may be rather slow once the limit is reached (Chapter 28), leading to the appearance of the maintenance of variation over moderate time scales.

26

Long-term Response: 2. Finite Population Size and Mutation

It is almost impossible with any brevity to exemplify the notion of adaptation. Fisher (1934)

Adaptation depends on how the various evolutionary processes shape variation in populations. Barton and Partridge (2000)

As we saw in Chapter 18, drift has significant short-term consequences for a population under selection, generating variation around the expected response. Through its role in changing allele frequencies, drift has even more impact over longer time scales. Given that most artificial selection experiments tend to have small effective population sizes, drift is a major factor in the limit of selection response and usually causes a population to plateau at levels below its genetic potential. Another evolutionary force entering into long-term response is mutation. After a sufficient amount of time, new response will be entirely driven by variation that was not present at the start of selection.

This chapter examines the roles of both finite population size and mutation in long-term response, considering how drift interacts with selection to change allele frequencies and how new mutations contribute to selection response. Throughout, we restrict attention to directional selection, deferring the consideration of stabilizing selection until Chapter 28. To distinguish between the contributions of initial variation and new mutation to selection responses, we use **long-term response** to refer to the gain due to variation in the population at the start of selection. As this initial variation is progressively exhausted, response from this component reaches a selection limit. The actual response, however, can continue well past this limit due to the input from new mutation, eventually (under constant selection) approaching an **asymptotic rate of response**, wherein the contribution to σ_A^2 from new mutation is balanced by its removal by drift and selection.

Much of the material here builds on results from Chapter 7 on the interaction between selection and drift, and it may prove useful to review that material before proceeding. This chapter is organized as follows. First, we review a few key results from Chapter 7, and then we expand on our brief discussion from Chapter 3 on the subtle (but important) effects of selection in decreasing the effective population size. We then turn to drift and mutation. Historically, long-term response theory focused solely on the effects of drift, and was only augmented later to include the role of mutation. Our discussion follows this same historical development by first considering Robertson's theory for the expected response in the presence of drift using only the initial variation.

Two important results emerge from Robertson's work. First, he obtained a *simple upper bound* of $2N_eR(1)$, twice the effective population size times the response in the first generation, for the selection limit. This simple, and rather general, result is in sharp contrast to the lack of such an analytical bound (based on easily quantified measures) under completely deterministic theory (Chapter 25). Second, there is an *optimal selection intensity*. If there is a fixed number of measured individuals, as we increase the selection intensity (and hence the short-term response), we do so at the expense of reducing the effective population size, thus reducing long-term response. We conclude our discussion of Robertson's theory by considering extensions allowing for linkage and various aspects of population structure (such as family selection, drift in the base population, and selection in a subdivided population).

Finally, we consider the effects of new mutations, both in terms of their (generally) minor role at the start of selection (except for rare mutations of large effect), and their

Evolution and Selection of Quantitative Traits. Bruce Walsh & Michael Lynch.
Published 2018 by Oxford University Press. © Bruce Walsh & Michael Lynch 2018.
DOI 10.1093/oso/9780198830870.001.0001

growing role as drift and selection erode the initial variance, which eventually leads to an asymptotic rate of response biased entirely on mutational input. Over sufficiently long time scales, response to continual directional selection is due to the fixation of a series of new mutations, an **adaptive walk**, which is examined in Chapter 27.

THE POPULATION GENETICS OF SELECTION AND DRIFT

In Chapter 7 we examined the interaction of selection and drift at a single locus, and we briefly review a few of the key results here. If the population size is finite, a favorable allele may become lost, and hence our interest is in u, the probability that an allele is fixed. In an infinite population, $u = 1$ for an allele favored by selection (provided it is not overdominant). In a finite population, $u < 1$, and its particular value depends on (among other things) its initial frequency, p_0, and the effective population size, N_e. Let $u(p_0)$ denote the probability that an allele starting at an initial frequency of p_0 becomes fixed. Recall from Chapter 2 that the fixation probability of a neutral allele depends only on its initial frequency, with

$$u(p_0) = p_0 \tag{26.1}$$

This independence from population size is not the case for an allele under selection. For additive selection (with fitnesses of $0 : s : 2s$)

$$u(p_0) \simeq \frac{1 - e^{-4N_e s p_0}}{1 - e^{-4N_e s}} \tag{26.2a}$$

$$\simeq p_0 + 2N_e s p_0(1 - p_0) \quad \text{when} \quad 2N_e|s| \leq 1 \tag{26.2b}$$

Equation 26.2a, from Kimura (1957), is derived using diffusion theory in Appendix 1, and values of $u(p_0)$ are plotted in Figure 26.1. Equation 26.2b, which is from Robertson (1960a), uses the approximations $e^{-x} \simeq 1 - x + x^2/2$ and $(1 - x)^{-1} \simeq 1 + x$ (both for $|x| \ll 1$) to simplify Kimura's result under weak selection. Equation 26.2b shows that

$$u(p_0) \simeq p_0 \qquad \text{if} \qquad 2N_e|\,s\,| \ll 1 \tag{26.2c}$$

Comparing this result to Equation 26.1 shows that an allele whose selection coefficient satisfies $2N_e|\,s\,| \ll 1$ behaves as if it were neutral over all allele frequencies, and it is called **effectively neutral** to reflect this fact. Equations 7.18–7.20 present corresponding expressions for the probability of fixation when dominance is present.

Even when an allele is strongly selected ($4N_e s \gg 1$), drift is important when its frequency is near zero or one. Taylor-expanding the numerator of Equation 26.2a yields

$$1 - e^{-4N_e s p_0} \simeq 1 - (1 - 4N_e s p_0) = 4N_e s p_0, \quad \text{or} \quad 2s\frac{N_e}{N} \quad \text{for} \quad p_0 = \frac{1}{2N}$$

Hence, the probability of fixation starting with a single copy, $p_0 = 1/(2N)$, of an advantageous allele is approximately $2s\,(N_e/N)$ when $4N_e s \gg 1$, implying that a favorable allele introduced as a single copy is usually lost by drift. Conversely, the fixation of a favored allele becomes almost certain as its frequency becomes sufficiently large. Equation 26.2a shows that if $p_0 \geq 1/(2N_e s)$, the probability of fixation exceeds 0.86, while if $p_0 \geq 1/(N_e s)$, the probability of fixation exceeds 0.98. Thus, if a favorable allele initially increases by drift, it can reach a threshold frequency above which deterministic selection dominates, which rapidly increases its frequency toward 1.0. As it approaches a frequency of 1.0, drift will again dominate, fixing the allele much more rapidly than expected under deterministic selection (Example 8.1).

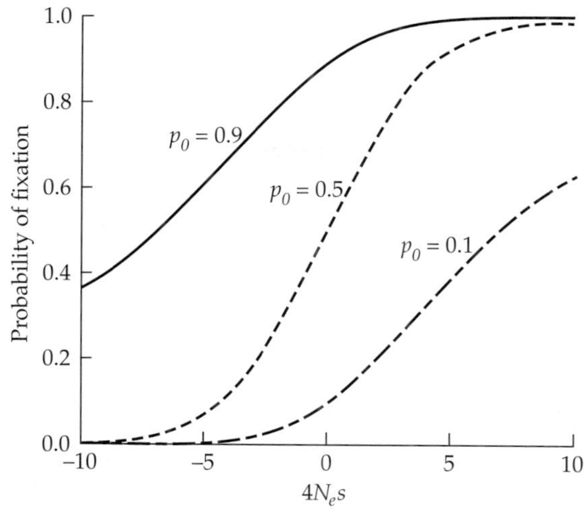

Figure 26.1 The probability of fixation of an additive allele as a function of $N_e s$ and its initial allele frequency, p_0 (from Equation 26.2a).

Finally, recall the Cohan effect (Example 7.4), which states that uniform selection can result in greater between-replicate divergence than drift. This occurs because the between-line divergence is maximized when the fixation probability of an allele is 0.5. Given that the fixation probability under neutrality is p_0, selection results in greater divergence than it does under neutrality when $u(p_0)[1 - u(p_0)] > p_0[1 - p_0]$. Figure 7.3 shows that the conditions for this to occur are not very restrictive under additive selection. As is discussed below, the Cohan effect has implications when crossing and then reselecting replicate lines.

Fixation Probabilities for Alleles at a QTL

We can translate the above results (and those from Chapter 7) for the fixation probability of an allele given its additive (s) and dominance (h) effects on *fitness* into the fixation probability for a favorable QTL allele as a function of its additive (a) and dominance (k) effects on the *trait* under selection. When the locus has only a small effect on the character, selection coefficients of $s = \bar{\imath}\, a/\sigma_z$ and $h = k$ (Equation 25.4) can be used in conjunction with Equation 7.18a to obtain fixation probabilities. If the allele displays no dominance in the character ($k = 0$), previous results imply that the probability of fixation exceeds 0.86 when $N_e s p_o \geq 1/2$, or

$$N_e \left(\bar{\imath}\, \frac{a}{\sigma_z} \right) p_0 \geq 1/2 \tag{26.3a}$$

For the fixation probability to exceed 0.86, the starting allele frequency must exceed

$$p_0 > \frac{\sigma_z}{a\, 2N_e\, \bar{\imath}} \tag{26.3b}$$

The fixation probability exceeds 98% when $N_e s p_0 > 1$, which corresponds to a critical allele frequency of twice that given by Equation 26.3b, namely,

$$p_0 > \frac{\sigma_z}{a\, N_e\, \bar{\imath}} \tag{26.4}$$

Note that if the product of initial allele frequency and its standardized effect, $p_0|a|/\sigma_z$, is sufficiently small, drift will dominate even if selection ($\bar{\imath}$) on the character is strong. With small values of $N_e\bar{\imath}$, only alleles of large effect or at moderate to high frequency are likely to be fixed by selection. As $N_e\bar{\imath}$ increases, favored alleles with smaller effects, or at lower frequencies, are increasingly more likely to become fixed. As a technical aside, the careful reader might recall from Chapter 14 that sampling causes fluctuations in $\bar{\imath}$ in a finite population. This additional complication need not overly concern us, as Hill (1969a, 1985) and Kojima (1961) showed that the error introduced by assuming a constant $\bar{\imath}$ is small.

Increased Recombination Rates Following Selection

We previously discussed the Hill-Robertson effect (Chapters 3, 7, and 8), wherein the effective population size is reduced in regions linked to a selected site, and in Chapter 4 we examined the evolution of the recombination rate. Otto and Barton (1997) showed that alleles at modifier loci that increase the recombination rate also increase the probability of fixation of favored alleles at selected loci linked to the modifier (also see Felsenstein 1974; Felsenstein and Yokoyama 1976). This can result in recombination modifiers hitchhiking along to fixation with the favored mutations. Such modifiers are favored because under low recombination, the effective selection coefficient on a particular mutation affecting a character depends on the selection coefficients at linked loci (e.g., Equation 7.42). Thus, the fate of a particular mutant is highly dependent upon the background in which it arose. As recombination increases, the fate of a mutation becomes increasingly uncoupled from the fate of its initial background. Rice and Chippendale (2001) experimentally demonstrated this by showing that the fixation probability of a mutation favored by artificial selection (white eye color in *Drosophilia*) increased with the recombination rate. Otto and Barton's theory makes the prediction that recombination rates may increase in selected populations relative to unselected controls, which is indeed observed in some experiments.

Example 26.1. Korol and Iliadi (1994) subjected a *Drosophila melanogaster* population to divergent selection for positive and negative geotaxis (an increased tendency to fly up and down, respectively). Recombination frequencies were scored on the unselected controls and the positively (geo^+) and negatively (geo^-) selected lines, with chromosomes II, III, and X scored after 36, 40, and 44 generations of selection. Over the scored regions (roughly 220 cM in the control), the map distance in the geo^+ line increased by a total of 78 cM (35%), while the geo^- line increased by 66 cM (30%). Presumably, these increases resulted from the increased probability of fixation of favorable mutations linked to modifiers increasing recombination frequencies. Other experiments in *Drosophila* also showed an increase in the recombination frequency following either directional or stabilizing selection (e.g., Thoday et al. 1964; Flexon and Rodell 1982; Zhuchenko et al. 1985; Rodell et al. 2004). A potentially related observation is that by Morran et al. (2009), who found that a collection of wild-type populations of the nematode *C. elegans* that were exposed to a bacterial pathogen had elevated rates of outcrossing compared with a set of controls that were not exposed. One concern with any such list of positive results is ascertainment bias, reporting a positive finding when it is present, but not when it is absent, thus inflating its apparent importance.

THE EFFECT OF SELECTION ON EFFECTIVE POPULATION SIZE

The simple act of selecting on a trait reduces the effective size, N_e, of a population below its actual size, N. Part of this is obvious, in that artificial selection proceeds by first choosing M (random) individuals to score and then selecting a fraction p of these to reproduce. Hence, $N_e \leq pM < N$, with stronger selection (smaller p) increasing the reduction in effective population size. As introduced in Chapter 3, there is a second, and much more subtle, effect that further reduces N_e for a selected population below that of an unselected control population with the same number of parents (i.e., $N_e < pM$). This additional reduction arises because selection inflates the variance in offspring number (more offspring are chosen from favorable families), reducing N_e. The initial realization of this phenomenon is often attributed to Morley (1954), who noted in sheep flocks exposed to selection that "the genetically superior individuals will tend to be most inbred," a result of a smaller (inbreeding) effective population size in the selected population relative to a control population of the same census size. However, Lush (1946) also very clearly understood this process, noting the "correlation between the fates of relatives" under selection and how this is expected to inflate the variance in offspring number.

While the effective size of a population under artificial selection can be retrospectively computed from either pedigree information or from the sampling variance in marker-allele frequencies (Chapter 4), predicting N_e *in advance* is more difficult. Its exact value depends on a variety of assumptions about both the family and population structure, and also on the underlying genetic model (the infinitesimal model is typically assumed). Theoretical investigations of the effects of selection on N_e were initiated by Robertson (1961), who presented simple approximations for both the single generation change and the asymptotic change following many generations of selection (Equations 3.29a and 3.29c).

Two different approaches have been used to examine the reduction in N_e from selection on unlinked loci. The first computes the expected variance in gene frequency for a neutral locus that is unlinked to any locus under selection (Robertson 1961; Nei and Murata 1966; Caballero 1994; Santiago and Caballero 1995), while the second computes the rate of inbreeding from the number of ancestors (Burrows 1984a, 1984b; Woolliams 1989; Verrier et al. 1990; Wray and Thompson 1990; Wray et al. 1990, 1994; Woolliams et al. 1993). The former corresponds to the variance effective population size, and the latter to the inbreeding effective population size. Strictly speaking, results based on diffusion theory require the usage of the variance effective size (as diffusion approximations use the sample variance in allele frequency). However, as discussed in Chapter 3, inbreeding and variance effective population sizes are usually equivalent unless the population size is changing over time. While all these treatments of the impact of selection on N_e consider the effective population size experienced by a neutral locus unlinked to loci influencing the traits under selection, the results should be very similar for selected loci under the infinitesimal model, as in this case drift (rather than selection) provides the major impetus for allele-frequency change.

The Expected Reduction in N_e from Directional Selection

Before proceeding, we will briefly review some results from Chapter 3. In the following discussion, N refers to the number of parents, pM. One of the assumptions of an ideal population (where the actual size, N, equals the effective size, N_e) is that all parents have an equal chance of contributing offspring. Equation 3.4 shows that variance, σ_k^2, in the number of offspring contributed by an individual reduces N_e, as $N_e = (N - 1/2)/(\sigma_k^2/4 + 1/2)$. If the number of offspring per parent follows a Poisson distribution, then $\sigma_k^2 = 2$ and $N_e = N - 1/2 \simeq N$. However, if some parents contribute a disproportionate number of offspring, then $\sigma_k^2 > 2$ and $N_e < N$. The more disproportionate the contribution, the larger is the offspring variance and the smaller is N_e.

As in Chapter 3, we use σ_k^2 to denote variation in offspring number for entirely nonheritable reasons. Here, the children of a parent contributing an excessive number of offspring are themselves no more likely to contribute an excessive number. But with selection on a heritable trait, this will no longer be true. A selected parent is likely to have offspring with favorable trait values that then *disproportionately* contribute to the next generation. Following Chapter 3, we separate these two sources of variance in offspring number, letting σ_k^2 refer to the variation in an unselected population and σ_w^2 to the additional among-family genetic variance in relative fitness (due to differential contribution from families with more favorable trait values). Because each family is assumed (on average) to contribute two offspring (in a population maintained at a stable size), the additional variance from selection is $2^2\sigma_w^2$, yielding an effective population size after a generation of selection of

$$N_e = \frac{4N}{\sigma_k^2 + 2 + 4\sigma_w^2} \tag{26.5}$$

Assuming $\sigma_k^2 = 2$ gives

$$N_e = \frac{N}{1+\sigma_w^2} \tag{26.6a}$$

a result from Robertson (1961). For truncation selection, Robertson (1961) and Milkman (1978) showed that $\sigma_w^2 \simeq \bar{\imath}^2\, t_{FS}$, where $t_{FS} = \text{Cov}(FS)/\sigma_z^2$ is the intraclass correlation of full sibs (LW Equation 17.3). Thus, the reduction in effective population size after a single generation of selection on unlinked loci becomes

$$N_e = \frac{N}{1+\bar{\imath}^2\, t_{FS}} \tag{26.6b}$$

There are two complications to consider when moving from this simple result for a single generation to the reduction in N_e over multiple generations of selection. First, as selection proceeds, genetic variances change, which in turn changes t_{FS}. The second complication, which was briefly noted in Chapter 3, is that continued selection has a *cumulative* effect. Robertson (1961) approximated this effect by noting that only half of the association between a neutral locus and unlinked loci under selection persists in each generation, yielding the among-family variance after τ generations of selection as

$$(1+1/2+1/4+\cdots+1/2^{\tau})^2\sigma_w^2 = Q_\tau^2\sigma_w^2$$

where Q_τ is the cumulative effect of τ generations of selection. Equation 26.6a thus becomes

$$N_{e,\tau} = \frac{N}{1+Q_\tau^2\,\sigma_w^2} = \frac{N}{1+Q_\tau^2\,\bar{\imath}^2\, t_{FS}} \tag{26.6c}$$

Q_τ approaches a limiting value of $Q = 2$, which yields Equation 3.29c, Robertson's long-term effective population size of $N/(1+4\sigma_w^2)$. As noted in Chapter 3, this result is approximate for several reason, most importantly because selection also reduces genetic variation, resulting in less than half of the value from the previous generation being passed along.

A more complete treatment was given by Santiago and Caballero (1995), who obtained a general expression for N_e under selection that allows for nonrandom mating. Assuming random mating, their expression for N_e after τ generations of selection is

$$\frac{N_{e,\tau}}{N} = \left[\frac{1-\gamma}{2} + \left(\frac{\sigma_k^2}{4} + Q_{*\tau}^2\sigma_w^2\right)(1+\gamma)\right]^{-1} \tag{26.6d}$$

where $\gamma = -1/(N-1)$ is a measure of the departure from Hardy-Weinberg due to finite population size (we generally assume that $\gamma \simeq 0$), and $Q_{*\tau}$ (given below; see Equation 26.9a) is a generalization of Robertson's Q_τ. For a single generation of selection, $Q_{*1} = 1$, implying that σ_w^2 is the effect of selection in the current generation, while $(Q_{*\tau}^2 - 1)\sigma_w^2$ is the cumulative effect of selection in previous generations. As will be shown, $Q_{*\tau}$ is a function of the selection intensity and heritability. Assuming that N is large (so $\alpha \simeq 0$) and a Poisson distribution of offspring in the absence of selection ($\sigma_k^2 = 2$), Equation 26.6d reduces to Equation 26.6c, but with $Q_{*\tau}$ replacing Q_τ. We are still, however, left with two complications—obtaining the value of t_{FS}, and finding $Q_{*\tau}$—which we address in turn.

Under the infinitesimal model, all of the selection-induced change in genetic variance is due to gametic-phase disequilibrium (Chapters 16 and 24). In Chapter 16 we showed that $\sigma_A^2 = \sigma_a^2 + d$, where σ_a^2 (the additive genic variance) is the additive genetic variance in the absence of disequilibrium and d is the disequilibrium contribution. Under the infinitesimal model, the within- and among-family contributions to the additive genetic variance differ, as (for unlinked loci) the within-family contribution ($\sigma_a^2/2$) is not influenced by disequilibrium, while the among-family variance ($\sigma_a^2/2 + d$) is (Chapter 16). Assuming the absence of both dominance and shared sib environmental effects, the intraclass correlation becomes

$$t_{FS} = \frac{\sigma_a^2/2+d}{\sigma_A^2+\sigma_E^2} = \frac{\sigma_a^2/2+d}{\sigma_a^2+d+\sigma_E^2} = \frac{h_0^2/2+d/\sigma_{z(0)}^2}{1+d/\sigma_{z(0)}^2} \tag{26.7a}$$

where h_0^2 and $\sigma^2_{z(0)} = \sigma^2_a + \sigma^2_E$ are, respectively, the heritability and phenotypic variance in the unselected base population ($d = 0$). Increasing either $\bar{\imath}^2$ or h^2 increases $\sigma^2_w = \bar{\imath}^2\, t_{FS}$, which in turn decreases N_e. Dominance variance does not change under the infinitesimal model (Chapter 16), so when it, or a common-family effect, occur, both appear as constants in the numerator and denominator of Equation 26.7a.

We now have all of the results needed to compute an improved expression for the single-generation reduction in N_e. Expressing the fraction of phenotypic variance in the selected parents as $\sigma^2_{z*}/\sigma^2_z = (1-\kappa)$, Equation 16.7a yields $d(1)/\sigma^2_{z(0)} = -\kappa\, h_0^4/2$ (Equation 16.11a gives κ as a function of $\bar{\imath}$). Substituting this value for $d(1)/\sigma^2_{z(0)}$ into Equation 26.7a,

$$t_{FS}(1) = \frac{h_0^2(1-\kappa h_0^2)/2}{1-\kappa h_0^4/2} \tag{26.7b}$$

Recalling that $Q_{*1} = 1$, Equation 26.6d (assuming $\gamma = 0, \sigma^2_k = 2$) yields a reduction in N_e from a single generation of selection of

$$N_{e,1} \simeq N\left(1 + \bar{\imath}^2\, \frac{(h_0^2/2)(1-\kappa h_0^2)}{1-\kappa h_0^4}\right)^{-1} \tag{26.8}$$

This result was first obtained by Robertson (1961), who did not include the $(1-\kappa h_0^4/2)$ term, which was subsequently added by Wray and Thompson (1990).

Turning to $Q_{*\tau}$, the cumulative effect of past selection, Santiago and Caballero (1995) showed that

$$Q_{*\tau} = 1 + \frac{G}{2}(1+\rho) + \cdots + \left[\frac{G}{2}(1+\rho)\right]^{\tau} = \sum_{i=0}^{\tau}\left[\frac{G}{2}(1+\rho)\right]^{i} \tag{26.9a}$$

where G is the fraction of genetic variance remaining after selection and ρ is the correlation between the selective values of mates ($\rho = -1/[N-1]$ under random mating). If $(1-\kappa)$ is the fraction of phenotypic variance after selection, then $G = 1 - \kappa\, h^2$ is the fraction of additive variance (Chapter 16). As with t_{FS}, G (and hence $Q_{*\tau}$) depends on a parameter (h^2) that changes under selection. However, recall from Chapter 16 that, under the infinitesimal model, h^2 quickly reaches its equilibrium value under directional selection (in roughly two or three generations). Thus, we typically use $G = 1 - \kappa\,\widehat{h}^2$, a function of the equilibrium heritability under the effects of selection and disequilibrium alone. In the limit ($\tau \to \infty$), the sum in Equation 26.9a converges to

$$\widetilde{Q}_* = \frac{2}{2-G(1+\rho)} \simeq \frac{2}{1+\kappa\,\widehat{h}^2} \quad \text{when } \rho \simeq 0 \tag{26.9b}$$

Robertson assumed a limiting value of $Q = 2$, but Equation 26.9b shows that this is an overestimate, which results in an underestimation of N_e. Substitution of Equation 26.9b into Equation 26.6c recovers Equation 3.29b (with the latter expressed in terms of $L = 1 - G$, where $L = \kappa h^2$ is the fractional loss of additive variance).

The general prediction that effective population size decreases in selected populations has been examined in a number of *Drosophila* experiments, where inbreeding is estimated directly from parental pedigrees. This prediction has generally been confirmed, with a reasonable fit to Robertson's theory (McBride and Robertson 1963; Jones 1969a, 1969b; López-Fanjul 1989). As expected from Equation 26.8, N_e is lowest in lines showing the greatest response to selection, as these lines have the highest realized heritabilities. Gallego and López-Fanjul (1983) tested a second prediction using selection on sternopleural bristles: because the reduction in N_e occurs from among-family selection (inflating the among-family variance, σ^2_k), no reduction in N_e is expected under within-family (full-sib) selection (Chapter 21). In accordance with theory, no reduction was observed.

As reviewed in Chapter 25, reproductive fitness often declines during long-term selection experiments. This can result in a further increase in the variance in fitness among individuals, which in turn further increases the variance in offspring number (as the latter is a measure of fitness). This increased variance can significantly decrease the effective population size below that predicted by Equation 26.8, which incorporates only the variance effects associated with artificial selection. Yoo (1980c) found that differences in fertility were more important in reducing effective population size than the effects of artificial selection during a long-term selection experiment for increased *Drosophila* bristle number.

Example 26.2. Consider directional truncation selection on a normally distributed character in which the uppermost 20% of the population ($p = 0.2$) is saved. From Example 16.3, this yields a selection intensity of $\bar{\imath} = 1.40$ and a reduction in variance of $\kappa = 0.781$. If we assume initial (before selection) values of $h_0^2 = 0.5$ and $\sigma^2_{z(0)} = 100$, Example 16.2 yields (under the infinitesimal model) equilibrium values of $\widehat{d} = -12.54$ and $\widehat{h}^2 = 0.428$. Hence

$$G = 1 - \kappa\,\widehat{h}^2 = 1 - 0.781 \cdot 0.428 = 0.665$$

Because we are assuming no dominance or common-family effects, the initial value of t_{FS} in the base population is $h_0^2/2 = 0.25$, while its equilibrium value becomes

$$\widehat{t}_{FS} = \frac{h_0^2/2 + \widehat{d}/\sigma^2_{z(0)}}{1 + \widehat{d}/\sigma^2_{z(0)}} = \frac{(0.5/2) - (12.54/100)}{1 - (12.54/100)} = 0.142$$

Hence, $\sigma^2_w = \bar{\imath}^2 \cdot \widehat{t}_{FS} = 1.4^2 \cdot 0.142 = 0.279$. Assuming $\rho = -1/(N-1) \simeq 0$, Equation 26.9b yields

$$\widetilde{Q}_* = \frac{2}{2 - G} = \frac{2}{2 - 0.665} = 1.498$$

Equation 26.6c yields an equilibrium effective population size of

$$N_e = \frac{N}{1 + \widetilde{Q}_*^2 \sigma_w^2} = \frac{N}{1 + 1.498^2 \cdot 0.279} = 0.615\,N$$

In contrast, using Robertson's approximation ($Q = 2$) in place of $\widetilde{Q}_*$ in the previous expresson returns a smaller value, $N_e = 0.473\,N$. Similar calculations using other p values (a smaller p equals stronger selection) yields

p	κ	$\widehat{d}$	$\widehat{t}_{FS}$	G	σ_w^2	$\widetilde{Q}_*$	N_e/N
0.50	0.64	-10.92	0.16	0.72	0.10	1.56	0.80
0.10	0.83	-13.05	0.14	0.65	0.42	1.48	0.52
0.05	0.86	-13.36	0.13	0.64	0.57	1.47	0.45
0.01	0.90	-13.76	0.13	0.62	0.93	1.45	0.34

Example 26.3. Cohan and Hoffmann (1986) examined the divergence between replicate lines of *Drosophila melanogaster* selected for increased resistance to ethanol. The selected lines had a higher among-line variance for characters associated with increased resistance than did the unselected control replicates. This could be explained by a reduction in effective population size due to selection or by the Cohan effect (Example 7.4). The reduction in effective population size, by increasing drift, is expected to increase the among-line variance in *any* character, selected or unselected. Conversely, the Cohan effect predicts that only characters under selection, or characters controlled by loci tightly linked to QTLs for these selected characters, should show increased divergence. Cohan and Hoffmann found no differences between the selected and control lines for three unselected characters, which suggested that the main cause of increased divergence was the Cohan effect.

DRIFT AND LONG-TERM SELECTION RESPONSE

Recall that in our distinction between long-term and asymptotic response, the former is attributable to the existing variation at the start of selection, while the latter is the expected eventual rate of response due to the input of new mutation. When the effective population size is small, essentially all of the observed response is due to the initial variation, with the population reaching an apparent limit until the appearance of new mutations allows for further response. In larger populations, these two components of response become more difficult to separate, and no limit may be observed when in fact all of the initial variation has been exhausted. Much of the initial theory of long-term response ignored mutation, and we examine this drift-only version first, as it provides a good description of how a population exhausts its initial variation.

Basic Theory

We expect the response to selection in very small populations to be significantly influenced by drift, showing less total response than in larger populations starting with the same initial genetic variance. A fairly extensive theory examining the effects of drift on long-term response (the utilization of the initial genetic variation) has been developed, starting with the extremely influential paper of Robertson (1960a). Most of this theory is based on summing over single-locus results, which we adhere to unless stated otherwise (this assumes that epistasis and linkage effects can be ignored).

As before, we first consider a single diallelic locus (indexed by i) where the genotypes $aa : Aa : AA$ have genotypic values (for the character under selection) of $0 : a(1+k) : 2a$. Let p_t denote the frequency of A at this locus in generation t, $\Delta_i(t)$ be the contribution to total response from this locus in generation t, and $u_i(p_0)$ be the probability that A is ultimately fixed at this locus, provided it starts at a frequency of p_0. The total response is obtained by summing over all loci, $R(t) = \sum_i \Delta_i(t)$. Under drift, both p_t and $\Delta_i(t)$ are random variables and (assuming that the genotypes are in Hardy-Weinberg proportions) are related by

$$\begin{aligned}\Delta_i(t) &= m_i(p_t) - m_i(p_0)\\ &= 2a\left[\, p_t - p_0 + k\left(\, p_t(1-p_t) - p_0(1-p_0)\right)\right] \qquad (26.10a)\end{aligned}$$

where $m_i(p)$, the expected contribution to the trait from locus i when the frequency of A is p, is given by Equation 25.1a. The expected contribution (at generation t) from this locus is

$$E[\,\Delta_i(t)\,] = 2a\left[\, E(p_t) - p_0 + k\left(\, E[\,p_t(1-p_t)\,] - p_0(1-p_0)\right)\right] \qquad (26.10b)$$

Because A is ultimately either fixed ($p_\infty = 1$) or lost ($p_\infty = 0$), $E(p_t)$ converges to

$$1 \cdot u_i(p_0) + 0 \cdot [\,1 - u_i(p_0)\,] = u_i(p_0)$$

while $E\,[\,p_t(1-p_t)\,]$ converges to zero. The limiting expected contribution from locus i becomes

$$E[\,\Delta_i(\infty)\,] = 2a\left[u_i(p_0) - p_0 - k\left(\, p_0(1-p_0)\right)\right] \qquad (26.11a)$$

Two cases of special interest are when A is additive ($k = 0$), yielding

$$E[\,\Delta_i(\infty)\,] = 2a\,[u_i(p_0) - p_0] \qquad (26.11b)$$

and when A is recessive ($k = -1$), in which case

$$E[\,\Delta_i(\infty)\,] = 2a\,\left[u_i(p_0) - p_0^2\,\right] \qquad (26.11c)$$

The variance (and indeed all higher moments) of the total response at the selection limit are easily computed, as, regardless of the value of k, $\Delta_i(\infty)$ takes on only two values,

$$\begin{aligned} \Delta_i(\infty) &= 2a - m_i(p_0) \quad \text{with probability } u_i(p_0) \\ &= \ \ 0 - m_i(p_0) \quad \text{with probability } 1 - u_i(p_0) \end{aligned} \tag{26.12}$$

In particular, the variance in the contribution from this locus over replicate selected lines is

$$\begin{aligned} \sigma^2\left[\,\Delta_i(\infty)\,\right] &= E\left[\,\Delta_i^2(\infty)\,\right] - \Big(E\left[\,\Delta_i(\infty)\,\right]\Big)^2 \\ &= 4a^2 u_i(p_0)\left[1 - u_i(p_0)\right] \end{aligned} \tag{26.13a}$$

With weak selection, $u_i(p_0) \simeq p_0$ (i.e., the allelic dynamics are largely governed by drift), implying

$$\sigma^2\left[\,R(\infty)\,\right] \simeq 4\sum a^2 p_0(1 - p_0) \tag{26.13b}$$

which is twice the initial additive-genetic variation (assuming that all loci are additive), and also is the expected among-line divergence under pure drift (Chapter 11). Under sufficiently strong selection, almost all favorable alleles will be fixed and the variance will be close to zero, as $u_i(p_0) \simeq 1$. When selection is moderate to weak, loci for which $u_i(p_0)[\,1 - u_i(p_0)\,] > p_0(1 - p_0)$ show a Cohan effect. If such loci are sufficiently frequent, selection *increases* the among-line variance relative to drift. This requires both weak selection and that most favored alleles be rare. The variance in response at the selection limit was considered in more detail by Hill and Rasbash (1986), Zeng and Cockerham (1990), and Zhang and Hill (2005a).

The variance in the selection limit across replicate lines has a direct bearing on whether further response can occur by crossing plateaued lines and then reselecting. If drift has played a significant role in the selection response, a line formed by crossing replicate plateaued lines should show further response to selection, as each line should be fixed for a considerable number of unfavorable alleles. In particular, with weakly selected loci, the Cohan effect can inflate the among-replicate variance over that expected under drift, increasing the potential for additional response when crossing weakly selected lines over that expected from crossing lines generated by drift alone.

Replicate lines at their selection limits usually show considerable genetic differences (reviewed by Cohan 1984a, 1984b). For example, Scowcroft (1965) used chromosomal analysis to show that three replicate *Drosophila* lines selected for increased scutellar bristles differed considerably in the amount of response attributable to each chromosome and the nature of interactions between chromosomes. Synthetic lines formed by crossing either replicate plateaued lines (Frankham et al. 1968b; Eisen 1975; Frankham 1980) or unrelated plateaued lines (Falconer and King 1953; Roberts 1967) generally respond to selection.

An interesting exception was revealed by Gallego and López-Fanjul (1983), who selected on sternopleural bristle number in *Drosophila*. Replicate lines showed a very rapid exhaustion of response, and crosses between lines did not result in further response. The authors interpreted these results as being consistent with a few alleles of large effect, which were initially at an intermediate frequency. These alleles rapidly go to fixation, with all lines being fixed for the same major alleles. Similarly, Skibinski and Shereif (1989) found that the among-line variance of lines selected for sternopleural bristle number decreased over time. The among-line variance is expected to increase over time if drift dominates (e.g., Equation 12.1b) or if there is weak selection on the underlying loci, but it is expected to decrease if the lines are fixed for the same few major genes.

Robertson's Theory of Selection Limits

Equations 26.10–26.13 are fairly general, assuming only Hardy-Weinberg, no epistasis, and that single-locus results can be added across loci. To proceed further, we need explicit expressions for $u_i(p_0)$ to describe the limit, and for both $E(p_t)$ and $E[\,p_t(1-p_t)\,]$ to describe the dynamics. The most complete description, by Robertson (1960a), is for additive alleles, where $E[\,\Delta_i(t)\,] = 2a[\,E(p_t) - p_0)\,]$. Recalling Equation 7.28a yields

$$E(p_t) \simeq p_0 + 2N_e\left(1 - e^{-t/2N_e}\right) s\,p_0(1-p_0)$$

as an approximate expression for the expected allele frequency, under the assumption that the allele has a small effect (i.e., is nearly neutral). For notational ease, we will drop the expectation notation, but the reader should keep in mind that we are examining the expected response. Recalling from Equation 25.4 that $s = a\bar{\imath}/\sigma_z = aS/\sigma_z^2$ (as $\bar{\imath} = S/\sigma_z$), substitution into Equation 7.28a yields an expected response from locus i after t generations of selection of

$$\Delta_i(t) = 2a[\,E(p_t) - p_0)\,] \simeq 2N_e\left(1 - e^{-t/2N_e}\right)\left(\frac{aS}{\sigma_z^2}\right) 2a\,p_0(1-p_0) \tag{26.14a}$$

This can by simplified further by noting that $2a^2p_0(1-p_0)$ is the initial additive variance contributed by the locus. Because we assumed no epistasis and no linkage disequilibrium, summing over all loci gives the cumulative response at generation t as

$$R(t) \simeq 2N_e\left(1 - e^{-t/2N_e}\right)\frac{S\sigma_A^2(0)}{\sigma_z^2} \tag{26.14b}$$

Note that $S\,\sigma_A^2(0)/\,\sigma_z^2 = S\,h^2(0) = R(1)$ is the expected response in the first generation, provided that the conditions for the breeder's equation hold. Equation 26.14b implies that

$$R(t) \simeq 2N_e\left(1 - e^{-t/2N_e}\right)R(1) \tag{26.15a}$$

returning an expected limiting total response of

$$R(\infty) \simeq 2N_e\,R(1) \tag{26.15b}$$

Because $R(1)/\sigma_z = h^2\,S/\sigma_z = h^2\,\bar{\imath}$, the expected limiting response in terms of phenotypic standard deviations is

$$R(\infty)/\sigma_z \simeq h^2(2N_e\,\bar{\imath}\,) \tag{26.15c}$$

Note that Equation 26.15a motivates the use of exponential regressions in Chapter 25 to estimate selection limits (Equation 25.10). The careful reader will note that we assumed that the phenotypic variance remains relatively constant over time, as would occur if h^2 were small (and hence a decrease in the heritability will have little impact on σ_z^2). Provided this assumption holds, the total expected response is simply $2N_e$ times the initial response, as first suggested by Dempster (1955b) and formally derived by Robertson (1960a). An alternative derivation of Equation 26.15a is as follows. Assuming the main force for allele-frequency change is drift, Equation 11.2 yields

$$\sigma_A^2(t) \simeq \sigma_A^2(0)[1 - 1/(2N_e)]^t \simeq \sigma_A^2(0)\exp\left[-t/(2N_e)\right] \tag{26.15d}$$

Writing the response in generation t as $h^2(t)S = \sigma_A^2(t)\,\bar{\imath}/\sigma_z$, summing over generations and applying Equation 7.28b recovers Equation 26.15a.

Equation 26.15b is an *upper limit* for the total response, which may seem somewhat counterintuitive because it was derived by assuming weak selection. The key to understanding this upper bound is that (everything else being equal) the initial response $R(1)$ when selection dominates is much larger than when drift dominates, so $2N_e$ (the time for drift to remove a significant amount of genetic variation) times the initial response overestimates the total response when selection dominates. To see this point, consider the maximal contribution, $\Delta_i^{max} = 2a(1-p_0)$, from a locus (which occurs when the favored allele is fixed) relative to the predicted contribution, $\Delta_i(\infty)$. From Equation 26.14a, it follows that

$\Delta_i(\infty) = 2N_e\, 2a^2 p_0(1-p_0) S/\sigma_z^2$. Substituting $S/\sigma_z^2 = \bar{\imath}/\sigma_z$ yields the ratio of maximum to expected contribution as

$$\frac{\Delta_i^{max}}{\Delta_i(\infty)} = \frac{1}{2N_e}\,\frac{2a(1-p_0)}{2a^2 p_0(1-p_0)\,\bar{\imath}/\sigma_z} = \frac{\sigma_z}{2N_e\,\bar{\imath}\,a\,p_0} \tag{26.16a}$$

Thus, $2N_e R(1)$ overestimates the ultimate limit ($\Delta_i(\infty) > \Delta_i^{max}$) when

$$2N_e\,\bar{\imath}\,a\,p_0 > \sigma_z \tag{26.16b}$$

implying that $2N_e R(1)$ overestimates the ultimate limit when this inequality is satisfied.

Recalling Equation 26.3b, the probability of fixation is greater than 86% when Equation 26.16b is satisfied. Increasing the effective population size above this threshold has little effect on increasing the selection limit, as $u_i(p_0) \simeq 1$ and, hence, the contribution from the ith locus is Δ_i^{max}. In contrast, when the inequality provided by Equation 26.16b fails, $\Delta_i^{max} > \Delta_i(\infty)$. However, in this case drift is expected to dominate (see Equation 26.3a), so we do not expect ot obtain the maximal possible response from each locus, as many favored loci will be lost, rather than fixed.

Another quantity of interest is the expected half-life of response, $t_{0.5}$, the time required to obtain half the final response. Recalling Equation 26.14a, and solving $1 - e^{-t_{0.5}/2N_e} = 1/2$, yields an expected half-life of

$$t_{0.5} = N_e \ln 2 \simeq 1.4N_e \tag{26.17}$$

Again, this is an *upper limit*, with the half-life decreasing as the product $N_e\bar{\imath}$ increases. An observed half-life considerably below that predicted by Equation 26.17 suggests that a large portion of the response is due to the fixation of favorable alleles by selection, as selection (when it dominates) changes allele frequencies much faster than drift.

Equations 26.14–26.17 rely on a number of assumptions besides additivity: no opposing natural selection, no linkage effects, two alleles per locus, and weak selection (on loci). Several authors have examined the robustness of these results. Hill and Rasbash (1986) found, for diallelic loci, that the distribution of allelic effects is relatively unimportant, but differences in allele frequencies can be critical. In particular, increasing the effective population size has much more of an effect on the selection limit when favored alleles are rare. This is expected, as the dynamics of common alleles at selected loci are largely governed by selection rather than drift (Equation 26.3b). Increasing population size lowers the critical allele-frequency threshold for selection to dominate, eventually capturing even rare alleles (Zhang and Hill 2005a). Latter and Novitski (1969) and Zeng and Cockerham (1990) examined the effects of multiple alleles, and found that the results for the expected limit (Equation 26.15b) and the half-life (Equation 26.17) are reasonable when selection is weak. As $N_e\bar{\imath}$ increases, $R(\infty)/R(1)$ becomes highly dependent on the number and frequencies of alleles at each locus (Chapter 25). In general, this ratio increases with the number of alleles, and decreases with increasing $N_e\bar{\imath}$, all the while remaining bounded by a factor of $2N_e$. Likewise, $t_{0.5}$ decreases as $N_e\bar{\imath}$ increases, but it is rather insensitive to the number of alleles.

With dominance, analytic results for the limit and half-life ($R[\infty]$ and $t_{0.5}$) are more complicated. Strictly recessive alleles have received the most study. In this case, the selection limit can considerably exceed $2N_e$ times the initial response when the character is controlled by a large number of rare recessives (Robertson 1960a). Additive genetic variance increases, often considerably, as these recessives increase in frequency, so this result should not be surprising (Chapter 25). With weak selection, the half-life from recessives varies from approximately N_e when $p \simeq 1$ to approximately $2N_e$ when $p \simeq 0$ (Robertson 1960a). Again, as $N_e\,\bar{\imath}$ increases, half-life decreases. Even with strictly additive loci, a temporary increase in the genetic variance (even in the face of genetic drift) can occur if there are a number of rare, but favored, alleles (Chapter 25). As these alleles increase in frequency, the additive variance also increases. If genetic drift strictly governs the dynamics of the additive variance, these rare alleles have only a small chance of increasing and do not significantly (on average) inflate the variance. However, if selection is of even modest importance to the dynamics

Table 26.1 Observed and predicted selection limits ($2N_e h^2 \bar{\imath}$, scaled in terms of σ_z; Equation 26.15c) and half-lives (scaled in terms of N_e) for a variety of characters in laboratory populations of mice. The Ratio column under Half-life is the fraction of the predicted upper limit for the half-life ($1.4N_e$; Equation 26.17), observed. (From Hanrahan et al. 1973; Eisen 1975; and Falconer 1977.)

Character	Direction of	Total Response			(Half-life)/N_e	
Selected	Selection	Observed	Predicted	Ratio	Observed	Ratio
Weight						
Strain N	Up	3.4	7.2	0.47	0.6	0.43
	Down	5.6	15.9	0.35	0.6	0.43
Strain Q	Up	3.9	15.8	0.27	0.2	0.14
	Down	3.6	9.6	0.38	0.4	0.29
Growth	Up	2.0	7.4	0.27	0.3	0.21
	Down	4.5	13.7	0.33	0.5	0.36
Litter Size	Up	1.2	2.3	0.52	0.5	0.36
	Down	0.5	7.7	0.06	0.5	0.36
Postweaning weight gain						
Line M4	Up	1.5	5.4	0.27	0.9	0.64
Line M8	Up	2.0	10.0	0.20	0.5	0.36
Line M16	Up	4.3	45.0	0.10	0.3	0.21

at any particular locus, as Chapter 25 highlights, the single-generation response is a very poor predictor of the long-term response.

James (1962), Verghese (1974), Nicholas and Robertson (1980), and Zeng and Hill (1986) extended Robertson's theory for various models of natural selection opposing artificial selection. Not surprisingly, the selection limit is reduced by the presence of opposing natural selection. In the absence of mutation, none of these models retain genetic variability, as drift eventually fixes all loci, even those displaying overdominance in fitness.

TESTS OF ROBERTSON'S THEORY OF SELECTION LIMITS

Robertson's theory applies to the expected response from the existing variation in the base population at the start of selection. Eventually, mutational input becomes important and will ultimately dominate the long-term response, a point we will develop in detail shortly. In the very small population sizes common in many selection experiments, the distinction between exhaustion of the initial variation and the additional response due to new mutation can be fairly clear, as the latter takes many more generations to become apparent than it takes to remove existing variation. For larger population sizes, the two sources of response become increasingly blurred. Hence, most tests of Robertson's theory use very small populations.

Observed limits and half-lives are usually considerably below the values predicted from Robertson's theory (reviewed in Roberts 1966; Kress 1975; Eisen 1980; Falconer and Mackay 1996). Table 26.1 gives various results from experiments with mice. These discrepancies between observation and theory are not unexpected. Robertson's theory assumes that the limit is reached as genetic variance is exhausted by fixation at all loci. As noted in Chapter 25, selection limits can occur despite significant additive genetic variance, often because natural and artificial selection are in conflict. Further, the selection limit of $2N_e R(1)$, and the half-life of $1.4N_e$, are expected *upper* limits that assume that drift largely dominates. An additional complication is that the effective population size is overestimated by taking N_e as the number of parents (Chapter 3). For example, variation in male mating success in *Drosophila* can decrease the effective population size to less than half the actual number of male parents (Crow and Morton 1955). Further, most experiments have not corrected for the expected reduction in N_e from the effects of artificial selection (Equation 26.6c).

Table 26.2 The cumulative response after 50 generations of selection for increased abdominal bristle number in *Drosophila melanogaster* as a function of the effective population size and the selection intensity. N_e is estimated as half the number of parents. None of the lines showed an apparent plateau, but the experiment was stopped after 50 generations. For fixed N_e, the response increases with $\bar{\imath}$ (compare entries within a column), while for fixed $\bar{\imath}$, response increases with N_e (compare entries across a row). (After Jones et al. 1968.)

N_e	$\bar{\imath}$	$R(50)$	N_e	$\bar{\imath}$	$R(50)$	N_e	$\bar{\imath}$	$R(50)$
10	1.6	16.3	20	1.7	20.3	40	1.7	31.7
10	1.3	11.2	20	1.4	14.7	40	1.4	18.8
10	0.9	8.1	20	1.0	12.2	40	1.0	16.4

A more direct test of Robertson's theory evaluates whether the selection limit increases, and the half-life decreases, as $N_e\bar{\imath}$ increases. In general, both of these predictions hold. For example, the estimated effective population sizes of lines M4, M8, and M16 in Table 26.1 were 7.7, 18.6, and 40.9, while each line experiences essentially the same value of $\bar{\imath}$ (Eisen 1975). For this data set, half-life decreases as $N_e\,\bar{\imath}$ increases, as predicted by theory. In a more extensive experiment, Jones et al. (1968) examined the effects of changing N_e or $\bar{\imath}$ on otherwise replicate lines of *Drosophila melanogaster*. Because all of their populations were still responding at the end of the experiment (50 generations), they did not estimate the limit or half-lives (although one could use their data with Equation 25.10a to do so). Nevertheless, their data (Table 26.2) are consistent with Robertson's qualitative predictions, as long-term response increases with $N_e\bar{\imath}$ (Figure 26.2).

Robertson's theory further predicts that when the effective population size is sufficiently large, further increases in N_e should not change the limit (provided mutational input can be ignored), as all favorable alleles that were initially present become fixed. This has yet to be observed, which is perhaps not surprising given that most experiments have value of N_e below 50. By designing ingenious devices to facilitate mass selection in *Drosophila melanogaster*, Weber (1990, 1996, 2004; Weber and Diggins 1990) were able to examine the consequences of larger population sizes. Selection experiments on wing-tip height (Weber 1990) and ethanol tolerance (Weber and Diggins 1990) had effective population sizes on the order of $N_e \simeq 200$–400. Both characters showed an increased response with increasing N_e. The data for wing-tip height are given in Figure 26.3A. Figure 26.3B summarizes the results of nine other experiments from previous studies, showing the ratio of response after 50 generations to the initial response. As predicted, this ratio generally increases with values of N_e. The implication is that there is additional "usable" genetic variation present in the base population that can be exploited by increasing the scaled strength of selection ($N_e\,\bar{\imath}$). In very small populations, only major alleles are influenced by selection (Equation 26.3). The observation that response continues to increase with N_e suggests a large pool of alleles of smaller effects, or at lower frequencies, or both. As $N_e\,\bar{\imath}$ increases, favorable alleles at these loci are more likely to become fixed, increasing response. Larger populations also provide a greater chance for recombination to remove deleterious linked combinations, which might be fixed in smaller populations, further increasing the potential for response.

One complication with Robertson's theory is that as population size increases, the contribution from mutational input becomes increasingly important over the time scales that it takes to remove the initial variation. We will address this point shortly. A second complication is that when the character value is influenced by inbreeding depression (as will occur if directional dominance is present), its effects are more dramatic in smaller populations. One test for whether inbreeding depression is reducing the selection response is to cross divergently selected lines and look for significant increases in the mean in the resulting F_1 population (e.g., Eisen 1975; Kownacki 1979).

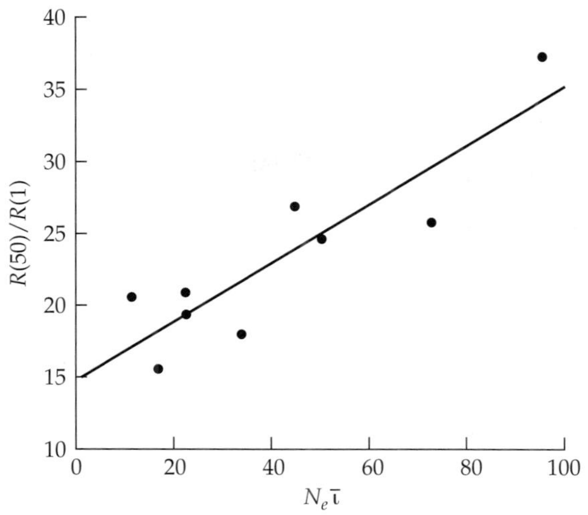

Figure 26.2 Cumulative response at generation 50 as a function of $N_e\bar{\iota}$ for selection on increased abdominal bristle number in *Drosophila melanogaster*. (After Jones et al. 1968.)

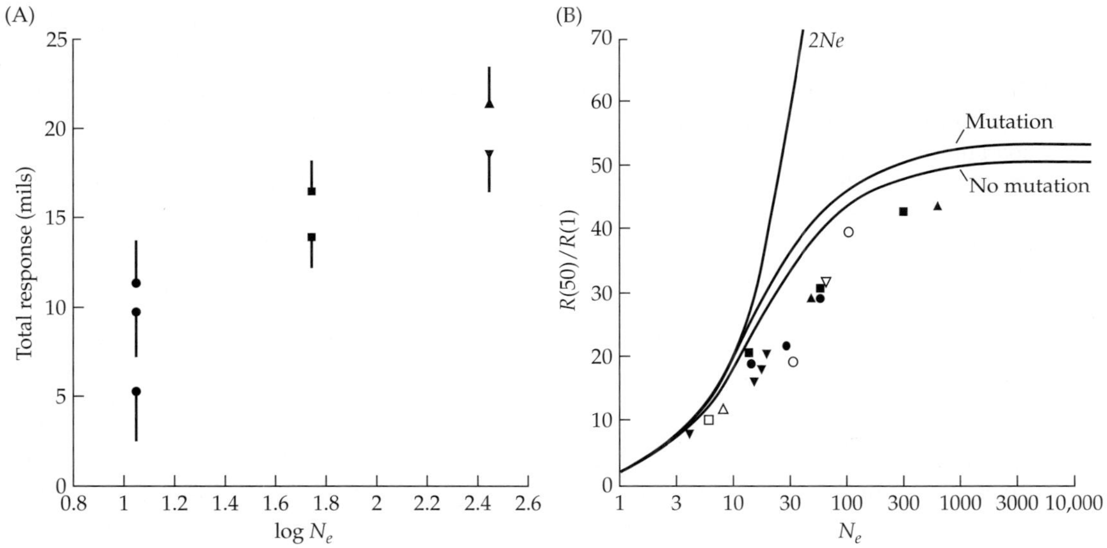

Figure 26.3 **A:** Selection for wing-tip height in *Drosophila melanogaster* (in mils = 0.001 inches). Three replicated selection lines (left to right) with estimated population sizes of 11 (3 replicates), 56 (2 replicates) and 280 (2 replicates) were used. Symbols represents the mean value, while the one-sided whisker shows its associated standard error (displayed in only one direction). Note that the response increases with N_e. (After Weber 1990.) **B:** The ratio of cumulative selection response at generation 50 to response in the first generation as a function of effective population size, for nine different experiments (represented by the different symbols). The lower sigmoidal curve is the prediction using Equation 26.15a; the upper sigmoidal curve is the prediction given by Equation 26.30c, which allows for response from new mutations (assuming $\sigma_m^2/\sigma_E^2 = 0.001$). The curve marked $2N_e$ is the expected limit under Robertson's additive model (Equation 26.15b). (After Weber and Diggins 1990.)

Weber's Selection Experiment on *Drosophila* Flight Speed

Perhaps the largest long-term artificial selection experiment (outside of microbes) is the heroic effort of Weber, which was introduced in Chapter 25. Weber (1996) scored a total

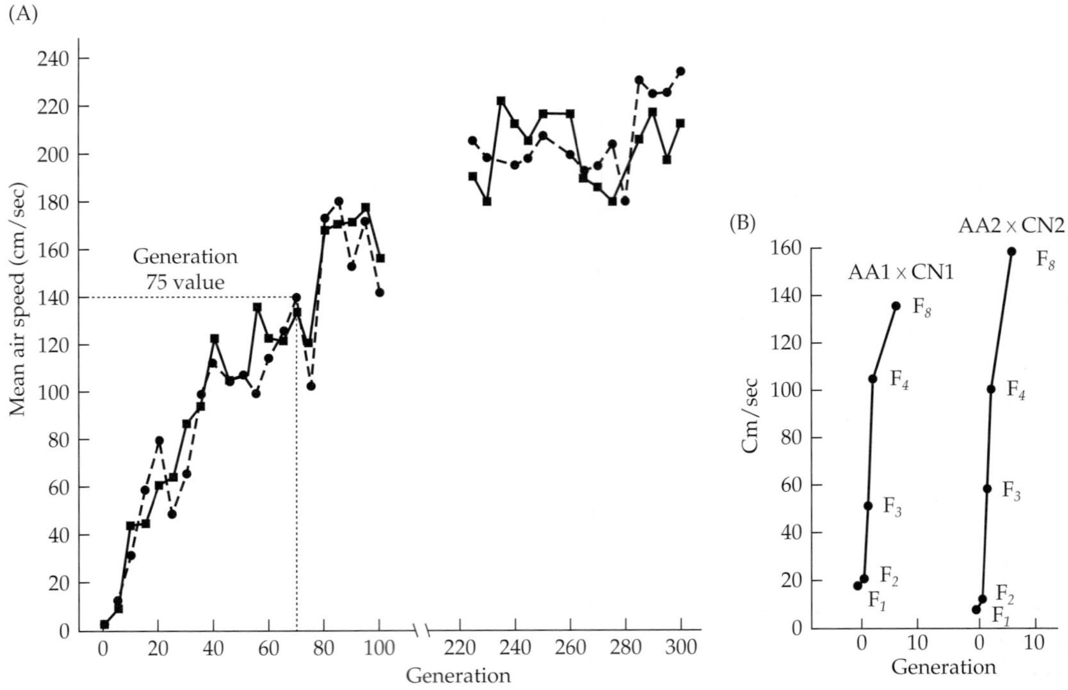

Figure 26.4 Weber's selection experiments for increased flight speed in *Drosophila*. **A:** Results of 300 generations of selection in two replicate lines (circles with dashed lines and squares with solid lines); also see Figure 25.9 for the first 100 generations. (After Weber 2004.) **B:** Response to selection in hybrid sublines formed by crossing two replicate generation 75 selection lines (AA1 and AA2) back to controls (CN1 and CN2). Selection started on the F_2 lines, with only six generations of selection (the F_8 lines) required to recover essentially all of the initial response of ~140 cm/sec. (After Weber 1996.)

of over 9,000,000 *Drosophila* for flight speed in two replicate lines subjected to 100 generations of selection (Figure 25.9). The resulting N_e was in the 500–1000 range, with a percent selected of $p = 0.045$ (for a selection intensity of $\bar{\imath} = 2.11$). The average speed before selection was around 2 cm/second, while the mean speed at generation 100 was 170 cm/sec. As shown in Figure 25.9, response continued in both lines for 100 generations but was diminishing with time, as indicated by a significant quadratic component in the response curve. Figure 26.3A shows the results for over 300 generations of selection from Weber (2004). As of this writing, the experiment is over 650 generations, with response, albeit diminishing, still occurring (Weber, pers. comm.).

Unlike in many artificial selection experiments, there was little slippage upon relaxation of selection and only a minimal loss in fitness relative to the control populations (fitness decreases of 6% and 7% at generations 50 and 85, respectively). Weber attributes this to the larger effective population size, which both reduces the level of inbreeding and allows for more efficient selection on modifiers. The latter can reduce deleterious pleiotropic effects that might accompany major alleles improving flight speed, as the weak second-order effects on modifiers are much easier to select for in larger populations. Larger population sizes also allow recombination to be more efficient, reducing the effects of deleterious alleles linked to alleles improving flight speed.

Weber gained some insight into the genetic nature of the response by examining the selection response in hybrid lines formed by crossing each replicate selection line at generation 75 (lines AA1 and AA2) back to control lines (CN1 and CN2). As Figure 26.4B shows, both the F_1 and F_2 were close to the control line values, indicating very strong dominance for reduced flight speed. Evidence for epistasis was more equivocal. From the theory of line-cross analysis (LW Chapter 9), an estimate of composite epistatic effects is provided by the linear contrast of means of the parental and first two filal populations, $4\,\overline{z}_{F_2} - 2\,\overline{z}_{F_1} - \overline{z}_{P_1} - \overline{z}_{P_2}$, but the resulting value was not significantly different from zero (-38.5 $\pm$ 37.5). Selection on both resulting F_2 lines required only six generations to recover essentially all of the response seen in the parental (75-generation) lines ($\sim$140 cm/sec).

THE EFFECTS OF LINKAGE ON THE SELECTION LIMIT

When QTLs are linked, we expect some reduction in the limit because selection on linked loci reduces the fixation probabilities of beneficial alleles (Hill and Robertson 1966; Birky and Walsh 1988; Barton 1995a). Simulation studies (Fraser 1957; Gill 1965a, 1965b, 1965c; Latter 1965a, 1966a, 1966b; Qureshi and Kempthorne 1968; Qureshi 1968; Qureshi et al. 1968) show that linkage has only a small effect unless loci are very close ($c \leq 0.05$). As mentioned in Chapter 7, most of these studies inflated the importance of linkage by assuming that all loci have equal effects. Simulation studies by McClosky and Tanksley (2013) found only modest reductions ($\simeq$ 10%) in short-term (less than 20 generations) response for populations with normal versus fully unconstrained levels of recombination.

An approximate analytic treatment of linkage was offered by Robertson (1970a, 1977a), and later by Hospital and Chevalet (1993, 1996) and Zhang and Hill (2005a), who relied on certain normality assumptions. In the absence of recombination, selection acts on an entire chromosome, and Robertson framed his results in terms of the response contributed by a single chromosome. Robertson considered three different limiting expected responses, L, corresponding to different amounts of recombination: L_f, the chromosomal limit with free recombination between all loci; L_0, the limit under complete linkage; and L_ℓ, the limit when the map length of the chromosome is ℓ (implying, for n loci, a recombination rate between the adjacent loci of approximately ℓ/n).

The completely additive model is assumed with loci starting in gametic-phase equilibrium. Let $\sigma^2_{A^*}$ be the initial additive genetic variance contributed by the focal chromosome and define $(h^*)^2 = \sigma^2_{A^*}/\sigma^2_z$ as the initial fraction of phenotypic variance attributable to this chromosome. The expected contribution from this chromosome following a single generation of selection is $S\sigma^2_{A^*}/\sigma^2_z = \bar{\imath}\, h^* \sigma_{A^*}$. When $N_e \bar{\imath}\, h^*$ is small, the expected limit for a chromosome with freely recombining loci is $2N_e$ times the initial response, yielding $L_f \simeq 2N_e \bar{\imath}\, h^* \sigma_{A^*}$ (Equation 26.15b, considering the response from a single chromosome). Assuming weak selection, Robertson (1970a) found that the ratio of the free-recombination limit to the complete-linkage limit (i.e., the best initial chromosome) is approximately

$$\frac{L_f}{L_0} \simeq 1 + \frac{2}{3}(N_e \bar{\imath}\, h^*)^2 \quad \text{when} \quad 2N_e \bar{\imath}\, h^* < 1 \tag{26.18}$$

Hence, for weak selection, complete linkage has only a trivial effect when the chromosome contains a large number of QTLs.

When selection is strong ($N_e \bar{\imath}\, h^* \gg 1$), the results are more complicated. Robertson assumed that there are n underlying loci, each with a frequency of p of the favored allele, which increases the character by $2a$ (the difference between the homozygotes). Under these assumptions, the additive variance contributed by this chromosome is $\sigma^2_{A*} = 2na^2 p(1-p)$. If selection is sufficiently strong, under free recombination all favored alleles will be fixed, and the total response becomes $L_f = 2na(1-p)$. Noting that $a = \sigma_{A*}\,/\sqrt{2np(1-p)}$, this can also be restated as

$$L_f = 2na(1-p) = \sigma_{A*}\sqrt{\frac{2n(1-p)}{p}} \tag{26.19}$$

On the other hand, with complete linkage the limit approaches twice the value of the best of the initial $2N$ chromosomes sampled (as this chromosome is ultimately fixed). The expected value for the best chromosome is given by the expected value of the largest order statistic (see Example 6 in LW Chapter 9). For a unit normal, this is expressed in terms of standard deviations (here σ_{A*}) above the mean, so that if x_{2N} is the standardized largest order statistic in a sample of $2N$ chromosomes, the limit is given by

$$L_0 = (x_{2N}\sqrt{2})\,\sigma_{A*} \tag{26.20a}$$

Robertson (1970a) showed, for $10 < N < 40$, that $x_{2N}\sqrt{2} \simeq 3$, so that $L_0/\sigma^2_{A^*} \simeq 3$. Hence, for these values of N,

$$\frac{L_f}{L_0} \simeq \frac{1}{3}\sqrt{\frac{2n(1-p)}{p}} \tag{26.20b}$$

The factor of 3 increases to 3.8 when $N = 80$ and to 4.6 when $N = 500$. For larger values of N, if we use the asymptotic approximation for the largest order statistic given by Kendall and Stuart (1977), the factor of 3 is replaced by

$$x_{2N}\sqrt{2} \simeq \frac{0.577}{\sqrt{\ln(2N)}} + 2\sqrt{\ln(2N)} \tag{26.20c}$$

Note that the increase in the selection limit is only weakly dependent on N, as the largest order statistic scales as $\sqrt{\ln(2N)}$. For example, for $N = 10^9$, $x_{2N}\sqrt{2} \simeq 9.4$.

Robertson suggested that as the number of loci, n, increases, the limit under free recombination approaches a value independent of n and p, namely, the infinitesimal limit, $L_f = 2N_e\bar{\imath}\,h^*\sigma_{A*}$. Thus, with strong selection and a large number of loci, Equation 26.20a implies that

$$L_f/L_0 \simeq \frac{2N_e\bar{\imath}\,h^*\sigma_{A*}}{(x_{2N}\sqrt{2})\,\sigma_{A*}} = \left(\frac{\sqrt{2}}{x_{2N}}\right) N_e\bar{\imath}\,h^* \qquad \text{when } N_e\,\bar{\imath}\,h^* > 5 \tag{26.20d}$$

Robertson also observed that for $N_e\bar{\imath}\,h^* > 5$, the half-life with no recombination is

$$t_{0.5} \simeq \frac{2}{\bar{\imath}\,h^*} \tag{26.20e}$$

generations, and that differences in response (relative to free recombination) only become apparent after this number of generations has passed.

Allowing for some recombination (at a rate of $\sim\ell/n$ between loci), Robertson found that the limit for a chromosome of length ℓ is

$$L_\ell/L_0 \simeq 1 + (N_e\,\ell/3) \qquad \text{when } N_e\ell \ll 1 \tag{26.21a}$$

To a poorer approximation, over the entire range of $N_e\,\ell$

$$L_\ell/L_0 \simeq 1 + \frac{KN_e\,\ell/3}{N_e\,\ell/3 + K} \tag{26.21b}$$

where $K = L_f/L_0$. Thus L_ℓ/L_0 approaches L_f/L_0 as $N_e\ell$ increases. Provided $L_f \gg L_0$, then L_ℓ is halfway between L_f and L_0 when $N_e\,\ell/3 = K = L_f/L_0$. Assuming moderate to large values of N_e, this result (together with Equation 26.20d) implies that if the amount of recombination satisfies $\ell > 2\bar{\imath}\,h^*$, then the selection response will be at least half that expected for free recombination. These expressions are approximate and assume that all loci have equal effects. Variation between loci in allelic effects reduces the effect of linkage (Hill and Robertson 1966; Robertson 1970a).

Experimental results generally confirm that the suppression of recombination has only a modest effect on the selection limit (Example 26.4). This is somewhat at odds with the increase in recombination rates seen during some artificial selection experiments (Example 26.1), although, as mentioned, this view may be tempered if there is significant ascertainment bias in the reporting of increased recombination following selection.

Robertson's result largely focused on the ultimate selection limit, while Hospital and Chevalet (1993, 1996) considered the dynamics of the approach to this limit. In particular, Hospital and Chevalet (1996) explicitly considered the effects of gametic-phase disequilibrium (also see Zhang and Hill 2005a). Initially, selection generates negative gametic-phase disequilibrium, which reduces the initial expressed additive-genetic variance and decreases the response. The tighter is the linkage, the more pronounced is this effect (Chapters 16 and 24). Surprisingly, Hospital and Chevalet (and also Zhang and Hill) showed that linkage can often result in an *increase* in the additive variation in later generations of selection. This seemingly counterintuitive result arises because selection increases the frequency of the gametes carrying the most favored alleles. Because any tightly linked alleles decreasing the trait are also dragged along, this reduces the ultimate selection limit (a phenomenon called **linkage drag**). On the other hand, rare recombination events among such gametes can result in the creation of new, even more favorable gametes, and generating a transient increase in the additive variance as these sweep through the population. Thus, the negative gametic-phase disequilibrium that suppresses the early response stores some genetic variation that can become released (via recombination) in later generations. This effect is most pronounced in larger populations, as in small populations, haplotypes can become fixed before such recombination events occur.

Example 26.4. By using the inversions *Curly* and *Moiré*, McPhee and Robertson (1970) were able to select for sternopleural bristles in *Drosophila* under conditions of suppressed recombination on chromosomes II and III. From previous work, $h^2 = 0.4$, with these chromosomes accounting for 1/3 and 1/2 (respectively) of the genetic variation in bristle number (and with the X chromosome accounting for the remaining 1/6). In lines that were suppressed for recombination on both chromosomes, the selection limit (on a transformed scale) was 0.166 ± 0.014 in up-selected lines and -0.134 ± 0.009 in down-selected lines, reductions of $28 \pm 8\%$ and $22 \pm 7\%$ relative to the limit obtained when normal recombination was allowed. For these studies, $N_e \simeq 10$ and $\bar{\imath} \simeq 1$, while $h^*_{II} = \sqrt{0.4/3} \simeq 0.37$ and $h^*_{III} = \sqrt{0.4/2} \simeq 0.45$. Thus, selection is strong on both chromosomes as $N_e\,\bar{\imath}\,h^*_{II} \simeq 3.7$ and $N_e\,\bar{\imath}\,h^*_{III} \simeq 4.5$.

Under these conditions, Robertson's theory predicts that the limiting contribution from each (recombination-suppressed) chromosome will be approximately $3\sigma_{A^*}$ (as $x_{2N}\,\sqrt{2} \sim 3$; Equation 26.20a). Given $\sigma_z = 0.059$ and $\sigma_{A^*} = 0.059 \cdot h^*$, the expected contributions to the selection limit from chromosomes II and III become $3 \cdot 0.059 \cdot 0.37 \simeq 0.065$ and $3 \cdot 0.059 \cdot 0.45 \simeq 0.080$, respectively, for a total absolute contribution of 0.145, consistent with the observed limits. Robertson's theory (Equation 26.20e) further predicts that the half-life in recombinationally suppressed lines is roughly $2/(\bar{\imath}\,h^*)$ generations, or $2/0.37 \simeq 5.4$ and $2/0.45 \simeq 4.4$ for chromosomes II and III, respectively, consistent with the observed half-life of 5 generations.

Other *Drosophila* experiments examined the consequences of suppressed recombination on selection response. Both Markow (1975) and Thompson (1977) used stocks with inversions while selecting for increased or decreased phototactic behavior. While Markow observed that recombination suppression reduced the selection limit, Thompson observed no differences. Markow did not use replicate lines, so the statistical significance of her results is unclear. However, she observed that the most recombinationally suppressed lines had the most reduced response, consistent with theory. In Thompson's experiments, $N_e \simeq 50, \bar{\imath} \simeq 1$, and $h^* \simeq 0.1$ (for both autosomes), yielding an expected (linkage) half-life of $2/(\bar{\imath}\,h^*) = 20$ generations (López-Fanjul 1989), as opposed to the value of $1.4 \cdot 50 = 70$ in the absence of linkage (Equation 26.17). Thompson's experiments were stopped at generation 21, so it is not surprising that he found no difference in response, as the reduction in total response from linkage is not readily apparent until well after the expected linkage half-life (Equation 26.20e).

Bourguet et al. (2003) commented that a potential flaw in these *Drosophilia* experiments is that balancer chromosomes were used to suppress recombination, which may have different levels of variation than their homologs in the base population. Using a more careful approach to suppress recombination, they observed no difference between normal and recombinationally suppressed lines in the response after 38 generations of selection for geotaxis. However, they noted that their experiment, like most others, suffered from low power.

Table 26.3 Differences in short-term versus long-term response as a function of the number of adults saved, N, when $M = 50$. Initially, $h^2 = 0.5$ and $\sigma_z^2 = 100$. The infinitesimal model is assumed with $N_e = N$. The selection intensity, $\bar{\imath}$, is obtained by using Equation 14.3a (corrected for finite population size). From Equation 13.6b, $R(1) = 5\,\bar{\imath}$, while from Equation 26.15b, $R(\infty) = 2N\,R(1)$.

N	p	$\bar{\imath}$	$R(1)$	$R(\infty)$
30	0.6	0.6	3.2	192
25	0.5	0.8	4.0	200
10	0.2	1.4	7.0	140
5	0.1	1.8	9.0	90

OPTIMAL SELECTION INTENSITIES FOR MAXIMIZING LONG-TERM RESPONSE

When a fixed number, M, of individuals is scored, there is a tradeoff between the intensity of selection ($\bar{\imath}$) and the amount of drift (N_e). If N individuals are allowed to reproduce (implying $p = N/M$ is the fraction saved), decreasing N (and hence p) increases $\bar{\imath}$ but also decreases N_e. Recalling Equation 26.15b, Robertson's selection limit can be expressed as

$$2N_e R(1) = N_e\,\bar{\imath}\left(\frac{2\sigma_A^2(0)}{\sigma_z}\right) \tag{26.22}$$

showing that the ultimate response (from the initial variation) depends on the product of N_e and $\bar{\imath}$. While decreasing p results in a larger short-term response due to increased $\bar{\imath}$, it also results in a decreased long-term response by decreasing N_e. Hence, the product $N_e\,\bar{\imath}$ decreases for sufficiently large or small values of p, suggesting that some intermediate value of p is optimal (see Equation 26.23). Table 26.3 and Figure 26.5B both illustrate this tradeoff. For example, while the single-generation response using $p = 0.50$ is less than half that for $p = 0.10$, it yields a selection limit over twice as large (200 vs. 90).

Supporting an earlier conjecture of Dempster (1955b), Robertson (1960a) found (for additive loci and normally distributed phenotypes) that the intensity of selection that yields the largest total response is $p = 0.5$, as $N_e\,\bar{\imath}$ is maximized for fixed M when half the population is saved. This can be seen directly for truncation selection on a normally distributed character. Recall from Equation 14.3a that $\bar{\imath} = \varphi(x_{[1-p]})/p$ (ignoring the correction for finite population size), where x_p satisfies $\Pr(U < x_{[p]}) = p$, and with U denoting a unit normal random variable and $\varphi(x)$ denotes the unit normal density function. Because the number saved is $N = Mp$, we have (following Hospital and Chevalet 1993)

$$\begin{aligned} R(t) &\simeq Mp\left(1 - e^{-t/2N_e}\right)\frac{\varphi(x_{[1-p]})\sigma_A^2(0)}{p\sigma_z} \\ &= \varphi(x_{[1-p]})\left[\frac{M\sigma_A^2(0)}{\sigma_z}\left(1 - e^{-t/2N_e}\right)\right] \end{aligned} \tag{26.23}$$

Because the term in brackets is independent of p, response (as a function of p) is maximized at the maximum value of $\varphi(x_{[1-p]})$, which occurs at $x = 0$, or a p value of 0.5.

As Figure 26.5A illustrates, the selection limit as a function of p becomes extremely flat-topped as M increases, so even fairly large deviations from $p = 0.50$ yield essentially the same limit. If we relax the assumption of normality, Cockerham and Burrows (1980) found that the optimal proportion for truncation selection is still near 0.50, unless the phenotypic distribution is extremely skewed. Hill and Robertson (1966), Robertson (1970a), and Hospital and Chevalet (1993) found that the optimal proportion increases to above $p = 0.50$ when linkage is important (recall from Chapter 24 that linkage disequilibrium generates skew in the genotypic distribution, causing it to depart from a normal).

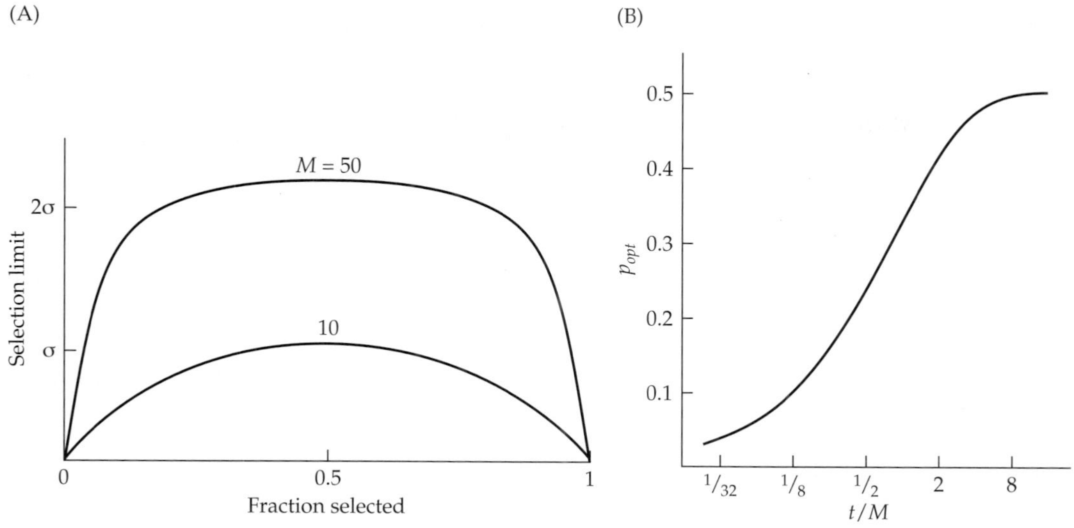

Figure 26.5 **A:** The selection limit as a function of the fraction selected (allowed to reproduce) for 10 and 50 individuals scored. (After Robertson 1960a.) **B:** The optimal proportion, p_{opt}, of individuals selected in each generation to maximize the selection advance over t generations, as a function of t/M, where M is the number of measured individuals. (After Robertson 1970b.) Both sets of curves can be generated using Equation 26.23, taking $M = N_e$.

Table 26.4 As selection intensity increases, the value of N_e becomes increasingly less than the actual number of parents ($N = pM$), further increasing drift. This additional reduction in effective population size due to selection is computed using the approach in Example 26.2. Parameters and assumptions are as in Table 26.3 ($M = 50$, $h^2 = 0.5$).

N	$\bar{\imath}$	N_e	N_e/N	$2N_e\,R(1)$
25	0.8	20.0	0.80	161
10	1.4	6.2	0.62	87
5	1.8	2.6	0.52	47

Robertson's prediction of the optimal selection intensity for long-term response is experimentally supported. Madalena and Robertson (1975) selected for decreased sternopleural bristle number in *Drosophila*. When the best 5 of 25 were chosen, the limit was 18.0 bristles, less extreme than the limit of 17.1 when the best 10 of 25 were chosen. Similar results were seen for increased abdominal bristle number in *Drosophila* (Jones et al. 1968), increased egg-laying in *Tribolium castaneum* (Ruano et al. 1975), and increased postweaning weight in mice (Hanrahan et al. 1973).

Using $N = pM$ as the effective population size is often a severe overestimate (Chapter 3), especially because, as Equations 26.6b–26.6d show, N_e/N decreases as selection intensity increases. Hence, increasing selection intensity increases drift by *both* reducing $N = pM$ *and* by further reducing the ratio of N_e to N. Table 26.4 illustrates this effect using the same parameters as Table 26.3. Without incorporating this further reduction in N_e, the ratio of expected limits when $p = 0.50$ versus $p = 0.10$ is $200/90 = 2.2$. When the reduction in N_e due to selection is accounted for, this ratio increases to $161/47 = 3.4$.

More generally, Robertson (1970b) obtained the optimal selection intensity when the goal is to maximize the total response (from the initial base population variation) at generation t. Robertson's derivation follows using Equation 26.15a. As Figure 26.5B shows, the optimal proportion is a function of t/M. Robertson assumed that the infinitesimal model held and that there were equal contributions from each sex. Jódar and López-Fanjul (1977) extended these results to unequal sex ratios, and found that the maximum response occurs when the number of individuals scored and the proportions that are selected are the same in each sex. This follows because effective population size is reduced as the sex ratio deviates from 1:1 (Equation 3.12), which increases the effects of drift. Hospital and Chevalet (1993) examined the effects of linkage and found that the amount by which the optimal value of p exceeds the predicted value (Figure 26.5B) increases with population size. In small populations, the value predicted from drift (for any particular t/M value) is close to the optimal value, while Robertson's value seriously underestimates the optimal p value in larger populations when linkage is present.

Ruano et al. (1975) and Frankham (1977) tested Robertson's predictions for the optimal response at a particular generation with selection experiments for egg-laying in *Tribolium* and for abdominal bristle number in *Drosophila*, respectively. The theory holds up well for $t/M \leq 0.2$, but both authors found discrepancies between the observed and predicted rank order of lines subjected to different selection intensities when $t/M > 0.2$. One explanation of these discrepancies could be the presence of major alleles, resulting in additive variance declining more rapidly than expected under the infinitesimal model. This results in the optimal proportions being larger than those predicted from Figure 26.5B. Frankham (1977) also suggested that not correcting for the additional decrease in N_e with increased selection intensity (e.g., Table 26.4) results in incorrect values of N_e, and hence incorrect optimal proportions. García-Dorado and López-Fanjul (1985) examined the consequences of unequal sex ratios using sternopleural bristle number in *Drosophila*. Equal sex ratios gave the highest response, and good agreement with the optimal values predicted by Jódar and López-Fanjul was seen when there were unequal sex ratios.

EFFECTS OF POPULATION STRUCTURE ON LONG-TERM RESPONSE

Our development of Robertson's theory of selection limits has made two assumptions regarding population structure: selection occurs in a large panmictic population, and the initial base population is infinite in size. This section relaxes these assumptions. We first examine the consequences of founder effects in the initial base population and of passing the population through bottlenecks during selection. We conclude by examining the expected limits when the population is subdivided and when selection is entirely within families.

Founder Effects and Population Bottlenecks

So far, we have been considering only the effects of drift due to selecting N adults in each generation from an initial base population that is assumed to be infinite. However, drift can also occur prior to selection if the base population itself was founded by sampling individuals from some larger population. By altering the starting additive variance, this initial sampling modifies the expected response, and (provided the founding event is severe), can have a significant impact on the selection response. Robertson (1966b), reporting on the unpublished thesis of Da Silva (1961), found that lines formed from a single parental pair underwent a decrease in the selection response of roughly 30% relative to a nonbottlenecked line from the base population (Figure 26.6A). Lines formed from taking single parental pairs for three consecutive generations showed only a modest further reduction in response, suggesting that most of the founder effect occurred in the first generation. Robertson's interpretation was that response in this population was due largely to alleles that were at an intermediate frequency, as alleles that are at low frequency are expected to be lost during the initial sampling. Segregating alleles present after this initial bottleneck of two individuals have intermediate frequencies ($1/4$, $1/2$, or $3/4$), which somewhat decreases their sensitivity to further sampling events.

Using this reasoning, Robertson (1960a) predicted that the effect of restricting population size after several generations of selection is expected to be small, as favored alleles are expected to be at intermediate to high frequencies. However, Jones et al. (1968) found that, even after many generations of selection, such bottlenecks can have a large effect. Sublines formed by taking ten pairs of adults from a parental line selected for 16

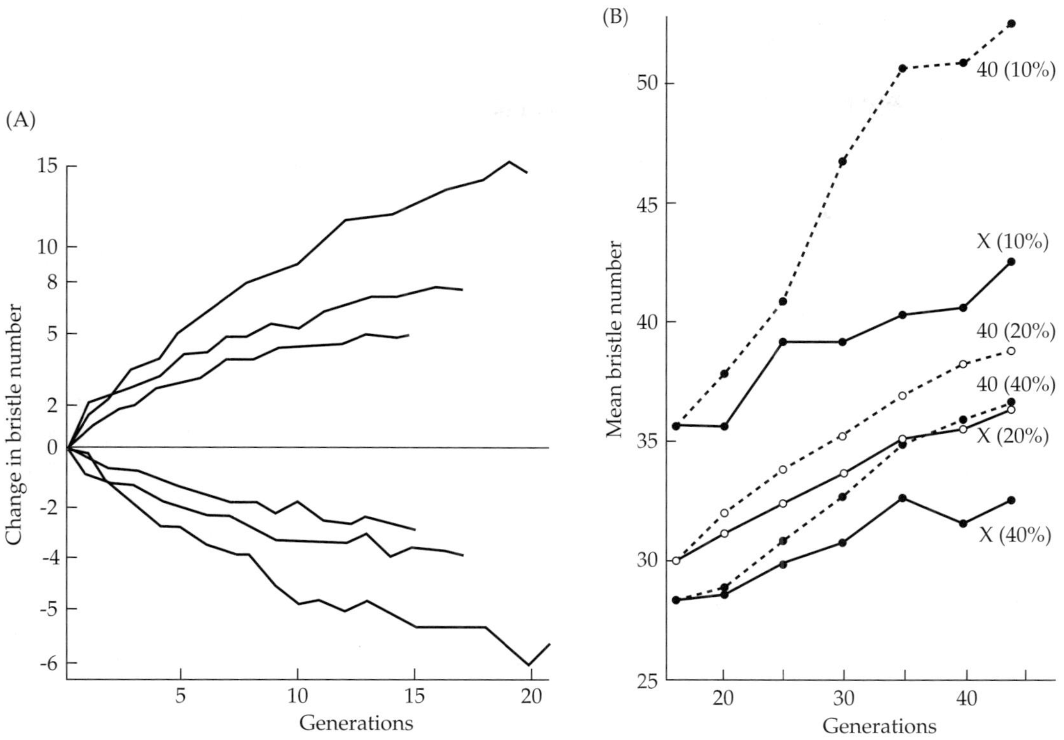

Figure 26.6 Effects of population bottlenecks on selection response. **A:** Selection for sternopleural bristle number in *Drosophila melanogaster*, with the most extreme 10 pairs out of 25 scored pairs selected in each generation. The outermost curves are responses using the base population. The middle curves are the responses for sublines formed from a single parental pair. The innermost curves correspond to sublines formed by single parental pairs for three consecutive generations prior to selection. (After Robertson 1966b.) **B:** Selection for abdominal bristle number in *Drosophila melanogaster*. The responses denoted by 40 (10%), 40 (20%), and 40 (40%) correspond to populations where the uppermost 40 pairs of adults are selected in each generation, with different selection intensities. For example, 200 pairs are scored and the uppermost 40 are chosen in the 40 (20%) population. Responses denoted by X (10%), X (20%), and X (40%) refer to lines split from the corresponding 40-pair lines after 16 generations of selection and selected thereafter at the same intensity with 10 pairs of parents per generation. Note that all X lines show reduced responses (being below their counterparts). (After Jones et al. 1968.)

generations showed reduced response relative to their parent lines (Figure 26.6B). One explanation for the results of Jones et al. is that there were still desirable alleles at low frequencies following 16 generations of selection. These alleles can be lost when the population passes through a bottleneck, reducing response. One source for these rare major alleles could be new mutations. Alternative explanations were considered by Frankham (1983b).

To present the theory for the impact of bottlenecks on selection response a bit more formally, results are developed for a single additive locus, and extended by assuming gametic-phase equilibrium and no epistasis. If N_0 is the number of founders, the initial expected additive- genetic variance in the founder population is $[\,1 - 1/(2N_0)]\,\sigma_A^2(0)$, with the expected response for the first generation of selection from a bottlenecked population being $[\,1 - 1/(2N_0\,)]$ times that for an initially infinite population (Jones 1970). The long-term effects of an initial bottleneck are more unpredictable, depending on initial allele frequencies and the relative strength of selection. When selection is weak at all loci (the infinitesimal model), the arguments leading to Equation 26.15a yield the expected response starting with a founder population of size N_0 as

$$R_{N_0}(t) = R(t)\left(1 - \frac{1}{2N_0}\right) \tag{26.24a}$$

where $R(t)$ is the response expected when the initial base population is infinite (Equation 26.15a). More generally, if two replicate populations of the same size are created using different numbers of founders (N_{01}, N_{02}) from a common, and large, base population, the ratio of the expected response at any generation is

$$\frac{R_{N_1}}{R_{N_2}} = \frac{1 - 1/(2N_{01})}{1 - 1/(2N_{02})} \tag{26.24b}$$

Thus, if selection at all loci is weak and all genetic variance is additive, the effect of a bottleneck depends only on the number of founders, N_0.

Founder effects are most serious when rare favorable alleles of large effect are present, but predicting the magnitude of the effect in any given population is difficult. When selection on a locus is strong ($2N_e s \gg 1$), the probability that a selected line formed from a bottlenecked base population will eventually become fixed for the favored allele converges to

$$u_{N_0}(p_0) = 1 - (1 - p_0)^{2N_0} \tag{26.25a}$$

where p_0 is the major-allele frequency in the population being sampled. This follows because if selection is sufficiently strong, the favored allele will become fixed if it is found in the initial sample, which occurs with a probability of $1-(1-p_0)^{2N_0}$. Using this approximation, the ratio of the expected limiting contribution from such a locus to the expected contribution when the founding population is infinite is

$$\frac{u_{N_0}(p_0) - p_0}{u(p_0) - p_0} \simeq \frac{1 - (1 - p_0)^{2N_0} - p_0}{1 - p_0} = 1 - (1 - p_0)^{2N_0 - 1} \simeq 1 - e^{-p_o(2N_0 - 1)} \tag{26.25b}$$

A more accurate measure would be to weight the fixation probability, $u(p)$, by the sampling probability given a starting allele frequency, $\sum_{i=1}^{2N_0} \Pr(i \,|\, p_0, 2N_0)\, u(i/[2N_0])$, where $\Pr(i \,|\, p_0, 2N_0)$ is the ith term in a binomial with parameters of p_0 and $2N_0$.

Because the initial frequencies of major alleles are unknown, the long-term effect of a bottleneck, even when all genetic variance is additive, is unpredictable. To see this, suppose that a rare ($p_0 \simeq 0$), but favorable (a is large), allele is initially present. Its contribution to the initial additive variance is $V = 2a^2 p_0(1 - p_0)$, while (if fixed), its contribution to the response is $R = 2a(1 - p_0)$. Hence, $R = V/(ap_0)$, so that if $ap_0 \ll 1$, but a is large, it makes a large contribution if it is fixed, but only a small contribution to the initial variance. If $p_0 \simeq 0$, an allele with a large effect can easily be lost by drift, with only a small effect on the additive variance, but leading to a large potential loss of response. Many artificial selection experiments examining the genetic architecture of a trait first start by breeding a wild-caught sample in the lab for many generations. This generates additional drift, and can result in rare (but important) alleles from the sampled population not being present at the start of artificial selection. Zhang and Hill (2005a) showed that a consequence of this sampling (coupled with selection-generated disequilibrium) is that a population with a significant number of rare alleles (and hence the potential for an accelerated response as rare alleles of large effect increase in frequency, increasing h^2; Chapter 25) often generates a response no different from that expected under an infinitesimal model.

Frankham (1980) examined founder effects in *Drosophila* populations that were selected for increased abdominal bristle number. As shown in Figure 26.7, the limit of bottlenecked populations formed from two founders was between 0.69 and 0.72 of that for nonbottlenecked populations, which is quite close to the value of $[\,1-1/(2N_0)\,] = 0.75$ that is predicted for additive loci under weak selection (Equation 26.24b). Frankham reported similar unpublished thesis results of Da Silva (1961) and Hammond (1973). However, while D. Robertson (1969, reported in James 1970) observed a decrease in response with decreasing number of

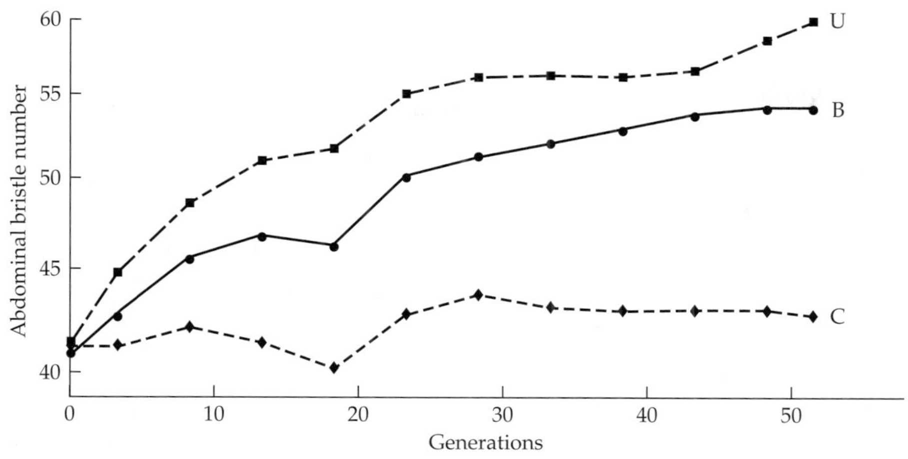

Figure 26.7 The effect of an initial bottleneck on selection for increased abdominal bristle number in *Drosophila*. Curve B corresponds to the response in bottlenecked populations formed from a single pair of parents ($N_0 = 2$), curve U to a nonbottlenecked population, and curve C corresponds to the response from the unselected control. All lines were maintained by using 20 pairs of parents in each generation. (After Frankham 1980.)

founders when the number of selected parents (N_e) was 10, there was no obvious effect when N_e was 40 (which is not unexpected because $1 - 1/80$ is negligible). We have been unable to find any reports of response increasing significantly when the population is passed through a bottleneck, as can occur if significant nonadditive variance is present (Chapter 11). Clearly, there is a need for further experiments.

An especially interesting experiment on founder effects was performed by Skibinski and Shereif (1989), who examined sternopleural bristle number in *Drosophila melanogaster*. Three initial lines were created from a large base population by taking parents from different parts of the distribution of bristle number to generate a high line, a low line, and a line from the central part of the distribution. The central line had the largest total response to divergent selection. Skibinski and Shereif suggested that these results were consistent with the assumption that a few major alleles underlay the trait, with the central line having higher heterozygosity at these loci (and hence more usable genetic variance) than the extreme lines. One caveat with this interpretation is that the central line had a larger initial population size than either extreme line.

Population Subdivision

Thus far, we have been considering the long-term response under mass selection in a single panmictic population. But how robust are these results if the total population is subdivided? Robertson (1960a) showed that when only additive variance is present, population structure has little effect on the selection limit. In particular, the expected limit for a population formed by crossing k (replicate) plateaued lines of size N is the same as for a single line of size Nk. Maruyama (1970) generalized this result by showing (for additive loci and ignoring linkage effects) that any subdivision of the population has the same limit, independent of when and how lines are crossed, provided there is no selection *among* lines. One caveat with this result is that breeders typically try to maximize gain under a set level of inbreeding, and Smith and Quinton (1993) showed that selecting and crossing sublines produces less total selection response for a fixed level of inbreeding than does selection in a single line.

Madalena and Hill (1972) further showed that linkage has only a minor effect on this conclusion. They also found (again assuming only additive variance) that while among-line selection (i.e., culling some of the lines) may increase short-term response, removing lines decreases the total genetic variance of the entire population, which decreases the limit. This reduction in the limit is most severe with free recombination, and it is negligible with tight linkage.

When significant nonadditive genetic variance is present, population subdivision may increase the selection limit. For example, when favorable rare recessives are present, subdividing the population and subsequently crossing these lines when they plateau and then reselecting yields a higher expected limit than using a single panmictic line of the same total size (Madalena and Hill 1972; Slatkin 1981b). The increased inbreeding in the sublines increases the frequency of homozygotes, which facilitates selection for favorable recessives.

Similarly, **Wright's shifting balance theory** (Wright 1931, 1951, 1978, 1982) asserts that local inbreeding due to population subdivision facilitates the accumulation of rare favorable epistatic combinations of loci. Crossing such fixed (or nearly fixed) lines increases the selection limit relative to a single panmictic population, much akin to what happens with rare recessives. Indeed, Enfield and Anklesaria (1986) found, in simulation studies, that when additive-by-additive epistatic variance is present, certain population subdivisions can result in greater short-term and long-term response than a single panmictic population.

There have been a number of contrasting views on the optimal population structure for evolution. Wright (1931, 1951, 1977, 1978, 1982) suggested that evolution is most rapid when the population is subdivided (henceforth, the **Wright structure**), while Fisher (1958) viewed a single large panmictic population (the **Fisher structure**) as the optimal structure. When mostly additive gene action is present, both the Wright and Fisher structures are expected to give comparable rates of evolution, although the Fisher structure may have a slight advantage when the effects of linkage are considered (in larger populations, the probability that a deleterious allele linked to a favorable allele will hitchhike to fixation is decreased, which increases the potential response). With nonadditive gene action, the optimal structure depends on the exact nature of gene action. With recessives, the Wright structure increases the response. With epistasis, this subdivision offers an advantage if epistatic combinations are such that their formation requires intermediate genotypes that are deleterious. Conversely, in other situations, the Fisher structure may offer an advantage in that it allows more gene combinations to be tested. There remains very significant debate over which structure is more relevant (Coyne et al. 1997, 2000; Peck et al. 1998, 2000; Wade and Goodnight 1998; Goodnight and Wade 2000).

Despite these concerns when nonadditive genetic variance is present, selection experiments with population subdivision (reviewed by Rathie and Nicholas 1980 and López-Fanjul 1989) generally have yielded results similar to those expected under the strictly additive model: subdivision usually has no effect on the selection limit. However, two experiments revealed exceptions to this trend. Madalena and Robertson (1975) selected for decreased sternopleural bristle number in *Drosophila melanogaster* under two different population structures: a single-cycle structure where sublines were crossed once, and a repeat-cycle structure where sublines were crossed multiple times. The limit under the single-cycle structure was essentially the same as for a panmictic population, regardless of whether among-line selection was practiced. The limit under the repeat-cycle structure was slightly more extreme than the panmictic population. These results are complicated by the presence in their lines of major alleles that are lethal as homozygotes but nevertheless suggest the presence of some favorable recessives initially at low frequency. The second exception was revealed an experiment by Katz and Young (1975), who selected for increased body weight in *Drosophila*. Populations that were subdivided with a small amount of migration among them gave a slightly larger response than the panmictic population.

One must keep in mind that the optimal population structure for maximizing response under one type of gene action may not be optimal for other types. In particular, many types of population structure that increase the probability of fixation of recessive or epistatic genes may retard the fixation of advantageous additive genes. Likewise, even structures that do not decrease the fixation probability may increase the fixation time, which reduces the rate of response.

Caballero et al. (1991) examined the types of mating schemes (following selection) that increase the fixation probability of recessive alleles while not significantly reducing the fixation probabilities or increasing the fixation times for additive genes. They found that mating full sibs wherever possible following selection increased the fixation probabilities for recessives (relative to random mating following selection), without any significant effect on additive alleles. The tradeoff here is a reduction in N_e (due to the increased inbreeding by full-sib mating following selection) versus the increased selection on recessives by inbreeding (compare Equations 7.19b and 7.20c). Recall from Equation 7.20c that the measure, f, of departures from Hardy-Weinberg frequencies enters into the selection coefficients. Caballero et al. showed that

$$f = \frac{N_{FS} - 1}{4N_{TM} - 3N_{FS} + 3} + f_r \tag{26.26a}$$

where N_{FS} is the number of full-sib matings, N_{TM} is the total number of matings, and f_r is the departure from Hardy-Weinberg genotype frequencies under random mating in a finite population, which is given by

$$f_r = -\left(\frac{1}{8N_f} + \frac{1}{8N_m} \right) \tag{26.26b}$$

where N_m and N_f are the numbers of reproducing males and females. Note that the negative sign implies that under random mating, there is a slight expected excessive of heterozygotes relative to the frequency expected from the allele frequencies alone. Caballero et al. (1991) noted that, under their random-mating scheme, the expected number of full-sib matings is close to one, so $N_{FS} - 1$ represents the excessive number of such matings.

Within-family Selection

The variance in the number of offspring contributed by each selected parent is an important determinant of the effective population size—the larger this variance, the smaller N_e (Equation 3.4). Exploiting this relationship, Toro and Nieto (1984) noted that deliberately assigning selected parents different probabilities of contributing offspring (according to a specific formula) results in populations with the same selection intensity but different effective population sizes relative to the situation in which the selected parents are randomly mated.

Suppose 20 individuals are measured ($M = 20$), and we wish the expected selection intensity to be $\bar{\imath} = 1.2$. This occurs if the best 5 individuals are chosen (using Equation 14.4b to correct $\bar{\imath}$ for finite population size) and each parent has an equal probability of contributing offspring. This same selection intensity, $\bar{\imath} = 1.2$, can be achieved by instead choosing the best 10 individuals and assigning these individuals unequal probabilities for contributing offspring (using effective selection differentials, which were introduced in Example 13.2; see Toro and Nieto [1984] for details). This latter scheme (while holding both selection intensity and the number of measured individuals, $M = 20$, constant) increases effective population size from 5.0 to 5.9, which in turn increases the long-term response.

The most extreme example of using a mating scheme to control N_e in a selected population occurs when selection is entirely *within families*: the best male and female are chosen from each full-sib family and mated at random between families. This doubles the effective population size compared to the result from selecting the same number of individuals independent of family structure. We remind the reader at this point of the important, but subtle, distinction between parents having an equal *probability* of contributing offspring versus parents contributing exactly the same *number* of offspring. In the former case, some parents will contribute no offspring and others will contribute more than one, generating a nonzero variance. In the latter case, recall from Equation 3.4 that if all parents contribute the same number of offspring, there will be no variance in offspring number and N_e will equal $2N$.

Thus, using only within-family selection results in a population with twice the effective size as one undergoing mass selection with the same number of individuals selected. However, as Robertson (1960a) noted, the usable additive genetic variance within full-sib families is only half that available under mass selection (see Chapter 21). This exactly cancels the advantage of a larger N_e, suggesting that both methods yield the same limit.

Dempflé (1975) pointed out that this conclusion relies critically on h^2 being low. Applying Equations 21.20 and 21.23, the response to a generation of within-family selection is (for full-sibs)

$$R_{wFS}(1) = \bar{\imath}\, h^2_{wFS}\, \sigma_{wFS}$$

where (with only additive genetic variance), the within-family heritability, the fraction of within-family differences due to differences in breeding values, is

$$h^2_{wFS} = \frac{\sigma^2_A/2}{\sigma^2_{wFS}}, \quad \text{where} \quad \sigma^2_{wFS} = \frac{\sigma^2_A}{2} + \sigma^2_{Es}$$

If the additive genetic variance is much larger than the within-family environmental variance (σ^2_{Es}), then $h^2_{wFS} \simeq 1$ and $\sigma^2_{wFS} \simeq \sigma^2_A/2$, which yields $R_{wFS}(1) \simeq \bar{\imath}\,\sigma_A/\sqrt{2}$. If the total environmental variance is much smaller than the additive variance, the expected response to individual selection will become $R(1) \simeq \bar{\imath}\,\sigma_A$. Thus, when additive genetic variance dominates, the ratio of expected limits is

$$\frac{4N R_{wFS}(1)}{2N R(1)} \simeq \sqrt{2}$$

and within-family selection increases the limit.

Three other factors can favor within-family selection:

1. *Retardation of the cumulative reduction in N_e from selection.* Recall that individual selection reduces N_e below the actual number of parents by inflating the among-family variance in offspring number when h^2 or $\bar{\imath}$ are large. This variance is zero under within-family selection ($Q = 0$ in Equation 26.6c), resulting in an effective population size greater than twice that for individual selection, so N_e(within-family) $> 2N_e$(individual).

2. *Significant among-family environmental variance.* If most of the environmental variance is due to among-family, rather than within-family, effects (i.e., if $\sigma^2_{Ec} > \sigma^2_{Es}$), within-family selection results in a larger single-generation response than individual selection (Chapter 21). Within-family selection is thus superior when the among-family component of environmental variance is sufficiently large, especially because this factor is in addition to its advantage from within-family selection generating a larger effective population size.

3. *Gametic-phase disequilibrium.* The presence of gametic-phase (linkage) disequilibrium also increases the effectiveness of within-family selection relative to individual selection. Under the assumptions of the infinitesimal model, the negative gametic-phase disequilibrium generated by directional selection reduces the among-family component of additive variance, while (for unlinked loci) the within-family component remains unchanged (Chapters 16 and 24). Hence, the usable additive variance in the mass-selection lines is decreased, while the usable additive variance in the within-family lines is unchanged. This effect is largely negligible unless selection is strong and heritability is high.

On the experimental side, von Butler et al. (1984) compared individual and within-family selection on 8-week body weight in mice. In one set of replicates, within-family selection initially showed a reduced response, but after 18 generations they had essentially the same response as the mass-selected lines. In another set of replicates (using a different base population), mass selection did better than within-family selection, but both populations were still responding after the experiment was stopped (after 18 generations). Because within-family selection is expected to show a longer period of response (due to a larger effective population size), the results for the second set of replicates are inconclusive.

ASYMPTOTIC RESPONSE DUE TO MUTATIONAL INPUT

As reviewed in Chapter 25 (and by Frankham 1980, 1983a; Weber and Diggins 1990; Weber 2004), there is strong evidence that new mutations contribute to selection response even during the relatively short time scales of many so-called "long-term" laboratory experiments. The limit resulting from drift and selection removing all initial genetic variation is thus an artifact of time scale, as it ignores ongoing mutational input. Even if an observed limit is due to a balance between natural and artificial selection, new mutations with less deleterious pleiotropic effects on fitness can arise, resulting in further response.

Confounding the issue of new mutations is the appearance of homozygotes involving recessive alleles that were initially present at a low frequency. If a recessive allele is present as a single copy, the expected time (conditional on it not being lost by drift) until the first appearance of a homozygote in a diploid population with an effective size of N_e is approximately $2N_e^{1/3}$ generations, with the appearance time following a nearly geometric distribution (Robertson 1978; Karlin and Tavaré 1980, 1981a, 1981b; Santago 1989). Because $N_e \leq 500$ for most selection experiments, any rare recessives that are initially present (and not lost by drift) will be expressed as homozygotes by around generation 15.

Our discussions of the nature of long-term response with mutational input largely follow Hill's pioneering treatment (1982a, 1982b). We start by assuming complete additivity. Recall from Chapter 11 (and LW Chapter 12) that one measure of mutational input is σ_m^2, the amount of new additive variance produced by mutation in each generation. Consider the ith locus, where each allele mutates to a new one with a per-generation rate of μ_i. The **incremental-mutation model** is assumed: when an allele A mutates to a new allele A', the genotypic values of AA' and $A'A'$ are $g_{AA} + \alpha$ and $g_{AA} + 2\alpha$, where g_{AA} is the genotypic value of AA. This model assumes that the genotypic value of the new mutant is the value of its parental allele plus an increment value, α. The distribution of α is assumed to be independent of the value of the parental allele, with $E[\alpha_i] = 0$ and $E[\alpha_i^2] = \sigma^2(\alpha_i)$. For n loci, the mutational variance for a diploid species becomes

$$\sigma_m^2 = 2\sum_{i=1}^{n} \mu_i \, \sigma^2(\alpha_i)$$

We first consider the infinitesimal model before examining a more general model and the consequences of dominance. An extensive discussion of different mutational models is given in Chapter 28.

Results for the Infinitesimal Model

We start by assuming complete additivity and ignore any effects of gametic-phase disequilibrium. From Equation 11.20b, the expected additive genetic variance at generation t is

$$\sigma_A^2(t) \simeq 2N_e\sigma_m^2 + \left[\sigma_A^2(0) - 2N_e\sigma_m^2\right] \exp(-t/2N_e) \tag{26.27}$$

Setting $\sigma_A^2(0) = 0$ gives the additive variance contributed entirely from mutation as

$$\sigma_{A,m}^2(t) \simeq 2N_e\sigma_m^2\left[\,1 - \exp(-t/2N_e)\,\right] \tag{26.28a}$$

Hence, the rate of response at generation t from mutational input is

$$r_m(t) = \bar{\imath}\,\frac{\sigma_{A,m}^2(t)}{\sigma_z} \simeq 2N_e\,\bar{\imath}\,\frac{\sigma_m^2}{\sigma_z}\left[\,1 - \exp(-t/2N_e)\,\right] \tag{26.28b}$$

where we have made the usual assumption that the phenotypic variance, σ_z^2, does not significantly change over time and that any disequilibrium is ignored. For $t \gg 2N_e$, the per-generation response approaches an asymptotic limit of

$$r_m(\infty) = 2N_e\,\bar{\imath}\,\frac{\sigma_m^2}{\sigma_z} \tag{26.29}$$

Assuming $\sigma^2_A(0) = 0$, Equation 26.28b shows that half this rate is achieved by $t \simeq 1.4N_e$, *independent* of the value of σ^2_m (Hill 1982a, 1982b). There are several ways to intuit the value of this asymptotic limit. From Robertson's theory, we expect the final response to be $2N_e$ times the initial response $R(1)$, which, for new mutants arising in any particular generation, is $R(1) = \bar{\imath}\,\sigma^2_m/\sigma_z$. Alternatively, recall (Equation 11.20c) that the equilibrium additive variance (assuming pure drift) is $2N_e\sigma^2_m$, which (upon recalling Equation 13.6b) recovers Equation 26.29. The assumption of the infinitesimal model implies vanishingly small selection coefficients at each underlying locus, which makes them effectively neutral.

Summing Equation 26.28b over generations (using the approximation given by Equation 7.28b) yields a cumulative response due to new mutation of

$$R_m(t) = \sum_{\tau=1}^{t} r_m(\tau) \simeq 2N_e\,\bar{\imath}\,\frac{\sigma^2_m}{\sigma_z}\Big(t - 2N_e[\,1 - \exp(-t/2N_e)\,]\Big) \tag{26.30a}$$

as found by Hill (1982a, 1990) and Weber and Diggins (1990). An approximation for genes of sufficiently large effect ($|a| \gg \sigma_z/N\,\bar{\imath}$) is to consider them as being essentially fixed instantaneously, in which case only the first term in the large parentheses in Equation 256.30a need be included, and the response approaches

$$R_m(t) = 2tN_e\,\bar{\imath}\,\frac{\sigma^2_m}{\sigma_z} \tag{26.30b}$$

as suggested by Hill (1982a). Note by comparison with Equation 26.29 that the instantaneous fixation assumption is equivalent to assuming that the asymptotic rate of response applies from generation 1.

Combining the mutational response with the response due to genetic variation that was originally in the base population (Equation 26.15a) yields an expected cumulative response of

$$R(t) = 2N_e\frac{\bar{\imath}}{\sigma_z}\left[t\,\sigma^2_m + \Big(1 - \exp(-t/2N_e)\Big)\Big(\sigma^2_A(0) - 2N_e\sigma^2_m\Big)\right] \tag{26.30c}$$

The $t\sigma^2_m$ term, which represents the asymptotic response, eventually dominates for sufficiently large t. The product term in the braces represents the transient effect of the initial additive variance, and it is zero if the population starts at the mutation-drift equilibrium (i.e., $\sigma^2_A(0) = 2N_e\sigma^2_m$).

Of considerable interest is the expected number of generations until the selection response from mutational input exceeds that contributed by the initial variation. Let t^* be the generation when the per-generation response from both sources is equal. At this value, the initial additive variance remaining equals the new additive variance cumulatively generated, or

$$\sigma^2_A(0)\exp(-t^*/2N_e) = 2N_e\,\sigma^2_m\,[\,1 - \exp(-t^*/2N_e)\,]$$

This equation has the solution

$$t^* = 2N_e\,\ln(1+\Psi) \quad \text{where} \quad \Psi = \frac{\sigma^2_A(0)}{2N_e\sigma^2_m} \tag{26.31a}$$

Denoting the initial heritability by h^2 and recalling that $\sigma^2_E = (1-h^2)\sigma^2_z$ yields

$$\frac{\sigma^2_A(0)}{\sigma^2_m} = \frac{h^2\sigma^2_z}{\sigma^2_m} = \frac{h^2}{\sigma^2_m/\sigma^2_z} = \frac{h^2}{(1-h^2)\sigma^2_m/\sigma^2_E} = \frac{h^2}{(1-h^2)h^2_m}$$

showing that

$$\Psi = \frac{h^2}{(1-h^2)\,2N_e\,h^2_m} \tag{26.31b}$$

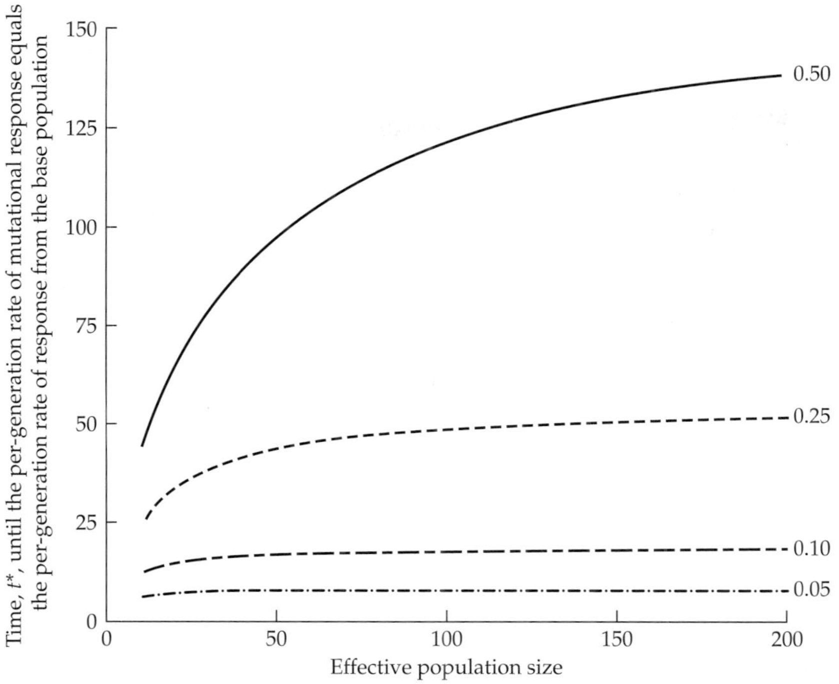

Figure 26.8 The expected generation, t^*, at which the per-generation response from mutational input equals the per-generation response from the initial variation in the base population (Equation 26.31a). We assume that $h_m^2 = \sigma_m^2/\sigma_E^2 = 0.005$, the average value in LW Table 12.1, from which Equation 26.31b yields $\Psi = 100h^2/[(1-h^2)\,N_e]$. The four curves correspond to initial heritabilities of 0.05, 0.10, 0.25 and 0.50. For sufficiently large values of N_e, t^* becomes independent of N_e and approaches the approximation given by Equation 26.31c. Note when h^2 is modest to large that Equation 26.31c significantly overestimate t^* for small N_e.

The average value of the mutational heritability, $h_m^2 = \sigma_m^2/\sigma_E^2$, is approximately 0.005 (LW Table 12.1). With this value, t^* is only rather weakly dependent on N_e (Figure 26.8). If $\Psi \ll 1$, meaning that the expected additive variance at the mutation-drift equilibrium exceeds the initial additive variance ($\sigma_A^2(0) \ll 2N_e\sigma_m^2$), the approximation $\ln(1+x) \simeq x$ for small values of $|x|$ yields

$$t^* \simeq 2N_e\Psi = \frac{h^2}{(1-h^2)h_m^2} \tag{26.31c}$$

Using $h_m^2 = 0.005$ yields $t^* \simeq 200h^2/(1-h^2)$. For h^2 values of 0.05, 0.10, and 0.25, respectively, this translates into 11, 22, and 67 generations until the per-generation response from mutational input exceeds that due to initial variation. For $h_m^2 = 0.001$, these values increase approximately five-fold to 52, 111, and 250 generations. Comparing these approximate results (from Equation 26.31c) with their exact values (Equation 26.31a) shows that Equation 26.31c tends to overestimate the true value of t^* when N_e is small (see Figure 26.8).

Recalling the discussion following Equation 26.27, it is important to stress that our expression for the half-life of selection response (from the initial genetic variation) assumes that drift dominates and tends to yield overestimates when selection is moderate to strong. Likewise, we expect that the infinitesimal model underestimates the changes in allele frequencies of new mutations under moderate to strong selection. Thus, Equation 26.31a is best considered as an upper bound for the number of generations after which mutation is expected to dominate.

Example 26.5. Yoo (1980a) observed a steady, and reasonably constant, increase in *Drosophila* abdominal bristle number over 80 generations of selection (Figure 25.8). In particular, an increase of about 0.3 bristles per generation was observed over generations 50 to 80. Assuming the infinitesimal model, how much of this response is due to mutational input? Yoo's base population had $\sigma_E^2 \simeq 4$, $\sigma_z^2 \simeq 5$, $h^2 \simeq 0.2$, and $\bar{\imath} \simeq 1.4$, with 50 pairs of parents chosen in each generation. Taking $\sigma_m^2 \simeq 0.001\sigma_E^2$ (the average for abdominal bristles in LW Table 26.1) gives $h_m^2 = 0.001$. Assuming $N_e \simeq 60$, Equation 26.31b yields

$$\Psi = \frac{0.2}{(1-0.2)\,2\cdot 60\cdot 0.001} = 2.083$$

Applying Equation 26.31a,

$$t^* = 2\cdot 60\,\ln(1+2.083) = 135$$

The approximation given by Equation 26.31c (which assumes that $\Psi \ll 1$) yields an overestimate of $t^* = 167$ generations. The expected asymptotic additive variance is

$$\widetilde{\sigma}_A^2 = 2N_e\sigma_m^2 = 2\cdot 60\cdot 0.004 = 0.48$$

yielding an expected asymptotic rate of response of

$$r = \bar{\imath}\,\frac{\widehat{\sigma}_A^2}{\widehat{\sigma}_z} = \bar{\imath}\,\frac{\widehat{\sigma}_A^2}{\sqrt{\widehat{\sigma}_A^2+\sigma_E^2}} = 1.4\cdot\frac{0.48}{\sqrt{0.48+4}} \simeq 0.32$$

While the observed rate of selection response (0.3) over generations 50 to 80 is close to the expected asymptotic rate, the expected time for half of the response to be from new mutations, $t^* = 135$, exceeds 80, showing that (under the infinitesimal assumptions) most of the response is still from the initial variation. Applying Equation 26.28b, the expected single-generation response from new mutational input at generation 60 has only reached a fraction

$$1-e^{-t/(2N_e)} = 1-e^{-60/120} \simeq 0.40$$

of its expected asymptotic rate, yielding $0.4\cdot 0.32 = 0.13$ as the expected response due to new mutants at $t = 60$. Assuming the phenotypic variance remains relatively constant, with $\sigma_z^2 \simeq 5$, the expected contribution at generation 60 from initial variation is

$$\bar{\imath}\,\frac{\sigma_{A,0}^2(t)}{\sigma_z} = \bar{\imath}\,\frac{h^2(0)\cdot\sigma_z^2\cdot e^{-t/(2N_e)}}{\sigma_z} = 1.4\cdot\frac{0.2\cdot 5\cdot e^{-60/120}}{\sqrt{5}} \simeq 0.38$$

Adding these two sources returns an expected total rate of response of 0.38 + 0.13 = 0.51 bristles per generation, 75% of which is due to the initial variation. While the predicted rate of 0.51 is larger than the observed rate, opposing natural selection likely slowed down the selection response in Yoo's lines, as evidenced by the rather sharp decay in response upon relaxation of selection, as well as the presence of segregating lethals within responding lines (Yoo 1980b).

A complication with applying this theory is that the presence of major alleles both decreases the time to lose initial variation (when they reside in the base population) and increases the expected response from new mutants (when they arise as mutations). Both of these factors result in a larger role for mutational input than predicted from the infinitesimal model (i.e., a much shorter value for t^*). Applying the approximation for mutations of large effect (Equation 26.30b) using the parameters in this example, the per-generation response from mutation is 0.32. Assuming that the initial variation decays according to the infinitesimal model gives a total rate of response (at generation 60) of 0.38 + 0.32 = 0.70, so mutation now accounts for a fraction, 0.32/0.70 = 0.46, of the total response. Further, when major alleles are present in the base population, the initial variation declines even faster than predicted by Equation 26.15a (as selection augments the amount of allele-frequency change expected under drift alone), suggesting that an even higher percentage of response may be due to new mutation.

Expected Asymptotic Response Under More General Conditions

The infinitesimal model assumes that allele-frequency changes are due entirely to drift. Clearly, selection can also change allele frequencies, and in this case other methods of analysis are required. One approach (Hill 1982a, 1982b) is to consider the expected contribution resulting from the eventual fixation by drift and selection of some of the new mutations that arise in each generation. Provided mutation and selection remain constant over time, at equilibrium the rate of response equals this expected per-generation contribution. Assuming M adults are measured, the frequency of a new mutant allele, A^*, is $1/(2M)$. To allow for dominance, assume that the genotypic values of AA^* and A^*A^* are, respectively, incremented by $\alpha(1+k)$ and 2α relative to the value of AA. As before, we assume that the joint distribution of α and k is independent of the genotypic value of the parental allele. Let $f(\alpha, k)$ denote this joint probability density function and let $\mu = \sum \mu_i$ be the total gametic mutation rate for the trait of interest. The expected contribution to the total response from a new mutant appearing as a single copy becomes $2\alpha \cdot u(1/[2M], \alpha, k)$, the change in genotypic value if the new allele is fixed times its probability of fixation (the latter can be obtained by Equation 7.18a, using the fitnesses given by Equation 25.4). Because $2M\mu$ new mutants appear each generation, the asymptotic rate of response is

$$\begin{aligned} r_m(\infty) &= 2M\mu\, E\left[\, 2\alpha \cdot u\left(\frac{1}{2M}, \alpha, k\right)\right] \\ &= 2M\mu \int_{-\infty}^{\infty}\int_{-\infty}^{\infty} 2\alpha \cdot u\left(\frac{1}{2M}, \alpha, k\right) f(\alpha, k)\, d\alpha\, dk \end{aligned} \tag{26.32}$$

Note that the expected asymptotic rate depends critically on the exact shape of the distribution of mutational effects (a point echoed in Chapter 28). Fortunately, some fairly general results emerge by using simple approximations for the probability of fixation (similar to Equations 7.19a and 7.19b; see Hill 1982a, 1982b for details).

Consider first the case where all new mutants are additive ($k = 0$). Hill (1982b) found that, provided major alleles are not common among new mutants,

$$r_m(\infty) \simeq 2N_e\,\bar{\imath}\,\mu\,\frac{E^+[\alpha^2]}{\sigma_z} = \frac{4N_e\,\bar{\imath}\,\sigma_m^2}{\sigma_z}\,\frac{E^+[\alpha^2]}{E[\alpha^2]} \tag{26.33a}$$

where

$$E^+[\alpha^2] = \int_0^{\infty} \alpha^2\, f(\alpha)\, d\alpha \tag{26.33b}$$

is the average squared increment of favorable alleles (i.e., those with $\alpha > 0$). If the distribution of mutational increments, $f(\alpha)$, is symmetric about zero, then $E^+[\alpha^2] = E[\alpha^2]/2$, as $\int_0^\infty f(\alpha)\, d\alpha = 1/2$, and the asymptotic response reduces to Equation 26.29. When major alleles are common among new mutants, correction terms involving $E^+[\alpha^3]$ appear; see Hill (1982b) for details. With divergent selection (the divergence between an up- and down-selected line; Chapter 25), effects due to asymmetry in $f(\alpha)$ cancel, and the asymptotic rate of divergence between high and low lines is simply twice the rate (for single-direction selection) predicted from the infinitesimal model, namely,

$$4N_e\bar{\imath}\,\frac{\sigma_m^2}{\sigma_z} \tag{26.33c}$$

independent of the shape of $f(\alpha)$. The effect of linkage on asymptotic response was examined by Keightley and Hill (1983, 1987), who found it to generally be small, with the relative effects of linkage increasing with σ_m^2 and N_e.

Hill and Keightley (1988) allowed for the possibility that new mutations are also influenced by natural selection. If both the trait and fitness effects of mutations are small, the distribution of α is symmetric, and natural selection effects are also symmetric in α (e.g., the change in fitness is a function only of $|\alpha|$), there is no change in the asymptotic rate of response. If these assumptions are violated, the asymptotic rate can be reduced.

To allow for dominance, we continue to assume the incremental mutation model. From LW Equation 4.12a, the additive variance contributed by a rare allele is

$$2p(1-p)\alpha^2[\,1+k(1-2p)\,]^2 \simeq 2p\alpha^2(1+k)^2$$

yielding a contribution to σ_A^2 from a single new mutation, where $p_0 = 1/(2M)$, of approximately

$$\alpha^2(1+k)^2/M$$

Because the expected number of new mutations per locus in any given generation is $2M\mu$, the expected additive variance contributed in each generation by new mutations at a given locus is

$$2M\mu\, E[\,\alpha^2(1+k)^2/M\,] = 2\mu\, E[\,\alpha^2(1+k)^2\,]$$

where the expectation is taken over the joint distribution of α and k values in new mutants. Summing over all loci, the expected new additive variance contributed each generation (in the absence of linkage disequilibrium) is

$$\sigma_m^2 = 2\sum_{i=1}^{n} \mu_i\, E[\,\alpha^2(1+k)^2\,] = 2\mu\, E[\,\alpha^2(1+k)^2\,] \tag{26.34a}$$

as obtained by Hill (1982b). The last equality assumes the distribution of mutational values and rates to be the same at each locus. When all mutations are additive ($k = 0$) and symmetric ($E[\alpha] = 0$), this reduces to our previous definition of σ_m^2. More generally, with complete additivity, but removing the assumption that $E[\alpha] = 0$, we have

$$\sigma_m^2 = 2\mu\, E[\,\alpha^2\,] \tag{26.34b}$$

while with complete dominance ($k = 1$),

$$\sigma_m^2 = 2\mu\, E[\,(2\alpha)^2\,] = 8\mu\, E[\,\alpha^2\,] \tag{26.34c}$$

For the same α and μ values, the mutational variance with complete dominance is four times larger than that for complete additivity (as the genotypic value of heterozygotes is doubled, which increases the variance by $2^2 = 4$).

For the case of complete dominance, Hill (1982b) found that the asymptotic rate of response is approximately

$$r_m(\infty) \simeq 16 N_e\, \bar{\imath}\, \mu\, E^+\left[\alpha^2\right]/\sigma_z \tag{26.35a}$$

where $E^+\left[\alpha^2\right]$ is defined by Equation 26.33b. With a symmetric distribution of mutational effects, Equation 26.35a reduces to

$$r_m(\infty) \simeq N_e\, \bar{\imath}\, \frac{\sigma_m^2}{\sigma_z} \tag{26.35b}$$

where σ_m^2 is given by Equation 26.34c. For the same values of σ_m^2, the response when all mutants are completely dominant is only half the expected response when alleles are additive (compare Equations 26.29 and 26.35b). However, for fixed values of μ and $E\left[\alpha^2\right]$, σ_m^2 is larger with complete dominance (compare Equations 26.34b and 26.34c), and the rate of response under dominance is twice as large as that expected for complete additivity.

If alleles are completely recessive, allelic effects are small, and the distribution of mutational effects is symmetric, the asymptotic response is approximately

$$r_m(\infty) \simeq 2N_e\,\bar{\imath}\,\mu\,E\left[\alpha^2\right]/\sigma_z \tag{26.36a}$$

(Hill 1982b). For recessives with large effects (cf. Equation 7.19b)

$$r_m(\infty) \simeq 2\,\mu\,E^+\left[\alpha^{3/2}\right]\sqrt{\frac{2N_e\,\bar{\imath}}{\pi\sigma_z}} \tag{26.36b}$$

Thus, the limiting response when all new mutations are recessive is not predictable from σ_m^2, even if mutational effects are symmetrically distributed. With recessive major alleles, the selection response scales as $\sqrt{N_e\,\bar{\imath}}$, and hence it increases much more slowly with $N_e\,\bar{\imath}$ than with complete dominance or additivity.

When loci are linked, the asymptotic response is reduced, but the effect is small unless linkage is tight, as might occur with a few small chromosomes (Keightley and Hill 1983). As mentioned previously, reduction in response also occurs if loci influencing the trait are linked to loci under natural selection.

Additional Models of Mutational Effects

A critical assumption in any analysis of mutational response is the mutational model. Given a current allelic effect of a, what can we say about the value, a^*, from a mutation in this allele? All of the above results make the incremental-mutation model assumption: $a^* = a + \alpha$, with the increment $\alpha \sim (0, \sigma_\alpha^2)$. This Brownian motion model (Appendix 1) implies that the additive variance (for neutral alleles) will be unbounded as N_e increases (Chapters 11 and 28).

As introduced in Chapter 11, the house-of-cards (HOC) is another potential mutation model. Here, each new allelic value is drawn from a constant distribution, *independent* of the current value of the parental allele, namely the HOC distribution: $a^* = \alpha$, with $\alpha \sim (0, \sigma_\alpha^2)$. Li and Enfield (1992) examined the long-term response under such a model. Starting with a population with no initial variation, they found that mutation increases the genetic variation up to some maximal value, after which it declines, with the time until this maximum is reached increasing with the number of loci. Li and Enfield only considered response over the first 120 generations, which was less than the smallest N_e value (150) in any of their simulations. Hence, the nature of any limit, or any asymptotic response, was not determined. The expectation under an HOC model is that an apparent selection limit is approached, although the population can still respond, but at an ever-diminishing rate, as further gains require random draws of ever-greater outliers from the HOC distribution of allelic-effects at a given locus. This view has connections with models of adaptive walks based on extreme-value theory, which are examined in the next chapter. A finite-value version of the HOC model, assuming that there are only k possible alleles at a locus, was examined by Zeng et al. (1989). As expected, the k-allele model results in an ultimate selection limit, as mutation cannot continue to generate better alleles indefinitely. In Chapter 11 we also introduced the Zeng-Cockerham model (Equation 11.23), $a^* = \tau a + \alpha$, which recovers Brownian motion when $\tau = 1$, and the HOC model when $\tau = 0$. To our knowledge, selection limits under the Zeng-Cockerham model have not been examined.

A second, very important, consideration is the role of pleiotropic fitness effects. These mutational models predict (for constant value of μ) that the equilibrium variance should linearly increase with N_e, at least when N_e is less than the reciprocal of the mutation rate (Chapter 11). However, even for modest N_e, the predicted equilibrium variances are too large to be comparable with observations (with heritabilities approaching 1.0, while most heritabilities in actual populations are below 0.5). This contradiction between theory and data as N_e increases is analogous to the limited observed range for molecular heterozygosity, which (assuming μ stays constant) should also approach one for large N_e (Chapter 2). If new mutations have pleiotropic fitness effects, the amount of usable variation will be overestimated at small N_e (the setting when σ_m^2 is measured). As detailed in Chapter 28, whether this results in a limiting value for $\widetilde{\sigma}_A^2$ as $N_e \to \infty$ depends on very delicate features of the joint distribution of (s, α) for values of s near zero.

Optimizing the Asymptotic Selection Response

Because the asymptotic response is a function of $N_e\,\bar{\imath}$, response is maximized by selection strategies that maximize this product. As was the case for maximizing long-term response (the total response using only the initial variation), there is a tradeoff in that the optimal short-term response (maximizing $\bar{\imath}$) is in conflict with the optimal asymptotic response (because increasing $\bar{\imath}$ decreases N_e). If our choice is simply the fraction of individuals to save, the previous discussion on the optimal selection intensity for long-term response also applies to considerations of the asymptotic response.

However, the breeder or experimentalist can use other design options beyond simply tuning the selection intensity. We have generally been assuming individual (or mass) selection, which is based solely on an individual's phenotype. There are, however, numerous other selection schemes, such as those incorporating information on the phenotypes of relatives (e.g., family-index and BLUP selection; Chapters 21 and 13, respectively). Schemes incorporating such information can improve the accuracy of an individual's breeding value estimate, and hence improve the accuracy of short-term response. This can be seen by recalling (Equation 13.11c) that the single-generation response, R, for any particular selection scheme is given by $R/(\bar{\imath}_x\,\sigma_A) = \rho(x,A)$, where selection occurs on some index, x, and $\rho(x,A)$ is the accuracy of the index (the correlation between an individual's index, x, and breeding values, A). Holding $\bar{\imath}$ constant, the single-generation response increases with the accuracy, $\rho(x,A)$, of the selection method. While different schemes can improve the short-term response over mass selection, what is their effect on asymptotic response? Once again, the answer is that schemes improving the short-term response usually do so at the expense of the asymptotic response.

Optimal asymptotic response occurs by maximizing the fixation probabilities of favorable QTLs, which amounts to maximizing $N_e s$, where s is the selection coefficient on the QTL. For an additive trait, Hill (1985) and Caballero et al. (1996) generalized Equation 25.4 to show that

$$s = \left(\bar{\imath}\,\frac{a}{\sigma_z}\right)\frac{\rho(x,A)}{h} \tag{26.37}$$

Note that $\rho(x,A) = h$ for individual selection (the index is simply the trait value, $x = z$), recovering Equation 25.4. Fixation probabilities under different selection schemes with the same selection intensities are thus functions of the product $N_e s$, which is proportional to $N_e \rho(x,A)$. The tradeoff is that increasing $\rho(x,A)$ typically decreases N_e by increasing the among-family variance in trait value (and hence in fitness). Thus, as was the case in our previous discussion on the optimal selection intensity, the optimal selection scheme for short-term response may differ from the optimal scheme for long-term response.

The accuracy, ρ, depends on the genetic variance, and hence can change over time as these variances change. As shown in Chapters 24 and 25, predicting long-term changes in variances can be extremely difficult. Once again, the analysis is greatly simplified by assuming the infinitesimal model. Under this model, the additive genetic variance eventually converges to a value of $\widetilde{\sigma}^{\,2}_A = 2N_e\sigma^2_m$. The effect of different selection schemes on the equilibrium additive variance (and ρ) is then entirely determined by the effective population size that each scheme generates. In comparing two different selection schemes (i and j) with the same selection intensity, Wei et al. (1996) showed that the ratio of asymptotic responses becomes

$$\frac{\widetilde{R}_i}{\widetilde{R}_j} = \frac{\widetilde{\rho}(i)\,\widetilde{\sigma}_A(i)}{\widetilde{\rho}(j)\,\widetilde{\sigma}_A(j)} = \frac{\widetilde{\rho}(i)}{\widetilde{\rho}(j)}\sqrt{\frac{N_e(i)}{N_e(j)}} \tag{26.38}$$

where a tilde denotes an equilibrium value and $\widetilde{\rho}(i)$ denotes the accuracy (at the equilibrium variances) of selection scheme i. The careful reader will note that the effect of N_e is twofold—there is a direct effect (the square root of the N_e ratio) and also an indirect effect through the ratio of the $\widetilde{\rho}$ (which is a function of $\widetilde{\sigma}_A$, and hence of N_e).

Example 26.6. Consider the asymptotic response to mass (m) versus within-family (w) selection. Under within-family (full-sib) selection, $N_{e(w)} \simeq 2N$, as the among-family variance is zero (Equation 3.4). In contrast, $N_{e(m)} < N$, with the difference between $N_{e(m)}$ and N increasing with the selection intensity and heritability (Equation 26.8), implying that

$$\sqrt{\frac{N_{e(w)}}{N_{e(m)}}} \geq \sqrt{2} \tag{29.39a}$$

The accuracy for mass selection is given by

$$\rho(z,A) = \frac{\sigma(z,A)}{\sigma_A \sigma_z} = \frac{\sigma_A^2}{\sigma_A \sigma_z} = \frac{\sigma_A^2}{\sqrt{\sigma_A^2\,(\sigma_A^2 + \sigma_E^2)}} \tag{29.39b}$$

yielding an asymptotic accuracy as

$$\widetilde{\rho}(m) = \frac{\widetilde{\sigma}_A^2}{\sqrt{\widetilde{\sigma}_A^2\,(\widetilde{\sigma}_A^2 + \sigma_E^2\,)}} = \frac{2N_{e(m)}\sigma_m^2}{\sqrt{2N_{e(m)}\sigma_m^2(\,2N_{e(m)}\sigma_m^2 + \sigma_E^2\,)}} \tag{29.39c}$$

as obtained by Wei et al. (1996).

Turning to within-family selection, let $\overline{z}_f$ denote the family mean. Selection decisions are based on the value of $z - \overline{z}_f$. Recalling our treatment of within-family selection from Chapter 21, the resulting accuracy for within-family (full-sib) selection becomes

$$\rho(w) = \rho(z - \overline{z}_f, A) \simeq \frac{\sigma(z - \overline{z}_f, A)}{\sqrt{\sigma^2(A)\sigma^2(z - \overline{z}_f)}} \simeq \frac{\sigma_A^2/2}{\sqrt{\sigma_A^2(\sigma_{Gw}^2 + \sigma_{E_s}^2)}} \tag{29.39d}$$

where the last step ignores the effect of the number of sibs (n) in each family by assuming that n is large (see Chapter 21 for expressions for when n is small). The within-family genetic variance, σ_{Gw}^2, equals $\sigma_A^2/2$ for a full-sib family with only additive effects, while the within-family environmental variance, $\sigma_{E_s}^2$, equals σ_E^2 under the assumption of no common-family effects (Chapter 21). We make these simplifying assumptions here, but more general expressions easily follow. At equilibrium

$$\widetilde{\rho}(w) = \frac{\widetilde{\sigma}_A^2/2}{\sqrt{\widetilde{\sigma}_A^2(\widetilde{\sigma}_A^2/2 + \sigma_E^2)}} = \frac{N_{e(w)}\sigma_m^2}{\sqrt{2N_{e(w)}\sigma_m^2(\,N_{e(w)}\sigma_m^2 + \sigma_E^2\,)}} \tag{29.39e}$$

(Wei et al. 1996). Applying Equation 26.39a along with Equations 26.39c and 26.39e yields

$$\frac{\rho_{(w,\infty)}}{\rho_{(m,\infty)}} \geq \frac{1}{\sqrt{2}}$$

Thus,

$$\frac{\widetilde{R}_w}{\widetilde{R}_m} = \left[\sqrt{\frac{N_{e(w)}}{N_{e(m)}}}\,\right]\left[\frac{\widetilde{\rho}_{(w)}}{\widetilde{\rho}_{(m)}}\right] \geq \sqrt{2}\,\frac{1}{\sqrt{2}} = 1$$

and hence $\widetilde{R}_w \geq \widetilde{R}_m$. That is, for the same selection intensity, the asymptotic response is greater under within-family selection than under mass selection.

The effects of different selection schemes on the effective population size can be seen by considering the general weighted index of within- and among-family information,

$$I = (z - \overline{z}_f) + \lambda(\overline{z}_f - \overline{z}) = (\text{within-family}) + \lambda\,(\text{among-family}) \tag{26.40}$$

where z is an individual's phenotypic value, $\overline{z}_f$ is the mean of its family, and $\overline{z}$ is the grand mean. A number of selection schemes can be represented (either exactly or to a good approximation) by this index (Chapter 21). For example, $\lambda = 1$ corresponds to individual selection (as $I = z - \overline{z}$), while $\lambda = 0$ corresponds to strict within-family selection ($I = z - \overline{z}_f$). The accuracy of selection using this index with an appropriately chosen value of λ is greater than the accuracy of individual selection ($\rho(I, A) > \rho(z, A)$; Equation 21.53b), and hence selection using the optimal index gives a greater short-term response than mass selection. To a first approximation, BLUP selection corresponds to this optimal index.

Because the effective population size is reduced by inflating the among-family variance, the larger the value of λ in Equation 26.40, the greater is the reduction in N_e. Larger values of λ place more weight on family information, resulting in more individuals from the best families being coselected. The reduction in N_e is greatest when heritability is small, as in these cases the index places the most weight on the among-family component. Yet, however, it is exactly this setting under which index and BLUP selection have the greatest short-term advantage over individual selection. Conversely, when care is taken to equalize the amount of inbreeding across methods, individual selection can produce a larger single-generation response than index selection or BLUP (Quinton et al. 1992; Andersson et al. 1998).

Can one balance this tradeoff between increased accuracy for short-term response using information from relatives versus inflation of the among-family variance (and the resulting reduction in the long-term response via reduction in N_e) that these schemes produce? Several authors have proposed schemes for reducing the among-family variance following selection. Toro and colleagues (Toro and Nieto 1984; Toro et al. 1988; Toro and Pérez-Enciso 1990) suggested that selected individuals be mated in ways that minimize the coancestry between them. A slightly different strategy, **compensatory mating**, was suggested by Grundy et al. (1994). Here, individuals from families that are overrepresented following selection are mated to individuals from underrepresented families. This has the effect of reducing the cumulative effect of selection (Q_T in Equation 26.6c) by reducing the variance in family contribution. Grundy et al. also suggested a more subtle approach. They noted that by using slightly biased selection parameters in the index (for example, using upwardly biased estimates of h^2 when computing the optimal λ), the slight reduction in the accuracy of the adjusted index from its optimal value is more than offset by a much smaller decrease in N_e. They suggested that this approach, combined with compensatory mating, provides a simple way for ameliorating the reduction in N_e. Verrier et al. (1993) also suggested that schemes placing slightly less emphasis on family information can, in small populations, give greater long-term response than BLUP selection. We examine the optimal control for inbreeding under BLUP in detail in Volume 3.

This tradeoff between optimal short-term versus optimal asymptotic response has economic consequences for breeders. While breeders are ultimately better off in the long run (in terms of total response) using selection schemes that are initially less accurate, competing breeders using the initially more accurate schemes will achieve a larger short-term response. Breeders must thus decide between staying in business over the short term versus experiencing a larger payoff (in terms of a greater response) over the long run.

27

Long-term Response: 3. Adaptive Walks

Real evolution may look less like an attempt to evolve uphill on a static landscape and more like an attempt to keep one's footing on an ever-changing landscape.
Orr (2009)

Our treatment of long-term selection response started by considering the role of existing genetic variation (Chapters 25 and 26). As such variation becomes exhausted through drift and selection, new mutations become increasingly more important (Chapter 26). Here we consider the logical conclusion of this second phase, the continual fixation of new mutations and their impact on long-term response over evolutionary time scales. In particular, our focus is on the nature of **adaptive walks**—the sequence of fixed mutational steps that underlie a specific adaptation. There is a rich (and growing) population-genetics literature on this subject which was nicely reviewed by Orr (2005a, 2005b). Much of this work starts with **Fisher's geometric model** (**FGM**) (Fisher 1930), an extremely influential model for the probability that a new mutation is adaptive (i.e., beneficial). A number of interesting results follow from this model, which assumes adaptation toward some optimal trait value (stabilizing selection). A second class of adaptive-walk models that we consider are based on extreme-value theory and focus on the *fitness effects* of new mutations (as opposed to the FGM focus on *trait values*). Surprisingly, both classes of models give the same general result: ***the effect-size distribution of genetic factors fixed along an adaptive walk is often approximately exponential.***

FISHER'S MODEL: THE ADAPTIVE GEOMETRY OF NEW MUTATIONS

Fisher (1930) offered a highly simplified, yet elegant and powerful, geometric argument suggesting that the probability that a new mutation increases fitness is a simple function of the size of the mutational effect relative to the distance of the original phenotype from the optimal trait value. Fisher believed that his model, while idealized, captured the "statistical requirements of the situation" of adaptation: one complex thing (the organism) must fit into another complex thing (the environment). Although Fisher envisioned that adaptation requires a highly multivariate phenotype meshing with a highly multivariate fitness function, Figure 27.1A shows the basic structure of his model in two dimensions. Stabilizing selection is assumed, with the current phenotypic value at a distance d from a fitness optimum, $\boldsymbol{\theta}$. The two traits have been scaled and rotated to be independent, with equal selection intensity on both (this can be accomplished using a transformation along the lines suggested by Equation A5.16).

As depicted in Figure 27.1A, we assume stabilizing selection acting on two independent traits. Let the vector $\mathbf{z}$, at distance d from the optimal value, $\boldsymbol{\theta}$, represent the current multivariate phenotype associated with a particular genotype. The phenotypic change by a new mutation is given by a vector of length r (the **effect size** of the mutation) extended from the current value ($\mathbf{z}$) in some random direction (i.e., the incremental model). Any mutation whose distance to $\boldsymbol{\theta}$ is less than d has increased fitness (and is said to be **beneficial** or **adaptive**), while a mutation whose distance from $\boldsymbol{\theta}$ is greater than d has lower fitness. Drawing a circle of radius d around $\boldsymbol{\theta}$ (a contour of equal fitnesses corresponding to that of the original phenotype), the probability that the new mutation is advantageous will be

Evolution and Selection of Quantitative Traits. Bruce Walsh & Michael Lynch.
Published 2018 by Oxford University Press. © Bruce Walsh & Michael Lynch 2018.
DOI 10.1093/oso/9780198830870.001.0001

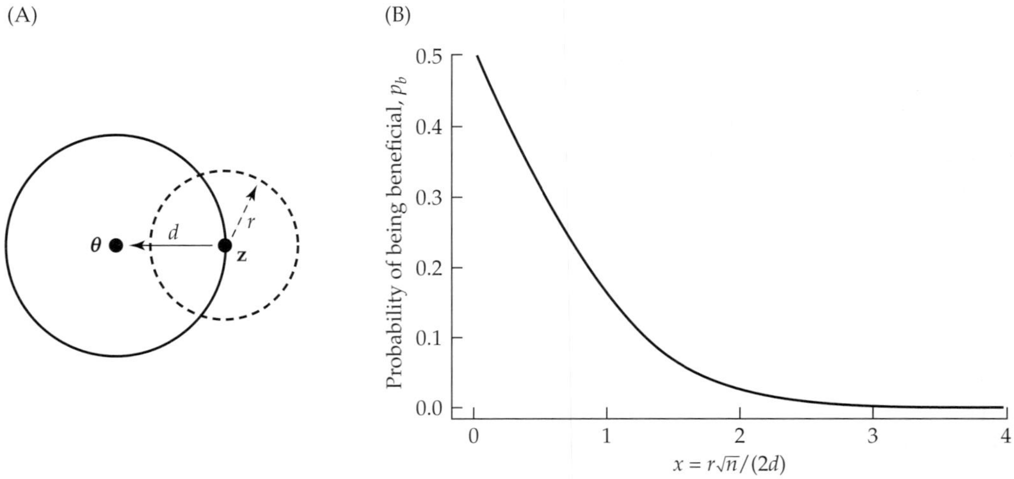

Figure 27.1 **A:** Fisher's (1930) model for the probability that a new mutation increases fitness for the simple case of two independent traits under stabilizing selection. The optimal fitness value occurs at $\boldsymbol{\theta}$, while the mean phenotypic value of the genotype about to experience a mutation is **z**, which is at distance d from the optimum. The solid circle denotes the fitness contour passing through the current value **z**, meaning that all points inside this circle have higher fitness. The effect of a new mutation is to move the expected phenotype by some distance (r) in a random direction around **z**, with the space of possible new phenotypes denoted by the dashed circle (the **isotropic** assumption of equal and independent mutational effects on both traits). The probability of increased fitness for a random mutation of effect r is the fraction of the circumference of the dashed circle that resides inside the solid circle (i.e., is closer to θ), namely, the fraction of all new mutations (whose range of possible phenotypes fall exactly along the circumference of the dashed circle) with a higher fitness than the original allele. More generally, under **anisotropic** conditions (correlated and/or unequal effects for selection, mutation, or both), these circles are replaced by ellipses in two dimensions, and the hyperspheres are replaced by hyperellipsoids in three or more dimensions. **B:** For an n-dimensional phenotype, the probability, p_b, that a new mutation is beneficial (has an increased fitness) is $1 - \Phi[r\sqrt{n}/(2d)]$, where $\Phi[x]$ is the cumulative density function of a unit normal (Equation 27.1b).

simply the probability that the mutation lies within this contour (i.e., is closer to $\boldsymbol{\theta}$).

To extend this simple two-dimensional model to n traits, let the vector **z** denote the current (multidimensional) phenotype of an allele, which is at a (Euclidean) distance of d from the optimum, $\boldsymbol{\theta}$. We express this condition as $||\mathbf{z}-\boldsymbol{\theta}|| = d$, where $||\mathbf{x}||$ denotes the length of the vector **x** (Equation A5.1a). Now suppose this allele mutates under the incremental model (Chapter 12), so that its new phenotype is $\mathbf{z}' = \mathbf{z}+\mathbf{m}$, where **m** denotes the vector of mutational increments. In particular, the new value for trait i is $z_i' = z_i+m_i$, where $E[m_i] = 0$. Further, we assume the mutational increments are uncorrelated and homoscedastic, so that we can write their covariance matrix as $\mathbf{M} = \sigma_m^2\mathbf{I}$. Fitness is increased if $||\mathbf{z}' - \boldsymbol{\theta}|| < ||\mathbf{z} - \boldsymbol{\theta}||$; namely, $\mathbf{z}'$ is closer to the optimum than **z**. Now consider a vector of random mutational inputs subject to the constraint that $r^2 = \sum m_i^2 = ||\mathbf{m}||^2$. Fisher's insight was that a single parameter, the **Fisher scaling parameter**,

$$x = \frac{r\sqrt{n}}{2d} \tag{27.1a}$$

is sufficient to compute the probability, p_b, that such a new mutation is beneficial, with

$$p_b = \frac{1}{\sqrt{2\pi}} \int_x^\infty \exp(-y^2/2)dy = 1 - \Phi(x) \tag{27.1b}$$

Here Φ is the cumulative density function of a unit normal. While Fisher did not present one, derivations of Equation 27.1b were provided by numerous authors (Kimura 1983; Leigh 1987; Rice 1990a; Hartl and Taubes 1996). Fisher's critical observation was that the probability (p_b) that a new mutation is beneficial is a decreasing function of the scaling parameter, x (Figure 27.1B). Increasing x decreases the chance that a mutation is beneficial, and conversely, decreasing x increases the chance it being beneficial. Indeed, as x approaches 0, the probability that a new mutation is beneficial approaches 0.5. The analogy Fisher used to explain this result was that of trying to improve the focus on a microscope. A very tiny change has close to a 50% chance of (very slightly) improving the focus, while a much larger change is far less likely to do so. This analogy led to the use of the term **Fisher's microscope** in the literature to describe this feature of the model.

Fisher's scaling factor (x) consists of two components. The first, r/d, is the ratio of the length of the vector of mutational increments to the distance from the optimum. Whether the mutational effect size (r) is regarded as large or small under Fisher's model depends on the distance of its progenitor allele from the optimum. Consider two mutations, both of which add a random vector of length 1 to the current phenotypic value. If $d = 50$, this is a mutation of small effect from the standpoint of x, while if $d = 2$, this is a large-effect mutation. Hence, the scaling of x accounts for the size of the mutation relative to the current distance from the origin. The second component of x, $\sqrt{n}/2$, accounts for the geometry of the fitness surface. As the dimensionality of the fitness surface increases, there are increasing numbers of constraints on a random mutation required for it to be beneficial. Hence, increasing the complexity of the phenotype (the number of independently selected traits, n, influenced by a new mutation) decreases the chance of mutations being beneficial. The original interpretation of Fisher's model was therefore twofold: small-effect mutations are the stuff of adaptation, and there is a "**cost of complexity**" (Orr 2000). As we will see, both of these initial assertions need significant refinement.

Orr (2006; also see Hartl and Taubes 1996, 1998; Waxman and Peck 1998) showed that Fisher's model provides a much richer description beyond simply predicting whether a mutation is beneficial—it can be used to generate the full distribution of fitness effects for the set of random mutations of a given size, r. Initially, this is performed under the assumption of equal and uncorrelated mutational and fitness effects (i.e., the distribution of trait displacements is a hypersphere of radius r and the fitness surface displays stabilizing selection with equal, and uncorrelated, selection along all trait dimensions). One can then integrate this result for a specific value of r over the assumed distribution of r values to obtain the unconditional distribution of fitness effects.

For the class of mutations that all have the same effect size, the fitness distribution is obtained by first computing the distribution of displacements toward the optimal value. Orr (198b) showed that this is approximately normal for large n, with a negative mean (on average, a random mutation moves the phenotype *away* from the optimum). Translating displacements into fitnesses under the assumption of Gaussian stabilizing selection, the resulting distribution of fitness effects of random mutations (again, all of the same size, r) is also approximately normal, with a mean less than the current fitness of the starting genotype. Martin and Lenormand (2006) and Tenaillon (2014) provided a more exact analysis, which yields a fitness distribution in the form of a modified gamma distribution (Appendix 2). Hence, most new mutations move a trait further away from its optimum, thereby lowering fitness. Only those mutations in the right tail of this displacement distribution (those with positive values, namely, those whose phenotypes following mutation are closer to $\boldsymbol{\theta}$) are advantageous. This important result foreshadows a rather different model of adaptation based on extreme-value theory (i.e., based on draws from the extreme right-hand tail of a fitness distribution), which will be examined in the second half of this chapter.

One might be concerned by the highly simplistic nature of Fisher's model: n independent traits with equal selection and mutation on each. However, these assumptions can be relaxed by replacing Fisher's assumed hyperspheres for fitness contours and mutational

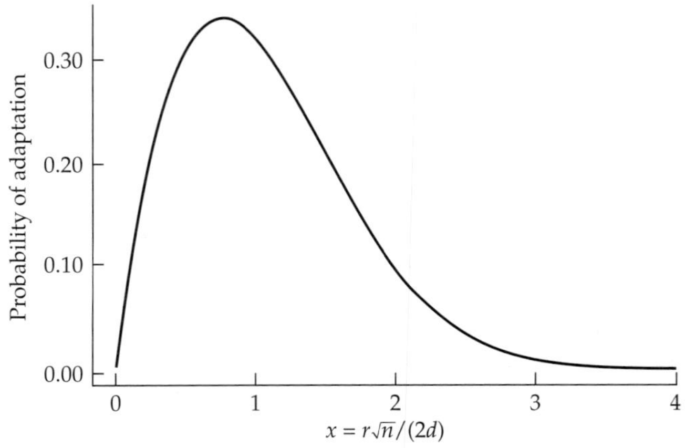

Figure 27.2 The probability of adaptation (the probability of being beneficial times the probability of fixation) for a new mutation with scaled effect $x = r\sqrt{n}/(2d)$. Figure 27.1B shows the probability, p_b, that such a mutation is *beneficial*, while the curve in this figure (which also incorporates the resulting selection coefficient) is the chance that it is beneficial and *fixed*.

effects (under the **isotropic assumption** of equal and uncorrelated effects) with hyperellipsoids, which allows for among-trait correlations and unequal traits effects in both fitness and mutation (e.g., Waxman and Welch 2005; Martin and Lenormand 2006, 2008; Waxman 2006, 2007). This **anisotropic assumption** of unequal and/or correlated pleiotropic mutational effects can be modeled by using a mutational covariance matrix, **M**, and unequal and/or correlated stabilizing selection (Chapter 30) can be accommodated with a semipositive-definite matrix, **S** (e.g., a matrix with no negative eigenvalues; Appendix 5). The resulting effective number of dimensions for x is given by $n/(1 + c_v^2)$, where c_v is the coefficient of variation of the eigenvalues of the matrix product of **MS**.

Fisher's model can also be derived from biological first principles. Martin (2014) showed that one can recover Fisher's model by starting with a complex metabolic network that influences a number of traits. Perhaps the best summary of Fisher's model is the comment by Fox (2014): "Sometimes, even really abstract mathematical models can make really good predictions about real-world biology. Further, they do so *because* of their abstractness, not *despite it*." Fisher's model is so powerful, not because it fully captures all of the biology, but rather because it *avoids* most of the biological details to make its critical points. Precisely because of this generality, there has been a recent resurgence of interest in Fisher's model as a framework for a number of evolutionary questions (reviewed by Tenaillon 2014).

Fisher-Kimura-Orr Adaptive Walks

A subtle feature, first noted by Kimura (1983), makes Fisher's smaller-is-better result misleading. An examination of Equation 27.1b and Figure 27.1B suggests that the majority of beneficial mutations are those with very small effects—those mutations whose length, r, satisfies $x \ll 1$, namely, $r \ll 2d/\sqrt{n}$. While this is indeed correct (under Fisher's model), Kimura noted that such small-effect mutations also have very small selective advantages, and therefore (while *beneficial*) are unlikely to be *fixed* by natural selection. Recall from Chapter 7 that the fixation probability of a favorable allele is $\simeq 2s$, so that the magnitude of s, not only its sign, is important in determining which factors become fixed.

Under Fisher's model, and conditioning on a mutation being beneficial, the expected value of s is linearly proportional to x, as the distance from θ is an indicator of fitness (Kimura 1983; Orr 1998b). This can be more compactly written as $E[s] \propto x$. Thus, the probability that a mutation will be *fixed* is a function of both its selective advantage *and* its chance of being beneficial, $(2s) \cdot p_b \propto (2x) \cdot [1 - \Phi(x)]$. We call this the **probability of adaptation**. As shown in Figure 27.2, the outcome is a dramatic shift in our interpretation of Fisher's result: adaptation favors mutations of *intermediate* effect. A comparison of the difference between Figures 27.1B and 27.2 highlights a critical distinction important throughout this chapter, namely, the difference between the distribution of effects (be they trait values or fitness) over *all* new beneficial mutations versus the distribution of effects restricted to *fixed* beneficial mutations.

There is an irony here in that although Fisher was a pan-selectionist, his initial analysis suggested that alleles of very small effect (i.e., nearly neutral) were more important for adaptation (i.e., comprise the majority of beneficial mutations). Conversely, Kimura, the founder of the neutral theory, showed that Fisher's model (when more carefully considered) argues for a role of stronger selection than envisioned by Fisher (i.e., fixing alleles of intermediate, rather than small, effect).

It is important to note that Figure 27.2 does *not* reflect the distribution of the x values for fixed mutations (i.e., the scaled lengths of fixed mutations). Rather, it describes the *filter of natural selection*, giving the chance that a mutation with effect x is fixed. The distribution of the x-scaled r values for fixed mutations depends on the product of the distribution of new mutational input (the distribution of r values among new alleles) and the chance that each is fixed (Figure 27.2).

To decouple the separate issues of the impact of selection (given a specific effect size) and the distribution of mutational effect sizes, both Kimura and Orr initially assumed a uniform distribution of effect sizes. This focuses discussion on the role of the filter of selection. The key finding, of a roughly exponential distribution of fixed trait effects, holds under rather general distributions of the effects of new mutations. For example, an exponential distribution of r, when compounded with an exponential distribution of fixed effects for a given value of r, returns an exponential. Orr's (1998b) simulations showed that the result of a roughly exponential distribution of the effect sizes of successful (i.e., fixed) mutations was remarkably robust to the assumed distribution of mutational effect sizes.

To examine the impact of selection on the pattern of successive fixations, we first consider the scaled length (x_1) of the initial successful (fixed) mutation at the start of a walk. The scaling from r into x accounts for both the initial distance from the optimum ($d_0 = ||\mathbf{z} - \boldsymbol{\theta}||$) and dimensionality ($n$) of the process (Equation 27.1a). Assuming a uniform distribution for the r values of new mutations, Kimura found that the expected value for $x_1 = r_1\sqrt{n}/[2d_0]$ of this first successful mutation is $E[x_1] \simeq 1.06$. Hence,

$$E[x_1] = \frac{E[r_1]\sqrt{n}}{2d_0} \simeq 1.06, \quad \text{or} \quad E[r_1] \simeq \frac{2.12\, d_0}{\sqrt{n}} \tag{27.2a}$$

where $E[r_1]$ is the expected value of r for the first-fixed mutation. For example, for $n = 10$, $E[r_1] = 0.67d_0$, while $E[r_1] = 0.3d_0$ for $n = 50$. Orr (1998b, 1999, 2000) noted that Kimura's result is simply the *first step* of the adaptive walk starting at distance d_0 from the optimum. Following this first fixation, the walk starts anew from distance $d_1 < d_0$ (Figure 27.3).

It is important to note that $E[r_1]$ is the expected total length of the vector of mutational increments (the distance between $\mathbf{z}$ and its value, $\mathbf{z}' = \mathbf{z} + \mathbf{m}$, after the mutation, with $r = ||\mathbf{m}||$). Our real interest is not this length, but rather in how much of this displacement is toward $\boldsymbol{\theta}$, meaning that the new phenotype is actually closer (and more adaptive) than $\mathbf{z}$, i.e., $||\mathbf{z}' - \boldsymbol{\theta}|| < ||\mathbf{z} - \boldsymbol{\theta}||$. A large value of r_1, by itself, does not imply a large jump closer to $\boldsymbol{\theta}$. Hence, to continue our analysis of the walk, we first need to compute the expected distance moved toward the optimum (i.e., the new distance, $d_1 = ||\mathbf{z}' - \boldsymbol{\theta}||$, from $\boldsymbol{\theta}$; Figure 27.3) after this first step of the walk (the first fixation event).

How much closer does fixing a mutation move us toward the optimum? The first fixed mutation almost never points exactly in the direction of the optimum, so that the actual amount of adaptation (distance moved toward θ) is less, and likely much less, than the length of the jump (Figure 27.3). Thus, instead of $E[x_1]$, we need to consider the length, $E[x_{p,1}]$, of the expected projection (Appendix 5) in the direction of the optimum. For large n, Orr (1998b) found that this projection is approximately

$$E[x_{p,1}] \simeq \frac{E[x_1]}{\sqrt{n}} \tag{27.2b}$$

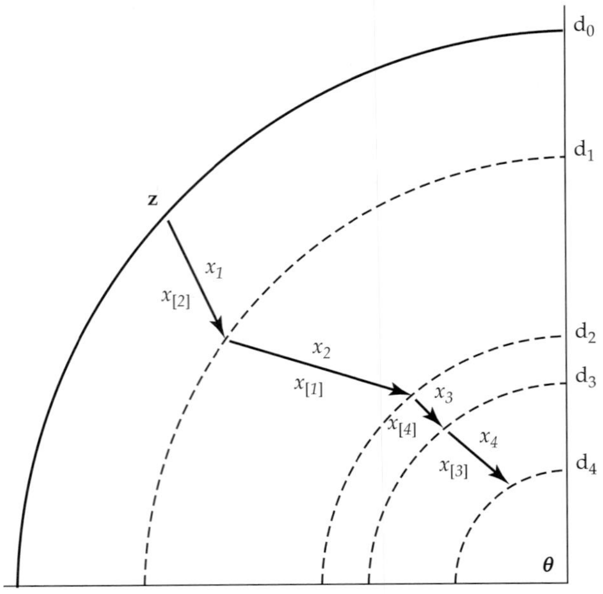

Figure 27.3 An example of an adaptive walk (in two dimensions), starting at **z** and moving toward the optimal value, $\boldsymbol{\theta}$. Here x_i denotes the scaled mutational size of the ith fixed mutation (with $r_i = 2d_0 x_i/\sqrt{n}$), while $x_{[i]}$ denotes the ith *largest* step (the ith largest x value over all mutations fixed during the walk). In this example, the largest jump, $x_{[1]}$, occurs at the second step (x_2), rather than at the first (x_1).

Scaling the phenotypes such that the initial distance from the optimum is one ($||\mathbf{z}-\boldsymbol{\theta}|| = d_0 = 1$), from Equation 27.1a the starting distance on the x scale is $\sqrt{n}/2$. Hence, following the first jump (the first fixation of a beneficial mutation), the population (on average) moves to a distance (in the x value scale) of

$$\sqrt{n}/2 - E[x_{p,1}] = \sqrt{n}/2 - E[x_1]/\sqrt{n}$$

which implies that the new distance to the optimum (on the x scale) following the first jump is a fraction,

$$\frac{\sqrt{n}/2 - E[x_1]/\sqrt{n}}{\sqrt{n}/2} = 1 - 2E[x_1]/n \tag{27.2c}$$

of the initial starting distance.

Orr's key insight was that the size distribution of the second jump (i.e., the distribution of x_2) is the same as that of the first jump, but with the distance to the optimum reduced an amount given by Equation 27.2c. Likewise, following the second jump, the distance to the optimum (as a fraction of the original distance) is the square of the value calculated by Equation 27.2c, and so forth, which generates a self-similar process for each step of the walk toward the optimum, for which $E[d_i] = (1 - 2E[x_1]/n)^i d_0$. It is this self-similar feature that generates the exponential distribution of effect sizes among fixed mutations. Orr (1998b) obtained the distribution for the Fisher-scaled effect size ($x = r\sqrt{n}/2$, taking $d = d_0 = 1$) of the new mutation (Equation 27.1a) during step k of the walk as

$$\psi(x, k) = C\, c_k^2\, x\, [1 - \Phi(c_k x)] \quad \text{where} \quad c_k = \left(1 - \frac{2E[x_1]}{n}\right)^{1-k} \tag{27.3}$$

where C is a normalization constant. This expression is for a uniform distribution of mutation effects (the r values). More generally, the $C\, c_k^2$ terms in Equation 27.3 are replaced by $C\, \phi(c_k x)\, c_k^2\, x$, where ϕ is the assumed distribution of mutational effects (Orr 1998b).

Assuming a uniform distribution of mutation effects (the r values), Orr showed that Equation 27.3 implies that the distribution of the $x_i = r_i\sqrt{n}/2$ is approximately exponential with parameter $\lambda \simeq 2.9$ (i.e., with a mean value of $1/\lambda \sim 0.34$). Simulation studies by Orr (1998b, 1999) showed that this result is very robust to different assumptions about the distribution of mutational inputs, especially when large-effect mutations are not common.

Orr's simulations also showed that the exponential approximation may break down for mutations near the very end of the walk, where the fixed values of x_i are very small as the process very closely approaches $\boldsymbol{\theta}$. In part, this behavior arises because of the drift barrier, discussed in Chapter 7. As the fitness effects of a mutation become sufficiently small ($4N_e|s| \ll 1$), the fixation probability approaches the neutral expectation (i.e., the initial allele frequency, $1/[2N]$), rather than $2s$. Hence, even in a constant environment, perfect adaptation does not occur. Rather, as the drift barrier is approached, there is a constant fixation of mutations of very small effect, which are random in their direction toward, and away from, the optimum. Hence, Orr's key finding that the distribution of the effects of fixed mutations during a walk is roughly exponential is tempered somewhat by departures for fixed mutations with very small effects. However, given that this class of fixed mutations with very small effects would be very difficult to detect, from an empirical standpoint, the bulk of any detected substitutions during a walk would follow an approximately exponential distribution.

A critical point noted by Orr (1998b), and shown in Figure 27.3, is that the *largest* "jump" in the walk, $x_{[1]}$ (the largest scaled length for *any* of the fixed mutations), does not necessarily correspond to the *first* jump, x_1 (the scaled length of the vector of effects for the first fixed mutation). The expected value of the largest jump is approximately

$$E[x_{[1]}] \simeq \frac{1}{\lambda}\left[\ln\left(\frac{\lambda n}{2}\right) + 0.5772\right] \tag{27.4a}$$

For example, taking $\lambda = 2.9$ and $n = 50$, Equation 27.4a yields the expected *largest* jump as $E[x_{[1]}] \simeq 1.68$, implying an expected total mutational size of $E[r_{[1]}] = 3.36d_0/\sqrt{n}$. Conversely, the expected value of the *first* jump is $E[x_1] \simeq 1.06$, or a mutation with $E[r_{[1]}] = 2.12d_0/\sqrt{n}$, which is ~60% of the expected size of the largest jump. More generally, the expected value of the kth largest jump is

$$E[x_{[k]}] \simeq \frac{1}{\lambda}\left[\ln\left(\frac{\lambda n}{2}\right) - \sum_{i=1}^{k-1}\frac{1}{i} + 0.5772\right] \tag{27.4b}$$

and Equation 27.1a yields $E[r_{[k]}] = (2d_0/\sqrt{n}\,)E[x_{[k]}]$. Using this result to consider the quantity $(E[x_{[k]}] - E[x_{[k+1]}])$ provides the expected difference in the scaled r values for the k and $k+1$ largest fixed mutations, which yields

$$E[x_{[k]}] - E[x_{[k+1]}] = E[\Delta_{[k]}] \simeq \frac{1}{\lambda}\left[\sum_{i=1}^{k}\frac{1}{i} - \sum_{i=1}^{k-1}\frac{1}{i}\right] = \frac{1}{\lambda k} = \frac{0.349}{k} \tag{27.4c}$$

for $\lambda \simeq 2.9$. For example, rearranging yields $E[x_{[2]}] = E[x_{[1]}] - 0.349 = 1.33$, and so on. This form of spacing will reappear in our discussion of walks modeled by extreme-value theory.

The careful reader might have noticed that Orr's finding of an approximate exponential distribution in the effects of fixed mutations refers to the lengths, r, of the vectors of these mutations. This is different from the actual phenotypic effect size for any *single trait*. Orr (1999) addressed this issue, and found that the projection of a favorable (but otherwise random with respect to direction) vector of length r onto any particular trait (say trait j) is roughly normally distributed, with an expected absolute value of

$$E[\,|z_j|\,] = \frac{2\,r}{\pi\,\sqrt{n}} \tag{27.5}$$

Hence, the distribution of the effect sizes for a single trait is the result of sampling from two distributions. First, the expected value of the effect size (Equation 27.5) is a function of the random variable, r, which is roughly exponentially distributed for fixed mutations. Second, there is sampling noise around the expected value given by Equation 27.5, namely, the roughly normal distribution for the projection from a given value of r onto the mutational effect on any given single trait. It is therefore not surprising that Orr's simulations showed that the effect-size distribution on any particular single trait also shows a strong exponential trend (with Gaussian sampling noise around the expected trend). Thus, while the total *lengths* of the mutational displacement vectors associated with fixed mutations are roughly exponentially distributed, so too are the resulting changes in the phenotypic values of fixed mutations for any particular *trait*. Consequently, if we could score the phenotypic effects on a trait over the set of fixed mutations leading to the current adaptation, we would expect them to be roughly exponential (except for those fixed mutations with effects on a trait that are extremely small, and therefore difficult to detect).

The view of a random walk as a series of discrete, fixed steps is one extreme vision, and essentially a hard-sweep view (Chapter 8). The other extreme view is of the constant generation of mutations of small effect, which mount a continuous infinitesimal response (soft or partial sweeps; Chapter 8). Lande's analysis (Chapter 25) showed that polygenic response (changes in the trait mean through allele-frequency changes, and not necessarily fixations, at a large number of loci of small effect) is favored when major genes have negative pleiotropic effects on other traits, unless there is strong selection on a focal trait. Hence, adaptive walks that follow Fisher's model may be biased toward cases of traits that are under strong selection.

A second assumption of the Fisher-Kimura-Orr adaptive walk model is weak mutation, with only a single beneficial mutation segregating at any one time. With multiple segregating mutations, Hill-Robertson effects (Chapters 3, 7, and 8) reduce the effectiveness of selection, the most extreme example being **clonal interference** in asexuals (Gerrish and Lenski 1998; Rozen et al. 2002). When considering the fixation of mutations in clonal (asexual) populations, a new beneficial mutation must first avoid stochastic loss when rare. While such surviving cosegregating beneficial mutations (which we denote as **competing**) may be fixed in a sexual population, in an asexual population they are fully linked to genomes that may experience additional beneficial mutations before the initial mutation becomes fixed. When $N\mu \ll 1$, the presence of multiple segregating beneficial mutations is unlikely (where μ is now the genomewide beneficial mutation rate). However, when $N\mu \gg 1$, clonal interference can occur, wherein the most fit genomes compete with each other and only one is ultimately fixed (Gerrish and Lenski 1998; Elena and Lenski 2003; Barrick and Lenski 2013; Levy et al. 2015). As noted by Barrett et al. (2006a), the winning genome often accrues *several* beneficial mutations on its way to fixation (outcompeting genomes with potentially more favorable initial mutations that did not accumulate additional beneficial mutations), which can result in an investigator overestimating the fitness effects under the assumption of a single beneficial mutation driving the fixation.

Example 27.1. Annual wild rice (*Oryza nivara*) is a photoperiod-insensitive selfer adapted to seasonally dry habits. It is thought to have evolved from a progenitor with properties of its sister species, *Oryza rufipogon*, a photoperiod-sensitive, perennial outcrosser found in deep-water swamps. As a potential model system for adaptation to new conditions, Grillo et al. (2009) crossed these two species and mapped QTLs for flowering time, mating system, and life-history traits. The resulting distribution of the values of QTL effects for these traits was roughly exponential, as predicted from Orr's adaptive walk. If directional selection was an important factor in the divergence of these two species, one would expect to find an excess of the detected QTL alleles in *nivara* in the direction of the adaptive phenotypes (the QTL sign test; Chapter 12). Indeed, for flowering time and duration, all six of the *nivara* QTL alleles act in the direction of reduced photoperiod sensitivity and longer flowering. Likewise, all seven of the QTLs for mating system traits in *nivara* acted in the direction of increased selfing.

One prediction from the Fisher-Kimura-Orr model is that fixations of large-effect mutations are more likely when the current population is far from its adaptive peak. Rogers et al.

(2012) compared threespine stickleback fish (*Gasterosteus aculeatus*) from two different lake types. This species has a marine origin, but it colonized freshwater lakes at the end of the last ice age. Based on the presence of a sculpin predator, one lake type (sculpin-present) was judged as being closer to the ancestral marine optimum than a second type (sculpin-absent). A higher frequency of large-effect QTLs for morphology (related to general predator defense) was observed in the sculpin-absent lake type, consistent with the fixation of larger-effect mutations when the optimum is further away.

A corollary to this prediction is that larger fixed-effect sizes are expected at loci underlying traits that experience continual and sudden shifts in their fitness surfaces. Louthan and Kay (2011) suggested that this often happens in traits under biotic selection. They investigated this hypothesis with a meta-analysis of 1950 QTLs (1721 controlled traits presumed to be under abiotic selection, and the rest under biotic selection) from 10 plant families (but mainly Brassicaceae). They found larger QTL effect sizes for traits under presumed biotic, as opposed to abiotic, selection.

The Cost of Pleiotropy

We now turn to the second implication from an initial analysis of Fisher's model, which is the impact of n on x, and hence on p_b. Equation 27.1a shows that the scaling parameter, x, increases with $\sqrt{n}$, the number of traits (i.e., the complexity of an organism). For the same r/d values, the probability that a mutation is adaptive is lower in more phenotypically complex organisms (those with larger n values), which has been termed the cost of complexity. When considering the *rate* of adaptation, Orr (2000; also see Welch and Waxman 2003) noted that the cost of complexity is actually far greater than suggested by a superficial analysis of Fisher's model. Consider a rate of adaptation defined by the rate at which a trait approaches its optimal value. This rate is proportional to the product of three components: (i) the probability that a mutation is beneficial; (ii) the probability that such a mutation is fixed; and (iii) the length of a move (or displacement) toward the optimal value (the increase in fitness) when such a mutation is fixed. These last two factors each scale as $1/\sqrt{n}$ (Equations 27.2b and 27.5), so together they scale as $1/n$. Given that the probability of an adaptive mutation also decreases with n (in a monotonic, but nonlinear, fashion, as $p_b = 1 - \phi[\sqrt{n}r/(2d)]$), the net result is that the decrease in the rate of adaptation is faster than $1/n$, which suggests a far higher cost than under Fisher's initial analysis.

What exactly is n and what forces determine this number? The original Fisher model effectively assumed **universal pleiotropy**, namely, that all new mutations essentially influence *every* trait in an organism. Under this assumption, the number of independent trait combinations under selection (the dimension of the fitness surface; technically, this is the number of nonzero eigenvalues of the stabilizing selection matrix, **S**; Appendix 5) in an organism is the factor that defines n. (The term universal pleiotropy first seems to have been used by Wright [1968], although, as noted by Wagner and Zhang [2011], Wright's usage appeared to be that every gene affects *more than one* trait, as opposed to the Fisherian notion that every gene affects *all* traits.) Under the Fisherian view of universal pleiotropy, the value of n is set by natural selection (n is the number of dimensions in the trait space that are under selection).

Given that Fisher's model is clearly biologically unrealistic, how valid is this concern of a cost of complexity (which, at face value, seems to imply that more complex organisms—those with more independent dimensions of trait space under selection—have a slower rate of evolution)? There are several potentially ameliorating factors. First, **mutational pleiotropy** is likely not universal, but rather more modular (the notion of **restricted pleiotropy**), wherein most mutations only influence traits within their developmental module (Wagner 1996b; Barton and Partridge 2000; Welch and Waxman 2003; Wagner et al. 2007; Wagner and Zhang 2011). Hence, the number of traits (on average) that a particular mutation influences is what determines n, not the overall dimension of the fitness surface. Notice that this shift in focus concerning the definition of n (from selection-driven to pleiotropy-driven) immediately suggests that n can evolve (just as one can envision selection to increase, or decrease, developmental modularity).

Under restricted pleiotropy, the focus shifts from the complexity of an organism to the complexity of a module, with a **cost of pleiotropy** (Wagner and Zhang 2011), as the rate of adaptation is decreased in modules of higher complexity (i.e., modules in which single mutations impact a larger number of traits). As reviewed by Wang et al. (2010) and Wagner and Zhang (2011), evidence from extensive QTL mapping and gene-knockout studies in yeast, mice, and nematodes suggests that most mutations show restricted pleiotropy. For example, Wang et al. (2010) examined $\sim$250 morphological traits in yeast and found that the median amount of pleiotropy for a given mutation was only 7 traits (2% of the traits examined) for the $\sim$2500 gene knockouts that were examined. Hill and Zhang (2012a, 2012b), however, have suggested that we exercise some caution before accepting restricted pleiotropy. They performed simulation studies using a model with highly pleiotropic alleles (in which new mutations influence a large number of traits), and found that low power in the ability to detect effects on any particular trait can create the appearance of restricted pleiotropy (see Wagner and Zhang 2012 for a reply). Corrections for multiple comparisons, given the large number of tested traits (Appendix 4), is expected to result in low power for most of these mutant screens.

A second mitigating factor in determining the cost of complexity is that the intensity of selection likely varies over traits. This reduces the actual number of traits, n, in a module to an effective number of traits, n_e, where n_e decreases as the variance in trait selection intensity increases (Rice 1990a; Orr and Coyne 1992; Orr 2000; Waxman and Welch 2005; Martin and Lenormand 2006, 2008). If just a few traits in a module face the majority of the selective pressure, this significantly lowers the cost (i.e., it increases the rate of adaptation). A similar argument applies if mutational effects are very uneven over traits. Finally, the presence of highly correlated mutational effects within a module (e.g., mutations tend to affect traits in the same, as opposed to random, directions) also reduces the effective n (Wang et al. 2010). The joint impact from all these factors suggests that the effective dimensionality of a module is less, and likely far less, than the actual number of selectively independent traits that are influenced by a given developmental module.

The final mitigating factor in any cost of complexity is more subtle, as it deals with the relationship (if any) between the average incremental effect, m_i, of a mutation on a specific trait and the total effect, $r = \sqrt{\sum m_i^2}$, of that mutation over all the traits that it affects, and how this scales with n. Under the **invariant total-effect model** (Wagner et al. 2008), the total effect (r) of any given mutation is independent of the number of traits it influences. In that case, a mutation influencing more and more traits will, on average, see no change in its r value. Under this assumption, as a mutation influences more traits, its average effect on a given trait decreases with n (with r constant, m_i decreases with n). The complementary assumption is **Euclidean superposition** (Wagner et al. 2008), wherein the effect of a mutation on any given *trait* is independent of n (i.e., independent of how many other traits that mutation impacts). In this case, a mutation's total effect scales as

$$r = \sqrt{\sum_{i=1}^{n} m_i^2} = \sqrt{n}\sqrt{m^2}$$

where m^2 is the average squared effect of a mutation on a random trait (with m^2 constant, r increases with n). Both of these models are special cases of the more general scaling, $r \simeq a\, n^b$, where $b = 0$ corresponds to the invariant model and $b = 0.5$ to the superposition model (Wang et al. 2010). With a large number of pleiotropic mutations in hand, one can regress their total effect over all traits, r, on their degree of pleiotropy, n, to estimate b. Surprisingly, studies in mice (Wagner et al. 2008) and yeast (Wang et al. 2010) suggest that the data are best fit by a model with $b > 0.5$, implying that the per-trait effect of a mutation *increases* with the amount of pleiotropy, n. Wang et al. showed that the consequence of $b > 0.5$ is that while the probability that a new mutation is advantageous decreases with increasing

n, its fixation probability, and its effect on fitness if fixed, both *increase* with n, resulting in the rate of adaptation being maximized at some intermediate value of n (akin to Kimura's observation of an intermediate value of x maximizing the probability of adaptation).

However, the observation of $b > 0.5$ needs to the treated with care. Hermisson and McGregor (2008) noted that QTL studies can result in inflated estimates of b in the presence of linked loci. While the results of Wang et al. (2010) avoid this concern (being based on single-gene knockouts), caution is still in order. Simulation studies by Hill and Zhang (2012b) found that a true value of $b = 0.5$ can easily generate data with b values greatly in excess of 0.5 if the traits and/or measurement errors are correlated.

Our conclusion from all of the above results is that any cost of complexity is based, not on organismal dimensionality per se, but rather on the amount of pleiotropy that a typical mutation experiences. There are reasons to suggest that this is often modest or small. Further, variances in the strength of selection, or in the distribution of mutational effects over individual traits, will lower the effective value of n, further reducing this cost. Finally, unresolved scaling issues (e.g., the relationship between total and individual-trait effects of new mutations) can further reduce this cost, provided they occur in the appropriate direction. While all of these issues can mitigate the cost of complexity, one may still remain. Cooper et al. (2007) used an extensive single-gene deletion dataset in yeast (*Saccharomyces cerevisiae*), where 501 morphological traits were measured for roughly 4800 single-gene deletions, along with the fitness effects for each deletion. There was a strong and significant negative correlation between the amount of pleiotropy (number of traits influenced) and fitness of the deletion (a squared correlation coefficient of roughly 0.18), meaning that organisms with greater numbers of pleiotropic mutations tend to be less fit.

Adaptive Walks Under a Moving Optimum

One critical assumption of Orr's analysis is that the change in the optimal value is sudden, large ($E[r] \ll d$), and effectively permanent. While this can happen following a sudden shift in the environment, an equally plausible scenario is that the optimal value slowly changes over time, e.g., $\theta(t) = \nu t$. In this setting, Bello and Waxman (2006), Collins et al. (2007), and Kopp and Hermisson (2007, 2009a, 2009b) found situations in which alleles of small to intermediate effect tend to be fixed before large-effect alleles. This is in contrast to the constant-optimum prediction that large-effect substitutions occur when the population is far from the optimum. Kopp and Hermisson (2009a, 2009b) made the key observation that the substitution pattern critically depends on the speed, ν, of environmental change relative to the genetic potential of the population. If the environmental change is sufficiently rapid, large-effect alleles are favored, as the population is waiting for the appearance of new mutations to help it keep up (and is said to be **genetically limited**). Conversely, when the environmental change is sufficiently slow, small-effect mutations tend to be fixed (and the population is said to be **environmentally limited**). For a single trait with a moving-optimum, Kopp and Hermisson (2009b) obtained a simple composite parameter

$$\gamma = \frac{\nu}{\omega^2 s N \mu \, \sigma_\alpha^2} \tag{27.6a}$$

to determine which region (environmentally, or genetically, limited) a population is in. Here μ is the haploid genomic mutation rate for the trait, while ν and ω^2 are, respectively, the rate of change in the optimum and the strength of stabilizing selection (larger values of ω^2 imply weaker selection; Equation 28.3b). Finally, σ_α^2 is the variance of the effects of new mutations (which are examined in detail in Chapter 28). The denominator in Equation 27.6a represents the genetic potential of the population (being a measure of the amount of new variation introduced into the population in each generation), while the numerator represents the speed of environmental change. When γ is small (as can happen in a large population), evolution is environmentally limited, while when γ is large, it is genetically limited.

Matuszewski et al. (2014) generalized this result to a multivariate vector of optima changing over time. The multivariate mutational structure is described by a pleiotropy mutation matrix, $\mathbf{M}$, while the multivariate nature of selection is described by a stabilizing selection matrix, $\mathbf{S}$, with the vector $\boldsymbol{\nu}$ showing the per-generation change in each component of $\boldsymbol{\theta}$. The basic results from the univariate optimum analysis hold, where the composite parameter determining which domain a population resides in (genetically or environmentally limited) is given by

$$\gamma = \frac{\sqrt{\boldsymbol{\nu}^T\,\mathbf{S}^{-1}\boldsymbol{\nu}}}{(N\mu)(\widetilde{s}/\widetilde{m})^{-3/2}}, \quad \text{where} \quad \widetilde{s}^2 = \sqrt[n]{\det(\mathbf{S})} \quad \text{and} \quad \widetilde{m}^2 = \sqrt[n]{\det(\mathbf{M})} \qquad (27.6b)$$

Recall that the determinant (det) of a matrix is the product of its eigenvalues (Appendix 5), and thus $\widetilde{s}^2$ and $\widetilde{m}^2$ are the geometric means of the eigenvalues of the stabilizing selection, $\mathbf{S}$, and mutation, $\mathbf{M}$, matrices, which can be thought of as the average widths of these processes. When both matrices are isotropic (a constant times the identity matrix), Equation 27.6b reduces to Equation 27.6a. Matuszewski et al. (2014) showed that in the environmentally limited range (where γ is sufficiently small), there is an optimal mutation size, with the distribution of the effects of fixed mutations being unimodal (like Kimura's distribution for the first-fixed factor; Figure 27.2), as opposed to Orr's exponential model.

There is also an interesting cost-of-complexity within the environmentally limited domain. Apparently at odds with Orr's result, Matuszewski et al. (2014) found that the size of the first fixed effect *increases* with the number of characters (n). *However*, a cost is imposed because the waiting time for a beneficial mutation that is destined to become fixed also increases with n. Because one has to wait longer for a beneficial mutation as complexity increases, the optimum has traveled further away, now making larger-effect mutations easier to fix. In the words of Matuszewski et al., "the moving-optimum model does not contradict the cost-of-complexity argument, but reveals yet another aspect of it."

WALKS IN SEQUENCE SPACE: MAYNARD-SMITH-GILLESPIE-ORR ADAPTIVE WALKS

Quantitative geneticists have historically been interested in the phenotypic effects of alleles, while the focus of population geneticists has been on their effect on a very particular trait, fitness. The power of Fisher's geometric model is that it addresses both of these concerns, as trait effect sizes can be translated into fitnesses (Martin and Lenormand 2006; Orr 2006; Tenaillon 2014). While the distribution of phenotypic values under Fisher's geometry is both unconstrained and continuous, a second class of adaptive walks, based on the much more constrained geometry of **sequence space**, has also been widely examined. The analysis of such models is strictly concerned with the sequence of fitness increases from beneficial alleles fixed during a walk, rather than their phenotypic effects (Maynard-Smith 1970; Gillepsie 1983, 1984b, 1994; Orr 2002, 2003a). Further, they focus on evolution at a *particular locus* (or more generally, a tightly linked region, such as an entire bacterial genome), while Fisher's model is concerned with the *genome-wide response* in a continuous trait, potentially involving a very large number of loci. Sequence-space walk models have several potentially very robust features, making them perhaps more general than FGM results.

One emerging feature in studies of molecular fitness spaces is the use of experimental tests of many of the assumptions and models. We will touch on this briefly here, and encourage the reader to work through the more technical sections that serve as leadins to experimental tests, as (is detailed below) much of this theory can be directly applied to model systems, especially in microbes. For example, the combination of high-throughput combinatorial DNA/RNA synthesis (tens of millions of base-pair combinations can be generated) coupled with high-throughput scoring (such as chip-based binding assays) is starting to open up the exploration of the fitness surfaces associated with single genes (reviewed in Conrad et al. 2011; de Vasser and Krug 2014). There is a rich literature—spanning statistical physics, computational theory, experimental molecular biology, and evolutionary biology—on mutational and fitness landscapes over sequence space beyond the models examined here. See Miller et al. (2011), Szendro et al. (2013), and de Vasser and Krug (2014) for brief overviews and an entreée into deeper modeling issues under more complex mutational-fitness surfaces.

SSWM Models and the Mutational Landscape

As first noted for protein sequences by Maynard-Smith (1962b, 1970), while phenotypic space is continuous, underlying evolutionary modifications involving changes in DNA sequences impose a discrete, and constrained, geometry. For a DNA sequence of length L, while there are 4^L possible states, each single point mutation can only move to one of $3L$ possible states. Under the strong selection ($Ns \gg 1$), weak mutation ($N\mu \ll 1$), assumption (the **SSWM**, or "swim," model), each new mutation will either be fixed or lost before the next one appears, meaning that double-mutations are not segregating. This shortened horizon in the sequence space leads to Gillepsie's (1983) **mutational landscape model (MLM)**, which constrains the possible geometry of evolutionary paths in sequence space to those that are one mutational step away (our discussion in Chapter 7 on mutations with contextual effects deals with some of the issues arising when the horizon is two, or more, mutations away). Given a starting allele, the set of "local" alleles are those $3L$ new sequences that are located one mutational step away. When one such adaptive mutation is fixed, there is now a new space of $3L - 1$ sequences ($3[L-1]$ sequences when ignoring the just-mutated site, which [ignoring the back mutation] can itself contribute 2 new sequences), and the walk continues until the most fit local allele is fixed and the walk stops. As we will detail shortly, Gillespie's (1983, 1984b) analysis of such models draws on the use of **extreme-value theory (EVT)**, which treats the distribution of extreme draws from some underlying distribution (Gumbel 1958; Leadbetter et al. 1989).

Extreme-Value Theory (EVT)

Under Gillespie's model, fitness values for new mutations are drawn from some unknown, and complex, underlying fitness distribution. However, the current allele is likely rather fit (even if moved to a new environment) relative to all of the possible $3L$ alleles that are one step (a single-base change) removed. Hence, its current fitness value is already in the right-hand tail of the fitness distribution, with new favorable alleles being drawn from values that are even more extreme (farther to the right); see Figure 27.4. In such settings, EVT states that one of three possible limiting distributions (referred to as **domains**) will occur (Example 27.2). Gillespie (1983, 1984b) assumed the **Gumbel domain**, which arises for a very wide range of underlying distributions. Joyce et al. (2008) examined the properties of adaptive walks under the two other EVT limiting distributions, the **Weibull** (truncated right tails) and **Fréchet** (tails heavier than exponential) **domains**. They did so by extending many of the results that we present below (which assume a Gumbel), under a generalized Pareto distribution as a function of its tail parameter (κ), which sets the domain (Example 27.2). Joyce et al. found that results based on the Gumbel ($\kappa = 0$) assumption are fairly robust for biologically realistic Fréchet distributions ($0 < \kappa < 0.5$), and for moderate Weibull distributions ($-0.5 \leq \kappa < 0$). However, for more extreme Weibull distributions ($\kappa < -0.5$), significant departures in behavior from Gumbel-assumption results can occur.

Example 27.2. The body of extreme-value theory that deals with the largest draws from a distribution provides very powerful machinery for the analysis of new beneficial mutations. Gillespie's (1983, 1984b) key insight was that the current alleles are likely rather fit (relative to all possible local alleles), and are therefore in the rightmost tail of the unknown, and likely very complex, distribution of potential fitnesses at a given locus (or over a very tightly linked region). *More* fit alleles represent even more extreme draws from this distribution (Figure 27.4), allowing many of their features to be derived from the extreme-value distribution of fitnesses at the target gene or region. There is another Fisherian irony here in that the field of EVT, which

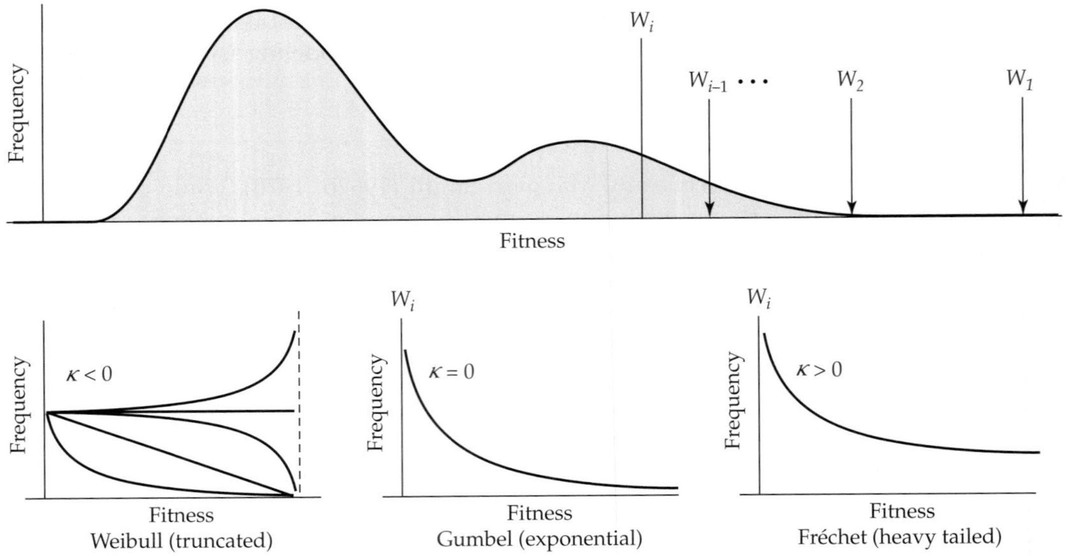

Figure 27.4 **Top:** Extreme-value theory applied to the distribution of fitness effects of new beneficial alleles. Assume some arbitrary, and unknown, fitness distribution for possible alleles at a given locus. Provided the current allele is fairly fit, it is a draw from the righthand tail (extreme upper values) of this distribution. Let W_i denote its current fitness, with i denoting the fitness rank of the current allele relative to the set that it can mutate to. W_1 is the most fit possible allele at this locus, W_2 is the next most fit, and so on. **Bottom:** The **trinity theorem** (Example 27.2) states that the limiting distribution of draws from the extreme tail of a distribution is one of only three possible types (or **domains**), depending upon a shape parameter, κ (see Equation 27.7). The figure displays the form of the density function for these three limiting cases. Under the **Gumbel domain**, the limiting extreme-value distribution is exponential. If there is a finite upper limit on the original distribution, its extreme-value distribution is from the **Weibull domain** (which, as depicted in the figure, can have a wide variety of shapes depending on the value of κ). Finally, if the tail of the original distribution is heavier (falls off more slowly) than an exponential, its limiting extreme-value distribution is in the **Fréchet domain**. (After Beisel et al. 2007.)

provides the basis of an alternative model to Fisher's geometric model, was first introduced by Fisher (Fisher and Tippett 1928). Leonard (L. H. C.) Tippett, a statistician working in the textiles industry, was a leading figure in the early development of quality-control methods and was also the first to publish a table of random numbers (Tippett 1927).

The critical result from EVT is the so-called **trinity** (or **Fisher–Tippett–Gnedenko**) **theorem**, which states that the distribution of draws of extremes (the **extreme-value distribution**) from any underlying distribution are given by the **generalized Pareto distribution** (Pickands 1975), a family of distributions determined by a scale parameter, τ, and shape parameter, κ (also called the **tail index**), which falls into one of three limiting types (or **domains**), depending on the value of κ (Equation 27.7). Because our interest is in the distribution of fitness values for new beneficial alleles, setting the fitness of the current allele to 1, then the fitness of a beneficial allele is 1 + X, where the fitness increase, X, is drawn from the tail of the underlying distribution to the right of zero. Following Beisel et al. (2007), the families of extreme-value distributions for such draws are given by

$$\Pr(X \le x \,|\, \tau, \kappa) = \begin{cases} 1 - (1 + \kappa x/\tau)^{1/\kappa}, & x \ge 0 \quad \text{if } \kappa > 0 \quad \text{(Fréchet)} \\ 1 - (1 + \kappa x/\tau)^{1/\kappa}, & 0 \le x < -\tau/\kappa \quad \text{if } \kappa < 0 \quad \text{(Weibull)} \\ 1 - \exp(-x/\tau), & x \ge 0 \quad \text{if } \kappa = 0 \quad \text{(Gumbel)} \end{cases} \tag{27.7}$$

Most common distributions (normal, gamma, etc.) have a **Gumbel** EV distribution ($\kappa = 0$), which has an exponential tail. This is the most commonly assumed EV distribution for beneficial mutations. When the underlying distribution is truncated to the right (the upper bound of $-\tau/\kappa > 0$, as $\kappa < 0$), the extreme-value distribution is in the **Weibull domain**.

Such an upper-truncated underlying distribution may hold for loci that are near their fitness optimum, as the fittest possible allele simply moves to the optimal value (Martin and Lenormand 2008). The final possible EV distribution (arising when $\kappa > 0$) is the **Fréchet domain**, which has much heavier tails than an exponential. Biologically, this implies that highly beneficial new mutations are rather likely to occur, and hence it is generally not used (Orr 2006), although some experimental data are consistent with the underlying fitness distribution being in the Fréchet domain (Table 27.1). By noting that κ determines the domain family, Beisel et al. (2007) developed a likelihood-ratio test for $\kappa = 0$. They found that the power to detect the Weibull domain alternative ($\kappa < 0$) versus a Gumbel null ($\kappa = 0$) is reasonable if ten or more beneficial mutations are scored.

One alternative approach to modeling walks is based upon the **Fisher–Kolmogorov equation** (Fisher 1937; Kolmogorov 1937), which was originally used to model the speed of the traveling wave of advance of an advantageous mutation over a linear cline (the analysis of allelic surfing is based on this approach; Chapters 8 and 9). As applied to an adaptive walk, one treats the "wave" of advance as the rate at which mean population fitness increases due to increases in the frequency of new advantageous mutations (see Hallatschek 2011 and Neher 2013 for reviews). The leading edge of this traveling wave of advance (the most fit alleles, namely, the extreme right tail of the fitness distribution) is often called the **nose** of the distribution, leading to the term **nose theory** being used to describe this modeling approach. The connection with EVT is that if the distribution of fitnesses in the nose (the extreme values of the distribution) falls off more quickly than an exponential, the distribution of fitness effects is in the Weibull domain, with the fitnesses of new advantageous alleles tending to be somewhat uniform (in the words of Neder 2013, their distribution "tends to be smooth"). Conversely, if the tail of the underlying fitness distribution declines at a rate slower than an exponential, extreme fitnesses are drawn from the Fréchet domain and the behavior of the walk can be dominated by a few large-effect mutations.

Structure of Adaptive Walks Under the SSWM Model

Given that the limiting EVT distribution is largely independent of the details of the underlying distribution generating it, several interesting results follow when the distribution of fitnesses for new mutations falls into the Gumbel domain. First, let s_j denote the fitness advantage of a new allele relative to the current one. Gillespie (1984b) showed that under strong selection and weak mutation (SSWM), the probability of allele j being the next one to be fixed during the walk is

$$\pi_j = \frac{s_j}{\sum_{\ell=1}^{k} s_\ell} \tag{27.8a}$$

where there are k alleles that are more fit than the current one (the set of potential alleles is typically taken as the set of all possible mutations that are one step away from the current allele). Hence, the chance that a mutation is fixed is proportional to its fitness advantage, but the most fit allele is by no means guaranteed of being the next allele fixed. Mutation rates do not appear in Equation 27.8a, as it makes the assumption that mutational change to any of the $3L$ adjacent sequences is equally likely (Equation 7.38 recovered a similar result for a somewhat different problem).

Gillespie (1983) also showed that the mean number of substitutions during a walk is small, with the mean number of steps until the most favorable allele is fixed, given the walk starts with the kth fittest, being

$$\frac{1}{2}\left(1 + \sum_{j=1}^{k-1} \frac{1}{j}\right) \tag{27.8b}$$

The simplification given in Equation 27.8b of Gillespie's original result is due to Orr (2002). Because Equation 27.8b scales as $\sim\ln(k)$, the mean number of steps (substitutions) typically ranges between two and five (Gillespie 1994). This sequential series of fixations until the best

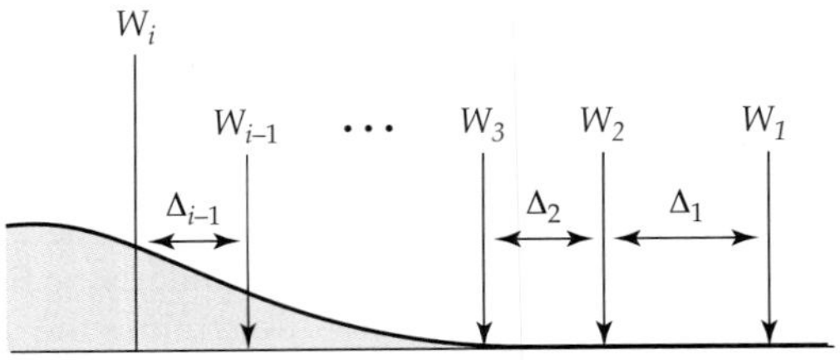

Figure 27.5 Let $\Delta_i = W_i - W_{i+1}$ denote the spacing (difference) in fitnesses between the genotype of fitness rank i and $i+1$. When the current genotype has a high fitness (its rank, i, is not too large) and the underlying fitness distribution is in the Gumbel domain, then the values of Δ_i follow an exponential distribution (Gillespie 1983; Orr 2003b).

allele is fixed yields a burst of substitutions followed by stasis (unless there is a significant change in the environment, and thus in the fitness function). The result is an overdispersed molecular clock (a higher variance than expected from the Poisson distribution for substitutions that is generated under drift). Remember that extreme-value theory walks focus on the substitution patterns in a single gene or tightly linked genetic region. Hence, many such walks may simultaneously be ongoing within a (recombining) genome. We return to the important issue of context-specific fitness (i.e., epistasis), and how it alters the walk, shortly.

Orr (2002) used Gillespie's SSWM process as an adaptive-walk model, making use of a key result from EVT on the **spacing** between extreme values. Let $\Delta_i = W_i - W_{i+1}$ denote the spacing between the fitness values for alleles of fitness ranks i and $i+1$ (Figure 27.5). When the fitness extreme-value distribution is in the Gumbel domain, then $E[\Delta_j] = E[\Delta_1]/j$ (Gumbel 1958; Weissman 1978), meaning that there is a regular pattern in the spacing that only depends on the fitness rank, j, of a given allele and on the expected spacing, $E[\Delta_1]$, for the most fit allele. (Recall that Equation 27.4c showed that a similar scaling of spacing, C/j, occurs between steps in an FGM walk.) This result leads to a dramatic simplification of Equation 27.8a. If the current allele has a fitness rank of i, the selection coefficient on a higher-ranked ($j < i$; i.e., more fit) allele is

$$s_j = \frac{\Delta_j}{W_i} = \frac{E[\Delta_1]/j}{W_i} = \frac{C_i}{j} \tag{27.9a}$$

where $C_i = E[\Delta_1]/W_i$. If we substitute $s_j = C_i/j$ into Equation 27.8a and note that the common C_i term in the numerator and denominator terms cancels, gives the transition probability (starting with an allele with a fitness rank of i) that the next allele to be fixed has a rank of $j < i$, as

$$\pi_{j,i} = \frac{1}{i-1} \sum_{k=j}^{i-1} \frac{1}{k} \quad \text{for} \quad j < i \tag{27.9b}$$

Using this expression, if the current sequence is the ith-fittest local allele, the mean fitness rank of the next fixed allele is

$$\sum_{j=1}^{i-1} j \cdot \pi_{j,i} = \sum_{j=1}^{i-1} j \left(\frac{1}{i-1} \sum_{k=j}^{i-1} \frac{1}{k} \right) = \frac{i+2}{4} \tag{27.9c}$$

Using Equation 27.9b, the corresponding variance in the fitness rank of this next fixed allele was obtained by Rokyta et al. (2006)

$$\frac{(i+2)(7i+6)}{144} \tag{27.9d}$$

For example, if the wild-type is the 10th-fittest (its rank is ten), Equation 27.9c calculates the expected rank following one substitution as three (the third-fittest), rather than the most fit (an allele of rank 1), as naively might be expected. From Equation 27.9d, the standard error on this expected rank is 2.5. Thus, while large jumps in fitness rank occur during the walk, these are typically not to the most fit allele. Orr (2002) noted that the adaptive process over this walk is the average of **perfect adaptation** (the best mutation, $i = 1$, is fixed in the first substitution; this is called **gradient adaptation** in the physics and computer science literature) and **random adaptation** (a randomly chosen fitter allele is fixed, for an average rank of $i/2$), as $(1 + i/2)/2 = (i + 2)/4$, recovering Equation 27.9c.

Unlike Fisher's model, Orr (2002) found that jumps under the SWMM model are few, and substantial, with over 30% of the total fitness change from the walk occurring on the first jump, and over 50% occuring from the largest jump. Orr also showed that the distribution of fixed selective effects over a number of such walks (ignoring the very smallest final steps) is again roughly exponential under certain conditions, a statement that we will refine below. Hence, adaptive walks over two very different geometries (continuous phenotypic vs. discrete sequence space), operating on different genetic scales (genomewide vs. a single gene or region), and treating rather different quantities (trait values vs. fitness effects), can both generate the same rough pattern of substitution effects. They differ in that a random walk in sequence space typically has only a few, mostly substantial, steps.

We have already discussed how the abstract nature of Fisher's model is perhaps its greatest strength, as it frees one from the tyranny of the idiosyncratic details for any particular trait. A similar feature occurs in sequence space through the use of EVT, to again obtain fairly general statements for an unknown distribution of the fitness effects of new alleles at a particular locus. What are some of the potential weaknesses of the sequence-space walk model? We can classify these as weaknesses as assumptions about mutation and assumptions about selection. An obvious issue concerning mutation is bias, in that the models assume that all (single-step) mutations are equally likely. However, the focus of these models is on how selection shapes the outcome under assumptions of equal mutation.

The first and foremost weakness with respect to assumptions about selection is that the fitness distribution may not be in the Gumbel domain, a point which will be discussed further below. Second, strong selection (the SS in SSWM) assumes that no slightly deleterious mutations can be fixed, while the weak-mutation assumption (the WM in SSWM) considers only two segregating alleles (the original and the mutation) at any point in time. Under these conditions, advantageous mutations that must be accessed through the fixation of a deleterious intermediate cannot be reached. A **selectively accessible** sequence is one that can be reached through a series of single-step substitutions, each of which is adaptive. This not need be the case when **sign epistasis** is present (epistasis that results in a change in the sign of fitness, meaning that an allele is deleterious in one background and favored in another). Here, one can have a mutation landscape with a series of fitness peaks that are inaccessible from one another through single mutations (Weinreich et al. 2005). As examined in Chapter 7, such peaks can potentially be reached (even under strong selection) when the weak mutation assumption is violated, such as in very large populations. This can occur via stochastic tunneling, wherein the second (advantageous) mutation can arise in the small fraction of segregating, initially deleterious, mutations. Finally, the weak mutation assumption ignores the important concern of clonal interference in asexual species.

The strong selection assumption also implies that no neutral (or even effectively neutral) alleles are fixed, but this is not a serious issue for the question being addressed by the model, namely the expected change in fitness from fixation of progressively more fit alleles (although it may be an important issue in terms of which adaptive peaks are accessible). For positively selected alleles, as selection coefficients become progressively smaller (but still positive), the drift barrier (Chapter 7) will be reached, and numerous substitutions of small (and effectively neutral) effect can occur. EVT theory ignores these as they all have essentially the same fitness (and hence are lumped into the same fitness class).

Example 27.3. Evolution in experimental microbial populations offers the opportunity to test certain features of the mutational landscape model (MLM), and in particular, the expected small number of steps during a walk and the prediction of rapidly diminishing fitness returns from new beneficial mutations (provided the initial genotype is itself reasonably fit). Holder and Bull (2001) examined the pattern of fitness increases (measured by doubling time) during adaptive walks for growth at high temperature in the bacteriophages ϕX174 and G4. The response seen for ϕX174 fit the pattern expected from MLM theory: the first mutation accounted for roughly 75% of the total gain in fitness, and the first two mutations accounted for 85%. A more complicated picture was seen for G4, which grows at lower temperatures than ϕX174. G4 showed no initial response for high-temperature (44°C) tolerance, and had to first be selected (for 50 generations) for adaptation to an intermediate temperature (41.5°C) before subjecting it to 44°C. While one of the first two mutations for growth at 44°C had a large fitness effect, there was otherwise no evidence of diminishing returns in the fitness in subsequently fixed beneficial mutations. In part this could be due to the essentially zero fitness of the initial genotype in the target environment (44°C), which violated the technical assumption that the starting value is itself from the extreme value distribution.

Adaptive walks in the fungus *Aspergillus nidulans* were studied by Schoustra et al. (2009) and Gifford et al. (2011). The observed number of steps in the walk was short (a mean of ~2), with diminishing fitness returns for fixed sites. By placing the original genotype in different environments with very different starting fitnesses, Gifford et al. were able to show that the length of the walk (which was still ~2 steps) was insensitive to the starting fitness.

Perhaps the most direct test of Orr's model was performed by Rokyta et al. (2005), who examined 20 single-step adaptations (for rapid replication under standard culture conditions) in the single-strand DNA bacteriophage ID11 (a relative of ϕX174). Genome sequencing showed that these 20 single-step events were comprised of nine different mutations. Based entirely on the fitness ranks of these mutations (with the wildtype having rank 10 and the most fit mutation having rank 1), Equation 27.9b gives a probability of 0.28 that the fitness-rank 1 mutation is fixed first. Hence, we expect that this will occur ~6 ($20 \cdot 0.28 = 5.6$) times in 20 random trails. However, it was only fixed once. Rokyta et al. used a multinomial distribution (Equation A2.37a) to assess whether where was a significant lack-of-fit between the observed probabilities for a mutation with fitness rank k being fixed first and the expected probability of this event under Equation 27.9b. The fit was poor ($p = 0.10$), but not significantly different from the null (which could simply reflect low power). The authors noted that the rank-1 mutation required a transversion mutation (these occur at lower rates than transition mutations, thus violating the equal-mutation assumption of Equation 27.8a). A much better fit ($p = 0.67$) was seen after correcting Equation 27.8a to allow for known differences in mutation rates.

Example 27.4. The examination of adaptive walks in sequence space between naturally occurring alternative sequences, and the possibility of local peaks that are not accessible through a series of single-step adaptive mutations, began with Malcolm et al. (1990). Game bird species (*Galliformes*) display one of two protein sequences for lysozymes, either Thr at position 40, Ile at position 55, and Ser at position 91 (the ancestral TIS sequence) or Ser 40, Val 55, Thr 91 (the derived SVT sequence). Malcolm et al. generated synthetic proteins representing all possible single-replacement pathways between these two states (e.g., TIS→SIS→SIT→SVT, TIS→TVS→SVS→SVT, and so on) and examined their thermostability (as a proxy for functional fitness). In all possible single-step pathways between TIS and SVT, an intermediate showed thermostability that was significantly outside the narrow range seen in either TIS or SVT, suggesting that these intermediate steps were deleterious.

A more direct study was performed by Weinreich et al. (2006), who examined penicillin resistance associated with the *E. coli* β-lactamase gene. Five point mutations (four nonsense and one 5′ noncoding) separate the resistant allele (TM^*) from its wild-type ancestor (TM^{wt}). By constructing the appropriate strains, Weinreich et al. examined the fitnesses of each intermediate sequence in all possible 5! = 120 single-step walks between TM^{wt} and TM^*. They found that 102 of the possible 120 trajectories (85%) were selectively inaccessible, involving transition through an intermediate of lower fitness, via sign epistasis. Using the estimated selection coefficients for each intermediate genotype, the authors applied Equation 27.8a to compute the expected probability of state transition (among those sites that are one mutational step away from the current state), and found that the 18 selectively accessible trajectories were not equally likely to occur, with just 6 accounting for half of all expected successful trajectories.

Table 27.1 Summary of several bacterial and viral experiments on the distribution of fitness effects among beneficial mutations. These studies either considered only *fixed* (Fixed = Yes) beneficial effects or considered *all* detected beneficials (Fixed = No). In order to detect all (as opposed to only fixed) beneficial mutations appearing in some interval during the walk, some experiments looked for rescue mutants that allow a strain to grow in a nonpermissive environment (Low = Yes). The Gillespie-Orr model predicts an exponential distribution (Gumbel domain of the EV distribution) when *all* beneficial mutations are considered (Fixed = No) *and* the genotype starts out at a high fitness (Low = No). Distributions indicated with the term "domain" used the formal likelihood-ratio approach of Beisel et al. (2007) to test the null (EV distribution shape parameter $\kappa = 0$; in the Gumbel domain) versus κ as significantly negative (Weibull domain) or significantly positive (Fréchet domain). The Barrett et al. study found a Weibull *distribution* of effects, which is unimodal, and *different* from the Weibull EV *domain* (indeed, if the underlying distribution is Weibull, its EV domain is actually Gumbel), and which is denoted by an asterisk to remind the reader of this difference.

Species	Fixed?	Low?	Beneficial Fitness Effects	Auhors
Escherichia coli	Yes	No	Exponential distribution	Imhof & Schlötterer 2001
	Yes	No	Normal distribution	Rozen et al. 2002
	No	Yes	Fréchet domain	Schenk et al. 2012
Pseudomonas fluorescens	No	Yes	Weibull distribution*	Barrett et al. 2006b
	No	No	Exponential distribution	Kassen & Bataillon 2006
	No	Yes	Weibull domain	Bataillon et al. 2011
	No	Yes	Normal distribution	McDonald et al. 2011
Pseudomonas aeruginosa	No	No	Exponential distribution	MacLean & Buckling 2009
	No	Yes	Weibull domain	MacLean & Buckling 2009
ID11 (ssDNA phage)	No	No	Weibull domain	Rokyta et al. 2008
ϕ6 (RNA phage)	No	Yes	Weibull domain	Rokyta et al. 2008
IAV (Influenza A virus)				
Oseltamivir absent	No	No	Weibull domain	Foll et al. 2014
Oseltamivir present	No	Yes	Fréchet domain	Foll et al. 2014
(the presence or absence of this drug changes the fitness conditions)				

The Fitness Distribution of Beneficial Alleles

Much of our discussion on walks (especially under Fisher's model) has been on the distribution of *fixed* beneficial mutations. We can use extreme-value theory to provide insight into a second issue: the distribution of effects for *all* beneficial mutations that appear at a given stage in the walk, not just those that are fixed (Orr 2003b, 2006, 2010; Martin and Lenormand 2008). When the fitness of an initial allele is high and the fitness distribution is in the Gumbel domain, the distribution of fitness values in new *beneficial* mutations is expected to be close to an exponential (Gillespie 1983; Orr 2003b). A remarkable feature about this distribution of effects is **Orr's invariance result**: at any stage of the walk, a new advantageous allele has the same mean fitness increase *independent* of the fitness of the current allele, as long as that allele's fitness is high (Orr 2003b).

As summarized in Table 27.1, a number of experiments with microbes have attempted to test the Gillespie-Orr prediction of an exponential distribution of fitness effects for all beneficial mutations. There are two important caveats when considering these results. First, the Gillespie-Orr prediction applies to the distribution of *all* newly beneficial mutations, *not* the distribution of those beneficial mutations that are *fixed* (although as mentioned above, the fixed distribution can itself be exponential in certain cases, a point that will be addressed shortly).

Second, the use of EVT assumes that the starting genotype already has a high fitness. A number of studies detect beneficial mutations by the rescue of deleterious mutations or by placing the genotype in a radically new, and very deleterious, environment. While this strategy provides a rapid screen for adaptive mutations (i.e., those that grow), the starting genotypes may not be sufficiently extreme in the fitness distribution for EVT to apply. As summarized in Table 27.1, all three EVT domains (Gumbel, Weibull, and Fréchet) are seen in the experimental data. As noted by Seetharaman and Jain (2014), there are qualitative differences in adaptive walks over these different domains. In particular, the expected fitness differences between successive substitutions are decreased under the Weibull domain, but they are increased under the Fréchet, and (as we have seen) they are constant under the Gumbel domain. When the selective effects are small (but still strong in a population-level sense, as we assume $4N_e s \gg 1$), the number of substitutions during a walk under the Fréchet domain is smaller than under the Weibull or Gumbel domains. However, when selective effects are large, the adaptive walk is shortest in the Gumbel domain.

Example 27.5. While detecting the vast majority of new beneficial mutations is an extremely challenging endeavor, it is not impossible, as an elegant study by Levy et al. (2015) demonstrated. By inserting $\sim$500,000 random DNA barcodes into a clonal founder yeast (*Saccharomyces cerevisiae*) population, they were able to track the appearance on new, favorable mutations by increases in the frequency of a specific barcode that was greater than expected from drift. In a population of $\sim 10^8$ cells descendant from this founding collection of random barcodes, the authors were able to ascertain that $\sim$25,000 lineage had gained a beneficial mutation within the roughly 170 (asexual) generations that they followed. By examining the rate of increase of a barcode, they were able to estimates its s value using the equations for allele-frequency change from Chapter 5. While there was a lower-limit in the s values that could be estimated (in particular, only those **established** mutants the escaped early stochastic loss were followed), this is still a remarkable system for tracking favorable mutations.

The findings from this study are at odds with much of the above theory. The distribution of detected beneficial fitness effects was not exponential. In fact, it was not even monotonic, but rather a multimodal distribution, with the bulk of (detected) mutations with modest effects ($0.02 < s < 0.05$), and then a rather flat (uniform) distribution for large effects, but with slight peaks around 0.07 and 0.10. Most beneficial mutations were lost, as their initial fitness advantage declined as the mean population fitness increased due to other clones acquiring beneficial mutations. This resulted in the small-effects mutations that dominated at the start of selection being overtaken by rarer mutations of large effect.

FISHER'S GEOMETRY OR EVT?

Fisher's geometric model (FGM) and the mutational landscape model (MLM) and its corresponding EVT results are, on the surface, very different views of adaptation. The former looks at the evolution of mutational-effect size for a trait under stabilizing selection, while the latter examines the evolution of fitness (a trait under strict directional selection) during a walk in a single linked region. However, at some fundamental level, we should be able to connect results from these models, as (ideally) the MLM is modeling the loci that result in phenotypes adapting under FGM. The results of Orr (2006) and, especially, Martin and Lenormand (2008), start to bridge this gap.

Recall for the FGM that Orr (2006) showed the distribution of fitness effects for *all* alleles (and not only the beneficial ones) to be roughly normal. As a result, under FGM, the underlying fitness distribution of new mutations is in the Gumbel domain (the EV domain associated with a normal distribution), and the fitness distribution for beneficial alleles (when they are rare, i.e., drawn from the extreme-value region of the distribution) is exponential. Thus, when beneficial mutations are rare, EVT results also apply to FGM. Provided the initial genotype starts at a sufficient distance from the optimum, so that many steps are needed in the walk, both the distribution of the phenotypic values of fixed mutations and the fitnesses of new beneficial mutations are exponential under FGM.

Likewise, the fitness distribution of *all* beneficial mutations is also exponential under a Gillespie-Orr walk when the Gumbel EV conditions hold (the starting genotype has high fitness and the fitness distribution is in the Gumbel domain). The distribution of fitness

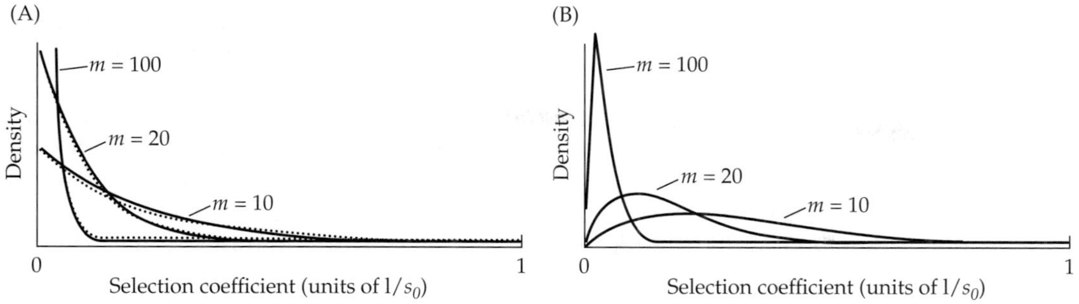

Figure 27.6 The distribution of the fitness effects for beneficial mutations at a locus underlying a phenotype under stabilizing selection, with the current allele close to the optimum. Let $s_0 \ll 1$ denote the maximal fitness advantage for any new mutation (which occurs when a mutation jumps the phenotype exactly to the optimum). **A:** The distribution of fitness effects among *all* beneficial mutations follows a beta distribution, Beta(1, m/2), where m is a measure of pleiotropy. The dotted curves show the corresponding exponential distributions, with the two distributions being very similar for values of m modest to large (>10). **B:** The distribution of fitness effects among *fixed* beneficial mutations also follows a beta distribution, but now with the first beta shape parameter changing from one to two, returning a unimodal distribution, Beta(2, m/2). Note that the exponential provides a very poor fit to this distribution unless m is very large and small values are ignored.

values among mutations that are *fixed* during the walk, however, departs from an exponential (Rozen et al. 2002; Barrett et al. 2006b; Martin and Lenormand 2008; Good et al. 2012). One reason is seen in Figure 27.2—while beneficial mutations with small fitness advantages are expected to be more common (given that they follow an exponential distribution), they also have lower fixation probabilities. Thus, the set of mutations surviving stochastic loss has a distribution that is shifted toward higher values (Figures 27.2 and 27.6).

Can we account for this difference in the distribution of fixed effects for trait values under FGM versus that for fitnesses under EVT? Likewise, what happens when the fitness distribution is not in the Gumbel domain? As Martin and Lenormand (2008) showed, these questions are connected. Recall that the trinity theorem (Example 27.2) states that when an underlying distribution has a limiting extreme-value distribution, it must reside in one of three domains. Most of our focus has been on results for the Gumbel domain. The Fréchet domain, with its heavier-than-exponential tails, has an excess of high fitness mutations (over the Gumbel domain). As such, Orr (2006) felt that fitness distributions were unlikely to fall in the extreme Fréchet domain (large values of κ). However, when the starting fitness is very low (such as when a genotype is very far from its optimal phenotype under FGM), there may indeed be an excess of high fitness mutations relative to the situation when the genotype is close to its optimal value.

Likewise, the Weibull domain is biologically quite feasible, simply requiring an upper limit on fitness (the underlying fitness distribution is truncated on the right). Indeed, this is exactly the domain that would be expected for a locus underlying a trait experiencing stabilizing selection, namely, the conditions for FGM. An allele at such a locus can have no higher fitness than that of a mutation that exactly jumps the genotype to the optimum, thus setting a strict upper limit on the fitness distribution. Let s_0 denote the selection coefficient for such a maximally beneficial mutation. Martin and Lenormand (2008) showed that when $s_0 \ll 1$ (the population is near an optimum), the distribution of fitness effects of *all* beneficial mutations is in the Weibull domain and is a beta (Equation A2.38), with

$$\frac{s}{s_0} \sim \text{Beta}(1, m/2) \quad \text{for} \quad 0 < s < s_0 \tag{27.10a}$$

where m is a measure of pleiotropy (m is the rank of the product of the mutational covariance matrix, **M**, and the stabilizing selection matrix, **S**). Figure 27.6A plots this distribution and shows that it is close to an exponential for m modest to large (indeed, a beta with a large m converges to an exponential). This convergence follows from the fact that the EVT shape parameter when Equation 27.10a holds is $\kappa = -2/m$, which approaches zero (and hence the Gumbel domain) for large m (see Equation 27.7). In this setting, the FGM generates a distribution of beneficial mutations that is roughly exponential, in accordance with EVT. This relationship also suggests an approach for estimating m, using the ML approach of Beisel et al. (2007) to estimate κ, with $\widehat{m} = -2/\widehat{\kappa}$.

Applying Equation A2.38b (the mean value for a beta distribution) yields a mean fitness value of all newly arising beneficial mutations of

$$E(s_b) = \frac{2s_0}{2+m} \tag{27.10b}$$

Note the cost of pleiotropy, with the expected fitness decreasing with m. Martin and Lenormand found that the distribution of selection coefficients for *fixed* beneficial alleles is also a beta, but now with a shape parameter that generates a unimodal distribution (see Figure 27.6B), with

$$\frac{s}{s_0} \sim \text{Beta}(2, m/2) \quad \text{for} \quad 0 < s < s_0 \tag{27.11a}$$

From Equation A2.38b, the mean fitness value of fixed beneficial mutations becomes

$$E(s_b) = \frac{4s_0}{4+m} \tag{27.11b}$$

which shows both a cost of pleiotropy among fixed mutations, as $E(s_b)$ decreases as m increases, and also that the mean of fixed mutations is larger than the mean of all beneficial mutations (as would be expected). Although this distribution is unimodal, if m is sufficiently large (a great number of weakly correlated traits are under selection) and small effects are ignored (the assumptions made by Orr), then it can be reasonably approximated by an exponential (Figure 27.6B). Outside of this range, however, the use of an exponential overestimates the fraction of fixed beneficial mutations with large fitness effects. Experimental support for Equation 27.11a was offered by the work of Sousa et al. (2012) on the distribution of fitness effects of compensatory mutations in *E. coli* that recover antibiotic resistance (also see Trindale et al. 2012).

THE IMPORTANCE OF HISTORY IN WALKS: LESSONS FROM LENKSI'S LTEE

The above theory of adaptive walks shows the importance of evolutionary history, with large steps in the walk typically occurring early and smaller steps typically occurring later (under constant selection). Empirically, the role of historical events can be clearly seen in Lenski's long-term evolution experiment, LTEE, which was introduced in Chapter 25 (Lenski et al. 1991; Lenski and Travisano 1994; Barrick et al. 2009; Wiser et al. 2013). The LTEE consists (as of this writing) of over 55,000 generations (the equivalent of over a million years of human evolution) of selection on 12 replicate populations of *Escherichia coli*. One unique feature of this experiment is that because material from every 500 generations is frozen, one can go back and "replay" evolution. We consider just three of the interesting features or examples that have come from this experiment. These empirical results highlight some of the nuances that were missed by the above theory of single-step mutational walks and also by the stochastic tunneling results for adaptations that require multiple steps to pass through deleterious intermediate genotypes (discussed in Chapter 7).

One feature is the general appearance of hypermutators (Elena and Lenski 2003; Barrick and Lenski 2013). This is not surprising given the lack of recombination in *E. coli*, as a hypermutator allele can generate a completely linked adaptive mutation, resulting in selection on this more favorable clone (as the adaptation counters the deleterious effects of the mutator, at least transiently). Once the beneficial site is fixed, the mutator is now deleterious, and it generates (weak) selection to reduce the mutation rate back toward more normal levels. Such a spike in mutation rates followed by a reversion to normal levels is routinely seen in the LTEE and other microbial experiments (reviewed in Barrick and Lenski 2013).

An interesting example from the LTEE is the evolution of the use of citrate as a carbon source, which serves as a potential model for the evolution of innovations (Blount et al. 2008, 2012). The growth medium for the LTEE contains abundant citrate, which was initially intended primarily as a chelating agent, and generally not accessible (under aerobic conditions) as a carbon source for *E. coli*. In one of the 12 replicates, a citrate-using strain (Cit^+) arose around generation 30,000 and became dominant around generation 33,000. Blount et al. (2008) found that the appearance of the initial Cit^+ mutation depended on one or more mutations in the population's history. When evolution was replayed, clones from the first 15,000 generations did not go on to produce Cit^+ mutations, while some clones chosen from later generations did. One or more potentiating events occurred at around generation 20,000, which led to conditions allowing for Cit^+ mutations to appear some 10,000 generations later, and these events did not result from a general increase in the mutation rate. Blount et al. (2012) partitioned the evolution of this trait into three phases: **potentiation, actualization**, and **refinement**. Events that occurred before generation 20,000 set up the conditions for the *potential* of Cit^+ mutations, which was *actualized* at around generation 30,000. Blount et al. (2012) detailed some of the molecular features leading to *refinement* from the initial extremely weak Cit^+ mutation, including duplication of *cit* genes.

A final example from the LTEE, which shows the full power of being able to "replay" evolution, is from Woods et al. (2011), who compared the fitnesses of four clones from generation 500. Two of these clones (EW1, EW2) were eventual winners, going on to become fixed, while two (EL1, EL2) were eventual losers. Surprisingly, when measured at generation 500, the fitness of the EL clones was significantly higher (by about 6%) than that of the EW clones. Indeed, these fitnesses predict that the EW clones would have gone extinct in roughly another 350 generations. Why, then, did the EW clones succeed? When evolution was replayed in several replicate experiments starting with these clones, eventually the majority of the EW clones went on to accrue higher fitnesses than the EL clones (presumably due to subsequent mutations), even though initially they had lower levels of fitness. The authors ruled out higher mutation rates among EW clones and determined instead that the key was a mutation in the EW clones in the *spoT* gene, a master regulator. In the EW background, this mutation confered a large beneficial fitness effect, but it was essentially neutral in the EL background. Thus, differences in fitness epistasis between the two types of clones (EL and EW) and the *spoT* gene resulted in the initially higher-fitness EL clones closing off access to a higher fitness peak via *spoT* mutations that the EW clones could achieve. In the very large population sizes of the LTEE, this inability of EL to access this fitness peak, while EW could do so, allowed secondary mutations to occur before the initial EW clones went extinct, which eventually led to an ancestor of the EW clones becoming fixed. This is an example of **second-order selection** (Tenaillon et al. 2001), whereby the fate of a new mutation is dependent on its ability to accrue secondary mutations, namely, the notion of stochastic tunneling, which is examined in Chapter 7.

28

Maintenance of Quantitative Genetic Variation

Empirical studies of quantitative genetic variation have revealed robust patterns that are observed both across traits and across species. However, these patterns have no compelling explanation, and some of the observations even appear to be mutually incompatible. Johnson and Barton (2005)

How wonderful that we have met with a paradox. Now we have some hope of making progress. Niels Bohr

Genetic variation is a ubiquitous feature of natural populations. The nature of the forces responsible for the maintenance of this variation, be it the distribution of allele frequencies, the level of heterozygosity, the amount of additive variation in a trait, or the joint distribution of allele frequencies and their effects, have long been of concern to both population and quantitative geneticists. The basic explanation is some balance of evolutionary forces: mutation/migration introducing new variation, which is removed by drift and/or selection against deleterious alleles. In some cases, selection by itself can maintain variation, such as when heterozygotes are advantageous. These various explanations are not mutually exclusive, and theorists have spent a great deal of effort in building models to examine the plausibility of each scenario. If the required parameter space to maintain variation is very narrow, a particular mechanism may account for the maintenance of variation in specific cases but is unlikely to be a general explanation.

Despite a wealth of possible explanations for the maintenance of variation, this is an area of some frustration among quantitative geneticists. At present, there are difficulties in reconciling most (some would say all) of the proposed explanations with estimates of observable parameters (such as the strength of the apparent stabilizing selection on a trait, its heritability, and the mutational variance). As this is a subject with a substantial body of complex theory, we present many of the derivational details in examples, which allows us to focus on the key results while still presenting the logic and assumptions behind the models. Reviews of the struggle to explain quantitative-genetic variation can be found in Nagylaki (1984), Turelli (1984, 1986, 1988), Barton and Turelli (1989), Bulmer (1989), Bürger (1998, 2000), Barton and Keightley (2002), Johnson and Barton (2005), Zhang and Hill (2005b, 2010), and Mitchell-Olds et al. (2007). Bürger (2000) is the standard reference for much of the theory developed here, and should be consulted by the more mathematically inclined reader.

OVERVIEW: THE MAINTENANCE OF VARIATION

Earlier chapters explored the roles of the major evolutionary forces (drift, mutation, and selection) and important modifiers (recombination and migration) in the maintenance of polymorphisms at individual loci. The effects of drift (removing variation) and mutation (generating variation) are straightforward (Chapter 2), while the effects of selection are more complicated, as it either retains or removes variation, depending on its nature (Chapter 5). With constant selection coefficients, overdominance (heterozygote advantage) retains variation in large populations, while all other constant-fitness schemes remove it (Figure 5.1). Selection can retain variation under a variety of circumstances when fitnesses vary, which we loosely lump together under the umbrella term of **balancing selection**. These conditions include frequency-dependent selection when rare alleles are favored, tradeoffs among different fitness components, sex-specific differences, and fitness changes over time and/or space (G $\times$ E). The conditions necessary to maintain variation can be rather delicate

Evolution and Selection of Quantitative Traits. Bruce Walsh & Michael Lynch.
Published 2018 by Oxford University Press. © Bruce Walsh & Michael Lynch 2018.
DOI 10.1093/oso/9780198830870.001.0001

for many of these strictly selective explanations.

The result of interactions between evolutionary forces can be straightforward—such as the mutation-drift equilibrium (Equation 2.24) or mutation-selection balance for deleterious alleles (Equation 7.6)—or they can be subtle and counterintuitive, such as the joint impact of selection, mutation, drift and recombination on the levels of variation under selective sweeps (Chapter 8). The goal in this chapter is build on these results in an attempt to explain the nature of the evolutionary forces maintaining quantitative-trait variation.

Maintaining Genetic Variation for Quantitative Traits

Most of our previous results on the maintenance of genetic variation were for population-genetic models, wherein the focus was solely on allele *frequencies*, and usually quantified by summary statistics such as the heterozygosity or the number of segregating alleles (Chapter 2). In this setting, the most complete equilibrium solution is given by the distribution of allele frequencies, such as Wright's result for a diallelic locus under mutation-selection-drift (Equation 7.31a) or the Watterson distribution for the site-frequency spectrum for mutation-drift balance under an infinite-sites model (Equation 2.34a). For quantitative traits, the allele-frequency distribution, by itself, is not sufficient to describe the equilibrium variation. Instead, one needs the full *joint* distribution of allele frequencies and their *effect sizes*, although we typically work with the additive-genetic variance as an appropriate summary statistic. Given the number of scenarios outlined above, it should not be surprising that a plethora of models have been proposed for the maintenance of genetic variation in quantitative traits. Figure 28.1 attempts to bring a little structure to this vast menagerie.

The simplest models are fully neutral: the trait, and its underlying loci, have no effects on fitness, leading to **mutation-drift models** (Chapters 11 and 12). Their problem is that they generate *too much* variation if the population size is modest to large. The most obvious correction is that there is some selection on the trait and/or on the underlying loci (independent of their effect on the focal trait). Models incorporating selection can be broken into two categories: those with at least some **direct selection** on the focal trait, and **pleiotropy models** that assume a neutral focal trait whose underlying loci have pleiotropic effects on fitness.

A central issue concerning direct-selection models is that stabilizing selection on a trait usually generates *underdominance* in fitness at its underlying loci, thus *removing* variation (Example 5.6). Hence, **strict stabilizing selection**, by itself, cannot account for quantitative-trait variance. This removal of variation could be countered by either mutation (**mutation-stabilizing selection balance**) or by selectively favored pleiotropic fitness effects. Under the latter scenario, loci underlying the trait under stabilizing selection are also under balancing selection for some other independent component of fitness (**balancing-stabilizing selection**). The central issue concerning mutation-selection balance is that the estimated strengths of stabilizing selection and polygenic mutation appear to be inconsistent with observed levels of heritability.

A critical question in the maintenance of genetic variation is just how much of observed stabilizing selection is actually real. Pleiotropic models can easily generate **apparent** (or **spurious**) **stabilizing selection** by returning a signature of stabilizing selection in a quadratic regression of fitness on the phenotypic value of a neutral trait (Chapters 29 and 30). Hence, it is possible that some (or perhaps much) of the observed stabilizing selection in nature is not real, but rather is instead due to pleiotropic fitness effects. Under pleiotropic models, the variation at the loci underlying a neutral trait is assumed to be maintained by either overdominant effects on fitness (**pleiotropic overdominance**) or because the underlying loci are slightly deleterious, but in mutation-selection equilibrium (**pleiotropic deleterious mutation-selection balance**). The problem with pleiotropy models is that the strength of selection on the underlying loci required to recover the observed strength of apparent stabilizing selection seen in nature is usually inconsistent with some other observable feature of the model. Various combinations of elements of these basic models have been proposed, as

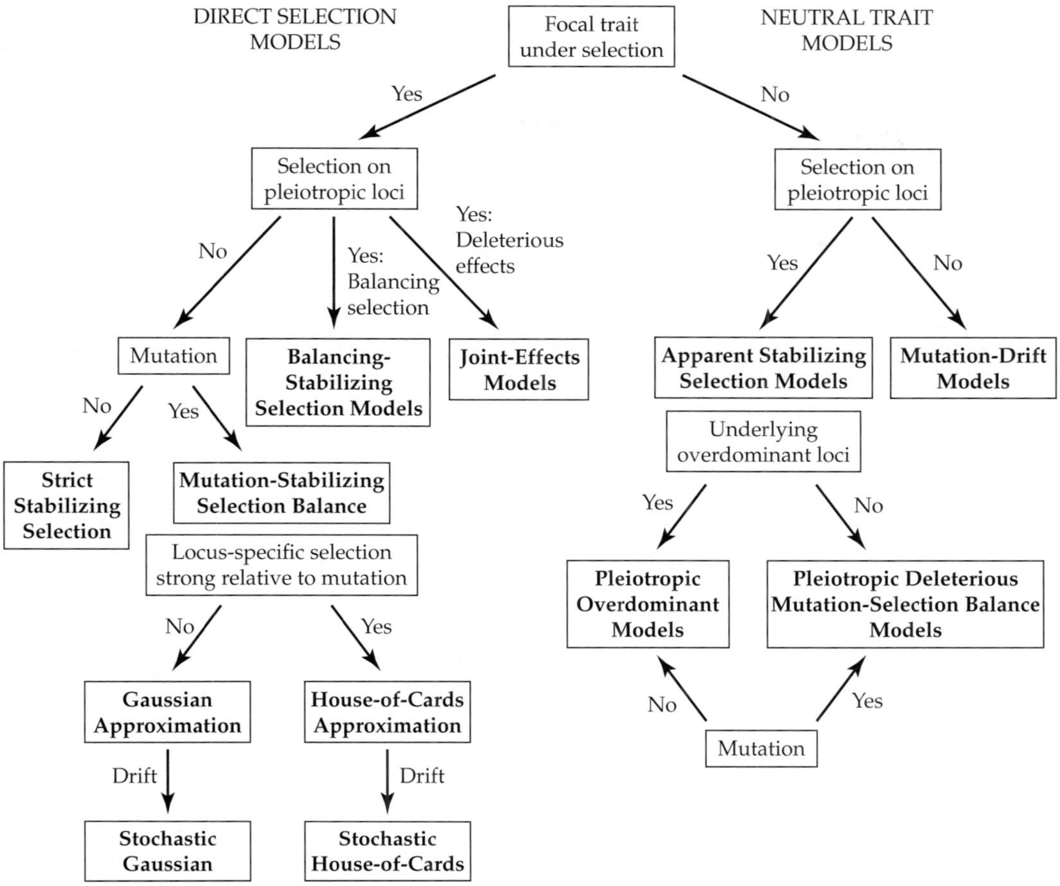

Figure 28.1 Flow chart of the various classes of models for the maintenance of quantitative-genetic variance. Roughly speaking, there are **direct-effect models** that assume that selection is acting on the phenotype of the focal trait (whose variation we are trying to explain) and models that assume that this trait is neutral. **Pleiotropic models** assume that loci underlying a trait have fitness effects *independent* of their impact on the focal trait (which is often assumed to be strictly neutral). As detailed below, models also vary in the importance assigned to mutation in countering the removal of genetic variation by selection and drift.

have refinements adding additional forces (such as drift), but most give inconsistent results when trying to simultaneously account for observed amounts of selection and variation.

Finally, differences in the assumed *granularity* of the underlying genetic architecture of a trait can significantly impact the results. If a few loci, each with only a few alleles, underlie a trait, the resulting genotypic values have a fairly granular distribution. The dynamics under stabilizing selection are different when one of these genotypic values matches the optimal stabilizing selection value compared to those of the situation when none do. Likewise, with just a few alleles at a few loci, the opportunity for independent selection on many traits is constrained. Conversely, under continuum-of-alleles (COA) models (Chapter 24), with their large number of alleles at each locus, there is a distribution of allelic effects and the potential for significantly more fine-turning. A key point of this chapter is that the relative strengths of the underlying evolutionary forces dictates which genetic architecture is more appropriate. If drift is strong relative to the other forces, then at most, only a few alleles at a locus are likely (beyond a constellation of very rare new mutations). The same is true when selection is strong relative to mutation. Conversely, when the strength of mutation is greater than the strength of selection or drift at a locus, we expect it to harbor a number of alleles in a large population. As we will see, differences in the strength of mutation relative to selection at a locus lead to qualitatively different results.

The (often fairly technical) analysis of the large number of models given in Figure 28.1

comprises the bulk of the chapter. There are several possible schemes by which to organize and discuss all of these alternative models. Our presentation is centered around increasing the complexity of evolutionary forces and their interactions. We start with drift interacting with neutral mutation, which serves as a useful baseline. We then consider models invoking only selection, either stabilizing selection on the focal trait or balancing selection on loci with pleiotropic effects on a strictly neutral focal trait.

These selection-only considerations provide the background for the major classes of models, those involving *both* selection and mutation. Much of the discussion on these models focuses on stabilizing selection countered by mutation, including the incorporation of drift. Most of the work on stabilizing selection has assumed either a Gaussian (Equation 28.3b) or quadratic (Equation 28.3a) fitness function. How these results translate to more general fitness functions with a stabilizing component remains a rather open question.

We conclude by discussing models in which a large fraction of the trait variance is assumed to result from pleiotropic effects of deleterious alleles, which are maintained by mutation-selection balance. Our analysis of this last class of models starts with a neutral focal trait, followed by **joint-effect models** allowing for *both* stabilizing selection on a focal trait *and* pleiotropic contributions from deleterious alleles.

To aid the more casual reader, Table 28.3 (near the end of the chapter) summarizes the major inconsistencies for each model, followed by an examination of the current data. This allows the reader to bypass the more technical discussions below, while still obtaining a general overview of the problem. The conclusion from this extensive analysis is that all of the models have significant inconsistencies with current estimates of strength of selection, mutational inputs, and amounts of standing genetic variation. The typical pattern seen is that for a model to accommodate one known feature (e.g., the observed strength of stabilizing selection), the required parameter values result in another known aspect (say, amount of standing variation) being inconsistent with observed values.

MUTATION-DRIFT EQUILIBRIUM

The most basic model for the maintenance of variation considers two universal (and counterbalancing) forces, drift and mutation. Chapter 2 examined the distribution of neutral allele frequencies and reviewed various resulting summary statistics under mutation-drift balance. At equilibrium, neutral allele frequencies are given by the Watterson distribution (Equation 2.34a), and the expected heterozygosity (for an infinite-alleles model) is $\widetilde{H} = \theta/(1+\theta)$, where $\theta = 4N_e\mu$ is the product of the effective population size and the mutation rate (Equation 2.24b). The problem with this expression, as noted by Lewontin (1974), is that heterozygosity should quickly approach one in large populations ($\theta \gg 1$), yet this is not seen. One possible explanation is that the mutation rate inversely scales with population size, so that θ is always $\ll 1$ (Chapter 4). Another, not necessarily exclusive, explanation is that selection at linked sites depresses variation by decreasing N_e (Chapters 3, 8, and 10). The impact on N_e from a pattern of recurrent sweeps is greatest in very large asexual populations, which otherwise would be predicted to have very high values of $\widetilde{H}$.

Mutational Models and Quantitative Variation

Chapters 11 and 12 developed the quantitative-genetic analog of $\widetilde{H}$ by considering the expected additive variance, $\widetilde{\sigma}^2_A$, that is maintained by neutral alleles in mutation-drift equilibrium. Two extensions, both concerning mutation, are required when moving from allelic frequencies to quantitative-trait variation. The first is that the **mutational variance** (the total amount of genetic variation arising in each generation), σ^2_m, replaces the mutation rate, μ (Chapter 11). The mutational variance contributed by (diploid) locus i is $2\mu_i\sigma^2_{\alpha_i}$, the product of its mutation rate and $\sigma^2_{\alpha_i}$, the **variance of mutational effects** (or **mutational-effects variance**). We use σ^2_α to denote an unspecified locus and $\sigma^2_{\alpha_i}$ to denote a specified one. With n equivalent loci, $\sigma^2_m = 2n\mu\sigma^2_\alpha$, while $\sigma^2_m = 2\sum_i \mu_i\sigma^2_{\alpha_i}$ when mutational effects vary over loci.

Table 28.1 Models for the effect of a new mutation on a quantitative trait. All make the infinite-alleles assumption that each new mutation creates a new allele. The effect, x', of this new allele is a function of its current value, x, and a random variable, $\alpha \sim (0, \sigma^2_\alpha)$. The incremental and house-of-cards (HOC) models are special cases of the Zeng-Cockerham regression model, corresponding to $\tau = 1$ and $\tau = 0$, respectively. Derivations can be found in Chapter 11, and in Zeng and Cockerham (1993).

Model	New Effect	$\widetilde{\sigma}^2_A$	$\widetilde{\sigma}^2_A$ as $N_e \to \infty$
Incremental, Random-walk, Brownian-motion	$x' = x + \alpha$	$4N_e\mu n\sigma^2_\alpha = 2N_e\sigma^2_m$	Unbounded
House-of-cards	$x' = \alpha$	$\dfrac{8N_e\mu n\sigma^2_\alpha}{1+4N_e\mu} = \dfrac{4N_e\sigma^2_m}{1+4N_e\mu}$	$2n\sigma^2_\alpha = \dfrac{\sigma^2_m}{\mu}$
Regression	$x' = \tau x + \alpha$	$\dfrac{8N_e\mu n\sigma^2_\alpha}{(1+\tau)[1+4N_e\mu(1-\tau)]} = \dfrac{4N_e\sigma^2_m/(1+\tau)}{1+4N_e\mu(1-\tau)}$	$\dfrac{2n\sigma^2_\alpha}{1-\tau^2} = \dfrac{\sigma^2_m}{\mu(1-\tau^2)}$

As reviewed in LW Chapter 12, the mutational variance can be estimated from the accumulation of additive variance in inbred lines. Such estimates are usually scaled by the environmental variance to yield the **mutational heritability**, $h^2_m = \sigma^2_m/\sigma^2_E$, and a typical value is $h^2_m = 10^{-3}$ (LW Table 12.1). Estimates of the component features of the mutational variance—the number of loci, n; the per-locus mutation rate, μ; and the variance of mutational effects, σ^2_α—are far more difficult to obtain. This is unfortunate, as many of the following models require the values of these components (n, μ, and σ^2_α), rather than their composite measure, σ^2_m. Some crude estimates follow from the widespread observation that h^2_m is typically on the order of 10^{-3}. If we assume that $\sigma^2_\alpha/\sigma^2_E = 1$, then the total trait mutation rate, $2n\mu$, will be on the order of 10^{-3}. For $n = 100$ loci, this implies a per-locus mutation rate (to new trait alleles) of $\mu = 5 \cdot 10^{-6}$. If the scaled variance of mutational effects is lower, then either the number of loci and/or the per-locus mutation rate must be correspondingly higher. Lyman et al. (1996) estimated a value of $\sigma^2_\alpha/\sigma^2_E \simeq 0.1$ for *Drosophila* bristle number mutations generated by *P*-element insertions. For $h^2_m = 10^{-3}$, this implies $2n\mu = 0.01$ (assuming that σ^2_α and μ for *P*-element insertions are representative of the wider mutational spectrum, which is unlikely).

The second required extension is some assumption relating the current effect of an allele, x, with its effect, x', after mutation (Table 28.1). (While we typically use a to denote allelic effects, given the close similarity to our use of α for the mutational effect, for clarity we will often use x in this chapter to denote an allelic effect.) The most widely used construct is the **incremental model** (also referred to as the **Brownian-motion** or **random-walk** model). Initially introduced by Clayton and Robertson (1955), and more formally by Crow and Kimura (1964) and Kimura (1965a), this model assumes that $x' = x + \alpha$, the pre-mutation value plus a random increment, where $\alpha \sim (0, \sigma^2_\alpha)$. When all mutations are additive, Equation 11.20c gives the (diploid population) mutation-drift equilibrium variance under this model as $\widetilde{\sigma}^2_A = 2N_e\sigma^2_m$. Equation 11.22a shows the expression for the additive variance when dominance is present. From Equation 11.21a, the expected equilibrium heritability becomes

$$\widetilde{h}^2 = \frac{2N_e h^2_m}{1 + 2N_e h^2_m} = 1 - \frac{1}{1 + 2N_e h^2_m} \tag{28.1}$$

Note the connection with the expression for neutral allelic heterozygosity, $\widetilde{H}$, as both are of the form $2N_e y/(1 + 2N_e y)$, with $y = h^2_m$ for heritability and $y = 2\mu$ for heterozygosity. As with $\widetilde{H}$, even modest values of N_e ($\sim$1000) return $\widetilde{h}^2$ values over 0.5, while larger values return heritabilities of close to one. For example, when $h^2_m = 0.001$, N_e is constrained to be in the range of 50–1200 in order to recover typical heritability values (0.1 to 0.6).

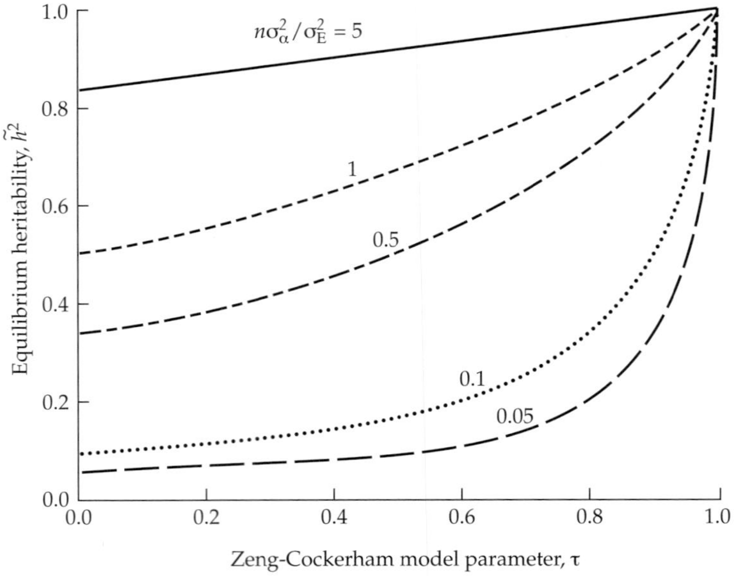

Figure 28.2 The expected heritability, $\widetilde{h}^2$, for large N_e, at mutation-drift equilibrium under the mutational regression model of Zeng and Cockerham (Equation 28.2a). This model includes the incremental ($\tau = 1$) and HOC ($\tau = 0$) models as special cases. Curves denote different values of $h_m^2/(2\mu) = n\sigma_\alpha^2/\sigma_E^2$, the ratio of the mutational heritability to the per-locus mutation rate.

As noted in Chapter 11, the incremental mutational model represents one extreme, wherein the value of the new mutation is closely tied to the evolutionary history (x) of its parental allele. The other extreme is the **house-of-cards (HOC)** model, which was formally developed by Kingman (1977, 1978; although also assumed by Wright 1948b, 1969). Under HOC, $x' = \alpha$, independent of an allele's starting value x, where again $\alpha \sim (0, \sigma_\alpha^2)$, so that past evolutionary history is completely irrelevant.

The incremental and HOC models present two extremes, one strongly influenced by evolutionary history and the other completely indifferent to it. Zeng and Cockerham (1993) proposed a more general **regression model**, $x' = \tau x + \alpha$, where $0 \leq \tau \leq 1$ and $\alpha \sim (0, \sigma_\alpha^2)$ (Table 28.1). The regression coefficient, τ, indicates the importance of past evolutionary history, recovering the incremental ($\tau = 1$) and HOC ($\tau = 0$) as special cases. This regression model is an Ornstein-Uhlenbeck process (Equation A1.33), as $E[\Delta x] = E[x' - x] = -(1 - \tau)x$. The parameter τ counters the diffusive effects of Brownian motion (the incremental random α) by exerting a restoring force toward the origin, and thus producing a bounded equilibrium distribution for $\widetilde{\sigma}_A^2$ (for $\tau < 1$). Under the regression model (provided $\tau \neq 1$), the equilibrium additive variance in a large population is bounded by $\sigma_m^2/[\mu(1 - \tau^2)]$, with a resulting heritability of

$$\widetilde{h}^2 = \frac{\sigma_m^2/[\mu(1 - \tau^2)]}{\sigma_m^2/[\mu(1 - \tau^2)] + \sigma_E^2} = 1 - \frac{1}{K + 1} \tag{28.2a}$$

where

$$K = \frac{h_m^2}{\mu(1 - \tau^2)} = \frac{2\mu n\sigma_\alpha^2/\sigma_E^2}{\mu(1 - \tau^2)} = \frac{2n\sigma_\alpha^2/\sigma_E^2}{1 - \tau^2} \tag{28.2b}$$

Figure 28.2 plots $\widetilde{h}^2$ as a function of τ and $h_m^2/(2\mu) = n\sigma_\alpha^2/\sigma_E^2$ (the scaled variance of mutational effects over all loci). The expected heritability increases as the role of past evolutionary history of an allele becomes increasingly important in predicting its mutated value (i.e., $\widetilde{h}^2$ increases with τ). Likewise, $\widetilde{h}^2$ increases with the total variance of mutational effects, $n\sigma_\alpha^2$. Assuming a typical value of $h_m^2 = 0.001$, an underlying per-locus mutation rate of $\mu = 10^{-3}$

(implying $2n\sigma_\alpha^2 = \sigma_E^2$) and a value of $\tau = 0.5$ gives $K = 2$ and $\widetilde{h}^2 = 0.67$. This decreases to 0.5 as we approach the HOC model ($\tau = 0$), and increases to one as we approach the incremental model ($\tau = 1$). Assuming that $h_m^2 = 0.001$ is a standard value for many traits, for large N_e this model requires a very high per-locus mutation rate ($\mu > h_m^2$ ~0.001; implying $K < 1$), otherwise the predicted heritabilities are too large. As with the incremental model, Equation 28.2a ignores the impact of deleterious mutations, and thus gives an upper limit on the equilibrium heritability.

MAINTENANCE OF VARIATION BY DIRECT SELECTION

As shown in Figure 28.1, a number of models for the maintenance of variation assume stabilizing selection on the focal trait. We start by examining stabilizing selection per se on both one, and n, traits. The conclusion is that only very limited amounts of genetic variation can be maintained in such settings, especially if a large number of genes, each of modest to small effect, underlie the trait. One potential countering selective force will arise if trait loci have overdominant pleiotropic effects on fitness, and this is discussed next. Such overdominance can arise when homozygotes have a higher environmental variance than heterozygotes for a trait under strict stabilizing selection. Fitness overdominance can also be generated when the underlying loci show G × E in the trait under selection, and we will examine both of these situations. Finally, the impacts of a changing optimum phenotypic value and frequency-dependent selection will be examined to see if these can help retain variation. As we detail, all of these models fall short in their attempt to account for observed levels of variation.

Fitness Models of Stabilizing Selection

Two standard fitness models of phenotypic stabilizing selection with trait value z appear in the literature: Wright's (1935a, 1935b) quadratic optimal model

$$W(z) = 1 - s(z-\theta)^2 \tag{28.3a}$$

and the Gaussian (or nor-optimal) model of Haldane (1954; also Weldon 1895)

$$W(z) = \exp\left[\frac{-(z-\theta)^2}{2\,\omega^2}\right] \tag{28.3b}$$

Recalling that $e^{-x} \simeq 1 - x$ for $|x| \ll 1$, the Gaussian reduces to the quadratic model under weak selection ($\omega^2 \gg 1$), as

$$W(z) \simeq 1 - \frac{(z-\theta)^2}{2\,\omega^2} \tag{28.3c}$$

As a result, these two models are used somewhat interchangeably, with $s \simeq 1/(2\omega^2)$. This is quite reasonable under the assumption of weak selection ($\omega^2 \gg 2\sigma_z^2$), but inappropriate under strong selection ($\omega^2 < 2\sigma_z^2$). While the Gaussian fitness function imposes no restrictions on the strength of stabilizing selection, the quadratic model does (to ensure that fitnesses are not negative at extreme values of z), which results in the two models showing very different behavior for loci under strong selection (Gimelfarb 1996b; see Example 5.11).

Discussions on the maintenance of variation often involve the mean fitness generated by a particular strength of selection. Under the quadratic model, this is a function of the mean and variance of z. If $z \sim (\overline{z}, \sigma_z^2)$, then

$$\overline{W}(z) = E[w(z)] = 1 - s\left(E[z^2] - 2\theta E[z] + \theta^2\right) = 1 - s\left[(\overline{z}-\theta)^2 + \sigma_z^2\right] \tag{28.3d}$$

where the last simplification follows from $E[z^2] = \overline{z}^2 + \sigma_z^2$. For Gaussian selection, if we assume that z is normal with $z \sim N(\overline{z}, \sigma_z^2)$, then

$$\begin{aligned}\overline{W} &= \frac{1}{\sqrt{2\pi\sigma_z^2}} \int \exp\left[\frac{-(z-\overline{z})^2}{2\,\sigma_z^2}\right] \exp\left[\frac{-(z-\theta)^2}{2\,\omega^2}\right] dz \\ &= \sqrt{\frac{\omega^2}{\omega^2+\sigma_z^2}}\, \exp\left[\frac{-(\overline{z}-\theta)^2}{2\,(\omega^2+\sigma_z^2)}\right]\end{aligned} \tag{28.3e}$$

(Kimura and Crow 1978). Equations 28.3d and 28.3e are special cases of our previous Equations 17.7b and 17.8a. An important application of Equation 28.3e is the expected fitness associated with a genotypic value of G. Assuming environmental effects are normally distributed around G, then $z|G \sim N(G, \sigma_E^2)$, and the resulting strength of stabilizing selection on G becomes

$$V_s = \omega^2 + \sigma_E^2 \tag{28.3f}$$

Larger values of V_s correspond to weaker selection, so (as expected) variation in the phenotype around a genotypic value weakens the strength of selection ($V_s > \omega^2$). V_s is a central parameter in the maintenance-of-variation literature, and it is usually scaled in units of σ_E^2, with $V_s = \omega^2/\sigma_E^2 + 1 \simeq \omega^2/\sigma_E^2$ under weak selection ($\omega^2 \gg \sigma_E^2$).

Assuming that the fitness function is given by Equation 28.3b, Equation 16.18a yields the phenotypic variance following selection as

$$\sigma_{z^*}^2 = \sigma_z^2 - \frac{\sigma_z^4}{\sigma_z^2 + \omega^2} \tag{28.3g}$$

When $\omega^2 \gg \sigma_z^2$ (weak selection), then for low heritability, $\sigma_z^2 + \omega^2 \simeq \sigma_E^2 + \omega^2 = V_s$, which rearranges to give an estimate of the strength of stabilizing selection as

$$\widehat{V}_s \simeq \frac{\sigma_z^4}{\sigma_z^2 - \sigma_{z^*}^2} \tag{28.3h}$$

This is a biased estimate in the presence of directional selection, which also reduces the phenotypic variance following selection (Equation 29.16a). Less biased estimates can be obtained from the quadratic term in the Pearson-Lande-Arnold fitness regression (Equation 29.29a),

$$w(z) = 1 + \beta(z - \mu_z) + \frac{\gamma}{2}\left[(z - \mu_z)^2 - \sigma_z^2\right] + e \tag{28.3i}$$

which adjusts for the reduction in variance from directional selection. Matching terms with Equation 28.3c, we find that $\gamma = -1/\omega^2$ (Keightley and Hill 1990). Under weak selection, $V_s = \omega^2 + \sigma_E^2 \simeq \omega^2$, returning an estimate of V_s as $\simeq -1/\gamma$.

Turelli (1984) suggested a typical value of $V_s/\sigma_E^2 \simeq 20$, which corresponds to $V_s/\sigma_A^2 \simeq 20$ when $h^2 = 0.5$. Under this strength of stabilizing selection (which implies that $V_s \simeq 10\sigma_z^2$), a phenotype two standard deviations from the mean has around 80% of the fitness at the phenotypic optimum. While Turelli's values are widely used in the maintenance-of-variation literature, more recent estimates (Kingsolver et al. 2001; summarized in Figure 30.5) are less clear. On the one hand, the average value of V_s among traits experiencing stabilizing selection (those with estimated negative γ values) is stronger than Turelli's assumed value, with a mean V_s of $\simeq 5\sigma_z^2$ ($\simeq 10\sigma_E^2$ when $h^2 \simeq 0.5$). Under this strength of selection, a phenotype two standard deviations from the mean has around 70% of the optimal fitness. On the other hand, Figure 30.5 shows that the distribution of estimated γ values from natural populations is largely symmetric around zero, implying that disruptive selection is as common as stabilizing selection. Although these results are colored by a lack of information on the statistical significance of many of the γ values plotted in Figure 30.5, they still raise the possibility that a typical trait may be under much weaker, or even nonexistent, stabilizing selection. Conversely, the long-term stasis of many traits over evolutionary time suggests that stabilizing selection is indeed a major force shaping evolution (Charlesworth et al. 1982; Maynard Smith 1983; Estes and Arnold 2007; Hunt 2007). Haller and Hendry (2013) discuss a variety of factors that might make stabilizing selection difficult to detect (Chapter 30).

An even larger issue, which frames much of the discussion on the maintenance of variance, is whether an observed amount of stabilizing selection on a trait is **real** or **apparent**. As we saw in Chapter 20 (and discuss extensively in Chapter 30), selection acting on a hidden feature correlated with the trait of interest will impart a signature of selection on that trait. Direct selection models assume that there is real selection on the focal trait. As we will see, their problem is that reasonable assumptions about the components of σ_m^2 predict heritabilities that are too small, given the observed values of V_s. Conversely, pleiotropic models that can account for the observed levels of heritability predict much larger apparent values of V_s (weaker selection) than are typically seen.

Stabilizing Selection on a Single Trait

In Chapter 5 we examined constant-fitness population-genetic models for alleles under strict selection (no mutation or drift) and showed that while heterozygote advantage can stably maintain both alleles at a diallelic locus, most forms of selection tend to remove variation. At first glance, an additive QTL for a trait under stabilizing selection seems to be an example of such a heterozygote advantage, as the heterozygote is intermediate in phenotype and an intermediate phenotype is preferred by selection. However, it is critical to recall that this is *not* the case. Example 5.6 showed that a QTL underlying a trait under stabilizing selection generally experiences selective *underdominance*, and hence the *removal*, rather than the maintenance, of variation by selection.

While Fisher (1930) was the first to suggest that stabilizing selection will remove, rather than retain, variation, the initial formal demonstration of this was due to Wright (1935a, 1935b) and Robertson (1956), and a vast literature has since followed. Assuming Gaussian stabilizing selection, and if the genotypes q_iq_i, Q_iq_i, and Q_iQ_i at locus i have effects of $-a_i$, 0, and a_i, respectively, then the dynamics for frequency p_i of allele Q_i are calculated by

$$\Delta p_i \simeq \frac{a_i}{V_s}\left[\frac{p_i(1-p_i)}{2}\right]\left[a_i(2p_i-1)+2(\theta-\overline{z})\right] \tag{28.4a}$$

(derived in Example 5.6). A useful way to understand these dynamics is to express them in the form of a weakly selected allele with additive effects (Equation 5.2), $\Delta p \simeq s_ip_i(1-p_i)$, where the selection coefficient becomes

$$s_i = \frac{a_i}{2V_s}\left[a_i(2p_i-1)+2(\theta-\overline{z})\right] \tag{28.4b}$$

The first term in the square brackets, $a_i(2p_i-1)$, represents stabilizing selection to reduce the variance generated by this locus, while the second term, $2(\theta-\overline{z})$, is the impact from directional selection. When $|\,\theta-\overline{z}\,| > a_i/2$, directional selection determines the dynamics. When this second term is negligible, selective underdominance occurs, as $\Delta p_i < 0$ for $p_i < 1/2$ and $\Delta p_i > 0$ for $p_i > 1/2$ (with $p = 1/2$ being an unstable equilibrium point). When $\overline{z} \simeq \theta$, the initial selection coefficient on a new allele ($p_i \simeq 0$) is

$$s_i \simeq -\frac{a_i^2}{2V_s} \tag{28.4c}$$

as found by Latter (1970), Kimura (1981), Bürger et al. (1989), and Houle (1989).

This is the crux of the problem with stabilizing selection per se—it drives allele frequencies towards fixation, *removing, rather than retaining*, variation at underlying loci (Robertson 1956). Additional analysis of single-locus models (ignoring linkage disequilibrium) showed that partial dominance (Kojima 1959; Lewontin 1964; Jain and Allard 1965; Singh and Lewontin 1966; Bulmer 1971a) or the presence of loci with unequal additive effects (Gale and Kearsey 1968; Kearsey and Gale 1968) can result in the maintenance of several polymorphic loci at equilibrium, although the parameter space for this to happen is extremely narrow for unlinked loci.

Analyses of two- and multiple-locus models (where LD is fully considered) again lead to the conclusion that selection removes variation for additive loci of equal effect. However, when selection is strong relative to recombination, multiple-locus polymorphisms can be maintained by stabilizing selection on a single trait when loci have unequal effects, or when dominance or epistasis is present in the trait under selection (Gimelfarb 1989, 1996b; Nagylaki 1989a; Zhivotovsky and Gavrilets 1992; Gavrilets and Hastings 1993, 1994a, 1994b).

Example 5.11 detailed Bürger and Gimelfarb's (1999) analysis of the general two-locus model under quadratic selection, and Willensdorfer and Bürger (2003) presented a similar analysis for Gaussian selection. The conditions under which stabilizing selection on a single trait can maintain polymorphisms at multiple loci are fairly stringent and generally result in high negative levels of disequilibrium, and hence small additive variances (Gimelfarb 1989; Zhivotovsky and Gavrilets 1992). Further, the genetic variance that can be maintained under such models generally decreases very rapidly with the number of loci, reflecting diminished selection coefficients on the individual loci (Bürger and Gimelfarb 1999). One subtle issue is the granularity of these models, in that if no genotype exists whose value exactly equals the optimal value under stabilizing selection, then small amounts of directional selection ($|\,\theta - \overline{z}\,| > 0$) can be present at equilibrium, and multilocus polymorphism (often with alleles at extreme values, and hence contributing little variation) can be maintained (Barton 1986).

Given that most traits seem be controlled by a moderate to large number of loci of moderate to small effect (Chapter 24), strong selection on *individual loci* (distinct from strong selection on the *trait*) is generally unlikely. Thus, the weak selection results suggest that, at best, only very modest amounts of additive variation are maintained by single-trait stabilizing selection in the absence of other forces.

Stabilizing Selection on Multiple Traits

The assumption that a gene only influences a single trait is biologically rather unrealistic, as it ignores the likely situation that ***the amount of standing variation at a given locus reflects the action of selection acting on multiple traits***. One model of such pleiotropic fitness effects assumes that a locus influences n independent traits, each under stabilizing selection. Hastings and Hom (1989) showed that when selection on individual loci is weak relative to recombination, at most k loci are polymorphic when k independent traits are under selection. Hence, under weak selection, the addition of pleiotropic stabilizing selection on a nonfocal trait does little to increase the amount of standing variation at a focal trait.

The effect of strong selection was examined by Gimelfarb (1986a, 1992, 1996a) and Hastings and Hom (1990). Gimelfarb (1986a) constructed a model with independent selection on two phenotypically uncorrelated traits (1 and 2, with phenotypic values of z_1 and z_2), which were determined by two additive loci with alleles A/a and B/b, whose joint allelic effects on the traits are $A : (z_1 = z_2 = 0)$, $a : (z_1 = z_2 = 1)$, $B : (z_1 = 0, z_2 = 1)$, and $b : (z_1 = 1, z_2 = 0)$, respectively. Fitness is assumed to be a function of the phenotypic values of each trait, and $W(z_1, z_2) = [1 - s_1(z_1 - \theta_1)][1 - s_2(z_2 - \theta_2)]$. Under this model there is pleiotropy (as the A locus influences both traits), and although Gimelfarb showed that, at equilibrium, both loci are polymorphic, the traits are phenotypically and genetically uncorrelated, and selection occurs independently on each. The result, in the words of Gimelfarb, is that "even if the investigator will be lucky enough to come across character 2, he is almost certain to discard it as having no biological connection with the character 1." The worrisome implications of this model foreshadow additional complications from pleiotropy, which are discussed below. While multiple-trait stabilizing selection can maintain variation at a number of loci, with selection strong relative to recombination, there is significant negative disequilibrium and often little additive variance in each trait (Gimelfarb 1992).

Barton (1990) raised several additional points on the limitations of multiple-trait stabilizing selection. First, simple genetic load arguments (the decrease in mean population fitness relative to the fittest possible genotype) place upper limits on the number of independently selected traits. Assume k traits, each under Gaussian selection with a common value of $V_s = \omega^2 + \sigma^2_E$. For populations at equilibrium ($\mu = \theta$), Equations 28.3e and 28.3f imply that genetic variation reduces the mean population fitness by $\sqrt{V_s/(V_s + \sigma^2_A)}$ for each trait. For $V_s \gg \sigma^2_A$ (weak selection), a Taylor-series argument shows that

$$\sqrt{\frac{V_s}{V_s + \sigma^2_A}} = \sqrt{\frac{1}{1 + \sigma^2_A/V_s}} \simeq 1 - \sigma^2_A/(2V_s) \simeq \exp\left[-\frac{\sigma^2_A}{2V_s}\right]$$

Assuming multiplicative fitnesses across the k independently selected traits, this yields a load of $\simeq \exp(-k\,\sigma_A^2/[2V_s])$. For $V_s = 20\sigma_A^2$, the mean fitness is around 8% of the highest fitness with $k = 100$ traits. For weaker selection, $V_s = 100\sigma_A^2$, this same load occurs for $k = 500$, while for stronger selection ($V_s = 5\sigma_A^2$) it occurs for $k = 25$. Thus, one quickly approaches an upper limit on the number of traits before the fitness load becomes unbearable.

As discussed in Chapter 7, such load arguments can be delicate because departures from the assumed multiplicative fitness model can either lessen the load (synergistic epistasis) or enhance it (diminishing-returns epistasis). However, the point remains that selection itself places a limit on the number of independent traits with segregating variance. There are also limits on the number of alleles at a given locus, again constraining the ability to evolve in an unlimited number of directions in phenotypic space (at least k alleles are required for a locus to evolve in independent directions at k traits).

Barton (1990) suggests there may be a modest number of phenotypic dimensions experiencing significant real stabilizing selection, which results in apparent stabilizing selection on any trait phenotypically correlated to one, or more, of these dimensions (Example 28.1). Further, we have shown that stabilizing selection per se, be it on a single or multiple traits, is unlikely to account for significant additive variance. Coupling these points suggests that stabilizing selection, by itself, is unlikely to explain more than a trivial amount of the genetic variance for a trait that appears to be under stabilizing selection, and that additional factors (such as mutation and pleiotropy) are critical. As succinctly stated by Barton "***heritable variation in any one trait is maintained as a side effect of polymorphisms which have nothing to do with selection on that trait***," an idea more fully explored throughout this chapter.

Example 28.1. As illustrated in Chapter 20, traits may show signs of directional selection (a covariance between trait value and fitness) without being the actual target of selection. The same is true for stabilizing selection, which appears as a negative covariance between the squared trait value and fitness (Chapters 29 and 30). Wagner (1996a) emphasized this point by considering two genetically uncorrelated traits, 1 and 2, that are phenotypically correlated through some shared environmental effect. Trait 1 is neutral (its trait value, z_1, has no effect on fitness), while trait 2 is under Gaussian stabilizing selection with a strength parameter of ω_2^2. Wagner showed that if ρ_z is the phenotypic correlation between the two traits, the expected fitness of phenotype z_1 is

$$W(z_1) = \exp\left(-\frac{z_1^2\,\rho_z\,\sigma_{z_2}^2}{2\sigma_{z_1}^2\,[\,\omega_2^2 + \sigma_{z_2}^2(1-\rho_z^2)]}\right) \tag{28.5a}$$

Matching terms with Equation 28.3b shows that trait 1 experiences apparent stabilizing (Gaussian) selection around an apparent optimum of 0 and with a strength of

$$\omega_1^2 = \frac{\sigma_{z_1}^2\,[\,\omega_2^2 + \sigma_{z_2}^2(1-\rho_z^2)]}{\rho_z\,\sigma_{z_2}^2} \tag{28.5b}$$

Note that $\omega_1^2 \to \infty$ (no selection) as $|\rho_z| \to 0$. Scaling both traits to have an environmental variance of one, then $\sigma_{z_i}^2 = \sigma_{G_i}^2 + 1$, where $\sigma_{G_i}^2$ is the (environmentally scaled) genetic variance of trait i. Using this scaling, Wagner rearranged Equation 28.5d to find a lower bound of

$$\omega_1^2 \geq 2\,\omega_2^2\,(\sigma_{G_1}^2 + 1)^2 \tag{28.5c}$$

This sets an upper limit on the strength of apparent stabilizing selection (as smaller ω_1^2 values imply stronger selection), with ω_1^2 increasing (selection becoming weaker) as the fraction of genetic variance in trait 1 increases. This is not surprising, as the apparent selection arises through the environmental component, which is decreased by increasing the genetic contribution.

What is surprising, however, is that the joint fitness for the genotypic values for both traits, (A_1, A_2), is

$$W(A_1, A_2) = \exp\left(-\frac{A_2^2}{2[\omega_2^2 + \sigma_{E_2}^2]}\right) \tag{28.6}$$

showing that there is no selection on the genotypic values of trait 1, which therefore evolves neutrally. Hence, the equilibrium heritability in trait 1 is entirely independent of the strength of the apparent selection on trait 1 (i.e., ω_1^2 does not not appear in this expression).

Stabilizing Selection Countered by Pleiotropic Overdominance

Extensions of direct-selection models to include pleiotropy assume that the loci underlying a trait under stabilizing selection also have independent effects on other fitness components. For example, an allele might influence the value of a trait under stabilizing selection (such as height), but might also influence fecundity, independent of any impact of height on fecundity. The motivation for this idea traces back to Lerner (1954), who suggested that "inheritance of metric traits may be considered, at least operationally, to be based on additively acting polygenic systems, while the totality of traits determining reproductive capacity and expressed as a single value (fitness) exhibits overdominance." While the support for overdominance has diminished over time (Lewontin 1974; Hedrick 2012; but see Manna et al 2011; Sellis et al. 2011; and Charlesworth 2015), a number of the initial pleiotropy models assumed overdominance (Robertson 1956; Lewontin 1964; Bulmer 1973; Gillespie 1984a). As we will see, such models can still be meaningful even in the absence of classical overdominance.

The basic structure of the **pleiotropic-overdominance-stabilizing-selection model** is as follows. For locus i, the genotypes $q_iq_i : Q_iq_i : Q_iQ_i$ have effects of $-a_i : 0 : a_i$ on a trait under stabilizing selection, and fitness effects of $1 : 1 + t_i : 1$ on an independent (and multiplicative) fitness component, with total fitness calculated as the product of $W(z)$ from stabilizing selection (e.g., Equation 28.3b) and the pleiotropic fitness of the genotype. Under this model, the change in allele frequency from weak overdominance alone is

$$\Delta p_i \simeq -t_i p_i(1-p_i)(2p_i - 1) \tag{28.7}$$

This form of selection maintains variation, as $\Delta p_i > 0$ when $p_i < 1/2$, while $\Delta p_i < 0$ when $p_i > 1/2$. Under the assumption of weak selection on the focal trait, we can add the change from stabilizing selection to obtain the approximate total allele-frequency change. Assuming Gaussian stabilizing selection, Equation 28.4a yields

$$\begin{aligned}\Delta p_i &\simeq -t_i p_i(1-p_i)[2p_i - 1] + \frac{a_i}{V_s}\left(\frac{p_i(1-p_i)}{2}\right)[\,a_i\,(2p_i - 1) + 2(\theta - \overline{z})\,] \\ &= p_i(1-p_i)\Big([2p_i - 1]\left[-t_i + a_i^2/(2V_s)\right] + [a_i(\theta - \overline{z})/V_s]\Big)\end{aligned} \tag{28.8}$$

which has a stable polymorphic equilibrium if $t_i > a_i^2/(2V_s)$, provided the population mean is close to the optimal trait value, θ (stability analyses are given by Gillespie 1984a; and Turelli and Barton 2004). Recalling Equation 28.4c, this condition can be restated as a stronger selection coefficient from overdominant selection than from stabilizing selection alone, namely, $t_i > a_i^2/(2V_s) = s_i$. If the phenotypic mean is sufficiently far away from the optimum value, then directional selection dominates (fixing Q_i if $\overline{z}$ is sufficiently below θ, and fixing q_i if $\overline{z}$ is sufficiently above θ). When $\overline{z} \simeq \theta$, balancing selection occurs, in which the net balance of the two selective forces maintains variation, resulting in intermediate allele frequencies at equilibrium.

While the preceding arguments are mathematically correct, the biological relevance of this model is less clear, especially given the difficulty of finding examples of loci that display classic fitness overdominance. However, there are several realistic settings involving stabilizing selection per se that also result in fitnesses that mimic heterozygote advantage. Zhivotovsky and Feldman (1992) noted that pleiotropic overdominance naturally arises when the environmental variance associated with a genotype decreases along with the number of heterozygous loci (Whitlock and Fowler 1999; Chapter 17). To see this, consider quadratic selection. The fitness associated with genotype g, where $z|g \sim (G, \sigma^2_{E(g)})$ is given from Equation 28.3d as

$$W(G) = 1 - s(G-\theta)^2 - s\sigma^2_{E(g)}$$

As the environmental variance, $\sigma^2_{E(g)}$, decreases, the fitness increases. This creates pleiotropic overdominance, as heterozygous individuals have a higher fitness than do more homozygous individuals with the same genotypic value (G) due to their smaller values of $\sigma^2_{E(g)}$ (also see Curnow 1964).

Gillespie and Turelli (1989, 1990) found that certain patterns of G $\times$ E (allelic effects change over environments, while the optimum phenotypic value, θ, remains unchanged) can also result in heterozygotes having higher fitnesses than homozygotes, which again recovers pleiotropic overdominance. However, Gimelfarb (1990) noted that the association between fitness and heterozygosity critically depends on strong G $\times$ E symmetry assumptions. A more general analysis of both spatial and temporal G $\times$ E models was provided by Turelli and Barton (2004), who found that a necessary condition for balancing selection to maintain polymorphisms in the face of stabilizing selection is that the coefficient of variation of allelic effects over environments exceed one. If the standard deviation of allelic effects over environments is less than their mean value, the loci are fixed. An interesting consequence of this condition is that sex-specific differences in allelic effects are not sufficient to maintain significant variation (i.e., more than one polymorphic locus) in polygenic models of stabilizing selection. While we showed, in Chapter 5, that antagonistic selection between the sexes can maintain variation in a single-locus model, moving to a polygenic model maintains no additional polymorphic loci.

Fluctuating and Frequency-dependent Stabilizing Selection

Balancing polymorphisms can potentially be maintained by **fluctuating selection**. The G $\times$ E models that we just considered assumed constant selection (θ fixed), with allelic effects changing over environments. In contrast, fluctuating stabilizing selection models assume constant allelic effects with the optimum value, θ, varying over time. Variation in θ can be random or include some periodicity. Starting with Dempster (1955a) and Haldane and Jayakar (1963), a large body of theoretical literature (reviewed by Felsenstein 1976; Hedrick 1986; Frank and Slatkin 1990a; Gillespie 1994; Lenormand 2002) showed that the conditions for temporal variation to retain a polymorphism at a single locus are delicate. Are the conditions any less restrictive with a polygenic trait under fluctuating stabilizing selection? Not substantially.

The simplest model involves random (uncorrelated) fluctuations in θ, and was considered by Ellner and Hairston (1994) and Ellner (1996). They showed that polymorphisms are maintained provided that $\gamma\sigma^2(\theta)/V_s > 1$, where $\sigma^2(\theta)$ is the temporal variance in θ and γ is a measure of the amount of population carryover when overlapping generations are present. Hence, rather large fluctuations are required. Are the conditions less restrictive when the change in θ is periodic? Bürger and Gimelfarb (2002) examined the impact of a fluctuating optimum under a model with built-in periodicity (the expected value of θ varied according to a sine function) plus additional stochasticity (the realization of θ at a particular time is its expected value plus a random increment). An autocorrelated moving optimum had little impact (relative to constant stabilizing selection) on maintaining genetic variation or increasing polymorphism. Further, the longer the periodicity of oscillation, the less was the impact on polymorphisms or on the level of genetic variation. As we will see later, when mutation is also allowed, fluctuating selection can significantly increase the amount of standing variation over models that assume a constant value of θ.

Spatial variation in θ can also maintain at least some variation. A simple example was given by Felsenstein (1977), who assumed a continuum-of-alleles model, with a Gaussian distribution of allelic effects at each locus (Chapter 24). Under Felsenstein's model, the optimal phenotypic value at position x along some linear cline (such as a river bank) is $\theta(x) = \beta x$. Individuals disperse along this cline with a mean distance of zero and a variance of σ_d^2. When selection is strong relative to migration ($V_s \ll \sigma_d^2$), the equilibrium additive variance is approximately $\beta^2\sigma_d^2$. When selection is weak relative to migration, the equilibrium variance is roughly $\beta(\sigma_d^2 V_s)^{1/2}$. More detailed analyses of this problem were presented by Tufto (2000) and Spichtig and Kawecki (2004).

Frequency-dependent selection is another possible mechanism for generating balancing selection. As discussed in Chapter 5, frequency dependence can maintain variation under selection alone (i.e., no other evolutionary forces need be invoked), and aspects of this process have been modeled by a number of researchers (Roughgarden 1972; Bulmer 1974b, 1980; Felsenstein 1977; Slatkin 1979; Clarke et al. 1988; Mani et al 1990; Kopp and Hermisson 2006). The most comprehensive analysis (in terms of maintenance of variation when stabilizing selection is occurring) is that of Bürger and Gimelfarb (2004). These authors assumed constant stabilizing selection on a trait that was also involved in intraspecific competition (as did Bulmer 1980). Individuals with increased differences in trait values from each other experienced reduced competition, and hence higher fitness, thus generating disruptive selection on the trait. Stabilizing selection on the focal trait was modeled by a quadratic fitness function with a selection effect of s (Equation 28.3a), whereas the amount of competition between phenotypes g and h also follows a quadratic, $1 - s_c(g - h)^2$. Assuming that these two components of fitness are multiplicative, Bürger and Gimelfarb found that the key parameter is $f = s_c/s$, the ratio of selection from competition to stabilizing selection. If f is below a critical value, the model essentially behaves like a standard model of stabilizing selection in removing variation. If f exceeds this critical value, however, there will be no stable monomorphic equilibria, and the genetic variance and amount of polymorphism will rapidly increase with f (since disruptive selection dominates).

Summary of Direct-selection Models

When the focal trait is under direct stabilizing selection, very little variation is maintained in the absence of other forces (such as mutation or countering selection). Likewise, stabilizing selection on multiple traits has little impact on increasing the amount of genetic variance that is at a focal trait, especially under weak selection (i.e., when selection on any given underlying locus is small relative to recombination). When the loci underlying a trait under stabilizing selection are also overdominant for an independent fitness component, sufficiently strong balancing selection can maintain significant variation. However, given the apparent scarcity of widespread fitness overdominance, this is an unlikely candidate to provide a general explanation for the maintenance of variation. Certain strictly stabilizing selection scenarios can mimic pleiotropic overdominance, such as environmental variances that decrease as a function of the total heterozygosity, or G $\times$ E when the genotypic values (but not the fitness optimum) change over time or space. A fluctuating optimum (a varying θ) is unlikely to retain significant variation by itself, but there are conditions under which density-dependent selection can maintain significant variation. As with any explanation presented here, demonstrating a potential to account for a pattern, even over a very wide parameter space, is not sufficient, as one also needs to have some idea about how common a particular mechanism actually is in nature.

NEUTRAL TRAITS WITH PLEIOTROPIC OVERDOMINANCE

In the preceding overdominance models, the removal of genetic variation for a trait under stabilizing selection is countered by advantageous pleiotropic fitness effects at the underlying loci. A natural extension of this idea is to imagine that there is *no* selection on a focal trait, but rather that trait variation is maintained *entirely* as a result of pleiotropic fitness effects at the underlying loci (e.g., Robertson 1956, 1967). These underlying polymorphisms could be maintained by advantageous fitness effects, such as overdominance or balancing selection, where the nature of selection is independent of the value of the focal trait. Another, more intriguing, possibility is that the underlying pleiotropic loci may have *deleterious* fitness effects, with variation now being maintained by mutation-selection balance. Given that strictly neutral models (i.e., in which none of the underlying loci are under any selection) maintain *too much* variation, perhaps making them slightly deleterious (for reasons other than their associated trait values) might allow them to generate the observed amounts of trait variation. Alas, however, as we will show later in the chapter, this is not the case.

An obvious concern that the careful reader might have with neutral-trait models is the widespread observation of apparent stabilizing selection on many traits. However, pleiotropic selection models can generate associations between the values at a neutral focal trait and fitness, thus generating false signals of stabilizing selection on that trait. In the case of underlying overdominant loci, more homozygous individuals have both lower fitness and more extreme trait values (Example 5.8). Likewise, under the pleiotropic deleterious mutation-selection balance model, individuals carrying more deleterious mutations also have more extreme trait values. In both settings, the neutral trait will show **apparent stabilizing selection** (Robertson 1956, 1967; Barton 1990; Kondrashov and Turelli 1992). Gavrilets and de Jong (1993) found that the conditions required for underlying pleiotropic loci to generate apparent stabilizing selection on a neutral trait are rather minimal. This has led to the suggestion that a significant fraction of apparent stabilizing selection on traits in natural populations is the result of selection on features other than the scored traits (e.g., Example 28.1; Gimelfarb 1996a). The limitation of pleiotropic-fitness models is that they cannot simultaneously account for the observed levels of variation (h^2) and the observed strengths of stabilizing selection (V_s). When one value matches the observations (say h^2), the corresponding value that the model generates for the other parameter (V_s) will be at odds with our current understanding of the data.

To see this last point, we turn to an analysis of Robertson's (1956, 1967) **pleiotropic overdominance** model, wherein loci under overdominant selection also have pleiotropic effects on a *neutral* focal trait (Example 5.8). This is in contrast to the previous pleiotropic overdominance model, in which the focal trait was under *stabilizing selection*, as opposed to being neutral. Consider the ith such pleiotropic locus, and assume that there are two alleles (the conditions for maintaining more than two alleles by overdominance at a locus are very delicate, so this is not an unreasonable assumption; Lewontin et al. 1978). Let the genotypes $Q_iQ_i : Q_iq_i : q_iq_i$ have fitnesses of $1 - s_i : 1 : 1 - t_i$, yielding (Example 5.4) an equilibrium frequency for Q_i of $\widetilde{p}_i = t_i/(s_i + t_i)$. Under an additive model in which the pleiotropic effects of this locus on the focal trait are $a_i : 0 : -a_i$, the equilibrium additive variance for the focal trait from this locus is $2a_i^2\, \widetilde{p}_i\, (1 - \widetilde{p}_i)$. When summed over n overdominant loci, the expected equilibrium additive variance is

$$\widetilde{\sigma}_A^2 = 2nE[\, a_i^2\, \widetilde{p}_i\, (1 - \widetilde{p}_i)\,] \tag{28.9}$$

The expectation is taken over all segregating overdominant loci influencing the trait. If homozygotes have rather similar fitnesses ($s_i \simeq t_i$), the equilibrium allele frequencies are intermediate ($\widetilde{p} \simeq 1/2$), resulting in $\widetilde{\sigma}_A^2 \simeq (n/2)E[\, a_i^2\,]$. If alternative homozygotes have very different fitnesses, the equilibrium frequencies will be close to zero or one, which results in drift quickly fixing one of the alleles (Figure 7.4). Consequently, under balancing selection models, segregating alleles are expected to be maintained at moderate frequencies.

Example 28.2. While pleiotropic overdominance models can maintain significant amounts of variation (Equation 28.9), they have limitations as a general explanation for quantitative-trait variation. The first problem is the scarcity of examples of actual overdominant selection in the wild (Lewontin 1974). However, one could argue that overdominance is widespread but overlooked, as very weak selection against both homozygotes still results in overdominance but would be difficult to detect in natural populations. Barton (1990) noted a second limitation:

the genetic load under a multilocus overdominance model constrains the expected response to artificial selection. We sketch this argument here.

Our starting point is Barton's (and Robertson's 1956) demonstration that the overdominance model with fitnesses of $1 - t_i : 1 : 1 - s_i$ (and phenotypic effects of $-a : 0 : a$) generates a strength of apparent stabilizing selection on the neutral focal trait of

$$V_s \simeq \sigma_A^2 / \overline{\ell} \tag{28.10a}$$

where σ_A^2 is the focal-trait additive variance, $\overline{\ell}$ is the average of the locus-specific segregation loads (the reduction in fitness from the optimal value) where

$$\ell_i = \frac{s_i t_i}{s_i + t_i} \simeq \frac{s_i}{2} \quad \text{when} \quad s_i \simeq t_i \tag{28.10b}$$

and $1 - \ell_i$ is the equilibrium mean fitness at locus i. Equation 28.10a implies that $\overline{\ell} = 0.05$ for a trait to have a typically assumed value of $V_s = 20\sigma_A^2$. Assuming n independent overdominant loci and multiplicative fitnesses, the expected mean population fitness becomes

$$\prod_{i=1}^{n}(1 - \ell_i) \simeq \exp(-\overline{\ell} n)$$

For $\overline{\ell} = 0.05$, around 20 such loci will result in the mean population fitness being about a third of its maximum possible value, so the number of such loci has to be modest for the load to remain reasonable. If loci have weaker effects ($\overline{\ell} < 0.05$), more polymorphisms can be maintained, but Equation 28.10a shows that the associated strength of apparent stabilizing selection on the neutral trait will be correspondingly weaker (V_s is larger).

Now consider the selection response (in the mean) when the focal trait is subjected to artificial directional selection that is strong enough to overpower any natural selection from overdominance. Assuming that the starting equilibrium allele frequencies (from overdominance) are $\widetilde{p}_i \simeq 1/2$ (i.e., $s_i \simeq t_i$), Equation 25.2a predicts that the fixation of all favorable alleles will result in an increase in the mean (measured in terms of standard deviations of the initial additive variance) of $\sqrt{2n}$ (this corrects the value given by Barton 1990). Hence, an observed response of R standard deviations requires $R^2/2$ such overdominant loci. Coupling this with the above load calculations suggests that for a population to show $5\sigma_A$ of response in a short-term selection experiment (a fairly typically result; Chapter 18), a lower bound of $5^2/2$ = 13 overdominant loci is required. Given a typical value of observed strength of (in this case, apparent) stabilizing selection of $\simeq 20\sigma_A^2$, $n = 13$ implies that the mean population fitness needs to be $(1-0.05)^{13}$, or roughly 50%, of the optimal fitness, to support such a response. For 10 standard deviations of response, the required reduction in fitness in the base population to support the required 50 overdominant loci is over 90% of the fitness of the optimal genotype.

Several factors can modify these results. First, as mentioned in Chapter 7, load can be diminished (and more loci can be maintained) under synergistic epistasis (as opposed to multiplicative fitnesses). Second, Equation 25.2a gives a lower bound on the required number of loci. The actual number of loci is much larger when their frequencies depart from 1/2 (homozygotes have unequal fitnesses), increasing the load. Finally, when homozygote fitnesses are rather unequal, the two alleles are maintained at more extreme frequencies (i.e., closer to 0 or 1), and thus potentially lost to drift in the small populations that characterize selection experiments (Chapter 26), resulting in an undercount of the number of overdominant loci.

MUTATION-STABILIZING SELECTION BALANCE: BASIC MODELS

Recurrent mutation can maintain at least some genetic variation even the face of strong selection (Chapter 7). For example, if μ is the mutation rate to a deleterious allele whose fitnesses are given by $1 : 1 - hs : 1 - s$, the infinite-population equilibrium frequency of the deleterious allele is $\widetilde{p} \simeq \mu/(hs)$ for $h \gg \sqrt{\mu/s}$ (Equation 7.6d) and $\widetilde{p} = \sqrt{\mu/s}$ for a recessive

($h = 0$; Equation 7.6d). While it is obvious that at least some variation can be maintained by the balance between stabilizing selection and mutation, the critical question is just how much. This apparently simple query has generated a huge amount of rather technical theory, with some surprising results.

We start our treatment by first considering the very different conclusions reached by Latter (1960) and Bulmer (1972) for diallelic models versus those by Kimura (1965a), Lande (1975, 1977a, 1980a, 1984a), and Fleming (1979) for continuum-of-alleles models. We show how these apparently disparate results are connected, with the different outcomes due, not to the number of assumed alleles per locus (two versus many), but rather to the relative strengths of mutation and selection (Turelli 1984). Given the rather complex nature of some of the theory, we have placed most of derivations, and many of the more technical details, in Examples 28.4–28.8 at the end of this section.

Latter-Bulmer Diallelic Models

While diallelic models of mutation and stabilizing selection trace back to Wright (1935a, 1935b), it was Latter (1960) and Bulmer (1972, 1980) who first considered the predicted equilibrium additive-genetic variance. To obtain their results, we start by slightly rewriting Equation 28.4a for the change in allele frequency due to Gaussian stabilizing selection as

$$\Delta p_i(\text{sel}) \simeq p_i(1-p_i)\,\frac{a_i^2(p_i-1/2)-a_i(\overline{z}-\theta)}{V_s} \tag{28.11a}$$

where a_i is the allelic-effect for locus i (assuming additive loci) and θ is the optimum phenotypic value. Assuming a simple diallelic model with equal mutation rates between alleles, the change from mutation becomes

$$\Delta p_i(\text{mut}) = -2\mu_i(p_i-1/2) \tag{28.11b}$$

Assuming that $\overline{z}=\theta$ at equilibrium (a subtle assumption that requires sufficient granularity in the allelic effects at the underlying loci; Barton 1986) and setting $\Delta p_i(\text{sel})+\Delta p_i(\text{mut})=0$ gives one equilibrium solution as

$$\widetilde{p}_i(1-\widetilde{p}_i)\,a_i^2 = 2\mu_i V_s \tag{28.11c}$$

The solutions to this quadratic equation are

$$\widetilde{p}_i = \frac{1}{2}\left(1\pm\sqrt{1-\frac{8\mu_i V_s}{a_i^2}}\,\right) \tag{28.11d}$$

An admissible solution ($0<\widetilde{p}<1$) requires that the strength of selection ($s_i = a_i^2/[2V_s]$; Equation 28.4c) on a locus be strong relative to mutation μ_i (Bulmer 1980; Slaktin 1987), namely, that

$$a_i^2 > 8\mu_i V_s, \quad \text{implying} \quad s_i > 4\mu_i \tag{28.11e}$$

Notice that the left-hand term in Equation 28.11c is just one half the additive variance contributed by the ith locus. Ignoring the contribution from linkage disequilibrium (which will be slightly negative; Chapter 16), summing over n loci yields an additive variance of

$$\widetilde{\sigma}_A^2 \simeq 4\,n\,\overline{\mu}\,V_s \tag{28.12a}$$

where $\overline{\mu}=n^{-1}\sum\mu_i$ is the average allelic mutation rate. Equation 28.12a is due to Latter (1960), who obtained it by a different approach. The surprising result is that the size of allelic effects (a_i) does not appear in $\widetilde{\sigma}_A^2$. This follows from Equation 28.11d, as increasing a_i results in a more extreme value of $\widetilde{p}$, and hence a smaller value for $\widetilde{p}(1-\widetilde{p})$; the two effects (larger effect size versus more extreme equilibrium frequencies) cancel, as is seen in Equation 28.11c.

If we consider the contribution from a single locus and then recall Equation 28.4c for the strength of selection against a new mutation ($2V_s = a_i^2/s_i$), Equation 28.11c yields the contribution from locus i to the additive variance as

$$\widetilde{\sigma}^2_{A(i)} = 2a_i^2\, \widetilde{p}_i[1 - \widetilde{p}_i] = (2\mu_i)(2V_s) = \frac{2\mu_i a_i^2}{s_i} = \frac{\sigma^2_{m_i}}{s_i} \tag{28.12b}$$

showing that the contribution from the ith locus is the ratio of its mutational variance, $\sigma^2_{m_i}$, and the strength of selection against new mutations, s_i; namely, the ratio of the rate of input of new variation to the rate of its removal (the analog of Equation 7.6b).

One interesting consequence of Equation 28.12a is that the mean fitness at equilibrium is independent of the strength of phenotypic selection, V_s. Substitution of Equation 28.12a into Equation 28.3e yields

$$\begin{aligned}\overline{W} &= \sqrt{\frac{V_s}{V_s + \widetilde{\sigma}^2_A}} = \sqrt{\frac{V_s}{V_s + 4\,n\,\overline{\mu}\,V_s}} \\ &= 1/\sqrt{1 + 4n\overline{\mu}} \simeq 1 - 2n\overline{\mu} \quad \text{for} \quad 4n\overline{\mu} \ll 1\end{aligned} \tag{28.12c}$$

This is another example of Haldane's principal (Chapter 7), namely, that the selective load is simply a function of the mutation rate, independent of the strength of selection.

Equation 28.12a ignores linkage disequilibrium, as it is simply the sum of the single-locus results. A more careful analysis by Bulmer (1980) accounting for gametic-phase disequilibrium (among unlinked loci) found that

$$\widetilde{\sigma}^2_A \simeq \frac{4\,n\,\overline{\mu}\,V_s}{1 - 8n\mu} \tag{28.12d}$$

which closely approximates Equation 28.12a unless the total mutation rate is large. More generally, Turelli (1984) found that the impact of linkage is typically small, unless it is very tight.

With $\widetilde{\sigma}^2_A = 4\,n\,\overline{\mu}\,V_s$, the equilibrium heritability becomes

$$\widetilde{h}^2 = \frac{\widetilde{\sigma}^2_A}{\widetilde{\sigma}^2_A + \sigma^2_E} = \frac{\widetilde{\sigma}^2_A/\sigma^2_E}{\widetilde{\sigma}^2_A/\sigma^2_E + 1} = \frac{4\,n\,\overline{\mu}\,(V_s/\sigma^2_E)}{4\,n\,\overline{\mu}\,(V_s/\sigma^2_E) + 1} \tag{28.12e}$$

Using Turelli's (1984) value of $V_s/\sigma^2_E \simeq 20$ (moderate selection), $n = 100$, and $\overline{\mu} = 10^{-5}$ returns an equilibrium heritability of 0.07. Increasing the per-locus mutation rate to 10^{-4} gives a value of 0.44. A total haploid mutation rate of $n\,\overline{\mu} = 0.0125$ is required to account for a heritability of 0.50 under $V_s/\sigma^2_E = 20$. Hence, unless stabilizing selection is weaker than it appears ($V_s/\sigma^2_E \gg 20$), the per-locus mutation rates are higher than expected ($\mu \gg 10^{-5}$), or the number, n, of loci is very large, the Latter-Bulmer model cannot account for the typically observed levels of heritability, a point made by Latter (1960).

A cautionary note on the Latter-Bulmer model was offered by Barton (1986). Due to the symmetry of the model (all loci have the same effect and heterozygote values equal the optimum value) and its diallelic nature, the above analysis assumes that the mean equals the optimum (set at $\theta = 0$) at equilibrium, such that there are an equal number of loci with equilibrium values of $\widetilde{p}$ and $1 - \widetilde{p}$, (contributing $2a(2\widetilde{p} - 1)$ and $2a(1 - 2\widetilde{p})$, respectively, to the overall mean). Barton showed that when the number of loci is large, equilibria exist at the underlying loci where the population mean *does not* equal the optimum, and in such settings the amount of additive variance exceeds the value predicted by Equation 28.12a, in some cases by a considerable amount. However, while such equilibria can indeed exist, they tend not to be reached, especially in the face of drift (Barton 1989; Hastings 1988, 1990d).

Turelli (1984) generalized the Latter-Bulmer result to a triallelic model, assuming additive effects and no epistasis, with Gaussian stabilizing selection occurring on n loci assumed to be in linkage equilibrium. At locus i, the alleles $A^{(i)}_{-1} : A^{(i)}_0 : A^{(i)}_1$ have values of $-a_i : 0 : a_i$, with the following mutational structure

$$A^{(i)}_{-1} \underset{\mu_i/2}{\overset{\mu_i}{\rightleftarrows}} A^{(i)}_0 \underset{\mu_i/2}{\overset{\mu_i}{\leftrightarrows}} A^{(i)}_1 \tag{28.13a}$$

This model also has a symmetry assumption, namely, that allele A_0 corresponds to the optimal value ($\theta = 0$). Provided that $\mu_i \ll a_i^2/V_s \ll 1$, the equilibrium allele frequencies are

$$\widetilde{p}_1^{\,(i)} = \widetilde{p}_{-1}^{\,(i)} \simeq \mu_i V_s / a_i^2 \tag{28.13b}$$

with $\widetilde{p}_0^{\,(i)} = 1 - 2\,\widetilde{p}_1^{\,(i)}$ (see Turelli 1984 for details). The resulting additive variance for locus i is

$$\widetilde{\sigma}^{\,2}_{A(i)} = 2\left[(-a_i)^2\,\widetilde{p}_{-1}^{\,(i)} + 0^2\,\widetilde{p}_0^{\,(i)} + a_i^2\,\widetilde{p}_1^{\,(i)}\right] = 4a_i^2\left(\mu_i V_s / a_i^2\right) = 4\mu_i V_s \tag{28.13c}$$

Under the assumption of linkage equilibrium, summing over loci recovers Equation 28.12a.

Kimura-Lande-Fleming Continuum-of-alleles Models

In contrast to the Latter-Bulmer two-allele model, starting with Kimura (1965a), a number of continuum-of-alleles models (Chapter 24) have been proposed that allow for a large number of alleles at a locus (Lande 1975, 1977a, 1980a, 1984a; Fleming 1979). Kimura's original analysis followed the distribution, $p_i(x)$, of allelic effects (x) at a given locus, i, assuming the incremental model of mutation (Table 28.1). As detailed in Example 28.4, by assuming small mutational effects, Kimura was able to use a Taylor-series approximation (Equation 28.22a) to show that the distribution of effects at an individual locus are normally distributed, with mean zero and variance $\sqrt{\mu_i \sigma^2_{\alpha_i} V_s}$. Kimura's result is for a haploid model, where $\sigma^2(a_i) = \sqrt{\mu_i \sigma^2_{\alpha_i} V_s}$ denotes the variance in allelic effects from locus i in a haploid gamete. Because of their similar notation, we remind the reader that $\sigma^2(a_i)$ denotes the equilibrium variance in allelic effects, while $\sigma^2_{\alpha_i}$ denotes the variance in mutational effects. Assuming additivity, the additive variance from locus i becomes $\sigma^2_{A(i)} = 2\sigma^2(a_i)$. Assuming no LD, Example 28.4 shows that summing over loci gives Kimura's expression for the additive variance with n equivalent underlying diploid loci as

$$\widetilde{\sigma}^{\,2}_A = \sqrt{2nV_s\sigma^2_m} \tag{28.14a}$$

When effects vary over loci, the above expression holds, with the effective number of loci

$$n_e = 2\left(\sum_{i=1}^{n}\sqrt{\mu_i\sigma^2_{\alpha_i}}\right)^2 \Big/ \sigma^2_m \tag{28.14b}$$

replacing n.

Lande (1975) extended Kimura's model to a full multilocus analysis to allow for linkage disequilibrium (Example 28.7). He did so by assuming that the vector of allelic effects for the n loci in a gamete is multivariate normal, and he obtained a slightly different expression for n equivalent underlying loci,

$$\widetilde{\sigma}^{\,2}_A = \sqrt{2n\sigma^2_m(V_s + n\sigma^2_m/2)} + n\sigma^2_m \tag{28.14c}$$

which essentially reduces to Kimura's result (Equation 28.14a) when $n\sigma^2_m \ll 1$. As with Equation 28.14a, when loci differ, n_e (Equation 28.14b) replaces n. Unlike Latter (1960),

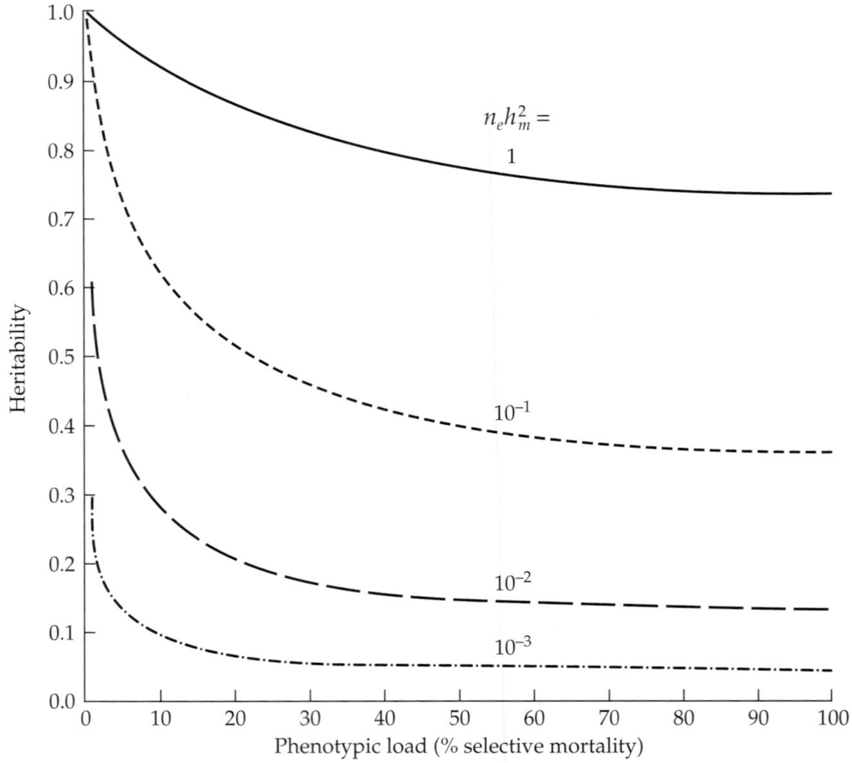

Figure 28.3 The equilibrium heritabilities expected under the Lande model (Equation 28.14c). Percent selective mortality is $100 \cdot (1 - \overline{W})$, where $\overline{W} = \sqrt{\omega^2/(\omega^2 + \sigma_E^2 + \widetilde{\sigma}_A^2)}$, assuming $\overline{z} = \theta$ (Equation 28.3e). Taking a typical value of $h_m^2 = 10^{-3}$, the plotted curves correspond (from top to bottom) to effective number of loci (n_e) values of 1000, 100, 10, and 1. Note that modest selection (low selective mortality) with a reasonable number of loci (n_e from 10 to 100, implying that $n_e h_m^2$ is 0.01 to 0.1) can account for the observed heritabilities in natural populations ($0.2 \leq h^2 \leq 0.7$). (After Lande 1975.)

Lande concluded that mutation-selection balance *could* indeed account for high levels of additive variation (Figure 28.3). Nagylaki (1984) and Turelli (1984) noted that with weak selection, Equation 28.14c slightly overestimates the genetic variance and is slightly less accurate than Equation 28.14a.

One can also recover Equation 28.14a using results from Chapter 24 on the Gaussian continuum-of-alleles model (which, as in Lande 1975, assumes the distribution of allelic effects at a locus to be normal). Equation 24.2a gave the dynamics for the change in genic variance, $\Delta\,\sigma_a^2(t)$, under the Gaussian COA model, which had an equilibrium value of $\widetilde{\sigma}_a^2 = 0$. However, adding a term, σ_m^2, for new mutation to Equation 24.2a, ignoring the effects of linkage disequilibrium (i.e., assuming $d = 0$, and hence the genic variance, σ_a^2, equals the additive-genetic variance σ_A^2), and setting $N_e = \infty$, yields

$$\Delta\,\widetilde{\sigma}_a^2 = 0 = -\frac{\kappa\,\widetilde{h}^2\,\widetilde{\sigma}_A^2}{2n} + \sigma_m^2, \quad \text{implying} \quad 2n\sigma_m^2 = \kappa\,\widetilde{h}^2\,\widetilde{\sigma}_A^2 \tag{28.14d}$$

Recall that κ is a measure of the strength of stabilizing selection, namely, the fraction by which the phenotypic variance is reduced following selection (Equation 16.10a). Because $\widetilde{h}^2 = \widetilde{\sigma}_A^2/\widetilde{\sigma}_z^2$, Equation 28.14d can be expressed as

$$\widetilde{\sigma}_A^4 = 2n\sigma_m^2(\widetilde{\sigma}_z^2/\kappa), \quad \text{yielding} \quad \widetilde{\sigma}_A^2 = \sqrt{2n\sigma_m^2(\widetilde{\sigma}_z^2/\kappa)} \tag{28.14e}$$

Because $\kappa \simeq \widetilde{\sigma}_z^2/V_s$ (Equation 16.18a), then $\widetilde{\sigma}_z^2/\kappa = V_s$, which recovers Equation 28.14a.

Fleming (1979) presented an improved (but still approximate) analysis of Kimura's model. He did so by scaling both the strengths of selection and mutation by a small parameter, ϵ, and then expressing the strength of selection as $(2V_s)^{-1} = \gamma\epsilon$ and the mutational-effects variance as $\sigma_\alpha^2 = \delta\epsilon$. This scaling (which Turelli notes implies a per-locus mutation rate $\mu \gg \sigma_\alpha^2/V_s$) assumes both selection and mutation are weak. By letting $\epsilon \to 0$, Fleming was able to express the equilibrium distribution of allelic effects in terms of zero and first-order expressions of ϵ, namely, $\phi(x) = \phi_0(x) + \epsilon\phi_1(x) + O(\epsilon^2)$. His zero-order term (i.e., the function $\phi_0[x]$) is a normal with a variance given by Equation 28.14a, and it is independent of the linkage map. The first-order expression (ϕ_1) has significant kurtosis, which shows that the distribution of segregating allelic effects departs from a Gaussian. When the mutational increment, α, is drawn from a normal distribution, Fleming's approximation yields

$$\widetilde{\sigma}_A^2 \simeq \sqrt{2nV_s\sigma_m^2}\left[1 + \left(1 - \frac{3}{16n\mu}\right)\sqrt{\frac{n\sigma_m^2}{2V_s}}\right] \tag{28.15}$$

Fleming (1979) and Bürger (1998a) present more general expressions, which allow for non-Gaussian kurtosis in the distribution of mutational effects. Simulation studies by Turelli (1984) found that Equation 28.15 is accurate over a much wider range of parameter values $(1 < \sigma_{\alpha_i}^2/(V_s\mu_i) < 10)$ than might be expected given the nature of the approximation. Applied mathematics aficionados are referred to Fleming's paper, and less technical discussions were provided by Nagylaki (1984) and Turelli (1984). By using methods from applied physics, Bürger (1986, 1988a, 1988c) obtained a number of conclusions regarding the solution to the general Kimura model, but as we now detail, most results are based on one of two different approximations of the equilibrium solution.

Gaussian Versus House-of-Cards Approximations for Continuum-of-alleles Models

Equations 28.12a and 28.14a offer very different predictions for the expected genetic variance under mutation-selection balance. Under Kimura's result (and Lande's extension), the effect of the number of loci and strength of phenotypic selection on the trait scale as $\sqrt{n}$ and $\sqrt{V_s}$, respectively, while under Latter's model, these scale as n and V_s. The Latter-Bulmer model (Equation 28.12a) simply requires the total mutation rate, $n\overline{\mu}$ (independent of the variance, σ_α^2, of mutational effects), while the Kimura-Lande-Fleming results (Equations 28.14a, 28.14c, and 28.15) are more pleasingly stated in terms of the mutational variance, σ_m^2, which is a more easily measured parameter than its components (μ, n, and σ_α^2). Further, the Latter-Bulmer model does not appear to maintain sufficient variation to account for observed h^2 values, while the Kimura-Lande-Fleming model does. Why is there this vast disparity, and which approach, if either, is correct?

Turelli (1984) showed that these rather different outcomes arise from different approximations of the complex integro-differential equation for the distribution of allelic effects for the general Kimura model (Equation 28.21c in Example 28.4). Kimura and Fleming obtained their approximate solutions by assuming that the variance of mutational effects at a locus (the allelic effects *given* that a mutation has occurred) to be much less that the current variance of allelic effects at that locus, $\sigma_{\alpha_i}^2 \ll \sigma_{A(i)}^2$, a point first stressed by Lande (1975). From Equation 28.14a, this condition implies that

$$\sigma_{\alpha_i}^2 \ll \sqrt{\mu_i\sigma_{\alpha_i}^2 V_s} \tag{28.16a}$$

which can be rearranged as

$$\mu_i \gg \frac{\sigma_{\alpha_i}^2}{V_s} \tag{28.16b}$$

If we recall Equation 28.4c, this condition is equivalent to $\mu_i \gg E[s_i]$, which shows that mutation is much stronger than selection at a given locus. Turelli (1984) referred to this as the **Gaussian approximation**, as the resulting equilibrium solution approaches a normal distribution of allelic effects at a locus (Example 28.5). Note that Lande (1975) *assumed* a Gaussian distribution of allelic effects in his multiple-locus treatment that accounted for linkage, whereas Kimura and Fleming *obtained* it following their assumption that $\sigma^2_{\alpha_i} \ll V_s\mu_i$. Kimura obtained exact normality with his solution, while normality was the zero-order term in Fleming's more careful analysis.

Turelli (1984) argued that the inequality given by Equation 28.16b is typically reversed, namely, $\mu_i \ll \sigma^2_{\alpha_i}/V_s$ (implying $\sigma^2_{\alpha_i} \gg \sigma^2_{A(i)}$), so that the Gaussian approximation is often inappropriate. His logic follows from the standard value of $\sigma^2_m = \sigma^2_E/10^3$, which implies $\sigma^2_m \simeq \sigma^2_A/10^3$ for a typical heritability ($0.3 \leq h^2 \leq 0.7$). Because both σ^2_m and σ^2_A are the sums of single-locus effects, with equivalent loci we can replace $\sigma^2_m \simeq \sigma^2_A/10^3$ by the single-locus contributions to each component to give $\mu_i\sigma^2_{\alpha_i} \simeq \sigma^2_{A(i)}/10^3$. Hence, the Gaussian approximation that $\sigma^2_{\alpha_i} \ll \sigma^2_{A(i)}$ (the variance of new mutations is much smaller than the standing variance) requires that $\mu_i \cdot 10^3 \gg 1$ or that $\mu_i \gg 10^{-3}$. This value is orders of magnitudes above traditional estimates of per-locus mutation rates.

Based on these concerns, Turelli considered Kimura's model when the inequality in Equation 28.16b is reversed

$$\mu_i \ll \frac{\sigma^2_{\alpha_i}}{V_s} \tag{28.17}$$

where now mutation is weak relative to selection ($\mu_i \ll E[s_i]$). Turelli's **house-of-cards approximation (HCA)** uses this assumption to obtain an equilibrium solution of the general Kimura equation (Example 28.5). The basis for Turelli's approximation follows from the HOC (house-of-cards) mutation model (Table 28.1), which assumes, at each locus, that the new mutational variance is likely to swamp any existing variance. (As a notational aside, we use HOC to refer to the mutational *model*, and HCA to refer to Turelli's *approximation* motivated by this model, to stress that these are *different*.) Under HOC mutation, the new allelic value, x', following mutation is independent of its current value, x (i.e., $x' = \alpha$; as opposed to the situation with the incremental model, where $x' = x + \alpha$). As shown in Example 28.5, the HCA gives

$$\widetilde{\sigma}^2_A \simeq 4V_s n\mu \tag{28.18a}$$

which is simply the Latter-Bulmer result (Equation 28.12a). The connection between the HCA and the Latter-Bulmer model follows because the latter requires $a_i^2 > 8\mu_i V_s$ (Equation 28.11e) in order to obtain Equation 28.12a, while the HCA requires that $\sigma^2_{\alpha_i} \gg \mu_i V_s$. The a_i^2 (mutational effects in a two-allele model) essentially equate to the mutational-effects variance $\sigma^2_{\alpha_i}$ under a continuum-of–alleles model. Under HCA conditions, selection is strong and the dominant (close to fixation) allele at a locus is expected to have a value close to the optimum. New mutations are thus deleterious, and tend to disappear quickly, resulting in most of the genetic variation being due to rare alleles with relatively large effects.

As with many of the results in this section, Equation 28.18a is simply the sum of single-locus results. Turelli and Barton (1990) examined the impact of linkage, finding that with n identical loci

$$\widetilde{\sigma}^2_A \simeq 4V_s n\mu \left[1 + \frac{2(n-1)\mu}{c_H}\right] \tag{28.18b}$$

where c_H is the harmonic mean of all pairwise recombination frequencies between all combinations of the underlying loci, or roughly 1/2 for loose linkage. As with the Gaussian approximation, the impact from linkage is small unless it is very tight.

As was discussed in Chapter 24, the kurtosis (given by $E[x^4]$ when $\mu_x = 0$; LW Chapter 2) provides one measure of departure from normality. The kurtosis for a normal equals $3\sigma_x^4$, suggesting two scaled measures of depature from normality. In Chapter 24, we used $\kappa_4 = (E[x^4] - 3\sigma_x^4)/\sigma_x^4$, which has a value of zero for a normal. Alternatively, we here use $k_4 = E[x^4]/(3\sigma_x^4)$, which equals one for a normal. Under the HCA, the resulting kurtosis for the distribution of allelic effects at locus i, where $E[\alpha_i] = 0$, is

$$k_{4,i} = \frac{E[\alpha_i^4]}{3E[\alpha_i^2]^2} \simeq \frac{2V_s\mu_i\sigma^2_{\alpha_i}}{3(2V_s\mu_i)^2} = \frac{\sigma^2_{\alpha_i}}{6V_s\mu_i} \tag{28.18c}$$

which is $\gg 1$ (highly leptokurtic, i.e., a heavier tail, and hence more outliers, than a Gaussian distribution) under the HCA (which follows from Equation 28.17, as $\sigma^2_{\alpha_i} \gg \mu_i V_s$). The resulting distribution of allelic effects thus departs significantly from a normal, with its leptokurtosis indicating the presence of rare alleles of large effect. Further, note that the (unscaled) kurtosis in the distribution of genotypic values (twice the haploid value) can be expressed as

$$2E[\alpha_i^4] = 4V_s\mu_i\sigma^2_{\alpha_i} = \widetilde{\sigma}^2_{A(i)}\sigma^2_{\alpha_i} \tag{28.18d}$$

with the last step following from Equation 28.11c. Recall from Example 24.11 that this expression for kurtosis has the same form as seen in the rare-alleles model (a constant times the second moment; Equation 24.32a), which is reasonable, as under HCA, most alleles are rare.

Kurtosis also influences the accuracy of Equation 28.18a, which is an upper bound. When the distribution of mutational effects is normal, the accuracy is quite good. As the distribution of mutational effects becomes increasing leptokurtic, the true variance (even under HCA conditions) can be significantly less than suggested by Equation 28.18a (Bürger and Hofbauer 1994; Bürger and Lande 1994).

Thus, we have Kimura-Lande-Fleming when $\mu_i \gg \sigma^2_{\alpha_i}/V_s$ (which fulfills the Gaussian assumption of that mutation is stronger than selection) and Latter-Bulmer when $\mu_i \ll \sigma^2_{\alpha_i}/V_s$ (the HCA assumption that selection is stronger than mutation). Extensive simulations by Turelli (1984) refined these domains. The Gaussian approximation overestimates the additive variance by less than 10% when $\mu_i \geq 5\sigma^2_{\alpha_i}/V_s$, while the HCA model gives a good fit when $\mu_i \leq 0.05\sigma^2_{\alpha_i}/V_s$. Bürger (1988a, 1988b) was able to obtain an upper bound for the equilibrium additive variance under a fairly general Kimura model (assuming symmetric mutations and quadratic fitnesses near the optimum). He found that the first-order bound is simply the HCA value, $\widetilde{\sigma}^2_A \leq 4\mu V_s$ (we remind the reader that σ^2_α is the variance of mutational effects, while σ^2_A refers to the additive variance). When Kimura's single-locus expression, $\sqrt{2V_s\sigma^2_m}$, exceeds this value, the Gaussian approximation has clearly failed, giving the restriction

$$\sqrt{2V_s\sigma^2_m} = \sqrt{2V_s\left(2\mu_i\sigma^2_{\alpha_i}\right)} = \sqrt{(4\mu_i V_s)\,\sigma^2_{\alpha_i}} \leq 4\mu_i V_s, \quad \text{or} \quad \sigma^2_{\alpha_i} \leq 4\mu_i V_s \tag{28.18e}$$

with the Gaussian approximation always failing when $\sigma^2_{\alpha_i} > 4\mu_i V_s$.

While the reader may perceive this difference between the Gaussian and HCA approximations as being a function of the assumed mutation *model*, it is rather a function of the *relative strengths* of selection to mutation at a locus. When mutation is strong, one expects a number of alleles at a locus, while when mutation is weak relative to selection, one expects very few segregating alleles (the rare-alleles model from Example 24.11). While both the Gaussian and HCA approximations follow from a continuum-of-alleles model, the transition from Gaussian to HCA behavior can be seen in models with a modest to small number of assumed alleles per locus. Equation 28.13c shows how the HCA variance follows from a triallelic model when Equation 28.17 holds.

An extension of Turelli's triallelic model provides further insight. Slatkin (1987a) assumed an unlimited number of alleles with a stepwise mutation model, with an allele mutating to a new effect with increment of α or $-\alpha$ (relative to its current value), with a mutation rate of $\mu/2$ for each step (a scheme also used by Narain and Chakraborty 1987), namely,

$$\cdots \quad -2\alpha \;\overset{\mu/2}{\underset{\mu/2}{\rightleftarrows}}\; -\alpha \;\overset{\mu/2}{\underset{\mu/2}{\rightleftarrows}}\; 0 \;\overset{\mu/2}{\underset{\mu/2}{\leftrightarrows}}\; \alpha \;\overset{\mu/2}{\underset{\mu/2}{\leftrightarrows}}\; 2\alpha \quad \cdots$$

As shown in Example 28.6, if selection is weak relative to mutation (such that many allelic states are present), this model reduces to Kimura's Gaussian result, while if selection is strong relative to mutation (meaning that a single major allele, whose value equals the phenotypic optimum, and two very minor alleles, each one step away, are present), this reduces to the HCA result (Turelli's triallelic model). Analyses of models assuming five alleles per locus further make this point (Turelli 1984; Slatkin 1987a). Example 28.8 presents Waxman's (2004) exact solution for the continuum-of-alleles model under a specific distribution of mutational effects, which recovers the HCA results for low mutation rates and the Gaussian for high rates, and shows the structure of the transition between these two domains.

Example 28.3. An interesting biological application of the house-of-cards versus Gaussian models is the work of Hodgins-Davis et al. (2015). These authors examined variation in gene expression in roughly 3500 yeast (*S. cerevisiae*) genes, 930 *C. elegans* genes, and 563 genes from *D. melanogaster*.

As discussed in Chapter 11, gene expression is thought to be under weak stabilizing selection. Hodgins-Davis et al. obtained estimates of σ_m^2 for expression data ($\simeq [10^{-5}$ to $10^{-4}]\sigma_E^2$, which tend to be much smaller than estimates for other traits), along with estimates of the number of genes (using both eQTL studies and extrapolation from single-gene knockout experiments). Coupling these values with estimates of μ, they estimated the variance of mutational effects, $\sigma_\alpha^2 = \sigma_m^2/(2n\mu)$. Next, given an estimate of σ_A^2, they estimated the potential strength of stabilizing selection from the standing variation in expression at each gene as either

$$V_s^G = \frac{\sigma_A^4}{2n\sigma_m^2} \quad \text{or} \quad V_s^{HC} = \frac{\sigma_A^2}{4n\mu}$$

which are obtained by rearranging either the Gaussian (Equation 28.14a) or HCA (Equation 28.18a) expressions.

With estimates of σ_α^2 and V_s (the latter is based on the Bayesian model-averaged values of V_s^G and V_s^{HC}) for each expression trait in hand, Hodgins-Davis et al. examined what fraction of traits fell within the HCA versus Gaussian domains. Recalling that $\sigma_\alpha^2 > 20\mu\, V_s$ is Turelli's bound for the HCA to hold, the authors plotted $20\mu\, V_s$ versus σ_α^2 for each expression trait. As shown in the following figure (after Hodgins-Davis et al. 2015), their finding was that the house-of-cards domain is satisfied by nearly all of the genes examined, with only a very few exhibiting any likelihood of being in the Gaussian domain.

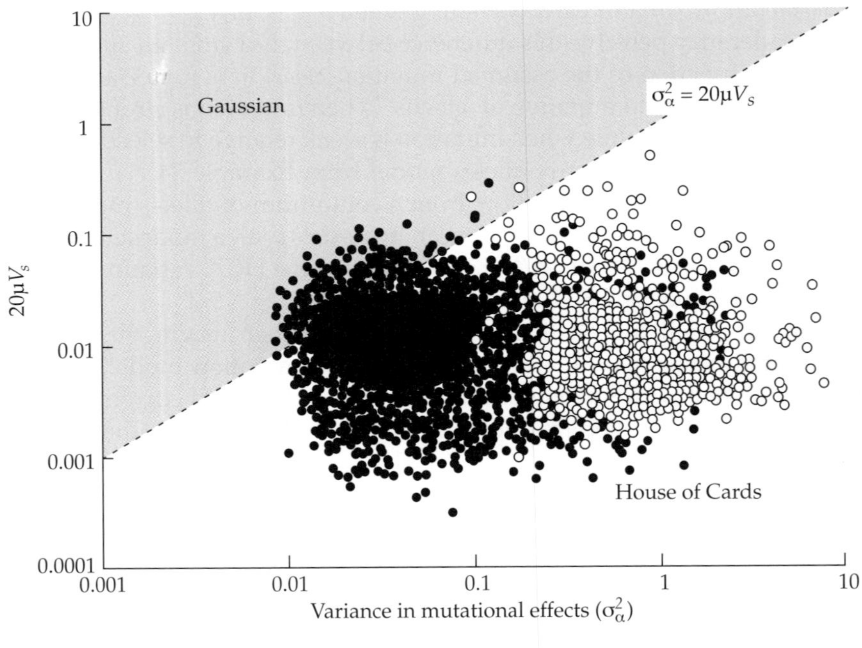

For example, the above figure plots values for expression levels for *D. melanogaster* (filled circles) and *C. elegans* (open circles) genes. Expression levels for most of the 563 measured *melanogaster* genes appear to evolve under the HCA framework (falling below the line $\sigma^2_\alpha = 20\mu V_s$), as do expression levels at essentially all of the 930 measured *C. elegans* genes.

Epistasis

Epistasis in models of stabilizing selection can act on several levels. **Fitness epistasis** naturally arises even for a completely additive trait under stabilizing selection, because the mapping from trait value to fitness is nonlinear (Examples 5.10 and 6.6). Likewise, the values for trait under stabilizing selection could themself show epistasis (**trait epistasis**). Hermisson et al. (2003) showed that trait epistasis reduces the amount of trait additive variation at equilibrium relative to purely additive models. While Gavrilets and de Jong (1993) found that certain models of fitness epistasis can both maintain a high amount of trait additive variation *and* show strong apparent stabilizing selection for that trait, this outcome results from assuming a neutral trait influenced by pleiotropic loci under balancing selection, a rather different setting from an epistatic trait itself being under stabilizing selection (see Lawson et al. 2011 for a potential example).

Tachida and Cockerham (1988) examined the expected amount of additive versus additive-by-additive variance in *fitness* for a trait under stabilizing selection. They found that additive-by-additive variance in fitness is larger than additive variance under conditions for the Gaussian approximation, but that the converse is true (additive variance is larger than nonadditive variance) under HCA conditions. In part, this likely arises because the rare-allele conditions under HCA implies that most of any genetic variation loads onto the additive component; see Example 24.11). Important caveats for this HCA result are that the number of loci per trait not be too large and that the number of traits that are influenced per locus (their amount of pleiotropy) be small. A review of the *Drosophila* fitness-components literature suggested more additive than additive-by-additive variance in fitness components, which led Tachida and Cockerham to suggest that the HCA domain might be more applicable in these cases. However, they also noted that this distinction between the two classes of approximation breaks down when the trait means depart from their optimal values.

Effects of Linkage and Mating Systems

The more diligent reader may recall situations in two-locus models wherein the effects of linkage disequilibrium were quite considerable (Chapter 5). This occurs in cases where selection is much stronger than recombination. In contrast, the analysis of polygenic models typically assumes that recombination is much stronger than selection at a given locus, resulting in linkage effects being much smaller, often to the point (depending on the problem) that they can be ignored as a good first approximation.

Most of the above analysis, under either the HCA or Gaussian approximations, extrapolates the additive variance by summing single-locus variances. Recalling Equation 16.2, the additive variance, σ^2_A, is the sum of the genic variance, σ^2_a (the additive variance in the absence of linkage disequilibrium), plus the disequilibrium contribution, d, with $\sigma^2_A = \sigma^2_a + d$. σ^2_A is often called the **expressed variation**, and $\sigma^2_a - \sigma^2_A = -d$ the **hidden variation** (the amount recovered upon decay of LD). Simply summing single-locus results (as we have done multiple times above) recovers the genic variance, σ^2_a, not the additive variance, σ^2_A, and we expect the genic variance to overestimate the additive variance (as $d < 0$ under stabilizing selection; Chapter 16). While the actual value of d can be considerable, simulations (Turelli 1984; Hastings 1989) and analytic results (Lande 1975; Fleming 1979; Nagylaki 1984; Bürger 1989) show that the relative error by ignoring d is generally small, negative, and increases (slowly) with n. Fleming found that the zero-order approximation of the distribution of allelic effects was independent of the recombination map, which entered as first-order terms. Assuming n equivalent loci ($\mu_i = \mu$, $\sigma^2_{\alpha_i} = \sigma^2_\alpha$), Turelli (1984) used the Lande and Fleming results to find that the relative error in using the linkage equilibrium

(LE) value in place of the true additive variance under the Gaussian approximation was

$$\frac{\widetilde{\sigma}^2_A(LE) - \widetilde{\sigma}^2_A}{\widetilde{\sigma}^2_A} \simeq \left(1 - \frac{1}{n}\right) \sqrt{\frac{n\sigma^2_m}{2V_s}} \tag{28.19}$$

A simulation study by Hastings (1989), essentially using the HCA approximation, found that the impact of LD is again small and scales with $n\mu$, the total (haploid) mutation rate. If $n\mu < 0.025$, the contribution from LD was small, less than 10% of the total variance. However, for $n\mu > 0.05$, the contribution can be considerable. Turelli and Barton (1990) also found that the impact of linkage under HCA scales with $n\mu$, see Equation 28.18b.

In an exact analysis of a two-locus model, Bürger (1989) found that the impact of linkage depends on the relative strengths of mutation and selection, namely, the HCA versus Gaussian assumptions. Under the Gaussian assumption, the genic variance, σ^2_a, remains constant, while the additive variation decreases as linkage becomes tighter (d becomes more negative). Under the HCA assumption, the genetic variance remains constant under linkage (as long as it is not too tight), while the genic variance increases with decreasing recombination (as seen in Equation 28.18b). If recombination is below a critical value, then the behavior is as for the Gaussian approximation.

A second issue of potential concern is the mating system. Thus far, we have been assuming random mating. However, previous chapters showed that inbreeding (Chapter 11) and assortative mating (Chapter 16) can both impact the additive-genetic variance. Given these observations, Lande (1977a) obtained the counterintuitive result that these departures from random mating have essentially no impact on the equilibrium additive variance for a Gaussian model with only additive effects. Inbreeding and assortative mating change the rate of approach to the equilibrium, but not its final value. Conversely, Turelli (1986) and Frank and Slatkin (1990b) found that inbreeding *does* change the equilibrium additive variance under HCA assumptions. Turelli suggested that the robustness of the Gaussian model to the mating system may be an artifact of the high mutation rate per locus required for this model to be accurate.

Spatial and Temporal Variation in the Optimum

As shown previously in this chapter, spatial and temporal variation in the optimum can maintain some variance under stabilizing selection in the absence of mutation, but the conditions necessary for a large effect are fairly restrictive. Does incorporating a variable optimum, θ, increase the additive variance when mutation is present? It does, and the increase can be substantial.

A number of authors have examined the impact of temporal variation in θ (Kirzhner et al. 1996a, 1996b; Kondrashov and Yampolsky 1996a, 1996b; Korol et al. 1996; Bürger 1999; Zhang 2012), with the most detailed treatment by Bürger and Gimelfarb (2002). When there is a periodic change in θ with a sufficiently long cycle time (>10 generations) and a sufficient amplitude ($>\sqrt{V_s}$), the amount of additive-genetic variation significantly will exceed the constant-θ value, often by at least an order of magnitude. When there are persistent directional shifts in the optimum, alleles that were initially rare and deleterious can become favorable and will be under directional selection to track the new optimum (Chapter 27). If the directional change persists for a sufficiently long period of time and the change in the trait mean while tracking θ is sufficiently large, significant allele-frequency change will occur, increasing the additive variance. The change in the optimum, however, must be ongoing, in that if θ stops at a new value, we return to a constant-θ model. The open question is not whether the optimum changes, as most ecologists would suspect that it does, but rather whether these changes are periodic and persistent enough to dramatically impact the additive variance. Changes in θ that are entirely random (i.e., with no short-term directional trend or positive autocorrelation) have little impact on the additive variance.

Table 28.2 Comparison of the Gaussian and house-of-cards (HCA) approximations for a trait under stabilizing selection. Here, V_s is the strength of selection on a genotypic value (Equation 28.3f), $\sigma^2_{A(i)}$ is the additive variation at locus i, n is the number of loci, $\sigma^2_{\alpha_i}$ and μ_i (respectively) are the variance of the effects of new mutations and the mutation rate at locus i, and $\sigma^2_m = \sum 2\mu_i\sigma^2_{\alpha_i}$ denotes the mutational variance. When mutational effects are constant over loci, we use σ^2_α and μ rather than retain the subscript i, with $\sigma^2_m = 2n\mu\sigma^2_\alpha$. See text for further details.

	Gaussian	HCA
Mutational input vs. standing variation	$\sigma^2_{\alpha_i} \ll \sigma^2_{A(i)}$ Smaller	$\sigma^2_{\alpha_i} \gg \sigma^2_{A(i)}$ Larger
Strength of mutation relative to selection	$\mu_i \gg \sigma^2_{\alpha_i}/V_s$ Stronger	$\mu_i \ll \sigma^2_{\alpha_i}/V_s$ Weaker
Domain of applicability (single trait, $N_e = \infty$)	$\sigma^2_{\alpha_i} \leq \mu_i V_s/5$	$\sigma^2_{\alpha_i} \geq 20\mu_i V_s$
Impact of drift on domain of applicability	Decreases domain	Little to no effect
Impact of pleiotropy on domain of applicability	Decreases domain	Increases domain
Equilibrium additive variance, $\widetilde{\sigma}^2_A$	$\sqrt{2nV_s\sigma^2_m}$	$4V_s n\mu$
Finite population $\widetilde{\sigma}^2_A$	$\sqrt{\left(\frac{nV_s}{2N_e}\right)^2 + 2n\sigma^2_m V_s} - \frac{nV_s}{2N_e}$	$\frac{4n\mu V_s}{1 + V_s/(N_e\sigma^2_\alpha)}$
Sensitivity to linkage map	Little unless $c_{ij} \simeq 0$	Little unless $c_{ij} \simeq 0$
Impact of mating system on $\widetilde{\sigma}^2_A$	Insensitive	Sensitive
Number of alleles/locus	Many	One major, few rare
Distribution of allelic effects	Normal. Many alleles at intermediate frequencies.	Leptokurtic. Rare alleles of large effect.
Impact of multiple-trait selection	None for uncorrelated traits.	Sensitive to uncorrelated traits.

The impact of spatial variation in the optimum under stabilizing selection has been examined by Felsenstein (1977), Slaktin (1978), and Barton (1999). We previously discussed Felsenstein's model, which assumed a linear gradient in the optimum, such that at location x on some linear cline (such as a river bank), $\theta(x) = \beta x$, with individuals randomly dispersing over some distance $d \sim N(0, \sigma^2_d)$. Felsenstein showed that this model can maintain at least some variation in the face of stabilizing selection without mutation, as migration effectively fills the role of generating variation. Slaktin (1978) and Barton (1999) extend Felsenstein's model to allow for mutation. Felsenstein and Slaktin both assumed a Gaussian distribution of mutational effects at a locus, which Barton showed was a good approximation even under HCA conditions. Slaktin found that the equilibrium additive variance becomes

$$\widetilde{\sigma}^2_A = 2Z\sqrt{V_s + Z^2} + 2Z^2 \tag{28.20a}$$

where

$$Z^2 = \sum_{i=1}^{n} \sqrt{\mu_i\sigma^2_{\alpha_i} + \beta^2\sigma^2_d} \tag{28.20b}$$

This is simply Lande's (1975) result (Equation 28.28f), with $\beta^2\sigma^2_d$ (a measure of how quickly selection changes relative to migration) augmenting the mutational variance. If this change is sufficiently large, namely, $\beta^2\sigma_d^2 > \mu_i\sigma_{\alpha_i}^2 = \sigma_m^2/(2n_e)$, then spatial differences in fitness (given by the variation in θ) dominate mutation, and $Z^2 \simeq n\beta\sigma_d$.

Summary: Implications of Gaussian Versus HCA Approximations

Table 28.2 summarizes the major features of the Gaussian and House-of-cards approximations and their differences in behavior (some of which are developed in later sections). While the reader might infer that the conditions for the Gaussian approximation to hold are unusual, Charlesworth (1993) and Bürger (2000) made the important point that this approximation might be highly relevant in asexual species or species with a large fraction of the genome in regions of low recombination or for species that undergo cyclical parthenogenesis (Lynch and Gabriel 1983). In these cases, the mutational size of what corresponds to a locus is much larger (equivalent to the entire genome), resulting in a higher mutation rate. We conclude this section with the derivations of many of the results given above, which can be skipped by the casual reader. A number of these results were also obtained by Zhang and Hill (2010), using the framework of the Price equation (Chapter 6), which offers the reader an independent set of derivations.

Example 28.4. Before proceeding with the derivation of Kimura's (1965a) result, recall the important distinction between an allele effect, a, and a mutational effect, α. Under the incremental model, the allelic effect following a new mutation is $a' = a + \alpha$. Our interest is in the variance of allelic effects, $\sigma^2(a)$, as this is half of the additive variance contributed by a locus (for a diploid and assuming additivity). As we will see, in many models, $\sigma^2(a)$ is a function of the variance in the *mutational effects*, σ_α^2. In what follows, because a and α are very similar in appearance, we use x for the allelic effect in much of the deviation, before returning to express $\sigma^2(x)$ as $\sigma^2(a)$ in the discussion. Finally, while our focus is for a particular locus, we suppress the subscript throughout much of the derivation to keep the notation simpler.

Kimura (1965a) considered a haploid continuum-of-alleles model, following the distribution, $p(x)$, of allelic effects at a locus with no linkage disequilibrium. He assumed a continuous-time model in Hardy-Weinberg equilibrium and under quadratic stabilizing selection (Equation 28.3a). Our treatment follows Bulmer (1989), who obtained the same result under Gaussian stabilizing selection (with the population mean set at θ). In this setting, Bulmer showed that the expected infinitesimal change in the distribution of allelic effects from selection is

$$\frac{\partial p(x)}{\partial t}(\text{sel}) = \frac{-p(x)[\,x^2 - \sigma^2(x)]}{2V_s} \tag{28.21a}$$

where $\sigma^2(x) = \sigma^2(a)$ is the variance in allelic effects at the focal locus.

Under the incremental mutational model, $\mu f(\alpha)$ is the probability that an allele mutates from value x to value $x+\alpha$, where $E(\alpha) = 0$ and $E(\alpha^2) = \sigma_\alpha^2$. The resulting rate of change from mutation becomes

$$\frac{\partial p(x)}{\partial t}(\text{mut}) = -\mu p(x) + \mu \int p(x-\alpha) f(\alpha) d\alpha \tag{28.21b}$$

The first term is the loss of alleles with an effect size of x due to mutation, and the second is the gain of such alleles from new mutation. The latter is expressed as the cumulative probability that alleles with some other effect mutate to an effect size of x.

Formally, if the life cycle is selection followed by mutation, then $p(x)$ in the Equation 28.21b is replaced by $p'(x)$, the postselection value. However, for weak selection and mutation, we can simply sum Equations 28.21a and 28.21b to yield the integro-differential equation for the distribution of allelic effects at time t. The equilibrium distribution is reached when

$$\frac{\partial p(x)}{\partial t} = \frac{\partial p(x)}{\partial t}(\text{sel}) + \frac{\partial p(x)}{\partial t}(\text{mut}) = 0$$

or

$$\frac{-p(x)[x^2 - \sigma^2(x)]}{2V_s} - \mu p(x) + \mu \int p(x-\alpha) f(\alpha) d\alpha = 0 \tag{28.21c}$$

which Bulmer (1989) called the **fundamental equation of the continuum-of-alleles model**. Most early workers assumed that such an equilibrium distribution exists and that it is unique, but this was not formally shown until the publications of Bürger (1986, 1988a, 1988b, 1988c, 1991b; Bürger and Bomze 1996). Notice that the solution to Equation 28.21c depends strongly on the choice of $f(\alpha)$, the assumed probability density function for new mutational effects.

The assumption of a Gaussian distribution of allelic effects breaks down in Equation 28.21b. Even assuming that a Gaussian allelic-effects distribution exists following selection *and* also that there is a Gaussian distribution of mutational effects, Equation 28.21b is a *weighted* sum of two Gaussians, with different variances, and hence is clearly *not* Gaussian. However, under certain conditions it can be close to normally distributed.

Different approximations have been used to proceed from Equation 28.21c to an explicit solution. Kimura assumed that mutational effects, α, are sufficiently small to approximate $p(x-\alpha)$ by a second-order Taylor series

$$p(x-\alpha) \simeq p(x) - \alpha\,\frac{\partial p(x)}{\partial x} + \frac{\alpha^2}{2}\,\frac{\partial^2 p(x)}{\partial^2 x} \tag{28.22a}$$

This approximation requires that $\sigma^2_\alpha \ll \sigma^2(x) = \sigma^2(a)$, namely, that the effect of a new mutation is much smaller than the current variance in allelic effects. Substituting this approximation into the integral in Equation 28.21c yields

$$\begin{aligned}
&\mu \int \left(p(x) - \alpha\frac{\partial p(x)}{\partial x} + \frac{\alpha^2}{2}\,\frac{\partial^2 p(x)}{\partial^2 x} \right) f(\alpha)d\alpha \\
&= \mu \left(p(x) \int f(\alpha)d\alpha - \frac{\partial p(x)}{\partial x} \int \alpha\, f(\alpha)d\alpha + \frac{1}{2}\frac{\partial^2 p(x)}{\partial^2 x} \int \alpha^2\, f(\alpha)d\alpha \right) \\
&= \mu \left(p(x) + 0 + \frac{1}{2}\frac{\partial^2 p(x)}{\partial^2 x}\sigma^2_\alpha \right)
\end{aligned} \tag{28.22b}$$

where we have used $\int \alpha f(\alpha)d\alpha = E[\alpha] = 0$. Using the approximation given by Equation 28.22a, the equilibrium solution of $p(x)$ satisfies the differential equation

$$\frac{-p(x)[x^2 - \sigma^2(x)]}{2V_s} + \frac{\mu\sigma^2_\alpha}{2}\,\frac{\partial^2 p(x)}{\partial^2 x} = 0 \tag{28.22c}$$

Kimura (1965a) showed that this equation is satisfied when $p(x)$ follows a normal distribution, with the parameters

$$p(x) \sim N\left(0, \sqrt{\mu\sigma^2_\alpha V_s}\right) \tag{28.22d}$$

In particular, the variance in allelic effects at locus i is given by

$$\sigma^2(a_i) = \sqrt{\mu_i \sigma^2_{\alpha_i} V_s} \tag{28.22e}$$

with the additive-genetic variance contributed by locus i being $\sigma^2(A_i) = 2\sigma^2(a_i)$.

This result of allelic effects at individual loci being normally distributed motivated the continuum-of-alleles models introduced in Chapter 24. Equation 28.22a is referred to as the **Gaussian approximation** because this weak selection assumption leads to a Gaussian distribution of effects at equilibrium (Equation 28.22d).

Ignoring LD, the additive variance is just twice (for the two alleles in a diploid) the sum of the locus-specific allelic variances

$$\widetilde{\sigma}^2_A = 2\sum_{i=1}^{n} \sigma^2(a_i) = 2\sqrt{V_s}\sum_{i=1}^{n}\sqrt{\mu_i\sigma^2_{\alpha_i}} \tag{28.22f}$$

With loci of equal effects, $\sigma^2_m = 2n\mu\sigma^2_\alpha$, implying $\mu\sigma^2_\alpha = \sigma^2_m/(2n)$. Substituting into Equation 28.22f recovers Equation 28.14a, as

$$\widetilde{\sigma}^2_A = 2\sqrt{V_s n}\sqrt{\sigma^2_m/(2n)} = \sqrt{2nV_s\sigma^2_m} \tag{28.22g}$$

Example 28.5. We now turn to the to the expected equilibrium additive variance under the house-of-cards model. As in Example 28.5, we use x (instead of a) to denote a random allelic effect to avoid confusion with α, the effect of a new mutation, and $p(x)$ and $f(\alpha)$, respectively, denote the probability density functions for allelic effects (x) and mutational increments (α). Under the house-of-cards mutational model, the allelic effect of a new mutation is independent of its current value, x (unlike the incremental model), and is drawn from a common distribution, so that the new allelic effect following a mutation is $x' = \alpha$. Under this model, the mutational input term in Equation 28.21c, $\mu \int p(x-\alpha) f(\alpha) d\alpha$, is replaced by $\mu f(x)$, yielding a much simpler equation for the equilibrium value of $p(x)$,

$$\frac{-p(x)[x^2 - \sigma^2(x)]}{2V_s} - \mu p(x) + \mu f(x) = 0 \tag{28.23a}$$

which has an immediate solution of

$$p(x) = \frac{2V_s \mu f(x)}{x^2 - \sigma^2(x) + 2V_s\mu} \tag{28.23b}$$

As noted by Bulmer (1989), the $\sigma^2(x)$ term—the variance in allelic effects, $\sigma^2(a)$—is a constant that can be found by noting that $\int p(x)dx = 1$, as $p(x)$ is a probability density function. Hence, for a given choice of $f(x)$, one integrates Equation 27.23b, and then solves for the value of $\sigma^2(x)$ that returns an integral of one.

If $x^2 \gg \sigma^2(x) + 2V_s\mu$, then

$$p(x) \simeq \frac{2V_s \mu f(x)}{x^2} \tag{28.23c}$$

Under this approximation, the expected value of x^k is

$$E[x^k] = \int x^k p(x) dx = \int x^k \frac{2V_s\mu f(x)}{x^2} dx = 2V_s \mu E[\alpha^{k-2}] \tag{28.23d}$$

namely, a function of the expected $k-2$ power of the mutational effects (α). Hence, setting k equal to 2 and 4, respectively, yields the equilibrium variance and kurtosis of allelic effects for locus i as

$$\sigma^2(a_i) \simeq 2V_s\mu_i E[\alpha_i^0] = 2V_s\mu_i \cdot 1 \quad \text{and} \quad E[x_i^4] \simeq 2V_s\mu_i E[\alpha_i^2] = 2V_s\mu_i\sigma^2_{\alpha_i}$$

Upon recalling that $\sigma^2_{A(i)} = 2\sigma^2(a_i)$, this first expression recovers Equation 28.18a, while the second expression yields Equation 28.18c.

Example 28.6. An intermediate model between Kimura's Gaussian and Turelli's HCA approximations was offered by Slatkin (1987), and our derivation here is based on his work, as well as that of Bulmer (1989). As with many of the above analyses, we start with a single-locus haploid model, which is extended to a diploid multilocus result by assuming additivity and no significant linkage effects. While our initial focus is on a single locus, the model now has multiple alleles, which we index by the subscript j. Further, our focus shifts from the distribution of allelic effects, $p(x)$ (from the previous example), to the frequency, p_j, of allele j. Again, the trait is scaled so that the optimum $\theta = 0$, and we assume that the current phenotypic mean resides at the optimum. Slaktin assumed a stepwise (as opposed to a continuum) series of alleles, where A_j mutates to either A_{j-1} or A_{j+1}, each with rate $\mu/2$ (independent of allelic state, j). Further, let us assume that allele A_j has effect $a \cdot j$. In that case Slaktin showed that the expected allele-frequency change from selection becomes

$$\frac{\partial p_j}{\partial t}(\text{sel}) = -\frac{p_j[\,a^2 j^2 - \sigma^2(x)\,]}{2V_s} \tag{28.24a}$$

where $\sigma^2(x)$ is the variance of allelic effects (which changes through time as the value of p_j change). The change from mutation is

$$\frac{\partial p_j}{\partial t}(\text{mut}) = -\mu p_j + \frac{\mu}{2}\left(p_{j-1} + p_{j+1}\right) \tag{28.24b}$$

Hence, at equilibrium,

$$-\frac{p_j[\,a^2j^2 - \sigma^2(x)\,]}{2V_s} + \frac{\mu}{2}\left(p_{j-1} - 2p_j + p_{j+1}\right) = 0 \tag{28.24c}$$

The mutation term in Equation 28.24c is a second-degree difference equation, which, in the limit, approaches a second derivative, as

$$\lim_{\delta\to 0}\left(\frac{f(x-\delta) - 2f(x) + f(x+\delta)}{\delta}\right)$$
$$= \lim_{\delta\to 0}\left(\frac{[f(x-\delta) - f(x)] - [f(x) - f(x+\delta)]}{\delta}\right) \to \frac{d^2f(x)}{dx^2}$$

Thus, if many alleles are segregating, the differences between the frequencies of adjacent allele are small, we can approximate the rightmost term in Equation 28.24c by the second derivation of p_j with respect to t, and Equation 28.24c becomes Kimura's Gaussian approximation (28.22c). Conversely, if selection is strong relative to mutation, there are typically only three alleles (one that is favored and the two single-step mutations), where p_0 is large and $p_{-1} = p_1$ are small. This is Turelli's triallelic model (Equation 28.13a), yielding $p_{-1} = p_1 = V_s\mu/a^2$ (Equation 28.13b), for a variance of $\sigma^2(x) = 2V_s\mu$, thus recovering the HCA results.

Example 28.7. A potential deficiency in Kimura's (1965a) mutation-selection balance model is that it is a one-locus haploid analysis extrapolated to n diploid loci by assuming no linkage effects. Lande (1975) attempted to remedy this by considering a model for a single trait with n underlying, potentially linked, loci under Gaussian stabilizing selection. (This paper is often cited as Lande 1976, as although his paper appeared in late 1975, the listed journal publication date was 1976.) In order to fully account for linkage effects, Lande followed the change over time in the covariances between the allelic effects at loci i and j in the maternal gamete and between those at i' and j' in the paternal gamete. We use the notation from Equation 16.1b, with $C_{ij} = \sigma\left(x^{(i)}, x^{(j)}\right)$ denoting the covariance between the effects of alleles at loci i and j. Random mating ensures that in each generation zygotes start with zero covariances between alleles residing on different gametes (i.e., $C_{ij'} = C_{i'j} = 0$). However, because of linkage disequilibrium, the corresponding covariances C_{ij} and $C_{i'j'}$ for loci on the *same* gamete are nonzero. Further, we expect selection to generate covariances between loci from different gametes, so

$$B_{ij}(t) = C_{ij'}(t)_s = C_{i'j}(t)_s \neq 0 \tag{28.25a}$$

where C_s denotes a covariance following selection. Assuming the incremental model, the change from mutation is

$$\Delta_m C_{ij} = \delta_{ij}\mu_i\sigma^2_{\alpha_i}, \quad \text{where} \quad \delta_{ij} = \begin{cases} 1 & i = j \\ 0 & i \neq j \end{cases} \tag{28.25b}$$

meaning that mutation changes the variances but not the covariances. Finally, let r_{ij} denote the recombination fraction between loci. Combining the joint actions of selection, recombination, and mutation (operating in that order) yields

$$C_{ij}(t+1) = (1 - r_{ij})C_{ij}(t)_s + r_{ij}B_{ij}(t) + \delta_{ij}\mu_i\sigma^2_{\alpha i} \tag{28.25c}$$

The last term accounts for mutation, while the first two account for recombination as follows. With probability $1 - r_{ij}$, loci i and j do not recombine, passing their covariance after selection, $C_{ij}(t)_s$, to their gametes, while with probability r_{ij} recombination does occur, with the covariance between i and j in a gamete equaling the between-gamete covariance following selection, $B_{ij}(t)$. (A notational aside is that, as in Chapter 24, we depart from using c_{ij} and instead use r_{ij} for recombination rates to avoid confusion with the C_{ij} terms.)

Because the $C_{ij}(t)$ determine the additive variance at generation t, let

$$2C_i(t) = 2\sum_{j=1}^{n} C_{ij}(t) \tag{28.25d}$$

denote the genetic variation that is due to locus i (the factor of two arises because C_{ij} is a covariance of single allelic effects, with both alleles contributing to the genetic variance; see Equation 16.1a). Recalling Equation 16.1a, the total additive variance at time t is

$$\sigma_A^2(t) = 2\sum_{i=1}^{n}\sum_{j=1}^{n} C_{ij}(t) = 2\sum_{i=1}^{n} C_i(t) \tag{28.25e}$$

In order to proceed, we need to compute the covariance, C_s, among alleles on the same gamete, and the covariance, B, among alleles on different gametes, after selection. Following Lande, we do so by considering the $n \times n$ matrices $\mathbf{C}_s(t), \mathbf{B}(t)$, and $\mathbf{C}(t)$ for the $C_{ij}(t)_s, B_{ij}(t)$ and $C_{ij}(t)$ elements. Lande's key assumption is that the joint distribution of allelic affects for the n loci in a gamete is multivariate normal (MVN) before selection. Under Gaussian stabilizing selection, it remains normal after selection. However, Equation 28.25c shows that the distribution of allelic effects following recombination is the weighted sum of two normals (with differing variances), which *is not* normal (Felsenstein 1977; Fleming 1979; Nagylaki 1984; Turelli 1984; Bürger 1986). Hence, the assumption of multivariate normality is an approximation, a point that Lande himself stressed. The same issue holds with mutation, where even if the mutational increments are Gaussian, Equation 28.25c again becomes a weighted sum of Gaussians, and hence is not strictly normal.

Under a MVN, the joint distribution of the vectors $\mathbf{x}$ and $\mathbf{x}'$ of maternal and paternal allelic effects in a newly formed zygote are also MVN, with covariance matrix

$$\mathbf{V}(t) = \begin{pmatrix} \mathbf{C}(t) & \mathbf{0} \\ \mathbf{0} & \mathbf{C}(t) \end{pmatrix} \tag{28.26a}$$

The matrix $\mathbf{0}$ of zeros on the off-diagonals corresponds to independent union of gametes (random mating), with the nonzero diagonal matrices, $\mathbf{C}(t)$, corresponding to the variances and LD structure (covariances) within each gamete. After Gaussian stabilizing selection, this covariance matrix becomes

$$\mathbf{K}(t) = \begin{pmatrix} \mathbf{C}_s(t) & \mathbf{B}(t) \\ \mathbf{B}(t) & \mathbf{C}_s(t) \end{pmatrix} \tag{28.26b}$$

and the task is to compute the elements of $\mathbf{K}(t)$. Under Gaussian stabilizing selection (with an optimal value of zero), the fitness of individuals with genotypic value $g = \sum(x_i + x_i')$ is

$$W(g) = \exp\left(-\left[\sum_{i=1}^{n}(x_i + x_i')\right]^2 \Big/ (2V_s) \right) = \exp\left(-\frac{\mathbf{x}\mathbf{1}\mathbf{x}^T + 2\mathbf{x}'\mathbf{1}\mathbf{x}^T + \mathbf{x}'\mathbf{1}(\mathbf{x}')^T}{2V_s} \right) \tag{28.26c}$$

where $\mathbf{1}$ is an $n \times n$ matrix of ones (i.e., $\mathbf{1}_{ij} = 1$).

The distribution of allelic effects after selection is proportional to the product of the maternal and paternal allelic-effect density functions (the independence of these follows because we assumed random mating) times their resulting fitness, yielding

$$p(\mathbf{x})p(\mathbf{x}')W\left[\sum(x_i + x_i')\right] \tag{28.26d}$$

Because all three terms contain exponentials of quadratic products, the resulting exponential term is the sum of the quadratic products. The quadratic product in W is given by Equation 28.26c, while the two quadratic products associated with the MVN density functions are of the form $[\mathbf{x}-\boldsymbol{\mu}]\mathbf{C}^{-1}[\mathbf{x}-\boldsymbol{\mu}]^T$ (LW Equation 8.24). Because the remaining terms are constants with respect to $\mathbf{x}$ and $\mathbf{x}'$, the product given by Equation 28.26d is proportional to $\exp(-F/2)$, where F equals the sum of the quadratic products

$$[\mathbf{x}-\boldsymbol{\mu}(t)]\mathbf{C}^{-1}(t)[\mathbf{x}-\boldsymbol{\mu}(t)]^T + [\mathbf{x}'-\boldsymbol{\mu}(t)]\mathbf{C}^{-1}(t)[\mathbf{x}'-\boldsymbol{\mu}(t)]^T + \frac{\mathbf{x}\mathbf{1}\mathbf{x}^T + 2\mathbf{x}'\mathbf{1}\mathbf{x}^T + \mathbf{x}'\mathbf{1}(\mathbf{x}')^T}{V_s} \tag{28.26e}$$

Because the resulting distribution of allelic effects after selection is also MVN with covariance matrix $\mathbf{K}$, it has an associated quadratic product of

$$\begin{pmatrix} \mathbf{x}-\boldsymbol{\mu}(t) \\ \mathbf{x}'-\boldsymbol{\mu}(t) \end{pmatrix} \mathbf{K}^{-1} \begin{pmatrix} \mathbf{x}-\boldsymbol{\mu}(t) \\ \mathbf{x}'-\boldsymbol{\mu}(t) \end{pmatrix}^T \tag{28.26f}$$

Our task is to find the value of $\mathbf{K}$ such that Equation 28.26f recovers Equation 27.26e. Matching terms yields

$$\mathbf{K}^{-1}(t) = \begin{pmatrix} \mathbf{C}^{-1}(t) + \mathbf{1}/V_s & \mathbf{1}/V_s \\ \mathbf{1}/V_s & \mathbf{C}^{-1}(t) + \mathbf{1}/V_s \end{pmatrix} \tag{28.26g}$$

Lande noted that because $\mathbf{K}\mathbf{K}^{-1} = \mathbf{I}$, Equations 28.26b and 28.26e imply

$$\begin{pmatrix} \mathbf{C}_s(t) & \mathbf{B}(t) \\ \mathbf{B}(t) & \mathbf{C}_s(t) \end{pmatrix} \begin{pmatrix} \mathbf{C}^{-1}(t) + \mathbf{1}/V_s & \mathbf{1}/V_s \\ \mathbf{1}/V_s & \mathbf{C}^{-1}(t) + \mathbf{1}/V_s \end{pmatrix} = \begin{pmatrix} \mathbf{I} & \mathbf{0} \\ \mathbf{0} & \mathbf{I} \end{pmatrix} \tag{28.26h}$$

Solving this system of equations gives

$$\mathbf{C}_s(t) = \frac{1}{2}\Big(\mathbf{C}^{-1}(t) + 2\cdot\mathbf{1}/V_s\Big)^{-1} + \frac{1}{2}\mathbf{C}(t) \tag{28.27a}$$

$$\mathbf{B}(t) = \mathbf{C}_s(t) - \mathbf{C}(t) \tag{28.27b}$$

Taking the inverse in 28.27a yields

$$\mathbf{C}_s(t) = \mathbf{C}(t) - \frac{\mathbf{C}(t)\,\mathbf{1}\,\mathbf{C}(t)}{V_s + \sigma_A^2(t)} \tag{28.27c}$$

The ijth term in the matrix product $\mathbf{C}(t)\,\mathbf{1}\,\mathbf{C}(t)$ is $C_i(t)\cdot C_j(t)$, and Equation 28.27c becomes

$$C_{ij}(t)_s = C_{ij}(t) - \frac{C_i(t)\,C_j(t)}{V_s + \sigma_A^2(t)} \tag{28.27d}$$

Recalling Equation 28.27b, this implies

$$B_{ij}(t) = C_{ij}(t)_s - C_{ij}(t) = -\frac{C_i(t)\,C_j(t)}{V_s + \sigma_A^2(t)} \tag{28.27e}$$

Substituting Equations 28.27d and 28.27e into Equation 28.25c yields the following set of recurrence equations

$$\Delta C_{ij}(t+1) = -\frac{C_i(t)\,C_j(t)}{V_s + \sigma_A^2(t)} - r_{ij}C_{ij}(t) + \delta_{ij}\mu_i\sigma_{\alpha_i}^2 \tag{28.28a}$$

where δ_{ij} is given by Equation 28.25b. At equilibrium, $\Delta C_{ij}(t) = 0$ or

$$\frac{\widetilde{C}_i\,\widetilde{C}_j}{V_s + \widetilde{\sigma}_A^2} + r_{ij}\widetilde{C}_{ij} = \delta_{ij}\mu_i\sigma_{\alpha_i}^2 \tag{28.28b}$$

For $i = j$, $r_{ij} = 0$ and Equation 28.28b reduces to

$$\widetilde{C}_i^2 = \mu_i\sigma_{\alpha_i}^2(V_s + \widetilde{\sigma}_A^2), \quad \text{hence} \quad \widetilde{C}_i = \sqrt{\mu_i\sigma_{\alpha_i}^2(V_s + \widetilde{\sigma}_A^2)} \tag{28.28c}$$

Likewise, the off-diagonal elements can be shown to have the solution

$$\widetilde{C}_{ij} = -\frac{\sqrt{\mu_i \sigma^2_{\alpha_i} \mu_j \sigma^2_{m_j}}}{r_{ij}} \quad \text{for} \quad i \neq j \tag{28.28d}$$

showing the presence of negative LD at equilibrium, as expected from Chapter 16. Note, however, that the $\widetilde{C}_{ij}$ values are independent of the strength of selection, V_s. Recalling Equation 28.25d, $\widetilde{C}_{ii} = \widetilde{C}_i - \sum_{j \neq i} \widetilde{C}_{ij}$, which yeilds

$$\widetilde{C}_{ii} = \sqrt{\mu_i \sigma^2_{\alpha_i} (V_s + \widetilde{\sigma}^2_A)} + \sqrt{\mu_i \sigma^2_{\alpha_i}} \sum_{j \neq i}^{n} \frac{\sqrt{\mu_j \sigma^2_{m_j}}}{r_{ij}} \tag{28.28e}$$

Finally, because the equilibrium additive variance can be expressed in terms of the $\widetilde{C}_i$, as $\widetilde{\sigma}^2_A = 2 \sum \widetilde{C}_i$, a little algebra yields

$$\widetilde{\sigma}^2_A = 2Z\sqrt{V_s + Z^2} + 2Z^2, \quad \text{where} \quad Z = \sum_{i=1}^{n} \sqrt{\mu_i \sigma^2_{\alpha_i}} \tag{28.28f}$$

For n equivalent loci, $\sigma^2_m = 2n\mu\sigma^2_\alpha$, reducing Z to $n\sqrt{\mu\sigma^2_\alpha}$, so $2Z^2 = 2n^2\mu\sigma^2_\alpha = n\sigma^2_m$, and we recover Equation 28.14c.

Example 28.8. Additional insight into the HCA versus Gaussian approximations was provided by the work of Waxman (2003) and Hermisson and Wagner (2004), who provided a solution (the former) and an approximation (the latter) ranging from the HCA result for low mutation rates to the Gaussian approximation for high mutation rates. As in Examples 28.4–28.6 we (mostly) denote allelic effects by x (instead of a) to avoid confusion with the mutational increment (α).

We again remind the reader that two different distributions appear in the fundamental equation of the continuum-of-alleles model (Equation 28.21c): the equilibrium distribution of allelic effects, $p(x)$, that we are trying to obtain, and the distribution of the assumed effects of new mutations, $f(\alpha)$, where the solution to $p(x)$ depends on the choice of $f(\alpha)$. Waxman (2003) made the clever observation that a closed-form solution of $p(x)$ can be obtained with a judicious choice of $f(\alpha)$. In particular, he assumed that the distribution of mutational effects is given by

$$f(\alpha) = \frac{\alpha}{\sigma^2_\alpha \sinh\left(\dfrac{\pi\alpha}{\sigma_\alpha\sqrt{2}}\right)} \tag{28.29a}$$

where $\sinh(y) = (1 - e^{-2y})/(2e^{-y})$ denotes the hyperbolic sine function, and σ^2_α is the mutational-effects variance. As Figure 28.4A shows, this distribution (solid curve) is a symmetric, unimodal function around its mean of zero and very close in appearance to a Gaussian (dashed curve). More important, when substituted into Equation 28.21c, it provides an *exact* solution to the equilibrium distribution of allelic effects, $p(x)$, with

$$p(x) = \frac{2^{\beta-3/2}}{\pi\sigma_\alpha} \cdot \frac{\left|\Gamma\left(\dfrac{\beta}{2} + i\dfrac{x}{\sigma_\alpha\sqrt{2}}\right)\right|^2}{\Gamma(\beta)} \tag{28.29b}$$

where Γ denotes the gamma function (Equation A2.26a) extended into the complex plane (with $i = \sqrt{-1}$). The parameter $\beta = 2\widetilde{\sigma}^2(a)/\sigma^2_\alpha$ is the ratio of the equilibrium variance in allelic effects, $\widetilde{\sigma}^2(a)$, to the variance of mutational increments, σ^2_α. Waxman found that

$$\widetilde{\sigma}^2(a) = \frac{\sigma^2_\alpha}{4}\left(\sqrt{1 + \frac{16\mu V_s}{\sigma^2_\alpha}} - 1\right) \tag{28.29c}$$

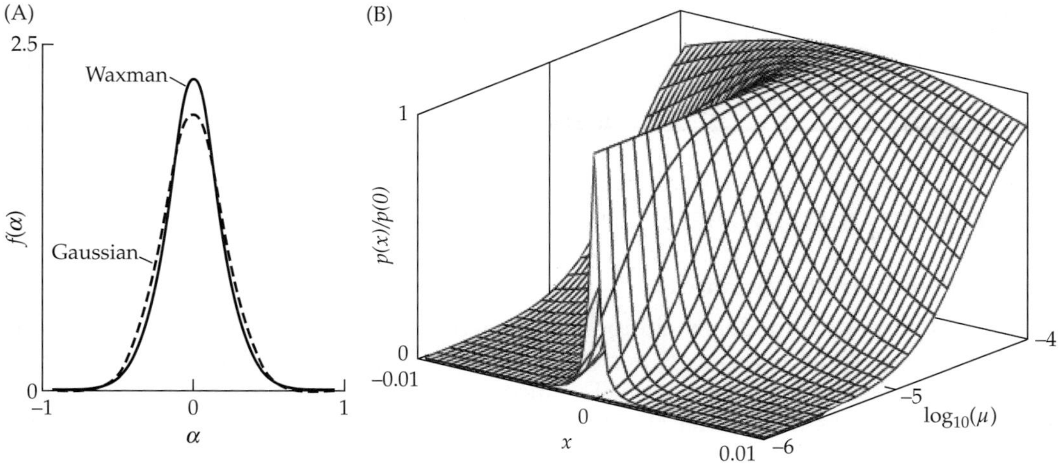

Figure 28.4 **A:** Comparison of Waxman's assumed distribution of the *mutational* effects, α (Equation 28.29a), and a Gaussian distribution. **B:** Use of the Waxman distribution for mutational effects leads to an exact solution of Equation 28.21c, showing the impact of varying mutation rates on the equilibrium distribution, $p(x)$, of *allelic* effects, x. The solution distribution is plotted as $p(x)/p(0)$, namely, the value of $p(x)$ scaled by its value at 0. In the front slice, the mutation rate is low and we recover the HCA result. As the mutation rate increases (moving toward the rear slices), the distribution becomes much more spread out, recovering the Gaussian approximation for the distribution of allelic effects. (After Waxman 2003.)

is the selection-mutation equilibrium variance in allele effects at a haploid locus (half the additive variance from that locus in a diploid, assuming additive effects). Equation 28.29a defines the **Waxman model** of mutational effects, with Equation 28.29b denoted as the **Waxman distribution** distribution of allelic effects at equilibrium, plotted in Figure 28.4B as a function of the mutation rate, μ.

For low mutation rates, Equation 28.29c recovers the HCA variance, while the Gaussian variance is recovered when mutation rates are high. To see this, let $y = 16\mu\, V_s/\sigma_\alpha^2$. For $y \ll 1$ (low mutation rates), $\sqrt{1+y}-1 \simeq (1+y/2)-1 = y/2$, yielding $\widetilde{\sigma}^{\,2}(a) = \sigma_\alpha^2\, y/8 = 2\mu\, V_s$, recovering the (haploid) HCA variance (Equation 28.12a). For sufficiently high mutation rates ($y \gg 1$), $\sqrt{1+y}-1 \simeq \sqrt{y}$, giving $\widetilde{\sigma}^{\,2}(a) = \sigma_\alpha^2\,\sqrt{y}/4 = \sqrt{u\,\sigma_\alpha^2 V_s}$, the single-locus haploid version of Kimura's Gaussian result (Equation 28.14a). Figure 28.4B shows more fully the impact of tuning the mutation rate in the Waxman solution. Under weak mutation (front slices of the graph), a sharply peaked distribution of allelic effects around zero occurs, while for stronger mutation (back slices of the graph) a much broader distribution occurs.

A complementary extension of Waxman's result is provided by the **House-of-Gauss** (**HG**) **model** of Hermisson and Wagner (2004). They assumed weak stabilizing selection around an optimum of $\theta = 0$, so that we can write the fitness as $W(z) = 1 - z^2/(2V_s)$. Under these assumptions, the equilibrium additive-genetic variance can be expressed in terms of cumulants (Chapter 24) of the distribution of allelic effects, with

$$\frac{2\widetilde{\sigma}^{\,4}(a) + \widetilde{K}_4}{2V_s} = \mu\sigma_\alpha^2 \tag{28.29d}$$

at equilibrium, representing the balance between the removal of variation by selection (left-hand side) and its introduction by new mutations (right-hand side). The Gaussian assumption is that the equilibrium fourth-order cumulant $\widetilde{K}_4 = \widetilde{\mu}_4(a) - 3\widetilde{\sigma}^{\,4}(a)$ (Equation 24.20) is small relative to $\widetilde{\sigma}^{\,4}(a)$, in which case solving Equation 28.29d recovers Kimura's haploid variance (Equation 28.14a). Under the HCA, Equation 28.18d shows that $\widetilde{K}_4$ is replaced by $\sigma_\alpha^2\widetilde{\sigma}^{\,2}(a)$ and the $\widetilde{\sigma}^{\,4}(a)$ term is ignored (the rare-alleles approximation discussed in Example 24.11), in which case Equation 28.29d recovers the haploid HCA variance (Equation 28.12a). The **House-of-Gauss approximation** (**HGA**) retains $\widetilde{\sigma}^4(a)$ in Equation 28.29d (the Gaussian approximation) while replacing $\widetilde{K}_4$ by its HCA, $\sigma_\alpha^2\sigma^2(a)$, which changes Equation 28.29d to

$$2\widetilde{\sigma}^{\,4}(a) + \sigma_\alpha^2\widetilde{\sigma}^{\,2}(a) = 2V_s\mu\sigma_\alpha^2 \tag{28.29e}$$

where $\widetilde{\sigma}^2(a)$ denotes the equilibrium variance of allelic effects and σ^2_α denotes the variance of mutational effects. It is remarkable that the solution using this approximation recovers Equation 28.29c, the equilibrium variance for the Waxman distribution.

MUTATION-STABILIZING SELECTION BALANCE: DRIFT

Impact on Equilibrium Variances

Because mutation-drift models yield too large a genetic variance, and mutation-selection models yield too small a variance, perhaps a mutation-selection model with drift might be just right. Alas, this is not the case. The incorporation of drift into mutation-selection balance models starts with Latter (1970) and Bulmer (1972). As might be expected, if the strength of selection is sufficiently weak (V_s is sufficiently large), the equilibrium variance approaches the pure-drift result (Equation 11.20c), while if the effects of drift are small (N_e sufficiently large), it approaches its deterministic value (e.g., Equation 28.18a under HCA).

Ignoring linkage disequilibrium (as above, by simply summing the single-locus results) and assuming Gaussian selection and the incremental mutational model with $\alpha \sim N(0, \sigma^2_\alpha)$, Bürger et al. (1989) obtained their **stochastic house-of-cards** (**SHC**) approximation

$$\widetilde{\sigma}^2_A(SHC) \simeq \frac{4n\mu V_s}{1 + V_s/(N_e\sigma^2_\alpha)} \tag{28.30a}$$

Bürger (1988a), Keightley and Hill (1988), Barton (1989), and Houle (1989) all obtained similar expressions using different approaches. As with the deterministic HCA model, linkage has little effect on this result, leading to only a slight overestimate (Bürger 1988a; Bürger et al. 1989; Bürger and Lande 1994). Equation 28.30a interpolates between the pure-selection HCA result (Equation 28.18a), denoted $\widetilde{\sigma}^2_A(HC)$, and the pure-drift (neutral) result (Equation 11.20), denoted $\widetilde{\sigma}^2_A(N)$. Following Bürger et al. (1989), a little algebra shows that

$$\widetilde{\sigma}^2_A(SHC) = \frac{\widetilde{\sigma}^2_A(HC) \cdot \widetilde{\sigma}^2_A(N)}{\widetilde{\sigma}^2_A(HC) + \widetilde{\sigma}^2_A(N)} \tag{28.30b}$$

which is simply half the harmonic mean of the pure selection and pure drift results. Analysis of Equation 28.30a gives the selection- and drift-dominated domains as

$$\widetilde{\sigma}^2_A(SHC) \simeq \begin{cases} \widetilde{\sigma}^2_A(N) & \text{when} \quad N_e\sigma^2_\alpha \ll V_s \\ \\ \widetilde{\sigma}^2_A(HC) & \text{when} \quad N_e\sigma^2_\alpha \gg V_s \end{cases} \tag{28.30c}$$

An alternative way to recover these domains is to recall that selection overpowers drift at a single locus when $|\,4N_e s\,| \gg 1$, while drift dominates when $|\,4N_e s\,| \ll 1$ (Chapter 7). Using Equation 28.4c, the expected selection coefficient for a new mutation (effect α_i under HOC) is

$$E(s_i) = \frac{E(\alpha_i^2)}{2V_s} = \frac{\sigma^2_{\alpha_i}}{2V_s} \tag{28.30d}$$

Hence, $|\,4N_e s\,| \gg 1$ implies $2N_e\sigma^2_{\alpha_i} \gg V_s$, while $|\,4N_e s\,| \ll 1$ implies $2N_e\sigma^2_{\alpha_i} \ll V_s$, thus recovering the selection- and drift-dominated domains given in Equation 28.30c.

An important caveat is that finite-population expressions for $\widetilde{\sigma}^2_A$ are *expected values*. Simulations show considerable spread around this expected value (Keightley and Hill 1988; Bürger 1989; Bürger et al. 1989; Bürger and Lande 1994). Barton (1989) found that the variation in the realizations is approximately

$$\sigma^2\left[\,\widetilde{\sigma}^2_A(SHC)\,\right] \simeq \left(\frac{\sigma^2_\alpha}{1 + \sigma^2_\alpha N_e/V_s}\right)\widetilde{\sigma}^2_A(SHC) \tag{28.31}$$

This reduces (to leading order) to Equation 11.25 as $V_s \to \infty$ (i.e., as the strength of selection approaches zero).

The Gaussian counterpart to the stochastic HCA can be obtained using the same logic leading to Equations 28.14d and 28.14e. Again, we start with Equation 24.2a, which assumes a Gaussian distribution of allelic effects. Adding a term σ_m^2 for new mutation and ignoring disequilibrium ($d = 0, \sigma_a^2 = \sigma_A^2$), then at equilibrium

$$\sigma_m^2 = \frac{\widetilde{\sigma}_A^2}{2N_e} + \left(1 - \frac{1}{N_e}\right)\frac{\widetilde{\sigma}_A^4}{2nV_s} \simeq \frac{\widetilde{\sigma}_A^2}{2N_e} + \frac{\widetilde{\sigma}_A^4}{2nV_s} \tag{28.32a}$$

where again we used the result that $\kappa h^2 = \sigma_A^2/V_s$ (note that our use of κ here and in Equations 28.14d and 28.14e follows from its use in Equation 16.10a to measure the reduction in variance following selection, and is distinct from κ_4, which is used above to denote a scaled measure of kurtosis; Equation 28.18c). This yields the quadratic equation

$$\widetilde{\sigma}_A^4 + \left(\frac{nV_s}{N_e}\right)\widetilde{\sigma}_A^2 - 2n\sigma_m^2 V_s = 0 \tag{28.32b}$$

whose solution is the **stochastic Gaussian** result

$$\widetilde{\sigma}_A^2 \simeq \sqrt{\left(\frac{nV_s}{2N_e}\right)^2 + 2n\sigma_m^2 V_s} - \left(\frac{nV_s}{2N_e}\right) \tag{28.32c}$$

Latter (1970), Keightley and Hill (1988), Houle (1989), Lynch and Lande (1993), and Santago (1998) all independently obtained slightly different versions of this expression. For sufficiently weak drift (large N_e), Equation 28.32c approaches Kimura's Gaussian result (Equation 28.14a). For sufficiently weak selection (large V_s), the $\widetilde{\sigma}_A^4$ term in Equation 28.32b can be ignored, which recovers the pure drift result (Equation 11.20c). Bürger (2000) found that the stochastic version of Fleming's Gaussian approximation (Equation 28.15) is also of the form of Equation 28.32c, with the $2n\sigma_m^2 V_s$ term (the square of Kimura's result; Equation 28.14a) replaced by the square of Fleming's result (Equation 28.15).

Using the stochastic version of the House-of-Gauss (HG) approximation (Example 28.8) of Hermisson and Wagner (2004) provides a unified treatment of the above results. Recall that the HG approximation recovers the HCA when the mutation rate is sufficiently low and the Gaussian when the mutation rate is sufficiently high. Adding a drift term, $-\widetilde{\sigma}^2(a)/N_e$, to the right-hand side of the deterministic version (Equation 28.29d) of the HG approximation yields

$$\frac{2\widetilde{\sigma}^4(a) + \sigma_\alpha^2 \widetilde{\sigma}^2(a)}{2V_s} = \mu\sigma_\alpha^2 - \frac{\widetilde{\sigma}^2(a)}{N_e} \tag{28.33a}$$

where $\widetilde{\sigma}^2(a)$ is the haploid, single-locus variance. Under linkage equilibrium, for n equivalent loci, $\widetilde{\sigma}^2(A) = 2n\widetilde{\sigma}^2(a)$. Using this result and solving Equation 28.33a, results in Hermisson and Wagner's **stochastic House-of-Gauss** expression

$$\widetilde{\sigma}^2(a) = 2n\sigma_\alpha^2 \frac{\gamma N_e + 1}{4\gamma N_e}\left(\sqrt{1 + 2\frac{\gamma N_e \cdot 4\mu N_e}{(\gamma N_e + 1)^2}} - 1\right), \quad \text{where} \quad \gamma = \frac{\sigma_\alpha^2}{2V_s} \tag{28.33b}$$

Finally, a more subtle effect of drift is that it can impact the domain of applicability of the Gaussian approximation. Houle (1989) noted that higher mutation rates than those necessary for the deterministic Gaussian approximation are required to compensate for the loss of alleles from drift, further restricting its domain of applicability.

Near Neutrality at the Underlying Loci?

Lande (1975) noted that with n loci, selection to move the mean to the optimum uses only a single degree of freedom (the sum of the allelic effects over all loci). He argued that the remaining $n-1$ degrees of freedom leaves ample opportunity for drift at the underlying loci, and an important role for historical events, as well as considerable genetic differentiation between populations, while still preserving the same mean and variance. The possibility of extensive neutral evolution at such loci was first examined by Kimura (1981), and later by Foley (1987, 1992), Hastings (1987b), Barton (1989) and Bürger et al. (1989). As we will see, Lande's initial suggestion of extensive neutrality is only partly correct.

A point in favor of Lande's suggestion is that the loci underlying a trait under stabilizing selection experience underdominant selection (heterozygote disadvantage; see Example 5.6), and such underdominant mutations are far more likely to become fixed than an unconditionally deleterious mutation with the same (initial) selection coefficient (Kimura 1981). Equation 28.4b shows that s_i varies with allele frequency, moving from negative values for $p < 1/2$ to positive values for $p > 1/2$. From Equation 28.4c, the initial selection against a new mutation (assuming that $\overline{z} \simeq \theta$) is $s_i = a_i^2/(2V_s)$, which decreases to zero (neutrality) as p_i approaches 1/2 (Equation 28.4b). Once the frequency drifts above 1/2, the allele is now favored, and increasingly so, as p_i approaches one. Kimura (1981) found that as a result of these frequency-dependent changes in s_i, extensive neutral evolution at the underlying loci is possible when $N_e s_i \ll 2$, a larger region than for a deleterious mutation with constant selection coefficient of the same value. Foley (1987) refined Kimura's result, showing that the expected substitution rate, λ, at loci underlying a trait under stabilizing selection is

$$\lambda \simeq \frac{\mu}{\sqrt{1+\sigma_m^2 N_e/V_s}} \tag{28.34a}$$

$$\simeq \begin{cases} \mu & \text{when } \sigma_m^2 N_e \ll V_s \quad \text{(effective neutrality)} \\ \mu\sqrt{\dfrac{V_s}{\sigma_m^2 N_e}} & \text{when } \sigma_m^2 N_e \gg V_s \quad \text{(strong constraint)} \end{cases} \tag{28.34b}$$

Kimura also suggested that underdominance results in a more U-shaped allele-frequency distribution (which has a larger probability mass near both zero and one) relative to a neutral diallelic locus with the same mutation rates. Foley (1992) obtained weak-selection approximations for the number of alleles and the frequency spectrum under the infinite-alleles model. These results showed that the Lande-Kimura notion of nearly neutral behavior at the loci of a trait undergoing stabilizing selection does not hold. Rather, their behavior is more akin to loci subjected to weak purifying selection (e.g., background selection; Chapters 3 and 8), and Barton (1989) noted that it was not possible to use the allele-frequency distribution to distinguish between stabilizing and weak purifying selection.

Finally, Bürger et al. (1989) examined the heterozygosity at the underlying loci through simulation studies. Generally, there was a reasonable fit between the fully neutral expectation of $\widetilde{H}_n = \theta/(1+\theta)$, where $\theta = 4N_e\mu$, and the observed value, $\widetilde{H}_o$, except under strong selection or a high variance of mutational effects. Foley (1992) found that a slightly better fit was obtained by replacing θ by

$$\theta_s = \frac{\theta}{\sqrt{1+\sigma_m^2 N_e/(2V_s)}} \tag{28.34c}$$

Both Foley and Bürger et al. noticed that heterozygosity is not necessarily highly correlated with the additive variance. In particular, Bürger et al. noted that the relationship often used for a diallelic locus to relate the equilibrium additive variance to the observed heterozygosity, namely $\widetilde{\sigma}_A^2 = n\sigma_\alpha^2 \widetilde{H}_o$ (e.g., Bulmer 1972), generally *does not* hold under the infinite-alleles assumption.

MUTATION-STABILIZING SELECTION BALANCE: PLEIOTROPY

Finally, we conclude our discussion of models assuming direct stabilizing selection on a focal trait balanced by mutation by considering the impact when mutations have pleiotropic effects (namely, influencing additional traits, beyond the focal one, that are under selection). The presence of pleiotropy introduces considerable complications. Many of the previous models depend on difficult-to-estimate quantities (n, μ, and σ_α^2). Pleiotropy adds additional, usually hidden, players that are even more difficult to detect and whose effects are essentially impossible to estimate with any precision. This is especially problematic, as seemingly very small differences in pleiotropy models can lead to qualitatively different outcomes. Johnson and Barton (2005) stressed that the lack of understanding of both the nature of pleiotropy, and how to robustly model it, are the main impediments to a deeper understanding of the maintenance of variation. This section considers the impact of adding pleiotropic effects to direct-selection models (wherein the trait itself is under selection), while pure pleiotropy models (wherein the focal trait is neutral) are examined in the final selection.

Gaussian Results

To model multiple-trait selection with pleiotropic mutations, we follow the standard approach of working with a single-locus haploid model, whose results are then extended to a diploid multilocus model by summing over loci (i.e., assuming additivity and ignoring linkage disequilibrium). In order to proceed, several conceptual extensions are required to move from a single- to a k-trait model. While our discussion is for a particular locus, i, we will often suppress the subscript for ease of presentation.

First, the single effect, a, of an allele on the focal trait is replaced by the vector of k allelic effects, $\mathbf{a}$, whose jth element is the allelic effect for trait j. As a result, the variance of allelic effects at a given locus is replaced by a variance-covariance matrix, $\mathbf{V}_a$, of effects on all pairs of traits influenced by pleiotropic mutations (that involve the focal trait), where

$$(\mathbf{V}_a)_{j,\ell} = \sigma\,(a_j, a_\ell) \qquad 1 \leq j, \ell \leq k \tag{28.35a}$$

namely, the covariance between the effects of an allele at the focal locus on traits i and j.

Second, under the incremental model, the vector of allelic effects following mutation becomes $\mathbf{a}' = \mathbf{a} + \boldsymbol{\alpha}$, whose jth element is $a_j' = a_j + \alpha_j$. As a result, the single-trait mutational effects variance, σ_α^2, is replaced by a **pleiotropic mutation matrix**, $\mathbf{V}_m$, whose elements are given by

$$(\mathbf{V}_m)_{j,\ell} = \sigma\,(\alpha_j, \alpha_\ell) \qquad 1 \leq j, \ell \leq k \tag{28.35b}$$

where α_j is the mutational increment to trait j. A critical point is that ***extensive pleiotropy can occur without any mutational covariance between traits***, namely, with $\sigma(\alpha_j, \alpha_\ell) = 0$ for all values of j and ℓ, a condition referred to as **hidden pleiotropy**. For example, consider a locus that influences two traits, with all mutations having pleiotropic effects, but comprising a random collection of $++$, $--$, $+-$ and $-+$ effects on the two traits. The between-trait covariance for the mutational effects is zero, even though there is complete pleiotropy (all mutations impact both traits). Short of actually measuring the joint effects of individual mutations, the presence of hidden pleiotropy would be difficult, if not impossible, to detect, yet has dramatic consequences for mutation-selection balance, and for multivariate evolution in general (the latter discussed in detail in Volume 3).

Finally, modeling selection requires a multivariate extension of Equation 28.3b. If $\mathbf{z}$ is the vector of k trait values, and $\boldsymbol{\theta}$ a vector of optimum values, then

$$W(\mathbf{z}) = \exp\left(-\frac{(\mathbf{z}-\boldsymbol{\theta})\mathbf{V}_\omega^{-1}(\mathbf{z}-\boldsymbol{\theta})^T}{2} \right) \tag{28.36a}$$

where $\mathbf{V}_\omega$ is a symmetric, positive-definite matrix (a matrix with all positive eigenvalues; see Chapter 30 and Appendix 5 for further details). For weak selection, expanding the quadratic product in the exponential gives

$$W(\mathbf{z}) \simeq 1 - \frac{1}{2}\sum_{j=1}^{k}\sum_{\ell=1}^{k}(z_j-\theta_j)(z_\ell-\theta_\ell)V_{j\ell} \tag{28.36b}$$

where $V_{j\ell}$ is the $j\ell$th element of $\mathbf{V}_\omega^{-1}$. From the assumed positive-definiteness of $\mathbf{V}_\omega$, outside of $\mathbf{z} = \boldsymbol{\theta}$, this double summation is always positive (Equation A5.17a), resulting in fitness being maximized at $\mathbf{z} = \boldsymbol{\theta}$ and declining quadratically in any direction around $\boldsymbol{\theta}$ (Chapter 30). Assuming environmental effects are multivariate normal, $\mathbf{z}|\mathbf{a} \sim \text{MVN}(\mathbf{a}, \mathbf{V}_E)$, the multivariate version of Equation 28.3e gives the fitness associated with $\mathbf{a}$ as $w(\mathbf{a}) \propto \exp[-(\mathbf{a}-\boldsymbol{\theta})\mathbf{V}_s^{-1}(\mathbf{a}-\boldsymbol{\theta})^T/2]$, where

$$\mathbf{V}_s = \mathbf{V}_\omega + \mathbf{V}_E \tag{28.36c}$$

is the multivariate extension of Equation 28.3f. Below, we will use the weak-selection approximation that $\mathbf{V}_\omega + \mathbf{V}_z \simeq \mathbf{V}_s$ and $\mathbf{V}_\omega + \mathbf{V}_a \simeq \mathbf{V}_s$.

Assuming that the vector of phenotypes is multivariate normal, $\mathbf{z} \sim \text{MVN}(\mathbf{0}, \mathbf{V}_z)$, the multivariate analog for the change by selection in the (univariate) phenotypic variance (Equation 28.3g) is given by the change in the phenotypic covariance matrix, $\mathbf{V}_z$, where

$$\Delta\mathbf{V}_z = -\mathbf{V}_z(\mathbf{V}_\omega + \mathbf{V}_z)^{-1}\mathbf{V}_z \simeq -\mathbf{V}_z\mathbf{V}_s^{-1}\mathbf{V}_z \tag{28.36d}$$

Similarly, when $\mathbf{a}$ is multivariate normal, the change in the covariances for the vector of allelic effects following selection becomes

$$\left(\Delta\mathbf{V}_a\right)_s = -\mathbf{V}_a\left(\mathbf{V}_s + \mathbf{V}_a\right)^{-1}\mathbf{V}_a \simeq -\mathbf{V}_a\,\mathbf{V}_s^{-1}\,\mathbf{V}_a \tag{28.36e}$$

Following Lande (1980), the change in $\mathbf{V}_a$ from the joint action of selection and mutation is

$$\Delta\mathbf{V}_a = \left(\Delta\mathbf{V}_a\right)_s + \mu_i\mathbf{V}_m \tag{28.37a}$$

which is zero at equilibrium. Note that $\mathbf{V}_m$ is the multivariate extension of σ^2_α, and hence is multiplied by the mutation rate at the ith locus, μ_i.

Recalling Equation 28.36e, the removal of genetic variances and covariances by selection balances the input from new mutation when

$$\widetilde{\mathbf{V}}_a\,\mathbf{V}_s^{-1}\,\widetilde{\mathbf{V}}_a = \mu_i\mathbf{V}_m \tag{28.37b}$$

which has the solution

$$\widetilde{\mathbf{V}}_a \simeq \mathbf{V}_s^{1/2}\left(\mu_i\mathbf{V}_s^{-1/2}\mathbf{V}_m\mathbf{V}_s^{-1/2}\right)^{1/2}\mathbf{V}_s^{1/2} \tag{28.37c}$$

where $\mathbf{V}^{1/2}$ denotes the square root of $\mathbf{V}$ (where $\mathbf{V}^{1/2}\mathbf{V}^{1/2} = \mathbf{V}$; see Equation A5.11b). If $\mathbf{V}_s$ and $\mathbf{V}_m$ are diagonal matrices (corresponding to no correlations in the fitness function and no pleiotropic covariance between mutational effects, respectively), Equation 28.37c gives the equilibrium variance for trait j at locus i as

$$\widetilde{\sigma}^2(a_{j,i}) \simeq \sqrt{\mu_i\,\sigma^2_{\alpha_{j,i}}V_{s,j}} \tag{28.38a}$$

where $\sigma^2_{\alpha_{j,i}}$ is the variance in the effects on trait j from new mutations at locus i. This is simply Kimura's result for a single trait (Equation 28.22d), and Lande's key finding: ***under the Gaussian assumption, the equilibrium additive variance of a trait is unaffected by selection on uncorrelated (both selectively and mutationally) traits***.

As noted by Turelli (1985), the condition for the multivariate Gaussian approximation to be reasonable is that

$$\mu_i \gg \frac{\sigma^2_{\alpha_{k,i}}}{V_{s,k}} \tag{28.38b}$$

for *all* loci and *all* traits. The presence of *any* locus-trait combination that violates this condition invalidates the Gaussian approximation. A second restriction on the plausibility of the Gaussian approximation is the realistic number of functionally distinct alleles that can be maintained at a locus. Turelli (1984) found that a locus with roughly 20 alleles can fairly closely match the continuum-of-alleles model for a single trait. However, with two traits, allowing over 100 alleles still did not provide sufficient granularity to capture the bivariate continuum-of-alleles structure. This problem becomes increasingly more acute as the number of traits that a pleiotropic mutant impacts grows.

Example 28.9. To be a bit more formal on the conditions required for selection to be uncorrelated over a set of traits, consider the two-trait versions of the matrices $\mathbf{V}_\omega$, $\mathbf{V}_E$, and $\mathbf{V}_s$. When the off-diagonal element of this last matrix is zero, the two traits are selectively uncorrelated. Because $\mathbf{V}_\omega$ is a symmetric, positive definite matrix, it is also a covariance matrix (Appendix 5), and hence we can write it as

$$\mathbf{V}_\omega = \begin{pmatrix} \omega_1^2 & \rho_\omega\,\omega_1\,\omega_2 \\ \rho_\omega\,\omega_1\,\omega_2 & \omega_2^2 \end{pmatrix}$$

A nonzero value of ρ_ω implies selection favoring a covariance between z_1 and z_2 (Chapter 30). Similarly expressing the covariance matrix of environmental effects as

$$\mathbf{V}_E = \begin{pmatrix} \sigma_{E_1}^2 & \rho_e\sigma_{E_1}\sigma_{E_2} \\ \rho_e\sigma_{E_1}\sigma_{E_2} & \sigma_{E_2}^2 \end{pmatrix} \quad \text{yields} \quad \mathbf{V}_s = \mathbf{V}_\omega + \mathbf{V}_E = \begin{pmatrix} V_{s,1} & C_s \\ C_s & V_{s,2} \end{pmatrix}$$

where

$$V_{s,i} = \omega_i^2 + \sigma_{E_i}^2 \quad \text{and} \quad C_s = \rho_e\sigma_{E_1}\sigma_{E_2} + \rho_\omega\,\omega_1\,\omega_2 \tag{28.39a}$$

The two traits are selectively uncorrelated when $C_s = 0$, which requires that the phenotypic selection (ρ_ω) and environmental (ρ_e) correlations are zero, or the unlikely event that $\rho_e = -\rho_\omega\,\omega_1\,\omega_2/(\sigma_{E_1}\sigma_{E_2})$. We can also write $\mathbf{V}_s$ as

$$\mathbf{V}_s = \begin{pmatrix} V_{s,1} & \rho_s\sqrt{V_{s,1}V_{s,2}} \\ \rho_s\sqrt{V_{s,1}V_{s,2}} & V_{s,2} \end{pmatrix}$$

where

$$\rho_s = \frac{C_s}{\sqrt{V_{s,1}V_{s,2}}} = \frac{\rho_e\sigma_{E_1}\sigma_{E_2} + \rho_\omega\,\omega_1\,\omega_2}{\sqrt{V_{s,1}V_{s,2}}} \tag{28.39b}$$

HCA Results

An encouraging feature of the single-trait house-of-cards analysis was its relative robustness to the underlying genetic model. Provided that Equation 28.17 (or its diallelic counterpart; Equation 28.11e) hold, the equilibrium additive-genetic variance (Equation 28.18a) is independent of many of the underlying genetic details, such as the number of alleles per locus. Unfortunately, this robustness vanishes when pleiotropy is introduced (Turelli 1985, 1986). Even more troubling, and unlike the Gaussian result just obtained (Equation 28.38a), selection acting on pleiotropically connected but *uncorrelated* traits influences the additive variance of a focal trait under the HCA (Turelli 1985, 1986, 1988; Wagner 1989; Slaktin and Frank 1990).

Turelli (1985) examined the simplest case of the HCA for a pleiotropic continuum-of-alleles model: two traits that are mutationally and selectively uncorrelated, $(\mathbf{V}_m)_{12} = 0$ and $\rho_s = 0$ (defined by Equation 28.39b). He found that the bivariate condition for the HCA approximation to be reasonable was more lenient than in the univariate condition. For two (uncorrelated) traits, the condition for locus i becomes

$$\mu_i \ll \sqrt{\frac{\sigma_{\alpha_{1,i}}^2\,\sigma_{\alpha_{2,i}}^2}{V_{s,1}\,V_{s,2}}} \tag{28.40}$$

which (unlike the Gaussian approximation) can be satisfied even when one of the loci does not itself satisfy the univariate HCA condition (Equation 28.17). Under the bivariate HCA, the equilibrium additive variance in trait 1 becomes

$$\widetilde{\sigma}^2_{A(i)} \simeq \frac{4\mu_i V_{s,1}}{1+\beta_i}, \quad \text{where} \quad \beta_i = \sqrt{\frac{\sigma^2_{\alpha 2,i} V_{s,1}}{\sigma^2_{\alpha 1,i} V_{s,2}}} \tag{28.41a}$$

Even if trait 1 is mutationally and selectively uncorrelated to trait 2, it is still impacted by selection on the latter when $\sigma^2_{\alpha 2,i} > 0$, namely, locus i experiences pleiotropic mutations influencing trait 2. Recalling Equation 28.30d, we see that β_i^2 is the ratio of the average selection coefficients for the two traits associated with a mutation at locus i (Turelli 1985), as

$$\beta_i^2 = \frac{\sigma^2_{\alpha 2,i}/(2V_{s,2})}{\sigma^2_{\alpha 1,i}/(2V_{s,1})} = \frac{E[s_{2,i}]}{E[s_{1,i}]} \tag{28.41b}$$

Under the HCA setting (wherein selection dominates mutation), both traits are near their optimum ($\overline{\mathbf{z}} \simeq \boldsymbol{\theta}$), meaning that a change in any direction is likely deleterious. Thus, any new mutations influencing trait 1 will also change trait 2, further lowering fitness. The stronger selection is on trait 2, the greater the additional reduction in fitness for a mutation that also affects trait 1. When most of the selection on new mutations at locus i is on trait 1 ($\beta_i \ll 1$), Equation 28.41a is close to the single-trait HCA value (Equation 28.18a). Conversely, when there is much stronger selection on trait 2 ($\beta_i \gg 1$), the amount of variation maintained for trait 1 is considerably below its single-trait HCA prediction. If a large number of traits are under multivariate Gaussian selection, one can easily construct a single synthetic trait (a linear combination of the remaining traits) to reduce this to a two-trait (focal plus synthetic) model (Example 28.10). Hence, for most traits we expect selection on the "other" trait to be larger, and likely considerably so, resulting in an overprediction of trait 1's genetic variance using the univariate HCA result (Equation 28.18a).

Finally, because the HCA assumes that selection at a locus is much stronger than mutation, a consequence of this additional selection on the nonfocal trait is to make the HCA approximation more plausible. Thus, pleiotropy expands the domain of applicability of the HCA while shrinking that of the Gaussian approximation (Equation 28.38b).

In addition to generating a dependency on hidden traits, pleiotropy has another, equally insidious, feature. Under single-trait HCA conditions, the equilibrium variance does not depend on the genetic details beyond the total mutation rate (with diallelic, triallelic, and continuum-of-alleles models all giving the same results). However, Turelli (1985) found that a five-allele model (the bivariate extension of his univariate triallelic model, now following five alleles, $A_{0,0}$ and $A_{i,j}$ for $i, j = \pm 1$, where allele A_{ij} has effects of $i \cdot a_1$ and $j \cdot a_2$ on traits 1 and 2, respectively) yielded

$$\widetilde{\sigma}^2_{A(i)} \simeq \frac{4\mu_i V_{s,1}}{1+\beta_i^2} \tag{28.42}$$

which is a different result from the continuum-of-alleles model (Equation 28.41a). Thus, additional genetic details (such as the number of alleles) seem to matter under pleiotropy.

In the univariate case, the qualitative difference in the additive variance under continuum-of-alleles (Kimura-Lande-Fleming) versus diallelic (Latter-Bulmer) models was due to the relative strengths of mutation and selection, *not* the number of alleles. Wagner (1989) suggested that something similar is behind the difference between Equations 28.41a and 28.42, with the amount of pleiotropic constraints among the effects of new mutations, rather than the number of alleles, accounting for the difference in equilibrium values. Turelli's five-allele model is highly constrained due to the limited number of alleles, while this is not the case for the continuum-of-alleles result. Wagner considered a model of constraints wherein the effect on trait j from a mutation of effect α_i at an underlying generator locus i is $b_{ij}\alpha_i$ (where the b_{ij} are constants). This structure implies that all of the mutational effects from a given locus are completely correlated, with the two-trait version recovering Turelli's five-allele result. Wagner suggested that differences in the amount of constraint on the pleiotropic nature of new mutations accounts for the difference between Equations 28.41a (little constraint) and 28.42 (significant constraint). Zhang and Hill (2003) showed this to be the case, recovering the Turelli five-allele result from a two-trait, continuum-of-alleles model when the correlation between the the pleiotropic effects of new mutations is high.

Finally, as in the univariate case, the radical differences in behavior between the Gaussian and HCA approximations in the presence of pleiotropy are the result of differences in the strength of selection relative to mutation, *not* the underlying mutational model. Slatkin and Frank (1990) verified this by considering a nine-allele model, whose mutational structure is given by the following lattice of trait values. The center allele (0,0) is at the bivariate optimum, and this allele is allowed to have a nonpleiotropic mutation for either trait, e.g., $(-a, 0), (a, 0), (0, -a)$, or $(0, a)$; or to jointly mutate for both traits in four possible directions, $(-a, -a), (-a, a), (a, -a), (a, a)$. This model offers a bit more granularity than Turelli's five-allele model. Slatkin and Frank found that, depending on the relative strengths of selection and mutation, this model can generate either the Gaussian behavior (weak to no impact from selection on uncorrelated traits) or the HCA behavior (strong impact from uncorrelated selection).

Example 28.10. Under the fitness function given by Equation 28.36a, Turelli (1985) and Zhang and Hill (2003) showed that if the population is at (or very close to) its optimal value ($\boldsymbol{\theta} = \mathbf{0}$), then the initial selection coefficient against a new mutation at locus i with effects vector $\boldsymbol{\alpha}^T = (\alpha_{1,i}, \alpha_{2,i}, \cdots, \alpha_{k,i})$ is approximately given by the quadratic product

$$s_i \simeq \frac{\boldsymbol{\alpha}^T \mathbf{V}_s^{-1} \boldsymbol{\alpha}}{2} \tag{28.43a}$$

This is the multivariate equivalent of Equation 28.4c. When $\mathbf{V}_s$ is diagonal (uncorrelated selection; Example 28.9), this reduces to

$$s_i \simeq \sum_{j=1}^{k} \frac{\alpha_{j,i}^2}{2V_{s,j}}, \quad \text{implying} \quad E[s_i] \simeq \sum_{j=1}^{k} \frac{\sigma_{\alpha_{j,i}}^2}{2V_{s,j}} = \sum_{j=1}^{k} E[s_{j,i}] \tag{28.43b}$$

namely, the sum of the average selection coefficients of a new mutation associated with each of the k traits. More generally, expanding the quadratic product given by Equation 28.43a yields

$$s_i \simeq \sum_{j=1}^{k} \sum_{\ell=1}^{k} \alpha_{j,i} \cdot V_{j,\ell} \cdot \alpha_{\ell,i}$$

where $V_{j,\ell}$ is the $j\ell$th element of $\mathbf{V}_s^{-1}$. Taking expectations yields

$$E[s_i] \simeq \sum_{j=1}^{k} \sum_{\ell=1}^{k} \left[V_{j,\ell} \cdot \sigma(\alpha_{k,i}, \alpha_{\ell i}) \right] \tag{28.43c}$$

showing that the average selection coefficient for a new mutation depends on the mutational covariances, $\sigma(\alpha_{k,i}, \alpha_{\ell i})$, in addition to the pattern and strength of stabilizing selection, $\mathbf{V}_s$.

Zhang and Hill (2003) noted that if all n traits impact locus i with roughly similar selection coefficients ($E[s_{j,i}] \simeq s_i$), then from the central limit theorem, the distribution of the s_i approaches a normal. Further, the coefficient of variation of this distribution goes to zero at an approximate rate of $\sqrt{(3\kappa_{4,s} - 1)/n}$, where $\kappa_{4,s} = E[(s_i - \overline{s}_i)^4] / \left\{3[\sigma^2(s_i)]^2\right\}$ is the scaled kurotsis of s_i. Thus, for a sufficiently large number of independent traits under selection, s_i is approximately a constant plus a small amount of normally distributed error.

MAINTENANCE OF VARIATION BY PLEIOTROPIC DELETERIOUS ALLELES

We conclude our discussion of theory with **deleterious-pleiotropy models**, in which the loci underlying a trait have pleiotropic effects on fitness, independent of their trait effects. This is the extension of the previous models, wherein the pleiotropic effects were on traits that were themselves under stabilizing selection. We start this section by assuming that the focal trait is *neutral*—its phenotypic value has no direct fitness consequences—but that its underlying loci are under selection (i.e., they have pleiotropic effects on both the trait and fitness), relaxing this neutral trait assumption later. We previously considered Robertson's model, wherein the underlying loci display fitness overdominance, which had a number of problems as a general explanation for the maintenance of trait variation (Example 28.2). We now turn to models in which the underlying loci are deleterious, and variation is maintained through selection-mutation balance. The logic behind such models is twofold. First, many new mutations are expected to be deleterious. Second, while purely neutral models (wherein all underlying loci have no fitness consequences) maintain too much variation in even modestly sized populations, perhaps introducing slightly deleterious underlying alleles allows the population to achieve the observed levels of variation.

This is the idea behind the **Hill-Keightley** (1988) (**HK**) model of **pleiotropic side-effects**, wherein the amount of additive-genetic variation for the neutral trait is determined by pleiotropic effects from deleterious alleles in mutation-selection balance. One observation motivating this model is that mutations that have major effects on a trait also tend to be deleterious (Chapters 25 and 26). Coupling this with the belief (and some observations; Chapter 27) that single mutations often influence multiple traits, and that many mutations are at least slightly deleterious, suggests that at least *some* of the variation for any trait is due to such deleterious alleles. As with much of the above analysis, the issue is whether such a model, by itself, can generate *both* sufficient variation *and* sufficiently strong apparent stabilizing selection to account for the observed values of these two features.

The Hill-Keightley Pleiotropic Side-effects Model

Organisms, and their underlying genetic systems, are expected to be highly integrated, with single genes and single traits unlikely to be isolated from others. Hence, pleiotropy and correlated selection are expected to be the norm, not the exception. We have previously considered one special case of this, namely a locus influencing a number of traits that are all under stabilizing selection. Is there a more general way to model this complex situation?

Hill and Keightley (1988) and Keightley and Hill (1990) suggested that one approach is to sweep all pleiotropic selective effects into a single fitness term, s, generally expected to be deleterious (e.g., Example 28.10). Their model assumes that each new mutation has *two effects*: (i) α, on the focal trait; and (ii) s, on fitness (measured as the fitness reduction in heterozygotes, as homozygotes are expected to be rare and hence make little contribution). From Chapter 7, the (infinite-population) equilibrium frequency of such an allele arising under recurrent mutation is $\widetilde{p} = \mu/s \ll 1$. Assuming additive trait effects, the contribution to the additive-genetic variance of the trait from this locus is

$$2\alpha^2\,\widetilde{p}\,(1-\widetilde{p}\,) \simeq 2\alpha^2\,\widetilde{p} \simeq \frac{2\alpha^2\mu}{s} \tag{28.44a}$$

Because new mutations show a distribution of both α and s values, the expected contribution from this locus becomes $2\mu E[\alpha^2/s]$. Some insight into this expectation is offered by using a Taylor-series approximation for the expected value of a ratio (LW Equation A1.19a). Recalling that $E[\alpha^2] = \sigma^2_\alpha$ (as we assumed that $E[\alpha] = 0$) and denoting $E[s]$ by $\overline{s}$, LW Equation A1.19a yields

$$E\left[\frac{\alpha^2}{s}\right] \simeq \frac{\sigma^2_\alpha}{\overline{s}}\left(1 + \frac{\sigma^2(\alpha^2)}{\sigma^2(s)} - \frac{\sigma(\alpha^2, s)}{\sigma^2_\alpha \cdot \overline{s}}\right) \tag{28.44b}$$

Hence, in addition to σ^2_α and $\overline{s}$, the equilibrium additive variance also depends on at least three other quantities. The first is the kurtosis of mutational trait effects, which enters through the $\sigma^2(\alpha^2) = E[\alpha^4] - \sigma^4(\alpha)$ term in Equation 28.44b. The second is $\sigma(\alpha^2, s)$, the covariance between the selective effect, s, of a new mutation and the squared effect, α^2, of that mutation on the focal trait. The third is $\sigma^2(s)$, the variance in pleiotropic selection coefficients. All three of these quantities influence the expected equilibrium variance. It is important to stress that knowledge of these quantities may not be sufficient, however, as the approximation given by Equation 28.44b can easily break down, making $E[\alpha^2/s]$ dependent on additional moments of the bivariate mutational distribution of (α^2, s).

Barton (1990) and Kondrashov and Turelli (1992) examined a simplified version of this model with n identical loci, in which all mutations have the same deleterious effect, s (Example 28.10 provides some justification for this assumption), while α can vary. This base model sidesteps delicate issues on the bivariate (α, s) distribution, such as the nature of the covariance $\sigma(\alpha^2, s)$, and the behavior of the distribution for values of s near zero. While these **constant-*s* models** offer some important insights, as we detail below they can also be misleading.

Barton (1990) assumed multiplicative fitnesses, with an individual heterozygous at k deleterious loci having a fitness of $(1-s)^k \simeq \exp(-sk)$, while Kondrashov and Turelli (1992) allowed for much more general fitness functions (including synergistic epistasis, and hence less of a selective load; Chapter 7). Both approaches yielded essentially the same conclusions. Let $\overline{k} = 2n\mu/s = 2n\widetilde{p}$ denote the average number of deleterious alleles per diploid individual (an average of $2\widetilde{p}$ deleterious alleles per locus). Assuming no linkage disequilibrium, summing Equation 28.44a over the contributions from the n loci yields

$$\widetilde{\sigma}^2_A \simeq 2\,n\,E[\alpha^2]\,\widetilde{p} = \frac{2n\mu\sigma^2_\alpha}{s} = \overline{k}\,\sigma^2_\alpha \tag{28.45a}$$

Because $\sigma^2_m = 2n\mu\sigma^2_\alpha$, Equation 28.45a can also be expressed as the ratio of the amount of variation introduced by mutation each generation to the rate of its removal by selection

$$\widetilde{\sigma}^2_A = \frac{\sigma^2_m}{s} \tag{28.45b}$$

Recall that this is an alternative expression for the equilibrium additive variance under the Latter-Bulmer (and hence HCA) model (Equation 28.12b), showing that these very different models (pure pleiotropy versus stabilizing selection) have some similar features. This should not be surprising, as under the HCA, wild-type alleles are near the optimum and thus nearly all new mutations are deleterious (Chapter 27). The conceptual distinction between these two models is that the reason that a mutation is deleterious is specified under the HCA (it impacts a trait under stabilizing selection), but unspecified under the deleterious-pleiotropy model.

Using the standard value of $\sigma^2_m \simeq 10^{-3}\sigma^2_E$ (LW Table 12.1) with $s \simeq 0.001$, Equation 28.45b yields $\widetilde{\sigma}^2_A \simeq \sigma^2_E$, and hence $\widetilde{h}^2 \simeq 0.5$. This result looks promising in that weakly deleterious alleles can maintain levels of additive variance similar to those seen in natural populations. This is perhaps not surprising, in that a strictly neutral model maintains extensive variation in a large population, so that a model where the underlying loci are somewhat close to neutral should also accommodate significant variation.

Because there is no assumed selection on the focal trait, can this model also generate sufficiently strong apparent stabilizing selection? Individuals carrying more deleterious alleles also tend to have more extreme (positive and negative) trait values, generating a quadratic relationship between trait value and fitness, and thus a spurious signature of stabilizing selection. For example, an individual with k deleterious alleles has an approximate fitness of $1 - sk$ (under Barton's model) and a squared trait value of $z^2 = \sum_i^k \alpha_i^2$, where $E[z^2 \mid k] = k\sigma^2_\alpha$ (as $E[\alpha_i] = 0$ and $E[\alpha_i^2] = \sigma^2_\alpha$). The apparent strength of stabilizing selection, $\widehat{V}_s$, follows from the quadratic term in Equation 28.3i, γ_{w,z^2}, the regression of relative fitness on the squared deviation from the mean (set here to zero for convenience),

$$\widehat{V}_s \simeq -\frac{1}{\gamma_{w,z^2}} = \frac{\sigma^2(z^2)}{2\sigma(w, z^2)} \tag{28.46a}$$

as obtained by Barton (1990), Kondrashov and Turelli (1992), Gavrilets and de Jong (1993), Bürger (2000), and Zhang et al (2002). By evaluating the variance and covariance term in Equation 28.46a, Barton (1990) showed that the resulting apparent strength of stabilizing selection becomes

$$\widehat{V}_s = \frac{\sigma^2_\alpha(3\kappa_4 + 2\overline{k}\,)}{2s} = 3\frac{\sigma^2_\alpha \kappa_4}{2s} + 2\frac{\sigma^2_\alpha n\mu}{s^2} = 3\frac{\sigma^2_\alpha \kappa_4}{2s} + \frac{\sigma^2_m}{s^2} \tag{28.46b}$$

with the middle step following from $\overline{k} = 2n\mu/s$. Here $\kappa_4 = E[\alpha^4]/(3\sigma^4_\alpha)$, the scaled kurtosis of trait mutational effects (Equation 28.18c), is greater than $1/3$ if there is any variation in the values of trait mutations (as $E[\alpha^4] \geq \sigma^4_\alpha$). For $\overline{k} \gg \kappa_4$, Equation 28.46b implies that

$$\widehat{V}_s \simeq \frac{\sigma^2_\alpha \overline{k}}{s} = \frac{\sigma^2_m}{s^2} = \frac{\widetilde{\sigma}^{\,2}_A}{s} \tag{28.46c}$$

For arbitrary values of $\overline{k}$, combining Equations 28.45b and 28.46b yields the relationship

$$\frac{\widetilde{\sigma}^{\,2}_A}{\widehat{V}_s} = \left(\frac{2n\mu\sigma^2_\alpha}{s}\right)\left(\frac{2s}{\sigma^2_\alpha[\,3\kappa_4 + 2\overline{k}\,]}\right) = \frac{4n\mu}{3\kappa_4 + 2\overline{k}} = \frac{4n\mu}{3\kappa_4 + 4n\mu/s} \tag{28.46d}$$

which can be rearranged to

$$\widetilde{\sigma}^{\,2}_A = \frac{4\widehat{V}_s n\mu}{3\kappa_4 + 2\overline{k}} < 4\widehat{V}_s n\mu \tag{28.46e}$$

where the last step follows because $3\kappa_4 \geq 1$. Hence, for the same apparent strength of stabilizing selection, less additive variation is maintained under the constant-s pleiotropy model than with the same amount of real stabilizing selection under HCA conditions ($4V_s n\mu$; Equation 28.18a).

In contrast to the emergence of $\widetilde{\sigma}^{\,2}_A = \sigma^2_m/s$ for both the HCA and pleiotropy models (Equations 28.12b and 28.45b), the ratio of the equilibrium additive-genetic variance to the apparent strength of stabilizing selection is rather different between models, as

$$\widetilde{\sigma}^{\,2}_A/\widehat{V}_s = \begin{cases} 4n\mu & \text{HCA} \\ 4n\mu/(3\kappa_4) & \text{Deleterious pleiotropy (with } \overline{k} \ll 1) \\ s & \text{Deleterious pleiotropy (with } \overline{k} \gg 1) \end{cases} \tag{28.47a}$$

Under direct selection, V_s determines s (Equation 28.12c), while under pleiotropy, s determines $\widehat{V}_s$ (Equation 28.46b).

Can the constant-s model account for *both* the observed levels of variation *and* the strengths of stabilizing selection? It can not. With $s = 0.001$ and $\sigma^2_m = \sigma^2_E/10^3$, Equation 28.45b yields $\widetilde{\sigma}^{\,2}_A = \sigma^2_E$. From Equation 28.47a, the induced apparent strength of stabilizing selection is $1000\sigma^2_E$ ($\widehat{V}_s = \widetilde{\sigma}^{\,2}_A/s = \sigma^2_E/0.001$), far too weak relative to estimates from natural populations of $\sim 20\sigma^2_E$. Conversely, taking observed values of V_s to be around $20\sigma^2_E$ and using a value of $\sigma^2_m = \sigma^2_E/10^3$ in Equation 28.46c yields $s = \widetilde{\sigma}^{\,2}_A/\widehat{V}_s = \sigma^2_E/(20\sigma^2_E) = 0.05$. With this value of s, Equation 28.45a yields $\widetilde{\sigma}^{\,2}_A = 10^{-3}\sigma^2_E/0.05 = 0.02\sigma^2_E$, for an equilibrium heritability of $\widetilde{h}^2 = 0.02/(1+0.02) \simeq 0.02$. Hence, using the typical estimate of the strength of stabilizing selection to estimate s yields too small a heritability.

The problem with the constant-s model is that it either does not produce enough additive variance (s is too large) or it gives apparent stabilizing selection that is too weak (s is too small, and the corresponding $\widehat{V}_s$ is too big). This failure follows because s influences both $\widetilde{\sigma}^{\,2}_A$ and $\widehat{V}_s$, imposing a constraint on their relationship (Barton 1990; Kondrashov and Turelli 1992; Gavrilet and de Jong 1993; Zhang et al 2002). From Equation 28.46d

$$\frac{\widehat{V}_s}{\widetilde{\sigma}^{\,2}_A} = \frac{3\kappa_4 + 4n\mu/s}{4n\mu} = \frac{3\kappa_4}{4n\mu} + \frac{1}{s} \geq \frac{1}{s} = \frac{\widetilde{\sigma}^{\,2}_A}{\sigma^2_m} \tag{28.47b}$$

Noting that $\sigma_A^2 = [h^2/(1-h^2)]\sigma_E^2$, Equation 28.47b can be expressed as

$$\frac{\widehat{V}_s}{\sigma_E^2} \geq \left[\frac{\widetilde{h}^2}{1-\widetilde{h}^2}\right]^2 \frac{\sigma_E^2}{\sigma_m^2} \tag{28.47c}$$

Using this expression, typical values for selection ($\widehat{V}_s/\sigma_E^2 = 20$) and mutational variance ($\sigma_E^2/\sigma_m^2 = 10^3$) imply an equilibrium heritability of less than 0.17. Thus, as with previous models, no constant value of s in the pure-pleiotropy model can produce *both* sufficiently strong apparent stabilizing selection *and* a moderate heritability.

While the constant-s model is mathematically tractable, it is also biologically unrealistic, as we expect s to vary and to be at least somewhat correlated with α, because mutations with large absolute effects are expected to be more deleterious. Does incorporation of these features resolve the inconsistencies between the equilibrium additive variance and strength of apparent stabilizing selection? The short answer is no, while the longer answer is that variation in s introduces additional complications.

When s (and α) vary over mutations, Equation 28.44a shows that the expected additive variation from a locus is $2\mu E[\alpha^2/s]$, and Equation 28.44b shows that in order to approximate $E[\alpha^2/s]$, one must (at least) specify both the correlation, ρ, between s and α^2, as well as the kurtosis of the distribution of trait mutational effects, $\sigma^2(\alpha^2)$. Further, different families of bivariate distributions that otherwise have the same values for ρ and $\sigma^2(\alpha^2)$ can give *very* different results, making the outcome *extremely* model-dependent (Hill and Keightley 1988; Caballero and Keightley 1990; Keightley and Hill 1990; Johnson and Barton 2005), see Example 28.12.

One immediate problem arises from ρ. If $\rho = 1$, the HK model simply recovers mutation-stabilizing selection balance with its inherent limitations (because mutational effects are completely correlated, the value of α^2 determines the value of s). Conversely, for variable s when $\rho < 1$, the additive variance can continue to increase without limit with N_e. This occurs because some small fraction of new mutations are effectively neutral, with the additive variance approaching the neutral result (Equation 11.20c), but with a lower mutation rate. Because the effectively neutral mutation rate decreases as N_e increases (for any value of s, a sufficiently large value of N_e satisfies $4N_e|s| \gg 1$), the result is a less than linear increase in additive variation with N_e, but the resulting variance is still unbounded under many joint distributions of s and α. As Johnson and Barton (2005) note, the conditional distribution of α for those values of s very near zero (near neutrality) determines whether the additive variance is unbounded in N_e, and very slight differences in the assumed joint distribution of α and s can result in dramatic differences in behavior.

Despite this impact of ρ on the model behavior for large values of N_e, a few general features of the HK model emerge from extensive simulations by Caballero and Keightley (1990) and Keightley and Hill (1990), as well as from analytic results assuming a general bivariate gamma distribution for α and s (Zhang et al. 2002):

1. Allowing s to vary increases both $\widetilde{\sigma}_A^2$ and V_s relative to a constant-s model (using $E[s]$ as the constant value), so that the strength of apparent stabilizing selection is generally too weak relative to observed values in nature, although abundant variation can potentially be maintained.

2. Dominance in *trait* mutations has little effect on the amount of trait variance maintained (this is not the case for fitness mutations, as we discuss shortly).

3. Increasing the correlation, ρ, between α^2 and s, decreases $\widetilde{\sigma}_A^2$, as does increasing the average strength of deleterious selection, $E[s]$.

4. The volume of mutations in the effectively neutral region ($0 \leq N_e|s| \leq 1$) significantly impacts the resulting genetic architecture.

5. Increasing the kurtosis (generating a thicker tail, and therefore more outliers relative to a normal) has opposite effects for trait and fitness mutations. The equilibrium

additive variance for the focal trait increases with the kurtosis of the fitness effects, s, (and can be much larger than that for a constant-s model), but decreases with the kurtosis of trait mutational effects, α. Increased kurtosis in the distribution of fitness effects of mutations implies more nearly neutral mutations (and hence higher equilibrium frequencies), while increased trait kurtosis implies a larger fraction of small trait-effect mutations (with a smaller variance contribution per mutation). Recall that Example 28.10 showed that if the pleiotropic effects are the result of stabilizing selection on a number of independent traits, the distribution of s values approaches a normal, so that the resulting fitness distribution is *not* leptokurtic and does not generate extra variation.

Example 28.11. While the correlation between the fitness and trait effects of new mutations is extremely difficult to directly measure, McGuigan and Blows (2012) used a clever mutation-accumulation (MA) design (LW Chapter 12) in *Drosophila serrata* to examine the genetic covariances between fitness and two traits (wing size and shape) due to new mutations. Their experiment consisted of 100 completely inbred MA lines, where a female was allowed to choose among five brothers, allowing for sexual selection based on mate choice (S lines) versus another MA experiment in which random brothers were used for mating (N lines). Both mating designs have the same effective population size (females were allowed to only mate once), so that any significant changes in trait values (relative to the control, which accounts for the effects of drift on both traits) are the consequences of selection for mate choice. Previous work demonstrated that these wing features were not involved in mate choice, so that any significant change in these traits is due to pleiotropic effects from loci under selection for mate choice.

In addition to contrasting the evolution of these wing features in S versus N lines, McGuigan and Blows also scored fitness from the extinction rate of the MA lines and from the productivity (the number of offspring) of the extant lines. In the N lines, both wing traits decreased in extant lines, with lines with larger trait values also showing greater productivity. However, these associations were not seen in the sexually selected S lines. Both observations suggest deleterious mutations (for either total fitness or sexual selection) also had pleiotropic effects on the wing traits scored.

Example 28.12. As mentioned, the HK model for the maintenance of variation critically depends on the fine details of the joint distribution of fitness and trait effects in new mutations. While parameters of the joint distribution of α and s for spontaneous mutations are extremely difficult to obtain, Mackay et al. (1992) were able to estimate these for bristle number using a set of spontaneous *P*-factor insertion-induced mutations in *Drosophila melanogaster*. The mean effect of an insertion on bristle number was around 0.4 phenotypic standard deviations (σ_z), the mean s effect was 0.2, and the haploid genome mutation rate was about 0.1. The distributions of both s and α were leptokurtic, with many mutations having little to no effect, and a few mutations having major effects. The correlation, ρ, between the selection coefficient and absolute mutational effect was around 0.4.

Caballero and Keightley (1994) used these *P*-factor mutational values to parameterize an (α-reflected) bivariate gamma distribution ($s \leq 0; -\infty < \alpha < \infty$), which was then used to generate an α and s value for a new mutation. These authors used $N_e = 10^4$, and hence an average number of $2N_e\mu = 2000$ new mutations per generation, with the bivariate effects of each new mutation drawn from the aforementioned distribution. These simulations returned an equilibrium heritability of 0.4. A closer look at the architecture of the mutations generated in the simulations showed that the vast majority (87%) of these were highly deleterious ($N_e s < -30$). The table below gives the expected number of mutations (each generation) in various (s, α) classes, and the percentage of total additive-genetic trait variance attributable to each class. For example, in their simulations, on average 7 slightly deleterious mutations ($-1 \leq N_e s \leq 0$) arose in each generation, with an effect of 0.125–0.25 standard deviations on the trait. Such mutations account for roughly 13% of the equilibrium additive-genetic variance.

	$-1 \leq N_e s \leq 0$		$-5 \leq N_e s < -1$	
α/σ_z	Num.	% σ^2_A	Num.	% σ^2_A
0–0.125	33	4	47	2
0.125–0.25	7	13	7	6
0.25–0.5	3	23	6	14
0.5–1	< 1	7	< 1	3
> 1	< 1	7		

The classes presented in this table accounted for 79% of the total additive variance, with the remaining 21% associated with highly deleterious mutations ($N_e s < -5$). The bulk, 56%, of additive-genetic trait variance is due to mildly to weakly deleterious alleles ($-5 \leq N_e s \leq 0$) of modest effect ($0.125 < \alpha/\sigma_z < 0.5$). On average, this contribution was due to only 1% (23/2000) of the new mutations that arose in the population each generation.

Caballero and Keightley (1994) also used a second, more crude, set of estimates of α and s from mutation-accumulation (MA) lines (and hence likely more reflective of the full spectrum of spontaneous mutations than the *P*-element values). The kurtosis and correlation of α and s were unknown for these MA data, but the haploid mutation rate was much higher ($\sim$1), while the average effects on fitness (s in the range of 0.01–0.02) and bristle number ($\sim 0.07\sigma_z$) were much lower than those estimated from the *P*-insertion data. The estimates of σ^2_m differed by roughly a factor of two between the *P* and MA datasets, yet both datasets gave roughly similar expected heritabilities in simulations. Thus, when judged by the two easiest-to-measure macroscopic parameters, σ^2_m and h^2, the simulated outcomes using the MA and *P*-based joint distributions of α and s appear rather similar. Their resulting genetic architectures, however, are radically different. The bulk of additive variation in simulations using the MA-line values is due to alleles with small α values (58% with α less than $0.125\sigma_z$), most of which are highly deleterious ($N_e s < -30$). These disparate results arose from assumed differences in the bivariate distribution of effects, and thus depend on microscopic parameters that are extremely challenging to measure, much less with any precision.

Example 28.13. Much of the analytic results for the pleiotropic side-effects model follow from the results of Wright (1938b), and especially Kimura (1969), on the expected time, $\phi(x)$, that a selected (additive) mutation under irreversible mutation spends at frequency x. From Equation 7.13b, this is given by

$$\phi(x|N, N_e, s) = \frac{2N_e(1 - \exp[-4N_e s(1-x)])}{Nx(1-x)[1 - \exp(-4N_e s)]} \quad \text{for} \quad \frac{1}{2N} \leq x \leq 1 - \frac{1}{2N}$$

Given that the additive variance contributed by an additive allele with effect α at frequency x is $2\alpha^2 x(1-x)$, the variance contributed by the flux of new alleles ($2N\mu$ per generation) becomes

$$2N\mu \int_{1/(2N)}^{1-1/(2N)} \int_{s,\alpha} 2\alpha^2 x(1-x) \cdot \phi(x|N, N_e, s) \cdot \varphi(s,\alpha)\, ds\, d\alpha\, dx$$

$$= 4N_e\mu \int_{1/(2N)}^{1-1/(2N)} \int_{s,\alpha} \frac{\alpha^2(1 - \exp[-4N_e s(1-x)])}{1 - \exp(-4N_e s)} \varphi(s,\alpha)\, ds\, d\alpha\, dx$$

where $\varphi(s,\alpha)$ is the joint distribution of s and α in new mutations (Keightley and Hill 1990; Zhang and Hill 2002; Eyre-Walker 2010). Caballero and Keightley (1994) and Zhang et al. (2004) present a more general version allowing for dominance in both the trait and fitness.

An interesting analysis using this approach was provided by Eyre-Walker (2010). For analytic tractability, he assumed that the trait effect, α, of a mutation was related to its (deleterious) selection coefficient, s, by

$$\alpha = \delta(4N_e s)^{\tau}(1+\epsilon)$$

where δ and ϵ are random variables, with δ either $+1$ or -1 with equal probability and $\epsilon \sim N(0, \sigma_\epsilon^2)$. Note that as $\sigma_\epsilon \rightarrow 0$, $|\alpha|$ and s are perfectly associated, while they become uncorrelated as $\sigma_\epsilon \rightarrow \infty$. The parameter τ measures how $|\alpha|$ increases with the strength of selection, for example, linearly when $\tau = 1$.

By assuming that the distribution of s was gamma, Eyre-Walker was able to obtain an analytic expression for the additive variance (in terms of the Hurwitz Zeta function, see his paper for details). Of particular interest was the fraction of the equilibrium additive-genetic variance contributed by alleles at frequency x, which turned out to be independent of σ_ϵ^2. Figure 28.5 plots this fraction for various assumed values of τ for strong ($4N_e\overline{s} = 3000$) and weak ($4N_e\overline{s} = 30$) average selection against new mutations. When $\tau = 0$, trait alleles are neutral, and the majority of the genetic variance is due to intermediate to high frequency (derived) alleles. Conversely, as τ increases, the majority of variation is due to alleles of large effect at very low frequency. For example, for $4N_e\overline{s} = 3000$ and $\tau = 1$, 96% of the variance is contributed by alleles with frequency <0.1%. As selection becomes weaker ($4N_e\overline{s}$ becomes smaller), the vast majority of the variation is due to alleles that are at low, rather than very low, frequencies.

The variance to allele-frequency relationship for purely neutral alleles is in sharp contrast to this observation that rare alleles account for most of the variation under the pleiotropic selection model. Because alleles are assumed neutral, there is no correlation between the effect of a new mutation, α, on a trait and its allele frequency. When expressed in terms of the minor-allele frequency (MAF), x, the folded Watterson distribution (Equation 2.34b) for neutral alleles at equilibrium is

$$\phi(x) = \frac{\theta}{x(1-x)} \quad \text{for} \quad \frac{1}{2N} \leq x \leq \frac{1}{2} \quad \text{where} \quad \theta = 4N_e\mu$$

Because the additive variance is given by $\sigma_A^2(x) = 2\alpha^2 x(1-x)$, we have $\phi(x) \cdot \sigma_A^2(x) = 2\alpha^2/\theta$ for all values of x. As shown in Figure 28.6, this implies a uniform distribution, over [0,1/2], for the fraction of equilibrium variation contributed by a given allele-frequency class, with a fraction $2x$ of the total additive variance due to alleles with MAF $\leq x$ (Visscher et al. 2012b; Robinson et al. 2014). For example, in this neutral setting, 80% of all variation is due to alleles with a minor-allele frequency between 0.1 and 0.5.

Example 28.14. An interesting example of apparent stabilizing selection was given by McGuigan and Blows (2009). Recall from Chapter 13 (e.g., Equation 13.26a) that the genetic variance-covariance matrix, **G**, associated with a vector, **z**, of trait values is critical to understanding multivariate evolution (discussed in great detail in Volume 3). The first principal component (the leading eigenvector) of this matrix, $\mathbf{g}_1$, represents the linear combination that accounts for the most genetic variation in this set of traits. In particular, the index $\mathbf{g}_1^T\mathbf{z} = \sum g_{1,i}z_i$ (where $g_{1,i}$ is the ith element of the leading eigenvector) has the largest genetic variance of any index of **z** (Appendix 5).

McGuigan and Blows classified male *Drosophila bunnanda* into those that were successful in a mate-choice experiment (high fitness) and those that were not (low fitness). Previous work from the Blows lab demonstrated that females favor males with a particular combination of cuticular hydrocarbons (CHC), which we can write as an index score, $I = \mathbf{b}^T\mathbf{z}$, where **b** is the vector of weights and **z** is the vector of CHC scores. This score has been under strong directional selection, and hence it is not surprising that there is very little genetic variation in this index (i.e., little variation along the direction of the CHC multivariate-trait space given by the vector **b** of weights). McGuigan and Blows compared the mating success of individuals based on a *different* CHC index, $H = \mathbf{g}_1^T\mathbf{z}$, namely, the index with the maximum genetic variance for CHC traits. Low-fitness males (poor mating success) tended to have extreme (high or low) H scores, while high-fitness males tended to have intermediate H values. Hence, a plot of mating success versus H shows an intermediate optimum, namely apparent stabilizing selection on H. However, there is no direct selection on H (indeed, it is almost orthogonal to the index, I, under selection by female choice), so the appearance of stabilizing selection likely arises from pleiotropic stabilizing selection on other fitness components that also influence CHC scores.

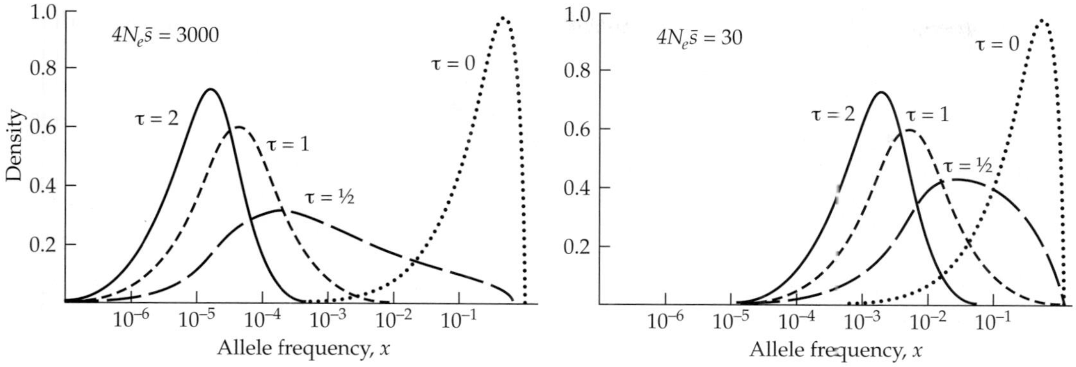

Figure 28.5 Results of the Eyre-Walker (2010) model for the maintenance of genetic variation for a neutral trait by deleterious mutations having pleiotropic effects. Here, $\overline{s}$ is the average strength of selection against a new allele, and τ is a measure connecting the trait effect, α, to s, with $\tau = 0$ indicating neutrality (see Example 28.13 for details). The figure plots the probability density of the amount of equilibrium additive-genetic variance accounted for by alleles at a specific allele frequency, x, for two different values of $4N_e\overline{s}$. Hence, the amount of variation attributable to alleles in a certain frequency range is simply the area under the curve for that range.

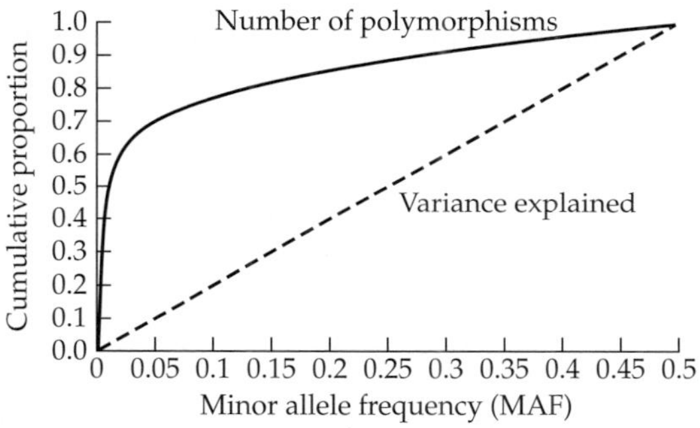

Figure 28.6 Expected number of polymorphisms, and the fraction of the trait variance explained, as a function of the minor allele frequency (MAF) under an equilibrium neutral model (i.e., the Watterson distribution). See Example 28.13 for details.

Sztepanacz and Rundle (2012) also observed this pattern in mate-choice experiments in the sister species *D. serrata*. Because $\mathbf{g}_1$ represents the direction of the most genetic variation, McGuigan et al. (2011) stated that it "is expected to capture a greater portion of the accumulated pleiotropic mutation in a set of traits, including mutations with pleiotropic effects on fitness. Consequently, strong stabilizing selection should be generated on $\mathbf{g}_1$, providing the opportunity to investigate the genetic basis of fitness using this simple summary statistic."

Deleterious Pleiotropy-stabilizing Selection (Joint-effects) Models

The final class of models relaxes the Hill-Keightley assumption that the focal trait is neutral, and instead allows it to be under stabilizing selection. This unified model is the most realistic, allowing for direct stabilizing selection, deleterious pleiotropic effects, and drift, but as such is also the most highly parameterized. It has all of the inherent complexity just seen for the HK model plus the additional complication of real stabilizing selection. While the most detailed analyses are by Zhang and Hill (Zhang and Hill 2002, 2003, 2005b; Zhang et al. 2004a), and indeed the term **joint-effects model** was coined by Zhang and Hill (2002), its roots trace back to brief comments by Kondrashov and Turelli (1992). These were expanded on by Tanaka (1996b, 1998), who considered a model with a constant pleiotropic selection coefficient and (what amounts to) a constant effect, α, on the trait from new mutations. First, we present these early results, the conclusions of which are substantially altered when either α or s varies (Zhang and Hill 2002).

For weak selection, Kondrashov and Turelli noted that the total selection coefficient, s_T, on a new mutation is approximately the sum of its (assumed constant) pleiotropic deleterious effect, s_p, and the deleterious effect, s_d, from direct stabilizing selection (Equation 28.4c), yielding

$$s_T \simeq s_p + s_d = s_p + \frac{\alpha^2}{2V_s} \tag{28.48a}$$

Kondrashov and Turelli suggested that even when fairly strong stabilizing selection occurs on a trait, most of the selection on its underlying loci is from pleiotropic effects. One way to see their point is to consider the average value of s_d, which (from Equation 28.30d) is $\sigma_\alpha^2/(2V_s)$. Note that this increases with the mutational effects variance, σ_α^2, and decreases with V_s (i.e., increases with the strength of stabilizing selection). To illustrate Kondrashov and Turelli 's conjecture, we bias our assumptions in favor of larger values of s_d by assuming a high value, $0.1\sigma_E^2$, for σ_α^2. By taking the standard value of $V_s \simeq 20\sigma_E^2$, Equation 28.30d yields $s_d \simeq 0.1/40 = 0.0025$. Using the Crow and Simmons (1983; also see Charlesworth 2015) estimate of $s_p = 0.02$ for deleterious mutations in *Drosophila* gives an eight-fold higher selection coefficient from pleiotropy, even under our assumption that was biased towards higher s_d values. Assuming an even larger mutational-effects variance ($\sigma_\alpha^2 = \sigma_E^2$, and hence $s_d = 0.025$) still leaves almost half of the selection from pleiotropic effects.

Given the value of s_T, Equation 28.45b suggests the equilibrium variance as

$$\widetilde{\sigma}_A^2 \simeq \frac{\sigma_m^2}{s_T} = \frac{\sigma_m^2}{s_p + \alpha^2/(2V_s)} \tag{28.48b}$$

Further, as an approximation for α^2, we can replace s_d by its average value (Equation 28.30d) to yield

$$\widetilde{\sigma}_A^2 \simeq \frac{\sigma_m^2}{s_p + \sigma_\alpha^2/(2V_s)} = \frac{2V_s\sigma_m^2}{2V_s s_p + \sigma_\alpha^2} \tag{28.48c}$$

This expression recovers the pure pleiotropy value of σ_m^2/s_p (Equation 28.45b) for sufficiently large V_s (very weak stabilizing selection), and the HCA value of $2V_s\sigma_m^2/\sigma_\alpha^2 = 4V_s n\mu$ (Equation 28.18a) for sufficiently small s_p (very weak pleiotropic effects).

Comparison of Equations 28.48b and 28.45b shows that adding any amount of real stabilizing selection on the focal trait results in a reduction of the equilibrium variance relative to the pure pleiotropy model value of σ_m^2/s_p, as

$$\widetilde{\sigma}_A^2 \simeq \frac{\sigma_m^2}{s_p + \alpha^2/(2V_s)} < \frac{\sigma_m^2}{s_p} \tag{28.48d}$$

This immediately resolves the delicate issue of additive variance increasing without limit as N_e increases under the HK model (when $\rho < 1$). Under the joint-effects model, as the effective population size increases, eventually $N_e|s_d| \gg 1$, and none of the trait mutations are effectively neutral, limiting the increase in additive variance as N_e increases.

Tanaka (1996b) and Zhang et al. (2004a) noted that the strength of *apparent* stabilizing selection, $\widehat{V}_s$, under the joint action of *real* stabilizing selection, V_s, and deleterious pleiotropic effects, s_p, is

$$\widehat{V}_s^{-1} = \widehat{V}_{s,p}^{-1} + V_s^{-1} \tag{28.49}$$

where $\widehat{V}_{s,p}$ is the induced strength of stabilizing selection from the pleiotropic effects alone (Equation 28.46b). Equation 28.49 implies $\widehat{V}_s \leq V_s$, giving the apparent strength as greater (as $\widehat{V}_s$ is smaller) that the true amount of actual stabilizing selection, V_s. As noted by Zhang and Hill (2002), incorporating real stabilizing selection partly breaks the constraint given by Equation 28.46d between $\widehat{V}_s$ and $\widetilde{\sigma}^2_A$ that prevents a pure pleiotropy model from generating *both* significant variance and strong apparent stabilizing selection. However, even here there is a tradeoff. Relative to a pure pleiotropy model, adding direct stabilizing selection ($s_d > 0$) increases the apparent strength of selection (Equation 28.49), but does so at the expense of decreasing the equilibrium variance (Equation 28.48d).

Zhang and Hill (2002) showed that the connection between the observed strength of apparent stabilizing selection, $\widehat{V}_s$, and any real stabilizing selection on the trait, V_s, is given by

$$\widehat{V}_s = V_s \left(\frac{\widetilde{\sigma}^4_A/V_s + \sigma^2_m - \sigma_p(w, z^2)}{\widetilde{\sigma}^4_A/V_s + \sigma^2_m} \right) \tag{28.50a}$$

where $\sigma_p(w, z^2)$ is the covariance between the relative fitness from pleiotropic effects and squared trait deviations from the mean (the latter is assumed to be at the optimum of zero). The bounds for this expression are

$$\widehat{V}^*_s \leq \widehat{V}_s \leq V_s \tag{28.50b}$$

The upper bound V_s (weakest apparent selection) occurs when the effects of pleiotropy are very small (Equation 28.50a converges to V_s as $\sigma_p[w, z^2] \to 0$). Conversely, the lower bound

$$\widehat{V}^*_s = \frac{\widetilde{\sigma}^4_A}{\widetilde{\sigma}^4_A/V_s + \sigma^2_m} \tag{28.50c}$$

corresponds to the strongest possible apparent selection under the joint-effects model. This lower bound is approached when the effects from pleiotropic selection dominate, so that $\sigma_p(w, z^2) \to \sigma^2_m$ (Zhang and Hill 2002; Zhang et al. 2002). When the lower bound holds, pleiotropy dominates, so $s_T \simeq s_p$, with Equation 20.45b yielding $\widetilde{\sigma}^2_A \simeq \sigma^2_m/s_p$, and Equation 28.50c rearranges to

$$\widehat{V}^*_s = \frac{(\sigma^2_m/s_p)^2/\sigma^2_m}{(\sigma^2_m/s_p)^2/(V_s\,\sigma^2_m) + 1} = \frac{(\sigma^2_m/s_p)/s_p}{(\sigma^2_m/s_p)/(V_s\,s_p) + 1} = \frac{\widetilde{\sigma}^2_A/s_p}{\widetilde{\sigma}^2_A/(V_s\,s_p) + 1} \tag{28.50d}$$

Recalling Equation 28.46c, $\widetilde{\sigma}^2_A/s_p$ is the apparent stabilizing selection, $\widehat{V}_{s,p}$, under pure pleiotropy, implying that the smallest (strongest) value for apparent stabilizing selection under the joint-effects model is

$$\widehat{V}^*_s = \frac{\widehat{V}_{s,p}}{\widehat{V}_{s,p}/V_s + 1} \tag{28.50e}$$

Because $\widehat{V}_{s,p} = \widetilde{\sigma}^2_A/s_p$ and $\widetilde{\sigma}^2_A = \sigma^2_m/s_p$ under the pure pleiotropy model (Equations 28.45b and 28.46c), when

$$\frac{\widehat{V}_{s,p}}{V_s} = \frac{(\widetilde{\sigma}^2_A/s_p)}{V_s} \simeq \frac{\sigma^2_m}{V_s\,s^2_p} > 1 \tag{28.50f}$$

then $\widehat{V}^*_s$ is less than half its constrained value under the pure pleiotropy model, $\widehat{V}_{s,p}$ (i.e., the apparent strength of stabilizing selection is twice as strong). Usually $\widehat{V}_{s,p}$ is fairly large when Equation 28.50f is satisfied, so reducing its value by half still leaves $\widehat{V}_s$ too large to account for typically assumed values of $\sim 20\sigma^2_E$.

While joint-effects models with α and s constant give some insight as to how pure pleiotropy and real stabilizing selection interact, they also miss important consequences when either (or both) vary. A simple example makes the point. Suppose there are two equally frequent classes of pleiotropic mutants. The first has $s_p = 0.001$, while the second has $s_p = 0.1$. Further suppose that in both cases $s_d = 0.001$, generating half the mutations with $s_T = 0.002$ and the other half with $s_T = 0.101$. Substituting their average, $\overline{s}_T = 0.0515$, into Equation 28.48b gives $\widetilde{\sigma}_A^2 = \sigma_m^2/0.0515 \simeq 19\sigma_m^2$. However, the correct value is the average of the variation generated by each class

$$\frac{\sigma_m^2/0.002 + \sigma_m^2/0.101}{2} \simeq 255\sigma_m^2$$

resulting in substantially more genetic variance. More generally, suppose there are k classes of mutations, the ith of which has a total selection coefficient of $s_{T,i}$, and contributes a fraction, π_i, of the total mutational variance, then

$$\widetilde{\sigma}_A^2 = \sigma_m^2 \sum_{i=1}^{k} \frac{\pi_i}{s_{T,i}} = \frac{\sigma_m^2}{(\overline{s_T})_H} \quad \text{where} \quad (\overline{s_T})_H = \left(\sum_{i=1}^{k} \frac{\pi_i}{s_{T,i}} \right)^{-1} \tag{28.51}$$

with s_T in Equation 28.48b replaced by its harmonic mean $(\overline{s_T})_H$ (weighted by the fraction of σ_m^2 accounted for by a particular fitness class), rather than its arithmetic mean. This same argument applies to the pure pleiotropy model, and is the reason why variation in s generates a higher equilibrium variance than a constant-s model with the same mean selection value.

When s_p is constant but α varies, then for $s_p \gg \overline{s}_d$, the strength of apparent stabilizing selection approaches the lower bound, $\widehat{V}_s^*$, given by Equation 28.50c. Zhang and Hill (2002) showed that the equilibrium additive variance is given by Equation 28.45b with

$$s_T \simeq s_p + 3\kappa_4\, \overline{s}_d = s_p + 3\kappa_4 \frac{\sigma_\alpha^2}{2V_s} \tag{28.52a}$$

where κ_4 is the scaled kurtosis of trait mutational effects (Equation 28.18c). κ_4 is bounded below by 1/3, with $\kappa_4 = 1$ if mutational effects are drawn from a normal distribution and $\kappa_4 > 1$ for a leptokurotic distribution. Equation 28.52a shows (as mentioned previously) that increasing the kurtosis of the trait mutational-effect values lowers the equilibrium additive variance. Alternatively, when $\overline{s}_d \gg s_p$, the apparent strength of stabilizing selection approaches V_s, and the equilibrium variance is given by the HCA (Equation 28.18a).

Finally, Zhang and Hill (2002) showed that when *both* α and s vary independently, these no longer need act in an additive fashion. By modeling α as normal and s as gamma, they found that

$$\overline{s}_T \simeq \overline{s}_d + \sqrt{\overline{s}_d \cdot \overline{s}_p} \tag{28.52b}$$

resulting in the average strength of direct selection, $\overline{s}_d$, being the primary driver for s_T, in contrast to the suggestion by Kondrashov and Turelli (1992), obtained by assuming constant values. Both types of selection (direct and pleiotropic) reduce the equilibrium additive variance, but the impact of pleiotropic selection now depends in part on the magnitude of direct selection. As Zhang and Hill noted, this unequal influence arises because a large value of s_d is always associated with large α^2 values, while the same is not true for s_p (as α, and hence s_d, and s are assumed to be uncorrelated in Equation 28.52b). For $\overline{s}_d \gg \overline{s}_p$, the equilibrium variance again reduces to the HCA result (Equation 28.18a), with $\widehat{V}_s \simeq V_s$. When $\overline{s}_d \ll \overline{s}_p$, $\widehat{V}_s \simeq \widehat{V}_s^*$, and $s_T \simeq \sqrt{\overline{s}_d \cdot \overline{s}_p}$, and substituting into Equation 28.45b yields

$$\widetilde{\sigma}_A^2 \simeq \frac{\sigma_m^2}{\sqrt{\overline{s}_d \cdot \overline{s}_p}} = \sqrt{\frac{\sigma_m^2}{\overline{s}_p}} \cdot \sqrt{\frac{\sigma_m^2}{\overline{s}_d}} \tag{28.52c}$$

Recalling (Equation 28.30d) that $\overline{s}_d = \sigma_\alpha^2/(2V_s)$ yields

$$\frac{\sigma_m^2}{\overline{s}_d} = \frac{2V_s\sigma_m^2}{\sigma_\alpha^2} = \frac{2V_s(2n\mu\sigma_\alpha^2)}{\sigma_\alpha^2} = 4n\mu V_s$$

resulting in

$$\widetilde{\sigma}^2_A \simeq \sqrt{4n\mu V_s \cdot \sigma^2_m / \,\overline{s}_p} \tag{28.52d}$$

as found by Zhang and Hill (2002), who noted that this is the geometric mean of the HCA and pure pleiotropy models. Note that $\overline{s}_d \ll \overline{s}_p$ can occur under even strong direct stabilizing selection (V_s small) if the mutation rate is sufficiently high, as $\overline{s}_d = \sigma^2_\alpha/(2V_s)$, and for fixed σ^2_m, increasing the total mutation rate, $n\mu$, decreases $\sigma^2_\alpha = \sigma^2_m/(2n\mu)$, and hence decreases $\overline{s}_d$.

The above analysis for the joint-effects model assumes that mutations are additive for both the trait and for any pleiotropic effect on fitness. Zhang et al. (2004) noted that this assumption is not supported by the data, which suggest significant dominance, especially in fitness. Further, fitness mutations of large effect tend to be more recessive (otherwise they would simply not be segregating in the population). Under the assumption that mutations are more likely to be recessive for fitness than for a trait, Zhang et al. showed that the joint-effects model generates considerably more variation than under the above additive assumptions. Most of the resulting additive variance comes from alleles with nearly neutral effects on fitness as heterozygotes, while alleles of large effect contribute the most to the apparent stabilizing selection. Thus, there appear to be regions of the parameter space under which the joint-effects model could account for both significant additive variation and sufficiently strong stabilizing selection. The unresolved issue is whether these regions are biologically realistic. There is also the secondary concern (from our previous load arguments) that strong direct stabilizing selection can act on only a limited number of traits, which suggests that weak true stabilizing selection is the norm, not the exception, significantly narrowing the size of these successful regions of the parameter space.

Example 28.15. As an application of the joint-effects model, suppose that

$$V_s = 100\sigma^2_E, \quad \overline{s}_p = 0.005, \quad \sigma^2_m = \sigma^2_E/10^3, \quad \text{and} \quad 2n\mu = 0.01$$

Thus, $\sigma^2_\alpha = \sigma^2_m/(2n\mu) = 0.1$ and $\overline{s}_d = \sigma^2_\alpha/(2V_s) = 0.1/(200) = 5 \cdot 10^{-5}$ (Equation 28.30d), so that $\overline{s}_d \ll \overline{s}_p$. Under HCA (Equation 28.18a), the equilibrium additive variance is $4n\mu V_s = 0.02 \cdot 100\,\sigma^2_E = 2\,\sigma^2_E$, while under pure pleiotropy (Equation 28.45b), it is $\sigma^2_m / \,\overline{s}_p = 10^{-3}\sigma^2_E/0.005 = 0.2\,\sigma^2_E$. Applying Equation 28.52d, the equilibrium variance under the joint-effects model becomes $\sqrt{2 \cdot 0.2} \cdot \sigma^2_E = 0.63\,\sigma^2_E$, for a heritability of 0.63/(1+0.63) = 0.39. Because $\overline{s}_d \ll \overline{s}_p$, the strength of apparent stabilizing selection is given by $\widehat{V}^*_s$ (Equation 28.50d)

$$\widehat{V}^*_s/\sigma^2_E = \frac{0.63/0.005}{0.63/(100 \cdot 0.005) + 1} = 55.7$$

Thus, a reasonable amount of genetic variation is maintained at a moderate apparent strength of stabilizing selection.

HOW WELL DO THE MODELS FIT THE DATA?

To aid the reader who either skimmed or skipped the preceding rather technical theory sections, the major results from this analysis are summarized in Table 28.3. The central conclusion is that essentially all of the models have issues, generally being unable to simultaneously generate *both* a high value of $\widetilde{\sigma}^2_A$ (heritability in the 0.2 to 0.6 range) *and* sufficiently strong apparent (or real) stabilizing selection ($\widehat{V}_s \leq 20\sigma^2_E$). This led Johnson and Barton (2005) to lament that "it is puzzling that levels of heritability are so pervasive, so

Table 28.3 Inconsistencies between model predictions and the observed amounts of genetic variation, $\widetilde{\sigma}^2_A$, and apparent strengths of natural selection, $\widehat{V}_s$. Figure 28.1 summarized the various models.

Neutral focal trait, no selection	
Mutation-drift	Does not account for apparent stabilizing selection. Additive variance increases without limits as $N_e \to \infty$ under the incremental mutational model (Table 28.1).
Neutral focal trait, selection on pleiotropic underlying loci	
Fitness overdominance	Required strength of selection at overdominant loci generates a very large genetic load.
Mutation-selection balance	Prevents a small $\widehat{V}_s$ without a small $\widetilde{\sigma}^2_A$.
Direct selection on focal trait	
Strict stabilizing selection	Fitness underdominance generated at underlying loci. Very little additive variance at equilibrium.
Stabilizing selection-mutation balance	Too much additive variance for the observed strengths of stabilizing selection.
Pleiotropic overdominance	Load and selection-response arguments (see Example 28.2).
Pleiotropic deleterious alleles (Joint-effects models)	Some parameter combinations allow for moderately strong apparent stabilizing selection with reasonable heritabilities.

high and roughly constant," and that "we are in the somewhat embarrassing position of observing some remarkably robust patterns . . . and yet seeing no compelling explanation for them." Before condemning the models, a more careful look at the data is in order.

Strength of Selection: Direct Selection on a Trait

Most of the above models can easily accommodate sufficient genetic variation. Indeed, a strictly neutral model generates *too much* variation. The more problematic issue is accounting for the observed strength of real (or apparent) stabilizing selection in nature, warranting a more careful look at the assumed literature values. If the strength of real or apparent stabilizing selection is weaker than is typically assumed, many of the apparent contradictions disappear, and a number of models can potentially account for the observations.

While Turelli's (1984) benchmark of $V_s \simeq 20\sigma^2_E$ is typically assumed, the data today are both more extensive, and more problematic, than when he extracted this value from the literature. The classic paper by Lande and Arnold (1983), which launched an entire cottage industry on the estimation of these parameters, appeared at essentially the same time as Turelli's analysis. We examine fitness estimation in detail in Chapters 29 and 30, noting here the basic conclusion that there is considerable uncertainty on the strength of natural selection on a typical trait. The relative constancy of many morphological phenotypes over evolutionary time is consistent with some form of stabilizing selection, as are the divergence data for gene-expression levels (Chapter 12). However, the *strength* of such selection is far less clear. The meta-analysis by Kingsolver et al. (2001) on the quadratic term, γ, of a Lande-Arnold fitness regression (Figure 30.5) shows that it is equally likely to be positive (disruptive selection) or negative (stabilizing selection). Conditioning on this value being negative, the mean strength is slightly stronger than Turelli's value ($\sim 10\sigma^2_E$). If correct, these higher estimates of V_s are *more* problematic for the previous models.

Besides the standard concerns of measurement error and power (especially with an inherently noisy trait like fitness), there are three issues that significantly obscure the actual strength of selection on a trait (Chapters 29 and 30). First, almost all fitness-trait regressions in the literature use a *component* of fitness (such as mating success, fecundity, or viability), *not* total fitness itself. Such component-based estimates can be very misleading,

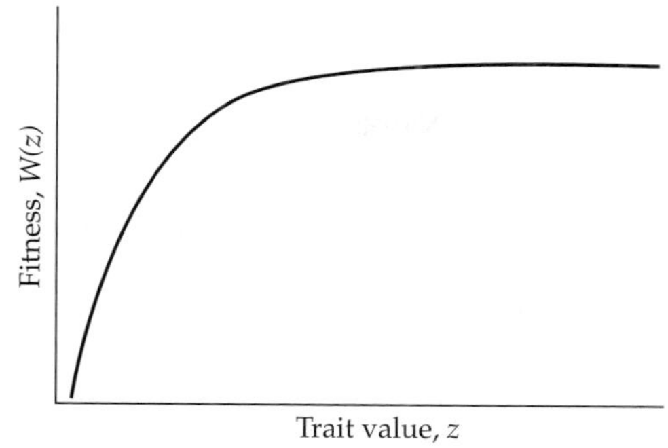

Figure 28.7 If the true fitness function has an asymptotic optimum, approximating it with a quadratic can be very misleading.

representing only a fraction of the total fitness (e.g., Johnston et al. 2013). Second, selection acting on phenotypically correlated characters obscures not just the actual strength of selection on a target trait, but more fundamentally can also disguise its true nature. For example, a neutral trait can show a strong signal of stabilizing selection if there is selection on phenotypically correlated traits (Example 28.1). The standard approach for dealing with this concern is a multivariate regression with a number of traits, in the hope that some of them are highly correlated with the actual targets of selection, so their inclusion acts as a covariate to reduce spurious associations. However, this approach is far from foolproof (Chapter 30). As Example 28.1 highlights, what matters for evolution is not the strength of selection on the phenotype, but rather the strength of selection on the *breeding value*. A highly heritable trait under strong apparent stabilizing selection can experience little to no selection on its breeding value if the target of selection is not the trait itself (Chapters 20, 29, and 30).

A more subtle issue is that most estimates of the strength of stabilizing selection are based on the quadratic term in a fitness regression (Equation 28.3i; Chapter 29). However, if Gaussian stabilizing selection or quadratic are poor models of the actual nature of nonadditive selection, these regression terms can be very misleading (see Figure 29.9). An alternative class of nonlinear candidate fitness functions are those that reach as asymptotic value (Figure 28.7). The quadratic approximations of such fitness functions are expected to be rather misleading, and (to our knowledge) little work has been done on the maintenance of quantitative-genetic variation under such a fitness function. The drift barrier (Chapter 7) arising from improved alleles eventually having too small a selective advantage to overpower drift could potentially result in considerable levels of variation at equilibrium.

Finally, there is also the issue of load discussed previously (e.g., Example 28.2), which suggests an upper bound on the number of independent traits under selection. Barton (1990) and Walsh and Blows (2009) suggested that strong selection is likely confined to a few dimensions in the multivariate trait space (i.e., a few indices of trait values), in which case selection impacts a very large number of traits, but each only weakly.

Example 28.16. Selection acts on multivariate phenotypes. When we examine the nature of selection on a trait-by-trait basis, we are examining the projection of this true multivariate selection surface onto a single dimension, which can be quite misleading. An excellent example of how to examine correlations between strength of stabilizing selection and genetic variation in a multivariate setting is found in Hunt et al. (2007b), which examined selection on mate calls in the cricket *Teleogryllus commodus*.

One advantage of this system is that the multivariate trait, call signal, can have any of its individual components artificially changed via computer software, and its impact on the measured fitness component (mate attraction) directly assessed through acoustic playback trials in natural populations (Brooks et al. 2005). Five call components were examined, whose heritabilities ranged from 17% to 72%. These components were also strongly genetically correlated, with values ranging from -0.65 to 0.40. Factor-analytic modeling (Hine and Blows 2006) gave strong support for three dimensions of the resulting genetic variance-covariance (**G**) matrix for these five traits, with the first three eigenvalues (λ_1 to λ_3) accounting for 90% of the total additive-genetic variation. Here we denote their associated eigenvectors by $\mathbf{g}_1$, $\mathbf{g}_2$, and $\mathbf{g}_3$. There was also close to significant support for the fourth eigenvalue, but no support for the final (fifth) dimension, as λ_5 was not significantly different from zero.

Using the methods of Chapter 30, the 5×5 matrix, $\boldsymbol{\gamma}$, of quadratic selection gradients of call components on mate attractiveness was estimated (Brooks et al. 2005). The diagonal elements (γ_{ii}) of $\boldsymbol{\gamma}$ correspond to the amount of quadratic selection on call component i, with a large negative value indicating strong stabilizing selection. This can be generalized to consider the amount of quadratic selection acting on some linear combination (i.e., an index) of trait values, $I = \sum a_i z_i = \mathbf{a}^T\mathbf{z}$, which is given by the quadratic product $\mathbf{a}^T\boldsymbol{\gamma}\,\mathbf{a}$ (Chapter 30 and Appendix 5). As expected, this generalization recovers the univariate result when the index $I = z_i$, as $\mathbf{a}^T\boldsymbol{\gamma}\,\mathbf{a} = \gamma_{ii}$ when $a_i = 1$ and all other elements of **a** are 0.

Using these results, Hunt et al. examined the strength of stabilizing (negative values) or disruptive (positive values) selection on the index of trait values, $\mathbf{g}_i^T\mathbf{z}$, whose weights are given by the ith eigenvector. This corresponds to the nature of selection along the ith-largest axis of genetic variation, where the resulting strength of selection is given by $\mathbf{g}_i^T\boldsymbol{\gamma}\,\mathbf{g}_i$.

	$\mathbf{g}_1$	$\mathbf{g}_2$	$\mathbf{g}_3$	$\mathbf{g}_4$	$\mathbf{g}_5$
λ_i, eigenvalue of **G**	0.930	0.468	0.235	0.125	0.065
% Genetic Variance	51.0	25.6	12.9	6.9	3.6
$\mathbf{g}_i^T\boldsymbol{\gamma}\,\mathbf{g}_i$	0.005	−0.012	−0.051	−0.097	−0.011

As the above table shows, there was very weak disruptive selection along the first eigenvector ($\mathbf{g}_1^T\boldsymbol{\gamma}\,\mathbf{g}_1 = 0.005$, which was not significantly different from zero), and increasingly strong stabilizing selection on eigenvectors 2 through 4. As the strength of stabilizing selection increased (-0.012, -0.051, -0.097), the amount of genetic variation in that direction decreased (25.6%, 12.9%, 6.9%). The eigenvector $\mathbf{g}_5$ breaks this pattern, but λ_5 was not significantly different from zero, and hence this direction of **G** could simply be residual noise (Volume 3 examines the estimation of the true dimensionality of **G** matrices in some detail). As summarized in the table, axes (combinations of trait values) with stronger stabilizing selection showed less genetic variation.

Another measure of the impact of selection is the expected within-generation change in **G** from selection, which is given by $\Delta\mathbf{G} = \mathbf{G}\boldsymbol{\gamma}\mathbf{G}$ (in the absence of directional selection; Volume 3). The amount of change in the additive-genetic variance of a composite trait, $\mathbf{a}^T\mathbf{z}$, is

$$\Delta\sigma^2_A(\mathbf{a}^T\mathbf{z}) = \mathbf{a}^T\Delta\mathbf{G}\mathbf{a} = \mathbf{a}^T\mathbf{G}\,\boldsymbol{\gamma}\,\mathbf{G}\mathbf{a}$$

so that the amount of change in the genetic variance along the direction given by eigenvector i becomes $\Delta\sigma^2_A(\mathbf{g}_i) = \mathbf{g}_i^T\Delta\mathbf{G}\mathbf{g}_i = \mathbf{g}_i^T\mathbf{G}\,\boldsymbol{\gamma}\,\mathbf{G}\mathbf{g}_i$. The resulting absolute and percentage change in the genetic variance along each of the eigenvalues were as follows:

	$\mathbf{g}_1$	$\mathbf{g}_2$	$\mathbf{g}_3$	$\mathbf{g}_4$	$\mathbf{g}_5$
$\Delta\sigma^2_A(\mathbf{g}_i)$	0.00400	−0.00283	−0.00285	−0.00212	−0.00009
% change in σ^2_A	6.6	−7.8	−11.0	−13.0	−3.8

The expected change in the additive-genetic variance along the directions given by eigenvectors 2, 3, and 4 are −8%, −11%, and −13%, respectively. Axes (trait combinations) with the strongest amounts of stabilizing selection are expected to show the fastest erosion of additive variation, and this is the pattern seen in the above table.

For the five call components considered here, when considered one trait at a time (but still correcting for the fitness correlation among traits), the heritabilities and strengths of selection on the individual traits were

Trait	*CPN*	*CIPD*	*TN*	*ICD*	*DF*
h^2	0.719	0.388	0.257	0.167	0.293
γ_{ii}	0.006	−0.006	−0.040	−0.070	−0.047

While the previous pattern of lower heritabilities for traits under stronger selection still holds, note that stabilizing selection along the direction given by the eigenvectors of **G** (i.e., on an index of trait values in this direction) is stronger than selection on any given trait (compare the magnitudes of γ_{ii} with those for $\mathbf{g}_i^T \gamma \, \mathbf{g}_i$). If the strength of stabilizing selection was estimated in a truly univariate fashion (computing γ_{ii} ignoring the other four call components), estimates of the strength of quadratic selection would be even more untrustworthy given the strong correlations among these traits.

The message here is that a full multivariate analysis gives a much more accurate picture than a series of univariate analyses focusing on single traits, which can be very misleading (Blows and Walsh 2009; Walsh and Blows 2009; Volume 3). Assessment of the success of a model for the maintenance of variation has typically been applied to one trait at a time, by attempting to reconcile the observed values of V_s ($\simeq -1/\gamma_{ii}$; Equation 28.3i) with their corresponding h_i^2 values. As this example shows, a more accurate comparison is to examine the additive-genetic variance for the composite traits, $y_i = \mathbf{g}_i^T \mathbf{z}$, based on the eigenvalues of **G**, and their corresponding multivariate measure of stabilizing selection, $\mathbf{g}_i^T \gamma \, \mathbf{g}_i$.

Strength of Selection: Persistence Times of New Mutants

One measure of the strength of selection on the breeding value of a trait is offered by the ratio of the additive-genetic to mutational variances, $\widetilde{\sigma}_A^2/\sigma_m^2$. As the equilibrium is reached when the variation introduced by mutation, σ_m^2, is balanced by its removal, this ratio is a measure of the strength of selection against new mutations (whatever the cause, be it direct selection on the trait and/or pleiotropic fitness effects). This ratio also corresponds to the average number of individuals affected by a mutation before its removal (Li and Nei 1972), which Crow (1979, 1993) called the **persistence time** (as $1/s$ is the time scale for selective removal). The weaker selection, the slower the removal and the longer the persistence time of a new mutation.

More formally, we can use this ratio to assign approximate selection coefficients. Under the deleterious pleiotropy model, Equation 28.45b gives this ratio as $\widetilde{\sigma}_A^2/\sigma_m^2 = 1/s$ when mutations have a fixed selective value. When s varies, it is replaced by their harmonic mean (Equation 28.51). Conversely, if the trait is under direct stabilizing selection, we can use the stochastic house of cards value for $\widetilde{\sigma}_A^2$ (Equation 28.30a) to rewrite the persistence time in terms of s as

$$\widetilde{\sigma}_A^2/\sigma_m^2 = \frac{1}{\sigma_m^2}\left[\frac{4n\mu V_s}{1+V_s/(N_e\sigma_\alpha^2)}\right] = \frac{N_e\sigma_\alpha^2}{2n\mu\sigma_\alpha^2}\left[\frac{4n\mu V_s}{N_e\sigma_\alpha^2+V_s}\right]$$

$$= \frac{2N_eV_s}{V_s+N_e\sigma_\alpha^2} = \frac{2N_e}{1+N_e\sigma_\alpha^2/V_s} = \frac{2N_e}{1+2N_es} \tag{28.53a}$$

with the last step, $\sigma_\alpha^2/(2V_s) = s$, following from Equation 28.30d. Hence

$$\widetilde{\sigma}_A^2/\sigma_m^2 = \frac{2N_e}{1+2N_es} \simeq \begin{cases} 2N_e & \text{for } N_es \ll 1 \\ s^{-1} & \text{for } N_es \gg 1 \end{cases} \tag{28.53b}$$

Thus, under both pleiotropy and direct selection, it is often the case that $\widetilde{\sigma}_A^2/\sigma_m^2 \simeq 1/s$. Akin to the use Robertson's secondary theorem of natural selection (Chapter 6 and 20) to examine the nature of selection acting directly on the breeding value of a trait, the persistence time measures the amount of selection (either directly or through pleiotropic effects) acting on the loci that underlie our focal trait.

Large values of this ratio (>1000) are more consistent with drift, smaller values with deleterious mutation-selection balance, be it pleiotropy or direct selection (Barton 1990). In a survey of l different organisms, Houle et al. (1996) found an average value of $\widetilde{\sigma}_A^2/\sigma_m^2 \simeq 50$ for life history traits and $\simeq 100$ for morphological traits, supporting some version of deleterious mutation-selection balance, with stronger selection (shorter persistence times) on life-history than on morphological traits. Houle et al. noted that these estimates raise a dilemma, in that if most of the genetic variation is associated with deleterious pleiotropic effects, it may have little bearing on adaptive evolution, which may largely be due to rare mutations with only weak pleiotropic side effects.

Number of Loci and Mutation Rates

One of the problematic issues with the maintenance of variation by stabilizing selection-mutation balance is that the haploid mutation rate, $n\mu$, must be sufficiently large to account for observed levels of variation. From Equation 28.12e, to achieve a heritability of h^2 when $V_s = K\sigma_E^2$ (i.e., a specific multiple, K) requires that

$$\frac{4V_s n\mu}{4V_s n\mu + \sigma_E^2} = \frac{4K\sigma_E^2 n\mu}{4K\sigma_E^2 n\mu + \sigma_E^2} = \frac{4Kn\mu}{4Kn\mu + 1} = h^2$$

or that $n\mu = h^2/[4K(1-h^2)]$. For Turelli's value ($K = 20$), $h^2 = 1/3$ requires $n\mu = 0.0065$, while $h^2 = 1/2$ requires $n\mu = 0.0125$. For a standard assumed per-locus (per-generation) mutation rate of 10^{-5}, this value of $n\mu$ requires over a thousand loci ($n = 1250$). This argument led Latter (1960) to conclude that stabilizing selection-mutation balance could not account for standing levels of variation, a point echoed by Turelli (1984), provided standard assumptions ($n < 100$, $\mu < 10^{-5}$) are correct. Have more recent data shifted this view? Gametic mutation rates for fitness components have been estimated to be in the 0.01 to 0.1 range (LW Chapter 12; Shaw et al. 2002; Halligan and Keightley 2009). The few estimates for nonfitness traits are also in this range, although none of these estimates are without problems (LW Chapter 12). What do we know about the components n and μ?

Consider the number of loci, n, first. Results from genome-wide association studies (GWASs) in humans typically find a large number of factors, each of very small effect (Chapter 24). The massive power loss in a typical GWAS due to conservative control over multiple comparisons ensures that the number of sites declared as significant is only a small fraction of the number of truly causative sites. This is one factor leading to the "problem" of "missing heritability" (Example 24.1). Thus, the notion that a typical trait may be influenced by hundreds of loci ($n > 500$) is less surprising that it once was. For example, Kemper et al. (2012) suggested that GWASs imply at least 1500 genes are involved in human height, while gene knock-out studies in mice suggested around 6000 loci for body size. At least several hundred loci are involved in maize plant height (Peiffer et al. 2014). Taken as a whole, the GWAS data have shifted the consensus to a much larger number of loci that can potentially influence a typical trait. Indeed, the method of genomic selection (Volume 3), which has rapidly been adopted by commercial breeders, rests on the assumption of a very large number of underlying loci, each of small effect (i.e., the infinitesimal genetics model; Chapter 24).

Turning to μ, as noted by Turelli (1984), the "typical" value of 10^{-6} to 10^{-5} for the mutation rate at a locus is based on alleles of large effect. One might easily imagine a higher mutation rate to alleles of smaller effect. Why is that the case? Under the view that much of quantitative-genetic variation is regulatory (as opposed to changes in amino-acid sequences), there is often a much larger, and far less granular, mutational target relative to a coding region, with many mutations likely resulting in very small regulatory changes. Other factors, such as the transposition of mobile elements (which often carry their own regulatory sequences) can potentially impact regulation at numerous sites far away from their point of insertion.

Assuming both a larger number of loci *and* a higher mutation rate per locus can account for the $n\mu$ values required for stabilizing selection-mutation balance to maintain sufficient variation, even in the face of fairly strong selection ($V_s = 20\sigma_E^2$). However, while one can certainly make a case for plausibility, it is also true that we are very uncertain about the estimates of key parameters (V_s, $n\mu$, σ_α^2). As a result, one can just as reasonably take values towards the lower end of their uncertainty distributions, retaining the claim that stabilizing selection-mutation balance cannot account for existing levels of variation. Further, as

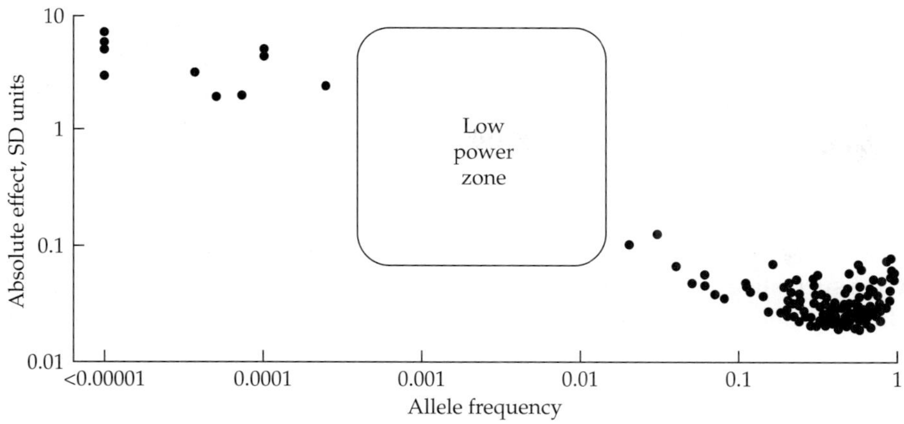

Figure 28.8 Plot of estimated allelic-effect size for known sites influencing human height versus allele frequency. The gap in the middle of the figure reflects a lack of power for either GWAS or linkage studies to detect genes in these regions. (After Kemper et al. 2012.)

we have noted above, load arguments imply that only a limited number of independent traits can be under stabilizing selection. Consequently, an observed V_s value for a focal trait is likely to be the projection (onto that trait) from some, likely very complicated, selection on a multivariate phenotype. Further, the strength of selection of interest is that on the *breeding value* of a trait, not that on its phenotype (Examples 28.1 and 28.16).

WHAT DOES GENETIC ARCHITECTURE TELL US?

A potential window into deciding which forces are predominantly responsible for quantitative variation is that the different models predict somewhat different genetic architectures. As noted by Kelly (2008), ideally such predictions are both **robust** and **exclusive**. Robust predictions imply that slight departures from model assumptions do not dramatically change the prediction, while exclusivity (predictions that are unique to a given model) is much more elusive.

One fairly robust prediction is that alleles in mutation-selection balance (MSB) should generally be at low frequencies. While this prediction is exclusive to MSB models, it does not distinguish between direct selection versus pleiotropic deleterious effects. In the case of direct stabilizing selection, alleles with larger trait effects have reduced fitness, generating a strong negative correlation between effect size and frequency. For trait alleles maintained by MSB due to pleiotropic deleterious fitness effects, the expected relationship between frequency and effect size is less clear. If there is a strong positive correlation between trait effect size and fitness, the same negative correlation is expected. Conversely, if there is a weak correlation, any such pattern would be greatly diminished.

These observations lead to the prediction of rare alleles of large effect under MSB scenarios (especially for direct stabilizing selection), while allele frequencies are expected to be more intermediate if balancing selection is involved. If trait alleles are largely neutral, (i.e., under selection, but only weakly so), then the distribution of allele frequencies is expected to be more L-shaped (approaching the Watterson distribution in an equilibrium population; Equation 2.34a), and (at best) show only a weak coupling between effect size and frequency. What do the data suggest? As we detail, the results from several independent lines of evidence are mixed.

As shown in Figure 28.8, the prediction under MSB of an inverse relationship between effect size and frequency clearly holds for human height. Alleles of large effect tend to be rare, although the poor resolution currently offered by mapping methods for genes of intermediate frequency and effect may temper this view somewhat. However, this observation leaves unresolved the issue of whether this pattern is due to direct stabilizing selection on height, pleiotropic fitness effects (especially for mutations of large effects), or both. It is worth mentioning in passing that the assumption of the basic form of genomic selection (GBLUP, RR-BLUP) of a roughly constant variance over sites implies an assumption that alleles of large effect are rare, so that their variances, $\simeq 2\alpha^2 p$, are roughly constant.

Accelerated Responses in Artificial Selection Experiments

If rare alleles of large effect are the norm, this would imply an increase in the additive variance when such alleles are favored by artificial selection (Barton and Turelli 1987; Maynard Smith 1989). While such accelerated selection responses are typically not seen (Chapter 25), their absence may not be very damning for the rare-alleles model (Keightley and Hill 1989; Zhang et al. 2004b; Zhang and Hill 2005a). Most experiments start with a small sample from a natural population is that bred at modest size in the laboratory for several generations before being subjected to selection, which is problematic for detecting alleles whose frequencies are on the order of $\sim\mu/s \ll 0.001$ in the base population. Under such conditions, with significant drift and founder effects, rare alleles will be lost (the majority of the time) or (rarely) increase to modest frequencies, in both cases diminishing the likelihood of generating an accelerated response (Zhang et al. 2004b). Keightley and Hill (1989) and Zhang and Hill (2005a) showed that the effects of such sampling, coupled with the effects of negative linkage disequilibrium generated by directional selection reducing the additive variance (Chapter 16), make the predicted short-term response under rare-alleles models very close to that from the classic infinitesimal model. Thus, the lack of accelerated response is not a fatal observation against rare-alleles models under many experimental designs.

However, as noted by Curtsinger and Ming (1997), using an appropriate design can significantly improve the chances of rare alleles being detected (also see simulations by Kelly 2008). Curtsinger and Ming constructed three replicate base populations with favorable alleles at low frequency. They did so by repeatedly backcrossing three different inbred lines to a line that was selected for high ethanol tolerance for over 50 generations. Specifically, the F_1s formed by crossing one of the inbred lines to the selected line was then backcrossed to the same inbred line, and then the resulting progeny again backcrossed to the same inbred line, and so on for five generations. Under this scheme, the frequencies of alleles from the increased tolerance line should be around 3% (0.5^5) in the base population for selection. They also constructed three control lines using the same general crossing scheme, but now backcrossing these same three inbred lines to an unselected population (the base population from which the tolerant line was selected). Thirty generations of selection for increased ethanol tolerance was performed using these six lines. All three lines constructed to contain favorable alleles (from the previously selected line) at low initial frequencies showed an acceleration in response around generation 15, while none of the control lines did. One key feature was large population size, with 1000 flies scored each generation and the top 20% used for the next generation.

Motivated by this "proof-of-concept" experiment, Kelly (2008) selected for both large and small flower size in *Mimulus guttatus*, using population sizes on the scale of the Curtsinger-Ming experiment. After accounting for potential scale effects (i.e., the variance increasing with the mean), Kelly found that the additive variance increased in the up-selected line, but decreased in the down-selected line. Such an asymmetric change in the variance is expected if rare alleles (presumably in MSB) disproportionately increase trait values. However, Kelly noted that such an asymmetric response could also occur with alleles at intermediate frequencies. He concluded that his results were, at best, only partly explained by the presence of rare alleles.

A related analysis by Nuzhdin et al. (1999) reached slightly different conclusions. They examined QTLs in high- and low-selected *Drosophila* lines for abdominal and sternopleural bristle number. While almost 30 QTLs were mapped, none was involved in the response in *both* the high- and low-selected lines. This suggests that none of the loci in the base population were segregating both positive and negative alleles at intermediate frequencies, indicating a pre-selection architecture closer to a rare-alleles model. This is somewhat surprising, as Long et al. (2000) found two intermediate-frequency polymorphisms in the *achaete scute* gene complex in natural *Drosophila* populations that generated significant variation for both types of bristle number. However, both polymorphisms resulted in a reduction in both abdominal and sternopleural bristles, suggesting that directional mutation bias at underlying loci (generating either largely positive or largely negative alleles) could have also generated the results observed by both Kelly and Nuzhdin et al.

A final selection-based test for the majority of standing variation being due to rare alleles is to compare the selection response using a bottlenecked versus a larger initial population (Robertson 1960a; James 1970; Frankham 1980). As reviewed in Chapter 26, these results are more consistent with intermediate-frequency alleles, but Zhang and Hill (2005a) cautioned that when linkage is considered, the tests may not have much discriminating power.

Kelly's Test for Rare Recessives

A related prediction from MSB is that deleterious alleles will not only be rare, but will also tend to be recessive (because additive alleles would be removed much more quickly). Further, there should be directional dominance (with heterozygotes being closer in fitness to wildtype, as opposed to mutant, homozygotes), leading to inbreeding depression (LW Chapter 10). Kelly (1999c) used this observation to construct a creative test for the presence of rare, recessive alleles. He noted that if genetic variation is due to such alleles, the ratio of the covariance of additive and dominance effects, $\sigma(a,d)$, to the additive variance, σ_A^2, should be greater than or equal to one (see the figure in Example 11.1). Recall that $\sigma(a,d) = \sigma_{ADI}/2$ (Table 11.1) appears in discussions of the covariance between inbred relatives (Chapter 11). Conversely, the ratio σ_{ADI}/σ_A^2 should be close to zero, or even negative, if most of standing variation is due to alleles at intermediate frequencies. Although Kelly assumed no epistasis, Charlesworth et al. (2007) discussed its impact on this test. Kelly (1999c, Kelly and Willis 2001) noted that while σ_{ADI} can be (rather imprecisely) estimated from covariances of inbred relatives (Chapter 11), a much cleaner estimate follows from a selection experiment. This is accomplished by contrasting the change in the mean, $\Delta\mu$, with the change in the coefficient, B, for inbreeding depression (Equation 23.1a). B (as well as μ) are measured over several generations of selection, and the ratio of their respective changes in computed (see Kelly 1999c for details). A value of $\Delta B/\Delta\mu$, which tracks σ_{ADI}/σ_A^2, that is greater than or equal to one is consistent with rare, recessive alleles. Negative values, consistent with intermediate alleles, were seen in three independent selection experiments on flower size in *Mimulus guttatus* (Kelly and Willis 2001), leading the authors to suggest that some form of balancing selection maintains flower size. Charlesworth et al. (2007) used Kelly's method to also find evidence of intermediate-frequency alleles underlying female fecundity in *Drosophila melanogaster* (also see Charlesworth 2015).

SUMMARY: WHAT FORCES MAINTAIN QUANTITATIVE-GENETIC VARIATION?

Over thirty years after Turelli (1984) lamented that more data and theory are needed to resolve the maintenance of variation conundrum, we have a flood of estimates of the strength of selection, rich GWAS and other genomic data, and considerably more theoretical development. Despite this, Turelli's main point (echoed by Johnson and Barton 2005) remains: ***There is still no clear resolution on the evolutionary forces responsible for the maintenance of variation***. Estimates of critical quantities, such as the strength of selection ($\widehat{V}_S$) and mutational parameters (n, μ, σ_α^2) are still sufficiently fuzzy to allow advocates of any particular model to proclaim that it largely fits the data, and opponents to insistent that it does not. Further, despite a wealth of sophisticated analysis of new models, most fail to jointly account for high apparent levels of stabilizing selection while maintaining sufficient variation. Parameter values allowing for sufficient variation result in too little apparent stabilizing selection, and vice-versa. Adding to the confusion, genetic architecture data suggest an important role for mutation-selection balance (MSB), namely, the negative association between trait effect and frequency, but also suggest intermediate frequency alleles are important for selection response, which is counter to predictions from MSB (at least in large populations).

What is clear? Barton's (1990) insight that much of the variation associated with a trait is likely maintained for reasons independent of that trait's fitness continues to gain traction, both empirically and theoretically. Load arguments suggest that stabilizing selection can only act on a relatively small set of independent dimensions in character space. Pleiotropy is clearly the norm for many or even most genes, although the number of traits that a new mutation impacts remains unclear, as does the critical question of the pleiotropic connection between alleles influencing a given trait and general fitness. Chapter 27 examined some of the theory on the issues.

One factor that is likely clouding our view is that much of the discussion of the maintenance of variation has been set in a univariate framework: given the apparent strength of selection on a specific trait, how can we account for its heritability? As Example 28.16 illustrates, and as we stress at length in Volume 3, selection and evolution act on integrated and connected multivariate phenotypes, not a series of independent univariate traits. One can have considerable heritabilities over a series of traits, yet have one (or more) indices based on these traits with heritabilities very close to zero. Hence, one could have selection based on one or more indices of trait values, where selection drives the heritability of the index to rather small values, while still having moderate heritabilities in the component traits. As illustrated in this chapter, the theory for the maintenance of variation based on single traits gives a very poor fit to the data, but it might be in better accord with the underlying selection index if we were able to extract such information from a multivariate analysis.

VII

Measuring Selection on Traits

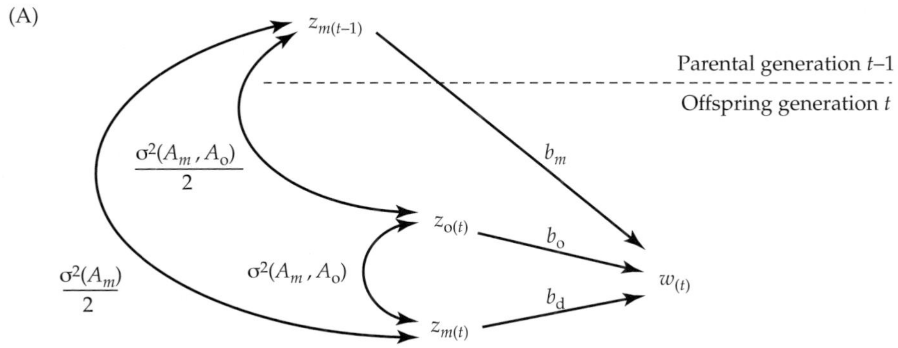

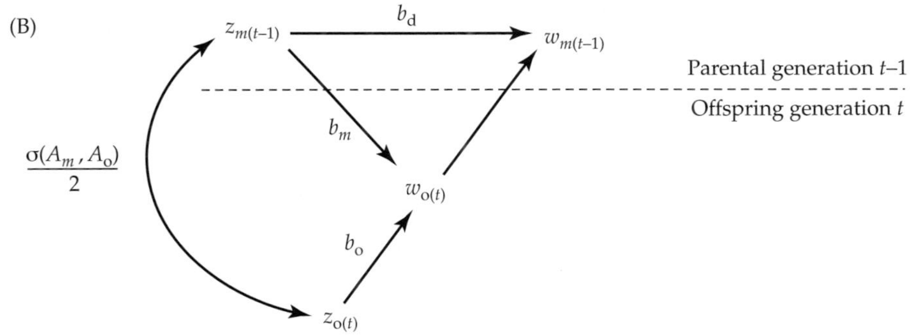

29

Individual Fitness and the Measurement of Univariate Selection

Natural selection is not evolution. R. A. Fisher (1930)

Previous chapters have examined the response to artificial selection, wherein the nature of selection is known to the investigator. Here we are concerned with the complementary issue of measuring how selection acts on particular phenotypes (**phenotypic selection**). This involves two related issues: measuring **individual fitness**, W, and determining if the phenotype of a particular character influences fitness (and, if so, the nature of this relationship). The latter question is often phrased in terms of estimating the generalized regression $W(z) = E[\, W \mid z]$, the expected fitness of an individual with character value z.

The first half of this chapter deals with various aspects of individual fitness, focusing on fitness components and measuring fitness over multiple episodes of selection. We review the properties of an especially useful statistic, the population variance in relative fitness (the **opportunity for selection**, I), which bounds the maximum possible within-generation change in the mean and variance of any character. We also examine issues regarding the measurement of **sexual selection** (fitness variation generated by mate choice).

The second half of this chapter considers the complementary problem of predicting the expected fitness of an individual given its phenotypic value. Our discussion here is concerned with selection acting exclusively on a single character. This admittedly unrealistic situation offers the advantage of allowing basic methodological points to be addressed without the additional complications inherent in a multivariate analysis. Chapter 30 extends these univariate ideas to the situation where the individual phenotypic value is a vector, $\mathbf{z}$, of trait values. The major complication in assigning fitnesses to traits is selection on phenotypically correlated characters. A within-generation change in the distribution of a trait may be due to **direct** selection on that trait, **indirect** effects of selection on phenotypically correlated (and usually unmeasured) characters, or both (Chapters 13, 20, and 30).

Excellent discussions on the detection of selection in natural populations can be found in Manly (1985), Endler (1986), Primack and Kang (1989), Brodie et al. (1995), and Kingsolver and Pfenning (2007), while summaries of estimates of the nature and strength of selection can also be found in Endler (1986), Conner (2001), Hoekstra et al. (2001), Kingsolver et al. (2001, 2012), Geber and Griffen (2003), Hereford et al. (2004), Kingsolver and Pfenning (2007), Siepielski et al. (2009, 2011, 2013), Kingsolver and Diamond (2011), and Morrissey and Hadfield (2012). We emphasize that these approaches are not restricted to "wild" populations. They can be used to study natural selection in domesticated populations, even those undergoing artificial selection. As noted by Stearns et al. (2010), these methods can also be used to study selection in contemporary human populations (Example 20.7), especially given the rich datasets that exist for a number of very large clinical cohorts.

Much of the literature on estimating fitness, fitness components, and fitness-trait relationships is dominated by ideas initiated by Pearson (1903) and greatly expanded by Lande and Arnold (1983) and Arnold and Wade (1984a, 1984b), which we largely follow. Indeed, almost all of the current estimates of fitness gradients are based on the approach of Lande and Arnold. Despite the vast reach and importance of these papers, the last few decades have seen movement away from some of their approaches and toward more statistically robust methods. Thus, we conclude with a discussion of some of these more recent developments, which continues into Chapter 30.

Evolution and Selection of Quantitative Traits. Bruce Walsh & Michael Lynch.
Published 2018 by Oxford University Press.
DOI 10.1093/oso/9780198830870.001.0001

EPISODES OF SELECTION AND THE ASSIGNMENT OF FITNESS

Selection can often be subdivided into discrete, nonoverlapping stages called **episodes of selection**. For example, a distinction is often made between **viability selection** (differences in survivorship) and **fertility selection** (differences in the number of offspring per mating). **Tradeoffs** may be found, wherein a trait that does well in one episode does poorly in another. For example, large body size is usually favored in adults of Darwin's medium ground finch (*Geospiza fortis*; Boag and Grant 1981; Price et al. 1984), while small body size is apparently favored in juveniles (Price and Grant 1984). The possibility of tradeoffs between **natural selection** and **sexual selection**, which was first suggested by Darwin (1859), has also received significant attention. It is interesting to note that Darwin's contemporary, and the co-proposer of evolution by natural selection, Alfred Wallace, did not believe that females would choose mates with lower viabilities, and hence discounted any such sexual selection versus natural selection tradeoffs (Moore and Pannell 2011). Sexual selection occurs when variance in the ability to acquire mates results in variance in the number of offspring, while natural selection results from variance in all other fitness components, such as viability and fertility differences, and variation in the quality in parental care.

Fitness Components

We start with two simplifying assumptions. The first is that generations are discrete and non-overlapping, where the actual *timing* of reproduction is unimportant. Second, we assume that the parental phenotypes (or more generally, the phenotypes of other individuals that interact with our focal individual) have no influence on that individual's fitness (Chapter 22). Some of the consequences when these assumptions fail will be considered shortly.

Under these simplifying assumptions, the **lifetime** (or **total**) **fitness** of an individual is the number of descendants that it leaves at the start of the next generation. When measuring the total fitness of an individual, care must be taken not to span generations or to overlook any stage of the life cycle in which selection acts. To accommodate these concerns, lifetime fitness is defined as the total number of zygotes (newly fertilized eggs) that an individual produces. Measuring total fitness from any other starting point in the life cycle (e.g., from adults in one generation to adults in the subsequent generation) can result in a very distorted picture of the true fitness of particular phenotypes (Prout 1965, 1969). In particular, if generations are spanned, measures of selection on a particular parental phenotype will be averages over both parental and offspring phenotypes, which can differ considerably. We revisit this strict view of not allowing fitness measures to cross generational boundaries later in this chapter (and also in Chapter 30 and in Volume 3), where we examine how fitnesses associated with parental care should be modeled.

Systems for measuring lifetime fitness have been especially well developed for laboratory populations of *Drosophila* (reviewed by Sved 1989). Measurement of *lifetime* fitness in nature is much more difficult and (not surprisingly) is rarely accomplished (although see Chapter 20). Attention instead is usually focused on particular *episodes* of selection or selection during particular phases of the life cycle. Fitness components for each episode of selection are usually defined to be **multiplicative**. For example, lifetime fitness (total number of offspring) can be partitioned as (probability of surviving to reproductive age)·(number of mates)·(number of zygotes per mating). Number of mates is a measure of sexual selection, while the viability and fertility components measure natural selection. A commonly measured fitness component is **reproductive success** (**RS**), the number of offspring per adult, which confounds natural selection (fertility) and sexual selection (the number of successful matings). Clutton-Brock (1988a) reviewed estimates of reproductive success from natural populations.

Fitness components can themselves be further decomposed. For example, fertility in plants might be decomposed as (seeds per plant) = (number of stems per plant)·(number of inflorescences per stem)·(average number of seed capsules per inflorescence)·(average number of seeds per capsule). This decomposition allows an investigator to ask questions of the form: do plants differ in the number of seeds mainly because some plants have more stems, or more flowers per stem, or are there tradeoffs between these? As this example suggests, the nature of fitness components chosen by an investigator is often set by ecological or behavioral, rather than evolutionary, concerns. Ideally, components should be *sequential*, with one episode finishing before the next begins, or (as with our plant example) they should represent nonoverlapping features. However, complications can occur. For example, the total number of mates (which is seemingly a measure of sexual selection) is equivalent to the product of the average number of mates per day and the number of days survived during the mating period. The former is a strict measure of sexual selection, whereas the latter (viability) is a measure of natural selection, resulting in the total number of mates being a compound measure of both natural and sexual selection.

Estimates of fitness can be obtained from either **longitudinal** studies or **cross-sectional** studies. A longitudinal study follows a cohort of individuals over time, while a cross-sectional study examines individuals at a single point (or a very narrow window) in time. Cross-sectional studies often generate only two fitness classes (e.g., dead versus living, mated versus unmated). Longitudinal studies are preferred, as the analysis of cross-sectional studies involves numerous assumptions (Lande and Arnold 1983; Arnold and Wade 1984b). Unfortunately, longitudinal studies usually require far more work and can be extremely challenging in many settings. As mentioned, age-structured populations pose further complications that will be considered in brief shortly, and in detail in Volume 3.

Finally, more complicated life cycles can be modeled as a path diagram (LW Appendix 2), generating a **life-history graph** (Figure 29.12). This more general structure allows for much more complex interactions among fitness components, which need not be sequential. One approach for the analysis of such graphs uses so-called **aster models** (Geyer et al. 2007; Shaw et al. 2008), a statistically rigorous framework that explicitly recovers the correct distribution of fitness effects after multiple episodes of selection. We examine this approach at the end of the chapter. Other approaches, such as those based on path analysis (e.g., latent variable analysis), are discussed in Chapter 30.

Assigning Fitness Components

We now turn to the task of partitioning measures of individual fitnesses in a longitudinal study into fitness components. A cohort of n individuals (indexed by $1 \leq r \leq n$) is followed through several discrete (nonoverlapping) episodes of selection. Let $W_j(r)$ be the fitness measure for the jth episode of selection for individual r. For example, if we are following viability, then W_j is either zero (dead) or one (alive) at the census period. **Relative fitness** components, $w_j(r) = W_j(r)/\overline{W}_j$ (mean-standardized fitnesses), will turn out to be especially useful (as $\overline{w}_j = 1$). At the start of a study, the frequency of each individual is $1/n$, yielding the mean fitness for the first (observed) episode of selection as

$$\overline{W}_1 = \frac{1}{n}\sum_{r=1}^{n} W_1(r) \tag{29.1a}$$

(We note the very real possibility that ***considerable selection on the focal trait may have already occurred prior to the life-cycle stages being examined***; Example 29.1.) Following this first episode, the new fitness-weighted frequency of the rth individual is $w_1(r)/n$, implying that the mean fitness for the second episode of selection is calculated by

$$\overline{W}_2 = \sum_{r=1}^{n} W_2(r) \cdot w_1(r) \cdot \left(\frac{1}{n}\right) \tag{29.1b}$$

In general, for the jth episode of selection,

$$\overline{W}_j = \sum_{r=1}^{n} W_j(r) \cdot w_{j-1}(r) \cdot w_{j-2}(r) \cdots w_1(r) \cdot \left(\frac{1}{n}\right) \tag{29.1c}$$

Note that if $W_j(r) = 0$, further fitness components for r are unmeasured. If we let $p_j(r)$ be the fitness-weighted frequency of individual r after j episodes of selection, it follows that $p_0(r) = 1/n$ and

$$p_j(r) = w_j(r) \cdot p_{j-1}(r) = \frac{1}{n} \prod_{i=1}^{j} w_i(r), \quad \text{where} \quad \sum_{r=1}^{n} p_j(r) = 1 \tag{29.2a}$$

The $1/n$ term represents the initial frequency of individual r, while the product is the total relative fitness for individual r following the j episodes of selection. It immediately follows that Equation 29.1c can be expressed as $\overline{W}_j = \sum W_j(r) \cdot p_{j-1}(r)$. Using these weights allows fitness-weighted moments to be calculated. For example, the fitness-weighted mean of a particular character following the jth episode is given by

$$\overline{z}_j = \sum_{r=1}^{n} z(r) \cdot p_j(r) \tag{29.2b}$$

where $z(r)$ is the value of the character in individual r. Likewise, the sample variance is

$$\text{Var}(z_j) = \frac{n}{n-1} \left[\sum_{r=1}^{n} z^2(r) \cdot p_j(r) - (\overline{z}_j)^2 \right] \tag{29.2c}$$

As discussed later in the chapter, the fitness distribution over a particular episode (or component) may closely correspond to a standard distribution, such as a Bernoulli random variable (with a value of zero or one) for survival or mating status, or a Poisson for the number of offspring or mates. However, the resulting compound distribution of fitness over several episodes is generally very complex. In particular, we usually expect a large probability mass at zero, representing individuals that did not reproduce. The resulting distribution generally does not correspond to any standard one, nor can it easily be transformed into one. As discussed at the end of the chapter, the aster-model approach can be used to construct the full unconditional distribution of total fitness.

Example 29.1. Mojica and Kelly (2010) presented an important cautionary study regarding the interpretation of fitness components. They reviewed a number of studies showing a distinct fitness advantage for larger flower size. To test this association, they planted seedlings of three distinct flower-size genotypes (High, Low, and Control) of yellow monkey flower (*Mimulus guttatus*) in the wild. These lines were the result of selection experiments for increased (High) and decreased (Low) flower size from the starting control population.

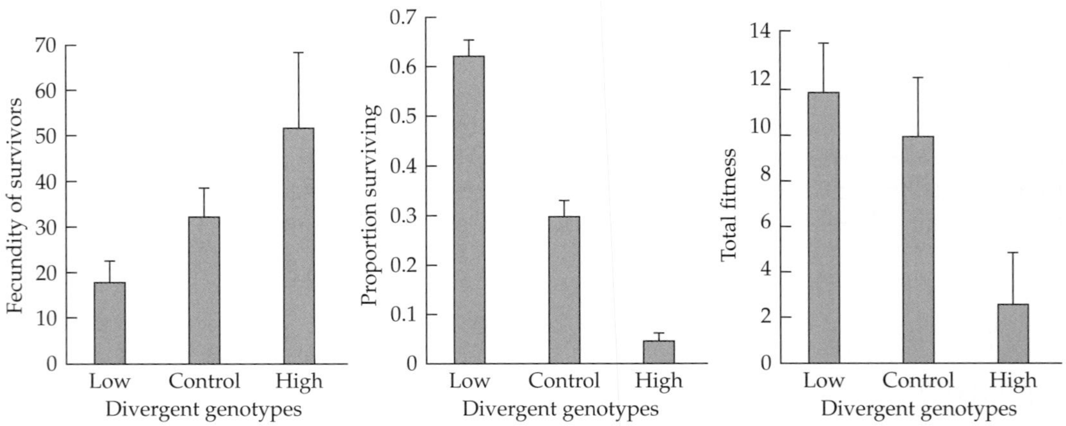

As the left-most figure shows, surviving individuals of the larger-flower (High) genotype had significantly higher fecundity than did the control or the small-flower (Low) genotype, consistent with the literature pattern of larger flowers having a selective advantage. However, the authors also found a massive (over tenfold) difference in survival (middle panel). The large-flower genotype required significantly more time to mature (flower) during a period when water stress becomes increasingly important. Total fitness (fecundity times survival; right panel) shows a very large advantage for the small-flower over the large-flower genotypes, which would be entirely missed in the absence of these transplant experiments. Note also that there is not a significant difference in total fitness between the control and small-flower genotypes, although the control genotype had significantly higher fecundity and significantly lower survival than the small-flower genotype, illustrating tradeoffs between life-history components.

Example 29.2. Total reproductive success ($W = W_1 \cdot W_2$) and its components, mating success (W_1) and fertility (W_2; eggs per successful mating), were measured in 38 male bullfrogs (*Rana catesbeiana*) in a longitudinal study conducted by Howard (1979). For illustrative purposes, we use part of this dataset to examine these fitness components in five males, along with their body size, z (in mm):

Male	z	W_1	W_2	w_1	p_1	w_2	p_2
1	145	1	25,820	0.714	0.143	1.628	0.233
2	128	1	22,670	0.714	0.143	1.429	0.204
3	148	0	0	0	0	0	0
4	138	2	7,230	1.429	0.286	0.456	0.130
5	141	3	15,986	2.143	0.428	1.008	0.433

Before selection, each male has frequency of $1/5$, resulting in

$$\overline{W}_1 = \frac{1}{5}\left(1+1+0+2+3\right) = \frac{7}{5} = 1.4 \quad \text{and } w_1 = \frac{W_1}{1.4}$$

While the *observed* frequencies of individuals have not changed after the first episode of selection (all are still present in the population), their *fitness-weighted* frequencies change due to differences in acquiring mates. For male 2, $p_1(2) = 0.714/5 = 0.143$ (e.g., 14.3% of the seven matings in the population involved this male), and the values for the other adults were similarly obtained. The mean fertility *per mating* is

$$\begin{aligned}\overline{W}_2 &= \sum_{r=1}^{n} W_2(r) \cdot p_1(r)\\ &= (25,820 \cdot 0.143) + (22,670 \cdot 0.143) + (7230 \cdot 0.286) + (15,986 \cdot 0.429) = 15,860\end{aligned}$$

If each reproducing male were weighted equally, average total fertility per individual (independent of the number of times each mates) would be (1/4)·(25,820 + 22,670 + 7230 + 15,986) = 17,927. The actual mean fertility *per mating*, $\overline{W}_2$ = 15,860, is lower because males 4 and 5 had much lower fertility than the other (successful) males. Using $w_2(r) = W_2(r)/(15{,}860)$ and recalling Equation 29.2a, the final fitness weighting for male 1 is (1/5)· 0.714· 1.628 = 0.233 (namely, 23.3% of all eggs produced by these five males were sired by this male). The remaining p_2 values in the table are similarly computed.

Using the body sizes for the males given above, the preselection mean and variance are $\overline{z}_0$ = 140.0 and Var(z_0) = 59.5. From Equation 29.2b, the fitness-weighted mean following the first episode of selection is

$$\overline{z}_1 = (145 \cdot 0.143) + (128 \cdot 0.143) + (148 \cdot 0) + (138 \cdot 0.286) + (141 \cdot 0.429) = 138.86$$

Thus, if we take a hypothetical offspring following the first episode of selection, the mean size of the adult that produced this randomly chosen offspring is 139. Similarly, we find that Var(z_1) = 30.51, $\overline{z}_2$ = 138.88, and Var(z_1) = 43.7.

Potential Issues With Assigning Discrete Fitness Values

In many studies, fitness cleanly falls into discrete categories. Often these are simply binary, such as alive/dead or mated/unmated. Zuk (1988) and Blanckenhorn et al. (1999) noted that biased sampling is not uncommon in such situations, with a tendency to oversample the rarer class. While such sampling can improve hypothesis testing (increasing the power to test whether a trait is a target of selection), if the investigator is careless in the weighting of the estimates from these two samples, such biased sampling generates biased estimates of phenotypic selection (Example 29.3).

Example 29.3. Suppose that 10% mortality occurs during an episode of selection and one is examining whether a focal trait influences fitness. The trait may be somewhat difficult to measure, making it tempting to oversample dead individuals in order to gain more power for testing whether a difference in trait value occurs between groups. Suppose the mean of the trait is 100 in surviving individuals and 50 in those showing mortality. With 10% mortality, the mean (before selection) is just $0.1 \cdot 50 + 0.9 \cdot 100 = 95$, giving a selection differential of $S = 100 - 95 = 5$. However, if one oversampled dead individuals to gain more precision in the mean of the trait in this group, say with 25% of the sample being dead individuals, then the (before-selection) sample mean becomes $0.25 \cdot 50 + 0.75 \cdot 100 = 87.5$, and the resulting selection differential is overestimated, with $S = 100 - 87.5 = 12.5$. If the investigator neglects to standardize the weights for these classes when computing the before-selection mean, oversampling of a rare class that is favored (survives or mates) relative to a more common unfavored class (dies or does not mate) results in an *underestimation* of selection differentials. Conversely, oversampling the rare class that is unfavored results in an *overestimation* of the selection differential.

A second, more subtle, issue with assigning fitness values to discrete classes was noted by Brodie and Janzen (1996). The absolute value of fitness assigned to a class can, in some cases, influence measures of selection. In the binary (Bernoulli) case, where fitness is scored as 0 and x (such as dead/living or unmated/mated), any (strictly positive) value can be chosen for x, as the relative fitnesses are independent of the choice of x. However, with three (or more) discrete fitness classes, this is no longer true. Brodie and Janzen presented an example in which laboratory survivorship was followed over a four-year period in turtles. One could code these data as simply 0 (do not survive) or 1 (survive). However, the data could also be scored as 0 through 4, depending on the latest year of survival. They also noted that it might make sense to double the assigned fitness value in year 4, as this is when reproduction typically starts in nature. These different weighting schemes give different relative fitness values for the surviving age classes, and hence different measures of selection. This problem arises because only one component of selection (viability) is measured while reproduction is also occurring, and the different proposed fitness-weighting schemes attempt to accommodate for this unmeasured component. As discussed later, using an aster-model approach with the appropriate life-history graph avoids this problem.

Assigning Components of Offspring Fitness to Their Mothers

When assigning a parent's reproductive fitness, offspring generally should be counted at the zygote (newly fertilized egg) stage. If counted later in life, offspring may have experienced selection based on their own, as opposed to their parents', phenotypes, thus confounding the targets of selection (Lande and Arnold 1983; Cheverud and Moore 1994). However, it is not uncommon for avian and mammalian evolutionary geneticists to count only "successful" offspring. For example, although eggs in the nest is a close measure of number of zygotes produced, the number of offspring from a parent is often scored as **hatchlings** (eggs that successfully hatch), **fledglings** (offspring that successfully leave the nest), or **recruits** (offspring recorded as breeding later in life). These measures move increasingly away from number of zygotes and can reflect selection on features of the offspring, rather than its parent. Many biologists counter that maternal care is critical, and thus the success of the offspring is a maternal, rather than offspring, trait (Grafen 1988). Where indeed does one draw the line? The strict barrier that is set at zygotes is formally correct *provided* parental phenotypes have no influence on offspring fitness. However, it is also clear that in species with significant parental investment in offspring care (such as birds and mammals, as well as seeds with significant endosperm contribution), the genotype and phenotype of the mother can influence the fitness of her offspring *independently* of the offspring phenotype.

Much of this apparent confusion arises from thinking about how to partition fitness over parents and offspring as a univariate selection problem. In reality, however, this assignment is a multiple-trait problem, with offspring survival potentially involving both the **direct** effect of an offspring on its own fitness and an **indirect** effect from the genotype/phenotype of the mother (Kirkpatrick and Lande 1989), see Figure 15.5. Recall that we have visited this issue before in Chapter 22, wherein the fitness of an individual is influenced by its own phenotype (its direct effect) and the **associative effects** of individuals (potentially parents, other relatives, or even unrelated individuals) around it. From the standpoint of assigning fitnesses, when the maternal phenotype has no effect on offspring fitness, assigning the fitness of the mother by counting her offspring *after* they may have experienced one or more episodes of selection (such as viability selection) is misleading.

Conversely, if the maternal phenotype *does* impact the survival of her offspring independent of their own direct effects, failure to include this also creates a misleading picture (Wolf and Wade 2001; Thomson and Hadfield 2017). In terms of correlated traits, one can view maternal performance as one trait and offspring performance as another. Both can influence the fitness of an offspring. If these are uncorrelated (i.e., there is no genetic or phenotypic correlation between the direct and maternal effects), maternal fitness should be assigned as the number of offspring *following* the episodes of selection in the offspring that are influenced by the maternal performance trait. For example, a trait may influence survival to the fledging or weaned stage but then have no future impact. Simply counting the number of eggs laid misses the maternal contribution. Conversely, even if the maternal contribution is important, if there are correlations between direct and maternal traits, then a misleading picture of selection can arise unless we uncouple them.

Given this uncertainty, how should one proceed? One suggestion is to perform separate analyses with several measures of reproductive success (based on counting offspring at different stages) and examine their consistency. Alternatively, as we saw with Equations 15.39a and 15.40, one can use a multiple-regression approach, including both offspring- and maternal-trait values (Kirkpatrick and Lande 1989; Thomson and Hadfield 2017). The limitation of this latter approach is that it is trait-based (one has to specify, and measure, maternal features influencing offspring fitness). We return to this issue in Chapter 30.

Concurrent Selective Episodes, Reproductive Timing, and Individual Fitness, λ_{ind}

The partitioning of selection into discrete episodes is, of course, an abstraction of the real world. Often this partitioning is simply done for the convenience of a researcher who wishes to measure individual fitness by considering just one or two episodes, potentially looking for tradeoffs between them. Selection episodes (e.g., viability and reproduction) often occur *concurrently*, rather than sequentially. For example, in the turtle data of Brodie and Janzen (1996), one has only information about viability (in the laboratory), and how to weight this information to account for the translation of viability into the timing of reproductive success raised issues in assigning fitness values. While methods to partition concurrent episodes of selection into their components have been proposed (e.g., Hamon 2005), the more general problem is assigning an appropriate measure of fitness that accounts for reproductive timing.

We have been stressing the use of **lifetime reproductive success** (**LRS**), but this is a *rate-insensitive* measure, which counts only the total number of offspring and not the actual *timing* of when offspring are produced. If reproduction occurs in a discrete window and the generations do not overlap, LRS provides a good measure of fitness. However, when generations do overlap, the actual timing of reproductive events can be more important than the total number of offspring, as earlier-produced offspring have a head start on contributing their own offspring to the population (Lande 1982). Consider three individuals, all of which produce 40 offspring during their lifespan, and hence each has the same fitness when measured by LRS. However, suppose individual 1 has 10 offspring each at ages 2, 3, 4, and 5; individual 2 has 20 offspring at both ages 4 and 5; and individual 3 has 20 offspring at ages 2 and 3. Clearly, individual 3 has a higher *rate* of offspring production than individuals 1 or 2 on an absolute time basis, even though it has the same LRS as the other two individuals. How do we account for this?

One solution is based on **age-projection** (or **Leslie**) **matrices** (Leslie 1945, 1948; Caswell 1989, 2001), an approach first proposed by Lenski and Service (1982). McGraw and Caswell (1996) forcefully argued for this approach as a measure of individual fitness in age-structured populations. The growth rate of an age-structured population can be determined from ℓ_x, the probability of surviving from age (or class/stage) x to $x+1$, and b_x, the birth rate (fecundity) in class x (the ℓ_x and b_x are collectively referred to as the **vital rates**). This information is expressed in a $k \times k$ matrix, where k is the upper age limit on reproduction. The first row of the Leslie matrix contains the fecundities, while the below-diagonal line consists of the survival values, ℓ_x,

$$\mathbf{L} = \begin{pmatrix} b_1 & b_2 & \cdots & b_{k-1} & b_k \\ \ell_1 & 0 & \cdots & 0 & 0 \\ 0 & \ell_2 & \cdots & 0 & 0 \\ \vdots & & \ddots & 0 & 0 \\ 0 & 0 & \cdots & \ell_{k-1} & 0 \end{pmatrix} \tag{29.3a}$$

If $\mathbf{n}(t)$ is a vector of the number of individuals in each age/stage class at time t, then $\mathbf{n}(t+1) = \mathbf{Ln}(t)$. The **asymptotic growth rate**, λ, for this population is the largest eigenvalue of $\mathbf{L}$, while its associated eigenvector is the stable age distribution. The Leslie matrix is just one type of life-history projection matrix. More generally, the life history of a species may be more accurately defined by **stages**, rather than *ages*. For example, a perennial plant could spend many years in a rosette stage before flowering. Life-history graphs are a more general approach for categorizing such stage-structured organisms. In an age-structured model, individuals increase in age at each step, but in a stage-structured model, an individual can *remain* in the same stage on the next step (e.g., stays as a rosette), generating a loop in the graph (an arrow that circles back to itself), and hence a nonzero diagonal element, L_{ii}, representing the chance of remaining in stage i in the next step (Caswell 1989, 2001).

Now consider a particular individual that last reproduced at age m. The resulting Leslie matrix for *this individual* is

$$\mathbf{L}_{ind} = \begin{pmatrix} f_1 & f_2 & \cdots & f_{m-1} & f_m \\ 1 & 0 & \cdots & 0 & 0 \\ 0 & 1 & \cdots & 0 & 0 \\ \vdots & & \ddots & 0 & 0 \\ 0 & 0 & \cdots & 1 & 0 \end{pmatrix} \tag{29.3b}$$

where $f_x = b_x/2$ is half the number of offspring that the individual produced at age x (using half accounts for both males and females contributing to the population, to avoid double-counting of the offspring). Note that LRS $= 2\sum_j^k f_j$. The individual-specific matrix, $\mathbf{L}_{ind}$, is very similar to a population growth matrix ($\mathbf{L}$), but in $\mathbf{L}_{ind}$, the fecundities are those that were *observed* for the focal individual (as opposed to *population averages* for each age class). In addition, because this individual survives to (at least) age m, we replace the ℓ_x elements in $\mathbf{L}$ (the average survival per age class) by ones, indicating the actual survival of the focal individual. The "growth" rate, λ_{ind}, for this individual is given by the largest eigenvalue of $\mathbf{L}_{ind}$, which can be obtained by solving the modified **Euler-Lokta equation**

$$\sum_{i=1}^{m} f_i \, \lambda_{ind}^{-i} = 1 \tag{29.3c}$$

Note from this equation that λ_{ind} can be thought of as an **age-discounted LRS**, a rate-sensitive measure of fitness, as opposed the rate-insensitive measure given by LRS.

McGraw and Caswell (1996) noted two sources of bias in using λ_{ind}. First, for any given phenotypic class, the particular realization of the matrix of vital rates ($\mathbf{L}_{ind}$) for an individual in that class (i.e., the expected survival, $\ell_x(z)$, and fecundity, $b_x(z)$, values for phenotype z) is a biased estimator of the matrix for that class. The reason is that random death prevents a proper estimation of the values of b_x, especially at later ages. The second source of bias, which was also a concern for Lenski and Service (1982), was that the average of the leading eigenvalue for each realization of $\mathbf{L}_{ind}$ does not generally equal the leading eigenvalue for the average matrix (i.e., the average of the values of λ_{ind} does not equal the population growth rate, λ). Despite these issues, McGraw and Caswell (1996) still favored the use of λ_{ind}, while Lenski and Service (1982) developed a resampled measure (based on $\mathbf{L}_{ind}$) with less bias. Alternatively, if the mean population growth rate (λ) is known, then an unbiased estimator for the relative fitness for individual j is

$$w(j) = \sum_{i=1}^{k} f_i(j) \; \lambda^{-i} \tag{29.3d}$$

where $f_i(j)$ denotes half of j's fecundity at age i (Lenski and Service 1982).

Example 29.4. In the case of overlapping generations, the timing of when reproduction occurs can be at least as important as the lifetime reproductive success (LRS). Consider the three hypothetical individuals mentioned previously, each with an LRS of 40 but with differences in the timing of reproduction. Recalling that the values of f_i in Equation 29.3b represent half the number of offspring, the Leslie matrices for these three individuals become

$$\mathbf{L}_1 = \begin{pmatrix} 0 & 5 & 5 & 5 & 5 \\ 1 & 0 & 0 & 0 & 0 \\ 0 & 1 & 0 & 0 & 0 \\ 0 & 0 & 1 & 0 & 0 \\ 0 & 0 & 0 & 1 & 0 \end{pmatrix}, \qquad \mathbf{L}_2 = \begin{pmatrix} 0 & 0 & 0 & 10 & 10 \\ 1 & 0 & 0 & 0 & 0 \\ 0 & 1 & 0 & 0 & 0 \\ 0 & 0 & 1 & 0 & 0 \\ 0 & 0 & 0 & 1 & 0 \end{pmatrix}, \qquad \mathbf{L}_3 = \begin{pmatrix} 0 & 10 & 10 \\ 1 & 0 & 0 \\ 0 & 1 & 0 \end{pmatrix}$$

The leading eigenvalues for these three matrices are $\lambda_1 = 2.77$, $\lambda_2 = 1.97$, and $\lambda_3 = 3.58$. Under this metric, individuals 1 and 2 are, respectively, 77% and 55% as fit as individual 3.

Example 29.4 shows that individuals with the same LRS can have very different λ_{ind} values. Indeed, the rankings of individuals based on LRS versus λ_{ind} can be rather nonconcordant. Hence, it should not be surprising that Brommer et al. (2002) noted several cases where inferences on selection were altered when λ_{ind} was used in place of LRS. An especially telling example in shown in Figure 29.1, which plots these two measures of fitness for female Ural owls (*Strix uralensis*). As the figure shows, there is a diminishing return in λ_{ind} with increasing values of LRS. This occurs because reproductive contributions later in life are increasingly down-weighted by λ_{ind} (Equation 29.3c).

While there are strengths to using λ_{ind}, it is not without problems. Brommer et al. (2002) made the important point that the stages *when* offspring are scored (e.g., eggs vs. hatchlings) is critical. Although LRS can also change given differences in the stage of scoring,

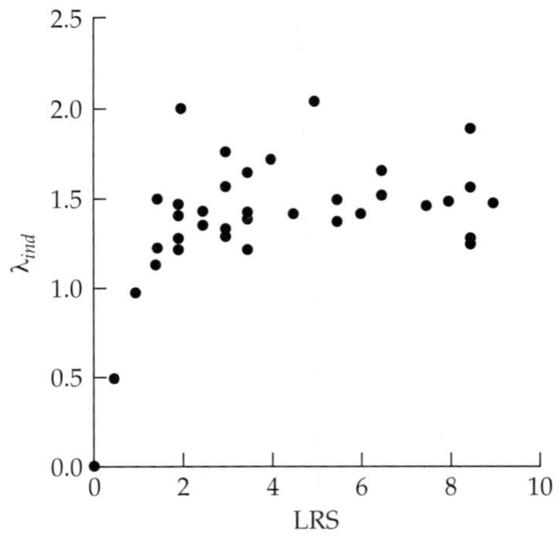

Figure 29.1 Lifetime reproductive success (LRS) versus λ_{ind} in 56 Ural owls (*Strix uralensis*) where offspring were measured as fledglings. Increasing LRS results in diminishing returns in λ_{ind}, which occurs because additional reproduction later in life is increasingly down-weighted by this rate-sensitive measure. (After Brommer et al. 2002.)

the effects can be much more dramatic for λ_{ind}, which differentially weights offspring produced at different ages. An interesting discussion on the merits of LRS versus λ_{ind} was given by Brommer et al. (2004), who compared these single-generation measures of fitness with the long-term genetic contribution of an individual. Using multigenerational population studies on Ural owls (a long-lived species) and shorter-lived collard flycatchers (*Ficedula albicollis*), they examined the correlation of both measures (LRS and λ_{ind}) with the long-term genetic contribution of an individual to future generations. The latter was assessed from known pedigrees as the expected number of gene copies left after more than two generations. Following the standard practice of ornithologists, the number of offspring from a parent was counted using two different measures of success: number of fledglings and number of recruits. This resulted in fitnesses measured by lifetime fledgling production (LFP) and lifetime recruit production (LRP) as the LRS analogs, and their rate-sensitive counterparts, $\lambda_{ind}(fp)$ and $\lambda_{ind}(rp)$. LFP had a higher correlation with the number of descendants (gene copies left after at least two generations) than did $\lambda_{ind}(fp)$. For flycatchers, the correlation between LFP and number of descendants was 0.35, while it was only 0.20 for $\lambda_{ind}(fp)$. In owls, these correlations were 0.54 for LFP and 0.19 for $\lambda_{ind}(fp)$. When fitness was based on recruits, LRS and $\lambda_{ind}(rp)$ performed equally well in both species in predicting the long-term contribution. Brommer et al. expressed concern that rate-sensitive measures such as λ_{ind} discounted later reproduction too harshly, and that rate-insensitive measures (such as LRS) might be a more robust choice.

In closing, we remark that there is a robust discussion in the literature on the appropriate measures of selection in age-structured populations (a partial list includes Charlesworth 1980, 1983, 1994a; Lande 1982; Lenski and Service 1982; Travis and Henrich 1986; Partridge and Harvey 1988; Henle 1991; Metz et al. 1992; Kozlowski 1993; de Jong 1994; Benton and Grant 1996, 2000; McGraw and Caswell 1996; Brommer 2000; Link et al. 2002; Brommer et al. 2003; Coulson et al. 2005; Metcalf and Pavard 2007; Horvitz et al. 2010; and Engen et al. 2011, 2012). Indeed, Stearns (1976) said it best: "Fitness: something everyone understands but no one can define precisely," and later (1992) noted that "all fitness definitions are tools invented by scientists to analyze natural selection." In part, this debate arises because different questions (e.g., whether a new phenotype can invade a population vs. determining the change in a trait value following selection) can result in different appropriate definitions of fitness.

Sensitivities and Elasticities of the Elements of L

Measures of how a small change in a vital rate (one of the elements of **L**) impact λ provide a connection between selection on particular traits and demography (van Tienderen 2000). An introduction to these metrics was offered by de Kroon et al (2000), with a more detailed treatment by Caswell (1989, 2001). The **sensitivity**, s_{ij}, of an element, L_{ij}, of a Leslie matrix is defined as the impact of a small change in that element on the total growth rate,

$$s_{ij} = \frac{\partial \lambda}{\partial L_{ij}} = \frac{v_i \cdot w_j}{\mathbf{v}^T \mathbf{w}} \tag{29.4a}$$

where **v** and **w** are the left and right eigenvectors associated with the dominant eigenvalue (λ) of **L** (Caswell 1978). The right-most identity follows from the spectral decomposition (Equation A5.9a) of **L**. The right eigenvector is the familiar $\mathbf{L}\mathbf{w} = \lambda\mathbf{w}$, while the **left eigenvector** satisfies $\mathbf{v}^T\mathbf{L} = \lambda\mathbf{v}^T$. Taking transposes on both sides shows that $\mathbf{L}^T\mathbf{v} = \lambda\mathbf{v}$, which means that **v** is the right eigenvalue of $\mathbf{L}^T$. Sensitivity provides the *absolute* change in the growth rate, λ, given an infinitesimal change in L_{ij}.

In contrast, the **elasticity**, e_{ij}, of element L_{ij} is the *proportional* change in λ given a *proportional* change in L_{ij},

$$e_{ij} = \frac{\partial \lambda / \lambda}{\partial L_{ij} / L_{ij}} = \frac{L_{ij}}{\lambda} \frac{\partial \lambda}{\partial L_{ij}} = \frac{L_{ij}}{\lambda} s_{ij} \tag{29.4b}$$

Recalling from the chain rule of calculus that $\partial \ln[f(x)]/\partial x = [1/f(x)] \cdot [\partial f(x)/\partial x]$ shows that elasticities can also be expressed as

$$e_{ij} = \frac{\partial \lambda / \lambda}{\partial L_{ij} / L_{ij}} = \frac{\partial \ln(\lambda)}{\partial \ln(L_{ij})} \tag{29.4c}$$

Caswell (1984) and de Kroon et al. (1986) showed that the elasticities of all of the elements of **L** sum to one, giving the result of de Kroon et al. (2000), namely, that

$$\lambda = \lambda \cdot 1 = \lambda \sum_{ij} e_{ij} = \lambda \sum_{ij} \frac{L_{ij}}{\lambda} s_{ij} = \sum_{ij} L_{ij} s_{ij} \tag{29.4d}$$

As will be discussed later in the chapter, elasticities provide a link between the impact of a trait on fitness components and the impact of those components on the growth rate (van Tienderen 2000). Volume 3 examines selection in age-structured populations in more detail.

VARIANCE IN INDIVIDUAL FITNESS

How do we compare the amount of selection acting on different populations? At first glance, one might consider using the variance-standardized selection differential (the selection intensity), $\bar{\imath} = S/\sigma$, to compare the relative strengths of individual selection between populations (Chapter 13). The limitation with $\bar{\imath}$ as a measure of *overall* selection on populations is that it is *character specific*. Hence, $\bar{\imath}$ may be appropriate when comparing the strength of selection on a particular *character* (one could also use mean-standardization, S/μ, instead; Chapters 13 and 30), but it is inappropriate for comparing the overall strength of selection in a *population*. This is because two populations may have the same $\bar{\imath}$ value for a given character, but if that character is tightly correlated with fitness in one population and only weakly correlated in the other, selection will be much stronger in the latter population.

Another issue with using $\bar{\imath}$ is that considerable selection can occur without changing the mean (e.g., stabilizing selection). Haldane (1954) and Van Valen (1965) proposed measures of the intensity of selection, which is suitable for stabilizing selection. Both contrast the mean fitness, $\overline{W}$, with W_o, the fitness of the optimal phenotype (usually estimated as the value of the mode of a stabilizing selection function, e.g., θ in Equation 28.3). Haldane suggested

$$I_H = \ln(W_o) - \ln(\overline{W})$$

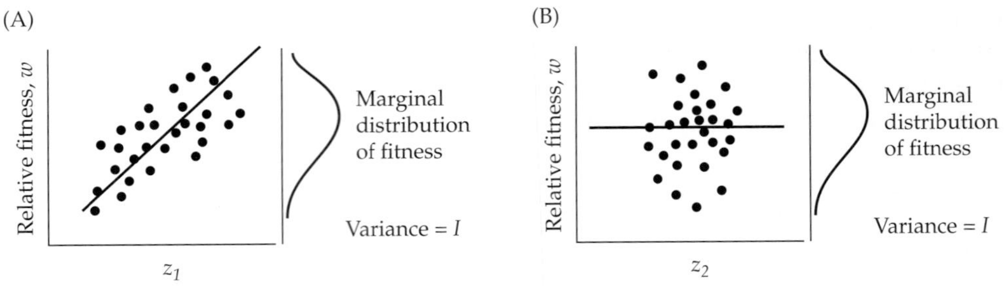

Figure 29.2 The fraction of I (the opportunity for selection) that is translated into a change in the mean depends on the correlation, $\rho(z, w)$, between relative fitness and the trait value. Traits z_1 and z_2 have the same marginal distribution of fitness, but only the regression of w on z_1 is significant. Thus, within a generation, selection changes the mean of z_1, but not of z_2.

while Van Valen proposed

$$I_{VV} = \frac{W_o - \overline{W}}{W_o} = 1 - \exp(-I_H)$$

For small values, note that $I_{VV} \simeq I_H$ (as $e^{-x} \simeq 1 - x$ for small x). These measures, however, also have limitations as comparative measures of the strength of selection. First, the value of W_o when selection is strictly directional is unclear, and second, like $\bar{\imath}$, these metrics are also trait-specific.

A much cleaner measure (which requires no assumptions about which traits are under selection or on the nature of that selection) is I, the **opportunity for selection**, which is defined as the variance in *relative* fitness

$$I = \sigma_w^2 = \frac{\sigma_W^2}{\overline{W}^2} \tag{29.5a}$$

This measure was introduced by Crow (1958; reviewed in 1989), who referred to it as the **index of total selection**, and was independently developed by O'Donald (1970). I is estimated by

$$\widehat{I} = \text{Var}(w) = \frac{n}{n-1}\left(\overline{w^2} - 1\right) \tag{29.5b}$$

Crow noted that if fitness is perfectly heritable ($h_W^2 = 1$), then $I = \Delta\overline{w}$, namely, the expected selection response in relative fitness. This follows from the Robertson-Price identity (Equation 6.10), $S = \sigma(z, w)$. When $z = W$, then (applying the breeder's equation with $h_W^2 = 1$) we have

$$R_W = h_W^2 S_W = S_W = \sigma(W, w) = \overline{W}\sigma(W/\overline{W}, w) = \overline{W}\sigma_w^2$$

implying that $R_W/\overline{W} = \Delta\overline{w} = \sigma_w^2 = I$. Following Arnold and Wade (1984a, 1984b), I is called the opportunity for selection, as any variation in individual fitness represents an *opportunity* for a within-generation change in a trait. The opportunity for selection provides an upper bound on $\bar{\imath}$. This follows by using: (i) the definition of a correlation, ρ, (ii) the Price-Robertson identity, $S = \sigma(z, w)$, and (iii) the fact that $|\rho| \leq 1$, to yield

$$|\,\rho_{z,w}\,| = \frac{|\sigma(z, w)|}{\sigma_z \sigma_w} = \frac{|S|}{\sigma_z\sqrt{I}} \leq 1 \tag{29.6a}$$

implying that

$$|\,\bar{\imath}\,| \leq \sqrt{I} \tag{29.6b}$$

Thus, the most that the mean of any trait can be shifted within a generation is $\sqrt{I}$ phenotypic standard deviations.

The usefulness of I as a bound of $\bar{\imath}$ depends on the correlation between relative fitness and the character being considered. From Equation 29.6a,

$$|\,\bar{\imath}\,| = |\rho(z,w)|\,\sqrt{I} \tag{29.6c}$$

Figure 29.2 shows scatterplots of relative fitness versus two characters (z_1 and z_2) measured in the same set of individuals. The marginal distributions of fitness are identical for both characters (because the same set of individuals, and hence the same set of fitnesses, were measured for each trait), and thus both have the same opportunity for selection. The association between relative fitness and z_1 is fairly strong, while there is no relationship between relative fitness and z_2, which means that z_1 realizes much of the opportunity for change, while z_2 realizes none of it.

Example 29.5. To estimate I for the bullfrog data used in Example 29.2, we first compute lifetime relative fitnesses, $w = w_1 \cdot w_2$, which yields values of 1.162, 1.020, 0, 0.652, and 2.160 for the five chosen males. Hence,

$$\overline{w^2} = (1/5)\left[1.162^2 + 1.020^2 + 0^2 + 0.652^2 + 2.160^2\right] = 1.496$$

giving

$$\widehat{I} = \frac{5}{4}(1.496 - 1) = 0.62$$

From Equation 29.6b, the most that selection can change the mean of any character within a generation is $\sqrt{I} \simeq 0.79$ standard deviations. The observed change in male body size (in standard deviations) from Example 29.2 is $(138.9 - 140)/\sqrt{59.5} = -0.143$, which is less than one-fifth of the maximum absolute change of 0.79.

Example 29.6. In many cases individual fitnesses are not recorded, and the average fitness for each phenotypic class is simply estimated. O'Donald (1970, 1971) analyzed the data of Dowdeswell (1961), who searched for signs of selection on hindwing eyespot number in the meadow brown butterfly (*Maniola jurtina*; see Brakefield 1984 for a review of the biology of this character). Dowdeswell compared the population distributions of eyespot number between a series of 471 wild-collected females (field) and a series of 237 females reared from larvae (lab). Presumably, any difference in distributions was due to selection on adults in nature. If p_i and p_i' denote the laboratory and field proportions for spot-class i, respectively, then from Equation 5.7a, the relative fitness of class i follows because $p_i' = w_i p_i$, and hence $w_i = p_i'/p_i$, assuming that the laboratory frequencies match the field frequencies before selection. The resulting data were as follows:

	Number		Proportion		
Eyespot No.	Lab	Field	Lab	Field	$w_i = p_i'/p_i$
0	124	294	0.523	0.624	1.1930
1	67	111	0.283	0.236	0.8336
2	34	53	0.143	0.113	0.7844
3	10	13	0.042	0.028	0.6541
4	2	0	0.008	0	0

Because the eyespot frequencies in the laboratory sample presumably represent the frequencies before selection, these are the ones we use, which yields

$$\overline{w^2} = \frac{1}{237}\left[(1.1930^2 \cdot 124) + (0.8336^2 \cdot 67) + (0.7844^2 \cdot 34) + (0.6541^2 \cdot 10)\right] \simeq 1.0475$$

and hence

$$\widehat{I} = \frac{237}{236}\left(1.0475 - 1^2\right) \simeq 0.0477$$

This is an *underestimate* of I, as to properly estimate I the distribution of *individual* fitnesses, rather than mean fitness for each phenotypic class (*class* fitnesses), is required. The Dowdeswell data only allow us to estimate the between-group variance in fitness (the variance in average fitness for the different eyespot classes) and neglects the additional within-group variance (the variance in fitness among individuals with the same number of eyespots).

Partitioning I Across Episodes of Selection

The total opportunity for selection can be partitioned into opportunities associated with each episode. Such a partitioning allows for comparisons of the relative strength of selection across episodes, as well as bounding the change in means and variances due to selection during any particular episode. Denote the opportunity of selection associated with the jth episode by I_j. By analogy with the definition of I, Arnold and Wade (1984a) suggested that the appropriate definition is the variance in the relative fitnesses of the jth fitness component

$$I_j = \sigma^2\left(w_j\right) = E(w_j^2) - 1 \tag{29.7a}$$

which is estimated by

$$\begin{aligned} \widehat{I}_j = \text{Var}(w_j) &= \frac{n}{n-1}\left(\overline{w_j^2} - 1\right) \\ &= \frac{n}{n-1}\left(\sum_{r=1}^{n} w_j^2(r)\, p_{j-1}(r) - 1\right) \end{aligned} \tag{29.7b}$$

where $p_j(r)$ is the fitness-weighted frequency of individual r after j episodes of selection (Equation 29.2a). Arnold and Wade showed that the partition for I over k episodes of selection is given by

$$I = \sum_{j=1}^{k} I_j + R \tag{29.8}$$

where the remainder term, R, represents a complex sum of covariances between fitness components (see Arnold and Wade 1984a for details). Webster et al. (1995) examined partitions of I involving both multiplicative and additive components of fitness. The latter can arise when fitness is influenced through two (or more) distinct paths, for example, mating success through both bonded-pair and extra-pair matings (see Whittingham and Dunn 2005 and Webster et al. 2007 for applications of the Webster approach).

Example 29.7. Compute the estimates of I_1 and I_2 using the data from Example 29.2. Using the relative fitnesses given in the table with $p_0(r) = 1/5$ yields

$$\overline{w_1^2} = \frac{1}{5}\left(0.714^2 + 0.714^2 + 0^2 + 1.429^2 + 2.143^2\right) \simeq 1.531$$

Likewise,

$$\begin{aligned} \overline{w_2^2} = \sum_{r=1}^{5} w_2^2(r)\, p_1(r) &= \left(1.628^2 \cdot 0.143\right) + 0 + \left(1.429^2 \cdot 0.143\right) \\ &\quad + \left(0.456^2 \cdot 0.286\right) + \left(1.008^2 \cdot 0.429\right) \simeq 1.165 \end{aligned}$$

Hence,

$$\widehat{I}_1 = \frac{5}{4}\,(1.531 - 1) \simeq 0.664 \quad \text{and} \quad \widehat{I}_2 = \frac{5}{4}\,(1.165 - 1) \simeq 0.206$$

Because $\widehat{I} = 0.62$ (Example 29.5), $\widehat{I}_1 + \widehat{I}_2 = 0.87 \neq \widehat{I}$. From Equation 29.8,

$$\widehat{R} = 0.62 - 0.87 = -0.25$$

which reflects the strong negative covariance within individuals between the first and second fitness components (in this dataset, individuals with high w_1 tend to have a low w_2, and vice versa).

Correcting Lifetime Reproductive Success for Random Offspring Mortality

As noted by Clutton-Brock (1988b), the variance in female lifetime reproductive success tends to increase with the age at which the offspring are counted. For example, in birds, I typically increases as we measure reproductive success by offspring at increasing ages: from eggs to hatchings and finally to fledglings. One might expect to see this pattern when maternal care is important, but it can also arise simply from random offspring mortality. Cabana and Kramer (1991) noted that a correction proposed by Crow and Morton (1955), originally suggested for allele-frequency change, can be used to adjust for random mortality. Suppose that the mean family size before an episode of selection is μ with a variance of σ_1^2. If offspring mortality is entirely random, with s being the probability of survival of a random individual, then the mean family size after selection is $s\mu$ and (from Crow and Morton) the variance σ_2^2 satisfies the relationship

$$\frac{\sigma_2^2/(s\mu) - 1}{s\mu} = \frac{\sigma_1^2/\mu - 1}{\mu} \tag{29.9a}$$

Because $s\mu$ and μ are the mean fitnesses, $\sigma_2^2/(s\mu)^2 = I_2$ and $\sigma_1^2/\mu^2 = I_1$, which yields

$$I_2 - \frac{1}{s\mu} = I_1 - \frac{1}{\mu}$$

which Cabana and Kramer rearrange to find that the opportunity for selection generated entirely from random offspring mortality is given by

$$I_2 = I_1 + \frac{1/s - 1}{\mu} \tag{29.9b}$$

Random offspring mortality always inflates the variance in reproductive success, and its effect is inversely proportional to the survival probability, s, and the mean family size, μ. Cabana and Kramer applied Equation 29.6b to 43 case studies of vertebrate reproductive success. In all cases, the mean decreased and the variance increased with the age at which the offspring were counted. They found that the predicted value of I_2 from random mortality alone (Equation 29.9b) was roughly 45% of the observed value of I_2 for a dataset of 16 bird species (measuring fitness as the number of fledglings per nesting attempt), while the predicted random-mortality value of I_2 was around 85% of the observed I_2 for a dataset of lifetime fledging and weaning success in a set of 8 birds and mammals. Thus, in many cases a large fraction of I can often be accounted for by this simple model of random mortality.

Caveats in Using the Opportunity for Selection

There are several subtle issues in the interpretation of I. To begin with, even though this variable appears to remove scaling effects due to different types of fitnesses, for estimates of I to

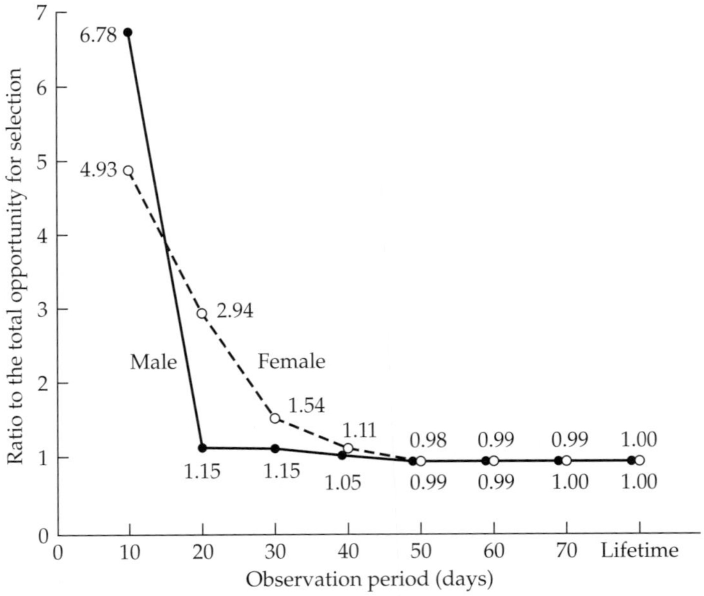

Figure 29.3 The ratio of the opportunity for selection on reproductive success to the lifetime opportunity for the coreid bug, *Colpula lativentris*, as a function of observation period. Males are given by the filled circles and females by the open circles. Note the inflation in I when very short time intervals are considered. (After Nishida 1989.)

be truly comparable, they must be based on consistent measures of fitness (Trial 1985). Consider the opportunity for selection based on number of *mates* per male (sexual selection), I_s, versus the opportunity for selection based on total male reproductive success (the number of *offspring* per male), I_{rs}. Total male reproductive success depends on both the number of mates and the fertility per mating. Recalling Equation 29.8, $I_{rs} = I_s + I_f + R$, where I_f is the opportunity for selection based on differences in male fertility per mating. Hence, I_{rs} is expected to exceed I_s unless there is sufficient negative covariance between the mating success and fertility components ($R < -I_f$).

A second point is that if the variance in fitness is not independent of $\overline{W}$, comparisons of I values between populations will be compromised. This dependency occurs in cross-sectional studies that measure sexual selection by simply counting the number of mating pairs (in such studies, an unequal sex ratio further biases comparisons of I between the sexes). If the time scale is such that only single matings are observed, the fitness of any individual is either 1 (mating) or 0 (not mating). The resulting fitnesses of randomly drawn individuals are binomially distributed with a mean of p (the mean copulatory success for the sex being considered) and variance $p(1-p)$, hence

$$I = \frac{p(1-p)}{p^2} \simeq \frac{1}{p} \quad \text{if } p \ll 1 \tag{29.10}$$

In this case, the mean and variance in individual fitness are not independent, and the opportunity for selection depends entirely on mean population fitness. As the time window for observing mating pairs decreases, fewer matings are seen and p decreases, which increases I. As the data plotted in Figure 29.3 illustrate, the opportunity for selection is often inflated if the observation period is short relative to the total mating period. Exactly the same problem occurs for viability selection (where p is now the probability of survival), but here p is expected to *decrease* (increasing I) as the window becomes longer.

Another example of the lack of independence between $\overline{W}$ and σ^2_W was given by Downhower et al. (1987). The Poisson distribution is a reasonable initial model for the number of mates under random mating. Indeed, Joshi et al. (1999) found that it provides an excellent fit for male mating success in laboratory populations of *Drosophila melanogaster*. Assuming that the number of mates for any given male follows a Poisson distribution, the variance in number of mates equals the mean number of mates, resulting in

$$I = \frac{\overline{W}}{\overline{W}^2} = \overline{W}^{-1}$$

where $\overline{W}$ is the mean number of mates per male. Thus, differences in I between populations do not necessarily indicate *biological* differences in male mating ability. For example, in a population of 100 males, if only 5 females mate, average male mating success is $\overline{W} = 0.05$, while if 50 females mate, $\overline{W} = 0.5$. For this example, differences in I come solely from variation in the number of mating females, *not* biological differences between males in their ability to acquire mates. Downhower et al. concluded from this example that comparing I values with the Poisson prediction ($I = 1/\overline{W}$), or some other appropriate random distribution, may help clarify the interpretation of I. Values of I less than the Poisson prediction indicate a more uniform distribution of fitness than expected if mate choice is random, while values in excess of this expectation indicate disproportionately high fitness among a limited set of individuals. In a study of four species with lek mating systems (where matings typically occur only in a specified area), Mackenzie et al. (1995) found that the combination of random mating and variation in male attendance at the lek accounted for ~40% of the total variation in male mating success. Thus, the variance in mating success was inflated by random factors, but still contains a substantial fraction of unexplained variance.

This comparison of I to the value expected under a Poisson distribution of individual fitness is an attempt to account for differences in opportunities for selection due to differences in mean fitness. In effect, this is a problem of stabilizing the variances (LW Chapter 11). Because I is the squared coefficient of variation in fitness, it is plagued by the same statistical problems as the Roginskii-Yablokov effect: a negative correlation is often expected between x/y and y, even when x and y are independent (LW Chapter 11). Thus, in most cases we expect I to be somewhat dependent on $\overline{W}$.

The Poisson mating example highlights that random variation (differences in individual fitness not attributable to intrinsic differences between individuals) reduces the correlation between phenotypic value and relative fitness. Although carefully controlled studies can reduce the error variance induced by chance (e.g., Houck et al. 1985), accounting for the inflation in the opportunity for selection by random effects remains problematic. It should be stressed that, despite these concerns, I still provides an upper bound on the strength of selection on any trait (over the measured episodes of selection). Accounting for random sources of variation that inflate I can tighten this limit.

MEASURING SEXUAL SELECTION

The idea of sexual selection, differences in reproductive success arising by differential mating success (**premating selection** or, in animals, **precopulatory selection**) and differential fertilization following mating (**postmating selection** or, in animals, **postcoupulatory selection**) began with Darwin's (1959) *Origins* and was more fully examined in his *The Descent of Man, and Selection in Relation to Sex* (Darwin 1872). As a product of his times, Darwin viewed females as essentially monogamous, and thus never considered postmating selection. He also assumed that sexual selection was mainly restricted to animals, because (as quaintly noted by Moore and Pannel 2011) organisms outside of animals "could surely not appreciate each other's beauty." Modern evolutionary biologists ascribe a much broader role (both across genders and across phyla) to Darwin's original concept. For example, starting with David Lloyd, many botanists have viewed sexual selection for increased siring ability (i.e., pollen dissemination) as one of the major evolutionary forces shaping floral evolution (Moore and Pannel 2011). Likewise, Darwin's original views on gender roles in sexual selection have been replaced with a more gender-neutral one, allowing for either, or both, males and females to be under strong sexual selection, depending on the biology of their situation.

Premating selection can be **intrasexual** (male-male or female-female competition for access to mates) or **intersexual** (female, or more rarely, male choice of mates). Postmating selection is possible when a female has mated multiple times. **Sperm competition** (or more generally, **gamete competition**) among the donations from different males within a single female is an example of intrasexual postmating selection, and **cryptic female choice** (preferential selection among male gametes by a female) is an example of intersexual postmating selection. By the nature of their reproductive biology, postmating selection may be rather common in the flowering plants. The growth of a pollen tube from a pollen grain down to the ovule of a flower has the potential for considerable selection (especially in the case of self-incompatible species). Given that a large fraction (often over 50%) of the genome is expressed during pollen tube growth (Mascarenhas 1990), such selection has underappreciated evolutionary implications. Our focus here is mainly concerned with premating selection and assessing which traits (if any) are under sexual selection. However, with the rise of molecular markers, the ability to examine the hidden postmating selection that may be ongoing within a female's reproductive system has now become accessible.

Sexual selection, and the evolution of mate choice, is an enormous (and growing) area of evolutionary biology, especially with the realization that sexual selection may be just as, if not more, intense as natural selection in some species (Hoekstra et al. 2001; Siepielski et al 2001; Kingsolver et. al. 2012). Jones and Ratterman (2009) provided a nice overview. A very incomplete list of more general reviews include those examining: the entire field (Arnold 1983a; Bradbury and Andersson 1987; Andersson 1994; Andersson and Iwasa 1996; Murphy 1998; Mead and Arnold 2004; Arnqvist and Rowe 2005; Kokko et al. 2006; Clutton-Brock 2007, 2009; Hunt et al. 2009; Tobias et al. 2012; Reid 2014), the evolution of mate choice and mating systems (Bateson 1983; Kokko et al. 2003; Shuster and Wade 2003; Neff and Pitcher 2005; Andersson and Simmons 2006; Kotiaho and Puurtinen 2007; Shuster 2009), postmating selection (Parker 1970, 2006; Smith 1984; Birkhead and Møller 1998; Birkhead and Pizzari 2002; Eberhard 2009), and sexual selection in plants (Willson and Burley 1983; Ashman and Morgan 2004; Delph and Ashman 2006; Moore and Pannell 2011).

Roughgarden et al. (2006) ignited a bit of a firestorm when they proclaimed that the notion of sexual selection is "always mistaken" and that it "needs to be replaced." One foundation for their argument was, as noted by Shuker (2010), that "various societal biases associated with gender have detrimentally influenced how scientists have thought about sexual selection." It is always appropriate to examine how societal biases may warp one's scientific viewpoint, but given that 40 prominent evolutionary biologists immediately, and vehemently, disagreed with Roughgarden et al. (Kavanagh 2006), there seems to be little traction for their radical reassessment. Additional debate on this issue can be found in Roughgarden and Akçay (2010a, 2010b), Clutton-Brock (2010), and Shuker (2010).

Bateman's Principles

In a classic paper, overlooked for many years (until it was rediscovered by Trivers 1972, who allegedly was directed to it by Ernst Mayr), Bateman (1948) used data from mating success and fecundity in laboratory populations of *Drosophila* to propose several principles regarding sexual selection. As much of the debate on the measure (and potential mismeasure) of sexual selection is centered around these principles, we discuss them here before considering how to quantify sexual selection. Ironically, Gowaty et al. (2012) recently suggested that Bateman's experimental design had a critical flaw (the genetic markers that were used influenced viability), producing a systematic bias in the estimates of number of offspring by sex. Despite this potential issue, Bateman's ideas have shaped most of the attempts to describe sexual selection, in large part due to Arnold's (1994) framing of the three principles from Bateman's experiment:

1. *Males show a greater variance in the number of* ***offspring*** *than do females*. The opportunity for selection, when measured by reproductive success (a composite measure of both

sexual and natural selection), is greater for males than females.

2. *Males show a greater variance in number of* ***mates*** *than do females.* The opportunity for selection, when measured by number of mates, is larger in males than in females. This suggests that variance in number of mates is a potential measure of sexual selection.

3. *The total number of offspring for males is an increasing function of the number of mates, but is largely independent of the number of mates in females.* Provided females have mated, their fecundity is largely independent of their number of mates. Thus, in females, the average fecundity per mating should be a *decreasing* function of number of mates, as their total fecundity, (number of mates)$\times$(fecundity per mating), is largely independent of the number of mates (provided she has mated at least once).

The essence of Bateman's observations was that female fitness is not increased by additional matings, while male fitness is, meaning that the third principle drives the first two. While these three principles served as a useful baseline for thinking about sexual selection, Arnold (1994) and Arnold and Duvall (1994) showed that each can be violated in different mating systems. If female fecundity increases with the number of matings (such as can occur when males provide nuptial gifts), we expect sex-specific differences in the opportunity for selection (based on either total reproductive success or the number of mates) to decrease. For example, Rodríguez-Muñoz et al. (2010) showed that female field crickets (*Gryllus campestris*), like their male counterparts, leave more offspring when they have more mates. A hypothetical (extreme) example arises in some mantids, wherein the female eats her mate, accruing additional nutrition. In such cases, "successful" males mate just once, while female fecundity could easily (in theory) be an increasing function of number of mates (or, in this case, meals).

It is well known that there are numerous exceptions to Bateman's assumed male-dominated sexual selection (e.g., Brown et al. 2009), but the focus on this issue misses the point that Arnold was trying to make. He stressed that the utility of Bateman's principles is that they *codify what needs to be measured in order to show that the* ***opportunity*** *for sexual selection exists*: variance in total fitness (reproductive success [RS]), variance in mating success (MS), and a relationship between increased MS and increased RS. Following Jones et al. (2002), we can restate Bateman's principles in a more gender-neutral fashion as follows:

1. *The sex experiencing the strongest sexual selection has a higher variance in reproductive success (RS).*

2. *The sex experiencing the strongest sexual selection has a higher variance in mating success (MS).*

3. *The slope of the RS-on-MS regression is larger for the sex experiencing the strongest sexual selection.*

It needs to be stressed that neither, or both, sexes could be experiencing strong sexual selection. The metric of the first principle is just the opportunity for selection (I) and the variance in mating success is the corresponding metric for the second principle. The critical principle is the third, which provides a direct connection between the number of matings and reproductive success. If both sexes experience strong sexual selection, both will show significant regressions of reproductive success on mating success.

Variance in Mating Success

The gender-neutral version of Bateman's second principle suggests that a natural measure for the *potential* of sexual selection is variance in mating success. Further, in many (but certainly not all) species, a higher variance is expected in males because some will fail to mate whereas essentially all females are (generally) assumed to have mated. This led Wade (1979) to suggest that the variance in male mating success was a measure of the potential for sexual selection. Arnold (1994, Arnold and Duvall 1994) codified this idea by considering the **opportunity for sexual selection,**

$$I_s = \frac{\sigma^2(MS)}{\overline{MS}^2} \tag{29.11a}$$

which is the variance in mating success (MS = number of mates per individual), scaled by the square of the average number of mates per male (with a corresponding definition for females). As this is just the variance, σ^2_{ms}, in **relative mating success** ($ms = MS/\overline{MS}$ being the mean-standardized mating success), it is the logical extension of the opportunity for selection ($I = \sigma^2_w$), which is based on relative fitness (Equation 29.5a). The importance of this measure is that if I_s (which also appears in the literature as I_{mates}) is nearly zero, there is little chance of sexual selection acting on any trait. However, a large value of I_s, by itself, does not guarantee that *any* sexual selection is occurring (i.e., it is a *necessary*, but not *sufficient*, condition). As we have seen above, chance variation in mating success can generate a significant values of I_s without any sexual selection occurring. Likewise, even if specific traits result in a strong mating advantage, if this does not translate into increased reproductive success (increased numbers of offspring), no sexual selection occurs.

For this reason, measures of sexual selection based entirely on I_s have been criticized (Banks and Thompson 1985; Sutherland 1985a, 1985b; Koenig and Albano 1986; Hubbell and Johnson 1987; Klug et al. 2010a; Jennions et al. 2012; Henshaw et al. 2016), and a number of alternative metrics have been suggested (reviewed by Kokko et al. 1999; Mills et al. 2007; Klug et al. 2010a; Henshaw et al. 2016). These attempt to adjust for the impact that sample size and mean mating success can have on I_s under random mating. The two metrics most widely appearing in the literature have their roots in ecological measures of competition and resource allocation. The use of **Morisita's index** (Morisita 1962),

$$I_\delta = N\left(\frac{\sum_{i=1}^N m_i^2 - \sum_{i=1}^N m_i}{(\sum_{i=1}^N m_i)^2 - \sum_{i=1}^N m_i}\right) \tag{29.11b}$$

where m_i is the number of mates for individual i ($1 \leq i \leq N$), was recommended by Fairbairn and Wilby (2001), as it adjusts the variance in mating success by the variance generated when all individuals have an equal chance of mating. An alternative measure is **Green's index of resource monopolization** (Green 1966),

$$Q = \frac{\sigma^2_m - \mu_m}{N\mu^2_m - \mu_m} \tag{29.11c}$$

where μ_m and σ^2_m are the mean and variance in number of mates. Q adjusts the observed variance in mating success with its maximal possible value, and was used by Ruzzante et al. (1996) and Blanckenhorn et al. (1998). These alternative indices are both attempts to generate a **null model**, namely, the variance expected under the mating system when there is no systematic mating bias, but rather only random encounters within the parameters set by the biology, such as the sex ratio or the "handling time" of a mating.

The **operational sex ratio** (**ORS**) of Emlen and Oring (1977), which is the ratio of sexually active males to fertilizable females, is also used as a metric for the potential for sexual selection (Kvarnemo and Ahnesjö 1996). A male-biased ORS (>1) offers the *possibility* of sexual selection on males, and a female-biased ORS (<1) offers the opportunity for sexual selection on females. One important issue with *any* measure is that sexual selection can be very context-specific (e.g., Fitze and Le Galliard 2011). For example, if there is a large pool of males, but only a small fraction have the capacity to mate, then an *observed* male-biased sex ratio could actually mask a female-biased *operational* sex ratio. As Klug et al. (2010b) stressed, it is not trivial deciding whom to include in measures of sexual selection. Finally, there is a density component to mate choice as well. When both sexes are scarce, individuals may be far less choosy about mates.

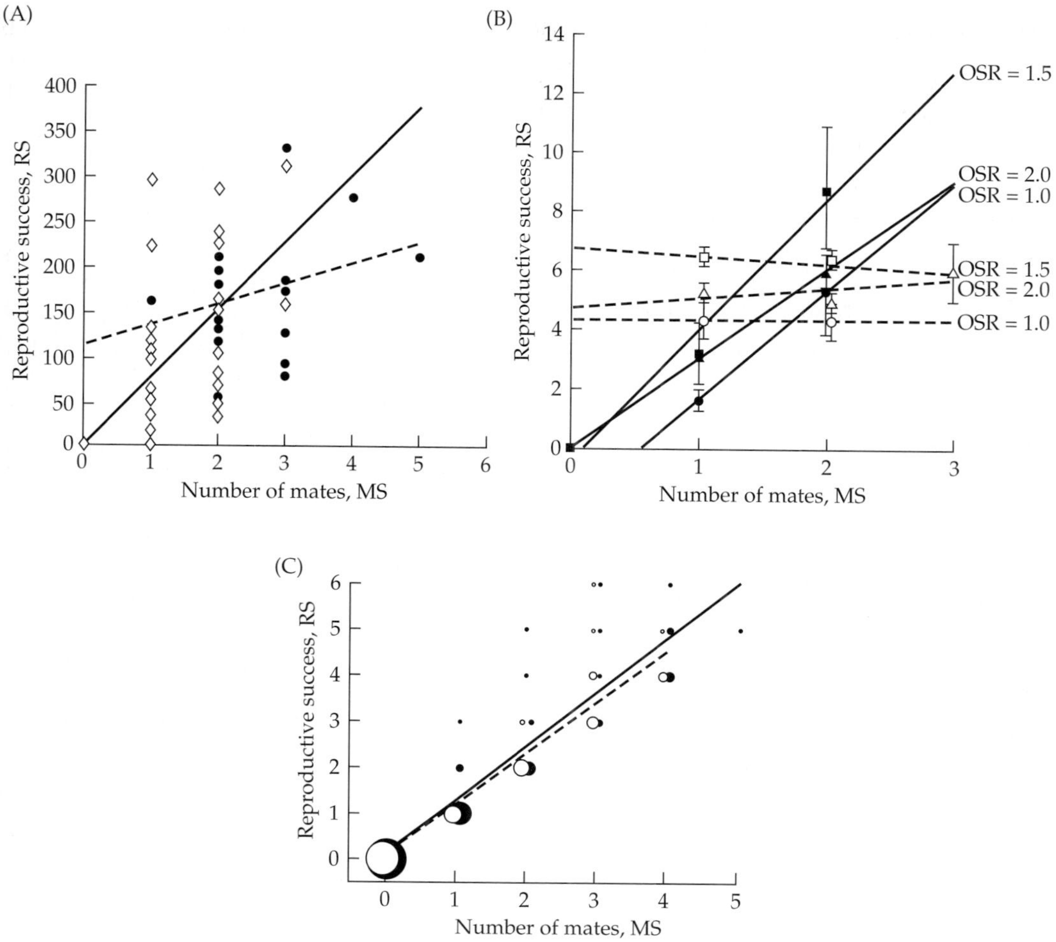

Figure 29.4 Examples of Bateman gradients, the regression of total offspring number (reproductive success [RS]) as a function of number of mates (mating success [MS]). **A:** Data for male (open diamonds and solid line) and female (filled circles and dashed line) rough-skinned newts (*Taricha granulosa*). The slope for males is significantly different from zero while the slope for females is not, and the sex-specific slopes are significantly different from each other. (After Jones et al. 2002.) **B:** Data for male (filled symbols and solid lines) and female (open symbols and dashed lines) bank voles (*Clethrionomys glareolus*), for three different operational sex ratios (ORS), indicated by squares, circles, and triangles. Again, the male slopes (at all three values of ORS) are significantly positive, while none of the female slopes are. (After Mills et al. 2007.) **C:** Gradients for a Québec population of eastern chipmunks (*Tamias striatus*). Males are open circles (regression given by the dashed line), and females are filled circles (solid line). The gradients on both sexes are positive and very similar, allowing for the possibility of strong sexual selection in both males and females. (After Bergeron et al. 2012.)

While the above alternative measures (I_δ, Q, ORS) may contain information to better inform a behavioral ecologist, the advantage of I_s is that it bounds the opportunity for sexual selection, no matter what the cause. It must be measured over an appropriate time window and in an appropriate setting, which are legitimate concerns. However, a dispassionate reading of the literature that is critical of the use of I_s shows that concern centers more around the reasonable objection that an investigator might be sloppy and *directly* relate I_s with *actual* sexual selection (rather than as a measure of its *opportunity*). From an evolutionary standpoint, I_s and two other measures to be introduced shortly—-the Bateman gradient (Equation 29.12) and Jones index (Equation 29.14a)—are the most natural measures of the *potential* for sexual selection (Jones et al. 2000, 2002; Wade and Shuster 2005; Mills et al. 2007;

Jones 2009; Krakauer et al. 2011). Nonetheless, as stressed by Croshaw (2010), while I_s is a natural measure of sexual selection potential, it should always be contrasted with its value under a null model. Croshaw observed high values of both I and I_s in marbled salamanders (*Ambystoma opacum*), but only I was significantly greater than its expected value under a null model.

The Sexual Selection, or Bateman, Gradient

Bateman's first two principles, by themselves, do not guarantee that sexual selection is occurring. Rather, it is his third principle, relating total reproductive success (RS) with mating success (MS), that indicates the potential for sexual selection to exist. Arnold and Duvall (1994) quantified this relationship by regressing total fecundity on number of mates,

$$RS_s = \text{const} + \beta_{ss} MS_s + e_s \tag{29.12}$$

where s indexes the sex (m for male or f for female), with the regression coefficient (β_{ss}), which they termed the **sexual selection gradient**, indicating the potential for sexual selection in that sex. Biologically, β_{ss} is the additional number of offspring gained per mating. As noted by Jones et al. (2002), the gradient is the final statistical path to fitness from all sources of sexual selection. One can have large variances in both RS and MS for a particular sex, but if the two variables are uncorrelated over individuals, there is no sexual selection. Andersson and Iwasa (1996) used the term **Bateman gradient** for β_{ss}, and this has generally replaced Arnold and Duvall's term in the literature. Examples of Bateman gradients are given in Figure 29.4. Bateman's original expectation was that males will show a positive slope, while females will show little to no slope (provided they have been mated at least once). The former (a positive male Bateman gradient; $\beta_{mm} > 0$) is routinely seen, but the assumption of a small female Bateman gradient ($\beta_{ff} \simeq 0$) is often incorrect. Gerlach et al. (2012) reviewed a wide range of species (ranging from invertebrates to fish, birds, reptiles, amphibians, and mammals) in which females showed a positive Bateman gradient, indicating the potential for sexual selection.

Although Bateman gradients are widely used for demonstrating the potential of sexual selection in one or both sexes (Jones et al. 2000, 2002; Mills et al. 2007), Kokko et al. (2012) made the point that the operational sex ratio (ORS) and Bateman gradients are complementary measures. The latter describes the fitness gain with increasing number of mates, while the former measures the potential difficulty in obtaining such matings. If one has a strong Bateman gradient on males (for example, estimated under controlled settings) but females are very scarce in nature (i.e., the ORS is small), much of this gain is never realized.

While least-squares linear regressions are typically used to model the RS-to-MS relationship, Arnold (1994) noted that the functional nature of this relationship may be one of four general forms. First, there could be **single-mating saturation**, as when a single mating allows a female to acquire sufficient male gametes to fertilize all of her eggs (females in Figure 29.4B). Alternatively, this relationship could be linear, with total reproductive output increasing linearly with the number of mates (as shown for all males in Figure 29.4 and for females in Figures 29.4A and 29.4C). More generally, because of additional costs and benefits of mating, the RS-to-MS functional relationship could show a **diminishing-returns** shape (e.g., increased mating yields more offspring but also puts the mating individual at additional risk). Finally, this relationship could exhibit an **intermediate optimum** number of matings. Given that most studies have insufficient numbers of individuals that have mated three or more times, attempts to fit nonlinear regressions are typically not performed.

Although Bateman gradients are usually framed for dioecious species, Arnold (1994) also considered monoecious (hermaphroditic) species (for example, many flowering plants), and a much more complete treatment was offered by Anthes et al. (2010), whose approach we follow. Provided one can track the offspring produced via male and female gametes from the same individual, the generalized Bateman gradients become

$$RS_m = \text{const} + \beta_{mm} MS_m + \beta_{mf} MS_f + e_m \tag{29.13a}$$

$$RS_f = \text{const} + \beta_{ff} MS_f + \beta_{fm} MS_m + e_f \tag{29.13b}$$

RS_m and RS_f correspond, respectively, to the male RS (offspring resulting from sperm or pollen donated by the focal individual fertilizing the eggs of others) and female RS (offspring resulting from eggs from the focal individual fertilized by the sperm or pollen of others). Likewise, MS_m and MS_f correspond to the number of matings that the focal individual was involved in as a sperm or pollen donor and as an egg donor, respectively.

The various regression coefficients in Equation 29.13 can be interpreted as follows. Consider the situation in which an individual gains an additional mating in which it contributes male gametes, while the number of times it served as an egg donor is held constant (MS_m increases by one and MS_f is held constant). From Equations 29.13a and 29.13b, we expect the reproductive success from mate gametes (RS_m) to increase by β_{mm} and reproductive success from female gametes (RS_f) to change by β_{fm}. A negative value of the latter suggests a tradeoff between male and female gamete production and mating. Likewise, β_{ff} and β_{mf} are the expected changes to RS_f and RS_m, respectively, caused by a one-unit change in MS_f with MS_m held constant. Finally, when individuals can self, a third system of gradients appears,

$$RS_{s_e} = \text{const} + \beta_{s_e m} MS_m + \beta_{s_e f} MS_f + e_{s_e} \tag{29.13c}$$

where now the index s_e refers to selfing events (Anthes et al. 2010).

A final composite measure of the potential for sexual selection is the **Jones Index**

$$I_J = \beta_{ss} \sqrt{I_s} \tag{29.14a}$$

(Jones 2009). This expression shows the impact of both the number of matings to fitness relationship (β_{ss}; Equation 29.12) and the variance in number of mates (I_s). If *either* one is small, there is very little sexual selection pressure on any trait. The original motivation for Equation 29.14a was to place an upper bound on the absolute change in any trait that could be generated by sexual selection. Using the same logic that led to Equation 29.6b, Jones showed that

$$|\bar{\imath}| \leq I_J \tag{29.14b}$$

Henshaw et al. (2016) suggested that the Jones index deserves a more prominent role in measures of the opportunity for sexual selection. Using data from five rather different species (one large and one small mammal, a bird, a beetle, and a fish) as a set of model organisms showing diverse mating systems, they simulated 500 biologically plausible mating systems and examined the correlation between measures of the potential sexual selection ($I_s, \beta_{ss}, I_J, I_\delta$, and Q) and the actual strength of sexual selection (the actual strength of selection on the simulated mating trait). They found that the Jones index performed the best (had the highest correlation with the actual strength), especially in cases with sex-specific differences. Surprisingly, I_s and β_{ss}, while still correlated with the actual strength of sexual selection, performed the poorest, with I_δ and Q having intermediate levels of success.

The above discussion has mainly dealt with the *potential* for sexual selection, without invoking any specific *traits*. While the Bateman gradient can be thought of as a special case of a trait-fitness regression (where the trait is number of mates), our analysis has otherwise been trait-free. This is both an advantage and a limitation. The advantage is that if the Bateman gradients for both sexes are essentially zero, there is no need to search for traits under sexual selection (Equation 29.14b). The limitation is that a large Jones index (which means that both β_{ss} and I_s are large) only indicates that sexual selection is occurring, and is otherwise uninformative as to the target trait or traits.

A more direct description of the amount of sexual selection on a candidate trait follows from the Robertson-Price identity (Equation 6.10). As above, let $ms = MS/\overline{MS}$ denote the *relative* mating success (the number of mates standardized by the mean number of mates) and z_{sd} denote the variance-standardized value of a candidate trait, namely, $\sigma^2(z_{sd}) = 1$. With these definitions, Jones (2009) defined the standardized strength of sexual selection on a specific trait, the analog of $\bar{\imath} = \sigma(z_{sd}, w)$, as

$$m' = \sigma(z_{sd}, ms) \tag{29.14c}$$

This is also called the **Jones mating differential**. Using this definition with Equation 6.10 yields the **Jones equation** (Jones 2009),

$$\bar{\imath} = \beta_{ss}\, m' \tag{29.14d}$$

which relates the standardized strength of selection on a trait (the selection intensity, $\bar{\imath}$) with the Bateman gradient (β_{ss}) and the standardized strength of sexual selection (m').

DESCRIBING PHENOTYPIC SELECTION: INTRODUCTORY REMARKS

Thus far, we have dealt with the fitness of individuals, independent of any knowledge of the phenotypes that must be the targets of selection. We conclude this chapter by examining this latter issue of detecting which traits are under selection and describing the nature of that selection. Our ultimate interest in a particular trait might be in predicting its change over time, which requires knowledge of both the nature of selection on that trait and its variance components (Chapter 13). Alternatively, we may simply wish to explore the trait's ecological implications by examining how expected fitness changes with the character value, independently (at least initially) of any concerns about its genetics. Further, we may wish to partition trait fitness across episodes of selection to enhance our understanding of the biological implications of a trait.

Example 29.2 showed how to weight individuals—and therefore how to weight measured traits in those individuals—based on their fitnesses. Hence, one simple approach for detecting selection on a character is to compare its fitness-weighted phenotypic distributions before and after some episode of selection. Such a seemingly reasonable comparison entails a number of significant caveats. We assume that any within-generation change following an episode of selection is *not* due to selection on unmeasured traits that are phenotypically correlated with our focal trait. A closely related assumption is that no spurious fitness-trait associations are generated via unmeasured environmental variables (namely, environmental factors that influence both fitness and our focal trait). Chapter 20 discussed some partial solutions to both of these problems, and we continue this discussion in Chapter 30.

A further caveat is that the phenotypic distribution of a trait can change over time for a myriad of reasons other than selection, such as growth or other ontogenetic changes, immigration, and environmental changes, and great care must be taken to account for these factors. Further, as discussed in Chapters 5 and (especially) 28, pleiotropy can generate fitness associations for a strictly neutral trait. Under the Hill-Keightley model (Chapter 28), trait variation is maintained by the pleiotropic effects on a neutral trait from alleles with deleterious fitness effects. Individuals carrying more mutations have more extreme trait values and lower fitness, which generates apparent stabilizing selection. Similarly, if the direction of new pleiotropic mutations is biased (e.g., on average, mutations lower trait value), then a spurious signal of directional selection can also be generated, as individuals with more mutations have (in this case) lower trait values and lower fitness.

Finally, as we have already seen in this chapter, there are a number of subtle issues in assigning fitnesses to phenotypes when the fitness of a focal individual is influenced by other interacting individuals (Chapter 22). Thoughtful reviews of some of these assignment issues were given by Grafen (1988), Cheverud and Moore (1994), Wolf and Wade (2001), Hadfield and Thomson (2017), and Thomson and Hadfield (2017).

Fitness Surfaces and Landscapes

The expected fitness, $W(z) = E[\,W \mid z\,]$, of an individual with a phenotypic value of z, describes a **fitness surface** (or equivalently, a **fitness function** or a **fitness profile**), relating fitness and character value. The **relative fitness surface**, $w(z) = W(z)/\overline{W}$, is often more convenient than $W(z)$, and we use the two variables somewhat interchangeably. The nature of selection on a character is determined by the local geometry of the individual fitness surface over the range of phenotypic values in the population (Figure 29.5). If fitness is increasing (or decreasing) over some range of phenotypes, a population having its mean value in this interval experiences **directional selection**.

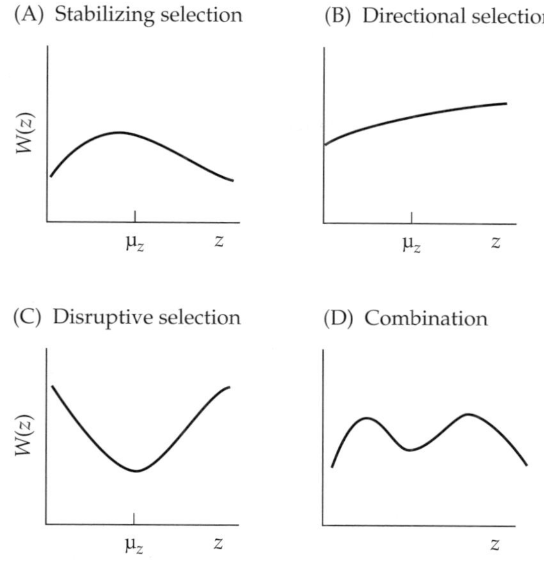

Figure 29.5 Phenotypic selection has historically been classified into three basic forms depending on the local geometry of the individual fitness surface, $W(z)$: stabilizing (**A**), directional (**B**), and disruptive (**C**). As **D** illustrates, populations can also simultaneously experience multiple forms of selection. Likewise, if the population mean (μ_z) in (**A**) does not correspond with the maximum value of $W(z)$, the population experiences directional selection as it is moved toward this value.

The curvature of the fitness surface is important as well. If $W(z)$ contains a local maximum at $z = \theta$, a population with members within an interval encompassing θ is said to experience **stabilizing selection**. If the population is distributed around a local minimum, **disruptive selection** occurs. More generally, we can have negative curvature (downward or **concave**; with a negative second derivative) without having a maximum over the observed range of phenotypes and thus we will not formally have stabilizing selection. Likewise, positive curvature (upward or **convex**; with a positive second derivative) can occur without having a minimum over the observed range of phenotypes and hence will not formally have disruptive selection. As illustrated in Figure 29.5D, when the local geometry of the fitness surface over the range spanned by a population is complex (e.g., multimodal), the simplicity of description offered by these types of quadratic selection surfaces (which can have only a single maximum or minimum) breaks down. Further, $W(z)$ can vary over genotypic and environmental backgrounds. In some situations, the fitness of a phenotype depends on the frequency of other phenotypes in the population (e.g., sexual selection, dominance hierarchies, truncation selection, and when predators have search images). In this case, fitnesses are said to be **frequency-dependent**.

A second important geometry describes how $\overline{W}$, the expected fitness of the *population*, varies as function of the distribution of phenotypes ($p[z|\boldsymbol{\Theta}]$; where $\boldsymbol{\Theta}$ is the vector of distribution parameters, such as the mean and variance) in that population,

$$\overline{W}(\boldsymbol{\Theta}) = \int W(z)\, p(z|\boldsymbol{\Theta})\, dz$$

Mean fitness is a function of both individual fitness, $W(z)$, and the parameters, $\boldsymbol{\Theta}$, of the phenotypic distribution of z, and we are interested in how changes in one or more of the distribution parameters change $\overline{W}$. For example, if z is normally distributed, we may wish to plot the function, $\overline{W}(\mu_z)$, of mean fitness as a function of μ_z for a given value of σ_z^2. More generally, one can consider the three-dimensional surface, $\overline{W}(\mu_z, \sigma_z^2)$, of how $\overline{W}$ jointly varies with μ_z and σ_z^2.

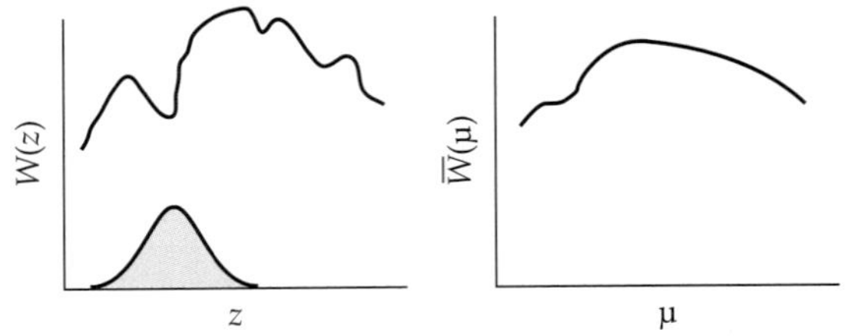

Figure 29.6 In this example, a small change in z is associated with a large change in the *individual* fitness surface, $W(z)$. However, because the mean population fitness, $\overline{W}(\mu_z)$, averages individual fitnesses over the phenotypic distribution (the filled curve in the left panel), small changes in μ_z result in only small changes in $\overline{W}(\mu_z)$. As a result, the corresponding fitness landscape is much smoother than the underlying individual fitness surface.

To stress this distinction between the $W(z)$ and $\overline{W}$ fitness geometries, the former is referred to as the **individual fitness surface**, and the latter as the **mean fitness surface** or (preferably) the **fitness landscape**. The term *landscape* traces back to Simpson (1944, 1952), who called $\overline{W}(\Theta)$ the **adaptive landscape** as an extension (to changes in trait means in phenotypic space) of Wright's notion of an adaptive topography for allele-frequency change in genotypic space (Equation 5.5b). Knowledge of the individual fitness surface allows us to compute the fitness landscape for any specified phenotypic distribution, $p(z)$, but the converse is not true. Given this important distinction between the two different geometries, Lande (1976), Arnold (2003), and Morrissey and Sakrejda (2013) suggested restricting the term *surface* to individual fitnesses, $W(z)$, and the term *landscape* to mean population fitnesses, $\overline{W}(\Theta)$, and we follow this convention here.

Populations, not *individuals*, evolve, and the fitness landscape provides the connection between the fitness of individuals and the evolution of the population. When the breeder's equation holds, the partial derivatives of $\overline{W}$ with respect to the first two phenotypic moments (μ_z and σ_z^2) describe the changes in mean and variance (Example 24.9). More generally, partials of $\overline{W}$ with respect to higher phenotypic moments describe the dynamics of selection in the Barton-Turelli selection-response equations (Equations 24.26 and 24.28). The individual fitness surface may have discontinuities and rough spots, for example, regions where very small changes in phenotypic values result in large changes in individual fitness. As shown in Figure 29.6, such regions are smoothed (at least to some extent) by the averaging of $W(z)$ over $p(z)$ in the fitness landscape, and this smoothing facilitates the existence of the various partials of mean fitness used in the Barton-Turelli equations.

DESCRIBING PHENOTYPIC SELECTION: CHANGES IN PHENOTYPIC MOMENTS

Selection for particular phenotypes changes the trait distribution (although it need not change all moments; for example, the mean may be unchanged). Thus, selection on a target trait can be detected by testing for differences between the distribution of phenotypes before and after some episode of selection. While complete distributions can be compared (e.g., using the d statistic from the DSD approach, as discussed below), the most common procedure for detecting selection acting on a trait is to test for changes in its first few phenotypic moments. Standard statistical tests for differences in means (t tests) and variances (F tests) can be used, but these tests rely on normality assumptions that are often violated, and nonparametric tests may be more appropriate. Differences in means can be tested using the Wilcoxon-Mann-Whitney test, while Conover's squared rank test (Conover 1999) can be used to test for changes in variances. Other nonparametric tests for changes in variance exist, but care must be exercised, as some (e.g., the Siegel-Tukey test) are quite sensitive to differences in means; see Conover (1999) and Sprent and Smeeton (2007). While these issues are important, the main problem in detecting whether selection is occurring on a particular trait is that changes in the moments may be due entirely to selection on phenotypically correlated characters (Chapters 20 and 30). Keeping this important caveat in mind, we now examine measures of selection for single characters.

Henshaw's Distributional Selection Differential (DSD)

While various nonparametric approaches can be used to compare a distribution of phenotypes before and after an episode of selection, simply finding a significant difference is not particularly informative. Henshaw (Henshaw et al. 2016; Henshaw and Zemel 2017) proposed a metric, the **distribution selection differential** (**DSD**, whose statistic is denoted by d), that uses differences in the shape of the entire distribution to quantify the nature and strength of selection. The original definition (Henshaw et al. 2016) was

$$d = \int_{-\infty}^{\infty} |F^*(z) - F(z)|\, dz, \quad \text{where} \quad F(z) = \int_{-\infty}^{z} p(x)dx \tag{29.15a}$$

namely, the total difference in the cumulative distribution functions of a trait before and after selection, $F(z)$ and $F^*(z)$, respectively. Empirically, this metric is estimated by

$$\widehat{d} = \sum_{i=1}^{n-1} \left[(z_{i+1} - z_i) \cdot \left| \sum_{j=1}^{i} \frac{1 - w_j}{n} \right| \right] \tag{29.15b}$$

where we have ordered phenotypes from smallest (z_1) to largest (z_n), with w_j indicating the relative fitness of the jth-ranked phenotypic value.

Henshaw and Zemel (2017) demonstrated that Equation 29.15a is equivalent to several other definitions, including a generalized Price-Robertson covariance expression (see their paper for details). They also noted two additional useful features of the DSD. First, under strictly directional selection, in which $W(z)$ is a monotonic function of z,

$$d = |\,\bar{\imath}\,| \tag{29.15c}$$

Hence, when computed using a variance-standardized trait, d equals the selection intensity (Equation 13.6a). This observation implies that

$$d_N = d - |\,\bar{\imath}\,| \tag{29.15d}$$

provides a measure of the amount of residual selection remaining after the strictly directional component has been removed (the so-called **nondirectional selection fraction**). Second, again for a variance-standardized trait, d is bounded by the variance, I, in relative fitness,

$$d \leq \sqrt{I} \tag{29.15e}$$

which shows that the bound from the opportunity for selection (Equation 29.6b) also extends to the DSD. See Henshaw and Zemel (2017) for a discussion of additional features and extensions of the DSD concept to multiple traits.

Directional Selection

Turning to moment-based comparisons, three measures of the within-generation change in phenotypic mean have been previously introduced: the **directional selection differential**, S (Equation 13.7); the variance-standardized directional selection differential (the **selection intensity**), $\bar{\imath} = S/\sigma$ (Equation 13.6a); and the **directional selection gradient**, $\beta = S/\sigma^2$ (Equation 13.8b). While these measures are interchangeable for selection acting on a single trait (using an appropriate scaling factor, as all have identical values when $\sigma_z^2 = 1$), the multivariate extension of the gradient (the vector $\boldsymbol{\beta}$) is the measure of choice when multiple characters are considered (Chapter 30). This is because the elements of $\boldsymbol{\beta}$ fully adjust for the effects of selection on phenotypically correlated traits *among the set of traits included in the regression*, thereby quantifying the amount of direct selection on a trait. In contrast, S and $\bar{\imath}$ confound these direct and indirect effects (Equation 30.3), and consequently, they are often called measures of **total selection** (the sum of the forces, direct and indirect, acting on a trait), while β is a measure of **direct selection**.

Quadratic Selection

Metrics can also be defined to quantify the change in variance. At first blush, this change seems best described by $\sigma^2_{z^*} - \sigma^2_z$, where $\sigma^2_{z^*}$ is the phenotypic variance following selection. The limitation of this metric as a measure of selection acting on the population variance of a trait is that strictly directional selection also reduces the variance. In particular, Lande and Arnold (1983) showed that

$$\sigma^2_{z^*} - \sigma^2_z = \sigma\left[w, (z-\mu_z)^2\right] - S^2 \tag{29.16a}$$

This result is derived in Example 30.2. Hence, directional selection *decreases* the phenotypic variance by S^2. With this in mind, Lande and Arnold suggested a corrected measure for the change in variance, their **stabilizing selection differential**,

$$C = \sigma^2_{z^*} - \sigma^2_z + S^2 \tag{29.16b}$$

which describes the nature of selection acting directly on the variance. Correcting for the effects of directional selection is important, as apparent signals of stabilizing selection based on a reduction in variance following selection could simply be the reduction in variance caused by directional selection. Similarly, signals of disruptive selection (an *inflated* postselection trait variance) can be masked by directional selection (e.g., Example 29.10).

The term *stabilizing selection differential* is actually a misleading description of C. One can have a negative value of C without having a fitness maximum over the range of phenotypes that is observed, and hence formally there will be no stabilizing selection. Because $C < 0$ ($C > 0$) is *consistent* with stabilizing (disruptive) selection, but is not *sufficient* in and of itself, following Phillips and Arnold (1989), we refer to C as the **quadratic selection differential**.

Analogous to S equaling the covariance between z and relative fitness, Equation 29.16a shows that C is the covariance between relative fitness and the squared deviation of a trait value from its mean,

$$C = \sigma\left[w, (z-\mu)^2\right] \tag{29.17}$$

A positive value of C indicates selection to increase the variance (convex selection, as would occur with disruptive selection), with fitness (on average) *increasing* for trait values further from the mean. Conversely, a negative value of C indicates selection to reduce the variance (concave selection, as would occur with stabilizing selection), with fitness (on average) *decreasing* as trait values move away from the mean.

As was the case with S, the opportunity for selection ($I = \sigma^2_w$) bounds the maximum possible within-generation change in variance (Arnold 1986). If we recall that $\sigma(x,y) = \rho_{xy}\sigma(x)\sigma(y)$, Equation 29.17 implies that $C = \rho_{w,(z-\mu)^2}\,\sigma_w\,\sigma[(z-\mu)^2]$. Because $\rho^2 \leq 1$,

$$C^2 \leq \sigma^2_w\,\sigma^2[(z-\mu)^2] = I \cdot \left(\mu_{4,z} - \sigma^4_z\right) \tag{29.18a}$$

where $\mu_{4,z} = E[(z-\mu)^4]$. The last step in Equation 29.18a follows if we recall that

$$\sigma^2[(z-\mu)^2] = E[(z-\mu)^4] - \left\{E[(z-\mu)^2]\right\}^2 = \mu_{4,z} - \left(\sigma^2_z\right)^2$$

Thus,

$$|C| \leq \sqrt{I\left(\mu_{4,z} - \sigma^4_z\right)} \tag{29.18b}$$

Furthermore, if z is normally distributed, then $\mu_{4,z} = 3\sigma^4_z$ (Kendall and Stewart 1977), implying that

$$\frac{|C|}{\sigma^2_z} \leq \sqrt{2I} \tag{29.18c}$$

Finally, the quadratic analog of β, the **quadratic** (or **stabilizing**) **selection gradient**, γ, was defined by Lande and Arnold (1983) as

$$\gamma = \frac{\sigma\left[w, (z-\mu)^2\right]}{\sigma^4_z} = \frac{C}{\sigma^4_z} \tag{29.19}$$

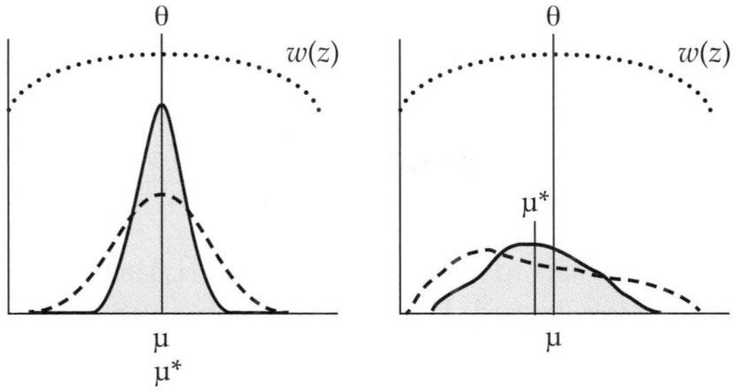

Figure 29.7 Even when a population is under strict stabilizing selection, the mean can change if the phenotypic distribution is skewed. One standard fitness function for stabilizing selection is Wright's quadratic, namely $W(z) = 1 - b(z - \theta)^2$ (Equation 28.3a). O'Donald (1968) found that when the population mean is at the optimum value ($\mu_z = \theta$), S is nonzero when the skew is nonzero ($\mu_{3,z} \neq 0$), as $S = -(b\mu_{3,z})/(1 - b\sigma_z^2)$. The dashed line and solid line (the latter with the corresponding filled curve), respectively, represent the preselection and postselection phenotypic distributions (both standarized to unit area). **Left**: If phenotypes are symmetrically distributed about the mean ($\mu_{3,z} = 0$), the distribution after selection has the same mean when $\mu_z = \theta$. **Right**: When the distribution is skewed, the longer tail experiences more selection than the shorter tail, which changes the mean.

As was the case for β, in its univariate form, γ appears as a simple rescaling of C, while its multivariate form (the matrix γ) accounts for the influence of phenotypic correlations among the measured traits (Chapter 30).

As with C, the interpretation of γ can be *extremely misleading* when the fitness surface is not well approximated by a quadratic over the range of scored phenotypes. We examine this issue in some detail both later in this chapter and in Chapter 30. A final complication in interpreting C is that if the phenotypic distribution is skewed, selection on the variance changes the mean (e.g., Equations 5.27b, 24.27, and 24.28a). This results in a nonzero value of S, which in turn inflates C (Figure 29.7).

Under Trait Normality, Gradients Describe the Local Geometry of the Fitness Landscape

A number of expressions equating β and γ to different gradient functions of fitness surfaces and landscapes appear in the literature. A potential source of confusion is that while these gradient functions are formally different (Geyer and Shaw 2008), they turn out to be equivalent when traits are normally distributed (Lande and Arnold 1983), and hence are used somewhat interchangeably in the literature.

We have frequently used the gradient of mean fitness with respect to the population mean,

$$\frac{\partial \ln \overline{W}}{\partial \mu_z} = \frac{1}{\overline{W}} \frac{\partial \overline{W}}{\partial \mu_z} \tag{29.20a}$$

We refer to gradients involving the mean fitness landscape as **landscape gradients**, and the particular gradient with respect to the population mean (Equation 29.20a) as the **landscape directional selection gradient**. Equations 24.26 and 24.27 show that landscape gradients with respect to higher moments of the phenotypic distribution appear in the general expression (free of normality assumptions) for selection response. When the trait is normally distributed and individual fitnesses are frequency-independent, the landscape directional selection gradient (Equation 29.20a) equals the directional selection gradient, $\beta = S/\sigma^2$ (Equation 13.8b); see Example A6.3 for the derivation.

A second class of gradients, based on the *individual* fitness surface, also appears in the literature. In particular, the **average directional selection gradient**,

$$\int \frac{\partial w(z)}{\partial z} p(z)\, dz = E_z \left[\frac{\partial w(z)}{\partial z} \right] \tag{29.20b}$$

is the average slope of the individual fitness surface, the expectation being taken over the phenotypic distribution for the population being studied. Similarly, the **average quadratic selection gradient**

$$\int \frac{\partial^2 w(z)}{\partial z^2}\, p(z)\, dz = E_z \left[\frac{\partial^2 w(z)}{\partial z^2} \right] \tag{29.20c}$$

is the average curvature of the individual fitness surface. Lande and Arnold (1983) showed that, provided z is normally distributed, Equation 29.20b equals β (the multivariate version is proved in Chapter 30) and Equation 29.20c equals γ (Equation 29.19).

Thus, β and γ provide measures of the geometry of the individual fitness surface averaged over the population being considered, *provided* z is normally distributed. When z is not Gaussian, Equations 29.20a and 29.20b need not equal β and Equation 29.20c need not equal γ. Indeed, unless the individual fitness function is of the form of a generalized exponential (Equation 29.34c), the postselection distribution is *expected* to depart from normality. Hence, after the first episode of selection, the fitness-weighted distribution is likely nonnormal, and these equalities can fail. This fact may be lost on an investigator examining selection over several episodes, as the *actual* distribution of phenotypes may not have changed, yet their *fitness-weighted* distributions may depart considerably from normality (for example, in a population before and after a episode of mate choice).

Under Trait Normality, Gradients Appear in Selection Response Equations

A final advantage of β and γ is that they appear as the only measure of phenotypic selection in the standard (i.e., normality-based) equations describing selection response. We have already seen (Equation 13.8c) that under the constraints of the breeder's equation, $R = \Delta\mu = \sigma_A^2\beta$, which is independent of any other measure of phenotypic selection. Similarly, under the infinitesimal model, the expected change in variance from a single generation of selection is given by Equation 16.7b,

$$\Delta\sigma_z^2 = \frac{h^4}{2}\,\delta_{\sigma_z^2} = \frac{\sigma_A^4}{2\,\sigma_z^4}\left(C - S^2\right) = \frac{\sigma_A^4}{2}\left(\gamma - \beta^2\right) \tag{29.21}$$

which decomposes the change in variance into the change due to selection on the variance (γ) and the change due to directional selection (β^2). More generally, when normality assumptions fail, the selection response equations for the variance involve various partials of landscape gradients (Equations 24.26 and 24.33c).

Again, while the distinction between differentials and gradients seems almost trivial in the univariate case (only a scale difference), their multivariate versions are *considerably different* from each other. As we will see in Chapter 30, gradients have the extremely important feature of removing the effects of phenotypic correlations among the set of *measured* traits, and the multivariate extensions of β and γ are what appear in the multivariate breeder's and Bulmer equations (Volume 3).

Partitioning Changes in Means and Variances Into Episodes of Selection

Suppose the total amount of within-generation selection is partitioned into k episodes of selection. Let μ_j and σ_j^2 be the (fitness-weighted) mean and variance after the jth episode of selection ($\mu = \mu_0$ and $\sigma^2 = \sigma_0^2$ are the baseline mean and variance before the first measured episode of selection). The definitions of S and C suggest that their appropriate counterparts for the jth episode of selection are calculated by

$$S_j = \mu_j - \mu_{j-1} \tag{29.22a}$$

$$C_j = \sigma_j^2 - \sigma_{j-1}^2 + S_j^2 \tag{29.22b}$$

(Arnold and Wade 1984a). The properties for S and C hold for each episode of selection, with

$$S_j = \sigma(w_j, z) \qquad \text{and} \qquad C_j = \sigma\left[w_j, (z - \mu_{j-1})^2\right] \tag{29.22c}$$

where w_j is the relative fitness for the jth episode of selection. Likewise, substituting $I_j = \sigma^2(w_i)$ for I in Equations 29.6 and 29.18c bounds the possible values of S_j and C_j.

How do these individual-episode measures relate to their total measure (i.e., over all episodes)? The partition of the total differentials, S_T and C_T, into those components are additive, with

$$S_T = \mu_k - \mu_0 = \sum_{j=1}^{k} (\mu_i - \mu_{i-1}) = \sum_{j=1}^{k} S_j \tag{29.22d}$$

and

$$C_T = \sigma_k^2 - \sigma_0^2 + S_T^2 = \sum_{j=1}^{k} \left(\sigma_i^2 - \sigma_{i-1}^2\right) + S_T^2$$

$$= \sum_{j=1}^{k} (C_j - S_j^2) + S_T^2 = \sum_{j=1}^{k} C_j + \left(S_T^2 - \sum_{j=1}^{k} S_j^2 \right) \tag{29.22e}$$

Example 29.8. For Example 29.2, following male body size over two episodes of selection, we find that $\widehat{S}_1 = \overline{z}_1 - \overline{z}_0 = -1.14$, $\widehat{S}_2 = \overline{z}_2 - \overline{z}_1 = 0.02$, and $\widehat{S}_T = -1.12$. Likewise,

$$\widehat{C}_1 = \text{Var}[z_1] - \text{Var}[z_0] + \widehat{S}_1^2 = 30.5 - 59.5 + (-1.12)^2 \simeq -27.7$$

Similar calculations yield $\widehat{C}_2 \simeq 13.19$ and $\widehat{C}_T \simeq -14.55$. Based on this limited dataset, there appears to be directional selection to reduce body size during the first episode. In addition, there appears to be selection to reduce the variance in body size during the first episode (mate choice) countered by selection to increase this variance during the second episode (fertility per mating).

Partitioning β and γ requires a little more care, as we have to account for changes in the phenotypic variance following each episode. From their definitions, the appropriate gradients for the jth episode of selection are

$$\beta_j = \frac{S_j}{\sigma_{j-1}^2} \quad \text{and} \quad \gamma_j = \frac{C_j}{\sigma_{j-1}^4} \tag{29.23a}$$

In their original paper, Arnold and Wade (1984a) initially defined gradients for each episode using the baseline variance, σ_0^2, throughout. With this definition, gradients are additive across episodes, as their numerators (being covariances) are additive and the denominators are unchanged. This approach is often called the **additive partition** (or more correctly, the **unweighted additive partition**), as $\beta_T = \sum \beta_j^0$, where $\beta_j^0 = S_j/\sigma_0^2$. Kalisz (1986) and Wade and Kalisz (1989) noted that the correct definition for an episodic gradient depends upon the trait variance at the *start* of each episode (Equation 29.23a), and the variance is expected to change during each episode. Writing $S_j = \beta_j \sigma_{j-1}^2$ gives

$$\beta_T = \frac{S_T}{\sigma_0^2} = \sum_{j=1}^{k} \frac{S_j}{\sigma_0^2} = \sum_{j=1}^{k} \frac{\beta_j \sigma_{j-1}^2}{\sigma_0^2} = \sum_{j=1}^{k} \beta_j \, a_j, \quad \text{where} \quad a_j = \frac{\sigma_{j-1}^2}{\sigma_0^2} \tag{29.23b}$$

Equation 29.23b defines the **variance-weighted additive partition** of gradients, where the total selection gradient is now a weighted sum of the individual gradients associated with each episode. The partition of γ follows in a similar manner, with

$$\gamma = \sum_{j=1}^{k} \gamma_j \, a_j^2 + \frac{1}{\sigma_0^4} \left[S_T^2 - \sum_{j=1}^{k} S_j^2 \right] \tag{29.23c}$$

where γ_j is calculated by Equation 29.23a. McGlothlin (2010) presented the multivariate version of Equation 29.23c and also revealed a key feature of Equation 29.23c. Note that γ can be nonzero even when *all* of the values of γ_j are zero. Further, when directional selection is occurring, the sign of γ can be different from the sign of the average value of γ_k.

Choice of the Reference Population: Independent Partitioning

Using the above partitioning schemes, selection differentials and gradients for a particular episode are based on their fitness-weighted values from the previous episode and are (with appropriate weighting) additive across episodes. Several authors have suggested that a slightly different approach, **independent partitioning**, may provide additional details about the nature of selection (Koenig and Albano 1987; Conner 1988; Koenig et al. 1991; Preziosi and Fairbairn 2000). Here, one uses the *observed* distribution of phenotypes as the reference population (as opposed to the *fitness-weighted* distribution) when computing statistics for each episode. While an independent partition provides a misleading picture of *evolutionary response* (by not weighting for previous selection), it may provide additional insight into the *nature of selection* in a particular episode. Further, if episodes are *not* sequential, the additive partition is not appropriate, while the independent partition can still be used to evaluate potential targets of selection. Fitness components estimated by the additive partition take into account any constraints imposed by prior selection, while independent partitioning ignores these prior constraints.

To see how using both partitions can provide additional insight into the nature of selection, consider the work of Koenig et al. (1991), who applied the additive and independent partitions to Howard's (1979) bullfrog dataset (Example 29.2). Howard concluded that larger frogs have more matings and produce more eggs and hatchlings. Fitness-component analysis (Example 29.2) using the additive partition showed that selection was largely on the number of mates, with no significant impact of body size on fertility per mate or hatching success (Arnold and Wade 1984b). Using an independent partition also found strong selection for body size via mating success, but, additionally, revealed strong selection for hatching success (offspring survivorship) as a function of body size. Thus, the independent partition suggested an additional interaction that was missed by the additive partition, thus generating new potential hypotheses for further testing (e.g., larger males may be better at defending eggs from predators or reside in better territories).

Example 29.9. To illustrate the difference between the additive and independent partitions, Koenig et al. (1991) presented the following hypothetical dataset for six individuals relating body size with natural (survival) and sexual (mating rate) selection.

Size z	Survivorship (Days alive), W_1	Mating rate (Mates/days alive), W_2	Total mates $W_1\,W_2$
11	1	3	3
12	1	2	2
13	1	1	1
11	10	1	10
12	10	2	20
13	10	3	30

Here, the mean body size before selection is 12. Focusing on total mates, the mean number of mates is 11, and the fitness-weighted trait mean (Equation 29.2b) becomes

$$\frac{1}{6}\left[11\left(\frac{3+10}{11}\right)+12\left(\frac{2+20}{11}\right)+13\left(\frac{1+30}{11}\right)\right]=12.3$$

Thus, the total selection differential on size due to differences in number of mates is 12.3 − 12 = 0.3. However, the selection differential for survivorship is 0 (the fitness-weighted trait mean is 12), and so the positive differential on total mates arises entirely via the fitness-weighted differences in mating rate (W_2).

Suppose, however, that one simply followed mating success per day and had no knowledge of survivorship. If one weights all six individuals equally (1/6) as the reference population, $\overline{W}_2 = 2$, which yields the mean after selection as

$$\frac{1}{6}\left[11\left(\frac{3+1}{2}\right)+12\left(\frac{2+2}{2}\right)+13\left(\frac{1+3}{2}\right)\right]=12$$

Thus, the selection differential on mating rate using this reference population is zero. If we ignore differences in survival, there are no effects of body size on mating rate. Despite the fact that both the survival and mating rate differentials are zero, the differential for total mates is not. This occurs because individuals with lower fitness in the first episode are given less weight when computing the additive partition, while they are equally weighted under an independent partition.

This example, along with Example 29.1, highlights the critical importance of accounting for all phases of selection. If this study followed just survivorship or just mates per day, no trait-fitness associations would have been seen. Grafen (1988) has coined the very appropriate term **invisible fraction** for that part of the population undergoing selection that is not seen by the investigator. Such missing data can significantly bias estimates of selection (e.g., Bennington and McGraw 1995; Hadfield 2008).

Standard Errors for Estimates of Differentials and Gradients

Because it is difficult to measure all of the individuals in a population, the effects of selection are usually estimated from a sample. This is also true in most longitudinal studies, with the cohort usually viewed as representative of phenotypes from the population. There are, however, occasional exceptions, such as when cohort members are intentionally chosen to include the most extreme phenotypes at much higher frequencies than their population levels.

There are a number of statistical issues involved in extrapolating from samples to the entire population, many of which still are unresolved. For example, individual fitness is usually measured with error. Further, there is generally a bias toward *underestimating* individual fitness—for example, marked individuals may not be recaptured and hence recorded as having zero fitness even if they survived. Likewise, the number of mates or offspring can be easily undercounted.

Assuming individual fitness is measured without error, the methods of LW Appendix 1 can be used to obtain approximate large-sample variances for estimators of differentials and gradients. The exact sampling variance for the directional selection differential is

$$\sigma^2\left(\widehat{S}_j\right)=\frac{\sigma_j^2}{n_j}+\frac{\sigma_{j-1}^2}{n_{j-1}} \tag{29.24a}$$

where n_j is the sample size for the jth episode. Using the delta-method approximation from LW Appendix 1, the large-sample variance for C is approximately

$$\sigma^2\left(\widehat{C}_j\right)\simeq 4S_j^2\,\sigma^2\left(\widehat{S}_j\right)+8S_j\left(\frac{\mu_{3,j}}{n_j}+\frac{\mu_{3,j-1}}{n_{j-1}}\right)$$
$$+\frac{\mu_{4,j}-\sigma_j^4}{n_j}+\frac{\mu_{4,j-1}-\sigma_{j-1}^4}{n_{j-1}} \tag{29.24b}$$

where μ_3 and μ_4 are the third and fourth central moments around the mean, $\mu_i = E[(z-\mu)^i]$. If phenotypes are normally distributed ($\mu_3 = 0, \mu_4 = 3\sigma^4$), this reduces to

$$\sigma^2\left(\widehat{C}_j\right)\simeq 4S_j^2\,\sigma^2\left(\widehat{S}_j\right)+2\left[\frac{\sigma_j^4}{n_j}+\frac{\sigma_{j-1}^4}{n_{j-1}}\right] \tag{29.24c}$$

If the scaled skewness, κ_3 (LW Equation 2.8), and kurtosis, κ_4 (LW Equation 2.12a), are small, this normal approximation is still appropriate.

Example 29.10. Boag and Grant (1981) observed intense natural selection in Darwin's medium ground finch (*Geospiza fortis*) during a severe drought on Daphne Major Island in the Galápagos. The estimated mean and variance of body weight in 642 adults before the drought were, respectively, 15.79 and 2.37, while the estimated mean and variance of 85 surviving adults after the drought were 16.85 and 2.43. Thus, $\widehat{S} = 16.85 - 15.79 = 1.06$ and Equation 29.24a returns the standard deviation of this estimate as

$$\text{SE}\left(\widehat{S}\right) \simeq \sqrt{\frac{2.37}{642} + \frac{2.43}{85}} \simeq \sqrt{0.0323} \simeq 0.180$$

implying that the directional selection differential on body size was significantly positive.

There appears to be very little selection on the variance when the uncorrected change in variance, $\text{Var}(z^*) - \text{Var}(z) = 2.43 - 2.37 = 0.06$, is used. However, using the quadratic selection differential to correct for the reduction in the variance from directional selection gives $\widehat{C} = 0.06 + 1.06^2 = 1.14$, consistent with selection to increase the variance (convex selection, whereby selection favors extreme individuals). From Equation 29.24c, assuming body size is normally distributed before and after the drought,

$$\text{SE}\left(\widehat{C}\right) \simeq \sqrt{4 \cdot (1.06)^2 \cdot 0.0323 + 2\left[\frac{(2.37)^2}{642} + \frac{(2.43)^2}{85}\right]} \simeq 0.549$$

Thus, $\widehat{C}$ is 2.08 standard errors above 0, suggesting that it is (at least) close to being significant.

DESCRIBING PHENOTYPIC SELECTION: INDIVIDUAL FITNESS SURFACES

The fitness (W) of an individual with a trait value of z can be decomposed into the sum of its **expected fitness**, $W(z) = E[\,W \mid z\,]$, plus a **residual deviation**, e_W,

$$W = E[\,W \mid z\,] + e_W = W(z) + e_W$$

The residual variance for a given z, $\sigma^2_{e_W}(z)$, measures the variance in fitness among individuals with a phenotypic value of z. Estimation of the individual fitness surface is thus a generalized regression problem, with the goal being to choose a candidate function for $W(z)$ that minimizes the average residual variance, $E_z[\sigma^2_{e_W}(z)]$. Because the total variance in fitness (σ^2_W) equals the sum of the within-group (individuals with the same trait value) and among-group variances in fitness,

$$\frac{\sigma^2_W - E_z[\,\sigma^2_{e_W}(z)\,]}{\sigma^2_W} = 1 - \frac{E_z[\,\sigma^2_{e_W}(z)\,]}{\sigma^2_W}$$

is the fraction of individual fitness variation accounted for by a particular estimate of $W(z)$, providing a metric for comparing different estimates of the fitness function.

There are at least two sources of variation that contribute to the fitness residual, e_W. First, there can be errors in measuring the actual fitness of an individual (these are almost always ignored, although they induce serious biases; see Hadfield 2008). Second, the *actual* (or **realized**) fitness of any particular individual is a random variable distributed around its *expected value*. Generally, these residual deviations are heteroscedastic, with the residual variance varying with z (Mitchell-Olds and Shaw 1987; Schluter 1988). As we will see later, aster models provide a general strategy for accounting for such errors structures.

Consider fitness measured by survival to a particular age. While $W(z) = p_z$ is the probability of survival for an individual with a character value of z, the fitness for a particular individual is a Bernoulli random variable, as it either survives (value of one) or does not (value of zero). The resulting residual has only two possible values, $e_W = 1 - p_z$ with probability p_z, and $e_W = -p_z$ with probability $1 - p_z$, giving

$$\sigma^2_{ew}(z) = E[(e_W)^2] = (1 - p_z)^2 p_z + p_z^2(1 - p_z) = p_z(1 - p_z)$$

Unless p_z is constant over z, the residuals are heteroscedastic. Note that after accounting for phenotypic-specific differences (here, the p_z values), there still remains substantial variance in individual fitness. This within-class variance is most extreme at intermediate survival rates ($p_z \simeq 1/2$).

Finally, inferences about the individual fitness surface are limited by the range of phenotypes in the population. Unless this range is very large, only a small region of the fitness surface can be explored with any precision. Estimates of the fitness surface at the tails of the current phenotypic distribution are extremely imprecise yet potentially very informative, as they suggest patterns of selection for populations at the margins of the observed range of phenotypes. A further complication is that the fitness surface can change as the environment changes, so that year-to-year estimates can differ (e.g., Kalisz 1986). Such change may require an investigator to choose between a time-averaged fitness surface (which can impart bias) versus a series of year-specific estimates (which suffer from lower power). Finally, as emphasized in Chapter 22, organisms often modify their environments as they evolve, implying that the biotic environment, and hence $W(z)$, also evolves.

Linear and Quadratic Approximations of $W(z)$

The individual fitness surface, $W(z)$, can be very complex, and a wide variety of functions may be chosen to approximate it. The simplest and most straightforward approach is to use a low-order polynomial (typically a linear or a quadratic; Lande and Arnold 1983). One justification for this approach is that in a sufficiently local region, even a highly nonlinear surface can be reasonably approximated by a quadratic function (this is simply a Taylor series approximation). Put another way, for a sufficiently small range of phenotypes, even a rather rough fitness surface can appear to be locally smooth.

Consider first the simple linear regression of *relative* fitness, w, as a function of phenotypic value, z. Because the directional selection gradient is defined as $\beta = S/\sigma_z^2 = \sigma(w, z)/\sigma_z^2$, it follows from regression theory (LW Equation 3.14b) that β is the slope of the least-squares linear regression of relative fitness on z,

$$w = a + \beta z + e \tag{29.25a}$$

which makes $\widehat{w}(z) = a + \beta z$ the best linear predictor of relative fitness (Geyer and Shaw 2008). Because the regression passes through the expected values of w and z (1 and μ_z, respectively), Equation 29.25a can be rewritten as

$$w = 1 + \beta(z - \mu_z) + e \tag{29.25b}$$

giving $\widehat{w}(z) = 1 + \beta(z - \mu_z)$.

If we assume that the fitness function (for the range of phenotypes scored) is well described by a linear regression, β is the expected change in relative fitness, given a unit change in z. Usually, **standardized regressions** are used, where the trait value is translated (shifted) and then variance-standardized to give it a mean of zero and a variance of one, with the standardized value, $z_{sd} = (z - \mu)/\sigma_z$, replacing z in the regression. The standardized version of the Lande-Arnold regression becomes

$$w = 1 + \beta_{sd} z_{sd} + e \tag{29.25c}$$

where β_{sd} gives the expected change in fitness given a standard deviation change in the trait. For example, a value of $\beta_{sd} = 0.2$ implies that a change of one standard deviation in the trait value increases relative fitness by 20%.

Note that fitness in the regression given by Equation 29.25c is *mean-standardized* (we use w in place of W), so our use of the term *standardized regression* is different from its use in many statistical packages, where *both* the dependent (w) and the predictor (z) variables are variance-standardized. Hence, it is best to directly feed in the transformed data (i.e., trait-standardized to a mean of zero and a variance of one, fitness-standardized to a mean of one) into a regression package, rather than worry about how a particular package will standardize the raw data before performing the regression.

As a consequence of standardization, we have

$$S_{sd} = \bar{\imath}_z \tag{29.25d}$$

showing that the selection differential for a standardized trait is simply the selection intensity, $\bar{\imath}_z$, of the trait on the original (untransformed) scale, allowing for an easy comparison of the relative strengths of selection among focal traits. Further, note (from the Robertson-Price identity) that because $S_{sd} = \sigma(z_{sd}, w)$, and hence $\beta_{sd} = \sigma(z_{sd}, w)/\sigma^2(z_{sd}) = S_{sd}$, yielding

$$\beta = \frac{\sigma(z,w)}{\sigma_z^2} = \frac{\sigma(z/\sigma_z, w)}{\sigma_z} = \frac{\sigma(z_{sd}, w)}{\sigma_z} = \frac{\beta_{sd}}{\sigma_z} \tag{29.25e}$$

From LW Equation 3.17, the fraction of variance in individual fitness accounted for by a linear regression in z is

$$r^2_{z,w} = \frac{[\sigma(z,w)]^2}{\sigma^2(z)\cdot\sigma^2(w)} = \widehat{\beta}^2\frac{\sigma^2(z)}{\widehat{I}} \tag{29.26a}$$

as obtained by Moorad and Wade (2013). For a standardized regression, this reduces to

$$r^2_{z_{sd},w} = \widehat{\beta}^2_{sd}/\,\widehat{I} \tag{29.26b}$$

Rearranging Equation 29.26a,

$$\widehat{I}r^2_{z,w} = \widehat{\beta}^2\sigma^2(z) = \frac{\sigma^2(z,w)}{\sigma^4(z)}\sigma^2(z) = \frac{\sigma(z,w)}{\sigma^2(z)}\sigma(z,w) = \widehat{\beta}S \tag{29.26c}$$

which shows that the amount of the variance in relative fitness (I) accounted for by directional selection on a focal trait is $\widehat{\beta}S$ (Moorad and Wade 2013).

If the individual fitness surface shows curvature, as expected if there is stabilizing or disruptive selection, a **quadratic regression** is more appropriate,

$$w = a + b_1 z + b_2(z-\mu_z)^2 + e \tag{29.27a}$$

Because the regression passes through the mean of all variables,

$$E[w] = 1 = a + b_1 E[z] + b_2 E[(z-\mu_z)^2] = a + b_1\mu_z + b_2\sigma_z^2$$

Solving for the intercept (a) and substituting back into Equation 29.26b yields

$$w = 1 + b_1(z-\mu_z) + b_2\left[(z-\mu_z)^2 - \sigma_z^2\right] + e \tag{29.27b}$$

The trait-standarized version ($\mu_{z_{sd}} = 0, \sigma^2(z_{sd}) = 1$) becomes

$$w = 1 + b_{sd,1} z_{sd} + b_{sd,2}\left(z_{sd}^2 - 1\right) + e \tag{29.27c}$$

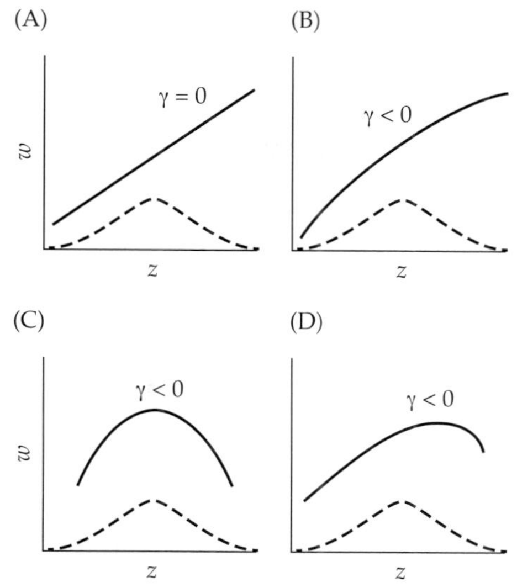

Figure 29.8 The relationship between γ and curvature of a quadratic fitness function (solid line). The dashed curve represents the distribution of z. **A:** *W(z)* is strictly linear, hence $\gamma = 0$. **B:** *W(z)* curves downward (is concave), but has no maximum over the range of z examined. Hence, $\gamma < 0$, implying stabilizing selection by the Lande-Arnold criterion, when in fact selection is entirely directional. **C:** Stabilizing selection only, as there is no change in the mean. **D:** A combination of directional and stabilizing selection, as the population mean is not at the optimal fitness value. (After Mitchell-Olds and Shaw 1987.)

which is also written as

$$w = a + b_{sd,1} z_{sd,1} + b_{sd,2} z_{sd}^2 + e \tag{29.27d}$$

A positive quadratic term (b_2 or $b_{sd,2} > 0$) indicates that the best-fitting quadratic approximation of the individual fitness surface has an upward (convex) curvature, while a negative value ($b_2 < 0$) implies that the curvature is downward (concave). Lande and Arnold (1983) suggested that $b_2 > 0$ indicates disruptive selection while $b_2 < 0$ indicates stabilizing selection. Their reasoning follows from elementary geometry in that a *necessary* condition for a local minimum is that a function must be convex over some interval, while a necessary condition for a local maximum is that the function is concave. However, Mitchell-Olds and Shaw (1987) and Schluter (1988) noted that this condition is not *sufficient* for the presence of a local extrema. Stabilizing selection is generally defined as the presence of a local maximum in $w(z)$ and disruptive selection by the presence of a local minimum, while b_2 indicates *curvature* rather than the presence of local extrema. As Figure 29.8B shows, a quadratic fitness function can curve downward without the population experiencing a local maximum. It can also curve upward without having a local minimum. Because of these issues, the current usage is to denote $b_2 < 0$ as concave selection and $b_2 > 0$ as convex selection.

Lande-Arnold Fitness Regressions

We solve for the regression coefficients b_1 and b_2 by transforming Equation 29.27a into a standard multiple regression problem by setting $x_1 = z$ and $x_2 = (z - \mu_z)^2$ and applying the methods of LW Chapter 8. To proceed, we need expressions for $\sigma(x_1, x_2)$, $\sigma(x_1, w)$, and $\sigma(x_2, w)$. From LW Equation A1.14, $\sigma(x_1, x_2) = \sigma\left(z, [z - \mu_z]^2\right) = \mu_{3,z}$, the skew of the phenotypic distribution before selection. Likewise, from Equations 13.7a and 29.17, $\sigma(x_1, w) = \sigma(z, w) = S$ and $\sigma(x_2, w) = \sigma([z - \mu_z]^2, w) = C$. Substituting these values into the results from LW Example 8.3 (which gives exact expressions for the partial regression coefficients of a bivariate regression), and note that $\sigma^2(\,[z-\mu_z]^2\,) = \mu_{4,z} - \sigma_z^4$, we obtain

$$b_1 = \frac{\sigma^2(x_2)\cdot\sigma(x_1,w) - \sigma(x_1,x_2)\cdot\sigma(x_2,w)}{\sigma^2(x_1)\cdot\sigma^2(x_2) - \sigma^2(x_1,x_2)} = \frac{(\mu_{4,z}-\sigma_z^4)\cdot S - \mu_{3,z}\cdot C}{\sigma_z^2\cdot(\mu_{4,z}-\sigma_z^4) - \mu_{3,z}^2} \tag{29.28a}$$

$$b_2 = \frac{\sigma^2(x_1)\cdot\sigma(x_2,w) - \sigma(x_1,x_2)\cdot\sigma(x_1,w)}{\sigma^2(x_1)\cdot\sigma^2(x_2) - \sigma^2(x_1,x_2)} = \frac{\sigma_z^2\cdot C - \mu_{3,z}\cdot S}{\sigma_z^2\cdot(\mu_{4,z}-\sigma_z^4) - \mu_{3,z}^2} \tag{29.28b}$$

The estimators of b_1 and b_2 are obtained by replacing $\mu_{k,z}$ with their sample estimates and using $\widehat{C}$ and $\widehat{S}$.

It is important to stress that the distributional assumptions leading to the solutions offered by Equation 29.28 are simply those of ordinarily least-squares (OLS), namely, homoscedastic and uncorrelated residuals (Chapter 18; LW Chapter 8). To this point, *no other assumptions* have been made about either the distribution of the trait values or the distributional form of the residuals (i.e., normality of the residuals is *not* required for Equation 29.28). Standard errors for parameter estimates follow from LW Equation 8.33b (again, requiring no assumptions on the distribution of residuals other that OLS). However, if we wish to turn standard errors into confidence intervals or to perform hypothesis testing, further distributional assumptions on residuals (such as normality) are required.

Provided z is normally distributed before selection, then $\mu_{3,z} = 0$ and $\mu_{4,z} - \sigma_z^4 = 2\sigma_z^4$. In this case, Equations 13.8b and 29.19 imply, respectively, that $b_1 = S/\sigma_z^2 = \beta$ and $b_2 = C/[2\sigma_z^4] = \gamma/2)$, giving the univariate version of the **Lande-Arnold regression** (1983)

$$w = 1 + \beta(z-\mu_z) + \frac{\gamma}{2}\left([z-\mu_z]^2 - \sigma_z^2\right) + e \tag{29.29a}$$

As this expresion was also motivated by Pearson (1903), we occasionally refer to Equation 29.29a as the **Pearson-Lande-Arnold regression**. In standarized form,

$$w = a + \beta_{sd}z_{sd} + \frac{\gamma_{sd}}{2}z_{sd}^2 + e \tag{29.29b}$$

Equations 29.29a and 29.29b provide a connection between selection gradients (directional and stabilizing, β and γ, respectively) and quadratic approximations of the individual fitness surface, *provided* z is normally distributed *and* residuals follow the OLS assumption.

An important point from Equation 29.28a is that if skew is present in the trait distribution ($\mu_{3,z} \neq 0$), then $b_1 \neq \beta$, and hence the slope (β) in the linear regression (the best *linear* fit) of $w(z)$ differs from the linear slope term (b_1) in the the best *quadratic* fit of $w(z)$. This inequality arises because the presence of skew generates a covariance between z and $(z-\mu_z)^2$. The biological significance of this inequality can be seen by reconsidering Figure 29.7, where the presence of skew in the phenotypic distribution results in a change in the mean of a population under strict stabilizing selection (defined as the population mean being at the optimum of the individual fitness surface). From the Robertson-Price identity (Equation 6.10), the within-generation change in the mean equals the covariance between phenotypic value and relative fitness. Because covariances measure the amount of *linear* association between variables, in describing the change in mean, the correct measure is the slope of the best *linear* fit of the individual fitness surface, which is unbiased by skew (Lande and Arnold 1983). If skew is present, using b_1 from the quadratic regression to describe the change in mean is incorrect, as this quadratic regression removes the effects on relative fitness from a linear change in z due to the correlation between z and $(z-\mu_z)^2$.

Observe from Equation 29.28b that the presence of skew (and kurtosis) also results in $2b_2$ being a biased estimator of γ. Despite this bias, $2b_2$ remains the standard estimate for γ in the literature. Shifting the trait to give it a mean of zero, a less-biased estimator of γ follows from a quadratic regression with no linear term,

$$w = a + bz^2 + e \tag{29.29c}$$

From standard regression theory (LW Chapter 3), the slope of Equation 29.29c is $b = \sigma(z^2, w)/\sigma^2(z^2)$, and using the covariances obtained for Equation 29.28 yields

$$b = \frac{\sigma(z^2, w)}{\sigma^2(z^2)} = \frac{C}{\mu_{4,z} - \sigma_z^4} \tag{29.29d}$$

Recall that the scaled kurtosis is $\kappa_4 = (\mu_4 - 3\sigma_z^4)/\sigma_z^4$ (LW Equation 2.12a). Therefore, rearranging yields $\mu_{4,z} - \sigma_z^4 = \sigma_z^4(\kappa_4 + 2)$, or

$$b = \frac{C}{\sigma^4(\kappa_4 + 2)} = \frac{\gamma}{\kappa_4 + 2} \tag{29.29e}$$

For a normal, $\kappa_4 = 0$, and $2E[b] = \gamma$. When kurtosis is present, using the sample estimate of $\widehat{\kappa}_4$ yields

$$\widehat{\gamma} = \widehat{b}\,(\widehat{\kappa}_4 + 2) \tag{29.29f}$$

which is an improved estimate of the quadratic gradient. Finally, despite these corrections, we should point out that bias induced by skewness and kurtosis is a matter of degree. Both of these moments have their own sampling problems, and using poor estimates of them to correct β and γ can be problematic.

The Geometry of Quadratic Fitness Functions

As we have seen, when traits are normally distributed, β and γ provide summary statistics of the geometry of the individual surface and mean fitness landscape, either by measuring the average slope or curvature of the individual fitness surface (average selection gradients; Equations 29.20b and 29.20c) or the slope or curvature (evaluated at the current mean, μ_z) of the fitness landscape (landscape selection gradients; Equation 29.20a). Table 30.1 summarizes these (and other) features for the multivariate versions of β and γ. ***These expressions hold for any form of w(z), provided*** z is normally distributed.

Moreover, when $w(z)$ is exactly a quadratic, similar expressions hold for *any* distribution of z. Taking the first two partial derivatives of Equation 29.27a yields

$$\frac{\partial w(z)}{\partial z} = b_1 + 2b_2(z - \mu_z) \quad \text{and} \quad \frac{\partial^2 w(z)}{\partial z^2} = 2b_2 \tag{29.30a}$$

Evaluating these derivatives at the current mean ($z = \mu_z$) returns

$$\left.\frac{\partial w(z)}{\partial z}\right|_{z=\mu_z} = b_1 \qquad \text{and} \qquad \left.\frac{\partial^2 w(z)}{\partial z^2}\right|_{z=\mu_z} = 2b_2 \tag{29.30b}$$

Hence, one interpretation of the regression coefficients in Equation 29.27a is that they describe the slope (b_1) and the curvature ($2b_2$) of a quadratic fitness surface at the population mean. A second interpretation (along the lines of Equation 29.20b and 29.20c) is that these coefficients also equal the average slope (b_1) and average curvature ($2b_2$) of the quadratic fitness surface over the entire distribution of z (i.e., they are average selection gradients). To see this, note that Equation 29.20b yields an average directional selection gradient of

$$E_z\left[\frac{\partial w(z)}{\partial z}\right] = E_z\left[b_1 + 2b_2(z - \mu_z)\right] = b_1 + 2b_2 E_z\left[z - \mu_z\right] = b_1 \tag{29.30c}$$

Likewise, from Equation 29.20, the average quadratic selection gradient is

$$E_z\left[\frac{\partial^2 w(z)}{\partial z^2}\right] = E_z\left[2b_2\right] = 2b_2 \tag{29.30d}$$

Even when the trait distribution (before selection) is normal, if $w(z)$ is not a quadratic, then the *average* slope and curvature need not equal the slope or the curvature of $w(z)$ that is evaluated at μ_z. In other words, the average of the function need not equal the value of the function evaluated at the average. Assuming a normally distributed trait, Phillips and Arnold (1989) performed Taylor-series expansions (around $z = \mu_z$) of Equations 29.20b and 29.20c, showing how the expected slope and curvature relate to the slope and curve at the mean,

$$\int \frac{\partial w(z)}{\partial z} p(z)\, dz = \left.\frac{\partial w(z)}{\partial z}\right|_{z=\mu_z} + \left.\left(\frac{\sigma_z^2}{2} \frac{\partial^3 w(z)}{\partial z^3} + \frac{\sigma_z^4}{8} \frac{\partial^5 w(z)}{\partial z^5} + \cdots \right)\right|_{z=\mu_z}$$

$$\int \frac{\partial^2 w(z)}{\partial z^2} p(z)\, dz = \left.\frac{\partial^2 w(z)}{\partial z^2}\right|_{z=\mu_z} + \left.\left(\frac{\sigma_z^2}{2} \frac{\partial^4 w(z)}{\partial z^4} + \frac{\sigma_z^4}{8} \frac{\partial^6 w(z)}{\partial z^6} + \cdots \right)\right|_{z=\mu_z} \tag{29.30e}$$

While trait normality ensures that the slope of the *mean fitness landscape*, $\overline{W}$, evaluated at the trait mean equals the average slope of $w(z)$, it *does not* ensure that the *average* slope or curvature of $w(z)$ equals the slope or curvature of $w(z)$ as is evaluated at the trait mean. As Equation 29.30e shows, this only happens when $w(z)$ is quadratic (and hence partials of order three and higher vanish). While this point may seem overly technical, it is important to stress because much of the elegant symmetry of the connections between β and γ and quantities of interest (average geometries of the fitness surface, regression coefficients, and important parameters in evolutionary equations for predicting response) only hold when the quadratic is an excellent approximation of the true fitness surface or when z is normal (before selection). See Geyer and Shaw (2008) for an extended discussion of this issue.

Finally, note that for a quadratic fitness function, for any distribution of z

$$\overline{W} = E_z \left[a + b_1 z + b_2 (z-\mu)^2 \right] = a + b_1\mu + b_2\sigma_2^2 \tag{29.30f}$$

which implies that

$$\frac{\partial \overline{W}}{\partial \mu} = b_1 \tag{29.30g}$$

It is important to stress that throughout, we have defined $\beta = S/\sigma_z^2$ (Equation 13.8b) and $\gamma = C/\sigma_z^4$ (Equation 29.19), making no assumptions about either the trait distribution or the individual fitness function. While alternative expressions (e.g., Equations 29.20 and 29.30) exist for β and γ under certain distributional (normality) or fitness function (quadratic) assumptions, these are simply equalities under certain conditions that follow given our (distributional and fitness function–independent) definitions of β and γ. The importance of β and γ is that they appear as the sole measures of selection in the breeder's equation (13.8c) and Bulmer equation (29.21) for predicting changes in means and variances. When the normality assumptions underlying the simple forms of these two selection-response (mean and variance, respectively) equations fail, the response depends, in complex ways, on the underlying genetic details and various partials of the mean fitness landscape (Equation 24.26).

Hypothesis Testing, Approximate Confidence Intervals, and Model Validation

While there is a large body of theory for testing the significance of regression coefficients, much of it assumes homoscedastic and normally distributed residuals, $e_w \sim N(0, \sigma_{e_w}^2)$, where the residual variance ($\sigma_{e_w}^2$) is a constant (i.e., independent of z). As mentioned above, these two assumptions are almost always violated with fitness data, invalidating standard tests for significance (Mitchell-Olds and Shaw 1987). Fortunately, a variety of resampling methods available for hypothesis testing are robust to heteroscedasticity and nonnormal residuals, and we briefly mention three of these procedures: jackknife confidence intervals, randomization tests of significance, and cross-validation. A fair amount of this material is presented for historical reasons, as the aster-model framework discussed at the end of the chapter provides a more statistically rigorous approach than the resampling approximations presented here.

Jackknife estimates were introduced by Tukey (1958) as a generalized statistical tool; see Miller (1974), Wu (1986), Shao and Tu (1996), and Manly (1997) for reviews, and Mitchell-Olds and Shaw (1987) and Mitchell-Olds and Bergelson (1990) for applications to fitness regressions. The idea is simple, **resampling**. Namely, base parameter estimates on the behavior of the distribution of estimates from subsamples of the original data. Consider the estimator of β for the linear regression given by Equation 29.25a. Denote by $\widehat{\beta}$ the standard least-squares estimate of β using the full dataset of n individuals, and we let $\widehat{\beta}_i$ denote the estimator using the complete dataset minus data for the ith individual. The resulting jackknife estimator is

$$\widehat{\beta}_{jack} = \frac{1}{n}\sum_{i=1}^{n} \phi_i = \overline{\phi} \quad \text{where} \quad \phi_i = n\widehat{\beta} - (n-1)\widehat{\beta}_i \tag{29.31a}$$

which has an approximate large-sample variance of

$$\text{Var}\left(\widehat{\beta}_{jack}\right) \simeq \frac{1}{n(n-1)}\sum_{i=1}^{n}(\phi_i - \overline{\phi})^2 \tag{29.31b}$$

Approximate large-sample confidence intervals follow using Equation 29.31b and the fact that $\widehat{\beta}_{jack}$ is approximately t-distributed with $n-1$ degrees of freedom. The jackknife estimator and its sampling variance are well behaved even when the residuals are heteroscedastic, thus allowing for valid hypothesis testing (Wu 1986). Wu presented a slightly improved jackknife estimator by weighting the ϕ_i values, but the difference between the weighted and unweighted estimates is usually small for large sample sizes.

Another resampling approach involves **randomization tests** for the significance of a regression. Again, the idea behind this class of tests is simple. A value of $\widehat{\beta}$ under the hypothesis of no association between fitness and trait value (z) is generated by assigning the n individual fitness values at random to the observed phenotypic values and estimating β for this scrambled (**randomized**) dataset. Repeating this resampling procedure several thousand times generates a distribution of regression coefficients under the null hypothesis of no association between individual fitness and character value. For example, suppose we obtain a standard least-squares estimate (assuming a linear regression) of $\widehat{\beta} = 1.25$, and upon subsequent randomization of the same dataset we find that only 14 out of 1000 randomized datasets have $|\,\widehat{\beta}\,|$ values in excess of 1.25, which suggests a p value of 0.014. Moore (1990) and Hews (1990) both illustrated applications of randomization tests to fitness data.

A final issue concerns the validity of the particular model chosen to fit $W(z)$. This is a difficult task because fitness data are inherently noisy—the residual variance can be rather large, even if we have perfectly fit $W(z)$. One approach for comparing models is not to contrast their goodness-of-fit (i.e., least-squares solutions), but rather to compare their *predictive* ability: if some of the data are ignored, how well are their fitnesses predicted using a model built from the remaining data? This **cross-validation** procedure (Snee 1977; Picard and Cook 1984) can be used both to assess how well a model predicts the data and as a model-selection approach. The latter follows by choosing the model (from a set of candidates) with the smallest prediction error (Chapter 30 further discusses issues in model selection). Different schemes for splitting the data into **training** (model-building) and **validation** (testing) sets have been proposed. At one extreme is **leave-one-out** cross-validation, wherein the first observation is left out, after which a model built from the remaining data and its fitness prediction of the removed value is assessed, proceeding in this fashion through all of the observations to yield a prediction error variance for a given model. **Generalized cross-validation** (Craven and Wahba 1979) estimates a smoothing parameter to yield a less rugged solution.

A computationally less intense approach is **k-fold** cross-validation, wherein data are randomly assigned to k groups (or **folds**), with the data from the first $k-1$ groups used to fit the model, and the data from the last group used to check its prediction accuracy. This procedure is repeated so that all k groups are used as test groups, typically with k somewhere between 4 and 10. The k-fold approach may be more appropriate when model fitting is computationally intense, as only $k \ll n$ models must be fitted. It is important to stress that fitting the model to existing data and predicting new data are quite different tasks, and that a model that fits the observed data well may not have good predictive ability.

Power

Another critical issue is the power of a regression to detect selection. As discussed in LW Appendix 5, power is the probability of detecting (i.e., declaring to be significant) an effect, given a preset significance level. Our main focus is on the power to detect a directional selection gradient (β) in a univariate regression. A convenient way to compute the power of this regression is to consider the correlation between the trait value and relative fitness,

$$\rho = \frac{\sigma(z, w)}{\sigma_z\, \sigma_w} = \frac{S}{\sigma_z\, \sigma_w} = \frac{\bar{\imath}}{\sqrt{I}} \tag{29.32a}$$

which is the ratio of the selection intensity to the square root of the opportunity for selection (Hersch and Phillips 2004). Note that we can alternatively express ρ in terms of either a unstandardized (β) or standardized directional selection gradient (β_{sd}), with

$$\rho = \frac{S}{\sigma_z\, \sigma_w} = \frac{S}{\sigma_z^2\, \sigma_w}\sigma_z = \beta\,\frac{\sigma_z}{\sigma_w} = \frac{\beta_{sd}}{\sigma_w} \tag{29.32b}$$

where the last identity follows from Equation 29.25e. Thus, the power to detect a directional selection gradient is a function of *both* the strength of selection (the selection intensity, $\bar{\imath} = \beta\sigma_z = \beta_{sd}$) *and* the variance in total fitness ($I = \sigma_w^2$). A strong amount of selection (a large value of $\bar{\imath}$) per se does not ensure high power: rather, it is the strength of selection *relative* to the total variance in fitness ($\bar{\imath}/\sqrt{I}$) that is critical. Hersch and Phillips (2004) worked with the adjusted sample correlation, which is defined as

$$r' = \sqrt{\frac{\text{Var}(w)}{\sigma_w^2}} \cdot \sqrt{\frac{\sigma_z^2}{\text{Var}(z)}} \cdot r \tag{29.32c}$$

where Var denotes a sample variance (for either z or w) and r denotes the sample correlation between z and w. Assuming fitness residuals are normal, r' can be scaled to follow a student-t distribution

$$\sqrt{n-1}\left(\frac{r' - \rho}{\sqrt{1-\rho^2}}\right) \sim t_{n-1} \tag{29.32d}$$

Using this result, power calculations follow from standard approaches, wherein for a given strength of selection (scaled as ρ), sample size (n), and level of significance (α), one can compute the probability of rejecting the null hypothesis of $\rho = 0$ (see similar examples in LW Appendix 5).

As shown in Figure 29.9, sample size typically needs to be in the hundreds to have significant power of detecting even modest selection on a trait. With zero/one coded fitness data (such as viability), the residuals follow binomial (rather than normal) distributions, and simulations by Hersch and Phillips (2004) showed that using Equation 29.32d *overestimates* the power to detect a trait-fitness association in this case. Values of n obtained from Equation 29.32d should thus be regarded as *lower bounds*. Further, because of the higher sampling variances with estimates of z^2, for a fixed value of n, the power to detect quadratic effects (such as γ) is expected to be much less than the power associated with a comparable value for a linear effect (β). Hence, if a large n is required to have any serious power to detect a linear effect, an even larger (and often *much* larger) value of n is required to detect a similar quadratic effect.

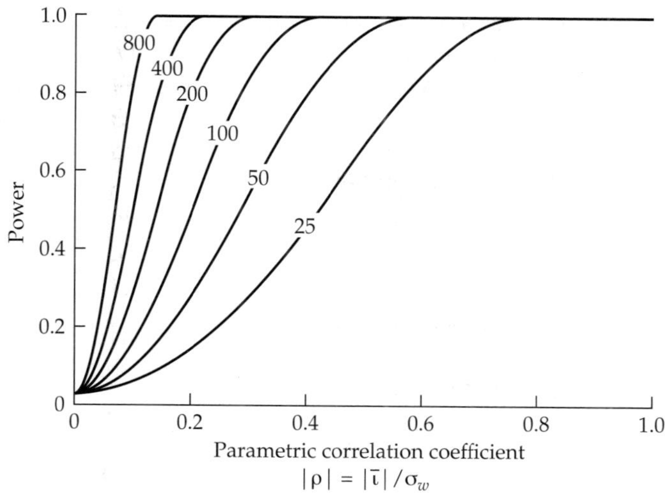

Figure 29.9 The power of a univariate regression to detect a directional selection gradient is a function of the correlation (ρ) between trait value and relative fitness, where $\rho = \bar{\imath}/\sigma_w = \beta\,\sigma_z/\sigma_w$. Power is plotted as a function of ρ with the curves for sample sizes starting at $n = 25$ and successively doubling until 800. Here power is the probability that the sample correlation is declared significantly different from zero using a test of significance of $\alpha = 0.05$. (After Hersch and Phillips 2004.)

Mean-standardized Gradients and Fitness Elasticities

Most traits of interest are scaled in the units of their measurement, which could be linear (lengths), areal (leaf surface), volumetric (mass), and even ordinal (traits such as shades of color). The evaluation of different traits across such a range of scales requires some sort of standardization to return a dimensionless quantity for comparison. Typically this is done by dividing a trait by its standard deviation (**variance-standardization**), or, less frequently, dividing it by its mean (**mean-standardization**). For example, Chapters 6 and 13 compared mean- and variance-standardization for the additive genetic variance and the selection response. As shown in Chapter 6, determining which standardization to use is not a trivial issue, as traits that are closely related to fitness have lower heritabilities (variance-standardized values) but higher coefficients of additive variance (mean-standardized values) relative to traits that are more distantly related to fitness.

This debate on the appropriate standardization also extends to selection gradients. Following Lande and Arnold (1983), much of the literature has reported variance-standardized gradients, $\beta_{sd} = \sigma_z\beta$ (also denoted as β_σ). There are, however, potential concerns with variance-standardization (Chapters 6 and 13). The genetic variance (which influences the total variance) is itself a product of past selection, implying that a trait with a history of strong selection (and hence a smaller value of σ_z^2, due to selection reducing the level of genetic variation) may be perceived as being under stronger current selection than the same trait in a population with a history of weaker past selection (and hence a larger value of σ_z^2). Further, the nature of the trait (whether ordinal, dichotomous, or continuous) can result in dramatic differences in variance. Finally, variance-standardization does not naturally lead to an unambiguous definition of the relative strength of selection. While β_{sd} is the expected change in relative fitness induced by one standard-deviation change in z, whether this is strong or weak depends, in part, on whether the standing variance is large or small.

An alternative is mean-standardization,

$$\beta_\mu = \mu\beta \tag{29.33a}$$

an approach used as early as Johnson et al. (1955), and championed by Morgan and Schoen (1997), van Tienderen (2000), Hereford et al. (2004), and Matsumura et al. (2012) as superior to variance-standardization. The motivation for this scaling is that if β represents the change in w from a unit change in z, then $\mu\beta$ represents the change in w for a change in z by an amount μ.

The relationship between the mean- and variance-standardizations is simply

$$\frac{\beta_\mu}{\mu} = \beta = \frac{\beta_{sd}}{\sigma}, \quad \text{hence} \quad \beta_\mu = \frac{\mu}{\sigma}\,\beta_{sd} \quad \text{and} \quad \beta_{sd} = \frac{\sigma}{\mu}\,\beta_\mu \tag{29.33b}$$

Mean-standardization offers three potential benefits, one obvious, the other two more subtle. First, it does not depend on the trait variance, and hence removes the influence that past selection (in shaping the current variance) has on the measure of current selection strength. Second, it provides a natural scale for the strength of selection, as $\beta_\mu = 1$ when the trait is fitness itself ($z = W$). To see this, first note that the relationship ($W = \overline{W} \cdot w$) between absolute (W) and relative (w) fitness implies, for $z = W$, that

$$\beta = \frac{\sigma(z,w)}{\sigma_z^2} = \frac{\sigma(W,w)}{\sigma^2(W)} = \frac{\overline{W}\sigma^2(w)}{\overline{W}^2\sigma^2(w)} = \frac{1}{\overline{W}} \tag{29.33c}$$

giving $\beta_\mu = \mu\beta = \overline{W}/\overline{W} = 1$. Mean-standardized gradients are also used to obtain the proportional selection response for a trait, with

$$\frac{R}{\mu_z} = \frac{\sigma_A^2}{\mu_z}\beta = \left(\frac{\sigma_A^2}{\mu_z^2}\right)\beta_u = I_A\beta_u \tag{29.33d}$$

where $I_A = (\sigma_A^2/\mu_z^2)$ is Hansen's (Hansen et al. 2003, 2011) metric of evolvability (the square of Houle's 1992 original definition); see Chapter 6.

A final advantage of mean-standardized gradients is that they measure the **elasticity of fitness** with respect to a trait, namely the proportional increase in fitness given a proportional increase in the trait (Morgan and Schoen 1997; van Tienderen 2000). Under the assumption that z is Gaussian distributed, Equation 29.20a implies that

$$\beta_u = \mu\beta = \mu\left(\frac{1}{\overline{W}}\frac{\partial\overline{W}}{\partial\mu}\right) = \left(\frac{\mu}{\partial\mu}\right)\left(\frac{\partial\overline{W}}{\overline{W}}\right) = \frac{\partial\ln(\overline{W})}{\partial\ln(\mu)} \tag{29.33e}$$

Comparison with Equation 29.4c shows that β_u is the elasticity of fitness (or more generally, of a fitness component) with respect to the trait. For example, if $\beta_\mu = 0.3$, increasing a trait value by μ increases fitness by 30%. Conversely, with variance-standardization, $\beta_{sd} = 0.3$ implies that increasing the trait value by one standard deviation results in a 30% increase in fitness. This latter value, however, is not an elasticity because it is not expressed as a function of the proportional change of the trait.

This interpretation of β_μ as an elasticity provides a natural connection between demography (the population growth rate, λ) and evolution (the rate of change in a trait, R). Because elasticities are both additive and also satisfy the chain rule of differentiation, the elasticity of a trait on the population growth rate (λ) can be computed as a function of the elasticity (e_i) of the ith (out of k) fitness component (W_i) of growth rate and the elasticity ($\beta_{\mu,i}$) of the trait on the ith fitness component, yielding

$$\frac{\partial\ln(\lambda)}{\partial\ln(\mu)} = \sum_{i=1}^{k}\left(\frac{\partial\ln(\lambda)}{\partial\ln(W_i)}\right)\left(\frac{\partial\ln(W_i)}{\partial\ln(\mu)}\right) = \sum_{i=1}^{k} e_i\beta_{\mu,i} \tag{29.33f}$$

Equation 29.33f, when paired with an appropriate path diagram connecting the trait to fitness components and also connecting those fitness components to λ, forms the basis of elasticity path analysis, which offers a very flexible approach for treating components of fitness (Chapter 30).

An example suggested by Hereford et al. (2004) illustrates this connection. In their meta-analysis, Hereford et al. found medians of the mean-standardized gradients, β_μ, over their sample of traits on viability and fecundity of 0.57 and 0.23, respectively. Likewise, using an avian dataset, Sæther and Bakke (2000) found average elasticities of viability and fecundity of $e = \partial \ln(\lambda)/\partial \ln(W_i) = 0.6$ and 0.25, respectively, on population growth. These values showed that a proportional change in viability had a far larger (0.6/0.25 = 2.4-fold) impact on the growth rate, λ, than did the same proportional change in fecundity.

Combining the median β_μ values for traits reported by Hereford et al. (2004) with the average elasticity values reported by Sæther and Bakke (2000), the elasticity of a trait on λ via viability selection is $0.6 \cdot 0.57 = 0.34$, while the elasticity through fecundity is $0.25 \cdot 0.23 = 0.06$, for a total elasticity of $0.34 + 0.06 = 0.40$, with the vast majority (86%) via its impact on viability. Hence, a doubling of the trait value results in a 40% increase in the growth rate, mostly through its impact on viability. This is not an isolated case. As summarized by Crone (2001), across a range of studies, adult viability usually had a much higher elasticity than fecundity. The exception was in **semelparous** plants (those with only a single episode of reproduction before death), where growth rate had the highest elasticity, followed by fecundity, and then adult viability. Hence, all else being equal, traits under viability selection likely have a larger impact on λ.

Despite their appeal, there are restrictions and caveats concerning mean-standardized gradients. First, they are only defined for traits where the origin on their measurement scale is not arbitrary. While many traits have a biologically motivated zero value (e.g., height, weight), this restriction excludes many interesting traits, such as seasonal timing (the origin of the calendar is arbitrary) and composite traits (e.g., principal components) (Kingsolver and Pfenning 2007). Second, there are delicate issues in the use of mean-standarization with dichotomous traits, such as the resistance fraction (f) and its complement, the susceptible fraction ($1 - f$). Because either fraction can be used as a measure of the strength of selection, they should both return the same standardized gradients, but the mean-standardized gradients for resistance and susceptibility are different unless $f = 1/2$ (Stinchcombe 2005). Finally, Kingsolver and Pfenning (2007) noted that traits with large β_μ values tend to have small coefficients of variation, a trend not expected if β_μ is a fair standardization.

MOVING AWAY FROM QUADRATIC FITNESS FUNCTIONS

Quadratic Surfaces Can Be Very Misleading

A potentially serious problem with quadratic regressions as estimators of $W(z)$ is that they allow for *only a single local maximum or minimum*. Fitness surfaces with multiple local maxima, or even sharp transitions, are thus poorly described by a quadratic. Figure 29.10A presents a particularly illustrative example, showing that a quadratic fit to a truncation-selection fitness function creates a spurious local minimum. Given this potential for a very misleading view of the fitness surface, why is there so much focus on quadratic regressions? There are two primary motivations. First, the quadratic is the simplest function that allows for curvature, and hence is the simplest estimate of any nonlinearity in the fitness surface. Second, and much more important, when the conditions for the breeder's equation hold (Chapters 6, 13, and 24), the sole measures of phenotypic change entering into the short-term selection response equations for mean (Equation 13.8c) and variance (Equation 16.7b) are the coefficients from the best linear (β from Equation 29.25b) and best quadratic (γ from Equation 29.29f) fit. Hence, even if the fitness surface is very poorly described by a quadratic, even to the point of being very misleading (Figure 29.10), one would still extract the coefficients for the response in the mean and variance by using the quadratic estimate. There is thus the potential for conflict in the use of quadratic regressions between ecologists (who wish to ascertain how traits influence fitness) and evolutionary biologists (who wish to examine how these traits might evolve). In reality, however, both viewpoints are correct. The more accurate the description of the fitness surface, the more ecological insight it provides into the trait, but this fitness surface also needs to be translated into the required parameters to predict the evolutionary dynamics of the trait.

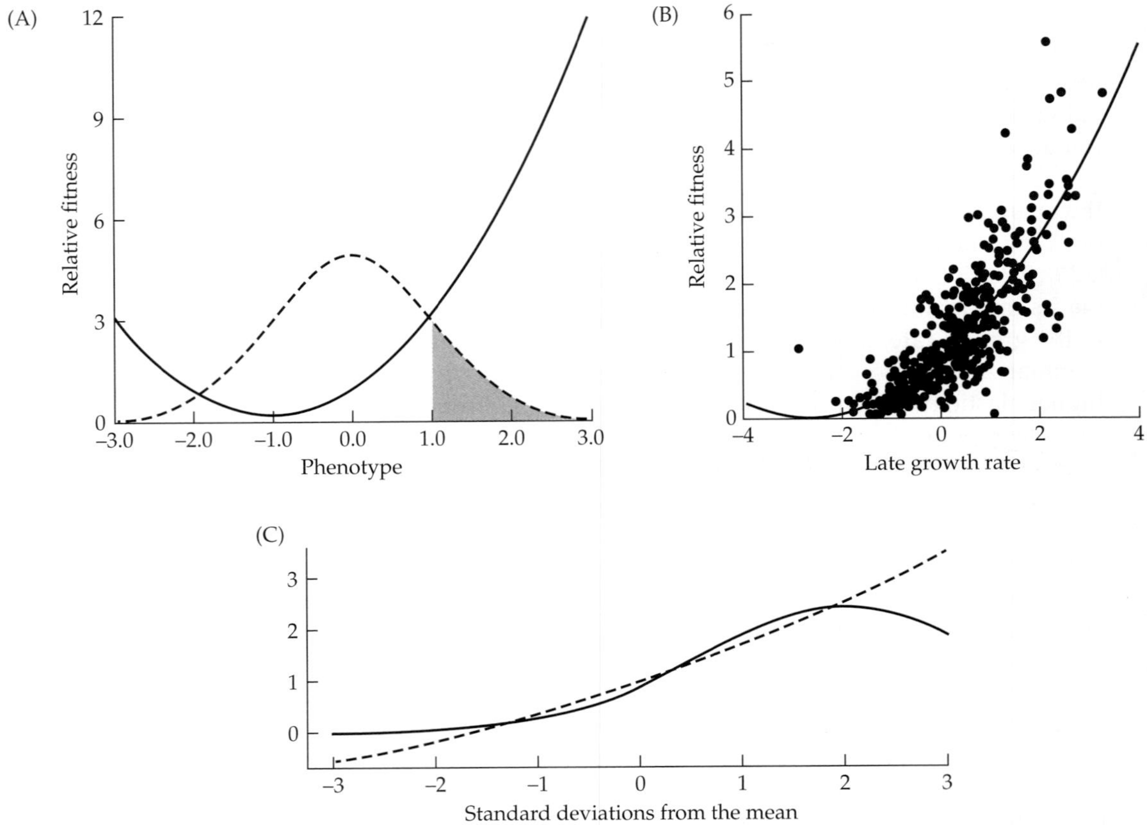

Figure 29.10 Examples of misleading approximations of *W(z)* resulting from using a quadratic regression. **A**: A hypothetical example wherein phenotypes are normally distributed with only individuals exceeding one phenotypic standard deviation surviving (i.e., threshold selection). The best quadratic regression erroneously suggests the presence of disruptive selection (by introducing a false minimum), rather than the strict directional selection that is actually occurring. In contrast (not shown), the same logic shows that if individuals *above* 1.0 are culled, then the quadratic surface would indicate stabilizing selection by introducing a false maximum. (After Schluter 1988.) **B**: Data from Mitchell-Olds and Bergelson (1990) on individual fitness as a function of late growth rate (z) for the annual plant *Impatiens capensis*. The data clearly depart from linearity, as they show curvature. The best-fitting quadratic (plotted) indicates a minimum in fitness (disruptive selection) around $z \simeq -2.4$. However, a better fit was given by a model of exponentially increasing fitness, with $w(z) + 0.5 = \exp(0.52 + 0.46 \cdot z - 0.002 \cdot z^2)$, suggesting that strict directional selection is acting on z, as this function monotonically increases over the character range being measured. (After Mitchell-Olds and Bergelson 1990.) **C**: When the optimal value under true stabilizing selection (solid curve; with the true *W(z)* given by Equation 29.34a) is sufficiently above the current population mean, the best-fitting quadratic (dashed curve) has a positive curvature ($\gamma > 0$). Here the population mean is at zero, and the optimum is two standard deviations above zero. If the mean were two standard deviations above the optimum, the best-fitting quadratic would have a negative curvature ($\gamma < 0$). (After Geyer and Shaw 2008; Shaw and Geyer 2010.)

Gaussian and Exponential Fitness Functions

If our sole focus is in describing the fitness surface (as opposed to extracting components of selection response), then other parametric forms besides the simple quadratic could be considered. One obvious candidate is a Gaussian fitness function

$$W(z) = a \cdot \exp\left(-\frac{(z-\theta)^2}{\omega^2}\right) \tag{29.34a}$$

Weldon (1901), in one of the first studies of selection on a quantitative trait in nature,

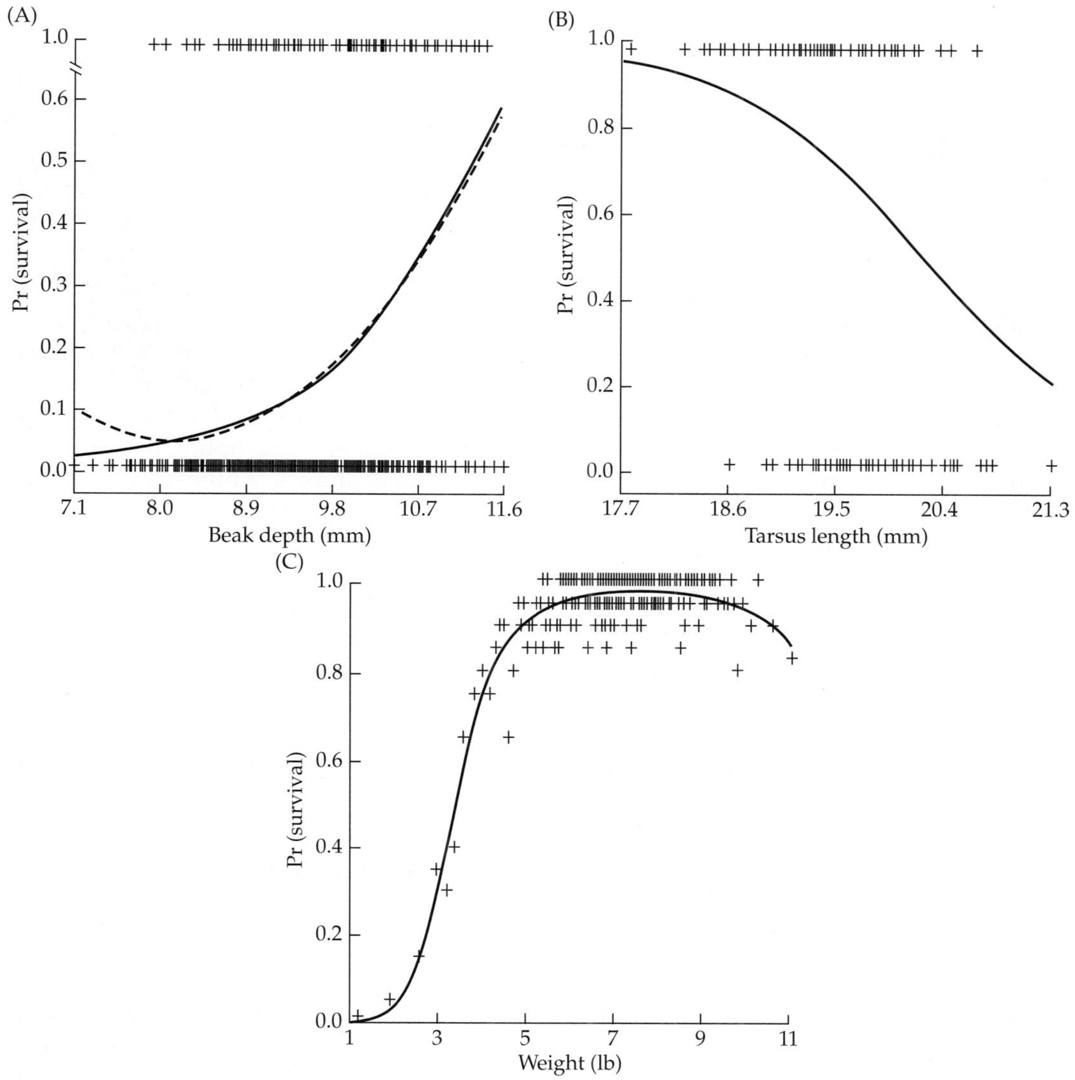

Figure 29.11 Examples of fitness surfaces generated using Schluter's cubic-spline estimate. The actual fitness values for individuals are indicated by +, and the solid curve indicates the cubic-spline estimate of *W(z)*. **A:** Probability of survival as a function of beak depth in Darwin's finch (*Geospiza fortis*). The dashed curve indicates the estimate of the surface by a quadratic regression, which generates a spurious minimum. **B:** Overwinter survival in juvenile female song sparrows (*Melospiza melodia*) as a function of tarsus length. **C:** Survival of male human infants as a function of birth weight. (After Schluter 1988.)

remarked that the Gaussian seems to be a good description of the fitness function (in his case, crab survival as a function of size), and it was used by Cavalli-Sforza and Bodmer (1971) to model the relationship between human birth weight and survival observed by Karn and Penrose (1951).

To model strictly directional selection, the exponential function,

$$W(z) = a \cdot \exp(bz) \tag{29.34b}$$

can be used, while the more general function,

$$W(z) = c \cdot \exp(a + bz + cz^2) \tag{29.34c}$$

allows for both directional and stabilizing or disruptive selection (Mitchell-Olds and Bergelson 1990; see Figure 29.10B). If a trait is normally distributed before selection, it remains normally distributed following selection when Equation 29.34c is the fitness function, of which Equations 29.34a and 29.34b are special cases (Felsenstein 1977).

Semiparametric Approaches: Schluter's Cubic-spline Estimate

In order to more reliably estimate the fitness function, Schluter (1988) developed a semiparametric method that makes no assumptions about the functional form of the fitness surface, but does require assumptions about the distribution of residuals. Schluter's approach fits the data with a series of **cubic splines** (a series of cubic polynomials that join smoothly together), and uses generalized cross-validation to generate a "best fit" criterion for model selection. In some sense, this is a nonparametric model, as it makes no assumptions about the functional form of the fitness surface. It is, however, parametric in the sense that model fitting requires assumptions about the distribution of the residuals as a function of phenotypic value, z. Schluter developed a program to estimate $W(z)$ when assuming either normal, binomial, or Poisson distributed residuals, using a resampling procedure to generate rough confidence intervals on estimates of $W(z)$. Examples of fitness surfaces that were estimated using this approach are given in Figure 29.11. When one of the classical examples of stabilizing selection, the data of Karn and Penrose (1951) relating survival of human infants to their birth weight, was reanalyzed using Schluter's method, the local maximum was not significant. Parametric tests of the significance of estimated local maxima or minima were discussed by Mitchell-Olds and Shaw (1987) for quadratic regressions, while nonparametric tests were discussed by Schluter (1988). Gimenez et al. (2006) extended Schluter's method to survival estimates based on capture-recapture data.

Janzen-Stern and Morrissey-Sakrejda Gradients: Calculating Average and Landscape Gradients Under General Fitness Functions

We have shown that when phenotypes are normally distributed before an episode of selection, $\beta = S/\sigma_z^2$ equals two different summary statistics of the geometry of selection,

$$\beta = \int \frac{\partial w(z)}{\partial z}\, p(z)dz = \frac{\partial \ln \overline{W}}{\partial \mu_z}$$

namely, the average selection gradient (the average slope of the individual fitness surface over the current distribution of phenotypes) and the landscape gradient (the gradient of log mean fitness with respect to the population mean). This relationship also holds for *any* phenotypic distribution when the individual fitness function is exactly a quadratic (Equations 29.30c and 29.30g). Outside of this special case for a quadratic $w(z)$ these equalities break down when the distribution of z is no longer normal distributed.

This raises the question of how to obtain these gradients when the trait does not follow a normal distribution and the individual fitness function is not a quadratic. For example, suppose the fitness surface is fitted by a nonquadratic function (such as Equations 29.34a or 29.34b) or by using Schluter's semiparametric regression. What should we take as the "gradient" of fitness in this case? The answer very much depends on the specific question that an investigator is asking. If one wishes to obtain a summary measure of the nature of selection, then the average gradients (the average slope or curvature; Equation 29.20b and 29.20c) might be a reasonable metric.

Janzen and Stern (1998) considered the case in which some general individual fitness surface, $w(z)$, was fit using n observations. Motivated by Equation 29.20b, they suggest using

$$\beta_{JS} = \int \frac{\partial w(z)}{\partial z}\, p(z)dz = E_z\left[\frac{\partial w(z)}{\partial z}\right] \simeq \frac{1}{n}\sum_{i=1}^{n} \left.\frac{\partial w(z)}{\partial z}\right|_{z=z_i} \tag{29.35a}$$

with the last step following by approximating the expectation by using the sampled values. We refer to this as a **Janzen-Stern selection gradient**. They used this approach to obtain a gradient when a logistic function is used to fit $w(z)$, where an explicit form of the required derivative is easily obtained. The same approach can be extended to the average curvature of the individual fitness surface by recalling Equation 29.20c, which calculates the corresponding Janzen-Stern (quadratic) gradient as

$$\gamma_{JS} = \int \frac{\partial^2 w(z)}{\partial z^2} \, p(z) \, dz \simeq \frac{1}{n} \sum_{i=1}^{n} \left. \frac{\partial^2 w(z)}{\partial z^2} \right|_{z=z_i} \tag{29.35b}$$

When the fitness surface is complex (for example, a fine-meshed grid with individual fitness values for each coordinate, such as would be returned by a semiparametric regression), numerical methods can be used to compute the required partials for Equations 29.35a and 29.35b. When $\delta \simeq 0$,

$$\begin{aligned} \left. \frac{\partial w(z)}{\partial z} \right|_{z=z_i} &\simeq \frac{w(z_i + \delta/2) - w(z_i - \delta/2)}{\delta} \\ \left. \frac{\partial^2 w(z)}{\partial z^2} \right|_{z=z_i} &\simeq \frac{w(z_i - \delta) - 2w(z_i) + w(z_i + \delta)}{\delta} \end{aligned} \tag{29.35c}$$

While average gradients of the individual fitness function provide reasonable summary statistics for discussions about the nature of selection, if the concern is the *evolution* of these traits over time, then landscape gradients are required. This follows from our extensive discussion in Chapter 24 of selection response when normality no longer holds. The generalized Barton-Turelli selection response equations (Equation 24.26) for changes in the phenotypic moments after a generation of selection requires estimates of various partials of the landscape gradient with respect to the moments of the phenotypic distribution. If the underlying genetics are such that the distribution of breeding values is nearly normal, then (Equation 24.30a) the only measure of selection entering into the equation for the change in mean is the landscape gradient with respect to the population mean. If the distribution of breeding values shows skew and/or kurtosis (specifically, a nonzero value of the scaled kurtosis, $\kappa_4 = [\mu_{4,G} - 3\sigma_A^4]/\sigma_A^4$), then partials of the landscape gradient with respect to the phenotypic variance and skew are required to predict the change in mean.

The requirement of landscape gradients to predict selection response motivates the definition of the **Morrissey-Sakrejda selection gradient** (Morrissey and Sakrejda 2013, 2014; Morrissey 2014a). Following Janzen and Stern, they suggest that this gradient (for a given estimated fitness function) can also be computed numerically. Here

$$\beta_{MS} = \frac{\partial \ln \overline{W}}{\partial \mu_z} = \frac{1}{\overline{W}} \frac{\partial \overline{W}}{\partial \mu_z}, \quad \text{where} \quad \overline{W} = \frac{1}{n} \sum_{i=1}^{n} W(z_i) \tag{29.35d}$$

A similar expression defines γ_{MS}. This approach works for general fitness surfaces, such as those estimated using a semiparametric regression, and nonnormal distributions.

MORE REALISTIC MODELS OF THE DISTRIBUTION OF FITNESS COMPONENTS

Placing the estimation of fitness components, total fitness, and fitness surfaces within a rigorous statistical framework requires more realistic models for the distribution of fitness residuals. We build toward this goal in two steps. First, we introduce a number of generalized linear models (a topic initially discussed in Chapter 14) that allow for more biologically realistic modeling of the distribution of fitness effects within a given episode of selection. We then turn to aster models that compute the much more complex distribution resulting from the compounding of component distributions over multiple episodes of selection.

Generalized Linear Models for Fitness Components

The data for total fitness are nonnegative integers, representing the number of offspring left by an individual. The components that make up total fitness are also integer data, and fall into two broad distributional categories. We first consider **Bernoulli random variables** (fitness components have a value of 1 with probability p, else they have a value of 0), such as occurs with viability selection (alive or dead) and mating status (mated or not mated).

Translating Bernoulli components of fitness into a fitness surface requires a model of how p varies with trait value, z (or more generally, a set of trait values represented by the vector, $\mathbf{z}$). A natural choice is the **logistic regression** model (Janzen and Stern 1998), which was introduced in Chapter 14. Equation 14.10a presented its simplest version,

$$W(z) = p(z) = \Pr(\text{survival} \mid z) = \frac{\exp(a + bz)}{1 + \exp(a + bz)} = \frac{1}{1 + \exp(-[a + bz])} \tag{29.36a}$$

Equation 29.36a is a model of directional selection, as $W(z)$ increases toward one as z increases when $b > 0$. Likewise, $W(z)$ decreases toward zero as z increases when $b < 0$. As discussed in Chapter 14, a logistic regression is equivalent to modeling $p(z)$ using the **logit** function,

$$\text{logit}[p(z)] = \ln\left[\frac{p(z)}{1 - p(z)}\right] = \begin{cases} a + bz & \text{(univariate case)} \\ a + \sum_i b_i z_i = a + \mathbf{b}^T\mathbf{z} & \text{(multivariate case)} \end{cases} \tag{29.36b}$$

The upper righthand expression in Equation 29.36b corresponds to modeling the logit of p with a simple linear regression, while the lower expression is a more general linear model, which could include quadratic terms (e.g., $b_1 z + b_2 z^2$), which allows for the possibility of either concave or convex selection. Most generally, the righthand expressions in Equation 29.36b could be a full mixed model. For example, Coulson (2012), modeled the impact of size, z, on survival in Soay sheep by assuming that

$$\text{logit}[p(z, t)] = a + bz - \eta N(t) + r(t)$$

where the fixed effect (η) accounts for density-dependent survival ($N[t]$ is the population size at time t), and the random effect, $r(t)$, accounts for the environmental effect at time t.

Logistic regressions naturally accommodate the heteroscedastic nature of residuals for viability (and other Bernoulli random variables), and maximum likelihood (LW Appendix 3) can be used to estimate model parameters. Modifications of logistic regressions can also be used to estimate fitness using capture-recapture data (Kingsolver and Smith 1995). Capture-recapture models were reviewed by Lebreton et al. (1992), while the analysis of the targets of selection using capture-recapture data was further discussed by Conroy et al. (2002), Zabel et al. (2005), and Gimenez et al. (2006).

The logit function takes an unconstrained linear model (with potential values over $-\infty$ to ∞) and maps it into the constrained parameter space for p, namely, between zero and one. As mentioned in Chapter 14, this is the basic structure of **generalized linear models**: the conditional expectation of an observed value, y, given some underlying variable, z, can be expressed as $E(y \mid z) = g(z)$ for some monotonic function g, where some inverse transformation returns a linear model,

$$g^{-1}[E(y \mid z)] = z = \mu + \sum \beta_k x_{k,i}$$

The inverse function, g^{-1}, is called the **link function**, as it transforms the conditional expectation into a linear model.

The basic idea of such models is that some underlying (and unobserved) **latent variable**, z, is transformed into an observed variable, y. Such a model can be thought of as having three distinct scales (de Villemereuil et al. 2016): (i) an underlying **latent scale**, the value of z; (ii) an **expected data scale**, $E(y \mid z) = g(z)$, which is the expected value of y for that value of z; and (iii) an **observed data scale**, the actual observed value of y, which is a function of its expectation plus a residual error, where the latter need not be normal (and indeed, usually is not). This approach is called a **generalized linear model** (**GLM**) if the latent variable, z, is modeled via a standard general linear model (involving only fixed effects), and a **generalized linear mixed model (GLMM)** if we model z as a mixed linear model (z is a mixture of both fixed and random effects; e.g., Coulson 2012). GLMs and GLMMs allow for fairly general error structures (e.g., Poisson, binomial, etc.) and are generally solved by either maximum-likelihood or Bayesian approaches (e.g., Hadfield 2010). GLMs and GLMMs were reviewed in detail by McCulloch et al. (2008) and Stroup (2012).

More general survival functions (beyond the simple logistic) can be found in the analysis of clinical trials (e.g., Klein and Moeschberger 1997; Lawless 2003) and time-to-failure analysis from industrial statistics (e.g., Kenett and Zacks 1998), both examples of **survival analysis**. Manly (1976) suggested the use of the **double exponential fitness function** for viability data, where

$$W(z) = \Pr(\text{survival} \mid z) = \exp(-\exp[a + bz]) \tag{29.37}$$

This is a special case of the classic **proportional-hazards model** (Cox 1972) for human clinical trails, wherein the probability of survival to time t is determined by a general risk for every individual in the population (specified by some monotonically nondecreasing function, $g(t)$, so that probability of survival declines with increasing values of t), and a specific (proportional) risk for the particular phenotype z, which yields

$$W(z,t) = \Pr(\text{survival} \mid z,t) = \exp(-\exp[ag(t) + bz]) \tag{29.38a}$$

One choice is $g(t) = \ln(t)$, giving the general risk as $\exp(-\exp[ag(t)]) = \exp(-\lambda t)$, namely the exponential distribution (a constant, i.e., age-independent, risk), weighted by the risk for phenotype z,

$$W(z,t) = \exp\left(-[\lambda t] \cdot \exp[bz]\right) \tag{29.38b}$$

If survival at stage j is measured, the proportional-hazards model can be written as

$$W(z,j) = \Pr(\text{survival} \mid z,j) = \exp(-\exp[h_j + bz]) \tag{29.38c}$$

One advantage of using proportional-hazards models (due to the memoryless feature of the exponential distribution) is that they can allow for certain types of missing, or **censored**, data (Little and Rubin 2002), such as the last time a marked individual was recovered.

Given that an individual has survived and mated, the second broad class of probability models for fitness components involves discrete counting distributions, which range over $0, 1, 2, \cdots$, such as number of mates or number of offspring per mating. One candidate distribution is the Poisson (Equation 14.16), which is solely a function of its mean value, θ. As with Bernoulli data, we need a model for how θ varies as a function of the trait value z, in which case the probability of j offspring (or j mates, or j of some other fitness component) is

$$\Pr(W = j \mid z) = e^{-\theta(z)}\, \frac{\theta(z)^j}{j!} \tag{29.39a}$$

From Equation 14.17b, the log-linear model is a candidate for translating z into a mean for the Poisson, where

$$\ln[\theta(z)] = a + b\,z \tag{29.39b}$$

Because $\theta(z)$ is constrained to be nonnegative, Equation 29.39b takes an unconstrained linear model and maps it into the range $(0, \infty)$, corresponding to the allowable range of means for a Poisson. More generally, for n traits we can use the log-linear model,

$$\ln[\theta(z)] = a + \sum_i^n b_i\, z_i = a + \mathbf{b}^T \mathbf{z} \tag{29.39c}$$

which again can include quadratic terms (e.g., $a + b_1 z + b_2 z^2$) and hence model concave or convex selection.

One limitation of the Poisson is that its mean equals its variance, whereas fitness data are often **overdispersed**, with the variance exceeding the mean (often, but not always, caused by an excess of zero values). In overdispersed settings, an alternative candidate model is the **negative binomial distribution**. This models the total number of successes (y) that occur before r failures for independent trials with a success probability of p,

$$\Pr(y = j \mid r, p) = \frac{(j + r - 1)!}{j!\,(r - 1)!}\, p^j (1 - p)^r \quad \text{for} \quad j = 0, 1, \cdots \tag{29.40a}$$

If a random variable, y, follows a negative-binomial distribution, denoted by $y \sim NB(r, p)$, then its mean and variance are

$$\mu_y = \frac{pr}{1 - p} \quad \text{and} \quad \sigma_y^2 = \frac{pr}{(1 - p)^2}. \quad \text{Hence} \quad \frac{\sigma_y^2}{\mu_y} = \frac{1}{1 - p} > 1 \tag{29.40b}$$

showing that the variance can exceed the mean by an arbitrary amount. Because of this feature, the negative-binomial is often used in place of a Poisson to allow for overdispersion (Wilson et al. 2005b; Yang et al. 2009). One approach for modeling fitness data using this distribution is to assume a constant r value for all trait values (estimated from the data), and then model $p(z)$ using the logit function (Equation 29.36b). The negative binomial is the sum of r geometric random variables, and hence is the discrete counterpart of the gamma distribution (which is the sum of r exponential random variables; Appendix 2).

Two other approaches for building discrete counting distributions for fitness components deal with the often problematic issue of handling an excess of zero values (Martin et al. 2005). The first is to use a **zero-truncated distribution**, whose values are *conditional* on being nonzero (i.e., they ignore zero values entirely). If we let $\phi(j \mid z) = \Pr(W = j \mid z)$ denote a discrete counting distribution (such as a Poisson, binomial, or negative binomial), then its zero-truncated version is given by

$$\Pr(W = j \mid z) = \frac{\phi(j \mid z)}{1 - \phi(0 \mid z)} \quad \text{for} \quad j \geq 1 \tag{29.41a}$$

where j equals number of offspring, mates, flower heads, and so forth. For example, the zero-truncated Poisson is

$$\Pr[W = j \mid \theta(z)] = \frac{\theta(z)^j e^{-\theta(z)}/j!}{1 - e^{-\theta(z)}} = \frac{e^{-\theta(z)}}{1 - e^{-\theta(z)}} \cdot \frac{\theta(z)^j}{j!} \quad \text{for} \quad j \geq 1 \tag{29.41b}$$

As above, the trait-specific mean $\theta(z)$ can be modeled as a log-linear model (Equations 29.39b and 29.39c). This can be generalized to a **k-truncated distribution**, wherein the first k states (i.e., 0 to a value of $k - 1$) are ignored

$$\Pr(W = j \mid z) = \frac{\phi(j \mid z)}{1 - \sum_{i=0}^{k-1} \phi(i \mid z)} \quad \text{for} \quad j \geq k \tag{29.41c}$$

and which occasionally appears in the selection literature (Shaw et al. 2008).

The other approach is to construct a **zero-inflated distribution**, which includes an additional point mass, $1 - q(z)$, at zero,

$$\Pr(W = j \mid z) = \begin{cases} 1 - q(z) + q(z)\phi(0 \mid z) & \text{for } j = 0 \\ q(z)\phi(j \mid z) & \text{for } j \geq 1 \end{cases} \tag{29.42}$$

where (as above) $\phi(j \mid z) = \Pr(W = j \mid z)$ is a discrete counting distribution and $q(z)$ is the probability that the observation given z is *not* zero-inflated. As above, we can model $q(z)$ using the logit function (Equation 29.36b). Alternatively, one can model the zero-inflated

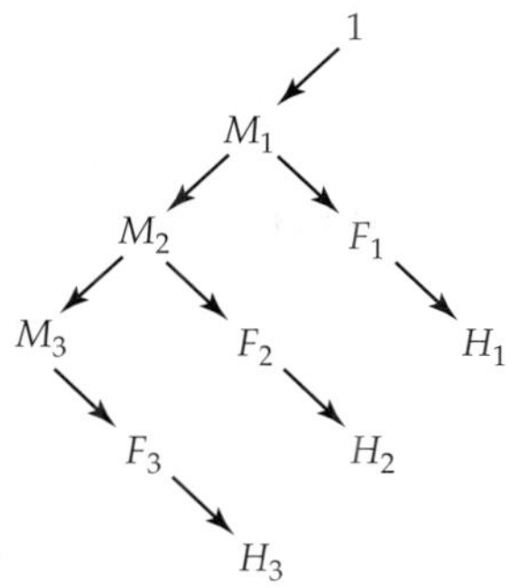

Figure 29.12 A life-history graph of potential fitness components in the aster *Echinacea angustifolia*. The distribution of lifetime fitness starts by considering the discrete number of offspring left by a single individual (the 1 at the upper right starting the graph). First, this individual must survive to year one (M_1), in which case it may flower (F_1), and if it flowers, it has some (discrete) number (H_1) of flower heads (taken as the surrogate for fitness). The same pattern continues (providing it survives) over years two and three, with lifetime fitness given by $H_1 + H_2 + H_3$. A life history can be constructed as a forest graph, as there are no loops and each node has only one predecessor. Further, if the value of a predecessor is zero, then all descendant values from that predecessor are also zero. For example, if an individual fails to survive to year two (M_2 is zero), then H_2 and H_3 are zero. If F_1 is zero, then H_1 is zero, but this (by itself) tells us nothing about H_2 or H_3, as they are not descendants of F_1. This graphical representation allows very complex life histories to be accurately modeled. (After Geyer et al. 2007.)

distribution as an **atom** (probability mass) at zero and a zero-truncated distribution if not zero.

Aster Models: Modeling the Distribution of Total Fitness

While the above section presented more realistic distributions for a given *component* of fitness, the distribution of *lifetime* fitness is much more complex, compounding a series of Bernoulli (viability, mating, etc.) and discrete counting distributions (number of mates, number of offspring, etc.). The resulting distribution incorporating all of these episodes has an excess of zero values and does not fit any standard distribution. Ideally, one would like an approach that allows an investigator to specify the distribution of each of the components and then arrive at the correct distribution of total fitness.

An example of this problem is presented in Figure 29.12, which shows the life history of the purple coneflower (*Echinacea angustifolia*) in the family Asteraceae (Geyer et al. 2007). A single individual may survive year one (M_1), in which case it may flower (F_1), in which case it has some discrete flower head–count status (H_1), with this pattern continuing in years two and three (assuming survival). Lifetime flower head production ($\sum H_i$) is used as the surrogate for total fitness. Even this fairly simple life history has a complex, very nonstandard, distribution for total fitness. One can model the M_i and F_i values as Bernoulli random variables, and the H_i values as zero-truncated Poissons (because we are conditioning on them flowering). Equivalently, conditioned on survival ($M_i = 1$), flower head status H_i could be directly modeled as a zero-inflated Poisson.

Motivated by the problem posed by Figure 29.12, Geyer et al. (2007; 2013; Shaw et al. 2008; Geyer 2010; Geyer and Shaw 2010a, 2010b) developed an approach they called **aster models** (a homage to the initial aster study) as a very flexible framework for modeling the distribution of total lifetime fitness given a life-history graph. As Geyer noted, the term aster models is far catchier than "forest graph exponential family conditional or unconditional canonical statistic models." Aster models start with a life-history graph (a **forest graph**, as there are no loops and each node has only one predecessor), and assume that the distribution to the next step of the graph (a fitness component) is a member of the exponential family (which includes Bernoulli, binomial, Poisson, and negative binomial distributions, as well as their truncated versions).

Shaw et al. (2008) noted that simple aster models correspond to previous methods of analysis. The simplest life-history graph is $1 \rightarrow X$. If X is normally distributed, this is a standard linear model (such as Lande-Arnold regression). If X is Bernoulli, this is a logistic regression (Equation 29.36), while if X is Poisson, this is the log-linear model (Equation 29.39c). The graph is $1 \rightarrow X \rightarrow Y$, where X is Bernoulli and Y is either a zero-truncated Poisson or a zero-truncated negative-binomial correspond to the zero-inflated versions of these distributions (Equation 29.42). Finally, the graph $1 \rightarrow X_1 \rightarrow X_2 \rightarrow \cdots \rightarrow X_n$ (where all the X_i are Bernoulli) corresponds to survival analysis. The power of aster models is that they use the specific life-history details of the organism of interest and reasonable candidate distributions for each fitness component to build up a proper distribution of lifetime fitness.

By modeling the parameter values for each fitness component distribution as functions of trait value z (e.g., Equations 29.36b and 29.39c), aster models can be used to construct fitness surfaces using the correct residual structure. This can be done using Geyer's `R` package `aster2`. We return to aster models and fitness surfaces in Chapter 30.

THE IMPORTANCE OF EXPERIMENTAL MANIPULATION

Several authors have stressed that regression approaches should be viewed as only the *preliminary step* in any analysis of the actual agents of selection, treating any suggested traits under selection as candidates to be further tested by experimental manipulation (Mitchell-Olds and Shaw 1987, 1990; Schluter 1988; Wade and Kalisz 1990; Kingsolver and Pfenning 2007). Further, as Example 29.1 highlights, the role of the **invisible fraction** (selection on a trait before it is expressed) is generally only accessible by direct manipulative experiments (Hadfield 2008). Recall that in Example 29.1, the very strong selection on the invisible fraction (here, viability selection on time to flowering) would have been missed if not for transplant experiments with known flower-size genotypes. A second interesting example is the work of Sinervo and McAdam (2008) on clutch size in side-blotched lizards (*Uta stansburiana*). Through the use of both hormonal control and surgical reduction in the number of ovaries, they showed a viability cost to increased clutch size that occurs before clutch size is expressed.

One can envision two (not mutually exclusive) experimental approaches for assessing trait-fitness relationships: **environmental manipulation** and **phenotypic manipulation**. The idea of assessing fitnesses in manipulated environments (be they physical or biotic alterations) traces back to the first attempts to measure natural selection on quantitative traits. Weldon (1895, 1899) observed that the frontal-length to carapace-length ratio in the European green crab (*Carcinus maenas*) declined over a five-year period in Plymouth Sound. Noting that a breakwater in the harbor had recently been competed, he hypothesized that increased silt accumulation might be a selective agent behind this change. To test this, he exposed crabs to silt conditions similar to that expected in Plymouth, and found that the survivors in this laboratory setting had reduced frontal length. He hypothesized that narrow frontal length had an impact on filtering out sediments.

A more recent example where the biotic, rather than physical, environment was altered involves the work of Breden and Wade (1989), who observed a positive relationship between group size and fitness in the imported willow leaf beetle. However, when predators were excluded, there was no such relationship, so that β is correlated with the environment (presence or absence of predators). Wade and Kalisz (1990) suggested computing fitness regressions in several different environments, and looking for correlations between β (and/or γ) and environmental features (Figure 29.13). Such correlations suggest specific environmental features for candidates as causal agents of selection.

Under phenotypic manipulation, the investigator artificially modifies the value of the focal trait in individuals and then assesses the fitness consequences in nature (e.g., Sinervo and Basolo 1996; Travis and Reznick 1998). A classic example of this approach is the work of Andersson (1982), which examined the impact of tail length in male mating success in a

Kenyan population of long-tailed widowbirds (*Euplectes progne*). Males with experimentally elongated tails showed greater mating success than males with normal or shortened tails. Further, shortened tails had no impact on the ability of males to control territories, indicating that female choice, rather than male-male competition, was the selective force acting on tail length.

One caveat about using artificial manipulations is that because they may produce variants far outside the normal range of the distribution of z, estimates of β and γ may be correspondingly spurious (Arnold 1988). Conversely, it has been argued that many traits may be near their optima and hence show a limited range of variation relative to the selective forces faced during their evolution. Given this potentially limited range of phenotypes, Lexer et al. (2003) suggested that experimental hybridization with closely related species (which is often easily to accomplish in plants) may provide important insight into the nature of selection. However, hybrids can have a number of significant fitness consequences outside the impact of specific traits on fitness, and hence caution is in order when using this approach.

Example 29.11. Grether (1996) examined male mating success and survival in a California population of the rubyspot damselfly (*Hetaerina americana*). Mature males of this species have a red spot at the base of their wings, and Grether was interested in whether variation in the size of this spot influences sexual and/or natural selection. Three fitness components were measured: reproductive life span (a measure of survival selection), mating rate (a measure of sexual selection), and lifetime mating success (a combination of both survival and sexual selection). All three showed significant selection gradients with red spot size, but not with body size. Thus, the Lande-Arnold regression suggested that larger spot size is favored by both sexual (mating rate) and natural (reproductive life span) selection. As a test of this, Grether examined these components in three additional groups: an "enlarged" group where red ink was used to increase spot size, a "sham" group where the same area was filled with clear ink, and an unmanipulated group. Mating success (mates per day) was significantly greater in the enlarged group, while unchanged in the sham versus control groups. Thus, experimental manipulation of this trait confirmed the role of red spot size in sexual selection. Surprisingly, males with enlarged spots had mortality rates that were 23% higher than the controls, while the sham and control groups showed no differences. Thus, contrary to what the fitness regression suggests, direct manipulation shows that increasing spot size decreases survival. Unmeasured traits (or other factors) generated an apparently strong positive selection gradient of survival on spot size that was not verified by phenotypic manipulation.

Performance and Eco-evolutionary Surfaces

The ultimate aim of detecting selection on a trait is not just to understand its nature (directional, convex, concave), but ideally its ecological *causes*. Phenotypic manipulation can verify that a trait is under selection, but may offer little direct insight into the forces underlying that selection. Functional morphology may offer increased insight, by considering the impact that a trait (or suite of traits) has on one (or more) composite **performance traits** that directly influence fitness (such as running, flying, or swimming speed; seed dispersal; or predator avoidance). The idea is that the performance value (h), not the trait value (z), is the direct target of selection, giving the causal (path) diagram as $z \to h \to w$, with more ecological insight garnered by considering the nature of selection acting on h. In exactly the same fashion that we constructed an individual fitness surface for the trait using regressions, Arnold (1983b, 2003) suggested the use of **performance surfaces**. Here linear or quadratic regressions are used to predict a performance value, h, given the observed traits,

$$h = a + \sum_i b_i z_i + \sum_j \sum_k c_{jk} z_j z_k + \epsilon \tag{29.43}$$

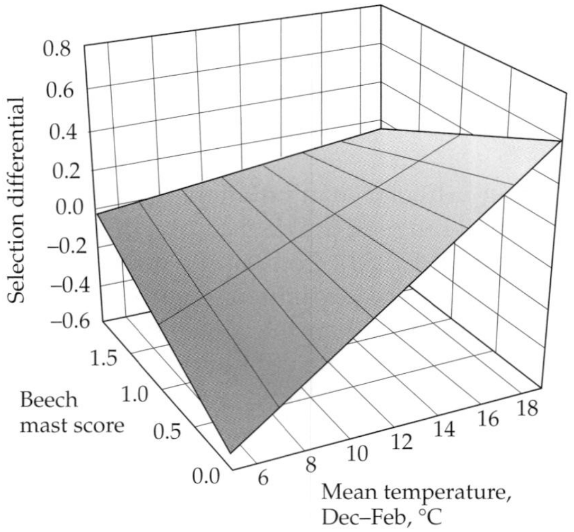

Figure 29.13 The eco-evolutionary surface for selection on fledging weight in great tits (*Parus major*), based on data from Wytham Woods, Oxford, England. The fitted surface plots the selection differential, S, for fledging weight as a function of biotic (food) and abiotic (temperature) variables. The abiotic variable was the average December to February temperature, the biotic variable the amount of beech tree (*Fagus sylvatica*) nuts, given by the mast score. (After MacColl 2011.)

Arnold suggested that performance surfaces can offer important insight, potentially bridging a conceptual gap between trait and fitness values through a functional intermediate that is more biologically interpretable (also see Geber and Griffen 2003 for a plant perspective). By analogy with Equation 29.29b, the regression coefficients (b_i and c_{jk}) in Equation 29.43 are often referred to as **performance gradients**.

Performance is occasionally used as a direct proxy for fitness (one measures performance, rather than directly measuring fitness), and the performance gradients subsequently taken as proxies for fitness gradients. Franklin and Morrissey (2017) cautioned that the conditions under which performance gradients (coefficients of z in the h on z regression) correspond to selection gradients (coefficients of z in the w on z regression) are very narrow. Specifically, the regression of fitness on the performance score (h) must be both linear and with an intercept of zero.

A complementary approach is that of MacColl (2011), who suggested that ecological insight can be gleaned by regressing selection differentials or gradients on potential biotic or abiotic factors, generating an **eco-evolutionary surface** (Figure 29.13), and thus generalizing the suggestion of Wade and Kalisz (1990). The idea is that one has trait-fitness data over a series of years along with candidate biotic and abiotic factors, fitting the regression

$$\beta = a + b_1 f_1 + b_2 f_2 + \cdots + b_k f_k + \epsilon \tag{29.44}$$

where f_i represent environmental and biotic values. This is a modification of the **factor regression** approach used by plant breeders to decompose G $\times$ E into sensitivities to specific environmental components (e.g., Baril 1992; Baril et al. 1995; Epinat-Le Signor 2001). Equation 29.44 generates a surface that attempts to predict trait selection as a function of ecological variables (regressing the selection gradient or differential for a trait as a function of ecological factors, as opposed to a fitness surfaces, which regresses fitness on trait values).

MacColl (2011) gave an example of selection on fledging weight in great tits (*Parus major*) as a function of December to February temperatures and the amount of nuts (the mast score) produced by beech trees (*Fagus sylvatica*). As shown in Figure 29.13, the resulting graph suggests selection for lower weight ($S < 0$) when food is scarce and temperatures are low, but selection for higher weight ($S > 0$) when temperatures are high. As with a

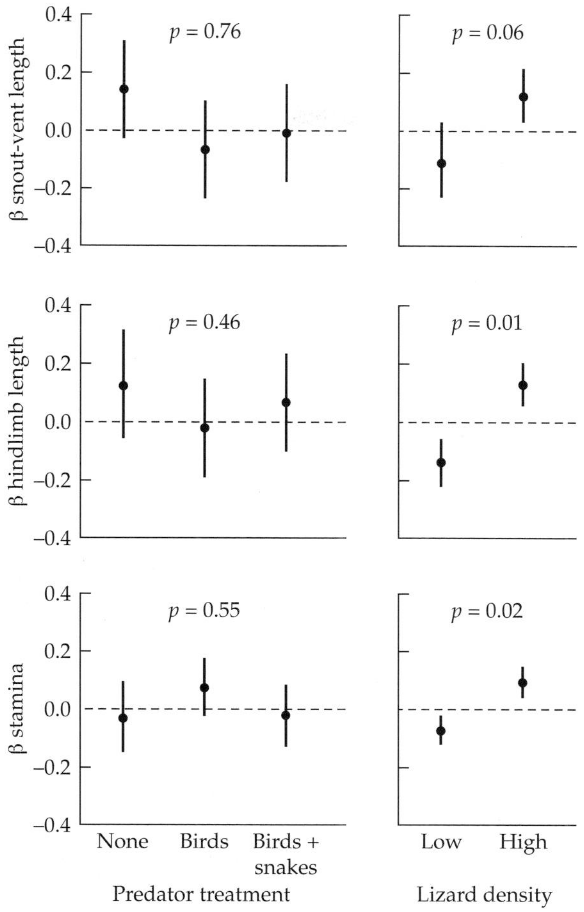

Figure 29.14 The impact of environmental manipulation of agents of possible selection on *Anolis* lizards in the Caribbean. The presence or absence of predators did not change the selection gradients on three traits (snout-vent length, limb length, running stamina), while lizard density had a significant impact ($p \leq 0.05$). See Example 29.12 for more details. (After Calsbeek and Cox 2010.)

fitness-trait regression, this approach is most appropriate when used to suggest hypotheses to be tested by manipulative experiments.

Example 29.12. Example 29.11 highlighted one type of manipulation: artificially changing trait values and examining the resulting impact on fitness. An alternative is to manipulate the *environment* and examine the impact on selection for particular traits, accomplishing experimentally what Equation 29.44 attempts to do statistically. A classic example of this is the work of Calsbeek and Cox (2010), on populations of the brown anole lizard (*Anolis sagrei*) on very small, environmentally manipulated, islands in the Caribbean. Some islands had no bird or snake predators, some had only bird predators, and some both birds and snakes. The presence of predators increased mortality, but did not significantly change the selection gradients on body size (snout-vent length), limb length, or running stamina (Figure 29.14). Conversely, altering the island lizard density resulted in significant differences in all three gradients (Figure 29.14), demonstrating that competition (or some other density-related effect, such as pathogen transmission) was much more important than predation as agents of selection on these traits.

30

Measuring Multivariate Selection

We have found that there are fundamental differences between the surviving birds and those eliminated, and we conclude that the birds which survived survived because they possessed certain structural characters, and that the birds which perished perished not through accident, but because they did not possess certain structural characters which would have enabled them to withstand the severity of the test imposed by nature; they were eliminated because they were unfit. Bumpus (1899)

While we have previously hinted at some of the features of multivariate selection (Chapters 13 and 20), we now start our formal discussion of selection on a vector of traits. As in Chapter 29, we distinguish between *selection*, the within-generation change in the (fitness-weighted) distribution of trait values and the *response to selection*, how the distribution of trait values changes across generations. *Phenotypic* correlations between traits within an individual influence within-generation changes, while *genetic* correlations (correlations among the breeding values for different traits within an individual) influence the between-generation response (Chapter 13). In particular, when a suite of traits is phenotypically correlated, simply observing the within-generation change in their means or variances is *not* sufficient to determine which ones are under selection.

To assess which traits are under selection, one must remove the effects of phenotypic correlations to separate **direct** selection from **correlated** selection. This is done by using partial regressions (linear for changes in the means and quadratic for changes in variances and covariances), and much of our focus here is on such fitness-trait regressions. Essentially all of the measures of phenotypic selection discussed here are multivariate extensions of the univariate measures introduced in Chapter 29. As these expressions are presented in terms of matrices and vectors, we rely rather heavily in places on matrix machinery (such as eigenvalues and eigenvectors, canonical decompositions, and, in a few places, matrix calculus). Appendix 5 discusses the mathematics of treating matrices as geometric objects, while Appendix 6 provides a brief refresher of important concepts and tools from the calculus of matrices (such as multidimensional derivatives and Taylor series).

The structure of this chapter is as follows: We first introduce the multivariate versions of selection differentials and gradients and their properties. Next, the geometry of quadratic regressions is examined in some detail, followed by discussions of multivariate nonparametric regressions and the strength of selection in natural populations. We conclude with some comments on unmeasured characters, the use of path analysis as an alternative (and often complementary) approach to the analysis of selection, and measures of multilevel selection.

As was the case for Chapter 29, much of the discussion here centers around the landmark paper of Lande and Arnold (1983) and the very long shadow it has cast for the past three decades in the phenotypic selection literature. There is a healthy debate as to the robustness of conclusions from this approach, especially in terms of measures of nonlinear selection. A second concern is that the ordinary least-squares (OLS) solutions of the Lande-Arnold regression assume that fitness residuals are normally distributed and homoscedastic, which is clearly incorrect. We introduced these issues in a univariate setting in Chapter 29, and here we continue their discussion in the multivariate setting.

SELECTION ON MULTIVARIATE PHENOTYPES: DIFFERENTIALS AND GRADIENTS

Chapter 29 described a variety of measures of univariate selection, with an emphasis on approximating the individual fitness function. In order to extend these methods to a vector

Evolution and Selection of Quantitative Traits. Bruce Walsh & Michael Lynch.
Published 2018 by Oxford University Press. © Bruce Walsh & Michael Lynch 2018.
DOI 10.1093/oso/9780198830870.001.0001

of characters, we need to account for phenotypic correlations. To do so, we follow the multiple regression approach of Lande and Arnold (1983), which was initially suggested by Pearson (1903). The phenotype of an individual is now a vector, $\mathbf{z} = (z_1, z_2, \cdots, z_n)^T$, of n character values. Suppose we denote the mean vector and covariance matrix of $\mathbf{z}$ before selection by $\boldsymbol{\mu}$ and $\mathbf{P}$, respectively, and by $\boldsymbol{\mu}^*$ and $\mathbf{P}^*$ after selection (but before reproduction). As an aside, we use this phrase *before reproduction*, which often appears in the literature. This does not mean that reproductive aspects of fitness are ignored; rather, it simply means that we ignore any complications arising from genetic transmission to the next generation. Formally, this means that the fitness-weight distribution of the parents (which reflects reproductive success; Chapter 29) is taken as the distribution after selection (as opposed to the trait distribution seen in their offspring, which confounds selection with the cross-generational transmission). To avoid additional complications, we examine only a single episode of selection. Partitions over multiple episodes follow as fairly straightforward extensions of the univariate partitions discussed in Chapter 29 (Arnold and Wade 1984a; Wade and Kalisz 1989; McGlothlin 2010).

Changes in the Mean Vector: The Directional Selection Differential Vector, S

The multivariate extension of the **directional selection differential** is the vector

$$\mathbf{S} = \boldsymbol{\mu}^* - \boldsymbol{\mu}$$

whose ith element is $S_i = \mu_i^* - \mu_i$, which is the differential for character z_i. (The fastidious reader might object to the nonstandard use of a capital bold letter, as opposed to the more standard bold lowercase letter, for a vector, but univariate selection theory uses S for the selection differential, and we keep that notation here.) As with the univariate case, the Robertson-Price identity (Equation 6.10) holds, with $\mathbf{S} = \sigma(\mathbf{z}, w)$, meaning that the elements of $\mathbf{S}$ represent the covariance between character value and relative fitness, $S_i = \sigma(z_i, w)$. This immediately implies (from Equation 29.6b) that the opportunity for selection, I (which is the population variance, σ_w^2, in relative fitness), bounds the range of S_i, and

$$\frac{|S_i|}{\sigma_{z_i}} \leq \sqrt{I} \tag{30.1a}$$

and

$$|\,\bar{\imath}_i\,| \leq \sigma_w \tag{30.1b}$$

Both show that selection intensity, $\bar{\imath}_i = S_i/\sigma_{z_i}$, for any trait (with $\sigma_{z_i}^2$ being the phenotypic variance of trait i) is bounded by the standard deviation of fitness.

As illustrated in Figure 30.1, $\mathbf{S}$ confounds the **direct effects** of selection on a focal trait with the **indirect effects** from selection on phenotypically correlated characters. Suppose character 1 is under direct selection to increase in value while character 2 is not directly selected. If z_1 and z_2 are uncorrelated, there is no within-generation change in μ_2 (the mean of z_2). However, if z_1 and z_2 are positively correlated, because individuals with large values of z_1 also tend to have large values of z_2, there will be a within-generation increase in μ_2 ($S_2 > 0$). Conversely, if z_1 and z_2 are negatively correlated, selection to increase z_1 will result in a within-generation decrease in μ_2 ($S_2 < 0$). Hence, a character not under selection can still experience a within-generation change resulting from selection on a phenotypically correlated trait (**indirect selection** or **correlated selection**). Fortunately, the **directional selection gradient, $\boldsymbol{\beta} = \mathbf{P}^{-1}\mathbf{S}$**, accounts for indirect selection resulting from phenotypic correlations (among the *measured* traits in the study), providing a less biased picture of the nature of directional selection acting on the component traits comprising $\mathbf{z}$.

The Directional Selection Gradient Vector, $\boldsymbol{\beta}$

As was discussed briefly in Chapters 13 and 20, $\boldsymbol{\beta}$, removes the effects of phenotypic correlations (among the set of traits being considered) because it is a vector of partial regression

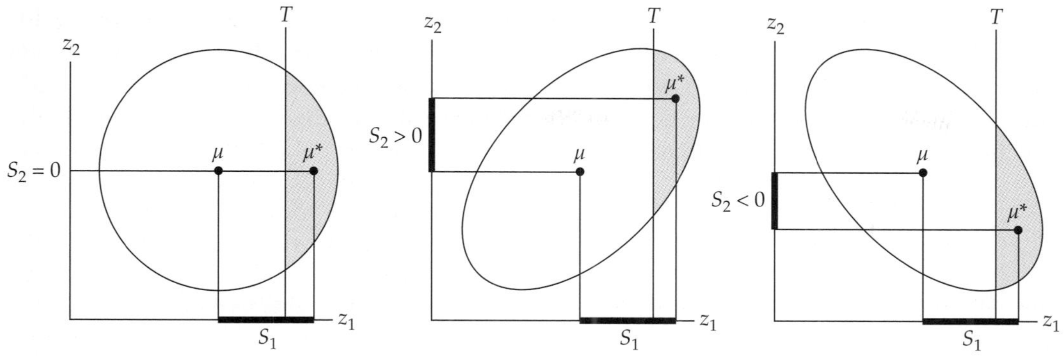

Figure 30.1 Selection on a character can result in a within-generation change in the mean of other phenotypically correlated characters that are not themselves under direct selection. Suppose we assume that character 1 is under simple truncation selection, so only individuals with values of $z_1 > T$ reproduce. There is no *direct* selection on trait 2. Here $\boldsymbol{\mu}$ and $\boldsymbol{\mu}^*$ are the population mean vector before and after selection, respectively. **Left:** When z_1 and z_2 are uncorrelated, $S_2 = 0$. **Center:** When z_1 and z_2 are positively correlated, $S_2 > 0$. **Right:** When z_1 and z_2 are negatively correlated, $S_2 < 0$.

coefficients. From multiple regression theory (LW Chapter 8), the vector of partial regression coefficients for predicting the value of w, given a vector of observations $\mathbf{z}$, is $\mathbf{P}^{-1}\boldsymbol{\sigma}(\mathbf{z}, w)$, where $\mathbf{P}$ is the covariance matrix of $\mathbf{z}$ and $\boldsymbol{\sigma}(\mathbf{z}, w)$ is the vector of covariances between the elements of $\mathbf{z}$ and fitness, with an ith element of $\sigma(z_i, w)$. Because $\mathbf{S} = \boldsymbol{\sigma}(\mathbf{z}, w)$, it immediately follows that

$$\mathbf{P}^{-1}\boldsymbol{\sigma}(\mathbf{z}, w) = \mathbf{P}^{-1}\mathbf{S} = \boldsymbol{\beta} \tag{30.2a}$$

is the vector of partial regression coefficients for the linear regression of relative fitness, w, on phenotypic value, $\mathbf{z}$, namely,

$$w(\mathbf{z}) = 1 + \sum_{j=1}^{n} \beta_j (z_j - \mu_j) + e = 1 + \boldsymbol{\beta}^T (\mathbf{z} - \boldsymbol{\mu}) + e \tag{30.2b}$$

Recalling the fact (LW Chapters 3 and 8) that the regression must pass through the means of both w and $\mathbf{z}$ (1 and $\boldsymbol{\mu}$, respectively), allows us to remove the intercept constant, a, in the regression. We could equivalently write Equation 30.2b as $w(\mathbf{z}) = a + \boldsymbol{\beta}^T\mathbf{z} + e$, where $a = 1 - \boldsymbol{\beta}^T\boldsymbol{\mu}$.

The partial regression coefficient, β_j, represents the change in w from a one-unit increase in z_j while holding all the other characters constant. It is important to note, however, that $\boldsymbol{\beta}$ accounts for the effects of phenotypic correlations only among the *measured* set of characters that comprise the elements of $\mathbf{z}$. Among the members of this set, a character under no directional selection has a value of $\beta_j = 0$.

Because $\mathbf{S} = \mathbf{P}\boldsymbol{\beta}$, we have

$$S_i = \sum_{j=1}^{n} \beta_j P_{ij} = \beta_i P_{ii} + \sum_{j \neq i}^{n} \beta_j P_{ij} \tag{30.3}$$

which illustrates that the directional selection differential for trait i confounds direct selection on that character (β_i) with indirect contributions due to selection on phenotypically correlated characters ($\beta_j P_{ij} \neq 0$). These contributions are given, respectively, by the first and second terms in Equation 30.3.

Operationally, fitness regressions are usually computed using **standardized variables**,

$$z_{sd,i} = \frac{z_i - \mu_i}{\sigma_i} \tag{30.4a}$$

with $z_{sd,i}$ having a mean of zero and unit variance (this is often denoted by z_i' in the literature). As mentioned in Chapter 29, one could also mean-standardize (z_i/μ_i), but when the term *standardized variable* is used in the evolutionary and ecological literature, it usually implies Equation 30.4a. The selection differential of the standardized variable is simply the selection intensity $\bar{\imath}$ (Equation 13.6a), as

$$S_{sd,i} = \sigma(z_{sd,i}, w) = \sigma(z_i/\sigma_i, w) = \frac{\sigma(z_i, w)}{\sigma_i} = \frac{S_i}{\sigma_i} = \bar{\imath}_i$$

The standardization given by Equation 30.4 can be expressed in matrix form as

$$\mathbf{z}_{sd} = \mathbf{D}^{-1}(\mathbf{z} - \boldsymbol{\mu}) \tag{30.4b}$$

where $\mathbf{D}$ is the diagonal matrix

$$\mathbf{D} = \begin{pmatrix} \sigma_1 & 0 & \cdots & 0 \\ 0 & \sigma_2 & \cdots & 0 \\ \vdots & & \ddots & \vdots \\ 0 & 0 & \cdots & \sigma_n \end{pmatrix} \tag{30.4c}$$

Equations 30.4b and 30.4c imply that the phenotypic covariance matrix ($\mathbf{P}_{sd}$) for the standardized variables is simply the matrix of all pairwise correlations. Following our notation from Chapter 29, we denote selection gradients using variance-standardized variables by $\beta_{sd,i}$, although they are often denoted by β_i' in the literature. When expressed with standardized variables, Equation 30.2b becomes

$$w(\mathbf{z}_{sd}) = 1 + \boldsymbol{\beta}_{sd}^T \mathbf{z}_{sd} + e, \quad \text{where} \quad \boldsymbol{\beta}_{sd} = \mathbf{P}_{sd}^{-1}\mathbf{S}_{sd}$$

Here $\beta_{sd,i}$ is the expected change in relative fitness given a change of one standard deviation in trait z_i when all of the other *measured* trait values (i.e., the elements of $\mathbf{z}$) are held constant.

The total strength of directional selection on the set of measured characters is a quantity of interest, and is given by the norm of $\boldsymbol{\beta}$, where (Equation A5.1a)

$$||\boldsymbol{\beta}|| = \sqrt{\boldsymbol{\beta}^T\boldsymbol{\beta}} = \sqrt{\sum \beta_i^2}$$

Morrissey (2014b) noted that $E[\,||\widehat{\boldsymbol{\beta}}||\,] \geq ||\boldsymbol{\beta}||$, as the values of β_i are measured with error, and such error carries over into the norm. Suppose k independent traits are measured, where $\widehat{\beta}_i \sim N(\beta, \sigma^2)$, so that each β_i is independent and identically distributed, with sampling error σ^2. In this case, $E\,[||\widehat{\boldsymbol{\beta}}||\,] = \sqrt{k(\beta + \sigma^2)} > ||\boldsymbol{\beta}|| = \sqrt{k\beta}$. Given that the standard error of an estimate of β is often of the same order as the estimate itself (Morrissey 2014b), the total strength of selection can be considerably overestimated by $||\widehat{\boldsymbol{\beta}}||$.

We also remark in passing that the Henshaw-Zemel (2017) distributional selection differential (a nonparametric measure of the total shift in a distribution following selection; Equation 29.15a) can be generalized into a multivariate version that also partitions trait selection into direct and indirect effects; see their paper for details.

Example 30.1. The original application of the Lande-Arnold regression was on a population of one-spotted stink bugs (*Euschistus variolarius*), collected along the shore of Lake Michigan after a storm (Lande and Arnold 1983), an insect equivalent of the classic Bumpus (1899) study. Of the 94 individuals collected (legend has it that some were deposited into an open adult beverage container, in a truly selfless act of science), 39 were alive. All individuals were measured for four characters: head (Hd) and thorax (Tx) width and scutellum (Sc) and forewing (Fw) length. The data were then logarithmically transformed to more closely approximate normality, and the resulting log-transformed variables were variance-standarized. Selection differentials were calculated as the difference between the trait means among those bugs that survived and the total sample (dead or alive). Because these selection differentials were scaled in standard deviations, they correspond to selection intensities ($\bar{\imath}$; Equation 13.6a). The resulting vector of standardized selection differentials and correlation matrix were

$$\mathbf{S}_{sd} = \begin{pmatrix} \bar{\imath}_{Hd} \\ \bar{\imath}_{Tx} \\ \bar{\imath}_{Sc} \\ \bar{\imath}_{Fw} \end{pmatrix} = \begin{pmatrix} -0.11 \\ -0.06 \\ -0.28^{*} \\ -0.43^{**} \end{pmatrix} \qquad \mathbf{P}_{sd} = \begin{pmatrix} 1.00 & 0.72 & 0.50 & 0.60 \\ 0.72 & 1.00 & 0.59 & 0.71 \\ 0.50 & 0.59 & 1.00 & 0.62 \\ 0.60 & 0.71 & 0.62 & 1.00 \end{pmatrix}$$

where $*$ and $**$ denote 5% and 1% significance for the elements in $\mathbf{S}_{sd}$. Hence,

$$\boldsymbol{\beta}_{sd} = (\mathbf{P}_{sd})^{-1}\,\mathbf{S}_{sd} = \begin{pmatrix} 1.00 & 0.72 & 0.50 & 0.60 \\ 0.72 & 1.00 & 0.59 & 0.71 \\ 0.50 & 0.59 & 1.00 & 0.62 \\ 0.60 & 0.71 & 0.62 & 1.00 \end{pmatrix}^{-1} \begin{pmatrix} -0.11 \\ -0.06 \\ -0.28 \\ -0.43 \end{pmatrix} = \begin{pmatrix} 0.02 \\ 0.53^{**} \\ -0.16 \\ -0.72^{**} \end{pmatrix}$$

The selection differentials for scutellum and wing length were both significantly different from zero, while the only significant gradients were for thorax and wing length. Hence, while the within-generation change in thorax length was negative (but not significantly different from zero), in reality there was strong direct selection to *increase* thorax length. An increase of one standard deviation in thorax length increases relative fitness by $\beta_{sd} = 0.53$ (i.e., 53%), while an increase of one standard deviation in wing length reduces fitness by 72%.

Directional Gradients, Fitness Surface Geometry, and Selection Response

Consider a scalar-valued function, $f(\mathbf{x})$, whose argument is a vector, $\mathbf{x} = (x_1, \cdots, x_n)^T$. Recall from vector calculus (Appendix 6) that the **gradient vector**, $\nabla_{\mathbf{x}} f(\mathbf{x})$, of f with respect to $\mathbf{x}$ is defined as

$$\nabla_{\mathbf{x}} f(\mathbf{x}) = \begin{pmatrix} \partial f/\partial x_1 \\ \partial f/\partial x_2 \\ \vdots \\ \partial f/\partial x_n \end{pmatrix}$$

namely, the vector of each partial of f with respect to x_i, for $1 \leq i \leq n$. Further, recall that the gradient vector of a multivariate function points in the direction of change that will give the greatest (local) increase in f.

When phenotypes are multivariate normal, $\boldsymbol{\beta} = \mathbf{P}^{-1}\mathbf{S}$ provides a convenient descriptor of the geometries, measured by gradients, of both the individual fitness surface, $w(z)$, and the mean population landscape, $\overline{W}$. Let us consider the landscape first. If we assume the $\mathbf{z}$ is multivariate normal (MVN), Example A6.3 shows that

$$\boldsymbol{\beta} = \nabla_{\boldsymbol{\mu}}[\ln \overline{W}(\boldsymbol{\mu})] = \overline{W}^{-1} \cdot \nabla_{\boldsymbol{\mu}}[\overline{W}(\boldsymbol{\mu})] \tag{30.5a}$$

which holds provided the fitnesses are frequency-independent (Lande 1976, 1979a). In this case, $\boldsymbol{\beta}$ is the gradient of *mean population* fitness with respect to the mean vector, $\boldsymbol{\mu}$, namely, the direction of steepest local increase in $\overline{W}$. Hence, given the current population mean, $\overline{W}$ increases most rapidly when the change in the vector of means ($\boldsymbol{\Delta\mu}$) is in the direction given by $\boldsymbol{\beta}$. If the fitnesses are frequency-dependent (individual fitnesses change as the population mean changes, so that $\nabla_{\boldsymbol{\mu}}[\,w(\mathbf{z})\,] \neq \mathbf{0}$), then provided $\mathbf{z}$ is multivariate-normally distributed, Example A6.3 further shows that

$$\boldsymbol{\beta} = \nabla_{\boldsymbol{\mu}}[\ln \overline{W}(\boldsymbol{\mu})] + \int \nabla_{\boldsymbol{\mu}}[\,w(\mathbf{z})\,]\,\varphi(\mathbf{z})\,d\mathbf{z} \tag{30.5b}$$

The integral in this expression accounts for the effects of frequency-dependence (the change in the *individual* fitness surface with respect to changes in the mean, $\boldsymbol{\mu}$) and φ is the MVN density function (Lande 1976). The vector $\boldsymbol{\beta}$ does not point in the direction of steepest increase in $\overline{W}$ unless the second integral is zero.

Equation 30.5a shows the connection between the fitness *landscape* (the mean fitness surface) and $\boldsymbol{\beta}$ when $\mathbf{z}$ is MVN. A similiar connection with $\boldsymbol{\beta}$ occurs (when $\mathbf{z}$ is also MVN) with the *individual fitness* surface. Here $\boldsymbol{\beta}$ as the average gradient of individual relative fitnesses (over the population distribution of phenotypes),

$$\boldsymbol{\beta} = E_{\mathbf{z}}\Big[\nabla_{\mathbf{z}}[\,w(\mathbf{z})\,]\Big] = \int \nabla_{\mathbf{z}}[\,w(\mathbf{z})\,]\,\varphi(\mathbf{z})\,d\mathbf{z} \tag{30.6}$$

which holds provided $\mathbf{z} \sim$ MVN (Lande and Arnold 1983). To see this, we use integration by parts to yield

$$\int_{\mathbf{a}}^{\mathbf{b}} \nabla_{\mathbf{z}}[\,w(\mathbf{z})\,]\,\varphi(\mathbf{z})\,d\mathbf{z} = w(\mathbf{z})\,\varphi(\mathbf{z})\Big|_{\mathbf{a}}^{\mathbf{b}} - \int_{\mathbf{a}}^{\mathbf{b}} \nabla_{\mathbf{z}}[\,\varphi(\mathbf{z})\,]\,w(\mathbf{z})\,d\mathbf{z}$$

If we take the limit as $\mathbf{a} \to -\infty$ and $\mathbf{b} \to \infty$, the first term on the righthand side vanishes as $\varphi(\mathbf{z}) \to 0$ when $\mathbf{z} \to \pm\infty$. If $\mathbf{z} \sim \text{MVN}(\boldsymbol{\mu}, \mathbf{P})$, then Equation A6.2a gives $\nabla_{\mathbf{z}}[\,\varphi(\mathbf{z})\,] = -\varphi(\mathbf{z})\,\mathbf{P}^{-1}\,(\mathbf{z}-\boldsymbol{\mu})$, implying

$$\begin{aligned}\int \nabla_{\mathbf{z}}[\,w(\mathbf{z})\,]\,\varphi(\mathbf{z})\,d\mathbf{z} = -\int \nabla_{\mathbf{z}}[\,\varphi(\mathbf{z})\,]\,w(\mathbf{z})\,d\mathbf{z} &= \int w(\mathbf{z})\,\varphi(\mathbf{z})\,\mathbf{P}^{-1}\,(\mathbf{z}-\boldsymbol{\mu})\,d\mathbf{z} \\ &= \mathbf{P}^{-1}\left(\int \mathbf{z}\,w(\mathbf{z})\,\varphi(\mathbf{z})\,d\mathbf{z} - \boldsymbol{\mu}\int w(\mathbf{z})\,\varphi(\mathbf{z})\,d\mathbf{z}\right) \\ &= \mathbf{P}^{-1}\,(\boldsymbol{\mu}^* - \boldsymbol{\mu}) = \mathbf{P}^{-1}\mathbf{S} = \boldsymbol{\beta}\end{aligned}$$

The first integral in the second line corresponds to the mean trait value weighted by relative fitness (μ^*), while the second integral is the average value of relative fitness (1). Note from this derivation that Equation 30.6 holds regardless of whether fitness is frequency-dependent or -independent. This follows because the gradient was taken with respect to $\mathbf{z}$, rather than with respect to the vector of means, $\boldsymbol{\mu}$.

Finally, while our focus has been on the role that $\boldsymbol{\beta}$ plays in measuring phenotypic selection, $\boldsymbol{\beta}$ also plays an important role in the response to selection. If we can assume that the breeder's equation holds, it is the only measure of phenotypic selection required to predict the response in the means, as the vector, $\mathbf{R}$, of response (changes in means) is calculated by $\mathbf{R} = \mathbf{G}\boldsymbol{\beta}$ (Equation 13.26a). Cheverud (1984b) made the important point that although it is often assumed that a set of phenotypically correlated traits responds to selection in a coordinated fashion, this is not necessarily the case. Because $\boldsymbol{\beta}$ removes the effects of phenotypic correlations, phenotypic characters will only respond as a group if they are all under direct selection or if they are *genetically* correlated, a point discussed in detail in Volume 3.

Changes in the Covariance Matrix: The Quadratic Selection Differential Matrix, C

Motivated by the univariate case wherein $C = \sigma[\,w, (z-\mu)(z-\mu)\,]$, the multivariate **quadratic selection differential** is defined as the square ($n \times n$) matrix, $\mathbf{C}$, whose elements are the covariances between all pairs of quadratic deviations, $(z_i - \mu_i)(z_j - \mu_j)$, and relative fitness, w, namely,

$$C_{ij} = \sigma[\,w, (z_i - \mu_i)(z_j - \mu_j)\,] \tag{30.7a}$$

As derived in Example 30.2, Lande and Arnold (1983) showed that

$$\mathbf{C} = \boldsymbol{\sigma}[w, (\mathbf{z}-\boldsymbol{\mu})(\mathbf{z}-\boldsymbol{\mu})^T] = \mathbf{P}^* - \mathbf{P} + \mathbf{S}\,\mathbf{S}^T \tag{30.7b}$$

where $\mathbf{P}$ and $\mathbf{P}^*$ are, respectively, the phenotypic covariance matrices before- and after-selection. If no quadratic selection is acting, the covariance between each quadratic deviation and fitness is zero, and with $\mathbf{C} = \mathbf{0}$. In this case, Equation 30.7b returns the within-generation change in $\mathbf{P}$ resulting from selection as

$$P^*_{ij} - P_{ij} = -S_i S_j \tag{30.7c}$$

This demonstrates that the $S_i S_j$ term corrects C_{ij} for the change in covariance caused by directional selection alone. This expression is the multivariate extension of the condition $\sigma^2(z^*) - \sigma^2(z) = P^* - P = -S^2$ for a single trait (Equation 29.16a); the latter expression follows as a special case ($i = j$) for Equation 30.7c.

Example 30.2. We wish to show that $\mathbf{P}^* - \mathbf{P} = \sigma[w(\mathbf{z}), (\mathbf{z}-\boldsymbol{\mu})(\mathbf{z}-\boldsymbol{\mu})^T] - \mathbf{S}\,\mathbf{S}^T$, which implies Equation 30.7b. From the definition of the variance-covariance matrix,

$$\mathbf{P} = E\left[\,(\mathbf{z}-\boldsymbol{\mu})\,(\mathbf{z}-\boldsymbol{\mu})^T\,\right] = \int (\mathbf{z}-\boldsymbol{\mu})(\mathbf{z}-\boldsymbol{\mu})^T\, p(\mathbf{z})\, d\mathbf{z} \tag{30.8a}$$

$$\mathbf{P}^* = E\left[\,(\mathbf{z}^*-\boldsymbol{\mu}^*)\,(\mathbf{z}^*-\boldsymbol{\mu}^*)^T\,\right] = \int (\mathbf{z}-\boldsymbol{\mu}^*)(\mathbf{z}-\boldsymbol{\mu}^*)^T\, p^*(\mathbf{z})\, d\mathbf{z} \tag{30.8b}$$

where $p^*(\mathbf{z}) = w(\mathbf{z})\,p(\mathbf{z})$ is the distribution of $\mathbf{z}$ after selection (but before reproduction). If we note that $\boldsymbol{\mu}^* = \boldsymbol{\mu} + \mathbf{S}$, the integrated expression in Equation 30.8b can be written as

$$\begin{aligned}(\mathbf{z}-\boldsymbol{\mu}^*)(\mathbf{z}-\boldsymbol{\mu}^*)^T &= (\mathbf{z}-\boldsymbol{\mu}-\mathbf{S})(\mathbf{z}-\boldsymbol{\mu}-\mathbf{S})^T \\ &= (\mathbf{z}-\boldsymbol{\mu}-\mathbf{S})(\,[\mathbf{z}-\boldsymbol{\mu}]^T - \mathbf{S}^T) \\ &= (\mathbf{z}-\boldsymbol{\mu})\,(\mathbf{z}-\boldsymbol{\mu})^T - (\mathbf{z}-\boldsymbol{\mu})\,\mathbf{S}^T - \mathbf{S}\,(\mathbf{z}-\boldsymbol{\mu})^T + \mathbf{S}\,\mathbf{S}^T \end{aligned} \tag{30.8c}$$

Because $\int \mathbf{z}\,p^*(\mathbf{z})\,d\mathbf{z} = \boldsymbol{\mu}^*$ and $\int p^*(\mathbf{z})\,d\mathbf{z} = 1$, then

$$\int (\mathbf{z}-\boldsymbol{\mu})\,\mathbf{S}^T\,p^*(\mathbf{z})\,d\mathbf{z} = \int\,[\mathbf{z}\,p^*(\mathbf{z})]\;\mathbf{S}^T\,d\mathbf{z} - \boldsymbol{\mu}\,\mathbf{S}^T \int p^*(\mathbf{z})\,d\mathbf{z} = (\boldsymbol{\mu}^*-\boldsymbol{\mu})\mathbf{S}^T = \mathbf{S}\,\mathbf{S}^T$$

$$\int \mathbf{S}(\mathbf{z}-\boldsymbol{\mu})^T\,p^*(\mathbf{z})\,d\mathbf{z} = \mathbf{S}(\boldsymbol{\mu}^*)^T - \mathbf{S}\boldsymbol{\mu}^T = \mathbf{S}(\boldsymbol{\mu}^*-\boldsymbol{\mu})^T = \mathbf{S}\,\mathbf{S}^T$$

$$\int \mathbf{S}\,\mathbf{S}^T\,p^*(\mathbf{z})\,d\mathbf{z} = \mathbf{S}\,\mathbf{S}^T$$

Substituting these results into Equation 30.8b yields

$$\begin{aligned}\mathbf{P}^* &= \int (\mathbf{z}-\boldsymbol{\mu})\,(\mathbf{z}-\boldsymbol{\mu})^T\,w(\mathbf{z})\,p(\mathbf{z})\,d\mathbf{z} - \mathbf{S}\,\mathbf{S}^T - \mathbf{S}\,\mathbf{S}^T + \mathbf{S}\,\mathbf{S}^T \\ &= E\left[\,w(\mathbf{z})\cdot(\mathbf{z}-\boldsymbol{\mu})\,(\mathbf{z}-\boldsymbol{\mu})^T\,\right] - \mathbf{S}\,\mathbf{S}^T \end{aligned} \tag{30.8d}$$

Because $E[\,w(\mathbf{z})\,] = 1$, we can write $\mathbf{P} = E[\,w(\mathbf{z})\,]\cdot\mathbf{P}$. Using the definition of $\mathbf{P}$,

$$\begin{aligned}\mathbf{P}^* - \mathbf{P} &= E\left[\,w(\mathbf{z})\cdot(\mathbf{z}-\boldsymbol{\mu})\,(\mathbf{z}-\boldsymbol{\mu})^T\,\right] - \mathbf{S}\,\mathbf{S}^T - E[\,w(\mathbf{z})\,]\cdot E\left[\,(\mathbf{z}-\boldsymbol{\mu})\,(\mathbf{z}-\boldsymbol{\mu})^T\,\right] \\ &= \boldsymbol{\sigma}\left[\,w(\mathbf{z}), (\mathbf{z}-\boldsymbol{\mu})(\mathbf{z}-\boldsymbol{\mu})^T\,\right] - \mathbf{S}\,\mathbf{S}^T \end{aligned} \tag{30.8e}$$

with the last equality following from the definition of a covariance, $\sigma(x,y) = E(x\cdot y) - E(x)\,E(y)$.

As was the case for **S**, the fact that C_{ij} is a covariance immediately allows us to bound its range using the opportunity for selection (Chapter 29). Because $\sigma^2(x,y) \leq \sigma^2(x)\,\sigma^2(y)$,

$$C_{ij}^2 \leq \sigma^2(w)\,\sigma^2[\,(z_i-\mu_i)(z_j-\mu_j)\,] = I\,\sigma^2[\,(z_i-\mu_i)(z_j-\mu_j)\,] \tag{30.9a}$$

When z_i and z_j are bivariate-normal, then (Kendall and Stuart 1983),

$$\sigma^2[\,(z_i-\mu_i)(z_j-\mu_j)\,] = P_{ij}^2 + P_{ii}\,P_{jj} = P_{ij}^2(1+\rho_{ij}^{-2}) \tag{30.9b}$$

where ρ_{ij} is the phenotypic covariance between z_i and z_j. Hence, for Gaussian-distributed phenotypes,

$$\left|\frac{C_{ij}}{P_{ij}}\right| \leq \sqrt{I}\,\sqrt{1+\rho_{ij}^{-2}} \tag{30.10}$$

which is a variant of the original bound based on I, as suggested by Arnold (1986). Note that when $i=j$, $\rho_{ii}=1$, and we recover Equation 29.18c.

The Quadratic Selection Gradient Matrix, γ

Like the directional selection differential vector, **S**, the quadratic selection differential, **C**, confounds the effects of direct selection with selection on phenotypically correlated characters. As was the case with **S**, these indirect effects can also be removed by a regression. Consider the quadratic regression of relative fitness as a function of phenotypic value,

$$w(\mathbf{z}) = a + \sum_{j=1}^{n} b_j\,z_j + \frac{1}{2}\sum_{j=1}^{n}\sum_{k=1}^{n} \gamma_{jk}\,(z_j-\mu_j)(z_k-\mu_k) \tag{30.11a}$$

$$= a + \mathbf{b}^T\mathbf{z} + \frac{1}{2}(\mathbf{z}-\boldsymbol{\mu})^T\,\boldsymbol{\gamma}\,(\mathbf{z}-\boldsymbol{\mu}) \tag{30.11b}$$

Using multiple regression theory, Lande and Arnold (1983) showed that when $\mathbf{z} \sim$ MVN, the matrix, $\boldsymbol{\gamma}$, of quadratic partial regression coefficients is given by

$$\boldsymbol{\gamma} = \mathbf{P}^{-1}\,\boldsymbol{\sigma}[\,w, (\mathbf{z}-\boldsymbol{\mu})(\mathbf{z}-\boldsymbol{\mu})^T\,]\,\mathbf{P}^{-1} = \mathbf{P}^{-1}\,\mathbf{C}\,\mathbf{P}^{-1} \tag{30.12}$$

This is the **quadratic selection gradient**, and (like $\boldsymbol{\beta}$) it removes the effects of phenotypic correlations (among the *measured* traits), thus providing a more accurate picture of how selection is operating on the multivariate phenotype.

As we saw in the univariate case (Chapter 29), the vector of linear coefficients (**b**) for the quadratic regression need not equal the vector of partial regression coefficients ($\boldsymbol{\beta}$) that is obtained by assuming only a *linear* regression (Equation 30.2b). Equation 29.28a showed (for the univariate case) that if the phenotypic distribution is skewed, the linear term (b) in the quadratic regression is a function of both S and C, while the linear term in a linear regression (β) is only a function of S. When phenotypes are multivariate normal, the skew is zero, and $\mathbf{b} = \boldsymbol{\beta}$ (Lande and Arnold 1983), which recovers the multivariate version of the Pearson-Lande-Arnold regression,

$$w(\mathbf{z}) = a + \boldsymbol{\beta}^T\mathbf{z} + \frac{1}{2}(\mathbf{z}-\boldsymbol{\mu})^T\boldsymbol{\gamma}\,(\mathbf{z}-\boldsymbol{\mu}) \tag{30.13a}$$

As with linear regression, one typically standardizes the trait values (Equation 30.4b), in which case Equation 30.13a can be written more compactly as

$$w(\mathbf{z}) = a + \boldsymbol{\beta}_{sd}^T\mathbf{z}_{sd} + \frac{1}{2}\mathbf{z}_{sd}^T\,\boldsymbol{\gamma}_{sd}\,\mathbf{z}_{sd} \tag{30.13b}$$

Because the elements, γ_{ij}, of the matrix $\boldsymbol{\gamma}$ (or its standarized counterpart, $\boldsymbol{\gamma}_{sd}$) are partial regression coefficients, they predict the change in expected fitness caused by changing the associated quadratic deviation while holding all other variables constant. Increasing

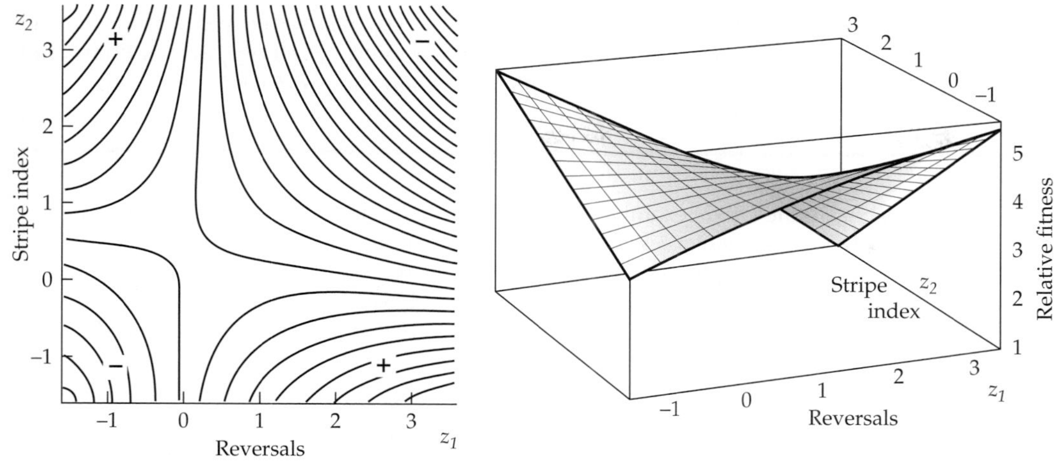

Figure 30.2 The fitness surface (measured as one-year survivorship) for number of reversals (z_1) and body stripe index (z_2) in the garter snake *Thamnophis ordinoides*. There is a significant correlational gradient between these two characters, and all other directional and quadratic gradients are nonsignificant. **Left:** A plot of contours of equal fitness, with peaks represented by $+$ and valleys by $-$, shows that the best-fitting quadratic fitness surface has a saddle point. **Right:** Three-dimensional representation of the best-fitting quadratic fitness surface. As will be discussed shortly, the eigenvalues of the $\boldsymbol{\gamma}$ matrix can be used to clarify the geometry of the associated quadratic fitness surface. Here, the eigenvalues of $\boldsymbol{\gamma}$ are 0.256 and -0.290, indicating roughly equal amounts of convex selection along one canonical axis (given by the index $0.77 \cdot z_1 - 0.64 \cdot z_2$) and concave selection along the other ($0.64 \cdot z_1 + 0.77 \cdot z_2$). (Based on data from Brodie 1992.)

the value of $(z_j - \mu_j)(z_k - \mu_k)$ by one unit in such a way as to hold all other variables and all other pairwise combinations of characters constant is expected to change relative fitness by γ_{jk} for $j \neq k$ and by $\gamma_{jj}/2$ if $j = k$ (the difference arises because $\gamma_{jk} = \gamma_{kj}$, so γ_{jk} appears twice in the regression unless $j = k$). The coefficients of $\boldsymbol{\gamma}$ thus describe the nature of selection on *quadratic* deviations from the mean for both single characters and pairwise combinations of characters. A value of $\gamma_{ii} < 0$ implies that fitness decreases as z_i moves away (in either direction) from its phenotypic mean. As was discussed in Chapter 29, this is a necessary, *but not sufficient*, condition for ensuring stabilizing selection on character i. As a result, the terms **concave selection** or **concave fitness surface** are often used to indicate this situation. The term *stabilizing selection* is restricted to situations where the fitness surface is concave *and* the population distribution is under a peak in the fitness surface. Similarly, $\gamma_{ii} > 0$ implies that fitness increases as i moves away from its mean (**convex selection** or **convex fitness surface**), which is again a necessary, but not sufficient condition, for disruptive selection. Turning to combinations of characters, nonzero values of γ_{jk} $(j \neq k)$ suggest the presence of **correlational selection**, with $\gamma_{jk} > 0$ suggesting selection for a positive phenotypic correlation between characters j and k, and $\gamma_{jk} < 0$ suggesting selection for a negative phenotypic correlation between those characters.. Although it appear to be straightforward to infer the overall nature of selection by looking at the various pairwise values of γ_{ij}, this can result in an extremely *misleading picture* about the geometry of the fitness surface (e.g., Figure 30.3). We will discuss this problem and its solution shortly.

Finally, as we did for directional selection differentials, $\mathbf{S}$ (Equation 30.3), we can partition changes in the quadratic selection differential, $\mathbf{C}$, into direct effects and indirect effects resulting from selection on phenotypically correlated traits. Solving for $\mathbf{C}$ by post- and pre-multiplying $\boldsymbol{\gamma}$ (Equation 30.12) by $\mathbf{P}$ gives $\mathbf{C} = \mathbf{P}\,\boldsymbol{\gamma}\,\mathbf{P}$, which yields

$$C_{ij} = \sum_{k=1}^{n} \sum_{\ell=1}^{n} \gamma_{k\ell}\, P_{ik}\, P_{\ell j} \tag{30.14}$$

showing that within-generation changes in phenotypic covariance between traits i and j, as measured by C_{ij}, are influenced by quadratic selection ($\gamma_{k\ell} \neq 0$) on pairs of characters, k and ℓ, that are correlated with i and j, specifically, when the product $P_{ik}\, P_{\ell j} \neq 0$.

Example 30.3. Brodie (1992) examined one-year survivorship in an Oregon population of garter snakes (*Thamnophis ordinoides*). Over a three-year period, 646 snakes were marked, 101 of which were eventually recaptured. Four morphological and behavioral characters were measured: overall stripedness of the body-color pattern (stripes), sprint speed, distance moved until an antipredator display was performed, and number of reversals of direction during flight from predators (reversals). None of the values of β_i or γ_{ii} were significant. However, there was a significant quadratic association between striping pattern and number of reversals, with $\gamma_{ij} = -0.268 \pm 0.097$ (confidence intervals were generated using the delete-one jackknife method of Mitchell-Olds 1989). As shown in Figure 30.2, the best-fitting quadratic regression of individual fitness has a saddle point, which means that concave selection (negative fitness surface curvature) occurs along one direction and convex selection (positive fitness surface curvature) along the other. Brodie suggested a biological explanation for selection favoring a negative correlation between these two characters. When the body pattern is banded, blotched, or spotted, the detection of movement by visual predators is enhanced. In such individuals, frequent reversals can disrupt a visual search. Conversely, the presence of body stripes makes it difficult for predators to judge the speed of the snake, so frequent reversals (and hence additional movement for predators to perceive) would be disadvantageous.

Quadratic Gradients, Fitness Surface Geometry, and Selection Response

As was the case for $\boldsymbol{\beta}$, when phenotypes are multivariate normal, $\boldsymbol{\gamma}$ also describes geometric features of both the individual fitness surface and the mean population fitness landscape. It provides a measure of the average curvature of the individual fitness surface, as

$$\boldsymbol{\gamma} = \int \mathbf{H_z}[\, w(\mathbf{z})\,]\, \varphi(\mathbf{z})\, d\mathbf{z} \tag{30.15a}$$

where $\mathbf{H_z}[\, f\,]$ denotes the Hessian matrix of f with respect to $\mathbf{z}$ (the matrix of all second partial derivatives; where $H_{ij} = \partial^2 f / \partial z_i\, \partial z_j$) and is a multivariate measure of the quadratic (local) curvature of a function (Appendix A6). This result, due to Lande and Arnold (1983) can be obtained by an integration-by-parts argument similar to that used to obtain Equation 30.6, and holds for both frequency-dependent and frequency-independent fitnesses.

When fitnesses are frequency-independent (again provided $\mathbf{z} \sim$ MVN), $\boldsymbol{\gamma}$ also provides a description of the curvature of the mean fitness landscape, with

$$\mathbf{H}_{\boldsymbol{\mu}}[\,\ln \overline{W}(\boldsymbol{\mu})\,] = \boldsymbol{\gamma} - \boldsymbol{\beta}\boldsymbol{\beta}^T \tag{30.15b}$$

This result is due to Lande (cited in Phillips and Arnold 1989), and it indicates that there are two sources for curvature in the mean fitness landscape: $-\boldsymbol{\beta}\boldsymbol{\beta}^T$ from directional selection and $\boldsymbol{\gamma}$ from quadratic selection.

Finally, when the breeder's equation holds, $\boldsymbol{\gamma}$ and $\boldsymbol{\beta}$ are sufficient to describe how phenotypic selection alters the additive-genetic covariance matrix. As we show in Volume 3, the additive-genetic covariance matrix, $\mathbf{G}^*$, following selection (but before reproduction) is calculated by

$$\mathbf{G}^* = \mathbf{G}\left(\boldsymbol{\gamma} - \boldsymbol{\beta}\boldsymbol{\beta}^T\right)\mathbf{G} + \mathbf{G} \tag{30.16}$$

Fitness Surface Curvature and Within-generation Changes in Variances and Covariances

Equations 30.15–30.16 provide some insight into the connection between within-generation changes in phenotypic and genetic variances and the curvature of the fitness surface. To see such connections, we ignore the complications introduced by either phenotypic or genetic correlations. First, consider the within-generation change in the phenotypic variance. Equations 30.7b and 30.14 imply that

$$C_{ii} = \sigma^2(z_i^*) - \sigma^2(z_i) + S_i^2 = \gamma_{ii}\, P_{ii}^2 = \gamma_{ii}\, \sigma^4(z_i) \tag{30.17a}$$

Hence, the within-generation change (denoted by $\delta[x]$) in the phenotypic variance for trait i is

$$\delta\left[\sigma^2(z_i)\right] = \gamma_{ii} P_{ii}^2 - S_i^2 \tag{30.17b}$$

Thus, concave selection ($\gamma_{ii} < 0$) reduces the phenotypic variance of a trait, while convex selection ($\gamma_{ii} > 0$) increases it. The net effect of directional selection ($S_i \neq 0$) is to always reduce the variance, which means that undetected directional selection (i.e., S was not measured) can mask the effects of convex selection (e.g., Example 29.10) and enhance the effects of concave selection. Likewise, from Equation 30.16 (and assuming there are no genetic correlations), the within-generation change in the additive variance is

$$\delta\left[\sigma^2(A_i)\right] = \left(\gamma_{ii} - \beta_i^2\right) \sigma^4(A_i) \tag{30.17c}$$

As with the phenotypic variance, both concave and directional selection reduce the additive variance, while convex selection increases it (Chapter 16). Note that Equation 30.17c (as well as Equation 30.16) is the *within-generation* change in the additive genetic variance. Recombination and segregation in the selected individuals will change the additive variance in the offspring generation by reducing the disequilibrium generated by selection and by adding segregation variance (Chapters 16 and 24).

What about the effect of correlational selection ($\gamma_{ij} \neq 0$)? Again, assuming all correlations (genetic and phenotypic) are (initially) zero, Equation 30.16 yields

$$C_{ij} = \sigma(z_i^*, z_j^*) - \sigma(z_i, z_j) + S_i S_j = 2\gamma_{ij} P_{ii} P_{jj} \tag{30.18a}$$

and the within-generation change in the phenotypic covariance becomes

$$\delta\left[\sigma(z_i, z_j)\right] = 2\gamma_{ij} P_{ii} P_{jj} - S_i S_j \tag{30.18b}$$

Positive values of γ_{ij} increase the phenotypic correlation, while negative values reduce it. Note that directional selection does not have a uniform effect: if both i and j are selected in the same direction, this decreases the phenotypic correlation, while if they are selected in opposite directions, this increases the correlation. Assuming no (initial) genetic correlations, Equation 30.16 gives the (within-generation) change in the genetic covariance as

$$\delta\left[\sigma(A_i, A_j)\right] = \left(2\gamma_{ij} - \beta_i \beta_j\right) \sigma^2(A_i)\sigma^2(A_j) \tag{30.18c}$$

Under the infinitesimal model, this change in the genetic covariance is due entirely to disequilibrium, which (for unlinked loci) is reduced by half in the offspring (Chapters 16 and 24).

The major features of linear and quadratic differentials and gradients discussed here are summarized in Table 30.1. Excellent overviews were also provided by Brodie et al. (1995) and Arnold et al. (2001).

MULTIDIMENSIONAL QUADRATIC FITNESS REGRESSIONS

As noted for univariate cases, approximating the individual fitness function by a quadratic can give a very distorted view of the true fitness surface (Figure 29.10). We expect this distortion to be even greater in a multivariate setting. With this caveat in mind, quadratic fitness surfaces are still quite useful. One advantage is that a quadratic is the simplest surface allowing for curvature. Further, when phenotypes are normally distributed, the coefficients

Table 30.1 Analogous features of directional and quadratic differentials and gradients. Details are in the text.

Changes in Means (Directional Selection)	**Changes in Covariances** (Quadratic Selection)
Differentials measure the covariance between relative fitness and phenotype	
$S_i = \sigma\,[\,w, z_i\,]$	$C_{ij} = \sigma\,[\,w, (z_i - \mu_i)(z_j - \mu_j)\,]$
The opportunity for selection bounds the differential	
$\dfrac{\lvert S_i\rvert}{\sigma(z_i)} \leq \sqrt{I}$ for any distribution of $\mathbf{z}$	$\left\lvert \dfrac{C_{ij}}{P_{ij}} \right\rvert \leq \sqrt{I}\,\sqrt{1+\rho_{ij}^{-2}}$ if $\mathbf{z} \sim$MVN
Differentials confound direct and indirect selection	
$\mathbf{S} = \boldsymbol{\mu}^* - \boldsymbol{\mu} = \mathbf{P}\boldsymbol{\beta}$	$\mathbf{C} = \mathbf{P}^* - \mathbf{P} + \mathbf{S}\mathbf{S}^T = \mathbf{P}\,\boldsymbol{\gamma}\,\mathbf{P}$
$S_i = \sum_{j=1}^{n} \beta_j\, P_{ij}$	$C_{ij} = \sum_{k=1}^{n}\sum_{\ell=1}^{n} \gamma_{k\ell}\, P_{ik}\, P_{\ell j}$
Gradients measure the amount of direct selection	
$\boldsymbol{\beta} = \mathbf{P}^{-1}\mathbf{S}$	$\boldsymbol{\gamma} = \mathbf{P}^{-1}\mathbf{C}\mathbf{P}^{-1}$
Gradients describe the slope and curvature of the log mean fitness landscape, provided $z \sim$ MVN and fitnesses are frequency-independent	
$\beta_i = \dfrac{\partial \ln \overline{W}(\boldsymbol{\mu})}{\partial \mu_i}$	$\gamma_{ij} = \dfrac{\partial^2 \ln \overline{W}(\boldsymbol{\mu})}{\partial \mu_i\,\partial \mu_j} + \beta_i\beta_j$
Gradients describe the average slope and average curvature of the individual fitness surface, provided $\mathbf{z} \sim$ MVN	
$\beta_i = \displaystyle\int \frac{\partial\, w(\mathbf{z})}{\partial z_i}\,\varphi(\mathbf{z})\,d\mathbf{z}$	$\gamma_{ij} = \displaystyle\int \frac{\partial^2\, w(\mathbf{z})}{\partial z_i\,\partial z_j}\,\varphi(\mathbf{z})\,d\mathbf{z}$
Gradients appear as coefficients in fitness regressions	
$w(\mathbf{z}) = 1 + \sum \beta_j\,(z_j - \mu_j)$	$w(\mathbf{z}) = a + \sum b_j\,(z_j - \mu_j) + \frac{1}{2}\sum_{j,k}\gamma_{jk}\,(z_j - \mu_j)(z_k - \mu_k)$
$w(\mathbf{z}) = 1 + \boldsymbol{\beta}^T(\mathbf{z} - \boldsymbol{\mu})$	$w(\mathbf{z}) = a + \mathbf{b}^T(\mathbf{z} - \boldsymbol{\mu}) + \frac{1}{2}(\mathbf{z} - \boldsymbol{\mu})^T\boldsymbol{\gamma}\,(\mathbf{z} - \boldsymbol{\mu})$
$w(\mathbf{z}) = 1 + \boldsymbol{\beta}_{sd}^T\mathbf{z}_{sd}$	$w(\mathbf{z}) = a + \mathbf{b}_{sd}^T\mathbf{z}_{sd} + \frac{1}{2}\mathbf{z}_{sd}^T\,\boldsymbol{\gamma}_{sd}\,\mathbf{z}_{sd}$
$\boldsymbol{\beta}$ = slope of the best linear fit	$\boldsymbol{\gamma}$ = the quadratic coefficient of the best quadratic fit. $\mathbf{b} = \boldsymbol{\beta}$ when $\mathbf{z} \sim$ MVN
Gradients appear as coefficients in evolutionary equations when $(\mathbf{z}, \mathbf{g}) \sim$ MVN	
$\Delta\boldsymbol{\mu} = \mathbf{G}\boldsymbol{\beta}$	$\mathbf{G}^* - \mathbf{G} = \mathbf{G}\left(\boldsymbol{\gamma} - \boldsymbol{\beta}\boldsymbol{\beta}^T\right)\mathbf{G}$

in the quadratic regression also appear as the coefficients of equations for predicting evolutionary change (Table 30.1). We will briefly review some statistical issues of fitting such regressions before examining the geometry of multivariate quadratic regressions.

Estimation, Hypothesis Testing, and Confidence Intervals

Even if we can assume that a best-fitting quadratic is a reasonable approximation of the individual fitness surface, we are still faced with a number of statistical issues. For k traits, the full quadratic regression (Equation 30.13) involves $k(k+3)/2$ parameters: $k(k+1)/2$ from $\boldsymbol{\gamma}$ (k γ_{ii} and $k[k-1]/2$ symmetric γ_{ij} terms) and k from $\boldsymbol{\beta}$. With 5, 10, and 25 characters, this corresponds to 20, 65, and 350 parameters, respectively. Hence, the number of observations should be $n \gg k(k+3)/2$ (ideally, by at least an order of magnitude) in order to estimate these parameters with any precision. Unless we test for, and confirm, trait multivariate normality (Appendix 5), $\boldsymbol{\beta}$ must be estimated from the best *linear* multiple regression, as the vector of linear slopes in a *quadratic* regression need not equal $\boldsymbol{\beta}$. Further, while the literature suggests that the estimate for $\boldsymbol{\gamma}$ is obtained from the best quadratic regression (Equation 30.13), Equation 29.28b showed that this estimate is biased by the presence of skew. Following Equation 29.29e, a strictly quadratic regression (i.e., with no linear terms),

$$w = a + \frac{1}{2}\, \mathbf{z}_{sd}^T\, \gamma_{sd}\, \mathbf{z}_{sd} + e$$

is not biased by skew, but is still a biased estimator for $\boldsymbol{\gamma}$ if the multivariate pattern of kurtosis differs from that for a multivariate normal (Equation 29.29f).

A second problem is **multicollinearity**—if many of the characters being measured are highly correlated, the phenotypic covariance matrix can be nearly singular, so even small errors in estimating $\mathbf{P}$ result in large differences in $\mathbf{P}^{-1}$. This, in turn, results in a very large sampling variance for the estimates of $\boldsymbol{\beta}$ and $\boldsymbol{\gamma}$ (which translates into their instability). A quick check for multicollinearity is to regress each trait on all the others. Subtracting the resulting model R^2 from 1.0 determines the **tolerance**, with very high R^2 or low tolerances indicating that multicollinearity is likely to be an issue. One possible solution is to use principal components (Appendix 5) to extract a subset of the characters (measured as PCs; namely, specific linear combinations of the characters) that explains most of the phenotypic variance of $\mathbf{P}$. Fitness regressions using the first few PCs as the characters can then be computed (Lande and Arnold 1983). This approach also reduces the problem of the number of parameters to estimate, but it risks the real possibility of removing the most important characters. Past strong selection may have eroded away much of the variation (Chapters 5, 16, 25, and 26), resulting in such traits being either excluded in a PC set or spread (with weak effects) over several indices of current traits. A further complication is that PCs—weighted indices of trait values—are often difficult to interpret biologically. While the first PC of $\mathbf{P}$ for morphological characters generally corresponds to a general measure of size (but see Somers 1986), the others are typically much more problematic to interpret. Finally, using PCs can spread the effects of direct selection on one character over several PCs, further complicating interpretation. While using the PCs of the phenotypic covariance matrix, $\mathbf{P}$, can be problematic, we will show that the PCs associated with the matrix of quadratic selection gradients, $\boldsymbol{\gamma}$, can provide considerable insight into the nature of selection.

A variety of additional concerns regarding fitness regressions were discussed in Chapter 29. Briefly, residuals of fitness regressions are expected by their nature to be poorly behaved, so using standard methods of confidence intervals on regression coefficients is often not appropriate. Mitchell-Olds and Shaw (1987) and Mitchell-Olds (1989) suggested using the delete-one jackknife method for approximating confidence intervals for coefficients in quadratic regressions when the residuals are not normal. Likewise, the discussions of randomization tests and cross-validation procedures in Chapter 29 extend to multivariate regressions in a straightforward manner. Multivariate tests of the presence of a single mode in the fitness surface were discussed by Mitchell-Olds and Shaw (1987), and we will introduce the Box-Hunter confidence volume for the stationary point on a quadratic fitness surface shortly (after first introducing some required matrix machinery).

Regression Packages and Coefficients of $\boldsymbol{\gamma}$

The coefficients of the elements of $\boldsymbol{\gamma}$ have a form that may be different from the output of a quadratic regression package. Suppose we have two variables with a mean of zero. Under a Lande-Arnold regression, the quadratic contribution to fitness, w, is

$$\left(\frac{\gamma_{11}}{2} z_1\right) + \left(\frac{\gamma_{22}}{2} z_2\right) + (\gamma_{12} \cdot z_1 \cdot z_2)$$

However, many regression packages output the quadratic coefficients as

$$(b_{11} \cdot z_1) + (b_{22} \cdot z_2) + (b_{12} \cdot z_1 \cdot z_2)$$

In such cases, $\gamma_{ii}/2 = b_{ii}$, or $\gamma_{ii} = 2b_{ii}$, while $\gamma_{ij} = b_{ij}$ for $i \neq j$. Failure to make this correction results in the reported γ_{ii} coefficients being only half their true value, thus *underestimating* the strength of quadratic selection on z_i (Lande and Arnold 1983; Stinchcombe et al. 2008). Just how widespread this mistake is in the literature remains unclear, but it may be the rule rather than the exception (at least for papers published before 2008). Indeed, Stinchcombe et al. found that almost 80% of the 33 studies they examined made this error.

Geometric Aspects

Despite their apparent simplicity, multivariate quadratic fitness regressions have a rather rich geometric structure. By adjusting the character values to give them a mean of zero, the general quadratic fitness regression can be written as

$$w(\mathbf{z}) = a + \sum_{i=1}^{n} b_1 z_i + \frac{1}{2} \sum_{i=1}^{n} \sum_{j=1}^{n} \gamma_{ij} z_i z_j = a + \mathbf{b}^T \mathbf{z} + \frac{1}{2} \mathbf{z}^T \boldsymbol{\gamma} \mathbf{z} \tag{30.19}$$

If $\mathbf{z} \sim$ MVN, then $\mathbf{b} = \boldsymbol{\beta}$ (the vector of coefficients of the best *linear* fit). Note that if we regard Equation 30.19 as a second-order Taylor series approximation (Equation A6.7) of $w(z)$, then $\mathbf{b}$ and $\boldsymbol{\gamma}$ can be interpreted as the gradient and Hessian, respectively, of individual fitness evaluated at the population mean (here $\boldsymbol{\mu} = \mathbf{0}$ by construction). Even though a quadratic is the simplest curved surface, its geometry can still be difficult to visualize (Phillips and Arnold 1989; Brodie et al. 1995). The key is that the nature of curvature of Equation 30.19 is determined by the eigenvalues of the $\boldsymbol{\gamma}$ matrix.

We start our exploration of this geometry by considering the gradient of this best-fitting quadratic fitness surface. Applying Equations A6.1b and A6.1c to Equation 30.19 yields

$$\nabla_{\mathbf{z}}[\, w(\mathbf{z})\,] = \mathbf{b} + \boldsymbol{\gamma}\mathbf{z} \tag{30.20a}$$

Hence, the direction of steepest ascent on the fitness surface (the direction in which to move in phenotype space to maximally increase local individual fitness) around $\mathbf{z}$ is the vector $\mathbf{b} + \boldsymbol{\gamma}\mathbf{z}$. If the true individual fitness surface is indeed a quadratic, the average gradient of individual fitness taken over the distribution of phenotypes is

$$\int \nabla_{\mathbf{z}}[\, w(\mathbf{z})\,]\, p(\mathbf{z})\, d\mathbf{z} = \mathbf{b} \int p(\mathbf{z})\, d\mathbf{z} + \boldsymbol{\gamma} \int \mathbf{z}\, p(\mathbf{z})\, d\mathbf{z} = \mathbf{b} \tag{30.20b}$$

as the last integral is $\boldsymbol{\mu}$ (which is zero by construction). Hence, if the true fitness function is quadratic, the average gradient of individual fitness is given by $\mathbf{b}$, independent of the distribution of $\mathbf{z}$.

Solving for $\nabla_{\mathbf{z}}[\, w(\mathbf{z})\,] = \mathbf{0}$, shows that a point, $\mathbf{z}_0$, that satisfies $\boldsymbol{\gamma}\,\mathbf{z}_0 = -\mathbf{b}$ is a candidate for a local extremum (also called a **stationary point**, as the gradient is zero). When $\boldsymbol{\gamma}$ is nonsingular,

$$\mathbf{z}_0 = -\,\boldsymbol{\gamma}^{-1}\mathbf{b} \tag{30.21a}$$

is the unique stationary point of this quadratic surface. Substituting into Equation 30.19, the expected individual fitness at this point is

$$w_0 = a + \frac{1}{2}\mathbf{b}^T \mathbf{z}_0 \tag{30.21b}$$

as obtained by Phillips and Arnold (1989). Because $\partial^2 w(\mathbf{z})/\partial z_i\, \partial z_j = \gamma_{ij}$, the Hessian of $w(\mathbf{z})$ is just $\boldsymbol{\gamma}$. Thus, $\mathbf{z}_0$ is a local minimum if $\boldsymbol{\gamma}$ is **positive-definite** (all eigenvalues are positive), a local maximum if γ is **negative-definite** (all eigenvalues are negative), or a saddle point if the eigenvalues differ in sign (see Equation A6.8b).

If γ is singular (i.e., it has at least one zero eigenvalue), then there is no unique stationary point. An example of this is seen in Figure 30.3B, where there is a ridge (rather than a single point) of phenotypic values having the highest fitness value. The consequence of a zero eigenvalue is that the fitness surface has no curvature along the axis that is defined by the associated eigenvector. If γ has k zero eigenvalues, then the fitness surface has no curvature along k dimensions. The remaining fitness space showing curvature has a single stationary point, which is given by Equation 30.21a for γ and $\mathbf{b}$ when reduced to the $n-k$ dimensions showing curvature.

A Brief Digression: Orthonormal and Diagonalized Matrices

We need some additional machinery on the geometry of matrices (from Appendix 5) to further our discussion of the geometry of the quadratic fitness surface. Matrix transformations (multiplying a vector by a matrix) consist of two basic operations: **rotations** (changes in the direction of the vector) and **scalings** (changes in its length). A transformation can be partitioned into these two basic operations by using **orthonormal** matrices. If we write a square matrix as $\mathbf{U} = (\mathbf{u}_1, \mathbf{u}_2, \cdots, \mathbf{u}_n)$, where each $\mathbf{u}_i$ is an n-dimensional column vector, $\mathbf{U}$ is said to be orthonormal if

$$\mathbf{u}_i{}^T \mathbf{u}_j = \begin{cases} 1 & \text{if } i = j \\ 0 & \text{if } i \neq j \end{cases}$$

In other words, each column of $\mathbf{U}$ is independent from every other column and has unit length. Matrices with this property are also referred to as **unitary** and satisfy $\mathbf{U}^T = \mathbf{U}^{-1}$, so

$$\mathbf{U}^T\,\mathbf{U} = \mathbf{U}\,\mathbf{U}^T = \mathbf{I} \tag{30.22}$$

The transformation induced by an orthonormal matrix has a very simple geometric interpretation in that it is a **rigid rotation** of the original coordinate system—all axes of the original coordinates are simply rotated by the same angle to create the new coordinate system (Appendix 5). The angle between any two vectors remains unchanged following their transformation by the same orthonormal matrix. If the angle between the vectors $\mathbf{x}_1$ and $\mathbf{x}_2$ is θ, then the angle between the transformed vectors $\mathbf{y}_1 = \mathbf{U}\mathbf{x}_1$ and $\mathbf{y}_2 = \mathbf{U}\mathbf{x}_2$ is also θ (Equation A5.5d).

A symmetric matrix $\mathbf{A}$ (such as a variance-covariance matrix) can be **diagonalized** as

$$\mathbf{A} = \mathbf{U}\boldsymbol{\Lambda}\mathbf{U}^T \tag{30.23}$$

where $\boldsymbol{\Lambda}$ is a diagonal matrix and $\mathbf{U}$ is an orthonormal matrix ($\mathbf{U}^{-1} = \mathbf{U}^T$). If λ_i and $\mathbf{e}_i$, respectively, denote the ith eigenvalue and its associated unit-length eigenvector of $\mathbf{A}$, then

$$\boldsymbol{\Lambda} = \text{diag}(\lambda_1, \lambda_2, \cdots, \lambda_n) = \begin{pmatrix} \lambda_1 & 0 & \cdots & 0 \\ 0 & \lambda_2 & \cdots & 0 \\ \vdots & & \ddots & \vdots \\ 0 & \cdots & \cdots & \lambda_n \end{pmatrix} \tag{30.24a}$$

and

$$\mathbf{U} = (\,\mathbf{e}_1, \mathbf{e}_2, \cdots, \mathbf{e}_n\,) \tag{30.24b}$$

Geometrically, $\mathbf{U}$ describes a rigid rotation of the original coordinate system while $\boldsymbol{\Lambda}$ shows the amounts that unit lengths in the original coordinate system are scaled in the transformed system. Using the unitary property of $\mathbf{U}$, premultiplying $\mathbf{A}$ by $\mathbf{U}^T$ and then postmultiplying by $\mathbf{U}$ results in a diagonal matrix whose elements are the eigenvalues of $\mathbf{A}$,

$$\begin{aligned} \mathbf{U}^T\mathbf{A}\mathbf{U} &= \mathbf{U}^T(\mathbf{U}\boldsymbol{\Lambda}\mathbf{U}^T)\mathbf{U} = (\mathbf{U}^T\mathbf{U})\boldsymbol{\Lambda}(\mathbf{U}^T\mathbf{U}) \\ &= \boldsymbol{\Lambda} \end{aligned} \tag{30.25}$$

The effect of using such a transformation is that (on this new scale) we remove all cross-product terms in a quadratic product (i.e., the $z_i z_j$ terms for $i \neq j$ in Equation 30.19 are absent). Put another way, ***on this new scale, there is no correlational selection***, as $\gamma_{ij} = 0$ for $i \neq j$. A few very useful results immediately follow from Equation 30.25. For $\mathbf{A}^{1/2}$ and $\mathbf{A}^{-1}$, the $\mathbf{U}$ matrix is unchanged, while the diagonal elements in the associated $\boldsymbol{\Lambda}$ matrix are given by the square root or inverse, respectively. Thus, $\mathbf{A}$, $\mathbf{A}^{1/2}$, and $\mathbf{A}^{-1}$ (provided the latter exists; i.e., no there are zero eigenvalues) all have the same eigenvectors and their eigenvalues are related as λ_i, $\lambda_i^{1/2}$, and λ_i^{-1}, respectively.

Example 30.4. Consider selection acting on two characters, z_1 and z_2. Suppose we find that $\gamma_{11} = -2$ and $\gamma_{22} = -1$, suggesting that the individual fitness surface has negative curvature in both z_1 and z_2. At first glance the picture this evokes is stabilizing selection on both z_1 and z_2, with the stabilizing selection surface perhaps rotated due to selection for correlations between z_1 and z_2. The first caveat, as mentioned in Chapter 29, is that negative curvature (concavity), by itself, does not imply a local maximum. Even if $\boldsymbol{\gamma}$ is negative definite (all $\lambda_i < 0$; Appendix 5), the location, $\mathbf{z}_0$, of the maximum in the quadratic surface may be *outside* of the observed range of population values and hence not currently applicable to the population being studied. A much more subtle point is that, as Figure 30.3 shows, the nature of the fitness surface is very much dependent on the amount of selection for correlations between z_1 and z_2. Figure 30.3 considers the surfaces associated with the same values for γ_{11} and γ_{22}, but three different values of γ_{12} under the assumption that $\mathbf{b} = 0$. Note that although in all three cases, $\gamma_{12} > 0$ (i.e., selection favors increased correlations between the phenotypic values of z_1 and z_2), the fitness surfaces are *qualitatively very different*. When $\gamma_{12} = 0.25$, the individual fitness surface indeed shows stabilizing selection in both characters. For $\gamma_{12} = \sqrt{2} \simeq 1.42$, the fitness surface has a ridge in one direction, with stabilizing selection in the other. When $\gamma_{12} = 4$, the fitness surface is a saddle, with convex selection along one axis and concave selection along the other. An especially troubling point is that if the standard error of γ_{12} is sufficiently large, we will not be able to distinguish between these very different types of selection even if we could show that $\gamma_{11}, \gamma_{22} < 0$, and $\gamma_{12} > 0$.

Canonical Transformation of $\boldsymbol{\gamma}$

While the curvature of a quadratic fitness surface is completely determined by $\boldsymbol{\gamma}$, it is easy to be misled about the actual nature of the fitness surface if one attempts to infer its multivariate structure from a simple inspection of the diagonal elements of $\boldsymbol{\gamma}$. As Figure 30.3 shows, even for two characters, visualizing the individual fitness surface is not trivial and can easily be extremely misleading. The problem is that the cross-product terms (γ_{ij} for $i \neq j$) make the quadratic form difficult to interpret geometrically. Removing these terms by a change of variables, so that the axes of new variables coincide with the axes of symmetry of the quadratic form (its **canonical axes**), greatly facilitates visualization of the fitness surface.

Motivated by this observation, Phillips and Arnold (1989) suggested using two slightly different versions of the canonical transformation of $\boldsymbol{\gamma}$ to clarify the geometric structure of the best fitting quadratic fitness surface. Applying Equation 30.25, if we consider the matrix, $\mathbf{U}$, whose columns are the eigenvectors of $\boldsymbol{\gamma}$, then the transformation $\mathbf{y} = \mathbf{U}^T\mathbf{z}$ (and hence $\mathbf{z} = \mathbf{U}\mathbf{y}$, because $\mathbf{U}^{-1} = \mathbf{U}^T$ as $\mathbf{U}$ is orthonormal) removes all the cross-product terms in the quadratic form, and returns

$$\begin{aligned} w(\mathbf{z}) &= a + \mathbf{b}^T\mathbf{U}\,\mathbf{y} + \frac{1}{2}\,(\mathbf{U}\mathbf{y})^T\boldsymbol{\gamma}(\mathbf{U}\mathbf{y}) = a + \mathbf{b}^T\mathbf{U}\,\mathbf{y} + \frac{1}{2}\,\mathbf{y}^T\left(\mathbf{U}^T\boldsymbol{\gamma}\mathbf{U}\right)\mathbf{y} \\ &= a + \mathbf{b}^T\mathbf{U}\,\mathbf{y} + \frac{1}{2}\,\mathbf{y}^T\boldsymbol{\Lambda}\,\mathbf{y} = a + \sum_{i=1}^{n}\theta_i\,y_i + \frac{1}{2}\sum_{i=1}^{n}\lambda_i\,y_i^2 \end{aligned} \tag{30.26}$$

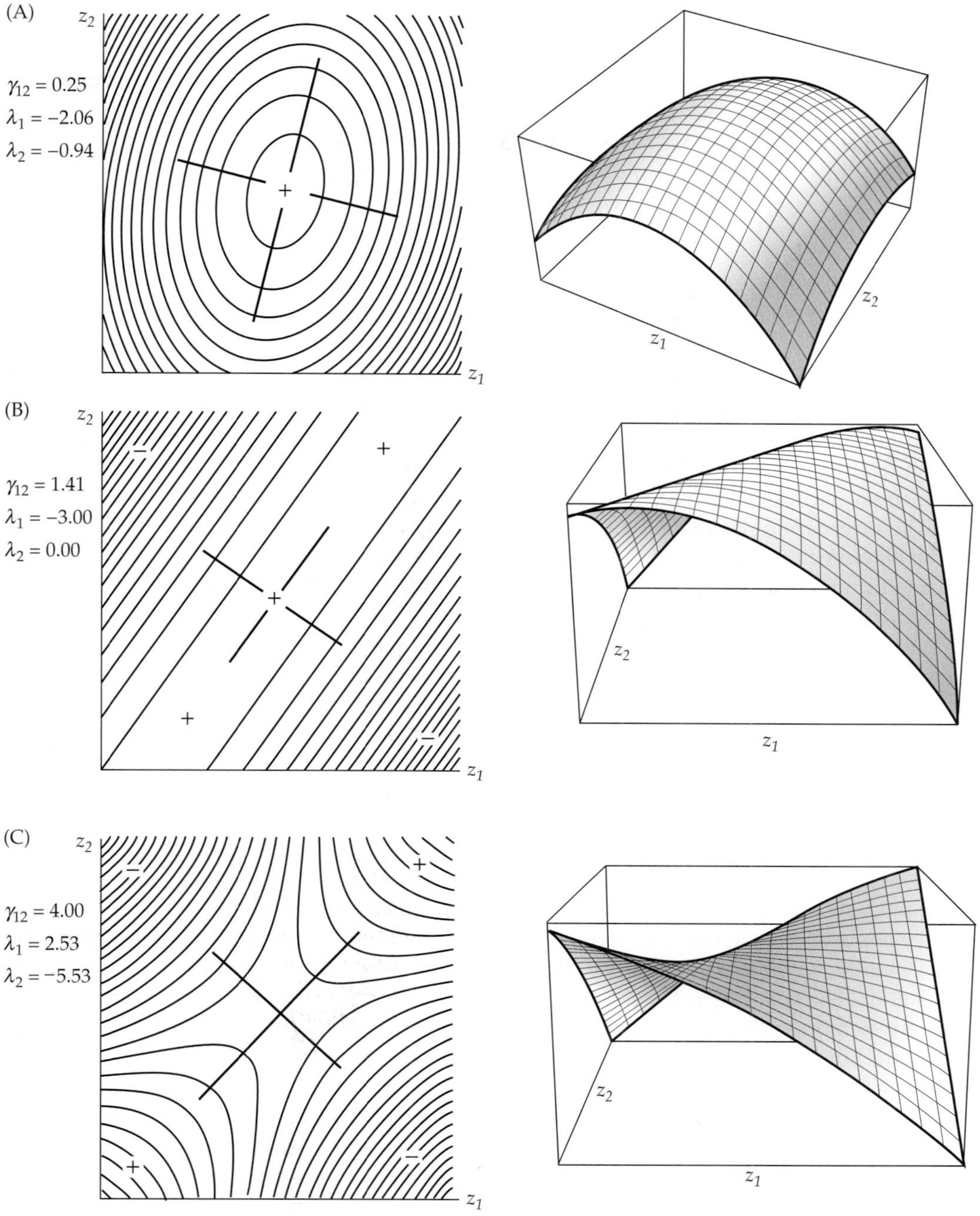

Figure 30.3 Three quadratic fitness surfaces, in all of which $\gamma_{11} = -2$, $\gamma_{22} = -1$, and $\mathbf{b} = \mathbf{0}$. On the left are fitness contour plots, with peaks represented by $+$ and valleys by $-$. **Axes of symmetry** of the surface (the **canonical**, **major**, or **principal axes** of γ; Equation 30.24b) are denoted by the thick lines. These axes correspond to the eigenvectors of γ (with their corresponding eigenvalues, λ_1 and λ_2, also shown in the figure). On the right are three-dimensional plots of individual fitness as a function of the phenotypic values of the characters z_1 and z_2. **A:** $\gamma_{12} = 0.25$. This corresponds to stabilizing selection on both characters, with fitness falling off more rapidly (as indicated by the shorter distance between contour lines) along the z_1 axis than along the z_2 axis. **B:** Here $\gamma_{12} = \sqrt{2} \simeq 1.41$, in which case γ is singular (as $\lambda_2 = 0$). The resulting fitness surface shows a ridge in one direction ($\mathbf{e}_2$, the eigenvector corresponding to $\lambda_2 = 0$) and strong stabilizing selection in the orthogonal direction (given by $\mathbf{e}_1$). **C:** When $\gamma_{12} = 4$, the fitness surface shows a saddle point, and there is concave selection along one of the canonical axes of the fitness surface ($\mathbf{e}_2$, corresponding to $\lambda_2 = -5.53$) and convex selection along the other ($\mathbf{e}_1$, corresponding to $\lambda_1 = 2.53$).

where $\theta_i = \mathbf{e}_i^T \mathbf{b}$ and $y_i = \mathbf{e}_i^T \mathbf{z}$, with λ_i and $\mathbf{e}_i$, respectively, representing the eigenvalues and associated unit eigenvectors of $\boldsymbol{\gamma}$. Alternatively, if a stationary point, $\mathbf{z}_0$, exists (e.g., $\boldsymbol{\gamma}$ is nonsingular, all $\lambda_i \neq 0$), the change of variables, $\mathbf{y} = \mathbf{U}^T(\mathbf{z} - \mathbf{z}_0)$, further removes all linear terms (Box and Draper 1987), so

$$w(\mathbf{z}) = w_0 + \frac{1}{2}\mathbf{y}^T \boldsymbol{\Lambda}\,\mathbf{y} = w_0 + \frac{1}{2}\sum_{i=1}^{n} \lambda_i\, y_i^2 \tag{30.27}$$

where $y_i = \mathbf{e}_i{}^T(\mathbf{z} - \mathbf{z}_0)$ and w_0 is shown by Equation 30.21b. Equation 30.26 is called the **A canonical form** and Equation 30.27, the **B canonical form** (Box and Draper 1987). Both forms represent a rotation of the original axis to the new set of coordinates (the canonical axes of $\boldsymbol{\gamma}$) that align them with axes of symmetry of the original quadratic surface. The B canonical form further shifts the origin to the stationary point, $\mathbf{z}_0$. Because the contribution to individual fitness from $\mathbf{b}^T\mathbf{z}$ is a hyperplane, its effect is to tilt the fitness surface. The B canonical form effectively levels this tilting, thus allowing us to focus entirely on the curvature (quadratic) aspects of the fitness surface.

The orientation (the **principal**, or **major, axes**) of the quadratic surface is determined by the eigenvectors $(\mathbf{e}_1, \cdots, \mathbf{e}_n)$ of $\boldsymbol{\gamma}$, while the eigenvalues $(\lambda_1, \cdots, \lambda_n)$ of $\boldsymbol{\gamma}$ determine the nature and amount of curvature of the surface along each canonical axis. Along the axis defined by $y_i = \mathbf{e}_i^T \mathbf{z}$, the individual fitness function has positive curvature (is convex) if $\lambda_i > 0$. It has negative curvature (is concave) if $\lambda_i < 0$, and no curvature (is a plane) if $\lambda_i = 0$. The amount of curvature is indicated by the magnitude of λ_i; the larger $|\lambda_i|$, the more extreme the curvature. An alternative way to envision the canonical transformation is that the original vector, $\mathbf{z}$, of n characters is transformed into a vector, $\mathbf{y}$, of n independent selection indices (Simms 1990). Directional selection on the index, y_i, is measured by θ_i, while quadratic selection on y_i is measured by λ_i.

Returning to Figure 30.3, we see that the axes of symmetry of the quadratic surface are the canonical axes of $\boldsymbol{\gamma}$. For $\gamma_{12} = 0.25$ (Figure 30.3A), $\lambda_1 = -2.06$ and $\lambda_2 = -0.94$, and so the fitness surface is concave along each canonical axis, with more extreme curvature along the y_1 axis. When $\gamma = \sqrt{2}$ (Figure 30.3B), one eigenvalue is zero while the other is -3, so the surface shows no curvature along one axis (it is a plane) but is strongly concave along the other. Finally, when $\gamma_{12} = 4$ (Figure 30.3C), the two eigenvalues differ in sign, as they are -5.53 and 2.53. This generates a saddle point, with a surface that is concave along one axis and convex along with other, and here with the concave curvature being more the extreme.

From Equation 30.26, we can see that the fitness change along a particular axis ($\mathbf{e}_i$) is $\theta_i\, y_i + (\lambda_i/2)\, y_i^2$. If $|\theta_i| \gg |\lambda_i| > 0$, the curvature of the fitness surface along this axis is dominated by the effects of linear (as opposed to quadratic) selection for modest values of $y_i = \mathbf{e}_i^T \mathbf{z}$. If $\lambda_i = 0$, the fitness surface along y_i has no curvature, so the fitness surface is a ridge along this axis. If $\theta_i > 0$, this is a rising ridge (fitness increases as y_i increases), whereas it is a falling ridge (fitness decreases as y_i increases) if $\theta_i < 0$, and it is flat if $\theta_i = 0$. Even if $\boldsymbol{\gamma}$ is not singular, it may be nearly so, with some of the eigenvalues being very close to zero. In this case, the fitness surface shows little curvature along the axes given by the eigenvectors associated with these nearly zero eigenvalues. Further issues relating to the visualization of multivariate fitness surfaces are discussed in Phillips and Arnold (1989), while Box and Draper (1987) review the statistical foundations of this approach.

Example 30.5. Consider the first two scenarios depicted in Figures 30.3A and 30.3B. The resulting $\boldsymbol{\gamma}$ and the component matrices for its diagonalization (which are easily obtained using the `eigen` function in `R`) are as follows:

$$\boldsymbol{\gamma}_1 = \begin{pmatrix} -2 & 0.25 \\ 0.25 & -1 \end{pmatrix}, \quad \mathbf{U}_1 = \begin{pmatrix} 0.230 & 0.973 \\ 0.973 & -0.230 \end{pmatrix}, \quad \boldsymbol{\Lambda}_1 = \begin{pmatrix} -2.06 & 0 \\ 0 & -0.94 \end{pmatrix}$$

and

$$\gamma_2 = \begin{pmatrix} -2 & 1.41 \\ 1.41 & -1 \end{pmatrix}, \quad \mathbf{U}_2 = \begin{pmatrix} 0.577 & 0.816 \\ 0.816 & -0.577 \end{pmatrix}, \quad \boldsymbol{\Lambda}_2 = \begin{pmatrix} 0 & 0 \\ 0 & -3.00 \end{pmatrix}$$

Because $\mathbf{U} = (\mathbf{e}_1 \;\; \mathbf{e}_2)$, the transformed variables, $y_i = \mathbf{e}_i^T\mathbf{z}$, for γ_1 (for Figure 30.3A) are

$$y_1 = \mathbf{e}_1^T\mathbf{z} = 0.230 \cdot z_1 + 0.973 \cdot z_2, \qquad y_2 = \mathbf{e}_2^T\mathbf{z} = 0.973 \cdot z_1 - 0.230 \cdot z_2$$

where the quadratic term now becomes

$$\frac{1}{2}\left(-2.06\, y_1^2 - 0.94\, y_2^2\right) = -1.03y_1^2 - 0.47y_2^2$$

For γ_2 (Figure 30.3B), there is a zero eigenvalue, corresponding to no curvature. This occurs in the direction of $y_1 = \mathbf{e}_1^T\mathbf{z} = 0.577 \cdot z_1 + 0.816 \cdot z_2$ (where $\mathbf{e}_1$ is the eigenvector associated with the zero eigenvalue), while the curvature in the direction of $y_2 = \mathbf{e}_2^T\mathbf{z} = 0.816 \cdot z_1 - 0.577 \cdot z_2$ is given by $-(3.00/2)y_2^2$.

Are Traits Based on Canonical Axes Meaningful?

While there clearly are significant benefits from using the canonical rotation of γ to infer those trait combinations that are under the strongest quadratic selection, this approach has also sparked a lively debate in the literature. Blows (2007a, 2007b) championed it as providing considerable insight into the nature of selection, while Conner (2007) suggested that "these advantages are usually outweighed by the disadvantage that the results are not very biologically interpretable," a point echoed by Hunt et al. (2007a). Basically, the concern these authors expressed parallels issues about the use of principal components: even though a specific combination of traits may account for most of the variation, their biological interpretation may be convoluted (at best). This argument raises two questions: What are the true targets of selection, and what is a trait?

The power of quantitative genetics is that *anything* we can measure can be regarded as a trait, no matter how strange or seemingly biologically unreasonable it may be. Clearly, ecologists and evolutionary biologists working on specific traits (such as clutch size or body weight) bring a wealth of empirical knowledge about these traits when considering possible targets of selection. In this sense, field biologists regard many traits as natural objects. While most would agree that some are (e.g., clutch size), other traits (such as body shape) are more problematic, as they can be defined and measured in a myriad of different ways. At a deeper level, it is the nature of the question that usually determines whether a biologist regards a particular trait as a natural object. Even when considering the same general features, a developmental biologist's view of natural traits may be quite different from an ecologist's view, and in turn both views may be different from those of an evolutionary biologist.

Thus, when a trait is not regarded as a natural object, but rather is some weighted combination of values of natural objects, biologists may feel that much of their intuitive and empirical knowledge about the individual components is diffused over some seemingly arbitrary combination of their values. This is not an unreasonable view. However, selection is not reasonable in that it does not care about how traits are defined; it simply acts on particular multivariate phenotypes. When selection is acting on a complex structure in a medium- or high-dimensional space, simply examining the fitness of projections from this space onto some subset of lower-dimensional traits can be extremely misleading (Walsh 2007; Blows and Walsh 2009). From the perspective of selection, the natural objects are the linear combinations of trait values that comprise the canonical axes of γ.

It is this connection with the axes of natural selection that imposes a very real difference between concerns about the interpretation of PCs for a phenotypic covariance matrix and the major axes of γ. The latter defines a real object of ecological importance, namely how selection views the traits under selection, while the former define axes of existing variation, which may (or may be) be attributable to selection. Using PCs from the phenotypic covariance matrix to define new traits can diffuse a true target of selection, as the PCs are used simply to deal with phenotypic correlations. In contrast, the canonical axes of γ specifically *highlight* the targets of selection. As Sewall Wright (1935b) insightfully noted: "It is the harmonious adjustment of all of the characertistics of the organism that is the object of selection, not the separate metrical 'characters.'"

Strength of Selection: γ_{ii} Versus λ

Recall from Equation 30.11a that $(\gamma_{ii}/2)(z_i - \mu_i)^2$ is the contribution toward relative fitness, w, from squared deviations of trait i from its mean. It is therefore natural to assume that if $\gamma_{ii} < 0$ (concave selection), this implies at least the potential of stabilizing selection on trait i. Figure 30.3 showed that using only the diagonal elements of γ can potentially give a very misleading picture of the nature of quadratic selection. However, the eigenvalues (λ) of γ provide a more exact description of the true nature of selection. Blows and Brooks (2003) stressed this point, and noted in an analysis of 19 studies that $|\,\gamma_{ii}\,|_{max} < |\,\lambda\,|_{max}$. Thus, studies that report weak values for quadratic selection are potentially biased if they use γ_{ii} values, rather than the full geometry of γ, as described by its eigenvalues. A further point (mentioned above) is that many published studies report only half the true value for λ_{ii} due to incorrect translation of the coefficient of the quadratic regression.

Blows and Brooks (2003) noted several advantages of focusing on estimation of the λ_i versus estimation of all of the γ_{ij}, noting that there are n eigenvalues, and $n(n+1)/2$ elements in γ. Further, given that many eigenvalues may be close to zero, a **subspace** of γ, such as the space spanned by the first few principal (i.e., canonical) components of γ may essentially capture all of the relevant information on the quadratic fitness surface. Following Simms (1990) and Simms and Rausher (1993), Blows and Brooks suggested that estimation and hypothesis testing can occur if we first obtain the eigenvectors of γ, and then use them to generate the transformed variables $\mathbf{y} = \mathbf{U}^T\mathbf{z}$ in the quadratic regression given by Equation 30.26. This approach is often referred to as a **double regression**, as one first uses the eigenvectors of γ to generate $\mathbf{y}$ and then fits a quadratic regression of w using $\mathbf{y}$. The quadratic terms in the regression correspond to the eigenvalues of γ (Equation 30.26), and confidence intervals and significance levels can be conducted within the standard GLM framework (LW Chapter 8). Bisgaard and Ankenman (1996) provided a formal statistical framework for generating standard errors for the estimated λ_i when using this procedure. However, Reynolds et al. (2010) noted that the initial transformation (to generate $\mathbf{y}$) biases tests of significance of the λ_i, and they suggested a permutation method to obtain correct type-I error rates.

Kruuk and Garant (2007) noted that the **Mercer-Mercer theorem** (2000) states that the magnitude of the largest eigenvalue of γ is as least as great as the largest magnitude of the diagonal elements of γ. Thus, it is "algebraically inevitable" that at least one combination of traits will show stronger quadratic selection than any of the original traits (provided all values of $\gamma_{ii} \neq 0$). Nevertheless, the biological issue here is whether λ_i is *significantly* greater than γ_{ii} (as opposed to only being slightly larger). Reynolds et al. (2010) found higher power for detecting curvature (nonzero eigenvalues of γ) using canonical analysis than when testing each value of γ_{ii} separately.

Example 30.6. Brooks and Endler (2001) examined four color traits in male guppies associated with sexual selection. The estimated γ matrix was

$$\gamma = \begin{pmatrix} 0.032 & -0.016 & -0.028 & 0.103 \\ -0.016 & 0.0001 & 0.066 & -0.131 \\ -0.028 & 0.066 & -0.022 & -0.099 \\ 0.103 & -0.131 & -0.099 & 0.060 \end{pmatrix}$$

The diagonal elements suggest evidence for weak convex selection ($\gamma_{44} = 0.060$, $\gamma_{11} = 0.032$) and some evidence for very weak concave selection ($\gamma_{33} = -0.022$). However, the eigenvalues of γ are 0.262, 0.012, -0.077, and -0.123. Of these eigenvalues, only the leading one (0.262) is significantly different from zero, with an amount of convex selection over four times that suggested from the largest γ_{ii} value (0.060). The take-home message is that simply relying upon a visual inspection of the diagonal elements of γ can depict a very misleading view of the nature of selection.

Significance and Confidence Regions for a Stationary Point

Recall from Equation 30.21a that when γ is nonsingular (i.e., it contains no zero eigenvalues), then $\mathbf{z}_0 = -\gamma^{-1}\mathbf{b}$ is the unique extremum (stationary point) in the quadratic regression. If γ contains k zero eigenvalues, there is no curvature along the trait combinations given by the k associated eigenvectors (the fitness surface is a k-dimensional hyperplane along these directions), while there is curvature (and a unique stationary point) in the remaining fitness space. As mentioned, this stationary value is a maximum if all of the eigenvalues are negative (and hence a test for significance for a maximum is that all of the eigenvalues should be significantly less than zero). Likewise, it is a local minimum if all the eigenvalues are positive (which can also be tested in the same manner). If γ contains both (significant) positive and negative eigenvalues, then $\mathbf{z}_0$ is a saddle point. Thus, tests for maxima or minima are straightforward. But what about the confidence region (or more correctly, a confidence *volume*) for the location of the stationary point?

A classical result for a quadratic regression is the **Box-Hunter confidence region**. Suppose we let $\mathbf{d_z}$ denote the gradient vector of the best quadratic regression of w on $\mathbf{z}$, which (Equation 30.20a) is calculated by

$$\mathbf{d_z} = \nabla_{\mathbf{z}}[\, w(\mathbf{z})\,] = \mathbf{b} + \gamma\mathbf{z}$$

Suppose there are k traits and n observations, and the residuals to the quadratic regression are independent and homoscedastic normal variables (as mentioned, for fitness data, this assumption is usually problematic). Box and Hunter (1954) showed that a $100\,(1-\alpha)\%$ confidence volume is given by those vectors $\mathbf{z}$ that satisfy the quadratic inequality

$$\mathbf{d_z}^T\,\mathbf{V}^{-1}\mathbf{d_z} = (\mathbf{b} + \gamma\mathbf{z})^T\,\mathbf{V}^{-1}\,(\mathbf{b} + \gamma\mathbf{z}) \le k\,F_{1-\alpha,k,n-p} \tag{30.28}$$

where F is the $(1-\alpha)$ value for an F distribution with k and $n-p$ degrees of freedom ($p = k(k+3)/2$ is the total number of estimated regression coefficients) and $\mathbf{V}$ is an estimate of the covariance matrix for $\mathbf{d_z}$. Values of $\mathbf{z}$ that satisfy Equation 30.28 are within the confidence volume for the stationary point. See Del Castillo and Cahya (2001) and Peterson et al. (2002) for further discussion and developments. It is critical to stress that these methods all make the strong assumption that fitness residuals are normally distributed, which typically fails. We strongly favor using aster models (Chapter 29), which correctly model the fitness distribution and for which Geyer and Shaw (2010a) developed likelihood-based approaches for hypothesis testing and confidence intervals.

Using Aster Models to Estimate Fitness Surfaces

Most of the approaches presented in this chapter (and in much of Chapter 29) make normality assumptions, which come in two different flavors: first, the assumed multivariate normality (MVN) of the distribution of *trait values*, $\mathbf{z}$; and second, the assumed MVN for the distribution of *residuals in the fitness regression*, $\mathbf{e} = w(\mathbf{z}) - \widehat{w}(\mathbf{z})$. We address these in turn.

As summarized in Table 30.1, when $\mathbf{z} \sim$ MVN, a variety of results hold that relate β and γ to measures of the geometries of the individual fitness surface and the mean fitness landscape. Further, as with the univariate case, for β and γ to equal, respectively, the linear and quadratic coefficients in a quadratic fitness regression also requires $\mathbf{z}$ to be MVN (lack of skew and the kurtosis of a normal, see Equations 29.28a and 29.28b).

Perhaps the more critical normality assumption involves the distributional behavior of the *fitness residuals*. As mentioned in Chapter 29, the assumption that $\mathbf{z} \sim$ MVN is *not* required to *fit* a regression, although when $\mathbf{z}$ is MVN, one can then equate estimated regression coefficients with the properties just mentioned. However, the assumption of multivariate normality of the fitness residuals is required, either directly (for hypotheses testing and constructing confidence intervals) or indirectly (the OLS assumption of homoscedastic residues is satisfied when residuals are MVN). Unfortunately, as mentioned in Chapter 29, most biologically realistic models for fitness components generate residuals that are both heteroscedastic and nonnormal.

As introduced in Chapter 29, Aster models (Geyer et al. 2007; Geyer and Shaw 2008, 2010a, 2010b; Shaw et al. 2008; Geyer 2010; Shaw and Geyer 2010) allow for a very wide range of distributional assumptions for the fitness-component residuals, and allow one to build up the distribution of total fitness by convoluting distributions across selection episodes. By fully, and correctly, accounting for the residua error structure, Aster models are far more statistically rigorous than Lande-Arnold estimation and can result in less bias when estimating the fitness surface. For example, in a simulated two-trait dataset fit by Shaw and Geyer (2010), Aster models correctly detected multivariate stabilizing selection (two negative eigenvalues for γ), while a Lande-Arnold regression of the same dataset suggested a saddle point (one positive and one negative eigenvalue). An OLS regression assumes that each value is equally weighted (the homoscedastic residual assumption), but when observations vary in their quality (such as heteroscedastic fitness residuals), weighting them properly (as done in Aster models) reduces bias.

While Aster models accommodate complex residual structures, they are not a panacea when it comes to estimating nonlinear fitness surfaces. In the fitness-estimation literature, *nonlinear* is often taken as synonymous with *quadratic*, but we have seen how misleading quadratic approximations can be (Figure 29.10). While the link functions used by the GLMs of an Aster model (transforming some underlying linear model into the expected data scale; Chapter 29) can induced nonlinearities (curvature), the transformation is also montonic (Geyer et al 2007; Geyer 2010, Shaw and Geyer 2010). Hence, while nonlinearities may be introduced, additional *peaks* are not. In their current form, Aster models assume a quadratic geometry (at most, a single extremum) when the underlying input is a quadratic function of trait values. Hence, many of the issues involved in estimating a fitness landscape by assuming a quadratic geometry remain when using the current Aster framework.

MULTIVARIATE SEMIPARAMETRIC FITNESS SURFACE ESTIMATION

As discussed in Chapter 29, using the best-fitting univariate quadratic can result in a very misleading picture when there are multiple peaks or sharp thresholds (such as truncation selection) in the individual fitness surface. This also holds in multivariate space, and hence regressions predicting fitness given a vector $\mathbf{z}$ of traits that are nonparameteric (free of any assumed functional form) certainly have some advantages. Because some assumption about the residual structure is typically needed to fit such regressions, they are formally called **semiparametric**, but we will still refer to them as nonparametric owing to the minimal number of assumptions they involve concerning about the functional form of $w(z)$.

Despite the advantage of having a minimal number of assumptions, these estimators also have significant disadvantages. We have already seen the difficulty in visualizing the multivariate fitness surface when only a simple quadratic function is assumed, but in this case the eigenvalues of γ provide significant help in interpretation. However, with general nonparametric methods, there is no corresponding metric for the fitness surface geometry, and thus one must resort to actual visualization of the fitness surface. This requires an examination of successive pairwise cross-sections of the fitness surface (projections onto two of the axes of the fitness function), generating a series of 3-D surfaces (one fitness axis and two trait axes; see Figure 30.4). Further, because these methods typically first construct new axes (linear combinations of the traits; e.g., $x_1 = \mathbf{a}_1^T\mathbf{z}$, $x_2 = \mathbf{a}_2^T\mathbf{z}$), and then use them to construct the

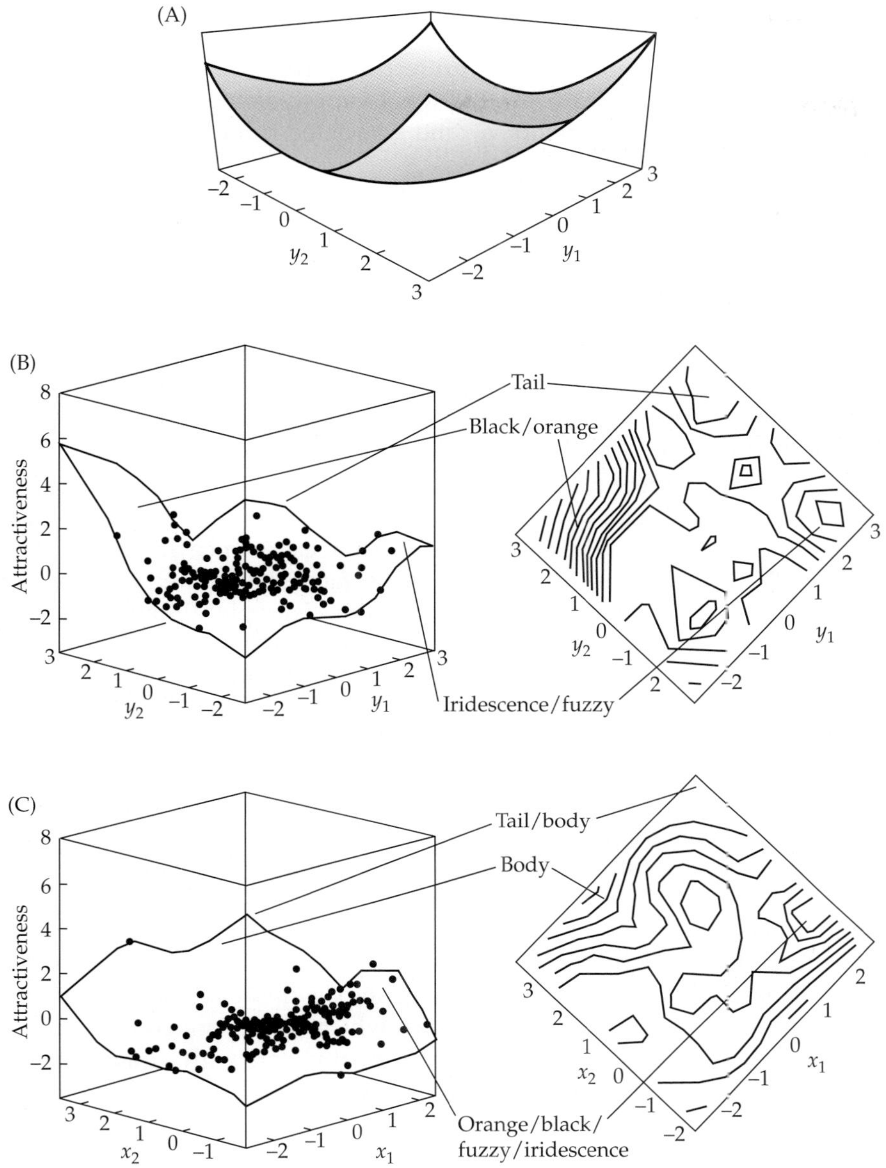

Figure 30.4 Visualization of the fitness surface for male attractiveness to female guppies as a function of color and size traits (for details see Example 30.7). **A:** The best-fitting quadratic regression. **B:** A thin-plate spline estimate of the fitness surface, using the axes given from the best quadratic regression ($y_i = \mathbf{e}_i^T \mathbf{z}$; Equation 30.26). **C:** The surface generated from projection-pursuit regressions (PPR) using the first two PPR axes ($x_i = \mathbf{a}_i^T \mathbf{z}$; Equation 30.29). While the thin-plate spline and PPR fitness surfaces look visually similar, it is important to recall that the trait loadings on their axes are different (see Example 30.7). The contour plots show the trait combinations that correspond to the different peaks. (After Blows et al. 2003.)

fitness surface, they display the pattern of selection on a series of composite traits, and not the actual traits themselves. This can make interpretation of the nature of selection on specific traits or combinations of traits problematic at best. Finally, the results from a nonparametric regression do not immediately provide coefficients to predict the response to selection, while (under normality assumptions) these follow automatically with a quadratic regression. Given these complementary strengths and weaknesses, researchers should use *both* quadratic and nonparametric regressions in the analysis of their selection data.

Projection-pursuit Regression and Thin-plate Splines

Schluter and Nychka (1994) extended Schluter's (1988) cubic spline univariate regressions (Chapter 29) to a vector of traits by using **projection-pursuit regression** (**PPR**; Friedman and Stuetzle 1981). The basic idea behind PPR is to approximate some complex function, $f(\mathbf{z})$, with a series of **projection vectors**, $\mathbf{a}_i$, and associated **ridge functions**, f_i,

$$\begin{aligned} f(\mathbf{z}) &\simeq f_1(\mathbf{a}_1^T\mathbf{z}) + f_2(\mathbf{a}_2^T\mathbf{z}) + \cdots + f_k(\mathbf{a}_k^T\mathbf{z}) \\ &= f_1(x_1) + f_x(x_2) + \cdots + f_k(x_k) \end{aligned} \tag{30.29}$$

Solutions require two numerically intensive steps: estimation of the best-fitting projection vectors $(\mathbf{a}_1, \cdots, \mathbf{a}_k)$, and then estimation of their associated ridge functions (Schluter and Nychka used cubic splines for the latter). Thus, one chooses optimality and smoothing criteria and then obtains the first projection vector while fitting a cubic spline to the data along this projection. One then moves on to fitting the second projection vector, and so on. The assumption is that a rather low-dimensional space captures most of the structure of the fitness surface, so the first few projection vectors are sufficient to approximate individual fitness. As with the eigenvectors of γ in a quadratic regression, the projection vectors $(\mathbf{a}_i)$ are those trait combinations (indices of trait values) that experience the strongest nonlinear selection. Visualization of the resulting complex surface is attempted by considering the pairwise projections on the first few projection vectors (Example 30.7 and Figure 30.4).

It is important to note that the projection vectors for a PPR do not necessarily correspond to the canonical axis of the γ matrix from a quadratic regression (Example 30.7). They do, however, correspond to the trait combinations under the strongest selection, but without the assumption of a quadratic fitness surface. As such, they represent a more general, and potentially natural, geometry for the axes of selection (Morrissey 2014b).

Another nonparametric approach is **thin-plate splines**, which are the two dimensional analog of univariate cubic splines. Using a projection onto two trait axes, thin-plate splines can be used to find the best-fitting 3-D surface on this reduced set of axes, thus employing another approach for visualizing complex surfaces.

The importance of allowing for more general fitness functions is highlighted by the work of Martin and Wainwright (2013), who examined fitness surfaces for a nascent adaptive radiation in a Bahamian clade of pupfish in the genus *Cyprinodon*. Using thin-plate splines, they found a fitness surface with multiple peaks, which corresponded to the morphology of existing populations in this radiation. Fitting the same data with a quadratic regression would have only resulted in a single peak (at most), which would have completely obscured important biological conclusions.

Example 30.7. Following up on the study of male color traits in guppies described in Example 30.6, Blows et al. (2003) examined six traits (two morphological and four color) for their role in female attractiveness. Both quadratic and projection-pursuit regressions were used to examine nonlinear selection on these traits. The trait loadings for the first two major axes of the quadratic regression of fitness on trait value (the eigenvectors $\mathbf{e}_1$ and $\mathbf{e}_2$, corresponding to the two leading eigenvectors of γ) and the first two projection vectors ($\mathbf{a}_1$ and $\mathbf{a}_2$) from projection-pursuit regression were as follows:

Trait	$\mathbf{e}_1$	$\mathbf{e}_2$	$\mathbf{a}_1$	$\mathbf{a}_2$
Body area	−0.065	0.157	0.185	0.551
Tail area	0.663	0.635	0.331	0.087
Black area	−0.245	0.645	0.430	−0.182
Fuzzy area	0.372	−0.173	0.101	−0.286
Iridescent area	0.436	−0.134	0.569	−0.757
Orange area	−0.411	0.333	0.581	−0.016

Note that these two sets of projection vectors from the different regressions are rather distinct. If we apply Equation A5.2b, the angle between $\mathbf{e}_1$ and $\mathbf{a}_1$ will be 88°, while the angle between $\mathbf{e}_2$ and $\mathbf{a}_2$ will be 87.5°. Thus, the major axes of the fitness surface are different (here, they are essentially orthogonal) if one uses a quadratic approximation as opposed to a projection-pursuit regression approximation. As Figure 30.4A shows, the best-fitting quadratic regression using $\mathbf{e}_1$ and $\mathbf{e}_2$ (where $y_1 = \mathbf{e}_1^T\mathbf{z}$ and $y_2 = \mathbf{e}_2^T\mathbf{z}$ are the trait values on this fitness surface for an individual with a trait vector, $\mathbf{z}$) shows disruptive selection, with fitness rising in all directions from a central minimum. If one uses these axes but then fits the data using thin-plate splines, the result is a rather different-looking fitness function (Figure 30.4B), but one that also shows multiple peaks corresponding to different combinations of the traits that females find attractive. When projection-pursuit regression is used (Figure 30.4C), the fitness surface (in rough appearance) is similar to that for thin-plate splines using the quadratic axes $(\mathbf{e}_1, \mathbf{e}_2)$. However, because the values for the two sets of axes ($x_1 = \mathbf{a}_1^T\mathbf{z}, x_2 = \mathbf{a}_2^T\mathbf{z}$ versus $y_1 = \mathbf{e}_1^T\mathbf{z}, y_2 = \mathbf{e}_2^T\mathbf{z}$) are rather different, this superficial visual appearance must be mapped into trait values (as indicated in Figure 30.4). Again, the result is that the fitness surface has multiple peaks.

Gradients for General Fitness Surfaces

Table 30.1 shows that when $\mathbf{z}$ is multivariate-normal, Lande-Arnold gradients describe the average geometry of an individual fitness surface and are the sole measure of selection in response equations (provided breeding values are Gaussian). When $\mathbf{z}$ is not multivariate normal, these features do not necessarily hold. We addressed the issue of how univariate generalized gradients are computed in such settings (and what they imply) in Chapter 29, and these same concepts easily extend to multivariate settings. Janzen-Stern gradients (Equations 29.35a and 29.35b) measure the average geometry of the individual fitness surface. Suppose we let $\mathbf{z}_i, \cdots, \mathbf{z}_n$ denote the n vectors of phenotypic observations. For trait i,

$$\beta_{JS,i} = E_{z_i}\left[\frac{\partial w(\mathbf{z})}{\partial z_i}\right] \simeq \frac{1}{n}\sum_{j=1}^{n} \left.\frac{\partial w(\mathbf{z})}{\partial z_i}\right|_{\mathbf{z}=\mathbf{z}_j} \tag{30.30a}$$

namely, the average value (over the n data vectors, $\mathbf{z}_1, \cdots, \mathbf{z}_n$) of the gradient with respect to the focal trait.

Morrissey-Sakrejda gradients (Equation 29.35d) are calculated based on the partials of the mean fitness landscape. Fitness-landscape partials are critical because these are what quantifies the nature of phenotypic selection in both the multivariate response equations when the distribution of breeding values is multivariate normal (Table 30.1) and also in the more general Barton-Turelli expression (Equation 24.26) for when normality is not assumed. The Morrissey-Sakrejda gradient for trait i is

$$\beta_{MS,i} = \frac{1}{\overline{\overline{W}}}\frac{\partial \overline{W}}{\partial \overline{z}_i}, \quad \text{where} \quad \overline{W} = \frac{1}{n}\sum_{j=1}^{n} W(\mathbf{z}_j) \tag{30.30b}$$

where, as above, there are n data vectors, with $W(\mathbf{z}_j)$ denoting the fitness associated with data vector $\mathbf{z}_j$. As in Chapter 29, these expressions apply to general fitness functions (such as those generated by PPR), and the required derivatives can be obtained numerically.

Calsbeek's Tensor Approach for Detecting Variation in Fitness Surfaces

When dealing with quadratic fitness surfaces, it is straightforward to test whether the regression coefficients vary among estimates from different temporal or spatial samples. How can this be accomplished when the fitness surface is generated by PPR, thin-plate splines, or some other semiparametric regression? Further, how does one best combine fitness surface estimates from a series of locations or times to display both an average fitness surface and also some measure of variability over samples?

A clever solution to both of these problems using **tensors** was suggested by Calsbeek (2012). The powerful tensor machinery, which is well known to physicists, has only been sporadically applied to quantitative-genetic problems (Rice 2002a, 2004b; Hine et al. 2009;

Aguirre et al. 2104). A matrix is a type of tensor with a table of items indexed by two subscripts. Now imagine a series of matrices stacked on top of each other to form a cube (think of a 3D chess board). This is a third-order tensor and is indexed by (i, j, k) (i.e., i for the ith matrix in the stack, whose elements are indexed by (j, k), so that the entry $(2, 4, 5)$ in this tensor (stack of matrices) is the value in row 4, column 5, of the second matrix in the stack. The analysis of fitness functions that may vary over time or space proceeds by considering each fitness-function estimate (corresponding to a specific location or time) as one of the matrices in the total stack of such matrices over all of the locations or times.

Calsbeek's analysis (for two traits) proceeds by constructing a suitable mesh (neither too fine nor too coarse) for the two trait values, say 50 equally spaced increments on both axes, giving a 50×50 matrix. The entry in $(i, 5, 7)$ is the estimated fitness from a semiparametric regression for sample i for the value in axis one corresponding to row five and the value in axis two corresponding to column seven. The stacked series of such estimates forms the tensor. Calsbeek computed several measures of dispersion and variation among the different matrices representing fitness surfaces from different samples. Just as one can decompose a matrix into a series of approximations (e.g., PC1, PC2, etc.) that represent major axes of variation, one can also decompose a tensor into a series of matrices presenting major lower-order dimensions of variation. The first such matrix (the tensor generalization of the PC1 vector) represents the average fitness surface over the sample. Parametric bootstrapping can be used for approximate hypothesis testing, such as for determining the significance of between-sample differences in some component of the fitness surface. See Calsbeek (2012) for details on both of these procedures.

Model Selection

Given the plethora of choices for modeling fitness surfaces (which traits to include, what method to use, what type of function to fit, etc.), a few comments on the delicate issue of **model selection** are in order (see Burham and Anderson 2002 for a complete treatment). If one has a candidate set of models, which should be used? If the models are *nested* (with one being a subset of the other), then standard likelihood-ratio tests can be used choose between them (LW Appendix 4). However, most models are not nested. In such cases, various informal statistics are used to compare models, and we focus on two, the AIC and BIC metrics. For both of these metrics, a smaller value means it is a better model. We stress that while both metrics are fully grounded in theoretical principles (Burham and Anderson 2004), comparing values for two different models is largely ad hoc in that there is no formal test for significance (i.e., no formal criterion for determining when one model is clearly better than the other).

The idea behind both metrics (as in a likelihood-ratio test) is reward goodness of fit (i.e., smaller values of $-2\ln[L]$ imply better fits, where L is the model likelihood), but also to penalize for the number of model parameters, k. One of the widely used model-comparison metrics is the **Akaike information criterion** (1973),

$$\text{AIC} = -2\ln(L) + 2k$$

which was adjusted for the sample size, n, by Sugiura (1978),

$$\text{AIC}_c = -2\ln(L) + 2k + \frac{2k(k+1)}{n-k-1} = -2\ln(L) + \frac{2kn}{n-k-1}$$

The latter was briefly introduced (Equation 12.25a) in Example 12.5. AIC_c should be used in place of AIC unless $n/k > 40$ (Burham and Anderson 2004).

The other widely used metric is the **Bayés information criterion**,

$$\text{BIC} = -2\ln(L) + \ln(n)k$$

which was introduced by Schwarz (1978), and thus is also known as the **Schwarz criterion**. While AIC and BIC are often used interchangeably, they are actually designed for slightly different purposes. When one of the models being compared is the true model, then BIC picks this model with a probability approaching one in large samples. Conversely, AIC considers the situation where *none* of the candidate models may be correct and then tries to pick among the best of these. As noted by Shaw and Geyer (2010), if one is just considering a few traits, BIC should be used, but with a large number of traits AIC (or AIC_c) is a better choice. Investigators often report both, as the two metrics often rank models differently.

THE STRENGTH AND PATTERN OF SELECTION IN NATURAL POPULATIONS

Just how strong is selection in natural populations? When phrased in this way, the question is ambiguous, as it is not clear if the focus is on individual variance in fitness or direct selection on particular traits. One can observe a very high variance in individual fitness, and hence much *potential* for phenotypic selection, but if all variation in fitness is random (character-independent), then (in the extreme) there will be no phenotypic selection. In focusing on the strength of selection for particular *traits*, Darwin (1859) felt that characters change very slowly, and hence selection on them is weak. Conversely, there are classical examples (e.g., insecticide resistance and industrial melanism) of rapid responses, and hence presumably strong selection, although such cases are often the result of sudden shifts in the environment (such as anthropogenic changes), and how representative they are remains unclear.

Attempts at measuring selection on quantitative traits in nature trace back to Bumpus (1899) and Weldon (1901). The classical book by Endler (1986), which was one of the first attempts to summarize the average strength of selection, found that strong selection "is not rare and may even be common," a conclusion that (at the time) was surprising to many. In 2001, Kingsolver and colleagues (Hoekstra et al. 2001; Kingsolver et al. 2001) started to harvest of the rich literature of Lande-Arnold fitness estimates that had been accumulating for almost two decades, and reached the conclusion that the average strength of selection was modest. As detailed shortly, these papers generated much discussion, with some proponents claiming they supported weak to modest selection and others claiming that they supported strong selection.

Meta-analysis

As reviewed in Appendix 4, **meta-analysis** is the field of statistics that deals with the analysis of trends over a large number of experiments (Hunter and Schmidt 2005; Borenstein et al. 2009a; Cooper et al. 2009; Harrison 2011; Koricheva et al. 2013). **Informal**, or **narrative**, **meta-analyses** are by far the most common approach in evolutionary biology. Here, one verbally summarizes the result of a number of studies. This is the type of analysis used by most of the early summaries of strengths of selection in the wild. However, by using the approaches in Appendix 4, one can combine p values over a number of experiments to reach a more global conclusion. While such global p values, and other summary statistics (such as empirical distributions by general trait type, and estimates of mean, or absolute, values), can appear in narrative meta-analyses, a **formal meta-analysis** is a much more effective way to use the data (Appendix 4).

Unfortunately, conducting a formal meta-analysis requires something that is very often lacking in published results, namely, *standard errors of estimates*. As detailed in Appendix 4, when standard errors are available, a meta-analysis can be placed into a very powerful fixed-effects or mixed-model framework that allows for a more rigorous, and detailed, analysis of the published data. In part, the focus on narrative approaches arises, not from a lack of statistical expertise from the authors of the meta-analyses, but rather from a lack of full transparency in the published literature, such as a failure to report (at a minimum) standard errors or (more ideally) making the entire dataset available for future analysis. In the words of Kingsolver et al. (2012), "we believe our general understanding of patterns of selection is most limited by lack of access to individual-level data (i.e., data on trait values and fitness measures for each individual) of most studies."

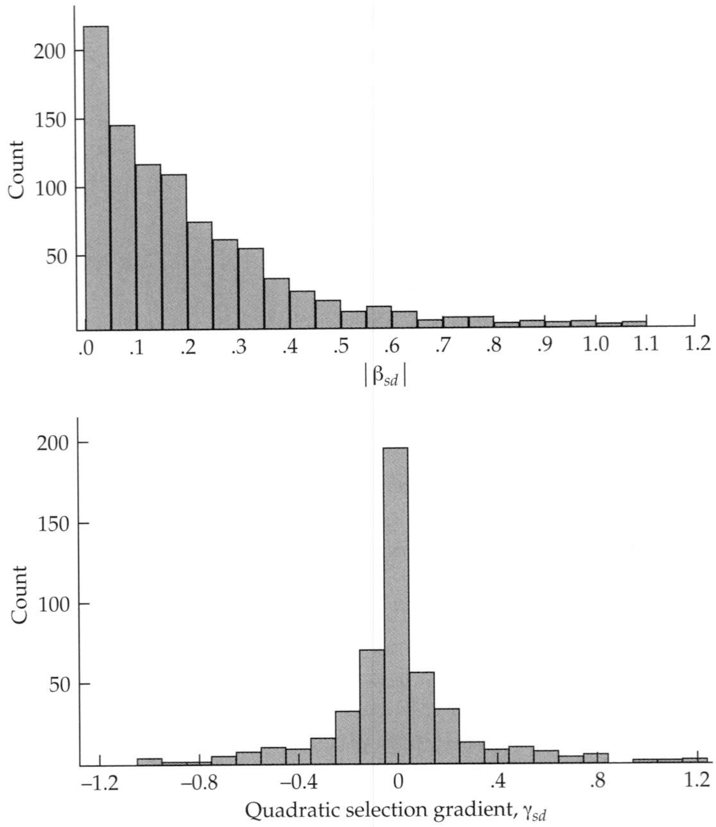

Figure 30.5 Summary of estimates of variance–standardized directional and quadratic gradients for natural populations. The data are from Kingsolver et al. (2001). **Top:** Plot of roughly 1000 estimated directional selection gradients in natural populations, with the median value of $|\,\beta_{sd}\,|$ being 0.16. The distribution of $|\,\beta_{sd}\,|$ was not significantly different from an exponential. **Bottom:** Plot of roughly 470 estimates of γ_{sd} from natural populations. The median value of $|\,\gamma_{sd}\,|$ was 0.10.

Kingsolver's Analysis

In a series of landmark papers, Kingsolver and colleagues (Hoekstra et al. 2001; Kingsolver et al. 2001) performed an informal meta-analysis on estimates of β and γ from 63 studies of natural populations published between 1984 and 1997. Their resulting frequency distributions of variance-standardized gradients ($\beta_{sd} = \sigma_z\beta$ and $\gamma_{sd} = \sigma_z^2\gamma$) are shown in Figure 30.5. These earlier findings have been updated using much larger datasets, but the basic conclusions from the initial 2001 papers remain the same, with narrative analyses by Geber and Griffen (2003), Kingsolver and Pfenning (2007), Cox and Calsbeek (2009), Siepielski et al. (2009, 2011, 2013), and Kingsolver and Diamond (2011) and more formal (mixed-model) analyses by Kingsolver et al. (2012), Morrissey and Hadfield (2012), and Morrissey (2016).

Kingsolver's initial analysis noted several trends. First, the distribution of absolute values of the variance-standardized directional selection gradients (β_{sd}) closely followed an exponential distribution, with a median (50% value) of 0.16. A β_{sd} of 0.16 implies that a change of one standard deviation in the trait changes relative fitness by 16%. Based on this observation, Kingsolver suggested that most directional selection in nature is fairly weak, although (as a result of the long tail of the exponential), there are a few large estimates (10% of estimates exceeded 0.5). Caution, however, is required with this initial estimate. It is well known that estimates of the expected *absolute* value, $|x|$, of a random variable, x, are inflated by sampling error (Equation A4.38a; Hereford et al. 2004; Morrissey 2016). For example, if

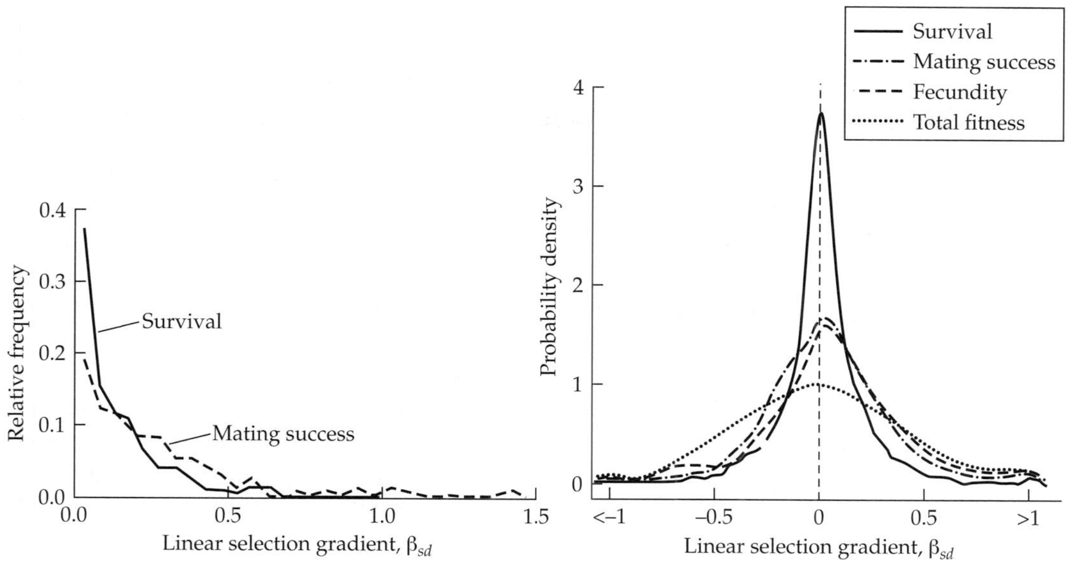

Figure 30.6 Distribution of variance-standardized directional selection gradients over different fitness components. **Left:** The distribution of $|\,\beta_{sd}\,|$ for both sexual and natural selection in the Kingsolver 2001 dataset. These gradients were measured by mating success and viability, respectively. The median value for viability was 0.153, but 0.250 for mating success. (After Hoekstra et al. 2001.) **Right:** Updated estimates of selection gradient using a larger dataset, showing the distribution of β_{sd} for traits influencing survival, mating success, fecundity, and total fitness. The latter three show a broader distribution of effects relative to survival. (After Kingsolver and Diamond 2011.)

$x \sim N(0,\sigma_x^2)$, then $E[|x|] = \sigma_x\sqrt{2/\pi}$ (Equation A4.37a), while if σ_e^2 is the error associated with estimating x using a sample value, $\widehat{x}$, then $E[|\widehat{x}|] = \sqrt{\sigma_x^2+\sigma_e^2}\sqrt{2/\pi} > E[|x|]$. Hence, the average true strength of directional selection, $E[|\beta|]$, is less than the estimated strength, $E[|\widehat{\beta}|]$. Indeed, Morrissey (2016) suggested that the initial Kingsolver value was roughly a twofold overestimate.

Two additional sources of bias can result in a further overestimation of the average strength of β, both of which are related to low power: Beavis effects (the overestimation of true effect size when using only values that are declared to be significant) and **publication bias** (the failure to publish nonsignificant results). Together, all these sources of bias strengthen the claim of weak selection. Consistent with this suggestion, most of the larger estimates of $|\,\beta_{sd}\,|$ occur in studies with small sample sizes, with most estimates below 0.1 occurring when the sample size was 1000 or greater. Hence, it is possible that some of the large β_{sd} values are simply a consequence of the larger estimation variance in smaller sample (the Beavis effect was discussed in LW Chapter 14 in the context of QTL mapping).

The discussion of β was focused on absolute values because their sign is likely to vary when considering a large set of traits and (without reference to a specific trait) the sign is usually not particularly relevant. Such is not the case with γ_{ii}, whose sign has an unambiguous meaning. Specifically, negative values indicate concave (and potentially stabilizing) selection, and positive values indicate convex (and potentially disruptive) selection. This leads to the second trend in selection data noted by Kingsolver, concerning the widespread belief that stabilizing selection is far more common than disruptive selection (Chapter 28). Rather than seeing this pattern (a skew toward negative γ_{ii} values), Kingsolver observed an essentially symmetric distribution of γ_{sd} values with a mean of zero (Figure 30.5), which implied that positively and negatively curved fitness surfaces are equally common. Further, the average strength of quadratic selection was weak, with an (again overestimated) median value of 0.10 for $|\,\gamma_{sd}\,|$. A more recent analysis of larger datasets reaffirmed this trend (Kingsolver and Diamond 2011; Kingsolver et al. 2012).

Kingsolver's third trend was that variance-standardized gradients, β_{sd}, for mating success and fecundity were larger than those for viability selection (Figure 30.6). This pattern of the weakest gradients often being associated with episodes of viability selection is seen in later (and larger) studies (Siepielski et al. 2011; Kingsolver and Diamond 2011). All of these authors noted that there is likely to be an intrinsic basis for mating traits in that most studies focused on measuring gradients for highly ornamental traits that were likely to be under strong sexual selection. However, no such apparent bias occurs in the choice of fecundity versus viability traits. This apparent pattern of higher variance-standardized gradients on fecundity than on viability provides some justification for the common tendency of studies of long-lived organisms to use annual fecundity (as opposed to annual survival) as a surrogate for total fitness.

However, this conclusion that fecundity is under stronger selection than viability is based on variance-standardized gradients. Recall our discussion from Chapter 29 on mean-standardized gradients, which measure fitness elasticity, and hence are a more natural metric for fitness effects. Crone (2001) found that most long-lived species had higher mean-standardized gradients for annual survival (viability) than for annual fecundity. Based on this result, Crone suggested that, contrary to the standard practice, annual viability may be a better surrogate for total fitness than annual fecundity. One notable exception to this pattern was in perennial semelparous plants (those with a single episode of reproduction), where the elasticity of growth rate was the largest, followed by fecundity, and then survival. For short-lived species, fecundity tends to have a higher elasticity than in longer-lived species. The most accurate approach for assessing the impact of these fitness components on total fitness is to weight the survival and fecundity gradients by the elasticity of each component on the population growth rate, λ; see Equation 29.33f.

Directional Selection: Strong or Weak?

Has the case for "weak" selection been made? As pointed out by Conner (2001), even so-called weak selection can be very effective at changing trait means. Consider Kingsolver's median value of $\beta_{sd} = 0.16$ ($\beta = 0.16 \cdot \sigma_z$). From Equation 13.22a, the single-generation expected change in the mean (in phenotypic standard deviations) is $h^2 \cdot 0.16$. With a typical heritability of 0.4, only 16 generations of selection are required to shift the population mean by one standard deviation, and only 80 generations are required to shift it by five standard deviations. Another way to think about the impact of a "weak" selection gradient of $\beta_{sd} = 0.16$ was mentioned by Hereford et al. (2004). They noted that even a population of modest size spans over four standard deviations of variation, implying a fitness range of $4 \cdot 0.16 = 0.64$, or a 64% variation in relative fitness among individuals within the population (assuming that the fitness-surface approximation holds over this span of phenotypic values).

Conversely, if concerns about publication bias and low power are correct, the true median value of $|\,\beta_{sd}\,|$ may be substantially less than the "weak" median value of 0.16 found by Kingsolver. Hersch and Phillips (2004) and Knapczyk and Conner (2007) examined the validity of these concerns. Figure 29.9 presented power curves for a univariate analysis (a single-trait fitness regression), which is a function of $\rho = \bar{\imath}/\sqrt{I}$. Hersch and Phillips showed that when multiple (potentially correlated) traits are considered, the adjusted version of the trait-fitness correlation, ρ, used in power calculations becomes

$$\rho^2 = \frac{\boldsymbol{\beta}_{sd}^T\,\mathbf{s}}{I} \tag{30.31}$$

Hence, $\bar{\imath}^{\,2}$ is replaced by $\boldsymbol{\beta}_{sd}^T\,\mathbf{s}$ (where $\mathbf{s}$ is the vector of variance-standardized selection differentials, i.e., a vector of selection intensities), which accounts for the correlations among the (measured) traits. Using Equation 30.31, Hersch and Phillips found that most of the studies summarized by Kingsolver for which power calculations could be performed (i.e., those that included estimates of I) were very underpowered, supporting Kingsolver's concern about the underreporting of small values and hence skewing the distribution of estimates of $|\,\beta_{sd}\,|$ upwards.

The impact of publication bias, which is also called the "**file-drawer effect**" (Rosenthal

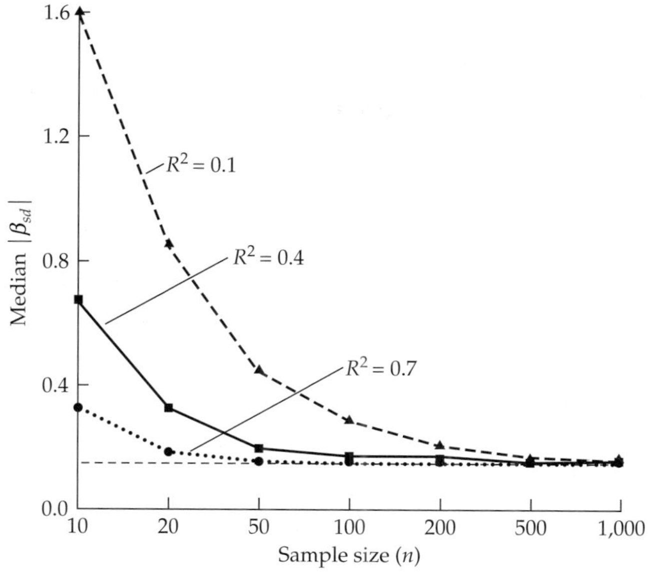

Figure 30.7 Overestimation of selection gradients due to publication bias. Hersch and Phillips (2004) simulated a model in which five values of β_{sd} were drawn from a distribution and only included in the final dataset if at least one of them was significant. The model coefficient of determination (R^2) is the total amount of variation in fitness explained by all of the traits (a small value of R^2 implies low power). The dashed line represents the true median value. Note the overestimation of the median β_{sd} values at small sample sizes and for small R^2 values.

1979; Rosenberg 2005), as nonsignificant results are not published and hence simply "left in a file," was also examined by Hersch and Phillips (2004). They simulated cases where results were only published when one (or more) of the gradients in a study were found to be significant. Although such a model does indeed allow for many nonsignificant values in the database, these are still biased because they were conditioned on at least one of the results being significant. This ascertainment scheme is certainly one reasonable model for quantifying publication bias. For each "publication," the simulation chose five "traits" by drawing five β_{sd} values at random. If one (or more) of the five tests was significant, all the values were retained, and if not, they all were discarded. As shown in Figure 30.7, this conditional sampling resulted in the median value of β being substantially overestimated when either the sample size or the amount of variation explained by the effects was small (low R^2 values), both of which are settings that result in low power. This is essentially another manifestation of the Beavis effect, overestimation when using statistics that are conditional on success (significance)

Knapczyk and Conner (2007) also considered sampling error and publication bias, but they arrived at the very different conclusion that "our understanding of selection is not strongly biased by these commonly invoked sources of error." To examine publication bias, they binned the datasets examined by Kingsolver based on sample size. If there was no publication bias, the distribution of $|\,\beta_{sd}\,|$ values should have a similar form over different sample sizes (following the exponential observed for the entire set by Kingsolver). They found departures from this pattern for studies with very small sample sizes (less than 40), but did not find departures for studies with larger sample sizes. Indeed, compared to the exponential distribution, they found an excess of weak selection gradients in the larger studies. Thus, they concluded that publication bias, except in the smallest studies, did not upwardly bias estimates of $|\,\beta_{sd}\,|$ in any appreciative way. Appendix 4 presents a number of metrics to assess publication bias.

Further complicating matters is the paper by Hereford et al. (2004), who found that using mean-standardized gradients ($\beta_\mu = \mu\beta$; Equation 29.33a) gave very different results. Recall that the mean-standardized gradient is $\beta_\mu = 1$ when the trait is fitness itself, providing

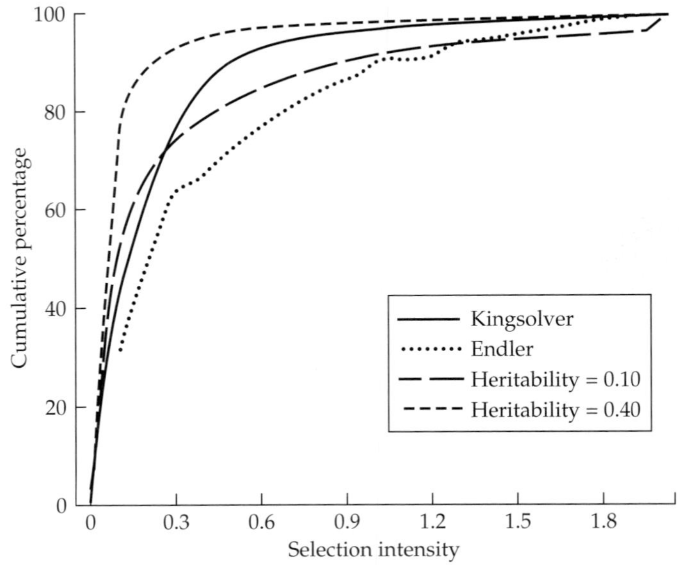

Figure 30.8 Four estimates of the distribution of selection intensities in nature. Two are direct estimates of selection from studies in nature (Endler 1986; Kingsolver et al. 2001), while the other two were estimated using observed rates of microevolution and then estimating $\bar{\imath}$ using two different assumed heritability values. (After Kinnison and Hendry 2001.)

a benchmark for the strength of selection on a trait (Equation 29.33c). Using a subset of the Kingsolver data (38 studies yielding 580 estimates) that included the required information for this standardization (i.e., those that reported the trait means), they found a median $|\,\beta_\mu\,|$ of 0.54, or 54% of the strength of selection on fitness. This is indeed strong selection. Even using a correction for the upward bias generated by sampling error (Equation A4.37a), which reduced the median value to 0.31, still left strong selection. The observed strengths of selection were so large that the authors suggested that the selection gradients were likely inflated by focusing on single episodes of selection rather than lifetime fitness. In essence, they suggested that tradeoffs resulted in the actual gradient of lifetime fitness for a trait being significantly less than the gradients based on single episodes of selection. However, as we discuss shortly, there appears to be little evidence for tradeoffs among the traits in the currently analyzed databases (Figure 30.9).

A final perspective on the strength of directional selection was offered by Hendry and Kinnison (1999) and Kinnison and Hendry (2001), who compiled data on over 2000 estimated rates of microevolution. While rates of divergence in trait means confound the strength of selection with the transmission genetics of the trait, an advantage of this approach is that such estimates focus on an average rate of change over time, thus smoothing out large values caused by brief episodes of strong selection. Such episodes are more likely to catch our attention, and considering only these episodes results in a biased view of the average strength of selection over time. Figure 30.8 plots the cumulative selection intensities obtained by Endler (1986) and Kingsolver et al. (2001) as well as the inferred average selection intensity values given the observed rates of microevolution, assuming heritabilities of 0.1 and 0.4. As can be seen, under either assumed heritability, there is an excess of weak selection (small $\bar{\imath}$) values relative to those seen by Kingsolver and Endler. While this suggests that weak selection is the norm, an argument for strong selection can also be made. While strong selection can certainly occur over brief episodes, if it continues over a sufficient amount of time, genetic variation for the selection response will be eroded and further response must wait for new variation (from mutation and immigration). Thus, with strong persistent selection, h^2 will likely become much smaller than 0.1. Conversely, if selection is episodic, selection to reduce h^2 will be smaller and genetic variation will erode less quickly.

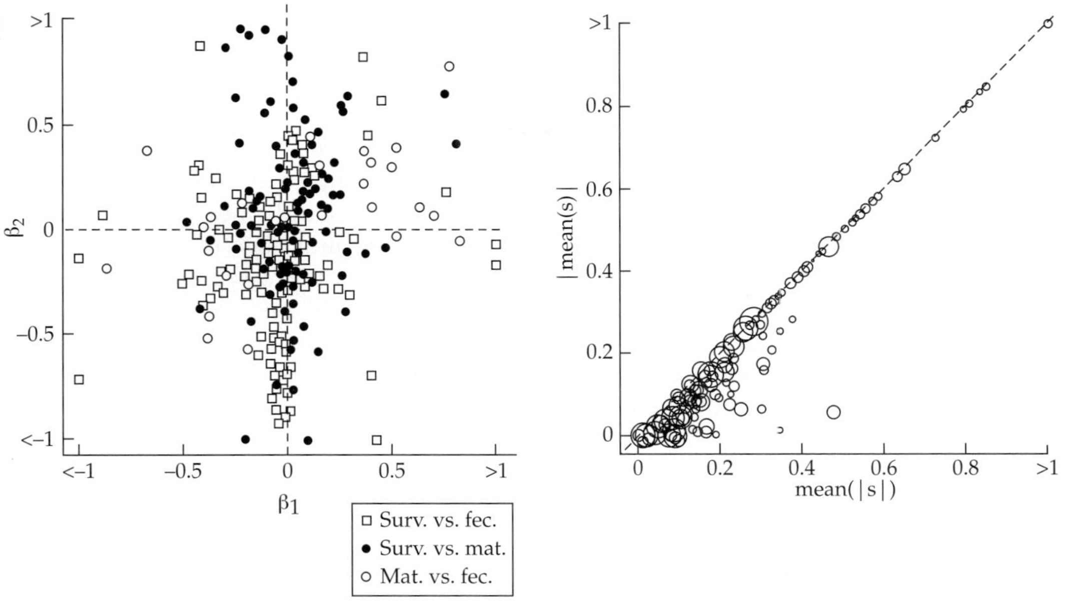

Figure 30.9 Additional results from the meta-analysis of Kingsolver and Diamond (2011). **Left:** Lack of evidence for fitness tradeoffs between survival (Surv.), mating (Mat.), and fecundity (Fec.). The scatterplot displays the variance-standardized linear gradients when a trait was measured for two or three of these components. Evidence of widespread tradeoffs would appear as an excessive number of values in the upper left ($\beta_{sd,1} < 0, \beta_{sd,2} > 0$) and lower right ($\beta_{sd,1} > 0, \beta_{sd,2} < 0$) quadrants. This is not seen. Rather, 57% of the data pairs have the same sign, which magnifies, rather than reduces, the total selective effect. **Right:** Temporal variation in standardized selection differentials ($s = S/\sigma = \bar{\imath}$) in samples with multiple time points. When sign reversals occur, $|\text{mean}(s)| < \text{mean}(|s|)$. The majority of values fall along the diagonal, indicating that sign reversal is not a common feature in this dataset (although values below the diagonal, and hence reversals, are seen). (After Kingsolver and Diamond 2011.)

An apparent excess of small selection intensities could reflect true low selection intensities when averaged over many generations, thus smoothing out strong episodes. However, it could equally likely reflect strong persistent selection quickly eroding heritability and hence reducing response.

Total Versus Direct Selection, Tradeoffs, and Temporal Variation

Three other evolutionary questions can be partly addressed through a meta-analysis of estimated gradients: the relative importance of correlated traits, the impact of fitness-component tradeoffs, and the magnitude of temporal variation.

The amount of **total selection** on a trait is given by S (and its variance-standardized counterpart, $s = \bar{\imath}$), which is the sum of the amount of direct (β, β_{sd}) and correlated selection (Equation 30.3). When β_{sd} is estimated from a multiple-trait regression, a plot of s versus β_{sd} can provide insight into the relative importance of direct versus correlated selection. A major caveat concerning this type of analysis is that it is only as good as the characters that are included in the regression. If causally important fitness traits that are correlated with a focal trait are not included, their effect can incorrectly be attributed to direct selection. Using this approach, Kingsolver and Diamond (2011) found that correlated selection had little impact, with s often falling very close to β_{sd} in value. An exception is size, where total selection is generally *less* than direct selection, which suggests negative correlations with other traits under selection. An analysis by Geber and Griffen (2003) of performance traits in plants found a different pattern than was seen by Kingsolver and Diamond. For these traits, they observed that $|\beta_{sd}|$ tended to be smaller than $|s|$, so not only were correlated effects important, but they also tended to be in the same direction as direct effects. Geber and Griffen estimated that ~40% of the amount of directional selection on a trait was the result of correlated effects. Moreover, they found the same pattern ($|\gamma_{sd}| < |C/\sigma^4|$) for quadratic selection, with correlated effects accounting for an average of ~60% of the total effect.

A second area of interest to evolutionary biologists is the relative importance of tradeoffs between different fitness components. For example, a trait that improves variability might do so at the expense of mating success or fecundity, so simply measuring the impact of a trait on one fitness component can give a very misleading view of its evolutionary potential (e.g., Example 29.1). Kingsolver and Diamond (2011) addressed this issue by plotting β_{sd} values for the same trait in those few studies that measured multiple episodes of selection. As shown in Figure 30.9, while examples of tradeoffs exist (the values of β_{sd} have different signs across episodes), there is no evidence that it is widespread. Indeed, in 57% of the comparisons, the two gradients had the *same* sign, thus reinforcing, rather than masking, selective effects.

Another potential tradeoff is, not between different episodes of selection, but rather between the sexes. Males and females may experience different amounts of selection for the same trait, and in the extreme may have opposite signs, or **sexually antagonistic selection** (**SA**). In part, the limit that any such SA may place on evolution depends on genetic correlation of a trait between the sexes. If this correlation is less than perfect (i.e., different sets of genes may be involved between the sexes), then we have a two-trait evolution question (Equation 13.26c), and whether the presence of different patterns of selection on the separate traits constrains their joint evolution depends as much on their genetic covariance structure (**G**) as it does on their strength of selection ($\boldsymbol{\beta}$). By ignoring this concern and instead focusing on selection itself, the potential importance of SA was examined in a narrative meta-analysis by Cox and Calsbeek (2009), who compared values of $|\beta_{male} - \beta_{female}|$ over traits, and obtained a median value of 0.13 (for variance-standardized gradients). Note that this is close to the initial Kingsolver value for absolute selection on traits, suggesting that between-sex differences are on par with the level of absolute selection on a trait. When it was broken down by fitness components, Cox and Calsbeek found that the median value of $|\beta_{male} - \beta_{female}|$ was smallest for viability and fecundity selection and largest for both sexual selection and total selection (a median value of around 0.2 to 0.3).

A strong caveat concerning these conclusions was noted by Morrissey (2016). As with $|\beta|$, estimates of an absolute difference, $|\beta_{male} - \beta_{female}|$, are inflated by sampling noise. Indeed, the expected value of $|x_i - x_j|$ for two random draws (x_i, x_j) from the same distribution is $2\sigma_x/\sqrt{\pi}$ (Nair 1936). Morrissey noted that such comparisons are best performed in a bivariate mixed-model framework (similar to the analysis used for Equations 20.26 through 20.29). His reanalysis of the Cox and Calsbeek data found that the correlation between male and female selection gradients was ~0.8, thus showing that selection gradients are highly positively correlated between the sexes and that SA is rare.

The final issue is the impact of temporal variation in selection strength and direction (Siepielski et al. 2009, 2011, 2013; Kingsolver and Diamond 2011). There are numerous ways to quantify temporal variation, such as measuring the standard deviation over a set of estimates (which compounds actual variation with sampling variation) or the fraction of times the sign changes (which can be high for a trait under very weak selection). A simple metric is to plot |mean(s)| versus mean($|s|$). If sign reversals are common, then |mean(s)| < mean($|s|$), while if the two are roughly equal, little sign reversal has occurred. While this metric obscures any variation in the strength of selection, it does address the impact of temporal variation in its direction. As shown in Figure 30.9, for the existing data, any directional variation is relatively modest, and it is greatest for traits under weak selection (mean[$|s|$] small), which could simply reflect sampling error.

These data were reexamined by Morrissey and Hadfield (2012), who performed a random-effects meta-analysis (Appendix 4) on a subset (consisting of those studies that reported standard errors, and hence allowed for a formal meta-analysis). If we let β_{ij} denote the jth temporal value from the ith study-trait combination, the model becomes

$$\beta_{ij} = \mu + u_i + t_{ij} + e_{ij}$$

where $\mu + u_i$ is the expected value for the gradient in the ith study-trait combination, t_{ij} is the residual due to temporal variance, and e_{ij} is the sampling error. The last three variables are treated as random effects, and one can estimate their variances in a mixed-model setting. Of interest is the ratio $\sigma_u^2/(\sigma_u^2 + \sigma_t^2)$, namely, the amount of variance due to among study-trait means relative to the trait plus temporal variance. Morrissey and Hadfield found this ratio was quite high, measuring 0.88 (with a 95% credible interval of 0.82 to 0.91), thus leading to the conclusion that selection over time is remarkably constant.

Siepielski et al. (2011) examined how the temporal strength of selection varied over different fitness components (viability, mating success, and fecundity). The temporal variation among variance-standardized gradients was highest for mating success and roughly equal for survival and fecundity. Given potential differences in sampling error over these different fitness components, this result could simply imply higher sampling variation in mating success over other components. They also found that reversals of signs were highest for survival, followed by mating, and then fecundity. They noted that frequent sign reversals could occur if a trait under viability stabilizing selection has its mean close to the optimum (θ). In such a setting, small fluctuations of the mean around θ result in sign variation in β.

To examine if this is the case, Siepielski et al. applied a result from Estes and Arnold (2007), who used β_{sd} and γ_{sd} to estimate the distance of a population mean from an optimum,

$$\frac{|\,\overline{z} - \theta\,|}{\sigma} \simeq \left| \frac{\beta_{sd}}{-\gamma_{sd}} \right| \tag{30.32}$$

This follows from Equation 30.21a, and it assumes a normally distributed trait with a fitness function that is well approximated (over the bulk of its phenotypic distribution) by a quadratic. Using Equation 30.32, Siepielski et al. found a value for the standardized difference of the mean from an optimum of 2.28 $\pm$ 0.54 for survival traits in their dataset, but values of 14.59 $\pm$ 9.14 for mating success and 11.62 $\pm$ 5.48 for fecundity. While Equation 30.32 is a crude metric, which is fraught with numerous pitfalls, it does suggest that survival traits might be closer to their optimum (provided it exists).

Directional Selection on Body Size and Cope's Law

Body size has several outlier features relative to other traits. As shown in Figure 30.10, while the distribution of β_{sd} values for morphological traits was symmetric around zero, the distribution for β_{sd} for body size was highly skewed toward positive values (Kingsolver and Pfenning 2004; Kingsolver and Diamond 2011). Kingsolver and Pfenning suggested that individual selection for larger body size underlies **Cope's law** (the tendency for the size of species within a lineage to increase over evolutionary time). This suggestion also presents a paradox: if it is true, why are most of the largest animals not present today? Kingsolver and Pfenning (2004, 2007) suggested that this is a consequence of differential extinction, pointing out that during the last widespread extinction of North American mammals, the largest species were the hardest hit. However, as we noted above, Kingsolver and Diamond (2011) also found that direct selection on size is greater than the total selection ($\beta_{size} > s_{size}$), suggesting the presence of selection against traits that are phenotypically correlated with body size, whose values tend to become more deleterious as body size increases. Further, they found that the distribution of γ, while symmetric for most classes of traits, was significantly negative for body size, suggesting that it is often under stabilizing selection. Figure 30.10 shows that another class of characters, phenological traits (season timing events such as breeding), also shows a skewed distribution, toward negative (i.e., earlier) values.

Quadratic Selection: Strong or Weak?

Discussions of whether the observed amount of quadratic selection is strong or weak are highly problematic. If estimates of β_{sd} are underpowered (and thus significant estimates are likely overvalued; i.e., the Beavis effect), the same is certainly true for quadratic estimates.

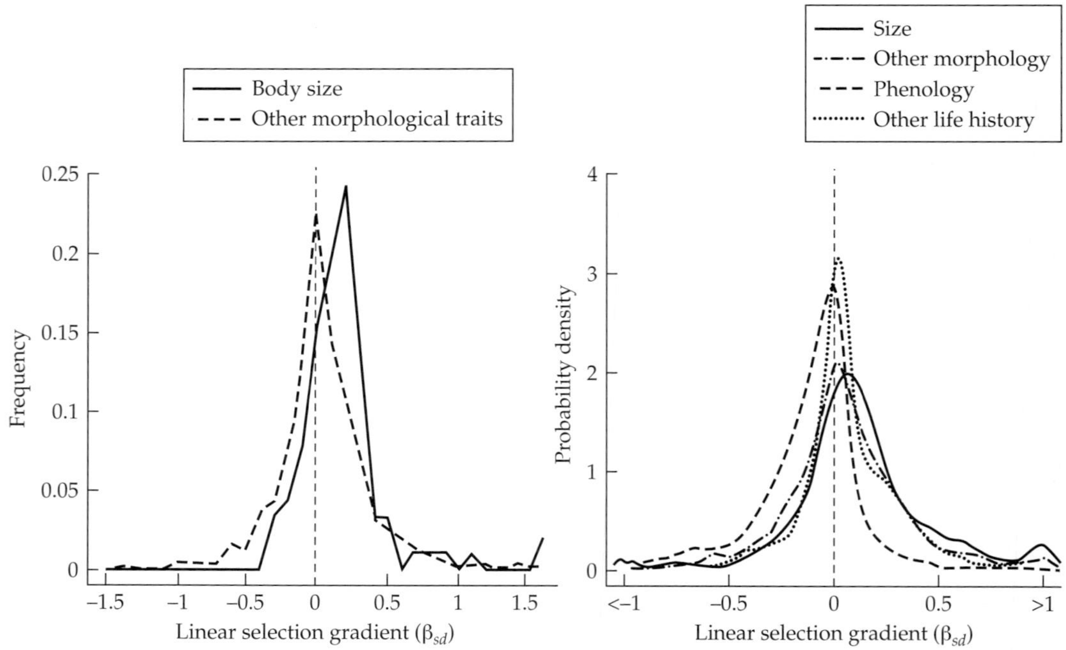

Figure 30.10 **Left:** Distribution of variance-standardized selection gradients, β_{sd}, for morphological traits and for size. While the distribution for morphological traits is largely symmetric around zero, the distribution of β_{sd} for body size is skewed toward positive values. (After Kingsolver and Pfenning 2004.) **Right:** A more recent analysis by Kingsolver and Diamond (2011) showing a positive-skew for size and a negative-skew for phenological traits.

Indeed, simulations by Haller and Hendry (2013) found that there is generally low power for detecting stabilizing selection due to small sample sizes. Simulations for traits under weak stabilizing selection typically resulted in a distribution of γ values similar to that seen in Figure 30.5, namely, largely symmetric about the origin (reflecting low power, and hence a roughly random distribution of estimates around zero). An important feature of the Haller-Hendry simulations was that they used a "**virtual ecologist**" approach, wherein simulated data from the quantitative-genetics model is combined with an **observer model** (i.e., a model for how the data would be ascertained in real-world settings, such as having living individuals that are not recaptured counted as dead). This mimics how real data would be "virtually" observed (Zurell et al. 2009), and hence more correctly reflects the actual power of collected data.

Balancing concerns about inflated estimates due to low power (Beavis effects), there are two reasons why reported estimates of quadratic gradients may be *undervalued*. As mentioned earlier, many published reports have made errors in obtaining γ_{ii} values from the output of standard regression packages, resulting in only half their true value being reported (Stinchcombe et al. 2008). Second, Blows and Brooks (2003) noted that $|\gamma_{ii}|$ likely significantly *underestimates* the strength of quadratic selection when multiple traits are considered (see Example 30.6). An analysis based on the eigenvalues (λ) of the γ matrix is much more informative as to the strength of quadratic selection in nature.

The much more problematic issue is whether γ is even a reasonable measure for nonlinear fitness surfaces. As Figure 29.10 illustrated, a value of γ can be entirely misleading when the fitness surface departs from a quadratic. As the above theory on canonical forms illustrates, a quadratic surface in k dimensions (traits) still has (at most) only a single extremum (Equation 30.21a). In this case, fitting a multimodal fitness surface with a quadratic will return a very misleading value of γ. Indeed, even with stabilizing Gaussian selection, if the mean is sufficiently above the optimum, the best quadratic fit will result in $\gamma > 0$, which falsely implies the presence of potentially disruptive selection (Figure 29.10).

Where Is All the Stabilizing Selection?

Perhaps the most striking observation from the meta-analysis of γ values is the lack of a clear trend toward negative values (except for body size), as would be expected if stabilizing selection were widespread. This is highly troubling given the historic view of most ecologists and evolutionary biologists that stabilizing selection is widespread and important over evolutionary time (Charlesworth et al. 1982; Maynard Smith 1983; Estes and Arnold 2007; Hunt et al. 2007a). One explanation is that γ does not capture the correct geometry in multimodal fitness surfaces. The second is the observation from the Haller-Hendry (2013) simulations of low power to detect stabilizing selection.

Haller and Hendry offered several other explanations for this "missing" stabilizing selection. The first is that selection may have already done its job in that the existing phenotypic variation is small relative to the width of the fitness peaks (Chapter 28). In their words, "selection 'erases its traces' once populations have adapted to a fitness peak." The second is the suggestion from Chapter 28 that strong stabilizing selection is likely to be confined to a few dimensions in the multivariate trait space (i.e., a few indices of trait values). This results in only a weak signal in any single component trait and low power of detection. Their final suggestion is that **squashed stabilizing selection** may be common, which again lowers power. Here competition results in disruptive selection among individuals whose trait values are near an adaptive peak, which flattens out (squashes) the fitness surface around the peak, making detection more difficult. An example of this phenomenon is the work of Morno-Rueda (2009) on selection on shell morphology in Spanish land snails (*Iberus gualtieranus*). Predation from black rats (*Rattus rattus*) results in distinctive markings on the shell, and using only such shells showed disruptive selection on shell height. Conversely, when shells lacking these markings were used in the analysis, stabilizing selection was seen. When all shells are considered as a single group, no significant quadratic term is observed.

A Plea to Fully Publish

In 1899, Herman Bumpus published a modest dataset involving 136 domestic sparrows (*Passer domesticus*) immobilized by an ice storm. Of these, 72 (21 females, 51 males) survived, while the remaining 64 (28 females, 36 males) perished. Despite the humble beginning, this has become the most widely analyzed study of selection (Harris 1911; Calhoun 1947; Grant 1972; Johnson et al. 1972; O'Donald 1973; Lande and Arnold 1983; Crespi and Bookstein 1988; Crespi 1990; Pugesek and Tomer 1996). In large part, this is because *Bumpus published all of his data*, allowing investigators to employ their own methods of analysis. In contrast, Kingsolver et al. (2001) lamented that many of the studies included in their meta-analysis did not even report standard errors, much less trait means (which are needed to allow for mean-standardized gradients to be computed), thus limiting their inclusion to only narrative meta-analyses. At a minimum, any published study should report standard errors for all estimates. More generally, studies should make their entire dataset available to the community. Indeed, such fully open access is a requirement for virtually all published molecular studies. As Kingsolver et al. (2012) pleaded "Recycle your hard-won, slightly used, still precious data today!"

UNMEASURED CHARACTERS AND OTHER BIOLOGICAL CAVEATS

Even if we are willing to assume that the best-fitting quadratic regression is a reasonable approximation of the individual fitness surface, there are still a number of important biological caveats to keep in mind (Chapters 20 and 29). For example, the fitness surface can change in both time and space, often over short spatial or temporal scales (e.g., Kalisz 1986; Stewart and Schoen 1987; Scheiner 1989; Jordan 1991; Garant et al. 2007; Siepielski et al. 2009, 2011, 2013; Bell 2010; but see Morrissey and Hadfield 2012), so one estimate of the fitness surface may be quite different from another for a different time or location (Figures 30.11 and 30.12). Hence, considerable care must be taken before pooling data from different times or sites to improve the precision of estimates. Conversely, fitness data are noisy,

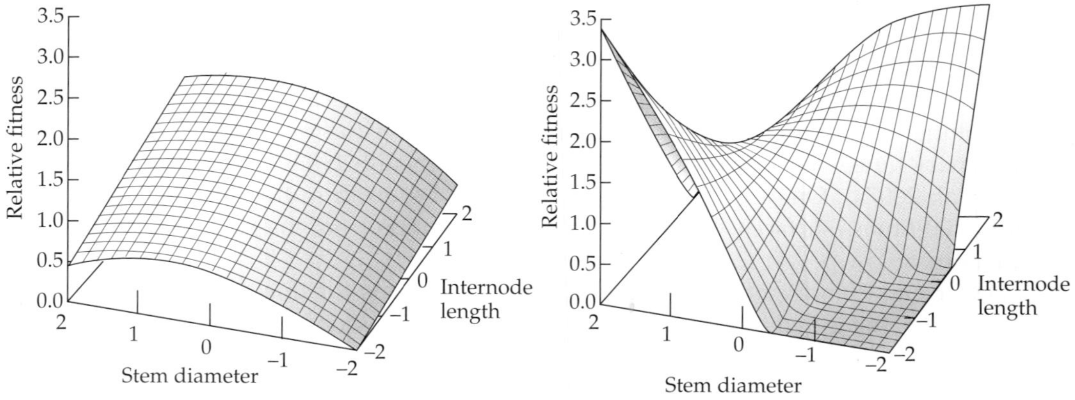

Figure 30.11 The selection surface can change over time. These two surfaces are the best-fitting quadratic surfaces for relative fitness as a function of stem diameter and internode length for a population of the annual plant *Diodia teres* (Rubiaceae) in North Carolina. **Left:** Estimate for June 1985 data. **Right:** Estimate for July 1985 data. (After Jordan 1991.)

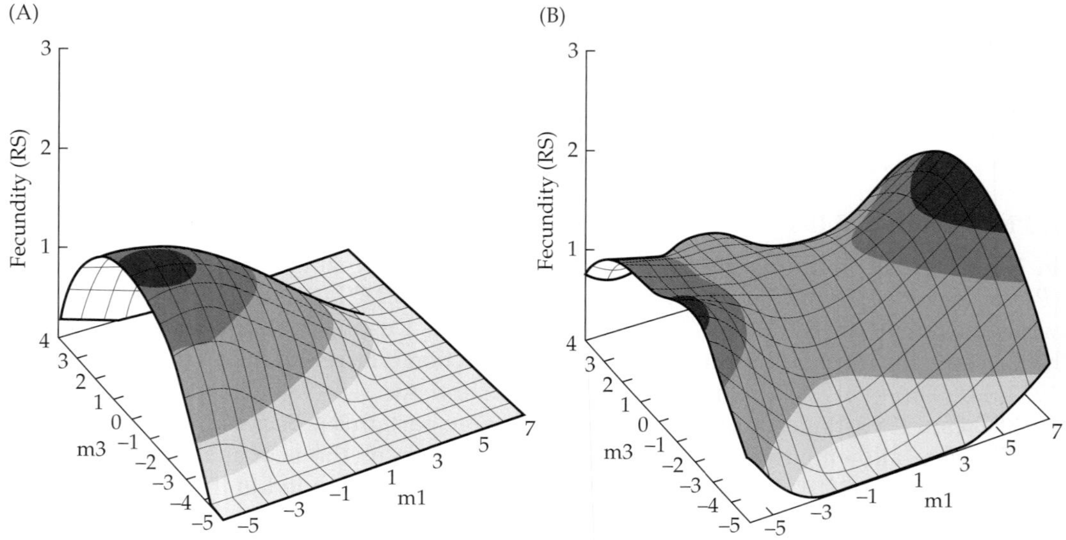

Figure 30.12 The fitness surface can change over space. The fecundity surface for great tits (*Parus major*) in two populations in Oxfordshire (U. K.) separated by less than four kilometers. Here, m1 and m3 correspond to the first and third canonical axes (weighted trait combinations) of the fitness surface. (After Garant et al. 2007.)

and the resulting surfaces have considerable uncertainty, so two, visually rather different-looking, surfaces might be not statistically significantly different. When the data are such that selection gradients can be estimated separately for different times or areas, interactions of space/time × gradient can be tested in a straightforward fashion (e.g., Mitchell-Olds and Bergelson 1990). Further, the evolution of a trait may result in a change in the biotic environment, which in turn may change the nature of selection on that trait (Chapters 20 and 22).

Population structure can also influence fitness surface estimation. If the population being examined has overlapping generations, fitness data must be adjusted to reflect this (Chapter 29). Likewise, if members of the population differ in their amount of inbreeding,

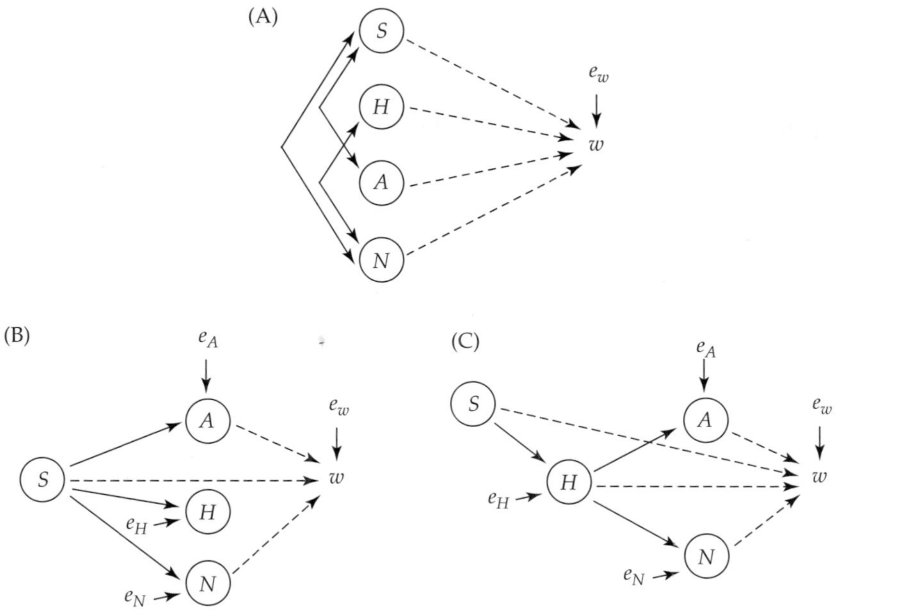

Figure 30.13 Four traits, seed weight (S), plant height (H), leaf area (A), and number of leaves (N), potentially influence fitness, w. Single-headed arrows indicate an assumed causality, while double-headed arrows indicate correlations (unspecified causality). Connections between a trait and fitness are shown as dashed arrows for visual clarity. Throughout, e represents a residual input (the variation not accounted for by the causal connections). **A:** The assumptions of a standard fitness regression. All four traits, some of which which may be correlated (doubled-headed arrows between trait combinations), potentially impact fitness. **B:** Seed weight impacts the three other traits, giving a direct impact on fitness ($S \rightarrow w$) and indirect effects on fitness through its impact on A, H, N and their subsequent impact on fitness ($S \rightarrow X \rightarrow w$). **C:** Seed weight impacts height directly, but not A or N. Height, in turn, impacts A and N. A key limitation (and also a strength) of a path analysis is that **B** and **C** assume paths (**causality structures**), which can result in rather different interpretations of the patterns of selection, and both (in turn) can depart from the pattern suggested from a standard regression (**A**). For example, when trait A has a large effect on fitness, the path in **B** may find that any correlation between S and w is entirely due to the impact of S on A, while the path in **C** may find that selection is from the impact of H on A.

measured characters and fitness may show a spurious correlation if both are affected by inbreeding depression, which in turn can inflate the apparent strength of directional and concave selection (Willis 1996). Finally, immigration must be accounted for (e.g., Garant et al. 2005), especially if there are different selection regimes within the area that encompasses the study population (e.g., Figure 30.12).

Despite these concerns, the most severe caveat for the regression approach of estimating $w(z)$ is unmeasured characters. Estimates of the amount of direct selection acting on a trait are biased if that traits is phenotypically correlated with unmeasured characters that are also under selection (Lande and Arnold 1983; Mitchell-Olds and Shaw 1987). Adding one or more of these unmeasured characters to the regression can change initial estimates of β and γ. Conversely, selection acting on unmeasured characters that are phenotypically *uncorrelated* with those being measured has no effect on estimates of β and γ.

PATH ANALYSIS AND FITNESS ESTIMATION

As we saw in Figure 29.12, complex life cycles can be represented as graphs, as can a set of cascading developmental traits (Figure 30.13), the elaborate interplay between fitness components (Figure 30.14), or the connection between traits, fitness components, and population growth rate (Figure 30.16). One way to explore these graphical structures is with **path analysis**, which requires a **path diagram**, which is a *hypothesis* about the structure of causality (LW Appendix 2; Kingsolver and Schemske 1991; Mitchell 1992; Shipley 1997). The strengths of relationships in a path diagram are represented by **path coefficients**, which are correlations among connected items in the path. As a very simple example, consider the impact that seed weight and plant height have on fitness.

One potential path diagram is seed weight → height → fitness, which states that seed weight only impacts fitness through its effect on plant height. A path analysis of this model first variance-standardizes all of the variables and then performs regressions (or partial regressions), generating correlations among the items in the path. Suppose the resulting coefficients are

$$\text{weight} \xrightarrow{0.35} \text{height} \xrightarrow{0.15} w$$

implying a correlation of $\rho = 0.35$ between seed weight and plant height, and a correlation of 0.15 between height and fitness. As drawn, this diagram implies that seed weight does not have a direct effect on fitness, but it does have an indirect effect through height. Because ρ^2 is the fraction of variation accounted for by a factor, we see that seed weight explains 13% ($0.35^2 = 0.13$) of the variance in plant height and that plant height accounts for 2% (0.15^2) of the variance in fitness. Further, the correlation between any two connected items in a path is the product of their path coefficients, giving a correlation between seed weight and fitness of $0.35 \cdot 0.15 = 0.0525$, showing that seed weight accounts for 0.3% (0.0525^2) of the variance in fitness.

Regression and path analysis offer *complementary approaches* for examining relationships between phenotypes and fitness. The purpose of a regression analysis is to predict fitness given character values, while path analysis provides a description of the biological nature of character covariances and how they interact with fitness. While regressions simply rely on the correlations among traits (Figure 30.13A), path analysis examines the *structure* of these correlations (Figures 30.13B and 30.13C), building on the biological intuition of the investigator. A number of authors have applied this approach to the analysis of natural selection (e.g., Arnold 1983b; Maddox and Antonovics 1983; Mitchell-Olds 1987; Crespi and Bookstein 1988; Crespi 1990; Jordan 1991; Weis and Kapelinski 1994; Conner 1996; Pugesek and Tomer 1996; Scheiner et al. 2000; van Tienderen 2000; Coulson et al. 2003; Latta and McCain 2009; Matsumura et al. 2012).

The analysis of selection on graphs falls into four categories. First, we have already seen the use of Aster models on life-history graphs to obtain a statistically rigorous distribution of fitness effects (Chapter 29). Second, path analysis can be used to represent a proposed causal structure among life-history components and traits to provide insights that a standard fitness-trait regression can miss. Third, selection could be on one or more **latent** (unobserved) features that are correlated with observed (and potentially unselected) traits. While the observed traits may show trait-fitness associations, these could all be indirect effects, resulting from the correlation with the latent traits. This setting can also create high collinearity, with many of the measured traits being highly correlated with the latent traits. Path-analysis models can be used in some cases in the analysis of such data. Finally, **elasticity path analysis** connects selection gradients of a trait on fitness components with the elasticity (Chapter 29) of those components on the population growth rate (λ) to provide the elasticity of that trait on λ (van Tienderen 2000). This approach provides a powerful connection between selection and demography. We address these last three topics in order, while Aster model analysis of life-history graphs were introduced in Chapter 29.

While powerful, path-analytic methods are not without significant caveats. First, all of the issues with non-Gaussian fitness residuals that are a concern with regression methods fully apply to path analysis (which, itself, is a modified regression method). Second, path analysis assumes that there are only linear interactions, and hence only model directional selection. Scheiner et al. (2000) described how to include quadratic factors by augmenting

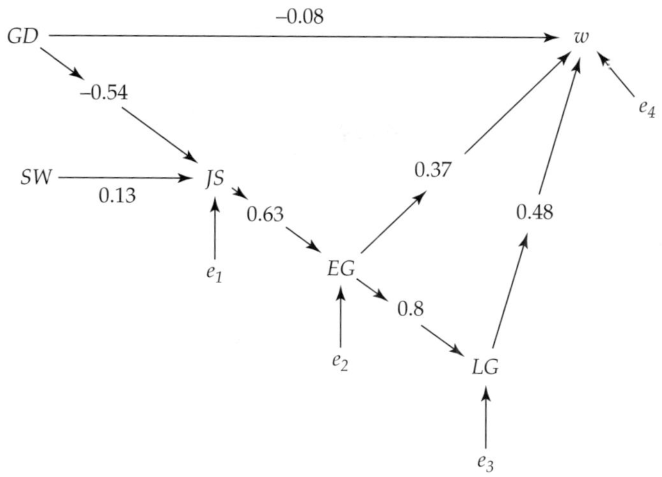

Figure 30.14 An example of the use of path analysis in describing multivariate selection from Mitchell-Olds and Bergelson (1990), who examined fitness (measured by adult size) for the annual plant *Impatiens capensis*. Germination date (GD), seed weight (SW), June size (JS), early growth rate (EG), and late growth rate (LG) were the measured characters, and e_i represents residual variance for trait i. Only significant paths, with their associated path coefficients (i.e., correlations) are shown. (After Mitchell-Olds and Bergelson 1990.)

each variable with its square and then performing a considerable amount of bookkeeping (see their paper for details). Finally, the results from a path analysis are *extremely* dependent on the assumed causality structure, so the interpretation of the results can be substantially biased if this structure is even slightly incorrect. Echoing the cautionary words of Kingsolver and Schemske (1991), "The uncritical application of path models to the analysis of selection in natural populations is likely to yield misleading and erroneous results."

Regressions Versus Path Analysis

The key distinction between a regression and a path analysis of selection is that regression attempts to statistically account for the covariance between fitness and the measured traits, while a path analysis further attempts to account for the *processes* generating correlations among the measured traits. Regressions assume that none of the traits are causal to any of the others, while Crespi (1990) noted that "path-analytic reasoning assumes that characters are correlated as a result of biological causes that should be used as information rather than adjusted away."

As an example, suppose traits are ordered in time, such as

$$\text{germination, } g \rightarrow \text{vegetative growth rate, } r \rightarrow \text{flowering time, } f$$

A regression analysis simply uses each trait's correlation with fitness without using the ordered structure inherent in the variables, while a path analysis explicitly uses this information. Examples of different casual assumptions for the same set of four traits are given in Figure 30.13. Each diagram displays a different causal structure that was assumed at the start of the analysis. Connections (arrows) that are nonsignificant or have very trivial values are usually left out of the final diagram, with the strength of connections often indicated by the line thickness of the arrows. Double-header arrows indicate correlations, namely, no assumption about the nature of causality between these variables, as would be the case for a standard regression (Figure 30.13A). A key limitation (and also strength) of a path analysis is that these different assumed path diagrams (causality structures) can result in rather different interpretations of the patterns of selection. If it is correct, the path diagram captures unique features missed by a regression (which *assumes* correlations, rather than attempting to *model how they arise*). If it is incorrect, a path analysis can introduce potentially serious biases.

Example 30.8. Mitchell-Olds and Bergelson (1990) measured fitness (using adult size as a proxy) in the annual plant *Impatiens capensis* as a function of five traits: seed weight, germination date, June size, early growth rate, and late growth rate. Figure 30.14 displays the significant paths between these variables and fitness. Note that the path diagram provides a description of the actual nature of the correlations, and in particular, the causal connections assumed between variables. From this diagram, we can examine the contribution to relative fitness from the direct effect of a character and from its indirect effects through its effect on other characters. For example, the direct effect of early growth is 0.37 (the path $EG \rightarrow w$), so this path accounts for 13% of the variation in fitness ($0.37^2 \simeq 0.13$). Early growth also influences fitness through an indirect path by influencing late growth rate, which in turn has a direct effect on fitness ($EG \rightarrow LG \rightarrow w$), and their product $0.8 \cdot 0.48 = 0.38$ accounts for an additional 15% (0.38^2) of the total variance in fitness. The total effect of early growth on fitness is the sum of the squared direct and indirect effects; here $0.37^2 + 0.38^2$, or 25%. Proceeding in a similar fashion, the direct and indirect effects for each trait on fitness are

Character	Direct Effect	Indirect Effect
Seed weight	0.04	0.10
Germination date	−0.08	−0.32
June size	−0.02	0.52
Early growth rate	0.37	0.38
Late growth rate	0.48	0.00

Negative coefficients imply that fitness increases with decreasing trait value. Observe that indirect effects are more important than direct effects for most characters.

While we have framed multiple-trait selection in the context of a number of traits in an individual and how these influence fitness, an equally common situation is when the "traits" are not properties of an individual, but rather biotic agents, such as the predators, competitors, or pollinators of the focal species. Path analysis allows one to examine the interactions among these multiple components and how they influence the fitness of the trait in the focal species.

Example 30.9. Weis and Kapelinski (1994) examined the nature of selective forces acting on gall size of a tephritid fly (*Eurosta solidadginis*), whose larvae makes a protective gall on goldrenrods. Larvae residing in small galls are susceptible to a parasitoid wasp (*Eurytoma gigantea*), whose ovipositor can reach the larvae through small, but not large, galls. Countering this, when galls become sufficiently large, they are preyed upon by insectivorous birds (mainly downy woodpeckers, *Picoides pubescens*). The result is selection pressure on both large and small galls, but through different biotic agents. Two other parasitoids, a second wasp and a beetle, also feed on larvae within galls, but here gall size does not appear to be a factor. However, galls attacked by these two insects appear to be distasteful, and the resulting parasitoids are not eaten by birds. Thus, this is a complex system with different agents and the possibility of frequency-dependent search images, all of which are acting to impart selection on the focal trait, which is gall size.

The authors examined two different path diagrams for selection intensity on gall size, one of which represented a conventional model of selection, with gall density and mean gall size included in the model. A second (search image-based) model replaced these two measures with the density of small galls and the density of large galls. Under the conventional model, there was a strong (but not significant) negative association (−0.43) between gall size and attacks by *Eurytoma* and a strong (and significant) association (0.37) between mean gall size and bird attacks. Conversely, under the frequency-dependent model, there was no association between small or large gall density and *Eurytoma* attacks, but a strong (and significant) association between density of small galls (−0.50) and density of large galls (0.46) and bird attacks.

As this example illustrates, a path-analysis model allows an investigator to examine the impact of several possible causality structures by leveraging known biological information (which does not enter into a standard regression analysis).

Morrissey's Extended Selection Gradient Vector, η

Consider the following two path diagrams connecting a focal trait (z), fitness (w), and some unmeasured factor (f):

path 1: $z \leftarrow f \rightarrow w$ path 2: $z \rightarrow f \rightarrow w$

Both paths generate a correlation between the focal trait and fitness, which results (in a regression analysis) in a nonzero gradient, β, when f is excluded (as z and w are correlated), and a β value of 0 when it is included (as, when it is conditioned on f, z and w are uncorrelated). Path 1 is the "missing trait" concern in a Lande-Arnold regression: if f is not included in the analysis, the regression falsely suggests a direct effect of z on fitness. Conversely, for path 2, z influences the unmeasured factor, f (e.g., some measure of performance; Arnold 1983b, 1988) which in turn influences fitness. While both paths generate a β value of zero when the factor f is included in the analysis, they reflect very different settings. In a regression analysis, we strive to avoid situations like those in path 1, while a nonzero value of β for path 2 when f is not in the analysis typically need not trouble us, as z has a causal influence on fitness (albeit an indirect one). Note that both paths can have identical correlation structures between all pairwise elements (and thus both would yield the exact same Lande-Arnold regression), which shows the importance of a causal diagram in the analysis of selection on correlated traits.

Motivated by situations like path 2, where z has an indirect causal impact on fitness but a β value of 0 if traits in the causal structure between z and w are included in the analysis (e.g., f in path 2), Morrissey (2014a) proposed the concept of an **extended selection gradient** vector, η. This variable measures the total selection caused by a trait and its downstream causative components. Recall (Equation 30.3) that the selection differential is measure of the total selection on a trait from both direct and indirect (correlated) selection, and that

$$S_i = \beta_i P_{ii} + \sum_{j \neq i} \beta_j P_{ij}$$

Morrissey's idea is that the extended selection gradient, η, is the *total selection* on a trait that is directly connected through its *causal* connections (as opposed to correlations) with w. For a given causal structure (path diagram), suppose we let $\beta_{pa,i}$ denote the selection gradient for the direct effect of a trait on fitness (i.e., for the single-arrow path $i \rightarrow w$). Likewise, let Φ_{ij} be the total impact over all connecting paths of the effect of i on trait j (see Example 30.10 for details). Then, by analogy with Equation 30.3,

$$\eta_i = \beta_{pa,i} \cdot 1 + \sum_{j \neq i} \beta_{pa,j} \Phi_{ij} \tag{30.33a}$$

As with S, the extended gradient decomposes into the direct effect from a trait (the path $i \rightarrow w$) plus the indirect effects from direct selection on all other traits that are causally associated with i ($i \rightarrow \cdots \rightarrow j \rightarrow w$).

One can compute the value of Φ_{ij} directly by taking the product of path coefficients for a path connecting i and j, and then summing over all such paths. However, Morrissey found a quicker approach for obtaining Φ_{ij} based on the matrix, $\mathbf{B}$, of path coefficients for a given causal structure. The elements of $\mathbf{B}$ are constructed as follows (see below for a worked example): if $i \rightarrow j$, then B_{ij} is the path coefficient between them, while $B_{ji} = 0$ because j does not influence i. Likewise, if there is not a *direct* connection between i and

j, then $B_{ij} = 0$, even if they are causally connected though intermediates; and finally, we define $B_{ii} = 0$. With $\mathbf{B}$ defined in this fashion, Morrissey found that the values of Φ_{ij} are calculated by

$$\boldsymbol{\Phi} = (\mathbf{I} - \mathbf{B})^{-1} \tag{30.33b}$$

From Equation 30.33a, we have

$$\boldsymbol{\eta} = \boldsymbol{\Phi}\boldsymbol{\beta}_{pa} = (\mathbf{I} - \mathbf{B})^{-1}\boldsymbol{\beta}_{pa} \tag{30.33c}$$

One very desirable feature of η_i is that if trait i has a causal effect (either direct or indirect) on fitness, then η_i is not changed by the addition (or subtraction) of new traits in the causal chain between i and w. In terms of our path 2 above ($z \to f \to w$), the value of η_z equals the value of β_z in a model where f is ignored ($z \to w$).

Example 30.10. Building on the work of Latta and McCain (2009), Morrissey (2014a) examined the relationship between a number of traits in wild oats (*Avena barbata*) and fitness, w (measured by number of reproductive spikes), using the following assumed causality structure. The values above the arrows are correlation coefficients, while the numbers 1 through 6 index the traits in the **B** matrix (as detailed below).

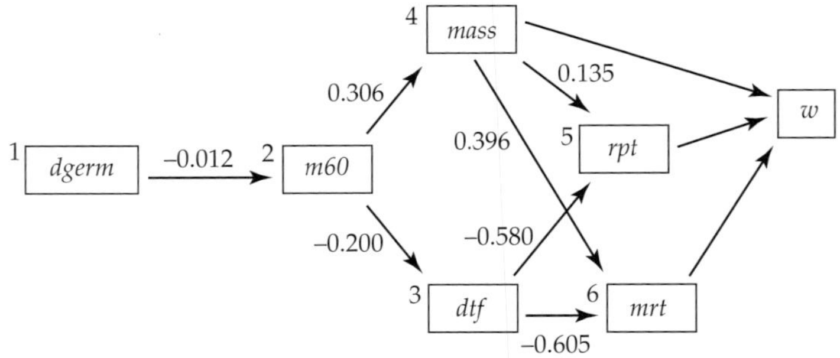

The six traits considered were days at germination (*dgerm*), mass at day 60 (*m60*), days to first flower (*dtf*), final total mass (*mass*), number of reproductive tillers (*rpt*), and mass of reproductive tillers (*mrt*). The (variance-standardized) directional selection gradients under Lande-Arnold ($\boldsymbol{\beta}_{LA}$, which is a multiple regression of w using all six traits) and under the path model ($\boldsymbol{\beta}_{pa}$) are quite different, which shows the impact of the assumed causal structure among the traits and with fitness:

$$\boldsymbol{\beta} = \begin{pmatrix} \beta_{dgerm} \\ \beta_{m60} \\ \beta_{dtf} \\ \beta_{mass} \\ \beta_{rpt} \\ \beta_{mrt} \end{pmatrix}, \quad \text{where} \quad \boldsymbol{\beta}_{LA} = \begin{pmatrix} 0.009 \\ 0.004 \\ -0.040 \\ -0.028 \\ 0.142 \\ 0.207 \end{pmatrix}, \quad \boldsymbol{\beta}_{pa} = \begin{pmatrix} 0 \\ 0 \\ 0 \\ -0.033 \\ 0.157 \\ 0.207 \end{pmatrix}$$

Note that only *mass, rpt*, and *mrt* directly impact fitness under the assumed path model.

The resulting matrix, **B**, of path coefficients is constructed as follows. The first row corresponds to those traits that are directly influenced by trait 1 (*dgerm*), which is only trait 2 (*m60*), as all the other entries are zero. The second row involves those traits that are directly influenced by trait 2, which are only traits 3 and 4, as all the other entries are zero. Continuing in this fashion, and noting that traits 5 and 6 influence no traits in the diagram (we are not considering w), yields

$$\mathbf{B} = \begin{pmatrix} 0 & -0.012 & 0 & 0 & 0 & 0 \\ 0 & 0 & -0.200 & 0.306 & 0 & 0 \\ 0 & 0 & 0 & 0 & -0.580 & -0.605 \\ 0 & 0 & 0 & 0 & 0.135 & 0.396 \\ 0 & 0 & 0 & 0 & 0 & 0 \\ 0 & 0 & 0 & 0 & 0 & 0 \end{pmatrix}$$

which in turn yields

$$\boldsymbol{\Phi} = (\mathbf{I} - \mathbf{B})^{-1} = \begin{pmatrix} 1 & -0.012 & 0.0240 & -0.0037 & -0.0018 & -0.0029 \\ 0 & 1 & -0.2000 & 0.3060 & 0.1573 & 0.2422 \\ 0 & 0 & 1 & 0 & -0.5800 & -0.6050 \\ 0 & 0 & 0 & 1 & 0.1350 & 0.3960 \\ 0 & 0 & 0 & 0 & 1 & 0 \\ 0 & 0 & 0 & 0 & 0 & 1 \end{pmatrix}$$

To see that **B** recovers the correct values of the Φ_{ij}, we compute a few of these elements by hand. The elements in the first row of $\boldsymbol{\Phi}$ correspond to the total causal effects from trait 1 (*dgerm*), which is -0.12 on trait 2 (*m60*), and $-0.012 \cdot 0.306 = -0.0037$ for trait 4 ($1 \rightarrow 2 \rightarrow 4$). A more complicated case is the casual connection of trait 1 on trait 5, as there are two paths, $1 \rightarrow 2 \rightarrow 4 \rightarrow 5$ and $1 \rightarrow 2 \rightarrow 3 \rightarrow 5$, for a total of $[-0.012 \cdot 0.306 \cdot 0.135] + [-0.012 \cdot (-0.200) \cdot (-0.580)] = -0.0018$.

Using $\boldsymbol{\Phi}$, the resulting vector of extended gradients becomes

$$\boldsymbol{\eta} = \begin{pmatrix} \eta_{dgerm} \\ \eta_{m60} \\ \eta_{dtf} \\ \eta_{mass} \\ \eta_{rpt} \\ \eta_{mrt} \end{pmatrix} = \boldsymbol{\Phi}\boldsymbol{\beta}_{path} = \boldsymbol{\Phi} \begin{pmatrix} 0 \\ 0 \\ 0 \\ -0.033 \\ 0.157 \\ 0.207 \end{pmatrix} = \begin{pmatrix} -0.001 \\ 0.065 \\ -0.216 \\ 0.070 \\ 0.157 \\ 0.207 \end{pmatrix}$$

The extended gradient ($\boldsymbol{\eta}$) gives the total amount of selection on each trait caused by all the paths that are causally connected to that trait. The values for *mrt* and *rpt* are the same as their β_{pa} values, as both of these traits have only a direct effect on fitness. Now consider the difference between β_{pa} for *mass* in the path diagram (-0.033) and its η value (0.070). While the direct effect, *mass*$\rightarrow w$, is negative, the *total* of the casual effects from *mass*, (*mass*$\rightarrow w$) + (*mass*$\rightarrow$ *rpt*$\rightarrow w$) + (*mass*$\rightarrow$ *mrt*$\rightarrow w$), is

$$-0.033 + (0.135 \cdot 0.157) + (0.396 \cdot 0.207) = 0.070$$

which is positive and larger in magnitude than the direct negative effect.

Selection on Latent Variables

Another application of path analysis arises when unmeasured traits that are under selection influence measured traits (Maddox and Antonovics 1983; Crespi and Bookstein 1988; Crespi 1990; Pugesek and Tomer 1996). This bane of a regression analysis can, *potentially*, be partly addressed with path analysis. Further, many of the measured traits may be highly correlated, raising issues of high collinearity (and instability in regression estimates). If this correlation structure is in large part due to one, or a few, underlying latent variables, then by focusing on these underlying variables, this problem may largely disappear.

For example, one common concern is that apparent selection on morphological traits may simply be a reflection of general selection on size (Figure 30.15). To this end, Crespi and Bookstein (1989) and Crespi (1990) suggested first extracting the leading principal component (PC 1) associated with the covariance matrix of the measured traits, which typically returns a generalized measure of size. Crespi and Bookstein then considered the univariate correlation of fitness of each size-adjusted trait, ignoring any further details about their correlation structure once size is removed.

Care must be taken, however, when the traits show any allometry (growth pattern) other than **isometric growth** (shape is independent of size; see LW Chapter 11). In such cases, PC 1 can contain both size *and* shape information. Jolicoeur (1963) and Somers (1986) suggested extracting PC 1 from the *correlation* matrix (as opposed to the covariance matrix)

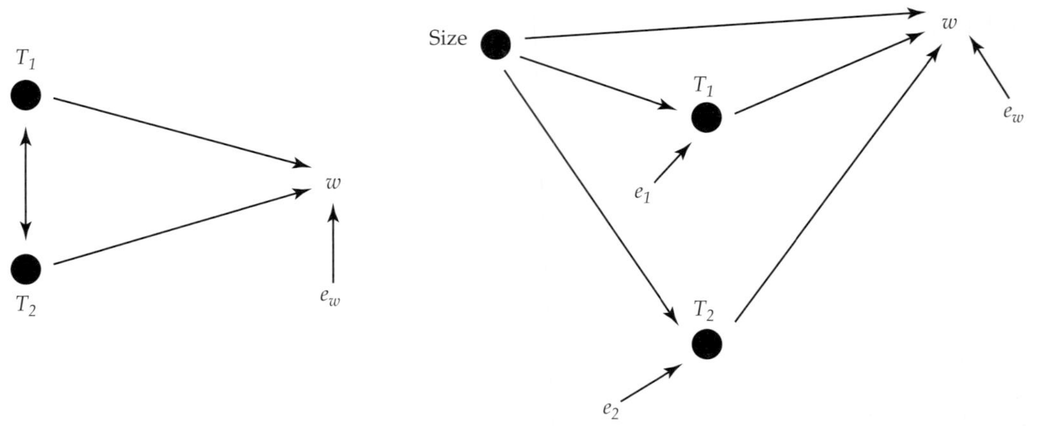

Figure 30.15 Morphological traits T_1 and T_2 being examined for their effects on fitness, w. **Left:** Under a standard Lande-Arnold regression, we allow for direct effects of both traits on fitness (w) as well as for indirect effects from correlations between these two traits (double-headed arrow). Here, e_w represents the additional variance in fitness that is not accounted for by the paths through T_1 and T_2. **Right:** Suppose size influences both traits as well as fitness. When we ignore size, it is possible that the correlation between trait value and fitness arises because the trait is serving as a surrogate for size. Because this is a missing trait (size was not included in the analysis), this effect would not be detected by a Lande-Arnold regression. However, by extracting a surrogate measure for size (such as PC 1) from the data, we can include it in the analysis. The path diagram allows for direct paths between size and fitness, and between both traits and fitness, after size effects have been removed ($T_1 \to w, T_2 \to w$).

for log-transformed traits. When isometric growth is present, the eigenvector corresponding to this first PC should have all traits weighted equally (Jolicoeur 1963), with (for n traits) each element in the vector being $1/\sqrt{n}$. This provides a check of whether the assumption of isometric growth is appropriate. Similarly, the latent trait could be some measure of performance (Arnold 1983b, 1988) that selection acts directly upon and that is influenced by the measured traits (Chapter 29).

Elasticity Path Diagrams

As we have seen, even very complex life histories can be represented as graphs (Figures 29.12 and 30.14). The impact of small demographic changes in the resulting transition matrix (or **projection matrix**) associated with this graph is measured by the elasticities of the elements in this matrix with respect to their impact on the population growth rate, λ (Chapter 29; Caswell 1989, 2001). If an element in the projection matrix has an elasticity of e, then a proportional change of f in that element results in a proportional change of $f \cdot e$ in λ (Equation 29.4b). Changes in elements with larger elasticities have a proportionally greater impact on λ. For example, recall that Sæther and Bakke (2000) found average elasticity values in birds of 0.6 and 0.25 for viability and fecundity, respectively, which showed that changes in viability result in a much greater proportional change in the growth rate.

Conner (1996) suggested that path analysis can be used to connect traits, their impact on fitness components, and the impact of those fitness components on the population growth rate, λ (namely, how changing the mean of a trait alters the mean population growth rate). In a traditional path-analysis setting, such an analysis would return correlations between these various elements (e.g., Example 30.8; Figure 30.14), and while these correlations have a straightforward biological interpretation, they are disconnected from our standard metric of selection gradients and population growth. In a classic (but often underappreciated) paper, van Tienderen (2000) showed that using elasticities in place of path coefficients (an **elasticity path analysis**, or **elastogram**) connects trait selection, fitness components, and the population growth rate. Equation 29.33f describes the elasticity of a trait on λ as the sum over all

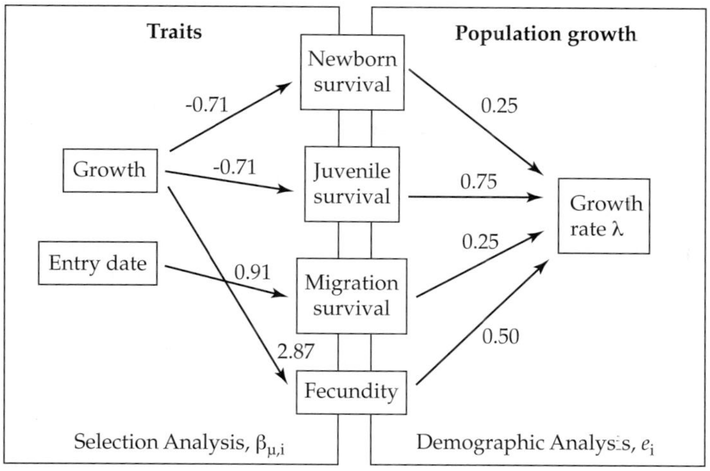

Figure 30.16 An example of the use of elastograms for problems arising in fishery management, showing an elastogram connecting two traits, annual juvenile growth rate and date of entry of a migrating individual into a river, with four fitness components influencing the population growth rate, λ. The impact of traits on the fitness components is assessed through their mean-standardized selection gradients (β_μ) on each fitness component, while a matrix projection model was constructed to obtain the elasticities (e_i) of these fitness components on λ. Details are presented in Example 30.11. (After Matsumura et al. 2012.)

fitness components of the product of its mean-standardized gradient ($\beta_{\mu,i}$) for fitness component, W_i, times the elasticity, $e_i = \partial\ln(\lambda)/\partial\ln(W_i)$, of that component on λ,

$$\frac{\partial\ln(\lambda)}{\partial\ln(\mu)} = \sum_{i=1} \frac{\partial\ln(\lambda)}{\partial\ln(W_i)}\frac{\partial\ln(W_i)}{\partial\ln(\mu)} = \sum_{i=1} e_i\beta_{\mu,i} \tag{30.34}$$

Example 30.11 and Figure 30.16 show the power of this approach, and a detailed example is given by Coulson et al. (2003).

Example 30.11. Matsumura et al. (2012) simulated a model examining the impact of traits on the success of spawning migrations of salmon. Two traits of interest were the annual juvenile growth rate and the date of river entry for migrating individuals. Three survival measures (newborn, juvenile, and migration) as well as fecundity were taken as the fitness components. The impact of a trait on each of these components was measured by mean-standardized selection gradients, which correspond to elasticities of these components with respect to the trait (Equation 29.33e). Likewise, a demographic analysis using a population projection matrix (Chapter 29) evaluated the elasticities of these fitness components with respect to the population growth rate λ (see their paper for details). Figure 30.16 presents the resulting elastogram. Consider the impact of entry date on λ. This trait impacts a single fitness component (migration survival) with an elasticity (mean-standardized gradient, β_μ) of 0.91, and migration survival has an elasticity (e_i) of 0.25 on λ, yielding (from Equation 30.34) an elasticity on λ as a function of this trait of $0.91 \cdot 0.25 = 0.23$. Hence, a 10% increase in mean entry date yields a 2.3% increase in the population growth rate. The second trait, juvenile growth rate, has significant effects on three fitness components. Its elasticity on λ through newborn survival is $-0.71 \cdot 0.25 = -0.18$; through juvenile survival is $-0.71 \cdot 0.75 = -0.53$, and through fecundity is $2.87 \cdot 0.50 = 1.44$; for a total elasticity (through all fitness components) of $-0.18 -0.53 + 1.44 = 0.725$. Thus, a 10% increase in mean juvenile growth rate results in a 7.3% increase in λ, mainly through its impact on fecundity (a 14% increase) and despite decreases of 1.8% and 5.3% in newborn and juvenile survival, respectively.

LEVELS OF SELECTION

Finally, while our focus (thus far) has been on predicting fitness given the phenotype of an *individual*, it is also possible that selection acts at levels above the individual, e.g., through **group-level effects** (Chapter 22). For example, Breden and Wade (1989) found that increasing larval group size in the imported willow leaf beetle (*Plagiodera versicolora*) increases individual survivorship, with each additional group member increasing fitness by ~7%. Here, we briefly examine the analysis of selection at multiple levels. While one can use Price's equation to decompose the selection differential of a trait into within-group and among-group components, as we will show, this can be misleading. Akin to the fact that selection differentials are unreliable indicators of the targets of selection when traits are correlated, this is also is the case when the traits have selective impacts at both the individual and group levels. As above, the amount of direct selection on a trait at a specific level of selection can be determined by using the appropriate multiple regression.

Contextual Analysis

Heisler and Damuth (1987) proposed that any effects of selection acting at some level above that of the individual can be estimated with the method of **contextual analysis**, which is widely used in the social sciences (e.g., Boyd and Iversen 1979). This regression-based approach amounts to simply adding group-level traits to a Lande-Arnold regression. Traits scored at the level of groups can be **aggregate characters** (simple functions of the individual phenotypes within a group, such as the group mean), or they can be **global** or **emergent characters** that can only be defined at the group level (such as number of group members or group density). As the following example illustrates, incorporating group-level traits into the Lande-Arnold fitness regression is straightforward.

Example 30.12. Aspi et al. (2003) examined the effects of selection on the riparian plant *Silene tatarica*, a threatened species from Finland in the family Caryophlleaceae. Plants tend to grow in patches and the authors envisioned that plant density within a patch might influence both pollinator visits and herbivory. The individual traits measured were plant height (z_1) and number of stems (z_2), while two aggregate traits were measured (the means $\overline{z}_1$ and $\overline{z}_2$ of each trait for the patch) along with the group-level trait of plant density, d. The resulting regression model for predicting the relative fitness of individual j in patch i becomes

$$w_{ij} = 1 + \beta_1 z_{1,ij} + \beta_2 z_{2,ij} + \beta_3 \overline{z}_{1,i} + \beta_4 \overline{z}_{2,i} + \beta_5 d_i + e_{ij}$$

The regression coefficients β_1 and β_2 correspond to direct selection on individual trait values, while β_3 and β_4 correspond to direct selection on the patch mean of each trait, and β_5 to direct selection on the density within a patch. Aspi et al. variance-standardized all variables, so a one standard deviation change in the variable of interest results in an expected change of β in fitness. For data collected in 1999, the estimated regression coefficients for a sample of 922 individuals were

	Height	Mean height	Stem No.	Mean stem No.	Density
β	0.589^{***}	0.653^{***}	0.187^{**}	-0.209^{***}	0.631^{***}

All of the β were significant, with ** denoting $p < 0.01$ and *** denoting $p < 0.001$. Note that (on a standardized scale) selection on group density is as strong as individual selection. Selection on height at the individual and group levels was in the same direction, which the authors suggested was due to pollinator attraction. However, selection on stems was in opposite directions, with individual selection increasing the number of stems, but patch-level selection decreasing them. The authors suggested that patch-level selection may be due to herbivory by reindeer.

A contextual analysis need not stop at two levels, as this approach easily extends to additional hierarchical levels of population organization, and hence allows for the potential for selection at these higher levels as well. Indeed, in some settings the individual level is ignored, and only higher levels of organization are contrasted. One example of this involves the work of Banschbach and Herbers (1996), who compared selection on nest- versus colony-level traits in a forest ant (*Myrmica punctiventris*), and found that fertility selection largely operates at the level of the nest, as opposed to the level of multiple-nest colonies.

In many settings, group composition is fairly obvious, due to a patchy distribution of individuals in space. However, in some situations, the appropriate set of individuals to include in a group is unclear. Indeed, for populations where individuals appear to be continuously distributed in space with few obvious breaks, group-level effects may be present when very close individuals are included yet largely disappear as the defined group becomes larger. In theory, one could assign different group sizes and use model-selection approaches to determine the group structure that gives the best fit of the fitness regression. However, care must be taken to distinguish a statistical fit from biological reality. A very reasonable biological definition of a group was offered by Uyenonyama and Feldman (1981), namely, the set of all individuals that influence the fitness of a focal individual.

An alternative to using group means is to model neighbor interactions that influence fitness (e.g., Equation 22.59a). In such cases, the fitness of a focal individual, i, can be written (for a single trait) as

$$w_i = 1 + \beta_1 z_i + \beta_2 \left(\sum_{i \neq j} z_j \right) + e_i \tag{30.35a}$$

Note that under this model, the group mean (which contains the value of the focal individual) is replaced by the interactions caused by the neighbors of the focal individual (the members in the sum). This model is easily connected to the model using the standard selection on group model by noting that

$$\sum_{i \neq j} z_j = n\overline{z} - z_i$$

when i interacts with $n - 1$ neighbors. Hence, we can express Equation 30.35a as

$$\begin{aligned} w_i &= 1 + \beta_1 z_i + \beta_2 \left(n\overline{z} - z_i \right) + e_i \\ &= 1 + (\beta_1 - \beta_2) z_i + \beta_2 \, n\overline{z} + e_i \end{aligned} \tag{30.35b}$$

The subtle distinction between the group-mean (Equation 30.35b) and neighborhood-fitness (Equation 30.35a) models is apparent in the case where only the neighbors influence the fitness of an individual. In this case, $\beta_1 = 0, \beta_2 \neq 0$, so that in the neighborhood-fitness model there is no weight on individual value. However, if a group-mean fitness model is fitted to these same data, there will be a nonzero regression slope on individual fitness (a value of $-\beta_2$), reflecting the input of an individual's value to its group mean. One could also weight the interactions in Equation 30.30a, for example, by replacing z_j with $f_{ij} z_j$, where f_{ij} is some measure of the amount of interaction between i and j, such as distance between plants or fraction of time observed interacting (Chapter 22).

Selection Can Be Antagonistic Across Levels

When selection is operating at both the individual and group levels, there is no a priori reason why its direction should be the same at the two levels. When the direction is the same and group-level effects are ignored, the effects of individual selection are overestimated. The more interesting case is that in which individual and group-level selection are **antagonistic**, working in opposite directions. Example 30.12 illustrates both cases: for height, selection at both levels was in the same direction, while for stem number, selection was in opposite directions at the individual and group levels. Antagonistic selection is commonly seen in the limited number of studies that have estimated multilevel selection components, although this might, in part, be due to a bias in initially choosing traits that are expected to be under differential individual, versus group, selection pressure.

One of the first applications of contextual analysis to levels-of-selection was by Stevens et al. (1995), using jewelweed (*Impatiens capensis*). They found that an overall measure of size was selected to decrease at the group level but to increase at the individual level. This pattern was observed when fitness was taken as either survival rate to two years or as seed production in open-pollinated cleistogamous flowers. The individual and group partial regression slopes (variance-standardized over trait values) for survival rate on overall size were $\beta_I = 1.74$ and $\beta_{group} = -3.03$. Hence, group selection was stronger (and of opposite sign) than individual selection. The partial regression slopes for cleistogamus seed production on size were $\beta_I = 0.51$ and $\beta_{group} = -0.52$. Similar observations were made by Weinig et al. (2007), who used a set of recombination inbred lines (RILs) in *Arabidopsis thaliana*, and found that two composite traits (size and elongation) were favored to increase under individual selection but to decrease under group selection. For both traits, individual selection was stronger than group selection (the variance-standardized β values were roughly twice as large). A final example is a study by Tsuji (1995), who worked with the Japanese queenless ant (*Pristomyrmex pungens*), an unusual species with no queens, whose workers are able to reproduce parthenogenetically. Larger size was favored at the individual level ($\beta_I = 0.07$), but selected against at the colony level ($\beta_{group} = -0.11$).

Group Selection Is Likely Density-Dependent

A number of workers (Goodnight et al. 1992; Donohue 2003, 2004) have suggested that at low density, group selection may be weak (or essentially absent), with individual selection dominating. As group density increases, so does competition, which may increase group effects. An empirical observation from plant ecology, the **law of constant final yield** (Harper 1977; Weiner and Freckleton 2010), has been offered as support for this view (Goodnight et al. 1992). Plant ecologists have noted that the total yield of a group initially increases with low density, but after a certain critical density is reached, the total biomass of the group remains roughly constant. Even though more plants are now in the group, their individual contribution has decreased. Goodnight et al. suggested that this law results from a balance between group and individual selection. One consequence of potential *density*-dependence at the group level is that an analysis of individual selection that ignores a group selection component can be mistaken for *frequency*-dependent selection (Uyenonyama and Feldman 1980, 1981; Damuth and Heisler 1988).

Working with the Great Lake sea rocket (*Cakile edentual*), a plant in the family Brassicaceae, Donohue (2004) did indeed observe that plant density influenced the relative strengths of individual and group selection. For plant height and stem mass, she observed antagonistic individual and group selection at varying densities, with individual selection favoring smaller plants with more stem mass, and group selection favoring larger plants with less stem mass. At all densities, individual selection remained significantly stronger than group selection. Surprisingly, group selection was strongest at the intermediate, as opposed to the high, density. Weinig et al. (2007) also found a density-dependent effect in *Arabidopsis thaliana*, where the density must be above some threshold before group selection becomes important.

Selection Differentials Can Be Misleading in Levels of Selection

Price (1972a) and Wade (1985) showed how to decompose the selection differential into individual and group components. They did so by using the Robertson-Price identity (Equation 6.10), which states that the within-generation change in a trait can be written as its covariance with relative fitness, $S = \sigma(w, z)$. This also holds more generally for levels of selection. Suppose we let w_{ij} and z_{ij} denote, respectively, the fitness and phenotype of the jth individual in group i. Using the definition of the covariance, Price showed that the total selection differential can be decomposed as

$$\sigma(z, w) = E_i\big[\sigma(w_{ij}, z_{ij})\big] + \sigma(\overline{z}_i, \overline{w}_i) \tag{30.36}$$

The first term is the covariance between individual phenotype and fitness within a group, and we take its average over all groups to obtain the within-group covariance between individual value and fitness. The second term is the covariance between the mean trait value, $\overline{z}_i$, and the mean fitness, $\overline{w}_i$, of the group. Equation 30.36 decomposes the selection differential into components from individual effects within groups and components from group effects.

Motivated by Equation 30.36, Price suggested that group selection implies a nonzero covariance between the mean trait value and mean fitness of groups, namely, $\sigma(\overline{z}_i, \overline{w}_i) \neq 0$. Under this definition, group selection cannot occur if there are no among-group differences in mean fitness. While this sounds reasonable, as we will see shortly, there are indeed situations where each group has the same mean fitness but there is still selection on the group means. Hence, group selection can occur even when the covariance between mean trait value and mean fitness is zero. Likewise, even when there is selection only on individual phenotypes, if group means differ, then (as we will show shortly) one can easily get a nonzero value for $\sigma(\overline{z}_i, \overline{w}_i)$.

Hence, this **covariance definition** (a nonzero group selection differential) for group selection is misleading, as one can have a group selection differential of zero when group selection *is* occurring, and a nonzero group selection differential when it *is not*. The root cause of this is a familiar problem: selection on phenotypically correlated traits can modify the selection differential of a focal trait, in the extreme showing an indirect response (nonzero S) when there is no direct selection on the trait. This also applies when considering selection differentials on group means. Indeed, by analogy to using a multiple regression to control for correlated traits, this was the motivation of Heisler and Damuth (1987; also Goodnight et al. 1992) for using contextual analysis. The partial regression coefficients in a contextual analysis control for any indirect effects of other correlated traits included *within the analysis*.

One consequence of indirect effects generating nonzero covariances (and hence nonzero selection differentials) is that an analysis restricted to one level of selection can be misleading. For example, suppose an investigator assumes that group selection might be important, and therefore only includes group means in the analysis. For a single trait, the investigator might consider the slope of a simple (i.e., univariate) regression of mean fitness on group trait mean,

$$\overline{w}_i = 1 + \beta\,\overline{z}_i + e_i \tag{30.37a}$$

If we recall (LW Chapter 3) that the slope of a univariate regression is the covariance divided by variance of the predictor variable, the slope becomes

$$\beta = \frac{\sigma(\overline{z}_i, \overline{w}_i)}{\sigma^2(\overline{z}_i)} \tag{30.37b}$$

Hence, a nonzero covariance implies a nonzero slope, and an investigator using this simple univariate regression would conclude that group selection is occurring on this trait. In reality, as we will see shortly, strict individual selection can generate a group-level covariance, and hence (in this case), a spurious assignment of a group-level effect to selection. By using a multiple regression that includes both individual and group mean values, a contextual analysis corrects for this concern (provided selection is limited to only these two levels).

Example 30.13. Recall the previous example, discussing the work of Aspi et al. (2003), who applied contextual analysis to height and stem number in *Silene tatarica*. They reported the within- and among-patch selection differentials as

Trait	Within-patch S	Among-patch S
Height	0.485***	0.025 ns
Stem number	0.855***	−0.006 ns

The use of selection differentials correctly determined the presence of positive directional selection on both individual height and stem number. However, the analysis also suggested that there were no among-patch effects (no selection on patch means), when in fact contextual analysis shows that these effects are highly significant for both traits (Example 30.12).

Hard, Soft, and Group Selection: A Contextual Analysis Viewpoint

Evolutionary biology has had an unfortunate history of, at times, being constrained by thinking about complex evolutionary processes in terms of simple catch phrases. We have seen examples of this with kin versus group selection, showing in Chapter 22 that these are special cases of the more general problem of selection with interacting individuals. Likewise, in this (and the previous) chapter, we have seen that the widely used terms of stabilizing and disruptive selection are special cases of the more general situation of a quadratic term (curvature) in the individual fitness surface. It should therefore not be surprising that there is some confusion with the exact meaning of group selection in the more general context of multilevel selection (reviewed by Okasha 2004, 2006). In the Price covariance framework, group selection is indicated by a nonzero covariance between the mean fitness and the mean trait value of a group, $\sigma(\overline{z}_i, \overline{w}_i) \neq 0$. This line of reasoning implies that across groups there must be variance in mean fitness, $\sigma^2(\overline{w}_i) \neq 0$, for group selection to exist. However, as stressed by Goodnight et al. (1992; Goodnight 2013, 2015), these covariance conditions for defining group selection are misleading. One can have group selection in the absence of variance in mean fitness across groups, and likewise, strict individual selection can still generate a covariance between group mean fitness and group mean trait value.

Just as the more general framework of selection in interacting populations removes much of the historical confusion between group selection and kin selection (Chapter 22), multilevel selection removes much of the confusion over whether group selection is occurring. Under multilevel selection, we simply have the fact that the fitness of a focal individual is influenced by other individuals, and hence selection has both individual and group components. The latter are very clearly defined by nonzero regression coefficients on group traits in a contextual analysis. As succinctly stated by Goodnight et al. (1992), "with contextual analysis, whether selection is acting at a particular level becomes a statistical rather than a philosophical question."

Wallace's (1968, 1976) distinction between hard and soft selection serves to highlight the differences between the covariance and multiple-regression definitions of group selection (Goodnight et al. 1992). **Hard selection** occurs when the fitness of a focal individual is solely determined by its phenotype and is unaffected by any group members. This results in the mean fitness of groups differing, with some groups leaving more offspring than others, as by chance, some groups are compromised of more fit individuals than others. Conversely, **soft selection** occurs when each group contributes equally to the next generation, which means that group mean fitness is a constant and there is no variance in mean fitness among groups. The law of constant yield is an example of soft selection, with fitness (number of seeds) being largely independent of patch density. Figure 30.17 shows the different consequences of hard, soft, and (strict) group selection on the within- and among-group regressions of phenotype on fitness.

As summarized in Table 30.2, hard, soft, and strict group selection can all be modeled as special cases of a contextual analysis regression. For simplicity, we focus on a single trait, and the resulting regression has the form

$$w_{ij} = 1 + \beta_1 z_{ij} + \beta_2 \overline{z}_i + e_{ij} \tag{30.38a}$$

We can also rewrite this regression in terms of selection on the within-group deviation plus selection on the group means,

$$w_{ij} = 1 + \beta_1^* (z_{ij} - \overline{z}_i) + \beta_2^* \overline{z}_i + e_{ij} \tag{30.38b}$$

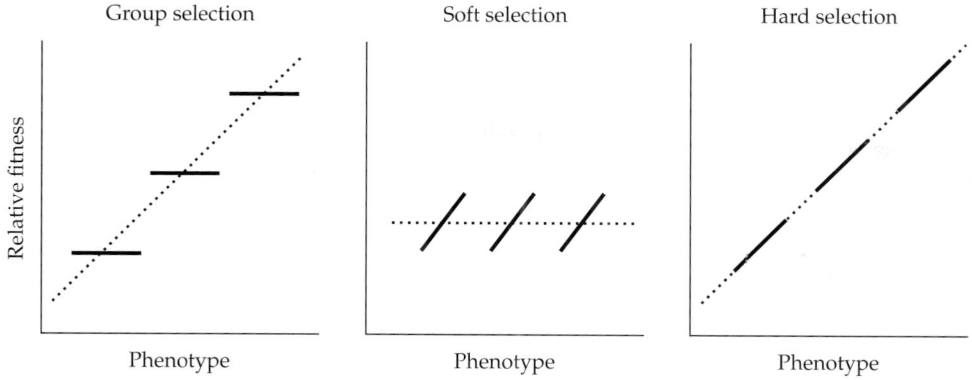

Figure 30.17 The regression of fitness on trait value within each group (solid line) and between the group means (dotted line). Under strict group selection (**left**), an individual's phenotypic value within a group has no predictive value on fitness (a within-group regression slope of zero), while the regression of fitness on group means is positive (a nonzero regression of individual fitness on group mean). Under soft selection (**center**), fitness is entirely determined by the within-group deviation. The result is a significant within-group regression but no among-group regression. Finally, under hard selection (**right**), fitness is entirely a function of phenotype, so that both the within-group and among-group regressions have the same slope (solid lines embedded with the dotted line with the same slope). (After Goodnight et al. 1992.)

Table 30.2 Hard, soft, and strict group selection can all be expressed as special cases of a more general contextual analysis. Model 1 presents the contextual regression coefficients (Equation 30.38a) on individual value (z_{ij}) and group mean $(\overline{z}_i)$. Model 2 presents the regression coefficients when the contextual analysis is framed in terms of within-group deviations and group means (Equation 30.38b). To contrast these results with those from the covariance criteria for group selection, the table also shows whether the type of selection results in a nonzero within-group (S_w) or among-group (S_b) selection differential and whether the among-group variance in mean fitness, $\sigma^2(\overline{w}_i)$, is nonzero.

	Model 1		Model 2				
Selection	z_{ij}	$\overline{z}_i$	$(z_{ij} - \overline{z}_i)$	$\overline{z}_i$	$S_w \neq 0$	$S_b \neq 0$	$\sigma^2(\overline{w}_i) > 0$
Hard	β	0	β	β	Yes	Yes	Yes
Soft	β	$-\beta$	β	0	Yes	No	No
Strict group	0	β	0	β	Yes	Yes	Yes

We use β_i^* to stress to the reader that the regression coefficients in Equation 30.38b can differ from those in Equation 30.38a. To see the connection, note that

$$w_{ij} = 1 + \beta_1^* z_{ij} + (\beta_2^* - \beta_1^*)\, \overline{z}_i + e_{ij} \tag{30.38c}$$

showing that β_1 is the same under both models ($\beta_1 = \beta_1^*$), while $\beta_2 = \beta_2^* - \beta_1^*$.

Hard selection occurs when selection is entirely on individual value, so that $\beta_2 = 0$ and hence there is no group component to multilevel selection. However, by chance, some groups will have more high-fitness individuals than others. This generates variance in group mean fitness and a nonzero covariance between group-mean trait values and mean fitness (Goodnight et al. 1992). Under this covariance criteria, one would infer group selection, even though it is absent. Here, a nonzero among-group selection differential arises as a correlated response to direct selection on individual value. In this simple case, the contextual analysis clearly shows no multilevel selection, but a simple univariate regression might suggest otherwise (e.g., Equation 30.37a).

The opposite conclusion arises with soft selection. Here selection is entirely based on within-group deviations (such as within-family selection; Chapter 21). Hence, $\beta_1 = -\beta_2$ as the regression of fitness becomes

$$w_{ij} = 1 + \beta\,(z_{ij} - \overline{z}_i) + e_{ij}$$
$$= 1 + \beta z_{ij} - \beta\,\overline{z}_i + e_{ij} \tag{30.39a}$$

Under soft selection, there is no variation in mean fitness across groups and $\sigma(\overline{z}_i, \overline{w}_i) = 0$. Hence, under Price's covariance definition, one might infer that there was no group selection. However, there is clearly multilevel selection occurring, as $\beta_2 \neq 0$ (cf. Equation 30.38a and Equation 30.39a). The lack of among-group variance in fitness shows that the effects of individual selection at the group level are exactly countered by selection at the group level.

Finally, consider strict group selection, when an individual's fitness is entirely a function of its group mean,

$$w_{ij} = 1 + \beta\,\overline{z}_i + e_{ij} \tag{30.39b}$$

while written here as a univariate regression, formally we have $\beta_1 = 0$ if the regression given by Equation 30.38a is fitted. If one simply fits an individual selection model ($w_{ij} = 1 + \beta\, z_{ij} + e_{ij}$), there is a nonzero regression slope if group effects are ignored, and hence $E_i\big[\sigma(w_{ij}, z_{ij})\big] \neq 0$ (Goodnight et al 1992). Again this occurs because individual value and group mean are correlated, as $\sigma(z_{ij}, \overline{z}_i) = \sigma^2_z/n$ (plus addition terms if group members are correlated; see Equation 22.25b). Hence, direct selection on one results in indirect selection (and hence a covariance) in the other. With contextual analysis, we reach the correct conclusion of selection only on group values.

Early Survival: Offspring or Maternal Fitness Component?

As mentioned in Chapters 15 and 22, a classic example of a levels-of-selection problem is maternal care, wherein features of both the mother and offspring may influence offspring survival. An area of some contention, especially among biologists working with species that display extensive maternal care (i.e., ornithologists and mammalogists), is whether the fitness associated with early survival should be assigned to the mother (acknowledging her maternal performance) or to the offspring (acknowledging features of their own that enhance survival). As mentioned in Chapter 29, the fitness of a mother is often scored as the number of her offspring remaining *after* some interval of potential early offspring mortality, rather than simply the total number of offspring that she produced. For birds, the number of hatchlings, fledglings, or recruits (offspring that have mated) are used as the measure of female fitness, rather than the number of eggs. In mammals, the number of weaned offspring or recruits is often used in place of number of births. This notion of fitness crossing generations has proponents (e.g., Clutton-Brock 1988) and opponents (Arnold 1983a; Lande and Arnold 1983; Cheverud 1984b; Thomson and Hadfield 2017), in part because different questions are being asked by the two sides. The implications of assigning fitnesses of offspring to their mothers (i.e., their early survival is part of their mother's fitness), and what potential biases this may induce, were examined by Wolf and Wade (2009; also see Thomson and Hadfield 2017), whose approach we closely follow.

Suppose the true fitness (such as early offspring survival) is a function of the offspring phenotype, $z_{o(t)}$, *and* the value of some trait in their mother, $z_{m(t-1)}$, where t indexes the current offspring generation and $t-1$ indexes a phenotype (in this case its mother) in the previous generation. This is an example of what Kirkpatrick and Lande (1989) called maternal selection (Chapter 15). The relative fitness of a male offspring is given by

$$w_{(t)}[\text{male}] = \alpha_{(t)} + b_m z_{m(t-1)} + b_o z_{o(t)} + \epsilon_{(t)} \tag{30.40a}$$

where b_o and b_m measure the strengths of selection on offspring value and maternal care, respectively. In female offspring, there is an additional source of potential selection, as a sex-limited maternal trait can have a cost. For example, a mother with a trait for increased

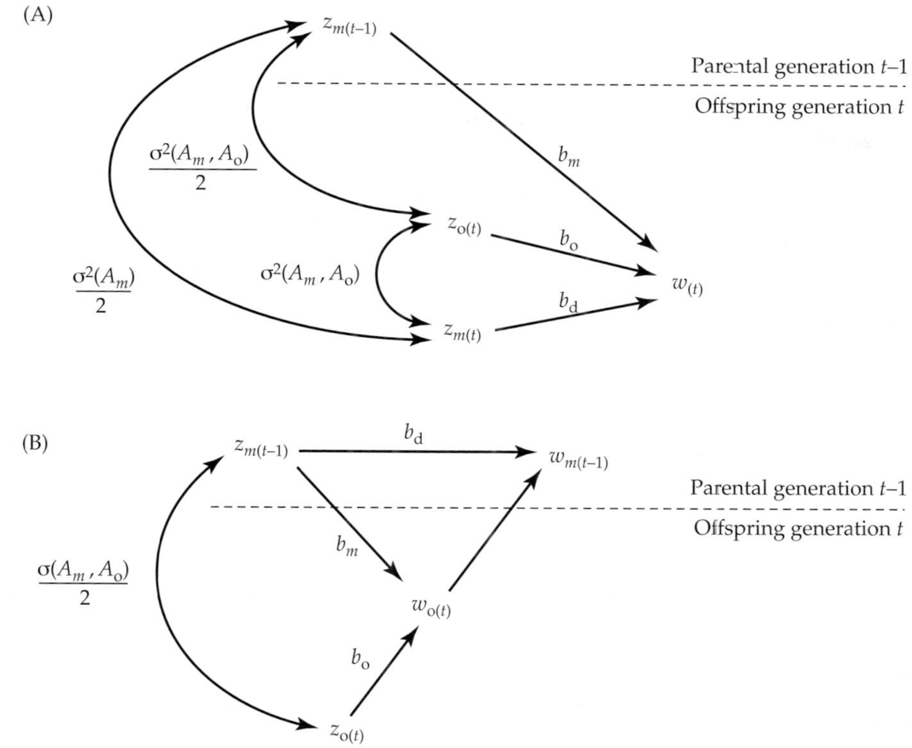

Figure 30.18 Alternate assignment of fitness, w, between an offspring and its mother. **A:** Suppose the true fitness is given by the **offspring-fitness model**. Here, the early survival of an offspring is a function of its phenotype, z_o, and the value of a maternal trait, $z_{m(t-1)}$, in its mother. The displayed path diagram is for female offspring, who have a potential extra component of fitness, as there may be a fitness cost when the offspring express the maternal trait, $z_{m(t)}$. This path through $z_{m(t)}$ is absent in males as they are assumed to not display this trait. **B:** The **mixed-fitness model**. Early offspring survival is now assigned to the mother (scoring maternal fitness as her number of fledglings or recruits). Under this model, we only follow the maternal trait value in the mother, $z_{m(t-1)}$, as opposed to the offspring-fitness model, where we also must follow the trait in the offspring, $z_{m(t)}$. As a result, there is only a single covariance (doubleheaded arrow) in the mixed-fitness model, connecting $z_{m(t-1)}$ and $z_{o(t)}$. In contrast, the offspring-fitness model (**A**) has two additional covariances: a cross-generational one between $z_{m(t-1)}$ and $z_{m(t)}$, and a within-generational one between $z_{m(t)}$ and $z_{o(t)}$. In (**A**), $z_{m(t)}$ is sex-limited and only impacts female offspring, while in both (**A**) and (**B**), all offspring have a mother and hence are impacted by $z_{m(t-1)}$. (After Wolf and Wade 2009.)

foraging enhances the survival of her offspring, but also increases her own risk of predation. Incorporating the *direct* cost (b_d) of such a trait in females yields

$$w_{(t)}[\text{female}] = \alpha_{(t)} + b_m z_{m(t-1)} + b_d z_{m(t)} + b_o z_{o(t)} + \epsilon_{(t)} \tag{30.40b}$$

Figure 30.18A presents the path diagram for this fitness model. If we assume that Equations 30.40a and 30.40b represents the correct fitness model, the selection differential, S_m, on the maternal trait value, z_m, can be computed from the Robertson-Price identity (Equation 6.10). Because $w(t)$ depends on whether the offspring is male or female, we let δ_f be a zero (for male) or one (for female) indicator variable. Equations 30.40a, 30.40b and 6.10 yield

$$\begin{aligned} S_m(\text{off}) &= \sigma(w_{(t)}, z_{m(t)}) \\ &= \sigma\big(b_m z_{m(t-1)} + \delta_f b_d z_{m(t)} + b_o z_{o(t)}, z_{m(t)}\big) \\ &= b_m \sigma(z_{m(t-1)}, z_{m(t)}) + \delta_f b_d \sigma(z_{m(t)}, z_{m(t)}) + b_o \sigma(z_{o(t)}, z_{m(t)}) \end{aligned} \tag{30.40c}$$

Because the total differential is the weighted average over both sexes (Equation 13.5),

$$\begin{aligned} S_m(\text{off}) &= b_m\sigma(z_{m(t-1)}, z_{m(t)}) + \frac{b_d}{2}\sigma(z_{m(t)}, z_{m(t)}) + b_o\sigma(z_{o(t)}, z_{m(t)}) \\ &= \frac{b_m}{2}\,\sigma^2(A_m) + \frac{b_d}{2}\,\sigma^2(z_m) + b_o\,\sigma(A_m, A_o) \end{aligned} \tag{30.40d}$$

This last step assumes that $\sigma(z_{o(t)}, z_{m(t)}) = \sigma(A_{o(t)} + e_{o(t)}, A_{m(t)} + e_{m(t)}) = \sigma(A_m, A_o)$, namely, that the environment effects between the maternal and direct effects are uncorrelated within an individual.

Now suppose that an investigator assumes the offspring fitness is assigned to its mother (Figure 30.18B), which Thomson and Hadfield (2017) call the **mixed-fitness model**. The mother's total fitness can be considered as the sum of two separate episodes: selection on early survival in her offspring, $w_{o(t)}$, and selection on her maternal trait, $w_{m(t-1)}$, where

$$w_{o(t)} = \alpha_{o(t)} + b_m z_{m(t-1)} + b_o z_{o(t)} + \epsilon_{(t)} \tag{30.41a}$$

$$w_{m(t-1)} = \alpha_{m(t-1)} + b_d z_{m(t-1)} + \epsilon_{(t-1)} \tag{30.41b}$$

giving her total fitness as

$$w_{o(t)} + w_{m(t-1)} = \big(\alpha_{o(t)} + \alpha_{m(t-1)}\big) + b_m z_{m(t-1)} + b_d z_{m(t-1)} + b_o z_{o(t)} + \big(\epsilon_{(t)} + \epsilon_{(t-1)}\big) \tag{30.41c}$$

The difference between this expression (mother-assigned offspring early-viability fitness) and Equation 30.40b (offspring assigned their own fitness) is in the direct fitness of the maternal effect. This component is weighted by $b_m z_{m(t-1)}$ in Equation 30.41c (the value of the mother), and by $b_m z_{m(t)}$ in Equation 30.40b (the cost to the offspring). Following the same logic leading to Equation 30.40d, the selection differential on the maternal trait becomes

$$\begin{aligned} S_m(\text{mother}) &= \sigma(w_{o(t)} + w_{m(t-1)}, z_{m(t-1)}) \\ &= \frac{1}{2}\left[(b_m + b_d)\sigma^2(z_m) + b_o\sigma(z_{o(t)}, z_{m(t-1)})\right] \\ &= \frac{1}{2}\left[(b_m + b_d)\sigma^2(z_m) + b_o\frac{\sigma^2(A_m, A_o)}{2}\right] \end{aligned} \tag{30.41d}$$

Subtracting Equation 30.41d from Equation 30.40d yields

$$S_m(\text{mother}) - S_m(\text{off}) = \frac{b_m}{2}\left[\sigma^2(z_m) - \sigma^2(A_m)\right] - \frac{3}{4}b_o\sigma(A_m, A_o) \tag{30.42}$$

The first term shows that maternal selection is *overestimated* by assigning fitness to mothers, as only a fraction (the additive variance, $\sigma^2[A_m]$) of the total maternal trait variation, $\sigma^2(z_m)$, is passed to the offspring. The second term either reduces this overestimation or amplifies it, depending the sign of the genetic covariance between the maternal trait and the offspring trait. Wolf and Wade (2001) stated that there is no reason to expect that these terms will largely cancel, making estimates of maternal selection when offspring fitness is assigned to mothers problematic in many cases.

Conversely, Equation 30.42 shows the conditions under which assigning fitnesses to mothers is unbiased. First, the maternal and offspring traits must be uncorrelated

$$\sigma\left(z_{o(t)}, z_{m(t)}\right) = 0$$

implying that $\sigma(A_m, A_o) = 0$, provided that $\sigma(e_o, e_m) = 0$. Second, and more problematic, the maternal trait must have a very high heritability, such that $\sigma^2(z_m) \simeq \sigma^2(A_m)$. As a result, as Wolf and Wade (2001) suggested, assigning offspring viability as a component of offspring fitness is less biased. Thomson and Hadfield (2017) echoed this point, noting that the conditions are even more restrictive for the mixed-fitness model to be unbiased if the parental trait influencing offspring fitness (upon which selection against the parent can occur) is not sex-limited. They also note that roughly 40% of fecundity estimates in the literature (mainly from birds and mammals) are based on mixed-fitness estimates (namely, weighted by some concept of offspring survival).

Both sets of authors pointed out that unless a fitness regression of the form given by Equation 30.40b is used, the levels-of-selection impact from a parent can easily be missed. Unfortunately, this requires one to either know, or at least have a good idea of, the potential parental traits that impact offspring survival. By using a list of such candidate traits, one can fit regressions of this form to test which traits (if any) have a significant impact on offspring fitness.

VIII

Appendices

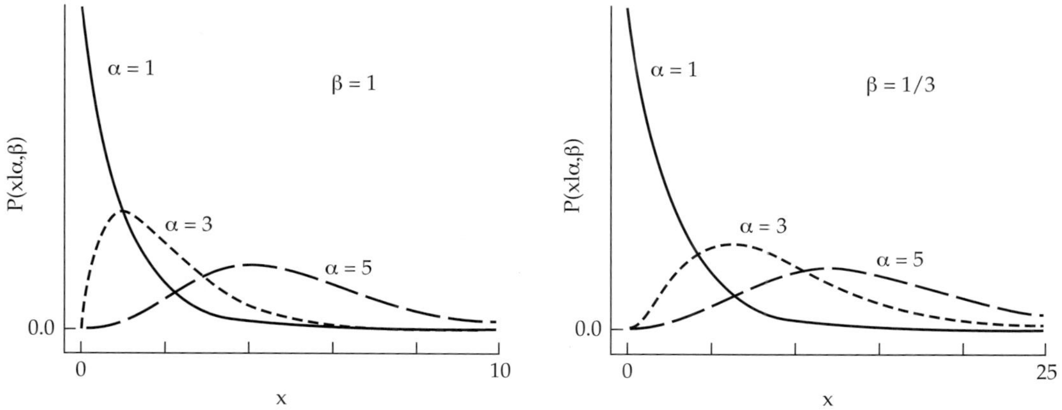

Appendix 1

Diffusion Theory

I believe that no one who is familiar, either with mathematical advances in other fields, or with the range of special biological conditions to be considered, would ever conceive that everything could be summed up in a single mathematical formula, however complex. Fisher (1932a)

Exact solutions of the dynamics of many random processes arising in population and quantitative genetics are either unknown or extremely cumbersome. Nevertheless, starting with Fisher (1922), Wright (1945), and Kimura (1955a, 1955b), the use of **diffusion approximations** in place of the exact dynamics of these processes has proven to be extremely powerful. Useful introductions to diffusion theory with special reference to genetics were given by Ewens (1979, 2004) and Karlin and Taylor (1981), with additional applications considered by Crow and Kimura (1970), Maruyama (1977), Kimura (1983), and Gale (1990). The goal of diffusion theory is to obtain expressions for $\varphi(x, t, p)$, the probability distribution for the random variable, x, at time t, given that the process starts at a value of p. It is often the case that for sufficiently large time, the probability distribution approaches a stationary value, $\varphi(x)$, independent of both time (t) and the starting value (p). Diffusion theory also provides approximations for a number of summary statistics of a particular process, such as probabilities and times to fixation for the various boundaries of the process (such as the loss or fixation of an allele). We will consider these issues in turn.

FOUNDATIONS OF DIFFUSION THEORY

Consider a continuous random variable, x_t, indexed by continuous time, t. If $\delta_x = x_{t+\delta_t} - x_t$ (the change in x_t over a very small time interval, δ_t) satisfies

$$E\left(\delta_x \,|\, x_t = x\right) = m(x)\delta_t + o(\delta_t) \tag{A1.1a}$$

$$\sigma^2(\delta_x \,|\, x_t = x) = v(x)\delta_t + o(\delta_t) \tag{A1.1b}$$

$$E\left(\,|\delta_x|^k\,\right) = o(\delta_t) \quad \text{for } k \geq 3 \tag{A1.1c}$$

then x_t is said to be a **diffusion process**, provided the additional technical restriction that x_t is a **Markov process** (the transition probabilities depend only on the current value of the process and no other aspects of its history) is satisfied. The functions $m(x)$ and $v(x)$ are defined shortly, while the notation $o(\delta_t)$ means that any remaining terms are of order δ_t^2 (or higher), and hence small relative to δ_t. Formally, $\lim_{\delta_t \to 0} o(\delta_t)/\delta_t = 0$, so we can ignore terms of order δ_t^2 (or higher) when δ_t itself is very small. Likewise, the notation $O(x)$ means terms of order x, and so for example, $O(x^2) = o(x)$.

The Infinitesimal Mean, *m*(*x*), and Variance, *v*(*x*)

The **infinitesimal mean**, $m(x)$, and **infinitesimal variance**, $v(x)$, correspond to the mean and variance of the process (given that it is at x) over a very small time interval. These are formally defined by

$$m(x) = \lim_{\delta_t \to 0} \frac{E\left(x_{t+\delta_t} - x_t \,|\, x_t = x\right)}{\delta_t} \tag{A1.2a}$$

$$v(x) = \lim_{\delta_t \to 0} \frac{E[\,(x_{t+\delta_t} - x_t)^2 \,|\, x_t = x\,]}{\delta_t} \tag{A1.2b}$$

In words, the diffusion assumption is that, over a very small time interval, the mean, $m(x)$, and variance, $v(x)$, around the current position, x, are sufficient to fully describe the process.

Evolution and Selection of Quantitative Traits. Bruce Walsh & Michael Lynch.
Published 2018 by Oxford University Press.
DOI 10.1093/oso/9780198830870.001.0001

In population genetics, diffusion approximations provide an elegant way to rescale a discrete space, discrete time, random variable (usually an allele frequency) to construct a new, continuously distributed random variable. For example, consider X_t, the number of copies of allele A in a discrete-generation population of N diploids at generation t. X_t takes on values of $0, 1, \ldots, 2N$, and time t is in units of discrete generations. Suppose we construct a new random variable $x_\tau^{(N)} = X_{(\tau)}/(2N)$, where $\tau = t/N$. A unit of time on this transformed scale (τ) corresponds to N generations on the original scale. If we take the limit as N approaches infinity, the limiting process x_τ is a continuous space (in x), continuous time (in τ) process that represents the allele frequency at time τ.

Example A1.1. Diffusion processes for allele frequencies are typically obtained by setting $v(x) = x(1-x)/(2N_e)$ (the per-generation variance in the change of allele frequencies due to drift; see Chapter 2), where x is the frequency of allele A. This is simply binomial sampling (Equation 2.1), and it is a function of the variance effective population size, N_e (Chapter 3). The deterministic (i.e., infinite-population) change in allele frequency is subsumed in $m(x)$. Because there is no deterministic change in allele frequency under pure drift (i.e., no selection or mutation), in this case the diffusion process is simply

$$m(x) = 0 \quad \text{and} \quad v(x) = \frac{x(1-x)}{2N_e} \tag{A1.3}$$

which is defined for $0 < x < 1$.

Now let us consider selection at a diallelic locus (with additive fitnesses of $1 : 1+s : 1+2s$ for the genotypes *aa*, *Aa*, and *AA*, respectively). Recalling Equation 5.2b, in an infinite population,

$$\Delta p = \frac{sp(1-p)}{1+2sp} = sp(1-p)[1+O(s)] = sp(1-p) + o(s)$$

The last step follows because $s\,O(s) = O(s^2) = o(s)$. Thus, ignoring terms of order $o(s)$ yields a diffusion approximation for the joint effects of drift and selection over $0 < x < 1$ of

$$m(x) = sx(1-x) \quad \text{and} \quad v(x) = \frac{x(1-x)}{2N_e} \tag{A1.4}$$

Finally, consider arbitrary selection with constant fitnesses, again with selection sufficiently weak such that $\overline{W} = 1 + O(s)$, and with forward and back mutation, where ν is the mutation rate from A to a and μ is the mutation rate from a to A. Applying Wright's formula (Equation 5.5b) and including mutation yields

$$m(x) = \frac{x(1-x)}{2}\frac{\partial \ln[\overline{W}(x)]}{\partial x} + (1-x)\mu - x\nu \quad \text{and} \quad v(x) = \frac{x(1-x)}{2N_e} \tag{A1.5}$$

We apply several of these expressions below.

The Kolmogorov Forward Equation

Given $m(x), v(x)$, and an initial frequency of p, the diffusion approximation for the dynamics of the probability density function for the realized value x at time t satisfies the **Kolmogorov forward equation** (or **KFE**)

$$\frac{\partial\, \varphi(x,t,p)}{\partial\, t} = \frac{1}{2}\frac{\partial^2\, v(x)\,\varphi(x,t,p)}{\partial\, x^2} - \frac{\partial\, m(x)\,\varphi(x,t,p)}{\partial\, x} \tag{A1.6}$$

where $\varphi(x, t, p)$ is the probability density for x at time t, given that the process starts at p, namely,

$$\Pr[\, c \leq x(t) \leq d \,|\, x(0) = p\,] = \int_c^d \varphi(x, t, p)\, dx$$

A derivation of the KFE is given below. When they can be found, closed-form solutions of $\varphi(x, t, p)$ are usually complicated (e.g., Example A1.2). The standard approach is to express the solution as a power series:

$$\varphi(x, t, p) = \sum_{i=1}^{\infty} f_i(x, p) e^{-\lambda_i t} \tag{A1.7}$$

Here, the values of λ_i are the **eigenvalues** associated with the partial differential equation given by Equation A1.6, and the functions, f_i, are the associated **eigenfunctions**. Notice that time, t, only appears as a multiplier of the eigenvalues, while the starting position, p, appears only in the eigenfunctions. Crow and Kimura (1970) presented exact solutions of $\varphi(x, t, p)$ for a number of population-genetic problems. As the next example shows, even in the simplest case (pure drift), the result are rather complex.

Example A1.2. From Equation A1.6, the KFE for the diffusion for pure drift, $m(x) = 0$, is

$$\frac{\partial\, \varphi(x, t, p)}{\partial\, t} = \frac{1}{2}\frac{\partial^2\, v(x)\, \varphi(x, t, p)}{\partial\, x^2}$$

Figure 2.4 plots solutions of this partial differential equation for several time points. As mentioned in Chapter 2, Kimura (1955b) obtained the solution for this particular KFE as

$$\varphi(x, t, p) = \sum_{i=1}^{\infty} p(1-p)i(i+1)(2i+1)g_i(p)g_i(x)e^{-\lambda_i t}$$

where

$$\lambda_i = \frac{i(i+1)}{4N_e} \quad \text{and} \quad g_i(x) = F(1-i, i+2, 2, x)$$

and where F is the hypergeometric function (Equation 15.1.1 in Abramowitz and Stegun 1972). In the notation of Equation A1.7, the eigenfunction associated with the i eigenvalue is

$$f_i(x, p) = p(1-p)i(i+1)(2i+1)F(1-i, i+2, 2, p)F(1-i, i+2, 2, x)$$

Although this solution is obtained using standard tools for the solution of partial differential equations (e.g., Farlow 1982), as one might expect given the form of the solution, this is not a trivial procedure.

While simple inspection of the above closed form solution provides only a small amount of insight into the process, it is clear that $\varphi(x, t, p) \to 0$ as $t \to \infty$ (because all $e^{-\lambda_i t} \to 0$ as $t \to \infty$). Thus, for large values of time, there is a vanishingly small probability mass in the open interval $0 < x < 1$. This occurs because drift results in an allele eventually either being fixed ($x = 1$) or being lost ($x = 0$). Other forces, such as mutation, are required for nonzero probability mass to remain in the interior space ($0 < x < 1$). For large spans of time, the solution is dominated by the first few eigenvalues. If we focus on just the eigenfunctions associated with the two leading eigenvalues, λ_1 and λ_2, the result is

$$\varphi(x, t, p) \simeq 6p(1-p)e^{-t/(2N_e)} + 30p(1-2p)(1-2x)e^{-3t/(2N_e)}$$

Notice that the eigenfunction for the dominant eigenvalue $(1/2N_e)$ is independent of x, so for large values of t, the probability density is a constant independent of x (i.e., a uniform over $0 < x < 1$), which decays to zero at a rate of $e^{-t/(2N_e)}$ (see Figure 2.4).

Boundary Behavior of a Diffusion

A critical feature of the density $\varphi(x, t, p)$ is its range of validity. Formally speaking, the diffusion approximation only applies within some *open* interval between two **boundary values**, a and b. The behavior *exactly* at the boundaries ($x = a$ and $x = b$) is beyond the realm of the approximation. In many cases, x_t does not change value once it reaches a boundary, in which case it is called **absorbing**. For example, in the absence of mutation and migration, once an allele frequency reaches either 0 or 1, it remains there, which means that both of these extreme states are absorbing boundaries. Further, a boundary is said to be **accessible** if it can be reached in finite time. When considering fixation probabilities, we are examining absorbing, accessible boundaries. Note that a finite boundary point may not be accessible. Suppose we consider a simple mutation-drift equilibrium. If the mutational pressure is sufficiently strong near a boundary (say for the loss of an allele), the resulting boundary may be **inaccessible**, with the population never being in a state where all copies of an allele are lost.

Derivation of the Kolmogorov Forward Equation

For completeness, we present a derivation of Equation A1.6, using the diffusion-approximation assumptions (Equations A1.1a–A1.1c). This section is a bit technical and can be skipped if readers desire. Consider the change from the probability distribution, $\varphi(x, t, p)$, at time t to a new distribution, $\varphi(x, t + \delta_t, p)$, after some very small time interval, δ_t. To arrive at x at time $t + \delta_t$, the frequency must have previously been at some value, $x - \delta_x$, and then moved by an amount, δ_x, over the interval δ_t. Let $\phi(\delta_x, x, \delta_t)$ be the probability of jumping by an amount, δ_x, over the time δ_t given the starting point, x. In our case, because we are assuming that there is a move from $x - \delta_x$ to x, we consider $\phi(\delta_x, x - \delta_x, \delta_t)$. Integrating over all possible jump values yields the **Chapman-Kolmogorov equation**

$$\varphi(x, t + \delta_t, p) = \int \varphi(x - \delta_x, t, p)\, \phi(\delta_x, x - \delta_x, \delta_t)\, d\delta_x \tag{A1.8a}$$

To simplify notation in the following derivation, we write $\varphi(x, t, p)$ as $\varphi(x, t)$, although the dependence on the initial value p should be kept in mind.

To solve the Chapman-Kolmogorov equation, we start by using the Taylor series approximation

$$f(x - a) = f(x) - a\frac{\partial f(x)}{\partial x} + \frac{a^2}{2}\frac{\partial^2 f(x)}{\partial x^2} - \frac{a^3}{6}\frac{\partial^3 f(x)}{\partial x^3} + o(a^3)$$

Expanding the function in the integral about $a = \delta_x$ yields

$$\begin{aligned}\varphi(x - \delta_x, t)\, \phi(\delta_x, x - \delta_x, \delta_t) &= \varphi(x, t)\, \phi(\delta_x, x, \delta_t) - \delta_x \frac{\partial\,[\,\varphi(x, t)\, \phi(\delta_x, x, \delta_t)\,]}{\partial x} \\ &+ \frac{\delta_x^2}{2}\frac{\partial^2[\,\varphi(x, t)\, \phi(\delta_x, x, \delta_t)\,]}{\partial x^2} - \frac{\delta_x^3}{6}\frac{\partial^3[\,\varphi(x, t)\, \phi(\delta_x, x, \delta_t)\,]}{\partial x^3} + o(\delta_x^3)\end{aligned}$$

If we ignoring terms of $o(\delta_x^3)$, i.e., of order δ_x^4 and higher, and integrate, the result is

$$\begin{aligned}\int \varphi(x - \delta_x, t)\, \phi(\delta_x, x - \delta_x, \delta_t)\, d\delta_x &\simeq \int \varphi(x, t)\, \phi(\delta_x, x, \delta_t)\, d\delta_x - \int \delta_x \frac{\partial[\,\varphi(x, t)\, \phi(\delta_x, x, \delta_t)\,]}{\partial x}\, d\delta_x \\ &+ \int \frac{\delta_x^2}{2}\frac{\partial^2[\,\varphi(x, t)\, \phi(\delta_x, x, \delta_t)\,]}{\partial x^2}\, d\delta_x - \int \frac{\delta_x^3}{6}\frac{\partial^3[\,\varphi(x, t)\, \phi(\delta_x, x, \delta_t)\,]}{\partial x^3}\, d\delta_x\end{aligned}$$

Because integration is done with respect to δ_x, partials with respect to x can be moved outside of the integrals, as can functions not involving δ_x, yielding

$$\int \varphi(x - \delta_x, t)\phi(\delta_x, x - \delta_x, \delta_t)d\delta_x = \varphi(x, t)\int \phi(\delta_x, x, \delta_t)d\delta_x - \frac{\partial}{\partial x}\left(\varphi(x, t)\int \delta_x\, \phi(\delta_x, x, \delta_t)d\delta_x\right)$$

$$+\frac{1}{2}\frac{\partial^2}{\partial x^2}\left(\varphi(x,t)\int \delta_x^2\,\phi(\delta_x,x,\delta_t)\,d\delta_x\right)-\frac{1}{6}\frac{\partial^3}{\partial x^3}\left(\varphi(x,t)\int \delta_x^3\,\phi(\delta_x,x,\delta_t)d\delta_x\right) \quad \text{(A1.8b)}$$

Because $\phi(\delta_x, x, \delta_t)$ is the probability distribution of moves of size δ_x over the time interval δ_t given we start at position x, we have

$$\int \phi(\delta_x,x,\delta_t)\,d\delta_x = 1 \quad \text{(A1.8c)}$$

$$\int \delta_x\,\phi(\delta_x,x,\delta_t)\,d\delta_x = E\left(\delta_x \mid x_t = x\right) = m(x)\delta_t + o(\delta_t) \quad \text{(A1.8d)}$$

The first identity follows from the fact that the integral over a probability distribution is equal to one, and the second is simply our first diffusion assumption (Equation A1.1a). Next, if we recall that $E(x^2) = \sigma_x^2 + [E(x)]^2$, we have

$$\int \delta_x^2\,\phi(\delta_x,x,\delta_t)d\delta_x = \sigma^2(\delta_x) + [E(\delta_x)]^2$$

which, using the diffusion approximation (Equations A1.1a and A1.1b), reduces to

$$\begin{aligned}\int \delta_x^2\,\phi(\delta_x,x,\delta_t)d\delta_x &= [v(x)\delta_t + o(\delta_t)] + [\,m(x)\delta_t + o(\delta_t)\,]^2 \\ &= v(x)\delta_t + o(\delta_t)\end{aligned} \quad \text{(A1.8e)}$$

The last step follows because the contribution from the squared change in the mean is $o(\delta_t)$, and we sweep all such terms into a single expression. Finally, under the diffusion assumption given by Equation A1.1c, moments of δ_x^3 and higher are negligible. Substituting the approximations given by Equations A1.8d and A1.8e into Equation A1.8b yields

$$\varphi(x,t+\delta_t) = \varphi(x,t) - \frac{\partial\,\{[\,m(x)\delta_t + o(\delta_t)\,]\,\varphi(x,t)\}}{\partial x} + \frac{1}{2}\frac{\partial^2\,\{[\,v(x)\delta_t + o(\delta_t)\,]\,\varphi(x,t)\}}{\partial x^2} \quad \text{(A1.9)}$$

Subtracting $\varphi(x,t)$ from both sides, dividing both by δ_t, and taking the limit at $\delta_t \to 0$ gives the left-hand size as

$$\lim_{\delta_t \to 0} \frac{\varphi(x,t+\delta_t) - \varphi(x,t)}{\delta_t} = \frac{\partial \varphi(x,t)}{\partial t}$$

Likewise, recalling that $\lim_{\delta_t \to 0} o(\delta_t)/\delta_t = 0$, the last two right-hand-side terms of Equation A1.9 simplify to

$$\lim_{\delta_t \to 0} \frac{1}{\delta_t}\left(\frac{\partial\,\{[\,m(x)\delta_t + o(\delta_t)\,]\,\varphi(x,t)\}}{\partial x}\right) = \frac{\partial[\,m(x)\varphi(x,t)\,]}{\partial x}$$

$$\lim_{\delta_t \to 0} \frac{1}{\delta_t}\left(\frac{\partial^2\{[\,v(x)\delta_t + o(\delta_t)\,]\,\varphi(x,t)\}}{\partial x^2}\right) = \frac{\partial^2[\,v(x)\varphi(x,t)\,]}{\partial x^2}$$

Together, these simplifications imply

$$\frac{\partial\,\varphi(x,t)}{\partial\,t} = -\frac{\partial[\,m(x)\,\varphi(x,t)\,]}{\partial\,x} + \frac{\partial^2[\,v(x)\,\varphi(x,t)\,]}{2\,\partial\,x^2}$$

which recovers the Kolmogorov forward equation (Equation A1.6), provided we again recall that $\varphi(x,t)$ is really $\varphi(x,t,p)$.

Stationary Distributions

At equilibrium, a probability density function does not change over time, namely,

$$\frac{\partial\,\varphi(x,t,p)}{\partial\,t} = 0 \quad \text{(A1.10a)}$$

When such a function exists, it is called the **stationary distribution** and is denoted by $\varphi(x)$. Stationary distributions are independent of the starting conditions: regardless of where the process starts in the interior of the diffusion, it converges to the same distribution. Hence, $\varphi(x,t,p)$ can be decomposed into a transient deviation (dependent on t and p) and a stationary expectation (independent of both t and p), $\varphi(x,t,p) = \varphi^*(x,t,p) + \varphi(x)$. The transient deviation satisfies $\lim_{t\to\infty} \varphi^*(x,t,p) = 0$, and the deviation from the equilibrium distribution decays to zero over time.

When Equation A1.10a is satisfied, Equation A1.6 becomes

$$\frac{\partial^2[\,v(x)\,\varphi(x)]}{2\,\partial x^2} = \frac{\partial[\,m(x)\,\varphi(x)]}{\partial x} \tag{A1.10b}$$

and integration of both sides reduces this to the simple differential equation,

$$\frac{\partial[\,v(x)\,\varphi(x)]}{\partial x} = 2m(x)\,\varphi(x) \tag{A1.10c}$$

Using standard methods for differential equations (e.g., Tenenbaum and Pollard 1963) yields the solution as

$$\varphi(x) = \frac{C}{v(x)\,G(x)} \tag{A1.11}$$

where G (which is called the **scale function** in the diffusion literature) is defined by the indefinite integral

$$G(x) = \exp\left[-2\int^x \frac{m(y)}{v(y)}\,dy\right] \tag{A1.12}$$

and C is a constant such that $\varphi(x)$ integrates to one, ensuring that Equation A1.11 is a proper probability density function. Stationary distributions for population-genetic processes were first obtained by Wright (1931, 1938b), who used a somewhat different approach. Note that $\int [v(x)\,G(x)]^{-1}\,dx$ may be infinite, in which case no stationary distribution exists. This happens, for example, in the absence of mutation and migration, where both boundaries (allele frequencies of 0 or 1) are absorbing.

Example A1.3. Consider the case in which random genetic drift is the only force influencing allele-frequency change. From Equation A1.3, $m(x) = 0$ and $v(x) = x(1-x)/(2N_e)$, which yields

$$G(x) = \exp\left[-4N_e\int^x \frac{0}{y(1-y)}\,dy\right] = e^{-4N_e 0} = 1 \tag{A1.13}$$

and from Equation A1.11,

$$\varphi(x) = \frac{2N_e C}{x(1-x)} \tag{A1.14}$$

The only valid equilibrium distribution is $\varphi(x) = 0$ (e.g., $C = 0$), as

$$\int_0^1 x^{-1}(1-x)^{-1}dx = \infty$$

As shown in Example A1.2, after sufficient time, and in the absence of any force to keep allele frequencies in the interior ($0 < x < 1$), all alleles eventually reach a frequency of zero or one. As a result, the equilibrium distribution on the open interval of $0 < x < 1$ is zero.

Example A1.4. Compute the stationary distribution for the frequency of an allele at a diallelic locus experiencing selection, mutation, and drift, where $\overline{W}(x)$ denotes the mean fitness as a function of the allele frequency, x. Using the definitions of $m(x)$ and $v(x)$ from Equation A1.5,

$$\int^x \frac{m(y)}{v(y)}\,dy = 2N_e \int^x \frac{y(1-y)d\ln[\overline{W}(y)]/(2\,dy) + (1-y)\mu - y\nu}{y(1-y)}\,dy$$
$$= N_e \int^x \frac{d\ln[\overline{W}(y)]}{dy}\,dy + 2N_e\mu \int^x \frac{1}{y}\,dy - 2N_e\nu \int^x \frac{1}{1-y}\,dy$$
$$= N_e \ln[\overline{W}(x)] + 2N_e\mu \ln(x) + 2N_e\nu \ln(1-x)$$

Hence

$$G(x) = \exp\left[-2\int^x \frac{m(y)}{v(y)}\,dy\right] = \overline{W}(x)^{-2N_e}\, x^{-4N_e\mu}\,(1-x)^{-4N_e\nu}$$

If we substitute this expression for $G(x)$ into Equation A1.11, it returns the stationary distribution as

$$\varphi(x) = C\overline{W}(x)^{2N_e}\, x^{4N_e\mu-1}\,(1-x)^{4N_e\nu-1} \quad \text{for } 0 < x < 1 \tag{A1.15}$$

a result first due to Wright (1931), obtained using a different approach.

Example A1.5. A particular application of Equation A1.15 is the case of a deleterious recessive allele maintained by mutation. In an infinite population (Chapter 7), the equilibrium frequency of this allele is approximately $\sqrt{\mu/s}$ (for $s \gg \mu$), where the fitness of the recessive homozygote is $1 - s$ and the mutation rate from normal to recessive allele is μ. Here, $\overline{W}(x) = 1 - sx^2$, and because $(1-x)^{N_e} \simeq e^{-xN_e}$, we have

$$[\,\overline{W}(x)\,]^{2N_e} = (1 - sx^2)^{2N_e} \simeq \exp(-2N_e s x^2)$$

resulting in an equilibrium distribution of

$$\varphi(x) = Ce^{-2N_e s x^2}\, x^{4N_e\mu-1}\,(1-x)^{4N_e\nu-1} \quad \text{for } 0 < x < 1$$

The Kolmogorov Backward Equation

While the KFE provides both the full solution of $\varphi(x,t,p)$ and, where appropriate, the equilibrium solution, we can obtain much simpler expressions for many quantities of interest when an equilibrium solution does not exist. For example, if one or both of the boundaries are accessible and absorbing, we can compute the fixation probabilities (the probability that the process eventually will reach a specified boundary), the time to reach that boundary (the time to loss or fixation), and the expected value of many other functions of interest. The key to all of these operations is the **Kolmogorov backward equation** (**KBE**), which is given by

$$\frac{\partial\,\varphi(x,t,p)}{\partial\,t} = \frac{1}{2}\,v(p)\,\frac{\partial^2\varphi(x,t,p)}{\partial\,p^2} + m(p)\,\frac{\partial\,\varphi(x,t,p)}{\partial\,p} \tag{A1.16}$$

The KBE is derived in a manner similar to the KFE. The KBE starts at the current time, t, and looks *backwards* in terms of how changes ($\partial/\partial p$) in the initial starting value, p, are influenced by the current position, x, hence the name. Conversely, the forward equation (Equation A1.6) examines how changes ($\partial/\partial x$) in the current position, x, are influenced by the starting value, p.

DIFFUSION APPLICATIONS IN POPULATION GENETICS

When no stationary distribution exists, useful summary statistics to describe the diffusion process are the probabilities of fixation of the various boundaries and the expected time to reach a specified boundary. We consider each in turn.

Probability of Fixation

When at least one boundary is absorbing (and accessible), no stationary distribution exists. In such cases, one important descriptor of the process is the probability of reaching one boundary before the other. One can show that the function $u(p,t)$, the probability of fixation by time t, given that we start at allele frequency p, satisfies the KBE. We are typically interested in the ultimate probability of fixation, $u(p) = \lim_{t\to\infty} u(p,t)$, in which case the partial derivative of $u(p)$ with respect to time is zero and, from Equation A1.16, $u(p)$ satisfies

$$0 = m(p)\,\frac{\partial\, u(p)}{\partial\, p} + \frac{1}{2}\,v(p)\,\frac{\partial^2 u(p)}{\partial\, p^2}$$

This has a solution of

$$u(p) = \frac{\int_0^p G(x)\,dx}{\int_0^1 G(x)\,dx} \tag{A1.17a}$$

where $G(x)$ is defined by Equation A1.12 (Kimura 1962). More generally, for any diffusion (regardless of the nature of the boundaries), the probability that the process reaches a value of b before a value of a, given it starts at p (where $A < a < p < b < B$, with the diffusion defined over $A < x < B$), is

$$u_{b,a}(p) = \frac{\int_a^p G(x)\,dx}{\int_a^b G(x)\,dx} \tag{A1.17b}$$

Example A1.6. Here we compute the probability of fixation of an allele under drift alone ($m(x) = 0$). In this setting $G(x) = 1$ (Equation A1.13), so Equation A1.17a reduces to

$$u(p) = \frac{\int_0^p 1\,dx}{\int_0^1 1\,dx} = \frac{p-0}{1-0} = p$$

In other words, the fixation probability of a neutral allele is simply its starting allele frequency (Chapter 2). This result makes intuitive sense, as there is no expected net change via drift alone, and hence the expected value of the allele frequency over a large number of replicates that have gone to fixation should be p. Likewise, Equation A1.17b shows the probability of reaching frequency b before frequency a (starting at a frequency p that satisfies $0 < a < p < b < 1$) as

$$u_{b,a}(p) = \frac{p-a}{b-a}$$

Now consider a diallelic locus experiencing additive selection and drift in a diploid population. From Equation A1.4 ,

$$m(x) = sx(1-x) \quad \text{and} \quad v(x) = \frac{x(1-x)}{2N_e}$$

which implies

$$G(x) = \exp\left[-4N_e s \int^x \frac{y(1-y)}{y(1-y)}dy\right] = e^{-4N_e sx}$$

Thus, the probability of fixation of an allele with initial frequency p is

$$u(p) = \frac{\int_0^p e^{-4N_e sx}\,dx}{\int_0^1 e^{-4N_e sx}\,dx} = \frac{1-e^{-4N_e sp}}{1-e^{-4N_e s}} \tag{A1.18a}$$

as obtained by Kimura (1957, 1962). A case of special interest is an initially rare allele ($p \ll 1$) that is under strong selection ($4N_e s \gg 1$). Here, the term in the denominator is essentially one, and because $1 - e^{-ax} = ax + o(x)$, we have

$$u(p) \simeq 4N_e sp \quad \text{for} \quad 4N_e s \gg 1 \quad \text{and} \quad p \ll 1$$

In particular, for a new mutation, $p = 1/(2N)$ and the fixation probability becomes

$$u\left(\frac{1}{2N}\right) = 4N_e s\,\frac{1}{2N} = 2s\frac{N_e}{N} \tag{A1.18b}$$

More generally, starting at p, the probability that we will reach b before reaching a is

$$u_{a,b}(p) = \frac{\int_a^p e^{-4N_e sx}\,dx}{\int_a^b e^{-4N_e sx}\,dx} = \frac{e^{-4N_e sa} - e^{-4N_e sp}}{e^{-4N_e sa} - e^{-4N_e sb}} \tag{A1.18c}$$

Finally, in allowing for dominance, we assign fitnesses as $1 : 1 + s(1+h) : 1 + 2s$. Under weak selection (ignoring terms of s^2 and higher),

$$m(x) = sx(1-x)[\,1 + h(1-2x)\,] \tag{A1.19a}$$

which yields

$$\begin{aligned} G(x) &= \exp\left[-4N_e s \int^x \frac{y(1-y)[\,1+h(1-2y)\,]}{y(1-y)}\,dy\right] \\ &= \exp\{-4N_e sx[\,1 + h(1-x)\,]\} \end{aligned} \tag{A1.19b}$$

and hence the fixation probability of the beneficial allele is

$$u(p) = \frac{\int_0^p \exp\{-4N_e sx[\,1+h(1-x)\,]\}\,dx}{\int_0^1 \exp\{-4N_e sx[\,1+h(1-x)\,]\}\,dx} \tag{A1.19c}$$

Example A1.7. An important special case of Equation A1.19c is a favored, but completely recessive, allele. In this case $h = -1$, and Equations A1.19a and A1.19b, respectively, become

$$m(x) = 2sx^2(1-x) \quad \text{and} \quad G(x) = \exp\left(-4N_e sx^2\right)$$

giving Equation A1.19c as

$$u(p) = \frac{\int_0^p e^{-4N_e sx^2}\,dx}{\int_0^1 e^{-4N_e sx^2}\,dx} \tag{A1.20a}$$

This integral can be expressed in terms of erf(x), the **error function** (Abramowitz and Stegun 1972), which is very closely related to the cumulative distribution of a unit normal, and is defined as

$$\text{erf}(x) = \frac{2}{\sqrt{\pi}} \int_0^x e^{-t^2}\,dt \tag{A1.20b}$$

With the change of variables, $t = x\sqrt{4N_e s}$, the numerator of Equation A1.20a becomes

$$\int_0^p e^{-4N_e sx^2}\,dx = \frac{1}{\sqrt{4N_e s}} \int_0^{p\sqrt{4N_e s}} e^{-t^2}\,dt = \frac{1}{4}\sqrt{\frac{\pi}{N_e s}}\,\text{erf}\left(p\sqrt{4N_e s}\right)$$

Hence

$$u(p) = \frac{\text{erf}\left(p\sqrt{4N_e s}\right)}{\text{erf}\left(\sqrt{4N_e s}\right)} \tag{A1.20c}$$

Now consider strong selection ($4N_e s \gg 1$), with the result that $\text{erf}\left(\sqrt{4N_e s}\right) \simeq 1$ (because $\text{erf}(x) \to 1$ for $x \gg 1$), and hence $u(p) \simeq \text{erf}\left(p\sqrt{4N_e s}\right)$. This expression can be further simplified by using the approximation that, for small values of x,

$$\text{erf}(x) \simeq \frac{2\,x}{\sqrt{\pi}}$$

(Abramowitz and Stegun 1972). Thus, when the recessive beneficial allele is initially rare ($x = p\sqrt{4N_e s} \ll 1$), its fixation probability is

$$u(p) \simeq \frac{2}{\sqrt{\pi}}\, p\sqrt{4N_e s} \tag{A1.20d}$$

which, for a newly arisen mutation, $p = 1/(2N)$, becomes

$$u\left(\frac{1}{2N}\right) \simeq \frac{1}{N}\sqrt{\frac{4N_e s}{\pi}} \tag{A1.20e}$$

This result was obtained by Kimura (1957), and for $N = N_e$ it reduces to $\sqrt{4s/N\pi} \simeq 1.13\sqrt{s/N}$. Previously, Haldane (1927) used the theory of branching processes to obtain an approximation of $\sqrt{2s/N} \simeq 1.41\sqrt{s/N}$, while Wright (1942) obtained an approximation of $\sqrt{s/N}$. Note that our expression is slightly different from the common form seen in the literature, as we used the general fitness model presented in Example A1.6 (the homozygote fitness is $1+2s$), while in models dealing with recessives, only the fitness is typically assigned as $1+s$ (e.g., Example A1.5).

Time to Fixation

The time for a process to reach a specified value (or values) is called the **sojourn time**. While we are often interested in the sojourn time to an absorbing boundary (the fixation time), the more general problem of the time to first reach a specific value within the interval over which the diffusion is defined is also tractable. Letting $\bar{t}_{a,b}$ denote the expected time that a diffusion spends in the interval (a, b) given it starts at frequency p, then

$$\bar{t}_{a,b} = \int_0^\infty \Pr\left(a \le x_t \le b\right) dt = \int_0^\infty \int_a^b \varphi(x,t,p)\, dx\, dt \tag{A1.21}$$

If a stationary distribution exists, this time is infinite and hence not really of much biological interest. However, if one or both boundaries are absorbing (and accessible), then $\bar{t}(p)$, the total time the diffusion spends in the interior, is obtained from Equation A1.21 by letting a and b be the lower and upper limits, respectively, of the diffusion (typically, $a = 0, b = 1$).

A very useful approach for solving Equation A1.21, and more general problems, is to consider $h(x,p)$, the expected amount of time that the process (starting at p) spends in the neighborhood of x before it is eventually absorbed. Formally, this is given by

$$h(x,p) = \int_0^\infty \varphi(x,t,p)\, dt \tag{A1.22a}$$

Thus, Equation A1.21 can be expressed as

$$\bar{t}_{a,b} = \int_a^b h(x,p)dx \tag{A1.22b}$$

$h(x,p)$ is called a **Green's function**, and it will prove very useful in solving a variety of problems. The general solution for $h(x,p)$, obtained by Maruyama (1977), is

$$h(x,p) = \begin{cases} \dfrac{2[1-u(p)]}{v(x)G(x)} \displaystyle\int_a^x G(y)\,dy & \text{for } a < x < p \\[2ex] \dfrac{2u(p)}{v(x)G(x)} \displaystyle\int_x^b G(y)\,dy & \text{for } p < x < b \end{cases} \tag{A1.22c}$$

where the fixation probability, $u(p)$, is provided by Equation A1.17a.

One can also obtain modified Green's functions for conditional processes, such as processes leading only to loss or processes leading only to fixation. For example, $\bar{t}_F$, the expected time to fix allele A (in those populations where it is fixed) is given by replacing $h(x,p)$ in Equation A1.22b by

$$h_1(x,p) = h(x,p)\,\frac{u(x)}{u(p)} \tag{A1.23a}$$

This follows from standard conditional probability arguments (see Ewens 1979, 2004), with $u(x)/u(p)$ correcting for the fact that we are only considering those sample paths over which the allele of interest (A) is fixed. This occurs with probability $u(p)$, and hence $u(x)/u(p)$ is the conditional density. Similarly, $\bar{t}_L$, the expected time to lose allele A, is obtained by replacing $h(x,p)$ in Equation A1.22b by

$$h_0(x,p) = h(x,p)\,\frac{1-u(x)}{1-u(p)} \tag{A1.23b}$$

Finally, we note that $\bar{t}$, $\bar{t}_F$, and $\bar{t}_L$ are related by

$$\bar{t}(p) = u(p)\,\bar{t}_F(p) + [1-u(p)]\,\bar{t}_L(p) \tag{A1.24}$$

That is, the expected time to loss or fixation is equal to the expected time to fixation multiplied by the probability of fixation plus the expected time to loss multiplied by the probability of loss.

Example A1.8. Compute the conditional and unconditional expected time to loss or fixation for a neutral allele. Under neutrality, $u(x) = x$, $v(x) = x(1-x)/(2N_e)$, and $G(x) = 1$. Hence, $\int_0^x G(y)dy = x$, $\int_x^1 G(y)dy = 1-x$, and Equation A1.22c simplifies considerably to

$$h(x,p) = \begin{cases} 4N_e(1-p)/(1-x) & \text{for } 0 < x < p \\[2ex] 4N_e p/x & \text{for } p < x < 1 \end{cases}$$

Thus, the expected amount of time that a neutral allele (with an initial frequency of p) remains polymorphic is

$$\begin{aligned} \bar{t}(p) &= \int_0^1 h(x,p)\,dx \\ &= 4N_e(1-p)\int_0^p \frac{dx}{1-x} + 4N_e p \int_p^1 \frac{dx}{x} \\ &= -4N_e[\,(1-p)\ln(1-p) + p\ln(p)\,] \end{aligned} \tag{A1.25}$$

Similarly, the conditional fixation times are obtained using Equations A1.23a and A1.23b, with

$$h_0(x,p) = \frac{1-x}{1-p}\,h(x,p) = \begin{cases} 4N_e & \text{for } 0 < x < p \\ 4N_e \dfrac{p(1-x)}{x(1-p)} & \text{for } p < x < 1 \end{cases} \tag{A1.26a}$$

for those paths leading to loss, and

$$h_1(x,p) = \frac{x}{p}\,h(x,p) = \begin{cases} 4N_e \dfrac{x(1-p)}{p(1-x)} & \text{for } 0 < x < p \\ 4N_e & \text{for } p < x < 1 \end{cases} \tag{A1.26b}$$

for those paths leading to fixation. The integration of Equation A1.22b then yields an expected conditional time to loss of

$$\begin{aligned} \bar{t}_L(p) &= \int_0^1 h_0(x,p)\,dx = 4N_e \int_0^p dx \,+\, 4N_e \frac{p}{1-p} \int_p^1 \frac{1-x}{x}\,dx \\ &= -4N_e \left(\frac{p}{1-p}\right) \ln(p) \end{aligned} \tag{A1.27}$$

and an expected conditional time to fixation of

$$\begin{aligned} \bar{t}_F(p) &= \int_0^1 h_1(x,p)\,dx = 4N_e \frac{1-p}{p} \int_0^p \frac{x}{1-x}\,dx \,+\, 4N_e \int_p^1 dx \\ &= -4N_e \left(\frac{1-p}{p}\right) \ln(1-p) \end{aligned} \tag{A1.28}$$

These results were first obtained by Kimura and Ohta (1969a, 1969b). As a check of consistency, we can apply Equation A1.24,

$$\begin{aligned} \bar{t} &= u(p)\,\bar{t}_F(p) + [1-u(p)]\,\bar{t}_L(p) \\ &= -4N_e \left[p\left(\frac{1-p}{p}\ln(1-p)\right) + (1-p)\left(\frac{p}{1-p}\ln(p)\right)\right] \\ &= -4N_e[\,(1-p)\ln(1-p) + p\ln(p)\,] \end{aligned}$$

which (as expected) recovers our expression (Equation A1.25) for the unconditional time $\bar{t}$.

For the special case of a neutral allele introduced as a single copy, $p = 1/(2N)$, which results in the approximations

$$(1-p)/p \simeq 1/p = 2N, \qquad p/(1-p) \simeq p = 1/(2N)$$

and

$$\ln(1-p) \simeq -p = -1/(2N), \qquad \ln(p) = -\ln(1/p) = -\ln(2N)$$

If we use these approximations, Equation A1.25 reduces to

$$\begin{aligned} \bar{t} &= -4N_e[\,(1-p)\ln(1-p) + p\ln(p)\,] \\ &\simeq -4N_e\big[-1/(2N) - \ln(2N)/(2N)\big] \simeq \frac{2N_e}{N}\ln(2N) \end{aligned}$$

Similarly, Equations A1.27 and A1.28 become

$$\bar{t}_L = -4N_e\left(\frac{p}{1-p}\right)\ln(p) \simeq \frac{2N_e}{N}\ln(2N)$$

and

$$\bar{t}_F = -4N_e \left(\frac{1-p}{p}\right) \ln(1-p) \simeq -4N_e \,(2N) \left(-\frac{1}{2N}\right) = 4N_e$$

Expectations of More General Functions

The expressions for sojourn times are relatively simple examples of computing expected values along a sample path of the diffusion process, and more complex functions can be evaluated in a similar manner. Let x_t, which resides in the open interval $(0,1)$, denote the values of our random variable along a particular sample path, and suppose that we wish to compute the integral (over all time) of some function f of x_t

$$I_f(p) = \int_0^\infty f(x_t)\, dt$$

Note that x_t is the realization at a particular time point, and hence the distribution of x_t (and thus ultimately I_f) is a function of p, the starting value of our process. Because x_t is a random variable, the integral $I_f(p)$ is also a random variable. Its expected value is given by

$$E[I_f(p)] = \int_0^\infty \int_0^1 f(x)\varphi(x,t,p)\, dt\, dx \qquad \text{(A1.29a)}$$

Fortunately, we do not have to solve for $\varphi(x,t,p)$, as (recalling Equation A1.22a) the general solution is obtained by using the Green's function, as described by Equation A1.22c, with

$$E[I_f(p)] = \int_0^1 f(x)\, h(x,p)\, dx \qquad \text{(A1.29b)}$$

This makes sense, as $h(x,p)dx$ is the expected amount of time that the process (starting at p) spends in the neighborhood of x before it is eventually lost or fixed. Thus, integration over all possible neighborhoods gives the expected value of the function over all sample paths.

Example A1.9. If we again assume an allele under drift only, consider two different functions, the total number of copies of the allele and total number of homozygotes involving this allele, with both expectations computed along only those paths that lose the allele. The relevant functions are, respectively, $f(x) = 2Nx$ for the total number of copies of the allele and $f(x) = Nx^2$ for the total number of homozygotes (assuming Hardy-Weinberg holds). To obtain the expected cumulative values of each function, we first apply Equation A1.26a, which gives the Green's function for pure drift on paths that are ultimately lost. This gives the general expression for any function, $f(x)$, as

$$\begin{aligned} E(f,p) &= \int_0^1 f(x)h_0(x,p)dx = 4N_e \int_0^p f(x)dx + 4N_e\frac{p}{1-p}\int_p^1 \frac{1-x}{x} f(x)dx \\ &= 4N_e \int_0^p f(x)dx + 4N_e\frac{p}{1-p}\left(\int_p^1 \frac{f(x)}{x}dx - \int_p^1 f(x)dx\right) \qquad \text{(A1.30)} \end{aligned}$$

Considering the total number of alleles, $f(x) = 2Nx$, the above expression becomes

$$E(2Nx) = 8N_eN\left[\int_0^p x\, dx + \frac{p}{1-p}\left(\int_p^1 dx - \int_p^1 x\, dx\right)\right]$$

The solution of the integrals in the square brackets reduces to

$$\frac{p^2}{2} + \frac{p}{1-p}\left((1-p) - \frac{1-p^2}{2}\right) = \frac{p}{2}$$

Hence, the expected total number of copies of the allele, starting at a frequency of p, during its entire sojourn to eventual loss is

$$8N_eNp/2 = 4N_eNp$$

Note that there are initially $2Np$ copies of the allele at the start of the process, so the total number of copies over the entire life of the conditional process leading to loss is simply $2N_e$ times the initial number. Hence, a single new mutation that is destined to be lost leaves an average of $2N_e$ copies over its lifetime. Of course, the distribution of the actual number of copies is heavily skewed, with a long right tail. Most mutations are quickly lost, but a few become very successful (and hence leave many copies) before eventually dying out.

If we turn to the expected total number of homozygotes, $f(x) = Nx^2$, which yields

$$\begin{aligned} E(Nx^2) &= 4N_eN\left[\int_0^p x^2 dx + \frac{p}{1-p}\left(\int_p^1 x\,dx - \int_p^1 x^2 dx\right)\right] \\ &= 4N_eN\left[\frac{p^3}{3} + \frac{p}{1-p}\left(\frac{1-p^2}{2} - \frac{1-p^3}{3}\right)\right] \\ &\simeq 4N_eN(p/6) = (2/3)N_eNp \quad (\text{for } p \ll 1) \end{aligned}$$

Again, if we start with a single copy, $E(Nx^2) \simeq N_e/3$.

APPLICATIONS IN QUANTITATIVE GENETICS

While we have focused on population-genetic applications of diffusion processes, this approach is also very useful for solving a number of problems in quantitative genetics. When attention shifts from individual alleles to a quantitative character, diffusions typically follow mean phenotypes instead of allele frequencies. Two well-studied diffusions, **Brownian motion** and the **Ornstein-Uhlenbeck process**, are especially useful (Chapter 12).

Brownian-motion Models

For Brownian motion (also called the **Wiener process**), the diffusion over the interval $-\infty < x < \infty$ is defined as

$$m(x) = a \quad \text{and} \quad v(x) = b \tag{A1.31a}$$

where $b > 0$. The general solution under Brownian motion starting at x_0 is that

$$x_t \sim \mathrm{N}(x_0 + at, bt) \tag{A1.31b}$$

namely, the distribution of x_t is normal, with a mean of $x_0 + at$ and a variance of $\sigma_t^2 = bt$. There is no equilibrium solution, as the process converges to a normal distribution with infinite variance (and, if $a \neq 0$, an infinite mean).

Example A1.10. Lande (1976) used the Brownian-motion model to approximate the change in the phenotypic mean of a neutral character with constant additive-genetic variance. In this case there is no directional force to change the mean, so $a = 0$. If we assume that the character has a strictly additive-genetic basis, the per-generation sampling variance in the mean is σ_A^2/N_e (Chapter 12), which is used for b in Equation A1.31a. Hence, at generation t, the distribution of phenotypic means is approximately normal, with expected an mean of μ_0 (the initial mean) and a variance of $\sigma_t^2 = t\sigma_A^2/N_e$.

One measure of how quickly phenotypic means drift is the minimum number of generations required for the mean of a random population to have at least a 50% probability of being more than K standard deviations from its initial mean. This is expressed as $\Pr(|x_t - \mu_0| \geq K\sigma_z) = 0.5$, where x_t is the mean of a randomly drawn replicate population and σ_z^2 is the phenotypic variance. Assuming Brownian motion, $(x_t - \mu_0)/\sigma_t$ is a unit normal random variable, and hence

$$\Pr(|x_t - \mu_0| \geq K\sigma_z) = \Pr\left[\frac{|x_t - \mu_0|}{\sigma_t} \geq \frac{K\sigma_z}{\sigma_t}\right] = \Pr\left[|U| \geq \frac{K\sigma_z}{\sigma_t}\right] = 0.5$$

For a unit normal U, $\Pr(|U| \geq 0.675) = 0.5$, which yields $K\sigma_z/\sigma_t = K\sigma_z/(\sigma_A\sqrt{t/N_e}) = 0.675$. Rearranging and substituting $h^2 = \sigma_A^2/\sigma_z^2$ yields

$$t = \frac{K^2\,N_e}{h^2\,0.675^2} \simeq 2\,N_e\,\frac{K^2}{h^2} \tag{A1.31c}$$

Thus, for $N_e = 10$, a neutral character with a constant heritability of $h^2 = 0.5$ requires $2 \cdot 10 \cdot 9/0.5 = 360$ generations until half the populations have phenotypic means of more than three standard deviations ($K = 3$) from their initial value. For $N_e = 10^5$, this time becomes 3.6 million generations.

The above analysis considers a single population over time (such as in a fossil sequence). Another situation of general interest involves an initial population that splits into two distinct, isolated populations. Here, the divergence between the two means is the sum of the two individual variances, meaning that if the initial difference is 0, then the distribution of the difference in mean after t generations is

$$d_t \sim \mathrm{N}\left[0, t \cdot \left(\frac{\sigma_{A_1}^2}{N_{e,1}} + \frac{\sigma_{A_2}^2}{N_{e,2}}\right)\right] \tag{A1.31d}$$

where σ_{Ai}^2 and $N_{e,i}$ are, respectively, the additive genetic variance and effective population size for population i.

Example A1.11. The careful reader may recall that drift also changes σ_A^2, with the assumption of a constant σ_A^2 being reasonable only for $t < N_e$ or when the population is at a mutation-drift equilibrium (Chapter 11). In Example A1.10, we fixed the genetic variance but allowed N_e to vary. However, if an ideal population (Chapter 2) has been at its current size for a sufficiently long time, the genetic variance for a neutral trait will be at its mutation-drift equilibrium value of $2N_e\sigma_m^2$, where σ_m^2 is the mutational variance (Equation 11.20c). The distribution of means now has an expected variance of

$$\sigma_t^2 = 2tN_e\sigma_m^2/N_e = 2t\sigma_m^2$$

From our previous example, Equation A1.31c calculates the expected number of generations until 50% of the means exceed K standard deviations as $K\sigma_z/\sigma_t = 0.675$. Using $\sigma_t^2 = 2t\sigma_m^2$ and solving for t yields

$$t = \frac{K^2\sigma_z^2}{2 \cdot 0.675^2 \cdot \sigma_m^2} = 1.1\,\frac{K^2\sigma_z^2}{\sigma_m^2} \simeq \frac{K^2\sigma_z^2}{\sigma_m^2} \tag{A1.31e}$$

Because $\sigma_z^2 = \sigma_A^2 + \sigma_E^2 = 2N_e\sigma_m^2 + \sigma_E^2$,

$$t \simeq K^2 \frac{\sigma_z^2}{\sigma_m^2} = K^2 \left(2N_e + \frac{1}{h_m^2} \right)$$

where $h_m^2 = \sigma_m^2 / \sigma_E^2$ is the mutational heritability.

According to LW Table 9.1, h_m^2 has an approximate average value of 0.006. Taking this value, and repeating the calculations from Example A1.10 (e.g., $N_e = 10$ and $K = 3$) yields $t = 9 \cdot (20 + 1/0.006) = 1680$ generations. The reason for the huge increase in time (1680 vs. 360) relative to the fixed variance example above is because the equilibrium additive variance is much smaller due to the small population size, and thus has an equilibrium heritability of

$$h^2 = \frac{2N_e\sigma_m^2}{2N_e\sigma_m^2 + \sigma_E^2} = \frac{2N_e\sigma_m^2/\sigma_E^2}{2N_e\sigma_m^2/\sigma_E^2 + 1} = \frac{2N_e h_m^2}{2N_e h_m^2 + 1} = \frac{0.12}{0.12 + 1} = 0.11$$

or roughly $1/5$ of the value used in Example A1.10. When $N_e = 10^5$, the time is 1.8 million generations, half the value obtained in Example A1.10, as now the equilibrium heritability is essentially one $(1200/1201)$, which is twice the value that was assumed in Example A1.10.

Example A1.12. The Brownian motion model easily extends to multiple traits. Suppose we let $\mathbf{x}$ denote a vector of traits, assume an infinitesimal mean change vector of zero, and denote the infinitesimal covariance matrix by $\mathbf{V}$ (the multivariate versions of $m(x)$ and $v(x)$, respectively). If the process starts at value $\mathbf{x}_0$, then

$$\mathbf{x}_t \sim \text{MVN}(\mathbf{x}_0, t\mathbf{V}) \tag{A1.32a}$$

where MVN denotes the multivariate normal. Lande (1979a) examined the change in a vector of trait means under drift and mutation. The per-generation sampling variance in the multivariate mean vector is $\mathbf{V} = \mathbf{G}/N_e$, the additive-genetic covariance matrix, $\mathbf{G}$ (Chapter 13), divided by the effective population size. If $\mathbf{u}_0$ is the starting vector of means, then

$$\mathbf{u}_t \sim \text{MVN}[\mathbf{u}_0, (t/N_e)\mathbf{G}] \tag{A1.32b}$$

A key feature of Equation A1.32b is that we expect the most divergence in population means to occur along the principal components of $\mathbf{G}$ (the eigenvectors corresponding to the leading eigenvalues; Appendix 5), as these are the axes with the most variance, and hence the most spread in the realizations of $\mathbf{u}_t$. Thus, if drift dominates, the divergence between two populations is largely in the direction of the first few principal components of $\mathbf{G}$ (a topic we return to in Volume 3).

Lande (1979a) assumed that over long periods of time, the genetic covariance matrix would approach its mutation-drift equilibrium value. Suppose we let $\mathbf{U}$ denote the matrix of mutational variances and covariances, which means that U_{ii} corresponds to the mutational (additive-genetic) variation in trait i generated in each generation (σ_m^2 in the univariate case), and U_{ij} is the mutational additive-genetic covariance between traits i and j (i.e., pleiotropy). For an effective population size of N_e, drift decays the current additive-genetic variance by an amount of $1/(2N_e)$ in each generation, but this is countered by new mutation, yielding an expected change in the average covariance matrix of

$$\Delta\mathbf{G} = \mathbf{G}/(2N_e) + \mathbf{U}$$

At equilibrium, $\Delta\widehat{\mathbf{G}} = \mathbf{0}$, or

$$\widehat{\mathbf{G}} = 2N_e\mathbf{U}$$

As expected, this reduces to the single-trait result of $2N_e\sigma_m^2$ (Equation 11.20c) when $\mathbf{U} = (\sigma_m^2)$. Hence, the sampling variance of the mean becomes $\widehat{\mathbf{G}}/N_e = 2\mathbf{U}$, which reduces Equation A1.32b to

$$\mathbf{u}_t \sim \text{MVN}(\mathbf{u}_0, 2t\mathbf{U}) \tag{A1.32c}$$

The neutral divergence between two populations now follows the leading eigenvectors of **U**, rather than those of the initial **G** matrix. This interesting finding requires the strong assumption that the population has been in mutation-drift equilibrium during the bulk of the divergence.

Ornstein-Uhlenbeck Models

The Ornstein-Uhlenbeck process is an extension of the Brownian-motion model to include a linear restoring force back to the origin. The one-dimensional version of this diffusion process for $-\infty < x < \infty$ is given by

$$m(x) = -ax \quad \text{and} \quad v(x) = b \tag{A1.33a}$$

with $a, b > 0$. As with Brownian motion, the distribution of x_t (given the starting condition x_0) is also normal, but now with a mean and a variance of

$$\mu_t = x_0 e^{-at} \quad \text{and} \quad \sigma_t^2 = \frac{b}{2a}\left(1 - e^{-2at}\right) \tag{A1.33b}$$

See Karlin and Taylor (1981) for a derivation. The resulting stationary distribution is normal with a mean of zero (on the assumed scale) and a variance of $b/(2a)$. More generally, if $m(x) = -a(x-\theta)$, so that the resorting force is toward the value $x = \theta$, then the variance will be unchanged, while

$$\mu_t = \theta + (x_0 - \theta)e^{-at} \tag{A1.33c}$$

Example A1.13. Lande (1976) examined the distribution of phenotypic means under drift and stabilizing selection, using the Gaussian fitness function (Equation 16.17) as a model of stabilizing selection. Here, the expected fitness of an individual with phenotypic value z is given by

$$W(z) = \exp\left(-\frac{z^2}{2\omega^2}\right)$$

The optimal phenotype is $z = 0$, and the strength of selection is given by the width, ω^2, with smaller values of ω^2 corresponding to stronger selection (fitness declining faster as one moves away from zero). Under nor-optimal selection, if phenotypes are normally distributed before selection with a mean of μ_t and a phenotypic variance of σ_z^2, they remain normal after selection, with a new mean of $\mu_t + S$, where

$$S = -\mu_t \frac{\sigma_z^2}{\sigma_z^2 + \omega^2}$$

and a new variance of

$$\sigma_z^2 - \frac{\sigma_z^4}{\sigma_z^2 + \omega^2}$$

We assume that selection is sufficiently weak ($\omega^2 \gg \sigma_z^4$) so that the variance remains effectively unchanged after selection. As expected, nor-optimal selection moves the mean in the direction of the optimum (decreasing the current mean if it is positive, and increasing it if negative). Assuming the breeder's equation ($R = h^2 S$), the change in the mean becomes

$$\Delta\mu = h^2 S = \frac{\sigma_A^2}{\sigma_z^2}\left(-\mu_t \frac{\sigma_z^2}{\sigma_z^2 + \omega^2}\right) = -\mu_t\left(\frac{\sigma_A^2}{\sigma_z^2 + \omega^2}\right)$$

Suppose we let $x_t = \mu_t$ be the mean in generation t of a randomly drawn replicate population. From Equation A1.33a, the resulting distribution of means can then be approximated by an Ornstein-Uhlenbeck process, with

$$a = \frac{\sigma_A^2}{\sigma_z^2 + \omega^2} \quad \text{and} \quad b = \frac{\sigma_A^2}{N_e}$$

where a follows from the change in mean and b is the drift variance under Brownian motion (Example A.10). From Equation A1.33b, the resulting distribution of phenotypic means in generation t is normal, with a mean of

$$\mu_t = \mu_0 \exp\left(-t \frac{\sigma_A^2}{\sigma_z^2 + \omega^2} \right) \tag{A1.34a}$$

and variance of

$$\sigma_t^2 = \frac{\sigma_z^2 + \omega^2}{2N_e} \left[1 - \exp\left(-2t \frac{\sigma_A^2}{\sigma_z^2 + \omega^2} \right) \right] \tag{A1.34b}$$

At equilibrium, the distribution of the means of replicate populations is normal, with mean of zero (i.e., centered at the fitness optimum) and a variance of $(\sigma_z^2 + \omega^2)/(2N_e)$.

Example A1.14. Example A1.13 shows how diffusion approximations yield a stationary distribution for the phenotypic mean under the joint action of drift and stabilizing selection. More general forms of selection can be considered as well. Suppose our concern is simply the change in the mean of a trait that is under arbitrary selection. Provided the phenotypes are normally distributed within a population, then, from Equation 13.27a the change in mean is

$$\Delta\mu = \sigma_A^2 \frac{\partial \ln[\overline{W}(\mu)]}{\partial \mu}$$

where the mean fitness, $\overline{W}(\mu)$, is a function of the population mean, μ. Because no assumptions were made about the nature of selection, any function for mean fitness can be used, *provided* the distribution of offspring phenotypes remains roughly normal. Hence, an approximating diffusion for the behavior of the mean under selection and drift is to let x equal the current mean and set

$$m(x) = \sigma_A^2 \frac{\partial \ln(\overline{W})}{\partial x} \quad \text{and} \quad v(x) = \frac{\sigma_A^2}{N_e} \tag{A1.35}$$

Substituting Equations A1.35 into Equation A1.12 yields

$$G(x) = \exp\left[-2N_e \int^x \frac{\partial \ln(\overline{W})}{\partial y} \partial y \right] = \exp[-2N_e \ln(\overline{W})] = \overline{W}(x)^{-2N_e}$$

Thus, Equation A1.11 gives the equilibrium distribution of population means as

$$\varphi(x) = \frac{C}{v(x)G(x)} = \frac{CN_e}{\sigma_A^2} \overline{W}(x)^{2N_e} \propto \overline{W}(x)^{2N_e} \tag{A1.36}$$

This result, which is from Lande (1976), shows that as the effective population size increases, the probability that the population mean is near a local maximum in fitness also increases. This follows because $\varphi(x)$ becomes increasingly peaked around local maxima relative to other parts of the fitness surface as we increase N_e. If $\overline{W}$ has multiple peaks, then so does $\varphi(x)$. Equations A1.35 and A1.36 rely on the assumptions that the phenotypic and additive-genetic variances remain constant as the mean changes and (as mentioned above) that the phenotypic distribution remains normal.

Appendix 2

Introduction to Bayesian Analysis

Now I've heard a Bayesian is half a man
With robustness not on his side,
But as the problems get tougher,
I just watch and wonder
As the frequentists run and hide!
— *Ain't Too Proud to Bayes*, B. Carlin (ISBA 2000)

The history of statistical methods in genetics closely parallels advances in computation. Before the widespread use of computers, **method-of-moments** approaches were common as they are relatively easy to obtain. Here, a summary statistic of the data is computed whose expected value is the parameter of interest (e.g., using the sample mean, $\overline{x}$, as an estimate of the true mean, μ_x, as $E[\overline{x}] = \mu_x$). In the mid-1970s, maximum-likelihood (ML) methods became much more common place, as they offer a very flexible platform for statistical analysis (estimation, determining precision, and hypothesis testing), but at the cost of numerically searching an often highly complex multidimensional likelihood surface (LW Appendix 4). Both these approaches typically return point estimators for the variables of interest, along with some measure of their uncertainty. As opposed to these classical (or **frequentist**) approaches, **Bayesian statistics** (which can be viewed as a natural extension of likelihood methods) is concerned with generating the full *distribution* for the parameters, $\boldsymbol{\Theta}$, given the data, $\mathbf{x}$, namely, obtaining the posterior distribution, $p(\boldsymbol{\Theta}\,|\,\mathbf{x})$. As such, Bayesian statistics provides a much more complete picture of the uncertainty in the estimation of the unknown parameters, especially after the confounding effects of nuisance parameters are removed.

Our treatment here is intentionally quite brief. A number of texts have presented excellent treatments of the statistical theory (e.g., Lindley 1965; Berger 1985; Carlin and Louis 2000; Lee 2012; Gelman et al. 2013). Blasco (2017) provided a very lucid introduction to applications in quantitative genetics, while Sorensen and Gianola (2002) offered a more comprehensive treatment. While very deep (and very subtle) differences in philosophy separate hard-core Bayesians from hard-core frequentists (Efron 1986; Glymour 1981), our treatment of Bayesian methods is motivated simply by their use as a powerful statistical tool. This appendix focuses on the basic theory, while the computational approaches that make these methods feasible are examined in Appendix 3.

WHY ARE BAYESIAN METHODS BECOMING MORE POPULAR?

In addition to providing a more formal framework for dealing with parameter uncertainty, two specific features have fueled the rapid growth of Bayesian approaches in genetics and genomics. First, under a Bayesian analysis, all parameters are random effects as opposed to fixed effects (Chapter 19). This has profound implications for degrees of freedom. Consider a gene expression study with 30,000 features (genes of interest), whose mRNA levels are contrasted over a set of 100 normal liver cells versus 100 cancerous ones. If we treat the differential expression level of any particular gene as a fixed effect (an unknown constant to be estimated) we will very quickly use all of the degrees of freedom, given the small sample size. Conversely, if these levels are treated as **random effects**, with the expression difference associated with a particular gene being a random variable drawn from some underlying (and unknown) distribution, then the only degrees of freedom lost will be those used to estimate the associated parameters for this underlying distribution. Further, prediction of the random realization that corresponds to a particular gene borrows information over all

Evolution and Selection of Quantitative Traits. Bruce Walsh & Michael Lynch.
Published 2018 by Oxford University Press.
DOI 10.1093/oso/9780198830870.001.0001

the genes. Thus, a Bayesian analysis can handle high-dimensional experiments in which the number of parameters, p, greatly exceeds the number of observations, n, in a framework that fully manages the uncertainty over all these estimates. Second, Bayesian methods are computationally feasible, as approaches such as MCMC (Appendix 3) allow high-dimensional datasets to be analyzed in a computationally efficient manner. In settings with a large number of nuisance parameters or a high-dimensional dataset, a Bayesian approach not only has considerable appeal, it may be the only approach that is even feasible.

BAYES' THEOREM

The foundation of Bayesian statistics is **Bayes' theorem**. Suppose we observe a random variable, x, and wish to make inferences about another random variable, θ, in the case where θ is drawn from some distribution, $\Pr(\theta)$. From the definition of conditional probability

$$\Pr(\theta \,|\, x) = \frac{\Pr(x, \theta)}{\Pr(x)} \tag{A2.1a}$$

where (for now) x and θ are discrete random variables. Again from the definition of conditional probability, we can express the joint probability by conditioning on θ to give

$$\Pr(x, \theta) = \Pr(x \,|\, \theta) \Pr(\theta) \tag{A2.1b}$$

Putting these together yields Bayes' theorem

$$\Pr(\theta \,|\, x) = \frac{\Pr(x \,|\, \theta) \Pr(\theta)}{\Pr(x)} \tag{A2.2a}$$

Bayes' theorem flips the conditioning variable, allowing us to move from $\Pr(x \,|\, \theta)$ to $\Pr(\theta \,|\, x)$. With k possible values of θ (θ_1 through θ_k), the discrete version of Bayes' theorem becomes

$$\Pr(\theta_j \,|\, x) = \frac{\Pr(x \,|\, \theta_j) \Pr(\theta_j)}{\Pr(x)} = \frac{\Pr(x \,|\, \theta_j) \Pr(\theta_j)}{\sum_{i=1}^{k} \Pr(\theta_i) \, \Pr(x \,|\, \theta_i)} \quad \text{for} \quad 1 \le j \le k \tag{A2.2b}$$

In Bayesian statistics, x represents an observable variable (the data), while θ represents a parameter describing the distribution of x. In this setting, $\Pr(\theta)$ is the **prior distribution** of possible parameter values, while $\Pr(\theta \,|\, x)$ is the subsequent **posterior distribution** of θ given the observed data x and the prior. In classical statistics, the unknown parameters are treated as fixed and the data are considered random, whereas under a Bayesian analysis, the data are considered fixed and the unknown parameters that generated the data are considered random.

All of the above statements also hold for continuous random variables, for which the probability density function, p, replaces the discrete probability value, Pr. In particular, the continuous multivariate version of Bayes' theorem is

$$p(\boldsymbol{\Theta} \,|\, \mathbf{x}) = \frac{p(\mathbf{x} \,|\, \boldsymbol{\Theta}) \, p(\boldsymbol{\Theta})}{p(\mathbf{x})} = \frac{p(\mathbf{x} \,|\, \boldsymbol{\Theta}) \, p(\boldsymbol{\Theta})}{\int p(\mathbf{x}, \boldsymbol{\Theta}) \, d\boldsymbol{\Theta}} \tag{A2.3}$$

where $\boldsymbol{\Theta} = (\theta_1, \theta_2, \cdots, \theta_n)$ is a vector of n (potentially) continuous variables. As with the univariate case, $p(\boldsymbol{\Theta})$ is the assumed prior distribution of the unknown parameters, while $p(\boldsymbol{\Theta} \,|\, \mathbf{x})$ is the posterior distribution given the prior, $p(\boldsymbol{\Theta})$, and the data, $\mathbf{x}$.

The origin of Bayes' theorem has a fascinating history (Stigler 1983). It is named after the Rev. Thomas Bayes, a priest who never published a mathematical paper during his lifetime. The paper in which the theorem appears was posthumously read before the Royal Society by his friend Richard Price in 1764. Stigler suggests it was first discovered by Nicholas

Saunderson, a blind mathematician and optician who, at age 29, became Lucasian Professor of Mathematics at Cambridge (the position held earlier by Issac Newton). This is an example of **Stigler's Law of Eponymy** (Stigler 1980), wherein no discovery or invention is named after its first discoverer (an **eponym**). As is fitting, Stigler's law is self-consistent, as this phenomenon was previously mentioned by Merton (1965).

Example A2.1. Consider a recessive color locus in cattle in which the genotypes *BB* and *Bb* are black, while *bb* is red. Two black-coated parents are crossed, and produce some red offspring, which implies that both parents must be *Bb*. A black-coated son of theirs is crossed to n red dams (*bb*), and all of his offspring are black. What is the posterior probability that he is *BB*?

To solve this problem using Bayes' theorem, we first define the indicator random variable

$$\theta = \begin{cases} 0 & \text{son is } Bb \\ 1 & \text{son is } BB \end{cases}$$

Given that both parents are *Bb*, the expected priors for their offspring are 1/4 for *BB* and 1/2 for *Bb*, resulting in a 3/4 prior for a black-coated offspring. Further, from conditional probability (Equation A2.1a), the prior that a black offspring is *BB* is

$$\Pr(BB \mid \text{Black}) = \frac{\Pr(BB, \text{Black})}{\Pr(\text{Black})} = \frac{\Pr(BB)}{\Pr(\text{Black})} = \frac{1/4}{3/4} = 1/3$$

where we used the fact that all *BB* are black, so that Pr(*BB*, Black) = Pr(*BB*). Hence, the prior becomes

$$\Pr(\theta) = \begin{cases} 0 & \text{is } 2/3 \\ 1 & \text{is } 1/3 \end{cases}$$

Further

$$\Pr(\text{all } n \text{ offspring are black} \mid \text{sire is } BB) = 1$$

$$\Pr(\text{all } n \text{ offspring are black} \mid \text{sire is } Bb) = (1/2)^n$$

and

$$\Pr(\text{all } n \text{ black}) = \Pr(\text{all black} \mid BB)^*\Pr(BB) + \Pr(\text{all black} \mid Bb)^*\Pr(Bb)$$

$$= 1 \cdot 1/3 + (1/2)^n \cdot (2/3)$$

If we combine the above values, Bayes' theorem yields

$$\Pr(\theta = 1 \mid n \text{ black offspring}) = \frac{\Pr(n \mid \theta = 1)\Pr(\theta = 1)}{\Pr(n)} = \frac{1 \cdot (1/3)}{1 \cdot (1/3) + (1/2)^n \cdot (2/3)}$$

which returns values of 0.5, 0.67, 0.8, 0.89, 0.94, and 0.998 for $n = 1, 2, 3, 4, 5$, and 10, respectively.

Example A2.2. Suppose a major gene (with alleles Q and q) underlies a character of interest. The distribution of phenotypic values for each major-locus genotype follows a normal distribution with a variance of 1 and means of 2.1, 3.5, and 1.3 for QQ, Qq, and qq, respectively. Suppose the frequencies of these genotypes for a random individual drawn from the population are 0.3, 0.2, and 0.5 (for QQ, Qq, and qq, respectively). If an individual from this population has a phenotypic value of 3, then what is the probability of it being QQ? Qq? qq? Let $\varphi(x \mid \mu, 1) = (2\pi)^{-1/2} e^{-(x-\mu)^2/2}$ denote the density function for a normal distribution with a mean of μ and variance of 1. To apply Bayes' theorem, note that the values for the priors and the conditionals are as follows:

Genotype, G	$\Pr(G)$	$p(\mathrm{x}\|G)$	$\Pr(G)\cdot p(\mathrm{x}\|G)$
QQ	0.3	$\varphi(3\,\|\,2.1, 1) = 0.266$	0.078
Qq	0.2	$\varphi(3\,\|\,3.5, 1) = 0.350$	0.070
qq	0.5	$\varphi(3\,\|\,1.3, 1) = 0.094$	0.047

Because $p(3) = \sum_G \Pr(G) \cdot p(3|G) = 0.195$, Bayes' theorem gives the posterior probabilities for the genotypes given the observed value of 3 as:

$$\Pr(QQ \mid x = 3) = 0.078/0.195 = 0.400$$

$$\Pr(Qq \mid x = 3) = 0.070/0.195 = 0.359$$

$$\Pr(qq \mid x = 3) = 0.047/0.195 = 0.241$$

Thus, there is a 40 percent chance that this individual has a genotype of QQ, a 36 percent chance it is Qq, and a 24 percent chance it is qq.

FROM LIKELIHOOD TO BAYESIAN ANALYSIS

The method of maximum likelihood (LW Appendix 4) and Bayesian analysis are closely related. Suppose $\ell(\boldsymbol{\Theta} \mid \mathbf{x})$ is the assumed likelihood function. Under ML estimation, we would compute the mode of the likelihood function (the maximal value of ℓ, as a function of $\boldsymbol{\Theta}$ given the data $\mathbf{x}$), and use the local curvature around the mode to construct confidence intervals. Hypothesis testing follows using likelihood-ratio (LR) statistics. The strengths of ML estimation rely on its *large-sample* properties, namely, that when the sample size is sufficiently large, we can assume both normality of the estimators and that most LR tests follow χ^2 distributions. These features, nice as they are, may not hold for small samples. Conversely, a Bayesian analysis is *exact* for any sample size, given a specified prior.

To transition from a likelihood to a Bayesian analysis, we start with some prior distribution, $p(\boldsymbol{\Theta})$, that captures our initial knowledge (or best guess) about the possible values of the unknown parameters. From Bayes' theorem, the data (likelihood) is combined with the prior to produce a posterior distribution,

$$p(\boldsymbol{\Theta} \mid \mathbf{x}) = \frac{1}{p(\mathbf{x})} \cdot p(\mathbf{x} \mid \boldsymbol{\Theta}) \cdot p(\boldsymbol{\Theta}) \tag{A2.4a}$$

$$= (\text{normalizing constant}) \cdot p(\mathbf{x} \mid \boldsymbol{\Theta}) \cdot p(\boldsymbol{\Theta}) \tag{A2.4b}$$

$$= \text{constant} \cdot \text{likelihood} \cdot \text{prior} \tag{A2.4c}$$

as $p(\mathbf{x} \mid \boldsymbol{\Theta}) = \ell(\boldsymbol{\Theta} \mid \mathbf{x})$ is simply the likelihood function (LW Appendix 4) and $1/p(\mathbf{x})$ is a constant (with respect to $\boldsymbol{\Theta}$). Consequently, the posterior distribution is often written as

$$p(\boldsymbol{\Theta} \mid \mathbf{x}) \propto \ell(\boldsymbol{\Theta} \mid \mathbf{x})\; p(\boldsymbol{\Theta}) \tag{A2.4d}$$

where the symbol $\propto$ means "proportional to" (equal up to a constant). Note that the constant $p(\mathbf{x})$ normalizes $p(\mathbf{x} \mid \boldsymbol{\Theta}) \cdot p(\boldsymbol{\Theta})$ to one, and hence can be obtained by integration

$$p(\mathbf{x}) = \int_{\boldsymbol{\Theta}} p(\mathbf{x} \mid \boldsymbol{\Theta}) \cdot p(\boldsymbol{\Theta})\, d\boldsymbol{\Theta} \tag{A2.5}$$

The dependence of the posterior on the prior (which can easily be assessed by trying different priors) provides an indication of how much information on the unknown parameter values is contained in the data (the curvature of the likelihood surface). If the posterior is highly dependent on the prior, then the data likely has little signal (a **flat likelihood surface**), while if the posterior is largely unaffected by different priors, then the data are likely highly informative (a sharply peaked likelihood surface). To see this, taking logs on Equation A2.4c yields

$$\log(\text{posterior}) = \log(\text{likelihood}) + \log(\text{prior}) + \text{constant} \tag{A2.6}$$

When the likelihood signal is strong, it largely dominates the prior in the resulting posterior, but when a likelihood is weak, the prior can dominate.

Marginal Posterior Distributions

Often only a subset of the unknown parameters is of concern to us, and the rest are **nuisance parameters** that are of no interest, but still must be fitted in the model. A strong feature of Bayesian analysis is that we can account for all the uncertainty introduced into the parameters of interest by any uncertainty in the values of nuisance parameters. This is accomplished by integrating the nuisance parameters out of the posterior distribution to generate a **marginal posterior distribution** for the parameters of interest. For example, suppose the mean and variance of data coming from a normal distribution are unknown, but our real interest is only in the variance. Estimating the mean introduces additional uncertainty into our variance estimate, which is not fully captured by standard classical approaches. Under a Bayesian analysis, the marginal posterior distribution for σ^2 is simply

$$p(\sigma^2 \,|\, \mathbf{x}) = \int p(\mu, \sigma^2 \,|\, \mathbf{x})\, d\mu$$

The resulting marginal posterior for σ^2 captures all of the uncertainty in the estimation of μ that influences the uncertainty in σ^2. This is an especially nice feature when a large number of nuisance parameters must be estimated.

The marginal posterior may involve several parameters (generating **joint marginal posteriors**). Suppose we write the vector of unknown parameters as $\boldsymbol{\Theta} = (\boldsymbol{\Theta}_1, \boldsymbol{\Theta}_{nu})$, where $\boldsymbol{\Theta}_{nu}$ is the vector of nuisance parameters. Integrating over $\boldsymbol{\Theta}_{nu}$ yields the desired marginal for the vector $\boldsymbol{\Theta}_1$ of parameters of interest as

$$p(\boldsymbol{\Theta}_1 \,|\, \mathbf{y}) = \int_{\boldsymbol{\Theta}_{nu}} p(\boldsymbol{\Theta}_1, \boldsymbol{\Theta}_{nu} \,|\, \mathbf{y})\, d\boldsymbol{\Theta}_{nu} \tag{A2.7}$$

While these complex integrals appear quite daunting (and indeed almost always are from an analytic standpoint), generating draws from the marginal distribution is usually very straightforward using MCMC methods (which are examined in Appendix 3).

SUMMARIZING THE POSTERIOR DISTRIBUTION

How do we extract a Bayesian estimator for some unknown parameter, θ? If our mindset is to use some sort of point estimator (as is usually done in classical statistics), then there are a number of candidates. We could follow maximum likelihood and use the **mode of the posterior distribution** (its maximal value)

$$\widehat{\theta} = \max_{\theta}\, [\, p(\theta \,|\, \mathbf{x})] \tag{A2.8a}$$

We could take the **expected value of** θ (its mean) given the posterior

$$\widehat{\theta} = E[\,\theta \,|\, \mathbf{x}\,] = \int \theta\, p(\theta \,|\, \mathbf{x}) d\theta \tag{A2.8b}$$

Another candidate is the **median of the posterior**, which is more robust than the mean to outliers. Here the estimator satisfies $\Pr(\theta > \widehat{\theta} \,|\, \mathbf{x}) = \Pr(\theta < \widehat{\theta} \,|\, \mathbf{x}) = 0.5$, hence

$$\int_{\widehat{\theta}}^{+\infty} p(\theta \,|\, \mathbf{x}) d\theta = \int_{-\infty}^{\widehat{\theta}} p(\theta \,|\, \mathbf{x}) d\theta = \frac{1}{2} \tag{A2.8c}$$

However, using any of the above estimators, or even all three simultaneously, loses the full power of a Bayesian analysis, as ***the full estimator is the entire posterior density itself***. If we cannot obtain the full form of the posterior distribution, then these estimates of general features of the distribution can be presented. However, as we will see in Appendix 3, we can generally obtain the full posterior by simulation using MCMC sampling, and hence the Bayesian estimate of a parameter is often presented as a frequency histogram (potentially smoothed) of the MCMC-generated samples from the posterior distribution (an **empirical posterior**).

Highest Density Regions (HDRs)

Given the posterior distribution, the construction of confidence intervals is straightforward. For example, a $100(1-\alpha)\%$ confidence interval is given by any $(L_{\alpha/2}, H_{\alpha/2})$ satisfying

$$\int_{L_{\alpha/2}}^{H_{\alpha/2}} p(\theta \,|\, \mathbf{x})\, d\theta = 1 - \alpha$$

To reduce the set of possible candidate intervals, one typically uses **highest density regions**, or **HDRs**, where, for a single parameter, the HDR $100(1-\alpha)$ region(s) are the shortest intervals giving an area of $(1-\alpha)$. More generally, if multiple parameters are being estimated, the HDR region(s) are those with the smallest *volume* in the parameter space. HDRs are also referred to as **Bayesian confidence intervals** or (better yet) **credible intervals**.

It is critical to note that there is a *profound difference* between a confidence interval (CI) from classical (frequentist) statistics and a Bayesian analysis. The interpretation of a classical confidence interval is that if we were to repeat the experiment a large number of times, and construct CIs in the same fashion, the fraction of the resulting collection of CIs that enclose the unknown parameter approaches $(1-\alpha)$. Thus, the frequentist CI is a measure of the *frequency* of occurrences in independent experiments in which the CI encloses the true value (and hence the term frequentist for this type of statistics). In contrast, with a Bayesian HDR, there is a probability of $(1-\alpha)$ that the interval contains the true value of the unknown parameter. While at first blush these two intervals appear to be essentially identical, they are not, and indeed they are fundamentally (but subtly) different. Often the CI and Bayesian intervals span essentially the same values, but again the interpretational difference remains. The key point is that the Bayesian prior allows us to make ***direct probability statements*** about θ, while under classical statistics we can only make statements about the behavior of the statistic if we consider ***repeating an experiment a large number of times***. Given the important conceptual difference between classical and Bayesian intervals, Bayesians typically avoid using the term *confidence interval*, using the term *credible interval* instead.

Bayes Factors and Hypothesis Testing

In the classical hypothesis-testing framework, we have two alternatives. The null hypothesis, H_0, that the unknown parameter, θ, belongs to some set or interval, Θ_0 ($\theta \in \Theta_0$), versus the alternative hypothesis, H_1, that θ belongs to the alternative set, Θ_1 ($\theta \in \Theta_1$). Θ_0 and Θ_1 contain no common elements ($\Theta_0 \cap \Theta_1 = \oslash$) and the union of Θ_0 and Θ_1 contains the entire space of values for θ (i.e., $\Theta_0 \cup \Theta_1 = \Theta$).

In the classical statistical framework of the frequentists, one uses the observed data to test the significance of a particular hypothesis, and (if possible) compute a p value (the probability, p, of observing a value equal to, or more extreme than, that of the test statistic if the null hypothesis is indeed correct). Initially, one would think that the idea of a hypothesis test is trivial in a Bayesian framework, as using the posterior distribution provides the expected p values directly, for example,

$$\Pr(\theta > \theta_0) = \int_{\theta_0}^{\infty} p(\,\theta \,|\, \mathbf{x})\, d\theta \qquad \text{and} \qquad \Pr(\theta_0 < \theta < \theta_1) = \int_{\theta_0}^{\theta_1} p(\,\theta \,|\, \mathbf{x})\, d\theta$$

The fault in this logic under a Bayesian framework is that we also have *prior information* and Bayesian hypothesis testing addresses whether, *given the data*, we are more or less inclined to believe the hypothesis than was suggested from the prior. Hence, the ***prior probabilities influence hypothesis testing***. To formalize this idea, let

$$p_0 = \Pr(\theta \in \Theta_0 \,|\, \mathbf{x}) \qquad \text{and} \qquad p_1 = \Pr(\theta \in \Theta_1 \,|\, \mathbf{x}) \tag{A2.9a}$$

denote the probabilities, given the observed data, **x**, that θ is in the null (p_0) and alternative (p_1) hypothesis sets. Note that these are *posterior* probabilities. Because $\Theta_0 \cap \Theta_1 = \emptyset$ and $\Theta_0 \cup \Theta_1 = \Theta$, it follows that $p_0 + p_1 = 1$. Likewise, for the *prior* probabilities we have

$$\pi_0 = \Pr(\theta \in \Theta_0) \qquad \text{and} \qquad \pi_1 = \Pr(\theta \in \Theta_1) \tag{A2.9b}$$

Thus the **prior odds** of H_0 versus H_1 are π_0/π_1, while the **posterior odds** are p_0/p_1.

The **Bayes factor**, $\mathbf{B}_0$, in favor of H_0 versus H_1 is calculated by the ratio of the posterior odds divided by the prior odds,

$$B_0 = \frac{p_0/p_1}{\pi_0/\pi_1} = \frac{p_0\pi_1}{p_1\pi_0} \tag{A2.10a}$$

The Bayes factor is loosely interpreted as the odds in favor of H_0 over H_1 as given by the data and our prior opinion. Because $\pi_1 = 1 - \pi_0$ and $p_1 = 1 - p_0$, we can also express this as

$$B_0 = \frac{p_0(1-\pi_0)}{\pi_0(1-p_0)} \tag{A2.10b}$$

By symmetry, note that the Bayes factor, B_1, in favor of H_1 versus H_0 is simply $B_1 = 1/B_0$.

Example A2.3. Suppose that the prior distribution of θ is such that $\Pr(\theta > \theta_0) = 0.10$, while for the posterior distribution $\Pr(\theta > \theta_0 \,|\, \mathbf{x}) = 0.05$. The latter is significant at the 5% level in a classical hypothesis-testing framework, but the data only doubles our confidence in the alternative hypothesis relative to our belief based on prior information. If $\Pr(\theta > \theta_0) = 0.50$ for the prior, then a 5% posterior probability would greatly increase our confidence in the alternative hypothesis. Consider the first case in this example, where the prior and posterior probabilities for the null were $\pi_0 = 0.1$ and $p_0 = 0.05$, respectively. The Bayes factor in favor of H_1 versus H_0 is

$$B_1 = \frac{\pi_0(1-p_0)}{p_0(1-\pi_0)} = \frac{0.1 \cdot 0.95}{0.05 \cdot 0.9} = 4.22$$

Similarly, for the second example, where the prior for the null was $\pi_0 = 0.5$,

$$B_1 = \frac{0.5 \cdot 0.95}{0.05 \cdot 0.5} = 19$$

Here, the data showed close to a 20-fold improvement (relative to the prior) in support of H_1. Bayes factors and p values represent fundamentally different approaches to an analysis and are not formally comparable. However, a *loose* interpretation is that a factor of 20 is akin to the level of support of a $p = 0.05$, and a factor of 100 to $p = 0.01$.

When the hypotheses are simple (i.e., single values), say $\Theta_0 = \theta_0$ vs. $\Theta_1 = \theta_1$, then

$$p_i \propto p(\theta_i)\, p(\mathbf{x} \,|\, \theta_i) = \pi_i\, p(\mathbf{x} \,|\, \theta_i) \quad \text{for} \quad i = 0, 1$$

Thus

$$\frac{p_0}{p_1} = \frac{\pi_0\, p(\mathbf{x} \,|\, \theta_0)}{\pi_1\, p(\mathbf{x} \,|\, \theta_1)} \tag{A2.11a}$$

and from Equation A2.10a, the Bayes factor (in favor of the null) reduces to

$$B_0 = \frac{p(\mathbf{x} \,|\, \theta_0)}{p(\mathbf{x} \,|\, \theta_1)} \tag{A2.11b}$$

which is simply a *likelihood ratio* (LW Appendix 4).

When hypotheses are **composite** (containing multiple elements), the situation is slightly more complicated. First, note that the prior distribution of θ conditioned on H_0 or H_1 is

$$p_i(\theta) = p(\theta)/\pi_i \qquad \text{for} \quad i = 0, 1 \tag{A2.12}$$

as the total probability $\theta \in \Theta_i = \pi_i$, so dividing by π_i normalizes the distribution to integrate to one. Thus,

$$\begin{aligned} p_i = \Pr(\theta \in \Theta_i \,|\, \mathbf{x}) &= \int_{\theta \in \Theta_i} p(\theta \,|\, \mathbf{x}) d\theta \\ &= \frac{1}{p(\mathbf{x})} \int_{\theta \in \Theta_i} p(\theta) p(\mathbf{x} \,|\, \theta) d\theta \\ &= \pi_i \int_{\theta \in \Theta_i} p(\mathbf{x} \,|\, \theta) p_i(\theta) d\theta \end{aligned} \tag{A2.13}$$

where the second step follows from Bayes' theorem, while the final step follows from Equation A2.12. The Bayes factor in favor of the null hypothesis becomes

$$B_0 = \left(\frac{p_0}{\pi_0}\right)\left(\frac{\pi_1}{p_1}\right) = \frac{\int_{\theta \in \Theta_0} p(\mathbf{x} \,|\, \theta) p_0(\theta) d\theta}{\int_{\theta \in \Theta_1} p(\mathbf{x} \,|\, \theta) p_1(\theta) d\theta} \tag{A2.14}$$

which is a ratio of the weighted likelihoods of Θ_0 and Θ_1.

THE CHOICE OF A PRIOR

Obviously, a critical feature of any Bayesian analysis is the choice of a prior. The key is that when the data have a sufficiently strong signal, even a poor choice of a prior will still not greatly influence the posterior. In a sense, it is an asymptotic (large-sample) property of Bayesian analysis in that all but pathological priors (those with zero probability where the true value lies) can be overcome by sufficient amounts of data. As mentioned above, one can check the impact of the prior by assessing the stability of posterior over a collection of diverse priors. The **location** of a parameter (mean or mode) and its **precision** (the reciprocal of the variance) of the prior is usually more critical than its actual shape in terms of conveying prior information. The shape (family) of the prior distribution is often chosen to facilitate calculation of the posterior, especially through the use of **conjugate priors** that, for a given likelihood function, return a posterior in the same distribution family as the prior (e.g., a gamma prior returns a gamma posterior when the likelihood is Poisson). We will return to conjugate priors, but first we will discuss other approaches for construction of priors.

Diffuse Priors

One of the most commonly used priors is the **flat** or **diffuse** (also called **uninformative** or **naive**) prior, which is simply a constant

$$p(\theta) = \frac{1}{b-a} \qquad \text{for} \qquad a \le \theta \le b \tag{A2.15a}$$

This conveys that we have no a priori reason to favor any particular parameter value over another. With a flat prior, the posterior is just a constant C times the likelihood

$$p(\theta \,|\, \mathbf{x}) = C\,\ell(\theta \,|\, \mathbf{x}) \tag{A2.15b}$$

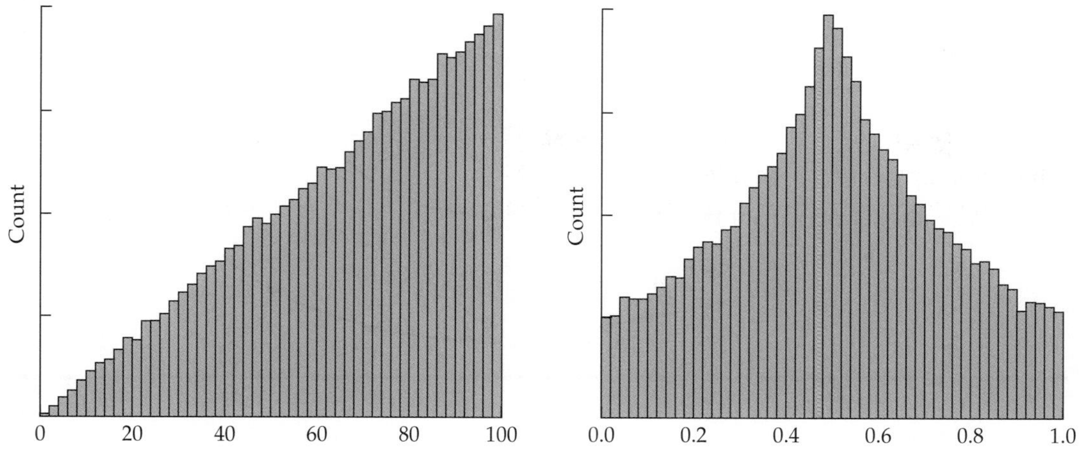

Figure A2.1 A uniform prior on one scale does not result in a flat prior on a transformed scale. Suppose a flat prior on (0,10000) is assumed for both the additive and residual variances. To mimic what happens under MCMC, we display these priors by using the resulting histograms generated from a large number of random draws, with a uniform expected to return a flat histogram. **Left:** The resulting prior for the standard deviation of either variance (the square root of a random draw). **Right:** The resulting prior for h^2, the ratio of a random draw for the additive variance divided by this value plus a random draw for the residual variance. Neither of these priors result in a uniform prior (namely, a flat histogram) on the transformed scale.

and we typically write that $p(\theta\,|\,\mathbf{x}) \propto \ell(\theta\,|\,\mathbf{x})$. In many cases, classical expressions from frequentist statistics are obtained by Bayesian analysis through assuming a flat prior.

If the variable (i.e., parameter) of interest ranges over $(0, \infty)$ or $(-\infty, +\infty)$, then, strictly speaking, a flat prior does not exist as, if the constant takes on any nonzero value, the integral does not exist. In such cases a flat prior (i.e., assuming $p[\theta\,|\,\mathbf{x}] \propto \ell[\theta\,|\,\mathbf{x}]$) is referred to as an **improper prior**, and care must be taken to ensure that the product of the prior and the likelihood results in a proper posterior (i.e., $\ell[\theta\,|\,\mathbf{x}]$ has a finite integral over the parameter range). This is by no means certain.

Another complication involved in using a uniform prior arises when the question of interest resides on a *different scale* than that used for the prior. A variable uniform on one scale may be far from uniform on a transformed scale. Figure A2.1 shows two examples based on the assumption that there was a flat prior on the variance. A uniform prior on the variance does *not* result in a uniform prior on the standard deviation (e.g., Van Dongen 2006). Likewise, if one assumes that the additive and residual variances have flat priors, this does not imply a flat prior for h^2, but rather a prior that is sharply peaked at 1/2. When assuming a flat prior, care must be taken that it is truly uninformative on the appropriate scale of biological interest. Otherwise, the choice of what superficially appears as an unbiased prior may instead create a bias that the signal in the data must overcome.

The Jeffreys Prior

Jeffreys (1961) proposed a general prior based on the Fisher information information, I, of the likelihood. Recall (LW Appendix 4) that

$$I(\theta\,|\,\mathbf{x}) = -E\left[\frac{\partial^2 \ln \ell(\theta\,|\,\mathbf{x})}{\partial\,\theta^2}\right]$$

The **Jeffreys prior** is as follows:

$$p(\theta) \propto \sqrt{I(\theta\,|\,\mathbf{x})} \tag{A2.16}$$

A full discussion, with derivation, can be found in Lee (2012).

When there are k parameters, $\mathbf{I}$ is the $k \times k$ Fisher information matrix of the expected second partials, where the elements of $\mathbf{I}$ are calculated by

$$\mathbf{I}(\,\boldsymbol{\Theta}\,|\,\mathbf{x}\,)_{ij} = -E_x\left[\frac{\partial^2 \ln \ell(\boldsymbol{\Theta}\,|\,\mathbf{x}\,)}{\partial\theta_i\,\partial\theta_j}\right]$$

In this case, the Jeffreys prior becomes

$$p(\boldsymbol{\Theta}) \propto \sqrt{\det[\mathbf{I}(\theta\,|\,\mathbf{x}\,)\,]} \tag{A2.17}$$

Example A2.4. Consider the likelihood of x successes in n independent draws from a binomial with a success parameter of θ,

$$\ell(\theta\,|\,\mathbf{x}) = C\theta^x(1-\theta)^{n-x}$$

where the constant C does not involve θ. Taking logs gives

$$L(\theta\,|\,\mathbf{x}) = \ln\left[\,\ell(\theta\,|\,\mathbf{x})\,\right] = \ln C + x\ln\theta + (n-x)\ln(1-\theta)$$

Thus

$$\frac{\partial L(\theta\,|\,\mathbf{x})}{\partial\theta} = \frac{x}{\theta} - \frac{n-x}{1-\theta}$$

and likewise

$$\frac{\partial^2 L(\theta\,|\,\mathbf{x})}{\partial\theta^2} = -\frac{x}{\theta^2} - (-1)\cdot(-1)\frac{n-x}{(1-\theta)^2} = -\left(\frac{x}{\theta^2} + \frac{n-x}{(1-\theta)^2}\right)$$

Because $E[\,x\,] = n\theta$, then

$$-E\left[\frac{\partial^2 \ln\ell(\theta\,|\,\mathbf{x}\,)}{\partial\,\theta^2}\right] = \frac{n\theta}{\theta^2} + \frac{n(1-\theta)}{(1-\theta)^2} = n\,\theta^{-1}(1-\theta)^{-1}$$

The resulting Jeffreys prior for this likelihood becomes

$$p(\theta) \propto \sqrt{\theta^{-1}(1-\theta)^{-1}} \propto \theta^{-1/2}(1-\theta)^{-1/2}$$

which is a U-shaped beta distribution with parameters $\alpha = \beta = 1/2$ (Equation A2.38a). This prior puts more weight on extreme values relative to assuming a uniform over (0,1), see Figure A2.3.

Example A2.5. Suppose our data consists of n independent draws from a normal distribution with an unknown mean and variance, μ and σ^2. In LW Appendix 4, we showed that the information matrix in this case is

$$\mathbf{I} = n\begin{pmatrix} \dfrac{1}{\sigma^2} & 0 \\ 0 & \dfrac{1}{2\sigma^4} \end{pmatrix}$$

Because the determinant of a diagonal matrix is the product of the diagonal elements, $\det(\mathbf{I}) \propto \sigma^{-6}$, giving the Jeffreys prior for μ and σ^2 as

$$p(\boldsymbol{\Theta}) \propto \sqrt{\sigma^{-6}} = \sigma^{-3}$$

Because the joint prior does not involve μ, this implies a flat prior for μ (i.e., $p[\mu] = c$). Note here that the prior distributions of μ and σ^2 are independent, as

$$p(\mu, \theta) = c \cdot \sigma^{-3} = p(\mu) \cdot p(\sigma^2)$$

POSTERIOR DISTRIBUTIONS UNDER NORMALITY ASSUMPTIONS

To introduce the basic ideas of Bayesian analysis, as well as treating a common assumption in quantitative genetics, consider the case where data are drawn from a normal (Gaussian) distribution, giving the likelihood function for the ith observation, x_i, as

$$\ell(\mu, \sigma^2 \,|\, x_i) = \frac{1}{\sqrt{2\pi\sigma^2}} \exp\left(-\frac{(x_i - \mu)^2}{2\sigma^2} \right) \tag{A2.18a}$$

If we assume independence, the resulting full likelihood for all n data points (with a sample mean of $\overline{x}$) is

$$\ell(\mu \,|\, \mathbf{x}) = \frac{1}{\sqrt{2\pi\sigma^2}} \exp\left(-\sum_{i=1}^{n} \frac{(x_i - \mu)^2}{2\sigma^2} \right) \tag{A2.18b}$$

$$= \frac{1}{\sqrt{2\pi\sigma^2}} \exp\left[-\frac{1}{2\sigma^2} \left(\sum_{i=1}^{n} x_i^2 - 2\mu n\overline{x} + n\mu^2 \right) \right] \tag{A2.18c}$$

The form of the posteriors given these normal likelihoods is a function of the assumed priors. By using the appropriate conjugate priors, these posteriors follow fairly standard distributions, and hence are easier to work with, as we now demonstrate.

Gaussian Likelihood With Known Variance and Unknown Mean

As a starting point, assume that the variance, σ^2, is known, while the mean, μ, is unknown. For a Bayesian analysis, it remains to specify the prior for μ, $p(\mu)$. Suppose we assume a Gaussian prior, $\mu \sim \mathrm{N}(\mu_0, \sigma_0^2)$, with

$$p(\mu) = \frac{1}{\sqrt{2\pi\sigma_0^2}} \exp\left(-\frac{(\mu - \mu_0)^2}{2\sigma_0^2} \right) \tag{A2.19}$$

The mean and variance of the prior, μ_0 and σ_0^2, are referred to as **hyperparameters**. Here, μ_0 specifies a prior location for the parameter (the unknown mean, μ), while σ_0^2 specifies our uncertainty in this prior location—the larger σ_0^2, the greater is our uncertainty. In the limit as $\sigma_0^2 \to \infty$, $p(\mu)$ approaches a flat (and in this case, improper) prior.

A useful device when calculating the posterior distribution is to ignore terms that are constants with respect to the unknown parameters. Suppose $\mathbf{x}$ denotes the data and $\boldsymbol{\Theta}_1$ is a vector of *known* model parameters, while $\boldsymbol{\Theta}_2$ is a vector of unknown parameters. If we can write the posterior as

$$p(\boldsymbol{\Theta}_2 \,|\, \mathbf{x}, \boldsymbol{\Theta}_1) = f(\mathbf{x}, \boldsymbol{\Theta}_1) \cdot g(\mathbf{x}, \boldsymbol{\Theta}_1, \boldsymbol{\Theta}_2) \tag{A2.20a}$$

then

$$p(\boldsymbol{\Theta}_2 \,|\, \mathbf{x}, \boldsymbol{\Theta}_1) \propto g(\mathbf{x}, \boldsymbol{\Theta}_1, \boldsymbol{\Theta}_2) \tag{A2.20b}$$

which follows since $f(\mathbf{x}, \boldsymbol{\Theta}_1)$ is constant with respect to $\boldsymbol{\Theta}_2$.

With the prior given by Equation A2.19, we can express the resulting posterior distribution as

$$p(\mu\,|\,\mathbf{x}) \propto \ell(\mu\,|\,\mathbf{x}) \cdot p(\mu)$$
$$\propto \exp\left[-\frac{(\mu-\mu_0)^2}{2\sigma_0^2} - \frac{1}{2\sigma^2}\left(\sum_{i=1}^{n} x_i^2 - 2\mu n\overline{x} + n\mu^2\right)\right] \tag{A2.21a}$$

We can factor out additional terms not involving μ to obtain

$$p(\mu\,|\,\mathbf{x}) \propto \exp\left(-\frac{\mu^2}{2\sigma_0^2} + \frac{\mu\,\mu_0}{\sigma_0^2} + \frac{\mu n\overline{x}}{\sigma^2} - \frac{n\mu^2}{2\sigma^2}\right) \tag{A2.21b}$$

Factoring in terms of μ, the term in the exponential becomes

$$-\frac{\mu^2}{2}\left(\frac{1}{\sigma_0^2} + \frac{n}{\sigma^2}\right) + \mu\left(\frac{\mu_0}{\sigma_0^2} + \frac{n\overline{x}}{\sigma^2}\right) = -\frac{\mu^2}{\sigma_*^2} + \frac{2\mu\mu_*}{2\sigma_*^2} \tag{A2.22a}$$

where

$$\sigma_*^2 = \left(\frac{1}{\sigma_0^2} + \frac{n}{\sigma^2}\right)^{-1} \quad \text{and} \quad \mu_* = \sigma_*^2\left(\frac{\mu_0}{\sigma_0^2} + \frac{n\overline{x}}{\sigma^2}\right) \tag{A2.22b}$$

Finally, by completing the square, we have

$$p(\mu\,|\,\mathbf{x}) \propto \exp\left[-\frac{(\mu-\mu_*)^2}{2\sigma_*^2} + f(\mathbf{x}, \mu_0, \sigma^2, \sigma_0^2)\right] \tag{A2.22c}$$

Recalling Equation A2.20b, we can ignore the second term in the exponential (as it does not involve μ), and the resulting posterior for μ (given the observed data **x**) becomes

$$p(\mu\,|\,\mathbf{x}) \propto \exp\left[-\frac{(\mu-\mu_*)^2}{2\sigma_*^2}\right] \tag{A2.23a}$$

demonstrating that the posterior density function for μ is a normal with a mean of μ_* and a variance of σ_*^2, namely,

$$\mu\,|\,(\mathbf{x}, \sigma^2) \sim \mathrm{N}\left(\mu_*, \sigma_*^2\right) \tag{A2.23b}$$

Notice that the posterior density is in the same form as the prior. This occurred because the prior **conjugated** with the likelihood function—the product of the prior and likelihood returned a distribution in the same family as the prior (but with different distribution parameters). The use of such **conjugate priors** associated with a given family of likelihood functions is a key concept in Bayesian analysis, and we will explore it more fully below.

We are now in a position to inquire about the relative importance of the prior versus the data. Under the assumed prior, the mean (and in this case, the mode as well) of the posterior distribution is

$$\mu_* = \mu_0\,\frac{\sigma_*^2}{\sigma_0^2} + \overline{x}\,\frac{\sigma_*^2}{\sigma^2/n} \tag{A2.24}$$

With a very diffuse prior on μ (i.e., $\sigma_0^2 \gg \sigma^2$), $\sigma_*^2 \to \sigma^2/n$ and $\mu_* \to \overline{x}$. Also note from Equation A2.22b that as we collect enough data (i.e., achieve a sufficiently large value of n), $\sigma_*^2 \to \sigma^2/n$ and again $\mu_* \to \overline{x}$, implying that primarily the data, rather than the prior, will influence the posterior when the value of n is sufficiently large.

Gamma, χ^2, Inverse-gamma, and χ^{-2} Distributions

Before examining the Gaussian likelihood with unknown variance, a brief aside is needed to develop the **inverse chi-square distribution**, denoted by χ^{-2}. We do this via the gamma and inverse-gamma distributions, as both χ^2 and χ^{-2} are special cases of these distributions.

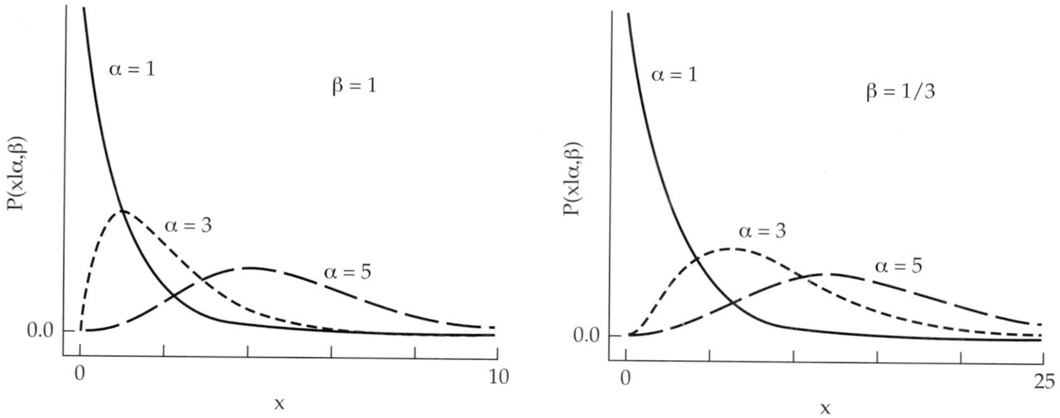

Figure A2.2 The effect of the **shape** (α) and **rate** ($\beta = 1/\lambda$, the inverse of the **scale**) **parameters** on the gamma distribution function. For $\alpha = 1$, the resulting distribution is the simple monotonically decreasing exponential, while for $\alpha > 1$, the distribution is unimodal. The effect of a change in the rate or scale is to keep the general shape but change the scaling with respect to x.

To motivate the **gamma distribution**, first consider the simple exponential waiting-time distribution, where β is the **rate** (the probability of a success in some small time unit, δ_t, is given by $\beta\,\delta_t$), then the **probability density function** (**pdf**) for the exponential is

$$p(x \mid \beta) = \beta e^{-\beta x} \quad \text{for} \quad 0 \leq x < \infty, \quad \beta > 0$$

Because the expected waiting time until a success is $\lambda = 1/\beta$, this can be reparameterized in terms of the **scale** (waiting time) **parameter** as

$$p(x \mid \beta) = \lambda^{-1} e^{-x/\lambda}$$

The sum of k exponentials with the same rate (or scale) parameter is called an **Erlang distribution**, and it was initially developed for certain problems in telephone queuing theory. Expressed in terms of the rate parameter, the resulting pdf becomes

$$p(x \mid k, \beta) = \frac{\beta^k}{(k-1)!}\, x^{k-1}\, e^{-\beta x} \quad \text{for} \quad 0 \leq x < \infty$$

where the integer k is called the **shape parameter**, with $k = 1$ recovering the exponential.

The gamma distribution follows by allowing the shape parameter to be any positive number, α, with $x \sim \text{Gamma}(\alpha, \beta)$ having its pdf defined by its shape (α) and rate (β) values,

$$p(x \mid \alpha, \beta) = \frac{\beta^\alpha}{\Gamma(\alpha)}\, x^{\alpha-1} e^{-\beta x} \quad \text{for } \alpha,\, \beta,\, x > 0 \tag{A2.25a}$$

Note that the factorial in the Erlang is replaced by the gamma function, $\Gamma(x)$, which is defined below (Equation A2.26a). Figure A2.2 shows how changes in these two parameters influence the shape of the distribution. Note that, as a function of x,

$$p(x \mid \alpha, \beta) \propto x^{\alpha-1} e^{-\beta x} \tag{A2.25b}$$

When expressed in terms of the scale ($\lambda = 1/\beta$) parameter, the pdf becomes

$$p(x \mid \alpha, \lambda) = \frac{\lambda^{-\alpha}}{\Gamma(\alpha)}\, x^{\alpha-1} e^{-x/\lambda}$$

Table A2.1 Summary of the functional forms (in terms of x) of various gamma-related distributions. See the text for further details.

Distribution	α	β	$p(x)$/constant
Gamma (α, β)			$x^{\alpha-1}e^{-\beta x}$
Chi-square, χ^2_n	$n/2$	$1/2$	$x^{n/2-1}e^{-x/2}$
Inverse-gamma (α, β)			$x^{-(\alpha+1)}e^{-\beta/x}$
Inverse chi-square, χ^{-2}_n	$n/2$	$1/2$	$x^{-(n/2+1)}e^{-1/(2x)}$
Scaled inverse chi-square, $\chi^{-2}_{(n,\sigma_0^2)}$	$n/2$	$\sigma_0^2/2$	$x^{-(n/2+1)}e^{-\sigma_0^2/(2x)}$

which yields

$$p(x \,|\, \alpha, \lambda) \propto x^{\alpha-1}e^{-x/\lambda} \tag{A2.25c}$$

Because both the rate and scale versions of the gamma distribution are widely used, take care to know which version your software package is using (for example, the default in R uses the scale parameter version). We can parameterize a gamma in terms of its mean and variance by noting that

$$\mu_x = \frac{\alpha}{\beta} = \alpha\,\lambda \quad \text{and} \quad \sigma_x^2 = \frac{\alpha}{\beta^2} = \alpha\,\lambda^2 \tag{A2.25d}$$

so that

$$\alpha = \frac{\mu_x^2}{\sigma_x^2} \quad \text{and} \quad \beta = \frac{\mu_x}{\sigma_x^2} \tag{A2.25e}$$

$\Gamma(\alpha)$, the **gamma function** evaluated at α (which normalizes the gamma distribution), is defined by

$$\Gamma(\alpha) = \int_0^\infty y^{\alpha-1}e^{-y}dy \qquad \text{for } \alpha > 0 \tag{A2.26a}$$

This is the generalization of the factorial function from the integers to any positive number. If n is an integer, then $\Gamma(n) = (n-1)!$ Using integration by parts, one can show that Γ satisfies the following identities

$$\Gamma(\alpha+1) = \alpha\Gamma(\alpha), \quad \Gamma(1) = 1, \quad \text{and} \quad \Gamma(1/2) = \sqrt{\pi} \tag{A2.26b}$$

The chi-square (χ^2) distribution is a special case of the gamma, as a χ^2 random variable with n degrees of freedom follows a gamma distribution with parameters $\alpha = n/2$ and $\beta = 1/2$ ($\lambda = 2$), namely, $\chi^2_n \sim \text{Gamma}(n/2, 1/2)$, giving the density function as

$$p(x \,|\, n) = \frac{2^{-n/2}}{\Gamma(n/2)}\, x^{n/2-1}e^{-x/2} \tag{A2.27a}$$

Hence for $x \sim \chi^2_n$,

$$p(x) \propto x^{n/2-1}e^{-x/2} \tag{A2.27b}$$

The **inverse-gamma** distribution will prove useful as a conjugate prior for Gaussian likelihoods with unknown variance. It is defined by the distribution of the random variable $y = x^{-1}$, where $x \sim \text{Gamma}(\alpha, \beta)$. The resulting density function is

$$p(x \,|\, \alpha, \beta) = \frac{\beta^\alpha}{\Gamma(\alpha)}\, x^{-(\alpha+1)}e^{-\beta/x} \quad \text{for } \alpha,\ \beta,\ x > 0 \tag{A2.28a}$$

The mean and variance for this distribution are only defined (i.e., finite) if α is sufficiently large, with

$$\mu_x = \frac{\beta}{\alpha-1} \quad \text{for } \alpha > 1 \quad \text{and} \quad \sigma_x^2 = \frac{\beta^2}{(\alpha-1)^2(\alpha-2)} \quad \text{for } \alpha > 2 \tag{A2.28b}$$

Note for the inverse gamma that

$$p(x \mid \alpha, \beta) \propto x^{-(\alpha+1)} e^{-\beta/x} \tag{A2.28c}$$

If $y \sim \chi_n^2$, then $x = 1/y$ follows an **inverse chi-square distribution**, which is denoted by $x \sim \chi_n^{-2}$. This is a special case of the inverse gamma, with (as for a normal χ^2) $\alpha = n/2$, $\beta = 1/2$. For $n > 4$ (i.e., $\alpha > 2$), the resulting density function is

$$p(x \mid n) = \frac{2^{-n/2}}{\Gamma(n/2)} x^{-(n/2+1)} e^{-1/(2x)} \tag{A2.29a}$$

with a mean and variance of

$$\mu_x = \frac{1}{n-2} \quad \text{and} \quad \sigma_x^2 = \frac{2}{(n-2)^2(n-4)} \tag{A2.29b}$$

The **scaled inverse chi-square distribution** is more typically used in a Bayesian analysis, where the rate parameter, β (which equals 1/2 under a chi-square), is replaced by $\beta = \sigma_0^2/2$, making the resulting pdf

$$p(x \mid n) \propto x^{-(n/2+1)} e^{-\sigma_0^2/(2x)} \tag{A2.30a}$$

where the $1/(2x)$ term in the exponential is replaced by a $\sigma_0^2/(2x)$ term. The scaled inverse chi-square distribution thus involves two parameters (σ_0^2 and n), and is denoted by $\chi^{-2}_{(n,\sigma_0^2)}$ or SI$-\chi^2(n, \sigma_0^2)$. Note that if

$$x \sim \chi^{-2}_{(n,\sigma_0^2)}, \quad \text{then} \quad \sigma_0^2 x \sim \chi_n^{-2} \tag{A2.30b}$$

which shows that σ_0^2 is a scaling factor on a standard ($\beta = 1/2$) inverse chi-square.

Gaussian Likelihood With Unknown Variance: Scaled Inverse-χ^2 Priors

Suppose data are drawn from a normal distribution with a known mean, μ, but unknown variance, σ^2. The resulting likelihood function can be expressed as

$$\ell(\sigma^2 \mid \mathbf{x}, \mu) \propto (\sigma^2)^{-n/2} \exp\left(-\frac{nS^2}{2\sigma^2}\right) \tag{A2.31a}$$

where

$$S^2 = \frac{1}{n} \sum_{i=1}^{n} (x_i - \mu)^2 \tag{A2.31b}$$

Notice that since we condition on **x** and μ (i.e., their values are known), S^2 is a constant. Further observe that, as a function of the unknown variance, σ^2, the likelihood is proportional to a scaled inverse χ^2 distribution (Equation A2.30a). If we take the prior for the unknown variance also as a scaled inverse χ^2 with hyperparameters ν_0 and σ_0^2, the posterior becomes

$$\begin{aligned} p(\sigma^2 \mid \mathbf{x}, \mu) &\propto (\sigma^2)^{-n/2} \exp\left(-\frac{nS^2}{2\sigma^2}\right) (\sigma^2)^{-\nu_0/2-1} \cdot \exp\left(-\frac{\sigma_0^2}{2\sigma^2}\right) \\ &= (\sigma^2)^{-(n+\nu_0)/2-1} \exp\left(-\frac{nS^2 + \sigma_0^2}{2\sigma^2}\right) \end{aligned} \tag{A2.32a}$$

Equation A2.30a shows the resulting posterior is also a scaled inverse χ^2 distribution with parameters $\nu_n = (n + \nu_0)$ and $\sigma_n^2 = (nS^2 + \sigma_0^2)$. Hence,

$$\text{the prior } \sigma^2 \sim \chi^{-2}_{\nu_0,\sigma_0^2} \quad \text{yields the posterior} \quad \sigma^2 \mid (\mathbf{x}, \mu) \sim \chi^{-2}_{\nu_n,\sigma_n^2} \tag{A2.32b}$$

Student's *t* Distribution

The final distribution needed for a Bayesian analysis of a Gaussian likelihood is the ***t*** (or **Student's *t***) distribution. Suppose that $x_i \sim N(\mu, \sigma^2)$, so for n independent draws, $\overline{x} \sim N(\mu, \sigma^2/n)$. This implies that $(\overline{x} - \mu)/\sqrt{\sigma^2/n} \sim U$, where $U \sim N(0,1)$ denotes a unit normal. Likewise, the sample variance, $\text{Var}(x)$, follows a scaled chi-square distribution, with $\text{Var}(x) \sim (n-1)\sigma^2\chi^2_{n-1}$ (LW Equation A5.14c). When the estimated variance, $\text{Var}(x)$, is used in place of the true variance, σ^2, the quantity $(\overline{x}-\mu)/\sqrt{\text{Var}(x)/n}$ follows a t distribution with $n-1$ degrees of freedom, giving rise to the very familiar ***t*-test**. Notice that

$$t_{n-1} = \left(\frac{\overline{x} - \mu}{\sigma/\sqrt{n}}\right)\left(\frac{1}{\sqrt{\text{Var(x)}/\sigma^2}}\right) = \frac{U}{\sqrt{\chi^2_{n-1}/(n-1)}}$$

Thus, a t_ν random variable follows the distribution of a unit normal divided by the square root of a chi-square with ν degrees of freedom,

$$t_\nu = \frac{U}{\sqrt{\chi^2_\nu/\nu}} \tag{A2.33a}$$

Note that $E(\chi^2_\nu) = \nu$, so $E(\chi^2_\nu/\nu) = 1$. Relative to a normal, a t distribution is more peaked and has heavier tails, and this kurtosis becomes more pronounced as ν decreases. Indeed, the tails fall off sufficiently slowly that a t random variable with two degrees of freedom has an infinite variance, while a t with four (or fewer) degrees of freedom has an infinite fourth moment. The coefficient of kurtosis (LW Equation 2.12a) for a t with $\nu > 4$ degrees of freedom is $k_4 = 6/(\nu-4)$, which approaches the value (zero) for a normal random variable for large values of ν. For $\nu > 30$, the t essentially becomes a unit normal distribution.

As with a unit normal, one can also add scale and location to a standard t_ν-distributed random variable, thus generating a three-parameter family of distributions,

$$t_\nu(\mu, \sigma) = \mu + \sigma \cdot t_\nu \tag{A2.33b}$$

The resulting mean and variance this distribution are

$$E[t_\nu(\mu,\sigma)] = \mu \quad \text{and} \quad \sigma^2[t_\nu(\mu,\sigma)] = \sigma^2\,\frac{\nu}{\nu-2} \quad \text{for } \nu > 2 \tag{A2.33c}$$

Hence, the choice of μ and σ control, respectively, the location and scale (uncertainty about the location), while ν controls the kurtosis, with heavy tails for values of ν that are small and little kurtosis for $\nu > 20$. The resulting probability density function thus becomes

$$p(x \,|\, \nu, \mu, \sigma) = \frac{\Gamma([\nu+1]/2)}{\Gamma(\nu/2)\sigma\,\sqrt{\pi\,\nu}}\left[1 + \frac{1}{\nu}\left(\frac{x-\mu}{\sigma}\right)^2\right]^{-(\nu+1)/2} \tag{A2.33d}$$

The role of the t distribution in Bayesian statistics is twofold. First, it is often used as a *more robust prior*, as its heavier tails may better account for outliers. Using a t distribution with low degrees of freedom (often $\nu = 5$) offers a prior that is similar to a normal but allows for more frequent extreme values. The second scenario is that the marginal posterior for μ of a Gaussian likelihood with a normal prior on the mean and an inverse chi-square prior on the variance is a t distribution. This arises after the joint posterior is integrated over all possible σ^2 values (i.e., over an inverse chi-square).

General Gaussian Likelihood: Unknown Mean and Variance

If we put all these pieces together, the posterior density for draws from a normal with unknown mean and variance is obtained as follows. First, we write the joint prior by conditioning on the variance,

$$p(\mu, \sigma^2) = p(\mu \,|\, \sigma^2) \cdot p(\sigma^2) \tag{A2.34a}$$

Table A2.2 Conjugate priors for common likelihood functions. If one uses the distribution family of the conjugate prior with its paired likelihood function, then the resulting posterior is in the same distribution family as the prior (albeit, of course, with different parameters).

Likelihood	Conjugate prior	Equation
Binomial	Beta	A2.38a
Multinomial	Dirichlet	A2.37b
Poisson	Gamma	A2.27a
Normal		
μ unknown, σ^2 known	Normal	A2.18a
μ known, σ^2 unknown	Inverse chi-square	A2.30a
Multivariate normal		
$\boldsymbol{\mu}$ unknown, **V** known	Multivariate normal	LW 8.24
$\boldsymbol{\mu}$ known, **V** unknown	Inverse-Wishart	A2.41

As above, we assume a scaled inverse chi-square distribution for the variance and, conditioned on the variance, a Gaussian prior for the mean with hyperparameters of μ_0 and σ^2/κ_0, namely,

$$\sigma^2 \sim \chi^{-2}_{\nu_0,\sigma_0^2} \quad \text{and} \quad \mu\,|\,\sigma^2 \sim \mathrm{N}\left(\mu_0, \frac{\sigma^2}{\kappa_0}\right) \tag{A2.34b}$$

We write the variance for the conditional mean prior in this way because σ^2 is known (as we condition on it) and we scale σ^2 by the hyperparameter, κ_0. The resulting marginal posterior becomes

$$\sigma^2\,|\,\mathbf{x} \sim \chi^{-2}_{\nu_n,\sigma_n^2} \quad \text{and} \quad \mu\,|\,\mathbf{x} \sim \mathrm{t}_{\nu_n}\left(\mu_n, \frac{\sigma_n^2}{\kappa_n}\right) \tag{A2.35}$$

where $\mathrm{t}_n(\mu, \sigma^2)$ denotes a t distribution with n degrees of freedom, mean μ, and scale parameter σ^2, and where

$$\nu_n = \nu_0 + n, \qquad \kappa_n = \kappa_0 + n \tag{A2.36a}$$

$$\mu_n = \mu_0\,\frac{\kappa_0}{\kappa_n} + \overline{x}\,\frac{n}{\kappa_n} = \mu_0\,\frac{\kappa_0}{\kappa_0 + n} + \overline{x}\,\frac{n}{\kappa_0 + n} \tag{A2.36b}$$

$$\sigma_n^2 = \frac{1}{\nu_n}\left(\nu_0\sigma_0^2 + \sum_{i=1}^{n}(x_i - \overline{x})^2 + \frac{\kappa_0 n}{\kappa_n}(\overline{x} - \mu_0)^2\right) \tag{A2.36c}$$

CONJUGATE PRIORS

The use of a prior density that conjugates the likelihood allows us to develop analytic expressions of the posterior density. As we will see in Appendix 3, this is critical in developing Gibbs samplers for problems of interest. Table A2.2 summarizes the conjugate priors for several common likelihood functions, with the various families of distributions discussed below.

The Beta and Dirichlet Distributions

With a binomial, each trial (observation) has two possible outcomes and the likelihood is a function of the sample size (number of trials), n, and a single success probability, p (as the two outcomes on any given trial have probabilities of p and $1-p$). The generalization of this model is the **multinomial distribution**, where now each trial has k possible outcomes, and which requires $k-1$ success probabilities to describe the likelihood. In particular, for a total of n observations, the probability that n_1 are in category 1, n_2 in category 2, $\cdots$, and n_k in category k is

$$p(n_1, \cdots n_k) = \frac{n!}{n_1!\,n_2!\cdots n_k!}\,p_1^{n_1}\cdots p_k^{n_k} \quad \text{where} \quad \sum_i n_i = n \quad \text{and} \quad \sum_i p_i = 1 \tag{A2.37a}$$

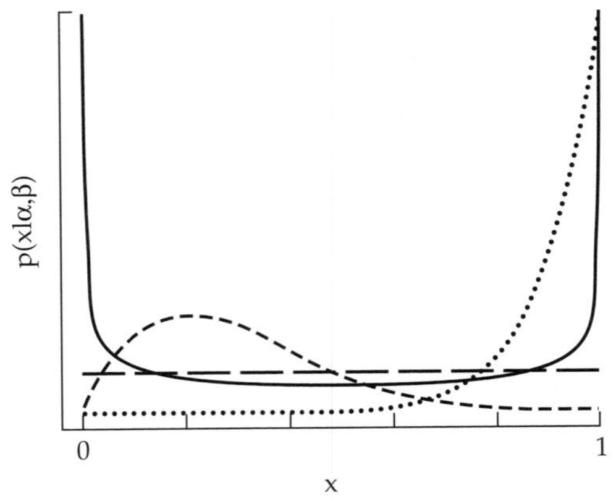

Figure A2.3 For $\alpha = \beta = 1$ (long-dashed curve), the beta distribution is simply the uniform distribution over $(0, 1)$. The pdf for the beta distribution can also be U-shaped ($\alpha = \beta = 0.5$; solid curve), unimodal ($\alpha = 2, \beta = 5$; short-dashed curve), or L-shaped ($\alpha = 10, \beta = 1$; dotted curve). Because the beta distribution is symmetric in α and β, switching their parameter values generates a distribution of the same shape translated about 0.5.

The conjugate prior for the multinomial likelihood is the **Dirichlet distribution**. If we let $\mathbf{x} = (x_1, x_2, \cdots, x_k)$ denote the k success probabilities, when pdf for $\mathbf{x} \sim$ Dirichlet$(\alpha_1, \cdots, \alpha_k)$ is

$$p(x_1, \cdots x_k \,|\, \alpha_1, \cdots, \alpha_k) = \frac{\Gamma(\alpha_0)}{\Gamma(\alpha_1) \cdots \Gamma(\alpha_k)}\, x_1^{\alpha_1 - 1} \cdots x_k^{\alpha_k - 1} \tag{A2.37b}$$

where

$$\alpha_0 = \sum_{i=1}^{k} \alpha_i \quad \text{with} \quad \alpha_i > 0, \quad \text{and} \quad \sum_{i=1}^{k} x_i = 1 \quad \text{with} \quad 0 \leq x_i \leq 1 \tag{A2.37c}$$

At first glance, this looks like the multinomial density function (with $\alpha_i - 1 = n_i$). The difference is that the multinomial is calculated over a set of discrete random variables (n_i), thus returning the expected probabilities for any vector of discrete numbers of counts (successes) in each category. Conversely, the Dirichlet treats an equivalent of the vector of outcomes (generalized to non-integers) as fixed and returns the continuous distribution for all possible configurations of the success parameters given this data, which means that the data (α_i) is fixed, and the success parameters (x_i) are random. A few key moments of this distribution are

$$\mu_{x_i} = \frac{\alpha_i}{\alpha_0}, \quad \sigma^2(x_i) = \frac{\alpha_i(\alpha_0 - \alpha_i)}{\alpha_0^2(\alpha_0 + 1)}, \quad \text{and} \quad \sigma(x_i, x_j) = -\frac{\alpha_i\, \alpha_j}{\alpha_0^2(\alpha_0 + 1)} \tag{A2.37d}$$

An important special case of the Dirichlet (for $k = 2$ classes) is the **beta distribution**, whose pdf is given by

$$p(x) = \frac{\Gamma(\alpha + \beta)}{\Gamma(\alpha)\Gamma(\beta)}\, x^{\alpha - 1}(1 - x)^{\beta - 1} \quad \text{for} \quad 0 \leq x \leq 1, \quad \alpha,\, \beta > 0 \tag{A2.38a}$$

which has a mean and a variance of

$$\mu = \frac{\alpha}{\alpha + \beta} \qquad \text{and} \qquad \sigma^2 = \frac{\alpha\beta}{(\alpha + \beta)^2(\alpha + \beta - 1)} \tag{A2.38b}$$

As Figure A2.3 illustrates, the beta distribution is *extremely flexible*, and can be flat, unimodal, U- , or L-shaped, depending on the choice of α and β.

Wishart and Inverse-Wishart Distributions

The **Wishart distribution** can be thought of as the multivariate extension of the χ^2 distribution. Suppose $\mathbf{x}_1, \cdots, \mathbf{x}_n$ are independent and identically distributed vectors, with $\mathbf{x}_i \sim \text{MVN}_k(\mathbf{0}, \mathbf{V})$. Using these n draws, and assuming that the mean is known to be zero, the resulting random ($k \times k$ symmetric, positive definite) sample covariance matrix, **W**, is given by

$$\mathbf{W} = \sum_{i=1}^{n} \mathbf{x}_i \, \mathbf{x}_i^T \sim \text{W}_n(\mathbf{V}) \tag{A2.39}$$

This sum is defined as a Wishart distribution with n degrees of freedom and a (matrix) parameter **V**. Recalling that the sum of n squared unit normals follows a χ^2_n distribution, the Wishart is the extension to the multivariate normal. Indeed, for $k = 1$ with $\mathbf{V} = (1)$, the Wishart is simply a χ^2_n distribution, as $\sum x_i^2 \sim \chi^2_n$, because $x_i \sim N(0, 1)$.

The Wishart is the sampling distribution for covariance matrices (just like the χ^2 is associated with the distribution of a sample variance for data drawn from a normal; Chapter 11). The pdf of the Wishart distribution is

$$p(\mathbf{W} \mid \mathbf{V}) = 2^{-nk/2} \pi^{-k(k-1)/k} \mid \mathbf{V} \mid^{-n/2} \mid \mathbf{W} \mid^{(n+k+1)/2} \frac{\exp\left(-\frac{1}{2}\text{tr}\left[\mathbf{V}^{-1}\mathbf{W}\right]\right)}{\prod_{i=1}^{k} \Gamma\left(\frac{n+1-i}{2}\right)} \tag{A2.40}$$

Recall that the trace (tr) of a matrix is just the sum of its diagonal elements, $\text{tr}(\mathbf{A}) = \sum A_{ii}$ (Appendix 5). Odell and Feiveson (1966) presenteds an algorithm for generating random draws from the Wishart.

If $\mathbf{Z} \sim \mathbf{W}_n(\mathbf{V})$, then $\mathbf{Z}^{-1} \sim \mathbf{W}_n^{-1}\left(\mathbf{V}^{-1}\right)$, where $\mathbf{W}^{-1}$ denotes the **inverse-Wishart distribution**. The density function for an inverse-Wishart distributed random matrix, **W**, is

$$p(\mathbf{W} \mid \mathbf{V}) = 2^{-nk/2} \pi^{-k(k-1)/k} \mid \mathbf{V} \mid^{n/2} \mid \mathbf{W} \mid^{-(n+k+1)/2} \frac{\exp\left(-\frac{1}{2}\text{tr}\left[\mathbf{V}\mathbf{W}^{-1}\right]\right)}{\prod_{i=1}^{k} \Gamma\left(\frac{n+1-i}{2}\right)} \tag{A2.41}$$

which is the distribution of the inverse of the sample covariance matrix.

Appendix 3

Markov Chain Monte Carlo and Gibbs Sampling

Far better an approximate answer to the right question, which is often vague, than an exact answer to the wrong question, which can always be made precise. Tukey (1962)

We are convinced nevertheless that Monte Carlo methods will one day reach an impressive maturity. Hammersley and Handscomb (1964)

The practice of MCMC is simple. Set up a Markov chain having the required invariant distribution, and run it on a computer. The folklore of simulation makes this seem more complicated that it really is. None of this folklore is justified by theory and none of it actually helps users do good simulations. Geyer (2011)

A historical impediment to the more widespread use of Bayesian methods was computational: obtaining the posterior distribution requires the integration of high-dimensional functions. This can be very difficult, but several analytic approximations short of direct integration have been proposed (reviewed by Smith 1991; Evans and Swartz 1995; Tanner 1996). Now, however, the widespread development of **Markov Chain Monte Carlo** (**MCMC**) methods has made obtaining very complex posteriors relatively easy from a conceptual standpoint (although they may still be rather computationally demanding). Indeed, MCMC allows Bayesian approaches to handle much higher-dimensional problems than other methods. MCMC approaches are so named because one uses the current sample value to randomly generate the next sample value, thus generating a **Markov chain**.

The realization in the early 1990s (Gelfand and Smith 1990) that one particular MCMC method, the **Gibbs sampler**, is widely applicable to a broad class of Bayesian problems sparked much of the current expansion in the use of Bayesian analysis. MCMC methods have their roots in the **Metropolis algorithm** (Metropolis and Ulam 1949; Metropolis et al. 1953), an attempt by physicists to compute complex integrals by expressing them as expectations of a much simpler distribution and then estimating this expectation by drawing samples from that distribution. While the Gibbs sampler had its origins outside of statistics, it is still somewhat surprising that the powerful machinery of MCMC had essentially no impact on the field of statistics until rather recently. MCMC methods were reviewed by Tanner (1996), Gammerman (1997), Chen et al. (2000), Robert and Casella (2004), and Brooks et al. (2011). The types of MCMC machinery used in quantitative genetics (often Gibbs samplers on linear mixed-models, e.g., Chapters 19 and 20) tend to be a bit different from those used in molecular population genetics (ABC methods for treating complex likelihoods, e.g., Chapters 9 and 10). Sorensen and Gianola (2002) extensively reviewed the former, while Marjoram and Tavaré (2006) provided a brief review of the latter. The roundtable discussion on practical issues of MCMC by Kass et al. (1998) is required reading for anyone using MCMC analysis, as is Geyer (2011). Finally, Robert and Casella (2011) presented a nice historical overview of the development of MCMC methods.

MONTE CARLO INTEGRATION

To better understand the historical roots of MCMC, we digress for a moment to discuss the original **Monte Carlo** approach, which used random-number generation to approximate

Evolution and Selection of Quantitative Traits. Bruce Walsh & Michael Lynch.
Published 2018 by Oxford University Press.
DOI 10.1093/oso/9780198830870.001.0001

integrals. Suppose we wish to compute a complex integral,

$$\int_a^b h(x)\,dx \tag{A3.1a}$$

If we can decompose $h(x)$ into the product of a function, $f(x)$, and a probability density function, $p(x)$, defined over the interval (a, b), then

$$\int_a^b h(x)\,dx = \int_a^b f(x)\,p(x)\,dx = E_{p(x)}[\,f(x)\,] \tag{A3.1b}$$

which means that the integral can be expressed as the expectation of $f(x)$ over the density $p(x)$. After sampling a large number of random variables, $(x_1, \cdots, x_n)$, from the density $p(x)$, we have that

$$\int_a^b h(x)\,dx = E_{p(x)}[\,f(x)\,] \simeq \frac{1}{n}\sum_{i=1}^{n} f(x_i) \tag{A3.1c}$$

This approach is referred to as **Monte Carlo integration**, and it can be used to approximate posterior (or marginal posterior) distributions required for a Bayesian analysis.

Consider the integral $I(y) = \int f(y\,|\,x)\,p(x)\,dx$, which we approximate by

$$\widehat{I}(y) = \frac{1}{n}\sum_{i=1}^{n} f(y\,|\,x_i) \tag{A3.1d}$$

where again x_i are draws from the density $p(x)$. Since we have framed this integral as an expectation, individual random draws have expected value equal to the integral. The spread of those values around their mean allows us to estimate a **Monte Carlo standard error** for our approximation, with

$$\text{SE}^2[\,\widehat{I}(y)\,] = \frac{1}{n}\left[\frac{1}{n-1}\sum_{i=1}^{n}\Big(f(y\,|\,x_i) - \widehat{I}(y)\Big)^2\right] \tag{A3.1e}$$

where the inner term estimates the sampling variance, σ^2. Notice from Equation A3.1e that the standard error scales as $\sigma/\sqrt{n}$, namely, with the square root of the sample size. Thus, a tenfold improvement in precision requires a hundredfold increase in the sample size.

Example A3.1. An interesting quantitative-genetic application of Monte Carlo integration was suggested by Ovaskainen et al. (2008) for comparing whether two **G** matrices are similar. As we detail in Volume 3, there are many proposed methods for comparing matrices, but Ovaskainen et al. suggested a (conceptually) simple approach. The **G** matrix for a given population is really a description of the distribution of breeding values, and, as such we can think of comparing **G** matrices as being akin to comparing two multivariate population distributions, whose densities are denoted by f and g. If $\mathbf{x}$ denotes a draw from one of these distributions, the probability that it originates from distribution g is simply $g(\mathbf{x})/[g(\mathbf{x}) + f(\mathbf{x})]$. Hence, the probability, q, that a random draw from f wil be incorrectly assigned to g is calculated as

$$q(f, g) = \int \frac{g(\mathbf{x})}{g(\mathbf{x}) + f(\mathbf{x})}\, f(\mathbf{x})\, d\mathbf{x} \tag{A3.2a}$$

If the two probability distributions are **essentially indistinquishable**, then $q(f, g) = 0.5$, while if they are **completely distinquishable** then $q(f, g) = 0$. Hence, $1 - 2q(f, g)$, which ranges from zero (indistinquishable) to one (fully distinquishable), provides a simple metric of the difference between them, and hence the difference between the two **G** matrices that comprise the distributions of f versus g. Ovaskainen et al. modified this idea further, suggesting the metric

$$d(f,g) = \sqrt{1 - 2q(f,g)} \tag{A3.2b}$$

While Equation A3.2a involves a complex integral, Equation A3.1d suggests that

$$\widehat{q}\,(f,g) = \frac{1}{n}\sum_{i=1}^{n} \frac{g(\mathbf{x}_i)}{g(\mathbf{x}_i) + f(\mathbf{x}_i)} \to q(f,g) \tag{A3.2c}$$

where $\mathbf{x}_1, \cdots, \mathbf{x}_n$ are random draws from f. For example, if f is a multivariate normal with a mean vector of **0** and a variance-covariance matrix of $\mathbf{G}_1$, then

$$f(\mathbf{x}) = (2\pi)^{-n/2} |\,\mathbf{G}_1\,|^{-1/2} \exp\left(- \frac{\mathbf{x}^T \mathbf{G}_1^{-1} \mathbf{x}}{2} \right)$$

where g is similarly defined, but with $\mathbf{G}_2$ replacing $\mathbf{G}_1$, which yields

$$\widehat{q}\,(f,g) = \frac{1}{n}\sum_{i=1}^{n} \frac{|\,\mathbf{G}_2\,|^{-1/2} \exp\left(-\mathbf{x}_i^T \mathbf{G}_2^{-1} \mathbf{x}_i/2\right)}{|\mathbf{G}_2|^{-1/2} \exp(-\mathbf{x}_i^T \mathbf{G}_2^{-1} \mathbf{x}_i/2) + |\mathbf{G}_1|^{-1/2} \exp(-\mathbf{x}_i^T \mathbf{G}_1^{-1} \mathbf{x}_i/2)} \tag{A3.2d}$$

The expression in the sum is not nearly as imposing as it appears. Both of the determinants, $|\,\mathbf{G}_1\,|$ and $|\,\mathbf{G}_2\,|$, are fixed constants, and they only need to be computed once. Likewise, $\mathbf{G}_1^{-1}$ and $\mathbf{G}_2^{-1}$ need only be inverted once and then can be stored. Monte Carlo integration proceeds by generating a set of random vectors, $(\mathbf{x}_1, \cdots, \mathbf{x}_n)$, from a MVN $\sim (\mathbf{0}, \mathbf{G}_1)$.

Importance Sampling

Consider any distribution whose probability density function, $p(x)$, has the same support as some target density, $q(x)$ (i.e., is nonzero over the same range of x values). Then,

$$\int f(x)\, q(x) dx = \int f(x) \left(\frac{q(x)}{p(x)} \right) p(x) dx = E_{p(x)} \left[f(x) \left(\frac{q(x)}{p(x)} \right) \right] \tag{A3.3a}$$

This forms the basis for the method of **importance sampling**, with

$$\int f(x)\, q(x) dx \simeq \frac{1}{n} \sum_{i=1}^{n} f(x_i) \left(\frac{q(x_i)}{p(x_i)} \right) \tag{A3.3b}$$

where again the x_i values are drawn from the distribution given by $p(x)$. For example, if we are interested in a marginal density as a function of y, $J(y) = \int f(y\,|\,x)\, q(x) dx$, we approximate this by

$$J(y) \simeq \frac{1}{n} \sum_{i=1}^{n} f(y\,|\,x_i) \left(\frac{q(x_i)}{p(x_i)} \right) \tag{A3.4}$$

where x_i are drawn from the approximating density p.

This method is so named in that a judicious choice of $p(x)$ ensures that values for which $f(x)$ is large (important) are appropriately sampled, which can result in a smaller Monte Carlo variance, and hence a more accurate estimate of the integral. Further, it can be faster to carry out in cases where p is more straightforward to sample from than q. Ideally, we would like $p(x)$ to be large when $f(x)q(x)$ is large, namely, $p(x) \propto f(x)q(x)$, although assessing when $q(x)$ is large may be problematic.

One issue that can arise is when $q(x)$ denotes a posterior distribution whose integration constant is unknown, and hence $q(x)$, as presented, will not be a formal probability distribution if $\int q(x) \neq 1$. If $C^{-1} = \int q(x)$, then $Cq(x)$ is formally a pdf. Fortunately, a simple trick allows us to use importance sampling to obtain C. Since $C\int q(x)dx = 1$, then

$$C^{-1} = \int q(x)dx \simeq \frac{1}{n}\sum_{i=1}^{n} w_i \quad \text{where} \quad w_i = \frac{q(x_i)}{p(x_i)} \tag{A3.5a}$$

with the x_i drawn from $p(x)$. Hence

$$C \cdot \int f(x)\, q(x)dx \simeq \widehat{I} = \sum_{i=1}^{n} w_i f(x_i) \Big/ \sum_{i=1}^{n} w_i \tag{A3.5b}$$

which has an associated Monte Carlo variance of

$$\text{Var}\left(\widehat{I}\right) = \sum_{i=1}^{n} w_i \left[f(x_i) - \widehat{I}\right]^2 \Big/ \sum_{i=1}^{n} w_i \tag{A3.5c}$$

INTRODUCTION TO MARKOV CHAINS

Before introducing two of the most common MCMC methods, the Metropolis-Hastings algorithm and the Gibbs sampler, a few introductory comments on Markov chains are in order. Let X_t denote the value of a random variable at time t, and let the **state space** refer to the range of possible X values. A random variable is said to follow a **Markov process** if the transition probabilities between different values in the state space depend only on the random variable's current state,

$$\Pr(X_{t+1} = s_j \mid X_0 = s_k, \cdots, X_t = s_i) = \Pr(X_{t+1} = s_j \mid X_t = s_i) \tag{A3.6}$$

Thus, the only information about the past needed for a Markov random variable to predict the future is the *current state* of the random variable. Knowledge of the values of earlier states does not influence the transition probability. A **Markov chain** refers to a sequence of random variables, $(X_0, \cdots, X_n)$, generated by a Markov process. A particular chain is defined by its **transition probabilities** (or the **transition kernel**), $P(i,j) = P(i \to j)$, which is the probability that a process at state space s_i moves to state s_j in a single step

$$P(i,j) = P(i \to j) = \Pr(X_{t+1} = s_j \mid X_t = s_i) \tag{A3.7a}$$

We use the notation $P(i \to j)$ to imply a move from i to j, as some texts define $P(i,j) = P(j \to i)$, so we will use the arrow notation to avoid any potential confusion. Let

$$\pi_j(t) = \Pr(X_t = s_j) \tag{A3.7b}$$

denote the probability that the chain is in state j at time t, and let $\boldsymbol{\pi}(t)$ denote the row vector of the state space probabilities at step (or time) t. We start the chain by specifying a starting row vector, $\boldsymbol{\pi}(0)$. Often all the elements of $\boldsymbol{\pi}(0)$ are zero except for a single element of 1, which corresponds to the process starting in that particular state. As the chain progresses, the probability values become more dispersed over the state space. The probability that the chain has a state value of s_j at time (or step) $t+1$ is obtained from the **Chapman-Kolomogrov (CK) equation**, which sums over the probability of being in a particular state at the current step value, t, and the transition probability from that state into state s_j

$$\begin{aligned}\pi_j(t+1) &= \Pr(X_{t+1} = s_j) \\ &= \sum_i \Pr(X_{t+1} = s_j \mid X_t = s_i) \cdot \Pr(X_t = s_i) \\ &= \sum_i P(i \to j)\, \pi_i(t) = \sum_i P(i,j)\, \pi_i(t)\end{aligned} \tag{A3.7c}$$

Successive iteration of the CK equations describes the evolution of the chain.

We can more compactly write the CK equations in matrix form, as follows: first we define the **probability transition matrix, P**, as the matrix whose ijth element is $P(i, j)$, the probability of moving from state i to state j, $P(i \rightarrow j)$. This implies that the rows of $\mathbf{P}$ sum to one, as when considering the ith row, $\sum_j P(i,j) = \sum_j P(i \rightarrow j) = 1$. In matrix form, the Chapman-Kolomogrov equation becomes

$$\boldsymbol{\pi}(t+1) = \boldsymbol{\pi}(t)\mathbf{P} \tag{A3.8a}$$

Hence,

$$\boldsymbol{\pi}(t) = \boldsymbol{\pi}(t-1)\mathbf{P} = \left[\boldsymbol{\pi}(t-2)\mathbf{P}\right]\mathbf{P} = \boldsymbol{\pi}(t-2)\mathbf{P}^2 \tag{A3.8b}$$

Continuing in this fashion yields the probability distribution in generation t as

$$\boldsymbol{\pi}(t) = \boldsymbol{\pi}(0)\mathbf{P}^t \tag{A3.8c}$$

Next we define the n-step transition probability, $p_{ij}^{(n)}$, as the probability that the process is in state j, given that it started in state i some n steps ago, namely,

$$p_{ij}^{(n)} = \Pr(X_{t+n} = s_j \mid X_t = s_i) \tag{A3.8d}$$

It immediately follows that $p_{ij}^{(n)}$ is simply the ijth element of $\mathbf{P}^n$, the nth power of the single-step transition matrix. The classic example of a Markov chain in genetics is the Wright-Fisher model of drift (Equations 2.1 and 2.2; Example 2.1).

Finally, a Markov chain is said to be **irreducibile** if there exists a positive integer, n_{ij}, such that $p_{ij}^{(n_{ij})} > 0$ for all ij. That is, all states **communicate** with each other, in that the process can always move from any one state to any other state (although this may require multiple steps). Likewise, a chain is said to be **aperiodic** when the number of steps required to move between two states (say x and y) is not always some multiple of a specific integer. Put another way, the chain is not forced into cycles of fixed length between certain states.

Example A3.2. Suppose the state space consists of three possible weather conditions (Rain, Sunny, Cloudy) and that weather patterns follow a Markov process (of course, they do not!). Under this assumption, the probability of tomorrow's weather simply depends on today's weather, and not on any other previous days. If this is the case, the observation that it has rained for three straight days does not alter the probability of tomorrow's weather compared to the situation where (say) it rained today but was sunny for the last week. Suppose the probability transitions given that today is rainy are

P(Rain tomorrow | Rain today) = 0.5
P(Sunny tomorrow | Rain today) = 0.25
P(Cloudy tomorrow | Rain today) = 0.25

This results in the first row of the transition probability matrix being $(0.5, 0.25, 0.25)$. Now suppose that the full transition matrix is

$$\mathbf{P} = \begin{pmatrix} 0.5 & 0.25 & 0.25 \\ 0.5 & 0 & 0.5 \\ 0.25 & 0.25 & 0.5 \end{pmatrix}$$

Note that this Markov chain is irreducible, as all states communicate with each other. Suppose today is sunny. What is the weather expected to be like two days from now? Seven days? Here $\boldsymbol{\pi}(0) = (\,0 \quad 1 \quad 0\,)$, which yeilds

$$\boldsymbol{\pi}(2) = \boldsymbol{\pi}(0)\mathbf{P}^2 = (\,0.375 \quad 0.25 \quad 0.375\,)$$

and

$$\boldsymbol{\pi}(7) = \boldsymbol{\pi}(0)\mathbf{P}^7 = (\,0.4 \quad 0.2 \quad 0.4\,)$$

Conversely, suppose today is rainy, so $\boldsymbol{\pi}(0) = (\,1 \quad 0 \quad 0\,)$. The expected future weather becomes

$$\boldsymbol{\pi}(2) = (\,0.4375 \quad 0.1875 \quad 0.375\,) \quad \text{and} \quad \boldsymbol{\pi}(7) = (\,0.4 \quad 0.2 \quad 0.4\,)$$

Note that after a sufficient amount of time, the expected weather is *independent of the starting value.* In other words, the chain has reached a stationary distribution, where the state space probability values are independent of the actual starting value.

As the above example illustrates, a Markov chain may reach a **stationary distribution**, $\boldsymbol{\pi}^*$, where the vector of probabilities of being in a particular state is independent of the initial starting distribution. The stationary distribution satisfies

$$\boldsymbol{\pi}^* = \boldsymbol{\pi}^*\mathbf{P} \tag{A3.9}$$

In other words, $\boldsymbol{\pi}^*$ is the **left eigenvector** associated with the eigenvalue $\lambda = 1$ of $\mathbf{P}$ (Appendix 5). Further, the spectral decomposition of $\mathbf{P}$ (Equation A5.9a) implies that the impact of the initial conditions after t steps decays as λ_2^t, where $\lambda_2 < 1$ is the second largest eigenvalue of $\mathbf{P}$. If this eigenvalue is very close to one, the impact of the initial conditions can persist for a substantial amount of time. Specifically, if $\lambda_2 = 1 - \delta$, then $\lambda_2^t = (1-\delta)^t \simeq \exp(-\delta t)$. One condition for the existence of a stationary distribution is that the chain must be irreducible and aperiodic. When a chain is periodic, it can cycle in a deterministic fashion between states and may never settle down to a stationary distribution (in effect, this cycling *is* the stationary distribution for this chain). Equation A5.11d can be used to show that if $\mathbf{P}$ has no eigenvalues equal to -1, it is aperiodic.

A sufficient condition for a unique stationary distribution is (for all i and j) that the **detailed balance equation** holds, namely,

$$P(j \rightarrow i)\,\pi_j^* = P(i \rightarrow j)\,\pi_i^* \tag{A3.10}$$

That is, at equilibrium, the amount of probability flux from state j to stage i is exactly matched by the probability flux in the opposite direction (for i to j), so a balance is reached, with no *net* flow of probability over the states. If Equation A3.10 holds for all values of i and j, the Markov chain is said to be **reversible**, and hence Equation A3.10 is also called the **reversibility condition**. Note that this condition implies that $\boldsymbol{\pi}^* = \boldsymbol{\pi}^*\mathbf{P}$, as the jth element of the row vector $\boldsymbol{\pi}^*\mathbf{P}$ is

$$(\boldsymbol{\pi}^*\mathbf{P})_j = \sum_i \pi_i^*\,P(i \rightarrow j) = \sum_i \pi_j^*\,P(j \rightarrow i) = \pi_j^* \sum_i P(j \rightarrow i) = \pi_j^*$$

where the key middle step follows from Equation A3.10, while the last step follows because rows of $\mathbf{P}$ sum to one.

The basic idea of a discrete-state Markov chain can be generalized to a continuous-state Markov process by replacing $\mathbf{P}$ with a probability kernel, $P(x, y)$, that satisfies

$$\int P(x, y)\, dy = 1$$

The continuous extension of the Chapman-Kolomogrov equation becomes

$$\pi_t(y) = \int \pi_{t-1}(x) P(x, y)\, dx \tag{A3.11a}$$

Finally, at equilibrium, the stationary distribution satisfies

$$\pi^*(y) = \int \pi^*(x)P(x,y)\,dx \tag{A3.11b}$$

THE METROPOLIS-HASTINGS ALGORITHM

The roots of MCMC methods trace back to a problem faced by mathematical physicists in applying Monte Carlo integration, namely, generating random samples from some complex distribution in order to apply Equation A3.1. Their solution was the Metropolis-Hastings algorithm (Metropolis and Ulam 1949; Metropolis et al. 1953; Hastings 1970), which was reviewed by Chib and Greenberg (1995).

Suppose our goal is to draw samples from some distribution, $p(\theta) = f(\theta)/K$, where the normalizing constant, K, may be very difficult to compute. Note that Bayesian posteriors are of this form (Equation A2.4b). The **Metropolis algorithm** (Metropolis and Ulam 1949; Metropolis et al. 1953), which generates a sequence of draws from this distribution, is as follows:

1. Start with any initial value, θ_0, satisfying $f(\theta_0) > 0$.
2. Using the current value of θ, sample a **candidate point**, θ^*, from some **jumping distribution**, $q(\theta_1, \theta_2) = \Pr(\theta_1 \to \theta_2)$, which is the probability of returning a value of θ_2 given a previous value of θ_1. This distribution is also referred to as the **proposal distribution** or **candidate-generating distribution**. The only restriction on the jump density in the Metropolis algorithm is that it is symmetric, with $q(\theta_1, \theta_2) = q(\theta_2, \theta_1)$.
3. Given the candidate point, θ^*, calculate the ratio of the density at the candidate and current, θ_{t-1}, points,

   $$\alpha = \frac{p(\theta^*)}{p(\theta_{t-1})} = \frac{f(\theta^*)}{f(\theta_{t-1})}$$

   Notice that because we are considering the ratio of $p(x)$ under two different values, the normalizing constant, K, cancels out.
4. If the jump increases the density ($\alpha > 1$), accept the candidate point (set $\theta_t = \theta^*$) and return to step 2. If the jump decreases the density ($\alpha < 1$), then with a probability of α we accept the candidate point, else we reject it and return to step 2. Note that α is a ratio of two likelihoods, so the probability of accepting a move is the ratio of the candidate to current likelihoods.

We can summarize Metropolis sampling as first computing

$$\alpha = \min\left(\frac{f(\theta^*)}{f(\theta_{t-1})}, 1\right) \tag{A3.12}$$

and then accepting a candidate value, θ^*, with probability α (the **probability of a move**). This generates a Markov chain, $(\theta_0, \theta_1, \cdots, \theta_k, \cdots)$, as the transition probabilities from θ_t to θ_{t+1} depend only on θ_t and not on $(\theta_0, \cdots, \theta_{t-1})$. Following a sufficient **burn-in period** (of, say, k steps), the chain approaches its stationary distribution and (as we will demonstrate shortly) samples from the vector $(\theta_{k+1}, \cdots, \theta_{k+n})$ are samples from $p(\theta)$.

Hastings (1970) generalized the Metropolis algorithm by using an arbitrary (as opposed to strictly symmetric) transition probability function, $q(\theta_1, \theta_2) = \Pr(\theta_1 \to \theta_2)$, and setting the acceptance probability for a candidate point as

$$\alpha = \min\left(\frac{f(\theta^*)\,q(\theta^*, \theta_{t-1})}{f(\theta_{t-1})\,q(\theta_{t-1}, \theta^*)}, 1\right) \tag{A3.13}$$

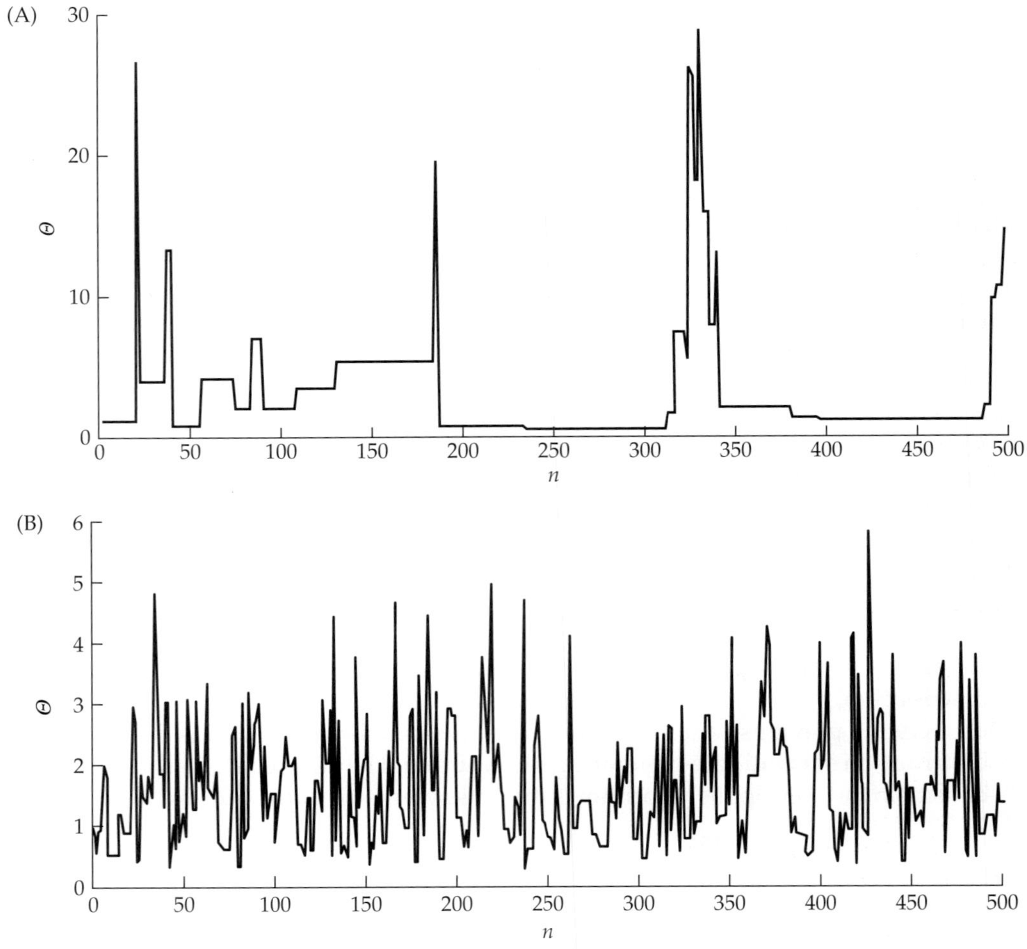

Figure A3.1 Traces for the samplers discussed in Example A3.3. **A:** A sample run when the candidate-generating distribution is a uniform over $(0, 100)$. **B:** A sample run when the candidate-generating distribution is a χ_1^2.

This is the **Metropolis-Hastings algorithm**, and Equation A3.13 is the **Hastings ratio**. When the proposal distribution is symmetric, $q(x, y) = q(y, x)$, we recover the original Metropolis algorithm.

Example A3.3. As a toy example, consider the scaled inverse-χ^2 distribution (Equation A2.30a),

$$p(x) = C \cdot x^{-(n/2+1)} \cdot \exp\left(\frac{-a}{2x}\right)$$

We wish to use the Metropolis algorithm to simulate draws from this distribution with $n = 3$ degrees of freedom and a scaling factor of $a = 4$. Here

$$f(x) = x^{-5/2} \exp[-4/(2x)\,] = x^{-2.5} \exp[-2/x]$$

Suppose we take as our candidate-generating distribution a uniform over $(0, 100)$. Clearly, there is probability mass above 100 for the scaled inverse-χ^2, but we assume this is sufficiently small that we can ignore it. Now let's run the algorithm. Take $\theta_0 = 1$ as our starting value, and suppose the uniform returns a candidate value of $\theta^* = 39.82$. Computing

α,

$$\alpha = \min\left(\frac{f(\theta^*)}{f(\theta_{t-1})}, 1\right) = \min\left(\frac{(39.82)^{-2.5} \cdot \exp(-2/39.82)}{(1)^{-2.5} \cdot \exp(-2/2 \cdot 1)}, 1\right) = 0.0007$$

Since $\alpha < 1$, the candidate value, θ^*, is accepted with a probability of 0.0007: we randomly draw U from a uniform $(0, 1)$ distribution and accept θ^* if $U \leq \alpha = 0.0007$. If $U > \alpha$, the candidate value is rejected, and we draw another candidate value from the proposal distribution and continue as above. A sample run resulting from first 500 values of θ is plotted in Figure A3.1A. Notice that there are long flat periods (corresponding to all candidate values, θ^*, being rejected). Such a chain is said to be **poorly mixing** (showing long periods between jumps) and is numerically inefficient for sampling the entire probability space defined by the target distribution. Conversely, balancing the poor mixing is the fact that when jumps occur, they tend to be large, thus exploring the extreme values of this distribution.

In contrast, suppose our proposal distribution is a χ_1^2. As this distribution is not symmetric, we must employ Metropolis-Hastings (see Example A3.5 for the details). A resulting Metropolis-Hastings sampling run is shown in Figure A3.1B. Note that the time series looks like **white noise**, and the chain is said to be well mixing. However, note that most jumps are small (< 5), with fewer extreme values sampled than under the uniform proposal distribution.

Example A3.4. This rather technical example shows that the Markov chain generated by the Metropolis-Hastings algorithm has a stationary distribution which corresponds to $p(x)$. Hence, draws from the stationary phase of this chain correspond to random draws from the distribution given by $p(x)$. We do so by showing that, under this chain, $p(x)$ satisfies the detailed balance equation (A3.10), and hence it must be the stationary distribution for the Markov chain.

Under Metropolis-Hasting, we sample from $q(x, y) = \Pr(x \to y \,|\, q)$ and then accept the move with probability $\alpha(x, y)$, so the transition probability kernel is given by

$$\Pr(x \to y) = q(x, y)\,\alpha(x, y) = q(x, y) \cdot \min\left[\frac{p(y)\,q(y, x)}{p(x)\,q(x, y)}, 1\right] \qquad \text{(A3.14a)}$$

If the Metropolis-Hasting kernel satisfies $P(x \to y)\,p(x) = P(y \to x)\,p(y)$, or

$$q(x, y)\,\alpha(x, y)\,p(x) = q(y, x)\,\alpha(y, x)\,p(y) \quad \text{for all } x, y \qquad \text{(A3.14b)}$$

then the stationary distribution from this kernel corresponds to draws from the target distribution, $p(x)$.

We show that the balance equation is indeed satisfied with this kernel by considering the three possible cases for any particular (x, y) pair:

1. $q(x, y)\,p(x) = q(y, x)\,p(y)$. Here $\alpha(x, y) = \alpha(y, x) = 1$ implying

$$P(x \to y)\,p(x) = q(x, y)\,p(x) \quad \text{and} \quad P(y \to x)p(y) = q(y, x)\,p(y)$$

Under our assumption that $q(x, y)\,p(x) = q(y, x)\,p(y)$, this reduces to $P(x \to y)\,p(x) = P(y \to x)\,p(y)$, showing that (for this case), the detailed balance equation holds.

2. $q(x, y)\,p(x) > q(y, x)\,p(y)$, in which case

$$\alpha(x, y) = \frac{p(y)\,q(y, x)}{p(x)\,q(x, y)} \quad \text{and} \quad \alpha(y, x) = 1$$

Hence,

$$\begin{aligned} P(x \to y)\,p(x) &= q(x, y)\,\alpha(x, y)\,p(x) \\ &= q(x, y)\,\frac{p(y)\,q(y, x)}{p(x)\,q(x, y)}\,p(x) \\ &= q(y, x)\,p(y) = q(y, x)\,\alpha(y, x)\,p(y) \\ &= P(y \to x)\,p(y) \end{aligned}$$

3. $q(x, y)\, p(x) < q(y, x)\, p(y)$. Here

$$\alpha(x, y) = 1 \quad \text{and} \quad \alpha(y, x) = \frac{q(x, y)\, p(x)}{q(y, x)\, p(y)}$$

Following the same logic just used yields

$$\begin{aligned} P(y \to x)\, p(y) &= q(y, x)\, \alpha(y, x)\, p(y) \\ &= q(y, x) \left(\frac{q(x, y)\, p(x)}{q(y, x)\, p(y)} \right) p(y) \\ &= q(x, y)\, p(x) = q(x, y)\, \alpha(x, y)\, p(x) \\ &= P(x \to y)\, p(x) \end{aligned}$$

which concludes the proof.

Burning-in the Sampler

A key issue in the successful implementation of Metropolis-Hastings, or any other MCMC sampler, is the number of steps (iterations) until the chain approaches stationarity (the length of the **burn-in period**), as initial values are somewhat dependent upon the starting position. Typically the first 1,000 to 50,000 (or more) values of the chain are discarded, and then various convergence tests (see below) are used to assess whether stationarity has indeed been reached.

A poor choice of either a proposal distribution or a starting value can greatly increase the required burn-in time, and an area of current research is whether an optimal starting point and proposal distribution can be found. One suggestion is to initiate the chain as close to the center of the distribution as possible, for example, taking a value close to the distribution's mode (such as using an approximate MLE as the starting value). Geyer (2011), however, offered the simple advice that "any point you don't mind having in a sample is a good starting point."

A chain is said to be **poorly mixing** if it stays in small regions of the parameter space for long periods of time, as opposed to a **well-mixing** chain, which seems to happily explore the entire space (albeit perhaps by very small steps, requiring very many draws to fully survey the entire space). A poorly mixing chain can arise because the target distribution is multimodal and our choice of starting values (or the chance steps that were initially taken) traps us near one of the modes (such multimodal posteriors can arise if we have a strong prior that is in conflict with the observed data). Two approaches have been suggested for situations where the target distribution may have multiple peaks. The most straightforward is to use widely dispersed initial values to start several different chains (Gelman and Rubin 1992). A less obvious approach, which we will discuss shortly, is to use **simulated annealing** on a single chain.

Even after an apparently successful burn-in, the real fear for the MCMC practitioner is that the chain is simply showing **pseudo-convergence**. It appears to have reached an equilibrium, when in fact it is simply exploring (albeit perhaps rather fully) some subset of the entire distribution. If the transition time between different parts of the full distribution is long relative to the sample size, the MCMC user will happily observe convergence and believe that the simulation is a fair representation of the true posterior. The unfortunate feature is that for *any* MCMC, the skeptic, or a hardened reviewer, can claim (with some justification) that convergence has not been demonstrated. The problem is akin to large-sample results in statistics (such as the large-sample features of ML; LW Appendix 4): the theory says that for a sufficiently large sample, certain desirable properties hold, but usually says rather little about how "large" is large. For a complex posterior, the required chain size might be far, far larger than an investigator envisions. The only real solution is to run the chain for as long as possible. In such a setting, Geyer (2011) argued, a burn-in is not required, and noted that "Burn-in is mostly harmless, which is perhaps why the practice persists. But everyone should understand that it is unnecessary, and those who do not use it are not thereby making an error."

Simulated Annealing

Simulated annealing was developed for finding the maximum of complex functions with multiple peaks where standard hill-climbing approaches may trap the algorithm on a less than optimal peak (Kirkpatrick et al. 1983). The idea is that when we initially start sampling the space, we will accept a reasonable probability of a down-hill move in order to explore the entire space. As the process proceeds, we decrease the probability of such down-hill moves. The analogy (and hence the term) is the annealing of a crystal as temperature decreases: initially there is a lot of movement, which gets smaller and smaller as the temperature cools. Simulated annealing is very closely related to Metropolis sampling, differing only in that the probability α of a move is given by

$$\alpha_{SA} = \min\left[1, \left(\frac{f(\theta^*)}{f(\theta_{t-1})}\right)^{1/T(t)}\right] = \alpha^{1/T(t)} \tag{A3.15a}$$

where the function $T(t)$ is called the **cooling schedule** (setting $T = 1$ recovers Metropolis sampling), and the particular value of T at any point in the chain is called the **temperature**. For example, suppose $f(\theta^*)/f(\theta_{t-1}) = 0.5$. If $T = 100$, $\alpha = 0.93$, while for $T = 1$, $\alpha = 0.5$, and for $T = 1/10$, $\alpha = 0.0098$. Hence, we start off with a high probability of a jump and then cool down to a very low jump probability. Replacing α with the Hastings ratio (Equation A3.13) allows us to apply simulated annealing to Metropolis-Hastings.

A function with geometric decline is typically used for $T(t)$. For example, to start at T_0 and cool down to a final temperature of T_f over n steps, we can set

$$T(t) = T_0 \left(\frac{T_f}{T_0}\right)^{t/n} \tag{A3.15b}$$

More generally, if we wish to cool off to T_f by time n and then keep the temperature constant at T_f for the rest of the run, we can take

$$T(t) = \max\left(T_0 \left(\frac{T_f}{T_0}\right)^{t/n}, T_f\right) \tag{A3.15c}$$

In particular, to cool down to Metropolis sampling (by step n), we set $T_f = 1$ and the cooling schedule becomes

$$T(t) = \max\left(T_0^{1-t/n}, 1\right) \tag{A3.15d}$$

Choosing a Jumping (Proposal) Distribution

The Metropolis sampler works with any symmetric proposal (or jumping) distribution, while Metropolis-Hastings allows for more general distributions. So how do we determine our best option for a proposal distribution? There are two general approaches: random walks and independent chain sampling. Under a sampler using a proposal distribution based on a **random-walk chain**, the new value, y, equals the current value, x, plus a random variable, namely, z,

$$y = x + z$$

In this case, $q(x, y) = g(y - x) = g(z)$, the density associated with the random variable z. If $g(z) = g(-z)$, i.e., the density for the random variable z is symmetric (as occurs with a normal or multivariate normal with a mean of zero, or a uniform centered around zero), then we can use Metropolis sampling, as $q(x, y)/q(y, x) = g(z)/g(-z) = 1$. The variance of the proposal distribution can be thought of as a **tuning parameter** that is adjusted to achieve better mixing.

Under a proposal distribution using an **independent chain**, the probability of jumping to point y is *independent* of the current position (x) of the chain, namely, $q(x, y) = g(y)$. The candidate value is simply drawn from a proposal distribution, independent of the current value (see Examples A3.3 and A3.5). Again, any number of standard distributions can be used for $g(y)$. Note that in this case, except for a uniform, the proposal distribution is generally not symmetric, as $g(x)$ is generally not equal to $g(y)$ (for $x \neq y$), and Metropolis-Hasting sampling must be used.

As mentioned, we can tune the proposal distribution to adjust the mixing, and in particular the acceptance probability, of the chain. This is generally accomplished by adjusting the standard deviation (SD) of the proposal distribution. This can be done by adjusting the variance for a normal or adjusting the eigenvalues of the covariance matrix for a multivariate normal. Likewise, one can increase or decrease the range, $(-a, a)$, if a uniform is used, or change the degrees of freedom (df) if a χ^2 is used (variance increases with the df). To increase the acceptance probability, one *decreases* the proposal distribution SD. There is a tradeoff in that if the SD is too large, jumps are potentially large (which is good), but are usually not accepted (which is a problem). This leads to high autocorrelation (see below) and very poor mixing, requiring much longer chains. On the other hand, if the proposal SD is too small, moves will generally be accepted (there is a high acceptance probability), but they will also be small, again generating high autocorrelations and poor mixing.

Example A3.5. Suppose we wish to use a χ^2 as our proposal distribution. Recall from Equation A2.27b, that for $x \sim \chi^2_n$,

$$g(x) \propto x^{n/2-1} e^{-x/2}$$

Thus, $q(x, y) = g(y) = C \cdot y^{n/2-1} e^{-y/2}$. Note that $q(x, y)$ is not symmetric, as $q(y, x) = g(x) \neq g(y) = q(x, y)$. Hence, we must use Metropolis-Hastings sampling, with an acceptance probability of

$$\alpha(x, y) = \min\left[\frac{p(y)\, q(y, x)}{p(x)\, q(x, y)}, 1\right] = \min\left[\frac{p(y)\, g(x)}{p(x)\, g(y)}, 1\right] = \min\left[\frac{p(y)\, x^{n/2-1} e^{-x/2}}{p(x)\, y^{n/2-1} e^{-y/2}}, 1\right]$$

If we use the same target distribution as in Example A3.3 (a scaled inverse-χ^2), $p(x) = C \cdot x^{-2.5}\, e^{-2/x}$, the rejection probability becomes

$$\alpha(x, y) = \min\left[\frac{(\,y^{-2.5}\, e^{-2/y}\,)\,(\,x^{n/2-1} e^{-x/2}\,)}{(\,x^{-2.5}\, e^{-2/x}\,)\,(\,y^{n/2-1} e^{-y/2}\,)}, 1\right]$$

Results for a run of the sampler under two different proposal distributions (a χ^2_2 and a χ^2_{10}) are plotted in Figure A3.2. The χ^2_2 has the smaller variance, and thus a higher acceptance probability. Further, it seems to explore more of the distribution space than a χ^2_{10} (compare the number of values above 4 in both traces).

Autocorrelation and Sample Size Inflation

We often expect adjacent members from an MCMC sequence to be positively correlated, and we can quantify the nature of this correlation by using an **autocorrelation function**. Consider

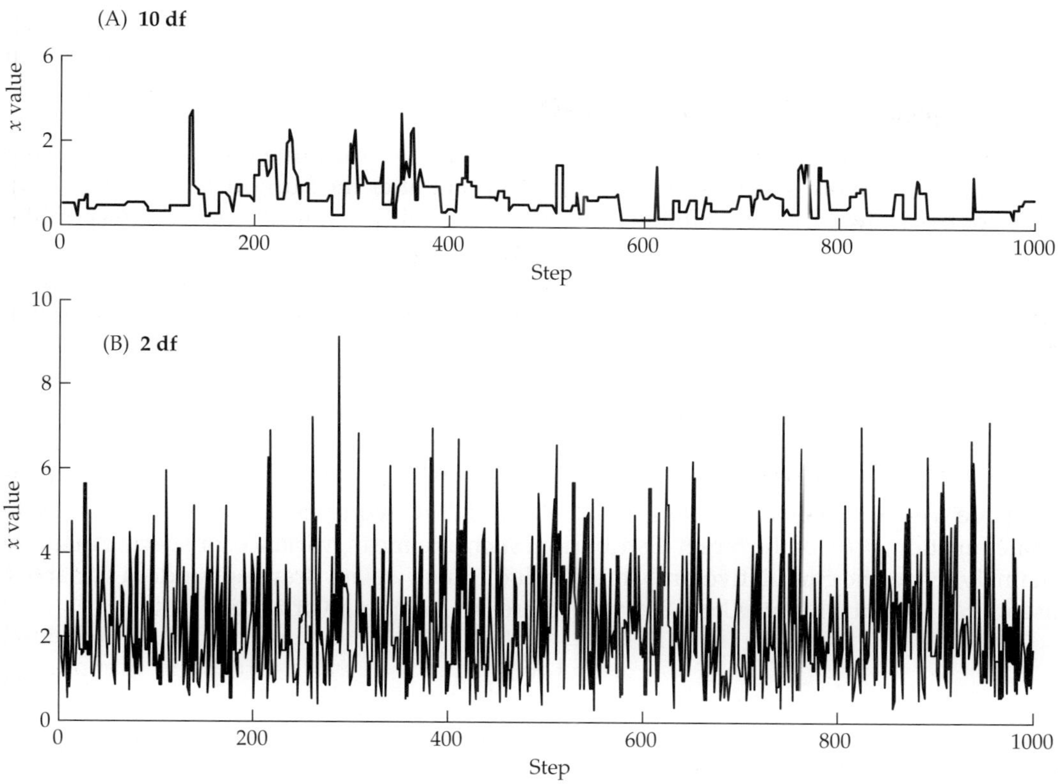

Figure A3.2 Trace plots for the two samplers discussed in Example A3.5. **A:** The proposal distribution is a χ^2_{10}. **B:** The proposal distribution is a χ^2_2.

a sequence, $(\theta_1, \cdots, \theta_n)$, of length n. Correlations can occur between adjacent members, $\rho(\theta_i, \theta_{i+1}) \neq 0$, and, more generally, between more distant members, $\rho(\theta_i, \theta_{i+k}) \neq 0$. The kth order autocorrelation, ρ_k, can be estimated by

$$\widehat{\rho}_k = \frac{\text{Cov}(\theta_t, \theta_{t+k})}{\text{Var}(\theta_t)} = \frac{\sum\limits_{t=1}^{n-k} \left(\theta_t - \overline{\theta}\right)\left(\theta_{t+k} - \overline{\theta}\right)}{\sum\limits_{t=1}^{n-k} \left(\theta_t - \overline{\theta}\right)^2}, \quad \text{with} \quad \overline{\theta} = \frac{1}{n}\sum_{t=1}^{n} \theta_t \tag{A3.16}$$

An important result from the theory of time series analysis is that if the values of θ_t are from a stationary (and correlated) process, correlated draws still provide an unbiased picture of the distribution ***provided the sample size is sufficiently large***. The reason for this large-sample stipulation is that, strictly speaking, Equation A3.16 applies only to a stationary process.

Some indication of the required sample size comes from the theory of a **first-order autoregressive process** (AR_1), where

$$\theta_t = \mu + \alpha(\theta_{t-1} - \mu) + \epsilon \tag{A3.17a}$$

where $|\alpha| < 1$ and ϵ is **white noise**, namely, $\epsilon \sim N(0, \sigma^2)$. Here $\rho_1 = \alpha$ and the kth order autocorrelation is given by $\rho_k = \rho_1^k = \alpha^k$. Under this process, $E(\overline{\theta}) = \mu$ with a standard error of

$$\text{SE}\left(\overline{\theta}\right) = \frac{\sigma}{\sqrt{n}}\sqrt{\frac{1+\rho}{1-\rho}} \tag{A3.17b}$$

The first ratio is the standard error for white noise, while the second ratio, $\sqrt{(1+\rho)/(1-\rho)}$, is the **sample size inflation factor** (**SSIF**), which shows how the autocorrelation inflates the sampling variance relative to independent samples. For $\rho = 0.5, 0.75, 0.9, 0.95$, and 0.99, the associated SSIF values become 3, 7, 19, 39, and 199, respectively. With an autocorrelation of 0.95 (which is not uncommon in a Metropolis-Hastings sequence), roughly 40 times as many iterations are required for the same precision as with an uncorrelated sequence.

One historical strategy for reducing autocorrelation is **thinning** (or **subsampling**) the output, which involves storing only every mth point after the burn-in period. Suppose a Metropolis-Hastings sequence follows an AR_1 model, with $\rho_1 = 0.99$. In this case, sampling every 50, 100, and 500 points gives the correlation between the thinned samples as 0.605 $(= 0.99^{50})$, 0.366, and 0.007, respectively, returning SSIF values of 2, 1.5, and 1.007. In addition to reducing autocorrelation, thinning the sequence also saves computer memory. However, this throws out a huge amount of the data. If computer memory is not an issue, Geyer (1992) and MacEachern and Berliner (1994) found that it is more efficient (yielding smaller Monte Carlo variances) to keep the entire chain than to use thinning.

The Monte Carlo Variance of a MCMC-Based Estimate

Suppose we are interested in using a burned-in MCMC sequence, $(\theta_1, \cdots, \theta_n)$, to estimate some function, $h(\theta)$, of the target distribution, such as a mean, variance, or specific quantile (namely, a specific cumulative probability value). Since we are drawing random variables, associated with the Monte Carlo estimate, there is a sampling variance of

$$\widehat{h} = \frac{1}{n} \sum_{i=1}^{n} h(\theta_i) \tag{A3.18}$$

A similar issue arose with the error in Monte Carlo integration, where we used the sample variance around $\widehat{h}$ to compute the Monte Carlo variance, dividing this by the number of iterations to yield a standard error (Equations A3.1e and A3.5c). This strategy exploited the fact that draws for Monte Carlo integration are *independent* and therefore uncorrelated. In stark contrast, we expect draws in an MCMC sequence to be highly *dependent*. A standard sample variance estimator ignores these correlations, potentially yielding a highly biased estimate of the Monte Carlo variance (see Equation A3.17b).

One direct approach is to run several chains (or extract subsamples from a very long chain) and use the between-chain variance in $\widehat{h}$ as our variance estimate. Specifically, if $\widehat{h}_j$ denotes the estimate for chain j $(1 \leq j \leq c)$, where each of the c chains has the same length, then the estimated variance of the Monte Carlo estimate is

$$\text{Var}\left(\widehat{h}\right) = \frac{1}{c-1} \sum_{j=1}^{c} \left(\widehat{h}_j - \widehat{h^*}\right)^2 \quad \text{where} \quad \widehat{h^*} = \frac{1}{c} \sum_{j=1}^{c} \widehat{h}_j \tag{A3.19}$$

Using only a single chain, an alternative approach is to use results from the theory of time series to account for the correlations among members in the chain. To start, we estimate the lag-k autocovariance associated with h by

$$\widehat{\gamma}(k) = \frac{1}{n-1} \sum_{i=1}^{n-k} \left[\left(h(\theta_i) - \widehat{h} \right) \left(h(\theta_{i+k}) - \widehat{h} \right) \right] \tag{A3.20}$$

This is natural generalization of the kth order autocorrelation (Equation A3.16) to the random variable generated by $h(\theta_i)$. Geyer (1992, 2011) obtained the resulting estimate of the Monte Carlo variance as

$$\text{Var}\left(\widehat{h}\right) = \frac{1}{n} \left(\widehat{\gamma}(0) + 2 \sum_{i=1}^{2\delta+1} \widehat{\gamma}(i) \right) \tag{A3.21}$$

Note that the zero-lag term, $\widehat{\gamma}(0)$, is simply the standard variance estimate when observations are uncorrelated, while δ is the largest positive integer satisfying $\widehat{\gamma}(2\delta) + \widehat{\gamma}(2\delta+1) > 0$, (i.e., the higher order, or lag, autocovariances above δ are zero).

One measure of the effects of autocorrelation between elements in the sampler is the **effective chain size,**

$$\widehat{n} = \frac{\widehat{\gamma}(0)}{\text{Var}\left(\widehat{h}\right)} \tag{A3.22}$$

where $\widehat{\gamma}(0)$ is the standard sample variance for uncorrelated observations. In the absence of autocorrelation between members, $\widehat{n} = n$.

CONVERGENCE DIAGNOSTICS

The careful reader will note that we have still not answered the fundamental, and vexing, question of how to determine whether the sampler has reached its stationary distribution. Many tests for stationarity have been proposed, and all can indeed show that stationarity has not yet been reached. However, it is important to stress that ***negative results from these tests are not conclusive evidence that the sampler has achieved stationarity.*** Comments from various published reviews on the subject stress this point: "all of the methods can fail to detect the sorts of convergence failure that they were designed to identify"(Cowles and Carlin 1996); "It is clear that the usefulness of convergence assessment methods are limited by their potential unreliability"(Brooks and Roberts 1998); "Diagnostics can only reliably be used to determine a lack of convergence and not detect convergence per se" (Brooks et al. 2003); and "no method works in every case "(El Adlouni et al. 2006). Despite this very serious concern, a variety of methods can certainly show a failure of convergence given the burn-in period used for a sampler. The greater question about trust is what we say when convergence diagnostics *do not* indicate a problem. We consider a few basic approaches here, while additional diagnostics were reviewed by Gelman and Rubin (1992), Geyer (1992), Raftery and Lewis (1992b), Cowles and Carlin (1996), Brooks and Roberts (1998), Kass et al. (1998), Mengerson et al. (1999), Robert and Casella (2004), El Adlouni et al. (2006), and Peltonen et al. (2009).

An extra level of complications occurs in models using **revisable jump MCMC,** wherein the chain actually jumps between difference *classes* of models. For example, suppose the number of QTLs, k, is a random variable. While the chain may come to a nearly equilibrium distribution for the parameters associated with a given value of k fairly quickly, one must also sample a sufficient number of jumps *between models* with different values of k to fully explore the model space. One might incorrectly assume that there appears to be convergence when the model quickly equilibrates for a given value of k, but jumps very slowly between different values of k so that only a small sample of the k space has been explored.

Visual Analysis

As shown in Figures A3.1 and A3.2, one should always examine the **time series trace**, the plot of the random variable being generated versus the number of iterations. In addition to showing evidence for poor mixing, such traces can also suggest a ***minimum*** burn-in period for some starting value. For example, suppose the trace moves very slowly away from the initial value to a rather different value (say after 5000 iterations), after which it appears to settle down. Clearly, the burn-in period is ***at least*** 5000 in this case. It must be cautioned that the actual burn-in time ***may be far longer*** than suggested by the trace. Nevertheless, the trace often indicates that more burn-in time is needed. A second approach is to run either several chains or subsample from a much longer chain, and then compare the traces. The expectation under stationarity is that these samples are drawn from the same distribution. Even if visual analysis suggests that two (or more) chains appear to be very similar and are both well-behaved, the concern is that both might be stuck in the same local maximum region of a complex posterior. In such cases, the impression of stationarity and consistency is presented, but significantly longer runs for each chain would show a different outcome.

Correlograms, which plot serial autocorrelations as a function of the time lag, are also very useful in accessing a run from an MCMC sampler. A plot of ρ_k versus k (the kth order autocorrelation versus the lag) should show geometric decay if the sampler series closely follows an AR_1 model. A plot of the **partial autocorrelations** as a function of lag is also useful. The kth partial autocorrelation is the excess correlation not accounted for by a $k-1$ order autogressive model (AR_{k-1}). Hence, if the first-order model fully fits the data, the second-order partial autocorrelation is zero, as the lagged autocorrelations are accounted for by the AR_1 model (i.e., $\rho_k = \rho_1^k$). Both of these autocorrelation plots can indicate an underlying correlation structure that may not be obvious from the time series trace.

More Formal Approaches

Gelman and Rubin's (1992) proposal was to first generate m separate chains at initial locations that are widely dispersed in the parameter space, with each returning a burned-in and thinned (such that autocorrelations between estimates are near zero) sequence of length n. Their approach is essentially an ANOVA, contrasting the estimated among-chain variance with its within-chain value. For the univariate case, let

$$B = \frac{n}{m-1}\sum_{i=1}^{m}(\overline{\theta}_{i\cdot} - \overline{\theta}_{\cdot\cdot})^2 \tag{A3.23a}$$

be the among-chain variance, where $\overline{\theta}_{i\cdot}$ is the mean parameter value for chain i, and $\overline{\theta}_{\cdot\cdot}$ is the grand mean over all chains. Likewise, for the within-chain variance, we define

$$W = \frac{1}{m}\left(\frac{1}{n-1}\sum_{i=1}^{m}\sum_{j=1}^{n}(\overline{\theta}_{ij} - \overline{\theta}_{i\cdot})^2\right) \tag{A3.23b}$$

A combined estimate of the variance is

$$\widehat{\sigma^2} = \frac{n-1}{n}W + \frac{1}{n}B \tag{A3.23c}$$

Gelman and Rubin defined the **potential scale reduction factor** (**PSRF**) as

$$R = \frac{\widehat{\sigma^2}}{W} = 1 + \frac{1}{n}\left(\frac{B}{W} - 1\right) \tag{A3.23d}$$

with values of R near one being consistent with convergence. A plot of R over time should converge to a value near one, if the chains are converging.

Because most MCMCs involve many variables, one could either look at the univariate R trace for each or use the multivariate version suggested by Brooks and Gelman (1998),

$$R = \frac{n-1}{n} + \left(1 + \frac{1}{m}\right)\lambda_1 \tag{A3.24}$$

where λ_1 is the leading eigenvalue of the matrix product $\mathbf{W}^{-1}\mathbf{B}/n$, and where $\mathbf{W}$ and $\mathbf{B}$ are the sample within- and among-covariance matrices for the parameters. Again, a value near 1 is consistent with convergence.

The **Geweke test** (Geweke 1992) splits the sample (after removing a burn-in period) into two parts. If the chain is at stationarity, the means of the two samples should be equal. A modified z-test can be used to compare the two subsamples, and the resulting test statistic is often referred to as a **Geweke z-score**. A value larger than 2 indicates that the mean of the series is still drifting and a longer burn-in is required before monitoring the chain (to extract a sampler) can begin.

A more informative approach is the **Raftery-Lewis test** (Raftery and Lewis 1992a). Here, one specifies a particular quantile, q, of the distribution of interest (typically testing both $q = 2.5\%$ and 97.5%, to give a 95% confidence interval), an accuracy, ϵ, of the quantile, and a power, $1-\beta$, for achieving this accuracy on the specified quantile. With these three parameter values set, the Raftery-Lewis test breaks the chain into a binary (1, 0) sequence, in which an element in the sequence takes on a value of 1 if $\theta_t \leq q$, otherwise it takes on a value of zero. This generates a two-state Markov chain, and the Raftery-Lewis test uses the sequence to estimate the transition probabilities between states. With these probabilities in hand, one can then estimate the number of additional burn-ins (if any) required to approach stationarity, the thinning ratio (how many points should be discarded for each sampled point), and the total chain length required to achieve the preset level of accuracy.

Practical MCMC: How Many Chains and How Long Should They Run?

One can either use a single long chain (Geyer 1992; Raftery and Lewis 1992b) or multiple chains, with each starting from different initial values (Gelman and Rubin 1992). Note that with parallel processing, using multiple chains may be computationally more efficient (given a fixed amount of analysis time) than a single long chain. One suggested approach (Kass et al. 1998) is to use five parallel chains with widely separated starting values, and to start the jth value for parameter θ_i at $\mu_i + (j-2)\sigma_i$, where μ_i and σ_i^2 are the prior values for θ_i. Alternatively, one could randomly sample j from $(-2, 1, 0, 1, 2)$ for each parameter to generate the vector of starting values.

Geyer (1992, 2011), however, forcefully argued that using a single long chain is the best approach. If the chain transition times between different parts of a distribution are long, then a series of shorter chains may entirely miss this problem, even if their starting values are widely dispersed. Theory basically tells us that, *provided* we run a chain for a long enough time, the results will be accurate. The best way to do so is to run a single, long chain. Convergent tests (still recognizing their problems mentioned previously) can be applied to subsections of a long chain as efficiently as they can be applied across shorter, parallel-running chains.

Given these concerns, what is the answer to the critical questor about how long to run a chain? **Geyer's dictums** (2011) are that (i) "the least one can do is to make an overnight run. What better way for your computer to spend its time?" and (ii) ideally, "one should start a run when the paper is submitted and keep running until the referees' reports arrive."

THE GIBBS SAMPLER

The **Gibbs sampler** (introduced in the context of image processing by Geman and Geman 1984), is a special case of Metropolis-Hastings sampling wherein the random value is always accepted (i.e., $\alpha = 1$). The key to the Gibbs sampler is that, at any point in the process, one only considers ***univariate*** conditional distributions—the distribution when all but one of the random variables are assigned fixed values. Such conditional distributions are far easier to simulate than complex joint distributions and usually have simple forms (often they are normals, inverse-χ^2, or other common prior distributions). An important use of conjugate priors (Table A2.2) is that they often result in relatively simple forms for conditional distributions. Thus, one simulates n random variables sequentially from the n univariate conditionals rather than generating a single n-dimensional vector in a single pass using the full joint distribution. Given their great efficiency (every value is accepted), it is often thought that one should strive to generate a Gibbs sampler whenever possible. However, one should consult Geyer (2011) for a different perspective.

To introduce the Gibbs sampler, consider a bivariate random variable (x, y), and suppose we wish to compute one or both of the marginals, $p(x)$ and $p(y)$. The idea behind the sampler is that it is far easier to consider a sequence of conditional distributions, $p(x \mid y)$ and $p(y \mid x)$, than it is to obtain the marginal by integration of the joint density $p(x, y)$, namely, $p(x) = \int p(x, y)dy$. The sampler starts with some initial value, y_0, for y and obtains x_0 by generating a random variable from the conditional distribution $p(x \mid y = y_0)$. The sampler then uses x_0 to generate a new value of y_1, drawing from the conditional distribution based on the value of x_0, $p(y \mid x = x_0)$. The sampler proceeds as follows:

$$x_i \sim p(x \mid y = y_{i-1}) \tag{A3.25a}$$

$$y_i \sim p(y \mid x = x_i) \tag{A3.25b}$$

Repeating this process k times generates a **Gibbs sequence** of length k, where a subset of points (x_j, y_j) for $1 \leq j \leq m < k$ are taken as our simulated draws from the full joint distribution. One iteration of all the univariate distributions is often called a **scan** of the sampler. To obtain the desired total of m sample values (here each value in the sampler is a vector of realizations of the two random variables), one samples the chain following a sufficient burn-in to remove the effects of the initial starting values, potentially thinning the sequence as well. The Gibbs sequence converges to a stationary (equilibrium) distribution that is independent of the starting values, and by construction this stationary distribution is the target distribution we are trying to simulate (Tierney 1994). As with any MCMC, a guarantee of convergence at some limit is no guarantee that our sample size is sufficiently large to justify the assumption of stationarity for a burned-in (and potentially thinned) sampler.

Example A3.6. Consider the following distribution from Casella and George (1992). Suppose the joint distribution of $x = 0, 1, \cdots n$ and $0 \leq y \leq 1$ is given by

$$p(x, y) = \frac{n!}{(n-x)!x!}\, y^{x+\alpha-1}\, (1-y)^{n-x+\beta-1}$$

Note that x is discrete and y continuous. While the joint density is complex, the conditional densities are simple distributions. To see this, first recall that a discrete binomial random variable z has a density proportional to

$$p(z \mid q, n) \propto \frac{q^z (1-q)^{n-z}}{z!(n-z)!} \quad \text{for} \quad 0 \leq z \leq n$$

where $0 < q < 1$ is the success parameter and n the number of traits, which we denote by $z \sim \text{B}(n, p)$. Likewise, recall the density for $z \sim \text{Beta}(a, b)$, a beta distribution (Equation A2.38a) with shape parameters a and b is given by

$$p(z \mid a, b) \propto z^{a-1}(1-z)^{b-1} \quad \text{for} \quad 0 \leq z \leq 1$$

Observe that the conditional distribution of x (treating y as a fixed constant) is $x \mid y \sim \text{B}(n, y)$, while $y \mid x \sim \text{Beta}(x + \alpha, n - x + \beta)$.

The power of the Gibbs sampler is that by computing a sequence of these univariate conditional random variables (a binomial and then a beta), we can compute any feature of either marginal distribution. Suppose $n = 10$ and $\alpha = 1, \beta = 2$. We start the sampler with (say) $y_0 = 1/2$ and then take it through three full iterations.

(i) x_0 is obtained by generating a random $\text{B}(n, y_0) = \text{B}(10, 1/2)$ random variable, returning $x_0 = 5$ in our simulation.

(ii) y_1 is obtained from a $\text{Beta}(x_0+\alpha, n-x_0+\beta) = \text{Beta}(5+1, 10-5+2)$ random variable, returning $y_1 = 0.33$.

(iii) x_1 is a realization of a $\text{B}(n, y_1) = \text{B}(10, 0.33)$ random variable, returning $x_1 = 3$.

(iv) y_2 is obtained from a $\text{Beta}(x_1+\alpha, n-x_1+\beta) = \text{Beta}(3+1, 10-3+2)$ random variable, returning $y_2 = 0.56$.

(v) x_2 is obtained from a $\text{B}(n, y_2) = \text{B}(10, 0.56)$ random variable, returning $x_2 = 7$.

Our particular realization of the Gibbs sequence after three iterations is thus (5, 0.5), (3, 0.33), (7, 0.56). We can continue this process to generate a chain of the desired length. Obviously, the initial values in the chain are dependent upon the value of y_0 chosen to start the chain. This dependence decays as the sequence length increases, so we typically start recording the sequence only after a sufficient number of burn-in iterations have occurred.

When more than two variables are involved, the sampler is extended in the obvious fashion. In particular, the value of the kth variable is drawn from the distribution $p(\theta^{(k)} \mid \boldsymbol{\Theta}^{(-k)})$, where $\boldsymbol{\Theta}^{(-k)}$ denotes a vector containing all of the variables but k. Thus, during the ith iteration of the sample, to obtain the value of $\theta_i^{(k)}$, we draw from the distribution

$$\theta_i^{(k)} \sim p(\theta^{(k)} \mid \theta^{(1)} = \theta_i^{(1)}, \cdots, \theta^{(k-1)} = \theta_i^{(k-1)}, \theta^{(k+1)} = \theta_{i-1}^{(k+1)}, \cdots, \theta^{(n)} = \theta_{i-1}^{(n)})$$

For example, if there are four variables, (w, x, y, z), the sampler becomes

$$\begin{aligned} w_i &\sim p(w \mid x = x_{i-1}, y = y_{i-1}, z = z_{i-1}) \\ x_i &\sim p(x \mid w = w_i, y = y_{i-1}, z = z_{i-1}) \\ y_i &\sim p(y \mid w = w_i, x = x_i, z = z_{i-1}) \\ z_i &\sim p(z \mid w = w_i, x = x_i, y = y_i) \end{aligned}$$

Gelfand and Smith (1990) illustrated the power of the Gibbs sampler to address a wide variety of statistical issues, while Smith and Roberts (1993) showed the natural marriage of the Gibbs sampler with Bayesian statistics for obtaining posterior distributions. A nice introduction to the sampler was presented by Casella and George (1992), while further details can be found in Tanner (1996), Besag et al. (1995), Sorensen and Gianola (2002), Robert and Casella (2004), and Lee (2013). Note that the Gibbs sampler can be thought of as a stochastic analog to the Expectation-Maximization (EM) approaches (LW Appendix 4) used to obtain likelihood functions when missing data are present. In the sampler, random sampling replaces the expectation and maximization steps.

Using the Gibbs Sampler to Approximate Marginal Distributions

Any feature of interest for the marginals can be computed from the m realizations of the Gibbs sequence. If $\theta_1, \cdots, \theta_m$ is an appropriately burned-in set of realizations from a Gibbs sampler, the expectation of any function, f, of the random variable θ is approximated by

$$E[f(\theta)]_m = \frac{1}{m} \sum_{i=1}^{m} f(\theta_i) \tag{A3.26a}$$

This is the **Monte-Carlo (MC) estimate** of $f(x)$, as $E[f(\theta)]_m \to E[f(\theta)]$ as $m \to \infty$. Likewise, the MC estimate for any function of n variables $(\theta^{(1)}, \cdots, \theta^{(n)})$ is calculated by

$$E[f(\theta^{(1)}, \cdots, \theta^{(n)})]_m = \frac{1}{m} \sum_{i=1}^{m} f(\theta_i^{(1)}, \cdots, \theta_i^{(n)}) \tag{A3.26b}$$

Example A3.7. Although the sequence of length 3 that was computed in Example A3.6 is too short (and too dependent on the starting value) to be a proper Gibbs sequence, for illustrative purposes we can use it to compute Monte-Carlo estimates. The MC estimates of the means of x and y are

$$\overline{x}_3 = \frac{5+3+7}{3} = 5, \quad \overline{y}_3 = \frac{0.5+0.33+0.56}{3} = 0.46$$

We use the subscript 3 throughout to remind the reader of the sample size of this particular Gibbs sampler. Similarly, $(\overline{x^2})_3 = 27.67$, $(\overline{y^2})_3 = 0.22$, and $(\overline{xy})_3 = 2.47$, returning the MC estimates of the variances of x and y as

$$\text{Var}(x)_3 = (\overline{x^2})_3 - (\overline{x}_3)^2 = 2.67, \qquad \text{Var}(y)_3 = (\overline{y^2})_3 - (\overline{y}_3)^2 = 0.25$$

and their covariance as

$$\text{Cov}(x,y)_3 = (\overline{xy})_3 - \overline{x}_3 \cdot \overline{y}_3 = 2.47 - 5 \cdot 0.46 = 0.16$$

While computing the MC estimate for any moment using the sampler is straightforward, computing the actual *shape* of the marginal density is slightly more involved. While one might use the empirical histogram of the Gibbs sequence as a rough approximation of the marginal distribution of x, this turns out to be inefficient, especially for obtaining the tails of the distribution. A better approach is to use the average of the conditional densities $p(x \mid y = y_i)$, as the function form of the conditional density contains more information about the shape of the entire distribution than the sequence of individual realizations, x_i (Gelfand and Smith 1990; Liu et al. 1991). Because

$$p(x) = \int p(x \mid y)\, p(y)\, dy = E_y\left[\, p(x \mid y)\,\right] \tag{A3.27a}$$

one can approximate the marginal density using

$$\widehat{p}_m(x) = \frac{1}{m} \sum_{i=1}^{m} p(x \mid y = y_i) \tag{A3.27b}$$

Example A3.8. Returning to the Gibbs sequence generated in Example A3.6, recall that the distribution of x given y is binomial, with $x \mid y \sim \text{B}(n, y)$. If we apply Equation A3.27b, the estimate (based on this sequence) of the marginal distribution of x is the weighted sum of three binomials with success parameters of 0.5, 0.33, and 0.56, yielding

$$p_3(x) = \frac{10!}{x!(10-x)!} \left[\frac{0.5^x(1-0.5)^{10-x} + 0.33^x(1-0.33)^{10-x} + 0.56^x(1-0.56)^{10-x}}{3} \right]$$

As Figure A3.3 shows, the resulting distribution (filled bars), although a weighted sum of binomials, departs substantially from the binomial based on the average (here 0.46) of the success parameter (open bars).

REJECTION SAMPLING AND APPROXIMATE BAYESIAN COMPUTATION (ABC)

In the words of Marjoram and Tavaré (2006), for many estimation problems in modern population genetics it is "far easier to simulate than to calculate." Consider the number of segregating sites in a sample of alleles from a population that went through a bottleneck of undetermined width at some unknown time in the past. A likelihood calculation running over all possible coalescent structures even in this simple case is extremely challenging. Conversely, if we specify the fractional reduction of the bottleneck and the time when this occurred, it is very easy to simulate a coalescent from this population structure and then randomly apply mutations to generate a sample of alleles. This logic forms the basis of an approach that Beaumont et al. (2002) called **approximate Bayesian computation (ABC)**, although its roots trace back to Tavaré et al. (1997), Weiss and von Haeseler (1998), and Pritchard

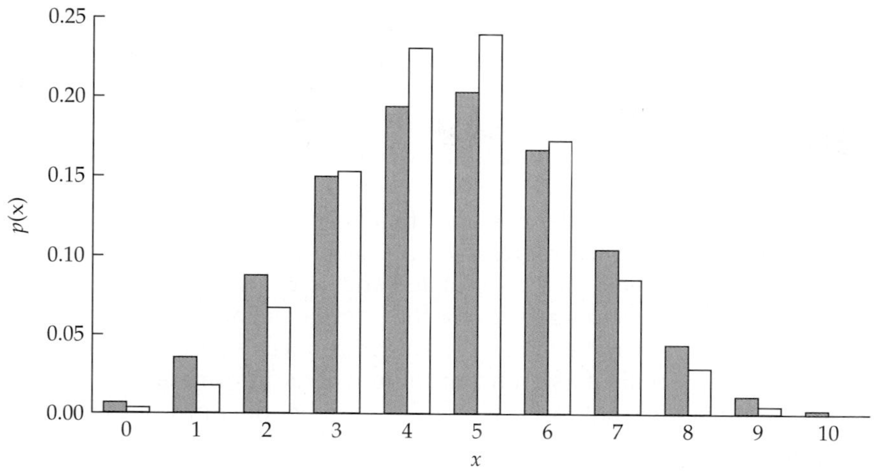

Figure A3.3 The approximation of the marginal posterior distribution for x for the distribution given in Example A3.6 using a Gibbs sequence. The open bars represent the approximation using a binomial with the Gibbs-sequence mean value of y as the success parameter, while the solid bars represent the approximation based on a weighted sum of binomials. See Example A3.8 for details.

et al. (1999). ABC is also referred to as **likelihood-free computation** in the literature (as the likelihood is obtained by simulation, rather than by direct calculation). Reviews of ABC were presented by Marjoram et al. (2003), Beaumont (2010), and Sisson and Fan (2011).

Although much of the early focus of ABC methods was on population-genetic problems, it has broad applicability in quantitative genetics. For example, Technow et al. (2015) used ABC approaches to combine crop growth models and genomic selection, while Csilléry et al. (2010) reviewed a number of applications in evolution and ecology. The use of ABC methods is a rapid growth area and, as such, our treatment is rather basic and likely dated.

A very simple example motivates ABC. Suppose we wish to calculate the posterior distribution for the success probability, p, given that a sample of size 10 has three successes and we assume a prior, $\phi(p)$, for the distribution of possible p values. We can obtain this posterior exactly without ever computing the likelihood as follows: first, we draw a random p value from the prior, $\phi(p)$, and then we generate a random draw from a binomial with $n = 10$ and the sampled value of p. If the binomial random variable generated equals 3, we record the p value (i.e., we accept it into the posterior). If not, we randomly draw a new p from $\phi(p)$ and then simulate using this new value of p, repeating this procedure until we have a sufficiently large number of accepted values to approximate the posterior. This is an example of **rejection sampling**.

More generally, one would draw a candidate vector, $\boldsymbol{\Theta}$, of model parameters from the prior, and then use these in a simulation to generate a dataset, D, which is then compared against the actual dataset, D^*. For our opening example, one would draw values of bottleneck width, time of the bottleneck, and scaled mutation rate from a joint prior. These values are then used to simulate a dataset, D, of polymorphism information for this simulated sample. If $D = D^*$, then we accept this value of $\boldsymbol{\Theta}$ into the posterior and draw again. If not, we reject it and draw again. This generates a value in the MCMC sample without ever having to compute a likelihood.

The complication involved in this basic idea obviously arises when datasets become even slightly complex, as the probability that $D = D^*$ becomes vanishingly small and the rejection probability is essentially one. In this case, one instead tries to match *summary statistics* from D and D^*. Ideally, these would be **sufficient statistics** for the parameters of interest. Recall that, conditioned on the value of a sufficient statistic, the distribution of the data does not depend on the parameter of interest—all of the information on that parameter in the data is provided by the sufficient statistic. Suppose $\mathbf{S}$ is a vector of sufficient statistics for the parameters of interest from the *simulated* dataset, D, and $\mathbf{S}^*$ is the analogous vector from D^* (the sufficient statistics based on the *original* data). A strict rejection method would accept the current draw from the prior if $\mathbf{S}$ equals $\mathbf{S}^*$. Here, again, the rejection probability is likely to be quite high. The approximate aspect of this approach arises when we don't require exact equality, but rather simply require that the two are "close," namely, that $|S - S^*| \leq \epsilon$ for a single sufficient statistic, or $m(\mathbf{S} - \mathbf{S}^*) \leq \epsilon$ for some distance metric, m (often this is Euclidean distance). In this case, the posterior that is generated is only an approximation. Beaumont et al. (2002) suggested regression approaches that attempt to adjust posterior values, given the distance between the simulated and actual data. A far more serious issue is when we do not use sufficient statistics for the parameters of interest (as these are often unknown, and, indeed, may not even exist). This introduces a second source of error that can be rather critical, especially if a significant amount of the information about an unknown parameter (or parameters) is not captured by $\mathbf{S}$. A number of approaches for dealing with this issue have been suggested (e.g., Joyce and Marjoram 2008; Nunes and Balding 2010; Jung and Marjoram 2011; Aeschbacher et al. 2012), but this is a very active research area with no current clean answer.

Example A3.9. Recall from Chapter 8 that an issue of much evolutionary interest concerns both the rate of selective sweeps and the average strength of selection for the swept allele. We can quantify the rate of sweeps by estimating the per-nucleotide adaptive evolution rate, λ, and quantify their strength by estimating the average selective advantage, s, of a swept allele. As discussed in Chapter 8, most standard estimates based on polymorphism and divergence data usually confound these two parameters. Jensen et al. (2008) developed an ABC approach to obtain separate estimates of λ and s. Their idea was to jointly consider the means and variances of several summary statistics measuring various sweep features from sequence polymorphism data. The features they included were nucleotide diversity, π, and number of segregating sites, S, for polymorphism levels; θ_H for departures in the site-frequency spectrum (Equations 9.27); and Z_{nS} for the local structure of LD (Equation 9.36b).

Their ABC approach proceeds as follows: First, values of λ and s are drawn from some joint prior, and then are used to generate a simulation of the sweep. The summary statistics of interest are then computed on this simulated data, and if they are sufficiently close to their observed values, the joint λ and s values are kept; otherwise, they are rejected and new values are drawn. This procedure is repeated to obtain (at least) several thousand accepted values to generate a joint empirical posterior distribution of λ and s values that is consistent with the observed data.

Appendix 4

Multiple Comparisons: Bonferroni Corrections, False-discovery Rates, and Meta-analysis

If your experiment needs statistics, you ought to have done a better experiment. Ernest Rutherford

Facts are stubborn things, but statistics are pliable. Mark Twain

Often one is faced with interpreting a list of p values (the probabilities of the observed outcomes under their null hypotheses), either from a set of independent experiments all testing the *same* hypothesis, or from a single experiment wherein a large number of *different* hypotheses are tested. Both of these are examples of **multiple comparisons**. In the former setting, the problem is how to best combine information from multiple, often rather disparate, studies to make more global statements. In the latter setting, our concern is controlling error over the entire *collection* of tests from a single experiment.

Issues with combining information across studies can be broken into methods for combining p values and more general methods for combining information on effect size. Given that methods for combining p values and issues related to multiple comparisons within a single experiment share some common features, we examine these first. We then turn to a more detailed look at the field of **meta-analysis** (**MA**), namely, how to most efficiently combine information on effect sizes over a series of studies.

The statistical analysis of a large dataset typically involves testing, not just a single hypothesis, but rather many hypothesis (often *very* many). This is especially true in the genomics era, wherein a single **high-dimensional experiment** may test tens of thousands of hypotheses (such as treatment-dependent expression over all of the genes in a genome). For any particular test, we may assign a preset probability, α, of a **Type-I error** (i.e., a **false positive**, declaring a test to be **significant**, namely, $p \leq \alpha$, when in fact the null hypothesis is true). Under this broad setting, there are two different strategies for controlling error. First, if we expect that the vast majority of the hypotheses are truly null, we are interested in controlling the **experiment-wide error rate**, the probability of a false positive over *all* of the tests (as the vast majority will be nulls). The standard approach in this setting is the classic **Bonferroni correction**—obtaining an experiment-wide error rate of π over a set of n comparisons by declaring a test to be significant when $p \leq \alpha = \pi/n$. However, this is usually far too stringent and results in an enormous loss of power.

We review sequential-testing schemes to improve this approach, but often the adjustment for multiple comparisons is best accomplished by a ***shift in thinking***. If we expect some reasonable number of the hypotheses to be false, then trying to avoid *any* false positives is not appropriate. Instead, controlling the *fraction* of false positives among the tests that are declared significant (**discoveries**) is a much better aim. This is especially true in large-scale exploratory experiments whose aim is to discover potential candidates for further study. In this setting, we attempt to control the **false-discovery rate** (**FDR**), namely, the fraction of all tests *declared to be significant* that are false positives, as opposed to the type I-error (false-positive) rate. The goal is to find a value, τ, such that the set of tests that is declared to be significant using $p \leq \tau$ has the desired false-discovery rate. As we will show below, the distinction between a false positive and a false discovery is very subtle but critically important. The former is simply the probability that a test under a true null will be declared significant, while the latter further depends on whether the null is indeed correct. If the null is wrong, then the probability of a false discovery is zero, while if the null is correct, then the false discovery rate equals the probability of a false positive.

Our treatment of these topics is organized as follows. First, we examine methods for

Evolution and Selection of Quantitative Traits. Bruce Walsh & Michael Lynch.
Published 2018 by Oxford University Press. © Bruce Walsh & Michael Lynch 2018.
DOI 10.1093/oso/9780198830870.001.0001

combining p values over independent tests. We then turn to controlling the overall false-positive rate for a collection of tests from a single experiment through the use of Bonferroni corrections and their extensions. Given that the decision to control false positives versus false discoveries hinges to a large extent on the fraction, π_0, of the hypotheses that are truly null, we then examine how π_0 can be estimated from the empirical distribution of the p values for a set of tests. Next, we then discuss approaches for controlling the false-discovery rate (settings where we expect some nontrivial fraction of the hypotheses to depart from the null). Finally, we turn to more formal issues in meta-analysis, developing a mixed-model framework for combining estimates across studies. We also examine metrics and graphical tools for assessing possible sources of bias.

COMBINING p VALUES OVER INDEPENDENT TESTS

Hypotheses of interest are often tested in multiple studies, and an important issue (the statistical field of meta-analysis—the analysis of analyses) is how best to combine the results from a set of such studies into a single global statement. The most obvious approach is simply to pool all of the data and perform a single test, but for a variety of reasons this is often not feasible. For example, different tests of the same hypothesis may involve different methodologies or very different settings. Further, published papers may not report the full dataset, but rather may just present a few summary statistics. In such settings, one straightforward approach is to consider the list of p values for the collection of experiments that all purport to test the same hypothesis and try to obtain a single global p value for this entire set. As in normal hypothesis testing, care must be taken in distinguishing between one- and two-tailed hypotheses. For example, we could get a significant p value for a new treatment over a series of tests because it is significantly *worse*, and we would draw an incorrect conclusion if we used combined p values for a two-tailed test when a one-tailed test (the treatment results in an improvement) is more appropriate.

This simple question of how best to combine p values from a set of experiments is potentially fraught with peril for several reasons. First, there is the issue of whether the different tests all really test the same hypothesis. The investigator must take care to assure that this is correct before proceeding. Second, there is concern about the so-called **file-drawer effect**, wherein nonsignificant results remain in the file drawer (i.e., are not published), leading to published results being biased toward smaller p values. One general trend seems to be a publication bias in studies with small sample sizes (where nonsignificant results are often not reported), but a reduction in this bias for larger samples (Easterbrook et al. 1991; Dickersin et al. 1992). The presumptive reason for this sample-size effect in publication bias is that small studies often lack power, so a nonsignificant result does not necessarily provide strong evidence that the null hypothesis is correct. Conversely, due to the higher power of larger studies, authors may feel more comfortable in publishing negative results. We examine several approaches that attempt to quantify the amount of any such **publication bias** in a meta-analysis at the end of this Appendix.

Fisher's χ^2 Method

Fisher (1932b) was among the first to offer a simple approach for combining p values (along with Tippett 1931), based on the important concept that ***the distribution of realized p values under the null follows a uniform distribution over (0,1)***. Further, if $p \sim$ Uniform(0,1), then $-2\ln(p) \sim \chi^2_2$ (Pearson 1938). Hence, under the null, twice the negative natural log of a p value follows a chi-square distribution with two degrees of freedom. If we have k independent tests, then the sum of their log-transformed p values is the sum of k chi-square variables. Such a sum is itself chi-square, with the degrees of freedom given by the sum of the degrees of freedom for the individual chi-squares (LW Appendix 5). These observations formed the motivation for **Fisher's combined probability test**: for k independent tests, where p_i denotes the p value for test i, the sum

$$X^2 = -2\sum_{i=1}^{k} \ln(p_i) \tag{A4.1a}$$

approximately follows a χ^2_{2k} distribution.

Example A4.1. Suppose five different groups collected data to test the same hypothesis, and these groups (perhaps using different methods of analysis) report p values of 0.10, 0.06, 0.15, 0.08, and 0.07. Notice that while none of these individual tests are significant, the trend is clearly that all are "close" to being significant ($\overline{p} = 0.09$). Fisher's statistic returns a value of

$$X^2 = -2\sum_{i=1}^{k} \ln(p_i) = 24.3921 \qquad \text{with} \qquad \Pr(\chi^2_{10} \geq 24.39) = 0.0066$$

Hence, when taken together, these five tests show a highly significant p value.

Rice (1990b; also see Whitlock 2005) noted that a problem with Fisher's method is that smaller p values are differentially weighted compared to complementary larger p values (e.g., p versus $1 - p$). Equation A4.1a can be rearranged to yield

$$X^2 = -2k \ln(\overline{p}_G) \tag{A4.1b}$$

where $\overline{p}_G$ is the geometric mean of the individual p values, which differentially weights smaller values. Under Fisher's method, an observed p value of (say) 0.001 receives more weight than a complementary value of 0.999, which is as extreme (with $-\ln$ weights of 6.9 vs. 0.001). However, under the Z-score transformation—which is obtained by solving $\Pr(U > Z) = p$, where $U \sim \text{N}(0, 1)$—the two complementary p values are of equal magnitude (Z scores of -3.09 and 3.09). This motivates the Z-score method, which we now consider.

Stouffer's Z Score

An alternative to Fisher's approach for combining p values was offered by Stouffer et al. (1949), who transformed the individual p values into Z scores. The sum of k independent unit normals is itself normal, with a mean of zero and a variance of k. These results lead to **Stouffer's Z score** method: assign a score of Z_i for test i by solving $\Pr(U > Z_i) = p_i$. Let Z_s denote the sum over the transformed p values of k tests, scaled by $k^{-1/2}$ to give it a variance of one, with

$$Z_s = \frac{\sum_{i=1}^{k} Z_i}{\sqrt{k}} \tag{A4.2a}$$

Because $Z_s \sim N(0, 1)$, the overall p value is obtained as

$$p = \Pr(U > Z_s) \tag{A4.2b}$$

As noted by Whitlock (2005), this test was first proposed in a footnote in Stouffer et al.'s sociological study of army life, making it one of the more obscure origins of a statistical method.

Example A4.2. Reconsider the data from Example A4.1. The Z_i values are easily obtained using `R`, as the command `qnorm(1-p)` returns Z satisfying $\Pr(U \leq Z) = 1 - p$, or (equivalently) $\Pr(U > Z) = p$. For example, Z_1 is calculated by `qnorm(1-0.1)`, or 1.281. Similarly computing the other Z_i values yields

$$\sum_{i=1}^{5} Z_i = 6.754, \qquad \text{hence} \qquad Z_s = \frac{6.754}{\sqrt{5}} = 3.020$$

Because $\Pr(U > 3.020) = 0.00126$, as in Example A4.1, the combined p value is highly significant.

Besides providing symmetric values for large and small p values (i.e., p and $1 - p$), a second advantage of the Z-score approach is that one can individually *weight* p values from different tests (Mosteller and Bush 1954; Liptak 1958), as the weighted sum of unit normals is itself a unit normal (while the weighted sum of χ^2 variables—the analog for Fisher's test—is considerably more complex). The resulting weighted version becomes

$$Z_w = \frac{\sum_{i=1}^{k} w_i Z_i}{\sqrt{\sum_{i=1}^{k} w_i^2}} \tag{A4.2c}$$

where $Z_w \sim N(0,1)$. As expected, Z_w (Equation A4.2c) reduces to Z_s (Equation A4.2a) when all the weights are equal. One can either weight by the degrees of freedom or by the reciprocal of the standard error of the estimate. Whitlock (2005) showed that the weighted Z-score method is superior to either X^2 or Z_s when sample size varies over the data. Z_w has higher power and also a higher correlation between its predicted p value and the actual p value obtained if one was able to merge all the samples. As noted by Whitlock, many studies in evolutionary biology examine whether a hypothesis consistently holds over a collection of species. In such cases, the number of species is the number of replicates, and weighting p values for individual species is inappropriate.

As detailed at the end of this Appendix, combining p values is one of the *least powerful* meta-analysis approaches, as it leaves much of the information from a collection of studies underutilized. A formal meta-analysis requires that studies report standard errors for their estimates. Unfortunately, many studies do not, and simply report p values instead, and in this setting the proceeding methods are the only type of meta-analysis available.

BONFERRONI CORRECTIONS AND THEIR EXTENSIONS

We now turn to the complementary problem of determining the significance level, α, for individual tests required to control the overall false-positive rate over a collection of n tests. The typical setting is that a single study or experiment has gathered data and a number of different tests, usually on *different* hypotheses, are performed using these data. Let π denote our desired experiment-wide false-positive rate—the probability of one (or more) false positives *over the entire collection* of n tests being no greater than π. The traditional approach for determining the appropriate α, given n and π, is to use **Bonferroni corrections**.

Standard Bonferroni Corrections

The probability of not making any Type-I errors (false positives) over n independent tests, each at level α, is $(1-\alpha)^n$. Hence, the probability, π, of having at least one false positive over the entire collection is simply one minus this or

$$\pi = 1 - (1-\alpha)^n \tag{A4.3a}$$

If we solve for the α value required for each test,

$$\alpha = 1 - (1-\pi)^{1/n} \tag{A4.3b}$$

This is often called the **Dunn-Ŝidák method.** If we note that $(1-\alpha)^n \simeq 1 - n\alpha$, we obtain the **Bonferroni method** by taking

$$\alpha = \pi/n \tag{A4.4}$$

Both Equations A4.3b and A4.4 are referred to as **Bonferroni corrections**. In the literature, π is the **family-wide error rate (FWER)**, while α is the **comparison-wise error rate (CWER)**, also referred to as the **point-wise significance level (PWSL)**.

Example A4.3. Suppose we have $n = 100$ independent tests and wish to obtain an overall π value of 0.05. What value of α should be used for each individual test to achieve an experiment-wide false-positive rate of 0.05? The Dunn-Ŝidák correction suggests

$$\alpha = 1 - (1 - 0.05)^{1/100} = 0.000512$$

while the Bonferroni correction is

$$\alpha = 0.05/100 = 0.0005$$

Note that using such small α values greatly reduces the power for any single test. For example, under a normal distribution, the 95% (two-sided) confidence interval(CI) for the true mean is $\overline{x} \pm 1.96\sqrt{\text{Var}}$, where Var denotes the variance of the sample mean. Moving to an α value of 0.0005 gives the associated CI as $\overline{x} \pm 3.48\sqrt{\text{Var}}$, as $\Pr(|z| \geq 3.48) = 0.0005$ for $z \sim N(0,1)$.

Sequential Bonferroni Corrections

Under a strict Bonferroni correction, only those tests whose associated p values are $\leq \pi/n$ are rejected (**declared significant**); all others are **accepted** (or more formally, **fail to be rejected**). This results in a considerable reduction in power if two or more of the hypotheses are actually false. When we reject a hypothesis, one fewer test remains, and the multiple comparison correction should reflect this, resulting in **sequential Bonferroni corrections**. Sequential approaches have increased power compared to standard Bonferroni corrections, as illustrated below in Example A4.4. Shaffer (1995) reviewed these and other approaches. The basic structure is that one has a collection of multiple tests, with $H(i)$ denoting the null hypothesis for test i—for example, the test that marker i has a nonzero effect, in which case $H(i)$ is the null hypothesis of no effect. In this case, rejecting $H(i)$ suggests evidence for a nonzero effect for marker i.

Holm's Method

The simplest of these sequential adjustments is **Holm's method** (Holm 1979). The first step is to order the p values for the n hypotheses being tested from smallest to largest, $p(1) \leq p(2) \leq \cdots \leq p(n)$, and let $H(i)$ be the hypothesis associated with $p(i)$. One proceeds with Holm's method as follows:

(i) If $p(1) > \pi/n$, accept all n null hypotheses (i.e., none are declared significant).

(ii) If $p(1) \leq \pi/n$, reject $H(1)$ [i.e., $H(1)$ is declared significant], and consider $H(2)$.

(iii) If $p(2) > \pi/(n-1)$, accept $H(i)$ (for $i \geq 2$).

(iv) If $p(2) \leq \pi/(n-1)$, reject $H(2)$ and move onto $H(3)$.

(v) Proceed with rejecting hypotheses until reaching the first i such that $p(i) > \pi/(n-i+1)$.

We can also apply Holm's method using Equation A4.3a—namely, $\alpha = 1-(1-\pi)^{1/n}$, the Dunn-Ŝidák correction—in place of $\alpha = \pi/n$.

Simes-Hochberg Method

With Holm's method, we stop once we fail to reject a hypothesis. An improvement on this approach is the **Simes-Hochberg correction** (Simes 1986; Hochberg 1988), which effectively starts backward, working with the largest p values first.

(i) If $p(n) \leq \pi$, then all hypothesis are rejected.

(ii) If not, $H(n)$ cannot be rejected, and we next examine $H(n-1)$.

(iii) If $p(n-1) \leq \pi/2$, then all $H(i)$ for $i \leq n-1$ are rejected.

(iv) If not, $H(n-1)$ cannot be rejected, and we compare $p(n-2)$ with $\pi/3$.

(v) In general, if $p(n-i) \leq \pi/(n-i+1)$, then all $H(i)$ for $i \leq n-i$ are rejected.

While the Simes-Hochberg approach is more powerful than that of Holm's (see Example A4.4), it is only strictly applicable when the group of tests being jointly considered are independent, whereas Holm's approach does not have this restriction. Hence, the general strategy is to use Holm's method if one is concerned about potential dependencies between tests, and the Simes-Hochberg's method if the tests are independent.

Hommel's Method

Hommel's method (1988) is slightly more complicated, but it is more powerful than the Simes-Hochberg correction (Hommel 1989). Under Hommel's method, we reject all hypotheses whose p values are less than or equal to π/k^*, where

$$k^* = \max_i \left(p(n-i+j) > \pi\frac{j}{i} \right) \quad \text{for all } j = 1, \cdots, i$$

Example A4.4 shows how all three of these methods are applied.

Example A4.4. Suppose for $n = 10$ tests, the (ordered) p values are as follows:

i	1	2	3	4	5	6	7	8	9	10
$p(i)$	0.0020	0.0045	0.0060	0.0080	0.0085	0.0090	0.0175	0.0250	0.1055	0.5350
$\frac{\pi}{n-i+1}$	0.0050	0.0056	0.0063	0.0071	0.0083	0.0100	0.0125	0.0167	0.0250	0.0500

For an experiment-wide level of significance of $\pi = 0.05$, the Bonferroni correction is $\alpha = 0.05/10 = 0.005$. Hence, using a strict Bonferrioni, we reject hypotheses 1 and 2, and we fail to reject (i.e., we accept) 3 through 10. To apply sequential methods, we use the associated $\pi/(n-i+1)$ values for $\pi = 0.05$, which are given above in the table. Under Holm's method, $p(i) \leq \pi/(n-i+1)$ for $i \leq 3$, and hence we reject *H*(1) through *H*(3) and accept the others. Under Simes-Hochberg, we fail to reject *H*(7) through *H*(10) [as $p(i) > \pi/(n-i+1)$], but because $p(6) = 0.009 \leq \pi/(n-i+1) = 0.010$, we reject *H*(6) through *H*(1).

To apply Hommel's method, we reject all hypotheses whose p values are less than or equal to π/k^*, where

$$k^* = \max_i \left(p(n-i+j) > \pi\frac{j}{i} \right) \quad \text{for all } j = 1, \cdots, i$$

Solving k^* requires an iterative approach, as follows. First, start with $i = 1$. Here, ($i = 1, j = 1$), $p(10) = 0.5350 > \pi \cdot (1/1) = 0.05$. Now let us try $i = 2$, which yields (for $j = 1, 2$), $p(9) = 0.1055 > \pi(1/2) = 0.025$ and (as above) $p(10) > \pi$. Hommel's condition still holds for $i = 3$, as $p(8) = 0.025 > \pi \cdot (1/3) = 0.0167$, $p(9) > \pi \cdot (2/3) = 0.033$, and $p(10) > \pi$. However, it fails for $i = 4$, as while it holds for $p(7) = 0.175 > \pi \cdot (1/4) = 0.0125$, it fails for ($i = 4, j = 2$) because $p(8) = 0.025 = \pi \cdot (1/2)$. Hence, $k^* = 3$ (because Hommel's condition holds for $k = 3$ but not for $k = 4$), and we reject all hypotheses whose p values are

$\leq 0.05/3 = 0.0167$, which means $H(1)$ through $H(6)$. Note that a strict Bonferroni rejected the fewest null hypotheses and Simes-Hochberg and Hommel's rejected the most null hypotheses (i.e., declared them to be significant), and all methods controlled the experiment-wide false-positive rate at 0.05.

Cheverud's Method and Other Approachs for Dealing with Dependence

When different tests share correlated data, it introduces dependency between the p values for these tests. How do we account for this? One approach (Cheverud 2001; Li and Ji 2005; Nyholt 2005) is to use the nature of the dependency structure of the data to estimate an **effective number of independent tests**, n_e. This value is then substituted for n in the above methods; e.g., Equation A4.3b becomes $\alpha = 1 - (1 - \pi)^{1/n_e}$. A classic application of this approach is correcting for correlations among tests of marker-trait associations over a set of linked markers in either a QTL mapping experiment or a GWAS (LW Chapters 15 and 16).

To proceed, we need to introduce a few facts about the eigenstructure of a *correlation matrix*, $\mathbf{C}$, whose eigenvalues are denoted (from largest to smallest) by $\lambda_1, \cdots, \lambda_n$. First, because $\mathbf{C}$ is a positive-semidefinite matrix, all $\lambda_i \geq 0$ (Appendix 5). Second, $\mathbf{C}$ is an $n \times n$ matrix with ones on its diagonal, which makes its trace (the sum of its diagonal elements; Appendix 5) equal to a value of n. The importance of this result is that the trace of a matrix equals the sum of its eigenvalues (Equation A5.8), which demonstrates that the average eigenvalue of $\mathbf{C}$ is

$$n^{-1} \sum_{i=1}^{n} \lambda_i = n^{-1} n = 1$$

When all of the underlying variables that generate $\mathbf{C}$ are uncorrelated, then $\lambda_1 = \cdots = \lambda_n = 1$, while when all of the observations are completely correlated, then $\lambda_1 = n$ and $\lambda_2 = \cdots = \lambda_n = 0$. These two cases represent the extremes of n independent tests (the former) and one independent test (the latter). As with principal components (Appendix 5), the spread of the eigenvalues tells us about dependency. One metric of this is the variance in the eigenvalues, $\sigma^2(\lambda)$. If all of the eigenvalues are equal, then $\sigma^2(\lambda) = 0$, while if one eigenvalue is nonzero, then $\sigma^2(\lambda) = n$.

Motivated by the above eigenstructure observations, **Cheverud's method** (2001) computes the effective number of independent tests as

$$n_{e,Cheverud} = n\left(1 - \frac{(n-1)\sigma^2(\lambda)}{n^2}\right), \qquad \text{where} \qquad \sigma^2(\lambda) = \frac{1}{n-1} \sum_{i=1}^{n} (\lambda_i - 1)^2 \quad \text{(A4.5a)}$$

This returns $n_e = n$ when $\sigma^2(\lambda) = 0$ and $n_e = 1$ when $\sigma^2(\lambda) = n$, which matches the expected results from the eigenvalue analysis for these extreme cases.

Li and Ji (2005) noted that Cheverud's method often returns an overly large value of n_e (and therefore, less power), especially when used with a large number of moderately correlated tests. While Cheverud's approach considered the two extreme cases (n vs. 1 independent test), Li and Ji noted that a set of c identical tests will result in an eigenvalue of $c > 1$, while tests that are only partially correlated with others will generate eigenvalues values < 1. Hence, they partitioned an eigenvalue into two parts, its integer value and the remainder, where the integer part implies identical tests (and hence is counted as contributing one independent test), while the remainder represents partial correlations. Hence, if an eigensequence is $4.1, 3.5, 1, 0.5, 0.1, \cdots$, then the first three eigenvalues correspond to independent tests, and with the total of their non-integer residuals $(0.1 + 0.5 + 0.5 + 0.1 = 1.2)$ adding one additional test, giving (for this part of the sequence) an effective number of four independent tests. Formally, the **Li-Ji method** is coded as

$$n_{e,Li-Ji} = \sum_{i=1}^{N} I(\lambda_i \geq 1) + \sum_{i=1}^{N} (\lambda_i - \text{floor}[\lambda_i]) \qquad \text{(A4.5b)}$$

where the indicator function $I(x \geq 1)$ returns a value of one when $x \geq 1$, and otherwise returns a value of zero. Hence, the first sum in Equation A4.5b is the number of eigenvalues of **C** that are ≥ 1. The floor$[x]$ function in the second term corresponds to the largest integer $\leq x$, so the second sum is all of the remainder terms (the effects of partial correlations among tests). Additional corrections for dealing with correlated tests have been proposed (e.g., Owen 2005; Efron 2007; Leek and Storey 2007, 2008; Li et al. 2012).

DETECTING AN EXCESS NUMBER OF SIGNIFICANT TESTS

While Bonferroni corrections (and their sequential counterparts) are widely used, when the number of tests is modest to large, their application significantly erodes power, leading to very high **Type-II errors** (failing to declare a test significant when the null is false). This tradeoff of Type-I versus Type-II errors applies to most statistical tests (LW Appendix 5), and the error of more concern to the investigator determines how to proceed. In many cases, our initial experiment is simply an enrichment method: we wish to take a large number of possible hypotheses and extract a subset showing the most support for their alternative hypotheses for further consideration. In such cases, we are often more concerned with Type-II errors, as the cost of rejecting a hypotheses that is truly null (i.e., improperly including it as significant; i.e., Type-I error) may be less than the cost of excluding a hypotheses from the alternative (failure to reject; i.e., a Type II-error). In such settings, we would like to calculate the number of hypotheses (n_0; or equivalently, the fraction $\pi_0 = n_0/n$) of true nulls among the n tested hypotheses. The first step toward doing so is to ask if the observed number of significant tests is excessive under the **global null hypotheses** (all hypotheses are truly nulls).

How Many False Positives?

Suppose we perform n independent tests, each with a Type-I error rate of α. If all hypotheses are truly null, the number, j, of false positives follows a binomial distribution, with a "success" probability (a false positive) of α, and n trials (the number of tests), yielding

$$\Pr(j \text{ false positives}) = \frac{n!}{(n-j)!\,j!}(1-\alpha)^{n-j}\,\alpha^j \tag{A4.6}$$

For n large and α small, this is closely approximated by the Poisson distribution, with Poisson parameter $n\alpha$ (the expected number of false positives), yielding

$$\Pr(j \text{ false positives}) \simeq \frac{(n\alpha)^j e^{-n\alpha}}{j!} \tag{A4.7}$$

Example A4.5. Suppose 250 independent tests are performed, each with $\alpha = 0.025$ (a 2.5% chance of declaring a result from the null hypothesis to be significant), and 15 tests are declared significant by this criteria. Is this number greater than expected by chance? The expected number of significant tests under the global null hypothesis is $n\alpha = 250 \cdot 0.025 = 6.25$. From Equation A4.5, the probability of observing 15 (or more) significant tests is

$$\sum_{j=15}^{250} \Pr(j \text{ false positives}) = \sum_{j=15}^{250} \frac{250!}{(250-j)!\,j!}(1-0.025)^{250-j}\,0.025^j$$

We could either sum this series directly or use the cumulative distribution function for a binomial. In `R`, the probability that a binomial with parameters n and p has a value of i or less is obtained by using the command `pbinom(i,n,p)`. The probability of 15 or greater is one minus the probability of 14 or less, or `1- pbinom(14,250,0.025)`, for which `R` returns 0.0018. A similar calculation can use the Poisson approximation (Equation A4.7), with `1-`

`ppois(14,6.25)` returing a value of 0.0021. Given that there is only a 0.2% chance of seeing this many significant tests under the global null, we expect that some of these significant tests are true **discoveries** (those whose associated null hypothesis is incorrect), not false positives. The critical question, of course, is which ones?

Testing for an excessive number of significant tests is a rather crude indicator of the actual number (n_0) of the n tests that are true nulls. It is very possible that $n_0 < n$ and yet we would not detect an excess of significant tests by the above method. Likewise, if an excessive number is detected, what really can we say about n_0 other than $n_0 < n$? For instance, Example A4.5 shows an excess of 9 significant tests (observed 15, expected 6), but clearly assuming $n_0 = n - 9$ is a bit naive. Finally, the outcome varies with our choice of α. One could easily imagine an excess of significant tests using $\alpha = 0.05$, but not when using $\alpha = 0.01$. Ideally, we would like to have an estimate for n_0 that is independent of the choice of α.

Such estimators readily follow from one of the key ideas in this Appendix, namely that ***if the null is correct, random draws of p values follow a uniform distribution over (0,1).*** A more careful examination of the empirical distribution of p values over our sample of tests, rather than simply how many we declare significant, is the key to obtaining estimates of n_0.

Schweder-Spjøtvoll plots

A simple graphical approach using the empirical distribution of p values was suggested by Schweder and Spjøtvoll (1982). If one rank-orders the p values from the smallest $p(1)$ to the largest $p(n)$, a plot of $p(i)$ versus i is a straight line under a uniform. Because our interest is usually in detecting an excessive number of small p values (as would be expected if $n_0 < n$), Schweder and Spjøtvoll suggest plotting $1 - p(i)$ values on the horizontal axis, and the ranks of these values (which are the reverse of the ranks of the $p[i]$) on the vertical axis. For example, the first point is $(1 - p[n], 1)$, the second $(1 - p[n-1], 2), \cdots$, and the nth $(1 - p[1], n)$. If all of the p values are indeed generated from null hypotheses, then these are drawn from a uniform, and the resulting plot will be a straight line (the solid triangles in Figure A4.1). Conversely, if some of the p values are drawn from hypotheses where the null is false, we expect an excess of small p values, and hence an over-abundance of $1 - p$ values near one (the open circles in Figure A4.1).

In addition to providing a quick visual check as to whether the p values follow a uniform, Schweder and Spjøtvoll suggest that these plots can also estimate n_0. One fits the best straight line until the upturn (i.e., inflection point) near one appears, extrapolating this line to obtain the n value for $1 - p = 1$ estimates the number of true null hypotheses, n_0. As shown in Figure A4.1, this gives a value very close to 80 (for the open circles), the correct number of true nulls used to generate this figure.

Estimating n_0: Subsampling From a Uniform Distribution

As suggested by the Schweder-Spjøtvoll plot, the distribution of p values offers insight into the number of truly null hypotheses, n_0. While this plot offers either a simple visual, or a more formal regression-based, estimator of n_0, it tends to overestimate the number of nulls. Further, it can be difficult to specify exactly where the upturn in the plotted values begins. A number of other estimators have been suggested, again based on a uniform distribution of p values for those tests under the null. Recall that the histogram from a sufficiently large number of draws from a uniform distribution is flat, as all values are equally likely (Figure A4.2A). However, if the null is false for at least some of the tests, then the distribution of p values is shifted away from uniform, and usually with a skew toward smaller values (Figure A4.2B), but potentially also skewed toward one (for example, if one-tailed tests are used when a two-tailed test is appropriate; Figure A4.2C).

If the collection of tests contains some alternative hypotheses mixed in with true nulls,

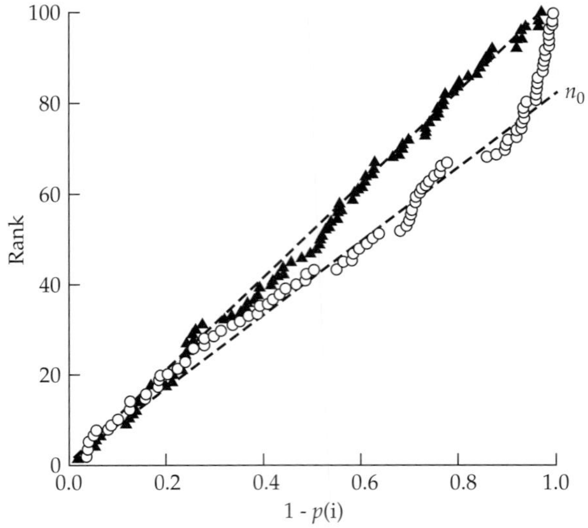

Figure A4.1 A Schweder-Spjøtvoll plot is one approach for detecting departures from a uniform distribution of p values. The p values are ordered from smallest, $p(1)$, to largest, $p(n)$, and one plots the rank of $1 - p(i)$ versus its value. These ranks are reversed from the ranks of $p(i)$, as the rank of $1 - p(n)$, being the smallest value, is 1. Under a uniform, the result is a straight line passing through the origin and the point $(1, n)$. The upper curve (solid triangles), generated by randomly sampling $n = 100$ values from a uniform (0,1), fits this pattern. The lower curve (open circles), generated by simulating p values for 80 true nulls and 20 tests where the alternative was correct, shows an inflation of p values near zero ($1-p$ values near one). This results in a strong departure from linearity near one. Ignoring this upturn and extrapolating the linear fit for the values below this inflection point gives an approximate value of 80 for the value of this projected line when $1 - p = 1$. This is the estimate of n_0.

we expect the distribution to be a mixture, with fraction, $\pi_0 = n_0/n$, consisting of draws from a uniform and the remaining fraction, $(1-n_0/n)$, from some other distribution. Figure A4.3 plots the empirical distribution of p values from a study by Mosig et al. (2001) on marker-trait associations. While the middle of the distribution appears to be consistent with random sampling around a flat average, there is a large excess of values near zero.

One simple approach for estimating n_0 is to use the average height for the middle range of the p-value histogram. Presumably, these p values are almost entirely drawn from null hypotheses, though this may not be the case for values near zero (and potentially one). Recall that the probability density function for a uniform over (0,1) has a very simple form

$$\phi_u(p) = 1 \quad \text{for} \quad 0 \leq p \leq 1 \tag{A4.8a}$$

If there are n_0 truly null tests, then the expected number of p values from these tests that fall within an interval $0 \leq a < b \leq 1$ is simply

$$n_0 \int_a^b \phi_u(p)\, dp = n_0 \int_a^b 1 \cdot dp = n_0(b-a) \tag{A4.8b}$$

Hence

$$\widehat{n}_0(a,b) = \frac{\text{Number of } p(i) \text{ values in } (a,b)}{b-a} \tag{A4.8c}$$

Likewise, an estimate for the fraction $\pi_0 = n_0/n$ of true nulls is

$$\widehat{\pi}_0(a,b) = \frac{\text{Number of } p(i) \text{ values in } (a,b)}{n(b-a)}$$

$$= \frac{\text{Fraction of } p(i) \text{ values in } (a,b)}{b-a} \tag{A4.8d}$$

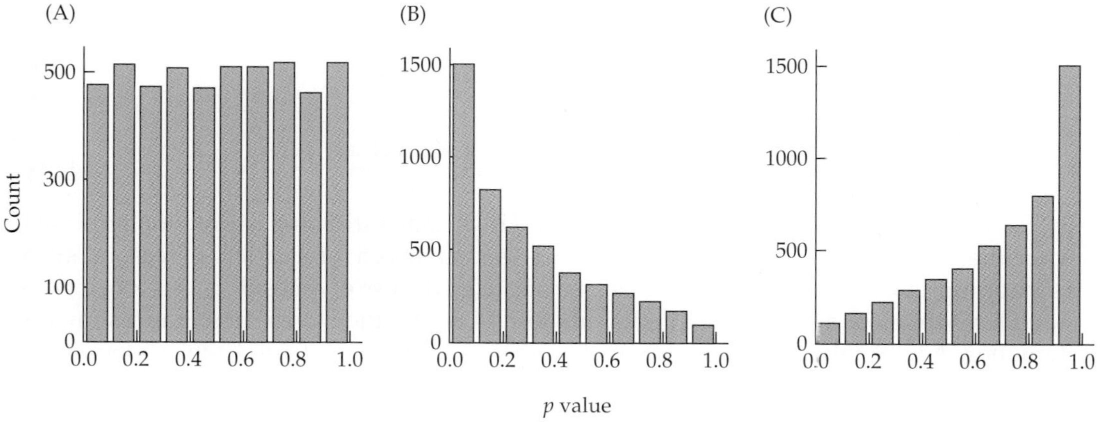

Figure A4.2 Simulated distribution of p values based on 5000 tests for samples of 25 draws from a normal distribution with a mean of μ and a variance of one. The null hypothesis is $H_0 : \mu \leq 0$. **A:** The distribution of p values when $\mu = 0$ (the null is correct) is uniform. **B:** The distribution when $\mu = 0.2$ is skewed toward an excess of values near zero. **C:** The distribution when $\mu = -0.2$ is skewed toward an excess of values near one.

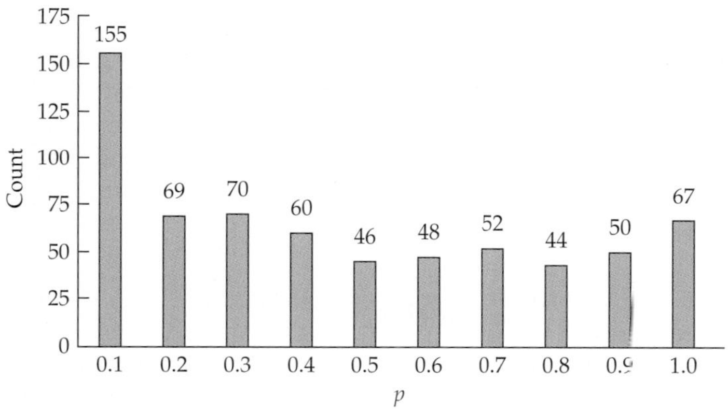

Figure A4.3 An empirical distribution of p values (for n = 644 tests) from Mosig et al. (2001). The number of p values in each of ten bins (of width 0.1) are given above the bars. Note the large excess of values near zero.

Example A4.6. According to the data in Figure A4.3, what is n_0? Consider the bins centered around $p = 0.5$. Based on the central three bins (0.4, 0.5, and 0.6), a total of 60 + 46 + 48 = 154 tests have p values in this interval. From Equation A4.8b, 154 = $n_0 \cdot$ 0.3 or n_0 = 154/0.3 = 513, and hence a fraction, $\pi_0 = n_0/n$ = 513/644 = 0.80, of the tests are true nulls. Using the bins from 0.3 to 0.8 yields n_0 = 322/0.6 = 537, or π_0 = 537/644 = 0 83. Hence, it appears that around 80% of the tests are consistent with true nulls. Mosig et al. (2001; also see Nettleton et al. 2006) used an iterative approach (also based on bin counts in the p-value histogram) and arrived at an estimate of n_0 = 500 (78%).

Storey and Tibshirani (2003) considered the number of p values exceeding some tunable parameter value, λ (taking $a = \lambda$ and $b = 1$ in Equation A4.8b), on the logic that for larger values of λ, most of these draws are from the uniform corresponding to draws from the null. Let $\widehat{\pi}_0(\lambda)$ denote the estimated fraction of truly null hypotheses based on using a tuning value of λ, then

$$\widehat{\pi}_0(\lambda) = \frac{\text{Number of } p(i) \text{ values } > \lambda}{n(1-\lambda)} \tag{A4.9a}$$

and

$$\widehat{n}_0(\lambda) = n \cdot \widehat{\pi}_0(\lambda) = \frac{\text{Number of } p(i) \text{ values } > \lambda}{1-\lambda} \tag{A4.9b}$$

By focusing on the interval $(\lambda, 1)$, the **Storey-Tibshirani estimator** is potentially biased when there are an excess of p values near one. This can happen for a variety of reasons, such as inappropriate assumptions for the test statistic (e.g., the use of one-sided tests when two-sided tests are more appropriate). Both Equation A4.8c and the Storey-Tibshirani estimator (Equation A4.9b) rely on tuning parameters (a, b, and λ, respectively) which define the region of the distribution of p values assumed to be drawn from a uniform (i.e., almost all p values in this interval are assumed to be generated under the null). Nettleton et al. (2006) reviewed these and other approaches for estimating n_0 from sampling parts of a presumed uniform and elaborated on their strengths and weaknesses.

One significant concern is that correlated tests can result in either an under- or over-dispersion of p values under the global null hypothesis, resulting in significant departure from a uniform distribution (Efron 2007; Hu et al. 2011; Leek and Storey 2011). This in turn compromises estimates of n_0.

Estimating n_0: Mixture Models

Allison et al. (2002) suggested that π_0 can be estimated by treating the distribution of p values as a mixture, a fraction π_0 of which comes from a uniform (and hence a uniform distribution function, ϕ_u), while the remainder $(1 - \pi_0)$ are from the distribution, $\phi_A(p)$, of p values when the alternative hypothesis is true (Figure A4.4). While the general form of $\phi_A(p)$ is unknown, a very flexible distribution to model it is by using the beta distribution (Appendix 2; Figure A2.3)

$$\phi_A(p) = \frac{\Gamma(a+b)}{\Gamma(a)\,\Gamma(b)}\, p^{a-1}(1-p)^{b-1} \tag{A4.10a}$$

Under the alternative hypothesis, we expect an increase in p values near zero, which occurs when $a < 1$. Likewise, the beta distribution can easily accommodate an increase in p values near one ($b < 1$). When $a = b = 1$, this simply reduces to a uniform.

Allison et al. (2002) suggested fitting the actual shape by using the data to obtain ML estimates of a and b, as well as our desired parameter, the fraction of true nulls, π_0. The resulting likelihood function for a single p value becomes

$$\ell(p) = (1-\pi_0)\,\phi_A(p) + \pi_0\,\phi_u(p) = (1-\pi_0)\frac{\Gamma(a+b)}{\Gamma(a)\,\Gamma(b)}\, p^{a-1}(1-p)^{b-1} + \pi_0 \tag{A4.10b}$$

with the resulting total likelihood over the n sampled p values (from independent tests) calculated by

$$\ell(\mathbf{p}) = \prod_{i=1}^{n} \ell(p_i) \tag{A4.10c}$$

Standard ML methods (LW Appendix 2) are used to solve for a, b, and π_0.

While hypothesis testing under a maximum likelihood framework is typically performed using the likelihood ratio (LR) test (LW Appendix 2), this is not appropriate for tests of the number of components in a mixture, as the LR does not approach a limiting χ^2 distribution, because π_0 is being tested against a boundary value, in this case a value of one (McLachlan 1987). While a modified LR test for mixtures can be constructed that behaves better (Chen et al. 2001), Allison et al. used a bootstrap approach (McLachlan 1987; Schork 1992). Here, one first uses the original distribution of p values to compute an LR test statistic for the null of a uniform versus the alternative of a mixture. One then generates **parametric**

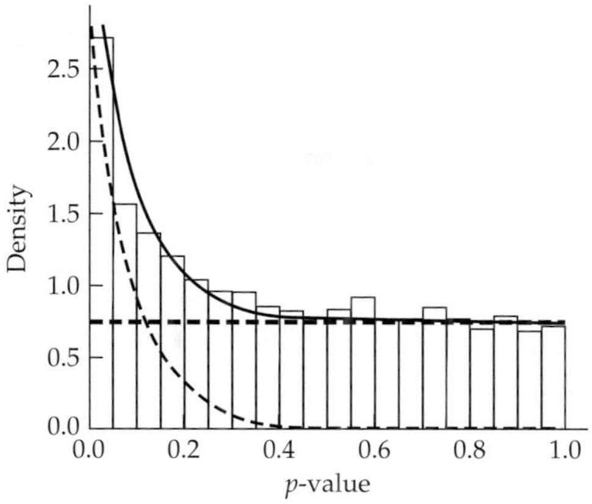

Figure A4.4 The empirical distribution of p values can be treated as a mixture model of a uniform plus a beta distribution (whose shape parameters, a and b, can be estimated via ML), see Equation A4.10b. In this hypothetical example, a weighted mixture of a uniform (horizontal dashed line) and a beta with ($a < 1, b = 1$; dashed curve), yields the mixture distribution (solid curve) that fits the empirical distribution of the p values.

bootstrap samples by drawing n values of p from the null distribution (here a uniform) and then using this simulated dataset to compute an LR test statistic for a mixture. This is done several thousand times to generate an approximate distribution of the LR statistic under the null, which is used to assess significance. For example, if only 0.25% of the bootstrap LR values are equal to (or exceed) the LR value for the original data, the significance is approximately 0.25%. Likewise, approximate standard errors for π_0 can be generated using a **conventional bootstrap** approach. One samples the *original* p values with replacement to generate a **bootstrap sample** of size n. This is then used to estimate π_0 (and the other parameters) under a standard ML framework. Several thousand bootstrap samples are generated, and the variation across estimates of π_0 (or any other parameter) over these samples provides an approximate estimate of the sampling variance for $\widehat{\pi}_0$.

Finally, while a beta (or weighted sum of betas) can be used as the functional form for ϕ_A, another approach is to use a nonparametric estimator for this unknown density function. This can be done using a **kernel density estimator**, whereby the form of an unknown density is estimated by using the observed number of counts within a series of bins spanning the distribution in conjunction with an appropriate smoothing function. This approach was used by Robin et al. (2007) and Guedj et al. (2009).

FDR: THE FALSE-DISCOVERY RATE

As mentioned, Bonferroni corrections (and their extensions) are appropriate when we expect that only a very few of the many null hypotheses are false. An alternate setting is that in which some substantial fraction of the null hypotheses is *expected* to be false. In such cases, even sequential Bonferroni corrections are likely to be too stringent, resulting in too many false negatives (Type-II errors; i.e., a failure to reject a false hypothesis). A different approach is required in these settings, most notably the **false-discovery rate** (**FDR**), introduced by Benjamini and Hochberg (1995).

The FDR is the fraction of false positives among all the tests that are declared to be significant. The motivation for using the FDR is that we may be conducting a very large number of tests, with those that are declared to be significant being subjected to further study. An example would be a search for differential expression over a huge set of genes. The goal of the initial analysis is to distill a large number of candidates down to a reduced set (for further analysis) that is highly enriched for true positives.

In such cases, we are more concerned with making sure all possible true alternatives are included in this reduced set, and we are willing to accept some false positives to accomplish this goal. However, we also don't want to be completely swamped with false positives. The goal is a statistical procedure that results in a significant *enrichment* of true positives (differentially expressed genes in our example), while controlling the fraction of false positives within this enriched set by specifying a value, δ, for the FDR. Choosing an FDR of 5% means that (on average) 5% of the genes that we *declare to be significant* are actually false positives. The flip side is that 95% of those genes (tests) that are declared to be significant do indeed have differential expression. Hence, screening genes with an FDR of 5% results in a significant enrichment of genes that are truly differentially expressed.

To formally motivate the concept of the FDR, suppose a total of n hypotheses are tested, S of which are judged significant (i.e., the p value for the test is $\leq$ some threshold value, τ). If we had complete knowledge, we would know that n_0 of the hypotheses have the null true and $n_1 = n - n_0$ have the alternative true, and we might find that F of the true nulls were called significant, while T of the alternative trues were called significant, yielding the following table

	Called significant	Called not significant	Total
Null true	F	$n_0 - F$	n_0
Alternative true	T	$n_1 - T$	n_1
Total	S	$n - S$	n

For this experiment, the false-discovery rate is the fraction of tests called significant that are actually true nulls, FDR = F/S. (The term **discovery** follows in that a significant result can be considered as a discovery for future work.) As a point of contrast, the normal Type-I error (which we can also call the **false-positive rate [FPR]**), which is the fraction of true nulls that are called significant, is F/n_0. Note the critical distinction between these two error rates. While the numerator of each is F, the denominators are considerably different—the total number, S, of tests called significant (for FDR), versus the number, n_0, of hypotheses that are truly null (FPR). As the threshold value (τ) for significance is changed, so too is the fraction F/S. To obtain a FDR of δ over our experiment, τ is adjusted to find its largest value such that some expectation of F/S is bounded above by δ. Finally, Gadbury et al. (2004) defined the **expected discovery rate** (**EDR**) as T/n_1 (the fraction of all true discoveries declared to be significant), which is the analog of statistical power in this setting.

Another way to see the distinction between the false-positive rate and the false-discovery rate is to consider them as probability statements for a single test involving hypothesis i. For the FDR, we condition on the test as being significant,

$$\text{FDR} = \Pr(i \text{ is truly null} \mid i \text{ is deemed significant}) = \delta \tag{A4.11a}$$

whereas for the false-positive rate, we condition on the hypothesis being null

$$\text{FPR} = \Pr(i \text{ is deemed significant} \mid i \text{ is truly null}) = \alpha \tag{A4.11b}$$

Morton's Posterior Error Rate (PER) and the FDR

Table A4.1 reminds the reader of the various test parameters that arise when multiple comparisons are considered. We now show how these are related. First, the relationship between α, π, and F is as follows. Suppose we have set the false-positive rate (i.e., the Type-I error rate) for an individual test at α. Such a p-value threshold only guarantees that the expected number of false positives is bounded above by $E[F] \leq \alpha \cdot n$. For n independent tests, a π-level experiment-wide false-positive error (setting $\alpha = \pi/n$; namely, the Bonferroni correction) implies that $\Pr(F \geq 1) \leq \pi$, i.e., the probability of at least one false positive is no greater than π. To show how α, β, π_0, and δ are related, we first need to introduce the concept of the posterior error rate.

Fernando et al. (2004) and Manly et al. (2004) both noted that FDR measures are closely

Table A4.1 Summary of the multiple comparisons parameters used in this Appendix. F denotes the number of false positives, namely, tests under the null that are declared significant.

Parameter	Definition
α	Comparison-wide Type-I error (false positive).
β	Type-II error (false negative); $1 - \beta$ = power.
π	Family-wide Type-I error; $\Pr(F > 0) = \pi$.
δ	False-discovery rate.
π_0	Fraction of all hypotheses that are truly null.
p	Probability of the test statistic under the null.
$p(k)$	kth smallest p value of the n tests.

related to Morton's (1955) **posterior error rate (PER)**, originally introduced in the context of linkage analysis in humans (this is also referred to as the **false positive report probability [FPRP]**; Wacholder et al. 2004). Morton's PER is simply the probability that a single significant test is a false positive,

$$\text{PER} = \Pr(F = 1 \,|\, S = n = 1) \tag{A4.12}$$

The connection between the FDR and the PER is that if we set the FDR to δ, then the PER for a randomly drawn significant test is also δ.

Framing tests in terms of the PER highlights the **screening paradox**: "Type-I error control may not lead to a suitably low PER" (Manly et al. 2004). For example, we might choose α = 0.05, but the PER may be far higher, which means that a test that is *declared to be significant* may have a much larger probability than 5% of being a false positive. The key is that because we are *conditioning on the test being significant* (as opposed to conditioning on *the hypothesis being a null*, as occurs with α), S may include either false positives or true positives. The relative fractions of each (and hence the probability of a false positive) is a function of the single test parameters, α and β, and the fraction, π_0, of hypotheses that are truly null. To see this, we apply Bayes' theorem (Equation A2.2a), which yields

$$\text{PER} = \Pr(F = 1 \,|\, S = n = 1) = \frac{\Pr(\text{false positive} \,|\, \text{null true}) \cdot \Pr(\text{null})}{\Pr(S = n = 1)} \tag{A4.13}$$

Consider the numerator of Equation A4.13 first. Let $\pi_0 = n_0/n$ be the fraction of all hypotheses that are truly null. The probability that a null is declared significant is simply the Type-I error, α, hence

$$\Pr(\text{false positive} \,|\, \text{null true}) \cdot \Pr(\text{null}) = \alpha \cdot \pi_0 \tag{A4.14a}$$

Turning to the denominator of Equation A4.13, what is the probability that a single (randomly chosen) test will be declared significant, $\Pr(S = n = 1)$? This event can occur because we choose to test a hypothesis that is truly null (π_0) and have a Type-I error (α), or because we choose to test a hypothesis that is truly false ($1 - \pi_0$) and avoid a Type-II error. For the latter, the power is simply T/n_1, the fraction of all tests under the alternative that is declared to be significant. If we write the power as $1 - \beta$ (β is the Type-II error), the resulting probability that a single (randomly drawn) test is significant is

$$\Pr(S = n = 1) = \alpha\pi_0 + (1 - \beta)(1 - \pi_0) \tag{A4.14b}$$

Substituting Equations A4.14a and A4.14b into Equation A4.13 yields

$$\text{PER} = \frac{\alpha \cdot \pi_0}{\alpha \cdot \pi_0 + (1 - \beta) \cdot (1 - \pi_0)} = \left(1 + \frac{(1 - \beta) \cdot (1 - \pi_0)}{\alpha \cdot \pi_0}\right)^{-1} \tag{A4.15a}$$

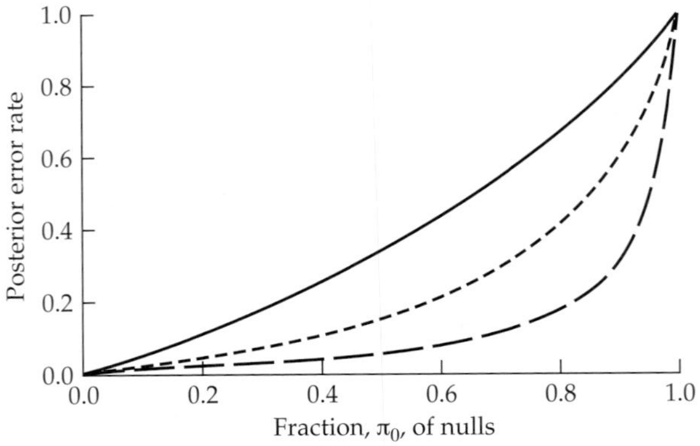

Figure A4.5 Plot of the posterior error rate (Equation A4.15a) for $\alpha = 0.05$, as a function of π_0 (the fraction of cases where the null hypothesis holds) and β (the Type-II error, which is one minus the power). The solid curve corresponds to $\beta = 0.9$ (10% power), the short-dashed curve corresponds to $\beta = 0.7$ (30% power), and the long-dashed (lower) curve corresponds to $\beta = 0$ (100% power).

Figure A4.5 plots Equation A4.15a for various values of π_0 and β.

Sham and Purcell (2014) noted that one can rearrange Equation A4.15a to find the α value to obtain a desired PER value of γ, with

$$\alpha = \left(\frac{\gamma}{1-\gamma}\right)\left(\frac{1-\pi_0}{\pi_0}\right)(1-\beta) \tag{A4.15b}$$

In particular, if there is complete power ($\beta = 0$) and only one of the n tested hypotheses departs from the null ($\pi_0 = n/[n+1]$), Equation A4.15c reduces to

$$\alpha = \left(\frac{\gamma}{1-\gamma}\right)\left(\frac{1}{n}\right) \simeq \frac{\gamma}{n} \tag{A4.15c}$$

which recovers the Bonferroni correction (Equation A4.4).

The Type-I error rate, α, of a *random* test, and the PER for a *significant* test, which are often assumed to be the same, are actually very different. In addition to α, the PER also depends on the power, β, of a test and the fraction, π_0, of tests that are truly null (as these latter parameters influence the probability that a test is declared to be significant). Manly et al. (2004) noted that the PER is acceptably low only if the fraction of alternative hypotheses $(1-\pi_0)$ is well above α.

Thinking in terms of the PER allows us to consider multiple comparisons in a continuum from Bonferroni-type corrections to using FDR to control the PER. If $\pi_1 = 1 - \pi_0$ is very small, most tests will truly be from the null hypothesis, and we can control the overall false-positive rate with a Bonferroni-type correction. However, if some fraction of the hypotheses is expected to be false (i.e., $1-\pi_0$ is at least of modest value), then using FDR corrections makes more sense for controlling the PER.

Example A4.7. In Morton's original application, because there are 23 pairs of human chromosomes, he argued that two randomly chosen genes had a $1/23 \simeq 0.05$ *prior probability of linkage*, namely, $1-\pi_0 = 0.05$, and thus $\pi_0 = 0.95$. Assuming a Type-I error rate of $\alpha = 0.05$ and 80% power to detect linkage ($\beta = 0.20$), applying Equation A4.15a yields a PER of

$$\frac{0.05 \cdot 0.95}{0.05 \cdot 0.95 + 0.80 \cdot 0.05} = 0.54$$

Hence, with a Type-I error control of $\alpha = 0.05$, a random test showing a significant result ($p \leq 0.05$) has a 54% chance of being a false positive. This occurs because most of the hypotheses are expected to be null—for example, if we draw 1000 random pairs of loci, 950 are expected to be unlinked and we expect $950 \cdot 0.05 = 47.5$ of these to show a false positive. Conversely, only 50 are expected to be linked, and we would declare $50 \cdot 0.80 = 40$ of these to be significant, so that $47.5/87.5 = 0.54$ of the significant results are due to false positives.

What value for α is needed under the above parameters to given a PER of 0.05? Solving for α in the expression

$$\frac{\alpha \cdot 0.95}{\alpha \cdot 0.95 + 0.80 \cdot 0.05} = 0.05$$

yields $\alpha = 0.0022$, and hence, setting this as the Type-I error gives a PER of 5%.

Example A4.8. Suppose we set $\alpha = 0.005$ for each test, and assume that the resulting power is essentially 1 (i.e., $\beta \simeq 0$). Consider 5000 tests under two different settings. First, suppose that the alternative is very rare, with $n_1 = 1$ ($\pi_0 = 0.9998$). Under this setting, we expect $4999 \cdot 0.005 = 24.995$ false positives and one true positive ($1 \cdot [1 - \beta] = 1$), yielding an expected PER of

$$\text{PER} = \frac{24.995}{24.995 + 1} = 0.961$$

Thus, a randomly chosen significant test has a 96.1% probability of being a false positive.

Now suppose that the alternative is not especially rare, for example $n_1 = 500$ ($\pi_0 = 0.9$). The expected number of false positives is $4500 \cdot 0.005 = 22.5$, while the expected number of true positives is 500, yielding a PER of

$$\text{PER} = \frac{22.5}{522.5} = 0.043$$

The PER is thus rather sensitive to π_0, the fraction of all tests that are truly from the null hypothesis. If π_0 is essentially 1, a PER of δ is obtained using the Bonferroni correction, $\alpha = \delta/n$. However, if π_0 departs even slightly from one (i.e., more than a few of the alternative hypotheses are correct), then the per-test level of α to achieve a desired PER rate is considerably larger (i.e., less stringent) than that given by the Bonferroni correction, namely, $\alpha(\delta) > \delta/n$. For example, for a 0.04 experiment-wide error rate, $\alpha = 0.04/5000 = 8 \cdot 10^{-6}$, which is roughly 625 times smaller than the value of $\alpha = 0.005$ required for a 4% FDR, highlighting the greatly increased power under the FDR framework. This increased power arises because the FDR approach acknowledges that some fraction of the tests are not from the null.

A Technical Aside: Different Definitions of False-discovery Rate

While the false-discovery rate for any experiment is simply F/S, there are several subtly different ways to formally define the expectation of this ratio. The original notion of a false-discovery rate is due to Benjamini and Hochberg (1995), with modifications suggested by a number of other workers, most notable Storey (2002) and Fernando et al. (2004); see Table A4.2.

While the technical distinction between these different false-discovery rates is important, when actually estimating a false-discovery rate from a collection of p values, one is usually left with an expression of the form $E(F)/E(S)$, which consists of the expected number of false positives divided by the expected number of significant tests. Strictly speaking, this is the **proportion of false positives (PFP)**.

The main distinction between the different false-discovery rates are: (i) the original method of Benjamini and Hochberg (1995), which assumes $n = n_0$ (all hypotheses are nulls); and (ii) all other estimators, which assume n_0 is not necessarily n, and thus attempt to estimate either π_0 or n_0, and then use either to estimate the false-discovery rate.

Table A4.2 Measures of false discovery. Here F is the number of false discoveries, S is the number of tests declared significant, and n is the total number of tests. (After Manly et al. 2004.)

	Name	Definition	Reference
FDR	False-discovery rate	$E(\frac{F}{S} \mid S > 0)\, \Pr(S > 0)$	Benjamini and Hochberg (1995)
pFDR	Positive false-discovery rate	$E(\frac{F}{S} \mid S > 0)$	Storey (2002)
PFP	Proportion of false positives	$E(F)/E(S)$	Fernando et al. (2004)
PER	Posterior error rate	$\Pr(F = 1 \mid S = n = 1)$	Morton (1955)
FPR	False-positive rate	$\Pr(F > 0)$	

The Benjamini-Hochberg FDR Estimator

The original estimator for the FDR was introduced by Benjamini and Hochberg (1995). Suppose we declare a test to be significant if its p value is at or below some threshold value, $\tau = p(k)$, in which case k of the hypotheses will be declared significant (as $p[k]$ is the kth smallest p value), and $S = k$. Likewise, if all n of the hypotheses are null, then the expected value of F (the number of false positives) is just $n\,p(k)$. The resulting fraction of all rejected hypotheses that are false discoveries becomes $F/S = n\,p(k)/k$. Hence, the false-discovery rate, δ_k, for hypothesis k is bounded by

$$\frac{np(k)}{k} \leq \delta_k \tag{A4.16a}$$

In particular, if we wish to obtain an FDR of δ for the entire experiment, then we reject (i.e., declare as significant) all hypotheses that satisfy

$$p(k) \leq \delta\,\frac{k}{n} \tag{A4.16b}$$

This simple (heuristic) derivation shows why the original Benjamini-Hochberg estimate of the FDR is conservative, as in those settings in which one applies the FDR criteria, the expectation is that some fraction of the hypotheses are not null, and so $n_0 < n$. The correct estimator of the expected number of rejected null hypotheses is $n_0 p(k)$, which leads to a more generalized estimate of the FDR, where $\widehat{n}_0$ (e.g., Equations A4.8–A4.10) replaces n. In this case, Equation A4.16a becomes

$$\widehat{\delta}_k = \frac{\widehat{n}_0\, p(k)}{k} \tag{A4.17}$$

Example A4.9. Consider again the 10 ordered p values from Example A4.4. Computing $n\,p(k)/k = 10\,p(k)/k$, where k denotes the test with the *k-th* smallest p value, yields the following table:

k	1	2	3	4	5	6	7	8	9	10
$p(k)$	0.0020	0.0045	0.0060	0.0080	0.0085	0.0090	0.0175	0.0250	0.1055	0.5350
$n\frac{p(k)}{k}$	0.0200	0.0225	0.0200	0.0200	0.0170	0.0150	0.0250	0.0313	0.1172	0.5350

Thus, if we wish an overall FDR value of $\delta = 0.05$, we would reject hypotheses when $n\,p(k)/k \leq \delta = 0.05$, which is satisfied by H(1) through H(8). Notice that this procedure rejects more hypotheses (i.e., returns more discoveries) than any of the sequential Bonferroni methods (Example A4.4).

A (Slightly More) Formal Derivation of the Estimated FDR

Following Storey and Tibshirani (2003), we consider the expected FDR for an experiment where we declare a hypothesis to be significant if its p value is less than or equal to some threshold value, τ. Obviously, as τ becomes smaller, the FDR is smaller (as significant nulls become increasingly less likely). However, if τ is set too small, we lose power (e.g., suppose we set $\tau = \pi/n$; namely, the Bonferroni correction). What we would like to do is to find the expected value of the FDR as a function of the chosen threshold parameter, τ, to allow us to optimally tune this parameter to obtain the desired FDR. If we have a large number of tested hypotheses,

$$E[FDR(\tau)] = E\left[\frac{F(\tau)}{S(\tau)}\right] \simeq \frac{E[F(\tau)]}{E[S(\tau)]} \tag{A4.18}$$

A simple estimate of $E[S(\tau)]$ is given by the observed number of significant tests when the threshold is τ.

To obtain an estimate for $E[F(\tau)]$, we again use the fact that the distribution of p values under the null follows a uniform $(0,1)$ distribution. Hence,

$$\Pr(p \leq \tau \,|\, \text{null hypothesis}) = \int_0^\tau \phi_u(p)dp = \tau \tag{A4.19}$$

where $\phi_u(p)$ is the uniform probability density function for p values under the null (Equation A4.8a). Hence, if n_0 of the n tests are truly null, then

$$E[F(\tau)] = n_0 \cdot \Pr(p \leq \tau \,|\, \text{null hypothesis}) \simeq n_0 \cdot \tau \tag{A4.20}$$

Hence,

$$E[FDR(\tau)] = \frac{n_0 \cdot \tau}{S(\tau)} \tag{A4.21}$$

Setting $\tau = p(k)$, then $S(\tau) = k$, and Equation A4.21 becomes $n_0\, p(k)/k$, recovering Equation A4.17. Using the Storey-Tibshirani estimator for n_0 (Equation A4.9b), an estimated value for the FDR using threshold value, τ (and based on the tuning parameter, λ, in the Storey-Tibshirani estimator), becomes

$$\widehat{FDR}(\tau) = n_0 \cdot \frac{\tau}{S(\tau)} = \left(\frac{N[\,p(i) \text{ values } > \lambda\,]}{1-\lambda}\right) \cdot \left(\frac{\tau}{N[\,p(i) \text{ values } \leq \tau\,]}\right) \tag{A4.22}$$

where $N[x]$ is the number of occurrences of event x. Ideally, over a reasonable range of λ values, we expect this estimate be stable. If λ is set too large, the likelihood that almost all values correspond to draws from a null will be countered by the much smaller sample size (and hence a larger sampling error) from using such a small fraction of the total data.

Under a mixture-model setting (e.g., Equation A4.10), the false-discovery rate for a given a significance threshold (τ) is simply the fraction of all true positives that are declared significant divided by the fraction of all tests that are declared significant (i.e., those tests for which $p \leq \tau$). This can be estimated directly from the parameters of the mixture distribution,

$$\text{FDR}(\tau) = \frac{\pi_0 \,\text{cdf}_U(\tau)}{\pi_0 \,\text{cdf}_U(\tau) + (1-\pi_0)\,\text{cdf}_A(\tau)} = \frac{\pi_0\, \tau}{\pi_0 \tau + (1-\pi_0)\,\text{cdf}_A(\tau)}, \tag{A4.23}$$

where cdf denotes the cumulative distribution function, with

$$\text{cdf}_U(x) = \int_0^x \phi_U(p)dp = x \qquad \text{cdf}_A(x) = \int_0^x \phi_A(p)dp$$

where ϕ_U is the uniform distribution under the null and ϕ_A is the distribution of p values under the alternative (for example, a fitted beta).

Storey's q Value

While we can control the FDR for an entire set of experiments, we would also like to have an indication of the FDR for any particular experiment (or test) within this family of tests. Intuitively, tests with smaller p values should also have smaller associated FDR values. To address this, Storey (2002; Storey and Tibshirani 2003) introduced the concept of a ***q* value** (as opposed to the p value) for a particular test, where q is the expected FDR rate for tests *within the current experiment* whose p values are at least as extreme as the test of interest. The estimated q value is a function of the p value for that test and the distribution of the entire set of p values from the family of tests being considered, namely,

$$\widehat{q}\,[p(i)] = \min_{\tau \geq p(i)} \widehat{FDR}(\tau) \tag{A4.24}$$

In words, the q value of a test is calculated by using the smallest FDR value over all significance threshold values (τ) such that this threshold is equal to, or greater than, the p value, $p(i)$ of the test.

To see why we used $\min_{\tau \geq p(i)}$ instead of simply setting $q_i = \widehat{FDR}[p(i)]$, recall Example A4.9. This example used the Benjamini-Hochberg estimator for FDR value (which differs from other FDR estimators by a constant, n_0/n). Notice that the smallest FDR occurs for hypothesis 6 (1.5%), and not for hypotheses with smaller p values. This reflects the tradeoff whereby increasing the threshold, τ, for significance results in the declaration of more tests as discoveries, so the ratio $\tau/S(\tau)$ need not monotonically increase as τ increases. As Example A4.9 shows, setting the threshold τ *above* the $p(i)$ value may actually result in a smaller q value, and hence Storey's definition.

Example A4.10. As an example of the interplay between the family-wide error rate (π), and the individual p and q values for a particular test, consider Storey and Tibshirani's (2003) analysis of a microarray dataset comparing *BRCA1* and *BRCA2* positive breast cancer tumors. A total of 3226 genes were examined. Setting a critical p value of $\alpha = 0.001$ detects 51 significant genes (i.e., those with differential expression between the two types of tumors). If we assume that the hypotheses being tested are independent (which is unlikely as expression can be highly correlated across sets of genes), the probability that there is at least one false positive is $\pi = 1 - (1 - .0001)^{3226} = 0.96$, while the expected number of false positives is 0.001 · 3226 = 3.2, or 6% (3.2/51) of the declared significant differences. After setting an FDR rate of $\delta = 0.05$, Storey and Tibshirani detected 160 genes that showed significant differences in expression. Of these 160, 8 (5%) are expected to be false positives. Compared to the Bonferroni correction (51 genes, 6% false positives), over three times as many genes were detected, and with a lower FDR rate. Further, Storey and Tibshirani estimated the fraction, π_0, of nulls (genes with no difference in expression) at 67%, which suggests that 33% (or roughly 1000 of the 3226 genes) are likely to be differentially expressed between the two tumor types.

This dramatic difference in performance between Bonferroni and FDR control arises because the former enforces very strict control over *any* false positives, resulting in a much smaller set of discoveries. Conversely, FDR is more concerned with the *fraction* of discoveries that are false, and by including more true discoveries, that fraction can be made smaller than under Bonferroni. Because the fraction of null hypotheses that are false in this study is rather substantial, a lower significance threshold includes more of these true discoveries, thus decreasing the FDR.

To contrast the distinction between the p and q values, consider the *MSH2* gene, which has a q value of 0.013 and a p value of $5.50 \cdot 10^{-5}$. This p value implies that the probability of seeing at least this level of difference in expression for a randomly drawn gene from the null hypothesis (no difference in expression) is $5.50 \cdot 10^{-5}$. Conversely, $q = 0.013$ indicates that, *for this experiment*, 1.3% of genes that show differences in expression that are as, or more, extreme (i.e., whose p values are at least as small) as that for *MSH2* are expected to be false positives.

Closing Caveats in Using the FDR

While controlling the FDR is a very powerful approach for many multiple-comparison problems, it is not a panacea. One concern is correlations among tests. As mentioned, in this case the null distribution of p values can significantly depart from a uniform, giving biased estimates of π_0 (and thus FDR). Further, recall that FDR control is accomplished by controlling the *expected* value of the FDR (or some closely related measure, such as the PFP). The *variance* in the FDR across independent experiments can be considerable, especially when the tests are correlated (Owen 2005; Leek and Storey 2011). One approach for treating these concerns is to use Leek and Storey's (2007, 2008) surrogate variable analysis to account for dependencies among the data before the actual p values for individual tests are obtained.

A second issue is a bit more subtle. Consider a standard QTL mapping experiment (LW Chapter 15) wherein a controlled cross is made between two lines (which are typically inbred) and one looks for marker-trait associations in the resulting F_2 (or other) progeny by scanning for linkage signals across a number of linked markers that span each chromosome. For each marker, the null hypothesis is that there is no linkage to a QTL influencing the trait, while the alternative is that the marker is linked to a QTL. As noted by Chen and Storey (2006), the linkage signal from a QTL influences essentially all the markers on the chromosome arm on which it resides, and so *as a group* they all satisfy the same hypothesis. Either all are nulls (unlinked to a QTL) or all are failures of the null (linked to a QTL, albeit with differing degrees of a linkage signal). As such, investigators can arbitrarily obtain any FDR level they desire by simply adding or subtracting linked markers, and FDR control is not appropriate for this setting (Chen and Storey 2006). To a much lesser extent, the same issue occurs in genome-wide association studies among sets of extremely tightly linked SNPs. However, because the linkage signal in these cases is the persistence of linkage disequilibrium (LD) over a large number of generations, any common signal is restricted to a set of very tightly linked markers rather than an entire chromosome, and control of the FDR *among* such clusters is appropriate.

FORMAL META-ANALYSIS

Another class of analysis involving multiple comparisons considers comparison *across* studies, rather than trying to adjust for multiple comparisons *within* a single study. Such an analysis of analyses, coined **meta-analysis** (**MA**) by Glass (1976), is a tool of expanding importance in quantitative genetics. While multiple-comparison corrections involve isolating the significance of *separate* variables from a single large study containing many factors, a meta-analysis *combines* all information from a set of studies in order to increase the power and insight over any single study. While the roots of MA trace back to Fisher (1932b) and Cochran (1937), much of the field was developed in the social sciences. General overviews can be found in Hunter and Schmidt (2005), Borenstein et al. (2009a), and Cooper et al. (2009), and reviews with a specific focus on issues that can arise in evolution (and ecology) can be found in Harrison (2011), Nakagawa and Santos (2012), and Koricheva et al. (2013).

Informal, or Narrative, Meta-analysis

Table A4.3 shows the canonical structure of the data for a meta-analysis: one has a number of studies, either published or unpublished, dealing with a specific question (such as the average strength of natural selection; Chapter 30). Study i reports an estimate, T_i, of an effect whose true (and unknown) value is denoted by θ_i. Unfortunately, however, many studies report only T_i and perhaps, p_i, although the latter is often simply reported in binary form (whether they are significant at some level or not) rather than as an actual value.

In settings where studies simply report T_i (and perhaps p_i), only an **informal**, or **narrative**, MA can be performed. Here, one simply presents the statistics on the collection of T_i values, such as a histogram of effects or some metric on their overall distribution (such as their mean, median, or variance). Often these summary statistics are compared over different

Table A4.3 The potential data available for a meta-analysis of k studies. One has reported values for the estimated effect, T_i, for each study. A study may also have m reported **moderator variables** (cofactors such as sex, laboratory, species, etc.). In the simplest setting, the study only reports a p value for whether the effect is significant. In such cases, an overall p-value can be obtained from the methods discussed earlier in this Appendix. If the study also reported the standard errors (SE) of the estimates, or (under the assumption of a constant error per observation over all studies) the sample size, then a formal (i.e., model-based) meta-analysis may be performed (as detailed in the text).

	Effect					
Study	Actual	Estimate	Sample size	SE	p value	Moderator variables
1	θ_1	T_1	n_1	s_1	p_1	$M_{11}, M_{12}, \cdots, M_{1m}$
2	θ_2	T_2	n_2	s_2	p_2	$M_{21}, M_{22}, \cdots, M_{2m}$
3	θ_3	T_3	n_3	s_3	p_3	$M_{31}, M_{32}, \cdots, M_{3m}$
$\vdots$	$\vdots$	$\vdots$	$\vdots$	$\vdots$	$\vdots$	$\vdots$
k	θ_k	T_k	n_k	s_k	p_k	$M_{k1}, M_{k2}, \cdots, M_{km}$

values of any **moderator variables** (cofactors) associated with the data. For example, a comparison of whether the distribution of reported selection gradient (β) values vary between the sexes or over different episodes of selection.

If the p values are only reported in a binary fashion (e.g., significant or nonsignificant), a **vote-counting** (Light and Smith 1971) scheme is used, which details the number of significant differences out of a collection of tests. Again, these results might be contrasted over different moderator variables, such as those over males versus females. If the actual p values are published, then the methods given earlier in this Appendix for combining the (presumably independent) p values can be used. Both the vote-counting and combined p value approaches are examples of the **null hypothesis significance testing** (**NHST**) framework, and represent the *least powerful* form of MA (Hedges and Olkin 1980). Stewart (2010) bemoaned the fact that these approaches have "a long history that lingers beyond its sell-by date." Unfortunately, many published studies in quantitative genetics only present p_i values (either actual values or values coded as significant or not significant), thus limiting the meta-analysist to these least-powerful approaches.

Furthermore, most MAs are not simply concerned about whether an effect is significant, as the usual assumption is that it is indeed. Rather, their motivation is in determining the *size of the effect* and whether it varies over cofactors of interest (for example, whether selection is stronger in males than females). If we focus only on p values, this information will be lost. When a study also reports standard errors for the test statistics (the s_i in Table A4.3), then a **formal MA** can proceed, which allows the meta-analyst to combine estimates of effect sizes, weighted by their precision, over the studies. This exploits the full power of an MA but critically depends on full transparency in reported studies (at a minimum, reporting values of s_i^2), something that is, surprisingly, often lacking.

Standardizing Effect Sizes

A formal meta-analysis proceeds by averaging over the standardized effect sizes for each study. As we will detail below, this weights each study by the strength (precision) of the evidence it provides. We will briefly review a few of the common standardizations here, pointing the reader to the general references cited at the start of this section, as well as to Nakagawa and Cuthill (2007), for more details.

The simplest setting is when study i estimates the mean of a quantity of interest, yielding

$$T_i = \overline{z}_i, \qquad \text{with} \qquad s_i^2 = \frac{1}{n_i - 1} \sum_{j=1}^{n_i} (z_{ij} - \overline{z}_i)^2 \tag{A4.25}$$

Another common setting is that in which the study contrasts the means of two different groups, which we will call the treatment (t) versus control (c),

$$T_i = \overline{z}_{i,t} - \overline{z}_{i,c} \tag{A4.26a}$$

Several slightly different standardizations exist in the literature for this case; they are of the form

$$D_i = \frac{T_i}{s_{i,p}} \tag{A4.26b}$$

where $s_{i,p}$ is the pooled standard error for study i. Different standardizations arise from slightly different estimates of this pooled standard error. **Cohen's *d* statistic** uses Equation A4.26b with

$$s_{i,p}^2 = \frac{(n_{i,t}-1)s_{i,t}^2 + (n_{i,c}-1)s_{i,c}^2}{n_{i,c}+n_{i,t}} \tag{A4.27}$$

Here, $n_{i,t}$ and $s_{i,t}$ (the latter from Equation A4.25 with $n_i = n_{i,t}$) are the sample size and standard error (respectively) for the treatment group, with similar expressions for the control group, c. This corresponds to the ML estimate of the pooled standard error (and hence is slightly biased; LW Chapter 27, LW Appendix 4).

Another common standardization in the literature is **Hedge's *g***, which uses

$$g_i = \frac{\overline{z}_{i,t} - \overline{z}_{i,c}}{s_{i,p}}, \quad \text{where} \quad s_{i,p}^2 = \frac{(n_{i,t}-1)s_{i,t}^2 + (n_{i,c}-1)s_{i,c}^2}{n_{i,c}+n_{i,t}-2} \tag{A4.28a}$$

which is the unbiased OLS estimate (and hence the REML estimate) of the pooled standard error. One further correction is often applied to g_i, in that the ratio given by Equation A4.28a is a biased estimator of $(\mu_t - \mu_c)/\sigma_p$, which can be adjusted by using

$$g_i\left(1 - \frac{3}{4(n_{i,c}+n_{i,t})-9}\right) \tag{A4.28b}$$

Besides comparisons of means, a meta-analysis often examines studies measuring the association between two variables, in which case the Pearson correlation coefficient, r_i, is the initial summary statistic. This is a slightly biased estimator of the true correlation coefficient (ρ), but a simple correction returns an unbiased estimate,

$$r_i^* = r_i + \frac{r_i(1-r_i^2)}{2(n_i-3)} \tag{A4.29a}$$

Fisher's *z*-transformation is used to both stabilize the variance and remove some of the skew:

$$z_{i,r} = \frac{1}{2}\ln\left(\frac{1+r_i}{1-r_i}\right), \quad \text{with} \quad s_i^2 = \frac{1}{n_i-3} \tag{A4.29b}$$

Typically the unbiased estimate (r^*) is used in place of r in the transformed data, and we use $T_i = z_{i,r}$ as the study effect (i.e., the value in Table A4.3). Note that the standard error of the transformed sample correlation is now simply a function of the sample size. Likewise, any estimates based on z can be back-converted to inferences of ρ by using Fisher's z-to-r transform

$$r = \frac{e^{2z_r}-1}{e^{2z_r}+1} \tag{A4.29c}$$

Other standardizations exist for other classes of comparisons, such as the log of the odds ratio.

Fixed-effects, Random-effects, and Mixed-model Meta-analysis

Once an appropriate summary statistic, along with its standard error, has been chosen, the next step is to decide if a fixed, random, or mixed meta-analysis should be used. For questions of interest to quantitative geneticists, a mixed-model analysis is likely the most appropriate. This is also the most general model, with the fixed-effects and random-effects models following as special cases. However, it will be useful to first consider the structure of these simpler models.

Under a **fixed-effects meta-analysis** (also called the **common-effect model**), we assume that the actual effect size is the *same* over all studies ($\theta_i = \theta$), which yields

$$T_i = \theta + e_i \tag{A4.30a}$$

where we assume that the residuals are independent but heteroscedastic, as $\sigma^2(e_i) = s_i^2$. Under the fixed-effects model, our interest is in combining studies to obtain a better estimate of the common (fixed) effect, θ. This simply involves generalized least-squares (GLS; LW Chapter 8), with the resulting meta-analysis global estimate of θ (given the k studies) being

$$\overline{T} = \frac{\sum_{i=1}^{k} w_i T_i}{\sum_{i=1}^{k} w_i}, \quad \text{where} \quad w_i = \frac{1}{s_i^2} \tag{A4.30b}$$

In other words, we use a weighted average, with each study weighted by its precision (studies with smaller standard errors receive larger weights). The meta-analysis standard error, $s_{\overline{T}}$, for the global estimate, $\overline{T}$, is

$$s_{\overline{T}}^2 = \frac{1}{\sum_{i=1}^{k} w_i} \tag{A4.30c}$$

For the situation where we assume that each individual observation in a given study has the same variance, so $\sigma^2(T_i) = \sigma^2/n_i$, then for k studies, each of size n,

$$\sigma^2(\overline{T}) = \frac{\sigma^2}{nk} \tag{A4.30d}$$

An obvious next line of inquiry is whether the assumption of a common effect over all studies is reasonable. This can be examined using the **Cochran *Q*** test of heterogeneity,

$$Q = \sum_{i=1}^{k} \frac{\left(T_i - \overline{T}\right)^2}{s_i^2} \tag{A4.31}$$

where (under the null of $\theta_1 = \cdots = \theta_k$, and assuming that the values of T_i are normally distributed), the distribution of Q is χ^2 with $(k-1)$ degrees of freedom.

One potential reason for a significant Q is that the study consists of different subsets of groups (say, males versus females), with a common effect that was the same in each group but differed among groups. In this case, we can extend the basic model by including a regression on moderator variables,

$$T_i = \theta + \sum_{j=1}^{m} b_j M_{ij} + e_i \tag{A4.32}$$

Often the values of M_{ij} are simply zero-one indicator variables (e.g., 0 for male, 1 for female), but they can be more general regression slopes as well. For example, M_{1j} could be the age of individuals within study j, with a significantly nonzero value of b_1 in Equation A4.32 indicating that the treatment mean varies with age. Again Equation A4.32 is simply a GLS regression, and one can test for moderator-variable effects ($b_j \neq 0$) in the standard regression fashion.

In most biological settings, the assumption of a single common value for the treatment mean over all studies is unrealistic. For example, in a meta-analysis of the strength of selection, we *expect* θ_i to vary over studies, and our interest shifts to the variance *among* the actual effects. This leads to the **random-effects meta-analysis** model

$$T_i = \mu + u_i + e_i \tag{A4.33a}$$

where $\mu_i \sim (0, \sigma_u^2)$. Typically, the effect sizes ($\theta_i = \mu + u_i$) are assumed to be drawn from a normal, $\theta_i \sim \text{N}(\mu, \sigma_u^2)$, and are independent of the residuals (which remain heteroscedastic). Under a random-effects analysis, our interest is the variation, σ_u^2, among the realized effects, in addition to their overall grand mean, μ. The estimate for the latter is also of the form of Equation A4.30b, but with a critical difference. Under a random-effects model, the weights are now given by

$$w_i = \frac{1}{s_i^2 + \widehat{\sigma}_u^2} \tag{A4.33b}$$

where $\widehat{\sigma}_u^2$ is the estimate of σ_u^2. One option for obtaining this variance is the **DerSimonian-Laird estimator**, which is based on Cochran's Q value (Equation A4.31),

$$\widehat{\sigma}_u^2 = \frac{Q - (k-1)}{S_1 - (S_2/S_1)}, \quad \text{where} \quad S_j = \sum_{i=1}^{k} s_i^{-2j} \tag{A4.33c}$$

which is set to zero if it is negative (DerSimonian and Laird 1986), although other approaches (e.g., REML) could also be used.

This difference in weighting schemes under fixed effects (Equation A4.30b) versus random effects (Equation A4.33b) can have profound implications (a nice review was presented by Borenstein et al. 2010b). In particular, the presence of $\widehat{\sigma}_u^2$ makes the random-effect weights more equal over studies, especially when $\widehat{\sigma}_u^2$ is of the same order as an average value of s_i^2. Under a fixed-effects setting, larger studies (i.e., those with smaller values of s_i^2) are given more weight than smaller studies. In a random-effects setting, this difference in weights is reduced such that larger studies lose influence and smaller studies gain influence. In the extreme when σ_u^2 is large relative to all of the s_i^2 values, all studies are given roughly equal weight, independent of their sample size.

To see why this occurs, assume we are using the same design that led to Equation A4.30d, namely, k studies all of size n, and all with a common residual variance (σ^2) for each observation. In this case, the variance of the estimate of μ becomes

$$\sigma^2(\overline{T}) = \frac{\sigma^2}{nk} + \frac{\sigma_u^2}{k} \tag{A4.34}$$

showing (unlike the fixed-effects case; Equation A4.30d) that the actual number (k) of studies is at least as important as the total number (nk) of observations over all studies. Under a fixed-effects design, all that matters is nk, as the only error is in estimating the common mean, with the contribution from any particular study being proportional to its sample size. Under a random-effects model, there is an *additional error* from the among-study variance, in that the mean realization for each study, $\theta_i = \mu + u_i$, differs, with each new study providing additional information on the among-study variance. This leads to the second important consideration for a random-effects model: power is as much a function of the number of studies (k) as it is of the precision of any particular study (n_i). Indeed, if θ_i is measured without error in any particular study, one still needs a reasonable number of θ_i values to estimate their variance with any precision. Hence, not surprisingly, random-effects models have lower power than fixed-effects models.

Given the lower power of a random-effects model, one might be tempted to start with a fixed-effects analysis and only considered random-effects when the fixed-effects model generates a significant Q value (Equation A4.31). Borenstein et al. (2010b) strongly caution against this approach. First, if the number of studies is small to moderate, Q can have poor power, so a failure to reject homogeneity among studies could simply be due to low power, and not an absence of among-study variation. Second, although there are costs (lower power) with a random-effects model, this arises because such models make a less stringent assumption: they allow the effect size to vary over studies. This is generally closer to reality, as the assumption of a constant θ value over all studies is unlikely to be correct.

Finally, in many settings, we might expect the grand mean to vary over different categories, as when the selection gradient differs between males and females. Similarly, we may wish to examine whether the strength of selection varies between life-history versus morphological traits. The potential of different means over different major categories can be accommodated in a meta-analysis model by the use of moderator variables (cofactors). These adjust the mean for a particular class, leading to a **mixed-model meta-analysis.** Suppose that there are $m \ll k$ moderators. The resulting mixed-model is

$$T_i = \mu + \sum_{j=1}^{m} b_j M_{ij} + u_i + e_i \tag{A4.35a}$$

where b_j is the effect of a moderator, j, which has a value of M_{ij} in study i. Equations A4.32 and A4.33a are special cases of Equation A4.35a, which we can write in general-linear-model form (LW Chapter 8) as

$$\mathbf{y} = \mathbf{Mb} + \mathbf{u} + \mathbf{e} \tag{A4.35b}$$

where $y_i = T_i$, and the ith row of the $k \times m$ matrix, $\mathbf{M}$, contains the values of the moderator variables associated with study i. The vectors, $\mathbf{u}$ and $\mathbf{e}$, of random effects are assumed uncorrelated, with $\mathbf{e} \sim (\mathbf{0}, \mathbf{R})$ and $\mathbf{u} \sim (\mathbf{0}, \mathbf{G})$, where $\mathbf{R}$ is a known diagonal matrix, diag$(s_1^2, s_2^3, \cdots, s_k^2)$, and $\mathbf{G} = \sigma_u^2 \mathbf{C}$, where $\mathbf{C}$ is a matrix of known constants.

While typically it is assumed that $\mathbf{C} = \mathbf{I}$ (i.e., effects are uncorrelated and homoscedastic), we can easily incorporate $\mathbf{C}$ matrices that account for phylogenetic relationships (i.e., correlations) when comparisons are made between species (Hadfield and Nakagawa 2010; Nakagawa and Santos 2012). In such settings, one could use either a Brownian-motion model (Equation A1.31) for the elements of $\mathbf{C}$ (divergence under drift) or an Ornstein-Uhlenbeck model (Equation A1.33) (divergence under drift and stabilizing selection). The bottom line is that our previous discussions on the great flexibility of mixed models (Chapters 19 and 20; LW Chapters 26 and 27) also hold when we conduct a meta-analysis using a mixed-model framework.

Publication and Other Sources in Bias

The most common concern of a meta-analysis is the issue of **publication bias**, whereby the chosen sample of studies is nonrandom with respect to all of the actual studies that have been done. Before addressing this, we need to stress that the most serious MA problem is simply ***poor data reporting*** (Gurevitch and Hedges 1999). The lack of inclusion of standard errors in many studies relegates their data to the lower status of a narrative MA. Ideally, *all* of the individual data, and not just summary statistics, should be readily available to the research community.

A second concern is **research bias** (Gurevitch and Hedges 1999). For example, in searching for signals of selection on traits, investigators are unlikely to choose a random set of characters. Rather, at least some of the traits will be explicitly chosen because an investigator feels that they are likely to be under selection. This introduces nonrandom sampling, and such a collection of estimates is not an unbiased sample of the nature of selection on a random trait. A related issue is **study bias**, in that human nature is such that investigators will often seek to make their task easier by nonrandomly choosing systems in which is it easier to measure quantities of interest.

These important concerns aside, meta-analysts worry about publication bias, such as the so-called **file-drawer effect**, whereby studies that do not achieve significance are not published but rather are simply left in the file drawer (Rosenthal 1979; Rosenberg 2005). Likewise, studies showing significant effects are more likely to be published in higher-profile

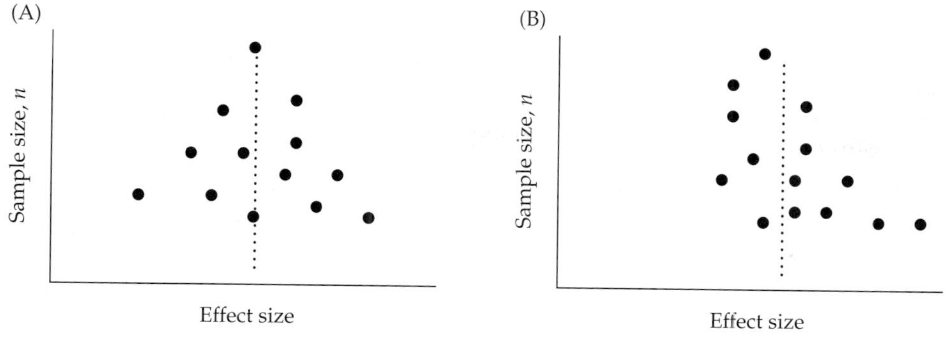

Figure A4.6 A funnel plot contrasts the estimated effect size (on the horizontal axis) of a study versus a measure of its precision (such as the sample size or $1/s_i^2$) on the vertical axis. **A:** An idealized funnel plot showing symmetry about the grand mean (dotted line). **B:** A funnel plot showing **asymmetry**, in this case an excess of larger effect-size estimates in studies with lower precision.

and more widely disseminated journals. Given this concern, a number of heuristic approaches for detecting the potential of publication bias have been proposed (a nice introduction can be found in Møller and Jennions 2001). Perhaps the best way to start any discussion of publication bias, and attempts to adjust for it, is with the comment by Copas and Shi (2000) that "correcting for publication bias is only possible if we are prepared to make unverifiable assumptions."

Methods for treating publication bias consist either of approaches that attempt to detect it (funnel plots, rank-correlation tests, Egger regressions) or that attempt to assess its impact (fail-safe numbers, trim-and-fill, model selection). Nice reviews of publication bias issues as they relate to evolutionary and ecological studies can be found in Møller and Jennions (2001) and Nakagawa and Santos (2012). While any serious meta-analysis should examine publication bias, it is often ignored. In a survey of 100 evolutionary and ecologically related meta-analyses, Nakagawa and Santos found that only 49% attempted to assess publication bias. Among those studies that did, only 45% searched for signals of bias, only 14% attempted to assess the impact of bias, and 41% attempted to do both.

A number of tests for the potential presence of publication bias are constructed around the notion of a **funnel plot** (Light and Pillemer 1984). As shown in Figure A4.6, for each study, one plots the estimated effect size on the horizontal axis and some measure of precision, such as s_i^{-2} or the sample size, on the vertical axis. Under a fixed-effects model, the scatter of points should be broad at the base (reflecting spread about the true mean due to larger standard errors in the estimate), and then narrowing as one moves vertically up in the plot (larger studies). This generates a plot that looks like an inverted funnel, hence the name. Under a fixed-effects model, data high on the vertical axis (i.e., studies with very large sample sizes) should show essentially no spread about the grand mean. Under a random-effects interpretation, even when sampling error is entirely removed (due to very large sample size), there will still be a spread of values around the grand mean, reflecting the random effect of sampling the mean for a particular realization. One will still see a funnel, but it will only narrow down (for large n) to a spread given by σ_u^2, namely, the funnel will taper up to a cylinder whose width is a function of σ_u^2.

Figure A4.6A shows an idealized setting, with estimates that are symmetrically distributed about the grand mean. Figure A4.6B shows a situation in which there is asymmetry in the funnel. In this case, there is an excess of large-effect estimates for studies with less precision (i.e., smaller samples). Publication bias can generate such an asymmetry, as studies whose estimated effects are smaller (and therefore either not significant, or only marginally so) are less likely to published. However, other sources of bias can also generate such an asymmetry, so that its presence does not automatically guarantee that publication

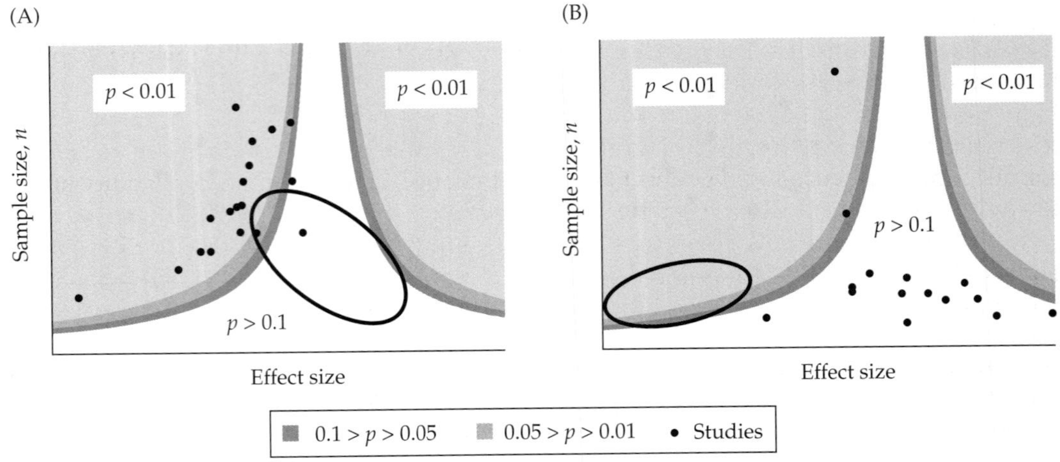

Figure A4.7 Contour-enhanced funnel plots help in the interpretation of asymmetry. Both of the plots show asymmetry, with the open ellipse showing the area of "missing" studies that generates the asymmetry. Such missing values would be added under the trim-and-fill method. The contours (showing regions of statistical significance) show the expected likelihood of study values under the null of no effect. **A:** The "missing" studies are in a region where they are likely to occur by chance ($p > 0.1$), suggesting they are absent due to publication bias. **B:** Here, these studies will all fall into regions unlikely to be seen under the null ($p < 0.05$), and hence may be missing for reasons other than publication bias. (After Peters et al. 2008.)

bias has occurred. For example, effect sizes could be correlated with their standard errors, as might occur in settings where the realization of a effect size is expected to be small (say, a selection gradient on what is regarded as a minor trait), and so the study collects a larger sample size in order to try to achieve sufficient power (Gurevitch and Hedges 1999).

While funnel asymmetry is essentially an informal visual judgment, the basic idea has been used for more formal tests. The **Begg-Mazumdar** (1994) **rank corrrelation test** looks for associations between the ranks of the standard errors (s_i) and study estimates (T_i). The **Egger regression test** (Egger et al. 1997) performs the regression

$$y_i = a + bs_i + e, \quad \text{where} \quad y_i = (T_i - \overline{T}) \tag{A4.36}$$

Under a symmetric funnel plot, the data are distributed symmetrically around zero, and the resulting intercept (a) should be zero. An a value that is significantly different from zero indicates asymmetry. Both of these tests suffer from low power, and they are not recommend when the number of studies is small (ten or fewer). Indeed, Sterne et al. (2011) recommended that these tests "should be used in only a minority of meta-analysis." Further, even a highly significant result does not imply publication bias, and a negative result does not imply that the study is free from such bias (especially if k is relatively small). Another indicator of the potential of publication bias is **time-lag bias**. Here, the initial studies report larger effects than are seen in follow-up studies. This can be tested by simply including the year of publication as a moderator variable in the original analysis (Nakagawa and Santos 2012).

While detecting publication bias can be problematic, *correcting* for it can be even more so. One approach is the **trim-and-fill** method of Duval and Tweedie (2000a, 2000b). Again, this method is based on funnel-plot asymmetry. In step one, smaller (i.e., lower-precision) studies are excluded in order to achieve a more symmetric plot, while step two replaces these excluded studies with "missing" studies whose values are imputed from the retained studies. Again, factors other than publication bias can result in asymmetric funnel plots, which compromises this approach.

The use of **contour funnel plots** (Figure A4.7) may provide some additional guidance (Peter et al. 2008). These are funnel plots enhanced by overlaying significance contours, which provides a visual test of whether "missing" studies occur in high- or low-probability regions. The former suggests bias, and the latter suggests simple sampling. Figure A4.7 shows two situations of asymmetry, and the region of the funnel plot with "missing" studies that might be imputed under a trim-and-fill approach. The contour plots show that in Figure A4.7A the missing studies were quite likely to have been seen had there been was no publication bias. However, in Figure A4.7B, the missing values are in regions of high significance (and hence could be absent simply by chance, given their low probability under the null), even with a full ascertainment of all studies.

A metric for assessing the impact of publication bias is the **fail-safe number** for an analysis (Rosenthal 1979; Rosenberg 2005). This is simply the number of additional studies (i.e., missed studies with no significance) that would have to be added to the analysis to invalidate (i.e., remove the significance of) the current analysis. The rough rule of thumb is that if this number exceeds $5k+10$, then the study is fairly robust (Rosenthal 1979; Rosenberg 2005). Despite often being reported, there are issues with this metric. The most obvious one is that a meta-analysis is usually concerned with the average effect size (or the variance in true effect sizes), rather than whether an effect is significant. Hence, even if the fail-safe number indicates that the current study is robust, it is simply robust to overall *significance* and not necessarily to overall *effect size*. Second, as we stressed above, a random effects meta-analysis is generally the most appropriate (which extends to a mixed model if moderator variables are added). Fail-safe numbers in random-effects setting are much smaller than their fixed-effects counterparts and are much more delicate to obtain. Rosenberg (2005) cited a meta-analysis based on 71 studies where the fixed-effects fail-safe number (based on slightly different methods) ranged from 7500 to 8500, while the fail-safe number under a random-effects model was approximately 9.

The most powerful, and also the most delicate, methods to both detect and adjust for publication bias are model-based approaches. These require a model of the ascertainment scheme. For example, Copas and Shi (2000) proposed an ascertainment model that is very much like the model for a threshold trait (Chapter 14), wherein some underlying latent variable determines whether a study is published or not. In their model, this variable is a function of the standardized effect size, which means that studies with smaller standardized effects, and therefore less significance, are less likely to be published. An excellent example of a model-based ascertainment scheme in quantitative-genetics was the study by Hersch and Phillips (2004). To obtain the bias in estimates of selection gradients, they simulated each study as a random draw of five gradients from a known distribution, for which the results were only reported when one (or more) of the values was significant. Figure 30.7 shows the results of their analysis, which found that such ascertained gradients were overestimated.

Bias When Estimating Magnitudes

Finally, consider a random-effects setting, in which the parameter of interest, θ_i, in a given study (i) is drawn from normal distribution with a mean of μ and a variance of σ_u^2, so $\theta_i \sim N(\mu, \sigma_u^2)$. Suppose we assume the observed effect, T_i, for this study is also normal, but with an additional sampling error, σ_e^2, so that $T_i \sim N(\mu, \sigma_u^2 + \sigma_e^2)$. This additional variance has no effect on our estimate of the desired mean, μ, as $E[T_i] = \mu$. However, in many setting in quantitative genetics (as well as ecology and evolution), our interest may be in the absolute magnitude, $|\theta|$, of the effect, rather than its mean value. An important example of this type of inquiry is the estimation of the average strength of selection, $|\beta|$, on a trait (Chapters 29 and 30). As noted by Hereford et al. (2004) and Morrissey (2016), $E[|T|] > E[|\theta|]$ when $\sigma_e^2 > 0$, which will result in an overestimate of the expected absolute value of the effect. In particular, if $\mu = 0$, then (Example A4.11),

$$E[|\theta|] = \sigma_u \sqrt{2/\pi}, \qquad \text{while} \qquad E[|T|] = \sqrt{\sigma_u^2 + \sigma_e^2} \cdot \sqrt{2/\pi} \tag{A4.37a}$$

which yields a relative error of

$$\frac{E[|T|] - E[|\theta|]}{E[|\theta|]} = \sqrt{\frac{\sigma_e^2}{\sigma_u^2} + 1} - 1 \tag{A4.37b}$$

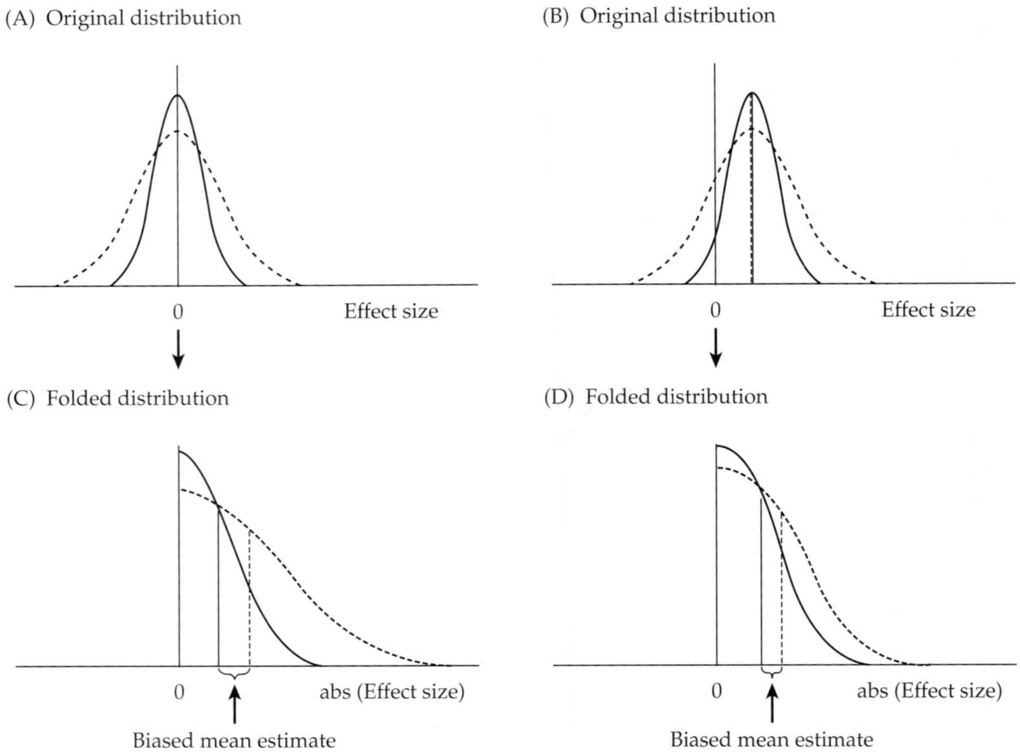

Figure A4.8 The consequences of considering the absolute value, $|\theta|$, of an effect. Here, the true size effect (θ) is normally distributed with a variance of σ^2, whose distribution is indicated by the solid curves. Of course, we do not observe the true value, but rather an estimate, T, which has the same mean as θ, but an additional sampling error, σ_e^2, so that while $\theta \sim N(\mu, \sigma^2)$, $T \sim N(\mu, \sigma^2 + \sigma_e^2)$, which (dashed curves) has a larger total variance. In (**A**) and (**C**), $\mu = 0$, while $\mu > 0$ in (**B**) and (**D**). The distribution of $|\theta|$ is given by **folding** the distribution about zero, with Pr($|\theta|$) = Pr(θ) + Pr($-\theta$) for $\theta > 0$, as shown in panels (**C**) and (**D**). While the sampling variance (σ_e^2) does not translate into bias when estimating the mean, as $E[T] = E[\theta]$, it introduces bias when the absolute value of θ is of interest, with $E[|T|] > E[|\theta|]$ when $\sigma_e^2 > 0$, as shown in (**C**) and (**D**). (After Nakagawa and Lagisz 2016.)

which can be a significant overestimation of the average magnitude of an effect if the error variance is large relative to the effect variance (Figure A4.8). Morrissey (2016) presented several additional examples where the concern of a meta-analysis is in some measure of dispersion about a mean, in which case the error variance associated with using an estimate (T_i) will result in an upwardly biased estimator. Morrissey showed how the use of appropriate mixed models helps to resolve this concern.

Example A4.11. Hereford et al. (2004) and Morrissey (2016) presented general expressions for $E[|x|]$ when $x \sim N(\mu, \sigma^2)$. As shown in Figure A4.8, the distribution of $|x|$ is given by a folded normal distribution, from which it follows that

$$E[|x|] = \sigma\sqrt{\frac{2}{\pi}}\exp\left(-\frac{\mu^2}{2\sigma^2}\right) + |\mu| \cdot \operatorname{erf}\left(\frac{|\mu|}{\sigma\sqrt{2}}\right) \tag{A4.38a}$$

where

$$\operatorname{erf}(x) = \frac{2}{\sqrt{\pi}}\int_0^x \exp(-t^2)\,dt$$

is the error function. Taking $\mu = 0$, Equation A4.38a reduces to Equation A4.37a. Likewise, when $\mu \gg \sigma^2$, Equation A4.38a reduces to

$$E[|x|] \sim |\mu| \tag{A4.38b}$$

which follows because for large values of x, $e^{-x} \rightarrow 0$ and $\text{erf}(x) \rightarrow 1$.

Appendix 5

The Geometry of Vectors and Matrices: Eigenvalues and Eigenvectors

Much of the presentation that follows is in matrix notation, and for this I offer no apology as this has rapidly become an essential tool of any serious student of animal breeding.
Henderson (1973)

The basic concepts of matrix algebra were introduced in LW Chapter 8 and LW Appendix A3, and we assume the reader has this level of understanding (which includes matrix multiplication, inverses, and determinants). If not, a quick review of LW Chapter 8 before proceeding will be helpful. A deeper understanding of multivariate issues in quantitative genetics requires an appreciation of matrix *geometry*. Our primary intent here is to introduce the reader to the idea of vectors and matrices as **geometric structures**, and thus viewing matrix operations as transformations converting one vector into another by a change in geometry (rotation and scaling), which is completely summarized by the **eigenvalues** (scaling), and their associated **eigenvectors** (rotation), of a matrix.

THE GEOMETRY OF VECTORS AND MATRICES

As there are numerous excellent texts on matrix algebra, we made little effort to prove most of the results given below. For statistical applications, concise introductions can be found in the chapters on matrix methods in Johnson and Wichern (1988) and Morrison (1976), while Dhrymes (1978) and Searle (1982) provided more extended treatments. Wilf's (1978) short chapter on matrix methods provides a very nifty review of methods useful in applied mathematics. Franklin (1968), Horn and Johnson (1985), and Gantmacher (1960), respectively, presented increasingly sophisticated treatments of matrix analysis.

Comparing Vectors: Lengths and Angles

As Figure A5.1A shows, a vector, $\mathbf{x}$, can be treated as a geometric object, consisting of an arrow leading from the origin to an n-dimensional point whose coordinates are given by the elements of $\mathbf{x}$. By changing coordinate systems, we change the resulting vector, potentially changing both its direction (**rotating** the vector) and length (**scaling** the vector). This geometric interpretation suggests several ways for comparing vectors, such as the **angle** between two vectors and the **projection** of one vector onto another.

Consider first the length (or **norm**) of a vector. The most common measure of length is the Euclidean distance of the vector from the origin, $||\mathbf{x}||$, defined as

$$||\mathbf{x}|| = \sqrt{x_1^2 + x_2^2 + \cdots + x_n^2} = \sqrt{\mathbf{x}^T\mathbf{x}} \tag{A5.1a}$$

For any scalar a, $||a\,\mathbf{x}|| = |a|\,||\mathbf{x}||$. Similarly, the squared Euclidean distance between the vectors $\mathbf{x}$ and $\mathbf{y}$ is

$$||\mathbf{x}-\mathbf{y}||^2 = \sum_{i=1}^{n}(x_i - y_i)^2 = (\mathbf{x}-\mathbf{y})^T(\mathbf{x}-\mathbf{y}) = (\mathbf{y}-\mathbf{x})^T(\mathbf{y}-\mathbf{x}) \tag{A5.1b}$$

Vectors can differ by length, direction, or both. The angle, θ, between two vectors ($\mathbf{x}$ and $\mathbf{y}$) provides a measure of how much they differ in direction (Figure A5.1C). If the vectors

Evolution and Selection of Quantitative Traits. Bruce Walsh & Michael Lynch.
Published 2018 by Oxford University Press. © Bruce Walsh & Michael Lynch 2018.
DOI 10.1093/oso/9780198830870.001.0001

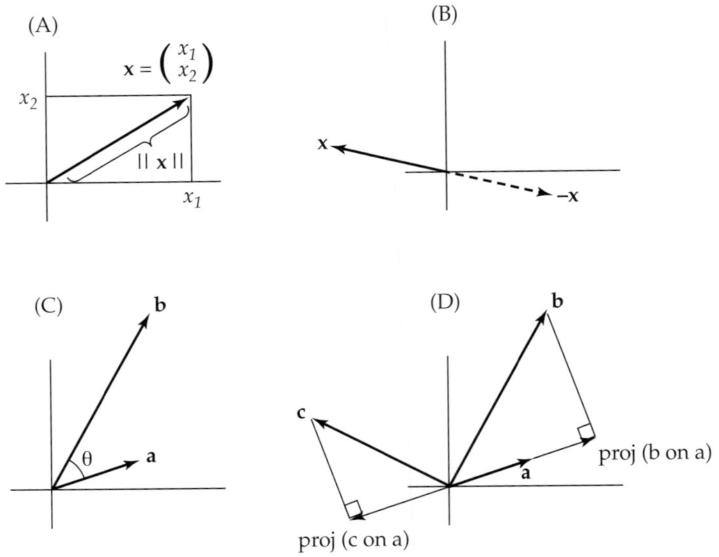

Figure A5.1 Some basic geometric concepts of vectors. While we use examples from two dimensions, these concepts easily extend to n dimensions. **A:** A vector $\mathbf{x}$ can be thought of as an arrow from the origin to a point in space whose coordinates are given by the elements of $\mathbf{x}$. **B:** Multiplying a vector by -1 results in a *reflection* about the origin. **C:** One measure of the difference in direction between two vectors is the angle (θ) between them. **D:** $\text{Proj}(\mathbf{b}\,\text{on}\,\mathbf{a})$ is the vector resulting from the projection of $\mathbf{b}$ onto $\mathbf{a}$. Note that the resulting projection vector is either in the same direction as $\mathbf{a}$ or in the direction of the reflection of $\mathbf{a}$, as seen for $\text{Proj}(\mathbf{c}\,\text{on}\,\mathbf{a})$.

satisfy $a\mathbf{x} = \mathbf{y}$, they both point in exactly the same direction ($\theta = 0$; they are **codirectional**) when $a > 0$. If $a < 0$, they are exactly 180 degrees apart and differ in direction only by a **reflection** about the origin (Figure A5.1B). At the other extreme, two vectors can be at right angles to each other ($\theta = 90^\circ$ or 270°), in which case they are said to be **orthogonal**. Orthogonal vectors of unit length are further said to be **orthonormal**. For any two n-dimensional vectors, θ satisfies

$$\cos(\theta) = \frac{\mathbf{x}^T\mathbf{y}}{||\mathbf{x}||\ ||\mathbf{y}||} = \frac{\mathbf{y}^T\mathbf{x}}{||\mathbf{x}||\ ||\mathbf{y}||} \tag{A5.2a}$$

Hence,

$$\theta = \cos^{-1}\left(\frac{\mathbf{y}^T\mathbf{x}}{||\mathbf{x}||\ ||\mathbf{y}||}\right) \tag{A5.2b}$$

If both $\mathbf{x}$ and $\mathbf{y}$ are of unit length, then $\theta = \cos^{-1}(\mathbf{y}^T\mathbf{x})$, which reveals the close connection between vector angles and inner products. Note that because $\cos(90^\circ) = \cos(270^\circ) = 0$, ***two vectors are orthogonal if, and only if, their inner product is zero,*** $\mathbf{x}^T\mathbf{y} = 0$.

Another way to compare two vectors is to consider the **projection** vector of one onto the other. $\text{Proj}(\mathbf{x}\,\text{on}\,\mathbf{y})$, the projection of $\mathbf{x}$ on $\mathbf{y}$, is a vector in the direction of $\mathbf{y}$, whose length is given by how much of the vector $\mathbf{x}$ lies along the direction of $\mathbf{y}$. For any two n-dimensional vectors, the projection of $\mathbf{x}$ on $\mathbf{y}$ is defined by

$$\text{Proj}(\mathbf{x}\,\text{on}\,\mathbf{y}) = \frac{\mathbf{x}^T\mathbf{y}}{\mathbf{y}^T\mathbf{y}}\,\mathbf{y} = \frac{\mathbf{x}^T\mathbf{y}}{||\mathbf{y}||^2}\,\mathbf{y} = \left(\cos(\theta)\,\frac{||\mathbf{x}||}{||\mathbf{y}||}\right)\,\mathbf{y} \tag{A5.3a}$$

The term in the parentheses (which follows from Equation A5.2a) is a scalar, representing the length that $\mathbf{x}$ projects in the direction of $\mathbf{y}$, which means that $\text{Proj}(\mathbf{x}\,\text{on}\,\mathbf{y})$ is a scaled version of the vector $\mathbf{y}$ onto which we are projecting. If $||\mathbf{y}|| = 1$, then

$$\text{Proj}(\mathbf{x}\,\text{on}\,\mathbf{y}) = (\mathbf{x}^T\mathbf{y})\,\mathbf{y} = (\cos(\theta)\,||\mathbf{x}||\,)\,\mathbf{y} \tag{A5.3b}$$

The vector resulting from the projection of $\mathbf{x}$ on $\mathbf{y}$ is in the same direction as $\mathbf{y}$ unless $90^\circ < \theta < 270^\circ$, in which case $\cos(\theta) < 0$ and the projection vector is in exactly the opposite direction (the reflection of $\mathbf{y}$ about the origin). The length of the projection vector is

$$||\text{Proj}(\mathbf{x}\,\text{on}\,\mathbf{y})|| = |\cos(\theta)|\;||\mathbf{x}|| \leq ||\mathbf{x}|| \qquad \text{(A5.3c)}$$

If two vectors lie in exactly the same direction ($\theta = 0$), the projection of one on the other simply recovers the vector (i.e., $\text{Proj}(\mathbf{x}\,\text{on}\,\mathbf{y}) = \mathbf{x}$). Conversely, if two vectors are orthogonal, the projection of one on the other yields a vector of length zero.

An important property of projection vectors is that if $\mathbf{y}_1, \mathbf{y}_2, \cdots, \mathbf{y}_n$ is any set of mutually orthogonal n-dimensional vectors, then any n-dimensional vector $\mathbf{x}$ can be represented as the sum of projections of $\mathbf{x}$ onto the members of this set, namely,

$$\mathbf{x} = \sum_{i=1}^{n} \text{Proj}(\mathbf{x}\,\text{on}\,\mathbf{y}_i) \qquad \text{(A5.4)}$$

One way to think about such a decomposition is as the transformation from one set of axes (or coordinates) into another (defined by the vectors, $\mathbf{y}_i$, that **span**, or completely cover, the vector space). We can also consider the projection of a vector into some **subspace** of a matrix (say $\mathbf{y}_1, \cdots, \mathbf{y}_k$, where $k < n$), namely, the projection onto some subset of the vectors that span the space of the original matrix. For example, one might consider the subspace of a covariance matrix imposed by (say) its three largest factors (eigenvalues). The notion of a subspace of the genetic covariance matrix $\mathbf{G}$ will prove useful in describing the constraints caused by the genetic covariance structure (Volume 3).

Matrices Describe Vector Transformations

When we multiply a vector, $\mathbf{x}$, by a matrix, $\mathbf{A}$, to create a new vector, $\mathbf{y} = \mathbf{Ax}$, $\mathbf{A}$ *rotates* and *scales* the original vector, $\mathbf{x}$, into the new vector, $\mathbf{y}$. $\mathbf{A}$ therefore describes a *transformation* of the original coordinate system of $\mathbf{x}$ into a new coordinate system, $\mathbf{y}$ (which has a different dimension from $\mathbf{x}$ unless $\mathbf{A}$ is square).

Example A5.1. Consider the Lande version of the multivariate breeder's equation, $\mathbf{R} = \mathbf{G}\boldsymbol{\beta}$ (Equation 13.26a). Here $\mathbf{R}$ is the change in the vector of phenotypic means resulting from selection, $\mathbf{G}$ is the covariance matrix of additive-genetic values (breeding values) of the characters, and $\boldsymbol{\beta}$ is the directional selection gradient (the direction of change in character means that results in the greatest increase in mean population fitness; Chapters 13 and 30). Suppose

$$\mathbf{G} = \begin{pmatrix} 4 & -2 \\ -2 & 2 \end{pmatrix} \quad \text{and} \quad \boldsymbol{\beta} = \begin{pmatrix} 1 \\ 3 \end{pmatrix}, \quad \text{yielding} \quad \mathbf{R} = \mathbf{G}\boldsymbol{\beta} = \begin{pmatrix} -2 \\ 4 \end{pmatrix}$$

The resulting direction of change in character means is different from that most favored by natural selection. Selection ($\boldsymbol{\beta}$) favors an increase in trait one (z_1), but when the genetic covariance structure is taken into account, the resulting change in the mean of z_1 is negative. If we take the appropriate inner products, we find $||\boldsymbol{\beta}|| = \sqrt{10}$, $||\mathbf{R}|| = \sqrt{20}$, and $\boldsymbol{\beta}^T\mathbf{R} = 10$. Equation A5.2a returns

$$\cos(\theta) = \frac{\boldsymbol{\beta}^T\mathbf{R}}{||\mathbf{R}||\;||\boldsymbol{\beta}||} = \frac{1}{\sqrt{2}}$$

The resulting angle between the selection gradient and response vector is $\cos^{-1}(1/\sqrt{2}) = 45^\circ$, implying that the constraints introduced by the genetic covariance matrix rotate the response vector considerably away from the direction most favored by natural selection (Figure A5.2).

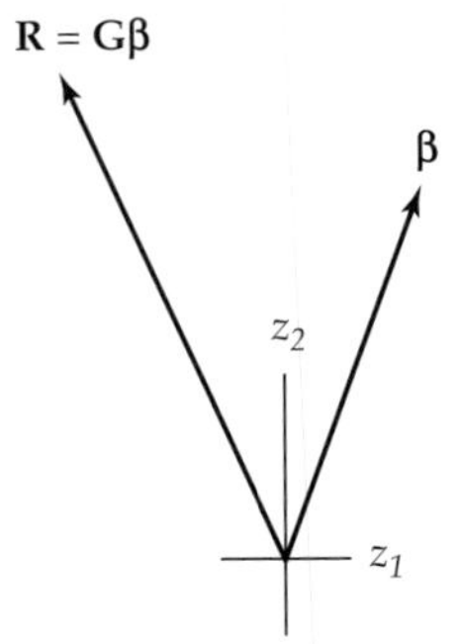

Figure A5.2 If we use the values of $\boldsymbol{\beta}$ and **G** from Example A5.1, observe that **G** translates the directional selection gradient vector ($\boldsymbol{\beta}$) ino the response vector (**R**) in a counterintuitive fashion. While $\boldsymbol{\beta}$ shows that fitness is maximized by increasing both traits 1 and 2, the resulting response vector, **R**, increases trait 2 but *decreases* trait 1. This behavior results from the strong negative additive-genetic covariance between z_1 and z_2, as will become more obvious shortly, when we consider the eigenvectors of **G** (Figure A5.3). As shown in Example A5.1, the angle between the vectors $\boldsymbol{\beta}$ and **R** is 45 degrees.

Orthonormal Matrices: Rigid Rotations

A key building block on our way to the partitioning of a matrix into its rotational and scaling components is the idea of an **orthonormal matrix**. Writing a square $n \times n$ matrix, **U**, as a row vector whose n elements are $1 \times n$ column vectors, $\mathbf{U} = (\mathbf{u}_1, \mathbf{u}_2, \cdots, \mathbf{u}_n)$, then **U** is said to be orthonormal if

$$\mathbf{u}_i^{\,T}\mathbf{u}_j = \begin{cases} 1 & \text{if } i = j \\ 0 & \text{if } i \neq j \end{cases}$$

Namely, each column of **U** is of unit length and is orthogonal to every other column. Matrices with this property are also referred to as **unitary** and satisfy

$$\mathbf{U}^T\mathbf{U} = \mathbf{U}\,\mathbf{U}^T = \mathbf{I} \tag{A5.5a}$$

As a result, the inverse of a unitary matrix is simply its transpose,

$$\mathbf{U}^T = \mathbf{U}^{-1} \tag{A5.5b}$$

The coordinate transformation induced by an orthonormal matrix has a very simple geometric interpretation: it is a **rigid rotation** of the original coordinate system—axes of the original coordinates are all rotated by the same angle to create the new coordinate system. To see this, first note that ***orthonormal matrices preserve all inner products***. Taking $\mathbf{y}_1 = \mathbf{U}\mathbf{x}_1$ and $\mathbf{y}_2 = \mathbf{U}\mathbf{x}_2$

$$\mathbf{y}_1^{\,T}\mathbf{y}_2 = \mathbf{x}_1^{\,T}(\mathbf{U}^T\mathbf{U})\mathbf{x}_2 = \mathbf{x}_1^{\,T}\mathbf{x}_2 \tag{A5.5c}$$

Thus, orthonormal matrices do not change (scale) the length of vectors, as $||\mathbf{y}_1|| = \mathbf{y}_1^{\,T}\mathbf{y}_1 = \mathbf{x}_1^{\,T}\mathbf{x}_1 = ||\mathbf{x}_1||$. Using these results, note that if θ is the angle between the vectors $\mathbf{x}_1$ and $\mathbf{x}_2$, then following transformation by an orthonormal matrix

$$\cos(\theta\,|\,\mathbf{y}_1, \mathbf{y}_2) = \frac{\mathbf{y}_1^{\,T}\mathbf{y}_2}{\sqrt{||\mathbf{y}_1||\;||\mathbf{y}_2||}} = \frac{\mathbf{x}_1^{\,T}\mathbf{x}_2}{\sqrt{||\mathbf{x}_1||\;||\mathbf{x}_2||}} = \cos(\theta\,|\,\mathbf{x}_1, \mathbf{x}_2) \tag{A5.5d}$$

which shows that the angle between the two vectors remains unchanged following their transformation by the same orthonormal matrix.

Eigenvalues and Eigenvectors

The eigenvalues, and their associated eigenvectors, of a square matrix describe its transformational geometry. Eigenvalues describe how the original coordinate axes are *scaled* in the new coordinate system that is described by the eigenvectors (i.e., how the original axes are *rotated*).

To more formally introduce eigenvalues and eigenvectors, suppose, for a square matrix $\mathbf{A}$, that the vector $\mathbf{y}$ satisfies the matrix equation

$$\mathbf{Ay} = \lambda\mathbf{y} \tag{A5.6}$$

for some scalar value, λ. Geometrically, this means that the new vector resulting from transformation of $\mathbf{y}$ by $\mathbf{A}$ points in the same direction as $\mathbf{y}$ (or is exactly reflected about the origin if $\lambda < 0$). For such vectors, the only action of the matrix transformation is to scale them by some amount, λ. These vectors thus represent the ***inherent axes associated with the transformation given by* A**, and the set of all such vectors, along with their corresponding scalar multipliers, completely describes the geometry of this transformation. Vectors that satisfy Equation A5.6 are referred to as **eigenvectors**, and their associated scaling factors are **eigenvalues**, and together they jointly describe the **eigenstructure** (the intrinsic geometry) of the square matrix, $\mathbf{A}$. If $\mathbf{y}$ is an eigenvector, then $a\mathbf{y}$ is also an eigenvector, as $\mathbf{A}(a\mathbf{y}) = a(\mathbf{Ay}) = \lambda(a\mathbf{y})$. Note, however, that the associated eigenvalue, λ, remains unchanged. Hence, we typically scale eigenvectors to be of unit length to yield **unit** or **normalized** eigenvectors. In particular, if $\mathbf{y}_i$ is any eigenvector associated with the ith eigenvalue, then the associated normalized eigenvector is $\mathbf{e}_i = \mathbf{y}_i/||\mathbf{y}_i||$.

The eigenvalues of an n-dimensional square matrix, $\mathbf{A}$, are solutions of Equation A5.6, which can be written as $(\mathbf{A} - \lambda\,\mathbf{I})\mathbf{y} = \mathbf{0}$. This implies that the determinant of $(\mathbf{A} - \lambda\,\mathbf{I})$ must equal zero, which gives rise to the **characteristic equation**, $|\mathbf{A} - \lambda\mathbf{I}| = 0$, whose solution yields the eigenvalues of $\mathbf{A}$. This equation can be also be expressed using the **Laplace expansion**,

$$|\mathbf{A} - \lambda\mathbf{I}| = (-\lambda)^n + S_1(-\lambda)^{n-1} + \cdots + S_{n-1}(-\lambda)^1 + S_n = 0 \tag{A5.7}$$

where $|\mathbf{A}|$ denotes the determinant of $\mathbf{A}$ and S_i is the sum of all **principal minors** (minors including diagonal elements of the original matrix) of order i (minors, which are subsets of the full matrix, were defined in LW Chapter 8). Finding the eigenvalues thus requires solving a polynominal equation of order n, implying that there are exactly n eigenvalues (some of which may be identical, i.e., **repeated**). In practice, for $n > 2$ this is accomplished numerically, and most statistical analysis packages offer routines to accomplish this task.

Two of these principal minors are easily obtained and provide information on the nature of the eigenvalues. The only principal minor having the same order of the matrix is the full matrix itself, which means that $S_n = |\mathbf{A}|$, the determinant of $\mathbf{A}$. S_1 is also related to an important matrix quantity, the **trace**. This is denoted by $\text{tr}(\mathbf{A})$, and is the sum of the diagonal elements of the matrix, namely,

$$\text{tr}(\mathbf{A}) = \sum_{i=1}^{n} A_{ii}$$

Observe that $S_1 = \text{tr}(\mathbf{A})$, as the only principal minors of order one are the diagonal elements themselves, the sum of which equals the trace. Both the trace and determinant can be expressed as functions of the eigenvalues, with

$$\text{tr}(\mathbf{A}) = \sum_{i=1}^{n} \lambda_i \qquad \text{and} \qquad |\mathbf{A}| = \prod_{i=1}^{n} \lambda_i \tag{A5.8}$$

Hence $\mathbf{A}$ ***is singular*** ($|\mathbf{A}| = 0$) ***if, and only if, at least one eigenvalue is zero***. As we will see, if $\mathbf{A}$ is a covariance matrix, then its trace (the sum of its eigenvalues) measures its total amount of variation, as the eigenvalues of a covariance matrix are nonnegative ($\lambda_i \geq 0$).

Let $\mathbf{e}_i$ be the (unit-length) eigenvector associated with eigenvalue λ_i. If the eigenvectors of $\mathbf{A}$ can be chosen to be mutually orthogonal, namely, $\mathbf{e}_i^T\mathbf{e}_j = 0$ for $i \neq j$, then we can express $\mathbf{A}$ as

$$\mathbf{A} = \lambda_1\mathbf{e}_1\mathbf{e}_1^T + \lambda_2\mathbf{e}_2\mathbf{e}_2^T + \cdots + \lambda_n\mathbf{e}_n\mathbf{e}_n^T \tag{A5.9a}$$

This is called the **spectral decomposition** of $\mathbf{A}$, and it is derived below in Equation A5.10d. Because $||\mathbf{e}_i|| = 1$, Equation A5.3b gives the projection of $\mathbf{x}$ on $\mathbf{e}_i$ as $(\mathbf{x}^T\mathbf{e}_i)\mathbf{e}_i$. Note that $\mathbf{e}_i(\mathbf{e}_i^T\mathbf{x}) = (\mathbf{e}_i^T\mathbf{x})\mathbf{e}_i = (\mathbf{x}^T\mathbf{e}_i)\mathbf{e}_i$, as $\mathbf{e}_i^T\mathbf{x}$ is a scalar, which implies that $\mathbf{e}_i^T\mathbf{x} = (\mathbf{e}_i^T\mathbf{x})^T = \mathbf{x}^T\mathbf{e}_i$. Hence, from Equation A5.3b, we have

$$\begin{aligned}\mathbf{Ax} &= \lambda_1\mathbf{e}_1\mathbf{e}_1^T\mathbf{x} + \lambda_2\mathbf{e}_2\mathbf{e}_2^T\mathbf{x} + \cdots + \lambda_n\mathbf{e}_n\mathbf{e}_n^T\mathbf{x} \\ &= \lambda_1\left(\mathbf{e}_1^T\mathbf{x}\right)\mathbf{e}_1 + \lambda_2\left(\mathbf{e}_2^T\mathbf{x}\right)\mathbf{e}_2 + \cdots + \lambda_n\left(\mathbf{e}_n^T\mathbf{x}\right)\mathbf{e}_n \\ &= \lambda_1\text{Proj}(\mathbf{x}\,\text{on}\,\mathbf{e}_1) + \lambda_2\text{Proj}(\mathbf{x}\,\text{on}\,\mathbf{e}_2) + \cdots + \lambda_n\text{Proj}(\mathbf{x}\,\text{on}\,\mathbf{e}_n)\end{aligned} \tag{A5.9b}$$

If we again apply Equation A5.3b, we can express this decomposition as

$$\mathbf{Ax} = ||\mathbf{x}||\sum_{i=}^{n}\left[\lambda_i\cdot\cos(\theta|\mathbf{x},\mathbf{e}_i)\right]\mathbf{e}_i \tag{A5.9c}$$

where $\theta|\mathbf{x},\mathbf{e}_i$ denotes the angle between the vectors $\mathbf{x}$ and $\mathbf{e}_i$. Thus, one can view a matrix as a series of vectors that form the **projection space** (the eigenvectors), so when a vector is multipled by this matrix, the resulting vector is the weighted (by the eigenvalues) sum of projections over all of the vectors (the $\mathbf{e}_i$) that span the space defined by the matrix.

Example A5.2. Determine the eigenstructure of the genetic covariance matrix $\mathbf{G}$ shown in Example A5.1. Writing the characteristic equation, and recalling the expression for the determinant of a 2×2 matrix (LW Equation 8.12a), yields

$$\begin{aligned}|\mathbf{G}-\lambda\mathbf{I}| &= \left|\begin{pmatrix} 4-\lambda & -2 \\ -2 & 2-\lambda\end{pmatrix}\right| \\ &= (4-\lambda)(2-\lambda)-(-2)^2 = \lambda^2 - 6\lambda + 4 = 0\end{aligned}$$

Alternatively, if we use the Laplace expansion (Equation A5.7), and note that tr($\mathbf{G}$) = 4+2 = 6 and $|\mathbf{G}|$= 4 · 2 - $(-2)^2$ = 4, we will also recover the characteristic equation, which has solutions

$$\lambda_1 = 3+\sqrt{5} \simeq 5.236 \qquad \lambda_2 = 3-\sqrt{5}\simeq 0.764$$

The associated unit eigenvectors (which as easily obtained, along with the eigenvectors, by using the `R` command `eigen`) are

$$\mathbf{e}_1 \simeq \begin{pmatrix}-0.851 \\ 0.526\end{pmatrix} \qquad \mathbf{e}_2 \simeq \begin{pmatrix}0.526 \\ 0.851\end{pmatrix}$$

These are orthogonal as $\mathbf{e}_1^T\mathbf{e}_2 = 0$.

The eigenstructure of $\mathbf{G}$ shows why the vector of responses, $\mathbf{R}$, is rotated away from the direction of the vector that corresponds to the direction of selection, $\boldsymbol{\beta}$. From Example A5.1, $||\boldsymbol{\beta}|| = \sqrt{10}$, while $\mathbf{e}_1^T\boldsymbol{\beta} \simeq 0.727$ and $\mathbf{e}_2^T\boldsymbol{\beta}\simeq 3.079$. Because $||\mathbf{e}_1|| = ||\mathbf{e}_2|| = 1$, Equation A5.2a simplifies to

$$\cos(\theta|\mathbf{e}_1,\boldsymbol{\beta}) \simeq \frac{0.727}{\sqrt{10}} \simeq 0.230 \qquad \text{and} \qquad \cos(\theta|\mathbf{e}_2,\boldsymbol{\beta}) \simeq \frac{3.079}{\sqrt{10}} \simeq 0.974$$

giving the angle between $\mathbf{e}_1$ and $\boldsymbol{\beta}$ as $\theta(\mathbf{e}_1,\boldsymbol{\beta}) \simeq 76.7^\circ$, while $\theta(\mathbf{e}_2,\boldsymbol{\beta})\simeq 13.2^\circ$. Applying Equation A5.3b, the corresponding scaled projections of $\boldsymbol{\beta}$ on these eigenvectors are

$$\begin{aligned}\lambda_1\text{Proj}(\boldsymbol{\beta}\ \text{on}\,\mathbf{e}_1) &= \lambda_1\cos(\theta|\mathbf{e}_1,\boldsymbol{\beta})\,||\boldsymbol{\beta}||\,\mathbf{e}_1 = \left(5.236\cdot 0.230\cdot\sqrt{10}\right)\mathbf{e}_1 \\ &= 3.803\begin{pmatrix}-0.851 \\ 0.526\end{pmatrix} = \begin{pmatrix}-3.236 \\ 2\end{pmatrix}\end{aligned}$$

$$\begin{aligned}\lambda_2\text{Proj}(\boldsymbol{\beta}\,\text{on}\,\mathbf{e}_2) &= \lambda_2\cos(\theta|\mathbf{e}_2,\boldsymbol{\beta})\,||\boldsymbol{\beta}||\,\mathbf{e}_2 = \left(0.764\cdot 0.974\cdot\sqrt{10}\right)\mathbf{e}_2 \\ &= 2.353\begin{pmatrix}0.526 \\ 0.851\end{pmatrix} = \begin{pmatrix}1.236 \\ 2\end{pmatrix}\end{aligned}$$

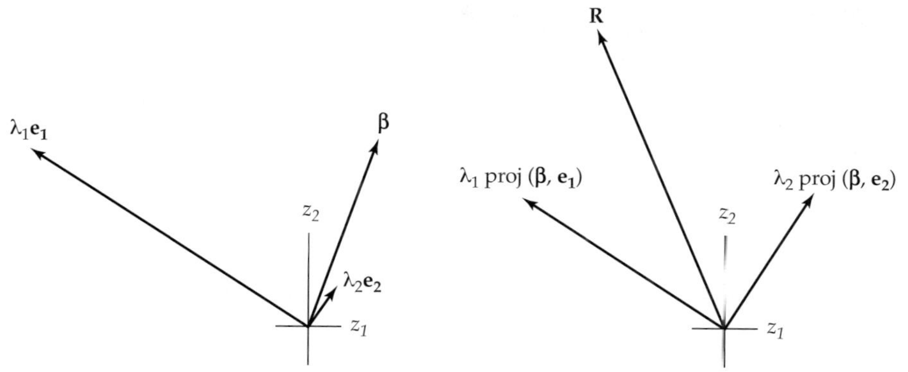

Figure A5.3 **Left:** The scaled eigenvectors associated with the covariance matrix, **G**, from Example A5.1, plotted along with the selection gradient, $\boldsymbol{\beta}$. Note that $\mathbf{e}_1$ and $\mathbf{e}_2$ are orthogonal and hence can be thought of as describing a new coordinate system. Because $\lambda_1 \gg \lambda_2$, the leading eigenvector, $\mathbf{e}_1$, largely dominates the transformation. **Right:** This is shown by taking the projections of $\boldsymbol{\beta}$ on each of these eigenvectors (shown here on a magnified scale relative to the left figure). Even though $\boldsymbol{\beta}$ is nearly parallel to $\mathbf{e}_2$ ($\theta|\mathbf{e}_1, \boldsymbol{\beta} = 13.2°$), the projection of $\boldsymbol{\beta}$ on $\mathbf{e}_1$ yields a vector of greater length than the projection of $\boldsymbol{\beta}$ on $\mathbf{e}_2$ (3.803 versus 2.353). From Equation A5.9b, the vector of responses to selection, **R**, is the sum of these two projections.

From Equation A5.9b, we can express the response, **R**, as the sum of the projections of $\boldsymbol{\beta}$ onto the eigenvalues of **G**, returning

$$\mathbf{R} = \mathbf{G}\boldsymbol{\beta} = \lambda_1 \text{Proj}(\boldsymbol{\beta} \text{ on } \mathbf{e}_1) + \lambda_2 \text{Proj}(\boldsymbol{\beta} \text{ on } \mathbf{e}_2)$$
$$= \begin{pmatrix} -3.236 \\ 2 \end{pmatrix} + \begin{pmatrix} 1.236 \\ 2 \end{pmatrix} = \begin{pmatrix} -2 \\ 4 \end{pmatrix}$$

As Figure A5.3 shows, the eigenstructure of **G** explains the unusual behavior of the selection response seen in Figure A5.2. The eigenvector associated with the **leading eigenvalue**, λ_1, accounts for most of the variation inherent in **G** (87%, as $\lambda_1/(\lambda_1 + \lambda_2) = 0.87$), and this eigenvector corresponds to a strong negative correlation between the additive-genetic values of z_1 and z_2. Hence, even though $\boldsymbol{\beta}$ points in very much the same direction as $\mathbf{e}_2$, because $\lambda_1 \gg \lambda_2$, the projection of $\boldsymbol{\beta}$ on $\mathbf{e}_1$ yields a vector of greater length than the projection of $\boldsymbol{\beta}$ on $\mathbf{e}_2$ (3.803 versus 2.353), and it is this $\mathbf{e}_1$ projection vector that results in the decrease in μ_{z_1}.

PROPERTIES OF SYMMETRIC MATRICES

Many of the matrices encountered in quantitative genetics are **symmetric**, satisfying $\mathbf{A} = \mathbf{A}^T$ (and therefore necessarily square). Examples include covariance matrices and the γ matrix of quadratic coefficients in the Pearson-Lande-Arnold fitness regression (Chapter 30). Symmetric matrices have a number of useful properties (proofs of which can be found in Dhrymes 1978; Horn and Johnson 1985; and Wilf 1978):

1. *If* **A** *is symmetric, then if* $\mathbf{A}^{-1}$ *exists, it is also symmetric.*
2. *The eigenvalues and eigenvectors of a symmetric matrix are all real.*
3. *For any n-dimensional symmetric matrix, a corresponding set of n orthonormal eigenvectors can be constructed,* namely, we can obtain a set of eigenvalues $\mathbf{e}_i$ for $1 \le i \le n$ that satisfies

$$\mathbf{e}_i{}^T\mathbf{e}_j = \begin{cases} 1 & \text{if } i = j \\ 0 & \text{if } i \ne j \end{cases}$$

   In particular, this guarantees that a spectral decomposition of **A** exists (Equation A5.9a).

4. A symmetric matrix $\mathbf{A}$ can be **diagonalized** as

$$\mathbf{A} = \mathbf{U}\boldsymbol{\Lambda}\mathbf{U}^T \tag{A5.10a}$$

where $\boldsymbol{\Lambda}$ is a diagonal matrix and $\mathbf{U}$ is an orthonormal matrix ($\mathbf{U}^{-1} = \mathbf{U}^T$). If λ_i and $\mathbf{e}_i$ are the ith eigenvalue and its associated unit eigenvector of $\mathbf{A}$, then

$$\boldsymbol{\Lambda} = \text{diag}(\lambda_1, \lambda_2, \cdots, \lambda_n) = \begin{pmatrix} \lambda_1 & 0 & \cdots & 0 \\ 0 & \lambda_2 & \cdots & 0 \\ \vdots & & \ddots & \vdots \\ 0 & \cdots & \cdots & \lambda_n \end{pmatrix} \tag{A5.10b}$$

and

$$\mathbf{U} = (\, \mathbf{e}_1, \mathbf{e}_2, \cdots, \mathbf{e}_n \,) \tag{A5.10c}$$

Geometrically, $\mathbf{U}$ is a unity matrix and thus describes a rigid rotation of the original coordinate system to a new coordinate system given by the eigenvectors of $\mathbf{A}$, while the diagonal elements of $\boldsymbol{\Lambda}$ give the amount by which vectors of unit length in the original coordinate system are scaled in the transformed system. If we use the decomposition $\boldsymbol{\Lambda} = \sum_{i=1}^{n} \boldsymbol{\Lambda}_i$, where $\boldsymbol{\Lambda}_i$ is a diagonal matrix whose elements are all zero, except for λ_i, then Equation A5.10a becomes

$$\mathbf{A} = \mathbf{U}\left(\sum_{i=1}^{n} \boldsymbol{\Lambda}_i\right)\mathbf{U}^T = \sum_{i=1}^{n} \mathbf{U}\boldsymbol{\Lambda}_i\mathbf{U}^T = \sum_{i=1}^{n} \lambda_i \mathbf{e}_i \mathbf{e}_i^T \tag{A5.10d}$$

recovering the spectral decomposition (Equation A5.9a). The last step in Equation A5.10d follows because $\mathbf{e}_i^T\mathbf{e}_j = 0$ for $i \neq j$. Because of this feature, Equation A5.10a is also called the **spectral factorization** or **eigendecomposition** of $\mathbf{A}$.

Using Equation A5.10a, it is easy to show that

$$\mathbf{A}^{-1} = \mathbf{U}\boldsymbol{\Lambda}^{-1}\mathbf{U}^T \tag{A5.11a}$$

To see this, note that

$$\mathbf{A}^{-1}\mathbf{A} = \left(\mathbf{U}\boldsymbol{\Lambda}^{-1}\mathbf{U}^T\right)\left(\mathbf{U}\boldsymbol{\Lambda}\mathbf{U}^T\right) = \mathbf{U}\boldsymbol{\Lambda}^{-1}\left(\mathbf{U}^T\mathbf{U}\right)\boldsymbol{\Lambda}\mathbf{U}^T = \mathbf{U}\boldsymbol{\Lambda}^{-1}\boldsymbol{\Lambda}\mathbf{U}^T = \mathbf{U}\mathbf{U}^T = \mathbf{I}$$

Similar logic yields

$$\mathbf{A}^{1/2} = \mathbf{U}\boldsymbol{\Lambda}^{1/2}\mathbf{U}^T \tag{A5.11b}$$

$$\mathbf{A}^{-1/2} = \mathbf{U}\boldsymbol{\Lambda}^{-1/2}\mathbf{U}^T \tag{A5.11c}$$

$$\mathbf{A}^{k} = \mathbf{U}\boldsymbol{\Lambda}^{k}\mathbf{U}^T \quad \text{for any integer } k \tag{A5.11d}$$

where the **square root matrix**, $\mathbf{A}^{1/2}$, satisfies $\mathbf{A}^{1/2}\mathbf{A}^{1/2} = \mathbf{A}$, and $\mathbf{A}^{-1/2}$ satisfies $\mathbf{A}^{-1/2}\mathbf{A} = \mathbf{A}\mathbf{A}^{-1/2} = \mathbf{A}^{1/2}$, as well as $\mathbf{A}^{-1/2}\mathbf{A}^{1/2} = \mathbf{A}^{1/2}\mathbf{A}^{-1/2} = \mathbf{I}$.

Because $\boldsymbol{\Lambda}$ is diagonal, the ith diagonal elements of $\boldsymbol{\Lambda}^{-1}$, $\boldsymbol{\Lambda}^{1/2}$, $\boldsymbol{\Lambda}^{-1/2}$, and $\boldsymbol{\Lambda}^{k}$ are λ_i^{-1}, $\lambda_i^{1/2}$, $\lambda_i^{-1/2}$, and λ_i^{k}, respectively, implying that if λ_i is an eigenvalue of $\mathbf{A}$, then λ_i^{-1}, $\lambda_i^{1/2}$, $\lambda_i^{-1/2}$, and λ_i^{k}, respectively, are eigenvalues of the matrices $\mathbf{A}^{-1}$, $\mathbf{A}^{1/2}$, $\mathbf{A}^{-1/2}$, and $\mathbf{A}^{k}$. Note that Equations A5.11a–A5.11d further imply that the matrices $\mathbf{A}$, $\mathbf{A}^{-1}$, $\mathbf{A}^{1/2}$, $\mathbf{A}^{-1/2}$, and $\mathbf{A}^{k}$ all have the same eigenvectors, namely the columns of $\mathbf{U}$. Finally, using Equation A5.10a, we see that premultiplying $\mathbf{A}$ by $\mathbf{U}^T$ and then postmultiplying by $\mathbf{U}$ gives a diagonal matrix whose elements are the eigenvalues of $\mathbf{A}$

$$\mathbf{U}^T\mathbf{A}\mathbf{U} = \mathbf{U}^T(\mathbf{U}\boldsymbol{\Lambda}\mathbf{U}^T)\mathbf{U} = (\mathbf{U}^T\mathbf{U})\boldsymbol{\Lambda}(\mathbf{U}^T\mathbf{U}) = \boldsymbol{\Lambda} \tag{A5.12}$$

5. The **Rayleigh-Ritz** theorem gives useful bounds on quadratic products associated with the symmetric matrix $\mathbf{A}$. It states that if the eigenvalues of $\mathbf{A}$ are ordered as $\lambda_{max} = \lambda_1 \geq \lambda_2 \geq \cdots \geq \lambda_n = \lambda_{min}$, then for any vector of constants $\mathbf{c}$ (for $||\mathbf{c}|| > 0$)

$$\lambda_1 \, ||\mathbf{c}|| \geq \mathbf{c}^T \mathbf{A} \mathbf{c} \geq \lambda_n \, ||\mathbf{c}|| \tag{A5.13a}$$

If $\mathbf{c}$ is of unit length, then all quadratic products are bounded by

$$\lambda_1 \geq \mathbf{c}^T \mathbf{A} \mathbf{c} \geq \lambda_n \tag{A5.13b}$$

The maximum and minimum quadratic products occur, respectively, when $\mathbf{c} = \mathbf{e}_1$ and $\mathbf{c} = \mathbf{e}_n$, the eigenvectors associated with λ_1 and λ_n. This is a useful result for bounding variances. Consider a univariate random variable, $y = \mathbf{c}^T \mathbf{x}$, formed by a linear combination of the elements of a random vector, $\mathbf{x}$. Recall from LW Equation 8.19 that the variance of a sum $y = \mathbf{c}^T \mathbf{x}$ is $\sigma^2(y) = \mathbf{c}^T \mathbf{V_x} \mathbf{c}$, where $\mathbf{V_x}$ is the covariance matrix for $\mathbf{x}$. If we apply Equation A5.13a we obtain

$$\lambda_1 ||\mathbf{c}||^2 \geq \sigma^2(y) \geq \lambda_n ||\mathbf{c}||^2 \tag{A5.14}$$

where λ_1 is the largest (leading or **dominant**) eigenvalue and λ_n is the smallest eigenvalue of the covariance matrix $\mathbf{V_x}$.

Example A5.3. Consider the additive-genetic covariance matrix $\mathbf{G}$ from Examples A5.1 and A5.2. Recalling the results from Example A5.2 and using Equation A5.10a, we can express $\mathbf{G}$ as $\mathbf{U} \boldsymbol{\Lambda} \mathbf{U}^T$, where

$$\boldsymbol{\Lambda} = \begin{pmatrix} 5.241 & 0 \\ 0 & 0.765 \end{pmatrix} \quad \text{and} \quad \mathbf{U} = (\, \mathbf{e}_1 \quad \mathbf{e}_2 \,) = \left(\begin{pmatrix} -0.851 \\ 0.526 \end{pmatrix} \quad \begin{pmatrix} 0.526 \\ 0.851 \end{pmatrix} \right)$$

From Equation A5.11a, the eigenvalues of $\mathbf{A}^{-1}$ are $(5.241)^{-1} \simeq 0.191$ and $(0.765)^{-1} \simeq 1.307$, while from Equation A5.11b, the eigenvalues of $\mathbf{A}^{1/2}$ are $\sqrt{5.241} \simeq 2.289$ and $\sqrt{0.765} \simeq 0.875$.

Correlations Can Be Removed by a Matrix Transformation

A powerful use of diagonalization is that it allows one to extract a set of n uncorrelated variables for any $n \times n$ nonsingular covariance matrix, $\mathbf{V_x}$. Consider the transformation

$$\mathbf{y} = \mathbf{U}^T \mathbf{x} \tag{A5.15a}$$

where $\mathbf{U} = (\mathbf{e}_1, \mathbf{e}_2, \cdots, \mathbf{e}_n)$ contains the normalized eigenvectors of $\mathbf{V_x}$. Because $\mathbf{U}$ is an orthonormal matrix, this transformation is a rigid rotation of the axes of the original $(x_1, \cdots, x_n)$ coordinate system to a new system given by $(e_1, \cdots, e_n)$. Applying LW Equation 8.21b and Equation A5.12, respectively, the covariance matrix for $\mathbf{y}$ is

$$\mathbf{V_y} = \mathbf{U}^T \mathbf{V_x} \mathbf{U} = \boldsymbol{\Lambda} \tag{A5.15b}$$

where $\boldsymbol{\Lambda}$ is a diagonal matrix whose elements are the eigenvalues of $\mathbf{V_x}$,

$$\sigma(y_i, y_j) = \begin{cases} \lambda_i & \text{if } i = j \\ 0 & \text{if } i \neq j \end{cases}$$

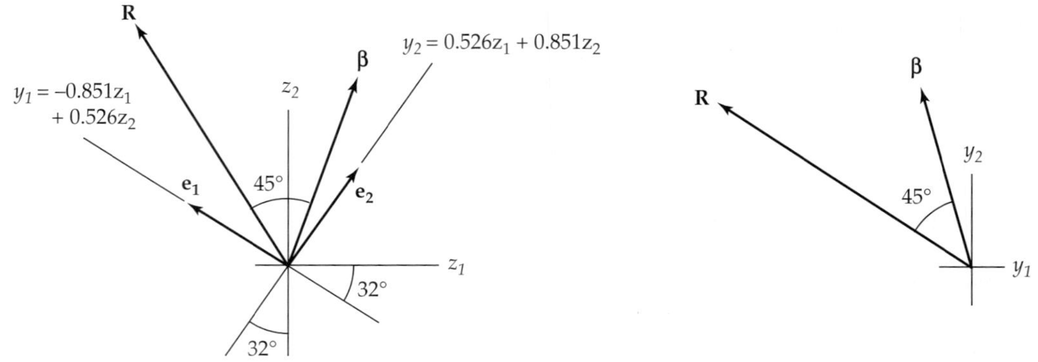

Figure A5.4 The transformation (Equation A5.15a) generating a set of independent variables for the covariance matrix **G** from Example A5.4 results in a rigid rotation of axes of the original traits onto the new, uncorrelated set. **Left:** The direction of the new axes are given by the eigenvectors $\mathbf{e}_1$ and $\mathbf{e}_2$. The angle between the new axis, $\mathbf{e}_1$, and the original $\mathbf{z}_1$ axis is given by the angle between $\mathbf{e}_1$ and $\mathbf{z}_1 = (1, 0)^T$. Here, $||\, \mathbf{e}_1 \,|| = ||\, \mathbf{z}_1 \,|| = 1$ and $\mathbf{e}_1{}^T \mathbf{z}_1 = 0.851$, giving $\theta = \cos^{-1}(0.851) \simeq 32°$. As this transformation is a rigid rotation, the angle between $\mathbf{e}_2$ and the $\mathbf{z}_2 = (0, 1)^T$ axis is also 32°. **Right:** On the (y_1, y_2) coordinates, the angle between **R** and $\boldsymbol{\beta}$ remains unchanged. See Example A5.4 for further details.

The rigid rotation introduced by **U** creates a set of n uncorrelated variables, the ith of which is

$$y_i = \mathbf{e}_i{}^T \mathbf{x} \tag{A5.15c}$$

Because the $\mathbf{e}_i$ are of unit length, from Equation A5.3b we have that $y_i = \mathbf{e}_i{}^T \mathbf{x}$ is the length of the projection of **x** onto the ith eigenvector of $\mathbf{V_x}$, which implies that the axes of the new coordinate system are given by the orthogonal set of eigenvectors of $\mathbf{V_x}$.

Defining the matrix **B** as

$$\mathbf{B} = \mathbf{U}\,\boldsymbol{\Lambda}^{-1/2} \tag{A5.15d}$$

the vector $\mathbf{y} = \mathbf{B}^T \mathbf{x}$ has a covariance matrix of $\mathbf{V_y} = \mathbf{I}$, which means that this transformation creates a set of uncorrelated variables, each with unit variance. To see this, note that

$$\begin{aligned} \mathbf{V_y} = \mathbf{B}^T \mathbf{V_x} \mathbf{B} &= \left(\mathbf{U}\boldsymbol{\Lambda}^{-1/2}\right)^T \left(\mathbf{U}\boldsymbol{\Lambda}\mathbf{U}^T\right) \left(\mathbf{U}\boldsymbol{\Lambda}^{-1/2}\right) \\ &= \boldsymbol{\Lambda}^{-1/2} \left(\mathbf{U}^T\mathbf{U}\right) \boldsymbol{\Lambda} \left(\mathbf{U}^T\mathbf{U}\right) \boldsymbol{\Lambda}^{-1/2} \\ &= \boldsymbol{\Lambda}^{-1/2}\, \boldsymbol{\Lambda}\, \boldsymbol{\Lambda}^{-1/2} = \mathbf{I} \end{aligned} \tag{A5.15e}$$

An alternative to Equation A5.15d is the **Cholesky decomposition**, $\mathbf{A} = \mathbf{C}^T\mathbf{C}$, of a square, symmetric matrix **A**, where **C** is an **lower triangular matrix** (all elements above the diagonal are zero). If **C** is the Cholesky decomposition for $\mathbf{V_x}$, then $\mathbf{y} = \mathbf{C}^{-1}\mathbf{x}$ also returns a covariance matrix of **I**.

Example A5.4. If we apply the change of variables suggested by Equation A5.15a to the vector, **z**, of characters with associated **G** matrix used in Example A5.1 and using the eigenvalues and vectors obtained in Example A5.2 yields

$$\begin{aligned} \mathbf{y} = \mathbf{U}^T \mathbf{z} &= \begin{pmatrix} \mathbf{e}_1{}^T \\ \mathbf{e}_2{}^T \end{pmatrix} \begin{pmatrix} z_1 \\ z_2 \end{pmatrix} \\ &= \begin{pmatrix} -0.851 & 0.526 \\ 0.526 & 0.851 \end{pmatrix} \begin{pmatrix} z_1 \\ z_2 \end{pmatrix} \\ &= \begin{pmatrix} -0.851 z_1 + 0.526 z_2 \\ 0.526 z_1 + 0.851 z_2 \end{pmatrix} \end{aligned}$$

From Equation A5.15b, $\mathbf{V_y} = \boldsymbol{\Lambda}$ as given in Example A5.3, showing that y_1 and y_2 are uncorrelated with $\sigma^2(y_1) = \lambda_1 = 5.241$ and $\sigma^2(y_2) = \lambda_2 = 0.765$. Hence, by considering the new coordinate system with

$$y_1 = \mathbf{e}_1{}^T\mathbf{z} = -0.851z_1 + 0.526z_2 \quad \text{and} \quad y_2 = \mathbf{e}_2{}^T\mathbf{z} = 0.526z_1 + 0.851z_2$$

we can transform the original coordinate system into a new system cn which there are no additive-genetic correlations between these new characters. Figure A5.4 shows that this transformation is simply a rigid rotation of the axes.

Likewise, from Equation A5.15d, the transformation that yields uncorrelated variables with unit variance is

$$\begin{aligned}\mathbf{y} = \boldsymbol{\Lambda}^{-1/2}\mathbf{U}^T\mathbf{z} &= \begin{pmatrix} 1/\sqrt{\lambda_1} & 0 \\ 0 & 1/\sqrt{\lambda_2} \end{pmatrix} \begin{pmatrix} \mathbf{e}_1{}^T \\ \mathbf{e}_2{}^T \end{pmatrix} \begin{pmatrix} z_1 \\ z_2 \end{pmatrix} \\ &= \begin{pmatrix} 1/\sqrt{5.236} & 0 \\ 0 & 1/\sqrt{0.764} \end{pmatrix} \begin{pmatrix} -0.851 & 0.526 \\ 0.526 & 0.851 \end{pmatrix} \begin{pmatrix} z_1 \\ z_2 \end{pmatrix} \\ &= \begin{pmatrix} -0.372 & 0.230 \\ 0.602 & 0.974 \end{pmatrix} \begin{pmatrix} z_1 \\ z_2 \end{pmatrix}\end{aligned}$$

Hence, the transformed variables $y_1 = -0.372z_1 + 0.230z_2$ and $y_2 = 0.602z_1 + 0.974z_2$ are uncorrelated, and each has unit variance.

An alternative set of uncorrelated random variables follows from the Cholesky decomposition, which can be compute in R using the `chol` command. (As an aside, `chol` returns the upper-triangular version of the decomposition, which is simply the transpose of the lower-triangular version). The resulting decomposition is

$$\mathbf{G} = \begin{pmatrix} 4 & -2 \\ -2 & 2 \end{pmatrix} = \mathbf{C}\,\mathbf{C}^T = \begin{pmatrix} 2 & 0 \\ -1 & 1 \end{pmatrix} \begin{pmatrix} 2 & -1 \\ 0 & 1 \end{pmatrix}$$

yielding

$$\mathbf{y} = \mathbf{C}^{-1}\mathbf{z} = \begin{pmatrix} 0.5 & 0 \\ 0.5 & 1 \end{pmatrix} \begin{pmatrix} z_1 \\ z_2 \end{pmatrix}$$

or

$$y_1 = z_1/2 \quad \text{and} \quad z_2 = z_1/2 + z_2$$

as a new set of uncorrelated variables , each with unit variance. One nice feature about using a Cholesky decomposition is that we can always isolate a given variable of interest (simply by putting first in the vector). Because $\mathbf{C}$ is lower-triangular, it always returns the first new uncorrelated variable as a scalar times the first original variable (rather than some linear combination of all the variables, as was the case for the first decomposition in this example).

Simultaneous Diagonalization

An extension of the notion of diagonalization is the **simultaneous diagonalization** of two symmetric matrices, $\mathbf{P}$ and $\mathbf{G}$, of the same dimension. There exists a matrix $\mathbf{T}$ such that

$$\mathbf{T}^T\mathbf{P}\mathbf{T} = \mathbf{I} \quad \text{and} \quad \mathbf{T}^T\mathbf{G}\mathbf{T} = \mathbf{D} \tag{A5.16}$$

where $\mathbf{D}$ is a diagonal matrix, whose elements are the eigenvalues of $\mathbf{P}^{-1}\mathbf{G}$. Hence, the same transformation simultaneously diagonalizes both $\mathbf{P}$ and $\mathbf{G}$. If one has a series of traits with both genetic ($\mathbf{G}$) and phenotypic ($\mathbf{P}$) covariances, they can be transformed to a scale where the new traits (based on linear combinations of the original traits) are genetically and phenotypically uncorrelated, where the elements of $\mathbf{D}$ correspond to the heritabilities of these new traits.

Example A5.5. To find the matrix, $\mathbf{T}$, that simultaneously diagonalizes both $\mathbf{P}$ and $\mathbf{G}$, we first use Equation A5.10a to write

$$\mathbf{P} = \mathbf{U}\boldsymbol{\Lambda}\mathbf{U}^T$$

where $\boldsymbol{\Lambda}$ is a diagonal matrix and $\mathbf{U}^T\mathbf{U} = \mathbf{U}\mathbf{U}^T = \mathbf{I}$. Defining $\mathbf{B} = \mathbf{U}\boldsymbol{\Lambda}^{-1/2}$, Equation A5.15e showed that $\mathbf{B}^T\mathbf{P}\mathbf{B} = \mathbf{I}$. Next, note for $\mathbf{M} = \mathbf{B}^T\mathbf{G}\mathbf{B}$, that $\mathbf{M} = \mathbf{M}^T$ (i.e., $\mathbf{M}$ is symmetric), as

$$\mathbf{M}^T = \left(\mathbf{B}^T\mathbf{G}\mathbf{B}\right)^T = \mathbf{B}^T\mathbf{G}^T\mathbf{B} = \mathbf{B}^T\mathbf{G}\mathbf{B} = \mathbf{M}$$

Hence, we can also diagonalize $\mathbf{M}$,

$$\mathbf{C}^T\mathbf{M}\mathbf{C} = \mathbf{D}$$

where $\mathbf{D}$ is a diagonal matrix and $\mathbf{C}^T\mathbf{C} = \mathbf{C}\mathbf{C}^T = \mathbf{I}$. Thus,

$$\mathbf{C}^T\mathbf{M}\mathbf{C} = \mathbf{C}^T\left(\mathbf{B}^T\mathbf{G}\mathbf{B}\right)\mathbf{C} = (\mathbf{B}\mathbf{C})^T\,\mathbf{G}\,(\mathbf{B}\mathbf{C}) = \mathbf{D}$$

Defining

$$\mathbf{T} = \mathbf{B}\mathbf{C} = \mathbf{U}\boldsymbol{\Lambda}^{-1/2}\mathbf{C}$$

we have from the previous expression that

$$\mathbf{T}^T\mathbf{G}\mathbf{T} = \mathbf{D}$$

Likewise,

$$\mathbf{T}^T\mathbf{P}\mathbf{T} = (\mathbf{B}\mathbf{C})^T\,\mathbf{P}\,(\mathbf{B}\mathbf{C}) = \mathbf{C}^T\left(\mathbf{B}^T\mathbf{P}\mathbf{B}\right)\mathbf{C} = \mathbf{C}^T\mathbf{C} = \mathbf{I}$$

showing that the matrix $\mathbf{T}$ satisfies Equation A5.16.

CANONICAL AXES OF QUADRATIC FORMS

The transformation $\mathbf{y} = \mathbf{U}^T\mathbf{x}$ given by Equation A5.15a applies to any symmetric matrix, and is referred to as its **canonical transformation**. This simplifies the interpretation of the quadratic form $\mathbf{x}^T\mathbf{A}\mathbf{x}$, as rotation of the original axes to align them with the eigenvectors of $\mathbf{A}$ removes all cross-product terms (x_ix_j for $i \neq j$) on this new coordinate system. Recall (Equation A5.5b) that $\mathbf{U}$ is a unitary matrix and hence $\mathbf{U}^T = \mathbf{U}^{-1}$. Thus,

$$\mathbf{U}\mathbf{y} = \mathbf{U}\mathbf{U}^T\mathbf{x} = \mathbf{x}$$

Applying Equations A5.15a and A5.12 transforms a quadratic form to one in which the square matrix is diagonal, which greatly simplifies the resulting quadratic product, as

$$\begin{aligned}\mathbf{x}^T\mathbf{A}\,\mathbf{x} &= (\mathbf{U}\mathbf{y})^T\mathbf{A}\mathbf{U}\mathbf{y} = \mathbf{y}^T(\mathbf{U}^T\mathbf{A}\mathbf{U})\mathbf{y} \\ &= \mathbf{y}^T\,\boldsymbol{\Lambda}\,\mathbf{y} \\ &= \sum_{i=1}^{n}\lambda_i\,y_i^2, \quad \text{with} \quad y_i = \mathbf{e}_i{}^T\mathbf{x}\end{aligned} \tag{A5.17a}$$

where λ_i and $\mathbf{e}_i$ are the eigenvalues and associated (normalized, i.e., $||\mathbf{e}_i|| = 1$) eigenvectors of $\mathbf{A}$. The new axes defined by the $\mathbf{e}_i$ vectors are the **canonical** (or **principal**)

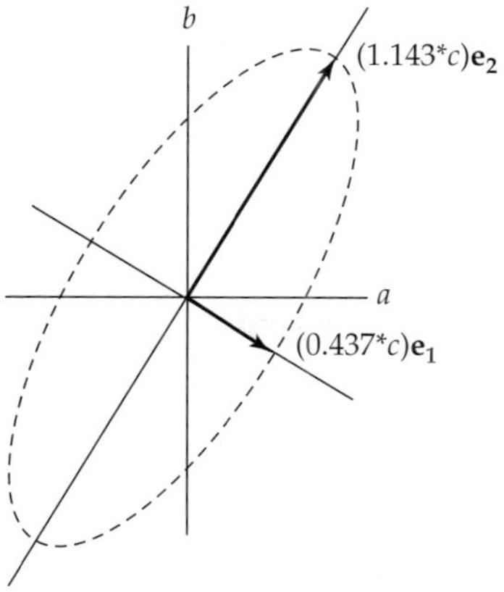

Figure A5.5 The general shape of surfaces of constant variance for the additive-genetic covariance matrix, **G**, given in Example A5.1. Defining a new composite character $y = az_1 + bz_2$, the rotated ellipse represents the set of weights (a, b) that give y the same additive-genetic variance, c^2. The major axis of the ellipse is along $\mathbf{e}_2$, the eigenvector associated with the smallest eigenvalue of **G**, where $\lambda_2 \simeq 0.765$, giving $1/\sqrt{\lambda_2} \simeq 1.143$. The minor axis of the ellipse is along $\mathbf{e}_1$, the eigenvector associated with the largest eigenvalue of **G**, where $\lambda_1 \simeq 5.241$, giving $1/\sqrt{\lambda_1} \simeq 0.437$.

axes of **A**. Because $y_i^2 \geq 0$, Equation A5.17a immediately shows the connection between the signs of the eigenvalues of a matrix and whether that matrix is positive definite, negative definite, or indefinite.

If all eigenvalues are positive (all $\lambda_i > 0$), then any quadratic form is always positive (unless all the y_i are zero) and hence **A** is **positive definite**. If one or more of the eigenvalues are zero, while the rest are positive, then **A** is said to be **positive semidefinite**, implying that quadratic products are either zero (corresponding to $\lambda_i = 0$) or positive. If all eigenvalues are negative (all $\lambda_i < 0$), then **A** is **negative definite** as any quadratic form is always negative, while **A** is said to be **negative semidefinite** if the eigenvalues are either zero or negative. If **A** has both positive and negative eigenvalues it is said to be **indefinite**, as quadratic products can be either positive or negative.

Equations of the form

$$\mathbf{x}^T \mathbf{A} \mathbf{x} = \sum_{i=1}^{n} \sum_{j=1}^{n} A_{ij} x_i x_j = c^2 \tag{A5.17b}$$

arise fairly frequently in quantitative genetics. For example, they describe surfaces of constant variance (Figure A5.5) or constant fitnesses in quadratic fitness regressions (Chapter 30). Solutions to Equation A5.17b describe **quadratic surfaces**—for two dimensions, these are the familiar conic sections (ellipses, parabolas, or hyperbolas). Equation A5.17a greatly simplifies the interpretation of these surfaces by removing all crcss product terms, yielding

$$\mathbf{x}^T \mathbf{A} \mathbf{x} = \sum_{i=1}^{n} \lambda_i y_i^2 = c^2 \tag{A5.17c}$$

Because $(y_i)^2$ and $(-y_i)^2$ have the same value, the canonical axes of **A** are also the **axes of symmetry** for the quadratic surface generated by quadratic forms involving **A**. When all eigenvalues of **A** are positive (as occurs with nonsingular covariance and other positive-definite matrices), Equation A5.17c describes an ellipsoid whose axes of symmetry are given by the eigenvectors of **A**. The distance from the origin to the surface along the $\mathbf{e}_i$ axis is $\lambda_i y_i^2 = c^2$ or $y_i = c\lambda_i^{-1/2}$, as can been seen by setting all the y_k equal to zero except for y_i, which yields

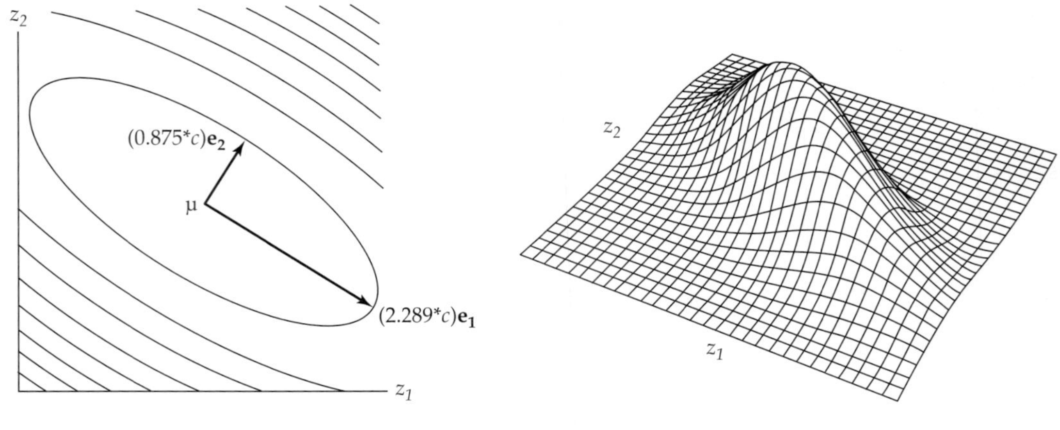

Figure A5.6 Surfaces for a multivariate normal (MVN) distribution. **Left:** Surfaces of equal probability assuming that the additive-genetic values associated with the characters z_1 and z_2 in Example A5.1 are $\sim$MVN($\boldsymbol{\mu}$,**G**). These surfaces are ellipses centered at μ, with the major axis of the ellipse along $\mathbf{e}_1$ and the minor axis along $\mathbf{e}_2$, whose lengths (for a fixed c) are, respectively, $\sqrt{\lambda_1} = 2.289$ and $\sqrt{\lambda_2} = 0.875$. **Right:** A plot of the associated probability density. Slicing along either the major or minor axis gives a normal curve. Because the variance in the major axis is greater, the curve is much broader along this axis. The covariance between the breeding values of z_1 and z_2 rotates the distribution so that the principal axes $(\mathbf{e}_1, \mathbf{e}_2)$ do not coincide with the original (z_1, z_2) axes.

$\mathbf{x}^T\,\mathbf{A}\,\mathbf{x} = \lambda_i y_i^2 = c^2$.

Consider a new variable (y) that is a weighted combination $y = ax_1 + bx_2 = \mathbf{b}^T\mathbf{x}$ of the original vector (**x**) of random variables, where $\mathbf{b}^T = (a, b)$. Its resulting variance is

$$\sigma^2(y) = a^2\sigma^2(x_1) + 2ab\sigma(x_1, x_2) + b^2\sigma^2(x_2) = \mathbf{b}^T\,\mathbf{V_x}\,\mathbf{b}$$

As shown in Figure A5.5, the collection of a, b values that result in the same variance (c^2) is the ellipse given by $c^2 = \mathbf{b}^T\,\mathbf{V_x}\,\mathbf{b}$. Variables with a large amount of variance require smaller weights to achieve the constant value (c^2) than do variables with lower variances. Thus, on a **constant-variance surface**, minor axes correspond to directions with *the most* variance, while major axes correspond to the directions with *the least* variability. This is in contrast to **surfaces of equal probability** (Figure A5.6), where major axes correspond to directions with the most variance. The reason for this reversal of roles is that constant-variance surfaces are functions of $\lambda_i^{-1/2}$, whereas constant-probability surfaces are functions of $\lambda_i^{1/2}$.

Implications for the Multivariate Normal Distribution

Recall the probability density function for the multivariate normal distribution (LW Chapter 8)

$$\phi(\mathbf{x}) = (2\pi)^{-n/2}\,|\mathbf{V_x}|^{-1/2}\,\exp\left[-\frac{1}{2}(\mathbf{x}-\boldsymbol{\mu})^T\,\mathbf{V_x^{-1}}\,(\mathbf{x}-\boldsymbol{\mu})\right] \tag{A5.18a}$$

Because only the quadratic product in the exponential varies with **x**, surfaces of equal probability for MVN distributed vectors satisfy

$$(\mathbf{x}-\boldsymbol{\mu})^T\,\mathbf{V_x^{-1}}\,(\mathbf{x}-\boldsymbol{\mu}) = c^2 \tag{A5.18b}$$

From the discussion following Equation A5.17c, these surfaces are n-dimensional ellipsoids centered at $\boldsymbol{\mu}$ whose axes of symmetry are given by the principal components (the eigenvectors) of the covariance matrix, $\mathbf{V_x}$. The length of the ellipsoid along the ith axis is $c\sqrt{\lambda_i}$ where λ_i is the eigenvalue associated with the eigenvector $\mathbf{e}_i$ (Figure A5.6).

Equation A5.18b motivates the **Mahalanobis distance**

$$D = \sqrt{(\mathbf{x}-\boldsymbol{\mu})^T\,\mathbf{V_x^{-1}}\,(\mathbf{x}-\boldsymbol{\mu})} \tag{A5.19}$$

which measures the distance of a point from its mean $\boldsymbol{\mu}$, correcting for its covariance structure, $\mathbf{V_X}$ (Mahalanobis 1938). As we detail shortly, D provides one metric for tests of multivariate normality.

Applying the canonical transformation (Equation A5.15a), we can change coordinate systems by a rigid rotation to remove any correlations between the variables in $\mathbf{x}$. If $\mathbf{x} \sim \text{MVN}(\boldsymbol{\mu}, \mathbf{V_X})$, then for $\mathbf{y} = \mathbf{U}^T(\mathbf{x} - \boldsymbol{\mu})$, it follows that

$$\mathbf{y} \sim \text{MVN}(\mathbf{0}, \boldsymbol{\Lambda}) \tag{A5.20a}$$

where $\boldsymbol{\Lambda}$ and $\mathbf{U}$ are the matrices defined by Equations A5.10b and A5.10bc for the diagonalization of $\mathbf{V_X}$. In particular,

$$y_i = \mathbf{e}_i^T(\mathbf{x} - \boldsymbol{\mu}) \qquad \text{where} \quad y_i \sim \text{N}(0, \lambda_i) \tag{A5.20b}$$

Note from Equation A5.20a that because the y_i are uncorrelated, they are also independent as the joint probability density is the product of n individual univariate normal densities. We can further transform the original vector by taking

$$z_i = \frac{\mathbf{e}_i^T(\mathbf{x} - \boldsymbol{\mu})}{\sqrt{\lambda_i}} \qquad \text{giving} \qquad z_i \sim \text{N}(0, 1) \tag{A5.20c}$$

Applying the transformation

$$\mathbf{z} = \boldsymbol{\Lambda}^{-1/2}\mathbf{U}^T(\mathbf{x} - \boldsymbol{\mu}) \tag{A5.20d}$$

results in $\mathbf{z} \sim \text{MVN}(\mathbf{0}, \mathbf{I})$, namely that the n elements of the vector $\mathbf{y}$ are each independent unit normal random variables.

Principal Components of the Variance-Covariance Matrix

We are often interested in how the variance of a random vector can be decomposed into independent components. For example, even though we may be measuring n variables, only one or two of these may account for the majority of the variation. If this is the case, we may wish to exclude those variables contributing very little variation from further analysis. More generally, if random variables are correlated, then certain **linear combinations** of the elements of $\mathbf{x}$ may account for most of the variance. The procedure of **principal component analysis (PCA)** extracts these combinations by decomposing the variance of $\mathbf{x}$ into the contributions from a series of orthogonal vectors, the first of which explains the most variation possible for any single vector, the second the next possible amount, and so on until we account for the entire variance of $\mathbf{x}$.

Consider Figure A5.5. Because the set of points comprising the ellipse represents the set of linear combinations (i.e., the set of weights) of the random variables of $\mathbf{z}$ that yield **equal** variance, a little thought shows that the closer a point on this curve is to the origin, the more variance there is in that direction. The points closest to the origin are those that lie along the axis defined by $\mathbf{e}_1$, while those furthest away lie along the axis defined by $\mathbf{e}_2$. Here $\mathbf{e}_1$ and $\mathbf{e}_2$ are the principal components of $\mathbf{G}$, with the first principal component accounting for most of the variation of $\mathbf{G}$. In particular, the ratio of additive variances for the characters $y_1 = \mathbf{e}_1^T\mathbf{z}$ and $y_2 = \mathbf{e}_2^T\mathbf{z}$ is $\sigma^2(y_1)/\sigma^2(y_2) = \sigma^2(\mathbf{e}_1^T\,\mathbf{z})/\sigma^2(\mathbf{e}_2^T\,\mathbf{z}) = \mathbf{e}_1^T\,\mathbf{G}\,\mathbf{e}_1/\mathbf{e}_2^T\,\mathbf{G}\,\mathbf{e}_2 = \lambda_1/\lambda_2 \simeq 5.241/0.765 \simeq 6.85$, so that a character in the direction of $\mathbf{e}_1$ has almost seven times as much additive variance as a character lying in the direction of $\mathbf{e}_2$.

In general, suppose we have an n-dimensional covariance matrix, $\mathbf{V_X}$. If we order the eigenvalues of $\mathbf{V_X}$ as $\lambda_1 \geq \lambda_2 \geq \cdots \geq \lambda_n$, then Equation A5.13b gives the maximum variance for any linear combination of the elements of $\mathbf{x}$ ($y = \mathbf{c}_1^T\mathbf{x}$, subject to the constraint that $||\mathbf{c}_1|| = 1$), as

$$\max\ \sigma^2(y) = \max_{||\mathbf{c}_1||=1} \sigma^2(\mathbf{c}_1^T\mathbf{x}) = \mathbf{c}_1^T\mathbf{V_X}\mathbf{c}_1 = \lambda_1$$

which occurs when $\mathbf{c}_1 = \mathbf{e}_1$ (the normalized eigenvector associated with the leading eigenvalue λ_1). This vector is the **first principal component** (often abbreviated as **PC1**), and accounts for the fraction $\lambda_1/\text{tr}(\mathbf{V_x})$ of the total variation in $\mathbf{x}$. We can partition the remaining variance in $\mathbf{x}$ after the removal of PC1 in a similar fashion. For example, the vector $\mathbf{c}_2$, that is orthogonal to PC1 ($\mathbf{c}_2^T\mathbf{c}_1 = 0$) and maximizes the remaining variance can be shown to be $\mathbf{e}_2$, which accounts for a fraction $\lambda_2/\text{tr}(\mathbf{V_x})$ of the total variation in $\mathbf{x}$ (e.g., Morrison 1976; Johnson and Wichern 1988). By proceeding in this fashion, we can see that the ith PC is given by $\mathbf{e}_i$, and that the amount of variation it accounts for is

$$\lambda_i \Big/ \sum_{k=1}^{n} \lambda_k = \frac{\lambda_i}{\text{tr}(\mathbf{V_x})} \tag{A5.21}$$

Hence $\sum \lambda_i = \text{tr}(\mathbf{V_x})$ is the total variance of the vector $\mathbf{x}$, while $\lambda_i/\text{tr}(\mathbf{V_x})$ is the fraction of that total variance explained by the linear combination $\mathbf{e}_i^T\mathbf{x}$.

Example A5.6. Again let us consider the additive-genetic covariance matrix, $\mathbf{G}$, as shown in Examples A5.1 and A5.2. Because $\lambda_1 \simeq 5.241$, $\lambda_2 \simeq 0.765$, and $\text{tr}(\mathbf{G}) = 4 + 2 = 6$, the first PC explains $5.241/6 \simeq 0.8735$, or 87% of the variance in $\mathbf{G}$. While the first PC accounts for the majority of variation over the entire space of the variables ($\mathbf{x}$), the amount of variation explained by PC1 for any *particular* weighted combination, $y = \mathbf{b}^T\mathbf{x}$, of the original variables depends on the projection of $\mathbf{b}$ onto PC1. For example, if $\mathbf{b} = \mathbf{e}_2$ (the weight vector corresponds to the second eigenvector), then the projection of $\mathbf{b}$ onto PC1 has a length of zero, because PC1 is orthogonal to $\mathbf{e}_2$, and hence PC1 explains none of the variation of this new variable.

Example A5.7 serves as a brief introduction to the important field of **morphometrics**, which is concerned with quantification and comparison of sizes and shapes of organisms. The reader is referred to Pimentel (1979), Reyment et al. (1984), Elewa (2004), Claude (2008) and especially Bookstein et al. (1985), Rohlf and Bookstein (1990), Reyment (1991), Bookstein (1997), Slice (2005), and Zelditch et al. (2012) for detailed treatments.

Example A5.7. Jolicoeur and Mosimann (1960) measured three carapace characters in 24 males of the painted turtle (*Chrysemys picta marginata*). Letting z_1 be the carapace length, z_2 be the maximun carapace width, and z_3 be the carapace height, the resulting sample covariance matrix ($\mathbf{S_z}$, the sample estimate of $\mathbf{V_z}$) for these data was found to be

$$\mathbf{S_z} = \begin{pmatrix} 138.77 & 79.15 & 37.38 \\ 79.15 & 50.04 & 21.65 \\ 37.38 & 21.65 & 11.26 \end{pmatrix}$$

Hence, $\text{tr}(\mathbf{S_z}) = 138.77 + 50.04 + 11.26 = 200.07$. Using R, the eigenvalues of $\mathbf{S_z}$ are found to be

$$\lambda_1 = 195.280, \qquad \lambda_2 = 3.687, \qquad \lambda_3 = 1.103$$

which (as expected) sum to the value of the trace, 200.07. The associated (normalized) eigenvectors are similarly found to be

$$\mathbf{e}_1 = \begin{pmatrix} 0.840 \\ 0.492 \\ 0.229 \end{pmatrix}, \qquad \mathbf{e}_2 = \begin{pmatrix} 0.488 \\ -0.870 \\ 0.079 \end{pmatrix}, \qquad \mathbf{e}_3 = \begin{pmatrix} 0.213 \\ 0.043 \\ -0.971 \end{pmatrix}$$

PC1 accounts for 97.6% of the variation (195.281/200.07 = 0.976), while PC2 and PC3 account for 1.84% and 0.55%, respectively. Jolicoeur and Mosimann interpret PC1 as measuring overall size, as the new variable

$$y_1 = \mathbf{e}_1^T\mathbf{z} = 0.840z_1 + 0.492z_2 + 0.229z_3$$

corresponds to a simultaneous change in all three variables in the same direction, as is expected as individuals change their overall size. Likewise, PC2 and PC3 are

$$y_2 = \mathbf{e}_2^T \mathbf{z} = 0.488 z_1 - 0.870 z_2 + 0.079 z_3$$
$$y_3 = \mathbf{e}_3^T \mathbf{z} = 0.213 z_1 + 0.043 z_2 - 0.971 z_3$$

which Jolicoeur and Mosimann interpreted as measures of shape. Because the coefficient on z_3 is small relative to the others in PC2, they interpret PC2 as measuring the tradeoff between length (z_1) and width (z_2). Thus, after removing the variation in size, 1.84% of the remaining variation can be accounted for by differences in the shape measured by length versus width. Likewise, because the PC3 coefficient for z_2 is very small, PC3 mainly measures shape differences due to length (z_1) versus height (z_3).

This example points out some of the advantages, and possible pitfalls, of using principal component analysis for dimensional reduction of the data. Namely, replacing the n-component vector $\mathbf{z}$ by an $m < n$ component vector $\mathbf{y}$ composed of linear combinations of the $\mathbf{z}$, i.e., $\mathbf{y}_{m\times 1} = \mathbf{M}_{m\times n}\mathbf{z}_{n\times 1}$, where $\mathbf{M} = (\mathbf{e}_1, \cdots, \mathbf{e}_m)^T$, with $y_i = \mathbf{e}_i^T\mathbf{z}$. Essentially all (over 97%) of the variance in the three measured characters is accounted for by variation in overall size, with the remaining variation accounted for by differences in shape. While the temptation is strong to simply consider overall size and ignore all shape information, it might be the case that selection is largely ignoring variation in size and instead focusing on (size-independent) shape differences. In this case, an analysis ignoring shape (as would occur if only the new character generated by PC1 were considered) would be very misleading. A further complication with principal component analysis is that it can often be difficult to give biological interpretations to the new characters resulting from the rotation of the coordinate system.

TESTING FOR MULTIVARIATE NORMALITY

Multivariate normality is often assumed in statistical procedures, but it is less often tested. In LW Chapter 11 we briefly discussed two approaches for testing univariate normality, one graphical and the other based on deviations of observed skewness and/or kurtosis from Gaussian expectations. As we now demonstrate, both of these approaches can be extended to testing for multivariate normality. Additional methods are reviewed by Malkovich and Afifi (1973), Gnanadesikan (1977), Cox and Small (1978), Seber (1984), Looney (1995), and Henze (2002).

Graphical Tests: Chi-square Plots

A fairly simple graphical test can be developed by extending the notion of the normal probability plot that is used to check univariate normality (LW Chapter 11), where observations were ranked and then plotted against their ranked expected values under normality. Departures from linearity signify departures from normality, and we can apply this same approach to check for multivariate normality. From Equation A5.20d, if $\mathbf{z} \sim \text{MVN}(\boldsymbol{\mu}, \mathbf{V_z})$, then each element of the vector

$$\mathbf{y} = \boldsymbol{\Lambda}^{-1/2}\mathbf{U}^T(\mathbf{z} - \boldsymbol{\mu})$$

is an independent unit normal, so that $\mathbf{y} \sim \text{MVN}(\mathbf{0}, \mathbf{I})$. Recalling that $\mathbf{U}^{-1} = \mathbf{U}^T$, we can rearrange this expression to yield

$$(\mathbf{z} - \boldsymbol{\mu}) = \mathbf{U}\boldsymbol{\Lambda}^{1/2}\mathbf{y}$$

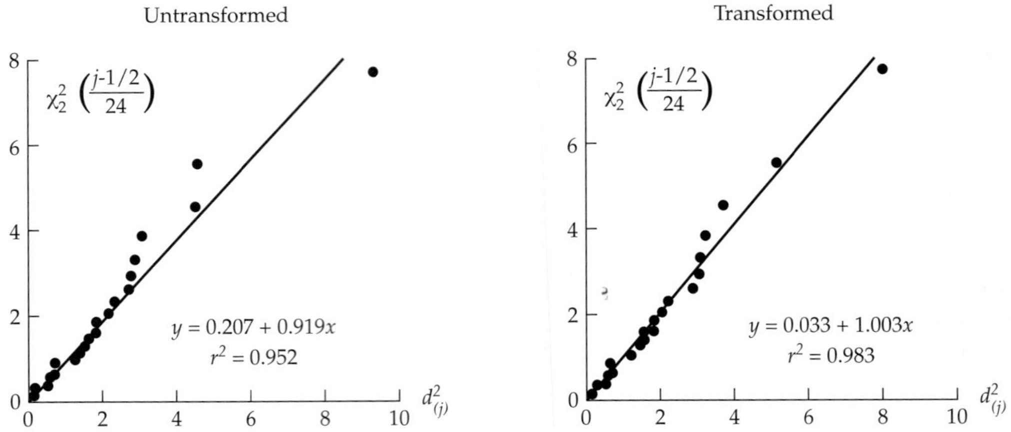

Figure A5.7 Plots of ranked distance data ($d^2_{(j)}$ being the jth smallest distance) versus the expected corresponding χ^2 value for the data of Jolicoeur and Mosimann from Example A5.8. **Left:** The untransformed data do not appear to depart significantly from linearity, although they depart slightly from the intercept (0) and slope (1) of the expected regression under multivariate normality. **Right:** Log-transforming the data gives a slightly better linear fit ($r^2 = 0.983$ versus $r^2 = 0.952$), with the best-fitting line passing through the origin as expected if the distance data follow a χ^2 distribution, and has a slope of essentially one. See Example A5.8 for more details.

Using this result and recalling Equation A5.11a, we have that

$$\begin{aligned}(\mathbf{z}-\boldsymbol{\mu})^T\mathbf{V}_{\mathbf{z}}^{-1}(\mathbf{z}-\boldsymbol{\mu}) &= \left(\mathbf{U}\boldsymbol{\Lambda}^{1/2}\mathbf{y}\right)^T\left(\mathbf{U}\boldsymbol{\Lambda}^{-1}\mathbf{U}^T\right)\left(\mathbf{U}\boldsymbol{\Lambda}^{1/2}\mathbf{y}\right)\\ &= \mathbf{y}^T\boldsymbol{\Lambda}^{1/2}\left(\mathbf{U}^T\mathbf{U}\right)\boldsymbol{\Lambda}^{-1}\left(\mathbf{U}^T\mathbf{U}\right)\boldsymbol{\Lambda}^{1/2}\mathbf{y}\\ &= \mathbf{y}^T\mathbf{y} = \sum_{i=1}^{n} y_i^2 \end{aligned} \tag{A5.22}$$

Thus if $\mathbf{z} \sim$ MVN, the quadratic form given by Equation A5.22 is the sum of n independent squared unit normal random variables. By definition, this sum is a χ^2 random variable with n degrees of freedom (LW Appendix 5), suggesting that one test for multivariate normality is to compare the goodness of fit of the scaled distances

$$d_i^2 = (\mathbf{z}_i - \overline{\mathbf{z}})^T\mathbf{S}_{\mathbf{z}}^{-1}(\mathbf{z}_i - \overline{\mathbf{z}}) \tag{A5.23}$$

to those generated by n (rank-ordered) draws from a χ^2_n. Here $\mathbf{z}_i$ is the vector of observations from the ith individual, $\overline{\mathbf{z}}$ the vector of sample means, and $\mathbf{S}_{\mathbf{z}}^{-1}$ the inverse of the sample covariance matrix. Note that the d_i are simply the squared Mahalanobis distances (Equation A5.19). We use the term *distance* because when $\mathbf{z}$ is transformed to $\mathbf{y}$, $\mathbf{V}_{\mathbf{y}} = \mathbf{I}$, giving the variance of the linear combination $\mathbf{c}^T\mathbf{y}$ as $\mathbf{c}^T\mathbf{V}_{\mathbf{y}}\mathbf{c} = \mathbf{c}^T\mathbf{I}\mathbf{c} = ||\mathbf{c}||^2$. Thus, regardless of orientation, any two $\mathbf{y}$ vectors having the same length also have the same variance, which equals their squared Euclidean distance.

The regression test for multivariate normality is based on *ordered* distances. Hence, we first order the distances generated by Equation A5.23 from smallest to largest,

$$d^2_{(1)} \leq d^2_{(2)} \leq \cdots \leq d^2_{(m)}$$

where m is the number of individuals sampled. Note that we use the subscription notation where $d^2_{(j)}$ denotes the jth smallest *distance* (the jth smallest value of Equation A5.23), whereas d_i^2 is the distance associated with the vector of observations for the ith *observation*.

Let $\chi_n^2(\alpha)$ correspond to the value of a chi-square random variable, X, with n degrees of freedom that satisfies Prob$[X \leq \chi_n^2(\alpha)] = \alpha$. Under multivariate normality, we expect the points

$$\left(d_{(i)}^2, \, \chi_n^2 \left[\frac{i - 1/2}{m} \right] \right) \qquad \text{for} \quad 1 \leq i \leq m$$

to fall along a line with a slope of one and an intercept of zero, as the ith ordered distance has i/m observations less than or equal to it (the factor of $1/2$ is added as a correction for continuity). As with univariate normal probability plots, departures from multivariate normality are indicated by departures from linearity. More formally, one can use a standard **Kolmogorov–Smirnov test** (Conover 1999) for comparing two distributions to compare the goodness-of-fit of these ordered distances with a χ_n^2.

Example A5.8. Consider again the data of Jolicoeur and Mosimann (1960) on carapace characters in 24 male turtles. Are the characters z_1 (carapace length) and z_2 (maximun carapace width) jointly bivariate normally distributed? Here $n = 2$ and $m = 24$ and

$$\overline{\mathbf{z}} = \begin{pmatrix} 113.13. \\ 88.29 \end{pmatrix}, \quad \mathbf{S_Z} = \begin{pmatrix} 138.77 & 79.15 \\ 79.15 & 50.04 \end{pmatrix}, \quad \mathbf{S_Z}^{-1} = \begin{pmatrix} 0.0737 & -0.1165 \\ -0.1165 & 0.2043 \end{pmatrix}$$

where $\mathbf{S_Z}$ is the sample covariance matrix. A partial list of the 24 vectors of observations is

$$\mathbf{z}_1 = \begin{pmatrix} 93 \\ 74 \end{pmatrix}, \quad \cdots, \quad \mathbf{z}_{11} = \begin{pmatrix} 113 \\ 88 \end{pmatrix}, \quad \cdots, \quad \mathbf{z}_{24} = \begin{pmatrix} 135 \\ 106 \end{pmatrix}$$

Applying Equation A5.23, these observations translate into the distances

$$d_1^2 = 4.45, \quad \cdots, \quad d_{11}^2 = 0.002, \quad \cdots, \quad d_{24}^2 = 9.277$$

After rank ordering, these correspond to $d_{(23)}^2$, $d_{(1)}^2$, and $d_{(24)}^2$, respectively. For $d_{(23)}^2$, the matching value when distances are χ^2-distributed is

$$\chi_2^2 \left(\frac{23 - 1/2}{24} \right) = \chi_2^2 \, (0.9375)$$

The R command `qchisq(0.9375,2)` returns a value of $x = 5.545$, which satisfies $\Pr(\chi_2^2 \leq x) = 0.9375$, and calculates the point generated from $\mathbf{z}_1$ as $(4.45, \, 5.545)$. Likewise, the χ^2 values for $d_{(1)}^2$ and $d_{(24)}^2$ are 0.043 and 7.742, respectively. Proceeding similarly for the other values, we obtain the regression plotted in Figure A5.7. This departs somewhat from linearity. Further, under the assumption of multivariate normality, the best-fitting linear regression is expected to have a slope of one and to pass through the origin, while the best linear fit of these data shows slight departures from these values. Transforming the data by taking logs results in a slightly better fit (Figure A5.7).

Mardia's Test: Multivariate Skewness and Kurtosis

As was the case for univariate normality, we can test for multivariate normality by examining the sample skewness and kurtosis. Mardia (1970, 1974) proposed multivariate extensions of skewness and kurtosis measures and suggested a large-sample test based on the asymptotic distribution of these statistics. If there are m vectors of observations (with each vector measuring n characters), then the multivariate skewness is estimated by

$$b_{1,n} = \frac{1}{m^2} \sum_{i=1}^{m} \sum_{j=1}^{m} \left[(\mathbf{z}_i - \overline{\mathbf{z}})^T \, \mathbf{S_Z}^{-1} \, (\mathbf{z}_j - \overline{\mathbf{z}}) \right]^3 \tag{A5.24a}$$

while the multivariate kurtosis is estimated by

$$b_{2,n} = \frac{1}{m} \sum_{i=1}^{m} \left[(\mathbf{z}_i - \overline{\mathbf{z}})^T \, \mathbf{S}_{\mathbf{Z}}^{-1} \, (\mathbf{z}_i - \overline{\mathbf{z}}) \right]^2 \qquad \text{(A5.24b)}$$

If $\mathbf{z} \sim$ MVN, then $b_{1,n}$ and $b_{2,n}$ have expected values 0 and $n(n+2)$. For large values of m, Mardia showed that the (scaled) multivariate skewness is asymptotically distributed as a chi-square random variable with f degrees of freedom, with

$$\frac{m}{6}\, b_{1,n} \sim \chi^2_f, \qquad \text{where } f = \frac{n(n+1)(n+2)}{6} \qquad \text{(A5.25a)}$$

Likewise for large values of m, the multivariate kurtosis (following appropriate scaling) is distributed as a unit-normal, with

$$\frac{b_{2,n} - n(n+2)}{\sqrt{8n(n+2)/m}} \sim N(0,1) \qquad \text{(A5.25b)}$$

If either Equation A5.25a or A5.25b is significant, then multivariate normality is rejected.

Example A5.9. Do the data considered in Example A5.8 display significant skewness or kurtosis? Here $n = 2$ and $m = 24$. Applying Equations A5.25a and A5.25b gives $b_{1,2} = 0.6792$ and $b_{2,2} = 7.6043$. Considering skewness first, from Equation A5.25a it follows that the value

$$\frac{m}{6}\, b_{1,2} = \frac{24}{6}\, 0.6792 = 2.717$$

is (under MVN) a draw from a chi-square distribution with $f = 2(2+1)(2+2)/6 = 4$ degrees of freedom. Because Prob$(\chi^2_4 \geq 2.717) \simeq 0.606$, this is not significant. Turning to kurtosis, Equation A5.25b yields

$$\frac{b_{2,n} - n(n+2)}{\sqrt{8n(n+2)/m}} = \frac{7.6043 - 8}{1.633} \simeq -0.2423$$

which is also not significant as Prob$(|N(0,1)| \geq 0.2423) \simeq 0.81$. Transforming the data by taking logs gives $b_{1,2} = 0.2767$ and $b_{2,2} = 7.1501$, and hence showing a slight decrease in skewness and and a slight increase in kurtosis relative to the untransformed data. Reyment (1971) presented a number of other biological examples using Mardia's test.

Appendix 6

Derivatives of Vectors and Vector-valued Functions

Mathematics is a collection of cheap tricks and dirty jokes. Lipman Bers

Quantitative genetics often deals with vector-valued functions, and here we provide a brief review of the calculus of such functions. In particular, we review common expressions for derivatives of vectors and vector-valued functions, introduce the gradient vector and Hessian matrix (for first and second partials, respectively), and use this machinery in multidimensional Taylor series for approximating functions around a specific value. We apply these results to several problems in selection theory and evolution.

DERIVATIVES OF VECTORS AND VECTOR-VALUED FUNCTIONS

Suppose we let $f(\mathbf{x})$ be a scalar (single-dimension) function of a column vector, $\mathbf{x} = (x_1, \cdots, x_n)^T$, of n variables. The **gradient** (or **gradient vector**) of f with respect to $\mathbf{x}$ is obtained by taking partial derivatives of the function with respect to each variable. In matrix notation, the gradient operator is denoted by

$$\nabla_{\mathbf{x}}[\, f\,] = \frac{\partial f}{\partial \mathbf{x}} = \begin{pmatrix} \dfrac{\partial f}{\partial x_1} \\ \vdots \\ \dfrac{\partial f}{\partial x_n} \end{pmatrix}$$

The gradient at a point, $\mathbf{x}_o$, corresponds to a vector indicating the direction of local steepest ascent of the function at that point (the multivariate slope of f at $\mathbf{x}_o$).

Example A6.1. For an $n \times 1$ column vector, $\mathbf{x}$, compute the gradient for

$$f(\mathbf{x}) = \sum_{i=1}^{n} x_i^2 = \mathbf{x}^T\mathbf{x}$$

Because $\partial f / \partial x_i = 2x_i$, the gradient vector is just $\nabla_{\mathbf{x}}[\, f(\mathbf{x})\,] = 2\mathbf{x}$. At the point $\mathbf{x}_o$, $\mathbf{x}^T\mathbf{x}$ locally increases most rapidly if we change $\mathbf{x}$ in the same the direction as the vector going from point $\mathbf{x}_o$ to point $\mathbf{x}_o + 2\delta\, \mathbf{x}_o = (1 + 2\delta)\, \mathbf{x}_o$, where δ is a small positive value.

Now consider an $m \times n$ matrix, $\mathbf{A}$, of constants. What is the derivative of the $n \times 1$ column vector $\mathbf{Ax}$ with respect to $\mathbf{x}$? Recall from the definition of matrix multiplication that the ith element of $\mathbf{Ax}$ is

$$(\mathbf{Ax})_i = \sum_{j=1}^{n} A_{ij} x_j, \quad \text{yielding} \quad \partial(\mathbf{Az})_i / \partial x_k = A_{ik}$$

Hence, the ith element of $\nabla_{\mathbf{x}}[\mathbf{Ax}]_k$ is $(\, A_{1k} \quad A_{2k} \quad \cdots \quad A_{mk}\,)$, namely the transpose of the kth column of $\mathbf{A}$, which yields $\nabla_{\mathbf{x}}[\mathbf{Ax}] = \mathbf{A}^T$.

Evolution and Selection of Quantitative Traits. Bruce Walsh & Michael Lynch.
Published 2018 by Oxford University Press.
DOI 10.1093/oso/9780198830870.001.0001

For a vector (**a**) and a matrix (**A**) of constants, using the same logic as in Example A6.1, it can be shown (e.g., Morrison 1976; Graham 1981; Searle 1982) that

$$\nabla_{\mathbf{x}}\left[\,\mathbf{a}^T\mathbf{x}\,\right] = \nabla_{\mathbf{x}}\left[\,\mathbf{x}^T\mathbf{a}\,\right] = \mathbf{a} \tag{A6.1a}$$

$$\nabla_{\mathbf{x}}\left[\,\mathbf{A}\mathbf{x}\,\right] = \mathbf{A}^T \tag{A6.1b}$$

Turning to quadratic forms, if **A** is symmetric ($\mathbf{A} = \mathbf{A}^T$), then

$$\nabla_{\mathbf{x}}\left[\,\mathbf{x}^T\mathbf{A}\mathbf{x}\,\right] = 2\,\mathbf{A}\mathbf{x} \tag{A6.1c}$$

$$\nabla_{\mathbf{x}}\left[\,(\mathbf{x}-\mathbf{a})^T\mathbf{A}(\mathbf{x}-\mathbf{a})\,\right] = 2\,\mathbf{A}(\mathbf{x}-\mathbf{a}) \tag{A6.1d}$$

$$\nabla_{\mathbf{x}}\left[\,(\mathbf{a}-\mathbf{x})^T\mathbf{A}(\mathbf{a}-\mathbf{x})\,\right] = -2\,\mathbf{A}(\mathbf{a}-\mathbf{x}) \tag{A6.1e}$$

Taking $\mathbf{A} = \mathbf{I}$, Equation A6.1c implies

$$\nabla_{\mathbf{x}}\left[\,\mathbf{x}^T\mathbf{x}\,\right] = \nabla_{\mathbf{x}}\left[\,\mathbf{x}^T\mathbf{I}\,\mathbf{x}\,\right] = 2\,\mathbf{I}\,\mathbf{x} = 2\,\mathbf{x} \tag{A6.1f}$$

as was found in Example A6.1. Two other useful identities follow from the chain rule of differentiation, namely,

$$\nabla_{\mathbf{x}}\left[\,\exp\{\,f(\mathbf{x})\,\}\,\right] = \exp[\,f(\mathbf{x})\,]\,\nabla_{\mathbf{x}}\left[\,f(\mathbf{x})\,\right] \tag{A6.1g}$$

$$\nabla_{\mathbf{x}}\left[\,\ln[\,f(\mathbf{x})\,]\,\right] = \frac{1}{f(\mathbf{x})}\cdot\nabla_{\mathbf{x}}\left[\,f(\mathbf{x})\,\right] \tag{A6.1h}$$

Finally, the product rule also applies to a gradient, with

$$\nabla_{\mathbf{x}}\left[\,f(\mathbf{x})\,g(\mathbf{x})\,\right] = \nabla_{\mathbf{x}}\left[\,f(\mathbf{x})\,\right]\,g(\mathbf{x}) + f(\mathbf{x})\,\nabla_{\mathbf{x}}\left[\,g(\mathbf{x})\,\right] \tag{A6.1i}$$

Example A6.2. The density function $\varphi(\mathbf{x}, \boldsymbol{\mu}, \mathbf{V})$ for a multivariate normal (MVN) distribution returns a scalar value and is a function of the data vector, **x**, the vector of means, $\boldsymbol{\mu}$, and the covariance matrix, **V**,

$$\varphi(\mathbf{x}, \boldsymbol{\mu}, \mathbf{V}) = a\exp\left(-\frac{1}{2}\cdot(\mathbf{x}-\boldsymbol{\mu})^T\,\mathbf{V}^{-1}\,(\mathbf{x}-\boldsymbol{\mu})\right)$$

where the constant $a = \pi^{-n/2}\,|\mathbf{V}|^{-1/2}$, and $|\mathbf{V}|$ denotes the determinant of **V**. To compute the gradient of the MVN with respect to the data vector, **x**, first apply Equation A6.1g to yield

$$\nabla_{\mathbf{x}}\left[\,\varphi(\mathbf{x}, \boldsymbol{\mu}, \mathbf{V})\,\right] = \varphi(\mathbf{x}, \boldsymbol{\mu}, \mathbf{V})\cdot\nabla_{\mathbf{x}}\left[\left(-\frac{1}{2}\right)\cdot(\mathbf{x}-\boldsymbol{\mu})^T\,\mathbf{V}_{\mathbf{X}}^{-1}\,(\mathbf{x}-\boldsymbol{\mu})\right]$$

Using this result along with Equation A6.1d returns

$$\nabla_{\mathbf{x}}\left[\,\varphi(\mathbf{x}, \boldsymbol{\mu}, \mathbf{V})\,\right] = -\varphi(\mathbf{x}, \boldsymbol{\mu}, \mathbf{V})\cdot\mathbf{V}^{-1}\,(\mathbf{x}-\boldsymbol{\mu}) \tag{A6.2a}$$

Similarly, Equation A6.1e implies that the gradient of the MVN with respect to the vector of means $\boldsymbol{\mu}$ is

$$\nabla_{\boldsymbol{\mu}}\left[\,\varphi(\mathbf{x}, \boldsymbol{\mu}, \mathbf{V})\,\right] = \varphi(\mathbf{x}, \boldsymbol{\mu}, \mathbf{V})\cdot\mathbf{V}^{-1}\,(\mathbf{x}-\boldsymbol{\mu}) \tag{A6.2b}$$

Example A6.3. Recall (Equations 13.27c and 30.5a) that when the distribution of phenotypes is multivariate normal, the directional selection gradient, $\boldsymbol{\beta} = \mathbf{P}^{-1}\mathbf{S}$, equals the gradient of log mean fitness with respect to the vector of trait means, $\nabla_{\boldsymbol{\mu}}[\ln \overline{W}(\boldsymbol{\mu})\,]$ (Lande 1979a). Hence, the increase in the mean population fitness is maximized if mean character values change in the same direction as the vector $\boldsymbol{\beta}$. To see this, first note that applying Equation A6.1h yields

$$\nabla_{\boldsymbol{\mu}}[\ln \overline{W}(\boldsymbol{\mu})\,] = \overline{W}^{\,-1}\,\nabla_{\boldsymbol{\mu}}[\,\overline{W}(\boldsymbol{\mu})\,] \tag{A6.2c}$$

If we write mean fitness as $\overline{W}(\boldsymbol{\mu}) = \int W(\mathbf{z})\,\varphi(\mathbf{z},\boldsymbol{\mu})\,d\mathbf{z}$ and take the gradient through the integral, we obtain

$$\nabla_{\boldsymbol{\mu}}[\ln \overline{W}(\boldsymbol{\mu})\,] = \overline{W}^{\,-1}\,\nabla_{\boldsymbol{\mu}}\left[\int W(\mathbf{z})\,\varphi(\mathbf{z},\boldsymbol{\mu})\,d\mathbf{z}\right] = \overline{W}^{\,-1}\int W(\mathbf{z})\,\nabla_{\boldsymbol{\mu}}\,[\,\varphi(\mathbf{z},\boldsymbol{\mu})\,]\,d\mathbf{z}$$

The last identity follows from the assumption that $W(\mathbf{z})$ is not a function of the vector of trait means, $\boldsymbol{\mu}$, that is, the fitnesses are *frequency-independent* ($\nabla_{\boldsymbol{\mu}}\,[\,W(\mathbf{z})\,] = 0$). If the trait vector $\mathbf{z} \sim \text{MVN}(\boldsymbol{\mu}, \mathbf{P})$, Equation A6.2b yields

$$\nabla_{\boldsymbol{\mu}}\,[\,\varphi(\mathbf{z},\boldsymbol{\mu})\,] = \varphi(\mathbf{z},\boldsymbol{\mu})\,\mathbf{P}^{-1}\,(\mathbf{z}-\boldsymbol{\mu}) \tag{A6.2d}$$

Hence,

$$\begin{aligned}
\overline{W}^{\,-1}\int W(\mathbf{z})\,\nabla_{\boldsymbol{\mu}}\,[\,\varphi(\mathbf{z},\boldsymbol{\mu})\,]\,d\mathbf{z} &= \int w(\mathbf{z})\,\varphi(\mathbf{z},\boldsymbol{\mu})\,\mathbf{P}^{-1}\,(\mathbf{z}-\boldsymbol{\mu})\,d\mathbf{z} \\
&= \mathbf{P}^{-1}\left(\int \mathbf{z}\,w(\mathbf{z})\,\varphi(\mathbf{z},\boldsymbol{\mu})\,d\mathbf{z} - \boldsymbol{\mu}\int w(\mathbf{z})\,\varphi(\mathbf{z},\boldsymbol{\mu})\,d\mathbf{z}\right) \\
&= \mathbf{P}^{-1}(\boldsymbol{\mu}^{*}-\boldsymbol{\mu}) = \mathbf{P}^{-1}\mathbf{S} = \boldsymbol{\beta}
\end{aligned} \tag{A6.2e}$$

which follows because the first integral (in the second line above) is the mean character value after selection, $\boldsymbol{\mu}^{*}$, while the second integral equals one by definition, as $E[w] = 1$.

If individual fitnesses are *frequency-dependent* (changing with $\boldsymbol{\mu}$), then, according to the product rule (Equation A6.1i), a second integral appears, and $\nabla_{\boldsymbol{\mu}}[\ln \overline{W}(\boldsymbol{\mu})\,]$ now becomes

$$\overline{W}^{\,-1}\int W(\mathbf{z})\,\nabla_{\boldsymbol{\mu}}\,[\,\varphi(\mathbf{z},\boldsymbol{\mu})\,]\,d\mathbf{z} + \overline{W}^{\,-1}\int \nabla_{\boldsymbol{\mu}}\,[\,W(\mathbf{z})\,]\,\varphi(\mathbf{z},\boldsymbol{\mu})d\mathbf{z}$$

which yields

$$\nabla_{\boldsymbol{\mu}}[\ln \overline{W}(\boldsymbol{\mu})\,] = \boldsymbol{\beta} + \overline{W}^{\,-1}\int \nabla_{\boldsymbol{\mu}}\,[\,W(\mathbf{z})\,]\,\varphi(\mathbf{z},\boldsymbol{\mu})d\mathbf{z} \tag{A6.2f}$$

Example A6.4. Consider the ordinary least-squares solution for the general linear model, $\mathbf{y} = \mathbf{X}\boldsymbol{\beta} + \mathbf{e}$, where $\boldsymbol{\beta}$ is the vector that minimizes the sum of squared residual errors, $\sum e_i^2$. In matrix form, this sum becomes

$$\begin{aligned}
\sum_{i=1}^{n} e_i^2 &= \mathbf{e}^T\mathbf{e} = (\mathbf{y}-\mathbf{X}\boldsymbol{\beta})^T(\mathbf{y}-\mathbf{X}\boldsymbol{\beta}) \\
&= \mathbf{y}^T\mathbf{y} - \boldsymbol{\beta}^T\mathbf{X}^T\mathbf{y} - \mathbf{y}^T\mathbf{X}\boldsymbol{\beta} + \boldsymbol{\beta}^T\mathbf{X}^T\mathbf{X}\boldsymbol{\beta} \\
&= \mathbf{y}^T\mathbf{y} - 2\boldsymbol{\beta}^T\mathbf{X}^T\mathbf{y} + \boldsymbol{\beta}^T\mathbf{X}^T\mathbf{X}\boldsymbol{\beta}
\end{aligned} \tag{A6.3a}$$

and the last step follows because the matrix product $\boldsymbol{\beta}^T\mathbf{X}^T\mathbf{y}$ yields a scalar, and hence equals its transpose,

$$\boldsymbol{\beta}^T\mathbf{X}^T\mathbf{y} = \left(\boldsymbol{\beta}^T\mathbf{X}^T\mathbf{y}\right)^T = \mathbf{y}^T\mathbf{X}\boldsymbol{\beta} \tag{A6.3b}$$

To find the vector $\boldsymbol{\beta}$ that minimizes $\mathbf{e}^T\mathbf{e}$, we take the derivative with respect to $\boldsymbol{\beta}$ and use Equations A6.1a–A6.1c, which yields

$$\nabla_{\boldsymbol{\beta}}\left[\,\mathbf{e}^T\mathbf{e}\,\right] = \frac{\partial\,\mathbf{e}^T\mathbf{e}}{\partial\,\boldsymbol{\beta}} = -2\mathbf{X}^T\mathbf{y} + 2\mathbf{X}^T\mathbf{X}\boldsymbol{\beta} \tag{A6.3c}$$

Setting this equal to zero yields $\mathbf{X}^T\mathbf{X}\boldsymbol{\beta} = \mathbf{X}^T\mathbf{y}$, which (provided the inverse of $\mathbf{X}^T\mathbf{X}$ exists) has a solution of

$$\boldsymbol{\beta} = \left(\mathbf{X}^T\mathbf{X}\right)^{-1}\mathbf{X}^T\mathbf{y} \tag{A6.3d}$$

More generally, if $\mathbf{X}^T\mathbf{X}$ is singular, we can still solve this equation by using a generalized inverse $\left(\mathbf{X}^T\mathbf{X}\right)^{-}$; see LW Appendix 3.

Example A6.5. Here we present one derivation of Henderson's mixed-model equations (Equation 19.4). Consider the mixed model $\mathbf{y} = \mathbf{X}\boldsymbol{\beta} + \mathbf{Z}\mathbf{u} + \mathbf{e}$, where $\mathbf{e} \sim \text{MVN}(0, \mathbf{R})$, $\mathbf{u} \sim \text{MVN}(0, \mathbf{G})$, and $\mathbf{e}$ and $\mathbf{u}$ are independent (Equation 19.1). If we recall the probability density function for a multivariate normal (Example A6.2), we have that

$$p(\mathbf{e}) \propto |\mathbf{R}|^{-1/2} \cdot \exp\left[-\frac{1}{2}\mathbf{e}^T\mathbf{R}^{-1}\mathbf{e}\right] \quad \text{and} \quad p(\mathbf{u}) \propto |\mathbf{G}|^{-1/2} \cdot \exp\left[-\frac{1}{2}\mathbf{u}^T\mathbf{G}^{-1}\mathbf{u}\right] \tag{A6.4a}$$

We can further note that the conditional distribution of $\mathbf{y}$ given $\mathbf{u}$ is

$$(\mathbf{y} - \mathbf{X}\boldsymbol{\beta} - \mathbf{Z}\mathbf{u})\,|\mathbf{u} = \mathbf{e} \sim \text{MVN}(0, \mathbf{R})$$

Hence,

$$p(\mathbf{y}, \mathbf{u}) = p(\mathbf{y}|\mathbf{u}) \cdot p(\mathbf{u}) = p(\mathbf{e}) \cdot p(\mathbf{u})$$

with the last step following because $\mathbf{e}$ and $\mathbf{u}$ are independent. From Equation A6.4a,

$$p(\mathbf{y}, \mathbf{u}) \propto$$

$$|\mathbf{R}|^{-1/2}|\mathbf{G}|^{-1/2} \cdot \exp\left[-\frac{1}{2}(\mathbf{y} - \mathbf{X}\boldsymbol{\beta} - \mathbf{Z}\mathbf{u})^T\mathbf{R}^{-1}(\mathbf{y} - \mathbf{X}\boldsymbol{\beta} - \mathbf{Z}\mathbf{u}) - \frac{1}{2}\mathbf{u}^T\mathbf{G}^{-1}\mathbf{u}\right] \tag{A6.4b}$$

Now consider the log of the density,

$$\ell = \ln\left[p(\mathbf{y}, \mathbf{u})\right] \propto$$

$$\left(-\frac{1}{2}\right)\left[\ln(|\mathbf{R}|) + \ln(|\mathbf{G}|) + (\mathbf{y} - \mathbf{X}\boldsymbol{\beta} - \mathbf{Z}\mathbf{u})^T\mathbf{R}^{-1}(\mathbf{y} - \mathbf{X}\boldsymbol{\beta} - \mathbf{Z}\mathbf{u}) + \mathbf{u}^T\mathbf{G}^{-1}\mathbf{u}\right] \tag{A6.4c}$$

We can expand the larger quadratic product to yield the last two terms of A6.4c as

$$-2\mathbf{y}^T\mathbf{R}^{-1}\mathbf{X}\boldsymbol{\beta} - 2\mathbf{y}^T\mathbf{R}^{-1}\mathbf{Z}\mathbf{u} + \boldsymbol{\beta}^T\mathbf{X}^T\mathbf{R}^{-1}\mathbf{X}\boldsymbol{\beta} + 2\boldsymbol{\beta}^T\mathbf{X}^T\mathbf{R}^{-1}\mathbf{Z}\mathbf{u} + \mathbf{u}^T\mathbf{Z}^T\mathbf{R}^{-1}\mathbf{Z}\mathbf{u} + \mathbf{u}^T\mathbf{G}^{-1}\mathbf{u} \tag{A6.4d}$$

Using Equations A6.1c through A6.1e to take the derivatives of ℓ with respect to $\boldsymbol{\beta}$ and $\mathbf{u}$ yields

$$\begin{pmatrix} \dfrac{\partial \ell}{\partial \boldsymbol{\beta}} \\ \dfrac{\partial \ell}{\partial \mathbf{u}} \end{pmatrix} = \begin{pmatrix} \mathbf{X}^T\mathbf{R}^{-1}\mathbf{y} - \mathbf{X}^T\mathbf{R}^{-1}\mathbf{X}\boldsymbol{\beta} - \mathbf{X}^T\mathbf{R}^{-1}\mathbf{Z}\mathbf{u} \\ \mathbf{Z}^T\mathbf{R}^{-1}\mathbf{y} - \mathbf{Z}^T\mathbf{R}^{-1}\mathbf{X}\boldsymbol{\beta} - \mathbf{Z}^T\mathbf{R}^{-1}\mathbf{Z}\mathbf{u} + \mathbf{G}^{-1}\mathbf{u} \end{pmatrix} \tag{A6.4e}$$

Denoting the value for $\boldsymbol{\beta}$ and $\mathbf{u}$ that return a zero vector for Equation A6.4e as $\widehat{\boldsymbol{\beta}}$ and $\widehat{\mathbf{u}}$ yields the following set of matrix equations

$$\begin{pmatrix} \mathbf{X}^T\mathbf{R}^{-1}\mathbf{y} \\ \mathbf{Z}^T\mathbf{R}^{-1}\mathbf{y} \end{pmatrix} = \begin{pmatrix} \mathbf{X}^T\mathbf{R}^{-1}\mathbf{X}\widehat{\boldsymbol{\beta}} + \mathbf{X}^T\mathbf{R}^{-1}\mathbf{Z}\widehat{\mathbf{u}} \\ \mathbf{Z}^T\mathbf{R}^{-1}\mathbf{X}\widehat{\boldsymbol{\beta}} + \mathbf{Z}^T\mathbf{R}^{-1}\mathbf{Z}\widehat{\mathbf{u}} + \mathbf{G}^{-1}\widehat{\mathbf{u}} \end{pmatrix} \tag{A6.4f}$$

which immediately yields Henderson's mixed-model equations,

$$\begin{pmatrix} \mathbf{X}^T\mathbf{R}^{-1}\mathbf{y} \\ \mathbf{Z}^T\mathbf{R}^{-1}\mathbf{y} \end{pmatrix} = \begin{pmatrix} \mathbf{X}^T\mathbf{R}^{-1}\mathbf{X} & \mathbf{X}^T\mathbf{R}^{-1}\mathbf{Z} \\ \mathbf{Z}^T\mathbf{R}^{-1}\mathbf{X} & \mathbf{Z}^T\mathbf{R}^{-1}\mathbf{Z} + \mathbf{G}^{-1} \end{pmatrix} \begin{pmatrix} \widehat{\beta} \\ \widehat{\mathbf{u}} \end{pmatrix} \tag{A6.4g}$$

Using the second equation (row two) of Equation A6.4g returns

$$\mathbf{Z}^T\mathbf{R}^{-1}\mathbf{X}\widehat{\beta} + \left(\mathbf{Z}^T\mathbf{R}^{-1}\mathbf{Z} + \mathbf{G}^{-1}\right)\widehat{\mathbf{u}} = \mathbf{Z}^T\mathbf{R}^{-1}\mathbf{y} \tag{A6.4h}$$

which can be rearranged to yields

$$\widehat{\mathbf{u}} = \left(\mathbf{Z}^T\mathbf{R}^{-1}\mathbf{Z} + \mathbf{G}^{-1}\right)^{-1}\mathbf{Z}^T\mathbf{R}^{-1}\left(\mathbf{y} - \mathbf{X}\widehat{\beta}\right) \tag{A6.4i}$$

as an alternative expression to Equation 19.3b for the BLUP of $\mathbf{u}$.

THE HESSIAN MATRIX, LOCAL MAXIMA/MINIMA, AND MULTIDIMENSIONAL TAYLOR SERIES

In univariate calculus, the local extrema of a function occur when its slope (first derivative) is zero. The multivariate extension is that the gradient vector is zero, so the slope of the function with respect to all variables is zero. A point $\mathbf{x}_e$ where this occurs is called a **stationary** or **equilibrium** point, and corresponds to either a local maximum, minimum, saddle point, or inflection point. As with the calculus of single variables, determining which of these cases is correct depends on the second derivative. With n variables, the appropriate generalization is the **Hessian** matrix

$$\mathbf{H}_{\mathbf{x}}[\,f\,] = \nabla_{\mathbf{x}}\left[\left(\nabla_{\mathbf{x}}[\,f\,]\right)^T\right] = \frac{\partial^2 f}{\partial\mathbf{x}\,\partial\mathbf{x}^T} = \begin{pmatrix} \dfrac{\partial^2 f}{\partial x_1^2} & \cdots & \dfrac{\partial^2 f}{\partial x_1\,\partial x_n} \\ \vdots & \ddots & \vdots \\ \dfrac{\partial^2 f}{\partial x_1\,\partial x_n} & \cdots & \dfrac{\partial^2 f}{\partial x_n^2} \end{pmatrix} \tag{A6.5}$$

Note that this is the **outer product** of $\nabla_{\mathbf{x}}[\,f\,]$ with itself. Recall for an n-dimensional column vector $\mathbf{a}_{n\times 1}$ that while the inner product, $\mathbf{a}_{1\times n}^T\mathbf{a}_{n\times 1} = \sum a_i$, returns a 1×1 matrix (a scalar), the outer product, $\mathbf{a}_{n\times 1}\mathbf{a}_{1\times n}^T$, returns an $n \times n$ matrix whose ijth element is $a_i a_j$, or (in our case)

$$\mathbf{H}_{ij} = \frac{\partial\left(f(\mathbf{x})/\partial x_i\right)}{\partial x_j} = \frac{\partial^2 f(\mathbf{x})}{\partial x_i\,\partial x_j}$$

This matrix is symmetric, as mixed partials are equal under suitable continuity conditions, and it measures the local multidimensional curvature of the function.

Example A6.6. Compute the Hessian matrix for the multivariate normal distribution with respect to the data vector $\mathbf{x}$. If we recall from Equation A6.2a that $\nabla_{\mathbf{x}}\left[\,\varphi(\mathbf{x},\boldsymbol{\mu})\,\right] = -\varphi(\mathbf{x},\boldsymbol{\mu})\cdot \mathbf{V}^{-1}(\mathbf{x}-\boldsymbol{\mu})$, we have

$$\begin{aligned}\mathbf{H}_{\mathbf{x}}\left[\,\varphi(\mathbf{x},\boldsymbol{\mu})\,\right] &= \nabla_{\mathbf{x}}\left[\left(\nabla_{\mathbf{x}}\left[\,\varphi(\mathbf{x},\boldsymbol{\mu})\,\right]\right)^T\right] = -\nabla_{\mathbf{x}}\left[\,\varphi(\mathbf{x},\boldsymbol{\mu})\cdot(\mathbf{x}-\boldsymbol{\mu})^T\mathbf{V}^{-1}\,\right]\\ &= -\nabla_{\mathbf{x}}\left[\,\varphi(\mathbf{x},\boldsymbol{\mu})\,\right]\cdot(\mathbf{x}-\boldsymbol{\mu})^T\mathbf{V}^{-1} - \varphi(\mathbf{x},\boldsymbol{\mu})\cdot\nabla_{\mathbf{x}}\left[\,(\mathbf{x}-\boldsymbol{\mu})^T\mathbf{V}^{-1}\,\right]\\ &= -\left[-\varphi(\mathbf{x},\boldsymbol{\mu})\cdot\mathbf{V}^{-1}(\mathbf{x}-\boldsymbol{\mu})\right]\cdot(\mathbf{x}-\boldsymbol{\mu})^T\mathbf{V}^{-1} - \varphi(\mathbf{x},\boldsymbol{\mu})\cdot\left[\mathbf{V}^{-1}\right]\\ &= \varphi(\mathbf{x},\boldsymbol{\mu})\cdot\left(\mathbf{V}^{-1}(\mathbf{x}-\boldsymbol{\mu})(\mathbf{x}-\boldsymbol{\mu})^T\mathbf{V}^{-1} - \mathbf{V}^{-1}\right)\end{aligned} \tag{A6.6a}$$

Here we have used the product rule (Equation A6.1i) and Equation A6.1b, respectively (recall that $\mathbf{V}$ is a symmetric matrix of constants). In a similar manner, the Hessian with respect to the vector of means, $\boldsymbol{\mu}$, is

$$\mathbf{H}_{\boldsymbol{\mu}}\left[\,\varphi(\mathbf{x},\boldsymbol{\mu})\,\right] = \varphi(\mathbf{x},\boldsymbol{\mu})\cdot\left(\mathbf{V}^{-1}(\mathbf{x}-\boldsymbol{\mu})(\mathbf{x}-\boldsymbol{\mu})^T\mathbf{V}^{-1} - \mathbf{V}^{-1}\right) \tag{A6.6b}$$

To see how the Hessian matrix determines the nature of equilibrium points, a slight digression on the **multidimensional Taylor series** is needed. Consider the (second-order) Taylor series of a scalar function of n variables, $f(x_1,\cdots,x_n)$, expanded about the point $\mathbf{x}_o$,

$$f(\mathbf{x}) \simeq f(\mathbf{x}_o) + \sum_{i=1}^{n}(x_i - x_{o,i})\frac{\partial f}{\partial x_i} + \frac{1}{2}\sum_{i=1}^{n}\sum_{j=1}^{n}(x_i - x_{o,i})(x_j - x_{o,j})\frac{\partial^2 f}{\partial x_i\,\partial x_j} + \cdots \tag{A6.7a}$$

where all partials are evaluated at $\mathbf{x}_o$. If we note that the first sum is the inner product of the gradient and $(\mathbf{x}-\mathbf{x}_o)$, and the double sum is a quadratic product involving the Hessian, we can express Equation A6.7a in matrix form as

$$f(\mathbf{x}) \simeq f(\mathbf{x}_o) + \nabla^T(\mathbf{x}-\mathbf{x}_o) + \frac{1}{2}(\mathbf{x}-\mathbf{x}_o)^T\mathbf{H}\,(\mathbf{x}-\mathbf{x}_o) \tag{A6.7b}$$

where ∇ and $\mathbf{H}$ are the gradient and Hessian of f with respect to $\mathbf{x}$ when evaluated at $\mathbf{x}_o$,

$$\nabla \equiv \nabla_{\mathbf{x}}[\,f\,]\big|_{\mathbf{x}=\mathbf{x}_o} \qquad \text{and} \qquad \mathbf{H} \equiv \mathbf{H}_{\mathbf{x}}[\,f\,]\big|_{\mathbf{x}=\mathbf{x}_o}$$

At an equilibrium point, $\widehat{\mathbf{x}}$, all first partials are zero, so $(\nabla_{\mathbf{x}}[\,f\,])^T$ is evaluated at $\widehat{\mathbf{x}}$ is a vector of length zero. Whether this point is a maximum or minimum is then determined by the quadratic product involving the Hessian when evaluated at $\widehat{\mathbf{x}}$. Consider a vector, $\mathbf{d}$, of a small change from the equilibrium point

$$f(\widehat{\mathbf{x}}+\mathbf{d}) - f(\widehat{\mathbf{x}}) \simeq \frac{1}{2}\cdot\mathbf{d}^T\,\mathbf{H}\,\mathbf{d} \tag{A6.8a}$$

Because $\mathbf{H}$ is a symmetric matrix, we can diagonalize it and apply a canonical transformation (Equation A5.17a) to simplify the quadratic product in Equation A6.8a, which returns

$$f(\widehat{\mathbf{x}}+\mathbf{d}) - f(\widehat{\mathbf{x}}) \simeq \frac{1}{2}\sum_{i=1}^{n}\lambda_i y_i^2 \tag{A6.8b}$$

where $y_i = \mathbf{e}_i^T \mathbf{d}$, with $\mathbf{e}_i$ and λ_i representing the eigenvectors and eigenvalues of the Hessian when evaluated at $\widehat{\mathbf{x}}$. Thus, if $\mathbf{H}$ is **positive-definite** (all eigenvalues of $\mathbf{H}$ are positive), f increases in all directions around $\widehat{\mathbf{x}}$, and hence $\widehat{\mathbf{x}}$ is a local minimum of f. If $\mathbf{H}$ is **negative-definite** (all eigenvalues of $\mathbf{H}$ are negative), f decreases in all directions around $\widehat{\mathbf{x}}$, and $\widehat{\mathbf{x}}$ is a local maximum of f. If the eigenvalues differ in sign ($\mathbf{H}$ is **indefinite**), $\widehat{\mathbf{x}}$ corresponds to a saddle point (to see this, suppose $\lambda_1 > 0$ and $\lambda_2 < 0$; any change along the vector $\mathbf{e}_1$ results in an increases in f, while any change along $\mathbf{e}_2$ results in a decrease in f).

Example A6.7. Consider the following demonstration (due to Lande 1979a) that the mean population fitness increases. A round of selection changes the current vector of means from $\boldsymbol{\mu}$ to $\boldsymbol{\mu} + \boldsymbol{\Delta\mu}$. Expanding the log of mean fitness in a Taylor series around the current population mean gives the change in mean population fitness as

$$\begin{aligned} \Delta \ln \overline{W}(\boldsymbol{\mu}) &= \ln \overline{W}(\boldsymbol{\mu} + \boldsymbol{\Delta\mu}) - \ln \overline{W}(\boldsymbol{\mu}) \\ &\simeq \Big(\nabla_{\boldsymbol{\mu}}[\ln \overline{W}(\boldsymbol{\mu})\,] \Big)^T \boldsymbol{\Delta\mu} + \frac{1}{2} \boldsymbol{\Delta\mu}^T\, \mathbf{H}_{\boldsymbol{\mu}}[\ln \overline{W}(\boldsymbol{\mu})\,]\, \boldsymbol{\Delta\mu} \end{aligned} \tag{A6.9a}$$

If we assume that second- and higher-order terms can be neglected (as would occur with weak selection and the population mean away from an equilibrium point), then Equation A6.9a simplifies to

$$\Delta \ln \overline{W}(\boldsymbol{\mu}) \simeq \Big(\nabla_{\boldsymbol{\mu}}[\ln \overline{W}(\boldsymbol{\mu})\,] \Big)^T \boldsymbol{\Delta\mu} \tag{A6.9b}$$

Further assuming that the joint distribution of phenotypes and additive genetic values is MVN, then substituting Equation A6.2e into Equation A6.9b yields

$$\Delta \ln \overline{W}(\boldsymbol{\mu}) \simeq \boldsymbol{\beta}^T \boldsymbol{\Delta\mu} \tag{A6.9c}$$

Because $\boldsymbol{\Delta\mu}$ is the response vector, $\mathbf{R}$, rearranging Equation 13.26a yields $\boldsymbol{\beta} = \mathbf{G}^{-1}\mathbf{R} = \mathbf{G}^{-1}\boldsymbol{\Delta\mu}$. Substituting this expression into Equation A6.9c yields

$$\Delta \ln \overline{W}(\boldsymbol{\mu}) \simeq (\mathbf{G}^{-1}\boldsymbol{\Delta\mu})^T\, \boldsymbol{\Delta\mu} = (\boldsymbol{\Delta\mu})^T\, \mathbf{G}^{-1}\, \boldsymbol{\Delta\mu} \geq 0 \tag{A6.9d}$$

The inequality follows because $\mathbf{G}$ is a variance-covariance matrix and hence is nonnegative-definite (all its eigenvalues are nonnegative). Under these conditions, mean population fitness never decreases, although because $\boldsymbol{\Delta\mu} \neq \nabla_{\boldsymbol{\mu}}[\ln \overline{W}(\boldsymbol{\mu})\,]$, the local increase in fitness does not necessarily improve in the fastest possible manner. Note that

$$\Delta \ln \overline{W}(\boldsymbol{\mu}) = \ln \overline{W}(\boldsymbol{\mu}[t+1]) - \ln \overline{W}(\boldsymbol{\mu}[t]) = \ln \left(\frac{\overline{W}(\boldsymbol{\mu}[t+1])}{\overline{W}(\boldsymbol{\mu}[t])} \right) \tag{A6.9e}$$

so $\Delta \ln \overline{W}(\boldsymbol{\mu}) > 0$ implies $\overline{W}(\boldsymbol{\mu}[t+1]) > \overline{W}(\boldsymbol{\mu}[t])$. Notice from Equation A6.2f that when fitnesses are frequency-dependent, $\nabla_{\boldsymbol{\mu}}[\ln \overline{W}(\boldsymbol{\mu})\,]$ has an extra term beyond $\boldsymbol{\beta}$, and the above proof does not hold (namely, mean fitness can decrease).

OPTIMIZATION UNDER CONSTRAINTS

Occasionally, we wish to find the maximum or minimum of a function subject to a constraint. The solution is to use **Lagrange multipliers**. Suppose we wish to find the extrema of a function, $f(\mathbf{x})$, subject to the constraint that $h(\mathbf{x}) = c$. We first construct a new function by considering

$$g(\mathbf{x}, \lambda) = f(\mathbf{x}) - \lambda[\, h(\mathbf{x}) - c\,] \tag{A6.10a}$$

Because $h(\mathbf{x}) - c = 0$, the extrema of $g(\mathbf{x}, \lambda)$ correspond to the extrema of $f(\mathbf{x})$ under the constraint. Local maxima and minima are obtained by solving the following set of equations:

$$\nabla_{\mathbf{x}}[\, g(\mathbf{x}, \lambda)\,] = \nabla_{\mathbf{x}}[\, f(\mathbf{x})\,] - \lambda \cdot \nabla_{\mathbf{x}}[\, h(\mathbf{x})\,] = \mathbf{0} \tag{A6.10b}$$

$$\frac{d\, g(\mathbf{x}, \lambda)}{d\lambda} = h(\mathbf{x}) - c = 0 \tag{A6.10c}$$

Observe that the second equation is satisfied by the constraint.

Example A6.8. A standard approach for selection on multiple traits is to use a **selection index**, I, which is a new (univariate) trait that is a linear combination of n characters $\mathbf{z} = (z_1, z_2, \cdots, z_n)^T$, with

$$I = \mathbf{b}^T \mathbf{z} = \sum_{k=1}^{n} b_k\, z_k \tag{A6.11a}$$

Notice that what matters for the weights are their *relative proportions*, as, if we were to multiply all of the weights by the same constant (e.g., $a\, b_k$), this new index would still choose the same individuals. Hence, to obtain a standardized index, we impose the constraint $\mathbf{b}^T\mathbf{b} = 1$. We denote the directional selection differential on the index by S_I, and observe that if $\mathbf{S}$ is the vector of directional selection differentials generated by selecting on I, then $S_I = \mathbf{b}^T\mathbf{S}$.

Smith (1936) and Hazel (1943) showed that a larger response in a target index is obtained by selecting on a *different* index. Suppose our aim is to maximize the response in the index,

$$J = \mathbf{a}^T \boldsymbol{\mu} = \sum a_k\, \mu_k \tag{A6.11b}$$

What are the standardized weights, $\mathbf{b}$, for the optimal **Smith-Hazel** index (for a fixed amount of selection, $S_I = s$)? If we assume that the conditions leading to the multivariate breeder's equation hold, the function to maximize is

$$f(\mathbf{b}) = \mathbf{a}^T \boldsymbol{\Delta\mu} = \mathbf{a}^T\mathbf{R} = \mathbf{a}^T\mathbf{G}\mathbf{P}^{-1}\mathbf{S} \tag{A6.11c}$$

under the associated constraint function

$$g(\mathbf{b}) - c = \mathbf{b}^T\mathbf{b} - 1 = 0 \tag{A6.11d}$$

Because $S_I = \mathbf{b}^T\mathbf{S}$, and we have the constraint that $\mathbf{b}^T\mathbf{b} = 1$, set $\mathbf{S} = s \cdot \mathbf{b}$, so that $S_I = \mathbf{b}^T\mathbf{S} = s \cdot \mathbf{b}^T\mathbf{b} = s$. If we use these functions, Equation A6.10a yields

$$\begin{aligned} \nabla_{\mathbf{b}}[\, g(\mathbf{b}, \lambda)\,] &= s \cdot \nabla_{\mathbf{b}}[\, \mathbf{a}^T\mathbf{G}\mathbf{P}^{-1}\mathbf{b}\,] - \lambda \cdot \nabla_{\mathbf{b}}[\, \mathbf{b}^T\mathbf{b}\,] \\ &= s \cdot \left(\mathbf{a}^T\mathbf{G}\mathbf{P}^{-1}\right)^T - (2\lambda) \cdot \mathbf{b} \end{aligned} \tag{A6.11e}$$

which is equal to zero when

$$\mathbf{b} = (2\lambda/s) \cdot \mathbf{P}^{-1}\mathbf{G}^T\mathbf{a} \tag{A6.11f}$$

Note that the constant $2\lambda/s$ simply standardizes the weights (so that $\mathbf{b}^T\mathbf{b} = 1$), and hence the Smith-Hazel weights are usually written as

$$\mathbf{b}_{SH} = \mathbf{P}^{-1}\mathbf{G}^T\mathbf{a} \tag{A6.11g}$$

which can be standardized by noting that

$$\mathbf{b} = \frac{\mathbf{b}_{SH}}{||\mathbf{b}_{SH}||} = \frac{\mathbf{b}_{SH}}{\sqrt{\mathbf{b}_{SH}^T\mathbf{b}_{SH}}}$$

Index selection is examined in detail in Volume 3.

Literature Cited

The numbers in brackets following each reference denote the chapters in which the reference is cited.

Aastveit, A. H., and K. Aastveit. 1990. Theory and application of open-pollination and polycross in forage grass breeding. *Theor. Appl. Genet.* 79: 618–624. [21]

Abbot, P., and 126 others. 2011. Inclusive fitness theory and eusociality. *Nature* 471: E1–E4. [22]

Abney, M., M. S. McPeek, and C. Ober. 2000. Estimation of variance components of quantitative traits in inbred populations. *Amer. J. Hum. Genet.* 66: 629–650. [11]

Abplanalp, H. 1962. Modification of selection limits for egg number. *Genet. Res.* 3: 210–225. [25]

Abramowitz, M., and I. A. Stegun. 1972. *Handbook of mathematical functions, with formulas, graphs, and mathematical tables.* Dover Publications, Inc., New York, NY. [A1, 2, 7]

Achaz, G. 2008. Testing for neutrality in samples with sequencing errors. *Genetics* 179: 1409–1424. [9]

Achaz, G. 2009. Frequency spectrum neutrality tests: one for all and all for one. *Genetics* 183: 249–258. [9]

Ackerman, M. S., and 6 others. 2017. Estimating seven coefficients of pairwise relatedness using population genomic data. *Genetics* 206: 105–118. [11, 20]

Ackermann, R. R., and J. M. Cheverud. 2004. Detecting genetic drift versus selection in human evolution. *Proc. Natl. Acad. Sci. USA* 101: 17946–17051. [12]

Acquaah, G. 2007. *Principles of plant genetics and breeding.* Blackwell Publishing, Malden, MA. [21]

Aeschbacher, S., M. A. Beaumont, and A. Futschik. 2012. A novel approach for choosing summary statistics in approximate Bayesian computation. *Genetics* 192: 1027–1047. [A3]

Agrawal, A. F., and J. R. Chasnov. 2001. Recessive mutations and the maintenance of sex in structured populations. *Genetics* 158: 913–917. [7]

Aguadé, M., N. Miyashita, and C. H. Langley. 1989. Reduced variation in the *yellow-achaete-scute* region in natural populations of *Drosophila melanogaster. Genetics* 122: 607–615. [8]

Aguirre, J. D., E. Hine, K. McGuigan, and M. W. Blows. 2014. Comparing G: multivariate analysis of genetic variation in multiple populations. *Heredity* 112: 21–30. [30]

Ahn, B. Y, and D. M. Livingston. 1986. Mitotic gene conversion lengths, coconversion patterns, and the incidence of reciprocal recombination in a *Saccharomyces cerevisiae* plasmid system. *Mol. Cell. Biol.* 6: 3685–3693. [4]

Aieri, C. G., and H. B. Fraser. 2014. Evolution at two levels of gene expression in yeast. *Genome Res.* 24: 411-421. [12]

Akaike, H. 1974. A new look at the statistical model identification. *Automatic Control, IEEE Transactions* 19: 716–723. [12, 30]

Akashi, H. 1995. Inferring weak selection from patterns of polymorphism and divergence at "silent" sites in *Drosophila* DNA. *Genetics* 139: 1067–1076. [8, 10]

Akashi, H. 1999. Within- and between-species DNA sequence variation and the 'footprint' of natural selection. *Genetics* 238: 39–51. [8]

Akashi, H., N. Osada, and T. Ohta. 2012. Weak selection and protein evolution. *Genetics* 192: 15–31. [8]

Akashi, H., and S. W. Schaeffer. 1997. Natural selection and the frequency distribution of "silent" DNA polymorphisms in *Drosophila. Genetics* 146: 295–307. [8]

Akçay, E., and J. Van Cleve. 2016. There is no fitness but fitness, and the lineage is its bearer. *Phil. Trans. Roy. Soc. Lond. B* 371: 20150085. [22]

Åkesson, M, S. Bensch, D. Hasselquist, M. Tarka, and B. Hansson. 2008. Estimating heritabilities and genetic correlations: comparing the 'animal model' with parent-offspring regression using data from a natural population. *PLoS One* 3: e1739. [20]

Akey, J. M. 2009. Constructing genomic maps of positive selection in humans: Where do we go from here? *Genome Res.* 19: 711–722. [9]

Akey, J. M., G. Zhang, K. Zhang, L. Jin, and M. D. Shriver. 2002. Interrogating a high-density SNP map for signatures of natural selection. *Genome Res.* 12: 1805–1814. [9]

Akey, J. M.. and 6 others. 2004. Population history and natural selection shape patterns of genetic variation in 132 genes. *PLoS Biol.* 2: e286. [8, 9]

Akey, J. M., and 7 others. 2010. Tracking footprints of artificial selection in the dog genome. *Proc. Natl. Acad. Sci. USA* 107: 1160–1165. [9]

Akin, E. 1979. *The geometry of population genetics*, Springer-Verlag, Berlin, DE. [5]

Akin, E. 1982. Cycling in simple genetic systems. *J. Math. Biol.* 13: 305–324. [5]

Alachiotis, N., A. Stamatakis, and P. Pavlidis. 2012. OmegaPlus: a scalable tool for rapid detection of selective sweeps in whole-genome datasets. *Bioinformatics* 23: 2274–2275. [9]

Alatalo, R. V., L. Gustafsson, and A. Lunberg. 1990. Phenotypic selection on heritable size traits: environmental variance and genetic response. *Am. Nat.* 135: 464–471. [20]

Albert, F. W., and L. Kruglyak. 2015. The role of regulatory variation in complex traits and disease. *Nat. Rev. Genet.* 16: 197–212. [12]

Albertson, R. C., J. T. Streelman, and T. D. Kocher. 2003. Directional selection has shaped the oral jaws of Lake Malawi cichlid fishes. *Proc. Natl. Acad. Sci. USA* 100: 5252–5257. [12]

Alemu, S. W., P. Berg, L. Janss, and P. Bijma. 2014. Indirect genetic effects and kin recognition: estimating IGEs when interactions differ between kin and strangers. *Heredity* 112: 197–206. [22]

Allaby, R. G., D. Q. Fuller, and T. A. Brown 2008. The genetic expectations of a protracted model for the origins of domesticated crops. *Proc. Natl. Acad. Sci. USA* 105: 13982–13986. [9]

Allan, M. F., E. J. Eisen, and D. Pomp. 2005. Genomic mapping of direct and correlated responses to long-term selection for rapid growth rate in mice. *Genet.* 170: 1863–1877. [25]

Allard, R. W. 1975. The mating system and microevolution. *Genetics* 79: 115–126. [23]

Allard, R. W. 1999. *Principles of plant breeding*, 2nd ed. Wiley, New York, NY. [21]

Allard, R. W., and J. Adams. 1969. The role of intergenotypic interactions in plant breeding. *Proc. XII Int. Congress Genetics* 3: 349–370. [22]

Allison, D. B., and 6 others. 2002. A mixture model approach for the analysis of microarray gene expression data. *Comput. Stat. Data Anal.* 39: 1–20. [A4]

Al-Murrani, W. K. and R. C. Roberts. 1974. Genetic variation in a line of mice selected to its limit for high body weight. *Anim. Prod.* 19: 273–289. [25]

Alves, I., A. S. Hanulova, M. Foll, and L. Excoffier. 2012. Genomic data reveal a complex making of humans. *PLoS Genet.* 8: e1002837. [8, 9]

Amato, R., G. Miele, A. Monticelli, and S. Cocozza. 2011. Signs of selective pressure on genetic variants affecting human height. *PLoS One* 6: e27588. [9]

Anderson, D. R., K. P. Burnham, and W. L. Thompson. 2000. Null hypothesis testing: problems, prevalence, and an alternative. *J. Wildlife Manage.* 64: 912–923. [12]

Anderson, E. C. 2005. An efficient Monte Carlo method for estimating N_e from temporally spaced samples using a coalescent-based likelihood. *Genetics* 170: 955–967. [4]

Anderson, E. C., and M. Slatkin. 2003. Orr's quantitative trait loci sign test under conditions of trait ascertainment. *Genetics* 165: 445–446. [12]

Anderson, E. C., E. G. Williamson, and E. A. Thompson. 2000. Monte Carlo evaluation of the likelihood for N_e from temporally spaced samples. *Genetics* 156: 2109–2118. [4]

Andersen, E. C., and 7 others. 2012. Chromosome-scale selective sweeps shape *Caenorhabditis elegans* genomic diversity. *Nat. Genet.* 44: 285–290. [8]

Anderson, W. W. 1973. Genetic divergence in body size among experimental populations of *Drosophila pseudoobscura* kept at different temperatures. *Evolution* 27: 278–284. [12]

Andersson, E. W., K. A. Spanos, T. J. Mullin, and D. Lindgren. 1998. Phenotypic selection can be better than selection for breeding value. *Scand. J. Forest Res.* 13: 7-11. [26]

Andersson, M. 1982. Female choice selects for extreme tail length in a widowbird. *Nature* 299: 818–820. [29]

Andersson, M. B. 1994. *Sexual selection.* Princeton University Press, Princeton, NJ. [29]

Andersson, M., and Y. Iwasa. 1996. Sexual selection. *Trends Ecol. Evol.* 11: 53–58. [29]

Andersson, M., and L. W. Simmons. 2006. Sexual selection and mate choice. *Trends Ecol. Evol.* 21: 296–302. [29]

Andersson, S., M. Ellmer, T. H. Jorgensen, and Anna Palmé. 2010. Quantitative genetic effects of bottlenecks: experimental evidence from a wild plant species, *Nigella degenii. J. Heredity* 101: 298–307. [11]

Andolfatto, P. 2001. Adaptive hitchhiking effects on genome variability. *Curr. Opin. Genet. Develop.* 11: 635–641. [8]

Andolfatto, P. 2005. Adaptive evolution of non-coding DNA in *Drosophila. Nature* 437: 1149-1152. [8, 10]

Andolfatto, P. 2007. Hitchhiking effects of recurrent beneficial amino acid substitutions in the *Drosophila melanogaster* genome. *Genome Res.* 17: 1755–1762. [8, 10]

Andolfatto, P. 2008. Controlling type-1 error of the McDonald-Kreitman test in genomewide scans for selection on noncoding DNA. *Genetics* 180: 1767–1771. [10]

Andolfatto, P., and M. Nordborg. 1998. The effect of gene conversion on intralocus associations. *Genetics* 148: 1397–1399. [4, 9]

Andolfatto, P., and M. Przeworski. 2000. A genome-wide departure from the standard neutral model in natural populations of *Drosophila. Genetics* 156: 257–268. [4]

Andolfatto, P., and M. Przeworski. 2001. Regions of lower crossing over harbor more rare variants in African populations of *D. melanogaster. Genetics* 158: 657–665. [8]

Andolfatto, P., J. D. Wall, and M. Kreitman. 1999. Unusual haplotype structure at the proximal breakpoint of *In(2L)t* in a natural population of *Drosophila melanogaster. Genetics* 153: 1297–1311. [9]

André, J. B., and B. Godelle. 2006. The evolution of mutation rate in finite asexual populations. *Genetics* 172: 611–626. [4]

Andrés, A. M., and 11 others. 2009. Targets of balancing selection in the human genome. *Mol. Biol. Evol.* 26: 2755–2764. [9]

Angerer, W. P. 2001a. A note on the evaluation of fluctuation experiments. *Mutat. Res.* 479: 207–224. [4]

Angerer, W. P. 2001b. An explicit representation of the Luria-Delbrück distribution. *J. Math. Biol.* 42: 145–174. [4]

Anisimova, M., J. P. Bielawski, and Z. Yang. 2001. Accuracy and power of the likelihood ratio test in detecting adaptive molecular evolution. *Mol. Biol. Evol.* 18: 1585–1592. [10]

Anisimova, M., J. P. Bielawski, and Z. Yang. 2002. Accuracy and power of Bayes prediction of amino acids sites under positive selection. *Mol. Biol. Evol.* 19: 950–958. [10]

Anisimova, M., and D. A. Liberles. 2007. The quest for natural selection in the age of comparative genomics. *Heredity* 99: 567–579. [10]

Anisimova, M., R. Nielsen, and Z. Yang. 2003. Effect of recombination on the accuracy of the likelihood method for detecting positive selection at amino acid sites. *Genetics* 164: 1229–1236. [10]

Anisimova, M., and Z. Yang. 2007. Multiple hypothesis testing to detect lineages under positive selection that affects only a few sites. *Mol. Biol. Evol.* 24: 1219–1228. [10]

Anthes, N., and 11 others. 2010. Bateman gradients in hermaphrodites: an extended approach to quantify sexual selection. *Am. Nat.* 176: 249–263. [29]

Aoki, K. 1981. Algebra of inclusive fitness. *Evolution* 35: 659–663. [22]

Aoki, K. 1982. Additive polygenic formulation of Hamilton's model of kin selection. *Heredity* 49: 163–169. [22]

Aquadro, C. F., and B. D. Greenberg. 1983. Human mitochondrial DNA variation and evolution: analysis of nucleotide sequences from seven individuals. *Genetics* 103: 287–312. [9]

Arango, J., I. Misztal, S. Tsuruta, M. Culbertson, and W. Herring. 2005. Estimation of variance components including competitive effects of Large White growing gilts. *J. Anim. Sci.* 83: 1241–1246. [22]

Arbeithuber, B., A. J. Betancourt, T. Ebner, and I. Tiemann-Boege. 2015. Crossovers are associated with mutation and biased gene conversion at recombination hotspots. *Proc. Natl. Acad. Sci. USA* 112: 2109–2114. [4]

Arbiza, L., and 6 others. 2013. Genome-wide inference of natural selection on human transcription factor binding sites. *Nat. Genet.* 45: 723–729. [10]

Arendt, J., and D. Reznick. 2008. Convergence and parallelism reconsidered: what have we learned about the genetics of adaptation? *Trends Ecol. Evol.* 23: 26–32. [8]

Aris-Brosou, S. 2006. Identifying sites under positive selection with uncertain parameter estimates. *Genome* 49: 767–776. [10]

Árnason, T. 2001. Trends and asymptotic limits for racing speed in standardbred trotters. *Livestock Prod. Sci.* 72: 135–145. [25]

Arnheim, N., P. Calabrese, and I. Tiemann-Boege. 2007. Mammalian meiotic recombination hot spots. *Ann. Rev. Genet.* 41: 369–399. [4]

Arnold, S. J. 1983a. Sexual selection: the interface of theory and empircism. *In* P. Bateson (ed.), *Mate choice*, pp. 67–107. Cambridge University Press, Cambridge, UK. [29, 30]

Arnold, S. J. 1983b. Morphology, performance and fitness. *Amer. Zool.* 23: 347–361. [29, 30]

Arnold, S. J. 1986. Limits on stabilizing, distruptive, and correlational selection set by the opportunity for selection. *Am. Nat.* 128: 143–146. [29]

Arnold, S. J. 1988. Behavior, energy and fitness. *Am. Zoo.* 28: 815-827. [29, 30]

Arnold, S. J. 1994. Bateman's principles and the measurement of sexual selection in plants and animals. *Am. Nat.* 144: S126–S149. [29]

Arnold, S. J. 2003. Performance surfaces and adaptive landscapes. *Integ. Comp. Biol.* 43: 367–375. [29]

Arnold, S. J., and D. Duvall. 1994. Animal mating systems: a synthesis based on selection theory. *Am. Nat.* 143: S317–348. [29]

Arnold, S. J., M. E. Pfrender, and A. G. Jones. 2001. The adaptive landscape as a conceptual bridge between micro- and macroevolution. *Genetica* 112-113: 9–32. [30]

Arnold, S. J., and M. J. Wade. 1984a. On the measurement of natural and sexual selection: theory. *Evolution* 38: 709–719. [13, 29, 30]

Arnold, S. J., and M. J. Wade. 1984b. On the measurement of natural and sexual selection: applications. *Evolution* 38: 720–734. [13, 29]

Arnqvist, G., and L. Rowe. 2005. *Sexual conflict*. Princeton Univ. Press, Princeton, NJ. [29]

Arora, V., and P. Lahiri. 1997. On the superiority of the Bayesian method over BLUP in small area estimation problems. *Statistica Sinica* 7: 1053–1063. [22]

Arunyawat, U., W. Stephan, and T. Städler. 2007. Using multilocus sequence data to assess population structure, natural selection, and linkage disequilibrium in wild tomatoes. *Mol. Biol. Evol.* 24: 2310–2322. [4]

Ashman, T.-L., and M. T. Morgan. 2004. Explaining phenotypic selection on plant attractive characters: male function, gender balance or ecological context? *Proc. Roy. Soc. Lond. B* 271: 553–559. [29]

Aspi, J., A. Jäkäläniemi, J. Tuomi, and P. Siikamäki. 2003. Multilevel phenotypic selection on morphological characters in a metapopulation of *Silene tatarica*. *Evolution* 57: 509–517. [30]

Asthana, S., S. Schmidt, and S. Sunyaev. 2005. A limited role for balancing selection. *Trends Genet.* 21: 30–32. [9]

Atkins, K. D., and R. Thompson. 1986. Predicted and realized response to selection for an index of bone length and body weight in Scottish Blackface sheep. 1. Responses in the index and component traits. *Anim. Prod.* 43: 421–435. [18, 19]

Atwood, K. C., L. K. Schneider, and F. J. Ryan. 1951a. Periodic selection in *Escherichia coli*. *Proc. Natl. Acad. Sci. USA* 37: 146–155. [8]

Atwood, K. C., L. K. Schneider, and F. J. Ryan. 1951b. Selective mechanisms in bacteria. *Cold Spring Harbor Symp. Quant. Biol.* 16: 345–355. [8]

Avalos, E. and W. G. Hill. 1981. An experimental check of an index combining individual and family measurements. *Anim. Prod.* 33: 1–5. [21]

Avery, P. J., and W. G. Hill. 1977. Variability in genetic parameters among small populations. *Genet. Res.* 29: 193–213. [11]

Axelsson, E., and H. Ellegren. 2009. Quantification of adaptive evolution of genes expressed in avian brain and population size effect on the efficacy of selection. *Mol. Biol. Evol.* 26: 1073–1079. [10]

Ayroles, J. F., and 7 others. 2015. Behavioral idiosyncrasy reveals genetic control of phenotypic variability. *Proc. Natl. Acad. Sci. USA* 112: 6706–6711. [17]

Azevedo, L., G. Suriano, B. van Asch, R. M. Harding, and A. Amorim. 2006. Epistatic interactions: how strong in disease and evolution? *Trends Genet.* 22: 581-585. [7]

Bachtrog, D. 2008. Similar rates of protein adaptation in *Drosophila miranda* and *D. melanogaster*, two species with different current effective population sizes. *BMC Evol. Biol.* 8: 334. [8, 10]

Baer, C. F. 1999. Among-locus variation in F_{ST}: fish, allozymes and the Lewontin-Krakauer test revisited. *Genetics* 152: 653–659. [9]

Baer, C. F., M. M. Miyamoto, and D. R. Denver. 2007. Mutation rate variation in multicellular eukaryotes: causes and consequences. *Nat. Rev. Genet.* 8: 619–631. [4]

Baer, C. F., and 8 others. 2006. Cumulative effects of spontaneous mutations for fitness in *Caenorhabditis*: role of genotype, environment and stress. *Genetics* 174: 1387–1395. [12]

Baguisi, A., and 19 others. 1999. Production of goats by somatic cell nuclear transfer. *Nat. Biotechnol.* 17: 456–461. [21]

Bailey, D. W. 1959. Rates of subline divergence in highly inbred lines of mice. *J. Heredity* 50: 26-30. [12]

Bailey, N. W. 2012. Evolutionary models of extended phenotypes. *Trends Ecol. Evol.* 27: 561–569. [22]

Baker, H. G. 1955. Self-compatibility and establishment after 'long-distance' dispersal. *Evolution* 9: 347–349. [23]

Baker, R. J. 1973. Assortative mating and artificial selection. *Heredity* 31: 231–238. [16]

Balding, D. G. 2003. Likelihood-based inference for genetic correlation coefficients. *Theor. Popul. Biol.* 63: 221–230. [2]

Baldwin-Brown, J. G., A. D. Long, and K. R. Thornton. 2014. The power to detect quantitative trait loci using resequenced, experimentally evolved populations of diploid, sexual organisms. *Mol. Biol. Evol.* 31: 1040–1055. [18]

Bamshad, M., and S. P. Wooding. 2003. Signatures of natural selection in the human genome. *Nat. Rev. Genet.* 4: 99–111. [9]

Bank, C., G. B. Ewing, A. Ferrer-Admettla, M. Foll, and J. D. Jensen. 2014. Thinking too positive? Revisiting current methods of population genetic selection inference. *Trends Genet.* 30: 540–546. [9]

Banks, M. J., and D. J. Thompson. 1985. Lifetime mating success in the damselfly *Coenagrion puella*. *Anim. Behav.* 33: 1175–1183. [29]

Banschbach, V. S., and J. M. Herbers. 1996. Complex colony structure in social insects. II. Reproduction, queen-worker conflict, and levels of selection. *Evolution* 50: 298–307. [30]

Baril, C. P. 1992. Factor regression for interpreting genotype-environment interaction in bread-wheat trails. *Theor. Appl. Genet.* 83: 1022–1026. [9, 29]

Baril, C. P., J.-B. Denis, R. Wustman, and F. A. van Eeuwijk. 1995. Analysing genotype by environment interaction in Dutch potato variety trials using factorial regression. *Euphytica* 82: 149–155. [29]

Barker, J. S. F., and L. J. Cummins. 1969. Disruptive selection for sternopleural bristle number in *Drosophila melanogaster*. *Genetics* 61: 697–712. [16]

Barreiro, L. B., G. Laval, H. Quach, E. Patin, and L. Quintana-Murci. 2008. Natural selection has driven population differentiation in modern humans. *Nat. Genet.* 40: 340–345. [9]

Barreiro, L. B., and 17 others. 2009. Evolutionary dynamics of human Toll-like receptors and their different contributions to host defense. *PLoS Genet.* 5: e1000562. [9]

Barrett, R. D. H., and H. E. Hoekstra. 2011. Molecular spandrels: tests of adaptation at the genetic level. *Nat. Rev. Genet.* 12: 767–780. [10]

Barrett, R. D. H., L. K. M'Gonigle, and S. P. Otto. 2006a. The distribution of beneficial mutant effects under strong selection. *Genetics* 174: 2071–2079. [27]

Barrett, R. D. H., R. C. MacLean, and G. Bell. 2006b. Mutations of intermediate effect are responsible for adaptation in evolving *Pseudomonas fluorescens* populations. *Biol. Lett.* 2: 236–238. [27]

Barrett, R. D. H., and D. Schluter. 2008. Adaptation from standing genetic variation. *Trend. Ecol. Evol.* 23: 38–44. [10]

Barrett, S. C. H. 2002. The evolution of plant sexual diversity. *Nat. Rev. Genet.* 3: 274–284. [23]

Barrett, S. C. H., R. Arunkumar, and S. I. Wright. 2014. The demography and population genomics of evolutionary transitions to self-fertilization in plants. *Phil. Trans. Roy. Soc. Lond. B* 369: 20130344. [23]

Barrick, J. E., and R. E. Lenski. 2013. Genome dynamics during experimental evolution. *Nat. Rev. Genet.* 14: 827–839. [27]

Barrick, J. E., and 7 others. 2009. Genome evolution and adaptation in a long-term experiment with *Escherichia coli*. *Nature* 461: 1243–1247. [25, 27]

Barton, N. H. 1979. Gene flow past a cline. *Heredity* 43: 333–339. [9]

Barton, N. H. 1986. The maintenance of polygenic variability by a balance between mutation and stabilizing selection. *Genet. Res.* 47: 209–216. [5, 7, 28]

Barton, N. H. 1989. The divergence of a polygenic system subject to stabilizing selection, mutation, and drift. *Genet. Res.* 47: 209–216. [7, 28]

Barton, N. H. 1990. Pleiotropic models of quantitative variation. *Genetics* 124: 773–782. [28]

Barton, N. H. 1993. The probability of fixation of a favored allele in a subdivided population. *Genet. Res.* 62: 149–157. [7]

Barton, N. H. 1995a. Linkage and the limits to natural selection. *Genetics* 140: 821–841. [3, 8, 26]

Barton, N. H. 1995b. A general model for the evolution of recombination. *Genet. Res.* 65: 123–145. [4]

Barton, N. H. 1998. The effect of hitch-hiking on neutral genealogies. *Genet. Res.*. 72: 123–133. [8]

Barton, N. H. 1999. Clines in polygenic traits. *Genet. Res.* 74: 223—236. [28]

Barton, N. H. 2000. Genetic hitchhiking. *Phil. Trans. Roy. Soc. Lond.* B 355: 1553–1562. [8]

Barton, N. H. 2010. Mutation and the evolution of recombination. *Philos. Trans. Roy. Soc. Lond. B Biol. Sci.* 365: 1281–1294. [4]

Barton, N. H., and B. Charlesworth. 1984. Genetic revolutions, founder effects, and speciation. *Ann. Rev. Ecol. Syst.* 15: 133–164. [11]

Barton, N. F., A. M. Etheridge, and A. Véber. 2017. The infinitesimal model: Definition, derivation, and implications. *Theor. Popul. Biol.* 118: 50–73. [24]

Barton, N. H., and P. D. Keightley. 2002. Understanding quantitative genetic variation *Nat. Rev. Genet.* 3: 11–21. [28]

Barton, N. H., and A. Navarro. 2002. Extending the coalescent to multilocus systems: the case of balancing selection. *Genet. Res.* 7: 129–140. [9]

Barton, N. H., and S. P. Otto. 2005. Evolution of recombination due to random drift. *Genetics* 169: 2353–2370. [4, 7]

Barton, N. H., and L. Partridge. 2000. Limits to natural selection. *BioEssays* 22: 1075–1084. [26]

Barton, N. H., and M. Turelli. 1987. Adaptive landscapes, genetic distances and the evolution of quantitiative characters. *Genet. Res.* 49: 157–173. [1, 5, 15, 24, 28]

Barton, N. H., and M. Turelli. 1989. Evolutionary quantitative genetics: how little do we know?. *Ann. Rev. Genet.* 23: 337–370. [28]

Barton, N. H., and M. Turelli. 2004. Effects of genetic drift on variance components under a general model of epistasis. *Evolution* 58: 2111–2132. [11]

Bataillon, T., T. Zhang, and R. Kassen. 2011. Cost of adaptation and fitness effects of beneficial mutations in *Pseudomonas fluorescens*. *Genetics* 189: 939–949. [27]

Bateman, A. J. 1948. Intra-sexual selection in *Drosophila*. *Heredity* 2: 349–368. [29]

Bateman, A. J., and K. Mather. 1951. The progress of inbreeding in barley. *Heredity* 5: 323–348. [11]

Bateson, P. (ed.) 1983. *Mate choice*. Cambridge University Press, Cambridge, UK. [29]

Baudry, E., and F. Depaulis. 2003. Effect of misoriented sites on neutrality tests with outgroup. *Genetics* 165: 1619–1622. [9]

Baxter, S. W., and 8 others. 2011. Parallel evolution of *Bacillus thuringiensis* toxin resistance in Lepidoptera. *Genetics* 189: 675–679. [25]

Bazin, E., K. J. Dawson, and M. A. Beaumont. 2010. Likelihood-free inference of population structure and local adaptation in a Bayesian hierarchical model. *Genetics* 185: 587–602. [9]

Beaulieu, J. M., D.-C. Jhwueng, C. Boettiger, and B. C. O'Meara. 2012. Modeling stabilizing selection: expanding the Ornstein-Uhlenbeck model of adaptive evolution. *Evol.* 66: 2369–2383. [12]

Beaumont, M. A. 2003. Estimation of population growth or decline in genetically monitored populations. *Genetics* 164: 1139–1160. [4]

Beaumont, M. A. 2005. Adaptation and speciation: what can F_{ST} tell us? *Trends Ecol. Evol.* 20: 435–440. [9]

Beaumont, M. A. 2010. Approximate Bayesian computation in evolution and ecology. *Ann. Rev. Ecol. Evol. Syst.* 41: 379–406. [A3]

Beaumont, M. A., and D. J. Balding. 2004. Identifying adaptive genetic divergence among populations from genome scans. *Mol. Ecol.* 13: 969–989. [9]

Beaumont, M. A., and R. A. Nichols. 1996. Evaluating loci for use in the genetic analysis of population structure. *Proc. Roy. Soc. Lond. B* 263: 1619–1626. [9]

Beaumont, M. A., W. Zhang, and D. J. Balding. 2002. Approximate Bayesian computation in population genetics. *Genetics* 162: 2025–2035. [A3]

Becker, W. A. 1992. *Manual of quantitative genetics*, Fifth Edition. Academic Enterprises, Pullman, Washington. [21]

Bedford, T., and D. L. Hartl. 2009. Optimization of gene expression by natural selection. *Proc. Natl. Acad. Sci. USA* 106: 1133–1138. [12]

Begg, C. B.. and M. Mazumda. 1994. Operating characteristics of a rank correlation test for publication bias. *Biometrics* 50: 1088–1101. [A4]

Begun, D. J., and C. F. Aquadro. 1991. Molecular population genetics of the distal portion of the *X* chromosome in *Drosophila*: Evidence for genetic hitchhiking of the *yellow-achaete* region. *Genetics* 129: 1147–1158. [8, 10]

Begun, D. J., A. J. Betancourt, C. H. Langley, and W. Stephan. 1999. Is the fast/slow allozyme variation at the *Adh* locus of *Drosophila melanogaster* an ancient balanced polymorphism? *Mol. Biol. Evol* 16: 1816–1819. [8]

Beisel, C. J., D. R. Rokyta, H. A. Wichman, and P. Joyce. 2007. Testing the extreme value domain of attraction for distributions of beneficial fitness effects. *Genetics* 176: 2441–2449. [27]

Beissinger, T. M., and 8 others. 2014. A genome-wide scan for evidence of selection in a maize population under long-term artificial selection for ear number. *Genetics* 196: 829–840. [25]

Bell, G. 2010. Fluctuating selection: the perpetual renewal of adaptation in variable environments. *Phil. Trans. Roy. Soc. Lond. B:* 365: 87–97. [30]

Bello, Y., and D. Waxman. 2006. Near-periodic substitution and the genetic variance induced by environmental change. *J. Theor. Biol.* 239: 152–160. [27]

Beniwal, B. K., I. M. Hastings, R. Thompson, and W. G. Hill. 1992a. Estimation of changes in genetic parameters in selected lines of mice using REML with an animal model. 1. Lean mass. *Heredity* 69: 352–360. [19]

Beniwal, B. K., I. M. Hastings, R. Thompson, and W. G. Hill. 1992b. Estimation of changes in genetic parameters in selected lines of mice using REML with an animal model. 2. Body weight, body composition and litter size. *Heredity* 69: 361–371. [19]

Benjamini, Y., and Y. Hochberg. 1995. Controlling the false discovery rate: a practical and powerful approach to multiple testing. *J. Royal Stat. Soc. B* 85: 289–300. [A4]

Benjamini, Y., and Y. Hochberg. 2000. On the adaptive control of the false discovery rate in multiple testing with independent statistics. *J. Educ. Behav. Stat.* 26: 60–83. [A4]

Bennington, C. C., and J. B. McGraw. 1995. Phenotypic selection in an artificial population of *Imatiens pallida:* the importance of the invisible fraction. *Evolution* 49: 317–324. [29]

Benton, T. G., and Grant, A. 1996. How to keep fit in the real world: elasticity analyses and selection pressures on life histories in a variable environment. *Am. Nat.* 147: 115–139. [29]

Benton, T. G., and Grant, A. 2000. Evolutionary fitness in ecology: comparing measures of fitness in stochastic, density-dependent environments. *Evol. Ecol. Res.* 2: 769–789. [29]

Bérénos, C., P. A. Ellis, J. G. Pilkington, and J. M. Pemberton. 2014. Estimating quantitative genetic parameters in wild populations: a comparison of pedigree and genomic approaches. *Mol. Ecol.* 23: 3434–3451. [20]

Berg, J. J., and G. Coop. 2014. A population genetic signal of polygenic adaptation. *PLoS Genet.* 10: e1004412. [8, 9, 12]

Berg, J. J., and G. Coop. 2015. A coalescent model for a sweep of a unique standing variant. *Genetics* 201: 707–725. [8]

Bergelson, J., G. Dwyer, and J. J. Emerson. 2001. Models and data on plant-enemy coevolution. *Ann. Rev. Genet.* 35: 469–499. [10]

Berger, J. O. 1985. *Statistical decision theory and Bayesian analysis.* Springer-Verlag, New York, NY. [A2]

Bergeron, P., P.-O. Montiglio, D. Réale, M. M. Humphries, and D. Garant. 2012. Bateman gradients in a promiscuous mating system. *Behav. Ecol. Sociobio.* 66: 1125–1130. [29]

Berglund, J., K. S. Pollard, and M. T. Webster. 2009. Hotspots of biased nucleotide substitutions in human genes. *PLoS Biol* : e1000026. [10]

Bergsma, R., E. Kanis, E. F. Knol, and P. Bijma. 2008. The contribution of social effects to heritable variation in finishing traits of domestic pigs (*Sus scrofa*). *Genetics* 178: 1559–1570. [22]

Bernardo, J. 1996. Maternal effects in animal ecology. *Am. Zoo.* 36: 83–105. [15]

Bernardo, R. 2010. *Breeding for quantitative traits in plants*, 2nd Ed. Stemma Press, Woodbury, MN. [13, 19, 21]

Berry, A. J., J. W. Ajioka, and M. Kreitman. 1991. Lack of polymorphism on the *Drosophila* fourth chromosome resulting from selection. *Genetics* 129: 1111–1117. [8]

Berthier, P., M. A. Beaumont, J. M. Cornuet, and G. Luikart. 2002. Likelihood-based estimation of the effective population size using temporal changes in allele frequencies: a genealogical approach. *Genetics* 160: 741–751. [4]

Besag, J., P. J. Green, D. Higdon, and K. L. M. Mengersen. 1995. Bayesian computation and stochastic systems (with discussion). *Stat. Sci.* 10: 3–66. [A3]

Betancourt, A. J., and D. C. Presgraves. 2002. Linkage limits the power of natural selection in *Drosophila*. *Proc. Natl. Acad. Sci. USA* 99: 13616–13620. [8]

Betancourt, A. J., J. J. Welch, and B. Charlesworth. 2009. Reduced effectiveness of selection caused by a lack of recombination. *Curr. Biol.* 19: 655–660. [8]

Bhargava, A., and F. F. Fuentes. 2010. Mutational dynamics of microsatellites. *Mol. Biotech.* 44: 250–266. [8]

Bierne, N. 2010. The distinctive footprints of local hitchhiking in a varied environment and global hitchhiking in a subdivided population. *Evolution* 64: 3254–3272. [8]

Bierne, N., and A. Eyre-Walker. 2004. The genomic rate of adaptive amino acid substitution in *Drosophila*. *Mol. Biol. Evol.* 21: 1350–1360. [10]

Bierne, N., D. Roze, and J. J. Welch. 2013. Pervasive selection or is it...? Why are F_{ST} outliers sometimes so frequent? *Mol. Ecol.* 22: 2061–2064. [9]

Bijma, P. 2010a. Fisher's fundamental theorem of inclusive fitness and the change in fitness due to natural selection when conspecifics interact. *J. Evol. Biol.* 23: 194–206. [22]

Bijma, P. 2010b. Multilevel selection 4: modeling the relationship of indirect genetic effects and group size. *Genetics* 186: 1029–1031. [22]

Bijma, P. 2010c. Estimating indirect genetic effects: precision of estimates and optimum designs. *Genetics* 186: 1013–1028. [22]

Bijma, P. 2011. A general definition of the heritable variation that determines the potential of a population to respond to selection. *Genetics* 189: 1347–1359. [22]

Bijma, P. 2014. The quantitative genetics of indirect genetic effects: a selective review of modelling issues. *Heredity* 112: 61–69. [22]

Bijma, P., W. M. Muir, and J. A. M. Van Arendonk. 2007a. Multilevel selection 1: quantitative genetics of inheritance and response to selection. *Genetics* 175: 277–288. [1, 22]

Bijma, P., W. M. Muir, E. D. Ellen, J. B. Wolf, and J. A. M. Van Arendonk. 2007b. Multilevel selection 2: Estimating the genetic parameters determining inheritance and response to selection. *Genetics* 175: 289–299. [22]

Bijma, P., and M. J. Wade. 2008. The joint effects of kin, multilevel selection and indirect genetic effects on response to genetic selection. *J. Evol. Biol.* 21: 1175–1188. [22]

Birch, J. 2014. Hamilton's rule and its discontents. *Br. J. Philos. Sci.* 65: 381–411. [22]

Birch, J., and S. Okasha. 2015. Kin selection and its critics. *BioScience* 65: 22–32. [22]

Birkhead, T. R., and A. P. Møller (eds). 1999. *Sperm competition and sexual selection.* Academic Press, San Diego, CA. [29]

Birkhead, T. R., and T. Pizzari. 2002. Postcopulatory sexual selection. *Nat. Rev. Genet.* 3: 262–273. [29]

Birky, C. W., and J. B. Walsh. 1988. Effects of linkage on rates of molecular evolution. *Proc. Natl. Acad. Sci. USA* 85: 6414–6418. [7, 26]

Bisgaard, S., and B. Ankenman. 1996. Standard errors for the eigenvalues in second-order response surface models. *Technometrics* 38: 238–246. [30]

Bishop, J. F., A. M. Dean, and T. Mitchell-Olds. 2000. Rapid evolution in plant chitinases: molecular targets of selection in plant-pathogen coevolution. *Proc. Natl. Acad. Sci. USA* 97: 5322–5327. [10]

Biswas, S., and J. M. Akey. 2006. Genomic insights into positive selection. *Trends Genet.* 22: 437-446. [9]

Bjöklund, M. 1991. Sexual dimorphisms and mating system in the grackles (*Quiscalus* spp.: Icterinae). *Evolution* 45: 608–621. [12]

Blair, H. T., and E. J. Pollak. 1984. Estimation of genetic trend in a selected population with and without the use of a control population. *J. Anim. Sci.* 58: 878–886. [19]

Blanckenhorn, W. U., J. W. A. Grant, and D. J. Fairbairn. 1998. Monopolization in a resource queue: water striders competing for food and mates. *Behav. Ecol. Sociobio.* 42: 63–70. [29]

Blanckenhorn, W. U., M. Reuter, P. L. Ward, and A. D. Barbour. 1999. Correcting for sampling bias in quantitative measures of selection when fitness is discrete. *Evolution* 53: 286–291. [29]

Blasco, A. 2001. The Bayesian controversy in animal breeding. *J. Anim. Sci.* 79: 2023–2046. [19]

Blasco, A. 2017. *Bayesian data analysis for animal scientists.* Springer, Gewerbestrasse, CH. [19, A2]

Blasco, A., D. Sorensen, and J. P. Bidanel. 1998. Bayesian inference of genetic parameters and selection response for litter size components in pigs. *Genetics* 149: 301–306. [19]

Bliss, F. A., and C. E. Gates. 1968. Directional selection in simulated populations of self-pollinated plants. *Aust. J. Biol. Sci.* 21: 705–720. [23]

Blomquist, G. E. 2010. Heritability of individual fitness in female macaques. *Evol. Ecol.* 24: 657–669. [6]

Blouin, M. S. 2003. DNA-based methods for pedigree reconstruction and kinship analysis in natural populations. *Trends Ecol. Evol.* 18: 503–511. [20]

Blount, Z. D., C. Z. Borland, and R. E. Lenski. 2008. Historical contingency and the evolution of a key innovation in an experimental population of *Escherichia coli*. *Proc. Natl. Acad. Sci. USA* 105: 7899–7906. [27]

Blount, Z. D., J. E. Barrick, C. J. Davidson, and R. E. Lenski. 2012. Genomic analysis of a key innovation in an experimental *Escherichia coli* population. *Nature* 489: 513–518. [27]

Blows, M. W. 2007a. A tale of two matrices: multivariate approaches in evolutionary biology. *J. Evol. Biol.* 20: 1–8. [30]

Blows, M. W. 2007b. Complexity for complexity's sake? *J. Evol. Biol.* 20: 39–44. [30]

Blows, M. W., and R. Brooks. 2003. Measuring nonlinear selection. *Am. Nat.* 162: 815–820. [30]

Blows, M. W., R. Brooks, and P. G. Kraft. 2003. Exploring complex fitness surfaces: multiple ornamentation and polymorphism in male guppies. *Evolution* 57: 1622–1630. [30]

Blows, M. W., S. F. Chenoweth, and E. Hine. 2004. Orientation of the genetic variance-covariance matrix and fitness surface for multiple male sexually selected traits. *Am. Nat.* 163: 329–340. [30]

Blows, M., and A. A. Hoffmann. 2005. A reassessment of genetic limits to evolutionary change. *Ecology* 86: 1371–1384. [25]

Blows, M., and B. Walsh. 2009. Spherical cows grazing in flatland: constraints to selection and adaptation. *In* J. van der Werf, H.U. Graser, R. Frankham, and C. Gondro (eds.) *Adaptation and fitness in animal populations: evolutionary and breeding perspectives on genetic resource management*, pp 83–101. Springer. Netherlands. [25, 28, 30]

Blumenstiel, J. P., X. Chen, M. He, and C. M. Bergman. 2014. An age-of-allele test of neutrality for transposable element insertions. *Genetics* 196: 523–538. [9]

Blyth, C. R. 1972. On Simpson's paradox and the surething principle. *J. Am. Stat. Assoc.* 67: 364--366. [10]

Boag, P. T., and P. R. Grant. 1981. Intense natural selection in a population of Darwin's finches (Geospizinae) in the Galápagos. *Science* 214: 82–84. [29]

Bodbyl Roels, S. A., and J. K. Kelly. 2011. Rapid evolution caused by pollinator loss in *Mimulus guttatus*. *Evolution* 65: 2541–2552. [23]

Bodmer, W. F. 1965. Differential fertility in population genetics models. *Genetics* 51: 411–424. [5]

Bohn, G. I., G. W. Friars, and I. McMillan. 1983. The variance of response to selection in *Tribolium castaneum*. *Can. J. Genet. Cytol.* 25: 44–46. [18]

Boichard, D., B. Bonaiti, A. Barbat, and S. Mattalia. 1995. Three methods to validate the estimation of genetic trend for dairy cattle. *J. Diary Sci.* 78: 431–437. [19]

Boitard, S., C. Schlötterer, and A. Futschik. 2009. Detecting selective sweeps: A new approach based on hidden Markov models. *Genetics* 181: 1567–1578. [9]

Bolker, B. M., and 6 others. 2009. Generalized linear mixed models: a practical guide for ecology and evolution. *Trends Ecol. Evol.* 24: 127–135. [14, 20]

Bollback, J. P., T. L. York, and, R. Nielsen. 2008. Estimation of $2N_e s$ from temporal allele frequency data. *Genetics* 179: 497–502. [4, 9]

Bonduriansky, R., A. J. Crean, and T. Day. 2012. The implications of nongenetic inheritance for evolution in changing environments. *Evol. Applications* 5: 192–201. [15]

Bonduriansky, R., and T. Day. 2009. Nongenetic inheritance and its evolutionary implications. *Ann. Rev. Ecol Evol. Syst.* 40: 103–125. [15]

Bonhomme, M., and 6 others. 2010. Detecting selection in population trees: the Lewontin and Krakauer test extended. *Genetics* 186: 241–262. [9]

Bonnin, I., J.-M. Prosperi, and I. Olivieri. 1996. Genetic markers and quantitative genetic variation in *Medicago truncatula* (Leguminosae): a comparative analysis of population structure. *Genetics* 143: 1795–1805. [12]

Bookstein, F. L. 1987. Random walk and the existence of evolutionary rates. *Paleobiology* 13: 446–464. [12]

Bookstein, F. L. 1988. Random walk and the biometrics of morphological characters. *Evol. Biol.* 23: 369–398. [12]

Bookstein, F. L. 1997. *Morphometric tools for landmark data: Geometry and biology.* Cambridge University Press, Cambridge, UK. [A5]

Bookstein, F.L., 2013. Random walk as a null model for high-dimensional morphometrics of fossil series: geometrical considerations. *Paleobiology* 39: 52–74. [12]

Bookstein, F. L., and 5 others. 1985. *Morphometrics in evolutionary biology.* Special publication 15, The Academy of Natural Sciences of Philadelphia. [A5]

Borenstein, M., L. V. Hedges, J. P. T. Higgins, and H. R. Rothstein. 2009a. *Introduction to meta-analysis.* Wiley, Chichester, UK. [30, A4]

Borenstein, M., L. V. Hedges, J. P. T. Higgins, and H. R. Rothstein. 2009b. A basic introduction to fixed-effect and random-effects models for meta-analysis. *Res. Syn. Meth.* 1: 97–111. [A4]

Bos, I., and P. Caligari. 1995. *Selection methods in plant breeding.* Chapman and Hall, London, UK. [13, 21]

Bos, M., and W. Scharloo. 1973a. The effects of disruptive and stabilizing selection on body size in *Drosophila melanogaster*. I. Mean values and variances. *Genetics* 75: 679–693. [16]

Bos, M., and W. Scharloo. 1973b. The effects of disruptive and stabilizing selection on body size in *Drosophila melanogaster*. II. Analysis of responses in the thorax selection lines. *Genetics* 75: 695–708. [16]

Bossert, W. H. 1963. Simulation of character displacement. Harvard University, Cambridge, MA (unpublished). [24]

Bourguet, D., J. Gair, M. Mattice, and M. C . Whitlock. 2003. Genetic recombination and adaptation to fluctuating environments: selection for geotaxis in *Drosophila melanogaster. Heredity* 91: 78–84. [26]

Bouwman, A. C., R. Bergsma, N. Duijvesteijn, and P. Bijma. 2010. Maternal and social genetic effects on average daily gain of piglets from birth until weaning. *J. Anim. Sci.* 88: 2883–2892. [22]

Bowcock, A. M., and 6 others. 1991. Drift, admixture, and selection in human evolution: a study with DNA polymorphisms. *Proc. Natl. Acad. Sci. USA* 1991 88: 839–843. [9]

Box, G. E. P. 1953. Non-normality and tests on variances. *Biometrika* 40: 318–335. [17]

Box, G. E. P., and N. R. Draper. 1987. *Empirical model-building and response surfaces.* John Wiley and Sons, New York, NY. [30]

Box, G. E. P., and J. S. Hunter. 1954. A confidence region for the solution of a set of simultaneous equations with an application to experimental designs. *Biometrika* 41: 190–199. [30]

Boyd, L. H., and G. R. Iversen. 1979. *Contextual analysis: concepts and statistical techniques.* Wadsworth, Belmont, Califorina. [30]

Boyd, R., and P. J. Richerson. 1980. Effect of phenotypic variation on kin selection. *Proc. Natl. Acad. Sci. USA* 77: 7506–7509. [22]

Boyd, R., and P. Richerson. 2005. *The origin and evolution of cultures.* Oxford University Press, Oxford, UK. [6]

Boyko, A. R., and 13 others. 2008. Assessing the evolutionary impact of amino acid mutations in the human genome. *PLoS Genet.* 4: e1000083. [10]

Bradbury, J. W., and M. B. Andersson. 1987. *Sexual selection: testing the alternatives.* John Wiley and Sons, New York, NY. [29]

Bradshaw, J. E. 1983. Estimating the predicted response to S_1 family selection. *Heredity* 51: 415–418. [23]

Bradshaw, J. E. 1984. Computer simulation of family selection schemes suitable for kale (*Brassica oleracea* L.), involving half-sibs, full-sib, and selfed families. *Theor. Appl. Genet.* 68: 503–508. [23]

Brakefield, P. M. 1984. The ecological genetics of quantitative characters of *Maniola jurtina* and other butterflies. *In* R. J. Vane-Wright and P. R. Ackery (eds.), *The biology of butterflies*, pp. 167–190. Princeton Univeristy Press, Princeton, NJ. [29]

Braverman, J. M., R. R. Hudson, N. L. Kaplan, C. H. Langley, and W. Stephan. 1995. The hitchhiking effect on the site frequency spectrum of DNA polymorphisms. *Genetics* 140: 783–796. [8]

Brawand, D., and 10 others. 2011. The evolution of gene expression levels in mammalian organs. *Nature* 478: 343–348. [12]

Breden, F. J., and M. J. Wade. 1989. Selection within and between kin groups of the imported willow leaf beetle. *Am. Nat.* 134: 35–50. [29, 30]

Breen, M. S., C. Kemena, P. K. Vlasov, C. Notredame, and F. A. Kondrashov. 2012. Epistasis as the primary factor in molecular evolution. *Nature* 490: 535-538. [7]

Breese, E. L. 1956. The genetical consequences of assortative mating. *Heredity* 10: 323–343. [16]

Brem, R. B., and L. Kruglyak. 2005. The landscape of genetic complexity across 5,700 gene expression traits in yeast. *Proc. Natl. Acad. Sci. USA* 102: 1572–1577. [12]

Brem, R. B., J. D. Storey, J. Whittle, and L. Kruglyak. 2005. Genetic interactions between polymorphisms that affect gene expression in yeast. *Nature* 436: 701–703. [12]

Brem, R. B., G. Yvert, R. Clinton, and L. Kruglyak. 2002. Genetic dissection of transcriptional regulation in budding yeast. *Science* 296: 752–755. [12]

Brennan, P., and D. E. Byth. 1979. Genotype × environment interactions in wheat yields and selection for widely adapted wheat genotypes. *J. Agric. Sci.* 30: 221–232. [21]

Bridges, W. C., Jr, S. J. Knapp, and P. L. Cornelius. 1991. Standard errors and confidence interval estimators for expected selection response. *Crop Sci.* 31: 253–255. [18]

Briggs, W. H., and I. L. Goldman. 2006. Genetic variation and selection response in model breeding populations of *Brassica rapa* following a diversity bottleneck. *Genetics* 172: 457-465. [11]

Brim, C. A., and C. C. Cockerham. 1961. Inheritance of quantitative characters in soybeans. *Crop Sci.* 1: 187–190. [23]

Brim, C. A., and C. W. Stuber. 1973. Application of genetic male sterility to recurrent selection schemes in soybeans. *Crop Sci.* 13: 528–530. [21]

Brinkman, M. A., and K. J. Frey. 1977. Yield component analysis of oat isolines that produce different grain yields. *Crop Sci.* 17: 165–168. [8]

Brinkmann, R. 1929. Statistisch-biostratigraphische Untersuchungen an mitteljurassischen Ammoniten über Artbegriff und Stammesenwicklung. *Abh., Ges. Wiss. Göttingen, Math-Phys Kl., N.F.* 13: 1–249. [12]

Brisbane, J. R., and J. P. Gibson. 1995. Balancing selection response and rate of inbreeding by including genetic relationships in selection decisions. *Theor. Appl. Genet.* 91: 421–431. [23]

Briscoe, D. A., and 6 others. 1992. Rapid loss of genetic variation in large captive populations of *Drosophila* flies: implications for the genetic management of captive populations. *Conserv. Biol.* 6: 416–425. [3]

Britten, H. B. 1996. Meta-analyses of the association between multilocus heterozygosity and fitness. *Evolution* 50: 2158–2164. [17]

Britten, R. J., and E. H. Davidson. 1969. Gene regulation for higher cells: a theory. *Science* 165: 349–357. [12]

Britten, R. J., and E. H. Davidson. 1971. Repetitive and non-repetitive DNA sequences and a speculation on the origins of evolutionary novelty. *Q. Rev. Biol.* 46: 111–138. [12]

Broadley, M. R., and 9 others. 2008. Evidence of neutral transcriptome evolution in plants. *New Phytol.* 180: 587–593. [12]

Brodie, E. D. III 1992. Correlational selection for color pattern and antipredator behavior in the garter snake *Thamnophis ordinoides*. *Evolution* 46: 1284–1308. [30]

Brodie, E. D. III, and F. J. Janzen. 1996. On the assignment of fitness values in statistical analysis of selection. *Evolution* 50: 437–442. [29]

Brodie, E. D. III, A. J. Moore, and F. J. Janzen. 1995. Visualizing and quantifying natural selection. *Trends Ecol. Evol. Biol.* 10: 313–318. [29, 30]

Brommer, J. E. 2000. The evolution of fitness in life-history theory. *Biol. Rev. Cambridge Phil. Sco.* 75: 377–404. [29]

Brommer, J. E. 2011. Whither P_{ST}? The approximation of Q_{ST} by P_{ST} in evolutionary and conservation biology. *J. Evol. Biol.* 24: 1160–1168. [12]

Brommer, J. E., L. Gustafsson, H. Pietiäinen, and J. Merilä. 2004. Single generation estimates of individual fitness as proxies for longterm genetic contribution. *Am. Nat.* 163: 505–517. [29]

Brommer, J. E., J. Merilä, and H. Kokko. 2002. Reproductive timing and individual fitness. *Ecol. Letters* 5: 802–810. [29]

Brooks, R., and J. A. Endler. 2001. Direct and indirect sexual selection and quantitative genetics of male traits in gupppies (*Poecilia reticulata*. *Evolution* 55: 1002 - 1015. [30]

Brooks, R., and 5 others. 2005. Experimental evidence for multivariate stabilizing sexual selection. *Evolution* 59: 871–880. [29]

Brooks, S., A. Gelman, G. Jones, and X.-L. Meng (eds). 2011. *Handbook of markov chain Monte Carlo.* CRC press, Boca Raton, FL. [A3]

Brooks, S. P., P. Giudici, and A. Philippe. 2003. Nonparametric convergence assessment for MCMC model selection. *J. Comput. Graph. Stat.* 12: 1–22. [A3]

Brooks, S. P., and G. O. Roberts. 1998. Convergence assessment techniques for Markov chain Monte Carlo. *Stat. Comput.* 8: 319–335. [A3]

Brown, G. R., G. P. Gill, R. J. Kuntz, C. H. Langley, and D. B. Neale. 2004. Nucleotide diversity and linkage disequilibrium in loblolly pine. *Proc. Natl. Acad. Sci. USA* 101: 15255–15260. [4]

Brown, G. R., K. N. Laland, and M. B. Mulder. 2009. Bateman's principles and human sex roles. *Trends Ecol. Evol.* 24: 297–304. [29]

Brown, P. O., and D. Botstein. 1999. Exploring the new world of the genome with DNA microarrays. *Nat. Genet.* 21: 33–37. [12]

Brown, W. M., and A. E. Bell. 1961. Genetic analysis of plateaued populations of *Drosophila*. *Genetics* 46: 407–425. [25]

Brown, W. M., and A. E. Bell. 1980. An experimental comparison of selection alternatives to plateaued response. *Genetics* 94: 477–496. [25]

Brumpton, R. J., H. Boughey, and J. L. Jinks. 1977. Joint selection for both extremes of mean performance and of sensitivity to a macro-environmental variable. *Heredity* 38: 219—226. [21]

Bryant, E. H., L. M. Combs, and S. A. McCommas. 1986a. Morphometric differentiation among experimental lines of the housefly in relation to a bottleneck. *Genetics* 114: 1213–1223. [12]

Bryant, E. H., S. A. McCommas, and L. M. Combs. 1986b. The effect of an experimental bottleneck upon quantitative genetic variation in the housefly. *Genetics* 114: 1191–1211. [11]

Bryant, E. H., and L. M. Meffert. 1996. Nonadditive genetic structuring of morphometric variation in relation to a population bottleneck. *Heredity* 77: 168–176. [11]

Bubb, K. L., and 11 others. 2006. Scan of human genome reveals no new loci under ancient balancing selection. *Genetics* 173: 2165–2177. [9]

Bull, J. J. 1987. Evolution of phenotypic variance. *Evolution* 41: 303–315. [17]

Bullard, J. H., Y. Mostovoy, S. Dudoit, and R. B. Brem. 2010. Polygenic and directional regulatory evolution across pathways in *Saccharomyces*. *Proc. Natl. Acad. Sci. USA* 107: 5058–5063. [12]

Bullaughey, K., M. Przeworski, and G. Coop. 2008. No effect of recombination on the efficacy of natural selection in primates. *Genome Res.* 18: 544–554. [8]

Bulmer, M. 1991. The selection-mutation-drift theory of synonymous codon usage. *Genetics* 129: 897–907. [1, 7, 8]

Bulmer, M. 2004. Did Jenkin's swamping argument invalidate Darwin's theory of natural selection? *Brit. J. Hist. Sci.* 37: 281–297. [1]

Bulmer, M. G. 1971a. The stability of equilibria under selection. *Heredity* 27: 157–162. [5, 28]

Bulmer, M. G. 1971b. The effect of selection on genetic variability. *Am. Nat.* 105: 201–211. [1, 15, 16, 19, 24]

Bulmer, M. G. 1972. The genetic variability of polygenic characters under optimizing selection, mutation and drift. *Genet. Res.* 19: 17–25. [7, 28]

Bulmer, M. G. 1973. The maintenance of the genetic variability of polygenic characters by heterozygous advantage. *Genet. Res.* 22: 9–12. [28]

Bulmer, M. G. 1974a. Linkage disequilibrium and genetic variability. *Genet. Res.* 23: 281–289. [1, 16, 24]

Bulmer, M. G. 1974b. Density-dependent selection and character displacement. *Am. Nat.* 108: 45–58. [28]

Bulmer, M. G. 1976a. The effect of selection on genetic variability: a simulation study. *Genet. Res.* 28: 101–117. [11, 16, 24]

Bulmer, M. G. 1976b. Regressions between relatives. *Genet. Res.* 28: 199–203. [16]

Bulmer, M. G. 1980. *The mathematical theory of quantitative genetics.* Oxford Univ. Press, Oxford, UK. [11, 14, 15, 16, 24, 28]

Bulmer, M. G. 1989. Maintenance of genetic variability by mutation-selection balance: a child's guide through the jungle. *Genome* 31: 761–767. [28]

Bumpus, H. C. 1899. The elimination of the unfit as illustrated by the introduced house sparrow, *Passer domesticus*. *Biol. Lectures, Marine Biol. Lab., Woods Hole*: 209-226. [1, 30]

Bünger, L., and G. Herrendörfer. 1994. Analysis of a long-term selection experiment with an exponential model. *J. Anim. Breed. Genet.* 111: 1–13. [25]

Bünger, L., and 9 others. 2001. Inbred lines of mice derived from long-term growth selected lines: unique resources for mapping growth genes. *Mamm. Genome* 12: 678–686. [25]

Bürger, R. 1986. On the maintenance of genetic variation: global analysis of Kimura's continuum-of-alleles model. *J. Math. Biol.* 24: 341–351. [28]

Bürger, R. 1988a. Mutation-selection balance and continuum-of-alleles models. *Math. Biosci.* 91: 67–83. [28]

Bürger, R. 1988b. The maintenance of genetic variation: A functional analysis approach to quantitative genetic models. *In* G. de Jong (ed.), *Population genetics and evolution*, pp. 63–72. Springer-Verlag, Berlin, DE. [28]

Bürger, R. 1988c. Perturbations of positive semigroups and applications to population genetics. *Math. Z.* 197: 259–272. [28]

Bürger, R. 1989. Linkage and the maintenance of heritable variation by mutation-selection balance. *Genetics* 121: 175–184. [7, 28]

Bürger, R. 1991a. Moments, cumulants, and polygenic dynamics. *J. Math. Biol.* 30: 199–213. [1, 24]

Bürger, R. 1991b. An integro-differential equation from population genetics and perturbations of differentiable semigroups in Fréchet spaces. *Proc. Roy. Soc. Edinb. A* 118: 63–73. [28]

Bürger, R. 1993. Predictions of the dynamics of a polygenic character under directional selection. *J. Theor. Biol.* 162: 487–513. [24]

Bürger, R. 1998. Mathematical properties of mutation-selection models. *Genetica* 102: 279–298. [28]

Bürger, R. 1999. Evolution of genetic variability and the advantage of sex and recombination in changing environments. *Genetics* 153: 1055–1069. [28]

Bürger, R. 2000. *The mathematical theory of recombination, selection, and mutation.* Wiley, New York, NY. [5, 7, 24, 28]

Bürger, R., and I. M. Bomze. 1996. Stationary distributions under mutation-selection balance: structure and properties. *Adv. Appl. Prob.* 28: 277–251. [28]

Bürger, R., and W. J. Ewens. 1995. Fixation probabilities of additive alleles in diploid populations. *J. Math. Biol.* 33: 557–575. [7]

Bürger, R., and A. Gimelfarb. 1999. Genetic variation maintained in multilocus models of additive quantitative traits under stabilizing selection. *Genetics* 152: 807–820. [5, 28]

Bürger, R., and A. Gimelfarb. 2002. Fluctuating environments and the role of mutation in maintaining quantitative genetic variation. *Genet. Res.* 80: 31–46. [28]

Bürger, R., and A. Gimelfarb. 2004. The effects of intraspecific competition and stabilizing selection on a polygenic trait. *Genetics* 167: 1425–1443. [28]

Bürger, R., and J. Hofbauer. 1994. Mutation load and mutation-selection-balance in quantitative genetic traits. *J. Math. Biol.* 32: 193–218. [28]11,

Bürger, R., and R. Lande. 1994. On the distribution of the mean and variance of a quantitative trait under mutation-selection-drift balance. *Genetics* 138: 901–912. [28]

Bürger, R., G. P. Wagner, and F. Stettinger 1989. How much heritable variation can be maintained in finite populations by mutation-selection balance? *Evol.* 43: 1748-1766. [28]

Buri, P. 1956. Gene frequency in small populations of mutant *Drosophila. Evolution* 10: 367–402. [2]

Burke, M. K. 2012. How does adaptation sweep through the genome? Insights from long-term selection experiments. *Proc. Roy. Soc. B* 279: 5029–5038. [25]

Burke, M. K., and 5 others. 2010. Genome-wide analysis of a long-term evolution experiment with *Drosophila. Nature* 467: 587–590. [25]

Burnham, K. P., and D. R. Anderson. 2002. *Model selection and multimodel inference: a practical information-theoretic approach*, 2nd Ed. Springer-Verlag, New York, NY. [30] .

Burnham, K. P., and D. R. Anderson. 2004. Multimodel inference: understanding AIC and BIC in model selection. *Sociological Methods & Research* 33: 261–304. [30]

Burrows, P. M. 1972. Expected selection differentials for directional selection. *Biometrics* 28: 1091–1100. [14]

Burrows, P. M. 1975. Variance of selection differentials in normal samples. *Biometrics* 31: 125–133. [14]

Burrows, P. M. 1984a. Inbreeding under selection from related families. *Biometrics* 40: 895–906. [26]

Burrows, P. M. 1984b. Inbreeding under selection from unrelated families. *Biometrics* 40: 357–366. [26]

Burton, G. W. 1974. Recurrent restricted phenotypic selection increases forage yields of Pensacola bahiagrass. *Crop Sci.* 14: 831–835. [13]

Burton, G. W. 1982. Improved recurrent restricted phenotypic selection increases bahiagrass forage yields.

Crop Sci. 22: 1058–1061. [13]

Burton, J. W., and C. A. Brim. 1981. Recurrent selection in soybeans. III. Selection for increased percent oil in seeds. *Crop Sci.* 21: 31–34. [21]

Burton, J. W., and B. F. Carver. 1993. Selection among S_1 families vs. selfed half-sib or full-sib families in autogamous crops. *Crop Sci.* 33: 21–28. [23]

Burton, J. W., L. H. Penny, A. R. Hallauer, and S. A. Eberhart. 1971. Evaluation of synthetic populations developed from a maize variety (BSK) by two methods of recurrent selection. *Crop Sci.* 11: 361–365. [18, 23]

Bustamante, C. D., J. Wakeley, S. A. Sawyer, and D. L. Hartl. 2001. Directional selection and the site-frequency spectrum. *Genetics* 19: 1779–1788. [10]

Bustamante, C. D., and 5 others. 2002. The cost of inbreeding in *Arabidopsis*. *Nature* 416: 531–534. [10]

Bustamante, C. D., and 13 others. 2005. Natural selection on protein-coding genes in the human genome. *Nature* 437: 1153–1157. [10]

Butcher, D. 1995. Muller's ratchet, epistasis and mutation effects. *Genetics* 141: 431–437. [4]

Butler, M. A., and A. A. King. 2004. Phylogenetic comparative analysis: a modeling approach for adaptive evolution. *Am. Nat.* 164: 683–695. [12]

Butruille, D. V., H. D. Silva, S. M. Kaeppler, and J. G. Coors. 2004. Response to selection and genetic drift in three populations derived from the golden glow maize population. *Crop Sci.* 44: 1527–1534. [18]

Caballero, A. 1994. Developments in the prediction of effective population size. *Heredity* 73: 657–679. [3, 26]

Caballero, A. 1995. On the effective size of populations with separate sexes, with particular reference to sex-linked genes. *Genetics* 139: 1007–1011. [3]

Caballero, A. 1996. A note on the change in gene frequency of a selected allele in partial full-sib mating populations. *Genetics* 142: 649–650. [7]

Caballero, A., and W. G. Hill. 1992a. A note on the inbreeding effective population size. *Evolution* 46: 1969-1972. [3]

Caballero, A., and W. G. Hill. 1992b. Effects of partial inbreeding on fixation rates and variation of mutant genes. *Genetics* 131: 493–507. [7, 8]

Caballero, A., and P. D. Keightley. 1994. A pleiotropic nonadditive model of variation in quantitative traits. *Genet.* 138: 883–900. [28]

Caballero, A., P. D. Keightley, and W. G. Hill. 1991. Strategies for increasing fixation probabilities of recessive mutations. *Genet. Res.* 58: 129–138. [7, 26]

Caballero, A., M. Wei, and W. G. Hill. 1996. Survival rates of mutant genes under artificial selection using individual and family information. *J. Genet.* 75: 63–80. [26]

Cabana, G., and D. L. Kramer. 1991. Random offspring mortality and variation in parental fitness. *Evolution* 45: 228–234. [29]

Cai, J. J., J. M. Macpherson, G. Sella, and D. A. Petrov. 2009. Pervasive hitchhiking at coding and regulatory sites in humans. *PLoS Genet.* 5: e1000336. [8]

Caicedo, A. L., and 11 others. 2007. Genome-wide patters of nucleotide polymorphism in domesticated rice. *PLoS Genet.* 3: e163. [9]

Calhoun, J. B. 1947. The role of temperature and natural selection in relation to the variations in the size of the English sparrow in the United States. *Am. Nat.* 81: 203–28. [30]

Calsbeek, B. 2012. Exploring variation in fitness surfaces over time or space. *Evolution* 66: 1126–1137. [30]

Calsbeek, R., and R. M. Cox. 2010. Experimentally assessing the relative importance of predation and competition as agents of selection. *Nature* 465: 613–616. [29, 30]

Cameron, N. D. 1997. *Selection indices and prediction of genetic merit in animal breeding*. CAB International, Wallingford, UK. [13]

Campbell, C. D., and 12 others. 2012. Estimating the human mutation rate using autozygosity in a founder population. *Nat. Genet.* 44: 1277-1281. [4]

Campbell, D. R. 1997. Genetic and environmental variation in life-history traits of a monocarpic perennial: a decade-long field experiment. *Evolution* 51: 373–382. [6]

Campbell, K. H. S., J. McWhir, W. A. Ritchie, and I. Wilmut. 1996. Sheep cloned by nuclear transfer from a cultured cell line. *Nature* 380 64–66. [21]

Campo, J. L., and M. Garcia Gill. 1993. Assortative mating and directional or stabilizing selection for a nonlinear function of traits in *Tribolium*. *J. Anim. Breed. Genet.* 110: 74–80. [16]

Campo, J. L., and M. Garcia Gill. 1994. The effects of assortative mating on the genetic change due to linear index selection in *Tribolium*. *J. Anim. Breed. Genet.* 111: 213–219. [16]

Campo, J. L., and P. Tagarro. 1977. Comparisons of three selection methods for pupal weight of *Tribolium castaneum*. *Ann. Genet. Sel. Anim.* 9: 259–268. [21]

Campos, J. L., K. Zeng, D. J. Parker, B. Charlesworth, and P. R. Haddrill. 2012. Codon usage bias and effective population sizes on the X chromosome versus autosomes in *Drosophila melanogaster*. *Mol. Biol. Evol.* 30: 811–823. [8, 10]

Cannings, C., E. A. Thompson, and M. H. Skolnick. 1978. Probability functions on complex pedigrees. *Adv. App. Prob.* 10: 26–61. [24]

Cantet, R. J. C., and E. P. Cappa. 2008. On identifiability of (co)variance components n animal models with competition effects. *J. Anim. Breed. Genet.* 125: 371–381. [22]

Cappa, E. P., and R. J. C. Cantet. 2006. Bayesian estimation of direct and competition additive (co) variance in individuals mixed models. *Proceedings of the 8th World Congress on Genetics Applied to Livestock Production*, contribution 28-06, Belo Horizonte, MG, Brazil. [22]

Cappa, E. P., and R. J. C. Cantet. 2008. Direct and competition additive effects in tree breeding: Bayesian estimation from an individual tree mixed model. *Silvae Genetica* 57: 45–56. [17]

Carangal, V. R., S. M. Ali, F. Koble, E. H. Rinke, and J. C. Sentz. 1971. Comparison of S_1 and testcross evaluation for recurrent selection in maize. *Crop Sci.* 11: 658–661. [23]

Carlborg, Ö., and C. S. Haley. 2004. Epistasis: too often neglected in complex trait studies? *Nat. Rev. Genet.* 8: 618–625. [25]

Carlborg, Ö, L. Jacobsson, P. Åhgren, P. Siegel, and L. Andersson. 2006. Epistasis and the release of genetic variation during long-term selection. *Nat. Genet.* 38: 418–420. [25]

Carlin, B. P., and T. A. Louis. 2000. *Bayes and empirical Bayes methods for data analysis,* 2nd Ed. Chapman and

Hall, New York, NY. [A2]

Carlson, C. S., and 6 others. 2005. Genomic regions exhibiting positive selection identified from dense genotype data. *Genome Res.* 15: 1553–1565. [9]

Carneiro, M., and 8 others. 2012. Evidence for widespread positive and purifying selection across the European rabbit (*Oryctolagus cuniculus*) genome. *Mol. Bio. Evol.* 29: 1837–1849. [8, 10]

Caporale, L. H. (ed.) 2006. *The implicit genome.* Oxford Univ. Press, Oxford, UK. [1]

Carr, R. N., and R. F. Nassar. 1970. Effects of selection and drift on the dynamics of finite populations. *Biometrics* 26: 41–49. [7]

Carroll, S. B. 2005. Evolution at two levels: on genes and form. *PLoS Biol.* 3: e245. [12]

Carroll, S. B. 2008. Evo-devo and an expanding evolutionary synthesis: a genetic theory of morphological evolution. *Cell* 134: 25–36. [9, 12]

Carson, H. L. 1968. The population flush and its genetic consequences. *In* R. C. Lewontin (ed.), *Population biology and evolution,* pp. 123–137. Syracuse Univ. Press, Syracuse, NY. [11]

Carson, H. L. 1975. The genetics of speciation at the diploid level. *Am. Nat.* 109: 83–92. [11]

Carson, H. L., and A. R. Templeton. 1984. Genetic revolutions in relation to speciation phenomena: the founding of new populations. *Ann. Rev. Ecol. Syst.* 15: 97–131. [11]

Carter, A. J. R., E. Osborne, and D. Houle. 2009. Heritability of directional asymmetry in *Drosophila melanogaster. Int. J. Evol. Biol.* Vol. 2009: 1–7 [18]

Carter, A. J., and G. P. Wagner. 2002. Evolution of functionally conserved enhancers can be accelerated in large populations: a population-genetic model. *Proc. Roy. Soc. Lond. B* 269: 953–960. [7]

Casa, A. M., and 6 others. 2006. Evidence for a selective sweep on chromosome 1 of cultivated sorghum. *Crop Sci.* 46: S27–S40. [9]

Casacuberta, E., and J. Gonzáez. 2013. The impact of transposable elements in environmental adaptation. *Mol. Ecol.* 22: 1503–1517. [8]

Casella, G., and E. I. George. 1992. Explaining the Gibbs sampler. *Am. Stat.* 46: 167–174. [A3]

Casellas, J., and J. F. Medrano. 2008. Within-generation mutation variance for litter size in inbred mice. *Genetics* 179: 2147–2155. [12]

Cash, W. S. 1977. An improved solution for the ultimate probability of fixation of a favorable allele. I. Ultimate probability of fixation of a favorable allele. *Biometrics* 33: 528–532. [7]

Casler, M. D., and E. C. Brummer. 2008. Theoretical expected genetic gains for among-and-within-family selection methods in perennial forage crops. *Crop Sci.* 48: 890–902. [21]

Caswell, H. 1978. A general formula for the sensitivity of population growth rate to changes in life history parameters. *Theor. Popul. Biol.* 14: 215–230. [29]

Caswell, H. 1984. Optimal life histories and age-specific costs of reproduction. *J. Theor. Biol.* 98: 519–529. [29]

Caswell, H. 1989. *Matrix population models.* Sinauer, Sunderland, MA. [29, 30]

Caswell, H. 2001. *Matrix population models, 2nd edition* Sinauer, Sunderland, MA. [29, 30]

Cavalli-Sforza, L. L. 1966. Population structure and human evolution. *Proc. Roy. Soc. Lond. B* 164: 362–379. [9]

Cavalli-Sforza, L. L., and W. F. Bodmer. 1971. *The genetics of human populations.* W. H. Freeman and Co., San Francisco, CA. [7, 29]

Cavalli-Sforza, L. L., and M. W. Feldman. 1978. Darwinian selection and "altruism." *Theor. Popul. Biol.* 14: 268–280. [22]

Chabora, A. J. 1968. Disruptive selection for sternopleural chaeta number in various strains of *Drosophila melanogaster. Am. Nat.* 102: 525–532. [16]

Chaix, R., M. Somel, D. P. Kreil, P. Khaitovich, and G. Lunter. 2008. Evolution of primate gene expression: drift and corrective sweeps? *Genetics* 180: 1379–1389. [12]

Chakraborty, R. 1987. Biochemical heterozygosity and phenotypic variability of polygenic traits. *Heredity* 59: 19–28. [17]

Chakraborty, R., and M. Nei. 1982. Differentiation of quantitative characters between populations or species. I. Mutation and random genetic drift. *Genet. Res.* 39: 303–314. [11, 12]

Chamary, J. V., J. L. Parmley, and L. D. Hurst. 2006. Hearing silence: non-neutral evolution at synonymous sites in mammals. *Nat. Rev. Genet* 7: 98–108. [10]

Chan, Y. F., and 5 others. 2012. Parallel selection mapping using artificially selected mice reveals body weight control loci. *Curr. Biol.* 22: 794–800. [9, 25]

Chang, S. M., and R. G. Shaw. 2003. The contribution of spontaneous mutation to variation in environmental response in *Arabidopsis thaliana:* responses to nutrients. *Evolution* 57: 984–994. [12]

Chapuis, E., G. Martin, and J. Goudet. 2008. Effects of selection and drift on **G** matrix evolution in a heterogeneous environment: a multivariate Q_{ST}-F_{ST} test with the freshwater snail *Galba truncatula. Genetics* 180: 2151–2161. [12]

Charlesworth, B. 1980. *Evolution in age-structured populations.* Cambridge University Press, Cambridge, UK. [29]

Charlesworth, B. 1983. Natural selection on multivariate traits in age-structured populations. *Proc. Roy. Soc. Lond. B* 251: 47–52. [29]

Charlesworth, B. 1984a. The evolutionary genetics of life histories. *In* B. Shorrocks (ed.), *Evolutionary ecology,* pp. 117–133. Blackwell Scientific, Oxford, UK. [6]

Charlesworth, B. 1984b. The cost of phenotypic evolution. *Paleobiol.* 10: 319-327. [1]

Charlesworth, B. 1987. The heritability of fitness. *In* J. W. Bradbury and M. B. Andersson (eds.), *Sexual selection: testing the alternatives,* pp. 21-40. Wiley, New York, NY. [6]

Charlesworth, B. 1990. Mutation-selection balance and the evolutionary advantage of sex and recombination. *Genet. Res.* 55: 199–221. [1, 4, 7, 16]

Charlesworth, B. 1992. Evolutionary rates in partially self-fertilizing species. *Am. Nat.* 140: 126–148. [8]

Charlesworth, B. 1993. Directional selection and the evolution of sex and recombination. *Genet. Res.* 61: 205–224. [28]

Charlesworth, B. 1994a. *Evolution in age-structured populations, 2nd edition.* Cambridge University Press, Cambridge, UK. [3, 29]

Charlesworth, B. 1994b. The effect of background selection against deleterious mutations on weakly selected, linked variants. *Genet. Res.* 63: 213–227. [7, 10]

Charlesworth, B. 1996. Background selection and patterns of genetic diversity in *Drosophila melanogaster*. *Genet. Res.* 68: 131–149. [8]

Charlesworth, B. 1998. Measures of divergence between populations and the effect of forces that reduce variability. *Mol. Biol. Evol.* 15: 538–543. [9]

Charlesworth, B. 2001. The effect of life-history and mode of inheritance on neutral genetic variability. *Genet. Res.* 77: 153–166. [3]

Charlesworth, B. 2009. Effective population size and patterns of molecular evolution and variation. *Nat. Rev. Genet.* 10: 195–205. [3, 8]

Charlesworth, B. 2010. Molecular population genomics: a short history. *Genet. Res.* 29: 397–411. [8]

Charlesworth, B. 2012. The effects of deleterious mutations on evolution at linked sites. *Genetics* 190: 5–22. [3, 8]

Charlesworth, B. 2015. Causes of natural variation in fitness: evidence from studies of *Drosophila* populations. *Proc. Natl. Acad. Sci. USA* 112: 1662–1669. [28]

Charlesworth, B., and D. Charlesworth. 1983. The population dynamics of transposable elements. *Genet. Res.* 42: 1–27. [1]

Charlesworth, B., and D. Charlesworth. 2000. The degeneration of Y chromosomes. *Phil. Trans. Roy. Soc. Lond. B* 355: 1563–1572. [1]

Charlesworth, B., R. Lande, and M. Slatkin. 1982. A neo-Darwinian commentary on macroevolution. *Evolution* 36: 474–498. [1, 25, 28, 30]

Charlesworth, B., and C. H. Langley. 1986. The evolution of self-regulated transposition of transposable elements. *Genetics* 112: 359–383. [1]

Charlesworth, B., T. Miyo, and H. Borthwick. 2007. Selection responses of means and inbreeding depression for female fecundity in *Drosophila melanogaster* suggest contributions from intermediate-frequency alleles to quantitative trait variation. *Genet. Res.* 89: 85–91. [11, 28]

Charlesworth, B., M. T. Morgan, and D. Charlesworth. 1993a. The effect of deleterious mutations on neutral molecular variation. *Genetics* 134: 1289–1303. [3, 8]

Charlesworth, D. 2006a. Balancing selection and its effects on sequences in nearby genome regions. *PLoS Genet.* 2: e64. [9]

Charlesworth, D. 2006b. Evolution of plant breeding systems. *Current Biology* 16: R726–R735. [23]

Charlesworth, D., N. H. Barton, and B. Charlesworth. 2017. The sources of adaptive variation. *Proc. Roy. Soc. Lond. B* 284: 2016.2864. [1]

Charlesworth, D., and B. Charlesworth. 1995. Quantitative genetics in plants: the effect of the breeding system on genetic variability. *Evolution* 49: 911–920. [23]

Charlesworth, D., B. Charlesworth, and M. T. Morgan. 1995. The pattern of neutral molecular variation under the background selection model. *Genetics* 141: 1619–1632. [8]

Charlesworth, D., M. T. Morgan, and B. Charlesworth. 1993b. Mutation accumulation in finite outbreeding and inbreeding populations. *Genet. Res.* 61: 39–56. [23]

Charlesworth, D., and J. H. Willis. 2009. The genetics of inbreeding depression. *Nat. Rev. Genet.* 10: 783–796. [23]

Charlesworth, J., and A. Eyre-Walker. 2006. The rate of adaptive evolution in enteric bacteria. *Mol. Biol. Evol.* 23: 1348–1356. [10]

Charlesworth, J., and A. Eyre-Walker. 2007. The other side of the nearly neutral theory, evidence of slightly advantageous back-mutations. *Proc. Natl. Acad. Sci. USA* 104: 16992–16997. [12]

Charlesworth, J., and A. Eyre-Walker. 2008. The McDonald-Kreitman test and slightly deleterious mutations. *Mol. Biol. Evol.* 25: 1007–1015. [10]

Charmantier, A., and D. Garant. 2005. Environmental quality and evolutionary potential: lessons from wild populations. *Proc. Roy. Soc. Lond. B* 272: 1415–1425. [20]

Charmantier, A., D. Garant, and L. E. B. Kruuk (eds). 2014. *Quantitative genetics in the wild.* University of Oxford Press, Oxford, UK [20]

Charmantier, A., L. E. B. Kruuk, J. Blondel, and M. M. Lambrechts. 2004. Testing for microevolution in body size in three blue tit populations. *J. Evol. Biol* 17: 732–743. [20]

Charmantier, A., C. Perrins, R. H. McCleery, and B. C. Sheldon. 2006. Evolutionary response to selection in clutch size in a long-term study of the mute swan. *Am. Nat.* 167: 453-465. [20]

Charmantier, A., and D. Réale. 2005. How do missassigned paternities affect the estimation of heritability in the wild? *Mol. Ecol.* 14: 2839–2850. [20]

Chen, C. T., Q. S. Chi, and S. A. Sawyer. 2008. Effects of dominance on the probability of fixation of a mutant allele. *J. Math. Biol.* 56: 413–434. [7]

Chen, C. Y., R. K. Johnson, S. D. Kachman, and L. D. Van Vleck. 2006. Estimation of variance components due to competition effects for selected lines of swine. *Proceedings of the 8th World Congress on Genetics Applied to Livestock Production*, contribution 17–03, Belo Horizonte, MG, Brazil. [22]

Chen, C. Y., R. K. Johnson, S. Newman, and L. D. Van Vleck. 2008. Estimation of genetic parameters for average daily gain using models with competition effects. *J. Anim. Sci.* 86: 2525–2530. [22]

Chen, C. Y., R. K. Johnson, S. Newman, S. D. Kachman, and L. D. Van Vleck. 2009. Effects of social interactions on empirical responses to selection for average daily gain of boars. *J. Anim. Sci.* 87: 844–849. [22]

Chen, C.-H., T.-J. Chuang, B.-Y. Liao, and F.-C. Chen. 2009. Scanning for signatures of positive selection for human-specific insertions and deletions. *Genome Biol. Evol.* 1: 415–419. [10]

Chen, H., J. Chen, and J. D. Kalbfleisch. 2001. A modified likelihood ratio test for homogeneity in finite mixture models. *Stat. Meth.* 63: 19–29. [A4]

Chen, H., P. L. Morrell, M. de la Cruz, and M. T. Clegg. 2008. Nucleotide diversity and linkage disequilibrium in wild avocado (*Persea americana* Mill.). *J. Heredity* 99: 382–389. [4]

Chen, H., N. Patterson, and D. Reich. 2010. Population differentiation as a test for selective sweeps. *Genome Res.* 20: 393–402. [9]

Chen, J. M., D. N. Cooper, N. Chuzhanova, C. Férec, and G. P. Patrinos. 2007. Gene conversion: mechanisms, evolution and human disease. *Nat. Rev. Genet.* 8: 762–775. [4]

Chen, L., and J. D. Storey. 2006. Relaxed significance criteria for linkage analysis. *Genetics* 173: 2371–2381. [9, A4]

Chen, M.-H., Q.-M. Shao, and J. G. Ibrahim. 2000. *Monte Carlo methods in Bayesian computation.* Springer-Verlag, New York, NY. [A3]

Cheng, J., S. Janssens, and N. Buys. 2009. Full sib pens of pigs are not suitable to identify variance component of associative effect: a simulation study using Gibbs Sampling. *BMC Genetics* 10: 9. [22]

Chenoweth, S. F., and M. Blows. 2008. Q_{ST} meets the *G* matrix: the dimensionality of adaptive divergence in multiple correlated quantitative traits. *Evolution* 62: 1437–1449. [12]

Cherry, L. M., S. M. Case, J. G. Kunkel, J. S. Wyles, and A. C. Wilson. 1982. Body shape metrics and organismal evolution. *Evolution* 36: 914–933. [12]

Chevalet, C. 1988. Control of genetic drift in selected populations. *In* B. S. Weir, E. J. Eisen, M. M. Goodman, and G. Namkoong (eds.), *Proceedings of the second international conference on quantitative genetics,* pp. 379–394. Sinauer Assocs., Sunderland, MA. [24]

Chevalet, C. 1994. An approximate theory of selection assuming a finite number of quantitative trait loci. *Genet. Sel. Evol.* 26: 379–400 [1, 24]

Cheverud, J. M. 1984a. Evolution by kin selection: A quantitative genetic model illustrated by maternal performance in mice. *Evolution* 38: 766–777. [15, 22]

Cheverud, J. M. 1984b. Quantitative genetics and developmental constraints on evolution by selection. *J. Theor. Biol.* 110: 155–171. [30]

Cheverud, J. M. 1985. A quantitative genetic model of altrustic selection. *Behav. Ecol. Sociobiol.* 16: 239–243. [22]

Cheverud, J. M. 2001. A simple correction for multiple comparisons in interval mapping genome scans. *Heredity* 87: 52–58. [A4]

Cheverud, J. M., and W. P. J. Dittus. 1992. Primate population studies at Polonnaruwa. 2. Heritability of body measurements in a natural population of Toque macaques (*Macaca sinica*). *Am. J. Primatology* 27: 145–156. [1, 20]

Cheverud, J. M., and A. J. Moore. 1994. Quantitative genetics and the role of the environment provided by relatives in the evolution of behavior. *In* C. R. B. Boake (ed.), *Quantitative genetic studies of behavioral evolution,* pp. 67–100. University of Chicago Press, Chicago, IL. [29]

Cheverud, J. M., and E. J. Routman. 1996. Epistasis as a source of increased additive genetic variance at population bottlenecks. *Evolution* 50: 1042–1051. [11]

Cheverud, J. M., and 7 others. 1999. Epistasis and the evolution of additive genetic variance in populations that pass through a bottleneck. *Evolution* 53: 1009-1018. [11]

Chevin, L.-M., S. Billiard, and F. Hospital. 2008. Hitchhiking both ways: effects of two interfering selective sweeps on linked neutral variation. *Genetics* 180: 301–316. [8]

Chevin, L.-M., and F. Hospital. 2008. Selective sweep at a quantitative trait locus in the presence of background genetic variation. *Genetics* 180: 1645–1660. [8]

Chevin, L.-M., and R. Lande. 2013. Evolution of discrete phenotypes from continuous norms of reaction. *Am. Nat.* 182: 13–27. [14]

Chiaromonte, F., and 5 others. 2003. The share of human genomic DNA under selection estimated from human–mouse genomic alignments. *Cold Spring Harb. Symp. Quant. Biol.* 68: 245–254. [10]

Chib, S., and E. Greenberg. 1995. Understanding the Metropolis-Hastings algorithm. *Am. Stat.* 49: 327–335. [A3]

Chimpanzee Sequencing and Analysis Consortium. 2005. Initial sequence of the chimpanzee genome and comparison with the human genome. *Nature* 437: 69–87. [10]

Choo, T. M., and L. W. Kannenberg. 1979a. Relative efficiencies of population improvement methods in corn: A simulation study. *Crop Sci.* 19: 179–185. [23]

Choo, T. M., and L. W. Kannenberg. 1979b. Changes in gene frequencies during mass, modified ear-to-row, and S_1 selection: A simulation study. *Crop Sci.* 19: 503–509. [23]

Choo, T. M., and L. W. Kannenberg. 1981. Comparison of predicted and simulated response to S_1 selection in a diploid, cross-fertilized species. *Can. J. Plant Sci.* 61: 9–15. [23]

Choy, S. C., and B. S. Weir. 1978. Exact inbreeding coefficients in populations with overlapping generations. *Genetics* 89: 591–614. [3]

Christiansen, F. B. 2000. *Population genetics of multiple loci.* Wiley, New York, NY. [5]

Chung, C. S., and A. B. Chapman. 1958. Comparisons of the predicted with actual gains from selection of parents of inbred progeny of rats. *Genetics* 43: 594–600. [18]

Churchill, G. A. 2002. Fundamentals of experimental design for cDNA microarrays. *Nat. Genet.* 32: 490–495. [12]

Civáň, P., H. Craig, C. J. Cox, and T. A. Brown. 2015. Three geographically separate domestications of Asian rice. *Nat. Plants* 1: 15164. [9]

Clark, A. G. 1998. Mutation-selection balance with multiple alleles. *Genetica* 102/103: 41–47. [7]

Clark, A. G., M. J. Hubisz, C. D. Bustamante, S. H. Williamson, and R. Nielsen. 2005. Ascertainment bias in studies of human genome-wide polymorphism. *Genet. Res.* 15: 1496–1502. [9]

Clark, D., J. W. Dudley, T. R. Rocheford, and J. R. LeDeaux. 2006. Genetic analysis of corn kernel chemical composition in the random mated 10 generation of the cross of generations 70 of IHO x ILO. *Crop Sci.* 46: 807–819. [25]

Clark, R. 1984. *J. B. S: The life and work of J. B. S. Haldane.* Oxford University Press, USA. [5]

Clark, R. M., E. Linton, J. Messing, and J. F. Doebley. 2004. Pattern of diversity in the genomic region near the maize domestication gene *tb1*. *Proc. Natl. Acad. Sci. USA* 101: 700–707. [9]

Clark, R. M., T. N. Wagler, P. Quijada, and J. Doebley, 2006. A distant upstream enhancer at the maize domestication gene *tb1* has pleiotropic effects on plant and inflorescent architecture. *Nat. Genet.* 38: 594-597. [9]

Clarke, B. C., P. R. Shelton, and G. S. Mani. 1988. Frequency-dependent selection, metrical characters and molecular evolution. *Phil. Trans. Roy Soc. Lond. B* 319: 631–640. [28]

Clarke, G. M., and J. A. McKenzie. 1987. Developmental stability of insecticide resistant phenotypes in blowfly: a result of canalizing natural selection. *Nature* 325: 345–346. [25]

Claude, J. 2008. *Morphometrics with R.* Springer, New York, NY. [A5]

Clay, J. S., W. E. Vinson, and J. M. White. 1979. Heterogeneity of daughter variances of sires for milk yield.

J. Dairy Sci. 62: 985–989. [17]

Clayton, G. A., J. A. Morris, and A. Robertson. 1957. An experimental check on quantitative genetical theory. I. Short-term responses to selection. *J. Genet.* 55: 131–151. [21]

Clayton, G. A., and A. Robertson. 1955. Mutation and quantitative variation. *Am. Nat.* 89: 151–158. [11, 28]

Clayton, G. A., and A. Robertson. 1957. An experimental check on quantitative genetical theory. II. The long term effects of selection. *J. Genet.* 55: 152–170. [18, 25]

Clément, V., and 6 others. 2001. Simulation analysis to test the influence of model adequacy and data structure on the estimation of genetic parameters for traits with direct and maternal effects. *Genet. Sel. Evol.* 33: 369–395. [19, 20]

Clune, J., and 5 others. 2008. Natural selection fails to optimize mutation rates for long-term adaptation on rugged fitness landscapes. *PLoS Comput. Biol.* 4: e1000187. [4]

Clutton-Brock, T. 2007. Sexual selection in males and females. *Science* 318: 1882–1885. [29]

Clutton-Brock, T. 2009. Sexual selection in females. *Anim. Behav.* 77: 3–11. [29]

Clutton-Brock, T. 2010. We do not need a sexual selection 2.0—nor a theory of genial selection. *Anim. Behav.* 79: e7–e10. [29]

Clutton-Brock, T. H. (ed.) 1988a. *Reproductive success: studies of individual variation in contrasting breeding systems.* University of Chicago Press, Chicago, IL. [29]

Clutton-Brock, T. H. 1988b. Reproductive success, *in* T. H. Clutton-Brock (ed.), *Reproductive success: studies of individual variation in contrasting breeding systems*, pp. 472–486. University of Chicago Press, Chicago, IL. [29, 30]

Clutton-Brock, T. H., F. E. Guinness, and S. D. Albon. 1982. *Red deer: behavior and ecology of the two sexes.* Univ. Chicago Press, Chicago, IL. [3]

Clutton-Brock, T., and B. C. Sheldon. 2010. Individuals and populations: the role of long-term, individual-based studies of animals in ecology and evolutionary biology. *Trends Ecol. Evol.* 25: 562–573. [20]

Cochran, W. G. 1937. Problems arising in the analysis of a series of similar experiments. *J. Roy. Stat. Soc.* 4: 102–118. [A4]

Cockerham, C. C. 1971. Higher order probability functions of identity of alleles by descent. *Genetics* 69: 235–246. [11]

Cockerham, C. C. 1973. Analyses of gene frequencies. *Genetics* 74: 679-700. [12]

Cockerham, C. C. 1983. Covariances of relatives from self-fertilization. *Crop Sci.* 23: 1177–1180. [11, 23]

Cockerham, C. C. 1984a. Covariances of relatives for quantitative characters with drift. *In* A. Charkravarti (ed.), *Human population genetics,* pp. 195–208. Van Nostrand Reinhold, New York, NY. [11]

Cockerham, C. C. 1984b. Additive by additive variance with inbreeding and linkage. *Genetics* 108: 487–500. [11]

Cockerham, C. C. 1984c. Drift and mutation with a finite number of allelic states. *Proc. Natl. Acad. Sci. USA* 81: 530–534. [2]

Cockerham, C. C., and P. M. Burrows. 1971. Populations of interacting autogenous components. *Am. Nat.* 105: 13–29. [22]

Cockerham, C. C., and P. M. Burrows. 1980. Selection limits and strategies. *Proc. Natl. Acad. Sci. USA* 77: 546–549. [26]

Cockerham, C. C., P. M. Burrows, S. S. Young, and T. Prout. 1972. Frequency-dependent selection in randomly mating populations. *Am. Nat.* 106: 493–515. [22]

Cockerham, C. C. and D. F. Matzinger. 1985. Selection response based on selfed progenies. *Crop. Sci.* 25: 483–488. [23]

Cockerham, C. C., and H. Tachida. 1987. Evolution and maintenance of quantitative genetic variation by mutations. *Proc. Natl. Acad. Sci. USA* 84: 6205–6209. [11, 12]

Cockerham, C. C., and H. Tachida. 1988. Permanency of response to selection for quantitative characters in finite populations. *Proc. Natl. Acad. Sci. USA* 85: 1563–1565. [11]

Cockerham, C. C., and B. S. Weir. 1968. Sib mating with two linked loci. *Genetics* 60: 629–640. [11]

Cockerham, C. C., and B. S. Weir. 1973. Descent measures for two loci with some applications. *Theor. Popul. Biol.* 4: 300–330. [11]

Cockerham, C. C., and B. S. Weir. 1977. Digenic descent measures for finite populations. *Genet. Res.* 30: 121–147. [11]

Cockerham, C. C., and B. S. Weir. 1983. Variance of actual inbreeding. *Theor. Popul. Biol.* 23: 85–109. [3, 11, 12]

Cockerham, C. C., and B. S. Weir. 1984. Covariances of relatives stemming from a population undergoing mixed self and random mating. *Biometrics* 40: 157–164. [11]

Cohan, F. M. 1984a. Genetic divergence under uniform selection. I. Similarity among populations of *Drosophila melanogaster* in their responses to artificial selection for modifiers of ci^D. *Evolution* 38: 55–71. [26]

Cohan, F. M. 1984b. Can uniform selection retard random genetic divergence between isolated populations? *Evolution* 38: 495–504. [1, 7, 26]

Cohan, F. M., and A. A. Hoffmann. 1986. Genetic divergence under uniform selection. II. Different responses to selection for knockdown resistance to ethanol among *Drosophila melanogaster* populations and their replicate lines. *Genetics* 114: 145–163. [26]

Collins, S., J. de Meaux, and C. Acquisti. 2007. Adaptive walks toward a moving optimum. *Genetics* 176: 1089–1099. [27]

Colosimo, P. F., and 9 others. 2005. Widespread parallel evolution in sticklebacks by repeated fixation of *Ectodysplasin* alleles. *Science* 307: 1928–1933. [8]

Coltman, D. W. 2005. Testing marker-based estimates of heritability in the wild. *Mol. Ecol.* 14: 2593–2599. [20]

Coltman, D. W., P. O'Donoghue, J. T. Hogg, and M. Festa-Bianchet. 2005. Selection and genetic (co)variance in bighorn sheep. *Evolution* 59: 1372–1383. [6, 20]

Coltman, D. W., and 5 others. 2003. Undesirable evolutionary consequences of trophy hunting. *Nature* 426: 655–658. [20]

Comeron, J. M., and T. B. Guthrie. 2005. Intragenic Hill-Robertson interference influences selection intensity on synonymous mutations in *Drosophila*. *Mol. Biol. Evol.* 22: 2519–2530. [8]

Comeron, J. M., and M. Kreitman. 2002. Population,

evolutionary and genomic consequences of interference selection. *Genetics* 161: 389–410. [8]

Comeron, J. M., M. Kreitman, and M. Aguadé. 1999. Natural selection on synonymous sites is correlated with gene length and recombination in *Drosophila*. *Genetics* 151: 239–249. [8]

Comeron, J. M., A. Williford, and R. M. Kliman. 2008. The Hill-Robertson effect: evolutionary consequences of weak selection and linkage in finite populations. *Heredity* 100: 19–31. [8]

Compton, W. A. 1968. Recurrent selection in self-pollinated crops without extensive crossing. *Crop Sci.* 8: 773. [23]

Compton, W. A., and K. Bahadur. 1977. Ten cycles of progress from modifed ear-to-row selection in corn. *Crop Sci.* 17: 378–380. [21]

Compton, W. A., and R. E. Comstock. 1976. More on modified ear-to-row selection in corn. *Crop Sci.* 16: 122. [21]

Comstock, R. E., and F. D. Enfield. 1981. Gene number estimation when multiplicative genetic effects are assumed—growth in flour beetles and mice. *Theor. Appl. Genet.* 59: 373–379. [25]

Comstock, R. E., and R. H. Moll. 1963. Genotype-environment interactions. *In* W. D. Hanson and H. F. Robinson (eds.), *Statistical genetics and plant breeding*, pp. 164–196. Natl. Acad. Sci., Natl. Res. Council Publ. No. 982, Washington, D.C. [21]

Comstock, R. E., and H. F. Robinson. 1948. The components of genetic variance in populations of biparental progenies and their use in estimating the average degree of dominance. *Biometrics* 4: 254–266. [21]

Conrad, T. M., N. E. Lewis, and B. Ø. Palsson. 2011. Microbial laboratory evolution in the era of genomescale science. *Mol. Systems. Biol* 7: 509. [27]

Conner, J. 1988. Field measurements of natural and sexual selection: applications. *Evolution* 42: 736–749. [29]

Conner, J. K. 1996. Understanding natural selection: an approach integrating selection gradients, multiplicative fitness components, and path analysis. *Ethol. Ecol. Evol.* 8: 387–397. [30]

Conner, J. K. 2001. How strong is natural selection? *Trends Ecol. Evol.* 16: 215–217. [29, 30]

Conner, J. K. 2007. A tale of two methods: putting biology before statistics in the study of phenotypic evolution. *J. Evol. Biol.* 20: 17–19. [30]

Conomos, M. P., A. P. Reiner, B. S. Weir, and T. A. Thornton. 2016. Model-free estimation of recent genetic relatedness. *Amer. J. Hum. Genet.* 98: 127–148. [20]

Conover, D. O., and E. T. Schultz. 1995. Phenotypic similarity and the evolutionary significance of countergradient variation. *Trends Ecol. Evol.* 10: 248–252. [20]

Conover, W. J. 1999. *Practical nonparametric statistics, 3rd edition*. Wiley and Sons, New York, NY. [9, 18, 28, 29, 30, A5]

Conrad, D. F., and 17 others. 2011. Variation in genome-wide mutation rates within and between human families. *Nat. Genet.* 43: 712–714. [4]

Conroy, M. J., J. C. Senar, and J. Doménech. 2002. Analysis of individual-and time-specific covariate effects on survival of *Serinus serinus* in north-eastern Spain. *J. Appl. Stat.* 29: 125–142. [29]

Cook, L. M., and D. A. Jones. 1996. The *medionigra* gene in the moth *Panaxia dominula*: the case for selection. *Phil. Trans. Roy. Soc. Lond. B* 351: 1623–1634. [9]

Cooke, F., P. D. Taylor, C. M. Francis, and R. F. Rockwell. 1990. Directional selection and clutch size in birds. *Am. Nat.* 136: 261–267. [20]

Coop, G., and M. Przeworski. 2007. An evolutionary view of human recombination. *Nat. Rev. Genet.* 8: 23–34. [9]

Coop, G., and P. Ralph. 2012. Patterns of neutral diversity under general models of selective sweeps. *Genetics* 192: 205–224. [8]

Coop, G., X. Wen, C. Ober, J. K. Pritchard, and M. Przeworski. 2008. High-resolution mapping of crossovers reveals extensive variation in fine-scale recombination patterns among humans. *Science* 319: 1395–1398. [4]

Coop, G., D. Witonsky, A. Di Rienzo, and J. K. Pritchard. 2010. Using environmental correlations to identify loci underlying local adaptation. *Genetics* 185: 1411–1423. [9, 12]

Cooper, H., L. V. Hedges, and J. C. Valentine 2009. *The handbook of research synthesis and meta-analysis*, 2nd Ed. Sage Foundation, New York, NY. [30, A4]

Cooper, T. F., E. A. Ostrowski, and M. Travisano. 2007. A negative relationship between mutation pleiotropy and fitness effect in yeast. *Evolution* 61: 495–499. [27]

Copas, J. and J. Q. Shi. 2000. Meta-analysis, funnel plots and sensitivity analysis. *Biostatistics* 1: 247–262. [A4]

Coque, M., and A. Gallias. 2006. Genomic regions involved in response to grain yield selection at high and low nitrogen fertilization in maize. *Theor. Appl. Genet.* 112: 1205–1220. [9]

Corbett-Detig, R. B., D. L. Hartl, and T. B. Sackton. 2015. Natural selection constrains neutral diversity across a wide range of species. *PLoS Biol.* 13: e1002112. [4, 8, 9]

Cornelius, P. L. 1975. Identity coefficients and covariances between relatives in a parent-offspring inbreeding system. *Theor. Appl. Genet.* 46: 201–212. [11]

Cornelius, P. L. 1988. Properties of components of covariance of inbred relatives and their estimates in a maize population. *Theor. Appl. Genet.* 75: 701–711. [11]

Cornelius, P. L., and J. W. Dudley. 1975. Theory of inbreeding and covariances between relatives under full-sib mating in diploids. *Biometrics* 31: 169–187. [11]

Cornelius, P. L., and D. A. Van Sanford. 1988. Quadratic components of covariance of inbred relatives and their estimation in naturally self-pollinated species. *Crop Sci.* 28: 1–7. [11]

Cornish, M. A. 1990a. Selection during a selfing programme. I. The effects of a single round of selection *Heredity* 65: 201–211. [23]

Cornish, M. A. 1990b. Selection during a selfing programme. II. The effects of two or more rounds of selection. *Heredity* 65: 213–220. [23]

Costa e Silva, J., B. M. Potts, P. Bijma, R. J. Kerr, and D. J. Pilbeam. 2013. Genetic control of interactions among individuals: contrasting outcomes of indirect genetic effects arising from neighbour disease infection and competition in a forest tree. *New Phytologist* 197: 631–641. [22]

Cotterman, C. W. 1940. A calculus for statistico-genetics. Ph. D. Thesis, Ohio State Univ., Columbus. [20]

Coulson, T. 2012. Integral projections models, their construction and use in posing hypotheses in ecology. *Oikos* 121: 1337–1350. [29]

Coulson, T., L. E. B. Kruuk, G. Tavecchia, J. M. Pemberton, and T. H. Clutton-Brock. 2003. Estimating selection on neonatal traits in red deer using elasticity path analysis. *Evolution* 57: 2879–2892. [30]

Coulson, T., and 5 others. 2006. Estimating individual contributions to population growth: evolutionary fitness in ecological time. *Proc. Roy. Soc. Lond. B.* 273: 547–555. [29]

Cowles, C. R., J. N. Hirschhorn, D. Altshuler, and E. S. Lander. 2002. Detection of regulatory variation in mouse genes. *Nat. Genet.* 32: 432–437. [12]

Cowles, M. K., and B. P. Carlin. 1996. Markov chain Monte Carlo diagnostics: A comparative review. *J. Am. Stat. Assoc.* 91: 883–904. [A3]

Cowling, W. A., and 7 others. 2015. Using the animal model to accelerate response to selection in a self-pollinating crop. *Genes Genomes Genet.* 5: 1419–1428. [23

Cox, D. R. 1972. Regression models and life tables. *J. Roy. Statist. Soc.* 34: 187–220. [29]

Cox, D. R., and N. J. H. Small. 1978. Testing multivariate normality. *Biometrika* 65: 263–272. [A5]

Cox, R. M., and R. Calsbeek. 2009. Sexually antagonistic selection, sexual dimorphism, and the resolution of intralocus sexual conflict. *Am. Nat.* 173: 176–187. [30]

Cox, T. S. 1995. Simultaneous selection for major and minor resistance genes. *Crop. Sci.* 35: 1337–1346. [25]

Coyne, J. A. 1987. Lack of response to selection for directional asymmetry in *Drosophila melanogaster*. *J. Heredity* 78: 119. [18]

Coyne, J. A. 1996. Genetics of a difference in male cuticular hydrocarbons between two sibling species, *Drosophila simulans* and *D. sechellia*. *Genetics* 143: 1689–1698. [12]

Coyne, J. A., N. H. Barton, and M. Turelli. 1997. Perspective: a critique of Sewall Wright's shifting balance theory of evolution. *Evolution* 51: 643–671. [26]

Coyne, J. A., N. H. Barton, and M. Turelli. 2000. Is Wright's shifting balance process important in evolution? *Evolution* 54: 306–317. [26]

Coyne, J. A., and E. Beecham. 1987. Heritability of two morphological characters within and among natural populations of *Drosophila melanogaster*. *Genetics* 117: 727–737. [13]

Coyne, J. A., and R. Lande 1985. The genetic basis for species difference in plants. *Am. Nat.* 126: 141–145. [25]

Craig, D. M. 1982. Group selection versis individual selection: an experimental analysis. *Evolution* 36: 271–282. [22]

Craig, J. V., and W. M. Muir. 1996. Group selection for adaptation to multiple-hen cages: beak-related mortalty, feathering, and body weight responses. *Poultry Sci.* 75: 294–302. [22]

Craven, P., and G. Wahba. 1979. Smoothing noisy data with spline functions: estimating the correct degree of smoothing by the method of generalized cross-validation. *Numerische Mathematik* 31: 377–403. [29]

Crespi, B. J. 1990. Measuring the effect of natural selection on phenotypic interaction systems. *Am. Nat.* 135: 32–47. [30]

Crespi, B. J., and F. L. Bookstein. 1988. A path-analytic model for the measurement of selection on morphology. *Evolution* 43: 18-28. [30]

Crisci, J. L., Y.-P. Poh, A. Bean, A. Simkin, and J. D. Jensen. 2012. Recent progress in polymorphism-based population genetic inference. *J. Heredity* 103: 287–296. [9]

Crisci, J. L., Y.-P. Poh, S. Mahajan, and J. D. Jensen. 2013. The impact of equilibrium assumptions on tests of selection. *Front. Genet.* 4: 235. [9]

Crnokrak, P., and D. A. Roff. 1995. Dominance variance: associations with selection and fitnesses. *Heredity* 75: 530–540. [6]

Crone, E. E. 2001. Is survivorship a better fitness surrogate than fecundity? *Evolution* 55: 2611–2614. [29, 30]

Croshaw, D. A. 2010. Quantifying sexual selection: a comparison of competing indices with mating system data from a terrestrially breeding salamander. *Biol. J. Linn. Soc.* 99: 73–83. [29]

Crossa, J., and C. O. Gardner. 1989. Predicted and realized grain yield response to full-sib family selection in CIMMYT maize (*Zea mays* L.) populations. *Theor. Appl. Genet.* 77: 33–38. [21]

Crow, J. F. 1948. Alternative hypotheses of hybrid vigor. *Genetics* 33: 477–487. [1]

Crow, J. F. 1954. Breeding structure of populations. II. Effective population number. *In* O. Kempthorne, T. A. Baneroft, J. W. Gowen, and J. L. Lush (eds.), *Statistics and mathematics in biology*, pp. 543–556. Iowa State College Press, Ames, IA. [3]

Crow, J. F. 1958. Some possibilities for measuring selection intensities in man. *Hum. Biol.* 30: 1–13. [29]

Crow, J. F. 1972. The origins of theoretical population genetics. *Amer. J. Hum. Genet.* 24: 608–609. [1]

Crow, J. F. 1979. Minor viability mutants in*Drosophila*. *Genetics* 92: s165–s172. [28]

Crow, J. F. 1992. Centennial: J. B. S. Haldane, 1892–1964. *Genetics* 130: 1–6. [5]

Crow, J. F. 1993. Mutation, mean fitness, and genetic load. *Oxford Surv. Evol. Biol.* 9: 3–42. [7, 28]

Crow, J. F. 2000. The origins, patterns and implications of human spontaneous mutation. *Nat. Rev. Genet.* 1: 40–47. [4]

Crow, J. F. 2002. Perspective: Here's to Fisher, additive genetic variance, and the fundamental theorem of natural selection. *Evolution* 56: 1313–1316. [6]

Crow, J. F. 2008. Maintaining evolvability. *J. Genet.* 87: 349–353. [24, 25]

Crow, J. F. 2010. On epistasis: why it is unimportant in polygenic directional selection. *Phil. Trans. Roy. Soc. Lond. B* 365: 1241–1244. [15, 25]

Crow, J. F., and K. Aoki. 1982. Group selection for a polygenic behavioral trait: a differential proliferation model. *Proc. Natl. Acad. Sci. USA* 79: 2628–2631. [22]

Crow, J. F., and C. Denniston. 1988. Inbreeding and variance effective population numbers. *Evolution* 42: 482–495. [3]

Crow, J. F., and M. Kimura 1964. The theory of genetic loads. *Proceed. of the XIth Inter. Congr. Genetics*, Vol. 2, 495–505. [7, 24, 28]

Crow, J. F., and M. Kimura. 1970. *An introduction to population genetics theory.* Harper and Row, New York, NY. [1, 2, 3, 5, A1]

Crow, J. F., and M. Kimura. 1979. Efficiency of truncation selection. *Proc. Natl. Acad. Sci. USA* 76: 396–299. [14]

Crow, J. F., and N. E. Morton. 1955. Measurement of gene frequency drift in small populations. *Evolution* 9: 202–214. [3, 26, 29]

Crow, J. F., and T. Nagylaki. 1976. The rate of change of a character correlated with fitness. *Am. Nat.* 110: 207–213. [6]

Crow, J. F., and M. J. Simmons. 1983. The mutational load in *Drosphilia*. *In* M Ashbruner, H. L. Carson, and J. N. Thompson (eds.), *The genetics and biology of Drosophila*, pp. 2–33.Academic Press, New York, NY. [28]

Crudent, D. 1949. The computation of inbreeding coefficients for closed populations. *J. Heredity* 40: 248–251. [11]

Cruz, A., and J. W. Wiley. 1989. The decline of an adaptation in the absence of a presumed selection pressure. *Evolution* 43: 55–62. [25]

Csilléry, K., M. G. B. Blum, O. E. Gaggiotti, and O, François. 2010. Approximate Bayesian computation (ABC) in practice. *Trends Ecol. Evol.* 25: 410–418. [A3]

Csilléry, K., and 7 others. 2006. Performance of marker-based relatedness estimators in natural populations of outbred vertebrates. *Genetics* 173: 2091–2101. [20]

Cullis, B. R., R. M. Thomson, J. A. Fisher, A. R. Gilmour, and R. Thompson. 1996. The analysis of the NSW wheat variety database. I. Modeling trail error variance. *Theor. Appl. Genet.* 92: 21–27. [17]

Curnow, R. N. 1961. The estimation of repeatability and heritability from records subject to culling. *Biometry* 17: 553–566. [19]

Curnow, R. N. 1964. The effect of continued selection of phenotypic intermediates on gene frequency. *Genet. Res.* 5: 341–353 [28]

Curnow, R. N., and L. H. Baker. 1968. The effect of repeated cycles of selection and regeneration in populations of finite size. *Genet. Res.* 11: 105–112. [7]

Curnow, R. N., and L. H. Baker. 1969. A correction to our earlier paper on "The effect of repeated cycles of selection and regeneration in populations of finite size". *Genet. Res.* 13: 105–106. [7]

Currat, M., and 6 others. 2006. Comment on "Ongoing adaptive evolution of *ASPM*, a brain size determinant in *Homo sapiens*" and "*Microcephalin*, a gene regulating brain size, continues to evolve adaptively in humans". *Science* 313: 172. [9, 10]

Curtsinger, J. W., and R. Ming. 1997. Non-linear selection response in *Drosophila*: a strategy for testing the rare-alleles model of quantitative genetic variability. *Genetica* 99: 59–66. [28]

Cutter, A. D., and B. A. Payseur. 2013. Genomic signatures of selection at linked sites: unifying the disparity among species. *Nat. Rev. Genet.* 14: 262–274. [8, 9]

da Fonseca, R. R., and 17 others. 2013. Influenza virus drug resistance: a time-sampled population genetics perspective. *PLoS Genet.* 10: e1004185. [9]

da Fonseca, R. R., and 19 others. 2015. The origin and evolution of maize in the Southwestern United States. *Nature Plants* 1: 1–5. [9]

Da Silva, J. M. P. 1961. Limits of response to selection. Ph. D. Thesis, University of Edinburgh, UK. [26]

da Silva, W. J., and J. H. Lonnquist. 1968. Genetic variances in population developed from full-sib and S_1 testcross progeny selection in an open-pollinated variety of maize. *Crop Sci.* 8: 201–204. [21]

Damerval, C., A. Maurice, J. M. Josse, and D. de Vienne. 1994. Quantitative trait loci underlying gene product variation: a novel perspective by analyzing regulation of genome expression. *Genetics* 137: 289–301. [12]

Damgaard, L. H. 2007. Technical note: How to use Winbugs to draw inferences in animal models. *J. Anim. Sci.* 85: 1363–1368. [19]

Darwin, C. 1859. *On the origin of species by means of natural selection, or the preservation of favoured races in the struggle for life.* John Murray, London, UK. [2, 29, 30]

Darwin, C. 1868. *The variation of animals and plants under domestication.* John Murray, London, UK. [9]

Darwin, C. 1871. *The descent of man and selection in relation to sex.* John Murray, London, UK. [20, 22, 29]

Darwin, C. 1876. *The effects of cross- and self- fertilisation in the vegetable kingdom.* John Murray, London, UK. [23]

Darwin, C. 1877. *The different forms of flowers on plants of the same species.* John Murray, London, UK. [23]

Daub, J. T., and 6 others. 2013. Evidence for polygenic adaptation to pathogens in the human genome. *Mol. Biol. Evol.* 30: 1544–1558. [9, 12]

David, F. N. 1981. *Order statistics*, 2nd Ed. Wiley, New York, NY. [14]

Dawson, K. J. 1997. Linkage disequilibrium and the infinitesimal limit. *Theor. Popul. Biol.* 52: 127–154. [24]

Dawson, K. J. 1999. The dynamics of infinitesimally rare alleles, applied to the evolution of mutation rates and the expression of deleterious mutations. *Theor. Popul. Biol.* 55: 1–22. [4]

Dawson, P. S. 1969. A conflict between darwinian fitness and population fitness in *Tribolium* "competition" experiments. *Genetics* 62: 413–419. [5, 22]

Day, T., and R. Bonduriansky. 2011. A unified approach to the evolutionary consequences of genetic and non-genetic inheritance. *Am. Nat.* 178: E18–E36. [15]

Day-Williams, A. G., J. Blangero, T. D. Dyer, K. Lange, and E. M. Sobel. 2011. Linkage analysis without defined pedigrees. *Genet. Epid.* 35: 360–370. [20]

de Boer, I. J. M., and I. Hoeschele. 1993. Genetic evaluation methods for populations with dominance and inbreeding. *Theor. Appl. Genet.* 86: 245–258. [11]

de Boer, I. J. M., and J. A. M. van Arendonk. 1992. Prediction of additive and dominance effects in selection or unselected populations with inbreeding. *Theor. Appl. Genet.* 84: 451–459. [19, 23]

DeGiorgio, M., K. E. Lohmueller, and R. Nielsen. 2014. A model-based approach for identifying signatures of ancient balancing selection in genetic data. *PLoS Genet.* 10: e1004561. [9]

de Jong, G. 1994. The fitness of fitness concepts and the description of natural selection. *Q. Rev. Biol.* 69: 3–29. [29]

De Koeyer, D. L., R. L. Phillips, and D. D. Stuthman. 2001. Allelic shifts and quantitative trait loci in a recurrent selection population of oat. *Crop Sci.* 41: 1228–1234. [9]

de Koning, D.-J., and C. S. Haley. 2005. Genetical genomics in humans and model organisms. *Trends Genet.* 21: 377–381. [12]

De Kovel, C. G. F. 2006. The power of allele frequency comparisons to detect the footprint of selection in natural and experimental populations. *Genet. Sel. Evol.* 38: 3–23. [9]

de Kroon, H., A. Plaisier, J. van Groenendael, and H. Caswell. 1986. Elasticity: the relative contribution of demographic parameters to population growth rate. *Ecology* 67: 1427–1431. [29]

de Kroon, H., J. van Groenendael, and J. Ehrlén. 2000. Elasticities: A review of methods and model limitations. *Ecology* 81: 607–618. [29]

De Lange, A. O. 1974. A simulation study of the effects of assortative mating on the response to selection. *1st World Congress Genet. Appl. Livestock Prod., Madrid, 7-11 October 1974* 3: 421–425. Editorial Garsi, Madrid. [16]

de Massy, B. 2003. Distribution of meiotic recombination sites. *Trends Genet.* 19: 514–522. [4]

De Mita, S., and 6 others. 2013. Detecting selection along environmental gradients: analysis of eight methods and their effectiveness for outbreeding and selfing populations. *Mol. Ecol.* 22: 1383–1399. [9]

de Silva, E., and M. P. H. Stumpf. 2004. HIV and the *CCR5*-Δ*32* resistance allele. *FEMS Microbiol. Letters* 241: 1–12. [9]

de Visser, J. A. G. M., and J. Krug. 2014. Empirical fitness landscapes and the predictability of evolution. *Nat. Rev. Genet.* 15: 480—490. [27]

de Villemereuil, P., É. Frichot, É. Bazin, O. François, and O. E. Gaggiotti. 2014. Genome scan methods against more complex models: when and how much should we trust them? *Mol. Ecol.* 23: 2006–2019. [9]

de Villemereuil, P., and O. E. Gaggiotti. 2015. A new F_{ST}-based method to uncover local adaptation using environmental variables. *Meth. Ecol. Evol.* 6: 1248–1258. [9]

de Villemereuil, P., O. Gimenez, and B. Doligez. 2013. Comparing parent–offspring regression with frequentist and Bayesian animal models to estimate heritability in wild populations: a simulation study for Gaussian and binary traits. *Meth. Ecol. Evol.* 4: 260–275. [20]

de Villemereuil, P, H. Schielzeth, S. Nakagawa, and M. Morrissey. 2016. General methods for evolutionary quantitative genetic inference from generalized mixed models. *Genetics* 204: 1281–1294. [14, 20, 29]

Decker, J. E., and 10 others. 2012. A novel analytical method, Birth Date Selection Mapping, detects response of the Angus (*Bos taurus*) genome to selection on complex traits. *BMC Genomics* 13: 606. [9]

Del Castillo, E., and S. Cahya. 2001. A tool for computing confidence regions on the stationary point of a response surface. *Am. Stat.* 55: 358–365. [30]

Delph, L. F., and T.-L. Ashman. 2006. Trait selection in flowering plants: how does sexual selection contribute? *Int. Comp. Biol.* 46: 465–472. [29]

Dempflé, L. 1975. A note on increasing the limit of selection through selection within families. *Genet. Res.* 24: 127–135. [21, 26]

Dempfle, L. 1990. Statistical aspects of design of animal breeding programs: a comparison among various selection schemes. *In* D. Gianola and K. Hammond (eds.), *Statistical methods for genetic improvement of livestock,* pp. 98–117. Springer-Verlag, NY. [21]

Dempster, E. R. 1955a. Maintenance of genetic heterogeneity. *Cold Spring Harbor Symp. Quant. Biol.* 20: 25–32. [4, 28]

Dempster, E. R. 1955b. Genetic models in relation to animal breeding. *Biometrics* 11: 535–536. [26]

Denamur, E., and I. Matic. 2006. Evolution of mutation rates in bacteria. *Mol. Microbiol.* 60: 820–827. [4]

Deng, W. Q., and G. Paré. 2011. A fast algorithm to optimize SNP prioritization for gene-gene and gene-environment interactions. *Genet. Epid.* 35: 729–738. [17]

Denniston, C., and J. F. Crow. 1990. Alternative fitness models with the same allele frequency dynamics. *Genetics* 152: 201–205. [5]

Denver, D. R., and 5 others. 2005. The transcriptional consequences of mutation and natural selection in *Caenorhabditis elegans. Nat. Genet.* 37: 544–548. [12]

Denver, D. R., and 5 others. 2012. Variation in base-substitution mutation in experimental and natural lineages of *Caenorhabditis* nematodes. *Genome Biol. Evol.* 4: 513–522. [4]

Depaulis, F., S. Mousset, and M. Veuille. 2001. Haplotype tests using coalescent simulations conditional on the number of segregating sites. *Mol. Biol. Evol.* 18: 1136–1138. [9]

Depaulis, F., S. Mousset, and M. Veuille. 2003. Power of neutrality tests to detect bottlenecks and hitchhiking. *J. Mol. Evol.* 57: S190–S200. [9]

Depaulis, F., S. Mousset, and M. Veuille. 2005. Detecting selective sweeps with haplotype tests: Hitchhiking and haplotype tests. *In* D. Numinsky (ed.), *Selective sweep*, pp. 34–54. Springer, New York, NY. [9]

Depaulis, F., and M. Veuille. 1998. Neutrality tests based on the distribution of haplotypes under an infinite-site model. *Mol. Biol. Evol.* 15: 1788–1790. [9]

Dermitzakis, E. T., and A. G. Clark. 2002. Evolution of transcription factor binding sites in Mammalian gene regulatory regions: conservation and turnover. *Mol. Biol. Evol.* 19: 1114–1121. [10]

DerSimonian, R. and N Laird. 1986. Meta-analysis in clinical trials. *Controlled Clinical Trails* 7: 177–188. [A4]

DeRose, M. A., and D. A. Roff. 1999. A comparison of inbreeding depression in life-history and morphological traits in animals. *Evolution* 53: 1288–1292. [6]

Desai, M. M., and D. S. Fisher. 2007. Beneficial mutation–selection balance and the effect of linkage on positive selection. *Genetics* 176: 1759–1798. [8]

Desai, M. M., and D. S. Fisher. 2011. The balance between mutators and nonmutators in asexual populations. *Genetics* 188: 997-1014. [4]

Desai, M. M., and J. B. Plotkin. 2008. The polymorphism frequency spectrum of finitely many sites under selection. *Genetics* 180: 2175–2191. [10]

Devaux, C., and R. Lande. 2009. Selection on variance in flowering time within and among individuals. *Evolution* 64: 1311-1320. [17]

Devaux, C., R. Lande, and E. Porcher. 2014. Pollination ecology and inbreeding depression control individual flowering phenologies and mixed mating. *Evolution* 68: 3051–3065. [23]

Devlin, B., and K. Roeder. 1999. Genomic control for association studies. *Biometrics* 55: 997–1004. [9]

Dhillon, B. S. 1991a. Recurrent mass selection based on selfed-plant evaluation in allogamous species. *Crop Sci.* 31: 1075–1077. [23]

Dhillon, B. S. 1991b. Alternate recurrent selection of S_1 and half-sib families for intrapopulation improvement. *Maydica* 36: 45–48. [23]

Dhrymes, P. J. 1978. *Mathematics for econometrics.* Springer-Verlag, New York, NY. [A5]

Diamond, J. 2002. Evolution, consequences and future of plant and animal domestication. *Nature* 418: 700–707. [9]

DiBattista, J. D., K. A. Feldheim, D. Garant, S. H. Gruber, and A. P. Hendry. 2009. Evolutionary potential of a large marine vertebrate: quantitative genetic parameters in a wild population. *Evolution* 63: 1051–1067. [20]

Dickersin, K., Y.-I. Min, and C. L. Meinert. 1992. Factors influencing publication of research results: follow-up of applications submitted to two institutional review boards. *J. Am. Med. Assoc.* 267: 374–378. [A4]

Dickerson, G. E. 1947. Composition of hog carcasses as influenced by heritable differences in rate and economy of gain. *Res. Bul. Iowa Agr. Exp. Sta.* 354: 489–524. [15, 22]

Dickerson, G. E. 1973. Inbreeding and heterosis in animals. *Proceedings Animal Breeding and Genetics Symposium in Honor of J. L. Lush*, pp. 54–77. American Society of Animal Science, Champaign, Il. [23]

Dickerson, G. E., and L. N. Hazel. 1944. Effectiveness of selection on progeny performance as a supplement to earlier culling in livestock. *J. Agric. Res.* 69: 459–476. [13]

Dickerson, G. E. and N. B. H. Lindhé. 1977. Potential uses of inbreeding to increase selection response. *In* E. Pollak, O. Kempthorne, and T. B. Bailey, Jr., (eds.), *Proceedings of the international conference on quantitative genetics*, pp. 323–342. Iowa State Univ. Press, Ames, IA. [23]

Dickinson, W. J. 2008. Synergistic fitness interactions and a high frequency of beneficial changes among mutations accumulated under relaxed selection in *Saccharomyces cerevisiae*. *Genetics* 178: 1571–1578. [12]

Dion, N., and F. Minvielle. 1985. The effects of alternated cycles of full-sib and random mating on selection for pupal weight in *Tribolium castaneum*. *Can. J. Genet. Cytol.* 27: 251–254. [23]

Doebley, J. 2004. The genetics of maize evolution. *Ann. Rev. Genet.* 38: 37–59. [9]

Doebley, J., A. Stec, and C. Gustus. 1995. *Teosinte branched1* and the origin of maize: evidence for epistasis and the evolution of dominance. *Genetics* 141: 333–346. [9]

Doebley, J. F., B. S. Gaut, and B. D. Smith. 2006. The molecular genetics of crop domestication. *Cell* 127: 1309–1321. [9]

Doggett, H. 1972. Recurrent selection in sorghum populations. *Heredity* 928: 9–29. [23]

Doggett, H., and S. A. Eberhart. 1968. Recurrent selection in sorghum. *Crop Sci.* 8: 119–121. [21]

Domínguez, A., J. Albornoz, and E. Santiago. 1987. Analysis of lethals in selected lines of *Drosophila melanogaster*. *Theor. Appl. Genet.* 74: 409–413. [25]

Dongen, S. V. 2006. Fluctuating asymmetry and developmental instability in evolutionary biology: past, present and future. *J. Evol. Biol.* 19: 1727–1743. [17]

Donnelly, P., and S. Tavaré. 1995. Coalescents and genealogical structure under neutrality. *Ann. Rev. Genet.* 29: 401–421. [2]

Donohue, K. 2003. The influence of neighbor relatedness on multilevel selection in the Great Lakes Sea Rocket. *Am. Nat.* 162: 77–92. [30]

Donohue, K. 2004. Density-dependent multilevel selection in the Great Lakes Sea Rocket. *Ecology* 85: 180–191. [30]

Doolittle, W. F. 2013. Is junk DNA bunk? A critique of ENCODE. *Proc. Natl. Acad. Sci. USA* 110: 5294–5300. [8]

Dowdeswell, W. H. 1961. Experimental studies on natural selection in the butterfly, *Maniola jurrina*. *Heredity* 16: 39–52. [29]

Downhower, J. F., L. S. Blumer, and L. Brown. 1987. Opportunity for selection: an appropriate measure for evaluating variation in the potential for selection? *Evolution* 41: 1395–1400. [29]

Downie, D. A. 2003. Effects of short-term spontaneous mutation accumulation for life history traits in grape phylloxera, *Daktulosphaira vitifoliae*. *Genetica* 119: 237–251. [12]

Drake, J. W. 1991. A constant rate of spontaneous mutation in DNA-based microbes. *Proc. Natl. Acad. Sci. USA* 88: 7160–7164. [4]

Drake, J. W., B. Charlesworth, D. Charlesworth, and J. F. Crow. 1998. Rates of spontaneous mutation. *Genetics* 148: 1667–1686. [4]

Drost, J. B., and W. R. Lee. 1995. Biological basis of germline mutation: comparisons of spontaneous germline mutation rates among *Drosophila*, mouse, and human. *Environ. Mol. Mutagen.* 25 Suppl 26: 48–64. [4]

Druger, M. 1967. Selection and the effect of temperature on scutellar bristle number in *Drosophila*. *Genetics* 56: 39–47. [21]

Du, F-X., I. Hoeschele, and K. M. Gage-Lahti. 1999. Estimation of additive and dominance variance components in finite polygenic models and complex pedigrees. *Genet. Res.* 74: 179–187. [24]

Duclos, A., and P. L. Crane. 1968. Comparative performance of topcross and S_1 progeny for improving populations of corn (*Zea mays* L.). *Crop Sci.* 8: 191–194. [23]

Dudley, J. W. 1977. 76 generations of selection for oil and protein percentage in maize. *In* E. Pollak, O. Kempthorne, and T. B. Bailey, Jr., (eds.), *Proceedings of the international conference on quantitative genetics*, pp. 459–473. Iowa State Univ. Press, Ames, IA. [25]

Dudley, J. W. 2007. From means to QTL: The Illinois long-term selection experiment as a case study in quantitative genetics. *Crop Sci.* 47: S20 - S31. [25]

Dudley, J. W., T. H. Busbice, and C. S. Levings. 1969. Estimates of genetic variance in 'Cherokee' alfalfa (Medicago sativa, L.). *Crop. Sci.* 9: 228–231. [21]

Dudley, J. W., D. Clark, T. R. Rocheford, and J. R. LeDeaux. 2007. Genetic analysis of corn kernel chemical composition in the random mated 7 generation of the cross of generations 70 of IHP x ILP. *Crop Sci.* 47: 45–57. [25]

Dudley, J. W., and R. J. Lambert. 1992. Ninety generations of selection for oil and protein in maize. *Maydica* 37: 81–87. [25]

Dudley, J. W., and R. J. Lambert. 2004. 100 generations of selection for oil and protein in maize. *Plant Breeding Reviews* 24: 79–110. [25]

Duforet-Frebourg, N., E. Bazin, and M. G. B. Blum. 2014. Genome scans for detecting footprints of local adaptation using a bayesian factor model. *Mol. Biol. Evol.* 31: 2483–2495. [9]

Duggan, D. J., M. Bittner, Y. Chen, P. Meltzer, and J. M. Trent. 1999. Expression profiling using cDNA microarrays. *Nat. Genet.* 21: 10–14. [12]

Dumont, B. L., and B. A. Payseur. 2008. Evolution of the genomic rate of recombination in mammals. *Evolution* 62: 276–294. [4]

DuMont, V. B., J. C. Fay, P. P. Calabrese, and C. F. Aquadro. 2004. DNA variability and divergence at the *Notch* locus in *Drosophila melanogaster* and *D. simulans:* A case of accelerated synonymous site divergence. *Genetics* 167: 171–185. [10]

Durand, E. Y., N. Patterson, D. Reich, and M. Slatkin. 2011. Testing for ancient admixture between closely related populations. *Mol. Biol. Evol.* 28: 2239–2252. [9]

Duret, L., and N. Galtier. 2009. Biased gene conversion and the evolution of mammalian genomic landscapes. *Ann. Rev. Genom. Hum. Genet.* 10: 285–311. [8]

Durrett, R., and J. Schweinsberg. 2004. Approximating selective sweeps. *Theor. Popul. Biol.* 66: 129–138. [8]

Duval, S., and R. Tweedie. 2000a. Trim and fill: a simple funnel-plot–based method of testing and adjusting for publication bias in meta-analysis. *Biometrics* 56: 455–462. [A4]

Duval, S., and R. Tweedie. 2000b. A nonparametric "trim and fill" method of accounting for publication bias in meta-analysis. *J. Am. Stat. Assoc.* 95: 89–98. [A4]

Duvick, D. N., 2001. Biotechnology in the 1930s: the development of hybrid maize. *Nat. Rev. Genet.* 2: 69–74. [1]

Duvick, D. N. 2005. The contribution of breeding to yield advances in maize (*Zea mays* L.). *Advan. Agron.* 86: 83–145. [1]

Dykhuizen, D. E. 1990. Experimental studies of natural selection in bacteria. *Ann. Rev. Ecol. Syst.* 21: 373–398. [8]

Easterbrook, P. J., R. Gopalan, J. A. Berlin, and D. R. Matthews. 1991. Publication bias in clinical research. *Lancet* 337: 867–872. [A4]

Eberhard, W. G. 2009. Postcopulatory sexual selection: Darwin's omission and its consequences. *Proc. Natl. Acad. Sci. USA* 196: 10025–10032. [29]

Eberhart, S. A. 1972. Techniques and methods for more efficient population improvement in Sorghum. *In* N. G. P. Rao and L. R. House (eds.), *Sorgum in the seventies*, pp. 197–213. Oxford and IBH Publ. Co, New Delhi, IN. [23]

Eberhart, S. A., R. H. Moll, H. F. Robinson, and C. C. Cockerham. 1966. Epistatic and other genetic variances in two varieties of maize. *Crop Sci.* 6: 275–280. [21]

Eckert, A. J., and 5 others. 2010. Back to nature: ecological genomics of loblolly pine (*Pinus taeda, Pinaceae*). *Mol. Ecol.* 19: 3789–3805. [9]

Edelaar, P., and M. Björklund. 2011. If F_{ST} does not measure neutral genetic differentiation, then comparing it with Q_{ST} is misleading. Or is it? *Mol. Ecol.* 20: 1805–1812. [12]

Edelaar, P., P. Burraco, and I. Gomez-Mestre. 2011. Comparisons between Q_{ST} and F_{ST}—how wrong have we been? *Mol. Ecol.* 20: 4830–4839. [12]

Edge, M. D., and N. A. Rosenberg. 2014. Upper bounds on F_{ST} in terms of the frequency of the most frequent allele and total homozygosity: The case of a specified number of alleles. *Theor. Popul. Biol.* 97: 20–34. [9]

Edmonds, C. A., A. S. Lillie, and L. L. Cavalli-Sforza. 2004. Mutations arising in the wave front of an expanding population. *Proc. Natl. Acad. Sci. USA* 101: 975–979. [9].

Edwards, A. W. F. 1990. Fisher, $\overline{W}$, and the fundamental theorem. *Theor. Popul. Biol.* 38: 276–284. [6]

Edwards, A. W. F. 1994. The fundamental theorem of natural selection. *Biol. Rev.* 69: 443–474. [6]

Edwards, A. W. F. 2000. Sewall Wright's equation $\Delta q = (q(1\text{-}q)\, \partial w / \partial q)/2w$. *Theor. Popul. Biol.* 57: 67–70. [5]

Edwards, J. W. 2008. Predicted genetic gain and inbreeding depression with general inbreeding levels in selection candidates and offspring. *Crop Sci.* 45: 2086–2096. [23]

Edwards, M. D., C. W. Stuber, and J. F. Wendel. 1987. Molecular-marker-facilitated investigations of quantitative-trait loci in maize. I. Numbers, genomic distributions and types of gene action. *Genetics* 116: 113–125. [17]

Efron, B. 1986. Why isn't everyone a Bayesian? *Am. Stat.* 40: 1–11. [19, A2]

Efron, B. 2007. Correlation and large-scale simultaneous significance testing. *J. Amer. Stats Assoc.* 102: 93–103. [A4]

Egea, R., S. Casillas, and A. Barbadilla. 2008. Standard and generalized McDonald–Kreitman test: a website to detect selection by comparing different classes of DNA sites. *Nucleic Acids Res.* 36: W157–W162. [10]

Egger, M., G. D. Smith, M .Schneider, and C. Minder. 1997. Bias in meta-analysis detected by a simple, graphical test. *BMJ* 315: 629–634. [A4]

Eisen, E. J. 1972. Long-term selection response for 12-day litter weight in mice. *Genetics* 72: 129–142. [25]

Eisen, E. J. 1975. Population size and selection intensity effects on long-term selection response in mice. *Genetics* 79: 305–323. [26]

Eisen, E. J. 1980. Conclusions from long-term selection experiments with mice. *Z. Tierzüchtg. Züchtgsbiol.* 97: 305–319. [26]

Eisen, E. J. 1989. Selection experiments for body composition in mice and rats: a review. *Livestock Prod. Sci.* 23: 17–32. [18]

Eisen, E. J., and J. P. Hanrahan. 1972. Selection for sexual dimorphism in body weight in mice. *Aust. J. Biol. Sci.* 25: 1015–1024. [21]

Eisen, E. J., and J. P. Hanrahan. 1974. Genetic drift and inbreeding depression measured from control populations of mice. *Can. J. Genet. Cytol.* 16: 91–104. [12]

Eitan, Y., and M. Soller. 2004. Selection induced genetic variation. *In* S. Wasser (ed.), *Evolutionary theory and processes: Modern horizons*, pp 153–176. Kluwer Academic Publishers, Dordrecht. [25]

Eklund, J., and G. E. Bradford. 1977. Genetic analysis of a strain of mice plateauted for litter size. *Genetics* 85: 529–542. [25]

El Adlouni, S., A.-C. Favre, and B. Bobée. 2006. Comparison of methodologies to assess the convergnce of Markov chain Monte Carlo methods. *Comp. Stat. Data Anal.* 50: 2685–2701. [A3]

Eldredge, N., and S. J. Gould. 1972. Punctuated equilibria: an alternative to phyletic gradualism. *In* J. M. T. Schopft (ed.), *Models in paleobiology*, pp 82–115. Freeman, San Francisco. [12, 25]

Eldredge, N., and 9 others. 2005. The dynamics of evolutionary stasis. *Paleobiology* 31: 133-145. [12]

Elena, S. F., and R. E. Lenski. 2003. Evolution experiments with microorganisms: The dynamics and genetic bases of adaptation. *Nat. Rev. Genet.* 4: 457–469. [27]

Elewa, A. M. T. 2004. *Morphometrics: Applications in biology and paleontology*. Springer, New York, NY. [A5]

Ellen, E. D., W. M. Muir, F. Teuscher, and P. Bijma. 2007. Genetic improvement of traits affected by interactions among individuals: sib selection schemes. *Genetics* 176: 489–499. [22]

Ellen, E. D., J. Visscher, J. A. M. van Arendonk, and P. Bijma. 2008. Survival of laying hens: Genetic parameters for direct and associative effects in three purebred layer lines. *Poulty Sci.* 87: 233–239. [22]

Ellner, S. 1996. Environmental fluctuations and the maintenance of genetic diversity in age or stage-structured populations. *Bull. Math. Biol.* 58: 103–127. [28]

Ellner, S., and N. G. Hairston Jr. 1994. Role of overlapping generations in maintaining genetic variation in a fluctuating environment. *Am. Nat.* 143 : 403–417. [28]

Emerson, J. J., and W.-H. Li. 2010. The genetic basis of evolutionary change in gene expression levels. *Phil. Trans. Roy. Soc. Lond. B* 365: 2581–2590. [12]

Emerson, J. J., and 7 others. 2010. Natural selection on *cis* and *trans* regulation in yeasts. *Genome Res.* 20: 826–836. [12]

Emigh, T. H., and E. Pollak. 1979. Fixation probabilities and effective population numbers in diploid populations with overlapping generations. *Theor. Popul. Biol.* 15: 86–107. [3]

Emik, L. O., and C. E Terrill. 1949. Systematic procedures for calculating inbreeding coefficients. *J. Heredity* 40: 51–55. [11]

Emlen, S. T., and L. W. Oring. 1977. Ecology, sexual selection, and the evolution of mating systems. *Science* 197: 215–223. [29]

Empig, L. T., C. O. Gardner, and W. A. Compton. 1972. Theoretical gains for different population improvement procedures. *Univ. Nebraska Agric. Exp. Stn Bull. Misc. Pub.* 26. [23]

Enard, D., P. W. Messer, and D. A. Petrov. 2014. Genome-wide signals of positive selection in human evolution. *Genome Res.* 24: 885–895. [8, 9]

Enard, W., and 7 others. 2002. Molecular evolution of FOXP2, a gene involved in speech and language. *Nature* 418: 869–872. [8]

Enattah, N. S., and 26 others. 2007. Evidence of still-ongoing convergence evolution of the lactase persistence T_{-13910} alleles in humans. *Am. J. Hum. Genet.* 81: 615–625. [8]

ENCODE Project Consortium. 2012. An integrated encyclopedia of DNA elements in the human genome. *Nature* 489: 57–74. [8, 10]

Endler, J. A. 1986. *Natural selection in the wild.* Princeton University Press, Princeton, NJ. [1, 29, 30]

Endo, T., K. Ikeo, and T. Gojobori. 1996. Large-scale search for genes on which positive selection may operate. *Mol. Biol. Evol.* 13: 685–690. [10]

Enfield, F. D. 1972. Patterns of response to 70 generations of selection for pupa weight in *Tribolium*. *Proc. 21st National Breeders Roundtable*, pp. 52–69. [25]

Enfield, F. D. 1980. Long term effects of selection: The limits to response. *In* A. Roberston, (ed.), *Selection experiments in laboratory and domestic animals*, pp. 69–86. Commonwealth Agricultural Bureaux, Slough, UK. [25]

Engels, W. R. 1983. Evolution of altrusitic behavior by kin selection: an alternative approach. *Proc. Natl. Acad. Sci. USA* 80: 515–518. [22]

Engen, S., R. Lande, and B.-E. Sæther. 2011. Evolution of a plastic quantitative trait in an age structured population in a fluctuating environment. *Evolution* 65: 2893–2906. [29]

Engen, S., B.- E. Sæther, T. Kvalnes, and H. Jensen. 2012. Estimating fluctuating selection in age-structured populations. *J. Evol. Biol.* 25: 1487–1499. [29]

Epinat-Le Signor, C., and 6 others. 2001. Interpretation of genotype x environment interactions for early maize hybrids over 12 years. *Crop Sci.* 41: 663–669. [29]

Escobar, J. S., and 9 others. 2011. Patterns of mating system evolution in hermaphroditic animals: correlations among selfing rate, inbreeding depression, and the timing of reproduction. *Evolution* 65: 1233–1253. [23]

Eshel, I., and M. W. Feldman. 1970. On the evolutionary effect of recombination. *Theor. Popul. Biol.* 1: 88–100. [4]

Estany, J., M. Baselga, A. Blasco, and J. Camacho. 1989. Mixed model methodology for the estimation of genetic response to selection in litter size of rabbits. *Livestock Prod. Sci.* 21: 67–75. [19]

Estes, S., and S. J. Arnold. 2007. Resolving the paradox of stasis: models with stabilizing selection explain evolutionary divergence on all timescales. *Am. Nat.* 169: 227–244. [12, 28, 30]

Etheridge, A., P. Pfaffelhuber, and A. Wakolbinger. 2006. An approximate sampling formula under genetic hitchhiking. *Ann. Appl. Prob.* 16: 685–729. [8]

Ethier, S. N., and R. C. Griffiths. 1990. On the two-locus sampling distribution. *J. Math. Biol.* 29: 131–159. [4]

Etterson, J. R., and R. G. Shaw. 2001. Constraint to adaptive evolution in response to global warming. *Science* 294: 151–154. [20]

Evans, M., and T. Swartz. 1995. Methods for approximating integrals in statistics with special emphasis on Bayesian integration problems. *Stat. Sci.* 10: 254–272. [A3]

Evans, P. D., J. R. Anderson, E. J. Vallender, S. S. Choi, and B. T. Lahn. 2004b. Reconstructing the evolutionary history of *microcephalin*, a gene controlling human brain size. *Hum. Mol. Genet.* 13: 1139–1145. [10]

Evans, P. D., and 6 others. 2004a. Adaptive evolution of *ASPM*, a major determinant of cerebral cortical size in humans. *Hum. Mol. Genet.* 13: 489–494 [10]

Evans, P. D., and 8 others. 2005. *Microcephalin*, a gene regulating brain size, continues to evolve adaptively in humans. *Science* 309: 1717–1720. [10]

Evenson, R. E., and D. Gollin. 2003. Assessing the impact of the Green Revolution, 1960 to 2000. *Science* 300: 758–762. [1]

Ewens, W J. 1969. A generalized fundamental theorem of natural selection. *Genetics* 63: 531–537. [6]

Ewens, W. J. 1972. The sampling theory of selectively neutral alleles. *Theor. Popul. Biol.* 3: 87–112. [2, 8, 9]

Ewens, W. J. 1976. Remarks on the evolutionary effect of natural selection. *Genetics* 83: 601–607. [6]

Ewens, W. J. 1979. *Mathematical population genetics.* Springer, Berlin, DE. [A1]

Ewens, W. J. 1989. An interpretation and proof of the fundamental theorem of natural selection. *Theor. Popul. Biol.* 36: 167–180. [6, 22]

Ewens, W. J. 1992. An optimizing principle of natural selection in evolutionary population genetics. *Theor. Popul. Biol.* 42: 333–346. [6, 22]

Ewens, W. J. 1994. The changing role of population genetics theory. *In* S. Levin (ed.), *Frontiers in Mathematical Biology*, pp. 186-197. Springer, New York, NY [5, 6]

Ewens, W. J. 2004. *Mathematical population genetics*, 2nd Edition. Springer-Verlag, New York, NY. [2, 5, 7, A1]

Ewens, W. J., and G. Thomson. 1970. Heterozygote selective advantage. *Ann. Hum. Genetics* 33: 365–376. [7]

Ewens, W. J., and G. Thomson. 1977. Properties of equilibria in multi-locus genetic systems. *Genetics* 87: 807–819. [6]

Ewing, G., and J. Hermisson. 2010. MSMS: a coalescent simulation program including recombination, demographic structure and selection at a single locus. *Bioinformatics* 26 : 2064–2065. [9]

Ewing, G., J. Hermisson, P. Pfaffelhuber, and J. Rudolf. 2011. Selective sweeps for recessive alleles and for other modes of dominance. *J. Math. Biol.* 63: 399–431 [8]

Excoffier, L., M. Foll, and R. J. Petit. 2009a. Genetic consequences of range expansions. *Ann. Rev. Ecol. Evol. Syst.* 40: 481—501. [9]

Excoffier, L., T. Hofer, and M. Foll. 2009b. Detecting loci under selection in a hierarchically structured population. *Heredity* 103: 285–298. [2, 9]

Excoffier, L., and N. Ray. 2008. Surfing during population expansions promotes genetic revolutions and structuration. *Trends Ecol. Evol.* 23: 347–351. [9]

Eyre-Walker, A. 2002. Changing effective population size and the McDonald-Kreitman test. *Genetics* 162: 2017–2024. [10]

Eyre-Walker, A. 2006. The genomic rate of adaptive evolution. *Trends Ecol. Evol.* 10: 569–575. [10]

Eyre-Walker, A. 2010. Genetic architecture of a complex trait and its implications for fitness and genome-wide association studies. *Proc. Natl. Acad. Sci. USA* 107: 1752–1756. [28]

Eyre-Walker, A., and P. D. Keightley. 2007. The distribution of fitness effects of new mutations. *Nat. Rev. Genet.* 8: 610–618. [10]

Eyre-Walker, A., and P. D. Keightley. 2009. Estimating the rate of adaptive molecular evolution in the presence of slightly deleterious mutations and population size change. *Mol. Biol. Evol.* 26: 2097–2108. [10]

Eyre-Walker, A., and P. D. Keightley. 2007. The distribution of fitness effects of new mutations. *Nat. Rev. Genet.* 8: 610–618. [12]

Eyre-Walker, A., M. Woolfit, and T. Phelps. 2006. The distribution of fitness effects of new deleterious amino acid mutations in humans. *Genetics* 173: 891–900. [10]

Fagny, M., and 5 others. 2014. Exploring the occurrence of classic selective sweeps in humans using whole-genome sequencing data sets. *Mol. Biol. Evol.* 31: 1850–1868. [9]

Fairbairn, D. J., and A. E. Wilby. 2001. Inequality of opportunity: measuring the potential for sexual selection. *Evol. Ecol. Res.* 3: 667–686. [29]

Fairfull, R. W., and W. M. Muir. 1996. Selection and breeding of laying hens: present and future solutions. *Proc. 20th World Poultry Congress*, New Delhi, pp. 395–415. [18]

Falconer, D. S. 1952. The problem of environment and selection. *Am. Nat.* 86: 293–298. [17]

Falconer, D. S. 1953. Selection for large and small size in mice. *J. Genet.* 51: 470–501. [21]

Falconer, D. S. 1954. Asymmetrical response in selection experiments. *In* A. A. Buzzati-Traverso (ed.) *Symposium on genetics of population structure* pp. 16–41. International Union Biol. Sci., Naples Series B, No. 15. [18]

Falconer, D. S. 1955. Patterns of response in selection experiments with mice. *Cold Spring Harbor Symp. Quant. Biol.* 20: 178–196. [25]

Falconer, D. S. 1957. Selection for phenotypic intermediates in *Drosophila*. *J. Genet.* 55: 651–561. [16]

Falconer, D. S. 1960a. Selection of mice for growth on high and low planes of nutrition. *Genet. Res.* 1: 91–113. [21]

Falconer, D. S. 1960b. The genetics of litter size in mice. *J. Cell Comp. Physiol.* 56 (Suppl 1): 153–167. [15, 21]

Falconer, D. S. 1965. Maternal effects and selection response. *In* S. J. Geerts (ed.) *Genetics today, proceedings of the XI international congress on genetics, Vol. 3* pp. 763–774. Pergamon, Oxford, UK. [1, 15, 22]

Falconer, D. S. 1971. Improvement of litter size in a strain of mice at a selection limit. *Genet. Res.* 17: 215–235. [25]

Falconer, D. S. 1973. Replicated selection for body weight in mice. *Genet. Res.* 22: 291–321. [18]

Falconer, D. S. 1977. Some results of the Edinburgh selection experiments with mice. *In* E. Pollak, O. Kempthorne, and T. B. Bailey, Jr., (eds.), *Proceedings of the international conference on quantitative genetics*, pp. 101–115. Iowa State Univ. Press, Ames, IA. [26]

Falconer, D. S. 1981. *Introduction to quantitative genetics.* 2nd Edition. Longman, New York, NY. [18]

Falconer, D. S. 1985. A note on Fisher's 'average effect' and 'average excess'. *Genet. Res.* 46: 337–347. [6]

Falconer, D. S. 1992. Early selection experiments. *Ann. Rev. Genet.* 26: 1–16. [18]

Falconer, D. S., and J. W. B. King. 1953. A study of selection limits in the mouse. *J. Genet.* 51: 561–581. [25, 26]

Falconer, D. S. and M. Latyszewski. 1952. The environment in relation to selection for size in mice. *J. Genet.* 51: 67–80. [21]

Falconer, D. S., and T. F. C. Mackay. 1996. *Introduction to quantitative genetics.* 4th Ed. Longman Sci. and Tech., Harlow, UK. [26]

Falke, K. C., and 5 others. 2007. Temporal changes in allele frequencies in two European F_2 flint maize populations under modified recurrent full-sib selection. *Theor. Appl. Genet.* 114: 765–776. [25]

Fariello, M. I., S. Boitard, H. Naya, M. SanCristobal, and B. Servin. 2013. Detecting signatures of selection through haplotype differentiation among hierarchically structured populations. *Genetics* 193: 929–941. [9]

Fariello, M.-I., and 7 others. 2014. Selection signatures in worldwide Sheep populations. *PLoS One* 9: e103813. [9]

Farlow, A., and 6 others. 2015. The spontaneous mutation rate in the fission yeast *Schizosaccharomyces pombe*. *Genetics* 201: 737–744. [4]

Farlow, S. J. 1982. *Partial differential equations for scientists and engineers*. Wiley, New York, NY. [A1]

Fay, J. C. 2011. Weighing the evidence for adaptation at the molecular level. *Trends Genet.* 27: 343–349. [10]

Fay, J. C., and P. J. Wittkopp. 2007. Evaluating the role of natural selection in the evolution of gene regulation. *Heredity* 100: 191–199. [12]

Fay, J. C., and C.-I. Wu. 2000. Hitchhiking under positive Darwinian selection. *Genetics* 155: 1405–1413. [8, 9]

Fay, J. C., G. J. Wyckoff, and C.-I Wu. 2001. Positive and negative selection on the human genome. *Genetics* 158: 1227–1234. [10]

Fay, J. C., G. J. Wyckoff, and C.-I Wu. 2002. Testing the neutral theory of molecular evolution with genomic data from *Drosophila*. *Nature* 415: 10024–1026. [10]

Fearnhead, P., and P. Donnelly. 2001. Estimating recombination rates from population genetic data. *Genetics* 159: 1299–1318. [4]

Feder, A. F., S. Kryazhimskiy, and J. B. Plotkin. 2014. Identifying signatures of selection in genetic time series. *Genetics* 196: 509–522. [9]

Fehr, W., and H. Hadley. 1980. *Hybridization of crop plants*. American Soceity of Agronomy, Madison, WI. [21]

Feinberg, A. P., and R. A. Irizarry. 2010. Stochastic epigenetic variation as a driving force of development, evolutionary adaptation, and disease. *Proc. Natl. Acad. Sci. USA* 107: 1757–1764. [17]

Feldman, M. W., S. P. Otto, and F. B. Christiansen. 1996. Population genetic perspectives on the evolution of recombination. *Ann. Rev. Genet.* 30: 261–295. [4]

Felsenstein, J. 1965. The effect of linkage on directional selection. *Genetics* 52: 349–363. [16]

Felsenstein, J. 1971. Inbreeding and variance effective numbers in populations with overlapping generations. *Genetics* 68: 581–597. [3, 4]

Felsenstein, J. 1974. The evolutionary advantage of recombination. *Genetics* 78: 737–756. [7, 8, 26]

Felsenstein, J. 1976. The theoretical population genetics of variable selection and migration. *Ann. Rev. Genet.* 10: 253–280. [28]

Felsenstein, J. 1977. Multivariate normal genetic models with a finite number of loci. *In* E. Pollak, O. Kempthorne, and T. B. Bailey, Jr., (eds.), *Proceedings of the international conference on quantitative genetics*, pp. 227–246. Iowa State Univ. Press, Ames, IA. [24, 28, 29]

Felsenstein, J. 1979. Excursions along the interface between disruptive and stabilizing selection. *Genetics* 93: 773–795. [24]

Felsenstein, J. 1981. Continuous-genotype models and assortative mating. *Theor. Popul. Biol.* 19: 341–357. [16]

Felsenstein, J. 1985. Phylogenies and the comparative method. *Am. Nat.* 125: 1–15. [10, 12]

Felsenstein, J. 1992. Estimating effective population size from samples of sequences: inefficiency of pairwise and segregating sites as compared to phylogenetic estimates. *Genet. Res.* 59: 139–147. [4]

Felsenstein, J. 2002. Contrasts for a within-species comparative method, *In* M. Slatkin and M. Veuille (eds.), *Modern developments in theoretical population genetics: the legacy of Gustave Malécot*, pp. 118-129. Oxford University Press, Oxford, UK. [9]

Felsenstein, J. 2004. *Inferring phylogenies*. Sinauer, Sunderland, MA [10, 12]

Felsenstein, J. 2008. Comparative methods with sampling error and within-species variation: contrasts revisited and revised. *Am. Nat.* 171: 713–725. [12]

Felsenstein J., and S. Yokoyama. 1976 The evolutionary advantage of recombination II. Individual selection for recombination. *Genet.* 83: 845–8759. [4, 26]

Fernández, A., M. A. Toro, and C. López-Fanjul. 1995. The effect of inbreeding on the redistribution of genetic variance of fecundity and viability in *Tribolium castaneum*. *Heredity* 75: 376–381. [11]

Fernando, R. L., and D. Gianola. 1986. Effect of assortative mating on genetic change due to selection. *Theor. Appl. Genet.* 72: 395–404. [16]

Fernando, R. L., and D. Gianola. 1990. Statistical inferences in populations undergoing selection or nonrandom mating. *In* D. Gianola and K. Hammond (eds.) *Statistical methods for genetic improvement of livestock*, pp. 437–453. Springer-Verlag, NY. [19]

Fernando, R. L., C. Stricker, and R. C. Elston 1994. The finite polygenic mixed model: an alternative formulation for the mixed model of inheritance. *Theor. Appl. Genet.* 88: 573–580. [24]

Fernando, R. L., and 5 others. 2004. Controlling the proportion of false positives in multiple dependent tests. *Genetics* 166: 611–619. [A4]

Ferrer-Admetlla, A., M. Liang, T. Korneliussen, and R. Nielsen. 2014. On detecting incomplete soft or hard selective sweeps using haplotype structure. *Mol. Biol. Evol.* 31: 1275-1291. [9]

Field, Y., and 10 others. 2016. Detection of human adaptation during the past 2000 years. *Science* 354: 760–764. [9, 12]

Fijarczyk, A., and W. Babik. 2015. Detecting balancing selection in genomes: limits and prospects. *Mol. Ecol.* 24: 3529–3545. [9]

Firth, J. A., J. D. Hadfield, A. W. Santure, J. Slate, and B. C. Sheldon. 2015. The influence of nonrandom extrapair paternity on heritability estimates derived from wild pedigrees. *Evolution* 69: 1336–1344. [20]

Fisher, R. A. 1918. The correlation between relatives on the supposition of Mendelian inheritance. *Trans. Royal Soc. Edinb.* 52: 399–433. [1, 16, 24, 25]

Fisher, R. A. 1922. On the dominance ratio. *Proc. Roy. Soc. Edinb.* 42: 321–341. [2, 5, A1]

Fisher, R. A. 1930. *The genetical theory of natural selection*. Clarendon, Oxford, UK. [6, 13, 26, 27, 28, 29]

Fisher, R. A. 1932a. The evolutionary modification of genetic phenomena. *Proc. 6th Int. Congress Genet.* 1:165–172. [A1]

Fisher, R. A. 1932b. *Statistical methods for research workers*. Oliver and Boyd, Edinburgh, UK. [A4]

Fisher, R. A. 1934. Adaptation and mutations. *School Science Review* 15: 294–301. [26]

Fisher, R. A. 1937. The wave of advance of advantageous genes. *Annals Eugenics* 7: 355–369. [8, 27]

Fisher, R. A. 1938. Presidential address to the first Indian statistical congress. *Sankhya* 4: 14–17. [18]

Fisher, R. A. 1941. Average excess and average effect of a gene substitution. *Ann. Eugen.* 11: 53–63. [6, 23]

Fisher, R. A. 1958. *The genetical theory of natural selection*. Dover Publ., NY. [13, 20, 26]

Fisher, R. A., and E. B. Ford. 1947. The spread of a gene in natural conditions in a colony of the moth *Panaxia dominuia* L. *Heredity* 1: 143–173. [9]

Fisher, R. A., and C. S. Stock. 1915. Cuénot on preadaptation: a criticism. *Eugenics Review* 7: 46–61. [1]

Fisher, R. A., and L. H. C. Tippett. 1928. Limiting forms of the frequency distribution of the largest and smallest member of a sample. *Proc. Camb. Phil. Soc.* 24: 180–190. [27]

Fitch, W. M., R. M. Bush, C. A. Vender, and N. J. Cox. 1997. Long-term trends in the evolution of H(3) HA1 human influenza type A. *Proc. Natl. Acad. Sci. USA* 94: 7712–7718. [10]

Fitze, P. S., and J.-F. Le Galliard. 2011. Inconsistency between different measures of sexual selection. *Am. Nat.* 178: 256–268. [29]

Fitzpatrick, M. J., E. Feder, L. Rowe, and M. B. Sokolowski. 2007. Maintaining a behavior polymorphism by frequency-dependent selection on a single gene. *Nature* 447: 210-213. [5]

Flachenecker, C., M. Frisch, K. C. Falke, and A. E. Melchinger. 2006. Genetic drift and selection effects of modified recurrent full-sib selection in two F_2 populations of European flint maize. *Theor. Appl. Genet.* 113: 1113–1120. [18]

Flatt, T. 2005. The evolutionary genetics of canalization. *Q. Rev. Biol.* 80: 287–316. [17]

Fleming, W. H. 1979. Equilibrium distributions of continuous polygenic traits. *SIAM J. Appl. Math.* 36, 148–168. [28]

Flexon, P. B., and C. F. Rodell. 1982. Genetic recombination and directional selection for DDT resistance in *Drosophila melanogaster*. *Nature* 298: 672–674. [26]

Flint, J., and T. F. C. Mackay. 2009. Genetic architecture of quantitative traits in mice, flies, and humans. *Genet. Res.* 19: 723–733. [24, 25]

Flowers, J. M., and 5 others. 2012. Natural selection in gene-dense regions shapes the genomic pattern of polymorphism in wild and domesticated rice. *Mol. Biol. Evol.* 29: 675–687. [8]

Flury, C., and 7 others. 2010. Effective population size of an indigenous Swiss cattle breed estimated from linkage disequilibrium. *J. Anim. Breed. Genet.* 127: 339–337. [4]

Flux, J. E. C., and M. M. Flux 1982. Artificial selection and gene flow in wild starlings, *Sturnus vulgaris*. *Naturwissenschaften* 69: 96–97. [18]

Foley, P. 1987. Molecular clock rates at loci under stabilizing selection. *Proc. Natl. Acad. Sci. USA* 84: 7996–8000. [28]

Foley, P. 1992. Small population genetic variability at loci under stabilizing selection. *Evolution* 46: 763–774. [28]

Foll, M., and O. Gaggiotti. 2008. A genome-scane method to identify selected loci appropriate for both dominant and codominant markers: A Bayesian perspective. *Genetics* 180: 977–993. [9]

Foll, M., H. Shim, and J. D. Jensen. 2015. WFABC: a Wright–Fisher ABC-based approach for inferring effective population sizes and selection coefficients from time-sampled data. *Mol. Ecol. Resources* 15: 87–98. [9]

Foll, M., and 17 others. 2014. Influenza virus drug resistance: a time-sampled population genetics perspective. *PLoS Genet.* 10: e1004185. [27]

Ford, M. J. 2002. Applications of selective neutrality tests to molecular ecology. *Mol. Ecol.* 11: 1245–1262. [9]

Forester, B. R., M. R. Jones, S. Joost, E. L. Landguth, and J. R. Lasky. 2016. Detecting spatial genetic signatures of local adaptation in heterogeneous landscapes. *Mol. Ecol.* 25: 104–120. [9]

Foulley, J. L. 1992. Prediction of selection response from threshold dichotomous traits. *Genetics* 132: 1187–1194. [14]

Foulley, J. L. 1993. Prediction of selection response for Poisson distributed traits. *Genet. Sel. Evol.* 25: 297–303. [14]

Foulley, J. L., D. Gianola, and R. Thompson. 1983. Prediction of genetic merit from data on binary and quantitative variates with an application to calving difficulty, birth weight, and pelvic opening. *Genet. Sel. Evol.* 15: 401–424. [14]

Foulley, J. L., and S. Im. 1993. A marginal quasi-likelihood approach to the analysis of Poisson variables with generalized linear mixed models. *Genet. Sel. Evol.* 25: 101–107. [14]

Foulley, J. L., and R. L. Quaas. 1995. Heterogeneous variances in Gaussian linear mixed models. *Genet. Sel. Evol.* 27: 211–228. [17]

Foulley, J. L., M. SanCristobal, D. Gianola, and S. Im. 1992. Marginal likelihood and Bayesian apporaches to the analysis of heterogeneous residual variances in mixed linear Gaussian models. *Comput. Stat. Data Anal.* 13: 291–305. [17]

Fourcade, Y., A. Chaput-Bardy, J. Secondi, C. Fleurant, and C. Lemaire. 2013. Is local selection so widespread in river organisms? Fractal geometry of river networks leads to high bias in outlier detection. *Mol. Ecol.* 22: 2065–2073. [9]

Force, A., M. Lynch, B. Pickett, A. Amores, Y.-L. Yan, and J. Postlethwait. 1999. Preservation of duplicate genes by complementary, degenerative mutations. *Genetics* 151: 1531–1545. [1]

Fowler, K., C. Semple, N. H. Barton, and L. Partridge. 1997. Genetic variation for total fitness in *Drosophila melanogaster. Proc. Roy. Soc. Lond.* B 264: 191–199. [6]

Fox, J. 2014. The fitness landscape of beer: Fisher's geometrical model vs. IPA. http://dynamicecology.wordpress.com/2014/05/12/fishers-geometrical-model-vs-beer/ [27]

Foxe, J. P., and 5 others. 2008. Selection on amino acid substitutions in *Arabidopsis. Mol. Biol. Evol.* 25: 1375–1383. [10]

François, O., H. Martins, K. Caye, and S. D. Schoville. 2016. Controlling false discoveries in genome scans for selection. *Mol. Ecol.* 25: 454–469. [9]

Frahm, R., and K.-I. Kojima. 1966. Comparison of selection response on body weight under divergent larval density conditions in *Drosophila pseudoobscura*. *Genetics* 54: 625–637. [25]

Frank, S. A. 1995. George Price's contributions to evolutionary genetics. *J. Theor. Biol.* 175: 373–388. [1, 6]

Frank, S. A. 1997. The Price equation, Fisher's fundamental theorem, kin selection, and causal analysis. *Evolution* 51: 1712–1729. [6]

Frank, S. A. 1998. *The foundations of social evolution*. Princeton University Press, Princeton, NJ. [6, 22]

Frank, S. A. 2007. Maladaptation and the paradox of robustness in evolution. *PLoS One* 2: e1021. [1]

Frank, S. A. 2012. Natural selection. IV. The Price equation. *J. Evol. Biol.* 25: 1002–1019. [6]

Frank, S.A. 2013. Natural selection. VII. History and interpretation of kin selection theory. *J. Evol. Biol.* 28: 1151–1184. [22]

Frank, S. A., and M. Slatkin. 1990a. Evolution in a variable environment. *Am. Nat.* 136: 244–260. [28]

Frank, S. A., and M. Slatkin. 1990b. The distribution of allelic effects under mutation and selection. *Genet. Res.* 55: 111–117. [28]

Frank, S. A., and M. Slatkin. 1992. Fisher's fundamental theorem of natural selection. *Trends Ecol. Evol.* 7: 92–95. [6]

Frankham, R. 1977. Optimum selection intensities in artificial selection programmes: an experimental evaluation. *Genet. Res.* 30: 115–119. [26]

Frankham, R. 1980. The founder effect and response to artificial selection in *Drosophila*. *In* A. Roberston, (ed.), *Selection experiments in laboratory and domestic animals*, pp. 87–90. Commonwealth Agricultural Bureaux, Slough, UK. [26, 28]

Frankham, R. 1982. Contribution of *Drosophila* research to quantitative genetics and animal breeding. *Proc. 2nd World Congress on Genetics Applied to Livestock Production* V: 43–56. [21]

Frankham, R. 1983a. Origin of genetic variation in selection lines. *Proceedings 32nd Annual National Breeders Roundtable*, pp. 56-68. [26]

Frankham, R. 1983b. The effects of population size restriction on selection response. *Proceedings 32nd Annual National Breeders Roundtable*, pp. 155-168. [26]

Frankham, R. 1988. Exchanges in the rRNA multigene family as a source of genetic variation. *In* B. S. Weir, E. J. Eisen, M. M. Goodman, and G. Namkoong (eds.), *Proceedings of the second international conference on quantitative genetics*, pp. 236–242. Sinauer Assoc., Sunderland, MA. [25]

Frankham, R. 1990. Are responses to artificial selection for reproductive fitness characters consistently asymmetrical? *Genet. Res.* 56: 35–42. [18]

Frankham, R. 1995. Effective population size/adult population size ratios in wildlife: a review. *Genet. Res.* 66: 95–107. [3, 4]

Frankham, R., D. A. Briscoe, and R. K. Nurthen. 1978. Unequal crossing over at the rRNA locus as a source of quantitative genetic variation. *Nature* 272: 80–81. [25]

Frankham, R., D. A. Briscoe, and R. K. Nurthen. 1980. Unequal crossing over at the rRNA tandem as a source of quantitative genetic variation in *Drosophila. Genetics* 95: 727–742. [25]

Frankham, R., L. P. Jones. and J. S. F. Barker. 1968a. The effects of population size and selection intensity in selection for a quantitative trait in *Drosophila*. I. Short-term response to selection. *Genet. Res.* 12: 237–243. [18]

Frankham, R., L. P. Jones and J. S. F. Barker. 1968b. The effects of population size and selection intensity in selection for a quantitative trait in *Drosophila*. III. Analysis of the lines. *Genet. Res.* 12: 267–283. [25, 26]

Frankham, R., and R. K. Nurthen. 1981. Forging links between population and quantitative genetics. *Theor. Appl. Genet.* 59: 251–263. [18, 25]

Frankham, R., B. H. Yoo, and J. S. F. Barker. 1988. Reproductive fitness and artificial selection in animal breeding: culling on fitness prevents a decline in reproductive fitness in lines of *Drosophila melanogaster* selected for increased inebriation time. *Theor. Appl. Genet.* 76: 909–914. [25]

Franklin, I. R. 1977. The distribution of the proportion of the genome which is homozygous by descent in inbred individuals. *Theor. Popul. Biol.* 11: 60–80. [3]

Franklin, J. N. 1968. *Matrix theory*. Prentice-Hall, Upper Saddle River, NJ. [A5]

Franklin, O. D., and M. B. Morrissey. 2017. Inference of selection gradients using performance measures as fitness proxies. *Methods Ecol. Evol.* 8: 663–677. [29]

Frary, A., and 10 others. 2000. *fw2.2:* A quantitative trait locus key to the evolution of tomato fruit size. *Science* 289: 85–88. [9]

Fraser, A. S. 1957. Simulation of genetic systems by automatic digital computers. II. Effects of linkage on rates of advance under selection. *Aust. J. Biol. Sci.* 10: 492–499. [7, 26]

Fraser, H. B. 2011. Genomewide approaches to the study of adaptive gene expression evolution. *Bioessays* 33: 469–477. [12]

Fraser, H. B. 2013. Gene expression drives local adaptation in humans. *Genome Res.* 23: 1089–1096. [8, 12]

Fraser, H. B., A. M. Moses, and E. E. Schadt. 2010. Evidence for widespread adaptive evolution of gene expression in budding yeast. *Proc. Natl. Acad. Sci. USA* 107: 2977–2982. [12]

Fraser, H. B., and E. E. Schadt. 2010. The quantitative genetics of phenotypic robustness. *PLoS One* 5: e8625. [17]

Fraser, H. B., and 6 others. 2011. Systematic detection of polygenic *cis*-regulatory evolution. *PLoS Genet.* 7: e1002023. [12]

Fredeen, H. T. 1984. Selection limits: have they been reached with pigs? *Can. J. Anim. Sci.* 64: 223–234. [25]

Frentiu, F. D., and 5 others. 2008. Pedigree-free animal models: the relatedness matrix reloaded. *Proc. Roy. Soc. Lond. B* 275: 639–648. [20]

Frichot, E., S. D. Schoville, G. Bouchard, and O. François. 2013. Testing for associations between loci and environmental gradients using latent factor mixed models. *Mol. Biol. Evol.* 30: 1687–1699. [9]

Friedman, J. A., and W. Stuetzle. 1981. Projection pursuit regression. *J. Am. Stat. Assoc.* 76: 817–823. [30]

Frisse, L., and 5 others. 2001. Gene conversion and different population histories may explain the contrast between polymorphism and linkage disequilibrium levels. *Am. J. Hum. Genet.* 69: 831–843. [4]

Fu, W., and J. M. Akey. 2013. Selection and adaptation in the human genome. *Ann. Rev. Genomics Hum. Genet.* 14: 467–489. [8, 9]

Fu, Y.-X. 1994a. A phylogenetic estimator of effective population size or mutation rate. *Genetics* 136: 685–692. [4]

Fu, Y.-X. 1994b. Estimating effective population size or mutation rate using the frequencies of mutations of various classes in a sample of DNA sequences. *Genetics* 138: 1375–1386. [4]

Fu, Y.-X. 1995. Statistical properties of segregating sites. *Theor. Popul. Biol.* 48: 172-197. [2, 4]

Fu, Y.-X. 1996. New statistical tests of neutrality for DNA samples from a population. *Genetics* 143: 557–570. [9]

Fu, Y.-X. 1997. Statistical tests of neutrality of neutrality against population growth, hitchhiking and background selection. *Genetics* 147: 915–925. [9]

Fu, Y.-X., and W.-H. Li. 1993a. Maximum likelihood estimation of population parameters. *Genetics* 134: 1261–1270. [4]

Fu, Y.-X., and W.-H. Li. 1993b. Statistical tests of neutrality of mutations. *Genetics* 133: 693–709. [4]

Fu, Y.-X., and W.-H. Li. 1999. Coalescing into the 21st century: an overview and prospects of coalescent theory. *Theor. Popul. Biol.* 56: 1–10. [2]

Fujimoto, A., T. Kado, H. Yoshimaru, Y. Tsumura, and H. Tachida. 2008. Adaptive and slightly deleterious evolution in a conifer, *Cryptomeria japonica*. *J. Mol. Evol.* 67: 201–210. [4]

Fumagalli, M., and 5 others. 2011. Signatures of environmental genetic adaptation pinpoint pathogens as the main selective pressure through human evolution. *PLoS Genet.* 7: e1002355. [8, 9, 12]

Gadbury, G. L., and 8 others. 2004. Power and sample size estimation in high dimensional biology. *Stat. Meth. Med. Res.* 13: 325–338. [A4]

Gadbury, G. L., and 5 others. 2008. Evaluating statistical methods using plasmode data sets in the age of massive public databases: An illustration using false discovery rates. *PLoS Genet.* 6: e1000098. [A4]

Gaffney, B., and E. P. Cunningham. 1998. Estimation of genetic trend in racing performance of thoroughbred horses. *Nature* 332: 722–724. [25]

Gale, J. S. 1990. *Theoretical population genetics.* Unwin Hyman, London, UK. [2, A1]

Gale, J. S., and M. J. Kearsey. 1968. Stable equilbria under stabilizing selection in the absence of dominance. *Heredity* 23: 553–561. [5, 28]

Gallais, A. 1975. Selection with truncation in autotetraploids: Comparison with diploids. *Theor. Appl. Genet.* 46: 386–394. [15]

Gallais, A. 1976. Effects of competition on means, variances and covariances in quantitative genetics with an application to general combiing ability selection. *Theor. Appl. Genet.* 47: 189–195. [22]

Gallais, A. 1989. Optimization of recurrent selection on the phenotypic value of doubled haploid lines. *Theor. Appl. Genet.* 77: 501–504. [23]

Gallais, A. 1990. *Théorie de la sélection en amélioration des plantes.* Masson, Paris, FR.

Gallais, A. 2003. *Quantitative genetics and breeding methods in autopolyploid plants.* INRA, Paris, FR. [13, 15, 21]

Gallavotti, A., and 7 others. 2004. The role of *barren stalk1* in the architecture of maize. *Nature* 432: 630–635. [9]

Gallego, A., and C. López-Fanjul. 1983. The number of loci affecting a quantitative trait in *Drosophila melanogaster* revealed by artificial selection. *Genet. Res.* 42: 137–149. [26]

Galtier, N., F. Depaulis, and N. H. Barton. 2000. Detecting bottlenecks and selective sweeps from DNA sequence polymorphism. *Genetics* 155: 981–987. [9]

Galtier, N., L. Duret, S. Glémin, and V. Ranwez. 2009. GC-biased gene conversion promotes the fixation of deleterious amino acid changes in primates. *Trends Genet.* 25: 1–5. [10]

Galtier, N., G. Piganeau, D. Mouchiroud, and L. Duret. 2001. GC-content evolution in mammalian genomes: the biased gene conversion hypothesis. *Genetics* 159: 907–911. [10]

Galvani, A. P., and J. Novembre. 2005. The evolutionary history of the *CCR5-Δ32* HIV-resistance mutation *Microbes & Infection* 7: 302–309. [9]

Gammerman, D. 1997. *Markov chain Monte Carlo.* Chapman and Hall, New York, NY.[A3]

Gantmacher, F. R. 1960. *The theory of matrices* (two volumes). Chelsea, New York, NY. [A5]

Garant, D., and L. E. B. Kruuk. 2005. How to use molecular marker data to measure evolutionary parameters in wild populations. *Mol. Ecol.* 14: 1843–1859. [20]

Garant, D., L. E. B. Kruuk, R. H. McCleery, and B. C. Sheldon. 2004. Evolution in a changing environment: a case study with great tit fledging mass. *Am. Nat.* 164: E115–E129. [20]

Garant, D., L. E. B. Kruuk, R. H. McCleery, and B. C. Sheldon. 2007. The effects of environmental heterogeneity on multivariate selection on reproductive traits in female great tits. *Evolution* 61: 1546–1559. [30]

Garant, D., L. E. B. Kruuk, T. A. Wilkin, R. H. McCleery, and B. C. Sheldon. 2005. Evolution driven by differential dispersal within a wild bird population. *Nature* 433: 60–64. [20, 30]

Garant, D., B. C. Sheldon, and L. Gustaffson. 2004. Climatic and temporal effects on the expression of secondary sexual characters: genetic and environmental components. *Evolution* 58: 634–644. [20]

García, N., C. López-Fanjul, and A. García-Dorado. 1994. The genetics of viability in *Drosophila melanogaster:* effects of inbreeding and artificial selection. *Evolution* 48: 1277-1285. [11]

Garcia, C., and L. Sanchez. 1992. Assortative mating and selection response in *Drosophila melanogaster. J. Anim. Breed. Genet.* 109: 161–167. [16]

García-Cortés, L. A., A. Legarra, and M. A. Toro. 2014. The coefficient of dominance is not (always) estimable with biallelic markers. *J. Anim. Breed. Genet.* 131: 97–104. [11]

García-Dorado, A., and C. López-Fanjul. 1983. Accumulation of lethals in highly selected lines of *Drosophila melanogaster. Theor. Appl. Genet.* 66: 221–223. [25]

García-Dorado, A. and C. López-Fanjul. 1985. Optimum selection strategies: studies with *Drosophila melanogaster. Genet. Res.* 46: 101–105. [26]

Garcia-Gonzalez, F., L. W. Simmons, J. L. Tomkins, J. S. Kotiaho, and J. P. Evans. 2012. Comparing evolvabilities: common errors surrounding the calculation and use of coefficients of additive genetic variation. *Evolution* 66: 2341–2349. [6]

Gardner, A. 2008. The Price equation. *Curr. Biol.* 18: R198–R202. [6]

Gardner, A. 2015. The genetical theory of multilevel selection. *J. Evol. Biol.* 28: 305–319. [22]

Gardner, A., S. A. West, and N. H. Barton. 2007. The relationship between multilocus population genetics and social evolution theory. *Am. Nat.* 169: 207–226. [22]

Gardner, C. O. 1961. An evaluation of effects of mass selection and seed irradiation with thermal neutrons on yields of corn. *Crop Sci.* 1: 241–245. [13]

Gardner, C. O. 1973. Evaluation of mass selection and of seed irradiation with mass selection for population improvement in maize. *Genetics* 74: S88–S89. [21]

Gardner, C. O., and S. A. Eberhart. 1966. Analysis and interpretation of the variety cross diallel and related populations. *Biometrics* 22: 439–452. [18]

Gardner, M. P., K. Fowler, N. H. Barton, and L. Partridge. 2005. Genetic variation for total fitness in *Drosophila melanogaster*: complex yet replicable patterns. *Genetics* 169: 1553–1571. [6]

Garland, T., A. W. Dickerman, C. M. Janis, and J. A. Jones. 1993. Phylogenetic analysis of covariance by computer simulation. *System. Biol.* 42: 265–292. [12]

Garland, T., Jr., and M. R. Rose. 2009. *Experimental evolution. Concepts, methods, and applications of selection experiments.* University of California Press, Berkeley, CA. [18]

Garreau, H., and 6 others. 2008. Results of four generations of a canalising selection for rabbit birth weight. *Livestock Sci.* 119: 55–62. [17]

Garud, N. R., P. W. Messer, E. O. Buzbas, and D. A. Petrov. 2015. Recent selective sweeps in North American *Drosophila melanogaster* show signatures of soft sweeps. *PLoS Genet.* 11: e1005004. [8, 9]

Garud, N. R., and N. A. Rosenberg. 2015. Enhancing the mathematical properties of new haplotype homozygosity statistics for the detection of selective sweeps. *Theor. Popul. Biol.* 92: 94–101. [8, 9]

Garwood, V. A., and P. C. Lowe. 1981. A comparison of combination and family selection in chickens. *Poulty Sci.* 60: 285–288. [21]

Garwood, V. A., P. C. Lowe, and B. B. Bohren 1980. An experimental test of the efficiency of family selection in chickens. *Theor. Appl. Genet.* 56: 6–9. [21]

Gates, C. E. 1954. *The consitution of genetic variances in self-fertilized crops assuming linkage.* Unpublished Ph.D. thesis, D. H. Hill Library, North Carolina State University, Raleigh, NC. [23]

Gates, C. E., R. E. Comstock, and H. F. Robinson. 1957. Generalized genetic variance and covariance formulae for self-fertilized crops assuming linkage. *Genetics* 42: 749–763. [23]

Gaut, B. S., C. M. Díez, and P. L. Morrell. 2015. Genomics and the contrasting dynamics of annual and perennial domestication. *Trends Genet.* 31: 709–719. [9]

Gautier, M. 2015. Genome-wide scan for adaptive divergence and association with population-specific covariates. *Genetics* 201: 1555–1579. [9]

Gautier, M., T. D. Hocking, and J.-L. Foulley. 2010. A Bayesian outlier criterion to detect SNPs under selection in large data sets. *PLoS One* 8: e11913. [9]

Gavrilets, S., and G. de Jong. 1993. Pleiotropic models of polygenic variation, stabilizing selection, and epistasis. *Genetics* 134: 609–625. [28]

Gavrilets, S., and A. Hastings. 1993. Maintenance of genetic variability under strong stabilizing selection: a two-locus model. *Genetics* 134: 377–386. [5, 28]

Gavrilets, S., and A. Hastings. 1994a. Dynamics of genetic variability in two-locus models of stabilizing selection. *Genetics* 138: 519–532. [5, 7, 24, 28]

Gavrilets, S., and A. Hastings. 1994b. Maintenance of multilocus variability under strong stabilizing selection. *J. Math. Bio.* 32: 287–302. [5, 28]

Gavrilets, S., and A. Hastings. 1994c. A quantitative-genetic model of selection on developmental noise. *Evolution* 48: 1478–1486. [17]

Gavrilets, S., and A. Hastings. 1995. Dynamics of polygenic variability under stabilizing selection, recombination, and drift. *Genet. Res.* 65: 63–74. [24]

Gay, L., M. Siol, and J. Ronfort. 2013. Pedigree-free estimates of heritability in the wild: promising prospects for selfing populations. *PLoS One* 8: e66983. [20]

Geber, M. A., and L. R. Griffen. 2003. Inheritance and natural selection on functional traits. *Int. J. Plant Sci.* 164: S21–S42. [29, 30]

Geiler-Samerotte, K. A., and 5 others. 2013. The details in the distributions: why and how to study phenotypic variability. *Curr. Opin. Biotech.* 24: 752–759. [17]

Geldermann, H., U. Pieper, and W. E. Weber. 1986. Effect of misidentification on the estimation of breeding value and heritability in cattle. *J. Anim. Sci.* 63: 1759–1768. [20]

Gelfand, A. E., and A. F. M. Smith. 1990. Sampling-based approaches to calculating marginal densities. *J. Am. Stat. Assoc.* 85: 398–409. [A3]

Gelman, A., J. B. Carlon, H. S. Stern, and D. B. Rubin. 2013. *Bayesian data analysis*, 3rd edition. Chapman and Hall, New York, NY. [A2]

Gelman, A., and D. B. Rubin. 1992. Inferences from iterative simulation using multiple sequences (with discussion). *Stat. Sci.* 7: 457–511. [A3]

Geman, S. and D. Geman. 1984. Stochastic relaxation, Gibbs distribution and Bayesian restoration of images. *IEE Transactions on Pattern Analysis and Machine Intelligence* 6: 721–741. [19, A3]

Genter, C. F. 1973. Comparison of S_1 and testcross evaluation after two cycles of recurrent selection in maize. *Crop Sci.* 13: 524–527. [23]

Gepts, P., T. C. Osborn, K. Rashka, and A. Bliss. 1986. Phaseolin-protein variability in wild forms and landraces of the common bean (*Phaseolus vulgris*): evidence for multiple centers of domestication. *Econ. Bot.* 40: 451–486. [9]

Gerlach, N. M., J. W. McGlothlin, P. G. Parker, and E. D. Ketterson. 2012. Reinterpreting Bateman gradients: multiple mating and selection in both sexes of a songbird species. *Behav. Ecol.* 23: 1078–1088. [29]

Gerrish, P. J., and R. E. Lenski. 1998. The fate of competing beneficial mutations in an asexual population. *Genetica* 102: 127–144. [27]

Geweke, J. 1992. Evaluating the accuracy of sampling-based approaches to the calculation of posterior moments. *In* J. M. Bernardo, J. O. Berger, A. P. Dawid, and A. F. M. Smith (eds.), *Bayesian Statistics 4*, pp. 169-193. Oxford University Press, Oxford, UK. [A3]

Geyer, C. J. 1992. Practical Markov chain Monte Carlo (with discussion). *Stat. Sci.* 7: 473–511. [19, A3]

Geyer, C. J. 2010. A philosophical look at aster models. Technical Report No. 676, Sch. of Statistics, Univ. of Minnesota, http://purl.umn.edu/57163. [29, 30]

Geyer, C. J. 2011. Introduction to markov chain Monte Carlo. *In* S. Brooks, A. Gelman, G. Jones, and X.-L. Meng (eds), *Handbook of markov chain Monte Carlo*, pp. 2–48. CRC press, Boca Raton, FL. [A3]

Geyer, C. J., C. E. Ridley, R. G. Latta, J. R. Etterson, and R. G. Shaw 2013. Local adaptation and genetic effects on fitness: calculations for exponential family models with random effects. *Ann. Appl. Stat.* 7: 1778–1795. [29]

Geyer, C. J., and R. G. Shaw. 2008. Commentary on Lande-Arnold analysis. Technical Report No. 670, Sch. of Statistics, Univ. of Minnesota, http://purl.umn.edu/56218. [29, 30]

Geyer, C. J., and R. G. Shaw. 2010a. Hypothesis tests and confidence intervals involving fitness landscapes fit by aster models (revised). Technical Report No. 674 (revised), Sch. of Statistics, Univ. of Minnesota, http://purl.umn.edu/56328. [29, 30]

Geyer, C. J., and R. G. Shaw. 2010b. Aster models and Lande-Arnold beta (revised). Technical Report No. 675 (revised), Sch. of Statistics, Univ. of Minnesota, http://purl.umn.edu/56394. [2, 30]

Geyer, C. J., S. Wagenius, and R. G. Shaw. 2007. Aster models for life history analysis. *Biometrika* 94: 415–426. [29, 30]

Gianola, D., and R. L. Fernando. 1986. Bayesian methods in animal breeding theory. *J. Anim. Sci.* 63: 217–246. [19]

Gianola, D., R. L. Fernando, S. Im, and J. L. Foulley. 1989. Likelihood estimation of quantitative genetic parameters when selection occurs: models and problems. *Genome* 31: 768–777. [19]

Gianola, D., J. L. Foulley, and R. L. Fernando. 1986. Prediction of breeding values when variances are not know. *Genet. Sel. Evol.* 18: 485–498. [19]

Gianola, D., S. Im, and R. L. Fernando. 1988. Prediction of breeding values under Henderson's selection model: a revisitation. *J. Dairy Sci.* 71: 2790–2798. [19]

Gianola, D., and G. J. M. Rosa. 2015. One hundred years of statistical developments in animal breeding. *Ann. Rev. Anim. Biosci.* 3: 19–56. [1, 19]

Gibson, G. 2009. Decanalization and the origin of complex disease. *Nat. Rev. Genet.* 10: 134–140. [17]

Gibson, G., and B. Weir. 2005. The quantitative genetics of transcription. *Trends Genet.* 21: 616–623. [12]

Gibson, J. B., and B. P. Bradley. 1974. Stabilising selection in constant and fluctuating environments. *Heredity* 33: 293–302. [16]

Gibson, J. B., N. Lewis, A. Adenca, and S. R. Wilson. 1979. Selection for ethanol tolerance in two populations of *Drosophila melanogaster* segregating alcohol dehydrogenase allozymes. *Aust. J. Biol. Sci.* 32: 387–398. [9]

Gibson, J. B., and J. M. Thoday. 1962. Isolation by disruptive selection. *Nature* 193: 1164 - 1166. [16]

Gibson, J. B., and J. M. Thoday. 1963. Effects of disruptive selection. VIII. Imposed quasi-random mating. *Heredity* 18: 513–524. [16]

Gienapp, P., E. Postma, and M. E. Visser. 2006. Why breeding time has not responded to selection for earlier breeding in a songbird population. *Evolution* 60: 2381–2388. [20]

Gienapp, P., C. Teplitsky, J. S. Alho, J. A. Mills, and J. Merilä. 2008. Climate change and evolution: disentangling environmental and genetic responses. *Mol. Ecol.* 17: 167–178. [20]

Gienapp, P., and 5 others. 2017. Genomic quantitative genetics to study evolution in the wild. *Trends Ecol. Evol.* 32: 897–908. [20]

Gifford, D. R., S. E. Schoustra, and R. Kassen. 2011. The length of adaptive walks is insensitive to starting fitness in *Aspergillus nidulans*. *Evolution* 65: 3070–3078. [27]

Gilad, Y., A. Oshlack, G. K. Smyth, T. P. Speed, and K. P. White. 2006. Expression profiling in primates reveals a rapid evolution of human transcription factors. *Nature* 440: 242–245. [12]

Gilad, Y., S. A. Rifkin, and J. K. Pritchard. 2008. Revealing the architecture of gene regulation: the promise of eQTL studies. *Trends Genet.* 24: 408–415. [12]

Gill, J. L. 1965a. Effects of finite size on selection advance in simulated genetic populations. *Aust. J. Biol. Sci.* 18: 599–617. [7, 26]

Gill, J. L. 1965b. A Monte Carlo evaluation of predicted selection response. *Aust. J. Biol. Sci.* 18: 999–1007. [7, 26]

Gill, J. L. 1965c. Selection and linkage in simulated genetic populations. *Aust. J. Biol. Sci.* 18: 1171–1187. [7, 26]

Gillespie, J. H. 1973. Natural selection with varying selection coefficients—a haploid model. *Genet. Res.* 21: 115–120. [5]

Gillespie, J. H. 1977. Natural selection for variances in offspring number: a new evolutionary principle. *Am. Nat.* 111: 1010–1014. [5]

Gillespie, J. H. 1983. A simple stochastic gene substitution model. *Theor. Popul. Biol.* 23: 202—215. [27]

Gillespie, J. H. 1984a. Pleiotropic overdominance and the maintenance of genetic variation in polygenic characters. *Genetics* 107: 321–300. [24, 28]

Gillespie, J. H. 1984b. Molecular evolution over the mutational landscape. *Evolution* 38: 1116–1129. [7, 27]

Gillespie, J. H. 1994. *The causes of molecular evolution.* Oxford Univ. Press, Oxford, UK. [7, 27, 28]

Gillespie, J. H. 1997. Junk ain't what junk does: neutral alleles in a selected context. *Gene* 205: 291–299. [8]

Gillespie, J. H. 2000. Genetic drift in an infinite population: the pseudohitchhiking model. *Genetics* 155: 909–919. [3, 8]

Gillespie, J. H., and Turelli, M. 1989. Genotype-environment interactions and the maintenance of polygenic variation. *Genetics* 121: 129–138. [28]

Gillespie, J. H., and Turelli, M. 1990. Genotype-environment interactions: a reply. *Genetics* 124: 447. [28]

Gillois, M. 1965. Relation d'identité en génétique I. Postulats et axiomes mendéliens. *Annales de l'institut Henri Poincaré (B) Probabilités et Statistiques* 2: 1–94. [11]

Gilmore, E. C., Jr. 1964. Suggested method of using reciprocal recurrent selection in some naturally self-pollinated species. *Crop Sci.* 4: 323–325. [21]

Gimelfarb, A. 1986a. Additive variation maintained under stabilizing selection: a two-locus model of pleiotropy for two quantitative characters. *Genetics* 112: 717–725. [28]

Gimelfarb, A. 1986b. Multiplicative genotype-environment interaction as a cause of reversed response to directional selection. *Genetics* 114: 333–343. [18]

Gimelfarb, A. 1989. Genotypic variation for a quantitative character maintained under stabilizing selection without mutations: epistasis. *Genetics* 123: 217–227. [28]

Gimelfarb, A. 1990. How much genetic variation can be maintained by genotype-environment interactions? *Genetics* 124: 443–445. [28]

Gimelfarb, A. 1992. Pleiotropy and multilocus polymorphisms. *Genetics* 130: 223–227. [28]

Gimelfarb, A. 1996a. Pleiotropy as a factor maintaining genetic variation in quantitative characters under stabilizing selection. *Genet. Res.* 68: 65–73. [28]

Gimelfarb. A. 1996b. Some additional results about polymorphisms in models of an additive quantitative trait under stabilizing selection. *J. Math. Biol.* 35: 88–96. [5, 28]

Gimelfarb, A., and J. H. Willis. 1994. Linearity versus nonlinearity of offspring-parent regression: An experimental study in *Drosophila melanogaster*. *Genetics* 138: 343–352. [13]

Gimenez, O., and 5 others. 2006. Nonparametric estimation of natural selection on a quantitative trait using mark-recapture data. *Evolution* 60: 460–466. [29]

Gingerich, P. D. 1974. Stratigraphic record of early Eocene *Hyopsodus* and the geometry of mammalian phylogeny. *Nature* 248: 107–109. [12]

Gingerich, P. D. 1979. The stratophenetic approach to phylogeny reconstruction in vertebrate paleontology. *In* J. Cracraft and N. Eldredge (eds.), *Phylogenetic analysis and paleontology*, pp . 41–77. Columbia University Press, New York, NY. [12]

Gingerich, P. D. 1993. Quantification and comparison of evolutionary rates. *Am. J. Sci.* 293: 453–478. [12]

Glass, G. V. 1976. Primary, secondary, and meta-analysis of research. *In* L. Shulman (ed.) *Review of research in education, Vol 5*, pp. 351–379. Peacock, Itasca, IL. [A4]

Glémin, S. 2007. Mating systems and the efficacy of selection at the molecular level. *Genet.* 177: 905–916. [23]

Glémin, S. 2012. Extinction and fixation times with dominance. *Theor. Popul. Biol.* 81: 310–316. [7, 8, 23]

Glinka, S., D. De Lorenzo, and W. Stephan. 2006. Evidence of gene conversion associated with a selective sweep in *Drosophila melanogaster. Mol. Biol. Evol.* 23: 1869–1878. [9]

Glymour, C. 1981. Why I am not a Bayesian, *In* C. Glymour (ed.) *Theory and evidence*, pp. 63–93. University of Chicago Press, Chicago, IL. [19, A2]

Gnanadesikan, R. 1977. *Methods for statistical data analysis of multivariate observations.* John Wiley and Sons, New York, NY. [A5]

Goddard, M. E. 2001. The validity of genetic models underlying quantitative traits. *Live. Prod. Sci.* 72: 117–127. [24]

Goddard, M. E., N. R. Wray, K. Verbyla, and P. M. Visscher. 2009. Estimating effects and making predictions from genome-wide marker data. *Stat. Sci.* 24: 517–529. [25]

Gojobori, J., H. Tang, J. M. Akey, and C.-I. Wu. 2007. Adaptive evolution in humans revealed by the negative correlation between the polymorphism and fixation phases of evolution. *Proc. Natl. Acad. Sci. USA* 104: 3907–3912. [10]

Gokhale, C. S., Y. Iwasa, M. A. Nowak, and A. Traulsen. 2009. The pace of evolution across fitness valleys. *J. Theor. Biol.* 259: 613–620. [7]

Golding, G. B. 1984. The sampling distribution of linkage disequilibrium. *Genetics* 108: 257–274. [4]

Golding, G. B. 1997. The effect of purifying selection on genealogies. *In* P. Donnelly and S. Tavare (eds.), *Progress in population genetics and human evolution*, pp. 271–285. Springer-Verlag, New York, NY. [8]

Golding, G. B., and C. Strobeck. 1980. Linkage disequilibrium in a finite population that is partially selfing. *Genetics* 94: 777–789. [2]

Goldman, I. L. 2004. The intellectual legacy of the Illinois long-term selection experiment. *Plant Breeding Reviews* 24: 61–78. [25]

Goldman, N., and Z. Yang. 1994. A codon-based model of nucleotide substitution for protein-coding DNA sequences. *Mol. Biol. Evol.* 11: 725–736. [10]

Goldringer, I., and T. Bataillon. 2004. On the distribution of temporal variations in allele frequency: consequences for the estimation of effective population size and the detection of loci undergoing selection. *Genetics* 168: 563–568. [4, 9]

Goldringer, I., P. Brabant, and A. Gallais. 1996. Theoretical comparison of recurrent selection methods for the improvement of self-pollinated crops. *Crop Sci.* 36: 1171–1180. [11, 23]

Goldschmidt, R. 1940. *The material basis of evolution.* Yale University Press, New Haven, CT. [1, 25]

Gomez-Raya, L., and E. B. Burnside. 1990. The effect of repeated cycles of selection on genetic variance, heritability, and response. *Theor. Appl. Genet.* 79: 568–574. [16]

Gompert, Z. 2016. Bayesian inference of selection in a heterogeneous environment from genetic time-series data. *Mol. Ecol.* 25: 121–134. [9]

Gomulkiewicz, R., R. D. Holt, M. Barfield, and S.L. Nuismer. 2010. Genetics, adaptation, and invasion in harsh environments. *Evol. Appl.* 3: 97–108. [25]

González, J., K. Lenkov, J. M. Macpherson, and D. A. Petrov. 2008. High rate of recent transposable element-induced adaptation in *Drosophila melanogaster. PLoS Bio.* 6: e251. [8]

Good, B. H., and M. M. Desai. 2014. Deleterious passengers in adapting populations. *Genetics* 198: 1183–1208. [8]

Good, B. H., I. M. Rouzine, D. J. Balick, O. Hallatschek, and M. M. Desai. 2012. Distribution of fixed beneficial mutations and the rate of adaptation in asexual populations. *Proc. Natl. Acad. Sci. USA* 109: 4950–4955. [27]

Good, I. J., and Y. Mittal. 1987. The amalgamation and geometry of two-by-two contingency tables. *Ann. Stat.* 15: 694–711. [10]

Goodale, H. D. 1938. A study of the inheritance of body weight in the albino mouse by selection. *J. Heredity* 29: 101–112. [25]

Goodman, S. J. 1997. R_{st} Calc: a collection of computer programs for calculating estimates of genetic differentiation from microsatellite data and determining their significance. *Mol. Ecol.* 6: 881–885. [2, 9, 12]

Goodnight, C. J. 1985. The influence of environmental variation on group and individual selection in a cress. *Evolution* 39: 545–558. [22]

Goodnight, C. J. 1988. Epistasis and the effect of founder events on the additive genetic variance. *Evolution* 42: 441–454. [11]

Goodnight, C. 2004. Gene interaction and selection. *Plant Breeding Reviews* 24: 269–292. [25]

Goodnight, C. J. 2005. Multilevel selection: the evolution of cooperation in non-kin groups. *Popul. Ecol.* 47: 3–12. [22]

Goodnight, C. 2013. On multilevel selection and kin selection: contextual analysis meets direct fitness. *Evolution* 67: 1539–1548. [22, 30]

Goodnight, C. J. 2015. Multilevel selection theory and evidence: a critique of Gardner, 2015. *J. Evol. Biol.* 28: 1734–1746. [22, 30]

Goodnight, C. J., J. M. Schwartz, and L. Stevens. 1992. Contextual analysis of models of group selection, soft selection, hard selection, and the evolution of altruism. *Am. Nat.* 140: 743–761. [30]

Goodnight, C. J., and L. Stevens. 1997. Experimental studies of group selection: what do they tell us about group selection in nature? *Am. Nat.* 150: S59–S79. [22]

Goodnight, C. J., and M. J. Wade. 2000. The ongoing synthesis: a reply to Coyne, Barton, and Turelli. *Evolution* 54: 317–324. [26]

Goodwillie, C., S. Kalisz, and C. G. Eckert. 2005. The evolutionary enigma of mixed mating systems in plants: occurrence, theoretical explanations, and empirical evidence. *Annu. Rev. Ecol. Evol. Syst.* 36: 47–79. [23]

Gordo, I., A. Navarro, and B. Charlesworth. 2002. Muller's ratchet and the pattern of variation at a neutral locus. *Genetics* 161: 835–848. [8]

Göring, H. H., J. D. Terwilliger, and J. Blangero. 2001. Large upward bias in estimation of locus-specific effects from genomewide scans. *Amer. J. Hum. Genet.* 69: 1357–1369. [25]

Gosler, A. G., and D. G. C. Harper. 2000. Assessing the heritability of body condition in birds: a challenge exemplified by the great tit *Parus major*. L. (Aves). *Biol. J. Linn. Soc.* 71: 103–117. [20]

Goss, P. J. E., and R. C. Lewontin. 1996. Detecting heterogeneity of substitution along DNA and protein sequences. *Genetics* 143: 589–602. [10]

Gossmann, T. I., D. Waxman, and A. Eyre-Walker.

2014. Fluctuating selection models and McDonald-Kreitman type analyses. *PLoS One* 9: e84540. [10]

Gossmann, T. I., and 7 others. 2010. Genome wide analysis reveal little evidence for adaptive evolution in many plant species. *Mol. Biol. Evol.* 27: 1822–1832. [10]

Gottlieb, L. D. 1984. Genetics and morphological evolution in plants. *Am. Nat.* 123: 681–709. [25]

Gottlieb, L. D. 1985. Reply to Coyne and Lande. *Am. Nat.* 126: 146–150. [25]

Goudet, J., and L. Büchi. 2006. The effects of dominance, regular inbreeding and sampling design on Q_{ST}, an estimator of population differentiation for quantitative traits. *Genetics* 172: 1337-1347. [12]

Goudet, J., and G. Martin. 2007. Under neutrality, $Q_{ST} \leq F_{ST}$ when there is dominance in an island model. *Genetics* 176: 1371–1374. [12]

Goulas, C. K., and J. H. Lonnquist. 1976. Combined half-sib and S_1 family selection in a maize composite population. *Crop Sci.* 16: 461–464. [23]

Gould, F. 1998. Sustainability of transgenic insecticidal cultivars: integrating pest genetics and ecology. *Ann. Rev. Entom.* 43: 701–726. [25]

Gould, S. J., and N. Eldredge. 1977. Punctuated equilibria: the tempo and mode of evolution reconsidered. *Paleobiology* 3: 115–151. [12]

Gould, S. J., and R. C. Lewontin. 1979. The spandrels of San Marco and the Panglossian paradigm: a critique of the adaptationist programme. *Proc. Roy. Soc. Lond. B* 205: 581–598. [10]

Gowaty, P. A., Y.-K. Kim, and W. W. Anderson. 2012. No evidence of sexual selection in a repetition of Bateman's classic study of *Drosophila melanogaster*. *Proc. Natl. Acad. Sci. USA* 109: 11740–11745. [29]

Gowe, R. S. 1983. Lessons from selection studies in poultry for animal breeders. *Proc 32nd Annual Breed. Roundtable*, pp. 22-50. [25]

Gowe, R. S., R. W. Fairfull, I. McMillan, and G. S. Schmidt. 1993. A strategy for maintaining high fertility and hatchability in a multiple-trait egg strock selection program. *Poultry Sci.* 72: 1433–1448. [25]

Gowe, R. S., A. Robertson, and B. D. H. Latter. 1959. Environment and poultry breeding problems. 5. The design of poultry control strains. *Poultry Sci.* 38: 462–471. [3]

Grafen, A. 1988. On the uses of data on lifetime reproductive success. *In* T. H. Clutton-Brock (ed.), *Reproductive success: studies of individual variation in contrasting breeding systems*, pp. 454-471. University of Chicago Press, Chicago, IL. [29]

Grafen, A. 2006. Optimization of inclusive fitness. *J. Theor. Biol.* 238: 541–563. [22]

Graham, A. 1981. *Kronecker products and matrix calculus with applications*. Halsted Press, New York, NY. [A6]

Graham, J. H., S. Raz, H. Hel-Or, and E. Nevo. 2010. Fluctuating asymmetry: method, theory and applications. *Symmetry* 2: 466–540. [17]

Grant, B., and L. E. Mettler. 1969. Disruptive and stabilizing selection on the "escape" behavior of *Drosophila melanogaster*. *Genetics* 62: 625–637. [16]

Grant, P. R. 1972. Centripetal selection and the house sparrow. *Syst. Zool.* 21: 23–30. [30]

Grant, P. R. 1990. Reproductive fitness in birds. *Trends Ecol. Evol.* 5: 379–380. [3]

Grant, P. R., and B. R. Grant. 1995. Predicting microevolutionary responses to directional selection on heritable variation. *Evolution* 49: 241–251. [20]

Grant, P. R., and B. R. Grant. 2002. Unpredictable evolution in a 30-year study of Darwin's finches. *Science* 296: 707–711. [20]

Grant, P. R., and B. R. Grant. 2006. Evolution of character displacement in Darwin's finches. *Science* 313: 224–226. [20]

Graser, H.-U., S. P. Smith, and B. Tier. 1987. A derivative-free approach for estimating variance components in animal models by restricted maximum likelihood. *J. Anim. Sci.* 64: 1362–1370. [19]

Grassini, P., K. M. Eskridge, and K. G. Cassman. 2013. Distinguishing between yield advances and yield plateaus in historical crop production trends. *Nature Commun.* 4: 2918. [25]

Graur, D., and W.-H. Li. 1991. Neutral mutation hypothesis test. *Nature* 354: 114-115. [10]

Graur, D., and W.-H. Li. 2000. *Fundamentals of molecular evolution*. Sinauer, Sunderland, MA [10]

Graur, D., and 5 others. 2013. On the immortality of television sets: "function" in the human genome according to the evolution-free gospel of ENCODE. *Genome Bio. Evol.* 5: 578–590. [8]

Greaves, J. H., R. Refern, P. B. Ayres and J. E. Gill. 1977. Warfin resistance: a balanced polymorphism in the Norway rat. *Genet. Res.* 30: 257–263. [25]

Green, A. J. 2001. Mass/length residuals: measures of body condition or generators of spurious results? *Ecology* 82: 1473–1483. [20]

Green, R. E., and 55 others. 2010. A draft sequence of the Neandertal genome. *Science* 328: 710–722. [9]

Green, R. H. 1966. Measurement of non-randomness in spatial distributions. *Res. Pop. Ecol.* 8: 1–7. [29]

Greenland, S. 1982. Interpretation and estimation of summary ratios under heterogeneity. *Stat. Med.* 1: 217–227. [10]

Grether, G. F. 1996. Sexual selection and survival selection on wing coloration and body size in the rubyspot damselfly *Hetaerina americana*. *Evolution* 50: 1939–1948. [29]

Griffing, B. 1960a. Theoretical consequences of truncation selection based on the individual phenotype. *Aust. J. Biol. Sci.* 13: 307–343. [5, 15, 16]

Griffing, B. 1960b. Accommodation of linkage in mass selection theory. *Aust. J. Biol. Sci.* 13: 501–526. [15, 16]

Griffing, B. 1967. Selection in reference to biological groups. I. Individuals and group selection applied to populations of unordered groups. *Aust. J. Biol. Sci.* 20: 127–139. [22]

Griffing, B. 1968a. Selection in reference to biological groups. II. Consequences of selection in groups of one size when evaluated in groups of a different size. *Aust. J. Biol. Sci.* 21: 1163–1170. [22]

Griffing, B. 1968b. Selection in reference to biological groups. III. Generalized results of individuals and group selection in terms of parent-offspring covariances. *Aust. J. Biol. Sci.* 21: 1171–1178. [22]

Griffing, B. 1969. Selection in reference to biological groups. IV. Application of selection index theory. *Aust. J. Biol. Sci.* 22: 131–142. [22]

Griffing, B. 1976a. Selection in reference to biological groups. V. Analysis of full-sib groups. *Genetics* 82: 703–722. [22]

Griffing, B. 1976b. Selection in reference to biological groups. VI. Use of extreme forms of non-random groups to increase selection efficiency. *Genetics* 82: 723–713. [22]

Griffing, B. 1977. Selection for populations of interaction genotypes. *In* E. Pollak, O. Kempthorne, and T. B. Bailey, Jr., (eds.), *Proceedings of the international conference on quantitative genetics*, pp. 413–434. Iowa State Univ. Press, Ames, IA. [22]

Griffiths, R. C. 1982. The number of alleles and segregating sites in a sample from the infinite-alleles model. *Adv. Appl. Prob.* 14: 225–239. [9]

Griffiths, R. C. 2003. The frequency spectrum of a mutation, and its age, in a general diffusion model. *Theor. Popul. Biol.* 64: 241–251. [9]

Griffiths, R. C., and S. Tavare. 1994a. Simulating probability distributions in the coalescent. *Theor. Popul. Biol.* 46: 131–159. [9]

Griffiths, R. C., and S. Tavare. 1994b. Sampling theory for neutral alleles in varying environments. *Phil. Trans. Roy. Soc. Lond. B* 403–410. [9]

Griffiths, R. C., and S. Tavare. 1998. The age of a mutation in a general coalescent tree. *Stoch. Models* 14:273—95. [9]

Grillo, M. A., and 6 others. 2009. Genetic architecture for the adaptive origin of annual wild rice, *Oryza nivara*. *Evolution* 63: 870–883. [27]

Griswold, C. K., and M. C. Whitlock. 2003. The genetics of adaptation: the roles of pleiotropy, stabilizing selection and drift in shaping the distribution of bidirectional fixed mutational effects. *Genetics* 165: 2181–2192. [12]

Groeters, F. R., and B. E. Tabashnik. 2000. Roles of selection intensity, major genes, and minor genes in evolution of insecticide resistance. *J. Econ. Ent.* 93: 1580–1587. [25]

Gronau, I., L. Arbiza, J. Mohammed, and A. Siepel. 2013. Inference of natural selection from interspersed genomic elements based on polymorphism and divergence. *Mol. Biol. Evol.* 30: 1159–1171. [10]

Gross, B. L., and K. M. Olsen. 2010. Genetic perspectives on crop domestication. *Trends Plant Sci.* 15: 529–537. [9]

Grossenbacher, D., R. B. Runquist, E. E. Goldberg, and Y. Brandvain. 2015. Geographic range size is predicted by plant mating system. *Ecol. Let.* 18: 706–713.

Grossman, S. R., and 12 others 2010. A composite of multiple signals distinguishes causal variants in regions of positive selection. *Science* 327: 883–886. [9]

Grossman, S. R., and 19 others. 2013. Identifying recent adaptations in large-scale genomic data. *Cell* 152: 703–713. [9]

Grundy, B., A. Caballero, E. Santiago, and W. G. Hill. 1994. A note on using biased parameter values and non-random mating to reduce rates of inbreeding in selection programs. *Anim. Prod.* 59: 465–468. [23, 26]

Grundy, B., B. Villanueva, and J. A. Woolliams. 1998. Dynamic selection for constrained inbreeding and their consequences for pedigree development. *Genet. Res.* 72: 159–168. [23]

Grundy, B., B. Villanueva, and J. A. Woolliams. 2000. Dynamic selection for maximizing response with constrained inbreeding in schemes with overlapping generations. *Anim. Sci.* 70: 373–382. [23]

Gu, X. 2004. Statistical framework for phylogenomic analysis of gene family expression profiles. *Genetics* 167: 531–542. [12]

Guedj, M., S. Robin, A. Celisse, and G. Nuel. 2009. Kerfdr: a semi-parametric kernel-based approach to local false discovery rate estimation. *BMC Bioinformatics* 10: 84. [A4]

Guinand, B., C. Lemaire, and F. Bonhomme. 2004. How to detect polymorphisms undergoing selection in marine fishes? A review of methods and case studies, including fishes. *J. Sea Res.* 51: 167–182. [9]

Guindon, S., M. Black, and A. Rodrigo. 2006. Control of the false discovery rate applied to the detection of positively selected amino acid sites. *Mol. Biol. Evol.* 23: 919–926. [10]

Gumbel, E. J. 1958. *Statistics of extremes.* Columbia Univ. Press, New York, NY. [27]

Günther, T., and G. Coop. 2013. Robust identification of local adaptation from allele frequencies. *Genetics* 195: 205–220. [9, 12]

Guo, S. W. 1996. Variation in genetic identity among relatives. *Hum. Hered.* 46: 61–70. [20]

Gurevitch, J., and L. V. Hedges. 1999. Statistical issues in ecological meta-analyses. *Ecology* 80: 1142–1149. [A4]

Gustafsson, L. 1986. Lifetime reproductive success and heritability: empirical support for Fisher's fundamental theorem. *Am. Nat.* 128: 761–764. [6]

Gutierrez, J. P., B. Neito, P. Piqueras, N. Ibáñez, and C. Salgado. 2006. Genetic parameters for canalisation analysis of litter size and litter weight traits at birth in mice. *Genet. Sel. Evol.* 38: 445–462. [17]

Guttman, D. S., and D. E. Dykhuizen. 1994. Detecting selective sweeps in naturally occurring *Escherichia coli*. *Genetics* 138: 993–1003. [8]

Haag, C. R., and D. Roze. 2007. Genetic load in sexual and asexual diploids: segregation, dominance and genetic drift. *Genetics* 176: 1663–1678. [7]

Haasl, R. J., and B. A. Payseur. 2016. Fifteen years of genomewide scans for selection: trends, lessons and unaddressed genetic sources of complication. *Mol. Ecol.* 25: 5–23. [9]

Haddrill, P. R., D. Bachtrog, and P. Andolfatto. 2008. Positive and negative selection on noncoding DNA in *Drosophila simulans*. *Mol. Bio. Evol.* 25: 1825–1834. [10]

Haddrill, P. R., D. L. Halligan, D. Tomaras, and B. Charlesworth. 2007. Reduced efficacy of selection in regions of the *Drosophila* genome that lack crossing over. *Genome Biol.* 8: R18. [8]

Haddrill, P. R., L. Loewe, and B. Charlesworth. 2010. Estimating the parameters of selection on nonsynonymous mutations in *Drosophila pseudoobscura* and *D. miranda*. *Genetics* 185: 1381–1396. [10]

Hadfield, J. D. 2008. Estimating evolutionary parameters when viability selection is operating. *Proc. Roy. Soc. Lond. B.* 275: 723–734. [20, 29]

Hadfield, J. D. 2010. MCMC methods for multi-response generalized linear mixed models: the MCMCglmm R package. *J. Stat. Software* 33: 1–22. [29]

Hadfield, J. D., D. S. Richardson, and T. Burke. 2006. Towards unbiased parentage assignment: combining genetic, behavioural and spatial data in a Bayesian framework. *Mol. Ecol.* 15: 3715–3730. [20]

Hadfield, J. D., and S. Nakagawa. 2010. General quantitative genetic methods for comparative biology: phylogenies, taxonomies and multitrait models for continuous and categorical characters. *J. Evol. Biol.* 23: 494–508. [A4]

Hadfield, J. D., and C. E. Thomson. 2017. Interpreting selection when individuals interact. *Methods Ecol.*

Evol. 8: 688–699. [29]

Hadfield, J. D., and A. J. Wilson. 2007. Multilevel selection 3: Modelling the effects of interacting individuals as a function of group size. *Genetics* 177: 667–668. [22]

Hadfield, J. D., A. J. Wilson, D. Garant, B. C. Sheldo, and L. E. B. Kruuk. 2010. The misuse of BLUP in ecology and evolution. *Am. Nat.* 175: 116–125. [20]

Hadfield, J. D., A. J. Wilson, and L. E. B. Kruuk. 2011. Cryptic evolution: does environmental deterioration have a genetic basis?. *Genetics* 187: 1099–1113. [20]

Hagenblad, J., and M. Nordborg 2002. Sequence variation and haplotype structure surrounding the flowering time locus *FRI* in *Arabidopsis thaliana*. *Genetics* 161: 289–298. [4]

Hagg, W. L., and R. R. Hill, Jr. 1974. Comparison of selection methods for autotetraploids. II. Selection for disease resistance in alfalfa. *Crop Sci.* 14: 591–593. [15]

Hahn, M. W. 2008. Toward a selection theory of molecular evolution. *Evolution* 62: 255–265. [1, 8]

Haldane, J. B. S. 1919. The combination of linkage values, and the calculation of distance between the loci of linked factors. *J. Genetics* 8: 299–309. [4]

Haldane, J. B. S. 1927. A mathematical theory of natural and artificial selection. Part V. Selection and mutation. *Proc. Cambridge Philos. Soc.* 23: 838–844. [7, A1]

Haldane, J. B. S. 1931. A mathematical theory of natural and artificial selection. Part VII. Selection intensity as a function of mortality rate. *Proc. Camb. Philos. Soc.* 27: 131–136. [5, 18]

Haldane, J. B. S. 1932a. *The causes of evolution.* Longmans, London, UK. [22]

Haldane, J. B. S. 1932b. A mathematical theory of natural and artifical selection. Part IX. Rapid selection. *Proc. Camb. Philos. Soc.* 28: 244–248. [5]

Haldane, J. B. S. 1937a. Some theoretical results of continued brother-sister mating. *J. Genet.* 34: 265–274. [3]

Haldane, J. B. S. 1937b. The effect of variation on fitness. *Am. Nat.* 71: 337–349. [7]

Haldane, J. B. S. 1949. Suggestions as to quantitative measurement of rates of evolution. *Evolution* 3: 51–56. [12]

Haldane, J. B. S. 1954. The measurement of natural selection. *Proc. IX Internal. Cong. Genet.* 1: 480–487. [13, 16, 20, 28, 29]

Haldane, J. B. S. 1955. Population genetics. *Penguin New. Bio.* 18: 34–51. [22]

Haldane, J. B. S. 1956. The estimation and significance of the logarithm of a ratio of frequencies. *Ann. Hum. Genet.* 20: 309–311. [10]

Haldane, J. B. S. 1964. A defense of beanbag genetics. *Perspect. Biol. Med.* 7: 343–359. [1]

Haldane, J. B. S., and S. D. Jayakar. 1963. Polymorphisms due to selection of varying direction. *Heredity.* 58: 237–242. [5, 28]

Hall, D. W., R. Mahmoudizad, A. W. Hurd, and S. B. Joseph. 2008. Spontaneous mutations in diploid *Saccharomyces cerevisiae:* another thousand cell generations. *Genet. Res.* 90: 229–241. [12]

Hallatschek, O. 2011. The noisy edge of traveling waves. *Proc. Natl. Acad. Sci. USA* 108: 1783–1787. [9, 37]

Hallatschek, O., P. Hersen, S. Ramanathan, and D. R. Nelson. 2007. Genetic drift at expanding frontiers promotes gene segregation. *Proc. Natl. Acad. Sci. USA* 104: 19926–19930. [9]

Hallatschek, O., and D. R. Nelson. 2008. Gene surfing in expanding populations. *Theor. Popul. Biol.* 73: 158–170. [9]

Hallatschek, O., and D. R. Nelson. 2009. Population genetics and range expansion. *Physics Today* 62: 42–48. [9]

Hallauer, A. R. 1981. Selection and breeding methods. *In* K. J. Frey (ed.), *Plant breeding II*, pp. 3–55. Springer-Verlag, New York, NY. [21]

Hallauer, A.R. 1985. Compendium of recurrent selection methods and their application. *Critical Reviews Plant Sci.* 3: 1–33. [21]

Hallauer, A. R., M. J. Carena, and J. B. Miranda Filho. 2010. *Quantitative genetics in maize breeding*, 3rd Ed. Springer-Verlag, New York, NY. [13, 21]

Hallauer, A. R. and J. B. Miranda. 1981. *Quantitative genetics in maize breeding*. Iowa State University Press, Ames, IA. [21, 23]

Hallauer, A. R., W. A. Russell, and K. R. Lamkey. 1988. Corn breeding. *In* G. F. Sprague and J. W. Dudley (eds.), *Corn and corn improvement, 3rd ed*, pp. 463–564. Springer-Verlag, New York, NY. [21]

Haller, B. C., and A. P. Hendry. 2013. Solving the paradox of stasis: squashed stabilizing selection and the limits of detection. *Evolution* 68: 483–500. [27]

Haller, B. C., and A. P. Hendry. 2013. Solving the paradox of stasis: squashed stabilizing selection and the limits of detection. *Evolution* 68: 483–500. [28]

Halliburton, R., and G. A. E. Gall. 1981. Disruptive selection and assortative mating in *Tribolium castaneum*. *Evolution* 35: 829–843. [16]

Halligan, D. L., and P. D. Keightley. 2006. Ubiquitous selective constraints in the *Drosophila* genome revealed by a genome-wide interspecies comparison. *Genome Res.* 16: 875–884. [8, 10]

Halligan, D. L., and P. Keightley. 2009. Spontaneous mutation accumulation studies in evolutionary genetics. *Ann. Rev. Ecol. Evol. Syst.* 40: 151—172. [28]

Halligan, D. L., F. Oliver, A. Eyer-Walker, B. Harr, and P. D. Keightley. 2010. Evidence for pervasive adaptive protein evolution in wild mice. *PLoS Genet.* 6: e1000825. [10]

Halpern, A. L., and W. J. Bruno. 1998. Evolutionary distances for protein-coding sequences: modeling site-specific residue frequencies. *Mol. Biol. Evol.* 15: 910–917. [10]

Hamblin, M. T., and 6 others. 2006. Challenges of detecting directional selection after a bottleneck: Lessons from *Sorhum bicolor*. *Genetics* 173: 953–964. [9]

Hamblin, M. T., and 5 others. 2005. Equilibrium processes cannot explain high levels of short- and medium-range linkage disequilibrium in the domesticated grass *Sorghum bicolor. Genetics* 171: 1247–1256. [4]

Hamilton, W. D. 1963. The evolution of altruistic behavior. *Am. Nat.* 97: 354–356. [22]

Hamilton, W. D. 1964a. The genetical evolution of social behavior. I. *J. Theor. Biol.* 7: 1–16. [22]

Hamilton, W. D. 1964b. The genetical evolution of social behavior. II. *J. Theor. Biol.* 7: 17–52. [22]

Hamilton, W. D. 1970. Selfish and spiteful behavious in an evolutionary model. *Nature* 228: 1218–1220. [22]

Hammersley, J. M., and D. C. Handscomb. 1964. *Monte Carlo methods* Springer, Dordrecht, NL. [A3]

Hammond, J. J., and C. O. Gardner. 1974. Modification of the variety cross diallel model for evaluating cycles of selection. *Crop Sci.* 14: 6–8. [18]

Hammond, K. 1973. Population size, selection response and variation in quantitative characters. Ph. D. Thesis, University of Sydney, AU. [26]

Hamon, T. R. 2005. Measurement of concurrent selection episodes. *Evolution* 59: 1096–1103. [29]

Hanchard, N. A., and 9 others. 2006. Screening for recently selected alleles by analysis of human haplotype similarity. *Am. J. Hum. Genet.* 78: 153–159. [9]

Hancock, A. M., G. Alkorta-Aranburu, D. B. Witonsky, and A. Di Rienzo. 2010b. Adaptations to new environments in humans: the role of subtle allele frequency shifts. *Phil. Trans. Roy. Soc. Lond. B* 365: 2459–2468. [8, 9, 12]

Hancock, A. M., and 10 others. 2010a. Human adaptations to diet, subsistence, and ecoregion are due to subtle shifts in allele frequency. *Proc. Natl. Acad. Sci. USA* 107: 8924–8930. [8, 9, 12]

Hancock, A. M., and 6 others. 2008. Adaptations to climate in candidate genes for common metabolic disorders. *PLoS Genet.* 4: e32. [9]

Hancock, A. M., and 9 others. 2011. Adaptations to climate-mediated selective pressures in humans. *PLoS Genet.* 7: e1001375. [12]

Hanrahan, J. P., E. J. Eisen, and J. E. Legates. 1973. Effects of population size and selection intensity on short-term response to selection for postweaning gain in mice. *Genetics* 73: 513–530. [18, 26]

Hansen, T. F. 1997. Stabilizing selection and the comparative analysis of adaptation. *Evolution* 51: 1341–1351. [12]

Hansen, T. F. 2006. The evolution of genetic architecture. *Ann. Rev. Eco. Evol. Syst.* 37: 123–157. [17]

Hansen, T. F. 2013. Why epistasis is important for selection and adaptation. *Evolution* 67: 3501–3511. [25]

Hansen, T. F., A. J. Carter, and C. Pélabon. 2006. On adaptive accuracy and precision in natural populations. *Am. Nat.* 168: 168–181. [17]

Hansen, T. F., and E. P. Martins. 1996. Translating between microevolutionary process and macroevolutionary patterns: the correlation structure of interspecific data. *Evolution* 50: 1404–1417. [12]

Hansen, T. F., C. Pélabon, W. S. Armbruster, and M. L. Carlson. 2003. Evolvability and genetic constraint in *Dalechampia* blossoms: components of variance and measures of evolvability. *J. Evol. Biol.* 16: 754–766. [13, 29]

Hansen, T. F., C. Pélabon, and D. Houle. 2011. Heritability is not evolvability. *Evol. Biol.* 38: 258–277. [6, 13, 29]

Hanson, M. A., and 6 others. 1996. Evolution of anthocyanin biosynthesis in maize kernels: The role of regulator and enzymatic loci. *Genetics* 143: 1395–1407. [9]

Hanson, W. D. 1963. Heritability. *In* W. D. Hanson and H. F. Robinson (eds.), *Statistical genetics and plant breeding*, pp. 125–140. Natl. Acad. Sci., Natl. Res. Council Publ. No. 982, Washington, DC. [21]

Harding, R. M., and 10 others. 2000. Evidence for variable selective pressures at MC1R. *Am. J. Hum. Genet.* 66: 1351–1361. [10]

Harman, O. 2011. *The price of altruism: George Price and the search for the origins of kindness.* Norton, New York, NY. [6]

Harper, J. L. 1977. *Population biology in plants.* Academic Press, New York, NY. [30]

Harr, B., M. Kauer, and C. Schlötterer. 2002. Hitchhiking mapping: a population-based fine mapping strategy for adaptive mutations in *Drosophila melanogaster*. *Proc. Natl. Acad. Sci. USA* 99: 12949–12954. [9]

Harris, D. L. 1964. Genotypic covariances between inbred relatives. *Genetics* 50: 1319–1348. [11]

Harris, D. L. 1982. Long-term response to selection I. Relation to breeding population size, intensity, and accuracy with additive gene action. *Genetics* 100: 511–532. [25]

Harris, E. E., and D . Meyer. 2006. The molecular signature of selection underlying human adaptations. *Yearbook of Phy. Antro.* 49: 89–130. [9]

Harris, J. A. 1911. A neglected paper on natural selection in the English sparrow. *Am. Nat.* 45: 314–318. [30]

Harris, K., and R. Nielsen. 2014. Error-prone polymerase activity causes multinucleotide mutations in humans. *Genome Res.* 24: 1445–1454. [4]

Harrison, F. 2011. Getting started with meta-analysis. *Methods Ecol. Evol.* 2: 1–10. [30, A4]

Harter, H. L. 1961. Expected values of normal order statistics. *Biometrika* 48: 151–166. [14]

Harter, H L. 1970a. *Order statistics and their use in testing and estimation. Volume 1: Tests based on range and studentized range of samples from a normal population.* U. S. Gov. Printing Off., Washington, DC. [14]

Harter, H. L. 1970b. *Order statistics and their use in testing and estimation. Volume 2: Estimates based on order statistics of samples from various populations.* U. S. Gov. Printing Off., Washington, DC. [14]

Hartfield, M., T Bataillon, S. Glémin. 2017. The evolutionary interplay between adaptation and self-fertilization. *Trends Genet.* 33: 420–431. [8, 23]

Hartfield, M., and S. P. Otto. 2011. Recombination and hitchhiking of deleterious alleles. *Evolution* 65: 2421–2434. [8]

Hartfield, M., S. P. Otto, and P. D. Keightley. 2010. The role of advantageous mutations in enhancing the evolution of a recombination modifier. *Genetics* 184: 1153–1164. [4]

Hartl, D. L., and R. B. Campbell. 1982. Allelic multiplicity of simple Mendelian disorders. *Am. J. Hum. Genet.* 34: 866–873. [7]

Hartl, D. L., D. E. Dykhuizen, and A. M. Dean. 1995. Limits of adaptation: the evolution of selective neutrality. *Genetics* 111: 655-674. [7]

Hartl, D. L., E. N. Moriyama, and S. A. Sawyer. 1994. Selection intensity for codon bias. *Genetics* 138: 227–234. [10]

Hartl, D. L., and C. H. Taubes. 1996. Compensatory nearly neutral mutations: selection without adaptation. *J. Theor. Biol.* 182: 303–309. [27]

Hartl, D. L., and C. H. Taubes. 1998. Towards a theory of evolutionary adaptation. *Genetica* 102/103: 525–533. [7, 27]

Harville, D. A. 1977. Maximum likelihood approaches to variance component estimation and to related problems. *J. Am. Stat. Assoc.* 72: 320–338. [19, 22]

Harville, D. A. 1990. BLUP (Best Linear Unbiased Prediction) and beyond. *In* D. Gianola and K. Hammond (eds.), *Advances in statistical methods for genetic improvement of livestock*, pp. 239–276. Springer-Verlag, New York, NY. [19]

Hasegawa, M., H. Kishino, and T. Yano. 1985 Dating of the human-ape splitting by a molecular clock of mitochondrial DNA. *J. Mol. Evol.* 22: 160–174. [10]

Hastings, A. 1981a. Stable cycling in discrete time genetic models. *Proc. Natl. Acad. Sci. USA* 78: 7224–7225. [5]

Hastings, A. 1981b. Disequilibrium, selection and recombination: limits in two-locus two-allele models. *Genetics* 98: 659–668. [5]

Hastings, A. 1985. Four simultaneously stable polymorphic equilibria in two-locus two-allele models. *Genetics* 109: 255–261. [5]

Hastings, A. 1986. Limits to the relationship among recombination, disequilibrium, and epistasis in two locus models. *Genetics* 113: 177–185. [5]

Hastings, A. 1987a. Monotonic change of the mean phenotype in two-locus models. *Genetics* 117: 583–585. [5]

Hastings, A. 1987b. Substitution rates under stabilizing selection. *Genetics* 116: 479–486. [28]

Hastings, A. 1988. Dependence of expected heterozygosity on locus number with stabilizing selection and drift. *J. Theor. Biol.* 134: 103–112. [28]

Hastings, A. 1989. Linkage disequilibrium and genetic variances under mutation-selection balance. *Genetics* 121: 856–860. [28]

Hastings, A. 1990a. Second-order approximations for selection coefficients at polygenic loci. *J. Math. Biol.* 28: 475–483. [5]

Hastings, A. 1990b. Determinstic multilocus population genetics: an overview. *In* A. Hastings (ed.), *Some mathematical questions in biology: Models in population biology*, pp. 27–54. American Mathematical Society, Providence, RI. [5]

Hastings, A. 1990c. The interaction between selection and linkage in plant populations. *In* M. J. Clegg, A. L. Kahlem, and B. S. Weir (eds.), *Plant population genetics, breeding and genetic resources*, pp. 163–180. Sinauer Associates, Sunderland, MA. [5]

Hastings, A. 1990d. Maintenance of polygenic variation through mutation-selection balance: bifurcation analysis of a biallelic model. *J. Math. Biol.* 28: 329–340. [28]

Hastings, A. 1992. Second-order approximations for selection coefficients at polygenic loci. II. Pleiotrophy. *J. Math. Biol.* 30: 379–388. [5]

Hastings, A., and C. L. Hom. 1989. Pleiotropic stabilizing selection limits the number of polymorphic loci to at most the number of characters. *Genetics* 122: 459–463. [5, 28]

Hastings, A., and C. L. Hom. 1990. Multiple equilibria and maintenance of additive genetic variance in a model of pleiotropy. *Evolution* 44: 1153–1163. [5, 28]

Hastings, W. K. 1970. Monte Carlo sampling methods using Markov Chains and their applications. *Biometrika* 57: 97–109. [19, A3]

Haubold, B., P. Pfaffelhuber, and M. Lynch. 2010. mlRho–A program for estimating the population mutation and recombination rates from shotgun-sequenced genomes. *Mol. Ecol.* 19, Suppl. 1: 277–284. [4]

Hayashi, T., and Y. Ukai. 1994. Change in genetic variance under selection in a self-fertilizing population. *Genetics* 136: 693–704. [23]

Hayes, B. J., P. M. Visscher, and M. E. Goddard. 2009. Increased accuracy of artificial selection by using the realized relationship matrix. *Genet. Res.* 9: 47–60. [20]

Hayes, B. J., P. M. Visscher, H. C. McPartlan, and M. E. Goddard. 2003. Novel multilocus measure of linkage disequilibrium to estimate past effective population size. *Genome Res.* 13: 635–643. [4]

Hayes, B. J., and 7 others. 2008. A genome map of divergent artificial selection between *Bos taurus* dairy cattle and *Bos taurus* beef cattle. *Anim. Genet.* 40: 176–184. [9]

Hazel, L. N. 1943. The genetic basis for constructing selection indexes. *Genetics* 28: 476–490. [22, 34, A6]

Hazel, W. N., R. Smock, and M. D. Johnson. 1990. A polygenic model for the evolution and maintenance of conditional strategies. *Proc. Roy. Soc. Lond. B* 242: 181–187. [14]

He, Z., and 9 others 2011. Two evolutionary histories in the genomes of rice: the roles of domestication genes. *PLoS Genet.* 7: e1002100. [9]

Heath, S. C., G. Bulfield, R. Thompson, and P. D. Keightley. 1995. Rates of change of genetic parameters of body weight in selected mouse lines. *Genet. Res.* 66: 19–25. [19]

Hedgecock, D. 1994. Does variance in reproductive success limit effective population size of marine organisms? *In* A. Beaumont (ed.), *Genetics and evolution of aquatic organisms*, pp. 122–134. Chapman and Hall, London, UK. [3]

Hedges, L. V., and I. Olkin. 1980. Vote-counting methods in research synthesis. *Psycholog. Bull.* 88: 359–369. [A4]

Hedrick, P. W. 1986a. Genetic polymorphisms in heterogeneous environments: a decade later. *Ann. Rev. Ecol. System.* 7: 1–33. [28]

Hedrick, P. W. 1986b. Average inbreeding or equilibrium inbreeding? *Am. J. Hum. Genet.* 38: 965–970. [3]

Hedrick, P. W. 1999. Perspective: highly variable loci and their interpretation in evolution and conservation. *Evolution* 53: 313–318. [9]

Hedrick, P. W. 2005. Large variance in reproductive success and the N_e/N ratio. *Evolution* 59: 1596–1599. [3, 4]

Hedrick, P. W. 2012. What is the evidence for heterozygote advantage selection? *Trends Ecol. Evol.* 27: 698–704. [28]

Hedrick, P. W., and C. C. Cockerham. 1986. Partial inbreeding: equilibrium heterozygosity and the heterozygosity paradox. *Evolution* 40: 856–861. [3]

Hein, J., M. H. Schierup, and C. Wiuf. 2005. *Gene genealogies, variation and evolution—a primer in coalescent theory*. Oxford Univ. Press, Oxford, UK. [2]

Heisler, I. L., and J. Damuth. 1987. A method of analyzing selection in hierachically structured populations. *Am. Nat.* 130: 582–602. [30]

Helanterä, H., and T. Uller. 2010. The Price equation and extended inheritance. *Philos. & Theory in Biol.* 2: e101. [6]

Hellmann, I., and 6 others. 2008. Population genetic analysis of shotgun assemblies of genomic sequences from multiple individuals. *Genome Res.* 18: 1020–1029. [8]

Helms, T. C., A. R. Hallauer, and O. S. Smith. 1989. Genetic drift and selection evaluated from recurrent selection programs in maize. *Crop. Sci.* 29: 602–607. [18]

Henderson, C. R. 1949. Estimates of changes in herd environment. *J. Dairy Sci.* 32: 706. [19]

Henderson, C. R. 1973. Sire evaluation and genetic trends. In *Proceeding of the animal breeding and genetics symposium in honor of Dr Jay L. Lush*, pp. 10-41. Champaign, IL. [A5]

Henderson, C. R. 1975. Best linear unbiased estimation and prediction under a selection model. *Biometrics* 31: 423–447. [1, 19]

Henderson, C. R. 1976. A simple method for the inverse of a numerator relationship matrix used in prediction of breeding values. *Biometrics* 32: 69–83. [19]

Henderson, C. R. 1984. *Applications of linear models in animal breeding*. University of Guelph, Guelph, CA. [13]

Henderson, C. R. 1988. Use of an average numerator relationship matrix for multiple-sire joinings. *J. Anim. Sci.* 66: 1614–1621. [20]

Henderson, C. R., O. Kempthorne, S. R. Searle, and C. M. von Krosigk. 1959. The estimation of environmental and genetic trends from records subject to culling. *Biometrics* 15: 192–218. [19]

Hendry, A. P. 2002. $Q_{ST} >=\neq< F_{ST}$? *Trends. Ecol. Evol.* 17: 502. [12]

Hendry, A. P., and M. T. Kinnison. 1999. The pace of modern life: measuring rates of contemporary microevolution. *Evolution* 53: 1637–1653. [30]

Henle, K. 1991. Some reflections on evolutionary theories with a classification of fitness. *Acta Biotheor.* 39: 91–106. [29]

Henshaw, J. M., A. T. Kahna, and K. Fritzschebe. 2016. A rigorous comparison of sexual selection indexes via simulations of diverse mating systems. *Proc. Natl. Acad. Sci. USA* 113: E300–E308. [29]

Henshaw, J. M., and Y. Zemel. 2016. A unified measure of linear and nonlinear selection on quantitative traits. *Methods Ecol. Evol.* 8: 604–614. [29, 30]

Henze, N. 2002. Invariant tests for multivariate normality: a critical review. *Stat. Papers* 43: 467–506. [A5]

Hereford, J., T. F. Hansen, and D. Houle. 2004. Comparing strengths of directional selection; how strong is strong? *Evolution* 58: 2133–2143. [29, 30, A4]

Hermisson, J. 2009. Who believes in whole-genome scans for selection? *Heredity* 103: 283–284. [9]

Hermisson, J., T. F. Hansen, and G. P. Wagner. 2003. Epistasis in polygenic traits and the evolution of genetic architecture under stabilizing selection. *Am. Nat.* 161: 708–734. [28]

Hermisson, J., and A. P. McGregor. 2008. Pleiotropic scaling and QTL data. *Nature* 456: E3–E3. [27]

Hermisson, J., and P. S. Pennings. 2005. Soft sweeps: population genetics of adaptation from standing genetic variation. *Genetics* 169: 2335–2352. [8]

Hermisson, J., and P. S. Pennings. 2017. Soft sweeps and beyond: understanding the patterns and probabilities of selection footprints under rapid adaptation. *Meth. Ecol. Evol.* 8: 700–716. [8]

Hermisson, J., and G. P. Wagner. 2004. The population genetic theory of hidden variation and genetic robustness. *Genetics* 168: 2271–2284. [28]

Hernandez, R. D., S. H. Williamson, and C. D. Bustamante. 2007. Context dependence, ancestral misidentification, and spurious signatures of natural selection. *Mol. Bio. Evol.* 24: 1782–1800. [9]

Hernandez, R. D., and 7 others. 2011. Classic selective sweeps were rare in recent human evolution. *Science* 331: 920–924. [8, 9]

Herr, A. J., and 7 others. 2011. Mutator suppression and escape from replication error-induced extinction in yeast. *PLoS Genet.* 7: e1002282. [4]

Herrendörfer, G., and L. Bünger. 1988. Estimation of the h2-function in long-term selection experiments, *In* D. Rasch, F. Pirchner, and J. Adam (eds.) *Proc. Internat. Conf. on Population Mathematics, Part III*, pp. 45–49. [25]

Hersch, E. I., and P. C. Phillips. 2004. Power and potential bias in field studies of natural selection. *Evolution* 58: 479–485. [29, 30, A4]

Hews, D. K. 1990. Examining hypotheses generated by field measures of sexual selection on male lizards, *Uta palmeri*. *Evolution* 44: 1956–1966. [29]

Hey, J. 1991. A multi-dimensional coalescent process applied to multi-allelic selection models and migration models. *Theor. Popul. Biol.* 39: 30–48. [2]

Hey, J., and R. M. Kliman. 2002. Interactions between natural selection, recombination and gene density in the genes of *Drosophila*. *Genetics* 160: 595–608. [8]

Hey, J., and J. Wakeley. 1997. A coalescent estimator of the population recombination rate. *Genetics*. 145: 833–846. [4]

Heywood, J. S. 1986. The effect of plant size variation on genetic drift in populations of annuals. *Am. Nat.* 127: 851–861. [3]

Heywood, J. S. 2005. An exact form of the breeder's equation for the evolution of a quantitative trait under natural selection. *Evolution* 59: 2287–2298. [6, 20]

Hicks, W. M., M. Kim, and J. E. Haber. 2010. Increased mutagenesis and unique mutation signature associated with mitotic gene conversion. *Science* 329: 82–85. [4]

Higgins, K., and M. Lynch. 2001. Metapopulation extinction due to mutation accumulation. *Proc. Natl. Acad. Sci. USA* 98: 2928–2933. [1]

Higgs, P. G. 1998. Compensatory neutral mutations and the evolution of RNA. *Genetica* 102/103: 91–101. [7]

Hill, R. R. Jr. 1971. Selection in autotetraploids. *Theor. Appl. Genet.* 41: 181–186. [15]

Hill, R. R., Jr., and W. L. Hagg. 1974. Comparison of selection methods for autotetraploids. I. Theoretical. *Crop Sci.* 14: 587–590. [15]

Hill, W. G. 1969a. On the theory of artificial selection in finite populations. *Genet. Res.* 13: 143–163. [7, 26]

Hill, W. G. 1969b. The rate of selection advance with nonadditive loci. *Genet. Res.* 13: 165–173. [7, 26]

Hill, W. G. 1971. Design and efficiency of selection experiments for estimating genetic parameters. *Biometrics* 27: 293–311. [18]

Hill, W. G. 1972a. Estimation of genetic change. I. General theory and design of control populations. *Anim. Breed. Abstracts* 40: 1–15. [12, 18]

Hill, W. G. 1972b. Estimation of genetic change. II. experimental evaluation of control populations. *Anim. Breed. Abstracts* 40: 193–213. [18]

Hill, W. G. 1972c. Estimation of realized heritabilities from selection experiments. I. Divergent selection. *Biometrics* 28: 747–765. [18]

Hill, W. G. 1972d. Estimation of realized heritabilities from selection experiments. II. Selection in one direction. *Biometrics* 28: 767–780. [18]

Hill, W. G. 1972e. Effective size of populations with overlapping generations. *Theor. Popul. Biol.* 3: 278–289. [3]

Hill, W. G. 1974a. Prediction and evaluation of response to selection with overlapping generations. *Anim. Prod.* 18: 117–139.

Hill, W. G. 1974b. Variability of response to selection in genetic experiments. *Biometrics* 30: 363–366. [18]

Hill, W. G. 1975. Linkage disequilibrium among multiple neutral alleles produced by mutation in finite population. *Theor. Popul. Biol.* 8: 117–126. [2, 4]

Hill, W. G. 1976. Order statistics of correlated variables and implications in genetic selection programmes. *Biometrics* 32: 889–902. [14, 21]

Hill, W. G. 1977a. Variation in response to selection. *In* E. Pollak, O. Kempthorne, and T. B. Bailey, Jr., (eds.), *Proceedings of the international conference on quantitative genetics*, pp. 343–365. Iowa State Univ. Press, Ames, IA. [18]

Hill, W. G. 1977b. Order statistics of correlated variables and implications in genetic selection programmes. II. Response to selection. *Biometrics* 33: 703–712. [14, 21]

Hill, W. G. 1979. A note on effective population size with overlapping generations. *Genetics* 92: 317–322. [3]

Hill, W. G. 1980. Design of quantitative genetic selection experiments. *In* A. Roberston, (ed.), *Selection experiments in laboratory and domestic animals*, pp. 1–13. Commonwealth Agricultural Bureaux, Slough, UK. [18]

Hill, W. G. 1981. Estimation of effective population size from data on linkage disequilibrium. *Genet. Res.* 38: 209–216. [4]

Hill, W. G. 1982a. Predictions of response to artificial selection from new mutations. *Genet. Res.* 40: 255–278. [26]

Hill, W. G. 1982b. Rates of change in quantitative traits from fixation of new mutations. *Proc. Natl. Acad. Sci. USA* 79: 142–145. [26]

Hill, W. G. (ed). 1984. *Quantitative Genetics, Part II: Selection. Benchmark Papers in Genetics, Volume 15.* Van Nostrand Reinhold, New York, NY. [18]

Hill, W. G. 1985. Fixation probabilities of mutant genes with artificial selection. *Genet. Sel. Evol.* 17: 351–358. [26]

Hill, W. G. 1986. Population size and design of breeding programmes. *Proceedings of the 3rd World Congress on Genetics Applied to Livestock Production*, Vol. 12: 245–256. University of Nebraska, Lincoln, NE. [18]

Hill, W. G. 1988. Selective breeding: Why aren't horses faster? *Nature* 332: 678. [25]

Hill, W. G. 1990. Mutation, selection and quantitative genetic variation. *In* N. Takahata and J. F. Crow (eds.) *Population biology of genes and molecules*, pp. 219–232. Baifukan, Tokyo. [26]

Hill, W. G. 2007. The variance of the variance—genetic analysis of environmental variability. *J. Anim. Breed. Genet.* 124: 1. [17]

Hill, W. G. 2010. Understanding and using quantitative genetic variation. *Phil. Trans. Roy. Soc. Lond. B* 365: 73–85. [24]

Hill, W. G. 2011. Can more be learned from selection experiments of value in animal breeding programmes? Or is it time for an obituary? *J. Anim. Breed. Genet.* 128: 87–94. [18]

Hill, W. G. 2014. Applications of population genetics to animal breeding, from Wright, Fisher and Lush to genomic prediction. *Genetics* 196: 1–16. [1]

Hill, W. G., N. H. Barton, and M. Turelli. 2006. Prediction of effects of genetic drift on variance components under a general model of epistasis. *Theor. Popul. Biol.* 70: 56–62. [11]

Hill, W. G., and A. Caballero. 1992. Artificial selection experiments. *Ann. Rev. Ecol. Syst.* 23: 287–310. [18]

Hill, W. G., A. Caballero, and L. Dempfle. 1996. Prediction of response to selection within families. *Genet. Sel. Evol.* 28: 379–383 [21]

Hill, W. G., M. E. Goddard, and P. M. Visscher. 2008. Data and theory point to mainly additive genetic variance for complex traits. *PLoS Genet.* 4: e1000008. [11, 15, 24, 25]

Hill, W. G., and P. D. Keightley. 1988. Interrelations of mutation, population size, artificial and natural selection. *In* B. S. Weir, E. J. Eisen, M. M. Goodman, and G. Namkoong (eds.), *Proceedings of the second international conference on quantitative genetics*, pp. 57–70. Sinauer Assocs., Sunderland, MA. [26, 28]

Hill, W. G., and M. Kirkpatrick. 2010. What animal breeding has taught us about evolution. *Ann. Rev. Ecol. Evol. System.* 41: 1–19. [1]

Hill, W. G., and T. F. C. Mackay (eds.). 1989 *Evolution and animal breeding*. C. A. B. International, Wallingford, UK. [18]

Hill, W. G., and S. H. Mbaga. 1998. Mutation and conflicts between artificial and natural selection for quantitative traits. *Genetica* 102: 171–181. [25]

Hill, W. G., and H. A. Mulder. 2010. Genetic analysis of environmental variation. *Genet. Res.* 92: 381–395. [17]

Hill, W. G., and J. Rasbash. 1986. Models of long term artificial selection in finite population. *Genet. Res.* 48: 41–50. [26]

Hill, W. G., and A. Robertson. 1966. The effects of linkage on limits to artificial selection. *Genet. Res.* 8: 269–294. [1, 2, 7, 8, 26]

Hill, W. G., and A. Robertson. 1968. Linkage disequilibrium in finite populations. *Theor. Appl. Genet.* 38: 226–231. [2, 9]

Hill, W. G., and B. S. Weir. 1988. Variances and covariances of squared linkage disequilibria in finite populations. *Theor. Popul. Biol.* 33: 54–78. [2]

Hill, W. G., and B. S. Weir. 2011. Variation in actual relationship as a consequence of Mendelian sampling and linkage. *Genet. Res.* 93: 47–64. [20]

Hill, W. G., and X.-S. Zhang. 2004. Effects on phenotypic variability of directional selection arising through genetic difference in residual variability. *Genet. Res.* 83: 121–132. [17]

Hill, W. G., and X.-S. Zhang. 2012a. On the pleiotropic structure of the genotype-phenotype map: the evolvability of complex organisms. *Genetics* 190: 1131–1137. [27]

Hill, W. G., and X.-S. Zhang. 2012b. Assessing pleiotropy and its evolutionary consequences: pleiotropy is not necessarily limited, nor need it hinder the evolution of complexity. *Nat. Rev. Genet.* 13: 296. [27]

Hilliker, A. J., and 5 others. 1994. Meiotic gene conversion tract length distribution within the *rosy* locus of *Drosophila melanogaster*. *Genetics* 137: 1019–1026. [4]

Hilu, K. 1983. The role of single-gene mutations in the evolution of flowering plants. *Evol. Biol.* 16: 97–128. [25]

Hine, E., and M. W. Blows. 2006. Determining the effective dimensionality of the genetic variance-covariance matrix. *Genetics* 173: 1135–1144. [28]

Hine, E., S. F. Chenoweth, H. D. Rundle, and M. W. Blows. 2009. Characterizing the evolution of genetic variance using genetic covariance tensors. *Phil. Tran. Roy. Soc. Lond. B*: 364: 1567–1578. [30]

Hoaglin, D. C., and R. E. Welsch. 1978. The Hat Matrix in regression and ANOVA. *Am. Stat.* 32: 17–22. [19]

Hoban, S., and 9 others. Finding the genomic basis of local adaptation: pitfalls, practical solutions, and future directions. *Am. Nat.* 188: 379–397. [9]

Hochberg, Y. 1988. A sharper Bonferroni procedure for multiple tests of significance. *Biometrika* 75: 800–802. [A4]

Hodgins-Davis, A., D. Rice, and J. Townsend. 2015. Gene expression evolves under a House-of-Cards model of stabilizing selection. *Mol. Biol. Evol.* 32: 2130–2140. [12, 27, 28]

Hoekstra, H. E., and J. A. Coyne. 2007. The locus of evolution: evo devo and the genetics of adaptation. *Evolution* 61: 995–1016. [12]

Hoekstra, H. E., and 7 others. 2001. Strength and tempo of directional selection in the wild. *Proc. Nat. Acad. Sci. USA* 98: 9157–9160 [29, 30]

Hoeschele, I., and P. M. Van Raden. 1991. Rapid inversion of dominance relationship matrices for noninbred populations by including sire by dam subclass effects. *J. Dairy Sci.* 74: 557–569. [19]

Hoeschele, I., and A. R. Vollema. 1993. Estimation of variance components with dominance and inbreeding in dairy cattle. *J. Anim. Breed. Genet.* 110: 93–104. [11]

Hofbauer, J., and K. Sigmund. 1998. *The theory of evolution and dynamical systems.* Cambridge University Press, Cambridge, UK. [5]

Hofer, A. 1998. Variance component estimation in animal breeding: a review. *J. Anim. Breed. Genet.* 115: 247–265. [19]

Hofer, T., N. Ray, D. Wegmann, and L. Excoffier. 2009. Large allele frequency differences between human continental groups are more likely to have occurred by drift during range expansions than by selection. *Ann. Hum. Genet.* 73: 95–08. [9]

Hoffmann, A. A., and J. Merilä. 1999. Heritable variation and evolution under favorable and unfavourable conditions. *Trends Ecol. Evol.* 14: 96–101. [20]

Hoffmann, A. A., and P. A. Parsons. 1991. *Evolutionary genetics and environmental stress.* Oxford University Press, Oxford, UK. [20]

Hoffmann, A. A., and P. A. Parsons. 1997a. *Extreme environmental change and evolution.* Cambridge University Press, Cambridge, UK. [20]

Hoffmann, A. A., and P. A. Parsons. 1997b. Consistent heritability changes under poor growth conditions. *Trends Ecol. Evol.* 12: 460–461. [20]

Hogben, L. 1974. The origins of theoretical population genetics. *Brit. J. Hist. Sci.* 7: 176–179. [1]

Hohenboken, W. D. 1985. The manipulation of variation in quantitative traits: a review of possible genetic strategies. *J. Anim. Sci.* 60: 101–110. [16, 17]

Holand, A. M., H. Jensen, J. Tufto, and R. Moe. 2011. Does selection or genetic drift explain geographic differentiation of morphological characters in house sparrows? *Genet. Res.* 93: 367–379. [12]

Holder, K. K., and J. J. Bull. 2001. Profiles of adaptation in two similar viruses. *Genet.* 159: 1393–1404. [27]

Holderegger, R., and 10 others. 2008. Land ahead: using genome scans to identify molecular markers of adaptive relevance. *Plant Ecol. Diver.* 1: 273–283. [9]

Holland, J. B., W. E. Nyquist, and C. T. Cervantes-Martínez. 2003. Estimating and interpreting heritability for plant breeding: An update. *Plant Breed. Rev.* 22: 9–112. [21]

Hollingdale, B. 1971. Analyses of some genes from abdominal bristle number selection lines in *Drosophila melanogaster*. *Theor. Appl. Genet.* 41: 292–301. [25]

Holloway, G. J., S. R. Povey, and R. M. Sibly. 1990. The effect of new environment on adapted genetic architecture. *Heredity* 64: 323–330. [20]

Holm, S. 1979. A simple sequential rejective multiple test procedure. *Scand. J. Stat.* 6: 65–70. [A4]

Holsinger, K. E. 1991. Mass-action models of plant mating systems: the evolutionary stability of mixed mating systems. *Am. Nat.* 138: 606–622. [23]

Holsinger, K. E., and B. S. Weir. 2009. Genetics in geographically structured populations: defining, estimating and interpreting F_{ST}. *Nat. Rev. Genet.* 10: 639–650. [12]

Holt, M., T. Meuwissen, and O. Vangen. 2005. Long-term responses, changes in genetic variances and inbreeding depression from 122 generations of selection on increased litter size in mice. *J. Anim. Breed. Genet.* 122: 199–209. [18, 19, 25]

Hommel, G. 1988. A stagewise rejective multiple test procedure on a modified Bonferroni test *Biometrika* 75: 383–386. [A4]

Hommel, G. 1989. A comparison of two modified Bonferroni procedures. *Biometrika* 76: 624–625. [A4]

Hopkins, C. G. 1899. Improvement in the chemical composition of the corn kernel. *Illinois Agr. Exp. Sta. Bull.* 55. [21, 25]

Hopkins, M. J., and S. Lidgard. 2012. Evolutionary mode routinely varies among morphological traits within fossil species lineages. *Proc. Natl Acad. Sci. USA* 109: 20520–20525. [12]

Hõrak, P., R. Mänd, and I. Ots. 1997. Identifying targets of selection: a multivariate analysis of reproductive traits in the great tit. *Oikos* 78: 592–600. [20]

Horn, R. A., and C. R. Johnson. 1985. *Matrix analysis.* Cambridge University Press, Cambridge, UK. [A5]

Horner, T. W. 1952. *Non-allelic gene interaction and the interpretation of quantitative genetics data.* Unpublished Ph.D. thesis, D. H. Hill Library, North Carolina State University, Raleigh, NC. [23]

Horner, T. W., and C. R. Weber. 1956. Theoretical and experimental study of self fertilized populations. *Biometrics* 12: 404–414. [11]

Horvat, S., and 6 others. 2000. Mapping of obesity QTLs in a cross between mouse lines divergently selected on fat content. *Mamm. Genom.* 11: 2–7. [25]

Horvitz, C. C., T. Coulson, S. Tuljapurkar, and D. W. Schemske. 2010. A new way to integrate selection when both demography and selection gradients vary over time. *Int. J. Plant Sci.* 171: 945–959. [29]

Hospital, F., and C. Chevalet. 1993. Effects of population size and linkage on optimal selection intensity. *Theor. Appl. Genet.* 86: 775–780. [26]

Hospital, F., and C. Chevalet. 1996. Interactions of selection, linkage and drift in the dynamics of polygenic characters. *Genet. Res.* 67: 77–87. [24, 26]

Hothorn, L. A., O. Libiger, and D. Gerhard. 2012. Model-specific tests on variance heterogeneity for detection of potentially interacting genetic loci. *BMC Genet.* 13: 59. [17]

Houck, L. D., S. J. Arnold, and R. A. Thisted. 1985. A statistical study of mate choice: Sexual selection in a plethodontid salamander (*Desmognathus ochrophaesus*). *Evolution* 39: 370–386. [29]

Houle, D. 1989. The maintenance of polygenic variation in finite populations. *Evolution* 43: 1767–1780. [28]

Houle, D. 1992. Comparing evolvability and variability of quantitative traits. *Genetics* 130: 195–204. [6, 13, 29]

Houle, D., B. Morikawa, and M. Lynch. 1996. Comparing mutational variabilities. *Genetics* 143: 1467–1483. [6, 28]

Houle, D., C. Pélabon, G. P. Wagner, and T. F. Hansen. 2011. Measurement and meaning in biology. *Quart. Rev. Biol.* 86: 3–34. [13]

Howard, R. D. 1979. Estimating reproductive success in natural populations. *Am. Nat.* 114: 221–231. [29]

Howard, R. D., J. A. DeWoody, and W. M. Muir. 2004. Transgenic male mating advantage provides opportunity for Trojan gene effect in a fish. *Proc. Natl. Acad. Sci. USA* 101: 2934–3938. [5]

Hsieh, W.-P., T.-M. Chu, R. D. Wolfinger, and G. Gibson. 2003. Mixed-model reanalysis of primate data suggests tissue and species biases in oligonucleotide-based gene expression profiles. *Genetics* 165: 747–757. [12]

Hsu, W. L., R. K. Johnson, and L. D. Van Vleck. 2010. Effects of pen mates on growth, backfat depth, and longissimus muscle area of swine. *J. Anim. Sci.* 88: 895–902. [22]

Hu, X., G. L. Gadbury, Q. Xiang, and D. B. Allison. 2011. Illustrations on using the distribution of a p-value in high dimensional data analysis. *Adv. Appl. Stat. Sci.* 1: 191–213. [A4]

Huang, W., and 7 others. 2015. Genetic basis of transcriptome diversity in *Drosophila melanogaster.*. *Proc. Natl. Acad. Sci. USA* 112: E6010–E6019. [17]

Huang, X., and 34 others. 2012. A map of rice genome variation reveals the origin of cultivated rice. *Nature* 490: 497–501. [8, 9]

Hubbell, S. P., and L. K. Johnson. 1987. Environmental variance in lifetime mating success, mate choice, and sexual selection. *Am. Nat.* 130: 91–112. [29]

Huber, C. D., M. DeGiorgio, I. Hellmann, and R. Nielsen. 2015. Detecting recent selective sweeps while controlling for mutation rate and background selection. *Mol. Ecol.* 25: 142–156. [9]

Huber, C. D., M. Nordborg, J. Hermisson, and I. Hellmann. 2014. Keeping it local: evidence for positive selection in Swedish *Arabidopsis thaliana*. *Mol. Biol. Evol.* 31: 3026–2039. [9]

Hudson, R. R. 1987. Estimating the recombination parameter of a finite population model without selection. *Genet. Res.* 50: 245–250. [4]

Hudson, R. R. 1990. Gene genealogies and the coalescent process. *Oxford Surv. Evol. Biol.* 7: 1–43. [2]

Hudson, R. R. 1993. Levels of DNA polymorphism and divergence yield important insights into evolutionary processes. *Proc. Natl. Acad. Sci. USA* 90: 7425–7426. [10]

Hudson, R. R. 1994. How can the low levels of DNA sequence variation in regions of the *Drosophila* genome with low recombination rates be explained? *Proc. Natl. Acad. Sci. USA* 91: 6815–6818. [8]

Hudson, R. R. 2001. Two-locus sampling distributions and their application. *Genetics* 159: 1805–1817. [4]

Hudson, R. R. 2002. Generating samples under a Wright-Fisher neutral model. *Bioinformatics* 18: 337–338. [2, 9]

Hudson, R. R., K. Bailey, D. Skarecky, J. Kwiatowski, and F. J. Ayala. 1994. Evidence for positive selection in the superoxide dismutase (*Sod*) region of *Drosophila melanogaster.* *Genetics* 136: 1329–1340. [9]

Hudson, R. R., and N. L. Kaplan. 1985. Statistical properties of the number of recombination events in the history of a sample of DNA sequences. *Genetics* 111: 147–164. [4, 8, 9]

Hudson, R. R., and N. L. Kaplan. 1988. The coalescent process in models with selection and recombination. *Genetics* 120: 831–840. [8]

Hudson, R. R., and N. L. Kaplan. 1994. Gene trees with background selection. *In* G. B. Golding (ed.), *Non-neutral evolution: theories and molecular data*, pp. 140–153. Chapman and Hall, New York, NY. [3]

Hudson, R. R., and N. L. Kaplan. 1995. Deleterious background selection with recombination. *Genetics* 141: 1605–1617. [3, 8]

Hudson, R. R., M. Kreitman, and M. Aguadé. 1987. A test of neutral molecular evolution based on nucleotide data. *Genetics* 116: 153–159. [10]

Huelsenbeck, J. P., and K. A. Dyer. 2004. Bayesian estimation of positively selected sites. *J. Mol. Evol.* 58: 661–672. [10]

Huerta-Sanchez, E., R. Durrett, and C. D. Bustamante. 2008. Population genetics of polymorphism and divergence under fluctuating selection. *Genetics* 178: 325–337. [10]

Hufford, K. M., P. Canaran, D. H. Ware, M. D. McMullen, and B. S. Gaut. 2007. Patterns of selection and tissue-specific expression among maize domestication and crop improvement loci. *Plant Physiol.* 144: 1642–1653. [9]

Hufford, M. B., and 24 others. 2012. Comparative population genomics of maize domestication and improvement. *Nat. Genet.* 44: 808–813. [9]

Hughes, A. L. 1999. *Adaptive evolution of genes and genomes.* Oxford University Press, Oxford, UK. [10]

Hughes, A. L. 2007. Looking for Darwin in all the wrong places: the misguided quest for positive selection at the nucleotide sequence level. *Heredity* 99: 364–373. [9, 10]

Hughes, A. L., and M. Nei. 1988. Pattern of nucleotide substitution at major histocompatibility complex class I loci reveals overdominant selection. *Nature* 335: 167–170. [10]

Hughes, A. L., and M. Nei. 1989. Nucleotide substitution at major histocompatibility complex class II loci: Evidence for overdominant selection. *Proc. Natl. Acad. Sci. USA* 86: 958–962. [10]

Hughes, A. L., and 5 others. 2003. Widespread purifying selection at polymorphic sites in human protein-coding loci. *Prod. Natl. Acad. Sci. USA* 100: 15754–15757. [10]

Huisman, J. 2017. Pedigree reconstruction from SNP data: parentage assignment, sibship clustering and beyond. *Mol. Ecol. Resources* 17: 1009—1024. [20]

Hull, F. H. 1945. Recurrent selection and specific combining ability in corn. *J. Am. Soc. Agron.* 37: 134–145. [21]

Hull, F. H. 1952. Recurrent selection and overdominance. *In* J. W. Gowen (ed.), *Heterosis,* pp. 451–473. Iowa State College Press, Ames, IA. [21]

Hume, A. N. 1919. Corn families of South Dakota. *South Dakota Agri. Exp. Stn. Coll.* 186. [21]

Hummel, S., D. Schmidt, B. Kremeyer, B. Herrmann, and M. Oppermann. 2005. Detection of the *CCR5-*Δ*32* HIV resistance gene in Bronze Age skeletons. *Genes & Immunity* 6: 371–374. [9]

Hunt, G. 2006. Fitting and comparing models of phyletic evolution: random walks and beyond. *Paleobiology*

32: 578–601. [12]

Hunt, G. 2007. The relative importance of directional change, random walks, and stasis in the evolution of fossil lineages. *Proc. Natl. Acad. Sci. USA* 104: 18404–18408. [12, 28, 30]

Hunt, G. 2008a. Gradual or pulsed evolution: when should punctuational explanations be preferred? *Paleobiology* 34: 360–377. [12]

Hunt, G. 2008b. Evolutionary patterns within fossil lineages: model-based assessment of modes, rates, punctuations and process. *In* P. H. Kelley and R. K. Bambach (eds.), *From evolution to geobiology: research questions driving paleontology at the start of a new century. Paleontological Society Papers* 14: 117–131. [12]

Hunt, G., and M. T. Carrano. 2010. Models and methods for analyzing phenotypic evolution in lineages and clades. *In* J. Alroy and G Hunt (eds.), *Quantitative methods in paleobiology*, pp. 245–269. [12]

Hunt, G., M. J. Hopkins, and S. Lidgard. 2015. Simple versus complex models of trait evolution and stasis as a response to environmental change. *Proc. Natl. Acad. Sci. USA* 112: 4885–4890. [12]

Hunt, H. R., C. A. Hoppert, and W. G. Erwin. 1944. Inheritance of susceptibility to caries in albino rats (*Mus norvegicus*). *J. Dent. Res.* 23: 385–401. [18]

Hunt, J., M. W. Blows, F. Zajitschek, M. D. Jennions, and R. Brooks. 2007b. Reconciling strong stabilizing selection with the maintenance of genetic variation in a natural population of black field crickets (*Teleogryllus commodus*). *Genetics* 177: 875–880. [28]

Hunt, J., C. J. Breuker, J. A. Sadowski, and A. J. Moore. 2009. Male-male competition, female mate choice and their interaction: determining total sexual selection. *J. Evol. Biol.* 22: 13–26. [29]

Hunt, J., J. B. Wolf, and A. J. Moore. 2007a. The biology of multivariate evolution. *J. Evol. Biol.* 20: 24–27. [30]

Hunter, J. E., and F. L. Schmidt. 2005. *Methods of meta-analysis: Correcting error and bias in research findings*. Sage, Thousand Oaks, CA. [30, A4]

Hunton, P. 1984. Selection limits: have they been reached in the poultry industry? *Can. J. Anim. Sci.* 64: 217–221. [25]

Hurst, H. E. 1951. Long-term storage capacity of reservoirs. *Trans. Amer. Soc. Civil Eng.* 116: 770–808. [12]

Hurst, L. D., and C. Pál. 2001. Evidence for purifying selection acting on silent sites in *BRCA1*. *Trends Genet.* 17: 62–65. [10]

Husband, B. C., and D. W. Schemske. 1996. Evolution of the magnitude and timing of inbreeding depression in plants. *Evolution* 50: 54–70. [23]

Husby, A., M. E. Visser, and L. E. B. Kruuk. 2011. Speeding up microevolution: the effects of increasing temperature on selection and genetic variance in a wild bird population. *PLoS Bio.* 9: e1000585. [20]

Ibáñez-Escriche, N., D. Sorensen, R. Waagepetersen, and A. Blasco. 2008b. Selection for environmental variance: a statistical analysis and power calculations to detect response. *Genetics* 180: 2209–2226. [17]

Ibáñez-Escriche, N., L. Varona, D. Sorensen, and J. L. Noquera. 2008c. A study of heterogeneity of environmental variance for slaughter weight in pigs. *Animal* 2: 19–26. [17]

Ibáñez-Escriche, and 5 others. 2008a. Genetic parameters related to environmental variability of weight traits in a selection experiment for weight gain in mice: signs of correlated response. *Genet. Sel. Evol.* 40: 279–293. [17]

Igic, B., and J. W. Busch. 2013. Is self-fertilization an evolutionary dead end? *New Phytol.* 198: 386–397. [23]

Igic, B., and J. R. Kohn. 2006. The distribution of plant mating systems: study bias against obligately outcrossing species. *Evolution* 60: 1098–1103. [23]

Igic, B., R. Lande, and J. R. Kohn. 2008. Loss of self-incompatibility and its evolutionary consequences. *Int. J. Plant Sci.* 169: 93–104. [23]

Iles, M. M., K. Walters, and C. Cannings. 2003. Recombination can evolve in large finite populations given selection on sufficient loci. *Genetics* 165: 2249–2258. [4]

Im, S., R. L. Fernando, and D. Gianola. 1989. Likelihood inferences in animal breeding under selection: a missing-data theory view point. *Genet. Sel. Evol.* 21: 399–414. [19, 20]

Imasheva, A. G., L. A. Zhivotovsky, and O. E. Lazebny. 1991. Effects of artifical stabilizing selecion on *Drosophila* populations subjected to directional selection for another trait. *Genetica* 83: 247–256. [25]

Imhof, M., and C. Schlötterer. 2001. Fitness effects of advantageous mutations in evolving *Escherichia coli* populations. *Proc. Natl. Acad. Sci. USA* 98: 1113–1117. [27]

Ingvarsson, P. K. 2004. Population subdivision and the Hudson-Kreitman-Aguade test: testing for deviations from the neutral model in organelle genomes. *Genet. Res.* 83: 31–39. [10]

Ingvarsson, P. K. 2010. Natural selection on synonymous and nonsynonymous mutations shapes patterns of polymorphism in *Populus tremula*. *Mol. Biol. Evol.* 27: 650–660. [8, 10]

Innan, H. 2006. Modified Hudson-Kreitman-Aguadé test and two-dimensional evaluation of neutrality tests. *Genetics* 173: 1725–1733. [10]

Innan, H., and Y. Kim. 2004. Pattern of polymorphism after strong artificial selection in a domestication event. *Proc. Natl. Acad. Sci. USA* 101: 10667–10672. [8]

Innan, H., and W. Stephan. 2000. The coalescent in an exponentially growing metapopulation and its application to *Arabidopsis thaliana*. *Genetics* 155: 2015–2019. [9]

Innan, H., and W. Stephan. 2001. Selection intensity against deleterious mutations in RNA secondary structures and rate of compensatory nucleotide substitutions. *Genetics* 159: 389–399. [7]

Innan, H., and W. Stephan. 2003. Distinguishing the hitchhiking and background selection models. *Genetics* 165: 2307–2312. [8]

Innan, H., and F. Tajima. 1997. The amounts of nucleotide variation within and between allelic classes and the reconstruction of the common ancestral sequence in a population. *Genetics* 147: 1431–1444. [8]

Innan, H., K. Zhang, P. Marjoram, S. Tavaré, and N. A. Rosenberg. 2005. Statistical tests of the coalescent model based on the haplotype frequency distribution and the number of segregating sites. *Genetics* 169: 1763–1777. [9]

Inukai, T., A. Sako, H.-Y. Hirano, and Y. Sano. 2000. Analysis of intragenic recombination at *wx* in rice: correlation between the molecular and genetic maps within the locus. *Genome* 43: 589–596. [9]

Irgang, R., E. U. Dillard, M. W. Tess, and O. W. Robison. 1985. Selection for weaning weight and postweaning weight in hereford cattle. II. response to selection. *J. Anim. Sci.* 60: 1142–1155. [18]

Ishii, T., and 11 others. 2013. *OsLG1* regulates a closed panicle trait in domesticated rice. *Nat. Genet.* 45: 462–465. [9]

Israel C., and J. I. Weller. 2000. Effect of misidentification on genetic gain and estimation of breeding value in dairy cattle populations. *J. Dairy Sci.* 83: 181–187. [20]

Iwasa, Y., F. Michor, and M. A. Nowak. 2004. Stochastic tunnels in evolutionary dynamics. *Genetics* 166: 1571–1579. [7]

Jablonka, E., and M. J. Lamb. 2005. *Evolution in four dimensions: genetic, epigenetic, behavioral, and symbolic variation in the history of life.* MIT Press, Cambridge, MA. [1]

Jacobsson, L., and 6 others. 2005. Many QTLs with minor additive effects are associated with a large difference in growth between two selection lines in chickens. *Genet. Res.* 86: 115–125. [25]

Jacquard, A. 1966. Logique du calcul des coefficients d'identité entre deux individus. *Population* 21: 751–776. [11]

Jacquard, A. 1974. *The genetic structure of populations.* Springer, Berlin, DE. [11]

Jaenicke-Després, V., and 6 others. 2003. Early allelic selection in maize as revealed by ancient DNA. *Science* 302: 1206–1208. [8, 9]

Jain, K., and A. Nagar. 2013. Fixation of mutators in asexual populations: the role of genetic drift and epistasis. *Evolution* 67: 1143–1154. [4]

Jain, S. K., and R. W. Allard. 1965. The nature and stability of equilibria under optimizing selection. *Proc. Natl. Acad. Sci. USA* 54: 1436–1443. [28]

Jakob, E. M., S. D. Marshall, and G. W. Uetz. 1996. Estimating fitness: A comparison of body condition indices. *Oikos* 77: 61–67. [20]

Jakobsson, M., M. D. Edge, and N. A. Rosenberg. 2013. The relationship between F_{ST} and the frequency of the most frequent allele. *Genetics* 193: 515–528. [9]

James, J. W. 1962. Conflict between directional and centripetal selection. *Heredity* 17: 487–499. [25, 26]

James, J. W. 1965. Response curves in selection experiments. *Heredity* 20: 57–63. [25]

James, J. W. 1970. The founder effect and the response to artificial selection. *Genet. Res.* 16: 241–250. [18, 26, 28]

James, J. W., and G. McBride. 1958. The spread of genes by natural and artificial selection in a closed poultry flock. *J. Genet.* 56: 55–62. [16]

Jan-Orn, J., C. O. Gardner, and W. M. Ross. 1976. Quantitative genetics studies of the NP3R random-mating grain sorghum population. *Crop Sci.* 16: 489–496. [23]

Jarne, P., and J. R. Auld. 2006. Animals mix it up too: the distribution of self-fertilization among hermaphroditic animals. *Evolution* 60: 1816–1824. [23]

Jansen, R. C., and J.-P. Nap. 2001. Genetical genomics: the added value from segregation. *Trends Genet.* 17: 388–391. [12]

Janssen, G. M., D. DeJong, E. N. G. Joose, and W. Scharloo. 1988. A negative maternal effect in springtails. *Evolution* 42: 828–834. [15]

Janzen, F. J., and H. S. Stern. 1998. Logistic regression for empirical studies of multivariate selection. *Evolution* 52: 1564–1571. [29]

Jeffreys, A. J., and 6 others. 2004. Meiotic recombination hot spots and human DNA diversity. *Phil. Trans. Roy. Soc. Lond. B* 359: 141–152. [4]

Jeffreys, H. 1961. *Theory of probability*, 3rd ed. Oxford University Press, Oxford, UK. [A2]

Jenkin, F. 1867. Review of *The origin of species. The North British Review* 46: 277–318. [1]

Jenkins, P. A., and Y. S. Song. 2009. Closed-form two-locus sampling distributions: accuracy and universality. *Genetics* 183: 1087–1103. [4]

Jennions, M.D., H. Kokko, and H. Klug. 2012. The opportunity to be misled in studies of sexual selection. *J. Evol. Biol.* 25: 591–598. [29]

Jensen, H., M. Szulkin, and J. Slate. 2014. Molecular quantitative genetics. *In* A. Charmantier, D. Garant, and L. E. B. Kruuk (eds.), *Quantitative genetics in the wild*, pp. 209–227. Oxford University Press, Oxford, UK. [20]

Jensen, J. D. 2014. On the unfounded enthusiasm for soft selective sweeps. *Nat. Commun.* 5: 5281. [8]

Jensen, J. D., and D. Bachtrog. 2010. Characterizing recurrent positive selection at fast-evolving genes in *Drosophila miranda* and *Drosophila pseudoobscura*. *Genome Biol. Evol.* 2: 371–378. [8]

Jensen, J. D., Y. Kim, V. B. DuMont, C. F. Aquadro, and C. D. Bustamante. 2005. Distinguishing between selective sweeps and demography using DNA polymorphism data. *Genetics* 170: 1401–1410. [9]

Jensen, J. D., K. R. Thornton, and P. Andolfatto. 2008. An approximate bayesian estimator suggests strong, recurrent selective sweeps in *Drosophila*. *PLoS Genet.* 4: e1000198. [8, A3]

Jensen, J. D., K. R. Thornton, C. D. Bustamante, and C. F. Aquadro. 2007. On the utility of linkage disequilibrium as a statistic for identifying targets of positive selection in nonequilibrium populations. *Genetics* 176: 2371–2379. [8, 9]

Jensen, J., C. S. Wang, D. A. Sorensen, and D. Gianola. 1994. Marginal inferences of variance and covariance components for traits influenced by maternal and direct genetic effects using the Gibbs sampler. *Acta Agric. Scand.* 44: 193–201. [19]

Jensen, M. A., B. Charlesworth, and M. Kreitman. 2002. Patterns of genetic variation at a chromosome 4 Locus of *Drosophila melanogaster* and *D. simulans*. *Genetics* 160: 493–507. [8]

Jewell, N. P. 1986. On the bias of commonly used measures of association for 2 x 2 tables. *Biometrics* 42: 351–358. [10]

Jeyaruban, M. G., and J. P. Gibson. 1996. Estimation of additive genetic variance in commercial layer poultry and simulated populations under selection. *Theor. Appl. Genet.* 92: 483–491. [19]

Jiang, J. 1996. REML estimation: asymptotic behavior and related topics. *Ann. Stats.* 24: 255–286. [11, 22]

Jimenez-Gomes, J. M., J. A. Corwin, B. Joseph, J. N. Maloff, and D. J. Kliebenstein. 2011. Genomic analysis of QTLs and genes altering natural variation in stochastic noise. *PLoS Genet.* 7: e1002295. [17]

Jódar, B., and C. López-Fanjul. 1977. Optimum proportions selected with unequal sex numbers. *Theor. Appl. Genet.* 50: 57–61. [26]

Johansson, A. M., M. E. Pettersson, P. B. Siegel, and Ö. Carlborg. 2010. Genome-wide effects of long-term divergent selection. *PLoS Genet.* 6: e1001188. [9, 25]

Johnson, D. L. 1977a. Variance-covariance structure of group means with overlapping generations. *In* E. Pollak, O. Kempthorne, and T. B. Bailey, Jr., (eds.), *Proceedings of the international conference on quantitative genetics*, pp. 851–858. Iowa State Univ. Press, Ames, IA. [18]

Johnson, D. L. 1977b. Inbreeding in populations with overlapping generations. *Genetics* 87: 581–591. [3]

Johnson, H. W., H. F. Robinson, and R. E. Comstock. 1955. Estimates of genetic and environmental variability in soybeans. *Agronomy J.* 47: 314–318. [29]

Johnson, N. L., A. W. Kemp, and S. Kotz. 2005. *Univariate discrete distributions.* John Wiley & Sons, Inc., Hoboken, NJ. [2]

Johnson, N. L., and S. Kotz. 1970a. *Continuous univariate distributions—1.* John Wiley & Sons, New York, NY. [14, 16, 24]

Johnson, N. L., and S. Kotz. 1970b. *Continuous univariate distributions—2.* John Wiley & Sons, New York, NY. [14]

Johnson, P. L., and M. Slatkin. 2008. Accounting for bias from sequencing error in population genetic estimates. *Mol. Biol. Evol.* 25: 199–-206. [4, 9]

Johnson, R. A., and D. W. Wichern. 1988. *Applied multivariate statistical analysis, 2nd ed.* Prentice-Hall, Upper Saddle River, NJ. [A5]

Johnson, R. K., D. M. Niles, and S. A. Rohwer. 1972. Hermon Bumpus and natural selection in the house sparrow *Passer domesticus. Evolution* 26: 22–31. [30]

Johnson, T. 1999a. The approach to mutation-selection balance in an infinite asexual population, and the evolution of mutation rates. *Proc. Roy. Soc. Lond. B* 266: 2389–2397. [4]

Johnson, T. 1999b. Beneficial mutations, hitchhiking and the evolution of mutation rates in sexual populations. *Genetics* 151: 1621–1631. [4]

Johnson, T., and N. Barton. 2005. Theoretical models of selection and mutation on quantitative traits. *Phil. Trans. Roy. Soc. Lond. B.* 360: 1411–1425. [28]

Johnston, S. E., and 6 others. 2013. Life history trade-offs at a single locus maintain sexually selected genetic variation. *Nature* 502: 93–95. [26]

Jolicoeur, P. 1963. The multivariate generalization of the allometry equation. *Biometrics* 19: 497–499. [30]

Jolicoeur, P., and J. E. Mosimann. 1960. Size and shape variation in the painted turtle. A principal component analysis. *Growth* 24: 339–354. [A5]

Jones, A. G. 2009. On the opportunity for sexual selection, the Bateman gradient and the maximum intensity of sexual selection. *Evolution* 63: 1673–1648. [29]

Jones, A. G., and W. R. Ardren. 2003. Methods for parentage analysis in natural populations. *Mol. Ecol.* 12: 2511–2523. [20]

Jones, A. G., J. R. Arguello, and S. J. Arnold. 2002. Validation of Bateman's principles: a genetic study of sexual selection and mating patterns in the rough–skinned newt. *Proc. Roy. Soc. Lond. B* 269: 2533–2539. [29]

Jones, A. G., S. J. Arnold, and R. Bürger. 2003. Stability of the G-matrix in a population experiencing pleiotropic mutation, stabilizing selection, and genetic drift. *Evolution* 57: 1747–1760. [1]

Jones, A. G., S. J. Arnold, and R. Büurger. 2007. The mutation matrix and the evolution of evolvability. *Evolution* 61: 727–745. [1]

Jones, A. G., R. Büurger, S. J. Arnold, P. A. Hohenlohe, and J. C. Uyeda. 2012. The effects of stochastic and episodic movement of the optimum on the evolution of the G-matrix and the response of the trait mean to selection. *J. Evol. Biol.* 25: 2210–2231. [1]

Jones, A. G., and N. L. Ratterman. 2009. Mate choice and sexual selection: what have we learned since Darwin? *Proc. Nat. Acad. Sci. USA* 106: 10001–10008. [29]

Jones, A. G., G. Rosenqvist, A. Berglund, S. J. Arnold, and J. C. Avise. 2000. The Bateman gradient and the cause of sexual selection in a sex–role–reversed pipefish. *Proc. Roy. Soc. Lond. B:* 267: 677–680. [29]

Jones, A. G., C. M. Small, K. A. Paczolt, and N. L. Ratterman. 2010. A practical guide to methods of parentage analysis. *Mol. Ecol. Resources* 10: 6–30. [20]

Jones, D. A. 1989. 50 years of studying the scarlet tiger moth. *Trends Ecol. Evol.* 4: 298–301. [9]

Jones, D. A., and J. Wakeley. 2008. The influence of gene conversion on linkage disequilibrium around a selective sweep. *Genetics* 180: 1251–1259. [8, 9]

Jones, L. P. 1969a. Effects of artificial selection on rates of inbreeding in populations of *Drosophila melanogaster.* I. Effect in early generations. *Aust. J. Biol. Sci.* 22: 143–155. [26]

Jones, L. P. 1969b. Effects of artificial selection on rates of inbreeding in populations of *Drosophila melanogaster.* II. Effect of previous selection on rates of inbreeding. *Aust. J. Biol. Sci.* 22: 157–169. [26]

Jones, L. P., R. Frankham, and J. S. F. Barker. 1968. The effects of population size and selection intensity in selection for a quantitative character in *Drosophila.* II. Long-term response to selection. *Genet. Res.* 12: 249–266. [26]

Joost, S., and 6 others. 2007. A spatial analysis method (SAM) to detect candidate loci for selection: towards a landscape genomics approach to adaptation. *Mol. Ecol.* 16: 3955–3969. [9]

Jordan, N. 1991. Multivariate analysis of selection in experimental populations derived from hybridization of two ecotypes of the annual plant *Diodia teres. Evolution* 45: 1760–1772. [30]

Jorjani, H. 1995. Assortative mating and selection in populations of various sizes: A review of the literature. Swedish University of Agricultural Sciences, Department of Animal Breeding and Genetics Publication No. 121. Uppsala. [16]

Jorjani, H., G. Engström, E. Strandberg, and L.-E. Lijedahl 1997a. Genetic studies of assortative mating—a simulation study. I. Characteristics of the control population. *Acta. Agric. Scand. A* 47: 65–73. [16]

Jorjani, H., G. Engström, E. Strandberg, and L.-E. Lijedahl. 1997b. Genetic studies of assortative mating—a simulation study. II. Assortative mating in unselected populations. *Acta. Agric. Scand. A* 47: 74–81. [16]

Jorjani, H., G. Engström, E. Strandberg, and L.-E. Lijedahl. 1997c. Genetic studies of assortative mating—a simulation study. III. Assortative mating in selected populations. *Acta. Agric. Scand. A* 47: 129–137. [16]

Jorjani, H., G. Engström, E. Strandberg, and L.-E. Lijedahl. 1998. Assortative mating and dominance. *6th World Cong. Genet. Appl. Anim. Prod.* University of New England, Armidale, Australia. 26: 45–48.

Joshi, A., M. H. Do, and L. D. Mueller. 1999. Poisson distribution of male mating success in laboratory populations of *Drosophila melanogaster. Genet. Res.* 73: 239–249. [29]

Joyce, P., and P. Marjoram. 2008. Approximately sufficient statistics and Bayesian computation. *Stat. Appl. Genet. Mol. Biol.* 7: a26. [A3]

Joyce, P., D. R. Rokyta, C. J. Beisel, and H. A. Orr. 2008. A general extreme value theory model for the adaptation of DNA sequences under strong selection and weak mutation. *Genetics* 180: 1627–1643. [27]

Judd, S. R., and T. D. Petes. 1988. Physical lengths of meiotic and mitotic gene conversion tracts in *Saccharomyces cerevisiae*. *Genetics* 118: 401–410. [4]

Judson, H. F. 1979. *The eighth day of creation: Makers of the revolution in biology*. Simon and Schuster, New York, NY. [1]

Juga, J., and R. Thompson. 1989. Estimation of variance components in populations selected over multiple generations. *Acta Agric. Scand.* 39: 79–89. [19]

Jung, H., and P. Marjoram. 2011. Choice of summary statistic weights in approximate Bayesian computation. *Stat. Appl. Genet. Mol. Biol.* 10: a45. [A3]

Kackar, R. N., and D. A. Harville. 1981. Unbiasedness of two-stage estimation and prediction for mixed linear models. *Comm. Stat. Theor. Meth.* A10: 1249–1261. [19]

Kaiser, V. B., and B. Charlesworth. 2009. The effects of deleterious mutations on evolution in non-recombining genomes. *Trends Genet.* 25: 9–12. [8]

Kalinka, A. T., and 8 others. 2010. Gene expression divergence recapitulates the developmental hourglass model. *Nature* 468: 811–814. [12]

Kalisz, S. 1986. Variable selection on the timing of germination in *Collinsia verna (Scrophulariaceae)*. *Evolution* 40: 479–491. [29, 30]

Kane, N. C., and L. H. Rieseberg. 2007. Selective sweeps reveal candidate genes for adaptation to drought and salt tolerance in common sunflower, *Helianthus annuus*. *Genetics* 175: 1823–1834. [9]

Kang, C. J., and P. Marjoram. 2011. Inference of population mutation rate and detection of segregating sites from next-generation sequence data. *Genetics* 189: 595–605. [4]

Kaplan, N. L., T. Darden, and R. R. Hudson. 1988. The coalescent process in models with selection. *Genetics* 120: 819–829. [8]

Kaplan, N. L., R. R. Hudson, and C. H. Langley. 1989. The "hitchhiking effect" revisited. *Genetics* 123: 887–899. [8, 9]

Karasov, T., P. W. Messer, and D. A. Petrov. 2010. Evidence that adaptation in *Drosophila* is not limited by mutation at single sites. *PLoS Genet.* 6: e1000924. [8]

Karhunen, M., J. Merilä, T. Leinonen, J. M. Cano, and O. Ovaskainen. 2013. DRIFTSEL: an R package for detecting signals of natural selection in quantitative traits. *Mol. Eco. Resources* 13: 746–754. [12]

Karhunen, M., O. Ovaskainen, G. Herczeg, and J. Merilä. 2014. Bringing habitat information into statistical tests of local adaptation in quantitative traits: a case study of ninespined sticklebacks. *Evolution* 68: 559–568. [12]

Karlin, S. 1975. General two-locus selection models: some objectives, results and interpretations. *Theor. Popul. Biol.* 7: 364–398. [5]

Karlin, S., and U. Liberman. 1979. Representation of nonepistatic selection models and analysis of multilocus Hardy-Weinberg equilibrium configurations. *J. Math. Biol.* 7: 353–374. [5]

Karlin, S., and J. L. McGregor. 1972. Addendum to a paper of W. Ewens. *Theor. Popul. Biol.* 3: 113–116. [2]

Karlin, S., and S. Tavaré. 1980. The detection of a recessive visible gene in finite populations. *Genet. Res.* 37: 33–46. [26]

Karlin, S., and S. Tavaré. 1981a. The detection of particular genotypes in finite populations. I. Natural selection effects. *Theor. Popul. Biol.* 19: 187–214. [26].

Karlin, S., and S. Tavaré. 1981b. The detection of particular genotypes in finite populations. I. The effects of partial penetrance and family structure. *Theor. Popul. Biol.* 19: 215–229. [26].

Karlin, S., and H. M. Taylor. 1981. *A second course in stochastic processes*. Academic Press, New York, NY. [A1]

Karn, M. N., and L. S. Penrose. 1951. Birth weight and gestation time in relation to maternal age, parity and infant survival. *Ann. Eugen.* 16: 147–164. [29]

Kass, R. E., B. P. Carlin, A. Gelman, and R. M. Neal. 1998. Markov chain Monte Carlo in practice: A roundtable discussion. *Am. Stat.* 52: 93–100. [A3]

Kassen, R., and T. Bataillon. 2006. Distribution of fitness effects among beneficial mutations before selection in experimental populations of bacteria. *Nat. Genet.* 38: 484–488. [27]

Katz, A. J., and S. S. Y. Young. 1975. Selection for high adult body weight in *Drosophila* populations with different structures. *Genetics* 81: 163–175. [26]

Kauer, M. O., D. Dieringer, and C. Schlötterer. 2003. A microsatellite variability screen for positive selection associated with the "Out of Africa" habitat expansion of *Drosophila melanogaster*. *Genetics* 165: 1137–1148. [9]

Kauffman, S. A. 1969. Metabolic stability and epigenesis in randomly constructed genetic nets. *J. Theor. Biol.* 22: 437–467. [12]

Kaufman, P. K., F. D. Enfield, and R. E. Comstock. 1977. Stabilizing selection for pupa weight in *Tribolium castaneum*. *Genetics* 87: 327–341. [16]

Kavanagh, E. (ed.). 2006. Debating sexual selection and mating strategies. *Science* 312: 689–697. [29]

Kavanaugh, C. M., and R. G. Shaw. 2005. The contribution of spontaneous mutation to variation in environmental responses of *Arabidopsis thaliana:* responses to light. *Evolution* 59: 266–275. [12]

Kawecki, T. J., and 5 others. 2012. Experimental evolution. *Trends Ecol Evol.* 27: 547–560. [18]

Kayser, M., S. Brauer, and M. Stoneking. 2003. A genome scan to detect candidate regions influenced by local natural selection in human populations. *Mol. Biol. Evol* 20: 893–900. [11]

Kearsey, M. J., and J. S. Gale. 1968. Stabilising selection in the absence of dominance: an additional note. *Heredity* 23: 617–620. [5, 28]

Keightley, P. D., and G. Bulfield. 1993. Detection of quantitative trait loci from frequency changes of marker alleles under selection *Genet. Res.* 62:195–203. [9]

Keightley, P. D., and A. Eyre-Walker. 2012. Estimating the rate of adaptive molecular evolution when the evolutionary divergence between species is small. *J. Mol. Evol.* 74: 61–68. [10]

Keightley, P. D., and D. L. Halligan. 2009. Analysis and implications of mutational variation. *Genetica* 136: 359–369. [7]

Keightley, P. D., and D. L. Halligan. 2011. Inference of site frequency spectra from high-throughput sequence data: quantification of selection on nonsynonymous and synonymous sites in humans. *Genetics* 188: 931–940. [4]

Keightley, P. D., and W. G. Hill. 1983. Effects of linkage on response to directional selection from new mutations. *Genet. Res.* 42: 193–206. [26]

Keightley, P. D., and W. G. Hill. 1987. Directional selection and variation in finite populations. *Genetics* 117: 573–582. [1, 16, 24, 26]

Keightley, P. D., and W. G. Hill. 1988. Quantitative genetic variability maintained by mutation-stabilizing selection balance in finite populations. *Genet. Res.* 52: 33–43. [28]

Keightley, P. D., and W. G. Hill. 1989. Quantitative genetic variability maintained by mutation-stabilizing selection balance: sampling variation and response to subsequent directional selection. *Genet. Res.* 54: 45–58. [11, 28]

Keightley, P. D., and W. G. Hill. 1990. Variation maintained in quantitative traits with mutation-selection balance: pleiotropic side-effects on fitness traits. *Proc. Roy. Soc. Lond. B.* 242: 95–100. [28]

Keightley, P. D., and M. Lynch. 2003. Toward a realistic model of mutations affecting fitness. *Evolution* 57: 683–685. [12]

Keightley, P. D., and S. P. Otto. 2006. Interference among deleterious mutations favours sex and recombination in finite populations. *Nature* 443: 89–92. [4]

Keightley, P. D., and 7 others. 2015. Estimation of the spontaneous mutation rate in *Heliconius melpomene*. *Mol. Biol. Evol.* 32: 239–243. [4]

Keith, N., and 11 others. 2015. High mutational rates of large-scale duplication and deletion in *Daphnia pulex*. *Genom. Res.* 26: 60–69. [4]

Keller, L. F., P. R. Grant, B. R. Grant, and K. Petren. 2001. Heritability of morphological traits in Darwin's finches: misidentified paternity and maternal effects. *Heredity* 87: 325–336. [20]

Kelly, J. K. 1997. A test of neutrality based on interlocus associations. *Genetics* 146: 1197–1206. [9]

Kelly, J. K. 1999a. Response to selection in partially self-fertilizing populations. I. selection on a single trait. *Evolution* 53: 336–349. [23]

Kelly, J. K. 1999b. Response to selection in partially self-fertilizing populations. II. selection on a multiple traits. *Evolution* 53: 350–357. [23]

Kelly, J. K. 1999c. An experimental method for evaluating the contribution of deleterious mutations to quantitative trait variation. *Genet. Res.* 73: 263–273. [11, 13, 28]

Kelly, J. K. 2008. Testing the rare-allele model of quantitative variation by artificial selection. *Genetica* 132: 187–198. [28]

Kelly, J. K., and H. S. Arathi. 2003. Inbreeding and the genetic variance in floral traits of *Mimulus guttatus*. *Heredity* 90: 77–83. [11]

Kelly, J. K., and M. J. Wade. 2000. Molecular evolution near a two-locus balanced polymorphism. *J. Theor. Biol.* 204: 83–101. [9]

Kelly, J. K., and S. Williamson. 2000. Predicting response to selection on a quantitative trait: a comparison between models for mixed-mating populations. *J. Theor. Biol.* 207: 37–56. [23]

Kelly, J. K., and J. H. Willis. 2001. Deleterious mutations and genetic variation for flower size in *Mimulus guttatus*. *Evolution* 55: 937–942. [11, 28]

Kemper, K. E., S. J. Saxton, S. Bolormaa, B. J. Hayes, and M. E. Goddard. 2014. Selection for complex traits leaves little or no classic signatures of selection. *BMC Genom.* 15: 246. [8, 9]

Kemper, K. E., P. M. Visscher, and M. E. Goddard. 2012. Genetic architecture of body size in mammals. *Genome Bio.* 13: 244. [24, 25, 28]

Kempthorne, O. 1954. The correlation between relatives in a random mating population. *Proc. Royal Soc. Lond. B* 143: 103–113. [1]

Kempthorne, O. 1957. *An introduction to genetic statistics*. Iowa State University Press, Ames, IA. [6, 15]

Kempthorne, O., and E. Pollak. 1970. Concepts of fitness in Mendelian populations. *Genetics* 64: 125–145. [5]

Kendall, M., and A. Stuart. 1977. *The advanced theory of statistics. Vol. 1. Distribution theory*. 4th Ed. Macmillan, New York, NY. [14, 24, 26]

Kendall, M., and A. Stuart. 1983. *The advanced theory of statistics. Volume 3: Design and analysis, and time-series*. 4th Ed. Griffin, London, UK. [29, 30]

Kenett, R., and S. Zacks. 1998. *Modern industrial statistics: the design and control of quality and reliability*. Cengage Learning, Boston, MA. [29]

Kennedy, B. W. 1984. Selection limits: have they been reached with the dairy cow? *Can. J. Anim. Sci.* 64: 207–215. [25]

Kennedy, B. W. 1990. Use of mixed model methodology in analysis of designed experiments. *In* D. Gianola and K. Hammond (eds.), *Statistical methods for genetic improvement of livestock*, pp. 77–97. Springer-Verlag, New York, NY. [19]

Kennedy, B. W., L. R. Schaeffer, and D. A. Sorensen. 1988. Genetic properties of animal models. *J. Dairy Sci.* 17: Supp. 2 17–26. [19]

Kennedy, B. W., and D. A. Sorensen. 1988. Properties of mixed-model methods for prediction of genetic merit. *In* B. S. Weir, E. J. Eisen, M. M. Goodman, and G. Namkoong (eds.), *Proceedings of the second international conference on quantitative genetics*, pp. 91–103. Sinauer Assoc., Sunderland, MA. [19]

Kennedy, B. W., and D. Trus. 1993. Considerations on genetic connectedness between management units under an animal model. *J. Anim. Sci.* 71: 2341–2352. [19]

Kern, A. D., and D. Haussler. 2010. A population genetic hidden Markov model for detecting genomic regions under selection. *Mol. Biol. Evol.* 27: 1673–1685. [9]

Kerr, B., and P. Godfrey-Smith. 2008. Generalization of the Price equation for evolutionary change. *Evolution* 63: 531–536. [6]

Kerr, M. K. 2003. Design considerations for efficient and effective microarray studies. *Biometrics* 59: 822–828. [12]

Kerr, M. K., and G. A. Churchill. 2001a. Experimental design for gene expression microarrays. *Biostatistics* 2: 183–201. [12]

Kerr, M. K., and G. A. Churchill. 2001b. Statistical design and the analysis of gene expression microarray data. *Genet. Res.* 77: 123–128. [12]

Keynes, J. M. 1921. *A treatise on probability*. Macmillan, London, UK. [19]

Khaitovich, P., S. Pääbo, and G. Weiss. 2005. Toward a neutral evolutionary model of gene expression. *Genetics* 170: 929–939. [12]

Khaitovich, P., and 8 others. 2004. A neutral model of transcriptome evolution. *PLoS Biol.* 2: 682–689. [12]

Kidwell, J. F., M. T. Clegg, F. M. Stewart, and T. Prout. 1977. Regions of stable equilibria for models of differential selection in the two sexes under random mating. *Genetics* 85: 171–183. [5]

Kiesselbach, T. A. 1916. Recent developments in our knowledge concerning corn. *Nebraska Corn Improver's Assoc. 7th Annu. Rep.* 15–42. [21]

Kiesselbach, T. A. 1922. Corn investigations. *Nebraska Agr. Exp. Sta. Res. Bull.* 20. [21]

Kim, S., and 8 others. 2007. Recombination and linkage disequilibrium in *Arabidopsis thaliana. Nat. Genet.* 39: 1151–1155. [4]

Kim, Y. 2004. Effects of strong directional selection on weakly selected mutations at linked sites: implications for synonymous codon usage. *Mol. Biol. Evol.* 21: 286–294. [8]

Kim, Y. 2006. Allele frequency distribution under recurrent selective sweeps. *Genetics* 172: 1967–1978. [8]

Kim, Y., and T. Maruki. 2011. Hitchhiking effect of a beneficial mutation spreading in a subdivided population. *Genetics* 189: 213–226. [9]

Kim, Y., and R. Nielsen. 2004. Linkage disequilibrium as a signature of selective sweeps. *Genetics* 167: 1513–1524. [8, 9]

Kim, Y., and W. Stephan. 2000. Joint effects of genetic hitchhiking and background selection on neutral variation. *Genetics* 155: 1415–1427. [3, 8]

Kim, Y., and W. Stephan. 2002. Detecting a local signature of genetic hitchhiking along a recombining chromosome. *Genetics* 160: 765–777. [8, 9]

Kim, Y., and W. Stephan. 2003. Selective sweeps in the presence of interference among partially linked loci. *Genetics* 164: 389–398. [3, 8]

Kimura, M. 1955a. Stochastic process and distribution of gene frequencies under natural selection. *Cold Spring Harbor Symp. Quant. Biol.* 20: 33–53. [1, A1]

Kimura, M. 1955b. Solution of a process of random genetic drift with a continuous model. *Proc. Natl. Acad. Sci. USA* 41: 144–150. [1, 2, A1]

Kimura, M. 1956. A model of a genetic system which leads to closer linkage by natural selection. *Evolution* 10: 278–287. [1, 5]

Kimura, M. 1957. Some problems of stochastic processes in genetics. *Ann. Math. Stat.* 28: 882–901. [1, 7, 26, A1]

Kimura, M. 1962. On the probability of fixation of mutant genes in a population. *Genetics* 47: 713–719. [1, A1]

Kimura, M. 1965a. A stochastic model concerning the maintenance of genetic variability in a quantitative character. *Proc. Natl. Acad. Sci. USA* 54: 731–736. [1, 5, 24, 28]

Kimura, M. 1965b. Attainment of quasi linkage equilibrium when gene frequencies are changing by natural selection. *Genetics* 52: 875–890. [6, 24]

Kimura, M. 1967. On the evolutionary adjustment of spontaneous mutation rates. *Genet. Res.* 9: 23–34. [4]

Kimura, M. 1968a. Genetic variability maintained in a finite population due to mutational production of neutral and nearly neutral isoalleles. *Genet. Res.* 11: 247–269. [2]

Kimura, M. 1968b. Evolutionary rate at the molecular level. *Nature* 217: 624–626. [1, 8]

Kimura, M. 1969. The number of heterozygous nucleotide sites maintained in a finite population due to steady flux of mutations. *Genetics* 61: 893–903. [7, 10, 28]

Kimura, M. 1981. Possibility of extensive neutral evolution under stabilizing selection with special reference to nonrandom usage of synonymous codons. *Proc. Natl. Acad. Sci. USA* 78: 5773–5777. [28]

Kimura, M. 1983. *The neutral theory of molecular evolution.* Cambridge Univ. Press, Cambridge, UK. [1, 2, 7, 8, 10, 12, 27, A1]

Kimura, M. 1985. The role of compensatory neutral mutations in molecular evolution. *J. Genet.* 64: 7–19. [7]

Kimura, M., and J. F. Crow. 1963a. On the maximum avoidance of inbreeding. *Genet. Res.* 4: 399–415. [3, 12]

Kimura, M., and J. F. Crow. 1963b. The measurement of effective population number. *Evolution* 17: 279–288. [3]

Kimura, M., and J. F. Crow. 1964. The number of alleles that can be maintained in a finite population. *Genetics* 49: 725–738. [1, 2, 11, 24]

Kimura, M., and J. F. Crow. 1978. Effect of overall phenotypic selection on genetic change at individual loci. *Proc. Natl. Acad. Sci. USA* 75: 6168–6171. [5, 14, 28]

Kimura, M., and T. Maruyama. 1966. The mutational load with epistatic gene interactions in fitness. *Genetics* 54: 1337–1351. [3, 7, 8]

Kimura, M., T. Maruyama, and J. F. Crow. 1963. The mutation load in small populations. *Genetics* 48: 1303–1312. [7]

Kimura, M., and T. Ohta. 1969a. The average number of generations until fixation of an individual mutant gene in a finite population. *Genetics* 61: 763–771. [7, A1]

Kimura, M., and T. Ohta. 1969b. The average number of generations until extinction of an individual mutant gene in a finite population. *Genetics* 63: 701–709. [7, A1]

Kimura, M., and T. Ohta. 1973. The age of a neutral mutation persisting in a finite population. *Genetics* 75: 199–212. [1, 2]

Kimura, R., A. Fujimoto, K. Tokunaga, and J. Ohashi. 2007. A practical genome scan for population-specific strong selective sweeps that have reached fixation. *PLoS One* 3: e286. [9]

Kindred, B. 1965. Selection for temperature sensitivity in scute *Drosophilia. Genetics* 52: 723–28. [21]

King, J. L., and T. H. Jukes. 1969. Non-Darwinian evolution. *Science* 164: 788–798. [8]

King, J. W. B. 1955. Observations on the mutant 'pygmy' in the house mouse. *J. Genet.* 53: 487–497. [25]

King, M.-C., and A. C. Wilson. 1975. Evolution at two levels in humans and chimpanzees. *Science* 188: 107–116. [9, 12]

Kinghorn, B. P., J. H .J. Van der Werf, and M. Ryan (Eds.) 2000. *Animal breeding: Use of new technologies.* The Post Graduate Foundation in Veterinarian Science of the University of Sydney, Australia. [13]

Kingman, J. F. C. 1961a. On an inequality in partial averages. *Q. J. Math.* 12: 78–80. [5]

Kingman, J. F. C. 1961b. A mathematical problem in population genetics. *Proc. Camb. Philos. Soc.* 57: 574–582. [5]

Kingman, J. F. C. 1977. On the properties of bilinear models for the balance between genetic mutation and selection. *Math. Proc. Camb. Phil. Soc.* 81: 443–453. [11, 28]

Kingman, J. F. C. 1978. A simple model for the balance between selection and mutation. *J. Appl. Prob.* 15: 1–12. [11, 28]

Kingman, J. F. C. 1982a. The coalescent. *Stoch. Proc. Appl.* 13: 235–248. [1, 2]

Kingman, J. F. C. 1982b. On the genealogy of large populations. *J. Appl. Prob.* 19: 27–43. [1, 2]

Kingman, J. F. C. 2000. Origins of the coalescent 1974–1982. *Genetics* 156: 1461–1463. [2]

Kingsolver, J. G., and S. E. Diamond. 2011. Phenotypic selection in natural populations: what limits directional selection? *Am. Nat.* 177: 346–357. [29, 30]

Kingsolver, J. G., S. E. Diamond, A. M. Siepielski, and S. M. Carlson. 2012. Synthetic analyses of phenotypic selection in natural populations: lessons, limitations and future directions. *Evol. Ecol.* 26: 1101–1118. [29, 30]

Kingsolver, J. G., and D. W. Pfenning. 2004. Individual-level selection as a cause of Cope's rule of phyletic size increase. *Evolution* 58: 1608–1612. [30]

Kingsolver, J. G., and D. W. Pfenning. 2007. Patterns and power of phenotypic selection in nature. *BioScience* 57: 561–572. [29, 30]

Kingsolver, J. G., and D. W. Schemske. 1991. Path analysis of selection. *Trends Ecol. Evol.* 6: 276–280. [30]

Kingsolver, J. G., and 8 others. 2001. The strength of phenotypic selection in natural populations. *Am. Nat.* 157: 245–261. [28, 29, 30]

Kinney, T. B. Jr., B. B. Bohren, J. V. Craig, and P. C. Lowe. 1970. Responses to individual, family or index selection for short term rate of egg production in chickens. *Poultry Sci.* 49: 1052–1064. [21]

Kinnison, M. T., and A. P. Hendry. 2001. The pace of modern life II: from rates of contemporary microevolution to pattern and process. *Genetica* 112-113: 145–164. [30]

Kirby, D. A., and W. Stephan. 1995. Haplotype test reveals departure from neutrality in a segment of the *white* gene of *Drosophila melanogaster*. *Genetics* 141: 1483–1490. [9]

Kirkpatrick, M., T. Johnson, and N. Barton. 2002. General models of multilocus evolution. *Genetics* 161: 1727–1750. [5]

Kirkpatrick, M., and R. Lande. 1989. The evolution of maternal characters. *Evolution* 43: 485–503. [15, 22, 29, 30]

Kirkpatrick, M., and R. Lande. 1992. The evolution of maternal characters: Errata. *Evolution* 46: 284. [15]

Kirkpatrick, S., C. D. Gelatt, and M. P. Vecchi. 1983. Optimization by simulated annealing. *Science* 220: 671–680. [A3]

Kirzhner, V. M., A. B. Korol, and E. Nevo. 1996a. Complex dynamics of multilocus systems subjected to cyclical selection. *Proc. Natl. Acad. Sci. USA* 93: 6532–6535. [28]

Kirzhner, V. M., A. B. Korol, and E. Nevo. 1996b. Complex limiting behaviour of multilocus genetic systems in cyclical environments. *J. Theor. Biol.* 190: 215–225. [28]

Klein, J. P., and M. L. Moeschberger. 1997. *Survival analysis: techniques for censored and truncated data.* Springer, New York, NY. [29]

Klein, J., 1980. Generation of diversity at MHC loci: implications for T-cell receptor repertoires, *In* M. Fougereau and J. Dausset (eds.), *Immunology 80*, pp. 239–253. Academic Press, London, UK. [9]

Klein, J., A. Sato, S. Nagl, and C. O'hUigín. 1998. Molecular trans-species polymorphism. *Ann. Rev. Ecol. Syst.* 29 1–21. [9]

Kliman, R. M., and J. Hey. 1993. Reduced natural selection associated with low recombination in *Drosophila melanogaster*. *Mol. Biol. Evol.* 10: 1239–1258. [8]

Klopfstein, S., M. Currat, and L. Excoffier. 2006. The fate of mutations surfing on the wave of a range expansion. *Mol. Biol. Evol.* 23: 482–490. [9]

Klug, H., J. Heuschele, M. D. Jennions, and H. Kokko. 2010a. The mismeasurement of sexual selection. *J. Evol. Biol.* 23: 447–462. [29]

Klug, H., K. Lindstrom, and H. Kokko. 2010b. Who to include in studies of sexual selection is no trivial matter. *Ecol. Lett.* 13: 1094–1102. [29]

Knapczyk, F. N., and J. K. Conner. 2007. Estimates of the average strength of natural selection are not inflated by samplng error or publication bias. *Am. Nat.* 170: 501–508. [30]

Knapp, S. J., W. C. Bridges, Jr., and M.-H. Yang. 1989. Nonparametric confidence interval estimators for heritability and expected selection response. *Genetics* 121: 891–898. [18]

Knight, G. R., and A. Robertson. 1957. Fitness as a measurable character in *Drosophila*. *Genetics* 42: 524–530. [25]

Knight, J. C. 2004. Allele-specific gene expression uncovered. *Trends Genet.* 20: 113–116. [12]

Koch, A. L 1974. The pertinence of the periodic selection phenomenon to prokaryote evolution. *Genetics* 77: 127–142. [8]

Kochert, G., and 5 others. 1996. RFLP and cytogenetic evidence on the origin and evolution of allotetraploid domesticated peanut, *Arachis hypogaea* (Leguminosae). *Am. J. Botany* 83: 1282–1291. [9]

Koenig, W. D., and S. S. Albano. 1986. On the measurement of sexual selection. *Am. Nat.* 127: 403–409. [29]

Koenig, W. D., and S. A. Albano. 1987. Lifetime reproductive success, selection, and the opportunity for selection in the white-tailed skimmer *Plathemis lydia* (Odonta: Libellulidae). *Evolution* 41: 22–36. [29]

Koenig, W. D., S. A. Albano, and J. L .Dickinson. 1991. A comparison of methods to partition selection via components of fitness: do larger male bullfrogs have greater hatching success? *J. Evol. Biol.* 4: 309–320. [29]

Kofler, R., and C Schlötterer. 2013. A guide for the design of evolve and resequencing studies. *Mol. Biol. Evol.* 31: 474–483.

Kohn, M. H., H.-J. Pelz, and R. K. Wayne. 2000. Natural selection mapping of the warfarin-resistance gene. *Proc. Natl. Acad. Sci. USA* 97: 7911-7915. [9]

Kojima, K. 1959. Stable equilibria for the optimum model. *Proc. Natl. Acad. Sci. USA* 45: 989—993. [28]

Kojima, K , and T. M. Kelleher. 1961. Changes of mean fitness in random-mating populations when epistasis and linkage are present. *Genetics* 46: 527–540. [5]

Kojima, K , and H. E. Schaffer. 1967. Survival process of linked mutant genes. *Evolution* 21: 518–531. [8]

Kojima, K.-I. 1961. Effects of dominance and size of population on response to mass selection. *Genet. Res.* 2: 177–188. [26]

Kokko, H., R. Brooks, M.l D. Jennions, and J. Morley. 2003. The evolution of mate choice and mating biases. *Proc. Roy Soc. Lond. B* 270: 653–664. [29]

Kokko, H., M. D. Jennions, and R. Brooks. 2006. Unifying and testing models of sexual selection. *Ann. Rev. Ecol. Evol. Syst.* 37: 43–66. [29]

Kokko, H., A. Mackenzie, J. D. Reynolds, J. Lindström, and W. J. Sutherland. 1999. Measures of inequality are not equal. *Am. Nat.* 154: 358–382. [29]

Kolmogorov, A. N. 1937. A study of the equation of diffusion with increase in the quantity of matter, and its application to a biological problem. *Moscow Univ. Bull. Math.* 1 :1–25. [27]

Komarova, N. L., A. Sengupta, and M. A. Nowak. 2003. Mutation-selection networks of cancer initiation: tumor suppresser genes and chromosomal instability. *J. Theor. Biol.* 223: 433–450. [7]

Kondrashov, A. S. 1984. Deleterious mutations as an evolutionary factor. 1. The advantage of recombination. *Genet. Res.* 44: 199–217. [7]

Kondrashov, A. S. 1988. Deleterious mutations and the evolution of sexual reproduction. *Nature* 336: 435–440. [1, 4, 7]

Kondrashov, A. S. 2003. Direct estimates of human per nucleotide mutation rates at 20 loci causing Mendelian diseases. *Hum. Mutat.* 21: 12–27. [4, 7]

Kondrashov, A. S., and J. F. Crow. 1993. A molecular approach to estimating the human deleterious mutation rate. *Hum. Mutat.* 2: 229–234. [7]

Kondrashov, A. S., S. Sunyaev, and F. A. Kondrashov. 2002. Dobzhansky-Muller incompatibilities in protein evolution. *Proc. Natl. Acad. Sci. USA* 99: 14878–14883. [7]

Kondrashov, A. S., and M. Turelli. 1992. Deleterious mutations, apparent stabilizing selection and the maintenance of quantitative variation. *Genetics* 132: 603–618. [28]

Kondrashov, A. S., and L. Y. Yampolsky. 1996a. High genetic variability under the balance between symmetric mutation and fluctuating stabilizing selection. *Genet. Res.* 68: 157–164. [28]

Kondrashov, A. S., and L. Y. Yampolsky. 1996b. Evolution of amphimixis and recombination under fluctuating selection in one and many traits. *Genet. Res.* 68: 165–173. [28]

Kong, A., and 20 others. 2012. Rate of *de novo* mutations and the importance of father's age to disease risk. *Nature* 488: 471-475. [4]

Konigsberg, L. K., and J. M. Cheverud. 1992. Uncertain paternity in primate quantitative genetic studies. *Am. J. Primatology* 27: 133–143. [20]

Konishi, S., and 6 others. 2006. An SNP caused loss of seed shattering during rice domestication. *Science* 312: 1293–1396. [9]

Kopp, M., and J. Hermisson. 2006. The evolution of genetic architecture under frequency-dependent disruptive selection. *Evolution* 60: 1537–1550. [28]

Kopp, M., and J. Hermisson. 2007. Adaptation of a quantitative trait to a moving optimum. *Genetics* 176: 715–719. [27]

Kopp, M., and J. Hermisson. 2009a. The genetic basis of phenotypic adaptation II: fixation of beneficial mutations in the moving optimum mode. *Genetics* 182: 233–249. [27]

Kopp, M., and J. Hermisson. 2009b. The genetic basis of phenotypic adaptation II: the distribution of adaptive substitutions in the moving optimum model. *Genetics* 183: 1453–1476. [27]

Koricheva, J., J. Gurevitch, and K. Mengersen (eds). 2013. *Handbook of meta-analysis in ecology and evolution*, Princeton University Press, Princeton, NJ. [30, A4]

Korol, A. B., and K. G. Iliad. 1994. Increased recombination frequencies resulting from directional selection for geotaxis in *Drosophila*. *Heredity* 72: 64–68. [26]

Korol, A. B., V. M. Kirzhner, Y. I. Ronin, and E. Nevo. 1996. Cyclical environmental changes as a factor maintaining genetic polymorphism. 2. Diploid selection for an additive trait. *Evolution* 50: 1432–1441. [28]

Korsgaard, I. R., A. H. Andersen, and J. Jensen. 2002. Prediction error variance and expected response to selection, when selection is based on the best predictor—for Gaussian and threshold characters, traits following a Poisson mixed model and survival traits. *Genet. Sel. Evol.* 34: 307–333. [14]

Kosakovsky Pond, S. L., and S. D. W. Forst. 2005. A genetic algorithm approach to detecting lineage-specific variation in selection pressure. *Mol. Biol. Evol.* 22: 478–485. [10]

Kosakovsky Pond, S. L., and 5 others. 2011. A random effects branch-site model for detecting episodic diversifying selection. *Mol. Biol. Evol.* 28: 3033–3043. [10]

Kosiol, C., and 6 others 2008. Patterns of positive selection in six mammalian genomes. *PLoS Genet.* 4: e1000144. [10]

Kotiaho, J. S., and M. Puurtinen. 2007. Mate choice for indirect genetic benefits: scrutiny of the current paradigm. *Func. Ecol.* 21: 638–644. [29]

Kousathanas, A., and P. D. Keightley. 2013. A comparison of methods to infer the distribution of fitness effects of new mutations. *Genetics* 193: 1197–1208. [10]

Kousathanas, A., F. Oliver, D. L. Halligan, and P. D. Keightley. 2011. Positive and negative selection on noncoding DNA close to protein-coding genes in wild house mice. *Mol. Biol. Evol.* 28: 1183–1191. [10]

Kouyos, R. D., O. K. Silander, and S. Bonhoeffer. 2007. Epistasis between deleterious mutations and the evolution of recombination. *Trends Ecol. Evol.* 22: 308–315. [7]

Kownacki, M. 1979. Effect of reciprocal crossing of selected lines of mice. *Theor. Appl. Genet.* 54: 169–175. [26]

Kozlowski, J. 1993. Measuring fitness in life-history studies. *Trends Ecol. Evol.* 8: 84–85. [29]

Kraft, P. 2008. Curses—winner's and otherwise—in genetic epidemiology. *Epidemiology* 19: 649–651. [10]

Krakauer, A. H., M. S. Webster, E. H. Duval, A. G. Jones, and S. M. Shuster. 2011. The opportunity for sexual selection: not mismeasured, just misunderstood. *J. Evol. Biol.* 24,: 2064–2071. [29]

Kramer, M. G., and 7 others. 1998. Genetic variation in body weight growth and composition in the intercross of large (LG/J) and small (SM/J) inbred strains of mice. *Genet. Mol. Biol.* 21: 211–218. [11]

Kreitman, M. 2000. Methods to detect selection in populations with applications to the human. *Ann. Rev. Genomics Hum. Genet.* 1: 539–559. [9]

Kreitman, M., and R. R. Hudson. 1991. Inferring the evolutionary histories of the *Adh* and *Adh-dup* Loci in *Drosophila melanogaster* from patterns of polymorphism and divergence. *Genetics* 127: 565–582. [8]

Kremer, A., and V. Le Corre. 2012. Decoupling of differentiation between traits and their underlying genes in response to divergent selection. *Heredity* 108: 375–385. [12]

Kress, D. D. 1975. Results for long-term selection experiments relative to selection limits. *In, Genetics Lectures 4*, pp. 253–271. Oregon State University Press, Corvallis, Oregon. [26]

Krimbas, C. B., and S. Tsakas. 1971. The genetics of *Dacus oleae.* V. Changes of esterase polymorphism in a natural population following insecticide control-selection or drift? *Evolution* 25: 454–460. [4]

Kristensen, T. N., and 5 others. 2005. A test of quantitative genetic theory using *Drosophila*–effects of inbreeding and rate of inbreeding on heritabilities and variance components. *J. Evol. Biol.* 18: 763–770. [11]

Kronholm, I., O. Loudet, and J. de Meaux. 2010. Influence of mutation rate on estimators of genetic differentiation-lessons from *Arabidopsis thaliana. BMC Genet.* 11: 33. [12]

Kruuk, L. E. B. 2004. Estimating genetic parameters in natural populations using the 'animal models.' *Phil. Trans. R. Soc. Lond B* 358: 873–890. [1, 19, 20]

Kruuk, L. E. B., T. Clutton-Brock, and J. M. Pemberton. 2014. Case study: quantitative genetics and sexual selection of weaponry in a wild ungulate. *In* A. D. Charmantier, D. Garant, and L. E. B. Kruuk (eds.), *Quantitative genetics in the wild*, pp.160–172. University of Oxford Press, Oxford, UK. [20]

Kruuk. L. E. B., and D. Garant. 2007. A wake-up call for studies of natural selection? *J. Evol. Biol.* 20: 30–33. [30]

Kruuk, L. E. B., and J. D. Hadfield. 2007. How to separate genetic and environmental causes of similarity between relatives. *J. Evol. Biol.* 20: 1890–1903. [20]

Kruuk, L. E. B., and W. G. Hill. 2008. Introduction. Evolutionary dynamics of wild populations: the use of long-term pedigree data. *Proc. R. Soc. B* 275: 593–596. [20]

Kruuk, L. E. B., J. Merilä, B. C. Sheldon. 2001. Phenotypic selection on a heritable size trait revisited. *Am. Nat.* 158: 557–571. [20]

Kruuk, L. E. B., J. Slate, and A. J. Wilson. 2008. New answers from old questions: The evolutionary quantitative genetics of wild animal populations. *Ann. Rev. Ecol. Evol. Syst*, 39: 525–548. [20]

Kruuk, L. E. B., and 5 others. 2000. Heritability of fitness in a wild mammal population. *Proc. Natl. Acad. Sci. USA* 97: 698–703. [6, 20]

Kruuk, L. E. B., and 5 others. 2002. Antler size in red deer: heritability and selection but no evolution. *Evolution* 56: 1683–1695. [20]

Kuhner, M. K. 2009. Coalescent genealogy samplers: windows into population history. *Trends Ecol. Evol.* 24: 86–93. [4]

Kuhner, M. K., J. Yamato, and J. Felsenstein. 2000. Maximum likelihood estimation of recombination rates from population data. *Genetics* 156: 1393–1401. [4]

Kulathinal, R. J., B. R. Bettencourt, and D. L. Hartl. 2004. Compensated deleterious mutations in insect genomes. *Science* 306: 1553–1554. [7]

Kuroda, R., B. Endo, M. Abe, and M. Shimiizu. 2009. Chiral blastomere arrangement dictates zygotic left–right asymmetry pathway in snails. *Nature* 462: 790–794. [15]

Kurtz, T. G. 1971. Limit theorems for sequence of jump Markov processes approximating ordinary differential equations. *J. Appl. Prob.* 8: 344–356. [8]

Kvarnemo, C., and I. Ahnesjö. 1996. The dynamics of operational sex ratios and competition for mates. *Trends Ecol. Evol.* 11: 404–408. [29]

Labate, J. A., K. R. Lamkey, M. Lee, and W. L. Woodman. 1999. Temporal changes in allele frequencies in two reciprocally selected maize populations. *Theor. Appl. Genet.* 99: 1166–1178. [9]

Lacerda, M., and C. Seoighe. 2014. Population genetics inference for longitudinally-sampled mutants under strong selection. *Genetics* 198: 1237–1250. [9]

Lachmann-Tarkhanov, M., and S. Sarkar. 1994. The alternative fitness sets which preserve allele trajectories: a general treatment. *Genetics* 138: 1323–1330. [5]

Lagercrantz, U., M. K. Osterberg, and M. Lascoux. 2002. Sequence variation and haplotype structure at the putative flowering-time locus *COL1* of *Brassica nigra. Mol. Biol. Evol.* 19: 1474–1482. [4]

Laidig, F., H.-P. Piepho, T. Drobek, and U. Meyer. 2014. Genetic and non-genetic long-term trends of 12 different crops in German official variety performance trials and on-farm yield trends. *Theor. Appl. Genet.* 127: 2599–2617. [19]

Laidig, F., and 5 others. 2017. Breeding progress, variation, and correlation of grain and quality traits in winter rye hybrid and population varieties and national on-farm progress in Germany over 26 years. *Theor. Appl. Genet.* 130: 981–998. [19]

Laland, K. N., and 7 others. 2015. The extended evolutionary synthesis: its structure, assumptions and predictions. *Proc. Roy. Soc. Lond. B* 282: 20151019. [1]

Laland, K., et al. 2014. Does evolutionary theory need a rethink? *Nature* 514: 161–164. [1]

Lande, R. 1975. The maintenance of genetic variability by mutation in a polygenic character with linked loci. *Genet. Res.* 26: 221–235. [5, 24, 28]

Lande, R. 1976. Natural selection and random genetic drift in phenotypic evolution. *Evolution* 30: 314–334. [1, 6, 11, 12, 13, 21, 29, 30, A1]

Lande, R. 1977a. The influence of the mating system on the maintenance of genetic variability in polygenic characters. *Genetics* 86: 485–498. [23, 24, 28]

Lande, R. 1977b. Statistical tests for natural selection on quantitative characters. *Evolution* 31: 442–446. [12]

Lande, R. 1977c. On comparing coefficients of variation. *Syst. Zoo.* 26: 214–217. [6]

Lande, R. 1978. Evolutionary mechanisms of limb loss in tetrapods. *Evolution* 32: 73–92. [1, 14]

Lande, R. 1979a. Quantitative genetic analysis of multivariate evolution, applied to brain:body size allometry. *Evolution* 33: 402–416. [12, 13, 30, A1, A6]

Lande, R. 1979b. Effective deme sizes during long-term evolution estimated from rates of chromosomal rearrangement. *Evolution* 33: 234–251. [7]

Lande, R. 1980a. The genetic covariance between characters maintained by pleiotropic mutations. *Genetics* 94: 203–215. [28]

Lande, R. 1980b. Genetic variation and phenotypic evolution during allopatric speciation. *Am. Nat.* 116: 463–479. [17]

Lande, R. 1982. A quantitative genetic theory of life history evolution. *Ecology* 63: 607–615. [29]

Lande, R. 1983. The response to selection on major and minor mutations affecting a metrical trait. *Heredity*

50: 47–65. [8, 24, 25]

Lande, R. 1984a. The genetic correlation between characters maintained by selection, linkage, and inbreeding. *Genet. Res.* 44: 309–320. [28]

Lande, R. 1984b. The expected fixation rate of chromosomal inversions. *Evolution* 38: 743–752. [7]

Lande, R. 1985. The fixation of chromosomal rearrangements in a subdivided population with local extinction and colonization. *Heredity* 54: 323–332. [7]

Lande, R. 1992. Neutral theory of quantitative genetic variance in an island model with local extinction and colonization. *Evolution* 46: 381–389. [11, 12]

Lande, R., and S. J. Arnold. 1983. The measurement of selection on correlated characters. *Evolution* 37: 1210–1226. [1, 13, 24, 28, 29, 30]

Lande, R., and G. F. Barrowclough. 1987. Effective population size, genetic variation, and their use in population management. *In* M. E. Soulé (ed.), *Viable populations for conservation*, pp. 87–124. Cambridge Univ. Press, Cambridge, UK. [3]

Lande, R., and M. Kirkpatrick. 1990. Selection response in traits with maternal inheritance. *Genet. Res.* 55: 189–197. [22]

Lande, R., and M. Kirkpatrick. 1990. Selection response in traits with maternal inheritance. *Genet. Res.* 55: 189–197. [15]

Lande, R., and E. Porcher. 2015. Maintenance of quantitative genetic variance under partial self-fertilization, with implications for evolution of selfing. *Genetics* 200: 891–906. [23]

Lande, R., and D. W. Schemske. 1985. The evolution of self-fertilization and inbreeding depression in plants. I. Genetic models. *Evolution* 39: 24–40. [23]

Lande, R., D. W. Schemske, and S. T. Schultz. 1994. High inbreeding depression, selective interference among loci, and the threshold selfing rate for purging recessive lethal mutations. *Evolution* 48: 965–978. [23]

Lande, R., and R. Thompson. 1990. Efficiency of marker-assisted selection in the improvement of quantitative traits. *Genetics* 124: 743–756. [1]

Lange, J. D., and J. E. Pool. 2016. A haplotype method detects diverse scenarios of local adaptation from genomic sequence variation. *Mol. Ecol.* 25: 3081–3100. [9]

Lange, K. 1978. Central limit theorems for pedigrees. *J. Math. Biol.* 6: 59–66. [24]

Lange, K. 1997. An approximate model of polygenic inheritance. *Genetics* 147: 1423–1430. [24]

Langley, C. H., B. P. Lazzaro, W. Phillips, E. Heikkinen, and J. M. Braverman. 2000. Linkage disequilibria and the site frequency spectra in the *su(s)* and $su(W^a)$ regions of *Drosophila melanogaster* X chromosome. *Genetics* 156: 1837–1852. [4, 8]

Lango Allen, H., and >100 others. 2010. Hundreds of variants clustered in genomic loci and biological pathways affect human height. *Nature* 467: 832–838. [24]

Larson, G., and J. Burger. 2013. A population genetics view of animal domestication. *Trends Genet.* 29: 197–205. [9]

Larson, G., and D. Q. Fuller. 2014. The evolution of animal domestication. *Ann. Rev. Ecol. Syst.* 45: 115–136. [9]

Larson, G., and 23 others. 2014. Current perspectives and the future of domestication studies. *Proc. Natl. Acad. Sci. USA* 111: 6139–6146. [9]

Larsson, K., H. P. van der Jeugd, I. T. van der Veen, P. Forslund. 1998. Body size declines despite positive selection on heritable size traits in a Barnacle Goose population. *Evolution* 52: 1169–1184. [20]

Lassalle, F., and 5 others. 2015. GC-content evolution in bacterial genomes: the biased gene conversion hypothesis expands. *PLoS Genet.* 11: e1004941. [10]

Latta, L. C. and 5 others. 2013. Genomic background and generation time influence deleterious mutation rates in *Daphnia*. *Genetics* 193: 539–544. [12]

Latta, R. G. 1998. Differentiation of allelic frequencies at quantitative trait loci affecting locally adaptive traits. *Am. Nat.* 151: 283–292. [12]

Latta, R. G. 2003. Gene flow, adaptive population divergence and comparative population structure across loci. *New Phytol.* 161: 51–58. [12]

Latta, R. G., and C. McCain. 2009. Path analysis of natural selection via survival and fecundity across contrasting environments in *Avena barbata*. *J. Evol. Biol.* 22: 2458–2469. [30]

Latter, B. D. H. 1959. Genetic sampling in a random mating control population of constant size and sex-ratio. *Aust. J. Biol. Sci.* 12: 500–505. [3]

Latter, B. D. H. 1960. Natural selection for an intermediate optimum. *Aust. J. Biol. Sci.* 13: 30–35. [25, 28]

Latter, B. D. H. 1965a. The response to artificial selection due to autosomal genes of large effect. I. Changes in gene frequency at an additive locus. *Aust. J. Biol. Sci.* 18: 585–598. [5, 14, 25, 26]

Latter, B. D. H. 1965b. The response to artificial selection due to autosomal genes of large effect. II. The effects of linkage on the limits to selection in finite populations. *Aust. J. Biol. Sci.* 18: 1009–1023. [7]

Latter, B. D. H. 1966a. The response to artificial selection due to autosomal genes of large effect. III. The effects of linkage on the rate of advance and approach to fixation in finite populations. *Aust. J. Biol. Sci.* 19: 131–146. [7, 26]

Latter, B. D. H. 1966b. The interaction between effective population size and linkage intensity under artificial selection. *Genet. Res.* 7: 313–323. [7, 26]

Latter, B. D. H. 1970. Selection in finite populations with multiple alleles. II. Centripetal selection, mutation, and isoallelic variation. *Genetics* 66: 165–186. [24, 28]

Latter, B. D. H., and C. E. Novitski. 1969. Selection in finite populations with multiple alleles I. Limits to directional selection. *Genetics* 62: 859–876. [26]

Latter, B. D. H., and A. Robertson. 1962. The effects of inbreeding and artificial selection on reproductive fitness. *Genet. Res.* 3: 110–138. [25]

Laurie, C. C., J. R. True, J. Liu, and J. M. Mercer. 1997. An introgression analysis of quantitative trait loci that contribute to a morphological difference between *Drosophila simulans* and *D. mauritiana*. *Genetics* 145: 339–348. [12]

Laurie, C. C., and 9 others. 2004. The genetic architecture of response to long-term artificial selection for oil concentration in the maize kernel. *Genetics* 168: 2141–2155. [25]

Lawless, J. F. 2003. *Statistical models and methods for lifetime data*, 2nd Ed. Wiley, New York, NY. [29]

Lawrance, A. J. 1976. On conditional and partial correlation. *Am. Stat.* 30: 146–149. [6]

Lawrie, D. S., P. W. Messer, R. Hershberg, and D. A. Petrov. 2013. Strong purifying selection at synonymous sites in *D. melanogaster*. *PLoS Genet.* 9: e1003527. [10]

Lawrie, D. S., D. A. Petrov, and P. W. Messer. 2011. Faster than neutral evolution of constrained sequences: the complex interplay of mutational biases and weak selection. *Genome Biol. Evol.* 3: 383–395. [10]

Lawson, H. A., and 5 others. 2011. Genetic effects at pleiotropic loci are context-dependent with consequences for the maintenance of genetic variation in populations. *PLoS Genet.* 7: e1002256. [28]

Layard, M. W. J. 1973. Robust large-sample tests for homogeneity of variances. *J. Amer. Stat. Assoc.* 68: 195–198. [17]

Le Corre, V. 2005. Variation at two flowering time genes within and among populations of *Arabidopsis thaliana*: comparison with markers and traits. *Mol. Ecol.* 14: 4181–4192. [12]

Le Corre, V., and A. Kremer. 2003. Genetic variability at neutral markers, quantitative trait loci and trait in a subdivided population under selection. *Genetics* 164: 1205–1219. [12]

Le Corre, V., and A. Kremer. 2012. The genetic differentiation at quantitative trait loci under local adaptation. *Mol. Eco.* 21: 1548–1566.[12]

Le Corre, V., F. Roux, and X. Rebound. 2002. DNA polymorphism at the *FRIGIDA* gene in *Arabidopsis thaliana:* Extensive nonsynonymous variation is consistent with local selection for flowering time. *Mol. Biol. Evol.* 19: 1261–1271. [10]

Le Cren, E. 1951. The length-weight relationship and seasonal cycle in gonad weight and condition in the perch (*Perca fluviatilis*). *J. Anim. Ecol.* 66: 1504–1512. [20]

Le Rouzic, A., and Ö. Carlborg. 2007. Evolutionary potential of hidden genetic variation. *Trend. Ecol. Evol.* 23: 33–37. [25]

Lea, D. E., and C. A. Coulson. 1949. The distribution of the number of mutants in bacterial populations. *J. Genet.* 28: 264–285. [4]

Leadbetter, M. R., G. Lindgren, and H. Rootzén. 1989. *Extremes and related properties of random sequences and processes*. Springer, New York, NY. [27]

Leamy, L. J., and C. P. Klingenberg. 2005. The genetics and evolution of fluctuating asymmetry. *Ann. Rev. Eco. Evol. System.* 36: 1–21. [17]

Lebreton, J.-D., K. P. Burnham, J. Clobert, and D. R. Anderson. 1992. Modeling survival and testing biological hypotheses using marked animals: a unified approach with case studies. *Ecol. Monogram.* 62: 67–118. [29]

Lee, C., and E. J. Pollak. 1997. Influence of sire misidentification on sire x year interaction variance and direct-maternal genetic covariance for weaning weight in beef cattle. *J. Anim. Sci.* 75: 2858–2863. [20]

Lee, P. M. 2013. *Bayesian statistics: An introduction*, 4th ed. Wiley, New York, NY. [A2, A3]

Lee, S. H., N. R. Wray, M. E. Goddard, and P. M. Visscher. 2011. Estimating missing heritability for disease from genome-wide association studies. *Amer. J. Hum. Genet.* 88: 294–305. [24]

Leek, J. T., and J. D. Storey. 2007. Capturing heterogeneity in gene expression studies by surrogate variables. *PLoS Genet.* 3: e161. [A4]

Leek, J. T., and J. D. Storey. 2008. A general framework for multiple testing dependence. *Proc. Natl. Acad. Sci. USA* 105: 18718–18723. [A4]

Leek, J. T., and J. D. Storey. 2011. The joint null criterion for multiple hypothesis tests. *Stat. Appl. Genet. Mol. Biol.* 10: Issue 1, Article 28. [A4]

Lees, D. R. 1981. Industrial melanisms: Genetic adaptation of animals to air pollution. *In* J. A. Bishop and L. M. Cook (eds.), *The genetic consequences of man made change*, pp. 129–176. Academic Press, London, UK. [25]

Lefebvre, J. F., and D. Labuda. 2008. Fraction of informative recombinations: a heuristic approach to analyze recombination rates. *Genetics* 178: 2069–2079. [4]

Leffler, E. M., and 7 others. 2012. Revisiting an old riddle: what determines genetic diversity levels within species. *PLoS Biol.* 10: e1001388. [4, 8]

Leffler, E. M., and 12 others. 2013. Multiple instances of ancient balancing selection shared between humans and chimpanzees. *Science* 339: 1578–1582 [9]

Lehmann, L., and L. Keller. 2006a. The evolution of cooperation and altruism—a general framework and a classification of models. *J. Evol. Biol.* 19: 1365–1376. [22]

Lehmann, L., and L. Keller. 2006b. Synergy, partner choice and frequency dependence: their integraton into inclusive fitness theory and their interpretation in terms of direct and indirect fitness effects. *J. Evol. Biol.* 19: 1426–1436. [22]

Lehmann, L., L. Keller, S. West, and D. Roze. 2007. Group and kin selection: two concepts but one process. *Proc. Natl. Acad. Sci. USA* 104: 6736–6739. [22]

Lehtonen, J. 2016. Multilevel selection in kin selection language. *Trends Ecol. Evol.* 31: 752–762. [22]

Leigh, E. G. 1987. Ronald Fisher and the development of evolutionary history. II. Influences of new variation on evolutionary process. *Oxford Surv. Evol. Biol.* 4: 212–264. [27]

Leigh, E. G., Jr. 1970. Natural selection and mutability. *Am. Nat.* 104: 301–305. [4]

Leigh, E. G., Jr. 1973. The evolution of mutation rates. *Genetics* 73: 1–18. [4]

Leinonen, T., J. M. Cano, H. Makinen, and J. Merilä. 2006. Contrasting patterns of body shape and neutral genetic divergence in marine and lake populations of threespine sticklebacks. *J. Evol. Biol.* 19: 1803–1812. [12]

Leinonen, T., R. J. S. McCairns, R. B. O'Hara, and J. Merilä. 2013. Q_{ST}–F_{ST} comparisons: evolutionary and ecological insights from genomic heterogeneity. *Nat. Rev. Genet.* 14: 179–190. [12]

Leinonen, T., R. B. O'Hara, J. M. Cano, and J. Merilä. 2008. Comparative studies of quantitative trait and neutral marker divergence: a meta-analysis. *J. Evol. Biol.* 21: 1-17. [12]

Lemos, B., G. Marroig, and R. Cerqueira. 2001. Evolutionary rates and stabilizing selection in large-bodied opossum skulls (Didelphimorphia: Didelphidae). *J. Zool.* 255: 181–189. [12]

Lemos, B., C. D. Meiklejohn, M. Cáceres, and D. L. Hartl. 2005. Rates of divergence in gene expression profiles of primates, mice, and flies: stabilizing selection and variability among functional categories. *Evolution* 59: 126-137. [12]

Lenormand, T. 2002. Gene flow and the limits to natural selection. *Trends Ecol .Evo.* 17: 183–189. [23]

Lenski, R. E., M. R. Rose, S. C. Simpson, and S. C. Tadler. 1991. Long-term experimental evolution in

Escherichia coli. I. Adaptation and divergence during 2,000 generations. *Am. Nat.* 138: 1315–1341. [25, 27]

Lenski, R. E., and P. M. Service. 1982. The statistical analysis of population growth rates calculated from schedules for survivorship and fecundity. *Ecology* 63: 655–662. [29]

Lenski, R. E., and M. Travisano. 1994. Dynamics of adaptation and diversification: a 10,000-generation experiment with bacterial populations. *Proc. Natl. Acad. Sci. USA* 91: 6808–6814. [25, 27]

Lerner, I. M. 1954. *Genetic Homeostasis*. Oliver and Boyd, Edinburgh, UK. [17, 25, 28]

Lerner, I. M., and E. R. Dempster. 1951. Attenuation of genetic progress under continued selection in poultry. *Heredity* 5: 75–94. [25]

Leroy, G., and 6 others. 2012. An ABC estimate of pedigree error rate: application in dog, sheep and cattle breeds. *Anim. Genet.* 43: 309–314. [20]

Lesecque, Y., P. D. Keightley, and A. Eyre-Walker. 2012. A resolution of the mutation load paradox in humans. *Genetics* 191: 1321–1330. [7]

Leslie, P. H. 1945. On the use of matrices in certain population mathematics. *Biometrika* 33: 183–212. [29]

Leslie, P. H. 1948. Some further notes on the use of matrices in certain population mathematics. *Biometrika* 35:213–245. [29]

Lessard, S. 1981. Is the between-population variance negligible in the total variance of heterozygosity? Case of a finite number of loci subject to the infinite-allele model in finite monoecious populations. *Theor. Popul. Biol.* 20: 394–410. [2]

Lessard, S. 1997. Fisher's fundamental theorem of natural selection revisited. *Theor. Popul. Biol.* 52: 119–136. [6]

Lessard, S., and A.-M. Castilloux. 1995. The fundamental theorem of natural selection in Ewens' sense (case of fertility selection). *Genetics* 141: 733–742. [6]

Levins, R. 1968. *Evolution in changing environments*. Princeton Univeristy Press, Princeton, NJ. [20]

Levins, R. 1970. Extinction. *Am. Math. Soc.* 2: 75–108.

Levy, S. F., and 5 others. 2015. Quantitative evolutionary dynamics using high-resolution lineage tracking. *Nature* 519: 181–186. [27]

Lewin, K. 1943. Psychology and the process of group living. *J. Social Psych.* 17: 113–131. [1]

Lewis, W. L., and E. J. Warwick. 1953. Effectiveness of selection for body weight in mice. *J. Heredity* 44: 233–238. [18]

Lewontin, R. C. 1964. The interaction of selection and linkage. II. Optimum models. *Genetics* 50: 757—782. [28]

Lewontin, R. C. 1974. *The genetic basis of evolutionary change*. Columbia University Press, New York, NY. [1, 4, 8, 10, 18, 28]

Lewontin, R. C., L. R. Ginzburg, and S. D. Tuljapurkar. 1978. Heterosis as an explanation for large amounts of genic polymorphism. *Genetics* 88: 149–169. [28]

Lewontin, R. C., and K. Kojima. 1960. The evolutionary dynamics of complex polymorphisms. *Evolution* 14: 458–472. [1, 5]

Lewontin, R. C., and J. Krakauer. 1973. Distribution of gene frequency as a test of the theory of selective neutrality of polymorphisms. *Genetics* 74: 175–195. [4, 9, 12]

Lewontin, R. C., J. A. Moore, W. B. Provine, and B. Wallace (eds.). 1981. *Dobzhansky's genetics of natural populations I-XLIII*. Columbia University, New York, NY. [9]

Lexer, C., R. A. Randell, and L. H. Rieseberg. 2003. Experimental hybridization as a tool for studying selection in the wild. *Ecology* 84: 1688–1699. [29]

Lexer, C., D. M. Rosenthal, O. Raymond, L. A. Donovan, and L. H. Rieseberg. 2005. Genetics of species differences in the wild annual sunflowers, *Helianthus annuus* and *H. petiolaris*. *Genetics* 169: 2225–2239. [12]

Li, C., A. Zhou, and T. Sang. 2006. Rice domestication by reducing shattering. *Science* 311: 1936–1939. [9]

Li, C. C. 1967. Fundamental theorem of natural selection. *Nature* 214: 505–506. [6]

Li, H., and W. Stephan. 2005. Maximum-likelihood methods for detecting recent positive selection and localizing the selected site in the genome. *Genetics* 171: 377–384. [8, 9]

Li, H., and W. Stephan. 2006. Inferring the demographic history and rate of adaptive substitution in *Drosophila*. *PLoS Genet.* 2: e166. [8, 9]

Li, J. and L. Ji. 2005. Adjusting multiple testing in multilocus analyses using the eigenvalues of a correlation matrix. *Heredity* 95: 221–227. [A4]

Li, J., and 5 others. 2012. Joint analysis of demography and selection in population genetics: where do we stand and where could we go? *Mol. Ecol.* 21: 28–44. [9]

Li, M. D., and F. D. Enfield. 1992. A computer simulation evaluation of the role of mutations in finite populations on the response to directional selection: the generations required to attain maximum genetic variance. *Theor. Appl. Genet.* 84: 995–1001. [26]

Li, M. X., J. M. Yeung, S. S. Cherny, and P. C. Sham. 2012. Evaluating the effective numbers of independent tests and significant p-value thresholds in commercial genotyping arrays and public imputation reference datasets. *Human Genet.* 131: 747–756. [A4]

Li, W.-H. 1976. Distribution of nucleotide differences between two randomly chosen cistrons in a subdivided population: the finite island model. *Theor. Popul. Biol.* 10: 303–308. [2, 11]

Li, W.-H. 1987. Models of nearly neutral mutations with particular implications for nonrandom usage of synonymous codons. *J. Mol. Evol.* 24: 337–345. [7, 8]

Li, W.-H. 2006. *Molecular evolution*. Sinauer, Sunderland, MA [10]

Li, W.-H., and Y.-X. Fu. 1999. Coalescent theory and its application in population genetics. *In* M. E. Halloran and S. Geisser (eds.), *Statistics in genetics*, pp. 45–79. Springer Verlag, New York, NY. [4]

Li, W.-H., and M. Nei. 1972. Total number of individuals affected by a single deleterious mutation in a finite population. *Am. J. Hum. Gene.* 24: 667–679. [7, 28]

Li, W.-H., and M. Nei. 1975. Drift variances of heterozygosity and genetic distance in transient states. *Genet. Res.* 25: 229–248. [2]

Li, Y. F., J. C. Costello, A. K. Holloway, and M. W. Hahn. 2008. "Reverse ecology" and the power of population genomics. *Evolution* 62: 2984–2994. [9, 10]

Liang, L., S. Zöllner, and G. R. Abecasis. 2007. GENOME: a rapid coalescent-based whole genome simulator. *Bioinformatics* 23: 1565–1567. [9]

Light, R. J., and D. B. Pillemer. 1971. *Summing up: the science of reviewing research*. Harvard Univ. Press. Cambridge, MA. [A4]

Light, R., and P. Smith. 1971. Accumulating evidence: procedures for resolving contradictions among different research studies. *Harv. Educ. Rev.* 41: 429–471. [A4]

Lin, K., H. Li, C. Schlötterer, and A. Futschik. 2011. Distinguishing positive selection from neutral evolution: Boosting the performance of summary statistics. *Genetics* 187: 229–244. [9]

Lindblad-Toh, K., and 46 others. 2005. Genome sequence, comparative analysis and haplotype structure of the domestic dog. *Nature* 438: 803–819. [10]

Lindblad-Toh, K., and 63 others. 2011. A high-resolution map of human evolutionary constraint using 29 mammals. *Nature* 47: 476–482. [10]

Lindgren, D., and J.-E. Nilsson 1985. Calculations concerning selection intensity. Report 5, Swedish University of Agricultural Sciences, Umea, Sweden. [14]

Lindgren, D., R.-P. Wei, and S. J. Lee. 1997. How to calculate optimum family number when starting a breeding program. *Forest Sci.* 43: 206–212. [21]

Lindley, D. V. 1965. *Introduction to probability and statistics from a Bayesian viewpoint* (2 Volumes), University Press, Cambridge, UK. [A2]

Link, W. A., E. G. Cooch, and E. Cam. 2002. Model-based estimation of individual fitness. *J. Appl. Stat.* 29: 207–224. [29]

Lipschutz-Powell, D., J. A. Woolliams, P. Bijma, and A. B. Doeschl-Wilson. 2012a. Indirect genetic effects and the spread of infectious disease: are we capturing the full heritable variation underlying disease prevalence? *PLoS One* 7: e39551. [22]

Lipschutz-Powell, D. and 5 others. 2012b. Bias, accuracy, and impact of indirect genetic effects in infectious diseases. *Front. Genet.* 3: 215. [22]

Liptak, T. 1958. On the combination of independent tests. *Magyar Tud. Akad. Mat. Kutato Int. Kozl.* 3: 171–197. [A4]

Little, R. J. A, and D. B. Rubin. 2002. *Statistical analysis with missing data*, 2nd ed. John Wiley and Sons, New York, NY. [20, 29]

Liu, A., and J. M. Burke. 2006. Patterns of nucleotide diversity in wild and cultivated sunflower. *Genetics* 173: 321–330. [4]

Liu, J., W. H. Wong, and A. Kong. 1991. Correlation structure and convergence rates of the Gibbs sampler (I): Application to the comparison of estimators and augmentation schemes. Technical Report 299, Dept. Statistics, University of Chicago, IL. [A3]

Liu, X., and Y.-X. Fu. 2008. Algorithms to estimate the lower bounds of recombination with or without recurrent mutations. *BMC Genomics* 9 (Suppl. 1): S24. [4]

Livshits, G., and E. Kobyliansky. 1985. Lerner's concept of developmental homeostasis and the problem of heterozygosity level in natural populations. *Heredity* 55: 341–353. [17]

Lloyd, D. G. 1979. Some reproductive factors affecting the selection of self-fertilization in plants. *Am. Nat.* 113: 67–79. [23]

Loewe, L., and B. Charlesworth. 2007. Background selection in single genes may explain patterns of codon bias. *Genetics* 175: 1381–1393. [8]

Lohmueller, K. E., and 11 others. 2008. Proportionally more deleterious genetic variation in European than in African populations. *Nature* 451: 994–997. [10]

Lohmueller, K. E., and 18 others. 2011. Natural selection affects multiple aspects of genetic variation at putatively neutral sites across the human genome. *PLoS Genet.* 7: e1002326. [8, 9]

Londo, J. P., Y.-C. Chiang, K.-H. Hung, T.-Y. Chiang, and B. A. Schaal. 2006. Phylogeography of Asian wild rice, *Oryza rufipogon*, reveals multiple independent domestications of cultivated rice, *Oryza sativa*. *Proc. Natl. Acad. Sci. USA* 103: 9578–9583. [9]

Long, A., G. Liti, A. Luptak, and O. Tenaillon. 2015. Elucidating the molecular architecture of adaptation via evolve and resequence experiments. *Nat. Rev. Genet.* 16: 567–582. [18]

Long, A. D., R. F. Lyman, A. H. Morgan, C. H. Langley, and T. F. C. Mackay. 2000. Both naturally occurring insertions of transposable elements and intermediate frequency polymorphisms at the *achaete-scute complex* are associated with variation in bristle number in *Drosophila melanogaster*. *Genetics* 154: 1255—1269. [28]

Long, H., M. Behringer, E. Williams, R. Te., and M. Lynch. 2016. Similar mutation rates but highly diverse mutation spectra in ascomycete and basidiomycete yeasts. *Genom. Bio. Evol.* 8: 3815–3821. [4]

Long, Q., and 17 others. 2013. Massive genomic variation and strong selection in *Arabidopsis thaliana* lines from Sweden.. *Nat. Genet.* 45: 884–890. [9]

Lonnquist, J. H. 1964. A modification of the ear-to-row procedutre for the improvement of maize populations. *Crop Sci.* 4: 227–228. [21]

Lonnquist, J. H., and M. F. Lindsey. 1964. Topcross versus S_1 line performance in corn. *Crop Sci.* 4: 580–584. [23]

Looney, S. W. 1995. How to use tests for univariate normality to assess multivariate normality. *Am. Stat.* 49: 64–70. [A5]

Lopes, M. S., and 6 others. 2013. Improved estimation of inbreeding and kinship in pigs using optimized SNP panels. *BMC Genet.* 14: 92. [20]

López-Fanjul, C. 1982. Some experimental evaluations of selection theory. *Proc. Second World Cong. on Genetics Applied to Livestock Production*, pp. 57–65. Madrid, Spain. [18]

López-Fanjul, C. 1989. Tests of theory by selection experiments. *In* W. G. Hill, and T. F. C. Mackay, (eds.), *Evolution and animal breeding*, pp. 129–133. C. A. B. International, Wallingford, UK. [26]

López-Fanjul, C., and M. L. Domínguez. 1982. Etude expérimentale de la variabilité de la réponse à la sélection chez *Drosophila melanogaster*. *Annales de Génétique et de Sélection Animale* 14: 213-224 [18]

López-Fanjul, C., A. Fernández, and M. A. Toro. 1999. The role of epistasis in the increase in the additive genetic variance after population bottlenecks. *Genet. Res.* 73: 45–59. [11]

López-Fanjul, C., A. Fernández, and M. A. Toro. 2002. The effect of epistasis on the excess of the additive and nonadditive variances after population bottlenecks. *Evolution* 56: 865–876. [11]

López-Fanjul, C., A. Fernández, and M. A. Toro. 2003. The effect of neutral nonadditive gene action on the quantitative index of population divergence. *Genetics* 164: 1627-1633. [12]

López-Fanjul, C., A. Fernández, and M. A. Toro. 2006. The effect of genetic drift on the variance/co-variance components generated by multilocus additive × additive epistatic systems. *J. Theor. Biol.* 239: 161-171. [12]

López-Fanjul, C., A. Fernández, and M. A. Toro. 2007. The effect of dominance on the use of the Q_{ST}–F_{ST}

contrast to detect natural selection on quantitative traits. *Genetics* 176: 725-727. [12]

López-Fanjul, C., J. Guerra, and A. García. 1989. Changes in the distribution of the genetic variance of a quantitative trait in small populations of *Drosophila melanogaster*. *Genet. Sel. Evol.* 21: 159–168. [12]

López-Fanjul, C., and B. Jódar. 1977. The genetic properties of egg laying of virgin females of *Tribolium castaneum*. *Heredity* 39: 251–258. [11]

López-Fanjul, C., and A. Villaverde. 1989. Inbreeding increases genetic variance for viability in *Drosophila melanogaster*. *Evolution* 43: 1800–1804. [11, 23]

Losos, J. B., T. R. Jackman, A. Larson, K. de Querioz, and L. Rodríguez-Schettino. 1998. Adaptive differentiation following experimental island colonization in *Anolis* lizards. *Nature* 387: 70–73. [20]

Lotterhos, K. E., and M. C. Whitlock. 2014. Evaluation of demographic history and neutral parameterization on the performance of F_{ST} outlier tests. *Mol. Ecol.* 23: 2178–2192. [9]

Lotterhos, K. E., and M. C. Whitlock. 2015. The relative power of genome scans to detect local adaptation depends on sampling design and statistical method. *Mol. Ecol.* 24: 1031–1046. [9]

Lotterhos, K. E., and 5 others. 2017. Composite measures of selection can improve the signal-to-noise ratio in genome scans. *Meth. Ecol. Evol.* 8: 717–727. [9]

Louthan, A. M., and K. M. Kay. 2011. Comparing the adaptive landscape across trait types: larger QTL effect size in traits under biotic selection. *BMC Evol. Biol.* 11: 60. [12, 27]

Lu, J., and 5 others. 2006. The accumulation of deleterious mutations in rice genomes: a hypothesis on the cost of domestication. *Trends Genet.* 22: 126–131. [8]

Luikart, G., and J. M. Cornuet. 1999. Estimating the effective number of breeders from heterozygote excess in progeny. *Genetics* 151: 1211–1216. [4]

Lujan, S. A., and 10 others. 2014. Heterogeneous polymerase fidelity and mismatch repair bias genome variation and composition. *Genome Res.* 24: 1751–1764. [4]

Luo, Z, W., J. A. Woolliams, and R. Thompson. 1995. Controlling inbreeding in dairy MOET nucleus schemes. *Anim. Sci.* 60: 379–387. [23]

Luque, V. J. 2017. One equation to rule them all: a philosophical analysis of the Price equation. *Biol. Philos.* 32: 97–125. [6]

Luria, S. E., and M. Delbrück. 1943. Mutations of bacteria from virus sensitivity to virus resistance. *Genetics* 28: 491–511. [4]

Lush, J. L. 1937. *Animal breeding plans.* Iowa State University Press, Ames, IA. [1, 6, 13]

Lush, J. L. 1945. *Animal breeding plans, 3rd Edition.* Iowa State University Press, Ames, IA. [6, 13, 16]

Lush, J. L. 1946. Chance as a cause of changes in gene frequencies within pure breeds of livestock. *Am. Nat.* 80: 318–342. [26]

Lush, J. L. 1947. Family merit and individual merit as bases for selection. *Am. Nat.* 81: 241–261, 362–379. [1, 21]

Lush, J. L. 1948. *The genetics of populations.* Animal Husbandry Department, Iowa State University, Ames, IA. [15]

LW. *See* Lynch, M., and B. Walsh. 1998.

Lyman, R. F., F. Lawrence, S. V. Nuzhdin, and T. F. C. Mackay. 1996. The effects of single *P* element insertions on bristle number and viability in *Drosophila melanogaster*. *Genetics* 143: 277–292. [28]

Lynch, C. B. 1980. Response to divergent selection for nesting behavior in *Mus musculus*. *Genetics* 96: 757–765. [21]

Lynch, M. 1984. The selective values of alleles underlying polygenic traits. *Genetics* 108: 1021–1033. [5]

Lynch, M. 1986. Random drift, uniform selection, and the degree of population differentiation. *Evolution* 40: 640–643. [1, 7]

Lynch, M. 1987. Evolution of intrafamilial interactions. *Proc. Nat. Acad. Sci. USA* 84: 8507–8511. [22]

Lynch, M. 1988a. Design and analysis of experiments on random drift and inbreeding depression. *Genetics* 120: 791–807. [11, 12]

Lynch, M. 1988b. The divergence of neutral quantitative characters among partially isolated populations. *Evolution* 42: 455–466. [11]

Lynch, M. 1988c. Estimation of relatedness by DNA fingerprinting. *Mol. Biol. Evol.* 5: 584–599. [20]

Lynch, M. 1990. The rate of morphological evolution in mammals from the standpoint of the neutral expectation. *Am. Nat.* 136: 727–741. [1, 12]

Lynch, M. 1991. Methods for the analysis of comparative data in evolutionary biology. *Evolution* 45: 1065–1080. [12]

Lynch, M. 1994. The neutral theory of phenotypic evolution, *In* L. Real (ed.) *Ecological genetics*, pp. 86–108. Princeton Univ. Press, Princeton, NJ. [11]

Lynch, M. 2006. The origins of eukaryotic gene structure. *Mol. Biol. Evol.* 23: 450–468. [4]

Lynch, M. 2007. *The origins of genome architecture.* Sinauer Assocs., Inc., Sunderland, MA. [3, 4, 8, 10, 11]

Lynch, M. 2008a. Estimation of nucleotide diversity, disequilibrium coefficients, and mutation rates from high-coverage genome-sequencing projects. *Mol. Biol. Evol.* 25: 2409–2419. [4]

Lynch, M. 2008b. The cellular, developmental, and population-genetic determinants of mutation-rate evolution. *Genetics* 180: 933–943. [4]

Lynch, M. 2009a. Estimation of allele frequencies from high-coverage genome-sequencing projects. *Genetics* 182: 295–301. [4]

Lynch, M. 2009b. Rate, molecular spectrum, and consequences of spontaneous mutations in man. *Proc. Natl. Acad. Sci. USA* 107: 961–968. [4]

Lynch, M. 2010a. Evolution of the mutation rate. *Trends Genet.* 26: 345–352. [4]

Lynch, M. 2010b. Scaling expectations for the time to establishment of complex adaptations. *Proc. Natl. Acad. Sci. USA* 38: 16577–16582. [1, 7]

Lynch, M. 2011. The lower bound to the evolution of mutation rates. *Genome Biol. Evol.* 3: 1107–1118. [1, 4]

Lynch, M. 2012a. The evolution of multimeric protein assemblages. *Mol. Biol. Evol.* 29: 1353-1366. [7]

Lynch, M. 2012b. The evolutionary layering of cellular functions and the limits to molecular perfection. *Proc. Natl. Acad. Sci. USA* 109: 18851-18856. . [1, 7]

Lynch, M. 2013. Evolutionary diversification of the multimeric states of proteins. *Proc. Natl. Acad. Sci. USA* 110: E2821–E2828. [1]

Lynch, M., and A. Abegg. 2010. The rate of establishment of complex adaptations. *Mol. Biol. Evol.* 27: 1404–1414. [7]

Lynch, M., L. M. Bobay, F. Catania, J.-F. Gout, and M. Rho. 2011. The repatterning of eukaryotic genomes by random genetic drift. *Ann. Rev. Genomics Hum. Genet.* 12: 347–366. [4]

Lynch, M., D. Bost, S. Wilson, and T. Maruki. 2014. Population-genetic inference from pooled-sequencing data. *Genome Biol. Evol.* 6: 1210–1218. [9]

Lynch, M., J. Conery, and R. Bürger. 1995a. Mutational meltdowns in sexual populations. *Evolution* 49: 1067–1080. [7]

Lynch, M., J. Conery, and R. Bürger. 1995b. Mutation accumulation and the extinction of small populations. *Am. Nat.* 146: 489–518. [7]

Lynch, M., and T. J. Crease. 1990. The analysis of population survey data on DNA sequence variation. *Mol. Biol. Evol.* 7: 377–394. [4]

Lynch, M., and A. Force. 2000. Gene duplication and the origin of interspecific genomic incompatibility. *Am. Nat.* 156: 590–605. [1]

Lynch, M., and W. Gabriel. 1983. Phenotypic evolution and parthenogenesis. *Am. Nat.* 122: 745–764. [28]

Lynch, M., and K. Hagner. 2014. Evolutionary meandering of intermolecular interactions along the drift barrier. *Proc. Natl. Acad. Sci. USA* 112: E30–E38. [1]

Lynch, M., and W. G. Hill. 1986. Phenotypic evolution by neutral mutation. *Evolution* 40: 915–935. [2, 11, 12]

Lynch, M., and R. Lande. 1993. Evolution and extinction in response to environmental change. *In* P. M. Kareiva, J. G. Kingsolver, and R. B. Huey (eds.), *Biotic interactions and global change*, pp. 234–250. Sinauer, Sunderland, MA [28]

Lynch, M., L. Latta, J. Hicks, and M. Giorgianni. 1998. Mutation, selection, and the maintenance of life-history variation in a natural population. *Evolution* 52: 727–733. [12]

Lynch, M., and B. Walsh. 1998. *Genetics and analysis of quantitative traits.* Sinauer Assocs., Inc., Sunderland, MA. [most chapters, as LW]

Lynch, M., and 6 others. 2016. Genetic drift, selection and the evolution of the mutation rate. *Nat. Rev. Genet.* 17: 704–714 [1, 4]

Lyttle, T. W. 1991. Segregation distorters. *Ann. Rev. Genet.* 25: 511–581. [25]

Lyttle, T. W. 1993. Cheaters sometimes prosper: distortion of mendelian segregation by meiotic drive. *Trends Genet.* 9: 205–210. [25]

Ma, Y., and 5 others. 2015. Properties of different selection signature statistics and a new strategy for combining them. *Heredity* 115: 426–436. [9]

Ma, X.-F., D. Hall, K. R. St Onge, S. Jansson, and P. K. Ingvarsson. 2010. Genetic differentiation, clinal variation and phenotypic associations with growth cessation across the *Populus tremula* photoperiodic pathway. *Genetics* 186: 1033–1044. [12]

MacArthur, J. W. 1949. Selection for small and large body size in the house mouse. *Genetics* 34: 194–209. [25]

MacColl, A. D. C. 2011. The ecological causes of evolution. *Trend. Ecol. Evol.* 26: 514–522. [29]

MacEachern, S. N., and L. M. Berliner. 1994. Subsampling the Gibbs sampler. *Amer. Stat.* 48: 188–190. [A3]

Mackay, T. F. C. 1985a. Transposable element-induced response to artificial selection in *Drosophila melanogaster*. *Genetics* 111: 351–374. [18]

Mackay, T. F. C. 1985b. A quantitative genetic analysis of fitness and its components in *Drosophila melanogaster. Genet. Res.* 47: 59–70. [6]

Mackay, T. F. C. 1988. Transposable element induced quantitative genetic variation in *Drosophila*. *In* B. S. Weir, E. J. Eisen, M. M. Goodman, and G. Namkoong (eds.), *Proceedings of the second international conference on quantitative genetics*, pp. 219–235. Sinauer, Sunderland, MA. [25]

Mackay, T. F. C., and R. F. Lyman. 2005. *Drosophila* bristles and the nature of quantitative genetic variation. *Phil. Trans. Roy. Soc. Lond. B* 360: 1513–1527. [17]

Mackay, T. F. C., R. F. Lyman, and M. S. Jackson. 1992. Effects of *P* element insertions on quantitative traits in *Drosophilia melanogaster. Genetics* 130: 315–332. [28]

Mackay, T. F. C, E. A. Stone, and J. F. Ayroles. 2009. The genetics of quantitative traits: challenges and prospects. *Nat. Rev. Genet.* 10: 565–577. [24, 25]

Mackenzia, A., J. D. Reynolds, V. J. Brown, and W. J. Sutherland. 1995. Variation in male mating success on leks. *Am. Nat.* 145: 633–652. [29]

MacLean, R. C., and A. Buckling. 2009. The distribution of fitness effects of beneficial mutations in *Pseudomonas aeruginosa. PLoS Genet.* 5: e1000406. [27]

Macnair, M. R. 1991. Why the evolution of resistance to anthropogenic toxins normally involves major gene changes: the limits to natural selection. *Genetica* 84: 213–219. [25]

MacNeil, M. D., D. D. Kress, A. E. Flower, R. P. Weber, and R. L. Blackwell. 1984. The effects of mating system in Japanese Quail. 2. Genetic parameters, response, and correlated response. *Theor. Appl. Genet.* 67: 407–412. [23]

Macpherson, J. M., G. Sella, J. C. Davis, and D. A. Petrov. 2007. Genomewide spatial correspondence between nonsynonymous divergence and neutral polymorphism reveals extensive adaptation in *Drosophila. Genetics* 177: 2083–2099. [8]

Macpherson, J. M., and 6 others. 2008. Nonadaptive explanations for signatures of partial sweeps in *Drosophila. Mol. Biol. Evol.* 25: 1025–1042. [9]

Madalena, F. E., and W. G. Hill. 1972. Population structure in artificial selection programmes: simulation studies. *Genet. Res.* 20: 75–99. [26]

Madalena, F. E., and A. Robertson. 1975. Population structure in artificial selection: studies with *Drosophila melanogaster*. *Genet. Res.* 24: 113–126. [25, 26]

Maddox, G. D., and J. Antonovics. 1983. Experimental ecological genetics in *Plantago*: a structural equation approach to fitness components in *P. aristata* and *P. patagonica*. *Ecology* 64: 1092–1099. [30]

Magnussen, S. 1993. Bias in genetic variance estimates due to spatial autocorrelation. *Theor. Appl Genet.* 86: 349–355. [20]

Mahalanobis, P. C. 1936. On the generalised distance in statistics. *Proc. Nat. Inst. Sci. India* 2: 49–55. [A5]

Maher, B. 2008. Personal genomes: The case of the missing heritability. *Nature* 456: 18–21. [24]

Majoram, P., J. Molito, V. Plagnol, and S. Tavaré. 2003. Markov chain Monte Carlo without likelihoods. *Proc. Natl. Acad. Sci. USA* 100: 15324–15328. [A3]

Majoram, P., and S. Tavaré. 2006. Modern computational approaches for analysing molecular genetic variation data. *Nat. Rev. Genet.* 7: 759–770. [A3]

Mäki-Tanila, A., and W. G. Hill. 2014. Influence of gene interaction on complex trait variation with multilocus models. *Genetics* 198: 355–367. [11, 15, 24, 25]

Mäki-Tanila, A., and B. W. Kennedy. 1986. Mixed model methodology under genetic models with a small number of additive and non-additive loci. *Proc. Third World Cong. on Genetics Applied to Livestock Production* XII: 443–448. Ames, IA. [19]

Malaspinas, A.-S. 2016. Methods to characterize selective sweeps using time serial samples: an ancient DNA perspective. *Mol. Ecol.* 25: 24–41. [9]

Malaspinas, A.-S., O. Malaspinas, S. N. Evans, and M. Slatkin. 2012. Estimating allele age and selection coefficient from time-serial data. *Genetics* 192: 599–607. [9]

Malcolm, B. A., K. P. Wilson, B. W. Matthews, J. F. Kirsch, and A. C. Wilson. 1990. Ancestral lysozymes reconstructed, neutrality tested and thermostability linked to hydrocarbon packing. *Nature* 345: 86–89. [27]

Malécot, G. 1948. *Les mathématiques de l'hérédité.* Masson, Paris, FR. [20]

Malkova, A., and J. E. Haber. 2012. Mutations arising during repair of chromosome breaks. *Ann. Rev. Genet.* 46: 455–473. [4]

Malkova, A., and 6 others. 2004. Gene conversion and crossing over along the 405-kb left arm of *Saccharomyces cerevisiae* chromosome VII. *Genetics* 168: 49–63. [4]

Malkovich, J. F., and A. A. Afifi. 1973. On tests for multivariate normality. *J. Am. Stat. Ass.* 68: 176–179. [A5]

Malmberg, R. L., and R. Mauricio. 2005. QTL-based evidence for the role of epistasis in evolution. *Genet. Res.* 86: 89–95. [25]

Maloney, M. A., J. C. Gilbreath and R. D. Morrison. 1963. Two-way selection for body weight in chickens. 1. The effectiveness of selection for twelve-week body weight. *Poultry Sci.* 42: 326–334. [18]

Mancera, E., R. Bourgon, A. Brozzi, W. Huber, and L. M. Steinmetz. 2008. High-resolution mapping of meiotic crossovers and non-crossovers in yeast. *Nature* 454: 479–485. [4]

Manel, S., and R. Holderegger. 2013. Ten years of landscape genetics. *Trends Ecol. Evol.* 28: 614–621. [9]

Manel, S., M. K. Schwartz, G. Luikart, and P. Taberlet. 2003. Landscape genetics: combining landscape ecology and population genetics. *Trends Ecol. Evol.* 18: 189–197. [9]

Manel, S., and 8 others. 2010. Perspectives on the use of landscape genetics to detect genetic adaptive variation in the field. *Mol. Ecol.* 19: 3760–3772. [9]

Mani, G. S., B. C. Clarke, and P. R. Shelton. 1990. A model of quantitative traits under frequency-dependent balancing selection. *Phil. Trans. Roy Soc. Lond. B* 240: 15–28. [28]

Manicacci, D., and 7 others. 2007. Maize *Sh2* gene is constrained by natural selection but escaped domestication. *J. Evol. Biol.* 20: 503–516. [9]

Manly, B. F. J. 1985. *The statistics of natural selection.* Chapman and Hall, London, UK. [29]

Manly, B. F. J. 1997. *Randomization, bootstrap and Monte Carlo methods in biology, 2nd edition.* Chapman and Hall, London, UK. [29]

Manly, K. F., D. Nettleton, and J. T. G. Hwang. 2004. Genomics, prior probability, and statistical tests of multiple hypotheses. *Genome Res.* 14: 997–1001. [A4]

Manna, F., G. Martin, and T. Lenormand. 2011. Fitness landscapes: an alternative theory for the dominance of mutation. *Genetics* 189: 923–937. [28]

Manolio, T. A., and 26 others. 2009. Finding the missing heritability of complex diseases. *Nature* 461: 747–753. [1, 24]

Mansai, S. P., and H. Innan H. 2010. The power of the methods for detecting interlocus gene conversion. *Genetics* 184: 517–527. [4]

Marais, G., D. Mouchiround, and L. Duret. 2001. Does recombination improve selection on codon usage? Lessons from nematode and fly complete genomes. *Proc. Natl. Acad. Sci. USA* 98: 5688–5692. [8]

Mardia, K. V. 1970. Measures of multivariate skewness and kurtosis with applications. *Biometrika* 57: 519–530. [A5]

Mardia, K. V. 1974. Applications of some measures of multivariate skewness and kurtosis in testing normality and robustness studies. *Sankhyā, Series B* 36: 115–128. [A5]

Markovtsova, L., P. Marjoram, and S. Taveré. 2001. On a test of Depaulis and Veuille. *Mol. Biol. Evol.* 18: 1132–1133. [9]

Markow, T. A. 1975. A genetic analysis of phototactic behavior in *Drosophila melanogaster*. I. Selection in the presence of inversions. *Genetics* 79: 527–534. [26]

Marrot, P., D. Garant, and A. Charmantier. 2017. Multiple extreme climatic events strengthen selection for earlier breeding in a wild passerine. *Phil. Trans. Roy. Soc. Lond. B* 372: 20160372. [20]

Martin, C. H., and P. C. Wainwright. 2013. Multiple fitness peaks on the adaptive landscape drive adaptive radiation in the wild. *Science* 339: 208–211. [30]

Martin, G. 2014. Fisher's geometrical model emerges as a property of complex integrated phenotypic networks. *Genetics* 197: 237–255. [26]

Martin, G., and T. Lenormand. 2006. A general multivariate extension of Fisher's geometrical model and the distribution of mutation fitness effects across species. *Evolution* 60: 893–907. [27]

Martin, G., and T. Lenormand. 2008. The distribution of beneficial and fixed mutation fitness effects close to an optimum. *Genetics* 179: 907–916. [27]

Martin, G., E. Chapuis, and J. Goudet. 2008. Multivariate Q_{ST}–F_{ST} comparisons: a neutrality test for the evolution of the *G* matrix in structured populations. *Genetics* 180: 2135–2149. [12]

Martin, T. G., and 7 others. 2005. Zero tolerance ecology: improving ecological inference by modelling the source of zero observations. *Ecol. Lett.* 8: 1235–1246. [29]

Martinez, V., L. Bünger, and W. G. Hill. 2000. Analysis of response to 20 generations of selection for body composition in mice: fit to the infinitesimal model assumptions. *Genet. Sel. Evol.* 32: 3–21. [19]

Martins, E. P. 1994. Estimating the rate of phenotypic evolution from comparative data. *Am. Nat.* 144: 193–209. [12]

Martins, E. P., and T. F. Hansen. 1997. Phylogenies and the comparative method: a general approach to incorporating phylogenetic information into the analysis of interspecific data. *Am. Nat.* 149: 646–667. [12]

Maruki, T., and M. Lynch. 2013. Genome-wide estimation of linkage disequilibrium from population-level high-throughput sequencing data. *Genetics* 197: 1303–1313. [9]

Maruyama, T. 1970. On the fixation probability of mutant genes in a subdivided population. *Genet. Res.* 15: 221–225. [7, 26]

Maruyama, T. 1971. An invariant property of subdivided populations. *Genet. Res.* 18: 81–84. [2]

Maruyama, T. 1974. The age of an allele in a finite population. *Genet. Res.* 23: 137–143. [2]

Maruyama, T. 1977. *Stochastic problems in population genetics.* Springer-Verlag, Berlin, DE. [7, 26, A1]

Maruyama, T., and C. W. Birky. 1991. Effects of periodic selection on gene diversityi n organelle genomes and other systems without recombination. *Genetics* 127: 449–451. [3]

Maruyama, T., and M. Kimura. 1974. A note on the speed of gene frequency changes in reverse direction in a finite population. *Evolution* 28: 161–163. [1, 7]

Maruyama, T., and M. Kimura. 1980. Genetic variability and effective population size when local extinction and recolonization of subpopulations are frequent. *Proc. Natl. Acad. Sci. USA* 77: 6710–6714. [3]

Mascarenhas, J. P. 1990. Gene activity during pollen development. *Ann. Rev. Plant Biol.* 41: 317–338. [29]

Maside, X., A. W. Lee, and B. Charlesworth. 2004. Selection on codon usage in *Drosophila americana. Curr. Biol.* 14: 150–154. [8]

Massey, S. E. 2008. The proteomic constraint and its role in molecular evolution. *Mol. Biol. Evol.* 25: 2557-2565. [4]

Mather, K. 1941. Variation and selection of polygenic characters. *J. Genet.* 41: 159–193. [25]

Mather, K. 1942. The balance of polygenic combinations. *J. Genet.* 43: 309–336. [25]

Mather, K. 1943. Polygenic inheritance and natural selection. *Biol. Rev.* 18: 32–64. [25]

Mather, K. 1983. Response to selection. *In* M. Ashburner, H. L. Carson, and J. N. Thompson Jr (eds.), *The genetics and biology of Drosophila, Vol 3.*, pp 155–221. Academic Press, New York, NY. [18]

Mather, K., and B. J. Harrison. 1949. The manifold effects of selection. *Heredity* 3: 1–52, 131–162. [25]

Mather, K. A., and 5 others. 2007. The extent of linkage disequilibrium in rice (*Oryza sativa* L.). *Genetics* 177: 2223–2232. [4]

Mathieson, I., and G. McVean. 2013. Estimating selection coefficients in spatially structured populations from time series data of allele frequencies. *Genetics* 193: 973–984. [9]

Mathieson, I., and 37 others. 2015. Genome-wide patterns of selection in 230 ancient Eurasians. *Nature* 538: 499–503. [8, 9, 12]

Matsumura, M. 1996. Genetic analysis of a threshold trait: density-dependent wing dimorphism in *Sogatella fucifera* (Horvath) (Hemiptera: Delphacidae), the whitebacked planthopper. *Heredity* 76: 229–237. [14]

Matsumura, S., R. Arlinghaus, and U. Dieckmann. 2012. Standardizing selection strengths to study selection in the wild: a critical comparison and suggestions for the future. *Bioscience* 62: 1039–1054. [29, 30]

Matsuoka, Y., and 5 others. 2002. A single domestication for maize shown by multilocus microsatellite genotyping. *Proc. Natl. Acad. Sci. USA* 99: 6080–6084. [9]

Matthysse, S., K. Lange, and D. K. Wagener. 1979. Continuous variation caused by genes with graduated effects. *Proc. Natl. Acad. Sci. USA* 76: 2862–2865. [24]

Matuszewski, S., J. Hermisson, and M. Kopp. 2014. Fisher's geometric model with a moving optimum. *Evolution* 68: 2571–2588. [27]

Mauricio, R., and L. E. Mojonnier. 1997. Reducing bias in the measurement of selection. *Trends Ecol. Evol.* 12: 433–436. [20]

Maynard Smith, J. 1962a. Disruptive selection, polymorphism and sympatric speciation. *Nature* 195: 60–62. [16]

Maynard Smith, J. 1962b. The limitations of molecular evolution. *In* I. J. Good (ed.). *The scientists speculates: an anthology of partly-baked ideas*, pp. 252–256. Basic Books, New York, NY. [27]

Maynard Smith, J. 1964. Group selection and kin selection. *Nature* 201: 1145–1147. [22]

Maynard Smith, J. 1966. Sympatric speciation. *Am. Nat.* 100: 637–650. [16]

Maynard Smith, J. 1970. Natural selection and the concept of a protein space. *Nature* 225: 563–564. [27]

Maynard Smith, J. 1976. Group selection. *Quart. Rev. Biol.* 51 277–283. [22]

Maynard Smith, J. 1983. The genetics of statis and punctuation. *Ann. Rev. Genet.* 17: 11–25. [28, 30]

Maynard Smith, J. 1989. *Evolutionary genetics.* Oxford University Press, London, UK. [28]

Maynard Smith, J., and J. Haigh. 1974. The hitch-hiking effect of a favorable gene. *Genet. Res.* 23: 23–35. [3, 8]

Maynard Smith, J., and G. R. Price. 1973. The logic of animal conflict. *Nature* 246: 15–18. [6]

Maynard Smith, J., and K. C. Sondhi. 1960. The genetics of a pattern. *Genetics* 45: 1039–1050. [18]

Mayo, O. 1987. *The theory of plant breeding, 2nd Edition.* Clarendon, Oxford, UK. [13, 21]

Mayr, E. 1959. Where are we? *Cold Spring Harbor Symp. Quant. Biol.* 124: 1–24. [1]

Mayr, E. 1963. *Animal species and evolution.* Belknap Press, Cambridge, MA. [1, 16]

Mayr, E. 1954. Change of genetic environment and evolution, *in* J. Huxley, A. C. Hardy, and E. B. Ford (eds.), *Evolution as a process*, pp. 157–180. Allen and Unwin, London, UK. [11]

McAdam, A. G., and S. Boutin. 2003. Variation in viability selection among cohorts of juvenile red squirrels (*Tamiasciurus hudsonicus*). *Evolution* 57: 1689–1697. [15]

McAdam, A. G., S. Boutin, D. Réale, and D. Berteaux. 2002. Maternal effects and the potential for evolution in a natural population of animals. *Evolution* 56: 846–851. [15]

McBride, G., and A. Robertson. 1963. Selection using assortative mating in *D. melanogaster. Genet. Res.* 4: 356–369. [16, 21, 26]

McCain, K. W. 2015. "Nothing as practical as a good theory" Does Lewin's Maxim still have salience in the applied social sciences? *Proc. Assoc. Inform. Sci. Technology* 52: 1–4. [1]

McCleery, R. H., and 5 others. 2004. Components of variance underlying fitness in a natural population of the great tit *Parus major. Am. Nat.* 164: E62–E72. [6]

McClosky, B., and S. D. Tanksley. 2013. The impact of recombination on short-term selection gain in plant breeding experiments. *Theor. Appl. Genet.* 126: 2299–2312. [26]

McCulloch, C. E., S. R. Searle, and J. M. Neuhaus. 2008. *Generalized, linear, and mixed models*, 2nd Ed. Wiley, Hoboken, NJ. [29]

McDonald, J. H. 1996. Detecting non-neutral heterogeneity across a region of DNA sequence in the ratio of polymorphisms to divergence. *Mol. Biol. Evol.* 13: 263–260. [10]

McDonald, J. H. 1998. Improved tests for heterogeneity across a region of DNA sequence in the ratio of polymorphisms to divergence. *Mol. Biol. Evol.* 15: 377–384. [10]

McDonald, J. H., and M. Kreitman. 1991a. Adaptive protein evolution at the *Adh* locus in *Drosophila*. *Nature* 351: 652–654. [10]

McDonald, J. H., and M. Kreitman. 1991b. Neutral mutation hypothesis test. *Nature* 354: 116. [10]

McDonald, M. J., T. F. Cooper, H. J. E. Beaumont, and P. B. Rainey. 2011. The distribution of fitness effects of new beneficial mutations in *Pseudomonas fluorescens*. *Bio. Lett.* 7: 98–100. [27]

McEvoy, B. P., J. E. Powell, M. E. Goddard, and P. M. Visscher. 2011. Human population dispersal "Out of Africa" estimated from linkage disequilibrium and allele frequencies of SNPs. *Genome Res.* 21: 821–829. [4]

McGill, C. B., B. K. Shafer, D. R. Higgins, and J. N. Strathern. 1990. Analysis of interchromosomal mitotic recombination. *Curr. Genet.* 18: 29–39. [4]

McGlothlin, J. W. 2010. Combining selective episodes to estimate lifetime nonlinear selection. *Evolution* 64: 1377–1385. [29, 30]

McGlothlin, J. W., and E. D. Brodie III. 2009. How to measure indirect genetic effects: the congruence of trait-based and variance-partitioning approaches. *Evolution* 63: 1785–1795. [22]

McGlothlin, J. W., A. J. Moore, J. B. Wolf, and E. D. Brodie III. 2010. Interacting phenotypes and the evolutionary process. III. Social evolution. *Evolution* 6: 2558–2574. [22]

McGlothlin, J. W., J. B. Wolf, E. D. Brodie, and A. J. Moore. 2014. Quantitative genetic versions of Hamilton's rule with empirical applications. *Phil. Trans. Roy. Soc. Lond. B:* 369: 20130358. [22]

McGraw, J. B., and H. Caswell. 1996. Estimation of individual fitness from life-history data. *Am. Nat.* 147: 47–64. [29]

McGuigan, K., and M. W. Blows. 2009. Asymmetry of genetic variation in fitnes srelated traits: apparent stabilizing selection on g_{max}. *Evolution* 63: 2838–2847. [28]

McGuigan, K., and M. W. Blows. 2012. Joint allelic effects on fitness and metric traits. *Evolution* 67: 1131–1142. [28]

McGuigan, K., L. Rowe, and M. W. Blows. 2011. Pleiotropy, apparent stabilizing selection and uncovering fitness optima. *Trend. Ecol. Evol.* 26: 22–29. [28]

McKay, J. K., and R. G. Latta. 2002. Adaptive population divergence: markers, QTL and traits. *Trends Ecol. Evol.* 17: 285–291. [12]

McKenzie, J. A. 2000. The character or the variation: The genetic analysis of the insectide-resistance phenotype. *Bull. Ent. Res.* 90: 3–7. [25]

McKenzie, J. A., A. G. Parker, and J. L. Yen. 1992. Polygenic and single gene responses to selection for resistance to diazinon in *Lucilia cuprina*. *Genet.* 130: 613–620. [25]

McLachlan, G. J. 1987. On bootstapping the likelihood ratio test statistic for the number of components in a normal mixture. *Appl. Stat.* 36: 318–324. [A4]

McPeek, M. S., and A. Strahs. 1999. Assessment of linkage disequilibrium by the decay of haplotype sharing, with application to fine-scale genetic mapping. *Am. J. Hum. Genet.* 65: 858–875. [9]

McPhee, C. P., and A. Robertson. 1970. The effect of suppressing crossing-over on the response to selection in *Drosophila melanogaster*. *Genet. Res.* 16: 1–16. [26]

McVean, G. 2007. The structure of linkage disequilibrium around a selective sweep. *Genetics* 175: 1395–1406. [8]

McVean, G., P. Awadalla, and P. Fearnhead. 2002. A coalescent-based method for detecting and estimating recombination from gene sequences. *Genetics* 160: 1231–1241. [4]

McVean, G. A. T., and B. Charlesworth. 1999. A population genetic model for the evolution of synonymous codon usage: patterns and predictions. *Genet. Res.* 74: 145–158. [7, 8, 10]

McVean, G. A. T., and B. Charlesworth. 2000. The effects of Hill-Robertson interference between weakly selected mutations on patterns of molecular evolution and variation. *Genetics* 155: 929–944. [8]

McVicker, G., D. Gordon, C. Davis, and P. Green. 2009. Widespread genomic signatures of natural selection in hominid evolution. *PLoS Genet.* 5: e1000471. [8]

Mead, L. S., and S. J. Arnold. 2004. Quantitative genetic models of sexual selection. *Trends Ecol. Evol.* 19: 264–271. [29]

Meader, S., C. P. Ponting, and G. Lunter. 2010. Massive turnover of functional sequence in human and other mammalian genomes. *Genome Res.* 20: 1335–1343. [10]

Meffert, L. M. 1995. Bottleneck effects on genetic variance for courtship repertoire. *Genetics* 139: 365–374. [11]

Meiklejohn, C. D., Y. Kim, D. L. Hartl, and J. Parsch. 2004. Identification of a locus under complex positive selection in *Drosophila simulans* by haplotype mapping and composite-likelihood estimation. *Genetics* 168: 265–279. [9]

Meirmans, P. G. 2012. The trouble with isolation by distance. *Mol. Ecol.* 21: 2839–2846. [9]

Meirmans, P. G., and P. W. Hedrick. 2011. Assessing population structure: F_{ST} and related measures. *Mol. Ecol. Resources* 11: 5–18. [9]

Mekel-Bobrov, N., and 7 others. 2005. Ongoing adaptive evolution of *ASPM*, a brain size determinant in *Homo sapiens*. *Science* 309: 1720–1722. [10]

Mekel-Bobrov, N., and 21 others. 2007. The ongoing adaptive evolution of *ASPM* and *microcephalin* is not explained by increased intelligence. *Hum. Mol. Genet.* 16: 600–608. [10]

Melchinger, A. E., and C. Flachenecker. 2006. An extension of the Smith model for quantitative genetic analysis of response under recurrent selection. *Plant Breed.* 125: 644–646. [18]

Mengerson, K. L., C. P. Robert, and C. Guihenneuc-Jouyaux. 1999. MCMC convergence diagnostics: A review. *In* J. M. Bernardo, J. O. Berger, A. P. Dawid, and A. F. M. Smith (eds.), *Bayesian Statistics 4*, pp. 415–440. Oxford University Press, Oxford, UK. [A3]

Mercer, A. M., and P. R. Mercer. 2000. Cauchy's interlace theorem and lower bounds for the spectral radius. *Inter. J. Mat. Math Sci.* 23: 563–566. [30]

Merilä, J. 1997. Quantitative trait and allozyme divergence in the greenfinch (*Carduelis chloris*, Aves: Fringillidae). *Biol. J. Linn. Soc.* 61: 243–266. [12]

Merilä, J., and P. Crnokrak. 2001. Comparison of genetic differentiation at marker loci and quantitative traits.

J. Evol. Biol. 14: 892–903. [12]

Merilä, J., L. E. B. Kruuk, and B. C. Sheldon. 2001a. Natural selection on the genetic component of variance in body condition in a wild bird population. *J. Evol. Biol.* 14: 98–929. [20]

Merilä, J., L. E. B. Kruuk, and B. C. Sheldon. 2001b. Cryptic evolution in a wild bird population. *Nature* 412: 76–79. [20]

Merilä, J., and B. C. Sheldon. 1999. Genetic architecture of fitness and nonfitness traits: empirical patterns and development of ideas. *Heredity* 83: 103–109. [6]

Merilä, J., and B. C. Sheldon. 2000. Lifetime reproductive success and heritability in nature. *Am. Nat.* 155: 301–310. [6]

Merilä, J., and B. C. Sheldon. 2001. Avian quantitative genetics. *Curr. Ornothol.* 16: 179–255. [20]

Merilä, J., B. C. Sheldon, and L. E. B. Kruuk. 2001c. Explaining stasis: microevolutionary studies in natural populations. *Genetica* 112-113: 199–222. [20]

Merton, R. K. 1965. *On the shoulders of giants*. Free Press, New York, NY. [A2]

Messer, P. W. 2009. Measuring the rates of spontaneous mutation from deep and large-scale polymorphism data. *Genetics* 182: 1219–1232. [4]

Messer, P. W., and R. A. Neher. 2012. Estimating the strength of selective sweeps from deep population diversity data. *Genetics* 181: 593–605. [8]

Messer, P. W., and D. A. Petrov. 2013a. Population genomics of rapid adaptation by soft selective sweeps. *Trends Genet.* 28: 659–669. [8]

Messer, P. W., and D. A. Petrov. 2013b. Frequent adaptation and the McDonald–Kreitman test. *Proc. Natl. Acad. Sci. USA* 110: 8615–8620. [10]

Messina, F. K. 1993. Heritability and 'evolvability' of fitness components in *Callosobruchus maculatus*. *Heredity* 71: 623–629. [6]

Meszaros, S. A., R. G. Banks, J. H. J. van der Werf, and M. Goddard. 1999. The use of genetic algorithms for optimizing age structure in breeding populations when inbreeding depresses genetic gain through effects on reproduction. *Anim. Sci.* 68: 449–457. [23]

Metcalf, C. J., and S. Pavard. 2007. Why evolutionary biologists should be demographers. *Trends Ecol. Evol.* 22: 205–212. [29]

Metropolis, N., A. W. Rosenbluth, M. N. Rosenbluth, A. Teller, and H. Teller. 1953. Equations of state calculations by fast computing machines. *J. Chem. Physics* 21: 1087–1091. [19, A3]

Metropolis, N., and S. Ulam. 1949. The Monte Carlo method. *J. Am. Stat. Assoc.* 44: 335–341. [19, A3]

Metz, J. A. J., R. M. Nisbet, and S. A. H. Green. 1992. How should we define 'fitness' for general ecological scenarios? *Trends Ecol. Evol.* 7: 198–202. [29]

Meuwissen, T. H. E. 1991. Reduction of selection differentials in finite populations with a nested full-half sib family structure. *Biometrics* 47: 195–203. [21]

Meuwissen, T. H. E. 1997. Maximizing the response of selection with a predefined rate of inbreeding. *J. Anim. Sci.* 75: 934–940. [23]

Meuwissen, T. H. E., B. J. Hayes, and M. E. Goddard. 2001. Prediction of total genetic value using genome-wide dense marker maps. *Genetics* 157: 1819–1829. [1]

Meuwissen, T. H. E., and A. K. Sonesson. 1998. Maximizing the response of selection with a predefined rate of inbreeding: overlapping generations. *J. Anim. Sci.* 76: 2575–2583. [23]

Meuwissen, T. H. E., and J. A. Woolliams. 1994. Effective sizes of livestock populations to prevent a decline in fitness. *Theor. Appl. Genet.* 89: 1019–1026. [23]

Meyer, H. H., and F. D. Enfield. 1975. Experimental evidence on limitations of the heritability parameter. *Theor. Appl. Genet.* 45: 268–273. [18]

Meyer, K. 1989. Approximate accuracy of genetic evaluation under an animal models. *Livest. Prod. Sci.* 21: 87–100. [20]

Meyer, K. 2008. Likelihood calculations to evaluate experimental designs to estimate genetic variances. *Heredity* 101: 212–221. [19]

Meyer, K., and W. G. Hill. 1991. Mixed model analysis of a selection experiment for food intake in mice. *Genet. Res.* 57: 71–81. [19]

Meyer, R. S., and M. D. Purugganan. 2013. Evolution of crop species: genetics of domestication and diversification. *Nat. Rev. Genet.* 14: 840–852. [9]

Michalakis, Y., and M. Slatkin. 1996. Interaction of selection and recombination in the fixation of negative-epistatic genes. *Genet. Res.* 67: 257–269. [7]

Michod, R. E. 1982. The theory of kin selection. *Ann. Rev. Ecol. Syst.* 13: 23–55. [22]

Michod, R. E., and R. Abugov. 1980. Adaptive topography in family-structured models of kin slection. *Science* 210: 667–669. [22]

Michod, R. E., and W. D. Hamilton. 1980. Coefficients of relatedness in sociobiology. *Nature* 288: 694–697. [22]

Milkman, R. 1978. Selection differentials and selection coefficients. *Genetics* 88: 391–403. [3, 5, 26]

Miller, C. R., P. Joyce, and H. A. Wichman. 2011. Mutational effects and population dynamics during viral adaptation challenge current models. *Genetics* 187: 185–202. [27]

Miller, D. E., and 12 others. 2012. A whole-chromosome analysis of meiotic recombination in *Drosophila melanogaster. Genes Genomes Genet.* 2: 249–260. [4]

Miller, J. R., M. C. Pugh, and M. B. Hamilton. 2006. A finite locus effect diffusion model for the evolution of a quantitative trait. *J. Math. Biol.* 52: 761–787. [24]

Miller, J. R., B. P. Wood, and M. B. Hamilton. 2008. F_{ST} and Q_{ST} under neutrality. *Genetics* 180: 1023–1037. [12]

Miller, R. G. 1974. The jackknife—a review. *Biometrika* 61: 1–15. [29]

Millicent, E., and J. M. Thoday. 1961. Effects of disruptive selection. IV. Gene flow and divergence. *Heredity* 16: 199–217. [16]

Milligan, B. N., D. Fraser, and D. L. Kramer. 2002 Within litter birth weight variation in the domestic pig and its relation to preweaning survival, weight gain, and variation in weaning weights. *Livestock Prod. Sci.* 76: 181–191. [17]

Mills, S. C., A. Grapputo, E. Koskela, and T. Mappes. 2007. Quantitative measure of sexual selection with respect to the operational sex ratio: a comparison of selection indices. *Proc. Roy. Soc. Lond. B* 274: 143–150. [29]

Milner, J. M., S. D. Albon, A. W. Illius, J. M. Pemberton, and T. H. Clutton-Brock. 1999. Repeated selection on morphometric traits in the Soay sheep on St. Kilda. *J. Anim. Ecol.* 68: 472–488. [19, 20]

Milner, J. M., J. M. Pemberton, S. Brotherstone, and S. D. Albon. 2000. Estimating variance components and heritabilities in the wild: a case study using the animal model approach. *J. Evol Biol.* 13: 804–813. [19]

Milot, E., and 5 others. 2011. Evidence for evolution in response to natural selection in a contemporary human population. *Proc. Natl. Acad. Sci. USA* 108: 17040–17045. [20]

Minawa, A., and A. J. Birley. 1978. The genetical response to natural selection by varied environments. *Heredity* 40: 39–50. [21]

Minvielle, F. 1980. A simulation study of truncation selection for a quantitative trait opposed by natural selection. *Genetics* 94: 989–1000. [25]

Misztal, I. 1997. Estimation of variance components with large-scale dominance models. *J. Dairy Sci.* 80: 965–974. [19]

Misztal, I. 2008. Reliable computing in estimation of variance components. *J. Anim. Breed. Genet.* 125: 363–370. [19]

Mitchell, R. J. 1992. Testing evolutionary and ecological hypotheses using path analysis and structural equation modelling. *Funct. Ecol.* 6: 123–129. [30]

Mitchell-Olds, T. 1987. Analysis of local variation in plant size. *Ecology* 68: 82–87. [30]

Mitchell-Olds, T. 1989. FreeStat Users' Manual. Technical Bulletin No. 101, University of Montana Herbarium. Missoula, MT. [30]

Mitchell-Olds, T., and J. Bergelson. 1990. Statistical genetics of *Impatiens capensis.* II. Natural selection. *Genetics* 124: 417–421. [29, 30]

Mitchell-Olds, T., and R. G. Shaw. 1987. Regression analysis of natural selection: statistical inference and biological interpretation. *Evolution* 41: 1149–1161. [24, 29, 30]

Mitchell-Olds, T., and R. G. Shaw. 1990. Comments on the causes of natural selection. *Evolution* 44: 2158. [29]

Mitchell-Olds, T., J. H. Willis, and D. B. Goldstein. 2007. Which evolutionary processes influence natural genetic variation for phenotypic traits? *Nat. Rev. Genet.* 8: 845–856. [28]

Mitton, J. B., and M. C. Grant. 1984. Associations among protein heterozygosity, growth rate and developmental homeostasis. *Ann. Rev. Ecol. Syst.* 15: 479–499. [17]

Miyashita, N., and C. H. Langley. 1988. Molecular and phenotypic variation of the *white* locus region in *Drosophila melanogaster. Genetics* 120: 199–212. [9]

Miyata, T., and T. Yasunaga. 1980. Molecular evolution of mRNA: A method for estimating evolutionary rates of synonymous and amino acid substitutions from homologous nucleotide sequences and its application. *J. Mol. Evol.* 16: 12–36. [10]

Mode, C. G., and H. F. Robinson. 1959. Pleiotropism and the genetic variance and covariance. *Biometrics* 15: 518–537. [1]

Modiano, D., and 10 others. 2001. Haemoglobin C protects against clinical *Plasmodium falciparum* malaria. *Nature* 414: 305–308. [9]

Mojica, J. P., and J. K. Kelly. 2010. Viability selection prior to trait expression is an essential component of natural selection. *Proc. Roy. Soc. Lond. B* 277: 2945–2950. [29]

Molina, J., and 11 others. 2011. Molecular evidence for a single evolutionary origin of domesticated rice. *Proc. Natl. Acad. Sci. USA* 108: 8351–8356. [9]

Moll, R. H., and O. S. Smith. 1981. Genetic variances and selection response in an advanced generation of a hybrid of widely divergent populations of maize. *Crop Sci.* 21: 387–391. [23]

Møller, A. P., and M. D. Jennions. 2001. Testing and adjusting for publication bias. *Trends Ecol. Evol.* 16: 580–586. [A4]

Montaldo, H. H. 1997. Optimization of selection response using artificial insemination and new reproductive technologies in diary cattle. Thesis, Department of Animal Science, University of Nebraska, Lincoln, NE. [14]

Montgomery, E. G. 1909. Experiments with corn. *Nebraska Agric. Exp. Stn. Bull.* 112. [21]

Montgomery, M. E., L. M. Woodworth, P. R. England, D. A. Briscoe, and R. Frankham. 2010. Widespread selective sweeps affecting microsatellites in *Drosophila* populations adapting to captivity: Implications for captive breeding programs. *Biol. Conserv.* 143: 1842–1849. [9]

Moody, D. E., D. Pomp, M. K. Nielsen, and L. D. Van Vleck. 1999. Identification of quantitative trait loci influencing traits related to energy balance in selection and inbred lines of mice. *Genetics* 152: 699–711. [25]

Moorad, J. A., and M. J. Wade. 2013. Selection gradients, the opportunity for selection, and the coefficient of determination. *Am. Nat.* 181: 291–300. [29]

Moore, A. J. 1990. The evolution of sexual dimorphism by sexual selection: the separate effects of intrasexual selection and intersexual selection. *Evolution* 44: 315–331. [29]

Moore, A. J., E. D. Brodie III, and J. B. Wolf. 1997. Interacting phenotypes and the evolutionary process. I. direct an indirect genetic effects of social interactions. *Evolution* 51: 1352–1362. [22]

Moore, A. J., and P. F. Kukuk. 2002. Quantitative genetic analysis of natural populations. *Nat. Rev. Genet.* 3: 971–978. [20]

Moore, J. C., and J. R. Pannell. 2011. Sexual selection in plants. *Curr. Biol.* 21: R176–R182. [29]

Moose, S. P., J. W. Dudley, and T. R. Rocheford. 2004. Maize selection passes the century mark: a unique resource for 21st century genomics. *Trends Plant Sci.* 9: 358–364. [25]

Moran, P. A. P. 1962. *The statistical processes of evolutionary theory.* Clarendon Press, Oxford, UK. [2]

Moreno-Rueda, G. 2009. Disruptive selection by predation offsets stabilizing selection on shell morphology in the land snail *Iberus g. gualtieranus. Evol. Ecol.* 23: 463–471. [30]

Morgan, A. D., R. W. Ness, P. D. Keightley, and N. Colegrave. 2014. Spontaneous mutation accumulation in multiple strains of the green alga, *Chlamydomonas reinhardtii. Evolution* 68: 2589–2602. [4]

Morgan, M. T., and D. J. Schoen. 1997. Selection on reproductive characters: floral morphology in *Asclepias syriaca. Heredity* 79: 433–441. [29]

Morgan, T. J., M. A. Evans, T. Garland, J. G. Swallow, and P. A. Carter. 2005. Molecular and quantitative genetic divergence among populations of house mice with known evolutionary histories. *Heredity* 94: 518–525. [12]

Morgante, F., P. Sørensen, D. A. Sorensen, C. Maltecca, and T. F. C. Mackay. 2015. Genetic architecture of micro-environmental plasticity in *Drosophila melanogaster. Sci. Reports* 5: 9785. [17]

Morisita, M. 1962. I_δ-index, a measure of dispersion of individuals. *Res. Pop. Ecol.* 4: 1–7. [29]

Morley, F. H. W. 1954. Selection for economic characters in Australian Merino sheep. IV. The effect of inbreeding. *Aust. J. Agric. Res.* 5: 305–316. [3, 26]

Morran, L. T., M. D. Parmenter, and P. C. Phillips. 2009. Mutation load and rapid adaptation favour outcrossing over self-fertilization. *Nature* 462: 350–352. [26]

Morrell, P. L., D. M. Toleno, K. E. Lundy, and M. T. Clegg. 2006. Estimating the contribution of mutation, recombination and gene conversion in the generation of haplotypic diversity. *Genetics* 173: 1705–1723. [4]

Morrison, D. F. 1976. *Multivariate statistical methods.* McGraw-Hill, New York, NY. [A5, A6]

Morrison, D. F. 1983. *Applied linear statistical methods.* Prentice-Hall, Upper Saddle River, NJ. [A5]

Morrissey, M. B. 2014a. Selection and evolution of causally covarying traits. *Evolution* 68: 1748–1761. [29, 30]

Morrissey, M. B. 2014b. In search of the best methods for multivariate selection analysis. *Methods Ecol. Evol.* 5: 1095–1109. [30]

Morrissey, M. B., 2015. Evolutionary quantitative genetics of nonlinear developmental systems. *Evolution* 69: 2050–2066. [14]

Morrissey, M. B. 2016. Meta-analysis of magnitudes, differences and variation in evolutionary parameters. *J. Evol. Biol.* 29: 1882–1904. [30, A4]

Morrissey, M. B., and J. D. Hadfield. 2012. Directional selection in temporally replicated studies is remarkably consistent. *Evolution* 66: 435–442. [20, 29, 30]

Morrissey, M. B., L. E. B. Kruuk, and A. J. Wilson. 2010. The danger of applying the breeder's equation in observational studies in natural populations. *J. Evol. Biol.* 23: 2277–2288. [20]

Morrissey, M. B., and K. Sakrejda. 2013. Unification of regression based methods for the analysis of natural selection. *Evolution* 67: 2094–2100. [29]

Morrissey, M. B., and K. Sakrejda. 2014. gsg: Calculation of selection coefficients. R package version 2.0. http://CRAN.R-project.org/package=gsg. [29]

Morrissey, M. B., and A. J. Wilson. 2010. PEDANTICS: An R package for pedigree-based genetic simulation and pedigree manipulation, characterization and viewing. *Mol. Ecol. Res.* 10: 711–719. [20]

Morrissey, M. B., A. J. Wilson, J. M. Pemberton, and M. M. Ferguson. 2007. A framework for power and sensitivity analysis of natural populations, and case studies in Soay sheep (*Ovis aries*). *J. Evol. Biol.* 20: 2309–2321. [20]

Morrissey, M. B., and 5 other. 2012. The prediction of adaptive evolution: Empirical application of the secondary theorem of selection and comparison to the breeder's equation. *Evolution* 66: 2399–2410. [20]

Morton, N. E. 1955. Sequential tests for the detection of linkage. *Am. J. Hum. Genet.* 7: 277–318. [A4]

Mosig, M. O., and 5 others. 2001. A whole genome scan for quantitative trait loci affecting milk protein percentage in Israeli-Holstein cattle, by means of selective milk DNA pooling in a daughter design, using an adjusted false discovery rate criterion. *Genetics* 157: 1683–1698. [A4]

Mostageer, A. 1971. A note on the use of inbred young bulls in progeny testing schemes. *Z. Tierz. Züchtungsbiol.* 88: 194–196. [23]

Mosteller, F., and R. R. Bush. 1954. Selected quantitative techniques. *In* G. Lindzer (ed.), *Handbook of social psychology*, Vol. 1, pp. 289–334. Addison-Wesley, Cambridge, MA. [A4]

Mousseau, T. A., and C. W. Fox (eds). 1998. *Maternal effects as adaptations.* Oxford University Press, Oxford, UK. [15]

Mousseau, T. A., and D. A. Roff. 1987. Natural selection and the heritability of fitness components. *Heredity* 59: 181–197. [6]

Mousseau, T. A.,T. Uller, E. Wapstra, and A. V. Badyaev. 2009. Evolution of maternal effects: past and present. *Phil. Trans. R. Soc. Lond. B* 364: 1035–1038. [15]

Mrode, R. A. 2014. *Linear models for the prediction of animal breeding values, 3rd Ed.* Cabi, Wallingford, UK. [13]

Mueller, J. P., and J. W. James. 1983. Effect on linkage disequilibrium of selection for a quantitative character with epistasis. *Theor. Appl. Genet.* 65: 25–30. [16, 24]

Mueller, L. D., B. A. Wilcox, P. R. Ehrlich, D. G. Heckel, and D. D. Murphy. 1985. A direct assessment of the role of genetic drift in determining allele frequency variation in populations of *Euphydryas editha*. *Genetics* 110: 495–511. [9]

Muir, C. D., J. B. Pease, and L. C. Moyle. 2014. Quantitative genetic analysis indicates natural selection on leaf phenotypes across wild tomato species (*Solanum* sect. *Lycopersicon*; Solanaceae). *Genetics* 198: 1629–1643. [12]

Muir, W. M. 1985. Relative efficiency of selection for performance of birds housed in colony cages based on production in single bird cages. *Poultry Sci.* 64: 2239–2447. [22]

Muir, W. M. 1986a. Estimation of response to selection and utilization of control populations for additional information and accuracy. *Biometrics* 42: 381–391. [18]

Muir, W. M. 1986b. Efficient design and analysis of selection experiments. *In* G. E. Dickerson and R. K. Johnson (eds.), *Proceedings of the 3rd world congress on genetics applied to livestock production.* Vol. 12: 269–282. Agric. Comms., Univ. Nebraska, Lincoln, NE. [18]

Muir, W. M. 1996. Group selection for adaptation to multiple-hen cages: selection program and direct responses. *Poultry Sci.* 75: 447–458. [22]

Muir, W. M. 2005. Incorporation of competitive effects in forest tree or animal breeding programs. *Genetics* 170: 1247–1259. [22]

Muir, W., P. Bijma, and A. Schinckel. 2013. Multilevel selection with kin and non-kin groups, experimental results with japanese quail (*Coturnix japonica*). *Evolution* 67: 1598–1606. [22]

Muir, W M., and J. V. Craig. 1998. Improving animal well-being through genetic selection. *Poultry Sci.* 77: 1781–1788. [22]

Muir, W M., and R. D. Howard. 1999. Possible ecological risks of transgenic organism release when transgenes affect mating success: sexual selection and the Trojan gene hypothesis. *Proc. Natl. Acad. Sci. USA* 24: 13853–13856. [5, 22]

Muir, W. M., and R. D. Howard. 2001. Fitness components and ecological risk of transgenic release: a model using Japanese Medaka (*Oryzias latipes*). *Am. Nat.* 158: 1-16. [5]

Muir, W. M., and A. Schinckel. 2002. Incorporation of competitive effects in breeding programs to improve productivity and animal well being. *Proc. 7th World*

Congr. Genet. Appl. Livestock Prod. Communication No. 14-07. Montpellier, FR. [22]

Mukherjee, S. 2016. *The gene: An intimate history*. Simon and Schuster, New York, NY. [1]

Mulambda, N. S., A. R. Hallauer, and O. S. Smith. 1983. Recurrent selection for grain yield in a maize population. *Crop Sci.* 23: 536–540. [23]

Mulder, H. A., P. Bijma, and W. G. Hill. 2007. Prediction of breeding values and selection response with genetic heterogeneity of environmental variance. *Genetics*: 175: 1895–1910. [17]

Mulder, H. A., P. Bijma, and W. G. Hill. 2008. Selection for uniformity in livestock by exploiting genetic heterogeneity of residual variance. *Genet. Sel. Evol*: 40: 37–60. [17]

Mulder, H. A., W. G. Hill, A. Vereijken, and R. F. Veerkamp. 2009. Estimation of genetic variation in residual variance in female and male broilers. *Animal* 3: 1673–1680. [17]

Muller, H. J. 1950. Our load of mutations. *Am. J. Hum. Genet.* 2: 111–176. [7]

Muller, H. J. 1964. The relation of recombination to mutational advance. *Mutat. Res.* 106: 2–-9. [8]

Murphy, C. G. 1998. Interaction-independent sexual selection and the mechanisms of sexual selection. *Evolution* 52: 8–18. [29]

Muse, S. V., and B. S. Gaut. 1994. A likelihood approach for comparing synonymous and nonsynonymous nucleotide substitution rates, with application to the chloroplast genome. *Mol. Biol. Evol.* 11: 715–724. [10]

Mutic, J. J., and J. B. Wolf. 2007. Indirect genetic effects from ecological interactions in *Arabidopsis thaliana*. *Mol. Ecol.* 16: 2371–2381. [22]

Myers, S., L. Bottolo, C. Freeman, G. McVean, and P. Donnelly. 2005. A fine-scale map of recombination rates and hotspots across the human genome. *Science* 310: 321–324. [4]

Myers, S. R., and R. C. Griffiths. 2003. Bounds on the minimum number of recombination events in a sample history. *Genetics* 163: 375–394. [4]

Nachman, M. W. 2004. Haldane and the first estimates of the human mutation rate. *Indian Acad. Sci.* 83: 231-233. [7]

Nachman, M. W., and S. L. Crowell. 2000. Estimate of the mutation rate per nucleotide in humans. *Genetics* 156: 297–304. [4]

Naciri-Graven, Y., and J. Goudet. 2003. The additive genetic variance after bottlenecks is affected by the number of loci involved in epistatic interactions. *Evolution* 57: 706–716. [11]

Nadot, R., and G. Vaysseix. 1973. Apparentement et identité. Algorithme du calcul des coefficients d'identité. *Biometrics* 29: 347–359. [11]

Nagylaki, T. 1976a. The evolution of one- and two-locus systems. *Genetics* 83: 583–600. [6]

Nagylaki, T. 1976b. A model for the evolution of self-fertilization and vegetative reproduction. *J. Theor. Biol.* 58: 55–58. [23]

Nagylaki, T. 1977a. *Selection in one- and two-locus systems.* Springer–Verlag, Berlin, DE. [5, 6]

Nagylaki, T. 1977b. The evolution of one- and two-locus systems. II. *Genetics* 85: 347–354. [6]

Nagylaki, T. 1982. Geographical invariance in population genetics. *J. Theor. Biol.* 99: 159–172. [2]

Nagylaki, T. 1984. Selection on a quantitative character. *In* A. Chakravarti (ed.), *Human population genetics: The Pittsburgh symposium*, pp. 275–306. Van Nostrand Reinhold, New York, NY. [5, 28]

Nagylaki, T. 1989a. The maintenance of genetic variability in two-locus models of stabilizing selection. *Genetics* 122: 235–248. [5, 28]

Nagylaki, T. 1989b. Rate of evolution of a character without epistasis. *Proc. Natl. Acad. Sci. USA* 86: 1910–1913. [5]

Nagylaki, T. 1991. Error bounds for the fundamental and secondary theorems of natural selection. *Proc. Natl. Acad. Sci. USA* 88: 2402–2406. [5, 6]

Nagylaki, T. 1992a. *Introduction to theoretical population genetics.* Springer-Verlag, Berlin, DE. [5, 7]

Nagylaki, T. 1992b. Rate of evolution of a quantitative character. *Proc. Natl. Acad. Sci. USA* 89: 8121–8124. [6]

Nagylaki, T. 1993. The evolution of multilocus systems under weak selection. *Genetics* 134: 627–647. [6]

Nagylaki, T. 1995. The inbreeding effective population number in dioecious populations. *Genetics* 139: 473–485. [3]

Nagylaki, T. 1998. The expected number of heterozygous sites in a subdivided population. *Genetics* 149: 1599–1604. [2]

Nagylaki, T. 2000. Geographical invariance and the strong-migration limit in subdivided populations. *J. Math. Biol.* 41: 123–142. [2, 11]

Nagylaki, T., J. Hofbauer, and P. Brunovsky. 1999. Convergence of multilocus systems under weak epistasis or weak selection. *J. Math. Biol.* 38: 103–133. [5]

Nair, U. S. 1936. The standard error of Gini's mean difference. *Biometrika* 28: 428–436. [30]

Nakagawa, S., and I. C. Cuthill. 2007. Effect size, confidence interval and statistical significance: a practical guide for biologists. *Biol. Rev.* 82: 591–605. [A4]

Nakagawa, S., and R. P. Freckleton. 2008. Missing inaction: the dangers of ignoring missing data. *Trends Ecol. Evol.* 23: 592–596. [20]

Nakagawa, S., and M Lagisz. 2016. Visualizing unbiased and biased unweighted metaanalyses. *J. Evol. Biol.* 29: 1914–1916. [A4]

Nakagawa, S., and E. S. A. Santos. 2012. Methodological issues and advances in biological meta-analysis. *Evol. Ecol.* 26: 1253–1274. [A4]

Namkoong, G. 1979. *Introduction to quantitative genetics in forestry.* Dept. of Agriculture Technical Bulletin No 1588. U. S. Governemnt Printing Office, Washington, DC. [13, 21]

Narain, P., and R. Chakraborty. 1987. Genetic differentiation of quantitative characters between populations or species II: optimal selection in infinite populations. *Heredity* 59: 199–212. [28]

Narum, S. R., and J. E. Hess. 2011. Comparison of F_{ST} outlier tests for SNP loci under selection. *Mol. Ecol. Resources* 11: 184–194. [9]

Nawa, N., and F. Tajima. 2008. Simple method for analyzing the pattern of DNA polymorphism and its application to SNP data of human. *Genes Genet. Syst.* 83: 353–360. [9]

Neff, B. D., and T. E. Pitcher. 2005. Genetic quality and sexual selection: an integrated framework for good genes and compatible genes. *Mol. Ecol.* 14: 19–38. [29]

Neher, R. A. 2013. Genetic draft, selective interference, and population genetics of rapid adaptation. *Ann. Rev. Ecol. Evol. Syst.* 44: 195–215. [8]

Neher, R. A., T. A. Kessinger, and B. I. Shraiman. 2013. Coalescence and genetic diversity in sexual populations under selection. *Proc. Natl. Acad. Sci. USA* 110: 15836–15841. [8]

Neher, R. A., and B. I. Shraiman. 2011. Genetic draft and quasi-neutrality in large facultatively sexual populations. *Genetics* 188: 975–996. [8]

Nei, M. 1969. Heterozygous effects and frequency changes of lethal genes in populations. *Genetics* 63: 669–680. [7]

Nei, M. 1971. Extinction time of deleterious mutant genes in large populations. *Theor. Popul. Biol.* 2: 419–425. [7]

Nei, M. 1978. Estimation of average heterozygosity and genetic distance from a small number of individuals. *Genetics* 89: 583–590. [4]

Nei, M. 1983. Genetic polymorphism and the role of mutation in evolution. *In* M. Nei and R. K. Koehn (eds.), *Evolution of genes and proteins,* pp. 165–190. Sinauer Assocs., Inc., Sunderland, MA. [4]

Nei, M. 1987. *Molecular evolutionary genetics.* Columbia Univ. Press, New York, NY. [12]

Nei, M., and R. K. Chesser. 1983. Estimation of fixation indices and gene diversities. *Ann. Hum. Genet.* 47: 253–259. [2]

Nei, M., and L. Jin. 1989. Variances of the average numbers of nucleotide substitutions within and between populations. *Mol. Biol. Evol.* 6: 290–300. [4]

Nei, M., and S. Kumar. 2000. *Molecular evolution and phylogenetics.* Oxford University Press, Oxford, UK. [2, 10]

Nei, M., and W.-H. Li. 1976. The transient distribution of allele frequencies under mutation pressure. *Genet. Res.* 28: 205–214. [2]

Nei, M., and T. Maruyama. 1975. Lewontin-Krakauer test for neutral genes. *Genetics* 80: 395. [9]

Nei, M., and M. Murata. 1966. Effective population size when fertility is inherited. *Genet. Res.* 8: 257–260. [26]

Nei, M., and A. K. Roychoudhury. 1973. Probability of fixation and mean fixation time of an overdominant mutation. *Genetics* 74: 371–380. [7]

Nei, M., and A. K. Roychoudhury. 1974. Sampling variances of heterozygosity and genetic distance. *Genetics* 76: 379–390. [4]

Nei, M., Y. Suzuki, and M. Nozawa. 2010. The neutral theory of molecular evolution in the genomic era. *Ann. Rev. Genom. Human Genet.* 11: 265–289. [8, 9]

Nei, M., and F. Tajima. 1981a. DNA polymorphism detectable by restriction endonucleases. *Genetics* 97: 145–163. [4]

Nei, M., and F. Tajima. 1981b. Genetic drift and estimation of effective population size. *Genetics* 98: 625–640. [4, 9]

Nei, M., and F. Tajima. 1983. Maximum likelihood estimation of the number of nucleotide substitutions from restriction sites data. *Genetics* 105: 207–217. [4]

Nei, M., and N. Takahata. 1993. Effective population size, genetic diversity, and coalescence time in subdivided populations. *J. Mol. Evol.* 37: 240–244. [2]

Nelson, R. M., M. E. Pettersson, and Ö. Carlborg. 2013. A century after Fisher: time for a new paradigm in quantitative genetics. *Trend. Genet.* 29: 669–676 [1, 11, 25]

Ness, R. W., A. D. Morgan, N. Colegrave, and P. D. Keightley. 2012. Estimate of the spontaneous mutation rate in *Chlamydomonas reinhardtii. Genetics* 192: 1447–1454. [4]

Nettleton, D., J. T. G. Hwang, R. A. Caldo, and R. P. Wise. 2006. Estimating the number of true null hypotheses from a histogram of *p* values. *J. Agr. Biol. Envir. Stats* 11: 337–356. [A4]

Neuhauser, C., and S. M. Krone. 1997. The genealogy of samples in models with selection. *Genetics* 145: 519–534. [8]

Newman, J. A., G. W. Rahnefeld, and H. T. Fredeen. 1973. Selection intensity and response to selection for yearling weight in beef cattle. *Can. J. Anim. Sci.* 53: 1–12. [18]

Newton, M., A. Noueiry, D. Sarkar, and P. Ahlquist. 2004. Detecting differential expression with a semiparametric hierarchical mixture method. *Biostats* 5: 155–176. [10]

Nguyen, H. T., and D. A. Sleper. 1983. Theory and application of half-sib matings in forage grass breeding. *Theor. Appl. Genet.* 64: 187–196. [21]

Nicholas, F. W. 1980. Size of population required for artificial selection. *Genet. Res.* 35: 85–105. [18]

Nicholas, F. W., and A. Robertson. 1980. The conflict between natural and artificial selection in finite populations. *Theor. Appl. Genet.* 56: 57–64. [25, 26]

Nicholson, G., and 5 others. 2002. Assessing population differentiation and isolation from single-nucleotide polymorphism data. *J. Roy. Stat. Soc. B.* 64: 695–715. [9]

Nicolas, F. W., and A. Robertson. 1976. The effect of selection on the standardized variance of gene frequency. *Theor. Appl. Genet.* 48: 263–266. [9]

Nielsen, B. V. H., and S. Andersen. 1987. Selection for growth on normal and reduced protein diets in mice. I. Direct and correlated responses for growth. *Genet. Res.* 50: 7–15. [21]

Nielsen, R. 2000. Estimation of population parameters and recombination rates from single nucleotide polymorphisms. *Genetics* 154: 931–942. [4]

Nielsen, R. 2001. Statistical tests of selective neutrality in the age of genomics. *Heredity* 86: 641–647. [9, 10]

Nielsen, R. 2005. Molecular signatures of natural selection. *Ann. Rev. Genet.* 39: 197–218. [9]

Nielsen, R. 2009. Adaptionism—30 years after Gould and Lewontin. *Evolution* 63: 2487–2490. [10]

Nielsen, R., M. J. Hubisz, and A. G. Clark. 2004. Reconstituting the frequency spectrum of ascertained single-nucleotide polymorphism data. *Genetics* 168: 237–2382. [9]

Nielsen, R., and Z. Yang. 1998. Likelihood models for detecting positively selected amino acid sites and application to the HIV-1 envelope gene. *Genetics* 148: 929–936. [10]

Nielsen, R., and 12 others. 2005a. A scan for positively selected genes in the genomes of humans and chimpanzees. *PLoS Bio.* 3: e170. [10]

Nielsen, R., and 5 others. 2005b. Genomic scans for selective sweeps using SNP data. *Genome Res.* 15: 1566–1575. [9]

Nielsen, R., and 12 others. 2009. Darwinian and demographic forces affecting human protein coding genes. *Genome Res.* 19: 838–849. [9]

Nishida, T. 1989. Is lifetime data always necessary for evaluating the "intensity" of selection? *Evolution* 43: 1826–1827. [29]

Nishino, J., and F. Tajima. 2004. Effect of dominance on heterozygosity and the fixation probability in a subdivided population. *Genes Genet. Syst.* 79: 41–48. [7]

Noël , E., and 6 others. 2016. Reduced mate availability leads to evolution of self-fertilization and purging of inbreeding depression in a hermaphrodite. *Evolution* 70: 625–640. [23]

Noël, E., and 6 others. 2017. Experimental evidence for the negative effects of self-fertilization on the adaptive potential of populations. *Curr. Biol.* 27: 237–242. [23]

Noor, M. A. F. 1999. Reinforcement and other consequences of sympatry. *Heredity* 83: 503–508. [16]

Nordborg, M. 1997. Structured coalescent processes on different time scales. *Genetics* 146: 1501–1514. [2]

Nordborg, M. 2000. Linkage disequilibrium, gene trees and selfing: an ancestral recombination graph with partial self-fertilization. *Genetics* 154: 923–929. [8, 9]

Nordborg, M. 2001. Introduction to coalescent theory *In* D. Balding, M. Bishop, and C. Cannings (eds.), *Handbook of statistical genetics*, pp. 179–212. Wiley, Chichester, UK. [2]

Nordborg, M., B. Charlesworth, and D. Charlesworth. 1996. The effect of recombination on background selection. *Genet. Res.* 67: 159–174. [3]

Nordskog, A. W., and J. Hardiman. 1980. Inbreeding depression and natural selection as factors limiting progress from selection in poultry. *In* A. Roberston, (ed.), *Selection experiments in laboratory and domestic animals*, pp. 91–99. Commonwealth Agricultural Bureaux, Slough, UK. [23]

Nordskog, A. W., and A. J. Wyatt. 1952. Genetic improvement as related to size of breeding operations. *Poultry Sci.* 31: 1062–1066. [14]

Norman, M. F. 1974. A central limit theorem for Markov processes that move by small steps. *Ann. Prob.* 2: 1065–1074. [8]

Notley-McRobb, L., S. Seeto, and T. Ferenci. 2002. Enrichment and elimination of *mutY* mutators in *Escherichia coli* populations. *Genetics* 162: 1055–1062. [4]

Novembre, J., and A. Di Rienzo. 2009. Spatial patterns of variation due to natural selection in humans. *Nat. Rev. Genet.* 10: 745–755. [9]

Novembre, J., A. P. Galvani, and M. Slatkin. 2005. The geographic spread of the *CCR5*-Δ*32* HIV-resistance allele. *PLoS Biol.* 3: e339. [9]

Novembre, J., and E. Han. 2012. Human population structure and the adaptive response to pathogen-induced selection pressures. *Phil. Trans. Roy. Soc. Lond. B* 367: 878–886. [8]

Nowak, M. A., C. E. Tarnita, and E. O. Wilson. 2010. The evolution of eusociality. *Nature* 466: 1057–1062. [22]

Nozawa, M., Y. Suzuki, and M. Nei. 2009. Reliabilities of identifying positive selection by branch-site and the site-prediction methods. *Proc. Natl. Acad. Sci. USA* 106: 6700–6705. [10]

Ntanos, D. A., and D. G. Roupakias. 2001. Comparative efficiency of two breeding methods for yield and quality in rice. *Crop. Sci.* 41: 345–350. [21]

Nunes, M. A., and D. J. Balding. 2010. On optimal selection of summary statistics for approximate Bayesian computation. *Stat. Appl. Genet. Mol. Biol.* 9: a34. [A3]

Nunney, L. 1993. The influence of mating system and overlapping generations on effective population size. *Evolution* 47: 1329–1341. [3]

Nunney, L. 1999. The effective size of a hierarchically structured population. *Evolution* 53: 1–10. [3]

Nunney, L. 2002. The effective size of annual plant populations: the interaction of a seed bank with fluctuating population size in maintaining genetic variation. *Am. Nat.* 160: 195–204. [3]

Nunney, L., and D. R. Elam. 1994. Estimating the effective size of conserved populations. *Conserv. Biol.* 8: 175–184. [3]

Nussbaum, R. L., R. R. McInnes, and H. F. Willard 2004. *Thompson & Thompson genetics in medicine*, 6th Edition. Elsevier, Amsterdam. [5]

Nuzhdin, S. V., C. L. Dilda, and T. F. C. Mackay. 1999. The genetic architecture of selection response: inferences from fine-scale mapping of bristle number quantitative trait loci in *Drosophila melanogaster*. *Genetics* 153: 1317–1331. [28]

Nuzhdin, S. V., P. D. Keightley, and E. G. Pasyukova. 1993. The use of retrotransposons as markers for mapping genes responsible for fitness differences between related *Drosophila melanogaster* strains. *Genet. Res.* 62: 125–133. [9]

Nuzhdin, S. V., and E. G. Pasyukova. 1991. New approach to polygene mapping. Location of genes controlling fitness level in related *Drosophila melanogaster* stocks. *Genetika* (Russ) 27: 849–859. [9]

Nuzhdin, S. V., M. L. Wayne, K. L. Harmon, and L. M. McIntyre. 2004. Common pattern of evolution of gene expression level and protein sequence in *Drosophila*. *Mol. Biol. Evol.* 21: 1308–1317. [12]

Nyholt, D. R. 2004. A simple correction for multiple testing for single-nucleotide polymorphisms in linkage disequilibrium with each other. *Am. J. Hum. Genet.* 74: 765–769. [A4]

Nyquist, W. E. 1991. Estimation of heritability and prediction of selection response in plant populations. *Critical Reviews Plant Sci.* 10: 235–322. [21]

O'Donald, P. 1968. Measuring the intensity of selection. *Nature* 220: 197–198. [24]

O'Donald, P. 1970. Change of fitness by selection for a quantitative character. *Theor. Popul. Biol.* 1: 219–232. [29]

O'Donald, P. 1971. Natural selection for quantitative characters. *Heridity* 27: 137–153. [29]

O'Donald, P. 1972. Natural selection for a quantitative character over several generations. *Nature* 237: 113–114. [24]

O'Donald, P. 1973. A further analysis of Bumpus' data: the intensity of natural selection. *Evolution* 27: 398–404. [30]

O'Hara, R. B. 2005. Comparing the effects of genetic drift and fluctuating selection on genotype frequency changes in the scarlet tiger moth. *Proc. Roy. Soc. Lond.* B 272: 211–217. [9]

O'Hara, R. B., J. M. Cano, O. Ovaskainen, C. Teplitsky, and J. S. Alho. 2008. Bayesian approaches in evolutionary quantitative genetics. *J. Evol. Biol.* 21: 949–957. [20]

O'Hara, R. B., and J. Merilä. 2005. Bias and precision in Q_{ST} estimates: Problems and some solutions. *Genetics* 171: 1331–1339. [12]

O'Reilly, P. F., E. Birney, and D. J. Balding. 2008. Confounding between recombination and selection, and the Ped/Pop method for detecting selection. *Genome Res.* 18: 1304–1313. [9]

O'Roak, B. J., and 15 others. 2011. Exome sequencing in sporadic autism spectrum disorders identifies severe *de novo* mutations. *Nat. Genet.* 43: 585–589. [4]

O'Roak, B. J., and 25 others. 2012. Multiplex targeted sequencing identifies recurrently mutated genes in autism spectrum disorders. *Science* 338: 1619–1622. [4]

O'Rourke, B. A., P. L. Greenwood, P. F. Arthur, and M. E. Goddard. 2012. Inferring the recent ancestry of *myostatin* alleles affecting muscle mass in cattle. *Anim. Genet.* 44: 86–90. [8]

Ochola, L. I., and 5 others. 2010. Allele frequency–based and polymorphism-versus-divergence indices of balancing selection in a new filtered set of polymorphic genes in *Plasmodium falciparum*. *Mol. Biol. Evol.* 27: 2344–251. [10]

Ødegård, J., and I. Olesen. 2011. Comparison of testing designs for genetic evaluation of social effects in aquaculture species. *Aquaculture* 317: 74–78. [22]

Odell, P. L., and A. H. Feiveson. 1966. A numerical procedure to generate a sample covariance matrix. *Am. Stat. Assoc. J.* 61: 198–203. [A2]

Ohta, T. 1973. Slightly deleterious mutant substitutions in evolution. *Nature* 246: 96–98. [8]

Ohta, T. 1982. Linkage disequilibrium due to random genetic drift in finite subdivided populations. *Proc. Natl. Acad. Sci. USA* 79: 1940–1944. [12]

Ohta, T. 1992. The nearly neutral theory of molecular evolution. *Ann. Rev. Ecol. Syst.* 23: 263–286. [8]

Ohta, T. 2002. Near-neutrality in evolution for genes and gene regulation. *Proc. Natl. Acad. Sci. USA* 99: 16134–16137. [8]

Ohta, T., and C. C. Cockerham. 1974. Detrimental genes with partial selfing and effects on a neutral locus. *Genet. Res.* 23: 191–200. [23]

Ohta, T., and M. Kimura. 1969a. Linkage disequilibrium due to random genetic drift. *Genet. Res.* 13: 47–55. [2]

Ohta, T., and M. Kimura. 1969b. Linkage disequilibrium at steady state determined by random genetic drift and recurrent mutation. *Genetics* 63: 229–238. [4]

Ohta, T., and M. Kimura. 1971. Linkage disequilibrium between two segregating nucleotide sites under the steady flux of mutations in a finite population. *Genetics* 68: 571–580. [2]

Ohta, T., and M. Kimura. 1975. The effect of selected linked locus on heterozygosity of neutral alleles (the hitch-hiking effect). *Genet. Res.* 25: 313-325. [8]

Okasha, S. 2004. Mutilevel selection and the partitioning of covariance: A comparison of three approaches. *Evolution* 58: 486–494. [30]

Okasha, S. 2006. *Evolution and levels of selection.* Oxford University Press, Oxford, UK. [6, 30]

Oleksyk, T. K., M. W. Smith, and S. J. O'Brien. 2010. Genome-wide scans for footprints of natural selection. *Phil. Trans. Roy. Soc. Lond. B* 365: 185–205. [9]

Oleksyk, T. K., and 5 others. 2008. Identifying selected regions from heterozygosity and divergence using a light-coverage genomic dataset from two human populations. *PLoS ONE* 3: e1712. [9]

Oliehoek, P. A., J. J. Windig, J. A. N. van Arendonk, and P. Bijma. 2006. Estimating relatedness between individuals in general populations with a focus on their use in conservation programs. *Genetics* 173: 483–496. [20]

Ollivier, L. 1988. Current principles and future prospects in selection of farm animals. *In* B. S. Weir, E. J. Eisen, M. M. Goodman, and G. Namkoong (eds.), *Proceedings of the second international conference on quantitative genetics*, pp. 438–450. Sinauer Assocs., Sunderland, MA. [13]

Ollivier, L. 1999. On the use of animal models in the analysis of selection experiments. *Genet. Sel. Evol.* 31: 135–148. [19]

Ollivier, L. 2008. Jay Lush: Reflections on the past. *Lohmann Information* 43: 3–12. [13].

Ollivier, L., L. A. Messer, M. F. Rothschild, and C. Legault. 1997. The use of selection experiments for detecting quantitative trait loci. *Genet. Res.* 69: 227–232. [9]

Olsen, K. M., and J. F. Wendel. 2013. A bountiful harvest: genomic insights into crop domestication phenotypes. *Ann. Rev. Plant Biol.* 64: 47–70. [9]

Olsen, K. M., and 5 others. 2006. Selection under domestication: evidence for a sweep in the rice *Waxy* genomic region. *Genetics* 173: 975–983. [9]

Ordas, B., R. A. Malvar, and W. G. Hill. 2008. Genetic variation and quantitative trait loci associated with developmental stability and the environmental correlation between traits in maize. *Genet. Res.* 90: 385–395. [17]

Orlove, M. J., and C. L. Wood. 1978. Coefficients of relationship and coefficients of relatedness in kin selection: a covariance form for the RHO formula. *J. Theor. Biol.* 73: 679–686. [22]

Ormond, L., M. Foll, G. B. Ewing, S. P. Pfeifer, and J. D. Jensen. 2015. Inferring the age of a fixed beneficial allele. *Mol. Ecol.* 25: 157–169. [8]

Orr, H. A. 1998a. Testing natural selection vs. genetic drift in phenotypic evolution using quantitative trait locus data. *Genetics* 149: 2099-2104. [12]

Orr, H. A. 1998b. The population genetics of adaptation: The distribution of factors fixed during adaptive evolution. *Evolution* 52: 935–949. [27]

Orr, H. A. 1999. The evolutionary genetics of adaptation: a simulation study. *Genet. Res.* 74: 207–214. [27]

Orr, H. A. 2000. Adaptation and the cost of complexity. *Evolution* 54: 13–20. [27]

Orr, H. A. 2002. The population genetics of adaptation: the adaptation of DNA sequences. *Evolution* 56: 1317–1330. [27]

Orr, H. A. 2003a. A minimum on the mean number of steps taken in adaptive walks. *J. Theor. Biol.* 220: 241–247. [27]

Orr, H. A. 2003b. The distribution of fitness effects among beneficial mutations. *Genetics* 163: 1519–1526. [27]

Orr, H. A. 2005a. The genetic theory of adaptation: A brief history. *Nat. Rev. Genet.* 6: 119–127. [27]

Orr, H. A. 2005b. Theories of adaptation: What they do and don't say. *Genetica* 123: 3–13. [26]

Orr, H. A. 2005c. The probability of parallel evolution. *Evolution* 59: 216–220. [8]

Orr, H. A. 2006. The distribution of fitness effects among beneficial mutations in Fisher's geometric model of adaptation. *J. Theor. Biol.* 238: 279–285. [27]

Orr, H. A. 2007. Absolute fitness, relative fitness, and utility. *Evolution* 61: 2997–3000. [5]

Orr, H. A. 2009. Fitness and its role in evolutionary genetics. *Nat. Rev. Genet.* 10: 351–359. [1, 27]

Orr, H. A. 2010. The population genetics of beneficial mutations. *Phil. Trans. R. Soc. Lond. B* 365: 1195–1201. [27]

Orr, H. A., and A. J. Betancourt. 2001. Haldane's sieve and adaptation from the standing genetic variation. *Genetics* 157: 875–884. [8]

Orr, H. A., and J. A. Coyne. 1992. The genetics of adaptation: a reassessment. *Am. Nat.* 140: 725–742. [25, 27]

Ortiz-Barrientos, D., A. Grealy, and P. Nosil. 2009. The genetics and ecology of reinforcement: Implications for the evolution of prezygotic isolation in sympatry and beyond. *Ann. N. Y. Acad. Sci.* 1168: 156–182. [16]

Osborne, R. 1957a. The use of sire and dam family averages in increasing the efficiency of selective breeding under a hierarchical mating system. *Heredity* 11: 93–116. [21]

Osborne, R. 1957b. Family selection in poultry: the use of sire and dam family averages in choosing male parents. *Proc. Roy. Soc. Edinb. B* 66: 374–393. [21]

Ossowski, S., and 7 others. 2010. The rate and molecular spectrum of spontaneous mutations in *Arabidopsis thaliana*. *Science* 327: 92–94. [4]

Ostrow, D., and 8 others. 2007. Mutational bias for body size in rhabditid nematodes. *Genetics* 176: 1653–1661. [12]

Otto, S. P. 2003. The advantages of segregation and the evolution of sex. *Genetics* 164: 1099–1118. [7]

Otto, S. P. 2004. Two steps forward, one step back: the pleiotropic effects of favoured alleles. *Proc. Roy. Soc. Lond. B* 271: 705–714. [25]

Otto, S. P., and N. H. Barton. 1997. The evolution of recombination: Removing the limits to natural selection. *Genetics* 147: 879–906. [8, 26]

Otto, S. P., and N. H. Barton. 2001. Selection for recombination in small populations. *Evolution* 55: 1921–1931. [1, 4]

Otto, S. P., and T. Lenormand. 2002. Resolving the paradox of sex and recombination. *Nat. Genet.* 3: 252–261. [4]

Otto, S. P., and M. C. Whitlock. 1997. The probability of fixation in populations of changing size. *Genetics* 146: 723–733. [7]

Ovaskainen, O., J. M. Cano, and J. Merilä. 2008. A Bayesian framework for comparative quantitative genetics. *Proc. Roy. Soc. B* 275: 669–678. [19, 20, A3]

Ovaskainen, O., M. Karhunen, C. Zheng, J. M. C. Arias, and J. Merilä. 2011. A new method to uncover signatures of divergent and stabilizing selection in quantitative traits. *Genetics* 189: 621–632. [12]

Owen, A. B. 2005. Variance of the number of false discoveries. *J. Roy. Stat. Soc. B* 67: 411–426. [A4]

Padhukasahasram, B., and B. Rannala. 2013. Meiotic gene-conversion rate and tract length variation in the human genome. *Eur. J. Hum. Genet.* 2013 Feb 27. [4]

Page, R. D. M. and E. C. Holmes. 1998. *Molecular evolution: a phylogenetic approach.* Wiley-Blackwell, New York, NY. [10]

Paigen, K., and 8 others. 2008. The recombinational anatomy of a mouse chromosome. *PLoS Genet.* 4: e1000119. [4]

Palaisa, K., M. Morgante, S. Tingey, and A. Rafalski. 2004. Long-range patterns of diversity and linkage disequilibrium surrounding the maize *Y1* gene are indicative of an asymmetric selective sweep. *Proc. Natl. Acad. Sci. USA* 101: 9885–9890. [9]

Pálsson, S. 2002. Selection on a modifier of recombination rate due to linked deleterious mutations. *J. Heredity* 93: 22–26. [4]

Pálsson, S., and P. Pamilo. 1999. The effects of deleterious mutations on linked, neutral variation in small populations. *Genetics* 153: 475-483. [3]

Palstra, F. P., and D. E. Ruzzante. 2008. Genetic estimates of contemporary effective population size: what can they tell us about the importance of genetic stochasticity for wild population persistence? *Mol. Ecol.* 17: 3428–3447. [4]

Pamilo, P., and S. L. Varvio-Aho. 1980. On the estimation of population size from allele frequency changes. *Genetics* 95: 1055–1057. [4]

Pandey, S., A. O. Diallo, T. M. T. Islam, and J. Deutsch. 1986. Progress from selection in eight tropical maize populations using international testing. *Crop Sci.* 26: 879–884. [21]

Pandey, S., A. O. Diallo, T. M. T. Islam, and J. Deutsch. 1987. Response to full-sib selection in four medium maturity maize populations. *Crop Sci.* 27: 617–622. [21]

Pannell, J. R. 2003. Coalescence in a metapopulation with recurrent local extinction and recolonization. *Evolution* 57: 949–961. [2]

Pannell, J. R., and B. Charlesworth. 1999. Neutral genetic diversity in a metapopulation with recurrent local extinction and recolonization. *Evolution* 53: 664–676. [3]

Pannell, J. R., and B. Charlesworth. 2000. Effects of metapopulation processes on measures of genetic diversity. *Phil. Trans. Roy. Soc. Lond. B* 355: 1851–1864. [2]

Pannell, J. R., and 18 others. 2015. The scope of Baker's law. *New Phytologist* 208: 656–667. [23]

Papaïx, J., and 5 others. 2010. Combining capture–recapture data and pedigree information to assess heritability of demographic parameters in the wild. *J. Evol. Biol.* 23: 2176–2184. [6, 20]

Paré, G., N. R. Cook, P. M. Ridker, and D. I. Chaseman. 2010. On the use of variance per genotype as a tool to identify quantitative trait interaction effects: a report from the women's genome health study. *PLoS Genet.* 6: e1000981. [17]

Parker, G. A. 1970. Sperm competition and its evolutionary consequences in the insects. *Biol. Rev.* 45: 525–567. [29]

Parker, G. A. 2006. Sexual conflict over mating and fertilization: an overview. *Phil. Trans. Roy. Soc. Lond. B* 361: 235–259. [29]

Partridge, L., and P. H. Harvey. 1988. The ecological context of life history evolution. *Science* 241: 1449–1455. [29]

Parts, L., and 13 others. 2011. Revealing the genetic structure of a trait by sequencing a population under selection. *Genome Res.* 21: 1131–1138. [25]

Paterniani, E. 1967. Selection among and within half-sib families in a Brazilian population of maize (*Zea mays* L.). *Crop Sci.* 7: 212–215. [21]

Patterson, H. D., V. Silvey, M. Talbot, and S. T. C. Weatherup. 1977. Variability of yields of cereal varieties in UK trials. *J. Agric. Sci.* 89: 238–245. [21]

Patterson, H. D., and R. Thompson. 1971. Recovery of interblock information when block sizes are unequal. *Biometrika* 58: 545–554. [19]

Pavlidis, P., J. D. Jensen, and W. Stephan. 2010. Searching for footprints of positive selection in whole-genome SNP data from nonequilibrium populations. *Genetics* 185: 907–922. [9]

Pavlidis, P., D. Metzler, and W. Stephan. 2012. Selective sweeps in multilocus models of quantitative traits. *Genetics* 192: 225–239. [8, 9]

Pavlidis, P., D. Živković, A. Stamatakis, and N. Alachiotis. 2013. SweeD: likelihood-based detection of selective sweeps in thousands of genomes. *Mol. Biol. Evol.* 30: 2224–2234. [9]

Pearson, E. S. 1938. The probability transformation for testing goodness of fit and combining independent tests of significance. *Biometrika* 30: 134–148. [A4]

Pearson, K. 1898. Mathematical contributions to the theory of evolution. On the law of ancestral heredity. *Proc. Roy. Soc. Lond.* 62: 386–412. [15]

Pearson, K. 1903. Mathematical contributions to the theory of evolution. XI. On the influence of natural selection on the variability and correlation of organs. *Phil. Trans. Roy. Soc. Lond. A* 200: 1–66. [1, 13, 16, 29, 30]

Peck, J. 1994. A ruby in the rubbish: beneficial mutations, deleterious mutations, and the evolution of sex. *Genetics* 137: 597–606. [7]

Peck, S. L., S. P. Ellner, and F. Gould. 1998. A spatially explicit stochastic model demonstrates the feasibility of Wright's shifting balance theory. *Evolution* 52: 1834–1839. [26]

Peck, S. L., S. P. Ellner, and F. Gould. 2000. Varying migration and deme size and the feasibility of the shifting balance. *Evolution* 54: 324–327. [26]

Pederson, D. G. 1969a. The prediction of selection response in a self-fertilizing species. I. Individual selection. *Aust. J. Biol. Sci.* 22: 117–129. [23]

Pederson, D. G. 1969b. The prediction of selection response in a self-fertilizing species. II. Family selection. *Aust. J. Biol. Sci.* 22: 1245–1257. [23]

Peiffer, J. A., and 10 others. 2014. The genetic architecture of maize height. *Genetics* 196: 1337–1356. [28]

Pélabon, C., and T. F. Hansen. 2008. On the adaptive accuracy of directional asymmetry in insect wing size. *Evolution* 62: 2855–2867. [18]

Pélabon, C., T. F. Hansen, A. J. R. Carter, and D. Houle. 2006. Response of fluctuating and directional asymmetry to selection on wing shape in *Drosophila melanogaster*. *J. Evol. Biol.* 19: 764–776. [18]

Pélabon, C., T. F. Hansen, A. J. R. Carter, and D. Houle. 2010. Evolution of variation and variability under fluctuating, stabilizing, and disruptive selection. *Evolution* 64: 1912–1925. 17

Peltonen, J., J. Venna, and S. Kaski. 2009. Visualizations for assessing convergence and mixing of Markov chain Monte Carlo simulations. *Compt. Stat. Data Anal.* 53: 4453–4470. [A3]

Pemberton, J. M. 2008. Wild pedigrees: the way forward. *Proc. Roy. Soc. Lond. B* 275: 613–621. [20]

Pemberton, J. M. 2010. Evolution of quantitative traits in the wild: mind the ecology. *Phil. Trans. Roy. Soc. Lond. B* 365: 2431–2438. [20]

Peng, Y., and 14 others. 2011. Genetic variations in Tibetan populations and high-altitude adaptation at the Himalayas. *Mol. Biol. Evol.* 28: 1075–1081. [9]

Pennings, P. S., and J. Hermisson. 2006a. Soft sweeps II—Molecular population genetics of adaptation from recurrent mutation or migration. *Mol. Biol. Evol.* 23: 1076–1084. [8]

Pennings, P. S., and J. Hermisson. 2006b. Soft sweeps III—The signature of positive selection from recurrent mutation. *PLoS Genet.* 2: e186. [8, 9]

Pepper, J. W. 2000. Relatedness in trait group models of social evolution. *J. Theor. Biol.* 206: 355–368. [22]

Pérez-Figueroa, A., M. J. García-Pereira, M. Saura, E. Rolán-Alvarez, and A. Caballero. 2010. Comparing three different methods to detect selective loci using dominant markers. *J. Evol. Biol.* 23: 2267–2276. [9]

Perlitz, M., and W. Stephan. 1997. The mean and variance of the number of segregating sites since the last hitchhiking event. *J. Math. Biol.* 36: 1–23. [8]

Perrins, C. M., and P. J. Jones. 1974. The inheritance of clutch size in the great tit (*Parus major*). *Condor* 76: 225–228. [20]

Perry, G. H., and 11 others. 2012. Comparative RNA sequencing reveals substantial genetic variation in endangered primates. *Genome Res.* 22: 602–610. [12]

Peter, B. M., E. Huerta-Sanchez, and R. Nielsen. 2012. Distinguishing between selective sweeps from standing variation and from a *de novo* mutation. *PLoS Genet.* 8: e1003011. [8]

Peters, J. L., A. J. Sutton, D. R. Jones, K. R. Abrams, and L. Rushton. 2008. Contour-enhanced meta-analysis funnel plots help distinguish publication bias from other causes of asymmetry. *J. Clinc. Epid.* 61: 991–996. [A4]

Peterson, J. J., S. Cahya, and E. del Castillo. 2002. A general approach to confidence regions for optimal factor levels of response surfaces. *Biometrics* 58: 422–431. [30]

Petes, T. D. 2001. Meiotic recombination hot spots and cold spots. *Nat. Rev. Genet.* 2: 360–369. [4]

Pfaffelhuber, P., B. Haubold, and A. Wakolbinger. 2006. Approximate genealogies under genetic hitchhiking. *Genetics* 174: 1995–2008. [8]

Pfaffelhuber, P., A. Lehert, and W. Stephan. 2008. Linkage disequilibrium under genetic hitchhiking in finite populations. *Genetics* 179: 527–537. [8]

Pfaffelhuber, P., and A. Studeny. 2007. Approximating genealogies for partially linked neutral loci under a selective sweep. *J. Math. Biol.* 55: 299–330. [8]

Pheasant, M., and J. S. Mattick. 2007. Raising the estimate of functional human sequences. *Genome Res.* 17: 1245–1253. [10]

Phillips, P. C. and S. J. Arnold. 1989. Visualizing multivariate selection. *Evolution* 43: 1209–1222. [29, 30]

Phillips, P. C., and N. A. Johnson. 1998. The population genetics of synthetic lethals. *Genetics* 150: 449–458. [25]

Picard, R. R., and R. D. Cook. 1984. Cross-validation of regression models. *Technometrics* 79: 575–583. [29]

Pickands, J. III. 1975. Statistical inference using extreme order statistics. *Ann. Stat.* 3: 119–131. [27]

Pickrell, J. K., and 10 others. 2009. Signals of recent positive selection in a worldwide sample of human populations. *Genome Res.* 19: 826–837. [9]

Piepho, H.-P., F. Laidig, T. Drobek, and U. Meyer. 2014. Dissecting genetic and non-genetic sources of long-term yield trend in German official variety trials. *Theor. Appl. Genet.* 127: 1009–1018. [19]

Piepho, H.-P., and J. Möhring. 2006. Selection in cultivar trials—is it ignorable? *Crop Sci.* 46: 192–201. [19]

Piepho, H.-P., J. Möhring, A. E. Melchinger, and A. Büchse. 2008. BLUP for phenotypic selection in plant breeding and variety testing. *Euphytica* 161: 209–228. [19]

Pigliucci, M., and J. Kaplan. 2000. The fall and rise of Dr. Pangloss: adaptationism and the spandrels paper

20 years later. *Trends Ecol. Evol.*15: 66–70. [1]

Pigliucci, M., and G. B. Müller. 2010. *Evolution: the extended synthesis.* MIT Press, Cambridge, MA. [1]

Pike, D. J. 1969. A comparison of two methods for predicting changes in the distribution of gene frequency when selection is applied repeatedly to a finite population. *Genet. Res.* 13: 117–126. [7]

Pimentel, R. A. 1979. *Morphometrics. The multivariate analysis of biological data.* Kendall and Hunt, Dubuque, IA. [A5]

Pirchner, F. 1983. *Population genetics in animal breeding*, 2nd Edition. Plenum, New York, NY. [13]

Plutynski, A. 2006. What was Fisher's fundamental theorem of natural selection and what was it for? *Stud. Hist. Phil. Biol. Biomed. Sci.* 37: 59–82. [6]

Pluzhnikov, A., and P. Donnelly. 1996. Optimal sequencing strategies for surveying molecular genetic diversity. *Genetics* 144: 1247–1262. [4]

Podlaha, O., D. M. Webb, P. K. Tucker, and J. Zhang. 2005. Positive selection for indel substitutions in the rodent sperm protein *catsper1*. *Mol. Biol. Evol.* 22: 1845–1852. [10]

Poh, Y.-P., V. S. Domingues, H. E. Hoekstra, and J. D. Jensen. 2014. On the prospect of identifying adaptive loci in recently bottlenecked populations. *PLoS One* 9: e110579. [9]

Pollak, E. 1983. A new method for estimating the effective population size from allele frequency changes. *Genetics* 104: 531–548. [4]

Pollinger, J. P., and 5 others. 2005. Selective sweep mapping of genes with large phenotypic effects. *Genome* 15: 1809–1819. [9]

Poncet, B. N., and 9 others. 2010. Tracking genes of ecological relevance using a genome scan in two independent regional population samples of *Arabis alpina*. *Mol. Ecol.* 19: 2896–2907. [9]

Pond, S. K., and S. V. Muse. 2005. Site-to-site variation of synonymous substitution rates. *Mol. Biol. Evol.* 22: 2375–2358. [10]

Pong-Wong, R., C. S. Haley, and J. A. Woolliams. 1999. Behaviour of the additive finite locus model. *Genet. Sel. Evol.* 31: 193–211. [24]

Pong-Wong, R., and J. A. Woolliams. 1998. Response to mass selection when an identified major gene is segregating. *Genet. Sel. Evol.* 30: 313–338. [25]

Poon, A., and S. P. Otto. 2000. Compensating for our load of mutations: freezing the meltdown of small populations. *Evolution* 54: 1467–1479. [7]

Porcher, E., T. Giraud, I. Goldringer, and C. Lavigne. 2004. Experimental demonstration of a causal relationship between heterogeneity of selection and genetic differentiation in quantitative traits. *Evolution* 58: 1434–1445. [12]

Porcher, E., T. Giraud, and C. Lavigne. 2006. Genetic differentiation of neutral markers and quantitative traits in predominantly selfing metapopulations: confronting theory and experiments with *Arabidopsis thaliana*. *Genet. Res.* 87: 1–12. [12]

Porcher, E., and R. Lande. 2005. The evolution of self-fertilization and inbreeding depression under pollen discounting and pollen limitation. *J. Evol. Biol.* 18: 497–508. [23]

Porcher, E., and R. Lande. 2013. Evaluating a simple approximation to modeling the joint evolution of self-fertilization and inbreeding depression. *Evolution* 67: 3628–3635. [23]

Postma, E. 2006. Implications of the differences between true and predicted breeding values for the study of natural selection and micro-evolution. *J. Evol. Biol.* 19: 309–320. [20]

Postma, E. 2014. Four decades of estimating heritabilities in wild vertebrate populations: improved methods, more data, better estimates. *In* A. Charmantier, D. Garant, and L. E. B. Kruuk (eds.), *Quantitative genetics in the wild*, pp. 16-33. Oxford University Press, Oxford, UK. [20]

Postma, E., and A. Charmantier. 2007. What 'animal models' can and cannot tell ornithologists about the genetics of wild populations. *J. Ornith.* 148.: 633–642. [20]

Postma, E., J. Visser, and A. J. van Noordwijk. 2007. Strong artificial selection in the wild results in predicted small evolutionary change. *J. Evol. Biol.* 20: 1823–1832. [20]

Powell, J. E., P. M. Visscher, and M. E. Goddard. 2010. Reconciling the analysis of IBD and IBS in complex trait studies. *Nat. Rev. Genet.* 11: 800–805. [20]

Powell, J. R., and E. M. Moriyama. 1997. Evolution of codon usage bias in *Drosophila. Proc. Natl. Acad. Sci. USA* 94: 7784–7790. [8]

Prasad, A., and 8 others. 2008. Linkage disequilibrium and signatures of selection on chromosomes 19 and 29 in beef and dairy cattle. *Anim. Genet.* 39: 597–605. [9]

Preston, C. R., and W. R. Engels. 1996. P-element-induced male recombination and gene conversion in *Drosophila. Genetics* 144: 1611–1622. [4]

Preziosi, R. F., and D. J. Fairbairn. 2000. Lifetime selection on adult body size and components of body size in a waterstrider: opposing selection and the maintenance of sexual size dimorphism. *Evolution* 54: 558–566. [29]

Price, G. R. 1970. Selection and covariance. *Nature* 227: 520–521. [1, 6, 13, 22]

Price, G. R. 1972a. Extension of covariance selection mathematics. *Ann. Hum. Genet.* 35: 485–490. [6, 13, 22, 30]

Price, G. R. 1972b. Fisher's 'fundamental theorem' made clear. *Ann. Hum. Genet.* 36: 129–140. [6, 22]

Price, T., M. Kirkpatrick, and S. J. Arnold. 1988. Directional selection and the evolution of breeding date in birds. *Science* 240: 798–799. [20]

Price, T., and L. Liou. 1989. Selection on clutch size in birds. *Am. Nat.* 134: 950–959. [20]

Price, T. D., and P. R. Grant. 1984. Life history traits and natural selection for small body size in a population of Darwin's finches. *Evolution* 38: 483–494. [29]

Price, T. D., P. R. Grant, H. L. Gibbs, and P. T. Boag. 1984. Recurrent patterns of natural selection in a population of Darwin's finches. *Nature.* 309: 787–789. [29]

Price, T. D., and D. Schluter. 1991. On the low heritability of life history traits. *Evolution* 45: 853–861. [6]

Primack, R. B., and H. Kang. 1989. Measuring fitness and natural selection in wild plant populations. *Annu. Rev. Ecol. Sys.* 20: 367–396. [29]

Pritchard, J. K., and A. Di Rienzo. 2010. Adaptation—not by sweeps alone. *Nat. Rev. Genet.* 11: 665–667. [8]

Pritchard, J. K., J. K. Prickrell, and G. Coop. 2010. The genetics of human adaptation: Hard sweeps, soft sweeps, and polygenic adaptation. *Current Biol.* 20: R208–R215. [7]

Pritchard, J. K., M. T. Seielstad, A. Perez-Lezaun, and M. W. Feldman. 1999. Population growth of human Y chromosomes: a study of Y chromosome microsatellites. *Mol. Bio. Evol.* 16: 1791–1798. [A3]

Prout, T. 1962a. The effects of stabilizing selection on the time of development in *Drosophila melanogaster*. *Genet. Res.* 3: 364–382. [16]

Prout, T. 1962b. The error variance of the heritability estimate obtained from selection response. *Biometrics* 18: 404–407. [18]

Prout, T. 1965. The estimation of fitness from genotypic frequencies. *Evolution* 19: 546–551. [29]

Prout, T. 1969. The estimation of fitness from population data. *Genetics.* 63: 949–967. [29]

Prout, T., and J. S. F. Barker. 1989. Ecological aspects of the heritability of body size in *Drosophila buzzatii*. *Genetics* 123: 803-813. [12]

Prout, T., and J. S. F. Barker. 1993. *F* statistics in *Drosophila buzzatii*: selection, population size and inbreeding. *Genetics* 134: 369–375. [12]

Provine, W. B. 1971. *The origins of theoretical population genetics*. University of Chicago Press, Chicago, IL. [1, 25]

Provine, W. B. 2001. *The origins of theoretical population genetics: With a new afterword*. University of Chicago Press, Chicago, IL. [1]

Przeworski, M. 2002. The signature of positive selection at randomly chosen loci. *Genetics* 160: 1179–1189. [8, 9]

Przeworski, M. 2003. Estimating the time since the fixation of a beneficial allele. *Genetics* 164: 1667–1676. [8]

Przeworski, M., B. Charlesworth, and J. D. Wall. 1999. Genealogies and weak purifying selection. *Mol. Biol. Evol.* 16: 246–252. [8]

Przeworski, M., G. Coop, and J. D. Wall. 2005. The signature of positive selection on standing genetic variation. *Evolution* 59: 2312–2323. [8]

Ptak, S. E., and M. Przeworski. 2002. Evidence for population growth in humans is confounded by fine-scale population structure. *Trends Genet.* 18: 559–563. [9]

Ptak, S. E., K. Voelpel, and M. Przeworski. 2004. Insights into recombination from patterns of linkage disequilibrium in humans. *Genetics* 167: 387–397. [4]

Pudovkin, A. I., D. V. Zaykin, and D. Hedgecock. 1996. On the potential for estimating the effective number of breeders from heterozygote–excess in progeny. *Genetics* 144: 383–387. [4]

Pugesek, B. H., and A. Tomer. 1996. The Bumpus house sparrow data: a reanalysis using structural equation models. *Evol. Ecol.* 10: 387–404. [30]

Pujol, B., A. J. Wilson, R. I. C. Ross, and J. R. Pannell. 2008. Are Q_{ST}–F_{ST} comparisons for natural populations meaningful? *Mol. Ecol.* 17: 4782–4785. [12]

Purcell, S. M., and >100 others. 2009. Common polygenic variation contributes to risk of schizophrenia and bipolar disorder. *Nature* 460: 748–752. [42]

Purnell, D. J., and J. N. Thompson, Jr. 1973. Selection for asymmetrical bias in a behavioral character in character in *Drosophila melanogaster*. *Heredity* 31: 401–405. [18]

Purugganan, M. D., A. L. Boyles, and J. I. Suddith. 2000. Variation and selection at the *CAULIFLOWER* floral homeotic gene accompanying the evolution of domesticated *Brassica oleracea*. *Genetics* 155: 855–862. [9]

Purugganan, M. D., and D. Q. Fuller. 2009. The nature of selection during plant domestication. *Nature* 457: 843–848. [9]

Pyhäjärvi, T., and 5 others. 2007. Demographic history has influenced nucleotide diversity in European *Pinus sylvestris* populations. *Genetics* 177: 1713–1724. [4]

Qin, H., W. B. Wu, J. M. Comeron, M. Kreitman, and W.-H. Li. 2004. Intragenic spatial patterns of codon usage bias in prokaryotic and eukaryotic genomes. *Genetics* 168: 2245–2260. [8]

Quaas, R. L. 1976. Computing the diagonal elements and inverse of a large numerator relationship matrix. *Biometrics* 32: 949–953. [19]

Quaas, R. L. 1988. Additive genetic model with groups and relationships. *J. Dairy Sci.* 71: 1338-1345. [12]

Quass, R. L., and E. J. Pollak. 1980. Mixed model methodology for farm and ranch beef cattle. *J. Anim. Sci.* 51: 1277–1287. [19]

Queller, D. C. 1991. Group selection and kin selection. *Trends. Ecol. Evol.* 6: 64. [22]

Queller, D. C. 1992a. Quantitative genetics, inclusive fitness, and group selection. *Am. Nat.* 139: 540–558. [20, 22]

Queller, D. C. 1992b. A general model for kin selection. *Evolution* 46: 376–380. [22]

Quinn, J. L., A. Charmantier, D. Garant and B. C. Sheldon. 2006. Data depth, data completeness, and their influence on quantitative genetic estimation in two contrasting bird populations. *J. Evol. Biol.* 19: 994–1002. [20]

Quinton, M., and C. Smith. 1995. Comparison of evaluation-selection schemes for maximizing genetic response at the same level of inbreeding. *J Anim. Sci.* 73: 2208–2212. [23]

Quinton, M., C. Smith, and M. E. Goddard. 1992. Comparison of selection methods at the same level of inbreeding. *J. Anim. Sci.* 70: 1060–1067. [23, 26]

Qureshi, A. W. 1968. The role of finite population size and linkage in response to continued trucation selection. II. Dominance and overdominance. *Theor. Appl. Genet.* 38: 264–270. [7, 26]

Qureshi, A. W., and O. Kempthorne. 1968. On the fixation of genes of large effects due to continued truncation selection in small populations of polygenic systems with linkage. *Theor. Appl. Genet.* 38: 229–255. [7, 26]

Qureshi, A. W., O. Kempthorne and L. N. Hazel. 1968. The role of finite population size and linkage in response to continued truncation selection. I. Additive gene action. *Theor. Appl. Genet.* 38: 256–263. [7, 26]

Raftery, A. E., and S. Lewis. 1992a. How many iterations in the Gibbs sampler? *In* J. M. Bernardo, J. O. Berger, A. P. Dawid, and A. F. M. Smith (eds.) *Bayesian statistics 4*, pp. 763–773. Oxford University Press, Oxford, UK. [19, A3]

Raftery, A. E., and S. Lewis. 1992b. Comment: One long run with diagnostics: Implementation strategies for Markov Chain Monte Carlo. *Stat. Sci.* 7: 493–497. [A3]

Ralph, P., and G. Coop. 2010. Parallel adaptation: One or many waves of advance of an advantageous allele? *Genetics* 186: 647–668. [8, 9]

Ramos-Onsins, S. E., S. Mousset, T. Mitchell-Olds, and W. Stephan. 2007. Population genetic inference using a fixed number of segregating sites: a reassessment. *Genet. Res.* 89: 231–244. [9]

Ramos-Onsins, S. E., and J. Rozas. 2002. Statistical properties of new neutrality tests against population growth. *Mol. Biol. Evol.* 19: 2092–2100. [9]

Rand, D. M., and L. M. Kann. 1996. Excessive amino acid polymorphism in mitochondrial DNA: Contrasts among genes from *Drosophila*, mice, and humans. *Mol. Biol. Evol.* 13: 735–748. [10]

Randhawa, I. A., M. S. Khatkar, P. C. Thomson, and H. W. Raadsma. 2014. Composite selection signals can localize the trait specific genomic regions in multi-breed populations of cattle and sheep. *BMC Genet.* 15: 34. [9]

Rands, C. M., S. Meader, C. P. Ponting, and G. Lunter. 2014. 8.2% of the human genome is constrained: variation in rates of turnover across functional element classes in the human lineage. *PLoS Genet.* 10: e1004525. [10]

Ranz, J. M., and C. A. Machado. 2006. Uncovering evolutionary patterns of gene expression using microarrays. *Trends Ecol. Evol.* 21: 29–37. [12]

Rao, B. R., M. L. Garg, and C. C. Li. 1968. Correlation between the sample variances in a singly truncated bivariate normal distribution. *Biometrika* 55: 433–436. [24]

Räsänen, K., and L. E. B. Kruuk. 2007. Maternal effects and evolution at ecological time-scales. *Func. Ecol.* 21: 408–421. [15]

Rathie, K. A., and J. S. F. Barker. 1968. Effectiveness of regular cycles of intermittent artificial selection for a quantitative character in *Drosophila melanogaster*. *Aust. J. Biol. Sci.* 21: 1187–1213. [25]

Rathie, K. A., and F. W. Nicholas. 1980. Artificial selection with differing population structures. *Genet. Res.* 36: 117–131. [26]

Ratnakumar, A., and 6 others. 2010. Detecting positive selection within genomes: the problem of biased gene conversion. *Phil. Trans. Roy. Soc. Lond. B* 365: 2571–2580. [10]

Raup, D.M., 1977. Stochastic models in evolutionary palaeontology. *Developments Palaeontology and Stratigraphy*5: 59–78. [12]

Raup, D. M., and R. E. Crick. 1981. Evolution of single characters in the Jurassic ammonite *Kosmoceras*. *Paleobiology* 7: 200–215. [12]

Rausher, M. D. 1992. The measurement of selection on quantitative traits: biases due to environmental covariances between traits and fitness. *Evolution* 46: 616–626. [20]

Rausher, M. D., and E. L. Simms. 1989. The evolution of resistance to heribivory in *Ipomoea purpurea*. I. Attempts to detect selection. *Evolution* 43: 563–572. [20]

Rawlings, J. O. 1976. Order statistics for a special class of unequally correlated multinormal variates. *Biometrics* 32: 875–887. [14, 21]

Ray, D. K., N. D. Mueller, P. C. West, and J. A. Foley. 2013. Yield trends are insufficient to double global crop production by 2050. *PLoS One* 8: e66428. [1]

Ray, D. K., N. Ramankutty, N. D. Mueller, P. C. West, and J. A. Foley. 2012. Recent patterns of crop yield growth and stagnation. *Nat. Comm.* 3: 1293. [1]

Réale, D., D. Berteaux, A. G. McAdam, and S. Boutin. 2003. Lifetime selection on heritable life-history traits in a natural population of red squirrels. *Evolution* 57: 2416–2423. [20]

Réale, D., M. Festa-Bianchet, and J. T. Jorgenson. 1999. Heritability of body mass varies with age and season in wild bighorn sheep. *Heredity* 83: 526–532. [20]

Reed, D. R., M. P. Lawler, and M. G. Tordoff. 2008. Reduced body weight is a common effect of gene knockout in mice. *BMD Genet.* 9: 4. [24]

Reed, F. A., and C. F. Aquadro. 2006. Mutation, selection and the future of human evolution. *Trends Genet.* 22: 479–84. [7]

Reeve, E. C. R. 1953. Studies on quantitative inheritance. III. Heritability and genetic correlation in progeny tests using different mating systems. *J. Genet.* 51: 520–542. [16]

Reeve, E. C. R., and F. W. Robertson. 1953. Studies in quantitative inheritance. II. Analysis of a strain of *Drosophila melanogaster* selected for long wings. *J. Genet.* 51: 276–316. [18, 25]

Rego, C., M. Matos, and M. Santos. 2006. Symmetry breaking in interspecific *Drosophila* hybrids is not due to developmental noise. *Evolution* 60: 746–761. [18]

Reich, D. E., and E. S. Lander. 2001. On the allelic spectrum of human disease. *Trends Genet.* 17: 502–510. [7]

Reid, J. M. 2014. Quantitative genetic approaches to understanding sexual selection and mating system evolution in the wild. *In* A. Charmantier, D. Garant, and L. B. Kruuk (eds.), *Quantitative genetics in the wild*, pp. 34–53. Oxford University Press, Oxford, UK. [29]

Reinhold, K. 2002. Maternal effects and the evolution of behavioral and morphological characters: a literature review indicates the importance of extended maternal care. *J. Heredity* 93: 400–405. [15]

Rellstab, C., F. Gugerli, A. J. Eckert, A. M. Hancock, and R. Holderegger. 2015. A practical guide to environmental association analysis in landscape genomics. *Mol. Ecol.* 24: 4348–4370. [9]

Rendel, J. M. 1943. Variations in the weights of hatched and unhatched duck's eggs. *Biometrika* 33: 48–58. [16]

Rendel, J. M. 1959. Optimum group size in half-sib family selection. *Biometrics* 15: 376–381. [21]

Rendel, J. M., and A. Robertson. 1950. Estimation of genetic gain in milk yield by selection in a closed herd of dairy cattle. *J. Genet.* 50: 1–8. [13]

Renne, U., and 9 others. 2013. Lifelong obesity in a polygenic mouse model prevents age-and diet-induced glucose intolerance—obesity is no road to late-onset diabetes in mice. *PLoS One* 8: e79788. [25]

Renner, S. S., and R.E. Ricklefs. 1995. Dioecy and its correlates in the flowering plants. *Am. J. Botany* 82: 596–606. [23]

Reyment, R. A. 1971. Multivariate normality in morphometric analysis. *Math. Geology* 3: 357–368. [A5]

Reyment, R. A. 1982. Phenotypic evolution in a Cretaceous foraminifer. *Evolution* 36: 1182–1199. [12]

Reyment, R. A. 1991. *Multidimensional palaeobiology*. Pergamon Press, Oxford, UK. [A5]

Reyment, R. A., R.E. Blackith, and N. A. Campbell. 1984. *Multivariate morphometrics*. Academic Press, New York, NY. [A5]

Reynolds, R. J., D. K. Childers, and N. M. Pajewski. 2010. The distribution and hypothesis testing of eigenvalues from the canonical analysis of the gamma matrix of quadratic and correlational selection gradients. *Evolution* 64: 1076–1085. [30]

Reznick, D., and J. Travis, 1996. The empirical study of adaptation in natural populations. *In* M. R. Rose and G. V. Lauder (eds.), *Adaptation*, pp. 243–289. Academic Press, San Diego, CA. [1]

Reznick, D. N., F. H. Shaw, F. H. Rodd, and R. G. Shaw. 1997. Evaluation of the rate of evolution in natural populations of guppies (*Poecilia retuculata*). *Science* 275: 1934–1937. [20]

Rhoné, B., R. Vitalis, I. Goldringer, and I. Bonnin. 2010. Evolution of flowering time in experimental wheat populations: a comprehensive approach to detect genetic signatures of natural selection. *Evolution* 64: 2110–2125. [12]

Rice, D. P., and J. P. Townsend. 2012. Resampling QTL effects in the QTL sign test leads to incongruous sensitivity to variance in effect size. *Genes, Genomes, Genetics* 2: 905–911. [12]

Rice, S. 1990a. A geometric model for the evolution of development. *J. Theor. Biol.* 143: 319–342. [27]

Rice, S. H. 2002a. A general population genetic theory for the evolution of developmental interactions. *Proc. Natl. Acad. Sci. USA* 99: 15518–15523. [30]

Rice, S. H. 2004a. *Evolutionary theory: mathematical and conceptual foundations.* Sinauer Assocates, Inc. Sunderland, MA. [6]

Rice, S. H. 2004b. Developmental associations between traits: covariance and beyond. *Genetics* 166: 513–526. [30]

Rice, W. R. 1990b. A consensus combined p-value test and the family-wide significance of component tests. *Biometrics* 46: 303–308. [A4]

Rice, W. R. 2002b. Experimental tests of the adaptive significance of sexual recombination. *Nat. Rev. Genet.* 3: 241–251. [7]

Rice, W. R., and A. K. Chippendale. 2001. Sexual recombination and the power of natural selection. *Science* 294: 555—559. [26]

Rich, S. S., A. E. Bell, D. A. Miles, and S. P. Wilson. 1984. An experimental study of genetic drift for two quantitative traits in *Tribolium. J. Heredity* 75: 191-195. [12]

Richardson, R. H., K. Kojima, and H. L. Lucas. 1968. An analysis of short-term selection experiments. *Heredity* 23: 493–506. [18]

Richey, F. D. 1922. The experimental basis for the present status of corn breeding. *J. Am. Soc. Agron.* 14: 1–17. [21]

Richman, A., 2000. Evolution of balanced genetic polymorphism. *Mol. Ecol.* 9: 1953–1963. [9]

Riebler, A., L. Held, and W. Stephan. 2008. Bayesian variable selection for detecting adaptive genomic differences among populations. *Genetics* 178: 1817–1829. [9]

Rieseberg, L. H., A. Widmer, A. M. Arntz, and J. M. Burke. 2002. Directional selection is the primary cause of phenotypic diversification. *Proc. Natl. Acad. Sci. USA* 99: 12242–12245. [12]

Rifkin, S. A., D. Houle, J. Kim, and K. P. White. 2005. A mutation accumulation assay reveals a broad capacity for rapid evolution of gene expression. *Nature* 438: 220–223. [12]

Rifkin, S. A., J. Kim, and K. P. White. 2003. Evolution of gene expression in the *Drosophila melanogaster* subgroup. *Nat. Genet.* 33: 138–144. [12]

Risch, N., and K. Lange. 1979. Application of a recombination model in calculating the variance of sib pair genetic identity. *Ann. Hum. Genet.* 43: 177–186. [20]

Risch, N., and 8 others. 1995. Genetic analysis of idiopathic torsion dystonia in Ashkenazi Jews and their recent descent from a small founder population. *Nat. Genet.* 9: 152–159. [9]

Riska, B., J. J. Rutledge, and W. R. Atchley. 1985. Covariance between direct and maternal genetic effects in mice with a model of persistent environmental influences. *Genet. Res.* 45: 287–297. [15]

Ritland, K. 2000. Marker-inferred relatedness as a tool for detecting heritability in nature. *Mol. Ecol.* 9: 1195–1204. [20]

Roach, D. A., and R. D. Wulff. 1987. Maternal effects in plants. *Ann. Rev. Ecol. Syst.* 18: 209–235. [15]

Roberge, C., H. Guderley, and L. Bernatchez. 2007. Genomewide identification of genes under directional selection: gene transcription Q_{ST} scan in diverging Atlantic salmon subpopulations. *Genetics* 177: 1011–1022. [12]

Robert, C. P., and G. Casella. 2004. *Monte Carlo statistical methods, 2nd Ed.* Springer Verlag, New York, NY. [A3]

Robert, C., and G. Casella. 2011. A short history of Markov chain Monte Carlo: Subjective recollections from incomplete data. *In* S. Brooks, A. Gelman, G. Jones, and X.-L. Meng (eds), *Handbook of markov chain Monte Carlo,* pp. 102-115. CRC press, Boca Raton, FL. [A3]

Roberts, R. C. 1966. The limits to artificial selection for body weight in the mouse. I. The limits attained in earlier experiments. *Genet. Res.* 8: 347–360. [25, 26]

Roberts, R. C. 1967. The limits to artificial selection for body weight in the mouse. III. Selection from crosses between previously selected lines. *Genet. Res.* 9: 73–85. [26]

Robertson, A. 1952. The effect of inbreeding on the variation due to recessive genes. *Genetics* 37: 189–207. [11]

Robertson, A. 1955a. Prediction equations in quantitative genetics. *Biometrics* 11: 95–98. [21]

Robertson, A. 1955b. Selection in animals: synthesis. *Cold Spring Harbor Symp. Quant. Biol.* 20: 225–229. [6]

Robertson, A. 1956. The effect of selection against extreme deviants based on deviation or homozygosis. *J. Genet.* 54: 236–248. [5, 25, 28]

Robertson, A. 1957. Optimum group size in progeny testing and family selection. *Biometrics* 13: 442–450. [21]

Robertson, A. 1960a. A theory of limits in artificial selection *Proc. Roy. Soc. Lond. B* 153: 234–249. [1, 7, 26, 28]

Robertson, A. 1960b. On optimum family size in selection programmes. *Biometrics* 14: 296–298. [21]

Robertson, A. 1961. Inbreeding in artificial selection programmes. *Genet. Res.* 2: 189–194. [3, 26]

Robertson, A. 1962. Selection for heterozygotes in small populations. *Genetics* 47: 1291–1300. [7, 10, 26]

Robertson, A. 1964. The effect of nonrandom mating within inbred lines on the rate of inbreeding. *Genet. Res.* 5: 164–167. [3]

Robertson, A. 1965. The interpretation of genotypic ratios in domestic animal populations. *Anim. Prod.* 7: 319–324. [4]

Robertson, A. 1966a. A mathematical model of the culling process in dairy cattle. *Anim. Prod.* 8: 95–108. [1, 6, 13, 20]

Robertson, A. 1966b. Artificial selection in plants and animals. *Proc. Roy. Soc. Lond. B:* 164: 341–349. [26]

Robertson, A. 1967. The nature of quantitative genetic variation. *In* R.A. Brink and E. D. Styles (eds.), *Heritage from Mendel,* pp. 265–280. Univ. Wisconsin Press, Madison, WI. [5, 28]

Robertson, A. 1968. The spectrum of genetic variation. *In* R. C. Lewontin (ed.), *Population biology and evolu-*

tion, pp. 5–16. Syracuse University Press, Syracuse, NY. [6, 20]

Robertson, A. 1970a. A theory of limits in artificial selection with many linked loci. *In*, K.-I. Kojima (ed.), *Mathematical topics in population genetics*, pp. 246–288. Springer-Verlag, New York, NY. [25, 26]

Robertson, A. 1970b. Some optimum problems in individual selection. *Theor. Popul. Biol.* 1: 120–127. [26]

Robertson, A. 1970c. A note on disruptive selection experiments in *Drosophila*. *Am. Nat.* 104: 561–569. [16]

Robertson, A. 1974. The validity of the optimum model. *Adv. Appl. Prob.* 6: 17–18. [28]

Robertson, A. 1975a. Remarks on the Lewontin-Krakauer test. *Genetics* 80: 396. [9]

Robertson, A. 1975b. Gene frequency distributions as a test of selective neutrality. *Genetics* 81: 775–785. [9]

Robertson, A. 1977a. Artificial selection with a large number of linked loci. *In* E. Pollak, O. Kempthorne, and T. B. Bailey, Jr., (eds.), *Proceedings of the international conference on quantitative genetics*, pp. 307–322. Iowa State Univ. Press, Ames, IA. [18, 26]

Robertson, A. 1977b. The effect of selection on the estimation of genetic parameters. *Z. Tier Zuchtungsbiol.* 94: 131–135. [19, 25]

Robertson, A. 1978. The time of detection of recessive visible genes in small populations. *Genet. Res.* 31: 255–264. [26]

Roberston, A. (ed). 1980. *Selection experiments in laboratory and domestic animals*. Commonwealth Agricultural Bureaux, Slough, UK. [18]

Robertson, A., and W. G. Hill. 1983. Population and quantitative genetics of many linked loci in finite populations. *Proc. Roy. Soc. Lond.* B 219: 253–264. [24]

Robertson, A., and E. C. R. Reeve. 1952. Heterozygosity, environmental variation and heterosis. *Nature* 170: 286. [17]

Robertson, D. E. 1969. Selection in small populations. Ph. D. Thesis, University of South Wales, Sydney, AU. [26]

Robertson, F. W. 1955. Selection response and the properties of genetic variation. *Cold Spring Harbor Symp. Quant. Biol.* 20: 166–177. [25]

Robertson, H. F., R. E. Comstock, and P. H. Harvey. 1955. Genetic variances in open pollinated varieties of corn. *Genetics* 40: 45–60. [21]

Robin, S., A. Bar-Hen, J. J. Daudin, and L. Pierre. 2007. A semi-parametic approach for mixture models: application to local false discovery rate estimation. *Comput. Stat. Data Anal.* 51: 5483–5493. [A4]

Robinson, M. R., N. R. Wray, and P. M. Visscher. 2014. Explaining additional genetic variation in complex traits. *Trend. Genet.* 30: 124–132. [24, 28]

Robinson, M. R., and 5 others. 2009. The impact of environmental heterogeneity on genetic architecture in a wild population of soay sheep. *Genetics* 181: 1639–1648. [20]

Robinson, M. R., and 42 others. 2015. Population genetic differentiation of height and body mass index across Europe. *Nat. Genet.* 47: 1357–1362. [8, 9, 12]

Rockman, M. V. 2012. The QTN program and the alleles that matter for evolution: all that's gold does not glitter. *Evolution* 66: 1–17. [8]

Rockman, M. V., M. W. Hahn, N. Soranzo, D. B. Goldstein, and G. A. Wray. 2003. Positive selection on a human-specific transcription factor binding site regulating *IL4* expresion. *Curr. Biol.* 13: 2118–2123. [9]

Rockman, M. V., and L. Kruglyak. 2006. Genetics of global gene expression. *Nat. Rev. Genet.* 7: 862–872. [12]

Rockman, M. V., S. S. Skrovanek, and L. Kruglyak. 2010. Selection at linked sites shapes heritable phenotypic variation in *C. elegans*. *Science* 330: 372–376. [8]

Rodell, C. F., M. R. Schipper, and D. K. Keenan. 2004. Modes of selection and recombination response in *Drosophila melanogaster*. *J. Heredity* 95: 70–75. [26]

Rodrigue, N., H. Philippe, and N. Lartillot. 2010. Mutation-selection models of coding sequence evolution with site-heterogeneous amino acid fitness profiles. *Proc. Natl. Acad. Sci. USA* 107: 4629–4634. [10]

Rodrigues-Motta, M., D. Gianola, B. Heringstad, G. J. M. Roza, and Y. M. Chang. 2007. A zero-inflated Poisson model for genetic analysis of number of mastitis cases in Norwegian Red cows. *J. Dairy Sci.* 90: 5306–5315. [14]

Rodriguez, M. C., M. Toro, and L. Silió. 1996. Selection on lean growth in a nucleus of Landrace pigs: an analysis using Gibbs sampling. *Anim. Sci.* 63: 243–253. [19]

Rodríguez-Muñoz, R., A. Bretman, J. Slate, C. A. Walling, and T. Tregenza. 2010. Natural and sexual selection in a wild insect population. *Science* 328: 1269–1272. [29]

Roesti, M., S. Gavrilets, A. P. Hendry, W. Salzburger, and D. Berner. 2014. The genomic signature of parallel adaptation from shared genetic variation. *Mol. Ecol.* 23: 3944–3956. [8]

Roff, D. A. 1996. The evolution of threshold traits in animals. *Q. Rev. Biol.* 71: 3–35. [14]

Roff, D. A., and K. Emerson. 2006. Epistasis and dominance: evidence for differential effects in life-history versus morphological traits. *Evolution* 60: 1981–1990. [6]

Roff, D. A., and T. A. Mousseau. 1987. Quantitative genetics and fitness: lessons from *Drosophila*. *Heredity* 58: 103–118. [6]

Rogell, B., and 7 others. 2012. Strong divergence in trait means but not in plasticity across hatchery and wild populations of sea-run brown trout *Salmo trutta*. *Mol. Ecol.* 21: 2963–2976. [12]

Rogers, A. R., and H. C. Harpending. 1983. Population structure and quantitative characters. *Genetics* 105: 985–1002. [12]

Rogers, S. M., and 6 others. 2012. Genetic signature of adaptive peak shift in threespine stickleback. *Evolution* 66: 2439–2450. [27]

Rohlf, F. J., and F. L. Bookstein. 1990. *Proceeding of the Michigan morphometrics workshop*. Special Publication No. 2, The University of Michigan Museum of Zoology, Ann Arbor, MI. [A5]

Rohlfs, R. V., P. Harrigan, and R. Nielsen. 2014. Modeling gene expression evolution with an extended Ornstein-Uhlenbeck process accounting for within-species variation. *Mol. Biol. Evol.* 31: 201–211. [12]

Rokyta, D. R., C. J. Beisel, and P. Joyce. 2006. Properties of adaptive walks on uncorrelated landscapes under strong selection and weak mutation. *J. Theor. Biol.* 243: 114–120. [27]

Rokyta, D. R., P. Joyce, S. B. Caudle, and H. A. Wichman. 2005. An empirical test of the mutational landscape model of adaptation using a single-stranded DNA virus. *Nat. Genet.* 37: 441–444. [27]

Rokyta, D. R., and 5 others. 2008. Beneficial fitness effects are not exponential for two viruses. *J. Mol. Evol.* 67: 368–376. [27]

Rolf, M. M., and 7 others. 2010. Impact of reduced marker set estimation of genomic relationship matrices on genomic selection for feed efficiency in Angus cattle. *BMC Genet.* 11: 24. [20]

Romero, I. G., I. Ruvinsky, and Y. Gilad. 2012. Comparative studies of gene expression and the evolution of gene regulation. *Nat. Rev. Genet.* 13: 505–516. [12]

Romiguier, J., and 20 others. 2014. Comparative population genomics in animals uncovers the determinants of genetic diversity. *Nature* 515: 261–263. [4, 8]

Ronald, J., and J. M. Akey. 2005. Genome-wide scans for loci under selection in humans. *Hum. Genomics* 2: 113–125. [9]

Rönnegård, L., and W. Valdar. 2011. Detecting major genetic loci controlling phenotypic variability in experimental crosses. *Genetics* 188: 435–447. [17]

Rönnegård, L., and W. Valdar. 2012. Recent developments in statistical methods for detecting genetic loci affecting phenotypic variability. *BMC Genet.* 13: 630. [17]

Roopnarine, P. D. 2001. The description and classification of evolutionary mode: a computational approach. *Paleobiology* 27: 446—465. [12]

Roopnarine, P. D., G. Byars, and P. Fitzgerald. 1999. Anagenetic evolution, stratophenetic patterns, and random walk models. *Paleobiology* 25: 41–57. [12]

Ros, M., and 6 others. 2004. Evidence for genetic control of adult weight plasticity in *Helix aspersa*. *Genetics* 168: 2089–2097. [17]

Rosa, G. J. M., J. P. Steibel, and R. J. Tempelman. 2005. Reassessing design and analysis of two-colour microarray experiments using mixed effects models. *Comp. Funct. Genom.* 6: 123–131. [12]

Rosche, W. A., and P. L. Foster. 2000. Determining mutation rates in bacterial populations. *Methods* 20: 4–17. [4]

Rose, M. 1982. Antagonistic pleiotropy, dominance and genetic variance. *Heredity* 48: 63–78. [6]

Roseman, C. C. 2004. Detecting interregionally diversifying natural selection on modern human cranial form by using matched molecular and morphometric data. *Proc. Natl. Acad. Sci. USA* 101: 12824–12829. [12]

Rosenberg, M. S. 2005. The file-drawer problem revisted: a general weighted method for calculating fail-safe numbers in meta-analysis. *Evolution* 59: 464–468. [30, A4]

Rosenberg, N. A. 2003. The shapes of neutral gene genealogies in two species: probabilities of monophyly, paraphyly, and polyphyly in a coalescent model. *Evolution* 57: 1465–1477. [9]

Rosenberg, N. A., and M. Nordborg. 2002. Genealogical trees, coalescent theory and the analysis of genetic polymorphisms. *Nat. Rev. Genet.* 3: 380–390. [2]

Rosenthal, R. 1979. The "file drawer problem" and tolerance for null results. *Physchol. Bull.* 86: 638–641. [30, A4]

Ross-Ibarra, J., P. L. Morrell, and B. S. Gaut. 2007. Plant domestication, a unique opportunity to identify the genetic basis of adaptation. *Proc. Natl. Acad. Sci. USA* 104: 8641–8648. [9]

Ross-Ibarra, J., and 7 others. 2008. Patterns of polymorphism and demographic history in natural populations of *Arabidopsis lyrata*. *PLoS One* 3: e2411. [9]

Rossiter, M. C. 1996. Incidence and consequences of inherited environmental effects. *Ann. Rev. Ecol. Syst.* 27: 451–476. [15]

Rothenberg, T. J. 1971. Identification in parametric models. *Econometrica* 39: 577–591. [11, 22]

Rothschild, M. F., C. R. Henderson, and R. L. Quaas. 1979. Effects of selection on variances and covariances of simulated first and second lactations. *J. Dairy Sci.* 62: 996–1002. [19]

Roughgarden, J. 1972. Evolution of niche width. *Am. Nat.* 106: 683–718. [1, 28]

Roughgarden, J., and E.l Akçay. 2010a. Do we need a sexual selection 2.0? *Anim. Behav.* 79: e1–e4. [29]

Routman, E., and J. Cheverud. 1997. Gene effects on a quantitative trait: two-locus epistatic effects measured at microsatellite markers and at estimated QTL. *Evolution* 51: 1654-1662. [11]

Roughgarden, J., and E. Akçay. 2010b. Final response: sexual selection needs an alternative *Anim. Behav.* 79: e18–e23. [29]

Roughgarden, J., M. Oishi, and E. Akçay. 2006. Reproductive social behavior: cooperative games to replace sexual selection. *Science* 311: 965–969. [29]

Rowe, D. E., and R. R. Hill, Jr. 1984. Theoretical improvement of autotetraploid crops: Interpopulation and intrapopulation selection. U. S. Dept. of Agriculture, Technical Bulletin No 1689. [15]

Rowe, S. J., J. M. S. White, S. Avendano, and W. G. Hill. 2006. Genetic heterogeneity of residual variance in broiler chickens. *Genet. Sel. Evol.* 38: 617–635. [17]

Rozas, J., M. Gullaud, G. Blandin, and M. Aguadé. 2001. DNA variation at the *rt49* geene region of *Drosophila simulans*: evolutionary inferences from an unusual haplotype structure. *Genetics* 158: 1147–1155. [9]

Rozas, J., C. Segarra, G. Ribó, and M. Aguadé. 1999. Molecular population genetics of the *rp49* gene region in different chromosomal inversions of *Drosophila subobscura*. *Genetics* 151: 189–202. [9]

Roze, D., and N. H. Barton. 2006. The Hill-Robertson effect and the evolution of recombination. *Genetics* 173: 1793–1811. [4]

Rozen, D E., J. A. G. M. de Visse, and P. J. Gerrish. 2002. Fitness effects of fixed beneficial mutations in microbial populations. *Curr. Biol.* 12: 1040–1045. [27]

Ruano, R G., F. Orozco, and C. López-Fanjul. 1975. The effect of different selection intensities on selection response in egg-laying of *Tribolium castaneum*. *Genet. Res.* 25: 17–27. [26]

Rubin, C.-J. and 17 others. 2010. Whole-genome resequencing reveals loci under selection during chicken domestication. *Nature* 464: 587–591. [9]

Rubin. C.-J. and 13 others. 2012. Strong signatures of selection in the domestic pig genome. *Proc. Natl. Acad. Sci. USA* 109: 19529–19536. [9]

Rudolph P. E., and G. Herrendörfer. 1995. Optimal experimental design and accuracy of parameter estimation for nonlinear regression models used in long-term selection. *Biomet. J.* 37: 183–190. [25]

Rukšć, A., P. L. Bell-Rogers, J. D. Smith, and M. D. Baker. 2008. Analysis of spontaneous gene conversion tracts within and between mammalian chromosomes. *J. Mol. Biol.* 377: 337–351. [4]

Russell, W. A., G. F. Sprague, and H. L. Penny. 1963. Mutations affecting quantitative characters in long-time inbred lines of maize. *Crop Sci.* 3: 175–178. [12]

Rutledge, J. J., E. J. Eisen, and J. E. Legates. 1973. An experimental evaluation of genetic correlation. *Genetics* 75: 709–726. [25]

Ruzzante, D. E., D. C. Hamilton, D. L. Kramer, and J. W. A. Grant. 1996. Scaling of the variance and the quantification of resource monopolization. *Behav. Ecol.* 7: 199–207. [29]

Sabeti, P. C., and 16 others. 2002. Detecting recent positive selection in the human genome from haplotype structure. *Nature* 419: 832–837. [9]

Sabeti, P. C., and 14 others. 2005. The case for selection at *CCR5-Δ32*. *PLoS Biol.* 3: e378. [9]

Sabeti, P. C., and 9 others. 2006. Positive natural selection in the human lineage. *Science* 312: 1614-1620. [9]

Sabeti, P. C., and 12 others. 2007. Genome-wide detection and characterization of positive selection in human populations. *Nature* 449: 913–919. [9]

Saccheri, I. J., R. A. Nichols, and P. M. Brakefield. 2001. Effects of bottlenecks on quantitative genetic variation in the butterfly *Bicyclus anynana*. *Genet. Res.* 77: 167–181. [11]

Sæther, S. A., and 7 others. 2007. Inferring local adaptation from Q_{ST}–F_{ST} comparisons: neutral genetic and quantitative trait variation in European populations of great snipe. *J. Evol. Biol.* 20: 1563–1576. [12]

Sæther, B.-E., and Ø. Bakke. 2000. Avian life history variation and contribution of demographic traits to the population growth rate. *Ecology* 81: 642–653. [29, 30]

Sakai, K. 1955. Competition in plants and its relation to selection. *Cold Spring Harbor Symp. Quant. Biol.* 20: 137–157. [22]

Saleh, M., and 10 others. 2004. Differential modulation of endotoxin responsiveness by human caspase-12 polymorphisms. *Nature* 429: 75–79. [8]

Sales, J., and W. G. Hill. 1976. Effect of sampling errors on efficiency of selection indices. 1. Use of information from relatives for single trait improvement. *Anim. Prod.* 22: 1–17. [21]

SanCristobal-Gaudy, M., L. Bodin, J.-M. Eisen, and C. Chevalet. 2001. Genetic components of litter size variability in sheep. *Genet. Sel. Evol.* 33: 249–271. [17]

SanCristobal-Gaudy, M., J.-M. Eisen, L. Bodin, and C. Chevalet. 1998. Prediction of the response to a selection for canalisation of a continuous trait in animal breeding. *Genet. Sel. Evol.* 30: 423–451. [17]

Sanders, D. C. 1981. The Bethlem lines: genetic selection for high and low rearing activity in rats. *Behavor. Genet.* 11: 491–503. [18]

Sandoval-Castellanos, E. 2010. Testing temporal changes in allele frequencies: a simulation approach. *Genet. Res.* 92: 309–320. [9]

Sang. T., and S. Ge. 2007. Genetics and phylogenetics of rice domestication. *Curr. Opin. Genet. Devel.* 17: 533–538. [9]

Sanghi, A. K. 1983. Genetic variances in maize composite "moti." *Indian J. Genet.* 43: 180–184. [21]

Sangster, T. A., and 6 others. 2008. HSP90 affects the expression of genetic variation and developmental stability in quantitative traits. *Proc. Natl. Acad. Sci. USA* 105: 2963–2968. [17]

Santiago, E. 1989. Time of detection of recessive genes: effects of system of mating and number of examined individuals. *Theor. Appl. Genet.* 77: 867–872. [26]

Santiago, E. 1998. Linkage and the maintenance of variation for quantitative traits by mutation–selection balance: an infinitesimal model. *Genet. Res.* 71: 161–170. [28]

Santiago, E., and A. Caballero. 1995. Effective size of populations under selection. *Genetics* 139: 1013–1030. [3, 26]

Santiago, E., and A. Caballero. 1998. Effective size and polymorphism of linked neutral loci in populations under directional selection. *Genetics* 149: 2105–2117. [3]

Santiago, E., and A. Caballero. 2005. Variation after a selective sweep in a subdivided population. *Genetics* 169: 475-483. [3, 8]

Santoyo, G., and D. Romero. 2005. Gene conversion and concerted evolution in bacterial genomes. *FEMS Microbiol. Rev.* 29: 169–183. [4]

Santure, A. W., and J. Wang. 2009. The joint effects of selection and dominance on the Q_{ST}–F_{ST} contrast. *Genetics* 181: 259–276. [12]

Santure, A. W., and 5 others. 2010. On the use of large marker panels to estimate inbreeding and relatedness: empirical and simulation studies of a pedigreed zebra finch population typed at 771 SNPs. *Mol. Ecol.* 19: 1439–1451. [20]

Sarhan, A. E. and B. G. Greenberg. 1962. *Contributions to order statistics*. Wiley, New York, NY. [14]

Sarkar, S., W. T. Ma, and G. H. Sandri. 1992. On fluctuation analysis: a new, simple and efficient method for computing the expected number of mutants. *Genetica* 85: 173–179. [4]

Sattah, S., E. Elyashiv, O. Kolodny, Y. Rinott, and G. Sella. 2011. Pervasive adaptive protein evolution apparent in diversity patterns around amino acid substitutions in *Drosophila simulans*. *PLoS Genet.* 7: e1001302. [8, 9]

Savalli, U. M. 1993. An application of the neutral model to the evolution of tail length in the genus *Euplectes* (Aves, Ploceidae). *Evolution* 47: 696–699. [12]

Sawyer, S. A., and D. L. Hartl. 1992. Population genetics of polymorphism and divergence. *Genetics* 132: 1161–1176. [10]

Sawyer, S. A., R. J. Kulathinal, C. D. Bustamante, and D. L. Hartl. 2003. Bayesian analysis suggests that most amino acid replacements in *Drosophila* are driven by positive selection. *J. Mol. Evol.* 57: S154–S164. [10]

Sawyer, S., J. Parsch, Z. Zhang, and D. L. Hartl. 2007. Prevalence of positive selection among nearly neutral amino acid replacements in *Drosophila*. *Proc. Natl. Acad. Sci. USA* 104: 6504–6510. [10]

Saxton, A. M. 1988. Further approximations for selection intensity. *Theor. Appl. Genet.* 76: 465–466. [14]

Schadt, E. E., and 13 others. 2003. Genetics of gene expression surveyed in maize, mouse and man. *Nature* 422: 297–302. [12]

Schaeffer, L. R., and H. Song. 1978. Selection bias and REML variance-covariance component estimation. *J. Dairy Sci.* 61: 91–92. [19]

Schaeffer, S. W. 2002. Molecular population genetics of sequence length diversity in the *Adh* region of *Drosophila pseudoobscura*. *Genet. Res.* 80: 163–175. [9]

Schaffer, H. E., D. Yardley, and W. W. Anderson. 1977. Drift or selection: A statistical test of gene frequency variation over generations. *Genetics* 87: 371–379. [9]

Schaffner, S. F., and 5 others. 2005. Calibrating a coalescent simulation of human genome sequence variation. *Genome Res.* 15: 1576–1583. [9]

Scharloo, W. 1964. The effects of disruptive and stabilizing selection on the expression of a *cubitus interruptus* mutant in *Drosophila*. *Genetics* 50: 553–562. [16]

Scharloo, W. 1971. Reproductive isolation by disruptive selection. *Am. Nat.* 105: 87–90. [16]

Scharloo, W., M. S. Hoogmoed, and A. Ter Kuile. 1967. Stabilizing and disruptive selection on a mutant character in *Drosophila*. I. The phenotypic variance and its components. *Genetics* 60: 373–388. [16]

Scharloo, W., A. Zweep, K. A. Schuitema, and J. G. Winjnstra. 1972. Stabilizing and disruptive selection on a mutant character in *Drosophila*. IV. Selection on sensitivity to temperature. *Genetics* 71: 551–566. [21]

Scheffler, K., and C. Seoighe. 2005. A Bayesian model comparison approach to inferring positive selection. *Mol. Biol. Evol.* 12: 2513–2540. [10]

Scheiner, S. M. 1989. Variable selection along a successional gradient. *Evolution* 43: 548–562. [30]

Scheiner, S. M., K. Donohue, L. A. Dorn, S. J. Mazer, and L. M. Wolfe. 2002. Reducing environmental bias when measuring natural selection. *Evolution* 56: 2156–2167. [20]

Scheiner, S. M., and R. F. Lyman. 1991. The genetics of phenotypic plasticity. II. Response to selection. *J. Evol. Biol.* 4: 23-50. [21]

Scheiner, S. M., R. J. Mitchell, and H. S. Callahan. 2000. Using path analysis to measure natural selection. *J. Evol. Biol.* 13: 423–433. [30]

Schemske, D.W. and Lande, R. 1985. The evolution of selffertilization and inbreeding depression in plants. II. Empirical observations. *Evolution* 39: 41–52. [23]

Schena, M., Dari S., R. W. Davis, and P. O. Brown. 1995. Quantitative monitoring of gene expression patterns with a complementary DNA microarray. *Science* 270: 467–470. [12]

Schenk, M. F., I. G. Szendro, J. Krug, and J. A. G. M. de Visser. 2012. Quantifying the adaptive potential of an antibiotic resistance enzyme. *PLoS Genet.* 8: e1002783. [27]

Schiffels, S., G. J. Szöllősi, V. Mustonen, and M. Lässig. 2011. Emergent neutrality in adaptive asexual evolution. *Genetics* 189: 1361–1375. [8]

Schlamp, F., and 5 others. 2016. Evaluating the performance of selection scans to detect selective sweeps in domestic dogs. *Mol. Ecol.* 25: 342–356. [9]

Schlötterer, C. 2002. A microsatellite-based multilocus screen for the identification of local selective sweeps. *Genetics* 160: 753–763. [9]

Schlötterer, C. 2003. Hitchhiking mapping—functional genomics from the population genetics perspective. *Trends Genet.* 19: 32–38. [9]

Schlötterer, C., C. Vogl, and D. Tautz. 1997. Polymorphism and locus-specific effects on polymorphism at microsatellite loci in natural *Drosophila melanogaster* populations. *Genetics* 146: 309–320. [9]

Schluter, D. 1988. Estimating the form of natural selection on a quantitiative trait. *Evolution* 42: 849–861. [29, 30]

Schluter, D., and L. Gustafsson. 1993. Maternal inheritance of condition and clutch size in the collared flycatcher. *Evolution* 47: 658–667. [15]

Schluter, D., and D. Nychka. 1994. Exploring fitness surfaces. *Am. Nat.* 143: 597–616. [30]

Schmalhausen, I. I. 1949. *Factors of evolution*. Blakiston, Philadelphia, PA. [17]

Schneider, A., B. Charlesworth, A. Eyre-Walker, and P. D. Keightley. 2011. A method for inferring the rate of occurrence and fitness effects of advantageous mutations. *Genetics* 189: 1427–1437. [8, 10]

Schnell, F. W. 1982. A synoptic study of the methods and categories of plant breeding. *Z. Pflanzenzüchtg.* 89: 1–18. [21]

Schoen, D. J., M. T. Morgan, and T. Bataillon. 1996. How does self-pollination evolve? Inferences from floral ecology and molecular genetic variation *Phil. Trans. Roy. Soc. Lond. B* 351: 1281–1290. [23]

Schöft, G., and C. Schlötterer. 2004. Patterns of microsatellite variability among X chromosomes and autosomes indicate a high frequency of beneficial mutations in non-African *D. simulans*. *Mol. Biol. Evol.* 21: 1384–1390. [9]

Schork, N. 1992. Bootstrapping likelihood ratios in quantitative genetics. *In* R. Lepage and L. Billard (eds.), *Exploring the limits of the bootstrap*, pp. 389–393. Wiley, New York, NY. [A4]

Schoustra, S. E., T. Bataillon, D. R. Gifford, and R. Kassen. 2009. The properties of adaptive walks in evolving populations of fungus. *PLoS Bio.* 7: e1000250. [27]

Schraiber, J. G., S. N. Evans, and M. Slatkin. 2016. Bayesian inference of natural selection from allele frequency time series. *Genetics* 203: 493–511. [9]

Schrider, D. R., D. Houle, M. Lynch, and M. W. Hahn. 2013. Rates and genomic consequences of spontaneous mutational events in *Drosophila melanogaster*. *Genetics* 194: 937–54. [4]

Schrider, D. R., J. N. Hourmozdi, and M. W. Hahn. 2011. Pervasive multinucleotide mutational events in eukaryotes. *Curr. Biol.* 21: 1051–1054. [4]

Schrider, D. R., F. K. Mendes, M. W. Hahn, and A. D. Kern. 2015. Soft shoulders ahead: spurious signatures of soft and partial selective sweeps result from linked hard sweeps. *Genetics* 200: 267–284. [8]

Schulte-Hostedde, A. I., B. Zinner, J. S. Millar, and G. J. Hickling. 2005. Restitution of mass-size residuals: validating body condition indices. *Ecology* 86: 155–163. [20]

Schultz, S. T., M. Lynch, and J. H. Willis. 1999. Spontaneous deleterious mutation in *Arabidopsis thaliana*. *Proc. Natl. Acad. Sci. USA* 96: 11393–11398. [12]

Schutz, W. M., C. A. Brim, and S. A. Usanis. 1968. Intergenotypic competition in plant populations. I. Feedback system with stable equilibriua in populations of autogamous homozygous lines. *Crop Sci.* 8: 61–66. [22]

Schutz, W. M., and C. C. Cockerham. 1966. The effect of field blocking on gain from selection. *Biometrics* 22: 843–863. [13]

Schutz, W. M., and S. A. Usanis. 1969. Inter-genotypic competition in plant populations. II. Maintenance of allelic polymorphisms with frequency-depedent selection and mixed selfing and random mating. *Genetics* 61: 875–891. [22]

Schwaegerle, K. E., and D. A. Levin. 1991. Quantitative genetics of fitness traits in a wild population of phlox. *Evolution* 45: 169–177. [6]

Schwartz, J. 2000. Death of an altruist. *Lingua Franca* 10: 51–61. [6]

Schwartz, L., and S. Wearden. 1959. A distribution-free asymptotic method of estimating, testing, and setting confidence limits for heritability. *Biometrics* 15: 227–235. [14]

Schwarz, G. 1978. Estimating the dimension of a model *Annals. Stat.* 6: 461–464. [30]

Schweder, T., and E. Spjøtvoll. 1982. Plots of *p*-values to evaluate many tests simultaneously. *Biometrika* 69: 493–502. [A4]

Scowcroft, W. R. 1965. Variation of scutellar bristles in *Drosophila*. X. Selection limits. *Aust. J. Biol. Sci.* 19: 807–820. [26]

Searle, S. S. 1982. *Matrix algebra useful for statistics*. John Wiley and Sons, New York, NY. [A5, A6]

Seber, G. A. F. 1984. *Multivariate observations*. John Wiley and Sons, New York, NY. [A5]

Seetharaman, S., and K. Jain. 2014. Adaptive walks and distribution of beneficial fitness effects. *Evolution* 68: 965–975. [27]

Seger, J. 1981. Kinship and covariance. *J. Theor. Biol.* 91: 191–213. [22]

Sehested, E., and I. L. Mao. 1992. Responses to selection on genotypic or phenotypic values in the presence of genes with major effects. *Theor. Appl. Genet.* 85: 403–406. [25]

Sella, G., and A. E. Hirsh. 2005. The application of statistical physics to evolutionary biology. *Proc. Natl. Acad. Sci. USA* 102: 9541–9546. [1, 7]

Sella, G., D. A. Petrov, M. Przeworski, and P. Andolfatto. 2009. Pervasive natural selection in the *Drosophila* genome? *PLoS Genet.* 5: e100049. [8, 10]

Sellis, D., B. J. Callahan, D. A. Petrov, and P. W. Messer. 2011. Heterozygote advantage as a natural consequence of adaptation in diploids. *Proc. Natl. Acad. Sci. USA* 108: 20666–20671. [28]

Semlitsch, R. D., and H. M. Wilbur. 1989. Artificial selection for paedomorphosis in the salamander *Ambystoma talpoideum*. *Evolution* 43: 105–112. [20]

Serero, A., C. Jubin, S. Loeillet, P. Legoix-Né, and A. G. Nicolas. 2014. Mutational landscape of yeast mutator strains. *Proc. Natl. Acad. Sci. USA* 111: 1897–1902. [4]

Serre, J. L., and 8 others. 1990. Studies of RFLP closely linked to the cystic fibrosis locus throughout Europe lead to new considerations in populations genetics. *Hum. Genet.* 84: 449–454. [9]

Servedio, M. R., and M. A. F. Noor. 2003. The role of reinforcement in speciation: Theory and data. *Ann. Rev. Ecol. Evol. Syst.* 34: 339–364. [16]

Seyfert, A. L., and 5 others. 2008. The rate and spectrum of microsatellite mutation in *Caenorhabditis elegans* and *Daphnia pulex*. *Genetics* 178: 2113–2121. [4]

Sgró, C. M., and A. A. Hoffmann. 2004. Genetic correlations, tradeoffs and environmental variation. *Heredity* 93: 241–248. [20]

Shaffer, J. P. 1995. Multiple hypothesis testing. *Ann. Rev. Psychol.* 46: 561–584. [A4]

Sham, P. C., and S. M. Purcell. 2014. Statistical power and significance testing in large-scale genetic studies. *Nat. Rev. Genet.* 15: 335–346. [A4]

Shao, J., and D. Tu. 1996. *The jackknife and bootstrap*. Springer, New York, NY. [29]

Shapiro, J. A. 2011. *Evolution: a View from the 21st Century*. FT Press Science, Upper Saddle River, NJ. [1]

Shaw, F. H., C. J. Geyer, and R. G. Shaw. 2002. A comprehensive model of mutations affecting fitness and inferences for *Arabidopsis thaliana.Evolution* 56: 453–463. [12, 28]

Shaw, F. H., and J. A. Woolliams. 1999. Variance component analysis of skin and weight data for sheep subjected to rapid inbreeding. *Genet. Sel. Evol.* 31: 43–59. [11]

Shaw, R. G. 1987. Maximum-likelihood approaches applied to quantitative genetics of natural populations. *Evolution* 41: 812–826. [1, 20]

Shaw, R. G., D. L. Byers, and E. Darmo. 2000. Spontaneous mutational effects on reproductive traits of *Arabidopsis thaliana*. *Genetics* 155: 369–738. [12]

Shaw, R. G., D. L. Byers, and F. H. Shaw. 1998. Genetic components of variation in *Nemophila menziesii* undergoing inbreeding: morphology and flowering time. *Genetics* 150: 1649–1661. [11, 23]

Shaw, R. G., and C. J. Geyer. 2010. Inferring fitness landscapes. *Evolution* 64: 2510–2520. [30]

Shaw, R. G., C. J. Geyer, S. Wagenius, H. H. Hangelbroek, and J. R. Etterson 2008. Unifying life history analyses for inference of fitness and population growth. *Am. Nat.* 172: E35–E47. [29, 30]

Sheets, H. D., and C. E. Mitchell. 2001. Why the null matters: statistical tests, random walks and evolution. *Genetica* 112-113: 105–125. [12].

Sheldon, B. 2014. Foward. *In* A. Charmantier, D. Garant, and L. E. B. Kruuk (eds.), *Quantitative genetics in the wild*, pp. v-vi. Oxford University Press, Oxford, UK. [1]

Sheldon, B. C., and H. Ellegren. 1999. Sexual selection resulting from extra-pair paternity in collared flycatchers. *Anim. Behav.* 57: 285–298. [20]

Sheldon, B. C., L. E. B. Kruuk, and J. Merilä. 2003. Natural selection and inheritance of breeding time and clutch size in the collared flycatcher. *Evolution* 57: 406–420. [20]

Sheldon, B. L. 1963. Studies in artificial selection of quantitative characters. I. Selection for abdominal bristles in *Drosophila melanogaster*. *Aust. J. Biol. Sci.* 16: 490–515. [18]

Shen, X., M. Pettersson, L. Rönnegård, and Ö. Carlborg. 2012. Inheritance beyond plan heritability: variance-controlling genes in *Arabidopsis thaliana*. *PLoS Genet.* 8: e1002839. [17]

Shepherd, R. K., and B. P. Kinghorn. 1994. A deterministic multi-tier model of assortative mating following selection. *Genet. Sel. Evol.* 26: 495–516. [16]

Sheridan, A. K. 1988. Agreement between estimated and realised genetic parameters. *Anim. Breed. Abstracts* 56: 877–889. [18]

Sherman, P. W. 1977. Nepotism and the evolution of alarm calls. *Science* 197: 1246–1253. [22]

Shipley, B. 1997. Exploratory path analysis with applications in ecology and evolution. *Am. Nat.* 149: 1113–1138. [30]

Shnol, E. E., and A. S. Kondrashov. 1993. The effect of selection on the phenotypic variance. *Genetics* 134: 995–996. [16]

Shuker, D.M. 2010. Sexual selection: endless forms or tangled bank? *Anim. Behav.* 79: e11–e17. [29]

Shuster, S. M. 2009. Sexual selection and mating systems. *Proc. Nat. Acad. Sci. USA* 106: 10009–10016. [29]

Shuster, S. M., and M. J. Wade. 2003. *Mating systems and strategies*. Princeton Univ. Press, Princeton, NJ. [29]

Siepielski, A. M., J. D. DiBattista, and S. M. Carlson. 2009. It's about time: the temporal dynamics of phenotypic selection in the wild. *Ecol. Let.* 12: 1261–1276. [20, 29, 30]

Siepielski, A. M., J. D. DiBattista, J. A. Evans, and Stephanie M. Carlson. 2011. Differences in the temporal dynamics of phenotypic selection among fitness components in the wild. *Proc. Roy. Soc. Lond. B.* 278: 1572–1580. [29, 30]

Siepielski, A. M., and 5 others. 2013. The spatial patterns of directional phenotypic selection. *Ecol. Lett.* 16: 1382–1392. [29, 30]

Siewerdt, F., E. J. Eisen, and J. D. Murray. 1999. Direct and correlated responses to short-term selection for 8-week body weight in lines of transgenic (oMt1a-oGH) mice. *In* J. D. Murray, G. B. Anderson, A. M. Oberbauer, and M. M. McGloughlin (eds.), *Transgenic animals in agriculture*, pp. 231–250. CAB, Wallingford, UK. [21]

Sillanpää, M. J. 2011. On statistical methods for estimating heritability in wild populations. *Mol. Ecol.* 20: 1324–1332. [20]

Silvela, L. 1980. Genetic changes with generations of artificial selection. *Genetics* 95: 769–782. [7]

Silvela, L., and R. Diez-Barra. 1985. Recurrent selection in autogamous species under forced random mating. *Euphytica* 34: 817–832. [23]

Silvela, L., R. Rodgers, A. Barrera, and D. E. Alexander. 1989. Effect of selection intensity and population size on percent oil in maize, *Zea mays* L. *Theor. Appl. Genet.* 78: 298–304. [18]

Simes, J. R. 1986. An improved Bonferroni procedure for multiple tests of significance. *Biometrika* 73: 75–754. [A4]

Simm, G. 1998. *Genetic improvement of cattle and sheep.* Farming press, Suffolk, UK. [13]

Simmonds, N. W. 1977. Approximations for *i*, intensity of selection. *Heredity* 38: 413–414. [14]

Simms, E. L. 1990. Examining selection on the multivariate phenotype: plant resistance to heribvores. *Evolution* 44: 1177–1188. [30]

Simms, E. L., and M. D. Rausher. 1993. Patterns of selection on phytophage resistance in *Ipomoea purpurea. Evolution* 47: 970–976 [30]

Simons, K. J., and 6 others. 2006. Molecular characterization of the major wheat domestication gene *Q. Genetics* 172: 547–555. [9]

Simonsen, K. L., G. A. Churchill, and C. F. Aquadro. 1995. Properties of statistical tests of neutrality for DNA polymorphism data. *Genetics* 141: 413–429. [9]

Simpson, E. H. 1951. The interpretation of interaction in contingency tables. *J. Roy. Stat. Soc. B* 13: 238–241. [10]

Simpson, G. G. 1944. *Tempo and mode in evolution.* Columbia University Press, New York, NY. [16, 29]

Simpson, G. G. 1953. *The major features of evolution.* Columbia Univ. Press, New York, NY. [29]

Sinervo, B., and A. L. Basolo. 1996. Testing adaptation using phenotypic manipulation. *In* M. R. Rose and G. V. Lauder (eds.), *Adaptation*, pp. 149–185. Academic Press, New York, NY. [29]

Sinervo, B., and A. G. McAdam. 2008. Maturational costs of reproduction due to clutch size and ontogenetic conflict as revealed in the invisible fraction. *Proc. Roy. Soc. Lond. B* 275: 629–638. [29]

Singh, M., and R. C. Lewontin. 1966. Stable equilibria under optimizing selection. *Proc. Natl. Acad. Sci. USA* 56: 1345–1348. [28]

Siol, M., S. I. Wright, and S. C. H. Barrett. 2010. The population genomics of plant adaptation. *New Phytol.* 188: 313–332. [9]

Sisson, S. A., and Y. Fan. 2011. Likelihood-free MCMC. *In* S. Brooks, A. Gelman, G. Jones, and X.-L. Meng (eds), *Handbook of markov chain Monte Carlo*, pp. 313–335. CRC press, Boca Raton, FL. [A3]

Skelly, D. A., J. Ronald, and J. M. Akey. 2009. Inherited variation in gene expression. *Ann. Rev. Genomics Hum. Genet.* 10: 313–332. [12]

Skibinski, D. O. F. 1986. Study of lethals in selection lines in *Drosophila melanogaster. J. Heredity* 77: 31–34. [25]

Skibinski, D. O. F., and N. A. K. Shereif. 1989. Directional selection in lines founded from different parts of the phenotypic distribution of sternopleural chaetae number in *Drosophila melanogaster. Theor. Appl. Genet.* 77: 409–415. [18, 26]

Slatkin, M. 1970. Selection and polygenic characters. *Proc. Natl. Acad. Sci. USA* 66: 87–93. [1]

Slatkin, M. 1977. Gene flow and genetic drift in a species subject to frequent local extinctions. *Theor. Popul. Biol.* 12: 253–262. [3]

Slatkin, M. 1978. Spatial patterns in the distributions of polygenic characters. *J. Theor. Biol.* 70: 213–228. [28]

Slatkin, M. 1979. Frequency- and density-dependent selection on a quantitative trait. *Genetics* 93: 755–771. [28]

Slaktin, M. 1981a. Population heritability. *Evolution* 35: 859–871. [22]

Slatkin, M. 1981b. Fixation probabilities and fixation times in a subdivided population. *Evolution* 35: 477–488. [7, 26]

Slatkin, M. 1987a. Heritable variation and heterozygosity under a balance between mutations and stabilizing selection. *Genet. Res.* 50: 53–62. [28]

Slatkin, M. 1987b. The average number of sites separating DNA sequences drawn from a subdivided population. *Theor. Popul. Biol.* 32: 42–49. [11]

Slatkin, M. 1991. Inbreeding coefficients and coalescent times. *Genet. Res.* 58: 167–175. [2]

Slatkin, M. 1994. An exact test for neutrality based on the Ewens sampling distribution. *Genet. Res.* 64: 71–74. [9]

Slatkin, M. 1995a. A measure of population subdivision based on microsatellite allele frequencies. *Genetics* 139: 457–462. [2, 9, 12]

Slatkin, M. 1995b. Hitchhiking and associative overdominance at a microsatellite locus. *Mol. Biol. Evol.* 12: 473–480. [8]

Slatkin, M. 1996. A correction to the exact test based on the Ewens sampling distribution. *Genet. Res.* 68: 259–260. [9]

Slatkin, M. 2000. Allele age and a test for selection on rare alleles. *Philos. Trans. Roy. Soc. Lond. B* 355: 1663–1668. [9]

Slatkin, M. 2008. A Bayesian method for jointly estimating allele age and selection intensity. *Genet. Res.* 90: 119–128. [9]

Slatkin, M., and G. Bertorelle. 2001. The use of intraallelic variability for testing neutrality and estimating population growth rate. *Genetics* 158: 865–687. [9]

Slatkin, M., and S. A. Frank. 1990. The quantitative genetic consequences of pleiotropy under stabilizing and directional selection. *Genetics* 125: 207–213. [28]

Slatkin, M., and B. Rannala. 1997. Estimating the age of alleles by the use of intraallelic variability. *Am. J. Hum Genet.* 60: 447–458. [2, 9]

Slatkin, M., and B. Rannala. 2000. Estimating allele age. *Ann. Rev. Genomics Hum. Genet.* 1: 225–249. [2, 9]

Slatkin, M., and L. Voelm. 1991. F_{ST} is a hierarchical island model. *Genetics* 127: 627–629. [2, 11]

Slaktin, M., and M. J. Wade. 1978. Group selection on a quantitative character. *Proc. Natl. Acad. Sci. USA* 75: 3531–3534. [22]

Slatkin, M., and T. Wiehe. 1998. Genetic hitch-hiking in a subdivided population. *Genet. Res.* 71: 155–160. [8]

Sleper, D. A., and J. M. Poehlman. 2006. *Breeding field crops, 5th Edition.* Blackwell Publishing, Malden, MA. [21]

Slice, D. E. 2005. *Modern morphometrics in physical anthropology.* Springer, New York, NY. [A5]

Slotte, T., J. P. Foxe, K. M. Hazzouri, and S. I. Wright. 2010. Genome-wide evidence for efficient positive and purifying selection in *Capsella grandiflora*, a plant species with a large effective population size. *Mol. Biol. Evol.* 27: 1813–1821. [10]

Smith, A. F. M. 1991. Bayesian computational methods. *Phil. Trans. Roy. Soc. Lond. A* 337: 369–386. [A3]

Smith, A. F. M., and G. O. Roberts. 1993. Bayesian computation via the Gibbs sampler and related Markov chain Monte-Carlo methods (with discussion). *J. Roy. Stat. Soc. B* 55: 3-23. [A3]

Smith, C. 1969. Optimum selection procedures in animal breeding. *Anim. Prod.* 11: 433–442. [14]

Smith, C., and M. Quinton. 1993. The effect of selection in sublines and crossing on genetic response and inbreeding. *J. Anim. Sci.* 71: 2631–2638. [26]

Smith, H. F. 1936. A discriminant function for plant selection. *Ann. Eugen.* 7: 240–250. [22, A6]

Smith, J., J. D. Van Dyken, and P. C. Zee. 2010. A generalization of Hamilton's rule for the evolution of microbial cooperation. *Science* 328: 1700–1703. [22]

Smith, L. H. 1908. Ten generations of corn breeding. *Illinois Agri. Exp. Stn. Bull.* 218. [21, 25]

Smith, L. H. 1909. The effect of selection upon certain physical characters in the corn plant. *Illinois Agri. Exp. Stn. Bull.* 132. 47–62 [21]

Smith, L. H., and A. M. Brunson. 1925. An experiment in selecting corn for yield by the method of the ear-to-row breeding plot. *Illinois Agri. Exp. Stn. Bull.* 271. [21]

Smith, N. G. C., and A. Eyre-Walker. 2002. Adaptive protein evolution in *Drosophila*. *Nature* 415: 1022–1024. [8, 10]

Smith, O. S. 1979a. A model for evaluating progress from recurrent selection. *Crop Sci.* 19: 223–226. [18]

Smith, O. S. 1979b. Application of a modified diallel analysis to evaluate recurrent selection for grain yield in maize. *Crop Sci.* 19: 819–822. [18]

Smith, O. S. 1983. Evaluation of recurrent selection in BSSS, BSCB1, and BS13 maize populations. *Crop Sci.* 23: 35–40. [18]

Smith, R. L. (ed.) 1984. *Sperm competition and the evolution of animal mating systems.* Academic Press, Orlando, FL. [29]

Smith, S. P., and A. Mäiki-Tanila. 1990. Genotypic covariance matrices and their inverses for models allowing dominance and inbreeding. *Genet. Sel. Evol.* 22: 65–91. [11, 19]

Smith, S. P., and H.-U. Graser. 1986. Estimating variance components in a class of mixed models by restricted maximum likelihood. *J. Dairy Sci.* 69: 1156–1165. [19]

Smith, S. P., and K. Hammond. 1987. Assortative mating and artificial selection: a second appraisal. *Genet. Sel. Evol.* 19: 181–196. [16]

Snee, R. D. 1977. Validation of regression models: methods and examples. *Technometrics* 19: 415–428. [29]

Sniegowski, P. D., P. J. Gerrish, T. Johnson, and A. Shaver. 2000. The evolution of mutation rates: separating causes from consequences. *Bioessays* 22: 1057–1066. [4]

Soliman, M. H. 1982. Directional and stabilizing selection for developmental time and correlated response in reproductive fitness in *Tribolium castaneum*. *Theor. Appl. Genet.* 63: 111–116. [16]

Somers, K. M. 1986. Multivariate allometry and removal of size with principal components analysis. *Syst. Zoo.* 35: 359–368. [30]

Sonesson, A. K., B. Grundy, J. A. Woolliams, and T. H. E. Meuwissen. 2000. Selection with control of inbreeding in populations with overlapping generations: a comparison of methods. *Anim. Sci.* 70: 1–8. [23]

Sonesson, A. K., and T. H. E. Meuwissen. 2000. Mating schemes for optimum contribution selection with constrained rates of inbreeding. *Genet. Sel. Evol.* 32: 231–248. [23]

Song, Y. S., and J. S. Song. 2007. Analytic computation of the expectation of the linkage disequilibrium coefficient r^2. *Theor. Pop. Biol.* 71: 49–60. [2, 4]

Sorensen, D., R. Fernando, and D. Gianola. 2001. Inferring the trajectory of genetic variance in the course of artificial selection. *Genet. Res.* 77: 83–94. [19]

Sorensen, D., and D. Gianola. 2002. *Likelihood, Bayesian and MCMC methods in quantitative genetics.* Springer, New York, NY. [19, A3]

Sorensen, D., and R. Waagepetersen. 2003. Normal linear models with genetically structured residual variance: a case study. *Genet. Res.* 82: 207–222. [17]

Sorensen, D., and R. Waagepetersen, and A. Blasco. 2008. Selection for environmental variance: a statistical analysis and power calculations to detect response. *Genetics* 180: 2209–2226. [17]

Sorensen, D. A. 1980. Effects of disruptive selection on genetic variability in *Drosophila*. *Heredity* 45: 142. [16]

Sorensen, D. A., B. Guldbrandtsen, and J. Jensen. 2003. On the need for a control line in selection experiments. A likelihood analysis. *Genet. Sel. Evol.* 35: 3–20. [19]

Sorensen, D. A., and W. G. Hill 1982. Effect of short term directional selection on genetic variability: experiments with *Drosophila melanogaster*. *Heredity* 48: 27–33. [16]

Sorensen, D. A., and W. G. Hill 1983. Effects of disruptive selection on genetic variance. *Theor. Appl. Genet.* 65: 173–180. [24]

Sorensen, D. A., and K. Johannsson. 1992. Estimation of direct and correlated response to selection using univariate animal models. *J. Anim. Sci.* 70: 2038–2044. [19]

Sorensen, D. A., and B. W. Kennedy. 1983. The use of the relationship matrix to account for genetic drift variances in the analysis of selection experiments. *Theor. Appl. Genet.* 66: 217–220. [19]

Sorensen, D. A., and B. W. Kennedy. 1984a. Estimation of response to selection using least-squares and mixed model methodology. *J. Anim. Sci.* 58: 1097–1106. [19]

Sorensen, D. A., and B. W. Kennedy. 1984b. Estimation of genetic variances from unselected and selected populations. *J. Anim. Sci.* 59: 1213–1223. [19]

Sorensen, D. A., and B. W. Kennedy. 1986. Analysis of selection experiments using mixed model methodology. *J. Anim. Sci.* 63: 245–258. [19]

Sorensen, D. A., C. S. Wang, J. Jensen, and D. Gianola. 1994. Bayesian analysis of genetic change due to selection using Gibbs sampling. *Genet. Sel. Evol.* 26: 333–360. [19]

Sørensen, P., G. de los Campos, F. Morgante, T. F. C. Mackay, and D. Sorensen. 2015. Genetic control of environmental variation of two quantitative traits of *Drosophila melanogaster* revealed by whole-genome sequencing. *Genetics* 201: 487–497. [17]

Sorrells, M. E., and S. E. Fritz. 1982. Application of dominant male-sterile allele to the improvement of self-pollinated crops. *Crop Sci.* 22: 1033–1035. [21]

Sousa, A., S. Magalhães, and I. Gordo. 2011. Cost of antibiotic resistance and the geometry of adaptation. *Mol. Biol. Evol.* 29: 1417–1428. [27]

Southwood, O. I., and B. W. Kennedy. 1991. Genetic and environmental trends for litter size in swine. *J. Anim. Sci.* 69: 3177–3182. [19]

Speed, D., and D. J. Balding. 2015. Relatedness in the post-genomic era: is it still useful?. *Nat. Rev. Genet.* 16: 33–44. [20]

Spicer, G. S. 1993. Morphological evolution of the *Drosophila virilis* species group as assessed by rate tests for natural selection on quantitative characters. *Evolution* 47: 1240–1254. [12]

Spichtig, M., and T. J. Kawecki. 2004. The maintenance (or not) of polygenic variation by soft selection in heterogeneous environments. *Am. Nat.* 164: 70–84. [28]

Spitze, K. 1993. Population structure in *Daphnia obtusa:* quantitative genetic and allozymic variation. *Genetics* 135: 367-374. [12]

Spofford, J. B. 1969. Heterosis and the evolution of duplications. *Am. Nat.* 103: 407–432. [9]

Spooner, D. M., K. McLean, G. Ramsay, R. Wuagh, and G. J. Bryan. 2005. A single domestication for potato based on multilocus amplified fragment length polymorphism genotyping. *Proc. Natl. Acad. Sci. USA* 102: 14694–14699. [9]

Sprague, G. F. 1955. Corn breeding. *In* G. F. Sprague (ed.), *Corn and corn improvement*, pp. 221–292. Academic Press, NY. [21]

Sprent. P., and N. C. Smeeton. 2007. *Applied nonparametric statistical methods, 4th edition.* Chapman and Hall, New York, NY. [29]

Stam, P. 1977. Selection response under random mating and under selfing in the progeny of a cross of homozygous parents. *Euphytica* 26: 169–184. [23]

Stam, P. 1980. The distribution of the fraction of the genome identical-by-descent in finite random mating populations. *Genet. Res.* 35: 131–155. [20]

Stamatoyannopoulos, J. A. 2004. The genomics of gene expression. *Genomics* 84: 449–457. [12]

Stapley, J., and 8 others. 2010. Adaptation genomics: the next generation. *Trends Ecol. Evol.* 25: 705–712. [25]

Stearns, S. C. 1976. Life-history tactics: a review of the idea. *Q. Rev. Biol.* 51: 3–47. [29]

Stearns, S. C. 1992. *The evolution of life histories.* Oxford University Press, Oxford, UK. [29]

Stearns, S. C., S. G. Byars, D. R. Govindaraju, and D. Ewbank. 2010. Measuring selection in contemporary human populations. *Nat. Rev. Genet.* 11: 611–622. [29]

Stebbins, G.L., 1957. Self fertilization and population variability in the higher plants. *Am. Nat.* 91: 337–354. [23]

Steinrücken, M., A. Bhaskar, and Y. S. Song. 2014. A novel spectral method for inferring general diploid selection from time series genetic data. *Ann. Appl. Stat.* 8: 2203–2222. [9]

Steinsland, I., C. T. Larsen, A. Roulin, and H. Jensen. 2014. Quantitative genetic modeling and inference in the presence of nonignorable missing data. *Evolution* 68: 1735–1747. [20]

Stephan, W. 1995. An improved method for estimating the rate of fixation of favorable mutations based on DNA polymorphism data. *Mol. Biol. Evol.* 12: 959–962. [8]

Stephan, W. 1996. The rate of compensatory evolution. *Genetics* 144: 419–426. [7]

Stephan, W. 2010a. Detecting strong positive selection in the genome. *Mol. Ecol. Res.* 10: 863–872. [8, 9]

Stephan, W. 2010b. Genetic hitchhiking versus background selection: the controversy and its implications. *Phil. Trans. Roy. Soc. Lond. B* 365: 1245–1253. [8]

Stephan, W. 2016. Signatures of positive selection: from selective sweeps at individual loci to subtle allele frequency changes in polygenic adaptation. *Mol. Ecol.* 25: 79–88. [9]

Stephan, W., B. Charlesworth, and G. McVean. 1999. The effect of background selection at a single locus on weakly selected, partially linked variants. *Genet. Res.* 73: 133–146. [7]

Stephan, W., and D. A. Kirby. 1993. RNA folding in *Drosophila* shows a distance effect for compensatory fitness interactions. *Genetics* 135: 97–103. [7]

Stephan, W., Y. S. Song, and C. H. Langley. 2006. The hitchhiking effect on linkage disequilibrium between linked neutral loci. *Genetics* 172: 2647–2663. [8]

Stephan, W., T. H. E. Wiehe, and M. W. Lenz. 1992. The effect of strongly selected substitutions on neutral polymorphisms: analytical results based on diffusion theory. *Theor. Popul. Biol.* 41: 237–254. [8]

Stephens, J C. 1986. On the frequency of undetectable recombination events. *Genetics* 112: 923–926. [4]

Stephens, J. C., and 38 others. 1998. Dating the origin of the *CCR5-Delta32* AIDS-resistance allele by the coalescence of haplotypes. *Am. J. Hum. Genet.* 62: 1507–1515. [2, 9]

Stephens, M. 2001. Inference under the coalescent *In* D. Balding, M. Bishop, and C. Cannings (eds.), *Handbook of statistical genetics*, pp. 213–228. Wiley, Chichester, UK. [2]

Stephens, M., and P. Scheet. 2005. Accounting for decay of linkage disequilibrium in haplotype inference and missing-data imputation. *Am. J. Hum. Genet.* 7: 449–462. [9]

Stephens, M., N. J. Smith, and P. Donnelly. 2001. A new statistical method for haplotype reconstruction from population data. *Am. J. Hum. Genet.* 68: 978–989. [9]

Sterne, J. A. C., and 18 others. 2011. Recommendations for examining and interpreting funnel plot asymmetry in meta-analyses of randomised controlled trials. *MBJ* 343: d4002 [A4]

Stevens, L., C. J. Goodnight, and S. Kalisz. 1995. Multilevel selection in natural populations of *Impatiens capensis. Am. Nat.* 145: 513–526. [30]

Stewart, F. M. 1976. Variability in the amount of heterozygosity maintained by neutral mutations. *Theor. Popul. Biol.* 9: 188–201. [2]

Stewart, G. 2010. Meta-analysis in applied ecology. *Biol. Lett.* 6: 78–81. [A4]

Stewart, S. C., and D. J. Schoen. 1987. Patterns of phenotypic variability and fecundity selection in a natural population of *Impatiens pallida*. *Evolution* 41: 1290–1301. [30]

Stigler, S. M. 1980. Stigler's law of eponymy. *Trans. NY Acad. Sci., Series II* 39: 147–158. [A2]

Stigler, S. M. 1983. Who discovered Bayes's theorem? *Am. Stat.* 37: 290–296. [A2]

Stinchcombe, J. R. 2005. Measuring natural selection on proportional traits: comparisons of three types of selection estimates for resistance and susceptibility to herbivore damage. *Evol. Ecol.* 19: 363–373. [29]

Stinchcombe, J. R., A. F. Agrawl, P. A. Hohenlohe, S. J. Arnold, and M. W. Blows. 2008. Estimating nonlinear selection gradients using quadratic regression coefficients: double or nothing? *Evolution* 62: 2435–2440. [30]

Stinchcombe, J. R., and H. E. Hoekstra. 2008. Combining population genomics and quantitative genetics: finding the genes underlying ecologically important traits. *Heredity* 100: 158–170. [9]

Stinchcombe, J. R., and 5 others. 2002. Testing for environmentally induced bias in phenotypic estimates of natural selection: theory and practice. *Am. Nat.* 160: 511–523. [22, 29]

Stoletzki, N., and A. Eyre-Walker. 2011. Estimation of the neutrality index. *Mol. Biol. Evol.* 28: 63–70. [10]

Stopher, K. V., and 7 others. 2012. Shared spatial effects on quantitative genetic parameters: accounting for spatial autocorrelation and home range overlap reduces estimates of heritability in wild red deer. *Evolution* 66: 2411–2426. [20]

Storey J.D. 2002. A direct approach to false discovery rates. *J. Roy. Stat. Soc. B* 64: 479–498. [A4]

Storey J.D. 2003. The positive false discovery rate: a Bayesian interpretation and the q-value. *Ann. Stat.* 31: 2013–2035. [A4]

Storey J.D., J. E. Taylor, and D. Siegmund. 2004. Strong control, conservative point estimation, and simultaneous conservative consistency of false discovery rates: A unified approach. *J. Roy. Stat. Soc. B* 66: 187–205. [A4]

Storey, J. D., and R. Tibshirani. 2003. Statistical significance for genomewide studies. *Proc. Natl. Acad. Sci. USA* 100: 9440–9445. [A4]

Storz, J. F. 2005. Using genome scans of DNA polymorphism to infer adaptive population divergence. *Mol. Ecol.* 14: 671-688. [9]

Storz, J. F., and C. W. Wheat. 2010. Integrating evolutionary and functional approaches to infer adaptation at specific loci. *Evolution* 64: 2489–2509. [10]

Stoskopf, N. C., D. T. Tomes, and B. R. Christie. 1993. *Plant breeding: Theory and practice.* Westview Press, Boulder, CO. [13, 21]

Stouffer, S. A., E. A. Suchman, L. C. DeVinney, S. A. Star, and R. M. Williams Jr. 1949. *The American soldier, Vol. 1: Adjustment during army life.* Princeton University Press, Princeton, NJ. [A4]

Strasburg, J. L., C. Scotti-Saintagne, I. Scotti, Z. Lai, and L. H. Rieseberg. 2009. Genomic patterns of adaptive divergence between chromosomally differentiated sunflower species. *Mol. Biol. Evol.* 26: 1341–1355. [10]

Stricker, C., R. L. Fernando, and R. C. Elston.1995. Linkage analysis with an alternative formulation for the mixed model of inheritance: the finite polygenic mixed model. *Genetics* 141: 1651–1656. [24]

Strobeck, C. 1983. Expected linkage disequilibrium for a neutral locus linked to a chromosomal arrangement. *Genetics* 103: 545–555. [8]

Strobeck, C. 1987. Average number of nucleotide differences in a sample from a single subpopulation: a test for population subdivision. *Genetics* 117: 149–153. [2, 9, 11]

Stroup, W. W. 2012. *Generalized linear mixed models: modern concepts, methods and applications*. CRC press, Boca Raton, FL. [29]

Struchlain, M. V., A. Dehghan, J. C. M. Witterman, C. van Duijn, and Y. S. Aulchenko. 2010. Variance heterogeneity analysis for detection of potentially interacting genetic loci: method and its limitations. *BMC Genet.* 11: 92. [17]

Stuber, C. W., and R. H. Moll. 1972. Frequency changes of isozyme alleles in a selection experiment for grain yield in maize (*Zea maize* L.). *Crop Sci.* 12: 337–340. [9]

Stuber, C. W., R. H. Moll, M. M. Goodman, H. E. Schaffer, and B. S. Weir. 1980. Allozyme frequency changes associated with selection for increased grain yield in maize (*Zea mays* L.) *Genetics* 95: 225–236. [9]

Studer, A., Q. Zhao, J. Ross-Ibarra, and J. Doebley. 2011. Identification of a functional transposon insertion in the maize domestication gene *tb1*. *Nat. Genet.* 43: 1160–1165. [8, 9, 12]

Stumpf, M. P. H, and H. M. Wilkinson-Herbots. 2004. Allelic histories: positive selection on a HIV-resistance allele. *Trends Ecol. Evol.* 19:166–168. [9]

Sturtevant, A. H. 1923. Inheritance of direction of coiling in Limnaea. *Science* 58: 269–270. [15]

Sturtevant, A. H. 1937. Essays on evolution. I. On the effects of selection on mutation rate. *Quart. Rev. Biol.* 12: 464–476. [4]

Su, G., P. Sørensen, and D. Sorensen. 1997. Inferences about variance components and selection response for body weight in chickens. *Genet. Sel. Evol.* 29: 413–425. [19]

Suarez, B. K., T. Reich, and P. M. Fishman. 1979. Variability in sib pair genetic identity. *Hum. Hered.* 29: 37–41. [20]

Subramanian, A. and 10 others. 2005. Gene set enrichment analysis: a knowledge-based approach for interpreting genome-wide expression profiles. *Proc. Natl. Acad. Sci. USA* 102: 15545–15550. [12]

Sugiura, N. Further analysis of the data by Akaike's information criterion and the finite corrections. *Comm. Stat. Theory and Methods* A7: 13–26. [30]

Sung, W., M. S. Ackerman, S. F. Miller, T. G. Doak, and M. Lynch. 2012. Drift-barrier hypothesis and mutation-rate evolution. *Proc. Natl. Acad. Sci. USA* 109: 18488–18492. [4]

Sutherland, T. M., P. E. Biondini and L. H. Haverland. 1968. Selection under assortative mating in mice. *Genet. Res.* 11: 171–178. [16]

Sutherland, W. J. 1985a. Chance can produce a sex difference in variance in mating success and explain Bateman's data. *Anim. Behav.* 33: 1349–1352. [29]

Sutherland, W. J. 1985b. Measures of sexual selection. *Oxford Surveys Evol. Bio.* 2: 90–101. [29]

Suzuki, Y., and T. Gojobori. 1999. A method for detecting positive selection at single amino acid sites. *Mol. Biol. Evol.* 16: 1315–1328. [10]

Suzuki, Y., and M. Nei. 2002. Simulation study of the reliability and robustness of the statistical methods for detecting positive selection at single amino acid sites. *Mol. Biol. Evol* 19: 1865–1869. [10]

Suzuki, Y., and M. Nei. 2004. False-positive selection identified by ML-based methods: Examples from the *Sig1* gene of the diatom *Thalassiosira weissflogii* and the *tax* gene of a human T-cell lymphotropic virus. *Mol. Biol. Evol.* 21: 914–921. [10]

Svardal, H., C . Rueffler, and J. Hermisson. 2011 Comparing environmental and genetic variance as adaptive response to fluctuating selection. *Evolution* 65: 2492–2513. [17]

Sved, J. A. 1971. Linkage disequilibrium and homozygosity of chromosome segments in finite populations. *Theor. Popul. Biol.* 2: 125–141. [2]

Sved, J. A. 1977. Opposition to artificial selection caused by natural selection at linked loci. *In* E. Pollak, O. Kempthorne, and T. B. Bailey, Jr., (eds.), *Proceedings of the international conference on quantitative genetics*, pp. 435–456. Iowa State Univ. Press, Ames, IA. [25]

Sved, J. A. 1989. The measurement of fitness in *Drosophila*. *In* W. G. Hill, and T. F. C. Mackay, (eds.), *Evolution and animal breeding*, pp. 113–120. CAB International, Wallingford, UK. [29]

Sved, J. A., and M. W. Feldman. 1973. Correlation and probability methods for one and two loci. *Theor. Popul. Biol.* 42: 129–132. [2]

Sved, J. A., A. F. McRae, and P. M. Visscher. 2008. Divergence between human populations estimated from linkage disequilibrium. *Am. J. Hum. Genet.* 83: 737–743. [4]

Swanson, M. R., J. W. Dudley, and S. G. Carmer. 1974. Simulated selection in autotetraploid populations. II: Effects of double reduction, population size, and selection intensity. *Crop. Sci.* 14: 630–636. [15]

Swanson-Wagner, R., and 7 others. 2012. Reshaping of the maize transcritptome by domestication. *Proc. Natl. Acad. Sci. USA* 109: 11878–11883. [9]

Szendro, I. G., M. F. Schenk, J. Franke, J. Krug, and J. A. G. M. de Visser. 2013. Quantitative analyses of empirical fitness landscapes. *J. Stat. Mech: Theory Exper* 13: P01005. [27]

Sztepanacz, J. L., and H. D. Rundle. 2012. Reduced genetic variance among high fitness individuals: Inferring stabilizing selection on male sexual displays in *Drosophila serrata*. *Evol.* 66: 3101–3110. [28]

Tabashnik, B. E., A. J. Gassmann, D. W. Crowder, and Y. Carriére. 2008. Insect resistance to Bt crops: evidence versus theory. *Nat. Biotech.* 26: 199–202. [25]

Tabery, J. G. 2004. The "Evolutionary Synthesis" of George Udny Yule. *J. History Biol.* 37: 73–101. [1]

Tachida, H. 2000. Molecular evolution in a multisite nearly neutral mutation model. *J. Mol. Evol.* 50: 69–81. [8]

Tachida, H., and C. C. Cockerham. 1988. Variance components of fitness under stabilizing selection. *Genet. Res.* 51: 47–53. [28]

Tachida, H., and C. C. Cockerham. 1989. Effects of identity disequilibrium and linkage on quantitative variation in finite populations. *Genet. Res.* 53: 63–70. [11]

Taft, H. R., and D. A. Roff. 2012. Do bottlenecks increase additive genetic variance? *Conser. Genet.* 13: 333–342. [11]

Tai, G. C. C. 1979. An interval estimation of expected response to selection. *Theor. Appl. Genet.* 54: 273–275. [18]

Tajima, F. 1983. Evolutionary relationship of DNA sequences in finite populations. *Genetics* 105: 437–460. [4]

Tajima, F. 1989. Statistical methods for testing the neutral mutation hypothesis by DNA polymorphism. *Genetics* 123: 585–595. [8, 9]

Tajima, F., and M. Nei. 1984. Note on genetic drift and estimation of effective population size. *Genetics* 106: 569–574. [4]

Takebayashi, N., and P. L. Morrell. 2001. Is self-fertilization an evolutionary dead end? Revisiting an old hypothesis with genetic theories and a macroevolutionary approach. *Am. J. Bot.* 88: 143–1150. [23]

Tallis, G. M 1987. Ancestral covariance and the Bulmer effect. *Theor. Appl. Genet.* 73: 815–820. [16]

Tallis, G. M., and P. Leppard. 1988a. The joint effects of selection and assortative mating on a single polygenic character. *Theor. Appl. Genet.* 75: 41–45. [16]

Tallis, G. M., and P. Leppard. 1988b. The joint effects of selection and assortative mating on multiple polygenic characters. *Theor. Appl. Genet.* 75: 278–281. [16]

Tallmon, D. A., G. Luikart, and M. A. Beaumont. 2004. Comparative evaluation of a new effective population size estimator based on approximate bayesian computation. *Genetics* 167: 977–988. [4]

Tamuri, A. U., M. dos Reis, and R. A. Goldstein. 2012. Estimating the distribution of selection coefficients from phylogenetic data using sitewise mutation-selection models. *Genetics* 190: 1101–1115. [10]

Tanaka, Y. 1996a. A quantitative genetic model of group selection. *Am. Nat.* 148: 660–683. [22]

Tanaka, Y. 1996b. The genetic variance maintained by pleiotropic mutation. *Theor. Popul. Biol.* 49: 211–231. [28]

Tanaka, Y. 1998. A pleiotropic model of phenotypic evolution. *Genetica* 102: 535–543. [28]

Tang, H., G. J. Wyckoff, J. Lu, and C.-I. Wu. 2004. A universal evolutionary index for amino acid changes. *Mol. Biol Evol.* 21: 1548–1556. [8]

Tang, K., K. R. Thornton, and M. Stoneking. 2007. A new approach for using genome scans to detect recent positive selection in the human genome. *PLoS Bio.* 5: e171. [9]

Tanner, A. H., and O. S. Smith. 1987. Comparison of half-sib and S_1 recurrent selection in the Krug Yellow Dent maize populations. *Crop Sci.* 27: 509–513. [23]

Tanner, M. A. 1996. *Tools for statistical inference*, 3rd ed. Springer-Verlag, New York, NY. [A3]

Tantawy, A. O., and E. C. R. Reeve. 1956. Studies in quantitative inheritance: IX. The effects of inbreeding at different rates in *Drosophila melanogaster*. *Z. indukt. Abst. Ver. bungslehre* 87: 648–667. [18]

Tantawy, A. O., and A. A. Tayel. 1970. Studies on natural populations of *Drosophila*. X. Effects of disruptive and stabilizing selection on wing length and the correlated response in *Drosophila melanogaster*. *Genetics* 65: 121–132. [16]

Tarone, R. E. 1981. On summary estimators of relative risk. *J. Chron. Dis.* 34: 463–468. [10]

Tavaré, S., D. J. Balding, R. C. Griffiths, and P. Donnelly. 1997. Inferring coalescence times from DNA sequence data. *Genetics* 145: 505–518. [A3]

Taylor, D. R., and P. K. Ingvarsson. 2003. Common features of segregation distortion in plants and animals. *Genetica* 117: 27–35. [25]

Taylor, M. F. J., Y. Shen, and M. E. Kreitman. 1995. A population genetic test of selection at the molecular level. *Science* 270: 1497–1499. [9]

Taylor, P. D., and S. A, Frank. 1996. How to make a kin selection model. *J. Theor. Biol.* 180: 27–37. [22]

Taylor, P. D., G. Wild, and A. Gardner. 2007. Direct fitness or inclusive fitness: how shall we model kin selection? *J. Evol. Biol.* 20: 301–309. [22]

Technow, F., C. D. Messina, L. R. Totir, and M . Cooper. 2015. Integrating crop growth models with whole genome prediction through approximate Bayesian computation. *PLoS One* 0130855 [A3]

Templeton, A. R. 1974. Analysis of selection in populations observed over a sequence of consecutive generations. *Theor. Appl. Genet.* 45: 179–191. [9]

Templeton, A. R. 1980. The theory of speciation via the founder principle. *Genetics* 94: 1011–1038. [11]

Templeton, A. R. 1987. Genetic systems and evolutionary rates. *In* K. S. W. Campbell and M. F. Day (eds.), *Rates of evolution*, pp. 218–234. Allen and Unwin, London, UK. [10]

Templeton, A. R. 1996. Contingency tests of neutrality using intra/interspecific gene trees: The rejection of neutrality for the evolution of the mitochondrial cytochrome oxidase II gene in the Hominoid primates. *Genetics* 144: 1263–1270. [10]

Templeton, A. R. 2000. Epistasis and complex traits. *In* J. B. Wolf, E. D. Brodie III, and M. J. Wade (eds.),*Epistasis and the evolutionary process,* pp. 41–57. Oxford Univ. Press, New York, NY. [11]

Templeton, A. R. 2006. *Population genetics and microevolutionary theory.* John Wiley & Sons, Inc., Hoboken, NJ. [3]

Tenaillon, M. I., J. U'Ren, O. Tenaillon, and B. S. Gaut. 2004. Selection versus demography: A multilocus investigation of the domestication process in maize. *Mol. Biol. Evol.* 21: 1214–1225. [4, 9]

Tenaillon, O. 2014. The utility of Fisher's geometric model in evolutionary genetics. *Ann. Rev. Ecol. Evol. Syst.* 45: 179–201. [27]

Tenaillon, O, F. Taddei, M. Radman, and I. Matic. 2001. Second-order selection in bacterial evolution: selection acting on mutation and recombination rates in the course of adaptation. *Res. Microbio.* 152: 11–16. [27]

Tenenbaum, M., and H. Pollard. 1963. *Ordinary differential equations: an elementary textbook for students of mathematics, engineering, and the sciences.* Harper and Row, New York, NY. [A1]

Tenesa, A., and 6 others. 2007. Recent human effective population size estimated from linkage disequilibrium. *Genome Res.* 17: 520–526. [4]

Teotónio, H., I. M. Chelo, M. Bradi, M. R. Rose, and A. D. Long. 2009. Experimental evolution reveals natural selection on standing genetic variation. *Nat. Genet.* 41: 251–257. [25]

Teplitsky, C., J. A. Mills, J. W. Yarrall, and J. Merilä. 2009. Heritability of fitness components in a wild bird population. *Evolution* 63: 716–726. [6]

Terhorst, J., C. Schlötterer, and Y. S. Song. 2015. Multilocus analysis of genomic time series data from experimental evolution. *PLoS Genet.* 11: e1005069. [9]

Teshmia, K. M., G. Coop, and M. Przeworski. 2006. How reliable are empirical genomic scans for selective sweeps? *Genome Res.* 16: 702–712. [9]

Teshima, K. M., and M. Przeworski. 2006. Directional positive selection on an allele of arbitrary dominance. *Genetics* 172: 713–718. [8]

Thériault, V., D. Garant, L. Bernatchez, and J. J. Dodson. 2007. Heritability of life-history tactics and genetic correlation with body size in a natural population of brook charr (*Salvelinus fontinalis*). *J. Evol. Biol.* 20: 2266–2277. [20]

Thoday, J. M. 1959. Effects of disruptive selection. I. Genetic flexibility. *Heredity* 13: 187–209. [16]

Thoday, J. M. 1972. Disruptive selection. *Proc. Roy. Soc. Lond. B* 182: 109–143. [16]

Thoday, J. M., and T. B. Boam. 1961. Regular responses to selection. I. Description of response. *Genet. Res.* 2: 161–176. [25]

Thoday, J. M., and J. B. Gibson. 1970. The probability of isolation by disruptive selection. *Am. Nat.* 104: 219–230. [16]

Thoday, J. M., J. B. Gibson, and S. G. Spickett. 1964. Regular responses to selection. II. Recombination and accelerated response. *Genet. Res.* 5: 1–19. [25, 26]

Thomas, M. G., and 5 others. 1998. Origins of old testament priests. *Nature* 395: 138–140. [9]

Thomas, S. C. 2005. The estimation of genetic relationships using molecular markers and their efficiency in estimating heritability in natural populations. *Phil. Trans. Roy. Soc. Lond. B* 360: 1457–1467. [20]

Thomas, S. C., D. W. Coltman, and J. M. Pemberton. 2002. The use of marker-based relationship information to estimate the heritability of body weight in a natural population: a cautionary tale. *J. Evol. Biol.* 15: 92–99. [20]

Thomas, S. C., and W. G. Hill. 2000. Estimating quantitative genetic parameters using sibships reconstructed from marker data. *Genetics* 155: 1961–1972. [20]

Thompson, N. J., and R. B. Cunningham. 1979. Genotype x environment interactions and evaluation of cotton cultivars. *Aust. J. Agric. Res.* 30: 105–112. [21]

Thompson, R. 1973. The estimation of variance and covariance components with an application when records are subject to culling. *Biometrics* 29: 527–550. [19]

Thompson, R. 1977. The estimation of heritability with unbalanced data. II. Data available on more than two generations. *Biometrics* 33: 496–504. [19]

Thompson, R. 1986. Estimation of realized heritability in a selected population using mixed model methods. *Genet. Sel. Evol.* 18: 475–484. [19]

Thompson, R. 2008. Estimation of quantitative genetic parameters. *Proc. Roy. Soc. Lond. B* 275: 679–686. [19]

Thompson, R., and K. D. Atkins. 1994. Sources of information for estimating heritability from selection experiments. *Genet. Res.* 63: 49–55. [19]

Thompson, R., S. Brotherstone, and I. M. White. 2005. Estimation of quantitative genetic parameters. *Phil. Trans. Roy. Soc. Lond. B* 360: 1469–1477. [19]

Thompson, R., and E. Mäntysaari. 2004. Prospects for statistical methods in animal breeding. *J. Ind. Soc. Agri. Stat.* 57: 15–25. [19]

Thompson, V. 1977. Recombination and response to selection in *Drosophila melanogaster*. *Genetics* 85: 125–140. [26]

Thomson, C. E., and J. D. Hadfield. 2017. Measuring selection when parents and offspring interact. *Methods Ecol. Evol.* 8: 678–687. [29, 30]

Thomson, G. 1977. The effect of a selected locus on linked neutral variation. *Genetics* 85: 753–788. [8]

Thornton, K. R., and J. D. Jensen. 2007. Controlling the false-positive rate in multilocus genome scans for selection. *Genetics* 175: 737–750. [9]

Thornton, K. R., J. D. Jensen, C. Becquest, and P. Andolfatto. 2007. Progress and prospects in mapping recent selection in the genome. *Heredity* 98: 340—348. [9]

Thornton, P. K. 2010. Livestock production: recent trends, future prospects. *Phil. Trans. Roy. Soc. Lond. B* 365: 2853–2867. [1]

Tian, F., N. M. Stevens, and E. S. Buckler. 2009. Tracking footprints of maize domestication and evidence for a massive selective sweep on chromosome 10. *Proc. Natl. Acad. Sci. USA* 106: 9979–9986. [9]

Tierney, L. 1994. Markov chains for exploring posterior distributions (with discussion). *Ann. Stat.* 22: 1701–1762. [A3]

Timothée, B., P. Wandeler, G. Camenisch, and E.Postma. 2017. Bigger is fitter? Quantitative genetic decomposition of selection reveals an adaptive evolutionary decline of body mass in a wild rodent population. *PLoS Biol.* 15: e1002592. [20]

Timpson, N., J. Heron, G. D. Smith, and W. Enard. 2007. Comment on papers by Evans *et al.* and Mekel-Bobrov *et al.* on evidence for positive selection of *MCPH1* and *ASPM*. *Science* 317: 1036. [10]

Tippett, L. H. C. 1927. *Random sampling numbers. Tracts for Computers* 15. Cambridge Univ. Press, London, UK. [27]

Tippett, L. H. 1931. *The methods of statistics.* Williams and Norgate, London, UK. [A4]

Tirosh, I., S. Reikhav, A. A. Levy, and N. Barkai. 2009. A yeast hybrid provides insight into the evolution of gene expression regulation. *Science* 324: 659–662. [12]

Tishkoff, S. A., and 18 others. 2007. Convergence adaptation of human lactase persistence in Africa and Europe. *Nat. Genet.* 39: 31–40. [8]

Tobias, J. A., R. Montgomerie, and B. E. Lyon. 2012. The evolution of female ornaments and weaponry: social selection, sexual selection and ecological competition. *Phil. Trans. Roy. Soc. Lond. B* 367: 2274–2293. [29]

Tong, Y. L. 1982. Some applications of inequalities for extreme order statistics to a genetic selection problem. *Biometrics* 38: 333–339. [21]

Toomajian, C., R. S. Ajioka, L. B. Jorde, J. P. Kushner, and M. Kreitman. 2003. A method for detecting recent selection in the human genome from allele age estimates. *Genetics* 165: 287–297. [9]

Toomajian, C., and 9 others. 2006. A nonparametric test reveals selection for rapid flowering in the *Arabidopsis* genome. *PLoS Bio.* 4: e137. [9]

Topa, H., Á. Jónás, R. Kofler, C. Kosiol, and A. Honkela. 2015. Gaussian process test for high-throughput sequencing time series: application to experimental evolution. *Bioinform.* 31: 1762–1770. [9]

Torgerson, D. G., and 10 others. 2009. Evolutionary processes acting on candidate cis-regulatory regions in humans inferred from patterns of polymorphism and divergence. *PLoS Genet.* 5: e1000592. [10]

Toro, M., R. Abugov, B. Charlesworth, and R. E. Michod. 1982. Exact versus heurtistic models of kin selection. *J. Theor. Biol.* 97: 699–713. [22]

Toro, M. A. 1993. A note on the use of inbreeding in progeny test schemes. *J. Anim. Breed. Genet.* 10: 234–240. [23]

Toro, M. A., and B. M. Nieto. 1984. A simple method for increasing the response to artificial selection. *Genet. Res.* 44: 347–349. [26]

Toro, M. A., B. Nieto, and C. Salgado. 1988. A note on minimization of inbreeding in small-scale selection programmes. *Livestock Prod. Sci.* 20: 317–323. [26]

Toro, M. A., and M. Pérez-Enciso. 1990. Optimization of selection response under restricted inbreeding. *Genet. Sel. Evol.* 22: 93–107. [23, 26]

Toro, M., and 5 others. 2002. Estimation of coancestry in Iberian pigs using molecular markers. *Conserv. Genet.* 3: 309–320. [20]

Tosh, J. J., and J. W. Wilton. 1994. Effects of data structure on variance of prediction error and accuracy of genetic evaluation. *J. Anim. Sci.* 72: 2568–2577. [20]

Tsuji, K. 1995. Reproductive conflicts and levels of selection in the ant *Pristomyrmex pungens*: contextual analysis and partitioning of covariance. *Am. Nat.* 146: 586–607. [30]

Trail, P. W. 1985. The intensity of selection: intersexual and interspecific comparisons require consistent measures. *Am. Nat.* 126: 434–439. [29]

Travis, J., and S. Henrich. 1986. Some problems in estimating the intensity of selection through fertility differences in natural and experimental populations. *Evolution* 40: 786–790. [29]

Travis, J., and D. N. Reznick. 1998. Experimental approaches to the study of evolution. *In* J. Resitartis and J. Bernardo (eds.), *Issues and perspectives in experimental biology*, pp. 437–459. Oxford, New York, NY. [29]

Travis, J. M. J., T. Münkemüller, O. J. Burton, C. Dytham, and K. Johst. 2007. Deleterious mutations can surf to high densities on the wave front of an expanding population. *Mol. Biol. Evol.* 24: 2334–2343. [9]

Trindade, S., A. Sousa, and I. Gordo. 2012. Antibiotic resistance and stress in the light of Fisher's model. *Evolution* 66: 3815–3824. [27]

Trivers, R. L. 1972. Parental investiment and sexual selection. *In* B. G. Campbell (ed.), *Sexual selection and the descent of man, 1871-1971*, pp. 136–179. Aldine, Chicago, IL. [29]

Tröbner, W., and R. Piechocki. 1984. Selection against hypermutability in *Escherichia coli* during long term evolution. *Mol. Gen. Genet.* 198: 177–178. [4]

True, J. R., J. Liu, L. F. Stam, Z.-B. Zeng, and C. C. Laurie. 1997. Quantitative genetic analysis of divergence in male secondary sexual traits between *Drosophila simulans* and *Drosophila mauritiana*. *Evolution* 51: 816–832. [12]

Tsakas, S., and C. B. Krimbas. 1976. Testing the heterogeneity of *F* values: A suggestion and a correction. *Genetics* 84: 399–401. [9]

Tufto, J. 2000. Quantitative genetic models for the balance between migration and stabilizing selection. *Genet. Res.* 76: 285–293. [28]

Tuinstra, E. J., G. De Jong, and W. Scharloo. 1990. Lack of response to family selection for directional asymmetry in *Drosophila melanogaster*: left and right are not distinguished in development. *Proc. Roy. Soc. Lond. B* 241: 146–152. [18]

Tukey, J. W. 1958. Bias and confidence in not quite large samples (Abstract). *Ann. Math. Stat.* 29: 614. [29]

Tukey, J. W. 1962. The future of data analysis. *Ann. Math. Stat.* 33: 1–67. [14, A3]

Turchin, M. C., and 6 others. 2012. Evidence of widespread selection on standing variation in Europe at height-associated SNPs. *Nat. Genet.* 44: 1015–1019. [8, 9, 12]

Turelli, M. 1982. *Cis-trans* effects induced by linkage disequilibrium. *Genetics* 72: 157–168. [5]

Turelli, M. 1984. Heritable genetic variation via mutation-selection balance: Lerch's zeta meets the abdominal bristle. *Theor. Popul. Biol.* 25: 138–193. [28]

Turelli, M. 1985. Effects of pleiotropy on predictions concerning mutation-selection balance for polygenic traits. *Genetics* 111: 165–195. [28]

Turelli, M. 1986. Gaussian versus non-gaussian genetic analysis of polygenic mutation-selection balance. *Evolutionary process and theory*, pp. 607–628. Academic Press, Orlando, FL. [28]

Turelli, M. 1988. Population genetic models for polygenic variation and evolution. *In* B. S. Weir, E. J. Eisen, M. M. Goodman, and G. Namkoong (eds.), *Proceedings of the second international conference on quantitative genetics*, pp. 601–618. Sinauer Assocs., Sunderland, MA. [1, 24, 28]

Turelli, M. 2017. Fishers infinitesimal model: A story for the ages. *Theor. Popul. Biol.* 118: 46–49. [24]

Turelli, M., and N. H. Barton. 1990. Dynamics of polygenic characters under selection. *Theor. Popul. Biol.* 38: 1–57. [1, 24, 28]

Turelli, M., and N. H. Barton. 1994. Genetic and statistical analyses of strong selection on polygenic traits: What, me normal? *Genetics* 138: 913–941. [24]

Turelli, M., and N. H. Barton. 2004. Polygenic variation maintained by balancing selection: pleiotropy, sex-dependent allelic effects and G x E interactions. *Genetics* 166: 1053–1079. [28]

Turelli, M., and N. H. Barton. 2006. Will population bottlenecks and multilocus epistasis increase additive genetic variance? *Evolution* 60: 1763–1776. [11]

Turelli, M., J. H. Gillespie, and R. Lande. 1988. Rate tests for selection on quantitative characters during macroevolution and microevolution. *Evolution* 42: 1085–1089. [12]

Turner, H. N., and S. S. Y. Young. 1969. *Quantitative genetics in sheep breeding*. Cornell University Press, Ithaca, NY. [13, 21]

Turner, J. R. G. 1981. Adaptation and evolution in *Heliconius*: a defense of NeoDarwinism. *Ann. Rev. Ecol. Syst.* 12: 99–121. [8]

Turner, T. L., A. D. Stewart, A. T. Fields, W. R. Rice, and A. M. Tarone. 2011. Population-based resequencing of experimentally evolved populations reveals the genetic basis of body size variation in *Drosophila melanogaster*. *PLoS Genet.* 7: e1001336. [18, 24, 25]

Uecker, H., and J. Hermisson. 2011. On the fixation process of a beneficial mutation in a variable environment. *Genetics* 188: 915–930. [7]

Uimari, P., and B. W. Kennedy. 1990. Mixed model methodology to estimate additive and dominance genetic values under complete dominance and inbreeding. *In* W. G. HIll, R. Thompson, and J. A. Wooliams (eds.), *Proc. 4th World Congress Genet. Appl. Livestock Prod.*, Vol 13, pp. 297–300. [23]

Utsunomiya, Y. T., and 7 others. 2013. Detecting loci under recent positive selection in dairy and beef cattle by combining different genome-wide scan methods. *PLoS One* 8: e64280. [9]

Uyeda, J. C., T. F. Hansen, S. J. Arnold, and J. Pienaar. 2011. The million-year wait for macroevolutionary bursts. *Proc. Natl. Acad. Sci. USA* 108: 15908–15913. [12]

Uyenoyama, M. K., and M. W. Feldman. 1980. Theories of kin and group selection: a populaion genetics perspective. *Theor. Pop. Biol.* 17: 380–414. [30]

Uyenoyama, M. K., and M. W. Feldman. 1981. On relatedness and adaptive topography in kin selection. *Theor. Popul. Biol.* 19: 87–123. [22, 30]

Van Buskirk, J., and Y. Willi. 2006. The change in quantitative genetic variation with inbreeding. *Evolution* 60: 2428-2434. [11]

van de Casteele, T., P. Galbusera, and E. Matthysen. 2001. A comparison of microsatellite-based pairwise relatedness estimators. *Mol. Ecol.* 10: 1539–1549. [20]

van der Werf, J. H. J. 1990. A note on the use of conditional models to estimate additive genetic variance in selected populations. *Proc. 4th World Congr. Genet. Appl. Livestock Prod. Edinburgh*, 13Z: 476–479. [19]

van der Werf, J. H. J., and I. J. M. de Boer. 1990. Estimation of additive genetic variance when base populations are selected. *J. Anim. Sci.* 68: 3124–3132. [19]

van der Werf, J. H. J., and R. Thompson. 1992. Variance decomposition in the estimation of genetic variance with selected data. *J. Anim. Sci.* 70: 2975–2985. [19]

Van Dongen, S. 2006. Prior specification in Bayesian statistics: three cautionary tales. *J. Theor. Biol.* 242: 90–100. [A2]

van Heerwaarden, B., Y. Willi, T. N. Kristensen, and A. A. Hoffmann. 2008. Population bottlenecks increase additive genetic variance but do not break a selection limit in rain forest *Drosophila*. *Genetics* 179: 2135–2146. [11]

van Noordwijk, A. J. 1988. Two-stage selection in which the first stage only reduces the environmental variation in body size in the great tit. *XIX International Ornithological Congress*, 1408–1415. [20]

van Tienderen, P. H. 2000. Elasticities and the link between demographic and evolutionary dynamics. *Ecology* 81: 666–679. [29, 30]

van Tienderen, P. H., and G. de Jong. 1994. A general model of the relation between phenotypic selection and genetic response. *J. Evol. Biol.* 7: 1–12. [20]

Van Valen, L. 1965. Selection in natural populations. III. Measurement and estimation. *Evolution* 19: 514–558. [29].

van Valen, L. 1973. A new evolutionary law. *Evol. Theory* 1: 1–30. [20]

Van Valen, L. 1976. Energy and evolution. *Evol. Theory* 1: 179–229. [3]

Van Valen, L. 1985. Why and how do mammals evolve unusually rapidly? *Evol. Theory* 7: 127–132. [12]

Van Vleck, L. D. 1968. Variation of milk records within paternal-sib groups. *J. Dairy Sci.* 51: 1465–1470. [17]

van Veelen, M. 2005. On the use of the Price equation. *J. Theor. Biol.* 237: 412–426. [6]

Van Veelen, M., J. García, M. W. Sabelis, and M. Egas. 2012. Group selection and inclusive fitness are not equivalent; the Price equation vs. models and statistics. *J. Theor. Biol.* 299: 64–80. [6, 22]

Van Vleck, L. D., and J. P. Cassady. 2005. Unexpected estimates of variance components with a true model containing genetic competition effects. *J. Anim. Sci.* 83: 68–74. [22]

Van Vleck, L. D., L. V. Cundiff, and R. M. Kock. 2007. Effect of competition on gain in feedlot bulls from Hereford selection lines. *J. Anim. Sci.* 85: 1625–1633. [22]

VanRaden, P. M. 2007. Genomic measures of relationship and inbreeding. *INTERBULL Bull.* 37: 33–36. [20]

VanRaden, P. M. 2008. Efficient methods to compute genomic predictions. *J. Dairy Sci.* 91: 4414–4423. [20]

van't Hof, A. E., and 8 others. 2016. The industrial melanism mutation in British peppered moths is a transposable element. *Nature* 534: 102–105. [8]

Vassilieva, L. L., A. M. Hook, and M. Lynch. 2000. The fitness effects of spontaneous mutations in *Caenorhabditis elegans*. *Evolution* 54: 1234–1246. [12]

Vatsiou, A. I., E. Bazin, and O. E. Gaggiotti. 2016. Detection of selective sweeps in structured populations: a comparison of recent methods. *Mol. Ecol.* 25: 89–103. [9]

Vazquez, A. I., D. Gianola, D. Bates, K. A. Weigel, and B. Heringstad. 2009. Assessment of Poisson, logit, and linear models for genetic analysis of clinical mastitis in Norwegian Red cows. *J. Dairy Sci.* 92: 739–748. [14]

Venn, O., and 5 others. 2014. Strong male bias drives germline mutation in chimpanzees. *Science* 344: 1272–1275. [4]

Verghese, M. W. 1974. Interaction between natural selection for heterozygotes and directional selection. *Genetics* 76: 163–168. [25, 26]

Verrier, E., J. J. Colleau, and J. L. Foulley. 1989. Effect of mass selection on the within-family genetic variance in finite populations. *Theor. Appl. Genet.* 77: 142–148. [16]

Verrier, E., J. J. Colleau, and J. L. Foulley. 1990. Predicting cumulated response to directional selection in finite panmictic populations. *Theor. Appl. Genet.* 79: 833–840. [26]

Verrier, E., J. J. Colleau, and J. L. Foulley. 1991. Methods for predicting response to selection in small populations under additive genetic models: a review. *Live. Prod. Sci.* 29: 93–114. [24]

Verrier, E., J. J. Colleau, and J. L. Foulley. 1993. Long-term effects of selection based on the animal model BLUP in a finite population. *Theor. Appl. Genet.* 87: 446–454. [26]

Via, S., and J. West. 2008. The genetic mosaic suggests a new role for hitchhiking in ecological speciation. *Mol. Ecol.* 17: 4334–4345. [9]

Victoria, J. K. 1972. Clutch characteristics and egg discrimination ability of the African Village Weaver *Ploceus cucullatus*. *Ibis* 114: 367–376. [25]

Vielle-Calzada, J.-P., and 10 others. 2009. The Palomero genome suggests metal effects on domestication. *Science* 326: 1078–1078. [9]

Vigouroux, Y., and 7 others. 2002. Identifying genes of agronomic importance in maize by screening microsatellites for evidence of selection during domestication. *Proc. Natl. Acad. Sci. USA* 99: 9650–9655. [9]

Vilas, A., A. Pérez-Figueroa, and A. Caballero. 2012. A simulation study on the performance of differentiation-based methods to detect selected loci using linked neutral markers. *J. Evol. Biol.* 25: 1364–1376. [9]

Villanueva, B., J. A. Woolliams, and G. Simm. 1994. Strategies for controlling rates of inbreeding in MOET nucleus schemes for beef cattle. *Genet. Sel. Evol.* 26: 517–535. [23]

Villanueva-Cañas, J. L., G. E Rech, M. A. Rodriguez de Cara, and J. Gonzáez. 2017. Beyond SNPs: how to detect selection on transposable element insertions. *Method. Ecol. Evol.* 8: 728–737. [8]

Visscher, P. M. 2016. Human complex trait genetics in the 21st century. *Genetics* 202: 377-379. [1]

Visscher, P. M., M. A. Brow., M. I. McCarthy, and J Yang. 2012a. Five years of GWAS discovery. *Amer. J. Hum. Genet.* 90: 7–24. [24]

Visscher, P. M., and M. E. Goddard. 2015. A general unified framework to assess the sampling variance of heritability estimates using pedigree or marker-based relationships. *Genetics* 199: 223–232. [20]

Visscher, P. M., M. E. Goddard, E. M. Derks, and N. R. Wray. 2012b. Evidence-based psychiatric genetics, AKA the false dichotomy between common and rare variant hypotheses. *Mol. Physch.* 17: 474–485. [28]

Visscher, P. M., and D. Posthuma. 2010. Statistical power to detect genetic loci affecting environmental sensitivity. *Behav. Genet.* 40: 728–733. [17]

Visscher, P., and R. Thompson. 1990. REML estimates of parameters for fat yield in pedigree herds in the U.K. using an individual animal model: male and female heritability estimates. *Proc. 4th World Congr. Genet. Appl. Livest. Prod. Edinburgh* 14: 484–487. [19]

Visscher, P., M., J. A. Woolliams, D. Smith, and J. L. Williams. 2002. Estimation of pedigree errors in the UK dairy population using microsatellite markers and the impact on selection. *J. Dairy Sci.* 85: 2368–2375. [20]

Visscher, P. M., and 7 others. 2006. Assumption free estimation of heritability from genome-wide identity-by descent sharing between full siblings. *PLoS Genet* 2: e41 [20]

Visscher, P. M., and 13 others. 2007. Genome partitioning of genetic variation for height from 11,214 sibling pairs. *Am. J. Hum. Genet.* 81: 1104–1110. [24]

Vitalis, R., K. Dawson, and P. Boursot. 2001. Interpretation of variation across marker loci as evidence of selection. *Genetics* 158: 181–1182. [9]

Vitalis, R., M. Gautier, K. J. Dawson, and M. A. Beaumont. 2014. Detecting and measuring selection from gene frequency data. *Genetics* 196: 799–817. [9]

Vitti, J. J., S. R. Grossman, and P. C. Sabeti. 2013. Detecting natural selection in genomic data. *Ann. Rev. Genet.* 47: 97–120. [9]

Vogel, K. P., and J. F. Pedersen. 1993. Breeding systems for cross-pollinated perennial grasses. *Plant Breed. Rev.* 11: 251–274. [21]

Vogler, D. W., and S. Kalisz. 2001. Sex among the flowers: the distribution of plant mating systems. *Evolution* 55: 202–204. [23]

Voight, B. F., S. Kudaravalli, X. Wen, and J. K. Pritchard. 2006. A map of recent positive selection in the human genome. *PLoS Bio.* 4: e72. [9]

Vøllestad, L. A., K. Hindar, and A. P. Møller. 1999. A meta-analysis of fluctuating asymmetry in relation to heterozygosity. *Heredity* 83: 206–218. [17]

von Butler, H. W., and F. Pirchner. 1984. Two-way within-family and mass selection for 8-week body weight in different mouse populations. *Genet. Res.* 43: 191–200.

von Butler, I., H. Willeke, and F. Pirchner. 1980. Preliminary results on antagonistic selection in mice. *In* A. Roberston, (ed.), *Selection experiments in laboratory and domestic animals*, pp. 114–115. Commonwealth Agricultural Bureaux, Slough, UK. [26]

Vucetich, J. A., T. A. Waite, and L. Nunney. 1997. Fluctuating population size and the ratio of effective to census population size. *Evolution* 51: 2017–2021. [3]

Vy, H. M. T., and Y. Kim. 2015. A composite-likelihood method for detecting incomplete selective sweep from population genomic data. *Genetics* 200: 633–649. [9]

Wacholder, S., S. Chanock, M. Garcia-Closas, L. El Ghormli, and N. l. Rothman. 2004. Assessing the probability that a positive report is false: an approach for molecular epidemiology studies. *J. Natl. Cancer Instit.* 6: 434–442. [A4]

Waddington, C. H. 1942. Canalizations of development and the inheritance of acquired characters. *Nature* 150: 563–565. [17]

Waddington, C. H. 1957. *The strategy of the genes.* MacMillan, New York, NY. [17]

Waddington, C. H. 1959. Canalization of development and genetic assimilation of acquired characters. *Nature* 183: 1654–1655. [17]

Waddington, C. H. 1960. Experiments on canalizing selection. *Genet. Res.* 1:140–150. [21]

Waddington, C. H., and E. Robertson 1966. Selection for developmental canalization. *Genet. Res.* 7: 303–12. [21]

Wade, M. J. 1976. Group selection among laboratory populations of *Tribolium. Proc. Natl. Acad. Sci. USA* 73: 4604–4607. [22]

Wade, M. J. 1977. An experimental study of group selection. *Evolution* 31: 134–153. [22]

Wade, M. J. 1978. A critical review of the models of group selection. *Quart. Rev. Bio.* 53: 101–14. [22]

Wade, M. J. 1979. Sexual selection and variance in reproductive success. *Am. Nat.* 114: 742–747. [29]

Wade, M. J. 1980. Kin selection: its components. *Science* 210: 665–667. [22]

Wade, M. J. 1985. Soft selection, hard selection, kin selection, and group selection. *Am. Nat.* 125: 61–73. [30]

Wade, M. J., P. Bijma, E. D. Ellen, and W. Muir. 2010. Group selection and social evolution in domesticated animals. *Evol. Appl.* 3: 453–465. [22]

Wade, M. J., and C. J. Goodnight. 1998. Perspective: the theories of Fisher and Wright in the context of metapopulations: when nature does many small experiments. *Evolution* 52: 1537–1553. [26]

Wade, M. J., and S. Kalisz. 1989. The additive partitioning of selection gradients. *Evolution* 43: 1567–1569. [29, 30]

Wade, M. J., and S. M. Shuster. 2005. Don't throw Bateman out with the bathwater. *Inter. Comp. Biol.* 45: 945–951. [29]

Wade, M. J., S. M. Shuster, and L. Stevens. 1996. Inbreeding: its effect on response to selection for pupal weight and the heritable variance in fitness in the flour beetle, *Tribolium castaneum. Evolution* 50: 723–733. [11]

Wagner, G. P. 1988. The influence of variation and of developmental constraints on the rate of multivariate phenotypic evolution. *J. Evol. Biol.* 1: 45–66. [26]

Wagner, G. P. 1989. Multivariate mutation-selection balance with constrained pleiotropic effects. *Genetics* 122: 223–234. [28]

Wagner, G. P. 1996a. Apparent stabilizing selection and the maintenance of neutral genetic variation. *Genetics* 143: 617–619. [28]

Wagner, G. P. 1996b. Homologues, natural kinds and the evolution of modularity. *Am. Zoo.* 36: 36–43. [27]

Wagner, G. P., G. Brooth, and H. Bagheri-Chaichian. 1997. A population genetic theory of canalization. *Evolution* 51: 329–347. [17]

Wagner, G. P., M. Pavlicev, and J. M. Cheverud. 2007. The road to modularity. *Nat. Rev. Genet.* 8: 921–931. [27]

Wagner, G. P., and J. Zhang. 2011. The pleiotropic structure of the genotype-phenotype map: the evolvability of complex organisms. *Nat. Rev. Genet.* 12: 204–213. [27]

Wagner, G. P., and J. Zhang. 2012. Universal pleiotropy is not a valid null hypothesis: reply to Hill and Zhang. *Nat. Rev. Genet.* 13: 296. [27]

Wagner, G. P., and 5 others. 2008. Pleiotropic scaling of gene effects and the cost of complexity. *Nature* 452: 470–473. [27]

Wahlberg, P., and 8 others. 2008. Genetic analysis of an F_2 intercross between two chicken lines divergently selected for body-weight. *BMG Genom.* 10: 248. [25]

Wakeley, J. 1997. Using the variance of pairwise differences to estimate the recombination rate. *Genet. Res.* 69: 45–48. [4]

Wakeley, J. 2003. Polymorphism and divergence for island-model species. *Genetics* 163: 411–420. [10]

Wakeley, J. 2006. *An introduction to coalescent theory.* Roberts & Co. Publ., Greenwood Village, CO. [2]

Wall, J. D. 1999. Recombination and the power of statistical tests of neutrality. *Genet. Res.* 74: 65–79. [9]

Wall, J. D. 2000. A comparison of estimators of the population recombination rate. *Mol. Biol. Evol.* 17: 156–163. [4]

Wall, J. D., and R. R. Hudson. 2001. Coalescent simulations and statistical tests of neutrality. *Mol. Biol. Evol.* 18: 1134–1135. [9]

Wallace, B. 1968. Polymorphism, population size, and genetic load. *In* R. C. Lewontin (ed.), *Population biology and evolution*, pp. 87–108. Syracuse University Press, Syracuse, NY. [30]

Wallace, B. 1975. Hard and soft selection revisited. *Evolution* 29: 465–473. [30]

Wallace, B. 1991. *Fifty years of genetic load: an odyssey.* Cornell Univ. Press, Ithaca, NY. [7]

Walling, C. A., J. M. Pemberton, J. D. Hadfield, and L. E. B. Kruuk. 2010. Comparing parentage inference software: reanalysis of a red deer pedigree. *Mol. Ecol.* 19: 1914–1928. [20]

Walsh, B. 2007. Escape from flatland. *J. Evol. Biol.* 20: 36–38. [30]

Walsh, B., and M. W. Blows 2009. Abundant genetic variation+ strong selection = multivariate genetic constraints: a geometric view of adaptation. *Ann. Rev. Ecol. Evol. Syst.* 40: 41–59. [13, 25, 28]

Walsh, J. B. 1982. Rate of accumulation of reproductive isolation by chromosome rearrangements. *Am. Nat.* 120: 510–532. [7]

Walsh, J. B. 1987. Sequence-dependent gene conversion: can duplicated genes diverge fast enough to escape conversion? *Genetics* 117: 543–557. [1]

Walsh, J. B. 1988. Unusual behaviour of linkage disequilibrium in two-locus gene conversion models. *Genet. Res.* 51: 55–58. [4]

Walsh, J. B. 1990. Inconsistencies in standard approximations for selection coefficients at loci affecting a polygenic character. *J. Math. Biol.* 28: 21–31. [5]

Walsh, J. B. 1995. How often do duplicated genes evolve new functions? *Genetics* 139: 421–428. [1, 7]

Walters, J. H., and H. Lehmann. 1956. Distribution of the S and C haemoglobin variants in two Nigerian communities. *Trans. Roy. Soc. Trop. Med. Hyg.* 50: 204–208. [9]

Wang, B., S. Sverdlov, and E. Thompson. 2017. Efficient estimation of realized kinship from single nucleotide polymorphism genotypes. *Genetics* 205: 1063–1078. [20]

Wang, C. S., J. J. Rutledge, and D. Gianola. 1993. Marginal inferences about variance components in a mixed linear model using Gibbs sampling. *Genet. Sel. Evol.* 21: 41–62. [19]

Wang, C. S., J. J. Rutledge, and D. Gianola. 1994b. Bayesian analysis of mixed linear models via Gibbs sampling with an application to litter size in Iberian pigs. *Genet. Sel. Evol.* 26: 91–115. [19]

Wang, C. S., and 5 others. 1994a. Response to selection for litter size in Danish landrace pigs: a Bayesian analysis. *Theor. Appl. Genet.* 88: 229–230. [19]

Wang, E. T., G. Kodama, P. Baldi, and R. K. Moyzis. 2006. Global landscape of recent inferred Darwinian selection for *Homo sapiens*. *Proc. Natl. Acad. Sci. USA* 103: 135–140. [9]

Wang, H., and 8 others. 2005. The origin of the naked grains of maize. *Nature* 436: 714–719. [9]

Wang, J. 1995. Exact inbreeding coefficient and effective size of finite populations under partial sib mating. *Genetics* 140: 357–363. [3]

Wang, J. 1996. Inbreeding and variance effective sizes for nonrandom mating populations. *Evolution* 50: 1786–1794. [3]

Wang, J. 1997. More efficient breeding systems for controlling inbreeding and effective size in animal populations. *Heredity* 79: 591–599. [3]

Wang, J. 2001. A pseudo–likelihood method for estimating effective population size from temporally spaced samples. *Genet. Res.* 78: 243–257. [4]

Wang, J. 2004. Sibship reconstruction from genetic data with typing errors. *Genetics* 166: 1963–1979. [20]

Wang, J. 2005. Estimation of effective population sizes from data on genetic markers. *Phil. Trans. Roy. Soc. Lond. B Biol. Sci.* 360: 1395–1409. [4]

Wang, J. 2009. A new method for estimating effective population sizes from a single sample of multilocus genotypes. *Mol. Ecol.* 18: 2148–2164. [4]

Wang, J., A. Caballero, P. D. Keightley, and W. G. Hill. 1998. Bottleneck effect on genetic variance: a theoretical investigation of the role of dominance. *Genetics* 150: 435–447. [11]

Wang, J., and A. W. Santure. 2009. Parentage and sibship inference from multilocus genotype data under polygamy. *Genetics* 181: 1579–1594. [20]

Wang, R.-L., A. Stec, J. Hey, L. Lukens, and J. Doebley. 1999. The limits of selection during maize domestication. *Nature* 398: 236–239. [8, 9]

Wang, X., W. E. Grus, and J. Zhang. 2006. Gene losses during human origins. *PLoS Biol.* 4: e52. [8, 9]

Wang, Z., M. Gerstein, and M. Snyder. 2009. RNA-Seq: a revolutionary tool for transcriptomics. *Nat. Rev. Genet.* 10: 57–63. [12]

Wang, Z., B.-Y. Liao, and J. Zhang. 2010. Genomic patterns of pleiotropy and the evolution of complexity. *Proc. Natl. Acad. Sci. USA* 107: 18034–18039. [27]

Waples, R. S. 1989a. A generalized approach for estimating effective population size from temporal changes in allele frequency. *Genetics* 121: 379–391. [4, 9]

Waples, R. S. 1989b. Temporal variation in allele frequencies: testing the right hypothesis. *Evolution* 43: 1236–1251. [9]

Waples, R. S., and M. Yokota. 2007. Temporal estimates of effective population size in species with overlapping generations. *Genetics* 175: 219–233. [4]

Warnefors, M., and A. Eyre-Walker. 2012. A selection index for gene expression evolution and its application to the divergence between humans and chimpanzees. *PLoS One* 7: e34935. [12]

Warwick, E. J., and W. L. Lewis. 1954. Increase in frequency of a deleterious recessive gene in mice. *J. Heredity* 45: 143–145. [25]

Watterson, G. A. 1975. On the number of segregation sites. *Theor. Popul. Biol.* 7: 256–276. [2, 4]

Watterson, G. A. 1976. Reversibility and the age of an allele. I. Moran's infinitely many neutral alleles model. *Theor. Popul. Biol.* 10: 239–253. [2, 9]

Watterson, G. A. 1977. Heterosis or neutrality? *Genetics* 85: 789–814. [9]

Watterson, G. A. 1978. The homozygosity test for neutrality. *Genetics* 38: 405–417. [9]

Watterson, G. A. 1982. Testing selection at a single locus. *Biometrics* 38: 323–331. [9]

Waxman, D. 2003. Numerical and exact solutions for continuum of alleles models. *J. Math. Biol.* 46: 225–240. [28]

Waxman, D. 2006. Fisher's geometrical model of evolutionary adaptation—beyond spherical geometry. *J. Theor. Biol.* 241: 887–895. [27]

Waxman, D. 2007. Mean curvature versus normality: A comparison of two approximations of Fisher's geometrical model. *Theor. Pop. Biol.* 71: 30–36. [27]

Waxman D., and J. R. Peck. 1998. Pleiotropy and the preservation of perfection. *Science* 279:1210–13. [27]

Waxman, D., and J. J. Welch. 2005. Fisher's microscope and Haldane's ellipse. *Am. Nat.* 166: 447–457. [27]

Webel, O. D., and J. H. Lonnquist. 1967. An evaluation of modified ear-to-row selection in a population of corn (*Zea mays* L.). *Crop Sci.* 7: 651–655. [21]

Weber, K. E. 1990. Increased selection response in larger populations. I. Selection for wing-tip height in *Drosophila melanogaster* at three population sizes. *Genetics* 125: 579–584. [26]

Weber, K. E. 1996. Large genetic change at small fitness cost in large populations of *Drosophila melanogaster* selected for wind tunnel flight: rethinking fitness surfaces. *Genetics* 144: 205–213. [25, 26]

Weber, K. E. 2004. Population size and long-term selection. *Plant Breeding Reviews* 24: 249–268. [25, 26]

Weber, K. E., and L. T. Diggins. 1990. Increased selection response in larger populations. II. Selection for ethanol vapor resistance in *Drosophila melanogaster* at two population sizes. *Genetics* 125: 585–597. [26]

Weber, W. E. 1982. Selection in segregating generations of autogamous species. I. Selection response for combined selection. *Euphytica* 3 493–502. [23]

Weber, W. E. 1984. Selection in segregating generations of autogamous species. II. Response to selection in two stages. *Euphytica* 33: 447–454. [23]

Webster, M. S., S. Pruett-Jones, D. F. Westneat, and S. J. Arnold. 1995. Measuring the effects of pairing success, extra-pair copulations and mate quality on the opportunity for sexual selection. *Evolution* 49: 1147–1157. [29]

Webster, M. S., K. A. Tarvin, E. M. Tuttle, and S. Pruett-Jones. 2007. Promiscuity drives sexual selection in a socially monogamous bird. *Evolution* 61: 2205–2211. [29]

Webster, M. T., and N. G. C. Smith. 2004. Fixation biases affecting humans SNPs. *Trends Genet.* 20: 122–126. [10]

Wei, M., A. Caballero, and W. G. Hill. 1996. Selection response in finite populations. *Genetics* 144: 1961–1974. [26]

Weiner, J., and R. P. Freckleton. 2010. Constant final yield. *Ann. Rev. Eco. Evol. Sys.* 41: 173–192. [30]

Weinig, C., J. A. Johnston, C. G. Willis, and J. N. Maloof. 2007. Antagonistic multilevel selection on size and architecture in variable density settings. *Evolution* 61: 58–67. [30]

Weinreich, D. M., and L. Chao. 2005. Rapid evolutionary escape by large populations from local fitness peaks is likely in nature. *Evolution* 59: 1175–1182. [7]

Weinreich, D. M., N. F. Delaney, M. A. DePristo, and D. L. Hartl. 2006. Darwinian evolution can follow only very few mutational paths to fitter proteins. *Science* 312: 111–114. [27]

Weinreich, D. M., and D. M. Rand. 2000. Contrasting patterns of nonneutral evolution in proteins encoded in nuclear and mitochondrial genomes. *Genetics* 156: 385–399. [10]

Weinreich, D. M., R. A. Watson, and L. Chao. 2005. Perspective: sign epistasis and genetic constraint on evolutionary trajectories. *Evolution* 59: 1165–1174. [27]

Weir, B. S. 1996. *Genetic data analysis II.* Sinauer Assocs., Inc., Sunderland, MA. [12]

Weir, B. S., A. D. Anderson, and A. B. Hepler. 2006. Genetic relatedness analysis: modern data and new challenges. *Nat. Rev. Genet* 7: 771–780. [20]

Weir, B. S., P. J. Avery, and W. G. Hill. 1980. Effect of mating structure on variation in inbreeding. *Theor. Popul. Biol.* 18: 396–429. [3, 11, 12]

Weir, B. S., and C. C. Cockerham. 1968. Pedigree mating with two linked loci. *Genetics* 61: 923–940. [11]

Weir, B. S., and C. C. Cockerham. 1969. Group inbreeding with two linked loci. *Genetics* 63: 711–742. [11]

Weir, B. S., and C. C. Cockerham. 1973. Mixed self and random mating at two loci. *Genet. Res.* 21: 247–262. [11]

Weir, B. S., and C. C. Cockerhman 1977. Two-locus theory in quantitative genetics. *In* E. Pollak, O. Kempthorne, and T. B. Bailey, Jr., (eds.), *Proceedings of the international conference on quantitative genetics*, pp. 247–262. Iowa State Univ. Press, Ames, IA. [11]

Weir, B. S., and C. C. Cockerham. 1984. Estimating *F*-statistics for the analysis of population structure. *Evolution* 38: 1358–1370. [2]

Weir, B. S., and W. G. Hill. 1980. Effect of mating structure on variation in linkage disequilibrium. *Genetics* 95: 477–488. [4, 11]

Weir, B. S., and W. G. Hill. 2002. Estimating *F*-statistics. *Ann. Rev. Genet.* 36: 721–750. [2]

Weis, A. E., and A. Kapelinski. 1994. Variable selection on *Eurosta*'s gall size. II. A path analysis of the ecological factors behind selection. *Evolution* 48: 734–746. [30]

Weiss, G., and A. von Haeseler. 1998. Inference of population history using a likelihood approach. *Genetics* 149: 1539–1546. [A3]

Weissman, D. B., and N. H. Barton. 2012. Limits to the rate of adaptive substitution in sexual populations. *PLoS Genet.* 8: e1002740. [8]

Weissman, D. B., M. M. Desai, D. S. Fisher, and M. W. Feldman. 2009. The rate at which asexual populations cross fitness valleys. *Theor. Pop. Biol.* 75: 286–300. [7]

Weissman, D. B., D. S. Fisher, and M. W. Feldman. 2010. The rate of fitness-valley crossing in sexual populations. *Genetics* 186: 1389–1410. [1, 7]

Weissman, I. 1978. Estimation of parameters and large quantiles based on the *k*th largest observations. *J. Am. Sta. Ass.* 73: 812–815. [27]

Welch, J. J. 2006. Estimating the genomewide rate of adaptive protein evolution in *Drosophila*. *Genetics* 173: 821–837. [10]

Welch, J. J. 2017. What's wrong with evolutionary biology? *Biol. Philos.* 32: 263–279. [1]

Welch, J. J., A. Eyre-Walker, and D. Waxman. 2008. Divergence and polymorphism under the nearly neutral theory of molecular evolution. *J. Mol. Evol.* 67: 418–426. [10]

Welch, J. J., and D. Waxman. 2003. Modularity and the cost of complexity. *Evolution* 57: 1723–1734. [27]

Weldon, W. F. R. 1895. An attempt to measure the death-rate due to the selective destruction of *Carcinus moenas* with respect to a particular dimension. *Proc. Roy. Soc. Lond.* 57: 360–379. [1, 16, 20, 28, 29]

Weldon, W. F. R. 1899. Presidential address. *British Association for the Advancement of Science. Transactions* D: 887–902. [1, 29]

Weldon, H. F. R. 1901. A first study of natural selection in *Clausilia laminata* (Montagu). *Biometrika* 1: 109–124. [29, 30]

Weller, A. M., C. Rödelsperger, G. Eberhardt, R. I. Molnar, and R. J. Sommer. 2014. Opposing forces of A/T-biased mutations and G/C-biased gene conversions shape the genome of the nematode *Pristionchus pacificus*. *Genetics* 196: 1145–1152. [4]

Weller, J. I. 1994. *Economic aspects of animal breeding*. Chapman and Hall, London, UK. [13]

Wellman, R., and J. Bennewtiz. 2001. The contribution of dominance to the understanding of quantitative genetic variation. *Genet. Res.* 93: 139–154. [24]

Wells, D. N., P. M. Misica, and H. R. Tervit. 1999a. Production of cloned calves following nuclear transfer with cultured adult mural granulosa cells. *Biol. Reprod.* 60: 996–1005. [21]

Wells, D. N., P. M. Misica, H. R. Tervit, and W. H. Vivanco. 1999b. Adult somatic cell nuclear transfer is used to preserve the last surviving cow of the Enderby Island cattle breed. *Reprod. Fert. Devel.* 10: 369–378. [21]

Wesselingh, R. A., and T. J. De Jong. 1995. Bidirectional selection on threshold size for flowering in *Cynoglossum officinale* (hound's-tough). *Heredity* 74: 415–424. [14]

West, S. A., and A. Gardner. 2013. Adaptation and inclusive fitness. *Curr. Biol.* 23 : R577–R584. [22]

West, S. A., A. S. Griffin, and A. Gardner. 2006. Social semantics: altruism, cooperation, mutualism, strong reciprocity and group selection. *J. Evol. Biol.* 20: 415–432. [22]

West, S. A., A. S. Griffin, and A. Gardner. 2008. Social semantics: how useful has group selection been? *J. Evol. Biol.* 21: 374–385. [22]

Westell, R. A., R. L. Quaas, and L. Dale Van Vleck. 1988. Genetic groups in an animal model. *J. Dairy Sci.* 71: 1310–1318. [12]

Wey, T., D. T. Blumstein, W. Shen, and F. Jordán. 2008. Social network analysis of animal behaviour: a promising tool for the study of sociality. *Anim. Behav.* 75: 333–344. [22]

Whitehead, A. N. 1920. *The concept of nature: Tarrner lectures delivered in Trinity College, November, 1919.* Cambridge University Press, Cambridge, UK.

Whitehead, A., and D. L. Crawford. 2006. Variation within and among species in gene expression: raw material for evolution. *Mol. Ecol.* 15: 1197–1211. [12]

Whitlock, M. C. 1999. Neutral additive genetic variance in a metapopulation. *Genet. Res.* 74: 215-221. [12]

Whitlock, M. C. 2003. Fixation probability and time in subdivided populations. *Genetics* 164: 767–779. [7]

Whitlock, M. C. 2005. Combining probability from independent tests: the weighted Z-method is superior to Fisher's approach. *J. Evol. Biol.* 18: 1368–1373. [A4]

Whitlock, M. C. 2008. Evolutionary inference from Q_{ST}. *Mol. Ecol.* 17: 1885-1896. [12]

Whitlock, M. C., and N. H. Barton. 1997. The effective size of a subdivided population. *Genetics* 146: 427–441. [3]

Whitlock, M. C., and K. Fowler. 1999. The changes in genetic and environmental variance with inbreeding in *Drosophila melanogaster*. *Genetics* 152: 345–353. [11, 16, 17, 28]

Whitlock, M. C., and K. J. Gilbert. 2012. Q_{ST} in a hierarchically structured population. *Mol. Ecol. Res.* 12: 481–483. [12]

Whitlock, M. C., and R. Gomulkiewicz. 2005. Probability of fixation in a heterogeneous environment. *Genetics* 171: 1407–1417. [7]

Whitlock, M. C., and F. Guillaume. 2009. Testing for spatially divergence selection: Comparing Q_{ST} to F_{ST}. *Genetics* 183: 1055–1063. [12]

Whitlock, M. C., and K. E. Lotterhos. 2015. Reliable detection of loci responsible for local adaptation: inference of a null model through trimming the distribution of F_{ST}. *Am. Nat.* 186: S24–S36 . [9]

Whitt, S. R., L. M. Wilson, M. I. Tenaillon, B. S. Gaut, and E. S. Buckler IV. 2002. Genetic diversity and selection in the maize starch pathway. *Proc. Natl. Acad. Sci. USA* 99: 12959–12962. [9]

Whittam, T. S., and M. Nei. 1991. Neutral mutation hypothesis test. *Nature* 354: 115-116. [10]

Whittingham, L. A., and P. O. Dunn. 2005. Effects of extra-pair and within-pair reproductive success on the opportunity for selection in birds. *Behav. Ecol.* 16: 138–144. [29]

Wiebe, G. A., F. C. Petr, and H. Stevens. 1963. Interplant competition between barley genotypes. *In* D. Hanson and H. F. Robinson (eds.), *Statistical genetics and plant breeding*, pp. 546–555. National Academy of Sciences, NRC No. 982. [22]

Wiehe, T. 1998. The effect of selective sweeps on the variance of the allele distribution of a linked multiallele locus: hitchhiking of microsatellites. *Theor. Popul. Biol.* 53: 272–283. [8]

Wiehe, T., V. Nolte, D. Zivkovic, and S. Schlötterer. 2007. Identification of selective sweeps using a dynamically adjusted number of linked microsatellites. *Genetics* 175: 207–218. [9]

Wiehe, T. H., and W. Stephan. 1993. Analysis of a genetic hitchhiking model, and its application to DNA polymorphism data from *Drosophila melanogaster*. *Mol. Biol. Evol.* 10: 842–854. [3, 8]

Wielgoss, S., and 9 others. 2013. Mutation rate dynamics in a bacterial population reflect tension between adaptation and genetic load. *Proc. Natl. Acad. Sci. USA* 110: 222–227. [4]

Wiener, P., and R. Pong-Wong. 2011. A regression-based approach to selection mapping. *J. Heredity* 102: 294–305. [9]

Wiener, P., and 8 others. 2003. Signatures of selection? Patterns of microsatellite diversity on a chromosome containing a selected locus. *Heredity* 90: 350–358. [8]

Wilf, H. S. 1978. *Mathematics for the physical sciences.* Dover, New York, NY. [A5]

Wilke, C. O., J. L. Wang, C. Ofria, R. E. Lenski, and C. Adami. 2001. Evolution of digital organisms at high mutation rates leads to survival of the flattest. *Nature* 412: 331–333. [4]

Willeke, H. 1982. Comparison of selection schemes for improving litter size in pigs—results of a simulation study. *Livestock Prod. Sci.* 8: 535–540. [21]

Willensdorfer, M., and R. Bürger. 2003. The two-locus model of Gaussian stabilizing selection. *Theor. Popul. Biol.* 64: 101–117. [5, 28]

Willham, R. L. 1963. The covariance between relatives for characters composed of components contributed by related individuals. *Biometrics* 19: 18–27. [1, 15, 22]

Willham, R. L. 1972. The role of maternal effects in animal breeding. III. Biometrical aspects of maternal effects in animals. *J. Anim. Sci.* 35: 1288–1293. [15, 22]

Williams, C. G., and F. A. Walton. 1915. Corn experiments. *Ohio Agric. Exp. Stn. Bull.* 282. [21]

Williams, G. C. 1966. *Adaptation and natural selection.* Princeton Univ. Press, Princeton, NJ. [22]

Williams, L. N., A. J. Herr, and B. D. Preston. 2013. Emergence of DNA polymerase ϵ antimutators that escape error-induced extinction in yeast. *Genetics* 193: 751-770. [4]

Williamson, E. G., and M. Slatkin. 1999. Using maximum likelihood to estimate population size from temporal changes in allele frequencies. *Genetics* 152: 755–761. [4]

Williamson, S., A. Fledel-Alon, and C. D. Bustamante. 2004. Population genetics of polymorphism and divergence for diploid selection models with arbitrary dominance. *Genetics* 168: 463–475. [10]

Williamson, S. H., and 5 others. 2007. Localizing recent adaptive evolution in the human genome. *PLoS Genet.* 3: e90. [9]

Willis, J. H. 1996. Measures of phenotypic selection are biased by partial inbreeding. *Evolution* 50: 1501–1511. [29, 30]

Willis, J. H., and H. A. Orr. 1993. Increased heritable variation following population bottlenecks: the role of dominance. *Evolution* 47: 949–956. [11]

Willson, M. F., and N. Burley. 1983. *Mate choice in plants: tactics, mechanisms, and consequences.* Princeton University Press, Princeton, NJ. [29]

Wilson, A. 2008. Why h^2 does not always equal VA/VP. *J. Evol. Bio.* 21: 647–650. [19, 20]

Wilson, A. J., U. Gelin, M.-C. Perron, and D. Réale. 2009. Indirect genetic effects and the evolution of aggression in a vertebrate system. *Proc. Roy. Soc. Lond. B* 276: 533–541. [22]

Wilson, A. J., and A. Rambaut. 2008. Breeding racehorses: what price good genes? *Biol. Let.* 4: 173–175. [25]

Wilson, A. J., and D. Réale. 2006. Ontogeny of additive and maternal genetic effects: Lessons from domestic mammals. *Am. Nat.* 167: E23–E38. [15]

Wilson, A. J., and 5 others. 2005a. Maternal genetic effects set the potential for evolution in a free-living vertebrate population. *J. Evol. Biol.* 18: 405–414. [15]

Wilson, A. J., and 6 others. 2005b. Selection on mothers and offspring: whose phenotype is it and does it matter? *Evolution* 59: 451–463. [15]

Wilson, A. J., and 6 others. 2006. Environmental coupling of selection and heritability limits evolution. *PLoS Biol.* 4: 1270–1275. [20]

Wilson, A. J., and 7 others. 2010. An ecologist's guide to the animal model. *J. Anim. Ecol.* 79: 13–26. [20]

Wilson, A. J., and 7 others. 2011. Indirect genetics effects and evolutionary constraint: an analysis of social dominance in red deer, *Cervus elaphus*. *J. Evol. Biol.* 24: 772–783. [22]

Wilson, B. A., D. A. Petrov, and P. W. Messer. 2014. Soft selective sweeps in complex demographic scenarios. *Genetics* 198: 669–684. [8]

Wilson, D. J., R. D. Hernandez, P. Andelfatto, and M. Przeworski. 2011. A population genetics-phylogenetics approach to inferring natural selection in coding sequences. *PLoS Genet.* 7: e1002395. [8]

Wilson, D. S. 1983. The group selection controversy: history and current status. *Ann. Rev. Ecol. Syst.* 14: 159–187. [22]

Wilson, D. S., and E. O. Wilson. 2007. Rethinking the theoretical foundation of sociobiology. *Quart. Rev. Biol.* 82: 327–348. [22]

Wilson, I. J., M. E. Weale, and D. J. Balding. 2003. Inferences from DNA data: population histories, evolutionary processes and forensic match probabilities. *J. Roy. Stat. Soc. A* 166: 155–201. [9]

Wilson, S. P. 1974. An experimental comparison of individual, family, and combination selection. *Genetics* 76: 823–836. [21]

Wilson, S. P. 1977. Selection experiments with laboratory animals. *Genetics Lectures* 3: 89–110. [18]

Wilson, S. P., H. D. Goodale, W. H. Kyle, and E. F. Godfrey. 1971. Long term selection for body weight in mice. *J. Heredity* 62: 228–234. [25]

Wilson, S. P., W. H. Kyle , and A. E. Bell. 1965. The effects of mating systems and selection on pupa weight in *Tribolium*. *Genet Res.* 6: :341–351. [16]

Wilson, S. R. 1980. Analyzing gene-frequency data when the effective population size is finite. *Genetics* 95: 489–502. [9]

Winn, A. A., and 10 others. 2011. Analysis of inbreeding depression in mixed-mating plants provides evidence for selective interference and stable mixed mating. *Evolution* 65: 3339–3359. [23]

Wiser, M. J., N. Ribeck, and R. E. Lenski. 2013. Long-term dynamics of adaptation in asexual populations. *Science* 342: 1364–1367. [25, 27]

Wisser, R. J., S. C. Murray, J. M. Kolkman, H. Ceballos, and R. J. Nelson. 2008. Selection mapping of loci of quantitative disease resistance in a diverse maize population. *Genetics* 180: 583–599. [9]

Wittkopp, P. J., B. K. Haerum, and A. G. Clark. 2004. Evolutionary changes in *cis* and *trans* gene regulation. *Nature* 430: 85–88. [12]

Wittkopp, P. J., B. K. Haerum, and A. G. Clark. 2008. Regulatory changes underlying expression differences within and between *Drosophila* species. *Nat. Genet.* 40: 346–350. [12]

Wiuf, C. 2006. Consistency of estimators of population scaled parameters using composite likelihood. *J. Math. Biol.* 53: 821–841. [9]

Wiuf, C., K. Zhao, H. Innan, and M. Nordborg. 2004. The probability and chromosomal extent of trans-specific polymorphism. *Genetics* 168: 2363–2372. [8, 9]

Wolc, A., I. M. S. White, S. Avendaon, and W. G. Hill. 2009. Genetic variability in residual variation of body weight and conformation scores in broiler chickens. *Poultry Sci.* 88: 1156–1161. [17]

Wolf, J. B. 2003. Genetic architecture and evolutionary constraint when the environment contains genetics. *Proc. Natl. Acad Sci. USA* 100: 4655–4660. [22]

Wolf, J. B., E. D. Brodie III, J. M. Cheverud, and A. J. Moore. 1998. Evolutionary consequences of indirect genetic effects. *Trends Ecol. Evol.* 13: 64–69. [22]

Wolf, J. B., E. D. Brodie III, and M. J. Wade (eds.) 2000. *Epistasis and the evolutionary process.* Oxford Univ. Press, New York, NY. [11]

Wolf, J. B., and A. J. Moore. 2010. Interacting phenotypes and indirect genetic effects. *In* D. Westneat and C. W. Fox (eds.), *Evolutionary behavioral ecology*, pp. 225–245. Oxford University Press, Ocford, UK. [22]

Wolf, J. B., J. J. Mutic, and P. X. Kover. 2011. Functional genetics of intraspecific ecological interactions in *Arabidopsis thaliana*. *Phil. Trans. Roy. Soc. Lond. B* 366: 1358–1367. [22]

Wolf. J. B., and M. J. Wade. 2001. On the assignment of fitness to parents and offspring: whose fitness is it and when does it matter? *J. Evol. Biol.* 14: 347–356. [15, 29]

Wolf, J. B., and M. J. Wade. 2009. What are maternal effects (and what are they not)? *Phil. Trans. Roy. Soc. Lond. B* 364: 1107–1115. [15, 30]

Wolf, J. B. W., A. Künstner, K. Nam, M. Jakobsson, and H. Ellegren. 2009. Nonlinear dynamics of nonsynonymous (d_N) and synonymous (d_S) substitution rates affects inference of selection. *Genom. Biol. Evol.* 1: 308–319. [10]

Wong, W. S. W., and R. Nielsen. 2004. Detecting selection in noncoding regions of nucleotide sequences. *Genetics* 167: 949–958. [10]

Wong, W. S. W., Z. Yang, N. Goldman, and R. Nielsen. 2004. Accuracy and power of statistical methods for detecting adaptive evolution in the protein coding sequences and for identifying positively selected sites. *Genetics* 168: 1041–1051. [10]

Wood, A. R., and >100 others. 2014. Defining the role of common variation in the genomic and biological architecture of adult human height. *Nat. Genet.* 46: 1173–1186. [24]

Woods, R. J., and 5 others. 2011. Second-order selection for evolvability in a large *Escherichia coli* population. *Science* 331: 1433–1436. [27]

Woolliams, J. A. 1989. Modifications to MOET nucleus breeding schemes to improve rates of genetic progress and decrease rates of inbreeding in dairy cattle. *Anim. Prod.* 49: 1-14. [13, 26]

Woolliams, J. A., N. R. Wray, and R. Thompson. 1993. Prediction of long-term contributions and inbreeding in populations undergoing mass selection. *Genet. Res.* 62: 231–242. [26]

Wootton, J. C., and 7 others. 2002. Genetic diversity and chloroquine selective sweeps in *Plasmodium falciparum*. *Nature* 418: 320–323. [9]

Wray, G. A. 2007. The evolutionary significance of *cis*-regulatory mutations. *Nat. Rev. Genet.* 8: 206–216. [12]

Wray, N. R. and W. G. Hill. 1989. Asymptotic rates of response from index selection. *Anim. Prod.* 49: 217–227. [21]

Wray, N. R., and R. Thompson. 1990. Prediction of rates of inbreeding in selected populations. *Genet. Res.* 55: 41–54. [26]

Wray, N. R., J. A. Woolliams and R. Thompson. 1990. Methods for predicting rates of inbreeding in selected populations. *Theor. Appl. Genet.* 80: 503–512. [26]

Wray, N. R., J. A. Woolliams and R. Thompson. 1994. Prediction of rates of inbreeding in populations undergoing index selection. *Theor. Appl. Genet.* 87: 878–892. [26]

Wricke, G. 1976. Comparison of selection based on yield of half-sib progenies and of I_1 lines *per se* on rye. *Theor. Appl. Genet.* 47: 265–269. [23]

Wricke, G., and W. E. Weber. 1986. *Quantitative genetics and selection in plant breeding*. Walter de Gruyter, New York, NY. [13, 15, 21, 23]

Wright, A. J. 1980. The expected efficiencies of half-sib, testcross, and S_1 progeny testing methods in single population improvement. *Heredity* 45: 361–376. [23]

Wright, A. J. 1986. Individual and group selection with competition. *Theor. Appl. Genet.* 72: 256–263. [22]

Wright, A. J. 1987. Covariance of inbred relatives with special reference to selfing. *Genetics* 115: 545–552. [11, 23]

Wright, A. J. 1988. Some applications of the covariances of relatives with inbreeding. *In* B. S. Weir, E. J. Eisen, M. M. Goodman, and G. Namkoong (eds.), *Proceedings of the Second International Conference on Quantitative Genetics*, pp. 10–20. Sinauer Assocs., Sunderland, MA. [11, 23]

Wright, A. J. and C. C. Cockerham. 1985. Selection with partial selfing. I. Mass selection. *Genetics* 109: 585–597. [23]

Wright, A. J. and C. C. Cockerham. 1986a. Selection with partial selfing. II. Family selection. *Crop Sci.* 26: 261–268. [11, 23]

Wright, A. J. and C. C. Cockerham. 1986b. Covariances of relatives and selection response in generations of selfing from an outcrossed base population. *Theor. Appl. Genet.* 72: 689–694. [23]

Wright, D. A., and F. H. Moyer. 1996. Parental influences on lactate dehydrogenase in the early development of hybrid frogs in the genus *Rana*. *J. Exper. Zoo.* 163: 215–229. [12]

Wright, S. 1921a. Systems of mating. *Genetics* 6: 111–178. [11]

Wright, S. 1921b. Systems of mating. II. The effects of inbreeding on the genetic composition of a population. *Genetics* 6: 124–143. [3, 12]

Wright, S. 1921c. Systems of mating. III. Assortative mating based on somatic resemblance. *Genetics* 6: 144–161. [16]

Wright, S. 1931. Evolution in Mendelian populations. *Genetics* 16: 97–159. [2, 3, 7, 26, A1]

Wright, S. 1932. The roles of mutation, inbreeding, crossbreeding and selection in evolution. *Proc. 6th Internat. Cong. Genetics* 1: 356–366. [7]

Wright, S. 1934a. An analysis of variability in number of digits in an inbred strain of guinea pigs. *Genetics* 19: 506–534. [1, 14]

Wright, S. 1934b. The results of crosses between inbred strains of guinea pigs, differing in number of digits. *Genetics* 19: 537–551. [1]

Wright, S. 1935a. The analysis of variance and the correlations between relatives with respect to deviations from an optimum. *J. Genet.* 30: 243–256. [5, 28]

Wright, S. 1935b. Evolution in populations in approximate equilibrium. *J. Genetics* 30: 257–266. [5, 28, 30]

Wright, S. 1937. The distribution of gene frequencies in populations. *Proc. Natl. Acad. Sci. USA* 23: 307–320. [5]

Wright, S. 1938a. Size of population and breeding structure in relation to evolution. *Science* 87: 430–431. [3]

Wright, S. 1938b. The distribution of gene frequencies under irreversible mutation. *Proc. Natl. Acad. Sci. USA* 24: 253–259. [7, 10, 28, A1]

Wright, S. 1939. Statistical genetics in relation to evolution. *Actualités Scientifiques et Industrielles, No. 802. Exposés de Biometrie et de Statistique Biologique. XIII.* Hermann and Cie, Paris, FR. [3]

Wright, S. 1940. Breeding structure of populations in relation to speciation. *Am. Nat.* 74: 232–248. [3]

Wright, S. 1942. Statistical genetics and evolution. *B. Am. Math. Soc.* 48: 223–246. [5, A1]

Wright, S. 1943. Isolation by distance. *Genetics* 28: 114–138. [2]

Wright, S. 1945. The differential equation of the distribution of gene frequencies. *Proc. Natl. Acad. Sci. USA* 31: 382–389. [A1]

Wright, S. 1948a. On the role of directed and random changes in gene frequency in the genetics of populations. *Evolution* 2: 279–294. [5, 9]

Wright, S. 1948b. Genetics of populations. *In Encyclopedia Britannica* Ed 14, Vol 10, pp 111-115. Encyclopedia Britannica, Chicago, IL. [28]

Wright, S. 1949. Adaptation and selection. *In* G. L. Jepson, G. G. Simpson, and E. Mayr (eds.), *Genetics, paleontology and evolution*, pp. 365–389. Princeton Univ. Press, Princeton, NJ. [7]

Wright, S. 1951. The genetical structure of populations. *Ann. Eugen.* 15: 323–354. [2, 9, 11, 12, 26]

Wright, S. 1955. Classification of the factors of evolution. *Cold Spring Harbor Symp. Quant. Bio.* 20: 16-24. [5]

Wright, S. 1968. *Evolution and the genetics of populations. I. Genetic and biometric foundations.* Univ. Chicago Press, Chicago, Il. [27]

Wright, S. 1969. *Evolution and the genetics of populations. II. The theory of gene frequencies.* Univ. Chicago Press, Chicago, IL. [3, 18, 28]

Wright, S. 1977. *Evolution and the genetics of populations. III. Experimental results and evolutionary deductions.* Univ. Chicago Press, Chicago, IL. [18, 25, 26]

Wright, S 1978. The relation of livestock breeding to theories of evolution. *J. Anim. Sci.* 46: 1192–1200. [26]

Wright, S. 1982. The shifting balance theory and macroevolution. *Ann. Rev. Genet.* 16: 1–20. [26]

Wright, S., and T. Dobzhansky. 1946. Genetics of natural populations. XII. Experimental reproduction of the changes caused by natural selection in certain populations of *Drosophila pseudoobscura*. *Genetics* 31: 125–156. [5]

Wright, S. I., and P. Andolfatto. 2008. The impact of natural selection on the genome: emerging patterns in *Drosophila* and *Arabidopsis*. *Ann. Rev. Ecol. Evol. Syst.* 39: 193–213. [10]

Wright, S. I., and B. Charlesworth. 2004. The HKA test revisited: A maximum-likelihood-ratio test of the standard neutral model. *Genetics* 168: 1071–1076. [10]

Wright, S. I., and B. S. Gaut. 2005. Molecular population genetics and the search for adaptive evolution in plants. *Mol. Bio. Evol.* 22: 506-519. [9]

Wright, S. I., S. Kalisz, and T. Slotte. 2013. Evolutionary consequences of self-fertilization in plants. *Proc. Roy. Soc. Lond. B* 280: 20130133. [23]

Wright, S. I., and 6 others. 2005. The effects of artificial selection on the maize genome. *Science* 308: 1310–1314 [9]

Wu, C. F. J. 1986. Jackknife, bootstrap and other resampling methods in regression analysis. *Ann. Stat.* 14: 1261–1295. [29]

Wyles, J. S., J. G. Kunkel, and A. C. Wilson. 1983. Birds, behavior, and anatomical evolution. *Proc. Natl. Acad. Sci. USA* 80: 4394-4397. [12]

Wynne-Edwards, V. C. 1962. *Animal dispersion in relation to social behavior.* Oliver and Boyd, Edinburgh, UK. [22]

Wynne-Edwards, V. C. 1963. Intergroup selection and the evolution of social systems. *Nature* 200: 623–626. [22]

Wynne-Edwards, V. C. 1986. *Evolution through group selection.* Blackwell, Oxford, UK. [22]

Xia, Q., and 57 others. 2009. Complete resequencing of 40 genomes reveals domestication events and genes in silkworm (*Bombyx*). *Science* 326: 433–446. [9]

Xiang-Yu, J., Z. Yang, K. Tang, and H. Li. 2016. Revisiting the false positive rate in detecting recent positive selection. *Quantit. Biol.* 4: 207–216. [9]

Xu, S. 2003. Theoretical basis of the Beavis effect. *Genetics* 16: 2259–2268. [25]

Xue, Y., and 13 others. 2006. Spread of an inactive form of Caspase-12 in humans due to recent positive selection. *Am. J. Hum. Genet.* 78: 659–670. [8]

Xue, Y., and 15 others. 2009. Human Y chromosome base–substitution mutation rate measured by direct sequencing in a deep–rooting pedigree. *Curr. Biol.* 19: 1453–1457. [4]

Yamasaki, M., and 7 others. 2005. A large-scale screen for artificial selection in maize identifies candidate agronomic loci for domestication and crop improvement. *Plant Cell* 17: 2859–2872. [9]

Yan, H., W. Yuan, V. E. Velculescu, B. Vogelstein, and K. W. Kinzler. 2002. Allelic variation in human gene expression. *Science* 297: 1143–1143. [12]

Yang, E., O. Christensen, and D. A. Sorensen. 2011. Analysis of a genetically structured variance heterogeneity model using the Box-Cox transformation. *Genet. Res.* 93: 33–46. [17]

Yang, J., S. H. Lee, M. E. Goddard, and P.M. Visscher. 2011a. GCTA: a tool for genome-wide complex trait analysis. *Am. J. Hum. Genet.* 88: 76–82. [20]

Yang, J., and 11 others. 2010. Common SNPs explain a large proportion of the heritability for human height. *Nat. Genet.* 42: 565-–569. [20, 24]

Yang, J., and 5 others. 2011b. Genome partitioning of genetic variation for complex traits using common SNPs. *Nat. Genet.* 6: 519–525. [24]

Yang, J., and 10 others. 2012. FTO genotype is associated with phenotypic variability of body mass index. *Nature* 490: 267–272. [17]

Yang, J., and 8 others. 2013. Ubiquitous polygenicity of human complex traits: genome-wide analysis of 49 traits in Koreans. *PLoS Genet.* 9: e1003355. [24]

Yang, S., and 7 others. 2012. Great majority of recombination events in *Arabidopsis* are gene conversion events. *Proc. Natl. Acad. Sci. USA* 109: 20992–20997. [4]

Yang, S., and 7 others. 2015. Parent-progeny sequencing indicates higher mutation rates in heterozygotes. *Nature* 523: 463–467 [4]

Yang, Y. H., and T. Speed. 2002. Design issues for cDNA microarray experiments. *Nat. Rev. Genet.* 3: 579–588. [12]

Yang, Z. 2006. *Computational molecular evolution.* Oxford University Press, Oxford, UK. [10]

Yang, Z., 2014. *Molecular evolution: a statistical approach.* Oxford University Press, Oxford, UK. [10]

Yang, Z., and J. P. Bielawski. 2000. Statistical methods for detecting molecular adaptation. *Trends Ecol. Evol.* 15: 496–503. [10]

Yang, Z., and M. dos Reis. 2011. Statistical properties of the branch-site test of positive selection. *Mol. Biol. Evol.* 28: 1217–1228. [10]

Yang, Z., J. W. Hardin, and C.L. Addy. 2009. Testing overdispersion in the zero-inflated Poisson model. *J. Stat. Planning Inference* 139: 3340–3353. [29]

Yang, Z., and R. Nielsen. 2002. Codon-substitution models for detecting molecular adaptation at individual sites along specific lineages. *Mol. Biol. Evol.* 19: 908–917. [10]

Yang, Z., and R. Nielsen. 2008. Mutation-selection models of codon substitution and their use to estimate selective strengths on codon usage. *Mol. Biol. Evol.* 25: 568–579. [10]

Yang, Z., R. Nielsen, N. Goldman, and A.-M. K. Pedersen. 2000. Codon-substitution models for heterogeneous selection pressures at amino acid sites. *Genetics* 155: 431–449. [10]

Yang, Z., W. S. W. Wong, and R. Nielsen. 2005. Bayes empirical Bayes inference of amino acid sites under positive selection. *Mol. Biol. Evol.* 22: 1107–1118. [10]

Yi, X., and 70 others. 2010. Sequencing of 50 human exomes reveals adaptation to high altitude. *Science* 329: 75–78. [9]

Yin, J., M. I. Jordan, and Y. S. Song. 2009. Joint estimation of gene conversion rates and mean conversion tract lengths from population SNP data. *Bioinformatics* 25: i231–i239. [4]

Yokoyama, S., and J. Felsenstein. 1978. A model of kin selection for an altruistic trait considered as a quantitative character. *Proc. Natl. Acad. USA* 75: 420–422. [22]

Yokoyama, S., T. Tada, H. Zhang, and L. Britt. 2008. Elucidation of phenotypic adaptations: molecular analysis of dim-light vision proteins in vertebrates. *Proc. Natl. Acad. Sci. USA* 105: 13480–13485. [10]

Yoo, B. H. 1980a. Long-term selection for a quantitative character in large replicate populations of *Drosophila melanogaster*. I: Response to selection. *Genet. Res.* 35: 1-17. [25, 26]

Yoo, B. H. 1980b. Long-term selection for a quantitative character in large replicate populations of *Drosophila melanogaster*. II: Lethals and visible mutants with large effects. *Genet. Res.* 35: 19-31. [25, 26]

Yoo, B. H. 1980c. Long-term selection for a quantitative character in large replicate populations of *Drosophila melanogaster*. V: The inbreeding effect of selection. *Aust. J. Biol. Sci.* 33: 713–723. [26]

Yu, F., and A. Etheridge. 2010. The fixation probability of two competing beneficial mutations. *Theor. Popul. Biol.* 78: 36–45. [9]

Yu, J., and 11 others. 2006. A unified mixed-model method for association mapping that accounts for multiple levels of relatedness. *Nat. Genet.* 38: 203–208. [9]

Yu, N., and 7 others. 2003. Low nucleotide diversity in chimpanzees and bonobos. *Genetics* 164: 1511–1518. [4]

Yule, G. U. 1902. Mendels laws and their probable relations to intra-racial heredity. *New Phytologist* 1: 193–207, 222—238. [1]

Yule, G. U. 1903. Notes on the theory of association of attributes in statistics. *Biometrika* 2: 121—134. [10]

Yule, G. U. 1906. On the theory of inheritance of quantitative compound characters on the basis of Mendels laws – a preliminary note. *Report of the Conference on Genetics* 140–142. [1]

Zabel, R. W., T. Wagner, J. L. Congleton, S. G. Smith, and J. G. Williams. 2005. Survival and selection of migrating salmon from capture-recapture models with individual traits. *Ecol. Appl.* 15: 1427–1439. [29]

Zeder, M. A., E. Emshwiller, B. D. Smith, and D. G. Bradley. 2006. Documenting domestication: the intersection of genetics and archaeology. *Trends Genet.* 22: 139–155. [9]

Zelditch, M., D. Swiderski, D. H. Sheets, and W. Fink. 2012. *Geometric morphometrics for biologists, 2nd ed.* Academic Press, New York, NY. [A5]

Zeng, K. 2010. A simple multiallele model and its application to identifying preferred–unpreferred codons using polymorphism data. *Mol. Bio. Evol.* 27 1327–1337. [8]

Zeng, K., and B. Charlesworth. 2009. Estimating selection intensity on synonymous codon usage in a nonequilibrium population. *Genetics* 183: 651–662. [8]

Zeng, K., and B. Charlesworth. 2010. Studying patterns of recent evolution at synonymous sites and intronic sites in *Drosophila melanogaster. J. Mol. Evol.* 70: 116–128. [8]

Zeng, K., S. Mano, S. Shi, and C.-I. Wu. 2007a. Comparisons of site-and haplotype-frequency methods for detecting positive selection. *Mol. Biol. Evol.* 24: 1562–1574. [9]

Zeng, K., S. Shi, and C.-I. Wu. 2007b. Compound tests for the detection of hitchhiking under positive selection. *Mol. Biol. Evol.* 24: 1898–1908. [9]

Zeng, K., Y.-X. Fu, S. Shi, and C.-I. Wu. 2006. Statistical tests for detecting positive selection by utilizing high-frequency variants. *Genetics* 174: 1431–1439. [9]

Zeng, Z.-B., 1987. Genotypic distribution at the limits of natural and artificial selection with mutation. *Theor. Popul. Biol.* 32: 90–113. [24]

Zeng, Z.-B., and C. C. Cockerham. 1990. Long-term response to artificial selection with multiple alleles—study by simulations. *Theor. Popul. Biol.* 37: 254–272. [26]

Zeng, Z. B., and C. C. Cockerham. 1991. Variance of neutral genetic variances within and between populations for a quantitative character. *Genetics* 129: 535–553. [11]

Zeng, Z. B., and C. C. Cockerham. 1993. Mutation models and quantitative genetic variation. *Genetics* 133: 729-736. [11, 12, 28]

Zeng, Z.-B., and W. G. Hill. 1986. The selection limit due to the conflict between truncation and stabilizing selection with mutation. *Genetics* 114: 1313–1328. [2, 26]

Zeng, Z. B., H. Tachida, and C. C. Cockerham. 1989. Effects of mutation on selection limits in finite populations with multiple alleles. *Genetics* 122: 977–984. [7, 26]

Zhai, W., R. Nielsen, and M. Slatkin. 2009. An investigation of the statistical power of neutrality tests based on comparative and population genetic data. *Mol. Biol. Evol.* 26: 273–283. [9, 10]

Zhang, G., L. J. Muglia, R. Chakraborty, J. M. Akey, and S. M. Williams. 2013. Signatures of natural selection on genetic variants affecting complex human traits. *Appl. Translational Genomics* 2: 78–94. [9]

Zhang, H., and 11 others. 2012. Diverse genetic basis of field-evolved resistance to Bt cotton in cotton bollworm from China. *Proc. Natl. Acad. Sci. USA* 109: 10275–10280. [25]

Zhang, J. 2003. Evolution of the human *ASPM* gene, a major determinant of brain size. *Genetics* 165: 2063–2070. [10]

Zhang, J. 2004. Frequent false detection of positive selection by the likelihood method with branch-site models. *Mol. Biol. Evol.* 21: 1332–1339. [10]

Zhang, J., R. Nielsen, and Z. Yang. 2005. Evaluation of an improved branch-site likelihood method for detecting positive selection at the molecular level. *Mol. Biol. Evol.* 22: 2472–2479. [10]

Zhang, L., and W.-H. Li. 2005. Human SNPs reveal no evidence of frequency positive selection. *Mol. Biol. Evol.* 22: 2504–2507. [10]

Zhang, X.-S. 2012. Fisher's geometrical model of fitness landscape and variance in fitness within a changing environment. *Evolution* 66: 2350–2368. [28]

Zhang, X.-S., and W. G. Hill. 2002. Joint effects of pleiotropic selection and stabilizing selection on the maintenance of quantitative genetic variation at mutation-selection balance. *Genetics* 162: 459–471. [28]

Zhang, X.-S., and W. G. Hill. 2003. Multivariate stabilizing selection and pleiotropy in the maintenance of quantitative genetic variation. *Evolution* 57: 1761–1775. [28]

Zhang, X.-S., and W. G. Hill. 2005a. Predictions of patterns of response to artificial selection in lines derived from natural populations. *Genetics* 168: 411–425. [24, 26]

Zhang, X.-S., and W. G. Hill. 2005b. Genetic variability under mutation selection balance. *Trends Ecol. Evol.* 20: 468–470. [28]

Zhang, X.-S., and W. G. Hill. 2010. Change and maintenance of variation in quantitative traits in the context of the Price equation. *Theor. Popul. Biol.* 77: 14–22. [28]

Zhang, X.-S., J. Wang, and W. G. Hill. 2002. Pleiotropic model of maintenance of quantitative genetic variation at mutation-selection balance. *Genetics* 161: 419–433. [28]

Zhang, X.-S., J. Wang, and W. G. Hill. 2004a. Influence of dominance, leptokurtosis and pleiotropy of deleterious mutations on quantitative genetic variation at mutation-selection balance. *Genetics* 166: 597–610. [28]

Zhang, X.-S., J. Wang, and W. G. Hill. 2004b. Redistribution of gene frequency and changes of genetic variation following a bottleneck in population size. *Genetics* 167: 1475–1492. [11, 28]

Zhao, O., A. L. Weber, M. D. McMullen, K. Guill, and J. Doebley. 2010. MADS-box genes in maize: frequent targets of selection during domestication. *Genet. Res.* 93: 65–75. [9]

Zhivotovsky, L. A., and M. W. Feldman. 1992. On models of quantitative genetic variability: a stabilizing selection-balance model. *Genetics* 130: 947–955. [28]

Zhivotovsky, L. A., and S. Gavrilets. 1992. Quantitative variability and multilocus polymorphism under epistatic selection. *Theor. Popul. Biol.* 42: 245–283. [28]

Zhou, D., and 10 others. 2011. Experimental selection of hypoxia-tolerant *Drosophila melanogaster*. *Proc. Natl. Acad. Sci. USA* 108: 2349–2354. [26]

Zhu, Y. O., M. L. Siegal, D. W. Hall, and D. A. Petrov. 2014. Precise estimates of mutation rate and spectrum in yeast. *Proc. Natl. Acad. Sci. USA* 111: E2310–E2318. [4]

Zhuchenko, A. A., A. B. Korol , and L. P. Kovtyukh. 1985. Change in the crossing-over frequency in *Drosophila* during selection for resistance to temperature fluctuations. *Genetica* 67: 73–78. [26]

Ẑivković, D., and T. Wiehe. 2008. Second-order moments of segregating sites under variable population size. *Genetics* 180: 341–357. [9]

Zohary, D. 1999. Monophyletic vs. polyphyletic origin of the crops on which agriculture was founded in the Near East. *Genet. Res. Crop Evol.* 46: 133-142. [9]

Zouros, E., and D. W. Foltz. 1987. The use of allelic isozyme variation for the study of heterosis. *Isozymes* 13: 1–59. [17]

Zuckerkandl, E. 1968. Hemoglobins, Haeckel's "Biogenic Law", and molecular aspects of development. *In* A. Rich, N. R. Davidson, and L. Pauling (eds.), *Structural chemistry and molecular biology*, pp. 256–274. Freeman, San Francisco. [9]

Zuidhof, M. J., B. L. Schneider, V. L. Carney, D. R. Korver, and F. E. Robinson. 2014. Growth, efficiency, and yield of commercial broilers from 1957, 1978, and 2005. *Poul. Sci.* 93: 1–13. [1]

Zuk, M. 1988. Parasite load, body size, and age of wild-caught male field crickets (Orthoptera: Grillidae): effects on sexual selection. *Evolution* 42: 696–976. [29]

Zurell, D., and 10 others. 2010. The virtual ecologist approach: simulating data and observers. *Oikos* 119: 622–635. [30]

Author Index

Organism and Trait Index

Subject Index